CALCULUS AND ANALYTIC GEOMETRY

CALCULUS AND ANALYTIC GEOMETRY

FOURTH EDITION

GEORGE B. THOMAS, JR.
Department of Mathematics
Massachusetts Institute of Technology

ADDISON-WESLEY PUBLISHING COMPANY
Reading, Massachusetts
Menlo Park, California · London · Don Mills, Ontario

This book is in the
Addison-Wesley Series in Mathematics

 Printed in the United States of America. Published simultaneously in Canada. Library of Congress Catalog Card No. 68-17568.

PREFACE

TO THE FOURTH EDITION

Calculus and Analytic Geometry is available in one complete volume or as two separate parts. The first part treats functions of one variable and analytic geometry in two and three dimensions. It includes some vector algebra and comprises the first twelve chapters of the complete book. The second part deals with linear algebra, functions of several variables, infinite series, vector analysis, and differential equations. It comprises Chapters 13 through 20 of the complete volume. An appendix on determinants appears in both the first and second parts. The first part is suitable for a one-year course: The complete book can be covered in three or four semesters, depending on the pace.

Although the applications of integration are richer than those of differentiation, the latter are not trivial, particularly if we include such topics as the derivation of Kepler's laws of motion from Newton's inverse-square gravitation law. Moreover, students generally find the algebra that is used in differential calculus relatively easy. This book begins with the traditional problem of the rate of change of a function. (It is not necessary to know all about conics and other topics from analytic geometry to begin the study of calculus. However, if an individual instructor prefers, he can begin with Chapter 10, which contains much of this material.) When we have formulated the rate problem in general terms, we are led to the study of limits. The theorems about limits of sums, products, and quotients that are proved in Chapter 2 are applied in Chapter 3 and later chapters to establish the standard differentiation formulas. In Chapter 4, applications of derivatives to problems in maxima and minima give the reader some experience of the power of the calculus.

The area problem is illustrated in Chapter 2 for a parabola. This early introduction has two purposes: (1) It provides another application of the theory of limits, and (2) it anticipates a more detailed treatment that comes in Chapter 5. Students who are concurrently studying physics and calculus may find it useful to see this early study of area, but should know that a fuller discussion follows.

The Riemann integral for continuous functions is defined in the usual way, and upper and lower Riemann sums are shown to be nearly equal for sufficiently small subinterval widths. In the present computer era, error analyses are almost as important as computational algorithms, and hence the estimation of an integral by a finite sum should also include some notion of the magnitude of their difference. The upper and lower Riemann sums give that kind of information, in addition to performing their usual roles in the theorem on the existence of the integral of a continuous function. Error estimates are included for the trapezoidal rule and for Simpson's rule. Although proofs are not given in this book, references are included.

The author of any book that has been in use for some time is aware that any deviation from standard approaches naturally raises questions in the minds of his colleagues. One such feature in this book is the inclusion of a brief review of trigonometry in Chapter 5. Why is it there? It could certainly have been left out because students of calculus have studied trigonometry. The only answer is the candid reply that some students have been confused by the mass of formulas they saw in trigonometry and don't remember those that we now consider to be essential. It doesn't take much space to derive those formulas from scratch: so, we do it. No further apology. Using those formulas, we quickly derive the differentiation and integration laws for the sine and cosine. The derivatives of the other trigonometric functions, the inverse trigonometric functions, the natural logarithm, and the exponential function are studied in Chapter 7. The systematic study of methods of integration in Chapter 9, plus Chapter 8 on hyperbolic functions and Chapter 18 on infinite series, complete the calculus of functions of one real variable. (The student might note that a brief table of integrals has been included on the endpapers of the book. This table will be fully useful to him after he has learned the techniques of integration presented in Chapter 9.)

The chapter on linear algebra is certainly no more than an introduction to the subject. Sets of linear equations provide the motivation for most of the theory. Because some, but not all, students who will use the text know something about determinants, that topic is handled in an appendix, in a traditional way that does not make use of the material presented in Chapter 13. The theory of determinants was once an important part of programs in algebra. In this text, determinants of order three are used as easy mnemonic devices in computing the cross product of two vectors, the curl of a vector field, and the triple scalar product of three vectors. For many students, and particularly for those who go on to differential equations or advanced calculus, it is more important that they have some knowledge of the necessary condition for the existence of a nontrivial solution of a system of n homogeneous linear equations in n unknowns. (The determinant of coefficients should be zero.) Hence the appendix on determinants, being a self-contained unit, can be used either for reference, or for a few lessons inserted at a convenient time.

Motion on a space curve and the calculus of functions of more than one real variable are studied, with frequent use of vectors, in Chapters 14, 15, and 16. Several topics included there, and in Chapter 19 on complex numbers and functions, are often postponed until a course in advanced calculus. However, there seem to be at least two reasons for including these in the first course in calculus, if time permits. First, they are useful to the student in his other studies in physics and in engineering, and should not be too long delayed. Second, the more difficult ideas need repeating. Even though the student of a first course may not master some of these ideas, he will profit from the preview, and will better understand the concepts when he next encounters them.

Not all of Chapter 17 is new, because the beginnings of vector analysis have been presented in the earlier editions. But many parts of the chapter are new, including the discussions of the divergence theorem, Green's theorem, and Stokes's theorem. These topics will be especially useful to electrical engineers and physicists, or others who study such subjects in the contexts of electromagnetic fields or fluid mechanics. Even theoretical mathematicians are likely to encounter analogous theorems in more advanced courses in mathematics; for example, Cauchy's integral theorem, in the theory of functions of complex variables, can be deduced from Green's theorem.

The treatment of differential equations in Chapter 20 is not exhaustive, but may serve as an introduction to a more systematic study of the important *linear* equations. It was this point of view which led, for example, to the choice of the method of variation of parameters for the solution of the nonhomogeneous linear equation with constant coefficients, even though the more restricted method of undetermined coefficients may be simpler to apply in some cases.

The reader is expected to know high-school algebra, geometry, and some trigonometry. The basic formulas from these subjects are collected for reference in an appendix at the end of the book but, in general, the more difficult formulas are reviewed in the text at the time they are needed. The author's experience, both as a student and as a teacher, indicates that one by-product of a good calculus course is likely to be a much firmer grasp of the fundamental skills of high-school mathematics. This is particularly true when the student supplies the missing steps in illustrative exercises and elsewhere.

Answers are provided for nearly all problems. Many of the Miscellaneous Exercises were generously furnished by Professors A. W. Tucker of Princeton and T. M. Apostol of the California Institute of Technology. Many of the answers were provided by Richard T. Bumby, Robert R. D. Kemp, Robert P. Walker, and C. Peter Wolk. The author is grateful for this help and for other valuable and friendly assistance.

Some students (and teachers) may like to have a source of review questions to help them go back over the theory and organize their thoughts. Such questions have been included (without answers) at the end of every chapter. It may be desirable to combine two or more chapters for review purposes, and this may easily be done, even if some chapters are taken out of their sequence.

After finishing this text, the reader should be well prepared for a course in differential equations or advanced calculus.

The most significant changes from the third edition are the following:

1. Three new chapters have been added. They are:

   Chapter 2, Limits,
   Chapter 13, Linear Algebra,
   Chapter 17, Vector Analysis.

2. Kepler's laws of motion for planetary orbits, and for earth satellites, have been deduced from Newton's inverse-square law of gravitational attraction.

3. An attempt has been made to improve the treatment of the definite integral and of the two Fundamental Theorems of Integral Calculus.

There has been no significant attempt to upgrade the general level of rigor of the book, except as noted above. One of the author's friends, Mr. Henry Roth, wrote that he feared that the new edition would be "rife with set theory." I believe that he, and others who have used the third edition, will find that only modest additions of set theory have been made. (For example, in Chapter 2, where neighborhoods are used to prove limit theorems, one needs to know that the intersection of two neighborhoods of a point c is a neighborhood of c.)

It is a pleasure to acknowledge the superb assistance in editing, design, composition, and illustration that the staffs of the Addison-Wesley Publishing Company and the Wolf Composition Company have given to the preparation of this edition. The author also acknowledges the many helpful suggestions of users of earlier editions of *Calculus and Analytic Geometry*.

Any errors that may appear are, of course, the responsibility of the author. He will very much appreciate having these brought to his attention.

Concord, Massachusetts
February 1968 G. B. T., Jr.

CONTENTS

113

THE RATE OF CHANGE OF A FUNCTION

CHAPTER 1

1.1 INTRODUCTION

Calculus is the mathematics of change and motion. Where there is motion or growth, where forces are at work producing acceleration, calculus is the right mathematical tool. This was true in the beginnings of the subject, and it is true today. Calculus is used to predict the orbits of earth satellites; in the design of inertial navigation systems, cyclotrons, and radar systems; to explore problems of space travel; and to test scientific theories about such things as the dynamics of the atmosphere, ocean currents, and even theories about economic, sociological, and psychological behavior. Of course the scientist needs a great deal more than mathematical competence, and he needs more mathematics than calculus. But calculus is a tool of great importance and usefulness and an understanding of its nature is prerequisite for further study in nearly all branches of higher mathematics.

One of the great mathematicians of the twentieth century, John von Neumann (1903–1957), has written:* "The calculus was the first achievement of modern mathematics, and it is difficult to overestimate its importance. I think it defines more unequivocally than anything else the inception of modern mathematics, and the system of mathematical analysis, which is its logical development, still constitutes the greatest technical advance in exact thinking."

The calculus is a branch of mathematics which provides methods for solving two large classes of problems. The first of these involves finding the rate at which a variable quantity is changing. For example, when a body, say a stone dropped from a tower, travels in a straight line, its distance from its starting point changes with time, and we may ask *how fast* it is moving at any given instant. *Differential calculus* is the branch of calculus which treats such problems.

The second type of problem the calculus deals with is that of finding a function when its rate of change is given. Thus, for example, if we are given the

---

* John von Neumann, "The Mathematician," in *The World of Mathematics,* Vol. 4, pp. 2053–2063.

velocity of a moving body at every instant of time, we may seek to find the distance it has moved as a function of time. This problem belongs to the domain of *integral calculus.*

Modern science and engineering use both branches of the calculus as a language for expressing physical laws in precise mathematical terms, and as a tool for studying the consequences of those laws. Thus Sir Isaac Newton (1642–1727) was able to explain the motion of the planets relative to the sun as a consequence of the physical assumption known today as the law of gravitational attraction. Some appreciation of the magnitude of his achievement may be gained from realizing that Kepler (1571–1630) spent some twenty years studying observational data and using empirical methods to discover the three laws now known as *Kepler's laws.* These laws may be stated as follows:

(a) Each planet describes an orbit about the sun which is an ellipse with the sun at one focus.

(b) The line joining the planet and the sun sweeps over equal areas in equal intervals of time.

(c) The squares of the periods of revolution of the planets are proportional to the cubes of their mean distances from the sun.

All three of these laws can be derived, using the calculus as the main mathematical tool, from the "inverse square" law of gravitational attraction and Newton's laws of motion. A large part of the subject matter of this branch of mathematics was developed by Newton as a tool to help him solve problems which arose in connection with his investigations in physics and astronomy. Credit for the invention of the calculus must also be shared, however, with a German mathematician and philosopher, Gottfried Wilhelm Leibniz (1646–1716), who independently and approximately concurrently also developed a large part of the subject, and whose notation has been almost universally adopted in preference to that used by Newton.

Analytic geometry, which forms a third division of the subject matter of this book, was the creation of several mathematicians.* Two French mathematicians, Pierre de Fermat (1601–1665) and René Descartes (1596–1650), are the chief claimants to the title of inventor of analytic geometry as we now know it. Descartes is credited with the important idea of locating a point in the plane by means of its distances from two perpendicular axes, and his name is commemorated in the terminology "Cartesian coordinates." We discuss coordinates in Article 1.2. The distinguishing characteristic of analytic geometry is that it uses algebraic methods and equations to study geometric problems. Conversely, it permits us to visualize algebraic equations in terms of geometric curves. This frequently adds clarity to abstract concepts. We shall see that most of the theory in calculus can be presented in geometrical terms, so that calculus and analytic geometry may profitably be united and studied as a whole.

1.2 COORDINATES

The basic idea in analytic geometry is the establishment of a one-to-one correspondence between the points of a plane on the one hand and pairs of numbers (x, y) on the other hand. This correspondence may be established in many ways, but the one most commonly used is as follows.

A horizontal line in the plane, extending indefinitely to the left and to the right, is chosen as the x-axis or axis of *abscissas.* A reference point O on this line and a unit of length are then chosen. The axis is scaled off in terms of this unit of length in such a way that the number zero is attached to O, the number $+a$ is attached to the point which is a units to the right of O, and $-a$ is attached to the symmetrically located point to the left of O. In this way, a one-to-one correspondence is established between the points of the x-axis and the set of all *real*

---

* See, for example, "Commentary on Descartes and Analytic Geometry," in *The World of Mathematics,* Vol. 1, pp. 235–237. Also Carl B. Boyer, "The Invention of Analytic Geometry," *Scientific American,* January, 1949.

numbers (that is, numbers which may be represented by terminating or nonterminating decimals).

Now through O take a second, vertical line in the plane, extending indefinitely up and down. This line becomes the y-axis, or axis of *ordinates*. The unit of length used to represent $+1$ on the y-axis need not be the same as the unit of length used to represent $+1$ on the x-axis. The y-axis is scaled off in terms of the unit of length adopted for it, with the positive number $+b$ attached to the point b units above O and negative number $-b$ attached to the symmetrically located point b units below O.

If a line parallel to the y-axis is drawn through the point marked a on the x-axis, and a line parallel to the x-axis is drawn through the point marked b on the y-axis, their point of intersection P is to be labeled $P(a, b)$. Thus, given the pair of real numbers a and b, we find one and only one point with abscissa a and ordinate b, and this point we denote by $P(a, b)$.

Conversely, if we start with any point P in the plane, we may draw lines through it parallel to the coordinate axes. If these lines intersect the x-axis at a and the y-axis at b, we then regard the pair of numbers (a, b) as corresponding to the point P. We say that the coordinates of P are (a, b).

The two axes divide the plane into four quadrants, called the first quadrant, second quadrant, and so on, and labeled I, II, III, IV in Fig. 1.1. Points in the first quadrant have both coordinates positive, and in the second quadrant the x-coordinate (abscissa) is negative and the y-coordinate (ordinate) is positive. The notations $(-, -)$ and $(+, -)$ in quadrants III and IV of Fig. 1.1 represent the signs of the coordinates of points in these quadrants.

Remark. There are times when it is quite obvious that there is no physical relationship between the unit of measurement of x and the unit of measurement of y. For example, if y is the total cost in dollars, to a certain shoe manufacturer, to produce x pairs of shoes per week, then "1" on the x-scale may stand for one pair of shoes, while "1" on the y-scale may stand for one dollar. Clearly, there is no reason for using the same scale on the two axes.

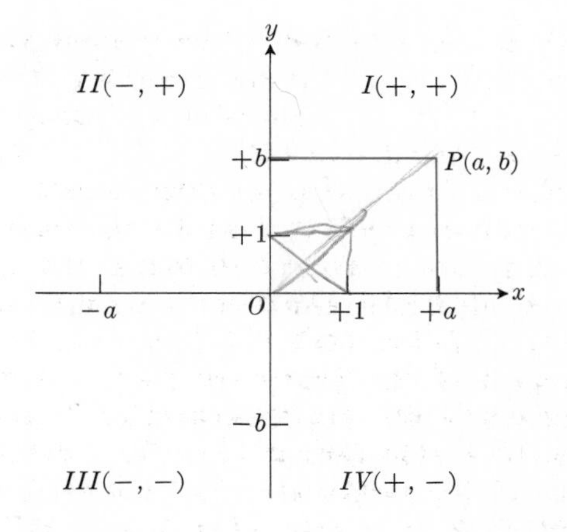

1.1

On the other hand, in surveying, one foot measured north-and-south should be the same as one foot measured east-and-west. In trigonometry, therefore, it is usually assumed that the units of length on the two axes are the same. This assumption is also made in analytic geometry.

In this book, if coordinates of points are given as pure numbers (i.e., without any physical units attached), it is to be assumed that the scales on the two axes are the same. In particular, this assumption is made wherever angles between lines or lengths of skew line segments are involved.

EXERCISES

In Exercises 1 through 15, plot the given point P and such of the following points as may apply.

(a) The point Q such that QP is perpendicular to the x-axis and is bisected by it. Give the coordinates of Q.

(b) The point R such that PR is perpendicular to and is bisected by the y-axis. Give the coordinates of R.

(c) The point S such that PS is bisected by the origin. Give the coordinates of S.

(d) The point T such that PT is perpendicular to and is bisected by the 45° line L through the origin bisecting the first and third quadrants. Give the coordinates of T, assuming equal units on the axes.

[*Note.* Q and P are symmetric with respect to the x-axis, R and P with respect to the y-axis, S and P with respect to the origin, and T and P with respect to L.] State rules for finding coordinates of Q, R, S, T in terms of the coordinates of P.

1. $(1, -2)$
2. $(2, -1)$
3. $(-2, 2)$
4. $(-2, 1)$
5. $(2, 2)$
6. $(-2, -2)$
7. $(0, 1)$
8. $(1, 0)$
9. $(-2, 0)$
10. $(0, -3)$
11. $(-1, -3)$
12. $(\sqrt{2}, -\sqrt{2})$
13. $(-\pi, -\pi)$
14. $(-1.5, 2.3)$
15. $(\sqrt{2}, 0)$

In Exercises 16 through 20, take the units of length on the two axes to be equal.

16. A straight line is drawn through the point $(0, 0)$ and the point $(1, 1)$. What acute angle does it make with the positive x-axis? Sketch.
17. A straight line is drawn through the point $A(1, 2)$ and the point $B(2, 4)$. Find $\tan \alpha$, where α is the acute angle that this line makes with the horizontal line through A. Sketch.
18. Show that the line through the pair of points $(1, 1)$ and $(2, 0)$ is parallel to the line through the pair of points $(-1, -1)$ and $(0, -2)$. Sketch.
19. The line through the pair of points $(2, 3)$ and $(1, 1)$ cuts the y-axis at the point $(0, b)$. Find b by using similar triangles. Sketch.
20. Does the line through the pair of points $(-1, -1)$ and $(1, 2)$ pass through the origin? Sketch. Give a reason, other than the evidence of your sketch, for your answer.

1.3 INCREMENTS

If a particle starts at a point $P_1(x_1, y_1)$ and goes to a new position $P_2(x_2, y_2)$, we say that its coordinates have received increments Δx (called *delta x*) and Δy (called *delta y*). For example, if a particle moves from $A(1, -2)$ to $B(6, 7)$, as in Fig. 1.2, then the increments received by its coordinates are

$$\Delta x = 5, \quad \Delta y = 9.$$

Observe that the increment in a coordinate is the net change given by

$$\Delta x = (x \text{ of terminal point}) - (x \text{ of initial point})$$

and

$$\Delta y = (y \text{ of terminal point}) - (y \text{ of initial point}).$$

So far as these increments are concerned, it doesn't matter at all what the positions of the particle may have been between its initial and terminal locations.

Definition. *If the initial position of a particle is $P_1(x_1, y_1)$ and its terminal position is $P_2(x_2, y_2)$, then the increments Δx and Δy are given by*

$$\Delta x = x_2 - x_1, \quad \Delta y = y_2 - y_1.$$

Remark. Either increment Δx or Δy can be any real number: positive, negative, or zero. In Fig. 1.3, for example, the increments from $C(-1, 5)$ to $D(2, -3)$ are

$$\Delta x = 2 - (-1) = 3, \quad \Delta y = -3 - 5 = -8.$$

Here, Δx is positive because x increases from C to D, and Δy is negative because the y coordinate decreases.

If the same unit of measurement is used on both axes, then we may also express all distances in the plane in terms of this fundamental unit. In particular, by applying the theorem of Pythagoras to the right triangle P_1RP_2 in Fig. 1.4, we readily find that the distance d between the two points $P_1(x_1, y_1)$ and $P_2(x_2, y_2)$ is simply

$$d = \sqrt{(\Delta x)^2 + (\Delta y)^2}$$
$$= \sqrt{(x_2 - x_1)^2 + (y_2 - y_1)^2}.$$

Example. If a particle starts from $P_1(-1, 2)$ and travels in a straight line to the point $P_2(2, -2)$, then its abscissa undergoes a change

$$\Delta x = x_2 - x_1 = 2 - (-1) = 3$$

and its ordinate changes by an amount

$$\Delta y = y_2 - y_1 = -2 - 2 = -4.$$

The distance the particle travels from P_1 to P_2 is

$$d = \sqrt{(\Delta x)^2 + (\Delta y)^2} = \sqrt{9 + 16} = 5.$$

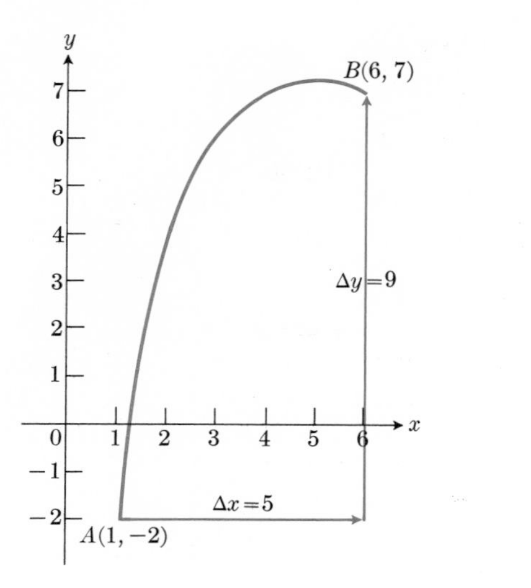

1.2

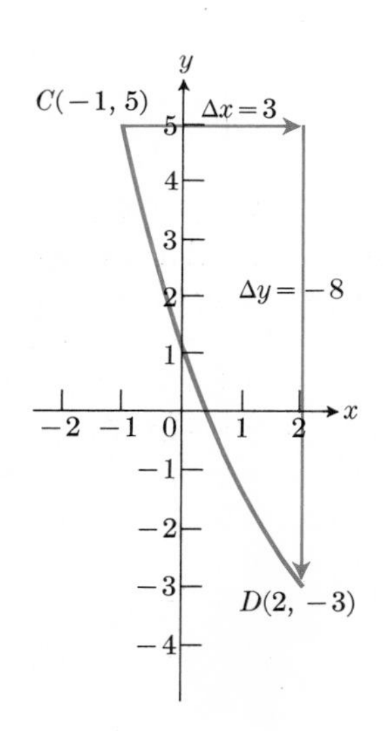

1.3

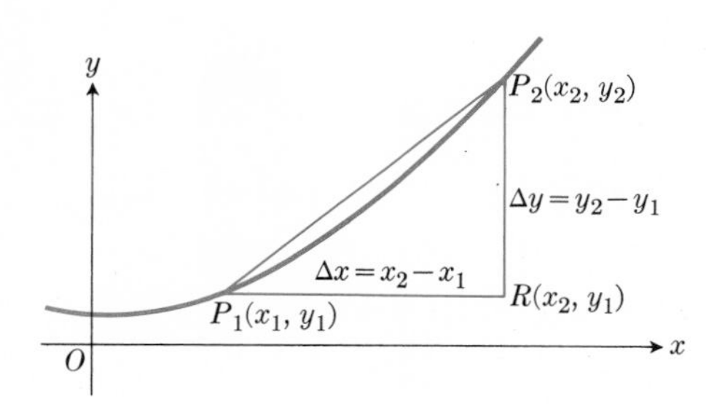

1.4

EXERCISES

In Exercises 1 through 10, points A and B are given and point C is to be found at the intersection of the horizontal line through A and the vertical line through B. Sketch, and find

(a) the coordinates of C, (b) Δx, (c) Δy,

(d) the length of AB under the assumption that the unit of length on the y-axis is the same as the unit of length on the x-axis.

1. $A(1, 1)$, $B(3, 4)$
2. $A(3, 4)$, $B(1, 1)$
3. $A(-1, 2)$, $B(2, -1)$
4. $A(2, -1)$, $B(-1, 2)$
5. $A(-1, -1)$, $B(1, 2)$
6. $A(1, 2)$, $B(-1, -1)$
7. $A(-3, 2)$, $B(-1, -2)$
8. $A(-1, -2)$, $B(-3, 2)$
9. $A(-3, -2)$, $B(-1, -1)$
10. $A(-1, -1)$, $B(-3, -2)$
11. What equation must be satisfied by the coordinates of a point $P(x, y)$ if P is on the circle of radius 5 with center at $C(h, k)$ if

    (a) $(h, k) = (0, 0)$, (b) $(h, k) = (5, 0)$,
    (c) $(h, k) = (3, 4)$, (d) $(h, k) = (h, k)$?
12. If a particle starts at $A(-2, 3)$ and its coordinates receive increments $\Delta x = 5$, $\Delta y = -6$, what will its new position be?
13. Find the starting position of a particle whose terminal position is $B(u, v)$ after its coordinates have received increments $\Delta x = h$, and $\Delta y = k$.
14. If a particle moves from the point $A(-2, 5)$ to the y-axis in such a way that $\Delta y = 3\,\Delta x$, what are its new coordinates?
15. A particle moves along the parabola $y = x^2$ from the point $A(1, 1)$ to the point $B(x, y)$. Show that

$$\frac{\Delta y}{\Delta x} = x + 1 \quad \text{if} \quad \Delta x \neq 0.$$

1.4 SLOPE OF A STRAIGHT LINE

Let L be a straight line which is not parallel to the y-axis. Let $P_1(x_1, y_1)$ and $P_2(x_2, y_2)$ be any two distinct points on L (Fig. 1.5a). Then we call

$$\Delta y = y_2 - y_1$$

the "rise" and

$$\Delta x = x_2 - x_1$$

the "run" along L from P_1 to P_2. (Note that either Δx or Δy, or both, may be negative.) We define the *slope of L* as the rate of rise per unit of run:

$$\text{slope} = \frac{\text{rise}}{\text{run}}. \tag{1a}$$

If we denote the slope by m, we have

$$m = \frac{\Delta y}{\Delta x} = \frac{y_2 - y_1}{x_2 - x_1}. \tag{1b}$$

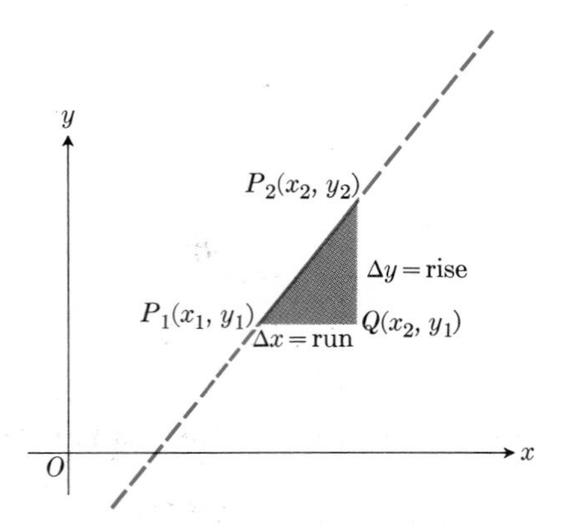

a | b

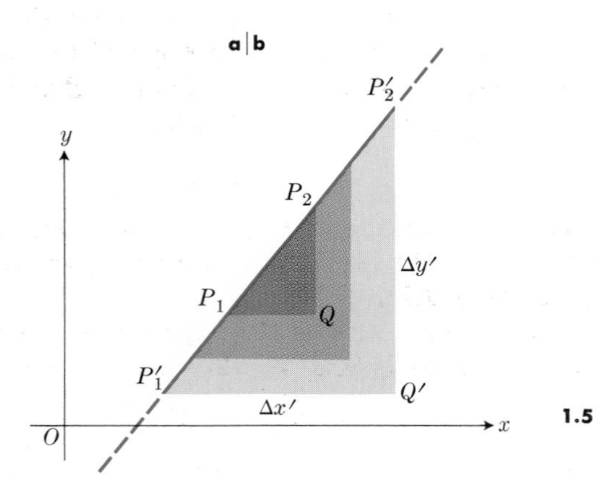

1.5

Remark 1. Suppose that instead of using the pair of points P_1 and P_2 to calculate the slope, Eq. (1a), we were to choose a different pair of points $P_1'(x_1', y_1')$, $P_2'(x_2', y_2')$ on L and calculate

$$m' = \frac{y_2' - y_1'}{x_2' - x_1'} = \frac{\Delta y'}{\Delta x'}.$$

Would we get the same answer for the slope? In other words, does

$$m' = m?$$

The answer is "yes." For in Fig. 1.5(b), the triangles P_1QP_2 and $P_1'Q'P_2'$ are similar, hence

$$m' = \frac{\Delta y'}{\Delta x'} = \frac{\Delta y}{\Delta x} = m.$$

In other words, Eq. (1b) relates the change in x and the change in y between *any* pair of points (x, y) and $(x + \Delta x, y + \Delta y)$ on L.

Remark 2. If we multiply the first two terms of Eq. (1b) by Δx, we obtain

$$\Delta y = m\,\Delta x. \tag{2}$$

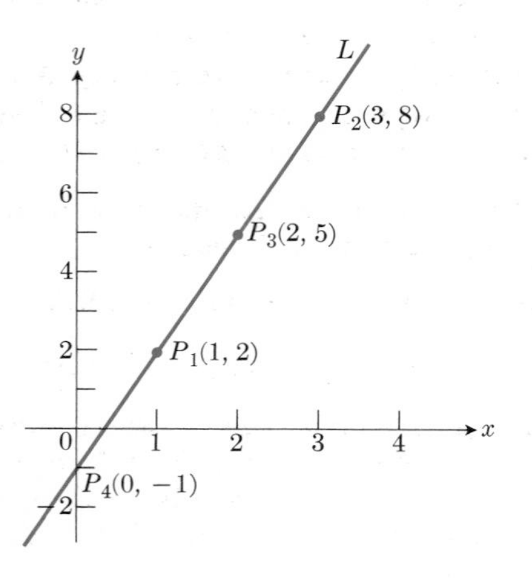

1.6

This means, in the light of the comment above, that as a particle moves along L, the change in y is proportional to the change in x, and the slope m is the proportionality factor.

For example, suppose L is the line determined by the points $P_1(1, 2)$ and $P_2(3, 8)$, Fig. 1.6. Then the slope of L is

$$m = \frac{8 - 2}{3 - 1} = 3.$$

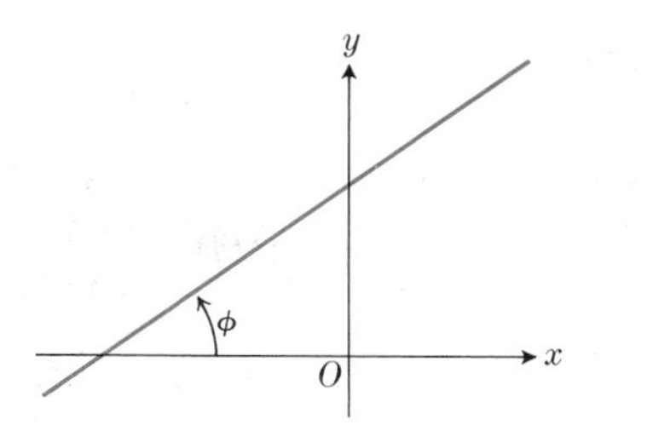

a | b

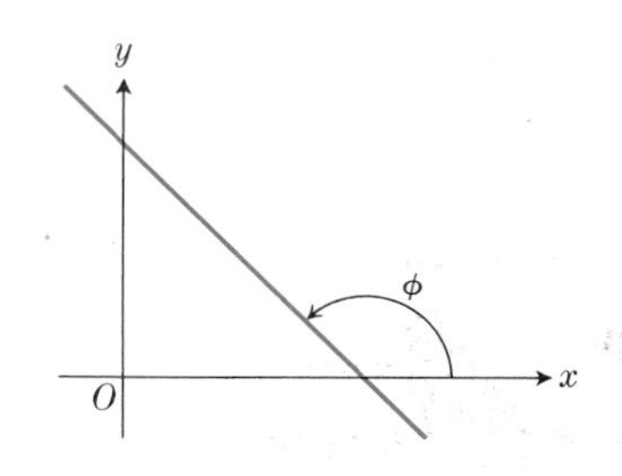

1.7

And, if $P(x, y)$, $Q(x + \Delta x, y + \Delta y)$ are any two points on L, then

$$\Delta y = 3\,\Delta x.$$

That is, for every unit of run, there are three units of rise. Thus, starting from $P_1(1, 2)$, if we increase x by 1 unit we should increase y by 3 units, which would bring us to the point $P_3(2, 5)$. Or, we could decrease x by 1 unit and y by 3 units and obtain the point $P_4(0, -1)$. [*Question:* At what point does L cross the x-axis?]

Remark 3. Interpolation. In using a table of logarithms, we may, for example, want to find log 3.1416. Suppose the table gives

$$\log 3.1410 = 0.49707,$$
$$\log 3.1420 = 0.49721,$$

and no intermediate values. Then, for any x between 3.1410 and 3.1420, we could approximate the graph of $y = \log x$ by the chord L through the two points $P_1(3.1410, 0.49707)$ and $P_2(3.1420, 0.49721)$. The slope of this chord is

$$m = \frac{\Delta y}{\Delta x} = \frac{14 \times 10^{-5}}{10 \times 10^{-4}} = 1.4 \times 10^{-1}.$$

Then, corresponding to

$$\Delta x = 3.1416 - 3.1410 = 6 \times 10^{-4},$$

we may take

$$\Delta y = 0.14\,\Delta x = 8.4 \times 10^{-5} \approx 8 \times 10^{-5}.$$

Hence, the ordinate of the point on L with abscissa 3.1416 is

$$0.49707 + 0.00008 = 0.49715.$$

That is, using the chord to approximate the curve, we find

$$\log 3.1416 = 0.49715.$$

Remark 4. If we take equal scales on the x- and y-axes, we may interpret the slope of the line L as the tangent of the angle of inclination that L makes with the positive x-axis. That is,

$$m = \tan \phi.$$

Here ϕ is the angle, $0 \le \phi < 180°$, that the line makes with the positively directed portion of the x-axis (Fig. 1.7).

For a line that slopes upward to the right (Fig. 1.7a), Δy is positive when Δx is. Also,

$$0 < \phi < 90°,$$

and $\tan \phi = \Delta y/\Delta x$ is positive.

For a line that slopes downward to the right (Fig. 1.7b), Δy is negative when Δx is positive. Also, $90° < \phi < 180°$, and $\tan \phi = \Delta y/\Delta x$ is negative.

For a horizontal line, Δy is zero for any Δx. Also $\phi = 0°$ and $\tan \phi = \Delta y/\Delta x$ is zero.

For a vertical line, Δx is zero for any Δy. Then $\Delta y/\Delta x$ is meaningless, since we cannot divide by zero. Also $\phi = 90°$, and $\tan 90°$ does not exist.

Example. Let us examine the slopes of lines whose inclination angles are near 90°, such as

$$\phi_1 = 89°59', \quad m_1 = \tan \phi_1 \approx 3437.7$$

and

$$\phi_2 = 90°01', \quad m_2 = \tan \phi_2 \approx -3437.7.$$

We see that such lines have slopes which are numerically very large. And by taking the angle still closer to 90°, we can make the slope numerically larger than any pre-assigned positive number N, no matter how large N may

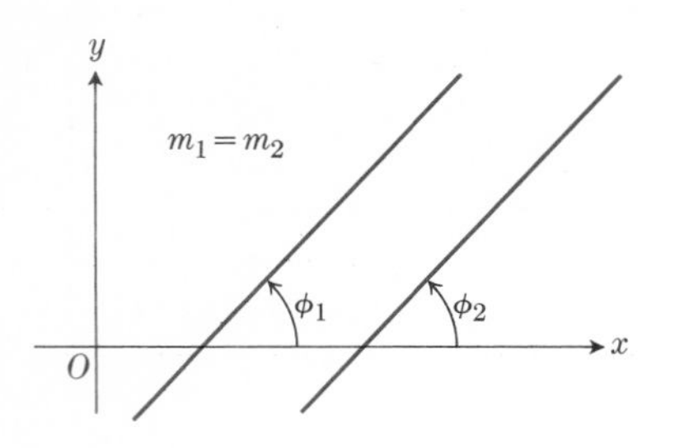

a|b

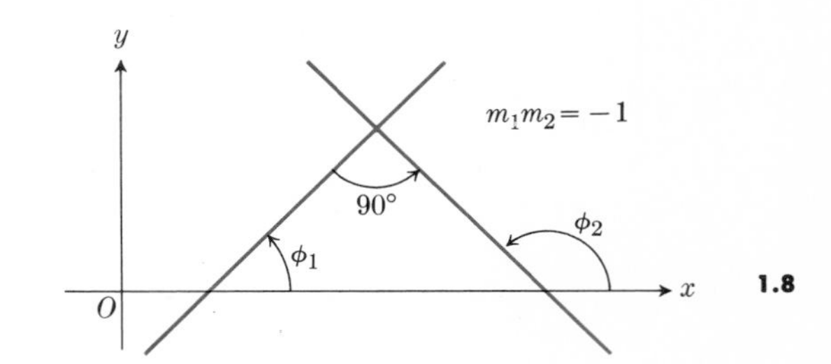

1.8

be. It is this fact which is sometimes summarized by the statements that "a vertical line has infinite slope," or "the slope of the line becomes infinite as its inclination angle approaches 90°."

The symbol ∞ is used to represent infinity, but we should not use this symbol in computations involving addition, subtraction, multiplication, or division in the way we use ordinary real numbers.

Two *parallel* lines (Fig. 1.8a) have equal angles of inclination and hence have equal slopes $m_1 = m_2$. Conversely, two lines having equal slopes

$$m_1 = \tan \phi_1, \quad m_2 = \tan \phi_2$$

also have equal inclinations since

$$\tan \phi_1 = \tan \phi_2,$$

together with

$$0 \leq \phi_1 < 180°, \quad 0 \leq \phi_2 < 180°,$$

implies that ϕ_1 and ϕ_2 are equal.

Next, consider two *perpendicular* lines (Fig. 1.8b). Their inclinations differ by 90°. Suppose, for example, that

$$\phi_2 = 90° + \phi_1.$$

Case 1. If one line is vertical, then the other must be horizontal.

Case 2. If neither line is vertical, then neither ϕ_1 nor ϕ_2 is 90° and

$$\begin{aligned} m_2 &= \tan \phi_2 = \tan (90° + \phi_1) \\ &= -\cot \phi_1 = -\frac{1}{\tan \phi_1} \\ &= -1/m_1. \end{aligned}$$

Thus the condition for *perpendicularity* in this case is

$$m_2 = -\frac{1}{m_1}. \tag{3}$$

Conversely, suppose that condition (3) is satisfied. Then one of the slopes is positive and the other is negative. With no loss of generality we may thus assume that m_1 is positive and m_2 is negative. (If not, renumber the lines.) Then

$$0 < \phi_1 < 90° < \phi_2 < 180°, \tag{4a}$$

and

$$\tan \phi_2 = -\cot \phi_1. \tag{4b}$$

Therefore

$$0 < (180° - \phi_2) < 90°, \tag{5a}$$

and

$$\tan (180° - \phi_2) = -\tan \phi_2 = \cot \phi_1. \tag{5b}$$

Equation (5b) and the inequalities (4a) and (5a) imply that ϕ_1 and $180° - \phi_2$ are complementary acute angles, as in Fig. 1.8b. Therefore, the two lines are perpendicular.

EXERCISES

In Exercises 1 through 15, plot the given points A and B and find the slope of the straight line determined by them and of a perpendicular line.

1. $A(1, 2)$, $B(2, -1)$
2. $A(-1, 2)$, $B(-2, -1)$
3. $A(-2, -1)$, $B(1, -2)$
4. $A(2, -1)$, $B(-2, 1)$
5. $A(1, 0)$, $B(0, 1)$
6. $A(-1, 0)$, $B(1, 0)$

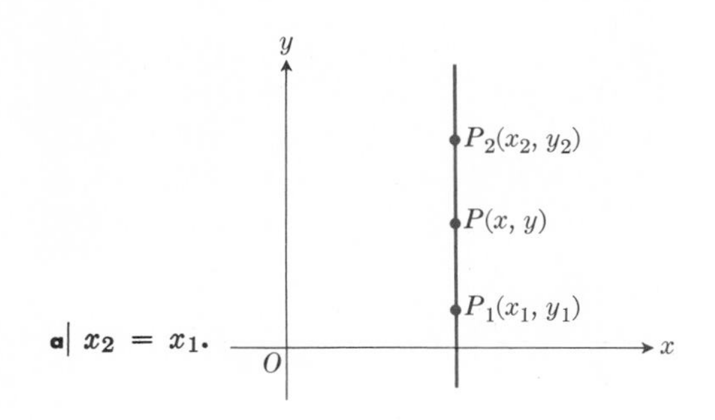

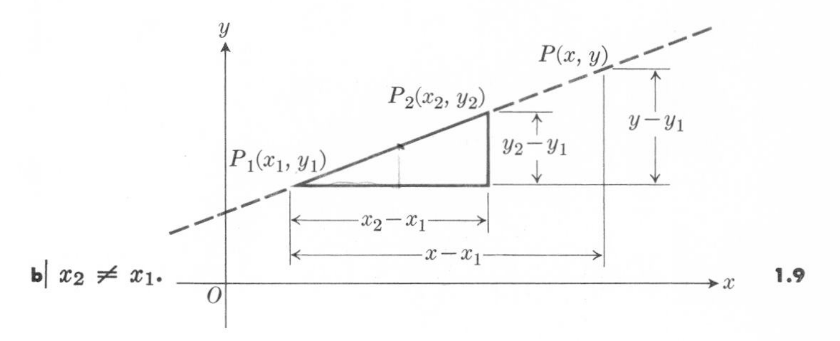

1.9

7. $A(2, 3)$, $B(-1, 3)$
8. $A(0, 3)$, $B(2, -3)$
9. $A(0, -2)$, $B(-2, 0)$
10. $A(1, 2)$, $B(1, -3)$
11. $A(\frac{1}{2}, 0)$, $B(0, -\frac{1}{3})$
12. $A(0, 0)$, $B(x, y)$ $(x \neq 0, y \neq 0)$
13. $A(0, 0)$, $B(x, 0)$ $(x \neq 0)$
14. $A(0, 0)$, $B(0, y)$ $(y \neq 0)$
15. $A(a, 0)$, $B(0, b)$ $(a \neq 0, b \neq 0)$

In Exercises 16 through 20, plot the given points A, B, C, and D. Determine in each case whether or not $ABCD$ is a parallelogram. Also indicate those parallelograms that are rectangles.

16. $A(0, 1)$, $B(1, 2)$, $C(2, 1)$, $D(1, 0)$
17. $A(3, 1)$, $B(2, 2)$, $C(0, 1)$, $D(1, 0)$
18. $A(-2, 2)$, $B(1, 3)$, $C(2, 0)$, $D(-1, -1)$
19. $A(1, -2)$, $B(2, -1)$, $C(2, 1)$, $D(1, 0)$
20. $A(-1, 0)$, $B(0, -1)$, $C(2, 0)$, $D(0, 2)$

21. Find the coordinates of a point $P(x, y)$ which is so located that the line L_1 through the origin and P has slope equal to $+2$, and the line L_2 through the point $A(-1, 0)$ and P has slope equal to $+1$.
22. A line L goes through the origin and has slope equal to $+3$. If $P(x, y)$ is a point other than the origin and on the line L, show that $y = 3x$. Sketch the line.

In Exercises 23 through 27, plot the given points and determine analytically whether or not they all lie on a straight line. Give reasons for your answers.

23. $A(1, 0)$, $B(0, 1)$, $C(2, -1)$
24. $A(-2, 1)$, $B(0, 5)$, $C(-1, 2)$
25. $A(-2, -1)$, $B(-1, 1)$, $C(1, 5)$, $D(2, 7)$
26. $A(-2, 3)$, $B(0, 2)$, $C(2, 0)$
27. $A(-3, -2)$, $B(-2, 0)$, $C(-1, 2)$, $D(1, 6)$
28. Let $P_1(x_1, y_1)$ and $P_2(x_2, y_2)$ be two points. Find the midpoint of the segment P_1P_2.
29. Given $A(0, -1)$, $B(4, 0)$, and $C(3, 4)$, show that ABC is a right triangle and find the center and radius of the circumscribed circle. Sketch.
30. Answer the question in Remark 2.

1.5 EQUATIONS OF A STRAIGHT LINE

In this article we shall discuss algebraic equations that represent straight lines. Let L be the line through the two points $P_1(x_1, y_1)$ and $P_2(x_2, y_2)$. We ask how the coordinates x and y are related when $P(x, y)$ is on L. It is desirable to distinguish between the cases (see Fig. 1.9)

$$x_2 = x_1 \quad \text{and} \quad x_2 \neq x_1.$$

Case 1. If $x_2 = x_1$, then the line L is vertical and all points on it have the same x-coordinate. Then $P(x, y)$ is on the line if and only if

$$x = x_1. \tag{1}$$

In this case the x- and y-coordinates of P are not related to each other. The ordinate y may have any value whatsoever and the point $P(x, y)$ will be on the line if and only if its abscissa x has the value x_1.

Case 2. If $x_2 \neq x_1$, we know the slope of the line is

$$m = \frac{y_2 - y_1}{x_2 - x_1}. \tag{2}$$

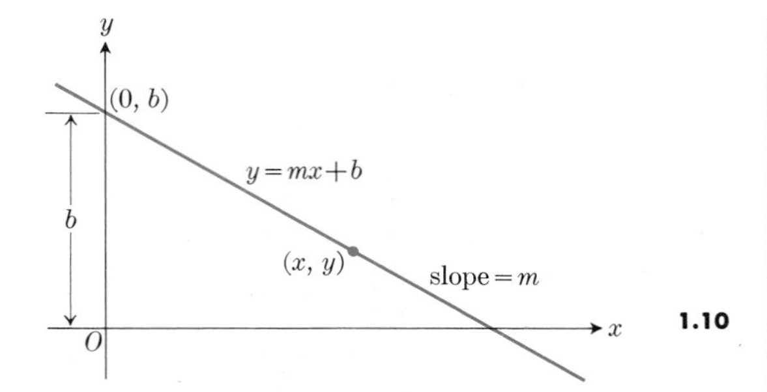

1.10

Then the point $P(x, y)$ is also on the line if and only if $P = P_1$ or $P \neq P_1$ and the slope of P_1P is the same as the slope of P_1P_2. If $P \neq P_1$, x and y must satisfy the equation

$$\frac{y - y_1}{x - x_1} = m,$$

or

$$y - y_1 = m(x - x_1), \qquad (3)$$

if P is on the line. Conversely, if x and y satisfy Eq. (3), then $P(x, y)$ is on the line. If P coincides with P_1, Eq. (3) is still satisfied, even though the slope of P_1P is then undefined.

Equation (3) is called the *point-slope* form of the equation of the line, since it gives the equation in terms of one point $P_1(x_1, y_1)$ on the line and the slope m. The three numbers x_1, y_1, and m are constants in Eq. (3), while x and y are variables. By a *variable* we mean a symbol, such as x, which may take any value in some set of numbers. In Eq. (3), x may take any real value whatever,

$$-\infty < x < +\infty,$$

and as x varies continuously from a negative value $-N$ to a positive value $+M$, the corresponding point $P(x, y)$ traces an unbroken portion of the line from left to right.

We may also write Eq. (3) in the equivalent form

$$y = mx + b, \qquad (4)$$

where m is the slope of the line and

$$b = y_1 - mx_1$$

is a constant. In fact, $(0, b)$ is the point where the line crosses the y-axis. Equation (4) is expressed in terms of the slope m and the *y-intercept* b, and is therefore called the *slope-intercept* equation of the line. Figure 1.10 displays this form of the linear equation.

Example. The equation

$$y = 2x + 3,$$

with values

$$m = 2, \quad b = 3,$$

may be compared with (4). This equation thus represents a straight line having y-intercept 3 and slope 2.

More generally, an equation

$$Ax + By + C = 0, \qquad (5)$$

where A, B, and C are constants with at least one of A and B different from zero, represents a *straight line*. For

$$\text{if} \quad B = 0, \quad \text{then} \quad Ax + C = 0,\ x = -\frac{C}{A},$$

and the line is a vertical line as in Eq. (1). On the other hand,

$$\text{if} \quad B \neq 0, \quad \text{then} \quad y = -\frac{A}{B}x - \frac{C}{B},$$

and this represents a straight line of the form of Eq. (4), with

$$m = -\frac{A}{B}, \quad b = -\frac{C}{B}.$$

An equation like (5) that contains only first powers of x and y is said to be "linear in x and in y." Thus we may summarize our discussion by saying that every straight line in the plane is represented by a linear equation and, conversely, every linear equation represents a straight line.

Problem 1. Find the slope of the line $2x + 3y = 5$.

Solution. Solving for y, we have

$$y = -\tfrac{2}{3}x + \tfrac{5}{3},$$

which compares with $y = mx + b$, with $m = -\frac{2}{3}$ and $b = \frac{5}{3}$. Hence the slope is $-\frac{2}{3}$.

Problem 2. Let L be a line with equation

$$Ax + By + C = 0.$$

Let $P_1(x_1, y_1)$ be a point not on L. Find a formula for the distance from P_1 to L.

Solution. The line L' through P_1 and perpendicular to L has equation $Bx - Ay + C' = 0$, with

$$C' = Ay_1 - Bx_1. \tag{6}$$

The point $P_2(x_2, y_2)$ at which L and L' intersect has coordinates that satisfy the simultaneous equations

$$Ax_2 + By_2 + C = 0, \quad Bx_2 - Ay_2 + C' = 0.$$

If we multiply the first of these by A, the second by B, and add, we get

$$(A^2 + B^2)x_2 + AC + BC' = 0.$$

Therefore

$$x_2 = -\frac{AC + BC'}{A^2 + B^2}. \tag{7a}$$

Similarly, if the first equation (for L) is multiplied by B and the second by $-A$ and the results added, we get

$$(B^2 + A^2)y_2 + BC - AC' = 0.$$

Hence

$$y_2 = \frac{AC' - BC}{A^2 + B^2}. \tag{7b}$$

From Eqs. (6) and (7a), we get

$$x_2 = -\frac{AC + B(Ay_1 - Bx_1)}{A^2 + B^2}, \tag{8}$$

so that

$$x_1 - x_2 = \frac{(A^2 + B^2)x_1 + AC + ABy_1 - B^2x_1}{A^2 + B^2}$$

$$= \frac{A}{A^2 + B^2}(Ax_1 + By_1 + C). \tag{9a}$$

Similarly, Eqs. (6) and (7b) lead to

$$y_2 = \frac{A(Ay_1 - Bx_1) - BC}{A^2 + B^2}$$

and

$$y_1 - y_2 = \frac{(A^2 + B^2)y_1 - A^2y_1 + ABx_1 + BC}{A^2 + B^2}$$

$$= \frac{B}{A^2 + B^2}(Ax_1 + By_1 + C). \tag{9b}$$

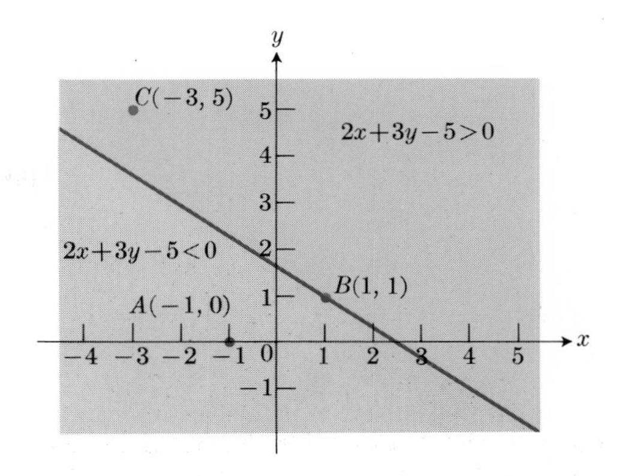

1.11 The line $2x + 3y - 5 = 0$ separates the plane into half-planes. Above L, $2x + 3y - 5$ is positive; below L, it is negative.

Note the common factor

$$h = \frac{Ax_1 + By_1 + C}{A^2 + B^2}$$

in Eqs. (9a) and (9b). From the distance formula, we get

$$d = \sqrt{(x_1 - x_2)^2 + (y_1 - y_2)^2}$$
$$= \sqrt{(Ah)^2 + (Bh)^2}$$
$$= |h|\sqrt{A^2 + B^2}.$$

The desired formula for the distance is therefore

$$d = \frac{|Ax_1 + By_1 + C|}{\sqrt{A^2 + B^2}}. \tag{10}$$

We apply this formula to find the distances from the points $A(-1, 0)$, $B(1, 1)$, and $C(-3, 5)$ to the line L: $2x + 3y - 5 = 0$ (see Fig. 1.11). The distances are

$$d_A = \frac{|2(-1) + 3(0) - 5|}{\sqrt{4 + 9}} = \frac{|-7|}{\sqrt{13}} = \frac{7}{\sqrt{13}},$$

$$d_B = \frac{|2(1) + 3(1) - 5|}{\sqrt{13}} = 0,$$

$$d_C = \frac{|2(-3) + 3(5) - 5|}{\sqrt{13}} = \frac{4}{\sqrt{13}}.$$

When the coordinates are substituted into the expression $2x + 3y - 5$, the results are -7, 0, and $+4$. These results indicate that point A is below L, B is on L, and C is above L. (See Exercise 28.) As Fig. 1.11 shows, the line $2x + 3y - 5 = 0$ separates the plane into half-planes. Above L, $2x + 3y - 5$ is positive; below L, it is negative.

EXERCISES

In Exercises 1 through 10, plot the given pair of points and find an equation for the line determined by them.

1. $(0, 0)$, $(2, 3)$
2. $(1, 1)$, $(2, 1)$
3. $(1, 1)$, $(1, 2)$
4. $(-2, 1)$, $(2, -2)$
5. $(-2, 0)$, $(-2, -2)$
6. $(1, 3)$, $(3, 1)$
7. $(a, 0)$, $(0, b)$ $(a \neq 0, b \neq 0)$
8. $(0, 0)$, $(1, 0)$
9. $(0, 0)$, $(0, 1)$
10. $(2, -1)$, $(-2, 3)$

In Exercises 11 through 20, find the slope of the given line.

11. $y = 3x + 5$
12. $2y = 3x + 5$
13. $x + y = 2$
14. $2x - y = 4$
15. $x - 2y = 4$
16. $3x + 4y = 12$
17. $4x - 3y = 12$
18. $x = 2y - 5$
19. $\dfrac{x}{a} + \dfrac{y}{b} = 1$ $(a, b \text{ constants} \neq 0)$
20. $x_1 x + y_1 y = 1$ $(x_1, y_1 \text{ constants} \neq 0)$
21. Find the line that passes through the point $(1, 2)$ and is parallel to the line $x + 2y = 3$.
22. (a) Find the line L through $A(-2, 2)$ and perpendicular to the line L': $2x + y = 4$.

    (b) Find the point B where the lines L and L' of part (a) intersect.

    (c) Using the result of part (b), find the distance from the point A to the line L' of part (a).
23. Find the line through $(1, 4)$ and having inclination angle $\phi = 60°$.
24. What is the inclination of the line $2x + y = 4$?
25. If A, B, C, C' are constants, show that

    (a) the lines
    $$Ax + By + C = 0,$$
    $$Ax + By + C' = 0$$
    are parallel, and

    (b) the lines
    $$Ax + By + C = 0,$$
    $$Bx - Ay + C' = 0$$
    are perpendicular.
26. (a) Let C and F denote, respectively, corresponding centigrade and Fahrenheit temperature readings. Given that the F vs. C graph is a straight line, find its equation from the following data: $(C = 0, F = 32)$ and $(C = 100, F = 212)$ are coordinates of two points on the graph.

    (b) Is there a temperature at which $C = F$? If so, what is it?
27. The perpendicular distance ON from the origin to line L is p, and ON makes an angle α with the positive x-axis. Show that L has equation
    $$x \cos \alpha + y \sin \alpha = p.$$
28. In connection with Problem 2 and Eq. (10), explain the rule: If $Ax_1 + By_1 + C$ agrees in sign with B, then the point $P_1(x_1, y_1)$ is above L; if the sign is opposite to the sign of B, then the point is below L. (Assume that $B \neq 0$. If $B = 0$, the line is parallel to, or coincides with, the y-axis. How should the rule be modified if $B = 0$?)
29. Use Eq. (10) to find the distances from each of the following points to the line $3y = 4x + 5$.

    (a) $(0, 0)$ (b) $(0, 4)$ (c) $(-2, 2)$ (d) $(1, 3)$

    Sketch the line and show that the given points follow the rule stated in Exercise 28.

1.6 FUNCTIONS AND GRAPHS

We have already mentioned that a *variable* is a symbol, such as x, that may take any value in some specified set of numbers. The set of numbers over which x may vary is called the *domain* of x. We also use the notation $x \in X$ to denote that x is an element of the set X. In most of our applications, the domains of our variables will be *intervals* of numbers such as the following:

1. The *open interval* consisting of all real numbers between the two fixed numbers a and b:
$$I_1 = \{x: a < x < b\}.$$

A shorthand notation for this open interval is (a, b), which is not to be confused with a point with abscissa a and ordinate b. The context will make clear whether an open interval is intended, or a point.

2. The *half-open interval* $[a, b)$ contains the left endpoint a as well as the numbers between a and b:

$$[a, b) = \{x: a \leq x < b\}.$$

3. The *half-open interval* $(a, b]$ contains the right endpoint:

$$(a, b] = \{x: a < x \leq b\}.$$

4. The *closed interval* $[a, b]$ contains both endpoints:

$$[a, b] = \{x: a \leq x \leq b\}.$$

Remark 1. We also have occasion to deal with various infinite intervals. For example, the set of all negative real numbers would be denoted by $(-\infty, 0)$. If zero is also to be included, we have the set $(-\infty, 0]$ of nonpositive real numbers. Similarly $(0, \infty)$ is used to denote the set of all positive real numbers, and $[0, \infty)$ is used to denote the set of all nonnegative reals. Neither $-\infty$ nor ∞ is a real number, and hence we do not use the square bracket ([) or (]) at the infinite end of these intervals. We shall either use R or $(-\infty, \infty)$ for the set of all real numbers.

Example. If both x and y are real variables and

$$y = \sqrt{1 - x^2},$$

then x^2 must not be greater than 1, because the square root of a negative number is imaginary. Thus the natural domain of x is the closed interval $[-1, 1]$. For

$$y = 1/\sqrt{1 - x^2},$$

the natural domain of x would be the open interval $(-1, 1)$, because we cannot divide by zero.

Function

In calculus we are most often interested in related variables. For example, the distance an object moves is related to its speed, a person's weight is related to his age, the amount of postage required to mail a letter is related to its weight, and so on. There is an especially important kind of relation between variables which characterizes a *function.* The key idea of a function is that as soon as we know the value of the first variable, the corresponding value of the second variable is determined.

Definition. *Let X and Y be nonempty sets. Let f be a collection of ordered pairs (x, y) with $x \in X$ and $y \in Y$. Then f is a function from X to Y if to every $x \in X$ there is assigned a unique $y \in Y$.*

Example 1. If

$$y = \tfrac{1}{2}x - 2,$$
$$X = R, \quad Y = R,$$

then to each x there corresponds exactly one y (Fig. 1.12).

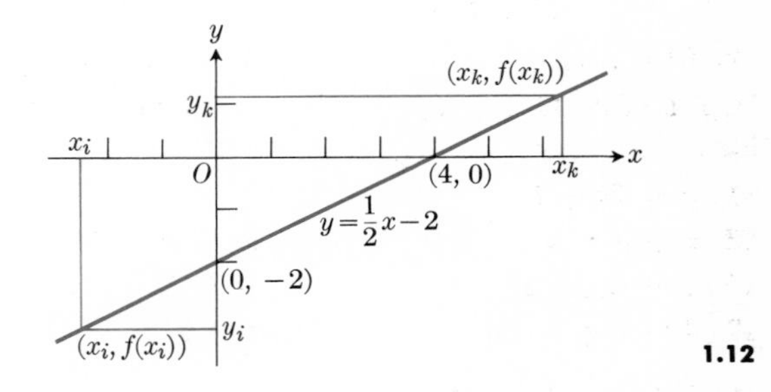

1.12

The collection of all first elements x of the pairs (x, y) in f is called the *domain* of f, and will be denoted by D_f. The *range* of f, denoted by R_f, is the set of all second elements y of the pairs (x, y) in f. The definition states that every x in X is the first entry of exactly one pair (x, y) in f.

Example 2. Let X be the set of all triangles and let Y be the set of all real numbers. Let f be the set of ordered pairs (x, y), where

x is any triangle, and

y is the area enclosed by that triangle, measured in appropriate units.

To each triangle x there corresponds a unique real number y that represents its area; so f is a function. The domain of f is all of X; the range consists of all positive real numbers.

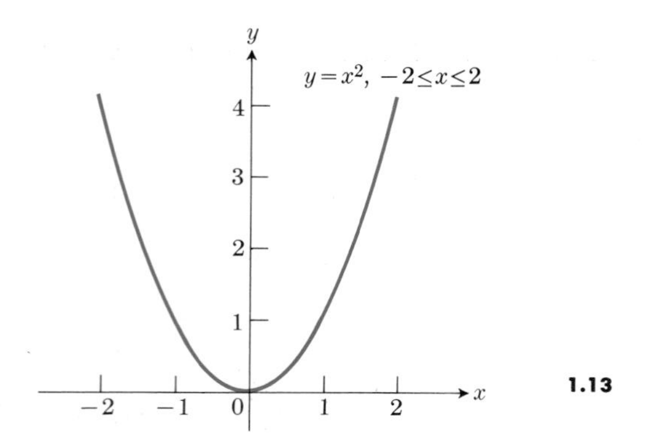

1.13

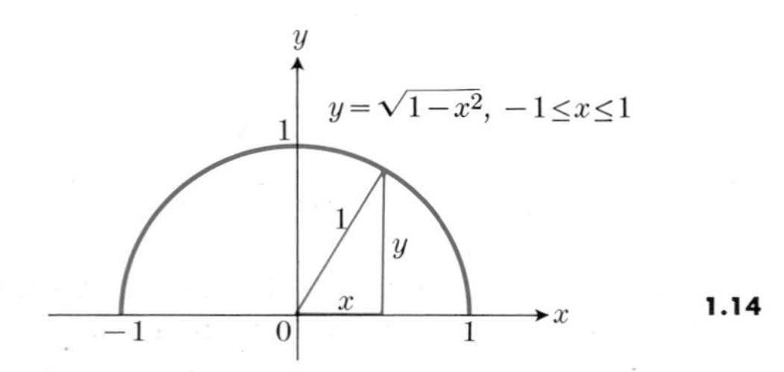

1.14

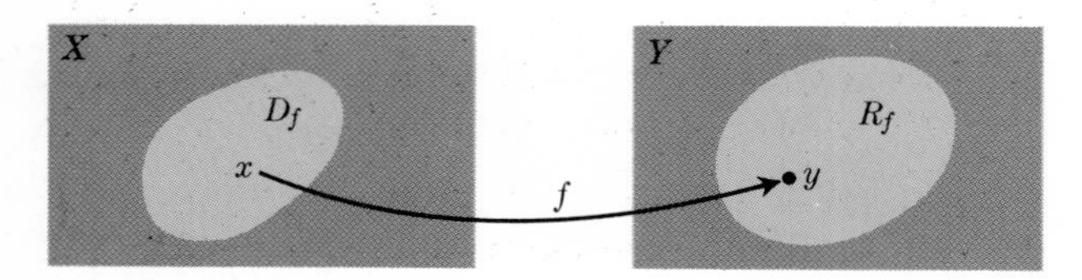

1.15 A function f maps the domain D_f onto the range R_f. The image of x is $y = f(x)$.

Example 3. Let X be the closed interval $[-2, 2]$. To each $x \in X$ assign the number x^2. This describes the function

$$f = \{(x, y): -2 \le x \le 2,\ y = x^2\}$$

with domain $D_f = [-2, 2]$ and range $R_f = [0, 4]$. The *graph* of the function is the set of points which correspond to the members of the function (see Fig. 1.13).

A function is often denoted by a single letter, say f or g, or F, or ϕ. Then if a is a particular number in the domain of f, and b is the value of y associated with a, we denote the fact that (a, b) is one of the pairs belonging to f by writing $b = f(a)$, and say that b is the *value* of the function f at a.

Example 4. Suppose with each number x in the closed interval $[-1, 1]$ we associate the number $y = \sqrt{1 - x^2}$. The domain is specified, and we have a rule which yields a unique number y when applied to any x in the domain. We imagine the set f of all possible ordered pairs of numbers of the form (x, y), with $-1 \le x \le 1$ and $y = \sqrt{1 - x^2}$. The value of this function at 0 is $f(0) = \sqrt{1 - 0^2} = 1$. Its value at $\frac{3}{5}$ is

$$f(\tfrac{3}{5}) = \sqrt{1 - (\tfrac{3}{5})^2} = \tfrac{4}{5},$$

and so on. It is not difficult to see that as x takes on all values from -1 to 0 inclusive, $y = \sqrt{1 - x^2}$ takes on all values from 0 to 1, and these are repeated in reverse order as x takes on the remaining values in its domain, namely from 0 to $+1$. The graph is the semicircle shown in Fig. 1.14, and the range of f is $[0, 1]$.

Definition. Independent and Dependent Variables. *The first variable x in the ordered pair (x, y) is often called the independent variable, or argument, of the function f. The second variable y is called the dependent variable.*

Mapping

We also speak of a function f from X to Y as a *mapping* that assigns to any element x in its domain a *unique* element y in its range such that the pair (x, y) belongs to f. This unique value of y that is thus associated with the given value of x is also expressed as $y = f(x)$, read "y equals f of x," or "y is the value of f at x." We say that f maps x onto $f(x)$, that it maps the domain D_f onto the range R_f, and that the image of x is $y = f(x)$ (see Fig. 1.15).

Definition. *A function f whose domain and range are sets of real numbers is said to be a real-valued function of a real variable.*

For the first seven chapters of this book, we shall usually mean a real-valued function of a real variable when we say merely *function*. However, we shall treat more general functions in later chapters.

Example 5. Let X and Y be the set of all real numbers. For each x in X let $f(x) = x^2$. Then

$$f = \{(x, y) : y = x^2, -\infty < x < \infty\}.$$

The domain of f is all of X, the range is the set of nonnegative real numbers. There are several ways of representing f:

1. By a table of corresponding values of x and x^2. This would be only a portion of f because such a table could not list all real values of x and their squares.

2. By corresponding number scales, as on a slide rule (Fig. 1.16a). This, too, is incomplete.

3. By the simple formula $f(x) = x^2$, which says, in effect, "choose any number x and square it." Thus

$$f(3) = 3^2 = 9, \quad f(-2) = (-2)^2 = 4,$$
$$f(a + h) = (a + h)^2 = a^2 + 2ah + h^2.$$

When the domain is understood to be the largest set of real numbers for which the formula makes sense, we shall often resort to such a shortened version as "the function $f(x) = x^2$" in place of a more exact statement in terms of ordered pairs.

4. By a graph of the equation $y = x^2$. This is a curve (Fig. 1.16b) consisting of points with coordinates (x, x^2). The domain of f is represented by the x-axis, the range by the nonnegative part of the y-axis. For any point P with abscissa x in the domain, we find the image by following the arrows, first from P up to $Q(x, x^2)$ on the curve, and then from Q over to the point R with ordinate x^2 on the y-axis.

Another notation, $f: x \to x^2$, is frequently used to represent the function f which maps x onto x^2. This notation is read "f sends x into x^2." It emphasizes the action of mapping, and, unless something is said to indicate otherwise, it is understood that the domain is the set of all real numbers to which the mapping can be applied. Thus, if

$$f: x \to x^2 + 2x - 3,$$

then the domain is

$$-\infty < x < \infty;$$

but if

$$f: x \to 1/x,$$

then the domain excludes 0.

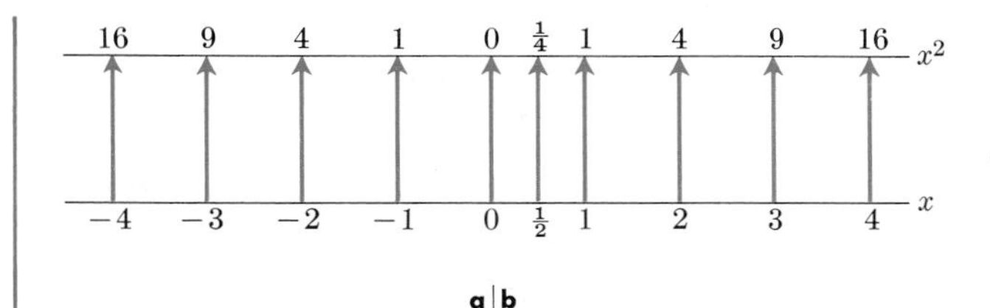

a | b

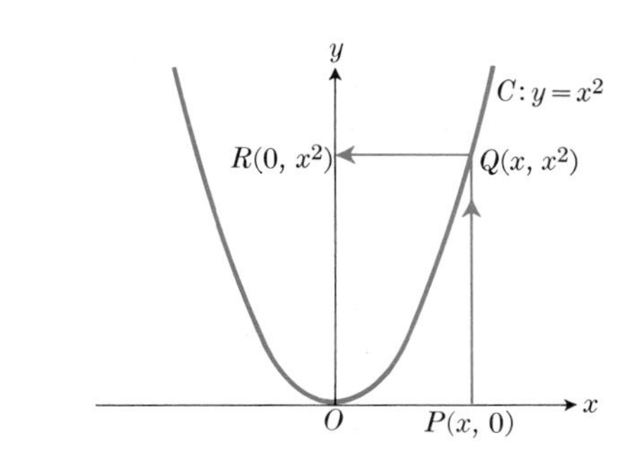

1.16

a | To each real number x on the lower scale there corresponds the number x^2 on the upper scale.

b | The path from P to Q to R gives the mapping $f: x \to x^2$.

It may happen that there is no formula to express the value of $f(x)$ in terms of algebraic operations on x. Sometimes special notation is invented, as in the next example.

Example 6. The *greatest-integer function* maps any real number x onto that unique integer which is the largest among all integers that are less than or equal to x. This, the *greatest integer in* x, is represented by putting x in square brackets:

$$[x] = \text{greatest integer that is} \leq x.$$

With this convention, we can represent the greatest-integer *function* as the set of ordered pairs

$$f = \{(x, [x]) : -\infty < x < \infty\}$$

or as the mapping

$$f: x \to [x].$$

A part of the graph of $y = [x]$ is shown in Fig. 1.17. It resembles a set of stairsteps without any risers, where the tread of each step starts at a point (n, n) with n an

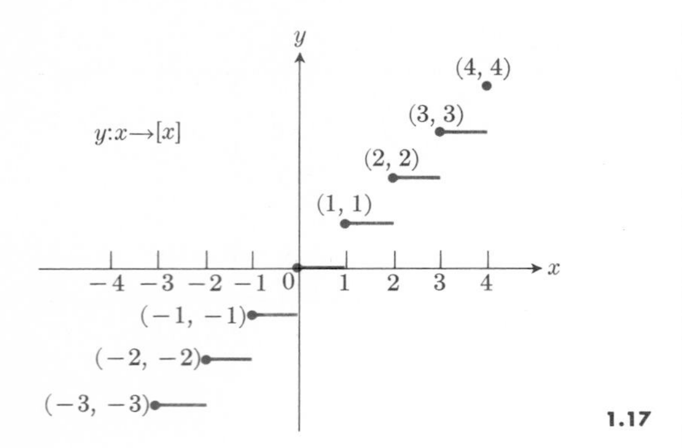

1.17

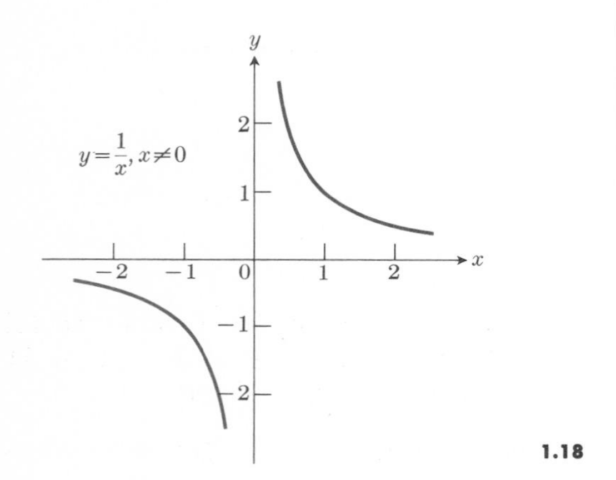

1.18

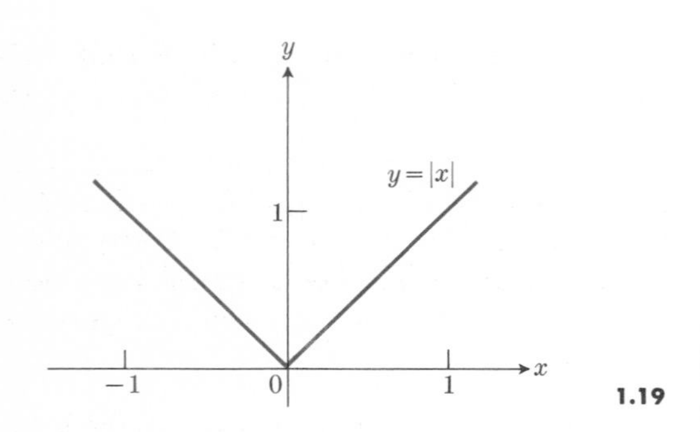

1.19

integer. For $x \in [n, n+1)$, we have $[x] = n$. Thus,

$$[-2] = -2, \quad [-1.8] = -2,$$

$$[0.2] = 0, \quad [3.99] = 3.$$

Each tread is a line segment one unit long, closed at its left end and open at its right end.

It is clear from the graph why such a function is called a "step function." There are many examples of step functions in everyday life. The following are typical:

1. The cash value y dollars of a \$100 Series E savings bond x years after the date of issue, $0 \leq x \leq 20$.
2. The postage y cents on a first-class letter weighing x ounces.
3. The toll charge y dollars on a long distance telephone call from Boston to Chicago lasting x minutes.

Such functions exhibit points of *discontinuity*, where the function suddenly jumps from one value to another without taking on any of the intermediate values. At $x = 2$ in Fig. 1.17, for example, as x approaches 2 from the left, y jumps from 1 to 2 without taking any of the intervening values.

Example 7. With each number $x \neq 0$ associate the number $y = 1/x$ (the reciprocal of x). This describes a function consisting of all pairs of numbers of the form $(x, 1/x)$, $x \neq 0$. Thus $(2, \frac{1}{2})$, $(5, \frac{1}{5})$, $(-\sqrt{2}, -1/\sqrt{2})$ are specific ordered pairs of numbers which belong to this function. The *image* of any number in the domain of the function is its reciprocal. The set of *all* these images, the *range*, is the set of all numbers different from zero, the same as the domain. Note that this function maps the finite interval $(0, 1]$ one-to-one onto the infinite interval $[1, \infty)$, which shows that there are as many points in the finite interval as there are in the infinite interval. The graph is shown in Fig. 1.18.

Example 8. The function represented in Fig. 1.19 is called the "absolute value of x." Its *domain* is the set of all real numbers x. To the number x it assigns the nonnegative number $\sqrt{x^2}$, which is represented by the symbol $|x|$. Therefore the value of the function at x is

$$|x| = \sqrt{x^2} = \begin{cases} x, & \text{if } x \geq 0, \\ -x, & \text{if } x < 0. \end{cases}$$

It maps every positive real number onto itself, and every negative real number onto its opposite, which is the corresponding positive number. Thus $|x|$ is never negative; it is like an electric current rectifier that converts either positive or negative current into positive.

Absolute value

The absolute value function discussed in Example 8 may also be given a geometrical interpretation as follows: $|x|$ measures the distance from the origin O to the point P that represents the number x on the scale of real numbers, regardless of whether x is positive or negative. Thus, in Fig. 1.20(a), x_1 is positive, P_1 is to the right of O, and the distance OP_1 is $|x_1|$; x_2 is negative, P_2 is to the left of O, and the distance P_2O is $|x_2|$.

If we want to say that a number x lies between -1 and $+1$, we can use the alternative expression that x is less than one unit away from zero; that is, $|x| < 1$. In other words, the statements $|x| < 1$ and $-1 < x < +1$ are identical.

If we wish to specify that the variable x is to be within 3 units, say, of a given number a (Fig. 1.20b), we may write

$$|x - a| < 3.$$

Since $|x - a|$ measures the distance between a and x, the inequality which says that this distance is less than 3 says, in effect, that x is between $a - 3$ and $a + 3$.

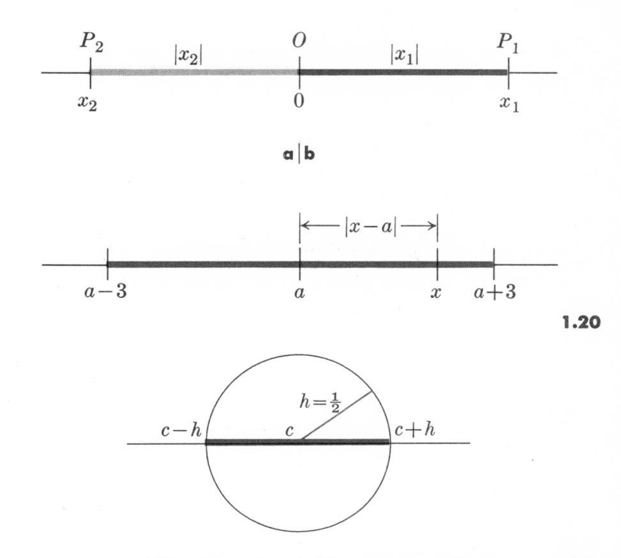

1.21 The neighborhood $N_h(c)$, centered at c, with radius h.

Neighborhood and deleted neighborhood

In the following chapter on limits, and in more advanced mathematics (in topology, for example), it is useful to be able to describe a set of numbers that are close to a fixed number c (*neighborhood*) and to describe such a set with c itself excluded (*deleted neighborhood*). To require x to be close to c is the same as requiring that $|x - c|$ be small. If x is not to be equal to c, then $|x - c| > 0$. Thus a *symmetric neighborhood* of c can be described by $|x - c| < h$ for some positive number h. To *delete c* from a symmetric neighborhood, we add the restriction that $|x - c| > 0$, and get $0 < |x - c| < h$.

Definition. *The symmetric neighborhood $N_h(c)$ of a point c is the open interval $(c - h, c + h)$, where h may be any positive number.*

For example, $N_h(c) = N_{1/2}(1)$ is the interval lying inside the circle with radius $h = \frac{1}{2}$ and centered at 1. We also say that $\frac{1}{2}$ is the *radius* of the interval (see Fig. 1.21).

A more general, possibly nonsymmetric, neighborhood of c is an open interval $(c - h, c + k)$ where h and k are any positive numbers.

Definition. *An open interval that contains c is called a neighborhood of c.*

Definition. *A neighborhood of c from which c itself has been removed is called a deleted neighborhood of c.*

For example, the open interval $(-1, 5)$ is a neighborhood of 2, of $-\frac{1}{5}$, of 0, of 4.9, and of every real number between -1 and 5. If any one of those numbers is removed, then the resulting union of two open intervals becomes a deleted neighborhood for that number.

Remark 1. If we wish to investigate the behavior of a quotient such as $(t^2 - 9)/(t + 3)$ for values of t near -3, then we restrict t to some deleted neighborhood of -3. We can't set t equal to -3 because $3^2 - 9 = 0$ and $-3 + 3 = 0$, and $0/0$ is meaningless. But if we restrict t to a small deleted neighborhood of -3, then $t + 3 \neq 0$ and the given quotient is exactly the same as $t - 3$ throughout that deleted neighborhood. Since $t - 3$ is close to -6 when t is close to -3, the given quotient must be very nearly equal to -6 in a sufficiently small deleted neighborhood of -3. This is the kind of idea that we shall investigate more thoroughly in the chapter on limits.

Remark 2. If h_1, h_2, k_1, and k_2 are all positive, and h is the minimum of h_1 and h_2, and k is the minimum of k_1 and k_2, then the intersection of the neighborhoods

$$(c - h_1, c + k_1) \quad \text{and} \quad (c - h_2, c + k_2)$$

is the neighborhood

$$(c - h, c + k).$$

Therefore the intersection of two neighborhoods of c is a neighborhood of c. Further, if we delete c from each of the original neighborhoods and from their intersection, then we get the deleted neighborhood

$$(c - h, c) \cup (c, c + k).$$

Thus the intersection of two neighborhoods of c is a neighborhood of c, and the intersection of two deleted neighborhoods of c is a deleted neighborhood of c.

For example, let $c = 2$, and consider the neighborhoods $A = (0, 5)$ and $B = (1, 7)$. If x is in both A and B, then it must satisfy $0 < x < 5$ and $1 < x < 7$. The requirements are met if and only if $1 < x < 5$. Therefore $A \cap B = (1, 5)$. If we delete 2 from both A and B, then we get the deleted neighborhoods

$$A^- = (0, 2) \cup (2, 5) \quad \text{and} \quad B^- = (1, 2) \cup (2, 7).$$

The intersection of these two sets is

$$A^- \cap B^- = (1, 2) \cup (2, 5)$$

which is identical with what we get by deleting $c = 2$ from $A \cap B$. In this example, $c = 2$, $c - h_1 = 0$, and $c + k_1 = 5$, and so $h_1 = 2$ and $k_1 = 3$. Also $c - h_2 = 1$ and $c + k_2 = 7$, and so $h_2 = 1$ and $k_2 = 5$. The minimum of 2 and 1 is $h = 1$, and the minimum of 3 and 5 is $k = 3$. Therefore, in the terminology of this remark, $(c - h, c + k) = (1, 5)$. But it is much easier to find the intersection using

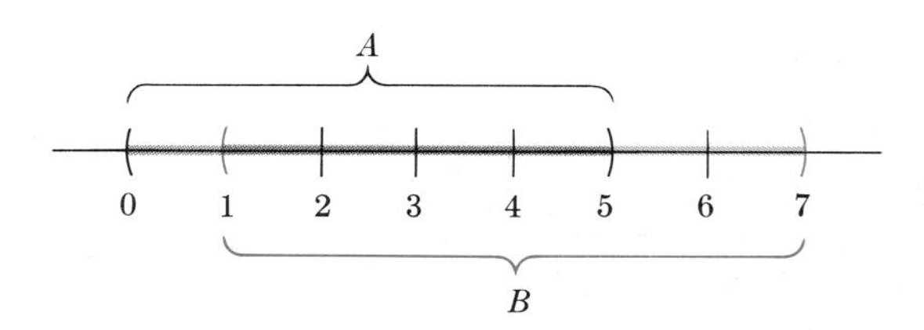

1.22. $A = (0, 5)$, $B = (1, 7)$, $A \cap B = (0, 5)$.

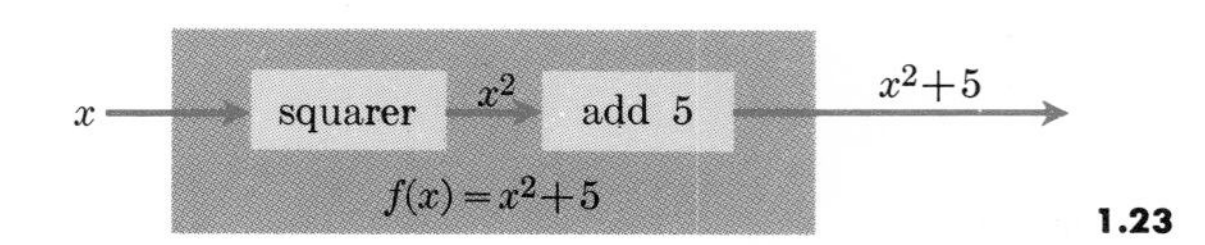

1.23

the inequalities as we did above, or by looking at the intervals A and B and taking the part that is in both of them. Figure 1.22 shows two neighborhoods, $A = (0, 5)$ and $B = (1, 7)$; one can easily see from the figure that $A \cap B = (1, 5)$.

Function machine

A function is determined by the *domain* and by any rule that tells what image in the range is to be associated with each element of the domain. For, once the domain and the rule are given, the set of all ordered pairs (x, y) can, at least in theory, be computed by a machine into which the elements of the domain are fed. The machine computes the value of y for the input x. Thus if the domain is the set of all real numbers and if the function associates the number

$$y = f(x) = x^2 + 5$$

with the number x, we think of a machine so constructed that when a real number x is fed into it, the machine produces $x^2 + 5$ as the result (Fig. 1.23). Whatever argument is substituted for x, the machine squares it and adds 5. For instance,

$$f(2) = 2^2 + 5 = 9,$$

and

$$\begin{aligned} f(a + 3) &= (a + 3)^2 + 5 \\ &= a^2 + 6a + 14. \end{aligned}$$

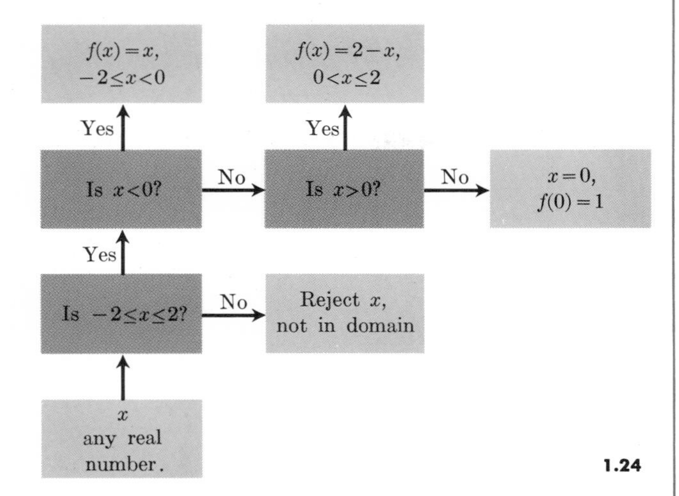

1.24

Remark 3. Another way that a machine can be used to give the value of a function at a particular x is to store in the machine's "memory" a complete table of the pairs (x, y) that constitute the function. Then tell the machine the value of x and ask it to produce the corresponding y. Or, instead of a complete table, we may store in the machine a partial set of function-values, and compute others from these by interpolation. In practice, the memory of a computer has limited capacity. Hence it is better to compute the function-values as needed, rather than storing them. But whether they are computed or stored in the form of a table, what is important is that there be just one "answer" y corresponding to any given x from the domain of the function. And when the values are given by a table, we may still say that there is a "rule" for determining the value of the function at a given x—the rule being, "look it up in the table." This is just what we do in evaluating logarithms, for example, though we shall later learn how logarithms are computed from series (Chapter 16).

Remark 4. The "function-as-a-machine" idea can be applied to functions defined by rules that are not simple formulas. For example, suppose f is a function whose domain is the interval $[-2, 2]$ with values given by the rule

$$f(x) = \begin{cases} x & \text{when } -2 \le x < 0, & \text{(5a)} \\ 1 & \text{when } x = 0, & \text{(5b)} \\ 2 - x & \text{when } 0 < x \le 2. & \text{(5c)} \end{cases}$$

A computer could be programmed to compute the values of $f(x)$, according to the flow diagram of Fig. 1.24. Start at the base and follow the arrows upward. The three blocks containing question marks are decision makers. They ask whether x is in the domain, and if so, whether $x < 0$ or $x = 0$ or $x > 0$. (If x is neither less than zero nor greater than zero, we conclude that $x = 0$.)

Since a function is determined by the domain and a rule, we shall often resort to such shorthand expressions as "consider the function $f(x) = x/(x - 2)$, $x \neq 2$" in place of the more precise expression "let f be a function whose domain is the set of all real numbers $x \neq 2$ and which associates with x the number $f(x) = x/(x - 2)$." Or again we may say, simply, "consider the function $y = x^2$," when we actually mean "the function defined on the domain of all real numbers which assigns the value x^2 to the function at x."

Two restrictions should always be kept in mind. First, we must **never divide by zero.** Thus if we see $y = x/(x - 2)$, we must always think "$x \neq 2$," even if this restriction is not stated explicitly. Second, we are dealing exclusively with real-valued functions. This means that the domain must be suitably restricted when we have square roots, or fourth roots, or other even roots. For example, if $y = \sqrt{4 - x^2}$, we should also think "x^2 must not be greater than 4; that is, the domain must not extend beyond the interval $-2 \le x \le +2$, or $|x| \le 2$."

Functions of more than one independent variable

Occasionally we have functions which depend upon several independent variables. For example, if r is the base radius and h is the altitude of a right circular cone (Fig. 1.25), then the volume

$$v = \frac{1}{3}\pi r^2 h$$

is uniquely determined when r and h are given definite values. The physical meaning requires that r, h, and v be positive. The set of all ordered triples (r, h, v) with

$$v = \frac{1}{3}\pi r^2 h, \quad r > 0, h > 0$$

is an example of a function of two independent variables (r and h). Its *domain* is the set of all pairs (r, h) with $r > 0$, $h > 0$. Its *range* is the set of positive numbers $v > 0$. The variables r and h are *independent* in the sense that the value assigned to either of them need not depend upon the other.

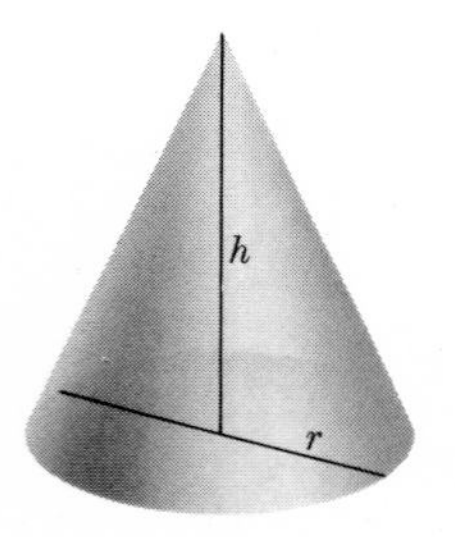

1.25

But the value of v is fixed by the other two variables, so we call v the *dependent* variable. It is also customary to be briefer, and somewhat less precise, and to say that "the volume of a right circular cone is a function of the radius and altitude," or, still more briefly, "v is a function of r and h." This means two things:

(a) r and h can be assigned values independently, and
(b) when r and h are given, v is uniquely determined.

Similarly, the set of all ordered triples (r, h, s) with

$$s = \pi r\sqrt{r^2 + h^2}, \quad r > 0, h > 0,$$

is another example of a function of the same two independent variables r and h. Here s is the lateral surface area of the cone. And again we might say, "the lateral surface area of a right circular cone is a function of the radius and altitude," or "s is a function of r and h."

More generally, suppose that some quantity y is uniquely determined by n other quantities $x_1, x_2, x_3, \ldots, x_n$. The set of all ordered $(n + 1)$-tuples $(x_1, x_2, x_3, \ldots, x_n, y)$ that can be obtained by substituting permissible values of the variables $x_1, x_2, \ldots, x_n$ and the corresponding values of y is a function whose domain is the set of all allowable n-tuples $(x_1, x_2, x_3, \ldots, x_n)$ and whose range is the set of all possible values of y corresponding to this domain. If values can be assigned independently to each of the x's, we call them independent variables and say that y is a function of the x's. We also write

$$y = f(x_1, x_2, x_3, \ldots, x_n)$$

to indicate that y is a function of the n x's, just as we write

$$y = f(x)$$

to indicate that y is a function of one independent variable x.

Our primary concern in this book will be with functions of a single independent variable.

Example 9. *Signum function.* As a final illustration, we define the signum function, sgn x. It derives its name from the Latin word for "sign." Its domain is $(-\infty, \infty)$, and the rule for assigning values is

$$\operatorname{sgn} x = \begin{cases} -1 & \text{if } x < 0, \\ 0 & \text{if } x = 0, \\ +1 & \text{if } x > 0. \end{cases}$$

Therefore the range is $\{-1, 0, 1\}$. In Exercise 13, you are asked to sketch the graph of this function.

EXERCISES

Solve each of the following equations for x in terms of y. On the basis of the assumption that x and y are real variables, discuss the possible sets of values of x and of y: the *domain* and the *range* of the functions defined by the formulas which here give y in terms of x.

1. $y = x^2$
2. $y = \dfrac{x+1}{x-1}$
3. $y = \dfrac{x^2}{x^2+1}$
4. $y = \sqrt{\dfrac{x}{x+1}}$
5. $y = x - \dfrac{1}{x}$

6. Separate the equation $x^2 + xy + y^2 = 3$ into two equations, each of which determines y as a function of x.

7. Given the function $f(x) = x^2 + (1/x)$; find $f(2)$, $f(-1)$, $f(1/x)$, $f(x + \Delta x)$.

8. Make a table of values corresponding to $x = 0, \frac{1}{2}, 1, \frac{3}{2}, 2$, and sketch a graph to represent the function

$$y = \begin{cases} x, & \text{when } 0 \le x \le 1, \\ 2 - x, & \text{when } 1 < x \le 2. \end{cases}$$

9. Describe the domain of the variable x without the use of the "absolute value" symbol in each of the following cases.

   (a) $|x| < 2$ (b) $|x| \ge 2$ (c) $|x - 1| \le 3$
   (d) $|x - 2| < 2$, where x must simultaneously satisfy the additional restriction $|x - 3| \ge 1$

10. Sketch a graph to represent the function $y = |4 - x^2|$ for $-3 \le x \le 3$.

11. When does $|1 - x|$ equal $1 - x$ and when does it equal $x - 1$?

12. Sketch a graph of the function $y = x - [x]$ for $-3 \le x \le 3$, where $[x]$ denotes the greatest integer in x.

13. (a) Sketch the graph of the signum function of Example 9.

   (b) Sketch the graph of $y = \text{sgn}\,(x^2 - 4)$ for $|x| < 4$.

Definition. Equality of Functions. *Two functions f and g are said to be equal if and only if they have the same domain* $D_f = D_g$, *and* $f(x) = g(x)$ *for all x in the domain.*

14. Use the foregoing definition of equality to determine which of the following functions are equal to the function

$$f: x \to \frac{x - 2x^2}{x}.$$

   (Use the natural domains in each instance.)

   (a) $g: x \to 1 - 2x$,
   (b) $h: x \to (x^2 - 2x^3)/x^2$,
   (c) $F: x \to \sqrt{1 - 4x + 4x^2}$,
   (d) $w: x \to \dfrac{(x^3 + x)(1 - 2x)}{x(1 + x^2)}$.

15. Are the three functions $f(x) = |x|$, $g(x) = \sqrt{x^2}$, and $h(x) = x \,\text{sgn}\, x$ all equal? Give reasons for your answers.

16. Sketch the graphs of the following functions to show that they are not the same.

   (a) $f_1(x) = x^2$, the natural domain,
   (b) $f_2(x) = x^2$, $x \ge 0$,
   (c) $f_3(x) = x^2$, $-2 \le x \le 2$,
   (d) $f_4(x) = x^2$, $-2 \le x \le 2$, where x is an integer.

17. State which of the following functions are equal to the function

$$r(x) = |x - 2|.$$

   Give reasons for your answers.

   (a) $s(x) = \sqrt{x^2 - 4x + 4}$
   (b) $t(x) = \sqrt{\dfrac{(x - 2)^3}{x - 2}}$
   (c) $u(x) = \dfrac{|x - 2|^2}{|x - 2|}$
   (d) $v(x) = \dfrac{|6x - 12|}{6}$
   (e) $w(x) = \left|\dfrac{x^2 + x - 6}{x + 3}\right|$

18. State which of the following functions are equal to the function

$$\phi(x) = \frac{\sin^2 x + 1}{x}.$$

   Give reasons for your answers.

   (a) $f(x) = \dfrac{\sin^2 x}{x} + \dfrac{1}{x}$
   (b) $g(x) = \dfrac{\sin^2 x}{x} + \dfrac{x - 1}{x^2 - x}$
   (c) $h(x) = \dfrac{2\sin^2 x + \cos^2 x}{x}$
   (d) $j(x) = \dfrac{2x - x\cos^2 x}{x^2}$

19. Sketch graphs of the following functions for

$$-3 \le x \le 3.$$

   (a) $h(x) = \dfrac{[x]}{x}$, $x \ne 0$ (b) $g(x) = [x] + [-x]$

20. Consider the function

$$V = f(l, w, h),$$

which gives the volume V of a rectangular solid with length l, width w, and height h.

(a) Name each independent variable.

(b) Name each dependent variable.

(c) What is the domain of each of the independent variables?

(d) What is the range of the function?

21. Let us define Δf as a function of two variables x and h such that

$$\Delta f(x, h) = f(x + h) - f(x).$$

(a) Name a sufficient condition for $\Delta f(x, h) = 0$.

(b) If $f(x) = x^2$, find

$$\Delta f(0, 0), \quad \Delta f(-2, 4), \quad \Delta f(0.0001, -0.0001).$$

(c) If $0 < h \leq 1$ and $|x| < 3$ determine the range of Δf for $f(x) = x^2$.

22. Suppose that A and B are two neighborhoods of c. Prove that $A \cup B$ is a neighborhood of c. How can you describe $A \cup B$ if $A = (a_1, a_2)$, $B = (b_1, b_2)$, $a = \min\{a_1, b_1\}$, and $b = \max\{a_2, b_2\}$?

23. Is the union of two deleted neighborhoods of c also a deleted neighborhood of c? Describe $A^- \cup B^-$ for A and B as in Exercise 22, and for A^- and B^-, the corresponding sets with c deleted.

24. Consider the infinitely many neighborhoods of 0 given by $I_1 = (-1, 1)$, $I_2 = (-\frac{1}{2}, \frac{1}{2}), \ldots, I_n = (-1/n, 1/n)$, $n = 1, 2, \ldots$ How many points are there in the intersection $\bigcap_{n=1}^{\infty} I_n$ of all these neighborhoods? Is this intersection a neighborhood of 0? Explain.

1.7 WAYS OF COMBINING FUNCTIONS

If f and g are two functions with domains D_f and D_g, then $f(x)$ and $g(x)$ exist for any x that is in both domains, and we can compute

$$f(x) + g(x), \quad f(x) - g(x), \quad \text{and} \quad f(x) \cdot g(x).$$

We can also find $f(x)/g(x)$, provided that $g(x) \neq 0$. The new functions so constructed are called the

$$\text{sum} \quad f + g: x \to f(x) + g(x), \tag{1a}$$

$$\text{product} \quad f \cdot g: x \to f(x) \cdot g(x), \tag{1b}$$

$$\text{difference} \quad f - g: x \to f(x) - g(x), \tag{1c}$$

$$\text{quotient} \quad \frac{f}{g}: x \to \frac{f(x)}{g(x)}. \tag{1d}$$

The domain of $f + g$, $f - g$, and $f \cdot g$ is the intersection of the domains of f and of g. The points at which $g(x) = 0$ must be excluded, however, to obtain the domain of f/g.

Example 1

$$f(x) = \sqrt{x}, \quad g(x) = \sqrt{1 - x}. \tag{2a}$$

The natural domains of f and g are

$$D_f = \{x : x \geq 0\}, \quad D_g = \{x : x \leq 1\}. \tag{2b}$$

The intersection of D_f and D_g is

$$D_f \cap D_g = \{x : 0 \leq x \leq 1\} = [0, 1]. \tag{3}$$

$$\text{sum} \quad f + g: x \to \sqrt{x} + \sqrt{1 - x},$$

$$\text{difference} \quad f - g: x \to \sqrt{x} - \sqrt{1 - x},$$

$$\text{product} \quad f \cdot g: x \to \sqrt{x(1 - x)},$$

$$\text{quotients} \quad \frac{f}{g}: x \to \sqrt{\frac{x}{1 - x}},$$

$$\frac{g}{f}: x \to \sqrt{\frac{1 - x}{x}}.$$

The domains of $f + g$, $f - g$, and $f \cdot g$ are all the same, $[0, 1]$. Since $g(x) = 0$ when $x = 1$, the domain of f/g is $[0, 1)$. Similarly, the domain of g/f is $(0, 1]$.

We can construct graphs of $f + g$, $f - g$, fg, f/g, or g/f from the separate graphs of f and g in the obvious way. For example, Fig. 1.26 shows a sketch of graphs of f, g, $f + g$, and $f \cdot g$.

Example 2. If $f(x) = x$ and $g(x) = 1/x$, then

$$D_f = (-\infty, \infty), \qquad D_g = \{x : x \neq 0\}.$$

The domain of $f \cdot g$ is the set of all nonzero real numbers. Hence, although $f(x) \cdot g(x) = 1$ when $x \neq 0$, we cannot say that $f \cdot g$ is the constant function 1, because the domain of that function is $(-\infty, \infty)$, and this is different from the domain of $f \cdot g$.

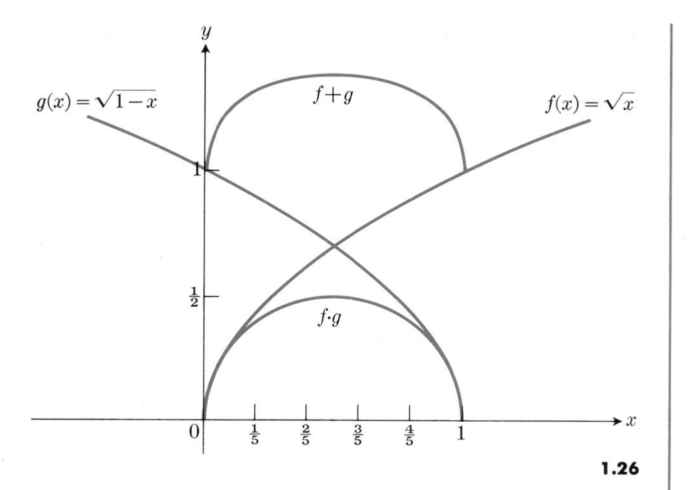

1.26

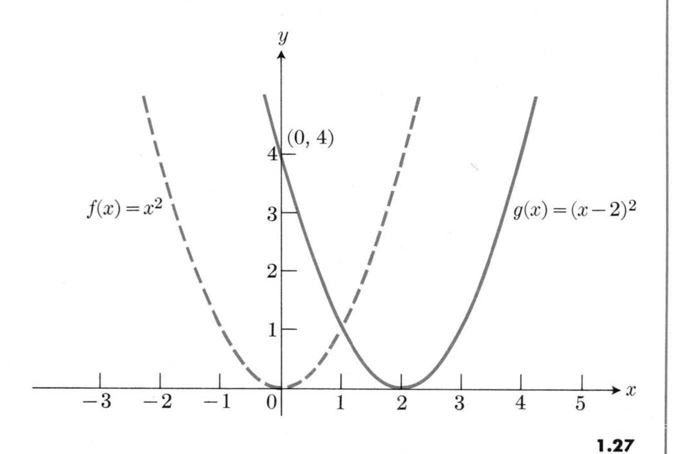

1.27

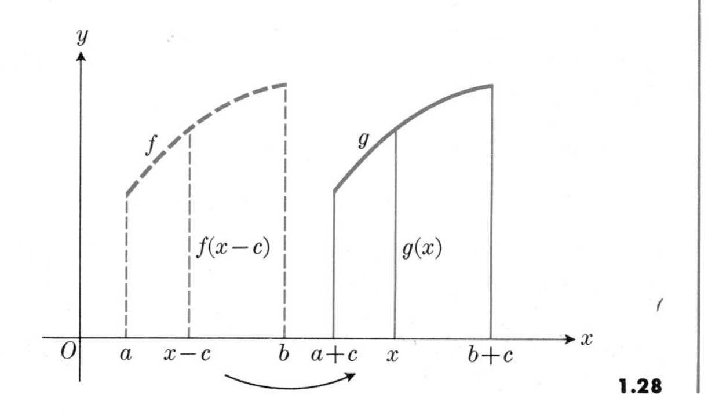

1.28

Translation and change of scale

It is well known that the graphs of

$$f(x) = x^2, \quad g(x) = (x - c)^2 \tag{4}$$

are very much alike, for any number c. Figure 1.27 shows the parabolas for the two functions f and g, with $c = 2$.

The formulas in Eq. (4) are related:

$$g(x) = f(x - c) = (x - c)^2. \tag{5}$$

If we look at the graphs in Fig. 1.27, we see that the graph of g is that of f *translated* or shifted two units to the right.

What can we say in general about the graphs of functions f and g that satisfy

$$g(x) = f(x - c) \tag{6}$$

for some number c? Assume that the domain of f is $[a, b]$. Then $f(x - c)$ is only defined when

$$a \leq x - c \leq b; \tag{7a}$$

or, adding c throughout, when

$$a + c \leq x \leq b + c. \tag{7b}$$

Thus

$$g(x) = f(x - c)$$

is defined on the domain

$$D_g = [a + c, b + c], \tag{8a}$$

while f itself has domain

$$D_f = [a, b]. \tag{8b}$$

Equations (8) show that the domain of g is the interval we get by shifting $[a, b]$ to the right c units, or to the left $|c|$ units if $c < 0$. Now take any x in the new interval $[a + c, b + c]$. Equation (6) shows that the value of g at that x is equal to the value of f at $x - c$. In other words, take the ordinate of f at $x - c$ and shift it to the right c units to get the ordinate of g at x. Figure 1.28 illustrates this situation for a positive value of c.

Problem. How is the graph of $y = |x|$ related to the graph of $y = |x - 2|$?

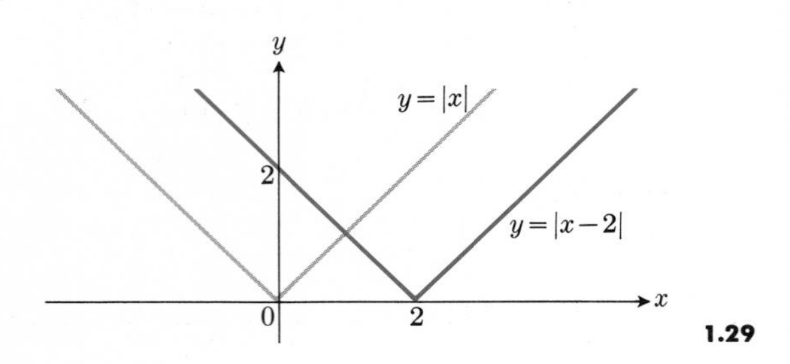

1.29

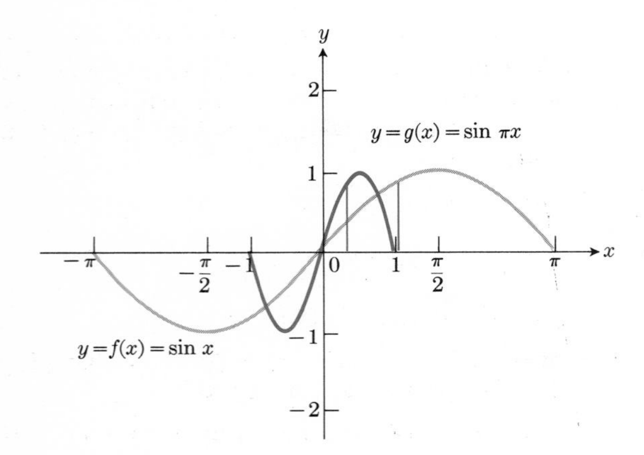

a | b

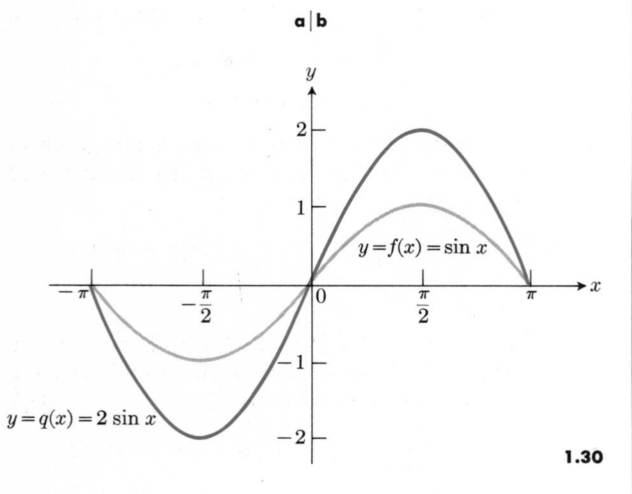

1.30

Solution. Figure 1.29 shows the two graphs. We just shift the graph of $|x|$ two units to the right to get the graph of $|x - 2|$.

A function can also be altered by a *change of scale* in the domain. This is illustrated by the example

$$\begin{aligned} f(x) &= \sin x, \\ g(x) &= \sin(\pi x). \end{aligned} \tag{9}$$

The natural domain is $(-\infty, \infty)$ for both functions f and g. But as x varies from 0 to 1, πx varies from 0 to π. Graphs of f, for $-\pi \leq x \leq \pi$, and of g, for $-1 \leq x \leq 1$, are shown in Fig. 1.30. The situation in Eq. (9) and Fig. 1.30(a) is that

$$g(x) = f(\pi x). \tag{10}$$

If we want the argument πx of f in Eq. (10) to cover the interval $[-\pi, \pi]$, which corresponds to one complete sine wave, then

$$-\pi \leq \pi x \leq \pi$$

corresponds to

$$-1 \leq x \leq 1.$$

We get the g curve in Fig. 1.30(a) by compressing the f curve horizontally in the ratio π to 1.

We can also alter a function by changing the scale in the range. Let us take, for example,

$$\begin{aligned} f(x) &= \sin x, \\ q(x) &= 2 \sin x. \end{aligned} \tag{11}$$

The graph of f and q for $-\pi \leq x \leq \pi$ is shown in Fig. 1.30(b). The range of f in Fig. 1.30(b) is $[-1, 1]$, but the range of q is $[-2, 2]$. We get the q curve by stretching the f curve vertically in the ratio of 2 to 1.

More generally, suppose k is any positive constant and that

$$g(x) = f(kx). \tag{12}$$

How is the graph of $g(x)$ related to the graph of $f(x)$? Figure 1.30(a) suggests that the answer is "compress or stretch the x-axis by shrinking (if $k > 1$) or stretching (if $k < 1$) every interval of length k on the x-axis into an interval of length 1."

Note. For example, if D_f, the domain of f, is

$$D_f = [a, b],$$

and if

$$g(x) = f(kx), \tag{13}$$

then g is defined by (13) if and only if

$$a \leq kx \leq b,$$

or, assuming $k > 0$,

$$\frac{a}{k} \leq x \leq \frac{b}{k}.$$

That is, D_g, the domain of g, is

$$D_g = \left[\frac{a}{k}, \frac{b}{k}\right], \tag{14}$$

and for any x in D_g, we find kx in D_f, and take

$$g(x) = f(kx).$$

Figure 1.30 illustrates

$$f(x) = \sin x \quad \text{and} \quad g(x) = \sin(\pi x) = f(\pi x)$$

for $k = \pi$.

Another example is illustrated by Fig. 1.31 which shows the effect for

$$f(x) = \sqrt{1 - x^2} \quad \text{and} \quad g(x) = f(2x) = \sqrt{1 - 4x^2}$$

for $k = 2$.

We can also look at the situation in Fig. 1.30(b) more broadly. Suppose k is a positive constant and that

$$q(x) = kf(x).$$

The figure suggests that we can obtain the graph of q from the graph of f in either of the following two equivalent ways:

1. Stretch the f curve vertically (if $k > 1$), or compress it (if $k < 1$).

2. Change the scale on the y-axis so that points labeled 1, 2, 3, . . . for the graph of f are relabeled k, $2k$, $3k$, . . . for the graph of q.

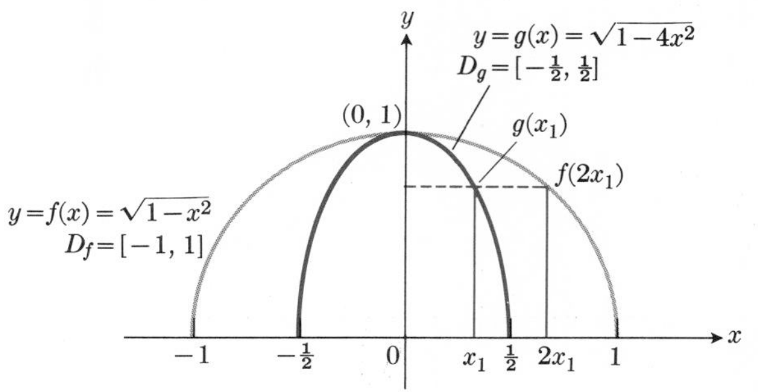

1.31

EXERCISES

For each of the following pairs of functions f and g in Exercises 1 through 3, specify the natural domains of f and g and the corresponding domains of $f + g$, $f \cdot g$, f/g, and g/f.

1. $f(x) = x$, $g(x) = \sqrt{x - 1}$
2. $f(x) = \dfrac{1}{x - 2}$, $g(x) = \dfrac{1}{\sqrt{x - 1}}$
3. $f(x) = \sqrt{x}$, $g(x) = \sqrt[4]{x + 1}$
4. Suppose $f(x) = |x|$ and $g(x) = f(x - 3)$. Graph f and g. Compare the graphs of f and g.
5. Suppose c is positive and $g(x) = f(x + c)$. If $D_f = [a, b]$, what is D_g? How can you get the graph of g from the graph of f? Illustrate with

$$f(x) = \sin x, \quad -\pi \leq x \leq \pi,$$

$$g(x) = f(x + \pi/2).$$

6. If $\phi(x) = \sin x$, $-\pi \leq x \leq \pi$, sketch the graph of each of the following and state the domain and range of each.

   (a) $p(x) = \phi\left(\dfrac{x}{2}\right)$ (b) $r(x) = \phi(3x)$

   (c) $q(x) = \phi(-x)$

7. If $f(x) = 2x^2 - 2[|x|]^2$, $-2 < x < 2$, sketch the graph of f and of each of the following and state the domain and range of each.

   (a) $g(x) = f\left(\dfrac{x}{4}\right)$ (b) $h(x) = f(\frac{3}{2}x)$

   (c) $k(x) = f(-2x)$

8. If $F(x) = 1 + 3\cos 2x$, $-\pi \leq x \leq \pi$, sketch the graph of F and of each of the following and state the domain and range of each.

   (a) $f(x) = \frac{1}{2}F(x)$ (b) $g(x) = F\left(\frac{x}{4}\right)$

   (c) $h(x) = F\left(x + \frac{\pi}{2}\right)$ (d) $H(x) = -F(x)$

9. Sketch the graph of

$$f(x) = \frac{1 + \operatorname{sgn}(x^2 - 4)}{3}, \quad -4 \leq x \leq 4,$$

   and state its range.

10. Sketch the graph of each of the following and state the domain and range of each, where $f(x) = \sqrt{1 - x^2}$ for $|x| \leq 1$.

    (a) $g(x) = \frac{3}{2}f(x)$ (b) $h(x) = f(\frac{2}{3}x)$

    (c) $F(x) = 3f(2x)$ (d) $H(x) = -f(-x)$

11. Sketch the graphs of

    (a) $f(x) = |x|$, (b) $g(x) = |2x|$,

    (c) $h(x) = |x^2 - 2x - 3|$,

    (d) $f(x) = |1 - x^2| - 2$,

    (e) $g(x) = \dfrac{2 + \operatorname{sgn}(-x)}{2}$,

    (f) $h(x) = 3 - [3x]$,

    (g) $f(x) = (\operatorname{sgn} x)(x - 1)^2 + x^2[\operatorname{sgn}(x - 1)]$.

1.8 BEHAVIOR OF FUNCTIONS

We know that the linear function

$$f(x) = 3x \tag{1}$$

has as its graph a straight line, through the origin, of slope $+3$. As x increases, the function values increase at a rate three times as fast, because a one-unit change in x produces a three-unit change in $f(x)$.

Example. If x_1 is an initial value of x, and $x_2 = x_1 + h$ is a new value obtained by increasing x by h units, then the corresponding increase in the value of the function is

$$f(x_1 + h) - f(x_1) = 3(x_1 + h) - 3x_1 = 3h, \tag{2}$$

which is exactly three times the increase in x. See Fig. 1.32.

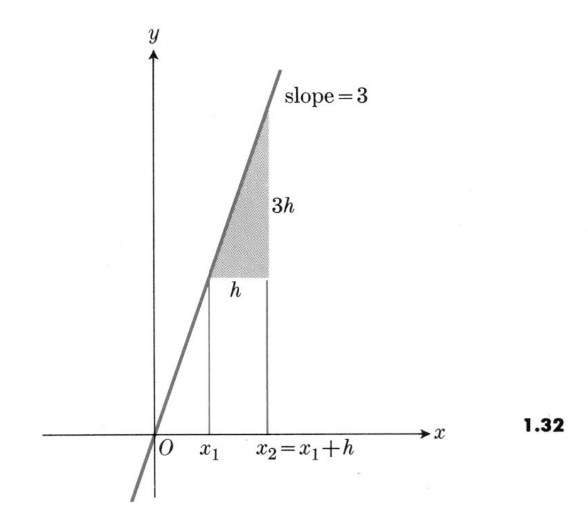

1.32

Remark 1. In Article 1.3, the notation Δx was introduced to represent the difference $x_2 - x_1$ between two values of x; that is,

$$\Delta x = x_2 - x_1. \tag{3}$$

In the preceding illustration, we located x_2, the new value of x, by increasing the initial value x_1 by h units:

$$x_2 = x_1 + h. \tag{4}$$

Note from (4) that

$$\Delta x = h = x_2 - x_1. \tag{5}$$

In other words Δx and h are different representations of the same thing, a change in the value of x. We shall use these two notations interchangeably.

Not all functions are as simple as the one just discussed. The function

$$f(x) = x^2 \tag{6}$$

increases over some intervals of x-values and decreases over others. Its rate of change, per unit change in x, is not constant. Substituting $x_1 + h$ and x_1 on both sides of Eq. (6) and subtracting, we get

$$\begin{aligned} f(x_1 + h) - f(x_1) &= (x_1 + h)^2 - x_1^2 \\ &= 2x_1h + h^2. \end{aligned} \tag{7}$$

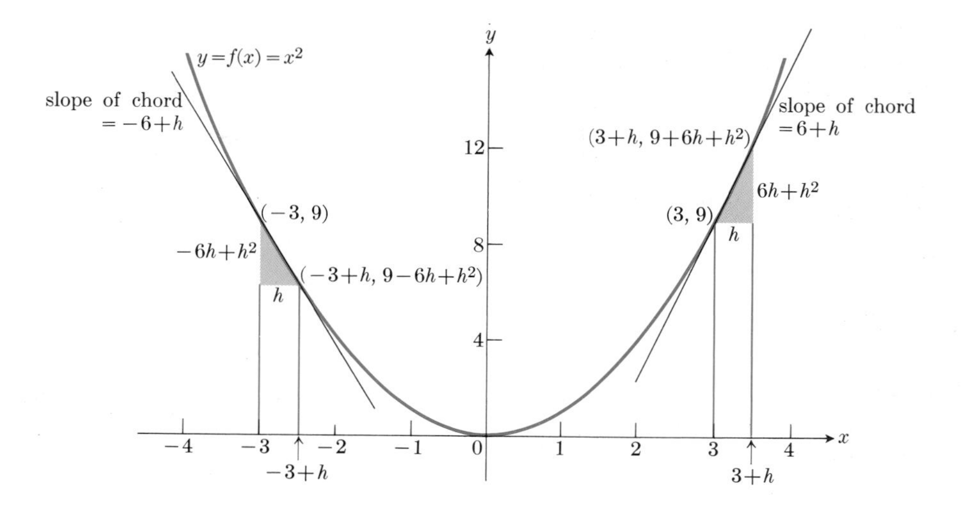

1.33

The increase in $f(x) = x^2$ depends upon x_1, the initial value of x, and also upon h, the amount of increase in x. We want to compare the increase in $f(x)$ with the increase in x. For the linear function $3x$ we saw that the increase in $f(x) = 3x$ was three times the increase in x. For the function $f(x) = x^2$, we see from Eq. (7) that we can write

$$f(x_1 + h) - f(x_1) = (2x_1 + h)h. \tag{8}$$

Here, h represents the increase in x from x_1 to $x_1 + h$. Therefore we can say, "the increase in x^2, as x increases from x_1 to $x_1 + h$, is $2x_1 + h$ times the increase in x." Alternatively we can divide the increase in $f(x)$ by the increase in x. This ratio,

$$\frac{f(x_1 + h) - f(x_1)}{(x_1 + h) - x_1} = \frac{f(x_1 + h) - f(x_1)}{h} = \frac{\text{change in } f(x)}{\text{change in } x}, \tag{9}$$

is called the *average rate of increase of $f(x)$, per unit of increase in x*, from x_1 to $x_1 + h$.

For the example $f(x) = x^2$, Eqs. (8) and (9) lead to the result

$$\frac{f(x_1 + h) - f(x_1)}{h} = 2x_1 + h \tag{10}$$

for the average rate of increase of x^2 per unit of increase in x. Thus, if $x_1 = 2$ and $h = 1$, the right side of Eq. (10) gives $2x_1 + h = 5$, so that x^2 increases five times as much as x does in going from $x_1 = 2$ to $x_1 + h = 3$. The following table shows how the average rate of increase of x^2, per unit of increase in x, varies for four different starting values, x_1, and several different values of $h > 0$.

	$2x_1 + h$			
h \ x_1	2	3	−2	−3
1	5	7	−3	−5
0.5	4.5	6.5	−3.5	−5.5
0.25	4.25	6.25	−3.75	−5.75
0.1	4.1	6.1	−3.9	−5.9
0.01	4.01	6.01	−3.99	−5.99
0.001	4.001	6.001	−3.999	−5.999

Two things are evident from the above table or from Eq. (10).

(a) If we fix x_1 and take smaller and smaller values of h, the average rate of change of x^2 per unit change in x tends to become almost equal to $2x_1$.

(b) When x_1 is negative, a small *increase* in x, from x_1 to $x_1 + h$, corresponds to a *negative increase* (that is, a *decrease*) in x^2.

Figure 1.33 illustrates some of these ideas. There, we can see that the slope of the secant line through

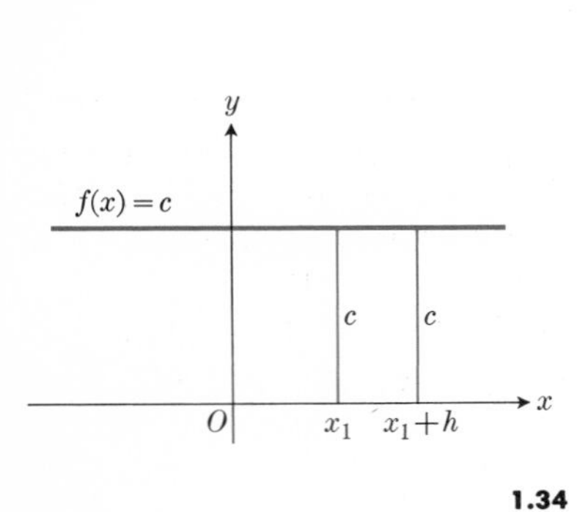

1.34

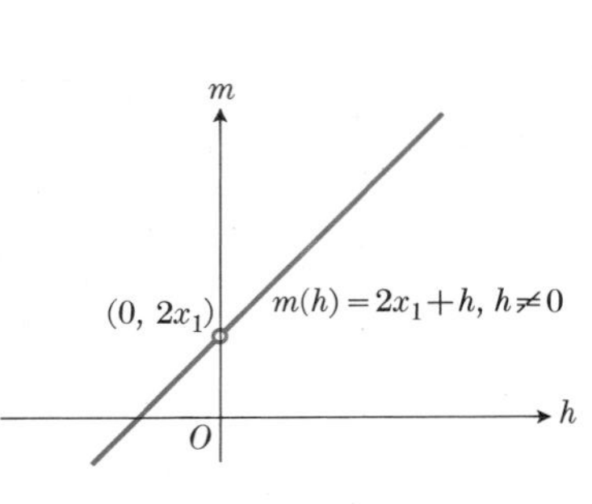

1.35

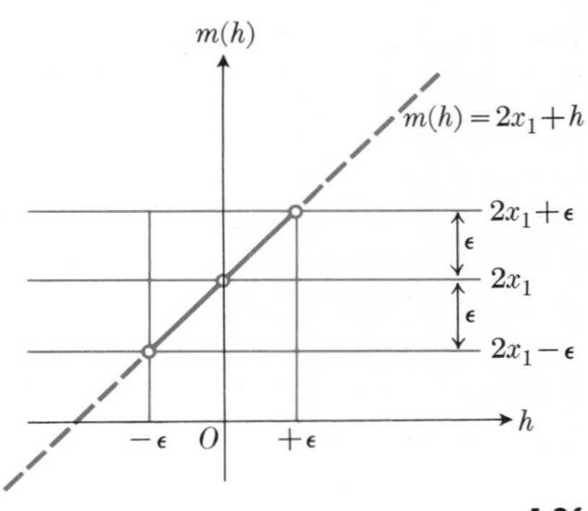

1.36

the points (x_1, x_1^2) and $((x_1 + h), (x_1 + h)^2)$ is $2x_1 + h$, for $x_1 = 3$ or -3.

For a *constant function*, such as

$$f(x) = 5,$$

of course, there is no change in the value of $f(x)$ when x increases from x_1 to $x_1 + h$, because

$$f(x_1 + h) = f(x_1) = 5.$$

For any such constant function, Eq. (9) leads to the result

$$\frac{f(x_1 + h) - f(x_1)}{h} = \frac{0}{h} = 0,$$

for any $h \neq 0$. We interpret this to say that "the average rate of increase of a constant function is zero," which is just what we would expect (see Fig. 1.34).

Remark 2. In the next chapters we shall develop systematic rules for calculating the rate of increase of a function $f(x)$ per unit change in x. *Differential calculus* (one of the two main branches of calculus—the other being integral calculus) is concerned with the *instantaneous* rate of increase, rather than with the average rate. For constant functions $f(x) = c$, and for linear functions $f(x) = mx + b$, the average rate of increase is constant and we need not distinguish between average rate and instantaneous rate. (By analogy, if an automobile is traveling at a steady rate of 45 miles per hour, its average speed and its instantaneous speed are the same.) But we saw that the square function does not have the same average rate of increase everywhere, because if $f(x) = x^2$, and $h \neq 0$, then

$$\frac{f(x_1 + h) - f(x_1)}{h} = 2x_1 + h. \tag{11}$$

Note the restriction $h \neq 0$: We cannot divide by 0 in Eq. (11). But we can consider h's in a deleted neighborhood of zero. Figure 1.35 shows a graph of the *slope function*

$$m(h) = 2x_1 + h \tag{12}$$

for a typical positive value of x_1 and h in a deleted neighborhood of zero (in other words, for h "near zero"). The point $(0, 2x_1)$, which would be on the graph if we allowed h to be equal to zero on the right side of Eq. (12), is *not included*, because h must be different from zero if we want to identify $m(h)$ from Eq. (12) with the difference quotient on the left-hand side of Eq. (11):

$$m(h) = \frac{f(x_1 + h) - f(x_1)}{h}. \tag{13}$$

But $m(h)$ is bounded between $2x_1 + 0.01$ and $2x_1 - 0.01$ when $|h|$ is, say, less than 0.01. In fact, if ϵ (Greek letter read "epsilon") is any positive number, and we choose $|h| < \epsilon$, then from Eq. (12) we see that $m(h)$ is between $2x_1 - \epsilon$ and $2x_1 + \epsilon$ because

$$-\epsilon < h < \epsilon \Rightarrow 2x_1 - \epsilon < 2x_1 + h < 2x_1 + \epsilon$$

(double arrow, $\Rightarrow$, read "implies"). We also say

that $m(h)$ is *approximated by* $2x_1$ *to within* ϵ when $0 < |h| < \epsilon$. Figure 1.36 illustrates the idea. The number $2x_1$ is called "the *limit* of $2x_1 + h$, as h approaches zero." This limit is also called the *instantaneous* rate of increase of $f(x) = x^2$, at $x = x_1$, and is indicated by the notation $f'(x_1)$, read "f prime at x_1." Formally, we make the following definition.

Definition. *We say $m(h)$ is approximated by L to within ϵ on an interval (a, b), provided*

$$L - \epsilon < m(h) < L + \epsilon$$

when

$$a < h < b.$$

Problem 1. Show $m(h) = 2h + 1$ is approximated by $L = 1$ to within $\epsilon = 0.5$ on the interval $(-\frac{1}{4}, \frac{1}{4})$.

Solution. We have

$$-\tfrac{1}{4} < h < \tfrac{1}{4}.$$

By using the properties of inequalities we get

$$2(-\tfrac{1}{4}) + 1 < 2h + 1 < 2(\tfrac{1}{4}) + 1.$$

Rewritten, we have

$$1 - 0.5 < m(h) < 1 + 0.5,$$

which demonstrates that the preceding definition applies with $L = 1$ and $\epsilon = 0.5$.

Problem 2. Find values of L and ϵ so that $m(h) = 3h + 2$ is approximated by L to within ϵ on the interval $(-0.5, 0.5)$.

Solution. A very large ϵ would always meet our conditions. We shall try to find a small ϵ. We know that

$$-0.5 < h < 0.5. \tag{14}$$

We want to find L and ϵ, so that

$$L - \epsilon < m(h) < L + \epsilon.$$

From (14) we have

$$3(-0.5) + 2 < 3h + 2 < 3(0.5) + 2,$$

or

$$2 - 1.5 < 3h + 2 < 2 + 1.5.$$

So we can conclude that if we take $\epsilon = 1.5$ and $L = 2$, all possible values of $m(h)$ over the given interval will be between $2 + 1.5$ and $2 - 1.5$.

EXERCISES

For the functions 1 through 6, write the expression for the slope function $m(h)$ as defined for f by Eq. (13). Compute $m(h)$ for the given values of x_1 and h.

1. $f(x) = 2x - 1, \quad x_1 = -1, \quad h = 4$
2. $g(x) = x^2 + x, \quad x_1 = 4, \quad h = 2$
3. $F(x) = \dfrac{1}{x}, \quad x_1 = 1, \quad h = \dfrac{1}{4}$
4. $f(x) = \dfrac{x}{x - 3}, \quad x_1 = 0, \quad h = \dfrac{1}{10}$
5. $g(x) = |x|, \quad x_1 = -1, \quad h = \frac{1}{2}$
6. $F(x) = \sin x, \quad x_1 = \dfrac{\pi}{3}, \quad h = \dfrac{\pi}{6}$

In Exercises 7 through 11, demonstrate the truth or falsity of "$m(h)$ is approximated by L to within ϵ" on the given interval (a, b) for the given values of L and ϵ. Sketch appropriate graphs, showing $m(h)$, L, $L - \epsilon$, $L + \epsilon$, and (a, b).

7. $m(h) = h^2 + 4$, $L = 4$, $\epsilon = 0.01$, $(a, b) = (-0.1, 0.1)$
8. $m(h) = 2h + 3$, $L = 3$, $\epsilon = 0.5$, $(a, b) = (-\frac{1}{4}, \frac{1}{4})$
9. $m(h) = 2h + 3$, $L = 4$, $\epsilon = 0.5$, $(a, b) = (-\frac{1}{4}, \frac{1}{4})$
10. $m(h) = 2h + 3$, $L = 3$, $\epsilon = 0.5$, $(a, b) = (-\frac{1}{2}, \frac{1}{2})$
11. $m(h) = 3 + (2/h)$, $L = 3$, $\epsilon = 1$, $(a, b) = (10, 20)$

In Exercises 12 through 16, find values of L and of ϵ so that it is true that $m(h)$ is approximated by L to within ϵ on (a, b). (Answers are not unique: One could always take ϵ very large and use many different values of L. Try to answer with small values of ϵ.)

12. $m(h) = 5 + 6h, \quad (a, b) = (-0.1, 0.1)$
13. $m(h) = 4 - 6h, \quad (a, b) = (-0.1, 0.1)$
14. $m(h) = \dfrac{3 + h}{2 - h}, \quad (a, b) = (-\frac{1}{2}, \frac{1}{2})$
15. $m(h) = (2 + h)(3 - h), \quad (a, b) = (-\frac{1}{2}, \frac{1}{2})$
16. $m(h) = (2 + h) + (5 - h^2), \quad (a, b) = (-0.1, 0.1)$

ok

1.9 SLOPE OF A CURVE

In Article 1.4, we discussed the slope of the straight line. In this article we shall define what we mean by the *slope of a curve* at a point $P(x, y)$ on the curve. This is the way in which Leibniz approached the "differential calculus."

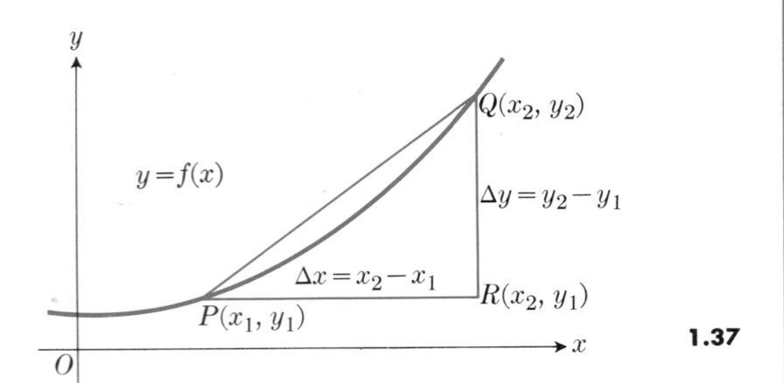

1.37

In Fig. 1.37, $P(x_1, y_1)$ is any point on the curve $y = f(x)$, and $Q(x_2, y_2)$ is another point on the same curve. Then the slope of the secant line joining the points P and Q is

$$m_{\text{sec}} = \frac{y_2 - y_1}{x_2 - x_1} = \frac{\Delta y}{\Delta x}. \tag{1}$$

Suppose now we hold P fixed and move Q along the curve toward P. As we do so, the slope of the secant line PQ will probably vary. But it may happen (and does for most curves encountered in practice) that as Q moves closer and closer to P, the slope of the secant line varies by smaller and smaller amounts and, in fact, approaches a constant limiting value. When this happens, as it does in the following example, we call this limiting value *the slope of the tangent to the curve at P* or, more briefly, *the slope of the curve at P.*

Remark. A geometric definition of tangent line can be given without reference to slope. One such definition is the following:

> Let C be a curve and P a point on C. If there exists a line L through P such that the measure of one of the angles between L and the secant line PQ approaches zero as Q approaches P along C, then L is said to be tangent to C at P.

One advantage of such a definition is that it clearly does not depend upon the particular choice of coordinate axes. It also permits the tangent line to be a vertical line (to which no real number can be assigned as slope). However, in most instances we start with a given equation, implying a choice of axes, and it is easiest to use the methods of calculus to find the slope and then write down an equation for the line tangent to the curve, using the point-slope formula

$$y - y_1 = m(x - x_1).$$

Example. Consider the curve

$$y = x^3 - 3x + 3. \tag{2}$$

If $P(x_1, y_1)$ is a point on this curve, then its coordinates must satisfy the equation

$$y_1 = x_1^3 - 3x_1 + 3. \tag{3a}$$

If $Q(x_2, y_2)$ is a second point on the curve, and if

$$\Delta x = x_2 - x_1, \quad \Delta y = y_2 - y_1,$$

then

$$x_2 = x_1 + \Delta x, \quad y_2 = y_1 + \Delta y$$

must also satisfy Eq. (2). That is,

$$\begin{aligned} y_1 + \Delta y &= (x_1 + \Delta x)^3 - 3(x_1 + \Delta x) + 3 \\ &= x_1^3 + 3x_1^2 \Delta x + 3x_1(\Delta x)^2 \\ &\quad + (\Delta x)^3 - 3x_1 - 3\,\Delta x + 3. \end{aligned} \tag{3b}$$

When we subtract the members of Eq. (3a) from Eq. (3b) we obtain

$$\Delta y = 3x_1^2 \Delta x + 3x_1(\Delta x)^2 + (\Delta x)^3 - 3\,\Delta x. \tag{4}$$

From this we readily find the slope of the secant line PQ to be

$$m_{\text{sec}} = \frac{\Delta y}{\Delta x} = 3x_1^2 + 3x_1\,\Delta x + (\Delta x)^2 - 3. \tag{5}$$

Now comes the most important step! When Q approaches P along the curve, Δx and Δy both approach zero. Thus the slope of PQ will be the ratio of two small numbers. This information is not helpful, since the ratio of two small numbers may be practically anything. But we have further information here, since we also know from Eq. (5) that

$$m_{\text{sec}} = (3x_1^2 - 3) + (3x_1 + \Delta x)\,\Delta x.$$

The right-hand side of this equation is the sum of two terms, one of which,

$$3x_1^2 - 3,$$

remains constant as Δx approaches zero, while the other,

$$(3x_1 + \Delta x)\,\Delta x,$$

becomes smaller and, in fact, itself approaches zero when Δx does. We summarize this by saying that

the limit of m_{sec} as Δx approaches zero is $3x_1^2 - 3$,

and by definition, this limit is the slope of the tangent to the curve, or the slope of the curve, at the point (x_1, y_1). Since (x_1, y_1) might have been *any* point on the curve of Eq. (2), we may delete the subscript 1 and say that

$$m = 3x^2 - 3 \tag{6}$$

gives the slope of the curve at any point $P(x, y)$ on it.

We can make use of the slope as well as the equation of the curve itself to learn more about the shape of the curve than we would be likely to discover if we did not have this additional information. For example, we see that the tangent to the curve is horizontal, that is, $m = 0$, when $x = \pm 1$. We also see that the slope is negative for x between -1 and $+1$, and that elsewhere it is positive. If we substitute values of x between -2 and $+2$, say, into Eq. (2) to find y and into Eq. (6) to find m, we obtain the following table of values.

x	y	m
-2	1	9
-1	5	0
0	3	-3
1	1	0
2	5	9

When these points are plotted and the tangent line at each point is drawn, we get a very good picture of the shape of the curve. Of course the curve itself is smooth and does not follow the tangent lines, but these lines do afford considerable assistance in sketching the curve (see Fig. 1.38).

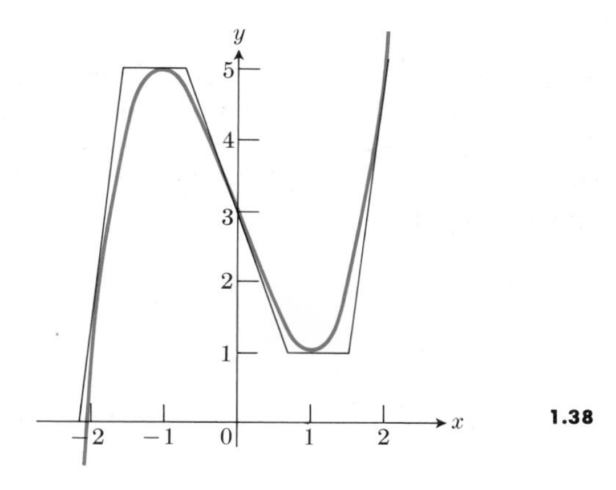

1.38

EXERCISES

Use the method illustrated in this article to find the slope of each of the following curves at a point (x, y) on the curve. Use the equation of the curve and the equation you get for its slope to make a table of values. Include in your table all points where the curve has a horizontal tangent. Sketch the curve, making use of all the information in your table of values.

1. $y = x^2 - 2x - 3$
2. $y = 2x^2 - x - 1$
3. $y = 4 - x^2$
4. $y = x^2 - 4x$
5. $y = x^2 - 4x + 4$
6. $y = x^2 + 4x + 4$
7. $y = 6 + x - x^2$
8. $y = 6 + 5x - x^2$
9. $y = x^2 + 3x + 2$
10. $y = 2 - x - x^2$
11. $y = 2x^3 + 3x^2 - 12x + 7$
12. $y = x^3 - 3x$
13. $y = x^3 - 12x$
14. $y = x^2(4x + 3) + 1$
15. $y = x^3 - 3x^2 + 4$

1.10 DERIVATIVE OF A FUNCTION

We now formulate the method for finding the slope of a curve representing the more general functional equation $y = f(x)$. Let $P(x_1, y_1)$ be a fixed point on the curve. The subscript 1 is used to emphasize that x_1 and y_1 are to be held constant throughout the following discussion. If $Q(x_1 + \Delta x, y_1 + \Delta y)$ is

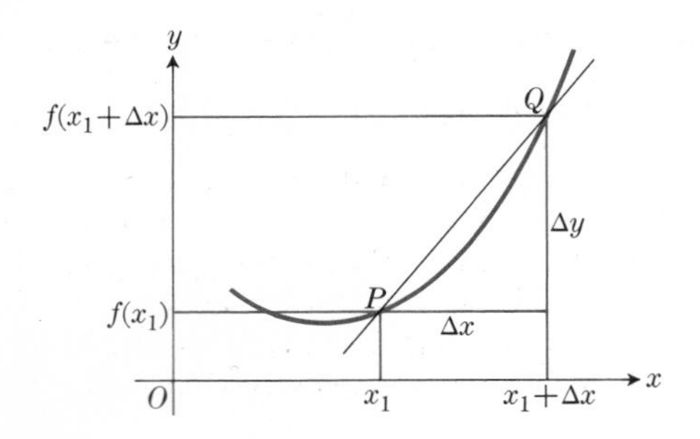

1.39. $\Delta y = f(x_1 + \Delta x) - f(x_1)$.

another point on the curve, then

$$y_1 + \Delta y = f(x_1 + \Delta x),$$

and from this we subtract

$$y_1 = f(x_1)$$

to obtain (Fig. 1.39)

$$\Delta y = f(x_1 + \Delta x) - f(x_1).$$

Then the slope of the secant line PQ is

$$m_{\text{sec}} = \frac{\Delta y}{\Delta x} = \frac{f(x_1 + \Delta x) - f(x_1)}{\Delta x}. \tag{1}$$

Let us note that the division in Eq. (1) can only be *indicated* when we are talking about a general function $f(x)$, but for any specific equation such as $f(x) = x^3 - 3x + 3$ in Eq. (2) of Article 1.9, this division of Δy by Δx is actually to be carried out [as we did in going from Eq. (4) to Eq. (5)] *before* we do the next operation.

Having performed the division indicated in Eq. (1), we now investigate what happens if we hold x_1 fixed and take Δx to be smaller and smaller, approaching zero. If m_{sec} approaches a constant value, we call this value its *limit* and define this to be the slope m_{tan} of the tangent to the curve at P. The mathematical symbols which summarize this discussion are

$$m_{\text{tan}} = \lim_{Q \to P} m_{\text{sec}} = \lim_{\Delta x \to 0} \frac{\Delta y}{\Delta x}$$
$$= \lim_{\Delta x \to 0} \frac{f(x_1 + \Delta x) - f(x_1)}{\Delta x}. \tag{2}$$

The symbol "lim" with "$\Delta x \to 0$" written beneath it is read "the limit, as Δx approaches zero, of . . ."

Chapter 2 will treat the theory of limits in greater depth. For the remainder of this chapter, the reader is encouraged to use the following intuitive idea: "L is the limit of $m(h)$ as h approaches zero" means that when h is very close to zero, but not equal to zero, then $m(h)$ is very close to L, and possibly equal to L.

The result of carrying out the operations represented by the last term in Eq. (2) produces a number which we have called m_{tan}, or the slope of the tangent to the curve $y = f(x)$ at the point $P(x_1, y_1)$. It is also customary to indicate that this number is related to the original function f by writing it as $f'(x_1)$ (read "f-prime at x_1"). Thus $f'(x_1)$ is *defined* by

$$f'(x_1) = \lim_{\Delta x \to 0} \frac{f(x_1 + \Delta x) - f(x_1)}{\Delta x}. \tag{3}$$

As mentioned above, the subscript 1 is used to emphasize the fact that x_1 is held constant while the various operations indicated on the right side of Eq. (3) are performed. The indicated limit may exist for some values of x_1 and fail to exist for other values. (We shall discuss this situation in more detail below.) At each point x_1 where the limit does exist, the function f is said to have a *derivative* (or to be *differentiable*) and the number $f'(x_1)$ is said to be the derivative of f at x_1.

The process of *finding the derivative of a function* is the fundamental operation of differential calculus. In fact we may now define *differential calculus* to be that branch of mathematics which is concerned with studying the following two general problems.

Problem 1. Given a function f, determine those values of x (in the domain of f) at which the function possesses a derivative.

Problem 2. Given a function f and an x at which the derivative exists, find $f'(x)$.

Example 1. In studying Eq. (2) of Article 1.9,

$$f(x) = x^3 - 3x + 3, \tag{4}$$

we found that for any x_1 whatever, the result of applying the operations on the right side of Eq. (3) gave us $3x_1^2 - 3$. That is, the function f in Eq. (4) possesses a derivative whose value at *any* x_1 is

$$f'(x_1) = 3x_1^2 - 3. \tag{5}$$

Thus the answer to Problem 1 above is: "The function f, Eq. (4), possesses a derivative at any x in the domain $-\infty < x < +\infty$," and the answer to Problem 2 is given by Eq. (5).

Now for most of the functions considered in this book, it will turn out that the answer to Problem 1 is that the derivative exists at all but a few values of x; that is, the places where the derivative fails to exist will be exceptional. Then the x_1 in Eq. (3) may be *any* of the nonexceptional values of x. Therefore, we might as well write x instead of x_1 in Eq. (3) if we remember that

x is to be held constant, while

Δx varies and approaches zero

in the calculation of the derivative

$$f'(x) = \lim_{\Delta x \to 0} \frac{f(x + \Delta x) - f(x)}{\Delta x}. \tag{6}$$

With this understanding, we shall henceforth omit the subscript 1 in talking about the derivative and shall use Eq. (6) to define the derivative of f, with respect to x, at any x in its domain for which the limit exists.

The derived function

Let x be a number in the domain of the function f. If the limit indicated in Eq. (6) exists, then it provides a rule for associating a number $f'(x)$ with the number x. The set of all pairs of numbers $(x, f'(x))$ that can be formed by this process is called the *derived function* f'. The *domain* of f' is a subset of the domain of f. It contains all numbers x in the domain of f such that the limit in (3) exists, but does not contain those exceptional values where the derivative fails to exist. The derived function f' is also called the *derivative* of f.

Example 2. Consider the absolute value function

$$f(x) = |x|, \quad -\infty < x < +\infty.$$

From the graph in Fig. 1.19 it is clear that the slope is $+1$ when x is positive and -1 when x is negative. But at $x_1 = 0$, the derivative fails to exist [the graph has no tangent at $(0, 0)$]. Hence the *domain* of the derived function does not include 0. It consists of all real numbers $x \neq 0$. The values of the derived function are given by the rule

$$f'(x) = \begin{cases} -1 & \text{if } x < 0, \\ +1 & \text{if } x > 0. \end{cases}$$

See Exercise 21.

Notations. In addition to $f'(x)$, various notations are used to denote the derivative of $y = f(x)$ with respect to x. The ones most commonly used are

$$y', \quad \frac{dy}{dx}, \quad D_x y.$$

The last two of these may be interpreted as

$$\frac{dy}{dx} = \frac{d}{dx}(y),$$

where $d/dx\,(\ldots)$ stands for the operation *derivative with respect to x of* (the expression following), and where

$$D_x y = D_x(y),$$

with $D_x \ldots$ meaning the same thing as $d/dx\,(\ldots)$, namely, *derivative with respect to x of* (the expression which follows it).

Problem 1. Find $f'(x)$ for the function

$$f(x) = x^2 + \frac{1}{x}, \quad x \neq 0.$$

Solution. We carry out the operations indicated on the right side of Eq. (6) in the following order.

1. $f(x + \Delta x) = (x + \Delta x)^2 + \dfrac{1}{x + \Delta x}$
$\qquad = x^2 + 2x\,\Delta x + (\Delta x)^2 + \dfrac{1}{x + \Delta x}.$

2. $f(x) = x^2 + \dfrac{1}{x}.$

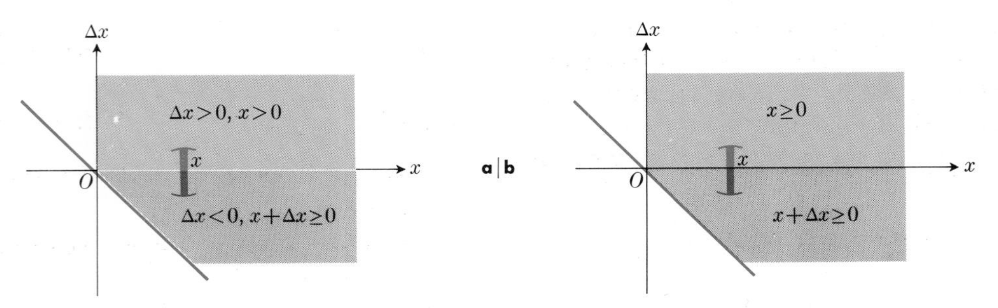

1.40 **a|** Domain of $g(x, \Delta x)$. Points on the x-axis are not included.
b| Domain of $h(x, \Delta x)$. The domain includes the positive x-axis.

Subtract (2) from (1):

$$3.\quad f(x+\Delta x) - f(x) = 2x\,\Delta x + (\Delta x)^2 + \frac{1}{x+\Delta x} - \frac{1}{x}$$

$$= 2x\,\Delta x + (\Delta x)^2 + \frac{x-(x+\Delta x)}{x(x+\Delta x)}$$

$$= \Delta x\left(2x+\Delta x - \frac{1}{x(x+\Delta x)}\right).$$

Divide by Δx:

$$4.\quad \frac{f(x+\Delta x)-f(x)}{\Delta x} = 2x+\Delta x - \frac{1}{x(x+\Delta x)}.$$

$$5.\quad f'(x) = \lim_{\Delta x\to 0}\frac{f(x+\Delta x)-f(x)}{\Delta x}$$

$$= \lim_{\Delta x\to 0}\left(2x+\Delta x-\frac{1}{x(x+\Delta x)}\right) \qquad \text{(a)}$$

$$= 2x+0-\frac{1}{x(x+0)} \qquad \text{(b)}$$

$$= 2x - \frac{1}{x^2}.$$

In a later article we shall discuss the validity of the various limit operations involved in going from (a) to (b) above. If we anticipate these results, we may say that *after* the division by Δx has been carried out and the expression has been reduced to a form [as in (a) above] which "makes sense" (that is, does not involve division by zero) when Δx is taken equal to zero, then the limit as Δx *approaches* zero does exist and may be found by simply replacing Δx by zero in this reduced form.

Remark. In Problem 2 below, we arrive at the following functions of two variables x and Δx:

$$g(x, \Delta x) = \frac{\sqrt{x+\Delta x}-\sqrt{x}}{\Delta x},$$

$$h(x, \Delta x) = \frac{1}{\sqrt{x+\Delta x}+\sqrt{x}}.$$

These algebraic expressions are equal on the intersection of their domains, but the two domains are not identical. For g, the domain is the set of points $(x, \Delta x)$ for which

$$x \geq 0,\quad \Delta x \neq 0,\quad x+\Delta x \geq 0.$$

This domain is shown in Fig. 1.40(a). The domain for h consists of the points $(x, \Delta x)$ for which

$$x \geq 0,\quad x+\Delta x \geq 0,\quad \sqrt{x}+\sqrt{x+\Delta x} \neq 0.$$

This domain is shown in the shaded portion of Fig. 1.40(b). Are there any points in D_g that are not in D_h? No. Are there any points in D_h that are not in D_g? Yes: all the points where $x > 0$ and $\Delta x = 0$. (Note that the origin $x = 0$, $\Delta x = 0$ is in neither D_g nor D_h.) Therefore

$$D_g \subset D_h,\quad D_g \cap D_h = D_g.$$

We now argue as follows: For any $x > 0$, we can put $\Delta x = 0$ in h and get

$$h(x, 0) = \frac{1}{\sqrt{x+0}+\sqrt{x}} = \frac{1}{2\sqrt{x}}.$$

Moreover it seems reasonable (and we shall prove it in Chapter 2) that $\sqrt{x+\Delta x}$ is very nearly equal to $\sqrt{x}$ when Δx is close to zero but different from zero. But when $\Delta x \neq 0$, the functions $g(x, \Delta x)$ and $h(x, \Delta x)$ are equal. Therefore we find it reasonable to expect that

$$g(x, \Delta x) \text{ is very nearly equal to } \frac{1}{2\sqrt{x}}$$

when $x > 0$ and

Δx is very nearly equal to 0 but different from 0,

because the corresponding statement for $h(x, \Delta x)$ seems reasonable. Thus, if the length of the colored segment in Fig. 1.40(a, b) is small, and the point $(x, \Delta x)$ is on that segment, and $\Delta x \neq 0$, then both g and h have values close to $1/(2\sqrt{x})$.

Problem 2. Find dy/dx if $y = \sqrt{x}$ and $x > 0$.

Solution. Let $y = \sqrt{x}$, $y + \Delta y = \sqrt{x + \Delta x}$, so that

$$\begin{aligned}\frac{\Delta y}{\Delta x} &= \frac{\sqrt{x+\Delta x} - \sqrt{x}}{\Delta x}\\ &= \frac{(\sqrt{x+\Delta x}-\sqrt{x})(\sqrt{x+\Delta x}+\sqrt{x})}{\Delta x(\sqrt{x+\Delta x}+\sqrt{x})} \qquad \text{(a)}\\ &= \frac{(x+\Delta x) - x}{\Delta x(\sqrt{x+\Delta x}+\sqrt{x})} = \frac{1}{\sqrt{x+\Delta x}+\sqrt{x}}.\end{aligned}$$

In step (a) we multiplied by a fraction equal to 1 and having a form such that it enabled us to get rid of the square roots in the numerator. Then in the next step we were able to carry out the division by Δx. Now we let Δx approach zero in the final form and have

$$\frac{dy}{dx} = \lim_{\Delta x \to 0} \frac{\Delta y}{\Delta x} = \lim_{\Delta x \to 0} \frac{1}{\sqrt{x+\Delta x}+\sqrt{x}} = \frac{1}{2\sqrt{x}}.$$

EXERCISES

For each of the following functions f, find the derivative $f'(x)$ by means of the definition in Eq. (6).

1. $f(x) = x^2$
2. $f(x) = x^3$
3. $f(x) = 2x + 3$
4. $f(x) = x^2 - x + 1$
5. $f(x) = \dfrac{1}{x}$
6. $f(x) = \dfrac{1}{x^2}$
7. $f(x) = \dfrac{1}{2x+1}$
8. $f(x) = \dfrac{x}{x+1}$
9. $f(x) = 2x^2 - x + 5$
10. $f(x) = x^3 - 12x + 11$
11. $f(x) = x^4$
12. $f(x) = ax^2 + bx + c$ (a, b, c constants)
13. $f(x) = x - \dfrac{1}{x}$
14. $f(x) = ax + \dfrac{b}{x}$ (a, b constants)
15. $f(x) = \sqrt{2x}$
16. $f(x) = \sqrt{x+1}$
17. $f(x) = \sqrt{2x+3}$
18. $f(x) = \dfrac{1}{\sqrt{x}}$
19. $f(x) = \dfrac{1}{\sqrt{2x+3}}$
20. $f(x) = \sqrt{x^2+1}$
21. In Example 2, $f(x) = |x|$. What is the *range* of the derived function f'? What is its graph? Compare the graphs of

$$y = f'(x) \quad \text{and} \quad y = |x|/x,\ x \neq 0.$$

1.11 VELOCITY AND RATES

When a body moves in a straight line, it is customary to represent the line of motion by a coordinate axis, select a reference point O on the line as origin, adopt a positive direction and a unit of distance on the line, and then describe the motion by means of an equation which gives the coordinate of the body as a function of the time t that has elapsed since the start of the motion.

Example 1. Thus for a freely falling body, the equation of motion is

$$s = \tfrac{1}{2}gt^2, \tag{1}$$

where g is the acceleration due to gravity and is approximately 32 (ft/sec^2), and s is the distance in feet that the body has fallen in t seconds from the start.

More generally, suppose the law of motion is given by a function f,

$$s = f(t), \tag{2}$$

and that we are required to find the *velocity* of the body at some instant of time t. First of all, how shall we *define* the instantaneous velocity of a moving body?

If we assume that *distance* and *time* are the fundamental physical quantities which we can measure, we may be led to reason as follows. At time t, the body is in the position

$$s = f(t), \tag{3}$$

and at time $t + \Delta t$ the body is in the position

$$s + \Delta s = f(t + \Delta t). \tag{4}$$

Hence, during the interval of time from t to $t + \Delta t$, it has undergone a displacement

$$\Delta s = f(t + \Delta t) - f(t). \tag{5}$$

The quantities in Eqs. (3), (4), and (5) are all physical quantities which can be measured (by clocks and tapelines, say).

We now define the *average velocity* to be Δs divided by Δt:

$$v_{\text{av}} = \frac{\Delta s}{\Delta t} = \frac{f(t + \Delta t) - f(t)}{\Delta t}. \tag{6}$$

Example 2. We are all familiar with applications of Eq. (6) in everyday life. For example, a sprinter runs 100 yards in 10 seconds. His average velocity is therefore

$$v_{\text{av}} = \frac{\Delta s}{\Delta t} = \frac{100 \text{ (yd)}}{10 \text{ (sec)}} = 10 \text{ (yd/sec)}.$$

To obtain the *instantaneous* velocity at time t, we take the average velocity over shorter and shorter intervals of time Δt. We are thereby led to the definition of instantaneous velocity, or what we call the velocity at time t, as the *limit* of v_{av} as Δt approaches zero; that is,

$$v = \lim_{\Delta t \to 0} \frac{\Delta s}{\Delta t} = \lim_{\Delta t \to 0} \frac{f(t + \Delta t) - f(t)}{\Delta t}. \tag{7}$$

We recognize that Eq. (7) is the same as the definition of the derivative

$$v = \frac{ds}{dt} = f'(t). \tag{8}$$

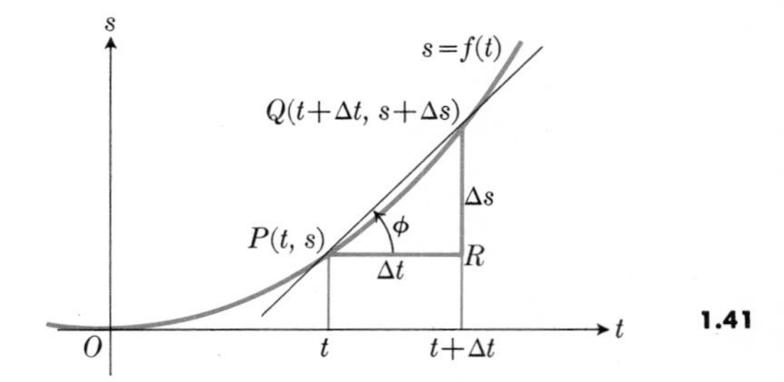

1.41

Example 3. For the motion governed by Eq. (1),

$$s = \tfrac{1}{2}gt^2,$$

we find, by the methods of the previous article, that the velocity at time t is

$$v = \frac{ds}{dt} = gt.$$

When we refer to Fig. 1.41, we see that

$$v_{\text{av}} = \frac{\Delta s}{\Delta t}$$

is the slope of the secant line PQ in the sense of measuring its rate of rise in units of s per unit of t. Here there is no reason to suppose that the unit of length used on the t-axis to represent one second is the same as the unit of length used on the s-axis to represent one foot. And certainly the geometrical inclination ϕ of the secant line PQ depends upon the scales used to represent the units of time and distance on the two axes. We can still say, however, that the average velocity equals the ratio of the opposite side to the adjacent side in the triangle PRQ, each with its proper units (and proper sign). Then the instantaneous velocity can also be interpreted geometrically as the rate of rise of the tangent line in units of s per unit of t.

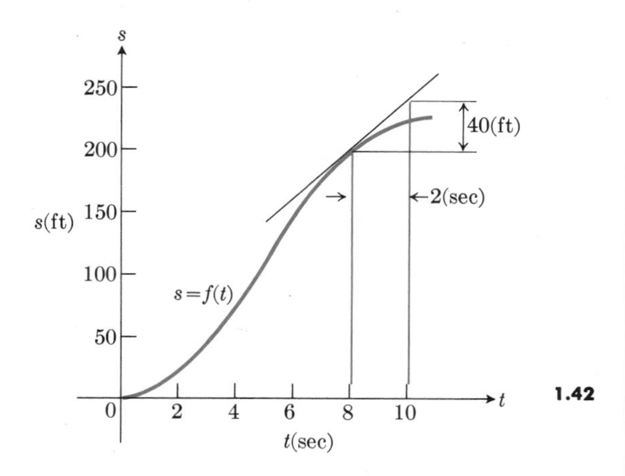

1.42

Example 4. This geometrical interpretation is sometimes used to estimate the velocity when the motion is given by means of a graph instead of by an equation $s = f(t)$. For example, in the graph of Fig. 1.42, s represents the coordinate of some point of a bicycle which a boy is riding along a straight road. At $t = 8$, the tangent line appears to be rising at the rate of 40 s-units in an interval of 2 t-units; that is, the velocity at time $t = 8$ is apparently

$$\frac{40\ \text{(ft)}}{2\ \text{(sec)}} = 20\ \text{(ft/sec)}.$$

We say the velocity is "apparently" 20 ft/sec because we cannot be sure that we have accurately constructed the tangent to the curve. Indeed, an accurate construction would require that we know the derivative $f'(t)$, exactly, at $t = 8$ and this is the very thing we are trying to estimate. On the other hand, we can make a fair approximation to the tangent line by inspection and this is what we have done in the graph.

There are many other applications of the notion of average rate and instantaneous rate. For example, the quantity of water Q (gal) in a reservoir at time t (min) is a function of t. Water may flow into or out of the reservoir. As it does so, suppose that Q changes by an amount ΔQ from time t to time $t + \Delta t$. Then the average rate of change of Q with respect to t is

$$\frac{\Delta Q}{\Delta t}\ \text{(gal/min)}$$

and the instantaneous rate of change of Q with respect to t is

$$\frac{dQ}{dt} = \lim_{\Delta t \to 0} \frac{\Delta Q}{\Delta t}\ \text{(gal/min)}.$$

Derivatives are important in economic theory, where they are usually indicated by the adjective "*marginal.*" Suppose, for example, that a manufacturer produces x tons of steel per week at a total cost of $y = f(x)$ dollars. This total cost includes such items as a proportionate part of the cost of building and maintaining the company's steel mills, salaries of executives, taxes, office maintenance, cost of raw materials, labor, etc. Suppose that in order to produce $x + \Delta x$ tons of steel weekly, it would cost $y + \Delta y$ dollars. The increase in cost per unit increase in output would be $\Delta y/\Delta x$. The limit of this ratio, as Δx tends to zero, is called the *marginal cost.* In other words, if total cost is y for weekly output x, then the *marginal* cost is just the derivative of y with respect to x. It gives the rate of increase of cost per unit increase of output from the level x.

The manufacturer is also interested in revenue and in profits. To sell his output of x tons per week he finds that he may charge a price $P = F(x)$ per ton. (Usually, at a higher price he would sell less; at a lower price he could sell more.) His *revenue* is then the product $xP = xF(x)$. The *marginal* revenue is the *derivative* of xP with respect to x. It gives the rate of increase of revenue per unit increase in output. The manufacturer's profit T is the difference between revenue and cost:

$$T = xP - y.$$

In Chapter 4 we shall see how the manufacturer should adjust his production to achieve maximum profit. The adjustment involves his *marginal* profit dT/dx, which is the rate of increase of profit per unit increase of production.*

* For a further discussion of applications to economics, see R. G. D. Allen, *Mathematical Analysis for Economists*, Macmillan Co., New York, 1939. In particular, marginal cost and marginal revenue are discussed on pp. 152 ff.

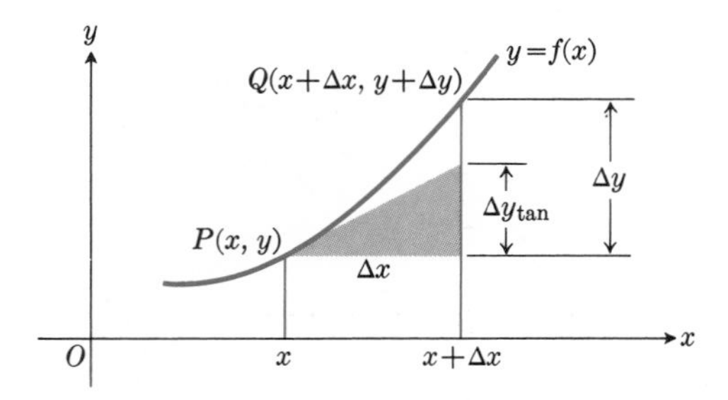

1.43. $\Delta y_{\tan} = f'(x)\,\Delta x.$

Any derivative may be interpreted as the instantaneous rate of change of one variable per unit change in the other. Thus, if the function defined by $y = f(x)$ has a derivative

$$f'(x) = \lim_{\Delta x \to 0} \frac{\Delta y}{\Delta x} = \lim_{\Delta x \to 0} \frac{f(x + \Delta x) - f(x)}{\Delta x},$$

we may interpret $\Delta y/\Delta x$ as *average* rate of change and

$$f'(x) = \lim_{\Delta x \to 0} \frac{\Delta y}{\Delta x}$$

as *instantaneous* rate of change of y with respect to x. Such a rate tells us the amount of change that would be produced in y a change of one unit in x, provided the rate of change remained constant.

The *average* rate of change of y per unit change in x, $\Delta y/\Delta x$, when multiplied by the number of units change in x, Δx, gives the actual change in y:

$$\Delta y = \frac{\Delta y}{\Delta x}\,\Delta x.$$

The *instantaneous* rate of change of y per unit change in x, $f'(x)$, multiplied by the number of units change in x, Δx, gives the change that would be produced in y if the point (x, y) were to move along the *tangent line* instead of moving along the curve; that is (Fig. 1.43)

$$\Delta y_{\tan} = f'(x)\,\Delta x. \tag{9}$$

One reason calculus is important is that it enables us to find quantitatively how a change in one of two related variables affects the second variable.

EXERCISES

1. If a, b, c are constants and
$$f(t) = at^2 + bt + c,$$
show that
$$f'(t) = \lim_{\Delta t \to 0} \frac{f(t + \Delta t) - f(t)}{\Delta t} = 2at + b.$$

In Exercises 2 through 10, the law of motion is used to express s as a function of t. Apply the *result* found in Exercise 1 above to write the velocity $v = ds/dt$ by inspection.

2. $s = 2t^2 + 5t - 3$
3. $s = \frac{1}{2}gt^2 + v_0t + s_0$ (g, v_0, s_0 constants)
4. $s = 4t + 3$
5. $s = t^2 - 3t + 2$
6. $s = 4 - 2t - t^2$
7. $s = (2t + 3)^2$
8. $s = (2 - t)^2$
9. $s = 3 - 2t^2$
10. $s = 64t - 16t^2$

11. The following data give the coordinate s of a moving body for various values of t. Plot s vs. t on coordinate paper and sketch a smooth curve through the given points. Assuming that this smooth curve represents the motion of the body, estimate the velocity (a) at $t = 1.0$, (b) at $t = 2.5$, (c) at $t = 2.0$.

s (ft)	10	38	58	70	74	70	58	38	10
t (sec)	0	0.5	1.0	1.5	2.0	2.5	3.0	3.5	4.0

12. A swimming pool is to be drained for cleaning. If Q represents the number of gallons of water in the pool t minutes after the pool has started to drain, and $Q = 200\,(30 - t)^2$, how fast is the water running out at the end of 10 minutes? What is the *average* rate at which the water flows out during the first 10 minutes?

13. (a) If the radius of a circle changes from r to $r + \Delta r$, what is the average rate of change of the area of the circle with respect to the radius?

    (b) Find the instantaneous rate of change of the area with respect to the radius.

14. The volume V (ft^3) of a sphere of radius r (ft) is $V = \frac{4}{3}\pi r^3$. Find the rate of change of V with respect to r.

15. The radius r and altitude h of a certain cone are equal at all times. Find the rate of change of the volume $V = \frac{1}{3}\pi r^2 h$ with respect to h.

REVIEW QUESTIONS AND EXERCISES

1. Define what is meant by the *slope* of a straight line. How would you find the slope of a straight line from its graph? From its equation?
2. In Fig. 1.44, the lines L_1, L_2, and L_3 have slopes m_1, m_2, m_3, respectively. Which slope is algebraically least? Greatest? Write the three slopes in order of increasing size with the symbols for *less than* or *greater than* correctly inserted between them.
3. Describe the family of lines $y - y_1 = m(x - x_1)$:

   (a) If (x_1, y_1) is fixed and different lines are drawn for different values of m.

   (b) If m and x_1 are fixed and different lines are drawn for different values of y_1.
4. Define *function*. What is the *domain* of a function? What is its *range?*
5. The domain of a certain function is the set $0 \leq x \leq 2$. The range of the function is the single number $y = 1$.

   (a) Sketch and describe the graph of the function.

   (b) Write an expression for the function.

   (c) If we interchange the axes in (a), does the new graph so obtained represent a function? Explain.
6. If $f(x) = 1/x$ and $g(x) = 1/\sqrt{x}$, what are the natural domains of f and g and the corresponding domains of $f - g$, $f \cdot g, f/g$, and g/f? If $h(x) = g(x + 4)$, what is the natural domain of h?
7. Give an example, different from those in the text, of a function whose domain could be the set of all real numbers, $-\infty < x < +\infty$.

Definition. *A* rational number *is one that can be expressed in the form p/q, where p and q are integers with no common divisor greater than one, and $q > 0$. A real number that is not rational is called* irrational.

8. With these definitions in mind, plot a few points on the graph of the function that maps the rational number p/q (reduced to lowest terms as above) into $1/q$, and maps each irrational number into zero. This function has been described as the "ruler function" because of the resemblance of its graph to the markings on an ordinary ruler showing inches, half-inches, quarters, eighths, and sixteenths by lines of different lengths. The edge of the ruler corresponds to the x-axis. Do you see why this terminology is appropriate?

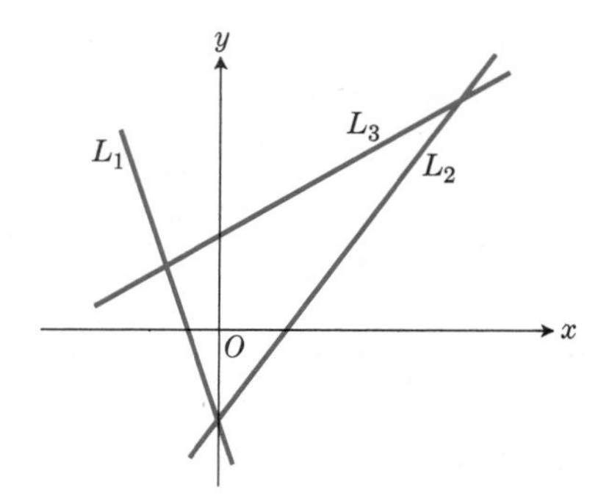

1.44 Lines L_1, L_2, and L_3 have slopes m_1, m_2, and m_3, respectively.

9. Is it appropriate to define a *tangent line* to a curve C at a point P on C as a line that has just the one point P in common with C? Illustrate your discussion with graphs.
10. Define, carefully, the concept of slope of a tangent to a curve at a point on the curve.
11. Define the concepts of *average velocity*, of *instantaneous velocity*.
12. What more general concept includes both the concept of slope of the tangent to a curve and the concept of instantaneous velocity?
13. Define the derivative of a function at a point in its domain. Illustrate your definition by applying it to the function f defined by $f(x) = x^2$, at the point $x = 2$.
14. A function f, whose domain is the set of all real numbers, has the property that

$$f(x + h) = f(x) \cdot f(h)$$

   for all x and h; and $f(0) \neq 0$.

   (a) Show that $f(0) = 1$. [*Hint.* Let $h = x = 0$.]

   (b) If f has a derivative at 0, show that f has a derivative at every real number x, and that

$$f'(x) = f(x) \cdot f'(0).$$

15. Give an example of a function, defined for all real x, that fails to have a derivative

   (a) at some point, (b) at several points,

   (c) at infinitely many points.

16. Suppose that f, g, and h are three functions whose domains include, respectively, the deleted neighborhoods A, B, and C of c. Let $N = A \cap B \cap C$. Prove this statement: If $f(t) \leq g(t) \leq h(t)$ for all $t \in N$, and if $f(t) \in (L - \epsilon, L + \epsilon)$ whenever $t \in A$ and if $h(t) \in (L - \epsilon, L + \epsilon)$ whenever $t \in C$, then $|g(t) - L| < \epsilon$ whenever $t \in N$. Illustrate with graphs.

17. Give an example, different from those mentioned in the text, of a step-function.

18. Sketch the graph of the function

$$f(x) = \max\{\cos x, \sin x\}, \quad -2\pi \leq x \leq 2\pi.$$

19. If $f(x) = \operatorname{sgn} x$, what is the domain of the derived function f'? Give an expression for $f'(x)$.

MISCELLANEOUS EXERCISES

1. (a) Plot the points $A(8, 1)$, $B(2, 10)$, $C(-4, 6)$, $D(2, -3)$, $E(4\frac{2}{3}, 6)$.

   (b) Find the slopes of the lines AB, BC, CD, DA, CE, BD.

   (c) Do four of the five points A, B, C, D, E form a parallelogram? (Why?)

   (d) Do three of the five points lie on a common straight line? (Why?)

   (e) Does the origin $(0, 0)$ lie on a straight line through two of the five points? (Why?)

   (f) Find equations of the lines AB, CD, AD, CE, BD.

   (g) Find the coordinates of the points in which the lines AB, CD, AD, CE, BD intersect the x- and y-axes.

2. Given the straight line $2y - 3x = 4$ and the point $(1, -3)$:

   (a) Find an equation of the straight line through the given point and perpendicular to the given line.

   (b) Find the shortest distance from the given point to the given line.

3. Plot the three points $A(6, 4)$, $B(4, -3)$, and $C(-2, 3)$.

   (a) Is triangle ABC a right triangle? Why?

   (b) Is it isosceles? Why?

   (c) Does the origin lie inside, outside, or on the boundary of the triangle? Why?

   (d) If C is replaced by a point $C'(-2, y)$ such that angle $C'BA$ is a right angle, find y, the ordinate of C'.

4. Find equations of the straight lines passing through the origin which are tangent to the circle of center $(2, 1)$ and radius 2.

5. Let $P_1(x_1, y_1)$ and $P_2(x_2, y_2)$ be any two points. Find the coordinates of the midpoint of the line segment P_1P_2.

6. The x- and y-intercepts of a line L are respectively a and b. Show that an equation of L is

$$(x/a) + (y/b) = 1, \quad \text{if} \quad ab \neq 0.$$

7. Given the line $L\colon ax + by + c = 0$. Find

   (a) its slope, (b) its y-intercept, (c) its x-intercept,

   (d) the line through the origin perpendicular to L.

8. Find the minimum and the maximum values of the function $f(x, y) = 3x + 4y$ on the domain S described by the following inequalities:

$$x + y \geq 2, \quad y \leq 3x + 2, \quad 5x \leq 10 - y.$$

   [*Suggested method:* Solve the problem graphically by first sketching S and then determining which lines of the family $3x + 4y = k$ have at least one point in common with S.] This is a simple exercise in linear programming. For more general information about such problems, see G. Hadley, *Linear Programming,* Addison-Wesley, Reading, Mass., 1962.*

9. How many circles can you find that are tangent to the three lines

$$L_1\colon x + y = 1, \quad L_2\colon y = x + 1, \quad L_3\colon x - 3y = 1?$$

   Give the center and radius of at least one such circle. You may use the formula for the distance from a point $P(x_1, y_1)$ to a line $ax + by + c = 0$,

$$d = \frac{|ax_1 + by_1 + c|}{\sqrt{a^2 + b^2}}.$$

10. Find, in terms of b, b', and m, the perpendicular distance between the parallel lines $y = mx + b$ and $y = mx + b'$.

---

* Also B. Noble, *Applications of Undergraduate Mathematics in Engineering,* Macmillan, New York, 1967.

11. Given the two lines

$$L_1: a_1x + b_1y + c_1 = 0, \quad L_2: a_2x + b_2y + c_2 = 0.$$

If k is a constant, what is the locus whose equation is

$$(a_1x + b_1y + c_1) + k(a_2x + b_2y + c_2) = 0?$$

12. Determine the coordinates of the point on the straight line $y = 3x + 1$ that is equidistant from $(0, 0)$ and $(-3, 4)$.
13. Find an equation of a straight line which is perpendicular to $5x - y = 1$ and is such that the area of the triangle formed by the x-axis, the y-axis, and the straight line is equal to 5.
14. Given the equation $y = (x^2 + 2)/(x^2 - 1)$. Express x in terms of y and determine the set of values of y for which x is real.
15. Express the area A and the circumference C of a circle as functions of the radius r. Express A as a function of C.
16. In each of the following functions, what is the largest domain of x and the corresponding range of y?

(a) $y = \dfrac{1}{1+x}$ (b) $y = \dfrac{1}{1+x^2}$ (c) $y = \dfrac{1}{1+\sqrt{x}}$

17. Without use of the absolute-value symbol, describe the domain of x for which $|x + 1| < 4$.
18. If $y = 2x + |2 - x|$, determine x as a function of y.
19. For what range of values of y does the equation $y = x + |2 - x|$ determine x as a single-valued function of y? Solve for x in terms of y on this range of values.
20. If $y = x + (1/x)$, express x in terms of y and determine the set of y for which x is real. Does this relation express x as a function of y? Why?
21. (a) If $f(x) = x^2 + 2x - 3$, find

$$f(-2), \quad f(-1), \quad f(x_1), \quad f(x_1 + \Delta x).$$

(b) If $f(x) = x - (1/x)$, show that

$$f(1/x) = -f(x) = f(-x).$$

22. Sketch the graph of both the following equations.

(a) $y = |x - 2| + 2$ (b) $y = x^2 - 1$

(c) Find the point on the curve in (b) where the tangent to the curve makes an angle of 45° with the positive x-axis.

23. Sketch a graph of the function $y = |x + 2| + x$ for the domain $-5 \leq x \leq 2$. What is the range?
24. Show that the expression

$$M(a, b) = (a + b)/2 + |a - b|/2$$

is equal to a when $a \geq b$ and is equal to b when $b \geq a$. In other words, $M(a, b)$ gives the larger of the two numbers a and b. Find a similar expression, $m(a, b)$, which gives the smaller of the two numbers. Note that $(a + b)/2$ is the midpoint, and $|a - b|/2$ is half the width, of the interval $[a, b]$ or $[b, a]$.

25. For each of the following expressions $f(x)$, sketch first the graph of $y = f(x)$, then the graph of $y = |f(x)|$, and finally the graph of

$$y = f(x)/2 + |f(x)|/2.$$

(a) $f(x) = (x - 2)(x + 1)$ (b) $f(x) = x^2$

(c) $f(x) = -x^2$ (d) $f(x) = 4 - x^2$

26. *Lagrange interpolation formula.* Let

$$(x_1, y_1), (x_2, y_2), \ldots, (x_n, y_n)$$

be n points in the plane, no two of them having the same abscissa. Find a polynomial, $f(x)$, of degree $(n - 1)$ which takes the value

$$y_1 \text{ at } x_1, \quad y_2 \text{ at } x_2, \quad \ldots, \quad y_n \text{ at } x_n;$$

that is,

$$f(x_i) = y_i \quad (i = 1, 2, \ldots, n).$$

Hint. $f(x) = y_1\phi_1(x) + y_2\phi_2(x) + \cdots + y_n\phi_n(x)$, where $\phi_k(x)$ is a polynomial which is zero at x_i $(i \neq k)$ and $\phi_k(x_k) = 1$.

27. Let $f(x) = ax + b$ and $g(x) = cx + d$. What condition must be satisfied by the constants a, b, c, and d in order that $f(g(x))$ and $g(f(x))$ shall be identical?
28. Let $f(x) = (ax + b)/(cx + d)$. If $d = -a$, show that $f(f(x)) = x$ for all x for which

$$(cx + d)(bc + d^2) \neq 0.$$

29. If $f(x) = x/(x - 1)$, find

(a) $f(1/x)$, (b) $f(-x)$, (c) $f(f(x))$, (d) $f(1/f(x))$.

30. Using the definition of the derivative, find $f'(x)$ if $f(x)$ is

(a) $(x - 1)/(x + 1)$, (b) $x^{3/2}$, (c) $x^{1/3}$.

31. Use the definition of the derivative to find

    (a) $f'(x)$ if $f(x) = x^2 - 3x - 4$,

    (b) $\dfrac{dy}{dx}$ if $y = \dfrac{1}{3x} + 2x$,

    (c) $f'(t)$ if $f(t) = \sqrt{t - 4}$.

32. (a) By means of the Δ-method, find the slope of the curve $y = 2x^3 + 2$ at the point $(1, 4)$.

    (b) At which point of the curve in (a) is the tangent to the curve parallel to the x-axis? Sketch the curve.

33. If $f(x) = 2x/(x - 1)$, find

    (a) $f(0), f(-1), f(1/x)$, (b) $\Delta f(x)/\Delta x$,

    (c) $f'(x)$, using the result of (b).

34. Given $y = 180x - 16x^2$. Using the method of Article 1.9 find the slope of the curve at the point (x_1, y_1). Sketch the curve. At what point does the curve have a horizontal tangent?

35. Find the velocity $v = ds/dt$ if the distance a particle moves in time t is given by $s = 180t - 16t^2$. When does the velocity vanish?

36. If a ball is thrown vertically upward with a velocity of 32 ft/sec, its height after t sec is given by the equation $s = 32t - 16t^2$. At what instant will the ball be at its highest point, and how high will it rise?

37. If the pressure P and volume V of a certain gas are related by the formula $P = 1/V$, find

    (a) the average rate of change of P with respect to V,

    (b) the rate of change of P with respect to V at the instant when $V = 2$.

38. The volume V (in^3) of water remaining in a leaking pail after t sec is $V = 2000 - 40t + 0.2t^2$. How fast is the volume decreasing when $t = 30$?

39. *Properties of inequalities.* If a and b are any two real numbers, we say a is less than b and write $a < b$ if (and only if) $b - a$ is positive. If $a < b$ we also say that b is greater than a $(b > a)$. Prove the following properties of inequalities.

    (a) If $a < b$, then $a \pm c < b \pm c$ for any real number c.

    (b) If $a < b$ and $c < d$, then $a + c < b + d$. Is it also true that $a - c < b - d$? If so, prove it; if not, give an example to support your contention.

    (c) If a and b are both positive (or both negative) and $a < b$, then $1/b < 1/a$.

    (d) If $a < 0 < b$, then $1/a < 0 < 1/b$.

    (e) If $a < b$ and $c > 0$, then $ac < bc$.

    (f) If $a < b$ and $c < 0$, then $bc < ac$.

40. *Properties of absolute values.*

    (a) Prove that $|a| < |b|$ if, and only if, $a^2 < b^2$.

    (b) Prove that $|a + b| \leq |a| + |b|$.

    (c) Prove that $|a - b| \geq \big||a| - |b|\big|$.

    (d) Prove by mathematical induction that

    $$|a_1 + a_2 + \cdots + a_n| \leq |a_1| + |a_2| + \cdots + |a_n|.$$

    (e) Using the result from (d), prove that

    $$|a_1 + a_2 + \cdots + a_n| \geq |a_1| - |a_2| - \cdots - |a_n|.$$

41. Give an example of a function f whose natural domain is the set of x for which

    (a) $x \geq 0$, (b) $x > 0$,

    (c) $-1 < x \leq 2$, (d) $|x| \geq 4$,

    (e) $|x| > 4$, (f) $x \in (-\infty, \infty)$.

LIMITS

CHAPTER 2

2.1 DEFINITION OF THE LIMIT OF A FUNCTION

The purpose of this chapter is to introduce the formal definition of *limit* as an operation applied to a function at a point, and to prove theorems about limits of sums, products, and quotients of functions. We have already seen in Article 1.11 that the average rate of increase of a function $f(x)$, when x goes from x_1 to $x_1 + h$, is

$$\frac{f(x_1 + h) - f(x_1)}{h}, \quad h \neq 0.$$

The *instantaneous* rate of increase of f at x_1 is the limit (if there is one) of this expression as h tends to zero. We cannot just put $h = 0$ in the quotient because $0/0$ is meaningless. Thus, it is desirable to study the theory of limits before proceeding. We shall apply the various limit theorems in Chapter 3 when we resume the study of the rate of change of a function.

Definition. *Suppose f is a function defined for values of x near a. (The domain of f need not include a, although it may.) We say that*

L is the limit of $f(x)$ as x approaches a,

and write

$$L = \lim_{x \to a} f(x), \tag{1}$$

provided that, for any $\epsilon > 0$ there corresponds a deleted neighborhood N of a such that

$$L - \epsilon < f(x) < L + \epsilon \tag{2}$$

whenever x is in N and in D_f.

Notation. When L is the limit of $f(x)$ as x approaches a, we may indicate this fact by writing

$$f(x) \to L \quad \text{as} \quad x \to a,$$

which we read "$f(x)$ approaches L as x approaches a." However, the verb "approaches" has a connotation of motion which is not always justified. For example, a constant function that has the value $f(x) = L$ for all values of x certainly fulfills the conditions of the definition. Note that the verb "is" conveys the message when we say that

$f(x)$ is in $(L - \epsilon, L + \epsilon)$ when x is in $N \cap D_f$.

There is no connotation of x moving toward a or $f(x)$ moving toward L in that statement. Nevertheless the idea of motion (for example, "point Q moves along a curve toward the point P") was, and is, also important in the development of the calculus.

Intuitively, what we are looking for is some L so that we can make "$f(x)$ as close to L as we please" by taking x in a "small enough" deleted neighborhood of a. The definition gives a precise mathematical formulation of the phrases enclosed in quotes in the previous sentence.

Problem. Find $\lim_{x\to 1} (5x - 3)$, and prove that your answer is correct.

Solution. Let

$$f(x) = 5x - 3 \quad \text{and} \quad a = 1. \tag{3}$$

We first guess an appropriate value of L by asking, what number is $f(x) = 5x - 3$ close to when x is close to 1? The obvious answer is

$$L = 2.$$

We try to satisfy the conditions in (2), assuming that $L = 2$ and ϵ is a given positive number. Thus we require that

$$2 - \epsilon < 5x - 3 < 2 + \epsilon. \tag{4a}$$

After adding 3 throughout, we see that (4a) is equivalent to

$$5 - \epsilon < 5x < 5 + \epsilon, \tag{4b}$$

or, after division by 5, to

$$1 - \frac{\epsilon}{5} < x < 1 + \frac{\epsilon}{5}. \tag{4c}$$

The inequalities in (4c) describe the open interval $(1 - \epsilon/5, 1 + \epsilon/5)$, which is a neighborhood of $a = 1$. If we delete 1 from that neighborhood, the resulting *deleted* neighborhood is described by

$$0 < |x - 1| < \frac{\epsilon}{5}. \tag{4d}$$

Figure 2.1 shows that if N_0 is this deleted neighborhood of 1, the inequalities in (4a) are satisfied whenever x is in N_0. Therefore, $\lim_{x\to 1} (5x - 3) = 2$.

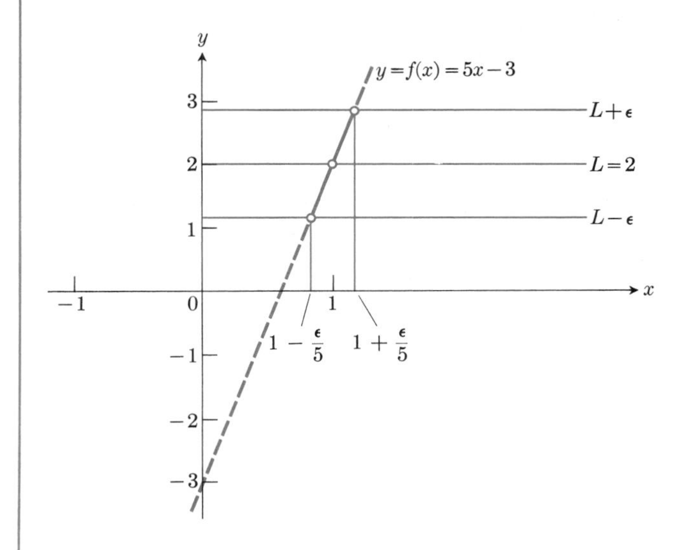

2.1 The deleted neighborhood N_0 extends from the line $x = 1 - \epsilon/5$ to the line $x = 1 + \epsilon/5$, but does not include the point $x = 1$.

Remark 1. Note that the half-width of the deleted neighborhood in (4d) is $\epsilon/5$. Thus the neighborhood gets smaller if ϵ is decreased.

Remark 2. The half-width $\epsilon/5$ is the largest that will work for the above example. But the deleted neighborhood N_0 is not unique: We could, for example, use a smaller neighborhood N_1 such as $0 < |x - 1| < \epsilon/10$ or use a nonsymmetric deleted neighborhood N_2 like the union of the intervals

$$(1 - \epsilon/6, 1) \quad \text{and} \quad (1, 1 + \epsilon/8).$$

Since $\epsilon/10$, $\epsilon/6$, and $\epsilon/8$ are all less than $\epsilon/5$, these other deleted neighborhoods lie inside N_0. Thus if x_1 is in N_2 it is also in N_0 and the value of f at x_1 is between $2 - \epsilon$ and $2 + \epsilon$.

Example 1. We can show that the function F (Fig. 2.2) defined by

$$F(t) = [t] = (\text{greatest integer in } t)$$

has no limit as t approaches 3. Our first guess might be that there is a limit, $L = 3$, since certainly the functional values of $F(t) = [t]$ are close to 3 when t is equal to or

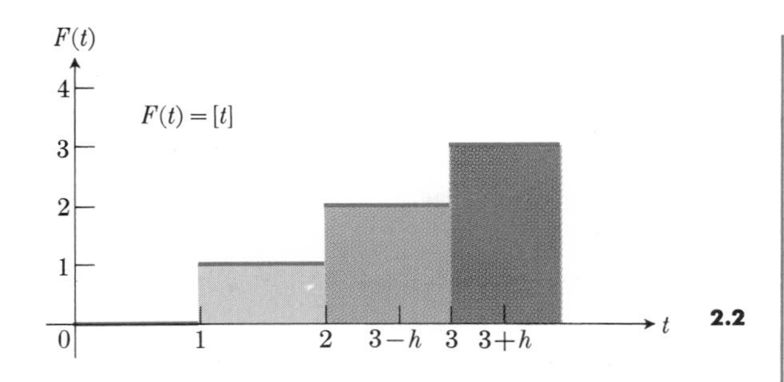

2.2

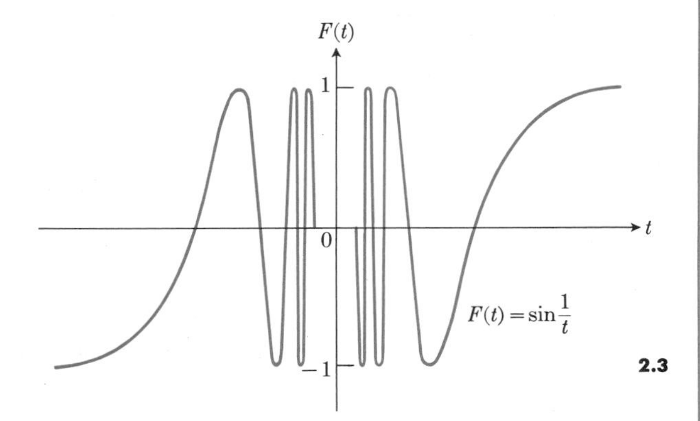

2.3

slightly *greater* than 3. But when t is slightly *less* than 3, say $t = 2.9999$, then $[t] = 2$. That is, if h is any positive number less than unity, $0 < h < 1$,

then $\qquad [t] = 2 \quad \text{if} \quad 3 - h < t < 3,$

while $\qquad [t] = 3 \quad \text{if} \quad 3 < t < 3 + h.$

Hence if we are challenged with a small positive ϵ, for example $\epsilon = 0.01$, we cannot find a deleted neighborhood N of 3 such that

$$|[t] - 3| < \epsilon \quad \text{when} \quad t \in N.$$

In fact, there is no number L that will work as the limit in this case, since when t is near 3 *some* of the functional values of $[t]$ are 2 while others are 3, and hence they are not all close to any one number L. That is,

$$\lim_{t \to 3} [t] \text{ does not exist.}$$

The so-called "right- and left-hand limits" do, however, exist. As the names imply, the right-hand limit $L+$ is a number such that the functional values of $F(t)$ are close to $L+$ when t is slightly greater than 3 (that is, to the right of 3), and the left-hand limit $L-$ is a number such that $F(t)$ is close to $L-$ when t is slightly less than 3.

In our example,

$$L+ = \lim_{t \to 3+} [t] = 3, \quad L- = \lim_{t \to 3-} [t] = 2.$$

The notation $t \to 3+$ may be read "t approaches 3 from above" (or "from the right," or "through values larger than 3") with an analogous meaning for $t \to 3-$.

Example 2. The function F, $F(t) = \sin(1/t)$, $t \neq 0$, possesses no limit as $t \to 0$. This is a consequence of the fact that in every neighborhood of $t = 0$ the function F takes all values between -1 and $+1$. Hence there is no single number L such that the functional values $F(t)$ are *all* close to L when t is close to zero. These remarks apply even when we restrict t to positive values or to negative values. In other words, this function does not even have a right-hand limit or a left-hand limit as t approaches zero (Fig. 2.3).

EXERCISES

Find an appropriate number L, and a deleted neighborhood N of a, such that given $\epsilon > 0$,

$$L - \epsilon < f(x) < L + \epsilon \quad \text{when } x \text{ is in } N.$$

Illustrate each of Exercises 1 through 7 with a graph similar to Fig. 2.1.

1. $f(x) = 5 - 3x, \quad a = 2$
2. $f(x) = 7, \quad a = -1$
3. $f(x) = 4/x, \quad a = 2$
4. $f(x) = \dfrac{x^2 - 4}{x - 2}, \quad a = 2$
5. $f(x) = \dfrac{x^2 + 6x + 5}{x + 5}, \quad a = 5$
6. $f(x) = \dfrac{3x^2 + 8x - 3}{2x + 6}, \quad a = -3$
7. $f(x) = \dfrac{(1/x) - (1/3)}{x - 3}, \quad a = 3$
8. In the next article we use the fact that if N_1 and N_2 are deleted neighborhoods of a, then so is their intersection $N_1 \cap N_2$. Prove this fact by telling what the intersection is if

$$N_1 = \{x \mid h_1 < x < a \quad \text{or} \quad a < x < k_1\},$$

$$N_2 = \{x \mid h_2 < x < a \quad \text{or} \quad a < x < k_2\}.$$

Illustrate with a sketch. [*Hint.* First set up conditions for h_3 and k_3 in terms of h_1 and h_2 and k_1 and k_2 respectively, such that for N_3, the intersection of N_1 and N_2, we have

$$N_3 = \{x \mid h_3 < x < a \quad \text{or} \quad a < x < k_3\}.$$

9. Find some deleted neighborhood of $t = 3$, that is, a domain $0 < |t - 3| < \delta$, such that when t is restricted to this domain, the difference between $t^2 + t$ and 12 will be numerically smaller than
(a) $\frac{1}{10}$, (b) $\frac{1}{100}$,
(c) ϵ, where ϵ may be any positive number.
10. Does the function $f(x) = x \sin(1/x)$, with domain all $x \neq 0$, have a limit as $x \to 0$? Prove your answer.
11. Does the function $g(x) = \operatorname{sgn} x$ have a limit as $x \to 0$? Explain.

2.2 THEOREMS ABOUT LIMITS

In Chapter 1 we saw how to combine functions by adding, multiplying, subtracting, and dividing. In this and the following article we shall prove some useful theorems about limits of such combinations of functions. Our first theorem, however, is suggested by the example of Article 2.1.

Theorem 1. *If m, b, and a are any real numbers,*

and
$$f(x) = mx + b, \tag{1}$$
then
$$\lim_{x \to a} f(x) = ma + b. \tag{2}$$

Proof. Let $\epsilon > 0$ be given, and let $L = ma + b$. We need to prove that there exists a deleted neighborhood N, of a, such that

$$(ma + b) - \epsilon < mx + b < (ma + b) + \epsilon \tag{3}$$

when x is in N. Subtract $(ma + b)$ from the three expressions in (3) and get

$$-\epsilon < mx - ma < \epsilon \tag{4a}$$

which is the same as

$$|m(x - a)| < \epsilon. \tag{4b}$$

In the statement of the theorem, m is an arbitrary real number, so it may be positive, negative, or zero.

Case 1. If m is positive, divide both sides of (4b) by m and get the equivalent inequality

$$|x - a| < \frac{\epsilon}{m}. \tag{5a}$$

Case 2. If m is negative, use the fact that

$$|m(x - a)| = |m|\,|x - a|$$

and divide both sides of (4b) by $|m|$ and get

$$|x - a| < \frac{\epsilon}{|m|}. \tag{5b}$$

Case 3. If $m = 0$, inequality (4b) is satisfied for all values of x because

$$|0(x - a)| = 0 < \epsilon.$$

For either Case 1 or Case 2 we can take, for the deleted neighborhood N, the set of all x's that satisfy

$$0 < |x - a| < \frac{\epsilon}{|m|}, \tag{6a}$$

and when $m = 0$ [$f(x) = b =$ constant], we can take N to be any deleted neighborhood, for example those x such that

$$0 < |x - a| < 1. \tag{6b}$$

Therefore, given m, b, a, and $\epsilon > 0$, we can always find a deleted neighborhood N of a such that the inequalities in (3) hold for all x in N. Therefore, $L = ma + b$ satisfies condition (2) in the definition of a limit, Article 2.1, and we conclude that

$$\lim_{x \to a} (mx + b) = ma + b. \ \square^*$$

Example 1.

$$\lim_{x \to 5} (3x - 7) = 15 - 7 = 8, \quad (m = 3, b = -7).$$

Example 2. $\lim_{x \to a} x = a, \quad (m = 1, b = 0).$

Example 3. $\lim_{x \to a} b = b, \quad (m = 0).$

---

* This symbol indicates that the proof has been completed.

Uniqueness of limit

Is it possible for a function to approach two different limits as x approaches a? The answer is "No," as the following theorem shows.

Theorem 2. *If $f(x)$ has a limit as x approaches a, that limit is unique.*

Proof (Indirect proof: by contradiction). Suppose L_1 and L_2 are two different numbers, and

$$\lim_{x\to a} f(x) = L_1, \quad \lim_{x\to a} f(x) = L_2. \tag{7}$$

Then $L_1 \neq L_2$ means that $|L_1 - L_2|$ is positive, and we let

$$\epsilon = \tfrac{1}{2}|L_1 - L_2|. \tag{8}$$

You will see why this choice is made later in the proof.

By definition of limit, applied to L_1, we know there is a deleted neighborhood N_1 of a (see Fig. 2.4) such that

$$|f(x) - L_1| < \epsilon \tag{9a}$$

when x is in N_1.* Likewise there is a deleted neighborhood N_2 of a such that

$$|f(x) - L_2| < \epsilon \tag{9b}$$

when x is in N_2. If

$$N_1 = \{x \mid h_1 < x < a \quad \text{or} \quad a < x < k_1\}, \tag{10a}$$

$$N_2 = \{x \mid h_2 < x < a \quad \text{or} \quad a < x < k_2\}, \tag{10b}$$

then the intersection of N_1 and N_2 is a deleted neighborhood N_3 such that

$$N_3 = \{x \mid h_3 < x < a \quad \text{or} \quad a < x < k_3\} \tag{10c}$$

where

$$h_3 = \max\{h_1, h_2\},$$

$$k_3 = \min\{k_1, k_2\}.$$

Figure 2.4 illustrates these ideas. N_1 goes from h_1 to k_1, N_2 from h_2 to k_2. Here points $\neq a$ between h_1 and k_2 are in both N_1 and N_2, and hence are in

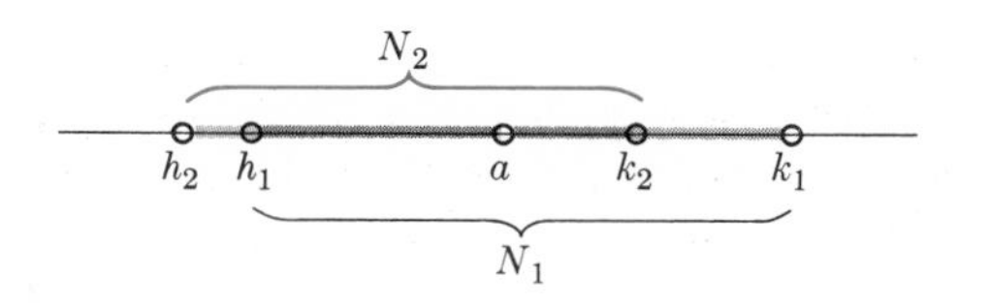

2.4

$N_3 = N_1 \cap N_2$, where $h_3 = h_1$ and $k_3 = k_2$. For any x in N_3, both (9a) and (9b) hold simultaneously, and lead to the false conclusion

$$\begin{aligned} |L_1 - L_2| &= |L_1 - f(x) + f(x) - L_2| \\ &\leq |L_1 - f(x)| + |f(x) - L_2| < \epsilon + \epsilon \\ &= 2\epsilon = |L_1 - L_2|. \end{aligned}$$

Since the assumption

$$L_1 \neq L_2$$

leads to the contradiction

$$|L_1 - L_2| < |L_1 - L_2|,$$

that assumption must be false. Therefore $L_1 = L_2$, so the limit is unique. □

Remark 1. The idea behind the foregoing result is quite simple and should not get lost in the fog: If we can make $f(x)$ very nearly equal to L_1 and also very nearly equal to L_2 by taking x close to a, then L_1 and L_2 must be close to each other. More precisely, if both

$$|f(x) - L_1| < \epsilon \quad \text{and} \quad |f(x) - L_2| < \epsilon,$$

then

$$|L_1 - L_2| < 2\epsilon.$$

However, L_1 and L_2 are fixed, and if they aren't equal, then we can't meet the ϵ-challenge for both L_1 and L_2 if ϵ *is not greater than* one-half the distance between them. In particular, we cannot meet the ϵ-challenge if $\epsilon = \frac{1}{2}|L_1 - L_2|$, as in the proof. Smaller starting values of ϵ, such as

$$\epsilon = \tfrac{1}{3}|L_1 - L_2|,$$

would obviously work. Hence we have shown that if $L_1 \neq L_2$, then there are *some* positive choices for ϵ such that condition (2) in the definition of limit cannot be satisfied for both. So $f(x)$ can't have two different limits as x approaches a.

---

* It is also to be understood that $x \in D_f$ whenever we refer to $f(x)$. Thus (9a) holds for $x \in N_1 \cap D_f$, and (9b) holds for $x \in N_2 \cap D_f$.

Remark 2. Theorem 2 shows that if we can guess an answer to the question of finding a limit L for a function $f(x)$ as x approaches a, and then prove that this number L satisfies the conditions in the definition of a limit, then it is *the* answer—there is no other.

Problem. Prove that $\lim_{x\to 3} (x^3 - 27)/(x - 3) = 27$.

Solution. Since

$$x^3 - 27 = (x - 3)(x^2 + 3x + 9),$$

we have

$$f(x) = \frac{x^3 - 27}{x - 3} = x^2 + 3x + 9 \quad \text{if } x \neq 3.$$

When x is *close* to (but not equal to) 3, $x^2 + 3x + 9$ is *close* to 27, and we guess that $L = 27$ is the answer. Now we can see if we can prove it. Suppose ϵ is a given positive number. Can we find a deleted neighborhood N of 3 such that

$$|f(x) - L| = |(x^2 + 3x + 9) - 27| < \epsilon \tag{11a}$$

when x is in N? We observe that

$$\begin{aligned} x^2 + 3x + 9 - 27 &= x^2 + 3x - 18 \\ &= (x - 3)(x + 6), \end{aligned}$$

so inequality (11a) is the same as

$$|x - 3|\,|x + 6| < \epsilon. \tag{11b}$$

When x is near 3, the factor $x - 3$ is near zero and $x + 6$ is near 9. Consider now the deleted neighborhood N_0,

$$0 < |x - 3| < 1. \tag{12}$$

If x is in N_0, then $-1 < x - 3 < 1$, or (by adding 9) $8 < x + 6 < 10$. Therefore $|x + 6| < 10$, and it follows that

$$|x - 3|\,|x + 6| < 10|x - 3|.$$

We now make

$$0 < 10|x - 3| < \epsilon$$

by keeping x in the deleted neighborhood N_1,

$$0 < |x - 3| < \frac{\epsilon}{10}\cdot \tag{13}$$

To satisfy both inequalities (12) and (13), we take for N the smaller of the two neighborhoods N_0 and N_1. If x is in N it follows that

$$\begin{aligned} |f(x) - 27| &= |(x^2 + 3x + 9) - 27| \\ &= |x^2 + 3x - 18| \\ &= |x + 6|\,|x - 3| \\ &< 10|x - 3| \quad \text{since } x \in N \subseteq N_0{}^{*} \\ &< 10\,\frac{\epsilon}{10} \qquad \text{since } x \in N \subseteq N_1 \\ &< \epsilon. \end{aligned}$$

Therefore

$$\lim_{x\to 3} f(x) = 27.$$

Remark 3. The foregoing example reduced to showing that

$$\lim_{x\to 3} (x^2 + 3x + 9) = 27.$$

If we could have said

$$\lim (x^2 + 3x + 9) = \lim (x^2) + \lim (3x + 9)$$

and

$$\lim (x^2) = (\lim x)(\lim x),$$

the problem would have been much simpler. These are, in fact, correct statements, as the next two theorems will show. (See Theorem 4, Article 2.3.)

Theorem 3. Limit of Sums.

If $\qquad \lim f(x) = L_1$ and $\lim g(x) = L_2$,

then $\qquad \begin{aligned} \lim \big(f(x) + g(x)\big) &= \lim f(x) + \lim g(x) \\ &= L_1 + L_2, \end{aligned}$

where all limits are taken as $x \to a$.

Proof. Suppose $\epsilon > 0$ is given. We want to find a deleted neighborhood N of a, such that

$$|f(x) + g(x) - (L_1 + L_2)| < \epsilon \tag{14}$$

when x is in N. We use the fact that

$$\begin{aligned} &|\big(f(x) + g(x)\big) - (L_1 + L_2)| \\ &\quad = |\big(f(x) - L_1\big) + \big(g(x) - L_2\big)| \\ &\quad \leq |(fx) - L_1| + |g(x) - L_2|. \end{aligned} \tag{15}$$

---

* We write $A \subseteq B$ to mean A is a subset of B which may be equal to B.

Because $\epsilon/2$ is positive and $f(x) \to L_1$, there is a deleted neighborhood N_1 of a such that

$$|f(x) - L_1| < \tfrac{1}{2}\epsilon \tag{16a}$$

when x is in N_1. Likewise there is a deleted neighborhood N_2 of a such that

$$|g(x) - L_2| < \tfrac{1}{2}\epsilon \tag{16b}$$

when x is in N_2. Let N be the intersection of N_1 and N_2:

$$N = N_1 \cap N_2.$$

Then the inequalities (16a) and (16b) both hold when x is in N, and so (14) also holds, since

$$\begin{aligned}|(f(x) + g(x)) - (L_1 + L_2)| &\le |f(x) - L_1| + |g(x) - L_2| \\ &< \tfrac{1}{2}\epsilon + \tfrac{1}{2}\epsilon = \epsilon. \ \square\end{aligned}$$

EXERCISES

Evaluate the limits indicated in Exercises 1 through 8.

1. $\lim_{x\to 2} 2x$
2. $\lim_{x\to 0} 2x$
3. $\lim_{x\to 4} 4$
4. $\lim_{x\to -2} 4$
5. $\lim_{x\to 1} (3x - 1)$
6. $\lim_{x\to 1/3} (3x - 1)$
7. $\lim_{y\to 2} \dfrac{y^2 + 5y + 6}{y + 2}$
8. $\lim_{y\to 2} \dfrac{y^2 - 5y + 6}{y - 2}$
9. Suppose $\lim_{x\to b} f(x) = 7$ and $\lim_{x\to b} g(x) = -3$. What is $\lim_{x\to b} [f(x) + g(x)]$?
10. Prove that

$$\lim_{x\to a} [p(x) + r(x) + s(x)] = \lim_{x\to a} p(x) + \lim_{x\to a} r(x) + \lim_{x\to a} s(x),$$

provided that the last three limits exist.

11. Suppose

$$\lim_{x\to -2} p(x) = 4, \quad \lim_{x\to -2} r(x) = 0, \quad \lim_{x\to -2} s(x) = -3.$$

What is $\lim_{x\to -2} [p(x) + r(x) + s(x)]$?

12. Prove that

(a) $\lim_{x\to 3} \dfrac{1}{x} = \dfrac{1}{3}$, (b) if $c \neq 0$, then $\lim_{x\to c} \dfrac{1}{x} = \dfrac{1}{c}$.

In Exercises 13 through 18, guess the limit, then prove that your guess is correct.

13. $\lim_{x\to 5} x^2$
14. $\lim_{x\to 2} x(2 - x)$
15. $\lim_{x\to 0} (x^2 - 2x + 1)$
16. $\lim_{x\to 5} (x^2 - 3x - 18)$
17. $\lim_{x\to 2} (x^3 - 4x + 1)$
18. $\lim_{x\to 4} \sqrt{x}$
19. Give an example of functions f and g such that $f(x) + g(x)$ approaches a limit as x approaches zero, even though $f(x)$ and $g(x)$, separately, do not approach limits as $x \to 0$.
20. Repeat Exercise 19 for $f(x) \cdot g(x)$ in place of $f(x) + g(x)$.

2.3 MORE THEOREMS ABOUT LIMITS

In Article 2.2, we proved a very useful theorem involving the limit of a sum. In this article we shall prove two theorems involving the limit of a product and the limit of a quotient.

Theorem 4. Limit of Products.

If $\lim f(x) = L_1$ *and* $\lim g(x) = L_2$,

then $\lim f(x)g(x) = (\lim f(x))(\lim g(x)) = L_1L_2$,

where all limits are taken as $x \to a$.

Proof. Let $\epsilon > 0$ be given. We need to show that there is a deleted neighborhood N of a such that

$$|f(x)g(x) - L_1L_2| < \epsilon \tag{1}$$

when x is in N.

We write

$$f(x) = L_1 + (f(x) - L_1),$$
$$g(x) = L_2 + (g(x) - L_2)$$

so that

$$f(x)g(x) - L_1L_2 = L_1(g(x) - L_2) + L_2(f(x) - L_1) + (f(x) - L_1)(g(x) - L_2).$$

Therefore

$$|f(x)g(x) - L_1L_2| \leq |L_1(g(x) - L_2)| + |L_2(f(x) - L_1)| + |(f(x) - L_1)(g(x) - L_2)|$$

and

$$|f(x)g(x) - L_1L_2| \leq |L_1| \cdot |g(x) - L_2| + |L_2| \cdot |f(x) - L_1| + |f(x) - L_1| \cdot |g(x) - L_2|. \qquad (2)$$

By hypothesis, there is a deleted neighborhood N_1 of a such that

$$|g(x) - L_2| < \frac{\epsilon}{3(1 + |L_1|)}$$

when x is in N_1, because $\lim g(x) = L_2$ and

$$\frac{\epsilon}{3(1 + |L_1|)}$$

is a positive number. Likewise there are deleted neighborhoods N_2, N_3, and N_4 such that

$$|f(x) - L_1| < \frac{\epsilon}{3(1 + |L_2|)} \quad \text{when } x \text{ is in } N_2,$$

$$|f(x) - L_1| < \sqrt{\epsilon/3} \quad \text{when } x \text{ is in } N_3,$$

$$|g(x) - L_2| < \sqrt{\epsilon/3} \quad \text{when } x \text{ is in } N_4.$$

[It will be apparent later why the particular positive numbers $\epsilon/3(1 + |L_1|)$, $\epsilon/3(1 + |L_2|)$, and $\sqrt{\epsilon/3}$ are chosen.] If N is the intersection of N_1, N_2, N_3, and N_4, the above inequalities all hold when x is in N, so that the right side of inequality (2) is less than

$$\frac{|L_1|\epsilon}{3(1 + |L_1|)} + \frac{|L_2|\epsilon}{3(1 + |L_2|)} + \sqrt{\frac{\epsilon}{3}} \cdot \sqrt{\frac{\epsilon}{3}}. \qquad (3)$$

Since

$$\frac{|L_1|}{1 + |L_1|} < 1 \quad \text{and} \quad \frac{|L_2|}{1 + |L_2|} < 1,$$

we have

$$\frac{|L_1|\epsilon}{3(1 + |L_1|)} < \frac{\epsilon}{3} \quad \text{and} \quad \frac{|L_2|\epsilon}{3(1 + |L_2|)} < \frac{\epsilon}{3}.$$

Therefore (3) is less than

$$\frac{\epsilon}{3} + \frac{\epsilon}{3} + \frac{\epsilon}{3} = \epsilon. \; \square$$

Remark 3. There is a reason for using $1 + |L_1|$ and $1 + |L_2|$ in the denominators in (3) instead of $|L_1|$ and $|L_2|$. If $L_1 = 0$, we do not want $|L_1|$ in the denominator. But $1 + |L_1|$ cannot be zero, and, as noted, $|L|/(1 + |L|)$ is always less than 1 because the denominator is greater than the numerator.

Corollary 4.1. *If* $\lim f(x) = L$ *and* k *is a constant, then* $\lim kf(x) = kL$, *where both limits are taken as* $x \longrightarrow a$.

Proof. Apply Theorem 4 with $g(x) = k$, so that

$$L_2 = \lim g(x) = k. \; \square$$

Corollary 4.2. *The* $\lim\limits_{x \to a} x^n = a^n$ *for any positive integer* n.

Proof (By induction on n). If $n = 1$, then

$$\lim_{x \to a} x = a,$$

by Theorem 1, with $m = 1$, $b = 0$.

If $n - 1$ is a positive integer such that

$$\lim_{x \to a} x^{n-1} = a^{n-1},$$

then

$$\lim_{x \to a} x^n = \lim_{x \to a} (x \cdot x^{n-1}) = a \cdot a^{n-1} = a^n,$$

by Theorem 4 with $f(x) = x$, $g(x) = x^{n-1}$, $L_1 = a$, and $L_2 = a^{n-1}$. $\square$

Corollary 4.3.

If $\quad f(x) = c_0x^n + c_1x^{n-1} + \cdots + c_n$

is a polynomial in x, *then*

$$\lim_{x \to a} f(x) = c_0a^n + c_1a^{n-1} + \cdots + c_n = f(a).$$

Proof (Again by induction on n). Since we have already proved it for $n = 1$ in Theorem 1, assume that the result holds for the polynomial

$$g(x) = c_1x^{n-1} + \cdots + c_n.$$

Then

$$\begin{aligned}\lim_{x\to a} (c_0x^n + g(x)) &= \lim_{x\to a} (c_0x^n) + \lim_{x\to a} g(x)\\ &= c_0a^n + g(a)\\ &= f(a),\end{aligned}$$

from Corollaries 4.1 and 4.2 with $k = c_0$. □

Example 1.

$$\lim_{x\to 2} (5x^2 + 4x - 10) = 20 + 8 - 10 = 18.$$

Quotients

Is it true that

$$\lim \frac{1}{g(x)} = \frac{1}{\lim g(x)}\ ?$$

Since division by zero is not valid, we must assume that

$$\lim g(x) \neq 0.$$

With this proviso, the answer is "yes."

Theorem 5. Limit of Quotients.

Let $$\lim_{x\to a} g(x) = L \neq 0.$$

Then $$\lim_{x\to a} \left(\frac{1}{g(x)}\right) = \frac{1}{L}. \tag{4}$$

Proof. Let $\epsilon > 0$ be given. We want a deleted neighborhood N of a such that

$$\left|\frac{1}{g(x)} - \frac{1}{L}\right| < \epsilon \tag{5}$$

when x is in N. Inequality (5) is the same as

$$\left|\frac{L - g(x)}{L \cdot g(x)}\right| < \epsilon. \tag{6}$$

The numerator on the left can be made as small as we like by taking x sufficiently near a, since $\lim_{x\to a} g(x) = L$. We apply Corollary 4.1 to the denominator:

$$\lim_{x\to a} L \cdot g(x) = L \lim_{x\to a} g(x) = L^2.$$

By definition of limit, with $\epsilon = L^2/2$, there is a deleted neighborhood N_1 of a such that, whenever x is in N_1,

$$L^2 - \epsilon < L \cdot g(x) < L^2 + \epsilon,$$

or, since $\epsilon = L^2/2$,

$$\frac{L^2}{2} < L \cdot g(x) < \frac{3L^2}{2}.$$

Hence $L \cdot g(x)$ is positive, and on taking reciprocals we get

$$\frac{2}{L^2} > \frac{1}{L \cdot g(x)}$$

when x is in N_1. Therefore

$$\left|\frac{L - g(x)}{L \cdot g(x)}\right| = \frac{|L - g(x)|}{L \cdot g(x)} < \frac{2}{L^2}|L - g(x)| \tag{7}$$

when x is in N_1. Finally, to satisfy (6), there is a deleted neighborhood N_2 such that

$$|L - g(x)| < \frac{\epsilon L^2}{2} \tag{8}$$

when x is in N_2. Now let N be the intersection of N_1 and N_2. When x is in N, inequalities (7) and (8) both hold, so that

$$\left|\frac{L - g(x)}{L \cdot g(x)}\right| < \frac{2}{L^2}|L - g(x)| < \frac{2}{L^2}\left(\frac{\epsilon L^2}{2}\right) = \epsilon.\ \square$$

Corollary 5.1.

If $\lim_{x\to a} f(x) = L_1$ *and* $\lim_{x\to a} g(x) = L_2$, $L_2 \neq 0$,

then $$\lim_{x\to a} \frac{f(x)}{g(x)} = \frac{L_1}{L_2} = \frac{\lim f(x)}{\lim g(x)}.$$

Proof. Combine Theorems 4 and 5.

$$\begin{aligned}\lim \frac{f(x)}{g(x)} &= \lim \left[f(x) \cdot \frac{1}{g(x)}\right]\\ &= \lim f(x) \cdot \lim \left(\frac{1}{g(x)}\right)\\ &= L_1 \cdot \frac{1}{L_2}.\ \square\end{aligned}$$

Corollary 5.2.

If $f(x) = b_0x^m + b_1x^{m-1} + \cdots + b_m$

and $g(x) = c_0x^n + c_1x^{n-1} + \cdots + c_n$

are polynomials and $g(a) \neq 0$, *then*

$$\lim_{x \to a} \frac{f(x)}{g(x)} = \frac{f(a)}{g(a)}.$$

Proof. Combine Corollaries 5.1 and 4.3 to get

$$\lim \frac{f(x)}{g(x)} = \frac{\lim f(x)}{\lim g(x)} = \frac{f(a)}{g(a)}. \ \square$$

Remark 4. In effect, Corollary 5.2 says that to find the limit of the quotient of two polynomials as *x approaches a*, substitute $x = a$ and compute $f(a)/g(a)$, provided the denominator turns out to be *different from zero*. But you cannot find

$$\lim_{x \to 3} \frac{x^2 - 9}{x - 3} = \lim_{x \to 3} (x + 3) = 6$$

by dividing $3^2 - 9 = 0$ by $3 - 3 = 0$.

Example 2. $$\lim_{x \to 1} \frac{x^2 - x + 2}{x^2 + x + 2} = \frac{1 - 1 + 2}{1 + 1 + 2} = \frac{1}{2}.$$

We conclude this article with a theorem which is used repeatedly in our later work.

Theorem 6. *Suppose*

$$f(x) \leq g(x) \leq h(x),$$

for all values of x in some deleted neighborhood N_1 *of c. Furthermore, suppose that the function values* $f(x)$ *and* $h(x)$ *approach a common limit L as x approaches c. Then* $g(x)$ *also approaches L as limit when x approaches c.*

Proof. By hypothesis, to any positive number ϵ there corresponds a deleted neighborhood N_2 of c such that both $f(x)$ and $h(x)$ lie between $L - \epsilon$ and $L + \epsilon$ when x is in N_2; that is

$$L - \epsilon < f(x) < L + \epsilon,$$

$$L - \epsilon < h(x) < L + \epsilon.$$

But, for x in the intersection $N = N_1 \cap N_2$, this also implies that

$$L - \epsilon < f(x) \leq g(x) \leq h(x) < L + \epsilon,$$

or

$$L - \epsilon < g(x) < L + \epsilon.$$

In other words, $|g(x) - L| < \epsilon$ when x is in N. This establishes the conclusion

$$\lim_{x \to c} g(x) = L. \ \square$$

Example 3. As an application of Theorem 6, we prove that

$$\lim_{\theta \to 0} \sin \theta = 0, \tag{9a}$$

$$\lim_{\theta \to 0} \cos \theta = 1, \tag{9b}$$

$$\lim_{\theta \to 0} \frac{\sin \theta}{\theta} = 1, \tag{9c}$$

provided that θ is measured in radians. [Recall that the radian measure of an angle can be defined by

$$\theta = s/r, \tag{10}$$

where s is the length of arc the angle intercepts on a circle of radius r when the center of the circle is the vertex of the angle. Figure 2.5(a) illustrates this definition.]

In Fig. 2.5(b), O is the center of a unit circle and θ is the radian measure of an acute angle AOP. Note that $s = \theta$ under these conditions. Now $\triangle APQ$ is a right triangle with legs of length

$$QP = \sin \theta, \quad AQ = 1 - \cos \theta.$$

From the theorem of Pythagoras, and the fact that $AP < \theta$, we get

$$\sin^2 \theta + (1 - \cos \theta)^2 = (AP)^2 < \theta^2. \tag{11}$$

Both terms on the left side of inequality (11) are positive, so each is smaller than their sum, and hence is less than θ^2:

$$\sin^2 \theta < \theta^2, \tag{12a}$$

$$(1 - \cos \theta)^2 < \theta^2. \tag{12b}$$

The inequalities (12a) and (12b) also imply

$$|\sin \theta| < |\theta|, \tag{13a}$$

$$|1 - \cos \theta| < |\theta|. \tag{13b}$$

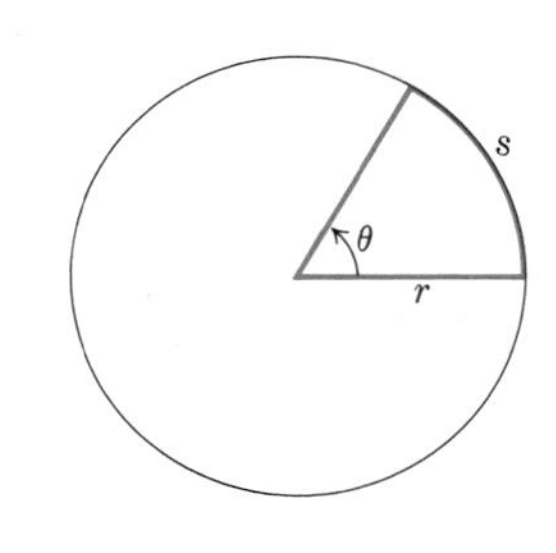

a | b

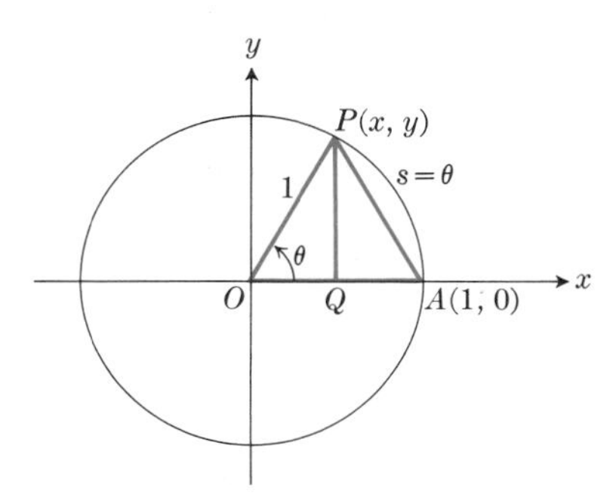

2.5 **a** | Radian measure.

b | Radian measure on a unit circle: $r = 1$, $s = \theta$.

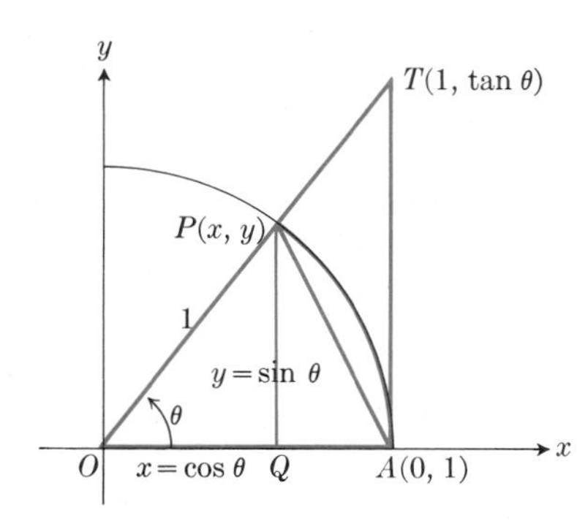

2.6 Area ΔAOP < area sector AOP < area ΔAOT.

If ϵ is any positive number we can take $h = \epsilon$ and conclude that

$$|\sin \theta - 0| < \epsilon$$

and

$$|1 - \cos \theta| < \epsilon$$

whenever $0 < |\theta| < h$. Therefore

$$\lim_{\theta \to 0} \sin \theta = 0 \quad \text{and} \quad \lim_{\theta \to 0} \cos \theta = 1.$$

To establish Eq. (9c), let us assume that θ is positive and less than $\pi/2$ in Fig. 2.6. We compare the areas of $\triangle AOP$, sector AOP, and $\triangle AOT$, and note that

$$\text{area } \triangle AOP < \text{area sector } AOP < \text{area } \triangle AOT.$$

The area of sector AOP is the same fraction of the area of a circle of radius 1 that its arc is of the total circumference*:

$$\frac{\text{area sector } AOP}{\text{area circle}} = \frac{\text{arc } AP}{\text{circumference}}.$$

Thus

$$\frac{\text{area sector } AOP}{\pi r^2} = \frac{r\theta}{2\pi r} = \frac{\theta}{2\pi},$$

and therefore

$$\text{area sector } AOP = \tfrac{1}{2} r^2 \theta \tag{14a}$$

$$= \tfrac{1}{2}\theta, \quad \text{when } r = 1. \tag{14b}$$

The altitude of $\triangle AOP$, with base $OA = 1$, is

$$y = QP = \sin \theta.$$

Hence

$$\text{area } \triangle AOP = \tfrac{1}{2} \sin \theta. \tag{15a}$$

The altitude of $\triangle AOT$, with base $OA = 1$, is

$$AT = \tan \theta,$$

and therefore

$$\text{area } \triangle AOT = \tfrac{1}{2} \tan \theta. \tag{15b}$$

But

$$\text{area } \triangle AOP < \text{area sector } AOP < \text{area } \triangle AOT,$$

and therefore

$$\tfrac{1}{2} \sin \theta < \tfrac{1}{2}\theta < \tfrac{1}{2} \tan \theta, \quad \text{if } 0 < \theta < \frac{\pi}{2}. \tag{16}$$

Dividing the three terms of inequality (10) by $\frac{1}{2} \sin \theta$, we get

$$1 < \frac{\theta}{\sin \theta} < \frac{1}{\cos \theta}, \quad \text{if } 0 < \theta < \frac{\pi}{2}. \tag{17a}$$

---

* We assume the results $A = \pi r^2$ and $C = 2\pi r$, for the area and circumference of a circle of radius r. These do, in fact, depend on limits but not on the trigonometric functions.

Recall that $\sin(-\theta) = -\sin\theta$, and $\cos(-\theta) = \cos\theta$, so (17a) also implies that

$$1 < \frac{-\theta}{\sin(-\theta)} < \frac{1}{\cos(-\theta)}, \quad \text{if } 0 < -\theta < \frac{\pi}{2},$$

or

$$1 < \frac{\theta}{\sin\theta} < \frac{1}{\cos\theta}, \quad \text{if } 0 > \theta > -\frac{\pi}{2}. \tag{17b}$$

We can combine (17a) and (17b) to say

$$1 < \frac{\theta}{\sin\theta} < \frac{1}{\cos\theta}, \quad \text{if } 0 < |\theta| < \frac{\pi}{2}. \tag{18}$$

Inequality (18) tells us about $\theta/(\sin\theta)$. We convert this to information about $(\sin\theta)/\theta$ by taking reciprocals and reversing the inequality signs, because

if a and b are positive and $a > b$;

then $(1/a) < (1/b)$.

Therefore $$1 > \frac{\sin\theta}{\theta} > \cos\theta$$

whenever $$0 < |\theta| < \frac{\pi}{2}.$$

Now apply Theorem 6. In this example, the two functions that are analogous to the functions $f(x)$ and $h(x)$ defined in Theorem 6 both approach 1 as common limit:

$$\lim_{\theta\to 0} 1 = 1, \quad \lim_{\theta\to 0} \cos\theta = 1.$$

Therefore we conclude that

$$\lim_{\theta\to 0} \frac{\sin\theta}{\theta} = 1.$$

Example 4

$$\lim_{x\to 0} \frac{\sin 3x}{x} = \lim_{3x\to 0} \frac{3\sin 3x}{3x} = 3\lim_{\theta\to 0}\left(\frac{\sin\theta}{\theta}\right) = 3.$$

Example 5

$$\lim_{x\to 0} \frac{\tan x}{x} = \lim_{x\to 0}\left(\frac{\sin x}{x}\,\frac{1}{\cos x}\right) = \left[\lim_{x\to 0}\left(\frac{\sin x}{x}\right)\right]\left[\lim_{x\to 0}\left(\frac{1}{\cos x}\right)\right] = (1)(1) = 1.$$

EXERCISES

1. Suppose $\lim_{x\to c} f(x) = 5$ and $\lim_{x\to c} g(x) = -2$.
   (a) What is $\lim_{x\to c} f(x)g(x)$?
   (b) What is $\lim_{x\to c} 2f(x)g(x)$?

Evaluate the limits indicated in Exercises 2 through 20.

2. $\lim_{x\to 1} [(2x-1)(x)]$
3. $\lim_{x\to 2} [3(2x-1)(x)]$
4. $\lim_{x\to 2} [3(2x-1)(x+1)]$
5. $\lim_{x\to 2} [3(2x-1)(x^2)]$
6. $\lim_{x\to 2} [3(2x-1)(x+1)^2]$
7. $\lim_{x\to -1} (x+3)$
8. $\lim_{x\to -1} (x+3)^2$
9. $\lim_{x\to -1} (x^2+6x+9)$
10. $\lim_{x\to -2} (x+3)^{171}$
11. $\lim_{x\to -4} (x+3)^{1967}$
12. $\lim_{x\to 1} (x^3+3x^2-2x-17)$
13. Suppose $\lim_{x\to c} f(x) = 2$ and $\lim_{x\to c} g(x) = 5$.
    (a) What is $\lim_{x\to c} \dfrac{f(x)}{g(x)}$?
    (b) What is $\lim_{x\to c} \dfrac{f(x)}{3g(x)}$?
14. $\lim_{t\to 2} \dfrac{t+3}{t+2}$
15. $\lim_{x\to 5} \dfrac{x^2-25}{x-5}$
16. $\lim_{x\to 5} \dfrac{x^2-25}{x+5}$
17. $\lim_{x\to 5} \dfrac{x-5}{x^2-25}$, if it exists.
18. $\lim_{x\to 5} \dfrac{x+5}{x^2-25}$, if it exists.
19. $\lim_{x\to 1} \dfrac{x^2-x-2}{x^2-1}$
20. $\lim_{x\to 0} \dfrac{5x^3+8x^2}{3x^4-16x^2}$
21. Give an example of functions f and g such that
    $$\lim_{x\to 0} f(x)/g(x)$$
    exists, but at least one of the functions f and g fails to have a limit as $x\to 0$.
22. As $x\to 0+$, the functions $1/x$, $1/x^2$, $1/\sqrt{x}$ all become very large. Which one increases most rapidly and which one least rapidly? [*Note.* $x\to 0+$ means

that x approaches 0 through positive values, that is, from the right.]

23. Suppose F is a function whose values are all less than or equal to some constant M: $F(t) \leq M$. Prove:

If $$\lim_{t \to c} F(t) = L,$$
then $$L \leq M.$$

Suggestion. An indirect proof may be used to show that $L > M$ is false. For if $L > M$, we may take $\frac{1}{2}(L - M)$ as a positive number ϵ, apply the definition of limit, and arrive at a contradiction.

24. Expand and simplify the left side of Eq. (11) and prove that $(1 - \cos\theta) < \theta^2/2$ for $0 < \theta < \pi/2$. Is $(1 - \cos\theta) < \theta^2/2$ true for $\theta = 0$? For $\theta = \pi/2$? For $0 < |\theta| < \pi/2$? For all $\theta \neq 0$?

Evaluate the following limits, or show that they do not exist.

25. (a) $\lim_{x \to 0} \dfrac{x}{\sin 2x}$ (b) $\lim_{x \to 0} \dfrac{\sin 7x}{4x}$

26. (a) $\lim_{\theta \to 0} \dfrac{\sec 2\theta \tan 3\theta}{5\theta}$ (b) $\lim_{t \to 0} \dfrac{15t}{\tan 6t}$

27. (a) $\lim_{x \to 0} \dfrac{\tan 2x}{\sin 7x}$ (b) $\lim_{x \to 0} \dfrac{\cot 4x}{\cot 3x}$

28. $\lim_{y \to \pi/2} \dfrac{\cos y}{\pi/2 - y}$ (Let $x = \pi/2 - y$.)

29. $\lim_{x \to 0} \dfrac{x \sin x}{1 - \cos x}$ (Multiply and divide by $1 + \cos x$.)

30. Use the trigonometric identity
$$\sin A - \sin B = 2 \sin \frac{A - B}{2} \cos \frac{A + B}{2}$$
to evaluate
$$\lim_{h \to 0} \frac{\sin (x + h) - \sin x}{h},$$
assuming that
$$\lim_{\theta \to 0} \cos (x + \theta) = \cos x.$$

31. Use the identity
$$\cos (A + B) = \cos A \cos B - \sin A \sin B$$
and appropriate theorems to prove that
$$\lim_{\theta \to 0} \cos (x + \theta) = \cos x.$$

2.4 INFINITY

The function
$$f(x) = \frac{1}{x}, \quad x \neq 0, \tag{1}$$
has as its domain the set of all real nonzero numbers. The graph is the curve shown in Fig. 2.7.

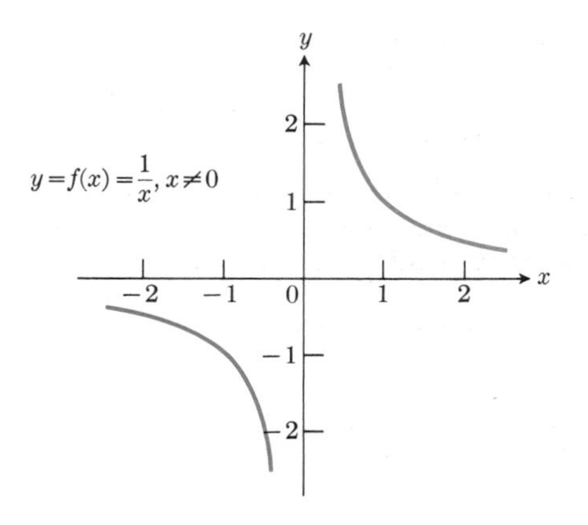

2.7

There are some facts about $1/x$ that are evident:

(a) When x is a small positive number, $1/x$ is a large positive number. For example,
$$1/0.001 = 1000.$$

(b) When x is a large number, $1/x$ is small. For example,
$$1/10{,}000 = 0.0001.$$

(c) When x is negative, $1/x$ is negative. For example,
$$1/(-0.001) = -1000$$
$$1/(-10{,}000) = -0.0001.$$

These facts are sometimes abbreviated by saying that

(a) $1/x$ approaches 0 as x tends to either $+\infty$ or $-\infty$,

(b) $1/x$ tends to $+\infty$ as x approaches 0 through positive values,

(c) $1/x$ tends to $-\infty$ as x approaches 0 through negative values.

The symbol ∞, is read *infinity*, but it does not represent any real number.

Note. In more advanced mathematics, a new system consisting of the real numbers and two additional elements, called $+\infty$ and $-\infty$, is sometimes used. It is called the *extended* real number system. However, special rules have to be set up, such as

$$a + (+\infty) = +\infty,$$
$$a - (+\infty) = -\infty,$$
$$a \cdot (+\infty) = +\infty \quad \text{if} \quad a > 0,$$
$$a \cdot (+\infty) = -\infty \quad \text{if} \quad a < 0,$$

for any real number a, and

$$(+\infty) \cdot (+\infty) = +\infty,$$
$$(+\infty) + (+\infty) = +\infty.$$

But such operations as $(+\infty) + (-\infty)$ and $0 \cdot (+\infty)$, or $(+\infty)/(+\infty)$ are not permitted because no consistent (unique) values can be assigned to such operations.

Mainly we shall avoid using the extended real number system in this book. And even in that system, division by zero has to be excluded if we hope to retain the usual laws of arithmetic and algebra.

Definition. *If $f(x)$ is defined for large values of x, we say that $f(x)$ approaches the real number L as limit as x tends to infinity, and write*

$$\lim_{x\to\infty} f(x) = L,$$

provided that to each positive number ϵ there corresponds a number M such that

$$L - \epsilon < f(x) < L + \epsilon \quad \textit{when } x > M. \tag{2}$$

Remark 1. Graphically, the inequalities in (2) mean that the curve $y = f(x)$ stays between the lines $y = L - \epsilon$ and $y = L + \epsilon$ for x sufficiently large. The part of the statement that reads

$$\text{when } x > M$$

can also be written

$$\text{for } M < x < \infty,$$

or

$$\text{for } x \text{ in the open interval } (M, \infty).$$

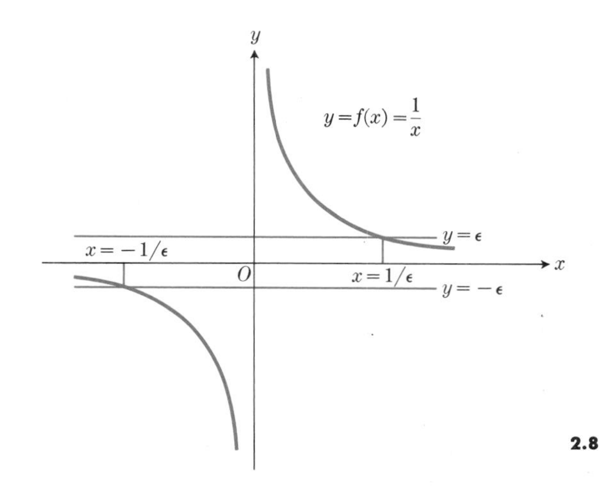

2.8

Such an open interval, consisting of all numbers greater than some fixed number M, is also called a *neighborhood* of ∞. Thus we could say that $f(x)$ stays inside the neighborhood $(L - \epsilon, L + \epsilon)$ when x is in the neighborhood (M, ∞).

Example 1. $$\lim_{x\to\infty} \frac{1}{x} = 0.$$

Example 2. $$\lim_{x\to\infty} \frac{x}{x+1} = \lim_{x\to\infty} \frac{1}{1 + (1/x)} = 1.$$

Example 3.

$$\lim_{x\to\infty} \frac{2x^2 - x + 3}{3x^2 + 5} = \lim_{x\to\infty} \frac{2 - (1/x) + (3/x^2)}{3 + (5/x^2)} = \frac{2}{3}.$$

For Example 1, $f(x) = 1/x$, we have

$$-\epsilon < \frac{1}{x} < \epsilon \tag{3a}$$

for any $\epsilon > 0$, provided that

$$|x| > \frac{1}{\epsilon}. \tag{3b}$$

Thus, $f(x)$ is in the neighborhood $(-\epsilon, \epsilon)$ of zero, when x is in the neighborhood $(1/\epsilon, \infty)$ of ∞; see Fig. 2.8.

See Fig. 2.9 for the example

$$f(x) = 2 + \frac{\sin x}{x}.$$

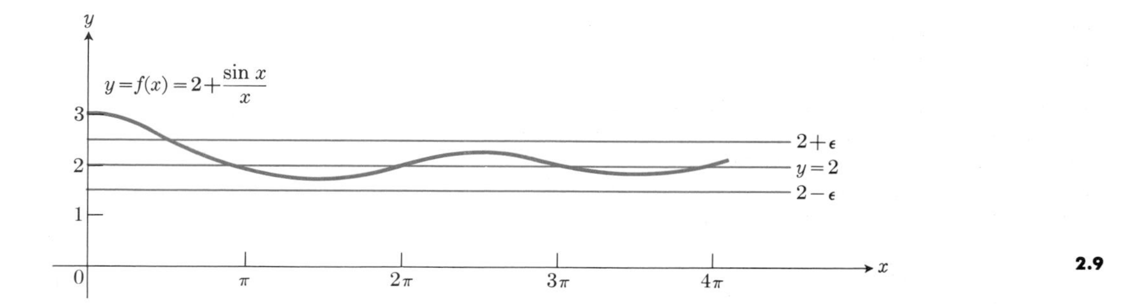

2.9

The graph of this function oscillates about the line $y = 2$. The amplitude of the oscillations decreases toward zero as $x \to \infty$, and the curve lies between $y = 2 + \epsilon$ and $y = 2 - \epsilon$ when $x > 1/\epsilon$. When x is large, $\sin x/x \to 0$ and $f(x) \to L = 2$.

The following theorems about limits of sums, differences, products, and quotients are analogous to the corresponding theorems for limits when $x \to a$.

Theorem 7.

If $\lim_{x\to\infty} f(x) = L_1$ *and* $\lim_{x\to\infty} g(x) = L_2$,

where L_1 and L_2 are (finite) real numbers, then:

$$\lim_{x\to\infty} [f(x) + g(x)] = L_1 + L_2, \tag{4a}$$

$$\lim_{x\to\infty} f(x)g(x) = L_1L_2, \tag{4b}$$

$$\lim_{x\to\infty} [f(x) - g(x)] = L_1 - L_2, \tag{4c}$$

$$\lim_{x\to\infty} \frac{f(x)}{g(x)} = \frac{L_1}{L_2}, \quad \textit{if} \quad L_2 \neq 0. \tag{4d}$$

To prove these theorems, one only needs to repeat the proofs for the corresponding theorems for limits as $x \to a$ with only minor changes (mainly, replacing neighborhoods of a by neighborhoods of ∞).

Example 4. $$\lim_{x\to\infty} x \sin\frac{1}{x} = \lim_{y\to 0} \frac{\sin y}{y} \quad \left(y = \frac{1}{x}\right)$$
$$= 1.$$

Finally, we sometimes want to say such things as

$$\lim_{x\to a} f(x) = \infty, \tag{5a}$$

$$\lim_{x\to a^+} f(x) = \infty, \tag{5b}$$

$$\lim_{x\to a^-} f(x) = \infty, \tag{5c}$$

$$\lim_{x\to\infty} f(x) = \infty. \tag{5d}$$

In every instance, the symbol ∞ means the same as

$$f(x) \to +\infty$$

or that for any real number M that is given, no matter how large M may be, the condition

$$f(x) > M$$

is satisfied for all values of x in some restricted set, usually depending upon M. The appropriate set in (5a) is

in some deleted neighborhood of a.

In (5b), $x \to a^+$ means that we consider numbers close to a but greater than a; that is, an interval of the form $(a, a + h)$ for some $h > 0$. In (5c), $x \to a^-$ means we restrict attention to values of x that are near a and less than a; an interval of the form $(a - h, a)$ with $h > 0$. In (5d), the neighborhood is of the form (c, ∞); that is, for x large, or for $x > c$ for some c.

The following theorems are stated without proof. The exercises ask for proofs of some of them. They involve various ideas about *bounded* functions. We shall first give some definitions and an example.

Definitions. *If there is a real number b such that for all x in the domain of g*

$$g(x) \geq b,$$

then $g(x)$ is bounded below.

If there is a real number b and a real number c such that for all x in the domain of g

$$g(x) \geq b \quad \text{for} \quad x > c,$$

then $g(x)$ is bounded below for x large.

If there is a real number a such that for all x in the domain of g

$$g(x) \leq a,$$

then $g(x)$ is bounded above.

If there are real numbers a, b, and c such that

$$a < g(x) < b \quad \text{when } x > c,$$

then $g(x)$ is bounded for x large.

Example 5.

$$g(x) = \frac{2x}{x+1} + \frac{1}{x}$$

is bounded for x large because

$$0 < g(x) < 2 \quad \text{when } x > 1.$$

Theorem 8. *If*

$$\lim_{x\to\infty} f(x) = \infty$$

and $g(x)$ is bounded below, for x large,

then
$$\lim_{x\to\infty} [f(x) + g(x)] = \infty.$$

Example 6.

$$\lim_{x\to\infty} x = \infty \quad \text{and} \quad \sin x \geq -1.$$

Hence

$$\lim_{x\to\infty} (x + \sin x) = \infty,$$

even though $\sin x$ does *not* have a limit as $x \to \infty$.

Theorem 9. *If*

$$\lim_{x\to\infty} f(x) = 0$$

and $g(x)$ is bounded for x large,

then
$$\lim_{x\to\infty} f(x)g(x) = 0.$$

Example 7.

$$\lim_{x\to\infty} (1/x) = 0 \quad \text{and} \quad -1 \leq \cos x \leq 1,$$

that is, $\cos x$ is bounded. Hence

$$\lim_{x\to\infty} \frac{\cos x}{x} = 0.$$

EXERCISES

Use Theorems 7, 8, and 9 to find the following limits.

1. $\lim\limits_{x\to\infty} \dfrac{2x+3}{5x+7}$
2. $\lim\limits_{x\to\infty} \dfrac{3x^3 + 5x^2 - 7}{10x^3 - 11x^2 + 5x}$
3. $\lim\limits_{x\to\infty} \dfrac{3x^2 - 6x}{4x - 8}$
4. $\lim\limits_{x\to\infty} \dfrac{7x - 28}{x^3}$
5. $\lim\limits_{x\to\infty} \dfrac{2x^2 + 3}{x^3 + x - 5} \operatorname{sgn} x$
6. $\lim\limits_{x\to\infty} \left(\dfrac{1}{x} + 1\right)\left(\dfrac{5x^2 - 1}{x^2}\right)$
7. $\lim\limits_{x\to\infty} \left(\sqrt{x} + \dfrac{1}{[x]}\right)$
8. $\lim\limits_{x\to\infty} [\tan x + \sec x(5 \cos x - \sin x)]$
9. $\lim\limits_{x\to\infty} \dfrac{\sqrt{x+1}}{\sqrt{4x - 1}}$
10. What is
$$\lim_{x\to\infty} \frac{f(x)}{g(x)}$$
if $f(x)$ is a polynomial of degree less than the degree of the polynomial $g(x)$?
11. What does $f(x)/g(x)$ tend to, as $x \to \infty$, if the degree of the polynomial $g(x)$ is less than the degree of the polynomial $f(x)$?
12. Prove Eq. (4a) of Theorem 7.
13. Use Eq. (4b) of Theorem 7 to prove the following corollary: If $\lim\limits_{x\to\infty} f(x) = L$ where L is finite,

then
$$\lim_{x\to\infty} [-f(x)] = -L.$$

14. Use the result of Exercise 13 to prove Eq. (4c) of Theorem 7.
15. Prove Theorem 8.
16. Prove Theorem 9.
17. (a) Is the polynomial $g(x) = x^2 - 3x + 4$ bounded below for x large?
    (b) Is it bounded below for all values of x?
    (c) Is it bounded for x large?
18. State and prove a theorem like Theorem 6 as x approaches ∞.
19. If
$$\frac{2x-3}{x} < f(x) < \frac{2x^2+5x}{x^2},$$
find $\lim_{x\to\infty} f(x)$.
20. Suppose that $f(x)$ and $g(x)$ are two polynomials, and that $g(x)$ is bounded below for x large. If the degree of $f(x)$ is less than the degree of $g(x)$, prove that $f(x) + g(x)$ is bounded below for x large.

2.5 LIMITS APPLIED TO AREAS

In Chapter 1, we met the limit concept in finding the slope of a line tangent to the graph of a function. This is one of the chief ways that limits are used in calculus. The second way is in connection with area, as Problem 1 below shows. In that problem, we need to know that the sum of the squares of the first n positive integers is given by the formula

$$1^2 + 2^2 + 3^2 + \cdots + n^2 = \frac{n(n+1)(2n+1)}{6}, \tag{1}$$

for any positive integer n. For the first few positive integers, we get from Eq. (1)

$$1^2 = \frac{1(1+1)(2+1)}{6}, \quad \text{or} \quad 1 = \frac{6}{6},$$

$$1^2 + 2^2 = \frac{2(2+1)(4+1)}{6}, \quad \text{or} \quad 5 = \frac{30}{6},$$

$$1^2 + 2^2 + 3^2 = \frac{3(3+1)(6+1)}{6}, \quad \text{or} \quad 14 = \frac{84}{6},$$

all of which are correct. No amount of such verification will establish the formula for all positive integers n, but the general method of *mathematical induction* will.

Principle of mathematical induction. Suppose that to each positive integer n there corresponds a proposition (or theorem, or formula) P_n which is either true or false. If P_1 is true, and for every positive integer k the truth of proposition P_k implies the truth of the following proposition P_{k+1}, then all of the propositions

$$P_1, P_2, P_3, \ldots, P_n, \ldots$$

are true. That is, P_n is true for every positive integer n.

[For a detailed discussion, see R. Courant and H. Robbins, *What is Mathematics?*, Oxford University Press, 1941, pp. 9–16, or E. P. Vance, *An Introduction to Modern Mathematics*, Addison-Wesley, 1963, pp. 246–251.] To apply this principle, we need, *first*, to show that P_1 is true, and, *second*, to show that $P_k \Rightarrow P_{k+1}$. (The symbol $\Rightarrow$ is read "implies.")

In the present application, let P_n be the proposition that Eq. (1) is correct for the integer n. We have verified the truth of P_1, P_2, and P_3 above. Such verification is necessary only for P_1, but the additional checking was easy and not harmful. Now we see if we can show that $P_k \Rightarrow P_{k+1}$ for any positive integer k. To test whether this is so, we assume that k is an integer for which P_k is true:

$$1^2 + 2^2 + \cdots + k^2 = \frac{k(k+1)(2k+1)}{6}.$$

If we add $(k+1)^2$ to both sides of this equation, we get

$$\begin{aligned} 1^2 + 2^2 + \cdots + k^2 + (k+1)^2 &= \frac{k(k+1)(2k+1)}{6} + (k+1)^2 \\ &= \frac{(k+1)}{6}[k(2k+1) + 6(k+1)] \\ &= \frac{(k+1)}{6}(2k^2 + 7k + 6). \end{aligned} \tag{2a}$$

Now P_{k+1} is the proposition that the sum of the squares on the left above should equal

$$n(n+1)(2n+1)/6$$

with $n = k + 1$; in other words, it should equal

$$(k+1)(k+2)(2k+3)/6.$$

Hence the truth of P_{k+1} follows from P_k provided it is true that

$$\frac{(k+1)}{6}(2k^2+7k+6) = \frac{(k+1)(k+2)(2k+3)}{6}. \tag{2b}$$

This equation is easily seen to be true because $(k+2)(2k+3) = 2k^2 + 7k + 6$. Thus the two steps needed to establish Eq. (1) for all positive integers n are complete. We shall apply the result in finding an area in the following problem.

Problem 1. Area. Find the area of the region AOB bounded above by the arc OB of the curve $y = x^2$, below by the segment OA of the x-axis, and on the right by the segment AB through the points $A(b, 0)$ and $B(b, b^2)$, $b > 0$. See Fig. 2.10, which shows the graph of $y = x^2$, $0 \leq x \leq b$, and $n = 8$ circumscribed rectangles.

Solution. Let n be a positive integer and divide the line segment OA into n subintervals, each of length $h = b/n$. In Fig. 2.10 we have used $n = 8$, but in the algebraic formulas below n can be any positive integer. With each subinterval we associate a rectangle having its base on that subinterval and its upper right vertex on the curve $y = x^2$. We number these rectangles $1, 2, \ldots, n$ in order from left to right. Thus

rectangle number	has its base on the subinterval	and has altitude
1	$0 \leq x \leq h$	h^2
2	$h \leq x \leq 2h$	$(2h)^2$
3	$2h \leq x \leq 3h$	$(3h)^2$
$\vdots$	$\vdots$	$\vdots$
n	$(n-1)h \leq x \leq nh$	$(nh)^2$

Let S_n denote the sum of the areas of these n circumscribed rectangles. Then

$$\begin{aligned} S_n &= h(h)^2 + h(2h)^2 + h(3h)^2 + \cdots + h(nh)^2 \\ &= h^3[1^2 + 2^2 + 3^2 + \cdots + n^2]. \end{aligned} \tag{3}$$

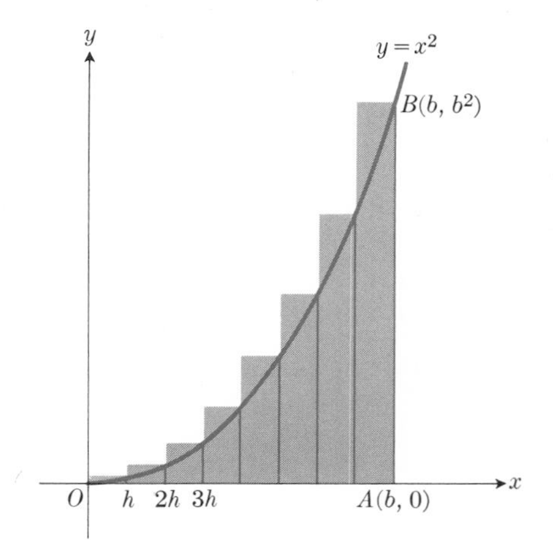

2.10 Circumscribed rectangles: $n = 8$.

But $h = b/n$, and the sum of squares in brackets in Eq. (3) is equal to $n(n+1)(2n+1)/6$, so we can also write S_n in the form

$$\begin{aligned} S_n &= \frac{b^3}{n^3}\,\frac{n(n+1)(2n+1)}{6} \\ &= \frac{b^3}{6}\left(\frac{n}{n}\right)\left(\frac{n+1}{n}\right)\left(\frac{2n+1}{n}\right). \end{aligned} \tag{4}$$

If S_n has a limit as $n \to \infty$, we define that limit to be the area of the region AOB. The last three quotients in Eq. (4) have the following limits:

$$\lim_{n\to\infty}\left(\frac{n}{n}\right) = 1, \quad \lim_{n\to\infty}\left(\frac{n+1}{n}\right) = 1,$$

$$\lim_{n\to\infty}\left(\frac{2n+1}{n}\right) = 2,$$

so

$$\text{area of } AOB = \lim S_n = \frac{b^3}{6}(1)(1)(2) = \frac{b^3}{3}. \tag{5}$$

The area of the parabolic region AOB is therefore $b^3/3$. Note that the area inside the triangle AOB is $b^3/2$. Thus the parabolic area is $\frac{2}{3}$ that of the corresponding triangle.

If we want the area of only a part of the region, say from a to b instead of from 0 to b, we can easily get it by subtraction. For example, suppose $0 < a < b$ and we want the area of the region over

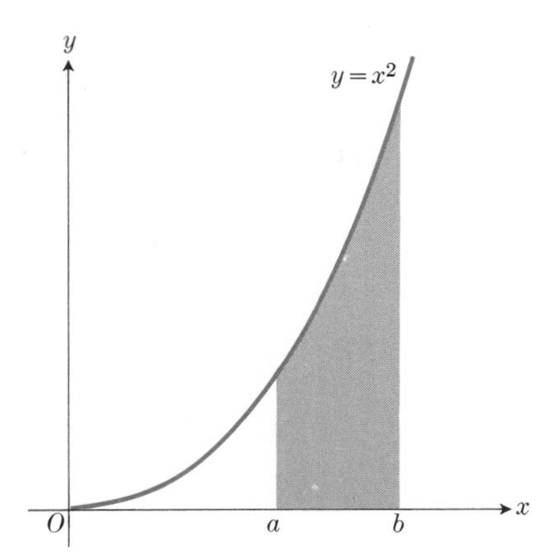

2.11 The area between $x = a$ and $x = b$, over the x-axis and under the curve $y = x^2$.

the interval $[a, b]$ and under the curve as shown in Fig. 2.11. The unshaded region to the left of $x = a$ has area

$$A_0^a = \frac{a^3}{3},$$

as given by Eq. (5), with a in place of b. The area between $x = a$ and $x = b$ is therefore just the difference

$$A_a^b = A_0^b - A_0^a = \frac{b^3}{3} - \frac{a^3}{3}. \tag{6}$$

Note that although we deduced Eq. (6) from Eq. (5) on the assumption that $a > 0$, it reduces to the correct result if $a = 0$, simply giving Eq. (5) back to us. It is also a correct formula when $a < 0$ because the curve $y = x^2$ is symmetric with respect to the y-axis and $-a^3/3$ corresponds to the area from $x = a$ to $x = 0$ when a is negative.

In calculus, the area of the parabolic region of Problem 1 is represented by $\int_a^b x^2\, dx$, which is read "the integral from a to b of $x^2\, dx$." Our result could therefore be expressed

$$\int_a^b x^2\, dx = \frac{b^3}{3} - \frac{a^3}{3}. \tag{7}$$

In most area problems, the algebra involved in a direct calculation using limits would be much more complicated. The great feature of calculus is that it gives a method for getting the answer without doing all that algebra. We shall come back to this in Chapter 5, where we shall develop the general theory. As a preview, here is a short outline of the program:

(a) Area under a curve is *defined* in terms of limits.

(b) If the graph of $y = f(t)$ is an unbroken curve lying above the t-axis over $[a, b]$, and $a < x < b$, then the area A_a^x of the region under the curve over $[a, x]$ is a differentiable function $F(x)$.

(c) The derivative of the area function F at x is the ordinate of the graph of f at x: $F'(x) = f(x)$.

(d) If G is any function whose derivative is f, then the area A_a^x is equal to $G(x) - G(a)$.

Definition. *A function G such that $G'(x) = f(x)$ is called a primitive of f, or an antiderivative of f, or an integral of f.*

Problem 2. If C is any constant and

$$G(x) = \frac{x^3}{3} + C,$$

show that G is a primitive of $f(x) = x^2$.

Solution. Let $\Delta x = h$ be different from zero.

Then $$G(x + \Delta x) = \frac{(x + h)^3}{3} + C,$$

and $$\frac{G(x + \Delta x) - G(x)}{\Delta x} = x^2 + xh + \frac{h^2}{3}.$$

When $\Delta x = h \to 0$, the right side of this equation has a limit x^2. Therefore, by definition of the derivative,

$$G'(x) = \lim_{\Delta x \to 0} \frac{G(x + \Delta x) - G(x)}{\Delta x} = x^2.$$

Remark. By item (d) of the outline above, the area from a to b under the parabola of Fig. 2.11 should be

$$G(b) - G(a) = \left(\frac{b^3}{3} + C\right) - \left(\frac{a^3}{3} + C\right) = \frac{b^3}{3} - \frac{a^3}{3},$$

as we also found by the direct limit approach.

EXERCISES

By the method of mathematical induction, prove the validity of statements 1 through 3, for all positive integers n.

1. $1 + 2 + 3 + \cdots + n = \dfrac{n(n+1)}{2}$

2. $1^3 + 2^3 + 3^3 + \cdots + n^3 = \dfrac{n^2(n+1)^2}{4}$

3. $1 \cdot 2 + 3 \cdot 4 + 5 \cdot 6 + \cdots + (2n-1)(2n)$
$$= \frac{n(n+1)(4n-1)}{3}$$

4. Prove that $n^3 - n$ is divisible by 3 for every positive integer n. [*Hint.* Use
$$(k+1)^3 - (k+1) = (k^3 - k) + 3(k^2 + k).]$$
Note. This is a special case of a theorem of Fermat that states that $n^p - n$ is divisible by p, if p is a prime.

5. Find the area of the region AOB of Problem 1, using *inscribed* rectangles instead of circumscribed rectangles. [*Check.* You should get
$$s_n = h^3[0^2 + 1^2 + 2^2 + \cdots + (n-1)^2]$$
in place of Eq. (3).]

6. Let S_n represent the sum of the areas of the n circumscribed rectangles used in Problem 1, and let s_n denote the sum of the areas of the corresponding inscribed rectangles used in Exercise 5. Show that
$$S_n - s_n = h^3n^2 = \frac{b^3}{n}.$$
Given b and $\epsilon > 0$, how large should N be in order to guarantee that $|S_n - s_n| < \epsilon$ whenever $n \geq N$? Would this also guarantee that both S_n and s_n differ from the area of the parabolic region AOB by no more than ϵ? Why?

7. Using the notation of Exercise 6, show that
$$s_n < \frac{b^3}{3} < S_n,$$
for every positive integer n.

8. Using the notation of Exercise 6, show that $\{s_n\}$ is a monotonically increasing sequence and that $\{S_n\}$ is a monotonically decreasing sequence. What is the geometric significance of these facts?

9. Show that the area of the shaded region in Fig. 2.11 is equal to $(b - a)c^2$ for some $c \in (a, b)$. Interpret this result geometrically as the area of a rectangle of altitude c^2.
Suggestion. Factor $b^3 - a^3$ and show that
$$3a^2 < a^2 + ab + b^2 < 3b^2 \quad \text{if} \quad 0 < a < b.$$

10. In Problem 1, relabel the horizontal axis as the t-axis and the curve as $y = t^2$. Then let $x > 0$ and let $F(x)$ be the area of the region under the curve and over the interval $[0, x]$. Equation (5) with $b = x$ then gives $F(x) = x^3/3$. Verify by direct calculation that the derivative of this function at x is the ordinate of the parabola $y = t^2$ at $t = x$.

In Exercises 11 through 13, let $f(x) = mx$ where $m > 0$. Let R be the triangular region bounded above by the line $y = mx$, below by the x-axis, and on the right by the line $x = b$, $b > 0$. Let n be a positive integer, and divide the interval $[0, b]$, mentally, into n subintervals of equal lengths $h = b/n$. With each subinterval associate a rectangle with its lower base on that subinterval and its upper base passing through a point of the corresponding segment of the line $y = mx$. Consider three possibilities:

(1) n circumscribed rectangles,

(2) n inscribed rectangles, and

(3) n rectangles with upper bases passing through the midpoints of the segments of $y = mx$ over the n subintervals.

11. Find s_n = the sum of areas of the inscribed rectangles, and $\lim_{n\to\infty} s_n$. Compare with the known formula for the area of R.

12. Find S_n = the sum of areas of the circumscribed rectangles, and $\lim_{n\to\infty} S_n$. Compare with the known formula for the area of R.

13. Find T_n = the sum of areas of the n "midpoint" rectangles, and $\lim_{n\to\infty} T_n$. Compare with the known area of the triangular region R.

14. If you solved Exercises 11 through 13, show that

$$T_n = \tfrac{1}{2}(s_n + S_n),$$

and obtain an expression for

$$S_n - T_n = T_n - s_n$$

in terms of n, b, and m. How large should n be to make this common difference less than ϵ, if $\epsilon > 0$?

15. (a) Show that

$$G(x) = \frac{mx^2}{2} + C$$

is an antiderivative of $f(x) = mx$, for C any constant.
(b) Let $0 < a < b$, and verify that the area of the trapezoidal region bounded above by the line $y = mx$, below by the x-axis, on the left by $x = a$, and on the right by $x = b$, is $G(b) - G(a)$, for G as in part (a) above.

16. Let D be the region bounded above by the line $y = mt$, below by the t-axis, on the left by the line $t = a$, and on the right by the line $t = x$. The area of this trapezoidal region is

$$F(x) = (x - a)\frac{ma + mx}{2} = \frac{m}{2}(x^2 - a^2).$$

Prove that the derivative is $F'(x) = mx =$ ordinate of line at $t = x$.

REVIEW QUESTIONS AND EXERCISES

1. Define what it means to say that

(a) $\lim_{x \to 2} f(x) = 5$, (b) $\lim_{x \to c} g(x) = k$.

2. Give two examples of deleted neighborhoods of the point $x = 2$, of which one is symmetric about 2 and the other is not.

3. Describe the intersection of the deleted neighborhoods

$$0 < |x - a| < 2 \quad \text{and} \quad (a - 1, a) \cup (a, a + 3).$$

4. What is wrong with the following "proof" of Theorem 5? Let

$$f_1(x) = g(x), \quad f_2(x) = 1/g(x),$$
$$L_1 = \lim_{x \to a} f_1(x), \quad L_2 = \lim_{x \to a} f_2(x).$$

Then, by Theorem 4,

$$\lim_{x \to a} f_1(x)f_2(x) = L_1L_2.$$

But $f_1(x)f_2(x) = 1$, so we have, by Theorem 1 with $m = 0$, $b = 1$,

$$1 = \lim_{x \to a} f_1(x)f_2(x) = L_1L_2.$$

Therefore

$$L_2 = 1/L_1.$$

5. Corollary 4.2, with $n = 2$, asserts that $\lim_{x \to a} x^2 = a^2$, but it does not say that $\lim_{x \to a} \sqrt{x} = \sqrt{a}$. Try to prove this result, for $a > 0$. (Do not assume the existence of the limit of $\sqrt{x}$. Work directly from the definition of limit.)

Evaluate the following limits.

6. $\lim_{x \to 1} \dfrac{x + 1}{x + 2}$

7. $\lim_{x \to 1} \dfrac{x - 1}{x + 2}$

8. $\lim_{x \to 2} \dfrac{x^2 + 1}{x^2 + 2}$

9. $\lim_{x \to 2} \dfrac{x^2 - x - 2}{x^2 - 4}$

10. $\lim_{x \to 2} \dfrac{x^2 - 4}{x^3 - 8}$

11. $\lim_{x \to 0} \dfrac{x^2 - 4}{x^3 - 8}$

12. $\lim_{x \to 0} \dfrac{\sin 3x}{5x}$

13. $\lim_{h \to 0} \dfrac{\sin 3h}{\sin 5h}$

14. $\lim_{\theta \to 0} \dfrac{\sin^2 \theta}{\theta}$

15. $\lim_{\theta \to 0} \dfrac{\sin^2 \theta}{\theta^2}$

16. (a) $\lim_{\theta \to 0} \dfrac{1 - \cos \theta}{\theta^2}$
Hint. Multiply and divide by $1 + \cos \theta$.

(b) $\lim_{\theta \to 0} \dfrac{1 - \cos \theta}{\theta}$

Determine which of the following indicated limits exist and evaluate those that do.

17. $\lim_{x \to \infty} \dfrac{2x + 3}{5x + 7}$

18. $\lim_{z \to \infty} \dfrac{4z^3 + 3z^2 + 2}{z^4 + 5z^2 + 1}$

19. $\lim_{x\to\infty} \frac{\sin x}{\sqrt{x}}$

20. $\lim_{x\to\infty} (1 - \cos x)$

21. $\lim_{x\to\infty} \frac{1 - \cos x}{x}$

22. $\lim_{x\to\infty} \sqrt{\frac{2 + 3x}{1 + 5x}}$

23. $\lim_{x\to\infty} \sqrt{\frac{2 + 3x}{1 - 5x}}$

24. $\lim_{x\to\infty} x^x$

25. $\lim_{x\to\infty} (x)^{1/x}$

[Try to guess an answer and argue for it even if you cannot prove it is correct. Use logarithms to evaluate $(1000)^{0.001}$.]

MISCELLANEOUS EXERCISES

Evaluate the following limits, or show that the limit does not exist.

1. $\lim_{x\to\infty} \frac{x + \sin x}{2x + 5}$

2. $\lim_{x\to\infty} \frac{1 + \sin x}{x}$

3. $\lim_{x\to 1} \frac{x^2 - 4}{x^3 - 8}$

4. $\lim_{x\to 0} \frac{x}{\tan 3x}$

5. $\lim_{x\to\infty} \frac{x \sin x}{x + \sin x}$

6. $\lim_{x\to a} \frac{x^2 - a^2}{x - a}$

7. $\lim_{x\to a} \frac{x^2 - a^2}{x + a}$

8. $\lim_{h\to 0} \frac{(x + h)^2 - x^2}{h}$

9. $\lim_{h\to 0} \frac{\sqrt{x + h} - \sqrt{x}}{h}$

10. $\lim_{\Delta x\to 0} \frac{1/(x + \Delta x) - 1/x}{\Delta x}$

11. $\lim_{x\to 0+} \frac{1}{x}$

12. $\lim_{x\to 1} \frac{1 - \sqrt{x}}{1 - x}$

13. $\lim_{x\to 1} \frac{(2x - 3)(\sqrt{x} - 1)}{2x^2 + x - 3}$

14. $\lim_{x\to\infty} (1 - x \cos x)$

The following exercises deal with limits of sequences. A function whose domain is the set of positive integers 1, 2, . . . is said to be a *sequence*. Instead of writing the value of the function at n as $f(n)$, it is customary to write the nth term of the sequence as a_n, or S_n, or something similar. A sequence a_n is said to converge to the limit L as $n \to \infty$ if and only if to each positive number ϵ there corresponds an integer N such that

$|a_n - L| < \epsilon$ for all values of n greater than N.

The following theorems about limits of sequences are analogous to theorems in this chapter about limits of other functions. Prove them.

15. If $\lim_{n\to\infty} a_n = A$ and $\lim_{n\to\infty} b_n = B$,

then $\lim_{n\to\infty} (a_n + b_n) = A + B.$

16. If $\lim_{n\to\infty} a_n = A$ and k is a number,

then $\lim_{n\to\infty} (ka_n) = kA.$

17. If $\lim a_n$ exists (is a finite real number), then the sequence a_n is bounded.

18. If $\lim_{n\to\infty} a_n = A$ and $\lim_{n\to\infty} b_n = B$,

then $\lim_{n\to\infty} (a_n b_n) = AB.$

19. If $\lim_{n\to\infty} b_n = B \neq 0,$

then $\lim_{n\to\infty} 1/b_n = 1/B.$

Evaluate the following limits.

20. $\lim_{n\to\infty} \frac{3n^2 - 5n + \sin n}{4n^2 + 7n + 6}$

Hint. Divide the numerator and denominator by n^2.

21. $\lim_{n\to\infty} \frac{2n^2 + 5n - 3}{n^3 + 2n}$

22. $\lim_{n\to\infty} (\sqrt{n^2 + n} - \sqrt{n^2 + 10})$

23. Suppose that $h_n = \sqrt[n]{n} - 1$. Then

$$n = (1 + h_n)^n = 1 + nh_n + \frac{n(n-1)}{2} h_n^2 + \cdots > \frac{n(n-1)}{2} h_n^2$$

when $n > 2$. From the above inequality, and the definition of h_n, deduce that

$$0 < h_n < \sqrt{\frac{2}{n - 1}} \quad \text{when} \quad n > 2.$$

Find (a) $\lim_{n\to\infty} h_n$ (b) $\lim_{n\to\infty} \sqrt[n]{n}$

The *right-hand limit* b of a function f at a point a is the limit of the function at a for a right-hand domain

$(a, a+\delta)$, $\delta > 0$. We write

$$\lim_{x \to a^+} f(x) = b.$$

Similarly, the *left-hand limit* c of a function f at a point a is the limit of the function at a for a left-hand domain $(a - \delta, a)$, $\delta > 0$. We write

$$\lim_{x \to a^-} f(x) = c.$$

For example, if $[x]$ is the greatest integer in x, then $\lim_{x \to 5^+} [x] = 5$, but $\lim_{x \to 5^-} [x] = 4$.

Find each of the following limits.

24. $\lim_{x \to 0^+} \dfrac{|x|}{x}$

25. $\lim_{x \to 0^-} \dfrac{|x|}{x}$

26. $\lim_{x \to 4^-} ([x] - x)$

27. $\lim_{x \to 4^+} ([x] - x)$

28. $\lim_{x \to 3^+} \dfrac{[x]^2 - 9}{x^2 - 9}$

29. $\lim_{x \to 3^-} \dfrac{[x]^2 - 9}{x^2 - 9}$

30. $\lim_{x \to 4^+} [[x]]$

31. $\lim_{x \to 0^+} \dfrac{\sqrt{x}}{\sqrt{4 + \sqrt{x}} - 2}$

32. Given $(x - 1)/(2x^2 - 7x + 5) = f(x)$. Find
(a) the limit of $f(x)$ as $x \to \infty$,
(b) the limit of $f(x)$ as $x \to 1$,
(c) $f(-1/x)$, $f(0)$, $1/f(x)$.

33. Compute the coordinates of the point of intersection of the straight lines

$$3x + 5y = 1, \quad (2 + c)x + 5c^2y = 1,$$

and determine the limiting position of this point as c tends to 1.

34. Find

(a) $\lim_{n \to \infty} (\sqrt{n^2 + 1} - n)$, (b) $\lim_{n \to \infty} (\sqrt{n^2 + n} - n)$.

35. Given $\epsilon > 0$, find $\delta > 0$ such that $\sqrt{t^2 - 1} < \epsilon$ when $0 < (t - 1) < \delta$.

36. Given $\epsilon > 0$, find M such that

$$\left| \frac{t^2 + t}{t^2 - 1} - 1 \right| < \epsilon \quad \text{when } t > M.$$

37. Show that $\lim_{t \to 0} t \sin (1/t)$ exists and is zero, even though $\sin (1/t)$ has no limit as t approaches zero.

38. Prove that if $f(t)$ is bounded [that is, $|f(t)| < M$ for some constant M] and $g(t)$ approaches zero as t approaches a, then $\lim_{t \to a} f(t)g(t) = 0$.

39. Prove that if $f(t)$ has a finite limit as t approaches a, then there exist numbers m, M, and $h > 0$ such that

$$m < f(t) < M \quad \text{if} \quad 0 < |t - a| < h.$$

DERIVATIVES OF ALGEBRAIC FUNCTIONS

CHAPTER 3

3.1 POLYNOMIAL FUNCTIONS AND THEIR DERIVATIVES

A single term of the form cx^n, where c is a constant and n is zero or a positive integer, is called a *monomial* in x. A function which is the sum of a finite number of monomial terms is called a *polynomial* in x. For example,

$$f(x) = x^3 - 5x + 7, \quad g(x) = (x^2 + 3)^3,$$
$$h(x) = 4x, \quad \phi(x) = 5$$

are special cases of polynomials in x, and

$$s = \tfrac{1}{2}gt^2, \quad v = v_0 + gt$$

are polynomials in t.

We shall now deduce some formulas which will enable us to find the derivatives of polynomial functions very easily. In every case we derive the formula by means of this basic definition:

Definition. *Let $y = f(x)$ define a function f. If the limit*

$$\frac{dy}{dx} = \lim_{\Delta x \to 0} \frac{\Delta y}{\Delta x}, \tag{1a}$$

meaning

$$f'(x) = \lim_{\Delta x \to 0} \frac{f(x + \Delta x) - f(x)}{\Delta x}, \tag{1b}$$

exists and is finite, we call this limit the derivative of y with respect to x and say that f is differentiable at x.

Theorem 1. *The derivative of a constant is zero.*

Proof. Let

$$y = c,$$

where c is a constant. Then when $x = x_1$ and again when $x = x_1 + \Delta x$, y has the same value c,

hence $$y = c$$

and $$y + \Delta y = c,$$

so that $$\Delta y = 0.$$

Then, dividing by Δx, we have

$$\frac{\Delta y}{\Delta x} = 0.$$

The limit of 0, as $\Delta x \to 0$, certainly exists, and

$$\frac{dy}{dx} = \lim_{\Delta x \to 0} \frac{\Delta y}{\Delta x} = 0. \square$$

Remark 1. The slope of the line $y = c$ (Fig. 3.1) is zero at every point on it, so Theorem 3.1 is not telling us anything new. The formal proof is given here for completeness, and because it gives us an easy start.

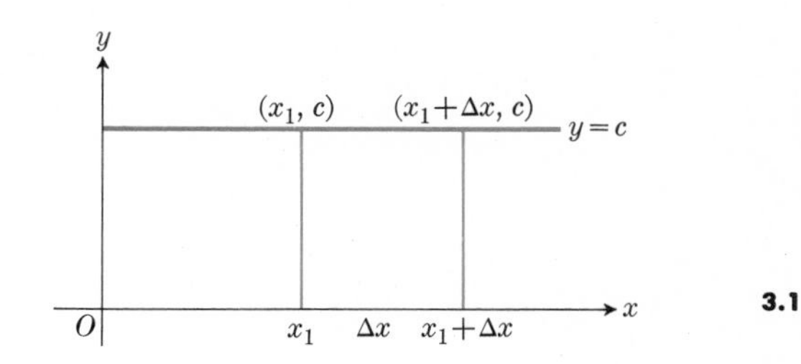

3.1

Theorem 2. *The derivative with respect to x of x^n is nx^{n-1} when n is any positive integer.*

Proof. Let $y = x^n$, where n may be any positive integer. Then, by the binomial theorem, we may express the sum $y + \Delta y$ as a polynomial. We have

$$y + \Delta y = (x + \Delta x)^n = \begin{cases} x + \Delta x & \text{if } n = 1, \\ x^2 + 2x\,\Delta x + (\Delta x)^2 & \text{if } n = 2, \\ x^3 + 3x^2\,\Delta x + 3x\,(\Delta x)^2 + (\Delta x)^3 & \text{if } n = 3, \\ x^n + nx^{n-1}\,\Delta x + (\text{terms in } x \text{ and } \Delta x) \cdot (\Delta x)^2 & \text{if } n > 3. \end{cases}$$

From this we subtract $y = x^n$ and obtain

$$\Delta y = \begin{cases} \Delta x & \text{if } n = 1, \\ 2x\,\Delta x + (\Delta x)^2 & \text{if } n = 2, \\ 3x^2\,\Delta x + (3x + \Delta x) \cdot (\Delta x)^2 & \text{if } n = 3, \\ nx^{n-1}\,\Delta x + (\text{terms in } x \text{ and } \Delta x) \cdot (\Delta x)^2 & \text{if } n > 3. \end{cases}$$

Dividing by Δx, we find next that

$$\frac{\Delta y}{\Delta x} = \begin{cases} 1 & \text{if } n = 1, \\ 2x + \Delta x & \text{if } n = 2, \\ 3x^2 + (3x + \Delta x) \cdot \Delta x & \text{if } n = 3, \\ nx^{n-1} + (\text{terms in } x \text{ and } \Delta x) \cdot \Delta x & \text{if } n > 3. \end{cases}$$

Finally, we let Δx approach zero and find

$$\frac{dy}{dx} = \lim_{\Delta x \to 0} \frac{\Delta y}{\Delta x} = \begin{cases} 1 & \text{if } n = 1, \\ 2x & \text{if } n = 2, \\ 3x^2 & \text{if } n = 3, \\ nx^{n-1} & \text{if } n > 3. \end{cases}$$

In particular, the case $n = 1$ tells us that

$$\frac{dx}{dx} = 1.$$

Since the results for $n = 1, 2, 3$ are simply special cases of the general result, we have

$$\frac{dy}{dx} = nx^{n-1} \quad \text{if} \quad y = x^n \tag{2}$$

and

$$n = \text{any positive integer.} \ \square$$

In words, this says: "*To find the derivative of x to the power n, simply multiply that power times x to a power one less.*"

Theorem 3. *If $u = f(x)$ is a differentiable function of x and c is a constant, then*

$$\frac{d(cu)}{dx} = c\,\frac{du}{dx}. \tag{3}$$

Proof. Let

$$y = cu, \tag{4a}$$

where

$$u = f(x).$$

Then if x is replaced by $x + \Delta x$, we have

$$y + \Delta y = c(u + \Delta u), \tag{4b}$$

where

$$u + \Delta u = f(x + \Delta x).$$

Subtracting (4a) from (4b), we obtain

$$\Delta y = c\,\Delta u$$

and dividing this by Δx, we have

$$\frac{\Delta y}{\Delta x} = c\,\frac{\Delta u}{\Delta x}. \tag{4c}$$

Since u has a derivative, the limit, when Δx approaches zero, of the right-hand side of Eq. (4c) exists:

$$\lim_{\Delta x \to 0} c\,\frac{\Delta u}{\Delta x} = c\,\frac{du}{dx}.$$

Since both sides of (4c) are equal, its left-hand side must have the same limit. Because this limit exists, it is also equal to dy/dx, from the definition of the derivative. All of this is summarized in the following equation, which follows by taking limits in Eq. (4c):

$$\frac{dy}{dx} = \lim_{\Delta x \to 0} \frac{\Delta y}{\Delta x} = \lim_{\Delta x \to 0} c\,\frac{\Delta u}{\Delta x} = c\,\frac{du}{dx}.$$

Since $y = cu$, this is the same as

$$\frac{d(cu)}{dx} = c\,\frac{du}{dx}. \ \square$$

Corollary. *If n is a positive integer and c is a number, then the function cx^n is differentiable for all x, and*

$$\frac{d}{dx}(cx^n) = cnx^{n-1}. \tag{5}$$

Proof. Let $u = x^n$, and apply Theorems 2 and 3. $\square$

Example 1. If

$$y = 7x^5,$$

then

$$\frac{dy}{dx} = 35x^4.$$

Theorem 4. *The derivative of the sum of a finite number of differentiable functions is equal to the sum of their derivatives.*

Proof. We first consider the sum of two terms,

$$y = u + v,$$

where we suppose that u and v are differentiable functions of x whose derivatives are

$$\frac{du}{dx} = \lim_{\Delta x \to 0} \frac{\Delta u}{\Delta x}, \qquad \frac{dv}{dx} = \lim_{\Delta x \to 0} \frac{\Delta v}{\Delta x}.$$

When x is replaced by $x + \Delta x$, the new values of the variables will satisfy the equation

$$y + \Delta y = (u + \Delta u) + (v + \Delta v).$$

We subtract

$$y = u + v$$

from this and obtain

$$\Delta y = \Delta u + \Delta v,$$

and hence

$$\frac{\Delta y}{\Delta x} = \frac{\Delta u}{\Delta x} + \frac{\Delta v}{\Delta x}.$$

By hypothesis, u and v are differentiable, so both $\Delta u/\Delta x$ and $\Delta v/\Delta x$ have limits as Δx approaches zero. Therefore their sum has a limit which is the sum of their limits. Hence

$$\frac{dy}{dx} = \lim_{\Delta x \to 0} \frac{\Delta y}{\Delta x} = \lim_{\Delta x \to 0} \left(\frac{\Delta u}{\Delta x} + \frac{\Delta v}{\Delta x}\right)$$
$$= \lim_{\Delta x \to 0} \frac{\Delta u}{\Delta x} + \lim_{\Delta x \to 0} \frac{\Delta v}{\Delta x},$$

and thus

$$\frac{d(u+v)}{dx} = \frac{du}{dx} + \frac{dv}{dx}.$$

This equation expresses the result that the derivative of the sum of two differentiable functions is the sum of their derivatives. □

Remark. We may proceed by induction to establish the result for the sum of any finite number of terms. For example, if

$$y = u_1 + u_2 + u_3,$$

where u_1, u_2, and u_3 are differentiable functions of x, then we may take

$$u = u_1 + u_2, \quad v = u_3,$$

and apply the result already established for the sum of two terms, namely,

$$\frac{dy}{dx} = \frac{d(u_1 + u_2)}{dx} + \frac{du_3}{dx}.$$

Since the first term is again a sum of two terms, we have

$$\frac{d(u_1 + u_2)}{dx} = \frac{du_1}{dx} + \frac{du_2}{dx},$$

so that

$$\frac{d(u_1 + u_2 + u_3)}{dx} = \frac{du_1}{dx} + \frac{du_2}{dx} + \frac{du_3}{dx}.$$

Finally, if it has been established for some integer n that

$$\frac{d(u_1 + u_2 + \cdots + u_n)}{dx} = \frac{du_1}{dx} + \frac{du_2}{dx} + \cdots + \frac{du_n}{dx},$$

and we let $y = u + v$, with

$$u = u_1 + u_2 + \cdots + u_n,$$
$$v = u_{n+1},$$

we find in the same way as above that

$$\frac{d(u_1 + u_2 + \cdots + u_{n+1})}{dx} = \frac{du_1}{dx} + \frac{du_2}{dx} + \cdots + \frac{du_{n+1}}{dx}.$$

This enables us to conclude that if the theorem is true for a sum of n terms it is also true for a sum of $(n + 1)$ terms, and since it is already established for the sum of two terms, we conclude that it is true for the sum of any finite number of terms.

Corollary. *Let f be a polynomial function*

$$f(x) = c_0x^n + c_1x^{n-1} + \cdots c_{n-1}x + c_n,$$

where n is a nonnegative integer and $c_0, c_1, \ldots, c_n$ are numbers. Then f is differentiable for all x, and

$$f'(x) = c_0nx^{n-1} + c_1(n-1)x^{n-2} + \cdots + c_{n-1}.$$

Proof. Apply Theorem 4 and the Corollary of Theorem 3. □

Problem 1. Find dy/dx if

$$y = x^3 + 7x^2 - 5x + 4.$$

Solution. By the results just established, we may find the derivatives of the separate terms of a sum and add the results. Thus

$$\frac{dy}{dx} = \frac{d(x^3)}{dx} + \frac{d(7x^2)}{dx} + \frac{d(-5x)}{dx} + \frac{d(4)}{dx}$$
$$= 3x^2 + 14x - 5.$$

Problem 2. What is the slope of the curve

$$y = x^3 - 6x + 2$$

where it crosses the y-axis?

Solution. The slope at the point (x, y) is equal to

$$\frac{dy}{dx} = 3x^2 - 6.$$

The curve crosses the y-axis when $x = 0$. Hence its slope at that point is $m = (3x^2 - 6)_{x=0} = -6$.

Second- and higher-order derivatives

The derivative with respect to x of $y' = dy/dx$ is called the *second derivative* of y with respect to x and is denoted by y''. That is,

$$y' = \frac{dy}{dx}$$

is the first derivative of y with respect to x, and its derivative,

$$y'' = \frac{dy'}{dx} = \frac{d}{dx}\left(\frac{dy}{dx}\right),$$

is the second derivative of y with respect to x. The operation of taking the derivative of a function twice in succession, denoted by

$$\frac{d}{dx}\left(\frac{d}{dx}\cdots\right),$$

is also indicated by the notation

$$\frac{d^2}{dx^2}(\cdots).$$

In this notation, we write the second derivative of y with respect to x as

$$\frac{d^2y}{dx^2}.$$

More generally, the result of differentiating a function $y = f(x)$ n times in succession is denoted by $y^{(n)}$, by $f^{(n)}(x)$, or by d^ny/dx^n.

Example 2. If $\quad y = x^3 - 3x^2 + 2,$

then $y' = \dfrac{dy}{dx} = 3x^2 - 6x, \quad y'' = \dfrac{d^2y}{dx^2} = 6x - 6,$

$$y''' = \frac{d^3y}{dx^3} = 6, \quad y^{(\text{iv})} = \frac{d^4y}{dx^4} = 0.$$

In mechanics, if $s = f(t)$ gives the position of a moving body at time t, then

(a) the first derivative ds/dt gives the *velocity*, and

(b) the second derivative d^2s/dt^2 gives the *acceleration*

of the body at time t.

Problem 3. A body moves in a straight line according to the law of motion

$$s = t^3 - 4t^2 - 3t.$$

Find its acceleration at each instant when the velocity is zero.

Solution. Here we find the velocity v and acceleration a are

$$v = \frac{ds}{dt} = 3t^2 - 8t - 3, \quad a = \frac{dv}{dt} = 6t - 8.$$

The velocity is zero when

$$3t^2 - 8t - 3 = (3t + 1)(t - 3) = 0,$$

that is, when

$$t = -\tfrac{1}{3} \quad \text{or} \quad t = 3.$$

The corresponding values of the acceleration are

$$t = -\tfrac{1}{3},\ a = -10; \quad t = 3,\ a = 10.$$

EXERCISES

In Exercises 1 through 5, s represents the position of a moving body at time t. Find the velocity

$$v = ds/dt$$

and the acceleration

$$a = dv/dt = d^2s/dt^2.$$

1. $s = t^2 - 4t + 3$
2. $s = 2t^3 - 5t^2 + 4t - 3$
3. $s = gt^2/2 + v_0t + s_0$ (g, v_0, s_0 constants)
4. $s = 3 + 4t - t^2$
5. $s = (2t + 3)^2$

Find both $y' = dy/dx$ and $y'' = dy'/dx$ in Exercises 6 through 15.

6. $y = x^4 - 7x^3 + 2x^2 + 5$
7. $y = 5x^3 - 3x^5$
8. $y = 4x^2 - 8x + 1$
9. $y = \dfrac{x^4}{4} - \dfrac{x^3}{3} + \dfrac{x^2}{2} - x + 3$
10. $y = 2x^4 - 4x^2 - 8$
11. $12y = 6x^4 - 18x^2 - 12x$
12. $y = 3x^7 - 7x^3 + 21x^2$
13. $y = x^2(x^3 - 1)$
14. $y = (x - 2)(x + 3)$
15. $y = (3x - 1)(2x + 5)$

Theorem 6. *At a point where $v \neq 0$, the derivative of the quotient*

$$y = \frac{u}{v}$$

of two differentiable functions u and v is given by

$$\frac{d}{dx}\left(\frac{u}{v}\right) = \frac{v\dfrac{du}{dx} - u\dfrac{dv}{dx}}{v^2}. \tag{4}$$

Proof. Consider a point $x = c$ where $v \neq 0$ and where u and v are differentiable. Let x be given an increment Δx, and let Δy, Δu, Δv be the corresponding increments in y, u, v. Then as $\Delta x \to 0$,

$$\lim (v + \Delta v) = \lim v + \lim \Delta v,$$

while

$$\lim \Delta v = \lim \frac{\Delta v}{\Delta x} \cdot \Delta x = \frac{dv}{dx} \cdot 0 = 0.$$

Therefore the value of $v + \Delta v$ is close to the value of v when $x + \Delta x$ is close to c, that is, when Δx is near zero. In particular, since $v \neq 0$ at $x = c$, it follows that $v + \Delta v \neq 0$ when Δx is *near* zero, say when $0 < |\Delta x| < h$. Let Δx be so restricted. Then $v + \Delta v \neq 0$ and

$$y + \Delta y = \frac{u + \Delta u}{v + \Delta v}.$$

From this we subtract

$$y = \frac{u}{v}$$

and obtain

$$\begin{aligned}\Delta y &= \frac{u + \Delta u}{v + \Delta v} - \frac{u}{v}\\ &= \frac{(vu + v\,\Delta u) - (uv + u\,\Delta v)}{v(v + \Delta v)}\\ &= \frac{v\,\Delta u - u\,\Delta v}{v(v + \Delta v)}.\end{aligned}$$

We divide this by Δx by dividing the numerator of the fraction on the right by Δx and have

$$\frac{\Delta y}{\Delta x} = \frac{v\dfrac{\Delta u}{\Delta x} - u\dfrac{\Delta v}{\Delta x}}{v(v + \Delta v)}.$$

When Δx approaches zero, we obtain

$$\lim \frac{\Delta u}{\Delta x} = \frac{du}{dx};$$

$$\lim \frac{\Delta v}{\Delta x} = \frac{dv}{dx};$$

$$\lim v(v + \Delta v) = \lim v \lim (v + \Delta v) = v^2 \neq 0;$$

and
$$\lim \frac{\Delta y}{\Delta x} = \frac{\lim \left(v\dfrac{\Delta u}{\Delta x} - u\dfrac{\Delta v}{\Delta x}\right)}{\lim v(v + \Delta v)},$$

or
$$\frac{dy}{dx} = \frac{v\dfrac{du}{dx} - u\dfrac{dv}{dx}}{v^2},$$

which establishes Eq. (4). □

Theorem 7. *At a point where $u = g(x)$ is differentiable and different from zero, the derivative of*

$$y = u^n$$

is given by

$$\frac{d(u^n)}{dx} = nu^{n-1}\frac{du}{dx} \tag{5}$$

if n is a negative integer.

Before we prove the theorem, we first note that formula (5) is the extension of (2) to the case where n is a *negative integer*.

Proof. To prove the theorem, we combine the results in Eqs. (2) and (4) as follows. Let

$$y = u^{-m} = \frac{1}{u^m},$$

where $n = -m$ is a negative integer, so that m is a positive integer. Then, using (4) for the derivative of a quotient, we have

$$\frac{dy}{dx} = \frac{d\left(\dfrac{1}{u^m}\right)}{dx} = \frac{u^m\dfrac{d(1)}{dx} - 1\dfrac{d(u^m)}{dx}}{(u^m)^2} \tag{6}$$

at any point where u is differentiable and different from zero. Now the various derivatives on the right side of (6) can be evaluated by formulas already proved, namely,

$$\frac{d(1)}{dx} = 0,$$

since 1 is a constant, and

$$\frac{d(u^m)}{dx} = mu^{m-1}\frac{du}{dx},$$

since m is a *positive integer*. Therefore

$$\frac{dy}{dx} = \frac{u^m \cdot 0 - 1 \cdot mu^{m-1}\dfrac{du}{dx}}{u^{2m}} = -mu^{-m-1}\frac{du}{dx}.$$

If $-m$ is replaced by its equivalent value n, the formula reduces to Eq. (5). □

Problem 1. Find the derivative of $y = x^2 + 1/x^2, x \neq 0$.

Solution. We may write

$$y = x^2 + x^{-2}.$$

Then

$$\begin{aligned}\frac{dy}{dx} &= 2x^{2-1}\frac{dx}{dx} + (-2)x^{-2-1}\frac{dx}{dx}\\ &= 2x - 2x^{-3}.\end{aligned}$$

Problem 2. Find the derivative of

$$y = \frac{x^2+1}{x^2-1}, \; x^2 \neq 1.$$

Solution. We apply the formula, Eq. (4), for the derivative of a quotient and have

$$\begin{aligned}\frac{dy}{dx} &= \frac{(x^2-1)\cdot 2x - (x^2+1)\cdot 2x}{(x^2-1)^2}\\ &= \frac{-4x}{(x^2-1)^2}.\end{aligned}$$

Problem 3. Find the derivative of

$$y = (x^2+1)^3(x^3-1)^2.$$

Solution. We could, of course, expand everything here and write y as a polynomial in x, but this is not necessary. Instead we use Eq. (1) for the derivative of a product.

$$\frac{dy}{dx} = (x^2+1)^3\frac{d}{dx}(x^3-1)^2 + (x^3-1)^2\frac{d}{dx}(x^2+1)^3.$$

The derivatives which are now to be evaluated are of the type in Eq. (2).

$$\begin{aligned}\frac{d}{dx}(x^3-1)^2 &= 2(x^3-1)\frac{d}{dx}(x^3-1)\\ &= 2(x^3-1)\cdot 3x^2 = 6x^2(x^3-1),\end{aligned}$$

and

$$\begin{aligned}\frac{d}{dx}(x^2+1)^3 &= 3(x^2+1)^2\frac{d}{dx}(x^2+1)\\ &= 3(x^2+1)^2\cdot 2x = 6x(x^2+1)^2.\end{aligned}$$

We substitute these into the earlier equation and have

$$\begin{aligned}\frac{dy}{dx} &= (x^2+1)^3 6x^2(x^3-1) + (x^3-1)^2 6x(x^2+1)^2\\ &= 6x(x^2+1)^2(x^3-1)[x(x^2+1) + (x^3-1)]\\ &= 6x(x^2+1)^2(x^3-1)(2x^3+x-1).\end{aligned}$$

EXERCISES

Find dy/dx in Exercises 1 through 8.

1. $y = x^3/3 - x^2/2 + x - 1$
2. $y = (x-1)^3(x+2)^4$
3. $y = (x^2+1)^5$
4. $y = (x^3-3x)^4$
5. $y = (x+1)^2(x^2+1)^{-3}$
6. $y = \dfrac{2x+1}{x^2-1}$
7. $y = \dfrac{2x+5}{3x-2}$
8. $y = \left(\dfrac{x+1}{x-1}\right)^2$

Find ds/dt in Exercises 9 through 15.

9. $s = \dfrac{t}{t^2+1}$
10. $s = (2t+3)^3$
11. $s = (t^2-t)^{-2}$
12. $s = t^2(t+1)^{-1}$
13. $s = \dfrac{2t}{3t^2+1}$
14. $s = (t+t^{-1})^2$
15. $s = (t^2+3t)^3$
16. With the book closed, state and prove the formula for the derivative of the product of two differentiable functions u and v.

17. With the book closed, state and prove the formula for the derivative of the quotient u/v of two differentiable functions u and v.
18. With the book closed, state and prove the formula for the derivative with respect to x of u^n, n a positive integer and u a differentiable function of x.
19. Prove the formula in Eq. (3) for the derivative of the product of n differentiable functions by induction on n (for $n \geq 2$).
20. Use the formula in Eq. (3) with $u_1, u_2, u_3, \ldots, u_n$ all equal to u to establish Eq. (2) for $n \geq 2$.

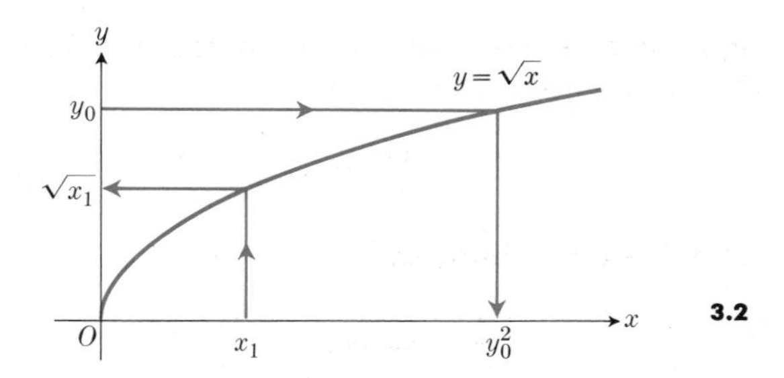

3.2

3.3 INVERSE FUNCTIONS AND THEIR DERIVATIVES

For some functions it is not only true that to each x in the domain there corresponds a unique y in the range, but also that each such y is the *image* of only one x. For example, the mapping

$$f: x \to \sqrt{x} \tag{1}$$

is such a one-to-one mapping from the domain $x \geq 0$ to the range $y \geq 0$. Figure 3.2 illustrates how the graph of $y = \sqrt{x}$ can be used to go from any nonnegative x to its nonnegative square root: Start at x, go up to the curve, and then go across to the y-axis to $\sqrt{x}$. Since every nonnegative y is the image of just one x under this mapping, we can reverse the procedure. That is, we can start with $y \geq 0$, go over to the curve, and then go down to y^2 on the x-axis. This reverse mapping

$$g: y \to y^2, \quad y \geq 0, \tag{2}$$

describes a new function (the squaring function) that is called the *inverse* of the function f (which in this example is the square-root function).

Remark 1. The mapping g, as written in Eq. (2), has y as its independent variable (the domain variable for g). Any other letter could equally well be used if we were studying only g: e. g., any of

$$g(t) = t^2, \quad g(a) = a^2, \quad g(x) = x^2,$$

with the understanding that the domain is the set of nonnegative real numbers.

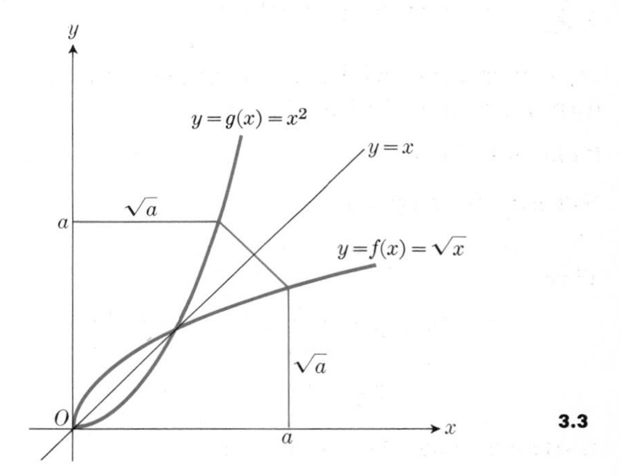

3.3

Remark 2. If the mapping g is expressed with x as independent variable, so that $g(x) = x^2$, the graph of g can be obtained by reflecting the graph of f in the 45° line $y = x$ as mirror (see Fig. 3.3). Any points $P(x_1, y_1)$ and $P'(y_1, x_1)$ are symmetric with respect to the line $y = x$. Thus $(a, \sqrt{a})$ and $(\sqrt{a}, a)$ are symmetric pairs. The first of these is on the graph of f, the second is on the graph of g.

If f and g are any pair of inverse functions, their graphs are mirror images of each other with respect to the 45° line $y = x$. Thus, if L_1 is the line tangent to the graph of f at (c, d) and L_2 is the mirror image of L_1 with respect to the 45° line, then it is reasonable to expect L_2 to be tangent to the graph of g at (d, c). (Figure 3.4 illustrates this idea.) Since (rise)/(run)

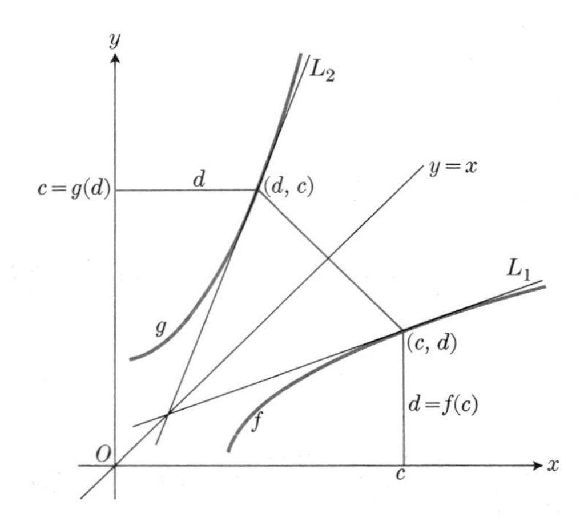

3.4 Graphs of inverse functions f and g and tangents L_1 and L_2 at corresponding points (c, d) and (d, c). Note that $d = f(c)$ and $c = g(d)$.

on L_2 corresponds to (run)/(rise) on L_1, we see that the slope of L_2 is just the reciprocal of the slope of $L_1: m' = 1/m$. Since L_1 is tangent to f at $(c, f(c))$,

$$m = f'(c).$$

And, if L_2 is tangent to the graph of g at $(d, g(d))$, then

$$m' = g'(d).$$

Hence the geometric evidence suggests that

$$g'(f(c)) = \frac{1}{f'(c)} \cdot \tag{3}$$

Naturally, Eq. (3) requires the existence of the derivative of f at c, and also requires that $f'(c) \neq 0$. In Fig. 3.4, f' is positive and f is a strictly increasing function.

The following two theorems are called *lemmas*. According to Webster, "lemma" means "a preliminary or auxiliary proposition or theorem demonstrated or accepted for immediate use in a demonstration of some other proposition." We shall use the two lemmas in establishing the validity of Eq. (3) when $f'(c)$ is positive. [The proof when $f'(c)$ is negative is only slightly different.] Briefly, what the first lemma says is that if dy/dx is positive at a point, then $\Delta y/\Delta x$ is also positive when Δx is near zero.

Lemma 1. *Let f be differentiable at c and $f'(c) > 0$. Then*

$$f(c + h) > f(c) \quad \textit{and} \quad f(c - h) < f(c) \tag{4}$$

for sufficiently small positive values of h.

Proof. Let $f'(c) = p$, where p is a positive number. When Δx is near zero, the difference quotient

$$\frac{f(c + \Delta x) - f(c)}{\Delta x} \tag{5}$$

is very nearly equal to p, by definition of $f'(c)$ as a limit. In particular, since p is positive, we can take $\epsilon = p/2$ and be sure that the above quotient is between $p - \epsilon$ and $p + \epsilon$ when Δx is in a sufficiently small deleted neighborhood of zero. Let N be such a deleted neighborhood. Then

$$\frac{p}{2} < \frac{f(c + \Delta x) - f(c)}{\Delta x} < \frac{3p}{2} \quad \text{when } \Delta x \in N.$$

Since the interval $(p/2, 3p/2)$ contains only positive numbers, the numerator and denominator in the difference quotient (5) must have the same sign, both positive when $\Delta x = h$ is positive and both negative when $\Delta x = -h$ is negative. That is,

$$f(c + h) - f(c) > 0 \quad \text{for } h > 0,\ h \in N$$

and

$$f(c - h) - f(c) < 0 \quad \text{for } h > 0,\ -h \in N.$$

These inequalities establish Lemma 1. □

Lemma 2. *If f is a differentiable function at x, and*

if $$\Delta y = f(x + \Delta x) - f(x),$$

then $$\lim \Delta y = 0 \quad \text{as} \quad \Delta x \to 0.$$

Proof. (This is left as an exercise. See the proof that $\Delta u \to 0$ as $\Delta x \to 0$ in the proof of Theorem 5.)

Theorem 8. *Let f be a one-to-one function with domain $[a, b]$ and range $[f(a), f(b)]$, whose derivative exists and is positive on (a, b). Then f has an inverse g, and if $y = f(x)$,*

then $$g'(y) = \frac{1}{f'(x)}, \quad \text{for } a < x < b. \tag{6}$$

Or, in another notation, if $y = f(x)$, then $x = g(y)$ and

$$\frac{dx}{dy} = \frac{1}{dy/dx}.$$

Proof. Because f is one-to-one from its domain D_f to its range R_f, the rule

$$g(y) = x \quad \text{if and only if} \quad y = f(x),\ x \in D_f$$

defines a function g whose domain D_g is the range of f:

$$D_g = R_f.$$

Figure 3.5 will help us to follow the remaining steps of the proof. Fix $y \in D_g$ and $x = g(y) \in D_f$. Then in Lemma 1 take $c = x$ and let N be a neighborhood of x contained in (a, b) and such that the inequalities (4) apply when $x + h$ and $x - h$ are in N and $h > 0$. This is possible because $a < x < b$. Also, let

$$y - k_1 = f(x - h), \quad y + k_2 = f(x + h).$$

Then k_1 and k_2 are positive numbers, so y is an inner point of the domain of g, and that domain contains the closed interval $[y - k, y + k]$, where

$$k = \min\{k_1, k_2\}.$$

For each $\Delta y \neq 0$ such that $|\Delta y| < k$, there exists $\Delta x \neq 0$ such that $|\Delta x| < h$ and

$$f(x + \Delta x) = y + \Delta y, \quad g(y + \Delta y) = x + \Delta x.$$

To prove that $g'(y)$ exists, we must show that the difference quotient

$$\frac{g(y + \Delta y) - g(y)}{\Delta y}$$

has a limit as $\Delta y \to 0$. But this is easy, because

$$g(y) = x, \qquad g(y + \Delta y) = x + \Delta x,$$
$$y = f(x), \qquad y + \Delta y = f(x + \Delta x),$$

so that

$$\frac{g(y + \Delta y) - g(y)}{\Delta y} = \frac{(x + \Delta x) - x}{f(x + \Delta x) - f(x)} = \frac{\Delta x}{f(x + \Delta x) - f(x)}, \tag{7}$$

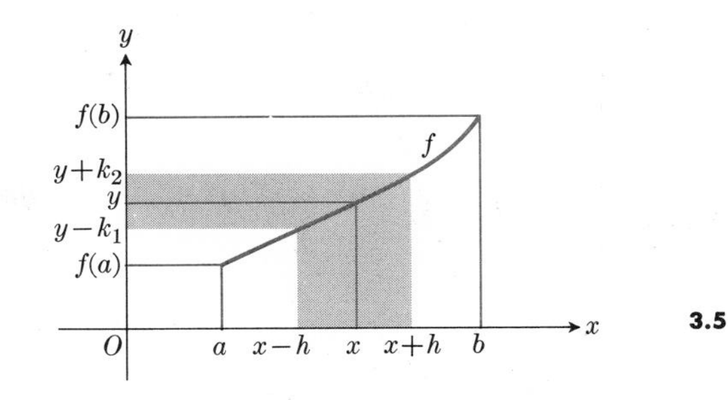

3.5

and

$$\lim_{\Delta x \to 0} \frac{f(x + \Delta x) - f(x)}{\Delta x} = f'(x). \tag{8a}$$

By hypothesis, $f'(x) \neq 0$. Therefore, taking reciprocals in Eq. (8a) we get

$$\lim_{\Delta x \to 0} \frac{\Delta x}{f(x + \Delta x) - f(x)} = \frac{1}{f'(x)}. \tag{8b}$$

By Lemma 2, $\Delta y \to 0$ when $\Delta x \to 0$. (The converse is also true because f and g are both one-to-one and continuous. Further discussion of continuity is postponed to the end of Chapter 4.) Therefore, if we let $\Delta x \to 0$ and use Eqs. (7) and (8b), we get

$$\lim_{\Delta y \to 0} \frac{g(y + \Delta y) - g(y)}{\Delta y} = \lim_{\Delta x \to 0} \frac{\Delta x}{f(x + \Delta x) - f(x)},$$

or

$$g'(y) = \frac{1}{f'(x)}. \quad \square$$

Example 1

Let $\quad a = 0, \quad b > 0, \quad f(x) = x^2.$

Then $\quad f'(x) = 2x > 0 \quad \text{for } a < x < b,$

and $\quad$ if $y = x^2 = f(x)$, then $x = \sqrt{y} = g(y)$

for $0 < x < b$ and $0 < y < b^2$. By Theorem 8,

$$g'(y) = \frac{1}{f'(x)} = \frac{1}{2x} = \frac{1}{2\sqrt{y}}.$$

For $y = f(x) = x^2$, $x \geq 0$, the inverse (expressed as a function of y) is $x = g(y) = y^{1/2}$. Similarly, if n is any positive integer and if

$$y = f(x) = x^n, \quad x \geq 0,$$

then the inverse (expressed as a function of y) is

$$x = g(y) = y^{1/n}, \quad y \geq 0.$$

We have just seen illustrated that

$$\frac{d}{dy}(y^{1/2}) = \tfrac{1}{2}y^{-1/2}, \quad \text{if} \quad y > 0.$$

The following corollary extends this result.

Corollary 8.1. *Let n be a positive integer, $y > 0$, and let*

$$g(y) = y^{1/n}. \tag{9a}$$

Then

$$g'(y) = \frac{1}{n}y^{(1/n)-1}. \tag{9b}$$

Proof. Let the inverse of g be f, defined for $x > 0$ such that $y = f(x)$, or $y = x^n$. Then by Theorem 8,

$$g'(y) = \frac{1}{f'(x)}, \quad \text{or} \quad g'(y) = \frac{1}{nx^{n-1}}.$$

Since $x^n = y$, we have

$$x = y^{1/n} \quad \text{and} \quad x^{n-1} = y^{(n-1)/n} = y^{1-(1/n)}.$$

Therefore

$$g'(y) = \frac{1}{ny^{1-(1/n)}} \quad \text{or} \quad g'(y) = \frac{1}{n}y^{(1/n)-1}. \ \square$$

Remark 3. In Theorem 8, we assumed that $f'(x) > 0$ for $a < x < b$. We could equally well have proved the theorem under the assumption $f'(x) < 0$ on (a, b). Under the first hypothesis f is a strictly increasing function, while the second hypothesis fits a strictly decreasing function. Only minor changes in the proof are needed in this alternative form. Indeed, one can achieve the desired result by applying Theorem 8 to the functions $f_1 = -f$ and $g_1 = -g$:

$$g_1'(y) = \frac{1}{f_1'(x)} \Rightarrow g'(y) = \frac{1}{f'(x)}.$$

Henceforth, when we use Theorem 8, we may do so under either of the assumptions

$$f'(x) > 0 \quad \text{for all } x \in (a, b),$$

and

$$f'(x) < 0 \quad \text{for all } x \in (a, b),$$

Corollary 8.2. *Let n be a negative integer and let*

$$g(y) = y^{1/n}, \ y > 0.$$

Then

$$g'(y) = \frac{1}{n}y^{(1/n)-1}.$$

Proof. For $x > 0$, let $f(x) = x^n$. Then

$$f'(x) = nx^{n-1} < 0 \quad \text{for } x > 0.$$

The functions f and g are inverses on $x > 0$, $y > 0$, because

$$y = x^n = f(x) \quad \text{if and only if} \quad x = y^{1/n} = g(y).$$

Hence

$$g'(y) = \frac{1}{f'(x)} = \frac{1}{nx^{n-1}} = \frac{1}{n}y^{(1/n)-1}. \ \square$$

Remark 4. These applications of Theorem 8 have been made easier by using the notation $g(y) = y^{1/n}$. However, it is also true that if $g(x) = x^{1/n}$, then $g'(x) = (1/n)x^{(1/n)-1}$, because there is nothing sacred about which letter of the alphabet we use for the variable in the definition of g. Moreover, instead of writing the exponent as $1/n$, with n a positive or negative integer, we could write the exponent as n and say that n is the reciprocal of a positive or negative integer. In other words, Corollaries 8.1 and 8.2 simply extend the familiar formula

$$\frac{d}{dx}(x^n) = nx^{n-1} \tag{10}$$

from its earlier applications, where n was a positive integer (and all values of x), or a negative integer (and $x \neq 0$), to the situation where n is the reciprocal of a positive or negative integer (and $x > 0$). We are now in a position to extend Eq. (10) to any *rational* exponent n, for $x > 0$.

Theorem 9. *Let n be a rational number. Let f be the function*

$$f(x) = x^n, \quad x > 0.$$

Then

$$f'(x) = nx^{n-1}. \tag{11}$$

Proof. Because n is rational, there are integers p and q with $p > 0$, such that $n = p/q$. Let

$$u = x^{1/q}, \quad y = x^{p/q} = u^p.$$

By Corollaries 8.1 and 8.2,

$$\frac{du}{dx} = \frac{1}{q}x^{(1/q)-1}.$$

By the corollary of Theorem 5,

$$\begin{aligned}\frac{dy}{dx} = pu^{p-1}\frac{du}{dx} &= pu^{p-1}\left(\frac{1}{q}x^{(1/q)-1}\right)\\ &= \frac{p}{q}(x^{(1/q)})^{p-1}x^{(1/q)-1}\\ &= \frac{p}{q}x^{((p/q)-(1/q)+(1/q)-1)}\\ &= nx^{n-1},\end{aligned}$$

because $p/q = n$. □

Example 2. If $\quad y = x^{2/3}, \quad x > 0,$

then $\quad dy/dx = \frac{2}{3}x^{-1/3}.$

Remark 5. The restriction $x > 0$ is not necessary in Example 2, because either positive or negative values of x have real cube roots and Theorem 2 could be extended to cover negative as well as positive values of x for all rational values of n for which x^n is real valued. The corresponding change in Corollary 8.2 would be to allow y to be negative (or positive) when n is an odd integer. The key feature there is that the equation $y = x^n$, for n an odd integer, establishes a one-to-one correspondence between the set of all nonzero real numbers x and the set of all nonzero real numbers y. (This is true for $n = 1, 3, 5, \ldots$ or $n = -1, -3, -5, \ldots$)

The next examples illustrate a technique known as *implicit differentiation.* (We shall not be able to justify the method rigorously because it involves more sophisticated ideas than we take up in this book. Nevertheless the method is easy to understand and to apply.) If a relation between x and y is expressed by an equation such as

$$x^2 + y^2 = 1, \quad xy = 1, \quad x^2 + xy + y^5 = 3,$$

it may be difficult (or impossible) to solve the equation explicitly for y in terms of x. Or, the solution may not express y as a *function* of x. The next example will clarify some of these points, and show the *method of* ***implicit differentiation.***

Example 3. If we solve the equation

$$x^2 + y^2 = 1 \tag{12}$$

for y, we get

$$y = \pm\sqrt{1 - x^2}.$$

For each $x \in (-1, 1)$, there are two corresponding values of y, not just one as is required for a function. But we can look at two different functions

$$f(x) = +\sqrt{1 - x^2}, \quad g(x) = -\sqrt{1 - x^2}, \quad |x| < 1.$$

Focus attention on $f(x)$. Call it u and *assume* that it is differentiable. (This will be proved later when we prove the chain rule.) Then u^2 is also differentiable and we know that

$$\frac{d}{dx}(u^2) = 2u\frac{du}{dx}.$$

Also, since $u = \sqrt{1 - x^2}$, it is clear that $u^2 + x^2$ is equal to the constant function 1 on the domain $|x| < 1$. Therefore

$$\frac{d}{dx}(u^2 + x^2) = \frac{d}{dx}(1) = 0 \quad \text{for } |x| < 1.$$

This implies that

$$2u\frac{du}{dx} + 2x = 0, \quad \text{so} \quad \frac{du}{dx} = \frac{-x}{u}.$$

Thus, if $u = \sqrt{1 - x^2}$ is differentiable, its derivative is

$$u' = \frac{-x}{u} \quad \text{for } |x| < 1.$$

Had we chosen the other function, say

$$v = g(x) = -\sqrt{1 - x^2},$$

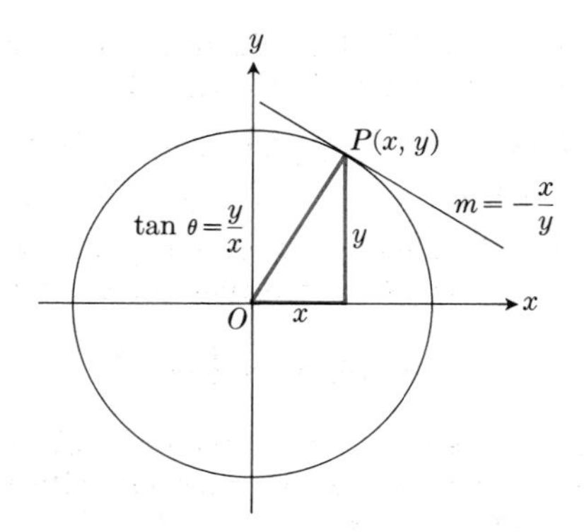

3.6 For the circle $x^2 + y^2 = 1$, the slope of the tangent at $P(x, y)$ is the negative reciprocal of the slope of OP: $dy/dx = -(x/y)$, $m_{OP} = y/x$.

the algebra would have been just the same:

$v^2 + x^2$ is the constant function 1 on $(-1, 1)$,

so, if v is differentiable, then

$$2vv' + 2x = 0, \quad v' = \frac{-x}{v}.$$

In practice, we would not have solved the original equation at all. We would have reasoned that if the original equation determines y as one or more functions of x, and if those functions are differentiable, then $x^2 + y^2$ is the sum of two differentiable functions and it is also the constant function 1 on $(-1, 1)$. Differentiating Eq. (12) implicitly with respect to x, we would get

$$\frac{d}{dx}(x^2 + y^2) = \frac{d}{dx}(1), \quad \text{or} \quad 2x + 2yy' = 0,$$

so

$$y' = -x/y.$$

See Fig. 3.6.

Example 4. If we differentiate both sides of the equation

$$xy = 1 \tag{13}$$

with respect to x, treating y as a differentiable function (which we happen in this case to know it is, for $x \neq 0$), then we get $xy' + y = 0$, or $y' = -y/x$. Here we could have done the problem differently:

$$y = x^{-1}, \quad y' = -1x^{-2}.$$

The answers agree, because $-y/x = -x^{-1}/x = -1x^{-2}$.

Example 5. In a more complicated equation, such as

$$x^2 + xy + y^5 = 3, \tag{14}$$

we cannot solve for y in terms of x. It is easy to verify that the point $(x_0, y_0) = (1, 1)$ is on the graph of Eq. (14). We proceed on faith, assuming that there is some neighborhood of x_0 throughout which $y = f(x)$ would satisfy the equation, with f a differentiable function at x_0. (There is nothing to be afraid of—go ahead and make the assumption.) Then y^5 is also differentiable and

$$\frac{d}{dx}(y^5) = 5y^4\frac{dy}{dx}.$$

Similarly xy is differentiable, and

$$\frac{d}{dx}(xy) = xy' + y.$$

Differentiating both sides of Eq. (14) implicitly with respect to x, we get

$$2x + (xy' + y) + 5y^4y' = 0.$$

This equation is linear in y', and can easily be solved to get

$$y' = -\frac{2x + y}{x + 5y^4}.$$

At the point $(x_0, y_0) = (1, 1)$, the slope of the curve is

$$y'(1, 1) = -\frac{2x + y}{x + 5y^4}\bigg]_{(1,1)} = -\frac{3}{6} = -\frac{1}{2}.$$

Example 6. If $\quad y = x^{p/q},$

then $\quad y^q = x^p,$

and implicit differentiation with respect to x yields the familiar results

$$qy^{q-1}\frac{dy}{dx} = px^{p-1},$$

or

$$\frac{dy}{dx} = \frac{p}{q}y^{1-q}x^{p-1} = \frac{p}{q}(x^{p/q})^{1-q}x^{p-1} = \frac{p}{q}x^{(p/q)-1}.$$

Remark 6. It is not easy to establish the validity of implicit differentiation without using partial derivatives. (See, for example: Olmsted, *Advanced Calculus*, Appleton-Century-Crofts, 1961, p. 289; or Fine, *Calculus*, Macmillan, 1937, p. 248.)

The foregoing examples were intended to convince you that it is all right to go ahead and differentiate implicitly. But if you find that you are trying to divide by zero: Stop.

EXERCISES

In Exercises 1 through 23, find dy/dx.

1. $x^2 + y^2 = 1$
2. $y^2 = \dfrac{x-1}{x+1}$
3. $x^2 + xy = 2$
4. $x^2y + xy^2 = 6$
5. $y^2 = x^3$
6. $x^{2/3} + y^{2/3} = 1$
7. $x^{1/2} + y^{1/2} = 1$
8. $x^3 - xy + y^3 = 1$
9. $x^2 = \dfrac{x-y}{x+y}$
10. $y = \dfrac{x}{\sqrt{x^2+1}}$
11. $y = x\sqrt{x^2+1}$
12. $y^2 = x^2 + \dfrac{1}{x^2}$
13. $2xy + y^2 = x + y$
14. $y = \sqrt{x} + \sqrt[3]{x} + \sqrt[4]{x}$
15. $y^2 = \dfrac{x^2-1}{x^2+1}$
16. $(x+y)^3 + (x-y)^3 = x^4 + y^4$
17. $(3x+7)^5 = 2y^3$
18. $y = (x+5)^4(x^2-2)^3$
19. $\dfrac{1}{y} + \dfrac{1}{x} = 1$
20. $y = (x^2 + 5x)^3$
21. $y^2 = x^2 - x$
22. $x^2y^2 = x^2 + y^2$
23. $y = \dfrac{\sqrt[3]{x^2+3}}{x}$
24. (a) By differentiating the equation $x^2 - y^2 = 1$ implicitly, show that $dy/dx = x/y$.

    (b) By differentiating both sides of the equation $dy/dx = x/y$ implicitly, show that

$$\frac{d^2y}{dx^2} = \frac{y - x\left(\dfrac{dy}{dx}\right)}{y^2} = \frac{y - \dfrac{x^2}{y}}{y^2} = \frac{y^2 - x^2}{y^3},$$

or, since $y^2 - x^2 = -1$ from the original equation, that

$$\frac{d^2y}{dx^2} = \frac{-1}{y^3}.$$

Use the method outlined in Exercise 24 to find dy/dx and d^2y/dx^2 in Exercises 25 through 28.

25. $x^2 + y^2 = 1$
26. $x^3 + y^3 = 1$
27. $x^{2/3} + y^{2/3} = 1$
28. $xy + y^2 = 1$
29. A particle of mass m moves along the x-axis. The velocity $v = dx/dt$ and position x satisfy the equation

$$m(v^2 - v_0^2) = k(x_0^2 - x^2),$$

where k, v_0, and x_0 are constants. Show, by implicitly differentiating this equation with respect to t, that whenever $v \neq 0$,

$$m\frac{dv}{dt} = -kx.$$

Find the lines that are respectively *tangent* and *normal* to the following curves at the points indicated as P_0. (A line is said to be *normal* to a curve at a point P_0 if it is perpendicular to the tangent at P_0.)

30. $x^2 + xy - y^2 = 1$, $P_0(2, 3)$
31. $x^2 + y^2 = 25$, $P_0(3, -4)$
32. $x^2y^2 = 9$, $P_0(-1, 3)$
33. $\dfrac{x-y}{x-2y} = 2$, $P_0(3, 1)$
34. $(y-x)^2 = 2x + 4$, $P_0(6, 2)$
35. (a) Sketch graphs of the curves $y = x^3$ and $y = x^{1/3}$ for $-2 \leq x \leq 2$, and sketch the lines tangent to them at the points $(1, 1)$ and $(-1, -1)$.

    (b) Which of these two functions fails to have a derivative at $x = 0$? What is the slope of the other curve at that point? What lines are tangent to the two curves at that point?
36. Consider the functions $f_n(x) = x^n$ on $[0, 1]$. Sketch graphs (rough approximations will do) for $n = 1$, 3, 10, $\frac{1}{3}$, and $\frac{1}{10}$. [For $n = 10$, you may estimate

$$(0.5)^{10} = (1/1024) = 0.001,$$

and for $n = \frac{1}{10}$,

$$(0.5)^{0.1} = \sqrt[10]{0.5} = 0.933.$$

These also mean that the tenth root of 0.001 is approximately $\frac{1}{2}$ and the tenth power of 0.933 is approximately $\frac{1}{2}$.]

37. (Continuation of Exercise 36.) What do you think happens to $f_n(x)$ for fixed $x \in [0, 1)$ as $n \to \infty$? To $f_n(1)$? Is there a limit for each $x \in [0, 1]$? If so, let

$$g(x) = \lim_{n\to\infty} f_n(x),$$

and describe the graph of $y = g(x)$ on $[0, 1]$.

38. Which of the five functions described in Exercise 36 is (are) differentiable on $(0, 1)$? If the defining equation makes sense for $x \in (-\infty, \infty)$ or for $x \in [0, \infty)$, let the domains be so extended. Of these new functions, which ones are differentiable for all x's in their domains? Where do some of the new functions fail to be differentiable? Discuss.

39. Of the functions of Exercise 38 (those of Exercise 36 with extended domains), which have inverses that are functions? Is the inverse of each of these five functions always one of the five? Discuss.

3.4 THE INCREMENT OF A FUNCTION

In this article, we shall learn how to estimate the change Δy produced in a function $y = f(x)$ when x changes by a small amount Δx. To be precise, let us focus our attention on a portion of the graph of $y = f(x)$ in the neighborhood of a point $P(x, y)$ where the function is differentiable. Then, if $Q(x + \Delta x, y + \Delta y)$ is a second point on the curve, with $\Delta x \neq 0$, we know that

$$\text{the slope of the secant line } PQ = \frac{\Delta y}{\Delta x}$$

approaches the limiting value

$$\frac{dy}{dx} = f'(x)$$

as Δx approaches zero. Therefore, the difference

$$\frac{\Delta y}{\Delta x} - \frac{dy}{dx}$$

is numerically small when $|\Delta x|$ is small. Let us denote this difference by ϵ:

$$\frac{\Delta y}{\Delta x} - \frac{dy}{dx} = \epsilon. \tag{1}$$

Then, saying that

$$\lim_{\Delta x\to 0} \frac{\Delta y}{\Delta x} = \frac{dy}{dx} = f'(x) \tag{2}$$

is equivalent to saying that

$$\lim_{\Delta x\to 0} \epsilon = 0. \tag{3}$$

In other words, we may deduce from Eqs. (1), (2), and (3) that

$$\frac{\Delta y}{\Delta x} = \frac{dy}{dx} + \epsilon, \tag{4a}$$

where

$$\epsilon \to 0 \quad \text{as} \quad \Delta x \to 0. \tag{4b}$$

If we multiply both sides of (4a) by Δx, we also have

$$\Delta y = \frac{dy}{dx}\,\Delta x + \epsilon\,\Delta x. \tag{4c}$$

Although Eq. (4c) was derived under the assumption that $\Delta x \neq 0$, it is still a true equation even when $\Delta x = 0$. In fact, if we define ϵ by Eq. (1) when $\Delta x \neq 0$, and take $\epsilon = 0$ when $\Delta x = 0$, then Eqs. (4b, c) hold whether $\Delta x = 0$ or not.

If we compare Fig. 3.7 with Fig. 1.43, Article 1.11, we see that the first term on the right-hand side of Eq. (4c) is equal to the change Δy_{tan} that would be produced in y by a change Δx in x if the point (x, y) were to move along the tangent line. This is called the *linear part* of Δy, and for small values of Δx it usually is large in comparison with the second term $\epsilon\,\Delta x$.

The significance of Eqs. (4b, c) is that

$$\begin{aligned}\Delta y &= \Delta y_{\text{tan}} + \epsilon\,\Delta x\\ &= \text{change in } y \textit{ along the curve}\end{aligned}$$

and

$$\begin{aligned}\Delta y_{\text{tan}} &= \frac{dy}{dx}\,\Delta x\\ &= \text{change in } y \textit{ along the tangent line}\end{aligned}$$

differ by an amount

$$\epsilon\,\Delta x,$$

which tends to zero more rapidly than Δx does when Δx approaches zero.

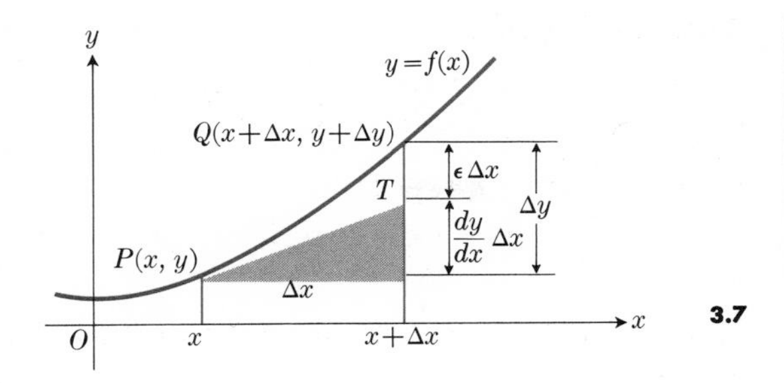

3.7

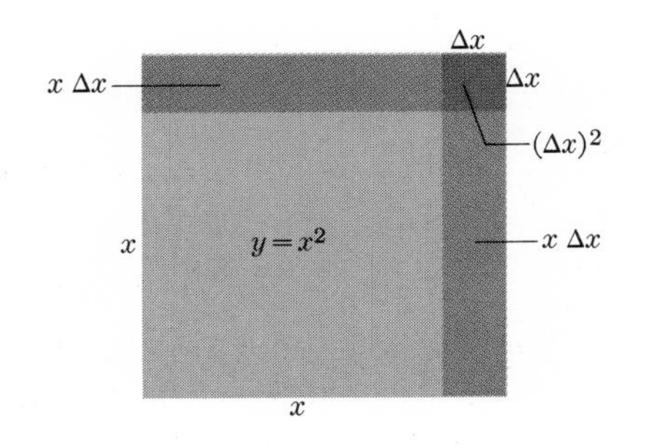

3.8

Example. If

$$y = x^2 = f(x),$$

then

$$\frac{dy}{dx} = 2x = f'(x),$$

and the linear part of Δy is

$$\frac{dy}{dx}\Delta x = 2x\,\Delta x.$$

On the other hand, the exact value of Δy is

$$\Delta y = 2x\,\Delta x + (\Delta x)^2.$$

We compare this with Eq. (4c) and see that for this particular example, the part which is denoted by $\epsilon\,\Delta x$ in (4c) is

$$\epsilon\,\Delta x = (\Delta x)^2.$$

When Δx is small, $(\Delta x)^2$ is the square of a small number and this is much smaller than Δx itself.

This example,

$$y = x^2, \quad (\text{linear part of } \Delta y) = 2x\,\Delta x,$$

also lends itself to a geometrical interpretation, as illustrated by Fig. 3.8. The original square, having sides x by x, has area $y = x^2$. When the sides are increased to $x + \Delta x$ by $x + \Delta x$, the area is increased by

$$\Delta y = 2x\,\Delta x + (\Delta x)^2.$$

When Δx is small in comparison with x, most of the change in y is given by the two rectangular pieces x by Δx and only a very small part is given by the small square Δx by Δx. Thus, if the original square has sides

$$x = 2 \text{ (in.)}$$

and the larger square has sides

$$x + \Delta x = 2.01 \text{ (in.)},$$

so that

$$\Delta x = +0.01 \text{ (in.)},$$

the area of the original square is

$$y = x^2 = 4 \text{ (in}^2\text{)}$$

and the area of the larger square is

$$\begin{aligned} y + \Delta y &= (x + \Delta x)^2 = x^2 + 2x\,\Delta x + (\Delta x)^2 \\ &= 4 + 0.04 + 0.0001 \text{ (in}^2\text{)}. \end{aligned}$$

The increment,

$$\Delta y = 0.04 + 0.0001 \text{ (in}^2\text{)},$$

has its linear part equal to

$$2x\,\Delta x = 0.04 \text{ (in}^2\text{)}$$

and its remainder

$$(\Delta x)^2 = 0.0001 \text{ (in}^2\text{)}.$$

If we were to use only the linear part of Δy as an approximation to Δy itself, we would thereby introduce an error of magnitude 0.0001 (in^2), which is just under one-fourth of one percent of Δy.

Since the linear part of Δy usually gives a good approximation to Δy itself (at least when Δx is small compared with dy/dx), it is customary to use the approximation

$$\Delta y \approx \frac{dy}{dx}\Delta x = \Delta y_{\tan} \tag{5}$$

in numerical calculations. (The symbol $\approx$ means "is approximately equal to.")

Problem 1. Use the approximation in Eq. (5) to approximate the change in $y = x^3$ when x changes from 3 to 2.98.

Solution. We take $x = 3$ and $x + \Delta x = 2.98$, so that $\Delta x = 2.98 - 3 = -0.02$. Since

$$\frac{dy}{dx} = 3x^2,$$

we find, using Eq. (5), that

$$\Delta y \approx 3x^2 \Delta x = 27(-0.02) = -0.54.$$

Problem 2. Using the approximation in Eq. (5), determine a reasonable approximation to $(2.98)^3$.

Solution. As in Problem 1, we consider

$$y = x^3.$$

Since 2.98 is near to 3 and $3^3 = 27$ is easily calculated, we think of starting from the point

$$(x = 3, \quad y = 27)$$

and trying to reach the point

$$(x + \Delta x = 2.98, \quad y + \Delta y = (2.98)^3 = ?).$$

Now instead of calculating $(2.98)^3$ exactly, we shall make use of the *approximation*

$$\Delta y \approx \frac{dy}{dx}\Delta x = 3x^2 \Delta x = -0.54.$$

Hence, adding this to $y = x^3 = 27$, we have

$$(2.98)^3 = y + \Delta y \approx y + \frac{dy}{dx}\Delta x = 27 - 0.54,$$

or

$$(2.98)^3 \approx 26.46.$$

Remark 1. Let f be a differentiable function in some neighborhood of $x = a$. Rewrite Eq. (4c) by setting

$$\Delta x = x - a, \quad \Delta y = f(x) - f(a),$$

and $dy/dx = f'(a)$. In the new equation, add $f(a)$ to both sides. You should get

$$f(x) = f(a) + f'(a)(x - a) + \epsilon(x - a). \tag{6a}$$

Condition (4b) translates into

$$\epsilon \to 0 \quad \text{as} \quad x \to a. \tag{6b}$$

The first two terms on the right-hand side of (6a) represent a linear function, called the *linearization of f at a:*

$$L_a(x) = f(a) + f'(a)(x - a). \tag{7}$$

The graph of this linear function goes through the point $(a, f(a))$ and is tangent to the graph of f at that point. Because the difference between $f(x)$ and $L_a(x)$ is the part $\epsilon\,(x - a)$ in (6a), and (6b) says that ϵ is small when x is near a, the linearization of f at a can be used to estimate with considerable accuracy the values of $f(x)$ for x's near a.

Problem 3. Estimate $1/(2.04)^2$.

Solution. Let $f(x) = 1/x^2$, $a = 2$, and use the linearization of f at a. Differentiating, we get

$$f'(x) = -2x^{-3} = -2/x^3.$$

Therefore

$$f(a) = f(2) = \tfrac{1}{4}, \quad f'(a) = f'(2) = -\tfrac{2}{8} = -\tfrac{1}{4},$$

and

$$L_a(x) = L_2(x) = \tfrac{1}{4} - \tfrac{1}{4}(x - 2),$$

so that

$$L_2(2.04) = 0.25 - \tfrac{1}{4}(0.04) = 0.24.$$

EXERCISES

For each of the following functions, find

(a) Δy,

(b) the linear part of Δy, namely, $\Delta y_{\tan} = (dy/dx)\,\Delta x$,

(c) the difference, $\Delta y - \Delta y_{\tan}$.

1. $y = x^2 + 2x$
2. $y = 2x^2 + 4x - 3$
3. $y = x^3 - x$
4. $y = x^4$
5. $y = x^{-1}$

In Exercises 6 through 10, define a suitable function f, choose an appropriate value of a, and find the corresponding linearization of f at a for the purpose of estimating the indicated number. Then use that linearization to estimate the number.

6. $\sqrt{4.5}$
7. $1/\sqrt{3.8}$
8. $(7.5)^{1/3}$
9. $(34)^{-1/5}$
10. $\dfrac{1.3}{1 + 1.3}$

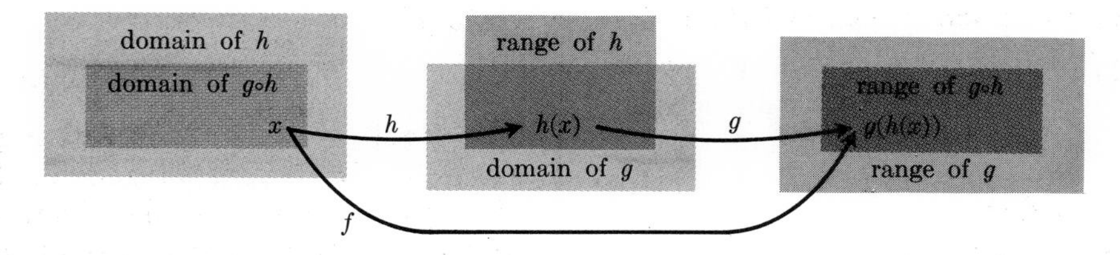

3.9 The composition of g with h: $f(x) = g(h(x))$, $f = g \circ h$.

11. A short table of squares is sometimes provided to be used with a small desk calculator. The instructions that come with it say something like this: To find $\sqrt{N}$, locate the nearest square in the table and compute $(N + a^2)/(2a)$. For instance, for $N = 7$, we might locate $(2.65)^2 = 7.0225$ and compute $14.0225/5.3$. (This division is easy to perform on the desk calculator. It comes out to approximately 2.64575. The CRC Mathematical Tables give $\sqrt{7} = 2.645751$.) Show that the above rule amounts to using the linearization of

$$f(x) = (x + a^2)^{1/2}$$

at 0 to estimate $\sqrt{N}$ with $x = N - a^2$.

12. The volume $y = x^3$ of a cube of edge x increases by an amount Δy when x increases by an amount Δx. Show that Δy may be represented geometrically as the sum of the volumes of

(a) three slabs of dimensions x by x by Δx,
(b) three bars of dimensions x by Δx by Δx,
(c) one cube of dimensions Δx by Δx by Δx.

Illustrate by a sketch.

3.5 COMPOSITE FUNCTIONS

Sometimes a function represents a mapping that can be achieved by performing two simpler mappings in succession. We illustrate this process of *composition of functions* with examples, and then give a general definition.

Example 1. The functions

$$f_1(x) = \sin(x^2), \quad -\infty < x < \infty, \tag{1a}$$

$$f_2(x) = \sin^2 x, \quad -\infty < x < \infty, \tag{1b}$$

are examples of composite functions. In Eq. (1a), we first square x, then find the sine of the result. Thus, if

$$g(x) = x^2 \tag{2a}$$

and

$$h(x) = \sin x, \tag{2b}$$

we can write Eq. (1a) in the form

$$f_1(x) = h(g(x)), \tag{3a}$$

or

$$f_1 = h \circ g, \tag{3b}$$

where (3a) is read "f_1 of x is equal to h of g of x," and (3b) is read "f_1 is the composition of h with g." The symbol $h \circ g$ is used to distinguish the composition of h with g from the ordinary product of h and g.

In the second equation, Eq. (1b), we first evaluate $\sin x$, then square the result. Using the notation of Eqs. (2a) and (2b), we can rewrite (1b) in the form

$$f_2(x) = g(h(x)), \tag{4a}$$

or, what amounts to the same thing,

$$f_2 = g \circ h. \tag{4b}$$

In both examples (1a) and (1b), the domains are the same—the set of all real numbers.

More generally, suppose we define the composition of g with h by the following rule:

$$f = g \circ h \tag{5a}$$

means that

$$f(x) = g(h(x)) \tag{5b}$$

for all values of x for which the right-hand side of Eq. (5b) is defined. Here g and h represent any two functions, no longer restricted as in (2a) and (2b). To evaluate (5b), we start with a value of x in the domain of h. (See Fig. 3.9, which illustrates the composition of g with h.) Then $h(x)$ is a number in the range of h. If this number $h(x)$ is also in the domain of g, then $g(h(x))$ is a number in the range of g. Thus, the domain of $g \circ h$ is a subset of the domain of h, and the range of $g \circ h$ is a subset of the range of g.

Example 2. Suppose that

$$g(x) = \sqrt{1 - x}, \quad x \leq 1, \tag{6a}$$

and

$$h(x) = \sqrt{x}, \quad x \geq 0. \tag{6b}$$

Then

$$\begin{aligned} g(h(x)) &= \sqrt{1 - h(x)}, \quad x \geq 0,\ h(x) \leq 1 \\ &= \sqrt{1 - \sqrt{x}}, \quad x \geq 0,\ \sqrt{x} \leq 1. \end{aligned}$$

From the restrictions $x \geq 0$, $\sqrt{x} \leq 1$, we find that the domain of

$$f = g \circ h$$

is the interval $0 \leq x \leq 1$. Figure 3.10(a, b) shows the graphs of the functions h and g, respectively. Fig. 3.10(c) emphasizes the interval [0, 1] of intersection of the domain of g and the range of h. Fig. 3.10(d) shows the graph of the composed function $f = g \circ h$ over the interval of intersection.

Definition 1. Composition of Functions. *Let g be a function with domain D_g and range R_g, and let h be a function with domain D_h and range R_h. The composition of g with h is the function f defined by*

$$f(x) = g(h(x))$$

with domain D_f consisting of those values x in D_h such that $h(x) \in D_g$.

Remark 1. Here is another way to look at the composite mapping $f = g \circ h$. Start with a number y in the intersection of D_g and R_h. Because this y is in the range of h, there is an x in D_h such that $h(x) = y$. And, because y is also in the domain of g, there is a number $z = g(y)$. Combining, we get

$$z = g(h(x)) = f(x).$$

Thus,

(x, y) is an element of the function h,

(y, z) is an element of the function g,

and

(x, z) is an element of the composite function $f = g \circ h$.

Therefore the ordered pair (x, z) belongs to the composite function $f = g \circ h$ if (and only if) there is a number y such that

(x, y) belongs to h and (y, z) belongs to g.

Remark 2. By using a three-dimensional (x, y, z)-coordinate system, we can get a graphical representation of the composition:

$$y = h(x), \quad z = g(y),$$
$$f(x) = g(h(x)).$$

Figure 3.10(e) shows such a representation for the example

$$y = h(x) = \sqrt{x}, \quad x \geq 0,$$
$$z = g(y) = \sqrt{1 - y},$$
$$f(x) = g(h(x)) = \sqrt{1 - \sqrt{x}}.$$

To find the value of $f(x)$ for a particular value of x, $0 \leq x \leq 1$, let Q be the point with abscissa x on the x-axis, and follow the diagram $Q \to R$, $R \to S$, $S \to T$, $T \to U$, $U \to P$. Here $R(x, y, 0)$ is on the graph of $y = h(x)$, $S(0, y, 0)$ has the same y-coordinate as R, and $T(0, y, z)$ is on the graph of $z = g(y)$. Because $U(0, 0, z)$ and $P(x, 0, z)$ have the same z-coordinates as T, the x- and z-coordinates of P satisfy the equations

$$z = g(y) = g(h(x)),$$

so that P is on the graph of $z = f(x)$ in the xz-coordinate plane (that is, $y = 0$). For the example shown in Fig. 3.10(d), the curve through A, P, and D is the graph of

$$z = \sqrt{1 - \sqrt{x}}, \quad 0 \leq x \leq 1.$$

See also Fig. 3.11, which illustrates the function $z = \sin(x^2)$ for $x \geq 0$.

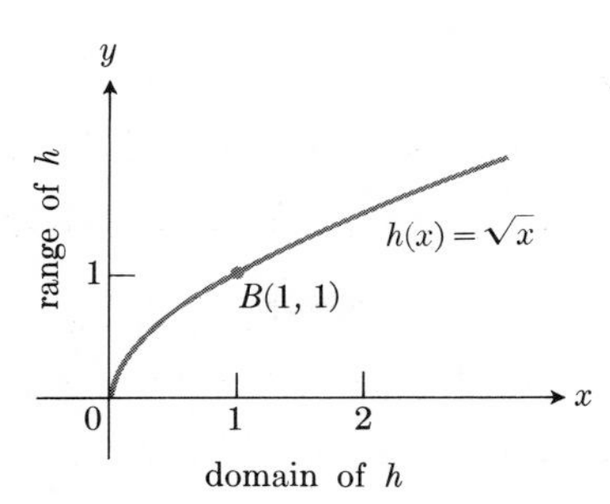

a| $y = h(x) = \sqrt{x}$.

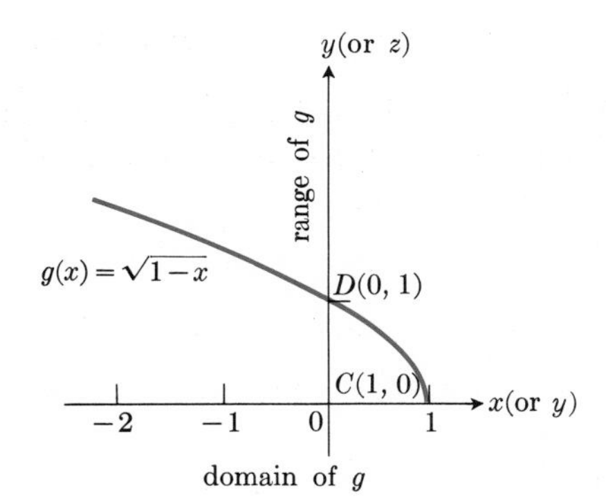

b| $y = g(x) = \sqrt{1-x}$ [or $z = g(y) = \sqrt{1-y}$].

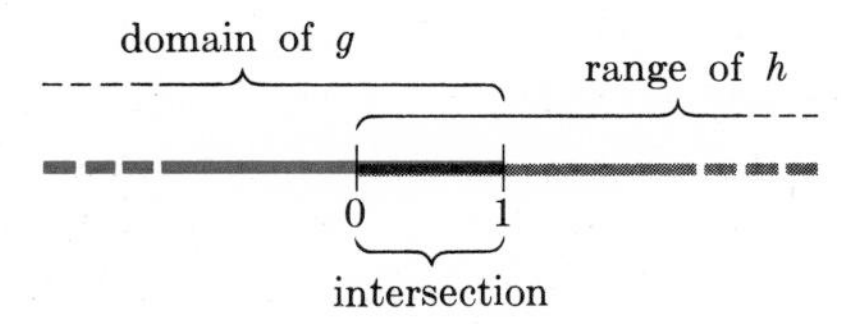

c| The intersection of the domain of g and the range of h.

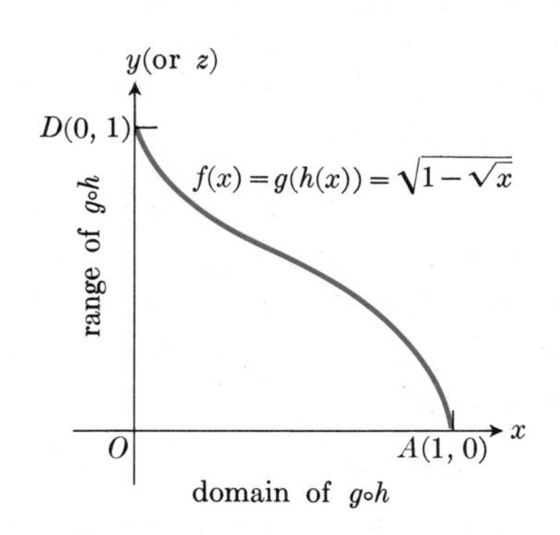

d| $y = f(x) = \sqrt{1-\sqrt{x}}$ [or $z = f(x) = \sqrt{1-\sqrt{x}}$].

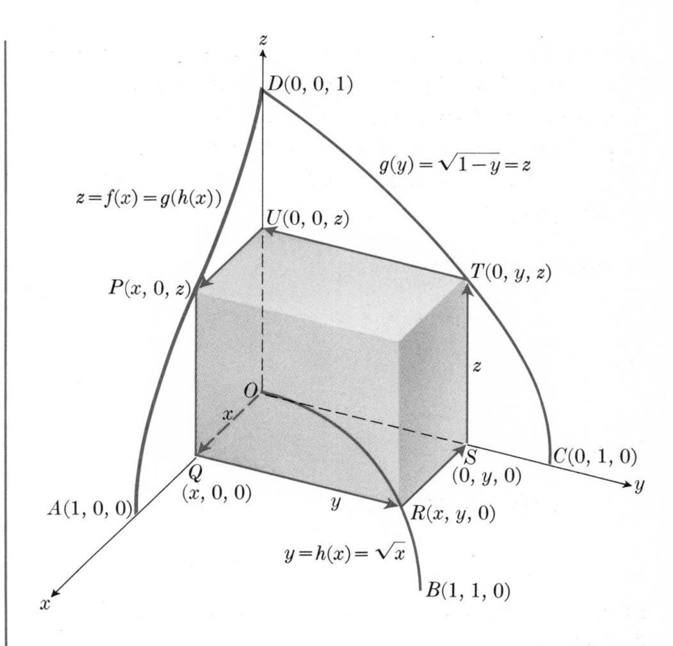

e| Representation of composition of functions in xyz-space:

$$y = \sqrt{x},$$
$$z = \sqrt{1-y}$$
$$= \sqrt{1-\sqrt{x}}.$$

3.10

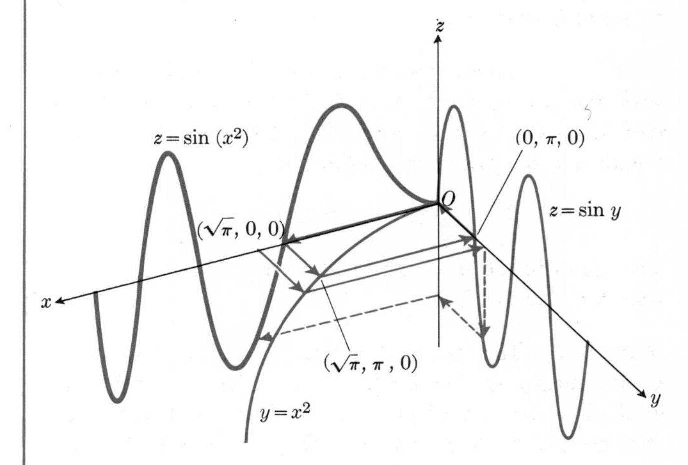

3.11 Composition:

$$y = x^2,$$
$$z = \sin y$$
$$= \sin (x^2).$$

EXERCISES

Find the compositions $f_1 = g \circ h$ and $f_2 = h \circ g$ for the following pairs of functions g and h. In each instance, indicate the domains of f_1 and f_2.

1. $g(x) = 2x + 1, \quad x \in (-\infty, \infty)$
   $h(x) = \sqrt{x}, \quad x \geq 0$
2. $g(x) = \dfrac{x}{x+1}, \quad x \neq -1$
   $h(x) = \dfrac{1}{x}, \quad x \neq 0$
3. $g(x) = \sqrt{x-1}, \quad x \geq 1$
   $h(x) = \dfrac{1}{x}, \quad x \neq 0$
4. $g(x) = \sin x, \quad x \in (-\infty, \infty)$
   $h(x) = 1/x, \quad x \neq 0$
5. $g(x) = \cos x, \quad x \in [-2\pi, 2\pi]$
   $h(x) = 1/\sqrt{x}, \quad x > 0$
6. $g(x) = |x|, \quad h(x) = \sin x, \quad x \in (-\infty, \infty)$
7. $g(x) = 2x + 4, \quad h(x) = |x|, \quad x \in (-\infty, \infty)$
8. $g\colon x \to \frac{1}{2}(x + |x|), \quad -\infty < x < \infty$
   $h\colon x \to \tan x, \quad x \in (-\pi/2, \pi/2)$
9. $g(x) = x^3, \quad h(x) = x^{1/3}, \quad x \in (-\infty, \infty)$
10. $g(x) = x^2, \quad h(x) = x^{1/2}$, on the natural domains
11. Figure 3.10(e) shows the graphs of $y = h(x)$ and $z = g(y)$ for a specific pair of functions h and g. Using these graphs (or otherwise), sketch graphs of $y = g(x)$, $z = h(y)$, and $z = h(g(x))$, for the natural domains of all the functions involved.
12. Figure 3.11 shows a portion of the graph of $z = \sin(x^2)$. Sketch the graph of $z = (\sin x)^2$ for $x \in [-2\pi, 2\pi]$.

3.6 DERIVATIVES OF COMPOSITE FUNCTIONS: THE CHAIN RULE

A derivative gives a rate of change. Thus, if

$$dy/dx = 5 \quad \text{and} \quad dx/dt = 3,$$

we might say "y changes five times as fast as x and x changes three times as fast as t," and conclude that "y changes fifteen times as fast as t." This suggests that

$$\frac{dy}{dt} = \frac{dy}{dx} \cdot \frac{dx}{dt}. \tag{1}$$

In this section, we shall see that Eq. (1), which is sometimes called the *chain rule* for derivatives, is indeed valid when properly interpreted, and under appropriate conditions.

Problem 1. A particle moves along the line

$$y = 5x - 2$$

in such a way that, at time t, its x-coordinate satisfies $x = 3t$. Show that Eq. (1) holds.

Solution. From the given data, we get

$$\frac{dy}{dx} = 5, \quad \frac{dx}{dt} = 3. \tag{2a}$$

On the other hand, we also have

$$y = 5x - 2 = 15t - 2,$$

so that

$$\frac{dy}{dt} = 15. \tag{2b}$$

The numbers in Eqs. (2a, b) satisfy Eq. (1).

Problem 2. A particle moves along the curve $y = x^2$. Its x-coordinate satisfies

$$x = 2t^2 + t + 1.$$

Show that Eq. (1) is satisfied when $t = 2$.

Solution. When $t = 2$, we have $x = 11$, $y = 121$, $dy/dx = 2x = 22$, and $dx/dt = 4t + 1 = 9$. Therefore

$$\left(\frac{dy}{dx}\right)\left(\frac{dx}{dt}\right) = (22)(9) = 198.$$

On the other hand, by substitution we get

$$\begin{aligned} y &= x^2 = (2t^2 + t + 1)^2 \\ &= 4t^4 + 4t^3 + 5t^2 + 2t + 1, \end{aligned}$$

so that

$$\frac{dy}{dt} = 16t^3 + 12t^2 + 10t + 2.$$

At $t = 2$, we therefore have

$$\frac{dy}{dt} = 128 + 48 + 20 + 2 = 198 = \frac{dy}{dx} \cdot \frac{dx}{dt}.$$

Remark 1. In Problem 2, we verified the chain rule at time $t = 2$. For the same problem, Eq. (1) holds for any value of t, because

$$\begin{aligned}\frac{dy}{dt} &= 16t^3 + 12t^2 + 10t + 2\\ &= 2(2t^2 + t + 1)(4t + 1)\\ &= 2x(4t + 1)\\ &= \left(\frac{dy}{dx}\right)\left(\frac{dx}{dt}\right).\end{aligned}$$

To prepare for a careful statement of the chain rule and its proof, we introduce some notation.

Let f be a function

$$f = \{(x, y): y = f(x),\ x \in D_f\}, \tag{3}$$

which is differentiable at an interior point x_0 of its domain D_f. Let g be a function

$$g = \{(t, x): x = g(t),\ t \in D_g\}, \tag{4a}$$

which is differentiable at an interior point t_0 of its domain such that

$$g(t_0) = x_0. \tag{4b}$$

Let $h = f \circ g$ be the composition of f with g:

$$h = \{(t, y): y = f(g(t)),\ t \in D_h\}. \tag{5}$$

Theorem 10. *Under the hypotheses described in the preceding paragraph and Eqs. (3), (4), (5), if t_0 is an inner point of the domain of h, then*

$$h'(t_0) = f'(x_0) \cdot g'(t_0). \tag{6}$$

Proof. Equations (3), (4), (5) say that, for any t in the domain of h, we have

$$y = h(t) = f(x),$$

provided

$$x = g(t).$$

By definition,

$$h'(t_0) = \lim_{t \to t_0} \frac{h(t) - h(t_0)}{t - t_0},$$

provided this limit exists. Now

$$\begin{aligned}h(t) - h(t_0) &= f(g(t)) - f(g(t_0))\\ &= f(x) - f(x_0),\end{aligned}$$

so that

$$\frac{h(t) - h(t_0)}{t - t_0} = \frac{f(x) - f(x_0)}{t - t_0}. \tag{7}$$

We would like to write the last term in Eq. (7) thus:

$$\begin{aligned}\frac{f(x) - f(x_0)}{t - t_0} &= \frac{f(x) - f(x_0)}{x - x_0} \cdot \frac{x - x_0}{t - t_0}\\ &= \frac{f(x) - f(x_0)}{x - x_0} \cdot \frac{g(t) - g(t_0)}{t - t_0},\end{aligned} \tag{8}$$

but Eq. (8) is valid only if $x \neq x_0$ and $t \neq t_0$. We cannot guarantee this. However, by Eq. (4c) of Article 3.4, we can write

$$f(x) - f(x_0) = \Delta y = \frac{dy}{dx}\Delta x + \epsilon\, \Delta x \tag{9a}$$

where

$$\frac{dy}{dx} = f'(x_0), \quad \Delta x = x - x_0, \tag{9b}$$

and

$$\epsilon \to 0 \quad \text{as} \quad \Delta x \to 0. \tag{9c}$$

Substitute from (9a, b) into the right-hand side of Eq. (7) and get

$$\begin{aligned}\frac{h(t) - h(t_0)}{t - t_0} &= (f'(x_0) + \epsilon)\frac{x - x_0}{t - t_0}\\ &= (f'(x_0) + \epsilon)\frac{g(t) - g(t_0)}{t - t_0}.\end{aligned} \tag{10}$$

Equation (10) brings us quickly to our goal, because $x = g(t)$ approaches $x_0 = g(t_0)$ when $t \to t_0$. (Why? You are asked to prove this statement in Exercise 11.) Therefore

$$\lim_{t \to t_0} \epsilon = \lim_{x \to x_0} \epsilon = 0$$

and

$$\lim_{t \to t_0} \frac{g(t) - g(t_0)}{t - t_0} = g'(t_0),$$

so that, from Eq. (10),

$$\begin{aligned}&\lim_{t \to t_0} \frac{h(t) - h(t_0)}{t - t_0}\\ &\quad = \lim_{t \to t_0} [f'(x_0) + \epsilon] \cdot \lim_{t \to t_0} \frac{g(t) - g(t_0)}{t - t_0}\\ &\quad = f'(x_0)g'(t_0).\end{aligned} \tag{11}$$

Equation (11) says that h has a derivative at t_0 and

$$h'(t_0) = f'(x_0)g'(t_0). \square$$

Remark 1. We have used subscripts on x_0 and t_0 to emphasize that these are held fixed while we take certain limits. The final result, Eq. (6), can equally well be written in either of the equivalent forms

$$h'(t) = f'(x)g'(t), \tag{12a}$$

or

$$\frac{dy}{dt} = \frac{dy}{dx}\,\frac{dx}{dt}. \tag{12b}$$

But it is understood that x is a function of t, so that dy/dx is also expressible as a function of t. When that has been done, the two sides of Eq. (12b) agree.

Problem 3. Find dy/dt

for $y = x^3 - 3x^2 + 5x - 4 = f(x)$,

where $x = t^2 + t = g(t)$.

Solution. From these equations, we find

$$\begin{aligned} f'(x) &= \frac{dy}{dx} = 3x^2 - 6x + 5 \\ &= 3(t^2 + t)^2 - 6(t^2 + t) + 5, \end{aligned}$$

$$g'(t) = \frac{dx}{dt} = 2t + 1,$$

so that Eq. (12b) becomes

$$\begin{aligned} \frac{dy}{dt} &= \frac{dy}{dx}\,\frac{dx}{dt} \\ &= [3(t^2 + t)^2 - 6(t^2 + t) + 5](2t + 1). \end{aligned}$$

If, on the other hand, we first substitute the value of x in terms of t into the equation for y, we have

$$y = (t^2 + t)^3 - 3(t^2 + t)^2 + 5(t^2 + t) - 4.$$

When we differentiate this with respect to t, we get

$$\begin{aligned} \frac{dy}{dt} &= 3(t^2 + t)^2(2t + 1) \\ &\quad - 6(t^2 + t)(2t + 1) + 5(2t + 1) \\ &= [3(t^2 + t)^2 - 6(t^2 + t) + 5](2t + 1), \end{aligned}$$

which agrees with the previous answer.

Remark 2. So far we have not defined dy and dx as separate entities, but have used the complete symbol dy/dx as the name for lim $(\Delta y/\Delta x)$ as $\Delta x \to 0$. Now that we have established Eq. (12b), we are in a position where we can (and we shall in the next article) define dy and dx as separate quantities whose ratio is the derivative of y with respect to x. Then Eq. (12b) is the form of the chain rule which is most easily remembered, since it suggests that dx may be cancelled from the two fractions on the right.

Remark 3. If $dx/dt \neq 0$, we may divide both sides of Eq. (12b) by dx/dt and write it in the equivalent form

$$\frac{dy}{dx} = \frac{dy/dt}{dx/dt}. \tag{13}$$

This form of the equation is particularly useful in dealing with the parametric equations

$$x = f(t), \quad y = g(t),$$

since it enables us to find the derivative of y with respect to x directly from these equations.

Example 1. The parametric equations

$$x = 2t + 3, \quad y = t^2 - 1$$

represent a curve whose equation in the form of $y = f(x)$ may be found by substituting

$$t = \frac{x - 3}{2}$$

from the first equation into the second; that is,

$$y = \tfrac{1}{4}(x - 3)^2 - 1.$$

If we work with the original equations and calculate

$$\frac{dx}{dt} = 2, \quad \frac{dy}{dt} = 2t,$$

then Eq. (13) gives

$$\frac{dy}{dx} = \frac{dy/dt}{dx/dt} = \frac{2t}{2} = t = \frac{x - 3}{2}.$$

On the other hand, from the equation for y in terms of x, we find

$$\frac{dy}{dx} = \frac{1}{2}(x-3)\frac{d}{dx}(x-3) = \frac{x-3}{2}.$$

Remark 4. The chain rule is also frequently expressed using other letters to represent the variables. Most common is the case where y is a function of u and u is a function of x,

$$y = F(u), \quad u = f(x).$$

Then the rule says

$$\frac{dy}{dx} = \frac{dy}{du}\frac{du}{dx}. \tag{14}$$

We have already had an illustration of this formula in the case of the derivative with respect to x of

$$y = u^n.$$

Then

$$\frac{dy}{du} = nu^{n-1}$$

and

$$\frac{dy}{dx} = \frac{dy}{du}\frac{du}{dx} = nu^{n-1}\frac{du}{dx}. \tag{15}$$

We have previously proved Eq. (15) for positive and negative integral exponents n. [See Article 3.2, Eqs. (2) and (5).] We can now extend it (at least for $u > 0$) to fractional exponents as well, for, in Article 3.3, Eq. (11), we established the result

$$\frac{d}{dx}(x^n) = nx^{n-1} \quad \text{for } n \text{ rational, } x > 0.$$

The same equation, written with u in place of x, says that

$$\frac{d}{du}(u^n) = nu^{n-1} \quad \text{for } n \text{ rational, } u > 0.$$

Combining this with the chain rule, Eq. (14), for u a differentiable function of x and n any rational number, we have the following theorem.

Theorem 11. *Let $u = f(x)$ be differentiable at $x = a$, and $f(a) > 0$. Let n be a rational number. Then $y = u^n$ is differentiable at $x = a$, and*

$$\frac{d}{dx}(u^n) = nu^{n-1}\frac{du}{dx}. \tag{16}$$

Example 2. If $y = (x^2 - a^2)^{1/2}$, where a is a number, then

$$\frac{dy}{dx} = \tfrac{1}{2}(x^2 - a^2)^{-1/2}(2x) = \frac{x}{y}.$$

EXERCISES

In Exercises 1 through 4, each pair of equations represents a curve in parametric form. In each case, find the equation of the curve in the form $y = F(x)$ by eliminating t, then calculate

$$\frac{dy}{dt}, \quad \frac{dy}{dx}, \quad \frac{dx}{dt}$$

and verify that these derivatives satisfy the chain rule, Eq. (12b).

1. $x = 3t + 1, \quad y = t^2$
2. $x = t^2, \quad y = t^3$
3. $x = \dfrac{t}{1-t}, \quad y = t^2$
4. $x = \dfrac{t}{1+t}, \quad y = \dfrac{t^2}{1+t}$
5. If a point traces the circle $x^2 + y^2 = 25$ and if $dx/dt = 4$ when the point reaches (3, 4), find dy/dt there.

In Exercises 6 through 10, find dy/dx

(a) by using the chain rule and also

(b) by first expressing y directly in terms of x, and then differentiating.

6. $y = u^2 + 3u - 7, \quad u = 2x + 1$
7. $y = \dfrac{u^2}{u^2+1}, \quad u = \sqrt{2x+1}$
8. $y = z^{2/3}, \quad z = x^2 + 1$
9. $y = w^2 - w^{-1}, \quad w = 3x$
10. $y = 2v^3 + \dfrac{2}{v^3}, \quad v = (3x+2)^{2/3}$
11. Prove that $x = g(t)$ approaches $x_0 = g(t_0)$ as $t \to t_0$ if g is differentiable at t_0.

3.7 THE DIFFERENTIALS dx AND dy

We are now prepared to define dx and dy in such a way that their ratio, when $dx \neq 0$, will be the same as the derivative of y with respect to x. This is very easy to do in the case where x is the *independent variable* and y is a function of x. When x is the independent variable and

$$y = f(x),$$

we shall adopt the following *definitions.*

Definition. *dx, called the "differential of x," may represent any real number whatever; that is, dx is another independent variable and its domain is*

$$-\infty < dx < +\infty.$$

Definition. *dy, called the "differential of y," represents the function of x and dx given by*

$$dy = f'(x)\,dx, \tag{1}$$

where $f'(x)$ is the derivative at x of the function f.

It is clear from (1) that the differentials dx and dy have the properties

(1) if $dx = 0$, then $dy = 0$, and

(2) if $dx \neq 0$, then

$$\frac{dy}{dx} = \frac{\text{the differential of } y \text{ in terms of } x \text{ and } dx}{\text{the differential of } x} = \frac{f'(x)\,dx}{dx} = f'(x) \tag{2}$$

is the derivative of y with respect to x.

There is certainly nothing mysterious about this last result, since we deliberately defined dy by Eq. (1) so that Eq. (2) would be true. In other words, we planned it that way!

Example 1. If $\quad y = x^3 = f(x),$

then $\quad f'(x) = 3x^2,$

and $\quad dy = f'(x)\,dx = 3x^2\,dx.$

We also have been accustomed to writing

$$\frac{dy}{dx} = 3x^2,$$

and if we treat this as a fraction and multiply both sides by dx, we obtain again

$$dy = 3x^2\,dx.$$

Again, if $\quad y = x = f(x),$

then $\quad f'(x) = 1,$

and $\quad dy = f'(x)\,dx = dx.$

That is, the *function* $f(x) = x$ has its differential, as given by Eq. (1), equal to the differential of the *independent variable* x. This is not a profound result; it simply shows that in the case where both (a) and (b) apply, the differential is the same whether given by (a) or by (b). That is, the two rules are *consistent.*

The geometric interpretation of dy becomes clear when we refer to Fig. 3.12, where in each part we indicate a curve $y = f(x)$ and a line segment PT tangent to the curve at P. The differential $dx = PR$ in each figure is considered positive if R is to the right of P [as in (a) and (c)], and negative when R is to the left of P [as in (b) and (d)]. The sign of dy is positive if T is above R [as in (a) and (d)] and negative if T is below R [as in (b) and (c)]. In any case, we find that

$$\frac{dy}{dx} = \frac{RT}{PR} = \tan\phi = f'(x)$$

gives the slope of the tangent to the curve at P, and

$$dy = f'(x)\,dx = (\tan\phi)(PR) = RT$$

is the *amount* of change in y along this tangent line that would be produced by a change dx in x. We thus have

(a) dy/dx is the *rate* of change of y per unit change in x, and

(b) dy is the *amount* of change of y for dx units of change in x along the *tangent* to the curve at P.

Therefore, if we take $dx = \Delta x$, then dy is the same as $\Delta y_{\tan}$ (Fig. 1.43), which is the principal part of Δy. In problems dealing with increments we therefore do take $dx = \Delta x$ and use dy as an approximation to Δy.

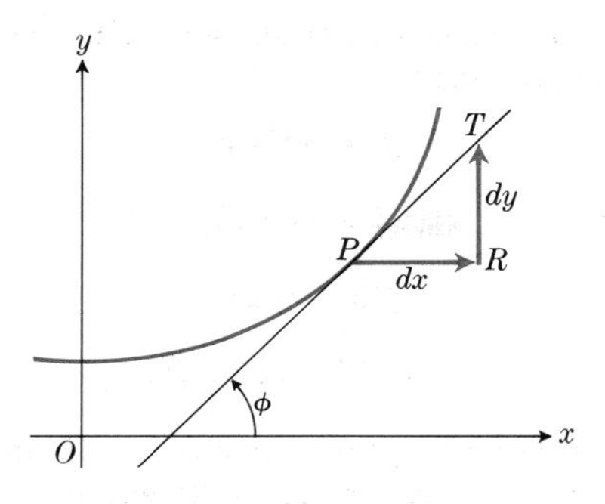

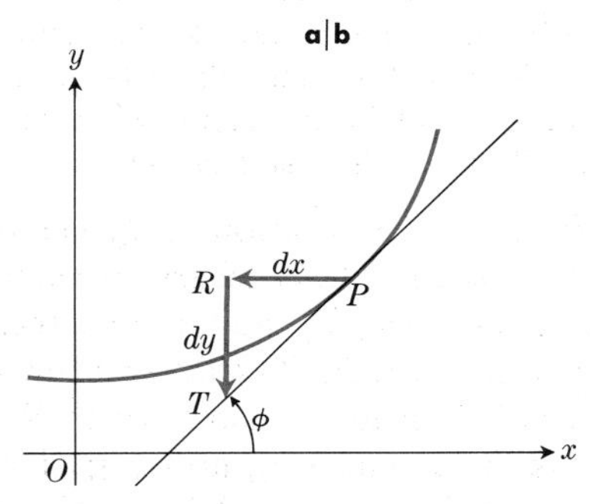

a|b

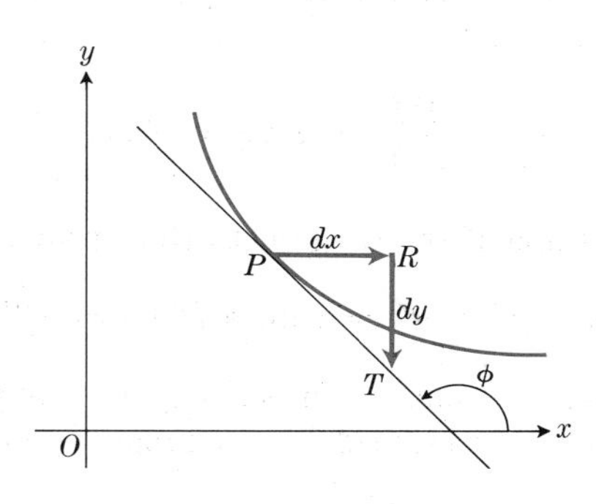

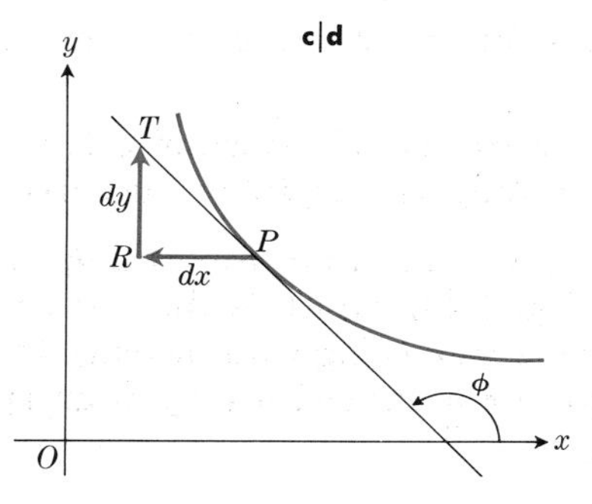

c|d

3.12

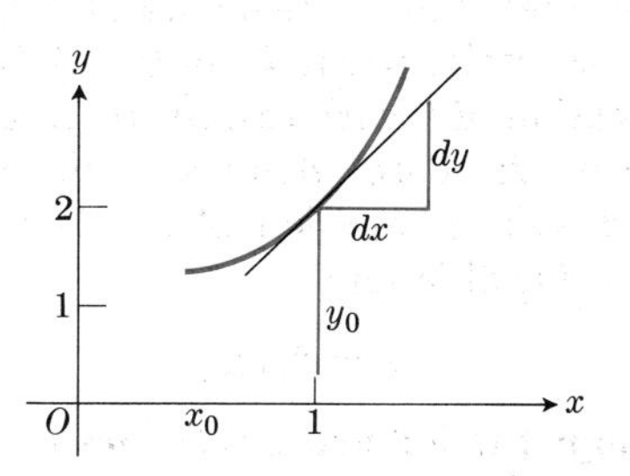

3.13

Remark 1. Equation (1) shows that at a point a where f is differentiable, the differential dy (or df) is directly proportional to dx. The proportionality coefficient is $f'(a)$. The function df is *linear* in dx. If we want to approximate the function f at x's near a, we can take $dx = \Delta x = x - a$, and use $df = f'(a)\,dx$ as the linear part of the increment (which it is). This is just another way of using the tangent line to the graph of f at a to estimate values of f for x's near a. The method is exactly the same as using the linearization of f, at a, as in Article 3.4, Eqs. (6) and (7), with a slight change in notation. You will find the differential notation widely used, as in the following problem.

Problem 1. Find a reasonable approximation to the value of

$$(1.001)^7 - 2(1.001)^{4/3} + 3.$$

Solution. At $x_0 = 1$, the function

$$f(x) = x^7 - 2x^{4/3} + 3$$

has the easily calculated value

$$y_0 = f(1) = 2.$$

The tangent to this curve $y = f(x)$ at $(1, 2)$ will remain near the curve for values of x close to x_0. See Fig. 3.13. As x changes from $x_0 = 1$ to $x_0 + dx = 1.001$, the change in y *along this tangent line will be*

$$dy = f'(x_0)\,dx.$$

Since

$$f'(x) = 7x^6 - \tfrac{8}{3}x^{1/3}$$

has the value

$$f'(x_0) = \tfrac{13}{3}$$

at $x_0 = 1$, when we take $dx = 0.001$, we have

$$dy = \tfrac{13}{3}(0.001) = 0.0043.$$

When this change in y is added to y_0, we have

$$y_0 + dy = 2.0043$$

as a good approximation to the value of y at the point on the curve for which $x = x_0 + dx$.

Suppose next that x is not the independent variable, but that x and y are both to be regarded as functions of t. Let

$$y = f(x), \quad x = g(t), \quad y = f(g(t)) = h(t), \qquad (3)$$

where t is now the independent variable for the functions g and h. If these functions f, g, and h satisfy the differentiability conditions stated in Theorem 10 of Article 3.6, then dt is a new independent variable on $(-\infty, \infty)$, and in terms of t and dt, we have

$$dx = g'(t)\,dt, \quad dy = h'(t)\,dt. \qquad (4a)$$

We can write Eq. (13) of Article 3.6 in the following form:

$$\frac{dy}{dx} = \frac{(dy/dt)}{(dx/dt)} = \frac{(dy/dt)\,dt}{(dx/dt)\,dt},$$

if

$$\frac{dx}{dt} \neq 0; \qquad (4b)$$

or

$$f'(x) = \frac{h'(t)\,dt}{g'(t)\,dt} = \frac{\text{differential of } y \text{ in terms of } t \text{ and } dt}{\text{differential of } x \text{ in terms of } t \text{ and } dt}. \qquad (4c)$$

That is, the derivative of y with respect to x is equal to

$$\frac{\text{the differential of } y}{\text{the differential of } x}, \quad \text{if} \quad dx \neq 0,$$

whether the differentials are in terms of x and dx or in terms of t and dt.

It is this property of the differentials that makes it possible to treat the derivative as the ratio of dy and dx, whether x is the independent variable or not, and enables us to take full advantage of the Leibniz notation.

Example 2. Let $y = t^2$ and $x = t^3$.

Then
$$\frac{dy}{dt} = 2t, \quad \frac{dx}{dt} = 3t^2,$$

$$dy = 2t\,dt, \quad dx = 3t^2\,dt,$$

$$\frac{dy}{dx} = \frac{2t\,dt}{3t^2\,dt} = \frac{2}{3t} \quad (\text{if } t \neq 0,\ dt \neq 0).$$

If we eliminate t before taking derivatives,

$$y = x^{2/3},$$

$$\frac{dy}{dx} = \frac{2}{3}x^{-1/3} = \frac{2}{3x^{1/3}} = \frac{2}{3t} \quad (\text{if } x \neq 0),$$

in agreement with the answer above.

The second derivative

The second derivative of y with respect to x has been interpreted as the result obtained from y by doing the operation "differentiate with respect to x" twice in succession; that is,

$$\frac{d^2y}{dx^2} = \frac{d}{dx}\left[\frac{d}{dx}(y)\right].$$

It is also possible to define a second differential, d^2y, in such a way that the second derivative of y with respect to x is equal to d^2y divided by $(dx)^2$. However, things get rather complicated in the case of parametric equations like Eqs. (3) above. In such cases, it is simplest to calculate second- and higher-order derivatives by using such rules as

$$\frac{dy}{dx} = y' = \frac{dy/dt}{dx/dt},$$

$$\frac{d^2y}{dx^2} = y'' = \frac{dy'}{dx} = \frac{dy'/dt}{dx/dt}.$$

It is important to note that to get the second derivative of y with respect to x, when the first derivative $dy/dx = y'$ is expressed in terms of t, we may differentiate y' with respect to t, provided we divide the result by dx/dt. (See Exercise 21 at the end of Article 3.8 for the general formula.)

Problem 2. Find d^2y/dx^2 if $x = t - t^2$ and $y = t - t^3$.

Solution.

$$y' = \frac{dy}{dx} = \frac{dy/dt}{dx/dt} = \frac{1 - 3t^2}{1 - 2t},$$

$$\frac{d^2y}{dx^2} = y'' = \frac{dy'}{dx} = \frac{dy'/dt}{dx/dt} = \frac{\frac{d}{dt}\left[\frac{1 - 3t^2}{1 - 2t}\right]}{(1 - 2t)}$$

$$= \frac{(1 - 2t)\cdot(-6t) - (1 - 3t^2)\cdot(-2)}{(1 - 2t)^3}$$

$$= \frac{2 - 6t + 6t^2}{(1 - 2t)^3}.$$

3.8 FORMULAS FOR DIFFERENTIATION REPEATED IN THE NOTATION OF DIFFERENTIALS

Earlier in this chapter we derived formulas for the derivatives listed below on the left. By multiplying each one by dx, we now obtain the corresponding formulas for differentials.

Derivatives	*Differentials*
I. $\frac{dc}{dx} = 0$	I′. $dc = 0$
II. $\frac{d(cu)}{dx} = c\frac{du}{dx}$	II′. $d(cu) = c\,du$
III. $\frac{d(u+v)}{dx} = \frac{du}{dx} + \frac{dv}{dx}$	III′. $d(u+v) = du + dv$
IV. $\frac{d(uv)}{dx} = u\frac{dv}{dx} + v\frac{du}{dx}$	IV′. $d(uv) = u\,dv + v\,du$
V. $\frac{d\left(\frac{u}{v}\right)}{dx} = \frac{v\frac{du}{dx} - u\frac{dv}{dx}}{v^2}$	V′. $d\left(\frac{u}{v}\right) = \frac{v\,du - u\,dv}{v^2}$
VI. $\frac{du^n}{dx} = nu^{n-1}\frac{du}{dx}$	VI′. $d(u^n) = nu^{n-1}\,du$
VIa. $\frac{dcx^n}{dx} = cnx^{n-1}$	VI′a. $d(cx^n) = cnx^{n-1}\,dx$

We collect these formulas primarily for future reference. Any problem involving differentials, say that of finding dy when y is a given function of x, may be handled in either of the following ways:

(a) by finding the derivative dy/dx and then multiplying by dx, or

(b) by direct use of formulas I′–VI′a.

Example. If we let $y = x^3 + 5x^2 - 7x + 4$,

then $$\frac{dy}{dx} = 3x^2 + 10x - 7,$$

and $$dy = (3x^2 + 10x - 7)\,dx,$$

or we may calculate directly:

$$\begin{aligned} dy &= d(x^3 + 5x^2 - 7x + 4) \\ &= d(x^3) + d(5x^2) + d(-7x) + d(4) && \text{(III′)} \\ &= 3x^2\,dx + 10x\,dx - 7\,dx + 0 && \text{(VI′a)} \\ &= (3x^2 + 10x - 7)\,dx. \end{aligned}$$

It should be noted that a *differential* on the left side of an equation, say dy, also calls for a *differential*, usually dx, on the right side of the equation. Thus we never have $dy = 3x^2$, but instead $dy = 3x^2\,dx$.

The term "differentiate" means either to find the derivative or to find the differential and either operation is referred to as "differentiation."

EXERCISES

In Exercises 1 through 8, find dy.

1. $y = x^3 - 3x^2 + 5x - 7$
2. $y^2 = (3x^2 + 1)^{3/2}$
3. $xy^2 + x^2y = 4$
4. $y = \frac{2x}{1 + x^2}$
5. $y = x\sqrt{1 - x^2}$
6. $y = \frac{x + 1}{x^2 - 2x + 4}$
7. $y = \frac{(1 - x)^3}{2 - 3x}$
8. $y = \frac{1 + x - x^2}{1 - x}$

Use differentials to obtain reasonable approximations to the following.

9. $\sqrt{145}$
10. $(2.1)^3$
11. $\sqrt[4]{17}$
12. $\sqrt[3]{0.126}$
13. $(8.01)^{4/3} + (8.01)^2 - \frac{1}{\sqrt[3]{8.01}}$

Given the following functions $x = f(t)$, $y = g(t)$:

(a) Express dx and dy in terms of t and dt.

(b) Use the results of (a) to find, at the point for which $t = 2$, the slope of the locus of the point (x, y).

(c) Use the results of (b) to find an equation of the line tangent to the curve at the point for which $t = 2$.

14. $x = t + \dfrac{1}{t}$, $y = t - \dfrac{1}{t}$

15. $x = \sqrt{2t^2 + 1}$, $y = (2t + 1)^2$

16. $x = t\sqrt{2t + 5}$, $y = (4t)^{1/3}$

17. $x = \dfrac{t-1}{t+1}$, $y = \dfrac{t+1}{t-1}$

18. $x = \dfrac{1}{t^2}$, $y = \sqrt{t^2 + 12}$

19. Find d^2y/dx^2 from the parametric equations in Exercise 14.

20. Find d^2y/dx^2 from the parametric equations in Exercise 17.

21. Given the parametric equations $x = f(t)$, $y = g(t)$. Show that

$$\frac{d^2y}{dx^2} = \frac{\dfrac{dx}{dt}\dfrac{d^2y}{dt^2} - \dfrac{d^2x}{dt^2}\dfrac{dy}{dt}}{(dx/dt)^3}.$$

22. (Franklin: *Treatise on Advanced Calculus.* Wiley, 1940, p. 110.) Let

$$y = F(x), \quad x = f(t), \quad F[f(t)] = g(t).$$

Using a subscript to denote the independent variable, define second differentials as follows:

$$(d^2y)_x = F''(x) \cdot (dx)^2,$$

$$(d^2y)_t = g''(t) \cdot (dt)^2, \quad (d^2x)_t = f''(t) \cdot (dt)^2.$$

Then show that

$$(d^2y)_t = (d^2y)_x + F'(x)(d^2x)_t.$$

Hint. By the chain rule, $g'(t) = F'(x)f'(t)$. From this deduce that

$$g''(t) = F'(x)f''(t) + F''(x)[f'(t)]^2$$

3.9 CONTINUITY

Before we take up applications of differentiation (Chapter 4), it is desirable to discuss the *continuity* of a function. We shall see that continuity is even more basic than differentiability.

Definition. *A function f which is defined in some neighborhood of c is said to be continuous at c provided that*

(a) *the function has a definite finite value $f(c)$ at c, and*

(b) *as x approaches c, $f(x)$ approaches $f(c)$ as limit:*

$$\lim_{x\to c} f(x) = f(c). \tag{1}$$

If a function is continuous at all points of an interval $a \le x \le b$ (or $a < x < b$, etc.), then it is said to be continuous on, or in, that interval.

Remark 1. The foregoing definition does not apply to certain situations. For instance,

$$\text{if} \qquad f(x) = \sqrt{x - x^2} \quad \text{for } x \in [0, 1], \tag{2}$$

then only the one-sided limits of $f(x)$ exist as x approaches the endpoint 0, or 1. We still say that f is continuous at these endpoints, however, because these one-sided limits are equal to the values of the function at the respective endpoints:

$$\lim_{x\to 0+} f(x) = \lim_{x\to 0+} \sqrt{x - x^2} = 0 = f(0), \tag{3a}$$

$$\lim_{x\to 1-} f(x) = \lim_{x\to 1-} \sqrt{x - x^2} = 0 = f(1). \tag{3b}$$

Remark 2. Sometimes the domain of a function has certain *isolated* points. For instance, the natural domain of the function

$$f(x) = \sqrt{-|x|}$$

is just the single point $x = 0$. For the function

$$g(x) = \sqrt{|x|(x - 1)},$$

the natural domain consists of $x = 0$ and all $x \ge 1$. How should we decide about the continuity of f and g at 0? The decision that has been made by a substantial number of mathematicians is that both f and g are *continuous* at $x = 0$. The reason they have made that choice is that they like the following alternative definition because it includes all possible points in the domain of f.

Alternative definition. *Let f be a function and let c be a number in the domain of f. Then f is* continuous *at c if and only if to each $\epsilon > 0$ there corresponds a $\delta > 0$ such that*

$$|f(x) - f(c)| < \epsilon \text{ whenever } |x - c| < \delta \text{ and } x \in D_f.$$

In spite of its lack of generality, we prefer the first definition because we can apply all of our limit theorems directly when we use it. [Also, because those same theorems would be true if each were modified to restrict x to D_f whenever $f(x)$ is mentioned, the continuity at endpoints of D_f, as in Eqs. (3a) and (3b), presents no problem.] For isolated points of D_f, we can just go along with the others and say, "All right, if you say so, f is continuous there too."

Remark 3. When a function fails to be continuous at a point of its domain, we say that it is *discontinuous* at that point. For instance, sgn x is discontinuous at $x = 0$ because sgn $0 = 0$, whereas

$$\lim \operatorname{sgn} x = -1, \quad \text{or} \quad +1,$$

when $x \to 0$ from the left, or from the right. Hence Eq. (1) is not satisfied.

Example 1. As an example of a continuous function, let us investigate

$$f(x) = x^2$$

at some fixed value $x = c$. Certainly requirement (a) of the definition is satisfied, since the function has the value

$$f(c) = c^2$$

at $x = c$. Furthermore, Eq. (1) is satisfied because

$$\lim_{x \to c} x^2 = c^2.$$

Therefore this function, $f(x) = x^2$, is continuous at *any* $x = c$; that is, it is continuous on $(-\infty, \infty)$.

The foregoing example is a special case of the following theorem.

Theorem 12. *Every polynomial function is continuous on $(-\infty, \infty)$.*

Proof. For any real number c,

$$\lim f(x) = f(c) \quad \text{when} \quad x \to c,$$

by Corollary 4.3, Chapter 2. □

The graph of a function which is continuous on an interval $a \leq x \leq b$ consists of an unbroken curve over that interval. This fact is of practical importance in using curves to represent functions. It makes it possible, for example, to sketch a curve by constructing a table of values (x, y), plotting relatively few points from this table, and then sketching a continuous (i.e., unbroken) curve through these points. For example, suppose we carry through the steps discussed for the simple case $y = x^2$.

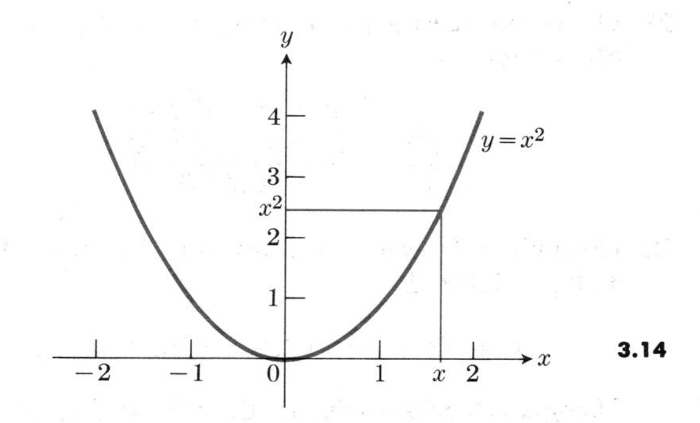

3.14

We first make a short table of values:

x	0	± 0.5	± 1	± 2
y	0	0.25	1	4

then plot these points and draw a continuous curve through them, as shown in Fig. 3.14. Then we might go one step further and use the curve itself to find the value of x^2 corresponding to an x different from those used in making the table of values. For example, in Fig. 3.14, $x = 1.6$ is indicated on the

x-scale and the corresponding reading on the y-scale gives $x^2 \approx 2.6$. Of course, the graph probably won't give *exact* values except at those points which have been plotted exactly. But it will give values which are near the exact values, if the calculated points aren't spread too far apart.

To appreciate the concept of continuity more fully, it is desirable to have some examples of the opposite behavior which some functions exhibit at certain points, that is, discontinuity. For example, the function F such that $F(t) = 1/t$ is continuous for $t \neq 0$, but is not defined at $t = 0$. Nor can it be defined in any way at $t = 0$ to satisfy Eq. (1), because it has no limit as t approaches zero. It is customary to say that this function (Fig. 3.15) has an infinite discontinuity at $t = 0$.

The function

$$F(t) = \text{greatest integer in } t = [t],$$

on the other hand, is discontinuous at $t = n$ for every integer $n = 0, \pm 1, \pm 2, \ldots,$ but its discontinuities are all finite. This function changes abruptly by unity when t decreases by ever so small an amount from 3 to $3 - \delta$, for example, no matter how small the positive number δ may be. (See Fig. 3.16.)

An extreme example of a discontinuous function is afforded by the function defined by the rule

$$f(x) = \begin{cases} 0, & \text{when } x \text{ is rational,} \\ 1, & \text{when } x \text{ is irrational.} \end{cases}$$

This function has a perfectly definite value for every x in the interval $0 \leq x \leq 1$. But no matter what c is chosen in this interval, there are points x which are arbitrarily close to c but having functional values $f(x)$ which differ from $f(c)$ by unity. That is,

$$|f(x) - f(c)|$$

is not arbitrarily small for all $x \neq c$ in some small neighborhood of c. This function is discontinuous at all points in the interval (0, 1). Its graph cannot be drawn, but it can be described by saying that it consists of all the rational points on the x-axis from 0 to 1, with the irrational points all pushed up one unit above the x-axis. (See Fig. 3.17.) Thus it consists of two parallel line segments, each of which is full of holes!

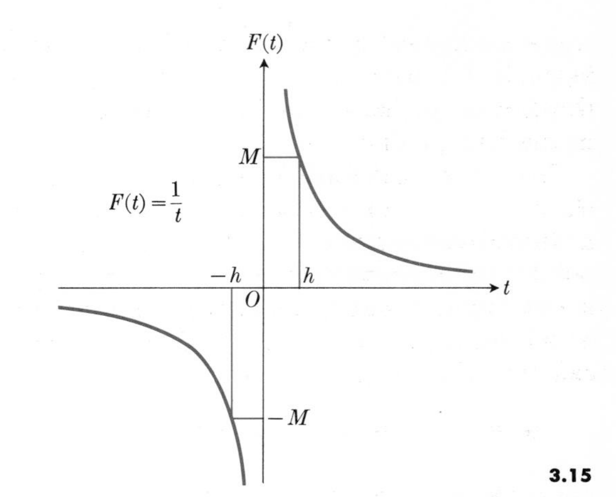

3.15

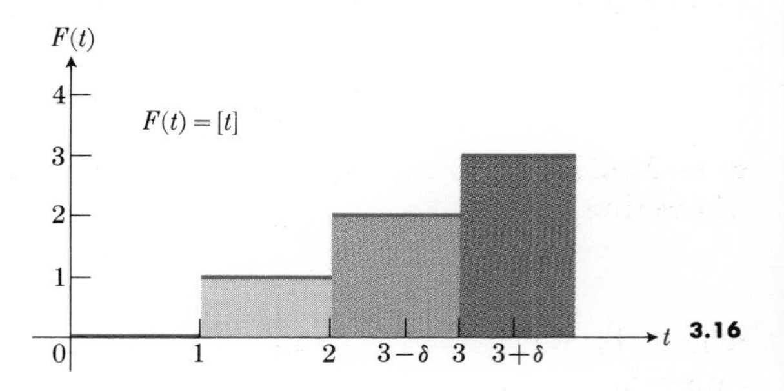

3.16

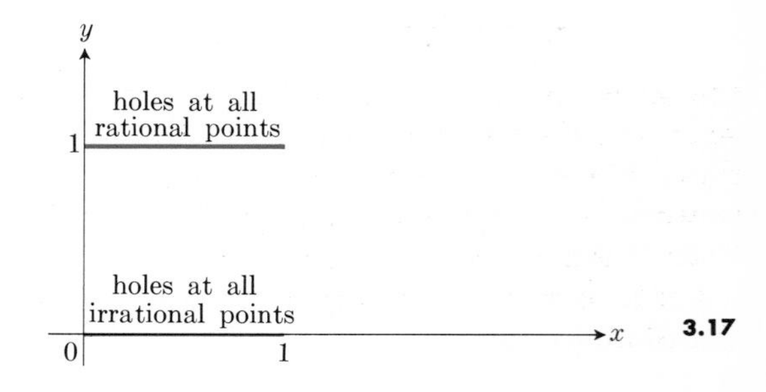

3.17

With these examples before us, the question arises as to how, in practice, we can tell whether or not a given function is continuous. When we sketch its graph, should we be sure to sketch a connected, unbroken curve, as for the graph of $y = x^2$, or must we be careful *not* to connect all of the points as, for example, in the case $y = 1/x$, where points in the third quadrant must not be connected with points in the first quadrant?

Since we have studied differentiation, the following theorem, which says that every function which has a derivative at a point is also continuous at that point, enables us to conclude that every polynomial is continuous for all x's, and every rational function is continuous for all x's in its domain. Thus, for example, the graphs of

$$y = x^3, \quad y = 2x^2 - x + 3, \quad y = \frac{x^2 - 1}{x^2 + 1}$$

are continuous, connected curves over the domain $-\infty < x < +\infty$. On the other hand, the graph of

$$y = \frac{x^2 + 1}{x^3 - 4x} \tag{4}$$

goes shooting off toward $\pm\infty$ as x approaches one of the three values

$$x = -2, \quad x = 0, \quad x = +2,$$

where the denominator vanishes. But, over the intervals

$$-\infty < x < -2, \quad -2 < x < 0,$$
$$0 < x < 2, \quad 2 < x < +\infty,$$

the graph consists of separate continuous curves, since the denominator does not vanish in any of these intervals. The reader will find it instructive to work out the graph of Eq. (4) for himself on the basis of the discussion above.

Let us now state and prove the theorem we mentioned above.

Theorem 13. *If the function f has a finite derivative*

$$f'(c) = \lim_{\Delta x \to 0} \frac{f(c + \Delta x) - f(c)}{\Delta x} \tag{5}$$

at $x = c$, then f is continuous at $x = c$.

Proof. For the limit in Eq. (5) to exist, it is necessary that $f(c)$ and $f(c + \Delta x)$ both exist, at least for all Δx near zero. We change the notation slightly by writing x in place of $c + \Delta x$. Then

$$x = c + \Delta x$$

is near c when

$$\Delta x = x - c$$

is near zero. Equation (5) now may be written in the form

$$f'(c) = \lim_{x \to c} \frac{f(x) - f(c)}{x - c}.$$

Suppose then that $f'(c)$ exists and is finite. We want to prove that $f(x) - f(c)$ tends to zero as $x \to c$. This follows at once from

$$\begin{aligned}\lim_{x \to c} [f(x) - f(c)] &= \lim_{x \to c} \left[(x - c)\frac{f(x) - f(c)}{x - c}\right] \\ &= \lim_{x \to c} (x - c) \cdot \lim_{x \to c} \frac{f(x) - f(c)}{x - c} \\ &= 0 \cdot f'(c) = 0. \ \square\end{aligned}$$

We have just shown that differentiability implies continuity. The converse, however, is not true, as is easily seen by considering the function

$$f(x) = |x|,$$

which is continuous at $x = 0$ but does not possess a derivative at that point. (See Fig. 3.18.)

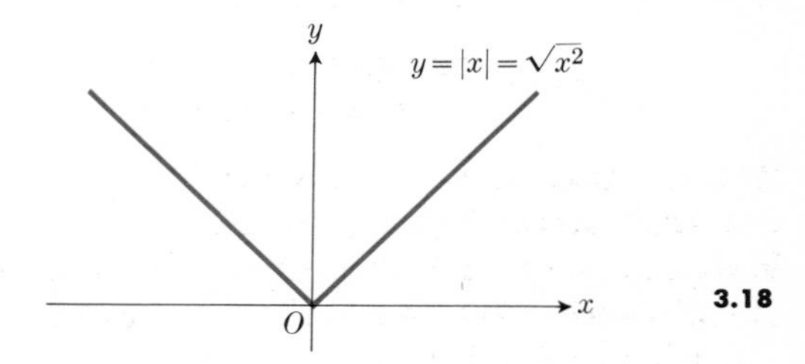

3.18

Note. The fact that $f(x) = |x|$ has no derivative at $x = 0$ follows from the fact that

$$\frac{f(0 + \Delta x) - f(0)}{\Delta x} = \frac{|\Delta x|}{\Delta x} = \begin{cases} +1, & \text{when } \Delta x > 0, \\ -1, & \text{when } \Delta x < 0, \end{cases}$$

and

$$\lim_{\Delta x \to 0} \frac{|\Delta x|}{\Delta x} \text{ does not exist.}$$

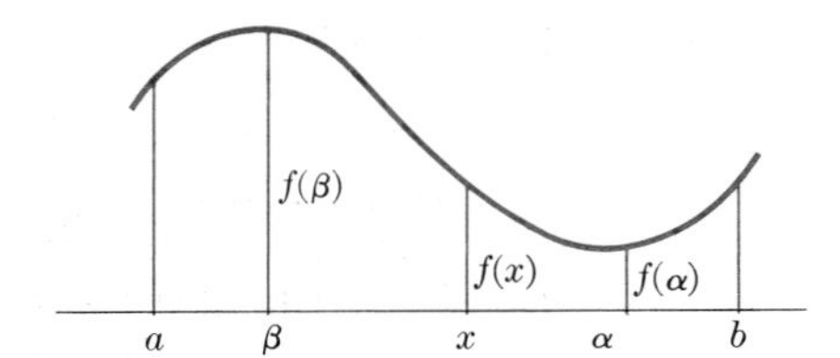

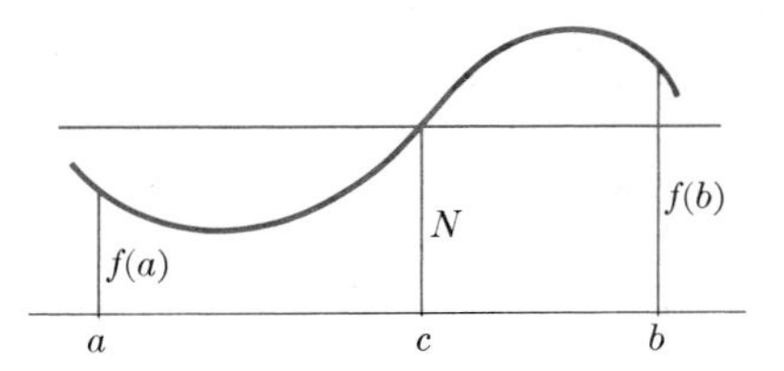

3.19

Until the middle of the 19th century, it was generally believed that a continuous function would at least have a derivative at most places, even though there might be isolated points (as at $x = 0$, in our example) where the derivative does not exist. However, the German mathematicians Riemann (1826–1866) and Weierstrass (1815–1897) gave examples of functions which are continuous but which fail to have derivatives anywhere. For a readable account of this, see *World of Mathematics*, Vol. 3, p. 1963.

We conclude this chapter with some theorems which we shall state without proofs. Some of these theorems follow rather simply from the corresponding theorems on limits. Others require, for rigorous proofs, a more penetrating investigation of the properties of the real number system than we are able to make at this time.

Theorem 14. *Suppose f and g are two functions that are continuous at $x = c$. Then the functions F_1, F_2, F_3 defined by*

$$F_1(x) = f(x) + g(x),$$

$$F_2(x) = kf(x), \quad k \text{ any number},$$

$$F_3(x) = f(x) \cdot g(x)$$

are also continuous at $x = c$. Furthermore, if $g(c)$ is not zero, then F_4,

$$F_4(x) = \frac{f(x)}{g(x)},$$

is also continuous at $x = c$.

Theorem 15. *A function f that is continuous for all x in the closed interval $a \leq x \leq b$ has the following properties (See Fig. 3.19):*

(i) *It has a minimum value m and a maximum value M; that is, there are numbers α and β in the interval such that $m = f(\alpha)$ and $M = f(\beta)$, and such that for all x in the interval, the condition*

$$m \leq f(x) \leq M$$

is satisfied.

(ii) *If N is any number between $f(a)$ and $f(b)$, then there is at least one number c between a and b such that $f(c) = N$.**

(iii) *Given any positive number ϵ there is a positive number δ (which may depend upon ϵ but does not depend upon x_1 or x_2) such that if x_1 and x_2 belong to the interval $a \leq x \leq b$ and $|x_1 - x_2| < \delta$, then $|f(x_1) - f(x_2)| < \epsilon$.*

Remark 4. A function that satisfies the condition (iii) above is said to be *uniformly continuous* for $a \leq x \leq b$, and the theorem says that a function that is continuous on a *closed* interval is automatically *uniformly* continuous there. The function $f(x) = 1/x$ is, on the other hand, continuous in the *open* interval $0 < x < 1$, but is not uniformly continuous there.

Example 2. The following physical interpretation of continuity may be of some value in throwing more light on the distinction between the notion of "continuous at every point of an interval" on the one hand and "uniformly continuous in an interval" on the other. We recall that a function is continuous at $x = c$ if $f(c)$ exists and $f(x) \to f(c)$ as $x \to c$. This means that if we are challenged with a positive "tolerance limit" ϵ and required to make

$$|f(x) - f(c)| < \epsilon \tag{6}$$

---

* This is known as the *intermediate value theorem.* Proofs of all parts of Theorem 15 can be found in Thomas, Moulton, Zelinka, *Elementary Calculus from an Advanced Viewpoint*, Addison-Wesley, 1967, Chapters 6 and 7.

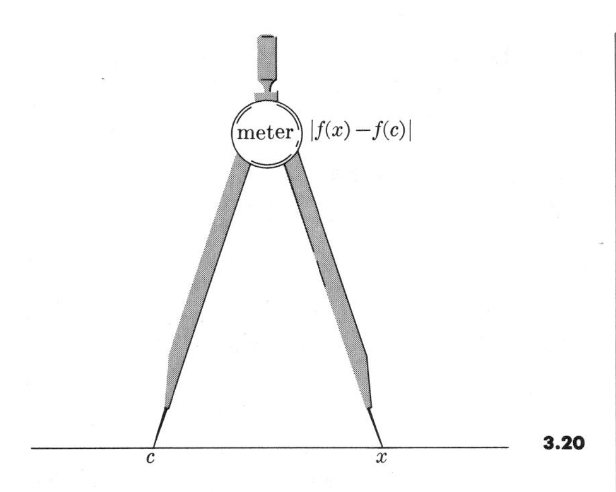

3.20

for *all* x sufficiently close to c, we must produce a positive number δ such that condition (3) is met whenever

$$|x - c| < \delta. \tag{7}$$

Let us translate this into physical terms.

Suppose we have a pair of dividers (Fig. 3.20) with a meter attached and having the property that when the points of the dividers are set at c and x respectively, the meter registers the numerical value of the difference in the functional values at c and x; that is,

the meter reads $|f(x) - f(c)|$
when the points touch x and c.

Suppose now we hold one prong of the dividers at c and move the other prong around near c, either to the left of c or to its right. Suppose that to *each* positive number ϵ we are able to assign an interval $(c - \delta, c + \delta)$ such that whenever the second prong is inside this interval, the meter reading is less than ϵ. Then the function is, by our definition, continuous at c.

Now the interval length 2δ will in general depend upon ϵ. That is, the smaller the tolerance limit ϵ that is required, the more narrowly must we restrict the range of x around c. Furthermore, the length 2δ will also, in general, depend upon c. That is, for a given function $f(x)$,

$$\delta = \delta(c, \epsilon)$$

is a function of both c and ϵ. But if the function $f(x)$ is continuous for all values of x in some domain, then with each x_0 in the domain and with each $\epsilon > 0$ we can associate a positive number $\delta = \delta(x_0, \epsilon)$, such that if x is in the domain and if $|x - x_0| < \delta(x_0, \epsilon)$, then

$$|f(x) - f(x_0)| < \epsilon.$$

Finally, as we vary x_0, it may happen that all of these numbers $\delta(x_0, \epsilon)$ are at least as large as some positive number $k = k(\epsilon)$. Then, given $\epsilon > 0$, we could use an interval of constant width $2k$ around *any* x_0 of the domain, and have

$$|f(x) - f(x_0)| < \epsilon$$

whenever x is in the domain and $|x - x_0| < k$. For the interval $(x_0 - k, x_0 + k)$ surely lies inside the interval $x_0 - \delta(x_0, \epsilon) < x < x_0 + \delta(x_0, \epsilon)$, since $\delta(x_0, \epsilon) \geq k$. In other words, if the two prongs of the dividers are spread apart any distance smaller than this k and the dividers are then placed so that they touch *any two points* x_1, x_2, in the domain of x, in the circumstances described above it will turn out that the meter reading is less than ϵ. This exemplifies what is meant by *uniform* continuity.

To summarize these remarks we may say that

(a) *ordinary continuity* at a point c means that the interval width depends not only upon the tolerance ϵ but also upon c, whereas

(b) *uniform continuity* means that with any positive ϵ we may associate *one* interval width that works for *all* points in the domain.

EXERCISES

1. For what values of x is each of the following functions discontinuous?

   (a) $f(x) = \dfrac{x}{x+1}$, (b) $f(x) = \dfrac{x+1}{x^2 - 4x + 3}$.

2. Is the function f defined by the rule

$$f(x) = \begin{cases} \dfrac{x^2 - 1}{x - 1}, & \text{when } x \neq 1, \\ 2, & \text{when } x = 1, \end{cases}$$

   continuous or discontinuous at $x = 1$? Prove your result.

3. The function $f(x) = |x|$ is continuous at $x = 0$. Given a positive number ϵ, how large may δ be in order that $|f(x) - 0| < \epsilon$ when $|x - 0| < \delta$?

4. What is the maximum of $f(x) = |x|$ for $-1 \leq x \leq 1$? What is the minimum? Sketch.

5. Does the function $f(x) = x^2$ have a maximum for $0 < x < 1$? Does it have a minimum? Give reasons for your answers.

6. A continuous function $y = f(x)$ is known to be negative at $x = 0$ and positive at $x = 1$. Why is it true that the equation $f(x) = 0$ has at least one root between $x = 0$ and $x = 1$? Illustrate with a sketch.

7. Is the function $f(x) = \sqrt[3]{x}$ continuous at $x = 0$? Is it differentiable there? Give reasons for your answers.

8. *Prove the following theorem.* Let f be continuous and positive at $x = c$. Prove that there is some interval around $x = c$, say $c - \delta < x < c + \delta$, throughout which $f(x)$ remains positive. Illustrate with a sketch. *Hint.* In the alternative definition of continuity, take $\epsilon = \frac{1}{2}f(c)$.

9. For $x \neq 2$, the function $f(x)$ is equal to

$$(x^2 + 3x - 10)/(x - 2).$$

What value should be assigned to $f(2)$ to make $f(x)$ continuous at $x = 2$?

10. Use the trigonometric identity

$$\sin x - \sin a = 2 \sin \frac{x - a}{2} \cos \frac{x + a}{2}$$

to prove that $\sin x$ is uniformly continuous on $(-\infty, \infty)$.

REVIEW QUESTIONS AND EXERCISES

1. Using the definition of the derivative, deduce the formula for the derivative of a product uv of two differentiable functions.

2. In the formula for the derivative of uv, let $v = u$, and thereby deduce a formula for the derivative of u^2. Repeat the process, with $v = u^2$, and get a formula for the derivative of u^3. Extend the result, by the method of mathematical induction, to deduce the formula for the derivative of u^n for every positive integer n.

3. Explain how the three formulas

(a) $\dfrac{d(x^n)}{dx} = nx^{n-1}$, (b) $\dfrac{d(cu)}{dx} = c\dfrac{du}{dx}$,

(c) $\dfrac{d(u + v)}{dx} = \dfrac{du}{dx} + \dfrac{dv}{dx}$

are sufficient to differentiate any polynomial.

4. What formula do we need, in addition to the three listed in Exercise 3 above, to differentiate rational functions?

5. Does the derivative of a polynomial function exist at every point of its domain? What is the largest domain the function can have? Does the derivative of a rational function exist at every point in its domain? What real numbers, if any, must be excluded from the domain of a rational function?

Definition. Algebraic Function. *Let $y = f(x)$ define a function of x such that every pair of numbers (x, y) belonging to f satisfy an irreducible equation of the form*

$$P_0(x)y^n + P_1(x)y^{n-1} + \cdots + P_{n-1}(x)y + P_n(x) = 0 \qquad (\alpha)$$

for some positive integer n, with coefficients

$$P_0(x), \ldots, P_n(x)$$

polynomials in x, and $P_0(x)$ not identically zero. Then f is said to be an algebraic function.

6. What technique of this chapter can be used to find the derivative of an algebraic function if the polynomial coefficients $P_0(x), \ldots, P_n(x)$ in its defining equation (α) are given?

7. Show that the function f defined by $f(x) = x^{2/3}$ is an algebraic function by finding an appropriate equation of the type (α) in the definition above. On what domain of values of x is this function defined? Where is it continuous? Where is it differentiable?

8. Show that every rational function is algebraic. Do you think the converse is also true? Explain.

9. It can be shown (though not very easily) that sums, quotients, products, powers, and roots of algebraic functions are algebraic functions. Thus

$$f(x) = x\sqrt{3x^2 + 1} + 5x^{4/3}/(3x + 2), \quad x \neq -\tfrac{2}{3},$$

defines an algebraic function. Find its derivative. What formulas of this chapter are used in finding derivatives of functions like this one?

10. Show that $y = |x|$ satisfies the equation $y^2 - x^2 = 0$. Is the absolute value function algebraic? What is its derivative? Where does the derivative exist? Where is the absolute value function continuous?

11. Define differentials. If $y = f(x)$ defines a differentiable function, how are dy and dx related? Give geometrical interpretations of dx and dy.

12. State the chain rule for derivatives. Prove it, with the book closed.

13. Show how the chain rule is used to prove that the derivative of y with respect to x is the ratio of dy to dx when x and y are both differentiable functions of t, and $dx \neq 0$.

14. Suppose that the mapping $f: x \to f(x)$, $a \leq x \leq b$, is one-to-one from $[a, b]$ to $[f(a), f(b)]$. Describe the inverse mapping. What is its domain? its range?

15. (Continuation of 14.) If $g(y) = x$ whenever $f(x) = y$, where f is one-to-one from $[a, b]$ to $[f(a), f(b)]$, then we also have

$$g(f(x)) = x \quad \text{for } x \in [a, b].$$

By applying the chain rule to the composite function $h = g \circ f$, deduce a general formula for the derivative of an inverse function.

Note. g is the inverse of f on a suitable domain.

16. Give an example of a function that is defined and bounded on $0 \leq t \leq 1$, continuous in the open interval $0 < t < 1$, and discontinuous at $t = 0$.

17. State and prove a theorem about the relationship between continuity and differentiability of a function, at a point in its domain.

18. Read the article "The Lever of Mahomet" by R. Courant and H. Robbins, and the accompanying "Commentary on Continuity" by J. R. Newman, in *World of Mathematics*, Vol. 4, pp. 2410–2413.

MISCELLANEOUS EXERCISES

Find dy/dx in Exercises 1 through 33.

1. $y = \dfrac{x}{\sqrt{x^2 - 4}}$
2. $x^2 + xy + y^2 - 5x = 2$
3. $xy + y^2 = 1$
4. $x^3 + 4xy - 3y^3 = 2$
5. $x^2y + y^2 = 6$
6. $y = (x+1)^2(x^2+2x)^{-2}$
7. $y = \dfrac{x}{\sqrt{1 - x^2}}$
8. $x^2 + 3xy^2 + y^3 = 2$
9. $y = \dfrac{x}{x+1}$
10. $y = \sqrt{2x+1}$
11. $y = x^2\sqrt{x^2 - a^2}$
12. $y = \dfrac{2x+1}{2x-1}$
13. $y = \dfrac{x^2}{1 - x^2}$
14. $y = (x^2 + x + 1)^3$
15. $y = \dfrac{1}{\sqrt{x^2 + x - 1}}$
16. $y = \dfrac{x^3 - 1}{x - 1}$
17. $y = \dfrac{(2x^2 + 5x)^{3/2}}{3}$
18. $y = \dfrac{3}{(2x^2 + 5x)^{3/2}}$
19. $xy^2 + \sqrt{xy} = 2$
20. $x^2 + y^2 = xy$
21. $x^{2/3} + y^{2/3} = a^{2/3}$
22. $x^{1/2} + y^{1/2} = a^{1/2}$
23. $xy = 1$
24. $\sqrt{xy} = 1$
25. $(x + 2y)^2 + 2xy^2 = 6$
26. $y = \sqrt{\dfrac{1 - x}{1 + x^2}}$
27. $y^2 = \dfrac{x}{x+1}$
28. $x^2y + xy^2 = 6(x^2 + y^2)$
29. $xy + 2x + 3y = 1$
30. $y = u^2 - 1,\ x = u^2 + 1$
31. $y = \sqrt{2t + t^2},\ t = 2x + 3$
32. $x = \dfrac{t}{1 + t^2},\ y = 1 + t^2$
33. $t = \dfrac{x}{1 + x^2},\ y = x^2 + t^2$
34. Find the slope of $y = x/(x^2 + 1)$ at the origin. Write an equation of the tangent line at the origin.
35. Write an equation of the tangent to the curve $x^2 - 2xy + y^2 + 2x + y - 6 = 0$ at $(2, 2)$.
36. Determine the constant c such that the straight line joining the points $(0, 3)$ and $(5, -2)$ is tangent to the curve $y = c/(x + 1)$.
37. What is the slope of the curve $y = 2x^2 - 6x + 3$ at the point on the curve where $x = 2$? What is an equation of the tangent line to the curve at this point?
38. Find the points on the curve

$$y = 2x^3 - 3x^2 - 12x + 20$$

where the tangent is parallel to the x-axis.

39. Find the derivatives of the following functions.

(a) $y = (x^2 + 2x)^5$ (b) $f(t) = \sqrt{3t^2 - 2t}$

(c) $f(r) = \sqrt{r^2 + 5} + \sqrt{r^2 - 5}$ (d) $f(x) = \dfrac{x^2 - 1}{x^2 + 1}$

40. Find an equation of the tangent to the curve

$$y = 2/\sqrt{x - 1}$$

at the point on the curve where $x = 10$.

41. Write an equation of the straight line passing through the point (1, 2) and normal to the curve $x^2 = 4y$.

42. Use the definition of the derivative to find dy/dx for $y = \sqrt{2x + 3}$ and then check the result by finding the same derivative by the power formula.

43. Find the value of

$$\lim_{\Delta x \to 0} \frac{[2 - 3(x + \Delta x)]^2 - [2 - 3x]^2}{\Delta x}$$

and specify the function $f(x)$ of which this is the derivative.

44. Find the slope of the curve $x^2y + xy^2 = 6$ at the point (1, 2).

45. A cylindrical can of height 6 in. and radius r in. has volume $V = 6\pi r^2$ in^3. What is the difference between ΔV and its linear part as r varies? What is the geometric significance of the linear part?

46. If a hemispherical bowl of radius 10 in. is filled with water to a depth of x in., the volume of water is given by $v = \pi[10 - (x/3)]x^2$. Find the rate of increase of the volume per inch increase of the depth.

47. A bus will hold 60 people. If the number x of persons per trip who use the bus is related to the fare charged (p nickels), by the law $p = [3 - (x/40)]^2$, write the function expressing the total revenue per trip received by the bus company. What is the number x_1 of people per trip that will make the marginal revenue equal to zero? What is the corresponding fare?

48. Prove Eq. (3), Article 3.2, by mathematical induction.

49. Given $y = x - x^2$, find the rate of change of y^2 with respect to x^2 (expressed in terms of x).

50. If $x = 3t + 1$ and $y = t^2 + t$, find dy/dt, dx/dt, and dy/dx. Eliminate t to obtain y as a function of x, and then determine dy/dx directly. Do the results check?

51. A particle projected vertically upward with a speed of a ft/sec reaches an elevation $s = at - 16t^2$ ft at the end of t sec. What must the initial velocity be in order for the particle to travel exactly 49 ft upward before it starts coming back down? [What is the velocity at the 49-ft level?]

52. Find the rate of change of $\sqrt{x^2 + 16}$ with respect to $x/(x - 1)$ at $x = 3$.

53. The circle $(x - h)^2 + (y - k)^2 = r^2$, center at (h, k), radius $= r$, is tangent to the curve $y = x^2 + 1$ at the point (1, 2).

(a) Find the locus of the point (h, k).

(b) If, also, the circle and the curve have the same second derivative at (1, 2), find h, k, and r. Sketch the curve and the circle.

54. If $y = x^2 + 1$ and $u = \sqrt{x^2 + 1}$, find dy/du.

55. If $x = y^2 + y$ and $u = (x^2 + x)^{3/2}$, find dy/du.

56. If $f'(x) = \sqrt{3x^2 - 1}$ and $y = f(x^2)$, find dy/dx.

57. If

$$f'(x) = \sin(x^2) \quad \text{and} \quad y = f\left(\frac{2x - 1}{x + 1}\right),$$

find dy/dx.

58. Find $\quad y' = dy/dx \quad$ and $\quad y'' = dy'/dx$

if $\quad y = x^2 - 3x + 5.$

59. If s represents the distance a body moves in time t, determine the acceleration

$$a = d^2s/dt^2$$

if $\quad s = 250 + 40t - 16t^2.$

60. If $\quad y = x\sqrt{2x - 3},$

find $\quad d^2y/dx^2.$

61. Find the value of d^2y/dx^2 in the equation $y^3 + y = x$ at the point (2, 1).

62. If $x = t - t^2$, $y = t - t^3$, find the values of dy/dx and d^2y/dx^2 at $t = 1$.

63. Prove Leibniz's rule:

(a) $\dfrac{d^2(uv)}{dx^2} = \dfrac{d^2u}{dx^2} \cdot v + 2\dfrac{du}{dx}\dfrac{dv}{dx} + u\dfrac{d^2v}{dx^2}$,

(b) $\dfrac{d^3(uv)}{dx^3} = \dfrac{d^3u}{dx^3} \cdot v + 3\dfrac{d^2u}{dx^2}\dfrac{dv}{dx} + 3\dfrac{du}{dx}\dfrac{d^2v}{dx^2} + u\dfrac{d^3v}{dx^3}$,

(c) $$\frac{d^n(uv)}{dx^n} = \frac{d^n u}{dx^n}\cdot v + n\frac{d^{n-1}u}{dx^{n-1}}\frac{dv}{dx} + \cdots + \frac{n(n-1)\cdots(n-k+1)}{k!}\frac{d^{n-k}u}{dx^{n-k}}\frac{d^k v}{dx^k} + \cdots + u\frac{d^n v}{dx^n}.$$

The terms on the right side of this equation may be obtained from the terms in the binomial expansion $(a+b)^n$ by replacing $a^{n-k}b^k$ by

$$(d^{n-k}u/dx^{n-k})\cdot(d^k v/dx^k)$$

for $k = 0, 1, 2, \ldots, n$, and interpreting d^0u/dx^0 as being u itself.

64. Find d^3y/dx^3 in the following cases.

(a) $y = \sqrt{2x-1}$ (b) $y = \dfrac{1}{3x+2}$

(c) $y = ax^3 + bx^2 + cx + d$

65. If $f(x) = (x-a)^n g(x)$, where $g(x)$ is a polynomial and $g(a) \neq 0$, show that

$$f(a) = 0 = f'(a) = \cdots = f^{(n-1)}(a);$$

but that $f^{(n)}(a) = n!g(a) \neq 0$.

66. If $y = 2x^2 - 3x + 5$, find Δy for $x = 3$ and $\Delta x = 0.1$. Approximate Δy by finding its linear part.

67. Find an approximate value of $\sqrt{26}$ by considering the function $y = \sqrt{x}$ when $x = 25$ and $\Delta x = 1$. Also find an approximate value of $\sqrt[3]{26}$. (Specify the function used for the purpose and the values of x and Δx.)

68. By means of differentials, find an approximate value of $\sqrt[10]{0.999}$.

69. To compute the height h of a lamppost, the length a of the shadow of a six-foot pole is measured. The pole is 20 ft from the lamppost. If $a = 15$ ft, with a possible error of less than one inch, find the height of the lamppost and estimate the error in height.

70. Find the differential dy in the following cases.

(a) $y = x^2/(1+x)$ (b) $x^2 - y^2 = 1$

(c) $xy + y^2 = 1$

71. Given a function f satisfying the following two conditions for all x and y:

(a) $f(x+y) = f(x)\cdot f(y)$,

(b) $f(x) = 1 + xg(x)$, where $\lim\limits_{x\to 0} g(x) = 1$.

Prove that

(a) the derivative $f'(x)$ exists, (b) $f'(x) = f(x)$.

72. Let $f(x) = x^2 + 1$. Given $\epsilon > 0$, find $\delta > 0$ such that $|f(x_1) - f(x_2)| < \epsilon$ whenever $|x_1 - x_2| < \delta$ and x_1, x_2 both lie in the closed interval $[-2, 2]$. State precisely what this means concerning the continuity of this function.

73. Given a function $f(x)$, defined for all real x, and a positive constant c such that $|f(x+h) - f(x)| \leq ch^2$ for all real h. Prove that

(a) $f(x)$ is uniformly continuous

(b) $f'(x) = 0$ for all x.

74. A function $f(x)$ is said to satisfy a Lipschitz condition of order m on the closed interval $[a, b]$ if there is a constant C such that

$$|f(x_2) - f(x_1)| \leq C|x_2 - x_1|^m$$

for all values of x_1, x_2 on $[a, b]$. Prove that a function which satisfies a Lipschitz condition of order $m > 0$ on $[a, b]$ is uniformly continuous there.

75. Suppose $[a, b]$ is the interval $[-1, 1]$ and

$$f(x) = \sqrt{1-x^2}.$$

Find appropriate values of C and of m to satisfy the conditions in Exercise 74.

Hint. Show that if

$$y_2 > y_1 > 0 \quad \text{and} \quad y_2 - y_1 = h,$$

then

$$|\sqrt{y_2} - \sqrt{y_1}| \leq \sqrt{h}.$$

APPLICATIONS

CHAPTER 4

4.1 INCREASING OR DECREASING FUNCTIONS: THE SIGN OF dy/dx

Some functions have the property that

$$f(x_1) > f(x_2) \quad \text{when} \quad x_1 > x_2. \tag{1}$$

For instance, a linear function $f(x) = mx + b$ with positive slope m has this property on $(-\infty, \infty)$. For other functions, the condition expressed by (1) may hold only for a part of the domain. For example, $f(x) = \sin x$ (Fig. 4.1) has the property when x_1 and x_2 are in $[-\pi/2, \pi/2]$ but not in $[\pi/2, 3\pi/2]$. On the latter interval, $\sin x$ is decreasing.

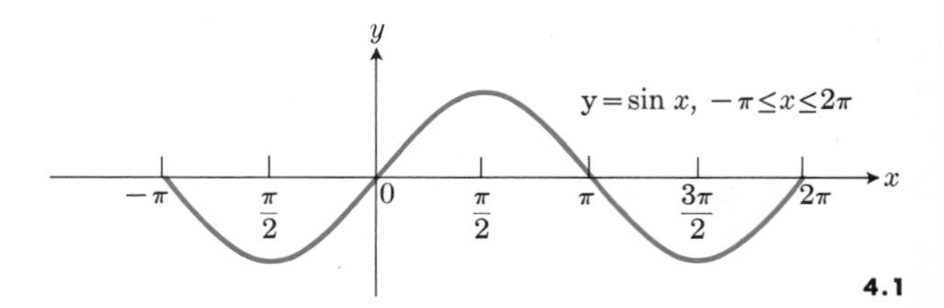

4.1

Definition. *If $f(x_1) > f(x_2)$ when $x_1 > x_2$ for all x_1 and x_2 in $[a, b]$, we say that f is an increasing function on $[a, b]$. If $f(x_1) < f(x_2)$ for $a \leq x_2 < x_1 \leq b$, we say that f is a decreasing function on $[a, b]$.*

Remark 1. Sometimes we consider functions that are increasing or decreasing on open intervals such as (a, b) and $(-\infty, 0)$, or half-open intervals such as $[0, 1)$.

Example 1. (See Fig. 4.2.)

$$f(x) = x^3 \text{ increases on } (-\infty, \infty).$$

Example 2. (See Fig. 4.3.)

$$f(x) = \tan x \text{ increases on } (-\pi/2, \pi/2).$$

Example 3. (See Fig. 4.4.)

$$f(x) = \frac{1}{x} \text{ decreases on } (-\infty, 0) \text{ and on } (0, \infty).$$

Example 4. Let $F(x)$ for $x > 0$ be the area of the shaded region shown in Figure 4.5. Then the area increases as x moves to the right, so F is an increasing function on $(0, \infty)$.

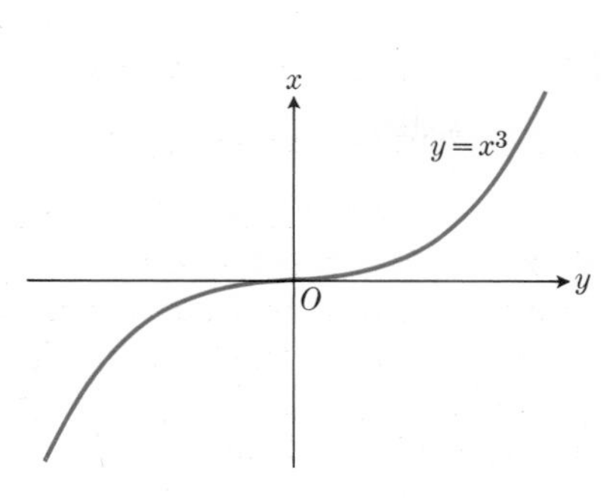

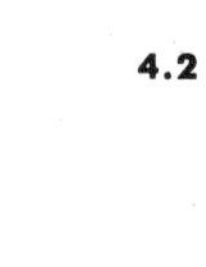

4.2

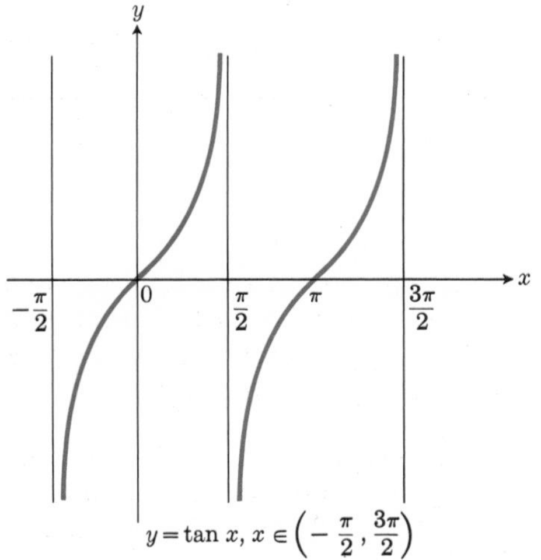

4.3

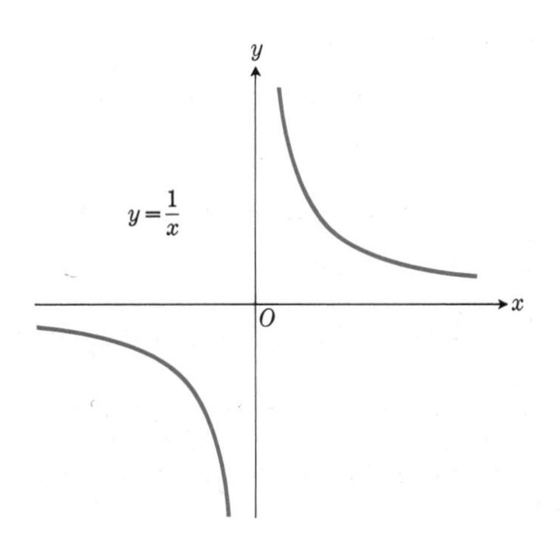

4.4

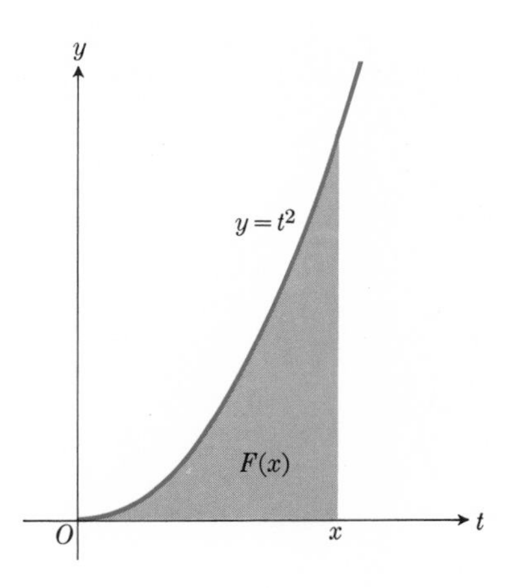

4.5 $F(x)$ = area of tinted region. The area gets larger as x moves to the right.

Remark 2. It is also customary to say that f is increasing at each point in (a, b) if f is increasing on (a, b). In fact, we say that f is *increasing at a point* c if there is some neighborhood N of c such that $f(x) > f(c)$ when $x > c$ and $f(x) < f(c)$ when $x < c$, for all $x \in N$. For example, $\operatorname{sgn} x$ is increasing at $c = 0$ because $\operatorname{sgn} 0 = 0$, $\operatorname{sgn} x = -1 < 0$ when $x < 0$, and $\operatorname{sgn} x = +1 > 0$ when $x > 0$.

Remark 3. A function may oscillate so fast near a point that it is neither increasing nor decreasing there (nor remaining constant).

Example 5. Let f be the function defined by

$$f(x) = \begin{cases} x \sin(1/x) & \text{for } x \neq 0, \\ 0 & \text{for } x = 0. \end{cases}$$

No matter how small a neighborhood of zero N may be, there are x's in N for which $f(x)$ is positive and those for which it is negative. This function oscillates infinitely often between positive and negative values in every neighborhood of $x = 0$. It is neither increasing nor decreasing at 0.

Sign of dy/dx

For $y = x^3$, $dy/dx = 3x^2$ is positive when $x \neq 0$. Therefore, by Lemma 1 of Article 3.3, we can now

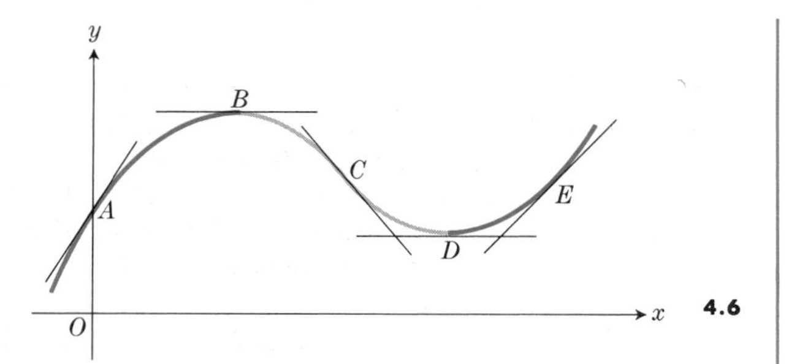

4.6

conclude that the function $f(x) = x^3$ is increasing at each $c \in (-\infty, 0)$ or in $(0, \infty)$. Because $f'(0) = 0$, the lemma does not apply at $c = 0$, although it is true that f is also increasing there, because x^3 is positive when x is greater than zero and is negative when x is less than zero.

At a point where $dy/dx > 0$, the tangent to the graph of $y = f(x)$ has positive slope, so the linearization of f at such a point is an increasing function. Since the tangent line stays close to the graph of f near the point of tangency, we also see geometrically the significance of Lemma 1 mentioned above: The curve rises as we go from left to right through this point. Of course, at a point where dy/dx is negative, the tangent line slopes downward and the curve falls as we go through the point of tangency from left to right.

Figure 4.6 shows a curve on which we have indicated certain points A, B, C, D, and E. At A and E, the tangents have positive slopes and the curve is rising. At C, the tangent has negative slope and the curve is falling. At B and D, the tangents have zero slope and there is a transition between rising and falling portions of the curve. [Note, however, that such a transition doesn't always occur when the derivative is zero: The curve may cross its horizontal tangent and keep on rising, as $y = x^3$ does at (0, 0).]

In relatively simple situations, $f'(x)$ is continuous and can only go from negative to positive values (or vice versa) by going through zero. [This is a consequence of the Intermediate Value Theorem, Theorem 15 (ii), Chapter 3.]

Example 6. We apply these ideas to the problem of sketching the curve $y = \frac{1}{3}x^3 - 2x^2 + 3x + 2$. The slope at (x, y) is

$$\frac{dy}{dx} = x^2 - 4x + 3 = (x - 1)(x - 3).$$

To determine where dy/dx is positive and where it is negative, we first determine where it is zero:

$$\frac{dy}{dx} = 0 \quad \text{when} \quad x = 1 \text{ or } x = 3,$$

since these points will mark the transition from positive to negative or from negative to positive slopes. The sign of dy/dx depends upon the signs of both factors $(x - 1)$ and $(x - 3)$, and since the sign of $(x - 1)$ is negative when x is to the left of 1 and positive to the right, we have the pattern of signs indicated in Fig. 4.7(a). Similarly the sign of $(x - 3)$ is as shown in Fig. 4.7(b), and the sign of $dy/dx = (x - 1)(x - 3)$ as shown in Fig. 4.7(c). We can get a rough idea of the shape of the curve just from this pattern of signs of its slope, if we sketch a curve which is rising, falling, and rising again for $x < 1$, $1 < x < 3$, and $x > 3$, respectively (Fig. 4.8).

To get a more accurate curve, we would construct a table for some range of values extending, say, from $x = 0$ to $x = 4$, which includes the transition points between rising and falling portions of the curve (Fig. 4.9).

The terms "rising" and "falling," or "increasing" and "decreasing," are always taken to apply to the behavior of the curve as the point that is tracing it moves from left to right or, in other words, relative to its behavior as x increases.

The same concepts apply to functions that vary with time. For example, Fig. 4.10 represents a rope running through a pulley at P, bearing a weight W at one end. The other end is held in a man's hand M at a distance of 5 feet above the ground as he walks in a straight line at the rate of 6 ft/sec. If, as in the figure, x represents the distance in feet of the man's hand away from the vertical line PW, then x is an increasing function of t and we are given the rate at which it changes with t, namely, 6 ft/sec. This would be stated in mathematical terms by saying $dx/dt = +6$ (ft/sec). On the other hand, if the man were to walk toward the line PW at the rate of 6 ft/sec, we would have $dx/dt = -6$ (ft/sec), because x would be then a decreasing function of t.

sign of $x-1$ | a

sign of $x-3$ | b

sign of $(x-1)(x-3)$ | c

4.7

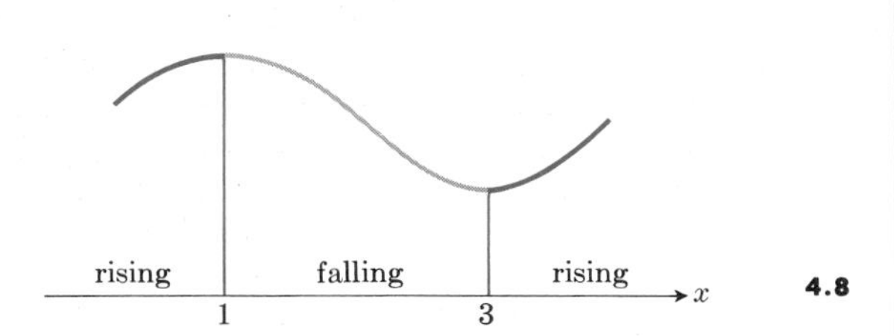

4.8

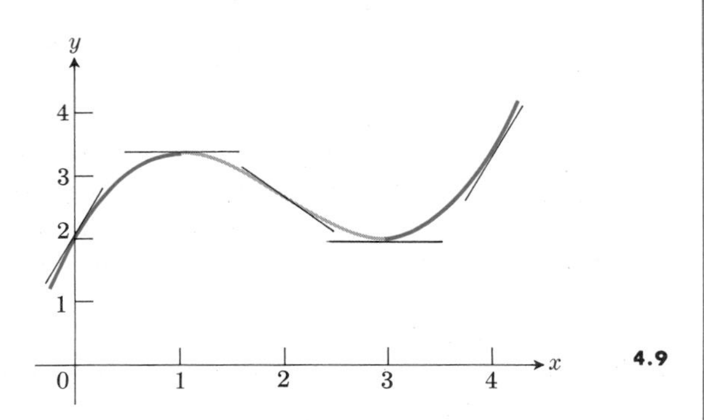

4.9

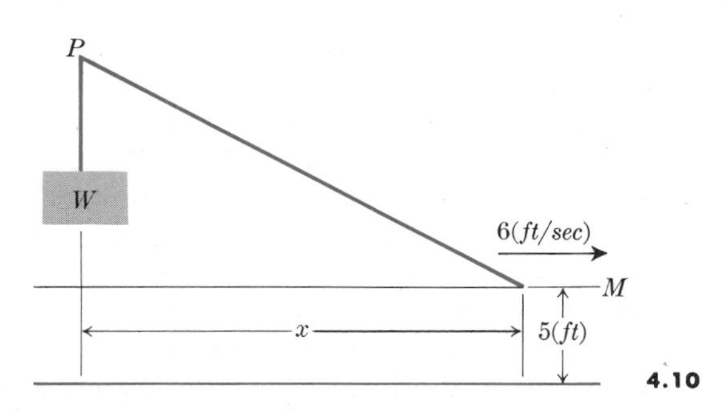

4.10

EXERCISES

In each of the following exercises, determine dy/dx and find the sets of values of x where the graph of y versus x is rising (to the right) and where it is falling. Sketch each curve, showing in particular the points of transition between falling and rising portions of the curve.

1. $y = x^2 - x + 1$
2. $y = \frac{x^3}{3} - \frac{x^2}{2} - 2x + \frac{1}{3}$
3. $y = 2x^3 - 3x^2 + 3$
4. $y = x^3 - 27x + 36$
5. $y = x^4 - 8x^2 + 16$
6. $y = (x + 1)(x - 2)^2$
7. $y = \frac{(x - 2)^2}{x + 1}$
8. $y = (x + 1)(x - 2)^{-2}$
9. $y = \sqrt{4 - x^2}$
10. $y = \sqrt{x^2 - 4}$
11. $y = x\,|x|$

4.2 RELATED RATES

In the example at the end of Article 4.1, suppose that the pulley is 25 ft above the ground, the rope is 45 ft long, and at a given instant the distance x is 15 ft and the man is walking away from the pulley. How fast is the weight being raised at this particular instant?

In Fig. 4.11, M represents the man's hand, P the pulley, W the weight, and OM is a horizontal line at the level of the man's hand which is 5 ft above the ground, so that P is 20 ft above O. The figure is drawn to illustrate the situation at any time t and

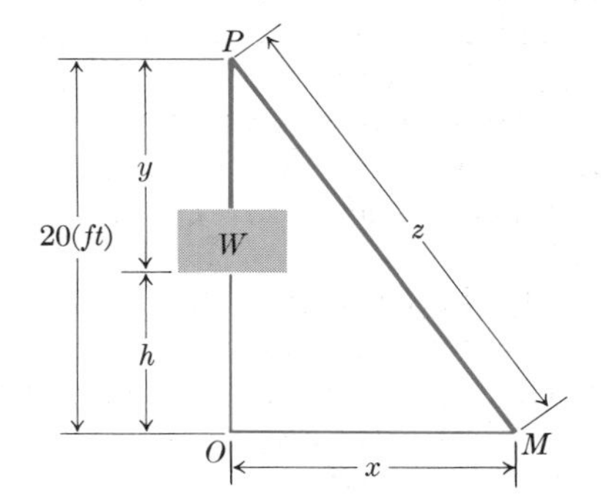

4.11

not just at the instant in question. The reason for this is that certain of the distances, namely, x, y, z, and h in the illustration, are quantities which vary and it is important to treat them as such and not as constants.

Having drawn a figure to illustrate the relationship between the various variable quantities involved in the problem, let us now state what is given and what the problem asks for in terms of these variables.

Given: (a) Relationships between the variables which are to hold for all instants of time:

$$y + z = 45,$$

$$h + y = 20,$$

$$20^2 + x^2 = z^2.$$

(b) At a given instant, which we may take to be $t = 0$:

$$x = 15, \qquad \frac{dx}{dt} = 6.$$

Find: $dh/dt = (?)$ at the instant $t = 0$.

Such a problem involving rates of related variables is called "a problem in related rates." It is typical of such problems that

(a) certain variables are related in a definite way for all values of t under consideration,

(b) the values of some or all of these variables and the rates of change of some of them are given at some particular instant, and

(c) it is required to find the rate of change of one or more of them at this instant.

The variables may then all be considered to be functions of time, and if the equations which relate them for all values of t are differentiated with respect to t, the new equations so obtained will tell how their rates of change are related. From this information it should be possible to answer the question posed by the problem.

In our problem, we shall modify the procedure outlined above by first obtaining a single equation relating the variable x (whose rate is given) and the variable h (whose rate is wanted) before we take derivatives. Using the equations given, we find successively:

$$z = 45 - y, \quad y = 20 - h,$$

$$z = 45 - (20 - h) = (25 + h),$$

$$20^2 + x^2 = z^2 = (25 + h)^2;$$

that is,

$$20^2 + x^2 = (25 + h)^2.$$

This last equation relates the variables x and h for all values of t under consideration (namely, in an interval of values of t near $t = 0$). Both sides of this equation are functions of t, and the equality says they represent the same function of t. Thus when we differentiate both sides of the equation with respect to t, we shall have another equation:

$$\frac{d}{dt}(20^2 + x^2) = \frac{d}{dt}(25 + h)^2,$$

$$0 + 2x\frac{dx}{dt} = 2(25 + h)\frac{dh}{dt}.$$

This may be solved for the rate we want, namely,

$$\frac{dh}{dt} = \frac{x}{25 + h}\frac{dx}{dt}.$$

Now at the given instant, we have

$$x = 15, \qquad \frac{dx}{dt} = 6,$$

and we find h at this instant from the equation

$$20^2 + 15^2 = (25 + h)^2, \quad 25 + h = 25.$$

When we substitute these values, we obtain

$$\frac{dh}{dt} = \frac{18}{5} = 3\tfrac{3}{5} \text{ (ft/sec)}$$

as the rate at which the weight is being raised at the instant in question.

As a second example, consider a ladder 26 ft long which leans against a vertical wall. At a particular instant, the foot of the ladder is 10 ft out from the base of the wall and is being drawn away from the wall at the rate of 4 ft/sec. How fast is the top of the ladder moving down the wall at this instant?

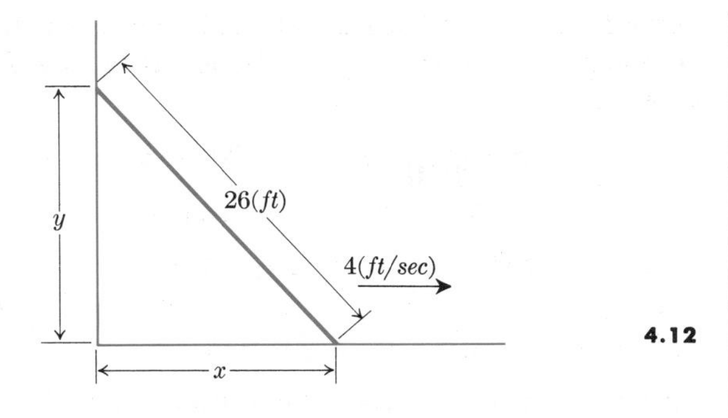

4.12

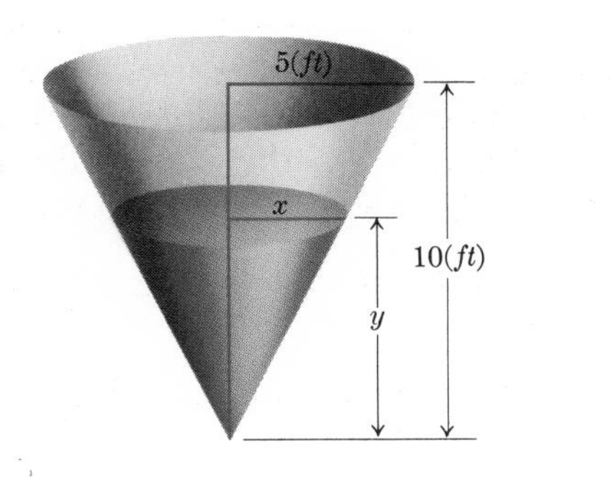

4.13

Figure 4.12 represents the position of the ladder, with the distance from the base of the wall to the foot of the ladder labeled x to indicate the fact that it is a variable, and the distance from the ground to the top of the ladder labeled y because it varies also. The statement of the problem may then be translated into mathematical terms as follows:

Given: $x^2 + y^2 = 26^2.$

Find: $\dfrac{dy}{dt} = (?)$

when

$$x = 10 \text{ (ft)} \quad \text{and} \quad \frac{dx}{dt} = 4 \text{ (ft/sec)}.$$

We differentiate the given equation implicitly with respect to t and get

$$2x\frac{dx}{dt} + 2y\frac{dy}{dt} = 0.$$

Hence

$$\frac{dy}{dt} = \frac{-x}{y}\frac{dx}{dt}.$$

When $x = 10$, $y = 24$, and $dx/dt = 4$, this leads to $dy/dt = -\frac{5}{3}$. That is, y is *decreasing* (the top of the ladder is moving *down*) at the rate of $\frac{5}{3}$ ft/sec.

As a final example, consider the conical reservoir, Fig. 4.13, into which water runs at the constant rate of 2 ft^3 per minute. How fast is the water level rising when it is 6 ft deep?

First, we must translate the problem into mathematical terms. To do this, we let

$v =$ volume (ft^3) of water in the tank at time t (min),

$x =$ radius (ft) of the section of the cone at the water line,

$y =$ depth (ft) of water in the tank at time t.

Then the statement that water runs into the tank at the rate of 2 ft^3/min becomes

$$\frac{dv}{dt} = 2.$$

The question we are to answer is

$$\frac{dy}{dt} = (?) \quad \text{when } y = 6.$$

The relationship between the volume of water v and depth y is expressed by the equation

$$v = \tfrac{1}{3}\pi x^2 y.$$

But this involves the additional variable x as well as v and y. However, we may eliminate x since, by similar triangles, we have

$$\frac{x}{y} = \frac{5}{10}, \quad \text{or} \quad x = \frac{1}{2}y.$$

Therefore

$$v = \tfrac{1}{12}\pi y^3,$$

and when we differentiate this with respect to t, we have

$$\frac{dv}{dt} = \frac{1}{4}\pi y^2\frac{dy}{dt}.$$

Hence

$$\frac{dy}{dt} = \frac{4\,dv/dt}{\pi y^2},$$

and when

$$\frac{dv}{dt} = 2 \quad \text{and} \quad y = 6,$$

this gives

$$\frac{dy}{dt} = \frac{2}{9\pi} \approx 0.071 \text{ ft/min.}$$

EXERCISES

1. Let A be the area of a circle of radius r. How is dA/dt related to dr/dt?
2. Let V be the volume of a sphere of radius r. How is dV/dt related to dr/dt?
3. Sand falls onto a conical pile at the rate of 10 ft^3/min. The radius of the base of the pile is always equal to one-half of its altitude. How fast is the altitude of the pile increasing when it is 5 ft deep?
4. Suppose that a raindrop is a perfect sphere. Assume that through condensation, the raindrop accumulates moisture at a rate proportional to its surface area. Show that the radius increases at a constant rate.
5. Point A moves along the x-axis at the constant rate of a ft/sec while point B moves along the y-axis at the constant rate of b ft/sec. Find how fast the distance between them is changing when A is at the point $(x, 0)$ and B is at the point $(0, y)$.
6. A spherical balloon is inflated with gas at the rate of 100 ft^3/min. Assuming that the gas pressure remains constant, how fast is the radius of the balloon increasing at the instant when the radius is 3 ft?
7. A boat is pulled in to a dock by means of a rope with one end attached to the bow of the boat, the other end passing through a ring attached to the dock at a point 4 ft higher than the bow of the boat. If the rope is pulled in at the rate of 2 ft/sec, how fast is the boat approaching the dock when 10 ft of rope are out?
8. A balloon is 200 ft off the ground and rising vertically at the constant rate of 15 ft/sec. An automobile passes beneath it traveling along a straight road at the constant rate of 45 mi/hr = 66 ft/sec. How fast is the distance between them changing one second later?
9. Water is withdrawn from a conical reservoir 8 ft in diameter and 10 ft deep (vertex down) at the constant rate of 5 ft^3/min. How fast is the water level falling when the depth of water in the reservoir is 6 ft?
10. A particle moves around the circle $x^2 + y^2 = 1$ with an x-velocity component $dx/dt = y$. Find dy/dt. Does the particle travel in the clockwise or counterclockwise direction around the circle?
11. A man 6 ft tall walks at the rate of 5 ft/sec toward a street light that is 16 ft above the ground. At what rate is the tip of his shadow moving? At what rate is the length of his shadow changing when he is 10 ft from the base of the light?
12. When air expands adiabatically, the pressure p and volume v satisfy the relationship $pv^{1.4}$ = constant. At a certain instant the pressure is 50 lb/in^2 and the volume 32 in^3 and is decreasing at the rate of 4 in^3/sec. How rapidly is the pressure changing at this instant?
13. A light is at the top of a pole 50 ft high. A ball is dropped from the same height from a point 30 ft away from the light. How fast is the shadow of the ball moving along the ground $\frac{1}{2}$ sec later? (Assume the ball falls a distance $s = 16t^2$ ft in t seconds.)
14. A boy flies a kite at a height of 300 ft, the wind carrying the kite horizontally away from the boy at a rate of 25 ft/sec. How fast must the boy pay out the string at the moment when the kite is 500 ft away from him?
15. A spherical iron ball 8 in. in diameter is coated with a layer of ice of uniform thickness. If the ice melts at the rate of 10 in^3 per minute, how fast is the thickness of the ice decreasing when it is 2 in. thick? At the same rate of melting, how fast is the outer surface area of ice decreasing?
16. Two ships A and B are sailing away from the point O along routes such that the angle $AOB = 120°$. How fast is the distance between them changing if, at a certain instant, $OA = 8$ mi, $OB = 6$ mi, ship A is sailing at the rate of 20 mi/hr, and ship B at the rate of 30 mi/hr?
Hint. Use the law of cosines.

4.3 SIGNIFICANCE OF THE SIGN OF THE SECOND DERIVATIVE

We have already seen how we may use the information given by dy/dx about the slope of the tangent to a curve to help in sketching it. We recall that in an interval where dy/dx is positive, the curve is rising to the right, while if dy/dx is negative the curve is falling to the right. Also it is evident that the regions of rise and fall are usually separated by high or low points where dy/dx is zero and the tangent is horizontal. However, it is possible for dy/dx to be zero at points which are neither high nor low points of the curve.

Example 1. If $\quad y = (x-2)^3 + 1,$

then
$$\frac{dy}{dx} = 3(x-2)^2,$$

and although
$$\frac{dy}{dx} = 0 \quad \text{at} \quad x = 2$$

for all values of x other than 2, the slope is positive and y is an increasing function of x (Fig. 4.14).

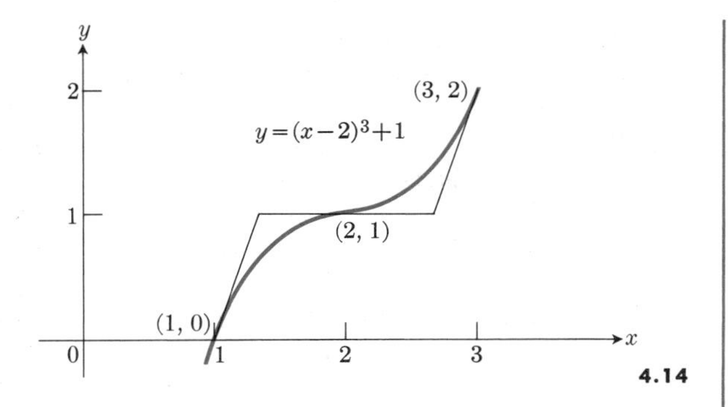

4.14

The regions of rise and fall may also be separated by points where the derivative fails to exist.

Example 2. If $y = x^{2/3}$, then

$$\frac{dy}{dx} = \frac{2}{3}x^{-1/3} = \frac{2}{3\sqrt[3]{x}}$$

is positive when x is positive, and negative when x is negative. At the transition point $x = 0$, dy/dx does not exist, but $dx/dy = \frac{3}{2}x^{1/3}$ is zero, which means that the tangent to the curve at $(0, 0)$ is vertical instead of horizontal (Fig. 4.15).

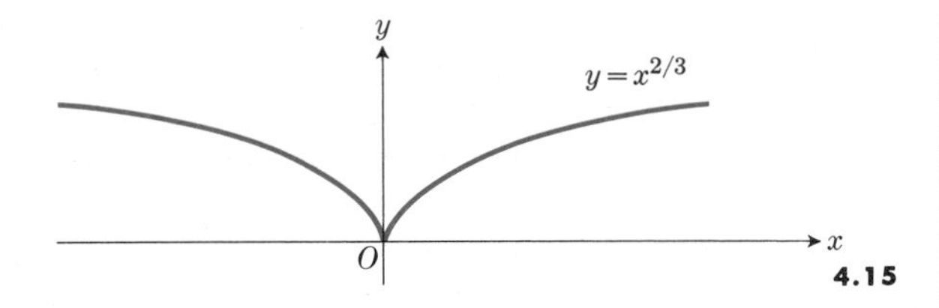

4.15

Next we shall see that the sign of the second derivative tells whether the graph of $y = f(x)$ is concave upward (y'' positive) or downward (y'' negative).

Consider, for example, the curve in Fig. 4.16(a). The first and second derivatives are represented by the curves in Fig. 4.16 (b) and (c). The arc ABC of the y-curve is concave upward; CDE is concave downward; EF is again concave upward. To focus attention, let us consider a section near A on the

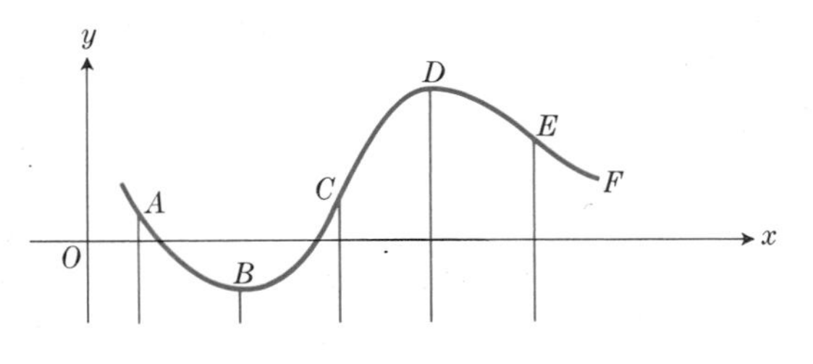

a | b

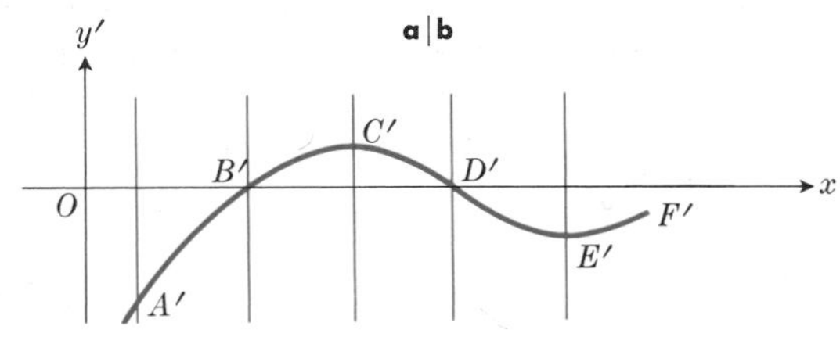

| c |

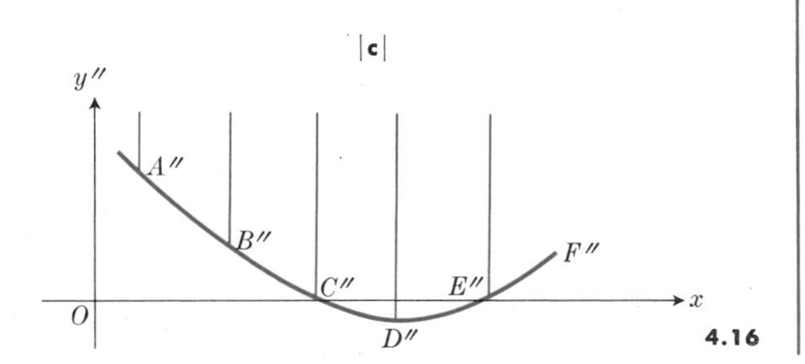

4.16

arc ABC. Here y' is negative and the y-curve slopes downward to the right. But as we travel through A, moving from left to right, we find the slope becomes less negative. That is, y' is an *increasing* function of x. Therefore the y'-curve slopes upward at A'. Hence its slope (that is, y'') is positive there. The same kind of argument applies at all points along the arc ABC; namely, y' is an increasing function of x, so its derivative (that is, y'') is positive. This is indicated by drawing the arc $A''B''C''$ of the y''-curve above the x-axis.

Similarly, where the y-curve is concave downward (along CDE), the y'-curve is falling, so its slope (that is, y'') is negative.

The direction of concavity is therefore

> *upward* if the second derivative is *positive*,
>
> *downward* if the second derivative is *negative.*

In the first case, we could also say the curve is cupped so as to "hold water"; in the second case it would "spill water."

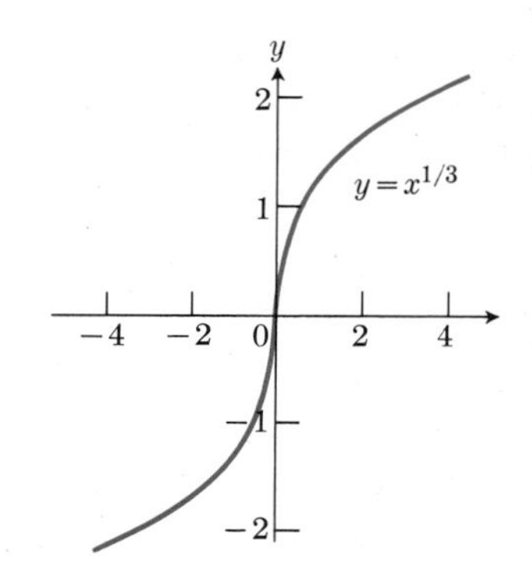

4.17

Points of inflection

A point where the curve changes the direction of its concavity from downward to upward or vice versa is called a *point of inflection.* Inflection points occur at C and E in Fig. 4.16(a), and are characterized by a change in the sign of d^2y/dx^2. Such a change of sign may occur where

(a) $d^2y/dx^2 = 0$, or

(b) d^2y/dx^2 fails to exist (for example, it may become infinite at such a point).

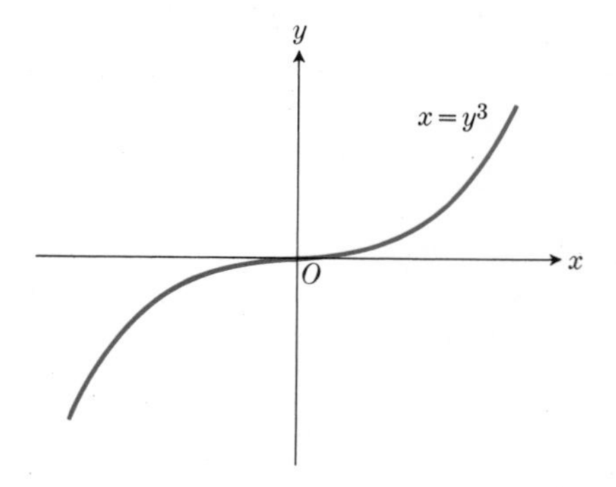

4.18

In Fig. 4.16, case (a) is shown. Case (b) is shown in Fig. 4.17, where the curve is

$$y = x^{1/3},$$

from which we find

$$y' = \tfrac{1}{3}x^{-(2/3)}, \quad y'' = -\tfrac{2}{9}x^{-(5/3)},$$

so that both y' and y'' become infinite when x approaches zero. On the other hand,

$$y' = \frac{1}{3x^{2/3}} = \frac{1}{3y^2}$$

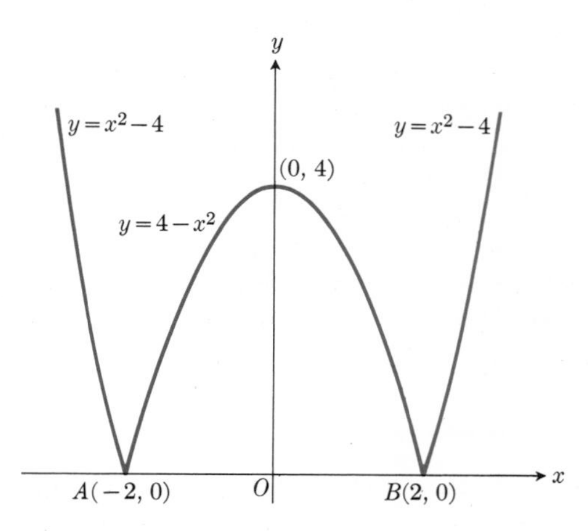

4.19 The graph of $y = |x^2 - 4|$.

is always positive for both positive and negative values of x, so the curve is always rising and has a vertical tangent at (0, 0). Also,

$$y'' = -\frac{2}{9}\left(\frac{1}{\sqrt[3]{x}}\right)^5 = \frac{-2}{9y^5}$$

is positive (the curve "holds water") when x is negative, and is negative (the curve "spills water") when x is positive, so that (0, 0) is a point of inflection.

Note. The illustration just given would have been much simpler to discuss if we had written the equation in the form $x = y^3$. Then

$$\frac{dx}{dy} = 3y^2, \qquad \frac{d^2x}{dy^2} = 6y,$$

so that we have

$$\frac{dx}{dy} = 0 \quad \text{and} \quad \frac{d^2x}{dy^2} = 0$$

at (0, 0). It is clear that dx/dy is positive for all y different from zero and d^2x/dy^2 has the same sign as y and therefore changes its sign at $y = 0$. If we interchange the x- and y-axes so as to take the x-axis to be vertical and the y-axis to be horizontal, the curve will be as shown in Fig. 4.18.

Remark. At a point of inflection, the curve should also have a unique tangent line, which the curve crosses there. Thus we rule out sharp corners, called *cusps*, such as points A and B on the graph of $y = |x^2 - 4|$, shown in Fig. 4.19. The curve to the left of A, and again to the right of B, has equation $y = x^2 - 4$, for which $y'' = +2$. Between A and B, the equation is $y = 4 - x^2$, for which $y'' = -2$. Thus the second derivative changes sign at A and again at B. However, there is not a (unique) tangent line to the graph at either A or B, and these points would not be called inflection points of the graph.

4.4 CURVE PLOTTING

The discussion in Article 4.3 may be summarized and stated in the following outline of procedure to apply in sketching the graph of an equation $y = f(x)$.

A. Calculate dy/dx and d^2y/dx^2.

B. Find the values of x for which dy/dx is positive and for which it is negative. Calculate y and d^2y/dx^2 at the points of transition between positive and negative values of dy/dx. These may give maximum or minimum points on the curve. (See Figs. 4.9, 4.15, 4.16, and 4.19.)

C. Find the values of x for which d^2y/dx^2 is positive and for which it is negative. Calculate y and dy/dx at the points of transition between positive and negative values of d^2y/dx^2. These may give points of inflection of the curve. (See Figs. 4.16 and 4.17.)

D. Plot a few additional points. In particular, points which lie between the transition points already determined or points which lie to the left and to the right of all of them will ordinarily be useful. The nature of the curve for large values of $|x|$ should also be indicated.

E. Sketch a smooth curve through the points found above, unless there are discontinuities in the curve or its slope. Have the curve pass through its points rising or falling as indicated by the sign of dy/dx, and concave upward or downward as indicated by the sign of d^2y/dx^2.

We shall now apply this technique to a few examples.

Problem 1. (See Fig. 4.20.) Sketch the curve of

$$y = \tfrac{1}{6}(x^3 - 6x^2 + 9x + 6).$$

Solution

$$\frac{dy}{dx} = \frac{1}{6}(3x^2 - 12x + 9) = \frac{1}{2}(x^2 - 4x + 3),$$

$$\frac{d^2y}{dx^2} = \frac{1}{2}(2x - 4) = x - 2.$$

In the factored form,

$$\frac{dy}{dx} = \frac{1}{2}(x - 1)(x - 3),$$

and the two factors change their signs at $x = 1$ and $x = 3$. (See Fig. 4.20a.)

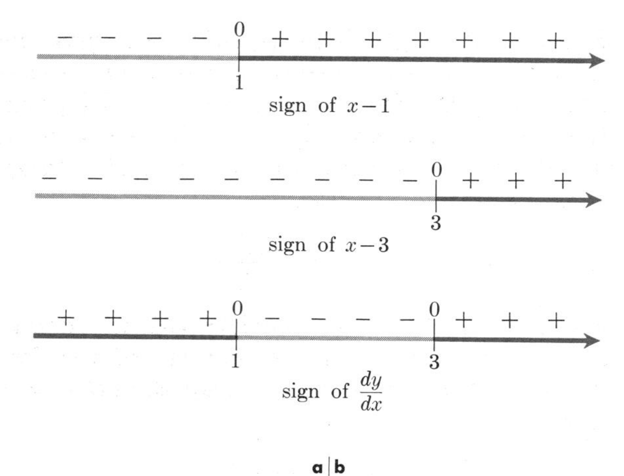

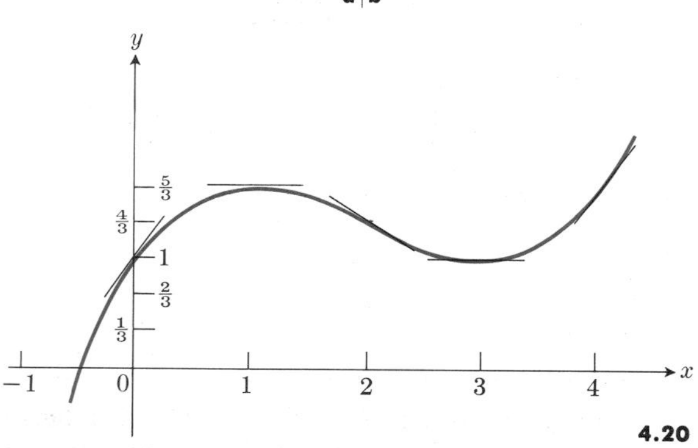

4.20

We readily see that $d^2y/dx^2 = x - 2$ is negative to the left of $x = 2$ and is positive to the right of $x = 2$, so that an inflection point occurs at $x = 2$. A fairly good sketch of the curve can now be made by using the information in the following table.

x	y	y'	y''	Remarks
-1	$-\frac{5}{3}$	$+$	$-$	Rising, concave down
0	1	$+\frac{3}{2}$	$-$	Rising, concave down
1	$\frac{5}{3}$	0	$-$	"Spills water", maximum
2	$\frac{4}{3}$	$-\frac{1}{2}$	0	Falling, point of inflection
3	1	0	$+$	"Holds water", minimum
4	$\frac{5}{3}$	$+\frac{3}{2}$	$+$	Rising, concave up

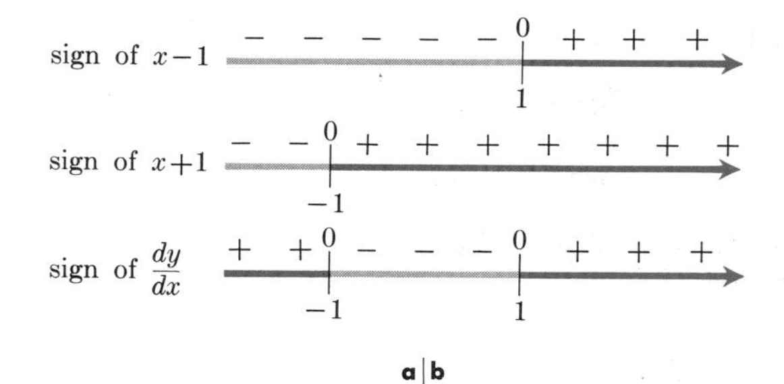

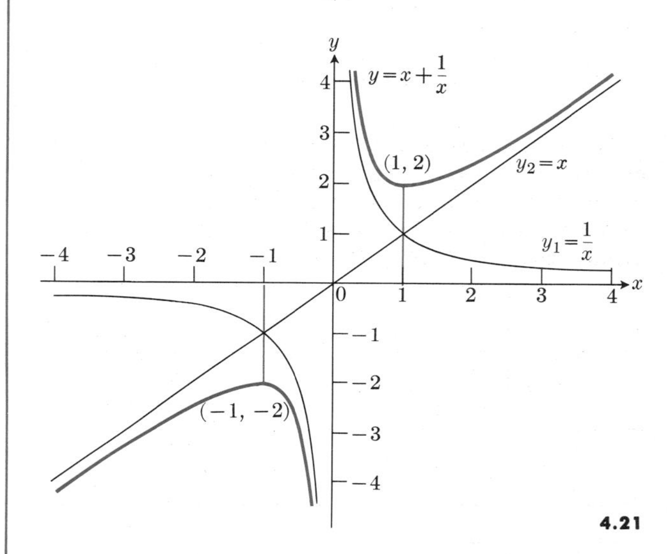

4.21

Problem 2. (See Fig. 4.21.) Sketch the curve of

$$y = x + \frac{1}{x} = x + x^{-1}.$$

Solution.

$$\frac{dy}{dx} = 1 - x^{-2} = 1 - \frac{1}{x^2} = \frac{x^2 - 1}{x^2},$$

$$\frac{d^2y}{dx^2} = 2x^{-3} = \frac{2}{x^3}.$$

The sign of dy/dx will be the same as the sign of

$$x^2 - 1 = (x - 1)(x + 1).$$

(See Fig. 4.21a.) But when x is near zero, $|y|$ will be large and y, dy/dx, d^2y/dx^2 all become infinite as x approaches zero. This curve is discontinuous at $x = 0$. Also, for large values of x, the $1/x$ term becomes small and $y \approx x$.

In fact,

$$\text{when } |x| \text{ is small,} \qquad y \approx \frac{1}{x},$$

$$\text{when } |x| \text{ is large,} \qquad y \approx x,$$

so that it is helpful to sketch (as can quickly be done) the curves

$$y_1 = \frac{1}{x}, \qquad y_2 = x,$$

as well as

$$y = x + \frac{1}{x} = y_1 + y_2.$$

Actually, replacing x by $-x$ merely changes the sign of y, so that it would suffice to sketch the portion of the curve in the first quadrant and then reflect the result with respect to the origin to get the portion in the third quadrant (Fig. 4.21b).

x	y_1	y_2	y	y'	y''	Remarks
-4	$-\frac{1}{4}$	-4	$-\frac{17}{4}$	$+$	$-$	Rising, concave down
-2	$-\frac{1}{2}$	-2	$-\frac{5}{2}$	$+$	$-$	Rising, concave down
-1	-1	-1	-2	0	-2	"Spills water," maximum
$-\frac{1}{2}$	-2	$-\frac{1}{2}$	$-\frac{5}{2}$	$-$	$-$	Falling, concave down
$-\frac{1}{4}$	-4	$-\frac{1}{4}$	$-\frac{17}{4}$	$-$	$-$	Falling, concave down
$+\frac{1}{4}$	4	$\frac{1}{4}$	$\frac{17}{4}$	$-$	$+$	Falling, concave up
$\frac{1}{2}$	2	$\frac{1}{2}$	$\frac{5}{2}$	$-$	$+$	Falling, concave up
1	1	1	2	0	$+2$	"Holds water", minimum
2	$\frac{1}{2}$	2	$\frac{5}{2}$	$+$	$+$	Rising, concave up
4	$\frac{1}{4}$	4	$\frac{17}{4}$	$+$	$+$	Rising, concave up

Remark. If $y = f(x)$ is the equation of a curve, and if $f(x) \to \infty$, or $f(x) \to -\infty$, as $x \to a$, or $x \to a-$, or $x \to a+$, then the graph of the curve will approach the line $x = a$ as an *asymptote*. For example, the line $x = 0$ (the y-axis) is an asymptote of the graph of $f(x) = x + x^{-1}$. When

$$f(x) = \frac{P(x)}{Q(x)},$$

where P and Q are polynomials with no common nonconstant factor, the asymptotes of the form $x = a$ are found by solving the equation $Q(x) = 0$. Thus

$$y = \frac{x^2 + 1}{x^3 - 4x}$$

has the lines

$$x = 0, \quad x = 2, \quad x = -2$$

as its vertical asymptotes.

EXERCISES

In Exercises 1 through 6, find intervals of values of x for which the curve is (a) rising, (b) falling, (c) concave upward, (d) concave downward. Sketch the curves, showing the high turning points M, low turning points m, points of inflection I, and vertical asymptotes.

1. $y = x^2 - 4x + 3$
2. $y = \dfrac{x}{x+1}$
3. $y = 4 + 3x - x^3$
4. $y = \dfrac{x^3}{3} - \dfrac{x^2}{2} - 6x$
5. $y = x + \dfrac{4}{x}$
6. $y = \dfrac{x^2}{x^2 - 1}$
7. Sketch a smooth curve $y = f(x)$, illustrating
$$f(1) = 0, \quad f'(x) < 0 \quad \text{for } x < 1,$$
$$f'(x) > 0 \quad \text{for } x > 1.$$
8. Sketch a smooth curve $y = f(x)$, illustrating
$$f(1) = 0, \quad f''(x) < 0 \quad \text{for } x < 1,$$
$$f''(x) > 0 \quad \text{for } x > 1.$$

Sketch curves 9 through 12, indicating high and low turning points and points of inflection

9. $y = 6 - 2x - x^2$
10. $y = 12 - 12x + x^3$
11. $y = x^4 - 32x + 48$
12. $y = x^3 - 3x^2 + 2$

For the curves 13 through 16, find vertical tangents and sketch the curves.

13. $x = y^3 + 3y^2 + 3y + 2$
14. $x = y^3 + 3y^2 - 9y - 11$
15. $x = y^4 - 2y^2 + 2$
16. $x = y^2 + \dfrac{2}{y}$
17. Sketch a continuous curve $y = f(x)$ having the following characteristics:

$$f(-2) = 8, \quad f(0) = 4, \quad f(2) = 0,$$
$$f'(x) > 0 \quad \text{for } |x| > 2,$$
$$f'(2) = f'(-2) = 0,$$
$$f'(x) < 0 \quad \text{for } |x| < 2,$$
$$f''(x) < 0 \quad \text{for } x < 0, \quad f''(x) > 0 \quad \text{for } x > 0.$$

18. Sketch a continuous curve $y = f(x)$ having

$$f'(x) > 0 \quad \text{for } x < 2, \quad f'(x) < 0 \quad \text{for } f > 2,$$

(a) if $f'(x)$ is continuous at $x = 2$,

(b) if $f'(x) \to 1$ as $x \to 2-$,
$f'(x) \to -1$ as $x \to 2+$,

(c) if $f'(x) = 1$ for all $x < 2$,
$f'(x) = -1$ for all $x > 2$.

19. Sketch a continuous curve $y = f(x)$ for $x > 0$ if

$$f(1) = 0, \quad \text{and} \quad f'(x) = 1/x \quad \text{for all } x > 0.$$

Is such a curve necessarily concave upward or concave downward?

20. Show that the sum of any positive real number and its reciprocal is at least 2.
21. Sketch the curve

$$y = 2x^3 + 2x^2 - 2x - 1$$

after locating its maximum, minimum, and inflection points. Then answer the following questions from your graph:

(a) How many times and approximately where does the curve cross the x-axis?

(b) How many times and approximately where would the curve cross the x-axis if $+3$ were added to all the y-values?

(c) How many times and approximately where would the curve cross the x-axis if -3 were added to all the y-values?

4.5 MAXIMA AND MINIMA: THEORY

A function f is said to have a relative, or local, *maximum* at $x = a$ if

$$f(a) \geq f(a + h)$$

for all positive and negative values of h sufficiently near zero. For a local *minimum* at $x = b$,

$$f(b) \leq f(b + h)$$

for values of h close to zero. The word *relative* or *local* is used to distinguish such a point from an *absolute* maximum or minimum that would occur if we could say, for example, that

$$f(a) \geq f(x)$$

for all x and not just for all x close to a. In Fig. 4.21, for example, the point (1, 2) is a *relative* minimum of the function

$$f(x) = x + \frac{1}{x}$$

because

$$x + \frac{1}{x} \geq 2$$

for all values of x close to 1. But certainly when x is negative, the inequality is no longer satisfied. Indeed, the function $f(x) = x + 1/x$ has no absolute minimum and no absolute maximum.

In Fig. 4.22 we indicate the graph of a function $y = f(x)$ defined only for the domain $[0, L]$. This function has a relative minimum at C, which is also an absolute minimum. It also has relative maxima at A and at B, and an absolute maximum at B.

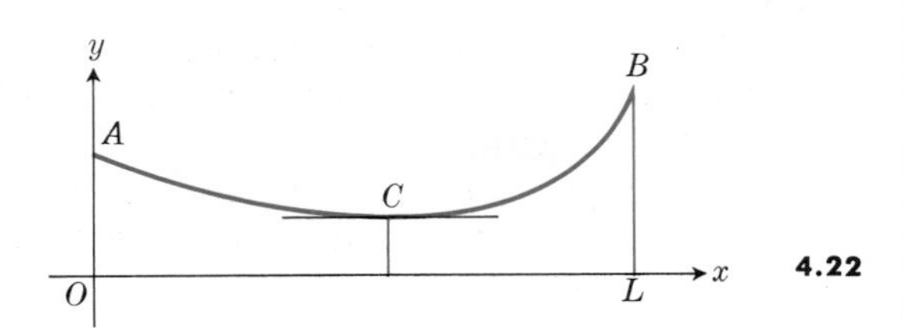

4.22

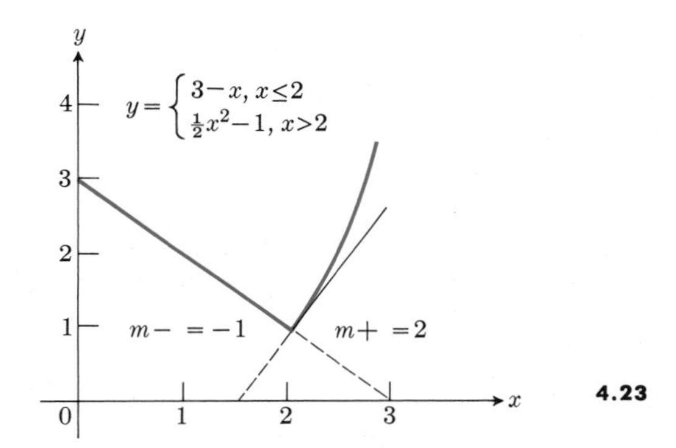

4.23

We have encountered relative maxima and minima in sketching curves and have observed that they occur at transition points between rising and falling portions of a curve. At such transition points we have also observed that dy/dx is usually zero, but may in exceptional cases become infinite (Fig. 4.15). We shall now prove the following theorem.

Theorem 1. *Let the function f be defined for $a \leq x \leq b$ and have a relative maximum or minimum at $x = c$, where $a < c < b$. If the derivative $f'(x)$ exists as a finite number at $x = c$, then*

$$f'(c) = 0. \tag{1}$$

Proof. By Lemma 1 of Article 3.3, we know that f would be increasing at c if $f'(c)$ were positive. But f is neither increasing nor decreasing at c because f has a local maximum or minimum at c. Hence $f'(c)$ cannot be positive. Likewise $f'(c)$ cannot be negative (because if it were, f would be decreasing at c). Therefore, $f'(c) = 0$. □

Remark 1. The reader is cautioned not to read into the theorem more than it says. It does not say what happens if a maximum or minimum occurs

(a) at a point c where the derivative fails to exist, or

(b) at an endpoint of the interval of definition of the function.

Neither does it say that the function necessarily does have a maximum or minimum at every place where the derivative is zero.

In Fig. 4.23, the graph represents the continuous function defined by

$$y = f(x) = \begin{cases} 3 - x & \text{for} \quad x \leq 2, \\ \frac{1}{2}x^2 - 1 & \text{for} \quad x > 2. \end{cases}$$

The slope is

$$\frac{dy}{dx} = \frac{d(3 - x)}{dx} = -1 \quad \text{for} \quad x < 2,$$

$$\frac{dy}{dx} = \frac{d(\frac{1}{2}x^2 - 1)}{dx} = x \quad \text{for} \quad x > 2.$$

At the point $x = 2$, the left-hand tangent has slope $m- = -1$ and the right-hand tangent has slope $m+ = 2$. Clearly, the curve is falling before $x = 2$ and rising after $x = 2$, and has a minimum at $x = 2$, $y = 1$. The derivative at $x = 2$ does not exist because $m+ \neq m-$. This shows that a minimum may occur where the derivative is not zero: where it doesn't even exist.

It is also easy to see from the proof of the theorem that when a maximum or minimum occurs at the *end* of a curve the derivative need not vanish at such a point. For example, suppose $g(x) = \sin x$ on $[-\pi/2, \pi/2]$ and that f is the inverse of g:

$$f(x) = \sin^{-1} x, \quad x \in [-1, 1].$$

The graphs of g and f are shown in Fig. 4.24. We know from elementary trigonometry that the minimum value of $\sin x$ is -1 and its maximum is $+1$, and that the sine is an increasing function on $[-\pi/2, \pi/2]$. Therefore its inverse function, $\sin^{-1} x$ (or, alternatively, $\arcsin x$), is an increasing function

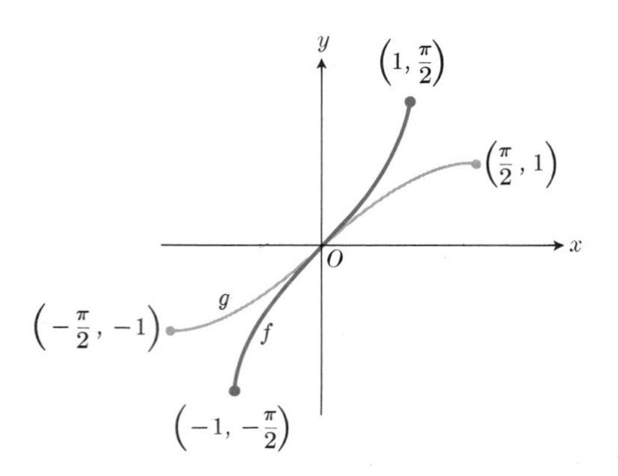

4.24 Graphs of $y = g(x) = \sin x$, $-\pi/2 \leq x \leq \pi/2$, and $y = f(x) = \sin^{-1} x$, $-1 \leq x \leq 1$.

on *its* domain $[-1, 1]$. Thus the minimum value of f is

$$f(-1) = -\pi/2,$$

and the maximum is $f(+1) = +\pi/2$. Although we don't yet have formulas for differentiating $\sin x$ and its inverse, it is clear from the graph that the slope of the inverse sine curve is not zero at its ends.

One-sided derivatives

Strictly, we can always say that $f'(a)$ and $f'(b)$ don't exist if $[a, b]$ is the domain of f, for in the definition of the derivative,

$$f'(x) = \lim_{h \to 0} \frac{f(x+h) - f(x)}{h}, \tag{2}$$

we are supposed to let h approach zero through either positive or negative values. If we don't get the same answer for h positive and negative, we say that the indicated limit doesn't exist. (See Fig. 4.23 for $x = 2$.) When $x = a$ is the left endpoint of the domain of f we can't put $h < 0$, because $f(a + h)$ doesn't make sense unless $a + h$ is in $[a, b]$. But it is useful to define the right-hand derivative of f at a, which we write $f'(a+)$, as follows:

$$f'(a+) = \lim_{h \to 0+} \frac{f(a+h) - f(a)}{h}, \tag{3a}$$

where $h \to 0+$ means that h approaches zero through positive values only. If this one-sided limit (3a) exists, it is customary to say that it is the (right-hand) derivative of f at a.

Example 1.

$$f(x) = x^{3/2}, \quad x \geq 0,$$

$$f'(x) = \tfrac{3}{2} x^{1/2}, \quad x > 0,$$

$$f'(0+) = \lim_{h \to 0+} \frac{h^{3/2} - 0}{h} = \lim_{h \to 0+} h^{1/2} = 0.$$

It is a little bit sloppy to say that f is differentiable for $x \geq 0$ and that the formula $f'(x) = \frac{3}{2}x^{1/2}$ holds for all $x \geq 0$, but lots of mathematicians would go ahead and say it anyway. [They expect us to be smart enough to know that they mean $f'(0+) = 0$, and charitable enough to them for not saying it exactly that way.]

At the right-hand endpoint of the domain $[a, b]$, we can similarly define a left-hand derivative:

$$f'(b-) = \lim_{h \to 0-} \frac{f(b+h) - f(b)}{h}, \tag{3b}$$

where the indicated limit is to be for h approaching zero through negative values, so that $b + h \in [a, b]$.

Remark 2. If f is defined on the domain $[a, b]$ and is differentiable on the open interval (a, b), and if $f'(a+)$ and $f'(b-)$ also exist, then we simply say that f is differentiable on $[a, b]$.

Example 2. $f(x) = (a^2 - x^2)^{3/2}$

is defined on the domain $[-a, a]$. For $x \in (-a, a)$, its derivative is

$$f'(x) = -3x(a^2 - x^2)^{1/2}. \tag{4}$$

The right-hand derivative at $-a$ is

$$\begin{aligned} f'(-a+) &= \lim \frac{f(x) - f(-a)}{x + a} \\ &= \lim \frac{(a - x)(a + x)(a^2 - x^2)^{1/2}}{a + x} \\ &= 0, \quad \text{where } x \to -a \text{ from the right.} \end{aligned}$$

Similarly, if we use the definition of the derivative $f'(a-)$, we find that

$$f'(a-) = 0.$$

The formula of Eq. (4) is also zero at $\pm a$. So we would say, in this example, that f is differentiable on $[-a, a]$ and that its derivative is given by Eq. (4).

Endpoint extrema

If the domain of f is the bounded, closed interval $[a, b]$ and if $f'(a+)$ and $f'(b-)$ exist, then it is easy to verify that

$$f \text{ has a local } \begin{Bmatrix} \text{maximum} \\ \text{minimum} \end{Bmatrix} \text{ at } a \text{ if } \begin{cases} f'(a+) < 0, \\ f'(a+) > 0, \end{cases} \tag{5a}$$

and

$$f \text{ has a local } \begin{Bmatrix} \text{minimum} \\ \text{maximum} \end{Bmatrix} \text{ at } b \text{ if } \begin{cases} f'(b-) < 0, \\ f'(b-) > 0. \end{cases} \tag{5b}$$

Summary

How does one find the maximum and minimum values of a function? Locate the points where

(1) the derivative is zero,

(2) the derivative fails to exist,

(3) the domain of the function has an end.

These are the candidates for values of the independent variable where f has an extremum. They are called the *critical points* for maximizing or minimizing f. By comparing the values of f at these points with each other and with the behavior at nearby points, we decide which of them, if any, give the local (or absolute) maxima and minima.

Problem. Let $f(x) = |4 - x^2|$ on $[-3, 3]$. Find the local (and absolute) maxima and minima.

Solution. We rewrite the formula for f:

$$f(x) = \begin{cases} 4 - x^2 & \text{for } |x| \le 2, \\ x^2 - 4 & \text{for } 2 < |x| \le 3. \end{cases}$$

The derivative is then given by

$$f'(x) = \begin{cases} -2x & \text{for } x \in (-2, 2) \\ 2x & \text{for } x \in (-3, -2) \text{ or } x \in (2, 3). \end{cases}$$

Hence the critical points are

(1) $x = 0$, where the derivative is zero;

(2) $x = -2$ and $x = +2$. Here the derivative fails to exist because $f'(2-) = -4$, whereas $f'(2+) = +4$, and $f'(-2-) = -4, f'(-2+) = +4$;

(3) $x = -3$ and $x = +3$, the endpoints of the domain.

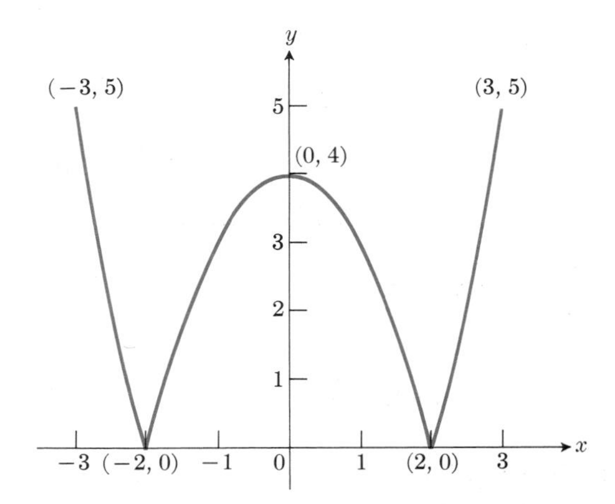

4.25 The graph of $y = |4 - x^2|, -3 \le x \le 3$.

The corresponding values of the function are

$$f(0) = 4, \quad f(2) = f(-2) = 0,$$
$$f(3) = f(-3) = 5.$$

There are minima (relative and absolute) at $x = \pm 2$, maxima (relative and absolute) at $x = \pm 3$, and a relative maximum at $x = 0$. The graph is shown in Fig. 4.25.

EXERCISES

Find the critical points for each of the following functions. For each critical point, determine whether the function has a local maximum or a local minimum there, or neither. If possible, find the absolute maximum and minimum values of the function on the indicated domain.

1. $f(x) = x - x^2$ on $[0, 1]$
2. $f(x) = x - x^3$ on $[0, 1]$
3. $f(x) = |x - x^2|$ on $[-2, 2]$
4. $f(x) = x(1 + x^2)^{-1}$ on $(-\infty, \infty)$
5. $f(x) = \dfrac{x}{1 + |x|}$ on $(-\infty, \infty)$
6. $f(x) = \dfrac{|x|}{1 + |x|}$ on $[0, \infty]$
7. $f(x) = x - [x]$ on $(-\infty, \infty)$, where $[x]$ is equal to the greatest integer in x

8. $f(x) = \operatorname{sgn}(\sin x)$ on $[-2\pi, 2\pi]$. Answer this question without using derivatives.
9. $f(x) = (x - x^2)^{-1}$ on $(0, 1)$
10. $f(x) = |x^3|$ on $[-2, 3]$. What would occur if the domain were changed to $[-2, 3)$?

4.6 MAXIMA AND MINIMA: PROBLEMS

The differential calculus is a powerful tool for solving problems that call for minimizing or maximizing a function. We shall illustrate how this is done in several instances, and then summarize the technique in a list of specific rules.

Problem 1. Find two positive numbers whose sum is 20 and such that their product is as large as possible.

Solution. If one of the numbers is x, the other is $(20 - x)$, and their product is

$$y = x(20 - x) = 20x - x^2. \tag{1}$$

Since both numbers are to be positive,

$$x > 0 \quad \text{and} \quad 20 - x > 0,$$

or

$$0 < x < 20.$$

From (1), we find

$$dy/dx = 20 - 2x = 2(10 - x),$$

which is

$$\text{positive when } x < 10,$$
$$\text{negative when } x > 10,$$
$$\text{zero when } x = 10.$$

Furthermore

$$\frac{d^2y}{dx^2} = -2$$

is always negative, and the curve representing (1) is concave downward at every point on it. The graph is the curve shown in Fig. 4.26, which has an absolute maximum at $x = 10$. The two numbers are thus $x = 10$, $20 - x = 10$.

Problem 2. A square sheet of tin a inches on a side is to be used to make an open-top box by cutting a small square of tin from each corner and bending up the sides. How large a square should be cut from each corner for the box to have as large a volume as possible?

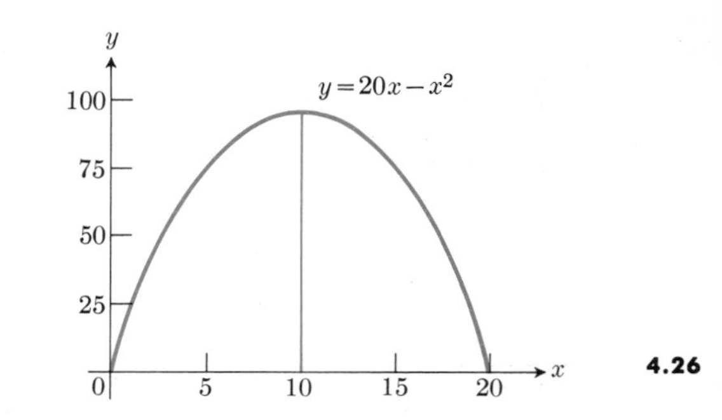

4.26

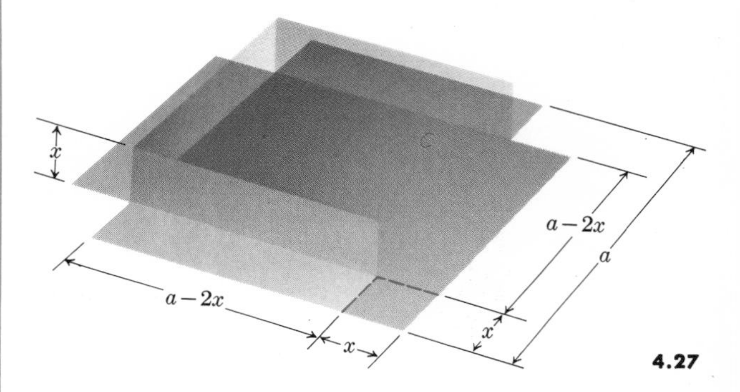

4.27

Solution. We first draw a figure to illustrate the problem (Fig. 4.27). In the figure, the side of the square cut from each corner is taken to be x inches and the volume of the box in cubic inches is then given by

$$y = x(a - 2x)^2, \quad 0 \le x \le a/2. \tag{2}$$

The restrictions placed on x in (2) are clearly those imposed by the fact that one can neither cut a negative amount of material from a corner, nor can one cut away an amount which is more than the total amount present. It is also evident that $y = 0$ when $x = 0$ or when $x = a/2$, so that the maximum volume y must occur at some value of x between 0 and $a/2$. The function in (2) possesses a derivative at *every* such point, hence

when y is a maximum, dy/dx must be zero.

From (2), we find that

$$y = a^2x - 4ax^2 + 4x^3,$$

$$\frac{dy}{dx} = a^2 - 8ax + 12x^2 = (a - 2x)(a - 6x).$$

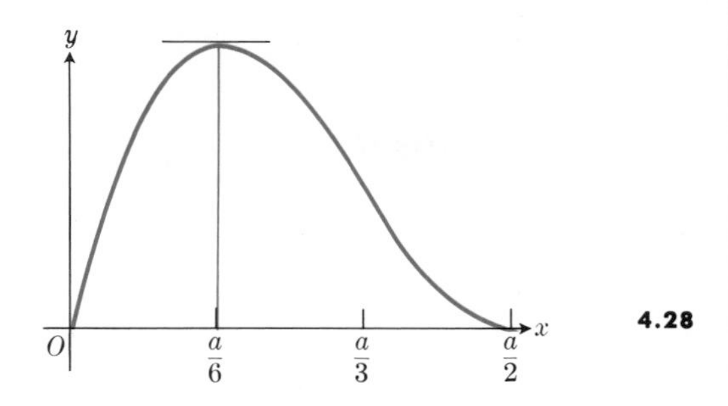

4.28

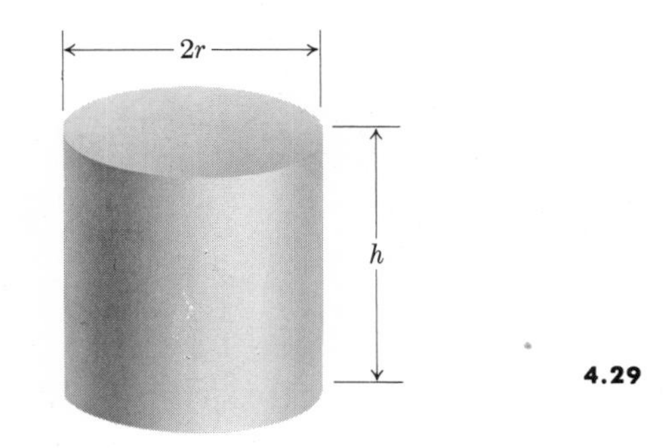

4.29

From the factored form of the equation for the derivative, it follows that

$$\frac{dy}{dx} = 0 \quad \text{when} \quad x = \frac{a}{2} \text{ or } \frac{a}{6}.$$

We gain additional information about the shape of the curve representing (2) from

$$\frac{d^2y}{dx^2} = 24x - 8a = 24\left(x - \frac{a}{3}\right),$$

which shows that the curve is

$$\text{concave downward} \quad \text{for } x < \frac{a}{3},$$

$$\text{concave upward} \quad \text{for } x > \frac{a}{3}.$$

In particular,

$$\begin{aligned} &\text{at } x = \frac{a}{6}: \quad \frac{dy}{dx} = 0, \quad \frac{d^2y}{dx^2} = -4a, \\ &\text{at } x = \frac{a}{2}: \quad \frac{dy}{dx} = 0, \quad \frac{d^2y}{dx^2} = +4a. \end{aligned} \tag{3}$$

From (3) we see that if we were to draw the graph of the curve (2) it would "spill water" at $x = (a/6)$, and this is the only point between $x = 0$ and $x = (a/2)$ where a relative maximum of the volume y may occur (Fig. 4.28). Each corner square should thus have dimensions $a/6$ by $a/6$ to produce a box of maximum volume.

Problem 3. An oil can is to be made in the form of a right circular cylinder and is to contain one quart of oil. What dimensions of the can will require the least amount of material?

Solution. Again we start with a figure to illustrate the problem (Fig. 4.29). Clearly, the requirement that the can hold a quart of oil is the same as

$$V = \pi r^2 h = a^3, \tag{4a}$$

if the radius r and altitude h are in inches and a^3 is the number of cubic inches in a quart ($a^3 = 57.75$). How shall we interpret the phrase "least amount of material"? A reasonable interpretation arises from neglecting the thickness of the material and the waste due to the manufacturing process. Then we ask for dimensions r and h that make the total surface area

$$A = 2\pi r^2 + 2\pi rh \tag{4b}$$

as small as possible while still satisfying (4a).

We are not quite ready to apply the methods used in Problems 1 and 2, because Eq. (4b) expresses A as a function of *two* variables, r and h, and our methods call for A to be expressed as a function of just *one* variable. However, Eq. (4a) may be used to express one of the variables r or h in terms of the other; in fact, we find

$$h = \frac{a^3}{\pi r^2}, \tag{4c}$$

or

$$r = \sqrt{\frac{a^3}{\pi h}}. \tag{4d}$$

The division in (4c) and (4d) is legitimate because neither r nor h can be zero, and only the positive square root is used in (4d) because the radius r can never be negative. If we substitute from (4c) into (4b), we have

$$A = 2\pi r^2 + \frac{2a^3}{r}, \quad 0 < r < \infty, \tag{4e}$$

and now we may apply our previous methods. A minimum of A can occur only at a point where

$$\frac{dA}{dr} = 4\pi r - 2a^3 r^{-2} \tag{4f}$$

is zero, that is, where

$$4\pi r = \frac{2a^3}{r^2}, \quad r = \frac{a}{\sqrt[3]{2\pi}}. \tag{4g}$$

At such a value of r, we shall have

$$\frac{d^2A}{dr^2} = 4\pi + 4a^3 r^{-3} = 12\pi > 0,$$

so the curve representing A as a function of r has $dA/dr = 0$ and d^2A/dr^2 positive ("holds water") at $r = a/\sqrt[3]{2\pi}$, which must therefore produce a relative minimum. Since the second derivative is *always positive* for $0 < r < \infty$, the curve is concave upward everywhere, and there can be no other relative minimum, so we have also found the *absolute* minimum. From (4g) and (4c), we find

$$r = a/\sqrt[3]{2\pi} = \sqrt[3]{V/2\pi},$$
$$h = 2a/\sqrt[3]{2\pi} = 2\sqrt[3]{V/2\pi}$$

as the dimensions of the can of volume V having minimal surface area. Figure 4.30 shows a curve representing A as a function of r as given by (4e).

There is an alternative method of solving problems of this type. Namely, instead of solving (4a) for h in terms of r, say, we differentiate both (4a) and (4b) with respect to r, remembering that h is a function of r. Thus, we have

$$V = \pi r^2 h,$$
$$A = 2\pi r^2 + 2\pi r h,$$

and because V is a constant,

$$dV/dr = 0.$$

Since we want to find when A will be a minimum, we put

$$\frac{dA}{dr} = 0$$

also. Hence

$$\frac{dV}{dr} = \pi\left(r^2\frac{dh}{dr} + 2rh\right) = 0, \quad \frac{dh}{dr} = -\frac{2h}{r}, \tag{4h}$$

$$\frac{dA}{dr} = 2\pi\left(2r + r\frac{dh}{dr} + h\right) = 2\pi(2r - h). \tag{4i}$$

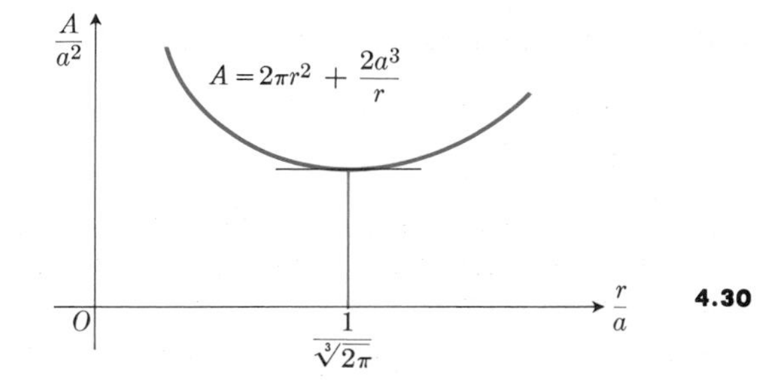

4.30

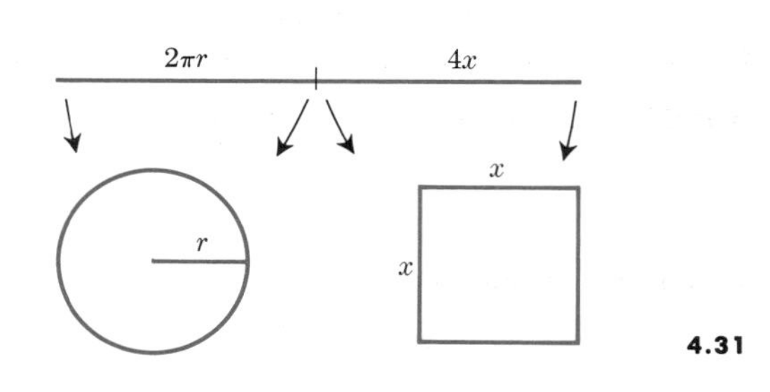

4.31

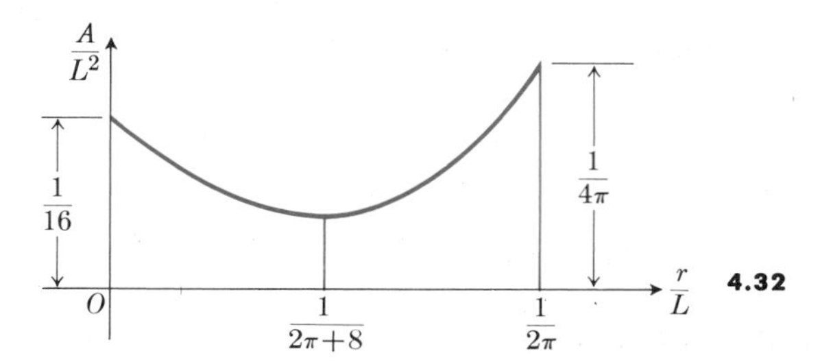

4.32

Therefore dA/dr is zero when

$$h = 2r. \tag{4j}$$

When this is substituted into (4a), we have

$$V = \pi r^2 h = 2\pi r^3,$$

so that

$$r = \sqrt[3]{V/2\pi}, \quad h = 2r = 2\sqrt[3]{V/2\pi}$$

give the critical dimensions for a cylindrical can of volume V.

We may use d^2A/dr^2 to test these critical values. We have

$$\frac{dA}{dr} = 2\pi(2r - h),$$

so that, if we take account of (4h), we find

$$\frac{d^2A}{dr^2} = 2\pi\left(2 - \frac{dh}{dr}\right) = 2\pi\left(2 + \frac{2h}{r}\right).$$

This is positive for all positive values of r and h, and hence A has a minimum at $r = r_{\text{crit}}$ where $dA/dr = 0$ and d^2A/dr^2 is positive.

Problem 4. A wire of length L is to be cut into two pieces, one of which is bent to form a circle and the other to form a square. How should the wire be cut if the sum of the areas enclosed by the two pieces is to be a maximum?

Solution. In the notation of Fig. 4.31, the sum of the combined areas is

$$A = \pi r^2 + x^2, \tag{5a}$$

where r and x must satisfy the equation

$$L = 2\pi r + 4x. \tag{5b}$$

We could solve (5b) for x in terms of r, but instead we shall differentiate both (5a) and (5b), treating A and x as functions of r for $0 \leq 2\pi r \leq L$. Then

$$\frac{dA}{dr} = 2\pi r + 2x\frac{dx}{dr}, \tag{5c}$$

and

$$\frac{dL}{dr} = 2\pi + 4\frac{dx}{dr} = 0, \quad \frac{dx}{dr} = -\frac{\pi}{2}, \tag{5d}$$

where dL/dr is zero because L is a constant. If we substitute dx/dr from (5d) into (5c), we get

$$\frac{dA}{dr} = \pi(2r - x) \tag{5e}$$

and, for future reference, we note that

$$\frac{d^2A}{dr^2} = \pi\left(2 - \frac{dx}{dr}\right) = \pi\left(2 + \frac{\pi}{2}\right) \tag{5f}$$

is a positive constant, so that the curve which represents A as a function of r is always concave upward (Fig. 4.32).

Now

$$\frac{dA}{dr} = 0 \quad \text{when} \quad x = 2r,$$

and when we substitute this into (5b),

$$L = 2\pi r + 4(2r) = (2\pi + 8)r,$$

we have

$$\frac{dA}{dr} = 0 \quad \text{when} \quad r = \frac{L}{2\pi + 8}, \quad x = \frac{L}{\pi + 4}.$$

But the fact that the second derivative is positive means that this value of r gives a *minimum* for A. The problem asks for the *maximum* of A.

Since r is limited to

$$0 \leq r \leq \frac{L}{2\pi},$$

we examine the values of A at the ends of this interval. When

$$r = 0: \qquad x = \frac{L}{4}, \qquad A = \frac{1}{16}L^2,$$

and when

$$r = \frac{L}{2\pi}: \qquad x = 0, \qquad A = \frac{1}{4\pi}L^2.$$

At the minimum,

$$r = \frac{1}{2}\frac{L}{\pi + 4}, \quad x = \frac{L}{\pi + 4}, \quad A = \frac{1}{4\pi + 16}L^2.$$

Using these values to make a rough sketch of A as a function of r (Fig. 4.32), we readily see that the maximum value of A occurs when $r = L/2\pi$, which means that the wire should not be cut at all, but all of it should be bent into the circle for maximum total area. Or, if we adopt the point of view that the wire *must* be cut, then there is no answer to the problem. For no matter how little of it is used for the square, we could always get a larger total area by using still less of it for the square.

Problem 5. Fermat's principle in optics states that light travels from a point A to a point B along that path for which the time of travel is a minimum. Let us find the path that a ray of light will follow in going from a point A in a medium where the velocity of light is c_1 to a point B in a second medium where the velocity of light is c_2, when both points lie in the xy-plane and the x-axis separates the two media. (See Fig. 4.33).

Solution. In either medium where the velocity of light remains constant, the light ray will follow a straight path, since then "shortest time" and "shortest distance" amount to the same thing. Hence the path will consist of a straight line segment from A to P in the first medium and another line segment PB in the second medium.

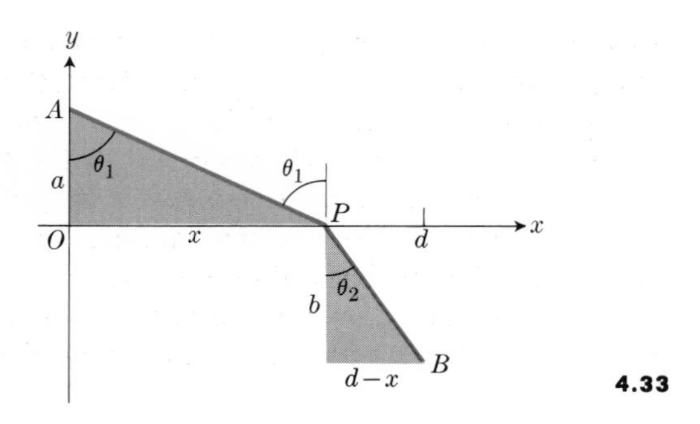

4.33

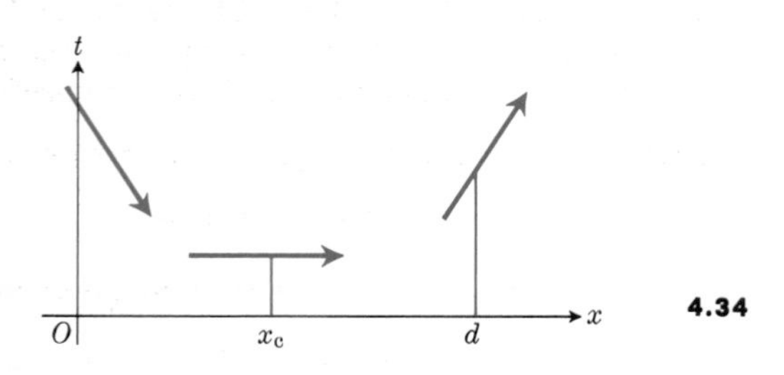

4.34

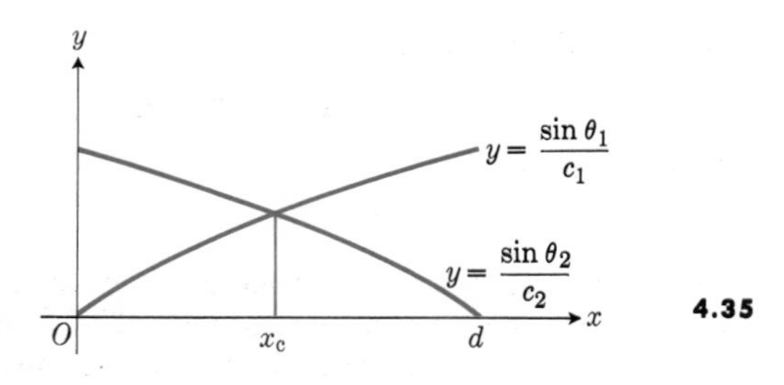

4.35

The time required for the light to travel from A to P is

$$t_1 = \frac{\sqrt{a^2+x^2}}{c_1},$$

and from P to B the time required is

$$t_2 = \frac{\sqrt{b^2+(d-x)^2}}{c_2}.$$

We therefore seek to minimize

$$t = t_1 + t_2 = \frac{\sqrt{a^2+x^2}}{c_1} + \frac{\sqrt{b^2+(d-x)^2}}{c_2}. \tag{6a}$$

We find

$$\frac{dt}{dx} = \frac{x}{c_1\sqrt{a^2+x^2}} - \frac{(d-x)}{c_2\sqrt{b^2+(d-x)^2}}, \tag{6b}$$

or

$$\frac{dt}{dx} = \frac{\sin\theta_1}{c_1} - \frac{\sin\theta_2}{c_2}, \tag{6c}$$

if we make use of the angles θ_1 and θ_2 in the figure.

If we restrict x to the interval $0 \le x \le d$, t has a negative derivative at $x = 0$ and a positive derivative at $x = d$, while at the value of x, say x_c, for which

$$\frac{\sin\theta_1}{c_1} = \frac{\sin\theta_2}{c_2}, \tag{6d}$$

dt/dx is zero. Figure 4.34 indicates the directions of the tangents to the curve giving t as a function of x, at these three points, and suggests that t will indeed be a minimum at $x = x_c$. If we want further information on this point, by referring to Fig. 4.33 we see that a decrease in x will cause P to move to the left, making θ_1 smaller, and hence also making $\sin\theta_1$ smaller, but will have the opposite effect on θ_2 and $\sin\theta_2$. Figure 4.35 shows that since $(\sin\theta_1)/c_1$ is an increasing function of x which is zero at $x = 0$, while $(\sin\theta_2)/c_2$ is a decreasing function of x which is zero at $x = d$, the two curves can cross at only one point, $x = x_c$, between 0 and d. To the right of x_c, the curve for $(\sin\theta_1)/c_1$ is above the curve for $(\sin\theta_2)/c_2$, but these roles are reversed to the left of x_c, so that

$$\frac{dt}{dx} = \frac{\sin\theta_1}{c_1} - \frac{\sin\theta_2}{c_2} \begin{cases} \text{is negative for } x < x_c, \\ \text{is zero for } x = x_c, \\ \text{is positive for } x > x_c, \end{cases}$$

and the minimum of t does indeed occur at $x = x_c$.

Instead of determining this value of x explicitly, it is customary to characterize the path followed by the ray of light by leaving the equation for $dt/dx = 0$ in the form (6d), which is known as the law of refraction, or Snell's law. (See Sears, *Optics*, 3rd edition, p. 27.)

The reader will very likely have noticed that the techniques summarized in the following rules have been used in the above problems.

First. When possible, draw a figure to illustrate the problem and label those parts that are important in the problem. Constants and variables should be clearly distinguished.

Second. Write an equation for the quantity that is to be a maximum or a minimum. If this quantity is denoted by y, it is desirable to express it in terms of a single independent variable x. This may require some algebraic manipulation to make use of auxiliary conditions of the problem.

Third. If $y = f(x)$ is the quantity to be a maximum or a minimum, find those values of x for which

$$\frac{dy}{dx} = f'(x) = 0.$$

Fourth. Test each value of x for which $f'(x) = 0$ to determine whether it provides a maximum or minimum or neither. The usual tests are:

(a) If $\frac{d^2y}{dx^2}$ is positive when $\frac{dy}{dx} = 0$,

then y is a minimum.

If $\frac{d^2y}{dx^2}$ is negative when $\frac{dy}{dx} = 0$,

then y is a maximum.

If $\frac{d^2y}{dx^2} = 0$ when $\frac{dy}{dx} = 0$,

then the test fails.

(b) If

$$\frac{dy}{dx} \text{ is } \begin{cases} \text{positive} & \text{for } x < x_c, \\ \text{zero} & \text{for } x = x_c, \\ \text{negative} & \text{for } x > x_c, \end{cases}$$

then a maximum occurs at x_c. But if dy/dx changes from negative to zero to positive as x advances through x_c, there is a minimum. If dy/dx does not change its sign, neither a maximum nor a minimum need occur.

Fifth. If the derivative fails to exist at some point, examine this point as possible maximum or minimum. (See Fig. 4.23.)

Sixth. If the function $y = f(x)$ is defined for only a limited range of values $a \leq x \leq b$, examine $x = a$ and $x = b$ for possible extreme values of y. (See Fig. 4.32.)

Remark. The following useful argument applies in many problems: It is often obvious from the formulas, or from physical conditions, that we have a continuous and everywhere-differentiable function that does not attain its maximum at an end point. Hence it has at least one maximum at an interior point, at which its derivative must be zero. So if we find just one zero for the derivative, we have the maximum without any appeal to second-derivative or other tests.

Problem 6. Suppose a manufacturer can sell x items per week at a price $P = 200 - 0.01x$ cents, and that it costs $y = 50x + 20{,}000$ cents to produce the x items. What is the production level for maximum profits?

Solution. The total revenue per week on x items is

$$xP = 200x - 0.01x^2.$$

The manufacturer's profit T is revenue minus cost:

$$\begin{aligned} T &= xP - y = (200x - 0.01x^2) - (50x + 20{,}000) \\ &= 150x - 0.01x^2 - 20{,}000. \end{aligned}$$

To maximize T, we find

$$dT/dx = 150 - 0.02x,$$

which is zero when

$$x = 7500.$$

We know that this production level gives maximum profit, since the second derivative $d^2T/dx^2 = -0.02$ is negative; this is also clear from the argument in the *Remark* above. To sell the 7500 items, the manufacturer should charge $1.25 per item.

EXERCISES

1. Show that the rectangle that has maximum area for a given perimeter is a square.
2. Find the dimensions of the rectangle of greatest area that can be inscribed in a semicircle of radius r.
3. Find the area of the largest rectangle with lower base on the x-axis and upper vertices on the curve

$$y = 12 - x^2.$$

4. An open rectangular box is to be made from a piece of cardboard 8 in. wide and 15 in. long by cutting a square from each corner and bending up the sides. Find the dimensions of the box of largest volume.

5. One side of an open field is bounded by a straight river. How would you put a fence around the other three sides of a rectangular plot in order to enclose as great an area as possible with a given length of fence?

6. An open storage bin with square base and vertical sides is to be constructed from a given amount of material. Determine its dimensions if its volume is to be a maximum. Neglect the thickness of the material and waste in construction.

7. A box with square base and open top is to hold 32 in^3. Find the dimensions that require the least amount of material. Neglect the thickness of the material and waste in construction.

8. A variable line through the point $(1, 2)$ intersects the x-axis at $A(a, 0)$ and the y-axis at $B(0, b)$. Find the area of the triangle AOB of least area if both a and b are positive.

9. A poster is to contain 50 in^2 of printed matter with margins of 4 in. each at top and bottom and 2 in. at each side. Find the overall dimensions if the total area of the poster is to be a minimum.

10. A right triangle of given hypotenuse is rotated about one of its legs to generate a right circular cone. Find the cone of greatest volume.

11. It costs a manufacturer c dollars each to manufacture and distribute a certain item. If he sells it at x dollars each, he estimates that the number he can sell is given by

$$n = a/(c - c) + b(100 - x),$$

where a and b are certain positive constants. What selling price will bring him a maximum profit?

12. One end of a cantilever beam of length L is built into a wall, while the other end is simply supported. If the beam weighs w lb per unit length, its deflection y at distance x from the built-in end satisfies the equation

$$48EIy = w(2x^4 - 5Lx^3 + 3L^2x^2),$$

where E and I are constants that depend on the material of the beam and the shape of its cross section. How far from the built-in end does the maximum deflection occur?

13. Determine the constant a so that the function

$$f(x) = x^2 + \frac{a}{x}$$

may have

(a) a relative minimum at $x = 2$,

(b) a relative minimum at $x = -3$,

(c) a point of inflection at $x = 1$.

Show that the function cannot have a relative maximum for any value of a.

14. Determine the constants a and b so that the function

$$f(x) = x^3 + ax^2 + bx + c$$

may have

(a) a relative maximum at $x = -1$ and a relative minimum at $x = 3$,

(b) a relative minimum at $x = 4$ and a point of inflection at $x = 1$.

15. A wire of length L is cut into two pieces, one being bent to form a square and the other to form an equilateral triangle. How should the wire be cut

(a) if the sum of the two areas is to be a minimum,

(b) if the sum of the areas is to be a maximum?

16. Find the points on the curve $5x^2 - 6xy + 5y^2 = 4$ that are nearest the origin.

17. The distance between the points (x_1, y_1) and (x_1, y_2) is

$$\sqrt{(x_2 - x_1)^2 + (y_2 - y_1)^2}.$$

Find the point on the curve $y = \sqrt{x}$ nearest the point $(c, 0)$

(a) if $c \geq \frac{1}{2}$, (b) if $c < \frac{1}{2}$.

18. Find the volume of the largest right circular cone that can be inscribed in a sphere of radius r.

19. Find the volume of the largest right circular cylinder that can be inscribed in a sphere of radius r.

20. Show that the volume of the largest right circular cylinder that can be inscribed in a given right circular cone is $\frac{4}{9} \times$ volume of the cone.

21. The strength of a rectangular beam is proportional to the product of its width and the square of its depth. Find the dimensions of the strongest beam that can be cut from a circular cylindrical log of radius r.

22. The stiffness of a rectangular beam is proportional to the product of its breadth and the cube of its

depth. Find the stiffest beam that can be cut from a log of given diameter.

23. The intensity of illumination at any point is proportional to the product of the strength of the light source and the inverse of the square of the distance from the source. If two sources of relative strengths a and b are a distance c apart, at what point on the line joining them will the intensity be a minimum? Assume the intensity at any point is the sum of intensities from the two sources.

24. A window is in the form of a rectangle surmounted by a semicircle. If the rectangle is of clear glass while the semicircle is of colored glass which transmits only half as much light per square foot as clear glass does, and the total perimeter is fixed, find the proportions of the window that will admit the most light.

25. Right circular cylindrical tin cans are to be manufactured to contain a given volume. There is no waste involved in cutting the tin that goes into the vertical sides of the can, but each end piece is to be cut from a square and the corners of the square wasted. Find the ratio of height to diameter for the most economical cans.

26. A silo is to be made in the form of a cylinder surmounted by a hemisphere. The cost of construction per square foot of surface area is twice as great for the hemisphere as for the cylinder. Determine the dimensions to be used if the volume is fixed and the cost of construction is to be a minimum. Neglect the thickness of the silo and waste in construction.

27. If the sum of the areas of a cube and a sphere is constant, what is the ratio of an edge of the cube to the diameter of the sphere when
   (a) the sum of their volumes is a minimum,
   (b) the sum of their volumes is a maximum?

*28. Two towns, located on the same side of a straight river, agree to construct a pumping station and filtering plant at the river's edge, to be used jointly to supply the towns with water. If the distances of the two towns from the river are a and b and the distance between them is c, show that the sum of the lengths of the pipe lines joining them to the pumping stations is at least as great as $\sqrt{c^2 + 4ab}$.

* You may find it easier to do these problems without calculus.

*29. Light emanating from a source A is reflected to a point B by a plane mirror. If the time required for the light to travel from A to the mirror and then to B is a minimum, show that the angle of incidence is equal to the angle of reflection.

30. Show that a manufacturer's profit is maximized (or minimized) at a level of production where his marginal revenue equals marginal cost.

31. Suppose the government imposes a tax of ten cents, for each item sold, on the product of Problem 6, but other features are unchanged. How much of the tax should the manufacturer absorb and how much should he pass on to the customer? Why? Compare his profits before and after the tax.

4.7 ROLLE'S THEOREM

There is strong geometrical evidence to support the belief that between two points where a smooth curve $y = f(x)$ crosses the x-axis there must be at least one point where it has a horizontal tangent (Fig. 4.36a). But such is not the case if the curve has a corner, as in Fig. 4.36(b), where the derivative fails to exist. More precisely, we have the following theorem.

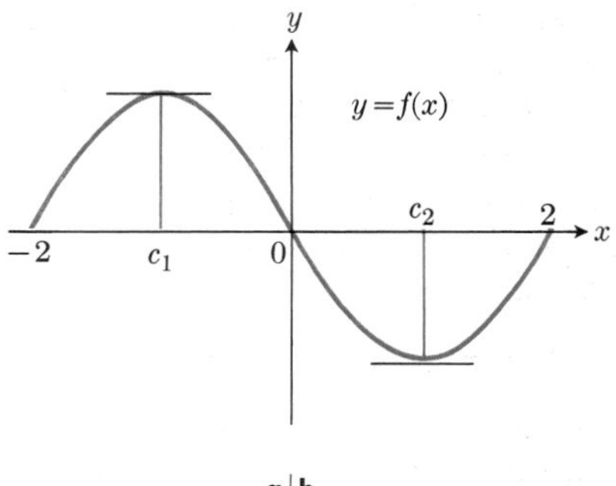

a | b

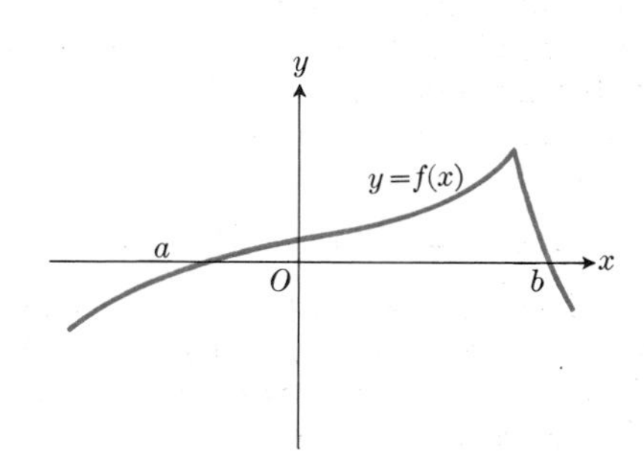

4.36

Rolle's Theorem.* *Let the function f be defined and continuous on the closed interval [a, b] and differentiable in the open interval (a, b). Furthermore, let*

$$f(a) = f(b) = 0.$$

Then there is at least one number c between a and b where $f'(x)$ is zero; that is,

$$f'(c) = 0 \quad \text{for some } c \text{ in } (a, b).$$

Proof. Either $f(x)$ is identically equal to zero for all x in $[a, b]$, or else $f(x)$ is different from zero for some values of x in this interval. In the former case, $f'(x)$ is also identically zero and the theorem is true for this case.

But if $f(x)$ is not zero everywhere between a and b, then either it is positive someplace, or negative someplace, or both. In any case, the function will then have a maximum positive value or a minimum negative value, or both. That is, it has an extreme value at a point c where $f(c)$ is negative (in the case of a minimum) or $f(c)$ is positive (in the case of a maximum). In either case, c is neither a nor b, since

$$f(a) = f(b) = 0, \qquad f(c) \neq 0.$$

Therefore c is between a and b, and Theorem 1 applies, showing that the derivative must be zero at $x = c$:

$$f'(c) = 0 \quad \text{for some } c, \quad a < c < b. \ \square$$

Remark 1. There may be more than one place between a and b where the derivative is zero. In Fig. 4.36(a), for example, dy/dx is zero at c_1 and at c_2, and both points lie between $a = -2$ and $b = +2$.

Example. The polynomial

$$y = x^3 - 4x = f(x)$$

is continuous and differentiable for all x, $-\infty < x < +\infty$. If we choose values

$$a = -2, \qquad b = +2,$$

the hypotheses of Rolle's Theorem are satisfied, since

$$f(-2) = f(+2) = 0.$$

Thus

$$f'(x) = 3x^2 - 4$$

must be zero at least once between -2 and $+2$. In fact, we find

$$3x^2 - 4 = 0$$

at

$$x = c_1 = -2\sqrt{3}/3 \quad \text{and} \quad x = c_2 = +2\sqrt{3}/3.$$

Remark 2. The reader may combine Rolle's Theorem with part (ii) of Theorem 15, Chapter 3 to obtain the following criterion for isolating the real roots of an equation $f(x) = 0$. Suppose a and b are two real numbers such that

(a) $f(x)$ is continuous on $[a, b]$ and its first derivative $f'(x)$ exists on (a, b),

(b) $f(a)$ and $f(b)$ have opposite signs, and

(c) $f'(x)$ is different from zero for all values of x in (a, b).

Then there is one and only one real root of the equation $f(x) = 0$ between a and b.

To see the significance of this, suppose the equation is

$$f(x) = x^3 + 3x + 1 = 0$$

(see Fig. 4.37). Then

$$f'(x) = 3x^2 + 3$$

is always positive, so that the equation $f'(x) = 0$ can have no real roots. Hence the original equation $f(x) = 0$ cannot have as many as two real roots. It does, in fact, have exactly one real root, since the graph of

$$y = x^3 + 3x + 1$$

crosses the x-axis between $x = -1$ and $x = 0$ but can never cross the x-axis at any other point. (If it did, the derivative would have to be zero at some point between.)

---

* Published in 1691 in *Méthode pour Résoudre les Egalités* by the French mathematician Michel Rolle. See D. E. Smith, *Source Book in Mathematics*, p. 253.

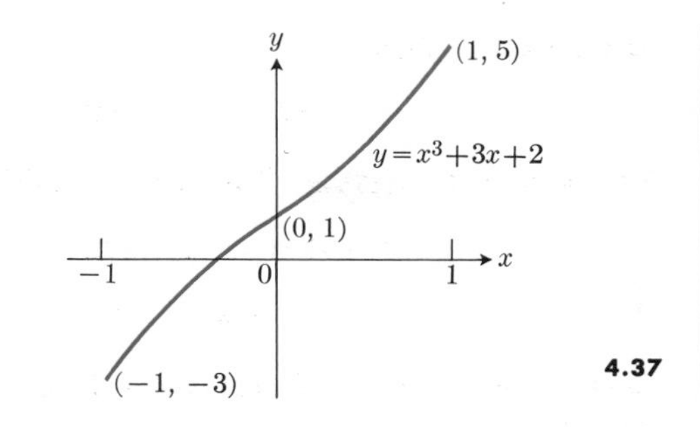

4.37

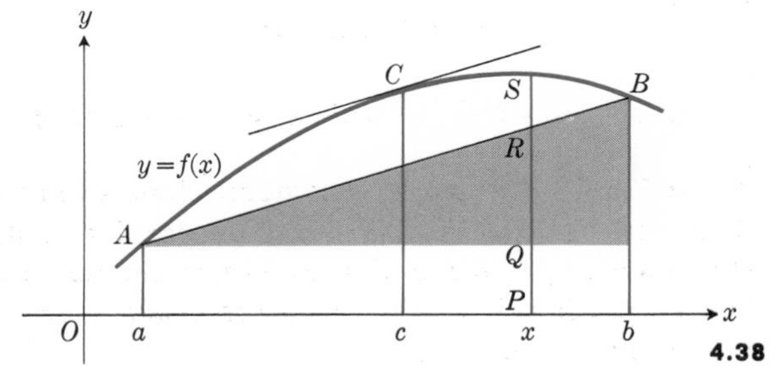

4.38

EXERCISES

Without trying to solve the equations exactly, show that the equation $f(x) = 0$ has one and only one real root between the numbers of the pair given for each of the following functions $f(x)$.

1. $x^4 + 3x + 1$, $(-2, -1)$
2. $x^4 + 2x^3 - 2$, $(0, 1)$
3. $2x^3 - 3x^2 - 12x - 6$, $(-1, 0)$
4. Let $f(x)$, together with its first two derivatives $f'(x)$ and $f''(x)$, be continuous for $a \leq x \leq b$. Suppose the curve $y = f(x)$ intersects the x-axis in at least three different places between a and b, inclusive. Using Rolle's Theorem, show that the equation $f''(x) = 0$ has at least one real root between a and b. Generalize this result.
5. Assume that $\sin x$ is differentiable. (This is true and will be proved later. You don't need the formula for the derivative to answer this question.) From elementary trigonometry we know that

$$\sin 0 = \sin \pi = 0,$$

and that the maximum of $\sin x$ is 1. What do you conclude about the value of c in Rolle's Theorem for $f(x) = \sin x$ on $[a, b] = [0, \pi]$?

4.8 THE MEAN VALUE THEOREM

In this article we shall prove the Mean Value Theorem, which is a generalization of Rolle's Theorem. Again we shall be concerned with a function $y = f(x)$ which is continuous on $[a, b]$ and which has a nonvertical tangent at each point between $A(a, f(a))$ and $B(b, f(b))$, although the tangent may be vertical at one or both of the end points A and B (Fig. 4.38). For example, the function might be

$$f(x) = \sqrt{a^2 - x^2} \quad \text{on} \quad [-a, a],$$

which represents a semicircle that fulfills the requirements above.

Demonstration

Geometrically, the Mean Value Theorem states that if the function f is continuous on $[a, b]$ and differentiable on (a, b), then there is at least one number c in (a, b) where the tangent to the curve is parallel to the chord through A and B. This is intuitively plausible, for if we consider displacing the chord AB in Fig. 4.38 upward, keeping it parallel to its original position, there will be a transition between a position where it cuts the curve in two nearby points and a position where it will fail to touch the curve, and this transition will take place at a point C where the line will be tangent to this curve. This transition will occur where the vertical distance between the chord AB and the curve is a maximum.

The analytic proof of the Mean Value Theorem has its key idea in this last statement. The vertical distance between the chord and the curve is measured by RS in Fig. 4.38, and

$$RS = PS - PR.$$

Now PS is simply the ordinate y on the curve $y = f(x)$, so that

$$PS = f(x).$$

On the other hand, PR is the ordinate on the chord AB, and its value is given by

$$g(x) = f(a) + m(x - a), \tag{1a}$$

from the equation of the straight line through $A(a, f(a))$ with slope

$$m = \frac{f(b) - f(a)}{b - a}. \tag{1b}$$

Therefore $\qquad RS = f(x) - g(x)$

measures the vertical displacement from the chord AB to the curve $y = f(x)$ for any x on $[a, b]$. Note that $f(a) - g(a) = 0$ and $f(b) - g(b) = 0$ because this vertical displacement is zero at the ends of the arc where the curve and chord meet. To complete this demonstration of the theorem, we apply Rolle's Theorem to the function

$$F(x) = f(x) - g(x). \tag{2}$$

Because both f and g are continuous on $[a, b]$, F is also continuous there. Likewise f and g are differentiable in (a, b), so F is also differentiable there. Hence F satisfies the hypotheses of Rolle's Theorem. From Rolle's theorem we therefore conclude that $F'(c) = 0$ for some $c \in (a, b)$. From Eq. (2),

$$F'(x) = f'(x) - g'(x),$$

so the conclusion is that

$$f'(c) = g'(c).$$

But from Eqs. (1a, b) we see that $g'(c) = m$. Therefore

$$f'(c) = \frac{f(b) - f(a)}{b - a} \quad \text{for some } c \in (a, b). \tag{3}$$

The result (3) is equivalent to stating that

$$f(b) - f(a) = f'(c)(b - a)$$

for some c,

$$a < c < b. \tag{4}$$

These results are summarized in the following theorem.

The Mean Value Theorem. *Let $y = f(x)$ be continuous on $[a, b]$ and be differentiable in the open interval (a, b). Then there is at least one number c between a and b such that*

$$f(b) - f(a) = f'(c)(b - a).$$

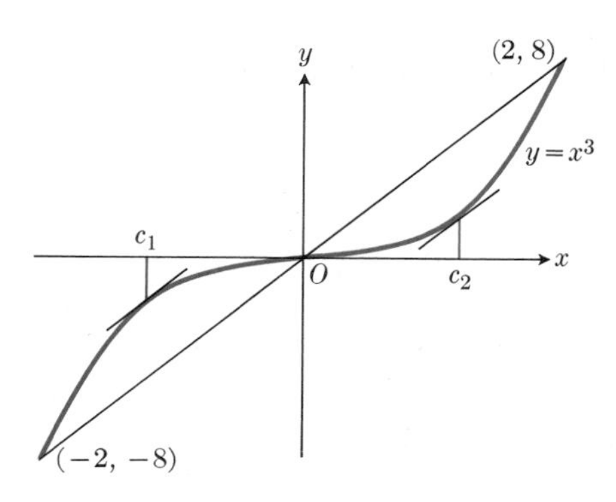

4.39

We note that Eq. (3b) states that the slope $f'(c)$ of the curve at $C(c, f(c))$ is the same as the slope $[f(b) - f(a)]/(b - a)$ of the chord joining the point $A(a, f(a))$ and $B(b, f(b))$; this is a form that is easily recalled.

Equation (4) contains a symbol c which is not very well defined except by this equation itself. For a specific function f and specific values of a and b, however, the equation can be used to find one or more values of c.

Example 1. (See Fig. 4.39.) Let

$$f(x) = x^3, \quad a = -2, \quad \text{and} \quad b = +2.$$

Then

$$f'(x) = 3x^2, \qquad f'(c) = 3c^2,$$

$$f(b) = 2^3 = 8, \qquad f(a) = (-2)^3 = -8,$$

$$\frac{f(b) - f(a)}{b - a} = \frac{8 - (-8)}{2 - (-2)} = \frac{16}{4} = 4,$$

so that

$$f'(c) = \frac{f(b) - f(a)}{b - a}$$

becomes

$$3c^2 = 4, \qquad c = \pm\tfrac{2}{3}\sqrt{3}.$$

There are thus two values of c, namely,

$$c_1 = -\tfrac{2}{3}\sqrt{3}, \qquad c_2 = +\tfrac{2}{3}\sqrt{3},$$

between $a = -2$ and $b = +2$, where the tangent to the curve $y = x^3$ is parallel to the chord through the points $(-2, -8)$ and $(+2, +8)$.

Example 2. (See Fig. 4.40.) Let

$$f(x) = x^{2/3}, \quad a = -8, \quad \text{and} \quad b = +8.$$

Then

$$f'(x) = \frac{2}{3}x^{-1/3} = \frac{2}{3\sqrt[3]{x}}$$

exists everywhere between a and b *except* at $x = 0$. We find that

$$\frac{f(b) - f(a)}{b - a} = \frac{(8)^{2/3} - (-8)^{2/3}}{8 - (-8)} = \frac{4 - 4}{16} = 0,$$

and that

$$f'(c) = \frac{2}{3\sqrt[3]{c}}$$

is not zero for any finite value of c. The result (3) need not hold, and does not hold, in this case due to the failure of the derivative to exist at a point, namely $x = 0$, between $a = -8$ and $b = +8$.

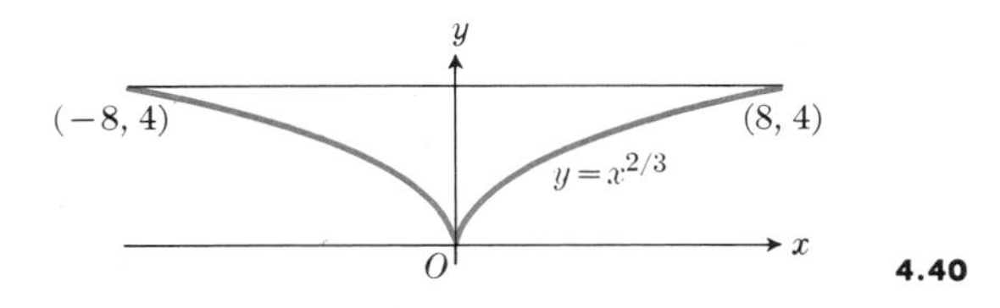

4.40

Remark. The Mean Value Theorem has the following interpretation when applied to an equation of motion $s = f(t)$. The quantity $\Delta s = f(b) - f(a)$ is the change in s corresponding to $\Delta t = b - a$, and the right-hand side of Eq. (3) is

$$\frac{\Delta s}{\Delta t} = \frac{f(b) - f(a)}{b - 1}$$

$$= \text{average velocity from } t = a \text{ to } t = b. \quad (5)$$

The equation then tells us that there is an instant $t = c$ between a and b at which the *instantaneous* velocity $f'(c)$ is equal to the *average* velocity. The theorem may therefore be paraphrased as follows: If a motorist makes a trip in which his *average* velocity is 30 mi/hr, then at least once during the trip his speedometer must also have registered precisely 30 mi/hr.

Corollary 1. *If a function F has a derivative which is equal to zero for all values of x in an interval (a, b), that is, if*

$$F'(x) \equiv 0 \quad \textit{for } x \in (a, b), \tag{6a}$$

then the function is constant throughout the interval:

$$F(x) \equiv \text{constant} \quad \textit{for } x \in (a, b). \tag{6b}$$

Proof. Suppose (6a) is satisfied. Let x_1 and x_2 be any two points in the interval with

$$a < x_1 < x_2 < b.$$

Then, since the function is differentiable for

$$x_1 \leq x \leq x_2,$$

it is also continuous in the same closed interval. Hence the Mean Value Theorem applies. That is, there is at least one number c, $x_1 < c < x_2$, such that

$$F(x_1) - F(x_2) = F'(c)(x_1 - x_2).$$

But $F'(c)$ is zero by hypothesis. Therefore

$$F(x_1) = F(x_2).$$

That is, the value of the function at any point x_1 is the same as its value at any other point x_2, for all x_1, x_2 in the interval (a, b). This is what is meant by Eq. (6b). □

Corollary 2. *If F_1 and F_2 are two functions each of which has its derivative equal to $f(x)$ for $a < x < b$, that is,*

$$\textit{if} \quad \frac{dF_1(x)}{dx} = \frac{dF_2(x)}{dx} = f(x) \quad \textit{for } a < x < b,$$

$$\textit{then} \quad F_1(x) - F_2(x) \equiv \text{constant} \quad \textit{in } (a, b).$$

Proof. Apply Corollary 1 to the function

$$F(x) = F_1(x) - F_2(x).$$

Corollary 3. *Let f be continuous on $[a, b]$ and differentiable on (a, b). If $f'(x)$ is positive throughout (a, b), then f is an increasing function on $[a, b]$, and if $f'(x)$ is negative throughout (a, b), then f is decreasing on $[a,b]$.*

Proof. Let x_1 and x_2 be any two numbers in $[a, b]$, such that $x_1 < x_2$. Apply the Mean Value Theorem to f on $[x_1, x_2]$:

$$f(x_2) - f(x_1) = f'(c)(x_2 - x_1) \quad \text{for some } c \in (x_1, x_2).$$

The sign of the right-hand side is the same as the sign of $f'(c)$ because $x_2 - x_1$ is positive. Therefore

$$f(x_2) > f(x_1) \quad \text{if} \quad f'(x) \text{ is positive on } (a, b),$$

and

$$f(x_2) < f(x_1) \quad \text{if} \quad f'(x) \text{ is negative on } (a, b). \ \square$$

EXERCISES

In Exercises 1 through 5, a, b, and c refer to the equation $f(b) - f(a) = (b - a)f'(c)$, which expresses the Mean Value Theorem. Given $f(x)$, a, and b, find c.

1. $f(x) = x^2 + 2x - 1, \quad a = 0, \quad b = 1.$
2. $f(x) = x^3, \quad a = 0, \quad b = 3.$
3. $f(x) = x^{2/3}, \quad a = 0, \quad b = 1.$
4. $f(x) = x + \dfrac{1}{x}, \quad a = \frac{1}{2}, \ b = 2.$
5. $f(x) = \sqrt{x - 1}, \quad a = 1, \ b = 3.$
6. Suppose you know that $f'(x)$ always has a value between -1 and $+1$. Show that
$$|f(x) - f(a)| \le |x - a|.$$
7. Let $P_1(x_1, y_1)$ and $P_2(x_2, y_2)$ be any two points on the parabola $y = ax^2 + bx + c$, and let $P_3(x_3, y_3)$ be the point on the arc P_1P_2 where the tangent is parallel to the chord P_1P_2. Show that
$$x_3 = (x_1 + x_2)/2.$$
8. Inequalities can often be established using the Mean Value Theorem. For example, if we know that $m < f'(x) < M$ when $a < x < b$, we can apply the Theorem to conclude that
$$f(a) + m(x - a) < f(x) < f(a) + M(x - a)$$
for all $x \in (a, b)$. Prove this result.
9. Suppose that f is a function that is differentiable for all $x \in (0, \infty)$, with $f'(x) = 1/x$. (You won't need any more information than this to answer the question. Such a function exists, even if we don't know yet what it is.) If $f(1) = 0$, prove that $0.5 < f(2) < 1.0$.
10. (a) If $f'(x) = 1/(x^2 + 1)$ for all x, and $f(0) = 0$, prove that $0.4 < f(2) < 2.0$.
(b) What better estimate of $f(2)$ could you give if you were given the further information that $f(1) = \pi/4$? (You should be able to prove that $0.2 < f(2) - f(1) < 0.5$.)

4.9 EXTENSION OF THE MEAN VALUE THEOREM

In Article 4.8 we established the existence of a number c between a and b, such that

$$f(b) - f(a) = f'(c)(b - a) \tag{1}$$

under suitable hypotheses on the function f. If the number c is replaced by a on the right side of Eq. (1), the equality must be changed to an approximation, which turns out to be the approximation we obtain by using the line tangent to the curve at $(a, f(a))$ to approximate the curve $y = f(x)$ at $x = b$. When b is close to a, we expect the approximation to be quite good. The following theorem tells us that if the function has a second derivative, as well as a first, then the difference between the tangent approximation and the function itself is approximately proportional to $(b - a)^2$.

Extended Mean Value Theorem. *Let $f(x)$ and its first derivative $f'(x)$ be continuous on the closed interval $[a, b]$, and suppose its second derivative $f''(x)$ exists in the open interval (a, b). Then there is a number c_2 between a and b such that*

$$f(b) = f(a) + f'(a)(b - a) + \tfrac{1}{2}f''(c_2)(b - a)^2. \tag{2}$$

Proof. Let K be the number defined by the equation

$$f(b) = f(a) + f'(a)(b - a) + K(b - a)^2. \tag{3}$$

Consider the function $F(x)$ that we get by replacing b by x in Eq. (3) and subtracting the right-hand side from the left:

$$F(x) = f(x) - f(a) - f'(a)(x - a) - K(x - a)^2. \tag{4}$$

Then, by substitution in (4), we find

$$F(a) = 0.$$

Also, from (3), we have

$$F(b) = 0.$$

Moreover, F and its first derivative are continuous on $[a, b]$, and

$$F'(x) = f'(x) - f'(a) - 2K(x - a). \tag{5}$$

Therefore F satisfies the hypotheses of Rolle's Theorem. Hence there is a number c_1 between a and b such that

$$F'(c_1) = 0.$$

And, by substitution in (5), we also have

$$F'(a) = 0.$$

The derived function F' satisfies Rolle's Theorem on the interval $[a, c_1]$. Hence there is a number c_2 between a and c_1 such that

$$F''(c_2) = 0.$$

We differentiate (5) and get

$$F''(x) = f''(x) - 2K. \tag{6}$$

If we put $x = c_2$ in (6), set the result equal to zero, and solve for K, we have

$$K = \tfrac{1}{2}f''(c_2).$$

When this is substituted into Eq. (3), we have Eq. (2). □

By a similar method, it is easy to prove a more general Extended Mean Value Theorem (see Miscellaneous Exercise 74 at the end of this chapter.)

Remark 1. Equation (2) can be used to estimate the size of the error that is made when we use the linearization of f at a to estimate $f(b)$, as in the following problem.

Problem 1. Use the linearization of $f(x) = \sqrt{x}$ at $a = 4$ to approximate $\sqrt{5}$, and estimate the size of the error in the approximation.

Solution. The linearization of f at a is

$$L_a(x) = f(a) + f'(a)(x - a),$$

as in Eq. (7), Article 3.4. For $f(x) = \sqrt{x}$ we have

$$f'(x) = 1/(2\sqrt{x}).$$

Hence $f(a) = \sqrt{4} = 2$ and

$$f'(a) = 1/(2\sqrt{4}) = 1/4,$$

so that

$$L_4(x) = 2 + \tfrac{1}{4}(x - 4). \tag{7a}$$

Substituting $x = 5$ in this gives $L(5) = 2.25$ as the approximation to $\sqrt{5}$ that we get by using Eq. (7a).

Now Eq. (2), with $a = 4$, $b = 5$, and $f(x) = \sqrt{x}$, says that

$$f(b) - L_a(b) = \tfrac{1}{2}f''(c_2)(b - a)^2, \tag{7b}$$

because the first two terms on the right side of Eq. (2) are just the same as $L_a(b)$. For the present problem,

$$f''(x) = -1/(4x^{3/2}) \tag{7c}$$

and $4 < c_2 < 5$. We don't know just what number to substitute for c_2, but it is clear that $f''(c_2)$ lies between $f''(4)$ and $f''(5)$, and we compute these by substituting into (7c):

$$f''(4) = -1/32, \qquad f''(5) = -1/(20\sqrt{5}).$$

Because $2 < \sqrt{5} < 2.25$, we estimate $20\sqrt{5}$ to be between 40 and 45; thus $f''(c_2)$ is between $-1/32$ and $-1/45$. Therefore, when $a = 4$ and $b = 5$, the right side of Eq. (7b) is between $-1/64$ and $-1/90$, which, in turn, are between -0.016 and -0.011. This means that the linear approximation 2.25 is too large by something between 0.01 and 0.02, so the error is less than one percent. As a practical matter, we would correct our estimate from 2.25 to 2.23 or 2.24. (A table in three decimals gives 2.236.)

Remark 2. If the hypotheses of the Extended Mean Value Theorem are satisfied on $[a, b]$, then they also hold on $[a, x]$ for any $x \in (a, b)$, and Eq. (2) can be rewritten with x in place of b, as follows:

$$f(x) = f(a) + f'(a)(x - a) + \tfrac{1}{2}f''(c)(x - a)^2 \quad \text{for some } c \in (a, x). \tag{8}$$

The first two terms on the right-hand side of Eq. (8) are the same as $L_a(x)$, the linearization of f at a.

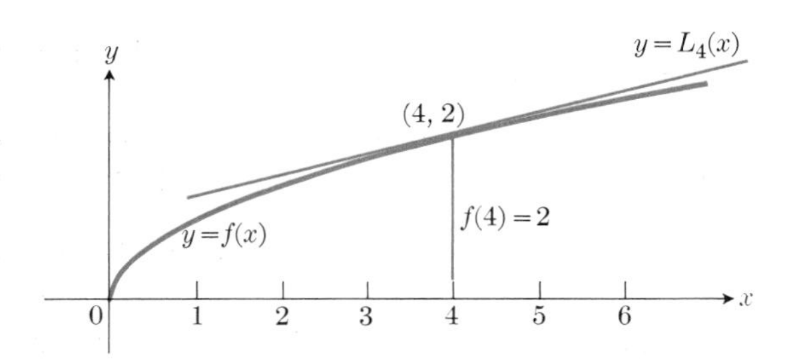

4.41 The graph of $f(x) = \sqrt{x}$ and its linearization at $a = 4$. Here $f(x) - L_4(x)$ is negative because $f''(x) < 0$.

The last term, therefore, shows that the difference

$$f(x) - L_a(x) = \tfrac{1}{2}f''(c)(x - a)^2 \tag{9}$$

is, as we announced just before the statement of the Extended Mean Value Theorem, approximately proportional to $(x - a)^2$. We have to say "approximately" because the coefficient $(\frac{1}{2})f''(c)$ is not constant when x varies. But if x stays close to a, then c also stays close to a, and the quadratic function

$$Q_a(x) = f(a) + f'(a)(x - a) + \tfrac{1}{2}f''(a)(x - a)^2 \tag{10}$$

has a graph which should stay close to the graph of f near a. (For a further discussion of this feature, see Article 18.3.)

Remark 3. If the second derivative is continuous and positive at $x = a$, then it is positive throughout a small neighborhood of a, and Eq. (8) shows that near a, the graph of f lies above its tangent (at a). This validates our earlier conclusion, and replaces our geometrical argument, that the graph of f is concave upward (holds water) when its second derivative is positive. Similarly, if $f''(a)$ is negative, and the second derivative is continuous at a, the graph of f is concave downward at a because it lies below the tangent to the curve. In Fig. 4.41, for example, we see that near $a = 4$ the graph of $f(x) = \sqrt{x}$ lies below the graph of $L_a(x)$. Also, from Eq. (9), we have a quantitative measure of the difference $f(x) - L_a(x)$. That difference, which we previously wrote in the form $\epsilon(x - a)$, with $\lim \epsilon = 0$ as $x \to a$, is now seen to be very nearly equal to $f''(a)(x - a)^2$ when x is near a.

EXERCISES

1. Use the linearization of $f(x) = \sqrt{x}$ at $a = 25$ to estimate $\sqrt{24}$, and use the Extended Mean Value Theorem to determine limits on the error.

2. If $f(x) = 3x^2 + 2x + 4$ and $a = 1$, show that the quadratic approximation of Eq. (10), and the exact expression of Eq. (8), are identical. Explain.

3. Use both the linear approximation $L_4(x)$ and the quadratic approximation $Q_4(x)$ of the function $\sqrt{x}$ to estimate

   (a) $\sqrt{4.4}$, (b) $\sqrt{2}$,
   (c) $\sqrt{3.6}$, (d) $\sqrt{5.3}$.

   (The five-decimal values are 2.09762, 1.41421, 1.89737, and 2.30217.)

4. In Fig. 4.41, the vertical displacement between the curve and the tangent line at the abscissa x is $|f(x) - L_4(x)|$. From the solution of Problem 1 in Remark 1, we saw that this is about $0.02(x - 4)^2$. In one figure, sketch portions of the graphs of the parabola $y = 0.02(x - 4)^2$, the curve $y = \sqrt{x}$, and the line $y = 2 + (\frac{1}{4})(x - 4)$, for $x \in [3, 5]$. (To three decimals, $\sqrt{3} = 1.732$ and $\sqrt{5} = 2.236$.)

5. If f and its first two derivatives are continuous functions in some neighborhood of $x = a$, what can you say about the existence and the values of the following limits? [$L_a(x)$ is the linearization of f at a.]

   (a) $\lim_{x \to a} |f(x) - f(a)|$ (b) $\lim_{x \to a} \dfrac{f(x) - f(a)}{x - a}$

   (c) $\lim_{x \to a} \dfrac{f(x) - L_a(x)}{x - a}$ (d) $\lim_{x \to a} \dfrac{f(x) - L_a(x)}{(x - a)^2}$

6. Use Eq. (10) to prove that if f and its first two derivatives are continuous in some neighborhood of $x = a$, and $f''(x)$ changes its sign when x increases through a, then the graph of f lies above the tangent line (at a) on one side of a and below it on the other side of a. (This is another way of saying that there is a *point of inflection* at a.)

7. Prove that a polynomial $f(x)$ of degree n may be written, precisely, in the form

$$f(x) = f(a) + f'(a)(x-a) + \frac{f''(a)}{2}(x-a)^2 + \cdots + \frac{f^{(n)}(a)}{n!}(x-a)^n.$$

Suggestion. Write the polynomial in the form

$$f(x) = a_0 + a_1(x-a) + a_2(x-a)^2 + \cdots + a_n(x-a)^n,$$

and evaluate the function and its derivatives at $x = a$.

8. Use the Extended Mean Value Theorem, Eq. (2), to prove the following: Let $f(x)$ be continuous and have continuous first and second derivatives. Suppose that $f'(a) = 0$. Then $f(x)$ has

   (a) a relative maximum at a if its second derivative is less than or equal to zero throughout some neighborhood of a, and

   (b) a relative minimum at a if its second derivative is greater than or equal to zero throughout some neighborhood of a.

9. The Extended Mean Value Theorem can be further extended as follows: Suppose that f, f', and f'' are continuous on $[a, b]$, and that f''' exists on (a, b). Let K be the number for which it is true that

$$f(b) = f(a) + f'(a)(b-a) + \tfrac{1}{2}f''(a)(b-a)^2 + K(b-a)^3.$$

By an argument like the one used in the proof of the Extended Mean Value Theorem, show that

$$K = \tfrac{1}{6}f'''(c_3) \quad \text{for some } c_3 \text{ in } (a, b).$$

REVIEW QUESTIONS AND EXERCISES

1. Discuss the significance of the signs of first and second derivatives. Sketch a small portion of a curve, illustrating how it looks near a point where

   (a) both y' and y'' are positive;

   (b) $y' > 0$, $y'' < 0$; (c) $y' < 0$, $y'' > 0$;

   (d) $y' < 0$, $y'' < 0$.

2. Define *point of inflection.* How do you find points of inflection from an equation of a curve?

3. How do you locate local maximum and minimum points of a curve? Discuss exceptional points, such as end points and points where the derivative fails to exist, in addition to the nonexceptional type. Illustrate with graphs.

4. Let n be a positive integer. For which values of n does the curve $y = x^n$ have

   (a) a local minimum at the origin,

   (b) a point of inflection at the origin?

5. Outline a general method of attack for solving "related rates" problems.

6. Outline a general method of attack for solving "max-min" problems.

7. What are the hypotheses of Rolle's theorem? What is the conclusion?

8. Is the converse of Rolle's theorem true?

9. With the book closed, state and prove the Mean Value Theorem. What is its geometrical interpretation?

10. We know that if $F(x) = x^2$, then $F'(x) = 2x$. If someone knows a function G such that $G'(x) = 2x$ but $G(x) \neq x^2$, what can be said about the difference $G(2) - G(1)$? Explain.

11. Use the Extended Mean Value Theorem to estimate $\sqrt[3]{9}$. Compare the answer with what you get using a linear approximation.

12. Read the article "Mathematics in Warfare" by F. W. Lanchester, *World of Mathematics*, Vol. 4, pp. 2138–2157, as a discussion of a practical problem in "related rates."

MISCELLANEOUS EXERCISES

In Exercises 1 through 14, find y' and y''. Determine in each case the sets of values of x for which

(a) y is increasing (as x increases),

(b) y is decreasing (as x increases),

(c) the graph is concave upward,

(d) the graph is concave downward.

Also sketch the graph in each case, indicating *high* and *low* turning points and points of inflection.

1. $y = 9x - x^2$
2. $y = x^3 - 5x^2 + 3x$
3. $y = 4x^3 - x^4$
4. $y = 4x + x^{-1}$
5. $y = x^2 + 4x^{-1}$
6. $y = x + 4x^{-2}$
7. $y = 5 - x^{2/3}$
8. $y = \dfrac{x-1}{x+1}$
9. $y = x - \dfrac{4}{x}$
10. $y = x^4 - 2x^2$
11. $y = \dfrac{x^2}{ax+b}$, $a > 0$, $b > 0$
12. $y = 2x^3 - 9x^2 + 12x$
13. $y = (x-1)(x+1)^2$
14. $y = x^2 - \frac{1}{6}x^3$
15. The slope of a curve at any point (x, y) is given by the equation

$$\frac{dy}{dx} = 6(x-1)(x-2)^2(x-3)^3(x-4)^4.$$

(a) For what value (or values) of x is y a maximum? Why?

(b) For what value (or values) of x is y a minimum? Why?

16. A particle moves along the x-axis with velocity $v = dx/dt = f(x)$. Show that its acceleration is $f(x)f'(x)$.
17. A meteorite entering the earth's atmosphere has velocity inversely proportional to $\sqrt{s}$ when at distance s from the center of the earth. Show that its acceleration is inversely proportional to s^2.
18. If the velocity of a falling body is $k\sqrt{s}$ at the instant when the body has fallen a distance s, find its acceleration.
19. The volume of a cube is increasing at a rate of 300 in^3/min at the instant when the edge is 20 in. Find the rate at which the edge is changing.
20. Sand falling at the rate of 3 ft^3/min forms a conical pile whose radius always equals twice the height. Find the rate at which the height is changing at the instant when the height is 10 ft.
21. The volume of a sphere is decreasing at the rate of 12π ft^3/min. Find the rates at which the radius and the surface area are changing at the instant when the radius is 20 ft. Also find approximately how much the radius and surface area may be expected to change in the following 6 sec.
22. At a certain instant, airplane A is flying a level course at 500 mi/hr. At the same time, airplane B is straight above airplane A and flying at the rate of 700 mi/hr on a slant course that intercepts A's course at a point C that is 4 mi from B and 2 mi from A.

(a) At the instant in question, how fast is the distance between the airplanes decreasing?

(b) What is the minimum distance between the airplanes, if they continue on the present courses at constant speed?

23. A point moves along the curve $y^2 = x^3$ in such a way that its distance from the origin increases at the constant rate of 2 units per second. Find dx/dt at $(2, 2\sqrt{2})$.
24. Refer to the triangle in Fig. 4.12. How fast is its area changing when $x = 17\sqrt{2}$?
25. Suppose the cone in Fig. 4.13 has a small opening at the vertex through which the water escapes at the rate of $0.08\sqrt{y}$ ft^3/min when its depth is y. Water is also running into the cone at a constant rate of c ft^3/min. When the depth is $6\frac{1}{4}$ ft it is observed to be increasing at the rate 0.02 ft/min. Under these conditions, will the tank fill? Give a reason for your answer.
26. A particle projected vertically upward from the surface of the earth with initial velocity v_0 has velocity $\sqrt{v_0^2 - 2gR[1 - (s/R)]}$ when it reaches a distance s from the *center* of the earth. Here R is the radius of the earth. Show that the acceleration is inversely proportional to s^2.
27. Given a triangle ABC. Let D and E be points on the sides AB and AC, respectively, such that DE is parallel to BC. Let the distance between BC and DE equal x. Show that the derivative, with respect to x, of the area $BCED$ is equal to the length of DE.
28. Points A and B move along the x- and y-axes, respectively, in such a way that the perpendicular distance r (inches) from the origin to AB remains constant. How fast is OA changing, and is it increasing or decreasing, when $OB = 2r$ and B is moving toward O at the rate of $0.3r$ in/sec?
29. Ships A and B start from O at the same time. Ship A travels due east at a rate of 15 mi/hr. Ship B travels in a straight course making an angle of 60° with the

path of ship A at a rate of 20 mi/hr. How fast are they separating at the end of 2 hr?

30. Water is being poured into an inverted conical tank (vertex down) at the rate of 2 ft^3/min. How fast is the water level rising when the depth of the water is 5 ft? The radius of the base of the cone is 3 ft and the altitude is 10 ft.

31. Divide 20 into two parts (not necessarily integers) such that the product of one part with the square of the other shall be a maximum.

32. Find the largest value of

$$f(x) = 4x^3 - 8x^2 + 5x \quad \text{for} \quad 0 \leq x \leq 2.$$

Give reasons for your answer.

33. Find two *positive* numbers whose sum is 36 and such that their product is as large as possible. Can the problem be solved if the product is to be as small as possible?

34. Determine the coefficients a, b, c, d so that the curve whose equation is

$$y = ax^3 + bx^2 + cx + d$$

has a maximum at $(-1, 10)$ and an inflection point at $(1, -6)$.

35. Find that number which most exceeds its square.

36. The perimeter p and area A of a circular sector ("piece of pie") of radius r and arc length s are given by

$$p = 2r + s \quad \text{and} \quad A = \tfrac{1}{2}rs,$$

respectively. If the perimeter is known to be 100 ft, what value of r will produce a maximum area?

37. If a ball is thrown vertically upward with a velocity of 32 ft/sec, its height after t sec is given by the equation $s = 32t - 16t^2$. At what instant will the ball be at its highest point, and how high will it rise?

38. A right circular cone has altitude 12 ft and radius of base 6 ft. A cone is inscribed with its vertex at the center of the base of the given cone and its base parallel to the base of the given cone. Find the dimensions of the cone of maximum volume that can be so inscribed.

39. An oil can is to be made in the form of a right circular cylinder to contain 16π in^3. What dimensions of the can will require the least amount of material to be used?

40. An isosceles triangle is drawn with its vertex at the origin, its base parallel to and above the x-axis and, the vertices of its base on the curve $12y = 36 - x^2$. Determine the area of the largest such triangle.

41. A tire manufacturer is able to make x (hundred) grade A tires and y (hundred) grade B tires per day, where

$$y = \frac{40 - 10x}{5 - x}, \quad 0 \leq x \leq 4.$$

If the profit on each grade A tire is twice the profit on a grade B tire, how many grade A tires per day should he make?

42. Find the points on the curve $x^2 - y^2 = 1$ which are nearest the point $P(a, 0)$ if

(a) $a = 4$, (b) $a = 2$,
(c) $a = \sqrt{2}$.

43. A motorist is stranded in a desert 5 mi from a point A, which is the point on a long straight road nearest to him. He wishes to get to a point B on the road. If he can travel at 15 mi/hr on the desert and 39 mi/hr on the road, find the point at which he must meet the road to get to B in the shortest possible time if

(a) B is 5 mi from A, (b) B is 10 mi from A,
(c) B is 1 mi from A.

44. Points A and B are ends of a diameter of a circle and C is a point on the circumference. Which of the following statements about triangle ABC is (or are) true?

(a) The area is a maximum when the triangle is isosceles.

(b) The area is a minimum when the triangle is isosceles.

(c) The perimeter is a maximum when the triangle is isosceles.

(d) The perimeter is a minimum when the triangle is isosceles.

45. The base and the perimeter of a triangle are fixed. Determine the remaining two sides if the area is to be a maximum.

46. The base b and the area k of a triangle are fixed. Determine the base angles if the angle at the vertex opposite b is to be a maximum.

47. A line is drawn through a fixed point (a, b) to meet the axes Ox, Oy in P and Q. Show that the minimum values of PQ, $OP + OQ$, and $OP \cdot OQ$ are respectively $(a^{2/3} + b^{2/3})^{3/2}$, $(\sqrt{a} + \sqrt{b})^2$, and $4ab$.

48. Find the smallest value of the constant m if

$$mx - 1 + \frac{1}{x}$$

is to be greater than or equal to zero for all positive values of x.

49. Let s be the distance from the fixed point $P_1(x_1, y_1)$ to a point $P(x, y)$ on the line $L: ax + by + c = 0$. Using calculus methods,

(a) show that s^2 is a minimum when P_1P is perpendicular to L, and

(b) show that the minimum distance is

$$|ax_1 + by_1 + c|/\sqrt{a^2 + b^2}.$$

50. A playing field is to be built in the shape of a rectangle plus a semicircular area at each end. A 440-yd race track is to form the perimeter of the field. Find the dimensions of the field if the rectangular part is to have as large an area as possible.

51. If $ax + (b/x) \geq c$ for all positive values of x, where a, b, and c are positive constants, show that

$$ab \geq c^2/4.$$

52. Prove that if

$$ax^2 + (b/x) \geq c$$

for all positive values of x, where a, b, and c are positive constants, then

$$27ab^2 \geq 4c^3.$$

53. Given $f(x) = ax^2 + 2bx + c$ with $a > 0$. By considering the minimum, prove that $f(x) \geq 0$ for all real x if, and only if,

$$b^2 - ac \leq 0.$$

54. In Exercise 53, take

$$f(x) = (a_1x + b_1)^2 + (a_2x + b_2)^2 + \cdots + (a_nx + b_n)^2,$$

and deduce Schwarz's inequality:

$$(a_1b_1 + a_2b_2 + \cdots + a_nb_n)^2 \leq (a_1^2 + a_2^2 + \cdots + a_n^2)(b_1^2 + b_2^2 + \cdots + b_n^2).$$

55. In Exercise 54, prove that equality can hold only in case there is a real number x such that $b_i = -a_i\, x$ for every $i = 1, 2, \ldots, n$.

56. If x is positive and m is greater than one, prove that

$$x^m - 1 - m(x - 1)$$

is not negative.

57. What are the dimensions of the rectangular plot of greatest area which can be laid out within a triangle of base 36 ft and altitude 12 ft? Assume that one side of the rectangle lies on the base of the triangle.

58. Find the width across the top of an isosceles trapezoid of base 12 in. and slant sides 6 in. if its area is a maximum.

59. A fence h ft high runs parallel to and w ft from a vertical wall. Find the length of the shortest ladder which will reach from the ground across the top of the fence to the wall.

60. Assuming that the cost per hour of running an ocean liner is $a + bv^n$, where a, b, and n are positive constants, $n > 1$, and v is the velocity through the water, find the speed for making the run from Liverpool to New York at minimum cost.

61. A flower bed is to be in the shape of a circular sector of radius r and central angle θ (i.e., like a piece of pie). Find r and θ if the area is fixed and the perimeter is a minimum.

62. A reservoir is to be built in the form of a right circular cone and the lateral area waterproofed. If the capacity of the reservoir is to be 72π ft^3 and one gallon of waterproofing material will cover 80 ft^2, how many gallons are required?

63. Given two concentric circles, C_1 of radius r_1 and C_2 of radius r_2, $r_2 > r_1 > 0$. Let A be the area between them.

(a) How fast is A increasing (or decreasing), when $r_1 = 4$ in. and is increasing at the rate of 0.02 in/sec while $r_2 = 6$ in. and is increasing at the rate of 0.01 in/sec?

(b) Suppose that at time $t = 0$, r_1 is 3 in. and r_2 is 5 in., and that for $t > 0$, r_1 increases at the constant rate of a in/sec and r_2 increases at the constant rate of b in/sec. If

$$\tfrac{3}{5}a < b < a,$$

find when the area A will be a maximum.

64. Given two concentric spheres, S_1 of radius r_1 and S_2 of radius r_2, $r_2 > r_1 > 0$. Let V be the volume between them. Suppose that at time $t = 0$, $r_1 = r$ in.

and $r_2 = R$ in., and that for $t > 0$, r_1 increases at the constant rate of a in/sec and r_2 increases at the constant rate of b in/sec. If

$$a > b > ar^2/R^2,$$

find when V will be a maximum.

65. The motion of a particle in a straight line is given by $s = \lambda t - (1 + \lambda^4)t^2$. Show that the particle moves forward initially when λ is positive but ultimately retreats. Show also that for different values of λ the maximum possible distance that the particle can move forward is $\frac{1}{8}$.

66. Let $h(x) = f(x)g(x)$ be the product of two functions that have first and second derivatives and are positive; that is,

$$f(x) > 0, g(x) > 0.$$

(a) Is it true, if f and g both have a relative maximum at $x = a$, that h has a relative maximum at $x = a$?

(b) Is it true, if f and g both have a point of inflection at $x = a$, that h has a point of inflection at $x = a$?

For both (a) and (b) either give a proof or construct a numerical example showing the statement false.

67. The numbers $c_1, c_2, \ldots, c_n$ are recorded in an experiment. It is desired to determine a number x with the property that

$$(c_1 - x)^2 + (c_2 - x)^2 + (c_3 - x)^2 + \cdots + (c_n - x)^2$$

shall be a minimum. Find x.

68. The 4 points

$$(-2, -\tfrac{1}{2}), \quad (0, 1), \quad (1, 2), \quad \text{and} \quad (3, 3)$$

are observed to lie more or less close to a straight line of equation $y = mx + 1$. Find m if the sum

$$(y_1 - mx_1 - 1)^2 + (y_2 - mx_2 - 1)^2 + (y_3 - mx_3 - 1)^2 + (y_4 - mx_4 - 1)^2$$

is to be a minimum, where $(x_1, y_1), \ldots, (x_4, y_4)$ are the coordinates of the given points.

69. The *geometric mean* of the n positive numbers $a_1, a_2, \ldots, a_n$ is the nth root of $a_1 a_2 \ldots a_n$ and the arithmetic mean is $(a_1 + a_2 + \cdots + a_n)/n$. Show that if $a_1, a_2, \ldots, a_{n-1}$ are fixed and $a_n = x$ is permitted to vary over the set of positive real numbers, the ratio of the arithmetic mean to the geometric mean is a minimum when x is the arithmetic mean of $a_1, a_2, \ldots, a_{n-1}$.

70. The curve $(y + 1)^3 = x^2$ passes through the points $(1, 0)$ and $(-1, 0)$. Does Rolle's Theorem justify the conclusion that dy/dx vanishes for some value of x in the interval $-1 \le x \le 1$? Give reasons for your answer.

71. If $a < 0 < b$ and $f(x) = x^{-1/3}$, show that there is no c that satisfies Eq. (3), Article 4.8. Illustrate with a sketch of the graph.

72. If $a < 0 < b$ and $f(x) = x^{1/3}$, show that there is a value of c that satisfies Eq. (3), Article 4.8, even though the function fails to have a derivative at $x = 0$. Illustrate with a sketch of the graph.

73. Show that the equation

$$f(x) = 2x^3 - 3x^2 + 6x + 6 = 0$$

has exactly one real root and find its value accurate to two significant figures. [*Hint.* $f(-1) = -5$, $f(0) = +6$, and $f'(x) > 0$ for all real x.]

74. General Extended Mean Value Theorem. Suppose $f(x)$ and its derivatives $f'(x), f''(x), \ldots, f^{(n-1)}(x)$ of order one through $n - 1$ are continuous on $a \le x \le b$, and $f^{(n)}(x)$ exists for $a < x < b$. If

$$F(x) = f(x) - f(a) - (x - a)f'(a) - \frac{(x-a)^2 f''(a)}{2!} - \cdots - \frac{(x-a)^{n-1} f^{(n-1)}(a)}{(n-1)!} - K(x-a)^n,$$

where K is chosen so that $F(b) = 0$, show that

(a) $F(a) = F(b) = 0$,

(b) $F'(a) = F''(a) = \cdots = F^{(n-1)}(a) = 0$,

(c) there exist numbers $c_1, c_2, c_3, \ldots, c_n$ such that

$$a < c_n < c_{n-1} < \cdots < c_2 < c_1 < b$$

and such that

$$F'(c_1) = 0 = F''(c_2) = F'''(c_3) = \cdots = F^{(n-1)}(c_{n-1}) = F^{(n)}(c_n).$$

(d) Hence, deduce that

$$K = \frac{f^{(n)}(c_n)}{n!}$$

for c_n as in (c); or, in other words, since $F(b) = 0$,

$$\begin{aligned} f(b) = f(a) &+ f'(a)(b - a) \\ &+ \frac{f''(a)}{2!}(b - a)^2 + \cdots \\ &+ \frac{f^{(n-1)}(a)}{(n - 1)!}(b - a)^{n-1} \\ &+ \frac{f^{(n)}(c_n)}{n!}(b - a)^n \end{aligned}$$

for some c_n, $a < c_n < b$. [*Amer. Math. Monthly*, Vol. 60 (1953), p. 415, James Wolfe.]

75. Suppose that it costs a manufacturer $y = a + bx$ dollars to produce x units per week. Assume that the price, P dollars per item, at which he can sell x items per week is $P = c - ex$.

(a) What level of production maximizes his profits?

(b) What is the corresponding price?

(c) What is his profit (per week) at this level of production?

(d) If a tax of t dollars per item sold is imposed on this product, and the manufacturer still wishes to maximize his profit, at what price should he sell each item? Comment on the difference between this price and the price before tax.

INTEGRATION

CHAPTER 5

5.1 INTRODUCTION

In the preceding chapters we have pursued one of the two main branches of the calculus, namely, *differential calculus.* We shall now turn our attention to the other main branch of the subject, *integral calculus.* Today, "to integrate" has two meanings when used in connection with calculus. The deeper and more fundamental meaning is nearly the same as the nontechnical definition: "to indicate the whole of; to give the sum or total of" (Webster). The mathematical meaning of the word in this sense will be amply illustrated in finding areas bounded by curves, volumes of various solids, lengths of curves, and centers of gravity, and in other applications.

The second mathematical meaning of the verb "to integrate" is "to find a function whose derivative is given." This is the aspect of integration that we shall discuss in the next two articles.

The two kinds of integration are called respectively *definite* and *indefinite,* and the connection between the two is given by a theorem which is called the *fundamental theorem* of integral calculus.

5.2 THE INDEFINITE INTEGRAL

Suppose that we are given the derivative dy/dx as a function

$$\frac{dy}{dx} = f(x), \quad x \in (a, b), \tag{1}$$

and are asked to find $y = F(x)$.

Example 1. We might be asked to find y as a function of x if

$$\frac{dy}{dx} = 2x.$$

From our experience with derivatives, we can guess one answer, namely,

$$y = x^2.$$

On the other hand, we realize that this is not the only answer, since

$$y = x^2 + 1,$$
$$y = x^2 - \sqrt{2},$$
$$y = x^2 + 5\pi$$

are also valid answers. Indeed,

$$y = x^2 + C$$

is an answer if C is any constant.

Definition 1. *An equation such as* (1), *which specifies the derivative as a function of* x (*or as a function of* x *and* y), *is called a differential equation.*

Example 2.
$$\frac{dy}{dx} = 2xy^2$$

is a differential equation. Second-, third-, and higher-order derivatives may also occur in differential equations, such as

$$\frac{d^2y}{dx^2} + 6xy\frac{dy}{dx} + 3x^2y^3 = 0,$$

and so on. For the time being we shall restrict attention to the special type of differential equation considered in Eq. (1). Differential equations of more general types will be considered in Chapter 20.

Definition 2. *A function* $y = F(x)$ *is called a solution of the differential equation* (1), *if over the domain* $a < x < b$, $F(x)$ *is differentiable and*

$$\frac{dF(x)}{dx} = f(x). \tag{2}$$

We also say, in these circumstances, that $F(x)$ *is an integral of* $f(x)$ *with respect to* x.

Remark 1. It is clear from this definition that if $F(x)$ is an integral of $f(x)$ with respect to x, then $F(x) + C$ is also such an integral when C is any constant whatever. For if Eq. (2) is satisfied, then we also have

$$\frac{d}{dx}[F(x) + C] = \frac{dF(x)}{dx} + \frac{dC}{dx}$$
$$= f(x) + 0 = f(x).$$

What is not clear, however, is whether there are other integrals of $f(x)$ not contained in this collection given by the formula $F(x) + C$. Specifically, we know that $y = x^2 + C$ is a solution, for any constant C, of the differential equation $dy/dx = 2x$. But are there any other solutions?

This question is answered by the second corollary of the Mean Value Theorem, Article 4.8. For if both $F_1(x)$ and $F_2(x)$ are integrals of $f(x)$,

then
$$\frac{dF_1(x)}{dx} = \frac{dF_2(x)}{dx} = f(x)$$

or
$$\frac{d[F_1(x) - F_2(x)]}{dx} = 0,$$

and hence
$$F_1(x) - F_2(x) = C,$$

where C is a constant. Hence, if we take

$$F_2(x) = F(x),$$

we have
$$F_1(x) = F(x) + C.$$

Therefore, if $y = F(x)$ is any solution whatever of Eq. (1), then *all* solutions are contained in the formula

$$y = F(x) + C,$$

where C is an arbitrary constant. This is indicated by writing

$$\int f(x)\,dx = F(x) + C, \tag{3}$$

where the symbol $\int$ is called an "integral sign," and Eq. (3) is read "The integral of $f(x)\,dx$ is $F(x)$ plus C." This is a standard notation. We may interpret it in either of two ways:

1. We may think of the symbol

$$\int \ldots dx \tag{4}$$

as meaning "integral, with respect to x, of . . ." The symbol (4) is then interpreted as the inverse of the symbol

$$\frac{d}{dx}\ldots,$$

which means "derivative, with respect to x, of . . ." In this interpretation the integral sign and the dx go together; the integral sign specifies the operation of integration, and the dx tells us that the "variable of integration" is x.

2. Or, we may think of Eq. (2) as written in *differential* form:

$$dF(x) = f(x)\,dx, \tag{5}$$

before the operation indicated by the integral sign is performed. Then, when we introduce the integral sign in Eq. (5) (that is, when we "integrate" both sides of the equation), we get

$$\int dF(x) = \int f(x)\, dx.$$

If we compare this with Eq. (3), we have

$$\int dF(x) = F(x) + C. \tag{6}$$

In other words, when we integrate the *differential* of a function we get that function plus an arbitrary constant. In this interpretation, we therefore think of the symbol $\int$ for integration (without absorbing the dx as part of the symbol) as meaning the operation which is the inverse of the operation denoted by the symbol d for differentiation. This is the interpretation which we shall adopt in this book.

Example 3. If we want to solve the differential equation

$$\frac{dy}{dx} = 3x^2,$$

we use the definition of the differential

$$dy = \frac{dy}{dx}\, dx$$

to change it to the differential form

$$dy = 3x^2\, dx.$$

Now we know, from past experience, that

$$d(x^3) = 3x^2\, dx.$$

Hence we have

$$y = \int 3x^2\, dx = \int d(x^3) = x^3 + C.$$

If both x and y occur in the differential equation, but in such a way that we can separate the variables so as to combine all y terms with dy and all x terms with dx, then we can integrate as in the following problem.

Problem 1. Solve the differential equation

$$\frac{dy}{dx} = x^2\sqrt{y}, \quad y > 0. \tag{7}$$

Solution. We change to differentials

$$dy = x^2\sqrt{y}\, dx$$

and then divide by $\sqrt{y}$ to obtain

$$y^{-(1/2)}\, dy = x^2\, dx,$$

in which the variables have been separated. The left side of this equation is

$$d(2y^{1/2}) = y^{-(1/2)}\, dy,$$

while the right side is

$$d(x^3/3) = x^2\, dx.$$

Therefore

$$d(2y^{1/2}) = d(x^3/3).$$

When we integrate this equation, we may write

$$2y^{1/2} + C_1 = \frac{x^3}{3} + C_2, \quad \text{or} \quad 2y^{1/2} = \frac{x^3}{3} + C, \tag{8a}$$

where we have combined the two constants C_1 and C_2 into a single constant

$$C = C_2 - C_1.$$

In fact, when we integrate the two sides of a differential equation, it always suffices to add the arbitrary constant C to just one side of the equation since, in any case, if we add constants to both sides of the equation they may always be combined into a single constant.

Remark 2. We readily verify that Eq. (8a) applies to a family of functions

$$y = F(x) = \frac{1}{4}\left(\frac{x^3}{3} + C\right)^2. \tag{8b}$$

To make $y > 0$, we would restrict the domain of F to

$$x^3 + 3C > 0 \quad \text{or} \quad x^3 + 3C < 0.$$

If we choose the former, then Eq. (7) is satisfied in the form

$$F'(x) = x^2\sqrt{F(x)}, \tag{8c}$$

because

$$2\left[\frac{1}{4}\left(\frac{x^3}{3} + C\right)\right]x^2 = x^2\sqrt{\frac{1}{4}\left(\frac{x^3}{3} + C\right)^2},$$

if

$$x^3 + 3C > 0.$$

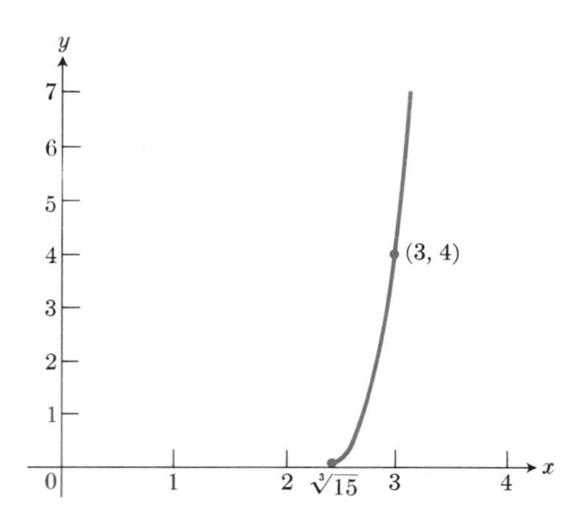

5.1 Solution of $dy/dx = x^2\sqrt{y}$ through (3, 4).

If, in addition to the differential equation (7), we want the graph of F to contain some given point, say (3, 4), we must choose C so that these coordinates satisfy Eqs. (8a, b):

$$2(4)^{1/2} = \frac{27}{3} + C, \quad \text{or} \quad C = -5.$$

The corresponding function F is defined by

$$F(x) = \frac{1}{4}\left(\frac{x^3}{3} - 5\right)^2, \quad x^3 - 15 > 0. \tag{9}$$

Figure 5.1 shows part of the graph of this function.

Remark 3. If in Eq. (7) we remove the restriction to positive values of y, and also allow $y = 0$, then we cannot divide when $y = 0$. However, we can see from the solution in Eq. (9), that the original form of the equation is satisfied in the sense of zero slope at $x = \sqrt[3]{15}$, $y = 0$. Observe also that the constant function $y = 0$ satisfies Eq. (7). Therefore, if we extend the graph to the left of $x = \sqrt[3]{15}$, not according to the algebraic formula of Eq. (9), but by following the x-axis, then we obtain a new function defined as

$$F(x) = \begin{cases} 0 & \text{for} \quad x \leq \sqrt[3]{15}, \\ \frac{1}{36}(x^3 - 15)^2 & \text{for} \quad x > \sqrt[3]{15}. \end{cases} \tag{10}$$

For every point (x, y) on the graph of (10), the slope satisfies the equation $dy/dx = x^2\sqrt{y}$.

Integration, as defined above, requires the ability to guess the answer. But the following formulas help to reduce the amount of guesswork in many cases. In these formulas, u and v denote differentiable functions of some independent variable (say of x) and a, n, and C are constants.

$$\int du = u + C, \tag{a}$$

$$\int a\,du = a\int du, \tag{b}$$

$$\int (du + dv) = \int du + \int dv, \tag{c}$$

$$\int u^n\,du = \frac{u^{n+1}}{n+1} + C, \quad (n \neq -1). \tag{d}$$

In words, these formulas say that:

(a) The integral of the differential of a function u is u plus an arbitrary constant C.

(b) A constant may be moved across the integral sign. [*Caution.* Variables must *not* be moved across the integral sign.]

(c) The integral of the sum of two differentials is the sum of their integrals. We note that formula (c) may be extended to the sum of any *finite* number of differentials:

$$\int (du_1 + du_2 + \cdots + du_n) = \int du_1 + \int du_2 + \cdots + \int du_n.$$

(d) If n is not equal to minus one, the integral of $u^n\,du$ is obtained by adding one to the exponent and dividing by the new exponent. For example,

$$\int x^{1/2}\,dx = \frac{x^{3/2}}{3/2} + C = \tfrac{2}{3}x^{3/2} + C.$$

Caution. One *must* have precisely du as well as u^n to use this formula.

Example 4. $$\int \sqrt{2x+1}\,dx$$

does not fit the formula if we let

$$u = 2x + 1, \qquad n = \tfrac{1}{2},$$

because then

$$du = \frac{du}{dx} \cdot dx = 2\,dx$$

is *not* present precisely. But only the constant factor 2 is missing, and this factor can be introduced after the integral sign provided we compensate for it by a factor of $\frac{1}{2}$ in front of the integral sign [by (b)]. Thus we write

$$\begin{aligned}\int \sqrt{2x+1}\,dx &= \frac{1}{2}\int \sqrt{2x+1}\cdot 2\,dx\\ &= \frac{1}{2}\int u^{1/2}\,du, \quad (u = 2x+1,\ du = 2\,dx)\\ &= \frac{1}{2}\,\frac{u^{3/2}}{3/2} + C\\ &= \frac{1}{3}(2x+1)^{3/2} + C.\end{aligned}$$

Remark 4. The word *antidifferentiation* is also used as a synonym for indefinite *integration.* Thus one *antiderivative* of $\sqrt{x}$ is $(\frac{2}{3})x^{3/2}$, and every antiderivative is obtained from this one by adding a constant. Another synonym for *antiderivative,* or *indefinite integral,* is *primitive.* Thus a function F of which f is the derivative is called a *primitive* of f.

Example 5. The derivative of $F(x) = \sqrt{x^2+1}$ is $f(x) = x/\sqrt{x^2+1}$. Therefore, for any constant C,

$$\sqrt{x^2+1} + C \quad \text{is a primitive of} \quad \frac{x}{\sqrt{x^2+1}}.$$

EXERCISES

Solve the following differential equations.

1. $\dfrac{dy}{dx} = x^2 + 1$
2. $\dfrac{dy}{dx} = \dfrac{1}{x^2} + x, \quad x > 0$
3. $\dfrac{dy}{dx} = \dfrac{x}{y}, \quad y > 0$
4. $\dfrac{dy}{dx} = \sqrt{xy}, \quad x > 0, \quad y > 0$
5. $\dfrac{dy}{dx} = \sqrt[3]{y/x}, \quad x > 0, \quad y > 0$
6. $\dfrac{dy}{dx} = 2xy^2,\ y > 0$
7. $\dfrac{dy}{dx} = 3x^2 - 2x + 5$
8. $\dfrac{ds}{dt} = 3t^2 + 4t - 6$
9. $\dfrac{dr}{dz} = (2z+1)^3$
10. $\dfrac{du}{dv} = 2u^2(4v^3 + 4v^{-3}), \quad v > 0, \quad u > 0$
11. $\dfrac{dx}{dt} = 8\sqrt{x}, \quad x > 0$
12. $\dfrac{dy}{dt} = (2t + t^{-1})^2, \quad t > 0$
13. $\dfrac{dy}{dz} = \sqrt{(z^2 - z^{-2})^2 + 4}, \quad z > 0$

Evaluate the following integrals.

14. $\displaystyle\int (2x+3)\,dx$
15. $\displaystyle\int (x^2 - \sqrt{x})\,dx$
16. $\displaystyle\int (3x-1)^{234}\,dx$
17. $\displaystyle\int (2-7t)^{2/3}\,dt$
18. $\displaystyle\int \sqrt{2+5y}\,dy$
19. $\displaystyle\int \frac{dx}{(3x+2)^2}$
20. $\displaystyle\int \frac{3r\,dr}{\sqrt{1-r^2}}$
21. $\displaystyle\int \sqrt{2x^2+1}\,x\,dx$
22. $\displaystyle\int t^2(1+2t^3)^{-(2/3)}\,dt$
23. $\displaystyle\int \frac{y\,dy}{\sqrt{2y^2+1}}$
24. $\displaystyle\int \left(\sqrt{x} + \frac{1}{\sqrt{x}}\right)dx$
25. $\displaystyle\int \frac{(z+1)\,dz}{\sqrt[3]{z^2+2z+2}}$

5.3 APPLICATIONS OF INDEFINITE INTEGRATION

Differential equations, such as Eq. (1) or (7) of Article 5.2, arise in chemistry, physics, mathematics, and all branches of engineering. Some of these applications will be illustrated in the examples that follow. Before proceeding with these, however, let us consider the meaning of the arbitrary constant C, which always enters when we integrate a differential equation. If we draw one of the integral

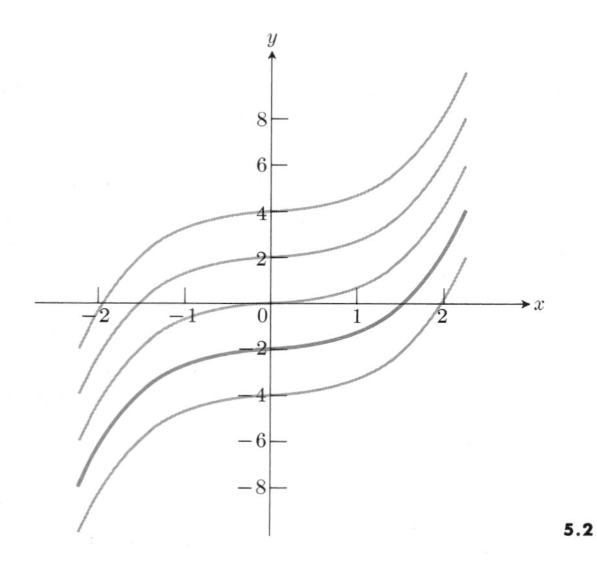

5.2

curves $y = F(x)$ (corresponding to taking $C = 0$), then any other integral curve $y = F(x) + C$ is obtained by simply shifting this curve through a vertical displacement C. Thus we obtain a family of "parallel" curves, as in Fig. 5.2. They are parallel in the sense that the slope of the tangent to any one of them, at the point of abscissa x, is $f(x)$, the same for all curves $y = F(x) + C$. Now clearly this family of parallel curves has the property that, given any point (x_0, y_0) with x_0 in the allowed domain of the independent variable x, there is one and only one curve of the family that passes through this particular point. For if that curve is to pass through the point, the equation must be satisfied by these particular coordinates. This uniquely specifies the value of C, namely,

$$C = y_0 - F(x_0).$$

With C thus determined, we get a definite function expressing y in terms of x.

The condition imposed that $y = y_0$ when $x = x_0$ is often referred to as an "initial condition." This terminology is particularly appropriate in connection with physical problems, where time is the independent variable and initial velocities or initial positions of moving bodies are specified.

Problem 1. The velocity, at time t, of a moving body is given by

$$v = at,$$

where a is a constant. If the body's coordinate is s_0 at time $t = 0$, find the distance s as a function of t.

Solution. The velocity v is the same as the derivative ds/dt. Hence we want to solve the problem that consists of

the differential equation $$\frac{ds}{dt} = at \qquad (1)$$

and the initial condition $$s = s_0 \quad \text{when} \quad t = 0. \qquad (2)$$

From Eq. (1), we have

$$ds = at\,dt,$$

$$s = \int at\,dt = a\frac{t^2}{2} + C.$$

The constant of integration may now be determined from the initial condition, which requires that $s_0 = C$. Hence, the solution of the problem is

$$s = a\frac{t^2}{2} + s_0.$$

Problem 2. (See Fig. 5.2.) Find the curve whose slope at the point (x, y) is $3x^2$ if the curve is also required to pass through the point $(1, -1)$.

Solution. In mathematical language, we have the following problem:

differential equation: $$\frac{dy}{dx} = 3x^2,$$

initial condition: $$y = -1 \quad \text{when} \quad x = 1.$$

First, we integrate the differential equation.

$$dy = 3x^2\,dx,$$

$$y = \int 3x^2\,dx = x^3 + C.$$

Then we impose the initial condition to evaluate the constant C:

$$-1 = 1^3 + C; \quad C = -2.$$

We substitute the value of C into the solution of the differential equation, and obtain the particular integral curve that passes through the given point, namely,

$$y = x^3 - 2.$$

EXERCISES

In Exercises 1 through 6, find the position s as a function of t from the given velocity $v = ds/dt$. Evaluate the constant of integration so as to have $s = s_0$ when $t = 0$.

1. $v = 3t^2$
2. $v = 2t + 1$
3. $v = (t+1)^2$
4. $v = (t^2+1)^2$
5. $v = (t+1)^{-2}$
6. $v = \sqrt{2gs}$ (g constant)

In Exercises 7 through 11, find the velocity v and position s as functions of t from the given acceleration $a = dv/dt$. Evaluate the constants of integration so as to have $v = v_0$ and $s = s_0$ when $t = 0$.

7. $a = g$ (constant)
8. $a = t$
9. $a = \sqrt[3]{2t+1}$
10. $a = (2t+1)^{-3}$
11. $a = (t^2+1)^2$
12. The gravitational attraction exerted by the earth on a particle of mass m at distance s from the center is given by $F = -mgR^2s^{-2}$, where R is the radius of the earth and F is negative because the force acts in opposition to increasing s. If a particle is projected vertically upward from the surface of the earth with initial velocity $v_0 = \sqrt{2gR}$, apply Newton's second law $F = ma$ with $a = v(dv/ds)$ to show that

$$v = v_0\sqrt{R/s} \quad \text{and} \quad s^{3/2} = R^{3/2}[1 + (3v_0t/2R)].$$

Remark. The initial velocity $v_0 = \sqrt{2gR}$ (approximately 7 miles per second) is known as the "velocity of escape," since the displacement s tends to infinity with increasing t provided the initial velocity is this large. Actually, a somewhat larger initial velocity is required for escape from the earth's gravitational attraction, due to the retardation effect of air resistance, which we have here neglected for the sake of simplicity.

Solve the following differential equations subject to the prescribed initial conditions.

13. $\dfrac{dy}{dx} = x\sqrt{y}$, $\quad x = 0,\ y = 1$
14. $\dfrac{dy}{dx} = 2xy^2$, $\quad x = 1,\ y = 1$
15. $\dfrac{dy}{dx} = x\sqrt{1+x^2}$, $\quad x = 0,\ y = -3$
16. $\dfrac{dy}{dx} = \dfrac{4\sqrt{(1+y^2)^3}}{y}$, $\quad x = 0,\ y = 1$

5.4 BRIEF REVIEW OF TRIGONOMETRY

Many natural phenomena are periodic; that is, they repeat after definite periods of time. Such phenomena are most readily studied through the use of the trigonometric functions, particularly sines and cosines. Our object in this article and the next is to apply the operations of the calculus to these functions, but before we do so, we shall review some of their properties.

When an angle of measure θ is placed in standard position at the center of a circle of radius r, as in Fig. 5.3, the trigonometric functions of θ are defined by the equations

$$\begin{aligned} \sin\theta &= \frac{y}{r}, \quad \cos\theta = \frac{x}{r}, \quad \tan\theta = \frac{y}{x}, \\ \csc\theta &= \frac{r}{y}, \quad \sec\theta = \frac{r}{x}, \quad \cot\theta = \frac{x}{y}. \end{aligned} \tag{1}$$

Observe that $\tan\theta$ and $\sec\theta$ are not defined for values of θ such that $x = 0$. In radian measure, this means that $\pi/2, 3\pi/2, \ldots, -\pi/2, -3\pi/2, \ldots$ are excluded from the domains of the tangent and the secant functions. Similarly $\cot\theta$ and $\csc\theta$ are not defined for values of θ corresponding to $y = 0$: that is, for $\theta = 0, \pi, 2\pi, \ldots, -\pi, -2\pi, \ldots$ For those values of θ where the functions are defined, it follows from Eqs. (1) that

$$\csc\theta = \frac{1}{\sin\theta}, \quad \sec\theta = \frac{1}{\cos\theta}, \quad \cot\theta = \frac{1}{\tan\theta}.$$

Since, by the theorem of Pythagoras, we have

$$x^2 + y^2 = r^2,$$

it follows that

$$\cos^2\theta + \sin^2\theta = 1. \tag{2}$$

It is also useful to express the coordinates of P in terms of r and θ as follows:

$$\begin{aligned} x &= r\cos\theta, \\ y &= r\sin\theta. \end{aligned} \tag{3}$$

When $\theta = 0$ in Fig. 5.3, we have $y = 0$ and $x = r$; hence, from the definitions (1), we obtain

$$\sin 0 = 0, \quad \cos 0 = 1.$$

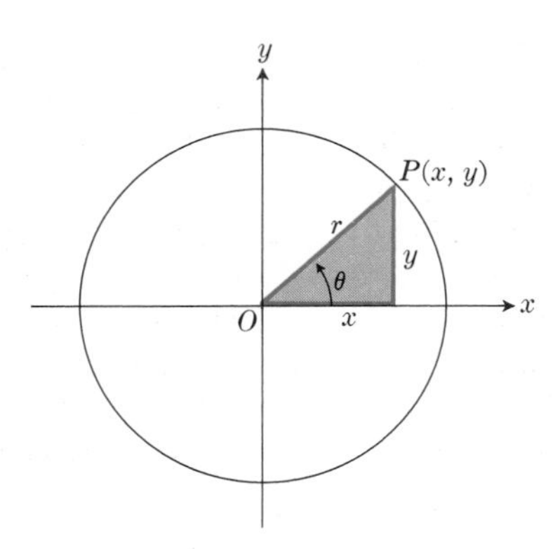

5.3 Angle θ in standard position.

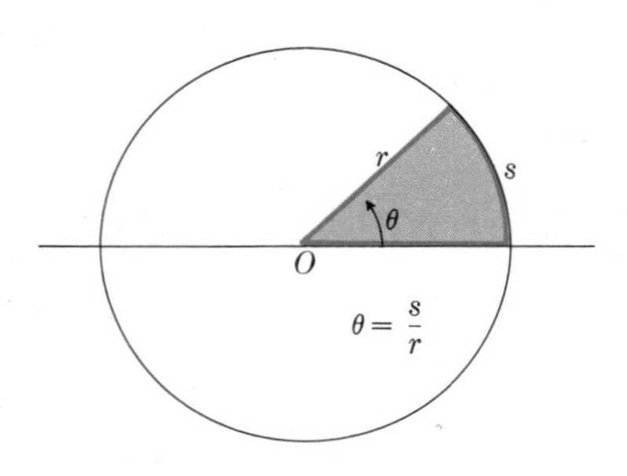

5.4 In radian measure, $\theta = s/r$.

Similarly, for a right angle, $\theta = \pi/2$, we have $x = 0$, $y = r$; hence

$$\sin\frac{\pi}{2} = 1, \qquad \cos\frac{\pi}{2} = 0.$$

Radian measure

In all work with the trigonometric functions in the calculus, it is desirable to measure the angle in *radians*. The number of radians θ in the angle in Fig. 5.4 is defined as the number of "radius units" contained in the arc s subtended by the central angle θ; that is

$$\theta \text{ (in radians)} = \frac{s}{r}. \tag{4a}$$

This also implies that

$$s = r\theta \qquad (\theta \text{ in radians}). \tag{4b}$$

Another useful interpretation of radian measure is easy to get if we take $r = 1$ in (4b). Then the central angle θ, in radians, is just equal to the arc s subtended by θ. We may imagine the circumference of the circle marked off with a scale from which we may read θ. We think of a number scale, like the y-axis shifted one unit to the right, as having been wrapped around the circle. The unit on this number scale is the same as the unit radius. We put the zero of the scale at the place where the initial ray crosses the circle, and then we wrap the positive end of the scale around the circle in the counterclockwise direction, and wrap the negative end around in the opposite direction (see Fig. 5.5). Then θ can be read from this curved s-"axis."

Two points on the s-axis that are exactly 2π units apart will map onto the same point on the unit circle when the wrapping is carried out. For example, if $P_1(x_1, y_1)$ is the point to which an arc of length s_1 reaches, then arcs of length $s_1 + 2\pi$, $s_1 + 4\pi$, and so on, will reach exactly the same point after going completely around the circle one, or two, or more, times. Similarly P_1 will be the image of points on the negative s-axis at $s_1 - 2\pi$, $s_1 - 4\pi$, and so on. Thus, from the wrapped s-axis, we could read

$$\theta_1 = s_1,$$

or

$$\theta_1 + 2\pi,\ \theta_1 + 4\pi,\ \ldots,\ \theta_1 - 2\pi,\ \theta_1 - 4\pi,\ \ldots$$

A unit of arc length $s = 1$ radius subtends a central angle of 57°18′ (approximately); so

$$1 \text{ radian} \approx 57°18'. \tag{5}$$

We find this, and other relations between degree measure and radian measure, by using the fact that the full circumference has arc length $s = 2\pi r$ and central angle 360°. Therefore

$$360° = 2\pi \text{ radians}, \tag{6a}$$

$$180° = \pi = 3.14159\ldots \text{ radians}, \tag{6b}$$

$$\left(\frac{360}{2\pi}\right)^{\circ} = 1 \text{ radian} \approx 57°17'44.8'', \tag{6c}$$

$$1° = \frac{2\pi}{360} = \frac{\pi}{180} \approx 0.01745 \text{ radian}. \tag{6d}$$

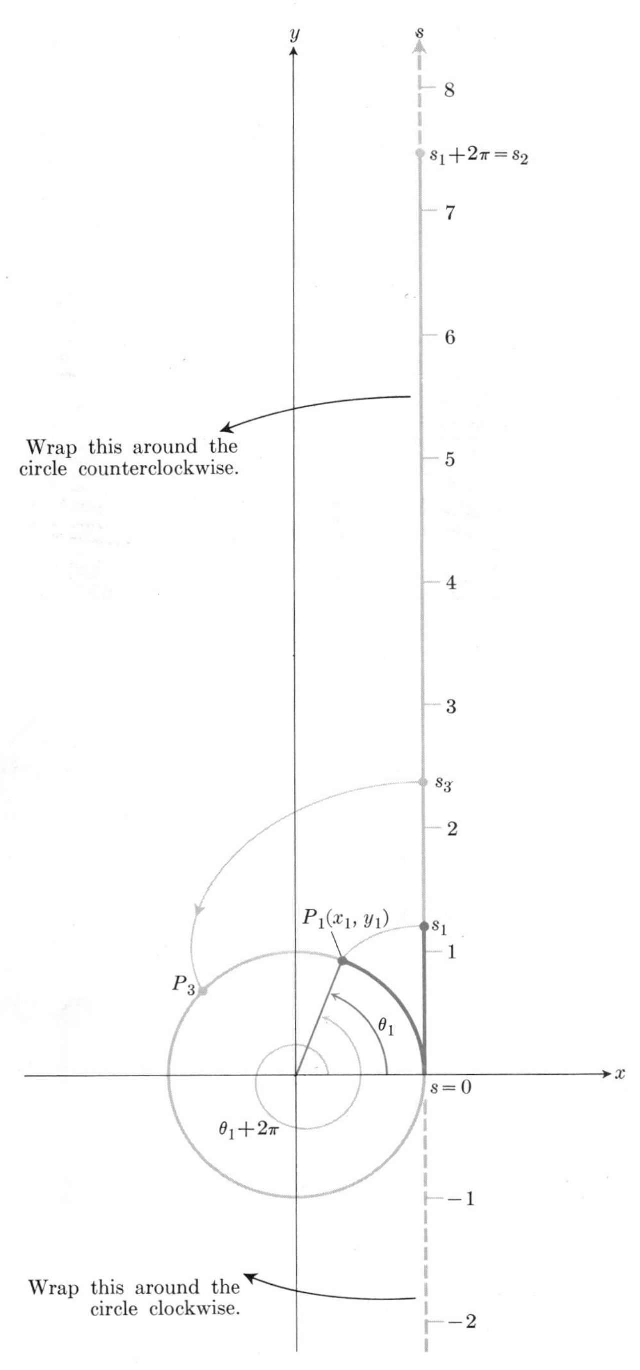

5.5 The curved s-"axis" wrapped around the unit circle.

It should be emphasized, however, that the radian measure of an angle is dimensionless, since r and s in Eqs. (4a, b) both represent lengths measured in identical units, for instance feet, inches, centimeters, or light-years. Thus $\theta = 2.7$ is to be interpreted as a pure number. The sine and cosine of 2.7 are the ordinate and abscissa, respectively, of the point $P(x, y)$ on a circle of radius r at the end of an arc of length 2.7 radii. For practical purposes we would convert 2.7 radians to 2.7 $(360/2\pi)$ degrees and say

$$\sin 2.7 = \sin\left[2.7\left(\frac{360}{2\pi}\right)^\circ\right] \approx \sin\,[154°41'55'']$$
$$\approx 0.42738.$$

We include here a short table of the angles most frequently used, their radian measures, and their sines and cosines.

Degrees	0°	30°	45°	60°	90°	180°	270°	360°
Radians	0	$\frac{\pi}{6}$	$\frac{\pi}{4}$	$\frac{\pi}{3}$	$\frac{\pi}{2}$	π	$\frac{3\pi}{2}$	2π
Sine	0	$\frac{1}{2}$	$\frac{\sqrt{2}}{2}$	$\frac{\sqrt{3}}{2}$	1	0	−1	0
Cosine	1	$\frac{\sqrt{3}}{2}$	$\frac{\sqrt{2}}{2}$	$\frac{1}{2}$	0	−1	0	1

Periodicity

The mapping from the real numbers s onto points $P(x, y)$ on the unit circle by the wrapping process described above and illustrated in Fig. 5.5 defines the coordinates as functions of s because Eqs. (1) apply, with $\theta = s$ and $r = 1$:

$$x = \cos\theta = \cos s, \qquad y = \sin\theta = \sin s.$$

Because $s + 2\pi$ maps onto the same point that s does, it follows that

$$\cos(\theta + 2\pi) = \cos\theta,$$
$$\sin(\theta + 2\pi) = \sin\theta. \tag{7}$$

Equations (7) are *identities;* that is, they are true for all real numbers θ. These identities would be

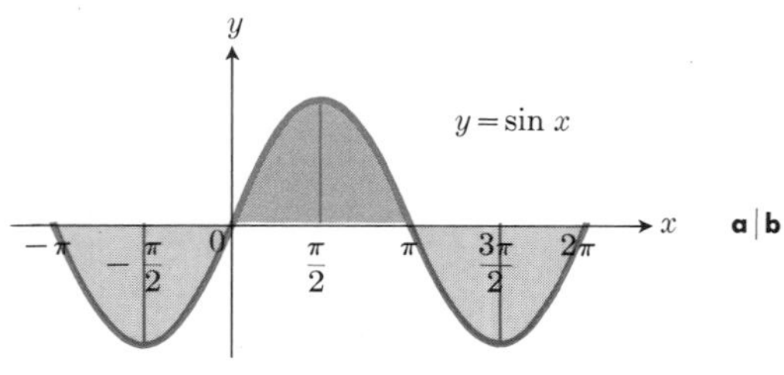

a | b

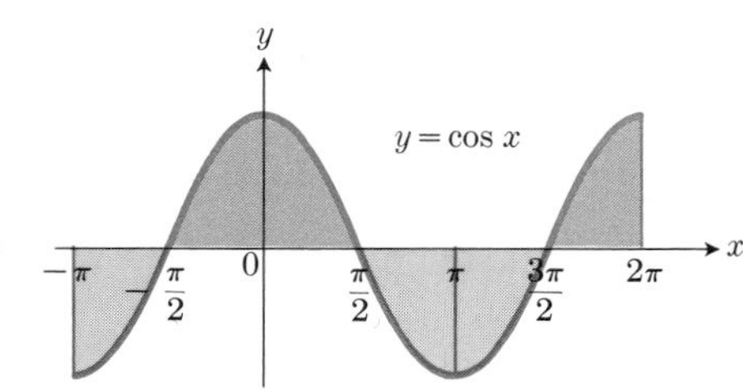

5.6

true for $\theta' = \theta + 2\pi$:

$$\cos\theta' = \cos(\theta' - 2\pi) \quad \text{and} \quad \sin\theta' = \sin(\theta' - 2\pi). \tag{8}$$

Equations (7) and (8) say that 2π can be added to or subtracted from the domain variable of the sine or cosine functions with no change in the function values. The same process could be repeated any number of times. Consequently

$$\begin{aligned} \cos(\theta + 2n\pi) &= \cos\theta, \\ \sin(\theta + 2n\pi) &= \sin\theta, \quad n = 0, \pm 1, \pm 2, \ldots \end{aligned} \tag{9}$$

Figure 5.6(a, b) shows graphs of the curves

$$y = \sin x \quad \text{and} \quad y = \cos x.$$

The portion of either curve between 0 and 2π is repeated endlessly to the left and to the right. We also note that the cosine curve is the same as the sine curve shifted to the left an amount $\pi/2$.

More trigonometric identities

Figure 5.7 shows two angles of opposite sign but of equal magnitude. By symmetry, the points $P(x, y)$ and $P'(x, -y)$, where the rays of the two angles θ and $-\theta$ intersect the circle, have equal abscissas and ordinates that differ only in sign. Hence we have

$$\sin(-\theta) = -\frac{y}{r} = -\sin\theta, \tag{10a}$$

$$\cos(-\theta) = \frac{x}{r} = \cos\theta. \tag{10b}$$

In particular,

$$\sin\left(-\frac{\pi}{2}\right) = -\sin\frac{\pi}{2} = -1,$$

$$\cos\left(-\frac{\pi}{2}\right) = \cos\frac{\pi}{2} = 0.$$

Both $\sin\theta$ and $\cos\theta$ are continuous functions of θ, since a small change in θ produces corresponding small changes in both x and y.

It will be helpful, for reasons which will soon be made apparent, to review the formulas

$$\sin(A + B) = \sin A\cos B + \cos A\sin B, \tag{11a}$$

$$\cos(A + B) = \cos A\cos B - \sin A\sin B, \tag{11b}$$

together with two formulas obtained from these by replacing B by $-B$ and recalling that

$$\sin(-B) = -\sin B, \quad \cos(-B) = \cos B, \tag{11c}$$

namely,

$$\sin(A - B) = \sin A\cos B - \cos A\sin B, \tag{11d}$$

$$\cos(A - B) = \cos A\cos B + \sin A\sin B. \tag{11e}$$

Equation (11e) may be established for all angles A and B by two applications of the formula for the

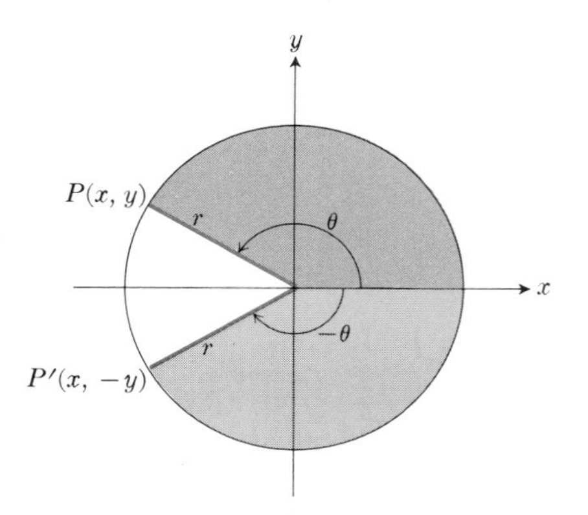

5.7 Angles of opposite sign.

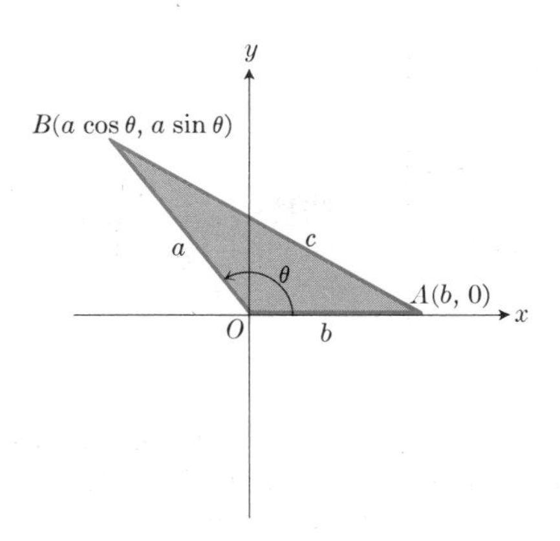

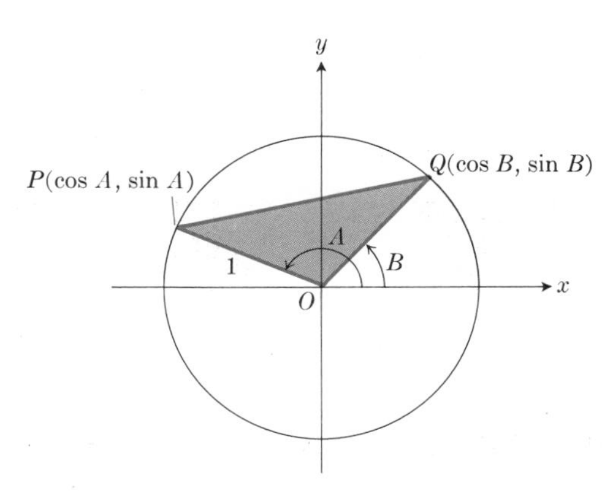

5.8 a| Law of cosines. b| Diagram for cos $(A - B)$.

distance between two points:

$$d = \sqrt{(x_2 - x_1)^2 + (y_2 - y_1)^2}.$$

The first application gives the law of cosines. The second then yields the identity (11e). The other formulas (11a, b, d) can be derived from (11e) as shown below.

Law of cosines. In Fig. 5.8(a), triangle OAB has been placed with one vertex O at the origin, and a second vertex A on the x-axis at $A(b, 0)$. The third vertex B has coordinates

$$x = a \cos \theta, \qquad y = a \sin \theta,$$

as given by Eq. (3) with $r = a$. The angle AOB has measure θ. By the formula for the distance between two points, the square of the distance c from A to B is

$$\begin{aligned} c^2 &= (a \cos \theta - b)^2 + (a \sin \theta)^2 \\ &= a^2 (\cos^2 \theta + \sin^2 \theta) + b^2 - 2ab \cos \theta, \end{aligned}$$

or

$$c^2 = a^2 + b^2 - 2ab \cos \theta. \qquad (12)$$

Equation (12) is called the *law of cosines.* In words, it says: "The square of any side of a triangle is equal to the sum of the squares of the other two sides minus twice the product of those two sides and the cosine of the angle between them." When the angle θ is a right angle, its cosine is zero, and Eq. (12) reduces to the theorem of Pythagoras. Equation (12) holds for a general angle θ, since it is based solely on the distance formula and on Eqs. (3) for the coordinates of a point. The same equation works with the exterior angle $(2\pi - \theta)$, or the opposite of $(2\pi - \theta)$, in place of θ, because

$$\cos (2\pi - \theta) = \cos (\theta - 2\pi) = \cos \theta.$$

It is still a valid formula when B is on the x-axis and $\theta = \pi$ or $\theta = 0$, as we can easily verify if we remember that $\cos 0 = 1$ and $\cos \pi = -1$. In these special cases, the right side of Eq. (12) becomes $(a - b)^2$ or $(a + b)^2$.

Addition formulas. Equation (11e) follows from the law of cosines applied to the triangle OPQ in Fig. 5.8(b). We take $OP = OQ = r = 1$. Then the coordinates of P are

$$x_P = \cos A, \qquad y_P = \sin A$$

and of Q,

$$x_Q = \cos B, \qquad y_Q = \sin B.$$

Hence the square of the distance between P and Q is

$$\begin{aligned} (PQ)^2 &= (x_Q - x_P)^2 + (y_Q - y_P)^2 \\ &= (x_Q^2 + y_Q^2) + (x_P^2 + y_P^2) - 2(x_Q x_P + y_Q y_P) \\ &= 2 - 2 (\cos A \cos B + \sin A \sin B). \end{aligned}$$

Next, imagine that a new set of axes is introduced in such a way that the x'-axis contains the point Q. In other words, rotate the original axes through an angle B to obtain the x'- and y'-axes. Relative to these axes, P has new coordinates

$$x' = \cos(A - B),$$
$$y' = \sin(A - B),$$

and Q has coordinates $(1, 0)$. The distance formula applies just as well in this coordinate system as in the original and gives the result

$$\begin{aligned}(PQ)^2 &= (OP)^2 + (OQ)^2 - 2(OP)(OQ)\cos(A - B)\\ &= 2 - 2\cos(A - B).\end{aligned}$$

When we equate these two expressions for $(PQ)^2$, we obtain

$$\cos(A - B) = \cos A \cos B + \sin A \sin B. \quad (11e)$$

We now deduce Eqs. (11a, b, d) from Eq. (11e). We shall also need the results

$$\begin{aligned}&\sin 0 = 0, \quad \sin(\pi/2) = 1, \quad \sin(-\pi/2) = -1,\\ &\cos 0 = 1, \quad \cos(\pi/2) = 0, \quad \cos(-\pi/2) = 0.\end{aligned} \quad (13)$$

1. In Eq. (11e), we put $A = \pi/2$ and use Eqs. (13) to get

$$\cos\left(\frac{\pi}{2} - B\right) = \sin B. \quad (14a)$$

In this equation, if we replace B by $\pi/2 - B$ and $\pi/2 - B$ by $\pi/2 - (\pi/2 - B)$, we get

$$\cos B = \sin\left(\frac{\pi}{2} - B\right). \quad (14b)$$

Equations (14a, b) express the familiar results that the sine and cosine of an angle are the cosine and sine, respectively, of the complementary angle.

2. We next put $B = -\pi/2$ in Eq. (11e) and use (13) to get

$$\cos\left(A + \frac{\pi}{2}\right) = -\sin A. \quad (14c)$$

3. We can get the formula for $\cos(A + B)$ from Eq. (11e) by substituting $-B$ for B everywhere:

$$\begin{aligned}\cos(A + B) &= \cos[A - (-B)]\\ &= \cos A \cos(-B) + \sin A \sin(-B)\\ &= \cos A \cos B - \sin A \sin B, \quad (11b)\end{aligned}$$

where the final equality uses Eqs. (11c), established earlier.

4. To derive formulas for $\sin(A \pm B)$, we use the identity (14a) with B replaced by

$$A + B \quad \text{or} \quad A - B.$$

Thus we have

$$\begin{aligned}\sin(A + B) &= \cos[\pi/2 - (A + B)]\\ &= \cos(\pi/2 - A - B)\\ &= \cos(\pi/2 - A)\cos B\\ &\qquad + \sin(\pi/2 - A)\sin B\\ &= \sin A \cos B + \cos A \sin B. \quad (11a)\end{aligned}$$

Equation (11d) follows from this if we replace B by $-B$.

These are the key results of analytic trigonometry, and all have been derived simply from the distance formula and the definitions of sine and cosine. The most important formulas to remember are the following ones:

$$\sin(A + B) = \sin A \cos B + \cos A \sin B, \quad (11a)$$
$$\cos(A + B) = \cos A \cos B - \sin A \sin B, \quad (11b)$$
$$\sin(-B) = -\sin B, \quad \cos(-B) = \cos B. \quad (11c)$$

If we let

$$\alpha = A + B \quad \text{and} \quad \beta = A - B,$$

so that

$$\begin{aligned}A &= \tfrac{1}{2}(\alpha + \beta),\\ B &= \tfrac{1}{2}(\alpha - \beta),\end{aligned}$$

and subtract Eq. (11d) from Eq. (11a), we obtain the further useful identity

$$\sin\alpha - \sin\beta = 2\cos\frac{\alpha + \beta}{2}\sin\frac{\alpha - \beta}{2}. \quad (15)$$

EXERCISES

In Exercises 1 through 6, sketch the graph of the given equation.

1. $y = 2 \sin x$
2. $y = 3 \sin 2x$
3. $y = A \sin (Bx + C)$, A, B, C constants $\neq 0$
4. $y = 2 \cos 3x$
5. $y = 3 \cos (2x + \pi/4)$
6. $y = 2 \tan x/3$
7. Show that the area of a sector of a circle having central angle θ and radius r is $\frac{1}{2}r^2\theta$, if θ is measured in radians.
8. Let $A(r, 0)$ be the point where the positive x-axis cuts a circle of radius r, center at the origin O. Let $P(r \cos \theta, r \sin \theta)$ be a point on the circle in the first quadrant, with angle $AOP = \theta$ radians. Let AT be tangent to the circle at A and suppose it intersects the line OP at T. By considering the areas of triangle AOP, sector AOP, and triangle AOT, prove the following inequality:
$$\sin \theta < \theta < \tan \theta \quad \text{if} \quad 0 < \theta < \pi/2.$$
9. In Eq. (11e) take $B = A$. Does the result agree with something else you know?
10. In Eq. (11d) take $B = A$. Does the result agree with something you already know?
11. Derive a formula for $\tan (A - B)$ from Eqs. (11d, e).
12. Derive a formula for $\tan (A + B)$ from Eqs. (11a, b).
13. Express all of the trigonometric functions of a general angle θ in terms of $\sin \theta$ and $\cos \theta$.
14. A function $f(\theta)$ is said to be
$$\text{an even function of } \theta \text{ if } f(-\theta) = f(\theta),$$
$$\text{an odd function of } \theta \text{ if } f(-\theta) = -f(\theta).$$
Which of the six trigonometric functions are even, and which are odd?
15. Deduce formulas for $\cos 2A$ and $\sin 2A$ from Eqs. (11a, b).
16. Let P and Q be points on a circle with radius $r = 1$, center at the origin O, and such that OP makes an angle $-B$ with the positive x-axis, OQ an angle A. Use the law of cosines to derive a formula for $\cos (A + B)$ directly from this configuration.

5.5 DIFFERENTIATION AND INTEGRATION OF SINES AND COSINES

We shall now apply the operations of the calculus to the sine and cosine functions. We need the results given by the following theorem.

Theorem. *Let θ be measured in radians. Then*

$$\lim_{\theta \to 0} \sin \theta = 0, \tag{1a}$$
$$\lim_{\theta \to 0} \cos \theta = 1, \tag{1b}$$
$$\lim_{\theta \to 0} \frac{\sin \theta}{\theta} = 1. \tag{1c}$$

These results were established in Article 2.3, Eqs. (9a, b, c).

Derivative of sin u

We now consider a function defined by

$$y = \sin u,$$

and calculate the derivative from the definition

$$\frac{dy}{du} = \lim_{\Delta u \to 0} \frac{\Delta y}{\Delta u}.$$

Let u be given an increment Δu and y a corresponding increment Δy. Then

$$y + \Delta y = \sin (u + \Delta u),$$

and hence

$$\Delta y = \sin (u + \Delta u) - \sin u \tag{a}$$
$$= 2 \cos \left(u + \frac{\Delta u}{2}\right) \sin \frac{\Delta u}{2}, \tag{b}$$

where we have made use of Eq. (15) of Article 5.4, with $\alpha = u + \Delta u$ and $\beta = u$, in going from (a) to (b). If we divide (b) by Δu, we have

$$\frac{\Delta y}{\Delta u} = 2 \cos \left(u + \frac{\Delta u}{2}\right) \frac{\sin (\Delta u/2)}{\Delta u}$$
$$= \cos (u + \theta) \frac{\sin \theta}{\theta}$$
$$= [\cos u \cos \theta - \sin u \sin \theta] \frac{\sin \theta}{\theta},$$

where $\theta = \Delta u/2$. We now let Δu and θ approach zero and make use of Eqs. (1a, b, c):

$$\lim_{\Delta u \to 0} \frac{\Delta y}{\Delta u} = [\cos u \cdot 1 - \sin u \cdot 0] \cdot 1 = \cos u.$$

But since $y = \sin u$, this means that

$$\frac{dy}{du} = \frac{d(\sin u)}{du} = \cos u.$$

If u is a differentiable function of x, we may apply the chain rule

$$\frac{dy}{dx} = \frac{dy}{du}\frac{du}{dx}$$

to this, with the result that we obtain

$$\frac{d(\sin u)}{dx} = \cos u \cdot \frac{du}{dx}. \qquad \text{(VII)}$$

Thus to find the derivative of the sine of a function, we take the cosine of the same function and multiply it by the derivative of the function.

Example 1. If

$$y = \sin 2x,$$

then

$$\frac{dy}{dx} = \cos 2x \cdot \frac{d(2x)}{dx} = 2\cos 2x.$$

Derivative of cos *u*

To obtain a formula for the derivative of $\cos u$, we make use of the identities

$$\cos u = \sin\left(\frac{\pi}{2} - u\right), \qquad \sin u = \cos\left(\frac{\pi}{2} - u\right).$$

Thus

$$\frac{d(\cos u)}{dx} = \frac{d\sin\left(\frac{\pi}{2} - u\right)}{dx}$$

$$= \cos\left(\frac{\pi}{2} - u\right)\frac{d\left(\frac{\pi}{2} - u\right)}{dx}$$

$$= \sin u \cdot -\frac{du}{dx},$$

or

$$\frac{d(\cos u)}{dx} = -\sin u\,\frac{du}{dx}. \qquad \text{(VIII)}$$

This equation tells us that the derivative of the cosine of a function is minus the sine of the same function, times the derivative of the function.

Example 2. Let

$$y = \cos(x^2).$$

Then

$$\frac{dy}{dx} = -\sin(x^2)\frac{d(x^2)}{dx} = -2x\sin(x^2).$$

The formulas (VII) and (VIII) may be combined with the formulas given previously, as in the following problems.

Problem 1. Find dy/dx if

$$y = \sin^2(3x) = u^2, \qquad u = \sin 3x.$$

Solution.

$$\frac{dy}{dx} = 2u\frac{du}{dx} = 2\sin 3x\,\frac{d(\sin 3x)}{dx}$$

$$= 2\sin 3x \cdot \cos 3x\,\frac{d(3x)}{dx}$$

$$= 6\sin 3x\cos 3x.$$

Problem 2. Find dy/dx if

$$y = \sec^2 5x = (\cos 5x)^{-2}.$$

Solution. First we must apply the formula for the derivative of a function to a power,

$$\frac{d(u^n)}{dx} = nu^{n-1}\frac{du}{dx}.$$

We then would have

$$\frac{dy}{dx} = -2(\cos 5x)^{-3}\frac{d(\cos 5x)}{dx}$$

$$= (-2\sec^3 5x)\left(-\sin 5x\,\frac{d(5x)}{dx}\right)$$

$$= +10\sec^3 5x\sin 5x.$$

Problem 3. Show that the function

$$f(x) = \tan x$$

is an increasing function of x at any point where

$$\cos x \neq 0.$$

Solution. (See Fig. 5.9.) To show this, we let

$$y = f(x) = \tan x, \quad \text{or} \quad y = \frac{\sin x}{\cos x}.$$

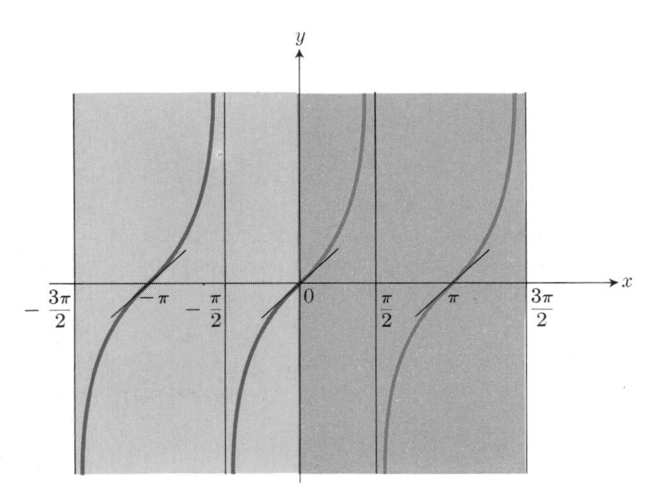

5.9. $y = \tan x$.

Then
$$\frac{dy}{dx} = \frac{\cos x \dfrac{d(\sin x)}{dx} - \sin x \dfrac{d(\cos x)}{dx}}{\cos^2 x}$$

$$= \frac{\cos x \cdot \cos x \dfrac{dx}{dx} - \sin x \left(-\sin x \dfrac{dx}{dx}\right)}{\cos^2 x}$$

$$= \frac{\cos^2 x + \sin^2 x}{\cos^2 x} = \frac{1}{\cos^2 x} = f'(x)$$

exists and is positive at any point where $\cos x \neq 0$. Since $f'(x)$ is positive, $f(x)$ is an increasing function of x. In fact, since $\cos^2 x \leq 1$, we observe that $dy/dx \geq 1$ at every point on the curve $y = \tan x$. (When $\cos x = 0$, $\tan x$ does not exist.)

Problem 4. Find the velocity and acceleration of a particle moving in a circle of radius r with constant angular velocity ω, $\omega > 0$.

Solution. (See Fig. 5.10.) If the position of the particle at time t is $P(x, y)$, then

$$x = r \cos \theta, \qquad y = r \sin \theta, \tag{2a}$$

and it is given that the angular velocity

$$d\theta/dt = \omega \tag{2b}$$

is constant. The velocity is a vector with components

$$v_x = \frac{dx}{dt} = \frac{dx}{d\theta}\frac{d\theta}{dt}, \qquad v_y = \frac{dy}{dt} = \frac{dy}{d\theta}\frac{d\theta}{dt} \tag{2c}$$

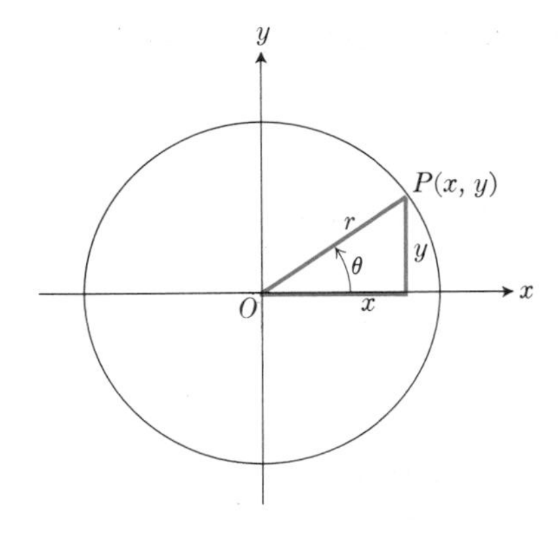

a | b

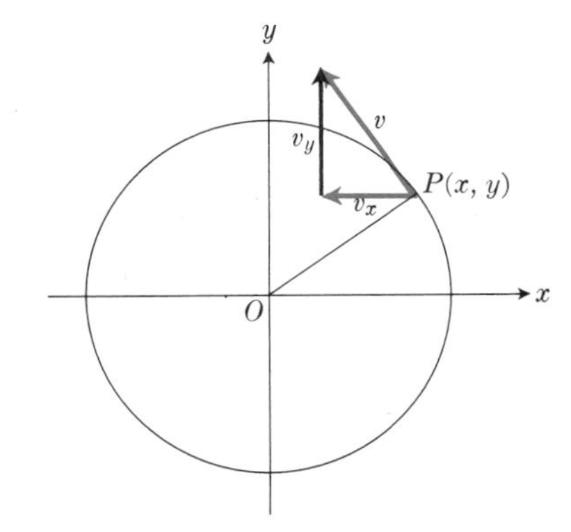

| c |

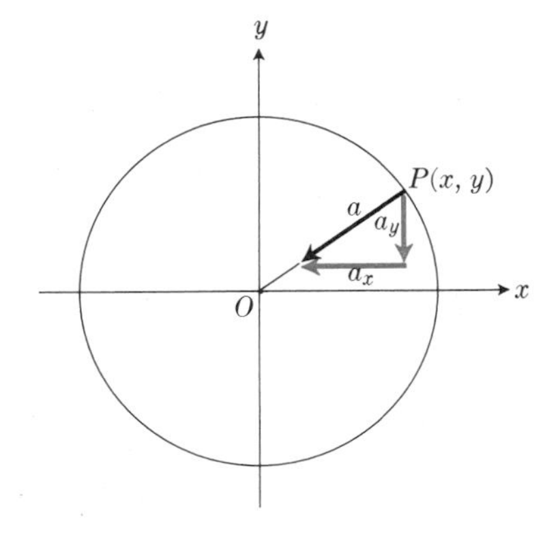

5.10

parallel to the x- and y-axes respectively. Also, the acceleration is a vector with components

$$a_x = \frac{dv_x}{dt}, \qquad a_y = \frac{dv_y}{dt} \tag{2d}$$

parallel to the coordinate axes. From (2a) and (2b), we find

$$\begin{aligned} v_x &= \frac{dx}{dt} = -r\sin\theta\,\frac{d\theta}{dt} = -\omega r\sin\theta = -\omega y, \\ v_y &= \frac{dy}{dt} = r\cos\theta\,\frac{d\theta}{dt} = \omega r\cos\theta = \omega x, \end{aligned} \tag{2e}$$

and from these we find in turn

$$\begin{aligned} a_x &= \frac{dv_x}{dt} = -\omega\frac{dy}{dt} = -\omega^2 x, \\ a_y &= \frac{dv_y}{dt} = \omega\frac{dx}{dt} = -\omega^2 y. \end{aligned} \tag{2f}$$

The velocity vector has both magnitude and direction. Its magnitude is

$$\sqrt{v_x^2 + v_y^2} = \sqrt{\omega^2 y^2 + \omega^2 x^2} = \omega\sqrt{y^2 + x^2} = \omega r,$$

which is the angular velocity ω times the radius of the circle. Its direction is specified, except for sense, by its slope, which is given by

$$\frac{v_y}{v_x} = \frac{dy/dt}{dx/dt} = \frac{dy}{dx},$$

which is the same as the slope of the tangent to the curve at P. In Fig. 5.10(b), the sense is seen graphically to be in the counterclockwise direction, since from (2e) the x-component has a sign opposite to the sign of y if, as we assume, ω is positive.

Similarly, the acceleration vector has magnitude

$$\sqrt{a_x^2 + a_y^2} = \omega^2\sqrt{x^2 + y^2} = \omega^2 r,$$

and its direction will be opposite to that of the vector from O to P, since the latter has components x and y, while (2f) shows that the acceleration has its corresponding components just $(-\omega^2)$ times these; that is,

$$\text{the acceleration vector } \mathbf{a} = -\omega^2\overrightarrow{OP}.$$

This shows that the acceleration is toward the center of the circle at each instant and has a constant magnitude $\omega^2 r$.

To summarize: If a particle moves in a circle of radius r with constant angular velocity ω, then its velocity vector is tangent to the circle and has magnitude ωr, while its acceleration vector points toward the center of the circle and has magnitude $\omega^2 r$. Hence (by Newton's second law) the force needed to keep a particle of mass m moving at constant speed ωr in a circle of radius r is $m\omega^2 r$, directed toward the center of the circle.

Integration of sines and cosines

Corresponding to the derivative formulas (VII) and (VIII), we also have the differential formulas

$$d(\sin u) = \cos u\,du, \tag{VII$'$}$$

$$d(\cos u) = -\sin u\,du, \tag{VIII$'$}$$

and the integration formulas

$$\begin{aligned} \int \cos u\,du &= \sin u + C, \\ \int \sin u\,du &= -\cos u + C. \end{aligned} \tag{3}$$

Example.
$$\begin{aligned} \int \cos 2t\,dt &= \tfrac{1}{2}\int \cos 2t \cdot 2\,dt \\ &= \tfrac{1}{2}\int \cos u\,du \quad (u = 2t) \\ &= \tfrac{1}{2}\sin 2t + C. \end{aligned}$$

Problem 5. Evaluate the integral

$$\int \frac{\cos 2x}{\sin^3 2x}\,dx.$$

Solution. Since

$$d(\sin 2x) = 2\cos 2x\,dx,$$

we recognize the numerator as being

$$\tfrac{1}{2}d(\sin 2x).$$

Hence we have

$$\begin{aligned} \int \frac{\cos 2x\,dx}{\sin^3 2x} &= \int (\sin 2x)^{-3}\cdot\tfrac{1}{2}d(\sin 2x) \\ &= \frac{1}{2}\int u^{-3}\,du \quad (u = \sin 2x) \\ &= \frac{1}{2}\,\frac{u^{-2}}{-2} + C \\ &= \frac{-1}{4\sin^2 2x} + C. \end{aligned}$$

EXERCISES

Evaluate limits 1 through 20 by making use of Eq. (1) together with appropriate trigonometric identities and theorems on limits.

1. $\lim_{\theta \to 0} \dfrac{\tan \theta}{\theta}$

2. $\lim_{\theta \to \pi} \dfrac{\sin \theta}{\pi - \theta}$
[*Hint.* Let $x = \pi - \theta$.]

3. $\lim_{\theta \to 0} \dfrac{\sin 2\theta}{\theta}$

4. $\lim_{x \to 0} \dfrac{\sin x}{3x}$

5. $\lim_{x \to 0} \dfrac{\sin 5x}{\sin 3x}$

6. $\lim_{x \to 0} \tan 2x \csc 4x$

7. $\lim_{\theta \to 0} \dfrac{\sin^2 \theta}{\theta}$

8. $\lim_{\theta \to 0} \dfrac{1 - \cos \theta}{\theta}$

$\left[\textit{Hint.} \text{ If } |\theta| < \pi, \text{ then } 1 - \cos \theta = \dfrac{\sin^2 \theta}{1 + \cos \theta} \cdot \right]$

9. $\lim_{\theta \to 0} \dfrac{1 - \cos \theta}{\theta^2}$

10. $\lim_{y \to 0} \dfrac{\tan 2y}{3y}$

11. $\lim_{u \to 0} \dfrac{3u}{\sin 5u}$

12. $\lim_{x \to \infty} x \sin \dfrac{1}{x}$
[*Hint.* Let $1/x = u$.]

13. $\lim_{y \to \infty} 2y \tan \dfrac{\pi}{y}$

14. $\lim_{x \to \pi/2} \dfrac{2x - \pi}{\cos x}$

15. $\lim_{\theta \to 0} \theta \cot 2\theta$

16. $\lim_{x \to 0} \dfrac{x^2 + 2x}{\sin 2x}$

17. $\lim_{x \to 0} \dfrac{\sin 2x}{2x^2 + x}$

18. $\lim_{h \to 0} \dfrac{\sin (a + h) - \sin a}{h}$

19. $\lim_{h \to 0} \dfrac{\cos (a + h) - \cos a}{h}$

20. $\lim_{h \to 2} \dfrac{\cos (\pi/h)}{h - 2}$

In Exercises 21 through 36, find dy/dx.

21. $y = \sin (3x + 4)$

22. $y = x \sin x$

23. $y = \dfrac{\sin x}{x}$

24. $y = \cos 5x$

25. $y = x^2 \sin 3x$

26. $y = \sqrt{2 + \cos 2x}$

27. $y = \sin^2 x + \cos^2 x$

28. $y = \dfrac{2}{\cos 3x}$

29. $y = 3 \sin 2x - 4 \cos 2x$

30. $y = 3 \cos^2 2x - 3 \sin^2 2x$

31. $y = 2 \sin x \cos x$

32. $y = \dfrac{1}{\sin x}$

33. $y = \cos^2 3x$

34. $y = \cot x$

35. $x \sin 2y = y \cos 2x$

36. $y^2 = \sin^4 2x + \cos^4 2x$

37. Using the Mean Value Theorem, prove that

$$|\sin b - \sin a| \le |b - a|.$$

38. Show that the curve

$$y = x + \sin x$$

has no relative maxima or minima even though it does have points where dy/dx is zero. Sketch the curve.

39. A particle moves on the curve

$$x = a \cos \omega t, \qquad y = b \sin \omega t,$$

where a, b, and ω are constants. Show that the acceleration components are

$$a_x = -\omega^2 x \quad \text{and} \quad a_y = -\omega^2 y.$$

Evaluate the integrals in Exercises 40 through 60.

40. $\int \sin 3x \, dx$

41. $\int \cos (2x + 4) \, dx$

42. $\int x \sin (2x^2) \, dx$

43. $\int (\cos \sqrt{x}) \dfrac{dx}{\sqrt{x}}$

44. $\int \sin 2t \, dt$

45. $\int \cos (3\theta - 1) \, d\theta$

46. $\int 4 \cos 3y \, dy$

47. $\int 2 \sin z \cos z \, dz$

48. $\int \sin^2 x \cos x \, dx$

49. $\int \cos^2 2y \sin 2y \, dy$

50. $\int (1 - \sin^2 3t) \cos 3t \, dt$

51. $\int \dfrac{\sin x \, dx}{\cos^2 x}$

52. $\int \dfrac{\cos x \, dx}{\sin^2 x}$

52. $\int \sqrt{2 + \sin 3t} \cos 3t \, dt$

54. $\int \dfrac{\sin 2t \, dt}{\sqrt{2 - \cos 2t}}$

55. $\int \sin^3 \dfrac{y}{2} \cos \dfrac{y}{2} \, dy$

56. $\int \dfrac{\sin \dfrac{z - 1}{3} \, dz}{\cos^2 \dfrac{z - 1}{3}}$

57. $\int \cos^2 \dfrac{2x}{3} \sin \dfrac{2x}{3} \, dx$

58. $\int (1 + \sin 2t)^{3/2} \cos 2t \, dt$

59. $\int (3 \sin 2x + 4 \cos 3x) \, dx$

60. $\int \sin t \cos t (\sin t + \cos t) \, dt$

5.6 AREA UNDER A CURVE

In geometry we learned how to find areas of certain polygons: rectangles, triangles, parallelograms, trapezoids. Indeed, the area of any polygon can be found by cutting it into triangles.

The area of a circle is easily computed from the familiar formula $A = \pi r^2$. But the idea behind this simple formula isn't so simple. In fact, it is the subtle concept of a *limit*, the area of the circle being *defined* as the limit of areas of inscribed (or circumscribed) regular polygons as the number of sides increases without bound. A similar idea is involved in the definition we now introduce for other plane areas.

Let $y = f(x)$ define a continuous function of x on the closed interval $[a, b]$. For simplicity, we shall also suppose that $f(x)$ is nonnegative over $[a, b]$. We consider the problem of calculating the area bounded above by the graph of the function, on the sides by vertical lines through $x = a$ and $x = b$, and below by the x-axis (Fig. 5.11).

We divide the area into n thin strips of uniform width $\Delta x = (b - a)/n$ by lines perpendicular to the x-axis through the end points $x = a$ and $x = b$ and many intermediate points, which we number as $x_1, x_2, \ldots, x_{n-1}$ (Fig. 5.12). We use an inscribed rectangle to approximate the area in each strip. For instance, in the figure, we approximate the area of the strip $aP_0P_1x_1$ by the shaded rectangle of altitude aP_0 and base $a \ldots x_1$. The area of this rectangle is

$$f(a) \cdot (x_1 - a) = f(a) \cdot \Delta x,$$

since the length of the altitude aP_0 is the value of f at $x = a$, and the length of the base is $x_1 - a = \Delta x$. Similarly the inscribed rectangle in the second strip

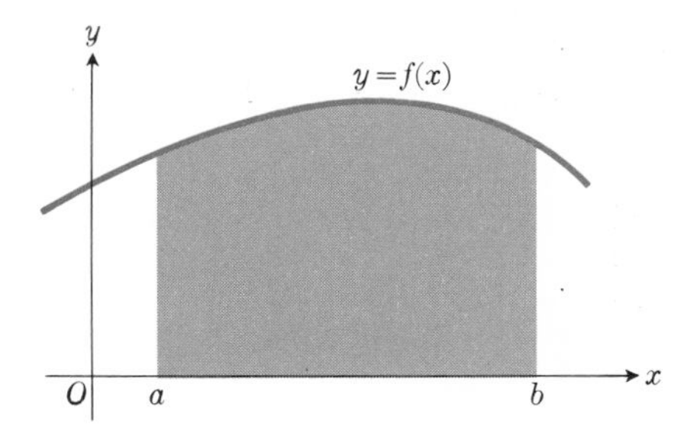

5.11 The area under a curve.

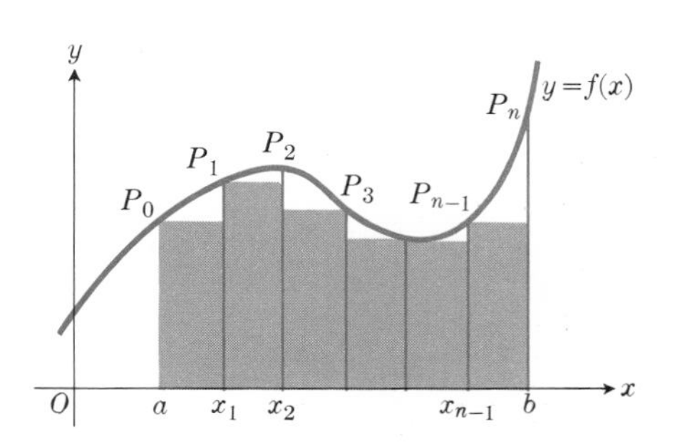

5.12 The area under a curve $y = f(x)$ divided into strips. Each strip is approximated by a rectangle.

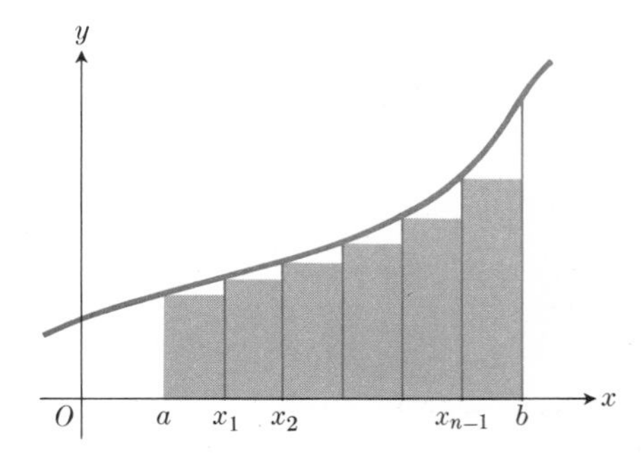

5.13 Rectangles under the graph of an increasing function.

has an area equal to

$$f(x_1) \cdot \Delta x.$$

Continuing in this fashion, we inscribe a rectangle in each strip.

In the special case where the function increases with x, as in Fig. 5.13, we always use the ordinate

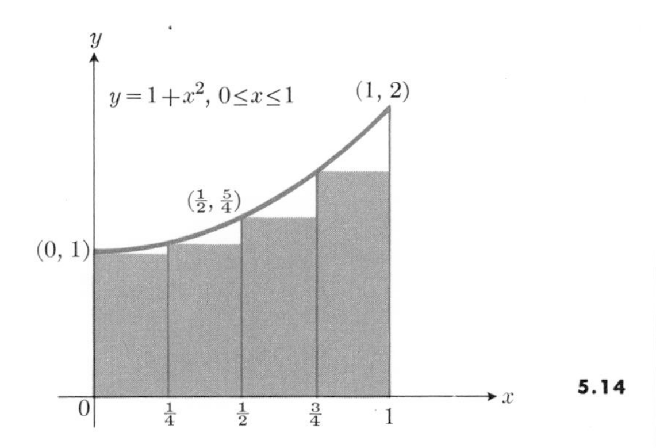

5.14

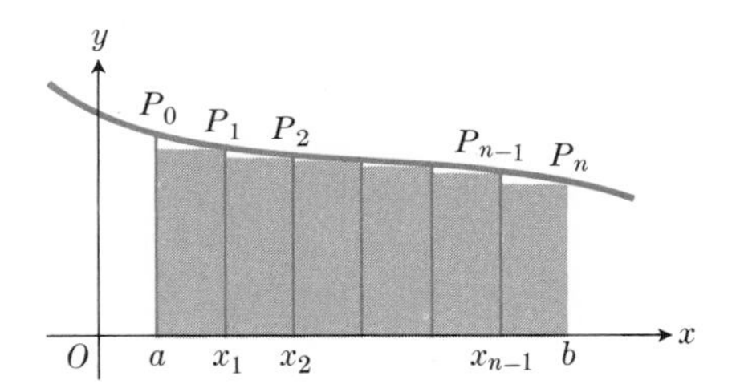

5.15 Rectangles under the graph of a decreasing function.

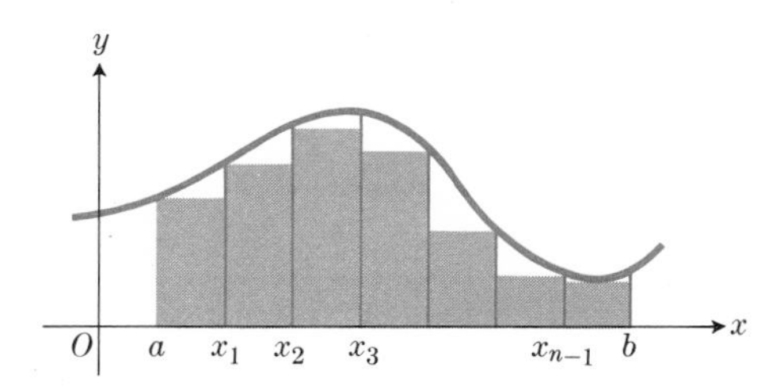

5.16 Rectangles under a curve that rises and falls between a and b.

at the left edge of the strip as the altitude of the corresponding rectangle. Then we have

$$
\begin{aligned}
\text{area of first rectangle} &= f(a) \cdot \Delta x, \\
\text{area of second rectangle} &= f(x_1) \cdot \Delta x, \\
\text{area of third rectangle} &= f(x_2) \cdot \Delta x, \\
&\vdots \\
\text{area of } n\text{th and last rectangle} &= f(x_{n-1}) \cdot \Delta x.
\end{aligned}
$$

Example. (See Fig. 5.14.) Suppose $f(x) = 1 + x^2$; $a = 0$, $b = 1$, and $n = 4$. There are $n - 1 = 3$ intermediate points, $x_1 = \frac{1}{4}$, $x_2 = \frac{1}{2}$, and $x_3 = \frac{3}{4}$, which divide the interval $[0, 1]$ into $n = 4$ subintervals, each of length $\Delta x = \frac{1}{4}$. The inscribed rectangles have areas

$$
\begin{aligned}
f(0) \cdot \Delta x &= 1 \cdot \tfrac{1}{4} = \tfrac{16}{64} \\
f(\tfrac{1}{4}) \cdot \Delta x &= \tfrac{17}{16} \cdot \tfrac{1}{4} = \tfrac{17}{64} \\
f(\tfrac{1}{2}) \cdot \Delta x &= \tfrac{5}{4} \cdot \tfrac{1}{4} = \tfrac{20}{64} \\
f(\tfrac{3}{4}) \cdot \Delta x &= \tfrac{25}{16} \cdot \tfrac{1}{4} = \tfrac{25}{64} \\
\text{Sum} &= \tfrac{78}{64} = 1.21875.
\end{aligned}
$$

Since the area under the curve is larger than the sum of the areas of these inscribed rectangles, we may expect it to be somewhat larger than 1.22. (In fact, by using methods we shall soon develop, we shall find that the area is exactly $\frac{4}{3}$.) Thus our estimate of 1.22 is about 8% too small. There are easy ways of improving the accuracy, for example by using trapezoids in place of rectangles to approximate each strip, but we defer that to a later time.

If the curve slopes downward, as in Fig. 5.15, the inscribed rectangles have areas as follows:

$$
\begin{aligned}
\text{first rectangle} \quad & f(x_1) \cdot \Delta x, \\
\text{second rectangle} \quad & f(x_2) \cdot \Delta x, \\
\text{third rectangle} \quad & f(x_3) \cdot \Delta x, \\
& \vdots \\
n\text{th and last} \quad & f(b) \cdot \Delta x.
\end{aligned}
$$

More generally, the curve may rise and fall between $x = a$ and $x = b$, as in Fig. 5.16. But there is a number c_1 between a and x_1 inclusive such that the first inscribed rectangle has area $f(c_1) \cdot \Delta x$; and a number c_2 in the second closed subinterval such that the area of the second inscribed rectangle is $f(c_2) \cdot \Delta x$; and so on. The number c_1 is the place between a and x_1 inclusive where f is minimized for the first subinterval. Similarly the minimum value of f for x in the second subinterval is attained at c_2, and so on.

The sum of the areas of these inscribed rectangles is

$$S_n = f(c_1) \cdot \Delta x + f(c_2) \cdot \Delta x + \cdots + f(c_n) \cdot \Delta x. \quad (1)$$

We may also write this in more abbreviated form by using the sigma notation:

$$S_n = \sum_{k=1}^{n} f(c_k) \cdot \Delta x. \tag{2}$$

The Greek letter $\sum$ (capital sigma) is used this way in mathematics to denote a sum. Note that each term of the sum in (1) is of the form $f(c_k) \cdot \Delta x$, with only the subscript on c changing from one term to another. We have indicated the subscript by k, but we could equally well have used i or j or any other symbol except a letter which is already in use for something else. In the first term in the sum on the right side of Eq. (1), the subscript is $k = 1$; in the second, $k = 2$; and so on to the last, or nth, in which $k = n$. We indicate this by writing $k = 1$ below the $\sum$ in (2), to say that the sum is to *start* with the term we get by replacing k by 1 in the expression that follows. The n above the sigma tells us where to *stop*. For instance, if $n = 4$, we have

$$\sum_{k=1}^{4} f(c_k) \cdot \Delta x = f(c_1) \cdot \Delta x + f(c_2) \cdot \Delta x + f(c_3) \cdot \Delta x + f(c_4) \cdot \Delta x.$$

The only thing that changes from one summand to the next is the numeral in the place indicated by k. First we replace k by 1, then by 2, then 3, then 4. Then we add.

Here we give a few other sums as examples of the $\sum$-notation:

(a) $\sum_{k=1}^{5} k^2 = 1^2 + 2^2 + 3^2 + 4^2 + 5^2,$

(b) $\sum_{k=1}^{3} \frac{k}{k+1} = \frac{1}{1+1} + \frac{2}{2+1} + \frac{3}{3+1},$

(c) $\sum_{j=0}^{2} \frac{j+1}{j+2} = \frac{0+1}{0+2} + \frac{1+1}{1+2} + \frac{2+1}{2+2},$

(d) $\sum_{i=1}^{4} x_i = x_1 + x_2 + x_3 + x_4,$

(e) $\sum_{k=1}^{4} x^k = x + x^2 + x^3 + x^4.$

We turn our attention once more to the area under a curve. We *define* it to be

the limit of the sums of the areas of inscribed rectangles as their number increases without bound. In symbols,

$$A = \lim_{n\to\infty} [f(c_1)\,\Delta x + f(c_2)\,\Delta x + \cdots + f(c_n)\,\Delta x] = \lim_{n\to\infty} \sum_{k=1}^{n} f(c_k)\,\Delta x. \tag{3}$$

Remark 1. The limit indicated in Eq. (3) exists, provided f is continuous. That is, by taking larger and larger values of n and computing the sum of the areas of inscribed rectangles for each n, we get answers which differ from one another (and from what we would intuitively call the area under the curve) by amounts that become arbitrarily small as n increases.

Remark 2. We could have used circumscribed instead of inscribed rectangles. We would then replace the c_k in Eq. (3) by other numbers, say e_k. These would be the places in the subintervals where the function takes on its *maximum* instead of minimum values. The corresponding sums would tend to overestimate the exact area, but in the limit we would get the same answer whether we used inscribed or circumscribed rectangles. The fact that the two kinds of sums of areas of rectangles give the same limit is a consequence of the *uniform continuity* of the function f over the domain $a \leq x \leq b$. It is a theorem which is usually proved in more advanced courses in mathematical analysis. A substantial part of the proof goes as follows: Let

$$m_k = f(c_k) = \min \{f(x) \text{ for } x \text{ in the } k\text{th subinterval}\},$$

$$M_k = f(e_k) = \max \{f(x) \text{ for } x \text{ in the } k\text{th subinterval}\}.$$

The expression $m_k\,\Delta x$ gives the area of the inscribed rectangle, and the expression $M_k\,\Delta x$ gives the area of the circumscribed rectangle.

The exact area A, then, is a number between the *lower sum*

$$L = m_1\,\Delta x + m_2\,\Delta x + \cdots + m_n\,\Delta x = \sum_{k=1}^{n} m_k\,\Delta x \tag{4a}$$

and the *upper sum*

$$U = M_1\,\Delta x + M_2\,\Delta x + \cdots + M_n\,\Delta x = \sum_{k=1}^{n} M_k\,\Delta x. \tag{4b}$$

By subtraction, we can see that since

$$L \le A \le U, \tag{5}$$

we have

$$|A - L| \le U - L \quad \text{and} \quad |U - A| \le U - L, \tag{6a}$$

where

$$U - L = (M_1 - m_1)\,\Delta x + (M_2 - m_2)\,\Delta x + \cdots + (M_n - m_n)\,\Delta x. \tag{6b}$$

The right-hand side of (6b) will be small if the differences $M_k - m_k$ are *all* small, for $k = 1, 2, \ldots, n$. Here is where the *uniform* continuity of f over $[a, b]$ is important. [See part (iii) of Theorem 15, Article 3.9.] Suppose that ϵ is an arbitrary positive number. Then $\epsilon' = \epsilon/(b - a)$ is also positive, and by the theorem cited there exists a number $\delta > 0$ such that for any $x, x' \epsilon [a, b]$,

$$|f(x) - f(x')| < \frac{\epsilon}{b - a} \quad \textit{if} \quad |x - x'| < \delta. \tag{7}$$

Let n be any positive integer large enough so that

$$\Delta x = \frac{b - a}{n} < \delta, \quad \text{that is,} \quad n > \frac{b - a}{\delta}. \tag{8}$$

Then each subinterval is so short that c_k and e_k are within δ-distance of each other for each $k = 1, 2, \ldots, n$. Hence, by (7),

$$|M_k - m_k| = |f(e_k) - f(c_k)| < \frac{\epsilon}{b - a} \quad \text{for } k = 1, 2, \ldots, n. \tag{9}$$

Therefore each of the n summands on the right-hand side of Eq. (6b) is less than $\epsilon\,\Delta x/(b - a)$. Consequently, whenever condition (8) is satisfied,

$$U - L < \frac{n \cdot \epsilon \cdot \Delta x}{b - a} = \epsilon \frac{n\,\Delta x}{b - a} = \epsilon \frac{b - a}{b - a} = \epsilon. \tag{10}$$

It is this inequality, and the inequalities in (6a), that justifies the conclusions

$$\lim_{\Delta x \to 0} U = A \quad \text{and} \quad \lim_{\Delta x \to 0} L = A.$$

Remark 3. Just as we use the simple formula $A = \pi r^2$ to find the area of a circle, rather than resorting to a calculation of the limit of areas of inscribed polygons, so with the area under a curve. We shall not actually compute many areas directly from the definition in Eq. (3). Rather, we shall develop from it a method for getting answers very quickly and simply. But first we need some definition of area, as given above, as a starting place.

Remark 4. Our definition of area as the limit of sums of areas of inscribed rectangles is less general than definitions you will find in more advanced books. For more sophisticated approaches, you may wish to consult one of the following references:

(1) Thomas, Moulton, and Zelinka, *Elementary Calculus from an Advanced Viewpoint*. Addison-Wesley Publishing Co., Inc. (1967), Chapter 12.

(2) R. C. Buck and E. F. Buck, *Advanced Calculus*. McGraw Hill (1965), Section 3.1.

(3) J. M. H. Olmsted, *Advanced Calculus*. Appleton-Century-Crofts, Inc. (1961), pp. 337–342.

For the types of regions considered here, the definition given in Eq. (3) and these more advanced definitions lead to the same numerical results.

EXERCISES

In Exercises 1 through 5, sketch the graph of the given equation over the interval $a \le x \le b$. Divide the interval into $n = 4$ subintervals each of length $\Delta x = (b - a)/4$.

(a) Sketch the inscribed rectangles and compute the sum of their areas.

(b) Do the same using the circumscribed in place of the inscribed rectangle on each subinterval.

1. $y = 2x + 1, \quad a = 0, \quad b = 1.$
2. $y = x^2, \quad a = -1, \quad b = 1.$

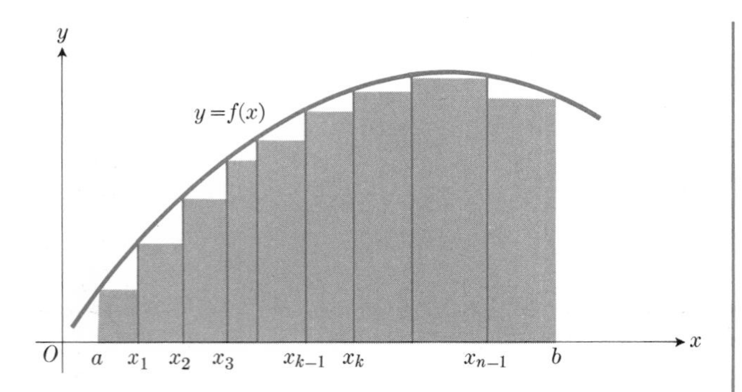

a | b

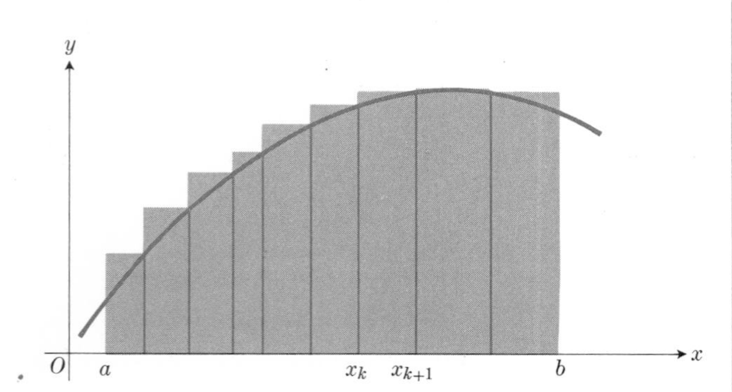

| c |

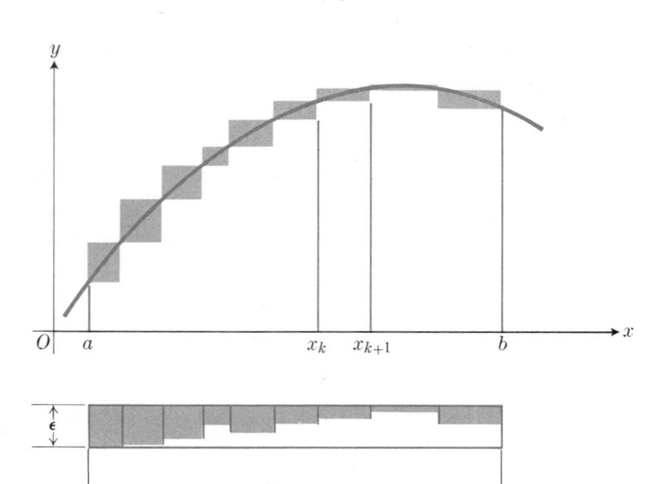

5.17

3. $y = \sin x, \quad a = 0, \quad b = \pi.$
4. $y = 1/x, \quad a = 1, \quad b = 2.$
5. $y = \sqrt{x}, \quad a = 0, \quad b = 4.$
6. Suppose that f is continuous and nonnegative over $[a, b]$, as in Fig. 5.17(a). By inserting points

$$x_1, x_2, \ldots, x_{k-1}, x_k, \ldots, x_{n-1}$$

as shown, divide $[a, b]$ into n subintervals of lengths $\Delta x_1 = x_1 - a, \Delta x_2 = x_2 - x_1, \ldots, \Delta x_n = b - x_{n-1}$, which need not be equal.

(a) If $m_k = \min \{f(x) \text{ for } x \text{ in the } k\text{th subinterval}\}$, explain the connection between the *lower sum*

$$L = m_1 \Delta x_1 + m_2 \Delta x_2 + \cdots + m_n \Delta x_n$$

and the shaded region in one of the parts of Figs 5.17.

(b) If $M_k = \max \{f(x) \text{ for } x \text{ in the } k\text{th subinterval}\}$, explain the connection between the *upper sum*

$$U = M_1 \Delta x_1 + M_2 \Delta x_2 + \cdots + M_n \Delta x_n$$

and the shaded region in one of the parts of Fig. 5.17.

(c) Explain the connection between $U - L$ and the shaded regions along the curve in Fig. 5.17(c).

(d) If $\epsilon > 0$ is given, why is it possible to make $U - L \leq \epsilon \cdot (b - a)$ by making the largest of the Δx_k's sufficiently small?

7. Suppose that the graph of f is always rising toward the right between $x = a$ and $x = b$, as in Fig. 5.18. Take $\Delta x = (b - a)/n$. Show, by reference to Fig. 5.18, that the difference between the upper and lower sums is representable graphically as the area $[f(b) - f(a)] \Delta x$ of the rectangle R. [*Hint.* The

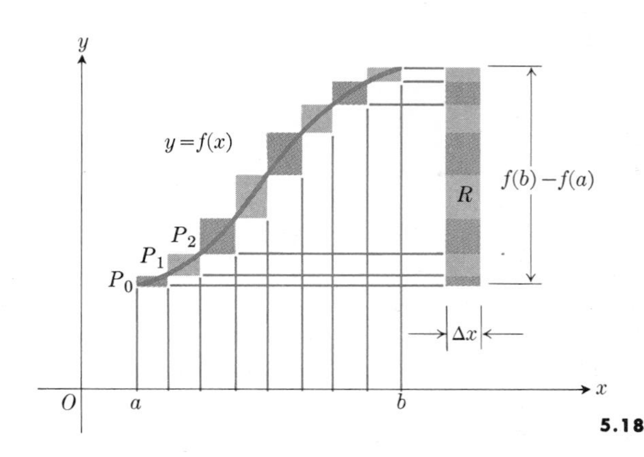

5.18

difference $U - L$ is the sum of areas of rectangles with diagonals P_0P_1, P_1P_2, and so forth along the curve, and there is no overlapping when these are all displaced horizontally into the rectangle R.]

8. Draw a figure to represent a continuous curve $y = f(x)$ which is always falling to the right between $x = a$ and $x = b$. Suppose the subdivisions Δx_k are again equal, $\Delta x_k = \Delta x = (b - a)/n$. Obtain an expression for the difference $U - L$ analogous to the expression in Exercise 7.
9. In Exercise 7 or 8, if the Δx_k's are not all equal, show that

$$U - L \leq |f(b) - f(a)|(\Delta x_{\max}),$$

where $\Delta x_{\max}$ is the largest of the Δx_k's for

$$k = 1, 2, \ldots, n.$$

5.7 COMPUTATION OF AREAS AS LIMITS

In Article 5.6, we defined the area under the graph of $y = f(x)$ over the closed interval $[a, b]$ as the limit of sums of areas of inscribed rectangles. We could have used circumscribed rectangles, as we did in the example worked out in Article 2.5. There we found the area under the graph of $y = x^2$ over an interval $[0, b]$ by computing the limit of sums of areas of circumscribed rectangles. At that time we used the formula

$$1^2 + 2^2 + 3^2 + \cdots + n^2 = \frac{n(n+1)(2n+1)}{6},$$

which we proved by the method of mathematical induction. (Recall that this consists in showing that the formula is true for $n = 1$, and then showing that if the formula is true for any integer n, then it is true for the next integer, $n + 1$. If we can demonstrate these two conditions, we can then conclude that the formula is correct for every positive integer n.) Although the procedure is called mathematical *induction*, it is strictly a *deductive* argument. Something more akin to the usual meaning of induction (Webster: "An instance of reasoning from particulars to generals . . .") is used in the next two paragraphs to show how such a formula might be discovered.

First, consider the sum of first powers:

$$F(n) = 1 + 2 + 3 + \cdots + n.$$

Here is a short table that shows how $F(n)$ increases with n. The last column exhibits $F(n)/n$, the ratio of $F(n)$ to n.

n	$F(n)$	$F(n)/n$
1	1	$1 = \frac{2}{2}$
2	$1 + 2 = 3$	$\frac{3}{2} = \frac{3}{2}$
3	$1 + 2 + 3 = 6$	$\frac{6}{3} = \frac{4}{2}$
4	$1 + 2 + 3 + 4 = 10$	$\frac{10}{4} = \frac{5}{2}$
5	$1 + 2 + 3 + 4 + 5 = 15$	$\frac{15}{5} = \frac{6}{2}$
6	$1 + 2 + 3 + 4 + 5 + 6 = 21$	$\frac{21}{6} = \frac{7}{2}$

The last column seems to indicate that the ratio $F(n)/n$ is equal to $(n + 1)/2$. At least, such is the case for all the entries in the table $(n = 1, 2, 3, 4, 5, 6)$. In other words, the formula

$$\frac{F(n)}{n} = \frac{n+1}{2},$$

or

$$1 + 2 + 3 + \cdots + n = \frac{n(n+1)}{2}, \tag{1}$$

is true for $n = 1, 2, 3, 4, 5, 6$. Suppose now that n is any integer for which (1) is known to be true (at the moment, n could be any integer from 1 through 6). Then if $(n + 1)$ were added to both sides of the equation, the new equation

$$1 + 2 + 3 + \cdots + n + (n + 1) = \frac{n(n+1)}{2} + (n + 1) \tag{2}$$

would also be true for that same n. But the right side of (2) is

$$\frac{n(n+1)}{2} + (n + 1) = \frac{(n+1)}{2}(n + 2) = \frac{(n+1)(n+2)}{2},$$

so that (2) becomes

$$1 + 2 + 3 + \cdots + n + (n + 1) = \frac{(n+1)[(n+1)+1]}{2},$$

which is just like Eq. (1) except that n is replaced by $n+1$. Thus if Eq. (1) is true for an integer n, it is also true for the next integer $n+1$. Hence we now know that it is true for $n+1=7$, since it was true for $n=6$. Then we can say it is true for $n+1=8$, since it is true for $n=7$. By the principle of mathematical induction, then, it is true for every positive integer n.

Now let's consider the squares. Let

$$Q(n) = 1^2 + 2^2 + 3^2 + \cdots + n^2$$

be the sum of the squares of the first n positive integers. Obviously this grows faster than the sum of first powers, but let us look at the ratio to compare them.*

n	$F(n)$	$Q(n)$	$Q(n)/F(n)$
1	1	$1^2 = 1$	$\frac{1}{1} = \frac{3}{3}$
2	3	$1^2+2^2 = 5$	$\frac{5}{3} = \frac{5}{3}$
3	6	$1^2+2^2+3^2 = 14$	$\frac{14}{6} = \frac{7}{3}$
4	10	$1^2+2^2+3^2+4^2 = 30$	$\frac{30}{10} = \frac{9}{3}$
5	15	$1^2+2^2+3^2+4^2+5^2 = 55$	$\frac{55}{15} = \frac{11}{3}$
6	21	$1^2+2^2+3^2+4^2+5^2+6^2 = 91$	$\frac{91}{21} = \frac{13}{3}$

We note how regular the last column is: $\frac{3}{3}$, $\frac{5}{3}$, $\frac{7}{3}$, and so on. In fact it is just

$$(2n+1)/3 \quad \text{for } n = 1, 2, 3, 4, 5, 6;$$

that is,

$$Q(n) = F(n) \cdot \frac{2n+1}{3}.$$

But from Eq. (1), $F(n) = n(n+1)/2$, and hence

$$Q(n) = 1^2 + 2^2 + 3^2 + \cdots + n^2 = \frac{n(n+1)(2n+1)}{6} \tag{3}$$

is true for the integers n from 1 through 6. To establish it for all other positive integers, we could proceed as before. [See Eqs. (2a, b), Article 2.5 for details.]

We now apply the formulas (1) and (3) to find areas under two graphs.

---

* The author is indebted to Professor Donald E. Richmond for suggesting this method of discovering the formula for $Q(n)$.

Example 1. (See Fig. 5.19.) Consider the line $y = mx$. Let b and a be any positive numbers, $b > a$. Let n be a positive integer, and divide the interval $[a, b]$ into n subintervals each of length $\Delta x = (b-a)/n$, by inserting the points

$$\begin{aligned} x_1 &= a + \Delta x, \\ x_2 &= a + 2\,\Delta x, \\ x_3 &= a + 3\,\Delta x, \\ &\vdots \\ x_{n-1} &= a + (n-1)\,\Delta x. \end{aligned}$$

The inscribed rectangles have areas

$$\begin{aligned} f(a)\,\Delta x &= ma \cdot \Delta x, \\ f(x_1)\,\Delta x &= m(a + \Delta x) \cdot \Delta x, \\ f(x_2)\,\Delta x &= m(a + 2\,\Delta x) \cdot \Delta x, \\ &\vdots \\ f(x_{n-1})\,\Delta x &= m[a + (n-1)\,\Delta x] \cdot \Delta x, \end{aligned}$$

whose sum is

$$\begin{aligned} S_n &= m[a + (a + \Delta x) + (a + 2\,\Delta x) + \cdots \\ &\qquad + (a + (n-1)\,\Delta x)] \cdot \Delta x \\ &= m[na + (1 + 2 + \cdots + (n-1))\,\Delta x]\,\Delta x \\ &= m\left[na + \frac{(n-1)n}{2}\,\Delta x\right]\Delta x \\ &= m\left[a + \frac{n-1}{2}\,\Delta x\right] n\,\Delta x, \quad \Delta x = \frac{b-a}{n}, \\ &= m\left[a + \frac{b-a}{2} \cdot \frac{n-1}{n}\right] \cdot (b-a). \end{aligned}$$

The area under the graph is defined to be the limit of S_n as $n \to \infty$. In the final form, the only place n appears

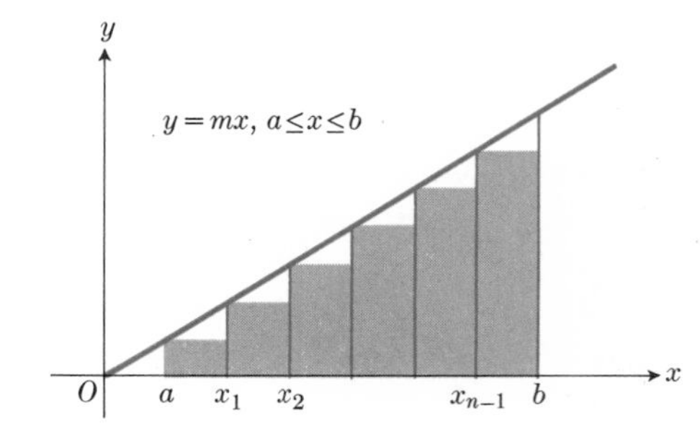

5.19

is in the fraction

$$\frac{n-1}{n} = 1 - \frac{1}{n},$$

and $1/n \to 0$ as $n \to \infty$, so

$$\lim \frac{n-1}{n} = 1.$$

Therefore

$$\lim S_n = m\left(a + \frac{b-a}{2}\right) \cdot (b-a)$$
$$= \frac{ma + mb}{2} \cdot (b-a).$$

This is easily interpreted as the area of a trapezoid, with "bases" (in this case vertical) ma and mb and with altitude $(b-a)$.

Example 2. (See Fig. 5.20.) Consider the graph of $y = x^2$, $0 \leq x \leq b$. Let n be a positive integer; divide the interval $0 \leq x \leq b$ into n subintervals each of length $\Delta x = b/n$, by inserting the points

$$x_1 = \Delta x, \quad x_2 = 2\,\Delta x, \quad x_3 = 3\,\Delta x, \quad \ldots,$$
$$x_{n-1} = (n-1)\,\Delta x.$$

The inscribed rectangles have areas

$$\begin{aligned} f(0)\,\Delta x &= 0, \\ f(x_1)\,\Delta x &= (\Delta x)^2\,\Delta x, \\ f(x_2)\,\Delta x &= (2\,\Delta x)^2\,\Delta x, \\ f(x_3)\,\Delta x &= (3\,\Delta x)^2\,\Delta x, \\ &\vdots \\ f(x_{n-1})\,\Delta x &= ((n-1)\,\Delta x)^2\,\Delta x. \end{aligned}$$

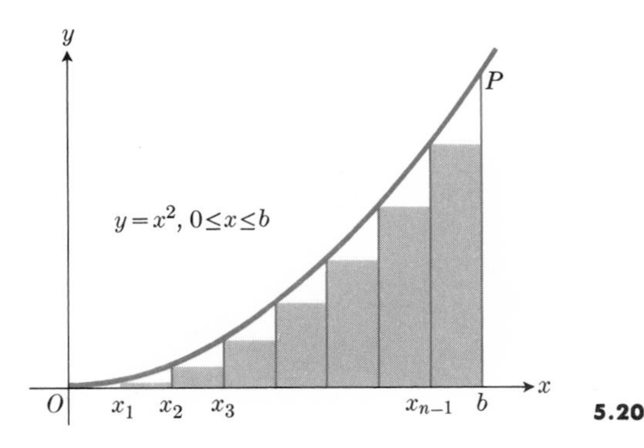

5.20

The sum of these areas is

$$\begin{aligned} S_n &= [1^2 + 2^2 + 3^2 + \cdots + (n-1)^2](\Delta x)^3 \\ &= \frac{(n-1)n(2n-1)}{6}\left(\frac{b}{n}\right)^3 \\ &= \frac{b^3}{6} \cdot \frac{n-1}{n} \cdot \frac{n}{n} \cdot \frac{2n-1}{n} \\ &= \frac{b^3}{6}\left(1 - \frac{1}{n}\right)\left(2 - \frac{1}{n}\right). \end{aligned}$$

To find the area under the graph, we let n increase without bound and get

$$A = \lim S_n = \frac{b^3}{3}.$$

EXERCISES

1. Verify the formula

$$\sum_{k=1}^{n} k^3 = 1^3 + 2^3 + \cdots + n^3 = \left[\frac{n(n+1)}{2}\right]^2$$

for $n = 1, 2, 3$. Then add $(n+1)^3$ and thereby prove by mathematical induction (as in the text) that the formula is true for all positive integers n.

2. Using the result of Exercise 1 and the method of Example 2 in the text, show that the area under the graph of $y = x^3$ over the interval $0 \leq x \leq b$ is $b^4/4$.

3. Find the area under the graph of $y = mx$ over the interval $a \leq x \leq b$ by using *circumscribed* rectangles in place of the inscribed rectangles of Example 1 in the text.

4. Do Exercise 2 by using circumscribed rectangles instead of inscribed rectangles.

5. Establish the formulas given below, for every positive integer n, by showing
   (a) that the formula is correct for $n = 1$,
   (b) if true for n, the formula is also true for $n + 1$.

$$\sum_{k=1}^{n} (2k-1) = 1 + 3 + 5 + \cdots + (2n-1) = n^2,$$

$$\sum_{k=1}^{n} \frac{1}{k(k+1)} = \frac{1}{1 \cdot 2} + \frac{1}{2 \cdot 3} + \cdots + \frac{1}{n \cdot (n+1)} = \frac{n}{n+1}.$$

6. The formula

$$\sin h + \sin 2h + \sin 3h + \cdots + \sin mh = \frac{\cos(h/2) - \cos(m + 1/2)h}{2\sin(h/2)}$$

may be assumed for this problem. Use it to find the area under the graph of $f(x) = \sin x$ over $[0, \pi/2]$, in two steps:

(1) Divide the interval $[0, \pi/2]$ into n equal subintervals and calculate the corresponding upper sum U; then

(2) find the limit of U as $n \to \infty$ and

$$\Delta x = (b - a)/n \to 0.$$

5.8 AREAS BY CALCULUS

In Article 5.6, we defined the area under a curve and showed how we could estimate it by computing sums of areas of rectangles. Nothing more than arithmetic is involved in these calculations, but we pay the price of getting only an estimate of the true area. On the other hand, in Article 5.7, we used algebraic techniques and actually computed *limits,* thus getting exact areas at the cost of fairly extensive algebraic preliminaries. In this article we shall follow the path blazed by Leibniz and Newton to show how exact areas can be easily computed by using calculus.

To begin we need some preliminary results. We consider a function f which is continuous and nonnegative over the domain $[a, b]$. Let A_a^c, A_c^b, A_a^b denote the areas under the graph and above the x-axis from a to c, from c to b, and from a to b, respectively (Fig. 5.21). If c is between a and b, we have

$$A_a^c + A_c^b = A_a^b. \tag{1}$$

And this is also true when $c = a$ if we define the area from a to a to be zero,

$$A_a^a = 0. \tag{2}$$

Equation (1) is also true if c is beyond b, provided we adopt some conventions about signed areas. Let us say that area above the x-axis is positive if we

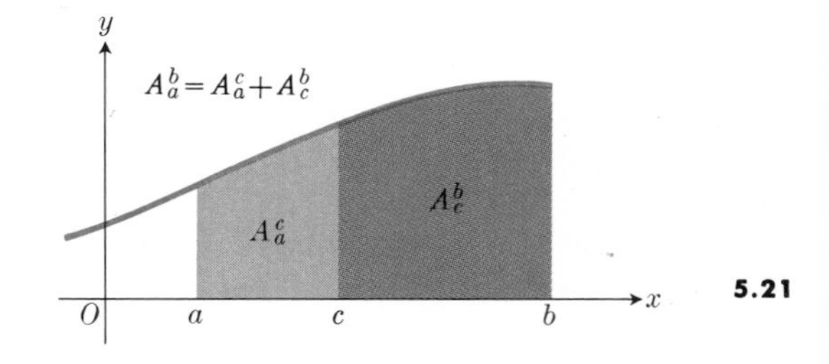

5.21

go from left to right, and negative if we go from right to left. Thus, in Fig. 5.21, A_a^c, A_c^b, and A_a^b are all positive, while

$$A_c^a = -A_a^c, \quad A_b^c = -A_c^b, \quad A_b^a = -A_a^b \tag{3}$$

are negative. So formula (1) is also true in the form

$$A_a^b + A_b^c = A_a^c, \tag{4}$$

because adding A_b^c to A_a^b just subtracts A_c^b from the latter, leaving A_a^c.

The following theorem will also be used.

Theorem 1. The Mean Value Theorem for Areas. *Let f be a nonnegative continuous function over the domain $[a, b]$. Let A_a^b denote the area under the graph of f over the domain. Then there is at least one number c between a and b such that*

$$A_a^b = f(c) \cdot (b - a). \tag{5}$$

Proof. Suppose m and M are respectively the minimum and maximum values of f over the domain $[a, b]$. Then

$$m(b - a) \leq A_a^b \leq M(b - a),$$

and therefore

$$m \leq \frac{A_a^b}{b - a} \leq M.$$

Since $A_a^b/(b - a)$ lies between the minimum and maximum values of $f(x)$ for $a \leq x \leq b$, there is at least one place c between a and b where $f(c)$ is equal to this number:

$$f(c) = \frac{A_a^b}{b - a} \qquad \text{[Theorem 15, Article 3.9].}$$

Equation (5) follows. □

Remark. The geometric interpretation of this Mean Value Theorem is this: A line drawn parallel to the x-axis at the right place between m and M units above the x-axis will serve as upper boundary of a rectangle having the same area as that under the curve, and it will surely cut the curve at least once between $x = a$ and $x = b$. The abscissa of that cut is a suitable value of c in Eq. (5) (see Fig. 5.22).

Theorem 2. *Let f be a nonnegative continuous function over $[a, b]$. Let P be a set of points*

$$P = \{a = x_0 < x_1 < x_2 < \cdots < x_{n-1} < x_n = b\}$$

that partitions $[a, b]$ into n subintervals, not necessarily of equal lengths. Let c_k be any number in the kth subinterval $[x_{k-1}, x_k]$, and let $\Delta x_k = x_k - x_{k-1}$, for $k = 1, 2, \ldots, n$. Then

$$\lim_{\max \Delta x_k \to 0} \sum_{k=1}^{n} f(c_k)\, \Delta x_k = A_a^b. \tag{6}$$

Proof. We apply the Mean Value Theorem for areas to each of the subintervals. If ΔA_k is the exact area under the curve on the kth subinterval, there exists a number d_k on the segment $[x_{k-1}, x_k]$ such that

$$\Delta A_k = f(d_k)\, \Delta x_k, \quad k = 1, 2, \ldots, n. \tag{7a}$$

Thus, if we choose $d_1, d_2, \ldots, d_n$ to satisfy Eq. (7a), and add, we get

$$A_a^b = \sum_{k=1}^{n} f(d_k)\, \Delta x_k. \tag{7b}$$

We compare this with the sum we would obtain if we used the numbers c_k, $k = 1, 2, \ldots, n$, as on the left-hand side of (6), by computing the difference

$$A_a^b - \sum_{k=1}^{n} f(c_k)\, \Delta x_k = \sum_{k=1}^{n} [f(d_k) - f(c_k)]\, \Delta x_k. \tag{8}$$

Because f is assumed to be continuous on the closed interval $[a, b]$ it is *uniformly* continuous there. Hence, if $\epsilon > 0$, there exists $\delta > 0$ such that

$$|f(d_k) - f(c_k)| < \frac{\epsilon}{b - a} \quad \text{if} \quad \max \Delta x_k < \delta. \tag{9}$$

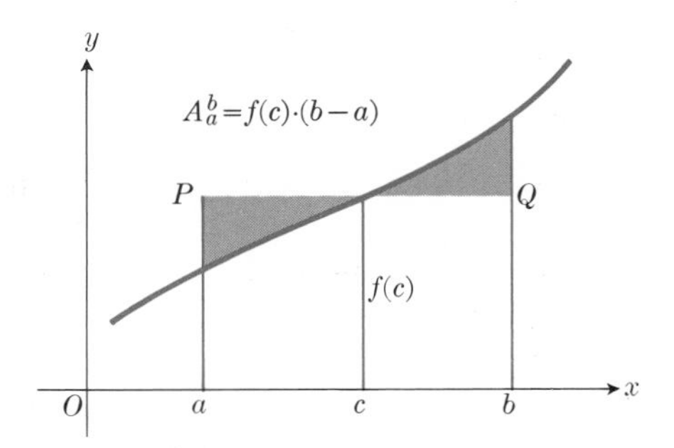

5.22 The area under the curve is equal to the area of the rectangle $aPQb$: the shaded regions above and below the curve have equal areas.

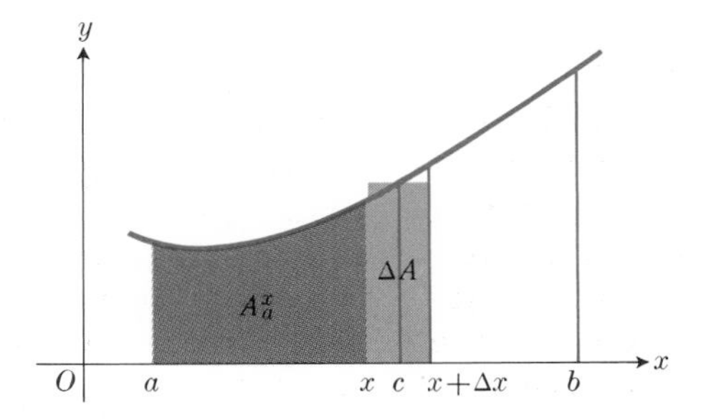

5.23 Increment of area function.

When $\max \Delta x_k \to 0$, therefore, conditions (9) are satisfied, and from Eq. (8) we see that

$$\left| A_a^b - \sum_{k=1}^{n} f(c_k)\, \Delta x_k \right| < \frac{\epsilon}{b - a} \sum_{k=1}^{n} \Delta x_k$$
$$= \frac{\epsilon}{b - a}(b - a) = \epsilon. \tag{10}$$

This inequality is just what the limit statement in Eq. (6) means. □

This theorem says that whether we use inscribed rectangles, circumscribed rectangles, or intermediate rectangles anywhere between, the sum of the areas $f(c_k)\, \Delta x_k$ approaches the exact area under the curve, provided we make the lengths of all the subintervals approach zero.

To *compute* the area A_a^b, we now consider any abscissa x between a and b, the area A_a^x, and the area $A_a^{x+\Delta x}$, where $\Delta x \neq 0$ (Fig. 5.23). We shall derive a differential equation for the area function A_a^x. The solution of this differential equation, with the initial condition

$$A_a^a = 0,$$

will enable us to compute the area from a to any x and, in particular, from a to b.

From Eq. (1) we have

$$A_a^x + A_x^{x+\Delta x} = A_a^{x+\Delta x},$$

so that

$$\Delta(A_a^x) = A_a^{x+\Delta x} - A_a^x = A_x^{x+\Delta x}. \tag{11a}$$

Then, by Eq. (5),

$$A_x^{x+\Delta x} = f(c) \cdot \Delta x, \tag{11b}$$

where c is a number between x and $x + \Delta x$, which must therefore approach x as Δx approaches zero. Combining (11a) and (11b) and dividing by Δx, we get

$$\frac{\Delta A_a^x}{\Delta x} = f(c), \tag{11c}$$

and therefore

$$\begin{aligned} \frac{dA_a^x}{dx} &= \lim_{\Delta x \to 0} \frac{\Delta A_a^x}{\Delta x} \\ &= \lim_{c \to x} f(c) \\ &= f(x), \end{aligned}$$

where the last line follows from the fact that f is continuous.

Therefore the area function A_a^x satisfies the differential equation

$$\frac{dA_a^x}{dx} = f(x) \tag{12a}$$

and the initial condition

$$A_a^a = 0. \tag{12b}$$

Thus if $F(x)$ is any integral of $f(x)\,dx$, we have

$$A_a^x = \int f(x)\,dx = F(x) + C, \tag{13a}$$

$$A_a^a = 0 = F(a) + C.$$

Hence $$C = -F(a)$$

and $$A_a^x = F(x) - F(a).$$

Finally, by taking $x = b$, we get

$$A_a^b = F(b) - F(a). \tag{13b}$$

Equations (13a) *and* (13b) *summarize the method for finding the area under a curve by integration. If the equation of the curve is* $y = f(x)$, *we integrate this function* f *to find*

$$F(x) + C = \int f(x)\,dx. \tag{14a}$$

If the interval is $a \leq x \leq b$, *we then compute*

$$A_a^b = F(x)]_a^b = F(b) - F(a). \tag{14b}$$

The standard notation $F(x)]_a^b$ in Eq. (14b) simply means: First replace x by the upper value b to calculate $F(b)$, and from this subtract the value $F(a)$ obtained by setting $x = a$. For example,

$$2x + 3]_1^5 = 13 - 5 = 8.$$

It is worth noting that the constant of integration may be omitted in evaluating Eq. (14b). For if we use $F(x) + C$ in place of $F(x)$ in (14b), we find

$$\begin{aligned} F(x) + C]_a^b &= [F(b) + C] - [F(a) + C] \\ &= F(b) - F(a) \\ &= F(x)]_a^b. \end{aligned}$$

Example 1. (See Fig. 5.19.) The area under the graph of $y = mx$, $a \leq x \leq b$, is

$$\begin{aligned} \int mx\,dx\Big]_a^b = \frac{mx^2}{2}\Big]_a^b &= \frac{mb^2}{2} - \frac{ma^2}{2} \\ &= \frac{mb + ma}{2} \cdot (b - a). \end{aligned}$$

Example 2. (See Fig. 5.20.) The area under the graph of

$$y = x^2, \quad 0 \leq x \leq b,$$

is

$$\int x^2\,dx\Big]_0^b = \frac{x^3}{3}\Big]_0^b = \frac{b^3}{3}.$$

Compare both the results and the methods with Examples 1 and 2 of Article 5.7.

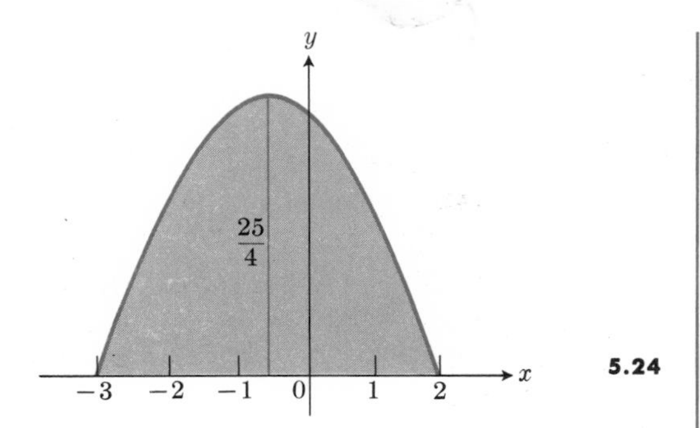

5.24

Problem 1. As another illustration of the method of finding areas, calculate the area bounded by the parabola $y = 6 - x - x^2$ and the x-axis.

Solution. We find where the curve crosses the x-axis by setting

$$y = 0 = 6 - x - x^2 = (3 + x)(2 - x),$$

which gives

$$x = -3 \quad \text{or} \quad x = 2.$$

The curve is sketched in Fig. 5.24.

According to Eqs. (14a, b), the area is

$$\begin{aligned} A_{-3}^{2} &= \int (6 - x - x^2)\,dx \Big]_{-3}^{2} \\ &= 6x - \frac{x^2}{2} - \frac{x^3}{3}\Big]_{-3}^{2} \\ &= (12 - 2 - \tfrac{8}{3}) - (-18 - \tfrac{9}{2} + \tfrac{27}{3}) \\ &= 20\tfrac{5}{6}. \end{aligned}$$

The curve in Fig. 5.24 is an arch of a parabola, and it is interesting to note that the area is exactly equal to two-thirds the base times the altitude:

$$\tfrac{2}{3}(5)(\tfrac{25}{4}) = \tfrac{125}{6} = 20\tfrac{5}{6}.$$

Problem 2. Show that the area under one arch of the curve $y = \sin x$ is 2.

Solution. One arch of the sine curve extends from $x = 0$ to $x = \pi$ (Fig. 5.6a). Therefore the area is

$$\begin{aligned} A_0^{\pi} &= \int \sin x \, dx \Big]_0^{\pi} = -\cos x \Big]_0^{\pi} \\ &= -\cos \pi + \cos 0. \end{aligned}$$

Since $\cos \pi = -1$ while $\cos 0 = +1$, this gives $A_0^{\pi} = 2$.

EXERCISES

In Exercises 1 through 10, find the area bounded by the x-axis, the given curve $y = f(x)$, and the given vertical lines.

1. $y = x^2 + 1, \quad x = 0, \quad x = 3.$
2. $y = 2x + 3, \quad x = 0, \quad x = 1.$
3. $y = \sqrt{2x + 1}, \quad x = 0, \quad x = 4.$
4. $y = \dfrac{1}{\sqrt{2x+1}}, \quad x = 0, \quad x = 4.$
5. $y = \dfrac{1}{(2x+1)^2}, \quad x = 1, \quad x = 2.$
6. $y = (2x + 1)^2, \quad x = -1, \quad x = 3.$
7. $y = x^3 + 2x + 1, \quad x = 0, \quad x = 2.$
8. $y = x\sqrt{2x^2 + 1}, \quad x = 0, \quad x = 2.$
9. $y = \dfrac{x}{\sqrt{2x^2+1}}, \quad x = 0, \quad x = 2.$
10. $y = \dfrac{x}{(2x^2+1)^2}, \quad x = 0, \quad x = 2.$
11. Find the area bounded by the coordinate axes and the line $x + y = 1$.
12. Find the area between the curve $y = 4 - x^2$ and the x-axis.
13. Find the area between the curve $y = 1/\sqrt{x}$, the x-axis, and the lines $x = 1$, $x = 4$.
14. Find the area between the curve $y = \sqrt{1 - x}$ and the coordinate axes.
15. Find the area between the curve $x = 1 - y^2$ and the y-axis.
16. Find the area contained between the x-axis and one arch of the curve $y = \cos 3x$.

*17. Take $B = A$ in Eq. (11b), Article 5.4, and show that

$$\cos 2A = \cos^2 A - \sin^2 A.$$

Combine this with the identity

$$1 = \cos^2 A + \sin^2 A$$

* The method used in this problem should be used in several later problems involving integration of squares of sines and cosines.

to show that

$$\cos^2 A = \tfrac{1}{2}(1 + \cos 2A), \quad \sin^2 A = \tfrac{1}{2}(1 - \cos 2A).$$

Make use of the last of these identities to find the area contained between the x-axis and one arch of the curve $y = \sin^2 3x$.

18. The graph of $y = \sqrt{a^2 - x^2}$ over $-a \le x \le a$ is a semicircle of radius a.

(a) Using this fact, explain why it is true that

$$\int \sqrt{a^2 - x^2}\, dx \Big]_{-a}^{a} = \frac{1}{2}\pi a^2.$$

(b) Evaluate $\int \sqrt{a^2 - x^2}\, dx]_0^a$.

19. The integral in Exercise 18 can be evaluated by using the substitution

$$x = a \cos \theta, \qquad \pi \ge \theta \ge 0,$$

to replace $\sqrt{a^2 - x^2}\, dx$ by $-a^2 \sin^2 \theta\, d\theta$. Combine this with Exercise 17 to show that

$$\int \sqrt{a^2 - x^2}\, dx \Big]_{-a}^{a} = \int -a^2 \sin^2 \theta\, d\theta \Big]_{\pi}^{0}$$
$$= -\frac{a^2}{2} \int (1 - \cos 2\theta)\, d\theta \Big]_{\pi}^{0},$$

and thus verify the result given in Exercise 18(a) by evaluating this final integral.

5.9 THE DEFINITE INTEGRAL AND THE FUNDAMENTAL THEOREMS OF INTEGRAL CALCULUS

In Articles 5.6, 5.7, and 5.8, we made a systematic study of the area problem. We have arrived at the following result.

If the function f is nonnegative and continuous over the domain $[a, b]$, then the area under its graph is

$$A_a^b = \lim \sum f(c_k)\, \Delta x = \int f(x)\, dx]_a^b$$
$$= F(x)]_a^b$$
$$= F(b) - F(a). \tag{1}$$

The first part of this equation is like the definition of the area as the limit of the sum of areas of constructed rectangles. The last part of the equation gives a short way of evaluating this limit, that is, by calculus. Therein lies one of the most powerful ideas of "modern" mathematics ("modern" in the sense of post-Renaissance); for the key idea is this: that the limit can be evaluated by integration. This, essentially, is what is known as the *First Fundamental Theorem* of integral calculus. It ties together the summation process (which Archimedes used over two thousand years ago for finding areas, volumes, and centers of gravity) and the differentiation process, from which one may find the tangent to a curve. It is a remarkable fact that the inverse of the "tangent problem" (that is, the inverse of differentiation) provides a ready tool for solving the summation problem. Its applications extend far beyond the finding of areas; they include finding volumes of solids, lengths of curves, areas of surfaces of revolution, centers of gravity, work done by a variable force, and gravitational and electrical potential.

While Eq. (1) is expressed in terms of area, and up until now has been restricted to nonnegative functions, the Fundamental Theorem is less restrictive.

First fundamental theorem of integral calculus. *Let f be a function which is continuous over the domain $[a, b]$. Let*

$$a, x_1, x_2, \ldots, x_{n-1}, b \tag{2}$$

be a set of numbers $a < x_1 < x_2 < \cdots < x_{n-1} < b$ that partition the interval $a \ldots b$ into n equal subintervals each of length

$$\Delta x = (b - a)/n. \tag{3}$$

Let $c_1, c_2, \ldots, c_n$ be a set of n numbers, one in each subinterval,

$$\begin{aligned} a &\le c_1 \le x_1, \\ x_1 &\le c_2 \le x_2, \\ &\vdots \\ x_{n-1} &\le c_n \le b. \end{aligned} \tag{4}$$

Let

$$S_n = f(c_1)\, \Delta x + f(c_2)\, \Delta x + \cdots + f(c_n)\, \Delta x$$
$$= \sum_{k=1}^{n} f(c_k)\, \Delta x. \tag{5}$$

Finally, let $F(x)$ be any primitive of $f(x)$, so that

$$f(x) = F'(x). \tag{6}$$

Then, as $n \to \infty$,

$$\lim S_n = \lim \sum f(c_k)\,\Delta x = F(b) - F(a). \tag{7}$$

Proof. We shall first prove (7) for the special set of numbers $c_1, c_2, \ldots, c_n$ that we get by applying the Mean Value Theorem of Article 4.8 to the function F in each subinterval. We can do this because F is differentiable and continuous. Thus, remembering that $F'(x) = f(x)$, we have

$$\begin{aligned}
F(x_1) - F(a) &= F'(c_1)\cdot(x_1 - a) = f(c_1)\,\Delta x,\\
F(x_2) - F(x_1) &= F'(c_2)\cdot(x_2 - x_1) = f(c_2)\,\Delta x,\\
F(x_3) - F(x_2) &= F'(c_3)\cdot(x_3 - x_2) = f(c_3)\,\Delta x,\\
&\;\;\vdots \qquad\qquad\qquad\qquad\qquad\qquad (8)\\
F(x_{n-1}) - F(x_{n-2}) &= F'(c_{n-1})\cdot(x_{n-1} - x_{n-2})\\
&= f(c_{n-1})\,\Delta x,\\
F(b) - F(x_{n-1}) &= F'(c_n)\cdot(b - x_{n-1})\\
&= f(c_n)\,\Delta x.
\end{aligned}$$

We add Eqs. (8), and note that

$$F(x_1), F(x_2), \ldots, F(x_{n-1})$$

all appear twice on the left side, once positive and once negative. Hence these terms cancel out, leaving only $F(b) - F(a)$ in the sum. Thus we get

$$F(b) - F(a) = f(c_1)\,\Delta x + f(c_2)\,\Delta x + \cdots + f(c_n)\,\Delta x. \tag{9}$$

Since the left side of this equation does not in any way involve n, it remains fixed as we let $n \to \infty$, thus establishing Eq. (7), for this particular way of choosing the numbers $c_1, c_2, \ldots, c_n$.

But the theorem states that the same answer is obtained no matter how the c's are chosen in the subintervals, so long as there is one c in each subinterval. To establish this final result, we recall that the function f is continuous on the closed interval $[a, b]$ and therefore is *uniformly* continuous there [Theorem 15 (iii), Chapter 3]. Hence, if ϵ is any positive number, there exists a positive number δ, depending only upon ϵ, such that

$$|f(c_k) - f(c'_k)| < \epsilon \tag{10}$$

whenever

$$|c_k - c'_k| < \delta. \tag{11}$$

And we can make $\Delta x = (b - a)/n < \delta$ by making

$$n > (b - a)/\delta. \tag{12}$$

For all sufficiently large n, (12) is satisfied. Now let c_1, c'_1 be two numbers in the first subinterval, c_2, c'_2 in the second, and so on. Form the sums

$$S_n = \sum_{k=1}^{n} f(c_k)\,\Delta x, \qquad S'_n = \sum_{k=1}^{n} f(c'_k)\,\Delta x.$$

Their difference is less than or equal to

$$|f(c_1) - f(c'_1)|\,\Delta x + |f(c_2) - f(c'_2)|\,\Delta x + \cdots + |f(c_n) - f(c'_n)|\,\Delta x. \tag{13a}$$

Every term in (13a) is less than $\epsilon \cdot \Delta x$, by (10), and there are n terms. Therefore

$$\begin{aligned}
|S_n - S'_n| < n\cdot(\epsilon\cdot\Delta x) &= \epsilon\cdot(n\,\Delta x)\\
&= \epsilon\cdot(b - a), \quad (13b)
\end{aligned}$$

provided condition (12) is satisfied. This inequality, (13b), says that the sums S_n and S'_n that we get from two different choices of the c's in the subintervals can be made to differ by as little as we please [no more than $\epsilon\cdot(b - a)$] by making n sufficiently large. But for the particular choice of the c's in Eqs. (8), we have

$$S_n = F(b) - F(a).$$

Therefore S'_n differs arbitrarily little from

$$F(b) - F(a)$$

when n is sufficiently large. This means that

$$\lim S'_n = F(b) - F(a). \tag{14}$$

Equation (14) completes the proof. □

Remark 1. The integral sign, $\int$, is a modified capital S (for sum), intended to remind us of the close connection between integration and summation.

Remark 2. The limit in Eq. (7),

$$\lim \sum f(c_k)\, \Delta x,$$

is called the *definite integral* of f from a to b. It is denoted by the symbol $\int_a^b f(x)\, dx$. The first Fundamental Theorem of integral calculus thus says that

$$\int_a^b F'(x)\, dx = F(b) - F(a). \tag{15}$$

Remark 3. It really isn't necessary to make the subdivisions of $[a, b]$ regular. We could equally well have put

$$\Delta x_1 = x_1 - a,$$
$$\Delta x_2 = x_2 - x_1, \ldots,$$
$$\Delta x_n = b - x_{n-1},$$

and

$$S = \sum_{k=1}^{n} f(c_k)\, \Delta x_k.$$

The uniform continuity of f on $[a, b]$ guarantees that if the largest Δx_k is made to approach zero, then all sums S of this form approach the same number as limit: namely, $F(b) - F(a)$, where F is any primitive of f. Since any two primitives of the same function f differ at most by a constant, and since

$$F(x) + C]_a^b = F(b) - F(a),$$

the same answer is obtained no matter which primitive is used.

Remark 4. If the function has only nonnegative values over $[a, b]$, then the definite integral

$$\int_a^b f(x)\, dx = F(x)]_a^b = F(b) - F(a) \tag{16}$$

represents the area under its graph, above the x-axis, between the ordinates at $x = a$ and $x = b$. If, on the other hand, the function f were everywhere negative between a and b, the summands $f(c_k)\, \Delta x$ would all be negative [assuming $b > a$ and $\Delta x = (b - a)/n$ is positive], and the integral in Eq. (16) would be negative.

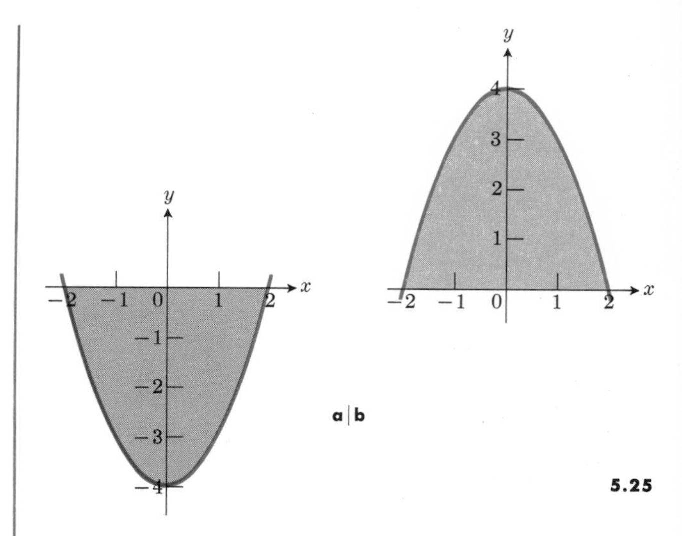

5.25

Example 1. For $|x| < 2$, $f(x) = x^2 - 4$ is negative (see Fig. 5.25a):

$$\int_{-2}^{2} f(x)\, dx = \frac{x^3}{3} - 4x \Big]_{-2}^{2} = \left(\tfrac{8}{3} - 8\right) - \left(-\tfrac{8}{3} + 8\right)$$
$$= -\tfrac{32}{3}.$$

The area between the curve and the x-axis, from $x = -2$ to $x = +2$, contains $\frac{32}{3}$ units of area. The sign is negative because the area is below the x-axis. Clearly, the graph in Fig. 5.25(b) of

$$y = g(x) = -f(x) = 4 - x^2, \quad -2 \le x \le 2,$$

is just the mirror image, with respect to the x-axis as mirror, of the curve in Fig. 5.25(a). The area between the graph of $y = g(x)$ and the x-axis is

$$\int_{-2}^{2} g(x)\, dx = 4x - \frac{x^3}{3}\Big]_{-2}^{2} = \tfrac{32}{3}.$$

The absolute value is the same as for the integral of $f(x)\, dx$ between the same limits. For $g(x)\, dx$ the sign is positive, since the area between the g-curve and the x-axis is above the axis.

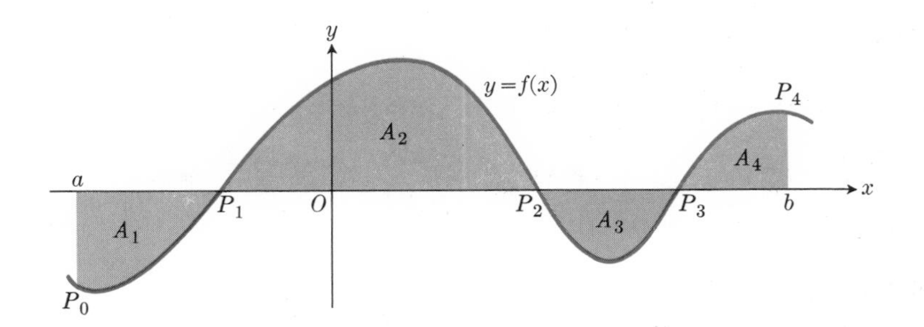

5.26

Remark 5. If the graph of $y = f(x)$, $a \leq x \leq b$, is partly below and partly above the x-axis, as in Fig. 5.26, then

$$\lim \sum f(c_k)\,\Delta x = \int_a^b f(x)\,dx = F(x)]_a^b = F(b) - F(a)$$

is the algebraic sum of *signed* areas, positive areas above the x-axis, negative areas below. For example, if the absolute values of the areas between the curve and the x-axis in Fig. 5.26 are A_1, A_2, A_3, A_4, then the definite integral of f from a to b is equal to

$$\int_a^b f(x)\,dx = -A_1 + A_2 - A_3 + A_4.$$

Thus, if we wanted the sum of the absolute values of these signed areas, that is

$$A = |-A_1| + A_2 + |-A_3| + A_4,$$

then we should need to find the abscissas s_1, s_2, s_3 of the points P_1, P_2, P_3 where the curve crosses the x-axis. We would then compute, separately,

$$-A_1 = \int_a^{s_1} f(x)\,dx, \qquad A_2 = \int_{s_1}^{s_2} f(x)\,dx,$$

$$-A_3 = \int_{s_2}^{s_3} f(x)\,dx, \qquad A_4 = \int_{s_3}^{b} f(x)\,dx,$$

and add their absolute values.

Problem 1. Find the total area bounded by the curve $y = x^3 - 4x$ and the x-axis.

Solution. The graph, shown in Fig. 5.27 lies above the x-axis from -2 to 0, below from 0 to $+2$. [The polynomial $x^3 - 4x$ factors as

$$x^3 - 4x = x(x - 2)(x + 2).$$

It is easy to determine the sign of the product from the signs of the three factors.]

$$A_1 = \int_{-2}^{0} (x^3 - 4x)\,dx = \frac{x^4}{4} - 2x^2\Big]_{-2}^{0} = 0 - (4 - 8) = +4,$$

$$-A_2 = \int_0^2 (x^3 - 4x)\,dx = \frac{x^4}{4} - 2x^2\Big]_0^2 = (4 - 8) - 0 = -4,$$

$$A_1 + |-A_2| = 4 + |-4| = 8.$$

Remark 6. The definite integral, as a limit of sums, provides a new way of defining functions. The integral of any continuous function $f(t)$ from $t = a$

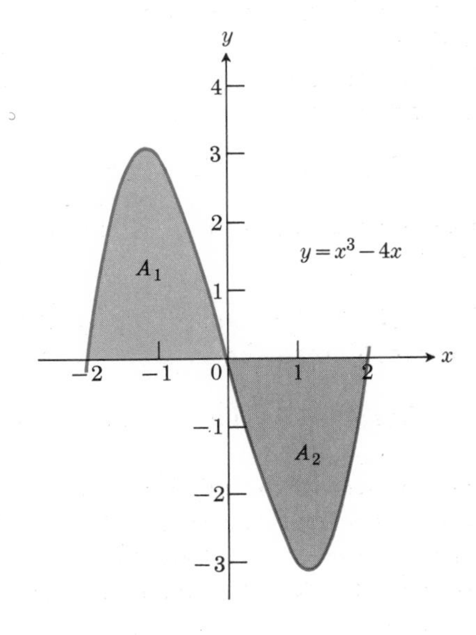

5.27

to $t = x$ defines a number

$$F(x) = \int_a^x f(t)\, dt,$$

which can be computed as a limit (at least in theory). The function F defined by this formula has certain rather obvious properties:

1. $F(a) = 0$.
2. F is an increasing function of x over intervals where f is positive.
3. When x changes from x_0 by a small amount, say when it changes to $x_0 + \Delta x$, $F(x)$ changes by an amount that is approximately $f(x_0)\,\Delta x$. These properties lead us to conjecture that F is differentiable and that its derivative is f. This conjecture is true (for continuous functions f). It is stated as the Second Fundamental Theorem of Integral Calculus.

Second fundamental theorem of integral calculus. *Let f be continuous on $[a, b]$, and let*

$$F(x) = \int_a^x f(t)\, dt \quad \text{for } x \in [a, b]. \tag{17a}$$

Then F is differentiable on (a, b), and

$$F'(x) = f(x) \quad \text{for } a < x < b. \tag{17b}$$

Proof. The same argument that led to Eq. (5), Article 5.8 can be applied to obtain the following:

$$F(x + \Delta x) - F(x) = \int_x^{x+\Delta x} f(t)\, dt = f(c)\,\Delta x$$

for some c between x and $x + \Delta x$. Divide by Δx and take limits:

$$\lim_{\Delta x \to 0} \frac{F(x + \Delta x) - F(x)}{\Delta x} = \lim_{c \to x} f(c),$$

or

$$F'(x) = f(x). \ \square$$

Example 2. Suppose

$$F(x) = \int_0^x \sqrt{1 - t^2}\, dt, \quad 0 < x < 1.$$

Then $\qquad F'(x) = \sqrt{1 - x^2}.$

Problem 2. Let $F(x) = \int_0^{\sqrt{x}} (t + 1)^{1/2}\, dt$ for $x > 0$. Find $F'(x)$.

Solution 1. By direct integration, we get

$$\int (t + 1)^{1/2}\, dt = \tfrac{2}{3}(t + 1)^{3/2} + C,$$

and $\qquad F(x) = \tfrac{2}{3}(\sqrt{x} + 1)^{3/2} - \tfrac{2}{3}.$

When we differentiate this we get

$$\begin{aligned} F'(x) &= (\sqrt{x} + 1)^{1/2} \frac{d}{dx}(\sqrt{x} + 1) \\ &= (\sqrt{x} + 1)^{1/2}(\tfrac{1}{2}x^{-1/2}). \end{aligned} \tag{18a}$$

Solution 2. In the original integral, call the upper limit u and apply the chain rule:

$$F(x) = G(u) = \int_0^u (t + 1)^{1/2}\, dt \quad \text{with } u = \sqrt{x}.$$

The Second Fundamental Theorem applies to $G'(u)$:

$$G'(u) = (u + 1)^{1/2}.$$

Now apply the chain rule:

$$F'(x) = G'(u)\frac{du}{dx} = (u + 1)^{1/2}(\tfrac{1}{2}x^{-1/2}). \tag{18b}$$

Eqs. (18a) and (18b) are equivalent because $u = \sqrt{x}$.

Remark 7. The First Fundamental Theorem is sometimes used as a tool for approximating sums. This is the reverse of what we did in Article 5.7. There we found formulas for sums of first powers, and of squares, of the positive integers 1 through n. Then we used these formulas to compute limits of sums of areas of inscribed rectangles for the graphs of $y = mx$, $a \le x \le b$, and of $y = x^2$, $0 \le x \le b$. In the following example, we work back from the definite integral (or, what amounts to the same thing, the area under the graph) to an approximation for the sum of the square roots of the integers 1 through n. The process is not completely reversible. We can go from exact formulas for sums to definite integrals by way of limits, but we cannot go back from the definite integral to an exact formula for the sum, because we don't know which terms went to zero in the limit process, and hence we cannot recover them.

Example 3. Consider the function defined by $f(x) = \sqrt{x}$, $0 \le x \le 1$. Let n be a positive integer, and $\Delta x = 1/n$. In Eq. (7), take $c_1 = x_1 = \Delta x$, $c_2 = x_2 = 2\,\Delta x, \ldots$, $c_n = b = 1 = n\,\Delta x$. Then

$$\begin{aligned} S_n &= \sqrt{c_1}\,\Delta x + \sqrt{c_2}\,\Delta x + \cdots + \sqrt{c_n}\,\Delta x \\ &= \frac{\sqrt{1} + \sqrt{2} + \cdots + \sqrt{n}}{n^{3/2}} \\ &\to \int \sqrt{x}\,dx\Big]_0^1 = \frac{x^{3/2}}{3/2}\Big]_0^1 = \frac{2}{3} \quad \text{as } n \to \infty. \end{aligned}$$

When n is large, S_n will be close to its limit $\frac{2}{3}$. This means that the numerator

$$\sqrt{1} + \sqrt{2} + \cdots + \sqrt{n}$$

is approximately equal to $\frac{2}{3}n^{3/2}$. For $n = 10$, the sum of the square roots is 22.5^-, while $\frac{2}{3}n^{3/2}$ is 21.1^-, so that the approximation is in error by about 6%.

EXERCISES

1. For each of the following cases, integrate the given function f to find a new function F, defined by $F(x) = \int f(x)\,dx$. Apply the Mean Value Theorem to this new function F to find an expression for c_k in terms of x_k and x_{k-1} such that

$$F(x_k) - F(x_{k-1}) = F'(c_k)(x_k - x_{k-1}).$$

(a) $f(x) = x$ (b) $f(x) = x^2$

(c) $f(x) = x^3$ (d) $f(x) = 1/\sqrt{x}$

2. Consider the function defined by $f(x) = x$, and take $a = 0$, $b > 0$. Show that by taking

$$c_k = \tfrac{1}{2}(x_k + x_{k-1})$$

in Eq. (5), the resulting sum has the constant value $S_n = \frac{1}{2}b^2$, independently of the value of n. (Let $a = x_0$ and $b = x_n$.)

3. Take $f(x) = x^2$, $a = 0$, $b > 0$ and form S_n, Eq. (5), using

$$c_k = \sqrt{\frac{x_k^2 + x_k x_{k-1} + x_{k-1}^2}{3}}.$$

Show that no matter what value n has, the resulting sum S_n has the constant value $S_n = \frac{1}{3}b^3$. (Let $a = x_0$ and $b = x_n$.) What is the limit of S_n as $n \to \infty$? Note that one should substitute $x_k - x_{k-1}$ for Δx and recognize that

$$\left(\frac{x_k^2 + x_k x_{k-1} + x_{k-1}^2}{3}\right)\Delta x = \frac{x_k^3 - x_{k-1}^3}{3}.$$

4. Take $f(x) = x^3$, $a = 0$, $b > 0$, and calculate S_n, Eq. (5), taking

$$c_k = \sqrt[3]{\frac{x_k^3 + x_k^2 x_{k-1} + x_k x_{k-1}^2 + x_{k-1}^3}{4}}.$$

Express the result in a form which is independent of the number of subdivisions n. (Let $a = x_0$ and $b = x_n$.)

5. Take $f(x) = 1/\sqrt{x}$, $a = 1$, $b > 1$. Use the intermediate values

$$c_k = \left(\frac{\sqrt{x_{k-1}} + \sqrt{x_k}}{2}\right)^2$$

and calculate S_n, Eq. (5). Express the result in a form which is independent of the number of subdivisions. (Let $a = x_0$ and $b = x_n$.)

Evaluate the definite integrals 6 through 16.

6. $\displaystyle\int_1^2 (2x + 5)\,dx$

7. $\displaystyle\int_0^1 (x^2 - 2x + 3)\,dx$

8. $\displaystyle\int_{-1}^1 (x + 1)^2\,dx$

9. $\displaystyle\int_0^2 \sqrt{4x + 1}\,dx$

10. $\displaystyle\int_0^\pi \sin x\,dx$

11. $\displaystyle\int_0^\pi \cos x\,dx$

12. $\displaystyle\int_{\pi/4}^{\pi/2} \frac{\cos x\,dx}{\sin^2 x}$ (Let $u = \sin x$.)

13. $\displaystyle\int_0^{\pi/6} \frac{\sin 2x}{\cos^2 2x}\,dx$

*14. $\displaystyle\int_0^\pi \sin^2 x\,dx$

*15. $\displaystyle\int_0^{2\pi/\omega} \cos^2(\omega t)\,dt$ (ω constant)

16. $\displaystyle\int_0^1 \frac{dx}{(2x + 1)^3}$

* See Exercise 17, Article 5.8.

17. (a) Express the area between the curves $y = x^2$, $y = 18 - x^2$ as a *limit of a sum* of areas of rectangles.
(b) Evaluate the area of (a) by Eq. (7).

18. Approximate the area under the curve $y = 1/x$ between $x = 1$ and $x = 2$ by five rectangles, each of base $\Delta x = 0.2$. Use the ordinate of the curve at the *midpoint* of each subinterval as the altitude of the approximating rectangle. The following short table of reciprocals may be used for convenience:

x	1.1	1.3	1.5	1.7	1.9
$1/x$	0.909	0.769	0.667	0.588	0.526

(Compare your answer with 0.693, which is the value $\int_1^2 dx/x$ to 3 decimals.)

In Exercises 19 through 25, use Eqs. (17a, b) to find $F'(x)$ for the given functions F:

19. $F(x) = \int_0^x \sqrt{1+t^2}\, dt$

20. $F(x) = \int_1^x \frac{dt}{t}$

21. $F(x) = \int_x^1 \sqrt{1-t^2}\, dt$

22. $F(x) = \int_0^x \frac{dt}{1+t^2}$

23. $F(x) = \int_1^{2x} \cos(t^2)\, dt$

Hint. Use the chain rule with $u = 2x$.

24. $F(x) = \int_1^{x^2} \frac{dt}{1+\sqrt{1-t}}$

25. $F(x) = \int_{\sin x}^0 \frac{dt}{2+t}$

5.10 THE TRAPEZOIDAL RULE FOR APPROXIMATING AN INTEGRAL

Any definite integral may be thought of as an area, or an algebraic sum of signed areas, as discussed in Article 5.9. We know how to evaluate a definite integral when an indefinite integral of the integrand is known. But there are simple integrands, for example $(\sin x)/x$, for which no simple indefinite integral is known. In such instances, if the integrand is continuous, the definite integral still exists and we may wish to evaluate it numerically. We can often obtain good accuracy by using numerical methods for approximating a definite integral. These methods are also important for computations done by machines. One of the simplest of these numerical methods is the *trapezoidal rule*, which we shall now derive. (See Fig. 5.28.)

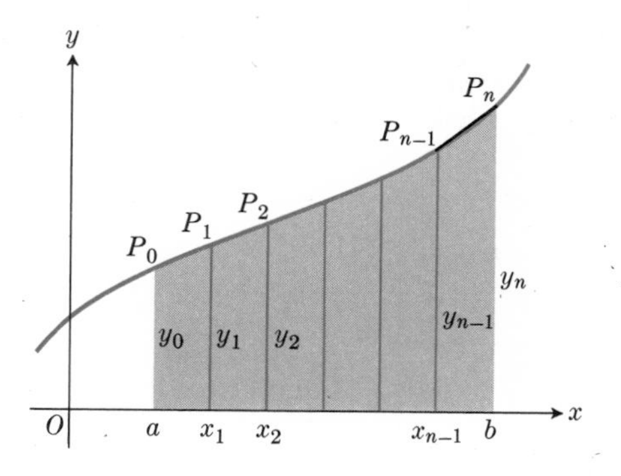

5.28 The sum of the areas of trapezoids approximates the area under the curve.

Suppose that the definite integral $\int_a^b f(x)\, dx$ is to be evaluated. We divide the interval $a \le x \le b$ into n subintervals, each of length $\Delta x = (b - a)/n$, in the usual way by inserting the points

$$
\begin{aligned}
x_1 &= a + \Delta x, \\
x_2 &= a + 2\,\Delta x, \\
&\vdots \\
x_{n-1} &= a + (n-1)\,\Delta x
\end{aligned}
$$

between $\quad x_0 = a \quad$ and $\quad x_n = b.$

The integral from a to b is just the sum of the integrals from a to x_1, from x_1 to x_2, and so on, and, finally, the integral from x_{n-1} to b:

$$
\begin{aligned}
\int_a^b f(x)\, dx &= \int_a^{x_1} f(x)\, dx + \int_{x_1}^{x_2} f(x)\, dx + \cdots \\
&\qquad + \int_{x_{n-1}}^b f(x)\, dx \\
&= \sum_{k=1}^{n} \int_{x_{k-1}}^{x_k} f(x)\, dx.
\end{aligned}
$$

The integral over the first subinterval is now *approximated* by the area of the trapezoid $aP_0P_1x_1$, which equals $\frac{1}{2}(y_0 + y_1)\,\Delta x$; over the second subinterval by the area of the trapezoid $x_1P_1P_2x_2$, which is $\frac{1}{2}(y_1 + y_2)\,\Delta x$; and so on. The *trapezoidal rule* is: To estimate the definite integral $\int_a^b f(x)\,dx$, use the *trapezoidal approximation* T given by

$$\begin{aligned} T &= \tfrac{1}{2}(y_0 + y_1)\,\Delta x + \tfrac{1}{2}(y_1 + y_2)\,\Delta x + \cdots \\ &\quad + \tfrac{1}{2}(y_{n-2} + y_{n-1})\,\Delta x + \tfrac{1}{2}(y_{n-1} + y_n)\,\Delta x \qquad \text{(1a)} \\ &= (\tfrac{1}{2}y_0 + y_1 + y_2 + \cdots + y_{n-1} + \tfrac{1}{2}y_n)\,\Delta x, \end{aligned}$$

where

$$\begin{gathered} y_0 = f(x_0), \\ y_1 = f(x_1),\ \ldots,\ y_{n-1} = f(x_{n-1}), \qquad \text{(1b)} \\ y_n = f(x_n). \end{gathered}$$

Problem 1. Use the trapezoidal rule with $n = 4$ to estimate $\int_1^2 x^2\,dx$ and compare this approximation with the exact value of the integral.

Solution. The exact value of this integral is

$$\int_1^2 x^2\,dx = \frac{x^3}{3}\bigg]_1^2 = \frac{7}{3}.$$

For the trapezoidal approximation we have

$$\begin{gathered} x_0 = a = 1, \quad x_n = b = 2, \quad n = 4, \\ \Delta x = (b - a)/n = (2 - 1)/4 = \tfrac{1}{4}, \end{gathered}$$

so that

$$\begin{aligned} x_0 &= a &&= 1, & y_0 &= f(x_0) = 1^2 = \tfrac{16}{16}, \\ x_1 &= a + \Delta x &&= \tfrac{5}{4}, & y_1 &= f(x_1) = (\tfrac{5}{4})^2 = \tfrac{25}{16}, \\ x_2 &= a + 2\,\Delta x &&= \tfrac{6}{4}, & y_2 &= f(x_2) = (\tfrac{6}{4})^2 = \tfrac{36}{16}, \\ x_3 &= a + 3\,\Delta x &&= \tfrac{7}{4}, & y_3 &= f(x_3) = (\tfrac{7}{4})^2 = \tfrac{49}{16}, \\ x_4 &= b &&= 2, & y_4 &= f(x_4) = (\tfrac{8}{4})^2 = \tfrac{64}{16}, \end{aligned}$$

and

$$\begin{aligned} T &= (\tfrac{1}{2}y_0 + y_1 + y_2 + y_3 + \tfrac{1}{2}y_4)\,\Delta x \\ &= (\tfrac{75}{8})\,\Delta x = \tfrac{75}{32} = 2.34375. \end{aligned}$$

Thus the approximation is too large by about 1 part in 233, or less than one-half of one percent.

The graph in Fig. 5.29 shows clearly that each approximating trapezoid contains slightly more area than the corresponding strip of area under the curve. But the polygonal line $P_0P_1P_2P_3P_4$ lies very close to the curve, so we would expect the area under the curve to be quite close to the area of the polygon $AP_0P_1P_2P_3P_4BA$, which is the trapezoidal approximation.

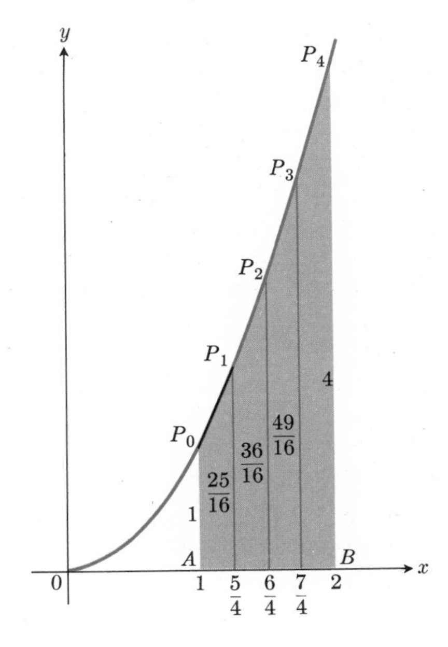

5.29 Trapezoidal approximation of the area under the curve.

Accuracy of the trapezoidal approximation

It is easy to demonstrate that as n increases and Δx approaches zero, the trapezoidal approximation approaches the exact value of the definite integral as limit. For we have

$$\begin{aligned} T &= (y_1 + y_2 + \cdots + y_n)\,\Delta x + \tfrac{1}{2}(y_0 - y_n)\,\Delta x \\ &= \sum_{k=1}^{n} f(x_k)\,\Delta x + \tfrac{1}{2}[f(a) - f(b)]\,\Delta x. \qquad (2) \end{aligned}$$

When n increases without bound and Δx approaches zero, the sum that is indicated by the sigma approaches $\int_a^b f(x)\,dx$ as limit, and the last term in (2) approaches zero. Therefore

$$\lim T = \int_a^b f(x)\,dx.$$

This means, of course, that by taking n sufficiently large the difference between T and the integral can be made as small as desired.

By methods studied in more advanced calculus (an extension of the Mean Value Theorem), it is possible to prove* that if f is continuous on

$$a \leq x \leq b,$$

and twice differentiable on $a < x < b$, then there is a number e between a and b such that

$$\int_a^b f(x)\,dx = T - \frac{b-a}{12} f''(e) \cdot (\Delta x)^2. \quad (3)$$

Thus as Δx approaches zero the "error," that is, the difference between the integral and the trapezoidal approximation, approaches zero as the *square* of Δx. By estimating the size of the second derivative between a and b, we can get a good estimate of how accurate T is as an approximation to the integral.

Example 2. In Problem 1 above, $f(x) = x^2$, $f''(x) = 2$, and the predicted error in (3) is

$$-\frac{b-a}{12} f''(e) \cdot (\Delta x)^2 = -\frac{2-1}{12} \cdot 2 \cdot \frac{1}{16} = -\frac{1}{96}.$$

This is precisely what we find when we subtract $T = \frac{75}{32}$ from $\int_1^2 x^2\,dx = \frac{7}{3}$, since $\frac{7}{3} - \frac{75}{32} = -\frac{1}{96}$. Here we are able to give the error *exactly*, since the second derivative of $f(x) = x^2$ is a constant and we have no uncertainty caused by not knowing e in the term $f''(e)$. Of course we are not always this lucky; in most cases the best we can do is *estimate* the difference between the integral and T.

Problem 2. Suppose the trapezoidal rule, with $n = 5$, is used to approximate $\int_1^2 (1/x)\,dx$. Estimate the error.

Solution. Here

$$f(x) = x^{-1}, \quad f'(x) = -x^{-2}, \quad f''(x) = 2x^{-3} = 2/x^3.$$

Since $f''(x)$ steadily decreases

$$\text{from} \quad f''(1) = 2 \qquad \text{to} \quad f''(2) = \tfrac{1}{4},$$

we can estimate the error only as lying between

$$-\tfrac{1}{12} \cdot 2 \cdot \tfrac{1}{25} = -\tfrac{1}{150} \approx -0.0067$$

and

$$-\tfrac{1}{12} \cdot \tfrac{1}{4} \cdot \tfrac{1}{25} = -\tfrac{1}{1200} \approx -0.0008.$$

Therefore the integral lies between

$$T - \tfrac{1}{150} \quad \text{and} \quad T - \tfrac{1}{1200},$$

so that it is slightly less than T, but differs from it by less than 0.01. In other words, we can get about two-decimal place accuracy for the integral by using the trapezoidal rule with $n = 5$, $\Delta x = \frac{1}{5}$. By taking $n = 10$, $\Delta x = \frac{1}{10}$, the error is cut by a factor of approximately 4. [The e of $f''(e)$ in the error may change when n does, and hence we cannot say that the error is cut precisely by a factor of 4.]

* See, for example, J. M. H. Olmsted, *Intermediate Analysis*, Appleton-Century-Crofts, 1956, p. 145.

EXERCISES

1. Interpret the meaning of the sign of the correction term in Eq. (3) if the graph of $y = f(x)$ is
   (a) concave upward over $a < x < b$,
   (b) concave downward.

   Illustrate both cases with sketches.

Test the accuracy of the trapezoidal approximation to the definite integral $\int_1^2 f(x)\,dx$ in Exercises 2 through 5.

2. $f(x) = x$, $a = 0$, $b = 2$, $n = 4$
3. $f(x) = x^3$, $a = 0$, $b = 2$, $n = 4$
4. $f(x) = \sqrt{x}$, $a = 0$, $b = 2$, $n = 4$
5. $f(x) = 1/x^2$, $a = 1$, $b = 2$, $n = 2$

5.11 SOME COMMENTS ON NOTATION

In each application of the definite integral it is reasonably easy to set up sums that approximate the answer to some physical problem. In general, it will even be possible (though not necessary) to find *particular* choices of the points c_k such that the *particular* sums

$$S_n = \sum_{k=1}^{n} f(c_k)\,\Delta x$$

give the *exact* value. Then

$$\lim_{n\to\infty} S_n = \int_a^b f(x)\,dx = F(x)]_a^b = F(b) - F(a)$$

also gives the *exact* value. But since all sums S_n give the *same* answer in the limit [provided that the function $f(x)$ is continuous for $a \leq x \leq b$; this condition will usually be satisfied by the functions we shall encounter], we could get the same limit from

$$S_n = \sum_{k=1}^{n} f(x_k)\,\Delta x. \tag{1a}$$

Finally, we modify this notation slightly by dropping the subscript k entirely and writing simply

$$S_a^b = \sum_{a}^{b} f(x)\,\Delta x, \tag{1b}$$

where we write a and b to indicate that we have a sum of expressions, each of the form $f(x)\,\Delta x$, extending over a set of subintervals from $x = a$ to $x = b$. Of course it is not the *sum* (1b) that we are interested in, but rather the *limit* of the sum as given by the definite integral, and the notation (1b) is most suggestive of the final form, namely,

$$\lim_{\Delta x \to 0} S_a^b = \lim_{\Delta x \to 0} \sum_{a}^{b} f(x)\,\Delta x = \int_a^b f(x)\,dx = F(x)\Big]_a^b. \tag{2}$$

As remarked in Article 5.9, it is the close relationship between sums and integrals which leads to the adoption of the symbol $\int$ (a modified S) to denote integration.

It should also be noted that *since it is the integral* that gives the *exact* answer in (2), and since the expression $f(x)\,dx$ after the integral sign is the same as the differential of $F(x)$,

$$\frac{dF(x)}{dx} = f(x), \tag{3a}$$

$$dF(x) = f(x)\,dx, \tag{3b}$$

we would arrive at the *correct final result* for the area under the curve $y = f(x)$ between $x = a$ and $x = b$, say, if we were to write

$$dA = f(x)\,dx, \qquad A = \int_a^b f(x)\,dx. \tag{4}$$

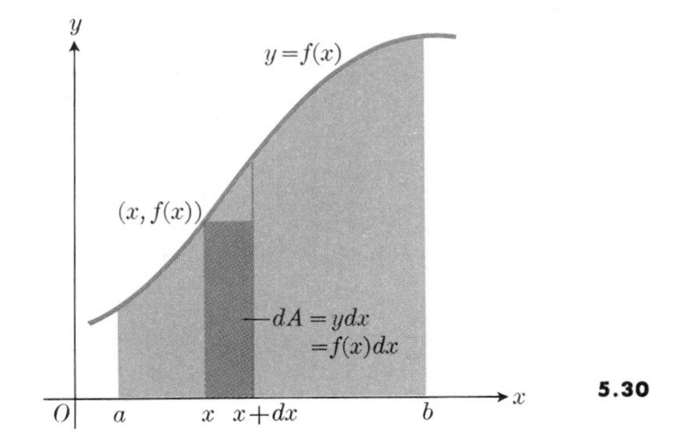

5.30

It is easy to attach a geometric interpretation to (4), but it is also easy to be misled by this interpretation into feeling that the integral in (4) gives only an approximation to the area, rather than giving its exact value. Enough has already been said on this latter point, however, so that the careful reader need not be a victim of this delusion. If we accept the interpretation simply as a short cut for setting up the *integral whose value gives the exact result,* we may thereby gain from its simplicity. Explicitly, if we again think of the area from the left boundary a to a right boundary x, this area is a function of x, with derivative $dA/dx = y = f(x)$, as we found in Article 5.8. When we multiply both sides of this equation by dx, we get the result $dA = f(x)\,dx$, as in (4). But if we think of dx as a small increment in x, we may also interpret this as the area of a small rectangle with base extending along the x-axis from x to $x + dx$, and with altitude $f(x) = y$ equal to the ordinate of the curve (see Fig. 5.30). Then this rectangle gives an *approximation* to that portion of the area under the curve lying between x and $x + dx$, and we think of the integral

$$A = \int_a^b f(x)\,dx$$

as adding together all of these small rectangles from $x = a$ to $x = b$ and then taking the *limit* of the sum so as to give the *exact value.*

It should be borne in mind that what has just been discussed should in no way obscure the fact that a *precise* formulation of the problem would involve a process of

(a) subdividing the interval (a, b),

(b) forming a sum $\sum_a^b f(x)\,\Delta x$,

(c) taking the limit $\lim_{\Delta x \to 0} \sum_a^b f(x)\,\Delta x$, and

(d) applying the fundamental theorem to evaluate this limit:

$$\lim_{\Delta x \to 0} \sum_a^b f(x)\,\Delta x = \int_a^b f(x)\,dx = F(x)]_a^b.$$

Insofar as the short cut described above leads to the same final answer, without leading to confusion or feelings of misgiving, it may be helpful and time-saving. But whenever there is doubt that the final answer may only be an approximation instead of a mathematically exact answer, a reconsideration should be made, along the lines which led up to the fundamental theorem. It is probably desirable, at least in the beginning, to go through the precise formulation first, and then it is but a matter of seconds to repeat the setup of the problem by the short-cut method.

5.12 SUMMARY

Much that we have done in this chapter is related to area. This is useful and gives us a way of interpreting sums like those in Eq. (5), Article 5.9, and Eq. (1a) in Article 5.11. But it is also good to express the main results in ways that don't depend upon area. In this article we shall summarize six results of major importance. Although we have appealed to the notion of area in presenting them in earlier sections, they can all be proved by purely analytical methods, with no reference to area. (See, for example, Apostol, *Modern Mathematical Analysis;* Buck, *Advanced Calculus;* or Rudin, *Principles of Analysis.*)

Once again we start with a function $f(x)$ defined on an interval $[a, b]$. We partition the interval into n subintervals by inserting points $x_1, x_2, \ldots, x_{n-1}$ between $a = x_0$ and $b = x_n$. We do not require that the points be uniformly spaced. The kth subinterval has length

$$\Delta x_k = x_k - x_{k-1},$$

which may vary with k. In the kth subinterval a number c_k is chosen arbitrarily, for each k from 1 through n. Then we form the sum

$$\sum_{k=1}^{n} f(c_k) \cdot (x_k - x_{k-1}). \tag{1}$$

This sum depends on the function f, on the points $x_0, x_1, x_2, \ldots, x_n$, and on the c's. However, for a given function f and interval $[a, b]$ it may happen that all sums like (1) are nearly equal to some constant, provided that the subintervals are all sufficiently short. If this is true, then that constant is called the *Riemann integral* of f from a to b, and is variously denoted by $R_a^b(f)$, or $\int_a^b f$, or $\int_a^b f(x)\,dx$, or $\int_a^b f(t)\,dt$, or $\int_a^b f(u)\,du$, and so on.

Definition of Riemann Integral. *Let f be a function whose domain includes the interval $[a, b]$, $a < b$. Let f have the property that there exists a number $R_a^b(f)$ so related to f that to each positive number ϵ there corresponds a positive number δ such that every sum* (1) *differs from $R_a^b(f)$ by less than ϵ whenever all subintervals have lengths Δx_k less than δ. Then:*

(a) *f is said to be Riemann-integrable* (*or, more briefly, integrable*) *over $[a, b]$, and*

(b) *$R_a^b(f)$ is called the Riemann integral* (*or, more briefly, the integral*) *of f from a to b.*

Theorem 3. *If f is Riemann-integrable over an interval $[a, b]$, then it is also integrable over any subinterval contained in $[a, b]$.*

Theorem 4. *Every function that is continuous on a closed bounded interval $[a, b]$ is Riemann-integrable there.*

Remark. Although we are mainly interested in continuous functions, they are not the only ones that are integrable. All bounded functions that are *piecewise*-continuous are also integrable. "Piecewise-continuous" means that the interval can be divided into a finite number of nonoverlapping open subintervals over each of which the function is continuous. "Bounded" means that, for some finite

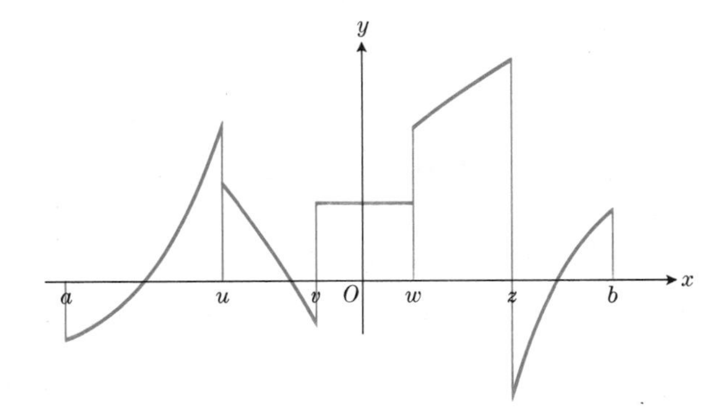

5.31 Graph of a piecewise-continuous function.

constant M, $|f(x)| \leq M$ for all values of x in the interval. Figure 5.31 shows the graph of a bounded piecewise-continuous function over $[a, b]$. For this example, the integral $R_a^b(f)$ would be computed as the sum

$$R_a^b(f) = R_a^u(f) + R_u^v(f) + R_v^w(f) + R_w^z(f) + R_z^b(f).$$

If $b < a$ and $R_b^a(f)$ exists, then we define $R_a^b(f)$ to be its negative:

$$R_a^b(f) = -R_b^a(f). \tag{2}$$

If $a = b$, we define $R_a^a(f)$ to be zero.

Theorem 5. *Let a, b, c be three numbers belonging to an interval over which the Riemann integral f exists. Then*

$$R_a^c(f) = R_a^b(f) + R_b^c(f). \tag{3}$$

Theorem 6. *Let f be Riemann-integrable over $[a, b]$. Let x be any number between a and b at which f is continuous. Then $R_a^x(f)$ is differentiable at x and its derivative there is $f(x)$:*

$$\frac{d}{dx} R_a^x(f) = f(x). \tag{4}$$

Theorem 7. The Mean Value Theorem for Integrals. *Let f be continuous on $[a, b]$. Then there exists at least one number c between a and b such that*

$$R_a^b(f) = f(c) \cdot (b - a). \tag{5}$$

Theorem 8. Fundamental Theorem of Integral Calculus. *Let f be Riemann-integrable over $[a, b]$. Let F be any continuous indefinite integral of f, that is,*

$$F'(x) = f(x) \quad \text{for } a < x < b.$$

Then

$$R_a^b(f) = F(b) - F(a). \tag{6}$$

EXERCISES

1. Rewrite Equations (2) through (6), Article 5.12, using the notation $\int_a^b f(t)\,dt$ in place of $R_a^b(f)$ for the Riemann integral of f from a to b.
2. Sketch graphs of the following piecewise-continuous functions on (0, 2) and compute $\int_0^2 f(x)\,dx$.

   (a) $f(x) = \begin{cases} x & \text{for } 0 \leq x < 1 \\ \sin(\pi x) & \text{for } 1 \leq x \leq 2 \end{cases}$

   (b) $f(x) = \begin{cases} \sqrt{1-x} & \text{for } 0 \leq x \leq 1 \\ (7x-6)^{-1/3} & \text{for } 1 < x \leq 2 \end{cases}$

3. Find $F'(x)$ for $x = \frac{1}{2}$ and for $x = \frac{3}{2}$

   if $$F(x) = \int_0^x f(t)\,dt$$

   and

   (a) f is defined as in Exercise 2(a) above,

   (b) f is defined as in Exercise 2(b) above, and

   (c) $f(t) = \dfrac{\sin(\pi t)}{1+t}$.

REVIEW QUESTIONS AND EXERCISES

1. The process of "indefinite integration" is sometimes called "antidifferentiation." Explain why the two terms should be synonymous.
2. Can there be more than one indefinite integral of a given function? If more than one exist, how are they related? What theorem of Chapter 4 is the key to this relationship?
3. If the acceleration of a moving body is given as a function of time (t), what further information do you need to find the law of motion, $s = f(t)$? How is s found?

4. Develop the law of cosines from the formula for the distance between two points.

5. Develop the formula for $\cos(A - B)$ from the law of cosines.

6. Write expressions for $\sin(A + B)$ and $\sin(A - B)$. Set $\alpha = A + B, \beta = A - B$ and develop a formula for $\sin\alpha - \sin\beta$.

7. Under what assumptions is it true that

$$\lim (\sin\theta/\theta) = 1?$$

Prove the result.

8. Write expressions for $\cos(A + B)$ and $\cos(A - B)$. Set $\alpha = A + B, \beta = A - B$ and develop a formula for $\cos\alpha - \cos\beta$. Then use this result to develop a formula for the derivative of $\cos x$, from the definition

$$\frac{d(\cos x)}{dx} = \lim_{\Delta x \to 0} \frac{\cos(x + \Delta x) - \cos x}{\Delta x}.$$

9. Let A_n be the area bounded by a regular n-sided polygon inscribed in a circle of radius r. Show that $A_n = (n/2)r^2 \sin(2\pi/n)$. Find $\lim A_n$ as $n \to \infty$. Does the result agree with what you know about the area of a circle?

10. A function G is defined by

$$G(x) = \int_0^x (1 + t^2)^{1/2}\, dt.$$

What is the largest possible domain for G? For which values of x in this domain is G continuous? Differentiable? For what values of x does $G'(x) = 3$? Where is $G'(x) = 0$? What is the minimum value of $G'(x)$?

11. Write a formula for the area bounded above by the semicircle $y = \sqrt{r^2 - t^2}$, below by the t-axis, on the left by the y-axis, and on the right by the line $t = x$. Without making any calculations, what do you know must be the derivative, with respect to x, of this expression? Explain.

12. What property of continuous functions is used in the proof of the Mean Value Theorem for Areas?

13. Is every Riemann-integrable function also differentiable? Give a reason for your answer.

14. Is the function f, defined on the domain $0 \leq x \leq 1$ by

$$f(x) = \begin{cases} 0 \text{ when } x \text{ is rational,} \\ 1 \text{ when } x \text{ is irrational,} \end{cases}$$

Riemann-integrable over (0, 1)? Give a reason for your answer.

15. If you could not find an indefinite integral of $f(x)\,dx$, how could you still calculate $\int_a^b f(x)\,dx$ to any specified number of decimal places of accuracy, assuming f is continuous over $a \leq x \leq b$? Illustrate for

$$f(x) = 1/x, \quad a = 1, \quad b = 2,$$

and one-decimal-place accuracy.

MISCELLANEOUS EXERCISES

Solve the differential equations in Exercises 1 through 5.

1. $\dfrac{dy}{dx} = xy^2$

2. $\dfrac{dy}{dx} = \sqrt{1 + x + y + xy}$

3. $\dfrac{dy}{dx} = \dfrac{x^2 - 1}{y^2 + 1}$

4. $\dfrac{dx}{dy} = \dfrac{y - \sqrt{y}}{x + \sqrt{x}}$

5. $\dfrac{dx}{dy} = \left(\dfrac{2 + x}{3 - y}\right)^2$

6. Solve each of the following differential equations subject to the prescribed initial conditions.

   (a) $\dfrac{dy}{dx} = x\sqrt{x^2 - 4}, \quad x = 2, \quad y = 3$

   (b) $\dfrac{dy}{dx} = xy^3, \quad x = 0, \quad y = 1$

7. Can there be a curve satisfying the following conditions? When $x = 0$, then $y = 0$ and $dy/dx = 1$, and also d^2y/dx^2 is everywhere equal to zero. Give a reason for your answer.

8. Find an equation of the curve whose slope at the point (x, y) is $3x^2 + 2$, if the curve is required to pass through the point $(1, -1)$.

9. A particle moves along the x-axis. Its acceleration is $a = -t^2$. At $t = 0$, the particle is at the origin. In the course of its motion, it reaches the point $x = b$, where $b > 0$, but no point beyond b. Determine its velocity at $t = 0$.

10. A particle moves with acceleration $a = \sqrt{t} - (1/\sqrt{t})$. Assuming that the velocity $v = 2$ and the distance $s = 5$ when $t = 0$, find

    (a) the velocity v in terms of t,

    (b) the distance s in terms of t.

11. A particle is accelerated with acceleration $3 + 2t$, where t is the time. At $t = 0$, the velocity is 4. Find the velocity as a function of time and the distance between the position of the particle at time zero and at time 4.

12. The acceleration of a particle moving along the x-axis is given by $d^2x/dt^2 = -4x$. If the particle starts from rest at $x = 5$, find the velocity when it first reaches $x = 3$.

13. Let $f(x)$, $g(x)$ be two continuously differentiable functions satisfying the relationships

$$f'(x) = g(x), \qquad f''(x) = -f(x).$$

Let

$$h(x) = f^2(x) + g^2(x).$$

If $h(0) = 5$, find $h(10)$.

14. The family of straight lines $y = ax + b$ (a, b arbitrary constants) can be characterized by the relation $y'' = 0$. Find a similar relation satisfied by the family of all circles $(x - h)^2 + (y - h)^2 = r^2$, where h and r are arbitrary constants. [*Hint.* Eliminate h and r from the set of three equations including the given one and two obtained by successive differentiation.]

15. Assume that the brakes of an automobile produce a constant deceleration of k ft/sec^2.

(a) Determine what k must be to bring an automobile traveling 60 mi/hr (88 ft/sec) to rest in a distance of 100 ft from the point where the brakes are applied.

(b) With the same k, how far would a car traveling 30 mi/hr travel before being brought to a stop?

16. Solve the differential equation $dy/dx = x\sqrt{1 + x^2}$ subject to the condition that $y = -2$ when $x = 0$.

17. The acceleration due to gravity is -32 ft/sec^2. A stone is thrown upward from the ground with a speed of 96 ft/sec. Find the height to which the stone rises in t sec. What is the maximum height reached by the stone?

18. Show that the following procedure will produce a continuous polygonal "curve" whose slope at the point (x_k, y_k) will be $f(x_k)$. First, sketch the auxiliary curve $C: y = xf(x)$. Then through the point $P_0(x_0, y_0)$, draw the ordinate $x = x_0$ intersecting C in

$$Q_0(x_0, x_0 f(x_0)).$$

Through P_0 draw a line segment P_0P_1 parallel to OQ_0. Then the slope of P_0P_1 is $f(x_0)$. Now take $P_1(x_1, y_1)$ on this segment to lie close to P_0. For example, take $x_1 = x_0 + h$, where h is small. Then find $Q_1(x_1, x_1 f(x_1))$ on C, and through P_1 draw a line segment P_1P_2 parallel to OQ_1. Continue the process by taking $P_2(x_2, y_2)$ close to P_1, with $x_2 = x_1 + h$; then find $Q_2(x_2, x_2 f(x_2))$ on C, and through P_2 draw a line segment P_2P_3 parallel to OQ_2; and so on. M. S. Klamkin, *Amer. Math. Monthly*, 1955.

19. (a) Apply the procedure of Exercise 18 to the case $f(x) = 1/x$ with $x_0 = 1$, $y_0 = 1$, and $h = \frac{1}{4}$. Continue the process until you reach the point $P_4(x_4, y_4)$. What is your value of y_4?

(b) Repeat the construction of part (a), but with $h = \frac{1}{8}$, and continue until you reach $x_8 = 2$. What is your value of y_8?

20. A body is moving with velocity 16 ft/sec when it is suddenly subjected to a deceleration. If the deceleration is proportional to the square root of the velocity, and the body comes to rest in 4 sec,

(a) how fast is the body moving 2 sec after it begins decelerating, and

(b) how far does the body travel before coming to rest?

Evaluate the integrals 21 through 33.

21. $\displaystyle\int \frac{x^3 + 1}{x^2}\,dx$

22. $\displaystyle\int y\sqrt{1 + y^2}\,dy$

23. $\displaystyle\int t^{1/3}(1 + t^{4/3})^{-7}\,dt$

24. $\displaystyle\int \frac{(1 + \sqrt{u})^{1/2}\,du}{\sqrt{u}}$

25. $\displaystyle\int \frac{dr}{\sqrt[3]{(7 - 5r)^2}}$

26. $\displaystyle\int \cos 4x\,dx$

27. $\displaystyle\int \sin^2 3x \cos 3x\,dx$

28. $\displaystyle\int \frac{\cos x\,dx}{\sqrt{\sin x}}$

29. $\displaystyle\int \cos(2x - 1)\,dx$

30. $\displaystyle\int \frac{y\,dy}{\sqrt{25 - 4y^2}}$

31. $\displaystyle\int \frac{dt}{t\sqrt{2t}}$

32. $\displaystyle\int (x^2 - \sqrt{x})\,dx$

33. $\displaystyle\int \frac{dx}{(2 - 3x)^2}$

Find dy/dx in Exercises 34 through 38.

34. $y = \cos(1 - 2x)$

35. $y = \dfrac{\cos x}{\sin x}$

36. $y = \sec^2(5x)$

37. $y = \sin^4 5x$

38. $y^3 = \sin^3 x + \cos^3 x$

39. Given $y = 3 \sin 2x$ and $x = u^2 + \pi$, find the value of dy/du when $u = 0$.

40. If $0 < x < \pi/2$, prove that $x > \sin x > 2x/\pi$.

41. If one side and the opposite angle of a triangle are fixed, prove that the area is a maximum when the triangle is isosceles.

42. A light hangs above the center of a table of radius r ft. The illumination at any point on the table is directly proportional to the cosine of the angle of incidence (i.e., the angle a ray of light makes with the normal) and is inversely proportional to the square of the distance from the light. How far should the light be above the table in order to give the strongest illumination at the edge of the table?

43. If A, B, C are constants, $AB \neq 0$, prove that the graph of the curve $y = A \sin(Bx + C)$ is always concave toward the x-axis and that its points of inflection coincide with its points of intersection with the x-axis.

44. Two particles move on the same straight line so that their distances from a fixed point O, at any time t, are

$$x_1 = a \sin bt \quad \text{and} \quad x_2 = a \sin[bt + (\pi/3)],$$

where a and b are constants, $ab \neq 0$. Find the greatest distance between them.

45. If the identity $\sin(x + a) = \sin x \cos a + \cos x \sin a$ is differentiated with respect to x, is the resulting equation also an identity? Does this principle apply to the equation $x^2 - 2x - 8 = 0$? Explain why or why not.

46. A revolving beacon light in a lighthouse $\frac{1}{2}$ mile offshore makes two revolutions per minute. If the shoreline is a straight line, how fast is the ray of light moving along the shore when it passes a point one mile from the lighthouse?

47. The coordinates of a moving particle are

$$x = a \cos^3 \theta, \qquad y = a \sin^3 \theta.$$

If a is a positive constant and θ increases at the constant rate of ω rad/sec, find the magnitude of the velocity vector.

48. The area bounded by the x-axis, the curve $y = f(x)$, and the lines $x = 1$, $x = b$ is equal to

$$\sqrt{b^2 + 1} - \sqrt{2} \quad \text{for all } b > 1.$$

Find $f(x)$.

49. Evaluate the following limits.

(a) $\displaystyle\lim_{x \to 0} \frac{2 \sin 5x}{3x}$

(b) $\displaystyle\lim_{x \to 0} \sin 5x \cot 3x$

(c) $\displaystyle\lim_{x \to 0} x \csc^2 \sqrt{2x}$

50. Find dy/dx and d^2y/dx^2 if

$$x = \cos 3t, \qquad y = \sin^2 3t.$$

51. Let $f(x)$ be a continuous function. Express

$$\lim_{n \to \infty} \frac{1}{n}\left[f\left(\frac{1}{n}\right) + f\left(\frac{2}{n}\right) + \cdots + f\left(\frac{n}{n}\right)\right]$$

as a definite integral.

52. Use the result of Exercise 51 to evaluate

(a) $\displaystyle\lim_{n \to \infty} \frac{1}{n^{16}}[1^{15} + 2^{15} + 3^{15} + \cdots + n^{15}]$,

(b) $\displaystyle\lim_{n \to \infty} \frac{\sqrt{1} + \sqrt{2} + \sqrt{3} + \cdots + \sqrt{n}}{n^{3/2}}$,

(c) $\displaystyle\lim_{n \to \infty} \frac{1}{n}\left[\sin\frac{\pi}{n} + \sin\frac{2\pi}{n} + \sin\frac{3\pi}{n} + \cdots + \sin\frac{n\pi}{n}\right]$.

53. Find

(a) $\displaystyle\lim_{h \to 0} \frac{1}{h} \int_x^{x+h} \frac{du}{u + \sqrt{u^2 + 1}}$,

(b) $\displaystyle\lim_{x \to x_1} \left[\frac{x}{x - x_1} \int_{x_1}^{x} f(t)\, dt\right]$.

54. Variables x and y are related by the equation

$$x = \int_0^y \frac{1}{\sqrt{1 + 4t^2}}\, dt.$$

Show that d^2y/dx^2 is proportional to y and find the constant of proportionality.

55. (a) Show that the perimeter P_n of an n-sided regular polygon inscribed in a circle of radius r is

$$P_n = 2nr \sin(\pi/n).$$

(b) Find the limit of P_n as $n \to \infty$. Is the answer consistent with what you know about the circumference of a circle?

APPLICATIONS OF THE DEFINITE INTEGRAL

CHAPTER 6

6.1 INTRODUCTION

In Chapter 5 we discovered a close connection between sums of the form

$$S_a^b = \sum_a^b f(x)\,\Delta x \tag{1}$$

and integration, the inverse of differentiation. When f is continuous on $[a, b]$, we found that the *limit* of S_a^b as Δx approaches zero is just $F(b) - F(a)$, where F is any integral or primitive of f:

$$F(x) = \int f(x)\,dx. \tag{2}$$

We applied this to the problem of computing the area between the x-axis and the graph of $y = f(x)$, $a \leq x \leq b$. In this chapter we shall extend the applications to the following topics: area between two curves, distance, volumes, lengths of curves, areas of surfaces of revolution, average value of a function, center of mass, centroid, work, and hydrostatic force. Further applications will be taken up in Chapter 16 in connection with double and triple integrals.

6.2 AREA BETWEEN TWO CURVES

Suppose that

$$y_1 = f_1(x), \qquad y_2 = f_2(x)$$

define two functions of x that are continuous on $[a,b]$, and furthermore suppose that

$$f_1(x) \geq f_2(x) \quad \text{for } a \leq x \leq b.$$

Then the y_1-curve is above the y_2-curve from a to b (Fig. 6.1), and we consider the problem of finding the area bounded above by the y_1-curve, below by the y_2-curve, and on the sides by the vertical lines $x = a$, $x = b$.

If the x-interval from a to b is divided into n equal subintervals, each of width

$$\Delta x = \frac{b-a}{n},$$

and a rectangle of width Δx and altitude extending from the y_2-curve to the y_1-curve is used to approximate that portion of the area between the curves

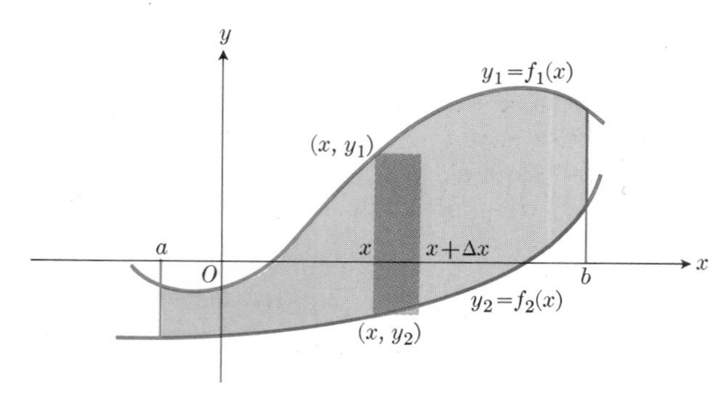

6.1

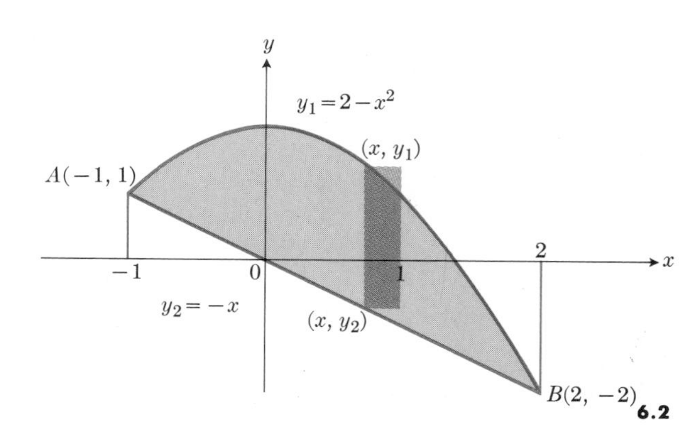

6.2

that lies between x and $x + \Delta x$, we find the area of such a rectangle is

$$(y_1 - y_2)\,\Delta x = [f_1(x) - f_2(x)]\,\Delta x.$$

The total area in question is *approximated* by adding together the areas of all such rectangles:

$$A \approx \sum_a^b [f_1(x) - f_2(x)]\,\Delta x.$$

Finally, if we let $\Delta x \to 0$, we obtain the exact area:

$$A = \lim_{\Delta x \to 0} \sum_a^b [f_1(x) - f_2(x)]\,\Delta x$$
$$= \int_a^b [f_1(x) - f_2(x)]\,dx. \qquad (1)$$

Problem. Find the area (Fig. 6.2) bounded by the parabola

$$y = 2 - x^2$$

and the straight line

$$y = -x.$$

Solution. We first find where the curves intersect by finding points that satisfy both equations simultaneously. That is, we solve

$$2 - x^2 = -x, \quad \text{or} \quad x^2 - x - 2 = 0:$$
$$(x - 2)(x + 1) = 0, \qquad x = -1, 2.$$

The points of intersection are thus $A(-1, 1)$ and $B(2, -2)$. For all values of x between -1 and $+2$, the curve $y_1 = 2 - x^2$

is above the line $y_2 = -x$

by an amount

$$y_1 - y_2 = (2 - x^2) - (-x) = 2 - x^2 + x.$$

This is the altitude of a typical rectangle used to approximate that portion of the area lying between x and $x + \Delta x$. The total area is approximated by

$$A_{-1}^{2} \approx \sum_{-1}^{2} (y_1 - y_2)\,\Delta x = \sum_{-1}^{2} (2 - x^2 + x)\,\Delta x,$$

and is given exactly by

$$A_{-1}^{2} = \lim_{\Delta x \to 0} \sum_{-1}^{2} (2 - x^2 + x)\,\Delta x$$
$$= \int_{-1}^{2} (2 - x^2 + x)\,dx = 4\tfrac{1}{2}.$$

EXERCISES

1. Make a sketch to represent a region that is bounded on the right by a continuous curve $x = f(y)$, on the left by a continuous curve $x = g(y)$, below by the line $y = a$, and above by the line $y = b$. Divide the region into n horizontal strips each of altitude $\Delta y = (b - a)/n$ and express the area of the enclosed region
   (a) as a limit of a sum of areas of rectangles, and
   (b) as an appropriate definite integral.

2. Find the area bounded by

   (a) the x-axis and the curve $y = 2x - x^2$,
   (b) the y-axis and the curve $x = y^2 - y^3$,
   (c) the curve $y^2 = x$ and the line $x = 4$,
   (d) the curve $y = 2x - x^2$ and the line $y = -3$,
   (e) the curve $y = x^2$ and the line $y = x$,
   (f) the curve $x = 3y - y^2$ and the line $x + y = 3$,
   (g) the curves $y = x^4 - 2x^2$ and $y = 2x^2$.

3. Find the area of the "triangular" shaped region in the first quadrant bounded by the y-axis and the curves $y = \sin x$, $y = \cos x$.

4. The area bounded by the curve $y = x^2$ and the line $y = 4$ is divided into two equal portions by the line $y = c$. Find c.

5. Find the area bounded by the curve $\sqrt{x} + \sqrt{y} = 1$ and the coordinate axes.

6. Make a sketch to illustrate the following alternative way of arriving at Eq. (1) for the area between two curves $y = f_1(x)$ and $y = f_2(x)$. Draw a line parallel to the x-axis and below both curves for $a \leq x \leq b$. With no loss of generality, we may take this line to be C units below the x-axis for some positive number C.

   (a) If this line is taken as a new x-axis, show that the ordinates on the curves are given by

   $$y = f_1(x) + C \quad \text{and} \quad y = f_2(x) + C. \tag{i}$$

   Let A_1 be the area under the curve $f_1(x) + C$ over $[a, b]$, and let A_2 be the corresponding area under the curve $f_2(x) + C$. By results of Chapter 5, we know that

   $$A_1 = \int_a^b [f_1(x) + C]\, dx$$

   and

   $$A_2 = \int_a^b [f_2(x) + C]\, dx.$$

   Since the area between the two curves is $A_1 - A_2$, (b) subtract the two integrals of (i) above and deduce Eq. (1).

   (c) Show that the technique described above is equivalent to shifting both original curves upward C units, and therefore does not change the area between the curves.

6.3 DISTANCE

As a second application of the basic principles involved in the use of the Fundamental Theorem of the calculus, we shall calculate the distance traveled by a body moving with velocity

$$v = f(t). \tag{1}$$

To simplify the discussion we shall assume that $f(t)$ is positive, as well as being continuous, for $a \leq t \leq b$. This means that the body moves only in one direction and does no backing up.

Now there are two ways in which we can calculate the distance traveled by the body between $t = a$ and $t = b$.

First method. If we can integrate the differential equation

$$ds = f(t)\, dt, \tag{2}$$

which we get by substituting ds/dt for v in Eq. (1), then we can determine the position s of the body as a function of t, say

$$s = F(t) + C. \tag{3}$$

The distance traveled by the body between $t = a$ and $t = b$ is then given by

$$s\Big]_{t=a}^{t=b} = F(t) + C\Big]_a^b = F(b) - F(a).$$

We recognize this, of course, as saying that the distance is given by the definite integral

$$s\Big]_{t=a}^{t=b} = \int_a^b f(t)\, dt = F(t)\Big]_a^b = F(b) - F(a). \tag{4}$$

Second method. In this method, we imagine the total time interval $a \leq t \leq b$ as divided into n subintervals, each of duration $\Delta t = (b - a)/n$. The velocity at the beginning of the first subinterval is

$$v_1 = f(t_1) = f(a).$$

If Δt is small, the velocity remains nearly constant throughout the time from a to $a + \Delta t$. Hence during the first subinterval of time, the body travels a distance Δs_1, which is approximately equal to $v_1\, \Delta t$:

$$\Delta s_1 \approx v_1\, \Delta t = f(t_1)\, \Delta t. \tag{5}$$

If, instead of using the velocity v_1 at time t_1, we were to use the *average* velocity

$$\overline{v_1} = \frac{\Delta s_1}{\Delta t}, \tag{6a}$$

we could write

$$\Delta s_1 = \overline{v_1}\,\Delta t \tag{6b}$$

exactly. Now, by the Mean Value Theorem, we know that there is some instant, say T_1, between t_1 and $t_1 + \Delta t$ where the instantaneous velocity is equal to the average velocity $\overline{v_1}$. In other words,

$$f(T_1) = \overline{v_1} \quad \text{for some } T_1,\ t_1 < T_1 < t_1 + \Delta t, \tag{6c}$$

and therefore

$$\Delta s_1 = f(T_1)\,\Delta t$$

exactly. Reasoning in the same manner for the second, third, ..., nth subintervals, we conclude that there are instants of time $T_2, T_3, \ldots, T_n$ in these respective intervals such that

$$\begin{aligned} \Delta s_2 &= f(T_2)\,\Delta t, \\ \Delta s_3 &= f(T_3)\,\Delta t, \\ &\vdots \\ \Delta s_n &= f(T_n)\,\Delta t, \end{aligned}$$

where $\Delta s_2, \Delta s_3, \ldots, \Delta s_n$ represent the distances traveled during these respective time subintervals. Therefore the total distance traveled between $t = a$ and $t = b$ is

$$\begin{aligned} s\Big]_{t=a}^{t=b} &= \Delta s_1 + \Delta s_2 + \Delta s_3 + \cdots + \Delta s_n \\ &= f(T_1)\,\Delta t + f(T_2)\,\Delta t + \cdots + f(T_n)\,\Delta t \\ &= \sum_{k=1}^{n} f(T_k)\,\Delta t. \end{aligned} \tag{7}$$

Let us now take finer and finer subdivisions Δt and let n increase without limit. For each n we select the appropriate instants of time $T_1, T_2, \ldots, T_n$, according to the method described above. Then the particular sums used in Eq. (7) tend to the definite integral

$$\lim_{n\to\infty} \sum_{k=1}^{n} f(T_k)\,\Delta t = \int_a^b f(t)\,dt \tag{8}$$

as limit, by virtue of the Fundamental Theorem. On the other hand, these sums all give the distance traveled by the body. Therefore the distance traveled is equal to the integral in Eq. (8) (or the one in Eq. 4).

Remark 1. If, instead of using the velocities at the instants of time $T_1, T_2, \ldots, T_n$ described above, we had used the velocities

$$v_1 = f(t_1), \quad v_2 = f(t_2), \quad \ldots, \quad v_n = f(t_n)$$

at the beginnings of the various subintervals, then we would have had n *approximations* like (5) with subscripts 1, 2, ..., n. Then we would have obtained an *approximation*

$$\Delta s_1 + \Delta s_2 + \cdots + \Delta s_n \approx f(t_1)\,\Delta t + f(t_2)\,\Delta t + \cdots + f(t_n)\,\Delta t,$$

or

$$s\Big]_{t=a}^{t=b} \approx \sum_{k=1}^{n} f(t_k)\,\Delta t. \tag{9}$$

Now since $f(t)$ is continuous, the sums (7) and (9) have the same *limit* as $n \to \infty$. In other words, the approximation (9) gets better as n increases, and the *limit* again gives the exact value, namely, the integral in Eq. (8).

Remark 2. Of course, the two methods give the same result, Eqs. (4) and (8). The second method is useful primarily

(a) because it shows how the simple formula

$$\text{distance} = \text{velocity} \times \text{time},$$

which applies only to the case of *constant* velocity, can be extended to the case of variable velocity, provided we apply it to short subintervals Δt; and

(b) because it can be used [in the form of the approximation (9)] to estimate the distance in cases where the velocity is given empirically by a table of values or a graph instead of by a formula $f(t)$ with known indefinite integral $F(t)$.

Remark 3. If the velocity changes sign during the interval (a, b), then the integral in Eq. (4) gives only the *net* change in s. This permits cancellation of distances traveled forwards and backwards. For example, it would give an answer of 2 miles if the body traveled forward 7 miles and backed up 5 miles.

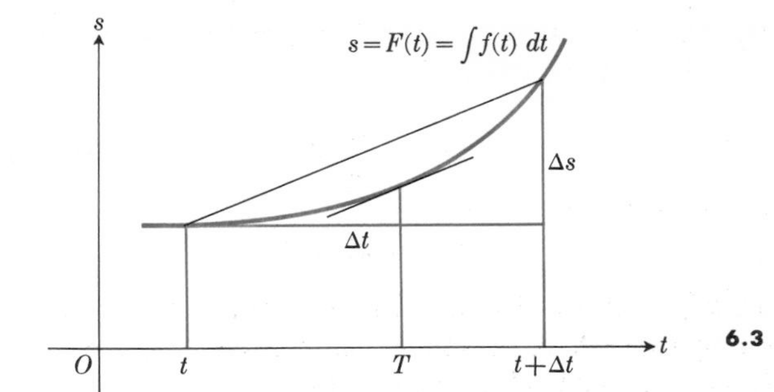

a|b 6.3

If we want to determine the actual total distance traveled, we should calculate the integral of the *absolute value* of the velocity,

$$\int_a^b |f(t)|\, dt. \tag{10}$$

This would be done by integrating separately over the intervals where v is positive and where v is negative and adding the absolute values of the results.

Remark 4. The preceding discussion shows that the distance traveled is given by precisely the same expression as that for the area bounded by the curve

$$v = f(t)$$

and the t-axis from a to b. All of the discussion above can be interpreted geometrically. For example, T is chosen between t and $t + \Delta t$ in such a way that the area of the rectangle of base Δt and altitude $f(T)$ (Fig. 6.3a) is precisely equal to the area under the velocity curve between t and $t + \Delta t$. This is also the place where the slope of the curve $s = F(t)$ (Fig. 6.3b) is equal to the slope of the chord.

EXERCISES

In Exercises 1 through 8, the function $v = f(t)$ represents the velocity v (ft/sec) of a moving body as a function of the time t (sec). Sketch the graph of v versus t and find that portion of the given time interval $a \le t \le b$ in which the velocity is (a) positive, and (b) negative. Then find the total distance traveled by the body between $t = a$ and $t = b$.

1. $v = 2t + 1, \quad 0 \le t \le 2$
2. $v = t^2 - t - 2, \quad 0 \le t \le 3$
3. $v = t - \dfrac{8}{t^2}, \quad 1 \le t \le 3$
4. $v = |t - 1|, \quad 0 \le t \le 2$
5. $v = 6 \sin 3t, \quad 0 \le t \le \dfrac{\pi}{2}$
6. $v = 4 \cos 2t, \quad 0 \le y \le \pi$
7. $v = \sin t + \cos t, \quad 0 \le t \le \pi$
8. $v = \sin t\sqrt{2 + 2\cos t}, \quad 0 \le t \le \pi$

In Exercises 9 through 13, the function $a = f(t)$ represents the acceleration (ft/sec^2) of a moving body and v_0 is its velocity at time $t = 0$. Find the *distance* traveled by the body between time $t = 0$ and $t = 2$.

9. $a = \sin t, \quad v_0 = 2$
10. $a = 1 - \cos t, \quad v_0 = 0$
11. $a = g$ (constant), $\quad v_0 = 0$
12. $a = \sqrt{4t + 1}, \quad v_0 = -4\frac{1}{3}$
13. $a = \dfrac{1}{\sqrt{4t + 1}}, \quad v_0 = 1$
14. Suppose water flows into a tank at the rate of $f(t)$ (gal/min), where f is a given, positive, continuous function of t. Let the amount of water in the tank at time $t = 0$ be Q_0 (gal). Apply the Fundamental Theorem to show that the amount of water in the tank at any later time $t = b$ is

$$Q = Q_0 + \int_0^b f(t)\, dt.$$

15. Use the trapezoidal rule, Article 5.10, to find (approximately) the distance traveled between $t = 0$ and $t = 2$ by a body whose velocity is given by the following table of values.

v (ft/sec)	2.2	2.5	3.0	3.8	5.0
t (sec)	0	0.5	1.0	1.5	2.0

Also find (approximately) the average velocity during the time $t = 0$ to $t = 2$.

6.4 VOLUMES

Method of slicing

The volumes of many solids can be found by an application of the *method of slicing*. Suppose, for example, that the solid is bounded by two parallel planes perpendicular to the x-axis at $x = a$ and $x = b$. Imagine the solid cut into thin slices of thickness Δx by planes perpendicular to the x-axis. Then the total volume V of the solid is the sum of the volumes of these slices (see Fig. 6.4).

Let ΔV be the volume of the representative slice between x and $x + \Delta x$. Then if A' and A'' are respectively the smallest and largest cross-sectional areas of the solid between x and $x + \Delta x$, it will be

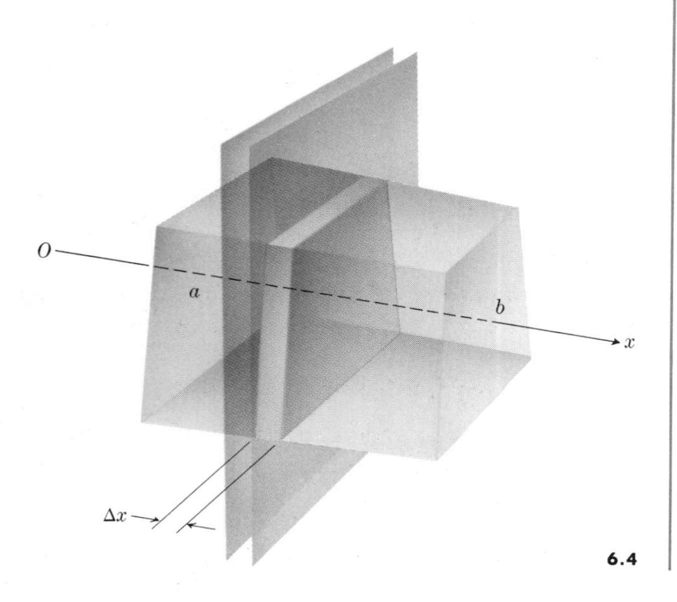

6.4

seen at once that

$$A' \,\Delta x \leq \Delta V \leq A'' \,\Delta x,$$

or

$$\Delta V = A(\bar{x}) \,\Delta x,$$

where $A(\bar{x})$ denotes an appropriate intermediate cross-sectional area of the solid at some $\bar{x}$ between x and $x + \Delta x$. Then

$$V = \sum_a^b A(\bar{x}) \,\Delta x$$

will give the total volume of the solid exactly. If, instead of choosing the exactly appropriate intermediate point $\bar{x}$ in the typical subinterval, we use the area $A(x)$ of the cross section at x, we have only an approximation

$$V \approx \sum_a^b A(x) \,\Delta x.$$

But now let $\Delta x \to 0$. Then, if the cross-sectional area is a continuous function of x, all these sums will approach the same limit. Since one sequence of sums always gives the value V, it will have this value as its limit. Hence the other sums will also give the value V in the limit, that is,

$$V = \lim_{\Delta x \to 0} \sum_a^b A(x) \,\Delta x = \int_a^b A(x) \,dx. \tag{1}$$

Volume of a solid of revolution

The solid generated by rotating a plane region about an axis on its plane is called a *solid of revolution*. To find the volume of such a solid, as shown in Fig. 6.5, we need only observe that the cross-sectional area $A(x)$ in Eq. (1) is the area of a circle of radius $r = y = f(x)$, so that

$$A(x) = \pi r^2 = \pi [f(x)]^2. \tag{2}$$

Example 1. Suppose the curve in Fig. 6.5 represents the graph of

$$y = \sqrt{x}$$

from (0, 0) to (4, 2). Then the volume of the representative slice is

$$\Delta V \approx \pi y^2 \,\Delta x = \pi (\sqrt{x})^2 \,\Delta x,$$

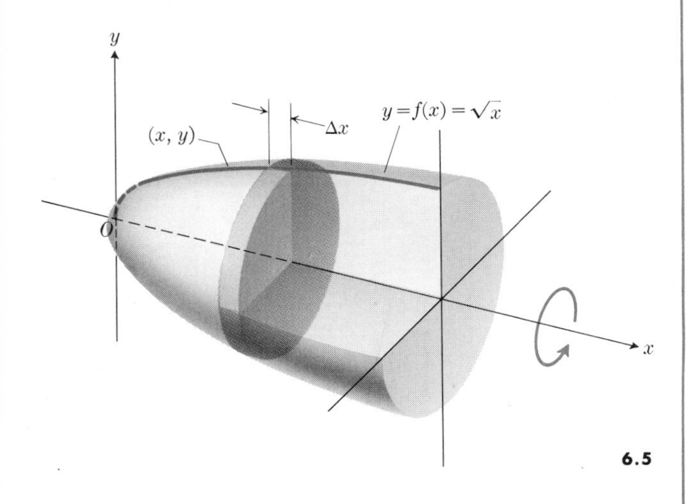

6.5

and the total volume is approximately

$$V \approx \sum_{0}^{4} \pi x \, \Delta x.$$

Exactly, we have

$$V = \lim_{\Delta x \to 0} \sum_{0}^{4} \pi x \, \Delta x = \int_{0}^{4} \pi x \, dx = \pi \frac{x^2}{2} \Big]_{0}^{4} = 8\pi.$$

Method of cylindrical shells

Suppose the region $PQRS$ in Fig. 6.6 is revolved around the y-axis. We can compute the generated volume in the following way. Consider a strip of area between the ordinate at x and the ordinate at $x + \Delta x$. When this strip is revolved around the y-axis, it generates a hollow, thin-walled shell of inner radius x, outer radius $x + \Delta x$, and volume ΔV. The base of this shell is a ring bounded by two concentric circles. The inner radius is

$$r_1 = x$$

and the outer radius is

$$r_2 = x + \Delta x.$$

The area of the ring is

$$\Delta A = \pi(r_2^2 - r_1^2) = 2\pi \left(\frac{r_2 + r_1}{2} \right) (r_2 - r_1)$$
$$= 2\pi r \, \Delta x.$$

Here
$$r = \frac{r_2 + r_1}{2} = x + \frac{\Delta x}{2}$$

is the radius of the circle midway between the inner and outer boundaries of the ring, and $2\pi r$ is its circumference.

Now if we had a cylindrical shell of constant altitude y standing on this base, its volume would be

$$\text{altitude} \times \text{base} = y \, \Delta A.$$

In the present case, the altitude of the shell varies between the minimum value of y and the maximum value of y between x and $x + \Delta x$. For an increasing function such as the one shown in Fig. 6.6, the volume therefore lies between $y \, \Delta A$ and $(y + \Delta y) \, \Delta A$.

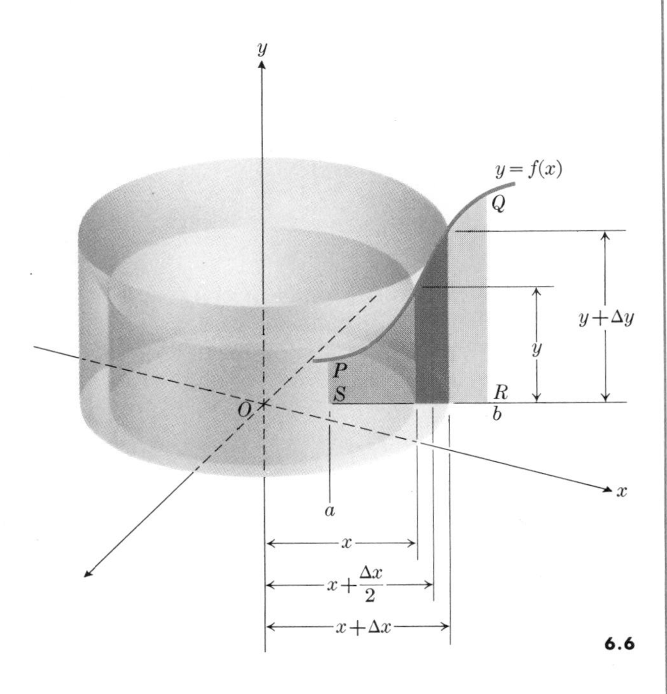

6.6

That is,

$$y\,\Delta A \le \Delta V \le (y + \Delta y)\,\Delta A,$$

or

$$y \cdot 2\pi r\,\Delta x \le \Delta V \le (y + \Delta y) \cdot 2\pi r\,\Delta x.$$

The total volume V is contained between the lower sum

$$s = \sum_a^b y \cdot 2\pi\left(x + \frac{\Delta x}{2}\right)\Delta x$$

and the upper sum

$$S = \sum_a^b (y + \Delta y) \cdot 2\pi\left(x + \frac{\Delta x}{2}\right)\Delta x.$$

As Δx approaches zero, both of these sums approach the following integral as limit (see Article 6.5). Hence this limit must also be equal to the volume V, namely,

$$V = \lim_{\Delta x \to 0} s = \lim_{\Delta x \to 0} S = \int_a^b y \cdot 2\pi x\,dx$$
$$= \int_a^b 2\pi x f(x)\,dx.$$

Remark. An easy way to visualize this result is to imagine that the volume element ΔV has been cut along a generator of the cylinder and that the shell has been rolled out flat like a thin sheet of tin. The sheet then has dimensions very nearly equal to $2\pi x$ by $y = f(x)$ by Δx. Hence

$$\Delta V \approx 2\pi x \cdot f(x) \cdot \Delta x,$$

and the total volume is given approximately by

$$V \approx \sum_a^b 2\pi x \cdot f(x) \cdot \Delta x.$$

As Δx approaches zero, we obtain a limit that is equal to the exact volume:

$$V = \lim_{\Delta x \to 0} \sum_a^b 2\pi x \cdot f(x) \cdot \Delta x$$
$$= \int_a^b 2\pi x f(x)\,dx.$$

Example 2. In Example 1, where the area under the curve $y = \sqrt{x}$ is revolved around the x-axis, we may

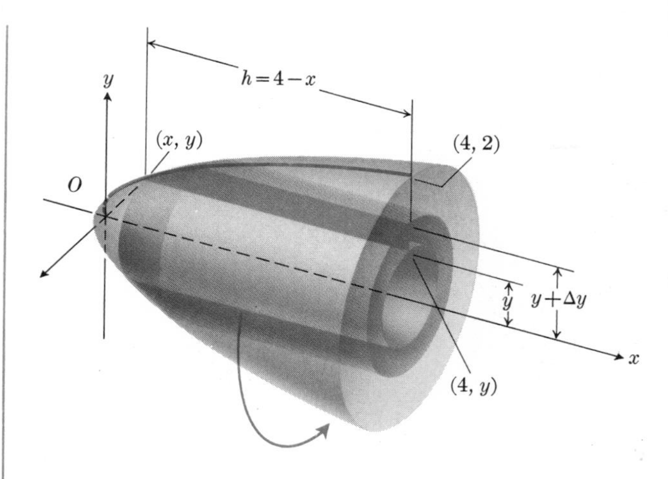

6.7 The solid of revolution generated by rotating the curve $y = \sqrt{x}$, [0, 4], about the x-axis.

take a horizontal strip of area between the lines at distances y and $y + \Delta y$ above the x-axis. The volume generated by revolving this strip around the x-axis is a hollow cylindrical shell of inner circumference $2\pi y$, inner length $4 - x$, and wall thickness Δy (see Fig. 6.7). Hence the total volume V is approximately

$$V \approx \sum_{y=0}^{2} 2\pi y(4 - x)\,\Delta y,$$

and is exactly equal to the limit of this sum as Δy approaches zero:

$$V = \lim_{\Delta y \to 0} \sum_{y=0}^{2} 2\pi y(4 - x)\,\Delta y$$
$$= \int_0^2 2\pi y(4 - y^2)\,dy$$
$$= 2\pi\left[2y^2 - \frac{y^4}{4}\right]_0^2 = 8\pi.$$

Illustrations

Various methods for finding volumes are illustrated in the following problems.

Problem 1. The region inside the circle

$$x^2 + y^2 = a^2$$

is rotated about the x-axis to generate a solid sphere. Find its volume.

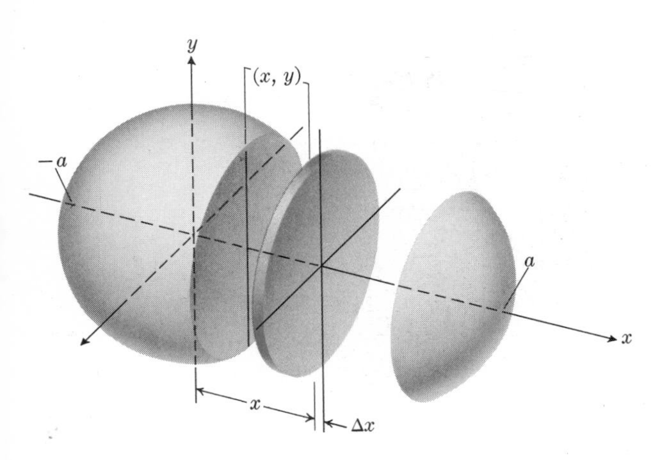

6.8 The sphere generated by rotating the circle $x^2 + y^2 = a^2$ about the x-axis.

Solution. We imagine the sphere cut into thin slices by planes perpendicular to the x-axis (Fig. 6.8). The volume of a typical slice between two planes at x and $x + \Delta x$ is approximately

$$\pi y^2 \,\Delta x = \pi(a^2 - x^2)\,\Delta x,$$

and the sum of all slices is approximately

$$V_{-a}^{a} \approx \sum_{-a}^{a} \pi(a^2 - x^2)\,\Delta x.$$

The exact volume is given by

$$\begin{aligned} V_{-a}^{a} &= \lim_{\Delta x \to 0} \sum_{-a}^{a} \pi(a^2 - x^2)\,\Delta x \\ &= \int_{-a}^{a} \pi(a^2 - x^2)\,dx \\ &= \pi\left[a^2 x - \frac{x^3}{3}\right]_{-a}^{a} = \tfrac{4}{3}\pi a^3. \end{aligned}$$

Problem 2. A hole of diameter a is bored through the center of the sphere of Problem 1. Find the remaining volume.

Solution. The volume in question could be generated by rotating about the y-axis the area inside the circle $x^2 + y^2 = a^2$ lying to the right of the line $x = a/2$ (Fig. 6.9a). The line and the circle intersect at the points $(a/2, \pm a\sqrt{3}/2)$. There are at least three methods for finding the required volume.

Method 1. Figure 6.9(b) shows an exploded view of the solid sphere with the core pulled out. Note that the core is a cylinder with a spherical cap at each end. Essentially, what we shall do is subtract the volume of this core from the volume of the sphere, but we shall simplify matters slightly by imagining that the two caps have been sliced off first, by planes perpendicular to the y-axis at $y = \pm a\sqrt{3}/2$. When these two caps have been removed, the truncated sphere has a volume V_1. From this truncated sphere we remove a right circular cylinder of radius $a/2$ and altitude $2(a\sqrt{3}/2) = a\sqrt{3}$. Then the volume of this flat-ended cylinder is

$$V_2 = \pi\left(\frac{a}{2}\right)^2 (a\sqrt{3}) = \frac{\sqrt{3}}{4}\pi a^3.$$

The desired volume is $V = V_1 - V_2$. We find V_1 by the same method we used for the spherical ball in Problem 1, but we keep in mind that the two caps have been removed at top and bottom. Imagine the solid as cut into slices by planes perpendicular to the y-axis. The slice between y and $y + \Delta y$ is approximated by a cylinder of altitude Δy and cross-sectional area

$$\pi x^2 = \pi(a^2 - y^2),$$

so that we have the approximation

$$V_1 \approx \sum_{-a\sqrt{3}/2}^{a\sqrt{3}/2} \pi(a^2 - y^2)\,\Delta y$$

and the exact value

$$V_1 = \int_{-a\sqrt{3}/2}^{a\sqrt{3}/2} \pi(a^2 - y^2)\,dy = \frac{3\sqrt{3}}{4}\pi a^3.$$

Thus

$$V = V_1 - V_2 = \frac{3\sqrt{3}}{4}\pi a^3 - \frac{\sqrt{3}}{4}\pi a^3 = \frac{\sqrt{3}}{2}\pi a^3.$$

Method 2. Instead of subtracting volumes, we may work directly with the volume required. Again, we imagine the solid to be cut into thin slices by planes perpendicular to the y-axis. Each slice (Fig. 6.9c) is now like a washer of thickness Δy, inner radius

$$r_1 = \frac{a}{2},$$

and outer radius

$$r_2 = x = \sqrt{a^2 - y^2}.$$

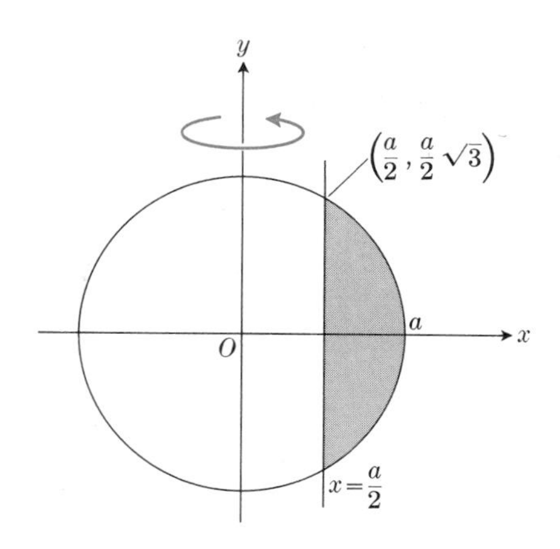

a|b

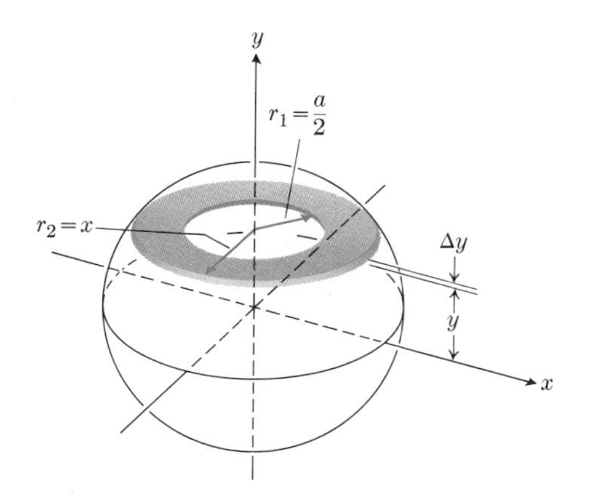

c|d

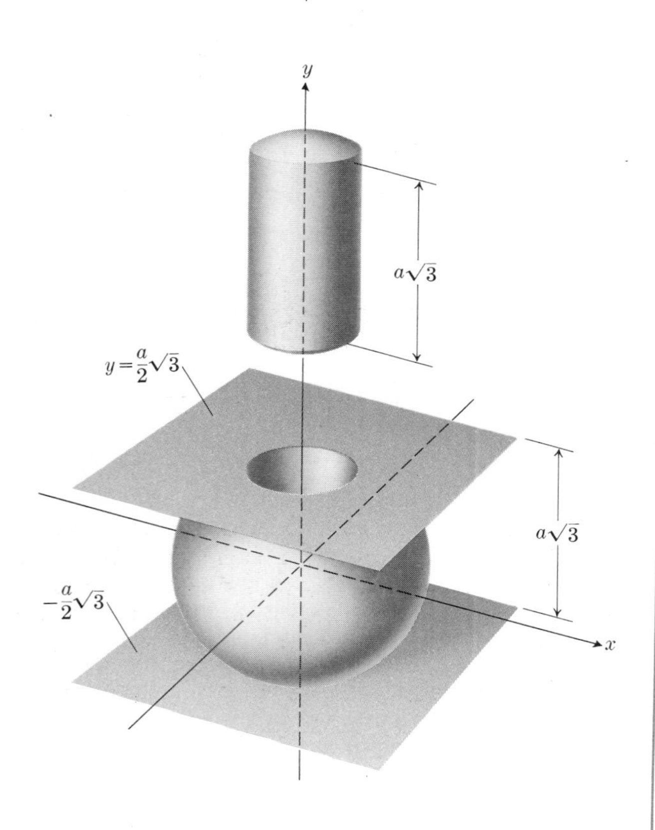

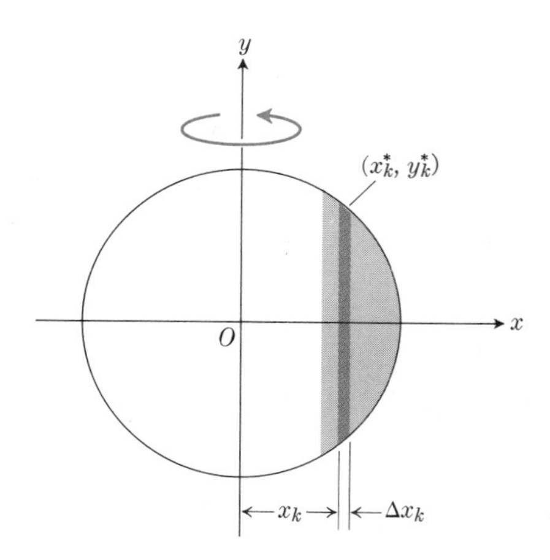

6.9 **a**| The tinted region generates the volume of revolution.
b| An exploded view showing the sphere with the core removed.
c| A phantom view showing a cross-sectional slice of the sphere with core removed.
d| The strip of area indicated by darker tint generates a cylindrical shell.

The area of a face of such a washer is

$$\pi r_2^2 - \pi r_1^2 = \pi(\tfrac{3}{4}a^2 - y^2)$$

and the volume of the solid is approximately

$$V \approx \sum_{-a\sqrt{3}/2}^{a\sqrt{3}/2} \pi(\tfrac{3}{4}a^2 - y^2)\,\Delta y,$$

while its exact value is

$$\begin{aligned} V &= \lim_{\Delta y \to 0} \sum_{-a\sqrt{3}/2}^{a\sqrt{3}/2} \pi(\tfrac{3}{4}a^2 - y^2)\,\Delta y \\ &= \int_{-a\sqrt{3}/2}^{a\sqrt{3}/2} \pi(\tfrac{3}{4}a^2 - y^2)\,dy \\ &= \frac{\sqrt{3}}{2}\pi a^2. \end{aligned}$$

Method 3. This time we shall use the cylindrical-shell approach (see Fig. 6.9a). We imagine the solid to be cut into a number of hollow cylindrical shells by cylinders having the y-axis as their common axis and having radii

$$x_1 = \frac{a}{2}, \quad x_2, \quad x_3, \quad \ldots, \quad x_n, \quad x_{n+1} = a.$$

The first shell has inner radius x_1 and outer radius x_2, the second shell has inner radius x_2 and outer radius x_3, and so on, the kth shell having inner radius x_k and outer radius x_{k+1}. This kth shell fills the region swept out when the dark strip in Fig. 6.9(d) is revolved about the y-axis. If we imagine this region to have been cut along a line parallel to the y-axis and flattened out, we see that the volume ΔV_k of the kth cylindrical shell is approximately

$$2\pi x_k^* \cdot \Delta x_k \cdot 2y_k^* = 4\pi x_k^* \sqrt{a^2 - (x_k^*)^2}\,\Delta x_k,$$

where (x_k^*, y_k^*), Fig. 6.9(d), is the point of abscissa

$$x_k^* = \tfrac{1}{2}(x_k + x_{k+1})$$

on the circle in the first quadrant. Then the entire volume will be approximated by

$$\begin{aligned} V &\approx \sum_{k=1}^{n} 4\pi x_k^* \sqrt{a^2 - (x_k^*)^2}\,\Delta x_k \\ &\approx \sum_{a/2}^{a} 4\pi x \sqrt{a^2 - x^2}\,\Delta x. \end{aligned}$$

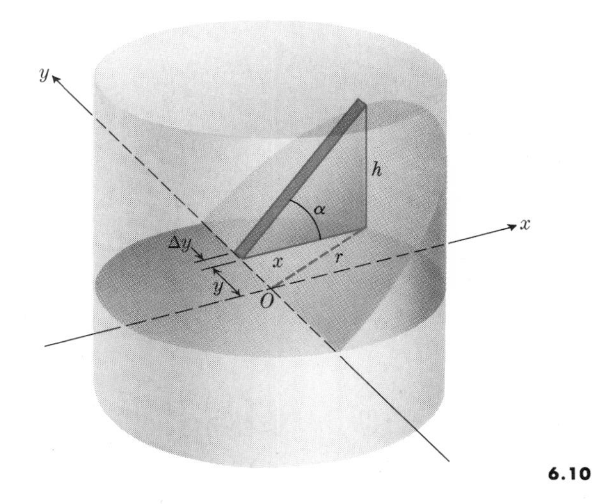

6.10

The exact value will be the *limit* of this sum, namely,

$$\begin{aligned} V &= \lim_{x \to 0} \sum_{a/2}^{a} 4\pi x \sqrt{a^2 - x^2}\,\Delta x \\ &= \int_{a/2}^{a} 4\pi x \sqrt{a^2 - x^2}\,dx. \end{aligned}$$

To evaluate this integral, we let

$$u = a^2 - x^2;$$

then

$$\begin{aligned} V &= 4\pi \int_{a/2}^{a} \sqrt{a^2 - x^2}\,(x\,dx) \\ &= 4\pi \int u^{1/2}(-\tfrac{1}{2}\,du) = -\frac{4\pi}{3}\Big[(a^2 - x^2)^{3/2}\Big]_{a/2}^{a} \\ &= \frac{\sqrt{3}}{2}\pi a^3. \end{aligned}$$

Problem 3. A wedge is cut from a right circular cylinder of radius r by two planes. One is perpendicular to the axis of the cylinder while the second makes an angle α with the first and intersects it at the center of the cylinder. Find the volume of the wedge.

Solution. The volume ΔV of the slice between y and $y + \Delta y$ in Fig. 6.10 is approximately

$$\Delta V \approx A(y)\,\Delta y,$$

where

$$A(y) = \tfrac{1}{2}xh$$

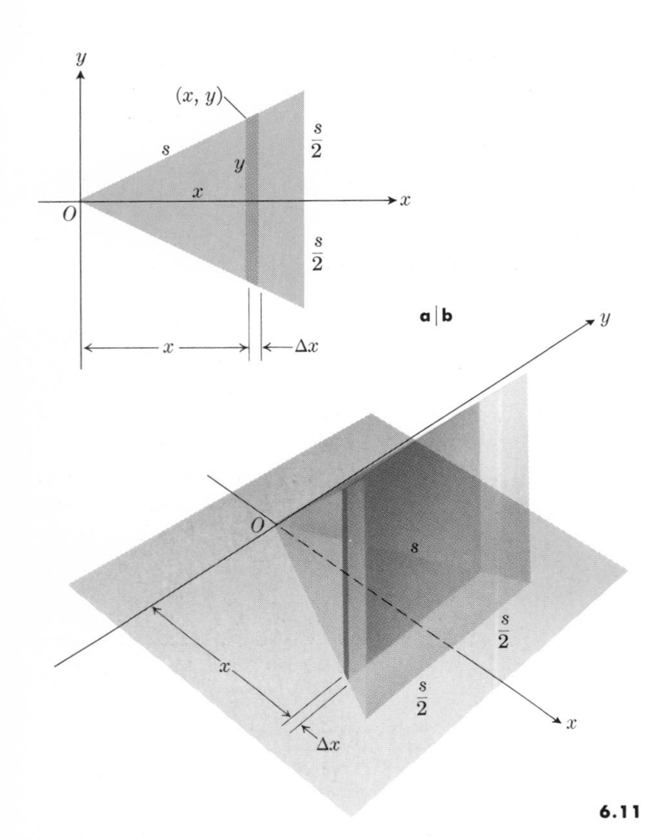

6.11

is the area of the triangle that forms one face of the slice and is to be expressed as a function of y. By trigonometry,

$$h = x \tan \alpha,$$

and by the theorem of Pythagoras,

$$x^2 + y^2 = r^2, \quad \text{or} \quad x^2 = r^2 - y^2.$$

Hence

$$A(y) = \tfrac{1}{2}x^2 \tan \alpha = \tfrac{1}{2} \tan \alpha \, (r^2 - y^2).$$

The total volume is given by

$$V = \lim_{\Delta y \to 0} \sum_{-r}^{r} A(y)\, \Delta y$$

$$= \int_{-r}^{r} \tfrac{1}{2} \tan \alpha (r^2 - y^2)\, dy.$$

Remembering that the factor $\frac{1}{2} \tan \alpha$ is a constant, we find

$$V = \tfrac{2}{3} r^3 \tan \alpha.$$

Problem 4. The base of a certain solid is an equilateral triangle of side s, with one vertex at the origin and an altitude along the x-axis. Each plane section perpendicular to the x-axis is a square, one side of which lies in the base of the solid. Find the volume of the solid.

Solution. Figure 6.11(a) represents the base of the solid with a strip of width Δx, which corresponds to a slice of the solid of volume

$$\Delta V \approx A(x)\, \Delta x,$$

where

$$A(x) = (2y)^2 = 4y^2$$

is the area of a face of the slice. The slice extends upward from the plane of the paper in Fig. 6.11(a). It is possible to find the required volume from the statement of the problem by reference to this figure and without visualizing the actual solid. A perspective view of the solid is, however, given in Fig. 6.11(b). It is required to find y in terms of x so that the area $A(x)$ will be given as a function of x. Since the base triangle is equilateral, its altitude h is given by

$$h^2 = s^2 - (s/2)^2 = \tfrac{3}{4}s^2,$$

or

$$h = \frac{s}{2}\sqrt{3}.$$

Then, by similar triangles, we find

$$\frac{2y}{x} = \frac{2}{h} = \frac{2}{\sqrt{3}}, \quad \text{or} \quad y = \frac{x}{\sqrt{3}}.$$

Then

$$V = \int_0^h A(x)\, dx = \int_0^{s\sqrt{3}/2} \tfrac{4}{3} x^2\, dx$$

$$= \frac{s^3}{2\sqrt{3}} = \frac{1}{3} h s^2.$$

EXERCISES

In Exercises 1 through 8, find the volumes generated when the areas bounded by the given curves and lines are rotated about the x-axis. [*Note.* $x = 0$ is the y-axis and $y = 0$ is the x-axis.]

1. $x + y = 2, \quad x = 0, \quad y = 0$
2. $y = \sin x, \quad y = 0 \quad (0 \leq x \leq \pi)$
   [See Exercise 17, Article 5.8.]

3. $y = x - x^2, \quad y = 0$
4. $y = 3x - x^2, \quad y = x$
5. $x = 2y - y^2, \quad x = 0$ (Use cylindrical shells.)
6. $y = x, \quad y = 1, \quad x = 0$
7. $y = x^2, \quad y = 4$
8. $y = 3 + x^2, \quad y = 4$
9. By integration find the volume generated by the triangle with vertices at $(0, 0)$, $(h, 0)$, (h, r) when it is rotated
   (a) about the x-axis, (b) about the y-axis.
10. (a) The area bounded by the curve $y = x^2$ and the line $y = x$ is revolved around the y-axis. Find the volume generated.
    (b) Find the volume generated if the area is revolved around the x-axis.
11. The area bounded by the curve $y = x^2$ and the line $y = 4$ generates various solids of revolution when rotated
    (a) about the y-axis, (b) about the line $y = 4$,
    (c) about the x-axis, (d) about the line $y = -1$,
    (e) about the line $x = 2$.
    Find the volume generated in each case.
12. (a) A hemispherical bowl of radius a contains water to a depth h. Find the volume of water in the bowl.
    (b) (Review exercise on related rates.) Water runs into a hemispherical bowl of radius 5 ft at the rate of 0.2 ft^3/sec. How fast is the water level in the bowl rising when the water is 4 ft. deep?
13. A football has a volume that is approximately the same as the volume generated by rotating the area inside the ellipse $b^2x^2 + a^2y^2 = a^2b^2$ (where a and b are constants) about the x-axis. Find the volume so generated.
14. The cross sections of a certain solid by planes perpendicular to the x-axis are circles with diameters extending from the curve $y = x^2$ to the curve $y = 8 - x^2$. The solid lies between the points of intersection of these two curves. Find its volume.
15. The base of a certain solid is the circle $x^2 + y^2 = a^2$. Each plane section of the solid cut out by a plane perpendicular to the x-axis is a square with one edge of the square in the base of the solid. Find the volume of the solid.
16. Two great circles, lying in planes that are perpendicular to each other, are marked on a sphere of radius a. A portion of the sphere is then shaved off in such a manner that any plane section of the remaining solid, perpendicular to the common diameter of the two great circles, is a square with vertices on these circles. Find the volume of the solid that remains.
17. The base of a certain solid is the circle $x^2 + y^2 = a^2$. Each plane section of the solid cut out by a plane perpendicular to the y-axis is an isosceles right triangle with one leg in the base of the solid. Find the volume.
18. The base of a certain solid is the region between the x-axis and the curve $y = \sin x$ between $x = 0$ and $x = \pi/2$. Each plane section of the solid perpendicular to the x-axis is an equilateral triangle with one side in the base of the solid. Find the volume.
19. A rectangular swimming pool is 30 ft wide and 50 ft long. The depth of water h (ft) at distance x (ft) from one end of the pool is measured at 5-ft intervals and found to be as follows:

x (ft)	0	5	10	15	20	25
h (ft)	6.0	8.2	9.1	9.9	10.5	11.0

x (ft)	30	35	40	45	50
h (ft)	11.5	11.9	12.3	12.7	13.0

    Use the trapezoidal rule to find (approximately) the volume of water in the pool.
20. The circle $x^2 + y^2 = a^2$ is rotated about the line $x = b$ $(b > a)$ to generate a torus. Find the volume generated.

    Hint. $$\int_{-a}^{a} \sqrt{a^2 - x^2}\, dx = \pi a^2/2,$$
    since it is the area of a semicircle of radius a.

6.5 APPROXIMATIONS

By now it is apparent that each application of the Fundamental Theorem that we have so far made has involved the following steps.

1. Select an independent variable, say x, such that the quantity U to be computed can be represented as a sum of pieces ΔU, where ΔU is that part of U that is associated with the subinterval $(x, x + \Delta x)$ of the domain $a \leq x \leq b$ of the variable x.

2. Subdivide the domain (a, b) into n subintervals. For the sake of simplicity we usually take the lengths all to be the same, namely, equal to $\Delta x = (b - a)/n$.

3. Approximate that portion ΔU of U that is associated with the subinterval $(x, x + \Delta x)$ by an expression of the form

$$\Delta U \approx f(X)\,\Delta x. \tag{1}$$

In this expression, X is to be some point in the subinterval $(x, x + \Delta x)$ and the function f is to be continuous over the entire domain $a \leq x \leq b$.

4. Observe that the total quantity U is given *approximately* by

$$U \approx \sum_a^b f(X)\,\Delta x. \tag{2}$$

5. Take the limit, as Δx approaches zero, of the sum in Eq. (2). This limit is the definite integral

$$\lim_{\Delta x \to 0} \sum_a^b f(X)\,\Delta x = \int_a^b f(x)\,dx. \tag{3}$$

Is this limit also an approximation to U or does it give U exactly?

In the case of area, distance, and volume we have seen that the *limit* in (3) does give U exactly. In order for us to be able to apply these methods to new situations as they arise, we need an answer to the following question.

Question. How accurate must the approximations in (1) and (2) be if we are to be able to say that all of the error is squeezed out in going to the limit (3), with the result that the exact value of U is given by the limit

$$U = \lim_{\Delta x \to 0} \sum_a^b f(X)\,\Delta x = \int_a^b f(x)\,dx? \tag{4}$$

Answer. Denote by $\alpha\,\Delta x$ the correction that must be added to the right side of Eq. (1) to give ΔU exactly. That is, suppose that

$$\Delta U = f(X)\,\Delta x + \alpha\,\Delta x, \tag{5}$$

exactly.

Case 1. Suppose these correction terms are no larger than a constant K times $(\Delta x)^2$:

$$|\alpha\,\Delta x| \leq K\,(\Delta x)^2$$

where K is a constant the same for all subintervals and all methods of subdivision. Then we have

$$\Delta U = f(X)\,\Delta x + \alpha\,\Delta x,$$

$$U = \sum_a^b f(X)\,\Delta x + \sum_a^b \alpha\,\Delta x;$$

and

$$\left|U - \sum_a^b f(X)\,\Delta x\right| = \left|\sum_a^b \alpha\,\Delta x\right| \leq \sum_a^b K\,(\Delta x)^2. \tag{6}$$

This last sum consists of n terms, each of which is equal to $K\,(\Delta x)^2$, where $\Delta x = (b - a)/n$. Therefore

$$\begin{aligned}\sum_a^b K\,(\Delta x)^2 &= nK(b - a)^2/n^2\\ &= K(b - a)^2/n = K(b - a)\,\Delta x,\end{aligned}$$

and (6) becomes

$$\left|U - \sum_a^b f(X)\,\Delta x\right| \leq K(b - a)\,\Delta x. \tag{7}$$

In other words, if the error in the approximation (1) to each individual ΔU is no more than a constant times the *square* of Δx, then the error in the approximation (2) to the *total* U is no more than a constant times the *first power* of Δx. Now let Δx approach zero in (7). The sum on the left becomes the definite integral (3), and the term on the right becomes zero. In other words, U is exactly equal to the definite integral in this case.

Case 2. Suppose the correction terms are numerically no larger than a constant K' times Δx times δy:

$$|\alpha\,\Delta x| \leq K'\,\Delta x\,\delta y,$$

where K' is the same for all subintervals and all methods of subdivision, and where δy is the oscillation in the interval $(x, x + \Delta x)$ of a function $y = \phi(x)$ that is continuous over the closed interval (a, b).

Then one finds in this case that the total error in the approximation (2) is no larger than

$$K'(b - a) \text{ times } (\max \delta y).$$

But the maximum δy also approaches zero when Δx does, so again we find that U is given exactly by the definite integral.

We have given criteria such as

$$|\alpha\, \Delta x| \leq K\,(\Delta x)^2 \quad \text{and} \quad |\alpha\, \Delta x| \leq K'\, \Delta x\, \delta y,$$

where K and K' are constants. Usually, in an application, we have a function of x in place of K or K' in these inequalities. But a function of x that is continuous over the closed interval $[a, b]$ has a maximum absolute value M on that interval, and we could take K (or K') equal to M in such a case. Also, the correction terms $\alpha\, \Delta x$ could involve finite combinations of the two cases discussed above, provided that the inequality

$$|\alpha\, \Delta x| \leq K\,(\Delta x)^2 + K'\, \Delta x\, \delta y$$

is satisfied for some choice of the constants K and K'.

Roughly speaking, we may say that in the approximation

$$\Delta U \approx f(X)\, \Delta x$$

we must include all first power Δx terms but we may omit higher power terms like $(\Delta x)^2$, $(\Delta x)^3$, and so on, or we may omit such mixed terms as$(\Delta x, \Delta y)$, $(\Delta x)^2(\Delta y)$, $(\Delta x)(\Delta y)^2$, and so on. If each separate piece ΔU is estimated to this degree of accuracy, then the total quantity U which is given approximately by the sum (2) will be given exactly by the limit of the sum, namely, by the integral (4).

In discussing the cylindrical shell method of finding a volume of revolution (Article 6.4), we found that we could estimate the volume ΔV in a hollow shell as follows:

$$2\pi\left(x + \frac{\Delta x}{2}\right) y\, \Delta x \leq \Delta V \leq 2\pi\left(x + \frac{\Delta x}{2}\right)(y + \Delta y)\, \Delta x. \tag{8}$$

Multiplied out, this becomes

$$2\pi xy\, \Delta x + \pi y\,(\Delta x)^2 \leq \Delta V,$$
$$\Delta V \leq 2\pi xy\, \Delta x + \pi y\,(\Delta x)^2 + 2\pi x\, \Delta x\, \Delta y + \pi\, \Delta y\,(\Delta x)^2.$$

If we ignore all except the first power terms in Δx, we have the approximation

$$\Delta V \approx 2\pi xy\, \Delta x,$$

with an error that involves combinations such as $\pi y\,(\Delta x)^2$, $2\pi x\, \Delta y\, \Delta x$, $\pi\, \Delta y\,(\Delta x)^2$. All such terms may be safely ignored when we go to the *limit* of the sum as Δx approaches zero:

$$V = \lim_{\Delta x \to 0} \sum_a^b 2\pi xy\, \Delta x$$
$$= \int_a^b 2\pi xy\, dx = \int_a^b 2\pi x f(x)\, dx.$$

Infinitesimals

A variable, such as Δx above, that approaches zero as a limit is called an *infinitesimal.* Terms that approach zero as the first power of Δx are said to be of the *same order* as Δx, but terms which approach zero as $(\Delta x)^2$ or as $\Delta y\, \Delta x$ are called infinitesimals of *higher order* than Δx. In setting up a definite integral, we must retain those infinitesimals that are of the same order as Δx, but we may omit those infinitesimals that are of higher order than Δx. Thus we may omit the term $\alpha\, \Delta x$, Eq. (5), if α is an infinitesimal of the same order as Δx (or higher order) or of the same order as δy (or higher order), and the *limit* in (4) will give the exact value of U.

As a further illustration of these ideas, we give here a proof of Bliss's theorem, a theorem which is often useful.

Bliss's Theorem.* Let f and g be continuous functions on $[a, b]$, where $a < b$, and a and b are finite real numbers. Let P be a partition of $[a, b]$ into n subintervals by points

$$a = x_0 < x_1 < x_2 < \cdots < x_n = b, \tag{9a}$$

and let

$$\delta = \max\,(x_i - x_{i-1}), \quad i = 1, 2, \ldots, n. \tag{9b}$$

---

* Olmsted, J. M. H., *Advanced Calculus*, Appleton-Century-Crofts, 1961, p. 116.

Let t_i and t'_i be in the subinterval $[x_{i-1}, x_i]$, and form the sum

$$S = \sum_{i=1}^{n} f(t_i)g(t'_i)\,\Delta x_i. \tag{10}$$

Then, as $n \to \infty$ and $\delta \to 0$,

$$\lim S = \int_a^b f(x)g(x)\,dx. \tag{11}$$

Proof. The result follows from the fact that f is bounded, g is uniformly continuous on $[a, b]$, and the sum

$$S^* = \sum_{i=1}^{n} f(t_i)g(t_i)\,\Delta x_i \tag{12}$$

approaches the integral in Eq. (11) when $n \to \infty$ and $\delta \to 0$. First we see that because f is bounded, there exists a positive number K such that

$$|f(x)| < K \quad \text{for } a \le x \le b. \tag{13}$$

Second, if $\epsilon > 0$, there exists $\delta' > 0$ such that

$$|g(t'_i) - g(t_i)| < \frac{\epsilon}{K(b-a)} \quad \text{if} \quad |t'_i - t_i| < \delta'.$$

Therefore, if $\delta \le \delta'$,

$$\begin{aligned} |S - S^*| &\le \sum_{i=1}^{n} |f(t_i)| \cdot |g(t'_i) - g(t_i)|\,\Delta x_i \\ &\le \sum_{i=1}^{n} K \cdot \frac{\epsilon}{K(b-a)}\,\Delta x_i \\ &= \frac{\epsilon}{b-a} \sum \Delta x_i = \epsilon. \end{aligned} \tag{14}$$

As $n \to \infty$ and $\delta \to 0$, therefore, S^* approaches the integral on the right side of Eq. (11); and $S - S^*$ approaches zero. Hence S approaches the integral, as required. □

Remark. Bliss's theorem can be extended to products of more than two functions. For instance, instead of $f(t_i)g(t'_i)$, as in Eq. (10), we might have three functions and three points in the ith subinterval: $f(t_i)g(t'_i)h(t''_i)$. If the three functions f, g, and h are all continuous on $[a, b]$, the limit of the corresponding sum would be the integral

$$\int_a^b f(x)g(x)h(x)\,dx.$$

Other combinations of functions than products, such as sums and quotients (denominators nonzero, of course), behave in a similar manner. For the arc length of a curve, we encounter a sum of the form

$$\sum \sqrt{[f(t_i)]^2 + [g(t'_i)]^2},$$

where f and g are continuous on $[a, b]$. Again, that sum has for its limit the same integral we would get if t'_i were replaced by t_i. In other words, the evaluation of one of the functions at t_i and the other at t'_i is of no consequence *in the limit* for continuous functions.

EXERCISES

1. In Fig. 6.6, Article 6.4, let V_a^x denote the volume generated by rotating about the y-axis the area under the curve $y = f(x)$ between the ordinate at a and the ordinate at x, $(x > a)$. By making use of the inequality (8) in this section, show that

$$\frac{dV_a^x}{dx} = \lim_{\Delta x \to 0} \frac{\Delta V_a^x}{\Delta x} = 2\pi x f(x).$$

From this, deduce that

$$V_a^b = \int_a^b 2\pi x f(x)\,dx.$$

2. Let U_a^x denote the amount of the quantity U associated with the interval (a, x), and let ΔU in Eq. (5) represent

$$U_a^{x+\Delta x} - U_a^x = U_x^{x+\Delta x}.$$

If the function $f(x)$ in Eq. (5) is continuous over the closed interval $a \le x \le b$ and $|\alpha\,\Delta x| \le K\,(\Delta x)^2$, where K is a constant, show that

$$\frac{dU_a^x}{dx} = \lim_{\Delta x \to 0} \frac{\Delta U}{\Delta x} = f(x).$$

From this, deduce that $U_a^b = \int_a^b f(x)\,dx$.

3. In Exercise 2, replace the condition $|\alpha\, \Delta x| \leq K(\Delta x)^2$ by the condition

$$|\alpha\, \Delta x| \leq |K'\, \Delta x\, \delta y|,$$

where K' is a constant and δy is the oscillation over the interval $(x, x + \Delta x)$ of a function $y = g(x)$ that is continuous over the closed interval $a \leq x \leq b$. Derive the same conclusion as in Exercise 2.

6.6 LENGTH OF A PLANE CURVE

Divide the arc AB, Fig. 6.12, into n pieces and join the successive points of division by straight lines. A representative line, such as PQ, will have length

$$PQ = \sqrt{(\Delta x_k)^2 + (\Delta y_k)^2}.$$

The length of the curve AB is approximately

$$L_A^B \approx \sum_{k=1}^{n} \sqrt{(\Delta x_k)^2 + (\Delta y_k)^2}.$$

When the number of division points is increased indefinitely and the lengths of the individual segments there tend to zero, we obtain

$$L_A^B = \lim_{n\to\infty} \sum_{k=1}^{n} \sqrt{(\Delta x_k)^2 + (\Delta y_k)^2}, \tag{1}$$

provided that the limit exists.* The sum on the right side of (1) is not in the standard form to which we can apply the Fundamental Theorem of the integral calculus, but it can be put into such a form, as follows.

Suppose that the function $y = f(x)$ is continuous and possesses a continuous derivative at each point of the curve from $A(a, f(a))$ to $B(b, f(b))$. Then, by the Mean Value Theorem, there is some point $P^*(x_k^*, y_k^*)$ between P and Q on the curve where the

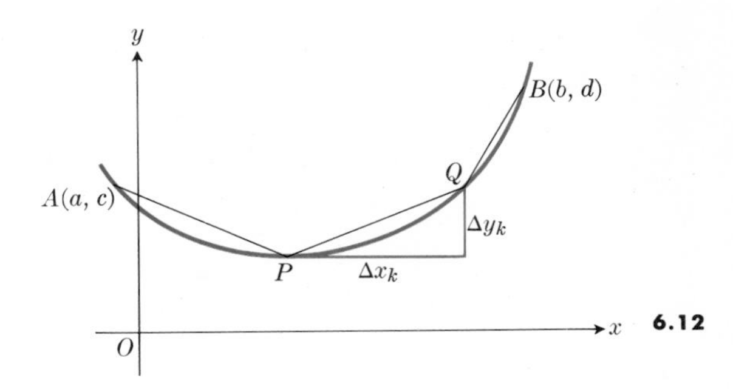

6.12

tangent to the curve is parallel to the chord PQ. That is,

$$f'(x_k^*) = \frac{\Delta y_k}{\Delta x_k}, \quad \text{or} \quad \Delta y_k = f'(x_k^*)\, \Delta x_k.$$

Hence (1) may also be written in the form

$$L_A^B = \lim_{n\to\infty} \sum_{k=1}^{n} \sqrt{(\Delta x_k)^2 + [f'(x_k^*)\, \Delta x_k]^2}$$

$$= \lim_{\Delta x \to 0} \left(\sum_a^b \sqrt{1 + [f'(x^*)]^2}\, \Delta x \right),$$

or

$$L_A^B = \int_a^b \sqrt{1 + \left(\frac{dy}{dx}\right)^2}\, dx, \tag{2a}$$

where we have written dy/dx for $f'(x)$.

We may remark that it is sometimes convenient, if x may be expressed as a single-valued function of y, to interchange the roles of x and y (indeed, the *length* of a curve does not depend on the particular axes used), and the length is also given by

$$L_A^B = \int_c^d \sqrt{1 + \left(\frac{dx}{dy}\right)^2}\, dy. \tag{2b}$$

In many applications, the curve is traced by a moving particle whose coordinates x and y are given as functions of some third variable, such as the time t. Let the equations of motion be

$$x = g(t), \qquad y = h(t), \tag{3}$$

and let t_k, t_{k+1} be the values of t at P and Q respectively. Suppose the arc AB is described just once by $P(x, y)$ as t goes from t_a at A to t_b at B. If

* For most smooth curves encountered in practice, the limit does exist. Such curves are called rectifiable. An example of a continuous curve that is not rectifiable is

$$y = \begin{cases} x \sin 1/x & \text{when } x \neq 0, \\ 0 & \text{when } x = 0, \end{cases} \quad \frac{-\pi}{2} \leq x \leq \frac{\pi}{2}.$$

the functions $g(t)$ and $h(t)$ are continuously differentiable for t between t_a and t_b inclusive, the Mean Value Theorem may be applied to Eq. (3) to give

$$\Delta x_k = x_{k+1} - x_k = g(t_{k+1}) - g(t_k) = g'(t'_k)\,\Delta t_k,$$
$$\Delta y_k = y_{k+1} - y_k = h(t_{k+1}) - h(t_k) = h'(t''_k)\,\Delta t_k,$$

where t'_k and t''_k are two suitably chosen values of t between t_k and t_{k+1}. Then (1) becomes

$$L_A^B = \lim_{n\to\infty} \sum_{k=1}^{n} \sqrt{[g'(t'_k)]^2 + [h'(t''_k)]^2}\,\Delta t_k$$
$$= \int_{t_a}^{t_b} \sqrt{\left(\frac{dx}{dt}\right)^2 + \left(\frac{dy}{dt}\right)^2}\,dt. \tag{4}$$

Clearly, nothing inherently depends on the fact that t stands for time in Eqs. (3) and (4). Any other variable could serve as well, and if θ is used instead of t in (3) in representing the curve, then we need only replace t by θ in (4) as well.

Example 1. The coordinates (x, y) of a point on a circle of radius r can be expressed in terms of the central angle θ, Fig. 6.13, as follows:

$$x = r\cos\theta, \qquad y = r\sin\theta.$$

The point $P(x, y)$ moves once around the circle as θ varies from 0 to 2π, so that the circumference of the circle is given by

$$C = \int_0^{2\pi} \sqrt{\left(\frac{dx}{d\theta}\right)^2 + \left(\frac{dy}{d\theta}\right)^2}\,d\theta.$$

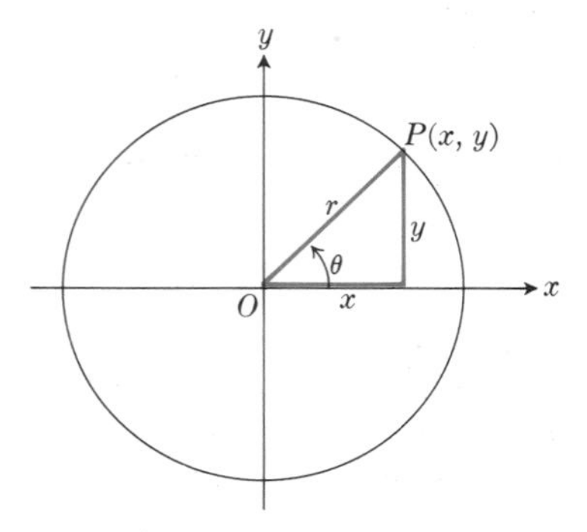

6.13

We find

$$\frac{dx}{d\theta} = -r\sin\theta, \qquad \frac{dy}{d\theta} = r\cos\theta,$$

so that

$$\left(\frac{dx}{d\theta}\right)^2 + \left(\frac{dy}{d\theta}\right)^2 = r^2(\sin^2\theta + \cos^2\theta) = r^2,$$

and hence

$$C = \int_0^{2\pi} r\,d\theta = r[\theta]_0^{2\pi} = 2\pi r.$$

It should be remarked that Eq. (4) is frequently written in terms of differentials in place of derivatives. This is done formally by writing $(dt)^2$ under the radical in place of the dt outside the radical, and then writing

$$(dx)^2 = \left(\frac{dx}{dt}\,dt\right)^2 = \left(\frac{dx}{dt}\right)^2 (dt)^2$$

and

$$(dy)^2 = \left(\frac{dy}{dt}\,dt\right)^2 = \left(\frac{dy}{dt}\right)^2 (dt)^2.$$

It is also customary to eliminate the parentheses in $(dx)^2$ and write dx^2 instead, so that Eq. (4) is written

$$L = \int \sqrt{dx^2 + dy^2}. \tag{5}$$

Of course, dx and dy must both be expressed in terms of one and the same variable, and appropriate limits must be supplied in (5) before the integration can be performed.

Note. A useful mnemonic device is also associated with (5). It is customary to write

$$ds = \sqrt{dx^2 + dy^2}, \tag{6}$$

and treat ds as the differential of arc length, which can be integrated (between appropriate limits) to give the total length of a curve. Figure 6.14(a) gives the exact interpretation of ds corresponding to Eq. (6). Figure 6.14(b) is not strictly accurate, but is to be thought of as a simplified version of Fig. 6.14(a), with the arc of the curve treated as the hypotenuse ds of a right triangle of sides dx and dy.

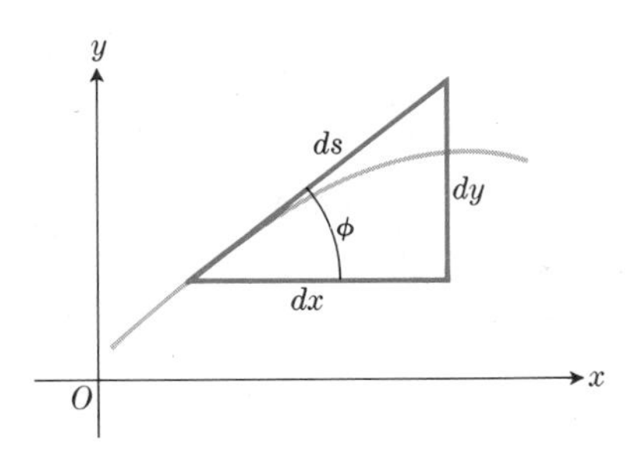

a | b

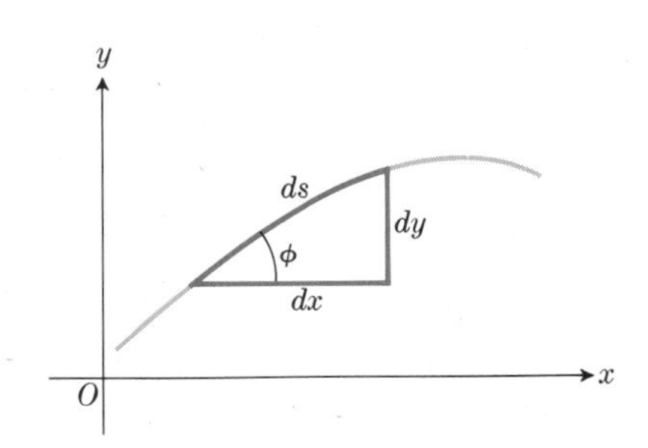

6.14

Problem 1. Find the length of the curve $y = x^{2/3}$ between $x = -1$ and $x = 8$.

Solution. From the equation of the curve, we find

$$\frac{dy}{dx} = \frac{2}{3}x^{-1/3}.$$

Since this becomes infinite at the origin (see Fig. 6.15), we use Eq. (2b) instead of (2a) to find the length of the curve. Then we need to find dx/dy. Since the equation of the curve,

$$y = x^{2/3},$$

can also be written as

$$x = \pm y^{3/2},$$

we find

$$\frac{dx}{dy} = \pm\frac{3}{2}y^{1/2}, \quad \text{or} \quad dx = \pm\frac{3}{2}y^{1/2}\,dy.$$

Then $\quad ds^2 = dx^2 + dy^2 = (\frac{9}{4}y + 1)\,dy^2,$

so that $\quad ds = \sqrt{\frac{9}{4}y + 1}\,dy.$

The portion of the curve between $A(-1, 1)$ and the origin has length

$$L_1 = \int_0^1 \sqrt{\tfrac{9}{4}y + 1}\,dy,$$

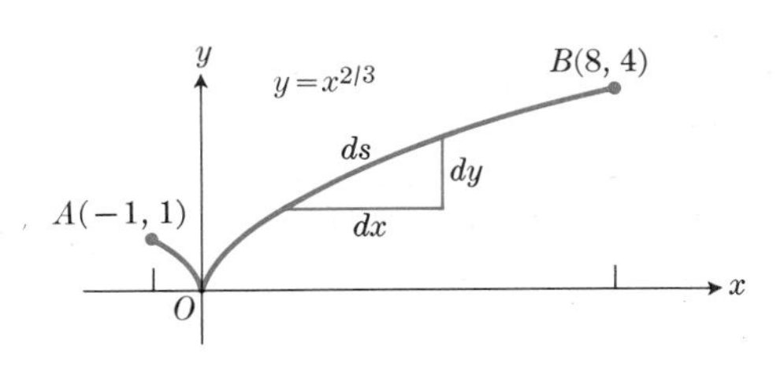

6.15

while the rest of the curve from the origin to $B(8, 4)$ has length

$$L_2 = \int_0^4 \sqrt{\tfrac{9}{4}y + 1}\,dy,$$

and the total length is

$$L = L_1 + L_2.$$

It is necessary to calculate the two lengths L_1 and L_2 as separate integrals, since $x = \pm y^{3/2}$ needs to be separated into two distinct functions of y. For the portion AO of the curve, we have $x = -y^{3/2}$, $0 \le y \le 1$; while on the arc OB, we have $x = +y^{3/2}$, $0 \le y \le 4$.

To evaluate the given integrals, we let

$$u = \tfrac{9}{4}y + 1;$$

then

$$du = \tfrac{9}{4}\,dy, \qquad dy = \tfrac{4}{9}\,du,$$

and

$$\int (\tfrac{9}{4}y + 1)^{1/2}\,dy = \tfrac{4}{9}\int u^{1/2}\,du = \tfrac{8}{27}u^{3/2}.$$

Therefore

$$L = \tfrac{8}{27}\left\{(\tfrac{9}{4}y + 1)^{3/2}\Big]_0^1 + (\tfrac{9}{4}y + 1)^{3/2}\Big]_0^4\right\}$$
$$= \tfrac{1}{27}(13\sqrt{13} + 80\sqrt{10} - 16) = 10.5.$$

As a check against gross errors, we calculate the sum of the lengths of the two inscribed chords:

$$AO + OB = \sqrt{2} + \sqrt{80} = 10.4.$$

The check appears to be satisfactory.

The curve in Fig. 6.15 has a cusp at (0, 0) where the slope becomes infinite. If we were to reconstruct the derivation of Eq. (2) for this particular curve, we would see that the crucial step that required an application of

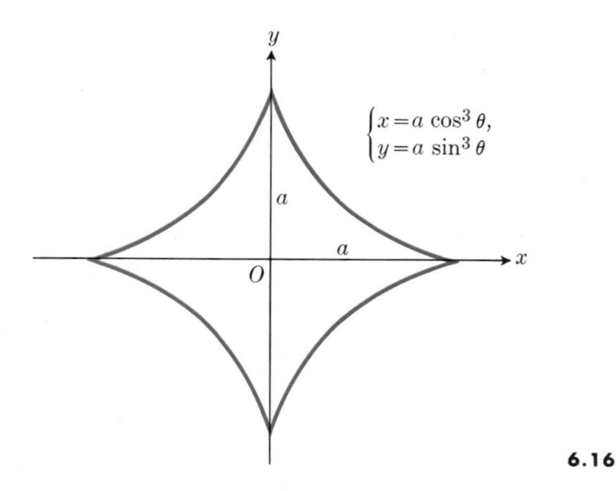

6.16

the Mean Value Theorem could not have been taken for the case of a chord PQ from a point P to the left of the cusp to a point Q to its right. For this reason, if for no other, when one or more cusps occur between the ends of a portion of a curve whose length is to be calculated, it is best to calculate the lengths of portions of the curve between cusps and add the results. We recall that the Mean Value Theorem is still valid even when the derivative becomes infinite at an extremity of the interval where it is to be applied, so that the derivation of Eq. (2) would be valid for the separate portions of a curve lying *between* cusps (or other discontinuities of dy/dx, such as occur at corners). Thus in the example just worked, we find the lengths L_1 from $A(-1, 1)$ to O (up to the cusp) and L_2 from O to $B(8, 4)$, then add the results to obtain $L = L_1 + L_2$. This was done by the sum of the two integrals. Had there been no point of discontinuity of dy/dx, it would not have been necessary to take two separate integrals, and Eq. (2a) could have been used.

Problem 2. The coordinates of the point $P(x, y)$ on the four-cusped hypocycloid (Fig. 6.16) are given by

$$x = a\cos^3\theta, \qquad y = a\sin^3\theta.$$

Find the total length of the curve.

Solution. When θ varies from 0 to $\pi/2$, P traces out the portion of the curve in the first quadrant. Also, from the fact that

$$\begin{aligned} x^{2/3} + y^{2/3} &= a^{2/3}(\cos^2\theta + \sin^2\theta) \\ &= a^{2/3}(1) = a^{2/3}, \end{aligned}$$

it will be seen that for every point (x, y) on the curve in the first quadrant, the corresponding points $(-x, y)$ $(-x, -y)$, and $(x, -y)$ in the other three quadrants are also on the curve, which is therefore symmetrical about both axes. The portion of the curve in the first quadrant lies *between* two cusps and is one-quarter of the total length of the curve. Therefore

$$L = 4\int_{\theta=0}^{\theta=\pi/2} \sqrt{dx^2 + dy^2}.$$

From the equation of the curve, we find

$$\frac{dx}{d\theta} = -3a\cos^2\theta\sin\theta, \quad dx = 3a\cos\theta\sin\theta\,(-\cos\theta)\,d\theta,$$

$$\frac{dy}{d\theta} = 3a\sin^2\theta\cos\theta, \quad dy = 3a\cos\theta\sin\theta\,(\sin\theta)\,d\theta,$$

and hence

$$\begin{aligned} ds^2 &= dx^2 + dy^2 \\ &= (3a\cos\theta\sin\theta)^2(\cos^2\theta + \sin^2\theta)(d\theta)^2, \end{aligned}$$

or

$$ds^2 = (3a\cos\theta\sin\theta\,d\theta)^2.$$

Hence, for θ between 0 and $\pi/2$,

$$ds = 3a\cos\theta\sin\theta\,d\theta$$

and

$$L = 4\int_0^{\pi/2} 3a\cos\theta\sin\theta\,d\theta.$$

To evaluate this integral, we let

$$u = \sin\theta, \qquad du = \cos\theta\,d\theta;$$

then

$$\int \cos\theta\sin\theta\,d\theta = \int u\,du = \tfrac{1}{2}u^2 = \tfrac{1}{2}\sin^2\theta,$$

so that

$$L = (12a)(\tfrac{1}{2}\sin^2\theta)\Big]_0^{\pi/2} = 6a.$$

EXERCISES

In Exercises 1 through 6, find the length of the segment of the curve described.

1. $y = \frac{1}{3}(x^2 + 2)^{3/2}$, from $x = 0$ to $x = 3$
2. $y = x^{3/2}$, from $(0, 0)$ to $(4, 8)$

3. $9x^2 = 4y^3$ from $(0, 0)$ to $(2\sqrt{3}, 3)$
4. $y = x^3/3 + 1/(4x)$, from $x = 1$ to $x = 3$
5. $x = y^4/4 + 1/(8y^2)$, from $y = 1$ to $y = 2$
6. $(y + 1)^2 = 4x^3$, from $x = 0$ to $x = 1$
7. Find the distance traveled between $t = 0$ and $t = \pi/2$ by a particle $P(x, y)$ whose position at time t is given by

$$x = a \cos t + at \sin t, \quad y = a \sin t - at \cos t,$$

where a is a positive constant.

8. Find the distance traveled by the particle $P(x, y)$ between $t = 0$ and $t = 4$ if the position at time t is given by

$$x = t^2/2, \qquad y = \tfrac{1}{3}(2t + 1)^{3/2}.$$

9. The position of a particle $P(x, y)$ at time t is given by

$$x = \tfrac{1}{3}(2t + 3)^{2/3}, \qquad y = t^2/2 + t.$$

Find the distance it travels between $t = 0$ and $t = 3$.

6.7 AREA OF A SURFACE OF REVOLUTION

Suppose that the curve in Fig. 6.17 is rotated about the x-axis. It will generate a surface in space. We approximate the curve by an inscribed polygon and also rotate it about the x-axis. The surface area generated by the curve is defined to be the limit of the surface areas generated by the inscribed polygons as their lengths approach the length of the arc of the curve; that is, as the number of segments approaches infinity and the lengths of the individual segments approach zero.

The inscribed polygonal line segments will generate inscribed frustums of cones. The sum of the lateral areas of these frustums will be an approximation to the area of the surface. To obtain an analytic expression for this approximation, we require the formula

$$A = \pi(r_1 + r_2)l \tag{1}$$

for the lateral area A of a frustum of slant height l, where r_1 and r_2 are the radii of its bases.

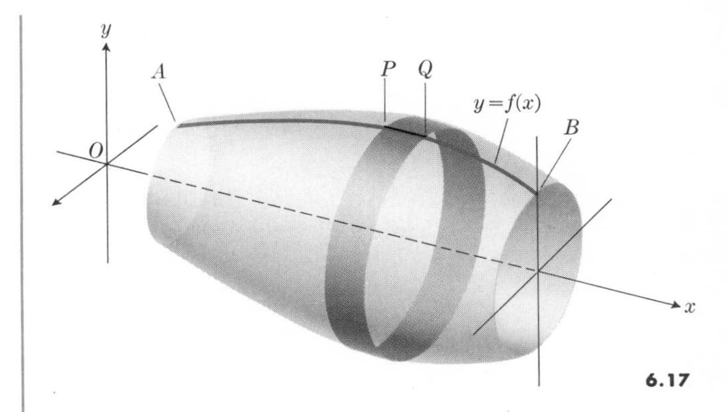

6.17

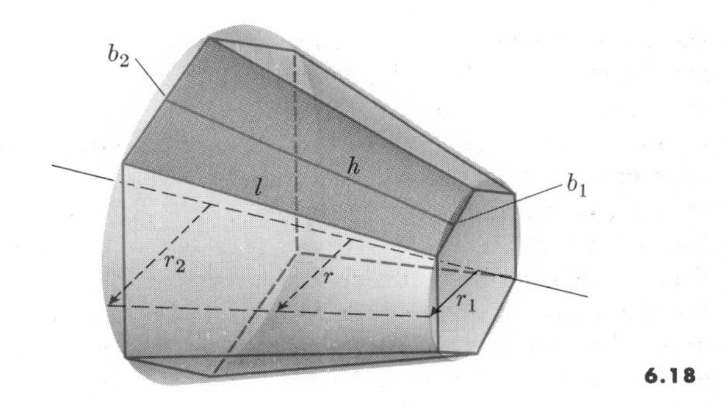

6.18

To establish Eq. (1), we may imagine the frustum of a cone to be the limit of inscribed frustums of pyramids (see Fig. 6.18). If each base of an inscribed frustum of a pyramid is a regular polygon of n sides, and b_1 is the length of one side in the upper face, while b_2 is the length of a side in the lower face, then the lateral area will consist of n trapezoids each having an area

$$\tfrac{1}{2}(b_1 + b_2)h.$$

Hence the n trapezoids have area

$$\tfrac{1}{2}(nb_1 + nb_2)h.$$

As n increases indefinitely, we have

$$\lim_{n\to\infty} nb_1 = 2\pi r_1, \qquad \lim_{n\to\infty} nb_2 = 2\pi r_2;$$

now

$$\lim_{n\to\infty} h = l,$$

so that

$$A = \lim_{n\to\infty} \tfrac{1}{2}(nb_1 + nb_2)h$$
$$= \tfrac{1}{2}(2\pi r_1 + 2\pi r_2)l = \pi(r_1 + r_2)l,$$

which establishes Eq. (1). Note that if $r_1 = r_2$, Eq. (1) reduces to the correct formula for the lateral area of a cylinder, and if $r_1 = 0$, it gives the correct result for the lateral area of a cone. It can also be put into the form

$$A = 2\pi rl,$$

where $r = \tfrac{1}{2}(r_1 + r_2)$ is the radius of the midsection of the frustum. In this form, (1) says that the lateral area of a frustum of a cone is equal to (circumference of mid-section) $\times$ (slant height).

We now consider the portion of the surface ΔS generated by the arc PQ, and its inscribed frustum of a cone generated by the chord PQ. Call the coordinates of P: (x, y), and of Q: $(x + \Delta x, y + \Delta y)$, and take

$$r_1 = y, \quad r_2 = y + \Delta y, \quad l = \sqrt{(\Delta x)^2 + (\Delta y)^2}$$

in Eq. (1). Then we have the approximation

$$S \approx \sum_{x=a}^{b} \pi(2y + \Delta y)\sqrt{(\Delta x)^2 + (\Delta y)^2},$$

or

$$S \approx \sum_{a}^{b} 2\pi\left(y + \frac{1}{2}\Delta y\right)\sqrt{1 + \left(\frac{\Delta y}{\Delta x}\right)^2}\,\Delta x.$$

In more formal notation, suppose that the curve is the graph of a function f that is continuous, together with its derived function f', on $[a, b]$. Let the segment $[a, b]$ be partitioned by points

$$a = x_0 < x_1 < x_2 < \cdots < x_n = b.$$

Then, by part (ii) of the Intermediate Value Theorem (Theorem 15, Article 3.9), there exists t_i on the ith subinterval such that

$$f(t_i) = y_i + \tfrac{1}{2}(\Delta y_i).$$

By the Mean Value Theorem (Article 4.8), there also exists t_i' on the ith subinterval such that

$$f'(t_i') = \frac{\Delta y_i}{\Delta x_i}.$$

Therefore, by the definition of S,

$$S = \lim_{n\to\infty} \sum_{i=1}^{n} 2\pi f(t_i)\sqrt{1 + [f_i'(t')]^2}\,\Delta x_i,$$

and, by Bliss's Theorem (Article 6.5), this leads to the result

$$S = \int_a^b 2\pi f(x)\sqrt{1 + [f'(x)]^2}\,dx,$$

or

$$S = \int_a^b 2\pi y\sqrt{1 + \left(\frac{dy}{dx}\right)^2}\,dx. \tag{2}$$

Our end result, Eq. (2), is easily remembered if we write

$$\sqrt{1 + \left(\frac{dy}{dx}\right)^2}\,dx = ds$$

and take

$$S = \int 2\pi y\,ds. \tag{3}$$

If we let

$$dS = 2\pi y\,ds,$$

so that

$$S = \int dS,$$

we may interpret dS as the product of

a circumference $2\pi y$ and a slant height ds.

Thus dS gives the lateral area of a frustum of a cone slant height ds if the point (x, y) is the midpoint of the element of arc length ds (Fig. 6.19).

Note. The following question often occurs to the alert student: Why not approximate the surface area S by inscribed *cylinders* and arrive at a result

$$S = \int 2\pi y\,dx,$$

having dx instead of ds, to replace Eq. (3)? This is a reasonable question. Since we know from our discussion on volume that inscribed cylinders work perfectly well for *volumes* of revolution, why not use them for *surfaces* of revolution also? The answer hinges on the fact that the approximation

$$\Delta V \approx \pi y^2\,\Delta x$$

for the volume of a slice involves, at worst, terms which are products like (Δx) (Δy) and higher powers of Δx, and

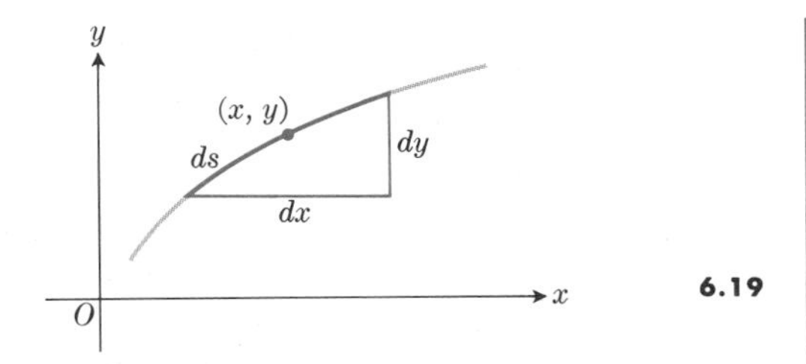

6.19

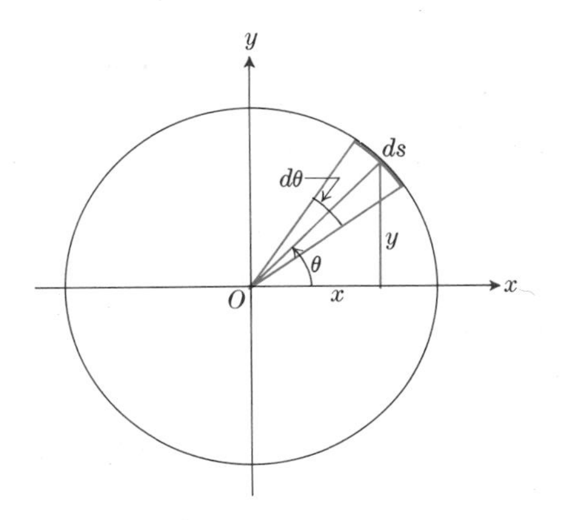

6.20

these contribute zero to the *limit* of the sum of volumes of the inscribed cylinders (see Article 6.5). But the approximations

$$\Delta S \approx 2\pi y\, \Delta x \tag{a}$$

and

$$\Delta S \approx 2\pi y \sqrt{(\Delta x)^2 + (\Delta y)^2} \tag{b}$$

for the surface area of a slice cannot *both* be this accurate, for the following reason: Their ratio is

$$\frac{2\pi y\sqrt{(\Delta x)^2 + (\Delta y)^2}}{2\pi y\, \Delta x} = \sqrt{1 + \left(\frac{\Delta y}{\Delta x}\right)^2},$$

which has the limiting value

$$\lim_{\Delta x \to 0} \sqrt{1 + \left(\frac{\Delta y}{\Delta x}\right)^2} = \sqrt{1 + \left(\frac{dy}{dx}\right)^2},$$

and this will be different from one (unless $dy/dx = 0$, which would not be generally true), whereas two approximations that differ only by terms involving products like $(\Delta y)(\Delta x)$ and higher powers of Δx have a ratio whose limiting value is unity. Since the approximations (a) and (b) above will usually lead to different answers when we pass to the corresponding definite integrals, they cannot both be correct, and we must abandon one or both of them. But the approximation (b) is clearly the one that corresponds to a natural way of defining the surface area of a surface of revolution, and it leads to (3).

If the axis of revolution is the y-axis, the corresponding formula that replaces (3) is

$$S = \int 2\pi x\, ds,$$

while more generally we may write

$$S = \int 2\pi \rho\, ds, \tag{4}$$

if ρ is the distance from the axis of revolution to the element of arc length ds. Both ρ and ds must be expressed in terms of some one variable and proper limits must be supplied to (4) in any particular problem.

Problem 1. The circle

$$x^2 + y^2 = r^2$$

is revolved about the x-axis (Fig. 6.20). Find the area of the sphere generated.

Solution. We write

$$dS = 2\pi y\, ds, \qquad ds = \sqrt{dx^2 + dy^2},$$

and use

$$x = r \cos \theta, \qquad y = r \sin \theta$$

to represent the circle. Then

$$dx = -r \sin \theta\, d\theta, \qquad dy = r \cos \theta\, d\theta,$$

so that

$$ds = r\, d\theta$$

and

$$\begin{aligned} dS &= 2\pi y\, ds = 2\pi(r \sin \theta) r\, d\theta \\ &= 2\pi r^2 \sin \theta\, d\theta. \end{aligned}$$

The top half of the circle generates the entire sphere, and the representative point (x, y) traces out this upper semicircle as θ varies from 0 to π. Hence

$$\begin{aligned} S &= \int_0^\pi 2\pi r^2 \sin \theta\, d\theta \\ &= 2\pi r^2 \Big[-\cos \theta\Big]_0^\pi = 4\pi r^2. \end{aligned}$$

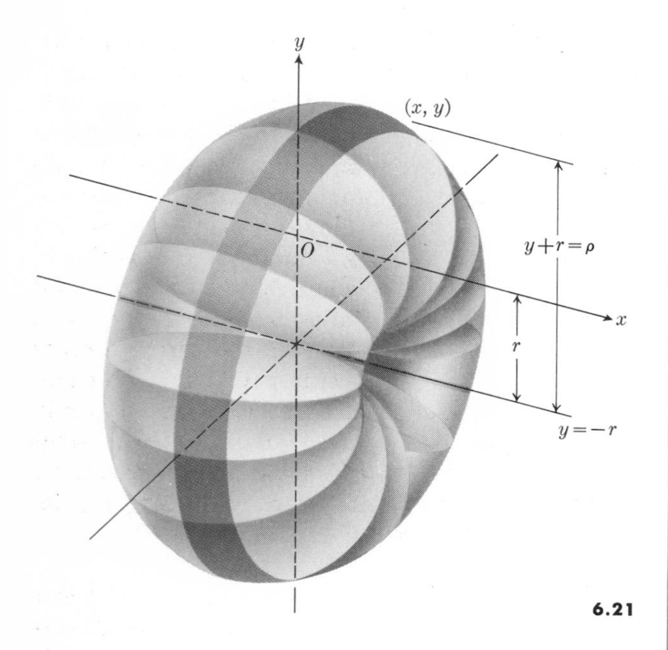

6.21

Problem 2. The circle in Problem 1 is revolved about the line $y = -r$, which is tangent to the circle at the point $(0, -r)$ (Fig. 6.21). Find the area of the surface generated.

Solution. Here it takes the whole circle to generate the surface, and in terms of the previous example, this means that we must let θ vary from 0 to 2π. The radius of rotation now becomes $\rho = y + r$, and we have

$$dS = 2\pi\rho\, ds = 2\pi(y + r)r\, d\theta = 2\pi(r \sin\theta + r)r\, d\theta.$$

Hence

$$\begin{aligned} S &= \int_0^{2\pi} 2\pi\,(\sin\theta + 1)r^2\, d\theta \\ &= 2\pi r^2\Big[-\cos\theta + \theta\Big]_0^{2\pi} = 4\pi^2 r^2. \end{aligned}$$

EXERCISES

1. Find the area of the surface generated by rotating the hypocycloid $x = a\cos^3\theta$, $y = a\sin^3\theta$ about the x-axis.
2. Find the area of the surface generated by rotating the portion of the curve $y = \frac{1}{3}(x^2 + 2)^{3/2}$ between $x = 0$ and $x = 3$ about the y-axis.
3. Find the area of the surface generated by rotating about the x-axis the arc of the curve $y = x^3$ between $x = 0$ and $x = 1$.
4. Find the area of the surface generated by rotating about the y-axis the arc of the curve $y = x^2$ between $(0, 0)$ and $(2, 4)$.
5. The arc of the curve $y = x^3/3 + 1/(4x)$ from $x = 1$ to $x = 3$ is rotated about the line $y = -1$. Find the surface area generated.
6. The arc of the curve $x = y^4/4 + 1/(8y^2)$ from $y = 1$ to $y = 2$ is rotated about the x-axis. Find the surface area generated.
7. The curve described by the particle $P(x, y)$,
$$x = t + 1, \qquad y = \frac{t^2}{2} + t,$$
from $t = 0$ to $t = 4$, is rotated about the y-axis. Find the surface area that is generated.
8. The loop of the curve $9x^2 = y(3 - y)^2$ is rotated about the x-axis. Find the surface area generated.

6.8 AVERAGE VALUE OF A FUNCTION

The process of finding the average value of a finite number of data is familiar to all students. For example, if $y_1, y_2, \ldots, y_n$ are the grades of a class of n students on a certain calculus quiz, then the class average on that quiz is

$$y_{\text{av}} = \frac{y_1 + y_2 + \cdots + y_n}{n}. \tag{1}$$

When the number of data is infinite, it is not feasible to use Eq. (1) (since it is likely to take on the meaningless form ∞/∞). This situation arises, in particular, when the data y are given by a continuous function

$$y = f(x), \quad a \le x \le b.$$

In this case, the average value of y, with respect to x, is defined to be

$$(y_{\text{av}})_x = \frac{1}{b - a}\int_a^b f(x)\, dx. \tag{2}$$

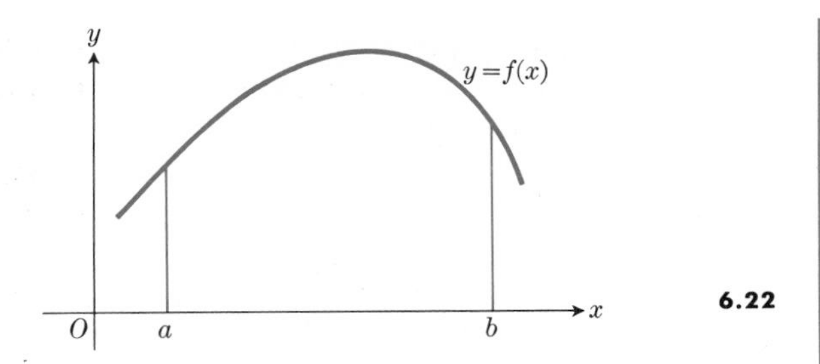

6.22

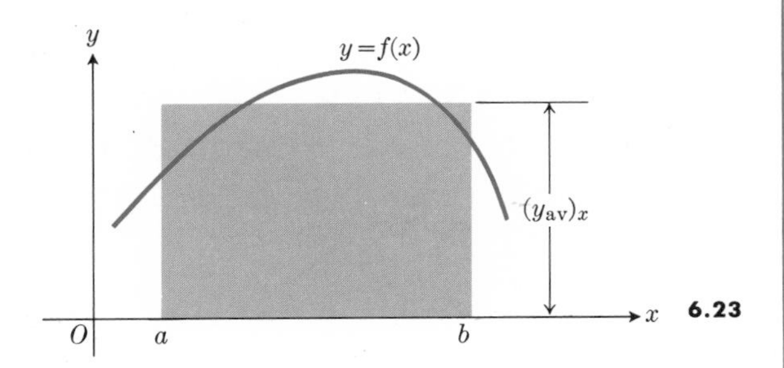

6.23

The curve in Fig. 6.22, for example, might represent temperature as a function of time over a twenty-four hour period. Equation (2) would then give the "average temperature" for the day.

Remark 1. It may be possible to express y as a function of x, or, alternatively, as a function of u. Then $(y_{\text{av}})_x$ and $(y_{\text{av}})_u$ need not be equal. For example, for a freely falling body starting from rest,

$$s = \tfrac{1}{2}gt^2, \qquad v = gt, \qquad v = \sqrt{2gs}.$$

Suppose we calculate the average velocity, first with respect to t and second with respect to s, from $t_1 = 0$, $s_1 = 0$ to $t_2 > 0$, $s_2 = \frac{1}{2}gt_2^2$. Then, by definition,

$$(v_{\text{av}})_t = \frac{1}{t_2 - 0}\int_0^{t_2} gt\,dt = \tfrac{1}{2}gt_2 = \tfrac{1}{2}v_2,$$

$$(v_{\text{av}})_s = \frac{1}{s_2 - 0}\int_0^{s_2} \sqrt{2gs}\,ds = \tfrac{2}{3}\sqrt{2gs_2} = \tfrac{2}{3}v_2.$$

Remark 2. If both sides of Eq. (2) are multiplied by $b - a$, we have

$$(y_{\text{av}})_x \cdot (b - a) = \int_a^b f(x)\,dx. \tag{3}$$

The right side of Eq. (3) represents the area bounded above by the curve $y = f(x)$, below by the x-axis, and on the sides by the ordinates $x = a$, $x = b$. The left side of the equation can be interpreted as the area of a rectangle of altitude $(y_{\text{av}})_x$ and of base $b - a$. Hence Eq. (3) provides a geometric interpretation of $(y_{\text{av}})_x$ as that ordinate of the curve $y = f(x)$ that should be used as altitude if one wishes to construct a rectangle whose base is the interval $a \leq x \leq b$ and whose area is equal to the area under the curve (Fig. 6.23). In the narrower sense of area, this statement is only valid when the curve lies above the x-axis. In case the curve lies partly or entirely below the x-axis, we would count areas below the x-axis as negative. In such a case there might well be a certain amount of canceling of positive and negative areas.

Remark 3. Equation (2) is a *definition*, and hence is not subject to proof. Nevertheless, some discussion may help to explain why this particular formula is used to define the average. One might arrive at it as follows. From the total "population" of x-values, $a \leq x \leq b$, we select a representative "sample," $x_1, x_2, \ldots, x_n$, uniformly distributed between a and b. Then, using Eq. (1), we calculate the average of the functional values

$$y_1 = f(x_1), \quad y_2 = f(x_2), \quad \ldots, \quad y_n = f(x_n)$$

associated with these representative x's. This gives us

$$\frac{y_1 + y_2 + \cdots + y_n}{n} = \frac{f(x_1) + f(x_2) + \cdots + f(x_n)}{n}. \tag{4}$$

Since we require that the x's be uniformly distributed between a and b, let us take the spacing to be Δx, with

$$x_2 - x_1 = x_3 - x_2 = \cdots = x_n - x_{n-1} = \Delta x,$$

$$\Delta x = \frac{b - a}{n}.$$

Then, in (4), let us replace the n in the denominator by $(b - a)/\Delta x$, thus obtaining

$$\frac{f(x_1) + f(x_2) + \cdots + f(x_n)}{(b-a)/\Delta x} = \frac{f(x_1)\,\Delta x + f(x_2)\,\Delta x + \cdots + f(x_n)\,\Delta x}{b-a}.$$

Now if n is very large (Δx small), the expression

$$f(x_1)\,\Delta x + f(x_2)\,\Delta x + \cdots + f(x_n)\,\Delta x = \sum_{k=1}^{n} f(x_k)\,\Delta x$$

is very nearly equal to $\int_a^b f(x)\,dx$. In fact, if we take limits, letting $n \to \infty$, we obtain precisely

$$\lim_{n\to\infty} \frac{f(x_1)\,\Delta x + f(x_2)\,\Delta x + \cdots + f(x_n)\,\Delta x}{b-a},$$

which is equal to

$$\frac{1}{b-a}\int_a^b f(x)\,dx.$$

This is the expression used to define the average value of y in Eq. (2).

EXERCISES

In Exercises 1 through 5, find the average value with respect to x, over the given domain, of the given function $f(x)$. In each case, draw a graph of the curve $y = f(x)$, and sketch a rectangle whose altitude is the average ordinate.

1. (a) $\sin x, \quad 0 \le x \le \pi/2$
   (b) $\sin x, \quad 0 \le x \le 2\pi$
2. (a) $\sin^2 x, \quad 0 \le x \le \pi/2$
   (b) $\sin^2 x, \quad \pi \le x \le 2\pi$
3. $\sqrt{2x+1}, \quad 4 \le x \le 12$
4. $\frac{1}{2} + \frac{1}{2}\cos 2x, \quad 0 \le x \le \pi$
5. $\alpha x + \beta; \quad a \le x \le b \quad (\alpha, \beta, a, b,$ constants)
6. Given a circle C of radius a, and a diameter AB of C. Chords are drawn perpendicular to AB, intercepting equal segments along AB. Find the limit of the average of the lengths of these chords, as the number of chords tends to infinity. [*Hint.* $\int_{-a}^{a} \sqrt{a^2 - x^2}\,dx$ is $\frac{1}{2}\pi a^2$, since it is the area of a semicircle of radius a.]
7. Solve Exercise 6 under the modified assumption that the chords intercept equal arcs along the circumference of C.
8. Solve Exercise 6 using the *squares* of the lengths, in place of the lengths of the chords.
9. Solve Exercise 7 using the *squares* of the lengths, in place of the lengths of the chords.

6.9 MOMENTS AND CENTER OF MASS

If several masses $m_1, m_2, \ldots, m_n$ are placed along the x-axis at distances $x_1, x_2, \ldots, x_n$, respectively, from the origin (Fig. 6.24), then their moment about the origin is defined by the following equality:

6.24

$$x_1m_1 + x_2m_2 + \cdots + x_nm_n = \sum_{k=1}^{n} x_km_k. \tag{1}$$

If all the mass

$$m_1 + m_2 + \cdots + m_n = \sum_{k=1}^{n} m_k \tag{2}$$

is concentrated at one point of abscissa $\overline{x}$, the total moment is

$$\overline{x}\left(\sum_{k=1}^{n} m_k\right).$$

The position of $\overline{x}$ for which this is the same as the total moment in (1) is called the *center of mass*. The condition

$$\overline{x}\sum_{k=1}^{n} m_k = \sum_{k=1}^{n} x_km_k$$

thus determines

$$\overline{x} = \frac{\sum_{k=1}^{n} x_km_k}{\sum_{k=1}^{n} m_k}. \tag{3}$$

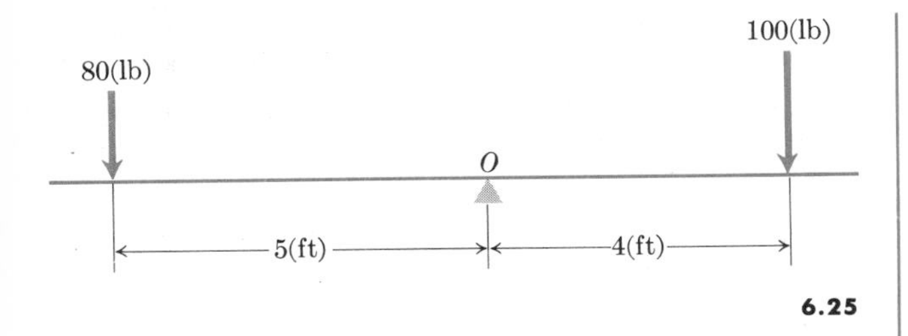

6.25

Example 1. The principle of a moment underlies the simple seesaw. For instance (Fig. 6.25), suppose one child weighs 80 lb and sits 5 ft from the point O, while the other child weighs 100 lb and is 4 ft from O. Then the child at the left end of the seesaw produces a moment of $80 \times 5 = 400$ (lb-ft), which tends to rotate the plank counterclockwise about O. The child at the right end of the plank produces a moment of $100 \times 4 = 400$ (lb-ft), which tends to rotate the plank clockwise around O. If we introduce coordinates $x_1 = -5$ and $x_2 = +4$, we find

$$\begin{aligned} x_1 m_1 + x_2 m_2 &= (-5)(80) + (4)(100) \\ &= -400 + 400 = 0 \end{aligned}$$

as the resultant moment about O. The same moment, zero, would be obtained if both children were at O, which is the center of their combined mass.

If, instead of being placed on the x-axis, the masses are located in the xy-plane at points

$$(x_1, y_1), \quad (x_2, y_2), \quad \ldots, \quad (x_n, y_n),$$

in that order, then we define their moments with respect to the y-axis and with respect to the x-axis as

$$M_y = x_1 m_1 + x_2 m_2 + \cdots + x_n m_n = \sum_{k=1}^{n} x_k m_k,$$

$$M_x = y_1 m_1 + y_2 m_2 + \cdots + y_n m_n = \sum_{k=1}^{n} y_k m_k.$$

The center of mass is the point $(\bar{x}, \bar{y})$,

$$\bar{x} = \frac{\sum x_k m_k}{\sum m_k}, \qquad \bar{y} = \frac{\sum y_k m_k}{\sum m_k}, \tag{4}$$

where the total mass could be concentrated and still give the same total moments M_y and M_x.

In space, three coordinates are needed to specify the position of a point. If the masses are located at the points

$$(x_1, y_1, z_1), \quad (x_2, y_2, z_2), \quad \ldots, \quad (x_n, y_n, z_n),$$

we define their moments with respect to the various coordinate planes as

$$M_{yz} = x_1 m_1 + x_2 m_2 + \cdots + x_n m_n = \sum_{k=1}^{n} x_k m_k,$$

$$M_{zx} = y_1 m_1 + y_2 m_2 + \cdots + y_n m_n = \sum_{k=1}^{n} y_k m_k,$$

$$M_{xy} = z_1 m_1 + z_2 m_2 + \cdots + z_n m_n = \sum_{k=1}^{n} z_k m_k.$$

(M_{yz} = moment with respect to the yz-plane, etc.)

The center of mass $(\bar{x}, \bar{y}, \bar{z})$ is the point where the total mass could be concentrated without altering these moments. Its coordinates therefore are given by

$$\bar{x} = \frac{\sum x_k m_k}{\sum m_k}, \qquad \bar{y} = \frac{\sum y_k m_k}{\sum m_k}, \qquad \bar{z} = \frac{\sum z_k m_k}{\sum m_k}. \tag{5}$$

Now most physical objects with which we deal are composed of enormously large numbers of molecules. It would be extremely difficult, and in most cases unnecessary, for us to concern ourselves with the molecular structure of a physical object, such as a pendulum, whose motion as a whole is to be studied. Instead, we make certain simplifying assumptions which we recognize as being only approximately correct. One such assumption is that the matter in a given solid is continuously distributed throughout the solid. Furthermore, if P is a point in the solid and ΔV is an element of volume which contains P, and if Δm is the mass of ΔV, then we assume that the ratio $\Delta m / \Delta V$ tends to a definite limit

$$\delta = \lim_{\Delta V \to 0} \frac{\Delta m}{\Delta V} \tag{6}$$

as the largest dimension of ΔV approaches zero. The limit δ is called the *density* of the solid at the point P. It is customary to write Eq. (6) in the alternative forms

$$\delta = dm/dV, \qquad dm = \delta\, dV. \tag{7}$$

Now if a solid is divided into small pieces ΔV of mass Δm and if $P(\tilde{x}, \tilde{y}, \tilde{z})$ is a point in ΔV and δ is the density at P, then

$$\Delta m \approx \delta\, \Delta V.$$

The moments of Δm with respect to the coordinate planes are not defined by what we have done thus far. But now we may think of replacing the Δm that fills the volume ΔV by an equal mass all of which is concentrated at the point P. The moments of this concentrated mass with respect to the coordinate planes are

$$\tilde{x}\, \Delta m, \qquad \tilde{y}\, \Delta m, \qquad \tilde{z}\, \Delta m.$$

Now we add the moments of all the concentrated masses in all the volume elements ΔV and take the limit as the ΔV's approach zero. This leads us to the following *definitions* of the moments M_{yz}, etc., for the mass as a whole:

$$M_{yz} = \lim_{\Delta m \to 0} \sum \tilde{x}\, \Delta m = \int \tilde{x}\, dm,$$

$$M_{zx} = \lim_{\Delta m \to 0} \sum \tilde{y}\, \Delta m = \int \tilde{y}\, dm,$$

$$M_{xy} = \lim_{\Delta m \to 0} \sum \tilde{z}\, \Delta m = \int \tilde{z}\, dm.$$

From these, we then deduce the equations

$$\bar{x} = \frac{\int \tilde{x}\, dm}{\int dm}, \qquad \bar{y} = \frac{\int \tilde{y}\, dm}{\int dm}, \qquad \bar{z} = \frac{\int \tilde{z}\, dm}{\int dm} \tag{8}$$

for the center of mass of the solid as a whole.

In theory, the element dm is approximately the mass of a volume element dV which has three small dimensions, as in Fig. 6.26. In a later chapter we shall see how to evaluate the integrals (8) which arise when this is done. In practice, however, we can usually take

$$dm = \delta\, dV,$$

with a volume element dV which has only one small dimension. Then we must interpret $\tilde{x}, \tilde{y}, \tilde{z}$ in the integrands in (8) as *the coordinates of the center of mass of the element dm.*

In many problems of practical importance, the density δ is constant and the solid possesses a plane of symmetry. Then it is easy to see that the center

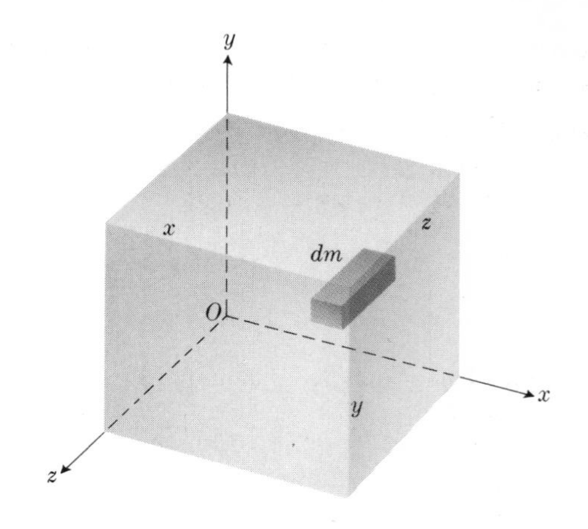

6.26

of mass lies in this plane of symmetry. For we may, with no loss of generality, choose our coordinate reference frame in such a way that the yz-plane is the plane of symmetry. Then for every element of mass Δm with a positive $\tilde{x}$ there is a symmetrically located element of mass with a corresponding negative $\tilde{x}$. These two elements have moments which are equal in magnitude and of opposite signs. The whole mass is made up of such symmetric pairs of elements and the sum of their moments about the yz-plane is zero. Therefore $\bar{x} = 0$; that is, the center of mass lies in the plane of symmetry. If there are two planes of symmetry, their intersection is an axis of symmetry and the center of mass must lie on this axis, since it lies in both planes of symmetry.

The most frequently encountered distributions of mass are

(a) along a thin wire or filament: $dm = \delta_1\, ds$,

(b) in a thin plate or shell: $dm = \delta_2\, dA$ or $\delta_2\, dS$,

(c) in a solid: $dm = \delta_3\, dV$,

where

ds = element of arc length,
δ_1 = mass per unit length of wire,
dA or dS = element of area,
δ_2 = mass per unit area of plate or shell,
dV = element of volume,
δ_3 = mass per unit volume of solid.

Problem 1. Find the center of mass of a thin homogeneous triangular plate of base b and altitude h.

Solution. Divide the triangle into strips of width dy parallel to the x-axis. A representative strip is shown in Fig. 6.27. The mass of the strip is approximately

$$dm = \delta_2\, dA,$$

where

$$dA = l\, dy$$

and l is the width of the triangle at distance y above its base. By similar triangles,

$$\frac{l}{b} = \frac{h-y}{h}, \quad \text{or} \quad l = \frac{b}{h}(h-y),$$

so that

$$dm = \delta_2 \frac{b}{h}(h-y)\, dy.$$

For the y-coordinate of the center of mass of the element dm, we have $\tilde{y} = y$. For the entire plate,

$$\bar{y} = \frac{\int y\, dm}{\int dm} = \frac{\int_0^h \delta_2 \frac{b}{h} y(h-y)\, dy}{\int_0^h \delta_2 \frac{b}{h}(h-y)\, dy} = \frac{1}{3}h.$$

Thus the center of mass lies above the base of the triangle at a distance one-third of the way toward the opposite vertex. By considering each side in turn as being a base of the triangle, this result shows that the center of gravity lies at the intersection of the medians.

Problem 2. A thin homogeneous wire is bent to form a semicircle of radius r (Fig. 6.28). Find its center of mass.

Solution. Here we take

$$dm = \delta_1\, ds,$$

where ds is an element of arc length of the wire and

$$\delta_1 = \frac{M}{L} = \frac{M}{\pi r}$$

is the mass per unit length of the wire. In terms of the central angle θ measured in radians (as usual), we have

$$ds = r\, d\theta$$

and

$$\tilde{x} = r \cos\theta, \qquad \tilde{y} = r \sin\theta.$$

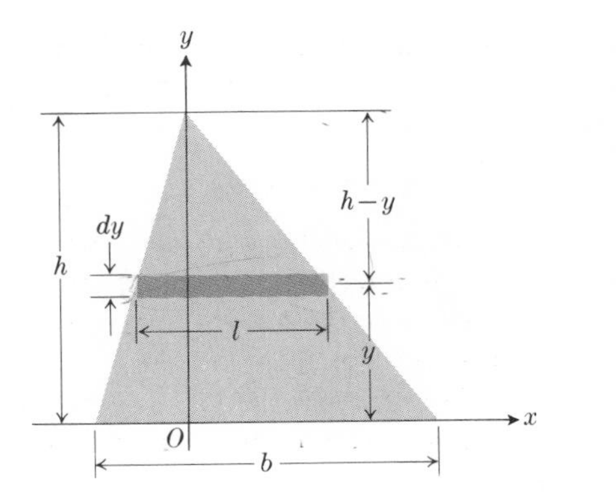

6.27

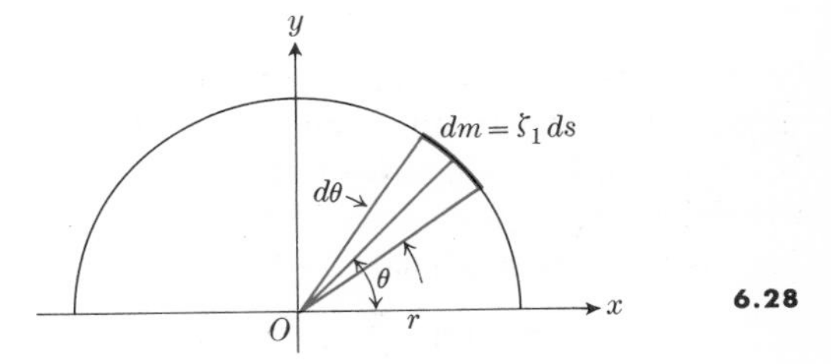

6.28

Hence

$$\bar{x} = \frac{\int_0^\pi r\cos\theta\, \delta_1 r\, d\theta}{\int_0^\pi \delta_1 r\, d\theta} = \frac{\delta_1 r^2 [\sin\theta]_0^\pi}{\delta_1 r[\theta]_0^\pi} = 0,$$

$$\bar{y} = \frac{\int_0^\pi r\sin\theta\, \delta_1 r\, d\theta}{\int_0^\pi \delta_1 r\, d\theta} = \frac{\delta_1 r^2 [-\cos\theta]_0^\pi}{\delta_1 r[\theta]_0^\pi} = \frac{2}{\pi} r.$$

The center of mass is therefore on the y-axis at a distance $2/\pi$ (roughly $\frac{2}{3}$) of the way up from the origin toward the intercept $(0, r)$.

Problem 3. Find the center of mass of a solid hemisphere of radius r if its density at any point P is proportional to the distance of P from the base of the hemisphere.

Solution. Imagine the solid cut into slices of thickness dy by planes perpendicular to the y-axis (Fig. 6.29), and take

$$dm = \delta_3\, dV,$$

where

$$dV = A(y)\, dy$$

is the volume of the representative slice at distance y above the base of the hemisphere, which is in the xz-plane, and where

$$\delta_3 = ky \quad (k = \text{constant})$$

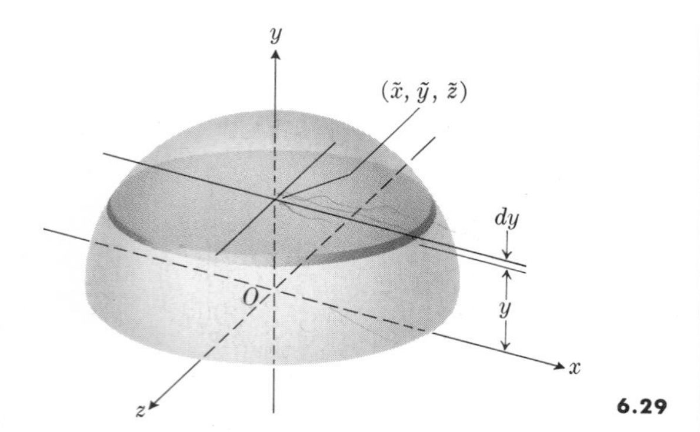

6.29

is the density of the solid in this slice. The area of a face of the slice dV is $A(y) = \pi x^2$, where $x^2 + y^2 = r^2$, that is,

$$dV = A(y)\,dy = \pi(r^2 - y^2)\,dy,$$

so that

$$dm = k\pi(r^2 - y^2)y\,dy.$$

The center of mass of the slice may be taken at its geometrical center, that is

$$(\tilde{x}, \tilde{y}, \tilde{z}) = (0, y, 0),$$

so that

$$\bar{x} = \bar{z} = 0$$

and

$$\bar{y} = \frac{\int y\,dm}{\int dm} = \frac{\int_0^r k\pi(r^2 - y^2)y^2\,dy}{\int_0^r k\pi(r^2 - y^2)y\,dy}$$

$$= \frac{k\pi\left[\frac{r^2y^3}{3} - \frac{y^5}{5}\right]_0^r}{k\pi\left[\frac{r^2y^2}{2} - \frac{y^4}{4}\right]_0^r} = \frac{8}{15}r.$$

EXERCISES

In Exercises 1 through 5, find the center of mass of a thin homogeneous plate covering the given portion of the xy-plane.

1. The first quadrant of the circle $x^2 + y^2 = a^2$.
2. The area bounded by the parabola $y = h^2 - x^2$ and the x-axis.
3. The "triangular" shaped area in the first quadrant between the circle $x^2 + y^2 = a^2$ and the lines $x = a$, $y = a$.
4. The area that lies between the x-axis and the curve $y = \sin x$ between $x = 0$ and $x = \pi$.
   Hint. Take $dA = y\,dx$ and $\tilde{y} = \frac{1}{2}y$.
5. The area that lies between the y-axis and the curve $x = 2y - y^2$.
6. Find the distance, from the base, of the center of mass of a thin triangular plate of base b and altitude h if its density varies as the square root of the distance from the base.
7. In Exercise 6, suppose that the density varies as the square of the distance.
8. Find the center of mass of a homogeneous solid right circular cone.
9. Find the center of mass of a solid right circular cone if the density varies as the distance from the base.
10. In Exercise 9, suppose that the density varies as the square of the distance.
11. In Problem 2 (see Fig. 6.28), suppose that the density is $\delta_1 = k \sin \theta$, k being constant. Find the center of mass.

6.10 THE CENTROID

If a mass has uniform density, Eqs. (8), Article 6.9, reduce to

$$\bar{x} = \frac{\int \tilde{x}\delta_3\,dV}{\int \delta_3\,dV} = \frac{\delta_3\int \tilde{x}\,dV}{\delta_3\int dV} = \frac{\int \tilde{x}\,dV}{\int dV},$$

$$\bar{y} = \frac{\int \tilde{y}\,dV}{\int dV}, \qquad \bar{z} = \frac{\int \tilde{z}\,dV}{\int dV} \tag{1}$$

for a solid, with similar equations having dA or ds in place of dV for a plate or a wire. Since these expressions involve only the geometric objects, namely, volumes, areas, and curves, we speak of the point $(\bar{x}, \bar{y}, \bar{z})$ in such cases as the *centroid* of the object.

Problem 1. Find the centroid of a solid hemisphere of radius r.

Solution. As in Problem 3, Article 6.9, imagine the solid (Fig. 6.29) cut into slices of thickness dy by planes perpendicular to the y-axis. The centroid of a slice is its geometrical center, which lies on the y-axis at $(0, y, 0)$. Its moment with respect to the xz-plane is

$$dM_{xz} = y\,dV = y\pi(r^2 - y^2)\,dy.$$

Hence

$$M_{xz} = \pi \int_0^r (r^2y - y^3)\,dy = \frac{\pi r^4}{4}.$$

Since the volume of the hemisphere is

$$V = \frac{2}{3}\pi r^3,$$

we have

$$\overline{y} = \frac{M_{xz}}{V} = \frac{3}{8}r.$$

Problem 2. Find the centroid of a thin hemispherical shell of inner radius r and thickness t.

Solution. We shall first solve this problem exactly (A), by using the results of the preceding example. Second (B), we shall see how the position of the centroid changes as we hold r fixed and let $t \to 0$. Finally (C), we shall solve the problem approximately without using the results of Problem 1, by considering the center of mass of an imaginary distribution of mass over the surface of the hemisphere of radius r, assuming that the thickness t is negligible in comparison with r (as, for example, the gold leaf covering the dome of the State House in Boston).

A. Let M_1, M_2 and M; V_1, V_2, and V denote respectively the moments with respect to the xz-plane and the volumes of the solid hemisphere of radius r, the solid hemisphere of radius $(r + t)$, and the hemispherical shell of thickness t and inner radius r. Since the moment of the sum of two masses is the sum of their moments and

$$V_1 + V = V_2,$$

we also have

$$M_1 + M = M_2, \quad \text{or} \quad M = M_2 - M_1.$$

But by Problem 1,

$$M_2 = \frac{\pi}{4}(r + t)^4, \qquad M_1 = \frac{\pi}{4}r^4,$$

so that

$$M = \frac{\pi}{4}[(r + t)^4 - r^4] = \frac{\pi}{4}[4r^3t + 6r^2t^2 + 4rt^3 + t^4],$$

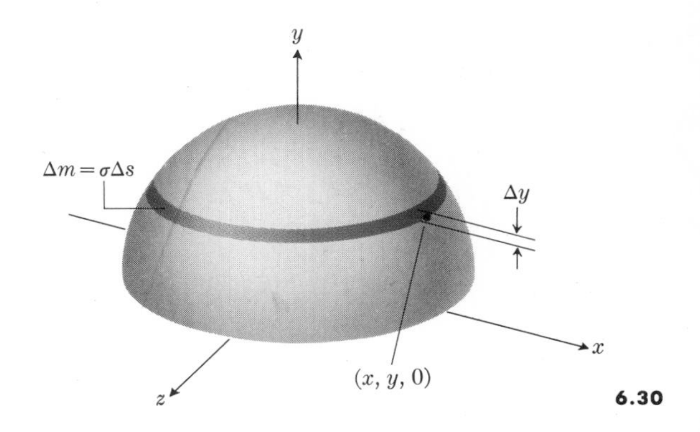

6.30

while

$$V = V_2 - V_1 = \tfrac{2}{3}\pi[(r + t)^3 - r^3] = \tfrac{2}{3}\pi(3r^2t + 3rt^2 + t^3);$$

hence

$$\overline{y} = \frac{M}{V} = \frac{1}{2}\,\frac{[r^3 + \frac{3}{2}r^2t + rt^2 + \frac{1}{4}t^3]}{[r^2 + rt + \frac{1}{3}t^2]}.$$

B. In the case of a shell where the thickness t is negligible in comparison with r, this reduces to $\overline{y} = \frac{1}{2}r$ if we write

$$\overline{y} = \frac{1}{2}\,\frac{r^3\left[1 + \frac{3}{2}\left(\frac{t}{r}\right) + \left(\frac{t}{r}\right)^2 + \frac{1}{4}\left(\frac{t}{r}\right)^3\right]}{r^2\left[1 + \left(\frac{t}{r}\right) + \frac{1}{3}\left(\frac{t}{r}\right)^2\right]},$$

and then let

$$(t/r) \to 0.$$

C. To solve the problem directly, without reference to Problem 1, consider the mass of the shell as being uniformly distributed over its surface so that the mass of any portion cut from the shell will be proportional to its (inner) surface area. The proportionality factor,

$$\sigma = \delta t,$$

where δ is the volume density and t the thickness of the shell, is frequently referred to as a surface density factor. (Thus in working with sheet metal stock, say aluminum of uniform thickness and density, we might speak of it as weighing so many pounds *per square foot*. This would be our σ, and A square feet of this particular stock would weigh σA pounds.) Then the slice of shell between planes perpendicular to the y-axis at distances y and $y + \Delta y$ respectively above the xz-plane has mass

$$\Delta m = \sigma\,\Delta S,$$

where ΔS is the surface area of the slice (see Fig. 6.30).

The moment of this slice with respect to the xz-plane then lies between

$$y\,\Delta m \quad \text{and} \quad (y + \Delta y)\,\Delta m,$$

and hence differs from $y\,\Delta m$ by a most $\Delta y\,\Delta m$. The sum of the moments of the slices has a limit, as Δy approaches zero, equal to $\int y\,dm$. We are thus led to

$$\bar{y} = \frac{\int y\,dm}{\int dm} = \frac{\int \sigma y\,dS}{\int \sigma\,dS} = \frac{\sigma \int y\,dS}{\sigma \int dS},$$

since σ is constant by our hypotheses. Here

$$dS = 2\pi x\,ds = 2\pi(r\cos\theta)r\,d\theta$$

if we use

$$x = r\cos\theta, \qquad y = r\sin\theta$$

to represent the circle $x^2 + y^2 = r^2$. Hence

$$\begin{aligned}\bar{y} &= \frac{\int_0^{\pi/2} r\sin\theta \cdot 2\pi(r\cos\theta)\cdot r\,d\theta}{\int_0^{\pi/2} 2\pi(r\cos\theta)r\,d\theta} \\ &= \frac{2\pi r^3 \int_0^{\pi/2} \sin\theta\cos\theta\,d\theta}{2\pi r^2 \int_0^{\pi/2} \cos\theta\,d\theta} \\ &= r\,\frac{[\frac{1}{2}\sin^2\theta]_0^{\pi/2}}{[\sin\theta]_0^{\pi/2}} = \frac{1}{2}r.\end{aligned}$$

Observe that the element of arc ds which, when rotated about the y-axis generates dS, need only move along the arc of the circle in the first quadrant, namely, $0 \le \theta \le \pi/2$, to give us the entire hemisphere.

By the symmetry of the shell, the centroid lies on the y-axis; that is,

$$(\bar{x}, \bar{y}, \bar{z}) = (0, \tfrac{1}{2}r, 0).$$

Remark. The *centroid* of a body is also called its *center of gravity*.

EXERCISES

In Exercises 1 through 5, find the centroid of the areas bounded by the given curves and lines.

1. The x-axis and the curve $y = c^2 - x^2$.
2. The y-axis and the curve $x = y - y^3$, $0 \le y \le 1$.
3. The curve $y = x^2$ and the line $y = 4$.
4. The curve $y = x - x^2$ and the line $x + y = 0$.
5. The curve $x = y^2 - y$ and the line $y = x$.
6. Find the center of gravity of a solid right circular cone of altitude h and base-radius r.
7. Find the center of gravity of the solid generated by rotating, about the y-axis, the area bounded by the curve $y = x^2$ and the line $y = 4$.
8. The area bounded by the curve $x = y^2 - y$ and the line $y = x$ is rotated about the x-axis. Find the center of gravity of the solid thus generated.
9. Find the center of gravity of a very thin right circular conical shell of base-radius r and altitude h.
10. Find the center of gravity of the surface area generated by rotating, about the line $x = -r$, the arc of the circle $x^2 + y^2 = r^2$ that lies in the first quadrant. [*Suggestion.* Use $x = r\cos\theta$, $y = r\sin\theta$ to represent the circle.]
11. Find the moment, about the x-axis, of the arc of the parabola $y = \sqrt{x}$ lying between the points $(0, 0)$ and $(4, 2)$.
12. Find the center of gravity of the arc length of one quadrant of a circle.

6.11 THE THEOREMS OF PAPPUS

skip

When a plane area, such as A in Fig. 6.31, is rotated about an axis in its plane that does not intersect the area, there is a useful formula that relates the volume swept out by the area to the path described by its centroid.

Theorem 1. *If a plane area is revolved about a line that lies in its plane but does not intersect the area, then the volume generated is equal to the product of the area and the distance traveled by its centroid.*

Proof. To prove this result, which is one of the Theorems of Pappus, let the x-axis coincide with the axis of revolution and divide the area into strips of width Δy by lines parallel to the x-axis. The entire volume is the sum of the volumes generated by these strips of area. Let $l = f(y)$ be the width of the area at distance y above the x-axis, $c \le y \le d$. By the same argument we used in Article 6.5 to justify the cylindrical-shell method of computing volumes, we have

$$V = \lim_{\Delta y \to 0} \sum 2\pi y l\,\Delta y = \int_c^d 2\pi y l\,dy. \tag{1}$$

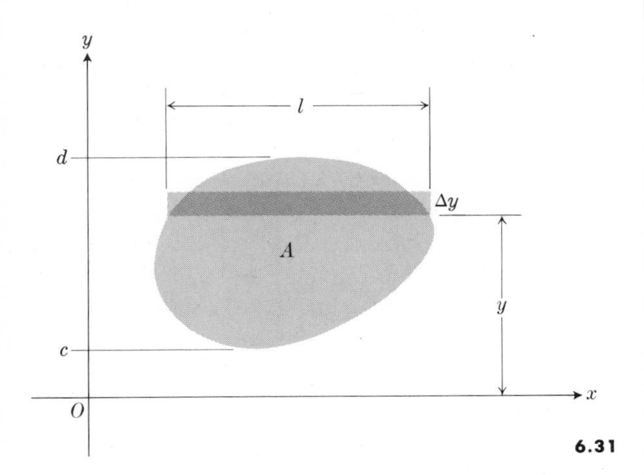

6.31

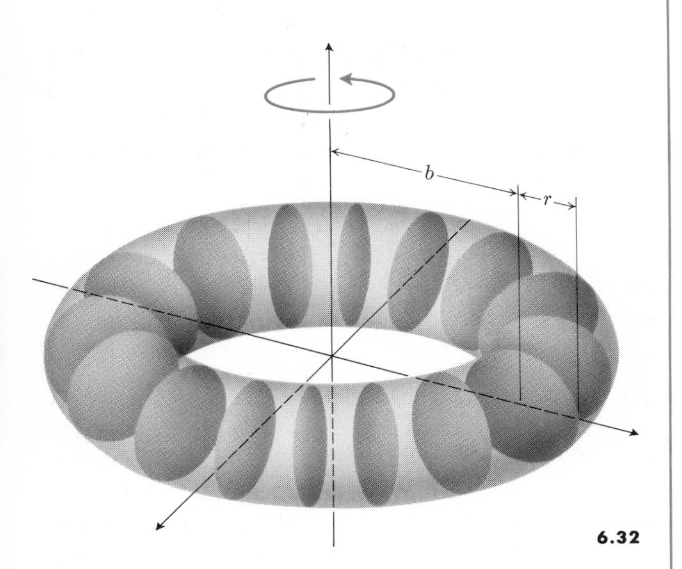

6.32

But the ordinate $\overline{y}_A$ of the centroid of the area A is given by

$$\overline{y}_A = \frac{\int y\,dA}{\int dA} = \frac{\int_c^d yl\,dy}{A},$$

so that

$$\int_c^d yl\,dy = A\overline{y}_A,$$

and hence Eq. (1) becomes

$$V = 2\pi\overline{y}_A A. \tag{2}$$

Since $2\pi\overline{y}_A$ is the distance traveled by the centroid of the area, the theorem is established. □

Problem 1. Find the volume of the torus (doughnut) generated by rotating a circle of radius r about an axis in its plane at a distance b from its center, $b > r$.

Solution. See Fig. 6.32. The center of the circle is its centroid, and this travels a distance $2\pi b$. The area of the circle is πr^2, so by Eq. (2) the volume of the torus is

$$V = (2\pi b)(\pi r^2) = 2\pi^2 br^2.$$

Theorem 2. *If an arc of a plane curve is revolved about a line that lies in its plane but does not intersect the arc, then the surface area generated by the arc is equal to the product of the length of the arc and the distance traveled by its centroid.*

Proof. In this, another Theorem of Pappus, the centroid is that of the curve (or of a fine homogeneous wire whose centerline coincides with the curve), and is generally different from the centroid of the enclosed plane area (if the curve is a closed curve). To prove the theorem, refer the curve to x- and y-axes in its plane and let the x-axis coincide with the axis of revolution (see Fig. 6.33). Then the surface area generated is

$$S = \int_{x=a}^{x=b} 2\pi y\,ds, \tag{3}$$

as we found in Article 6.7. But the centroid of the curve has ordinate

$$\overline{y}_c = \frac{\int y\,ds}{\int ds} = \frac{\int y\,ds}{L}, \quad \text{where} \quad L = \int ds$$

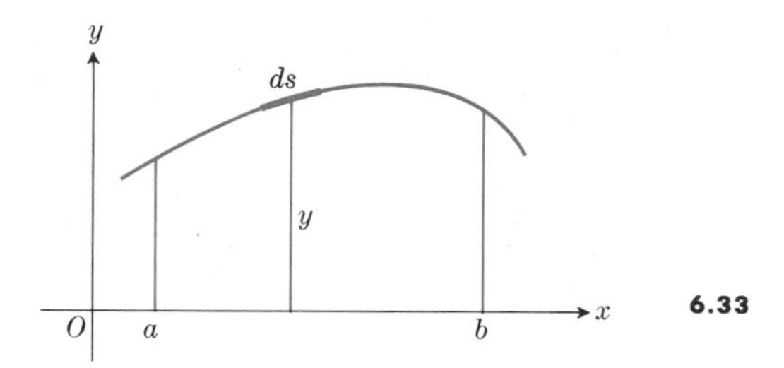

6.33

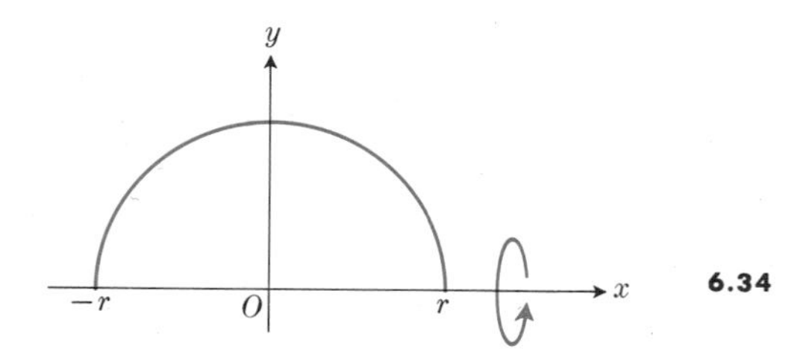

6.34

is the length of the arc. Hence

$$S = 2\pi \int_{x=a}^{x=b} y\, ds = 2\pi \bar{y}_c L. \; \square \tag{4}$$

Example 1. The surface area of the torus in Problem 1 is given by

$$S = (2\pi b)(2\pi r) = 4\pi^2 br.$$

The two Theorems of Pappus are useful in determining volumes or surface areas of solids of revolution when the centroids are known. They are equally useful in determining the centroids when the volume and surface area are known.

Problem 2. Use the first Theorem of Pappus and the fact that the volume of a sphere of radius r is $V = \frac{4}{3}\pi r^3$ to find the center of gravity of the area of a semicircle.

Solution. Note that we generate a sphere by rotating a semicircle about its diameter (Fig. 6.34). Then the axis of revolution does not intersect the area, and since

$$A = \tfrac{1}{2}\pi r^2,$$

while

$$V = 2\pi \bar{y} A,$$

we find

$$\bar{y} = \frac{V}{2\pi A} = \frac{\frac{4}{3}\pi r^3}{2\pi \cdot \frac{1}{2}\pi r^2} = \frac{4}{3\pi} r.$$

EXERCISES

1. Use the Theorems of Pappus to find the lateral surface area and the volume of a right circular cone.
2. Use the second Theorem of Pappus and the fact that the surface area of a sphere of radius r is $4\pi r^2$ to find the centroid of the semicircular arc
$$y = \sqrt{r^2 - x^2}, \quad -r \le x \le r.$$
3. The semicircular arc
$$y = \sqrt{r^2 - x^2}, \quad -r \le x \le r,$$
is rotated about the line $y = r$. Use the second Theorem of Pappus together with the answer to the preceding exercise to find the surface area generated.
4. The centroid of the area bounded by the x-axis and the semicircle $y = \sqrt{r^2 - x^2}$ is at $[0, (4/3\pi)r]$ (see Problem 2). Find the volume generated when this area is rotated about the line $y = -r$.
5. The area of the preceding exercise is rotated about the line $y = x - r$. Find the volume generated.
6. Use the answer to Exercise 2 above to find the surface area generated by rotating the semicircular arc
$$y = \sqrt{x^2 - r^2}, \quad -r \le x \le r,$$
about the line $y = x - r$.
7. Find the moment about the x-axis of the area in the semicircle of Fig. 6.34, Example 1. (If you use results already known, you won't need to integrate.)
8. Find the moment about the line $y = -r$ of the area in the semicircle of Fig. 6.34.
9. Find the moment about the line $y = x - r$ of the area in the semicircle of Fig. 6.34.

6.12 HYDROSTATIC PRESSURE

If a flat-bottomed container is filled with water to a depth h, the resulting force due to the pressure of the liquid it contains is

$$F = whA, \tag{1}$$

where w is the weight-density, which is nearly 62.5 lb/ft^3, and A is the area of the bottom of the container. Obviously, the units in Eq. (1) must be compatible, say h in feet, A in square feet, and w in pounds per cubic foot, giving F in pounds. It is a remarkable fact that this force does not depend on the shape of the sides of the vessel. The force on the bottom of the beaker and the flask in Fig. 6.35, for example, is the same if both containers have the same area at their bases and both have the same "head" h. (See Sears, *Mechanics, Heat, and Sound*, Chapter 16.) The *pressure*, or *force per unit area*, at the bottom of the container is therefore

$$p = wh. \tag{2}$$

Next consider any body of water, such as the water in a reservoir or behind a dam. According to Pascal's principle, the pressure $p = wh$ at depth h in such a body of water is the same *in all directions*. For a flat plate submerged *horizontally*, the downward *force* acting on its upper face due to this liquid pressure is the same as that given by Eq. (1). If the plate is submerged *vertically*, however, then the pressure against it will be different at different depths and Eq. (1) no longer is usable in that form because we would have different h factors for points at different depths. We get around this difficulty in a manner which is now familiar, by dividing the plate into many narrow strips, the representative strip having its upper edge at depth h and its lower edge at depth $h + \Delta h$ below the surface of the water. If the area of this strip is called ΔA, and the force on one side of it ΔF, then Eq. (1) suggests that

$$wh\,\Delta A \leq \Delta F \leq w(h + \Delta h)\,\Delta A,$$

since the pressure will vary from wh to $w(h + \Delta h)$ in this strip. When we add the forces for all strips, then take the limit of such sums as $\Delta h \to 0$, we may disregard the infinitesimals of higher order, such as $\Delta h\,\Delta A$, or we may apply Bliss's theorem, and we are led to

$$F = \lim_{\Delta h \to 0} \sum_a^b wh\,\Delta A,$$

or

$$F = \int_{h=a}^{h=b} wh\,dA = \int_a^b whl\,dh, \tag{3}$$

where $dA = l\,dh$ (see Fig. 6.36).

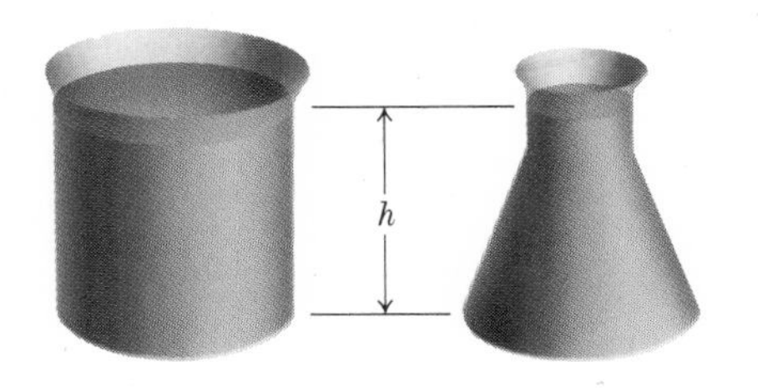

6.35

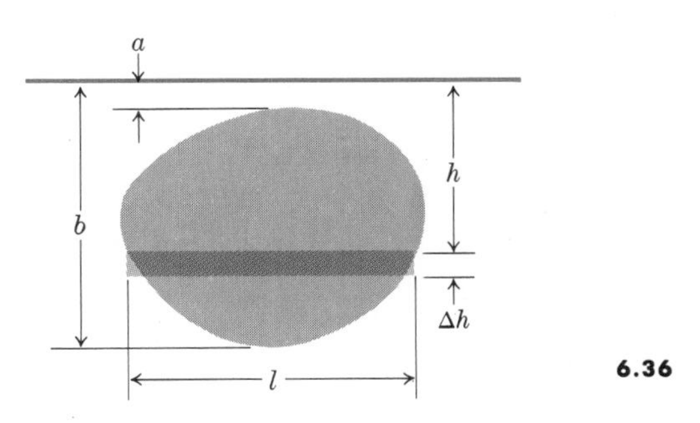

6.36

As a corollary of Eq. (3), if we denote the depth of the centroid of the area A by $\bar{h}$, then

$$\bar{h} = \frac{\int h\,dA}{\int dA}, \quad \text{or} \quad \int h\,dA = \bar{h}A.$$

Since w is a constant which may be moved across the integral sign in (3), we have

$$F = w\bar{h}A. \tag{4}$$

This states that the total force of the liquid pressing against one face of the plate (an equal and opposite force presses against the other face unless the plate forms part of a wall of the container) is the same as it would be according to (1) if all of the area A were at the depth $\bar{h}$ below the surface.

Equation (4) is the working tool used most frequently by the engineer in finding such hydrostatic forces as we have discussed. He can refer to a handbook to obtain the centroid of simple plane areas and from this he quickly finds $\bar{h}$. Of course, the location of the centroid which he finds in his handbook was calculated by someone who performed an integration equivalent to evaluating the integral in Eq. (3). It is recommended that the reader solve

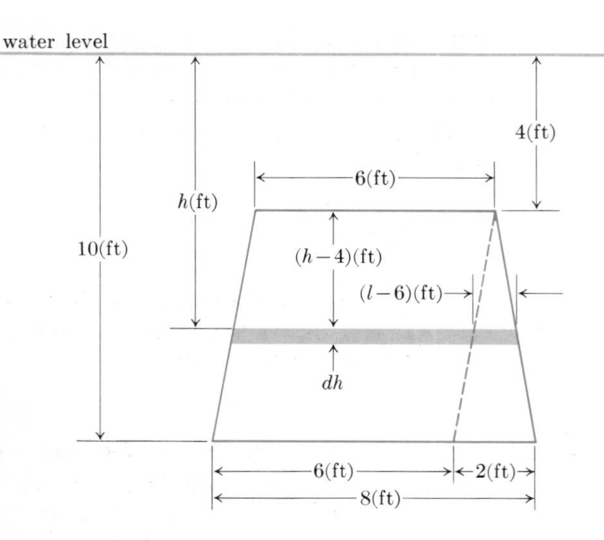

6.37 The trapezoid is submerged vertically beneath the water level.

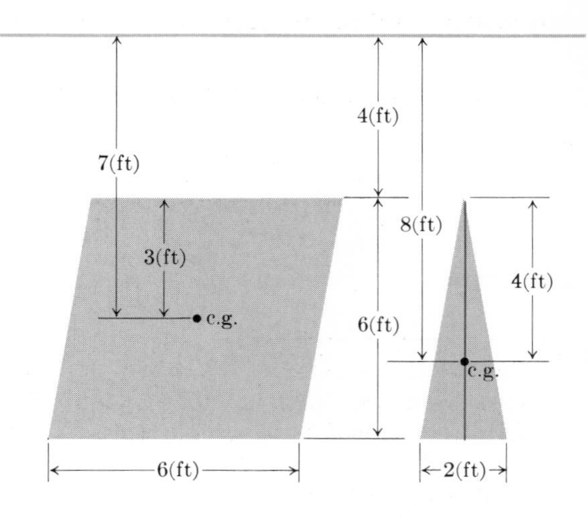

6.38 Centroids at points marked "c.g." (center of gravity).

problems of this type by thinking through the steps which lead up to Eq. (3) by integration, and that he then check his results, when he can conveniently do so, by Eq. (4).

Problem 1. A trapezoid is submerged vertically in water with its upper edge 4 ft below the surface and its lower edge 10 ft below the surface. If the upper and lower edges are respectively 6 ft and 8 ft long, find the total force on one face of the trapezoid.

Solution. In Fig. 6.37, we have, by similar triangles,

$$\frac{l-6}{2} = \frac{h-4}{6},$$

so that

$$l = \frac{h+14}{3}.$$

Then the force is

$$F = \int_4^{10} wh\left(\frac{h+14}{3}\right) dh = 300w.$$

Since

$$w = 62.5\ (\text{lb/ft}^3) = \tfrac{1}{32}\ (\text{ton/ft}^3),$$

we have

$$F = 9\tfrac{3}{8}\ \text{tons}.$$

Now we check by using Eq. (4). The trapezoid can be resolved into a parallelogram plus a triangle (Fig. 6.38). For the parallelogram,

$$\bar{h}_1 = 7 \quad \text{and} \quad A_1 = 36, \qquad F_1 = 252w.$$

For the triangle,

$$\bar{h}_2 = 8 \quad \text{and} \quad A_2 = 6, \qquad F_2 = 48w.$$

For the trapezoid,

$$F = F_1 + F_2 = 300w.$$

Problem 2. Find the moment produced about the lower edge of the trapezoid by the forces in the preceding example.

Solution. The force on the representative strip may be denoted by

$$dF = wh\,dA = wh\left(\frac{h+14}{3}\right) dh.$$

For the moment of this force about the lower edge of the trapezoid, we must multiply this force by the moment arm,

$$r = 10 - h,$$

to give the moment:

$$dM = r\,dF = wh(10-h)\,\frac{h+14}{3}\,dh.$$

The total moment due to forces on all the strips is

$$M = \frac{w}{3}\int_4^{10} (140h - 4h^2 - h^3)\, dh = 732w,$$

or

$$M = 22\tfrac{7}{8} \text{ (ft-tons)}.$$

EXERCISES

1. The vertical ends of a water trough are isosceles triangles of base 4 ft and altitude 3 ft. Find the force on one end if the trough is full of water weighing 62.5 lb/ft^3.
2. Find the force in the previous problem if the water level in the trough is lowered one foot.
3. A triangular plate ABC is submerged in water with its plane vertical. The side AB, 4 ft long, is 1 ft below the surface, while C is 5 ft below AB. Find the total force on one face of the plate.
4. Find the force on one face of the triangle ABC of Exercise 3 if AB is one foot below the surface as before, but the triangle is rotated 180° about AB so as to bring the vertex C 4 ft above the surface.
5. A semicircular plate is submerged in water with its plane vertical and its diameter in the surface. Find the force on one face of the plate if its diameter is 2 ft.
6. The face of a dam is a rectangle, $ABCD$, of dimensions $AB = CD = 100$ ft, $AD = BC = 26$ ft. Instead of being vertical, the plane $ABCD$ is inclined as indicated in Fig. 6.39, so that the top of the dam is 24 ft higher than the bottom. Find the force due to water pressure on the dam when the surface of the water is level with the top of the dam.
7. Find the moment, about AB, of the force in Exercise 6.

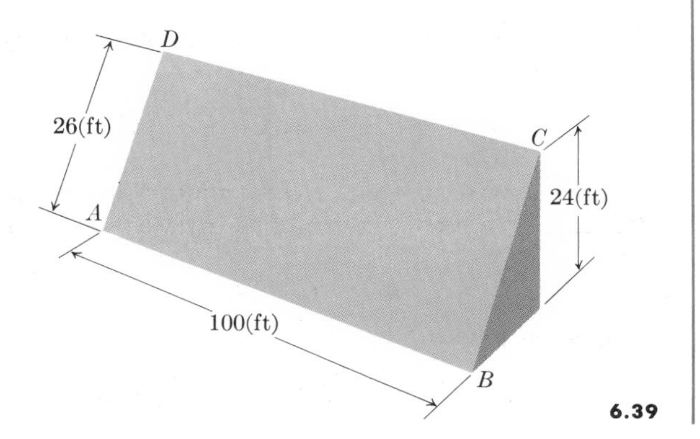

6.39

6.13 WORK

When a constant force F (pounds) acts throughout a distance s (feet), the *work* done (in foot-pounds) is the product of force and distance:

$$W = Fs. \tag{1}$$

When the force is not constant, as for instance in stretching or compressing a spring, then Eq. (1) cannot be used directly to give the work done. The law in (1) can be used, however, to find approximately the work done over a *short* interval Δs if the force is a continuous function of s. The integral process then enables us to extend the law in (1) to find the total work done.

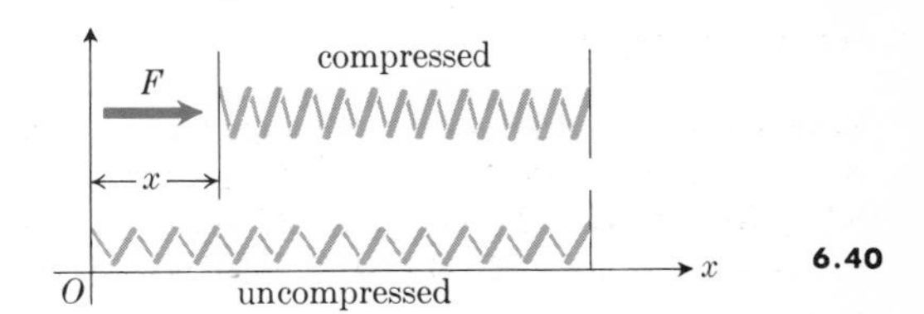

6.40

To illustrate, suppose we consider the work done in compressing a spring from its natural unstressed length L to a length $\frac{3}{4}L$ (Fig. 6.40) if the force required to hold it under compression is

$$F = cx, \tag{2}$$

where x is the amount it has been compressed and c is a proportionality factor (called the "spring constant"). Thus to compress the spring by an amount $L/4$, the force must be increased from

$$F_0 = c \times 0 = 0$$

to

$$F_1 = c\frac{L}{4},$$

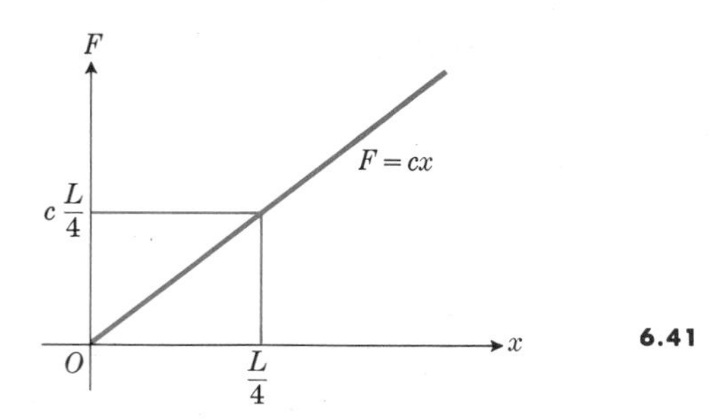

6.41

and as it does so the point of application of the force will move from $x = 0$ to $x = L/4$ (Fig. 6.41). How then shall we find the *work* done during the process? We imagine the x-interval from 0 to $L/4$ to be divided into a large number of subintervals, each of length Δx. As the spring is compressed an amount Δx, so that its left end moves across the representative subinterval from x to $x + \Delta x$, the *force* will vary from cx to $c(x + \Delta x)$, and since it acts through a *distance* Δx, the work done for this small compression will lie between

$$cx \, \Delta x \quad \text{and} \quad c(x + \Delta x) \, \Delta x,$$

so that the total work will be given approximately by

$$W \approx \sum_{x=0}^{L/4} cx \, \Delta x,$$

or

$$W = \sum_{x=0}^{L/4} c\tilde{x} \, \Delta x,$$

for appropriate choices of the $\tilde{x}$ between x and $x + \Delta x$. In the limit as Δx tends to zero, we have

$$W = \lim_{\Delta x \to 0} \sum_{x=0}^{L/4} cx \, \Delta x = \int_0^{L/4} cx \, dx$$
$$= \frac{cL^2}{32} = \frac{1}{2} \frac{cL}{4} \frac{L}{4}.$$

In this last form, the factor $\frac{1}{2}c(L/4)$ is one-half the final value reached by F when the spring has been compressed to $\frac{3}{4}$ of its original length, and the factor $L/4$ is the total distance through which the variable force has acted. This suggests

$$W = \overline{F}s \tag{3}$$

as a suitable modification of Eq. (1) when the force is variable, where $\overline{F}$ represents the *average* value of the variable force throughout the total displacement. However, the determination of $\overline{F}$ itself involves an integration in the general case, so that it is usually as easy to apply the principles illustrated in the example just considered as it is to apply Eq. (3). In fact, Eq. (3) may be interpreted as *defining* $\overline{F}$.

In a manner entirely analogous to that in the above example, it is easily seen that

$$W = \int_a^b F \, ds \tag{4}$$

gives the work done by a variable force (which, however, always acts along a given direction) as the point of application undergoes a displacement from $s = a$ to $s = b$. This leads to the following interesting theorem of mechanics.

Theorem 3. *Let F denote the resultant of all forces acting on a particle of mass m. Let the direction of F remain constant. Then whether the magnitude of F is constant or variable, the work done on the particle by the force F is equal to the change in the kinetic energy of the particle.*

The ingredients required for a proof of this theorem are:

(a) Eq. (4), which gives the work,

(b) the definition of kinetic energy: K.E. $= \frac{1}{2}mv^2$,

(c) Newton's second law: $F = m(dv/dt)$.

The fact that

$$v = \frac{ds}{dt} \quad \text{and} \quad \frac{dv}{dt} = \frac{dv}{ds}\frac{ds}{dt} = \frac{dv}{ds}v,$$

together with Newton's second law, enables us to write Eq. (4) in the form

$$W = \int_{s=a}^{s=b} mv \frac{dv}{ds} \, ds = \int_{v=v_a}^{v=v_b} mv \, dv,$$

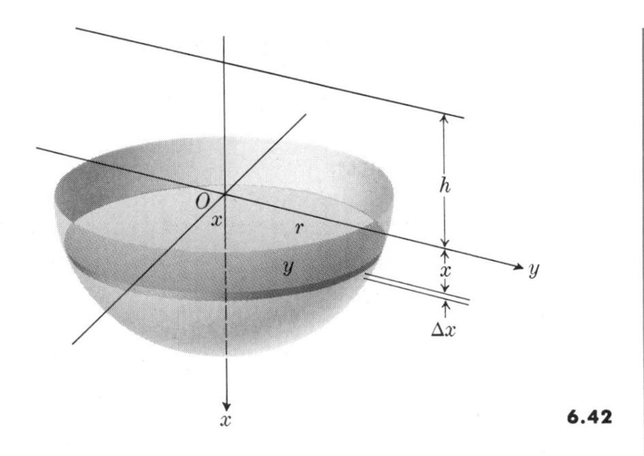

6.42

which leads to

$$W = \frac{1}{2}\, mv^2 \Big]_{v_a}^{v_b} = \frac{1}{2}\, mv_b^2 - \frac{1}{2}\, mv_a^2.$$

That is, the work done by the force F is the kinetic energy at b minus the kinetic energy at a or, more simply, the change in kinetic energy:

$$W = \Delta(\text{K.E.}).$$

As a second example of work in a situation where the simple Eq. (1) cannot be applied to the *total* but can be applied to a *small piece*, we consider the problem of pumping all the water from a full hemispherical bowl of radius r (feet) to a distance h (feet) above the top of the bowl (see Fig. 6.42).

We introduce coordinate axes as shown in Fig. 6.42 and imagine the bowl to be divided into a large number of thin slices by planes perpendicular to the x-axis between $x = 0$ and $x = r$. The representative slice between the planes at x and $x + \Delta x$ has volume ΔV which, if we neglect only infinitesimals of higher order than Δx, is given approximately by

$$\Delta V \approx \pi y^2 \, \Delta x = \pi(r^2 - x^2)\, \Delta x.$$

To the same order of approximation, the force F required to lift *this slice* is equal to its weight:

$$w\, \Delta V \approx \pi w(r^2 - x^2)\, \Delta x,$$

where w is the weight of a cubic foot of water. Finally, the *distance* through which this force must act lies between

$$h + x \quad \text{and} \quad h + x + \Delta x,$$

so that the work ΔW done in *lifting this one slice* is approximately

$$\Delta W \approx \pi w(r^2 - x^2)(h + x)\, \Delta x,$$

where we again have suppressed all higher powers of Δx. This is justified in the *limit* of the sum, and the total work is

$$\begin{aligned} W &= \lim_{\Delta x \to 0} \sum_0^r \pi w(r^2 - x^2)(h + x)\, \Delta x \\ &= \int_0^r \pi w(h + x)(r^2 - x^2)\, dx \\ &= hw \int_0^r \pi(r^2 - x^2)\, dx + w \int_0^r \pi x(r^2 - x^2)\, dx \\ &= hwV + \bar{x}wV. \end{aligned}$$

Here (wV) is the weight of the whole bowlful of water of volume V, and the second integral may be interpreted physically as giving the work required in pumping *all* the water from the depth of the center of gravity of the bowl to the level $x = 0$, while the first integral gives the work done in pumping the whole bowlful of water from the level $x = 0$ up a distance of h feet. The actual evaluation of the integrals leads to

$$W = \tfrac{2}{3}\pi r^3 w(h + \tfrac{3}{8}r).$$

Can you prove the following fact to yourself? No matter what the shape of the container in Fig. 6.42, the total work is the sum of two terms: One of these is

$$W_1 = hw \int dV,$$

and represents the total work done in lifting a bowlful of water a distance h, while the other is

$$W_2 = w \int x\, dV$$

and represents the work done in lifting a bowlful of water a distance equal to the depth of the center of gravity of the bowl.

EXERCISES

1. If the spring in Fig. 6.40 has natural length $L = 18$ in., and a force of 10 lb is sufficient to compress it to a length of 16 in., what is the value of the "spring constant" c for the particular spring in question? How much work is done in compressing it from a length of 16 in. to a length of 12 in.?
2. Answer the questions of Exercise 1 if the law of force is $F = c \sin(\pi x/2L)$ rather than Eq. (2).
3. Two electrons repel each other with a force inversely proportional to the square of the distance between them. Suppose one electron is held fixed at the point (1, 0) on the x-axis. Find the work required to move a second electron along the x-axis from the point $(-1, 0)$ to the origin.
4. If two electrons are held stationary at the points $(-1, 0)$ and $(1, 0)$ on the x-axis, find the work done in moving a third electron from (5, 0) to (3, 0) along the x-axis.
5. If a straight hole could be bored through the center of the earth, a particle of mass m falling in this hole would be attracted toward the center of the earth with force $mg\,r/R$ when it is at distance r from the center. (R is the radius of earth; g is the acceleration due to gravity at the surface of the earth.) How much work is done on the particle as it falls from the surface to the center of the earth?
6. A bag of sand originally weighing 144 lb is lifted at a constant rate of 3 ft/min. The sand leaks out uniformly at such a rate that half of the sand is lost when the bag has been lifted 18 ft. Find the work done in lifting the bag this distance.
7. Gas in a cylinder of constant cross-sectional area A expands or is compressed by the motion of a piston. If p is the pressure of the gas in pounds per square inch and v is its volume in cubic inches, show that the work done by the gas as it goes from an initial state (p_1, v_1) to a second state (p_2, v_2) is

$$W = \int_{(p_1,v_1)}^{(p_2,v_2)} p\,dv \text{ (inch-pounds).}$$

[*Hint.* Take the x-axis perpendicular to the face of the piston. Then $dv = A\,dx$ and $F = pA$.]

8. Use the result of Exercise 7 to find the work done on the gas in compressing it from $v_1 = 243\text{ in}^3$ to $v_2 = 32\text{ in}^3$ if the initial pressure $p_1 = 50\text{ lb/in}^2$ and the pressure and volume satisfy the law for adiabatic change of state, $pv^{1.4} = \text{constant}$.
9. Find the work done in pumping all the water out of a conical reservoir of radius 10 ft at the top, altitude 8 ft, to a height of 6 ft above the top of the reservoir.
10. Find the work done in Exercise 9 if at the beginning the reservoir is filled to a depth of 5 ft and the water is pumped just to the top of the reservoir.

REVIEW QUESTIONS AND EXERCISES

1. List nine applications of the definite integral presented in this chapter.
2. How do you define the area bounded above by a curve $y = f_1(x)$, below by a curve $y = f_2(x)$, and on the sides by $x = a$ and $x = b$, $a < b$?
3. How do you define the volume generated by rotating a plane area about an axis in its plane and not intersecting the area?
4. How do you define the length of a plane curve? Can you extend this to a curve in three-dimensional space?
5. How do you define the surface area of a sphere? Of other surfaces of revolution?
6. How do you define the average value of a function over an interval?
7. How do you define center of mass?
8. How do you define work done by a variable force?
9. How do you define the hydrostatic force on the face of a dam?

MISCELLANEOUS EXERCISES

1. Sketch the graphs of the equations $y = 2 - x^2$ and $x + y = 0$ in one diagram, and find the area bounded by them.
2. Find the maximum and minimum points of the curve $y = x^3 - 3x^2$ and find the total area bounded by this curve and the x-axis. Sketch.

In Exercises 3 through 15, find the area bounded by the given curves and lines. Use the trapezoidal rule, Article 5.10, to approximate any integral you cannot evaluate by the Fundamental Theorem. A sketch is usually helpful.

3. $y = x, \quad y = \dfrac{1}{x^2}, \quad x = 2$

4. $y = x, \quad y = \dfrac{1}{\sqrt{x}}, \quad x = 2$

5. $y = x + 1, \quad y = 3 - x^2$

6. $y = 2x^2, \quad y = x^2 + 2x + 3$

7. $x = 2y^2, \quad x = 0, \quad y = 3$

8. $4x = y^2 - 4, \quad 4x = y + 16$

9. $x^{1/2} + y^{1/2} = a^{1/2}, \quad x = 0, \quad y = 0$

10. $y = \sqrt{9 + x^2}, \quad y = 0, \quad x = 0, \quad x = 4$

11. $y^2 = 9x, \quad y = \dfrac{3x^2}{8}$

12. $y = \dfrac{1}{1 + x^2}, \quad y = 0, \quad x = 0, \quad x = 1$

13. $y = x\sqrt{2x^2 + 1}, \quad x = 0, \quad x = 2$

14. $y^2 = 4x$ and $y = 4x - 2$

15. $y = 2 - x^2$ and $y = x^2 - 6$

16. The function $v = 3t^2 + 18$ represents the velocity v (ft/sec) of a moving body as a function of the time t (sec).

 (a) Find that portion of the time interval $0 \leq t \leq 3$ in which the velocity is positive.

 (b) Find that portion where v is negative.

 (c) Find the total distance traveled by the body from $t = 0$ to $t = 3$.

17. The area from 0 to x under a certain graph is given to be

$$A = (1 + 3x)^{1/2} - 1, \quad x \geq 0.$$

 (a) Find the *average* rate of change of A with respect to x as x increases from 1 to 8.

 (b) Find the *instantaneous* rate of change of A with respect to x at $x = 5$.

 (c) Find the ordinate (height) y of the graph as a function of x.

 (d) Find the average value of the ordinate (height) y, with respect to x, as x increases from 1 to 8.

18. A solid is generated by rotating, about the x-axis, the area bounded by the curve $y = f(x)$, the x-axis, and the lines $x = a$, $x = b$. Its volume, for all $b > a$, is $b^2 - ab$. Find $f(x)$ (a = constant).

19. Find the volume generated by rotating the area bounded by the given curves and lines about the line indicated:

 (a) $y = x^2$, $y = 0$, $x = 3$; about the x-axis,

 (b) $y = x^2$, $y = 0$, $x = 3$; about the line $x = -3$.

 (c) $y = x^2$, $y = 0$, $x = 3$; about the y-axis, first integrating with respect to x and then integrating with respect to y.

 (d) $x = 4y - y^2$, $x = 0$; about the y-axis.

 (e) $x = 4y - y^2$, $x = 0$; about the x-axis.

20. A solid is generated by rotating

$$f(x), \quad 0 \leq x \leq a,$$

about the x-axis. Its volume for all a is $a^2 + a$; find $f(x)$.

21. The area bounded by the curve $y^2 = 4x$ and the straight line $y = x$ is rotated about the x-axis. Find the volume generated.

22. Sketch the area bounded by the curve $y^2 = 4ax$, the line $x = a$, and the x-axis. Find the respective volumes generated by rotating this area in the following ways:

 (a) about the x-axis, (b) about the line $x = a$,

 (c) about the y-axis.

23. The area bounded by the curve $y = x/\sqrt{x^3 + 8}$, the x-axis, and the line $x = 2$ is rotated about the y-axis, generating a certain volume. Set up the integral that should be used to evaluate the volume and then evaluate the integral.

24. Find the volume of the solid generated by rotating the larger area bounded by $y^2 = x - 1$, $x = 3$, and $y = 1$ about the y-axis.

25. The area bounded by the curve $y^2 = 4ax$ and the line $x = a$ is rotated about the line $x = 2a$. Find the volume generated.

26. A twisted solid is generated as follows: We are given a fixed line L in space, and a square of side s in a plane perpendicular to L. One vertex of the square is on L. As this vertex moves a distance h along L, the square turns through a full revolution, with L

as the axis. Find the volume generated by this motion. What would the volume be if the square had turned through two full revolutions in moving the same distance along L?

27. Two circles have a common diameter and lie in perpendicular planes. A square moves in such a way that its plane is perpendicular to this diameter and its diagonals are chords of the circles. Find the volume generated.

28. Find the volume generated by rotating about the x-axis the area bounded by the x-axis and one arch of the curve $y = \sin 2x$.

29. A round hole of radius $\sqrt{3}$ ft is bored through the center of a solid sphere of radius 2 ft. Find the volume cut out.

30. The cross section of a certain solid in any plane perpendicular to the x-axis is a circle having diameter AB with A on the curve $y^2 = 4x$ and B on the curve $x^2 = 4y$. Find the volume of the solid lying between the points of intersection of the curves.

31. The base of a solid is the area that is bounded by $y^2 = 4ax$ and $x = a$. Each cross section perpendicular to the x-axis is an equilateral triangle. Find the volume of the solid.

32. Find the length of the curve $y = (\frac{2}{3})x^{3/2} - (\frac{1}{2})x^{1/2}$ from $x = 0$ to $x = 4$.

33. Find the surface area generated when the curve of Exercise 32 is rotated about the y-axis.

34. Find the length of the curve $x = (\frac{3}{5})y^{5/3} - (\frac{3}{4})y^{1/3}$ from $y = 0$ to $y = 1$.

35. Find the surface area generated when the curve of Exercise 34 is rotated about the line $y = -1$.

36. Find the average value of y with respect to x for that part of the curve $y = \sqrt{ax}$ between $x = a$ and $x = 3a$.

37. Find the average value of y^2 with respect to x for the curve $ay = b\sqrt{a^2 - x^2}$ between $x = 0$ and $x = a$. Also find the average value of y with respect to x^2 for $0 \leq x \leq a$.

38. Consider the curve $y = f(x)$, $x \geq 0$ such that $f(0) = a$. Let $s(x)$ denote the arc length along the curve from $(0, a)$ to $(x, f(x))$.

 (a) Find $f(x)$ if $s(x) = Ax$. (What are the permissible values of A?)

 (b) Is it possible for $s(x) = x^n$, $n > 1$? Give a reason for your answer.

39. A point moves in a straight line during the time from $t = 0$ to $t = 3$, according to the law $s = 120t - 16t^2$.

 (a) Find the average value of the velocity, with respect to time, for these three seconds. [Compare with the "average velocity," Article 1.11, Eq. (6).]

 (b) Find the average value of the velocity with respect to the distance s during the three seconds.

40. Sketch a smooth curve through the points $A(1, 3)$, $B(3, 5)$, $C(5, 6)$, $D(7, 6)$, $E(9, 7)$, $F(11, 10)$. The area bounded by this curve, the x-axis, and the lines $x = 1$, $x = 11$ is rotated about the x-axis to generate a solid. Use the trapezoidal rule to approximate the volume generated. Also determine approximately the average value of the circular cross-sectional area (with respect to x).

41. Determine the center of mass of a thin homogeneous plate covering the area enclosed by the curves $y^2 = 8x$ and $y = x^2$.

42. Find the center of mass of a homogeneous plate covering the area in the first quadrant bounded by the curve $4y = x^2$, the y-axis, and the line $y = 4$.

43. Find the center of gravity of the area bounded by the curve $y = 4x - x^2$ and the line $2x - y = 0$.

44. Find the center of mass of a thin homogeneous plate covering the portion of the xy-plane, in the first quadrant, bounded by the curve $y = x^2$, the x-axis, and the line $x = 1$.

45. Consider a thin metal plate of area A and constant density and thickness. Show that if its first moment about the y-axis is M, its moment about the line $x = b$ is $M - bA$. Indicate why this result actually shows that the center of gravity is a physical property of the body, independent of the coordinate system used for finding its location.

46. Find the center of mass of a thin plate covering the region bounded by the curve $y^2 = 4ax$ and the line $x = a$, $a =$ positive constant, if the density at (x, y) is directly proportional to

 (a) x, (b) $|y|$.

47. Find the centroid of the area, in the first quadrant, bounded by two concentric circles and the coordinate axes, if the circles have radii a and b, $b > a > 0$, and their centers are at the origin. Also find the limits of the coordinates of the centroid as a approaches b, and discuss the meaning of the result.

48. (a) Find the centroid of the arc of the curve

$$x = a\cos^3\phi, \qquad y = a\sin^3\phi$$

that is in the first quadrant.

(b) Find the centroid of the surface generated by rotating the arc of part (a) about the y-axis.

49. A triangular corner is cut from a square 1 ft on a side. The area of the cutoff triangle is 36 in^2. If the centroid of the remaining area is 7 in. from one side of the original square, how far is it from the remaining sides?

50. A triangular plate ABC is submerged in water with its plane vertical. The side AB, 4 ft long, is 6 ft below the surface of the water, while the vertex C is 2 ft below the surface. Find the force of liquid pressure on one face of the plate.

51. A dam is in the form of a trapezoid, with its two horizontal sides 200 and 100 ft, respectively, the longer side being at the top; the height is 20 ft. What is the force of pressure on the dam when the water is level with the top of the dam?

52. The center of pressure on a submerged plane area is defined to be the point at which the total force could be applied without changing its total moment about any axis in the plane. Find the depth to the center of pressure

(a) on a vertical rectangle of height h and width b if its upper edge is in the surface of the water,

(b) on a vertical triangle of height h and base b if the vertex opposite b is a ft, and the base b is $a + h$ ft, below the surface of the water.

53. A container is filled with two nonmixing liquids with respective densities d_1 and d_2, $d_1 < d_2$. Find the force on one face of a square $ABCD$, $6\sqrt{2}$ feet on a side, immersed in the liquids with the diagonal AC normal to the free surface, if the highest point A of the square is 2 ft below the free surface and BD lies on the surface separating the two liquids.

54. A particle of mass M starts from rest at time $t = 0$ and is moved with constant acceleration a from $x = 0$ to $x = h$ against a variable force $F(t) = t^2$. Find the work done.

55. When a particle of mass M is at $(x, 0)$ it is attracted toward the origin with a force whose magnitude is k/x^2. If the particle starts from rest at $x = b$ and is acted upon by no other forces, find the work done on it by the time it reaches $x = a$, $0 < a < b$.

56. Below the surface of the earth the force of its gravitational attraction is directly proportional to the distance from its center. Find the work done in lifting an object, whose weight at the surface is w lb, from a distance r ft below the earth's surface to the surface.

57. A storage tank is a right circular cylinder 20 ft long and 8 ft in diameter with its axis horizontal. If the tank is half full of oil weighing w lb/ft^3, find the work done in emptying it through a pipe that runs from the bottom of the tank to an outlet that is 6 ft above the top of the tank.

TRANSCENDENTAL FUNCTIONS

CHAPTER 7

7.1 THE TRIGONOMETRIC FUNCTIONS

We have learned how to differentiate and integrate polynomials and certain other algebraic functions, including some rational functions and fractional powers. By definition, y is an *algebraic* function of x if it is a function that satisfies an irreducible algebraic equation of the form

$$P_0(x)y^n + P_1(x)y^{n-1} + \cdots + P_{n-1}(x)y + P_n(x) = 0,$$

with n a positive integer and with coefficients

$$P_0(x),\ P_1(x),\ \ldots$$

that are polynomials in x. For instance, $y = \sqrt{x}$, $x > 0$, defines an algebraic function whose elements (x, y) satisfy the irreducible equation

$$y^2 - x = 0.$$

Sums, products, quotients, powers, and roots of algebraic functions are in turn algebraic functions.

A function that is not algebraic is called *transcendental.* The class of transcendental functions includes the trigonometric, logarithmic, exponential, and inverse trigonometric functions, and many more which are less familiar.

We have learned how to differentiate and integrate two transcendental functions—the sine and cosine. In this chapter we shall add to this list the derivatives of the remaining trigonometric functions, their inverse functions, and the logarithmic and exponential functions.

At this stage it is well to recall the list of formulas for derivatives and differentials given in Chapter 3, which we present in expanded form here. The first six pairs of functions, (I, I′) through (VI, VI′), by themselves enable us to differentiate the *algebraic* functions, which include polynomials, ratios of polynomials, and roots or powers of either of these types. VII, VII′, VIII, and VIII′ are discussed in Article 5.5.

When we combine formulas V, VII, and VIII and apply them to the trigonometric identities

$$\tan u = \frac{\sin u}{\cos u}, \quad \cot u = \frac{\cos u}{\sin u},$$
$$\sec u = \frac{1}{\cos u}, \quad \csc u = \frac{1}{\sin u}, \tag{1}$$

Derivatives	*Differentials*
I. $\dfrac{dc}{dx} = 0.$	I′. $dc = 0.$
II. $\dfrac{d(cu)}{dx} = c\dfrac{du}{dx}\cdot$	II′. $d(cu) = c\,du.$
III. $\dfrac{d(u+v)}{dx} = \dfrac{du}{dx} + \dfrac{dv}{dx}\cdot$	III′. $d(u+v) = du + dv.$
IV. $\dfrac{d(uv)}{dx} = u\dfrac{dv}{dx} + v\dfrac{du}{dx}\cdot$	IV′. $d(uv) = u\,dv + v\,du.$
V. $\dfrac{d\left(\dfrac{u}{v}\right)}{dx} = \dfrac{v\dfrac{du}{dx} - u\dfrac{dv}{dx}}{v^2}\cdot$	V′. $d\left(\dfrac{u}{v}\right) = \dfrac{v\,du - u\,dv}{v^2}\cdot$
VI. $\dfrac{d(u^n)}{dx} = nu^{n-1}\dfrac{du}{dx}\cdot$	VI′. $d(u^n) = nu^{n-1}\,du.$
VII. $\dfrac{d\,(\sin u)}{dx} = \cos u\dfrac{du}{dx}\cdot$	VII′. $d\,(\sin u) = \cos u\,du.$
VIII. $\dfrac{d\,(\cos u)}{dx} = -\sin u\dfrac{du}{dx}\cdot$	VIII′. $d\,(\cos u) = -\sin u\,du.$
IX. $\dfrac{d\,(\tan u)}{dx} = \sec^2 u\dfrac{du}{dx}\cdot$	IX′. $d\,(\tan u) = \sec^2 u\,du.$
X. $\dfrac{d\,(\cot u)}{dx} = -\csc^2 u\dfrac{du}{dx}\cdot$	X′. $d\,(\cot u) = -\csc^2 u\,du.$
XI. $\dfrac{d\,(\sec u)}{dx} = \sec u\tan u\dfrac{du}{dx}\cdot$	XI′. $d\,(\sec u) = \sec u\tan u\,du.$
XII. $\dfrac{d\,(\csc u)}{dx} = -\csc u\cot u\dfrac{du}{dx}\cdot$	XII′. $d\,(\csc u) = -\csc u\cot u\,du.$

it is a simple matter to complete the list of formulas for differentiating the trigonometric functions and, doing so, we obtain the formulas that appear as the pairs (IX, IX′) through (XII, XII′). The proofs of these eight formulas should be carried through by the reader.

These formulas are in the forms that are customarily used, though there are alternate forms, such as the formula

$$\frac{d\,(\sec u)}{dx} = \frac{\sin u}{\cos^2 u}\frac{du}{dx},$$

which are obtained from the expressions given by using some trigonometric identities. One reason for preferring the forms above is that they are quite easy to remember if we note that *the derivative of every cofunction* (cos, cot, csc) *can be obtained from the derivative of the corresponding function* (sin, tan, sec) by

(a) *introducing a minus sign, and*

(b) *replacing each function by its cofunction.*

Apply this, for example, to the formula

$$\frac{d\,(\sec u)}{dx} = \sec u \tan u \frac{du}{dx}.$$

Replace sec u and tan u by their *co*functions on *both sides* of the equation and put in a *minus* sign; the result is

$$\frac{d\,(\csc u)}{dx} = -\csc u \cot u \frac{du}{dx}.$$

Thus it is really only necessary to "memorize" the two new formulas

$$\frac{d\,(\tan u)}{dx} = \sec^2 u \frac{du}{dx},$$

$$\frac{d\,(\sec u)}{dx} = \sec u \tan u \frac{du}{dx};$$

the above rule produces the other two.

The formulas IX′–XII′ for differentials immediately produce the four new integration formulas

$$\begin{aligned}\int \sec^2 u\,du &= \tan u + C,\\ \int \csc^2 u\,du &= -\cot u + C,\\ \int \sec u \tan u\,du &= \sec u + C,\\ \int \csc u \cot u\,du &= -\csc u + C.\end{aligned} \tag{2}$$

Problem 1. Find dy/dx if $y = \tan^2 3x$.

Solution. First we apply the power formula VI:

$$\frac{dy}{dx} = 2 \tan 3x \frac{d\,(\tan 3x)}{dx},$$

and then IX:

$$\frac{d\,(\tan 3x)}{dx} = \sec^2 3x \frac{d(3x)}{dx},$$

and then II:

$$\frac{d(3x)}{dx} = 3\frac{dx}{dx} = 3.$$

When these are combined, we get

$$\frac{dy}{dx} = 6 \tan 3x \sec^2 3x.$$

Problem 2. Integrate

$$\int \tan^2 3x\,dx.$$

Solution. At first sight, this seems to be impossible with the formulas we have for integration. However, we see that the trigonometric identity

$$\sin^2 3x + \cos^2 3x = 1$$

becomes

$$\tan^2 3x + 1 = \sec^2 3x$$

when we divide both sides by $\cos^2 3x$. Hence,

$$\tan^2 3x = \sec^2 3x - 1,$$

and

$$\begin{aligned}\int \tan^2 dx\,dx &= \int (\sec^2 3x - 1)\,dx\\ &= \int \sec^2 3x\,dx - \int 1\,dx\\ &= \tfrac{1}{3}\int \sec^2 3x\,d(3x) - \int 1\,dx\\ &= \tfrac{1}{3}\tan 3x - x + C.\end{aligned}$$

EXERCISES

Sketch the graphs of the equations in Exercises 1 through 10.

1. $y = \sin x$
2. $y = \cos x$
3. $y = \tan x$
4. $y = \cot x$
5. $y = \sec x$
6. $y = \csc x$
7. $y = 3 \sin 2x$
8. $y = 4 \cos (x/3)$
9. $y = \sin^2 x$
10. $y = \frac{1}{2} + \frac{1}{2}\cos 2x$
11. (a) Derive formula IX. (b) Derive formula XI.
12. (a) Derive formula X. (b) Derive formula XII.

Find dy/dx in Exercises 13 through 33.

13. $y = \tan(3x^2)$
14. $y = \tan^2(\cos x)$
15. $y = \cot(3x + 5)$
16. $y = \sqrt{\cot x}$
17. $y = \sec^2 x - \tan^2 x$
18. $y = \frac{1}{3}\sec^3 x$
19. $y = 3x + \tan 3x$
20. $y = 3\tan 2x$
21. $y = x - \tan x$
22. $y = \sec^4 x - \tan^4 x$
23. $y = (\csc x + \cot x)^2$
24. $y = 2\sin^2 3x$
25. $y = \frac{1}{2}x - \frac{1}{4}\sin 2x$
26. $y = \frac{1}{2}x + \frac{1}{4}\sin 2x$
27. $y = \sin^3 \dfrac{x}{3} - 3\sin\dfrac{x}{3}$
28. $y = 3\cos^2 \dfrac{x}{2}$
29. $y = 2\sin\frac{1}{2}x - x\cos\frac{1}{2}x$
30. $x^2 = \sin y + \sin 2y$
31. $y = \sin^3 x - \sin 3x$
32. $y = \frac{1}{3}\sin^3 5x - \sin 5x$
33. $x + \tan(xy) = 0$

34. Find the maximum height of the curve
$$y = 6\cos x - 8\sin x$$
above the x-axis.
35. Find the maximum and minimum values of the function $y = 2\sin x + \cos 2x$ for $0 \le x \le \pi/2$.
36. Show that the function $f(x) = x + \sin x$ has no relative maxima or minima.
37. Show that the curve $y = x\sin x$ is tangent to the line $y = x$ whenever $\sin x = 1$, and is tangent to the line $y = -x$ whenever $\sin x = -1$.
38. A revolving light 3 miles from a straight shoreline makes 8 revolutions per minute. Find the velocity of the beam of light along the shore at the instant when it makes an angle of 45° with the shoreline.
39. A rope with a ring in one end is looped over two pegs in a horizontal line. The free end, after being passed through the ring, has a weight suspended from it, so that the rope hangs taut. If the rope slips freely through the ring, the weight will descend as far as possible. Find the angle formed at the bottom of the loop. (Assume that the length of the rope is at least four times the distance between the pegs.)

Evaluate the following integrals.

40. $\displaystyle\int \sin 3t\, dt$
41. $\displaystyle\int \sec^2 2\theta\, d\theta$
42. $\displaystyle\int \tan^3 x \sec^2 x\, dx$
43. $\displaystyle\int \sec^3 x \tan x\, dx$
44. $\displaystyle\int \sec\frac{x}{2}\tan\frac{x}{2}\, dx$
45. $\displaystyle\int \cos^2 y\, dy$
46. $\displaystyle\int \frac{d\theta}{\cos^2\theta}$
47. $\displaystyle\int \frac{d\theta}{\sin^2(\theta/3)}$

48. Find the area bounded by the curve
$$y = \tan x \sec^2 x,$$
the x-axis, and the line $x = \pi/3$.
49. The area bounded above by the curve $y = \tan x$, below by the x-axis, and on the right by the line $x = \pi/4$ is revolved about the x-axis. Find the volume generated.
50. Prove that $\tan x > x$ for $0 < x < \pi/2$ by applying the Mean Value Theorem to $f(t) = \tan t - t$ on $[0, x]$.
51. Prove that (a) $t\sec^2 t - \tan t > 0$, $0 < t < \pi/2$, (b) $(\tan x)/x$ is an increasing function on $(0, \pi/2)$.

7.2 THE INVERSE TRIGONOMETRIC FUNCTIONS

The equation
$$x = \sin y \tag{1}$$
determines infinitely many real values of y for each x in the range $-1 \le x \le 1$.

Example. If $x = \frac{1}{2}$, then we ask for all angles y such that $\sin y = \frac{1}{2}$. The two angles 30° and 150° in the first and second quadrants occur to us immediately. But any whole multiple of 360° may be either added to or subtracted from these and the sine of the resulting angle will still be $\frac{1}{2}$. (See Fig. 7.1.) Expressing these facts in mathematical language and using radian measure for the angles, we have $\sin y = \frac{1}{2}$ if

$$y = \begin{cases} \dfrac{\pi}{6} + 2n\pi, \\[2mm] \dfrac{5\pi}{6} + 2n\pi, \end{cases} \qquad n = 0, \pm 1, \pm 2, \ldots$$

Similarly, we find that $\sin y = -\frac{1}{2}$ if

$$y = \begin{cases} -\dfrac{\pi}{6} + 2n\pi, \\[2mm] -\dfrac{5\pi}{6} + 2n\pi, \end{cases} \qquad n = 0, \pm 1, \pm 2, \ldots$$

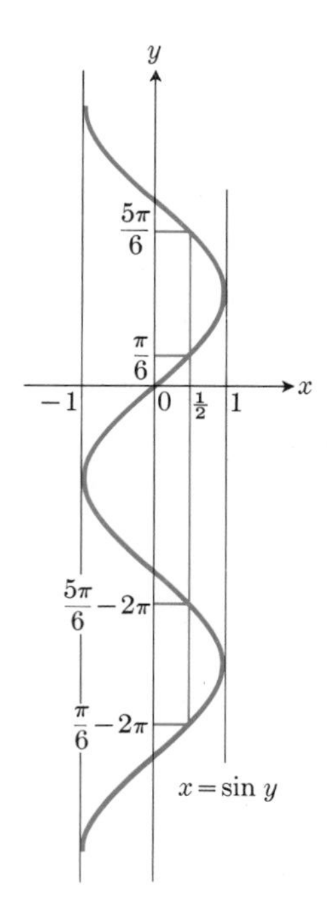

7.1

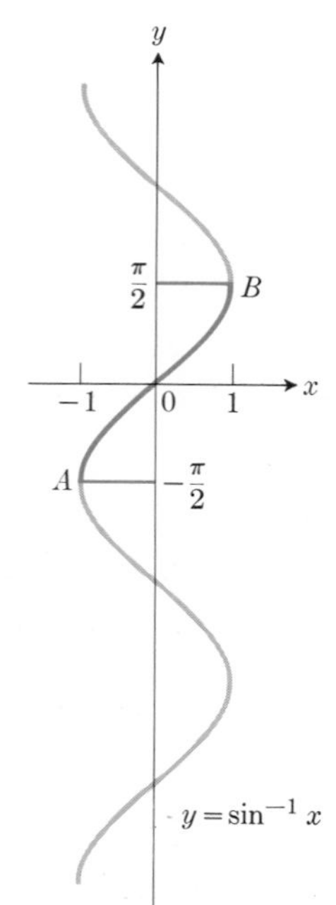

7.2 The principle values of the function $y = \sin^{-1} x$ lie on the portion of the curve printed in darker color.

Now it is desirable, for some of our later applications, to adopt a rule for picking a *principal value* of y. We shall require that the rule

(a) give one and only one value of y for each x, $-1 \leq x \leq 1$, such that $\sin y = x$; and

(b) give two "nearly equal" values of y for two "nearly equal" values of x.

To state requirement (b) more specifically, we require that if x_1 and x_2 are two values of x both between -1 and $+1$, and y_1 and y_2 are the corresponding values of y such that $\sin y = x$, then $y_2 - y_1$ should tend to zero if $x_2 - x_1$ does. The portion AOB of the curve, shown in darker color in Fig. 7.2, fulfills these requirements. For any x between -1 and $+1$ there is one and only one corresponding point (x, y) on this portion of the curve, and the y-value satisfies $\sin y = x$. Furthermore, this is a continuous curve, so that if (x_1, y_1) and (x_2, y_2) are two points on it and $(x_2 - x_1) \to 0$, then also $(y_2 - y_1) \to 0$.

To be sure that the point (x, y) stays on the portion AOB of the curve, we need only restrict y:

$$\sin y = x, \quad -\frac{\pi}{2} \leq y \leq \frac{\pi}{2}. \tag{2}$$

The restrictions placed on y give a continuous function

$$y = \sin^{-1} x, \quad -\frac{\pi}{2} \leq y \leq \frac{\pi}{2}, \tag{3}$$

called the "inverse sine of x," which is defined on the domain $-1 \leq x \leq 1$ and has values on the range $-\pi/2 \leq y \leq \pi/2$.

Notation. The minus one in Eq. (3), although written as an exponent, does *not* mean

$$(\sin x)^{-1} = \frac{1}{\sin x} = \csc x.$$

Actually, Eq. (3) merely gives a name to a new function that is defined implicitly by Eq. (2). The symbol $f^{-1}(x)$ is often used to denote the function y

defined implicitly by $f(y) = x$; that is,

$$y = f^{-1}(x).$$

The notation "arcsin" x is also used to represent the function that we have denoted by $\sin^{-1} x$, but the latter notation is now more commonly used. Thus,

$$\sin^{-1}\left(\frac{1}{2}\right) = \frac{\pi}{6}, \qquad \sin^{-1}\left(-\frac{1}{2}\right) = -\frac{\pi}{6}.$$

Note. In terms of angles referred to standard position in the various quadrants, Eq. (2) and the restriction $-\pi/2 \le y \le \pi/2$ mean that the answer to the question: "What is the angle y whose sine is x?" is always an angle in the *first* or *fourth* quadrant, but if it is in the *fourth* quadrant, it is to be expressed as the *negative* of an angle between 0 and $\pi/2$ and not as a positive angle $\ge 3\pi/2$, since the latter would violate the restriction in (2).

At the risk of appearing to labor a point which is already clear, we also note that the inverse sine is an *odd* function of x. This means that

$$\sin^{-1}(-x) = -\sin^{-1}(x).$$

Keeping this fact firmly in mind will prevent us from using values of y corresponding to points on the discontinuous curve composed of the portions OB and CD on the curve in Fig. 7.3. The objection to using these is that for a small positive value of x, say $x_1 = +0.01$, the point is near O and $y_1 \approx 0.01$; while for a small negative value of x, say $x_2 = -0.01$, the point is near D and $y_2 \approx 2\pi - 0.01$. The difference $y_2 - y_1$ is nearly $2\pi = 6.28+$, whereas the difference $x_1 - x_2$ is, in this case, only 0.02. This difficulty is not encountered if we restrict the value of y as in (3), which confines us to the *continuous* curve AOB of Fig. 7.2.

To each of the other trigonometric functions there also corresponds an *inverse* trigonometric function. In each case, we restrict the angle in such a way that the function is *single-valued.*

The same reasoning that led us to adopt the particular range

$$-\frac{\pi}{2} \le y \le \frac{\pi}{2}$$

for the principal value of $y = \sin^{-1} x$ prompts us to adopt a similar range for the principal value of y satisfying the equation $x = \tan y$ (see Fig. 7.4).

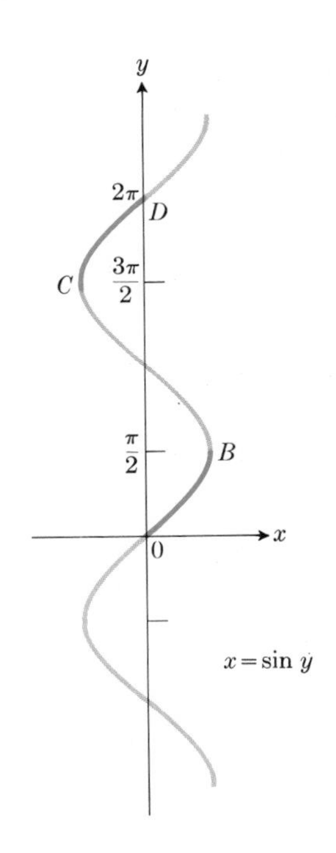

7.3

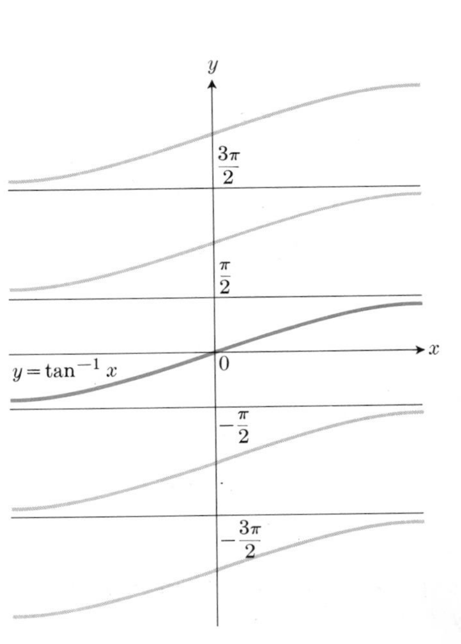

7.4

Thus we write

$$y = \tan^{-1} x, \quad -\frac{\pi}{2} < y < \frac{\pi}{2} \tag{4}$$

to mean the same as

$$\tan y = x, \quad -\frac{\pi}{2} < y < \frac{\pi}{2}. \tag{5}$$

The inverse tangent function, (4), is defined on the *domain* $-\infty < x < +\infty$, that is, for all real x. Its *range* is $-\pi/2 < y < +\pi/2$. The values $y = \pm\pi/2$ are to be excluded, since the tangent becomes infinite as the angle approaches $\pm\pi/2$.

When we come to the inverse cosine function, we simply *define* it by the equation

$$\cos^{-1} x = \frac{\pi}{2} - \sin^{-1} x. \tag{6}$$

Since

$$-\frac{\pi}{2} \leq \sin^{-1} x \leq \frac{\pi}{2},$$

the range of principal values for (6) is

$$0 \leq \cos^{-1} x \leq \pi. \tag{7}$$

The *domain* of the inverse cosine function is

$$-1 \leq x \leq 1.$$

Its *range* is $0 \leq y \leq \pi$. The principal value is thus represented by the heavily marked portion of the curve $x = \cos y$ shown in Fig. 7.5.

The reason for defining $\cos^{-1} x$ by Eq. (6) is as follows. In the right triangle in Fig. 7.6, the acute angles α and β are complementary, that is

$$\alpha + \beta = \frac{\pi}{2}.$$

But we also have

$$\sin \alpha = x = \cos \beta,$$

so that

$$\alpha = \sin^{-1} x, \qquad \beta = \cos^{-1} x,$$

and the relation $\alpha + \beta = \pi/2$ is the same as $\beta = \pi/2 - \alpha$. That is,

$$\cos^{-1} x = \frac{\pi}{2} - \sin^{-1} x.$$

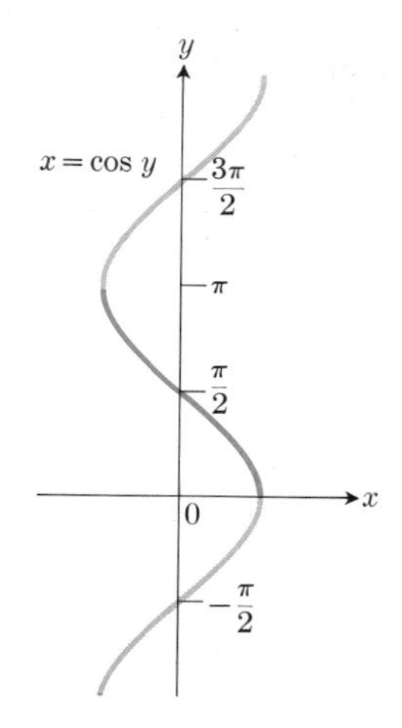

7.5

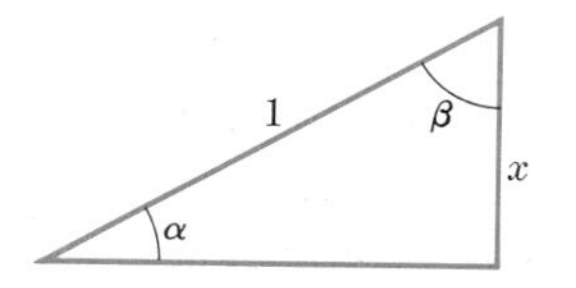

7.6

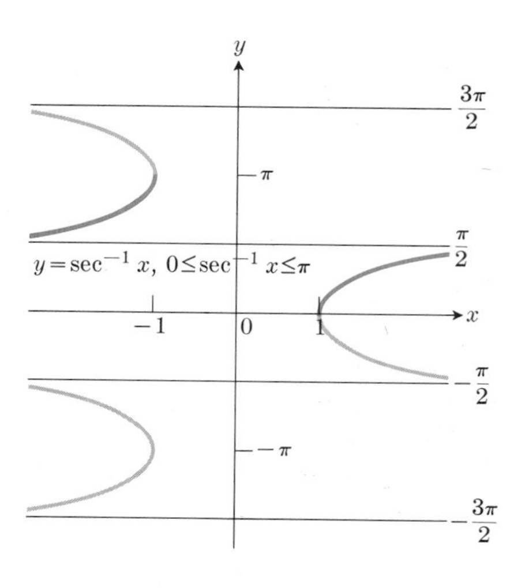

7.7

Thus Eq. (6) expresses the fact that the angle whose cosine is x is the complement of the angle whose sine is x.

For exactly analogous reasons, we also define the inverse cotangent to be

$$\cot^{-1} x = \frac{\pi}{2} - \tan^{-1} x, \tag{8}$$

and we thus restrict the principal value to

$$0 < \cot^{-1} x < \pi. \tag{9}$$

The inverse secant and inverse cosecant functions will be single-valued functions of x on the domain $|x| \geq 1$ if we adopt the definitions

$$\sec^{-1} x = \cos^{-1}\left(\frac{1}{x}\right), \tag{10a}$$

$$\csc^{-1} x = \sin^{-1}\left(\frac{1}{x}\right), \tag{10b}$$

with their principal values confined to the ranges

$$0 \leq \sec^{-1} x \leq \pi, \tag{11a}$$

$$-\frac{\pi}{2} \leq \csc^{-1} x \leq \frac{\pi}{2}. \tag{11b}$$

(Figure 7.7 illustrates the principle values of $\sec^{-1} x$.) We adopt the definition (10a) for the following reason:

$$\text{If} \quad \sec y = x, \qquad \text{then} \quad \cos y = \frac{1}{x}.$$

We simply ask that the principal values of y be the same in these two cases, that is, that

$$y = \sec^{-1} x = \cos^{-1}\left(\frac{1}{x}\right).$$

A similar reason explains the choice of (10b) and (11b).

Remark. From the graphs of the inverse sine, inverse tangent, and inverse secant, all of which are increasing functions on the domains specified, we expect that their derivatives will be positive. This is a consequence of the relation

$$dy/dx = (dx/dy)^{-1}$$

that connects the derivative of a function with its inverse. For example, the secant function has a nonnegative derivative on $[0, \pi]$, excluding $\pi/2$; hence it has an inverse there, and that inverse also has a nonnegative derivative. It should be noted, however, that some writers choose the principal value of the inverse secant of x to lie between 0 and $\pi/2$ when x is positive, and between $-\pi$ and $-\pi/2$ (hence, as a negative angle in the third quadrant) when x is negative. This latter method has an advantage in simplifying the formula for the derivative of $\sec^{-1} x$, but has the disadvantage of failing to satisfy the relationship expressed in Eq. (10a) when x is negative. Either method leads to a function which is discontinuous, as is inherent in the nature of the secant function itself, since to pass from a point where the secant is negative to a point where the secant is positive requires crossing one of the discontinuities of the curve.

EXERCISES

1. Given that $\alpha = \sin^{-1} \frac{1}{2}$, find $\cos\alpha$, $\tan\alpha$, $\sec\alpha$, $\csc\alpha$.
2. Given that $\alpha = \cos^{-1}(-\frac{1}{2})$, find $\sin\alpha$, $\tan\alpha$, $\sec\alpha$, $\csc\alpha$.
3. Evaluate $\sin^{-1}(1) - \sin^{-1}(-1)$.
4. Evaluate $\tan^{-1}(1) - \tan^{-1}(-1)$.
5. Evaluate $\sec^{-1}(2) - \sec^{-1}(-2)$.
6. A picture a feet high is placed on a wall with its base b feet above the level of an observer's eye. If the observer stands x feet from the wall, show that the angle of vision α subtended by the picture is given by

$$\alpha = \cot^{-1}\frac{x}{a+b} - \cot^{-1}\frac{x}{b}.$$

7. Prove
   (a) $\tan^{-1}(-x) = -\tan^{-1} x$,
   (b) $\cos^{-1}(-x) = \pi - \cos^{-1} x$,
   (c) $\sec^{-1}(-x) = \pi - \sec^{-1}(x)$.
8. Simplify each of the following expressions.
   (a) $\sin(\sin^{-1} 0.735)$
   (b) $\cos(\sin^{-1} 0.8)$
   (c) $\sin(2\sin^{-1} 0.8)$
   (d) $\tan^{-1}(\tan\pi/3)$
   (e) $\cos^{-1}(-\sin\pi/6)$
   (f) $\sec^{-1}[\sec(-30°)]$

7.3 DERIVATIVES OF THE INVERSE TRIGONOMETRIC FUNCTIONS

Note the formulas for derivatives and differentials at the bottom of the page. To illustrate how these are derived, we shall derive formulas XIII and XVII. Let

$$y = \sin^{-1} u, \quad -\frac{\pi}{2} \le y \le \frac{\pi}{2}.$$

Then

$$\sin y = u,$$

$$\cos y \frac{dy}{dx} = \frac{du}{dx}, \qquad \frac{dy}{dx} = \frac{1}{\cos y}\frac{du}{dx}.$$

To express the right-hand side in terms of u, we use the fact that

$$\cos y = \pm\sqrt{1 - \sin^2 y} = \pm\sqrt{1 - u^2}.$$

When $-\pi/2 \le y \le \pi/2$, $\cos y$ is not negative; hence

$$\cos y = +\sqrt{1 - u^2}$$

and

$$\frac{dy}{dx} = \frac{1}{\sqrt{1 - u^2}}\frac{du}{dx},$$

which establishes XIII.

Next, let us take

$$y = \sec^{-1} u, \quad 0 \le y \le \pi.$$

Then

$$\sec y = u, \qquad \sec y \tan y \frac{dy}{dx} = \frac{du}{dx},$$

$$\frac{dy}{dx} = \frac{1}{\sec y \tan y}\frac{du}{dx}.$$

Substituting

$$\sec y = u,$$

$$\tan y = \pm\sqrt{\sec^2 y - 1} = \pm\sqrt{u^2 - 1},$$

we have

$$\frac{dy}{dx} = \frac{1}{u(\pm\sqrt{u^2 - 1})}\frac{du}{dx}.$$

The ambiguous sign is determined by the sign of $\tan y$, and hence is

$+$ if $0 < y < \frac{\pi}{2}$; that is, if u is $+$,

$-$ if $\frac{\pi}{2} < y < \pi$; that is, if u is $-$.

Derivatives	*Differentials*
XIII. $\dfrac{d(\sin^{-1} u)}{dx} = \dfrac{1}{\sqrt{1 - u^2}}\dfrac{du}{dx}.$	XIII′. $d(\sin^{-1} u) = \dfrac{1}{\sqrt{1 - u^2}}\,du.$
XIV. $\dfrac{d(\cos^{-1} u)}{dx} = -\dfrac{1}{\sqrt{1 - u^2}}\dfrac{du}{dx}.$	XIV′. $d(\cos^{-1} u) = -\dfrac{1}{\sqrt{1 - u^2}}\,du.$
XV. $\dfrac{d(\tan^{-1} u)}{dx} = \dfrac{1}{1 + u^2}\dfrac{du}{dx}.$	XV′. $d(\tan^{-1} u) = \dfrac{1}{1 + u^2}\,du.$
XVI. $\dfrac{d(\cot^{-1} u)}{dx} = -\dfrac{1}{1 + u^2}\dfrac{du}{dx}.$	XVI′. $d(\cot^{-1} u) = -\dfrac{1}{1 + u^2}\,du.$
XVII. $\dfrac{d(\sec^{-1} u)}{dx} = \dfrac{1}{\lvert u\rvert\sqrt{u^2 - 1}}\dfrac{du}{dx}.$	XVII′. $d(\sec^{-1} u) = \dfrac{1}{\lvert u\rvert\sqrt{u^2 - 1}}\,du.$
XVIII. $\dfrac{d(\csc^{-1} u)}{dx} = \dfrac{-1}{\lvert u\rvert\sqrt{u^2 - 1}}\dfrac{du}{dx}.$	XVIII′. $d(\csc^{-1} u) = \dfrac{-1}{\lvert u\rvert\sqrt{u^2 - 1}}\,du.$

In other words,

$$\frac{dy}{dx} = \frac{1}{u\sqrt{u^2 - 1}} \frac{du}{dx} \quad \text{if} \quad u > 1$$

$$= \frac{1}{-u\sqrt{u^2 - 1}} \frac{du}{dx} \quad \text{if} \quad u < -1;$$

both of these are summarized in the single formula

$$\frac{dy}{dx} = \frac{1}{|u|\sqrt{u^2 - 1}} \frac{du}{dx}.$$

Remark 1. In some calculus books and handbooks of mathematical formulas (such as *CRC Standard Mathematical Tables*, 14th edition, pp. 504, 305) the principal value of $\sec^{-1} x$, or arcsec x is defined to be between 0 and $\pi/2$ for $x \geq 1$, and between $-\pi$ and $-\pi/2$ for $x \leq -1$. When this convention is adopted the absolute value signs, as used in formula XVII, are not required in the formula for the derivative.

The formulas XIII to XVIII are used directly as given, but even more important are the integration formulas obtained immediately from the differential formulas XIII′ to XVIII′, namely,

$$\int \frac{du}{\sqrt{1 - u^2}} = \sin^{-1} u + C,$$

$$\int \frac{du}{1 + u^2} = \tan^{-1} u + C, \qquad (1)$$

$$\int \frac{du}{u\sqrt{u^2 - 1}} = \int \frac{d(-u)}{(-u)\sqrt{u^2 - 1}}$$

$$= \sec^{-1} |u| + C.$$

Note that $\sec^{-1} u$ and $\sec^{-1} |u|$ are meaningless unless $|u| \geq 1$ (see Fig. 7.7).

Example

$$\int_0^1 \frac{dx}{1 + x^2} = \tan^{-1} x \Big]_0^1$$

$$= \tan^{-1} 1 - \tan^{-1} 0 = \frac{\pi}{4}.$$

Remark 2. When u is positive, both of the commonly used definitions of $\sec^{-1} u$ agree that

$$0 \leq \sec^{-1} u < \pi/2$$

and

$$d\,(\sec^{-1} u) = \frac{du}{u\sqrt{u^2 - 1}}.$$

Therefore, for $u > 1$,

$$\int \frac{du}{u\sqrt{u^2 - 1}} = \sec^{-1} u + C. \qquad (2a)$$

Although the definitions of $\sec^{-1} u$ disagree when u is negative, the integral in (2a) can be evaluated for $u < -1$ by making the simple substitution $u = -v$:

$$\int \frac{du}{u\sqrt{u^2 - 1}} = \int \frac{-dv}{-v\sqrt{v^2 - 1}} \quad (v > 1)$$

$$= \int \frac{dv}{v\sqrt{v^2 - 1}}$$

$$= \sec^{-1} v + C. \qquad (2b)$$

But, since $v = -u$, Eq. (2b) becomes

$$\int \frac{du}{u\sqrt{u^2 - 1}} = \sec^{-1} (-u) + C \quad \text{when } u < -1. \qquad (2c)$$

We can combine Eqs. (2a) and (2c) into one equation if we introduce $|u|$:

$$\int \frac{du}{u\sqrt{u^2 - 1}} = \sec^{-1} |u| + C, \quad \text{if} \quad u^2 > 1, \qquad (3)$$

because $|u| = u$ if u is positive and $|u| = -u$ if u is negative.

Problem. Sketch a portion of the graph of

$$f(x) = \frac{1}{x\sqrt{x^2 - 1}}, \quad 1 < |x| < 3,$$

and compare

$$\int_{-2}^{-\sqrt{2}} f(x)\, dx \quad \text{and} \quad \int_{\sqrt{2}}^{2} f(x)\, dx.$$

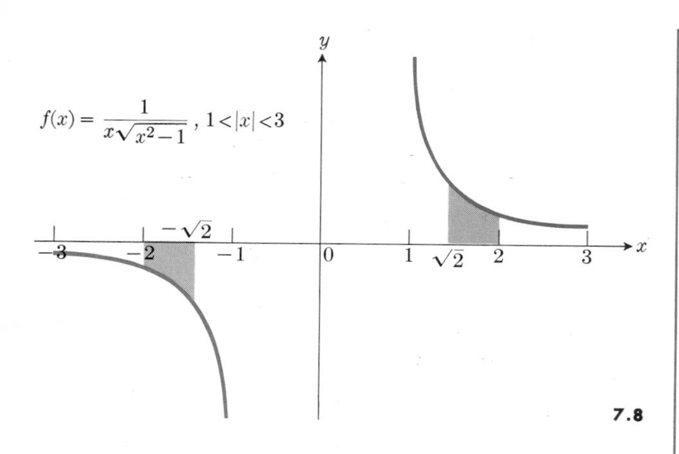

7.8

Solution. Figure 7.8 shows that the graph is below the x-axis for $x < -1$ and above the x-axis for $x > 1$. For the two integrals, we have

$$I_1 = \sec^{-1} |x|]_{-2}^{-\sqrt{2}} = \sec^{-1} \sqrt{2} - \sec^{-1} 2 = \frac{\pi}{4} - \frac{\pi}{3} = -\frac{\pi}{12}$$

and

$$I_2 = \sec^{-1} |x|]_{\sqrt{2}}^{2} = \sec^{-1} 2 - \sec^{-1} \sqrt{2} = \frac{\pi}{3} - \frac{\pi}{4} = \frac{\pi}{12}.$$

The two tinted regions shown in Fig. 7.8 have equal areas, $\pi/12$.

EXERCISES

1. Derive formula XIV.
2. Derive formula XV.
3. Derive formula XVI.
4. Derive formula XVIII.

Find dy/dx in Exercises 5 through 14.

5. $y = \sin^{-1} \frac{x}{2}$
6. $y = \frac{1}{3} \tan^{-1} \frac{x}{3}$
7. $y = \sec^{-1} 5x$
8. $y = \cos^{-1} 2x$
9. $y = \cot^{-1} \frac{2}{x} + \tan^{-1} \frac{x}{2}$
10. $y = \sin^{-1} \frac{x-1}{x+1}$
11. $y = \tan^{-1} \frac{x-1}{x+1}$
12. $y = x \sin^{-1} x + \sqrt{1-x^2}$
13. $y = x (\sin^{-1} x)^2 - 2x + 2\sqrt{1-x^2} \sin^{-1} x$
14. $y = x \cos^{-1} 2x - \frac{1}{2}\sqrt{1-4x^2}$
15. How far from the wall should the observer stand in order to maximize the angle of vision α in Exercise 6, Article 7.2?

Evaluate the following integrals.

16. $\int_0^{1/2} \frac{dx}{\sqrt{1-x^2}}$
17. $\int_{-1}^{1} \frac{dx}{1+x^2}$
18. $\int_{\sqrt{2}}^{2\sqrt{3}} \frac{dx}{x\sqrt{x^2-1}}$
19. $\int_{-2}^{-2/\sqrt{3}} \frac{dx}{x\sqrt{x^2-1}}$
20. $\int \frac{dx}{\sqrt{1-4x^2}}$
21. $\int_{1/\sqrt{3}}^{1} \frac{dx}{x\sqrt{4x^2-1}}$

7.4 THE NATURAL LOGARITHM

We have so far studied functions which are fairly familiar. Polynomials, rational functions, and other algebraic functions result from the familiar operations of arithmetic and algebra. The trigonometric functions can be identified with coordinates of points on a unit circle and with their ratios and reciprocals. The inverse trigonometric functions are probably less familiar, but neverthless they can be understood without any knowledge of the calculus. But we are now going to study a function, the natural logarithm, which depends on calculus for its very definition.

Definition. *The natural logarithm* of x, which is indicated by the notation $\ln x$, is defined, for positive x, as the integral*

$$\ln x = \int_1^x \frac{1}{t}\, dt. \tag{1}$$

* The first discovery of logarithms is credited to a Scottish nobleman, John Napier (1550–1617). For a biographical sketch of Napier, see the *World of Mathematics*, Vol. 1, "The Great Mathematicians," by H. W. Turnbull, pp. 121–125.

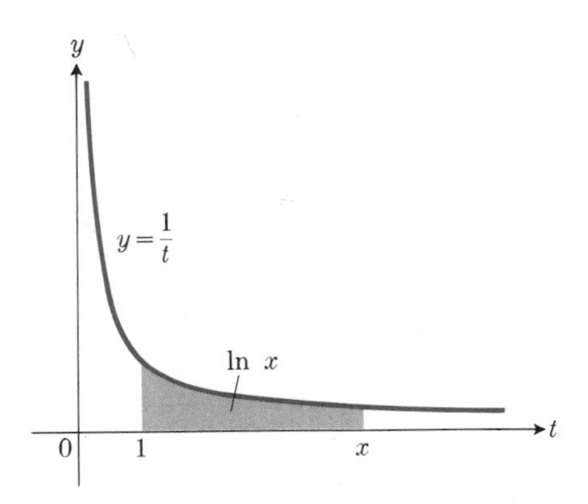

7.9 The tinted area represents $\ln x$.

For any x greater than 1, this integral represents the area bounded above by the curve $y = 1/t$, below by the t-axis, on the left by the line $t = 1$, and on the right by the line $t = x$ (see Fig. 7.9).

If $x = 1$, the left and right boundaries of the area are identical and the area is zero:

$$\ln 1 = \int_1^1 \frac{1}{t}\,dt = 0 \tag{2}$$

If x is less than 1, then the left boundary is the line $t = x$ and the right boundary is $t = 1$. In this case,

$$\ln x = \int_1^x \frac{1}{t}\,dt = -\int_x^1 \frac{1}{t}\,dt \tag{3}$$

is the negative of the area under the curve between x and 1.

In all cases, if x is any positive number, the value of the definite integral in Eq. (1) can be calculated to as many decimal places as desired by using inscribed rectangles, circumscribed rectangles, or trapezoids to approximate the appropriate area. (See Chapter 5. Another method for computing natural logarithms, by means of series, will also be discussed in Chapter 17. Also see Exercise 8 below.) In any event, Eq. (1) defines a function that is computable over the domain $(0, \infty)$, since the integrand $1/t$ is continuous over this domain. We study its range in Article 7.7.

EXERCISES

Obtain approximations to the natural logarithms, given in Exercises 1 through 4. Use trapezoids of altitude $\Delta t = 0.1$ to approximate the appropriate areas.

1. $\ln 1.2$
2. $\ln 1.4$
3. $\ln 2$
4. $\ln 0.5$

5. (a) Suppose h is small and positive. Does the approximation $\ln(1 + h) \approx h$ correspond to using an inscribed rectangle or a circumscribed rectangle to approximate the area from 1 to $1 + h$ under the curve $y = 1/t$?
   (b) Answer the same question in case h is negative.

6. Approximate $\ln(1 + h)$ by using the line L that is tangent to the curve $y = 1/t$ at $A(1, 1)$ as upper boundary of a trapezoid of altitude $|h|$. Sketch. Show that this leads to the approximation
$$\ln(1 + h) \approx h - \frac{h^2}{2}.$$
   Use this result to approximate
   (a) $\ln 1.04$, (b) $\ln 0.96$.

7. Let $aACBb$ be the polygon whose vertices are
$$a(1, 0),\ A(1, 1),\ C(\tfrac{4}{3}, \tfrac{2}{3}),\ B(2, \tfrac{1}{2}),\ \text{and}\ b(2, 0).$$
   Show that AC and CB are tangent to the hyperbola $y = 1/t$ at A and at B, respectively. Use the area of the polygon to find an approximation to $\ln 2$. Sketch.

8. (a) Show, by long division or otherwise, that
$$\frac{1}{1 + u} = 1 - u + u^2 - \frac{u^3}{1 + u}.$$
   hand in
   (The division could be continued. We stop here only for convenience.)
   (b) In Eq. (1), make the substitution
$$t = 1 + u, \qquad dt = du,$$
   and make the corresponding change in the limits of integration, thus obtaining
$$\ln x = \int_0^{x-1} \frac{du}{1 + u}.$$

(c) Combine the results of (a) and (b) to obtain

$$\ln x = \int_0^{x-1} \left(1 - u + u^2 - \frac{u^3}{1+u}\right) du,$$

or $\ln x = (x-1) - \frac{1}{2}(x-1)^2 + \frac{1}{3}(x-1)^3 - R$,

where

$$R = \int_1^{x-1} \frac{u^3}{1+u}\, du.$$

(d) Show that, if $x > 1$ and $0 \le u \le x - 1$, then

$$u^3/(1+u) \le u^3.$$

Deduce that therefore

$$R \le \int_0^{x-1} u^3\, du = \frac{(x-1)^4}{4}.$$

(e) Combining the results of (c) and (d), show that the approximation

$$\ln x \approx (x-1) - \tfrac{1}{2}(x-1)^2 + \tfrac{1}{3}(x-1)^3$$

tends to overestimate the value of $\ln x$, but with an error not greater than $(x-1)^4/4$.

(f) Use the result of (e) to estimate $\ln 1.2$.

9. Explain how Theorem 4, Article 5.12 ensures the existence of the integral that defines $\ln x$

(a) for $x > 1$, (b) for $0 < x < 1$.

7.5 THE DERIVATIVE OF $\ln x$

Since the function $F(x) = \ln x$ is defined by the integral

$$F(x) = \int_1^x \frac{1}{t}\, dt \quad (x > 0) \tag{1}$$

(see Fig. 7.10), it follows at once, from the Second Fundamental Theorem of integral calculus (Article 5.9) that $F'(x) = 1/x$. That is,

$$d(\ln x)/dx = 1/x. \tag{2}$$

We can obtain a slightly more general formula by considering $\ln u$, where u is a positive, differentiable, function of x. By the chain rule for derivatives,

$$\frac{d \ln u}{dx} = \frac{d \ln u}{du}\frac{du}{dx} = \frac{1}{u}\frac{du}{dx}, \tag{XIX}$$

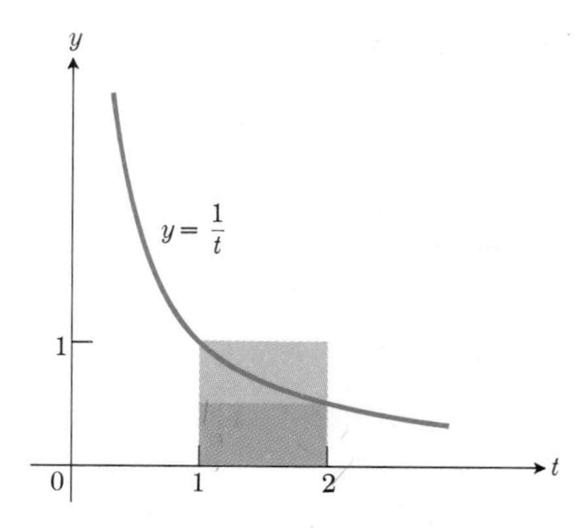

7.10 Area of inscribed rectangle $= \frac{1}{2}$, area of circumscribed rectangle $= 1$; $\frac{1}{2} < \ln 2 < 1$.

or, in terms of differentials,

$$d \ln u = \frac{du}{u}. \tag{XIX$'$}$$

Problem 1. Find $\dfrac{dy}{dx}$ if $y = \ln(3x^2 + 4)$.

Solution.

$$\frac{dy}{dx} = \frac{1}{3x^2+4}\frac{d(3x^2+4)}{dx} = \frac{6x}{3x^2+4}.$$

Formula XIX′ leads at once to the integration formula

$$\int \frac{du}{u} = \ln u + C, \tag{3a}$$

provided that u is positive. On the other hand, if u is negative, then $-u$ is positive and

$$\int \frac{du}{u} = \int \frac{d(-u)}{-u} = \ln(-u) + C. \tag{3b}$$

The two results (3a, b) can be combined into a single result, namely,

$$\int \frac{du}{u} = \begin{cases} \ln u + C & \text{if } u > 0, \\ \ln(-u) + C & \text{if } u < 0, \end{cases}$$

or

$$\int \frac{du}{u} = \ln |u| + C. \tag{4}$$

We recall that the formula

$$\int u^n\,du = \frac{u^{n+1}}{n+1} + C, \quad n \neq -1$$

failed to cover one case, namely, $n = -1$. Now Eq. (4) covers this case for us, since it tells us that

$$\int u^{-1}\,du = \int \frac{du}{u} = \ln|u| + C.$$

It is important, of course, to remember that

$$\int \frac{dx}{x^2} = \int x^{-2}\,dx = \frac{x^{-1}}{-1} + C$$

is *not* a logarithm.

Problem 2. Integrate

$$\int \frac{\cos\theta\,d\theta}{1+\sin\theta}.$$

Solution. In Eq. (4), let

$$u = 1 + \sin\theta.$$

Then

$$du = \frac{du}{d\theta}\,d\theta = \cos\theta\,d\theta,$$

so that

$$\int \frac{\cos\theta\,d\theta}{1+\sin\theta} = \int \frac{du}{u} = \ln|u| + C$$
$$= \ln|1+\sin\theta| + C.$$

Remark. Because $1 + \sin\theta$ is never negative, we can also write $\ln(1+\sin\theta)$ in place of $\ln|1+\sin\theta|$. Either answer is correct.

EXERCISES

Find dy/dx in Exercises 1 through 18.

1. $y = \ln(x^2 + 2x)$
2. $y = (\ln x)^3$
3. $y = \ln(\cos x)$
4. $y = \ln(\tan x + \sec x)$
5. $y = \ln(x\sqrt{x^2+1})$
6. $y = \ln(3x\sqrt{x+2})$
7. $y = x\ln x - x$
8. $y = x^3 \ln(2x)$
9. $y = \frac{1}{2}\ln\frac{1+x}{1-x}$
10. $y = \frac{1}{3}\ln\frac{x^3}{1+x^3}$
11. $y = \ln\frac{x}{2+3x}$
12. $y = \ln(x^2+4) - x\tan^{-1}\frac{x}{2}$
13. $y = \ln x - \frac{1}{2}\ln(1+x^2) - \frac{\tan^{-1}x}{x}$
14. $y = x(\ln x)^3$
15. $y = x[\sin(\ln x) + \cos(\ln x)]$
16. $y = x\sec^{-1}x - \ln(x + \sqrt{x^2-1}), \quad (x > 1)$
17. $y = x\ln(a^2+x^2) - 2x + 2a\tan^{-1}\frac{x}{a}$
18. $y = \ln(\ln x)$

Evaluate the integrals in Exercises 19 through 30.

19. $\int \frac{dx}{2x+3}$
20. $\int \frac{dx}{2-3x}$
21. $\int \frac{x\,dx}{4x^2+1}$
22. $\int \frac{\sin x\,dx}{2-\cos x}$
23. $\int \frac{\cos x\,dx}{\sin x}$
24. $\int \frac{2x-5}{x}\,dx$
25. $\int \frac{x\,dx}{x+1}$
26. $\int \frac{x^2\,dx}{4-x^3}$
27. $\int \frac{x\,dx}{1-x^2}$
28. $\int \frac{dx}{\sqrt{x}\,(1+\sqrt{x})}$
29. $\int (\ln x)^2\frac{dx}{x}$
30. $\int \frac{dx}{(2x+3)^2}$

7.6 PROPERTIES OF NATURAL LOGARITHMS

In this article we shall establish the following properties of the natural logarithm:

$$\ln ax = \ln a + \ln x, \tag{1}$$

$$\ln\frac{x}{a} = \ln x - \ln a, \tag{2}$$

$$\ln x^n = n\ln x, \tag{3}$$

provided x and a are positive and n is a rational number.

The demonstrations of these results are all based on the fact that

$$y = \ln x$$

satisfies the differential equation

$$\frac{dy}{dx} = \frac{1}{x} \quad \text{for all} \quad x > 0,$$

plus the fact that if two functions have the same derivative for all $x > 0$, then the two functions can differ only by a constant. That is, if

$$\frac{dy_1}{dx} = \frac{dy_2}{dx} \quad \text{for all} \quad x > 0,$$

then

$$y_1 = y_2 + \text{constant} \quad \text{for all} \quad x > 0.$$

Now to prove Eq. (1), let

$$y_1 = \ln ax, \qquad y_2 = \ln x.$$

Then

$$\frac{dy_1}{dx} = \frac{1}{ax}\frac{d(ax)}{dx} = \frac{a}{ax} = \frac{1}{x} = \frac{dy_2}{dx}. \tag{4}$$

Therefore we have

$$\ln ax = \ln x + C. \tag{5}$$

To evaluate C it suffices to substitute $x = 1$:

$$\ln a = \ln 1 + C = 0 + C,$$

which gives

$$C = \ln a.$$

Hence, by (5)

$$\ln ax = \ln x + \ln a. \ \square$$

To demonstrate Eq. (2), we first put $x = 1/a$ in Eq. (1) and recall that $\ln 1 = 0$:

$$0 = \ln 1 = \ln a + \ln (1/a),$$

so that

$$\ln \frac{1}{a} = -\ln a. \tag{6}$$

Now apply Eq. (1) with a replaced by $1/a$ and $\ln a$ replaced by $\ln (1/a) = -\ln a$. The result is Eq. (2).

To demonstrate Eq. (3), let

$$y_1 = \ln x^n.$$

Then

$$\frac{dy_1}{dx} = \frac{1}{x^n} \cdot nx^{n-1} = \frac{n}{x} = \frac{d}{dx}(n \ln x).$$

Hence $y_1 = \ln x^n$ and $y_2 = n \ln x$ have equal derivatives, so they differ at most by a constant:

$$\ln x^n = n \ln x + C. \tag{7}$$

But by taking $x = 1$, and remembering that $\ln 1 = 0$, we find $C = 0$, which gives Eq. (3). In particular, if $n = 1/m$, where m is a positive integer, we have

$$\ln \sqrt[m]{x} = \ln x^{1/m} = \frac{1}{m} \ln x. \tag{8}$$

EXERCISES

Express the following logarithms in terms of the two given logarithms $a = \ln 2$, $b = \ln 3$. [For example: $\ln 1.5 = \ln \frac{3}{2} = \ln 3 - \ln 2 = b - a$.]

1. $\ln 16$
2. $\ln \sqrt[3]{9}$
3. $\ln 2\sqrt{2}$
4. $\ln 0.25$
5. $\ln \frac{4}{9}$
6. $\ln 12$
7. $\ln \ln \frac{9}{8}$
8. $\ln 36$
9. $\ln 4.5$
10. $\ln \sqrt{13.5}$

7.7 GRAPH OF $y = \ln x$

The slope of the curve

$$y = \ln x \tag{1}$$

is given by

$$\frac{dy}{dx} = \frac{1}{x}. \tag{2}$$

Two conclusions can be drawn immediately from Eq. (2).

1. The natural logarithm is a *differentiable* function, and therefore is *continuous*, on $(0, \infty)$.

2. The graph of $y = \ln x$ steadily rises from left to right because the slope is positive.

Also, since the derivative is continuous the curve has a continuously turning tangent. The second derivative,

$$\frac{d^2y}{dx^2} = -\frac{1}{x^2}, \tag{3}$$

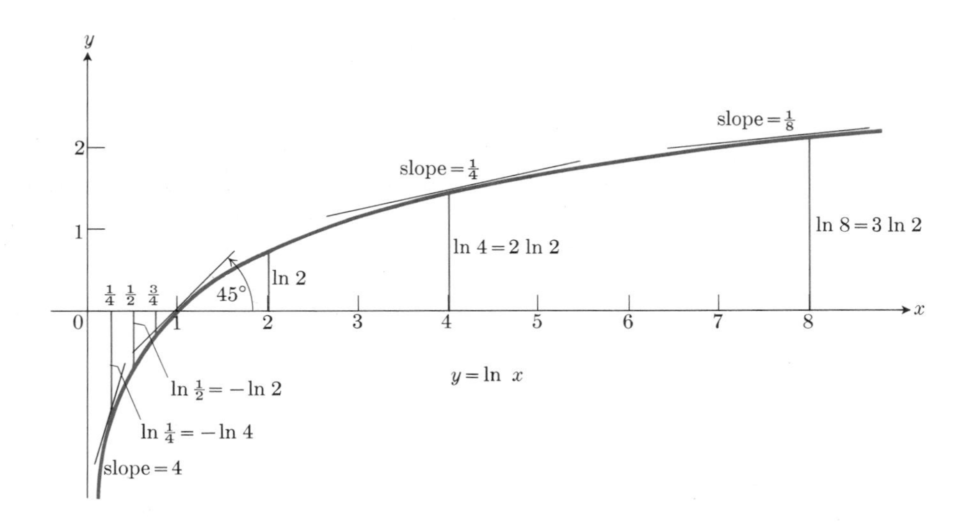

7.11

is always negative, so the curve (1) is everywhere concave downward.

The curve passes through the point (1, 0), since $\ln 1 = 0$. At this point its slope is $+1$, so the tangent line at this point makes an angle of 45° with the x-axis (if we use equal units on the x- and y-axes).

If we refer to the definition of ln 2 as an integral,

$$\ln 2 = \int_1^2 \frac{1}{t}\, dt,$$

we see that it may be interpreted as the area in Fig. 7.10 with $x = 2$. By considering the areas of rectangles of base 1 and altitudes 1 or $\frac{1}{2}$, respectively circumscribed over or inscribed under the given area, we see that

$$0.5 < \ln 2 < 1.0.$$

In fact, by more extensive calculations, the value of ln 2 to 5 decimal places is found to be

$$\ln 2 \approx 0.69315.$$

By Eq. (3) of Article 7.6, we have

$$\ln 4 = \ln 2^2 = 2 \ln 2 \approx 1.38630,$$
$$\ln 8 = \ln 2^3 = 3 \ln 2 \approx 2.07945,$$
$$\ln \tfrac{1}{2} = \ln 2^{-1} = -\ln 2 \approx -0.69315,$$
$$\ln \tfrac{1}{4} = \ln 2^{-2} = -2 \ln 2 \approx -1.38630, \ldots$$

We now plot the points that correspond to $x = \frac{1}{4}$, $\frac{1}{2}$, 1, 2, 4, 8 on the curve $y = \ln x$ and connect them with a smooth curve. The curve we draw should have slope $1/x$ at the point of abscissa x and should everywhere be concave downward. The curve is shown in Fig. 7.11.

Since ln 2 is greater than 0.5 and $\ln 2^n = n \ln 2$, it is clear that

$$\ln 2^n > 0.5n,$$

and hence $\ln x$ increases without limit as x does. That is,

$$\ln x \to +\infty \quad \text{as} \quad x \to +\infty. \tag{4}$$

On the other hand, as x approaches zero through positive values, $1/x$ tends to plus infinity. Hence, taking Eq. (6) of Article 7.6 into account, we have

$$\ln x = -\ln \frac{1}{x} \to -\infty \quad \text{as} \quad x \to 0+. \tag{5}$$

Summary

The natural logarithm, $y = \ln x$, is a function with the following properties.

1. Its *domain* is the set of positive real numbers, $x > 0$.
2. Its *range* is the set of all real numbers,

$$-\infty < y < +\infty.$$

3. It is a *continuous, increasing* function of x everywhere on its domain:

$$\text{If} \quad x_1 > x_2 > 0, \quad \text{then} \quad \ln x_1 > \ln x_2.$$

It provides a one-to-one mapping from its domain to its range.

4. *Multiplication* of numbers in the domain corresponds to *addition* of numbers in the range. That is,

$$\text{if} \quad \begin{cases} x_1 > 0, & y_1 = \ln x_1, \\ x_2 > 0, & y_2 = \ln x_2, \end{cases}$$

$$\text{then} \quad x_1 x_2 > 0 \quad \text{and} \quad y_1 + y_2 = \ln (x_1 x_2).$$

5. The *derivative* of $y = \ln x$ is $dy/dx = 1/x$.

EXERCISES

1. Take $\ln 3 = 1.09861$ as known. Plot the points corresponding to $x = \frac{1}{9}, \frac{1}{3}, 1, 3$, and 9 on the curve $y = \ln x$. Also, through each of these points construct a line segment tangent to the curve. Then sketch the curve itself. From the curve, read off $\ln 2$.
2. Let x_0 be the abscissa of the point $(x_0, 1)$ in which the line $y = 1$ intersects the curve $y = \ln x$. Show that the curve $y = x \ln x$ has a minimum at
$$(1/x_0, \; -1/x_0)$$
and sketch the curve for $x > 0$. Does this curve possess a point of inflection? If so, find it (or them), and if not, state why not. (See Exercise 5.)
3. Sketch the curve $x = \ln y$ for $y > 0$. What is its slope at the point (x, y)? Is it always concave upward, or downward, or does it have inflection points? Give reasons for your answers.
4. The curve $y = \ln x$ in Fig. 7.11 has the property that y becomes infinite when x does. But the ratio $(\ln x)/x$ approaches zero as x becomes infinite. Prove this latter statement by showing, first, that for $x > 1$,
$$\ln x = \int_1^x \frac{1}{t}\, dt \le \int_1^x \frac{1}{\sqrt{t}}\, dt = 2(\sqrt{x} - 1).$$
Hint. Compare the areas under the curves $y_1 = 1/t$ and $y_2 = 1/\sqrt{t}$ for $1 \le t \le x$.
5. From the result of Exercise 4, that
$$\lim_{x \to +\infty} \frac{\ln x}{x} = 0,$$
deduce that $\lim_{x \to 0+} x \ln x = 0$.
Hint. Let $x = 1/u$, $u \to +\infty$.
6. Show that the curve
$$y = \ln x - (x - 1) + \tfrac{1}{2}(x - 1)^2$$
has a point of inflection and a horizontal tangent at $(1, 0)$. Sketch the curve for $0 < x < 2$. (Compare Exercise 8, Article 7.4.)
7. Suppose you had a table of natural logarithms giving $\ln x$ for each $x > 0$. How could you use this to find $\log_{10} N$, where $\log_{10} N = b$ means $N = 10^b$?

7.8 THE EXPONENTIAL FUNCTION

Does the natural logarithmic function have an inverse, and, if so, what are its main properties? The answers are most easily understood in terms of the graphs of the curves

$$y = \ln x, \qquad x > 0 \tag{1a}$$

$$x = \ln y, \qquad y > 0. \tag{1b}$$

These two graphs are shown in Fig. 7.12. We know quite a bit about the graph of Eq. (1a).

1. Its slope at the point $(x, \ln x)$ is $1/x$.

2. Because the derivative of $\ln x$ is positive, the graph constantly rises from left to right.

3. The function $\ln x$ is positive for $x > 1$, zero for $x = 1$, negative for $x < 1$.

4. Since
$$\ln (2^n) = n \ln 2 \approx (0.693)n,$$
the points having abscissas
$$x_n = 2^n, \quad n = 1, 2, 3, \ldots,$$
have ordinates y_n that increase without bound as $n \to \infty$. Combining this with the fact that $\ln x$ is continuous and monotone increasing, we see that the portion of the domain where
$$1 \le x < \infty$$

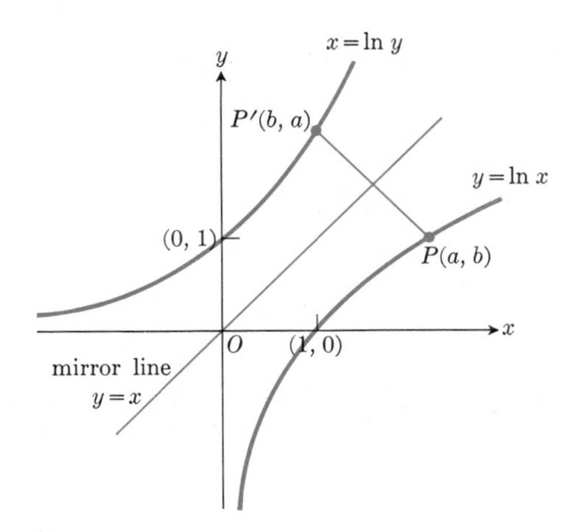

7.12 Each curve is the reflection of the other with respect to the line $y = x$. The points $P(a, b)$, with $b = \ln a$, and $P'(b, a)$ are symmetric with respect to the mirror line.

maps one-to-one onto

$$0 \leq y < \infty$$

under the natural logarithmic mapping $y = \ln x$.

5. Because $\ln (1/x) = -\ln x$, the last-named property means that the part of the domain where

$$0 < x \leq 1$$

maps one-to-one onto

$$-\infty < y \leq 0.$$

Thus, as x increases through the domain between 0 and $+\infty$, $\ln x$ increases between $-\infty$ and $+\infty$. Every real number y is the natural logarithm of precisely one positive real number x. The *inverse relation* that we get from ln by interchanging x and y in the set of ordered pairs

$$\ln = \{(x, y) : y = \ln x, \quad x > 0\}$$

is, therefore, a function—the *inverse natural logarithmic function*, which we shall temporarily denote by $\ln^{-1}$. Thus

$$y = \ln^{-1}(x) \tag{2a}$$

means just the same as

$$x = \ln y, \qquad y > 0, \qquad -\infty < x < \infty. \tag{2b}$$

The *domain* of $\ln^{-1}$ is the *range* of ln, hence is the set of all real numbers, $-\infty < x < \infty$. Thus, Eqs. (2a) and (2b) both correspond to the upper curve in Fig. 7.12.

The number whose natural logarithm is 1 plays a special role, and is the famous number e (for Euler):

$$e = \ln^{-1}(1), \qquad 1 = \ln e. \tag{3}$$

Equations (3) *define* e. In a later chapter we shall find that e can be found from an infinite series. Using that series, it is possible to compute as many significant places in the decimal expansion of e as we want. In the early 1960's, it was computed to 100,000 decimal places by Daniel Shanks and John W. Wrench, Jr., using an IBM 7090 computer. The computation took 2.5 hours, as compared with 8.7 hours for π. See Shanks and Wrench, "Calculation of π to 100,000 Decimals," in *Mathematics of Computation*, **16,** No. 77, January 1962, p. 78. To 15 places,

$$e = 2.7\ 1828\ 1828\ 45\ 90\ 45. \tag{4}$$

Without using the infinite series, we can approximate e by finding numbers $a, b, c, \ldots$ such that

$$\ln a + \ln b + \ln c + \cdots = 1.00000. \tag{5}$$

Assuming, for example, that $\ln 2 = 0.69315$, we can find logarithms of the square root, fourth root, eighth root, . . . of 2 by successive divisions of ln 2. The arithmetic is easy:

$$\begin{aligned}
\ln 2 &\approx 0.69315; & & \\
\ln 2^{1/2} &\approx 0.34658, & 2^{1/2} &\approx 1.414214; \\
\ln 2^{1/4} &\approx 0.17329, & 2^{1/4} &\approx 1.189207; \\
\ln 2^{1/8} &\approx 0.08664, & 2^{1/8} &\approx 1.090508; \\
\ln 2^{1/16} &\approx 0.04332, & 2^{1/16} &\approx 1.044274.
\end{aligned}$$

To find which of these logarithms have a sum close to 1.00000, we subtract from 1 the largest logarithm we can without getting a negative result (0.69315, in this case), and then repeat the subtraction process. This leads to

$$\begin{aligned}
1.00000 = {} & 0.69315 + 0.17329 \\
& + 0.08664 + 0.04332 + R,
\end{aligned}$$

where the remainder R is

$$R = 0.00360 < \ln 2^{1/160} \approx 0.00433.$$

This leads to the estimation

$$2 \cdot 2^{1/4} \cdot 2^{1/8} \cdot 2^{1/16} < e < 2 \cdot 2^{1/4} \cdot 2^{1/8} \cdot 2^{1/16} \cdot 2^{1/160} \tag{6}$$

We get further information about $\ln^{-1}$ by studying rational powers of e. Because

$$\ln (e^x) = x \ln e \quad \text{for } x \text{ rational},$$

and

$$\ln e = 1,$$

we have

$$\ln e^x = x \quad \text{for } x \text{ rational}. \tag{7}$$

When we compare Eq. (7) with Eqs. (2b) and (2a) we see that

$$e^x = \ln^{-1} (x) \quad \text{for all rational values of } x. \tag{8}$$

The right-hand side of Eq. (8) has a perfectly well-defined meaning for *any real number* x (not just for the rationals). When x happens to be *rational*, both sides of Eq. (8) are well defined, and are equal. When x is *irrational*, we use Eq. (8) to *define* e^x.

Definition 1. *For any real number x, we define e^x to be the same as $\ln^{-1}(x)$:*

$$e^x = \ln^{-1}(x). \tag{9}$$

Notation. The notation exp (exponential function) is also used to represent the inverse of the natural logarithm:

$$\exp (x) = \ln^{-1} (x), \quad x \text{ real}. \tag{10}$$

Thus Eq. (9) becomes

$$e^x = \exp (x).$$

The use of the exp-notation is particularly helpful when working with composite functions, such as

$$e^{(\sin^{-1} x)} = \exp (\sin^{-1} x),$$

or other expressions leading to complicated typography.

Summary

So far, we have

1. defined ln and studied it,
2. defined $\exp = \ln^{-1}$,
3. shown that

$$e^x = \exp (x) \quad \text{for } x \text{ rational},$$

4. *defined* e^x for x irrational to be $\exp (x)$.

Result. The exponential function (base e) is the inverse of the natural logarithmic function. All of its properties are derived from that fact.

Theorem 1. *For every real x,*

$$\frac{d}{dx} (e^x) = e^x, \tag{11a}$$

or

$$\exp' (x) = \exp (x). \tag{11b}$$

Proof. From the general theory of inverse functions, Theorem 8, Article 3.3, we can write

$$y = e^x \Leftrightarrow x = \ln y$$

and

$$\frac{dy}{dx} = \frac{1}{dx/dy} = \frac{1}{(1/y)} = y. \square$$

Corollary. *If u is a differentiable function of x, then*

$$\frac{d}{dx} (e^u) = e^u \frac{du}{dx}; \tag{XX}$$

in terms of differentials,

$$d(e^u) = e^u \, du. \tag{XX$'$}$$

Proof. For formula XX, apply the chain rule and Theorem 1:

$$\frac{d}{dx} (e^u) = \frac{d}{du} (e^u) \frac{du}{dx} = e^u \frac{du}{dx}.$$

The formula for the differential follows from this if we multiply both sides of the equation by dx and use the definition

$$dy = \frac{dy}{dx} dx. \square$$

Problem 1. Let $P(x, y)$ be a point on the graph of $y = e^x$. Let Q be the point where the tangent to the graph at P intersects the x-axis. Find the length of the segment QP.

Solution. The slope of the tangent at P is equal to the ordinate of P because

$$\frac{dy}{dx} = e^x = y.$$

Since the slope of QP is also rise/run, and the rise is y, the run must equal 1. Therefore

$$|QP| = \sqrt{y^2 + 1}.$$

See Fig. 7.13.

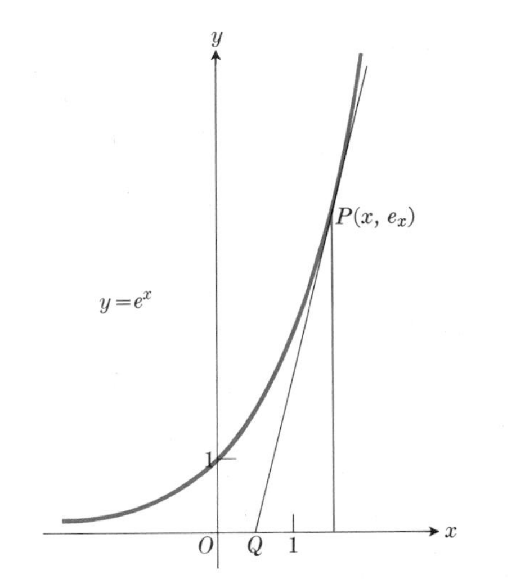

7.13

Problem 2. Find dy/dx if $y = e^{\tan^{-1} x}$.

Solution.

$$\frac{dy}{dx} = e^{\tan^{-1} x} \frac{d \tan^{-1} x}{dx}$$

$$= e^{\tan^{-1} x} \cdot \frac{1}{1 + x^2} \cdot \frac{dx}{dx}$$

$$= \frac{e^{\tan^{-1} x}}{1 + x^2}.$$

The differential formula XX′ leads to the integration formula

$$\int e^u \, du = e^u + C. \tag{12}$$

Problem 3. Find the area under the curve $y = e^{-x}$ from $x = 0$ to $x = b(>0)$ and show that this area *remains finite* as $b \to +\infty$.

Solution. The area from $x = 0$ to $x = b$ is given by

$$A_0^b = \int_0^b y \, dx = \int_0^b e^{-x} \, dx.$$

To evaluate this integral, we compare it with our standard forms and see that it is almost, but not exactly, like

$$\int e^u \, du = e^u + C.$$

To bring it into precisely this form, we let

$$u = -x, \qquad du = -dx,$$

or

$$dx = -du,$$

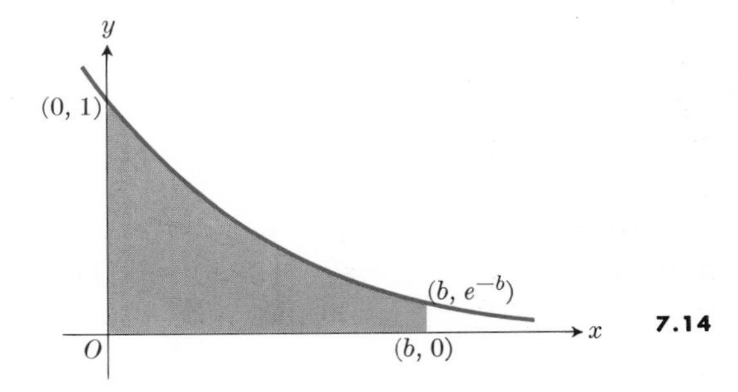

7.14

so that

$$\int e^{-x} \, dx = \int e^u(-du) = -\int e^u \, du$$

$$= -e^u + C = -e^{-x} + C.$$

Therefore

$$A_0^b = -e^{-x}\Big]_0^b = -e^{-b} + e^0 = 1 - e^{-b}.$$

From this we see that the number of square units of area in the tinted region of Fig. 7.14 is somewhat less than one for b large and positive. Moreover, as b increases without bound, so does e^b. Hence

$$\lim_{b \to \infty} e^{-b} = \lim_{b \to \infty} (1/e^b) = 0,$$

and

$$\lim_{b \to \infty} A_0^b = \lim_{b \to \infty} (1 - e^{-b}) = 1.$$

We summarize this by writing

$$\int_0^\infty e^{-x}\,dx = \lim_{b\to\infty}\int_0^b e^{-x}\,dx = 1.$$

The area from 0 to b remains finite and approaches 1 as $b \to \infty$.

Some properties of the function

$$y = e^x$$

are worthy of additional comment.

Theorem 2. *For all real x_1 and x_2,*

$$e^{x_1}e^{x_2} = e^{x_1+x_2}, \tag{13a}$$

or

$$\exp(x_1)\cdot\exp(x_2) = \exp(x_1+x_2). \tag{13b}$$

Proof. Let $\quad y_1 = e^{x_1}, \quad y_2 = e^{x_2}.$

Then, by definition,

$$x_1 = \ln y_1, \quad x_2 = \ln y_2,$$

and

$$x_1 + x_2 = \ln y_1 + \ln y_2 = \ln(y_1y_2).$$

Therefore

$$y_1y_2 = \ln^{-1}(x_1+x_2) = \exp(x_1+x_2),$$

so that, by substitution,

$$e^{x_1}e^{x_2} = \exp(x_1+x_2) = e^{(x_1+x_2)}. \ \square$$

Corollary. *For every real x,*

$$e^{-x} = \frac{1}{e^x},$$

or

$$\exp(-x) = [\exp(x)]^{-1}. \tag{14}$$

Proof. Let $x_1 = -x$ and $x_2 = x$ in Eq. (13a) or (13b):

$$e^{-x}e^x = e^{-x+x} = e^0 = \ln^{-1}(0) = 1.$$

Therefore, by division,

$$e^{-x} = \frac{1}{e^x}. \ \square$$

Remark. In Exercise 4, Article 7.7, it is shown that

$$\lim_{x\to\infty}\frac{\ln x}{x} = 0.$$

From this it also follows that for any real number n,

$$\lim_{x\to\infty}\frac{\ln(x^n)}{x} = \lim_{x\to\infty}\frac{n\ln x}{x} = 0. \tag{15}$$

If n is a positive integer and

$$u = x^n, \qquad x = \sqrt[n]{u},$$

where both x and u are positive, then Eq. (15) implies

$$\lim_{u\to\infty}\frac{\ln u}{u^{1/n}} = 0. \tag{16}$$

Equation (16) says that the logarithm function increases more slowly than the nth root function, for any positive integer n. It can also be made to give information about the rate of growth of the exponential function. In Eq. (16), let

$$u^{1/n} = e^y,$$

or

$$y = (\ln u)/n.$$

Because the limit of a product of a finite number of factors is the product of their limits, we also have

$$\lim_{u\to\infty}\frac{(\ln u)^n}{u} = \left[\lim_{u\to\infty}\frac{\ln u}{u^{1/n}}\right]^n = 0^n = 0 \tag{17}$$

for any positive integer n. This form is one that easily gives us a comparison between the rates of growth of the exponential function and the nth power function. In Eq. (17), we let

$$\ln u = y \quad \text{and} \quad u = e^y,$$

and get

$$\lim_{y\to\infty}\frac{y^n}{e^y} = 0 \quad \text{for any positive integer } n. \tag{18}$$

The limit result (18) says that as $y \to \infty$, e^y increases more rapidly than any fixed positive integral power of y.

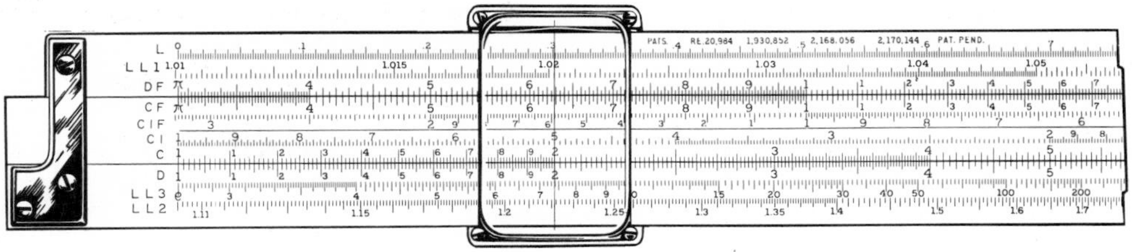

7.15

Problem 4. Find a number c such that $e^y > y^{10}$ for $y > c$.

Solution. The condition to be met is equivalent to the condition that $y > 10 \ln y$. Consider the function $g(y) = y - 10 \ln y$ for $y > 0$. Its derivative is

$$g'(y) = 1 - 10/y,$$

which is positive for $y > 10$. Because we know something about ln 2, we look for $y = 2^k$, such that $2^k > 10$ and

$$2^k > 10 \ln 2^k = (10k) \ln 2 = 6.9315k.$$

By trial and error, we see that $k = 5$ is too small but $k = 6$ is large enough. In other words,

$$\begin{aligned} g(2^6) &= g(64) = 64 - 10 \ln 64 \\ &= 64 - 60 \ln 2 \\ &> 64 - 60(0.694) > 0. \end{aligned}$$

Therefore $c = 64$ is sufficiently large to guarantee that $g(y)$ is positive for $y > c$, because

$$g(64) > 0 \quad \text{and} \quad g'(y) > 0 \quad \text{for } y > 10$$

together imply that

$$g(y) > 0 \quad \text{for all } y > 64.$$

As a check, observe that $e > 2$, so that

$$e^{64} > 2^{64} \quad \text{and} \quad 64^{10} = 2^{60} < 2^{64}.$$

Theorem 3. *For every real x and rational r,*

$$(e^x)^r = e^{rx}, \tag{19a}$$

or

$$[\exp(x)]^r = \exp(rx). \tag{19b}$$

Proof. Let

$$y = e^x, \quad \text{or} \quad x = \ln y.$$

Then, for any rational r,

$$rx = r \ln y = \ln(y^r),$$

so that

$$y^r = \ln^{-1}(rx) = \exp(rx) = e^{rx}.$$

But $y = e^x$. Therefore

$$(e^x)^r = e^{rx}. \; \square$$

Powers of e are tabulated in many mathematical tables, such as *CRC* and Burington's. They may also be read directly from the LL-scales of a log-log slide rule. For example, if the number x on the D-scale in Fig. 7.15 is $x = 2$, then the corresponding readings on the LL-scales are:

$$\begin{aligned} e^2 &= 7.4 && \text{on LL 3,} \\ e^{0.2} &= 1.2215 && \text{on LL 2,} \\ e^{0.02} &= 1.0202 && \text{on LL 1.} \end{aligned}$$

The value of e^{-2} can be read from the LL00-scale on the reverse side of the slide rule opposite the 2 on the right half of the A-scale (the reading gives $e^{-2} = 0.135$). It is also possible to use

$$e^{-2} = \frac{1}{e^2} = \frac{1}{7.4} = 0.135$$

to arrive at the same result. The use of the LL-scales is described in the manual of instructions that comes with the log-log slide rule.

A table of e^x and e^{-x} is included at the end of the book for convenience and to give an appreciation of the rapid rate of increase of the exponential function for positive values of x.

EXERCISES

1. Simplify each of the following expressions.

(a) $e^{\ln x}$ (b) $\ln (e^x)$ (c) $e^{-\ln (x^2)}$

(d) $\ln (e^{-x^2})$ (e) $\ln (e^{1/x})$ (f) $\ln (1/e^x)$

(g) $e^{\ln (1/x)}$ (h) $e^{-\ln (1/x)}$ (i) $e^{\ln 2 + \ln x}$

(j) $e^{2 \ln x}$ (k) $\ln (e^{x-x^2})$ (l) $\ln (x^2 e^{-2x})$

(m) $e^{x + \ln x}$ (n) $\exp (\ln x - 2 \ln y)$

Find dy/dx in Exercises 2 through 19.

2. $y = x^2 e^x$

3. $y = e^{2x}(2 \cos 3x + 3 \sin 3x)$

4. $y = \ln \dfrac{e^x}{1 + e^x}$

5. $y = \frac{1}{2}(e^x - e^{-x})$

6. $y = \frac{1}{2}(e^x + e^{-x})$

7. $y = \dfrac{e^x - e^{-x}}{e^x + e^{-x}}$

8. $y = e^{\sin^{-1} x}$

9. $y = (1 + 2x)e^{-2x}$

10. $y = (9x^2 - 6x + 2)e^{3x}$

11. $y = \dfrac{ax - 1}{a^2} e^{ax}$

12. $y = e^{-x^2}$

13. $y = x^2 e^{-x^2}$

14. $y = e^x \ln x$

15. $y = \tan^{-1} (e^x)$

16. $y = \sec^{-1} (e^{2x})$

17. $e^{2x} = \sin (x + 3y)$

18. $y = e^{1/x}$

19. $\tan y = e^x + \ln x$

20. Let h be a number between $\frac{1}{2}$ and 1. (For example, $h = \ln 2$ would qualify.) Consider the set S of numbers

$$\begin{aligned} a_1 &= h, \\ a_2 &= h/2, \\ a_3 &= h/4, \ \ldots, \\ a_n &= h/2^{n-1}, \ \ldots \end{aligned}$$

Given any $\epsilon > 0$, you could find a finite subset of S whose sum would lie between $1 - \epsilon$ and $1 + \epsilon$. Explain how this could be done if you had access to a computer.

21. Recall that $\sqrt{a}$ can be approximated by the following procedure: From a table of squares, locate a number x such that x^2 is reasonably near a. For example, for $a = 2$ you might use the squares $1.4^2 = 1.96$ and $1.5^2 = 2.25$ to give the estimate $x = 1.4$. Then get an improved estimate

$$x' = \tfrac{1}{2}(x + a/x).$$

These operations are easy for a computer, and the process can be repeated with x replaced by the improved estimate to get a still better estimate. When $a = 1 + h$ and h is small, you could begin with the approximation $x = 1 + h/2$. Using these ideas, explain how you could compute good approximations, successively, for the square root, fourth root, eighth root, and so on, of 2, if you had access to a computer.

22. (a) Show that $y = Ce^{ax}$ is a solution of the differential equation $dy/dx = ay$ for any choice of the constant C.

(b) Using the result of (a), find a solution of the differential equation $dy/dt = -2y$ satisfying the initial condition $y = 3$ when $t = 0$.

23. Show that $y = e^{-ax} u$ is a solution of the differential equation

$$\frac{d^2y}{dx^2} + 2a\frac{dy}{dx} + (a^2 + b^2)y = 0$$

if and only if u satisfies the equation

$$\frac{d^2u}{dx^2} + b^2 u = 0,$$

provided that a and b are constants.

Evaluate the integrals in Exercises 24 through 31.

24. $\displaystyle\int e^{2x}\, dx$

25. $\displaystyle\int xe^{x^2}\, dx$

26. $\displaystyle\int e^{\sin x} \cos x\, dx$

27. $\displaystyle\int e^{x/3}\, dx$

28. $\displaystyle\int \frac{4\, dx}{e^{3x}}$

29. $\displaystyle\int \frac{e^x\, dx}{1 + 2e^x}$

30. $\displaystyle\int_e^{e^2} \frac{dx}{x \ln x}$

31. $\displaystyle\int \frac{(e^x - e^{-x})\, dx}{(e^x + e^{-x})}$

32. If a particle moves along the x-axis so that its position at time t is given by

$$x = ae^{\omega t} + be^{-\omega t},$$

where a, b, ω, are constants, show that it is repelled from the origin with a force proportional to the displacement. Assume that

$$\text{force} = \text{mass} \times \text{acceleration}.$$

33. Find a positive number p such that $e^x > x^{20}$ for $x > p$.
34. Find a positive number b such that $e^u > 20u^{10}$ for $u > b$. [*Suggestion.* Observe that $u^{11} > 20u^{10}$ when u is sufficiently large.]

7.9 THE FUNCTIONS a^u AND $\log_a u$

If a is any positive real number and

$$b = \ln a, \tag{1a}$$

then

$$a = e^b. \tag{1b}$$

The two equations (1a, b) may be combined to give the result

$$a = e^{\ln a}: \tag{2}$$

> $\ln a$ *is the power to which the base* e *must be raised to give* a.

Suppose we now raise both sides of Eq. (2) to the power u, where u is any real number. We would like to have the law of exponents

$$(e^{\ln a})^u = e^{u \ln a}$$

be true for all real values of u. (We proved that it is true if $u = r$ is rational. See Theorem 3, Article 7.8, with $x = \ln a$, and $u = r$.) Hence we now make the following definition.

Definition 1. *Let a be a positive number. Then, for any real number u, we define a^u to be* $\exp(u \ln a)$:

$$a^u = \exp(u \ln a), \tag{3a}$$

or

$$a^u = e^{u \ln a}. \tag{3b}$$

Since the right sides of Eqs. (3a) and (3b) are well defined for all positive numbers a and all real numbers u, we have certainly defined the general exponential function (base a, $a > 0$) for the domain $-\infty < u < \infty$. It comes as no surprise, but is nevertheless good news, that for any *rational* exponent u, say

$$u = m/n, \qquad m, n \text{ integers}, \quad n \neq 0,$$

the earlier definition,

$$a^u = \sqrt[n]{a^m},$$

and the new definition, (3a), agree. In particular, the following results are immediate consequences of Eqs. (3a) and (3b):

$$a^1 = \exp(1 \ln a) = \exp(\ln a) = \ln^{-1}(\ln a) = a,$$

$$a^0 = \exp(0 \ln a) = \exp(0) = \ln^{-1}(0) = 1,$$

$$\begin{aligned} a^u a^v &= \exp(u \ln a) \cdot \exp(v \ln a) \\ &= \exp[(u + v) \ln a] \\ &= a^{u+v}, \end{aligned}$$

$$\begin{aligned} (a^{m/n})^n &= \left[\exp\left(\frac{m}{n} \ln a\right)\right]^n \\ &= \exp(m \ln a) \qquad \text{if } n \text{ is rational (by Theorem 3, Article 7.8)} \\ &= a^m \qquad \text{(by Eq. 3a).} \end{aligned}$$

Also,

$$(ab)^u = a^u b^u \quad \text{if} \quad a > 0,\ b > 0,$$

because

$$\begin{aligned} (ab)^u &= \exp[u \ln(ab)] = \exp[u \ln a + u \ln b] \\ &= \exp(u \ln a) \cdot \exp(u \ln b) = a^u \cdot b^u. \end{aligned}$$

Remark 1. In Eq. (3b), let a be a fixed positive number different from 1. Then the exponent $u \ln a$ takes all real values between $-\infty$ and $+\infty$ when u does. The mapping

$$u \to a^u, \quad -\infty < u < \infty,$$

is one-to-one from the domain of all real numbers u onto the range of positive real numbers a^u. That is, for each positive real number x there exists exactly one real number u (given $a > 0$, $a \neq 1$) such that

$$x = a^u. \tag{4a}$$

The exponent u for which Eq. (4a) is true is called the *logarithm of x to the base a*:

$$\log_a x = u \Leftrightarrow x = a^u. \tag{4b}$$

Comparing Eqs. (3b) and (4b), we see also that

$$\begin{aligned} \log_a x = u \Leftrightarrow x &= e^{u \ln a} \\ &\Updownarrow \\ \ln x &= u \ln a. \end{aligned} \tag{4c}$$

The statements (4c) lead to the conclusion

$$u = \log_a x = \frac{\ln x}{\ln a}. \tag{4d}$$

Definition 2. *Let a be a positive number not equal to one. The logarithmic function, to the base a, is defined on the domain $x > 0$ by Eqs.* (4b) *or* (4d):

$$\log_a x = \frac{\ln x}{\ln a}.$$

Example. From a table of natural logarithms, we get

$$\log_{10} 2 = \frac{\ln 2}{\ln 10} \approx \frac{0.6931}{2.3025} \approx 0.3010.$$

Derivatives

With the definition of a^u now before us, we are ready to calculate its derivative in case u is a differentiable function of x. We apply formula XX, Article 7.8, and remembering that $\ln a$ is constant when a is, we obtain

$$\frac{da^u}{dx} = \frac{de^{u \ln a}}{dx} = e^{u \ln a} \cdot \frac{du}{dx} \cdot \ln a,$$

$$\frac{da^u}{dx} = a^u \cdot \frac{du}{dx} \cdot \ln a. \tag{5a}$$

If u is a differentiable positive function of x, we may differentiate both sides of the defining equation

$$\log_a u = \frac{\ln u}{\ln a}$$

and get

$$\frac{d \log_a u}{dx} = \frac{1}{u \ln a} \frac{du}{dx}. \tag{5b}$$

It is clear that this formula is simpler for logarithms to the base e than for any other base. The only bases of practical importance are $a = e$ (natural logarithms) and $a = 10$ (common logarithms). Since $\ln 10 = 2.30259\ldots$, Eq. (5b) is certainly less attractive for the case $a = 10$ than it is for the case $a = e$. This is the primary reason why natural logarithms are to be preferred over common logarithms in applications of the calculus.

Note. Logarithms to the base 10 are used primarily as aids in performing arithmetical calculations such as multiplication, division, and raising numbers to powers. The use of slide rules, desk calculators, and new high-speed computers is rapidly making this use of logarithms less and less important. Natural logarithms, on the other hand, are not in any immediate danger of becoming obsolete, since they occur naturally when some common kinds of differential equations are solved. (See Problems 1 and 2, Article 7.10.)

Equation (5a) gives us a formula for the derivative of any positive constant a raised to a variable power. There is very little reason to memorize the result, since it may always be derived very quickly by the method of logarithmic differentiation, which will presently be discussed. It illustrates the fact that the base e is the most desirable one to encounter, since $\ln e = 1$ and Eq. (5a) is the same as formula XX in case $a = e$. It also enables us to write down a formula for the integral of $a^u\,du$. For, if we multiply both sides of Eq. (5a) by dx to change it to differential form:

$$da^u = \ln a \cdot a^u\,du,$$

and then divide by $\ln a$, provided $a \neq 1$, we see that

$$\int a^u\,du = \int \frac{1}{\ln a}\,da^u = \frac{a^u}{\ln a} + C, \qquad (a > 0,\ a \neq 1). \tag{6}$$

Logarithmic differentiation

The properties

$$\begin{aligned} &\text{(a)}\ \ln uv = \ln u + \ln v, \\ &\text{(b)}\ \ln \frac{u}{v} = \ln u - \ln v, \\ &\text{(c)}\ \ln u^n = n \ln u, \\ &\text{(d)}\ \ln a^u = u \ln a \end{aligned} \tag{7}$$

can be used to advantage in calculating derivatives of products, quotients, roots, and powers, as illustrated in the problems that follow. The method, known as *logarithmic differentiation,* is to take the natural logarithm of both sides of an equation

$$y = f(x),$$
$$\ln y = \ln f(x),$$

simplify $\ln f(x)$ as much as possible by making use of the properties in Eq. (7), and then differentiate implicitly with respect to x:

$$\frac{d}{dx} \ln y = \frac{1}{y} \frac{dy}{dx}.$$

Problem 1. Derive Eq. (5a).

Solution. Let $\quad y = a^u.$

Then

$$\ln y = \ln a^u = u \ln a,$$
$$\frac{1}{y}\frac{dy}{dx} = \frac{du}{dx} \ln a,$$
$$\frac{dy}{dx} = y \frac{du}{dx} \ln a,$$

or, since $y = a^u$,

$$\frac{da^u}{dx} = a^u \frac{du}{dx} \ln a.$$

Problem 2. If n is a real constant and u is a positive differentiable function of x, show that

$$\frac{du^n}{dx} = nu^{n-1} \frac{du}{dx}.$$

Solution. Let $\quad y = u^n.$

Then

$$\ln y = \ln u^n = n \ln u,$$
$$\frac{1}{y}\frac{dy}{dx} = n \frac{1}{u}\frac{du}{dx},$$
$$\frac{dy}{dx} = n \frac{y}{u}\frac{du}{dx} = n \frac{u^n}{u}\frac{du}{dx}$$
$$= nu^{n-1}\frac{du}{dx}.$$

Problem 3. Find dy/dx if $y = x^x, x > 0$.

Solution.

$$\ln y = \ln x^x = x \ln x,$$
$$\frac{1}{y}\frac{dy}{dx} = x \cdot \frac{1}{x} + \ln x = 1 + \ln x,$$
$$\frac{dy}{dx} = y(1 + \ln x) = x^x(1 + \ln x).$$

Problem 4. Find dy/dx if

$$y^{2/3} = \frac{(x^2+1)(3x+4)^{1/2}}{\sqrt[5]{(2x-3)(x^2-4)}} \quad (x > 2).$$

Solution.

$$\ln y^{2/3} = \tfrac{2}{3} \ln y$$
$$= \ln (x^2+1) + \tfrac{1}{2} \ln (3x+4)$$
$$-\tfrac{1}{5} \ln (2x-3) - \tfrac{1}{5} \ln (x^2-4), \quad \text{(a)}$$

and

$$\frac{2}{3} \cdot \frac{1}{y}\frac{dy}{dx} = \frac{2x}{x^2+1} + \frac{1}{2} \cdot \frac{3}{3x+4}$$
$$- \frac{1}{5} \cdot \frac{2}{2x-3} - \frac{1}{5} \cdot \frac{2x}{x^2-4},$$

so that

$$\frac{dy}{dx} = \frac{3y}{2}\left[\frac{2x}{x^2+1} + \frac{\frac{3}{2}}{3x+4} - \frac{\frac{2}{5}}{2x-3} - \frac{2x/5}{x^2-4}\right].$$

Note. The restriction that $x > 2$ ensures that all the quantities whose logarithms are indicated in (a) above are positive. If we want dy/dx at a point where y is negative, then we may multiply our equation by -1 before taking logarithms; for if y is negative, then $-y$ is positive and $\ln(-y)$ is well defined.

Economists define the ratio of "marginal cost" to "average cost" as the "elasticity of total cost." That is, if the total cost of producing x units of a given product is $y = f(x)$ dollars, then the "average cost" is y/x dollars per unit. The "marginal cost," which is, roughly speaking, the additional cost of producing one more unit, is dy/dx. Then, by definition,

$$\text{elasticity of total cost} = \frac{dy/dx}{y/x}$$
$$= \frac{dy/y}{dx/x} = \frac{d\,(\ln y)}{d\,(\ln x)}.$$

When the elasticity of total cost equals one, then marginal cost equals average cost (that is, $dy/dx = y/x$) and average cost has a stationary value [that is, $d(y/x)/dx = 0$]. (For further discussion, see R. G. D. Allen, *Mathematical Analysis for Economists*, Macmillan, 1939, p. 260 ff.)

EXERCISES

Use the method of logarithmic differentiation to find dy/dx in Exercises 1 through 8.

1. $y^2 = x(x+1) \quad (x > 0)$
2. $y = \sqrt[3]{\dfrac{x+1}{x-1}} \quad (x > 1)$
3. $y = \dfrac{x\sqrt{x^2+1}}{(x+1)^{2/3}} \quad (x > 0)$
4. $y = \sqrt[3]{\dfrac{x(x+1)(x-2)}{(x^2+1)(2x+3)}} \quad (x > 2)$
5. $y = x^{\sin x} \quad (x > 0)$
6. $y = (\sin x)^{\tan x} \quad (\sin x > 0)$
7. $y = 2^{\sec x}$
8. $y = x^{\ln x} \quad (x > 0)$
9. Show that $(ab)^u = a^u b^u$ if a and b are any positive numbers and u is any real number.
10. Show that the derivative of average cost is zero when marginal cost and average cost are equal.

Evaluate the integrals in Exercises 11 through 15.

11. $\displaystyle\int_0^{\ln 2} e^{-2x}\,dx$
12. $\displaystyle\int_0^{1.2} 3^x\,dx$
13. $\displaystyle\int_1^{\sqrt{2}} x\,2^{-x^2}\,dx$
14. $\displaystyle\int_0^1 5^{2t-2}\,dt$
15. $\displaystyle\int_0^{\pi/6} (\cos\theta)4^{-\sin\theta}\,d\theta$
16. Let a be a number greater than one. Prove that the graph of $y = a^x$ has the following characteristics:
    (a) If $x_1 > x_2$, then $a^{x_1} > a^{x_2}$.
    (b) The graph is everywhere concave upward.
    (c) The graph lies entirely above the x-axis.
    (d) The slope at any point is proportional to the ordinate there, and the proportionality factor is the slope at the y-intercept of the graph.
    (e) The curve approaches the negative x-axis as $x \to -\infty$.
17. If the graph of $y = a^x$ of Exercise 16 is reflected in the 45°-line $y = x$, the new curve so obtained is the graph of $x = a^y$ or $y = \log_a x$. Describe this curve, assuming that $a > 1$.

Because $y = \log_a x \Leftrightarrow x = a^y$, we can say that "the logarithm of x to the base a is the power y to which a must be raised to give x." Using this fact, determine the logarithms in Exercises 18 and 19.

18. (a) $\log_4 16$ (b) $\log_8 32$
    (c) $\log_5 0.04$ (d) $\log_{0.5} 4$
19. (a) $\log_2 4$ (b) $\log_4 2$
    (c) $\log_8 16$ (d) $\log_{32} 4$
20. Find x if $3^x = 2^{x+1}$.
21. Find x if $3^{\log_3 7} + 2^{\log_2 5} = 5^{\log_5 x}$.
22. Show that $\log_b u = \log_a u \cdot \log_b a$ if a, b, and u are positive numbers, $a \neq 1$, $b \neq 1$.
23. Use the equation
$$\log_a u = \frac{\ln u}{\ln a}, \quad a > 0, \quad a \neq 1,$$
    to derive the following properties of the logarithm function to base a.
    (a) $\log_a uv = \log_a u + \log_a v$
    (b) $\log_a \dfrac{u}{v} = \log_a u - \log_a v$
    (c) $\log_a u^v = v \log_a u$
24. Given $\ln 2 = 0.69315$, $\ln 10 = 2.30259$, $\log_{10} 2 = 0.30103$. Find
    (a) $\log_{10} 20$, $\log_{10} 200$, $\log_{10} 0.2$, $\log_{12} 0.02$,
    (b) $\ln 20$, $\ln 200$, $\ln 0.2$, $\ln 0.02$.
25. Use the table of natural logarithms at the end of the book to evaluate the following logarithms to two significant figures.
    (a) $\log_3 5$ (b) $\log_5 3$ (c) $\log_{1.4} 2.3$
26. Use the tables at the end of the book to evaluate the following to two significant figures.
    (a) $(1.6)^{1.5}$ (b) $\dfrac{\ln 0.7}{\ln 1.3}$
27. Find the derivatives of the following equations.
    (a) $y = 3^{\tan x}$ (b) $y = (x^2+1)^{\ln x}$
    (c) $s = 2^{-t^2}$ (d) $r = \theta e^{-2\theta}$

7.10 DIFFERENTIAL EQUATIONS

Suppose a quantity changes at a rate that at any instant of time t is proportional to the amount x present at that instant. This is expressed by the differential equation

$$\frac{dx}{dt} = kx, \tag{1}$$

where k is the proportionality constant. If the amount present at time $t = 0$ is x_0, the initial condition

$$x = x_0 \quad \text{when } t = 0 \tag{2}$$

enables us to evaluate the constant of integration that enters when we integrate Eq. (1). The method of solution will be illustrated by a problem.

Problem. As a result of leakage, an electrical condenser discharges at a rate proportional to the charge. If the charge Q has the value Q_0 at the time $t = 0$, find Q as a function of t.

Solution. The mathematical statement of the problem consists of the differential equation $dQ/dt = -kQ$ and the initial condition $t = 0$, $Q = Q_0$. Then

$$\frac{dQ}{Q} = -k\,dt,$$

$$\int \frac{dQ}{Q} = -k \int dt,$$

$$\ln Q = -kt + C. \tag{3}$$

The constant of integration C is evaluated by using the initial condition:

$$t = 0,\ Q = Q_0: \qquad \ln Q_0 = C.$$

Substituting this value of C back into (3) and transposing it to the left of the equation, we get

$$\ln Q - \ln Q_0 = -kt,$$

or

$$\ln \frac{Q}{Q_0} = -kt.$$

Hence, $-kt$ is the power to which e must be raised to give Q/Q_0, that is,

$$\frac{Q}{Q_0} = e^{-kt}, \quad \text{or} \quad Q = Q_0 e^{-kt}.$$

According to this equation, Q is never zero; that is, some charge would always remain. But when $kt = 10$, say, then $Q = Q_0 e^{-10}$ so that the charge Q has by then diminished to less than 5/1000 of 1 percent of its original value Q_0.

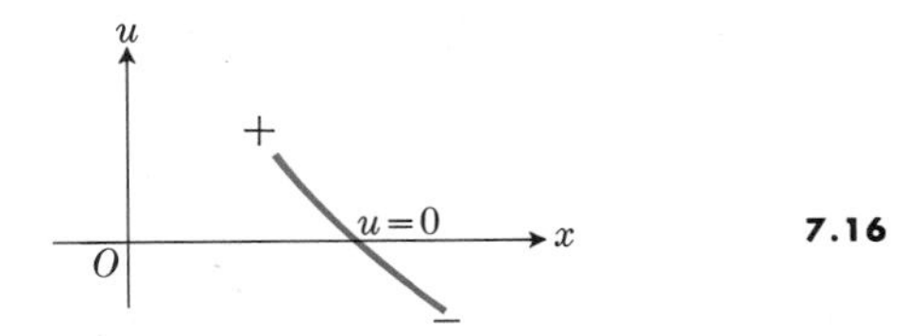

7.16

Remark. In this problem, the function integrated was positive, so that we were able to omit the absolute value symbols and write simply $\ln Q$ in place of $\ln |Q|$. Occasionally we encounter a problem where the quantity whose logarithm we require is an unknown function which presumably could be either positive or negative. To change sign from one to the other, however, would mean that the quantity would become zero at the transition point, provided it is a *continuous* function (Fig. 7.16). An integral of the form

$$\int \frac{du}{u} \tag{4}$$

between points where u changes continuously from positive to negative or from negative to positive is a so-called *improper intgeral,* because the integrand $1/u$ becomes infinite at $u = 0$. We shall discuss improper integrals in more detail in a later chapter. At present we shall assume that any physical problems we encounter that lead to integrals of the form of (4) arise from physical situations in which infinities are excluded. Then if we know the sign of u at one point of the interval of integration, we shall assume that it always has the same sign throughout the whole interval.

Example. Ohm's law, $E = Ri$, requires modification in a circuit containing self-inductance, as in a coil. The modified form is

$$L\frac{di}{dt} + Ri = E \tag{5}$$

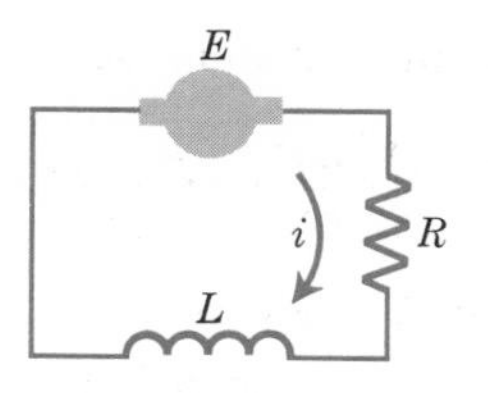

7.17

for the series circuit shown in Fig. 7.17, where R denotes the resistance of the circuit in ohms, L is the self-inductance in henries, E is the impressed electromotive force (emf) in volts, i is the current in amperes, and t is the time in seconds. (See Sears, *Electricity and Magnetism*, Chapter 13.) We shall solve the differential equation (5) under the assumption that E is constant and

$$i = 0 \quad \text{when} \quad t = 0. \tag{6}$$

To integrate (5), we first separate the variables in order to have the terms involving the variable current i combined with di for purposes of integration. Thus

$$L\,di = (E - Ri)\,dt$$

and

$$\int \frac{L\,di}{E - Ri} = \int dt. \tag{7}$$

We let

$$u = E - Ri; \tag{8}$$

then

$$du = -R\,di, \quad di = -\frac{1}{R}\,du,$$

so that (7) becomes

$$-\frac{L}{R}\int \frac{du}{u} = \int dt. \tag{9}$$

From (8) and the initial conditions (6) we see that $u = E - Ri$ has the positive value $u_0 = E$ at $t = 0$. We therefore assume that it remains positive, so that $|u| = u$ and the integral in (9) then gives

$$-\frac{L}{R}\ln u = t + C,$$

or

$$-\frac{L}{R}\ln(E - Ri) = t + C. \tag{10}$$

The initial conditions $t = 0$, $i = 0$ enable us to evaluate the constant of integration:

$$C = -\frac{L}{R}\ln E.$$

When we substitute this back into (10) and transpose it to the left side of the equation, we obtain

$$\frac{L}{R}[\ln E - \ln(E - Ri)] = t,$$

$$\ln\frac{E}{E - Ri} = \frac{Rt}{L},$$

$$\frac{E}{E - Ri} = e^{Rt/L},$$

$$\frac{E - Ri}{E} = e^{-Rt/L},$$

$$1 - \frac{R}{E}i = e^{-Rt/L},$$

$$i = \frac{E}{R}(1 - e^{-Rt/L}). \tag{11}$$

We see from this that the current (for $t > 0$) is always less than, but increases toward, the steady state value

$$i_{\text{s.s.}} = \frac{E}{R}, \tag{12}$$

which is the current that would flow in the circuit if either $L = 0$ (no inductance) or $di/dt = 0$ (steady current, i = constant) in Eq. (5). Indeed, if $L \to 0$, then for all $t > 0$, the exponent (in Eq. 11)

$$-\frac{Rt}{L} \to -\infty$$

and thus

$$i \to \frac{E}{R}.$$

A graph of the current versus time relation (11) is shown in Fig. 7.18.

Incidentally, in problems involving electrical circuits, the letter e is usually used to denote an emf and the letter ϵ is then used to denote the constant which we have

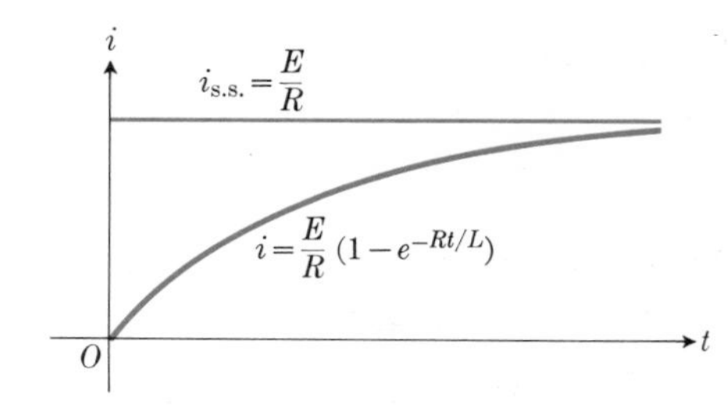

7.18

called (and shall continue to call) e. Mathematicians have become accustomed to the use of ϵ to denote an arbitrarily small positive number, so they are not likely to change their notation; nor are the electrical engineers likely to change theirs. However, this lack of uniformity should not cause confusion, since the meaning of the symbols will usually be clear from the nature of the context.

EXERCISES

1. Suppose a body of mass m moving in a straight line with velocity v encounters a resistance proportional to the velocity, and that this is the only force acting on the body. If the body starts with velocity v_0, how far does it travel in time t? Assume $F = d(mv)/dt$.
2. A radioactive substance disintegrates at a rate proportional to the amount present. How much of the substance remains at time t if the initial amount is Q_0?
3. If the bacteria in a culture increase continuously at a rate proportional to the number present, and the initial number is N_0, find the number at time t.

In Exercises 4 through 7, substitute $y = e^{rx}$ and find values of the constant r for which $y = e^{rx}$ is a solution of the given differential equation.

4. $\dfrac{d^2y}{dx^2} - 4\dfrac{dy}{dx} + 3y = 0$
5. $\dfrac{d^2y}{dx^2} - \dfrac{dy}{dx} - 2y = 0$
6. $\dfrac{d^3y}{dx^3} + 6\dfrac{d^2y}{dx^2} + 5\dfrac{dy}{dx} = 0$
7. $\dfrac{d^4y}{dx^4} - 13\dfrac{d^2y}{dx^2} + 36y = 0$
8. By substitution, verify that

$$y = (c_1 + c_2x)e^{r_1x}$$

satisfies the differential equation

$$\frac{d^2y}{dx^2} - 2r_1\frac{dy}{dx} + r_1^2y = 0$$

for all values of the constants c_1, c_2, and r_1.

9. By substitution, verify that

$$y = e^{ax}(c_1 \cos bx + c_2 \sin bx)$$

satisfies the differential equation

$$\frac{d^2y}{dx^2} - 2a\frac{dy}{dx} + (a^2 + b^2)y = 0$$

for all values of the constants c_1, c_2, a, and b.

10. Sketch $y = e^{-x} \cos 2x$ for $x \geq 0$. Show that it is tangent to the curves $y = \pm e^{-x}$ when $\cos 2x = \pm 1$.

REVIEW QUESTIONS AND EXERCISES

1. Review the formulas for derivatives of sums, products, quotients, powers, sines, and cosines. Using these, develop formulas for derivatives of the other trigonometric functions.
2. Define the inverse sine, inverse cosine, inverse tangent, and inverse secant functions. What is the *domain* of each? What is its *range*? What is the derivative of each? Sketch the graph of each.
3. Define the natural logarithm function. What is its domain? What is its range? What is its derivative? Sketch its graph.
4. What is the inverse function of the natural logarithm called? What is the domain of this function? What is its range? What is its derivative? Sketch its graph.
5. Define the term "algebraic function." Define "transcendental function." To which class (algebraic or transcendental) do you think the "greatest integer" step-function (Fig. 1.17) belongs?

[It is often difficult to prove that a particular function belongs to one class or the other. Numbers are also classified as algebraic and transcendental. *References:* (1) W. J. Le Veque, *Topics in Number Theory*, Addison-Wesley, Vol. II, Chap. 5, pp. 161–200; (2) I. M. Niven, *Irrational Numbers*, Carus monograph number 11 of the Mathematical Association of America (1956). Both π and e are transcendental numbers, and proofs may be found in the references cited. That π is transcendental was first proved by Ferdinand Lindemann in 1882. One result that follows from the transcendence of π is that it is impossible to construct, with unmarked straightedge and compass alone, a square whose area is equal to the area of a given circle. The first proof that e is transcendental was published by Charles Hermite in 1873.

You may also enjoy reading about a related feature of the graph of $y = \ln x$ in the article on logarithms on p. 301 of Vol. 14 of the *Encyclopedia Britannica* (1956 edition).]

6. Let n be an integer greater than 1. Divide the interval $1 \leq x \leq n$ into n equal subintervals and write an expression for the trapezoidal approximation to the area under the graph of $y = 1/x$, $1 \leq x \leq n$, based on this subdivision. Is this approximation less than, equal to, or greater than $\ln n$? Give a reason for your answer.

7. Prove that
$$\lim_{x\to\infty} \frac{\ln x}{x} = 0,$$
starting with the definition of $\ln x$ as an integral.

8. The *Encyclopedia Britannica* article on logarithms begins with the statement: "By shortening processes of computation, logarithms have doubled the working speed of astronomers and engineers." What property or properties of logarithms do you think the author of the article had in mind when he made that statement?

9. What is the meaning of the differential equation
$$\frac{dy}{dt} = ky?$$
What is the solution of this equation that satisfies the initial condition $y = y_0$ when $t = 0$?

10. Give the definitions that lead from one function to the next in the following sequence:
$$\ln x,\ e^x,\ a^x,\ \log_a x.$$

11. Prove
$$\lim_{h\to 0} \frac{\ln(1+h)}{h} = 1,$$
either from the definition of the logarithm as an integral, or by considering the definition and value of $f'(1)$ for $f(x) = \ln x$.

12. Use the result of Exercise 11 and the continuity of the logarithmic and exponential functions to prove that
$$\lim_{h\to 0} (1+h)^{1/h} = e.$$
(This limit is sometimes used as the definition of e.)

MISCELLANEOUS EXERCISES

In Exercises 1 through 25,
(a) find dy/dx, and (b) sketch the curve.

1. $y = \sin 2x$
2. $y = 4\cos(2x + \pi/4)$
3. $y = 2\sin x + \sin 2x$
4. $y = \dfrac{\sin x}{1 + \cos x}$
5. $x = \cos y$
6. $x = \tan y$
7. $y = 4\sin\left(\dfrac{x}{2} + \pi\right)$
8. $y = 1 - \sin x$
9. $y = x - \sin x$
10. $y = \dfrac{\sin x}{x}$ (What happens as $x \to 0$? As $x \to \infty$?)
11. $y = \dfrac{\sin x}{x^2}$ (What happens as $x \to 0$? As $x \to \infty$?)
12. $y = x \sin \dfrac{1}{x}$ (What happens as $x \to \infty$? As $x \to 0$?)
13. $y = x^2 \sin \dfrac{1}{x}$ (What happens as $x \to \infty$? As $x \to 0$?)
14. $x = \tan \dfrac{\pi y}{2}$
15. (a) $y = \frac{1}{2}(e^x + e^{-x})$ (b) $y = \frac{1}{2}(e^x - e^{-x})$
 (c) $y = \dfrac{e^x - e^{-x}}{e^x + e^{-x}}$
16. $y = xe^{-x}$
17. (a) $y = x \ln x$ (b) $y = \sqrt{x}\ln x$
18. (a) $y = \dfrac{\ln x}{x}$ (b) $y = \dfrac{\ln x}{\sqrt{x}}$
 (c) $y = \dfrac{\ln x}{x^2}$
19. $y = e^{-x}\sin 2x$
20. (a) $y = 2x - \frac{1}{2}e^{2x}$ (b) $y = \exp(2x - \frac{1}{2}e^{2x})$
21. (a) $y = x - e^x$ (b) $y = \exp(x - e^x)$
22. $y = \ln(x + \sqrt{x^2 + 1})$
23. $y = \dfrac{1}{2}\ln\dfrac{1+x}{1-x}$
24. $y = \ln(x^2 + 4)$
25. $y = x\tan^{-1}\dfrac{x}{2}$

Find dy/dx in Exercises 26 through 42.

26. $y = \dfrac{e^{2x} - e^{-2x}}{e^{2x} + e^{-2x}}$

27. $y = \ln \dfrac{\sec x + \tan x}{\sec x - \tan x}$

28. $y = x^2 e^{2x} \sin 3x$

29. $y = \sin^{-1}(x^2) - xe^{x^2}$

30. $y = \ln\left(\dfrac{x^4}{1+x^3}\right) + 7^{x^{2/3}}$

31. $y = (x^2+2)^{2-x}$

32. $y = \dfrac{\ln x}{e^x}$

33. $y = \ln \dfrac{x}{\sqrt{x^2+1}}$

34. $y = \ln(4 - 3x)$

35. $y = \ln(3x^2 + 4x)$

36. $y = x(\ln x)^3$

37. $y = x \ln(x^3)$

38. $y = x^3 \ln x$

39. $y = \ln e^x$

40. $y = x^2 e^x$

41. $x = \ln y$

42. $y = \ln(\ln x)$

43. Find dy/dx by logarithmic differentiation for

(a) $y = \dfrac{x}{x^2+1}$ $(x > 0)$,

(b) $y = \sqrt[3]{\dfrac{x(x-2)}{x^2+1}}$ $(x > 2)$.

44. If $f(x) = x + e^{4x}$, find $f(0)$ and $f'(0)$, and find an approximation for $f(0.01)$.

45. Find dy/dx for each of the following.

(a) $y = a^{x^2-x}$

(b) $y = \ln \dfrac{e^x}{1+e^x}$

(c) $y = x^x$

(d) $y = x^{1/x}$

46. Sketch the graphs of $y = \ln(1-x)$ and $y = \ln(1/x)$.

47. Sketch (in a common figure) the graphs of $y = e^{-x}$ and $y = -e^{-x}$.

48. Solve for x: $\tan^{-1} x - \cot^{-1} x = \pi/4$.

49. If $\dfrac{dy}{dx} = \dfrac{e^x - e^{-x}}{e^x + e^{-x}}$, find $\dfrac{d^2y}{dx^2}$ and y.

50. If $dy/dx = 2/e^y$ and $y = 0$ when $x = 5$, find y as a function of x.

51. $P(x_1, y_1)$ and $Q(x_2, y_2)$ are any two points (in the first quadrant) lying on the hyperbola $xy = R$ (R positive). Show that the area bounded by the arc PQ, the lines $x = x_1$, $x = x_2$, and the x-axis is equal to the area bounded by the arc PQ, the lines $y = y_1$, $y = y_2$, and the y-axis.

52. By the trapezoidal rule, find $\int_0^2 e^{-x^2}\,dx$, using $n = 4$.

53. In a condenser discharging electricity the rate of change of the voltage in volts per second is proportional to the voltage, being numerically equal to minus one-fortieth of the voltage. Express the voltage as a function of the time. In how many seconds will the voltage decrease to 10 percent of its original value?

54. (a) Sketch the graph $2y = (e^x + e^{-x})$. Observe that whenever $P_1(x_1, y_1)$ is on the graph, so is the point $P_1'(-x_1, y_1)$. This implies a certain symmetry property of the graph. What is it?

(b) Find the length of the curve from its lowest point to an arbitrary point $P_1(x_1, y_1)$ on it.

55. The area bounded by the x-axis and one arch of the curve $y = \sin x$ is revolved around the x-axis. Find the volume of the solid of revolution thus generated. Compare its volume with that of two inscribed cones.

56. A curve $y = f(x)$ goes through the points $(0, 0)$ and (x_1, y_1). It divides the rectangle $0 \le x \le x_1$, $0 \le y \le y_1$ into two regions: A above the curve and B below. Find the curve if the area of A is twice the area of B for all choices of $x_1 > 0$ and $y_1 > 0$.

57. Find a curve passing through the origin and such that the length s of the curve between the origin and any point (x, y) of the curve is given by

$$s = e^x + y - 1.$$

58. A particle starts at the origin and moves along the x-axis in such a way that its velocity at the point $(x, 0)$ is given by the formula $dx/dt = \cos^2 \pi x$. How long will it take to cover the distance from the origin to the point $x = \frac{1}{4}$? Will it ever reach the point $x = \frac{1}{2}$? Why?

59. A particle moves in a straight line with acceleration $a = 4/(4-t)^2$. If when $t = 0$ the velocity is equal to 2, find how far the particle moves between $t = 1$ and $t = 2$.

60. The velocity of a certain particle moving along the x-axis is proportional to x. At time $t = 0$ the particle is located at $x = 2$ and at time $t = 10$ it is at $x = 4$. Find its position at $t = 5$.

61. Solve the differential equation

$$dy/dx = y^2 e^{-x}$$

if $y = 2$ when $x = 0$.

62. It is estimated that the population of a certain country is now increasing at a rate of 2% per year. Assuming that this (instantaneous) rate will continue indefinitely, estimate what the population N will be t years from now, if the population now is N_0. How many years will it take for the population to double?

63. If
$$y = \frac{(e^{2x} - 1)}{(e^{2x} + 1)},$$
show that
$$dy/dx = 1 - y^2.$$

64. Determine the inflection points of the curve
$$y = e^{-(x/a)^2},$$
where a is a positive constant. Sketch the curve. (This is closely related to the normal curve used in statistics.)

65. Find the volume generated when the area bounded by $y = e^{-x}$, $x = 0$, $x = \ln 2$, $y = 0$ is rotated about the x-axis.

66. Find the volume generated when the area bounded by $x = \sec y$, $x = 0$, $y = 0$, $y = \pi/3$ is rotated about the y-axis.

67. (a) Find dy/dx if $y = \ln(\sec x + \tan x)$.
(b) Find the length of arc of the curve $y = \ln \sec x$ from $x = 0$ to $\pi/4$, using, at some stage, the result of part (a).

68. Find the length of the curve
$$\frac{x}{a} = \left(\frac{y}{b}\right)^2 - \frac{1}{8}\frac{b^2}{a^2}\ln\frac{y}{b}$$
from $y = b$ to $y = 3b$, assuming a and b to be positive constants.

69. Find the volume generated by rotating about the x-axis the area bounded by $y = e^x$, $y = 0$, $x = 0$, $x = 2$.

70. Find the area bounded by the curve
$$y = (a/2)(e^{x/a} + e^{-x/a}),$$
the x-axis, and the lines $x = -a$ and $x = +a$.

71. The portion of a tangent to a curve included between the x-axis and the point of tangency is bisected by the y-axis. If the curve passes through (1, 2), find its equation.

72. Use the trapezoidal rule with $n = 5$ to approximate the value of
(a) $\ln 0.5$, (b) $\ln 3$.

73. (a) If $y = x \ln x - x$, find y'.
(b) Use the results of 73(a) and 72(b) to evaluate $\int_1^3 \ln x\, dx$ approximately.

Evaluate the integrals in Exercises 74 through 76.

74. (a) $\displaystyle\int \frac{dx}{4 - 3x}$ (b) $\displaystyle\int \frac{5\, dx}{x - 3}$

75. (a) $\displaystyle\int_0^2 \frac{x\, dx}{x^2 + 2}$ (b) $\displaystyle\int_0^2 \frac{x\, dx}{(x^2 + 2)^2}$

76. (a) $\displaystyle\int \frac{x + 1}{x}\, dx$ (b) $\displaystyle\int \frac{x}{2x + 1}\, dx$

77. Given $\ln 2 = 0.6932$ and $\ln 5 = 1.6094$. Use the properties of logarithms to find numerical values for $\ln 0.1$, $\ln 0.25$, $\ln 10$, and $\ln 20$.

78. Find the volume generated by rotating about the x-axis the area in the first quadrant bounded by the curve $xy^2 = 1$, the x-axis, the line $x = 1$, and the line $x = 4$.

79. Find the volume generated if the area in Exercise 78 is rotated about the y-axis.

80. Evaluate

(a) $\displaystyle\int_0^1 \frac{x^2\, dx}{2 - x^3}$, (b) $\displaystyle\int_0^3 x(e^{x^2-1})\, dx$,

(c) $\displaystyle\int_1^3 \frac{dx}{x}$, (d) $\displaystyle\int_0^5 \frac{x\, dx}{x^2 + 1}$,

(e) $\displaystyle\int_0^1 (e^x + 1)\, dx$.

81. In the inversion of raw sugar the time rate of change of the amount of raw sugar present varies as the amount of raw sugar remaining. If after 10 hr 1000 lb of raw sugar has been reduced to 800 lb, how much raw sugar will remain after 24 hr?

82. A cylindrical tank of radius 10 ft and height 20 ft, with its axis vertical, is full of water but has a leak at the bottom. Assuming that water escapes at a rate proportional to the depth of water in the tank and that 10% escapes during the first hour, find a formula for the volume of water left in the tank after t hr.

83. Let p be a positive integer ≥ 2. Show that

$$\lim_{n\to\infty}\left(\frac{1}{n+1}+\frac{1}{n+2}+\cdots+\frac{1}{p\cdot n}\right) = \ln p.$$

84. (a) If $(\ln x)/x = (\ln 2)/2$, does it necessarily follow that $x = 2$?
(b) If $(\ln x)/x = -2 \ln 2$, does it necessarily follow that $x = \frac{1}{2}$? Give reasons for your answers.

85. Given that $\lim_{x\to\infty} (\ln x)/x = 0$ (see Article 7.7, Exercise 4). Prove that
(a) $\lim_{x\to\infty} (\ln x)/x^h = 0$ if h is any positive constant,
(b) $\lim_{x\to+\infty} x^n/e^x = 0$ if n is any constant.

86. Show that $\lim_{h\to 0} (e^h - 1)/h = 1$, by considering the definition of the derivative of e^x at $x = 0$.

87. Prove that if x is any positive number,

$$\lim_{n\to\infty} n(\sqrt[n]{x} - 1) = \ln x.$$

Hint. Take $x = e^{nh}$ and apply Exercise 86.

Remark. This result provides a method for finding $\ln x$ (to any desired finite number of decimal places) using nothing fancier than repeated use of the operation of extracting square roots. For we may take $n = 2^k$, and then $\sqrt[n]{x}$ is obtained from x by taking k successive square roots.

88. Show that

$$(x^2/4) < x - \ln(1+x) < (x^2/2) \text{ if } 0 < x < 1.$$

Hint. Let

$$f(x) = x - \ln(1+x)$$

and show that $(x/2) < f'(x) < x$.

89. Prove that the area under the graph of $y = 1/x$ over the interval $a \leq x \leq b$ $(a > 0)$ is the same as the area over the interval $ka \leq x \leq kb$, for any $k > 0$.

90. Find the limit, as $n \to \infty$, of

$$\frac{e^{1/n} + e^{2/n} + \cdots + e^{(n-1)/n} + e^{n/n}}{n}.$$

91. If interest on money is compounded n times per year at a rate of r percent per year, the amount (principal plus interest) at the end of t years is

$$A = A_0\left(1+\frac{r}{100n}\right)^{nt},$$

where A_0 is the initial amount. For fixed r and t, let $h = r/(100n)$, and use the result of Review Exercise 12 to show that

$$A \to A_0 e^{rt/100} \quad \text{as } n \to \infty.$$

(We interpret this to mean that if the interest is "continuously compounded" at a rate of r percent per year, so that $dA/dt = (r/100)A$, the amount at time t years is $A_0e^{rt/100}$.)

92. Use a geometric argument based on areas to prove that

$$\ln(ab) - \ln a = \ln b \quad \text{if } a > 0,\ b > 0.$$

Suggestion.

$$\ln b = \int_1^b \left(\frac{1}{t}\right) dt$$

and

$$\ln(ab) - \ln a = \int_a^{ab} \left(\frac{1}{t}\right) dt.$$

Approximate the first of these integrals by a Riemann sum

$$\sum_{k=1}^{n} \left(\frac{1}{t_k}\right)(t_k - t_{k-1}),$$

and show that each summand also represents the area of an approximating rectangle of altitude $1/(at_k)$ and base $[at_{k-1}, at_k]$ for an appropriate way of subdividing the interval $[a, ab]$ and setting up a Riemann sum for the second integral. Match each subdivision of $[1, b]$ with a corresponding subdivision of $[a, ab]$ and use the definition of each integral as a limit of Riemann sums.

HYPERBOLIC FUNCTIONS

CHAPTER 8

8.1 INTRODUCTION*

In this chapter we shall consider certain combinations of the exponentials e^x and e^{-x} which are called "hyperbolic functions." There are two reasons why we study these functions. One reason is that they are used in solving certain engineering problems. For example, the tension at any point in a cable suspended by its ends and hanging under its own weight, such as an electric transmission line, may be computed in terms of hyperbolic functions. We shall investigate the hanging cable in some detail in Article 8.6. A second reason for studying the hyperbolic functions is that they are useful in connection with differential equations.

8.2 DEFINITIONS AND IDENTITIES

The combinations $\frac{1}{2}(e^u + e^{-u})$ and $\frac{1}{2}(e^u - e^{-u})$ occur with sufficient frequency that it has been found convenient to give special names to them. It may not be clear at this particular time why the names about to be introduced are especially appropriate. But it will become more apparent as we proceed that these functions have many properties analogous, respectively, to $\cos u$ and $\sin u$. And just as $\cos u$ and $\sin u$ are easily identified with the point (x, y) on the unit circle $x^2 + y^2 = 1$, where by properly defining u we may take $x = \cos u$ and $y = \sin u$, so it is also possible to identify

$$\begin{aligned} \cosh u &= \tfrac{1}{2}(e^u + e^{-u}), \\ \sinh u &= \tfrac{1}{2}(e^u - e^{-u}), \end{aligned} \tag{1}$$

with the coordinates of the point (x, y) on the "unit hyperbola" $x^2 - y^2 = 1$.

Equations (1) are the *definitions* of the *hyperbolic cosine* of u ($\cosh u$, which is often pronounced to rhyme with "gosh you") and of the *hyperbolic sine* of u ($\sinh u$, which is pronounced as though it were spelled "cinch u").

Suppose we check the statement that the point (x, y) with $x = \cosh u$ and $y = \sinh u$ lies on the unit hyperbola. We simply substitute the defining

* This chapter can be omitted without loss of continuity.

relations (1) into the equation of the hyperbola, and see whether or not these coordinates do satisfy the equation:

$$x^2 - y^2 = 1,$$

$$\cosh^2 u - \sinh^2 u \stackrel{?}{=} 1,$$

$$\tfrac{1}{4}(e^{2u} + 2 + e^{-2u}) - \tfrac{1}{4}(e^{2u} - 2 + e^{-2u}) \stackrel{?}{=} 1,$$

$$\tfrac{1}{4}(e^{2u} + 2 + e^{-2u} - e^{2u} + 2 - e^{-2u}) \stackrel{?}{=} 1,$$

$$\tfrac{1}{4}(4) \stackrel{?}{=} 1. \quad \text{(Yes!)}$$

Actually, if we let

$$\begin{aligned} x &= \cosh u = \tfrac{1}{2}(e^u + e^{-u}), \\ y &= \sinh u = \tfrac{1}{2}(e^u - e^{-u}), \end{aligned} \tag{2}$$

then when u varies from $-\infty$ to $+\infty$, the point $P(x, y)$ describes the right-hand branch of the hyperbola $x^2 - y^2 = 1$. The sense in which the curve is described is indicated by the arrows in Fig. 8.1. Since e^u is always positive and $e^{-u} = 1/e^u$ is also positive, it follows that $x = \cosh u = \frac{1}{2}(e^u + e^{-u})$ is positive for all real values of u, $-\infty < u < +\infty$. Hence the point (x, y) remains always to the right of the y-axis.

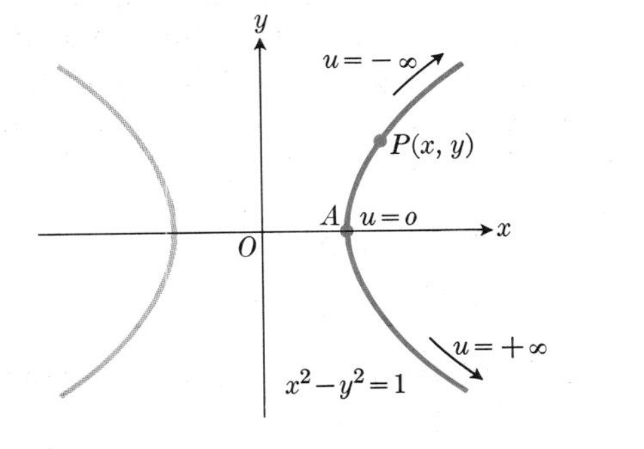

8.1

The bit of hyperbolic trigonometry that we have just established is the basic identity

$$\cosh^2 u - \sinh^2 u = 1. \tag{3}$$

This is analogous to, but not the same as, the ordinary trigonometric identity, $\cos^2 u + \sin^2 u = 1$. We shall now investigate additional points of analogy between these two types of functions.

The remaining hyperbolic functions are *defined* in terms of sinh u and cosh u as follows:

$$\begin{aligned} \tanh u &= \frac{\sinh u}{\cosh u} = \frac{e^u - e^{-u}}{e^u + e^{-u}}, \\ \coth u &= \frac{\cosh u}{\sinh u} = \frac{e^u + e^{-u}}{e^u - e^{-u}}, \\ \operatorname{sech} u &= \frac{1}{\cosh u} = \frac{2}{e^u + e^{-u}}, \\ \operatorname{csch} u &= \frac{1}{\sinh u} = \frac{2}{e^u - e^{-u}}. \end{aligned} \tag{4}$$

If we divide the identity (3) by $\cosh^2 u$, we get

$$1 - \tanh^2 u = \operatorname{sech}^2 u, \tag{5a}$$

and if we divide it by $\sinh^2 u$, we get

$$\coth^2 u - 1 = \operatorname{csch}^2 u. \tag{5b}$$

Since from (1) we find that

$$\cosh u + \sinh u = e^u, \tag{6a}$$

$$\cosh u - \sinh u = e^{-u}, \tag{6b}$$

it is apparent that *any* combination of the exponentials e^u and e^{-u} can be replaced by a combination of sinh u and cosh u, and conversely. Also, since e^{-u} is positive, (6b) shows that cosh u is always greater than sinh u. But for large values of u, e^{-u} is small, and $\cosh u \approx \sinh u$.

The graphs of the hyperbolic functions are shown in Fig. 8.2 (a–d).

At $x = 0$, $\cosh x = 1$ and $\sinh x = 0$, so the hyperbolic functions all have the same values at 0 that the corresponding trigonometric functions have. The hyperbolic cosine is an *even function*, that is,

$$\cosh(-x) = \cosh x, \tag{7}$$

and the hyperbolic sine is an *odd function*, that is,

$$\sinh(-x) = -\sinh x; \tag{8}$$

so one curve is symmetric about the y-axis and the other is symmetric with respect to the origin. Here again the hyperbolic functions behave like the ordinary trigonometric (or circular) functions.

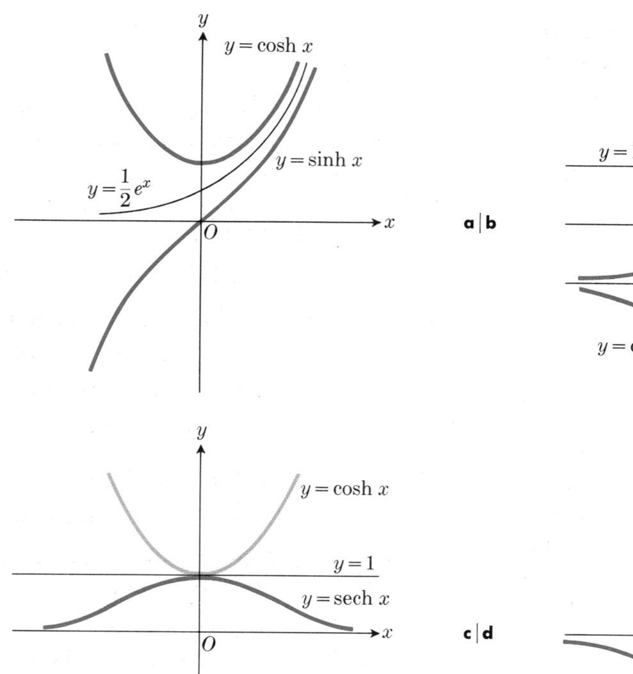

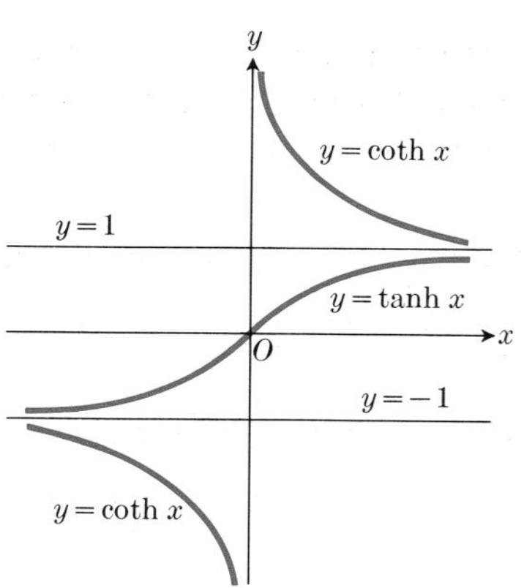

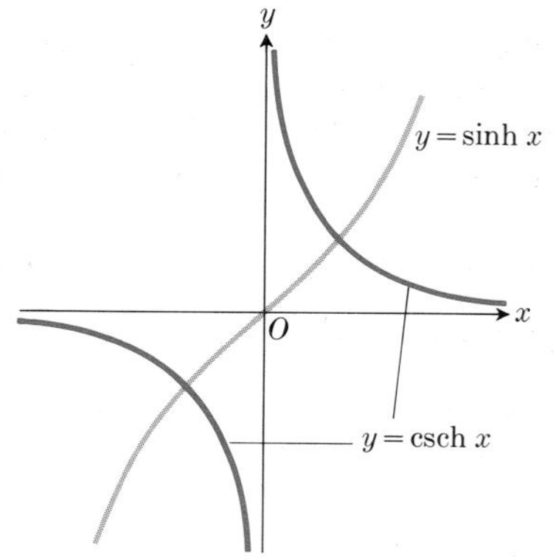

a|b c|d

8.2

The hyperbolic functions are so important and useful that their numerical values have been calculated and tabulated just as have the circular functions, at least for $\sinh x$, $\cosh x$, and $\tanh x$. The values of the other functions can be expressed in terms of these, so it is unnecessary to tabulate them too. Some slide rules also have $\sinh x$ and $\cosh x$ scales, and the values of these functions may be read directly in the same way that e^x may be read from the LL-scales.

Certain major differences between the hyperbolic and the circular functions should be noted. For example, the circular functions are periodic:

$$\sin(x + 2\pi) = \sin x, \quad \tan(x + \pi) = \tan x, \quad \text{etc.}$$

But the *hyperbolic functions are not periodic*. Again, they differ greatly in the range of values they assume:

$\sin x$ varies between -1 and $+1$, oscillates;
$\sinh x$ varies from $-\infty$ to $+\infty$, steadily increases;

$\cos x$ varies between -1 and $+1$, oscillates;
$\cosh x$ varies from $+\infty$ to $+1$ to $+\infty$;

$|\sec x|$ is never less than unity;
$\operatorname{sech} x$ is never greater than unity, is always positive;

$\tan x$ varies between $-\infty$ and $+\infty$;
$\tanh x$ varies between -1 and $+1$.

Another difference is exhibited in the behavior of the functions as $x \to \pm\infty$. In the case of the circular functions, $\sin x$, $\cos x$, $\tan x$, and so on, we can say nothing very specific about their behavior for large values of x. On the other hand, the hyperbolic functions behave very much like $e^x/2$, $e^{-x}/2$, unity, or zero, as follows:

For x large and positive:

$$\cosh x \approx \sinh x \approx \tfrac{1}{2}e^x,$$
$$\tanh x \approx \coth x \approx 1,$$
$$\operatorname{sech} x \approx \operatorname{csch} x \approx 2e^{-x} \approx 0.$$

For x negative, $|x|$ large:

$$\cosh x \approx -\sinh x \approx \tfrac{1}{2}e^{-x},$$
$$\tanh x \approx \coth x \approx -1,$$
$$\operatorname{sech} x \approx -\operatorname{csch} x \approx 2e^{x} \approx 0. \tag{9}$$

Additional analogies will be apparent when we study the calculus of hyperbolic functions in the next article. We conclude this article with certain formulas which the reader may readily verify. The following identities require only some algebraic calculations combined with the definitions (1), namely,

$$\sinh(x+y) = \sinh x \cosh y + \cosh x \sinh y,$$
$$\cosh(x+y) = \cosh x \cosh y + \sinh x \sinh y. \tag{10}$$

These in turn give

$$\sinh 2x = 2 \sinh x \cosh x, \tag{11a}$$

$$\cosh 2x = \cosh^2 x + \sinh^2 x, \tag{11b}$$

when we take $y = x$. The second of these leads to certain useful "half-angle" formulas when we combine it with the basic identity

$$1 = \cosh^2 x - \sinh^2 x. \tag{3}$$

For if we add (11b) and (3), we have

$$\cosh 2x + 1 = 2\cosh^2 x, \tag{12a}$$

while if we subtract (3) from (11b), we get

$$\cosh 2x - 1 = 2\sinh^2 x. \tag{12b}$$

Thus practically all the circular trigonometric identities have hyperbolic analogies. The formulas (3) and (10) are easily verified in terms of the definitions (1). The others follow from these by straightforward algebraic manipulations.

EXERCISES

1. Show that $x = -\cosh u$, $y = \sinh u$ represents a point on the left-hand branch of the hyperbola $x^2 - y^2 = 1$.
2. Using the definitions of $\cosh u$ and $\sinh u$ given by Eq. (1), show that
$$\cosh(-u) = \cosh u \quad \text{and} \quad \sinh(-u) = -\sinh u.$$
3. Verify Eqs. (10) for $\sinh(x+y)$ and $\cosh(x+y)$.
4. Show that $(\cosh x + \sinh x)^n = \cosh nx + \sinh nx$.
5. Let L be the line tangent to the hyperbola
$$x^2 - y^2 = 1$$
at the point $P_1(x_1, y_1)$, where
$$x_1 = \cosh u, \qquad y_1 = \sinh u.$$
Show that L cuts the x-axis at the point $(\operatorname{sech} u, 0)$ and the y-axis at $(0, -\operatorname{csch} u)$.

In Exercises 6 through 11, one of the six hyperbolic functions of u is given; determine the remaining five.

6. $\sinh u = -\frac{3}{4}$
7. $\cosh u = \frac{17}{15}$
8. $\tanh u = -\frac{7}{25}$
9. $\coth u = \frac{13}{12}$
10. $\operatorname{sech} u = \frac{3}{5}$
11. $\operatorname{csch} u = \frac{5}{12}$
12. Show that the distance r from the origin O to the point $P(\cosh u, \sinh u)$ on the hyperbola $x^2 - y^2 = 1$ is $r = \sqrt{\cosh 2u}$.
13. Show that the line tangent to the hyperbola at its vertex A in Fig. 8.1 intersects the line OP in the point $(1, \tanh u)$. This gives a geometric representation of $\tanh u$.
14. If θ lies in the interval $-\pi/2 < \theta < \pi/2$ and $\sinh x = \tan\theta$, show that
$$\cosh x = \sec\theta, \quad \tanh x = \sin\theta, \quad \coth x = \csc\theta,$$
$$\operatorname{csch} x = \cot\theta, \quad \text{and} \quad \operatorname{sech} x = \cos\theta.$$

8.3 DERIVATIVES AND INTEGRALS

Let u be a differentiable function of x and differentiate

$$\sinh u = \tfrac{1}{2}(e^u - e^{-u}),$$
$$\cosh u = \tfrac{1}{2}(e^u + e^{-u}), \tag{1}$$

with respect to x. Applying the formulas

$$\frac{de^u}{dx} = e^u \frac{du}{dx},$$

$$\frac{de^{-u}}{dx} = e^{-u}\frac{d(-u)}{dx} = -e^{-u}\frac{du}{dx},$$

we get

$$\frac{d(\sinh u)}{dx} = \cosh u \frac{du}{dx}, \qquad \text{(XXI)}$$

$$\frac{d(\cosh u)}{dx} = \sinh u \frac{du}{dx}. \qquad \text{(XXII)}$$

Then, if we let

$$y = \tanh u = \frac{\sinh u}{\cosh u}$$

and differentiate as a fraction, we get

$$\frac{d(\tanh u)}{dx} = \frac{\cosh u \dfrac{d(\sinh u)}{dx} - \sinh u \dfrac{d(\cosh u)}{dx}}{\cosh^2 x}$$

$$= \frac{\cosh^2 u \dfrac{du}{dx} - \sinh^2 u \dfrac{du}{dx}}{\cosh^2 u}$$

$$= \frac{(\cosh^2 u - \sinh^2 u)\dfrac{du}{dx}}{\cosh^2 u}$$

$$= \frac{1}{\cosh^2 u}\frac{du}{dx}$$

$$= \operatorname{sech}^2 u \frac{du}{dx}.$$

In a similar manner, we may establish the rest of the formulas in the following list.

$$\frac{d(\tanh u)}{dx} = \operatorname{sech}^2 u \frac{du}{dx} \qquad \text{(XXIII)}$$

$$\frac{d(\coth u)}{dx} = -\operatorname{csch}^2 u \frac{du}{dx} \qquad \text{(XXIV)}$$

$$\frac{d(\operatorname{sech} u)}{dx} = -\operatorname{sech} u \tanh u \frac{du}{dx} \qquad \text{(XXV)}$$

$$\frac{d(\operatorname{csch} u)}{dx} = -\operatorname{csch} u \coth u \frac{du}{dx} \qquad \text{(XXVI)}$$

Note that aside from the pattern of algebraic signs, these formulas are the exact analogs of the formulas for the corresponding circular functions. Again, exactly half of them have minus signs, but we no longer attach the minus signs to the derivatives of the cofunctions, but rather note that the first three, $\sinh u$, $\cosh u$, and $\tanh u$, have positive derivatives and the last three have negative derivatives.

Each of these derivative formulas has a matching differential formula. These in turn may be integrated at once to produce the following integration formulas.

$$\begin{aligned}
&(1)\ \int \sinh u\, du = \cosh u + C\\
&(2)\ \int \cosh u\, du = \sinh u + C\\
&(3)\ \int \operatorname{sech}^2 u\, du = \tanh u + C\\
&(4)\ \int \operatorname{csch}^2 u\, du = -\coth u + C\\
&(5)\ \int \operatorname{sech} u \tanh u\, du = -\operatorname{sech} u + C\\
&(6)\ \int \operatorname{csch} u \coth u\, du = -\operatorname{csch} u + C
\end{aligned} \qquad (2)$$

Problem 1. Show that $y = a \cosh (x/a)$ satisfies the differential equation

$$\frac{d^2y}{dx^2} = \frac{w}{H}\sqrt{1 + \left(\frac{dy}{dx}\right)^2}, \qquad (3)$$

provided that $a = H/w$, where H and w are constants.

Solution. By differentiating $y = a \cosh (x/a)$, we find

$$\frac{dy}{dx} = a \sinh \frac{x}{a}\cdot\frac{1}{a} = \sinh\frac{x}{a}, \qquad \frac{d^2y}{dx^2} = \cosh\frac{x}{a}\cdot\frac{1}{a}.$$

We substitute these into (3) and obtain

$$\frac{1}{a}\cosh\frac{x}{a} \stackrel{?}{=} \frac{w}{H}\sqrt{1 + \sinh^2\frac{x}{a}} \qquad \text{(a)}$$

$$= \frac{w}{H}\sqrt{\cosh^2\frac{x}{a}} \qquad \text{(b)}$$

$$= \frac{w}{H}\cosh\frac{x}{a}, \qquad \text{(c)}$$

which is a true equation provided $a = H/w$.

Remark 1. In going from (a) to (b) above, we used the fundamental identity

$$\cosh^2 u - \sinh^2 u = 1$$

in the form

$$1 + \sinh^2 u = \cosh^2 u.$$

In going from (b) to (c), we used the fact that $\cosh u$ is always positive, and hence

$$\sqrt{\cosh^2 u} = |\cosh u| = \cosh u.$$

Remark 2. The differential equation (3) expresses the condition for equilibrium of forces acting on a section AP of a hanging cable (Fig. 8.3). We imagine the rest of the cable as having been removed and the section AP from the lowest point A to the representative point $P(x, y)$ as being in equilibrium under the forces

(1) H = horizontal tension pulling on the cable at A,

(2) T = tangential tension pulling on the cable at P,

(3) $W = ws$ = weight of s feet of the cable at w pounds per foot of length from A to P.

Then equilibrium of the cable requires that the horizontal and vertical components of T balance H and W respectively:

$$T \cos \phi = H, \qquad T \sin \phi = W = ws. \tag{4}$$

By division, we get

$$\frac{T \sin \phi}{T \cos \phi} = \tan \phi = \frac{W}{H}$$

or

$$\frac{dy}{dx} = \frac{ws}{H}, \tag{5}$$

since $\tan \phi = dy/dx$. The arc length s in Eq. (5) would be found by integrating

$$ds = \sqrt{1 + \left(\frac{dy}{dx}\right)^2}\, dx$$

from A to P. But instead of doing so, we may differentiate Eq. (5) with respect to x:

$$\frac{d^2y}{dx^2} = \frac{w}{H}\frac{ds}{dx},$$

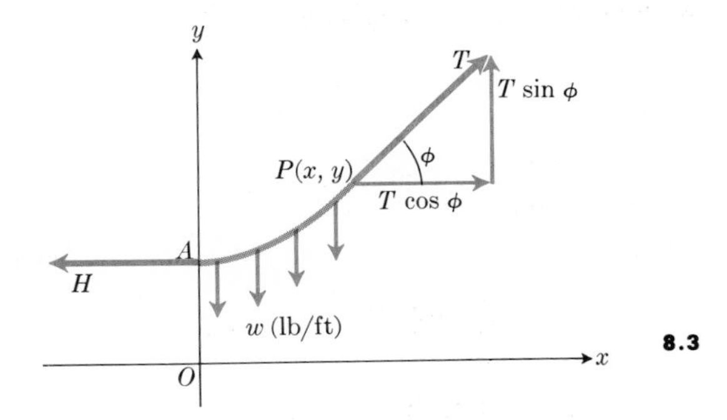

8.3

and then substitute $\sqrt{1 + (dy/dx)^2}$ in place of ds/dx to obtain Eq. (3) above.

Remark 3. In Problem 1, we pulled the equation $y = a \cosh (x/a)$ out of the hat, so to speak, and showed by substitution that it satisfies the differential equation (3). In Article 8.6, we shall adopt the more straightforward approach of simply solving Eq. (3) subject to the initial conditions

$$\frac{dy}{dx} = 0 \quad \text{and} \quad y = y_0$$

$$\text{when} \quad x = 0.$$

By choosing the origin so that

$$y_0 = a = \frac{H}{w},$$

we shall find that the solution is indeed the one given above, namely,

$$y = a \cosh \frac{x}{a} \quad \text{with } a = \frac{H}{w}. \tag{6}$$

Problem 2. Show that the tension in the cable at $P(x, y)$ in Fig. 8.3 is

$$T = wy.$$

Solution. We make use of the fact that

$$\frac{dy}{dx} = \tan \phi,$$

since T acts along the tangent, and

$$T = \frac{H}{\cos \phi} = H \sec \phi,$$

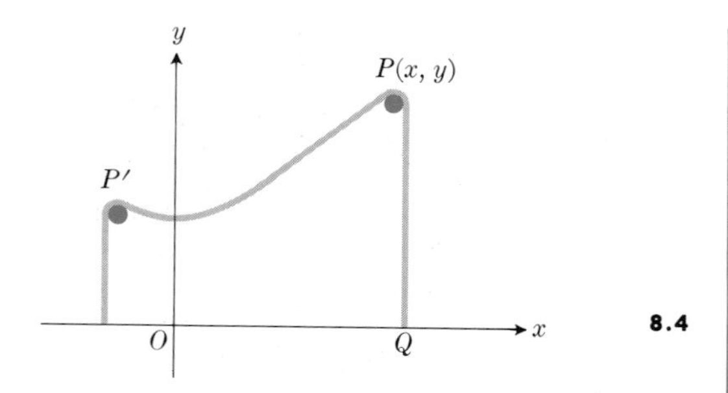

8.4

by Eq. (4a). Then, differentiating (6), we have

$$\tan \phi = \frac{dy}{dx} = \sinh \frac{x}{a}$$

and

$$\sec \phi = \sqrt{\sec^2 \phi} = \sqrt{1 + \tan^2 \phi} = \sqrt{1 + \left(\frac{dy}{dx}\right)^2}$$

$$= \sqrt{1 + \sinh^2 \frac{x}{a}} = \sqrt{\cosh^2 \frac{x}{a}} = \cosh \frac{x}{a}.$$

Therefore

$$T = H \sec \phi = H \cosh \frac{x}{a}, \tag{7a}$$

where

$$a = \frac{H}{w}, \quad \text{or} \quad H = wa. \tag{7b}$$

Combining (7a) and (7b), we have $T = wa \cosh (x/a)$, or, when we take account of Eq. (6),

$$T = wy.$$

This means that the tension at P is equal to the weight of y feet of the cable. Thus if the end of the cable to the right of P is allowed to hang down over a smooth peg while the cable is held at P so that it does not slip, and if the cable is then cut off at the point Q where it crosses the x-axis (Fig. 8.4), it may then be released at P and the weight wy of the section of cable PQ will be just sufficient to prevent the cable from slipping. If this is carried out at two different points P and P', the cable may be draped over two smooth pegs without slipping, provided the free ends reach just to the x-axis. The curve, $y = a \cosh (x/a)$, is called a *catenary* from the Latin word *catena*, meaning chain. The x-axis is called the directrix of the catenary.

EXERCISES

1. Establish the validity of the formulas XXIV through XXVI.

Find dy/dx in Exercises 2 through 9.

2. $y = \sinh 3x$
3. $y = \cosh^2 5x$
4. $y = \cosh^2 5x - \sinh^2 5x$
5. $y = \tanh 2x$
6. $y = \coth (\tan x)$
7. $y = \operatorname{sech}^3 x$
8. $y = 4 \operatorname{csch} (x/4)$
9. $\sinh y = \tan x$

Integrate each of the following.

10. $\int \cosh (2x + 1)\, dx$
11. $\int \tanh x\, dx$
12. $\int \frac{\sinh x}{\cosh^4 x}\, dx$
13. $\int \frac{4\, dx}{(e^x + e^{-x})^2}$
14. $\int \frac{e^x - e^{-x}}{e^x + e^{-x}}\, dx$
15. $\int \tanh^2 x\, dx$
16. $\int \frac{\sinh \sqrt{x}}{\sqrt{x}}\, dx$
17. $\int \cosh^2 3x\, dx$
18. $\int \sqrt{\cosh x - 1}\, dx$
19. Find the area of the hyperbolic sector AOP bounded by the arc AP and the lines OA, OP through the origin, in Fig. 8.1.
20. Show that the straight line $y = x/2 + 1$ and the catenary $y = \cosh x$ intersect at the two points (0, 1) and (0.930, 1.465). (Consult tables for $\cosh x$.)
21. Two successive poles supporting an electric power line are 100 ft apart, the supporting members being at the same level. If the wire dips 25 ft at the center,

    (a) find the length of the wire between supports, and

    (b) find the tension in the wire at its lowest point if its weight is $w = 0.3$ lb/ft.

    Hint. First find a from the equation

    $$25/a + 1 = \cosh 50/a,$$

    which can be related to Exercise 20 with $x = 50/a$.
22. The equation $\sinh x = \tan \theta$, $-\pi/2 < \theta < \pi/2$, defines θ as a function of x:

    $$\theta = \tan^{-1} (\sinh x),$$

to which the name "gudermannian of x" has been attached, written

$$\theta = \text{gd } x.$$

Show that

$$\frac{d\theta}{dx} = \frac{1}{\cosh x} = \frac{1}{\sec \theta},$$

so that $\sec \theta \, d\theta = dx$ and

$$\int \sec \theta \, d\theta = \text{gd}^{-1} \theta + C = \sinh^{-1} (\tan \theta) + C.$$

23. Sketch the curve $\theta = \text{gd } x$ (see Exercise 22) by the following procedure. First sketch the curves

$$y = \sinh x, \quad y = \tan \theta, \quad -\pi/2 < \theta < \pi/2,$$

on separate xy- and θy-planes. Starting in the xy-plane with any value of x, $-\infty < x < +\infty$, determine the corresponding value of y from the curve $y = \sinh x$. Transfer this y reading to the y-axis in the θy-plane and determine the corresponding value of θ from the curve $y = \tan \theta$. Use this value of θ as ordinate and the original value of x as abscissa to plot a point on the curve $\theta = \text{gd } x$.

24. Show that the curve $\theta = \text{gd } x$ (see Exercises 22 and 23) has the lines $\theta = \pm\pi/2$ as horizontal asymptotes, is always rising (from left to right), has a point of inflection at the origin, and is symmetric with respect to the origin.

8.4 GEOMETRIC MEANING OF THE HYPERBOLIC RADIAN

We are now in a position to illustrate the meaning of the variable u in the equations

$$x = \cosh u, \qquad y = \sinh u \tag{1}$$

as they relate to the point $P(x, y)$ on the unit hyperbola

$$x^2 - y^2 = 1. \tag{2}$$

Before we do so, however, we shall find the analogous meaning of the variable θ in the equations

$$x = \cos \theta, \qquad y = \sin \theta \tag{3}$$

as they relate to the point $P(x, y)$ on the unit circle

$$x^2 + y^2 = 1. \tag{4}$$

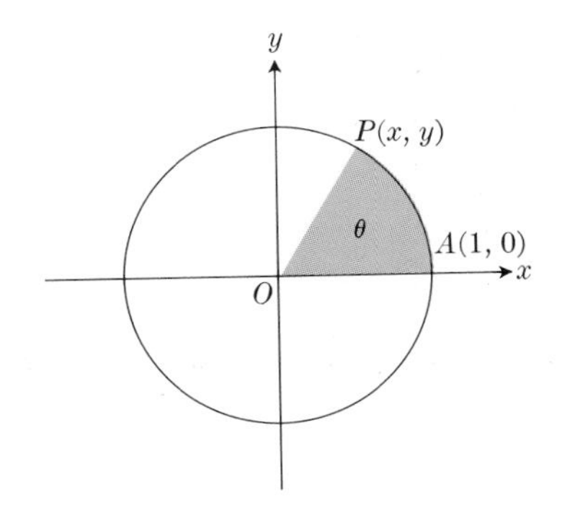

8.5

The most familiar interpretation of θ is, of course, that it is the radian measure of the angle AOP in Fig. 8.5, that is

$$\theta = \frac{\text{arc } AP}{\text{radius } OA}.$$

But we also recall that the area of a circular sector of radius r and central angle θ (radians) is given by $(\frac{1}{2})r^2\theta$. Since this is a unit circle,

$$\text{area of sector } AOP = \tfrac{1}{2}\theta$$

or, if we solve this for θ,

$$\theta = \text{twice the area of sector } AOP. \tag{5}$$

Of course we must realize that θ is a pure (dimensionless) number and that Eq. (5) really says that the *value* of θ which Eqs. (3) associate with the point $P(x, y)$ on the unit circle is twice the *number* of square units of area that the radius vector OP sweeps out as P moves along the circle from A to its final position P. Thus when the area of the sector AOP is one-half the area of a square having OA as a side, then $\theta = 1$ and the coordinates of P represent cos 1 and sin 1. Negative values of θ would be interpreted as corresponding to areas swept over in a clockwise rotation of OA.

Now for the unit hyperbola

$$x^2 - y^2 = 1,$$

we shall find an analogous interpretation for the variable u in the equations

$$x = \cosh u, \qquad y = \sinh u.$$

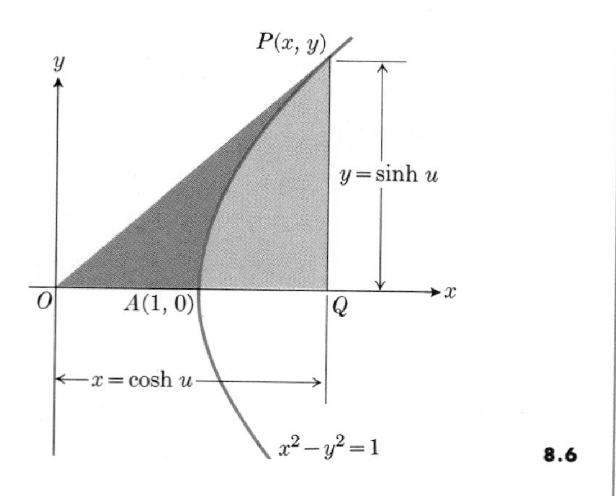

8.6

To see that this is indeed the case, we shall calculate the area of the sector AOP in Fig. 8.6. This area is clearly equal to the area of the triangle QOP minus the area QAP bounded above by the curve, below by the x-axis, and on the right by the vertical line QP. But this area is simply

$$\text{area } AQP = \int_A^P y\,dx = \int_A^P \sinh u\,d(\cosh u)$$
$$= \int_A^P \sinh^2 u\,du.$$

By Eq. (126), Article 8.2, we then have

$$\text{area } AQP = \tfrac{1}{2}\int_A^P (\cosh 2u - 1)\,du$$
$$= \tfrac{1}{2}[\tfrac{1}{2}\sinh 2u - u]_{A(u=0)}^{P(u=u)}$$
$$= \tfrac{1}{4}\sinh 2u - \tfrac{1}{2}u$$
$$= \tfrac{1}{2}\sinh u \cosh u - \tfrac{1}{2}u.$$

Hence the area of the sector AOP is equal to

$$\text{area of } OQP - \text{area of } AQP;$$

in terms of hyperbolic functions, this is

$$\tfrac{1}{2}\sinh u \cosh u - (\tfrac{1}{2}\sinh u \cosh u - \tfrac{1}{2}u) = \tfrac{1}{2}u.$$

Solving for u, we have

$$u = \text{twice the area of the sector } AOP. \tag{6}$$

As for the circle, a positive value of u is associated with an area above the x-axis and a negative value with an area below the x-axis, and areas are to be measured in terms of the unit square having OA as side. The term *hyperbolic radian* is sometimes used in connection with the variable u in Eq. (6), but here again, u is just a dimensionless real number. For example, cosh 2 and sinh 2 may be interpreted as the coordinates of P when the area of the sector AOP is just equal to the area of a square having OA as side.

8.5 THE INVERSE HYPERBOLIC FUNCTIONS

If we start with

$$x = \sinh y, \tag{1}$$

then as y varies continuously from $-\infty$ to $+\infty$, x does likewise. Graphically, this means that we may start with any real value on the y-axis in Fig. 8.7 and draw a horizontal line to the curve. Then a vertical line to the x-axis locates exactly one value of x such that the point (x, y) is on the curve. On the other hand, we could equally well reverse these steps and start with any real value x on the x-axis, go along a vertical line to the curve, then on a horizontal line to the y-axis. This latter procedure gives us y as a function of x, and the notation we use is

$$y = \sinh^{-1} x. \tag{2}$$

Here we have no problem about principal values, as we did in the case of the inverses of the circular functions, since the correspondence between the real numbers x and the real numbers y in Eq. (1) is one-to-one. Equation (1) says exactly the same thing as Eq. (2):

$$y = \sinh^{-1} x \quad \text{means} \quad x = \sinh y.$$

The inverse hyperbolic cosine, however, is double-valued, and it is desirable to prescribe a principal branch. For if we start with the equation

$$x = \cosh y, \tag{3}$$

then y and $-y$ both give the same value of x; that is, the correspondence between y-values and x-values is two-to-one, so that if x is considered the independent variable, there are *two* corresponding values of y. This is analogous to the situation surrounding

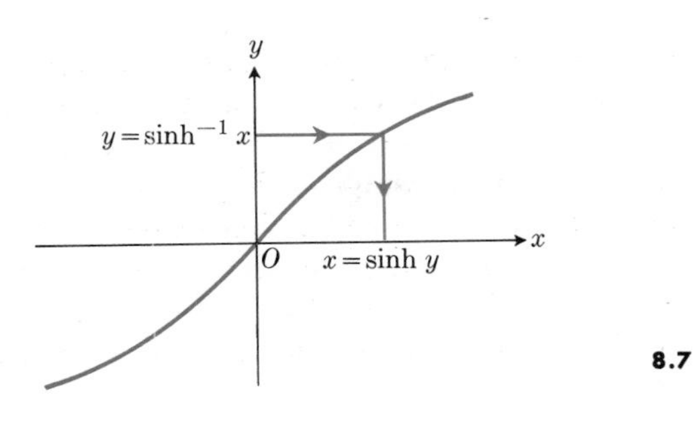

8.7

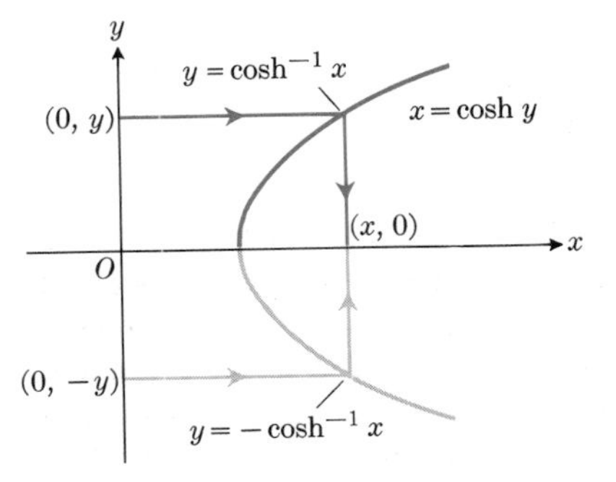

8.8

the equation $x = y^2$, which defines x as a single-valued function of y; but when x is given and we ask for y, the result is $y = \pm\sqrt{x}$. In the case of the inverse hyperbolic cosine, we take the positive values of y as the principal branch:

$$y = \cosh^{-1} x \quad \text{means} \quad x = \cosh y,$$

where

$$y \geq 0 \quad \text{and} \quad x \geq 1. \tag{4}$$

Thus in Fig. 8.8, the equation $x = \cosh y$ represents the entire curve, but $y = \cosh^{-1} x$ represents only that portion above the x-axis, while the portion below the x-axis is given by $y = -\cosh^{-1} x$.

Reference to Fig. 8.2 shows that the only other double-valued inverse is the inverse hyperbolic secant, and again we select the positive branch. That is,

$$y = \operatorname{sech}^{-1} x, \quad y > 0, \quad 0 < x \leq 1 \tag{5}$$

defines the principal branch and the numbers x and y satisfy

$$x = \operatorname{sech} y. \tag{6}$$

Since

$$\operatorname{sech} y = \frac{1}{\cosh y},$$

Eq. (6) is equivalent to

$$\cosh y = \frac{1}{x}$$

and the restriction $y > 0$ defines the same branch as in (5), so that

$$y = \cosh^{-1} \frac{1}{x}.$$

That is,

$$\operatorname{sech}^{-1} x = \cosh^{-1} \frac{1}{x}.$$

Similarly, we see that $y = \operatorname{csch}^{-1} x = \sinh^{-1}(1/x)$ means

$$\operatorname{csch} y = x. \tag{7}$$

Also,

$$y = \tanh^{-1} x \quad \text{means} \quad x = \tanh y, \tag{8a}$$

$$y = \coth^{-1} x \quad \text{means} \quad x = \coth y. \tag{8b}$$

The graphs of the various inverse hyperbolic functions are given in Figs. 8.7, 8.8, and 8.9. Numerical values of $\sinh^{-1} x$, $\cosh^{-1} x$ and $\tanh^{-1} x$ may be read from the tables of $\sinh x$, $\cosh x$, and $\tanh x$ respectively in the same way that antilogarithms are read from tables of logarithms.

Example 1. Suppose it is required to find $\tanh^{-1} 0.25$. If we let

$$x = \tanh^{-1} 0.25,$$

then

$$\tanh x = 0.25.$$

We consult a table of hyperbolic tangents and we find the following entries.

x	$\tanh x$	$\tanh x$ (rounded off)
0.25	0.24492	0.24
0.26	0.25430	0.25

Since we want the entry which corresponds to

$$\tanh x = 0.25,$$

we may simply round off the tabular entries to two significant figures and read $x = 0.26$. Or else we may

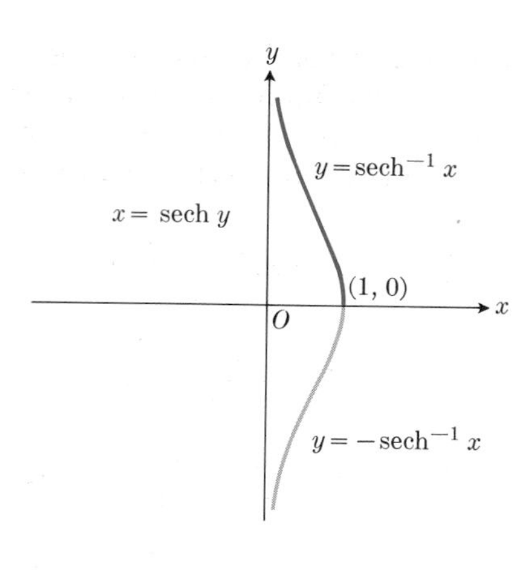

a|b

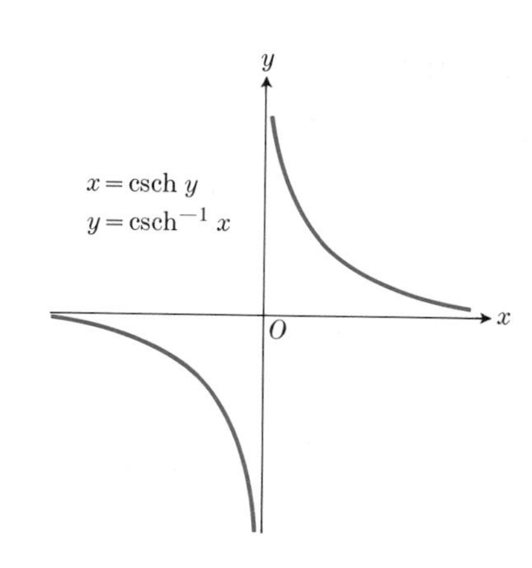

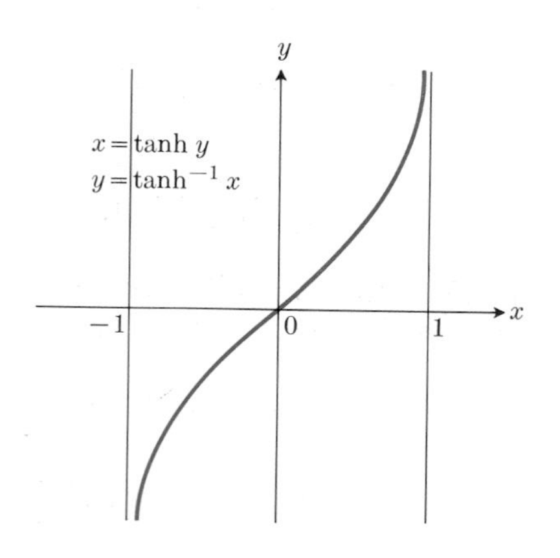

c|d

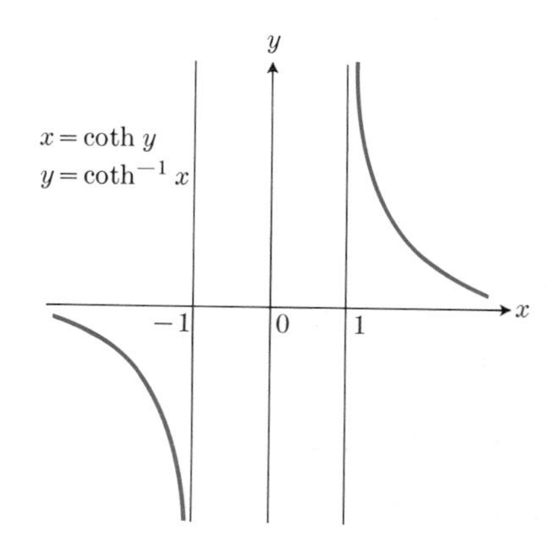

8.9

take $\tanh x = 0.25000$ and interpolate between the readings listed above. Using ordinary proportional parts, we have

$$x_1 = 0.25, \qquad \tanh x_1 = 0.24492,$$

$$x = (?), \qquad \tanh x = 0.25000,$$

$$x_2 = 0.26, \qquad \tanh x_2 = 0.25430;$$

$$\frac{x - x_1}{x_2 - x_1} = \frac{\tanh x - \tanh x_1}{\tanh x_2 - \tanh x_1},$$

$$\frac{x - x_1}{0.01} = \frac{508}{938} = 0.54,$$

$$x = 0.2554 = \tanh^{-1} 0.25000.$$

An alternative method for evaluating the inverse hyperbolic functions is to express each of them in terms of logarithms, as illustrated below for $\tanh^{-1} x$. Let

$$y = \tanh^{-1} x;$$

then

$$\tanh y = x,$$

or

$$x = \frac{\sinh y}{\cosh y} = \frac{\frac{1}{2}(e^y - e^{-y})}{\frac{1}{2}(e^y + e^{-y})}$$

$$= \frac{e^y - \dfrac{1}{e^y}}{e^y + \dfrac{1}{e^y}} = \frac{e^{2y} - 1}{e^{2y} + 1}.$$

We now solve this equation for e^{2y}:

$$xe^{2y} + x = e^{2y} - 1,$$

or

$$1 + x = e^{2y}(1 - x),$$

and

$$e^{2y} = \frac{1 + x}{1 - x}.$$

Hence

$$y = \tanh^{-1} x = \frac{1}{2} \ln \frac{1 + x}{1 - x}, \quad |x| < 1. \tag{9}$$

The variable x in Eq. (9) is restricted to the domain $|x| < 1$, since $x = \tanh y$ lies in this interval for all real values of y, $-\infty < y < +\infty$.

Example 2

$$\begin{aligned} \tanh^{-1} 0.25 &= \frac{1}{2} \ln \frac{1.25}{0.75} = \frac{1}{2} \ln \frac{5}{3} \\ &= \tfrac{1}{2}(\ln 5 = \ln 3) = 0.25542. \end{aligned}$$

The expressions for the other inverse hyperbolic functions in terms of logarithms are found in a similar manner. They are:

$$\begin{aligned} \sinh^{-1} x &= \ln (x + \sqrt{x^2 + 1}), \quad -\infty < x < \infty, \\ \cosh^{-1} x &= \ln (x + \sqrt{x^2 - 1}), \quad x \geq 1, \\ \operatorname{sech}^{-1} x &= \ln \left(\frac{1 + \sqrt{1 - x^2}}{x} \right) \\ &= \cosh^{-1} \left(\frac{1}{x} \right), \quad 0 < x \leq 1, \\ \operatorname{csch}^{-1} x &= \ln \left(\frac{1}{x} + \frac{\sqrt{1 + x^2}}{|x|} \right) \\ &= \sinh^{-1} \left(\frac{1}{x} \right), \quad x \neq 0, \\ \coth^{-1} x &= \frac{1}{2} \ln \frac{x + 1}{x - 1} = \tanh^{-1} \left(\frac{1}{x} \right), \quad |x| > 1. \end{aligned} \tag{10}$$

It should be evident that the logarithmic expressions are, on the whole, rather cumbersome, and that the inverse hyperbolic functions do provide a useful shorthand wherever these expressions arise.

The chief merit of the inverse hyperbolic functions lies in their usefulness in integration. This will easily be understood after we have derived the following formulas for their derivatives.

$$\frac{d(\sinh^{-1} u)}{dx} = \frac{1}{\sqrt{1 + u^2}} \frac{du}{dx}, \tag{XXVII}$$

$$\frac{d(\cosh^{-1} u)}{dx} = \frac{1}{\sqrt{u^2 - 1}} \frac{du}{dx}, \tag{XXVIII}$$

$$\frac{d(\tanh^{-1} u)}{dx} = \frac{1}{1 - u^2} \frac{du}{dx}, \quad |u| < 1, \tag{XXIX}$$

$$\frac{d(\coth^{-1} u)}{dx} = \frac{1}{1 - u^2} \frac{du}{dx}, \quad |u| > 1, \tag{XXX}$$

$$\frac{d(\operatorname{sech}^{-1} u)}{dx} = \frac{-1}{u\sqrt{1 - u^2}} \frac{du}{dx}, \tag{XXXI}$$

$$\frac{d(\operatorname{csch}^{-1} u)}{dx} = \frac{-1}{|u|\sqrt{1 + u^2}} \frac{du}{dx}. \tag{XXXII}$$

The proofs of these all follow the same method. We illustrate for the case of $\cosh^{-1} u$.

Let

$$y = \cosh^{-1} u;$$

then

$$\cosh y = u, \qquad \sinh y \frac{dy}{dx} = \frac{du}{dx},$$

and

$$\frac{dy}{dx} = \frac{1}{\sinh y} \frac{du}{dx}.$$

But

$$\cosh^2 y - \sinh^2 y = 1,$$

$$\cosh y = u,$$

so that

$$\begin{aligned} \sinh y &= \pm\sqrt{\cosh^2 y - 1} \\ &= \pm\sqrt{u^2 - 1} \end{aligned}$$

and

$$\frac{dy}{dx} = \frac{1}{\pm\sqrt{u^2 - 1}} \frac{du}{dx}.$$

The ambiguous sign will be + if we restrict attention to the principal value, $y = \cosh^{-1} y$, $y \geq 0$, for then $\sinh y \geq 0$ and the ambiguous sign is the same as the sign of $\sinh y$. Thus XXVIII is established. The identities (5a) and (5b), Article 8.2, with y in place of u, will be found to be useful in proving formulas XXIX to XXXII. The reader should have no difficulty in deriving these results.

Remark. The restrictions $|u| < 1$ and $|u| > 1$ in XXIX and XXX respectively are due to the fact that

if $\quad y = \tanh^{-1} u,$

then $\quad u = \tanh y,$

and since $\quad -1 < \tanh y < 1,$

this means $|u| < 1$. Similarly,

$$y = \coth^{-1} u, \qquad u = \coth y$$

requires $|u| > 1$. The distinction becomes important when we invert the formulas to get integration formulas, since otherwise we would be unable to tell whether we should write $\tanh^{-1} u$ or $\coth^{-1} u$ for

$$\int \frac{du}{1 - u^2}.$$

The following integration formulas follow at once from the differential formulas XXVII′ to XXXII′, that are obtained by multiplying both sides of XXVII to XXXII by dx:

(a) $\displaystyle \int \frac{du}{\sqrt{1 + u^2}} = \sinh^{-1} u + C,$

(b) $\displaystyle \int \frac{du}{\sqrt{u^2 - 1}} = \cosh^{-1} u + C,$

(c) $$\int \frac{du}{1 - u^2} = \begin{cases} \tanh^{-1} u + C & \text{if } |u| < 1 \\ \coth^{-1} u + C & \text{if } |u| > 1 \end{cases}$$

$$= \frac{1}{2} \ln \left| \frac{1 + u}{1 - u} \right| + C, \tag{11}$$

(d) $$\int \frac{du}{u\sqrt{1 - u^2}} = -\operatorname{sech}^{-1} |u| + C = -\cosh^{-1}\left(\frac{1}{|u|}\right) + C,$$

(e) $$\int \frac{du}{u\sqrt{1 + u^2}} = -\operatorname{csch}^{-1} |u| + C = -\sinh^{-1}\left(\frac{1}{|u|}\right) + C.$$

EXERCISES

1. Solve the equation $x = \sinh y = \frac{1}{2}(e^y - e^{-y})$ for e^y in terms of x, and thus show that
$$y = \ln (x + \sqrt{1 + x^2}).$$
(This equation expresses $\sinh^{-1} x$ as a logarithm.)
2. Express $\cosh^{-1} x$ in terms of logarithms by using the method of Exercise 1.
3. Establish formula XXVII.
4. Establish formula XXIX.
5. Establish formula XXXI.

Find dy/dx in Exercises 6 through 10.

6. $y = \sinh^{-1} (2x)$
7. $y = \tanh^{-1} (\cos x)$
8. $y = \cosh^{-1} (\sec x)$
9. $y = \coth^{-1} (\sec x)$
10. $y = \operatorname{sech}^{-1} (\sin 2x)$

Evaluate the following integrals.

11. $\displaystyle \int \frac{dx}{\sqrt{1 + 4x^2}}$
12. $\displaystyle \int \frac{dx}{\sqrt{4 + x^2}}$
13. $\displaystyle \int_0^{0.5} \frac{dx}{1 - x^2}$
14. $\displaystyle \int_{5/4}^{2} \frac{dx}{1 - x^2}$
15. $\displaystyle \int \frac{dx}{x\sqrt{4 + x^2}}$
16. If a body of mass m falling from rest under the action of gravity encounters an air resistance proportional to the square of the velocity, then the velocity v at time t satisfies the differential equation
$$m(dv/dt) = mg - kv^2,$$
where k is a constant of proportionality and $v = 0$ when $t = 0$. Show that
$$v = \sqrt{\frac{mg}{k}} \tanh\left(\sqrt{\frac{gk}{m}}\, t\right),$$
and hence deduce that the body approaches a "limiting velocity" $= \sqrt{mg/k}$ as $t \to \infty$.

8.6 THE HANGING CABLE

We conclude this chapter by deriving the solution of the differential equation

$$\frac{d^2y}{dx^2} = \frac{w}{H}\sqrt{1 + \left(\frac{dy}{dx}\right)^2}, \tag{1}$$

which is the equation of equilibrium of forces on a hanging cable discussed in Problem 1, Article 8.3. Since Eq. (1) involves the second derivative, we shall require two conditions to determine the constants of integration. By choosing the y-axis to be the vertical line through the lowest point of the cable, one condition becomes

$$\frac{dy}{dx} = 0 \quad \text{when } x = 0. \tag{2a}$$

Then we may still move the x-axis up or down to suit our convenience. That is, we let

$$y = y_0 \quad \text{when } x = 0, \tag{2b}$$

and we may choose y_0 so as to give us the simplest form in our final answer. (See Fig. 8.10.)

It is customary, when solving an equation such as (1), to introduce a single letter to represent dy/dx. The letter p is often used. Thus we let

$$\frac{dy}{dx} = p, \tag{3a}$$

and then we may write

$$\frac{d^2y}{dx^2} = \frac{dp}{dx}, \tag{3b}$$

so that Eq. (1) takes the form

$$\frac{dp}{dx} = \frac{w}{H}\sqrt{1 + p^2}.$$

We may now separate the variables to get

$$\frac{dp}{\sqrt{1 + p^2}} = \frac{w}{H}\,dx,$$

or

$$\int \frac{dp}{\sqrt{1 + p^2}} = \frac{w}{H}x + C_1. \tag{4a}$$

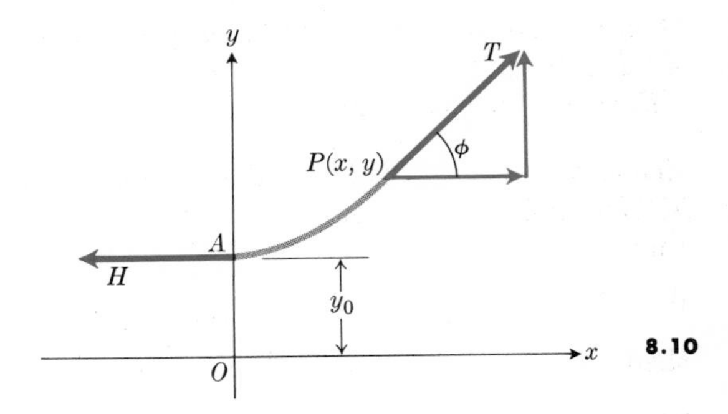

8.10

The integral on the left is of the same form, with p in place of u, as (11a), Article 8.5. Hence (4a) becomes

$$\sinh^{-1} p = \frac{w}{H}x + C_1. \tag{4b}$$

Since $p = dy/dx = 0$ when $x = 0$, we determine the constant of integration $\sinh^{-1} 0 = C_1$, from which (see Fig. 8.7) we have $C_1 = 0$. Therefore Eq. (4) becomes

$$\sinh^{-1} p = \frac{w}{H}x,$$

or

$$p = \sinh\left(\frac{w}{H}x\right). \tag{5a}$$

We substitute $p = dy/dx$, multiply by dx, and have

$$dy = \sinh\left(\frac{w}{H}x\right)dx,$$

or

$$\begin{aligned} y &= \int \sinh\left(\frac{w}{H}x\right)dx \\ &= \frac{H}{w}\int \sinh\left(\frac{w}{H}x\right)d\left(\frac{w}{H}x\right) \\ &= \frac{H}{w}\cosh\left(\frac{w}{H}x\right) + C_2. \end{aligned} \tag{5b}$$

The condition $y = y_0$ when $x = 0$ determines C_2:

$$y_0 = \frac{H}{w}\cosh 0 + C_2,$$

$$C_2 = y_0 - \frac{H}{w}, \tag{5c}$$

and hence

$$y = \frac{H}{w} \cosh\left(\frac{w}{H} x\right) + y_0 - \frac{H}{w}. \tag{5d}$$

Clearly, this equation will have a simpler form if we choose y_0 so that

$$y_0 = \frac{H}{w}.$$

We do so. The answer then takes the form

$$y = \frac{H}{w} \cosh\left(\frac{w}{H} x\right),$$

or

$$y = a \cosh \frac{x}{a} \quad \text{with } a = \frac{H}{w}. \tag{6}$$

EXERCISES

1. Find the length of arc of the catenary

$$y = a \cosh x/a$$

from $A(0, a)$ to $P_1(x_1, y_1)$, $x_1 > 0$.

2. Show that the area bounded by the x-axis, the catenary $y = a \cosh x/a$, the y-axis, and the vertical line through $P_1(x_1, y_1)$, $x_1 > 0$, is the same as the area of a rectangle of altitude a and base s, where s is the length of the arc from $A(0, a)$ to P_1. (See Exercise 1.)

3. The catenary $y = a \cosh x/a$ is revolved about the x-axis. Find the surface area generated by the portion of the curve between the points $A(0, a)$ and $P_1(x_1, y_1)$, $x_1 > 0$.

4. Find the center of gravity of the arc of the catenary $y = a \cosh x/a$ between two symmetrically located points $P_0(-x_1, y_1)$ and $P_1(x_1, y_1)$.

5. Find the volume generated when the area of Exercise 2 is revolved about the x-axis.

6. (a) The length of the arc AP (Fig. 8.10) is

$$s = a \sinh x/a.$$

(See Exercise 1.) Show that the coordinates of $P(x, y)$ may be expressed as functions of the arc length s as follows:

$$x = a \sinh^{-1} \frac{s}{a}, \qquad y = \sqrt{s^2 + a^2}.$$

(b) Calculate dx/ds and dy/ds from (a) and verify that $(dx/ds)^2 + (dy/ds)^2 = 1$.

7. A cable 32 feet long and weighing 2 pounds per foot has its ends fastened at the same level to two posts 30 feet apart.

(a) Show that the constant a in Eq. (6) must satisfy the equation

$$\sinh u = \frac{16}{15} u, \qquad u = \frac{15}{a}.$$

(See Exercise 6a.)

(b) Sketch graphs of the curves $y_1 = \sinh u$, $y_2 = \frac{16}{15}u$ and show (by consulting tables) that they intersect at $u = 0$ and $u = \pm 0.6$ (approximately).

(c) Using the results of (b), find the dip in the cable at its center.

(d) Using the results of (b), find the tension in the cable at its lowest point.

REVIEW QUESTIONS AND EXERCISES

1. Define each of the hyperbolic functions.
2. State three trigonometric identities [such as formulas for $\sin(A + B)$, $\cos(A - B)$, $\cos^2 A + \sin^2 A = 1$, and so forth]. What are the corresponding hyperbolic identities? Verify them.
3. Develop formulas for derivatives of the six hyperbolic functions.
4. What is the domain of the hyperbolic sine? What is its range?
5. What are the domain and range of the hyperbolic cosine? Of the hyperbolic tangent?
6. State some differences between the graphs of the trigonometric functions and their hyperbolic counterparts (for example, sine and sinh, cosine and cosh, tangent and tanh).
7. If $y = A \sin(at) + B \cos(at)$, then $y'' = -a^2 y$. What is the corresponding differential equation satisfied by $y = A \sinh(at) + B \cosh(at)$?
8. Define $\sinh^{-1}$ and $\cosh^{-1}$ functions. What are their domains? What are their ranges? What are their derivatives?

MISCELLANEOUS EXERCISES

1. Prove the hyperbolic identity

$$\cosh 2x = \cosh^2 x + \sinh^2 x.$$

2. Verify that $\tanh x = \sinh 2x/(1 + \cosh 2x)$.
3. Sketch the curves $y = \cosh x$ and $y = \sinh x$ in one diagram. To each positive value of x corresponds a point P on $y = \sinh x$ and a point Q on $y = \cosh x$. Calculate the limit of the distance PQ as x becomes infinitely large.
4. If $\cosh x = \frac{5}{4}$, find $\sinh x$ and $\tanh x$.
5. If $\operatorname{csch} x = -\frac{9}{40}$, find $\cosh x$ and $\tanh x$.
6. If $\tanh x > \frac{5}{13}$, show that

$$\sinh x > 0.4 \quad \text{and} \quad \operatorname{sech} x < 0.95.$$

7. Let $P(x, y)$ be a point on the curve $y = \tanh x$ (Fig. 8.2b). Let AB be the vertical line segment through P with A and B on the asymptotes of the curve. Let C be a semicircle with AB as diameter. Let L be a line through P perpendicular to AB and cutting C in a point Q. Show that $PQ = \operatorname{sech} x$.
8. Prove that $\sinh 3u = 3 \sinh u + 4 \sinh^3 u$.
9. Find equations of the asymptotes of the hyperbola represented by the equation $y = \tanh(\frac{1}{2} \ln x)$.
10. A particle moves along the x-axis according to one of the following laws:

(a) $\quad x = a \cos kt + b \sin kt,$

(b) $\quad x = a \cosh kt + b \sinh kt.$

In both cases, show that the acceleration is proportional to x, but that in the first case it is always directed toward the origin while in the second case it is directed away from the origin.

11. Show that $y = \cosh x$, $\sinh x$, $\cos x$, and $\sin x$ all satisfy the relationship $d^4y/dx^4 = y$.

Find dy/dx in Exercises 12 through 21.

12. $y = \sinh^2 3x$
13. $\tan x = \tanh^2 y$
14. $\sin^{-1} x = \operatorname{sech} y$
15. $\sinh y = \sec x$
16. $\tan^{-1} y = \tanh^{-1} x$
17. $y = \tanh(\ln x)$
18. $x = \cosh(\ln y)$
19. $y = \sinh(\tan^{-1} e^{3x})$
20. $y = \sinh^{-1}(\tan x)$
21. $y^2 + x \cosh y + \sinh^2 x = 50$

Evaluate the integrals in Exercises 22 through 31.

22. $\displaystyle\int \frac{d\theta}{\sinh \theta + \cosh \theta}$
23. $\displaystyle\int \frac{\cosh \theta \, d\theta}{\sinh \theta + \cosh \theta}$
24. $\displaystyle\int \sinh^3 x \, dx$
25. $\displaystyle\int e^x \sinh 2x \, dx$
26. $\displaystyle\int \frac{e^{2x} - 1}{e^{2x} + 1} \, dx$
27. $\displaystyle\int_0^1 \frac{dx}{4 - x^2}$
28. $\displaystyle\int_3^5 \frac{dx}{4 - x^2}$
29. $\displaystyle\int \frac{e^t \, dt}{\sqrt{1 + e^{2t}}}$
30. $\displaystyle\int \frac{\sin x \, dx}{1 - \cos^2 x}$
31. $\displaystyle\int \frac{\sec^2 \theta \, d\theta}{\sqrt{\tan^2 \theta - 1}}$

Sketch the loci in Exercises 32 through 34.

32. $y = \dfrac{1}{2} \ln \dfrac{1 + \tanh x}{1 - \tanh x}$
33. $y = \tan\left(\dfrac{\pi}{2} \tanh x\right)$
34. $\cosh y = 1 + \dfrac{x^2}{2}$
35. If the arc s of the catenary $y = a \cosh(x/a)$ is measured from the lowest point, show that

$$dy/dx = s/a.$$

36. A body starting from rest falls under the attraction of gravity but encounters resistance proportional to the square of its velocity. Show that if the body could continue to fall indefinitely under these same conditions, its velocity would approach a limiting value. Find the distance it would fall in time t.
37. Evaluate the limit, as $x \to \infty$, of $\cosh^{-1} x - \ln x$.
38. Evaluate

$$\lim_{x \to \infty} \int_1^x \left(\frac{1}{\sqrt{1 + t^2}} - \frac{1}{t} \right) dt.$$

METHODS OF INTEGRATION

CHAPTER 9

9.1 BASIC FORMULAS

Since indefinite integration is defined as the inverse of differentiation, the problem of evaluating an integral

$$\int f(x)\,dx \tag{1}$$

is equivalent to finding a function F such that

$$dF(x) = f(x)\,dx. \tag{2}$$

At first sight, this may seem like a hopeless task when we think of adopting a "trial-and-error" approach and realize that it is simply impossible to try *all functions* as F in Eq. (2) hoping that we find one that does the job.* In fact, it is even possible to write down fairly simple integrals, such as

$$\int e^{-x^2}\,dx, \tag{3}$$

which cannot be expressed in terms of finite combinations of the so-called "elementary functions" which we have studied thus far. We shall leave a discussion of such integrals as (3) to a later time, and say simply that the words "finite combinations" must give way and an integral like (3) may then be expressed in terms of an infinite series.

To cut down on the amount of trial-and-error involved in integration, it is useful to build up a table of standard types of integral formulas by inverting formulas for differentials as we have done in the previous chapters. Then we try to match any integral that confronts us against one of the standard types. This usually involves a certain amount of algebraic manipulation.

A table of integrals, more extensive than the list of basic formulas given here, probably is, or will be,

---

* If f is continuous, then an integral function F satisfying Eq. (2) always exists. For example, for $F(x)$ we could take the area A_a^x discussed in Article 5.8. Theoretically, this, together with the fact that such an area can be calculated as a limit of sums of areas of rectangles, gives us a rule for computing $F(x)$. However, we are interested in avoiding these arithmetical calculations whenever possible. We prefer to find simple closed-form expressions for the integral whenever possible.

a part of the reader's library. To use such a table intelligently, it is necessary to become familiar with certain basic techniques that one can apply to reduce a given integral to a form that matches an entry in the tables. The examples and problems in this book should serve to develop our skill with these techniques. So that we may concentrate on the techniques without becoming entangled in a mass of algebra, these problems and examples have been kept fairly simple. This in turn means that *these particular* problems can frequently be solved immediately by consulting an integral table. But the reader should realize that if he adopts such a course, he will defeat the intended purpose of developing his own power. And it is this power which is important, rather than the specific answer to any given problem.

Perhaps a good way to develop the skill we are aiming for is for each student to build his own table of integrals. He may, for example, make a notebook in which the various sections are headed by the standard forms such as $\int u^n\,du$, $\int du/u$, $\int e^u\,du$, and so on, and then under each heading he may include several examples to illustrate the range of application of the particular formula.

Making such a notebook probably has educational value. But once it is made, it should rarely be necessary to refer to it!

The student's success in integration hinges on his ability to spot the particular part of the integrand that he ought to call u. If he picks u skillfully, then he will have du so that he can apply a formula that he already knows. This means that the *first requirement* for skill in integration is a thorough mastery of the formulas for differentiation. If the student will look on the following page, he will see that we have listed certain formulas for differentials that have been derived in previous chapters, together with their integral counterparts.

At present, let us just consider this list a handy reference and *not* a challenge to our memories! A bit of examination will probably convince us that we are already rather familiar with the first twelve formulas for *differentials* and, so far as *integration* is concerned, part of our training in technique will show how we can get the *integrals* in the last six cases without memorizing the formulas for the differentials of the inverse trigonometric functions.

Just what types of functions can we integrate directly by use of this short table of integrals?

Powers $\quad \int u^n\,du, \quad \int \frac{du}{u}$

Exponentials $\quad \int e^u\,du, \quad \int a^u\,du$

Trigonometric functions

$$\int \sin u\,du, \quad \int \cos u\,du.$$

Algebraic functions

$$\int \frac{du}{\sqrt{1-u^2}}, \quad \int \frac{du}{1+u^2}, \quad \int \frac{du}{u\sqrt{u^2-1}}$$

Of course, there are additional combinations of trigonometric functions, such as $\int \sec^2 u\,du$, which are integrable accidentally, so to speak.

What common types of functions are *not* included in the table?

Logarithms $\quad \int \ln u\,du, \quad \int \log_a u\,du$

Trigonometric functions

$$\int \tan u\,du, \quad \int \cot u\,du, \quad \int \sec u\,du, \quad \int \csc u\,du$$

Algebraic functions

$$\int \frac{du}{a^2+u^2}, \quad \text{etc., with } a^2 \neq 1$$

$$\int \sqrt{a^2 \pm u^2}\,du, \quad \int \sqrt{u^2-a^2}\,du, \quad \text{etc.}$$

Inverse functions

$$\int \sin^{-1} u\,du, \quad \int \tan^{-1} u\,du, \quad \text{etc.}$$

We shall eventually see how to handle all of these and some others, but there are no methods which will solve *all* integration problems in terms of elementary functions.

Differentials	*Integrals*
1. $du = \frac{du}{dx}\,dx$	1. $\int du = u + C$
2. $d(au) = a\,du$	2. $\int a\,du = a\int du$
3. $d(u + v) = du + dv$	3. $\int (du + dv) = \int du + \int dv$
4. $d(u^n) = nu^{n-1}\,du$	4. $\int u^n\,du = \frac{u^{n+1}}{n+1} + C, \quad n \neq -1$
5. $d\,(\ln u) = \frac{du}{u}$	5. $\int \frac{du}{u} = \ln \lvert u \rvert + C$
6. (a) $d(e^u) = e^u\,du$	6. (a) $\int e^u\,du = e^u + C$
(b) $d(a^u) = a^u \ln a\,du$	(b) $\int a^u\,du = \frac{a^u}{\ln a} + C$
7. $d\,(\sin u) = \cos u\,du$	7. $\int \cos u\,du = \sin u + C$
8. $d\,(\cos u) = -\sin u\,du$	8. $\int \sin u\,du = -\cos u + C$
9. $d\,(\tan u) = \sec^2 u\,du$	9. $\int \sec^2 u\,du = \tan u + C$
10. $d\,(\cot u) = -\csc^2 u\,du$	10. $\int \csc^2 u\,du = -\cot u + C$
11. $d\,(\sec u) = \sec u \tan u\,du$	11. $\int \sec u \tan u\,du = \sec u + C$
12. $d\,(\csc u) = -\csc u \cot u\,du$	12. $\int \csc u \cot u\,du = -\csc u + C$
13. $d\,(\sin^{-1} u) = \frac{du}{\sqrt{1 - u^2}}$ 14. $d\,(\cos^{-1} u) = \frac{-du}{\sqrt{1 - u^2}}$	13. and 14. $\int \frac{du}{\sqrt{1 - u^2}} = \begin{cases} \sin^{-1} u + C \\ -\cos^{-1} u + C' \end{cases}$
15. $d\,(\tan^{-1} u) = \frac{du}{1 + u^2}$ 16. $d\,(\cot^{-1} u) = \frac{-du}{1 + u^2}$	15. and 16. $\int \frac{du}{1 + u^2} = \begin{cases} \tan^{-1} u + C \\ -\cot^{-1} u + C' \end{cases}$
17. $d\,(\sec^{-1} u) = \frac{du}{\lvert u \rvert \sqrt{u^2 - 1}}$ 18. $d\,(\csc^{-1} u) = \frac{-du}{\lvert u \rvert \sqrt{u^2 - 1}}$	17. and 18. $\int \frac{du}{u\sqrt{u^2 - 1}} = \begin{cases} \sec^{-1} \lvert u \rvert + C \\ -\csc^{-1} \lvert u \rvert + C \end{cases}$

EXERCISES

Evaluate each of the following integrals by reducing the integrand to one of the standard forms 1 through 18. In each case, indicate what you have called u and refer by number to the standard formula used.

1. $\int \sqrt{2x+3}\,dx$
2. $\int \frac{dx}{3x+5}$
3. $\int \frac{dx}{(2x-7)^2}$
4. $\int \frac{(x+1)\,dx}{x^2+2x+3}$
5. $\int \frac{\sin x\,dx}{2+\cos x}$
6. $\int \tan^3 2x \sec^2 2x\,dx$
7. $\int \frac{x\,dx}{\sqrt{1-4x^2}}$
8. $\int x^{1/3}\sqrt{x^{4/3}-1}\,dx$
9. $\int \frac{x\,dx}{(3x^2+4)^3}$
10. $\int x^2\sqrt{x^3+5}\,dx$
11. $\int \frac{x^2\,dx}{\sqrt{x^3+5}}$
12. $\int \frac{x\,dx}{4x^2+1}$
13. $\int e^{2x}\,dx$
14. $\int \sin x\, e^{\cos x}\,dx$
15. $\int \frac{dx}{e^{3x}}$
16. $\int \frac{e^{\sqrt{x+1}}}{\sqrt{x+1}}\,dx$
17. $\int \cos^2 x \sin x\,dx$
18. $\int \frac{\cos x\,dx}{\sin^3 x}$
19. $\int \cot^3 x \csc^2 x\,dx$
20. $\int \tan 3x \sec^2 3x\,dx$
21. $\int \frac{e^{2x}+e^{-2x}}{e^{2x}-e^{-2x}}\,dx$
22. $\int \sin 2x \cos^2 2x\,dx$
23. $\int (1+\cos\theta)^3 \sin\theta\,d\theta$
24. $\int te^{-t^2}\,dt$
25. $\int \frac{\cos x\,dx}{\sin x}$
26. $\int \frac{\cos x\,dx}{1+\sin x}$
27. $\int \sec^3 x \tan x\,dx$
28. $\int \frac{\sin\theta\,d\theta}{\sqrt{1+\cos\theta}}$
29. $\int \sec^2 3x\, e^{\tan 3x}\,dx$
30. $\int \cos 2t\sqrt{4-\sin 2t}\,dt$
31. $\int \frac{1+\cos 2x}{\sin^2 2x}\,dx$
32. $\int \frac{\sin^2 2x}{1+\cos 2x}\,dx$
33. $\int \frac{\csc^2 2t}{\sqrt{1+\cot 2t}}\,dt$
34. $\int e^{3x}\,dx$
35. $\int \frac{e^{\tan^{-1}2t}}{1+4t^2}\,dt$
36. $\int xe^{-x^2}\,dx$
37. $\int 3^x\,dx$
38. $\int 10^{2x}\,dx$
39. Each of the following integrals may be easily evaluated for a particular numerical value of n. Choose this value and integrate. For example,

$$\int x^n \cos(x^2)\,dx$$

is easily evaluated for $n = 1$:

$$\int x\cos(x^2)\,dx = \tfrac{1}{2}\sin(x^2)+C.$$

(a) $\int x^n \ln x\,dx$ (b) $\int x^n e^{x^3}\,dx$ (c) $\int x^n \sin\sqrt{x}\,dx$

40. The integral $\int_a^\infty e^{-x^2}\,dx$ arises in statistics and elsewhere. It is defined to mean the same thing as

$$\lim_{b\to+\infty}\int_a^b e^{-x^2}\,dx.$$

Show that if $x > a \geq 1$, then $e^{-x^2} < e^{-ax}$. Hence, by comparing the integral from a to ∞ of e^{-x^2} with the integral of e^{-ax} and evaluating the latter integral, show that

$$\int_a^\infty e^{-x^2}\,dx < \frac{1}{a}e^{-a^2} \quad \text{if } a \geq 1.$$

9.2 POWERS OF TRIGONOMETRIC FUNCTIONS

The formula

$$\int u^n\,du = \begin{cases} \dfrac{u^{n+1}}{n+1}+C, & n \neq -1, \\ \ln|u|+C, & n = -1 \end{cases}$$

may be used to evaluate certain integrals involving powers of the trigonometric functions, as illustrated by the examples that follow. The same methods work for powers of other functions. The reader should pay attention to the *methods* rather than trying to remember specific results.

Problem 1. $\int \sin^n ax \cos ax\, dx.$

Solution. If we let $u = \sin ax$, then

$$du = \cos ax\, d(ax) = a \cos ax\, dx,$$

so that we need only multiply the integral by unity in the form of a times $1/a$. Since a and $1/a$ are constants, we may write a inside the integral sign and $1/a$ in front of the integral sign, so that we have

$$\begin{aligned}\int \sin^n ax \cos ax\, dx &= \frac{1}{a}\int (\sin ax)^n (a \cos ax\, dx) \\ &= \frac{1}{a}\int u^n\, du \\ &= \begin{cases} \dfrac{1}{a}\dfrac{u^{n+1}}{n+1} + C, & n \neq -1, \\ \dfrac{1}{a}\ln|u| + C, & n = -1; \end{cases}\end{aligned}$$

that is,

$$\int \sin^n ax \cos ax\, dx = \frac{\sin^{n+1} ax}{(n+1)a} + C, \quad n \neq -1, \tag{1a}$$

and if $n = -1$, we get

$$\int \cot ax\, dx = \frac{1}{a}\ln|\sin ax| + C. \tag{1b}$$

Note that the success of the method depended upon having $\cos ax$ to go with the dx as part of du.

Problem 2. $\int \sin^3 x\, dx.$

Solution. The method of the previous example does not work because there is no $\cos x$ to go with dx to give du if we try letting $u = \sin x$. But if we write

$$\sin^3 x = \sin^2 x \cdot \sin x = (1 - \cos^2 x) \cdot \sin x$$

and let

$$u = \cos x, \qquad du = -\sin x\, dx,$$

we have

$$\begin{aligned}\int \sin^3 x\, dx &= \int (1 - \cos^2 x) \cdot \sin x\, dx \\ &= \int (1 - u^2) \cdot (-du) = \int (u^2 - 1)\, du \\ &= \tfrac{1}{3}u^3 - u + C = \tfrac{1}{3}\cos^3 x - \cos x + C.\end{aligned}$$

This *method* may be applied whenever an *odd* power of $\sin x$ or $\cos x$ is to be integrated. For example, any positive odd power of $\cos x$ has the form

$$\begin{aligned}\cos^{2n+1} x = \cos^{2n} x \cdot \cos x &= (\cos^2 x)^n \cdot \cos x \\ &= (1 - \sin^2 x)^n \cdot \cos x,\end{aligned}$$

with n an integer ≥ 0. Then, if we let $u = \sin x$, $du = \cos x\, dx$, we have

$$\begin{aligned}\int \cos^{2n+1} x\, dx &= \int (1 - \sin^2 x)^n \cdot \cos x\, dx \\ &= \int (1 - u^2)^n \cdot du.\end{aligned}$$

The expression $(1 - u^2)^n$ may now be expanded by the binomial theorem and the result evaluated as a sum of individual integrals of the type $\int u^m\, du$.

Problem 3. $\int \sec x \tan x\, dx.$

Solution. Of course this is a standard form already, so there is no real problem in finding an answer. But in trigonometry we often express all trigonometric functions in terms of sines and cosines, and we now investigate what this does to the integral in question.

$$\int \sec x \tan x\, dx = \int \frac{1}{\cos x}\frac{\sin x}{\cos x}\, dx = \int \frac{\sin x\, dx}{\cos^2 x}.$$

Taking a clue from the previous examples (keeping in mind du as well as u), we let

$$u = \cos x, \qquad du = -\sin x\, dx,$$

and then

$$\begin{aligned}\int \frac{\sin x\, dx}{\cos^2 x} &= \int \frac{-du}{u^2} \\ &= -\int u^{-2}\, du = \frac{-u^{-1}}{-1} + C = \frac{1}{u} + C \\ &= \frac{1}{\cos x} + C = \sec x + C.\end{aligned}$$

Problem 4. $\int \tan^4 x\, dx.$

Solution. This does not lend itself readily to the use of sines and cosines, since both of them occur to even powers. We say: $u = \tan x$ would require $du = \sec^2 x\, dx$. Is there some way to include $\sec^2 x$? Yes; there is an identity involving tangents and secants. How does it go? Since

$$\sin^2 x + \cos^2 x = 1,$$

if we divide through by $\cos^2 x$ we get

$$\tan^2 x + 1 = \sec^2 x \quad \text{or} \quad \tan^2 x = \sec^2 x - 1.$$

Then

$$\begin{aligned}\int \tan^4 x\,dx &= \int \tan^2 x \cdot \tan^2 x\,dx \\ &= \int \tan^2 x \cdot (\sec^2 x - 1)\,dx \\ &= \int \tan^2 x \sec^2 x\,dx - \int \tan^2 x\,dx.\end{aligned}$$

The first of these is all set. But how about the $\tan^2 x$? Oh, yes,

$$\tan^2 x = \sec^2 x - 1.$$

So

$$\begin{aligned}\int \tan^4 x\,dx &= \int \tan^2 x \sec^2 x\,dx - \int \tan^2 x\,dx \\ &= \int \tan^2 x \sec^2 x\,dx - \int (\sec^2 x - 1)\,dx \\ &= \int \tan^2 x \sec^2 x\,dx - \int \sec^2 x\,dx + \int dx.\end{aligned}$$

In the first two of these, we let

$$u = \tan x, \qquad du = \sec^2 x\,dx$$

and we have

$$\int u^2\,du - \int du = \tfrac{1}{3}u^3 - u + C'.$$

The other is a standard form, so

$$\int \tan^4 x\,dx = \tfrac{1}{3}\tan^3 x - \tan x + x + C.$$

The *method* works for any *even* power of $\tan x$, but what is still better is a *reduction formula*, derived as follows:

$$\begin{aligned}\int \tan^n x\,dx &= \int \tan^{n-2} x\,(\sec^2 x - 1)\,dx \\ &= \int \tan^{n-2} x \sec^2 x\,dx - \int \tan^{n-2} x\,dx \\ &= \frac{\tan^{n-1} x}{n-1} - \int \tan^{n-2} x\,dx.\end{aligned}$$

This reduces the problem of integrating $\tan^n x\,dx$ to the problem of integrating $\tan^{n-2} x\,dx$. Since this decreases the exponent on $\tan x$ by 2, a repetition with the same formula will reduce the exponent by 2 again, and so on. Applying this to the problem above, we have

$$\begin{aligned}n = 4:&\quad \int \tan^4 x\,dx = \frac{\tan^3 x}{3} - \int \tan^2 x\,dx, \\ n = 2:&\quad \int \tan^2 x\,dx = \frac{\tan x}{1} - \int \tan^0 x\,dx, \\ n = 0:&\quad \int \tan^0 x\,dx = \int 1\,dx = x + C.\end{aligned}$$

Therefore

$$\begin{aligned}\int \tan^4 x\,dx &= \tfrac{1}{3}\tan^3 x - (\tan x - x + C) \\ &= \tfrac{1}{3}\tan^3 x - \tan x + x + C'.\end{aligned}$$

This reduction formula works whether the original exponent n is even or odd, but if the exponent is odd, say $2m + 1$, after m steps it will be reduced by $2m$, leaving

$$\begin{aligned}\int \tan x\,dx = \int \frac{\sin x}{\cos x}\,dx &= -\int \frac{d(\cos x)}{\cos x} \\ &= -\ln|\cos x| + C \qquad (2)\end{aligned}$$

as the final integral to be evaluated.

These examples illustrate how it is possible, by using the trigonometric identities

$$\sin^2 x + \cos^2 x = 1, \qquad \tan^2 x + 1 = \sec^2 x,$$

and others readily derived from these, to evaluate the integrals of

(a) *odd* powers of $\sin x$ or $\cos x$,

(b) *any* integral powers of $\tan x$ (or $\cot x$), and

(c) *even* powers of $\sec x$ (or $\csc x$).

The even powers of $\sec x$, say $\sec^{2n} x$, can all be reduced to powers of $\tan x$ by employing the substitution $\sec^2 x = 1 + \tan^2 x$, and then using the reduction formula for integrating powers of $\tan x$ after expanding $\sec^{2n} x = (1 + \tan^2 x)^n$ by the binomial theorem. But it is even simpler to use the method below:

$$\begin{aligned}\int \sec^{2n} x\,dx &= \int \sec^{2n-2} x \sec^2 x\,dx\\ &= \int (\sec^2 x)^{n-1} \sec^2 x\,dx\\ &= \int (1+\tan^2 x)^{n-1} \sec^2 x\,dx\\ &= \int (1+u^2)^{n-1}\,du \qquad (u = \tan x).\end{aligned}$$

When $(1+u^2)^{n-1}$ is expanded by the binomial theorem, the resulting polynomial in u may be integrated term by term.

Example

$$\begin{aligned}\int \sec^6 x\,dx &= \int \sec^4 x \cdot \sec^2 x\,dx\\ &= \int (1+\tan^2 x)^2 \cdot \sec^2 x\,dx\\ &= \int (1+2u^2+u^4)\,du \quad (u = \tan x)\\ &= u + \frac{2u^3}{3} + \frac{u^5}{5} + C\\ &= \tan x + 2\frac{\tan^3 x}{3} + \frac{\tan^5 x}{5} + C.\end{aligned}$$

Problem 5. $\int \sec x\,dx.$

Solution. This is hard to evaluate unless one has seen the following trick!

$$\begin{aligned}\sec x &= \frac{\sec x\,(\tan x + \sec x)}{\sec x + \tan x}\\ &= \frac{\sec x \tan x + \sec^2 x}{\sec x + \tan x}.\end{aligned}$$

In this form the numerator is the derivative of the denominator. Therefore

$$\begin{aligned}\int \sec x\,dx &= \int \frac{\sec x \tan x + \sec^2 x}{\sec x + \tan x}\,dx\\ &= \int \frac{du}{u} \qquad (u = \sec x + \tan x)\\ &= \ln|u| + C.\end{aligned}$$

That is,

$$\int \sec x\,dx = \ln|\sec x + \tan x| + C. \tag{3}$$

EXERCISES

1. $\int \sin t\sqrt{1+\cos t}\,dt$
2. $\int \frac{\sin\theta\,d\theta}{2-\cos\theta}$
3. $\int \frac{\sec^2 2x\,dx}{1+\tan 2x}$
4. $\int \tan 3x\,dx$
5. $\int \cos^3 x\,dx$
6. $\int \tan^2 4\theta\,d\theta$
7. (a) $\int \sin^3 x \cos^2 x\,dx$ (b) $\int \frac{\sin^3 x\,dx}{\cos^2 x}$
8. $\int \sec^n x \tan x\,dx$
9. $\int \tan^n x \sec^2 x\,dx$
10. $\int \sin^n x \cos x\,dx$
11. $\int \cos^n x \sin x\,dx$
12. $\int \sin^2 3x \cos 3x\,dx$
13. $\int \cos^3 2x \sin 2x\,dx$
14. $\int \sec^4 3x \tan 3x\,dx$
15. $\int \sec^4 3x\,dx$
16. $\int \cos^3 2x\,dx$
17. $\int \tan^3 2x\,dx$
18. $\int \tan^3 x \sec x\,dx$
19. $\int \sin^3 x\,dx$
20. $\int \frac{\cos x\,dx}{(1+\sin x)^2}$
21. $\int \frac{\sec^2 x\,dx}{2+\tan x}$
22. $\int \frac{\cos^3 t\,dt}{\sin^2 t}$
23. $\int \frac{e^x\,dx}{1+e^x}$
24. $\int (\ln ax)^n \frac{dx}{x}$
25. $\int \frac{dx}{x \ln 3x}$
26. $\int \cot^3 x\,dx$
27. $\int \csc^3 2t \cot 2t\,dt$
28. $\int \csc^4 x\,dx$
29. Derive a reduction formula for $\int \cot^n ax\,dx$ and use the result to evaluate $\int \cot^4 3x\,dx$.

The following integrals require some knowledge of Chapter 8.

30. $\int \tanh u\,du$
31. $\int \tanh^2 3x\,dx$
32. $\int \operatorname{sech} 5x\,dx$
33. $\int \sinh^2 at\,dt \quad (a \neq 0)$
34. $\int \sinh u\sqrt{1+\sinh^2 u}\,du$

9.3 EVEN POWERS OF SINES AND COSINES

In Article 9.2, we saw how we could evaluate integrals of odd powers of sines and cosines. Indeed, any integral of the form

$$\int \sin^m x \cos^n x \, dx \tag{1}$$

in which at least one of the exponents m and n is a positive odd integer may be evaluated by these methods.

Problem. $\int \cos^{2/3} x \sin^5 x \, dx.$

Solution. Here we have $\sin x$ to an *odd* power. So we put one factor of $\sin x$ with dx and the remaining sine factors, namely $\sin^4 x$, can be expressed in terms of $\cos x$ without introducing any square roots, as follows:

$$\sin^4 x = (\sin^2 x)^2 = (1 - \cos^2 x)^2.$$

The $\sin x$ goes well with dx when we take

$$u = \cos x, \qquad du = -\sin x \, dx$$

and evaluate the integral as follows:

$$\begin{aligned}\int \cos^{2/3} x \sin^5 x \, dx &= \int \cos^{2/3} x(1 - \cos^2 x)^2 \sin x \, dx \\ &= \int u^{2/3}(1 - u^2)^2(-du) \\ &= -\int (u^{2/3} - 2u^{8/3} + u^{14/3}) \, du \\ &= -[\tfrac{3}{5}u^{5/3} - \tfrac{6}{11}u^{11/3} + \tfrac{3}{17}u^{17/3}] + C,\end{aligned}$$

so that we have

$$-\cos^{5/3} x[\tfrac{3}{5} - \tfrac{6}{11}\cos^2 x + \tfrac{3}{17}\cos^4 x] + C.$$

If both the exponents m and n in the integral (1) are even integers, the method illustrated above won't work. So we use one or both of the following trigonometric identities:

$$\sin^2 A = \tfrac{1}{2}(1 - \cos 2A), \tag{2a}$$

$$\cos^2 A = \tfrac{1}{2}(1 + \cos 2A). \tag{2b}$$

These identities may be derived very quickly by adding or subtracting the equations

$$\cos^2 A + \sin^2 A = 1,$$
$$\cos^2 A - \sin^2 A = \cos 2A,$$

and dividing by two.

Example 1

$$\begin{aligned}\int \cos^4 x \, dx &= \int (\cos^2 x)^2 \, dx = \int \tfrac{1}{4}(1 + \cos 2x)^2 \, dx && \text{(a)} \\ &= \tfrac{1}{4}\int (1 + 2\cos 2x + \cos^2 2x) \, dx && \text{(b)} \\ &= \tfrac{1}{4}\int [1 + 2\cos 2x + \tfrac{1}{2}(1 + \cos 4x)] \, dx \\ &= \tfrac{3}{8}x + \tfrac{1}{4}\sin 2x + \tfrac{1}{32}\sin 4x + C.\end{aligned}$$

In (a), we used Eq. (2b) with $A = x$, and in (b), we used Eq. (2b) with $A = 2x$. The final integrations involve

$$\int \cos ax \, dx = \frac{1}{a}\int \cos ax \, d(ax) = \frac{1}{a}\sin ax + C'.$$

Remark. An integral such as $\int \sin^2 x \cos^4 x \, dx$, which involves even powers of both $\sin x$ and $\cos x$, can be changed to a sum of integrals each of which involves only powers of one of them. Then these may be handled by the method illustrated above.

Example 2

$$\begin{aligned}\int \sin^2 x \cos^4 x \, dx &= \int (1 - \cos^2 x) \cos^4 x \, dx \\ &= \int \cos^4 x \, dx - \int \cos^6 x \, dx.\end{aligned}$$

We evaluated $\int \cos^4 x \, dx$ above, and

$$\begin{aligned}\int \cos^6 x \, dx &= \int (\cos^2 x)^3 \, dx = \tfrac{1}{8}\int (1 + \cos 2x)^3 \, dx \\ &= \tfrac{1}{8}\int (1 + 3\cos 2x \\ &\qquad + 3\cos^2 2x + \cos^3 2x) \, dx.\end{aligned}$$

We now know how to handle each term of this integral. The reader may supply the details and show that the result is

$$\int \cos^6 x \, dx = \tfrac{5}{16}x + \tfrac{1}{4}\sin 2x + \tfrac{3}{64}\sin 4x - \tfrac{1}{48}\sin^3 2x + C.$$

EXERCISES

Evaluate the following integrals.

1. $\int \sin^2 x \cos^3 x \, dx$
2. $\int \frac{\sin^3 x \, dx}{\cos^2 x}$
3. $\int \sin^2 2t \, dt$
4. $\int \cos^2 3\theta \, d\theta$

5. $\int \sin^4 ax\, dx$

6. $\int \sin^2 y \cos^2 y\, dy$

7. $\int \frac{dx}{\cos^2 x}$

8. $\int \frac{dx}{\sin^4 x}$

9. $\int \frac{\cos 2t\, dt}{\sin^4 2t}$

10. $\int \sin^6 x\, dx$

The following integrals require some knowledge of Chapter 8.

11. $\int \sinh^2 2x \cosh^3 2x\, dx$

12. $\int \frac{\sinh^3 3t\, dt}{\cosh^2 3t}$

13. $\int \frac{dx}{\cosh^4 ax}$ $(a \neq 0)$

14. $\int \operatorname{sech}^2 u \tanh^{2/3} u\, du$

15. $\int \tanh^3 5z\, dz$

9.4 INTEGRALS WITH TERMS $\sqrt{a^2 - u^2}$, $\sqrt{a^2 + u^2}$, $\sqrt{u^2 - a^2}$, $a^2 + u^2$, $a^2 - u^2$

Sometimes an integral with one of these terms may be treated directly by an integral formula from the list in Article 9.1 if we change the integral around so that $a = 1$. For example, we have

$$\int \frac{du}{1 + u^2} = \tan^{-1} u + C, \tag{1}$$

and to evaluate

$$\int \frac{du}{a^2 + u^2} \tag{2}$$

we factor a^2 out of the denominator and proceed as follows:

$$\begin{aligned}
\int \frac{du}{a^2 + u^2} &= \int \frac{du}{a^2\left[1 + \left(\frac{u}{a}\right)^2\right]} \\
&= \frac{1}{a^2} \int \frac{a\,d\left(\frac{u}{a}\right)}{1 + \left(\frac{u}{a}\right)^2} = \frac{1}{a} \int \frac{dz}{1 + z^2} \\
&= \frac{1}{a} \tan^{-1} z + C \\
&= \frac{1}{a} \tan^{-1} \frac{u}{a} + C \qquad \left(z = \frac{u}{a}\right).
\end{aligned}$$

That is,

$$\int \frac{du}{a^2 + u^2} = \frac{1}{a} \tan^{-1} \frac{u}{a} + C. \tag{3}$$

The essential feature here was the introduction of a new variable by means of the substitution

$$z = \frac{u}{a}, \quad \text{or} \quad u = az.$$

Such a substitution allows the a-terms to be brought outside the integral sign, and the resulting integral may then match one of the inverse trigonometric formulas.

An alternative approach will show, however, that one may bypass the job of learning these formulas. In addition, this method, which we shall soon illustrate, permits the evaluation of many additional integrals.

The method leans primarily upon the following identities:

$$\begin{aligned}
1 - \sin^2 \theta &= \cos^2 \theta, \\
1 + \tan^2 \theta &= \sec^2 \theta, \\
\sec^2 \theta - 1 &= \tan^2 \theta.
\end{aligned} \tag{4}$$

These identities may be multiplied by a^2, with the result that the substitutions listed below have the following effects:

if $u = a \sin \theta$,

then $a^2 \cos^2 \theta$ replaces $a^2 - u^2$; (5a)

if $u = a \tan \theta$,

then $a^2 \sec^2 \theta$ replaces $a^2 + u^2$; (5b)

if $u = a \sec \theta$,

then $a^2 \tan^2 \theta$ replaces $u^2 - a^2$. (5c)

Thus we see that corresponding to each of the *binomial* expressions $a^2 - u^2$, $a^2 + u^2$, and $u^2 - a^2$, we have a substitution that replaces the binomial by a single squared term. The particular substitution to use depends on the form of the integrand. In the problems that follow, a is a positive constant.

Problem 1. $\displaystyle\int \frac{du}{\sqrt{a^2 - u^2}}, \quad a > 0.$

Solution. We try the substitution (5a):

$$u = a \sin\theta,$$
$$du = a \cos\theta \, d\theta,$$
$$\theta = \sin^{-1}(u/a).$$

Thus

$$a^2 - u^2 = a^2(1 - \sin^2\theta) = a^2 \cos^2\theta.$$

Then

$$\int \frac{du}{\sqrt{a^2 - u^2}} = \int \frac{a\cos\theta\, d\theta}{\sqrt{a^2\cos^2\theta}} = \int \frac{a\cos\theta\, d\theta}{\pm a\cos\theta},$$

the sign depending on the sign of $\cos\theta$; continuing,

$$\int \frac{du}{\sqrt{a^2 - u^2}} = \pm\int d\theta = \pm(\theta + C).$$

Using only the *principal value* of $\sin^{-1} u/a$ means that θ will lie between $-\pi/2$ and $\pi/2$; hence $\cos\theta \geq 0$ and the ambiguous sign is $+$. That is,

$$\int \frac{du}{\sqrt{a^2 - u^2}} = \sin^{-1}\frac{u}{a} + C. \tag{6}$$

Problem 2. $\displaystyle\int \frac{du}{\sqrt{a^2 + u^2}}, \quad a > 0.$

Solution. This time we try

$$u = a\tan\theta,$$
$$du = a\sec^2\theta\, d\theta,$$
$$\theta = \tan^{-1}(u/a).$$

Thus

$$a^2 + u^2 = a^2(1 + \tan^2\theta) = a^2\sec^2\theta.$$

Then

$$\int \frac{du}{\sqrt{a^2 + u^2}} = \int \frac{a\sec^2\theta\, d\theta}{\sqrt{a^2\sec^2\theta}} = \pm\int \sec\theta\, d\theta,$$

the sign depending on the sign of $\sec\theta$.

By Eq. (3), Article 9.2, we know that

$$\int \sec\theta\, d\theta = \ln|\sec\theta + \tan\theta| + C.$$

If we take $\displaystyle\theta = \tan^{-1}\frac{u}{a}, \quad -\frac{\pi}{2} < \theta < \frac{\pi}{2},$

then $\sec\theta$ is positive, and

$$\begin{aligned}\int \frac{du}{\sqrt{a^2 + u^2}} &= \int \sec\theta\, d\theta \\ &= \ln|\sec\theta + \tan\theta| + C \\ &= \ln\left|\frac{\sqrt{a^2 + u^2}}{a} + \frac{u}{a}\right| + C \\ &= \ln|\sqrt{a^2 + u^2} + u| + C',\end{aligned} \tag{7}$$

where

$$C' = C - \ln a.$$

Problem 3. $\displaystyle\int \frac{du}{\sqrt{u^2 - a^2}}, \quad |u| > a > 0.$

Solution. We try the substitution

$$u = a\sec\theta,$$
$$du = a\sec\theta\tan\theta\, d\theta,$$
$$\theta = \sec^{-1}(u/a).$$

Thus

$$u^2 - a^2 = a^2(\sec^2\theta - 1) = a^2\tan^2\theta.$$

Then

$$\begin{aligned}\int \frac{du}{\sqrt{u^2 - a^2}} &= \int \frac{a\sec\theta\tan\theta\, d\theta}{\sqrt{a^2\tan^2\theta}} \\ &= \pm\int \sec\theta\, d\theta,\end{aligned}$$

the sign depending on the sign of $\tan\theta$. If we take

$$\theta = \sec^{-1}\frac{u}{a}, \quad 0 \leq \theta \leq \pi,$$

then

$$\tan\theta \text{ is positive} \quad \text{if} \quad 0 < \theta < \pi/2,$$
$$\tan\theta \text{ is negative} \quad \text{if} \quad \pi/2 < \theta < \pi,$$

and, from Eq. (3), Article 9.2,

$$\pm\int \sec\theta\, d\theta = \pm\ln|\sec\theta + \tan\theta| + C.$$

When $\tan\theta$ is positive, we must use the plus sign; when $\tan\theta$ is negative, the minus sign. Moreover,

$$\sec\theta = \frac{u}{a}, \qquad \tan\theta = \pm\frac{\sqrt{u^2 - a^2}}{a}.$$

So we have

$$\int \frac{du}{\sqrt{u^2 - a^2}} = \pm\ln\left|\frac{u}{a} \pm \frac{\sqrt{u^2 - a^2}}{a}\right| + C$$

$$= \begin{cases} \ln\left|\dfrac{u}{a} + \dfrac{\sqrt{u^2-a^2}}{a}\right| + C \\ \text{or} \\ -\ln\left|\dfrac{u}{a} - \dfrac{\sqrt{u^2-a^2}}{a}\right| + C. \end{cases}$$

But the two forms are actually equal, because

$$-\ln\left|\frac{u}{a} - \frac{\sqrt{u^2-a^2}}{a}\right|$$

$$= \ln\left|\frac{a}{u - \sqrt{u^2 - a^2}}\right|$$

$$= \ln\left|\frac{a(u + \sqrt{u^2 - a^2})}{(u - \sqrt{u^2-a^2})(u + \sqrt{u^2 - a^2})}\right|$$

$$= \ln\left|\frac{a(u + \sqrt{u^2 - a^2})}{a^2}\right|$$

$$= \ln\left|\frac{u + \sqrt{u^2-a^2}}{a}\right|.$$

Therefore

$$\int \frac{du}{\sqrt{u^2 - a^2}} = \ln|u + \sqrt{u^2 - a^2}| + C', \tag{8}$$

where we have replaced $C - \ln a$ by C' in this final form.

Problem 4. $\displaystyle\int \frac{du}{a^2 + u^2}, \quad a > 0.$

Solution. $u = a \tan\theta, \qquad du = a \sec^2\theta\, d\theta,$

$$a^2 + u^2 = a^2(1 + \tan^2\theta) = a^2\sec^2\theta.$$

Then

$$\int \frac{du}{a^2 + u^2} = \int \frac{a\sec^2\theta\, d\theta}{a^2\sec^2\theta} = \frac{1}{a}\int d\theta = \frac{1}{a}\theta + C.$$

Since $u = a\tan\theta$, we have

$$\tan\theta = \frac{u}{a}, \qquad \theta = \tan^{-1}\frac{u}{a},$$

and

$$\int \frac{du}{a^2 + u^2} = \frac{1}{a}\tan^{-1}\frac{u}{a} + C. \tag{9}$$

Problem 5. $\displaystyle\int \frac{x^2\, dx}{\sqrt{9 - x^2}}.$

Solution. If we substitute

$$x = 3\sin\theta, \quad -\frac{\pi}{2} < \theta = \sin^{-1}\left(\frac{x}{3}\right) < \frac{\pi}{2},$$

$$dx = 3\cos\theta\, d\theta,$$

$$9 - x^2 = 9(1 - \sin^2\theta) = 9\cos^2\theta,$$

then we find

$$\int \frac{x^2\, dx}{\sqrt{9 - x^2}} = \int \frac{9\sin^2\theta \cdot 3\cos\theta\, d\theta}{3\cos\theta} = 9\int \sin^2\theta\, d\theta.$$

This is considerably simpler than the integral we started with. To evaluate it, we make use of the identity

$$\sin^2\theta = \tfrac{1}{2}(1 - \cos 2\theta).$$

Then

$$\int \sin^2\theta\, d\theta = \tfrac{1}{2}\int (1 - \cos 2\theta)\, d\theta$$
$$= \tfrac{1}{2}[\theta - \tfrac{1}{2}\sin 2\theta] + C$$
$$= \tfrac{1}{2}[\theta - \sin\theta\cos\theta] + C.$$

Substituting this above, we get

$$\int \frac{x^2\, dx}{\sqrt{9 - x^2}} = \frac{9}{2}[\theta - \sin\theta\cos\theta] + C$$
$$= \frac{9}{2}\left[\sin^{-1}\frac{x}{3} - \frac{x\sqrt{9 - x^2}}{9}\right] + C.$$

The trigonometric substitutions (a), (b), and (c) of Eq. (5) can be easily remembered by thinking of the theorem of Pythagoras and taking a and u as two of the sides and θ as an angle in a right triangle. Thus $\sqrt{a^2 - u^2}$ suggests a as hypotenuse and u as a leg, $\sqrt{a^2 + u^2}$ suggests a and u as the legs, and $\sqrt{u^2 - a^2}$ suggests u as hypotenuse and a as a leg. These situations are shown in Fig. 9.1.

The trigonometric identities, Eqs. (4), are simply equivalent expressions of the theorem of Pythagoras applied to the right triangles in Fig. 9.2. The triangles in Figs. 9.2(b, c) are obtained from the basic triangle in Fig. 9.2(a) by dividing its sides all by $\cos\theta$ and by $\sin\theta$, respectively, to obtain similar triangles.

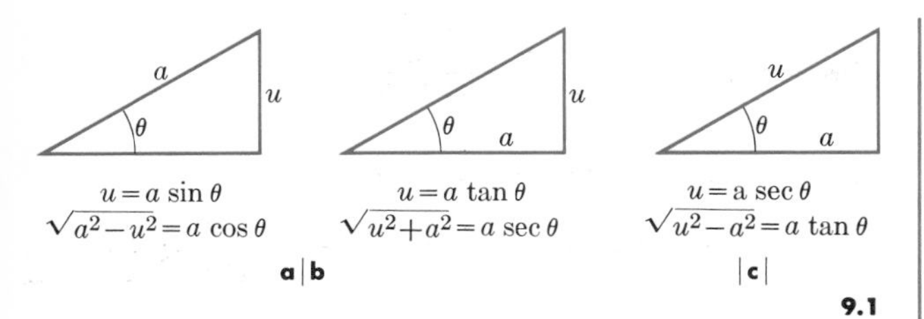

9.1

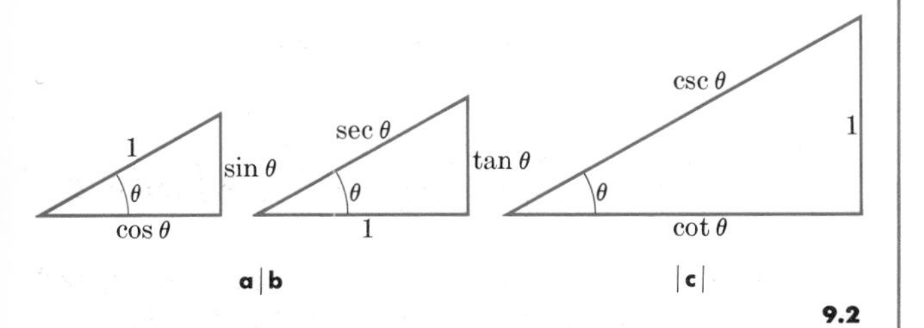

9.2

Remark (*for readers of Chapter 8*). Just as we have made use of the identities of Eqs. (4), we may also make use of the hyperbolic identities

$$\begin{aligned} \cosh^2\theta - \sinh^2\theta &= 1, \\ 1 - \tanh^2\theta &= \operatorname{sech}^2\theta, \\ \cosh^2\theta - 1 &= \sinh^2\theta, \\ 1 + \sinh^2\theta &= \cosh^2\theta. \end{aligned} \tag{10}$$

The last three of these identities follow directly from the first. If they are multiplied by a^2, we see that the substitutions listed below have the following effects:

if $\quad u = a \tanh\theta,$

then $\qquad a^2 \operatorname{sech}^2\theta$ replaces $a^2 - u^2$; (11a)

if $\quad u = a \sinh\theta,$

then $\qquad a^2 \cosh^2\theta$ replaces $a^2 + u^2$; (11b)

if $\quad u = a \cosh\theta,$

then $\qquad a^2 \sinh^2\theta$ replaces $u^2 - a^2$. (11c)

Problem 6. Use the substitutions

$$u = a\sinh\theta, \qquad du = a\cosh\theta\, d\theta, \qquad \theta = \sinh^{-1}(u/a)$$

to evaluate

$$\int \frac{du}{\sqrt{a^2 + u^2}}.$$

(The solution provides an alternate form for the answer to the integral of Problem 2.)

Solution. The indicated substitutions lead to

$$\begin{aligned} \int \frac{a\cosh\theta\, d\theta}{a\cosh\theta} &= \int d\theta = \theta + C \\ &= \sinh^{-1}(u/a) + C. \end{aligned}$$

We therefore have the alternate form

$$\int \frac{du}{\sqrt{a^2 + u^2}} = \sinh^{-1}(u/a) + C. \tag{12}$$

EXERCISES

Evaluate the following integrals.

1. $\displaystyle\int_0^{0.6a} \frac{x\,dx}{\sqrt{a^2 - x^2}}$
2. $\displaystyle\int \frac{dx}{\sqrt{1 - 4x^2}}$
3. $\displaystyle\int_0^a \sqrt{a^2 - x^2}\,dx$
4. $\displaystyle\int \frac{dx}{\sqrt{4 - (x-1)^2}}$
5. $\displaystyle\int \sec 2t\,dt$
6. $\displaystyle\int_0^2 \frac{dx}{\sqrt{4 + x^2}}$
7. $\displaystyle\int_0^1 \frac{dx}{\sqrt{4 - x^2}}$
8. $\displaystyle\int \frac{x\,dx}{\sqrt{4 + x^2}}$
9. $\displaystyle\int_0^2 \frac{x\,dx}{4 + x^2}$
10. $\displaystyle\int_0^2 \frac{dx}{4 + x^2}$
11. $\displaystyle\int_0^1 \frac{dx}{4 - x^2}$
12. $\displaystyle\int \csc u\,du$
13. $\displaystyle\int \frac{dx}{x\sqrt{a^2 + x^2}}$
14. $\displaystyle\int \frac{x + 1}{\sqrt{4 - x^2}}\,dx$
15. $\displaystyle\int \frac{dx}{\sqrt{2 - 5x^2}}$
16. $\displaystyle\int \frac{\sin\theta\,d\theta}{\sqrt{2 - \cos^2\theta}}$

17. $\displaystyle\int \frac{dx}{x\sqrt{x^2 - a^2}}$

18. $\displaystyle\int \frac{dx}{x\sqrt{a^2 - x^2}}$

19. $\displaystyle\int \frac{dx}{(a^2 - x^2)^{3/2}}$

20. $\displaystyle\int \frac{dx}{(a^2 + x^2)^2}$

9.5 INTEGRALS WITH $ax^2 + bx + c$

The general quadratic

$$f(x) = ax^2 + bx + c, \quad a \neq 0, \tag{1}$$

can be reduced to the form $a(u^2 + B)$ by completing the square:

$$\begin{aligned} ax^2 + bx + c &= a\left(x^2 + \frac{b}{a}x\right) + c \\ &= a\left(x^2 + \frac{b}{a}x + \frac{b^2}{4a^2}\right) + c - \frac{b^2}{4a} \\ &= a\left(x + \frac{b}{2a}\right)^2 + \frac{4ac - b^2}{4a}, \end{aligned}$$

and then substituting

$$u = x + \frac{b}{2a}, \qquad B = \frac{4ac - b^2}{4a^2}. \tag{2}$$

When the integrand involves the square root of

$$f(x) = ax^2 + bx + c,$$

we restrict attention to the case where $f(x)$ is not negative. If a is negative and B is positive, the square root is imaginary. We disregard this case and consider

$$\sqrt{a(u^2 + B)}$$

(1) when a is positive, and

(2) when a and B are both negative.

In the first case, the relation

$$\sqrt{a(u^2 + B)} = \sqrt{a}\sqrt{u^2 + B}$$

allows us to reduce the problem to a consideration of $\sqrt{u^2 + B}$, which we handle by the methods of Article 9.4. In the second case, when a and B are both negative, then $-a$ and $-B$ are positive and

$$\begin{aligned} \sqrt{a(u^2 + B)} &= \sqrt{-a(-B - u^2)} \\ &= \sqrt{-a}\sqrt{-B - u^2} \\ &\qquad (-a > 0, \quad -B > 0), \end{aligned}$$

and taking $-B = A^2$, say, we consider $\sqrt{A^2 - u^2}$ as in Article 9.4.

When the integrand does not involve an even root of $f(x)$, there need be no restriction on the signs of a and B. In these cases also, we try to make use of the trigonometric substitutions of the previous section, or we simply apply the integration formulas derived there.

Problem 1. $\displaystyle\int \frac{dx}{\sqrt{2x - x^2}}.$

Solution. The algebraic transformations proceed as follows:

$$\begin{aligned} \sqrt{2x - x^2} &= \sqrt{-(x^2 - 2x)} \\ &= \sqrt{-(x^2 - 2x + 1) + 1} \\ &= \sqrt{1 - u^2} \qquad (u = x - 1). \end{aligned}$$

Then, with $u = x - 1$,

$$du = dx$$

and

$$\int \frac{dx}{\sqrt{2x - x^2}} = \int \frac{du}{\sqrt{1 - u^2}} = \sin^{-1} u + C;$$

that is,

$$\int \frac{dx}{\sqrt{2x - x^2}} = \sin^{-1}(x - 1) + C.$$

Problem 2. $\displaystyle\int \frac{dx}{4x^2 + 4x + 2}.$

Solution. Again we start with the algebraic transformations:

$$\begin{aligned} 4x^2 + 4x + 2 &= 4(x^2 + x) + 2 \\ &= 4(x^2 + x + \tfrac{1}{4}) + (2 - \tfrac{4}{4}) \\ &= 4u^2 + 1 \qquad (u = x + \tfrac{1}{2}). \end{aligned}$$

Then we let

$$u = x + \tfrac{1}{2}, \qquad du = dx$$

and perform the integration as follows:

$$\int \frac{dx}{4x^2 + 4x + 2} = \int \frac{du}{4u^2 + 1} = \frac{1}{4}\int \frac{du}{u^2 + \frac{1}{4}} = \frac{1}{4}\int \frac{du}{u^2 + a^2} \quad (a = \tfrac{1}{2})$$

$$= \frac{1}{4}\left(\frac{1}{a}\tan^{-1}\frac{u}{a} + C\right) = \frac{1}{2}\tan^{-1}(2x + 1) + C.$$

Problem 3. Integrate the expression $\displaystyle\int \frac{(x + 1)\,dx}{\sqrt{2x^2 - 6x + 4}}$.

Solution. The quadratic part may be reduced as follows:

$$2x^2 - 6x + 4 = 2(x^2 - 3x) + 4 = 2(x^2 - 3x + \tfrac{9}{4}) + 4 - \tfrac{9}{2} = 2(u^2 - a^2), \quad \text{with} \quad u = x - \tfrac{3}{2},\ a = \tfrac{1}{2}.$$

Then

$$x = u + \tfrac{3}{2}, \qquad dx = du, \qquad x + 1 = u + \tfrac{5}{2},$$

and

$$\int \frac{(x + 1)\,dx}{\sqrt{2x^2 - 6x + 4}} = \int \frac{(u + \frac{5}{2})\,du}{\sqrt{2(u^2 - a^2)}} = \frac{1}{\sqrt{2}}\int \frac{u\,du}{\sqrt{u^2 - a^2}} + \frac{5}{2\sqrt{2}}\int \frac{du}{\sqrt{u^2 - a^2}}.$$

In the first term on the right-hand side of the last equation, we let

$$z = u^2 - a^2, \qquad dz = 2u\,du, \qquad u\,du = \tfrac{1}{2}\,dz,$$

and to the second we apply Eq. (8), Article 9.4:

$$\frac{1}{\sqrt{2}}\int \frac{u\,du}{\sqrt{u^2 - a^2}} = \frac{1}{2\sqrt{2}}\int \frac{dz}{\sqrt{z}} = \frac{1}{2\sqrt{2}}\int z^{-(1/2)}\,dz = \frac{1}{2\sqrt{2}}\cdot \tfrac{1}{2}\,z^{1/2} + C_1 = \sqrt{\frac{u^2 - a^2}{2}} + C_1$$

and

$$\frac{5}{2\sqrt{2}}\int \frac{du}{\sqrt{u^2 - a^2}} = \frac{5}{2\sqrt{2}}\ln|u + \sqrt{u^2 - a^2}| + C_2,$$

so that

$$\int \frac{(x + 1)\,dx}{\sqrt{2x^2 - 6x + 4}} = \sqrt{\frac{u^2 - a^2}{2}} + \frac{5}{2\sqrt{2}}\ln|u + \sqrt{u^2 - a^2}| + C$$

$$= \sqrt{\frac{x^2 - 3x + 2}{2}} + \frac{5}{2\sqrt{2}}\ln\left|x - \frac{3}{2} + \sqrt{x^2 - 3x + 2}\right| + C.$$

EXERCISES

1. $\displaystyle\int_1^3 \frac{dx}{x^2 - 2x + 5}$
2. $\displaystyle\int \frac{x\,dx}{\sqrt{x^2 - 2x + 5}}$
3. $\displaystyle\int \frac{(x + 1)\,dx}{\sqrt{2x - x^2}}$
4. $\displaystyle\int \frac{(x - 1)\,dx}{\sqrt{x^2 - 4x + 3}}$
5. $\displaystyle\int \frac{x\,dx}{\sqrt{5 + 4x - x^2}}$
6. $\displaystyle\int \frac{dx}{\sqrt{x^2 - 2x - 8}}$
7. $\displaystyle\int \frac{(1 - x)\,dx}{\sqrt{8 + 2x - x^2}}$
8. $\displaystyle\int \frac{x\,dx}{\sqrt{x^2 + 4x + 5}}$
9. $\displaystyle\int \frac{x\,dx}{x^2 + 4x + 5}$
10. $\displaystyle\int \frac{(2x + 3)\,dx}{4x^2 + 4x + 5}$

9.6 INTEGRATION BY THE METHOD OF PARTIAL FRACTIONS

In algebra we learned how to combine fractions over a common denominator. In integration, it is desirable to reverse the process and split a fraction into a sum of fractions having simpler denominators. The technique of doing this is known as the *method of partial fractions.*

Example

$$\frac{2}{x+1}+\frac{3}{x-3}=\frac{2(x-3)+3(x+1)}{(x+1)(x-3)}=\frac{5x-3}{(x+1)(x-3)}.$$

The reverse process consists in finding constants A and B such that

$$\frac{5x-3}{(x+1)(x-3)}=\frac{A}{x+1}+\frac{B}{x-3}.$$

(Pretend, for a moment, that we don't know that $A = 2$, $B = 3$ will work. Then we call A and B *undetermined coefficients.*) Clearing of fractions, we have

$$\begin{aligned}5x-3 &= A(x-3)+B(x+1)\\ &= (A+B)x-3A+B.\end{aligned}$$

This will be an identity in x if and only if coefficients of like powers of x on the two sides of the equation are equal:

$$A+B=5, \qquad -3A+B=-3.$$

These two equations in two unknowns determine

$$A=2, \qquad B=3.$$

More generally, suppose we wish to separate a rational function

$$\frac{f(x)}{g(x)} \tag{1}$$

into a sum of partial fractions. Success in doing so hinges upon two things:

1. The degree of $f(x)$ should be less than the degree of $g(x)$. If this is not the case, we first perform a long division, then work with the remainder term. This remainder can always be put into the required form.

2. The factors of $g(x)$ should be known. Theoretically, any polynomial $g(x)$ with real coefficients can be expressed as a product of real linear and quadratic factors. In practice, it may be difficult to perform the factorization.

Let us assume that these two conditions prevail. Let $x - r$ be a linear factor of $g(x)$. Suppose $(x - r)^m$ is the highest power of $x - r$ that divides $g(x)$. Then, to this factor, assign the sum of m partial fractions as follows:

$$\frac{A_1}{x-r}+\frac{A_2}{(x-r)^2}+\cdots+\frac{A_m}{(x-r)^m}.$$

Next, let $x^2 + px + q$ be a quadratic factor of $g(x)$. Suppose

$$(x^2+px+q)^n$$

is the highest power of this factor that divides $g(x)$. Then, to this factor, assign the sum of n partial fractions:

$$\frac{B_1x+C_1}{x^2+px+q}+\frac{B_2x+C_2}{(x^2+px+q)^2}+\cdots+\frac{B_nx+C_n}{(x^2+px+q)^n}.$$

Do this for each of the distinct linear and quadratic factors of $g(x)$. Then set the original fraction $f(x)/g(x)$ equal to the sum of all these partial fractions. Clear the resulting equation of fractions and arrange the terms in decreasing powers of x. Equate the coefficients of corresponding powers of x, and solve the resulting equations for the undetermined coefficients.

Problem 1. Express

$$\frac{4-2x}{(x^2+1)(x-1)^2}$$

as a sum of partial fractions.

Solution. Let

$$\frac{-2x+4}{(x^2+1)(x-1)^2}=\frac{Ax+B}{x^2+1}+\frac{C}{x-1}+\frac{D}{(x-1)^2}.$$

Then we have

$$\begin{aligned}-2x+4 &= (Ax+B)(x-1)^2+C(x-1)(x^2+1)\\ &\quad + D(x^2+1)\\ &= (A+C)x^3+(-2A+B-C+D)x^2\\ &\quad + (A-2B+C)x+(B-C+D).\end{aligned}$$

If this is to be an identity in x, it is both necessary and sufficient that the coefficient of each power of x that occurs should be the same on the left-hand side of the equation as it is on the right-hand side. Imposing this condition successively on the coefficients of x^3, x^2, x, and x^0, respectively, we have

$$\begin{aligned}0 &= A+C,\\ 0 &= -2A+B-C+D,\\ -2 &= A-2B+C,\\ 4 &= B-C+D.\end{aligned}$$

If we subtract the second equation from the fourth, we can solve for A:

$$2A = 4, \qquad A = 2.$$

Then from the first equation, we have

$$C = -A = -2.$$

Knowing A and C, we find B from the third equation:

$$B = 1.$$

Finally, from the fourth equation, we have

$$D = 4 - B + C = 1.$$

Hence

$$\frac{-2x+4}{(x^2+1)(x-1)^2} = \frac{2x+1}{x^2+1} - \frac{2}{x-1} + \frac{1}{(x-1)^2}.$$

Problem 2. Evaluate

$$\int \frac{x^5 - x^4 - 3x + 5}{x^4 - 2x^3 + 2x^2 - 2x + 1}\,dx.$$

Solution. The integrand is a fraction, but not a proper fraction. Hence we divide first, thus obtaining

$$\frac{x^5 - x^4 - 3x + 5}{x^4 - 2x^3 + 2x^2 - 2x + 1} = x + 1 + \frac{-2x+4}{x^4 - 2x^3 + 2x^2 - 2x + 1}. \tag{2}$$

The denominator factors as follows:

$$x^4 - 2x^3 + 2x^2 - 2x + 1 = (x^2+1)(x-1)^2.$$

By the result of Example 2, we have for the remainder term

$$\frac{-2x+4}{(x^2+1)(x-1)^2} = \frac{2x+1}{x^2+1} - \frac{2}{x-1} + \frac{1}{(x-1)^2}. \tag{3}$$

Hence, substituting from (3) into (2), multiplying by dx, and integrating, we have

$$\begin{aligned}&\int \frac{x^5 - x^4 - 3x + 5}{x^4 - 2x^3 + 2x^2 - 2x + 1}\,dx\\ &= \int \left[x + 1 + \frac{2x+1}{x^2+1} - \frac{2}{x-1} + \frac{1}{(x-1)^2}\right] dx\\ &= \frac{x^2}{2} + x + \ln(x^2+1) + \tan^{-1} x - 2\ln|x-1|\\ &\quad - \frac{1}{x-1} + K.\end{aligned}$$

In theory, any rational function of x can be integrated by the method of partial fractions. Once the necessary algebra has been done, the problem reduces to evaluating integrals of the following two types:

$$\int \frac{dx}{(x-r)^h}, \tag{4a}$$

$$\int \frac{(ax+b)\,dx}{(x^2+px+q)^k}. \tag{4b}$$

The first of these becomes simply $\int u^{-h}\,du$ when we let $u = x - r$. In the second type it is preferable to complete the square in the denominator:

$$x^2 + px + q = \left(x + \frac{p}{2}\right)^2 + q - \frac{p^2}{4},$$

and then let

$$u = x + \frac{p}{2}, \qquad c^2 = q - \frac{p^2}{4}.$$

Then

$$\begin{aligned}ax + b &= a\left(u - \frac{p}{2}\right) + b = au + b',\\ b' &= b - a\frac{p}{2},\end{aligned}$$

and we consider

$$\int \frac{au + b'}{(u^2 + c^2)^k}\,du = \frac{a}{2}\int (u^2 + c^2)^{-k}(2u\,du) + b'\int \frac{du}{(u^2 + c^2)^k}. \tag{5}$$

The first integral on the right in (5) readily yields to the substitution $z = u^2 + c^2$. In the second integral, we let

$$u = c\tan\theta, \quad \text{or} \quad \theta = \tan^{-1}\frac{u}{c},$$
$$du = c\sec^2\theta\,d\theta,$$
$$u^2 + c^2 = c^2\sec^2\theta,$$

and obtain

$$\int \frac{du}{(u^2 + c^2)^k} = c^{1-2k}\int \cos^{2k-2}\theta\,d\theta. \tag{6}$$

In the next article we shall obtain a reduction formula for $\int \cos^n\theta\,d\theta$, which may be used to evaluate the integral in (6). The trigonometric identity

$$\cos^2\theta = \frac{1 + \cos 2\theta}{2} \tag{7}$$

may also be used to advantage.

Problem 3. $\displaystyle\int \frac{dx}{1 - x^2}.$

Solution. By partial fractions,

$$\frac{1}{1 - x^2} = \frac{A}{1 - x} + \frac{B}{1 + x},$$
$$1 = A(1 + x) + B(1 - x)$$
$$= A + B + (A - B)x,$$
$$A + B = 1, \qquad A - B = 0,$$
$$A = B = \tfrac{1}{2}.$$

Therefore

$$\int \frac{1}{1 - x^2}\,dx = \int\left(\frac{\frac{1}{2}}{1 - x} + \frac{\frac{1}{2}}{1 + x}\right)dx$$
$$= -\tfrac{1}{2}\ln|1 - x| + \tfrac{1}{2}\ln|1 + x| + C$$
$$= \tfrac{1}{2}\ln\left|\frac{1 + x}{1 - x}\right| + C. \tag{8}$$

Problem 4. $\displaystyle\int \sec\theta\,d\theta.$

Solution. We did this in Problem 5, Article 9.2, by a trick. We can now do it in a more straightforward manner by writing $\sec\theta$ as $1/\cos\theta$ and proceeding as follows:

$$\int \sec\theta\,d\theta = \int \frac{d\theta}{\cos\theta} = \int \frac{\cos\theta\,d\theta}{\cos^2\theta}$$
$$= \int \frac{dx}{1 - x^2} \qquad (x = \sin\theta)$$
$$= \tfrac{1}{2}\ln\left|\frac{1 + x}{1 - x}\right| + C \quad \text{(Eq. 8)}$$
$$= \ln\sqrt{\frac{1 + \sin\theta}{1 - \sin\theta}} + C. \tag{9}$$

This answer is equivalent to Eq. (3), Article 9.2, as may be seen by multiplying numerator and denominator by $1 + \sin\theta$:

$$\sqrt{\frac{1 + \sin\theta}{1 - \sin\theta}} = \sqrt{\frac{(1 + \sin\theta)^2}{1 - \sin^2\theta}}$$
$$= \left|\frac{1 + \sin\theta}{\cos\theta}\right|$$
$$= |\sec\theta + \tan\theta|.$$

EXERCISES

1. $\displaystyle\int \frac{x\,dx}{x^2 + 4x - 5}$
2. $\displaystyle\int \frac{x\,dx}{x^2 - 2x - 3}$
3. $\displaystyle\int \frac{(x + 1)\,dx}{x^2 + 4x - 5}$
4. $\displaystyle\int \frac{x^2\,dx}{x^2 + 2x + 1}$
5. $\displaystyle\int \frac{dx}{x(x + 1)^2}$
6. $\displaystyle\int \frac{dx}{(x + 1)(x^2 + 1)}$
7. $\displaystyle\int \frac{dx}{x(x^2 + x + 1)}$
8. $\displaystyle\int \frac{\sin\theta\,d\theta}{\cos^2\theta + \cos\theta - 2}$
9. $\displaystyle\int \frac{e^t\,dt}{e^{2t} + 3e^t + 2}$
10. $\displaystyle\int \frac{dx}{(x^2 + 1)^2}$
11. $\displaystyle\int \frac{x^4\,dx}{(x^2 + 1)^2}$

9.7 INTEGRATION BY PARTS

There are really just two general methods of integration. One of these is the method of substitution, which we have illustrated in Articles 9.2 through 9.5. The method of partial fractions is really not a method of *integration* so much as it is a method of *algebraic transformation* of a rational function into an integrable form. The second general method of integration, called *integration by parts,* depends on the formula for the differential of a product:

$$d(uv) = u\,dv + v\,du \quad \text{or} \quad u\,dv = d(uv) - v\,du.$$

When this is integrated, we have

$$\int u\,dv = uv - \int v\,du + C. \tag{1}$$

By way of comment on the formula (1) for integration by parts, we observe that it expresses one integral $\int u\,dv$ in terms of a second integral $\int v\,du$. If, by proper choice of u and dv, the second integral is simpler than the first, we may be able to evaluate it quite simply and thus arrive at an answer.

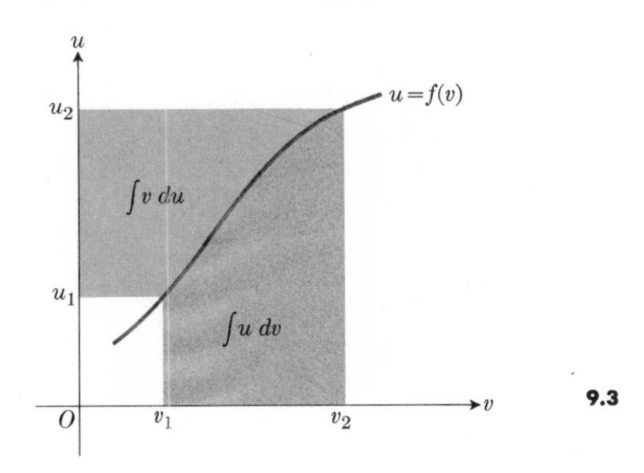

9.3

Note. In a definite integral, appropriate limits must be supplied. We may then interpret the formula for integration by parts,

$$\int_{(1)}^{(2)} u\,dv = uv\Big]_{(1)}^{(2)} - \int_{(1)}^{(2)} v\,du,$$

geometrically in terms of areas (see Fig. 9.3).

Problem 1. $\displaystyle\int \ln x\,dx.$

Solution. If we try to match $\ln x\,dx$ with $u\,dv$, we may take

$$u = \ln x \quad \text{and} \quad dv = dx.$$

Then, to use (1), we see that we also require

$$du = d(\ln x) = \frac{dx}{x},$$

$$v = \int dv = \int dx = x + C_1,$$

so that

$$\begin{aligned}\int \ln x\,dx &= (\ln x)(x + C_1) - \int (x + C_1)\frac{dx}{x} + C_2\\ &= x\ln x + C_1\ln x - \int dx - \int C_1\frac{dx}{x} + C_2\\ &= x\ln x + C_1\ln x - x - C_1\ln x + C_2\\ &= x\ln x - x + C_2.\end{aligned}$$

Note that the first constant of integration C_1 does not appear in the final answer. This is generally true, for if we write $v + C_1$ in place of v in the right side of (1), we obtain

$$\begin{aligned}&u(v + C_1) - \int (v + C_1)\,du\\ &\qquad = uv + C_1u - \int v\,du - \int C_1\,du = uv - \int v\,du,\end{aligned}$$

where it is understood that a constant of integration must still be added in the final result. It is therefore customary to drop the first constant of integration when determining v as $\int dv$. On the other hand, there is occasionally an advantage in taking C_1 equal to some specific constant. For instance, if we consider

$$\int \ln(x + 1)\,dx$$

and let

$$u = \ln(x + 1) \quad \text{and} \quad dv = dx,$$

then

$$du = \frac{dx}{x + 1} \quad \text{and} \quad v = x + C_1.$$

Equation (1) then gives

$$\int \ln(x + 1)\,dx = (x + C_1)\ln(x + 1) - \int \frac{x + C_1}{x + 1}\,dx.$$

Now taking $C_1 = 1$ greatly simplifies the second integral and leads to the result

$$\int \ln(x+1)\,dx = (x+1)\ln(x+1) - x + C_2.$$

Problem 2. $\int \tan^{-1} x\,dx.$

Solution. This is typical of the inverse trigonometric functions. Let

$$u = \tan^{-1} x, \qquad dv = dx.$$

Then

$$du = \frac{dx}{1+x^2} \quad \text{and} \quad v = x,$$

so that

$$\int \tan^{-1} x\,dx = x\tan^{-1} x - \int \frac{x\,dx}{1+x^2}$$
$$= x\tan^{-1} x - \tfrac{1}{2}\ln(1+x^2) + C.$$

Sometimes an integration by parts must be repeated to obtain an answer, as in the following example.

Problem 3. $\int x^2 e^x\,dx.$

Solution. Let $u = x^2, \quad dv = e^x\,dx.$

Then $du = 2x\,dx$ and $v = e^x$,

so that $\int x^2 e^x\,dx = x^2 e^x - 2\int xe^x\,dx.$

The integral on the right is similar to the original integral, except that we have reduced the power of x from 2 to 1. If we could now reduce it from 1 to 0, we could see success ahead. In $\int xe^x\,dx$, we therefore let

$$U = x, \qquad dV = e^x\,dx,$$

so that

$$dU = dx \quad \text{and} \quad V = e^x.$$

Then

$$\int xe^x\,dx = xe^x - \int e^x\,dx = xe^x - e^x + c,$$

and

$$\int x^2 e^x\,dx = x^2 e^x - 2xe^x + 2e^x + C.$$

Problem 4. Obtain a reduction formula for

$$J_n \equiv \int \cos^n x\,dx.$$

Solution. We may think of $\cos^n x$ as $\cos^{n-1} x \cdot \cos x$. Then we let

$$u = \cos^{n-1} x, \qquad dv = \cos x\,dx,$$

so that

$$du = (n-1)\cos^{n-2} x(-\sin x\,dx) \quad \text{and} \quad v = \sin x.$$

Hence

$$J_n = \cos^{n-1} x \sin x + (n-1)\int \sin^2 x \cos^{n-2} x\,dx$$
$$= \cos^{n-1} x \sin x + (n-1)\int (1-\cos^2 x)\cos^{n-2} x\,dx,$$

or

$$\int \cos^n x\,dx = \cos^{n-1} x \sin x + (n-1)\int \cos^{n-2} x\,dx - (n-1)\int \cos^n x\,dx.$$

The last integral on the right may now be transposed to the left to give

$$[1 + (n-1)]J_n = n\,J_n.$$

We then divide by n, and the final result is

$$\int \cos^n x\,dx = \frac{\cos^{n-1} x \sin x}{n} + \frac{n-1}{n}\int \cos^{n-2} x\,dx. \tag{2}$$

This allows us to reduce the exponent on $\cos x$ by 2, and is a very useful formula. When n is a positive integer, we may apply the formula repeatedly until the remaining integral is either

$$\int \cos x\,dx = \sin x + C$$

or

$$\int \cos^0 x\,dx = \int dx = x + C.$$

For example, with $n = 4$, we get

$$\int \cos^4 x\,dx = \frac{\cos^3 x \sin x}{4} + \frac{3}{4}\int \cos^2 x\,dx,$$

and with $n = 2$,

$$\int \cos^2 x\,dx = \frac{\cos x \sin x}{2} + \frac{1}{2}\int dx.$$

Therefore

$$\int \cos^4 x\,dx = \frac{\cos^3 x \sin x}{4} + \frac{3}{4}\left(\frac{\cos x \sin x}{2} + \frac{1}{2}x\right) + C.$$

The reader may find it instructive to derive the companion formula:

$$\int \sin^n x\,dx = -\frac{\sin^{n-1} x \cos x}{n} + \frac{n-1}{n}\int \sin^{n-2} x\,dx. \tag{3}$$

The integral in the following example occurs in electrical engineering problems. Its evaluation requires two integrations by parts, followed by solving for the unknown integral in a method analogous to that used above in finding reduction formulas.

Problem 5. $\int e^{ax} \cos bx\,dx.$

Solution. Let $u = e^{ax}$, $dv = \cos bx\,dx$.

Then $du = ae^{ax}\,dx$ and $v = \frac{1}{b}\sin bx,$

so that

$$\int e^{ax}\cos bx\,dx = \frac{e^{ax}\sin bx}{b} - \frac{a}{b}\int e^{ax}\sin bx\,dx.$$

The second integral is like the first except that it has $\sin bx$ in place of $\cos bx$. If we apply integration by parts to it, letting

$$U = e^{ax} \quad\text{and}\quad dV = \sin bx\,dx,$$

then

$$dU = ae^{ax}\,dx \quad\text{and}\quad V = -\frac{1}{b}\cos bx,$$

so that

$$\int e^{ax}\cos bx\,dx = \frac{e^{ax}\sin bx}{b} - \frac{a}{b}\left[-\frac{e^{ax}\cos bx}{b} + \frac{a}{b}\int e^{ax}\cos bx\,dx\right].$$

Now the unknown integral appears on the left with a coefficient of unity and on the right with a coefficient of $-a^2/b^2$. Transposing this term to the left and dividing by the new coefficient

$$1 + \frac{a^2}{b^2} = \frac{a^2+b^2}{b^2},$$

we have

$$\int e^{ax}\cos bx\,dx = e^{ax}\left(\frac{b\sin bx + a\cos bx}{a^2+b^2}\right) + C. \tag{4}$$

EXERCISES

1. $\int x \ln x\,dx$
2. $\int x^n \ln ax\,dx \quad (n \neq -1)$
3. $\int x \tan^{-1} x\,dx$
4. $\int \sin^{-1} ax\,dx$
5. $\int x \sin ax\,dx$
6. $\int x^2 \cos ax\,dx$
7. $\int x \sec^2 ax\,dx$
8. $\int e^{ax}\sin bx\,dx$
9. $\int \sin(\ln x)\,dx$
10. $\int \cos(\ln x)\,dx$
11. $\int \ln(a^2 + x^2)\,dx$
12. $\int x\cos(2x+1)\,dx$
13. $\int x \sin^{-1} x\,dx$
14. $\int_1^2 x \sec^{-1} x\,dx$
15. $\int_1^4 \sec^{-1}\sqrt{x}\,dx$
16. $\int x^2 \tan^{-1} x\,dx$

Derive the following reduction formulas and apply each one to the specific problem given.

17. (a) $\int x^m(\ln x)^n\,dx = \frac{x^{m+1}(\ln x)^n}{m+1} - \frac{n}{m+1}\int x^m(\ln x)^{n-1}\,dx$

    (b) $\int x^3(\ln x)^2\,dx$

18. (a) $\int \sin^n x\,dx = -\frac{\sin^{n-1} x\cos x}{n} + \frac{n-1}{n}\int \sin^{n-2} x\,dx$

    (b) $\int_0^{\pi/6} \sin^4 3x\,dx$

19. (a) $\int x^n e^x\,dx = x^n e^x - n\int x^{n-1}e^x\,dx,$

    (b) $\int x^3 e^x\,dx$

20. (a) $\int \sec^n x\,dx = \frac{\sec^{n-2} x\tan x}{(n-1)} + \frac{n-2}{n-1}\int \sec^{n-2} x\,dx$

    (b) $\int \sec^3 x\,dx$ [See Eq. (3), Article 9.2.]

    (c) $\int \sqrt{a^2+x^2}\,dx$ [*Hint.* Let $x = a\tan\theta$.]

21. Find the second-degree polynomial $P(x)$ that has the following properties:
   (a) $P(0) = 1$, (b) $P'(0) = 0$,
   (c) the indefinite integral
$$\int \frac{P(x)\,dx}{x^3(x-1)^2}$$
   is a rational function (that is, no logarithmic terms occur in the answer).

9.8 INTEGRATION OF RATIONAL FUNCTIONS OF sin x AND cos x, AND OTHER TRIGONOMETRIC INTEGRALS

It has been discovered that the substitution

$$z = \tan \frac{x}{2} \tag{1}$$

enables one to reduce the problem of integrating any rational function of $\sin x$ and $\cos x$ to a problem involving a rational function of z. This in turn can be integrated by the method of partial fractions discussed in Article 9.6. Thus the substitution (1) is a very powerful tool. This method is cumbersome, however, and is used only when the simpler methods outlined previously have failed.

To see the effect of the substitution, we calculate

$$\cos x = 2\cos^2\frac{x}{2} - 1 = \frac{2}{\sec^2\frac{x}{2}} - 1$$

$$= \frac{2}{1+\tan^2\frac{x}{2}} - 1 = \frac{2}{1+z^2} - 1,$$

or

$$\cos x = \frac{1-z^2}{1+z^2}; \tag{2a}$$

and

$$\sin x = 2\sin\frac{x}{2}\cos\frac{x}{2} = 2\frac{\sin\frac{x}{2}}{\cos\frac{x}{2}}\cdot\cos^2\frac{x}{2}$$

$$= 2\tan\frac{x}{2}\cdot\frac{1}{\sec^2\frac{x}{2}} = \frac{2\tan\frac{x}{2}}{1+\tan^2\frac{x}{2}},$$

or

$$\sin x = \frac{2z}{1+z^2}. \tag{2b}$$

Finally,

$$x = 2\tan^{-1} z,$$

so that

$$dx = \frac{2\,dz}{1+z^2}. \tag{2c}$$

Example 1.
$$\int \sec x\,dx = \int \frac{dx}{\cos x}$$

becomes

$$\int \frac{2\,dz}{1+z^2}\cdot\frac{1+z^2}{1-z^2} = \int \frac{2\,dz}{1-z^2}.$$

To this we apply the method of partial fractions:

$$\frac{2}{1-z^2} = \frac{A}{1-z} + \frac{B}{1+z},$$
$$2 = A(1+z) + B(1-z)$$
$$= (A+B) + (A-B)z,$$

which requires

$$A + B = 2, \qquad A - B = 0.$$

Hence

$$A = B = 1$$

and

$$\int \frac{2\,dz}{1-z^2} = \int \frac{dz}{1-z} + \int \frac{dz}{1+z}$$
$$= -\ln|1-z| + \ln|1+z| + C$$
$$= \ln\left|\frac{1+z}{1-z}\right| + C$$
$$= \ln\left|\frac{1+\tan\frac{x}{2}}{1-\tan\frac{x}{2}}\right| + C$$
$$= \ln\left|\frac{\tan\frac{\pi}{4}+\tan\frac{x}{2}}{1-\tan\frac{\pi}{4}\tan\frac{x}{2}}\right| + C$$
$$= \ln\left|\tan\left(\frac{\pi}{4}+\frac{x}{2}\right)\right| + C.$$

That is,

$$\int \sec x\,dx = \ln\left|\tan\left(\frac{\pi}{4}+\frac{x}{2}\right)\right| + C \tag{3}$$

is an alternative form, which may be used in place of Eq. (3), Article 9.2, or Eq. (9), Article 9.6.

Example 2. $$\int \frac{dx}{1+\cos x}$$

becomes

$$\int \frac{2\,dz}{1+z^2}\,\frac{1+z^2}{2} = \int dz = z + C = \tan\frac{x}{2} + C.$$

Example 3. $$\int \frac{dx}{2+\sin x}$$

becomes

$$\begin{aligned}
\int \frac{2\,dz}{1+z^2}\left[\frac{1+z^2}{2+2z+2z^2}\right] &= \int \frac{dz}{z^2+z+1} \\
&= \int \frac{dz}{(z+\frac{1}{2})^2+\frac{3}{4}} \\
&= \int \frac{du}{u^2+a^2} \qquad \left[\begin{aligned} u &= z+\tfrac{1}{2}, \\ a &= \sqrt{3}/2\end{aligned}\right] \\
&= \frac{1}{a}\tan^{-1}\frac{u}{a} + C \\
&= \frac{2}{\sqrt{3}}\tan^{-1}\frac{2z+1}{\sqrt{3}} + C \\
&= \frac{2}{\sqrt{3}}\tan^{-1}\frac{1+2\tan\frac{x}{2}}{\sqrt{3}} + C.
\end{aligned}$$

We have already made repeated use of trigonometric identities to aid us in evaluating integrals. The following types of integrals:

$$\begin{aligned}
&\int \sin mx \sin nx\,dx, \\
&\int \sin mx \cos nx\,dx, \qquad (4) \\
&\int \cos mx \cos nx\,dx,
\end{aligned}$$

arise in connection with alternating-current theory, heat transfer problems, bending of beams, cable stress analysis in suspension bridges, and many other places where trigonometric series (or Fourier series) are applied to problems in mathematics, science, and engineering. The integrals in (4) can be evaluated by the method of integration by parts, but two such integrations are required in each case. A simpler way to evaluate them is to exploit the trigonometric identities

$$\sin mx \sin nx = \tfrac{1}{2}[\cos (m-n)x - \cos (m+n)x], \tag{5a}$$

$$\sin mx \cos nx = \tfrac{1}{2}[\sin (m-n)x + \sin (m+n)x], \tag{5b}$$

$$\cos mx \cos nx = \tfrac{1}{2}[\cos (m-n)x + \cos (m+n)x]. \tag{5c}$$

These identities follow at once from

$$\begin{aligned}
\cos (A+B) &= \cos A \cos B - \sin A \sin B, \\
\cos (A-B) &= \cos A \cos B + \sin A \sin B,
\end{aligned} \tag{6a}$$

and

$$\begin{aligned}
\sin (A+B) &= \sin A \cos B + \cos A \sin B, \\
\sin (A-B) &= \sin A \cos B - \cos A \sin B.
\end{aligned} \tag{6b}$$

If, for example, we add the two equations in (6a) and then divide by 2, we obtain (5c) by taking $A = mx$ and $B = nx$. The identity in (5a) is obtained in a similar fashion by subtracting the first equation in (6a) from the second equation. Finally, if we add the two equations in (6b) we are led to the identity in (5b).

Example 4

$$\begin{aligned}
\int \sin 3x \cos 5x\,dx &= \tfrac{1}{2}\int [\sin (-2x) + \sin 8x]\,dx \\
&= \tfrac{1}{2}\int (\sin 8x - \sin 2x)\,dx \\
&= -\frac{\cos 8x}{16} + \frac{\cos 2x}{4} + C.
\end{aligned}$$

EXERCISES

1. $\displaystyle\int_0^{\pi} \frac{dx}{1+\sin x}$

2. $\displaystyle\int_{\pi/2}^{\pi} \frac{dx}{1-\cos x}$

3. $\displaystyle\int \frac{dx}{1-\sin x}$

4. $\displaystyle\int_0^{\pi/2} \frac{dx}{2+\cos x}$

5. $\int \frac{\cos x\, dx}{2 - \cos x}$

6. $\int_0^{\pi/2} \cos 3x \sin 2x\, dx$

7. $\int_{-\pi}^{\pi} \sin 3x \sin 2x\, dx$

8. $\int_{-\pi}^{\pi} \sin^2 3x\, dx$

9. Two functions f and g are said to be *orthogonal* on an interval $[a, b]$ if

$$\int_a^b f(x)\, g(x)\, dx = 0.$$

(a) Prove that $\sin mx$ and $\sin nx$ are orthogonal on any interval of length 2π provided that m and n are integers such that $m^2 \neq n^2$.

(b) Prove the same for $\sin mx$ and $\cos nx$.

(c) Prove the same for $\cos mx$ and $\cos nx$.

9.9 FURTHER SUBSTITUTIONS

Some integrals involving fractional powers of the variable x may be simplified by substituting $x = z^n$, where n is the least common multiple of the denominators of the exponents.

Example.

$$\int \frac{\sqrt{x}\, dx}{1 + \sqrt[4]{x}}$$

may be simplified by taking

$$x = z^4, \qquad dx = 4z^3\, dz.$$

This leads to

$$\begin{aligned}
\int \frac{\sqrt{x}\, dx}{1 + \sqrt[4]{x}} &= \int \frac{z^2 \cdot 4z^3\, dz}{1 + z} \\
&= 4\int \left(z^4 - z^3 + z^2 - z + 1 - \frac{1}{z+1}\right) dz \\
&= 4\left[\frac{z^5}{5} - \frac{z^4}{4} + \frac{z^3}{3} - \frac{z^2}{2} + z - \ln|z + 1|\right] + C \\
&= \tfrac{4}{5}x^{5/4} - x + \tfrac{4}{3}x^{3/4} - 2x^{1/2} + 4x^{1/4} \\
&\qquad - 4\ln|1 + \sqrt[4]{x}| + C.
\end{aligned}$$

In $\int x^3\sqrt{x^2 + a^2}\, dx$, on the other hand, the substitutions

$$z^2 = x^2 + a^2, \quad \text{or} \quad x^2 = z^2 - a^2,$$

and

$$2x\, dx = 2z\, dz$$

lead to

$$\begin{aligned}
\int x^3\sqrt{x^2 + a^2}\, dx &= \int x^2\sqrt{x^2 + a^2} \cdot x\, dx \\
&= \int (z^2 - a^2) \cdot z^2 \cdot dz \\
&= \int (z^4 - a^2z^2)\, dz \\
&= \frac{(x^2 + a^2)^{5/2}}{5} - \frac{a^2(x^2 + a^2)^{3/2}}{3} + C \\
&= \frac{3x^2 - 2a^2}{15}(x^2 + a^2)^{3/2} + C.
\end{aligned}$$

Even when it is not clear at the start that a substitution will work, it is advisable to try one that seems reasonable and pursue it until it either gives results or appears to make matters worse. In the latter case, try something else! Sometimes a chain of substitutions $u = f(x)$, $v = g(u)$, $z = h(v)$, and so on, will produce results when it is by no means obvious that this will work. The criterion of success is whether the new integrals so obtained appear to be simpler than the original integral. Here it is handy to remember that any rational function of x can be integrated by the method of partial fractions and that any rational function of $\sin x$ and $\cos x$ can be integrated by using the substitution $z = \tan x/2$. If we can reduce a given integral to one of these types, we then know how to finish the job.

Even in these cases, however, it is frequently simpler to use special methods suggested by the particular integrand rather than to use the general methods. For instance, one would hardly let $z = \tan x/2$ to evaluate $\int \sin x\, dx = -\cos x + C$. And reference to Eqs. (5) and (6), Article 9.6 and the accompanying discussion shows that a trigonometric substitution is used to evaluate certain of the integrals that arise in the method of partial fractions. It may be more convenient to use the same substitution *before* going through the algebraic reductions instead of *after* doing so.

Example. Consider the integral

$$\int_{-1}^{+1} \sqrt{\frac{1 + x}{1 - x}}\, dx, \tag{1}$$

which appears in the so-called "lifting line theory" in aerodynamics. If we let

$$\frac{1+x}{1-x} = z^2, \tag{2}$$

we shall, of course, get rid of the radical. Whether this is really a good substitution or not depends upon how complicated the expression is for dx in terms of z and dz. To settle this question, we solve for x as follows:

$$1 + x = z^2 - xz^2,$$

$$x(z^2 + 1) = z^2 - 1,$$

$$x = \frac{z^2 - 1}{z^2 + 1}. \tag{3}$$

Hence

$$dx = \frac{4z\,dz}{(z^2+1)^2}$$

and the integral becomes

$$\int \sqrt{\frac{1+x}{1-x}}\,dx = \int \frac{4z^2\,dz}{(z^2+1)^2}.$$

Now the integrand on the right is a rational function of z and we could proceed by the method of partial fractions. But the $z^2 + 1$ in the denominator also lends itself to the substitution

$$z = \tan\theta. \tag{4}$$

We try this, and have

$$dz = \sec^2\theta\,d\theta,$$

$$z^2 + 1 = \tan^2\theta + 1 = \sec^2\theta,$$

so that

$$\int \frac{4z^2\,dz}{(z^2+1)^2} = \int \frac{4\tan^2\theta\cdot\sec^2\theta\,d\theta}{\sec^4\theta}$$

$$= 4\int \frac{\sin^2\theta}{\cos^2\theta}\cdot\frac{1}{\sec^2\theta}\,d\theta$$

$$= 4\int \sin^2\theta\,d\theta,$$

and the last integral can now be evaluated at once to give

$$4\int \sin^2\theta\,d\theta = 2\left[\theta - \frac{\sin 2\theta}{2}\right] + C. \tag{5}$$

We now have two alternative courses of action for evaluating the *definite integral* (1):

(a) We may reverse the substitutions (3) and (4) and replace (5) by its equivalent in terms of x. Then the definite integral (1) could be evaluated by substituting the limits of integration. Or,

(b) we may determine new limits of integration corresponding to the new variable of integration.

We illustrate the second line of attack. In the original integral (1), x varies from -1 to $+1$. As this happens, the left side of (2) varies from 0 to $+\infty$, and since we have tacitly taken

$$z = \sqrt{\frac{1+x}{1-x}}$$

as the *positive* square root, we see that then z varies from 0 to $+\infty$. Finally, from (4), we have $\theta = \tan^{-1} z$, so that θ varies from 0 to $\pi/2$ as z varies from 0 to $+\infty$. Combining these results, we have

$$\int_{-1}^{+1} \sqrt{\frac{1+x}{1-x}}\,dx = \int_0^{\infty} \frac{4z^2\,dz}{(z^2+1)^2}$$

$$= 4\int_0^{\pi/2} \sin^2\theta\,d\theta$$

$$= 2\left[\theta - \frac{\sin 2\theta}{2}\right]_0^{\pi/2} = \pi.$$

The reader will find it instructive to evaluate the integral in (1) by making the alternative substitution

$$x = \cos 2t,$$

$$dx = -2\sin 2t\,dt = -4\sin t\cos t\,dt,$$

which is particularly well adapted to the integrand in question, since

$$1 + \cos 2t = 2\cos^2 t,$$

$$1 - \cos 2t = 2\sin^2 t.$$

But the shortest way of all is to notice that for

$$-1 < x < 1,$$

we have

$$\sqrt{\frac{1+x}{1-x}} = \frac{1+x}{\sqrt{1-x^2}},$$

and hence

$$\int \sqrt{\frac{1+x}{1-x}}\,dx = \int \frac{1+x}{\sqrt{1-x^2}}\,dx$$

$$= \sin^{-1} x - \sqrt{1-x^2} + C. \tag{6}$$

EXERCISES

1. $\int \frac{1-\sqrt{x}}{1+\sqrt{x}}\,dx$

2. $\int \frac{dx}{a+b\sqrt{x}}$ (a, b constant)

3. $\int x^2\sqrt{x+a}\,dx$

4. $\int \frac{\sqrt{x+a}\,dx}{x+b}$ $(b > a)$

5. $\int \sqrt{\frac{a+x}{b+x}}\,dx$ $(b \neq a)$

6. $\int \frac{dx}{x+\sqrt{x^2+a^2}}$ $(a > 0)$

7. $\int_0^1 \frac{x^3\,dx}{(x^2+1)^{3/2}}$

8. $\int \frac{dx}{x(ax^n+c)}$ $(anc \neq 0)$

9. $\int_3^8 \frac{(t+2)\,dt}{t\sqrt{t+1}}$

10. $\int \frac{dx}{x-x^{2/3}}$

11. $\int_0^3 \frac{\sqrt{x+1}-1}{\sqrt{x+1}+1}\,dx$

12. $\int \frac{dx}{x(1-\sqrt[4]{x})}$

13. $\int \frac{y^{2/3}\,dy}{y+1}$

14. $\int \frac{z^5\,dz}{\sqrt{1+z^3}}$

15. $\int \frac{\sqrt{t^3-1}}{t}\,dt$

16. $\int_0^{\ln 2} \frac{dw}{(1+e^w)}$

17. $\int_0^{\pi/2} \frac{\sin 2\theta\,d\theta}{2+\cos\theta}$

18. $\int \frac{\sin x\,dx}{\tan x+\cos x}$

19. $\int \frac{dx}{x^3+1}$

20. $\int \frac{dy}{y^{1/3}-y^{1/2}}$

9.10 IMPROPER INTEGRALS

An integral of the type discussed in the previous article,

$$\int_{-1}^{+1} \sqrt{\frac{1+x}{1-x}}\,dx,$$

is called an *improper* integral because the function

$$f(x) = \sqrt{\frac{1+x}{1-x}}$$

which appears after the integral sign becomes infinite at one of the limits of integration, in this case at $x = +1$. In general, integrals of the following two types are called improper integrals:

(a) $$\int_a^b f(x)\,dx,$$

where $f(x)$ becomes infinite at a value of x in $[a, b]$,

(b) $$\int_a^{\infty} f(x)\,dx \quad \text{or} \quad \int_{-\infty}^{b} f(x)\,dx,$$

where one or both of the limits of integration are infinite.

In the Example, Article 9.9, we had

$$\int_{-1}^{1} \sqrt{\frac{1+x}{1-x}}\,dx = \int_0^{\infty} \frac{4z^2\,dz}{(z^2+1)^2}$$

for

$$z = \sqrt{\frac{1+x}{1-x}},$$

where the integral on the left is an improper integral of the first type and the integral on the right is an improper integral of the second type.

Figure 9.4 (a) illustrates the case where the function becomes infinite at one of the limits of integration. If we interpret the integral in terms of area under the curve $y = f(x)$, say from -1 to $+1$ as in the figure, we see that the curve extends to infinity as x approaches 1 and the area from -1 to $+1$ is not well defined. Nevertheless, we can certainly define the area from $x = -1$ to $x = b$, where b is any positive number *less* than one; for example, we can take $b = 0.999$. This defines a function of b:

$$g(b) = \int_{-1}^{b} f(x)\,dx.$$

If this function has a finite limit as b approaches $+1$ from the left, we *define* this limit to be the value of the improper integral

$$\int_{-1}^{+1} f(x)\,dx = \lim_{b\to 1-} \int_{-1}^{b} f(x)\,dx.$$

In this case we also say that the improper integral *converges*. On the other hand, we say that the inte-

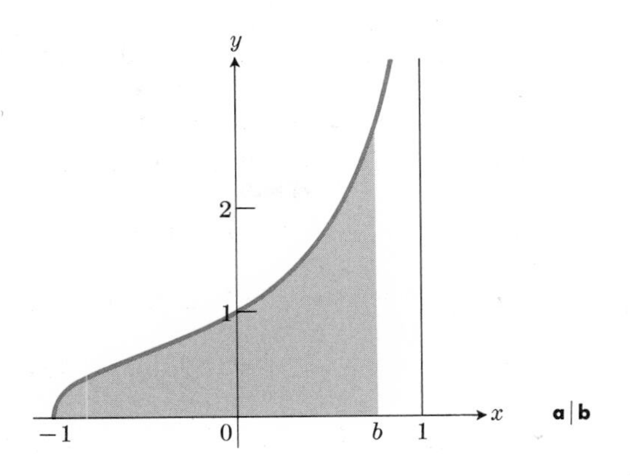

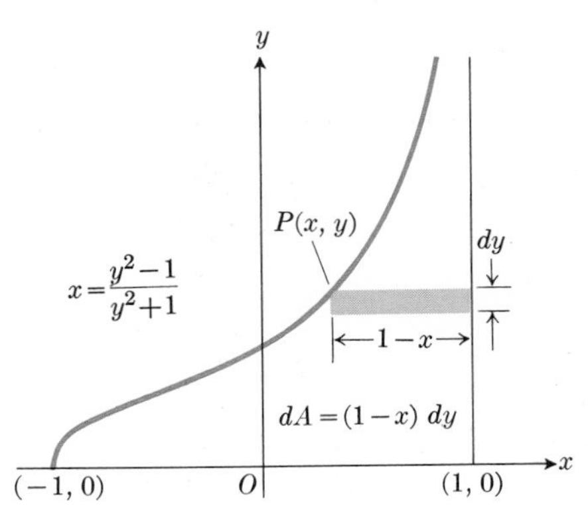

a | b

9.4

gral *diverges* if the function g has no definite finite limit as $b \to 1-$.

Example 1. In the particular integral under discussion, we find from Eq. (6) of Article 9.9 that

$$\int_{-1}^{b} \sqrt{\frac{1+x}{1-x}}\,dx = \sin^{-1} x - \sqrt{1-x^2}\Big]_{-1}^{b}$$
$$= \sin^{-1} b - \sqrt{1-b^2} + \frac{\pi}{2}.$$

This is the number denoted above by $g(b)$. When b is slightly less than $+1$, $\sin^{-1} b$ is slightly less than $\pi/2$ and $\sqrt{1-b^2}$ is nearly zero. Hence

$$\lim_{b\to 1-} \int_{-1}^{b} \sqrt{\frac{1+x}{1-x}}\,dx = \lim_{b\to 1-}\left[\sin^{-1} b - \sqrt{1-b^2} + \frac{\pi}{2}\right]$$
$$= \frac{\pi}{2} - 0 + \frac{\pi}{2} = \pi.$$

The given integral therefore converges and its value is π.

The same result is obtained if, instead of finding the area by summing vertical elements (Fig. 9.4a), we sum horizontal elements (Fig. 9.4b). Then we find

$$A = \int_{y=0}^{\infty} (1-x)\,dy = \int_{0}^{\infty} \frac{2}{y^2+1}\,dy.$$

This time, the integral to be evaluated is an improper integral of the second type because the range of integration extends to infinity. In this case we investigate the integral from $y = 0$ to $y = c$ for large values of c:

$$\int_0^c \frac{2\,dy}{y^2+1} = 2\tan^{-1} c.$$

We then *define* the integral from 0 to ∞ as the limit of this as $c \to \infty$ (if the limit exists); that is,

$$\int_0^{\infty} \frac{2\,dy}{y^2+1} = \lim_{c\to\infty} \int_0^c \frac{2\,dy}{y^2+1}$$
$$= \lim_{c\to\infty} (2\tan^{-1} c) = \pi.$$

The limit does exist, since $\tan^{-1} c$ approaches $\pi/2$ as c increases indefinitely.

Example 2. As a second example, we consider

$$\int_0^1 \frac{dx}{x}.$$

The function f in this case is defined by

$$f(x) = \frac{1}{x},$$

which becomes infinite at $x = 0$. We cut off the point $x = 0$ and start our integration at some positive number $b < 1$. (See Fig. 9.5). That is, we consider the integral

$$\int_b^1 \frac{dx}{x} = \ln x\Big]_b^1 = \ln 1 - \ln b = \ln\frac{1}{b},$$

and investigate its behavior as b approaches zero from the right. Since

$$\lim_{b\to 0+} \int_b^1 \frac{dx}{x} = \lim_{b\to 0+}\left(\ln\frac{1}{b}\right) = +\infty,$$

we say that the integral from $x = 0$ to $x = 1$ *diverges*.

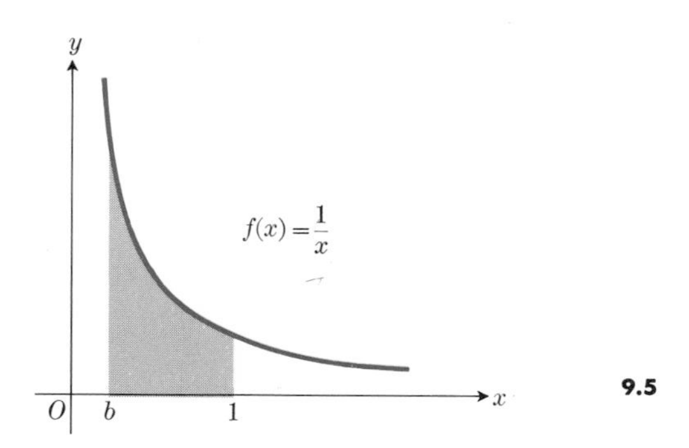

9.5

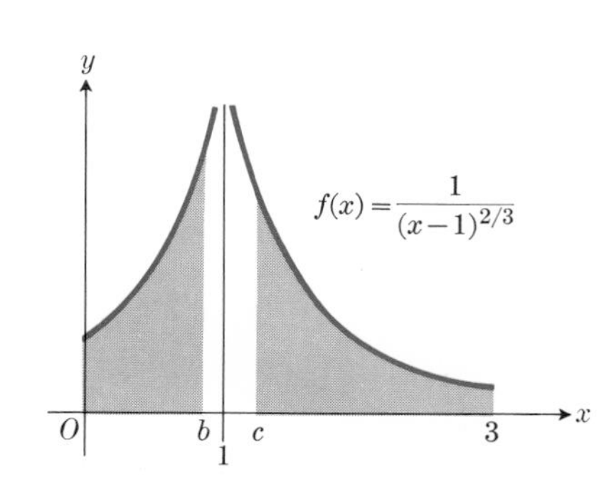

9.6

The method to be used when the function f becomes infinite at an interior point of the range of integration is illustrated in the following example.

Example 3. $$\int_0^3 \frac{dx}{(x-1)^{2/3}}.$$

The function

$$f(x) = \frac{1}{(x-1)^{2/3}}$$

becomes infinite at $x = 1$, which lies between the limits of integration 0 and 3. In such a case, we again cut out the point where $f(x)$ becomes infinite. This time we integrate from 0 to b, where b is slightly less than 1, and start again on the other side of 1 at c and integrate from c to 3 (Fig. 9.6). Then we get the two integrals

$$\int_0^b \frac{dx}{(x-1)^{2/3}} \quad \text{and} \quad \int_c^3 \frac{dx}{(x-1)^{2/3}}$$

to investigate. If the first of these has a definite limit as $b \to 1-$, and the second also has a definite limit as $c \to 1+$, then we say that the improper integral converges and that its value is given by

$$\int_0^3 \frac{dx}{(x-1)^{2/3}} = \lim_{b\to1-} \int_0^b \frac{dx}{(x-1)^{2/3}} + \lim_{c\to1+} \int_c^3 \frac{dx}{(x-1)^{2/3}}.$$

If either limit fails to exist, we say that the given improper integral *diverges*. For this example, first,

$$\lim_{b\to1-} \int_0^b (x-1)^{-2/3}\,dx = \lim_{b\to1-} [3(b-1)^{1/3} - 3(0-1)^{1/3}] = +3$$

and for the second limit,

$$\lim_{c\to1+} \int_c^3 (x-1)^{-2/3}\,dx = \lim_{c\to1+} [3(3-1)^{1/3} - 3(c-1)^{1/3}] = 3\sqrt[3]{2}.$$

Since both limits exist and are finite, the original integral is said to converge and its value is $3 + 3\sqrt[3]{2}$.

Example 4. As further examples of the second type of improper integral, we consider

$$\int_1^\infty \frac{dx}{x} \quad \text{and} \quad \int_1^\infty \frac{dx}{x^2}.$$

The two curves

$$y_1 = \frac{1}{x} \quad \text{and} \quad y_2 = \frac{1}{x^2}$$

both approach the x-axis as $x \to \infty$ (Fig. 9.7). In the first case,

$$\int_1^b \frac{dx}{x} = \ln x\Big]_1^b = \ln b.$$

If we now let b take on larger and larger positive values, the logarithm of b increases indefinitely and

$$\lim_{b\to\infty} \int_1^b \frac{dx}{x} = \infty.$$

We therefore say that

$$\int_1^\infty \frac{dx}{x} = \infty$$

and that the integral *diverges*.

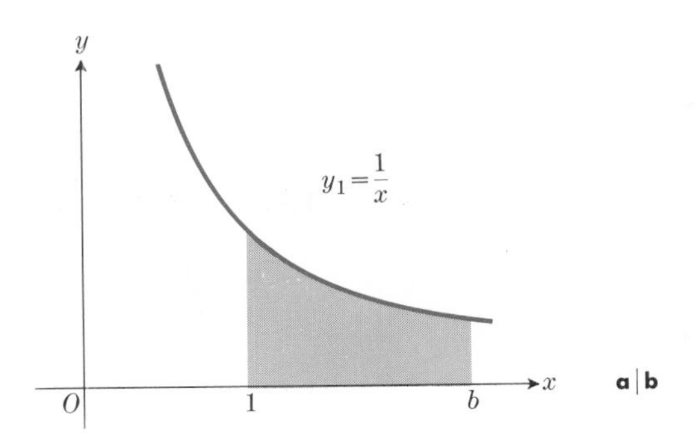

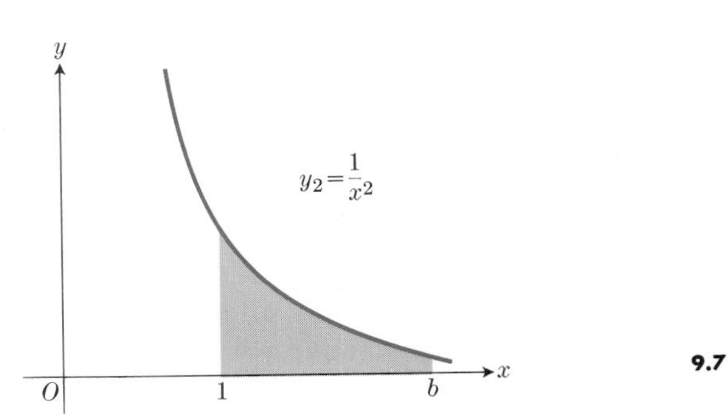

a | b 9.7

In the second case,

$$\int_1^b \frac{dx}{x^2} = -\frac{1}{x}\Big]_1^b = 1 - \frac{1}{b}.$$

This does have a definite limit, namely 1, as b increases indefinitely, and we say that the integral from 1 to ∞ converges and that its value is 1. That is,

$$\int_1^\infty \frac{dx}{x^2} = \lim_{b\to\infty} \int_1^b \frac{dx}{x^2} = \lim_{b\to\infty}\left(1 - \frac{1}{b}\right) = 1.$$

Sometimes we can determine whether a given integral converges or diverges by comparing it with a simpler integral.

Example 5. Even though we cannot find any simpler expression for

$$I(b) \equiv \int_1^b e^{-x^2}\,dx,$$

we can show that

$$\lim_{b\to\infty} I(b)$$

exists and is finite. For the function $I(b)$ represents the area between the x-axis and the curve

$$y = e^{-x^2}$$

between $x = 1$ and $x = b$. Clearly, this is an increasing function of b, so that there are two alternatives: Either

(a) $I(b)$ becomes infinite as $b \to \infty$, or

(b) $I(b)$ has a finite limit as $b \to \infty$.

We show that the first alternative cannot be the true one. We do this by comparing the area under the given curve $y = e^{-x^2}$ with the area under the curve $y = e^{-x}$ (Fig. 9.8). The latter area, from $x = 1$ to $x = b$, is given by the integral

$$\int_1^b e^{-x}\,dx = -e^{-x}\Big]_1^b = e^{-1} - e^{-b},$$

which approaches the finite limit e^{-1} as $b \to \infty$. Since e^{-x^2} is less than e^{-x} for all x greater than one, the area under the given curve is certainly no greater than e^{-1} no matter how large b is.

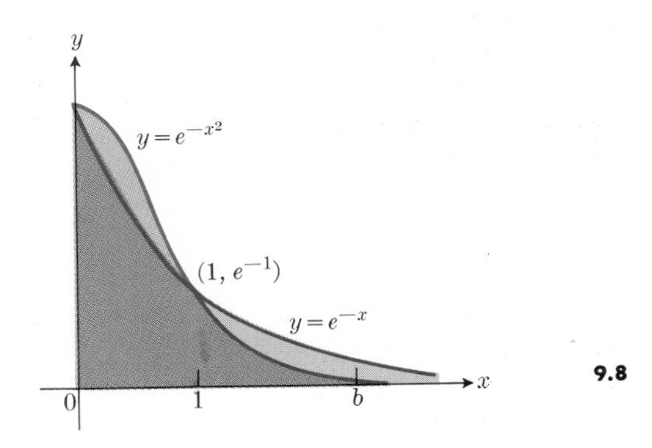

9.8

The discussion above is summarized in the following inequalities:

$$I(b) = \int_1^b e^{-x^2}\,dx \le \int_1^b e^{-x}\,dx = e^{-1} - e^{-b} \le e^{-1},$$

that is,

$$I(b) \le e^{-1} < 0.37.$$

Therefore $I(b)$ does not become infinite as $b \to \infty$, so alternative (a) is ruled out and alternative (b) must hold; that is,

$$\int_1^\infty e^{-x^2}\,dx = \lim_{b\to\infty}\int_1^b e^{-x^2}\,dx$$

converges to a definite finite value. We have not calculated what this limit is, but we know that it exists and is less than 0.37.

It should be pointed out that an improper integral may diverge without becoming infinite.

Example 6. The integral

$$\int_0^b \cos x\,dx = \sin b$$

takes all values between -1 and $+1$ as b varies between $2n\pi - \pi/2$ and $2n\pi + \pi/2$, where n is any integer. Hence,

$$\lim_{b\to\infty}\int_0^b \cos x\,dx$$

does not exist. We might say that this integral "diverges by oscillation."

EXERCISES

Show that the following improper integrals converge and evaluate them.

1. $\displaystyle\int_0^\infty \frac{dx}{x^2+1}$
2. $\displaystyle\int_0^1 \frac{dx}{\sqrt{x}}$
3. $\displaystyle\int_{-1}^1 \frac{dx}{x^{2/3}}$
4. $\displaystyle\int_1^\infty \frac{dx}{x^{1.001}}$
5. $\displaystyle\int_0^4 \frac{dx}{\sqrt{4-x}}$
6. $\displaystyle\int_0^1 \frac{dx}{\sqrt{1-x^2}}$
7. $\displaystyle\int_0^\infty e^{-x}\cos x\,dx$
8. $\displaystyle\int_0^1 \frac{dx}{x^{0.999}}$

Determine whether each of the following improper integrals converges or diverges.

9. $\displaystyle\int_1^\infty \frac{dx}{\sqrt{x}}$
10. $\displaystyle\int_1^\infty \frac{dx}{x^3}$
11. $\displaystyle\int_1^\infty \frac{dx}{x^3+1}$
12. $\displaystyle\int_0^\infty \frac{dx}{x^3}$
13. $\displaystyle\int_0^\infty \frac{dx}{x^3+1}$
14. $\displaystyle\int_0^\infty \frac{dx}{1+e^x}$
15. $\displaystyle\int_0^{\pi/2} \tan x\,dx$
16. $\displaystyle\int_{-1}^1 \frac{dx}{x^2}$
17. $\displaystyle\int_{-1}^1 \frac{dx}{x^{2/5}}$
18. $\displaystyle\int_0^\infty \frac{dx}{\sqrt{x}}$
19. $\displaystyle\int_0^\infty \frac{dx}{\sqrt{x+x^4}}$

Hint. Compare the integral with $\int dx/\sqrt{x}$ for x near zero and with $\int dx/x^2$ for large x.

9.11 NUMERICAL METHODS FOR APPROXIMATING DEFINITE INTEGRALS

Certain important integrals that arise in physical and mathematical problems cannot be expressed in terms of the so-called elementary functions. For example, the length of one arch of the curve $y = \sin x$, Fig. 9.9, is given by

$$L = \int_0^\pi \sqrt{1+\cos^2 x}\,dx.$$

The integral cannot be evaluated by any of the methods we have studied thus far. Nevertheless it can be evaluated numerically to any required number of decimal places by several methods. The trapezoidal rule is one. We indicate another below.

Any definite integral $\int_a^b f(x)\,dx$ can be interpreted as an area or as a combination of areas, some added and some subtracted. Such an integral can therefore be approximated by any method of approximating these areas. Instruments called *planimeters* are useful in this connection, but they have two drawbacks:

(a) the curve $y = f(x)$ must be drawn with some care, and

(b) the accuracy of the planimeter is limited to approximately the order of 1%.

Simpson's method, which we shall now study, avoids both of these drawbacks.

This method is based on the simple formula

$$A_p = \frac{h}{3}(y_0 + 4y_1 + y_2)$$

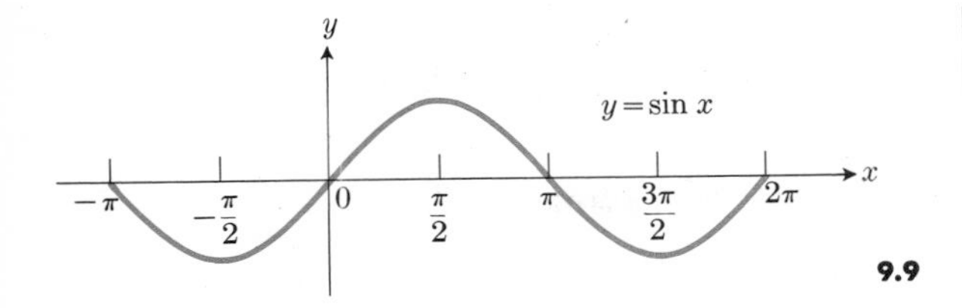

9.9

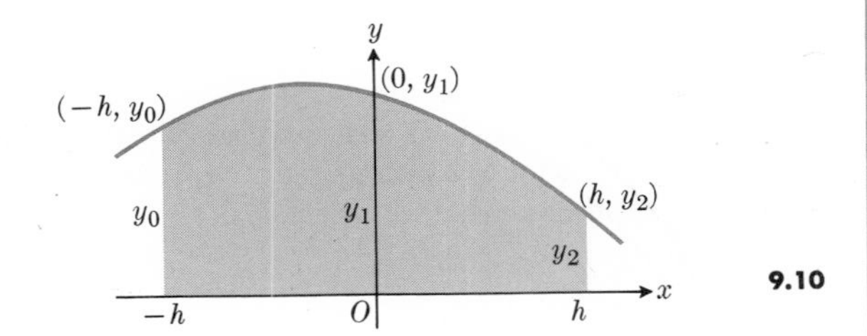

9.10

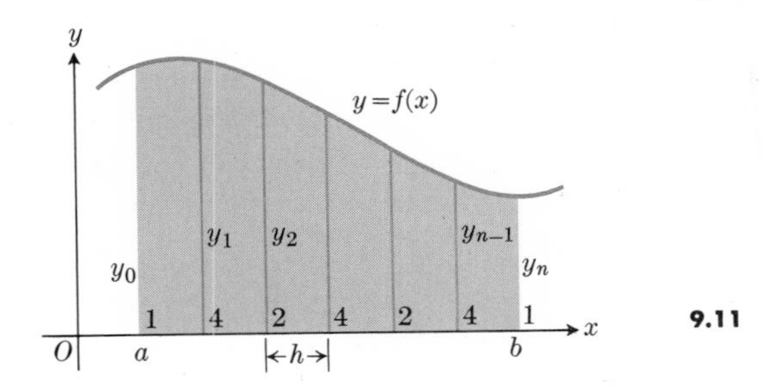

9.11

for the area under the arc of the parabola

$$y = Ax^2 + Bx + C$$

between $x = -h$ and $x = +h$. (See Fig. 9.10.*)

This formula is readily established as follows: In the first place, we have

$$\begin{aligned} A_p &= \int_{-h}^{h} (Ax^2 + Bx + C)\, dx \\ &= \frac{2Ah^3}{3} + 2Ch. \end{aligned}$$

Since the curve passes through the three points

$$(-h, y_0), \quad (0, y_1), \quad \text{and} \quad (h, y_2),$$

we also have

$$\begin{aligned} y_0 &= Ah^2 - Bh + C, \qquad y_1 = C, \\ y_2 &= Ah^2 + Bh + C, \end{aligned}$$

from which there follows

$$\begin{aligned} C &= y_1, \\ Ah^2 - Bh &= y_0 - y_1, \\ Ah^2 + Bh &= y_2 - y_1, \\ 2Ah^2 &= y_0 + y_2 - 2y_1. \end{aligned}$$

Hence, expressing the area A_p in terms of the ordinates y_0, y_1, and y_2, we have

$$A_p = \frac{h}{3}(2Ah^2 + 6C) = \frac{h}{3}[(y_0 + y_2 - 2y_1) + 6y_1]$$

or

$$A_p = \frac{h}{3}(y_0 + 4y_1 + y_2). \tag{1}$$

Simpson's rule follows from applying this result to successive pieces of the curve $y = f(x)$ between $x = a$ and $x = b$. Each separate piece of the curve, covering an x-subinterval of width $2h$, is approximated by an arc of a parabola through its ends and its mid-point. The area under each parabolic arc is then given by an expression like Eq. (1) and the results are added to give

$$\begin{aligned} A_S = \frac{h}{3}[y_0 + 4y_1 + 2y_2 + 4y_3 + 2y_4 + \cdots \\ + 2y_{n-2} + 4y_{n-1} + y_n], \end{aligned} \tag{2}$$

which is taken as the approximate value of $\int_a^b f(x)\, dx$. In Eq. (2), $y_0, y_1, y_2, \ldots, y_n$ are the ordinates of the curve $y = f(x)$ at the points of abscissas

$$\begin{aligned} x_0 &= a, \\ x_1 &= a + h, \\ x_2 &= a + 2h, \\ &\vdots \\ x_n &= a + nh = b \end{aligned}$$

corresponding to a subdivision of the interval $a \le x \le b$ into n equal subintervals each of width $h = (b - a)/n$. (See Fig. 9.11.) The number n of subdivisions must be an *even* integer if we are to apply the method. The application of the rule, Eq. (2), does not require a graph of the curve

* We choose the interval $-h \le x \le h$ to simplify the algebra. There is no loss in generality in so doing, because the area does not depend on the location of the y-axis.

$y = f(x)$ nor a determination of the approximating parabolic arcs. All the work is strictly arithmetical. For a continuous function f, the approximations generally get better as n increases and h gets smaller. By an extension of the Mean Value Theorem, it is possible to prove* that if f is continuous on $[a, b]$ and four times differentiable on (a, b), then there is a number c between a and b such that

$$\int_a^b f(x)\,dx = A_S - \frac{b-a}{180} f^{(4)}(c) \cdot h^4.$$

Problem 1. Use Simpson's rule with four subdivisions to find an approximate value of $\ln 2 = \int_1^2 (1/x)\,dx$.

Solution. We arrange the work in tabular form.

x	$y = 1/x$	Weighting factor	Product
1.00	1.00000	1	1.00000
1.25	0.80000	4	3.20000
1.50	0.66667	2	1.33333
1.75	0.57143	4	2.28572
2.00	0.50000	1	0.50000
		Sum =	8.31905

Since $h = \frac{1}{4}$, we must multiply the last sum by $h/3 = \frac{1}{12}$. Hence $A_S = \frac{1}{12}(8.31905) = 0.69325$. The approximation $\ln 2 \approx 0.69235$ is in error by 1 part in 6900 or about one-seventieth of one percent. Greater accuracy may be obtained by taking twice as many subdivisions.

EXERCISES

Find approximate numerical values for the following definite integrals by using Simpson's rule (a) with $n = 2$, and (b) with $n = 4$. Where possible, compare your answer with (c) the exact value.

1. $\int_0^2 x^2\,dx$ 2. $\int_0^\pi \sin x\,dx$ 3. $\int_0^\pi \sqrt{\sin x}\,dx$

4. $\int_0^\pi \frac{dx}{2 + \cos x}$ 5. $\int_0^\pi \frac{\sin x}{x}\,dx$ 6. $\int_0^1 \frac{dx}{x^2 + 1}$

7. Show that

$$\int_3^\infty e^{-3x}\,dx = \tfrac{1}{3}e^{-9} = 0.000041,$$

and hence that $\int_3^\infty e^{-x^2}\,dx < 0.000041$. Show further that $\int_0^\infty e^{-x^2}\,dx$ can be replaced by $\int_0^3 e^{-x^2}\,dx$ without introducing any errors in the first three decimal places of the answer. Evaluate this last integral by Simpson's rule with $n = 6$. (This illustrates one method by which a convergent improper integral may be approximated numerically.)

* See, for example, J. M. H. Olmsted, *Intermediate Analysis*, Appleton-Century-Crofts, p. 146.

REVIEW QUESTIONS AND EXERCISES

1. What are some of the general methods presented in this chapter for finding an indefinite integral?
2. What substitution(s) would you consider trying if the integrand contained the following:

   (a) $\sqrt{x^2 + 9}$, (b) $\sqrt{x^2 - 9}$,
   (c) $\sqrt{9 - x^2}$, (d) $\sin^3 x \cos^2 x$,
   (e) $\sin^2 x \cos^2 x$, (f) $\dfrac{1 + \sin\theta}{2 + \cos\theta}$?

3. What method(s) would you try if the integrand contained

   (a) $\sin^{-1} x$, (b) $\ln x$,
   (c) $\sqrt{1 + 2x - x^2}$, (d) $x \sin x$,
   (e) $\dfrac{2x + 3}{x^2 - 5x + 6}$, (f) $\sin 5x \cos 3x$,
   (g) $\dfrac{1 - \sqrt{x}}{1 + \sqrt[4]{x}}$, (h) $x\sqrt{2x + 3}$?

4. What numerical methods do you know for approximating a definite integral? How can an approximation usually be improved upon to give greater accuracy?
5. Discuss two types of improper integral. Define convergence and divergence of each type. Give examples of convergent and divergent integrals of each type.

MISCELLANEOUS EXERCISES

Try A lot of these

Evaluate the following integrals.

1. $\int \frac{\cos x\,dx}{\sqrt{1+\sin x}}$

2. $\int \frac{\sin^{-1} x\,dx}{\sqrt{1-x^2}}$

3. $\int \frac{\tan x\,dx}{\cos^2 x}$

4. $\int \frac{dx}{1-\sin x}$

5. $\int e^{\ln\sqrt{x}}\,dx$

6. $\int \frac{\cos\sqrt{x}}{\sqrt{x}}\,dx$

7. $\int \frac{dx}{\sqrt{x^2+2x+2}}$

8. $\int \frac{(3x-7)\,dx}{(x-1)(x-2)(x-3)}$

9. $\int x^2 e^x\,dx$

10. $\int \sqrt{x^2+1}\,dx$

11. $\int \frac{e^t\,dt}{1+e^{2t}}$

12. $\int \frac{dx}{e^x+e^{-x}}$

13. $\int \frac{dx}{1+\sqrt{x}}$

14. $\int \frac{dx}{\sqrt{1+\sqrt{x}}}$

15. $\int t^{2/3}(t^{5/3}+1)^{2/3}\,dt$

16. $\int \frac{\cot x\,dx}{\ln(\sin x)}$

17. $\int \frac{dt}{\sqrt{e^t+1}}$

18. $\int \frac{dt}{\sqrt{1-e^{-t}}}$

19. $\int \frac{\sin x\, e^{\sec x}}{\cos^2 x}\,dx$

20. $\int \frac{\cos x\,dx}{1+\sin^2 x}$

21. $\int \frac{dx}{\sqrt{2x-x^2}}$

22. $\int \frac{\sin x\,dx}{1+\cos^2 x}$

23. $\int \frac{\cos 2t}{1+\sin 2t}\,dt$

24. $\int \frac{dx}{\sin x\cos x}$

25. $\int \sqrt{1+\sin x}\,dx$

26. $\int \sqrt{1-\sin x}\,dx$

27. $\int \frac{dx}{\sqrt{(a^2-x^2)^3}}$

28. $\int \frac{dx}{\sqrt{(a^2+x^2)^3}}$

29. $\int \frac{\sin x\,dx}{\cos^2 x-5\cos x+4}$

30. $\int \frac{e^{2x}\,dx}{\sqrt[3]{1+e^x}}$

31. $\int \frac{dx}{x(x+1)(x+2)\cdots(x+m)}$

32. $\int \frac{dx}{x^6-1}$

33. $\int \frac{dy}{y(2y^3+1)^2}$

34. $\int \frac{x\,dx}{1+\sqrt{x}}$

35. $\int \frac{dx}{x(x^2+1)^2}$

36. $\int \ln\sqrt{x-1}\,dx$

37. $\int \frac{dx}{e^x-1}$

38. $\int \frac{d\theta}{1-\tan^2\theta}$

39. $\int \frac{(x+1)\,dx}{x^2(x-1)}$

40. $\int \frac{x\,dx}{x^2+4x+3}$

41. $\int \frac{du}{(e^u-e^{-u})^2}$

42. $\int \frac{4\,dx}{x^3+4x}$

43. $\int \frac{dx}{5x^2+8x+5}$

44. $\int \frac{\sqrt{x^2-a^2}}{x}\,dx$

45. $\int e^x\cos 2x\,dx$

46. $\int \frac{dx}{x(3\sqrt{x}+1)}$

47. $\int \frac{dx}{x(1+\sqrt[3]{x})}$

48. $\int \frac{\cot\theta\,d\theta}{1+\sin^2\theta}$

49. $\int \frac{z^5\,dz}{\sqrt{1+z^2}}$

50. $\int \frac{e^{4t}\,dt}{(1+e^{2t})^{2/3}}$

51. $\int \frac{dx}{x^{1/5}\sqrt{1+x^{4/5}}}$

52. $\int x\sec^2 x\,dx$

53. $\int x\sin^{-1} x\,dx$

54. $\int \frac{(x^3+x^2)\,dx}{x^2+x-2}$

55. $\int \frac{x^3+1}{x^3-x}\,dx$

56. $\int \frac{x\,dx}{(x-1)^2}$

57. $\int \frac{(2e^{2x}-e^x)\,dx}{\sqrt{3e^{2x}-6e^x-1}}$

58. $\int \frac{(x+1)\,dx}{(x^2+2x-3)^{2/3}}$

59. $\int \frac{dy}{(2y+1)\sqrt{y^2+y}}$

60. $\int \frac{dx}{x^2\sqrt{a^2-x^2}}$

61. $\int (1-x^2)^{3/2}\,dx$

62. $\int \ln(x+\sqrt{1+x^2})\,dx$

63. $\int x\tan^2 x\,dx$

64. $\int \frac{\tan^{-1} x}{x^2}\,dx$

65. $\int x\cos^2 x\,dx$

66. $\int x^2 \sin x \, dx$

67. $\int x \sin^2 x \, dx$

68. $\int \frac{dt}{t^4 + 4t^2 + 3}$

69. $\int \frac{du}{e^{4u} + 4e^{2u} + 3}$

70. $\int x \ln \sqrt{x + 2} \, dx$

71. $\int (x + 1)^2 e^x \, dx$

72. $\int \sec^{-1} x \, dx$

73. $\int \frac{8 \, dx}{x^4 + 2x^3}$

74. $\int \frac{x \, dx}{x^4 - 16}$

75. $\int_0^{\pi/2} \frac{\cos x \, dx}{\sqrt{1 + \cos x}}$

76. $\int \frac{\cos x \, dx}{\sin^3 x - \sin x}$

77. $\int \frac{du}{(e^u + e^{-u})^2}$

78. $\int \frac{x \, dx}{1 + \sqrt{x} + x}$

79. $\int \frac{\sec^2 t \, dt}{\sec^2 t - 3 \tan t + 1}$

80. $\int \frac{dt}{\sec^2 t + \tan^2 t}$

81. $\int \frac{dx}{1 + \cos^2 x}$

82. $\int e^{2t} \cos (e^t) \, dt$

83. $\int \ln \sqrt{x^2 + 1} \, dx$

84. $\int x \ln (x^3 + x) \, dx$

85. $\int x^3 e^{x^2} \, dx$

86. $\int \frac{\cos x \, dx}{\sqrt{4 - \cos^2 x}}$

87. $\int \frac{\sec^2 x \, dx}{\sqrt{4 - \sec^2 x}}$

88. $\int x^2 \sin (1 - x) \, dx$

89. $\int \frac{dx}{1 + \sin x}$

90. $\int \frac{dx}{1 + 2 \sin x}$

91. $\int \frac{dx}{\sin^3 x}$

92. $\int \frac{dx}{\cot^3 x}$

93. $\int (\sin^{-1} x)^2 \, dx$

94. $\int x \ln \sqrt[3]{3x + 1} \, dx$

95. $\int \frac{x^3 \, dx}{(x^2 + 1)^2}$

96. $\int \frac{x \, dx}{\sqrt{1 - x}}$

97. $\int x\sqrt{2x + 1} \, dx$

98. $\int \ln (x + \sqrt{x^2 - 1}) \, dx$

99. $\int \ln (x - \sqrt{x^2 - 1}) \, dx$

100. $\int \frac{dt}{t - \sqrt{1 - t^2}}$

101. $\int e^{-x} \tan^{-1} (e^x) \, dx$

102. $\int \sin^{-1} \sqrt{x} \, dx$

103. $\int \ln (x + \sqrt{x}) \, dx$

104. $\int \tan^{-1} \sqrt{x} \, dx$

105. $\int \ln (x^2 + x) \, dx$

106. $\int \ln (\sqrt{x} + \sqrt{1 + x}) \, dx$

107. $\int \cos \sqrt{x} \, dx$

108. $\int \sin \sqrt{x} \, dx$

109. $\int \tan^{-1} \sqrt{x + 1} \, dx$

110. $\int \sqrt{1 - x^2} \sin^{-1} x \, dx$

111. $\int x \sin^2 (2x) \, dx$

112. $\int \frac{\tan x \, dx}{\tan x + \sec x}$

113. $\int \frac{dt}{\sqrt{e^{2t} + 1}}$

114. $\int \frac{dx}{(\cos^2 x + 4 \sin x - 5) \cos x}$

115. $\int \frac{dt}{a + be^{ct}}, \quad abc \neq 0$

116. $\int \sqrt{\frac{1 - \cos x}{\cos \alpha - \cos x}} \, dx, \quad \alpha$ constant, $0 < \alpha < x < \pi$

117. $\int \frac{dx}{\sin x - \cos x}$

118. $\int \ln \sqrt{1 + x^2} \, dx$

119. $\int \ln (2x^2 + 4) \, dx$

120. $\int \frac{x^3}{\sqrt{1 - x^2}} \, dx$

121. $\int \frac{dx}{x(2 + \ln x)}$

122. $\int \frac{\cos 2x - 1}{\cos 2x + 1} \, dx$

123. $\int \frac{dx}{x^3 + 1}$

124. $\int \frac{e^{2x} \, dx}{\sqrt[4]{e^x + 1}}$

125. $\int \frac{e^x \, dx}{1 + e^{2x}}$

126. $\int e^{\sqrt{t}} \, dt$

127. $\int \sin \sqrt{x + 1} \, dx$

128. $\int \cos \sqrt{1 - x} \, dx$

Evaluate the limits in Exercises 129 through 135 by identifying each one with an appropriate definite integral and evaluating the latter.

129. $\lim_{n\to\infty}\left(\frac{1}{n+1}+\frac{1}{n+2}+\cdots+\frac{1}{2n}\right)$

130. $\lim_{n\to\infty}\left(\frac{1}{\sqrt{n^2}}+\frac{1}{\sqrt{n^2+n}}+\frac{1}{\sqrt{n^2+2n}}+\cdots+\frac{1}{\sqrt{n^2+(n-1)n}}\right)$

131. $\lim_{n\to\infty}\left(\frac{\sin 0+\sin\frac{\pi}{n}+\sin\frac{2\pi}{n}+\cdots+\sin\frac{(n-1)\pi}{n}}{n}\right)$

132. $\lim_{n\to\infty}\left(\frac{1+\sqrt[n]{e}+\sqrt[n]{e^2}+\sqrt[n]{e^3}+\cdots+\sqrt[n]{e^{n-1}}}{n}\right)$

133. $\lim_{n\to\infty}\left(\frac{n}{n^2+0^2}+\frac{n}{n^2+1^2}+\frac{n}{n^2+2^2}+\cdots+\frac{n}{n^2+(n-1)^2}\right)$

134. $\lim_{n\to\infty}\sum_{k=1}^{n}\ln\sqrt[n]{1+\frac{k}{n}}$

135. $\lim_{h\to\infty}\sum_{k=0}^{n-1}\frac{1}{\sqrt{n^2-k^2}}$

136. Show that $\int_0^\infty x^3e^{-x^2}\,dx$ is a convergent integral, and evaluate it.

137. Show that $\int_0^1 \ln x\,dx$ is a convergent integral and find its value. Sketch the integrand $y = \ln x$ for $0 < x \leq 1$.

138. Assuming that $|\alpha| \neq |\beta|$, prove that

$$\lim_{T\to\infty}\frac{1}{T}\int_0^T \sin\alpha x \sin\beta x\,dx = 0.$$

139. Evaluate $\displaystyle\lim_{h\to 0}\frac{1}{h}\int_2^{2+h} e^{-x^2}\,dx.$

140. At points of the curve $y^2 = 4px$, lines of length $h = y$ are drawn perpendicular to its plane. Find the area of the surface formed by these lines at points of the curve between $(0, 0)$ and $(p, 2p)$.

141. A plane figure is bounded by a 90° arc of a circle of radius r and a straight line. Find its area and its centroid.

142. Find the coordinates of the center of gravity of the area bounded by the curves $y = e^x$, $x = 1$, and $y = 1$.

143. At points of a circle of radius r, perpendiculars to its plane are erected, the perpendicular at each point P being of length ks, where s is the arc of the circle from a fixed point A to P. Find the area of the surface formed by the perpendiculars along the arc beginning at A and extending once around the circle.

144. A plate in the first quadrant, bounded by the curves $y = e^x$, $y = 1$, and $x = 4$, is submerged vertically in water with its upper corner on the surface. The surface of the water is given by the line $y = e^4$. Find the total force on one side of the plate if the weight of water is 62.5 lb/ft³, and if the units of x and y are also measured in feet.

145. The area under the curve $xy = 1$, for $x \geq 1$, is rotated about the x-axis. If A is the area under the curve, V is the volume generated, and S is the surface area of this volume,

(a) is A finite? (b) is V finite? (c) is S finite?

Give reasons for your answers.

146. Answer the questions of Exercise 145 if the curve given there is replaced by the curve $y = e^{-x}$. Give reasons for your answers.

147. Find the centroid of the area bounded by $y = e^x$, $y = 0$, $x = 0$, $x = 1$.

148. Find the centroid of the area bounded by $y = \ln x$, $x = 1$, $y = 1$.

149. Find the length of arc of $y = \ln x$ from $x = 1$ to $x = e$.

150. The arc of Exercise 149 is rotated about the y-axis. Find the surface area generated. Check your answer by comparing it with the area of a frustum of a suitably related cone.

151. Find the length of arc of the curve $y = e^x$ from $x = 0$ to $x = 2$.

152. The arc of Exercise 151 is rotated about the x-axis. Find the surface area generated. Compare this area with the area of a suitably related frustum of a cone to check your answer.

153. One arch of the curve $y = \cos x$ is rotated about the x-axis. Find the surface area generated.

154. Use Simpson's rule to compute, approximately, the length of arc of $y = \cos x$ from $x = -\pi/2$ to $x = \pi/2$. Check your answer by consulting a table of elliptic integrals, if one is available.

155. A thin wire is bent into the shape of one arch of the curve $y = \cos x$. Find its center of mass (see Exercises 153, 154).

156. The area between the graph of $y = \ln(1/x)$, the x-axis, and the y-axis is revolved around the x-axis. Find the volume of the solid it generates.

157. Find the total perimeter of the curve

$$x^{2/3} + y^{2/3} = a^{2/3}.$$

158. Find the length of arc of the curve $y = \ln(\cos x)$ from $x = 0$ to $x = \pi/3$.

159. Find the area of the surface generated by rotating the curve

$$x^{2/3} + y^{2/3} = a^{2/3}$$

about the x-axis.

160. A thin homogeneous wire is bent into the shape of one arch of the curve $x^{2/3} + y^{2/3} = a^{2/3}$. Find its center of mass (see Exercises 157, 159).

161. Determine whether the integral

$$\int_1^\infty \ln x \frac{dx}{x^2}$$

converges or diverges.

162. Sketch the curves

(a) $y = x - e^x$, (b) $y = e^{(x - e^x)}$,

and show the behavior of each for large $|x|$.

Show that the following integrals converge and compute their values.

(c) $\displaystyle\int_{-\infty}^{b} e^{(x-e^x)}\,dx$ (d) $\displaystyle\int_{-\infty}^{\infty} e^{(x-e^x)}\,dx$

163. The gamma function $\Gamma(x)$ is defined, for $x > 0$, by the definite integral

$$\Gamma(x) = \int_0^\infty t^{x-1}e^{-t}\,dt.$$

(a) Sketch graphs of the integrand $y = t^{x-1}e^{-t}$ against t, $t > 0$, for the three typical cases $x = \frac{1}{2}$, 1, 3. Find maxima, minima, and points of inflection (if they exist).

(b) Show that the integral converges if $x > 0$.

(c) Show that the integral diverges if $x \leq 0$.

(d) Using the integration by parts, demonstrate that $\Gamma(x + 1) = x\Gamma(x)$, for $x > 0$.

(e) Using the result of part (d), show that, if n is a positive integer, $\Gamma(n) = (n - 1)!$ where $0! = 1$ (by definition) and if m is a positive integer, $m!$ is the product of the positive integers from 1 through m inclusive.

(f) Discuss how one might compute a table of values of $\Gamma(x)$ for the domain $1 \leq x \leq 2$, say. Consult such a table, if available, and sketch the graph of $y = \Gamma(x)$ for $0 < x \leq 3$.

164. Determine whether the integral $\int_0^\infty t^{x-1}(\ln t)e^{-t}\,dt$ converges or diverges for fixed $x > 0$. Sketch the integrand $y = t^{x-1}\ln t\, e^{-t}$ vs. t, $t > 0$, for the two cases $x = \frac{1}{2}$ and $x = 3$.

165. Solve the differential equation $d^2x/dt^2 = -k^2x$, subject to the initial conditions $x = a$, $dx/dt = 0$ when $t = 0$.

Hint. Let $$dx/dt = v,$$

$$d^2x/dt^2 = dv/dt = v\,dv/dx.$$

166. Find $\displaystyle\lim_{n\to\infty}\int_0^1 \frac{ny^{n-1}}{1+y}\,dy.$

167. Prove that if n is a positive integer, or zero, then

$$\int_{-1}^{1}(1 - x^2)^n\,dx = \frac{2^{2n+1}(n!)^2}{(2n+1)!},$$

with $0! = 1$; $n! = 1 \cdot 2 \cdots n$ for $n \geq 1$.

PLANE ANALYTIC GEOMETRY

CHAPTER 10

10.1 CURVES AND EQUATIONS

We use analytic geometry, which combines algebra and calculus with geometry, in two types of problem:

(a) to study the geometric properties of a curve when its equation is given, and

(b) to find an equation of the curve when its geometric properties are known.

Locus

An equation of the form

$$F(x, y) = 0 \tag{1}$$

describes a set which is the locus of all and only those points $P(x, y)$ whose coordinates satisfy the given equation. If the locus for example, is a curve, every point on the curve must satisfy the equation and every point that satisfies the equation lies on the curve. We have already had some experience in analyzing a given equation to find the turning points and points of inflection of the curve it represents. We shall now consider certain other important characteristics that we may find by investigating a given equation.

In this chapter we shall assume that the scales on the x- and y-axes are the same, unless we specifically note otherwise.

Symmetry

If the equation of the curve is unaltered when x is replaced by $-x$, that is, if

$$F(x, y) = F(-x, y),$$

then the points (x_1, y_1) and $(-x_1, y_1)$ are both on the curve when either of them is. Then the curve is symmetric with respect to the y-axis. (See Fig. 10.1a.) In particular, an equation which contains only even powers of x represents such a curve.

Other symmetries that are easily detected are:

Symmetry about the x-axis, if the equation is unaltered when y is replaced by $-y$; in particular, if only even powers of y occur. (See Fig. 10.1b.)

Symmetry about the origin, if the equation is unaltered when x and y are replaced by $-x$ and $-y$ respectively. (See Fig. 10.1c.)

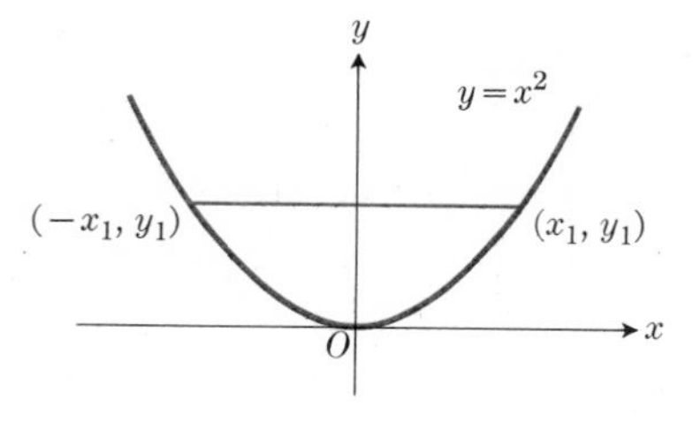

a|b

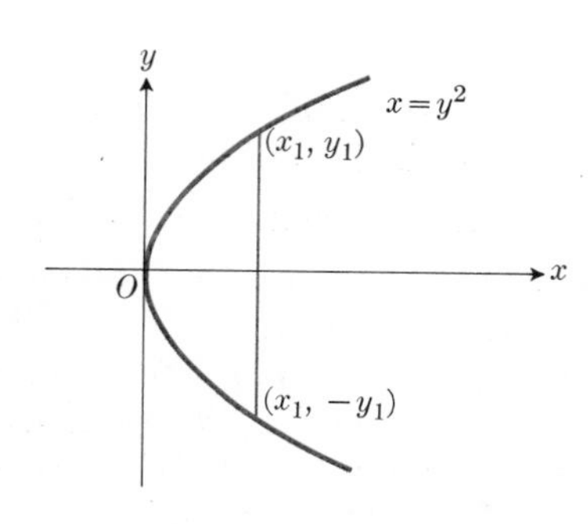

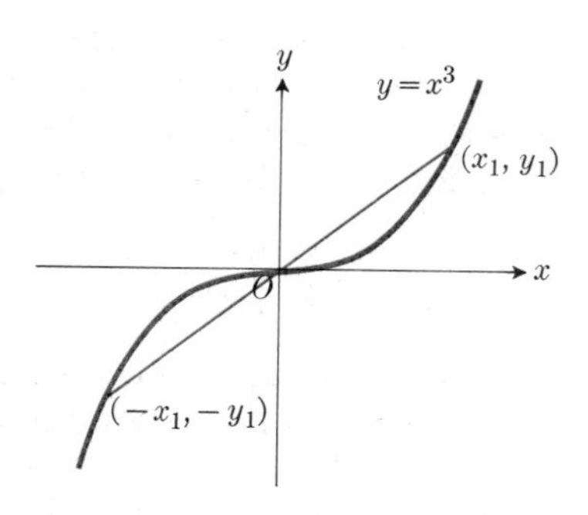

c|d

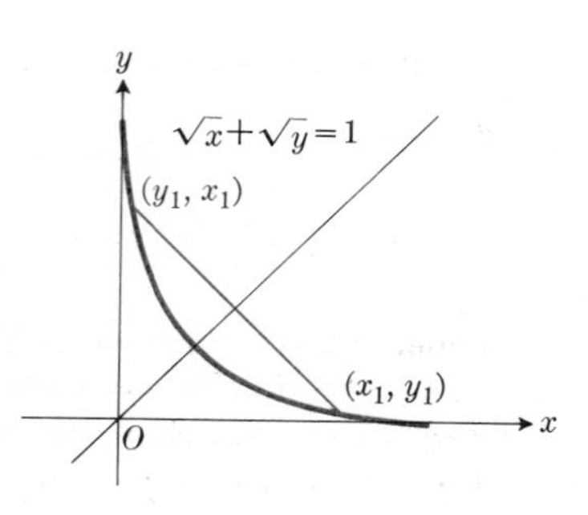

10.1

Symmetry about the 45° line $y = x$ if the equation is unaltered when x and y are interchanged. (See Fig. 10.1d.)

Example 1. $x^2 + y^2 = 1.$

This curve is symmetric about both axes, about the origin, and about the line $y = x$.

Example 2. $x^2 - y^2 = 1.$

This curve is symmetric about both axes and the origin. Not symmetric about the line $y = x$.

Example 3. $xy = 1.$

This curve is not symmetric about either axis. Symmetric about the origin and about the line $y = x$.

Extent

Only real values of x and y are considered in determining the points (x, y) whose coordinates satisfy a given equation. When even powers of a variable appear in the equation, the solution for that variable may involve square roots (or other even roots). The *extent* of the curve may then be limited by the condition that negative numbers do not have real square roots.

Example 4. $x^2 + y^2 = 1.$

When this equation is solved for y, we get

$$y = \pm\sqrt{1 - x^2}.$$

The quantity under the radical is negative if $|x| > 1$. The extent of the curve in the x-direction is therefore limited to the interval $-1 \leq x \leq 1$. By symmetry about the line $y = x$, the curve is also limited in the y-direction to $-1 \leq y \leq 1$.

Example 5. $x^2 - y^2 = 1.$

Solving for y, we get

$$y = \pm\sqrt{x^2 - 1}.$$

In this case, y is imaginary unless $|x| \geq 1$. There is no curve between the two lines $x = -1$ and $x = +1$. Solving for x, we get

$$x = \pm\sqrt{1 + y^2}.$$

The quantity under the radical is positive (actually it is greater than or equal to one) for all real values of y, $-\infty < y < +\infty$. The extent of this curve is not limited in the y-direction.

Intercepts

The point or points where a given curve crosses the x-axis can be found by setting $y = 0$ in the equation and solving for x. These abscissas are called the *x-intercepts*. The y-intercepts are found in an analogous way, by setting $x = 0$. If the labor involved in finding the intercepts is not excessive, they may be determined and will give specific points on the curve.

Example 6. $x^2 + y^2 = 1$.

Setting $y = 0$, we have $x^2 = 1$. Therefore $(1, 0)$ and $(-1, 0)$ give the x-intercepts. By symmetry, $(0, 1)$ and $(0, -1)$ give the y-intercepts.

Example 7. $x^2 - y^2 = 1$.

Setting $y = 0$, we find the x-intercepts at $(1, 0)$ and at $(-1, 0)$. Setting $x = 0$, we obtain $y^2 = -1$, which has no real solutions. The curve does not have intercepts on the y-axis (see Example 5).

Example 8. $xy = 1$.

Setting either $x = 0$ or $y = 0$ leads to the equation $0 = 1$, which has no solutions. There are no x- or y-intercepts.

Asymptotes

As the point $P(x, y)$ on a given curve moves farther and farther away from the vicinity of the origin, it may happen that the distance between P and some fixed line tends to zero. In such a case the line is called an *asymptote* of the curve. For example, if the equation of the curve has the form

$$y = \frac{N(x)}{D(x)}, \tag{2}$$

where $N(x)$ and $D(x)$ are polynomials without any factors in common, and if $x = c$ is a root of the equation

$$D(x) = 0, \tag{3}$$

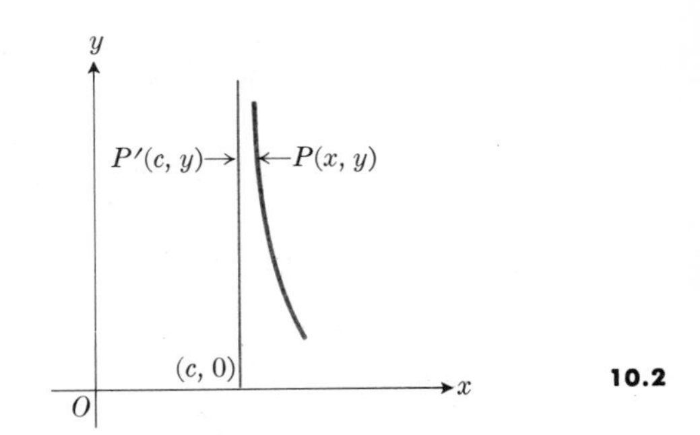

10.2

then as the x-coordinate of the tracing point $P(x, y)$ approaches c, two things will occur:

1. $y \to \infty$. Hence the distance $OP \to \infty$, and
2. $(x - c) \to 0$. That is, the horizontal distance $P'P$ between the vertical line $x = c$ and the curve tends to zero (see Fig. 10.2). In other words, *the line $x = c$ is an asymptote of the curve in Eq. (2) if $x = c$ makes the denominator, $D(x)$, vanish.* Such asymptotes are thus found by solving the equation explicitly for y in terms of x; then, if the result is a fraction, we set the denominator of the fraction equal to zero and solve for the numerical values of x. A similar procedure with the roles of x and y reversed shows that values of y which cause the denominator to vanish in the expression

$$x = \frac{f(y)}{g(y)} \tag{4}$$

may give horizontal asymptotes of the curve. An alternate method for finding such asymptotes is to let $x \to \pm\infty$ (provided the extent of the curve does not prohibit this) in Eq. (2) and find the limiting values of y.

Example 9. $y^2(x^2 - x) = x^2 + 1. \quad (5)$

Solving for y, we obtain

$$y = \pm\sqrt{\frac{x^2 + 1}{x(x - 1)}}. \tag{6}$$

The expression under the radical must not be negative, so no portion of the curve lies between the lines $x = 0$ and $x = 1$ (see Fig. 10.3). But all values of $x > 1$ and all negative values of x, that is, $x < 0$, are permissible.

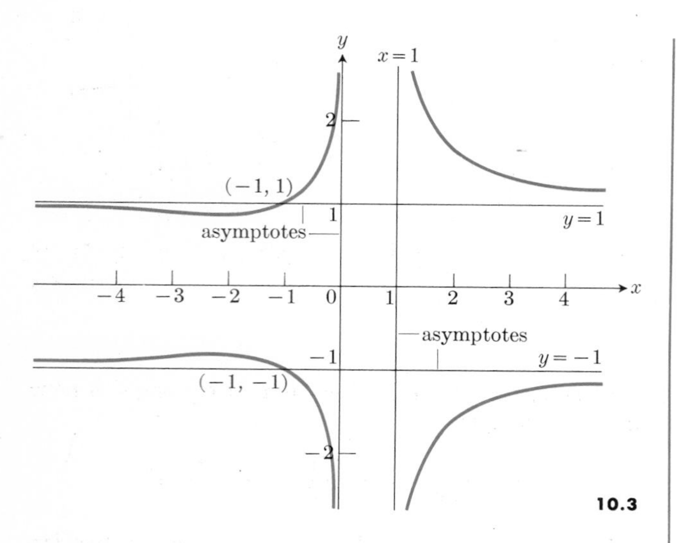

10.3

As x approaches zero *from the left*, $y \to \pm\infty$; and as x approaches one *from the right*, $y \to \pm\infty$. The lines $x = 0$ and $x = 1$ are asymptotes of the curve. Since arbitrarily large values of $|x|$ are permitted, we also investigate the behavior of the curve as $x \to -\infty$ and again as $x \to +\infty$. The expression

$$\frac{\infty^2 + 1}{\infty(\infty - 1)},$$

which appears if we substitute $x = \infty$ directly into (6), is meaningless, but if we write the equation of the curve in the equivalent form

$$y^2 = \frac{x^2 + 1}{x^2 - x} = \frac{1 + \dfrac{1}{x^2}}{1 - \dfrac{1}{x}}, \tag{7}$$

we readily see that $y^2 \to 1$ as $x \to \infty$ or as $x \to -\infty$. Thus the lines $y = 1$ and $y = -1$ are also asymptotes of the curve. We now scrutinize Eq. (7) more closely to see whether or not y^2 is necessarily always greater than one, or always less than one, or may be equal to one. Certainly when x tends to $+\infty$, the numerator in (7) is greater than one, while the denominator is less than one, and hence the fraction is larger than one. That is,

$$y^2 > 1 \quad \text{and} \quad y^2 \to 1 \quad \text{as } x \to +\infty. \tag{8}$$

On the other hand, when x tends to $-\infty$, both $1/x^2$ and $-1/x$ are positive, so that both the numerator and denominator in (7) are greater than one. However, $1/x^2$ will be less than $-1/x$, so that the numerator is less than the denominator. That is,

$$y^2 < 1 \quad \text{and} \quad y^2 \to 1 \quad \text{as } x \to -\infty. \tag{9}$$

To determine whether or not we may ever have $y^2 = 1$, we try it, say in (5), and find

$$y^2 = 1 \quad \text{when} \quad x = -1. \tag{10}$$

We now have quite a bit of information about the curve represented by Eq. (5). It is immediately evident that the curve is symmetric about the x-axis, because y may be replaced by $-y$ without changing the equation. There is no curve between $x = 0$ and $x = 1$. The lines

$$x = 0, \quad x = 1, \quad y = 1, \quad y = -1$$

are asymptotes of the curve. It crosses $y = 1$ and $y = -1$ at $x = -1$. There is no x-intercept, because putting $y = 0$ in Eq. (5) requires $0 = x^2 + 1$, which has only imaginary roots. Using this information, we may sketch the curve with a fair degree of accuracy (see Fig. 10.3). In fact, it becomes apparent that the curve has a minimum (and a symmetrically located maximum) somewhat to the left of $x = -1$, and that there is also a point of inflection to the left of this turning point. To find these points accurately requires that we determine dy/dx and d^2y/dx^2. The turning points will be found to occur at

$$\left(-1 - \sqrt{2}, \pm\sqrt{2\sqrt{2} - 2}\right).$$

The algebra involved in finding the points of inflection is excessive.

Direction at a point

The three curves

(a) $\qquad y = x, \quad$ for which $\dfrac{dy}{dx} = 1,$

(b) $\qquad y = x^2, \quad$ for which $\dfrac{dy}{dx} = 2x,$

(c) $\qquad y = \sqrt{x}, \quad$ for which $\dfrac{dy}{dx} = \dfrac{1}{2\sqrt{x}},$

all pass through the origin and the point (1, 1). But their behavior for small values of x is radically different. The individual nature of any one of them in the vicinity of (0, 0) is better indicated if the

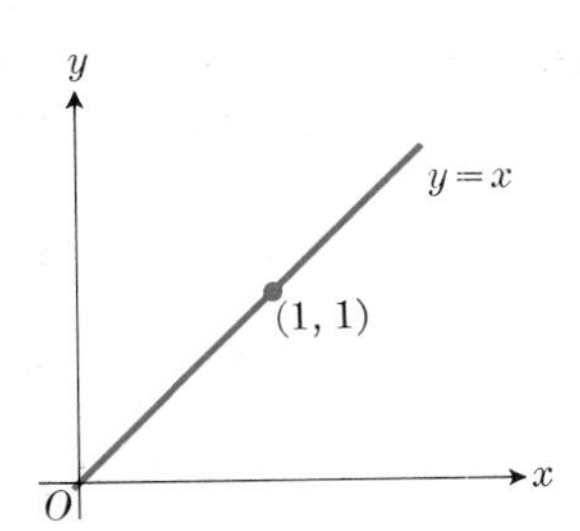

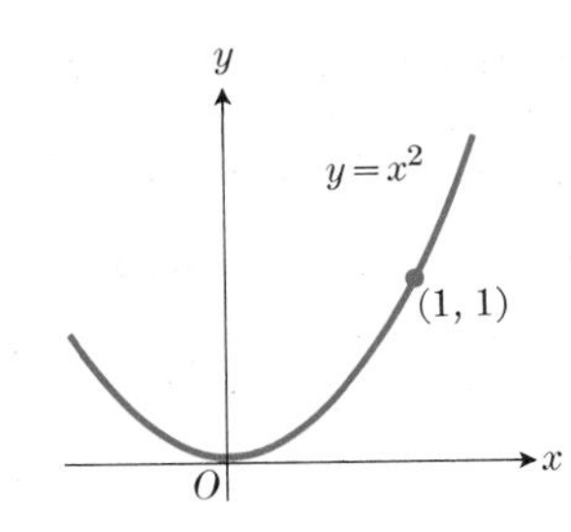

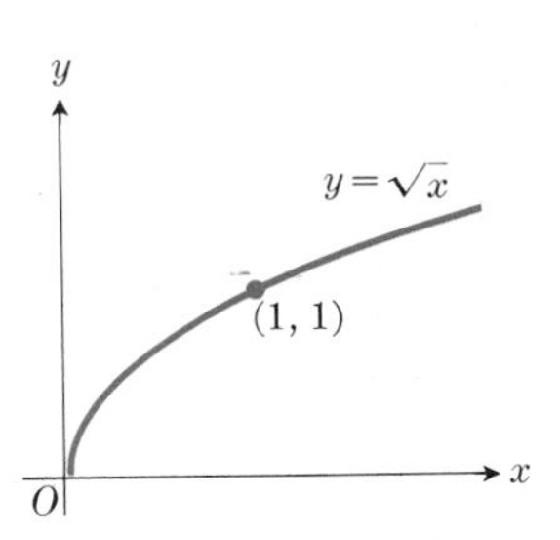

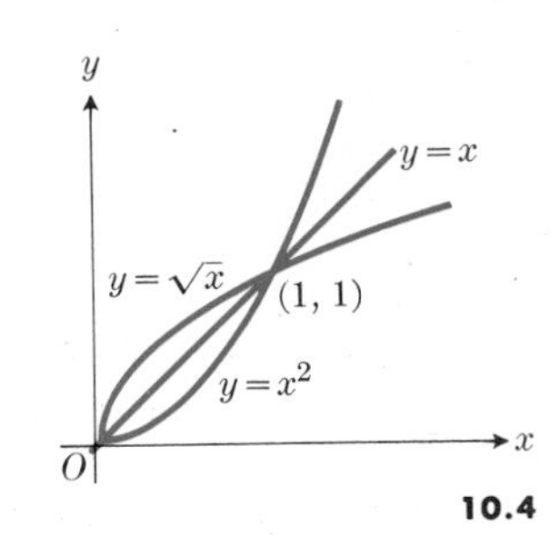

10.4

direction of the curve at that point is also given. This, of course, means that it is desirable to find the slope of each of them at (0, 0). We find the inclination angles are respectively 45°, 0°, and 90° (see Fig. 10.4).

Example 10. $y^2 = x^2(1 - x^2)$.

This curve has symmetry with respect to both axes and the origin (see Fig. 10.5). Its extent in the x-direction is limited to $|x| \leq 1$. Differentiating implicitly, we find

$$y' = \frac{x - 2x^3}{y}.$$

At one x-intercept, (1, 0), the slope is infinite. At the origin, which is another intercept, the expression we have for y' becomes 0/0, which is meaningless. However, if we replace y in the denominator by its equivalent expression in terms of x, we obtain

$$y' = \frac{x(1 - 2x^2)}{\pm x\sqrt{1 - x^2}} = \pm \frac{1 - 2x^2}{\sqrt{1 - x^2}}.$$

As $x \to 0$, this gives $y' = \pm 1$. The same result can be obtained by considering the limit, as P approaches the origin O, of the slope of the secant line through O and a second point $P(x, y)$ on the curve. Denoting this slope by m_{sec}, we have

$$m_{\text{sec}} = \frac{\text{rise}}{\text{run}} = \frac{y - 0}{x - 0} = \pm\sqrt{1 - x^2}.$$

When $P \to O$, $x \to 0$ and $m_{\text{sec}} \to \pm 1$. The double sign means, of course, that if P approaches O along the branch of the curve given by

$$y = x\sqrt{1 - x^2},$$

then the limit of the slope of the secant is $+1$, while the branch given by

$$y = -x\sqrt{1 - x^2}$$

has slope -1 at the origin (Fig. 10.5). We also find turning points at $x = \pm\sqrt{\frac{1}{2}}$, $y = \pm\frac{1}{2}$. Note that we may determine the extent of the curve in the y-direction by comparing the equation of the curve in the form

$$x^4 - x^2 + y^2 = 0$$

with the quadratic equation

$$az^2 + bz + c = 0,$$

whose solutions are

$$z = \frac{-b \pm \sqrt{b^2 - 4ac}}{2a}.$$

Taking $z = x^2$, $a = 1$, $b = -1$, and $c = y^2$, this leads to

$$x^2 = \frac{1 \pm \sqrt{1 - 4y^2}}{2},$$

and it is seen that there is no portion of the curve above the line $y = \frac{1}{2}$ or below the line $y = -\frac{1}{2}$.

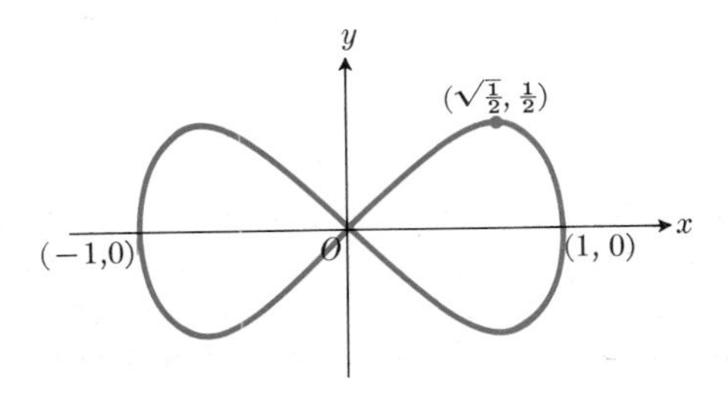

10.5

We observe in Example 10 that the equation is of the second degree in y and of the fourth degree in x. A line parallel to the y-axis has the equation $x =$ constant. Solving the equation of the curve and the equation of the line simultaneously will therefore produce at most two values of y; that is, such a line will cut the curve in at most two points. Similarly, because $y^2 = x^2(1 - x^2)$ is of the fourth degree in x, a line parallel to the x-axis will cut the curve in at most four points. The line $y = c$ and the curve will, in fact, have

(a) no intersections if $|c| > \frac{1}{2}$,
(b) two intersections if $|c| = \frac{1}{2}$,
(c) four intersections if $0 < |c| < \frac{1}{2}$,
(d) three intersections if $c = 0$.

In cases (b) and (d), the results may be interpreted as limiting cases of (c) with some of the intersections coinciding.

EXERCISES

Analyze each of the equations below to investigate the following properties of the curve:

(a) symmetry, (b) extent,
(c) intercepts, (d) asymptotes,
(e) slope at the intercepts.

Locate a few points and sketch the curve, taking into account the information discovered above.

1. $y^2 = x(x - 2)$
2. $y^2 = \dfrac{x}{x - 2}$
3. $x^4 + y^4 = 1$
4. $x^2 = \dfrac{1 + y^2}{1 - y^2}$
5. $x^2 = \dfrac{y^2 + 1}{y^2 - 1}$
6. $y = x + \dfrac{1}{x}$
7. $y = x^2 + 1$
8. $y = \dfrac{1}{x^2 + 1}$
9. $y = x^2 - 1$
10. $y = \dfrac{1}{x^2 - 1}$

10.2 TANGENTS AND NORMALS

In Chapter 1, we learned that a straight line of slope m which passes through the point (x_1, y_1) is represented by an equation

$$y - y_1 = m(x - x_1). \tag{1}$$

The equation of a line *tangent* to a given curve $y = f(x)$ at a given point (x_1, y_1) on the curve can be found at once by taking

$$m = \left(\frac{dy}{dx}\right)_{(x_1, y_1)} = f'(x_1)$$

in Eq. (1), which then becomes

$$y - y_1 = f'(x_1)(x - x_1). \tag{2}$$

In this equation, x_1, y_1, and $f'(x_1)$ are all *constants*; the variables x and y appear only linearly.

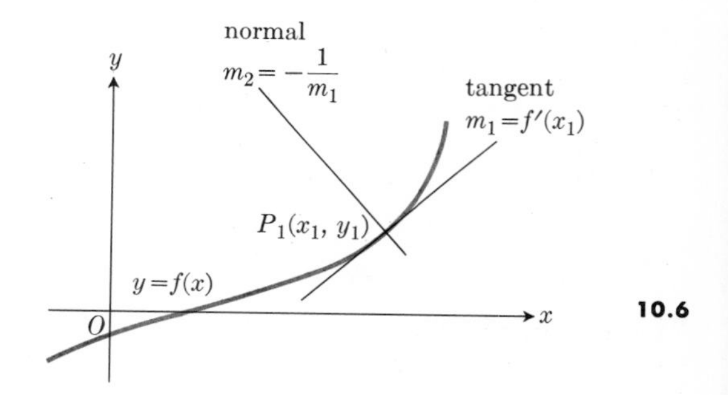

10.6

The line through $P_1(x_1, y_1)$ and perpendicular to the tangent line is called the *normal* to the curve at P_1 (see Fig. 10.6). Its slope is the negative reciprocal of the slope of the tangent, that is,

$$m_2 = -\frac{1}{m_1} = \frac{-1}{f'(x_1)},$$

and the equation of the *normal* is

$$y - y_1 = \frac{-1}{f'(x_1)}(x - x_1). \tag{3}$$

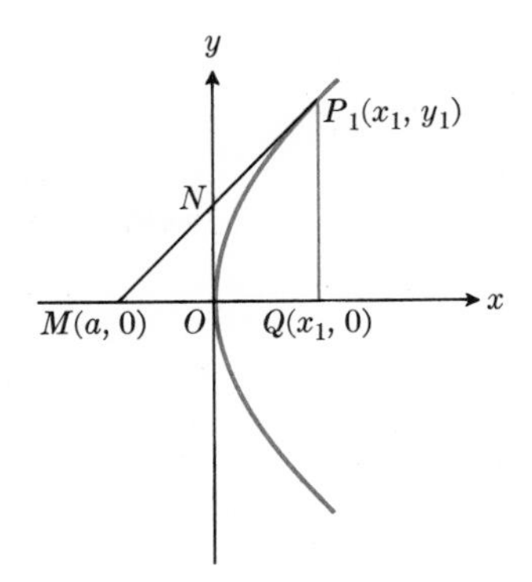

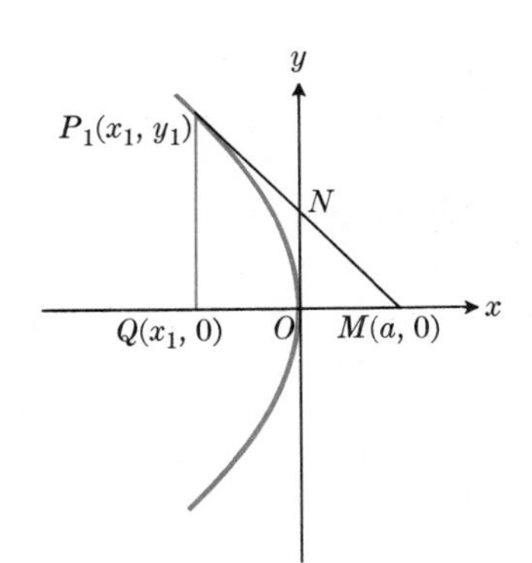

10.7

Problem 1. The tangent to the curve

$$y^2 = 4px \tag{4}$$

at the point $P_1(x_1, y_1)$ intersects the x-axis at M and the y-axis at N. Show that N is the midpoint of the segment MP_1.

Solution. The curve is symmetric about the x-axis and lies entirely to the right of the y-axis if p is positive, or entirely to the left of the y-axis if p is negative (see Fig. 10.7).

Its slope at any point is found from

$$\frac{dy}{dx} = \frac{2p}{y}.$$

The tangent to the curve at the origin is vertical and for this point the problem becomes meaningless. But if P_1 is any point on the curve (4) other than the origin, then the equation of the line tangent to the curve at P_1 is

$$y - y_1 = \frac{2p}{y_1}(x - x_1).$$

The point $M(a, 0)$ where this line cuts the x-axis *lies on the line*, and hence

$$-y_1 = \frac{2p}{y_1}(a - x_1),$$

which gives

$$a = x_1 - \frac{y_1^2}{2p}. \tag{5}$$

The point (x_1, y_1) is *on the curve*, and hence

$$y_1^2 = 4px_1,$$

so that

$$\frac{y_1^2}{2p} = 2x_1,$$

which reduces (5) to

$$a = -x_1. \tag{6}$$

Equation (6) tells us that the origin bisects the line segment MQ joining $M(-x_1, 0)$ and $Q(x_1, 0)$. Hence, by similar triangles, point N bisects the segment MP_1.

Problem 2. Show that the normal to the curve

$$x^2 + y^2 = a^2 \tag{7}$$

at any point (x_1, y_1) on it passes through the origin.

Solution. Differentiating (7) implicitly and solving for dy/dx, we find

$$\frac{dy}{dx} = -\frac{x}{y}$$

as the slope of the *tangent* at (x, y). The slope of the *normal* at (x_1, y_1) is therefore

$$m = \frac{y_1}{x_1},$$

and the equation of the normal is

$$y - y_1 = \frac{y_1}{x_1}(x - x_1). \tag{8}$$

When we clear of fractions, (8) reduces to

$$x_1 y = y_1 x.$$

Clearly, the origin lies on this line, since the coordinates $x = 0$, $y = 0$ satisfy the equation.

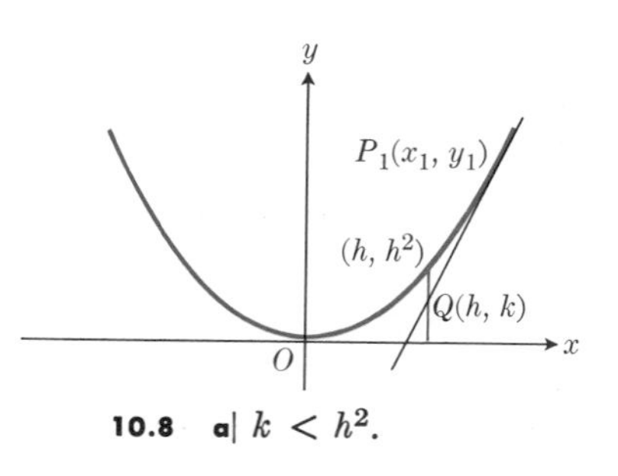

10.8 a| $k < h^2$.

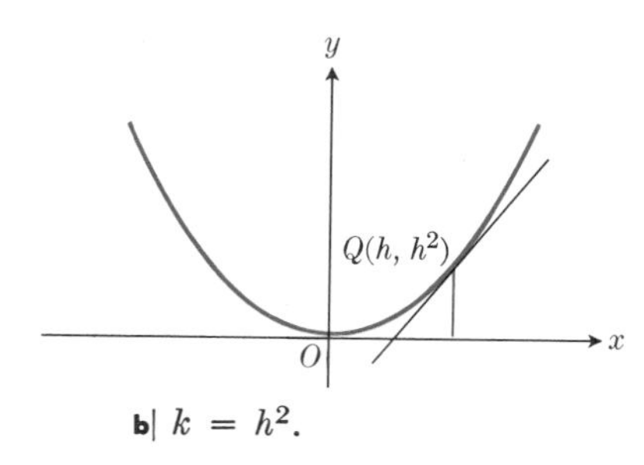

b| $k = h^2$.

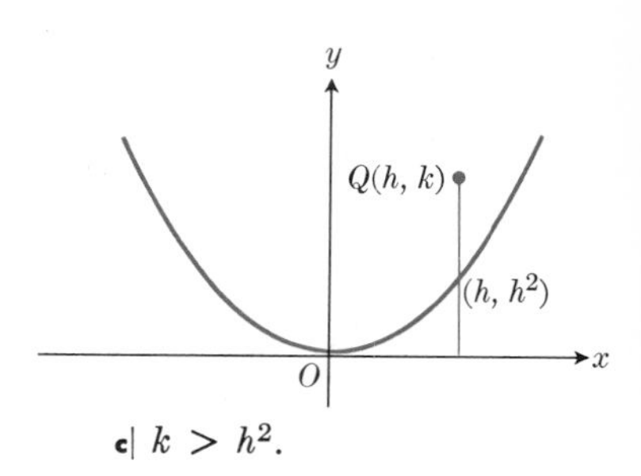

c| $k > h^2$.

Problem 3. Show that the number of tangent lines that can be drawn from the point $Q(h, k)$ to the curve

$$y = x^2 \tag{9}$$

is two, one, or zero accordingly as k is less than h^2, equal to h^2, or greater than h^2, respectively.

Solution. Figure 10.8 represents the curve and each of the three cases $k < h^2$, $k = h^2$, $k > h^2$. From the fact that $d^2y/dx^2 = 2$ is always positive, so that the curve is concave upward, we see that the results stated are equivalent to saying that two tangents can be drawn from a point on the convex side of the curve, no tangents can be drawn from a point on the concave side of the curve, and only one tangent can be drawn from a point on the curve. This is geometrically evident, but we shall adopt an analytical approach as follows.

Suppose that a tangent *can* be drawn to the curve from Q touching the curve at some point $P_1(x_1, y_1)$ which is at present not known. Then, using Eq. (2), we find the equation of this tangent line to be

$$y - y_1 = 2x_1(x - x_1).$$

Since $Q(h, k)$ is *on this line*, its coordinates must satisfy the equation

$$k - y_1 = 2x_1(h - x_1). \tag{10a}$$

Also, since (x_1, y_1) is *on the curve*, its coordinates must satisfy the equation

$$y_1 = x_1^2. \tag{10b}$$

Now in reality, h and k are known to us but x_1 and y_1 are not. We therefore have two equations, (10a) and (10b), for two unknowns, x_1 and y_1. We eliminate y_1 between them and obtain

$$k - x_1^2 = 2hx_1 - 2x_1^2,$$

or

$$x_1^2 - 2hx_1 + k = 0.$$

If we solve this for x_1 by means of the quadratic formula, we obtain

$$x_1 = h \pm \sqrt{h^2 - k}. \tag{10c}$$

Only real values of x_1 correspond to points P_1 on the curve; hence no tangent may be drawn if $h^2 - k$ is negative, which is the same condition as saying k is greater than h^2. If $h^2 = k$, then the two roots in (10c) coincide and give only one point, namely $x_1 = h$, $y_1 = k = h^2$, to which the tangent can be drawn. But if $k < h^2$, the two roots in (10c) are distinct and each root gives a value of x_1 which with $y_1 = x_1^2$ determines a point P_1 such that the line QP_1 is tangent to the curve, so that two tangents may be drawn.

Angle between two curves

Let C_1 and C_2 be two curves which intersect at a point P. Then the angle between C_1 and C_2 at P is, by definition, the angle β between their tangents at P (Fig. 10.9). If these tangents are, respectively, L_1 and L_2, they have slopes

$$m_1 = \tan \phi_1, \quad m_2 = \tan \phi_2.$$

Then $\tan \beta$ can be found from m_1 and m_2 as follows. Since

$$\beta = \phi_2 - \phi_1,$$

we have

$$\tan \beta = \tan(\phi_2 - \phi_1) = \frac{\tan \phi_2 - \tan \phi_1}{1 + \tan \phi_2 \tan \phi_1} = \frac{m_2 - m_1}{1 + m_2 m_1}. \tag{11}$$

Here β is measured in the counterclockwise direction from L_1 to L_2.

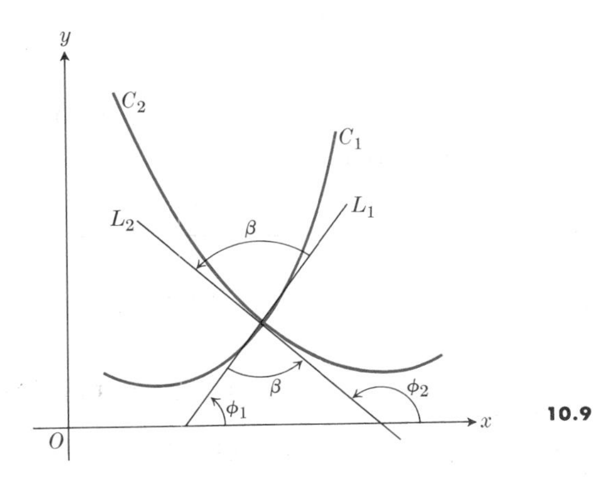

10.9

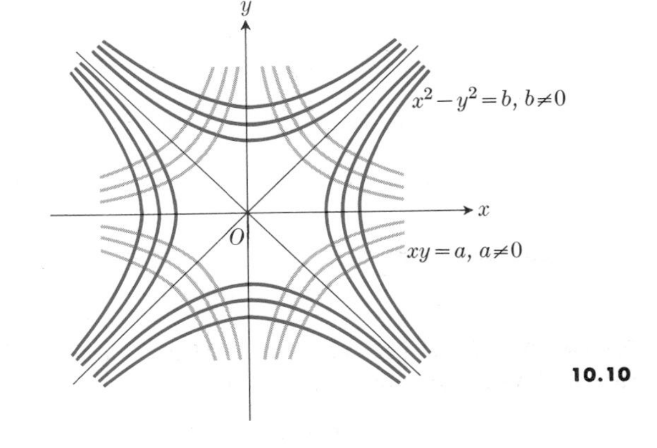

10.10

We recall that the lines L_1 and L_2 are perpendicular if $m_2 = -1/m_1$. In this case we say that the curves C_1 and C_2 are *orthogonal.*

Problem 4. Show that every curve of the family

$$xy = a, \quad a \neq 0 \tag{12a}$$

is orthogonal to every curve of the family

$$x^2 - y^2 = b, \quad b \neq 0. \tag{12b}$$

Solution. The two families of curves are sketched in Fig. 10.10. At a point $P(x, y)$ on any curve of (12a), the slope is

$$\frac{dy}{dx} = -\frac{y}{x}, \tag{13a}$$

and on any curve of (12b) the slope is

$$\frac{dy}{dx} = \frac{x}{y}. \tag{13b}$$

At a point of intersection, the values of x and y in (13b) are the same as in (13a), and the two curves are orthogonal because these slopes are negative reciprocals of each other. The cases $x = 0$ or $y = 0$ cannot occur in Eqs. (13) if (x, y) is a point of intersection of any curve (12a) and a curve (12b), since a and b are restricted to be constants different from zero. That every curve in (12a) does in fact intersect every curve in (12b) follows from the fact that the equation

$$x^2 - \frac{a^2}{x^2} = b,$$

which results from eliminating y between (12a) and (12b), has real roots for every pair of nonzero real constants a and b.

The curves in (12b) are called ***orthogonal trajectories*** of the curves in (12a). Such mutually orthogonal systems of curves are of particular importance in physical problems related to electrical potential, where the curves in one family correspond to lines of flow and those in the other family correspond to lines of constant potential. They also occur in hydrodynamics and in heat flow problems.

EXERCISES

1. Find the equation of the line tangent to the curve
$$y^2 - 2x - 4y - 1 = 0$$
at $(-2, 1)$.
2. Find the equation of the line normal to the curve
$$xy + 2x - 5y - 2 = 0$$
at $(3, 2)$.
3. Find the equations of the lines tangent to the curve
$$y = x^3 - 6x + 2$$
and parallel to the line $y = 6x - 2$.
4. Find the equations of the lines normal to the curve
$$xy + 2x - y = 0$$
and parallel to the line $2x + y = 0$.

5. Show that the two lines that are drawn from $(\frac{3}{2}, 0)$ tangent to the curve

$$x^2 - 4y + 4 = 0$$

are perpendicular.

6. Show that the two lines that are drawn from any point on the line $x = -p$ tangent to the curve $y^2 = 4px$ are perpendicular (p = constant).

7. Does the line that is tangent to the curve $y = x^3$ at the point (1, 1) intersect the curve at any other point? If so, find the point.

8. A tangent to the curve $y = x^3$ is drawn from $P_0(1, 5)$ to a point $P_1(x_1, y_1)$ on it. Find the coordinates of P_1. Is there more than one possibility for the point P_1? Why?

9. Find all lines that can be drawn normal to the curve $x^2 - y^2 = 5$ and parallel to the line $2x + 3y = 10$. Sketch the curve and the lines.

10. Find all lines that can be drawn tangent to the curve $4xy = 1$ from the point $P(-1, 2)$. Sketch the curve and the lines.

11. The line that is normal to the curve $y = x^2 + 2x - 3$ at (1, 0) intersects the curve at what other point?

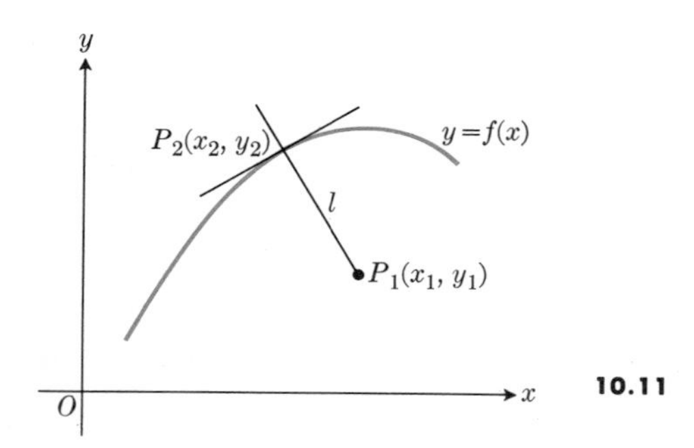

10.11

12. If l represents the distance from a fixed point $P_1(x_1, y_1)$ to a variable point $P(x, y)$ on a curve $y = f(x)$ (Fig. 10.11), then

$$l^2 = (x - x_1)^2 + [f(x) - y_1]^2.$$

Show that the derivative $d(l^2)/dx$ is equal to zero when $x = x_2$ if and only if the line P_1P_2 is normal to the curve at $P_2(x_2, f(x_2))$.

13. Show that the curve $2x^2 + 3y^2 = 5$ and the curve $y^2 = x^3$ intersect at right angles.

14. Let $P_1(x_1, y_1)$ and $P_2(x_2, y_2)$ be any two points on the curve $y = ax^2 + bx + c$, $a \neq 0$. If a line is drawn tangent to the curve at (x_0, y_0) and is parallel to the chord P_1P_2, show that $x_0 = (x_1 + x_2)/2$.

15. For what values of b is the line $y = 12x + b$ tangent to the curve $y = x^3$?

16. For what values of m is the line $y = mx$ tangent to the curve $y^2 + x^2 - 4x + 3 = 0$?

17. Find the interior angles of the triangle whose vertices are $A(1, 1)$, $B(3, -1)$, and $C(5, 2)$.

18. Find the slope of the line that bisects the angle ACB in the previous problem.

19. Let A, B, C be the interior angles of a triangle in which no angle is a right angle. By calculating $\tan(A + B + C)$ and observing that

$$A + B + C = 180°,$$

show that the equation

$$\tan A + \tan B + \tan C = \tan A \tan B \tan C$$

must be satisfied.

20. Calculate the interior angles of the triangle whose vertices are $A(1, 2)$, $B(2, -1)$, and $C(-1, 1)$, and check your results by showing that they satisfy the equation

$$\tan A + \tan B + \tan C = \tan A \tan B \tan C.$$

(See Exercise 19.)

21. Find the angles between the following pairs of curves.
(a) $3x + y = 5$, $2x - y = 4$
(b) $y = x^2$, $xy = 1$
(c) $x^2 + y^2 = 16$, $y^2 = 6x$
(d) $x^2 + xy + y^2 = 7$, $y = 2x$

10.3 NEWTON'S METHOD FOR APPROXIMATING ROOTS OF EQUATIONS

Suppose we want to find a real root of the equation $f(x) = 0$. If the equation is linear or quadratic, we know simple rules to use. There are formulas for solving third and fourth degree equations, too, though they are less familiar. But there are no algebraic formulas for solving equations of the fifth degree or higher. Moreover, even the quadratic formula doesn't immediately give the answer in decimal form. Often we must look up the square root in a table or compute it. Newton's method, however, has these advantages:

(a) It applies to an equation of any degree. Indeed, it even applies to nonpolynomial equations such as $2 \sin x - x = 0$.

(b) It gives the answer in numerical form. We continue the computations until we have the degree of accuracy we want.

Its disadvantage is that the arithmetic becomes involved unless a calculating machine is used.

Newton's method proceeds as follows. If the equation we want to solve is $f(x) = 0$, we make a rough graph of $y = f(x)$. We estimate the root r, where the curve crosses the x-axis. This estimate is our first approximation; call it x_1. The next approximation, x_2, is the place where the *tangent* to the curve at $(x_1, f(x_1))$ crosses the x-axis (see Fig. 10.12). The equation of this tangent line is

$$y - f(x_1) = f'(x_1)(x - x_1). \tag{1}$$

We assume that this line is not parallel to the x-axis, that is, $f'(x_1) \neq 0$. (If we started at a place where the slope is zero, we must start over again at a nearby place where the slope isn't zero.) Then the line crosses the x-axis at a point with coordinates $x = x_2$, $y = 0$ which satisfy (1). Hence

$$0 - f(x_1) = f'(x_1)(x_2 - x_1)$$

and

$$x_2 = x_1 - \frac{f(x_1)}{f'(x_1)}. \tag{2}$$

Equation (2) provides the basis for an iterative process, since we may now replace x_1 by x_2 on the right-hand side and call the new result x_3 instead of x_2. In general, having arrived at an approximation x_n, we replace x_1 by x_n on the right-hand side of (2) and we replace x_2 by x_{n+1}, thus:

$$x_{n+1} = x_n - \frac{f(x_n)}{f'(x_n)}. \tag{3}$$

A computing machine can be programmed with feedback to keep repeating the process of Eq. (3) automatically, as indicated in Fig. 10.13.

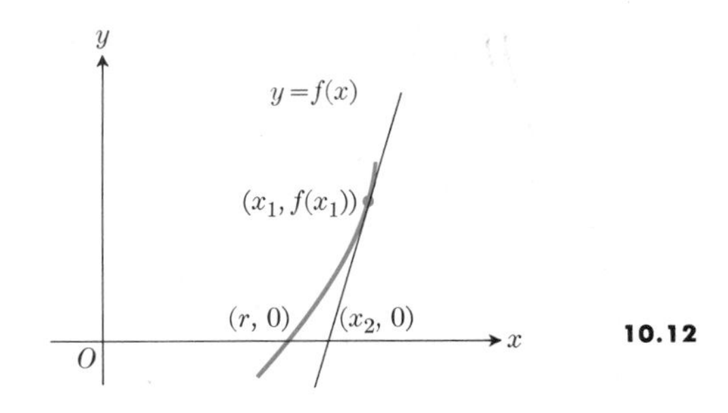

10.12

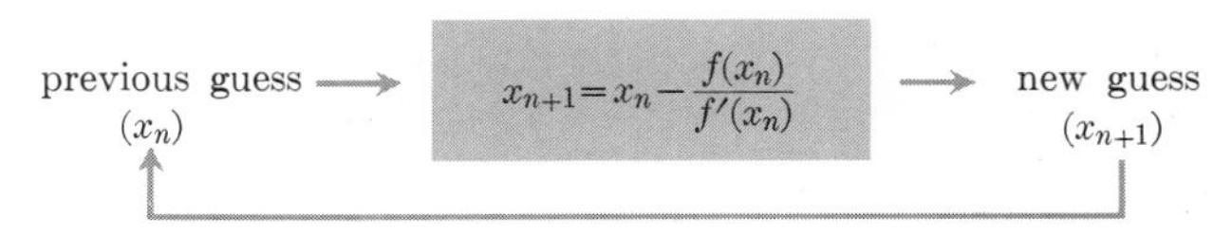

10.13 Feedback mechanism for iterating Newton's method.

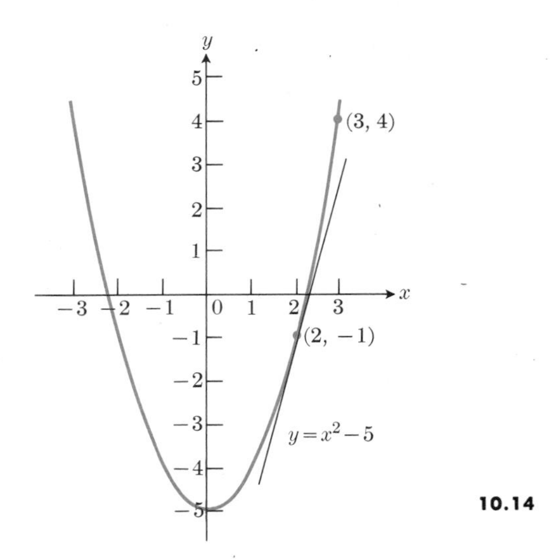

10.14

Example 1. We may use Newton's method to solve $x^2 = 5$ to two decimals. We take $f(x) = x^2 - 5$, and sketch the curve

$$y = f(x) = x^2 - 5.$$

A root of the equation $x^2 = 5$ corresponds to a place where this curve crosses the x-axis. From Fig. 10.14, we see that one crossing is between 2 and 3, and is closer to 2 than to 3. We therefore take $x_1 = 2$ as our first approximation. Then we compute x_2 from Eq. (2).

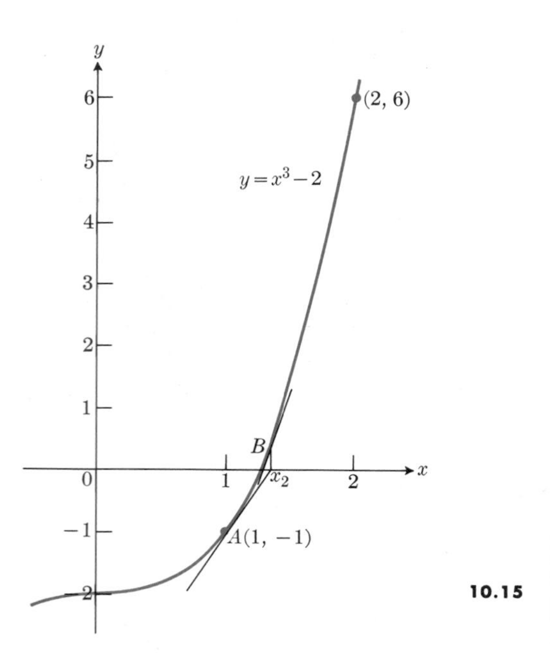

10.15

Here we have the following information:

$$f(x) = x^2 - 5, \qquad f(2) = -1,$$
$$f'(x) = 2x, \qquad f'(2) = 4.$$

Hence

$$x_2 = 2 - (-\tfrac{1}{4}) = \tfrac{9}{4} = 2.25.$$

To get the next approximation, we use Eq. (3) with $n = 2$:

$$x_3 = x_2 - \frac{f(x_2)}{f'(x_2)}$$

with

$$f(x_2) = f(\tfrac{9}{4}) = (\tfrac{9}{4})^2 - 5 = \tfrac{1}{16}$$
$$f'(x_2) = f'(\tfrac{9}{4}) = 2(\tfrac{9}{4}) = \tfrac{9}{2}.$$

Hence

$$x_3 = \frac{9}{4} - \frac{\frac{1}{16}}{\frac{9}{2}} = \frac{161}{72} = 2.236^+.$$

Since

$$(2.236)^2 = 4.999696 \quad \text{and} \quad (2.237)^2 = 5.004169,$$

it is clear that the positive root is between 2.236 and 2.237, hence is 2.24 to two decimal places.

Example 2. Newton's method may be used to find cube roots, fourth roots, and so on. We shall illustrate the method for finding the cube root of 2. That is, we shall try to find a root of the equation $x = \sqrt[3]{2}$ or $x^3 = 2$. We first put $f(x) = x^3 - 2$ and sketch the graph of $y = f(x)$ (see Fig. 10.15). Since $f(1) = -1$ and $f(2) = 6$, we let our first approximation be $x_1 = 1$. Then Eq. (2) becomes

$$\begin{aligned} x_2 &= x_1 - \frac{f(x_1)}{f'(x_1)} \\ &= x_1 - \frac{(x_1^3 - 2)}{3x_1^2} \\ &= \tfrac{2}{3}x_1 + \tfrac{1}{3}(2/x_1^2). \end{aligned} \tag{4}$$

We have written Eq. (4) in a form that shows x_2 is a "weighted average" of x_1 (with coefficient $\frac{2}{3}$) and of $2/x_1^2$ (with coefficient $\frac{1}{3}$). The sum of these coefficients is 1, but we "weight" x_1 in our average twice as heavily as we do $2/x_1^2$. This same procedure may of course be repeated with x_2 in place of x_1 on the right-hand side of (4) and x_3 on the left. In general, we have

$$x_{n+1} = \tfrac{2}{3}x_n + \tfrac{1}{3}(2/x_n^2). \tag{5}$$

If at any stage the numbers x_n and $2/x_n^2$ are equal, then x_{n+1} is also equal to this common value, and it is $\sqrt[3]{2}$,

since
$$x_n = 2/x_n^2$$
implies
$$x_n^3 = 2.$$

Starting with $x_1 = 1$, we find the first three approximations are

$$x_1 = 1, \quad x_2 = \tfrac{4}{3} = 1.33^+, \quad x_3 = \tfrac{91}{72} = 1.264^-.$$

A table of cube roots gives $\sqrt[3]{2} = 1.259921$ to six decimal places. Our third approximation is too large by about 0.004, but 1.26 is correct to two decimal places.

Remark 1. The accuracy of Newton's method for functions f which possess continuous first and second derivatives can be estimated by applying the Mean Value Theorem. (See Exercise 13.) One estimate that we can obtain by this theorem is

$$|x_2 - r| \le (x_1 - r)^2(M/m), \tag{6}$$

where M is the maximum absolute value of the second derivative and m the minimum absolute value of the first derivative in an interval that contains the root r and the estimates x_1 and x_2. The error decreases if the first guess is accurate enough to make

$$\left|(x_1 - r)\frac{M}{m}\right| < 1. \tag{7}$$

In Example 1, for instance, we had

$$f(x) = x^2 - 5, \quad f'(x) = 2x, \quad f''(x) = 2,$$

with $r = \sqrt{5}$, and we had $x_1 = 2$ and all other approximations x_n between 2 and $\frac{9}{4}$. Hence in (6) we could take $M = 2$, $m = 4$ and have

$$|x_2 - \sqrt{5}| \le \tfrac{1}{2}(x_1 - \sqrt{5})^2.$$

The same inequality applies to any two successive approximations x_{n+1} and x_n:

$$|x_{n+1} - \sqrt{5}| \le \tfrac{1}{2}(x_n - \sqrt{5})^2.$$

This inequality shows that the number of decimal places of accuracy doubles with each successive iteration.

In Example 2, we had

$$f(x) = x^3 - 2, \quad f'(x) = 3x^2, \quad f''(x) = 6x,$$

with $r = \sqrt[3]{2} = 1.26^-$; and $x_1 = 1$, $x_2 = \frac{4}{3}$, and all other approximations lie between x_1 and x_2. Hence in (6) we may take $M = 6(\frac{4}{3}) = 8$, $m = 3$, and get

$$|x_2 - \sqrt[3]{2}| \le (\tfrac{8}{3})(x_1 - \sqrt[3]{2})^2.$$

The condition (7) is satisfied since the left side of the inequality (7) is actually less than 0.7.

Remark 2. We can get a good picture of how the Newton process converges in a favorable situation such as that shown in Fig. 10.16. Here x_1 is to the left of r. The tangent to the curve at $A(x_1, f(x_1))$ crosses the x-axis at x_2 to the right of r. The second tangent line, through $B(x_2, f(x_2))$ crosses at x_3, much nearer r. And the third tangent line, through $C(x_3, f(x_3))$, is practically indistinguishable from the curve itself in the vicinity of the root r.

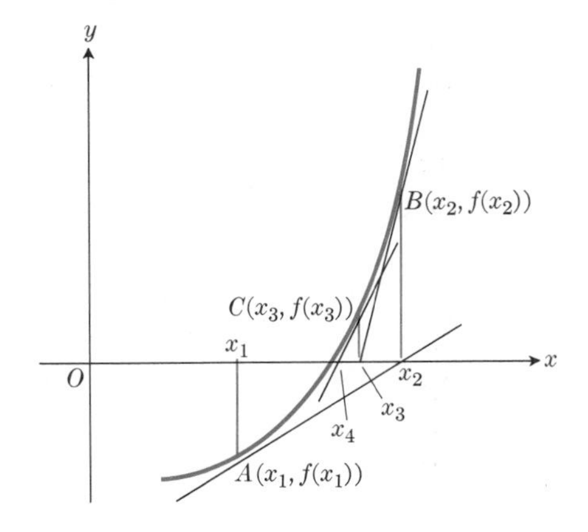

10.16

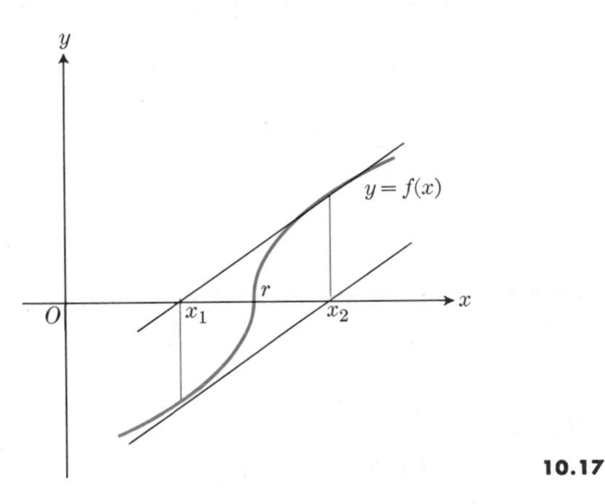

10.17

Remark 3. In some cases, Newton's method may not converge. For instance, if the graph of $y = f(x)$ is like Fig. 10.17, we may have

$$x_1 = r - h, \qquad x_2 = r + h,$$

and successive approximations go back and forth between these two values alternately. A function such as

$$f(x) = \begin{cases} \sqrt{x - r} & \text{for } x \ge r \\ -\sqrt{r - x} & \text{for } x \le r \end{cases} \tag{8}$$

has this property. No amount of iteration of the process will bring us any closer to the root r than our initial guess. (Also see Exercise 12.)

EXERCISES

Sketch the graph of $y = f(x)$ in Exercises 1 through 6. Show that $f(a)$ and $f(b)$ have opposite signs, and use Newton's method to estimate the root of the equation $f(x) = 0$ between a and b. One might use $x_1 = (a + b)/2$ as first approximation.

1. $f(x) = x^2 + x - 1, \quad a = 0, \quad b = 1$
2. $f(x) = x^3 + x - 1, \quad a = 0, \quad b = 1$
3. $f(x) = x^4 + x - 3, \quad a = 1, \quad b = 2$
4. $f(x) = x^4 - 2, \quad a = 1, \quad b = 2$
5. $f(x) = 2 - x^4, \quad a = -1, \quad b = -2$
6. $f(x) = \sqrt{2x+1} - \sqrt{x+4}, \quad a = 2, \quad b = 4$
7. Suppose our first guess is lucky, in the sense that x_1 is a root of $f(x) = 0$. What happens to x_2 and later approximations?
8. In Example 2, take $x_3 = 1.26$ and compute x_4 to five decimal places. Compare with the six-place $\sqrt[3]{2}$ given in the text. How does your result check with the statement that "when one approximation is good to two decimal places, the next Newton's-method approximation will be good to about four decimals"?
9. To find $x = \sqrt[q]{a}$, we apply Newton's method to $f(x) = x^q - a$. Here we assume that a is a positive real number and q is a positive integer. Show that x_2 is a "weighted average" of x_1 and a/x_1^{q-1}, and find the coefficients m_1, m_2 such that
$$x_2 = m_1 x_1 + m_2(a/x_1^{q-1})$$
for
$$m_1 > 0, \quad m_2 > 0 \quad m_1 + m_2 = 1.$$
What conclusion would you reach if x_1 and a/x_1^{q-1} were equal? What would be the value of x_2 in that case? (You may also wish to read the article by J. P. Ballantine, "An Averaging Method of Extracting Roots," *American Mathematical Monthly*, Vol. 63, 1956, pp. 249–252, where more efficient ways of averaging are discussed. Also see J. S. Frame, "The Solution of Equations by Continued Fractions," ibid., Vol. 60, 1953, pp. 293–305.)
10. In Example 2, let $e_1 = x_1 - r$ and $e_2 = x_2 - r$ be the errors in our first two approximations of $r = \sqrt[3]{2}$. Show algebraically that $e_2 = e_1^2(2x_1 + r)/(3x_1^2)$. Similar equations relate the errors in any two successive approximations. All our approximations in Example 2 lie between $x_1 = 1$ and $x_2 = \frac{4}{3}$, inclusive. Deduce that the factor $(2x_n + r)/(3x_n^2)$ is not greater than $(\frac{8}{3} + r)/3$, which in turn is less than the quantity $(2.67 + 1.26)/3 = 1.31$. Therefore, $e_{n+1} \le 1.31e_n^2$.
11. Show that Newton's method applied to $f(x)$ in Eq. (8) leads to $x_2 = r + h$ if $x_1 = r - h$, and to $x_2 = r - h$ if $x_1 = r + h$, $h > 0$. Interpret the result geometrically.
12. (See Remark 3.) Is it possible that successive approximations actually get "worse," in that x_{n+1} is farther away from the root r than x_n is? Can you find such a "pathological" example?
Hint. Try cube roots in place of square roots in Eq. (8).
13. Using the following hints, derive the inequality (6).

(1) By the Mean Value Theorem there is a number a between x_1 and r such that
$$f(x_1) - f(r) = f'(a)(x_1 - r).$$
If the minimum of the absolute value of $f'(x)$ in an interval containing x_1 and r is m, and $m \ne 0$, then $f'(a) \ne 0$. We assume that r is a root of $f(r) = 0$. Hence
$$x_1 - r = f(x_1)/f'(a). \tag{a}$$
(2) By Eq. (2),
$$x_2 - x_1 = -f(x_1)/f'(x_1). \tag{b}$$
(3) Add (a) and (b) to get
$$x_2 - r = \frac{f(x_1)}{f'(a)f'(x_1)}[f'(x_1) - f'(a)]$$
$$= \frac{x_1 - r}{f'(x_1)}[f'(x_1) - f'(a)]. \tag{c}$$
(4) By the Mean Value Theorem, there is a number b between x_1 and a such that
$$f'(x_1) - f'(a) = f''(b)(x_1 - a). \tag{d}$$
(5) From (c) and (d), deduce that
$$x_2 - r = (x_1 - r)(x_1 - a)\frac{f''(b)}{f'(x_1)}, \tag{e}$$
where a and b are numbers between x_1 and r. In particular, therefore,
$$|x_1 - a| \le |x_1 - r|. \tag{f}$$
Now deduce (6) from (e) and (f).

10.4 DISTANCE BETWEEN TWO POINTS: EQUATIONS OF LOCI

It is at once apparent from Fig. 10.18 and the theorem of Pythagoras that the distance d between the two points $P_1(x_1, y_1)$ and $P_2(x_2, y_2)$ is given by

$$d = \sqrt{(x_2 - x_1)^2 + (y_2 - y_1)^2}. \tag{1}$$

This formula is particularly useful in finding the equation of a curve when its geometric character depends on one or more distances.

Problem. Find the locus of $P(x, y)$ if the distance from P to the origin is equal to the distance from P to the line $L: x = 4$.

Solution. (See Fig. 10.19.) The distance from P to L is the perpendicular distance PQ from P to $Q(4, y)$, where

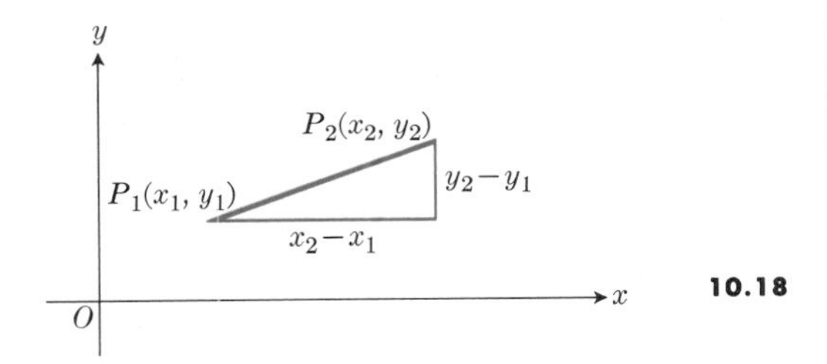

10.18

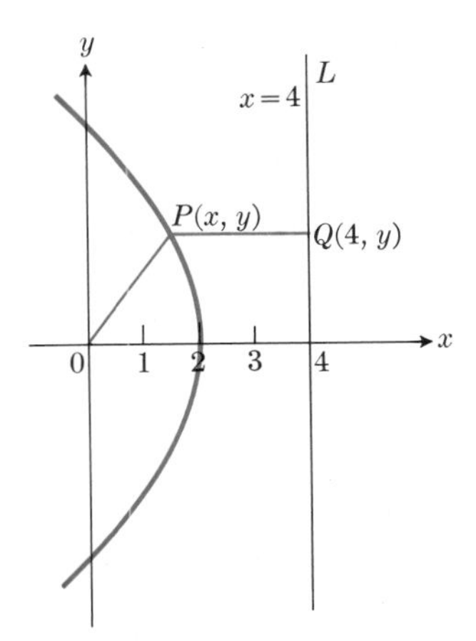

10.19 The locus of P when P is equidistant from the origin and from the line L.

Q has the same ordinate as P. Thus

$$PQ = \sqrt{(4 - x)^2 + (y - y)^2} = |4 - x|.$$

The distance OP is

$$OP = \sqrt{x^2 + y^2}.$$

The condition of the locus is $OP = PQ$, or

$$\sqrt{x^2 + y^2} = |4 - x|. \tag{1}$$

If (1) holds, so does the equation we get by squaring:

$$x^2 + y^2 = 16 - 8x + x^2,$$

or

$$y^2 = 16 - 8x. \tag{2}$$

That is, if a point belongs to the locus, then its coordinates must satisfy Eq. (2). The converse is also true, for if Eq. (2) holds, then

$$\sqrt{x^2 + y^2} = \sqrt{x^2 + (16 - 8x)} = \sqrt{(x - 4)^2}$$
$$= |x - 4| = |4 - x|,$$

and hence

$$OP = PQ.$$

Therefore Eq. (2) expresses both the necessary and the sufficient condition on the coordinates of $P(x, y)$ for P to belong to the locus.

EXERCISES

In Exercises 1 through 8, use the distance formula to derive the equation of the locus of points $P(x, y)$ that satisfy the specified conditions. Analyze the equation and sketch its graph.

1. P is equidistant from the two points $A(-2, 1)$ and $B(2, -3)$.
2. The distance from P to $F_1(-1, 0)$ is twice its distance to $F_2(2, 0)$.
3. The product of its distances from $F_1(-2, 0)$ and $F_2(2, 0)$ is 4.
4. The sum of the distances from P to $F_1(1, 0)$ and $F_2(0, 1)$ is constant and the curve passes through the origin.
5. The distance of P from the line $x = -2$ is 2 times its distance from the point $(2, 0)$.

6. The distance of P from the point $(-3, 0)$ is 4 more than its distance from the point $(3, 0)$.
7. The distance of P from the line $y = 1$ is 3 less than its distance from the origin.
8. P is 3 units from the point $(2, 3)$.
9. Find a point that is equidistant from the three points $A(0, 1)$, $B(1, 0)$, $C(4, 3)$. What is the radius of the circle through A, B, and C?
10. Find the distance from the point $P_1(x_1, y_1)$ to the straight line $Ax + By + C = 0$.

10.5 THE CIRCLE

Definition. *A circle is the locus of points in a plane that are at a given distance from a given point.*

The equation of a circle

Let $C(h, k)$ be the given point, the center of the circle. Let r be the given distance, the radius of the circle. Let $P(x, y)$ be a point on the circle. Then

$$CP = r, \tag{1}$$

or

$$\sqrt{(x-h)^2 + (y-k)^2} = r,$$

or

$$(x-h)^2 + (y-k)^2 = r^2. \tag{2}$$

If (1) is satisfied so is (2), and conversely if $r > 0$. Therefore (2) is the equation of the locus. (See Fig. 10.20.)

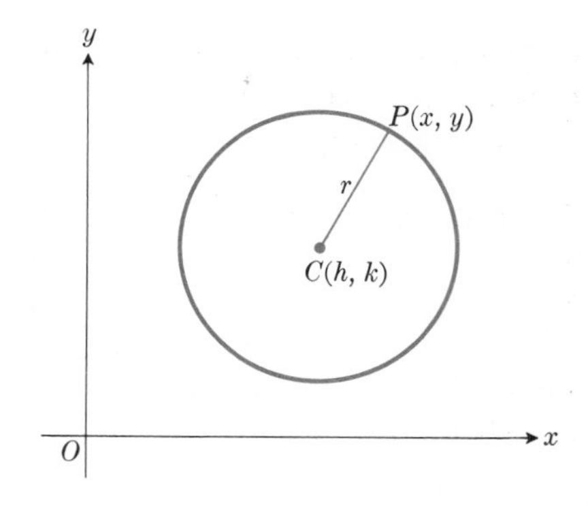

10.20 Circle: $(x-h)^2 + (y-k)^2 = r^2$.

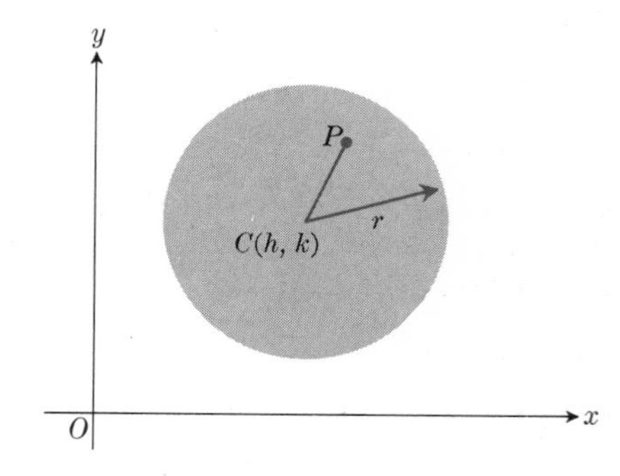

10.21 The region $(x-h)^2 + (y-k)^2 < r^2$, the interior of the circle with center $C(h, k)$ and radius r.

Problem 1. Find the equation of the circle with center at the origin and with radius r.

Solution. If $h = k = 0$, Eq. (2) becomes

$$x^2 + y^2 = r^2. \tag{3}$$

Problem 2. Find the circle through the origin with center at $C(2, -1)$.

Solution. Its equation, (2), is of the form

$$(x-2)^2 + (y+1)^2 = r^2.$$

Since it goes through the origin, $x = y = 0$ must satisfy the equation. Hence $(0-2)^2 + (0+1)^2 = r^2$, or $r^2 = 5$. The equation is

$$(x-2)^2 + (y+1)^2 = 5.$$

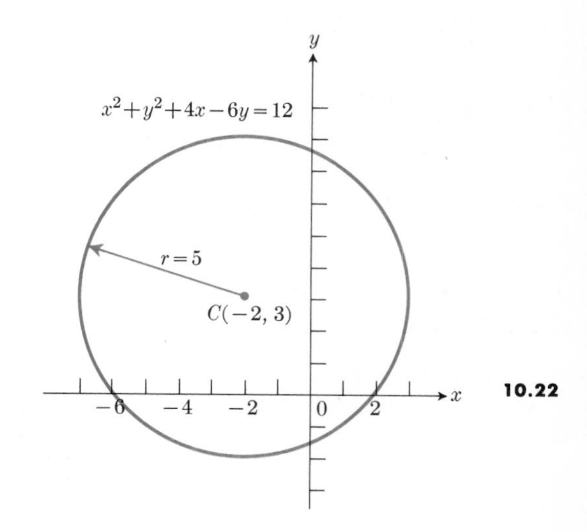

10.22

Problem 3. What is the locus of points $P(x, y)$ whose coordinates satisfy the inequality

$$(x - h)^2 + (y - k)^2 < r^2? \tag{4}$$

Solution. The left side of (4) is the square of the distance CP from $C(h, k)$ to $P(x, y)$. The inequality is satisfied if and only if

$$CP < r,$$

that is, if and only if P lies inside the circle of radius r with center at $C(h, k)$. (See Fig. 10.21.)

Problem 4. Analyze the equation

$$x^2 + y^2 + 4x - 6y = 12.$$

Solution. We complete the squares in the x terms and y terms and get

$$(x^2 + 4x + 4) + (y^2 - 6y + 9) = 12 + 4 + 9,$$

or

$$(x + 2)^2 + (y - 3)^2 = 25.$$

This is of the same form as Eq. (2); therefore it represents a circle with center $C(-2, 3)$ and radius $r = 5$. Its graph is shown in Fig. 10.22.

Remark 1. An equation of the form

$$Ax^2 + Ay^2 + Dx + Ey + F = 0, \quad A \neq 0, \tag{5}$$

can often be reduced to the form of Eq. (2) by completing the squares as we did in Example 4. More specifically, we may divide (5) by A and write

$$\left(x^2 + \frac{D}{A}x\right) + \left(y^2 + \frac{E}{A}y\right) = -\frac{F}{A}. \tag{6}$$

To complete the squares for x, we must add

$$(D/2A)^2 = D^2/(4A^2);$$

and for y we add

$$(E/2A)^2 = E^2/(4A^2).$$

Of course, we must add to both sides of Eq. (6), thus obtaining

$$\left(x + \frac{D}{2A}\right)^2 + \left(y + \frac{E}{2A}\right)^2 = -\frac{F}{A} + \frac{D^2 + E^2}{4A^2} = \frac{D^2 + E^2 - 4AF}{4A^2}. \tag{7}$$

Equation (7) is like Eq. (2), with

$$r^2 = \frac{D^2 + E^2 - 4AF}{4A^2}, \tag{8}$$

provided this expression is positive. Then (5) represents a circle with center at $(-D/2A, -E/2A)$ and radius $r = \sqrt{(D^2 + E^2 - 4AF)/(4A^2)}$.

If Eq. (8) is equal to zero, the locus reduces to a single point, and if it is negative, there are no points (with real coordinates) that satisfy Eq. (7) or Eq. (5).

It is recommended that the reader apply the method of Problem 4, rather than using formulas, to handle problems of this type. The thing to remember is that an equation like (5), which is *quadratic in x and in y, with equal coefficients of x^2 and y^2, and with no xy-term, represents a circle* (or a single point, or no real locus).

Remark 2. Equation (5) can be divided by A and be replaced by an equation of the form

$$x^2 + y^2 + C_1x + C_2y + C_3 = 0, \tag{9}$$

where C_1, C_2, C_3 are three constants. The three coefficients in (9) can often be determined so as to satisfy three prescribed conditions: for example, that the circle go through three given (noncollinear) points; or be tangent to three given nonconcurrent lines; or be tangent to two lines and pass through a given point not on either line.

Problem 5. Find the circle through the three points $A(1, 0)$, $B(0, 1)$, and $C(2, 2)$.

Solution 1. Let Eq. (9) be the equation of the circle. Then substitute for x and y the coordinates of A, B, and C, since these points are to be on the circle.

Point	$x^2 + y^2 +$	$C_1x +$	$C_2y +$	$C_3 = 0$	
$A(1, 0)$	1	$+ C_1$		$+ C_3 = 0$	
$B(0, 1)$	1		$+ C_2$	$+ C_3 = 0$	(10)
$C(2, 2)$	$4 + 4$	$+ 2C_1$	$+ 2C_2$	$+ C_3 = 0$	

Equations (10) are three equations for the three unknowns. Subtracting the second equation from the first, we get

$$C_2 = C_1.$$

Substitute C_1 for C_2 in the third equation:

$$8 + 4C_1 + C_3 = 0.$$

Subtract the first equation from this:

$$7 + 3C_1 = 0.$$

Hence

$$C_2 = C_1 = -\tfrac{7}{3}$$

and

$$C_3 = -1 - C_1 = \tfrac{4}{3}.$$

Therefore the equation is

$$x^2 + y^2 - \tfrac{7}{3}x - \tfrac{7}{3}y + \tfrac{4}{3} = 0,$$

or

$$3x^2 + 3y^2 - 7x - 7y + 4 = 0.$$

Of course this example could equally well have been solved starting with Eq. (2), which contains the three unknowns h, k, and r. Or we could have found the center Q of the circle by locating the point of intersection of the perpendicular bisectors of the segments AB and BC. The radius is the distance from the center Q to any one of the three points A, B, or C.

EXERCISES

In Exercises 1 through 6, find the equation of the circle having the given center $C(h, k)$ and radius r.

1. $C(0, 2)$, $r = 2$
2. $C(-2, 0)$, $r = 3$
3. $C(3, -4)$, $r = 5$.
4. $C(1, 1)$, $r = \sqrt{2}$
5. $C(-2, -1)$, $r = \sqrt{6}$
6. $C(-4, 2)$, $r = 4$

In Exercises 7 through 12, find the center and radius of the given circle.

7. $x^2 + y^2 - 2y = 3$
8. $x^2 + y^2 + 2x = 8$
9. $3x^2 + 3y^2 + 6x = 1$
10. $2x^2 + 2y^2 + x + y = 0$
11. $x^2 + y^2 + 2x - 4y + 5 = 0$
12. $x^2 + y^2 + 4x + 4y + 9 = 0$
13. The center of a circle is $C(2, 2)$. The circle goes through the point $A(4, 5)$. Find its equation.
14. The center of a circle is $C(-1, 1)$. The circle is tangent to the line $x + 2y = 4$. Find its equation.
15. A circle passes through the points $A(2, -2)$ and $B(3, 4)$. Its center is on the line $x + y = 2$. Find its equation.
16. Show geometrically that the lines that are drawn from the exterior point $P_1(x_1, y_1)$ tangent to the circle
$$(x - h)^2 + (y - k)^2 = r^2$$
have length l given by
$$l^2 = (x_1 - h)^2 + (y_1 - k)^2 - r^2.$$
17. Find an equation of the circle that passes through the three points $A(2, 3)$, $B(3, 2)$, and $C(-4, 3)$.
18. Find the locus of the point $P(x, y)$ if the sum of the squares of its distances from the two points $(-5, 2)$ and $(1, 4)$ is always 52. Identify and sketch the curve.
19. Is the point $(0.1, 3.1)$ inside, outside, or on the circle $x^2 + y^2 - 2x - 4y + 3 = 0$? Why?
20. If the distance from $P(x, y)$ to the point $(6, 0)$ is twice its distance from the point $(0, 3)$, show that the locus is a circle and find the center and radius.
21. Find the circle inscribed in the triangle whose sides are the lines
$$4x + 3y = 24,$$
$$3x - 4y = 18,$$
$$4x - 3y + 32 = 0.$$
Hint. The distance from the point (h, k) to the line $ax + by + c = 0$ is
$$\frac{|ah + bk + c|}{\sqrt{a^2 + b^2}},$$
by Eq. (10), Article 1.5.
22. Let P be a point outside a given circle C. Let PT be tangent to C at T. Let the line PN from P through the center of C intersect C at M and N. Prove that $PM \cdot PN = (PT)^2$.
23. It is known that any angle inscribed in a semicircle is a right angle. Prove the converse. If for every choice of the point $P(x, y)$ on a curve C joining O and A, the angle OPA is a right angle, then the curve is a circle or a semicircle having OA as diameter.
24. Suppose that Eqs. (2) and (9) represent the same circle.
    (a) Express C_1, C_2, and C_3 in terms of h, k, and r.
    (b) Express h, k, and r in terms of C_1, C_2, and C_3.

10.6 THE PARABOLA

Definition. *A parabola is the locus of points in a plane equidistant from a given point and a given line. The given point is called the focus of the parabola and the given line is called the directrix.*

The equation of a parabola

We choose the y-axis through the focus F and perpendicular to the directrix L, and take the origin halfway between F and L. If the distance between F and L is $2p$, we may assign F coordinates $(0, p)$ and the equation of L is $y = -p$, as in Fig. 10.23. Then $P(x, y)$ is on the parabola if and only if the distances PF and PQ are equal:

$$PF = PQ, \tag{1}$$

where $Q(x, -p)$ is the foot of the perpendicular from P to L. From the distance formula,

$$PF = \sqrt{x^2 + (y-p)^2} \quad \text{and} \quad PQ = \sqrt{(y+p)^2}.$$

We equate these two expressions, square, and simplify; then

$$x^2 = 4py. \tag{2}$$

This equation must be satisfied by any point on the locus. Conversely, if (2) is satisfied, then

$$\begin{aligned} PF &= \sqrt{x^2 + (y-p)^2} \\ &= \sqrt{4py + (y^2 - 2py + p^2)} \\ &= \sqrt{(y+p)^2} = PQ \end{aligned}$$

and $P(x, y)$ is on the locus. In other words, Eq. (2) is both a necessary and a sufficient condition on the coordinates x and y of points on the parabola.

Discussion. In Eq. (2), assume p is positive. Then y cannot be negative for real x, and the curve lies above the x-axis. It is symmetric about the y-axis, since x appears only to an even power.

The axis of symmetry of the parabola is also called the "axis of the parabola." The point on this axis midway between the focus and the directrix is on the parabola, since it is equidistant from the focus and the directrix. It is called the *vertex* of the parabola. The origin is the vertex of the parabola in Fig. 10.23. The tangent to a parabola at its vertex is parallel to the directrix. From Eq. (2) we find the slope

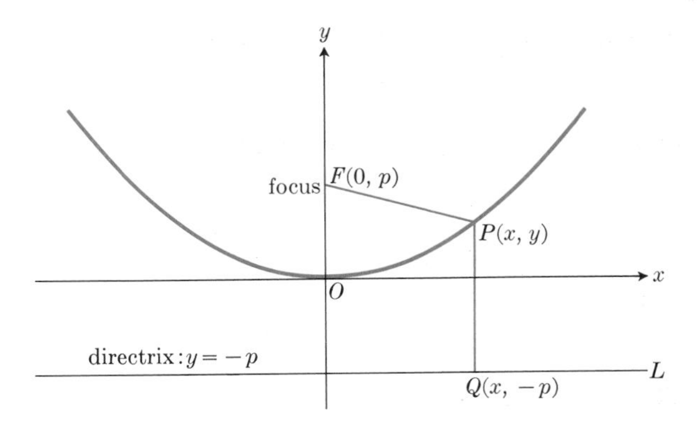

10.23 Parabola: $x^2 = 4py$.

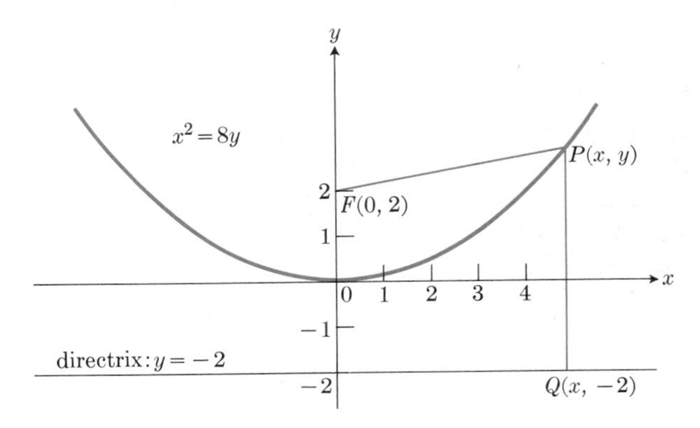

10.24

of the tangent at any point is $dy/dx = x/2p$, and this is zero at the origin. The second derivative is $d^2y/dx^2 = 1/2p$, which is positive, so the curve is concave upward. A geometric method for constructing points on a parabola (in contrast to using its equation to locate points) is described in Exercise 24 below.

Problem 1. Find the focus and directrix of the parabola

$$x^2 = 8y. \tag{3}$$

Solution. Equation (3) matches (2) if we take $4p = 8$, $p = 2$. The focus is on the axis of symmetry (the y-axis), p units from the vertex, that is, at $F(0, 2)$. The directrix is the line $y = -2$. Every point on the graph of (3) is equidistant from $F(0, 2)$ and from the line $y = -2$. See Fig. 10.24.

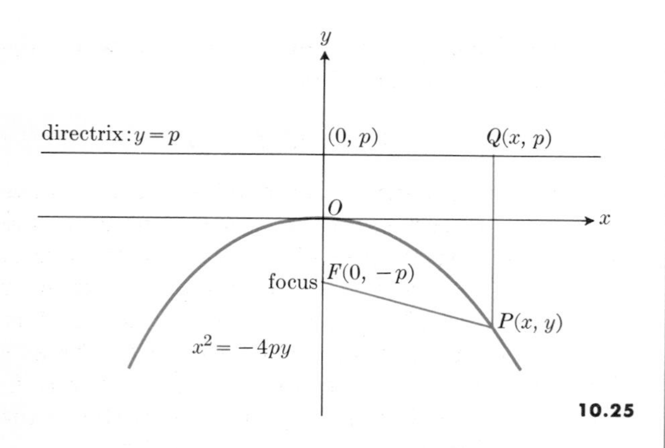

10.25

Remark 1. Suppose the parabola opens downward, as in Fig. 10.25, with its focus $F(0, -p)$ and directrix the line $y = p$. The effect is to change the sign of p in Eq. (2), which now becomes

$$x^2 = -4py. \tag{4}$$

We may also interchange the roles of x and y in Eqs. (2) and (4). The resulting equations

$$y^2 = 4px \tag{5a}$$

and

$$y^2 = -4px \tag{5b}$$

also represent parabolas, but now they are symmetric about the x-axis because y appears only to an even power. The *vertex* is still at the origin. The directrix is perpendicular to the axis of symmetry, and p units from the vertex. The focus is on the axis of symmetry, also p units from the vertex, and "inside" the curve. If we assume p is positive in Eqs. (5), then (5a) opens toward the right, because x must be greater than or equal to zero, while (5b) opens toward the left. Figure 10.26 shows graphs of parabolas of these two types.

Translation of axes

If the vertex of the parabola is at the point $V(h, k)$, Eqs. (2), (4), and (5) no longer apply in those forms. However, it is easy to determine what the appropriate equation is, by considering a new coordinate system, with its origin O' at V, and axes parallel to the original axes (see Fig. 10.27). Every point P in the plane then has two sets of coordinates, say x and y in the original system, and x', y' in the new. To go from O to P, we have a horizontal displacement x and a vertical displacement y. The abscissa may be resolved into two horizonal displacements: h from O to O', and x' from O' to P. Similarly, the ordinate is the resultant of two vertical displacements: k from O to O', and y' from O' to P. Thus

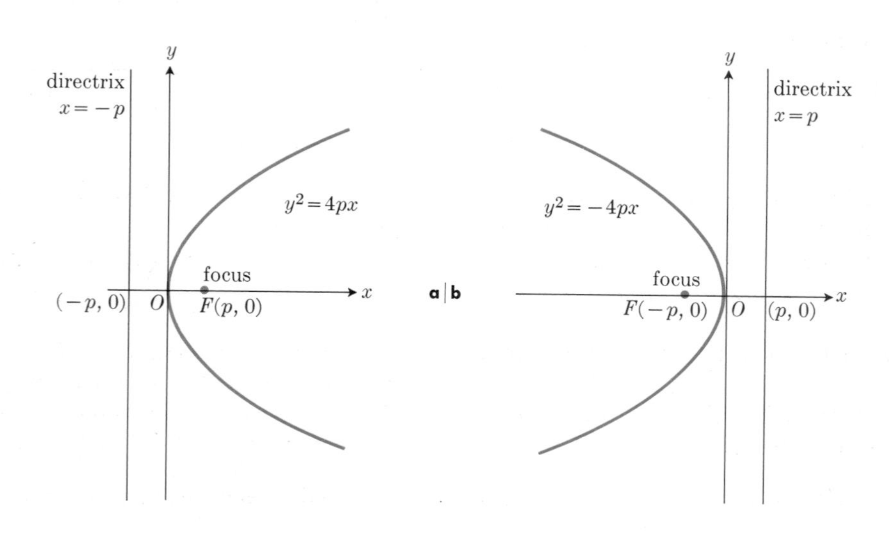

10.26

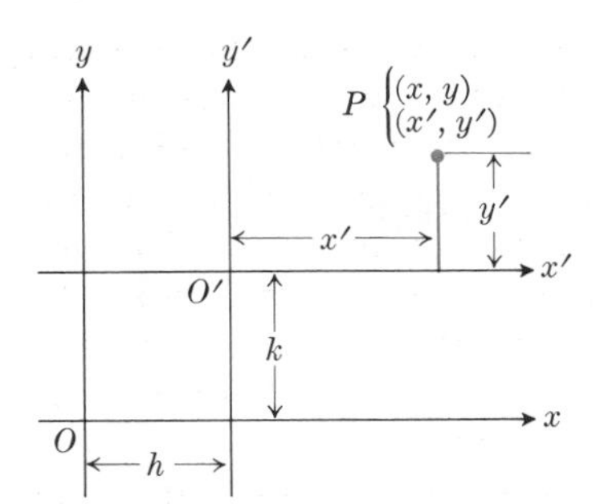

10.27 Translation of axes.

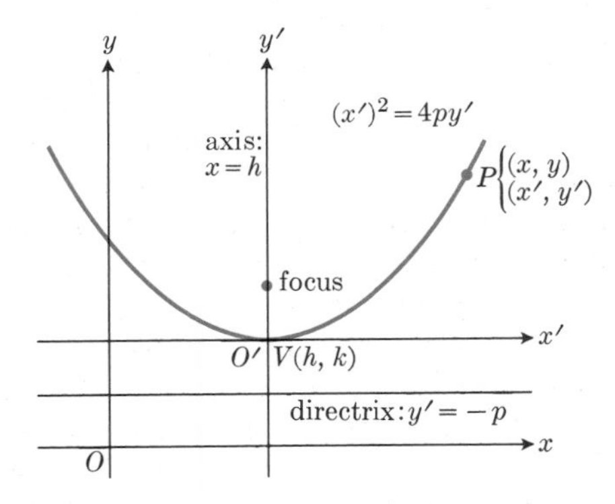

10.28 Parabola with vertex at $V(h, k)$.

the two sets of coordinates are related as follows:

$$(x = x' + h, \quad y = y' + k), \tag{6a}$$

or

$$(x' = x - h, \quad y' = y - k). \tag{6b}$$

Equations (6) are called the equations for *translation of axes,* because the new coordinate axes may be obtained by moving the old axes to the position of the new ones in a motion known as a pure translation without rotation.

Suppose we now consider a parabola with vertex $V(h, k)$ and opening upward, as in Fig. 10.28. In terms of $x'y'$-coordinates, Eq. (2) provides us with the equation of the parabola in the form

$$(x')^2 = 4py'. \tag{7}$$

By using Eqs. (6), we may express this in xy-coordinates by the equation

$$(x - h)^2 = 4p(y - k). \tag{8a}$$

The axis of symmetry of the parabola in (8a) is the line $x = h$. Observe that this corresponds to setting the quadratic term $(x - h)^2$ in (8a) equal to zero. When p is positive, $y - k$ must be greater than or equal to zero in (8a), for real $x - h$, and therefore the graph opens upward. The focus is on the axis of symmetry, p units above the vertex at $x = h$, $y = k + p$. The directrix is p units below the vertex and perpendicular to the axis of symmetry.

Other forms of equations of parabolas are

$$(x - h)^2 = -4p(y - k), \tag{8b}$$

$$(y - k)^2 = 4p(x - h), \tag{8c}$$

$$(y - k)^2 = -4p(x - h). \tag{8d}$$

Equation (8b) has a graph symmetric about $x = h$, opening downward; (8c), symmetric about $y = k$ and opening to the right $(x \geq h)$; (8d), symmetric about $y = k$ and opening to the left $(x \leq h)$.

Problem 2. Discuss the parabola

$$y = x^2 + 4x. \tag{9}$$

Solution. We complete the square in the x terms by adding 4 to both sides of Eq. (9):

$$\begin{aligned} y + 4 &= x^2 + 4x + 4 \\ &= (x + 2)^2. \end{aligned}$$

This is of the form

$$(x - h)^2 = 4p(y - k)$$

with

$$h = -2, \quad k = -4,$$

$$4p = 1, \quad p = \tfrac{1}{4}.$$

The vertex of the parabola is $V(-2, -4)$; its axis of symmetry $(x + 2)^2 = 0$; and it opens upward because $y \geq -4$ for real x. The graph is shown in Fig. 10.29. The focus is on the axis of symmetry, $\frac{1}{4}$-unit above the vertex, at $F(-2, -3\frac{3}{4})$. The directrix is parallel to the x-axis, $\frac{1}{4}$-unit below the vertex. Its equation is $y = -4\frac{1}{4}$.

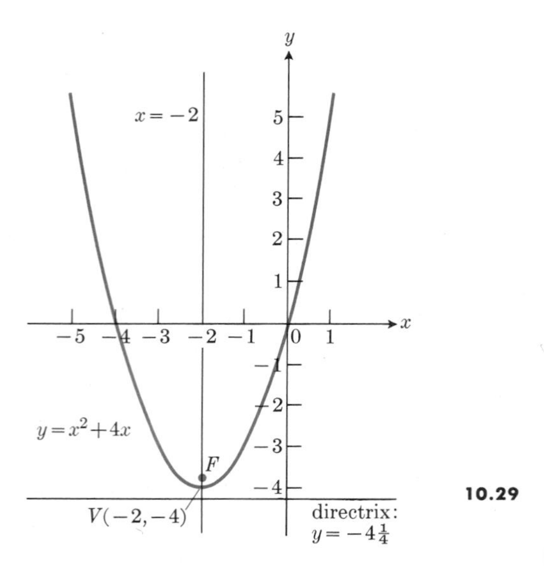

10.29

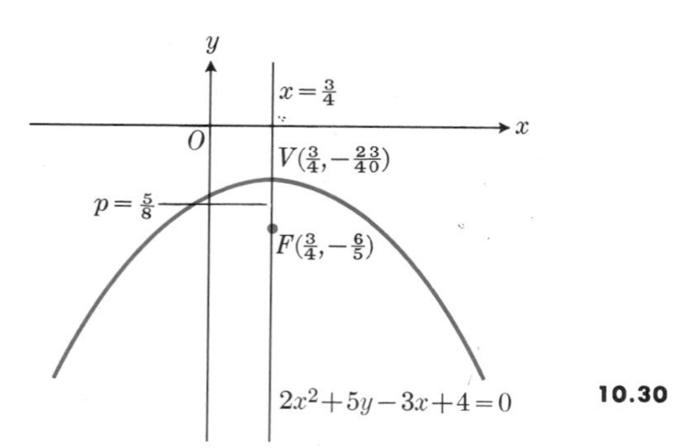

10.30

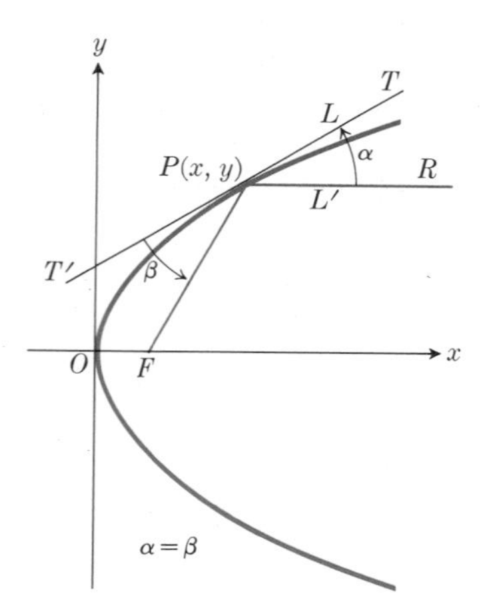

10.31 In a parabolic reflector, the angles α and β are equal.

Remark 2. The clue to an equation of a parabola is that it is quadratic in one of the coordinates and linear in the other. Whenever we have such an equation we may reduce it to one of the standard forms (8a, b, c, d) by completing the square in the coordinate that appears quadratically. We then put the linear terms in the form

$$\pm 4p(x - h) \quad \text{or} \quad \pm 4p(y - k).$$

Then the information about vertex, distance from vertex to focus, axis of symmetry, and direction the curve opens can all be read from the equation in this standard form.

Problem 3. Discuss the equation

$$2x^2 + 5y - 3x + 4 = 0.$$

Solution. This equation is quadratic in x, linear in y. We divide by 2, the coefficient of x^2, and collect all the x terms on one side of the equation:

$$x^2 - \tfrac{3}{2}x = -\tfrac{5}{2}y - 2.$$

Now we complete the square by adding $(-\frac{3}{4})^2 = \frac{9}{16}$ to both sides:

$$(x - \tfrac{3}{4})^2 = -\tfrac{5}{2}y - 2 + \tfrac{9}{16} = -\tfrac{5}{2}y - \tfrac{23}{16}.$$

To get the y terms in the form $-4p(y - k)$, we factor out $-\frac{5}{2}$, and write

$$(x - \tfrac{3}{4})^2 = -\tfrac{5}{2}(y + \tfrac{23}{40}).$$

This has the form

$$(x - h)^2 = -4p(y - k)$$

with

$$h = \tfrac{3}{4}, \quad k = -\tfrac{23}{40}, \quad 4p = \tfrac{5}{2}.$$

Hence the vertex is $V(\frac{3}{4}, -\frac{23}{40})$. The axis of symmetry is $(x - \frac{3}{4})^2 = 0$, or $x = \frac{3}{4}$; and the distance from the vertex to the focus is

$$p = \tfrac{5}{8}.$$

Since $y - k$ must here be ≤ 0 for real x, the curve opens downward, and the focus is p units below the vertex at $F(\frac{3}{4}, -\frac{6}{5})$. The graph is shown in Fig. 10.30.

Remark 3. In Fig. 10.31 the line L through $P(x, y)$ is tangent to the parabola, F is the focus, and the line L' through $P(x, y)$ is parallel to the axis of the parabola. The angles $\alpha = \angle RPT$ and $\beta = \angle T'PF$ are equal (see Exercise 31). This accounts for the property of a parabolic reflector that rays originating from the focus are reflected parallel to the axis. Or, rays coming into the reflector parallel to the axis are reflected to the focus. These properties are used in parabolic mirrors of telescopes and in parabolic radar antennas.

EXERCISES

In Exercises 1 through 6, the vertex V and focus F of a parabola are given. Find the equation of the parabola and of its directrix. Sketch the graph showing the focus, vertex, and directrix.

1. $V(0, 0)$, $F(0, 2)$
2. $V(0, 0)$, $F(-2, 0)$
3. $V(-2, 3)$, $F(-2, 4)$
4. $V(0, 3)$, $F(-1, 3)$
5. $V(-3, 1)$, $F(0, 1)$
6. $V(1, -3)$, $F(1, 0)$

In Exercises 7 through 12, the vertex V and directrix L of a parabola are given. Find the equation of the parabola and its focus. Sketch the graph showing the focus, vertex, and directrix.

7. $V(2, 0)$, L the y-axis
8. $V(1, -2)$, L the x-axis
9. $V(-3, 1)$, L the line $x = 1$
10. $V(-2, -2)$, L the line $y = -3$
11. $V(0, 1)$, L the line $x = -1$
12. $V(0, 1)$, L the line $y = 2$

In Exercises 13 through 22, find the vertex, axis of symmetry, focus, and directrix of the given parabola. Sketch the curve, showing these features.

13. $x^2 + 8y - 2x = 7$
14. $x^2 - 2y + 8x + 10 = 0$
15. $y^2 + 4x = 8$
16. $x^2 - 8y = 4$
17. $x^2 + 2x - 4y - 3 = 0$
18. $y^2 + x + y = 0$
19. $4y^2 - 8y + 3x - 2 = 0$
20. $y^2 + 6y + 2x + 5 = 0$
21. $3x^2 - 8y - 12x = 4$
22. $3x - 2y^2 - 4y + 7 = 0$

23. What is the locus of points whose coordinates satisfy the inequality $x^2 < 8y$? Sketch.
24. Explain why the following method of locating points on the graph of a parabola is valid.

(a) Construct a family of lines parallel to the directrix.

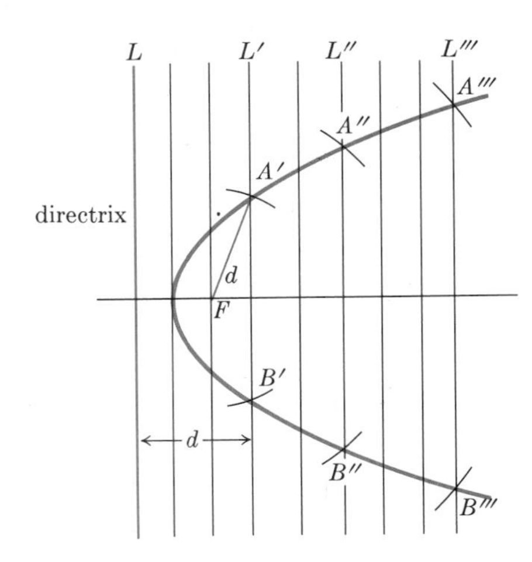

10.32 Construction of points on a parabola with focus F and directrix L.

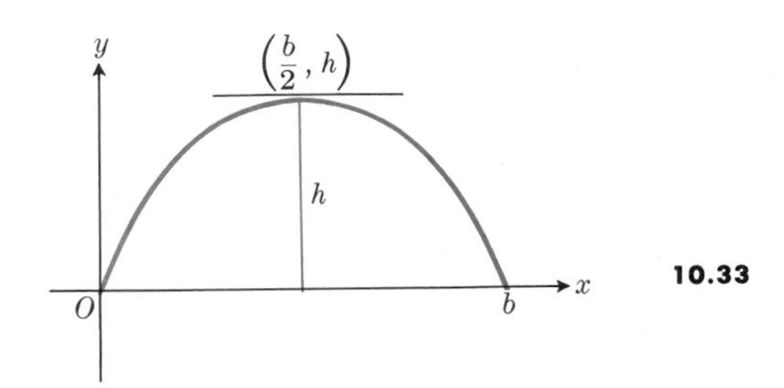

10.33

(b) Set a compass to measure the distance d from the directrix to one of these lines L'. With the focus as center, construct an arc of radius d intersecting L' in points A' and B', as in Fig. 10.32.

(c) Repeat the process with other lines L'', L''', and so on, constructing points A'', B'', A''', B''', and so on. The points so constructed lie on a parabola with focus F, directrix L.

25. Use the method of Exercise 24 to construct the graph of the parabola with focus $F(2, 2)$ and directrix the line $x + y = 0$.
26. Find the equation of the parabola of Exercise 25. *Hint.* The distance from a point (x, y) to the line $Ax + By + C = 0$ is
$$\frac{|Ax + By + C|}{\sqrt{A^2 + B^2}}.$$
27. Find the equation of the parabolic arch of base b and altitude h in Fig. 10.33.

28. Given the three points $(-1, 2)$, $(1, -1)$, and $(2, 1)$.
(a) Find a parabola passing through the given points and having its axis parallel to the x-axis.
(b) Find a parabola passing through the given points and having its axis parallel to the y-axis.

29. Suppose a and b are positive numbers. Sketch the parabolas
$$y^2 = 4a^2 - 4ax$$
and
$$y^2 = 4b^2 + 4bx$$
in the same diagram. Show that they have a common focus, the same for any a and b. Show that they intersect at $(a - b, \pm 2\sqrt{ab})$, and that each "a-parabola" is orthogonal to every "b-parabola." (Using different values of a and b, we obtain families of confocal parabolas. Each family is a set of orthogonal trajectories of the other family. See Article 10.2.)

30. What is the locus of the equation
$$(2x + y - 3)(x^2 + y^2 - 4)(x^2 - 8y) = 0?$$
Give a reason for your answer.

31. Prove that the angles α and β in Fig. 10.31 are equal.

32. Prove that the tangent to the parabola $y^2 = 4px$ at $P_1(x_1, y_1)$ intersects the axis of symmetry x_1 units to the left of the vertex. (This provides a simple method for constructing the tangent to a parabola at any point on it.)

33. Show that the area of a parabolic segment of altitude h and base b is $\frac{2}{3}bh$ (see Exercise 27).

34. Show that the volume generated by rotating the area bounded by the parabola $y = (4h/b^2)x^2$ and the line $y = h$ about the y-axis is equal to one and one-half times the volume of the corresponding inscribed cone.

35. The condition for equilibrium of the section OP of a cable that supports a weight of w pounds per foot measured along the horizontal (Fig. 10.34) is
$$\frac{dy}{dx} = \frac{wx}{H}\left(= \frac{T \sin \phi}{T \cos \phi}\right),$$
where the origin O is taken at the low point of the cable and H is the horizontal tension at O. Show that the curve in which the cable hangs is a parabola.

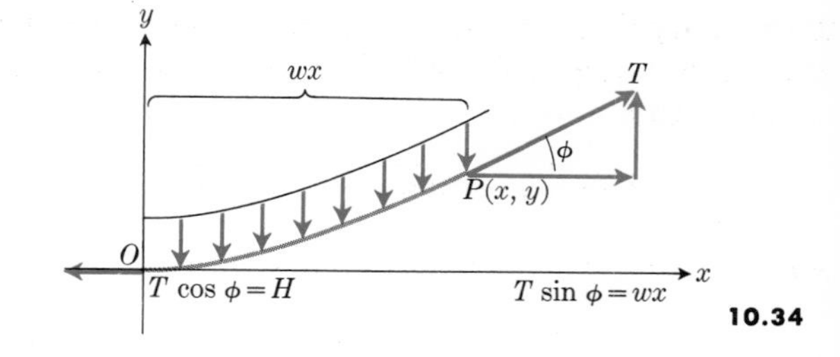

10.34

36. Assume, from optics, that when a ray of light is reflected by a mirror the angle of incidence is equal to the angle of reflection. If a mirror is formed by rotating a parabola about its axis and silvering the resulting surface, show that a ray of light emanating from the focus of the parabola is reflected parallel to the axis.

10.7 THE ELLIPSE

Definition. *An ellipse is the locus of points $P(x, y)$ the sum of whose distances from two fixed points is constant.*

The equation of an ellipse

If the two fixed points, called *foci*, are taken at $F_1(-c, 0)$ and $F_2(c, 0)$, and the sum of the distances $PF_1 + PF_2$ is denoted by $2a$ (see Fig. 10.35), then the coordinates of P must satisfy the equation
$$\sqrt{(x + c)^2 + y^2} + \sqrt{(x - c)^2 + y^2} = 2a.$$

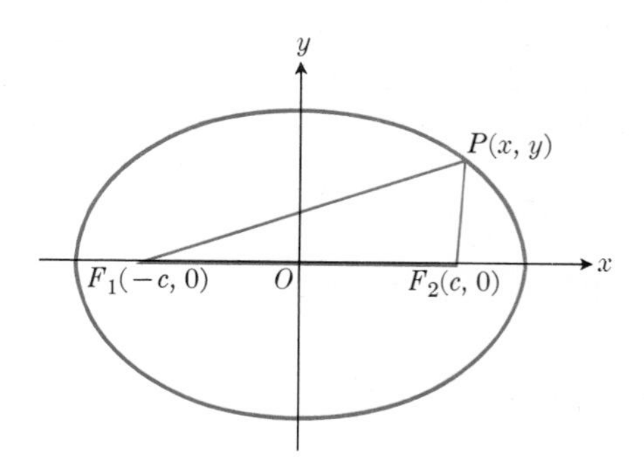

10.35 Ellipse: $\dfrac{x^2}{a^2} + \dfrac{y^2}{b^2} = 1.$

To simplify this expression, we transpose the second radical to the right side of the equation, square, and simplify, to obtain

$$a - \frac{c}{a}x = \sqrt{(x-c)^2 + y^2}.$$

Again we square and simplify, and obtain

$$\frac{x^2}{a^2} + \frac{y^2}{a^2 - c^2} = 1. \tag{1}$$

Since the sum $PF_1 + PF_2 = 2a$ of two sides of the triangle F_1F_2P is greater than the third side $F_1F_2 = 2c$, the term $(a^2 - c^2)$ in (1) is positive and has a real positive square root, which we denote by b:

$$b = \sqrt{a^2 - c^2}. \tag{2}$$

Then (1) takes the more compact form

$$\frac{x^2}{a^2} + \frac{y^2}{b^2} = 1, \tag{3}$$

from which it is readily seen that the curve is symmetric about both axes and lies inside the rectangle bounded by the lines $x = a$, $x = -a$, $y = b$, $y = -b$. The intercepts of the curve are at $(\pm a, 0)$ and $(0, \pm b)$. The curve intersects each axis at an angle of 90°, since

$$\frac{dy}{dx} = \frac{-b^2x}{a^2y}$$

is zero at $x = 0$, $y = \pm b$, and $dx/dy = 0$ at $y = 0$, $x = \pm a$.

We have shown that the coordinates of P must satisfy (1) if P satisfies the geometric condition $PF_1 + PF_2 = 2a$. Conversely, if x and y satisfy the algebraic equation (1) with $0 < c < a$, then

$$y^2 = (a^2 - c^2)\frac{a^2 - x^2}{a^2},$$

and substituting this in the radicals below, we find that

$$PF_1 = \sqrt{(x+c)^2 + y^2} = \left|a + \frac{c}{a}x\right| \tag{4a}$$

and

$$PF_2 = \sqrt{(x-c)^2 + y^2} = \left|a - \frac{c}{a}x\right|. \tag{4b}$$

Since x is restricted to the domain $-a \le x \le a$, the value of $(c/a)x$ lies between $-c$ and c, and thus both $a + (c/a)x$ and $a - (c/a)x$ are positive, both being between $a + c$ and $a - c$. Hence the absolute values in (4a) and (4b) yield

$$PF_1 = a + \frac{c}{a}x, \quad PF_2 = a - \frac{c}{a}x. \tag{5}$$

Adding these, we see that $PF_1 + PF_2$ has a value $2a$ independent of the position of P on the curve. Thus the *geometric property* and *algebraic equation* are equivalent.

In Eq. (3), $b^2 = a^2 - c^2$ is less than a^2. The *major axis* of the ellipse is the segment of length $2a$ between the x-intercepts $(\pm a, 0)$. The *minor axis* is the segment of length $2b$ between the y-intercepts $(0, \pm b)$. The numbers a and b are also referred to respectively as semimajor axis and semiminor axis. If these semiaxes are $a = 4$, $b = 3$, then Eq. (3) is

$$\frac{x^2}{16} + \frac{y^2}{9} = 1. \tag{6a}$$

On the other hand, if we interchange the roles of x and y in (6a), we get the equation

$$\frac{x^2}{9} + \frac{y^2}{16} = 1, \tag{6b}$$

which must also represent an ellipse, but one with its major axis vertical rather than horizontal. Graphs of Eqs. (6a) and (6b) are given in Figs. 10.36(a) and 10.36(b).

There is never any need for confusion in analyzing equations like (6a) and (6b). We simply find the intercepts on the axes of symmetry; then we know which way the major axis runs. The foci are always on the major axis. And if we use the letters a, b, and c to represent the lengths of semimajor axis, semiminor axis, and half-distance between foci, then Eq. (2) tells us that

$$b^2 = a^2 - c^2,$$

or

$$a^2 = b^2 + c^2. \tag{7}$$

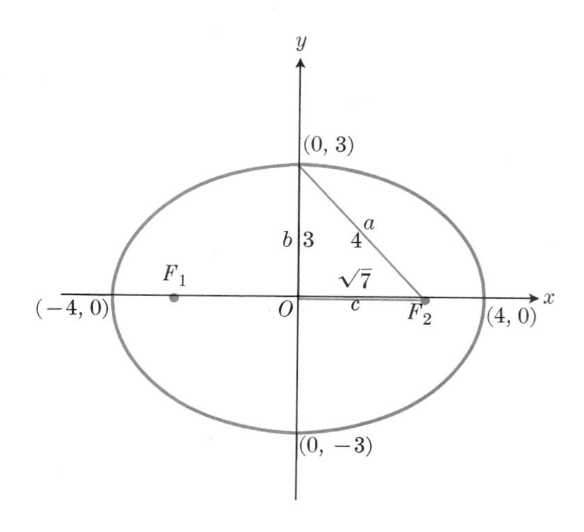

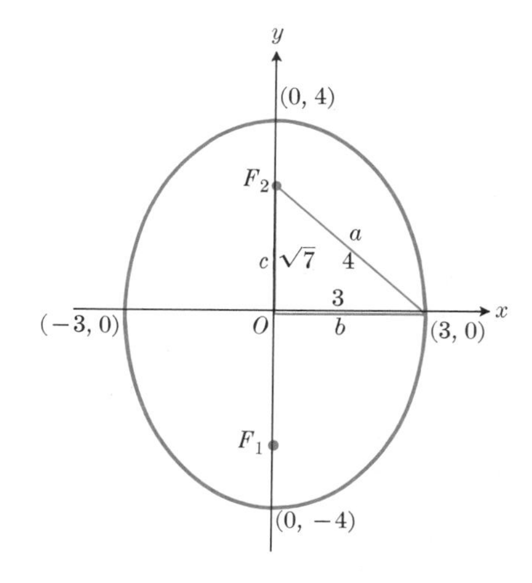

10.36 a| The major axis of $\dfrac{x^2}{16}+\dfrac{y^2}{9}=1$ is horizontal.

b| The major axis of $\dfrac{x^2}{9}+\dfrac{y^2}{16}=1$ is vertical.

Hence a is the hypotenuse of a right triangle of sides b and c, as in Fig. 10.36. When we start with an equation like (6a) or (6b), we can read off a^2 and b^2 from it at once. Then Eq. (7) determines c^2 as their difference. So in either of Eqs. (6a) and (6b), we have

$$c^2 = 16 - 9 = 7.$$

Therefore the foci are $\sqrt{7}$ units from the center of the ellipse, as shown.

Center not at the origin

The *center* of an ellipse is defined as the point of intersection of its axes of symmetry. If the center is at $C(h, k)$, and the axes of the ellipse are parallel to the x- and y-axes, then we may introduce new coordinates

$$x' = x - h, \quad y' = y - k, \tag{8}$$

using C as origin O' of $x'y'$-coordinates. The equation of the ellipse in the new coordinates is either

$$\frac{x'^2}{a^2} + \frac{y'^2}{b^2} = 1 \tag{9a}$$

or

$$\frac{x'^2}{b^2} + \frac{y'^2}{a^2} = 1, \tag{9b}$$

depending on which way the major axis runs.

Problem 1. Analyze the equation

$$9x^2 + 4y^2 + 36x - 8y + 4 = 0.$$

Solution. To complete the squares, we collect the x terms and the y terms separately:

$$9(x^2 + 4x) + 4(y^2 - 2y) = -4.$$

We then complete the square in each set of parentheses, obtaining

$$9(x^2 + 4x + 4) + 4(y^2 - 2y + 1) = -4 + 36 + 4.$$

We now divide both sides by 36 and write

$$\frac{(x+2)^2}{4} + \frac{(y-1)^2}{9} = 1.$$

Setting

$$x' = x + 2, \quad y' = y - 1,$$

we see that the new origin $x' = 0$, $y' = 0$ is the same as the point $x = -2$, $y = 1$. In terms of the new coordinates, the equation is

$$\frac{x'^2}{4} + \frac{y'^2}{9} = 1,$$

which represents an ellipse with intercepts at $(0, \pm 3)$ on the y'-axis and $(\pm 2, 0)$ on the x'-axis. To locate the foci, we use the relation

$$c = \sqrt{a^2 - b^2}.$$

Here $\quad a^2 = 9, \quad b^2 = 4,$

so $\quad c = \sqrt{5}.$

The foci are at the points $(0, \pm\sqrt{5})$ on the y'-axis or at $(-2, 1 \pm \sqrt{5})$ in terms of the original coordinates. (See Fig. 10.37.)

We recall (Fig. 10.35) that the essential *geometric* property of an ellipse is that the sum of the distances from any point on it to the two foci be a constant:

$$PF_1 + PF_2 = 2a.$$

The essential *algebraic* property of its equation when written in the form of a quadratic without a cross-product term is that the x^2- and y^2-terms have the same sign.

To discuss the properties of the ellipse in more detail, we shall assume that its equation has been reduced to the form

$$\frac{x^2}{a_2} + \frac{y^2}{b^2} = 1, \quad a > b > 0. \tag{10}$$

Although the distance c from the center of the ellipse to a focus does not appear in its equation, we may still determine c as in the examples above from the equation

$$c^2 = a^2 - b^2, \tag{11}$$

which is another way of writing Eq. (7).

A simple geometric method of finding c is shown in Fig. 10.38. (The figure also indicates $c = ae$ where $e = c/a$ is called the *eccentricity*.) This method for finding c is based on the following considerations. Since the sum of the distances

$$PF_1 + PF_2$$

is equal to $2a$ for *every point* P on the ellipse, then certainly this must hold when P is taken to be at the y-intercept B. But the y-axis is the perpendicular bisector of the line segment F_1F_2, and hence every point on it is equidistant from the foci F_1 and F_2. We therefore have

$$BF_1 + BF_2 = 2a \quad \text{and} \quad BF_1 = BF_2;$$

hence

$$BF_1 = BF_2 = a.$$

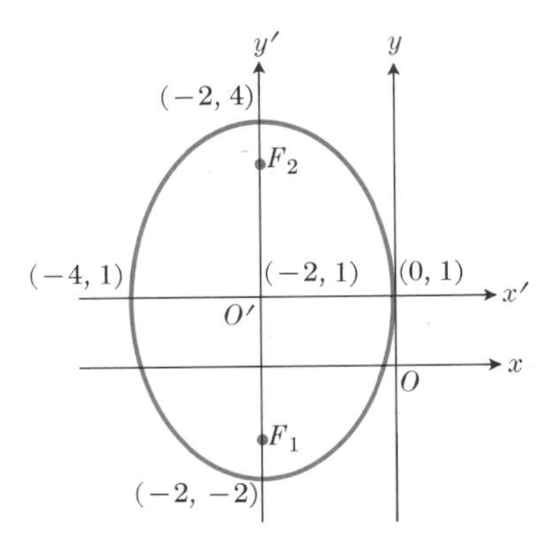

10.37 The ellipse $9x^2 + 4y^2 + 36x - 9y + 4 = 0$.

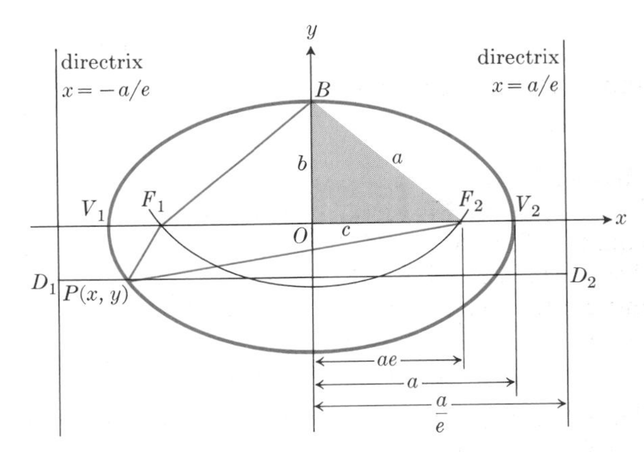

10.38

We may therefore use B as the center of a circular arc of radius a. It will cut the x-axis in the two foci F_1 and F_2. Equation (11) then follows at once from applying the Theorem of Pythagoras to the right triangle OBF_2 in the figure.

Eccentricity

If we keep a fixed and vary c over the range $0 \leq c \leq a$, the resulting ellipses will vary in shape, being circular when $c = 0$ and becoming flatter as c increases, until in the extreme case $c = a$, the "ellipse" reduces to the line segment F_1F_2 joining the two foci (Fig. 10.39). The ratio

$$e = \frac{c}{a}, \tag{12}$$

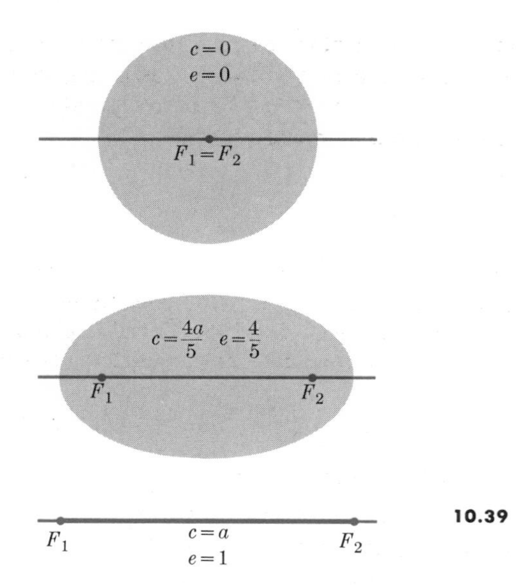

10.39

called the *eccentricity* of the ellipse, varies from 0 to 1 and indicates the degree of departure from circularity.

We recall that a parabola has one focus and one directrix. Each ellipse has two foci and two directrices. The *directrices* are lines perpendicular to the major axis of the ellipse at distances $\pm a/e$ from its center. The *parabola* has the property that

$$PF = 1 \cdot PD \tag{13}$$

for any point P on it, where F is the focus and D is the point nearest P on the directrix. For an *ellipse*, it is not difficult to show (see Exercise 21) that the equations that take the place of (13) are

$$PF_1 = e \cdot PD_1, \quad PF_2 = e \cdot PD_2. \tag{14}$$

Here e is the eccentricity, P is any point on the ellipse, F_1 and F_2 are the foci, and D_1 and D_2 are the points nearest P on the two directrices. In Eq. (14), the corresponding directrix and focus must be used; that is, if one uses the distance from P to the focus F_1, one must also use the distance from P to the directrix at the same end of the ellipse (see Fig. 10.38). We thus associate the directrix $x = -a/e$ with the focus $F_1(-c, 0)$, and the directrix $x = a/e$ with the focus $F_2(c, 0)$. In terms of the semimajor axis a and eccentricity $e < 1$, as one goes away from the center along the major axis, one finds successively

a *focus* at distance ae from the center,

a *vertex* at distance a from the center,

a *directrix* at distance a/e from the center.

An interesting feature of the "focus-and-directrix" property is that it furnishes a common bond uniting the parabola, ellipse, and hyperbola: If a point $P(x, y)$ is such that its distance PF from a fixed point (the focus) is proportional to its distance PD from a fixed line (the directrix), that is, if

$$PF = e \cdot PD, \tag{15}$$

where e is a constant of proportionality, then the locus of P is

(a) a *parabola* if $e = 1$,

(b) an *ellipse* of eccentricity e if $e < 1$, and

(c) a *hyperbola* of eccentricity e if $e > 1$.

Constructions

There are several methods of constructing an ellipse. One of these simply uses the definition directly. The two ends of a string of length $2a$ are held fixed at the foci F_1 and F_2 and a pencil traces the curve as it is held taut against the string (see Fig. 10.40a).

A second method makes use of a straightedge AB of length $a + b$. Place point A on the y-axis and B on the x-axis, and on the graph paper, make a dot at $P(x, y)$ at distance a from A (see Fig. 10.40b). In terms of the angle θ that line AB makes with the (negative) x-axis, we have

$$x = a \cos \theta, \quad y = b \sin \theta, \tag{16}$$

and hence

$$\frac{x^2}{a^2} + \frac{y^2}{b^2} = \cos^2 \theta + \sin^2 \theta = 1.$$

Therefore $P(x, y)$ is on the ellipse, Eq. (10).

A third method is to construct two concentric circles of radii a and b (see Fig. 10.40c). A line making an angle θ with the horizontal is drawn from the center cutting the two circles in points A and B respectively. The vertical line through A and the

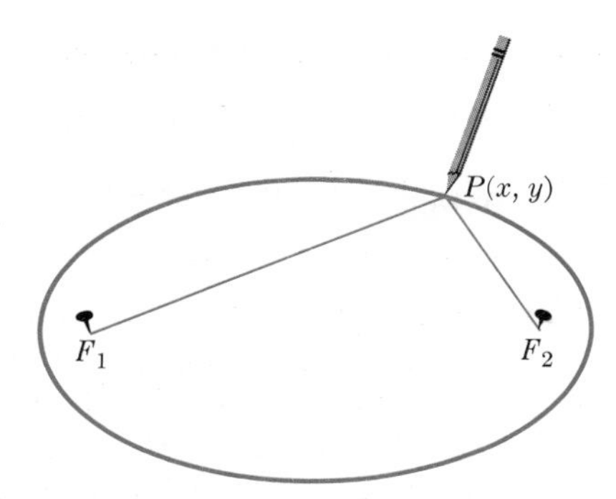

a| $PF_1 + PF_2 = 2a$.

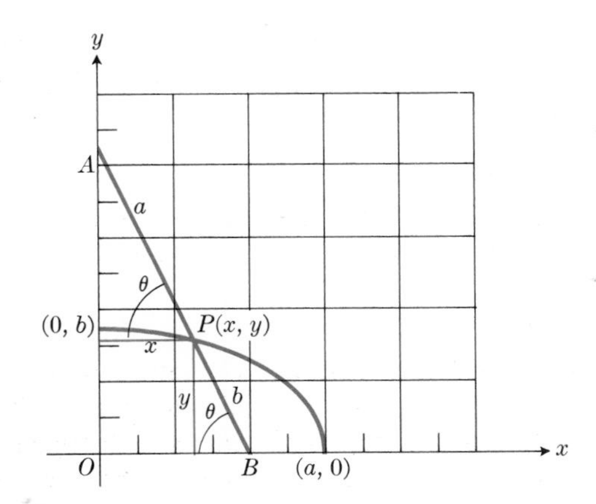

b| $x = a\cos\theta,\ y = b\sin\theta;\ \dfrac{x^2}{a^2} + \dfrac{y^2}{b^2} = 1.$

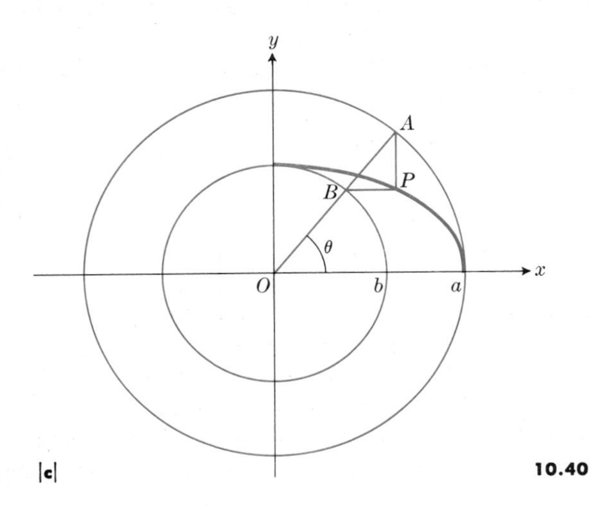

|c| 10.40

horizontal line through B intersect in a point P whose coordinates also satisfy (16). By varying the angle θ from 0 to 360°, as many points as desired may be obtained on the ellipse.

EXERCISES

In Exercises 1 through 5, find the equation of an ellipse having the given center C, focus F, and semimajor axis a. Sketch the graph and give the eccentricity of each ellipse.

1. $C(0, 0)$, $F(0, 2)$, $a = 4$
2. $C(0, 0)$, $F(-3, 0)$, $a = 5$
3. $C(0, 2)$, $F(0, 0)$, $a = 3$
4. $C(-3, 0)$, $F(-3, -2)$, $a = 4$
5. $C(2, 2)$, $F(-1, 2)$, $a = \sqrt{10}$
6. The end points of the major and minor axes of an ellipse are (1, 1), (3, 4), (1, 7), and (—1, 4). Sketch the ellipse, give its equation, and find its foci.
7. Find the center, vertices, and foci of the ellipse $25x^2 + 9y^2 - 100x + 54y - 44 = 0$. Sketch the curve.
8. Sketch the following ellipses.
   (a) $9x^2 + 4y^2 = 36$
   (b) $4x^2 + 9y^2 = 144$
   (c) $\dfrac{(x-1)^2}{16} + \dfrac{(y+2)^2}{4} = 1$
   (d) $4x^2 + y^2 = 1$
   (e) $16(x-2)^2 + 9(y+3)^2 = 144$
9. Find the equation of the ellipse that passes through the origin and has foci at (—1, 1) and (1, 1).
10. Find the eccentricity and the directrices of the ellipse $x^2/7 + y^2/16 = 1$.
11. Find the volume generated by rotating an ellipse of semiaxes a and b $(a > b)$ about its major axis.
12. Set up the integrals that give
    (a) the area of a quadrant of the circle $x^2 + y^2 = a^2$,
    (b) the area of a quadrant of the ellipse
    $$b^2x^2 + a^2y^2 = a^2b^2.$$
    Show that the integral in (b) is b/a times the integral in (a) and deduce the area of the ellipse from the known area of the circle.

13. An ellipsoid is generated by rotating an ellipse about its major axis. The inside surface of the ellipsoid is silvered to produce a mirror. Show that a ray of light emanating from one focus will be reflected to the other focus. (Sound waves also follow such paths; it is this property of ellipsoids that accounts for phenomena in certain "whispering galleries.")
14. Find the length of the chord perpendicular to the major axis of the ellipse $b^2x^2 + a^2y^2 = a^2b^2$ and passing through a focus. (This chord is called the "latus rectum" of the ellipse.)
15. Find the equation of an ellipse of eccentricity $\frac{2}{3}$ if the line $x = 9$ is one directrix and the corresponding focus is at $(4, 0)$.
16. Find the values of the constants A, B, and C if the ellipse $4x^2 + y^2 + Ax + By + C = 0$ is to be tangent to the x-axis at the origin and to pass through the point $(-1, 2)$.
17. Show that the line tangent to the ellipse

$$(x^2/a^2) + (y^2/b^2) = 1$$

at the point $P_1(x_1, y_1)$ on it is

$$\frac{xx_1}{a^2} + \frac{yy_1}{b^2} = 1.$$

18. Graph the locus of $P(x, y)$ if $9x^2 + 16y^2 < 144$.
19. Graph the locus of $P(x, y)$ if

$$(x^2 + 4y)(2x - y - 3)(x^2 + y^2 - 25) \times (x^2 + 4y^2 - 4) = 0.$$

20. Graph the locus of $P(x, y)$ if

$$(x^2 + y^2 - 1)(9x^2 + 4y^2 - 36) < 0.$$

21. Deduce Eqs. (14) from Eqs. (4a, b).

10.8 THE HYPERBOLA

Definition. *A hyperbola is the locus of $P(x, y)$ if the difference of its distances from two fixed points is constant.*

The equation of a hyperbola

Taking the fixed points at $F_1(-c, 0)$ and $F_2(c, 0)$ and the constant equal to $2a$ (see Fig. 10.41), we

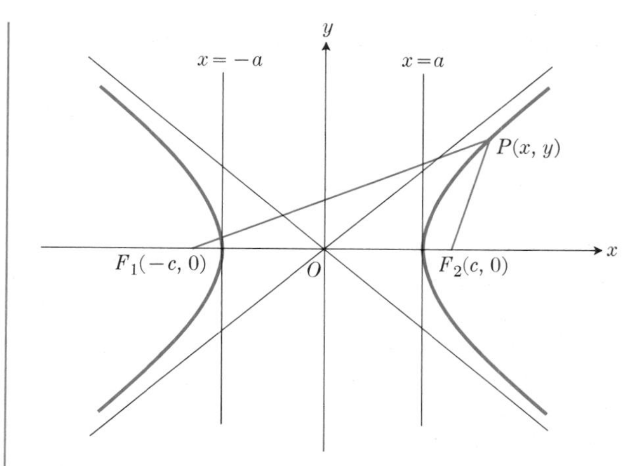

10.41 Hyperbola: $\dfrac{x^2}{a^2} - \dfrac{y^2}{b^2} = 1$.

have the condition

$$\sqrt{(x + c)^2 + y^2} - \sqrt{(x - c)^2 + y^2} = 2a$$

or

$$\sqrt{(x - c)^2 + y^2} - \sqrt{(x + c)^2 + y^2} = 2a.$$

The second equation is like the first, with $2a$ replaced by $-2a$. Hence we write the first one with $\pm 2a$, transpose one radical to the right-hand side of the equation, square, and simplify. One radical still remains. We isolate it and square again. We then obtain the equation

$$\frac{x^2}{a^2} + \frac{y^2}{a^2 - c^2} = 1. \tag{1}$$

So far, this is just like the equation for an ellipse. But now $a^2 - c^2$ is negative, because the *difference* in two sides of the triangle F_1F_2P is less than the third side:

$$2a < 2c.$$

So in this case $c^2 - a^2$ is positive and has a real positive square root which we call b:

$$b = \sqrt{c^2 - a^2}, \tag{2a}$$

or

$$a^2 - c^2 = -b^2. \tag{2b}$$

The equation of the hyperbola now becomes

$$\frac{x^2}{a^2} - \frac{y^2}{b^2} = 1, \tag{3}$$

which is analogous to the equation of an ellipse; the only differences are the minus sign in the equation of the hyperbola, and the new relation among a, b, and c given by Eq. (2b). The hyperbola, like the ellipse, is symmetric with respect to both axes and the origin, but it has no real y-intercepts and in fact no portion of the curve lies between the lines $x = a$ and $x = -a$.

If we start with a point $P(x, y)$ whose coordinates satisfy Eq. (3), the distances PF_1 and PF_2 will be given (see Exercise 14) by

$$PF_1 = \sqrt{(x+c)^2 + y^2} = \left|a + \frac{c}{a}x\right|, \tag{4a}$$

$$PF_2 = \sqrt{(x-c)^2 + y^2} = \left|a - \frac{c}{a}x\right|, \tag{4b}$$

as for the ellipse. But now c is greater than a, and P is either to the right of the line $x = a$, that is, either

$$x > a;$$

or else P is to the left of the line $x = -a$ and

$$x < -a.$$

The absolute values in Eqs. (4) work out to be

$$\left.\begin{aligned} PF_1 &= a + \frac{c}{a}x \\ PF_2 &= \frac{c}{a}x - a \end{aligned}\right\} \text{ if } x > a \tag{5a}$$

and

$$\left.\begin{aligned} PF_1 &= -\left(a + \frac{c}{a}x\right) \\ PF_2 &= a - \frac{c}{a}x \end{aligned}\right\} \text{ if } x < -a. \tag{5b}$$

Thus, when P is to the right of the line $x = a$, the condition $PF_1 - PF_2 = 2a$ is satisfied, while if P is to the left of $x = -a$, the condition

$$PF_2 - PF_1 = 2a$$

is fulfilled (Fig. 10.41). In either case, *any point P that satisfies the geometric conditions must satisfy the algebraic equation and, conversely, any point that satisfies the algebraic equation does also satisfy the geometric conditions.*

Asymptotes

The left-hand side of (3) can be factored and the equation written in the form

$$\left(\frac{x}{a} - \frac{y}{b}\right)\left(\frac{x}{a} + \frac{y}{b}\right) = 1,$$

or

$$\frac{x}{a} - \frac{y}{b} = \frac{ab}{bx + ay}. \tag{6a}$$

Analysis of (3) shows that one branch of the curve lies in the first quadrant and has infinite extent. If the point P moves along this branch so that x and y both become infinite, then the right-hand side of (6a) tends to zero; hence the left-hand side must do likewise. That is,

$$\lim_{\substack{x\to\infty \\ y\to\infty}} \left(\frac{x}{a} - \frac{y}{b}\right) = 0, \tag{6b}$$

which leads us to speculate that the straight line

$$\frac{x}{a} - \frac{y}{b} = 0 \tag{7a}$$

may be an asymptote of the curve. To see that this is definitely so, we investigate the vertical distance $(y_2 - y_1)$ between the curve and the line where we take

$$y_2 = \frac{b}{a}x$$

on the line, and

$$y_1 = \frac{b}{a}\sqrt{x^2 - a^2}$$

on the curve (Fig. 10.42). We then multiply both sides of Eq. (6b) by b, and see that

$$\lim_{x\to\infty} (y_2 - y_1) = 0.$$

Since this vertical distance tends to zero, certainly the perpendicular distance from the line to the curve

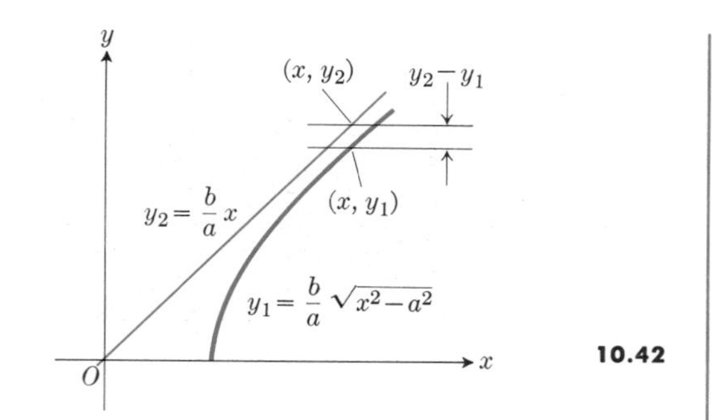

10.42

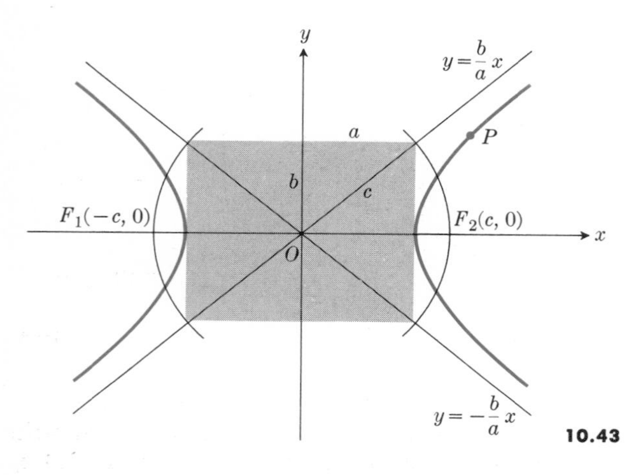

10.43

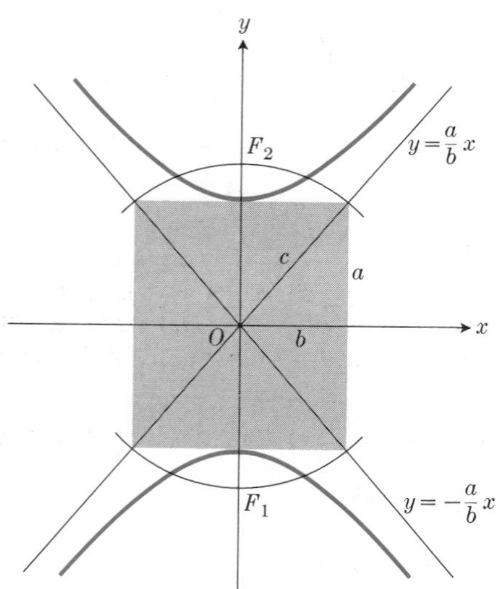

10.44 Hyperbola: $\frac{y^2}{a^2} - \frac{x^2}{b^2} = 1.$

also approaches zero, and the line in Eq. (7a) is an asymptote of the hyperbola.

By symmetry, the line

$$\frac{x}{a} + \frac{y}{b} = 0 \tag{7b}$$

is also an asymptote of the hyperbola. Both asymptotes may be obtained by simply replacing the "one" on the right side of (3) by a zero and then factoring. In sketching a hyperbola (Fig. 10.43), it is convenient to mark off distances a to the right and to the left of the origin along the x-axis and distances b above and below the origin along the y-axis, and to construct a rectangle with sides passing through these points, parallel to the coordinate axes. The diagonals of this rectangle, extended, are then the asymptotes of the hyperbola. The semidiagonal

$$c = \sqrt{a^2 + b^2}$$

can also be used as the radius of a circle that will cut the x-axis in two points, $F_1(-c, 0)$ and $F_2(c, 0)$, which are the foci of the hyperbola.

If we interchange x and y in Eq. (3), the new equation

$$\frac{y^2}{a^2} - \frac{x^2}{b^2} = 1 \tag{8}$$

represents a hyperbola with foci on the y-axis. Its graph is shown in Fig. 10.44.

Center not at the origin

The *center* of a hyperbola is the point of intersection of its axes of symmetry. If the center is $C(h, k)$, we may introduce a translation to new coordinates

$$x' = x - h, \quad y' = y - k \tag{9}$$

with origin O' at the center. In terms of the new coordinates, the equation of the hyperbola is either

$$\frac{x'^2}{a^2} - \frac{y'^2}{b^2} = 1, \tag{10a}$$

or

$$\frac{y'^2}{a^2} - \frac{x'^2}{b^2} = 1. \tag{10b}$$

Problem 1. Analyze the equation

$$x^2 - 4y^2 + 2x + 8y - 7 = 0.$$

Solution. We complete the squares in the x and y terms separately and reduce to standard form:

$$(x^2 + 2x) - 4(y^2 - 2y) = 7,$$

$$(x^2 + 2x + 1) - 4(y^2 - 2y + 1) = 7 + 1 - 4,$$

$$\frac{(x+1)^2}{4} - (y-1)^2 = 1.$$

The translation of axes

$$x' = x + 1, \quad y' = y - 1$$

reduces the equation to

$$\frac{x'^2}{4} - \frac{y'^2}{1} = 1,$$

which represents a hyperbola with center at $x' = 0$, $y' = 0$, or $x = -1$, $y = 1$, and having

$$a^2 = 4, \quad b^2 = 1, \quad c^2 = a^2 + b^2 = 5,$$

and asymptotes

$$\frac{x'}{2} - y' = 0, \qquad \frac{x'}{2} + y' = 0.$$

The foci have coordinates $(\pm\sqrt{5}, 0)$ relative to the new axes or, since

$$x = x' - 1, \quad y = y' + 1,$$

the coordinates relative to the original axes are

$$(-1 \pm \sqrt{5}, 1).$$

The curve is sketched in Fig. 10.45.

Problem 2. Analyze the equation

$$x^2 - 4y^2 - 2x + 8y - 2 = 0.$$

Solution. Proceeding as before, we obtain

$$(x-1)^2 - 4(y-1)^2 = 2 + 1 - 4 = -1.$$

The standard form requires a plus one on the right side of the equation, so we change signs and have

$$4(y-1)^2 - (x-1)^2 = 1.$$

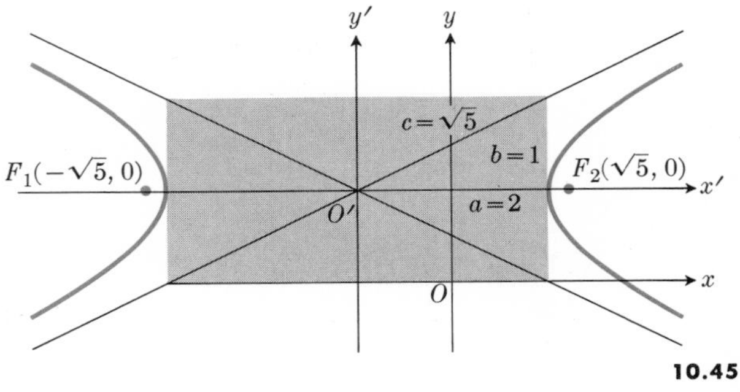

10.45

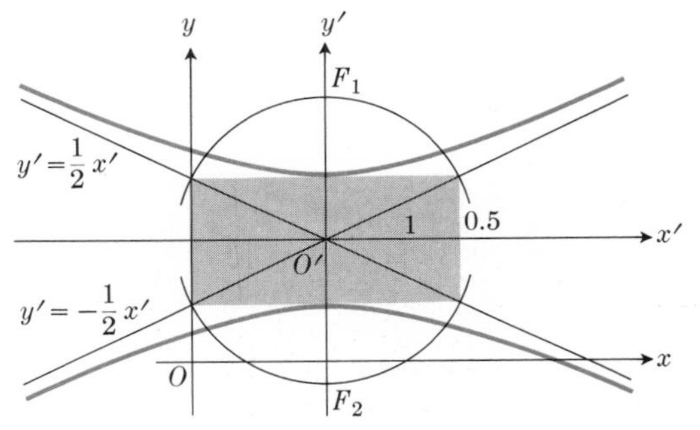

10.46

Comparison with (10) indicates that we should write the first term of this equation as $(y-1)^2$ divided by 0.25:

$$\frac{(y-1)^2}{0.25} - \frac{(x-1)^2}{1} = 1.$$

The translation $x' = x - 1$, $y' = y - 1$ replaces this by

$$\frac{y'^2}{0.25} - \frac{x'^2}{1} = 1,$$

which represents a hyperbola with center at $x' = y' = 0$, or $x = y = 1$. The curve has intercepts at $(0, \pm 0.5)$ on the y'-axis but does not cross the x'-axis. Here (10b) applies, with

$$a^2 = 0.25, \quad b^2 = 1, \quad c^2 = a^2 + b^2 = 1.25.$$

The lines (Fig. 10.46)

$$\frac{y'}{0.5} - x' = 0, \qquad \frac{y'}{0.5} + x' = 0$$

are the asymptotes, while the foci are at $(0, \pm\sqrt{1.25})$ on the y'-axis.

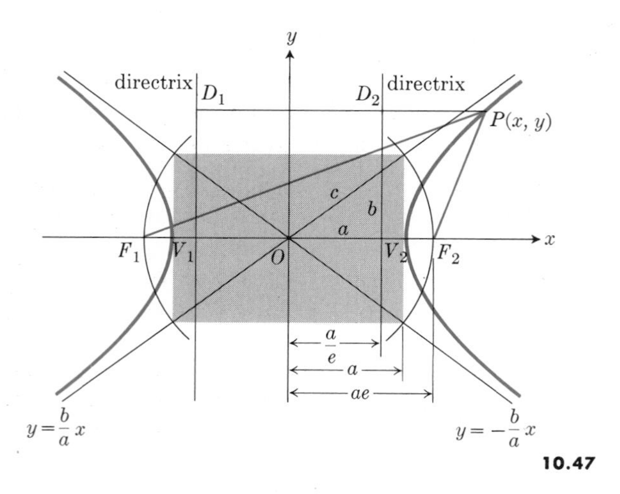

10.47

It is worth noting that there is no restriction $a > b$ for the hyperbola as there is for the ellipse, and the direction in which the hyperbola opens is controlled by the *signs* rather than by the relative *sizes* of the coefficients of the quadratic terms.

In our further discussion of the hyperbola, we shall assume that it has been referred to axes through its center and that its equation has the form

$$\frac{x^2}{a^2} - \frac{y^2}{b^2} = 1. \tag{11}$$

Then

$$c^2 = a^2 + b^2. \tag{12}$$

As for the ellipse, we define the *eccentricity* e to be

$$e = \frac{c}{a},$$

and since $c \geq a$, the eccentricity of a hyperbola is never less than unity. The lines

$$x = \frac{a}{e}, \qquad x = -\frac{a}{e}$$

are the *directrices*.

We shall now verify that a point $P(x, y)$ whose coordinates satisfy Eq. (11) also has the property that

$$PF_1 = e \cdot PD_1 \tag{13a}$$

and

$$PF_2 = e \cdot PD_2, \tag{13b}$$

where $F_1(-c, 0)$ and $F_2(c, 0)$ are the foci while

$$D_1\left(-\frac{a}{e}, y\right) \quad \text{and} \quad D_2\left(\frac{a}{e}, y\right)$$

are the points nearest P on the directrices.

We shall content ourselves with establishing the results (13a, b) for any point P on the right branch of the hyperbola; the method is the same when P is on the left branch. Reference to Eqs. (5a) then shows that

$$\begin{aligned} PF_1 &= \frac{c}{a}x + a = e\left(x + \frac{a}{e}\right) \\ PF_2 &= \frac{c}{a}x - a = e\left(x - \frac{a}{e}\right), \end{aligned} \tag{14a}$$

while we see from Fig. 10.47 that

$$\begin{aligned} PD_1 &= x + \frac{a}{e}, \\ PD_2 &= x - \frac{a}{e}. \end{aligned} \tag{14b}$$

These results combine to establish the "focus-and-directrix" properties of the hyperbola expressed in Eqs. (13a, b). Conversely, if Eqs. (14a) are satisfied, it is also true that

$$PF_1 - PF_2 = 2a.$$

That is, P satisfies the requirement that the difference of its distances from the two foci is constant.

It is possible to devise various schemes for sketching a hyperbola. In one method (see Fig. 10.48) we exploit the equation

$$PF_1 = 2a + PF_2$$

by constructing a circle of radius r with center at F_2 and another circle of radius $2a + r$ with center at F_1. The points P, P' where the two circles intersect are points on the hyperbola. By varying r, as many points on the hyperbola as desired may be obtained. Interchanging the roles of the two foci F_1 and F_2 allows us to obtain points on the other branch of the hyperbola.

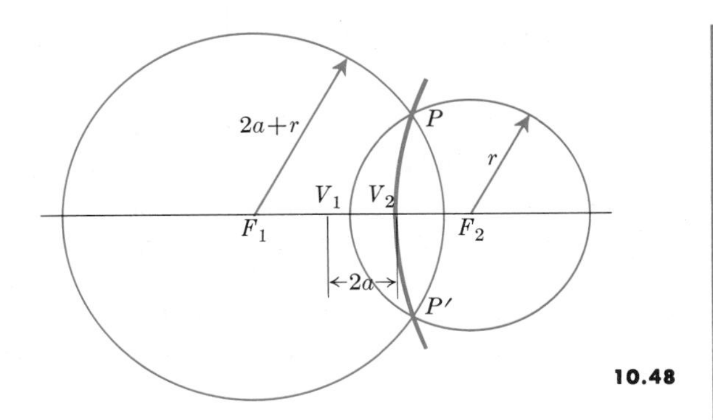

10.48

EXERCISES

1. Sketch the following hyperbolas.

(a) $\frac{x^2}{9} - \frac{y^2}{16} = 1$ (b) $\frac{x^2}{16} - \frac{y^2}{9} = 1$

(c) $\frac{y^2}{9} - \frac{x^2}{16} = 1$ (d) $\frac{x^2}{9} - \frac{y^2}{16} = -1$

In Exercises 2 through 8, find the center, vertices, foci, and asymptotes of the given hyperbola. Sketch the curve.

2. $9(x - 2)^2 - 4(y + 3)^2 = 36$
3. $4(x - 2)^2 - 9(y + 3)^2 = 36$
4. $4(y + 3)^2 - 9(x - 2)^2 = 1$
5. $5x^2 - 4y^2 + 20x + 8y = 4$
6. $4x^2 = y^2 - 4y + 8$ 7. $4y^2 = x^2 - 4x$
8. $4x^2 - 5y^2 - 16x + 10y + 31 = 0$
9. Show that the line tangent to the hyperbola

$$b^2x^2 - a^2y^2 = a^2b^2$$

at a point $P(x_1, y_1)$ on it has an equation that may be written in the form $b^2xx_1 - a^2yy_1 = a^2b^2$.

10. Find the volume generated when the area bounded by the hyperbola $b^2x^2 - a^2y^2 = a^2b^2$ and the line $x = c$, through its focus $(c, 0)$, is rotated about the y-axis.
11. Show that the equation

$$\frac{x^2}{9 - C} + \frac{y^2}{5 - C} = 1$$

represents

(a) an ellipse if C is any constant less than 5,
(b) a hyperbola if C is any constant between 5 and 9,
(c) no real locus if C is greater than 9.

Show that each ellipse in (a) and each hyperbola in (b) has foci at the two points $(\pm 2, 0)$, independent of the value of C.

12. Find the equation of the hyperbola with foci at $(0, 0)$ and $(0, 4)$ if it is required to pass through the point $(12, 9)$.
13. One focus of a hyperbola is located at the point $(1, -3)$ and the corresponding directrix is the line $y = 2$. Find the equation of the hyperbola if its eccentricity is $\frac{3}{2}$.
14. Deduce Eqs. (4a, b) from Eq. (1).

10.9 SECOND-DEGREE CURVES

The circle, parabola, ellipse, and hyperbola are curves whose equations are all special cases of the following general equation of the second degree:

$$Ax^2 + Bxy + Cy^2 + Dx + Ey + F = 0. \qquad (1)$$

For example, the circle

$$(x - h)^2 + (y - k)^2 = r^2$$

may be obtained from Eq. (1) by taking

$$A = C = 1, \quad B = 0, \quad D = -2h,$$
$$E = -2k, \quad F = h^2 + k^2 - r^2,$$

and the parabola

$$x^2 = 4py$$

is obtained by taking

$$A = 1, \quad E = -4p, \quad B = C = D = F = 0.$$

In fact, even the straight line is a special case of (1) with $A = B = C = 0$, but this reduces (1) to a *linear* equation instead of maintaining its status as a second-degree equation. The terms Ax^2, Bxy, and Cy^2 are the second degree, or quadratic terms, and we shall presently investigate the nature of the

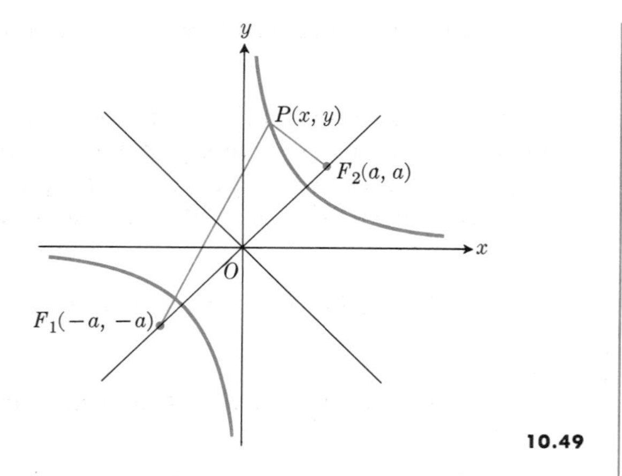

10.49

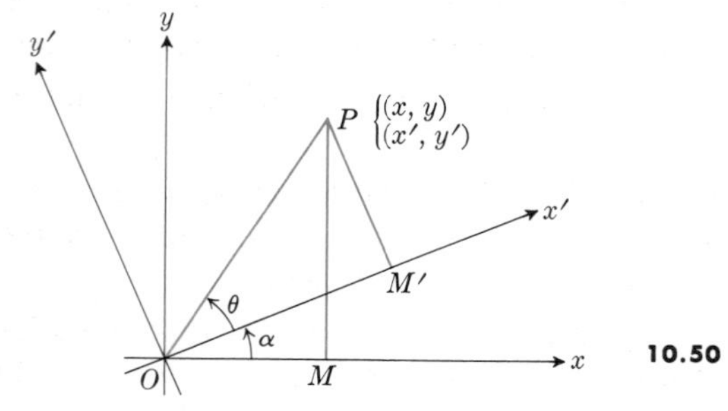

10.50

curve represented by Eq. (1) when at least one of these quadratic terms is present.

The so-called "cross-product" term Bxy has not appeared in the equations we found in Articles 10.5 through 10.8. This results from the way in which we chose coordinate axes, namely, such that at least one of them is parallel to an axis of symmetry of the curve in question. However, suppose we seek to find the equation of a hyperbola with foci at $F_1(-a, -a)$ and $F_2(a, a)$ for example, and with $|PF_1 - PF_2| = 2a$ (Fig. 10.49). Then

$$\sqrt{(x+a)^2 + (y+a)^2} - \sqrt{(x-a)^2 + (y-a)^2} = \pm 2a,$$

and when we transpose one radical, square, solve for the radical that still appears, and square again, this reduces to

$$2xy = a^2, \tag{2}$$

which is a special case of Eq. (1) in which the cross-product term is present. The asymptotes of the hyperbola in Eq. (2) are the x- and y-axes, and the transverse axis of the hyperbola (which is the axis of symmetry on which the foci lie) makes an angle of 45° with the coordinate axes. In fact, the cross-product term is present only in some similar circumstance, where the axes have been "tilted."

We shall now see how the equation of a curve is modified if it is referred to $x'y'$-axes by a rotation through an angle α in the counterclockwise direction. Referring to Fig. 10.50, we have

$$\begin{aligned} x &= OM = OP \cos(\theta + \alpha), \\ y &= MP = OP \sin(\theta + \alpha), \end{aligned} \tag{3a}$$

while

$$\begin{aligned} x' &= OM' = OP \cos\theta, \\ y' &= M'P = OP \sin\theta. \end{aligned} \tag{3b}$$

Using the relationships

$$\cos(\theta + \alpha) = \cos\theta\cos\alpha - \sin\theta\sin\alpha,$$
$$\sin(\theta + \alpha) = \sin\theta\cos\alpha + \cos\theta\sin\alpha,$$

in (3a) and taking account of (3b), we find

$$\begin{aligned} x &= x'\cos\alpha - y'\sin\alpha, \\ y &= x'\sin\alpha + y'\cos\alpha, \end{aligned} \tag{4}$$

which are known as the equations for *rotation of axes*.

To illustrate their application, suppose that we take $\alpha = 45°$ and find the equation of the hyperbola of (2) in terms of the new coordinates. Since $\cos 45° = \sin 45° = \sqrt{\frac{1}{2}}$, we substitute

$$x = \frac{x' - y'}{\sqrt{2}}, \quad y = \frac{x' + y'}{\sqrt{2}}$$

into Eq. (2) and obtain

$$(x')^2 - (y')^2 = a^2,$$

which is like Eq. (3), Article 10.8, with $b = a$.

If we apply the general rotation of axes of equations (4) to the general quadratic equation (1), we obtain a new quadratic equation of the form

$$A'x'^2 + B'x'y' + C'y'^2 + D'x' + E'y' + F' = 0, \tag{5}$$

with new coefficients related to the old as follows:

$$\begin{aligned}
A' &= A\cos^2\alpha + B\cos\alpha\sin\alpha + C\sin^2\alpha,\\
B' &= B(\cos^2\alpha - \sin^2\alpha) + 2(C - A)\sin\alpha\cos\alpha,\\
C' &= A\sin^2\alpha - B\sin\alpha\cos\alpha + C\cos^2\alpha,\\
D' &= D\cos\alpha + E\sin\alpha,\\
E' &= -D\sin\alpha + E\cos\alpha,\\
F' &= F.
\end{aligned} \tag{6}$$

If we start with an equation in which the cross-product term is present ($B \neq 0$), then we can always find an angle of rotation α such that the new cross-product term is eliminated. To find the angle α that does this, we simply put $B' = 0$ in the second equation of (6) and solve for α. It is easier to do this if we note that

$$\cos^2\alpha - \sin^2\alpha = \cos 2\alpha$$

and

$$2\sin\alpha\cos\alpha = \sin 2\alpha,$$

so that

$$B' = B\cos 2\alpha + (C - A)\sin 2\alpha.$$

Hence B' will vanish if we choose α so that

$$\cot 2\alpha = \frac{A - C}{B}. \tag{7}$$

Example. The curve whose equation is

$$x^2 + xy + y^2 = 3$$

has $A = B = C = 1$. Choosing α according to Eq. (7),

$$\cot 2\alpha = 0, \quad 2\alpha = 90°, \quad \alpha = 45°,$$

and substituting

$$x = \frac{x' - y'}{\sqrt{2}}, \quad y = \frac{x' + y'}{\sqrt{2}}$$

according to (4), we obtain

$$3(x')^2 + (y')^2 = 6.$$

This may be identified as an ellipse with its foci on the new y'-axis.

Since axes may be rotated to eliminate the cross-product term, there is no loss in generality in assuming that this has been done. Then the quadratic equation (5), with $B' = 0$, will look like Eq. (1), with $B = 0$:

$$Ax^2 + Cy^2 + Dx + Ey + F = 0. \tag{8}$$

Eq. (8) represents

(a) a straight line, if $A = C = 0$ and not both D and E vanish.

(b) a circle, if $A = C \neq 0$. In special cases the locus may reduce to a single point, or no real locus.

(c) a parabola, if (8) is quadratic in one variable, linear in the other.

(d) an ellipse, if A and C are both positive or both negative. In special cases the locus may reduce to a single point, or no real locus.

(e) a hyperbola, if A and C are of opposite signs, both different from zero. In special cases the locus may reduce to a pair of intersecting straight lines (for example, $x^2 - y^2 = 0$).

We have already seen how to reduce (8) to the standard form for the equation of a circle, parabola, ellipse, or hyperbola by completing the squares (as needed) and translating to new axes.

Summary

Any second-degree equation in x and y represents a circle, parabola, ellipse, or hyperbola (except for certain special cases in which the locus may reduce to a point, a line, a pair of lines, or fail to exist, as noted above). Conversely, any of these curves has an equation of the second degree. To find the curve, given its equation,

(1) rotate axes (if necessary) to eliminate the cross-product term, and

(2) translate axes (if desired) to reduce the equation to a recognizable standard form.

EXERCISES

1. Show that the equation $x^4 + 6x^2y^2 + y^4 = 32$ becomes $x'^4 + y'^4 = 16$ under a 45° rotation of axes. Sketch the curve and the two sets of axes.
2. Use the definition of an ellipse to find the equation of an ellipse with foci at $F_1(-1, 0)$ and $F_2(0, \sqrt{3})$ if it passes through the point $(1, 0)$. Through what angle α should the axes be rotated to eliminate the cross-product term from the equation found?
3. Show that the equation

$$x^2 + y^2 = r^2 \quad \text{becomes} \quad x'^2 + y'^2 = r^2$$

for every choice of the angle α in the equations for rotation of axes.
4. Show that $A' + C' = A + C$ for every choice of the angle α in Eqs. (6).
5. Show that $B'^2 - 4A'C' = B^2 - 4AC$ for every choice of the angle α in Eqs. (6).
6. Show that a rotation of the axes through 45° will eliminate the cross-product term from Eq. (1) whenever $A = C$.
7. Find the equation of the curve

$$x^2 + 2xy + y^2 = 1$$

after a rotation of axes that makes $A' = 0$ in Eq. (6).

By a rotation of axes, transform each equation into an equation that has no cross-product term.

8. $3x^2 + 2xy + 3y^2 = 19$
9. $x^2 - 3xy + y^2 = 5$
10. $3x^2 + 4\sqrt{3}xy - y^2 = 7$

10.10 INVARIANTS AND THE DISCRIMINANT

It is sometimes useful to apply directly to the equation

$$Ax^2 + Bxy + Cy^2 + Dx + Ey + F = 0 \qquad (1)$$

a criterion that will tell whether the curve is a parabola, an ellipse, or a hyperbola, without first performing a rotation of axes to eliminate the cross-product term. Our discussion has shown that a rotation of axes through an angle α determined by

$$\cot 2\alpha = \frac{A - C}{B} \qquad (2)$$

will transform the equation to the equivalent form

$$A'x'^2 + B'x'y' + C'y'^2 + D'x' + E'y' + F' = 0 \qquad (3)$$

with new coefficients $A', \ldots, F'$ related to the old as in Eq. (6), Article 10.9, and with $B' = 0$ for the particular choice of α satisfying Eq. (2) above.

Now if a curve has an equation (3) except that it has no cross-product term, it is

(a) *a parabola* if A' or $C' = 0$ (but not both) and if both x' and y' appear in the equation;

(b) *an ellipse* (or in exceptional cases, a single point, or empty) if A' and C' have the same sign, that is, if $A'C' > 0$;

(c) *a hyperbola* (or in exceptional cases, a pair of intersecting lines) if A' and C' have opposite signs, that is, if $A'C' < 0$.

However, it has been discovered that the coefficients A, B, C and A', B', C' satisfy the condition

$$B^2 - 4AC = B'^2 - 4A'C' \qquad (4)$$

for *any* rotation of axes. [This can easily be verified by use of Eq. (6), Article 10.9.] That is, the quantity $B^2 - 4AC$ is *invariant* under a rotation of axes. But when the particular rotation is performed that makes $B' = 0$, the right-hand side of (4) becomes simply $-4A'C'$. The criteria above, expressed in terms of A' and C', can now be expressed in terms of the *discriminant:*

$$\text{discriminant} = B^2 - 4AC. \qquad (5)$$

Thus we can say that the curve is

(a) *a parabola* (or in certain special cases, a pair of parallel lines, or one line, or an empty locus), if $B^2 - 4AC = 0$;

(b) *an ellipse* (or in exceptional cases, a single point, or empty) if $B^2 - 4AC < 0$;

(c) *a hyperbola* (or in exceptional cases, a pair of intersecting lines) if $B^2 - 4AC > 0$.

The following problem illustrates some of the special cases of (a).

Problem 1. What is the locus represented by the equation $x^2 - 4xy + 4y^2 + F = 0$ if

(a) $F = 0$, (b) $F = 1$, (c) $F = -1$?

Solution. The discriminant is

$$B^2 - 4AC = (-4)^2 - 16 = 0.$$

However, the locus is not a parabola because it is the same as $(x - 2y)^2 = -F$. This is a single line $x = 2y$ if $F = 0$; it is empty if $F = 1$; and it is two parallel lines $x - 2y = \pm 1$ if $F = -1$.

Another invariant associated with Eqs. (1) and (3) is the sum of the coefficients of the squared terms. For it is evident from Eq. (6), Article 10.9, that

$$A' + C' = A\ (\cos^2 \alpha + \sin^2 \alpha) + C\ (\sin^2 \alpha + \cos^2 \alpha),$$

or

$$A' + C' = A + C, \tag{6}$$

since

$$\sin^2 \alpha + \cos^2 \alpha = 1$$

for any angle α.

The two invariants (4) and (6) may be used as a check against numerical errors in performing a rotation of axes of a quadratic equation. They may also be used to find the new coefficients of the quadratic terms

$$A'x'^2 + B'x'y' + C'y'^2,$$

with

$$B' = 0,$$

as in the following problem.

Problem 2. Determine the equation to which

$$x^2 + xy + y^2 = 1$$

reduces when the axes are rotated to eliminate the cross-product term.

Solution. From the original equation we find

$$B^2 - 4AC = -3, \quad A + C = 2.$$

Then, taking $B' = 0$, we have, from (4) and (6),

$$-4A'C' = -3, \quad A' + C' = 2.$$

Substituting $C' = 2 - A'$ from the second of these into the first, we obtain the quadratic equation

$$4A'^2 - 8A' + 3 = 0,$$

which factors into

$$(2A' - 3)(2A' - 1) = 0$$

and gives

$$A' = \tfrac{3}{2} \quad \text{or} \quad A' = \tfrac{1}{2}.$$

The corresponding values of C' are

$$C' = \tfrac{1}{2} \quad \text{or} \quad C' = \tfrac{3}{2}.$$

The equation therefore is

$$\tfrac{3}{2}x'^2 + \tfrac{1}{2}y'^2 = 1,$$

or

$$\tfrac{1}{2}x'^2 + \tfrac{3}{2}y'^2 = 1$$

in the new coordinates. Hence the curve is an ellipse.

It should be noted that when no first power terms are present in the original equation, they will also be absent in the new equation. This is due to the fact that a rotation of axes preserves the algebraic degree of each term of the equation; or we may refer to Eq. (6), Article 10.9, which shows that D' and E' are both zero if D and E are.

EXERCISES

Use the discriminant to classify each of the following second-degree equations as representing a circle, an ellipse, a parabola, or a hyperbola.

1. $x^2 + y^2 + xy + x - y = 3$
2. $2x^2 - y^2 + 4xy - 2x + 3y = 6$
3. $x^2 + 4xy + 4y^2 - 3x = 6$
4. $x^2 + y^2 + 3x - 2y = 10$
5. $xy + y^2 - 3x = 5$
6. $3x^2 + 6xy + 3y^2 - 4x + 5y = 12$
7. $x^2 - y^2 = 1$
8. $2x^2 + 3y^2 - 4x = 7$
9. $x^2 - 3xy + 3y^2 + 6y = 7$

10. When $B^2 - 4AC$ is negative, the equation

$$Ax^2 + Bxy + Cy^2 = 1$$

represents an ellipse. If the semiaxes have lengths a and b, the area of the ellipse is πab. Show that the area of the ellipse given above is $2\pi/\sqrt{4AC - B^2}$.

11. Show, by reference to Eq. (6), Article 10.9 that

$$D'^2 + E'^2 = D^2 + E^2$$

for every angle of rotation α.

12. If $C = -A$ in Eq. (1), show that there is a rotation of axes for which $A' = C' = 0$ in the resulting Eq. (3). Find the angle α that makes $A' = C' = 0$ in this case.
Hint. Since $A' + C' = 0$, one need only make the further requirement that $A' = 0$ in Eq. (3).

10.11 SECTIONS OF A CONE

The circle, parabola, ellipse, and hyperbola are known as *conic* sections because each may be obtained by cutting a cone by a plane. If the cutting plane is perpendicular to the axis of the cone, the section is a circle.

In general, suppose the cutting plane makes an angle α with the axis of the cone and let the generating angle of the cone be β (see Fig. 10.51). Then the section is

(i) a circle, if $\alpha = 90°$,
(ii) an ellipse, if $\beta < \alpha < 90°$,
(iii) a parabola, if $\alpha = \beta$,
(iv) a hyperbola, if $0 \leq \alpha < \beta$.

The connection between these curves as we have defined them and the sections from a cone is readily made by reference to Fig. 10.52. The figure is drawn to illustrate the case of an ellipse, but the argument holds for the other cases as well.

A sphere is inscribed tangent to the cone along a circle C, and tangent to the cutting plane at a point F. Point P is any point on the conic section. We shall see that F is a focus, and that the line L, in which the cutting plane and the plane of the circle C intersect, is a directrix of the curve. To this end let Q be the point where the line through P parallel

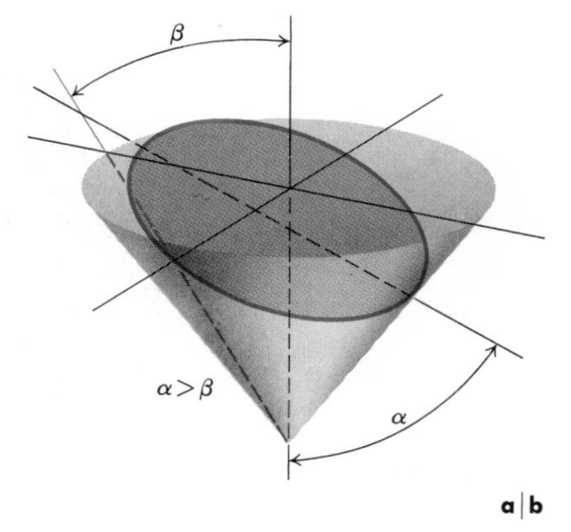

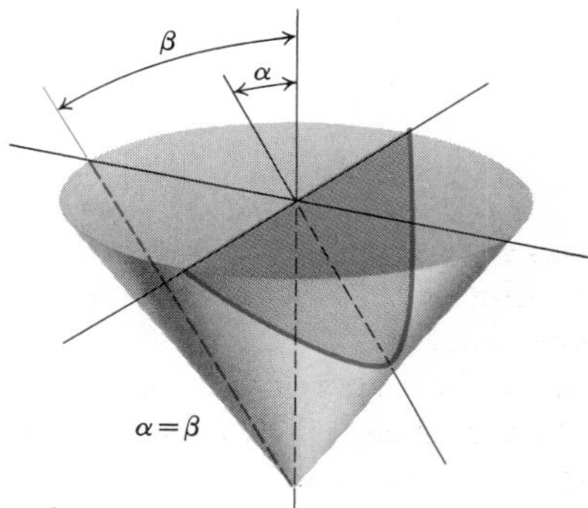

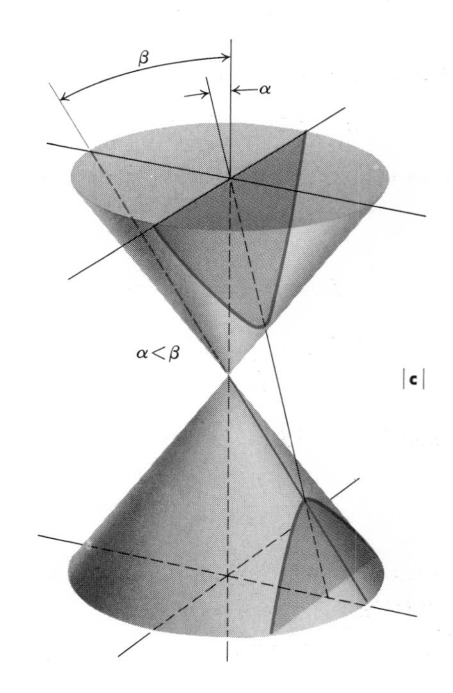

10.51

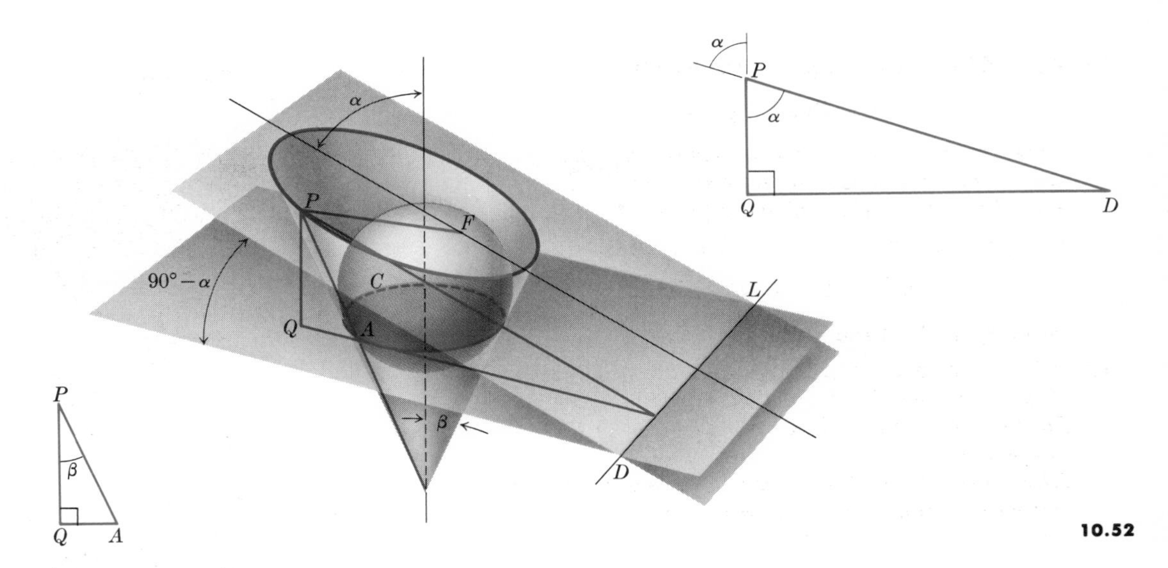

10.52

to the axis of the cone intersects the plane of C, let A be the point where the line joining P to the vertex of the cone touches C, and let PD be perpendicular to line L at D. Then PA and PF are two lines tangent to the same sphere from a common point P and hence have the same length:

$$PA = PF.$$

Also, from the right triangle PQA, we have

$$PQ = PA \cos \beta;$$

and from the right triangle PQD, we find that

$$PQ = PD \cos \alpha.$$

Hence

$$PA \cos \beta = PD \cos \alpha,$$

or

$$\frac{PA}{PD} = \frac{\cos \alpha}{\cos \beta}.$$

But since $PA = PF$, this means that

$$\frac{PF}{PD} = \frac{\cos \alpha}{\cos \beta}. \tag{1}$$

Since α and β are constant for a given cone and a given cutting plane, Eq. (1) has the form

$$PF = e \cdot PD.$$

This characterizes P as belonging to a parabola, an ellipse, or a hyperbola, with focus at F and directrix L, accordingly as $e = 1$, $e < 1$, or $e > 1$ respectively, where

$$e = \frac{\cos \alpha}{\cos \beta}$$

is thus identified with the eccentricity.

EXERCISES

1. Sketch a figure similar to Fig. 10.52 where the conic section is a parabola, and carry through the argument of Article 10.11 on the basis of such a figure.
2. Sketch a figure similar to Fig. 10.52 where the conic section is a hyperbola, and carry through the argument of Article 10.11 on the basis of such a figure.
3. Which parts of the construction described in Article 10.11 become impossible when the conic section is a circle?
4. Let one directrix be the line $x = -p$ and take the corresponding focus at the origin. Using Eq. (15) of Article 10.7, derive the equation of the general conic section of eccentricity e. If e is neither 0 nor 1, show that the center of the conic section has coordinates

$$\left(\frac{pe^2}{1 - e^2}, 0\right).$$

REVIEW QUESTIONS AND EXERCISES

1. Discuss criteria for symmetry of a curve with respect to

   (a) the x-axis, (b) the y-axis,
   (c) the origin, (d) the line $y = x$.

2. Define *asymptote*. How do you find vertical and horizontal asymptotes of a curve if its equation is given in the form $y = f(x)/g(x)$?
3. How do you find the extent of a curve (or, alternatively, strips of the plane from which the curve is excluded)?
4. Name the conic sections.
5. What kind of equation characterizes the conic sections?
6. If the equation of a conic section is given, and it contains no xy-term, how can you tell by inspection whether it is a parabola, circle, ellipse, or hyperbola? How can you tell what the curve is if there is an xy-term in the equation?
7. What are the equations of transformation of coordinates

   (a) for a translation of axes?
   (b) for a rotation of axes?

   Illustrate with diagrams.
8. What two quantities that are associated with the equation of a conic section remain invariant under a rotation of axes?
9. Sketch a parabola and label its vertex, focus, axis, and directrix. What is the definition of a parabola? What is the equation of your parabola?
10. Sketch an ellipse and label its vertices, foci, axes, and directrices. What is the definition of an ellipse? What is the equation of your ellipse?
11. Sketch a hyperbola and label its vertices, foci, axes, asymptotes, and directrices. What is the definition of a hyperbola? What is the equation of your hyperbola?
12. A ripple tank is made by bending a strip of tin around the perimeter of an ellipse for the wall of the tank and soldering a flat bottom onto this. An inch or two of water is put in the tank and the experimenter pokes a finger into it, right at one focus of the ellipse. Ripples radiate outward through the water, reflect from the strip around the edge of the tank, and (in a short time) a drop of water spurts up at the second focus. Why?

MISCELLANEOUS EXERCISES

In Exercises 1 through 9, determine the following properties of the curves whose equations are given: (A) symmetry, (B) extent, (C) intercepts, (D) asymptotes, (E) slope at intercepts. Use this information in sketching the curves.

1. (a) $y^2 = x(4 - x)$; (b) $y^2 = x(x - 4)$;
   (c) $y^2 = \dfrac{x}{4 - x}$
2. (a) $y = x + \dfrac{1}{x^2}$; (b) $y^2 = x + \dfrac{1}{x^2}$; (c) $y = x^2 + \dfrac{1}{x}$
3. (a) $y = x(x + 1)(x - 2)$; (b) $y^2 = x(x + 1)(x - 2)$
4. (a) $y = \dfrac{8}{4 + x^2}$; (b) $y = \dfrac{8}{4 - x^2}$; (c) $y = \dfrac{8x}{4 + x^2}$
5. (a) $xy = x^2 + 1$; (b) $y = \dfrac{x^2}{x - 1}$
6. (a) $y^2 = x^4 - x^2$; (b) $y^2 = \dfrac{x - 1}{x - 2}$
7. $x^2y - y = 4(x - 2)$
8. $y = \dfrac{x^2 + 1}{x^2 - 1}$
9. $x^2 + xy + y^2 = 3$
10. Let C be the curve in Exercise 9. Let $P(x, y)$ be a point on C. Let $P'(kx, ky)$ be a point on the line OP from the origin to P. If k is held constant, what is the equation of the locus described by P' as P traces out the curve C?
11. Sketch the graph whose equation is
$$(y - x + 2)(2y + x - 4) = 0.$$
12. A certain graph has an equation of the form
$$ay^2 + by = \frac{cx + d}{ex^2 + fx + g},$$

where a, b, c, d, e, f, and g are constants whose value in each case is either 0 or 1. From the following information about the graph, determine the constants and give a reason for your choice in each case.

Extent. The curve does not exist for $x < -1$. All values of y are permissible.

Symmetry. The curve is symmetric about the x-axis.

Intercepts. No y-intercept; x-intercept at $(-1, 0)$.

Asymptotes. Both axes; no others.

Sketch the graph.

13. Each of the following inequalities describes one or more regions of the xy-plane. Sketch first the locus obtained by replacing the inequality sign by an equal sign; then indicate the region that contains the points whose coordinates satisfy the given inequality.

(a) $x < 3$ (b) $x < y$
(c) $x^2 < y$ (d) $x^2 + y^2 > 4$
(e) $x^2 + xy + y^2 < 3$ (f) $x^2 + xy + y^2 > 3$
(g) $y^2 < \dfrac{x}{4 - x}$

14. Write an equation of the tangent to the curve

$$x^2 - 2xy + y^2 + 2x + y - 6 = 0.$$

at the point $(2, 2)$.

15. Sketch the curves $xy = 2$ and $x^2 - y^2 = 3$ in one diagram, and show that they intersect orthogonally.

16. Find equations of the lines that are tangent to the curve $y = x^3 - 6x + 2$ and are parallel to the line $y = 6x - 2$.

17. Prove that if a line is drawn tangent to the curve $y^2 = kx$ at a point $P(x, y)$ not at the origin, then the portion of the tangent that lies between the x-axis and P is bisected by the y-axis.

18. Through the point $P(x, y)$ on the curve $y^2 = kx$, lines are drawn parallel to the axes. The rectangular area bounded by these two lines and the axes is divided into two portions by the given curve.

(a) If these two areas are rotated about the y-axis, show that they generate two solids whose volumes are in the ratio of four to one.

(b) What is the ratio of the volumes of the solids generated when these areas are rotated about the x-axis?

19. Show that the curves

$$2x^2 + 3y^2 = a^2 \quad \text{and} \quad ky^2 = x^3$$

are orthogonal for all values of the constants a and k ($a \neq 0$, $k \neq 0$). Sketch the four curves corresponding to $a = 2$, $a = 4$, $k = \frac{1}{2}$, $k = -2$ in one diagram.

20. Show, analytically, that an angle inscribed in a semicircle is a right angle.

21. Two points P, Q are called symmetric with respect to a circle if P and Q lie on the same ray through the center and if the product of their distances from the center is equal to the square of the radius. Given that Q describes the straight line $x + 2y - 5 = 0$, find the locus of the point P that is symmetric to Q with respect to the circle $x^2 + y^2 = 4$.

22. A point $P(x, y)$ moves so that the ratio of its distances from two fixed points is a constant k. Show that the locus is a circle if $k \neq 1$, and is a straight line if $k = 1$.

23. Show that the centers of all chords of the parabola $x^2 = 4py$ with slope m lie on a straight line and find its equation.

24. The line through the focus F and the point $P(x_1, y_1)$ on the parabola $y^2 = 4px$ intersects the parabola in a second point $Q(x_2, y_2)$. Find the coordinates of Q in terms of y_1 and p. If O is the vertex and PO cuts the directrix at R, prove that QR is parallel to the axis of the parabola.

25. Find the point (or points) on the curve $x^2 = y^3$ nearest the point $P(0, 4)$. Sketch the curve and the shortest line from P to the curve.

26. Prove that every line through the center of the circle $(x - h)^2 + (y - k)^2 = r^2$ is orthogonal to the circle.

27. Find all points on the curve $x^2 + 2xy + 3y^2 = 3$, where the tangent line is perpendicular to the line $x + y = 1$.

28. A line PT is drawn tangent to the curve $xy = x + y$ at the point $P(-2, \frac{2}{3})$. Find equations of two lines that are normal to the curve and perpendicular to PT.

29. Sketch the loci described by the following equations.

(a) $(x + y)(x^2 + y^2 - 1) = 0$
(b) $(x + y)(x^2 + y^2 - 1) = 1$

Hint. In (b), consider intersections of the locus with the line $x + y = k$ for different values of the constant k.

30. Find the center and radius of the circle through the two points $A(2, 0)$ and $B(6, 0)$ and tangent to the curve $y = x^2$.

31. Find the center of the circle that passes through the point $(0, 1)$ and is tangent to the curve $y = x^2$ at $(2, 4)$.

32. Let L_1, L_2, L_3 be three straight lines, no two of which are parallel. Let $L_i = a_i x + b_i y + c_i = 0$, $i = 1, 2, 3$, be the equation of the line L_i.

(a) Describe the locus whose equation is

$$L_1 L_2 + h L_2 L_3 + k L_1 L_3 = 0,$$

assuming h and k are constants.

(b) Use the method of part (a) and determine h and k so that the equation represents a circle through the points of intersection of the lines

$$x + y - 2 = 0, \quad x - y + 2 = 0, \quad y - 2x = 0.$$

(c) Find a parabola, axis vertical, through the points of intersection of the lines in (b).

33. A comet moves in a parabolic orbit with the sun at the focus. When the comet is 4×10^7 miles from the sun, the line from the sun to it makes an angle of 60° with the axis of the orbit (drawn in the direction in which the orbit opens). Find how near the comet comes to the sun.

34. Sketch in one diagram the curves

$$y^2 = 4x + 4, \quad y^2 = 64 - 16x,$$

and find the angles at which they intersect.

35. Find an equation of the curve such that the distance from any point $P(x, y)$ on the curve to the line $x = 3$ is the same as its distance to the point $(4, 0)$. Sketch the curve.

36. Two radar stations lying along an east-west line are separated by 20 mi. Choose a coordinate system such that their positions are $(-10, 0)$ and $(10, 0)$. A low-flying plane traveling from west to east is known to have a speed of v_0 mi/sec. At $t = 0$ a signal is sent from the station at $(-10, 0)$, bounces off the plane, and is received at $(10, 0)$ $30/c$ sec later (c is the velocity of the signal). When $t = 10/v_0$, another signal is sent out from the station at $(-10, 0)$, reflects off the plane, and is once again received $30/c$ sec later by the other station. What is the position of the plane when it reflects the second signal, providing that one assumes $v_0 \ll c$ (that is, v_0 is much less than c)?

37. A line is drawn tangent to the parabola $y^2 = 4px$ at a point $P(x, y)$ on the curve. Let A be the point where this tangent line crosses the axis of the parabola, let F be the focus, and let PD be the line parallel to the axis of the parabola and intersecting the directrix at D. Prove that $AFPD$ is a rhombus.

38. Find the equation of the locus of a point $P(x, y)$ if the distance from P to the vertex is twice the distance from P to the focus of the parabola $x^2 = 8y$. Name the locus.

39. Prove that the tangent to a parabola at a point P cuts the axis of the parabola at a point whose distance from the vertex equals the distance from P to the tangent at the vertex.

40. Discuss the locus of the equation

$$x^4 - (y^2 - 9)^2 = 0$$

and sketch its graph.

41. Show that the curve $C: x^4 - (y^2 - 9)^2 = 1$ approaches part of the curve $x^4 - (y^2 - 9)^2 = 0$ as the point $P(x, y)$ moves farther and farther away from the origin. Sketch. Do any points of C lie inside the circle $x^2 + y^2 = 9$? Give a reason for your answer.

42. The ellipse $(x^2/a^2) + (y^2/b^2) = 1$ divides the plane into two regions; one inside the ellipse, the other outside. Show that points in one of these regions have coordinates that satisfy the inequality

$$(x^2/a^2) + (y^2/b^2) < 1,$$

while in the other, $(x^2/a^2) + (y^2/b^2) > 1$. (Consider the effect of replacing x, y in the given equation by $x' = kx$, $y' = ky$, with $k < 1$ in one case and $k > 1$ in the other.)

43. Find an equation of an ellipse with foci at $(1, 0)$ and $(5, 0)$, and one vertex at the origin.

44. Let $F_1 = (3, 0)$, $F_2 = (0, 5)$, $P = (-1, 3)$.

(a) Find the distances F_1P and F_2P.

(b) Does the origin O lie inside or outside the ellipse that has F_1 and F_2 as its foci and that passes through the point P? Why?

45. Find the greatest area of a rectangle inscribed in the ellipse $(x^2/a^2) + (y^2/b^2) = 1$, with sides parallel to the coordinate axes.

46. Show that the line $y = mx + c$ is tangent to the conic section $Ax^2 + y^2 = 1$ if and only if the constants A, m, and c satisfy the condition

$$A(c^2 - 1) = m^2.$$

47. Starting from the general equation for the conic, find the equation of the conic with the following properties:

(a) It is symmetric with respect to the origin.

(b) It passes through the point $(1, 0)$.

(c) The tangent to it at the point $(-2, 1)$ on it is the line $y = 1$.

48. Find an ellipse with one vertex at the point $(3, 1)$, the nearer focus at the point $(1, 1)$, and eccentricity $\frac{2}{3}$.

49. By a suitable rotation of axes, show that the equation $xy - x - y = 1$ represents a hyperbola. Sketch.

50. Find an equation of a hyperbola with eccentricity equal to $\sqrt{2}$ and with vertices at the points $(2, 0)$ and $(-2, 0)$.

51. Sketch the conic $\sqrt{2}y - 2xy = 3$. Locate its center and find its eccentricity.

52. If c is a fixed positive constant, then

$$\frac{x^2}{t^2} + \frac{y^2}{t^2 - c^2} = 1 \quad (c^2 < t^2)$$

defines a family of ellipses, any member of which is characterized by a particular value of t. Show that every member of the family

$$\frac{x^2}{t^2} - \frac{y^2}{c^2 - t^2} = 1 \quad (t^2 < c^2)$$

intersects any member of the first family at right angles.

53. Sketch the locus $|x| + |y| = 1$ and find the area it encloses.

54. Show that if the tangent to a curve at a point $P(x, y)$ passes through the origin, then $dy/dx = y/x$ at the point. Hence show that no tangent can be drawn from the origin to the hyperbola $x^2 - y^2 = 1$.

55. (a) Find the coordinates of the center and the foci, the lengths of the axes, and the eccentricity of the ellipse

$$x^2 + 4y^2 - 4x + 8y - 1 = 0.$$

(b) Do likewise with the hyperbola

$$3x^2 - y^2 + 12x - 3y = 0,$$

and in addition find equations of its asymptotes.

56. If the ends of a line segment of constant length move along perpendicular lines, show that a point P on the segment, at distances a and b from the ends, describes an ellipse.

57. Sketch the loci

(a) $(9x^2 + 4y^2 - 36)(4x^2 + 9y^2 - 36) = 0$,
(b) $(9x^2 + 4y^2 - 36)(4x^2 + 9y^2 - 36) = 1$.

Is the curve in (b) bounded or does it extend to points arbitrarily far from the origin? Give a reason for your answer.

58. Let p, q be positive numbers such that $q < p$. If r is a third number, prove that the equation

$$\left(\frac{x^2}{p - r}\right) + \left(\frac{y^2}{q - r}\right) = 1$$

represents

(a) an ellipse if $r < q$,
(b) a hyperbola if $q < r < p$,
(c) nothing if $p < r$.

Prove that all these ellipses and hyperbolas have the same foci, and find these foci.

59. Find the eccentricity of the hyperbola $xy = 1$.

60. On a level plane the sound of a rifle and that of the bullet striking the target are heard at the same instant. What is the locus of the hearer?

61. Show that any tangent to the hyperbola $xy = a^2$ determines with its asymptotes a triangle of area $2a^2$.

62. Given the hyperbola

$$9x^2 - 4y^2 - 18x - 16y + 29 = 0.$$

Find the coordinates of the center and foci, and the equations of the asymptotes. Sketch.

63. By an appropriate rotation, eliminate the xy-term from the equation

$$7x^2 - 8xy + y^2 = 9.$$

64. Show that the tangent to the conic section

$$Ax^2 + Bxy + Cy^2 + Dx + Ey + F = 0$$

at a point (x_1, y_1) on it has an equation that may be written in the form

$$Axx_1 + B\left(\frac{x_1y + xy_1}{2}\right) + Cyy_1 + D\left(\frac{x + x_1}{2}\right) + E\left(\frac{y + y_1}{2}\right) + F = 0.$$

65. (a) Find the eccentricity and center of the conic

$$x^2 + 12y^2 - 6x - 48y + 9 = 0.$$

(b) Find the vertex of the conic

$$x^2 - 6x - 12y + 9 = 0.$$

(c) Sketch the conics in one diagram.

66. Find an equation of the circle passing through the three points common to the conics of Exercise 65.

67. Two vertices A, B of a triangle are fixed and the vertex $C(x, y)$ moves in such a way that

$$\angle A = 2(\angle B).$$

Find the locus of C.

68. Find the equation into which $x^{1/2} + y^{1/2} = a^{1/2}$ is transformed by a rotation of axes through 45°, and elimination of radicals.

69. Show that $dx^2 + dy^2$ is invariant under any rotation of axes about the origin.

70. Show that $x\,dy - y\,dx$ is invariant under any rotation of axes about the origin.

71. Graph the locus $x^{2n} + y^{2n} = a^{2n}$ for the following values of n:

(a) 1, (b) 2, (c) 100.

In each instance find where the curve cuts the line $y = x$.

POLAR COORDINATES

CHAPTER 11

11.1 THE POLAR COORDINATE SYSTEM

We know that a point can be located in a plane by giving its abscissa and ordinate relative to a given coordinate system. Such x- and y-coordinates are called *Cartesian* coordinates, in honor of the French mathematician-philosopher René Descartes* (1596–1650), who is credited with discovering this method of fixing the position of a point in a plane.

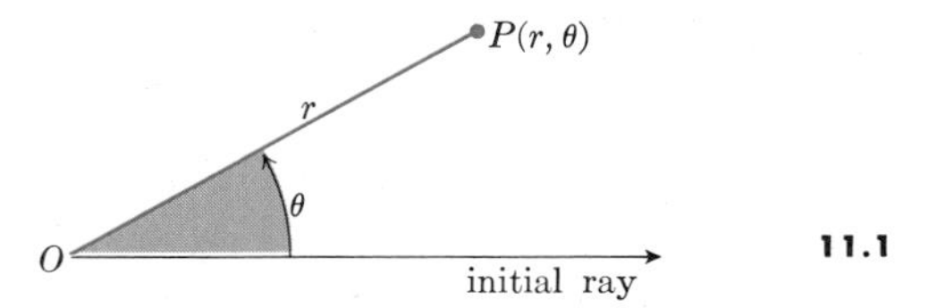

11.1

Another useful way to locate a point in a plane is by *polar coordinates* (see Fig. 11.1). First, we fix an *origin* O and an *initial ray*† from O. The point P has polar coordinates r, θ, with

$$r = \text{directed distance from } O \text{ to } P, \tag{1a}$$

and

$$\theta = \text{directed angle from initial ray to } OP. \tag{1b}$$

As in trigonometry, the angle θ is *positive* when measured counterclockwise and negative when measured clockwise (Fig. 11.1). But the angle associated with a given point is not unique (Fig. 11.2). For instance, the point 2 units from the origin, along the ray $\theta = 30°$, has polar coordinates $r = 2$, $\theta = 30°$. It also has coordinates $r = 2$, $\theta = -330°$, or $r = 2$, $\theta = 390°$.

There are occasions when we wish to allow r to be negative. That's why we say "directed distance"

---

* For an interesting biographical account together with an excerpt from Descartes' own writings, see *World of Mathematics*, Vol. 1, pp. 235–253.

† A *ray* is a half-line consisting of a vertex and points of a line on one side of the vertex. For example, the origin and positive x-axis is a ray. The points on the line $y = 2x + 3$ with $x \geq 1$ is another ray; its vertex is $(1, 5)$.

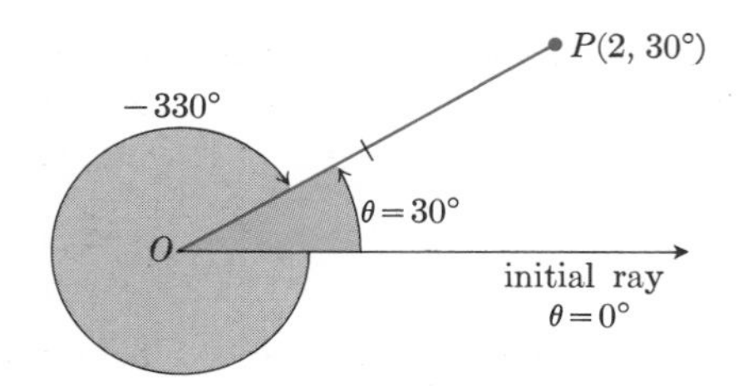

11.2 The ray $\theta = 30°$ is the same as the ray $\theta = -330°$.

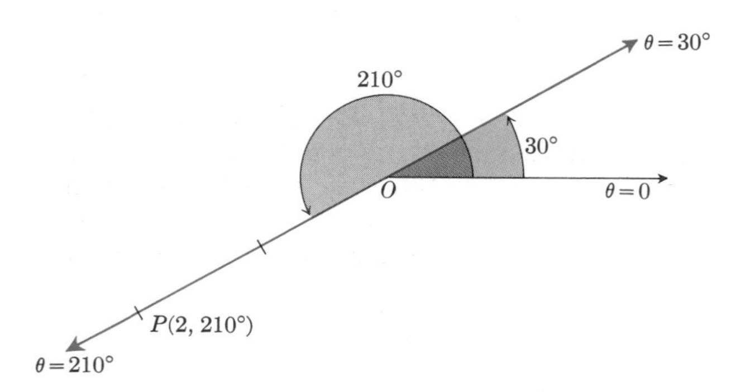

11.3 The rays $\theta = 30°$ and $\theta = 210°$ make a line.

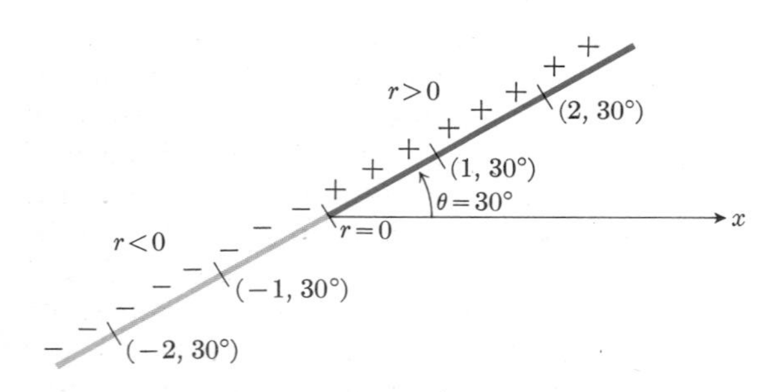

11.4 The terminal ray $\theta = \pi/6$ and its negative.

in Eq. (1a). The ray $\theta = 30°$ and the ray $\theta = 210°$ together make up a complete line through O (see Fig. 11.3). The point $P(2, 210°)$ 2 units from O on the ray $\theta = 210°$ has polar coordinates $r = 2$, $\theta = 210°$. It can be reached by a person standing at O and facing out along the initial ray, if he first turns 210° counterclockwise, and then goes forward 2 units. He would reach the same point by turning only 30° counterclockwise from the initial ray and then going *backward* 2 units. So we say that the point also has polar coordinates $r = -2$, $\theta = 30°$.

Whenever the angle between two rays is 180°, the rays actually make a straight line. We then say that either ray is the negative of the other. Points on the ray $\theta = \alpha$ have polar coordinates (r, α) with $r \geq 0$. Points on the negative ray, $\theta = \alpha + 180°$, have coordinates (r, α) with $r \leq 0$. The origin is $r = 0$. (See Fig. 11.4 for the ray $\theta = 30°$ and its negative. A word of caution: The "negative" of the ray $\theta = 30°$ is the ray $\theta = 30° + 180° = 210°$ and *not* the ray $\theta = -30°$. "Negative" refers to the directed distance r.)

There is a great advantage in being able to use both polar and Cartesian coordinates at once. To do this, we use a common origin and take the initial ray as the positive x-axis, and take the ray $\theta = 90°$ as the positive y-axis. The coordinates, shown in Fig. 11.5, are then related by the equations

$$x = r \cos \theta, \quad y = r \sin \theta. \tag{2}$$

These are the equations that define $\sin \theta$ and $\cos \theta$ when r is positive. They are also valid if r is negative, because

$$\cos (\theta + 180°) = -\cos \theta,$$
$$\sin (\theta + 180°) = -\sin \theta,$$

so positive r's on the $(\theta + 180°)$-ray correspond to negative r's associated with the θ-ray. When $r = 0$, then $x = y = 0$, and P is the origin.

If we impose the condition

$$r = a \quad (a \text{ constant}), \tag{3}$$

then the locus of P is a circle with center O and radius a, and P describes the circle once as θ varies from 0 to 360° (see Fig. 11.6). On the other hand, if we let r vary and hold θ fixed, say

$$\theta = 30°, \tag{4}$$

the locus of P is the straight line shown in Fig. 11.4.

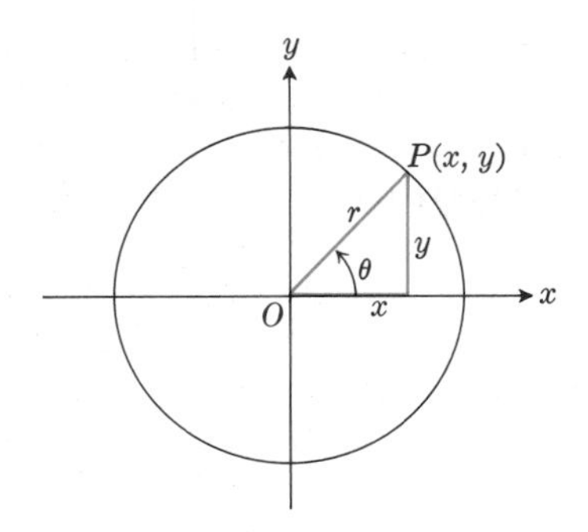

11.5 Polar and Cartesian coordinates.

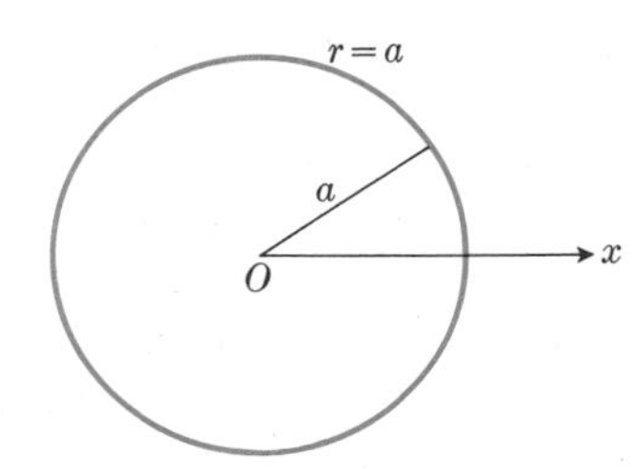

11.6 The circle $r = a$ is the locus P.

We adopt the convention that r may be any real number, $-\infty < r < \infty$. Then $r = 0$ corresponds to $x = 0$, $y = 0$ in Eqs. (2), regardless of θ. That is,

$$r = 0, \quad \theta \text{ any value}, \tag{5}$$

is the origin, $x = 0$, $y = 0$.

The same point may be represented in several different ways in polar coordinates. For example, the point $(2, 30°)$, or $(2, \pi/6)$, has the following representations: $(2, 30°)$, $(2, -330°)$, $(-2, 210°)$, $(-2, -150°)$. These and all others are summarized in the two formulas

$$\left.\begin{array}{r}(2, 30° + n\,360°),\\ (-2, 210° + n\,360°),\end{array}\right\} \quad n = 0, \pm 1, \pm 2, \ldots;$$

or, if we represent the angles in radians, in the two formulas

$$\left.\begin{array}{r}(2, \tfrac{1}{6}\pi + 2n\pi),\\ (-2, \tfrac{7}{6}\pi + 2n\pi),\end{array}\right\} \quad n = 0, \pm 1, \pm 2, \ldots$$

The fact that the same point may be represented in several different ways in polar coordinates makes added care necessary in certain situations. For example, the point $(2a, \pi)$ is on the curve

$$r^2 = 4a^2 \cos\theta \tag{6}$$

even though its coordinates as given do not satisfy the equation, because the same point is represented by $(-2a, 0)$ and these coordinates do satisfy the equation. The same point $(2a, \pi)$ is on the curve

$$r = a(1 - \cos\theta), \tag{7}$$

and hence this point should be included among the points of intersection of the two curves represented by Eqs. (6) and (7). But if we solve the equations simultaneously by first substituting $\cos\theta = r^2/4a^2$ from (6) into (7) and then solving the resulting quadratic equation

$$\left(\frac{r}{a}\right)^2 + 4\left(\frac{r}{a}\right) - 4 = 0$$

for

$$\frac{r}{a} = -2 \pm 2\sqrt{2}, \tag{8}$$

we do *not* obtain the point $(2a, \pi)$ as a point of intersection. The reason is simple enough: The point is not on the curves "simultaneously" in the sense of being reached at the "same time," since it is reached in the one case when $\theta = 0$ and in the other case when $\theta = \pi$. It is as though two ships describe paths that intersect at a point, but the ships do not collide because they reach the point of intersection at different times! The curves represented by Eqs. (6) and (7) are shown in Fig. 11.9(c). They are seen to intersect at the four points

$$(0, 0), \quad (2a, \pi), \quad (r_1, \theta_1), \quad (r_1, -\theta_1), \tag{9a}$$

where

$$\begin{gathered} r_1 = (-2 + 2\sqrt{2})a, \\ \cos\theta_1 = 1 - \frac{r_1}{a} = 3 - 2\sqrt{2}. \end{gathered} \tag{9b}$$

Only the last two of the points (9a) are found from the simultaneous solution; the first two are disclosed only by the graphs of the curves.

EXERCISES

1. Plot the following points, given in polar form, and find *all* polar coordinates of each point.

   (a) $(3, \pi/4)$ (b) $(-3, \pi/4)$
   (c) $(3, -\pi/4)$ (d) $(-3, -\pi/4)$

2. Graph the locus of points $P(r, \theta)$ whose polar coordinates satisfy the given equation, inequality, or inequalities.

   (a) $r = 2$ (b) $r < 2$
   (c) $r > 1$ (d) $1 < r < 2$
   (e) $0° \leq \theta \leq 30°$ $r \geq 0$ (f) $\theta = 120°$, $r \leq -2$
   (g) $\theta = 60°$, $-1 \leq r \leq 3$ (h) $\theta = 495°$, $r \geq -1$

3. Find Cartesian coordinates of the points in Ex. 1.

4. Graph the loci.

   (a) $r \cos \theta = 2$ (b) $r \sin \theta = -1$
   (c) $r \cos (\theta - 60°) = 3$
   Hint. Rotate axes so that $\theta' = \theta - 60°$.
   (d) $r \sin (\theta + 45°) = 4$ (e) $r \cos (30° - \theta) = 0$

5. Show that $(2, \frac{3}{4}\pi)$ is on the curve $r = 2 \sin 2\theta$.
6. Show that $(\frac{1}{2}, \frac{3}{2}\pi)$ is on the curve $r = -\sin (\theta/3)$.
7. Show that these equations represent the same curve: $r = \cos \theta + 1$, $r = \cos \theta - 1$.

Find some intersections of the following pairs of curves (a = constant).

8. $r^2 = 2a^2 \sin 2\theta$, $r = a$
9. $r = a \sin \theta$, $r = a \cos \theta$
10. $r = a(1 + \cos \theta)$, $r = a(1 - \sin \theta)$
11. $r = a(1 + \sin \theta)$, $r = 2a \cos \theta$
12. $r = a \cos 2\theta$, $r = a(1 + \cos \theta)$

11.2 GRAPHS OF POLAR EQUATIONS

The graph of an equation $F(r, \theta) = 0$ consists of all those points whose coordinates (in some form) satisfy the equation. Frequently the equation gives r explicitly in terms of θ, as

$$r = f(\theta).$$

As many points as desired may then be obtained by substituting values of θ and calculating the corresponding values of r. In particular, it is desirable to plot the points where r is a maximum or a minimum and to find the values of θ when the curve passes through the origin, if that occurs.

Certain types of *symmetry* are readily detected. For example, the curve is:

(a) symmetric about the origin, if the equation is unchanged when r is replaced by $-r$;

(b) symmetric about the x-axis, if the equation is unchanged when θ is replaced by $-\theta$;

(c) symmetric about the y-axis, if the equation is unchanged when θ is replaced by $\pi - \theta$.

These and certain other tests for symmetry are readily verified by considering the symmetrically located points in Fig. 11.7.

[*Question.* What kind of symmetry occurs if the equation is unaltered when r and θ are replaced by $-r$ and $-\theta$, respectively?]

To illustrate how symmetries are used in analyzing equations, we shall discuss and sketch three curves.

Example 1. Consider

$$r = a(1 - \cos \theta),$$

where a is a positive constant. Since

$$\cos (-\theta) = \cos \theta,$$

the equation is unaltered when θ is replaced by $-\theta$; hence the curve is symmetric about the x-axis (Fig. 11.7b). Also, since

$$-1 \leq \cos \theta \leq 1,$$

the values of r vary between 0 and $2a$. The minimum value $r = 0$ occurs at $\theta = 0$, and the maximum value $r = 2a$ occurs at $\theta = \pi$. Moreover, as θ varies from 0 to π, $\cos \theta$ decreases from 1 to -1, hence $1 - \cos \theta$ increases from 0 to 2; that is, r increases from 0 to $2a$ as the radius vector OP swings from $\theta = 0$ to $\theta = \pi$. We mark the points from the following table.

θ	0	60°	90°	120°	180°
r	0	$\frac{a}{2}$	a	$\frac{3a}{2}$	$2a$

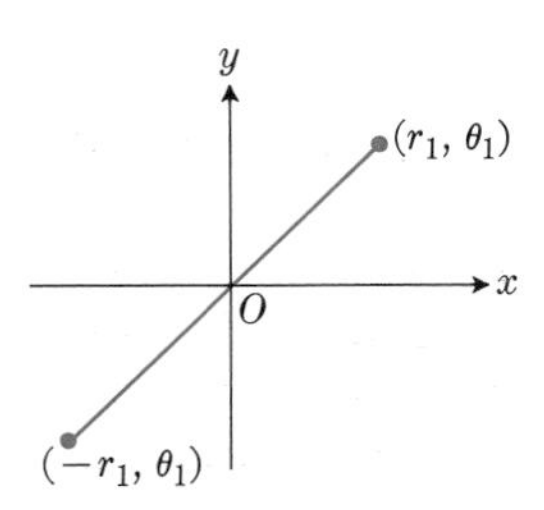

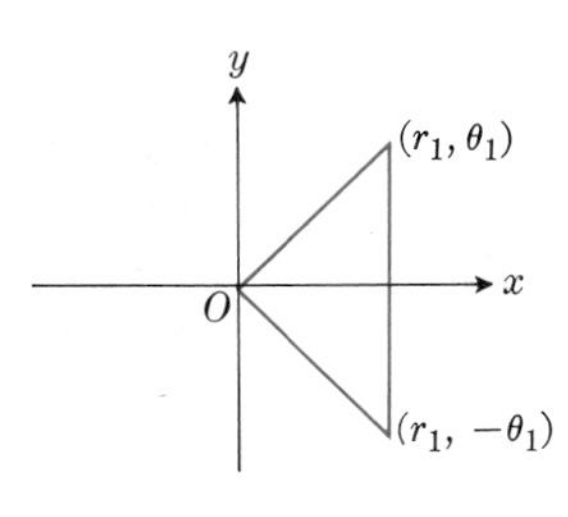

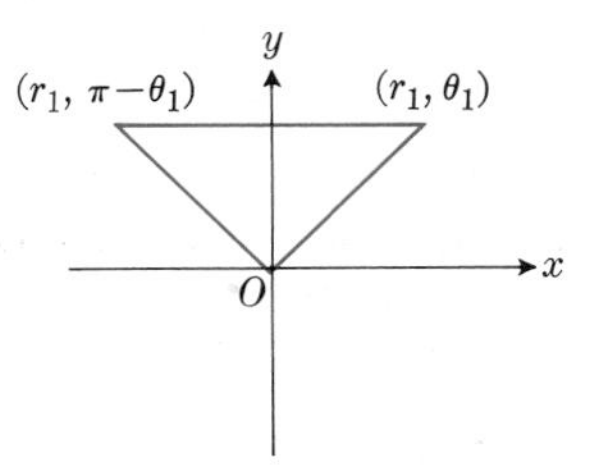

11.7 **a|** Symmetry about the origin. **b|** Symmetry about the x-axis. **c|** Symmetry about the y-axis.

We then sketch a smooth curve through these points (see Fig. 11.8) in such a way that r increases as θ increases. (Our reason for sketching it in this way is that

$$dr/d\theta = a \sin \theta$$

is positive for $0 < \theta < \pi$.) Then we exploit the symmetry of the curve and reflect this portion across the x-axis. The result is the curve shown in Fig. 11.9, which is called a *cardioid* because of its heart-shaped appearance. The behavior of this curve at the origin and the angles between its tangents and the coordinate axes at the other intercepts are more easily discussed at a later time. However, we may investigate the slope of the cardioid at the origin as follows: Let P be a point on the curve in the first quadrant, where P is destined to approach O along the curve (see Fig. 11.10a). Then, as $P \to O$, the slope of OP ($=\tan\theta$) approaches the slope of the tangent at O. But $P \to O$ as $\theta \to 0$, and hence

$$\lim_{\theta\to 0} (\text{slope of } OP) = \lim_{\theta\to 0} \tan\theta = 0.$$

That is, the slope of the tangent to the curve at the origin is zero.

In fact, for any curve that passes through the origin, say when $\theta = \theta_0$, the discussion above would be modified only to the extent of saying that $P \to O$ along the curve as $\theta \to \theta_0$, and hence that

$$\left(\frac{dy}{dx}\right)_{\theta=\theta_0} = \lim_{\theta\to\theta_0} (\tan\theta) = \tan\theta_0.$$

But $(dy/dx)_{\theta=\theta_0}$ is also the tangent of the angle between the x-axis and the curve at this point. Hence the line $\theta = \theta_0$ is tangent to the curve at the origin. In other words, whenever a curve passes through the origin for a value θ_0 of θ, it does so *tangent* to the line $\theta = \theta_0$. See Fig. 11.10(b).

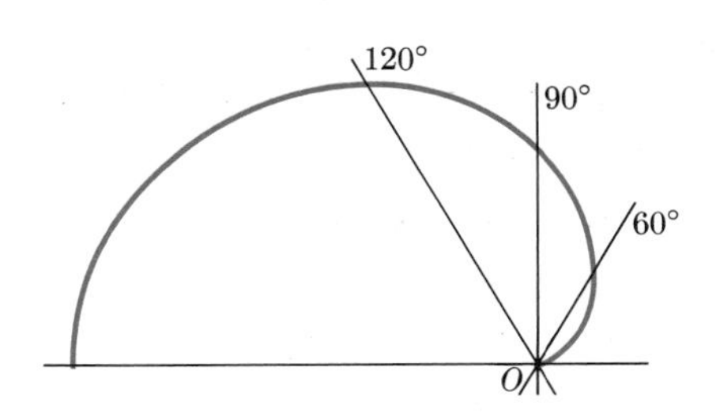

11.8

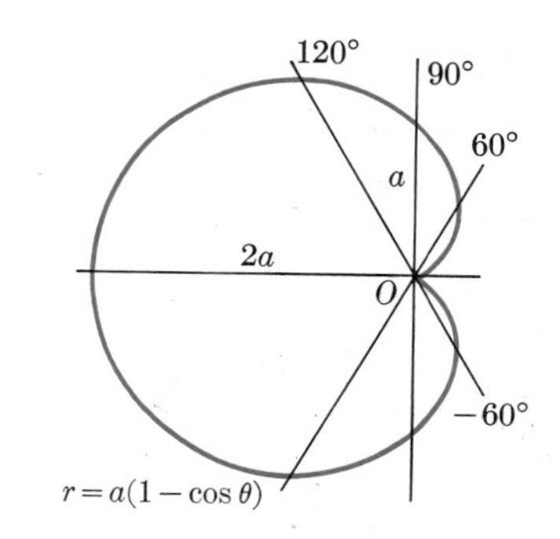

11.9

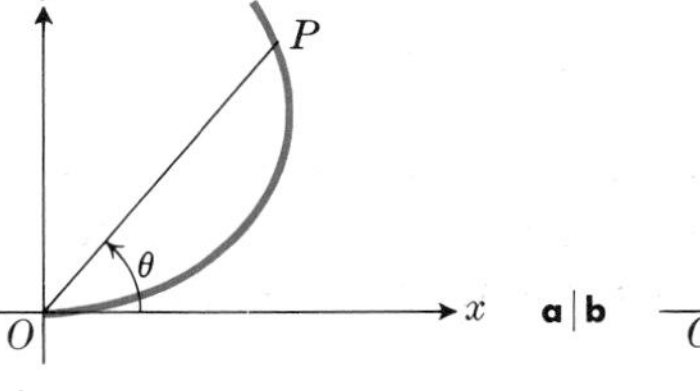

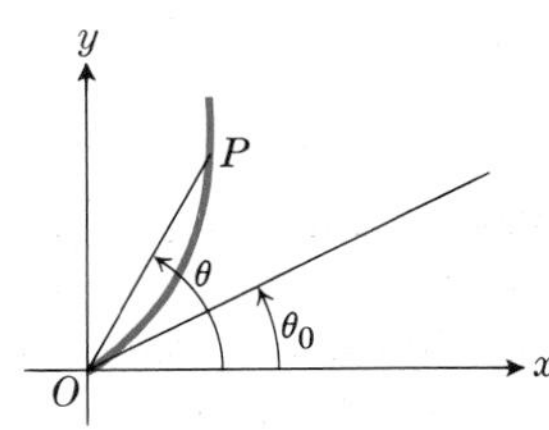

a|b **11.10**

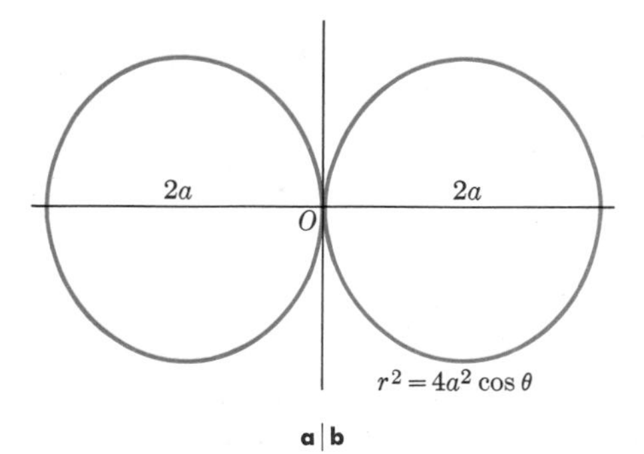

a | b

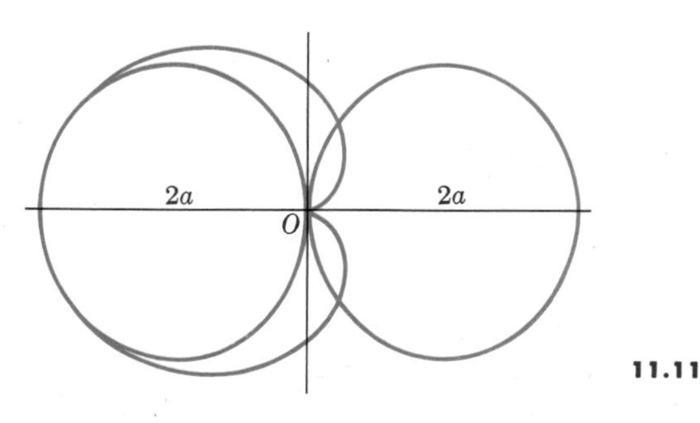

11.11

Example 2. Consider

$$r^2 = 4a^2 \cos\theta.$$

This curve is symmetric about the origin. Two values,

$$r = \pm 2a\sqrt{\cos\theta},$$

correspond to each value of θ for which $\cos\theta > 0$, namely,

$$-\frac{\pi}{2} < \theta < \frac{\pi}{2}.$$

Furthermore, the curve is symmetric about the x-axis, since θ may be replaced by $-\theta$ without altering the value of $\cos\theta$. The curve passes through the origin at $\theta = \pi/2$ and is tangent to the y-axis at this point. Since $\cos\theta$ never exceeds unity, the maximum value of r is $2a$, which occurs at $\theta = 0$. As θ increases from 0 to $\pi/2$, $|r|$ decreases from $2a$ to 0. The curve is sketched in Fig. 11.11(a) and again in Fig. 11.11(b) with the cardioid of Example 1.

Example 3. Consider $r\theta = a$, a a positive constant. When $\theta = 0$ the equation becomes $0 = a$, which is not true. That is, there is no point on the curve for $\theta = 0$. However, suppose θ is a *small* positive angle. Then from $r = a/\theta$, we see that r is large and positive. Consider the situation in Fig. 11.12. No matter how small the positive angle θ may be, if r is sufficiently large the point P may be far above the x-axis. In fact, we need to see what happens to

$$y = r\sin\theta = \frac{a}{\theta}\sin\theta$$

for small positive values of θ. We know, of course, that

$$\lim_{\theta\to 0}\frac{\sin\theta}{\theta} = 1,$$

and hence that

$$\lim_{\theta\to 0} y = \lim_{\theta\to 0} a\frac{\sin\theta}{\theta} = a.$$

This shows that the line $y = a$ is an asymptote of this curve. So we think of the curve being traced by a point P that starts far out near the line $y = a$ for $\theta = 0+$ and moves in the direction indicated by the arrows in Fig. 11.13(a) as θ increases and r decreases. As the radius vector OP continues to rotate about the origin, it shrinks in length, and P describes a spiral which coils around the origin with r tending to zero as the angle θ increases indefinitely.

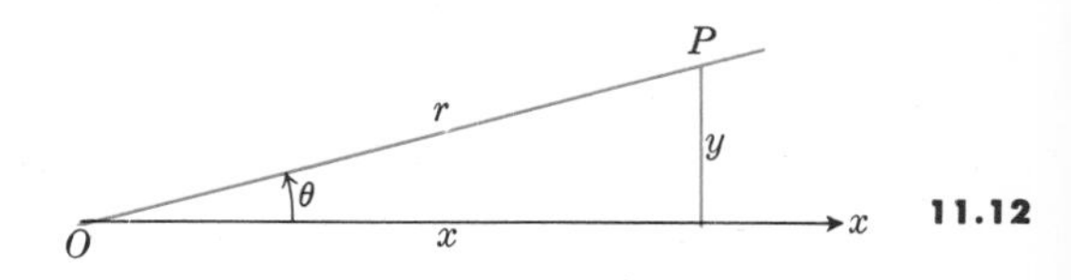

11.12

When r and θ are replaced by $-r$ and $-\theta$ respectively, the equation is unaltered. Hence for every point $(r_1\ \theta_1)$ on the curve in Fig. 11.13(a), there is a point $(-r_1, -\theta_1)$ symmetrically located with respect to the y-axis also on the curve $r\theta = a$. We therefore reflect the curve in Fig. 11.13(a) in the y-axis, obtaining another spiral which coils around the origin in the clockwise sense as θ approaches minus infinity. The complete curve is indicated in Fig. 11.13(b). It is called a *hyperbolic spiral*, the adjective "hyperbolic" being used in this connection because the equation $r\theta = a$ is analogous to the equation $xy = a$, which represents a hyperbola in Cartesian coordinates.

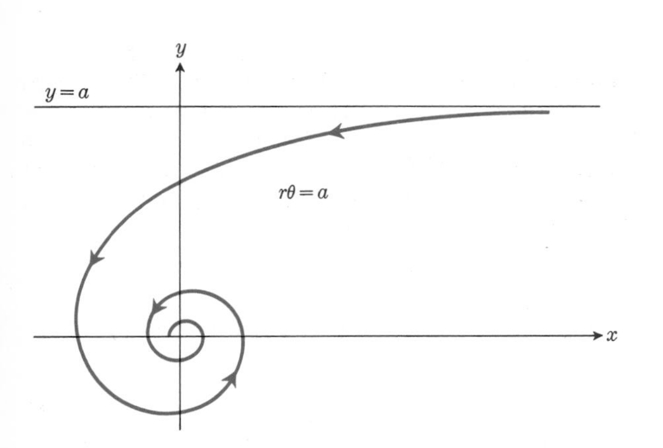

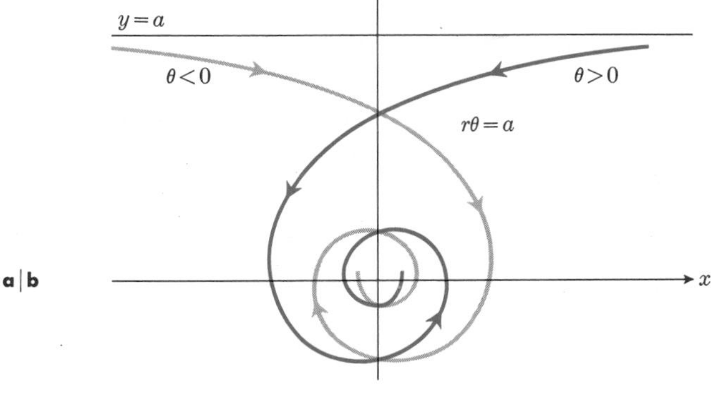

a|b

11.13 Hyperbolic spiral.

EXERCISES

1. Find the polar form of the equation of the line $3x + 4y = 5$.

2. If the polar coordinates (r, θ) of a point P satisfy the equation $r = 2a \cos \theta$, what equation is satisfied by the Cartesian coordinates (x, y) of P?

Discuss and sketch each of the following curves.

3. $r = a(1 + \cos \theta)$
4. $r = a(1 - \sin \theta)$
5. $r = a \sin 2\theta$
6. $r^2 = 2a^2 \cos 2\theta$
7. $r = a(2 + \sin \theta)$
8. $r = a(1 + 2 \sin \theta)$
9. $r = \theta$
10. $r = a \sin (\theta/2)$

11. Find the points on the curve $r = a(1 + \cos \theta)$ where the tangent is
    (a) parallel to the x-axis,
    (b) parallel to the y-axis.

    Hint. Express x and y in terms of θ. Then calculate dy/dx.

12. Sketch the curves

$$r = a(1 + \cos \theta), \qquad r = 3a \cos \theta$$

in one diagram and find the angle between their tangents at the point of intersection that lies in the first quadrant.

SKIP

11.3 POLAR EQUATIONS OF THE CONIC SECTIONS AND OTHER CURVES

The relationships given in Article 11.1 between Cartesian and polar coordinates enable us to change any Cartesian equation into a polar equation for the same curve.

Example. The circle

$$x^2 + y^2 - 2ax = 0, \tag{1}$$

with center at $(a, 0)$ and radius a, becomes

$$(r \cos \theta)^2 + (r \sin \theta)^2 - 2a(r \cos \theta) = 0,$$

which reduces to

$$r(r - 2a \cos \theta) = 0.$$

The locus obtained by putting the first factor equal to zero, $r = 0$, is just one point, the origin. The other factor vanishes when

$$r = 2a \cos \theta. \tag{2}$$

This includes the origin among its points and hence represents the entire locus given by Eq. (1).

The example illustrates one method of finding the polar equation for a curve, namely, by transforming its Cartesian equation into polar form. An alternative method is to derive the polar equation directly from some geometric property. For example,

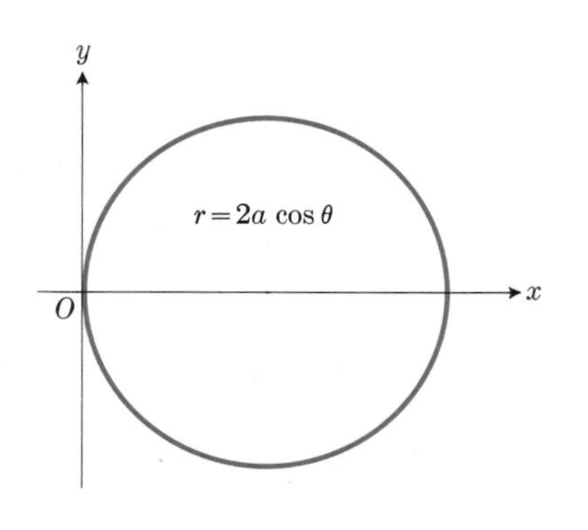

a | b

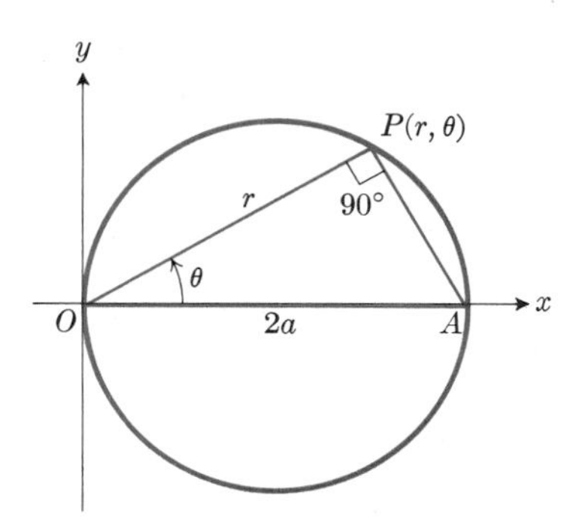

11.14

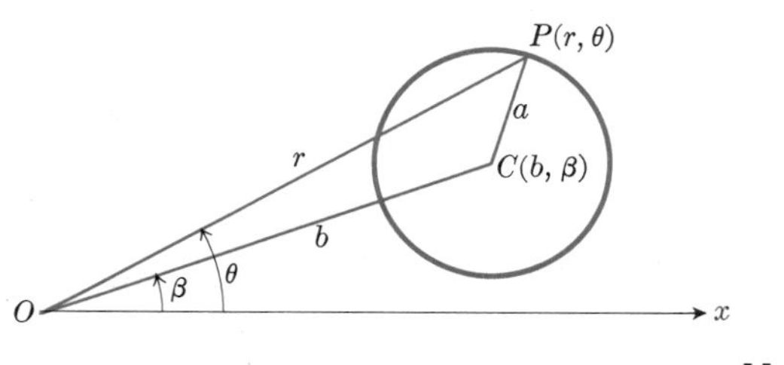

11.15

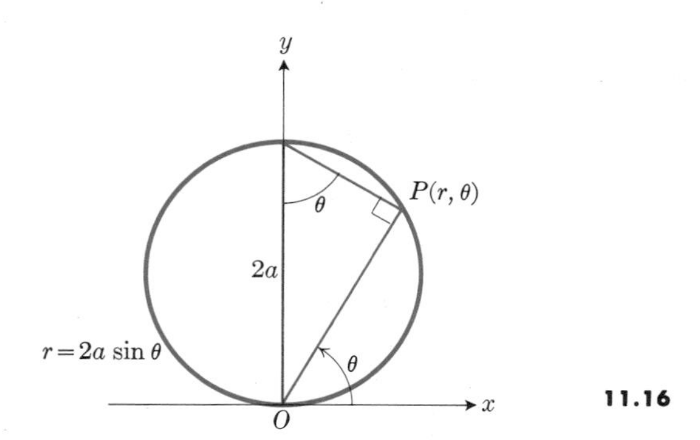

11.16

take the circle in Fig. 11.14 and let $P(r, \theta)$ be a representative point on the circle. Then angle OPA is a right angle (why?) and from the right triangle OPA, we read

$$\frac{r}{2a} = \cos\theta, \quad \text{or} \quad r = 2a\cos\theta,$$

which is the same equation obtained above.

We shall apply this second method to obtain equations of various curves in the following problems.

Problem 1. Find the polar equation of the circle of radius a with center at (b, β).

Solution. We let $P(r, \theta)$ be a representative point on the circle and apply the law of cosines to the triangle OCP (Fig. 11.15) to obtain

$$a^2 = b^2 + r^2 - 2br\cos(\theta - \beta). \tag{3}$$

If the circle passes through the origin, then $b = a$ and the equation takes the simpler form

$$r[r - 2a\cos(\theta - \beta)] = 0,$$

or

$$r = 2a\cos(\theta - \beta). \tag{4}$$

In particular, if $\beta = 0$, Eq. (4) reduces to the result we have obtained before, while if $\beta = 90°$, so that the center of the circle lies on the y-axis (Fig. 11.16), Eq. (4) reduces to

$$r = 2a\sin\theta. \tag{5}$$

Problem 2. The normal from the origin to the line L intersects L at the point $N(p, \beta)$. Find the polar equation of L.

Solution. We let $P(r, \theta)$ be a representative point on the line L in Fig. 11.17, and from the right triangle ONP we read the result

$$r\cos(\theta - \beta) = p. \tag{6}$$

This equation is simply a more general form of the equation

$$r\cos\theta = p, \tag{7}$$

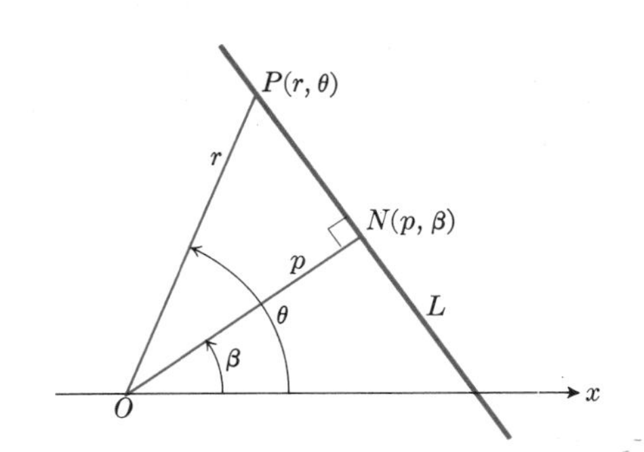

11.17

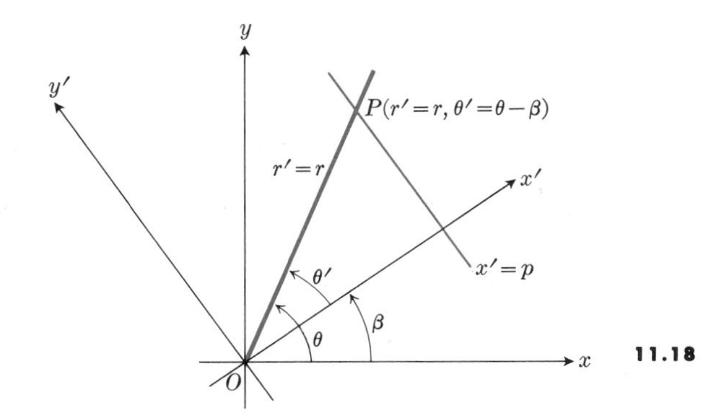

11.18

which is the polar form of the line

$$x = p.$$

In fact, if we perform a rotation of axes as in Fig. 11.18, the new polar coordinates (r', θ') are related to the old polar coordinates as follows:

$$r' = r, \quad \theta' = \theta - \beta. \tag{8}$$

If we apply this rotation to Eq. (6), we get

$$r' \cos \theta' = p.$$

But

$$r' \cos \theta' = x',$$

so the equation is the same as

$$x' = p,$$

which represents a straight line p units from and parallel to the y'-axis.

Problem 3. Find the polar locus of P if the product of its distances to the two points $F_1(a, \pi)$ and $F_2(a, 0)$ is a constant, say b^2.

Solution. We let $P(r, \theta)$ be a representative point on the locus and determine the equation satisfied by the coordinates r and θ in order to fulfill the requirement

$$d_1 d_2 = b^2, \tag{9}$$

where $d_1 = PF_1$ and $d_2 = PF_2$ (Fig. 11.19). We apply the law of cosines twice: once to the triangle OPF_2:

$$d_2^2 = r^2 + a^2 - 2ar \cos \theta,$$

and again to the triangle OPF_1:

$$d_1^2 = r^2 + a^2 - 2ar \cos (\pi - \theta).$$

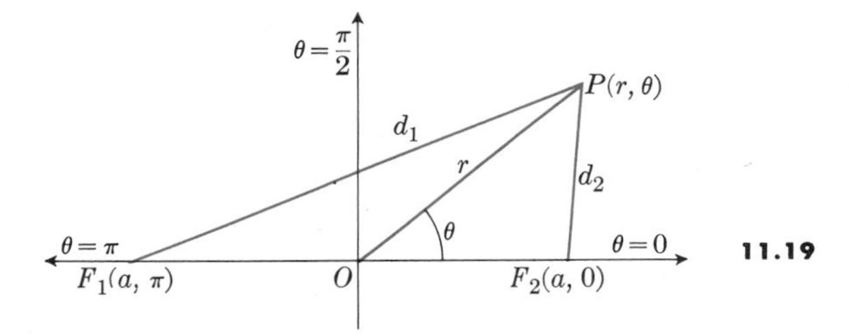

11.19

But $\qquad \cos (\pi - \theta) = -\cos \theta,$

so that $\qquad d_1^2 = r^2 + a^2 + 2ar \cos \theta.$

Hence $\qquad d_1^2 d_2^2 = (r^2 + a^2)^2 - (2ar \cos \theta)^2,$

or $\qquad b^4 = a^4 + r^4 + 2a^2r^2(1 - 2 \cos^2 \theta).$

The trigonometric identity

$$\cos 2\theta = 2 \cos^2 \theta - 1$$

enables us to put our equation in the form

$$b^4 = a^4 + r^4 - 2a^2r^2 \cos 2\theta. \tag{10}$$

One special case allows the locus to pass through the origin, namely, if $b = a$. Then the equation simplifies still further to that of the *lemniscate*

$$r^2 = 2a^2 \cos 2\theta. \tag{11}$$

The graphs of the curves represented by (10) for different values of the ratio b/a are shown in Fig. 11.20(a) through (c). The curves in both (a) and (b) are called *lemniscates*, and the curve in (c) consists of two separate closed portions known as "ovals of Cassini."

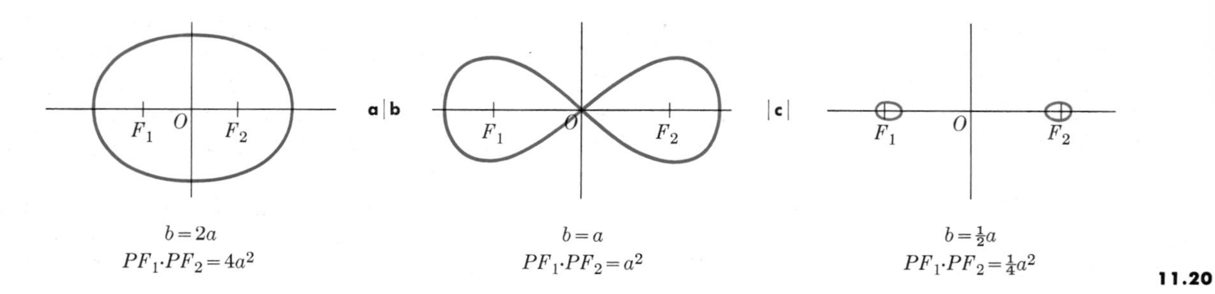

11.20

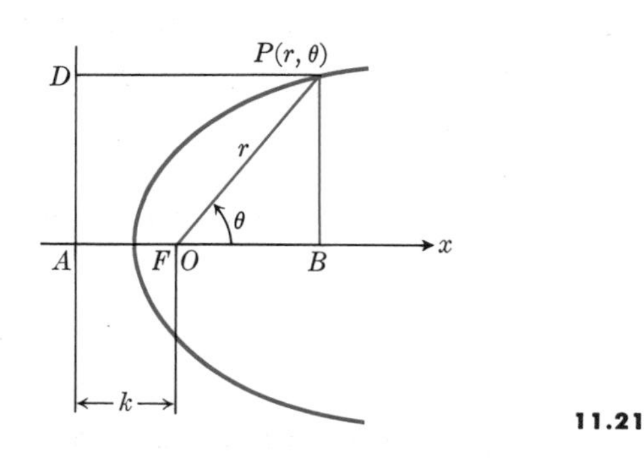

11.21

Problem 4. Find the polar equation of the conic section of eccentricity e if the focus is at the origin and the associated directrix is the line $x = -k$.

Solution. We adopt the notation of Fig. 11.21 and use the focus-and-directrix property

$$PF = e \cdot PD, \tag{12}$$

which allows us to handle the parabola, ellipse, and hyperbola all at the same time.

By taking the origin at the focus F, we have

$$PF = r,$$

while

$$PD = AB$$

and

$$AB = AF + FB = k + r\cos\theta.$$

Then Eq. (12) is the same as

$$r = e(k + r\cos\theta).$$

If we solve this equation for r, we get

$$r = \frac{ke}{1 - e\cos\theta}. \tag{13}$$

Typical special cases of (13) are obtained by taking

$$e = 1, \quad r = \frac{k}{1 - \cos\theta} = \frac{k}{2}\csc^2\frac{\theta}{2}, \tag{14a}$$

which represents a *parabola;*

$$e = \tfrac{1}{2}, \quad r = \frac{k}{2 - \cos\theta}, \tag{14b}$$

which represents an *ellipse;* and

$$e = 2, \quad r = \frac{2k}{1 - 2\cos\theta}, \tag{14c}$$

which represents a *hyperbola.*

It is worth noting that the denominator in Eq. (14b) for the ellipse can never vanish, so that r remains finite for all values of θ. But r becomes infinite as θ approaches 0 in (14a) and as θ approaches $\pi/3$ in (14c).

By replacing the constant ke in Eq. (13) by its equivalent value

$$ke = a(1 - e^2)$$

when $e < 1$, we may let $e \to 0$ and $ke \to a$ and get

$$r = a$$

in the limit. That is, the circle of radius a and center at O is a limiting case of the curves represented by Eq. (13).

EXERCISES

1. A line segment of length $2a$ slides with its ends on the x- and y-axes. Find the polar equation of the locus described by the point $P(r, \theta)$ at which the perpendicular from the origin intersects the moving line. Sketch the curve.

2. OA is a diameter of a circle of radius a, AC is tangent to the circle, and OC intersects the circle at B. On OC the point $P(r, \theta)$ is found such that $OP = BC$. With O as origin, OA as x-axis, and angle AOP as θ, find the polar equation of the locus of P and sketch the curve. Show that $x = 2a$ is an asymptote. The curve is called a "cissoid," meaning "ivy-like."

In Exercises 3 through 7, determine the polar equation and sketch the given curve.

3. $x^2 + y^2 - 2ay = 0$
4. $(x^2 + y^2)^2 + 2ax(x^2 + y^2) - a^2y^2 = 0$
5. $x \cos \alpha + y \sin \alpha = p$ (α, p constants)
6. $y^2 = 4ax + 4a^2$
7. $(x^2 + y^2)^2 = x^2 - y^2$

In Exercises 8 through 13, determine the Cartesian equation and sketch the given curve.

8. $r = 4 \cos \theta$
9. $r = 6 \sin \theta$
10. $r = \sin 2\theta$
11. $r^2 = 2a^2 \cos 2\theta$
12. $r = 8/(1 - 2 \cos \theta)$
13. $r = a(1 + \sin \theta)$

14. Sketch the following loci.

    (a) $r = 2 \cos (\theta + 45°)$
    (b) $r = 4 \csc (\theta - 30°)$
    (c) $r = 5 \sec (60° - \theta)$
    (d) $r = 3 \sin (\theta + 30°)$
    (e) $r = a + a \cos (\theta - 30°)$
    (f) $0 \leq r \leq 2 - 2 \cos \theta$

15. (a) How can the angle β and the distance p (see Fig. 11.17 and Eq. 6) be determined from the Cartesian equation $ax + by = c$ for a straight line? (b) Specifically, find p, β, and the polar equation for the straight line $\sqrt{3}x + y = 6$.

16. The focus of a parabola is at the origin and its directrix is the line $4 = r \cos (\theta + \pi/3)$. Find the polar equation of the parabola.

17. One focus of a hyperbola of eccentricity $\frac{5}{4}$ is at the origin and the corresponding directrix is the line $r \cos \theta = 9$. Find the polar coordinates of the second focus. Also determine the polar equation of the hyperbola and sketch.

18. From the origin a line is drawn perpendicular to a tangent of the circle $r = 2a \cos \theta$. Find the locus of the point of intersection and sketch the curve.

19. Find the maximum and minimum values of r that satisfy Eq. (10), first in terms of a and b, and then specifically for the cases $b = 2a$, $b = a$, and $b = a/2$.

20. Determine the slopes of the lines that are tangent to the lemniscate $r^2 = 2a^2 \cos 2\theta$ at the origin.

21. Equation (14b), $r = k/(2 - \cos \theta)$, represents an ellipse with one focus at the origin. Sketch the curve and its directrices for the case $k = 2$ and locate the center of the ellipse.

22. In the case $e > 1$, Eq. (13) represents a hyperbola. From the polar form of the equation, determine the slopes of the asymptotes of the hyperbola of eccentricity e.

23. Show, by reference to a figure, that the distance from the line
$$r \cos (\theta - \beta) = p$$
to the point $P_1(r_1, \theta_1)$ is
$$|r_1 \cos (\theta_1 - \beta) - p|.$$

11.4 THE ANGLE ψ BETWEEN THE RADIUS VECTOR AND THE TANGENT LINE

In Cartesian coordinates, when we want to discuss the direction of a curve at a point, we use the angle ϕ from the positive x-axis to the tangent line. In polar coordinates, it is more convenient to make use of the angle ψ (psi) from the *radius vector* to the tangent line. Then the relationship

$$\phi = \theta + \psi, \tag{1}$$

which can be read from Fig. 11.22, makes it a simple matter to find ϕ if that angle is desired instead of ψ.

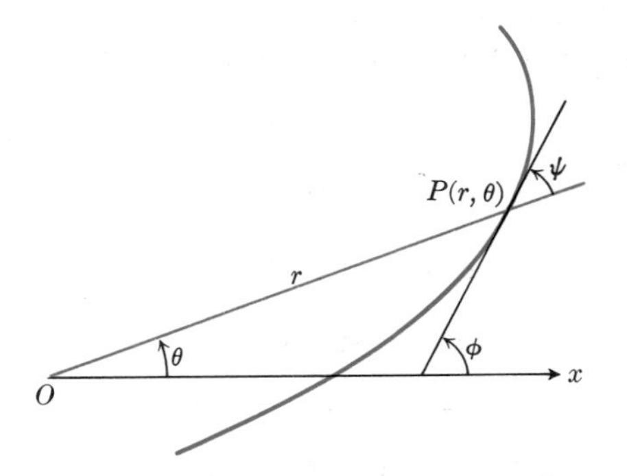

11.22 The angle ψ between the tangent vector and the radius vector.

Suppose the equation of the curve is given in the form $r = f(\theta)$, where $f(\theta)$ is a differentiable function of θ. Then from

$$x = r \cos \theta, \qquad y = r \sin \theta, \tag{2}$$

we see that x and y are differentiable functions of θ with

$$\begin{aligned} \frac{dx}{d\theta} &= -r \sin \theta + \cos \theta \frac{dr}{d\theta}, \\ \frac{dy}{d\theta} &= r \cos \theta + \sin \theta \frac{dr}{d\theta}. \end{aligned} \tag{3}$$

Since $\psi = \phi - \theta$ from (1),

$$\tan \psi = \tan (\phi - \theta) = \frac{\tan \phi - \tan \theta}{1 + \tan \phi \tan \theta},$$

while

$$\tan \phi = \frac{dy/d\theta}{dx/d\theta}, \qquad \tan \theta = \frac{y}{x}.$$

Hence

$$\tan \psi = \frac{\dfrac{dy/d\theta}{dx/d\theta} - \dfrac{y}{x}}{1 + \dfrac{y}{x}\dfrac{dy/d\theta}{dx/d\theta}} = \frac{x\dfrac{dy}{d\theta} - y\dfrac{dx}{d\theta}}{x\dfrac{dx}{d\theta} + y\dfrac{dy}{d\theta}}. \tag{4}$$

The numerator in the last expression in Eq. (4) is found by substitution from Eqs. (2) and (3) to be

$$x\frac{dy}{d\theta} - y\frac{dx}{d\theta} = r^2.$$

Similarly, the denominator is

$$x\frac{dx}{d\theta} + y\frac{dy}{d\theta} = r\frac{dr}{d\theta}.$$

When we substitute these into Eq. (4), we obtain the very simple final result

$$\tan \psi = \frac{r}{dr/d\theta}. \tag{5}$$

Note. This formula for $\tan \psi$ is much simpler than the formula

$$\tan \phi = \frac{r \cos \theta + (\sin \theta)\, dr/d\theta}{-r \sin \theta + (\cos \theta)\, dr/d\theta} \tag{6}$$

that one obtains by calculating

$$\frac{dy}{dx} = \frac{dy/d\theta}{dx/d\theta}$$

from (3). This is why it is usually preferable to work with the angle ψ rather than with ϕ in the case of polar coordinates.

One may also obtain a simple expression for the differential element of arc length ds by squaring and adding the differentials

$$\begin{aligned} dx &= -r \sin \theta \, d\theta + \cos \theta \, dr, \\ dy &= r \cos \theta \, d\theta + \sin \theta \, dr. \end{aligned}$$

We find that

$$ds^2 = dx^2 + dy^2 = r^2 \, d\theta^2 + dr^2.$$

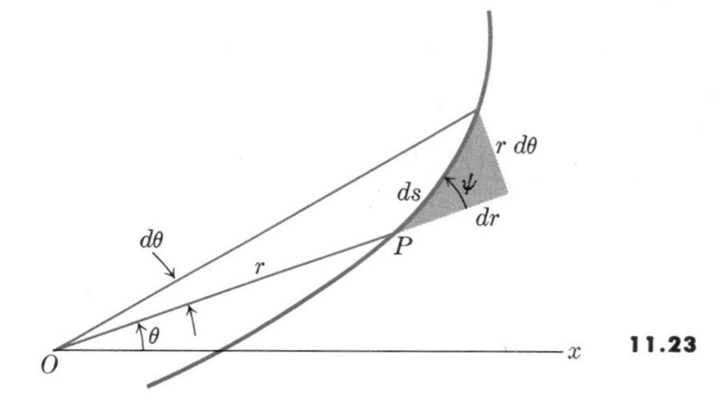

11.23

This result and the result for $\tan \psi$ are both easily remembered in the form

$$\tan \psi = \frac{r \, d\theta}{dr}, \tag{7a}$$

$$ds^2 = r^2 \, d\theta^2 + dr^2, \tag{7b}$$

if we refer to the "differential triangle" shown in Fig. 11.23. We simply treat dr and $r\,d\theta$ as the two legs and ds as the hypotenuse of an ordinary right triangle with the angle ψ opposite the side $r\,d\theta$. If we realize that certain terms of higher order are being neglected, we may think of dr as the component of displacement along the radius vector and $r\,d\theta$ as the component at right angles to this produced by the displacement ds along the curve. The relationships (7a) and (7b) may be read at once from this triangle. We should realize, of course, that the *proof* of these equations did not depend on any such "differential triangle" and that the latter represents only a mnemonic device.

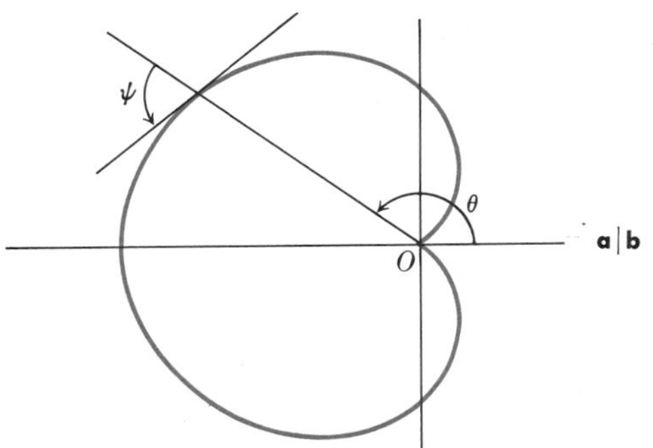

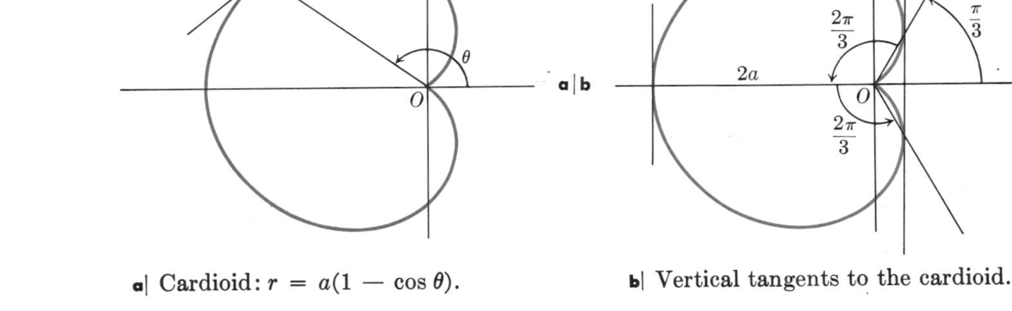

11.24 a| Cardioid: $r = a(1 - \cos\theta)$. b| Vertical tangents to the cardioid.

To find the angle ψ, we use (7a); and to find the length of a polar curve, we find ds from (7b) and integrate between appropriate limits as illustrated in the following problems.

Problem 1. Find the angle ψ for the cardioid (Fig. 11.24a)

$$r = a(1 - \cos\theta). \tag{8a}$$

Solution. From the equation of the curve, we get

$$dr = a \sin\theta \, d\theta, \tag{8b}$$

so that

$$\tan\psi = \frac{r\,d\theta}{dr} = \frac{a(1 - \cos\theta)\,d\theta}{a\sin\theta\,d\theta} = \tan\frac{\theta}{2}.$$

As θ varies from 0 to 2π, the angle ψ varies from 0 to π according to the equation $\psi = \theta/2$. Thus at the y-intercepts the tangent line makes an angle of 45° with the vertical, at the origin the curve is tangent to the x-axis, and at $(2a, \pi)$ the tangent line makes an angle of 90° with the negative x-axis.

Suppose we ask for the points on the cardioid where the tangent line is vertical (see Fig. 11.24b). Denoting the inclination angle of the tangent line by ϕ and recalling that

$$\phi = \psi + \theta \quad \text{and} \quad \psi = \tfrac{1}{2}\theta,$$

we get $\phi = \frac{3}{2}\theta$. Since the tangent is vertical when

$$\phi = \frac{\pi}{2} + n\pi,$$

the values of θ satisfy

$$\theta = \frac{2}{3}\phi = \frac{\pi}{3} + \frac{2}{3}n\pi.$$

Taking $n = 0, -1, 1, -2, 2, \ldots$, we obtain

$$\theta = \frac{\pi}{3}, \quad \frac{-\pi}{3}, \quad \pi, \quad -\pi, \quad \frac{5}{3}\pi, \quad \ldots$$

These lead to three distinct points, namely,

$$\left(\frac{a}{2}, \pm\frac{\pi}{3}\right) \quad \text{and} \quad (2a, \pi),$$

whose radius vectors are evenly spaced 120° apart.

Problem 2. Find the length of the cardioid

$$r = a(1 - \cos\theta).$$

Solution. Substituting dr and r from (8b, a) into (7b), we have

$$\begin{aligned} ds^2 &= a^2\,d\theta^2\,[\sin^2\theta + (1 - \cos\theta)^2] \\ &= 2a^2\,d\theta^2(1 - \cos\theta) \end{aligned}$$

and

$$ds = a\sqrt{2}\,\sqrt{1 - \cos\theta}\,d\theta.$$

To integrate the expression on the right, we recall the trigonometric identity

$$1 - \cos\theta = 2\sin^2\frac{\theta}{2},$$

and have

$$ds = 2a\left|\sin\frac{\theta}{2}\right|\,d\theta.$$

Since $\sin\theta/2$ is not negative when θ varies from 0 to 2π, we obtain

$$s = \int_0^{2\pi} 2a\sin\frac{\theta}{2}\,d\theta = -4a\cos\frac{\theta}{2}\bigg]_0^{2\pi} = 8a.$$

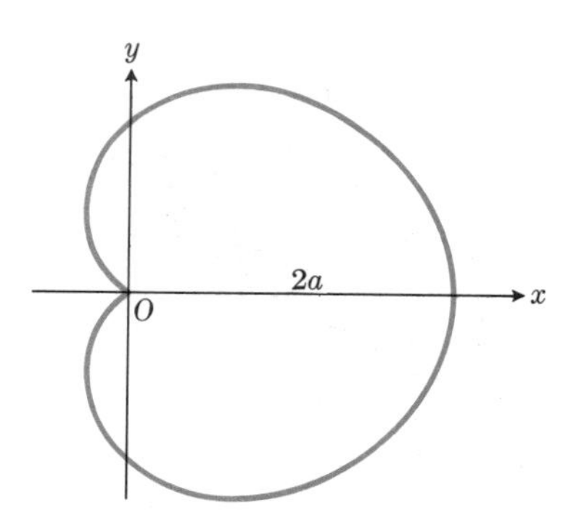

11.25 Cardioid: $r = a(1 + \cos\theta)$.

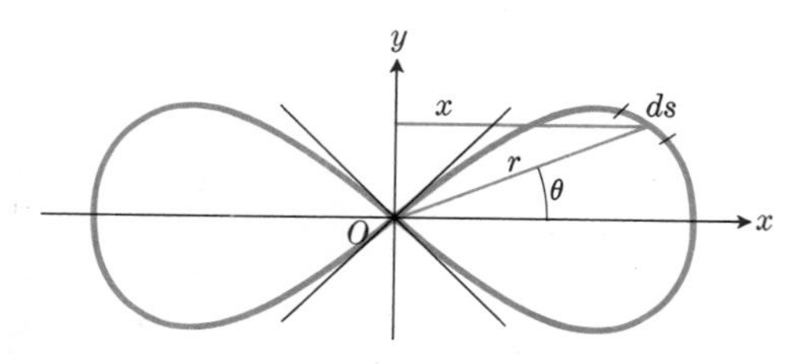

11.26 Lemniscate: $r^2 = 2a^2 \cos 2\theta$.

We observe in passing that in letting θ range from 0 to 2π, we start at the cusp at the origin, go once around the smooth portion of the cardioid and return to the cusp. In doing this we do not pass *across* the cusp (see Fig. 11.24a). If, on the other hand, we take the cardioid $r = a(1 + \cos\theta)$, the appropriate procedure for avoiding the cusp (Fig. 11.25) is to let θ increase from $-\pi$ to π, or we may use the symmetry of the curve and calculate half of the total length by letting θ vary from 0 to π.

Problem 3. The lemniscate

$$r^2 = 2a^2 \cos 2\theta$$

is revolved about the y-axis. Find the area of the surface generated.

Solution. An element of arc length ds (Fig. 11.26) generates a portion of surface area

$$dS = 2\pi x\, ds,$$

where

$$x = r\cos\theta, \quad ds = \sqrt{dr^2 + r^2\, d\theta^2}.$$

That is,

$$dS = 2\pi r \cos\theta \sqrt{dr^2 + r^2\, d\theta^2} = 2\pi \cos\theta \sqrt{r^2\, dr^2 + r^4\, d\theta^2}.$$

From the equation of the curve, we get

$$r\, dr = -2a^2 \sin 2\theta\, d\theta.$$

Then

$$(r^2\, dr^2 + r^4\, d\theta^2) = (2a^2\, d\theta)^2 (\sin^2 2\theta + \cos^2 2\theta)$$

and

$$dS = 4\pi a^2 \cos\theta\, d\theta.$$

The total surface area is generated by the loop of the lemniscate to the right of the y-axis between $\theta = -\pi/4$ and $\theta = +\pi/4$, so that

$$S = \int_{-\pi/4}^{\pi/4} 4\pi a^2 \cos\theta\, d\theta = 4\pi a^2 \sqrt{2}.$$

EXERCISES

1. For the hyperbolic spiral $r\theta = a$, show that $\psi = 135°$ when $\theta = 1$ radian, and that $\psi \to 90°$ as the spiral winds around the origin. Sketch the curve and indicate ψ for $\theta = 1$ radian.
2. Show, by reference to a figure, that the angle β between the tangents to two curves at a point of intersection may be found from the formula
$$\tan\beta = \frac{\tan\psi_2 - \tan\psi_1}{1 + \tan\psi_2 \tan\psi_1}.$$
When will the two curves intersect orthogonally?
3. Find a point of intersection of the parabolas
$$r = \frac{1}{1 - \cos\theta}, \quad r = \frac{3}{1 + \cos\theta},$$
and find the angle between the tangents to these curves at this point.
4. Find points on the cardioid $r = a(1 + \cos\theta)$ where the tangent line is horizontal.
5. Find the length of the cardioid $r = a(1 + \cos\theta)$. *Hint.*
$$\int \sqrt{1 + \cos\theta}\, d\theta = \int \sqrt{2}\, |\cos\theta/2|\, d\theta.$$
6. A thin, uniform wire is bent into the shape of the cardioid $r = a(1 + \cos\theta)$. Find its center of gravity $(\bar{x}, \bar{y})$.
Hint. $\int \cos\theta \cos\theta/2\, d\theta$ can be evaluated by substituting $\cos\theta = 1 - 2\sin^2\theta/2$ and then letting $u = \sin\theta/2$.

7. The lemniscate $r^2 = 2a^2 \cos 2\theta$ is rotated about the x-axis. Find the area of the surface generated.
8. Find the length of the curve $r = a \sin^2 \theta/2$ from $\theta = 0$ to $\theta = \pi$. Sketch the curve.
9. Find the length of the parabolic spiral $r = a\theta^2$ between $\theta = 0$ and $\theta = \pi$. Sketch the curve.
10. Find the length of the curve $r = a \sin^3 \theta/3$ between $\theta = 0$ and $\theta = \pi$. Sketch the curve.
11. Find the surface area generated by rotating the curve $r = 2a \cos \theta$ about the line $\theta = \pi/2$. Sketch.

11.5 PLANE AREAS IN POLAR COORDINATES

The area AOB in Fig. 11.27 is bounded by the rays $\theta = \alpha$, $\theta = \beta$, and the curve $r = f(\theta)$. We imagine the angle AOB as being divided into n parts

$$\Delta\theta = \frac{\beta - \alpha}{n},$$

and we approximate the area in a typical sector POQ by the area of a *circular* sector of radius r and central angle $\Delta\theta$, that is

$$\text{area of } POQ \approx \tfrac{1}{2}r^2 \,\Delta\theta,$$

and hence the entire area AOB is approximately

$$\sum_{\theta=\alpha}^{\beta} \tfrac{1}{2}r^2 \,\Delta\theta.$$

We might note that if in fact the function $r = f(\theta)$ which represents the polar curve is a continuous function of θ for $\alpha \leq \theta \leq \beta$, then there is a θ_k between θ and $\theta + \Delta\theta$ such that the circular sector of radius

$$r_k = f(\theta_k)$$

and central angle $\Delta\theta$ gives the *exact* area of POQ (Fig. 11.28). Then the entire area is given exactly by

$$A = \sum \tfrac{1}{2}r_k^2 \,\Delta\theta = \sum \tfrac{1}{2}[f(\theta_k)]^2 \,\Delta\theta.$$

If we let $\Delta\theta \to 0$, we may apply the Fundamental Theorem of the integral calculus and obtain

$$A = \lim_{\Delta\theta\to 0} \sum \tfrac{1}{2}[f(\theta_k)]^2 \,\Delta\theta = \tfrac{1}{2}\int_\alpha^\beta [f(\theta)]^2 \,d\theta.$$

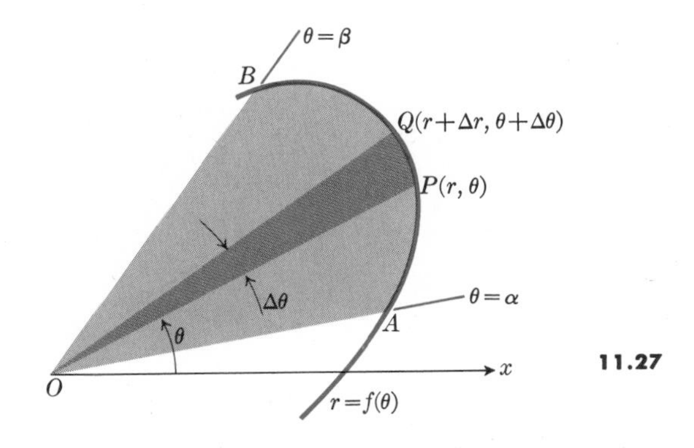

11.27

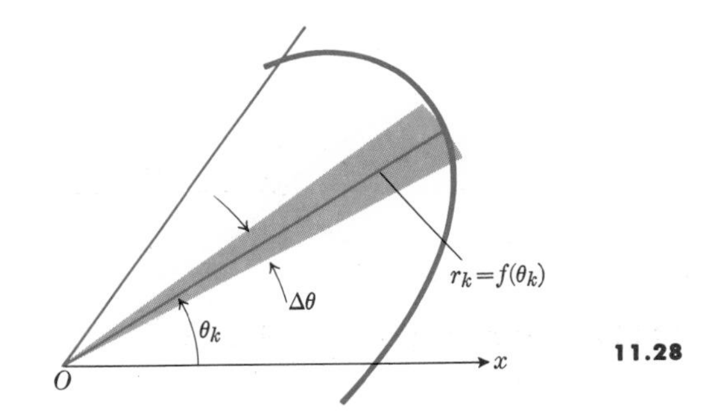

11.28

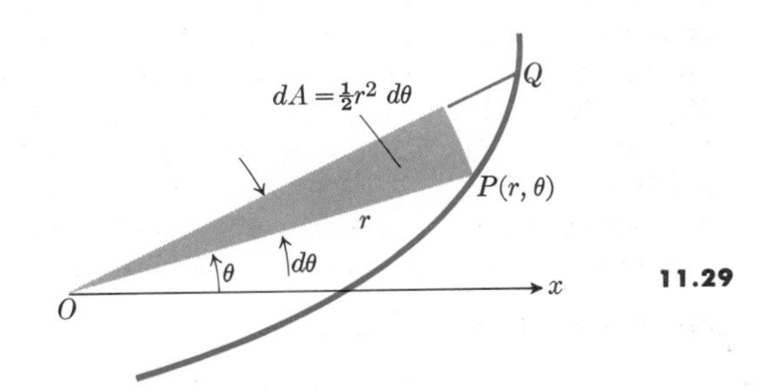

11.29

This is equivalent to the integral

$$A = \int_\alpha^\beta \tfrac{1}{2}r^2 \,d\theta. \tag{1}$$

This result may also be remembered as the integral of the differential element of area (Fig. 11.29),

$$dA = \tfrac{1}{2}r^2 \,d\theta,$$

taken between the appropriate limits on θ.

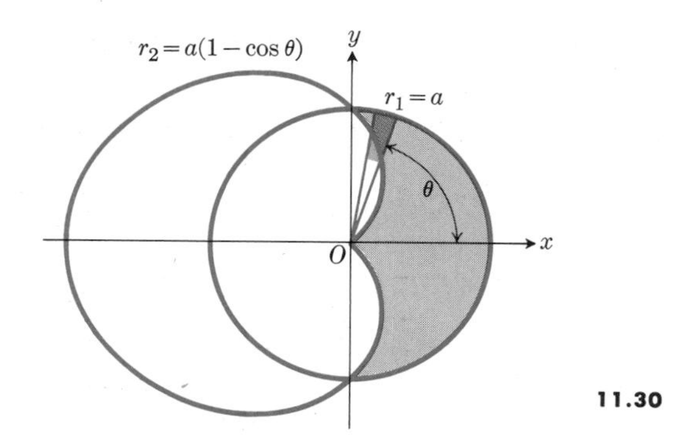

11.30

Problem. Find the area that is inside the circle $r = a$ and outside the cardioid $r = a(1 - \cos\theta)$. (See Fig. 11.30.)

Solution. We take a representative element of area

$$dA = dA_1 - dA_2,$$

where

$$dA_1 = \tfrac{1}{2}r_1^2\,d\theta, \quad dA_2 = \tfrac{1}{2}r_2^2\,d\theta$$

with

$$r_1 = a, \quad r_2 = a(1 - \cos\theta).$$

Such elements of area belong to the region inside the circle and outside the cardioid, provided that θ lies between

$$-\frac{\pi}{2} \text{ and } +\frac{\pi}{2},$$

where the curves intersect. Hence

$$\begin{aligned} A &= \int_{-\pi/2}^{\pi/2} \tfrac{1}{2}(r_1^2 - r_2^2)\,d\theta \\ &= \int_{-\pi/2}^{\pi/2} \frac{a^2}{2}(2\cos\theta - \cos^2\theta)\,d\theta \\ &= a^2\int_{-\pi/2}^{\pi/2}\cos\theta\,d\theta - \frac{a^2}{2}\int_{-\pi/2}^{\pi/2}\frac{1+\cos 2\theta}{2}\,d\theta \\ &= a^2\left(2 - \frac{\pi}{4}\right). \end{aligned}$$

As a check against gross errors, we observe that this is roughly 80% of the area of a semicircle of radius a, and this seems reasonable when we look at the tinted area in Fig. 11.30.

EXERCISES

Note. The trigonometric identities

$$\sin^2\theta = \tfrac{1}{2}(1 - \cos 2\theta), \quad \cos^2\theta = \tfrac{1}{2}(1 + \cos 2\theta)$$

should be used to evaluate $\int \sin^2\theta\,d\theta$ and $\int \cos^2\theta\,d\theta$ in certain of the problems that follow.

1. Find the total area inside the cardioid
$$r = a(1 + \cos\theta).$$
2. Find the total area inside the circle $r = 2a\sin\theta$.
3. Find the total area inside the lemniscate
$$r^2 = 2a^2\cos 2\theta.$$
4. Find that portion of the area inside the lemniscate $r^2 = 2a^2\cos 2\theta$ that is not included in the circle $r = a$. Sketch.
5. Find the area inside the curve $r = a(2 + \cos\theta)$.
6. Find the area common to the circles $r = 2a\cos\theta$ and $r = 2a\sin\theta$. Sketch.
7. Find the area inside the circle $r = 3a\cos\theta$ and outside the cardioid $r = a(1 + \cos\theta)$.
8. Since the center of gravity of a triangle is located on a median at a distance $\frac{2}{3}$ of the way from a vertex to the opposite base, the lever arm for the moment about the x-axis of the area of triangle POQ in Fig. 11.27 is $\frac{2}{3}r\sin\theta + \epsilon$, where $\epsilon \to 0$ as $\Delta\theta \to 0$. Deduce that the center of gravity of the area AOB in the figure is given by
$$y = \frac{\int \frac{2}{3}r\sin\theta \cdot \frac{1}{2}r^2\,d\theta}{\int \frac{1}{2}r^2\,d\theta}$$
and similarly,
$$x = \frac{\int \frac{2}{3}r\cos\theta \cdot \frac{1}{2}r^2\,d\theta}{\int \frac{1}{2}r^2\,d\theta},$$
with limits $\theta = \alpha$ to $\theta = \beta$ on all integrals.
9. Use the results of Exercise 8 to find the center of gravity of the area bounded by the cardioid
$$r = a(1 + \cos\theta).$$
10. Use the results of Exercise 8 to find the center of gravity of the area of a semicircle of radius a.

REVIEW QUESTIONS AND EXERCISES

1. Make a diagram to show the standard relations between Cartesian coordinates (x, y) and polar coordinates (r, θ). Express each set of coordinates in terms of the other kind.
2. If a point has polar coordinates (r_1, θ_1), what other polar coordinates represent the same point?
3. What is the expression for area between curves in polar coordinates?
4. What is the expression for length of arc of a curve in polar coordinates? For area of a surface of revolution?
5. What are some criteria for symmetry of a curve if its polar coordinates satisfy the equation $r = f(\theta)$? Illustrate your discussion with specific examples.
6. An artificial satellite is in an orbit that passes over the North and South Poles of the earth. When it is over the North Pole it is at the highest point of its orbit, 1000 miles above the earth's surface. Above the South Pole it is at the lowest point of its orbit, 300 miles above the earth's surface.

   (a) Assuming that the orbit (with reference to the earth) is an ellipse with one focus at the center of the earth, find its eccentricity. (Take the diameter of the earth to be 8000 miles.)

   (b) Using the north-south axis of the earth as polar axis, and the center of the earth as origin, find a polar equation of the orbit.

MISCELLANEOUS EXERCISES

Discuss and sketch each of the curves in Exercises 1 through 13 (where a is a positive constant).

1. $r = a\theta$
2. $r = a(1 + \cos 2\theta)$
3. (a) $r = a \sec \theta$ (b) $r = a \csc \theta$ (c) $r = a \sec \theta + a \csc \theta$
4. $r = a \sin\left(\theta + \dfrac{\pi}{3}\right)$
5. $r^2 + 2r(\cos \theta + \sin \theta) = 7$
6. $r = a \cos \theta - a \sin \theta$
7. $r \cos \dfrac{\theta}{2} = a$
8. $r^2 = a^2 \sin \theta$
9. $r^2 = 2a^2 \sin 2\theta$
10. $r = a(1 - 2 \sin 3\theta)$
11. (a) $r = \cos 2\theta$ (b) $r^2 = \cos 2\theta$
12. (a) $r = 1 + \cos \theta$ (b) $r = \dfrac{1}{1 + \cos \theta}$
13. (a) $r = \dfrac{2}{1 - \cos \theta}$ (b) $r = \dfrac{2}{1 + \sin \theta}$
14. Sketch and discuss the graph whose equation in polar coordinates is
$$r = 1 - \tan^2 \theta.$$
Show that it has no vertical asymptotes.

Sketch each of the pairs of curves in Exercises 15 through 20, and find all points of intersection and the angle between their tangents at each point of intersection.

15. $r = a, \quad r = 2a \sin \theta$
16. $r = a, \quad r = a(1 - \sin \theta)$
17. $r = a \sec \theta, \quad r = 2a \sin \theta$
18. $r = a \cos \theta, \quad r = a(1 + \cos \theta)$
19. $r = a(1 + \cos 2\theta), \quad r = a \cos 2\theta$
20. $r^2 = 4 \cos 2\theta, \quad r^2 = \sec 2\theta$
21. Find the equation, in polar coordinates, of a parabola whose focus is at $r = 0$ and whose vertex is at $r = 1, \theta = 0$.
22. Find the polar equation of the straight line with intercepts a and b on the lines $\theta = 0, \theta = \pi/2$.
23. Find the equation of a circle with center on the line $\theta = \pi$, of radius a, and passing through the origin.
24. Find the polar equation of a parabola with focus at the origin and vertex at $(a, \pi/4)$.
25. Find the polar equation of an ellipse with one focus at the origin, the other at $(2, 0)$, and a vertex at $(4, 0)$.
26. Find the polar equation of a hyperbola with one focus at the origin, center at $(2, \pi/2)$, and vertex at $(1, \pi/2)$.
27. Three loran stations are located (in polar coordinates) at $(a, 0)$, $(0, 0)$, and $(a, \pi/4)$. Radio signals are sent out from the three stations simultaneously. A plane receiving the signals notes that the signals from the second and third stations arrive $a/2\nu$ sec later than that from the first. If ν is the velocity of a radio signal, what is the location of the plane in polar coordinates?

End 112 and 113

28. Show that the parabolas

$$r = a/(1 + \cos\theta), \quad r = b/(1 - \cos\theta)$$

are orthogonal at each point of intersection ($ab \neq 0$).

29. Find the angle between the line $\theta = \pi/2$ and the cardioid $r = a(1 - \cos\theta)$ at their intersection.

30. Find the angle between the line $r = 3\sec\theta$ and the curve $r = 4(1 + \cos\theta)$ at one of their intersections.

31. Find the slope of the tangent line to the curve

$$r = a\tan(\theta/2)$$

at $\theta = \pi/2$.

32. Check that the two curves $r = 1/(1 - \cos\theta)$ and $r = 3/(1 + \cos\theta)$ intersect at the point $(2, \pi/3)$. Find the angle between the tangents to these curves at this point.

33. The equation $r^2 = 2\csc 2\theta$ represents a curve in polar coordinates.

(a) Sketch the curve.

(b) Find the equation of the curve in rectangular coordinates.

(c) Find the angle at which the curve intersects the line $\theta = \pi/4$.

34. A given curve cuts all rays θ = constant at the constant angle α.

(a) Show that the area bounded by the curve and two rays $\theta = \theta_1$, $\theta = \theta_2$, is proportional to $r_2^2 - r_1^2$, where (r_1, θ_1) and (r_2, θ_2) are polar coordinates of the ends of the arc of the curve between these rays. Find the factor of proportionality.

(b) Show that the length of the arc of the curve in (a) is proportional to $r_2 - r_1$ and find the proportionality constant.

35. The cardioid $r = a(1 - \cos\theta)$ is rotated about the initial line.

(a) Find the area of the surface generated. *Hint.* You may use the facts

$$\sin\theta = 2\sin(\theta/2)\cos(\theta/2),$$
$$1 - \cos\theta = 2\sin^2(\theta/2)$$

to evaluate your integral.

(b) Set up the definite integral (or integrals) that would be used to find the centroid of the area in (a).

(c) Find the centroid of the area in (a).

36. Let P be a point on the hyperbola $r^2 \sin 2\theta = 2a^2$. Show that the triangle formed by OP, the tangent at P, and the initial line is isosceles.

37. Verify that the formula for finding the length of a curve in polar form gives the correct result for the circumference of the circles

(a) $r = a$, (b) $r = a\cos\theta$,
(c) $r = a\sin\theta$.

38. If $r = a\cos^3(\theta/3)$, show that $ds = a\cos^2(\theta/3)\,d\theta$ and determine the perimeter of the curve.

39. Find the area that lies inside the curve

$$r = 2a\cos 2\theta$$

and outside the curve $r = a\sqrt{2}$.

40. Sketch the curves

$$r = 2a\cos^2(\theta/2), \quad r = 2a\sin^2(\theta/2),$$

and find the area they have in common.

Find the total area enclosed by each of the curves in Exercises 41 through 47.

41. $r^2 = a^2\cos 2\theta$
42. $r = a(2 - \cos\theta)$
43. $r = a(1 + \cos 2\theta)$
44. $r = 2a\cos\theta$
45. $r = 2a\sin 3\theta$
46. $r^2 = 2a^2\sin 3\theta$
47. $r^2 = 2a^2\cos^2(\theta/2)$

48. Find the area that is inside the cardioid

$$r = a(1 + \sin\theta)$$

and outside the circle $r = a\sin\theta$.

VECTORS AND PARAMETRIC EQUATIONS

CHAPTER 12

12.1 PARAMETRIC EQUATIONS IN KINEMATICS

In Newtonian mechanics, the motion of a particle in a plane is usually described by means of a pair of differential equations expressing Newton's second law of motion,*

$$\mathbf{F} = \frac{d(m\mathbf{v})}{dt}. \tag{1}$$

Here $\mathbf{F}$ is the vector force acting on the particle of mass m at time t and $\mathbf{v}$ is its vector velocity at time t. When the x- and y-components of Eq. (1) are separated, the equivalent equations are

$$F_x = \frac{d(mv_x)}{dt}, \qquad F_y = \frac{d(mv_y)}{dt}. \tag{2}$$

In addition to these laws of motion, we require a knowledge of the position and velocity of the particle at some given instant. Then the position of the particle at all later instants is determined by the prescribed forces. The result of integrating the second-order differential equations (2) subject to the given initial conditions usually yields equations that determine the x- and y-coordinates of the particle as functions of t:

$$x = f(t), \quad y = g(t). \tag{3}$$

These equations are called *parametric equations* of the path of the particle and t is called the *parameter*. Equations (3) contain more information about the motion of the particle than does the *Cartesian equation*

$$y = F(x) \tag{4}$$

that one obtains from (3) by eliminating t. The parametric equations tell *where* the particle goes and *when* it gets to any given place, whereas the Cartesian equation tells only the curve along which the particle travels.

---

* Vectors are indicated by bold-faced Roman letters. In handwritten work, it is customary to indicate them by a small arrow over each letter which represents a vector.

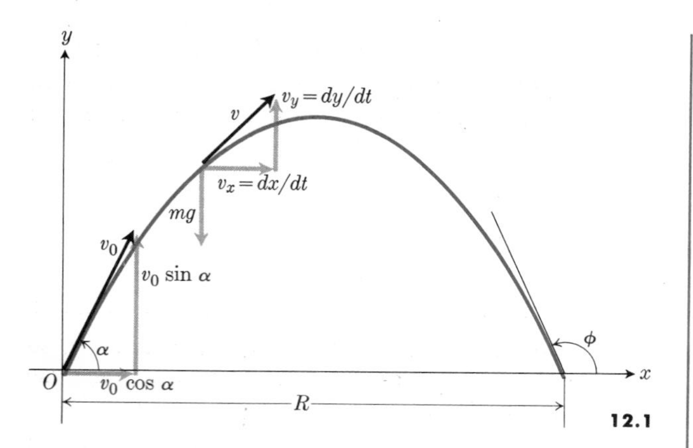

12.1

Example. We shall illustrate these ideas by determining the path of a projectile. Suppose the projectile is fired with an initial velocity v_0 at an angle of elevation α. Assuming that gravity is the only force acting on the projectile, we shall find its motion.

We introduce coordinate axes with origin at the point where the projectile begins its motion, as in Fig. 12.1. The initial conditions may then be taken to be

$$t = 0: \qquad x = 0, \qquad y = 0;$$

$$\frac{dx}{dt} = v_0 \cos \alpha, \qquad \frac{dy}{dt} = v_0 \sin \alpha. \tag{5}$$

The force components at time t, with appropriate sign, are

$$F_x = 0, \quad F_y = -mg,$$

so that the differential equations (2) become

$$0 = m\frac{d^2x}{dt^2}, \tag{6a}$$

$$-mg = m\frac{d^2y}{dt^2}. \tag{6b}$$

Each of Eqs. (6) requires two integrations. This introduces a total of four constants of integration. These may be evaluated by using the initial conditions (5). Thus by integration one obtains from (6a)

$$\frac{d^2x}{dt^2} = 0, \quad \frac{dx}{dt} = c_1, \quad x = c_1t + c_2, \tag{7a}$$

and from (6b)

$$\frac{d^2y}{dt^2} = -g, \qquad \frac{dy}{dt} = -gt + c_3,$$
$$y = -\tfrac{1}{2}gt^2 + c_3t + c_4. \tag{7b}$$

From the initial conditions, we find

$$c_1 = v_0 \cos \alpha, \quad c_2 = 0,$$
$$c_3 = v_0 \sin \alpha, \quad c_4 = 0. \tag{7c}$$

The position of the projectile at time t (seconds, say, after firing) is given by

$$x = (v_0 \cos \alpha)t, \tag{8a}$$

$$y = -\tfrac{1}{2}gt^2 + (v_0 \sin \alpha)t. \tag{8b}$$

For a given angle of elevation α and a given muzzle velocity v_0, the position of the projectile at any time may be determined from the parametric equations (8). The equations may be used to answer such questions as the following:

1. How high does the projectile rise vertically?
2. How far does the projectile travel horizontally?
3. What is the range of the projectile (horizontally) in terms of the angle of elevation? What angle produces the maximum range?

The projectile will attain maximum height when its y velocity component is zero; that is, when

$$\frac{dy}{dt} = -gt + v_0 \sin \alpha = 0$$

or when

$$t = t_m = \frac{v_0 \sin \alpha}{g}.$$

Then

$$y_{\max} = -\tfrac{1}{2}g(t_m)^2 + (v_0 \sin \alpha)t_m = \frac{(v_0 \sin \alpha)^2}{2g}.$$

To find the horizontal range of the projectile, we find the time when it strikes the ground, that is, when $y = 0$. From (8b), we have

$$y = t(-\tfrac{1}{2}gt + v_0 \sin \alpha) = 0$$

when

$$t = 0 \quad \text{or} \quad t = \frac{2v_0 \sin \alpha}{g} = 2t_m.$$

Clearly, $t = 0$ corresponds to the instant when the projectile is fired, so that $t = 2t_m$ gives the range $x = R$:

$$R = v_0 \cos\alpha \, \frac{2v_0 \sin\alpha}{g}$$
$$= \frac{v_0^2}{g} \sin 2\alpha.$$

For a given muzzle velocity, this shows that the maximum range is obtained when $\sin 2\alpha = 1$, that is, when $\alpha = 45°$.

The slope of the path of the projectile at any point may also be found directly from Eqs. (8):

$$\frac{dy}{dx} = \frac{dy/dt}{dx/dt} = \frac{-gt + v_0 \sin\alpha}{v_0 \cos\alpha}.$$

At $y = 0$, $x = R$, we found

$$t = \frac{2v_0 \sin\alpha}{g},$$

and hence

$$\frac{dy}{dx} = \frac{-2v_0 \sin\alpha + v_0 \sin\alpha}{v_0 \cos\alpha}$$
$$= -\tan\alpha.$$

This shows that the angle ϕ (Fig. 12.1) is $\pi - \alpha$, so that the projectile returns to the earth at the same angle as that at which it was fired.

Finally, the Cartesian equation of the path is readily obtained from Eqs. (8); for we need only substitute

$$t = \frac{x}{v_0 \cos\alpha}$$

from (8a) into (8b) to eliminate t and obtain

$$y = -\left(\frac{g}{2v_0^2 \cos^2\alpha}\right)x^2 + (\tan\alpha)x. \tag{9}$$

Since this equation is linear in y and quadratic in x, it represents a *parabola*. Thus the path of a projectile *in vacuo* (neglecting air resistance) is a parabola.

When air resistance is taken into account, the differential equations obtained are too complicated for an analytic solution, but they can be integrated numerically using high-speed computers. Against a moving target, in particular, the *time* is also of great importance, so that equations or tables which give x and y in terms of t are preferred over the Cartesian form.

EXERCISES

In Exercises 1 through 4, the projectile is assumed to obey the laws of motion discussed above, in which air resistance is neglected.

1. Find two values of the angle of elevation that will enable a projectile to reach a target on the same level as the gun and 25,000 feet distant from it if the initial velocity is 1000 ft/sec. Determine the times of flight corresponding to these two angles.
2. Show that doubling of the initial velocity of a projectile multiplies both the maximum height and the range by a factor of four.
3. Show that a projectile attains three-quarters of its maximum height in one-half the time required to reach that maximum.
4. Suppose a target moving at the constant rate of a ft/sec is level with and b ft away from a gun at the instant the gun is fired. If the target moves in a horizontal line directly away from the gun, show that the muzzle velocity v_0 and angle of elevation α must satisfy the equation

$$v_0^2 \sin 2\alpha - 2a\, v_0 \sin\alpha - bg = 0,$$

if the projectile is to strike the target.

In Exercises 5 through 7, find parametric equations and sketch the curve described by the point $P(x, y)$ for $t \geq 0$ if its coordinates satisfy the given differential equations and initial conditions.

5. $\dfrac{dx}{dt} = x, \quad \dfrac{dy}{dt} = -x^2; \quad t = 0, x = 1, y = -4$
6. $\dfrac{dx}{dt} = y, \quad \dfrac{dy}{dt} = y^2; \quad t = 0, x = 0, y = 1$
7. $\dfrac{dx}{dt} = \sqrt{1 - x^2}, \quad \dfrac{dy}{dt} = x^2;$
   $t = 0, x = 0, y = 1$

12.2 PARAMETRIC EQUATIONS IN ANALYTIC GEOMETRY

The solutions of differential equations of motion are not the only ways in which parametric equations arise. For example, we have frequently had occasion to use equations such as

$$x = a \cos \theta, \quad y = a \sin \theta \tag{1}$$

to represent a circle. Here θ is the parameter and Eqs. (1) are parametric equations of the circle (Fig. 12.2). Similarly

$$x = a \cos \phi, \quad y = b \sin \phi \tag{2}$$

are parametric equations of an ellipse whose Cartesian equation is

$$\frac{x^2}{a^2} + \frac{y^2}{b^2} = 1.$$

Example 1. The parabola

$$y^2 = 4px \tag{3}$$

can be parametrized in several ways. One method is to use as parameter the slope

$$t = dy/dx$$

of the tangent to the curve at (x, y) (see Fig. 12.3a). Since

$$2y \frac{dy}{dx} = 4p, \quad \text{or} \quad \frac{dy}{dx} = \frac{2p}{y},$$

the parametric equations in this case are

$$y = \frac{2p}{t}, \quad x = \frac{p}{t^2}. \tag{4}$$

If we use the parameter defined (Fig. 12.3b) by $m = y/x$, we have

$$y^2 = m^2x^2 \quad \text{and} \quad y^2 = 4px,$$

which lead to

$$x = \frac{4p}{m^2}, \quad y = \frac{4p}{m} \tag{5}$$

as the parametric equations.

Sometimes the parametric equations of a locus and the Cartesian equation are not coextensive.

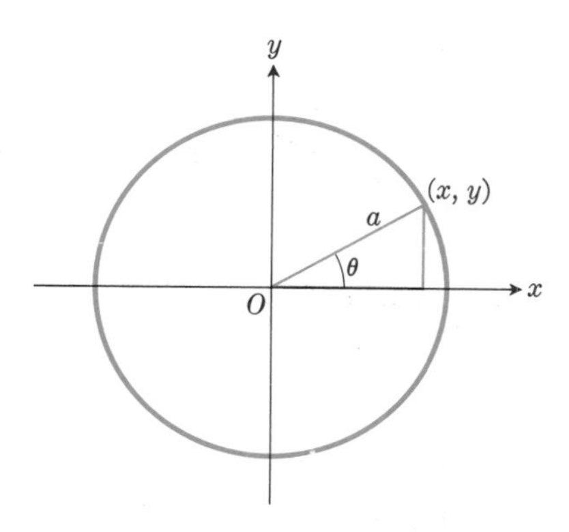

12.2

Example 2. Suppose the parametric equations of a curve are

$$x = \cosh \theta, \quad y = \sinh \theta. \tag{6}$$

Then the hyperbolic identity

$$\cosh^2 \theta - \sinh^2 \theta = 1$$

enables us to eliminate θ and write

$$x^2 - y^2 = 1 \tag{7}$$

as the Cartesian equation of the curve. Closer scrutiny, however, shows that Eq. (7) *includes too much.* For $x = \cosh \theta$ is never less than unity, so the parametric equations represent a curve lying wholly to the right of the y-axis, whereas the Cartesian equation (7) represents both the right- and left-hand branches of the hyperbola (Fig. 12.4). The left-hand branch could be excluded by taking only positive values of x. That is,

$$x = \sqrt{1 + y^2} \tag{8}$$

does represent the same curve as (6).

Example 3. Consider the curve whose parametric equations are

$$x = \cos 2\theta, \quad y = \cos \theta. \tag{9}$$

This curve is certainly restricted to lie within the square that is determined by the conditions

$$-1 \leq x \leq 1, \quad -1 \leq y \leq 1.$$

But we eliminate θ as follows:

$$x = \cos 2\theta = 2 \cos^2 \theta - 1 = 2y^2 - 1.$$

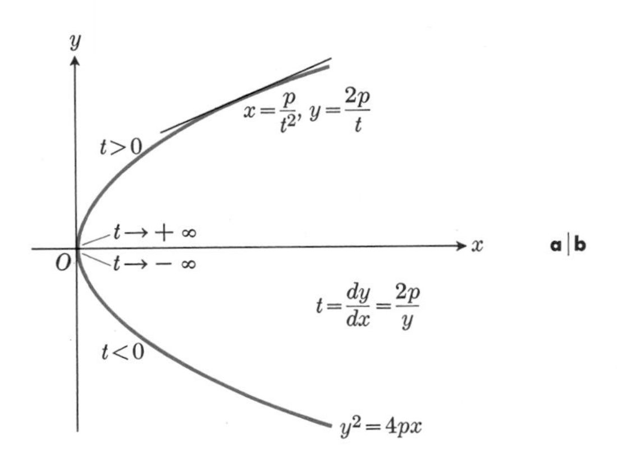

a | b

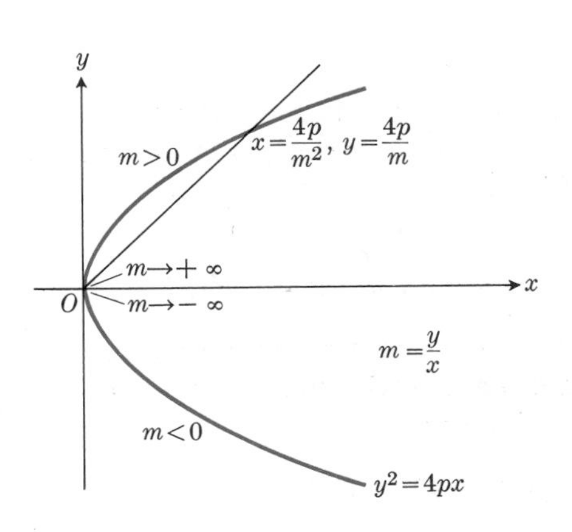

12.3

Thus every point on the locus described by Eqs. (9) also lies on the curve given by

$$x = 2y^2 - 1 \tag{10}$$

and, in addition, is restricted by

$$|x| \leq 1, \quad |y| \leq 1.$$

If we omit these restrictions, Eq. (10) represents the complete parabola

$$y^2 = \tfrac{1}{2}(x + 1)$$

with vertex at $(-1, 0)$ and opening to the right, as shown in Fig. 12.5. The parametric equations, on the other hand, represent only the arc ABC. Referring to Eqs. (9), we readily see that the point starts at $A(1, 1)$ when $\theta = 0$, moves along AB to $B(-1, 0)$ as θ varies from 0 to $\pi/2$, and continues to $C(1, -1)$ as θ increases to π. As θ varies from π to 2π, the point traverses the arc CBA back to A. Since both x and y are periodic functions of θ of period 2π, further variations in θ result in retracing the same portion of the parabola.

The following problem illustrates further how parametric equations may be derived from geometric properties of a curve. Suppose that a wheel of radius a rolls along a horizontal straight line without slipping: Find the locus described by a point on a spoke of the wheel at distance b from its center. Such a curve is called a *trochoid*. When $b = a$, the tracing point is on the circumference of the wheel, and the trochoid in this case is called a *cycloid*.

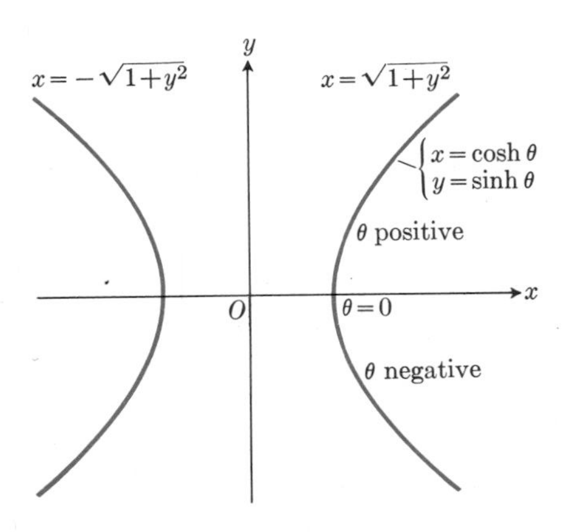

12.4

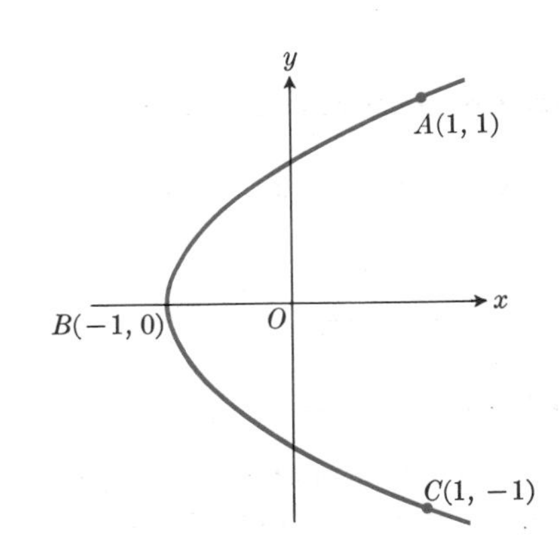

12.5

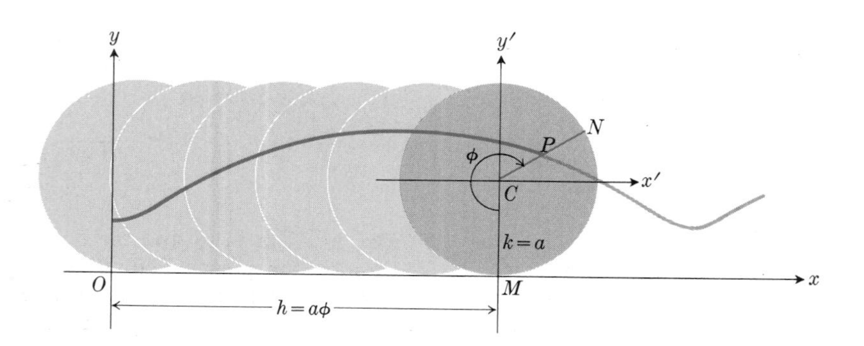

12.6

In Fig. 12.6, we take the x-axis as the line the wheel rolls along, with the y-axis through a low point of the trochoid. Point P describes the curve in question. It is customary to use the angle ϕ through which CP has rotated as the parameter. We introduce $x'y'$-axes parallel to the xy-axes but having their origin at the center of the circle C (Fig. 12.7). If (h, k) are the coordinates of C relative to the xy-axes, then the xy-coordinates of P are related to its $x'y'$-coordinates as follows:

$$x = h + x', \quad y = k + y'. \tag{11}$$

From the fact that the circle rolls along the x-axis without slipping, we see that the distance OM that the wheel has moved horizontally is just equal to the arc $MN = a\phi$. (Roll the wheel back; then N will fall at the origin O.) The xy-coordinates of C are therefore

$$h = a\phi, \quad k = a. \tag{12}$$

From Fig. 12.7, we may immediately read

$$x' = b \cos \theta, \quad y' = b \sin \theta,$$

or, since

$$\theta = \frac{3\pi}{2} - \phi,$$

$$x' = -b \sin \phi, \quad y' = -b \cos \phi. \tag{13}$$

We substitute these results and Eqs. (12) into Eqs. (11) and obtain

$$x = a\phi - b \sin \phi, \quad y = a - b \cos \phi \tag{14}$$

as parametric equations of the trochoid. The cycloid (Fig. 12.8a)

$$x = a(\phi - \sin \phi), \quad y = a(1 - \cos \phi), \tag{15}$$

obtained from (14) by taking $b = a$, is the most important special case of a trochoid.

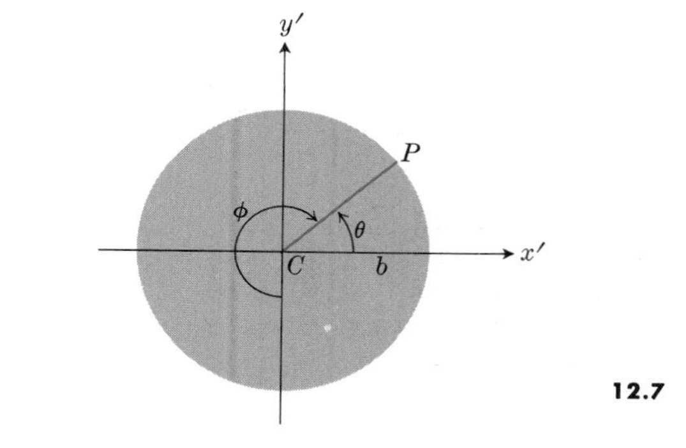

12.7

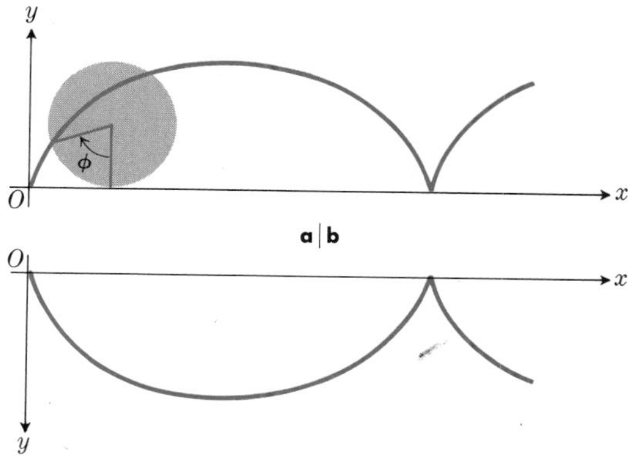

12.8 Cycloid: $x = a(\phi - \sin \phi)$, $y = a(1 - \cos \phi)$.

Note. If we reflect the cycloid and the y-axis across the x-axis, Eqs. (15) still apply and the resulting curve (Fig. 12.8b) has several interesting properties, one of which we shall now discuss without proof. The proofs belong to a branch of mathematics known as the calculus of variations. Much of the fundamental theory of this subject is attributable to the Bernoulli brothers, John and James, who were friendly rivals and stimulated each other with mathematical problems in the form of challenges. One of these, the brachistochrone problem, was: Among all smooth curves joining two given points, find that one along which a bead might slide, subject only to the force of gravity, *in the shortest time.* (*Brachistochrone* is derived from two Greek words that together mean "shortest time.")

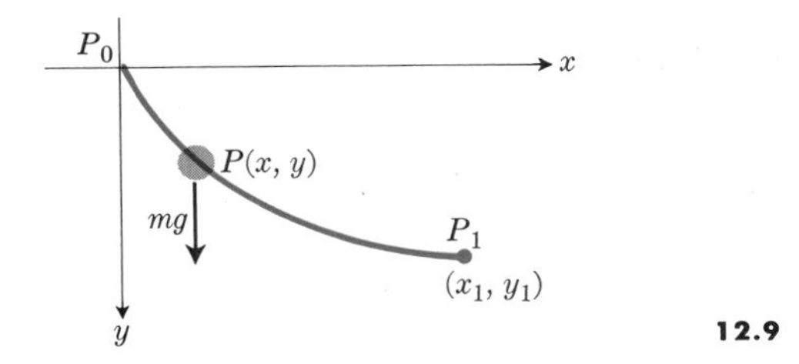

12.9

The two points, labeled P_0 and P_1 in Fig. 12.9, may be taken to lie at the origin and at (x_1, y_1) respectively in a vertical plane. We can formulate the problem in mathematical terms as follows. The kinetic energy of the bead at the start is zero, since its velocity is zero. The work done by gravity in moving the particle from $(0, 0)$ to any point (x, y) is mgy and this must be equal to the change in kinetic energy; that is,

$$mgy = \tfrac{1}{2}mv^2 - \tfrac{1}{2}m(0)^2.$$

Thus the velocity

$$v = ds/dt$$

that the particle has when it reaches $P(x, y)$ is

$$v = \sqrt{2gy}.$$

That is,

$$\frac{ds}{dt} = \sqrt{2gy},$$

or

$$dt = \frac{ds}{\sqrt{2gy}} = \frac{\sqrt{1 + \left(\frac{dy}{dx}\right)^2}\,dx}{\sqrt{2gy}}.$$

The time t_1 required for the bead to slide from P_0 to P_1 depends on the particular curve $y = f(x)$ along which it moves; it is given by

$$t_1 = \int_0^{x_1} \sqrt{\frac{1 + (f'(x))^2}{2gf(x)}}\,dx. \tag{16}$$

The problem is *to find the curve* $y = f(x)$ that passes through the points $P_0(0, 0)$ and $P_1(x_1, y_1)$ and minimizes the value of the integral in Eq. (16).

At first sight, one might guess that the straight line joining P_0 and P_1 would also yield the shortest time, but a moment's reflection will cast some doubt on this conjecture. For there may be some gain in time by having the particle start to fall vertically at first, thereby building up its velocity more quickly than if it were to slide along an inclined path. With this increased velocity, one may be able to afford to travel over a longer path and still reach P_1 in a shorter time. The solution of the problem is beyond the present book, but the brachistochrone curve is actually an arc of a cycloid through P_0 and P_1, having a cusp at the origin. [The interested reader is referred to the book *Calculus of Variations* by G. A. Bliss, The Mathematical Association of America (1925), for a discussion of this problem and others belonging to this branch of mathematics.]

If we write Eq. (16) in the equivalent form

$$t_1 = \int \sqrt{\frac{dx^2 + dy^2}{2gy}}$$

and then substitute Eqs. (15) into this, we obtain

$$t_1 = \int_{\phi=0}^{\phi_1} \sqrt{\frac{a^2(2 - 2\cos\phi)}{2ga(1 - \cos\phi)}}\,d\phi = \phi_1\sqrt{\frac{a}{g}}$$

as the time required for the particle to slide from P_0 to P_1. The time required to reach the bottom of the arc is obtained by taking $\phi_1 = \pi$. Now it is a remarkable fact, which we shall soon demonstrate, that the time required to slide along the cycloid from $(0, 0)$ to the lowest point $(a\pi, 2a)$ is the same as the time required for the particle, starting from rest, to slide from *any intermediate point* of the arc, say (x_0, y_0), to $(a\pi, 2a)$. For the latter case, one has

$$v = \sqrt{2g(y - y_0)}$$

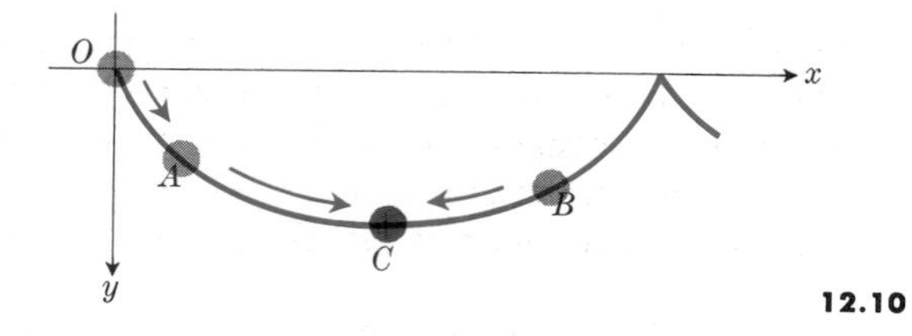

12.10

as the velocity at $P(x, y)$, and the time required is

$$T = \int_{\phi_0}^{\pi} \sqrt{\frac{a^2(2 - 2\cos\phi)}{2ag(\cos\phi_0 - \cos\phi)}}\, d\phi$$

$$= \sqrt{\frac{a}{g}} \int_{\phi_0}^{\pi} \sqrt{\frac{1 - \cos\phi}{\cos\phi_0 - \cos\phi}}\, d\phi$$

$$= \sqrt{\frac{a}{g}} \int_{\phi_0}^{\pi} \sqrt{\frac{2\sin^2\frac{\phi}{2}}{\left(2\cos^2\frac{\phi_0}{2} - 1\right) - \left(2\cos^2\frac{\phi}{2} - 1\right)}}\, d\phi$$

$$= 2\sqrt{\frac{a}{g}} \left[-\sin^{-1} \frac{\cos\frac{\phi}{2}}{\cos\frac{\phi_0}{2}} \right]_{\phi_0}^{\pi}$$

$$= 2\sqrt{\frac{1}{g}}\,(-\sin^{-1} 0 + \sin^{-1} 1) = \pi\sqrt{\frac{a}{g}}.$$

Since this answer is independent of the value of ϕ_0, it follows that the same length of time is required to reach the lowest point on the cycloid no matter where on the arc the particle is released from rest. Thus, in Fig. 12.10, three particles that start at the same time from O, A, and B will reach C simultaneously. In this sense, the cycloid is also a tautochrone (meaning "the same time") as well as being a brachistochrone.

EXERCISES

In Exercises 1 through 20, sketch the graph of the curve described by the point $P(x, y)$ as the parameter t varies over the domain given. Also determine the Cartesian equation of the curve in each case.

1. $x = \cos t, \quad y = \sin t, \quad 0 \le t \le 2\pi$
2. $x = \cos 2t, \quad y = \sin t, \quad 0 \le t \le 2\pi$
3. $x = \sec t, \quad y = \tan t, \quad -\pi/2 < t < \pi/2$
4. $x = 2 + 4\sin t, \quad y = 3 - 2\cos t, \quad 0 \le t \le 2\pi$
5. $x = 2t + 3, \quad y = 4t^2 - 9, \quad -\infty < t < \infty$
6. $x = \cosh t, \quad y = \sinh t, \quad 0 \le t < \infty$
7. $x = 2 + 1/t, \quad y = 2 - t, \quad 0 < t < \infty$
8. $x = t + 1, \quad y = t^2 + 4, \quad 0 \le t < \infty$
9. $x = t^2 + t, \quad y = t^2 - t, \quad -\infty < t < \infty$
10. $x = 3 + 2\,\text{sech}\, t, \quad y = 4 - 3\tanh t,$ $-\infty < t < \infty$
11. Find parametric equations of the semicircle
$$x^2 + y^2 = a^2, \quad y > 0,$$
using as parameter the slope $t = dy/dx$ of the tangent to the curve at (x, y).
12. Find parametric equations of the semicircle
$$x^2 + y^2 = a^2, \quad y > 0,$$
using as parameter the variable θ defined by the equation $x = a\tanh\theta$.
13. Find parametric equations of the circle
$$x^2 + y^2 = a^2,$$
using as parameter the arc length s measured counterclockwise from the point $(a, 0)$ to the point (x, y).
14. Find parametric equations of the catenary
$$y = a\cosh x/a,$$
using as parameter the length of arc s from the point $(0, a)$ to the point (x, y), with the sign of s taken to be the same as the sign of x.
15. If a string wound around a fixed circle is unwound while held taut in the plane of the circle, its end traces an *involute* of the circle. Let the fixed circle be located with its center at the origin O and have radius a. Let the initial position of the tracing point P be $A(a, 0)$ and let the unwound portion of the string PT be tangent to the circle at T. Derive parametric equations of the involute, using the angle AOT as the parameter ϕ.
16. When a circle rolls externally on the circumference of a second, fixed circle, any point P on the circumference of the rolling circle describes an *epicycloid*. Let the fixed circle have its center at the origin O and have radius a. Let the radius of the rolling circle be b and let the initial position of the tracing

point P be $A(a, 0)$. Determine parametric equations of the epicycloid, using as parameter the angle θ from the positive x-axis to the line of centers.

17. When a circle rolls on the inside of a fixed circle any point P on the circumference of the rolling circle describes a *hypocycloid*. Let the fixed circle be $x^2 + y^2 = a^2$, let the radius of the rolling circle be b, and let the initial position of the tracing point P be $A(a, 0)$. Use the angle θ from the positive x-axis to the line of centers as parameter and determine parametric equations of the hypocycloid. In particular, if $b = a/4$, show that

$$x = a \cos^3 \theta, \quad y = a \sin^3 \theta.$$

18. Find the length of one arch of the cycloid

$$x = a(\phi - \sin \phi), \quad y = a(1 - \cos \phi).$$

19. Show that the slope of the cycloid

$$z = a(\phi - \sin \phi), \quad y = a(1 - \cos \phi)$$

is $dy/dx = \cot \phi/2$. In particular, the tangent to the cycloid is vertical when ϕ is 0 or 2π.

20. Show that the slope of the trochoid

$$x = a\phi - b \sin \phi, \quad y = a - b \cos \phi$$

is always finite if $b < a$.

21. There exists a bell-shaped curve that can be constructed as follows: Let C be a circle of radius a having its center at $(0, a)$ on the y-axis. The variable line OA through the origin O intersects the line $y = 2a$ in the point A and intersects the circle in the point B. A point P on the curve is now located by taking the intersection of lines through A and B parallel to the y- and x-axes respectively. Find parametric equations of the curve, using as parameter the angle θ from the x-axis to the line OA. Also find the Cartesian equation.

12.3 VECTOR COMPONENTS AND THE UNIT VECTORS i AND j

Some physical quantities are completely determined when their magnitudes, in terms of specific units, are given. Such quantities are called *scalars*, and are exemplified by mass and length. Other quantities, like forces and velocities, in which the direction as well as the magnitude is important, are called *vectors*. It is customary to represent a vector by a directed line segment whose direction represents the direction of the vector and whose length (in terms of some chosen unit of length) represents the magnitude.

The most satisfactory algebra of vectors is based on a representation of each vector in terms of its components parallel to the axes of a Cartesian coordinate system. This is accomplished by using the same unit of length on both the x- and y-axes, with unit vectors along these axes used as basic components in terms of which every vector in the plane may be expressed. Thus, in Fig. 12.11, the vector from $(0, 0)$ to $(1, 0)$ is the unit vector $\mathbf{i}$, while $\mathbf{j}$ is the unit vector from $(0, 0)$ to $(0, 1)$. Then $a\mathbf{i}$, a being a scalar, represents a vector parallel to the x-axis, having magnitude $|a|$ and pointing to the right if a is positive, and to the left if a is negative. Similarly, $b\mathbf{j}$ is a vector parallel to the y-axis and having the same sense as $\mathbf{j}$ or the opposite sense, depending on the sign of b.

We shall ordinarily deal with "free vectors," meaning that a vector is free to move about under parallel displacements. We say that two vectors are *equal* provided they have the same direction and the same magnitude. The same condition may be expressed algebraically by saying that

$$a\mathbf{i} + b\mathbf{j} = a'\mathbf{i} + b'\mathbf{j} \quad \text{if and only if} \quad a = a', b = b'.$$

That is, two vectors are equal if and only if their corresponding components are equal. Thus, in Fig. 12.11, the vector $\overrightarrow{AB}$ is equal to $a\mathbf{i}$, and the vector $\overrightarrow{OP}$ from $(0, 0)$ to the point $(a, 0)$ on the x-axis is also equal to $a\mathbf{i}$.

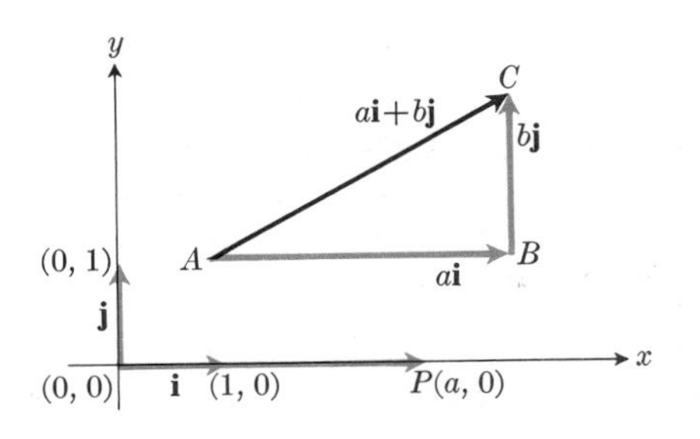

12.11

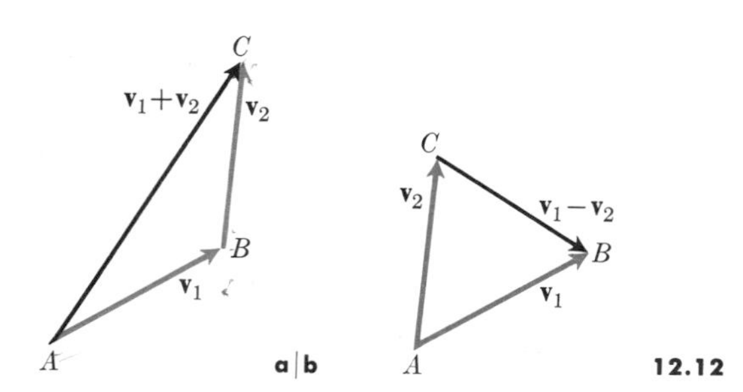

a | b 12.12

Addition

Two vectors $\mathbf{v}_1$ and $\mathbf{v}_2$ are *added* by drawing a vector $\mathbf{v}_1$, say from A to B in Fig. 12.12(a), and then a vector equal to $\mathbf{v}_2$ starting from the terminal point of $\mathbf{v}_1$; thus $\mathbf{v}_2 = \overrightarrow{BC}$ in Fig. 12.12(a). The sum $\mathbf{v}_1 + \mathbf{v}_2$ is then the vector from the starting point A of $\mathbf{v}_1$ to the terminal point C of $\mathbf{v}_2$:

$$\mathbf{v}_1 = \overrightarrow{AB}, \quad \mathbf{v}_2 = \overrightarrow{BC},$$

$$\mathbf{v}_1 + \mathbf{v}_2 = \overrightarrow{AB} + \overrightarrow{BC} = \overrightarrow{AC}.$$

If we apply this principle to the vectors $a\mathbf{i}$ and $b\mathbf{j}$ in Fig. 12.11, we see that $a\mathbf{i} + b\mathbf{j}$ is the vector hypotenuse of a right triangle whose vector sides are $a\mathbf{i}$ and $b\mathbf{j}$ respectively. If the vectors $\mathbf{v}_1$ and $\mathbf{v}_2$ are given in terms of components

$$\mathbf{v}_1 = a_1\mathbf{i} + b_1\mathbf{j},$$

$$\mathbf{v}_2 = a_2\mathbf{i} + b_2\mathbf{j},$$

then

$$\mathbf{v}_1 + \mathbf{v}_2 = (a_1 + a_2)\mathbf{i} + (b_1 + b_2)\mathbf{j} \tag{1}$$

has x- and y-components obtained by adding the x- and y-components of $\mathbf{v}_1$ and $\mathbf{v}_2$.

Subtraction

To *subtract* one vector $\mathbf{v}_2$ from another vector $\mathbf{v}_1$ geometrically, we draw them both from a common initial point and then draw the vector from the tip of $\mathbf{v}_2$ to the tip of $\mathbf{v}_1$. Thus in Fig. 12.12(b), we have

$$\mathbf{v}_1 = \overrightarrow{AB}, \quad \mathbf{v}_2 = \overrightarrow{AC},$$

and

$$\mathbf{v}_1 - \mathbf{v}_2 = -\mathbf{v}_2 + \mathbf{v}_1 = \overrightarrow{CA} + \overrightarrow{AB} = \overrightarrow{CB}.$$

In terms of components, vector subtraction follows the simple algebraic law

$$\mathbf{v}_1 - \mathbf{v}_2 = (a_1 - a_2)\mathbf{i} + (b_1 - b_2)\mathbf{j},$$

which says that corresponding components are subtracted.

Multiplication by scalars

The algebraic operation of multiplying a vector $\mathbf{v} = a\mathbf{i} + b\mathbf{j}$ by a scalar c is also simple:

$$c(a\mathbf{i} + b\mathbf{j}) = (ca)\mathbf{i} + (cb)\mathbf{j}.$$

Geometrically, $c\mathbf{v}$ is a vector whose length is $|c|$ times the length of $\mathbf{v}$ and whose direction is the same as that of $\mathbf{v}$ if c is positive, and is opposite to that of $\mathbf{v}$ if c is negative. Thus $3\mathbf{v}$ has the same sense as $\mathbf{v}$ and is three times as long; while $-2\mathbf{v}$ is twice as long as $\mathbf{v}$ but is directed in the opposite sense (Fig. 12.13).

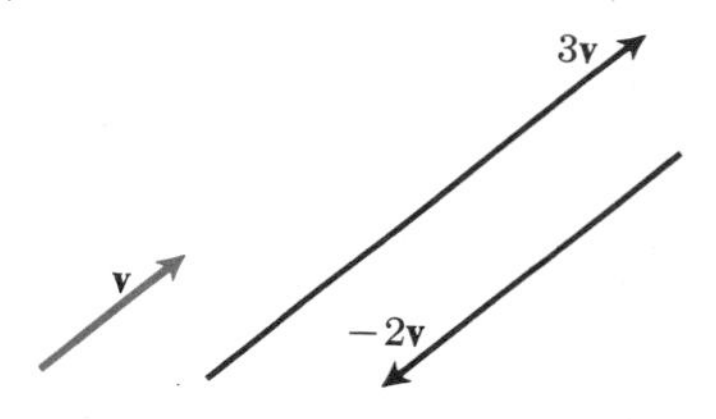

12.13

The unit vectors $\mathbf{i}$ and $\mathbf{j}$ serve the useful purpose of allowing us to keep the components separated from one another when we operate on the vectors algebraically.

The length of the vector $\mathbf{v}$ is usually denoted by $|\mathbf{v}|$, which is read "magnitude of v." Reference to Fig. 12.11 shows that $\mathbf{v} = a\mathbf{i} + b\mathbf{j}$ is the hypotenuse of a right triangle whose legs have lengths $|a|$ and $|b|$ respectively. Hence we may apply the theorem of Pythagoras to obtain

$$|a\mathbf{i} + b\mathbf{j}| = \sqrt{a^2 + b^2}. \tag{2}$$

Zero vector

Any vector whose length is zero is called the *zero vector*, $\mathbf{0}$. The vector

$$a\mathbf{i} + b\mathbf{j} = \mathbf{0},$$

if and only if

$$a = b = 0.$$

Unit vector

Any vector $\mathbf{u}$ whose length is equal to the unit of length used along the coordinate axes is called a *unit vector*. If $\mathbf{u}$ is a unit vector obtained by rotating the unit vector $\mathbf{i}$ through an angle θ in the positive direction, then (see Fig. 12.14) $\mathbf{u}$ has a horizontal component

$$u_x = \cos\theta$$

and a vertical component

$$u_y = \sin\theta,$$

so that

$$\mathbf{u} = \mathbf{i}\cos\theta + \mathbf{j}\sin\theta. \tag{3}$$

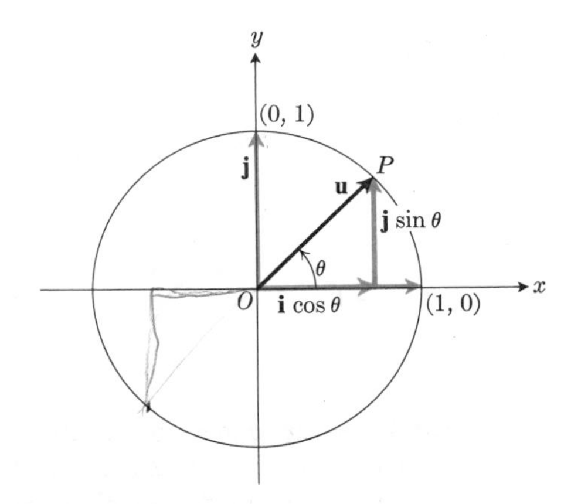

12.14

If we allow the angle θ in Eq. (3) to vary from 0 to 2π, say, then the point P traces out the unit circle $x^2 + y^2 = 1$ precisely once in the counterclockwise direction.

Example. As an illustration of how vector methods may be applied to problems in geometry, we shall derive parametric equations of the involute of a circle. This is the locus described by the endpoint P of a string that is held taut as it is unwound from the circle. In Fig. 12.15, the origin is taken at the center of the circle whose radius is a. The x-axis passes through the point A where P started when the string was all wound onto the circle. The figure illustrates the position of P after a length of string

$$TP = \text{arc } AT = a\theta$$

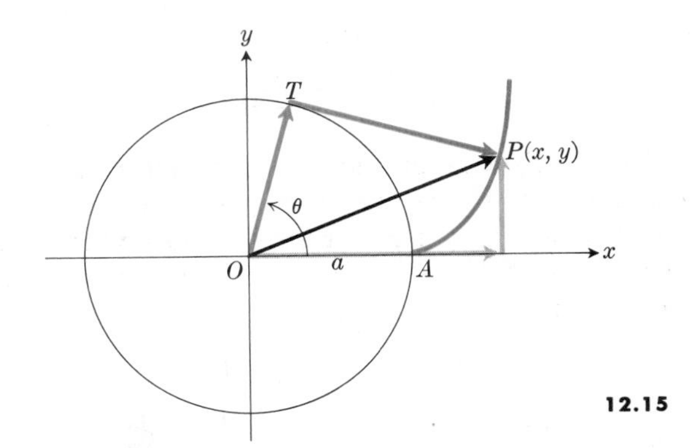

12.15

has been unwound. The line TP is drawn tangent to the circle, since the string is held taut as it is unwound. Now the vector from the origin O to the point P is simply

$$\overrightarrow{OP} = \mathbf{i}x + \mathbf{j}y. \tag{4}$$

On the other hand, the same vector is also given by

$$\overrightarrow{OP} = \overrightarrow{OT} + \overrightarrow{TP}, \tag{5}$$

and if we express the right-hand side in terms of the parameter θ, we may equate the two right-hand sides of (4) and (5) and separate the components to obtain x and y in terms of θ. Now

$$\overrightarrow{OT} = a\mathbf{u}_1 \quad \text{and} \quad \overrightarrow{TP} = (a\theta)\mathbf{u}_2,$$

where $\mathbf{u}_1$ and $\mathbf{u}_2$ are unit vectors making angles θ and $\theta + 3\pi/2$ respectively with the x-axis. (See Fig. 12.16.)

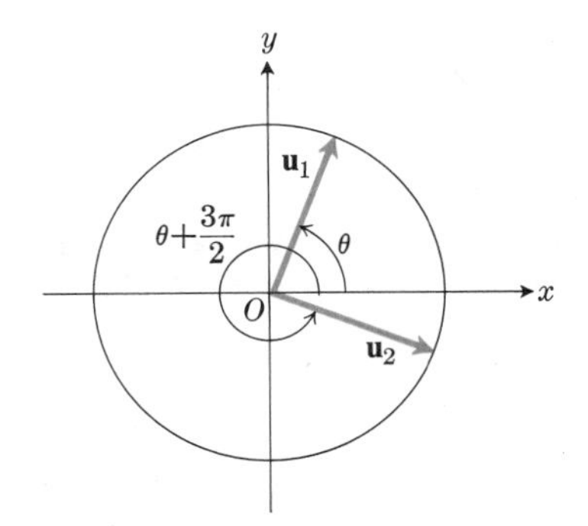

12.16

Then, applying (3), we have

$$\mathbf{u}_1 = \mathbf{i}\cos\theta + \mathbf{j}\sin\theta$$

and

$$\mathbf{u}_2 = \mathbf{i}\cos\left(\theta + \frac{3\pi}{2}\right) + \mathbf{j}\sin\left(\theta + \frac{3\pi}{2}\right)$$
$$= \mathbf{i}\sin\theta - \mathbf{j}\cos\theta.$$

Therefore

$$\begin{aligned}\overrightarrow{OP} &= a\mathbf{u}_1 + (a\theta)\mathbf{u}_2 \\ &= a(\mathbf{i}\cos\theta + \mathbf{j}\sin\theta) + a\theta(\mathbf{i}\sin\theta - \mathbf{j}\cos\theta) \\ &= a(\cos\theta + \theta\sin\theta)\mathbf{i} + a(\sin\theta - \theta\cos\theta)\mathbf{j}.\end{aligned}$$

We equate this with $x\mathbf{i} + y\mathbf{j}$ and, since corresponding components must be equal, we obtain the parametric equations

$$x = a(\cos\theta + \theta\sin\theta),$$
$$y = a(\sin\theta - \theta\cos\theta). \tag{6}$$

EXERCISES

In Exercises 1 through 10, express each of the vectors in the form $a\mathbf{i} + b\mathbf{j}$. Indicate all quantities graphically.

1. $\overrightarrow{P_1P_2}$, if P_1 is the point (1, 3) and P_2 is the point (2, —1)
2. $\overrightarrow{OP_3}$, if O is the origin and P_3 is the midpoint of the vector $\overrightarrow{P_1P_2}$ joining $P_1(2, -1)$ and $P_2(-4, 3)$
3. The vector from the point $A(2, 3)$ to the origin
4. The sum of the vectors $\overrightarrow{AB}$ and $\overrightarrow{CD}$, given the four points $A(1, -1)$, $B(2, 0)$, $C(-1, 3)$, and $D(-2, 2)$
5. A unit vector making an angle of 30° with the positive x-axis
6. The unit vector obtained by rotating $\mathbf{j}$ through 120° in the clockwise direction
7. A unit vector having the same direction as the vector $3\mathbf{i} - 4\mathbf{j}$
8. A unit vector tangent to the curve $y = x^2$ at the point (2, 4)
9. A unit vector normal to the curve $y = x^2$ at the point $P(2, 4)$ and pointing from P toward the concave side of the curve (that is, an "inner" normal)
10. A unit vector tangent to the involute of a circle whose parametric equations are given in Eq. (6)

Find the lengths of each of the following vectors and the angle that each makes with the positive x-axis.

11. $\mathbf{i} + \mathbf{j}$
12. $2\mathbf{i} - 3\mathbf{j}$
13. $\sqrt{3}\,\mathbf{i} + \mathbf{j}$
14. $-2\mathbf{i} + 3\mathbf{j}$
15. $5\mathbf{i} + 12\mathbf{j}$
16. $-5\mathbf{i} - 12\mathbf{j}$
17. Use vector methods to determine parametric equations for the trochoid of Fig. 12.6, by taking

$$\mathbf{R} = \overrightarrow{OP} = \overrightarrow{OM} + \overrightarrow{MC} + \overrightarrow{CP}.$$

18. Let A, B, C, D be the vertices, in order, of a quadrilateral. Let A', B', C', D' be the midpoints of the sides AB, BC, CD, and DA, in order. Prove that $A'B'C'D'$ is a parallelogram.
Hint. First show that $\overrightarrow{A'B'} = \overrightarrow{D'C'} = \frac{1}{2}\overrightarrow{AC}$.
19. Using vectors, show that the diagonals of a parallelogram bisect each other.
Method. Let A be one vertex and let M and N be the midpoints of the diagonals. Then show that $\overrightarrow{AM} = \overrightarrow{AN}$.

12.4 SPACE COORDINATES

Cartesian coordinates

In Fig. 12.17, a system of mutually orthogonal coordinate axes, Ox, Oy, and Oz, is indicated. The system is called *right-handed* if a right-threaded screw pointing along Oz will advance when the blade of the screw driver is twisted from Ox to Oy through an angle, say, of 90°. In the right-handed system shown, the y- and z-axes lie in the plane of the paper and the x-axis points out from the paper. The Cartesian coordinates of a point $P(x, y, z)$ in space may be read from the scales along the coordinate axes by passing planes through P perpendicular to each axis. All points on the x-axis have their y- and z-coordinates both zero; that is, they have the form $(x, 0, 0)$. Points in a plane perpendicular to the z-axis, say, all have the same value for their z-coordinate. Thus, for example, $z = 5$ is an equation satisfied by every point $(x, y, 5)$ lying in

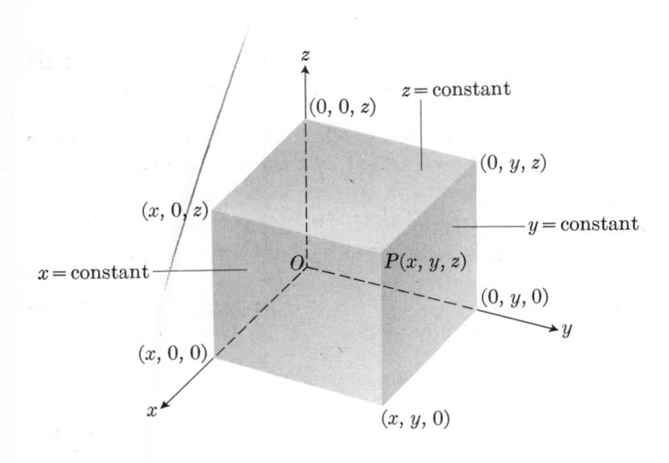

12.17 Cartesian coordinates.

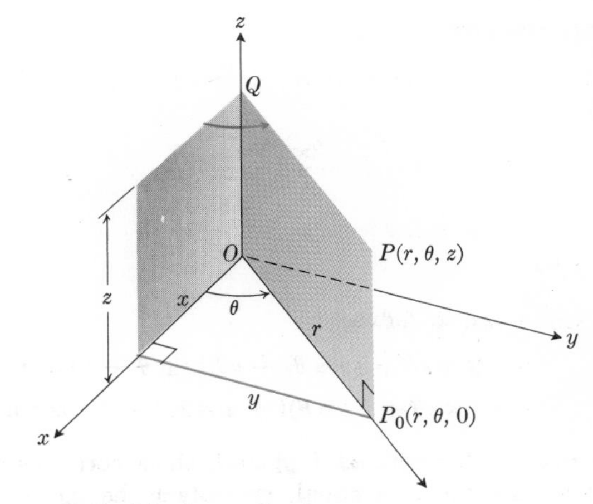

12.18 Cylindrical coordinates.

a plane perpendicular to the z-axis and 5 units above the xy-plane. The three planes

$$x = 2, \quad y = 3, \quad z = 5$$

intersect in the point $P(2, 3, 5)$. The yz-plane is characterized by $x = 0$. The three coordinate planes $x = 0, y = 0, z = 0$ divide the space into eight cells, called *octants.* That octant in which the points (x, y, z) have all three coordinates positive is called the *first octant,* but there is no conventional numbering of the remaining seven octants.

Cylindrical coordinates

It is frequently convenient to use cylindrical coordinates (r, θ, z) to locate a point in space. In particular, cylindrical coordinates are convenient when there is an *axis of symmetry* in a physical problem. Essentially, cylindrical coordinates are just the polar coordinates (r, θ), used instead of (x, y) in the plane $z = 0$, coupled with the z-coordinate (see Fig. 12.18). Cylindrical and Cartesian coordinates are related by the familiar equations

$$\begin{aligned} x &= r\cos\theta, \quad r^2 = x^2 + y^2, \\ y &= r\sin\theta, \quad \tan\theta = y/x, \\ z &= z. \end{aligned} \tag{1}$$

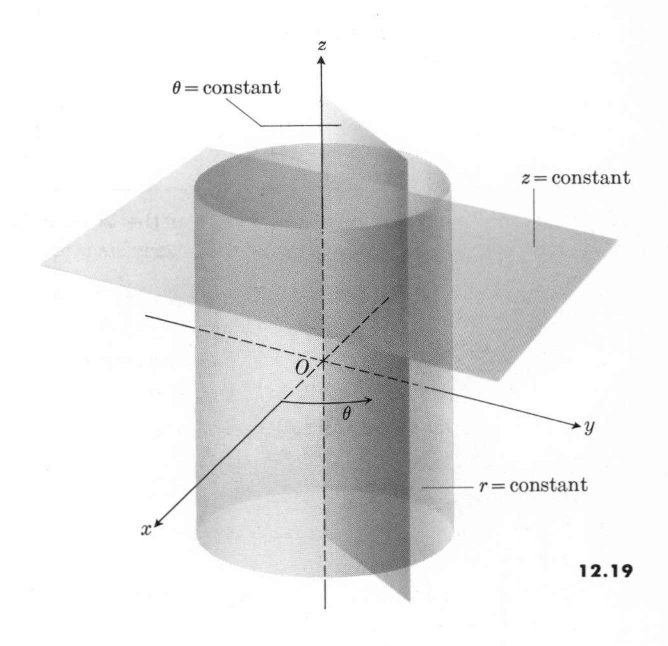

12.19

If we hold r = constant and let θ and z vary, the locus of $P(r, \theta, z)$ is then a right circular cylinder of radius r and axis along Oz. The locus $r = 0$ is just the z-axis itself. The locus θ = constant is a plane containing the z-axis and making an angle θ with the xz-plane (Fig. 12.19).

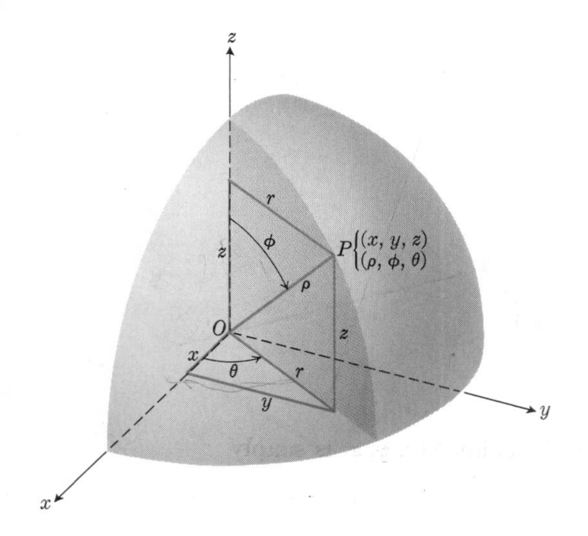

12.20 Spherical coordinates.

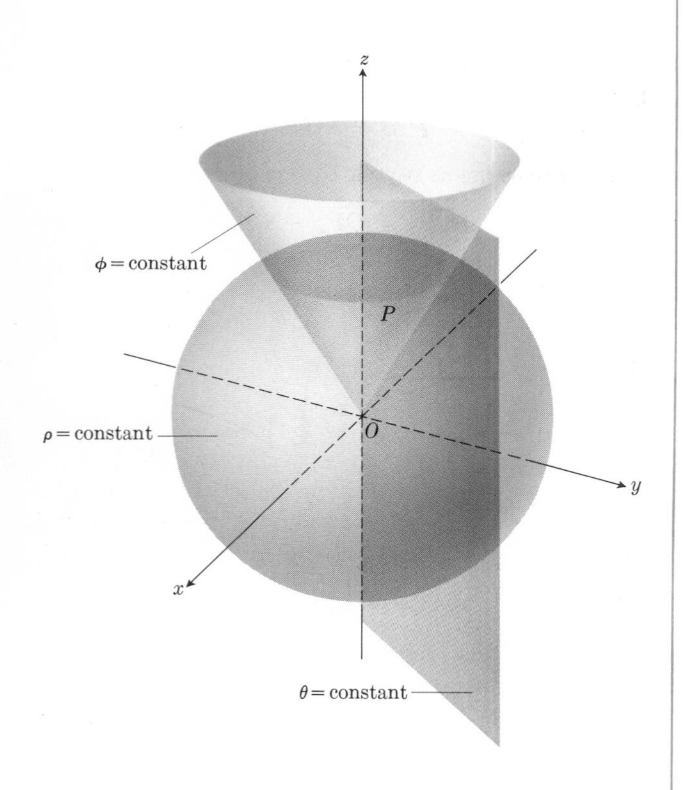

12.21 Loci of spherical coordinates.

Spherical coordinates

Just as cylindrical coordinates are convenient to use when there is an axis of symmetry in a physical problem, so are spherical coordinates useful when there is a point which is a *center* of symmetry. In this case, we would take this center as the origin. The *spherical coordinates* (ρ, ϕ, θ) are illustrated in Fig. 12.20. In the first place,

$$\rho = |OP|$$

is simply the distance from the origin to the point P and is always taken to be greater than or equal to zero. The locus of points $\rho =$ constant (Fig. 12.21) is the surface of a sphere of radius ρ with center at O. The second spherical coordinate, ϕ, is the angle measured down from the z-axis to the line OP. We say that the locus of points $\phi =$ constant is a cone with vertex at O, axis Oz, and generating angle ϕ. (Note that we must broaden our interpretation of the word "cone" in this case to include the xy-plane for which $\phi = \pi/2$ as well as cones with generating angles greater than $\pi/2$.) The third spherical coordinate, θ, is the same as the angle θ in cylindrical coordinates; it is the angle from the xz-plane to the plane through P and the z-axis.

We may read the following relationships between the Cartesian, cylindrical, and spherical coordinate systems from Fig. 12.20:

$$\begin{aligned} r &= \rho \sin \phi, & x &= r \cos \theta, & x &= \rho \sin \phi \cos \theta, \\ z &= \rho \cos \phi, & y &= r \sin \theta, & y &= \rho \sin \phi \sin \theta, \\ \theta &= \theta, & z &= z, & z &= \rho \cos \phi. \end{aligned} \tag{2}$$

Every point in the whole space can be given spherical coordinates restricted to the ranges

$$\rho \geq 0, \quad 0 \leq \phi \leq \pi, \quad 0 \leq \theta < 2\pi. \tag{3}$$

Because of the analogy between the surface of a sphere and the earth's surface, the z-axis is sometimes called the *polar axis*, while ϕ is referred to as *colatitude* and θ is referred to as *longitude*. One also speaks of *meridians*, *parallels*, and the *northern* and *southern hemispheres*.

EXERCISES

In Exercises 1 through 4, describe the locus of points $P(x, y, z)$ that satisfy the given pairs of simultaneous equations. Sketch.

1. $x = \text{constant}, \quad y = \text{constant}$
2. $y = x, \quad z = 5$
3. $x^2 + y^2 = 4, \quad z = -2$
4. $x = 0, \quad \dfrac{y^2}{a^2} + \dfrac{z^2}{b^2} = 1$

In Exercises 5 through 8, describe the locus of points $P(r, \theta, z)$ whose cylindrical coordinates satisfy the given pairs of simultaneous equations. Sketch.

5. $r = 2, \quad z = 3$
6. $\theta = \pi/6, \quad z = r$
7. $r = 3, \quad z = 2\theta$
8. $r = 2\theta, \quad z = 3\theta$

In Exercises 9 through 12, describe the locus of points $P(\rho, \phi, \theta)$ whose spherical coordinates satisfy the given pairs of simultaneous equations. Sketch.

9. $\rho = 5, \quad \theta = \pi/4$
10. $\rho = 5, \quad \phi = \pi/4$
11. $\theta = \pi/4, \quad \phi = \pi/4$
12. $\theta = \pi/2, \quad \rho = 4 \cos \phi$

Translate each of the following equations from the given coordinate system (Cartesian, cylindrical, or spherical) into the forms that are appropriate to the other two systems.

13. $x^2 + y^2 + z^2 = 4$
14. $x^2 + y^2 + z^2 = 4z$
15. $z^2 = r^2$
16. $\rho = 6 \cos \phi$

Describe the following space loci.

17. $x \geq 0$
18. $3 \leq \rho \leq 5$
19. $r \geq 2, \quad \rho \leq 5$
20. $0 \leq \theta \leq \pi/4, \quad 0 \leq \phi \leq \pi/4, \quad \rho \geq 0$
21. $4x^2 + 9y^2 \leq 36$

12.5 VECTORS IN SPACE

The study of solid analytic geometry is greatly facilitated by the use of vectors. The vectors from the origin to the points whose Cartesian coordinates are (1, 0, 0), (0, 1, 0), and (0, 0, 1), respectively, are the basic *unit* vectors, which we denote by **i**, **j**, and **k**. In this notation, the vector from the origin O to the point $P(x, y, z)$ is simply

$$\mathbf{R} = \overrightarrow{OP} = \mathbf{i}x + \mathbf{j}y + \mathbf{k}z. \tag{1}$$

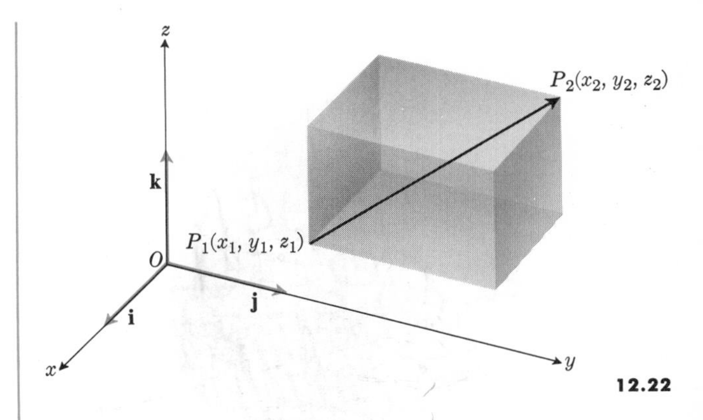

12.22

If $P_1(x_1, y_1, z_1)$ and $P_2(x_2, y_2, z_2)$ are two points in space (Fig. 12.22), then the vector from P_1 to P_2 is the vector sum

$$\overrightarrow{P_1P_2} = \overrightarrow{P_1O} + \overrightarrow{OP_2}.$$

Since

$$\overrightarrow{P_1O} = -\overrightarrow{OP_1},$$

this is the same as saying that

$$\overrightarrow{P_1P_2} = \overrightarrow{OP_2} - \overrightarrow{OP_1},$$

or

$$\overrightarrow{P_1P_2} = \mathbf{i}(x_2 - x_1) + \mathbf{j}(y_2 - y_1) + \mathbf{k}(z_2 - z_1). \tag{2}$$

The length of any vector

$$\mathbf{A} = a\mathbf{i} + b\mathbf{j} + c\mathbf{k}$$

is readily determined by applying the theorem of Pythagoras twice; once to the diagonal of a face of a rectangular box (Fig. 12.23) and then to the diagonal of the box. In the right triangle ABC in Fig. 12.23,

$$|\overrightarrow{AC}| = |a\mathbf{i} + b\mathbf{j}| = \sqrt{a^2 + b^2},$$

and in the right triangle ACD,

$$|\overrightarrow{AD}| = \sqrt{|\overrightarrow{AC}|^2 + |\overrightarrow{CD}|^2} = \sqrt{(a^2 + b^2) + c^2};$$

that is,

$$|a\mathbf{i} + b\mathbf{j} + c\mathbf{k}| = \sqrt{a^2 + b^2 + c^2}. \tag{3}$$

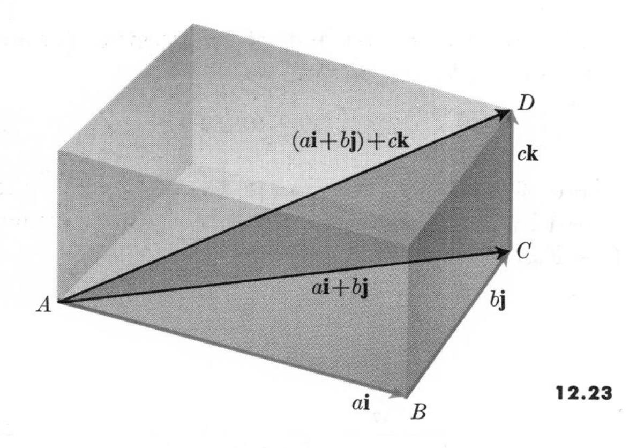

12.23

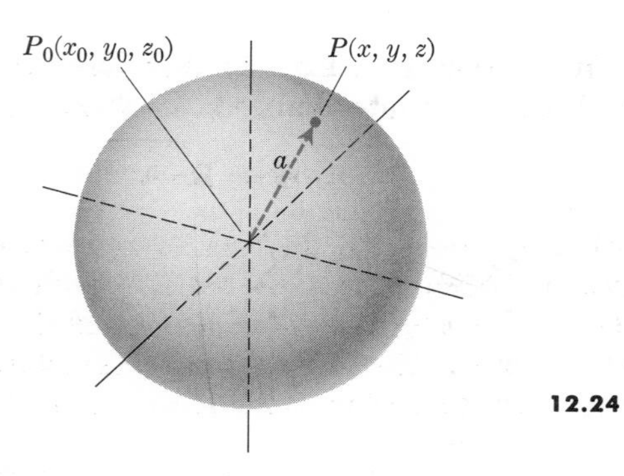

12.24

If we apply this result to the vector $\overrightarrow{P_1P_2}$ of Eq. (2), we obtain a formula for the distance between two points:

$$|\overrightarrow{P_1P_2}| = \sqrt{(x_2 - x_1)^2 + (y_2 - y_1)^2 + (z_2 - z_1)^2}. \tag{4}$$

Since a sphere is the locus of points in space that are equidistant from one fixed point at the center, Eq. (4) may be used to determine the equation satisfied by every point P on a sphere of radius a with center at $P_0(x_0, y_0, z_0)$ (see Fig. 12.24). P is on the sphere if and only if

$$|\overrightarrow{P_0P}| = a,$$

or, what amounts to the same thing, if and only if

$$(x - x_0)^2 + (y - y_0)^2 + (z - z_0)^2 = a^2. \tag{5}$$

Example 1. If we complete the squares in the equation

$$x^2 + y^2 + z^2 + 2x - 4y = 0,$$

thus obtaining

$$(x + 1)^2 + (y - 2)^2 + z^2 = 5,$$

we may identify the locus of points that satisfy the equation as a sphere with center at $(-1, 2, 0)$ and radius $\sqrt{5}$.

Direction

For any nonzero vector $\mathbf{A}$, we obtain a unit vector called *the direction of* $\mathbf{A}$ by dividing $\mathbf{A}$ by its own length*:

$$\text{direction of } \mathbf{A} = \frac{\mathbf{A}}{|\mathbf{A}|}. \tag{6}$$

Example 2. If $\quad \mathbf{A} = 2\mathbf{i} - 3\mathbf{j} + 7\mathbf{k},$

then its length is $\sqrt{4 + 9 + 49} = \sqrt{62}$, and

$$\text{direction of } (2\mathbf{i} - 3\mathbf{j} + 7\mathbf{k}) = \frac{2\mathbf{j} - 3\mathbf{j} + 7\mathbf{k}}{\sqrt{62}}.$$

EXERCISES

Find the center and radius of the spheres in Exercises 1 through 4.

1. $x^2 + y^2 + z^2 + 4x - 4z = 0$
2. $2x^2 + 2y^2 + 2z^2 + x + y + z = 9$
3. $x^2 + y^2 + z^2 - 2az = 0$
4. $3x^2 + 3y^2 + 3z^2 + 2y - 2z = 9$
5. What is the distance from the point $P(x, y, z)$ to
   (a) the x-axis, (b) the y-axis,
   (c) the z-axis, (d) the xy-plane?
6. The distance from $P(x, y, z)$ to the origin is d_1 and the distance from P to $A(0, 0, 3)$ is d_2. Find the locus of P if
   (a) $d_1 = 2d_2$, (b) $d_1 + d_2 = 6$,
   (c) $|d_1 - d_2| = 2$.

* The author credits Professor Arthur P. Mattuck with this definition.

Find the length of each of the following vectors.

7. $2\mathbf{i} + \mathbf{j} - 2\mathbf{k}$
8. $3\mathbf{i} - 6\mathbf{j} + 2\mathbf{k}$
9. $\mathbf{i} + 4\mathbf{j} - 8\mathbf{k}$
10. $9\mathbf{i} - 2\mathbf{j} + 6\mathbf{k}$
11. Find the direction of $4\mathbf{i} + 3\mathbf{j} + 12\mathbf{k}$.
12. Find the vector from the origin O to the point of intersection of the medians of the triangle whose vertices are the three points

$$A(1, -1, 2), \quad B(2, 1, 3), \quad C(-1, 2, -1).$$

13. A bug is crawling straight up the side of a rotating right circular cylinder of radius 2 ft. At time $t = 0$, he is at the point $(2, 0, 0)$ relative to a fixed set of xyz-axes. The axis of the cylinder lies along the z-axis. Assume that the bug travels along a generator (a line parallel to the z-axis) at the rate of c ft/sec, and that the cylinder rotates (counterclockwise as viewed from above) at the rate of b radians/sec. If $P(x, y, z)$ is the bug's position at the end of t seconds, show that

$$\overrightarrow{OP} = \mathbf{i}(2 \cos bt) + \mathbf{j}(2 \sin bt) + \mathbf{k}(ct).$$

12.6 THE SCALAR PRODUCT OF TWO VECTORS

So far, in our work with vectors, we have not defined what meaning, if any, is to be attached to the product of two vectors **A** and **B**. There are three kinds of multiplication of two vectors which have been defined and found to have significance in physical applications.

1. *The* scalar *product is denoted by* $\mathbf{A} \cdot \mathbf{B}$. *This product is also called the dot product because of the dot symbol used to denote it. The result is a scalar.*

2. *The* vector *product, or* cross *product, is denoted by* $\mathbf{A} \times \mathbf{B}$. *The result is a vector.*

3. *The* dyadic *product is denoted simply by* **AB** *with no sign written between the factors. The result is a new mathematical quantity called a dyad.*

(The study of dyads has largely been superseded by the subject of matrices. We shall not be concerned with dyads in this book.)

The *scalar* or *dot* product of two vectors **A** and **B** is defined by the equation

$$\mathbf{A} \cdot \mathbf{B} = |\mathbf{A}|\,|\mathbf{B}| \cos \theta, \tag{1}$$

where θ measures the angle $(0 \leq \theta \leq \pi)$ determined by **A** and **B** when their initial points coincide (see Fig. 12.25).

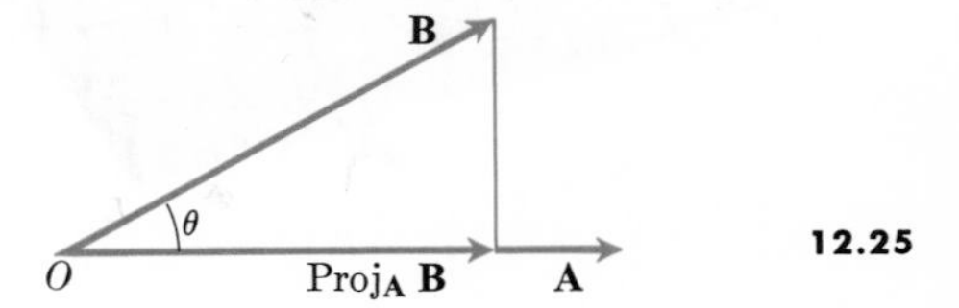

12.25

It is clear from Eq. (1) that the two factors **A** and **B** may be interchanged. That is,

$$\mathbf{A} \cdot \mathbf{B} = \mathbf{B} \cdot \mathbf{A}. \tag{2}$$

When an operation, such as addition or multiplication of ordinary numbers, satisfies such an equation as $a + b = b + a$, or $ab = ba$, we call that operation *commutative*. Equation (2) says, therefore, that the operation of scalar multiplication of two vectors is commutative.

The vector that we get by projecting **B** onto **A** is called the "*vector* projection" of **B** onto **A**. We shall denote it by $\text{proj}_\mathbf{A}\mathbf{B}$, as in Fig. 12.25. The component of **B** in the direction of **A** is plus or minus the length of this vector projection. The sign is to be plus if $\text{proj}_\mathbf{A}\mathbf{B}$ has the same sense as $+\mathbf{A}$, and it is to be minus if it has the same sense as $-\mathbf{A}$. Then the component of **B** in the direction of **A** is equal to $|\mathbf{B}| \cos \theta$. Hence the dot product, Eq. (1), may be interpreted geometrically as follows:

$$\begin{aligned}\mathbf{A} \cdot \mathbf{B} &= |\mathbf{A}|\,(|\mathbf{B}| \cos \theta)\\ &= (\text{length of } \mathbf{A}) \text{ times } (\mathbf{B}\text{-component in the } \mathbf{A}\text{-direction}).\end{aligned}$$

Of course, we may interchange the roles of $|\mathbf{A}|$ and $|\mathbf{B}|$ and write the dot product in the alternative form

$$\begin{aligned}\mathbf{A} \cdot \mathbf{B} &= |\mathbf{B}|\,(|\mathbf{A}| \cos \theta)\\ &= (\text{length of } \mathbf{B}) \text{ times } (\mathbf{A}\text{-component in the } \mathbf{B}\text{-direction}).\end{aligned}$$

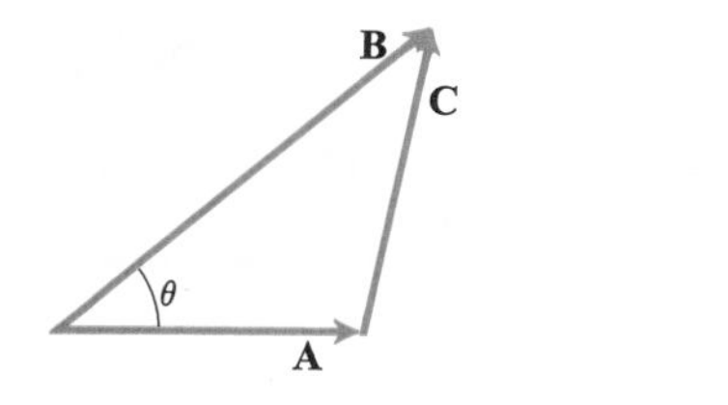

12.26

To calculate the scalar product $\mathbf{A} \cdot \mathbf{B}$ in terms of the components of the vectors, we let

$$\begin{aligned} \mathbf{A} &= a_1\mathbf{i} + a_2\mathbf{j} + a_3\mathbf{k}, \\ \mathbf{B} &= b_1\mathbf{i} + b_2\mathbf{j} + b_3\mathbf{k}, \end{aligned} \tag{3}$$

and

$$\begin{aligned} \mathbf{C} &= \mathbf{B} - \mathbf{A} \\ &= (b_1 - a_1)\mathbf{i} + (b_2 - a_2)\mathbf{j} + (b_3 - a_3)\mathbf{k}. \end{aligned}$$

Then we apply the law of cosines to the triangle whose sides represent the vectors **A**, **B**, and **C** (Fig. 12.26), and obtain

$$|\mathbf{C}|^2 = |\mathbf{A}|^2 + |\mathbf{B}|^2 - 2|\mathbf{A}|\,|\mathbf{B}| \cos\theta,$$

$$|\mathbf{A}|\,|\mathbf{B}| \cos\theta = \frac{|\mathbf{A}|^2 + |\mathbf{B}|^2 - |\mathbf{C}|^2}{2}. \tag{4}$$

The left side of this equation is the same as $\mathbf{A} \cdot \mathbf{B}$, and we may calculate all terms on the right-hand side of (4) by applying Eq. (3), Article 12.5, to find the lengths of the vectors **A**, **B**, and **C**. The result of this algebra is the formula

$$\mathbf{A} \cdot \mathbf{B} = a_1b_1 + a_2b_2 + a_3b_3, \tag{5}$$

which expresses the scalar product in terms of the components of the vectors. Thus, to find the scalar product of two given vectors, we simply multiply their *corresponding* components together and add the results.

Problem 1. Vectors are drawn from the origin to the two points $A(1, -2, -2)$ and $B(6, 3, -2)$. Find the angle AOB.

Solution. Let

$$\begin{aligned} \mathbf{A} &= \overrightarrow{OA} = \mathbf{i} - 2\mathbf{j} - 2\mathbf{k}, \\ \mathbf{B} &= \overrightarrow{OB} = 6\mathbf{i} + 3\mathbf{j} - 2\mathbf{k}, \end{aligned}$$

and denote the angle AOB by θ. Then

$$\mathbf{A} \cdot \mathbf{B} = 6 - 6 + 4 = 4,$$

from Eq. (5), while

$$\mathbf{A} \cdot \mathbf{B} = |\mathbf{A}|\,|\mathbf{B}| \cos\theta,$$

from Eq. (1). Since

$$|\mathbf{A}| = \sqrt{1 + 4 + 4} = 3, \quad |\mathbf{B}| = \sqrt{36 + 9 + 4} = 7,$$

we have

$$\cos\theta = \frac{\mathbf{A} \cdot \mathbf{B}}{|\mathbf{A}|\,|\mathbf{B}|} = \frac{4}{21},$$

$$\theta = \cos^{-1}\frac{4}{21} \approx 79°.$$

From Eq. (5), it is readily seen that if

$$\mathbf{C} = c_1\mathbf{i} + c_2\mathbf{j} + c_3\mathbf{k}$$

is any third vector, then

$$\begin{aligned} \mathbf{A} \cdot (\mathbf{B} + \mathbf{C}) &= a_1(b_1 + c_1) \\ &\quad + a_2(b_2 + c_2) + a_3(b_3 + c_3) \\ &= (a_1b_1 + a_2b_2 + a_3b_3) \\ &\quad + (a_1c_1 + a_2c_2 + a_3c_3) \\ &= \mathbf{A} \cdot \mathbf{B} + \mathbf{A} \cdot \mathbf{C}. \end{aligned}$$

Hence scalar multiplication obeys the *distributive* law:

$$\mathbf{A} \cdot (\mathbf{B} + \mathbf{C}) = \mathbf{A} \cdot \mathbf{B} + \mathbf{A} \cdot \mathbf{C}. \tag{6}$$

If we combine this with the commutative law, Eq. (2), it is also evident that

$$(\mathbf{A} + \mathbf{B}) \cdot \mathbf{C} = \mathbf{A} \cdot \mathbf{C} + \mathbf{B} \cdot \mathbf{C}. \tag{7}$$

Equations (6) and (7) together permit us to multiply sums of vectors according to the familiar laws of elementary algebra. For example,

$$(\mathbf{A} + \mathbf{B}) \cdot (\mathbf{C} + \mathbf{D}) = \mathbf{A} \cdot \mathbf{C} + \mathbf{A} \cdot \mathbf{D} + \mathbf{B} \cdot \mathbf{C} + \mathbf{B} \cdot \mathbf{D}. \tag{8}$$

Orthogonal vectors

It is clear from Eq. (1) that the dot product is zero when the vectors are perpendicular, since $\cos 90° = 0$. Conversely, if $\mathbf{A} \cdot \mathbf{B} = 0$ then one of the vectors is zero or else the vectors are perpendicular. The zero vector has no specified direction, and we might adopt the convention that it is perpendicular to any vector. Then we could say that $\mathbf{A} \cdot \mathbf{B} = 0$ if and only if the vectors $\mathbf{A}$ and $\mathbf{B}$ are perpendicular. Perpendicular vectors are also said to be *orthogonal.*

If the scalar product is negative, then $\cos \theta$ is negative and the angle between the vectors is greater than 90°.

If $\mathbf{B} = \mathbf{A}$, then $\theta = 0$ and $\cos \theta = 1$, so that

$$\mathbf{A} \cdot \mathbf{A} = |\mathbf{A}|^2.$$

Problem 2. Resolve the vector $\mathbf{B}$ into components $\mathbf{B}_1$ and $\mathbf{B}_2$, where $\mathbf{B}_1$ is to be parallel to a given vector $\mathbf{A}$ while $\mathbf{B}_2$ is to be perpendicular to $\mathbf{A}$ (see Fig. 12.27).

Solution. Let $\qquad \mathbf{B} = \mathbf{B}_1 + \mathbf{B}_2,$

with $\qquad \mathbf{B}_1 = c\mathbf{A} \quad \text{and} \quad \mathbf{B}_2 \cdot \mathbf{A} = 0.$

Then, substituting for $\mathbf{B}_1$, we have

$$\mathbf{B} = c\mathbf{A} + \mathbf{B}_2,$$

and the scalar c is determined by the equation

$$\begin{aligned} 0 &= \mathbf{B}_2 \cdot \mathbf{A} = (\mathbf{B} - c\mathbf{A}) \cdot \mathbf{A} \\ &= \mathbf{B} \cdot \mathbf{A} - c(\mathbf{A} \cdot \mathbf{A}), \end{aligned}$$

or

$$c = \frac{\mathbf{B} \cdot \mathbf{A}}{\mathbf{A} \cdot \mathbf{A}}.$$

Then

$$\mathbf{B}_2 = \mathbf{B} - \mathbf{B}_1 = \mathbf{B} - c\mathbf{A} = \mathbf{B} - \frac{\mathbf{B} \cdot \mathbf{A}}{\mathbf{A} \cdot \mathbf{A}}\mathbf{A}$$

is perpendicular to $\mathbf{A}$, because of the way in which c was determined. For example, if

$$\mathbf{B} = 2\mathbf{i} + \mathbf{j} - 3\mathbf{k} \quad \text{and} \quad \mathbf{A} = 3\mathbf{i} - \mathbf{j},$$

then

$$c = \frac{\mathbf{B} \cdot \mathbf{A}}{\mathbf{A} \cdot \mathbf{A}} = \frac{6 - 1}{9 + 1} = \frac{1}{2},$$

and

$$\mathbf{B}_1 = \tfrac{1}{2}\mathbf{A} = \tfrac{3}{2}\mathbf{i} - \tfrac{1}{2}\mathbf{j}$$

is parallel to $\mathbf{A}$, while

$$\mathbf{B}_2 = \mathbf{B} - \mathbf{B}_1 = \tfrac{1}{2}\mathbf{i} + \tfrac{3}{2}\mathbf{j} - 3\mathbf{k}$$

is perpendicular to $\mathbf{A}$.

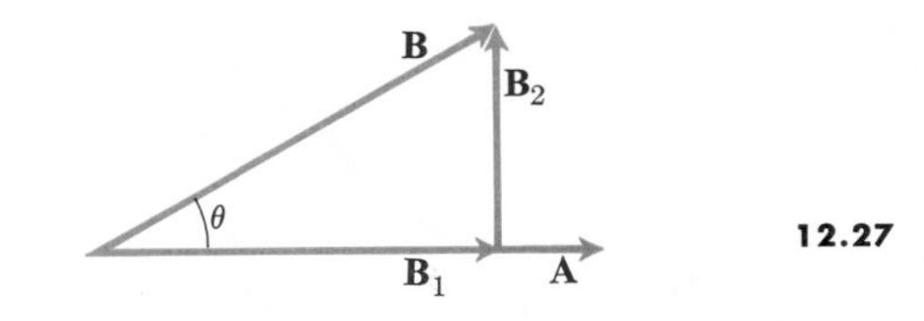

12.27

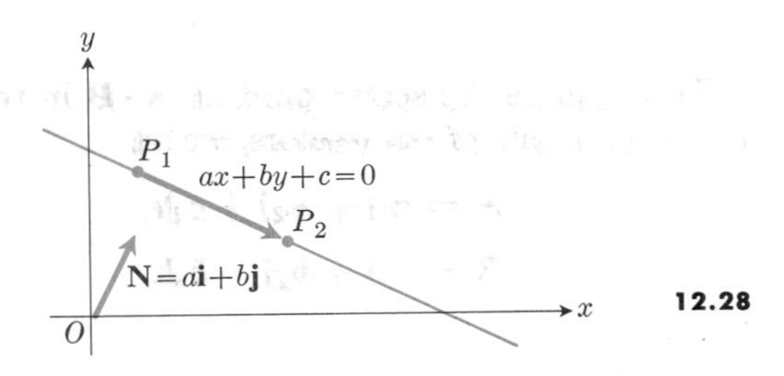

12.28

Problem 3. Show that the vector $\mathbf{N} = a\mathbf{i} + b\mathbf{j}$ is perpendicular to the line $ax + by + c = 0$ in the xy-plane (see Fig. 12.28).

Solution. Let $P_1(x_1, y_1)$ and $P_2(x_2, y_2)$ be any two points on the line; that is,

$$ax_1 + by_1 + c = 0, \qquad ax_2 + by_2 + c = 0.$$

By subtraction, we eliminate c and obtain

$$a(x_2 - x_1) + b(y_2 - y_1) = 0,$$

or

$$(a\mathbf{i} + b\mathbf{j}) \cdot [(x_2 - x_1)\mathbf{i} + (y_2 - y_1)\mathbf{j}] = 0. \tag{9}$$

Now $(x_2 - x_1)\mathbf{i} + (y_2 - y_1)\mathbf{j} = \overrightarrow{P_1P_2}$ is a vector joining two points on the line, while $\mathbf{N} = a\mathbf{i} + b\mathbf{j}$ is the given vector. Equation (9) says that either

$$\mathbf{N} = \mathbf{0}, \quad \text{or} \quad \overrightarrow{P_1P_2} = \mathbf{0},$$

or else $\mathbf{N} \perp \overrightarrow{P_1P_2}$. Now $ax + by + c = 0$ is assumed to be an honest equation of a straight line, so that a and b are not both zero, and hence that $\mathbf{N} \neq \mathbf{0}$. Furthermore, we may surely choose P_2 different from P_1 on the line, and hence $\overrightarrow{P_1P_2} \neq \mathbf{0}$. Therefore $\mathbf{N} \perp \overrightarrow{P_1P_2}$; that is, $\mathbf{N}$ is perpendicular to the line. For example, if the equation of the line is $2x - 3y - 5 = 0$, then the vector $\mathbf{N} = 2\mathbf{i} - 3\mathbf{j}$ is normal to the line.

Problem 4. Using vector methods, find the distance of the point (4, 3) from the line $x + 3y - 6 = 0$ (see Fig. 12.29).

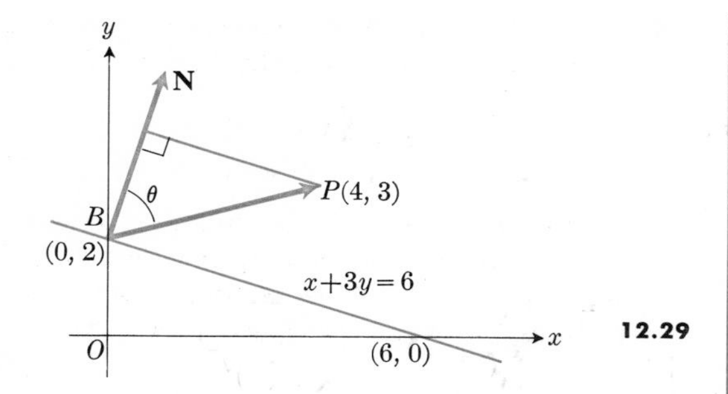

12.29

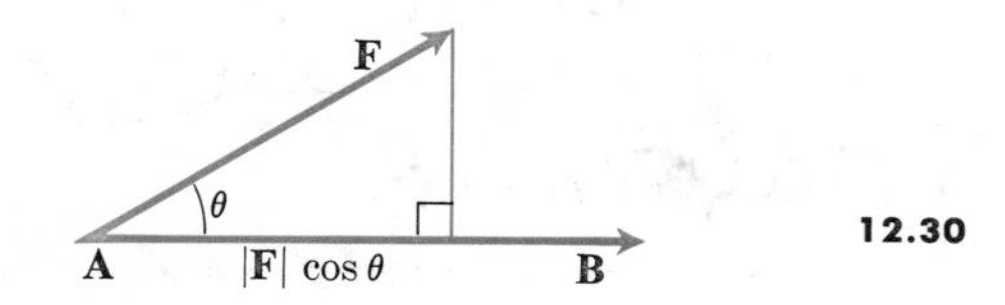

12.30

Solution. The line cuts the y-axis at $B(0, 2)$. At B, draw the vector

$$\mathbf{N} = \mathbf{i} + 3\mathbf{j}$$

normal to the line (see Problem 3). Then the distance from the line to P is $d = |\overrightarrow{BP}| \cos \theta$. This reminds one of the dot product

$$\mathbf{N} \cdot \overrightarrow{BP} = |\mathbf{N}|\,|\overrightarrow{BP}| \cos \theta = |\mathbf{N}| d,$$

from which it follows that

$$d = \frac{\mathbf{N} \cdot \overrightarrow{BP}}{|\mathbf{N}|}.$$

One readily finds $\overrightarrow{BP} = 4\mathbf{i} + \mathbf{j}$, and hence

$$d = \frac{4 + 3}{\sqrt{1 + 9}} = \frac{7}{10}\sqrt{10}.$$

In addition to such geometrical applications as indicated above, the scalar product is useful in mechanics, where it is used in calculating the work done by a force $\mathbf{F}$ when the point of application of $\mathbf{F}$ undergoes a displacement $\overrightarrow{AB}$. If the force remains constant in direction and magnitude, then (see Fig. 12.30) the work is given by

$$\begin{aligned} \text{work} &= (|\mathbf{F}| \cos \theta)\, |\overrightarrow{AB}| \\ &= \mathbf{F} \cdot \overrightarrow{AB}. \end{aligned}$$

The concept of work also enters into the study of electricity and magnetism, and the scalar product again plays a basic role. (See Sears, *Electricity and Magnetism*, Chapter 3.)

Remark. Observe that if we set $\mathbf{B} = \mathbf{i}$ in Eq. (3), and set $b_1 = 1$ and $b_2 = b_3 = 0$ in Eq. (5), then we get

$$\mathbf{A} \cdot \mathbf{i} = a_1.$$

Similarly,

$$\mathbf{A} \cdot \mathbf{j} = a_2, \quad \mathbf{A} \cdot \mathbf{k} = a_3.$$

The numbers a_1, a_2, and a_3 are just the components of $\mathbf{A}$ in the directions $\mathbf{i}$, $\mathbf{j}$, $\mathbf{k}$ respectively. More generally, if $\mathbf{u}$ is any unit vector, then the component of $\mathbf{A}$ in the $\mathbf{u}$-direction is just the scalar product $\mathbf{A} \cdot \mathbf{u}$, because

$$\begin{aligned} \mathbf{A} \cdot \mathbf{u} &= |\mathbf{A}|\,|\mathbf{u}| \cos \theta \\ &= |\mathbf{A}| \cos \theta \\ &= \mathbf{A}\text{-component in the } \mathbf{u}\text{-direction.} \end{aligned} \tag{10}$$

If $\mathbf{B}$ is any nonzero vector, we can let $\mathbf{u} = \mathbf{B}/|\mathbf{B}|$ in Eq. (10). Because the direction of $\mathbf{B}$ is then the $\mathbf{u}$-direction, we can also see that

$$(\mathbf{A}\text{-component in the } \mathbf{B}\text{-direction}) = \mathbf{A} \cdot \left(\frac{\mathbf{B}}{|\mathbf{B}|}\right). \tag{11}$$

The *vector projection* of $\mathbf{A}$ on $\mathbf{B}$ is

(**A**-component in the **B**-direction)
times
(direction of **B**),

which is equal to

$$\frac{\mathbf{A} \cdot \mathbf{B}}{|\mathbf{B}|} \frac{\mathbf{B}}{|\mathbf{B}|}, \quad \text{or} \quad \left(\frac{\mathbf{A} \cdot \mathbf{B}}{\mathbf{B} \cdot \mathbf{B}}\right) \mathbf{B}. \tag{12}$$

EXERCISES

1. Suppose it is known that $\mathbf{A} \cdot \mathbf{B}_1 = \mathbf{A} \cdot \mathbf{B}_2$, and $\mathbf{A}$ is not zero, but nothing more is known about the vectors $\mathbf{B}_1$ and $\mathbf{B}_2$. Is it permissible to cancel $\mathbf{A}$ from both sides of the equation? Give a reason for your answer.

2\. (a) Express the vector projection of **B** onto **A** in a vector form which is convenient for calculation.

(b) Find the vector projection of $\mathbf{B} = \mathbf{i} + 3\mathbf{j} + 4\mathbf{k}$ onto the vector $\mathbf{A} = 10\mathbf{i} + 11\mathbf{j} - 2\mathbf{k}$.

3\. Find the interior angles of the triangle ABC whose vertices are the points $A(-1, 0, 2)$, $B(2, 1, -1)$, and $C(1, -2, 2)$.

4\. Find the point $A(a, a, 0)$ on the line $y = x$ in the xy-plane such that the vector $\overrightarrow{AB}$ is perpendicular to the line OA. Here O is the origin and B is the point $(2, 4, -3)$.

5\. Find the scalar projection of the vector

$$\mathbf{A} = 2\mathbf{i} + 2\mathbf{j} + \mathbf{k}$$

onto the vector $\mathbf{B} = 2\mathbf{i} + 10\mathbf{j} - 11\mathbf{k}$.

6\. Find the angle between the diagonal of a cube and one of its edges.

7\. Find the angle between the diagonal of a cube and a diagonal of one of its faces.

8\. Find the angle between the vectors **A** and **B** of Exercise 5.

9\. How many lines through the origin make angles of 60° with both the y- and z-axes? What angles do they make with the positive x-axis?

10\. If $a = |\mathbf{A}|$ and $b = |\mathbf{B}|$, show that the vector

$$\mathbf{C} = \frac{a\mathbf{B} + b\mathbf{A}}{a + b}$$

bisects the angle between **A** and **B**.

11\. With the same notation as in Exercise 10, show that the vectors $a\mathbf{B} + b\mathbf{A}$ and $\mathbf{A}b - \mathbf{B}a$ are perpendicular.

12\. If **R** is the vector from the origin O to $P(x, y, z)$ and **k** is the unit vector along the z-axis, show geometrically that the equation

$$\frac{\mathbf{R} \cdot \mathbf{k}}{|\mathbf{R}|} = \cos 45°$$

represents a cone with vertex at the origin and generating angle of 45°. Express the equation in Cartesian form.

13\. Find the work done by a force $\mathbf{F} = -w\mathbf{k}$ as its point of application moves from the point $P_1(x_1, y_1, z_1)$ to a second point $P_2(x_2, y_2, z_2)$ along the straight line P_1P_2.

14\. Using vector methods, show that the distance d from the point (x_1, y_1) to the line $ax + by + c = 0$ is

$$d = \frac{|ax_1 + by_1 + c|}{\sqrt{a^2 + b^2}}.$$

15\. *Direction cosines.* If the vector $\mathbf{A} = a\mathbf{i} + b\mathbf{j} + c\mathbf{k}$ makes angles α, β, and γ, respectively, with the positive x-, y-, and z-axes, then $\cos\alpha$, $\cos\beta$, $\cos\gamma$ are called its *direction cosines*. Show that

$$\cos\alpha = \frac{a}{\sqrt{a^2 + b^2 + c^2}}, \quad \cos\beta = \frac{b}{\sqrt{a^2 + b^2 + c^2}},$$

$$\cos\gamma = \frac{c}{\sqrt{a^2 + b^2 + c^2}}.$$

Also show that

$$\cos^2\alpha + \cos^2\beta + \cos^2\gamma = 1,$$

and that

$$\mathbf{u} = \mathbf{i}\cos\alpha + \mathbf{j}\cos\beta + \mathbf{k}\cos\gamma$$

is a unit vector having the same direction as **A**.

16\. Let **A** and **B** be vectors, $\mathbf{B} \neq \mathbf{0}$. Show that

$$(\mathbf{A} - \text{proj}_{\mathbf{B}}\mathbf{A})$$

is orthogonal to **B**. Do this in two ways:

(1) Use the expressions (12) to find $\text{proj}_{\mathbf{B}}\mathbf{A}$, and use the distributive law to find the scalar product.

(2) (Geometrical method.) Use the geometric meaning of $\text{proj}_{\mathbf{B}}\mathbf{A}$.

Illustrate specifically for

$$\mathbf{A} = 2\mathbf{i} + 5\mathbf{j} \quad \text{and} \quad \mathbf{B} = 3\mathbf{i} - 4\mathbf{j}.$$

Use the results to express **A** in the form $c_1\mathbf{B} + c_2\mathbf{B}'$, where $\mathbf{B}'$ is orthogonal to **B** and c_1 and c_2 are appropriate scalars.

12.7 THE VECTOR PRODUCT OF TWO VECTORS

The two vectors **A** and **B** may be subjected to parallel displacements, if necessary, to bring their initial points into coincidence. Consider this as having been done and again let the angle from **A** to **B** be θ, with $0 \leq \theta \leq \pi$. Then, unless **A** and **B** are parallel, they now determine a plane. Let **n**

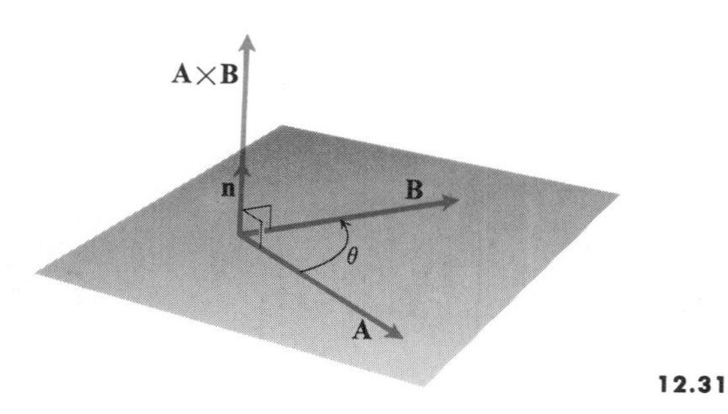

12.31

be a unit vector perpendicular to this plane and pointing in the direction in which a right-threaded screw advances when its head is rotated from **A** to **B** through the angle θ (see Fig. 12.31). The *vector product*, or *cross product*, of **A** and **B**, in that order, is then defined by the equation

$$\mathbf{A} \times \mathbf{B} = \mathbf{n}|\mathbf{A}|\,|\mathbf{B}| \sin\theta. \tag{1}$$

(Applications of the cross product to electricity and magnetism are discussed in Sears, *Electricity and Magnetism*, Chapters 9 and 11.)

If **A** and **B** are parallel, $\theta = 0$ or $180°$ and $\sin\theta = 0$, so that $\mathbf{A} \times \mathbf{B} = \mathbf{0}$. In this case, the direction of **n** is not determined, but this is immaterial, since the zero vector has no specific direction. In all other cases, however, **n** is determined and the cross product is a vector having the same direction as **n** and having magnitude equal to the area, $|\mathbf{A}|\,|\mathbf{B}| \sin\theta$, of the parallelogram determined by the vectors **A** and **B** (see Fig. 12.32).

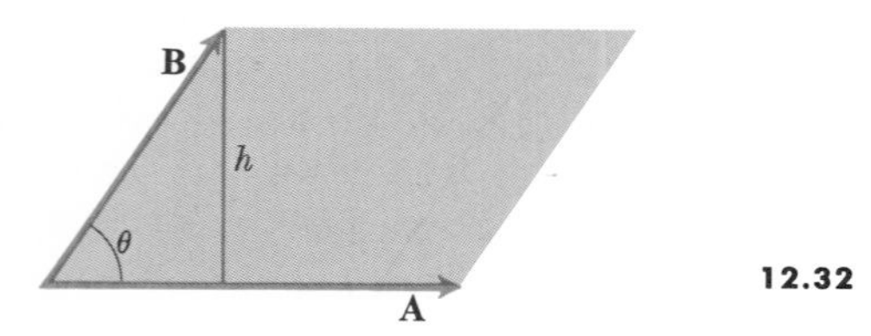

12.32

If the factors **A** and **B** are reversed in the above definition, the vector **n** is replaced by $-\mathbf{n}$ with the result that

$$\mathbf{B} \times \mathbf{A} = -\mathbf{A} \times \mathbf{B}. \tag{2}$$

This kind of multiplication is not commutative, and we must pay attention to the order of the factors.

When the definition is applied to the unit vectors **i**, **j**, and **k**, one readily finds that

$$\begin{aligned} \mathbf{i} \times \mathbf{j} &= -\mathbf{j} \times \mathbf{i} = \mathbf{k}, \\ \mathbf{j} \times \mathbf{k} &= -\mathbf{k} \times \mathbf{j} = \mathbf{i}, \\ \mathbf{k} \times \mathbf{i} &= -\mathbf{i} \times \mathbf{k} = \mathbf{j}, \end{aligned} \tag{3}$$

while

$$\mathbf{i} \times \mathbf{i} = \mathbf{j} \times \mathbf{j} = \mathbf{k} \times \mathbf{k} = \mathbf{0}.$$

Our next objective is to obtain a formula that will express $\mathbf{A} \times \mathbf{B}$ in terms of the components of **A** and **B**. First, we note that the associative law,

$$(r\mathbf{A}) \times (s\mathbf{B}) = rs(\mathbf{A} \times \mathbf{B}), \tag{4}$$

follows from the geometrical meaning of the cross product. Second, we adopt a geometric argument to establish the distributive law,

$$\mathbf{A} \times (\mathbf{B} + \mathbf{C}) = \mathbf{A} \times \mathbf{B} + \mathbf{A} \times \mathbf{C}. \tag{5}$$

Note. To see that Eq. (5) is valid, we interpret the cross product $\mathbf{A} \times \mathbf{B}$ in a slightly different way. The vectors **A** and **B** are drawn from the common point O, and a plane M is constructed perpendicular to **A** at O (see Fig. 12.33). Vector **B** is now projected orthogonally onto M, yielding a vector $\mathbf{B}'$ whose length is $|\mathbf{B}| \sin\theta$. The vector $\mathbf{B}'$ is then rotated 90° about **A** in the positive sense to produce a vector $\mathbf{B}''$. Finally, $\mathbf{B}''$ is multiplied by the length of **A**. The resulting vector $|\mathbf{A}|\,\mathbf{B}''$ is equal to $\mathbf{A} \times \mathbf{B}$, since $\mathbf{B}''$ has the same direction as **n** by its construction and

$$|\mathbf{A}|\,|\mathbf{B}''| = |\mathbf{A}|\,|\mathbf{B}'| = |\mathbf{A}|\,|\mathbf{B}| \sin\theta = |\mathbf{A} \times \mathbf{B}|.$$

Now each of these three operations,

(1) projection onto M,
(2) rotation about **A** through 90°,
(3) multiplication by the scalar $|\mathbf{A}|$,

when applied to a triangle will produce another triangle. If we start with the triangle whose sides are **B**, **C**, and $\mathbf{B} + \mathbf{C}$, as in Fig. 12.34, and apply these three steps, we successively obtain:

(1) a triangle whose sides $\mathbf{B}'$, $\mathbf{C}'$, and $(\mathbf{B} + \mathbf{C})'$ satisfy the vector equation

$$\mathbf{B}' + \mathbf{C}' = (\mathbf{B} + \mathbf{C})',$$

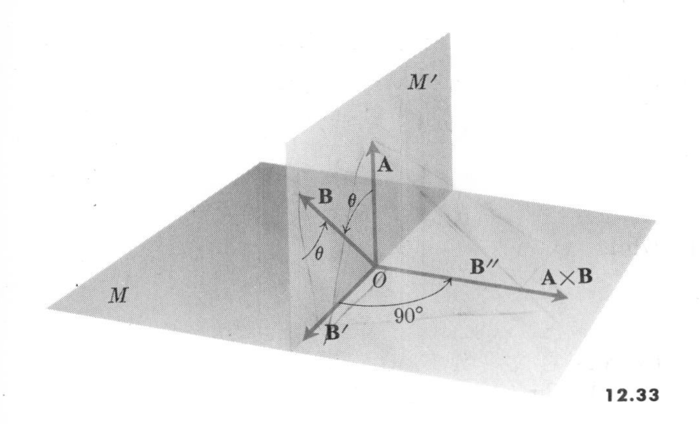

12.33

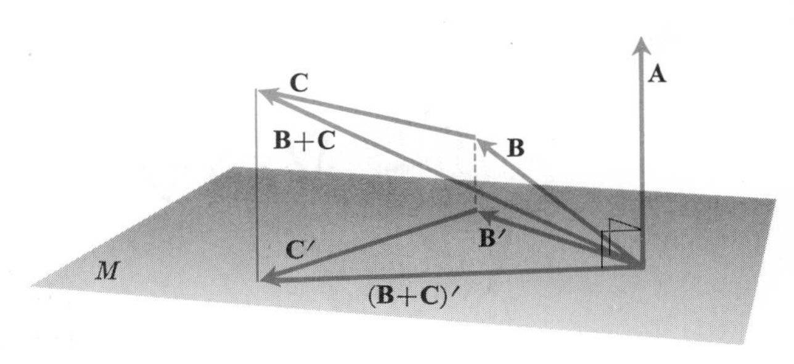

12.34

(2) a triangle whose sides $\mathbf{B}''$, $\mathbf{C}''$, and $(\mathbf{B}+\mathbf{C})''$ satisfy the vector equation

$$\mathbf{B}'' + \mathbf{C}'' = (\mathbf{B} + \mathbf{C})''$$

(the double-prime on each vector has the same meaning as in Fig. 12.33), and, finally,

(3) a triangle whose sides $|\mathbf{A}|\mathbf{B}''$, $|\mathbf{A}|\mathbf{C}''$, and $|\mathbf{A}|(\mathbf{B}+\mathbf{C})''$ satisfy the vector equation

$$|\mathbf{A}|\mathbf{B}'' + |\mathbf{A}|\mathbf{C}'' = |\mathbf{A}|(\mathbf{B} + \mathbf{C})''. \tag{6}$$

When we use the equations

$$|\mathbf{A}|\mathbf{B}'' = \mathbf{A} \times \mathbf{B},$$
$$|\mathbf{A}|\mathbf{C}'' = \mathbf{A} \times \mathbf{C}$$
$$|\mathbf{A}|(\mathbf{B} + \mathbf{C})'' = \mathbf{A} \times (\mathbf{B} + \mathbf{C})$$

which result from our discussion above, Eq. (6) becomes

$$\mathbf{A} \times \mathbf{B} + \mathbf{A} \times \mathbf{C} = \mathbf{A} \times (\mathbf{B} + \mathbf{C}),$$

which is the distributive law, Eq. (5), which we wanted to establish.

The companion law,

$$(\mathbf{B} + \mathbf{C}) \times \mathbf{A} = \mathbf{B} \times \mathbf{A} + \mathbf{C} \times \mathbf{A}, \tag{7}$$

now follows at once from Eq. (5) if we multiply both sides of it by -1 and take account of the fact that interchanging the two factors in a cross product changes the sign of the result.

From Eqs. (4), (5), and (7), we may conclude that multiplication of two vectors according to the cross product law follows the ordinary laws of algebra, *except that the order of the factors is not reversible.* If we apply these results to calculate $\mathbf{A} \times \mathbf{B}$ with

$$\mathbf{A} = a_1\mathbf{i} + a_2\mathbf{j} + a_3\mathbf{k},$$
$$\mathbf{B} = b_1\mathbf{i} + b_2\mathbf{j} + b_3\mathbf{k},$$

we obtain

$$\begin{aligned}\mathbf{A} \times \mathbf{B} &= (a_1\mathbf{i} + a_2\mathbf{j} + a_3\mathbf{k}) \times (b_1\mathbf{i} + b_2\mathbf{j} + b_3\mathbf{k}) \\ &= a_1b_1\mathbf{i}\times\mathbf{i} + a_1b_2\mathbf{i}\times\mathbf{j} + a_1b_3\mathbf{i}\times\mathbf{k} \\ &\quad + a_2b_1\mathbf{j}\times\mathbf{i} + a_2b_2\mathbf{j}\times\mathbf{j} + a_2b_3\mathbf{j}\times\mathbf{k} \\ &\quad + a_3b_1\mathbf{k}\times\mathbf{i} + a_3b_2\mathbf{k}\times\mathbf{j} + a_3b_3\mathbf{k}\times\mathbf{k} \\ &= \mathbf{i}(a_2b_3 - a_3b_2) + \mathbf{j}(a_3b_1 - a_1b_3) \\ &\quad + \mathbf{k}(a_1b_2 - a_2b_1),\end{aligned} \tag{8}$$

where Eqs. (3) have been used to evaluate the products $\mathbf{i}\times\mathbf{i} = \mathbf{0}$, $\mathbf{i}\times\mathbf{j} = \mathbf{k}$, and so on. The six terms on the right-hand side of Eq. (8) are precisely the same as the six terms in the expansion of the third-order determinant* in Eq. (9), so that the cross product may conveniently be calculated from the equation

$$\mathbf{A} \times \mathbf{B} = \begin{vmatrix} \mathbf{i} & \mathbf{j} & \mathbf{k} \\ a_1 & a_2 & a_3 \\ b_1 & b_2 & b_3 \end{vmatrix}. \tag{9}$$

* See Appendix I for a treatment of the theory of determinants.

Problem 1. Find the area of the triangle whose vertices are $A(1, -1, 0)$, $B(2, 1, -1)$, and $C(-1, 1, 2)$.

Solution. Two sides of the given triangle are represented by the vectors

$$\mathbf{a} = \overrightarrow{AB} = (2-1)\mathbf{i} + (1+1)\mathbf{j} + (-1-0)\mathbf{k}$$
$$= \mathbf{i} + 2\mathbf{j} - \mathbf{k},$$
$$\mathbf{b} = \overrightarrow{AC} = (-1-1)\mathbf{i} + (1+1)\mathbf{j} + (2-0)\mathbf{k}$$
$$= -2\mathbf{i} + 2\mathbf{j} + 2\mathbf{k}.$$

The vector

$$\mathbf{c} = \mathbf{a} \times \mathbf{b} = \begin{vmatrix} \mathbf{i} & \mathbf{j} & \mathbf{k} \\ 1 & 2 & -1 \\ -2 & 2 & 2 \end{vmatrix} = 6\mathbf{i} + 6\mathbf{k}$$

has magnitude $|\mathbf{c}| = \sqrt{36+36} = 6\sqrt{2}$, which is equal to the area of a parallelogram of which the given triangle is exactly one-half. Hence the area of the triangle is $\frac{1}{2}(|\mathbf{a} \times \mathbf{b}|) = 3\sqrt{2}$.

Problem 2. Find a unit vector perpendicular to both of the vectors $\mathbf{A} = 2\mathbf{i} + \mathbf{j} - \mathbf{k}$ and $\mathbf{B} = \mathbf{i} - \mathbf{j} + 2\mathbf{k}$.

Solution. The vector $\mathbf{N} = \mathbf{A} \times \mathbf{B}$ is perpendicular to both $\mathbf{A}$ and $\mathbf{B}$; hence $\mathbf{u} = c(\mathbf{A} \times \mathbf{B})$ is also such a vector for any scalar $c \neq 0$, and we may choose c so as to make $\mathbf{u}$ have unit length. We thus find

$$\mathbf{N} = \mathbf{A} \times \mathbf{B} = \mathbf{i} - 5\mathbf{j} - 3\mathbf{k},$$
$$\mathbf{u} \cdot \mathbf{u} = c^2\mathbf{N} \cdot \mathbf{N} = 35c^2 = 1,$$
$$c = \pm 1/\sqrt{35},$$

and

$$\mathbf{u} = \pm \frac{\mathbf{i} - 5\mathbf{j} - 3\mathbf{k}}{\sqrt{35}}.$$

EXERCISES

1. Find $\mathbf{A} \times \mathbf{B}$ if $\mathbf{A} = 2\mathbf{i} - 2\mathbf{j} - \mathbf{k}$, $\mathbf{B} = \mathbf{i} + \mathbf{j} + \mathbf{k}$.
2. Find a vector $\mathbf{N}$ perpendicular to the plane determined by the three points $A(1, -1, 2)$, $B(2, 0, -1)$, and $C(0, 2, 1)$.
3. Find the area of the triangle ABC of Exercise 2.
4. Find the distance from the origin to the plane ABC of Exercise 2 by projecting $\overrightarrow{OA}$ onto the normal vector $\mathbf{N}$.
5. Find a vector that is perpendicular to both the vectors $\mathbf{A} = \mathbf{i} + \mathbf{j} + \mathbf{k}$ and $\mathbf{B} = \mathbf{i} + \mathbf{j}$.
6. Vectors from the origin to the points A, B, and C are given by

$$\mathbf{A} = \mathbf{i} - \mathbf{j} + \mathbf{k},$$
$$\mathbf{B} = 2\mathbf{i} + 3\mathbf{j} - \mathbf{k},$$
$$\mathbf{C} = -\mathbf{i} + 2\mathbf{j} + 2\mathbf{k}.$$

Find all points $P(x, y, z)$ that satisfy the following requirements: $\overrightarrow{OP}$ is a unit vector perpendicular to $\mathbf{C}$, and P lies in the plane determined by $\mathbf{A}$ and $\mathbf{B}$.

7. Using vector methods, find the distance between the line L_1 determined by the two points $A(1, 0, -1)$ and $B(-1, 1, 0)$ and the line L_2 determined by the two points $C(2, 1, -1)$ and $D(4, 5, -2)$. The distance is to be measured along a line perpendicular to both L_1 and L_2.
8. The vector $\mathbf{A} = 3\mathbf{i} + \mathbf{j} - \mathbf{k}$ is normal to a plane M_1 and the vector $\mathbf{B} = 2\mathbf{i} - \mathbf{j} + \mathbf{k}$ is normal to a second plane M_2.

(a) Find the angle between the two normals.

(b) Do the two planes necessarily intersect if they are both extended indefinitely? Give a reason for your answer.

(c) If the two planes do intersect, find a vector which is parallel to their line of intersection.

12.8 EQUATIONS OF LINES AND PLANES

Lines

Suppose L is a line in space that passes through a given point $P_1(x_1, y_1, z)$ and is parallel to a given nonzero vector

$$\mathbf{v} = A\mathbf{i} + B\mathbf{j} + C\mathbf{k}.$$

Then L is the locus of all points $P(x, y, z)$ such that the vector $\overrightarrow{P_1P}$ is parallel to the given vector $\mathbf{v}$ (see Fig. 12.35).

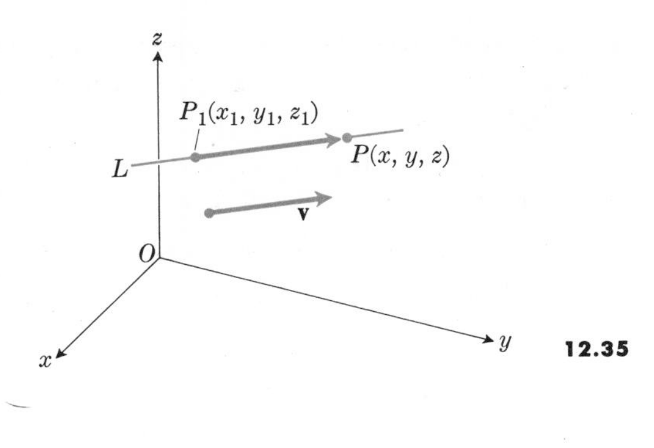

12.35

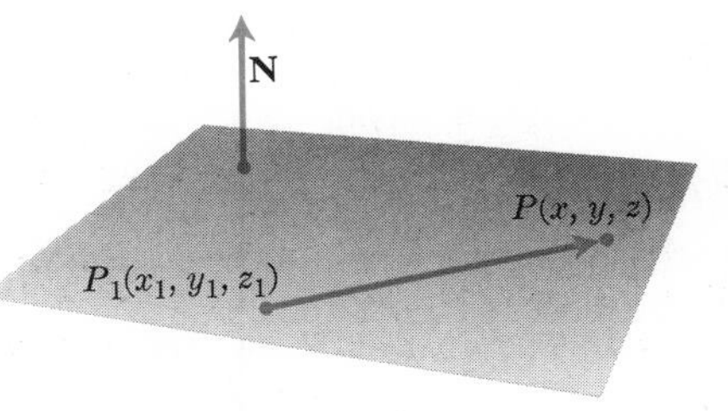

12.36

That is, P is on the line L if and only if there is a scalar t such that

$$\overrightarrow{P_1P} = t\mathbf{v}. \tag{1}$$

When we separate the components in Eq. (1), we have

$$x - x_1 = tA, \quad y - y_1 = tB, \quad z - z_1 = tC. \tag{2}$$

If we allow t to vary from $-\infty$ to $+\infty$, the point $P(x, y, z)$ given by Eqs. (2) will traverse the infinite line L through P_1. These equations may be interpreted as parametric equations of the line. To obtain its Cartesian equations, we may eliminate t, thereby obtaining

$$\frac{x - x_1}{A} = \frac{y - y_1}{B} = \frac{z - z_1}{C}. \tag{3}$$

If any one of the constants A, B, or C is zero in a denominator in Eq. (3), then the corresponding numerator must also be zero. This follows at once from the parametric form, Eqs. (2), which shows, for example, that

$$x - x_1 = tA \quad \text{and} \quad A = 0$$

together imply

$$x - x_1 = 0.$$

Thus, when one of the denominators in Eq. (3) vanishes, we interpret the equations to say that the corresponding numerator must vanish.

Planes

To obtain the equation of a *plane*, we suppose that a point $P_1(x_1, y_1, z_1)$ on the plane and a nonzero normal vector

$$\mathbf{N} = A\mathbf{i} + B\mathbf{j} + C\mathbf{k} \tag{4}$$

perpendicular to the plane are given (see Fig. 12.36). Then the point $P(x, y, z)$ will lie in the plane if and only if the vector $\overrightarrow{P_1P}$ is perpendicular to $\mathbf{N}$; that is, if and only if

$$\mathbf{N} \cdot \overrightarrow{P_1P} = 0,$$

or

$$A(x - x_1) + B(y - y_1) + C(z - z_1) = 0. \tag{5}$$

This equation may also be put in the form

$$Ax + By + Cz = D, \tag{6}$$

where D is the constant $Ax_1 + By_1 + Cz_1$. Conversely, if we start from any linear equation such as (6), we may find a point $P_1(x_1, y_1, z_1)$ whose coordinates do satisfy it, that is, whose coordinates are such that

$$Ax_1 + By_1 + Cz_1 = D.$$

Then, by subtraction, we may put the given Eq. (6) into the form of Eq. (5) and factor it into the dot product

$$\mathbf{N} \cdot \overrightarrow{P_1P} = 0,$$

with $\mathbf{N}$ as in Eq. (4). This says that the constant vector $\mathbf{N}$ is perpendicular to the vector $\overrightarrow{P_1P}$ for every pair of points P_1 and P whose coordinates satisfy the equation. Hence the locus of points

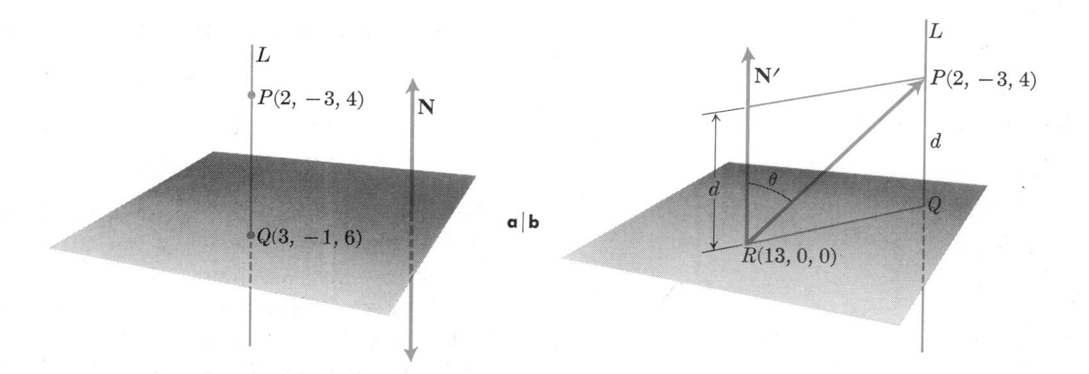

12.37

$P(x, y, z)$ whose coordinates satisfy such a linear equation is a plane, and the vector $A\mathbf{i} + B\mathbf{j} + C\mathbf{k}$, having the same coefficients that x, y, and z have in the given equation, is normal to the plane.

Problem 1. Find the distance from the point $P(2, -3, 4)$ to the plane $x + 2y + 2z = 13$.

Solution 1. The vector $\mathbf{N} = \mathbf{i} + 2\mathbf{j} + 2\mathbf{k}$ is normal to the given plane (see Fig. 12.37a). Now the line L defined by

$$\frac{x-2}{1} = \frac{y+3}{2} = \frac{z-4}{2}$$

goes through P and is parallel to $\mathbf{N}$. Hence L is normal to the plane. We may denote the common ratio in the equation for L by t:

$$\frac{x-2}{1} = \frac{y+3}{2} = \frac{z-4}{2} = t.$$

We then have

$$x = t + 2, \quad y = 2t - 3, \quad z = 2t + 4$$

as parametric equations of the line in terms of the parameter t. Substituting these into the equation of the plane, we obtain

$$(t + 2) + 2(2t - 3) + 2(2t + 4) = 13,$$

or $t = 1$ at the point of intersection of the plane and the line L. That is, $Q(3, -1, 6)$ is the point of intersection. Then the distance from the point to the plane is simply the distance between $P(2, -3, 4)$ and $Q(3, -1, 6)$. Hence

$$d = \sqrt{(3-2)^2 + (-1+3)^2 + (6-4)^2} = 3.$$

Solution 2. Let R be any point of the plane and project the skew segment $\overrightarrow{RP}$ onto a line normal to the plane (see Fig. 12.37b). This will give the distance d from the point P to the plane. The plane intersects the x-axis at $(13, 0, 0)$ which we take as point R on the plane. Now $\mathbf{N} = \mathbf{i} + 2\mathbf{j} + 2\mathbf{k}$ is normal to the plane, and hence $-\mathbf{N}$ is also normal to the plane. Let us consider the normal $\mathbf{N}' = +\mathbf{N}$ or $-\mathbf{N}$, which makes an angle $\theta < 90°$ with $\overrightarrow{RP}$. If we project the vector $\overrightarrow{RP} = -11\mathbf{i} - 3\mathbf{j} + 4\mathbf{k}$ onto this normal $\mathbf{N}'$, then we may obtain the distance d from the equation

$$d = |\overrightarrow{RP}| \cos\theta = \frac{|\mathbf{N}'|\,|\overrightarrow{RP}|\cos\theta}{|\mathbf{N}'|} = \frac{\mathbf{N}' \cdot \overrightarrow{RP}}{|\mathbf{N}'|}.$$

Taking $\mathbf{N}' = \pm\mathbf{N}$ (the sign to be decided later), we find that

$$\mathbf{N}' \cdot \overrightarrow{RP} = \pm(-11 - 6 + 8) = \pm(-9),$$
$$|\mathbf{N}'| = \sqrt{(\pm 1)^2 + (\pm 2)^2 + (\pm 2)^2} = 3,$$

and

$$d = \frac{\pm(-9)}{3}.$$

Clearly, we should take the ambiguous sign to be minus, that is, $\mathbf{N}' = -\mathbf{N}$; we then find $d = 3$.

Problem 2. Find the angle between the two planes $2x + y - 2z = 5$ and $3x - 6y - 2z = 7$.

Solution. Clearly the angle between two planes shown in Fig. 12.38 is the same as the angle between their normals. (Actually there are two angles in each case,

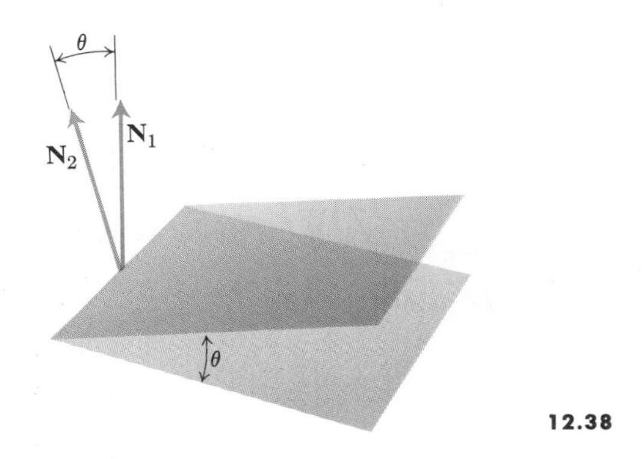

12.38

namely θ and $180° - \theta$.) From the equations of the planes we may read off their normal vectors:

$$\mathbf{N}_1 = 2\mathbf{i} + \mathbf{j} - 2\mathbf{k}, \quad \mathbf{N}_2 = 3\mathbf{i} - 6\mathbf{j} - 2\mathbf{k}.$$

Then

$$\cos\theta = \frac{\mathbf{N}_1 \cdot \mathbf{N}_2}{|\mathbf{N}_1|\,|\mathbf{N}_2|} = \frac{4}{21}, \quad \theta = \cos^{-1}\left(\frac{4}{21}\right) \approx 79°.$$

Problem 3. Find a vector parallel to the line of intersection of the two planes of Problem 2.

Solution. The requirements are met by the vector

$$\mathbf{v} = \mathbf{N}_1 \times \mathbf{N}_2 = \begin{vmatrix} \mathbf{i} & \mathbf{j} & \mathbf{k} \\ 2 & 1 & -2 \\ 3 & -6 & -2 \end{vmatrix} = -14\mathbf{i} - 2\mathbf{j} - 15\mathbf{k}.$$

In connection with the last two problems, it should be noted that any two intersecting planes determine a straight line in space. The equations of the two planes are satisfied simultaneously only by points on the line of intersection. Hence a pair of simultaneous linear equations may be interpreted as representing a line, namely, the line of intersection of the two planes represented by the individual equations. For example, recall Eq. (3),

$$\frac{x - x_1}{A} = \frac{y - y_1}{B} = \frac{z - z_1}{C}, \tag{3}$$

which we found for the line L through the point $P_1(x_1, y_1, z_1)$ and parallel to the vector

$$\mathbf{v} = A\mathbf{i} + B\mathbf{j} + C\mathbf{k}.$$

This is equivalent to the three simultaneous equations

$$B(x - x_1) = A(y - y_1), \tag{7a}$$

$$C(x - x_1) = A(z - z_1), \tag{7b}$$

$$C(y - y_1) = B(z - z_1). \tag{7c}$$

Each of these equations represents a plane. Any pair of them represents the line of intersection of the corresponding pair of planes. There are three such pairs of planes, namely 1st and 2nd, 1st and 3rd, 2nd and 3rd. But the three lines of intersection so determined are all identical; that is, there is just *one* line of intersection. To see that this is so, we consider three separate cases.

Case 1. If any two of the three coefficients A, B, and C are zero and the third one is different from zero, then one of the three equations (7) reduces to $0 = 0$, which imposes no restriction on (x, y, z), while the other two equations represent two planes that intersect in a common line.

Case 2. If only one of the coefficients is zero, say $A = 0$ and $BC \neq 0$, then the first two Eqs. (7) say simply that $x = x_1$. These two equations thus represent just one plane, and the intersection of this plane with the plane

$$C(y - y_1) = B(z - z_1)$$

is the line L.

Case 3. If $A \neq 0$, we may multiply Eq. (7a) by C/A and Eq. (7b) by B/A and subtract one from the other to obtain Eq. (7c). Thus we might just as well ignore the third equation, since it contains no new information.

In all cases, we see that the three Eqs. (7) reduce to just two independent equations, and the three planes intersect in one straight line.

Taken as the equation of a line, (3) is said to be in *standard form.* The line determined by the two planes in Problem 2 may be given in standard form by finding any one point, for example (9, 1, 7), on both planes, and then A, B, and C may be read from the coefficients of $\mathbf{v} = \mathbf{N}_1 \times \mathbf{N}_2$ (or, from $-\mathbf{v}$). Thus we may obtain

$$\frac{x - 9}{14} = \frac{y - 1}{2} = \frac{z - 7}{15}$$

as the "standard form" equation of the line.

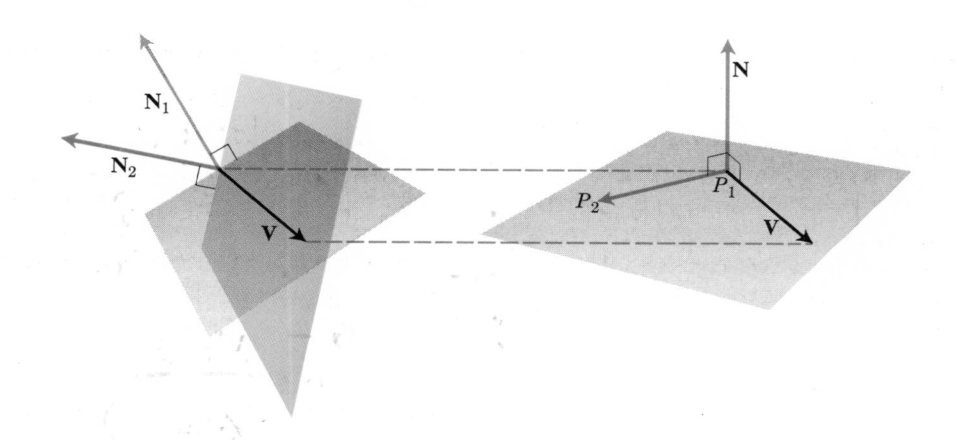

12.39

Problem 4. Find an equation of the plane that passes through the two points

$$P_1(1, 0, -1) \quad \text{and} \quad P_2(-1, 2, 1)$$

and is parallel to the line of intersection of the planes $3x + y - 2z = 6$ and $4x - y + 3z = 0$.

Solution. Our main problem is to find a vector

$$\mathbf{N} = \overrightarrow{P_1P_2} \times \mathbf{V},$$

normal to the plane in question (see Fig. 12.39). The line of intersection of the two given planes is parallel to the vector

$$\mathbf{V} = \mathbf{N}_1 \times \mathbf{N}_2 = \begin{vmatrix} \mathbf{i} & \mathbf{j} & \mathbf{k} \\ 3 & 1 & -2 \\ 4 & -1 & 3 \end{vmatrix} = \mathbf{i} - 17\mathbf{j} - 7\mathbf{k},$$

where $\mathbf{N}_1$ and $\mathbf{N}_2$ are normals to the two given planes. The vector $\overrightarrow{P_1P_2} = -2\mathbf{i} + 2\mathbf{j} + 2\mathbf{k}$ is to lie in the required plane. Now we may also slide $\mathbf{V}$ parallel to itself until it also lies in the required plane (since the plane is to be parallel to $\mathbf{V}$). Hence we may take

$$\mathbf{N} = \overrightarrow{P_1P_2} \times \mathbf{V} = 20\mathbf{i} - 12\mathbf{j} + 32\mathbf{k}$$

as a vector normal to the plane. Actually,

$$\tfrac{1}{4}\mathbf{N} = 5\mathbf{i} - 3\mathbf{j} + 8\mathbf{k}$$

serves just as well. From this normal vector, we may substitute

$$A = 5, \quad B = -3, \quad C = 8$$

in Eq. (5), together with $x_1 = 1$, $y_1 = 0$, $z_1 = -1$, since $P_1(1, 0, -1)$ is to lie in the plane. The required plane is therefore

$$5(x - 1) - 3(y - 0) + 8(z + 1) = 0,$$

or

$$5x - 3y + 8z + 3 = 0.$$

EXERCISES

1. Find the coordinates of the point P in which the line

$$\frac{x - 1}{2} = \frac{y + 1}{-1} = \frac{z}{3}$$

intersects the plane $3x + 2y - z = 5$.

2. Find parametric and Cartesian equations of the line joining the points $A(1, 2, -1)$ and $B(-1, 0, 1)$.

3. Show, by vector methods, that the distance from the point $P_1(x_1, y_1, z_1)$ to the plane

$$Ax + By + Cz - D = 0$$

is

$$\frac{|Ax_1 + By_1 + Cz_1 - D|}{\sqrt{A^2 + B^2 + C^2}}.$$

4. (a) Define what is meant by the angle between a line and a plane.

(b) Find the acute angle between the line

$$\frac{x + 1}{2} = \frac{y}{3} = \frac{z - 3}{6}$$

and the plane $10x + 2y - 11z = 3$.

5. Find a plane that passes through the point $(1, -1, 3)$ and is parallel to the plane $3x + y + z = 7$.
6. Show that the equations obtained by substituting different values for the constant D in the equation

$$2x + 3y - 6z = D$$

represent a family of parallel planes. What is the distance between two of these planes, one corresponding, say, to $D = D_1$ and the other to $D = D_2$?
7. Prove that the line

$$\frac{x-1}{2} = \frac{y+1}{3} = \frac{z-2}{4}$$

is parallel to the plane $x - 2y + z = 6$.
8. Find a plane through the points $A(1, 1, -1)$, $B(2, 0, 2)$, and $C(0, -2, 1)$.
9. Let $P_i(x_i, y_i, z_i)$, $i = 1, 2, 3$, be three points. What is the locus described by the equation

$$\begin{vmatrix} x & y & z & 1 \\ x_1 & y_1 & z_1 & 1 \\ x_2 & y_2 & z_2 & 1 \\ x_3 & y_3 & z_3 & 1 \end{vmatrix} = 0?$$

10. Find a plane through $A(1, -2, 1)$ and perpendicular to the vector from the origin to A.
11. Find a plane through $P_0(2, 1, -1)$ and perpendicular to the line of intersection of the planes

$$2x + y - z = 3, \qquad x + 2y + z = 2.$$

12. Find a plane through the points

$$P_1(1, 2, 3), \quad P_2(3, 2, 1),$$

and perpendicular to the plane $4x - y + 2z = 7$.
13. Find the distance from the origin to the line

$$\frac{x-2}{3} = \frac{y-1}{4} = \frac{2-z}{5}.$$

14. (a) Prove that three points A, B, C are collinear if and only if $\overrightarrow{AC} \times \overrightarrow{AB} = \mathbf{0}$.
(b) Are the points

$$A(1, 2, -3), \quad B(3, 1, 0), \quad C(-3, 4, -9)$$

collinear?
15. Prove that four points A, B, C, D are coplanar if and only if $\overrightarrow{AD} \cdot (\overrightarrow{AB} \times \overrightarrow{BC}) = 0$.
16. Show that the line of intersection of the planes

$$x + 2y - 2z = 5, \qquad 5x - 2y - z = 0$$

is parallel to the line

$$\frac{x+3}{2} = \frac{y}{3} = \frac{z-1}{4}.$$

Find the plane determined by these two lines.
17. Show that the lines

$$\frac{x-2}{1} = \frac{y-2}{3} = \frac{z-3}{1},$$

$$\frac{x-2}{1} = \frac{y-3}{4} = \frac{z-4}{2}$$

intersect. Find the plane determined by these two lines.
18. Find the direction cosines (Article 12.6, Exercise 15) of the line

$$2x + y - z = 5, \qquad x - 3y + 2z = 2.$$

19. The equation $\mathbf{N} \cdot \overrightarrow{P_1P} = 0$ represents a plane through P_1 perpendicular to $\mathbf{N}$. What locus does the inequality $\mathbf{N} \cdot \overrightarrow{P_1P} > 0$ represent? Give a reason for your answer.
20. The unit vector $\mathbf{u}$ makes angles α, β, γ, respectively, with the positive x-, y-, z-axes. Find the plane that is normal to $\mathbf{u}$ and goes through $P_0(x_0, y_0, z_0)$.

12.9 PRODUCTS OF THREE OR MORE VECTORS

Certain products involving three or more vectors arise in physical and engineering problems. For example (see Sears, *Electricity and Magnetism*, p. 286), the electromotive force $\overrightarrow{dE}$ induced in an element of a conducting wire $\overrightarrow{dl}$ moving with velocity $\mathbf{v}$ through a magnetic field at a point where the flux density is $\mathbf{B}$ is given by $\overrightarrow{dE} = (\mathbf{B} \times \overrightarrow{dl}) \cdot \mathbf{v}$. Here the factor in parentheses is a vector, and the result of forming the scalar product of this vector and $\mathbf{v}$ is a scalar. It is a real economy in thinking to represent the result in the compact vector form which removes the necessity of carrying factors such as the sine of the angle between $\mathbf{B}$ and $\overrightarrow{dl}$ and the cosine of the angle between the normal to their plane and the velocity vector $\mathbf{v}$. All these are automatically

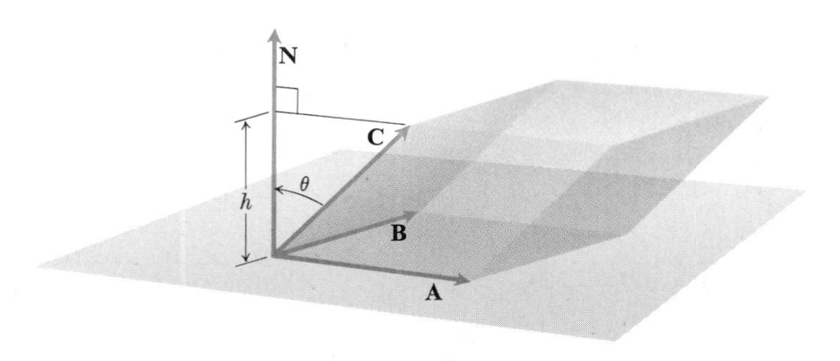

12.40

taken account of by the given product of three vectors.

Triple scalar product

The product $(\mathbf{A} \times \mathbf{B}) \cdot \mathbf{C}$, called the *triple scalar product,* has the following geometrical significance. The vector $\mathbf{N} = \mathbf{A} \times \mathbf{B}$ is normal to the base of the parallelepiped determined by the vectors **A**, **B**, and **C** (see Fig. 12.40). The magnitude of **N** equals the area of the base determined by **A** and **B**. Thus

$$(\mathbf{A} \times \mathbf{B}) \cdot \mathbf{C} = \mathbf{N} \cdot \mathbf{C} = |\mathbf{N}|\,|\mathbf{C}| \cos \theta$$

is, except perhaps for sign, the *volume of a box* of edges, **A**, **B**, and **C**, since

$$|\mathbf{N}| = |\mathbf{A} \times \mathbf{B}| = \text{area of base}$$

and

$$|\mathbf{C}| \cos \theta = \pm h = \pm \text{ altitude of box.}$$

If **C** and $\mathbf{A} \times \mathbf{B}$ lie on the same side of the plane determined by **A** and **B**, the triple scalar product will be positive. But if the vectors **A**, **B**, and **C** are so oriented as to form a left-handed system, then $(\mathbf{A} \times \mathbf{B}) \cdot \mathbf{C}$ is negative. By successively considering the plane of **B** and **C**, then the plane of **C** and **A**, as the base of the box, it is readily seen that

$$(\mathbf{A} \times \mathbf{B}) \cdot \mathbf{C} = (\mathbf{B} \times \mathbf{C}) \cdot \mathbf{A} = (\mathbf{C} \times \mathbf{A}) \cdot \mathbf{B}. \quad (1)$$

Since the dot product is commutative, we also have

$$(\mathbf{B} \times \mathbf{C}) \cdot \mathbf{A} = \mathbf{A} \cdot (\mathbf{B} \times \mathbf{C}),$$

so that Eq. (1) gives the result

$$(\mathbf{A} \times \mathbf{B}) \cdot \mathbf{C} = \mathbf{A} \cdot (\mathbf{B} \times \mathbf{C}). \quad (2)$$

Equation (2) says that the dot and the cross may be interchanged in the triple scalar product, provided only that the multiplications are performed in a way that "makes sense." Thus $(\mathbf{A} \cdot \mathbf{B}) \times \mathbf{C}$ is excluded on the ground that $(\mathbf{A} \cdot \mathbf{B})$ is a scalar and we never "cross" a scalar and a vector.

The triple scalar product in Eq. (2) is conveniently expressed in determinant form:

$$\mathbf{A} \cdot (\mathbf{B} \times \mathbf{C}) = \begin{vmatrix} a_1 & a_2 & a_3 \\ b_1 & b_2 & b_3 \\ c_1 & c_2 & c_3 \end{vmatrix}. \quad (3)$$

We may verify this by first considering the determinant representation for $\mathbf{B} \times \mathbf{C}$. Let us interpret the dot product of $\mathbf{A} = a_1\mathbf{i} + a_2\mathbf{j} + a_3\mathbf{k}$ and this vector $\mathbf{B} \times \mathbf{C} = B_1\mathbf{i} + B_2\mathbf{j} + B_3\mathbf{k}$ as being formed by replacing the unit vectors **i**, **j**, and **k** in the latter equation by the coefficients a_1, a_2, and a_3 respectively. Then this replacement may as well be carried out before the determinant for $\mathbf{B} \times \mathbf{C}$ is expanded. The result will be the one given in Eq. (3).

A product which involves three vectors but is much simpler than the triple scalar product is $(\mathbf{A} \cdot \mathbf{B})\mathbf{C}$. Here the scalar, $s = \mathbf{A} \cdot \mathbf{B}$, multiplies **C** and we have simply the result $s\mathbf{C}$.

Triple vector product

The triple vector products

$$(\mathbf{A} \times \mathbf{B}) \times \mathbf{C}$$

and

$$\mathbf{A} \times (\mathbf{B} \times \mathbf{C})$$

are, in general, not equal. Each of them can be expressed rather simply in terms of the vectors involved, by formulas which we shall now derive.

To focus our attention, let us consider the vector product $(\mathbf{A} \times \mathbf{B}) \times \mathbf{C}$. We shall now show that this is given by

$$(\mathbf{A} \times \mathbf{B}) \times \mathbf{C} = (\mathbf{A} \cdot \mathbf{C})\mathbf{B} - (\mathbf{B} \cdot \mathbf{C})\mathbf{A}. \tag{4}$$

Case 1. In the special case where any one of the vectors is the zero vector, Eq. (4) is certainly true, since both sides of it are then zero.

Case 2. If none of the vectors is zero, but if $\mathbf{B} = s\mathbf{A}$ for some scalar s, then again both sides of Eq. (4) are zero.

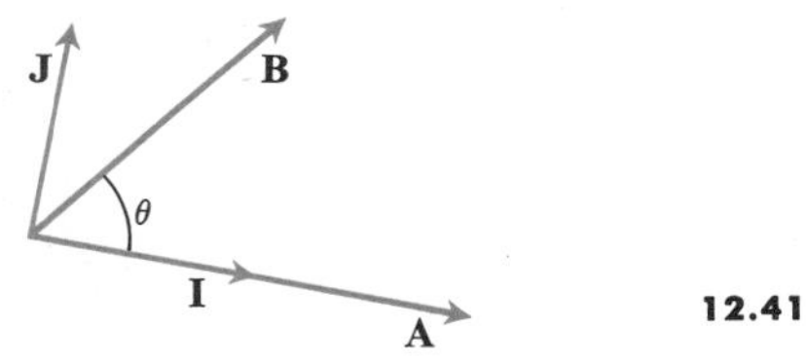

12.41

Case 3. Now we consider the case where none of the vectors is zero and where $\mathbf{A}$ and $\mathbf{B}$ are not parallel. The vector on the left in Eq. (4) is parallel to the plane determined by $\mathbf{A}$ and $\mathbf{B}$, so that it is possible to find scalars m and n such that

$$(\mathbf{A} \times \mathbf{B}) \times \mathbf{C} = m\mathbf{A} + n\mathbf{B}. \tag{5}$$

For easy calculation of m and n, we introduce orthogonal unit vectors $\mathbf{I}$ and $\mathbf{J}$ in the plane of $\mathbf{A}$ and $\mathbf{B}$, with $\mathbf{I} = \mathbf{A}/|\mathbf{A}|$ (see Fig. 12.41). We also introduce a third unit vector, $\mathbf{K} = \mathbf{I} \times \mathbf{J}$, and write all our vectors in terms of the unit vectors $\mathbf{I}$, $\mathbf{J}$, and $\mathbf{K}$:

$$\begin{aligned} \mathbf{A} &= a_1\mathbf{I}, \\ \mathbf{B} &= b_1\mathbf{I} + b_2\mathbf{J}, \\ \mathbf{C} &= c_1\mathbf{I} + c_2\mathbf{J} + c_3\mathbf{K}. \end{aligned} \tag{6}$$

Then

$$\mathbf{A} \times \mathbf{B} = a_1b_2\mathbf{K}$$

and

$$(\mathbf{A} \times \mathbf{B}) \times \mathbf{C} = a_1b_2c_1\mathbf{J} - a_1b_2c_2\mathbf{I}. \tag{7}$$

Comparing this with the right-hand side of Eq. (5), we have

$$m(a_1\mathbf{I}) + n(b_1\mathbf{I} + b_2\mathbf{J}) = a_1b_2c_1\mathbf{J} - a_1b_2c_2\mathbf{I}.$$

This is equivalent to the pair of scalar equations

$$\begin{aligned} ma_1 + nb_1 &= -a_1b_2c_2, \\ nb_2 &= a_1b_2c_1. \end{aligned}$$

If b_2 were equal to zero, $\mathbf{A}$ and $\mathbf{B}$ would be parallel, contrary to hypothesis. Hence b_2 is not zero and we may solve the last equation for n. We find

$$n = a_1c_1 = \mathbf{A} \cdot \mathbf{C}.$$

Then, by substitution,

$$\begin{aligned} ma_1 &= -nb_1 - a_1b_2c_2 \\ &= -a_1c_1b_1 - a_1b_2c_2, \end{aligned}$$

and since $|\mathbf{A}| = a_1 \neq 0$, we may divide by a_1 and obtain

$$m = -(b_1c_1 + b_2c_2) = -(\mathbf{B} \cdot \mathbf{C}).$$

When these values are substituted for m and n in Eq. (5), we obtain the result given in Eq. (4).

The identity

$$(\mathbf{B} \times \mathbf{C}) \times \mathbf{A} = (\mathbf{B} \cdot \mathbf{A})\mathbf{C} - (\mathbf{C} \cdot \mathbf{A})\mathbf{B} \tag{8a}$$

follows from Eq. (4) by a simple interchange of the letters $\mathbf{A}$, $\mathbf{B}$, and $\mathbf{C}$. If we now interchange the factors $\mathbf{B} \times \mathbf{C}$ and $\mathbf{A}$, we must also change the sign on the right-hand side of the equation. This gives the following identity, which is a companion of Eq. (4):

$$\mathbf{A} \times (\mathbf{B} \times \mathbf{C}) = (\mathbf{A} \cdot \mathbf{C})\mathbf{B} - (\mathbf{A} \cdot \mathbf{B})\mathbf{C}. \tag{8b}$$

The identities in Eqs. (4) and (8b) make it possible to reduce expressions involving the multiplication of three or more vectors to, at most,

(scalars) times (cross products of two factors).

Problem 1. Verify the result of Eq. (4) by calculating $(\mathbf{A} \times \mathbf{B}) \times \mathbf{C}$ in two ways, where

$$\begin{aligned} \mathbf{A} &= \mathbf{i} - \mathbf{j} + 2\mathbf{k}, \\ \mathbf{B} &= 2\mathbf{i} + \mathbf{j} + \mathbf{k}, \\ \mathbf{C} &= \mathbf{i} + 2\mathbf{j} - \mathbf{k}. \end{aligned}$$

Solution. One method of calculation is to use Eq. (4). This way, we find

$$\mathbf{A} \cdot \mathbf{C} = -3, \quad \mathbf{B} \cdot \mathbf{C} = 3,$$

and hence

$$(\mathbf{A} \times \mathbf{B}) \times \mathbf{C} = (\mathbf{A} \cdot \mathbf{C})\mathbf{B} - (\mathbf{B} \cdot \mathbf{C})\mathbf{A} = -3\mathbf{B} - 3\mathbf{A} = -9(\mathbf{i} + \mathbf{k}).$$

Another way is to calculate

$$\mathbf{A} \times \mathbf{B} = \begin{vmatrix} \mathbf{i} & \mathbf{j} & \mathbf{k} \\ 1 & -1 & 2 \\ 2 & 1 & 1 \end{vmatrix} = -3\mathbf{i} + 3\mathbf{j} + 3\mathbf{k}$$

and then

$$(\mathbf{A} \times \mathbf{B}) \times \mathbf{C} = \begin{vmatrix} \mathbf{i} & \mathbf{j} & \mathbf{k} \\ -3 & 3 & 3 \\ 1 & 2 & -1 \end{vmatrix} = -9\mathbf{i} - 9\mathbf{k}.$$

Problem 2. Use Eqs. (4) and (8b) to express

$$(\mathbf{A} \times \mathbf{B}) \times (\mathbf{C} \times \mathbf{D})$$

in terms involving at most multiplication of scalars with cross products of no more than two factors.

Solution. Write, for convenience,

$$\mathbf{C} \times \mathbf{D} = \mathbf{V}.$$

Then we use Eq. (4) to evaluate

$$(\mathbf{A} \times \mathbf{B}) \times \mathbf{V} = (\mathbf{A} \cdot \mathbf{V})\mathbf{B} - (\mathbf{B} \cdot \mathbf{V})\mathbf{A},$$

or

$$(\mathbf{A} \times \mathbf{B}) \times (\mathbf{C} \times \mathbf{D}) = (\mathbf{A} \cdot \mathbf{C} \times \mathbf{D})\mathbf{B} - (\mathbf{B} \cdot \mathbf{C} \times \mathbf{D})\mathbf{A}.$$

The result, as written, expresses the answer as a scalar times **B** minus a scalar times **A**. One could also represent the answer as a scalar times **C** minus a scalar times **D**. Geometrically, the vector is parallel to the line of intersection of the **A**, **B**-plane and the **C**, **D**-plane.

Problem 3. Let

$$\mathbf{A} = \overrightarrow{PQ}, \quad \mathbf{B} = \overrightarrow{PS}$$
$$\mathbf{A}' = \overrightarrow{P'Q'}, \quad \mathbf{B}' = \overrightarrow{P'S'}$$

be sides of parallelograms $PQRS$ and $P'Q'R'S'$ that are related in such a way that PP', QQ', RR', and SS' are parallel to one another and to the unit vector **n** (see Fig. 12.42). Show that

$$(\mathbf{A} \times \mathbf{B}) \cdot \mathbf{n} = (\mathbf{A}' \times \mathbf{B}') \cdot \mathbf{n}, \tag{9}$$

and discuss the geometrical meaning of this identity.

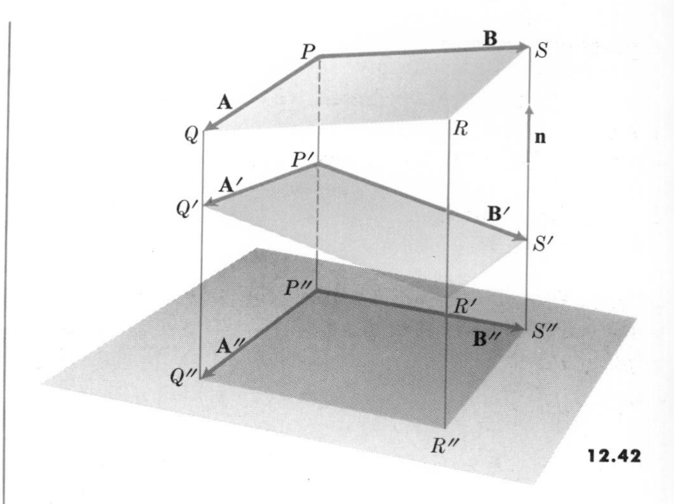

12.42

Solution. From the way the parallelograms are related, it follows that

$$\begin{aligned}\mathbf{A} = \overrightarrow{PQ} &= \overrightarrow{PP'} + \overrightarrow{P'Q'} + \overrightarrow{Q'Q} \\ &= \overrightarrow{P'Q'} + (\overrightarrow{PP'} - \overrightarrow{QQ'}) \\ &= \mathbf{A}' + s\mathbf{n}\end{aligned}$$

for some scalar s, since both $\overrightarrow{PP'}$ and $\overrightarrow{QQ'}$ are parallel to **n**. Similarly,

$$\mathbf{B} = \mathbf{B}' + t\,\mathbf{n}$$

for some scalar t. Hence

$$\begin{aligned}\mathbf{A} \times \mathbf{B} &= (\mathbf{A}' + s\mathbf{n}) \times (\mathbf{B}' + t\,\mathbf{n}) \\ &= \mathbf{A}' \times \mathbf{B}' + t(\mathbf{A}' \times \mathbf{n}) \\ &\quad + s(\mathbf{n} \times \mathbf{B}') + st(\mathbf{n} \times \mathbf{n}).\end{aligned} \tag{10}$$

But $\mathbf{n} \times \mathbf{n} = 0$, while $\mathbf{A}' \times \mathbf{n}$ and $\mathbf{n} \times \mathbf{B}'$ are both perpendicular to **n**. Therefore, when we dot both sides of (10) with **n**, we get Eq. (9).

Geometrical meaning of Equation (9)

The result (9) says that when the parallelograms $PQRS$ and $P'Q'R'S'$ are any two plane sections of a prism with sides parallel to **n**, then the box determined by **A**, **B**, and **n** has the same volume as the box determined by **A**′, **B**′, and **n**. Thus, in particular, we may replace the right side of (9) by $(\mathbf{A}'' \times \mathbf{B}'') \cdot \mathbf{n}$, where **A**″ and **B**″ are sides of a *right* section $P''Q''R''S''$, as in Fig. 12.42. Then $\mathbf{A}'' \times \mathbf{B}''$ is parallel to $\mathbf{n}$, and

$$\mathbf{A}'' \times \mathbf{B}'' = (\text{area right section})\ \mathbf{n}$$

and

$$(\mathbf{A}'' \times \mathbf{B}'') \cdot \mathbf{n} = \text{area right section.}$$

Therefore, by Eq. (9), we have the following interpretation:

$(\mathbf{A} \times \mathbf{B}) \cdot \mathbf{n}$ *is the area of the orthogonal projection of the parallelogram determined by* $\mathbf{A}$ *and* $\mathbf{B}$ *onto a plane whose unit normal is* $\mathbf{n}$.* (11)

In particular,

$$(\mathbf{A} \times \mathbf{B}) \cdot \mathbf{k} = \text{area of projection in the } xy\text{-plane,} \tag{12a}$$

$$(\mathbf{A} \times \mathbf{B}) \cdot \mathbf{j} = \text{area of projection in the } xz\text{-plane,} \tag{12b}$$

$$(\mathbf{A} \times \mathbf{B}) \cdot \mathbf{i} = \text{area of projection in the } yz\text{-plane.} \tag{12c}$$

EXERCISES

In Exercises 1 through 3, take

$$\mathbf{A} = 4\mathbf{i} - 8\mathbf{j} + \mathbf{k},$$
$$\mathbf{B} = 2\mathbf{i} + \mathbf{j} - 2\mathbf{k},$$
$$\mathbf{C} = 3\mathbf{i} - 4\mathbf{j} + 12\mathbf{k}.$$

1. Find $(\mathbf{A} \cdot \mathbf{B})\mathbf{C}$ and $\mathbf{A}(\mathbf{B} \cdot \mathbf{C})$.
2. Find the volume of the box having **A**, **B**, **C** as three coterminous edges.
3. (a) Find $\mathbf{A} \times \mathbf{B}$, and use the result to find

$$(\mathbf{A} \times \mathbf{B}) \times \mathbf{C}.$$

   (b) Find $(\mathbf{A} \times \mathbf{B}) \times \mathbf{C}$ by another method.

4. Prove that any vector **A** satisfies the identity

$$\mathbf{A} = \tfrac{1}{2}[\mathbf{i} \times (\mathbf{A} \times \mathbf{i}) + \mathbf{j} \times (\mathbf{A} \times \mathbf{j}) + \mathbf{k} \times (\mathbf{A} \times \mathbf{k})].$$

5. Express the product $\mathbf{R} = (\mathbf{A} \times \mathbf{B}) \times (\mathbf{C} \times \mathbf{D})$ in the form $a\mathbf{C} + b\mathbf{D}$, with a and b scalars.
6. Find the volume of the tetrahedron with vertices at (0, 0, 0), (1, −1, 1), (2, 1, −2), and (−1, 2, −1).
7. Use Eq. (3) to show that

   (a) $\mathbf{A} \cdot (\mathbf{C} \times \mathbf{B}) = -\mathbf{A} \cdot (\mathbf{B} \times \mathbf{C})$,
   (b) $\mathbf{A} \cdot (\mathbf{A} \times \mathbf{B}) = 0$,
   (c) $(\mathbf{A} + \mathbf{D}) \cdot (\mathbf{B} \times \mathbf{C}) = \mathbf{A} \cdot (\mathbf{B} \times \mathbf{C}) + \mathbf{D} \cdot (\mathbf{B} \times \mathbf{C})$.

   Interpret the results geometrically.
8. Explain the statement in the text that $(\mathbf{A} \times \mathbf{B}) \times \mathbf{C}$ is parallel to the plane determined by **A** and **B**. Illustrate with a sketch.
9. Explain the statement, at the end of Problem 2, that $(\mathbf{A} \times \mathbf{B}) \times (\mathbf{C} \times \mathbf{D})$ is parallel to the line of intersection of the **A**, **B**-plane and the **C**, **D**-plane. Illustrate with a sketch.
10. Find a line in the plane of $P_0(0, 0, 0)$, $P_1(2, 2, 0)$, and $P_2(0, 1, -2)$ that is perpendicular to the line

$$\frac{x+1}{3} = \frac{y-1}{2} = 2z.$$

11. Let $P(1, 2, -1)$, $Q(3, -1, 4)$, and $R(2, 6, 2)$ be three vertices of a parallelogram $PQRS$.
    (a) Find the coordinates of S.
    (b) Find the area of $PQRS$.
    (c) Find the area of the projection of $PQRS$ in the xy-plane; in the yz-plane; in the xz-plane.
12. Show that the area of a parallelogram in space is the square root of the sum of the squares of the areas of its projections on any three mutually orthogonal planes.

* This assumes that $\mathbf{A} \times \mathbf{B}$ and $\mathbf{n}$ lie on the same side of the plane $PQRS$. If they are on opposite sides, take the absolute value to get the area.

12.10 LOCI IN SPACE: CYLINDERS

In this article and the next, we shall consider some extensions of analytic geometry to space of three dimensions. We shall begin with the notion of a surface.

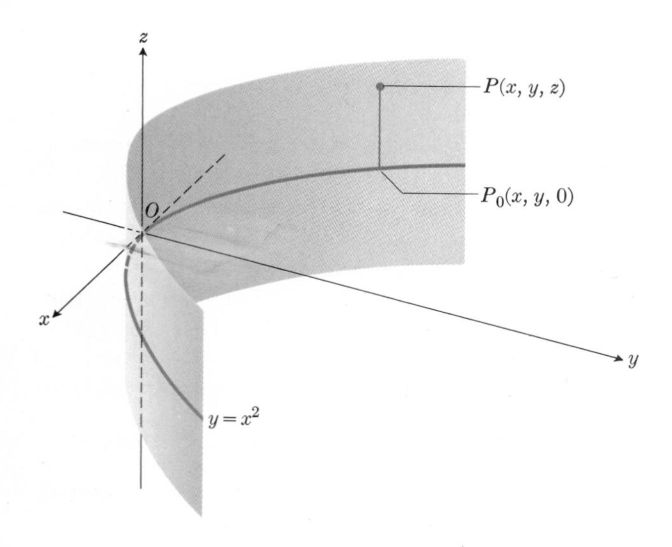

12.43 Parabolic cylinder.

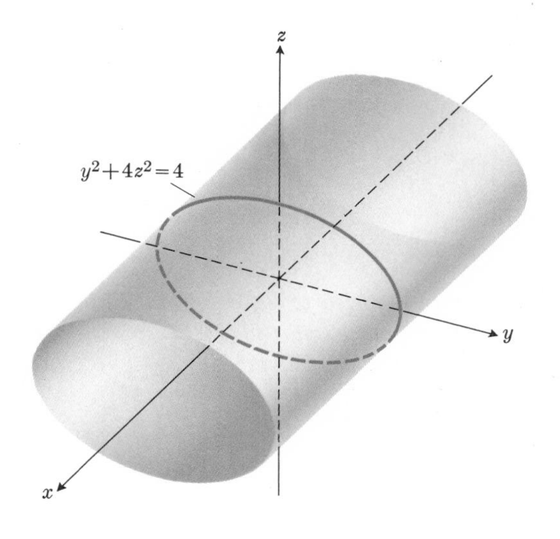

12.44 Elliptic cylinder.

The locus of points $P(x, y, z)$ that satisfy an equation

$$F(x, y, z) = 0 \tag{1}$$

may be interpreted in a broad sense as being a surface. The simplest surfaces are planes, and we have already seen that the equation of a plane is a linear equation; that is, it involves only first powers of the variables x, y, and z.

Next after planes in order of simplicity are *cylinders*. In general, a *cylinder* is a surface generated by a straight line that moves parallel to a given line and passes through a given curve. For example, the given curve may be the curve

$$f(x, y) = 0 \tag{2}$$

in the xy-plane and the generator of the cylinder may always be parallel to the z-axis. Then if the point $P_0(x, y, 0)$ lies on the curve (2), the point $P(x, y, z)$ having the same values of x and y and any value of z also lies on the surface. In other words, if its x- and y-coordinates satisfy the equation $f(x, y) = 0$, then the point P lies on the cylinder, regardless of the value of its z-coordinates. Conversely, if the point $P(x, y, z)$ is on the cylinder, then the point $P_0(x, y, 0)$ is on the curve in the xy-plane, and hence the x- and y-coordinates of P satisfy Eq. (2). Thus, if we interpret the equation $f(x, y) = 0$ as the equation of a space locus (rather than the equation of a plane curve), then the locus is a cylinder with elements parallel to the z-axis (the "missing letter" in the equation) and having the curve

$$f(x, y) = 0 \quad \text{in the plane} \quad z = 0$$

as a typical cross section.

Example 1. (See Fig. 12.43.) The cylinder

$$y = x^2$$

has elements parallel to the z-axis and a parabolic cross section in the plane $z = 0$. It is called a *parabolic cylinder.*

Clearly, the discussion above could be carried through for cylinders with elements parallel to the other coordinate axes. We may summarize the result by saying that *an equation in Cartesian coordinates, from which one letter is missing, represents a cylinder in space, with elements parallel to the axis associated with the missing letter.*

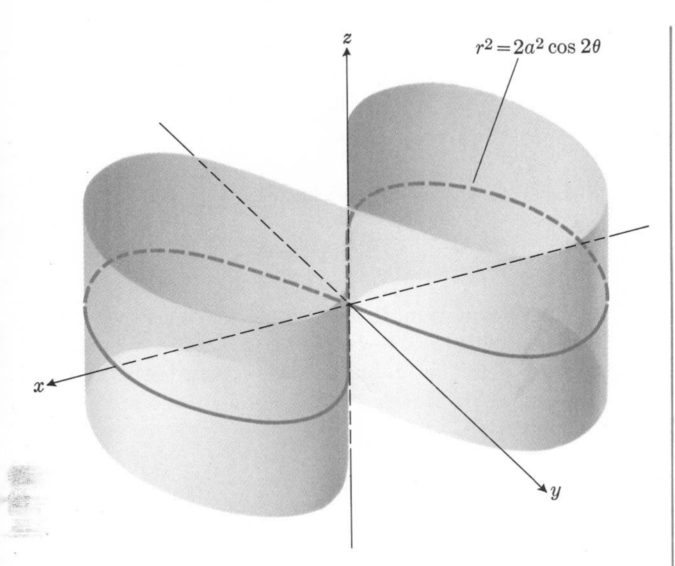

12.45 A cylinder whose cross section is a lemniscate.

Example 2. (See Fig. 12.44.) The surface

$$y^2 + 4z^2 = 4$$

is an elliptic cylinder with elements parallel to the x-axis. It extends indefinitely in both the negative and positive directions along the x-axis, which in this case is also called the axis of the cylinder, since it passes through the centers of the elliptical cross sections of the cylinder.

Example 3. (See Fig. 12.45.) The surface

$$r^2 = 2a^2 \cos 2\theta$$

in cylindrical coordinates is a cylinder with elements parallel to the z-axis. Each section perpendicular to the z-axis is a lemniscate. The cylinder extends indefinitely in the positive and negative directions along the z-axis.

EXERCISES

Describe and sketch each of the following surfaces [(r, θ, z) are cylindrical coordinates].

1. $x^2 + y^2 = a^2$
2. $x^2 - z^2 = 1$
3. $r = 2a \cos \theta$
4. $r = a(1 + \cos \theta)$
5. $y^2 + z^2 - 4z = 0$
6. $x^2 + 4z^2 - 4z = 0$

12.11 QUADRIC SURFACES

A surface whose equation is a quadratic in the variables x, y, and z is called a *quadric* surface. We shall not make an exhaustive study of such surfaces, but we shall briefly indicate how the simpler ones which are most frequently encountered may be recognized from their equations. Just as we have plane curves which are parabolas, ellipses, and hyperbolas, we here have surfaces called paraboloids, ellipsoids, and hyperboloids. We often analyze an equation to determine the nature of the curves that are obtained by cutting the given surface by planes

$$x = \text{constant}, \quad y = \text{constant}, \quad z = \text{constant}$$

perpendicular to the coordinate axes. This is usually sufficient to enable us to visualize the surface.

The *sphere*

$$(x - h)^2 + (y - k)^2 + (z - m)^2 = a^2 \tag{1}$$

with center at (h, k, m) and radius a has already been mentioned in Article 12.5. Similarly, the various *cylinders*

$$Ax^2 + Bxy + Cy^2 + Dx + Ey + F = 0 \tag{2}$$

with elements parallel to the z-axis, and others with elements parallel to the other coordinate axes, are familiar and will not be further discussed. In the examples that follow, we shall refer the surfaces discussed to coordinate axes that yield simple forms of the equations. For example, we take the origin to be at the center of the ellipsoid in Example 1 below. If the center were at (h, k, m) instead, the equation would simply have $x - h$, $y - k$, and $z - m$ in place of x, y, z respectively. We take a, b, and c to be positive constants in every case.

Example 1. The ellipsoid

$$\frac{x^2}{a^2} + \frac{y^2}{b^2} + \frac{z^2}{c^2} = 1 \tag{3}$$

cuts the coordinate axes at $(\pm a, 0, 0)$, $(0, \pm b, 0)$, and $(0, 0, \pm c)$. It is limited in extent to lie inside the rectangular box where

$$|x| \leq a, \quad |y| \leq b, \quad |z| \leq c.$$

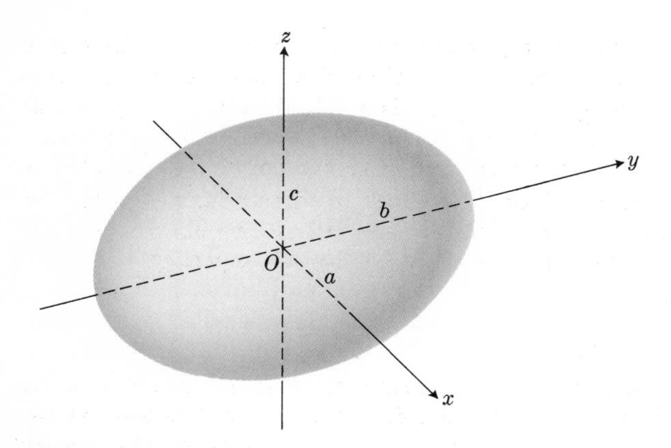

12.46 Ellipsoid.

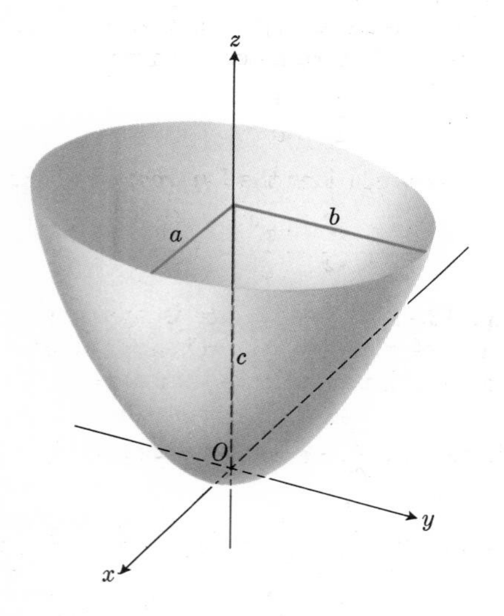

12.47 Elliptic paraboloid.

Since only even powers of x, y, and z occur in the equation, this surface is symmetric with respect to each of the coordinate planes. The sections cut out by the coordinate planes are ellipses; for example, we have

$$\frac{x^2}{a^2} + \frac{y^2}{b^2} = 1 \quad \text{when } z = 0.$$

Each section cut out by a plane

$$z = z_1, \quad |z_1| < c$$

is an ellipse

$$\frac{x^2}{a^2\left(1 - \dfrac{z_1^2}{c^2}\right)} + \frac{y^2}{b^2\left(1 - \dfrac{z_1^2}{c^2}\right)} = 1$$

with center on the z-axis and having semiaxes

$$\frac{a}{c}\sqrt{c^2 - z_1^2} \quad \text{and} \quad \frac{b}{c}\sqrt{c^2 - z_1^2}.$$

With this information, it is a simple matter to visualize the surface, which is sketched in Fig. 12.46. When two of the three semiaxes a, b, and c are equal, the surface is an ellipsoid of revolution, and when all three are equal, it is a sphere.

Example 2. Consider *the elliptic paraboloid*

$$\frac{x^2}{a^2} + \frac{y^2}{b^2} = \frac{z}{c}. \tag{4}$$

shown in Fig. 12.47. The surface is symmetrical with respect to the planes $x = 0$ and $y = 0$. The only intercept on the axes is at the origin. Since the left-hand side of the equation is nonnegative, the surface is limited to the region $z \geq 0$; that is, it lies above the xy-plane. The section cut out from the surface by the yz-plane is

$$x = 0, \quad y^2 = \frac{b^2}{c} z,$$

which is a parabola with vertex at the origin and opening upward. Similarly, one finds that

$$\text{when} \quad y = 0, \qquad \text{then} \quad x^2 = \frac{a^2}{c} z,$$

which also represents such a parabola. When $z = 0$, the cut reduces to the single point $(0, 0, 0)$. Each plane $z = z_1 > 0$ perpendicular to the z-axis cuts the surface in an ellipse of semiaxes

$$a\sqrt{z_1/c} \quad \text{and} \quad b\sqrt{z_1/c}.$$

These semiaxes increase in magnitude as z_1 increases. The paraboloid extends indefinitely upward. When $a = b$ the paraboloid is a paraboloid of revolution. In this case one may also give its equation very simply in cylindrical coordinates as

$$\frac{r^2}{a^2} = \frac{z}{c}. \tag{5}$$

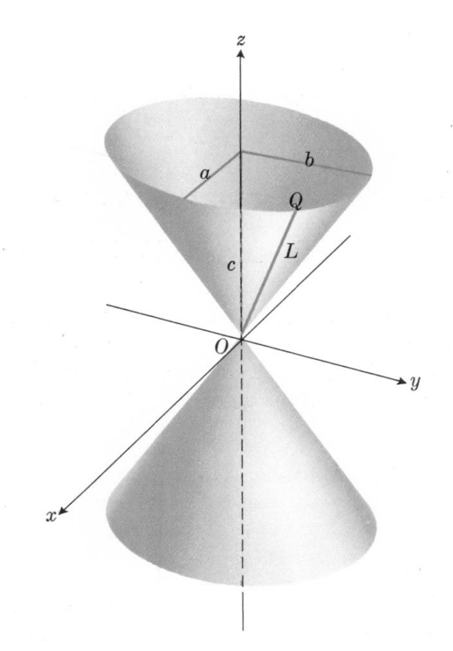

12.48 Elliptic cone.

Example 3. *The elliptic cone*

$$\frac{x^2}{a^2}+\frac{y^2}{b^2}=\frac{z^2}{c^2}, \tag{6}$$

shown in Fig. 12.48, is symmetrical with respect to all three coordinate planes. The plane $z = 0$ cuts the surface in the single point $(0, 0, 0)$. The plane $x = 0$ cuts it in the two intersecting straight lines

$$x = 0, \qquad \frac{y}{b} = \pm\frac{z}{c} \tag{7}$$

and the plane $y = 0$ gives

$$y = 0, \qquad \frac{x}{a} = \pm\frac{z}{c}. \tag{8}$$

The section cut out by a plane $z = z_1 > 0$ is an ellipse with center on the z-axis and vertices lying on the straight lines (7) and (8). In fact, the whole surface is generated by a straight line L passing through the origin and a point Q on the ellipse

$$z = c, \qquad \frac{x^2}{a^2}+\frac{y^2}{b^2} = 1.$$

As the point Q traces out the ellipse, the infinite line L generates the surface, which is a cone with elliptic cross sections. To see that this is indeed the case, suppose that $Q(x_1, y_1, z_1)$ is a point on the surface and t is any scalar. Then the vector from O to the point $P(tx_1, ty_1, tz_1)$ is simply t times $\overrightarrow{OQ}$, so that as t varies from $-\infty$ to $+\infty$ the point P traces out the infinite line L. But since Q is assumed to be on the surface, the equation

$$\frac{x_1^2}{a^2}+\frac{y_1^2}{b^2}=\frac{z_1^2}{c^2}$$

is satisfied. Multiplying both sides of this equation by t^2, we see that the point $P(tx_1, ty_1, tz_1)$ is also on the surface. This establishes the validity of the remark that the surface is a cone generated by the line L through O and the point Q on the ellipse.

If $a = b$, the cone is a right circular cone and its equation in cylindrical coordinates is simply

$$\frac{r}{a} = \frac{z}{c}. \tag{9}$$

Example 4. We now consider the *hyperboloid of one sheet*

$$\frac{x^2}{a^2}+\frac{y^2}{b^2}-\frac{z^2}{c^2} = 1, \tag{10}$$

shown in Fig. 12.49. This surface is symmetric with respect to each of the three coordinate planes. The sections cut out by the coordinate planes are

$$\text{the hyperbola } \frac{y^2}{b^2}-\frac{z^2}{c^2} = 1, \quad \text{for } x = 0, \tag{11a}$$

$$\text{the hyperbola } \frac{x^2}{a^2}-\frac{z^2}{c^2} = 1, \quad \text{for } y = 0, \tag{11b}$$

$$\text{the ellipse } \frac{x^2}{a^2}+\frac{y^2}{b^2} = 1, \quad \text{for } z = 0. \tag{11c}$$

The plane $z = z_1$ cuts the surface in an ellipse with center on the z-axis and vertices on the hyperbolas in (11). The surface is *connected*, by which we mean that it is possible to travel from any point on it to any other point on it without leaving the surface. For this reason, it is said to have *one* sheet, in contrast to the next example, which illustrates a surface of *two* sheets. In the special case where $a = b$, the surface is a hyperboloid of revolution with equation given in cylindrical coordinates by

$$\frac{r^2}{a^2}-\frac{z^2}{c^2} = 1. \tag{12}$$

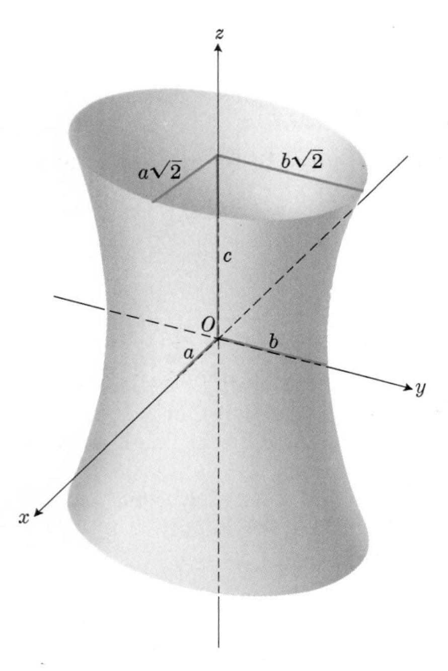

12.49 Hyperboloid of one sheet.

12.50 Hyperboloid of two sheets.

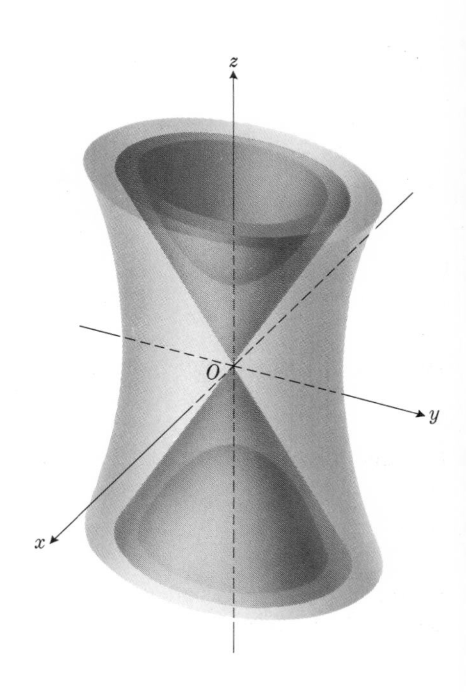

12.51 Cone asymptotic to hyperboloid of one sheet and hyperboloid of two sheets.

Example 5. Next we consider the *hyperboloid of two sheets*

$$\frac{z^2}{c^2} - \frac{x^2}{a^2} - \frac{y^2}{b^2} = 1, \tag{13}$$

shown in Fig. 12.50. The surface is symmetric with respect to the three coordinate planes. The plane $z = 0$ does not intersect the surface; in fact, one must have

$$|z| \geq c$$

for real values of x and y in Eq. (13). The hyperbolic sections

$$\frac{z^2}{c^2} - \frac{y^2}{b^2} = 1, \quad \text{for } x = 0,$$

$$\frac{z^2}{c^2} - \frac{x^2}{a^2} = 1, \quad \text{for } y = 0,$$

have their vertices and foci on the z-axis. The surface is separated into two portions, one above the plane $z = c$ and the other below the plane $z = -c$. This accounts for the name of the surface.

It is worth noting that Eqs. (10) and (13) differ in the number of negative terms that each contains on the left-hand side when the right-hand side is $+1$. The number of negative signs is the same as the number of sheets of the hyperboloid. If we compare these two equations with Eq. (6), we see that replacing the unity on the right-hand side of either Eq. (10) or (13) by zero gives the equation of a cone. This cone is, in fact, asymptotic to both of the hyperboloids (10) and (13) in the same way that the lines

$$\frac{y^2}{b^2} - \frac{z^2}{c^2} = 0$$

are asymptotic to the two hyperbolas

$$\frac{y^2}{b^2} - \frac{z^2}{c^2} = \pm 1$$

in the yz-plane (see Fig. 12.51).

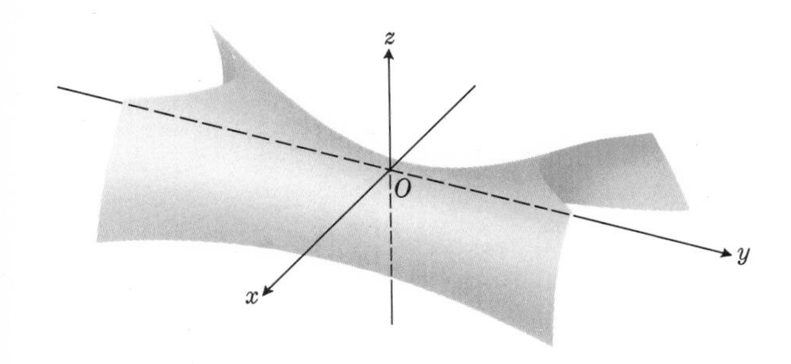

12.52 Hyperbolic paraboloid.

Example 6. *The hyperbolic paraboloid*

$$\frac{y^2}{b^2} - \frac{x^2}{a^2} = \frac{z}{c}, \tag{14}$$

shown in Fig. 12.52, has symmetry with respect to the planes $x = 0$ and $y = 0$. The sections in these planes are

$$y^2 = b^2 \frac{z}{c}, \quad \text{for } x = 0, \tag{15a}$$

$$x^2 = -a^2 \frac{z}{c}, \quad \text{for } y = 0, \tag{15b}$$

which are parabolas. In the plane $x = 0$, the parabola opens upward and has vertex at the origin. The parabola in the plane $y = 0$ has the same vertex, but it opens downward. If we cut the surface by a plane $z = z_1 > 0$, the section is a hyperbola,

$$\frac{y^2}{b^2} - \frac{x^2}{a^2} = \frac{z_1}{c}, \tag{16}$$

whose focal axis is parallel to the y-axis and which has its vertices on the parabola in (15a). If, on the other hand, z_1 is negative in Eq. (16), then the focal axis of the hyperbola is parallel to the x-axis, and its vertices lie on the parabola in (15b). Near the origin the surface is shaped very much like a saddle. To a person traveling along the surface in the yz-plane, the origin looks like a minimum. To a person traveling in the xz-plane, on the other hand, the origin looks like a maximum. Such a point is called a *minimax* or *saddle point* of a surface. We shall discuss maximum and minimum points on surfaces in the next chapter.

If $a = b$ in Eq. (14), then the surface is not a surface of revolution, but it is possible to express the equation in the alternative form

$$\frac{2x'y'}{a^2} = \frac{z}{c} \tag{17}$$

if we refer it to $x'y'$-axes obtained by rotating the xy-axes through 45°.

EXERCISES

Describe and sketch the following surfaces [(r, θ, z) are cylindrical coordinates].

1. $x^2 + y^2 + 4x - 6y = z$
2. $x^2 + y^2 + z^2 + 4x - 6y = 3$
3. $x^2 + 4y^2 + z^2 - 8y = 0$
4. $x^2 + 4y^2 + 4z^2 - 8y = 0$
5. $4x^2 + 4y^2 + 4z^2 - 8y = 0$
6. $x^2 - y^2 + z^2 + 4x - 6y = 9$
7. $x^2 - y^2 - z^2 + 4x - 6y = 9$
8. $z^2 = 4x$
9. $z^2 = 4xy$
10. $z = 4xy$
11. $z = r^2$
12. $z = r$
13. $z^2 = r$
14. $z^2 = x^2 + 4y^2$
15. $z^2 = x^2 - 4y^2$
16. $z^2 = 4y^2 - x^2$
17. $z^2 = x^2 + 4y^2 - 2x + 8y + 4z$
18. $z^2 = x^2 + 4y^2 - 2x + 8y + 4z + 1$
19. $x^2 + 4z^2 = 4$
20. $x = y^2 + 4z^2 + 1$
21. $z = r \cos \theta$
22. $z = r \sin \theta$
23. $z = \sin \theta \quad (0 \le \theta \le \pi/2)$
24. $z = \cosh \theta \quad (0 \le \theta \le \pi/2)$
25. (a) Express the area $A(z_1)$ of the cross section cut from the ellipsoid

$$\frac{x^2}{a^2} + \frac{y^2}{b^2} + \frac{z^2}{c^2} = 1$$

by the plane $z = z_1$ as a function of z_1. (The area of an ellipse of semiaxes A and B is πAB.)

(b) By integration, find the volume of the ellipsoid of (a). Consider slices made by planes perpendicular to the z-axis. Does your answer give the correct volume of a sphere if $a = b = c$?

26. By integration, prove that the volume of the segment of the elliptic paraboloid

$$\frac{x^2}{a^2} + \frac{y^2}{b^2} = \frac{z}{c}$$

cut off by the plane $z = h$ is equal to one-half the area of its base times its altitude.

27. (a) By integration, find the volume between the plane $z = 0$ and the plane $z = h$, enclosed by the hyperboloid of one sheet (Eq. 10).

(b) Express your answer to (a) in terms of the altitude h and the areas A_0 and A_h of the plane ends of the segment of the hyperboloid.

(c) Verify that the volume of (a) is also given exactly by the prismoid formula

$$V = h(A_0 + 4A_m + A_h)/6,$$

where A_0 and A_h are the areas of the plane ends of the segment of the hyperboloid and A_m is the area of its midsection cut out by the plane $z = h/2$.

28. If the hyperbolic paraboloid

$$\frac{y^2}{b^2} - \frac{x^2}{a^2} = \frac{z}{c}$$

is cut by the plane $y = y_1$, the resulting curve is a parabola. Find its vertex and focus.

29. What is the nature, in general, of a surface whose equation in spherical coordinates has the form $\rho = F(\phi)$? Give reasons for your answer.

Describe and sketch the following surfaces, which are special cases of Exercise 29.

30. $\rho = a \cos \phi$

31. $\rho = a(1 + \cos \phi)$

REVIEW QUESTIONS AND EXERCISES

1. When are two vectors equal?
2. How are two vectors added? Subtracted?
3. If a vector is multiplied by a scalar, how is the result related to the original vector? In your discussion include all possible values of the scalar: positive, negative, and zero.
4. In a single diagram, show the Cartesian, cylindrical, and spherical coordinates of an arbitrary point P, and write the expressions for each set of coordinates in terms of the other two kinds.
5. What is the locus, in space, described by

(a) x = constant,
(b) r = constant,
(c) θ = constant,
(d) ρ = constant,
(e) ϕ = constant,
(f) $ax + by + cz = d$,
(g) $ax^2 + by^2 + cz^2 = d$?

6. What is the length of the vector $a\mathbf{i} + b\mathbf{j} + c\mathbf{k}$? On what theorem of plane geometry does this result depend?
7. Define *scalar product* of two vectors. Which algebraic laws (commutative, associative, distributive) are satisfied by the operations of addition and scalar multiplication of vectors? Which of these laws is (are) not satisfied? Explain. When is the scalar product equal to zero?
8. Suppose that $\mathbf{i}, \mathbf{j}, \mathbf{k}$ is one set of mutually orthogonal unit vectors and that $\mathbf{i}', \mathbf{j}', \mathbf{k}'$ is another set of such vectors. Suppose that all the scalar products of a unit vector from one set with a unit vector from the other set are known. Let

$$\mathbf{A} = a\mathbf{i} + b\mathbf{j} + c\mathbf{k} = a'\mathbf{i}' + b'\mathbf{j}' + c'\mathbf{k}'$$

and express a, b, c in terms of a', b', c'; and conversely. (Expressions involve $\mathbf{i} \cdot \mathbf{i}'$, $\mathbf{i} \cdot \mathbf{j}'$, $\mathbf{i} \cdot \mathbf{k}'$, and so forth.)

9. List four applications of the scalar product.
10. Define *vector product* of two vectors. Which algebraic laws (commutative, associative, distributive) are satisfied by the vector product operation (combined with addition), and which are not? Explain. When is the vector product equal to zero?
11. Derive the formula for expressing the vector product of two vectors as a determinant. What is the effect of interchanging the order of the two vectors and the corresponding rows of the determinant?
12. How may vector and scalar products be used to find the equation of a plane through three given points?
13. With the book closed, develop equations for a line

(a) through two given points,
(b) through one point and parallel to a given line.

14. With the book closed, develop the equation of a plane
(a) through a given point and normal to a given vector,
(b) through one point and parallel to a given plane,
(c) through a point and perpendicular to each of two given planes.
15. What is the geometrical interpretation of

$$\mathbf{A} \cdot (\mathbf{B} \times \mathbf{C})?$$

When is this triple scalar product equal to zero?
16. What is the meaning of
(a) $\mathbf{A} \times (\mathbf{B} \times \mathbf{C})$, (b) $\mathbf{A} \times (\mathbf{B} \cdot \mathbf{C})$,
(c) $\mathbf{A} \cdot (\mathbf{B} \cdot \mathbf{C})$?
17. Given a parallelogram $PQRS$ in space, how could you find a vector normal to its plane and with length equal to its area?
18. What is the space locus of an equation of the form
(a) $f(x, y) = 0$, (b) $f(z, r) = 0$,
(c) $z = f(\theta)$, $0 \le \theta < 2\pi$?
19. Define *quadric surface*. Name and sketch six different quadric surfaces and indicate their equations.

MISCELLANEOUS EXERCISES

In Exercises 1 through 10, find parametric equations of the locus of $P(x, y)$ for the data given.

1. $\dfrac{dx}{dt} = x^2$, $\dfrac{dy}{dt} = x$; $t = 0, x = 1, y = 1$
2. $\dfrac{dx}{dt} = \cos^2 x$, $\dfrac{dy}{dt} = x$; $t = 0, x = \dfrac{\pi}{4}, y = 0$
3. $\dfrac{dx}{dt} = e^t$, $\dfrac{dy}{dt} = xe^x$; $t = 0, x = 1, y = 0$
4. $\dfrac{dx}{dt} = 6 \sin 2t$, $\dfrac{dy}{dt} = 4 \cos 2t$; $t = 0, x = 0, y = 4$
5. $\dfrac{dx}{dt} = 1 - \cos t$, $\dfrac{dy}{dt} = \sin t$; $t = 0, x = 0, y = 0$
6. $\dfrac{dx}{dt} = \sqrt{1 + y}$, $\dfrac{dy}{dt} = y$; $t = 0, x = 0, y = 1$
7. $\dfrac{dx}{dt} = \operatorname{sech} x$, $\dfrac{dy}{dt} = x$; $t = 0, x = 0, y = 0$
8. $\dfrac{dx}{dt} = \cosh \dfrac{t}{2}$, $\dfrac{dy}{dt} = x$; $t = 0, x = 2, y = 0$
9. $\dfrac{dx}{dt} = y$, $\dfrac{dy}{dt} = -x$; $t = 0, x = 0, y = 4$
10. $\dfrac{d^2x}{dt^2} = -\dfrac{dx}{dt}$, $\dfrac{dy}{dt} = x$; $t = 0, x = 1, y = 1, \dfrac{dx}{dt} = 1$
11. A particle is projected with velocity v at an angle α to the horizontal from a point that is at the foot of a hill inclined at an angle ϕ to the horizontal, where

$$0 < \phi < \alpha < (\pi/2).$$

Show that it reaches the ground at a distance

$$\frac{2v^2 \cos \alpha}{g \cos^2 \phi} \sin (\alpha - \phi)$$

measured up the face of the hill. Hence show that the greatest range achieved for a given v is when $\alpha = (\phi/2) + (\pi/4)$.
12. A wheel of radius 4 in. rolls along the x-axis with angular velocity 2 rad/sec. Find the locus described by a point on a spoke and 2 in. from the center of the wheel if it starts from the point $(0, 2)$ at time $t = 0$.
13. OA is the diameter of a circle of radius a. AN is tangent to the circle at A. A line through O making angle θ with diameter OA intersects the circle at M and tangent line at N. On ON a point P is located so that $OP = MN$. Taking O as origin, OA along the y-axis, and angle θ as parameter, find parametric equations of the locus described by P.
14. Let a line AB be the x-axis of a system of rectangular coordinates. Let the point C be the point $(0, 1)$. Let the line DE through C intersect AB at F. Let P and P' be the points on DE such that $PF = P'F = a$. Find parametric equations of the loci of P and P' in terms of the angle $\theta = \angle CFB$.
15. For the curve

$$x = a(t - \sin t), \quad y = a(1 - \cos t),$$

find the following quantities.
(a) The area bounded by the x-axis and one loop of the curve
(b) The length of one loop

(c) The area of the surface of revolution obtained by rotating one loop about the x-axis

(d) The coordinates of the centroid of the area in (a)

16. In the accompanying figure, D is the midpoint of side AB and E is one-third of the way between C and B. *Using vectors*, prove that F is midpoint of the line CD.

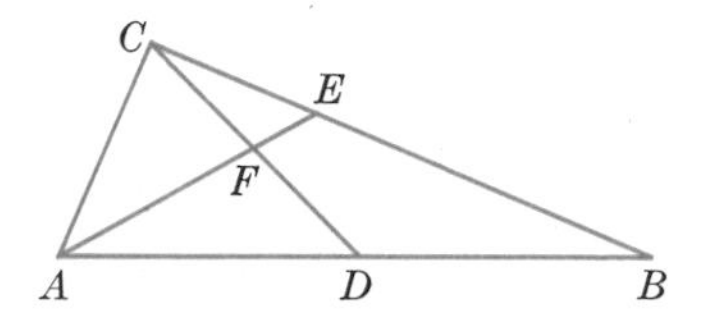

17. The vectors $2\mathbf{i} + 3\mathbf{j}$, $4\mathbf{i} + \mathbf{j}$, and $5\mathbf{i} + y\mathbf{j}$ have their initial points at the origin. Find the value of y so that the vectors terminate on one straight line.

18. $\mathbf{A}$ and $\mathbf{B}$ are vectors from the origin to the two points A and B. The point P is determined by the vector $\overrightarrow{OP} = x\mathbf{A} + y\mathbf{B}$, where x and y are positive quantities, neither of which is zero, and whose sum is equal to one. Prove that P lies on the line segment AB.

19. Using vector methods, prove that the segment joining the midpoints of two sides of a triangle is parallel to, and half the length of, the third side.

20. Let ABC be a triangle and let M be the midpoint of AB. Let P be the point on CM that is two-thirds of the way from C to M. Let O be any point in space.

(a) Show that

$$\overrightarrow{OP} = \left(\frac{\overrightarrow{OA} + \overrightarrow{OB} + \overrightarrow{OC}}{3}\right).$$

(b) Show how the result in (a) leads to the conclusion that the medians of a triangle meet in a point.

21. A, B, C are the vertices of a triangle and a, b, c are the midpoints of the opposite sides. Show that

$$\overrightarrow{Aa} + \overrightarrow{Bb} + \overrightarrow{Cc} = \mathbf{0}.$$

Interpret the result geometrically.

22. Vectors are drawn from the center of a regular polygon to its vertices. Show that their sum is zero.

23. Let $\mathbf{A}$, $\mathbf{B}$, $\mathbf{C}$ be vectors from a common point O to points A, B, C.

(a) If A, B, C are collinear, show that three constants x, y, z (not all zero) exist such that

$$x + y + z = 0 \quad \text{and} \quad x\mathbf{A} + y\mathbf{B} + z\mathbf{C} = \mathbf{0}.$$

(b) Conversely, if three constants x, y, z (not all zero) exist such that

$$x + y + z = 0 \quad \text{and} \quad x\mathbf{A} + y\mathbf{B} + z\mathbf{C} = \mathbf{0},$$

show that A, B, C are collinear.

24. Find the vector projection of $\mathbf{B}$ onto $\mathbf{A}$ if

$$\mathbf{A} = 3\mathbf{i} - \mathbf{j} + \mathbf{k}, \quad \mathbf{B} = 2\mathbf{i} + \mathbf{j} - 2\mathbf{k}.$$

25. Find the cosine of the angle between the line

$$(1 - x)/4 = y/3 = -z/5$$

and the vector $\mathbf{i} + \mathbf{j}$.

26. Given two noncollinear vectors $\mathbf{A}$ and $\mathbf{B}$. Given also that $\mathbf{A}$ can be expressed in the form $\mathbf{A} = \mathbf{C} + \mathbf{D}$, where $\mathbf{C}$ is a vector parallel to $\mathbf{B}$, and $\mathbf{D}$ is a vector perpendicular to $\mathbf{B}$. Express $\mathbf{C}$ and $\mathbf{D}$ in terms of $\mathbf{A}$ and $\mathbf{B}$.

27. The curve whose vector equation is

$$\mathbf{r} = (t^4 + 2t^2 + 1)\mathbf{i} + (1 + 4t - t^4)\mathbf{j}$$

intersects the line $x + y = 0$, $z = 0$. Find the cosine of the angle which the acceleration vector makes with the radius vector at the point of intersection.

28. *Using vectors*, prove that for any four numbers a, b, c, d we have the inequality

$$(a^2 + b^2)(c^2 + d^2) \geq (ac + bd)^2.$$

Hint. Consider $\mathbf{A} = a\mathbf{i} + b\mathbf{j}$ and $\mathbf{B} = c\mathbf{i} + d\mathbf{j}$.

29. Find a vector parallel to the plane $2x - y - z = 4$ and perpendicular to the vector $\mathbf{i} + \mathbf{j} + \mathbf{k}$.

30. Find a vector which is normal to the plane determined by the points

$$A(1, 0, -1), \quad B(2, -1, 1), \quad C(-1, 1, 2).$$

31. Given vectors

$$\mathbf{A} = 2\mathbf{i} - \mathbf{j} + \mathbf{k},$$
$$\mathbf{B} = \mathbf{i} + 2\mathbf{j} - \mathbf{k},$$
$$\mathbf{C} = \mathbf{i} + \mathbf{j} - 2\mathbf{k},$$

find a *unit* vector in the plane of $\mathbf{B}$ and $\mathbf{C}$ that is $\perp\mathbf{A}$.

32. By forming the cross product of two appropriate vectors, derive the trigonometric identity

$$\sin(\alpha - \beta) = \sin\alpha\cos\beta - \cos\alpha\sin\beta.$$

33. Find a vector *of length two* parallel to the line

$$x + 2y + z - 1 = 0, \quad x - y + 2z + 7 = 0.$$

34. Given a tetrahedron with vertices O, A, B, C. A vector is constructed normal to each face, pointing outwards, and having a length equal to the area of the face. Using cross products, prove that the sum of these four outward normals is the zero vector.

35. What angle does the line of intersection of the two planes

$$2x + y - z = 0, \quad x + y + 2z = 0$$

make with the x-axis?

36. Let $\mathbf{A}$ and $\mathbf{C}$ be given vectors in space, with $\mathbf{A} \neq \mathbf{0}$ and $\mathbf{A} \cdot \mathbf{C} = 0$, and let d be a given scalar. Find a vector $\mathbf{B}$ that satisfies both equations $\mathbf{A} \times \mathbf{B} = \mathbf{C}$ and $\mathbf{A} \cdot \mathbf{B} = d$ simultaneously. The answer should be given as a formula involving $\mathbf{A}$, $\mathbf{C}$, and d.

37. Given any two vectors

$$\mathbf{A} = a_1\mathbf{i} + a_2\mathbf{j}, \quad \mathbf{B} = b_1\mathbf{i} + b_2\mathbf{j}$$

in the plane, define a new vector, $\mathbf{A} \otimes \mathbf{B}$, called their "circle product," as follows:

$$\mathbf{A} \otimes \mathbf{B} = (a_1b_1 - a_2b_2)\mathbf{i} + (a_1b_2 + a_2b_1)\mathbf{j}.$$

This product satisfies the following algebraic laws.

(a) $\mathbf{A} \otimes \mathbf{B} = \mathbf{B} \otimes \mathbf{A}$
(b) $\mathbf{A} \otimes (\mathbf{B} \otimes \mathbf{C}) = (\mathbf{A} \otimes \mathbf{B}) \otimes \mathbf{C}$
(c) $\mathbf{A} \otimes (\mathbf{B} + \mathbf{C}) = (\mathbf{A} \otimes \mathbf{B}) + (\mathbf{A} \oplus \mathbf{C})$
(d) $|\mathbf{A} \otimes \mathbf{B}| = |\mathbf{A}|\,|\mathbf{B}|$

Prove (a) and (d).

38. Find the equations of the straight line that passes through the point (1, 2, 3) and makes an angle of 30° with the x-axis and an angle of 60° with the y-axis.

39. The line L, whose equations are

$$x - 2z - 3 = 0, \quad y - 2z = 0,$$

intersects the plane

$$x + 3y - z + 4 = 0.$$

Find the point of intersection P and find the equation of that line in this plane that passes through P and is perpendicular to L.

40. Find the distance of the point (2, 2, 3) from the plane $2x + 3y + 5z = 0$.

41. Given the two parallel planes

$$Ax + By + Cz + D_1 = 0, \quad Ax + By + Cz + D_2 = 0,$$

show that the distance between them is given by the formula

$$\frac{|D_1 - D_2|}{|A\mathbf{i} + B\mathbf{j} + C\mathbf{k}|}.$$

42. Consider the straight line through the point (3, 2, 1) and perpendicular to the plane

$$2x - y + 2z + 2 = 0.$$

Compute the coordinates of the point of intersection of that line and that plane.

43. Consider the space curve whose parametric equations are

$$x = t, \quad y = t, \quad z = \tfrac{2}{3}t^{3/2}.$$

Compute the equation of the plane that passes through the point $(1, 1, \frac{2}{3})$ of this curve, and is perpendicular to the tangent of this curve at the same point.

44. Find an equation of the plane parallel to the plane $2x - y + 2z + 4 = 0$ if the point $(3, 2, -1)$ is equidistant from both planes.

45. Given the four points

$$A = (-2, 0, -3), \quad B = (1, -2, 1),$$
$$C = (-2, -\tfrac{13}{5}, \tfrac{26}{5}), \quad D = (\tfrac{16}{5}, -\tfrac{13}{5}, 0).$$

(a) Find the equation of the plane through AB that is parallel to CD.

(b) Compute the shortest distance between the lines AB and CD.

46. Find an equation of the plane passing through the endpoints of the three vectors

$$\mathbf{A} = 3\mathbf{i} - \mathbf{j} + \mathbf{k},$$
$$\mathbf{B} = \mathbf{i} + 2\mathbf{j} - \mathbf{k},$$
$$\mathbf{C} = \mathbf{i} + \mathbf{j} + \mathbf{k},$$

supposed to be drawn from the origin.

47. Show that the plane through the three points

$$(x_1, y_1, z_1), \quad (x_2, y_2, z_2), \quad (x_3, y_3, z_3)$$

is given by

$$\begin{vmatrix} x_1 - x & y_1 - y & z_1 - z \\ x_2 - x & y_2 - y & z_2 - z \\ x_3 - x & y_3 - y & z_3 - z \end{vmatrix} = 0.$$

48. Given the two straight lines

$$x = a_1t + b_1, \quad y = a_2t + b_2, \quad z = a_3t + b_3,$$

$$x = c_1\tau + d_1, \quad y = c_2\tau + d_2, \quad z = c_3\tau + d_3,$$

where t and τ are parameters. Show that the necessary and sufficient condition that the two lines either intersect or are parallel is

$$\begin{vmatrix} a_1 & c_1 & b_1 - d_1 \\ a_2 & c_2 & b_2 - d_2 \\ a_3 & c_3 & b_3 - d_3 \end{vmatrix} = 0.$$

49. Given the vectors

$$\mathbf{A} = \mathbf{i} + \mathbf{j} - \mathbf{k},$$

$$\mathbf{B} = 2\mathbf{i} + \mathbf{j} + \mathbf{k},$$

$$\mathbf{C} = -\mathbf{i} - 2\mathbf{j} + 3\mathbf{k},$$

evaluate

(a) $\mathbf{A} \cdot (\mathbf{B} \times \mathbf{C})$, (b) $\mathbf{A} \times (\mathbf{B} \times \mathbf{C})$.

50. Given four points

$$A = (1, 1, 1), \quad B = (0, 0, 2),$$

$$C = (0, 3, 0), \quad D = (4, 0, 0),$$

find the volume of the tetrahedron with vertices at A, B, C, D, and find the angle between the edges AB and AC.

51. Prove or disprove the formula

$$\mathbf{A} \times [\mathbf{A} \times (\mathbf{A} \times \mathbf{B})] \cdot \mathbf{C} = -|\mathbf{A}|^2 \mathbf{A} \cdot \mathbf{B} \times \mathbf{C}.$$

52. If the four vectors $\mathbf{A}$, $\mathbf{B}$, $\mathbf{C}$, $\mathbf{D}$ are coplanar, show that

$$(\mathbf{A} \times \mathbf{B}) \times (\mathbf{C} \times \mathbf{D}) = \mathbf{0}.$$

53. Prove the following identities in which $\mathbf{i}$, $\mathbf{j}$, $\mathbf{k}$ are three mutually perpendicular unit vectors, and $\mathbf{A}$, $\mathbf{B}$, $\mathbf{C}$ are any vectors.

(a) $\mathbf{A} \times (\mathbf{B} \times \mathbf{C}) + \mathbf{B} \times (\mathbf{C} \times \mathbf{A}) + \mathbf{C} \times (\mathbf{A} \times \mathbf{B}) = \mathbf{0}$.

(b) $\mathbf{A} \times \mathbf{B} = [\mathbf{A} \cdot (\mathbf{B} \times \mathbf{i})]\mathbf{i} + [\mathbf{A} \cdot (\mathbf{B} \times \mathbf{j})]\mathbf{j} + [\mathbf{A} \cdot (\mathbf{B} \times \mathbf{k})]\mathbf{k}$.

54. Show that

$$(\mathbf{a} \times \mathbf{b}) \cdot (\mathbf{c} \times \mathbf{d}) = \begin{vmatrix} \mathbf{a} \cdot \mathbf{c} & \mathbf{b} \cdot \mathbf{c} \\ \mathbf{a} \cdot \mathbf{d} & \mathbf{b} \cdot \mathbf{d} \end{vmatrix}.$$

55. Sketch the surfaces

(a) $(x - 1)^2 + 4(y^2 + z^2) = 16$,

(b) $z = r^2$ (cylindrical coordinates),

(c) $\rho = a \sin \phi$ (spherical coordinates).

56. Find an equation of the locus of those points in space whose distance from the point $(2, -1, 3)$ is twice their distance from the xy-plane. Name the surface and find its center of symmetry.

57. Find an equation of the sphere that has the two planes

$$x + y + z - 3 = 0, \quad x + y + z - 9 = 0$$

as tangent planes, if the two planes

$$2x - y = 0, \quad 3x - z = 0$$

pass through the center of the sphere.

58. The two cylinders $z^3 - x = 0$ and $x^2 - y = 0$ intersect in a curve C. Find an equation of a cylinder parallel to the x-axis which passes through C. This cylinder traces out a curve C' in the yz-plane. Rotate C' about the y-axis and obtain the equation of the surface so generated.

LINEAR ALGEBRA: VECTORS IN n-SPACE

CHAPTER 13

13.1 VECTORS IN EUCLIDEAN n-SPACE

Many problems of engineering, economics, and physics involve more than three variables. An engineer may encounter an electrical network having branches that carry currents $i_1, i_2, \ldots, i_n$, for example, and an economist may study exports and imports of many different products. A physicist may assign three coordinates of position and three components of velocity to each of N particles, thus introducing 6N variables. In sports, the record of wins, losses, and ties for each baseball team in the major leagues also involves many variables. Not all of these examples fit the mathematical model we are about to introduce. Many do, however, and those that do not will fit models that are similar to this one in many respects.

Vectors as ordered n-tuples

The vectors $a_1\mathbf{i} + a_2\mathbf{j}$ or $a_1\mathbf{i} + a_2\mathbf{j} + a_3\mathbf{k}$ could equally well be represented by another notation: (a_1, a_2) or (a_1, a_2, a_3). Algebraically, there is no reason why we cannot consider more general objects, say

$$(a_1, a_2, a_3, a_4) \quad \text{or} \quad (a_1, a_2, \ldots, a_n).$$

Hence we now define n-dimensional Euclidean vector space.

Definition. *Let n be a positive integer. An ordered n-tuple of real numbers*

$$(a_1, a_2, \ldots, a_n) \tag{1}$$

is called a vector in n-space. The set of all such vectors, a_i real, under operations to be explained, is called Euclidean n-space, E^n.

Definitions

Equality. *Two vectors,*

$$\begin{aligned} \mathbf{A} &= (a_1, a_2, \ldots, a_n), \\ \mathbf{B} &= (b_1, b_2, \ldots, b_n) \end{aligned} \tag{2}$$

are said to be equal if and only if their components are all equal:

$$a_1 = b_1, \quad a_2 = b_2, \quad \ldots, \quad a_n = b_n. \tag{3}$$

Addition. *Vectors are added by adding corresponding components:*

$$\mathbf{A} + \mathbf{B} = (a_1 + b_1, a_2 + b_2, \ldots, a_n + b_n). \tag{4}$$

In particular,

$$\mathbf{A} + \mathbf{A} = (a_1 + a_1, a_2 + a_2, \ldots, a_n + a_n)$$
$$= (2a_1, 2a_2, \ldots, 2a_n).$$

Naturally we write this as

$$2\mathbf{A} = 2(a_1, a_2, \ldots, a_n).$$

Multiplication by scalars. *We define multiplication of a vector* $\mathbf{A}$ *by a scalar (real number)* c *by the formula*

$$c\mathbf{A} = (ca_1, ca_2, \ldots, ca_n). \tag{5}$$

Inner product. *The inner product of two vectors* $\mathbf{A}$ *and* $\mathbf{B}$ *is the direct analog of the dot product in two and three dimensions. We shall also denote it by* $\mathbf{A} \cdot \mathbf{B}$. *Thus*

$$\mathbf{A} \cdot \mathbf{B} = a_1b_1 + a_2b_2 + \cdots + a_nb_n. \tag{6}$$

Length. *The length of a vector* $\mathbf{A}$, *written* $|\mathbf{A}|$, *is defined to be the square root of* $\mathbf{A} \cdot \mathbf{A}$. *Thus*

$$|\mathbf{A}| = \sqrt{\mathbf{A} \cdot \mathbf{A}} = \sqrt{a_1^2 + a_2^2 + \cdots + a_n^2}. \tag{7}$$

The zero vector is the only vector with zero length:

$$\mathbf{0} = (0, 0, 0, \ldots, 0).$$

Observe that $a_i^2 \geq 0$, *so the right side of Eq. (7) is positive, and therefore* $|\mathbf{A}| > 0$, *unless every component of* $\mathbf{A}$ *is zero.*

All these definitions are straightforward extensions of corresponding notions mentioned in Chapter 12. There is no cross product, however, for $n \neq 3$.

The following theorem summarizes the chief properties of addition, scalar multiplication, and inner products in E^n. Proofs of parts of the theorem are called for in the exercises, but all the results are immediate consequences of the definitions and the corresponding properties of the real numbers. The proofs are easy.

Theorem 1. *Let* $\mathbf{A}$, $\mathbf{B}$, $\mathbf{C}$, $\mathbf{D}$ *be vectors in Euclidean n-space, and let* a, b, c, d *be real numbers. Let* $\mathbf{0}$ *be the zero vector. Then*

$$\begin{aligned}
\mathbf{A} + \mathbf{0} &= \mathbf{0} + \mathbf{A} = \mathbf{A}, \\
\mathbf{A} + (-1)\mathbf{A} &= \mathbf{0}, \\
\mathbf{A} + \mathbf{B} &= \mathbf{B} + \mathbf{A}, \\
\mathbf{A} + (\mathbf{B} + \mathbf{C}) &= (\mathbf{A} + \mathbf{B}) + \mathbf{C}, \\
\mathbf{A} \cdot \mathbf{B} &= \mathbf{B} \cdot \mathbf{A}, \\
\mathbf{A} \cdot (\mathbf{B} + \mathbf{C}) &= \mathbf{A} \cdot \mathbf{B} + \mathbf{A} \cdot \mathbf{C}, \\
(c + d)\mathbf{A} &= c\mathbf{A} + d\,\mathbf{A}, \\
a(b\mathbf{C}) &= (ab)\mathbf{C}, \\
|c\mathbf{A}| &= |c|\,|\mathbf{A}|.
\end{aligned}$$

Projections

For $n > 3$, we can and do still use geometric language, but we must be careful that our proofs are algebraic. Thus, for example, we speak of two vectors being *orthogonal* when their inner product is zero:

> $\mathbf{A}$ and $\mathbf{B}$ are orthogonal
> $\Leftrightarrow a_1b_1 + a_2b_2 + \cdots + a_nb_n = 0.$

We can also define the *direction* of a nonzero vector $\mathbf{B}$ as the unit vector $\mathbf{B}/|\mathbf{B}|$. Also, if $|\mathbf{B}| \neq 0$, we can always write $\mathbf{A}$ as the sum of two vectors $\mathbf{A}_1$ and $\mathbf{A}_2$, such that $\mathbf{A}_1$ is the projection of $\mathbf{A}$ on $\mathbf{B}$ and $\mathbf{A}_2$ is orthogonal to $\mathbf{B}$:

$$\mathbf{A} = \mathbf{A}_1 + \mathbf{A}_2, \tag{8a}$$

$$\mathbf{A}_1 = c\mathbf{B}, \quad \mathbf{A}_2 \cdot \mathbf{B} = \mathbf{0}. \tag{8b}$$

For $n = 2$ or 3, the results of Article 12.6 apply to yield

$$\mathbf{A}_1 = \left(\frac{\mathbf{A} \cdot \mathbf{B}}{\mathbf{B} \cdot \mathbf{B}}\right)\mathbf{B}. \tag{9a}$$

Certainly this has the form $c\mathbf{B}$, because

$$c = \frac{\mathbf{A} \cdot \mathbf{B}}{\mathbf{B} \cdot \mathbf{B}} \tag{9b}$$

is a real number (a scalar) if $|\mathbf{B}| \neq 0$. If we define $\mathbf{A}_1$ by (9a) and let

$$\mathbf{A}_2 = \mathbf{A} - \mathbf{A}_1 = \mathbf{A} - c\mathbf{B}, \tag{9c}$$

then we find that $\mathbf{A}_2$ is orthogonal to $\mathbf{B}$, because

$$\begin{aligned}\mathbf{A}_2 \cdot \mathbf{B} &= (\mathbf{A} - c\mathbf{B}) \cdot \mathbf{B} \\ &= \mathbf{A} \cdot \mathbf{B} - c(\mathbf{B} \cdot \mathbf{B}) \\ &= \mathbf{A} \cdot \mathbf{B} - \frac{\mathbf{A} \cdot \mathbf{B}}{\mathbf{B} \cdot \mathbf{B}}(\mathbf{B} \cdot \mathbf{B}) \\ &= \mathbf{A} \cdot \mathbf{B} - \mathbf{A} \cdot \mathbf{B} = 0.\end{aligned}$$

Remark 1. We see that Eqs. (8) and (9) give a way of resolving $\mathbf{A}$ into two components, one lying along $\mathbf{B}$ and the other being orthogonal to $\mathbf{B}$. The first of these, $\mathbf{A}_1$ as given by Eq. (9a), is also called the orthogonal *projection* of $\mathbf{A}$ on $\mathbf{B}$. The second,

$$\mathbf{A}_2 = \mathbf{A} - \mathbf{A}_1,$$

is called *the component of* $\mathbf{A}$ *orthogonal to* $\mathbf{B}$, or *the orthogonal complement of* $\mathbf{A}$ *relative to* $\mathbf{B}$.

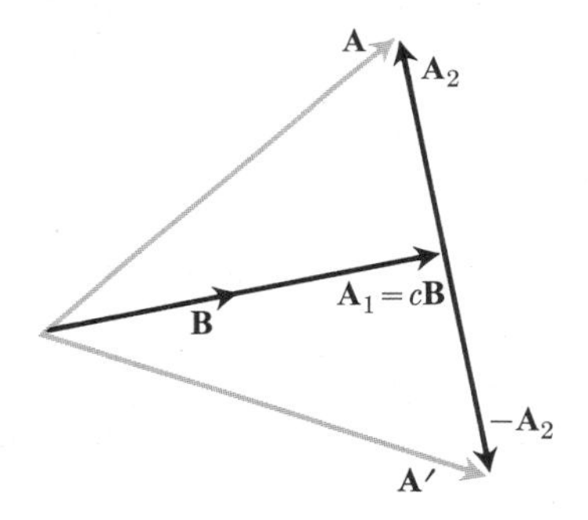

13.1 The vector $\mathbf{A}$ with components $\mathbf{A}_1$ along $\mathbf{B}$ and $\mathbf{A}_2$ and $-\mathbf{A}_2$ orthogonal to $\mathbf{B}$.

Reflection

In Fig. 13.1, suppose we want the reflection of the vector $\mathbf{A}$ across the line of $\mathbf{B}$. It appears that the appropriate vector is $\mathbf{A}_1 - \mathbf{A}_2$. If we *define* a new vector $\mathbf{A}'$ to be

$$\mathbf{A}' = \mathbf{A}_1 - \mathbf{A}_2$$

with

$$\mathbf{A}_1 = \left(\frac{\mathbf{A} \cdot \mathbf{B}}{\mathbf{B} \cdot \mathbf{B}}\right)\mathbf{B}, \qquad \mathbf{A}_2 = \mathbf{A} - \mathbf{A}_1,$$

then

$$\mathbf{A}' = \mathbf{A}_1 - (\mathbf{A} - \mathbf{A}_1) = 2\mathbf{A}_1 - \mathbf{A}.$$

Definition. *The reflected image of the vector* $\mathbf{A}$ *with respect to a nonzero vector* $\mathbf{B}$ *is the vector*

$$\mathbf{A}' = 2\left(\frac{\mathbf{A} \cdot \mathbf{B}}{\mathbf{B} \cdot \mathbf{B}}\right)\mathbf{B} - \mathbf{A}. \tag{10}$$

Problem 1. In E^2, find the reflected image of the vector $\mathbf{A} = (a, b)$ with respect to the vector $\mathbf{B} = (1, 1)$.

Solution. Since $\mathbf{A} \cdot \mathbf{B} = a + b$ and $\mathbf{B} \cdot \mathbf{B} = 2$, Eq. (10) yields

$$\begin{aligned}\mathbf{A}' &= 2\left(\frac{a+b}{2}\right)(1, 1) - (a, b) \\ &= (a + b, a + b) - (a, b) \\ &= (b, a).\end{aligned}$$

This result is what we naturally expect, because the vector from the origin to the point (1, 1) in the xy-plane lies on the line $y = x$. The reflection of (a, b) relative to this 45° line is just (b, a).

Problem 2. Let

$$n = 4, \quad \mathbf{A} = (1, -1, 0, 1), \quad \text{and} \quad \mathbf{B} = (1, 0, 1, 0).$$

Find the projection of $\mathbf{A}$ on $\mathbf{B}$, the orthogonal complement of $\mathbf{A}$ with respect to $\mathbf{B}$, and the reflected image of $\mathbf{A}$ with respect to $\mathbf{B}$.

Solution.

$$\begin{aligned}\mathbf{A} \cdot \mathbf{B} &= 1 + 0 + 0 + 0 = 1, \\ \mathbf{B} \cdot \mathbf{B} &= 1 + 0 + 1 + 0 = 2.\end{aligned}$$

Therefore the *projection* of $\mathbf{A}$ on $\mathbf{B}$ is

$$\mathbf{A}_1 = \left(\frac{\mathbf{A} \cdot \mathbf{B}}{\mathbf{B} \cdot \mathbf{B}}\right)\mathbf{B} = \tfrac{1}{2}\mathbf{B} = (\tfrac{1}{2}, 0, \tfrac{1}{2}, 0),$$

and the component of $\mathbf{A}$ orthogonal to $\mathbf{B}$ is

$$\mathbf{A}_2 = \mathbf{A} - \mathbf{A}_1 = (\tfrac{1}{2}, -1, -\tfrac{1}{2}, 1).$$

As a check, we note that

$$\mathbf{A}_1 + \mathbf{A}_2 = (1, -1, 0, 1) = \mathbf{A}$$

and

$$\mathbf{A}_2 \cdot \mathbf{B} = \tfrac{1}{2} + 0 - \tfrac{1}{2} + 0 = 0.$$

The *reflected image* of $\mathbf{A}$ with respect to $\mathbf{B}$ is

$$\mathbf{A}' = \mathbf{A}_1 - \mathbf{A}_2 = (0, 1, 1, -1).$$

As a check, we compute $\mathbf{A} - \mathbf{A}'$:

$$\begin{aligned}\mathbf{A} - \mathbf{A}' &= (1, -2, -1, 2) \\ &= 2(\tfrac{1}{2}, -1, -\tfrac{1}{2}, 1) = 2\mathbf{A}_2.\end{aligned}$$

The Pythagorean theorem

Is there a theorem that says that

$$|\mathbf{A}+\mathbf{B}|^2 = |\mathbf{A}|^2 + |\mathbf{B}|^2 \tag{11}$$

when $\mathbf{A}\cdot\mathbf{B} = 0$? Yes; and it is easily verified, because

$$\begin{aligned}|\mathbf{A}+\mathbf{B}|^2 &= (\mathbf{A}+\mathbf{B})\cdot(\mathbf{A}+\mathbf{B})\\ &= \mathbf{A}\cdot(\mathbf{A}+\mathbf{B}) + \mathbf{B}\cdot(\mathbf{A}+\mathbf{B})\\ &= \mathbf{A}\cdot\mathbf{A} + \mathbf{A}\cdot\mathbf{B} + \mathbf{B}\cdot\mathbf{A} + \mathbf{B}\cdot\mathbf{B}\\ &= |\mathbf{A}|^2 + 0 + 0 + |\mathbf{B}|^2\\ &= |\mathbf{A}|^2 + |\mathbf{B}|^2, \quad \text{if} \quad \mathbf{A}\cdot\mathbf{B} = 0.\end{aligned}$$

Triangle inequality

Another theorem from elementary geometry says that the length of any side of a triangle is less than the sum of the lengths of the other two sides. In terms of vectors (see Fig. 13.2), this would become

$$|\mathbf{A}+\mathbf{B}| < |\mathbf{A}| + |\mathbf{B}|.$$

However, such an inequality is false if $\mathbf{A}$ and $\mathbf{B}$ are collinear, for then we could have $\mathbf{B} = 2\mathbf{A}$, $\mathbf{A}+\mathbf{B} = 3\mathbf{A}$, and $|\mathbf{A}+\mathbf{B}| = |\mathbf{A}| + |\mathbf{B}| = 3|\mathbf{A}|$. But the modified triangle inequality for vectors

$$|\mathbf{A}+\mathbf{B}| \leq |\mathbf{A}| + |\mathbf{B}|$$

is correct, as we shall now prove.

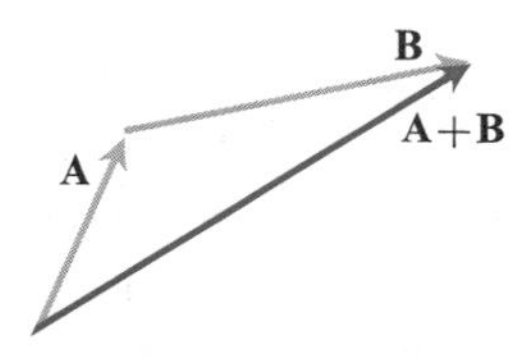

13.2 $\mathbf{A}$, $\mathbf{B}$, and $\mathbf{A}+\mathbf{B}$ as sides of a triangle: $|\mathbf{A}+\mathbf{B}| \leq |\mathbf{A}| + |\mathbf{B}|$.

Theorem 2. *Let* $\mathbf{A}$ *and* $\mathbf{B}$ *be vectors in* E^n. *Then*

$$|\mathbf{A}+\mathbf{B}| \leq |\mathbf{A}| + |\mathbf{B}|. \tag{12a}$$

Here equality holds if and only if one of the vectors is zero, or if the two vectors have the same direction; that is, if

$$\frac{\mathbf{A}}{|\mathbf{A}|} = \frac{\mathbf{B}}{|\mathbf{B}|}. \tag{12b}$$

Proof. Obviously, if $\mathbf{A} = \mathbf{0}$ or $\mathbf{B} = \mathbf{0}$, (12a) is correct as an equality:

$$|\mathbf{A}+\mathbf{0}| = |\mathbf{A}| + |\mathbf{0}| = |\mathbf{A}|$$

or

$$|\mathbf{0}+\mathbf{B}| = |\mathbf{0}| + |\mathbf{B}| = |\mathbf{B}|.$$

Next, suppose $\mathbf{B} \neq \mathbf{0}$. Then we can resolve $\mathbf{A}$ into components $\mathbf{A}_1$ and $\mathbf{A}_2$, as in Eqs. (8) and (9):

$$\mathbf{A} = \mathbf{A}_1 + \mathbf{A}_2,$$

$$\mathbf{A}_1 = c\mathbf{B}, \quad \text{with } c = \frac{\mathbf{A}\cdot\mathbf{B}}{\mathbf{B}\cdot\mathbf{B}},$$

$$\mathbf{A}_2 = \mathbf{A} - c\mathbf{B}.$$

Then $\quad |\mathbf{A}_2| > 0 \quad$ unless $\mathbf{A}_2 = \mathbf{0}$.

Case 1. If $\quad \mathbf{A}_2 = \mathbf{0},$

then $\quad \mathbf{A} = c\mathbf{B}$

and $\quad |\mathbf{A}+\mathbf{B}| = |(c+1)\mathbf{B}| = |c+1|\,|\mathbf{B}|.$

If $\quad c < 0,$

then $\quad c+1 < 1 < 1+|c|$

and $\quad -c-1 < -c = |c| < 1+|c|,$

so $\quad |c+1| < 1+|c|.$

Hence, if $\quad c < 0,$

then $\quad |\mathbf{A}+\mathbf{B}| < (1+|c|)|\mathbf{B}| = |\mathbf{B}| + |c\mathbf{B}| = |\mathbf{B}| + |\mathbf{A}|.$

If $\quad c > 0,$

then $\quad 1+c > 0$

and $\quad |1+c| = 1+c,$

so that

$$|\mathbf{A}+\mathbf{B}| = |c+1|\,|\mathbf{B}| = (c+1)|\mathbf{B}| = c|\mathbf{B}| + |\mathbf{B}| = |\mathbf{A}| + |\mathbf{B}|.$$

If $\quad c = 0,$

then $\quad \mathbf{A} = \mathbf{0}$

and $\quad |\mathbf{A}+\mathbf{B}| = |\mathbf{0}+\mathbf{B}| = |\mathbf{0}| + |\mathbf{B}|.$

Thus, if $\mathbf{A} = c\mathbf{B}$, inequality (12a) holds and

$$|\mathbf{A}+\mathbf{B}| = |\mathbf{A}| + |\mathbf{B}| \Leftrightarrow c \geq 0.$$

Case 2. If

$$\mathbf{A}_2 \neq \mathbf{0},$$

then

$$\begin{aligned}0 < |\mathbf{A}_2|^2 &= |\mathbf{A} - c\mathbf{B}|^2 \\ &= (\mathbf{A} - c\mathbf{B}) \cdot (\mathbf{A} - c\mathbf{B}) \\ &= \mathbf{A} \cdot \mathbf{A} - 2c\mathbf{A} \cdot \mathbf{B} + c^2\mathbf{B} \cdot \mathbf{B} \\ &= |\mathbf{A}|^2 - 2\left(\frac{\mathbf{A} \cdot \mathbf{B}}{\mathbf{B} \cdot \mathbf{B}}\right)(\mathbf{A} \cdot \mathbf{B}) + \frac{(\mathbf{A} \cdot \mathbf{B})^2}{(\mathbf{B} \cdot \mathbf{B})^2}(\mathbf{B} \cdot \mathbf{B}) \\ &= |\mathbf{A}|^2 - \frac{(\mathbf{A} \cdot \mathbf{B})^2}{|\mathbf{B}|^2}.\end{aligned}$$

Therefore, adding

$$\frac{(\mathbf{A} \cdot \mathbf{B})^2}{|\mathbf{B}|^2},$$

we get

$$\frac{(\mathbf{A} \cdot \mathbf{B})^2}{|\mathbf{B}|^2} < |\mathbf{A}|^2,$$

or

$$(\mathbf{A} \cdot \mathbf{B})^2 < |\mathbf{A}|^2|\mathbf{B}|^2. \tag{13}$$

Inequality (13) also implies

$$|\mathbf{A} \cdot \mathbf{B}| < |\mathbf{A}|\,|\mathbf{B}|, \tag{14}$$

because

$$x^2 < y^2 \Leftrightarrow |x| < |y|.$$

From (14), since $\mathbf{A} \cdot \mathbf{B}$ is just a real number whose absolute value is the maximum of $\mathbf{A} \cdot \mathbf{B}$ and $-(\mathbf{A} \cdot \mathbf{B})$, we can further deduce that both

$$\mathbf{A} \cdot \mathbf{B} < |\mathbf{A}|\,|\mathbf{B}| \tag{15a}$$

and

$$-\mathbf{A} \cdot \mathbf{B} < |\mathbf{A}|\,|\mathbf{B}|. \tag{15b}$$

Don't despair—we're almost there. Now we multiply both sides of (15a) by 2 and add

$$|\mathbf{A}|^2 = \mathbf{A} \cdot \mathbf{A} \quad \text{and} \quad |\mathbf{B}|^2 = \mathbf{B} \cdot \mathbf{B}$$

to get

$$\mathbf{A} \cdot \mathbf{A} + 2\mathbf{A} \cdot \mathbf{B} + \mathbf{B} \cdot \mathbf{B} < |\mathbf{A}|^2 + 2|\mathbf{A}|\,|\mathbf{B}| + |\mathbf{B}|^2.$$

This is the same as

$$(\mathbf{A} + \mathbf{B}) \cdot (\mathbf{A} + \mathbf{B}) < (|\mathbf{A}| + |\mathbf{B}|)^2,$$

or

$$|\mathbf{A} + \mathbf{B}|^2 < (|\mathbf{A}| + |\mathbf{B}|)^2.$$

Finally we take nonnegative square roots and get

$$|\mathbf{A} + \mathbf{B}| < |\mathbf{A}| + |\mathbf{B}|. \ \square$$

Remark 2. In component form, the triangle inequality (12a) reads

$$\sqrt{(a_1 + b_1)^2 + (a_2 + b_2)^2 + \cdots + (a_n + b_n)^2} \le \sqrt{a_1^2 + a_2^2 + \cdots + a_n^2} + \sqrt{b_1^2 + b_2^2 + \cdots + b_n^2}, \tag{16a}$$

and strict inequality holds

unless $a_1 = a_2 = \cdots = a_n = 0,$

or $b_1 = b_2 = \cdots = b_n = 0,$

or $a_1 = cb_1, \quad a_2 = cb_2, \quad \ldots, \quad a_n = cb_n$

for some positive real number c.

Remark 3. The inequality given by (14), or by (15a, b), says:

$$|a_1b_1 + a_2b_2 + \cdots + a_nb_n| \le \sqrt{a_1^2 + a_2^2 + \cdots + a_n^2} \times \sqrt{b_1^2 + b_2^2 + \cdots + b_n^2}. \tag{16b}$$

This is known as the *Cauchy-Schwarz inequality.*

Inequalities (14), (15a), and (15b) can also be made to tell us that if

$$|\mathbf{A}| \neq 0, \quad |\mathbf{B}| \neq 0, \quad \mathbf{A} \neq c\mathbf{B},$$

then

$$-1 < \frac{\mathbf{A} \cdot \mathbf{B}}{|\mathbf{A}|\,|\mathbf{B}|} < 1.$$

Hence there exists a real θ between 0 and π such that

$$\cos\theta = \frac{\mathbf{A} \cdot \mathbf{B}}{|\mathbf{A}|\,|\mathbf{B}|}, \tag{17a}$$

or

$$\mathbf{A} \cdot \mathbf{B} = |\mathbf{A}|\,|\mathbf{B}|\cos\theta. \tag{17b}$$

The number θ defined by (17a) is called *the measure of the angle between the vectors* **A** *and* **B**. If either **A** or **B** is the zero vector, we do not define an angle between them. Finally,

if $\quad \mathbf{A} = c\mathbf{B}$ and $c \neq 0,$

then $\quad \theta = \pi$ if $c < 0, \quad |\mathbf{A}| \neq 0, |\mathbf{B}| \neq 0,$

and $\quad \theta = 0$ if $c > 0, \quad |\mathbf{A}| \neq 0, |\mathbf{B}| \neq 0.$

Equation (17a) is then a valid equation for all nonzero vectors, and (17b) is to be interpreted as $0 = 0$ (but θ undefined) if either $\mathbf{A}$ or $\mathbf{B}$ is the zero vector.

Problem 3. Verify the triangle inequality for the following pairs of vectors in E^4.

(a) $\mathbf{A} = (1, -1, 2, 1)$, $\mathbf{B} = (0, 1, -1, 2)$
(b) $\mathbf{A} = (1, 2, 3, 4)$, $\mathbf{B} = (-1, -2, -3, -4)$
(c) $\mathbf{A} = (2, 0, 4, 6)$, $\mathbf{B} = (1, 0, 2, 3)$

Solutions

(a)
$$|\mathbf{A}| = \sqrt{1+1+4+1} = \sqrt{7},$$
$$|\mathbf{B}| = \sqrt{1+1+4} = \sqrt{6},$$
$$|\mathbf{A}+\mathbf{B}| = \sqrt{(1+0)^2+(-1+1)^2+(2-1)^2+(1+2)^2} = \sqrt{1+0+1+9} = \sqrt{11},$$
and
$$\sqrt{7}+\sqrt{6} > 2.6+2.4 = 5.0 > \sqrt{11}.$$

(b)
$$|\mathbf{A}| = \sqrt{1+4+9+16} = \sqrt{30} = |\mathbf{B}|,$$
$$|\mathbf{A}+\mathbf{B}| = |0| = 0,$$
and
$$2\sqrt{30} > 0.$$

(c)
$$|\mathbf{A}| = \sqrt{4+16+36} = \sqrt{56} = 2\sqrt{14},$$
$$|\mathbf{B}| = \sqrt{1+4+9} = \sqrt{14},$$
$$|\mathbf{A}+\mathbf{B}| = \sqrt{(2+1)^2+0^2+(4+2)^2+(6+3)^2} = \sqrt{9+36+81} = \sqrt{126} = \sqrt{9}\sqrt{14} = 3\sqrt{14}.$$
Here
$$|\mathbf{A}|+|\mathbf{B}| = 2\sqrt{14}+\sqrt{14} = 3\sqrt{14} = |\mathbf{A}+\mathbf{B}|,$$
and
$$\mathbf{A} = 2\mathbf{B}.$$

Problem 4. Find the cosines of the angles between the pairs of vectors of Problem 3 above.

Solutions. In (c), $\mathbf{A} = 2\mathbf{B}$, so $\theta = 0$ and $\cos\theta = 1$. In (b), $\mathbf{A} = -\mathbf{B}$, so $\theta = \pi$ and $\cos\theta = -1$. In (a), $\mathbf{A}\cdot\mathbf{B} = 0 - 1 - 2 + 2 = -1$, so
$$\cos\theta = \frac{\mathbf{A}\cdot\mathbf{B}}{|\mathbf{A}|\,|\mathbf{B}|} = \frac{-1}{\sqrt{7}\sqrt{6}} = \frac{-1}{\sqrt{42}}.$$

The following problem illustrates how the notions of vector, inner product, and orthogonal projection can be extended to functions. We shall use the interval $[0, 1]$ for convenience, but the ideas work equally well for any finite interval $[a, b]$.

Problem 5. Let f and g be functions that are continuous on $[0, 1]$. Two such functions can be added to give another such function; and a continuous function can be multiplied by a real constant to give another. We define the *inner product,* which we here write as $\langle f, g\rangle$, as follows:
$$\langle f, g\rangle = \int_0^1 f(x)g(x)\,dx.$$
If $f(x) = x$ and $g(x) = x^2$, find functions f_1 and f_2 such that $f(x) = f_1(x) + f_2(x)$, $f_1(x) = cg(x)$ for some constant c, and $\langle f_2, g\rangle = 0$.

Solution. As in Eqs. (9b, c), let
$$c = \frac{\langle f, g\rangle}{\langle g, g\rangle} = \frac{\int_0^1 x^3\,dx}{\int_0^1 x^4\,dx} = \frac{5}{4},$$
and take
$$f_1(x) = \tfrac{5}{4}x^2, \quad f_2(x) = x - \tfrac{5}{4}x^2.$$
Then
$$\langle f_2, g\rangle = \int_0^1 \left(x^3 - \frac{5}{4}x^4\right) dx = \frac{x^4}{4} - \frac{5}{4}\frac{x^5}{5}\Big]_0^1 = \tfrac{1}{4} - \tfrac{1}{4} = 0.$$

Remark. The functions f_2 and g are said to be *orthogonal* on $[0, 1]$.

EXERCISES

1. Prove that $c(\mathbf{A}+\mathbf{B}) = c\mathbf{A} + c\mathbf{B}$ for any scalar c and vectors $\mathbf{A}$ and $\mathbf{B}$ in E^n.
2. Prove that $\mathbf{A}\cdot(\mathbf{B}+\mathbf{C}) = \mathbf{A}\cdot\mathbf{B} + \mathbf{A}\cdot\mathbf{C}$ for any vectors $\mathbf{A}$, $\mathbf{B}$, $\mathbf{C}$ in E^n.
3. Prove that $\mathbf{A}\cdot\mathbf{B} = \mathbf{B}\cdot\mathbf{A}$ for all vectors in E^n.
4. Prove that $|\mathbf{A}+\mathbf{B}|^2 = |\mathbf{A}|^2 + 2\mathbf{A}\cdot\mathbf{B} + |\mathbf{B}|^2$ for all vectors in E^n.

In Exercises 5 through 7 let $n = 4$, with

$$\mathbf{A} = (-1, 1, 0, 2),$$
$$\mathbf{B} = (1, -1, 0, 1),$$
$$\mathbf{C} = (0, 0, 1, 1).$$

5. Compute $|\mathbf{A}|$, $|\mathbf{B}|$, $\mathbf{A} \cdot \mathbf{B}$, and $\cos \theta$, where θ measures the angle between $\mathbf{A}$ and $\mathbf{B}$. Find the orthogonal projection of $\mathbf{A}$ on $\mathbf{B}$ and the component of $\mathbf{A}$ that is orthogonal to $\mathbf{B}$. Comment.
6. Find the orthogonal projection of $\mathbf{B}$ on $\mathbf{C}$ and the component of $\mathbf{B}$ that is orthogonal to $\mathbf{C}$.
7. Find the orthogonal projections of $\mathbf{A}$ on $\mathbf{C}$, of $\mathbf{B}$ on $\mathbf{C}$, and of $(\mathbf{A} + \mathbf{B})$ on $\mathbf{C}$. How are they related? Does this agree with the result you would expect from reasoning geometrically about the analogous two- and-three-dimensional cases?
8. Suppose $\mathbf{A}$, $\mathbf{B}$, $\mathbf{C}$ are three vectors in E^n, and that $\mathbf{C} \neq \mathbf{0}$. Prove that the orthogonal projection of $(\mathbf{A} + \mathbf{B})$ on $\mathbf{C}$ is the sum of the orthogonal projections; that is,

$$\text{proj}_{\mathbf{C}}\,(\mathbf{A} + \mathbf{B}) = \text{proj}_{\mathbf{C}}\,(\mathbf{A}) + \text{proj}_{\mathbf{C}}\,(\mathbf{B}).$$

Do *not* use a geometric argument for your proof, but interpret the result geometrically after you have proved it.

In Exercises 9 through 11, use the ideas of Problem 5, and the notion *inner product* as there defined, to find the orthogonal projection f_1 of f on g, and also the component, f_2, of f that is orthogonal to g, for each of the following pairs f and g.

9. $f(x) = x, \quad g(x) = 1$
10. $f(x) = x^2, \quad g(x) = x$
11. $f(x) = x^3, \quad g(x) = x^2$

In Exercises 12 through 15, find the reflected image of the vector $\mathbf{A} = (a, b, c)$ in E^3 with respect to the nonzero vector $\mathbf{B}$ given, and comment on your answer.

12. $\mathbf{B} = (a, b, 0)$
13. $\mathbf{B} = (a, 0, c)$
14. $\mathbf{B} = (0, b, c)$
15. $\mathbf{B} = \left(\dfrac{a+b}{2}, \dfrac{a+b}{2}, c\right)$

16. (a) If a, b, c are numbers such that $ax^2 + bx + c \geq 0$ for all real x, prove that $b^2 - 4ac \leq 0$. [Observe that if the discriminant is positive, the graph of $y = ax^2 + bx + c$ crosses the x-axis at two distinct points.]

(b) Apply the result of (a) to deduce that

$$|\mathbf{A} \cdot \mathbf{B}| \leq |\mathbf{A}|\,|\mathbf{B}|,$$

from the fact that $|\mathbf{A}x + \mathbf{B}|^2 \geq 0$ for all real x. *Suggestion.* Expand

$$|\mathbf{A}x + \mathbf{B}|^2 = (\mathbf{A}x + \mathbf{B}) \cdot (\mathbf{A}x + \mathbf{B})|.$$

13.2 MATRICES AND SIMULTANEOUS LINEAR EQUATIONS: NOTATION

The two linear equations

$$2x + 3y - 4z = 5, \tag{1a}$$

$$3x - y + 2z = 2 \tag{1b}$$

represent planes in 3-space. The line of intersection consists of the set of points $P(x, y, z)$ whose coordinates simultaneously satisfy both equations. In Article 12.8, we developed methods for finding a vector (a, b, c) and a point $P_0(x_0, y_0, z_0)$ such that we could write the equation of the line (1) in the parametric form

$$(x, y, z) = (x_0, y_0, z_0) + t(a, b, c). \tag{2}$$

In this article we shall start afresh with this problem, and then extend the method we develop to m simultaneous equations in n unknowns.

Problem 1. Find all solutions of the two simultaneous equations (1a, b).

Solution. Multiply the second equation by 3 and add the result to the first equation to eliminate y:

$$11x + 2z = 11.$$

Solve this equation for x in terms of z:

$$x = 1 - \tfrac{2}{11}z. \tag{3a}$$

Substitute this into Eq. (1b) and solve for y:

$$\begin{aligned} y &= 3x + 2z - 2 \\ &= 3(1 - \tfrac{2}{11}z) + 2z - 2 \\ &= 1 + \tfrac{16}{11}z. \end{aligned} \tag{3b}$$

Therefore, if the vector (x, y, z) is a solution of (1a, b), it must be of the form

$$(x, y, z) = (1 - \tfrac{2}{11}z, 1 + \tfrac{16}{11}z, z),$$

or

$$(x, y, z) = (1, 1, 0) + z(-\tfrac{2}{11}, \tfrac{16}{11}, 1). \tag{4}$$

Observe that Eq. (4) matches Eq. (2), with

$$x_0 = 1, \quad y_0 = 1, \quad z_0 = 0,$$
$$t = z, \quad a = -\tfrac{2}{11}, \quad b = \tfrac{16}{11}, \quad c = 1.$$

We should also show, by direct substitution or otherwise, that every vector (x, y, z) given by (4) also represents a solution of both (1a) and (1b). By substitution, we have

$$\begin{aligned} 2x + 3y - 4z &= 2(1 - \tfrac{2}{11}z) + 3(1 + \tfrac{16}{11}z) - 4z \\ &= 2 - \tfrac{4}{11}z + 3 + \tfrac{48}{11}z - \tfrac{44}{11}z \\ &= 5 \end{aligned}$$

and

$$\begin{aligned} 3x - y + 2z &= 3(1 - \tfrac{2}{11}z) - (1 + \tfrac{16}{11}z) + 2z \\ &= 3 - \tfrac{6}{11}z - 1 - \tfrac{16}{11}z + \tfrac{22}{11}z \\ &= 2. \end{aligned}$$

Remark 1. There is a more compact way of writing Eqs. (1a, b). In *matrix form,* we write

$$\begin{bmatrix} 2 & 3 & -4 \\ 3 & -1 & 2 \end{bmatrix} \begin{bmatrix} x \\ y \\ z \end{bmatrix} = \begin{bmatrix} 5 \\ 2 \end{bmatrix}. \tag{5}$$

The matrix

$$\mathbf{A} = \begin{bmatrix} 2 & 3 & -4 \\ 3 & -1 & 2 \end{bmatrix} \tag{6a}$$

is called the *coefficient matrix,* or matrix of coefficients, of the system (1a, b). It has two rows:

$$\mathbf{R}_1 = [2 \quad 3 \quad -4], \tag{6b}$$

$$\mathbf{R}_2 = [3 \quad -1 \quad 2], \tag{6c}$$

each corresponding to a vector in E^3. We know, in fact, that $\mathbf{R}_1$ is normal to the first plane and $\mathbf{R}_2$ is normal to the second plane. The unknowns have been written as a *column vector,*

$$\mathbf{X} = \begin{bmatrix} x \\ y \\ z \end{bmatrix}, \tag{6d}$$

for reasons to be explained shortly. Of course, it would take less space to write (x, y, z) as an ordered row-triple, but *matrix multiplication* will be defined in the conventional way so that Eqs. (1a, b), or Eq. (5), become just

$$\mathbf{AX} = \mathbf{B}, \tag{7}$$

with $\mathbf{A}$ given by Eq. (6a), $\mathbf{X}$ by Eq. (6d), and

$$\mathbf{B} = \begin{bmatrix} 5 \\ 2 \end{bmatrix},$$

as in Eq. (5).

Look again at the left side of Eq. (5). Take the usual inner product of the first row, $\mathbf{R}_1$ of Eq. (6b), with the vector $\mathbf{X}$; but write it in the following matrix form:

$$\mathbf{R}_1 \cdot \mathbf{X} = [2 \quad 3 \quad -4] \begin{bmatrix} x \\ y \\ z \end{bmatrix} = 2x + 3y - 4z.$$

Similarly

$$\mathbf{R}_2 \cdot \mathbf{X} = [3 \quad -1 \quad 2] \begin{bmatrix} x \\ y \\ z \end{bmatrix} = 3x - y + 2z.$$

The rule is: Take the inner product of the ith row of $\mathbf{A}$ with the vector $\mathbf{X}$ and set the result equal to the element in the ith row of $\mathbf{B}$, for $i = 1, 2$. This simply gets us back to our original Eq. (1a, b). So far, we aren't trying to prove anything; we're just getting acquainted with the matrix notation.

Problem 2. Write the following system of equations in matrix form $\mathbf{AX} = \mathbf{B}$:

$$\begin{aligned} 3x_1 + 2x_2 - x_3 + x_4 &= 0, \\ x_1 + 5x_2 + 2x_3 - 3x_4 &= 2. \end{aligned}$$

Solution. By analogy with Problem 1, we write

$$\begin{bmatrix} 3 & 2 & -1 & 1 \\ 1 & 5 & 2 & -3 \end{bmatrix} \begin{bmatrix} x_1 \\ x_2 \\ x_3 \\ x_4 \end{bmatrix} = \begin{bmatrix} 0 \\ 2 \end{bmatrix}.$$

Here the column of x's represents a vector in E^4, written as a column but still called a *vector*. The coefficient matrix has two rows and four columns. Each *row* contains the coefficients in one of the original equations, and each *column* corresponds to one of the components of the vector **X**. If we follow the rule for multiplication, the product of the *first row* of the coefficient matrix **A** with the column vector **X** is

$$3x_1 + 2x_2 - x_3 + x_4.$$

If we set this equal to the element in the *first row* of the column

$$\mathbf{B} = \begin{bmatrix} 0 \\ 2 \end{bmatrix},$$

we get back the first of the given equations. Similarly the product of the *second* row of **A** with **X** is

$$x_1 + 5x_2 + 2x_3 - 3x_4,$$

and the equation $\mathbf{AX} = \mathbf{B}$ means that this product should equal 2, the element in the *second* row in **B**.

Extension to *m* equations in *n* unknowns

When the number of equations is m and the number of unknows is n, we write the equations as

$$\begin{aligned} a_{11}x_1 + a_{12}x_2 + \cdots + a_{1n}x_n &= b_1 \\ a_{21}x_1 + a_{22}x_2 + \cdots + a_{2n}x_n &= b_2 \\ \vdots \qquad\qquad &\quad \vdots \\ a_{m1}x_1 + a_{m2}x_2 + \cdots + a_{mn}x_n &= b_m. \end{aligned} \tag{8}$$

The double subscripts i and j on a_{ij} signify the position of the coefficient of x_j in the ith equation. For example, a_{21} is the coefficient of x_1 in the second equation. We write the *coefficient matrix* as $\mathbf{A} = [a_{ij}]$, with i going from 1 through m (the number of equations) and j going from 1 through n (the number of unknowns, or variables). Then we write the equations (8) in matrix form as

$$\begin{bmatrix} a_{11} & a_{12} & \cdots & a_{1n} \\ a_{21} & a_{22} & \cdots & a_{2n} \\ \vdots & \vdots & & \vdots \\ a_{m1} & a_{m2} & \cdots & a_{mn} \end{bmatrix} \begin{bmatrix} x_1 \\ x_2 \\ \vdots \\ x_n \end{bmatrix} = \begin{bmatrix} b_1 \\ b_2 \\ \vdots \\ b_m \end{bmatrix}, \tag{9}$$

or, much more compactly, as

$$\mathbf{AX} = \mathbf{B}, \tag{9'}$$

where **X** is the column vector of x's, and **B** is the column of constants from the right-hand sides of Eqs. (8). We can now define the product of the matrix **A** and the vector **X**.

Definition. Product of matrix A and vector X. *Let us form the inner product of the ith row of* **A** *with the vector* **X**, *and then set the result equal to* b_i, $i = 1, 2, \ldots, m$. *This is the product of matrix* **A** *and vector* **X**. *In symbols, using the summation sign,*

$$\sum_{j=1}^{n} a_{ij}x_j = b_i, \quad i = 1, \ldots, m. \tag{10}$$

This is how Eq. (10) is to be interpreted:

(a) Select any integer i from 1 through m.

(b) With i fixed, multiply the coefficient a_{ij} from the ith row and the jth column of **A** by x_j, and sum on j.

(c) Set this sum equal to b_i, the element in the ith row of **B**.

Example 1. Here is a numerical example of matrix multiplication.

$$\begin{bmatrix} 3 & 1 \\ 4 & -2 \\ 5 & -3 \end{bmatrix} \begin{bmatrix} -2 \\ 1 \end{bmatrix} = \begin{bmatrix} 3\cdot(-2) + 1\cdot 1 \\ 4\cdot(-2) + (-2)\cdot 1 \\ 5\cdot(-2) + (-3)\cdot 1 \end{bmatrix} = \begin{bmatrix} -5 \\ -10 \\ -13 \end{bmatrix}$$

Remark 2. This row-by-column kind of multiplication is also used in certain flow diagrams. For example, suppose that a system has three possible states S_1, S_2, S_3 (see Fig. 13.3). (We use three states for simplicity only.) When the system is in

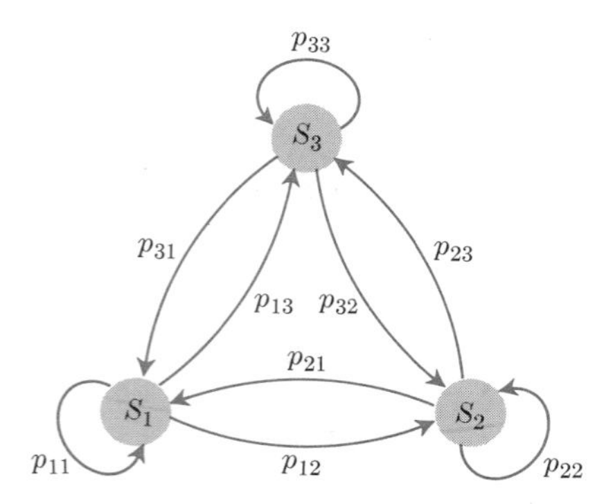

13.3 Flow diagram for a three-state system S_1, S_2, S_3.

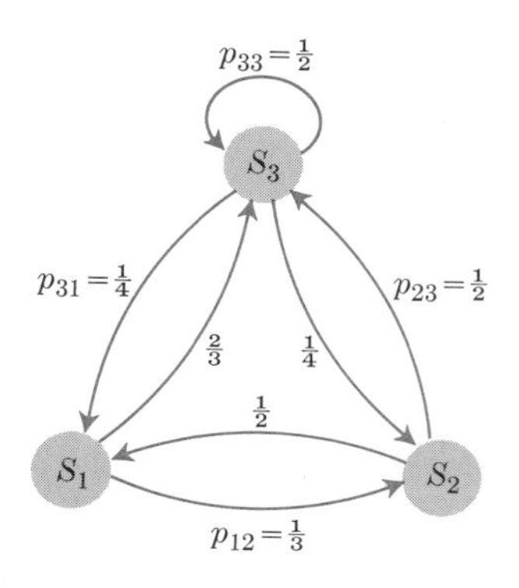

13.4 Flow diagram for the matrix of Eq. (12).

any given state S_i, there is a probability p_{ij} that it will, one time unit later, be in state S_j. The matrix

$$\mathbf{P} = \begin{bmatrix} p_{11} & p_{12} & p_{13} \\ p_{21} & p_{22} & p_{23} \\ p_{31} & p_{32} & p_{33} \end{bmatrix} \tag{11}$$

summarizes these one-step transition probabilities for a *Markov chain.* Suppose, for example, that the probabilities are

$$\mathbf{P} = \begin{bmatrix} 0 & \frac{1}{3} & \frac{2}{3} \\ \frac{1}{2} & 0 & \frac{1}{2} \\ \frac{1}{4} & \frac{1}{4} & \frac{1}{2} \end{bmatrix}. \tag{12}$$

The first row of this matrix tells us the transition probabilities from state S_1. These probabilities are

$$\text{prob}\,(S_1 \to S_1) = 0,$$
$$\text{prob}\,(S_1 \to S_2) = \tfrac{1}{3},$$
$$\text{prob}\,(S_1 \to S_3) = \tfrac{2}{3}.$$

In words, we say that the system always goes from S_1 to one of the other two states, and that it is twice as likely to go to state S_3 as to S_2. The second row shows that from S_2 it goes to S_1 or to S_3, with equal probabilities. Finally, the third row indicates that

$$\text{prob}\,(S_3 \to S_1) = \text{prob}\,(S_3 \to S_2) = \tfrac{1}{4},$$
$$\text{prob}\,(S_3 \to S_3) = \tfrac{1}{2}.$$

Note that the elements in any row sum to 1 because that is the probability that the system, in one step, goes from S_i to *some* state:

$$\sum_{j=1}^{3} p_{ij} = 1 \quad \text{for } i = 1, 2, 3.$$

The flow diagram for the matrix of Eq. (12) would be like Fig. 13.4.

Example 2. A coin is tossed repeatedly. The system is in state S_1 so long as no heads have appeared; in S_2 when one head has appeared; and in S_3 when two or more heads have appeared. For instance, in a particular experiment, these results were obtained:

$$T,\ H,\ T,\ T,\ H,\ H,\ T,\ \ldots.$$

We can describe this by saying that the successive states of the system were

$$S_1 \text{ (start)},\ S_1,\ S_2,\ S_2,\ S_2,\ S_3,\ S_3,\ S_3,\ \ldots$$

Another trial of the experiment resulted in

$$H,\ H,\ \ldots,$$

and the successive states were

$$S_1 \text{ (start)},\ S_2,\ S_3,\ S_3,\ S_3,\ \ldots$$

Fig. 13.5 is a flow diagram for this experiment. We see that the matrix of one-step transition probabilities is

$$\mathbf{P} = \begin{bmatrix} \frac{1}{2} & \frac{1}{2} & 0 \\ 0 & \frac{1}{2} & \frac{1}{2} \\ 0 & 0 & 1 \end{bmatrix}. \tag{13}$$

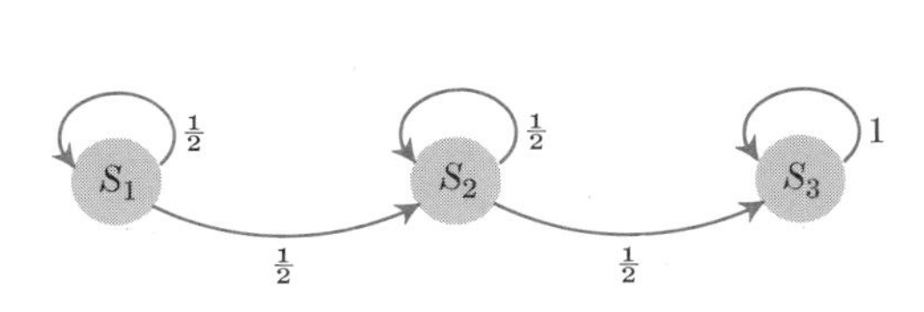

13.5 Flow diagram for the coin example, Eq. (13).

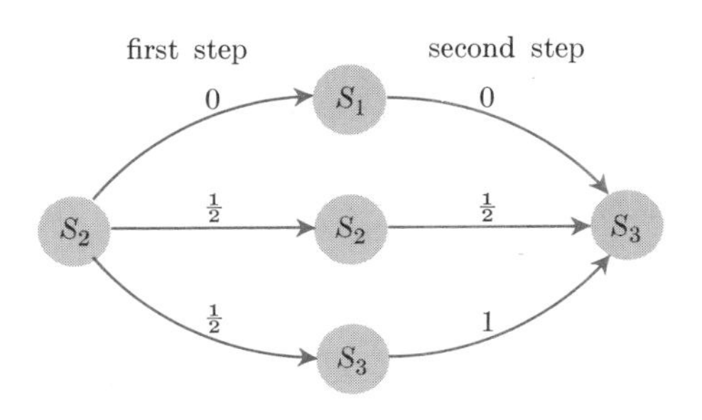

13.6 Transitions from S_2 to S_3 in two steps.

Note that it is impossible to go from state S_1 to S_3 in *one* step. But in *two* steps, it would be possible, by getting two heads in a row. The corresponding probability is

$$\text{prob } (S_1 \to S_2 \to S_3) = (\tfrac{1}{2})(\tfrac{1}{2}) = \tfrac{1}{4}.$$

Of course, at the end of two steps the system might still be in state S_1 (if the coin fell TT), or it might be in state S_2 (on TH or HT). The probabilities are

$$\text{prob } (S_1 \to S_1 \to S_1) = (\tfrac{1}{2})(\tfrac{1}{2}) = \tfrac{1}{4}$$

and

$$\text{prob } (S_1 \to S_2 \to S_2 \quad \text{or} \quad S_1 \to S_1 \to S_2) = \tfrac{2}{4}.$$

Now observe that these numbers, and other two-step transition probabilities, are also obtained by the following matrix multiplication:

$$\mathbf{P} \times \mathbf{P} = \begin{bmatrix} \frac{1}{2} & \frac{1}{2} & 0 \\ 0 & \frac{1}{2} & \frac{1}{2} \\ 0 & 0 & 1 \end{bmatrix} \begin{bmatrix} \frac{1}{2} & \frac{1}{2} & 0 \\ 0 & \frac{1}{2} & \frac{1}{2} \\ 0 & 0 & 1 \end{bmatrix} = \begin{bmatrix} \frac{1}{4} & \frac{2}{4} & \frac{1}{4} \\ 0 & \frac{1}{4} & \frac{3}{4} \\ 0 & 0 & 1 \end{bmatrix}.$$

How are these numbers computed, and what do they mean?

First, as to how we compute them: Multiply the elements of the ith *row* from the *first* matrix by the corresponding elements of the jth *column* of the *second* matrix, sum, and enter this sum as the ijth element of the product. For instance, with $i = 1$ and $j = 2$,

$$[\tfrac{1}{2} \;\; \tfrac{1}{2} \;\; 0] \begin{bmatrix} \frac{1}{2} \\ \frac{1}{2} \\ 0 \end{bmatrix} = \tfrac{1}{4} + \tfrac{1}{4} + 0 = \tfrac{2}{4};$$

and with $i = 2$ and $j = 3$,

$$[0 \;\; \tfrac{1}{2} \;\; \tfrac{1}{2}] \begin{bmatrix} 0 \\ \frac{1}{2} \\ 1 \end{bmatrix} = 0 + \tfrac{1}{4} + \tfrac{1}{2} = \tfrac{3}{4}.$$

Next, the interpretation: For $i = 2$ and $j = 3$, we computed

$$p_{21}p_{13} + p_{22}p_{23} + p_{23}p_{33},$$

or

$$\sum_{k=1}^{3} p_{2k}p_{k3}.$$

Now the product $p_{2k}p_{k3}$ is the probability of going from state S_2 to state S_k in *one* step and then going from S_k to S_3 in the *next* step. Hence $p_{2k}p_{k3}$ is the probability of going from S_2 to S_3 in two steps by way of S_k. The mutually exclusive and exhaustive possible cases are $k = 1, 2, 3$, as shown in Fig. 13.6. Hence the sum over k, as indicated, is the total probability of being in S_3 two steps after being in S_2.

We include another numerical example of matrix multiplication.

Problem 3. If

$$\mathbf{A} = \begin{bmatrix} 1 & -1 \\ 0 & 1 \end{bmatrix}, \quad \mathbf{B} = \begin{bmatrix} 1 & 0 \\ -1 & 1 \end{bmatrix},$$

$$\mathbf{C} = \begin{bmatrix} 0 & 1 \\ 1 & -1 \end{bmatrix},$$

find $\qquad \mathbf{AB}, \quad \mathbf{BC}, \quad (\mathbf{AB})\mathbf{C}, \quad \mathbf{A}(\mathbf{BC}).$

Solution. The reader should verify that

$$\mathbf{AB} = \begin{bmatrix} 1\cdot 1 + (-1)\cdot(-1) & 1\cdot 0 + (-1)\cdot 1 \\ 0\cdot 1 + 1\cdot(-1) & 0\cdot 0 + 1\cdot 1 \end{bmatrix} = \begin{bmatrix} 2 & -1 \\ -1 & 1 \end{bmatrix},$$

$$\mathbf{BC} = \begin{bmatrix} 0 & 1 \\ 1 & -2 \end{bmatrix}, \qquad \mathbf{A}(\mathbf{BC}) = \begin{bmatrix} -1 & 3 \\ 1 & -2 \end{bmatrix},$$

$$(\mathbf{AB})\mathbf{C} = \begin{bmatrix} 2 & -1 \\ -1 & 1 \end{bmatrix}\begin{bmatrix} 0 & 1 \\ 1 & -1 \end{bmatrix} = \begin{bmatrix} -1 & 3 \\ 1 & -2 \end{bmatrix}.$$

Matrix addition and multiplication by scalars

So far we have seen how matrices are multiplied, but we have not discussed addition. This is much easier: Just add corresponding elements. Thus, if $\mathbf{A} = [a_{ij}]$ and $\mathbf{B} = [b_{ij}]$ and both $\mathbf{A}$ and $\mathbf{B}$ are $m \times n$ matrices, then

$$\mathbf{A} + \mathbf{B} = [a_{ij}] + [b_{ij}] = [a_{ij} + b_{ij}],$$

where it is understood that i goes from 1 through m and j goes from 1 through n.

Example 3

$$\begin{bmatrix} 1 & -1 \\ 0 & 2 \end{bmatrix} + \begin{bmatrix} 2 & 3 \\ 4 & 5 \end{bmatrix} = \begin{bmatrix} 3 & 2 \\ 4 & 7 \end{bmatrix}$$

Example 4

If
$$\mathbf{A} = \begin{bmatrix} a & b & c \\ d & e & f \end{bmatrix},$$
then
$$\mathbf{A} + \mathbf{A} = \begin{bmatrix} 2a & 2b & 2c \\ 2d & 2e & 2f \end{bmatrix}.$$

In Example 4, we naturally want to write $\mathbf{A} + \mathbf{A}$ as $2\mathbf{A}$. The result of addition shows that this should mean that each element of $2\mathbf{A}$ is twice the corresponding element of $\mathbf{A}$. To make this consistent with the general definition of $c\mathbf{A}$, for c any scalar (number), we adopt the following definition: Each element of $c\mathbf{A}$ is c times the corresponding element of $\mathbf{A}$. In abbreviated matrix notation,

$$c\mathbf{A} = c[a_{ij}] = [ca_{ij}].$$

Example 5

$$\frac{3}{10}\begin{bmatrix} 2 & -1 \\ 1 & -2 \end{bmatrix} = \begin{bmatrix} 0.6 & -0.3 \\ 0.3 & -0.6 \end{bmatrix}.$$

What properties of the algebra of real numbers (commutative, associative, distributive) also apply to the algebra of matrices? The following theorem is a summary of properties that follow from the foregoing definitions of matrix addition and scalar multiplication.

Theorem 3. *Let* $\mathbf{A}$, $\mathbf{B}$, $\mathbf{C}$, *and* $\mathbf{0}$ *be* $m \times n$ *matrices whose entries are real numbers, and all entries of* $\mathbf{0}$ *are zeros. Let* s *and* t *be scalars. Then*

$$\begin{aligned}
&\text{(i)} & \mathbf{A} + \mathbf{0} &= \mathbf{0} + \mathbf{A} = \mathbf{A}, \\
&\text{(ii)} & \mathbf{A} + \mathbf{B} &= \mathbf{B} + \mathbf{A}, \\
&\text{(iii)} & \mathbf{A} + (-1)\mathbf{A} &= \mathbf{0}, \\
&\text{(iv)} & \mathbf{A} + (\mathbf{B} + \mathbf{C}) &= (\mathbf{A} + \mathbf{B}) + \mathbf{C}, \\
&\text{(v)} & s(\mathbf{A} + \mathbf{B}) &= s\mathbf{A} + s\mathbf{B}, \\
&\text{(vi)} & (s + t)\mathbf{A} &= s\mathbf{A} + t\mathbf{A}, \\
&\text{(vii)} & s(t\mathbf{A}) &= (st)\mathbf{A}.
\end{aligned}$$

Remark. The proof in each instance is almost an immediate consequence of the corresponding property of real numbers. For example, to prove (vi), we observe that the matrix $(s + t)\mathbf{A}$ on the left-hand side is obtained by multiplying each element of $\mathbf{A}$ by $s + t$. But

$$(s + t)a_{ij} = sa_{ij} + ta_{ij} \quad \text{for} \quad \begin{cases} i = 1, \ldots, m, \\ j = 1, \ldots, n, \end{cases}$$

by the distributive law for real numbers. The right-hand side of this equation is the sum of the element in the ith row and jth column of $s\mathbf{A}$ and the corresponding element of $t\mathbf{A}$. In terms of the compact matrix notation, the proof takes two lines:

$$\begin{aligned}
(s + t)\mathbf{A} &= (s + t)[a_{ij}] = [(s + t)a_{ij}] = [sa_{ij} + ta_{ij}] \\
&= [sa_{ij}] + [ta_{ij}] = s\mathbf{A} + t\mathbf{A}.
\end{aligned}$$

As another example of the compact notation:

$$\begin{aligned}\mathbf{A}+\mathbf{B} &= [a_{ij}]+[b_{ij}] = [a_{ij}+b_{ij}] = [b_{ij}+a_{ij}] \\ &= [b_{ij}]+[a_{ij}] = \mathbf{B}+\mathbf{A}.\end{aligned}$$

For further practice you are asked to prove some of the other properties in the exercises. The proofs are easy once you understand the notation.

Equality of matrices

As you will have noted above, whenever we write an equality between two matrices it means that they have the same dimensions (say both are $m \times n$), and that the entries in corresponding positions are equal. Thus, if G is a matrix with r rows and c columns and H is a matrix with m rows and n columns, and if $G = H$, then $r = m$, $c = n$, and $g_{ij} = h_{ij}$ for $i = 1, \ldots, m$ and $j = 1, \ldots, n$.

EXERCISES

In Exercises 1 through 4, compute the indicated products using **A**, **B**, and **C**, as in Problem 3 above.

1. **BA** 2. **AC** 3. **(BA)C** 4. **B(AC)**

5. If
$$\mathbf{A} = \begin{bmatrix} a & b \\ c & d \end{bmatrix} \quad \text{and} \quad \mathbf{I} = \begin{bmatrix} 1 & 0 \\ 0 & 1 \end{bmatrix},$$
show that $\mathbf{AI} = \mathbf{IA} = \mathbf{A}$.

6. If
$$\mathbf{A} = \begin{bmatrix} a & b \\ c & d \end{bmatrix} \quad \text{and} \quad \mathbf{B} = \begin{bmatrix} d & -b \\ -c & a \end{bmatrix},$$
show that
$$\mathbf{AB} = \mathbf{BA} = \begin{bmatrix} ad-bc & 0 \\ 0 & ad-bc \end{bmatrix} = (ad-bc)\begin{bmatrix} 1 & 0 \\ 0 & 1 \end{bmatrix}.$$

7. Let **A** be an $n \times n$ square matrix,
$$\mathbf{A} = [a_{ij}], \quad i = 1, \ldots, n \quad \text{and} \quad j = 1, \ldots, n.$$
Let $\mathbf{I} = [\delta_{ij}]$ also be an $n \times n$ matrix with entries δ_{ij} defined as follows:
$$\delta_{ij} = \begin{cases} 1, & \text{if } i = j, \\ 0, & \text{if } i \neq j, \end{cases} \quad \begin{matrix} i = 1, \ldots, n \\ j = 1, \ldots, n. \end{matrix}$$
(The symbol δ_{ij} is called *Kronecker's delta.*) Thus **I**, the $n \times n$ *identity matrix*, has 1's on the main diagonal and zeros elsewhere. Show that
$$\mathbf{AI} = \mathbf{IA} = \mathbf{A}.$$
Suggestion. Let $\mathbf{B} = \mathbf{AI} = [b_{ij}]$. Then by definition,
$$b_{ij} = \sum_{k=1}^{n} a_{ik}\,\delta_{kj},$$
and the only possible nonzero term in the sum is the one for which $k = j$. Similarly let $\mathbf{C} = \mathbf{IA} = [c_{ij}]$ with
$$c_{ij} = \sum_{k=1}^{n} \delta_{ik} a_{kj}.$$

8. Prove that
$$\mathbf{A} + (\mathbf{B} + \mathbf{C}) = (\mathbf{A} + \mathbf{B}) + \mathbf{C}$$
for $m \times n$ matrices **A**, **B**, **C**.

9. Prove that $s(\mathbf{A} + \mathbf{B}) = s\mathbf{A} + s\mathbf{B}$. (If you give a one- or two-line proof, support each equality by saying to yourself the reason for it: a definition, or some property of real numbers.)

10. If **A** is an $n \times n$ matrix and c is a scalar, what is the relation between the *determinant* of $c\mathbf{A}$ and the *determinant* of **A**? Why? (See Appendix I for properties of determinants.)

13.3 MATRICES AND SIMULTANEOUS LINEAR EQUATIONS: COMPUTATIONAL TECHNIQUES

In this article, we shall first prove the *associative law* for matrix multiplication, and then show how it is applied to solve simultaneous linear equations.

Theorem 4. *Let* **A** *be a matrix with* m *rows and* n *columns (an* $m \times n$ *matrix), let* **B** *be an* $n \times p$ *matrix, and let* **C** *be a* $p \times q$ *matrix. Then* (**AB**)**C** *and* **A**(**BC**) *are* $m \times q$ *matrices, and*

$$(\mathbf{AB})\mathbf{C} = \mathbf{A}(\mathbf{BC}). \tag{1}$$

Proof. First let us consider the matrix

$$\mathbf{AB} = \mathbf{D} = [d_{ij}.] \tag{2a}$$

This has as many rows as there are rows in **A** and as many columns as there are columns in **B**. To get the entry d_{ij} in the ith row and the jth column, we must form the scalar product of the ith row of **A** and the jth column of **B**:

$$\begin{aligned} d_{ij} &= [a_{i1}a_{i2} \ldots a_{in}] \begin{bmatrix} b_{1j} \\ b_{2j} \\ \vdots \\ b_{nj} \end{bmatrix} \\ &= a_{i1}b_{1j} + a_{i2}b_{2j} + \cdots + a_{in}b_{nj} \\ &= \sum_{k=1}^{n} a_{ik}b_{kj}. \end{aligned} \tag{2b}$$

Note that this product is well defined because the two vectors are in E^n: Each row of **A** has n elements (**A** has n columns), and each column of **B** has n elements (**B** has n rows). Since the product **AB** has as many rows as **A** and as many columns as **B**, $\mathbf{D} = \mathbf{AB}$ is an $m \times p$ matrix.

The left-hand side of Eq. (1) is $(\mathbf{AB})\mathbf{C} = \mathbf{DC}$. We let

$$\mathbf{DC} = \mathbf{E} = [e_{it}].$$

Since **D** is an $m \times p$ matrix and **C** is a $p \times q$ matrix, there are m rows and q columns in the product **DC**. The element in the ith row and tth column is the inner product of the ith row of **D** with the tth column of **C**:

$$\begin{aligned} e_{it} &= [d_{i1}d_{i2} \ldots d_{ip}] \begin{bmatrix} c_{1t} \\ c_{2t} \\ \vdots \\ c_{pt} \end{bmatrix} \\ &= d_{i1}c_{1t} + d_{i2}c_{2t} + \cdots + d_{ip}c_{pt} \\ &= \sum_{j=1}^{p} d_{ij}c_{jt}. \end{aligned} \tag{2c}$$

If we combine Eqs. (2b) and (2c), we get

$$e_{it} = \sum_{j=1}^{p} \left(\sum_{k=1}^{n} a_{ik}b_{kj} \right) c_{jt}. \tag{2d}$$

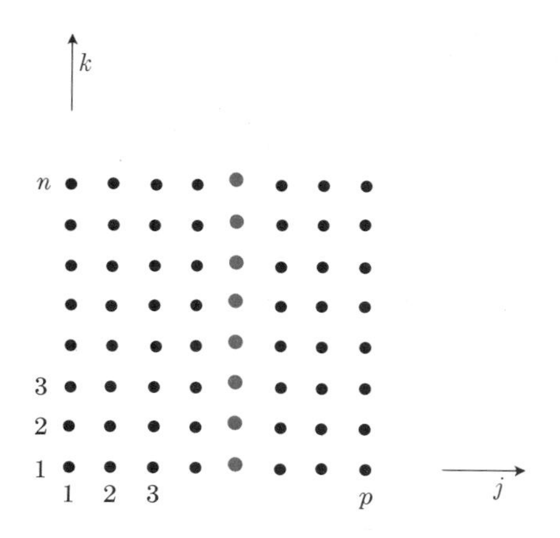

13.7 Lattice points (j, k), with $1 \leq j \leq p$, $1 \leq k \leq n$.

This is equivalent to a sum over a set of lattice points in a rectangular array, as illustrated in Fig. 13.7. On the right-hand side of Eq. (2d), i and t do not vary during the summation. To compute the sum, choose j from 1 through p inclusive, and take the summation of the products

$$a_{ik}b_{kj}c_{jt},$$

with k going from 1 through n. This corresponds to summing over the jth *column* of lattice points (as indicated by the colored dots in Fig. 13.7). Next, the leftmost summation in (2d) calls for summing these column totals as j goes from 1 through p. Now we get the same grand total if we first sum to get row subtotals [the inner sigma in Eq. 3(a)] and then sum these. That is, the left-hand side of the equation

$$e_{it} = \sum_{k=1}^{n} \left(\sum_{j=1}^{p} a_{ik}b_{kj}c_{jt} \right)$$

is equivalent to

$$\sum_{k=1}^{n} a_{ik} \left(\sum_{j=1}^{p} b_{kj}c_{jt} \right). \tag{3a}$$

The terms a_{ik} do not involve j, so they can be moved across the j-summation sigma, as we have done. (This follows from the distributive law for real numbers, because a_{ik} is a common factor in each of the summands.) Now we reverse steps:

$$\sum_{j=1}^{p} b_{kj}c_{jt} = b_{k1}c_{1t} + b_{k2}c_{2t} + \cdots + b_{kp}c_{pt}$$

$$= [b_{k1} \quad b_{k2} \quad \cdots \quad b_{kp}] \begin{bmatrix} c_{1t} \\ c_{2t} \\ \vdots \\ c_{pt} \end{bmatrix}$$

$$= f_{kt} \tag{3b}$$

is the element in the kth row and tth column of the product

$$\mathbf{BC} = \mathbf{F} = [f_{kt}].$$

Hence, combining (3a) and (3b), we have

$$e_{it} = \sum_{k=1}^{n} a_{ik}f_{kt} = a_{i1}f_{1t} + a_{i2}f_{2t} + \cdots + a_{in}f_{nt}$$

$$= [a_{i1} \quad a_{i2} \quad \cdots \quad a_{in}] \begin{bmatrix} f_{1t} \\ f_{2t} \\ \vdots \\ f_{nt} \end{bmatrix}$$

$$= g_{it}, \tag{3c}$$

where $\mathbf{G} = [g_{it}]$ is the product $\mathbf{AF}$:

$$\mathbf{G} = [g_{it}] = \mathbf{AF}.$$

The number of rows in $\mathbf{G}$ is m, the same as the number of rows in $\mathbf{A}$. The number of columns in $\mathbf{G}$ is q, the same as the number of columns in $\mathbf{F}$, or in $\mathbf{C}$, because $\mathbf{F} = \mathbf{BC}$. Hence $\mathbf{G}$, like $\mathbf{E}$, is an $m \times q$ matrix. Also, Eq. (3c) says that the entries in the two matrices are the same:

$$e_{it} = g_{it} \quad \text{for } i = 1, \ldots, m \quad \text{and} \quad t = 1, \ldots, q.$$

This is the definition of matrix equality:

$$\mathbf{E} = \mathbf{G}.$$

That is,

$$\mathbf{E} = \mathbf{DC} = (\mathbf{AB})\mathbf{C}$$

and

$$\mathbf{G} = \mathbf{AF} = \mathbf{A}(\mathbf{BC})$$

are equal. This completes the proof. □

Example 1. Let

$$\mathbf{A} = \begin{bmatrix} 1 & -2 & -1 \\ 2 & 1 & 0 \end{bmatrix}, \quad \text{a } 2 \times 3 \text{ matrix;}$$

$$\mathbf{B} = \begin{bmatrix} 0 & -1 \\ 1 & 2 \\ -2 & 1 \end{bmatrix}, \quad \text{a } 3 \times 2 \text{ matrix;}$$

$$\mathbf{C} = \begin{bmatrix} 1 & -1 & 2 & -2 \\ -1 & 0 & 1 & 2 \end{bmatrix}, \quad \text{a } 2 \times 4 \text{ matrix.}$$

Then

$$\mathbf{AB} = \begin{bmatrix} 0 & -6 \\ 1 & 0 \end{bmatrix}, \qquad (\mathbf{AB})\mathbf{C} = \begin{bmatrix} 6 & 0 & -6 & -12 \\ 1 & -1 & 2 & -2 \end{bmatrix}.$$

Next we compute $\mathbf{BC}$:

$$\mathbf{BC} = \begin{bmatrix} 1 & 0 & -1 & -2 \\ -1 & -1 & 4 & 2 \\ -3 & 2 & -3 & 6 \end{bmatrix}.$$

The element in the first row and third column of $\mathbf{A}(\mathbf{BC})$ is, for example,

$$[1 \quad -2 \quad -1] \begin{bmatrix} -1 \\ 4 \\ -3 \end{bmatrix} = -1 - 8 + 3 = -6;$$

this agrees with the corresponding entry in $(\mathbf{AB})\mathbf{C}$. In Exercise 8, the reader is asked to complete the check that

$$\mathbf{A}(\mathbf{BC}) = \begin{bmatrix} 6 & 0 & -6 & -12 \\ 1 & -1 & 2 & -2 \end{bmatrix}.$$

Application to systems of linear equations: elementary row operations

Example 2. When we "solved" the first system of equations of Article 13.2, we replaced the equivalent of

$$\begin{bmatrix} 2 & 3 & -4 \\ 3 & -1 & 2 \end{bmatrix} \begin{bmatrix} x \\ y \\ z \end{bmatrix} = \begin{bmatrix} 5 \\ 2 \end{bmatrix} \tag{4a}$$

by the equivalent of

$$\begin{bmatrix} 1 & 0 & \frac{2}{11} \\ 0 & 1 & -\frac{16}{11} \end{bmatrix} \begin{bmatrix} x \\ y \\ z \end{bmatrix} = \begin{bmatrix} 1 \\ 1 \end{bmatrix}. \tag{4b}$$

[See Eqs. (1a, b) and (3a, b) of Article 13.2.] To do this, we could have multiplied the second row of the augmented matrix

$$\mathbf{A}_1^* = \left[\begin{array}{ccc|c} 2 & 3 & -4 & 5 \\ 3 & -1 & 2 & 2 \end{array}\right]$$

by 3 and added it to the first row, thus getting the new augmented matrix

$$\mathbf{A}_2^* = \left[\begin{array}{ccc|c} 11 & 0 & 2 & 11 \\ 3 & -1 & 2 & 2 \end{array}\right].$$

The fourth column in $\mathbf{A}_1^*$, the one following the rule, is the column of constants from the right-hand side of the Eqs. (4a). When we modify the equations by combining them to form equivalent new equations, we modify both the coefficients and the column of constants. Thus $\mathbf{A}_2^*$ corresponds to a new system of equations that we get from (4a) by adding (3 × the second equation) to the first. But instead of working with the actual equations, we can work with the so-called *augmented matrix*

$$\mathbf{A}^* = [\mathbf{A} \mid \mathbf{B}]$$

that we get by adjoining to the coefficient matrix $\mathbf{A}$ the column of constants $\mathbf{B}$ from the right-hand side of the equation

$$\mathbf{AX} = \mathbf{B}.$$

Next, we divide the first row of $\mathbf{A}_2^*$ by 11 and get

$$\mathbf{A}_3^* = \left[\begin{array}{ccc|c} 1 & 0 & \frac{2}{11} & 1 \\ 3 & -1 & 2 & 2 \end{array}\right].$$

The first row of this matrix corresponds to the equation

$$x + 0y + \tfrac{2}{11}z = 1,$$

which can easily be solved for x in terms of z. We now modify the second row of $\mathbf{A}_3^*$ by adding to it (−3 × its first row). The result is

$$\mathbf{A}_4^* = \left[\begin{array}{ccc|c} 1 & 0 & \frac{2}{11} & 1 \\ 0 & -1 & \frac{16}{11} & -1 \end{array}\right],$$

and, to make this exactly like Eq. (4b) above, we multiply the last row by −1 to get

$$\mathbf{A}_5^* = \left[\begin{array}{ccc|c} 1 & 0 & \frac{2}{11} & 1 \\ 0 & 1 & -\frac{16}{11} & 1 \end{array}\right].$$

The augmented matrix $\mathbf{A}_5^*$ corresponds to the pair of equations

$$\begin{aligned} x + \tfrac{2}{11}z &= 1, \\ y - \tfrac{16}{11}z &= 1, \end{aligned} \tag{5}$$

which we see are the same as matrix equation (1b).

Equations (4b) and (5) are equivalent to (4a) in the following sense: Every vector (x, y, z) that satisfies one set satisfies the other set because all our operations are reversible. We could start from (4b), or from the augmented matrix $\mathbf{A}_5^*$, and reverse the steps back through $\mathbf{A}_4^*$, $\mathbf{A}_3^*$, and $\mathbf{A}_2^*$ to $\mathbf{A}_1^*$. The advantage that Eqs. (5) and (4b) have over the original set (4a) is the ease with which we can find *all* solutions in the parametric form

$$(x, y, z) = (1, 1, 0) + t(-\tfrac{2}{11}, \tfrac{16}{11}, 1), \tag{6}$$

in which t can be any real number.

The 2 × 2 identity matrix is

$$\mathbf{I} = \begin{bmatrix} 1 & 0 \\ 0 & 1 \end{bmatrix}. \tag{7a}$$

For reasons to be explained later, let us do to $\mathbf{I}$ the same things we did in going from $\mathbf{A}_1^*$ to $\mathbf{A}_5^*$. First, add (3 × the second row) to the first row and get

$$\mathbf{I}_2^* = \begin{bmatrix} 1 & 3 \\ 0 & 1 \end{bmatrix}.$$

Next, divide the first row by 11:

$$\mathbf{I}_3^* = \begin{bmatrix} \frac{1}{11} & \frac{3}{11} \\ 0 & 1 \end{bmatrix}.$$

Add (−3 × the first row) to the second row:

$$\mathbf{I}_4^* = \begin{bmatrix} \frac{1}{11} & \frac{3}{11} \\ -\frac{3}{11} & \frac{2}{11} \end{bmatrix}.$$

Finally, multiply the last row by −1:

$$\mathbf{I}_5^* = \begin{bmatrix} \frac{1}{11} & \frac{3}{11} \\ \frac{3}{11} & -\frac{2}{11} \end{bmatrix}.$$

Why have we done this? To show (as we now shall) that if we multiply the original augmented matrix A_1^* by I_5^*, then we get A_5^*:

$$\begin{bmatrix} \frac{1}{11} & \frac{3}{11} \\ \frac{3}{11} & -\frac{2}{11} \end{bmatrix} \left[\begin{array}{ccc|c} 2 & 3 & -4 & 5 \\ 3 & -1 & 2 & 2 \end{array}\right] = \left[\begin{array}{ccc|c} 1 & 0 & \frac{2}{11} & 1 \\ 0 & 1 & -\frac{16}{11} & 1 \end{array}\right].$$

For the general $m \times n$ matrix $\mathbf{A}$, we would start with an $m \times m$ identity matrix

$$\mathbf{I} = \begin{bmatrix} 1 & 0 & \cdots & 0 \\ 0 & 1 & \cdots & 0 \\ \vdots & & & \vdots \\ 0 & 0 & \cdots & 1 \end{bmatrix} m \text{ rows}$$
$$m \text{ columns}$$

By a sequence of elementary row operations (adding a multiple of one row to a different row, or interchanging two rows, or multiplying a row by a nonzero constant), we convert $\mathbf{I}$ to a matrix $\mathbf{P}$. Such a matrix $\mathbf{P}$ is said to be *row-equivalent* to the original matrix. In the example above, we have $m = 2$ and we could take $\mathbf{P} = \mathbf{I}_5^*$. We would then look for a sequence of row operations that converts the original coefficient matrix $\mathbf{A}$, or augmented matrix $\mathbf{A}^*$, to a row-equivalent *echelon form* $\mathbf{E}$ in which

(1) the first nonzero entry in each row is a 1, and the column entries above and below this 1 are all zeros,

(2) the number of zeros at the beginning of a row increases as we read down the rows.

Problem 1. By a sequence of row operations, reduce

$$\mathbf{A} = \begin{bmatrix} 1 & 2 & 0 \\ 2 & 1 & 2 \\ -1 & 3 & 0 \end{bmatrix}$$

to echelon form $\mathbf{E}$. Simultaneously, apply those operations to the 3×3 identity matrix and call the resulting matrix $\mathbf{P}$. Then verify that $\mathbf{PA} = \mathbf{E}$.

Solution. Multiply row one by -2 and add to row two:

$$\mathbf{A}_1 = \begin{bmatrix} 1 & 2 & 0 \\ 0 & -3 & 2 \\ -1 & 3 & 0 \end{bmatrix}, \quad \mathbf{I}_1 = \begin{bmatrix} 1 & 0 & 0 \\ -2 & 1 & 0 \\ 0 & 0 & 1 \end{bmatrix}.$$

Add row one to row three:

$$\mathbf{A}_2 = \begin{bmatrix} 1 & 2 & 0 \\ 0 & -3 & 2 \\ 0 & 5 & 0 \end{bmatrix}, \quad \mathbf{I}_2 = \begin{bmatrix} 1 & 0 & 0 \\ -2 & 1 & 0 \\ 1 & 0 & 1 \end{bmatrix}.$$

Interchange rows two and three and divide the new second row by 5:

$$\mathbf{A}_3 = \begin{bmatrix} 1 & 2 & 0 \\ 0 & 1 & 0 \\ 0 & -3 & 2 \end{bmatrix}, \quad \mathbf{I}_3 = \begin{bmatrix} 1 & 0 & 0 \\ \frac{1}{5} & 0 & \frac{1}{5} \\ -2 & 1 & 0 \end{bmatrix}.$$

Add ($-2 \times$ row two) to row one and ($3 \times$ row two) to row three, which combines two steps in one (note that row two is not changed):

$$\mathbf{A}_4 = \begin{bmatrix} 1 & 0 & 0 \\ 0 & 1 & 0 \\ 0 & 0 & 2 \end{bmatrix}, \quad \mathbf{I}_4 = \begin{bmatrix} \frac{3}{5} & 0 & -\frac{2}{5} \\ \frac{1}{5} & 0 & \frac{1}{5} \\ -\frac{7}{5} & 1 & \frac{3}{5} \end{bmatrix}.$$

Divide row three by two:

$$\mathbf{A}_5 = \mathbf{E} = \begin{bmatrix} 1 & 0 & 0 \\ 0 & 1 & 0 \\ 0 & 0 & 1 \end{bmatrix}, \quad \mathbf{I}_5 = \mathbf{P} = \begin{bmatrix} \frac{3}{5} & 0 & -\frac{2}{5} \\ \frac{1}{5} & 0 & \frac{1}{5} \\ -\frac{7}{10} & \frac{1}{2} & \frac{3}{10} \end{bmatrix}.$$

Verification:

$$\mathbf{PA} = \begin{bmatrix} \frac{3}{5} & 0 & -\frac{2}{5} \\ \frac{1}{5} & 0 & \frac{1}{5} \\ -\frac{7}{10} & \frac{1}{2} & \frac{3}{10} \end{bmatrix} \begin{bmatrix} 1 & 2 & 0 \\ 2 & 1 & 2 \\ -1 & 3 & 0 \end{bmatrix}$$
$$= \begin{bmatrix} \frac{5}{5} & 0 & 0 \\ 0 & \frac{5}{5} & 0 \\ 0 & 0 & \frac{2}{2} \end{bmatrix} = \mathbf{E}.$$

Remark. In this example, the echelon matrix $\mathbf{E}$ is the identity matrix $\mathbf{I}$, and we have found a matrix $\mathbf{P}$ such that $\mathbf{PA} = \mathbf{I}$. It can also be verified that $\mathbf{AP} = \mathbf{I}$, and we indicate this by calling $\mathbf{P}$ the *inverse* of $\mathbf{A}$. We write this as

$$\mathbf{P} = \mathbf{A}^{-1}.$$

Problem 2. Use the results of Problem 1 to solve the equations

$$\begin{aligned} x + 2y &= 1, \\ 2x + y + 2z &= -1, \\ -x + 3y &= 4. \end{aligned}$$

Solution. The equations have the form $\mathbf{AX} = \mathbf{B}$, with $\mathbf{A}$ as in Problem 1 above. Premultiplying both sides of this equation by $\mathbf{P}$, we get

$$\mathbf{P}(\mathbf{AX}) = \mathbf{PB},$$

or, using the associative law,

$$(\mathbf{PA})\mathbf{X} = \mathbf{PB}.$$

Since $\qquad \mathbf{PA} = \mathbf{I} \quad \text{and} \quad \mathbf{IX} = \mathbf{X}$

(see Exercise 7 below), we have

$$\mathbf{X} = \mathbf{PB} = \mathbf{A}^{-1}\mathbf{B}.$$

Therefore

$$\begin{bmatrix} x \\ y \\ z \end{bmatrix} = \begin{bmatrix} \frac{3}{5} & 0 & -\frac{2}{5} \\ \frac{1}{5} & 0 & \frac{1}{5} \\ -\frac{7}{10} & \frac{1}{2} & \frac{3}{10} \end{bmatrix} \begin{bmatrix} 1 \\ -1 \\ 4 \end{bmatrix} = \begin{bmatrix} -1 \\ 1 \\ 0 \end{bmatrix}.$$

It is easy to verify that $x = -1$, $y = 1$, $z = 0$ satisfies the original equations.

EXERCISES

In Exercises 1 through 6, given the matrix $\mathbf{A}$, find the row-equivalent echelon matrix $\mathbf{E}$ and related matrix $\mathbf{P}$ such that $\mathbf{PA} = \mathbf{E}$. Verify this second result by matrix multiplication. If $\mathbf{A}$ is an $n \times n$ matrix and $\mathbf{E}$ is the identity matrix, also verify that $\mathbf{AP} = \mathbf{E}$ by direct matrix multiplication.

1. $\mathbf{A} = \begin{bmatrix} 1 & 2 & 3 \\ 2 & -1 & 0 \end{bmatrix}$

2. $\mathbf{A} = \begin{bmatrix} 2 & 1 \\ 1 & 2 \end{bmatrix}$

3. $\mathbf{A} = \begin{bmatrix} 0 & 1 \\ 1 & 1 \end{bmatrix}$

4. $\mathbf{A} = \begin{bmatrix} 1 & 0 & 1 \\ 0 & 1 & 0 \\ 1 & 0 & 1 \end{bmatrix}$

5. $\mathbf{A} = \begin{bmatrix} 1 & a & b \\ 0 & 1 & a \\ 0 & 0 & 1 \end{bmatrix}$

6. $\mathbf{A} = \begin{bmatrix} 1 & 0 & 0 \\ -1 & 2 & 0 \\ 2 & 1 & 1 \end{bmatrix}$

7. If $\mathbf{I}$ is an $m \times m$ identity matrix (a matrix with 1's on the main diagonal and zeros elsewhere), and if $\mathbf{C}$ is any $m \times p$ matrix, verify that $\mathbf{IC} = \mathbf{C}$.

8. Using the product $\mathbf{BC}$ as computed in Example 1, complete the calculation of $\mathbf{A}(\mathbf{BC})$ and thus verify that $\mathbf{A}(\mathbf{BC}) = (\mathbf{AB})\mathbf{C}$ in that example.

9. For the matrices $\mathbf{A}$, $\mathbf{B}$, and $\mathbf{C}$ of Example 1, compute the product $\mathbf{BA}$. How many rows and how many columns are there in this product? Is it true that $(\mathbf{BA})\mathbf{C} = \mathbf{B}(\mathbf{AC})$? Comment.

10. What is the augmented matrix for the system of equations

$$\begin{aligned} x - 2y + 3z &= 4, \\ 2x + y - 4z &= -1, \\ y + 2z &= 3? \end{aligned}$$

By a sequence of elementary row operations, reduce this augmented matrix to echelon form and thereby solve the equations. Check your solution by direct substitution into the original equations. (This step often shows up little slips in the calculations.)

11. In Problem 1, for the given matrix $\mathbf{A}$ and $\mathbf{P} = \mathbf{I}_5$, verify that $\mathbf{AP} = \mathbf{I}$.

12. Write the following equations in matrix form $\mathbf{AX} = \mathbf{B}$ and verify that $\mathbf{A}$ is the same as the matrix $\mathbf{A}$ in Problem 1. Then use $\mathbf{A}^{-1}$ from Problem 1 to solve these equations:

$$\begin{aligned} x + 2y &= 2, \\ 2x + y + 2z &= -3, \\ -x + 3y &= 1. \end{aligned}$$

13. Write the following equations in matrix form $\mathbf{AX} = \mathbf{B}$, and solve for $\mathbf{X}$ by finding $\mathbf{A}^{-1}$:

$$\begin{aligned} x + y - 2z &= 3, \\ 2x - y + z &= 0, \\ 3x + y - z &= 8. \end{aligned}$$

14. (For readers with a knowledge of determinants.) Let $\mathbf{A}$ be an $n \times n$ matrix. By deleting the ith row and jth column from $\mathbf{A}$, one obtains an $(n-1) \times (n-1)$ matrix $\mathbf{M}_{ij}$. The number $A_{ij} = (-1)^{i+j} \det(\mathbf{M}_{ij})$ is called the *cofactor of* a_{ij} in $\mathbf{A}$.

Then $\displaystyle\sum_{k=1}^{n} a_{ik}A_{ik} = \det(\mathbf{A}), \quad i = 1, \ldots, n$

and $\displaystyle\sum_{k=1}^{n} a_{kj}A_{kj} = \det(\mathbf{A}), \quad j = 1, \ldots, n$

are equations for the expansion of the determinant of $\mathbf{A}$ in terms of cofactors of elements of the ith row, or of the jth column. But if the elements of any row are multiplied by the cofactors of the corresponding elements of a different row, then their sum is zero:

$$\sum_{k=1}^{n} a_{ik}A_{jk} = 0 \quad \text{if} \quad i \neq j.$$

Similarly, for columns,

$$\sum_{k=1}^{n} a_{ki}A_{kj} = 0 \quad \text{if} \quad i \neq j.$$

Use these results to show that

$$\begin{bmatrix} a_{11} & a_{12} & \cdots & a_{1n} \\ a_{21} & a_{22} & \cdots & a_{2n} \\ \vdots & & & \vdots \\ a_{n1} & a_{n2} & \cdots & a_{nn} \end{bmatrix} \begin{bmatrix} A_{11} & A_{21} & \cdots & A_{n1} \\ A_{12} & A_{22} & \cdots & A_{n2} \\ \vdots & \vdots & & \vdots \\ A_{1n} & A_{2n} & \cdots & A_{nn} \end{bmatrix}$$

is the matrix $(\det \mathbf{A})\mathbf{I}$, where $\mathbf{I}$ is the $n \times n$ identity matrix. (Note that the cofactors of the *rows* of $\mathbf{A}$ are written as the *columns* of the second factor. This gives the *transposed* matrix of cofactors and is called the *adjoint of* $\mathbf{A}$, written "adj $\mathbf{A}$.")

15. Use the result of Exercise 14 to find adj $\mathbf{A}$ for

$$\mathbf{A} = \begin{bmatrix} 1 & -2 & 1 \\ 2 & 1 & -3 \\ 0 & 1 & 1 \end{bmatrix}.$$

Divide each element of adj $\mathbf{A}$ by det $\mathbf{A}$ and show that the resulting matrix is the inverse of $\mathbf{A}$:

$$\mathbf{A}^{-1} = \frac{\text{adj}\,\mathbf{A}}{\det \mathbf{A}}.$$

13.4 LINEAR INDEPENDENCE AND LINEAR DEPENDENCE OF VECTORS

Suppose $\mathbf{V}_1, \mathbf{V}_2, \ldots, \mathbf{V}_s$ are vectors in E^m. If $c_1, c_2, \ldots, c_s$ are scalars (real numbers), we say that

$$c_1\mathbf{V}_1 + c_2\mathbf{V}_2 + \cdots + c_s\mathbf{V}_s \tag{1}$$

is a *linear combination* of $\mathbf{V}_1, \ldots, \mathbf{V}_s$.

Example 1. The equations we looked at in Article 13.2,

$$\begin{aligned} 2x + 3y - 4z &= 5, \\ 3x - y + 2z &= 2, \end{aligned} \tag{2}$$

can be interpreted as posing the following problem:

Find all ways of expressing the vector

$$\begin{bmatrix} 5 \\ 2 \end{bmatrix}$$

as

$$x\begin{bmatrix} 2 \\ 3 \end{bmatrix} + y\begin{bmatrix} 3 \\ -1 \end{bmatrix} + z\begin{bmatrix} -4 \\ 2 \end{bmatrix}.$$

Here we have four vectors in E^2:

$$\mathbf{V}_1 = \begin{bmatrix} 2 \\ 3 \end{bmatrix}, \quad \mathbf{V}_2 = \begin{bmatrix} 3 \\ -1 \end{bmatrix}, \quad \mathbf{V}_3 = \begin{bmatrix} -4 \\ 2 \end{bmatrix},$$

and

$$\mathbf{W} = \begin{bmatrix} 5 \\ 2 \end{bmatrix}. \tag{3}$$

The question is whether $\mathbf{W}$ is a linear combination of $\mathbf{V}_1, \mathbf{V}_2, \mathbf{V}_3$, and, if so, how to find the coefficients $c_1 = x$, $c_2 = y$, $c_3 = z$ that satisfy

$$\mathbf{W} = c_1\mathbf{V}_1 + c_2\mathbf{V}_2 + c_3\mathbf{V}_3. \tag{4}$$

We already know that there are infinitely many ways of satisfying Eq. (4):

$$c_1 = 1 - \tfrac{2}{11}t, \quad c_2 = 1 + \tfrac{16}{11}t, \quad c_3 = t \tag{5}$$

for all t, $-\infty < t < \infty$. [See Eqs. (1a, b) and (4) of Article 13.2.] Here we are not so much concerned with computational techniques as with a new interpretation and with the notions of linear combination, linear dependence, and linear independence, and so we press on.

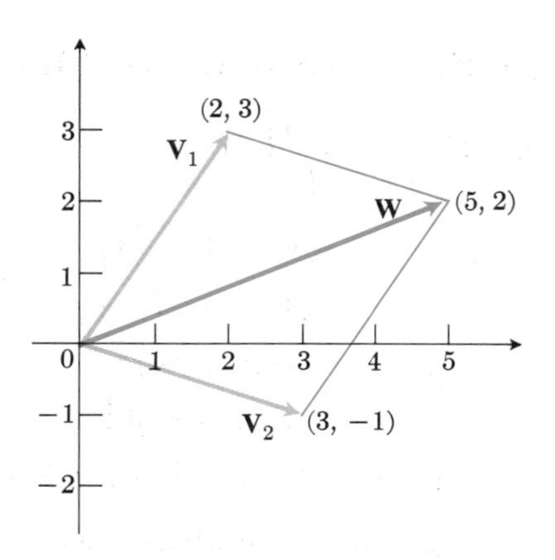

13.8 $\mathbf{W} = \mathbf{V}_1 + \mathbf{V}_2$.

Let us look at Eqs. (2) from the viewpoint of vectors in 2-space, as in Eqs. (3), (4), and (5). Figure 13.8 shows $\mathbf{V}_1$, $\mathbf{V}_2$, and $\mathbf{W}$. As it happens,

$$\mathbf{W} = \mathbf{V}_1 + \mathbf{V}_2, \tag{6}$$

because

$$\begin{bmatrix} 5 \\ 2 \end{bmatrix} = \begin{bmatrix} 2 \\ 3 \end{bmatrix} + \begin{bmatrix} 3 \\ -1 \end{bmatrix}.$$

Thus, Eq. (4) is satisfied, in particular, with

$$c_1 = 1, \quad c_2 = 1, \quad c_3 = 0.$$

This corresponds to taking $t = 0$ in Eqs. (5).

Now it is clear from the figure that we could introduce new coordinate axes, say along $\mathbf{V}_2$ and orthogonal to $\mathbf{V}_2$. The orthogonal projection of $\mathbf{V}_1$ on $\mathbf{V}_2$ is

$$\mathbf{V}_1' = \left(\frac{\mathbf{V}_1 \cdot \mathbf{V}_2}{\mathbf{V}_2 \cdot \mathbf{V}_2}\right)\mathbf{V}_2 = \frac{3}{10}\mathbf{V}_2 = \frac{3}{10}\begin{bmatrix} 3 \\ -1 \end{bmatrix} = \begin{bmatrix} 0.9 \\ -0.3 \end{bmatrix},$$

and the component of $\mathbf{V}_1$ that is orthogonal to $\mathbf{V}_2$ is

$$\mathbf{V}_1'' = \mathbf{V}_1 - \mathbf{V}_1' = \mathbf{V}_1 - \frac{3}{10}\mathbf{V}_2 = \begin{bmatrix} 2 \\ 3 \end{bmatrix} - \frac{3}{10}\begin{bmatrix} 3 \\ -1 \end{bmatrix}$$
$$= \begin{bmatrix} \frac{11}{10} \\ \frac{33}{10} \end{bmatrix} = \frac{11}{10}\begin{bmatrix} 1 \\ 3 \end{bmatrix}.$$

The vectors $(3, -1)$ and $(1, 3)$ in E^2 are orthogonal, and have *directions* (unit vectors)

$$\gamma_1 = \left(\frac{3}{\sqrt{10}}, -\frac{1}{\sqrt{10}}\right) \tag{6a}$$

$$\gamma_2 = \left(\frac{1}{\sqrt{10}}, \frac{3}{\sqrt{10}}\right). \tag{6b}$$

It is evident on geometric grounds (and can be verified algebraically) that *every vector* in E^2 can be expressed as a linear combination of these two orthogonal unit vectors γ_1 and γ_2, which are like $\mathbf{i}$ and $\mathbf{j}$ so rotated that

$$\mathbf{i} \to \gamma_1, \qquad \mathbf{j} \to \gamma_2.$$

Indeed, if β is any vector in E^2, we need only resolve β into its components β_1 along γ_1 and β_2 orthogonal to γ_1. Thus

$$\beta = \beta_1 + \beta_2,$$

where

$$\beta_1 = \frac{\beta \cdot \gamma_1}{\gamma_1 \cdot \gamma_1}\gamma_1, \qquad \beta_2 = \beta - \beta_1. \tag{7a}$$

In this case, we can also take

$$\beta_2 = \frac{\beta \cdot \gamma_2}{\gamma_2 \cdot \gamma_2}\gamma_2. \tag{7b}$$

Problem 2. Express both of the vectors $(1, 0)$ and $(0, 1)$ as linear combinations of γ_1 and γ_2 of Eqs. (6a, b).

Solution. Let

$$\alpha_1 = (1, 0), \qquad \alpha_2 = (0, 1).$$

Then the components of α_1 in the directions γ_1 and γ_2 are

$$\alpha_1 \cdot \gamma_1 = \frac{3}{\sqrt{10}}, \qquad \alpha_1 \cdot \gamma_2 = \frac{1}{\sqrt{10}},$$

so that

$$\alpha_1 = \frac{3}{\sqrt{10}}\gamma_1 + \frac{1}{\sqrt{10}}\gamma_2. \tag{8a}$$

Similarly

$$\alpha_2 \cdot \gamma_1 = -\frac{1}{\sqrt{10}}, \qquad \alpha_2 \cdot \gamma_2 = \frac{3}{\sqrt{10}},$$

$$\alpha_2 = -\frac{1}{\sqrt{10}}\gamma_1 + \frac{3}{\sqrt{10}}\gamma_2. \tag{8b}$$

We verify Eqs. (8a, b):

$$\alpha_1 = \begin{bmatrix}1\\0\end{bmatrix} \stackrel{?}{=} \frac{3}{\sqrt{10}}\begin{bmatrix}\frac{3}{\sqrt{10}}\\ \frac{-1}{\sqrt{10}}\end{bmatrix} + \frac{1}{\sqrt{10}}\begin{bmatrix}\frac{1}{\sqrt{10}}\\ \frac{3}{\sqrt{10}}\end{bmatrix}$$

$$= \begin{bmatrix}\frac{9}{10}\\ -\frac{3}{10}\end{bmatrix} + \begin{bmatrix}\frac{1}{10}\\ \frac{3}{10}\end{bmatrix} = \begin{bmatrix}1\\0\end{bmatrix} \quad \text{(Yes)},$$

and

$$\alpha_2 = \begin{bmatrix}0\\1\end{bmatrix} \stackrel{?}{=} -\frac{1}{\sqrt{10}}\begin{bmatrix}\frac{3}{\sqrt{10}}\\ \frac{-1}{\sqrt{10}}\end{bmatrix} + \frac{3}{\sqrt{10}}\begin{bmatrix}\frac{1}{\sqrt{10}}\\ \frac{3}{\sqrt{10}}\end{bmatrix}$$

$$= \begin{bmatrix}-\frac{3}{10}\\ \frac{1}{10}\end{bmatrix} + \begin{bmatrix}\frac{3}{10}\\ \frac{9}{10}\end{bmatrix} = \begin{bmatrix}0\\1\end{bmatrix} \quad \text{(Yes)}.$$

Remark 1. Any vector (a, b) in E^2 can be written as $(a, 0) + (0, b)$, or

$$\begin{bmatrix}a\\b\end{bmatrix} = a\begin{bmatrix}1\\0\end{bmatrix} + b\begin{bmatrix}0\\1\end{bmatrix}. \tag{9a}$$

Equations (8a) and (8b), in turn, show that

$$\begin{bmatrix}1\\0\end{bmatrix} = a_{11}\gamma_1 + a_{12}\gamma_2, \tag{9b}$$

$$\begin{bmatrix}0\\1\end{bmatrix} = a_{21}\gamma_1 + a_{22}\gamma_2 \tag{9c}$$

for

$$\mathbf{A} = \begin{bmatrix}a_{11} & a_{12}\\ a_{21} & a_{22}\end{bmatrix} = \frac{1}{\sqrt{10}}\begin{bmatrix}3 & 1\\ -1 & 3\end{bmatrix}.$$

Symbolically we can also rewrite Eqs. (8a, b) in the matrix form (note the transposed order)

$$[\alpha_1 \quad \alpha_2] = [\gamma_1 \quad \gamma_2]\begin{bmatrix}a_{11} & a_{21}\\ a_{12} & a_{22}\end{bmatrix}. \tag{10}$$

Then any vector (a, b) in E^2 can be expressed thus:

$$\begin{aligned} a\alpha_1 + b\alpha_2 &= [\alpha_1 \quad \alpha_2]\begin{bmatrix}a\\b\end{bmatrix} \\ &= [\gamma_1 \quad \gamma_2]\begin{bmatrix}a_{11} & a_{21}\\ a_{12} & a_{22}\end{bmatrix}\begin{bmatrix}a\\b\end{bmatrix} \\ &= [\gamma_1 \quad \gamma_2]\begin{bmatrix}a_{11}a + a_{21}b\\ a_{12}a + a_{22}b\end{bmatrix} \\ &= (a_{11}a + a_{21}b)\gamma_1 + (a_{12}a + a_{22}b)\gamma_2. \end{aligned}$$

This is the algebraic verification that

> every vector in E^2 is a linear combination of γ_1 and γ_2.

Summary

We have shown that the particular set of equations (2), (3), and (4) have infinitely many solutions. We showed that there is *one solution* with $c_1 = c_2 = 1$, $c_3 = 0$. We then showed that every vector in E^2 is a linear combination of γ_1 and γ_2, and hence is a linear combination of $\mathbf{V}_1$ and $\mathbf{V}_2$, because γ_1 and γ_2 are linear combinations of $\mathbf{V}_1$ and $\mathbf{V}_2$. This means, in particular, that $\mathbf{V}_3$ is a linear combination of $\mathbf{V}_1$ and $\mathbf{V}_2$:

$$\mathbf{V}_3 = k_1\mathbf{V}_1 + k_2\mathbf{V}_2, \tag{11}$$

for some scalars k_1 and k_2. Or, we can rewrite Eq. (11) in the form

$$k_1\mathbf{V}_1 + k_2\mathbf{V}_2 + k_3\mathbf{V}_3 = \mathbf{0}, \tag{12}$$

with $k_3 = -1$. Note now that we could add the zero vector given by Eq. (12) to any solution of

$$c_1\mathbf{V}_1 + c_2\mathbf{V}_2 + c_3\mathbf{V}_3 = \mathbf{W}$$

and have a new solution

$$(c_1 + k_1)\mathbf{V}_1 + (c_2 + k_2)\mathbf{V}_2 + (c_3 + k_3)\mathbf{V}_3,$$

which is equal to

$$\mathbf{W} + \mathbf{0} = \mathbf{W}.$$

There is a technical term, *linear dependence*, that describes a situation such as Eq. (12).

Definition 1. **Linear dependence.** *A set of vectors* $\mathbf{V}_1, \mathbf{V}_2, \ldots, \mathbf{V}_s$ *is said to be linearly dependent if and only if there exist scalars* $k_1, k_2, \ldots, k_s$, *not all zero, such that*

$$k_1\mathbf{V}_1 + k_2\mathbf{V}_2 + \cdots + k_s\mathbf{V}_s = \mathbf{0}. \tag{13}$$

Example 2. The vectors $\mathbf{V}_1$, $\mathbf{V}_2$, and $\mathbf{W}$ of Eq. (3) are linearly dependent because, as we saw above,

$$\mathbf{W} = \mathbf{V}_1 + \mathbf{V}_2,$$

and hence

$$k_1\mathbf{V}_1 + k_2\mathbf{V}_2 + k_3\mathbf{W} = \mathbf{0},$$

with

$$k_1 = 1, \quad k_2 = 1, \quad k_3 = -1.$$

If there is *no* linear combination like the left-hand side of Eq. (13) that is $\mathbf{0}$ *except when*

$$k_1 = k_2 = \cdots = k_s = 0,$$

then the vectors are said to be *linearly independent.*

Problem 2. Are the vectors

$$\mathbf{V}_1 = (3, 1), \quad \mathbf{V}_2 = (-1, 3)$$

linearly dependent or linearly independent?

Solution. Suppose

$$k_1\mathbf{V}_1 + k_2\mathbf{V}_2 = \mathbf{0}. \tag{14}$$

Then

$$k_1(3, 1) + k_2(-1, 3) = (3k_1 - k_2, k_1 + 3k_2) = (0, 0),$$

so that

$$3k_1 - k_2 = 0, \tag{15a}$$

and

$$k_1 + 3k_2 = 0. \tag{15b}$$

Equation (15a) implies $k_2 = 3k_1$. We substitute this into Eq. (15b) and get

$$k_1 + 9k_1 = 0, \quad \text{or} \quad 10k_1 = 0.$$

Hence Eq. (14) implies

$$k_1 = 0, \quad \text{and} \quad k_2 = 3k_1 = 0.$$

Therefore the given vectors are linearly *independent,* because

$$k_1\mathbf{V}_1 + k_2\mathbf{V}_2 = \mathbf{0} \Rightarrow k_1 = k_2 = 0.$$

Remark 2. The vectors of Problem 2 above are *orthogonal* because

$$\mathbf{V}_1 \cdot \mathbf{V}_2 = 3 \cdot (-1) + 1 \cdot (3) = 0.$$

We can use this fact to give a particularly easy proof that $\mathbf{V}_1$ and $\mathbf{V}_2$ are linearly independent: As in Eq. (14), assume that

$$k_1\mathbf{V}_1 + k_2\mathbf{V}_2 = \mathbf{0} \quad \text{and} \quad \mathbf{V}_1 \cdot \mathbf{V}_2 = 0,$$

whereas

$$|\mathbf{V}_1|^2 = \mathbf{V}_1 \cdot \mathbf{V}_1 = \mathbf{V}_2 \cdot \mathbf{V}_2 = 10.$$

Take the dot product of both sides of Eq. (14), first with $\mathbf{V}_1$ and then with $\mathbf{V}_2$, and get

$$k_1\mathbf{V}_1 \cdot \mathbf{V}_1 + k_2\mathbf{V}_2 \cdot \mathbf{V}_1 = \mathbf{0} \cdot \mathbf{V}_1 = 0,$$

so

$$10k_1 = 0, \quad \text{and} \quad k_1 = 0;$$

$$k_1\mathbf{V}_1 \cdot \mathbf{V}_2 + k_2\mathbf{V}_2 \cdot \mathbf{V}_2 = \mathbf{0} \cdot \mathbf{V}_2 = 0,$$

so

$$10k_2 = 0, \quad \text{and} \quad k_2 = 0. \;\square$$

There is a general theorem, which we now state and prove, about sets of orthogonal vectors: It says that if they are all nonzero, then they are independent.

Theorem 5. *Let* $\mathbf{V}_1, \mathbf{V}_2, \ldots, \mathbf{V}_s$ *be a set of nonzero, mutually orthogonal vectors:*

$$\mathbf{V}_i \cdot \mathbf{V}_j = 0 \quad \text{for } i \neq j, \quad 1 \leq i \leq s, 1 \leq j \leq s \tag{16a}$$

and

$$\mathbf{V}_i \cdot \mathbf{V}_i > 0 \quad \text{for } i = 1, \ldots, s. \tag{16b}$$

Then these vectors are linearly independent.

Proof. Suppose some linear combination of the given vectors is $\mathbf{0}$:

$$k_1\mathbf{V}_1 + k_2\mathbf{V}_2 + \cdots + k_s\mathbf{V}_s = \mathbf{0}. \tag{17}$$

Let i be any integer from 1 through s, and form the inner product of both sides of Eq. (17) with $\mathbf{V}_i$. On the left-hand side the distributive law applies and all these terms are zero except for

$$k_i\mathbf{V}_i \cdot \mathbf{V}_i;$$

and on the right,

$$\mathbf{0} \cdot \mathbf{V}_i = 0.$$

Hence

$$k_i \mathbf{V}_i \cdot \mathbf{V}_i = 0.$$

But, by hypothesis, Eq. (16b),

$$|\mathbf{V}_i|^2 = \mathbf{V}_i \cdot \mathbf{V}_i \neq 0.$$

Therefore

$$k_i = 0 \quad \text{for every } i,\ 1 \leq i \leq s. \tag{18}$$

Since the *only* linear combination of $\mathbf{V}_1, \ldots, \mathbf{V}_s$ that satisfies Eq. (17) is one for which

$$k_1 = k_2 = \cdots = k_s = 0,$$

the vectors are linearly *independent.* □

Problem 3. Show that the following vectors are mutually orthogonal:

$$\mathbf{V}_1 = (1, 0, 1, 0),$$
$$\mathbf{V}_2 = (1, 0, -1, 0),$$
$$\mathbf{V}_3 = (0, 2, 0, 3),$$
$$\mathbf{V}_4 = (0, 3, 0, -2).$$

Solution

$$\mathbf{V}_1 \cdot \mathbf{V}_2 = 1 - 1 = 0,$$
$$\mathbf{V}_1 \cdot \mathbf{V}_3 = \mathbf{V}_1 \cdot \mathbf{V}_4 = 0,$$
$$\mathbf{V}_2 \cdot \mathbf{V}_3 = \mathbf{V}_2 \cdot \mathbf{V}_4 = 0,$$
$$\mathbf{V}_3 \cdot \mathbf{V}_4 = 6 - 6 = 0.$$

Also, none of these vectors has zero length. Therefore, by Theorem 5, they form a *linearly independent set.*

Problem 4. Let $\mathbf{V}_1, \mathbf{V}_2, \mathbf{V}_3, \mathbf{V}_4$ be as they are in Problem 3, and let

$$\mathbf{V}_5 = (1, 0, 0, 0).$$

Show that $\mathbf{V}_1, \mathbf{V}_2, \mathbf{V}_3, \mathbf{V}_4, \mathbf{V}_5$ are *linearly dependent.*

Solution. By an easy inspection,

$$\mathbf{V}_1 + \mathbf{V}_2 = (2, 0, 0, 0) = 2\mathbf{V}_5.$$

Hence

$$1\mathbf{V}_1 + 1\mathbf{V}_2 + 0\mathbf{V}_3 + 0\mathbf{V}_4 - 2\mathbf{V}_5 = \mathbf{0},$$

and the coefficients

$$k_1 = 1, \quad k_2 = 1, \quad k_3 = 0, \quad k_4 = 0, \quad k_5 = -2$$

are *not* all zero. Hence, by Definition 1, the vectors are *linearly dependent.*

Remark 3. In Problem 4, the vectors $\mathbf{V}_1, \mathbf{V}_2, \mathbf{V}_5$ form a linearly dependent *subset* of the larger set $\mathbf{V}_1, \mathbf{V}_2, \mathbf{V}_3, \mathbf{V}_4, \mathbf{V}_5$. Whenever a set S contains a nonempty linearly dependent subset T, the set S is also linearly dependent, as we shall now show.

Theorem 6. *Let S be a set of vectors $\mathbf{V}_1, \mathbf{V}_2, \ldots, \mathbf{V}_s$ that contains a nonempty subset $T = \{\mathbf{V}_1', \mathbf{V}_2', \ldots, \mathbf{V}_t'\}$ of linearly dependent vectors. Then S is linearly dependent.*

Proof. By hypothesis, there are scalars $c_1, c_2, \ldots, c_t$ that are not all zero and such that

$$c_1\mathbf{V}_1' + c_2\mathbf{V}_2' + \cdots + c_t\mathbf{V}_t' = \mathbf{0}. \tag{19}$$

Since T is a subset of S, the vectors $\mathbf{V}_1, \ldots, \mathbf{V}_t$ are some (possibly all) of $\mathbf{V}_1, \ldots, \mathbf{V}_s$. Then we have

$$k_1\mathbf{V}_1 + k_2\mathbf{V}_2 + \cdots + k_s\mathbf{V}_s = \mathbf{0}, \tag{20}$$

provided we take $k_j = 0$ when $\mathbf{V}_j$ is not a vector in T and provided we take the other k's to be equal to the c's that appear in Eq. (19). Not all are zero. Hence (20) is satisfied with some of the k's not zero. Hence S is a linearly dependent set. □

Corollary. *Any set of vectors that contains the zero vector is linearly dependent.*

Proof. Let $T = \{\mathbf{0}\}$. Then T is linearly dependent because

$$k_1\mathbf{0} = \mathbf{0} \quad \text{for } k_1 = 1 \neq 0.$$

Hence S is dependent if $T \subset S$, by Theorem 6. □

Spanning set

If $\mathbf{V}_1, \mathbf{V}_2, \ldots, \mathbf{V}_s$ are vectors in some vector space $\mathbf{X}$, the collection $\mathbf{F}$ of *all vectors* $\mathbf{W}$ that can be expressed as linear combinations

$$\mathbf{W} = k_1\mathbf{V}_1 + k_2\mathbf{V}_2 + \cdots + k_s\mathbf{V}_s$$

is closed with respect to addition and with respect to multiplication by scalars. It forms what is called a *vector subspace* of $\mathbf{X}$. We also say that $\mathbf{F}$ is *spanned by* $\mathbf{V}_1, \mathbf{V}_2, \ldots, \mathbf{V}_s$.

Example 3. The vectors $\mathbf{V}_1 = (1, 2)$ and $\mathbf{V}_2 = (0, 1)$ span all of E^2, because any vector $\mathbf{W} = (a, b)$ in E^2 can be expressed as $k_1\mathbf{V}_1 + k_2\mathbf{V}_2$, with $k_1 = a$ and $k_2 = b - 2a$:

$$\begin{aligned}(a, b) &= a(1, 2) + (b - 2a)(0, 1) \\ &= (a, 2a) + (0, b - 2a).\end{aligned}$$

Example 4. In the space $E^3 = \{(x, y, z) : x, y, z \text{ real}\}$, the vectors $\mathbf{V}_1 = (1, 0, 0)$ and $\mathbf{V}_2 = (0, 1, 0)$ span the xy-plane, because all vectors of the form $(x, y, 0)$ are equal to $x\mathbf{V}_1 + y\mathbf{V}_2$. The xy-plane is a two-dimensional vector subspace of E^3.

Basis

Definition. *If a set of vectors*

$$S = \{\mathbf{V}_1, \mathbf{V}_2, \ldots, \mathbf{V}_s\}$$

in some vector space **X** *is a linearly independent set, then it is said to be a basis for the subspace* **F** *that it spans. Any other set of linearly independent vectors that span the same space* **F** *is also called a basis for* **F**.

Example 5. In E^2, the vectors $(1, 0)$ and $(0, 1)$ are linearly independent and span E^2. The same is true of the two vectors $(-1, 1)$ and $(1, 1)$; these two vectors are orthogonal and are therefore *independent* in E^2, and they *span* E^2 since

$$\begin{aligned}(a, b) &= a(1, 0) + b(0, 1) \\ &= \frac{a}{2}[(1, 1) - (-1, 1)] + \frac{b}{2}[(1, 1) + (-1, 1)] \\ &= \frac{a + b}{2}(1, 1) + \frac{b - a}{2}(-1, 1).\end{aligned}$$

The foregoing example illustrates an important feature of E^2: Each basis considered contains exactly two vectors. This is a special instance of this broader theorem.

Theorem 7. *Let* **F** *be a vector subspace of a vector space* **X**, *and suppose that*

$$S = \{\mathbf{V}_1, \mathbf{V}_2, \ldots, \mathbf{V}_s\},$$

$$T = \{\mathbf{W}_1, \mathbf{W}_2, \ldots, \mathbf{W}_t\}$$

are two bases for **F**. *Then $s = t$; that is, any two bases for* **F** *contain the same number of vectors.*

Proof. Because S is a basis, it spans **F**. Because $\mathbf{W}_1 \in \mathbf{F}$, we have

$$\mathbf{W}_1 = c_1\mathbf{V}_1 + c_2\mathbf{V}_2 + \cdots + c_s\mathbf{V}_s,$$

so the vectors

$$\mathbf{W}_1, \mathbf{V}_1, \mathbf{V}_2, \ldots, \mathbf{V}_s \tag{21}$$

are linearly dependent. Therefore there exist constants

$$k_1, k_2, \ldots, k_{s+1} \quad \text{not all zero}$$

such that

$$k_1\mathbf{W}_1 + k_2\mathbf{V}_1 + k_3\mathbf{V}_2 + \cdots + k_{s+1}\mathbf{V}_s = \mathbf{0}. \tag{22}$$

We could, for example, take

$$k_1 = -1, \quad k_2 = c_1,$$

$$k_3 = c_2, \ldots, k_{s+1} = c_s, \quad k_1 \neq 0.$$

Now comes a clever idea (from Birkhoff and MacLane, *A Survey of Modern Algebra*): In Eq. (22), let k_{j+1} be the *last* nonzero coefficient, so that all coefficients later than k_{j+1} (if $j \neq s$) are zero, and so that Eq. (22) is therefore the same as

$$k_1\mathbf{W}_1 + k_2\mathbf{V}_1 + \cdots + k_{j+1}\mathbf{V}_j = 0. \tag{23}$$

Since $k_{j+1} \neq 0$, we know that Eq. (23) can be multiplied by k_{j+1}^{-1} and solved for $\mathbf{V}_j$:

$$\mathbf{V}_j = -k_{j+1}^{-1}(k_1\mathbf{W}_1 + k_2\mathbf{V}_1 + \cdots + k_j\mathbf{V}_{j-1}). \tag{24}$$

Incidentally, $j > 0$ because k_1 cannot be the *last* nonzero coefficient in (22), since $\mathbf{W}_1 \neq \mathbf{0}$.

Now Eq. (24) says that in the list of vectors (21), the vector $\mathbf{V}_j$ is a linear combination of earlier vectors. Hence any linear combination of all the vectors from that list (including $\mathbf{V}_j$) can be replaced by a linear combination of the vectors in the set S' that we obtain by deleting $\mathbf{V}_j$:

$$S' = \{\mathbf{W}_1, \mathbf{V}_1, \mathbf{V}_2, \ldots, \mathbf{V}_s\} - \{\mathbf{V}_j\}. \tag{25}$$

The new set S' still has s vectors, and it also *spans* **F** because the set of vectors in the list (21) spans **F**.

Essentially, what the above argument has demonstrated is that we may substitute a vector $\mathbf{W}_1$ from the second basis set T for some $\mathbf{V}_j$ in the first basis set S.

Now, of course, if $t = 1$, we would be able to conclude that $s \geq t$, because S' contains s vectors and it also contains $\mathbf{W}_1$, so $s \geq 1$. What we shall prove, more generally, is that $s \geq t$. Therefore we continue to feed $\mathbf{W}$ vectors into the system from the left. We argue, as above, that some $\mathbf{V}$-vector is a linear combination of earlier vectors and that we can therefore remove it and still have a spanning set. To formalize this argument, let us suppose that we have inserted r of the vectors from T and removed r of the vectors from S, and now have a spanning set

$$U = \{\mathbf{W}_r, \mathbf{W}_{r-1}, \ldots, \mathbf{W}_1, \mathbf{V}_{i_1}, \mathbf{V}_{i_2}, \ldots, \mathbf{V}_{i_{s-r}}\},$$

which contains r $\mathbf{W}$-vectors and $s - r$ V-vectors. If $t = r$, all the $\mathbf{W}$-vectors are in the set U, and $s \geq r$, and therefore $s \geq t$. But if not all the $\mathbf{W}$-vectors are in the set U, then we argue this way:

> $\mathbf{W}_{r+1}$ is in $\mathbf{F}$, so it is a linear combination of vectors in the set U.

The new set

$$\{\mathbf{W}_{r+1}, \mathbf{W}_r, \ldots, \mathbf{W}_1, \mathbf{V}_{i_1}, \ldots, \mathbf{V}_{i_{s-r}}\} \tag{26}$$

is a set that spans $\mathbf{F}$ and is linearly dependent. Some one of its members can therefore be expressed as a linear combination of those appearing *before* it on the list. (The argument for this is like the one in the first paragraph of the proof.) However, no $\mathbf{W}$-vector is a linear combination of other $\mathbf{W}$-vectors because T is a *basis* (and is therefore a linearly independent set). So some $\mathbf{V}$-vector is still present in the list (26) *and* has the property that it is a linear combination of earlier vectors in that list. Delete it. The new set has s vectors, and still spans $\mathbf{F}$. We conclude that

$$\text{if } \quad r < t, \quad \text{then} \quad s \geq r + 1.$$

The logical conclusion of this argument (for $r = t - 1$) is that

$$s \geq t. \tag{27}$$

Equation (27) is almost our goal. To complete the proof we note that the whole argument could be repeated (it isn't necessary actually to do so), interchanging the roles of the sets S and T. That would lead to the conclusion that

$$t \geq s. \tag{28}$$

But (27) and (28) together imply

$$t = s. \;\square$$

Since all bases for $\mathbf{F}$ contain the same number of elements, the following definition of *dimension* makes sense.

Definition. Dimension. *If a vector space can be spanned by a finite set of vectors, we say that it is a finite-dimensional vector space, and its dimension is the number of vectors in any basis for it.*

Remark 4. Although we have restricted our discussion mainly to Euclidean spaces, there are other types of finite-dimensional vector spaces to which the foregoing theorems and definitions apply.

Example 6. Consider the set of polynomials containing the zero polynomial and all polynomials

$$f(x) = a_0 + a_1 x + a_2 x^2 \text{ of degree} \leq 2,$$

with a_0, a_1, a_2 real. The polynomials

$$p_1(x) = 1, \quad p_2(x) = x, \quad p_3(x) = x^2$$

form a basis for a vector space of such polynomials (which can be imbedded in a larger, infinite-dimensional vector space of all polynomials). This vector space has dimension 3, the number of elements in the basis

$$\{p_1, p_2, p_3\}.$$

Remark 5. In more extensive treatments of linear algebra, it is shown that

(a) any *spanning set* of vectors *contains* a basis for the space spanned; and

(b) any linearly *independent set* of vectors is *contained in* some basis.

For (a), the key idea is to delete vectors that can be expressed as linear combinations of other vectors in the spanning set, so long as there are any. One stops when the set is linearly independent. What's left is the basis.

For (b), the key idea is to augment the given independent set by appending a list of vectors that *is* a basis. This bigger set is now dependent, and some element of it is a linear combination of vectors that come earlier in the list. Delete one such vector, and continue until the remaining set is linearly independent. What's left is a basis, and it contains the original vectors because none of them is expressible as a linear combination of vectors preceding it in the list.

Problem 5. The vectors

$$\mathbf{V}_1 = (1, 1, 1) \quad \text{and} \quad \mathbf{V}_2 = (1, 1, 0)$$

are linearly independent. Find a basis for E^3 that contains both $\mathbf{V}_1$ and $\mathbf{V}_2$.

Solution. Let

$$\mathbf{W}_1 = (1, 0, 0), \quad \mathbf{W}_2 = (0, 1, 0), \quad \mathbf{W}_3 = (0, 0, 1).$$

These are a basis for E^3 (the *standard* basis), because

$$(x, y, z) = x\mathbf{W}_1 + y\mathbf{W}_2 + z\mathbf{W}_3.$$

Now

$$\mathbf{V}_1, \mathbf{V}_2, \mathbf{W}_1, \mathbf{W}_2, \mathbf{W}_3$$

are linearly dependent:

$$\mathbf{V}_1 = \mathbf{W}_1 + \mathbf{W}_2 + \mathbf{W}_3.$$

Therefore

$$-\mathbf{V}_1 + \mathbf{W}_1 + \mathbf{W}_2 + \mathbf{W}_3 = \mathbf{0}$$

or

$$\mathbf{W}_3 = \mathbf{V}_1 - \mathbf{W}_1 - \mathbf{W}_2.$$

so $\mathbf{W}_3$ is not needed to span E^3; the vectors

$$\{\mathbf{V}_1, \mathbf{V}_2, \mathbf{W}_1, \mathbf{W}_2\}$$

form a spanning set. But they are linearly dependent too:

$$\mathbf{V}_2 = \mathbf{W}_1 + \mathbf{W}_2, \quad \text{or} \quad -\mathbf{V}_2 + \mathbf{W}_1 + \mathbf{W}_2 = \mathbf{0}.$$

Therefore

$$\mathbf{W}_2 = \mathbf{V}_2 - \mathbf{W}_1,$$

and we don't need $\mathbf{W}_2$ either. The set

$$\{\mathbf{V}_1, \mathbf{V}_2, \mathbf{W}_1\}$$

spans E^3. Are these vectors linearly independent? We test them: Suppose

$$c_1\mathbf{V}_1 + c_2\mathbf{V}_2 + c_3\mathbf{W}_1 = \mathbf{0},$$

that is,

$$(c_1, c_1, c_1) + (c_2, c_2, 0) + (c_3, 0, 0) = (0, 0, 0).$$

Then, from the third component, c_1 must be zero, and

$$(c_2, c_2, 0) + (c_3, 0, 0) = (0, 0, 0).$$

Look at the second component:

$$c_2 + 0 = 0.$$

Therefore $c_2 = 0$ and

$$(c_3, 0, 0) = (0, 0, 0),$$

and, finally, c_3 must also be zero. Therefore $\{\mathbf{V}_1, \mathbf{V}_2, \mathbf{W}_1\}$ is a *basis* for E^3.

Remark 6. One consequence of the results (a) and (b) stated in Remark 5 is this:

If a vector space $\mathbf{X}$ has dimension r, then any set of r vectors that *either* is linearly independent *or* spans $\mathbf{X}$ is a basis for $\mathbf{X}$.

(a) Suppose S is a set of r *linearly independent* vectors in $\mathbf{X}$. Then S is contained in some basis for $\mathbf{X}$, say, T. But every basis contains exactly r elements since the dimension of $\mathbf{X}$ is r. Therefore $T = S$; so S is a basis.

(b) Suppose S is a set of r vectors that *spans* $\mathbf{X}$. Then S contains a subset, say U, that is a basis for $\mathbf{X}$. But every basis for $\mathbf{X}$ contains r elements. Therefore $U = S$; so S is a basis.

Problem 6. Show that the vectors

$$\mathbf{V}_1 = (1, 0, -1, 1) \quad \text{and} \quad \mathbf{V}_2 = (0, 1, 1, 0)$$

are linearly independent, and find a basis for E^4 that contains $\mathbf{V}_1$ and $\mathbf{V}_2$.

Solution. If $c_1\mathbf{V}_1 + c_2\mathbf{V}_2 = \mathbf{0}$, then

$$(c_1, 0, -c_1, c_1) + (0, c_2, c_2, 0) = (0, 0, 0, 0),$$

and so $c_1 = c_2 = 0$. Therefore $\mathbf{V}_1$ and $\mathbf{V}_2$ are linearly independent. For variety, we shall find vectors that are orthogonal to $\mathbf{V}_1$ and $\mathbf{V}_2$. Let $\mathbf{A} = (a_1, a_2, a_3, a_4)$ be orthogonal to $\mathbf{V}_1$ and to $\mathbf{V}_2$:

$$\mathbf{V}_1 \cdot \mathbf{A} = 0 = a_1 - a_3 + a_4,$$

$$\mathbf{V}_2 \cdot \mathbf{A} = 0 = a_2 + a_3.$$

From the second equation, we get

$$a_3 = -a_2,$$

and from the first we get

$$a_4 = a_3 - a_1 = -a_2 - a_1.$$

Hence

$$\begin{aligned}\mathbf{A} &= (a_1, a_2, a_3, a_4) = (a_1, a_2, -a_2, -a_1 - a_2)\\ &= a_1(1, 0, 0, -1) + a_2(0, 1, -1, -1).\end{aligned}$$

The vectors we get when

$$a_2 = 0 \quad \text{and} \quad a_1 = 1$$

or

$$a_1 = 0 \quad \text{and} \quad a_2 = 1$$

are

$$\mathbf{W}_1 = (1, 0, 0, -1),$$

$$\mathbf{W}_2 = (0, 1, -1, -1).$$

We now have four vectors, $\mathbf{V}_1$, $\mathbf{V}_2$, $\mathbf{W}_1$, $\mathbf{W}_2$, in E^4. If they span E^4 *or* are linearly independent, they form a basis. We shall show that they are *linearly independent.*

Suppose

$$c_1\mathbf{V}_1 + c_2\mathbf{V}_2 + c_3\mathbf{W}_1 + c_4\mathbf{W}_2 = \mathbf{0}.$$

Then

$$\begin{aligned}&(c_1, 0, -c_1, c_1) + (0, c_2, c_2, 0)\\ &\quad + (c_3, 0, 0, -c_3) + (0, c_4, -c_4, -c_4) = (0, 0, 0, 0).\end{aligned}$$

That is,

$$c_1 + c_3 = 0; \quad c_3 = -c_1,$$

$$c_2 + c_4 = 0; \quad c_4 = -c_2,$$

$$-c_1 + c_2 - c_4 = 0; \quad -c_1 + 2c_2 = 0; \quad c_1 = 2c_2,$$

$$c_1 - c_3 - c_4 = 0; \quad 2c_1 + c_2 = 0; \quad 5c_2 = 0.$$

The result $5c_2 = 0$ implies

$$c_2 = 0, \; c_1 = 2c_2 = 0,$$

$$c_3 = -c_1 = 0, \quad \text{and} \quad c_4 = -c_2 = 0.$$

Therefore the vectors are *linearly independent* (and we need not verify that they also span E^4).

Remark 7. In Problem 6, the vectors $\mathbf{V}_1$ and $\mathbf{V}_2$ span a two-dimensional subspace of E_4, and so do the vectors $\mathbf{W}_1$ and $\mathbf{W}_2$. Every vector in the second subspace is orthogonal to every vector in the first subspace because, by the distributive law,

$$\begin{aligned}(a\mathbf{W}_1 + b\mathbf{W}_2) \cdot (c\mathbf{V}_1 + d\mathbf{V}_2)&\\ = ac(\mathbf{W}_1 \cdot \mathbf{V}_1) &+ ad(\mathbf{W}_1 \cdot \mathbf{V}_2)\\ &+ bc(\mathbf{W}_2 \cdot \mathbf{V}_1) + bd(\mathbf{W}_2 \cdot \mathbf{V}_2),\end{aligned}$$

and

$$\mathbf{W}_1 \cdot \mathbf{V}_1 = \mathbf{W}_1 \cdot \mathbf{V}_2 = \mathbf{W}_2 \cdot \mathbf{V}_1 = \mathbf{W}_2 \cdot \mathbf{V}_2 = 0.$$

We say that the subspace spanned by $\mathbf{W}_1$ and $\mathbf{W}_2$ is the *orthogonal complement*, in E_4, of the subspace spanned by $\mathbf{V}_1$ and $\mathbf{V}_2$.

EXERCISES

1. Find all (if there are any) ways of expressing

$$\begin{bmatrix} -1 \\ 1 \end{bmatrix}$$

as a linear combination of

$$\mathbf{V}_1 = \begin{bmatrix} 2 \\ -1 \end{bmatrix}, \quad \mathbf{V}_2 = \begin{bmatrix} 1 \\ 2 \end{bmatrix}, \quad \mathbf{V}_3 = \begin{bmatrix} 1 \\ 0 \end{bmatrix}.$$

2. Express both

$$\mathbf{V}_1 = \begin{bmatrix} 1 \\ 0 \end{bmatrix} \quad \text{and} \quad \mathbf{V}_2 = \begin{bmatrix} 0 \\ 1 \end{bmatrix}$$

as linear combinations of

$$\mathbf{W}_1 = \begin{bmatrix} 1 \\ 1 \end{bmatrix} \quad \text{and} \quad \mathbf{W}_2 = \begin{bmatrix} -1 \\ 1 \end{bmatrix}.$$

3. Find two orthogonal unit vectors $\mathbf{V}_1$ and $\mathbf{V}_2$ that span the subspace of E_4 that is spanned by

$$\mathbf{W}_1 = (1, 1, 1, 1), \quad \mathbf{W}_2 = (1, 1, 1, 0).$$

Suggestion. Start with $\mathbf{V}_1$ in the direction of $\mathbf{W}_1$.

4. Let $\mathbf{A} = (1, 2, -1, 0)$ and $\mathbf{B} = (2, -1, 0, 1)$. Find all vectors $\mathbf{X} = (x, y, z, u)$ in E^4 that are orthogonal to both $\mathbf{A}$ and $\mathbf{B}$.

5. (Continuation of Exercise 4.) Find a basis for E^4 that contains $\mathbf{A}$ and $\mathbf{B}$ and two vectors $\mathbf{V}_1$ and $\mathbf{V}_2$ that are orthogonal to each other and to $\mathbf{A}$ and to $\mathbf{B}$.

The following procedure can be used to tell whether vectors

$$\begin{aligned} \mathbf{A}_1 &= (a_{11}, a_{12}, \ldots, a_{1n}), \\ \mathbf{A}_2 &= (a_{21}, a_{22}, \ldots, a_{2n}), \\ &\vdots \\ \mathbf{A}_m &= (a_{m1}, a_{m2}, \ldots, a_{mn}) \end{aligned}$$

are linearly independent or linearly dependent.

(a) First write the given vectors as the rows of an $m \times n$ matrix $\mathbf{A} = [a_{ij}]$.
(b) Then perform elementary (reversible) row operations on the rows of $\mathbf{A}$ to convert it into an echelon form $\mathbf{E}$.
(c) *Conclusion.* If $\mathbf{E}$ contains a row that is all zeros, the given vectors are linearly dependent; otherwise, they are linearly independent.

6. Explain why the foregoing procedure works. In your explanation, take account of the following example.

Example

$$\mathbf{A} = \begin{bmatrix} 1 & -2 & 3 & 4 \\ 2 & -1 & 0 & -2 \\ 1 & -5 & 9 & 14 \end{bmatrix}$$

Add $(-2 \times \text{row one})$ to row two, and $(-1 \times \text{row one})$ to row three. Don't change row one.

$$\mathbf{A}_1^* = \begin{bmatrix} 1 & -2 & 3 & 4 \\ 0 & 3 & -6 & -10 \\ 0 & -3 & 6 & 10 \end{bmatrix}.$$

Add row two to row three:

$$\mathbf{A}_2^* = \begin{bmatrix} 1 & -2 & 3 & 4 \\ 0 & 3 & -6 & -10 \\ 0 & 0 & 0 & 0 \end{bmatrix}.$$

We need not complete the echelon reduction, which consists of dividing row two by 3, and then adding $(2 \times \text{row two})$ to row one. The last row of $\mathbf{A}_2^*$ is all zeros, so we conclude that the vectors

$$\mathbf{A}_1 = (1, -2, 3, 4),$$

$$\mathbf{A}_2 = (2, -1, 0, -2),$$

$$\mathbf{A}_3 = (1, -5, 9, 14)$$

are linearly dependent. *Check.* $3\mathbf{A}_1 - \mathbf{A}_2 = \mathbf{A}_3$.

In Exercises 7 through 10, use the method illustrated for Exercise 6 to determine whether the following sets of vectors are linearly dependent or independent.

7. $(1, -2, 1, 1)$, $(2, 1, 0, 1)$, $(1, 0, 1, 0)$
8. $(-2, 1, 0, 1)$, $(1, 2, 1, 0)$, $(0, 5, 2, 1)$
9. $(1, 3, -2)$, $(2, 0, 1)$, $(0, 6, -5)$
10. $(1, 3, -2)$, $(2, 0, 1)$, $(1, 5, 3)$

The following procedure, known as the Gram-Schmidt process, leads to an orthogonal basis of unit vectors ("orthonormal basis") for the space spanned by the given nonzero vectors $\mathbf{V}_1, \mathbf{V}_2, \ldots, \mathbf{V}_r$.

(a) First, test the vectors for linear independence. If they are linearly independent, then proceed to step (b). If they are linearly dependent, then some one of them can be expressed in terms of those appearing before it in the list. Remove such a vector: The remaining vectors still span the same space. Test these for independence. Continue eliminating vectors (if any) that can be expressed in terms of those earlier in the list. Ultimately you have a set which can be relabeled

$$\{\mathbf{V}_1, \mathbf{V}_2, \ldots, \mathbf{V}_k\},$$

in which the vectors are linearly independent. Now go on to step (b).

(b) Let $\gamma_1 = \mathbf{V}_1/|\mathbf{V}_1|$. Then γ_1 is a unit vector.

(c) Find the orthogonal projection of $\mathbf{V}_2$ on γ_1:

$$\mathbf{V}_2' = \frac{\mathbf{V}_2 \cdot \gamma_1}{\gamma_1 \cdot \gamma_1}\gamma_1,$$

or

$$\mathbf{V}_2' = (\mathbf{V}_2 \cdot \gamma_1)\gamma_1 \quad \text{since} \quad \gamma_1 \cdot \gamma_1 = 1.$$

(d) Let
$$\mathbf{V}_2'' = \mathbf{V}_2 - \mathbf{V}_2'$$
be the orthogonal complement of $\mathbf{V}_2$ with respect to γ_1. This is not the zero vector because
$$\mathbf{V}_2 \neq \mathbf{V}_2' = (\mathbf{V}_2 \cdot \gamma_1)\frac{\mathbf{V}_1}{|\mathbf{V}_1|},$$
since $\mathbf{V}_1$ and $\mathbf{V}_2$ are independent. Let
$$\gamma_2 = \mathbf{V}_2''/|\mathbf{V}_2''|.$$
Now
$$\gamma_1, \gamma_2, \mathbf{V}_3, \ldots, \mathbf{V}_k$$
span the same space as the original vectors. They are linearly independent, and γ_1, γ_2 are orthogonal unit vectors.

(e) To continue, suppose we have arrived at a stage with j orthogonal unit vectors and $(k - j)$ of the original $\mathbf{V}$'s:
$$\gamma_1, \gamma_2, \ldots, \gamma_j, \mathbf{V}_{j+1}, \ldots, \mathbf{V}_k.$$
The orthogonal *projection* of $\mathbf{V}_{j+1}$ on the unit vectors $\gamma_1, \ldots, \gamma_j$ is
$$\mathbf{V}_{j+1}' = c_1\gamma_1 + c_2\gamma_2 + \cdots + c_j\gamma_j,$$
with
$$c_1 = \mathbf{V}_{j+1} \cdot \gamma_1,$$
$$c_2 = \mathbf{V}_{j+1} \cdot \gamma_2, \quad \ldots,$$
$$c_j = \mathbf{V}_{j+1} \cdot \gamma_j.$$
The component of $\mathbf{V}_{j+1}$ that is orthogonal to $\gamma_1, \ldots, \gamma_j$ is
$$\mathbf{V}_{j+1}'' = \mathbf{V}_{j+1} - \mathbf{V}_{j+1}'.$$
It isn't zero (by the linear independence of $\gamma_1, \ldots, \gamma_j, \mathbf{V}_{j+1}$), and we take
$$\gamma_{j+1} = \mathbf{V}_{j+1}''/|\mathbf{V}_{j+1}''|.$$

(f) Repeat step (e) until $j + 1 = k$. Then
$$\gamma_1, \gamma_2, \ldots, \gamma_k$$
is the desired orthonormal basis.

11. Explain in some detail just why it is that the Gram-Schmidt process leads to an orthonormal basis. In your explanation, take account of the following example.

Example.
$$\mathbf{V}_1 = (1, -1, 0, 1),$$
$$\mathbf{V}_2 = (1, 1, 1, 1),$$
$$\mathbf{V}_3 = (1, 0, 0, 2).$$
To test for independence we use echelon reduction (see Exercise 6) of the matrix $\mathbf{A}$, with rows $\mathbf{V}_1, \mathbf{V}_2, \mathbf{V}_3$:
$$\mathbf{A} = \begin{bmatrix} 1 & -1 & 0 & 1 \\ 1 & 1 & 1 & 1 \\ 1 & 0 & 0 & 2 \end{bmatrix} \sim \begin{bmatrix} 1 & -1 & 0 & 1 \\ 0 & -2 & 1 & 0 \\ 0 & 1 & 0 & 1 \end{bmatrix}$$
$$\sim \begin{bmatrix} 1 & -1 & 0 & 1 \\ 0 & -1 & \frac{1}{2} & 0 \\ 0 & 0 & \frac{1}{2} & 1 \end{bmatrix} \sim \begin{bmatrix} 1 & -1 & 0 & 1 \\ 0 & -1 & 0 & -1 \\ 0 & 0 & \frac{1}{2} & 1 \end{bmatrix}$$
$$\sim \begin{bmatrix} 1 & 0 & 0 & 2 \\ 0 & 1 & 0 & 1 \\ 0 & 0 & 1 & 2 \end{bmatrix}.$$
(The symbol "$\sim$" means "is row-equivalent to.") This shows that the rows are independent. Then
$$\mathbf{V}_1 \cdot \mathbf{V}_1 = 1 + 1 + 0 + 1 = 3,$$
so
$$\gamma_1 = \frac{1}{\sqrt{3}}(1, -1, 0, 1).$$
The component of $\mathbf{V}_2$ along γ_1 is
$$\mathbf{V}_2 \cdot \gamma_1 = \frac{1}{\sqrt{3}}(1 - 1 + 0 + 1) = \frac{1}{\sqrt{3}},$$
so
$$\mathbf{V}_2' = \frac{1}{\sqrt{3}}\gamma_1 = \frac{1}{3}(1, -1, 0, 1)$$
and
$$\mathbf{V}_2'' = \mathbf{V}_2 - \mathbf{V}_2' = (\tfrac{2}{3}, \tfrac{4}{3}, 1, \tfrac{2}{3})$$
$$= \tfrac{1}{3}(2, 4, 3, 2).$$
The second unit vector is
$$\gamma_2 = \frac{\mathbf{V}_2''}{|\mathbf{V}_2''|} = \frac{1}{\sqrt{4 + 16 + 9 + 4}}(2, 4, 3, 2)$$
$$= \frac{1}{\sqrt{33}}(2, 4, 3, 2).$$

Finally, the projections of $\mathbf{V}_3$ on γ_1 and γ_2 are

$$c_1 = \mathbf{V}_3 \cdot \gamma_1 = \frac{1}{\sqrt{3}}(1+0+0+2) = \sqrt{3},$$

$$c_2 = \mathbf{V}_3 \cdot \gamma_2 = \frac{1}{\sqrt{33}}(2+0+0+4) = \frac{6}{\sqrt{33}},$$

and so

$$\mathbf{V}_3' = c_1\gamma_1 + c_2\gamma_2 = \frac{\sqrt{3}}{\sqrt{3}}(1, -1, 0, 1) + \frac{6}{33}(2, 4, 3, 2)$$

$$= \frac{1}{11}(15, -3, 6, 15).$$

The component of $\mathbf{V}_3$ that is orthogonal to both γ_1 and γ_2 is

$$\mathbf{V}_3'' = \mathbf{V}_3 - \mathbf{V}_3' = (1, 0, 0, 2) - \frac{1}{11}(15, -3, 6, 15)$$

$$= \frac{1}{11}(-4, 3, -6, 7),$$

and the unit vector γ_3 is

$$\gamma_3 = \frac{\mathbf{V}_3''}{|\mathbf{V}_3''|} = \frac{1}{\sqrt{16+9+36+49}}(-4, 3, -6, 7)$$

$$= \frac{1}{\sqrt{110}}(-4, 3, -6, 7).$$

12. Show that these vectors are mutually orthogonal:

$$\mathbf{W}_1 = (1, -1, 0, 1), \quad \mathbf{W}_2 = (2, 4, 3, 2),$$
$$\mathbf{W}_3 = (-4, 3, -6, 7).$$

13. Express *each* of the vectors $\mathbf{V}_1$, $\mathbf{V}_2$, and $\mathbf{V}_3$ of the example following Exercise 11 as linear combinations of the vectors $\mathbf{W}_1$, $\mathbf{W}_2$, and $\mathbf{W}_3$ of Exercise 12. *Suggestion.* Let

$$\mathbf{V}_3 = c_{31}\mathbf{W}_1 + c_{32}\mathbf{W}_2 + c_{33}\mathbf{W}_3,$$

and take inner products with $\mathbf{W}_1$, then with $\mathbf{W}_2$, then with $\mathbf{W}_3$, and use orthogonality of the $\mathbf{W}$'s.

14. Let $\mathbf{AX} = \mathbf{B}$ be a system of m simultaneous linear equations in n unknowns. Suppose a *particular vector* $\mathbf{X}^*$ is known such that $\mathbf{AX}^* = \mathbf{B}$. (That is, $\mathbf{X}^*$ gives *one* solution of the system.) Show that

(a) if $\mathbf{Y}$ satisfies $\mathbf{AY} = \mathbf{0}$, then $\mathbf{X} = \mathbf{X}^* + \mathbf{Y}$ satisfies $\mathbf{AX} = \mathbf{B}$; and

(b) if $\mathbf{X}$ is any solution of $\mathbf{AX} = \mathbf{B}$, then

$$\mathbf{Y} = \mathbf{X} - \mathbf{X}^*$$

satisfies $\mathbf{AY} = \mathbf{0}$.

15. In Exercise 14, let

$$\mathbf{A} = \begin{bmatrix} 1 & -1 & 2 & -2 \\ 2 & 0 & 1 & 0 \end{bmatrix}, \quad \mathbf{B} = \begin{bmatrix} 1 \\ -1 \end{bmatrix}.$$

(a) Show that

$$\mathbf{X}^* = \begin{bmatrix} -\frac{1}{2} \\ -\frac{3}{2} \\ 0 \\ 0 \end{bmatrix}$$

is a particular solution of $\mathbf{AX}^* = \mathbf{B}$.

(b) Find *all* solutions of $\mathbf{AY} = \mathbf{0}$.

(c) Show that $\mathbf{X} = \mathbf{X}^* + \mathbf{Y}$ gives *all* solutions of $\mathbf{AX} = \mathbf{B}$.

If $\mathbf{A}$ is an $m \times n$ matrix, its rows are vectors in E^n and its columns are vectors in E^m. The subspace of E^n that is spanned by its rows is called the *row space* of $\mathbf{A}$, and the subspace of E^m that is spanned by its columns is called its *column space*. For each of the matrices given in Exercises 16 through 20, find a basis for the *row* space.

16. $\begin{bmatrix} 1 & 1 \\ 0 & 1 \end{bmatrix}$

17. $\begin{bmatrix} 0 & -1 & 2 \\ 0 & 1 & -2 \end{bmatrix}$

18. $\begin{bmatrix} -1 & 1 & 0 & 2 \\ 1 & 0 & 1 & -2 \end{bmatrix}$

19. $\begin{bmatrix} 0 & -1 & 0 & 1 \\ 1 & 0 & 2 & 0 \\ 1 & -1 & 2 & 1 \end{bmatrix}$

20. $\begin{bmatrix} 1 & -2 \\ 2 & 1 \\ 0 & 1 \\ 1 & 0 \end{bmatrix}$

As Exercises 21, 22, 23, 24, and 25, find a basis for the *column* space for each of the matrices, Exercises 16, 17, 18, 19, and 20, respectively. (Exercise 21 goes with Exercise 16, Exercise 22 goes with Exercise 17, and so on.)

26. Find a basis for the orthogonal complement of the row space of the matrix of Exercise 17.

27. Find a basis for the orthogonal complement of the column space of the matrix of Exercise 20.

28. What is the orthogonal complement of the row space of the matrix of Exercise 16? Of the column space of Exercise 19?

29. Suppose that the vectors $\mathbf{V}_1, \mathbf{V}_2, \ldots, \mathbf{V}_n$ are nonzero, mutually orthogonal vectors in E^n. Why do they form a basis for E^n? If $\mathbf{B}$ is any vector in E^n, how can one easily determine the coefficients $x_1, x_2, \ldots, x_n$ in the equation

$$x_1\mathbf{V}_1 + x_2\mathbf{V}_2 + \cdots + x_n\mathbf{V}_n = \mathbf{B}?$$

30. Show that the vectors

$$\mathbf{V}_1 = (1, 0, 2, 0), \quad \mathbf{V}_2 = (-2, 0, 1, 0),$$
$$\mathbf{V}_3 = (0, 1, 0, 4), \quad \mathbf{V}_4 = (0, -4, 0, 1)$$

are orthogonal, and express $\mathbf{B} = (a, b, c, d)$ as a linear combination of $\mathbf{V}_1, \mathbf{V}_2, \mathbf{V}_3, \mathbf{V}_4$.

31. In Problem 6, the vectors $\mathbf{W}_1$ and $\mathbf{W}_2$ are orthogonal to $\mathbf{V}_1$ and $\mathbf{V}_2$, but the vectors $\mathbf{V}_1$ and $\mathbf{V}_2$ are not orthogonal, nor are $\mathbf{W}_1$ and $\mathbf{W}_2$. But it is true that $\mathbf{V}_2 = \mathbf{V}_2' + \mathbf{V}_2''$, where $\mathbf{V}_2'$ is the orthogonal projection of $\mathbf{V}_2$ on $\mathbf{V}_1$. Similarly

$$\mathbf{W}_2 = \mathbf{W}_2' + \mathbf{W}_2'',$$

with $\mathbf{W}_2'$ the orthogonal projection of $\mathbf{W}_2$ on $\mathbf{W}_1$.

(a) Show that the vectors $\mathbf{V}_1, \mathbf{V}_2'', \mathbf{W}_1, \mathbf{W}_2''$ form an orthogonal basis for E^4. Find these vectors explicitly.

(b) Express x, y, z, u in terms of a, b, c, d if

$$x\mathbf{V}_1 + y\mathbf{V}_2 + z\mathbf{W}_1 + u\mathbf{W}_2 = a\mathbf{V}_1 + b\mathbf{V}_2'' + c\mathbf{W}_1 + d\mathbf{W}_2''.$$

(c) First find a, b, c, d to make the right side of the last equation equal to $(2, 1, -1, -2)$, and then find x, y, z, u as in (b).

[These three steps suggest a procedure for solving simultaneous equations by replacing the columns $\mathbf{V}_1, \mathbf{V}_2, \mathbf{W}_1, \mathbf{W}_2$ of a matrix by related orthogonal columns, exploiting the orthogonality to solve easier equations, and then backtracking to the original equations.]

32. If S is a basis for a subspace $\mathbf{X}$ of E^n and T is a basis for the orthogonal complement of $\mathbf{X}$ in E^n, show that $S \cup T$ is a basis for E^n. (Thus every vector $\mathbf{Z}$ in E^n is the sum of a vector $\mathbf{Z}'$ in $\mathbf{X}$ and a vector $\mathbf{Z}''$ in the orthogonal complement of $\mathbf{X}$.)

13.5 MATRICES AND LINEAR TRANSFORMATIONS

In this article, we look once more at a system of equations, say

$$\begin{aligned} 2x_1 + x_2 - x_3 &= b_1 \\ x_1 - 2x_2 + x_3 &= b_2. \end{aligned} \tag{1}$$

However, instead of specifying numerical values for b_1 and b_2, we interpret Eqs. (1) as equivalent to a machine into which a vector (x_1, x_2, x_3) can be fed and which then produces a vector (b_1, b_2) as its output. (See Fig. 13.9.) From this point of view, Eqs. 1, or the matrix equation

$$\mathbf{AX} = \mathbf{B}, \tag{2}$$

represents a *function* whose *domain* is E^3 and whose *range* is a subspace of E^2. (See Fig. 13.10.)

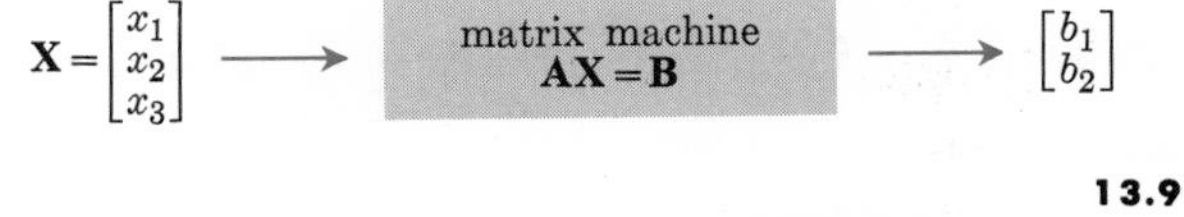

13.9

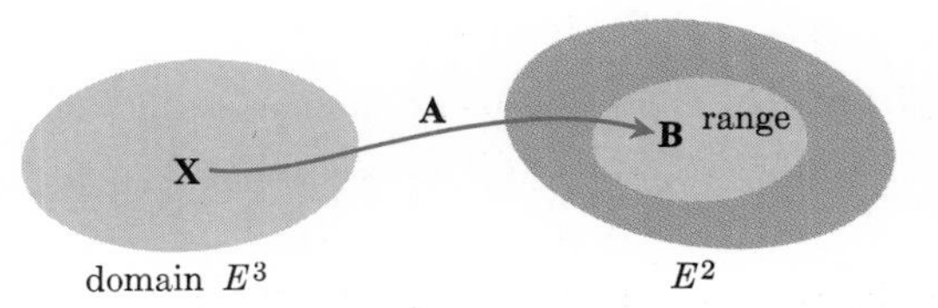

13.10 $\mathbf{B} = \mathbf{AX}$ as a mapping from E^3 to E^2.

Problem 1. Find a basis for the *range* of the mapping given by Eqs. (1).

Solution. Every vector in the domain E^3 is a linear combination of the vectors

$$\mathbf{V}_1 = (1, 0, 0), \quad \mathbf{V}_2 = (0, 1, 0), \quad \mathbf{V}_3 = (0, 0, 1).$$

The images of these vectors are

$$\mathbf{W}_1 = (2, 1), \quad \mathbf{W}_2 = (1, -2), \quad \text{and} \quad \mathbf{W}_3 = (-1, 1).$$

Now every vector $\mathbf{B}$ in the *range* is the image of some vector

$$\mathbf{X} = (x_1, x_2, x_3) = x_1\mathbf{V}_1 + x_2\mathbf{V}_2 + x_3\mathbf{V}_3,$$

so that we may write **B** as follows:

$$\mathbf{B} = \mathbf{A}\begin{bmatrix} x_1 \\ x_2 \\ x_3 \end{bmatrix} = \mathbf{A}\left\{x_1\begin{bmatrix} 1 \\ 0 \\ 0 \end{bmatrix} + x_2\begin{bmatrix} 0 \\ 1 \\ 0 \end{bmatrix} + x_3\begin{bmatrix} 0 \\ 0 \\ 1 \end{bmatrix}\right\}$$

$$= x_1\mathbf{A}\begin{bmatrix} 1 \\ 0 \\ 0 \end{bmatrix} + x_2\mathbf{A}\begin{bmatrix} 0 \\ 1 \\ 0 \end{bmatrix} + x_3\mathbf{A}\begin{bmatrix} 0 \\ 0 \\ 1 \end{bmatrix}$$

$$= x_1\mathbf{W}_1 + x_2\mathbf{W}_2 + x_3\mathbf{W}_3.$$

Therefore $\{\mathbf{W}_1, \mathbf{W}_2, \mathbf{W}_3\}$

spans the range. Since

$$\mathbf{W}_1 \cdot \mathbf{W}_2 = 0$$

and

$$\mathbf{W}_1 \neq \mathbf{0}, \qquad \mathbf{W}_2 \neq \mathbf{0},$$

the vectors

$$\mathbf{W}_1, \mathbf{W}_2$$

are independent. Indeed, they span E^2, and

$$\mathbf{W}_3 = c_1\mathbf{W}_1 + c_2\mathbf{W}_2,$$

with

$$c_1 = \frac{\mathbf{W}_3 \cdot \mathbf{W}_1}{\mathbf{W}_1 \cdot \mathbf{W}_1} = \frac{-2+1}{4+1} = -\frac{1}{5},$$

$$c_2 = \frac{\mathbf{W}_3 \cdot \mathbf{W}_2}{\mathbf{W}_2 \cdot \mathbf{W}_2} = \frac{-1-2}{1+4} = -\frac{3}{5}.$$

Check. $-\frac{1}{5}(2, 1) - \frac{3}{5}(1, -2) = (-\frac{5}{5}, \frac{5}{5}) = \mathbf{W}_3.$

Therefore $\mathbf{W}_1$ and $\mathbf{W}_2$ form a *basis* for the *range* of the mapping given by Eqs. (1).

Remark 1. In Problem 1, the range is all of E^2, so any basis for E^2 would be a basis for the range. However, the method shown is one that works, in general, when **A** is an $m \times n$ matrix. The method is this: Let $\mathbf{V}_1, \mathbf{V}_2, \ldots, \mathbf{V}_n$ be a basis for E^n, and let

$$\mathbf{W}_1 = \mathbf{A}\mathbf{V}_1, \quad \mathbf{W}_2 = \mathbf{A}\mathbf{V}_2, \ldots, \quad \mathbf{W}_n = \mathbf{A}\mathbf{V}_n.$$

Then $\{\mathbf{W}_1, \mathbf{W}_2, \ldots, \mathbf{W}_n\}$ *spans the range* of the mapping $\mathbf{AX} = \mathbf{B}$, and it contains a basis for the range.

Problem 2. Find a basis for the range of the mapping

$$\begin{bmatrix} 2 & -1 & 0 & 1 \\ 1 & 0 & -1 & -2 \end{bmatrix}\begin{bmatrix} x_1 \\ x_2 \\ x_3 \\ x_4 \end{bmatrix} = \begin{bmatrix} b_1 \\ b_2 \end{bmatrix}.$$

Solution. With $x_1 = 1$ and $x_2 = x_3 = x_4 = 0$, we get

$$\mathbf{W}_1 = \begin{bmatrix} 2 \\ 1 \end{bmatrix}.$$

Proceeding in a similar fashion with other vectors from the standard unit basis vectors for E^4, we next put $x_1 = 0$, $x_2 = 1$, $x_3 = x_4 = 0$, and get

$$\mathbf{W}_2 = \begin{bmatrix} -1 \\ 0 \end{bmatrix}.$$

Since $\mathbf{W}_1$ and $\mathbf{W}_2$ are obviously independent, they span a two-dimensional subspace in the range, but the range itself is in E^2, and hence $\mathbf{W}_1$ and $\mathbf{W}_2$ form a basis for the range. (We don't need to find $\mathbf{W}_3$ and $\mathbf{W}_4$ and then delete them.)

Properties of linear mappings

If **A** is an $m \times n$ matrix, the equation

$$\mathbf{AX} = \mathbf{B}$$

maps $\mathbf{X} \in E^n$ onto $\mathbf{B} \in E^m$. The *domain* is all of E^n; the range is a subspace of E^m. The mapping is also called a *linear transformation* because it has the linearity properties

$$\mathbf{A}(c\mathbf{X}) = c(\mathbf{AX}), \quad \mathbf{A}(\mathbf{X} + \mathbf{Y}) = \mathbf{AX} + \mathbf{AY}. \tag{3}$$

The zero vector in E^n goes into

$$\mathbf{A0} = \mathbf{0}. \tag{4}$$

There may be other vectors $\mathbf{X} \neq \mathbf{0}$ such that

$$\mathbf{AX} = \mathbf{0}. \tag{5}$$

The set of *all vectors* **X** in E^n that map onto **0** in E^m is called the *kernel*, or the *null space*, of the mapping. The dimension of the kernel is called the *nullity* of the mapping.

Problem 3. Find the kernel and the nullity of the mapping of Problem 2.

Solution. We seek all solutions of

$$\begin{bmatrix} 2 & -1 & 0 & 1 \\ 1 & 0 & -1 & -2 \end{bmatrix} \begin{bmatrix} x_1 \\ x_2 \\ x_3 \\ x_4 \end{bmatrix} = \begin{bmatrix} 0 \\ 0 \end{bmatrix}.$$

We modify this by adding ($-2 \times$ row two) to row one:

$$\begin{bmatrix} 0 & -1 & 2 & 5 \\ 1 & 0 & -1 & -2 \end{bmatrix} \begin{bmatrix} x_1 \\ x_2 \\ x_3 \\ x_4 \end{bmatrix} = \begin{bmatrix} 0 \\ 0 \end{bmatrix}.$$

This is equivalent to

$$x_2 = 2x_3 + 5x_4,$$
$$x_1 = x_3 + 2x_4,$$

with x_3 and x_4 arbitrary real numbers. The solution vector can be written

$$\begin{aligned}(x_1, x_2, x_3, x_4) &= (x_3 + 2x_4, 2x_3 + 5x_4, x_3, x_4) \\ &= x_3(1, 2, 1, 0) + x_4(2, 5, 0, 1).\end{aligned}$$

Hence a *basis for the kernel* is $\{\mathbf{V}_1, \mathbf{V}_2\}$, with

$$\mathbf{V}_1 = (1, 2, 1, 0), \quad \mathbf{V}_2 = (2, 5, 0, 1).$$

The dimension of the subspace of E^4 spanned by $\mathbf{V}_1$ and $\mathbf{V}_2$ is 2; therefore, the *nullity* of $\mathbf{A}$ is 2.

Remark 2. In Problem 3, let $\mathbf{V}_1, \mathbf{V}_2, \mathbf{V}_3, \mathbf{V}_4$ be a basis for all of E^4, with $\mathbf{V}_1$ and $\mathbf{V}_2$ as a basis for the kernel. Every vector in the range is $\mathbf{AX}$ for some

$$\mathbf{X} = c_1\mathbf{V}_1 + c_2\mathbf{V}_2 + c_3\mathbf{V}_3 + c_4\mathbf{V}_4.$$

Hence

$$\mathbf{W}_3 = \mathbf{AV}_3 \quad \text{and} \quad \mathbf{W}_4 = \mathbf{AV}_4$$

together span the range, because

$$\begin{aligned}\mathbf{AX} &= c_1\mathbf{AV}_1 + c_2\mathbf{AV}_2 + c_3\mathbf{AV}_3 + c_4\mathbf{AV}_4 \\ &= \mathbf{0} + \mathbf{0} + c_3\mathbf{W}_3 + c_4\mathbf{W}_4.\end{aligned}$$

Problem 4. In Problem 3, the vectors

$$\mathbf{V}_1 = (1, 2, 1, 0), \quad \mathbf{V}_2 = (2, 5, 0, 1)$$

were found to be orthogonal to the vectors

$$\mathbf{A}_1 = (2, -1, 0, 1), \quad \mathbf{A}_2 = (1, 0, -1, -2).$$

This suggests that we might take $\mathbf{V}_3 = \mathbf{A}_1$ and $\mathbf{V}_4 = \mathbf{A}_2$ to get a basis $\mathbf{V}_1, \mathbf{V}_2, \mathbf{V}_3, \mathbf{V}_4$ for E^4. Verify this possibility, and find a *basis for the range* of the mapping.

Solution. We write a matrix with

$$\mathbf{V}_1, \mathbf{V}_2, \mathbf{V}_3 = \mathbf{A}_1, \quad \mathbf{V}_4 = \mathbf{A}_2$$

as its rows; and start reducing it toward row-echelon form.

$$\begin{bmatrix} 1 & 2 & 1 & 0 \\ 2 & 5 & 0 & 1 \\ 2 & -1 & 0 & 1 \\ 1 & 0 & -1 & -2 \end{bmatrix} \sim \begin{bmatrix} 1 & 2 & 1 & 0 \\ 0 & 1 & -2 & 1 \\ 0 & -5 & -2 & 1 \\ 0 & -2 & -2 & -2 \end{bmatrix}$$
$$\sim \begin{bmatrix} 1 & 2 & 1 & 0 \\ 0 & 1 & -2 & 1 \\ 0 & 0 & -12 & 6 \\ 0 & 0 & -6 & 0 \end{bmatrix} \sim \begin{bmatrix} 1 & 0 & 5 & -2 \\ 0 & 1 & -2 & 1 \\ 0 & 0 & 0 & 6 \\ 0 & 0 & -6 & 0 \end{bmatrix}.$$

We can see that the rows are linearly independent, so we stop:

$$\mathbf{V}_1, \mathbf{V}_2, \mathbf{A}_1, \mathbf{A}_2$$

form a basis for E^4, because there are four independent vectors in the set. The *range* of the mapping is spanned by

$$\mathbf{W}_3 = \mathbf{AV}_3 = \begin{bmatrix} 2 & -1 & 0 & 1 \\ 1 & 0 & -1 & -2 \end{bmatrix} \begin{bmatrix} 2 \\ -1 \\ 0 \\ 1 \end{bmatrix} = \begin{bmatrix} 6 \\ 0 \end{bmatrix},$$

$$\mathbf{W}_4 = \mathbf{AV}_4 = \begin{bmatrix} 2 & -1 & 0 & 1 \\ 1 & 0 & -1 & -2 \end{bmatrix} \begin{bmatrix} 1 \\ 0 \\ -1 \\ -2 \end{bmatrix} = \begin{bmatrix} 0 \\ 6 \end{bmatrix}.$$

Note that $(6, 0) = 6(1, 0)$ and $(0, 6) = 6(0, 1)$, so that the range also has for its basis $\{(1, 0), (0, 1)\}$, the standard basis for E^2.

Remark 3. The foregoing problems suggest that in general we might do the following: Given an $m \times n$ matrix $\mathbf{A}$ and the mapping $\mathbf{AX} = \mathbf{B}$ from E^n to E^m. Let s be the nullity of $\mathbf{A}$. (If $s = 0$, only the zero vector in E^n is in the kernel.) If $s \geq 1$, let

$$\mathbf{V}_1, \mathbf{V}_2, \ldots, \mathbf{V}_s$$

be a basis for the kernel of $\mathbf{A}$. Extend this to a basis for all E^n:

$$\mathbf{V}_1, \mathbf{V}_2, \ldots, \mathbf{V}_s, \quad \mathbf{V}_{s+1}, \ldots, \mathbf{V}_n. \tag{6a}$$

The image of this basis spans the range, because every vector in the range is $\mathbf{AX}$ for some

$$\mathbf{X} = c_1\mathbf{V}_1 + c_2\mathbf{V}_2 + \cdots c_n\mathbf{V}_n.$$

The images of $\mathbf{V}_1, \ldots, \mathbf{V}_s$ are all $\mathbf{0}$, so

$$\begin{aligned} \mathbf{W}_{s+1} &= \mathbf{AV}_{s+1}, \\ \mathbf{W}_{s+2} &= \mathbf{AV}_{s+2}, \quad \ldots, \\ \mathbf{W}_n &= \mathbf{AV}_n \end{aligned} \tag{6b}$$

span the range. (*Exception.* If $s = n$, the range is just the zero vector in E^m, because the domain and kernel are the same. We bypass this exception temporarily.) Now it happens that these $\mathbf{W}$-vectors in Eqs. (6b) are also linearly independent, as we shall next prove. Then it will follow that they are a basis for the range and that the dimension of the range is $n - s$, where

$$s = \text{dimension of the kernel},$$

$$n = \text{dimension of the domain}.$$

To prove that the $\mathbf{W}$-vectors listed in (6b) are linearly independent, suppose that

$$k_1(\mathbf{AV}_{s+1}) + k_2(\mathbf{AV}_{s+2}) + \cdots + k_r(\mathbf{AV}_{s+r}) = \mathbf{0}, \tag{6c}$$

where $s + r = n$. Then, by the distributive law for matrix multiplication, we have

$$\mathbf{A}(k_1\mathbf{V}_{s+1} + k_2\mathbf{V}_{s+2} + \cdots + k_r\mathbf{V}_{s+r}) = \mathbf{0}. \tag{7}$$

Equation (7) says that

$$\mathbf{X} = k_1\mathbf{V}_{s+1} + k_2\mathbf{V}_{s+2} + \cdots + k_r\mathbf{V}_{s+r} \tag{8}$$

is in the kernel of the mapping (by definition of *kernel*). But

$$\mathbf{V}_1, \mathbf{V}_2, \ldots, \mathbf{V}_s$$

is a basis for the kernel, so

$$\mathbf{X} = c_1\mathbf{V}_1 + c_2\mathbf{V}_2 + \cdots + c_s\mathbf{V}_s. \tag{9}$$

Now subtract (8) from (9) and get

$$\begin{aligned} c_1\mathbf{V}_1 + c_2\mathbf{V}_2 + \cdots + c_s\mathbf{V}_s - k_1\mathbf{V}_{s+1} \\ - k_2\mathbf{V}_{s+2} - \cdots - k_r\mathbf{V}_n = \mathbf{0} \end{aligned}$$

However, $\mathbf{V}_1, \ldots, \mathbf{V}_s, \ldots, \mathbf{V}_n$ are a *basis* for E^n and must therefore be linearly independent. Hence

$$c_1 = 0, \; c_2 = 0, \; \ldots,$$

$$k_1 = 0, \; k_2 = 0, \; \ldots, \; k_r = 0.$$

Since the assumption in Eq. (6c) holds only when

$$k_1 = k_2 = \cdots = k_r = 0,$$

the vectors in the list (6b) are linearly independent.

We summarize by stating these results in the following theorem.

Theorem 8. *Let $\mathbf{A}$ be an $m \times n$ matrix corresponding to a linear mapping $\mathbf{AX} = \mathbf{B}$ from E^n to E^m. Let K be the kernel of this mapping:*

$$K = \{\mathbf{X}: \mathbf{AX} = \mathbf{0}, \; \mathbf{X} \in E^n\}.$$

Let R be the range of the mapping:

$$R = \{\mathbf{B}: \mathbf{AX} = \mathbf{B}, \; \mathbf{X} \in E^n\}.$$

Then

$$\dim(R) + \dim(K) = \dim(E^n) = n. \tag{10}$$

Remark 4. We proved Eq. (10) for

$$0 < \dim(K) = s < n.$$

There are two exceptional cases.

Case 1. If $s = 0$, then $K = \{\mathbf{0}\}$ and the argument from Eq. (7) can be modified by putting $s = 0$ and $s + r = n$. in Eq. (7). The right-hand side of Eq. (8) is then $\mathbf{0}$, and therefore $k_1 = k_2 = \cdots = k_n = 0$, and the vectors

$$\mathbf{W}_1 = \mathbf{AV}_1, \quad \mathbf{W}_2 = \mathbf{AV}_2, \quad \ldots, \quad \mathbf{W}_n = \mathbf{AV}_n$$

are linearly independent. They also span the range.

Hence Eq. (10) is correct in the form

$$\dim(R) = n, \quad \dim(K) = 0$$

when $K = \{\mathbf{0}\}$.

Case 2. If $s = n$, then $K = E^n$, and all of E^n is mapped into $\mathbf{0}$ in E^m:

$$R = \{\mathbf{0}\}, \quad \dim(R) = 0.$$

Therefore Eq. (10) is valid in the form

$$\dim(R) = 0, \quad \dim(K) = \dim(E^n) = n.$$

Example 1. The equations for rotation of axes in the plane (Article 10.9, Eq. 4) can be written as

$$\begin{bmatrix} x \\ y \end{bmatrix} = \begin{bmatrix} \cos\alpha & -\sin\alpha \\ \sin\alpha & \cos\alpha \end{bmatrix} \begin{bmatrix} x' \\ y' \end{bmatrix}. \tag{11a}$$

We may interpret this as a mapping from E^2 to E^2:

$$(x', y') \to (x, y).$$

To make this notation look more like our present notation, we consider the inverse mapping

$$\begin{bmatrix} x' \\ y' \end{bmatrix} = \begin{bmatrix} \cos\alpha & \sin\alpha \\ -\sin\alpha & \cos\alpha \end{bmatrix} \begin{bmatrix} x \\ y \end{bmatrix}. \tag{11b}$$

Birkhoff and MacLane point out that there is a conceptual difference between the kind of transformation in which a vector (or point) is moved to a new location, and the change-of-coordinates kind of transformation in which the points stay put, but the axes move. They use the term "*alibi*" when "the point goes *someplace else*," and the term "alias" when "the point has a *new name*." A matrix mapping can often be interpreted as corresponding to either of these conceptions. For the moment, we choose to ignore the change-of-coordinates aspect, and consider Eq. (11b) as a linear mapping $\mathbf{X}' = \mathbf{AX}$ from E^2 to E^2.

Now it is geometrically apparent that the range is all of E^2, so it must also be true that the kernel is $\{\mathbf{0}\}$. Let us verify this.

The kernel consists of those vectors (x, y) such that

$$\begin{bmatrix} \cos\alpha & \sin\alpha \\ -\sin\alpha & \cos\alpha \end{bmatrix} \begin{bmatrix} x \\ y \end{bmatrix} = \begin{bmatrix} 0 \\ 0 \end{bmatrix}. \tag{12}$$

We premultiply both sides of Eq. (12) by

$$P = \begin{bmatrix} \cos\alpha & -\sin\alpha \\ \sin\alpha & \cos\alpha \end{bmatrix}$$

and apply the associative law to get

$$\begin{bmatrix} \cos^2\alpha + \sin^2\alpha & 0 \\ 0 & \sin^2\alpha + \cos^2\alpha \end{bmatrix} \begin{bmatrix} x \\ y \end{bmatrix} = \begin{bmatrix} 0 \\ 0 \end{bmatrix},$$

or

$$\begin{bmatrix} 1 & 0 \\ 0 & 1 \end{bmatrix} \begin{bmatrix} x \\ y \end{bmatrix} = \begin{bmatrix} x \\ y \end{bmatrix} = \begin{bmatrix} 0 \\ 0 \end{bmatrix}.$$

Hence $K = \{\mathbf{0}\}$.

The following example shows how matrices may be used to represent mappings induced by differentiation. The inverse mapping corresponds to integration.

Example 2. Consider the set of all functions that are linear combinations of

$$\mathbf{V}_1 = e^{ax}\cos bx, \quad \mathbf{V}_2 = e^{ax}\sin bx,$$

where a and b are numbers different from zero.

If
$$f(x) = c_1\mathbf{V}_1 + c_2\mathbf{V}_2, \tag{13a}$$

then
$$f'(x) = (ac_1 + bc_2)\mathbf{V}_1 + (-bc_1 + ac_2)\mathbf{V}_2. \tag{13b}$$

Let any function $k_1\mathbf{V}_1 + k_2\mathbf{V}_2$ be considered as a vector (k_1, k_2). The mapping

$$f = \begin{bmatrix} c_1 \\ c_2 \end{bmatrix} \to f' = \begin{bmatrix} a & b \\ -b & a \end{bmatrix} \begin{bmatrix} c_1 \\ c_2 \end{bmatrix}$$

carries any vector in the space spanned by

$$\{e^{ax}\cos x, \quad e^{ax}\sin bx\}$$

into a vector in the same space. The mapping has an inverse in this space, because the matrix

$$\mathbf{A} = \begin{bmatrix} a & b \\ -b & a \end{bmatrix}$$

has the inverse

$$\mathbf{A}^{-1} = \frac{1}{(a^2 + b^2)} \begin{bmatrix} a & -b \\ b & a \end{bmatrix}.$$

Thus, a function $g = (k_1, k_2)$ has an inverse image

$$\frac{1}{(a^2+b^2)}\begin{bmatrix} a & -b \\ b & a \end{bmatrix}\begin{bmatrix} k_1 \\ k_2 \end{bmatrix} = \begin{bmatrix} \dfrac{ak_1 - bk_2}{a^2+b^2} \\ \dfrac{bk_1 + ak_2}{a^2+b^2} \end{bmatrix}.$$

In more familiar terms,

$$\int (k_1 e^{ax}\cos bx + k_2 e^{ax}\sin bx)\,dx = \frac{1}{a^2+b^2}[(ak_1 - bk_2)e^{ax}\cos bx + (bk_1 + ak_2)e^{ax}\sin bx].$$

If we set $k_1 = 1$ and $k_2 = 0$, then we get

$$\int e^{ax}\cos bx\,dx = \frac{e^{ax}}{a^2+b^2}[a\cos bx + b\sin bx], \quad (14a)$$

and if we set

$$k_1 = 0 \quad \text{and} \quad k_2 = 1,$$

then we get

$$\int e^{ax}\sin bx\,dx = \frac{e^{ax}}{a^2+b^2}[-b\cos bx + a\sin bx]. \quad (14b)$$

Remark 5. If we consider a larger class of functions, for example, those that are linear combinations of

$$1, \quad e^{ax}\cos bx \quad \text{and} \quad e^{ax}\sin bx,$$

and apply the differentiation operator, we have a kernel consisting of all constant functions:

$$k_1 \cdot 1 + k_2 e^{ax}\cos bx + k_3 e^{ax}\sin bx,$$

with $k_2 = k_3 = 0$. Antidifferentiation is no longer unique (in this new space). Thus arbitrary constants can be added to the right sides of Eqs. (14a, b) when our function space is enlarged from the space spanned by

$$\{e^{ax}\cos bx,\ e^{ax}\sin bx\}$$

to that spanned by

$$\{1,\ e^{ax}\cos bx,\ e^{ax}\sin bx\}.$$

Example 3. Lest one have the idea that linear mappings from $E^n \to E^m$ are only valid when $n \geq m$, we conclude with an example of a valid linear mapping from $E^1 \to E^3$:

$$\begin{bmatrix} 2 \\ -1 \\ 1 \end{bmatrix}[x_1] = \begin{bmatrix} b_1 \\ b_2 \\ b_3 \end{bmatrix}.$$

This maps a vector $[x_1]$ in E^1 onto the vector

$$(b_1, b_2, b_3) = (2x_1, -x_1, x_1)$$

in E^3. What is the kernel? Obviously, **0**. What spans the range? Obviously, $(2, -1, 1)$. We have

$$\dim K = 0, \quad \dim R = 1, \quad \dim E^1 = 1,$$

and the basic equation

$$\dim(\text{kernel}) + \dim(\text{range}) = \dim(\text{domain})$$

is satisfied since $0 + 1 = 1$.

Remark 6. In Example 3, note that the image in E^3 of the one-dimensional space E^1 is a *line:*

$$R = \{x_1(2, -1, 1)\}.$$

This is also a one-dimensional space, even though it lies in E^3. The fundamental equation

$$\dim(\text{kernel}) + \dim(\text{range}) = \dim(\text{domain})$$

shows that

$$\dim(\text{range}) \leq \dim(\text{domain}):$$

For a *linear* mapping the *range* never has dimension greater than that of the domain.

Remark 7. The dimension of the range of a linear mapping is also called the *rank* of the mapping. Because the columns of the matrix **A** span the range (a result which you are asked to verify in one of the exercises), the rank of the mapping is also called the *column rank* of the matrix. There is also a *row rank*, which is the dimension of the space spanned by the rows of the matrix, and there is a determinant rank. A theorem that states that these three ranks are all equal is proved in more advanced books. (See Perlis, *Theory of Matrices*, Addison-Wesley, 1952, p. 56, on the equality of row rank and column rank.)

EXERCISES

1. For the mapping $\mathbf{AX} = \mathbf{B}$ considered in Problem 2, find the images of these vectors.

   (a) $\mathbf{X} = (1, 1, 0, -1)$ (b) $\mathbf{X} = (0, -1, 1, 2)$

2. Show that any linear mapping $\mathbf{AX} = \mathbf{B}$ from E^n to E^m has the following two properties (which constitute the characteristics of *linearity*).

   (a) $\mathbf{A}(\mathbf{X} + \mathbf{Y}) = \mathbf{AX} + \mathbf{AY}$,

   (b) $\mathbf{A}(t\mathbf{X}) = t\mathbf{AX}$ for any real t.

3. In this exercise we do not start with a matrix, but show how different matrices might correspond to the same linear transformation, depending on the basis in E^n and the basis in E^m. We say that a transformation T from a vector space U to a vector space W is linear if and only if

$$T(\mathbf{X} + \mathbf{Y}) = T\mathbf{X} + T\mathbf{Y} \quad \text{and} \quad T(c\mathbf{X}) = c(T\mathbf{X})$$

   for all vectors $\mathbf{X}$ and $\mathbf{Y}$ and scalars c.

   (a) Show that T is linear if and only if

$$T(c_1\mathbf{X}_1 + c_2\mathbf{X}_2) = c_1(T\mathbf{X}_1) + c_2(T\mathbf{X}_2)$$

   for all vectors $\mathbf{X}_1$ and $\mathbf{X}_2$ and all scalars c_1 and c_2. Extend the result to all linear combinations of vectors in E^n.

   (b) Let $\mathbf{V}_1, \mathbf{V}_2, \ldots, \mathbf{V}_n$ be any basis for U and let $\{\mathbf{Z}_1, \mathbf{Z}_2, \ldots, \mathbf{Z}_m\}$ be any basis for W. Then any vector in U is a linear combination of the $\mathbf{V}$'s, and its image in W, after being transformed by T, is a linear combination of the $\mathbf{Z}$'s. Thus, if

$$\mathbf{X} = c_1\mathbf{V}_1 + c_2\mathbf{V}_2 + \cdots + c_n\mathbf{V}_n,$$

   linearity of T gives

$$\begin{aligned} T(\mathbf{X}) &= T(c_1\mathbf{V}_1 + c_2\mathbf{V}_2 + \cdots + c_n\mathbf{V}_n) \\ &= c_1T(\mathbf{V}_1) + c_2T(\mathbf{V}_2) + \cdots + c_nT(\mathbf{V}_n). \end{aligned} \tag{15}$$

   Because $T(\mathbf{V}_j) \in W$, there exist scalars

$$a_{1j}, a_{2j}, \ldots, a_{mj}$$

   such that

$$T(\mathbf{V}_j) = a_{1j}\mathbf{Z}_1 + a_{2j}\mathbf{Z}_2 + \cdots + a_{mj}\mathbf{Z}_m = \sum_{i=1}^{m} a_{ij}\mathbf{Z}_i. \tag{16}$$

   Suppose we now form a matrix $\mathbf{A}$ with the scalars from Eq. (16) in its jth column:

$$\mathbf{A} = \begin{bmatrix} a_{11} & a_{12} & \cdots & a_{1j} & \cdots & a_{1n} \\ a_{21} & a_{22} & \cdots & a_{2j} & \cdots & a_{2n} \\ \vdots & \vdots & & \vdots & & \vdots \\ a_{m1} & a_{m2} & \cdots & a_{mj} & \cdots & a_{mn} \end{bmatrix}. \tag{17}$$

   This is called the *matrix* of the transformation T *relative to the basis* $\mathbf{V}_1, \ldots, \mathbf{V}_n$ for U and *the basis* $\mathbf{Z}_1, \ldots, \mathbf{Z}_m$ for W. Now, for practice: Let T be a linear transformation from E^2 to E^3 that maps $(1, 0)$ onto $(-1, 1, 1)$ and maps $(0, 1)$ onto $(1, 0, 0)$. Find the matrix for T relative to the standard bases for E^2 and E^3; that is, with

$$\mathbf{V}_1 = (1, 0), \quad \mathbf{V}_2 = (0, 1),$$

$$\mathbf{Z}_1 = (1, 0, 0), \quad \mathbf{Z}_2 = (0, 1, 0), \quad \mathbf{Z}_3 = (0, 0, 1).$$

4. Let T be the same linear transformation from E^2 to E^3 as in Exercise 3, but suppose that

$$\mathbf{V}_1 = (1, 1), \quad \mathbf{V}_2 = (-1, 1),$$

$$\mathbf{Z}_1 = (0, 1, 0), \quad \mathbf{Z}_2 = (1, 1, 0), \quad \mathbf{Z}_3 = (1, 1, 1).$$

   Find the matrix for T relative to these new bases for E^2 and E^3.

5. Use Eqs. (15), (16), and (17) to show that if T is a linear transformation from U to W, and if

$$\mathbf{X} = c_1\mathbf{V}_1 + c_2\mathbf{V}_2 + \cdots + c_n\mathbf{V}_n$$

   and

$$T(\mathbf{X}) = d_1\mathbf{Z}_1 + d_2\mathbf{Z}_2 + \cdots + d_m\mathbf{Z}_m,$$

   then

$$\begin{bmatrix} d_1 \\ d_2 \\ \vdots \\ d_m \end{bmatrix} = \begin{bmatrix} a_{11} & a_{12} & \cdots & a_{1n} \\ a_{21} & a_{22} & \cdots & a_{2n} \\ \vdots & \vdots & & \vdots \\ a_{m1} & a_{m2} & \cdots & a_{mn} \end{bmatrix} \begin{bmatrix} c_1 \\ c_2 \\ \vdots \\ c_n \end{bmatrix}. \tag{18}$$

   [The scalars $c_1, c_2, \ldots, c_n$ are called the coordinates of $\mathbf{X}$ relative to the basis $\{\mathbf{V}_1, \mathbf{V}_2, \ldots, \mathbf{V}_n\}$, and the scalars $d_1, d_2, \ldots, d_m$ are the coordinates of $T(\mathbf{X})$ relative to the basis $\{\mathbf{Z}_1, \mathbf{Z}_2, \ldots, \mathbf{Z}_m\}$. Note that Eq. (18) shows how the before-and-after coordinates are related, using the appropriate matrix $\mathbf{A}$.

In the treatment in the body of the text, we have tacitly assumed throughout that the standard bases were being used in both E^n and E^m, and we used $x_1, x_2, \ldots, x_n$ for the coordinates of $\mathbf{X}$ and $b_1, b_2, \ldots, b_m$ for the coordinates of $T(\mathbf{X})$, relative to those bases. With this understanding, you will see that Eq. (18) is the same as Eq. (2).]

6. Verify by direct differentiation that Eq. (13a) implies Eq. (13b) for the vectors $\mathbf{V}_1$ and $\mathbf{V}_2$ of Example 2.
7. If $\mathbf{A}$ is the matrix of a linear transformation $\mathbf{AX} = \mathbf{B}$, show that the columns of $\mathbf{A}$ span the range of the mapping.

MISCELLANEOUS EXERCISES

Definition. Transposed matrices. *Given an $m \times n$ matrix $\mathbf{A}$, we can form a new matrix $\mathbf{A}^T$, called the transpose of $\mathbf{A}$, by writing the rows of $\mathbf{A}$ as the columns of $\mathbf{A}^T$. Thus the element b_{ij} in the ith row and jth column of $\mathbf{B} = \mathbf{A}^T$ is $b_{ij} = a_{ji}$, for $1 \le i \le n$ and $1 \le j \le m$.*

1. (a) Write the transpose of the matrix

$$\mathbf{A} = \begin{bmatrix} 2 & 1 & 0 \\ 1 & 0 & -2 \end{bmatrix}.$$

(b) With $\mathbf{A}$ as in (a), let

$$\mathbf{X} = \begin{bmatrix} 1 \\ -2 \\ 1 \end{bmatrix}$$

and compute $\mathbf{B} = \mathbf{AX}$.

(c) With $\mathbf{A}$, $\mathbf{X}$, and $\mathbf{B}$ as in (b), show that

$$|\mathbf{B}|^2 = \mathbf{B}^T\mathbf{B} = \mathbf{X}^T(\mathbf{A}^T\mathbf{A})\mathbf{X}$$

by computing separately $|\mathbf{B}|^2$ = inner product of $\mathbf{B}$ with $\mathbf{B}$, the matrix product $\mathbf{B}^T\mathbf{B}$, the matrix product $\mathbf{A}^T\mathbf{A}$, and the matrix products $(\mathbf{A}^T\mathbf{A})\mathbf{X}$ and $\mathbf{X}^T[(\mathbf{A}^T\mathbf{A})\mathbf{X}]$.

2. If $\mathbf{A}$ is an $m \times n$ matrix and $\mathbf{B}$ is an $n \times p$ matrix, prove that the transpose of the product of $\mathbf{A}$ and $\mathbf{B}$ is the product of the transposes in the reverse order:

$$(\mathbf{AB})^T = \mathbf{B}^T\mathbf{A}^T.$$

Definition. Orthogonal matrices. *An $n \times n$ square matrix $\mathbf{A}$ is said to be orthogonal if and only if $\mathbf{A}^T\mathbf{A} = \mathbf{A}\mathbf{A}^T = \mathbf{I}$ is the $n \times n$ identity matrix.*

3. Show that this means that $\mathbf{A}$ is an orthogonal matrix if and only if the rows of $\mathbf{A}$ are orthogonal unit vectors, and the columns of $\mathbf{A}$ are also orthogonal unit vectors, in E^n.
4. Refer to the definition in Exercise 3 and show that the 2×2 matrix in Eq. (11a), Article 13.5, is orthogonal.
5. Prove that the transpose of an orthogonal matrix is also orthogonal.

Suggestion. $\qquad (\mathbf{A}^T)^T = \mathbf{A}.$

6. A *linear transformation* is said to be orthogonal if and only if its matrix relative to bases composed of orthogonal unit vectors is an orthogonal matrix. Prove that every orthogonal linear transformation T from E^n to E^n preserves lengths:

$$|T(\mathbf{X})| = |\mathbf{X}| \quad \text{for every } \mathbf{X} \in E^n.$$

7. Prove that the product of two orthogonal $n \times n$ matrices $\mathbf{A}$ and $\mathbf{B}$ is an orthogonal matrix.

Definition. Quadratic forms. *A quadratic form in two variables is an algebraic expression of the form*

$$Q(x, y) = ax^2 + 2bxy + cy^2.$$

8. Show that the above expression can be written as the matrix product

$$Q(x, y) = [x, y]\begin{bmatrix} a & b \\ b & c \end{bmatrix}\begin{bmatrix} x \\ y \end{bmatrix}, \quad \text{or} \quad Q(x, y) = \mathbf{X}^T\mathbf{A}\mathbf{X},$$

where

$$\mathbf{X} = \begin{bmatrix} x \\ y \end{bmatrix}, \quad \mathbf{A} = \begin{bmatrix} a & b \\ b & c \end{bmatrix}.$$

(The matrix $\mathbf{A}$ is called the *matrix of the quadratic form Q*. Note that it is always a *symmetric matrix:* $\mathbf{A}^T = \mathbf{A}$. We achieve this by splitting the coefficient of xy equally between a_{12} and a_{21}.)

Exercises 9 through 16 require the definition given before Exercise 8. For each quadratic form given, write down the associated matrix.

9. $2x^2 - 4xy$
10. $x^2 + 2xy - y^2$
11. $x^2 + y^2$
12. $x^2 - 4y^2$
13. $2xy$
14. $2x^2 - 3xy + 5y^2$

15. The matrix

$$\mathbf{P} = \begin{bmatrix} \cos\theta & \sin\theta \\ -\sin\theta & \cos\theta \end{bmatrix}$$

is used to effect a change of variables $\mathbf{X} = \mathbf{PX}'$. If the quadratic form Q is written in terms of the new coordinates as $(\mathbf{X}')^T\mathbf{BX}'$, with symmetric matrix $\mathbf{B}$, show that $\mathbf{B} = \mathbf{P}^T\mathbf{AP}$.

16. (Continuation.) It is nice if $\mathbf{P}$ is chosen so that $\mathbf{B}$ is a diagonal matrix

$$\mathbf{B} = \begin{bmatrix} d_1 & 0 \\ 0 & d_2 \end{bmatrix}$$

and, at the same time, $\mathbf{P}$ is an orthogonal matrix. Can this be done? How?

17. (Continuation.) For a quadratic form in three variables,

$$Q(x, y, z) = [x \quad y \quad z] \begin{bmatrix} a & b & c \\ b & d & e \\ c & e & f \end{bmatrix} \begin{bmatrix} x \\ y \\ z \end{bmatrix}$$

is also written $\mathbf{X}^T\mathbf{AX}$, where $\mathbf{A}$ is the 3×3 symmetric matrix above. In more advanced books (for example, Birkhoff and MacLane, *A Survey of Modern Algebra*, third edition, Macmillan, 1965, Theorem 21, p. 259) it is shown that a suitable change of variables $\mathbf{X} = \mathbf{PX}'$ exists, with $\mathbf{P}$ orthogonal, such that $\mathbf{P}^T\mathbf{AP} = \mathbf{D}$, where $\mathbf{D}$ is a diagonal matrix

$$\mathbf{D} = \begin{bmatrix} d_1 & 0 & 0 \\ 0 & d_2 & 0 \\ 0 & 0 & d_3 \end{bmatrix}.$$

An equivalent form of the equation $\mathbf{P}^T\mathbf{AP} = \mathbf{D}$ is $\mathbf{AP} = \mathbf{PD}$, because $\mathbf{PP}^T = \mathbf{P}^T\mathbf{P} = \mathbf{I}$ is a 3×3 identity matrix when $\mathbf{P}$ is orthogonal. Temporarily, let $\mathbf{P}_1$ be the first column of $\mathbf{P}$, $\mathbf{P}_2$ the second column, and $\mathbf{P}_3$ the third column. These columns should be mutually orthogonal unit vectors such that

$$\mathbf{AP}_1 = d_1\mathbf{P}_1, \quad \mathbf{AP}_2 = d_2\mathbf{P}_2, \quad \mathbf{AP}_3 = d_3\mathbf{P}_3.$$

Verify that if $\mathbf{P}_1$, $\mathbf{P}_2$, and $\mathbf{P}_3$ are mutually orthogonal unit vectors that satisfy these equations for some scalars d_1, d_2, and d_3, and if these vectors are used as columns of a matrix $\mathbf{P}$, then $\mathbf{P}^T\mathbf{AP} = \mathbf{D}$ is diagonal, as claimed.

18. (Continuation.) Let $\mathbf{A}$ be a square matrix, $\mathbf{X}$ a nonzero vector, and d a scalar such that $\mathbf{AX} = d\mathbf{X}$. Refer to the treatment of homogeneous linear equations in Appendix I, or elsewhere, to show that d must be a root of the equation

$$\det(\mathbf{A} - d\mathbf{I}) = 0.$$

(Cramer's rule shows that $\mathbf{X}$ would have to be the zero vector if this determinant were not zero.)

19. (Continuation.) Use the result of Exercise 18 to find a scalar d and a nonzero unit vector $\mathbf{X}$ such that $\mathbf{AX} = d\mathbf{X}$ if

$$\mathbf{A} = \begin{bmatrix} 5 & 3 \\ 3 & -3 \end{bmatrix}.$$

20. (Continuation.)

(a) Show that

$$\mathbf{P} = \frac{1}{\sqrt{10}} \begin{bmatrix} 3 & 1 \\ 1 & -3 \end{bmatrix}$$

is an orthogonal matrix, and find the matrix $\mathbf{P}^T\mathbf{AP}$ with this $\mathbf{P}$ and with $\mathbf{A}$ as in Exercise 19 above.

(b) What change of coordinates in E^2 will convert the quadratic form

$$Q(x, y) = 5x^2 + 6xy - 3y^2$$

to diagonal form

$$Q'(x', y') = d_1(x')^2 + d_2(y')^2,$$

and what are the numerical values of d_1 and d_2?

(c) What curve is represented by the equation

$$5x^2 + 6xy - 3y^2 = k,$$

(i) if k is a positive constant,
(ii) if k is a negative constant,
(iii) if $k = 0$?

Remark. The numbers d_1 and d_2 are called *characteristic roots* of the matrix $\mathbf{A}$, and the associated nonzero vectors $\mathbf{X}_1$ and $\mathbf{X}_2$, such that

$$\mathbf{AX}_i = d_i\mathbf{X}_i, \quad i = 1 \text{ or } 2,$$

are called *characteristic vectors* of $\mathbf{A}$. The same terminology is used for $n \times n$ matrices for $n \geq 3$, and the characteristic roots of $\mathbf{A}$ must satisfy $\det(\mathbf{A} - d\mathbf{I}) = 0$. (See Birkhoff and MacLane, *op. cit.*, pp. 231–235.)

21. If $\mathbf{D}$ is a diagonal matrix with entries $d_{ij} = 0$ for $i \neq j$ and $d_{ii} = k_i$ for $i = 1, 2, \ldots, n$, show that $\mathbf{D}^2$ is also a diagonal matrix and find its diagonal elements. What can you say about higher powers $\mathbf{D}^3$, $\mathbf{D}^4$, and so on?

22. (Continuation.) Suppose that $\mathbf{A}$, $\mathbf{P}$, $\mathbf{D}$ are $n \times n$ matrices, such that $\mathbf{P}$ has an inverse (we then say that $\mathbf{P}$ is *nonsingular*) and that
$$\mathbf{P}^{-1}\mathbf{A}\mathbf{P} = \mathbf{D}.$$
Prove that

(i) $$\mathbf{D}^r = (\mathbf{P}^{-1}\mathbf{A}\mathbf{P})^r = \mathbf{P}^{-1}(\mathbf{A}^r)\mathbf{P},$$

(ii) $$\mathbf{A}^r = \mathbf{P}(\mathbf{D}^r)\mathbf{P}^{-1}.$$

23. (Continuation.) Suppose that $\mathbf{A}$, $\mathbf{P}$, $\mathbf{D}$ are $n \times n$ matrices with the properties ascribed to them in Exercise 22, and that $\mathbf{D}$ is a *diagonal* matrix. Describe a relatively quick way to compute $\mathbf{A}^r$ for positive integral exponents r. Apply your method with
$$\mathbf{A} = \begin{bmatrix} 5 & 3 \\ 3 & -3 \end{bmatrix}, \quad \mathbf{P} = \frac{1}{\sqrt{10}}\begin{bmatrix} 3 & 1 \\ 1 & -3 \end{bmatrix},$$
the appropriate matrix $\mathbf{D}$, and $r = 3$.

24. In the linear mapping $\mathbf{AX} = \mathbf{B}$, suppose that $\mathbf{A}$ is the matrix corresponding to a rotation of axes in $\mathbf{E}^2$ through an angle of measure θ. What does $\mathbf{A}^2$ represent? $\mathbf{A}^{-1}$?

Theorem. Determinants. *In Birkhoff and MacLane, A Survey of Modern Algebra, Theorem 4, pp. 285–286, you will find that the determinant of the product of two $n \times n$ matrices is the product of their determinants:*
$$\det(AB) = (\det A)(\det B). \tag{19}$$
[*To understand their proof, you also need to read Section 8 of Chapter VIII (pp. 215–218) of that book. You can read both sections in about half an hour, and it is well worth the time. Of course, for a mastery of the material, you would have to spend more time and solve some of the problems.*]

25. By direct calculation of the matrix $\mathbf{AB}$ as well as its determinant and $\det \mathbf{A}$ and $\det \mathbf{B}$, verify that Eq. (19) is satisfied by all *two* $\times$ *two* matrices
$$\mathbf{A} = \begin{bmatrix} a & b \\ c & d \end{bmatrix}, \quad \mathbf{B} = \begin{bmatrix} e & f \\ g & h \end{bmatrix}.$$

Definition. Groups. *An algebraic system G of elements A, B, ... is said to form a group under the operation of multiplication if and only if the closure, associative, identity, and inverse laws are satisfied.*

(i) Closure. *$AB \in G$ for all A and B in G.*

(ii) Associative.
$$A(BC) = (AB)C \text{ for all } A, B, C, \text{ in } G.$$

(iii) Identity. *There exists an element I in G such that $AI = IA = A$ for all A in G.*

(iv) Inverses. *For each A in G, there exists A^{-1} in G such that $AA^{-1} = A^{-1}A = I$, where I is the identity element of G.*

26. Prove that the set consisting of the two matrices
$$\mathbf{A} = \begin{bmatrix} 1 & 0 \\ 0 & 1 \end{bmatrix}, \quad \mathbf{B} = \begin{bmatrix} 0 & 1 \\ 1 & 0 \end{bmatrix}$$
is a group under matrix multiplication. What is the identity element? What is $\mathbf{B}^{-1}$?

27. Using Eq. (19), prove that the set G of all $n \times n$ matrices $\mathbf{A}$ (with n a fixed positive integer), such that $|\det \mathbf{A}| = 1$, is a group. (Thus $\mathbf{G}$ is the set of *all* matrices of fixed dimensions $n \times n$ with determinant either 1 or -1.)

28. (Continuation.) In Exercise 27, take $n = 2$, and let H be the set of all matrices that represent *rotations* in the plane. Prove that H is a group that is a proper subset of G, where G is as described in Exercise 27 for $n = 2$.

29. *Rotations and reflections in E^2.* Recall that an orthogonal matrix $\mathbf{A}$ satisfies $\mathbf{A}^T\mathbf{A} = \mathbf{A}\mathbf{A}^T = \mathbf{I}$. Prove that the determinant of an orthogonal matrix is either 1 or -1. Consider those 2×2 orthogonal matrices of determinant 1. Prove that they form a *group* under multiplication, and that each matrix in this group corresponds to some *rotation* in the plane. (We say that the rotation group is a subgroup of the orthogonal group: In the *orthogonal* group the determinant may be 1 or -1, but in the *rotation* group the determinant is 1.)

30. Let
$$\mathbf{B} = \begin{bmatrix} 0 & 1 \\ 1 & 0 \end{bmatrix}.$$
Is $\mathbf{B}$ orthogonal? Does it represent a rotation? Which rotation? If it does not represent a rotation,

what does the change of coordinates $\mathbf{X} = \mathbf{BX}'$ represent geometrically? Prove that every 2×2 orthogonal matrix $\mathbf{A}$ either represents a rotation or is the product of $\mathbf{B}$ (as above) and a matrix $\mathbf{C}$ that represents a rotation:

$$\mathbf{A} = \mathbf{C} \quad \text{or} \quad \mathbf{A} = \mathbf{BC}.$$

(You may assume the results of Exercise 29 as given.)

31. *Effect of linear transformation on area.* Suppose that T is a linear transformation from E^2 to E^2 with matrix (relative to the standard basis)

$$\mathbf{A} = \begin{bmatrix} a & b \\ c & d \end{bmatrix}.$$

Consider a parallelogram with one vertex at the origin and vectors

$$\mathbf{V}_1 = \begin{bmatrix} x_1 \\ y_1 \end{bmatrix}, \quad \mathbf{V}_2 = \begin{bmatrix} x_2 \\ y_2 \end{bmatrix}$$

as two coterminous sides. The area of the region enclosed by this parallelogram is the absolute value of the determinant of the matrix having $\mathbf{V}_1$ and $\mathbf{V}_2$ as its columns (or the transpose of that matrix, since the determinants are equal).

(a) Sketch the parallelogram and show that every point in or on it is the tip of a vector, from the origin, of the form $c_1\mathbf{V}_1 + c_2\mathbf{V}_2$, with $0 \leq c_i \leq 1$, $i = 1, 2$.

(b) Because T is a linear transformation,

$$T(c_1\mathbf{V}_1 + c_2\mathbf{V}_2) = c_1T(\mathbf{V}_1) + c_2T(\mathbf{V}_2).$$

Deduce from this and from (a) that the image under T of the parallelogram and its interior is also a parallelogram and its interior. The vertices of this parallelogram are at the origin and at the tips of the vectors $T(\mathbf{V}_1)$, $T(\mathbf{V}_2)$, and $T(\mathbf{V}_1 + \mathbf{V}_2)$.

(c) Let $\mathbf{W}_1 = \mathbf{AV}_1$ and $\mathbf{W}_2 = \mathbf{AV}_2$. Show that the matrix having $\mathbf{W}_1$ and $\mathbf{W}_2$ as its columns is $\mathbf{A}$ times the matrix that has $\mathbf{V}_1$ and $\mathbf{V}_2$ as its columns. Using this fact and Eq. (19) (which follows Exercise 24 above), show that the area of the transformed parallelogram is

$$|\det \mathbf{A}| \times \text{area of original parallelogram.}$$

Remark. Analogous results apply in E^n for $n \geq 3$. A linear mapping from E^3 to E^3 transforms a parallelepiped into a parallelepiped with volume $|\det \mathbf{A}|$ times the original volume. The reasoning is almost identical with that sketched above for E^2: One expresses the before-and-after volumes as third-order determinants and applies Eq. (19) to see how they are related.

32. Suppose that $\mathbf{V}_1$ and $\mathbf{V}_2$ are linearly independent vectors in E^2 (or in E^3). Consider the parallelogram having $\mathbf{V}_1$ and $\mathbf{V}_2$ as one pair of coterminous edges. Let

$$\mathbf{V}_2' = c\mathbf{V}_1, \quad \text{with } c = \frac{\mathbf{V}_1 \cdot \mathbf{V}_2}{\mathbf{V}_1 \cdot \mathbf{V}_1},$$

and let

$$\mathbf{V}_2'' = \mathbf{V}_2 - \mathbf{V}_2'.$$

(Thus $\mathbf{V}_2''$ is the component of $\mathbf{V}_2$ that is orthogonal to $\mathbf{V}_1$.)

(a) Show that

$$|\mathbf{V}_2 - c\mathbf{V}_1|^2 = |\mathbf{V}_2|^2 - c^2|\mathbf{V}_1|^2 = \mathbf{V}_2 \cdot \mathbf{V}_2 - \frac{(\mathbf{V}_1 \cdot \mathbf{V}_2)^2}{(\mathbf{V}_1 \cdot \mathbf{V}_1)}.$$

(b) Use the fact that the area enclosed by the parallelogram is the product of the length of $\mathbf{V}_2''$ and the length of $\mathbf{V}_1$, and the result of (a) above, to show that the square of that area is equal to the determinant of the matrix

$$\begin{bmatrix} \mathbf{V}_1 \cdot \mathbf{V}_1 & \mathbf{V}_1 \cdot \mathbf{V}_2 \\ \mathbf{V}_2 \cdot \mathbf{V}_1 & \mathbf{V}_2 \cdot \mathbf{V}_2 \end{bmatrix}.$$

(c) Show that the matrix exhibited in (b) is the product of $\mathbf{A}$ and $\mathbf{A}^T$, if $\mathbf{A}$ is the matrix having $\mathbf{V}_1$ as its first row and $\mathbf{V}_2$ as its second row.

(d) In E^2, let $\mathbf{V}_1 = (a_1, b_1)$ and $\mathbf{V}_2 = (a_2, b_2)$. Use the results of (b) and (c) and Eq. (19) to deduce that the area enclosed by the parallelogram is the absolute value of the determinant

$$\begin{vmatrix} a_1 & b_1 \\ a_2 & b_2 \end{vmatrix}.$$

VECTOR FUNCTIONS AND THEIR DERIVATIVES

CHAPTER 14

14.1 INTRODUCTION

In this chapter, we shall focus our attention on functions from E^1 to E^n. We shall be primarily interested in the motion of a particle in two- or three-dimensional space. Thus, for most applications, we shall take $n = 2$ or 3. We shall also use **i**, **j**, **k** for the unit vectors along the three coordinate axes x, y, z, respectively. When the motion is in a plane, we shall ordinarily assume that coordinates are chosen to make the plane of motion the xy-plane. This allows us to consider motion in the xy-plane as a special case of motion in 3-space, the special condition being that the z-coordinate of the particle is zero.

Problem 1. The position of a particle in the xy-plane at time t is given by

$$x = e^t, \quad y = te^t.$$

Let

$$\mathbf{R}(t) = \mathbf{i}x + \mathbf{j}y = \mathbf{i}e^t + \mathbf{j}te^t.$$

Where is the particle at time $t = 0$, and how fast is it moving and in what direction?

Solution. At $t = 0$, we have $x = 1$ and $y = 0$, so

$$\mathbf{R}(0) = \mathbf{i}.$$

This is the vector from the origin to the position of the particle at time $t = 0$. Next, how can we find the speed and direction of motion? If you have studied particle mechanics from a vector viewpoint, then you probably know the answer. If not, think where the particle might be a short time after $t = 0$: Both x and y will have increased somewhat, and there will be a vector to the new position. The vector between the two positions is

$$\Delta\mathbf{R} = \mathbf{i}\,\Delta x + \mathbf{j}\,\Delta y.$$

For a small Δt, this vector gives, approximately, the direction of motion. Also, $|\Delta\mathbf{R}/\Delta t|$ gives the speed, approximately. Now for small values of Δt, we know that

$$\Delta x \approx \frac{dx}{dt}\Delta t \quad \text{and} \quad \Delta y \approx \frac{dy}{dt}\Delta t.$$

Therefore

$$\Delta\mathbf{R} \approx \mathbf{i}\frac{dx}{dt}\Delta t + \mathbf{j}\frac{dy}{dt}\Delta t = \left(\mathbf{i}\frac{dx}{dt} + \mathbf{j}\frac{dy}{dt}\right)\Delta t,$$

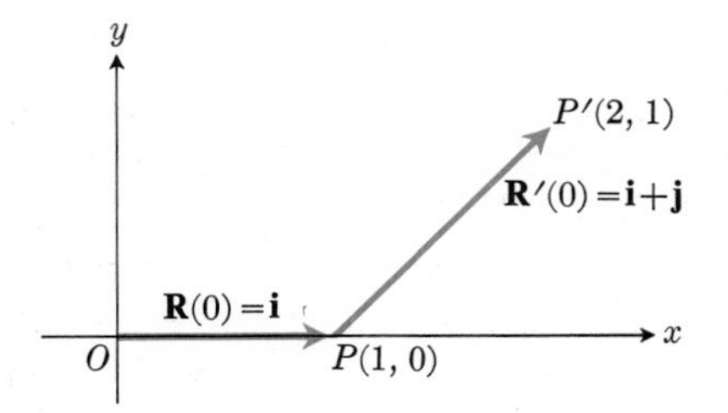

14.1 At $t = 0$, the position vector is $\mathbf{R} = \mathbf{i}$, and the velocity vector is $\mathbf{R}' = \mathbf{i} + \mathbf{j}$.

which leads to the conclusion that

$$\frac{\Delta \mathbf{R}}{\Delta t} \approx \mathbf{i}\frac{dx}{dt} + \mathbf{j}\frac{dy}{dt}.$$

From the above line of reasoning (which we shall make more precise in the next article), it seems reasonable to use

$$\mathbf{i}\frac{dx}{dt} + \mathbf{j}\frac{dy}{dt} = \mathbf{i}(e^t) + \mathbf{j}(te^t + e^t),$$

evaluated at $t = 0$, as the vector whose magnitude gives the speed, and whose direction gives the direction of motion at the instant $t = 0$. The vector is $\mathbf{i} + \mathbf{j}$, so

$$\text{speed} = |\mathbf{i} + \mathbf{j}| = \sqrt{2},$$

$$\text{direction} = \frac{\mathbf{i} + \mathbf{j}}{\sqrt{2}}.$$

Naturally, the vector $\mathbf{i} + \mathbf{j}$ is called the *velocity vector*. Its components are just the derivatives of the components of the *position vector* $\mathbf{R} = \overrightarrow{OP}$. (See Fig. 14.1.)

Remark 1. It is clear that we shall want to define the derivative of a vector function $\mathbf{R}(t)$. This in turn means that we need to define a limit such as

$$\lim_{h \to a} \mathbf{F}(h),$$

where $\mathbf{F}(h)$ is a *vector function* of h. For this definition, we assume that $\mathbf{F}$ has n components:

$$\mathbf{F} = (f_1, f_2, \ldots, f_n), \tag{1}$$

where each component is a function whose domain includes some deleted neighborhood of $h = a$. We then restrict h to the intersection of these deleted neighborhoods and say that $\mathbf{F}$ has a limit, as $h \to a$, if and only if each component of $\mathbf{F}$ has a limit. If, as $h \to a$,

$$\lim f_1(h) = L_1, \ \lim f_2(h) = L_2, \ \ldots, \ \lim f_n(h) = L_n, \tag{2a}$$

then we *define the limit* of $\mathbf{F}(h)$ to be $(L_1, L_2, \ldots, L_n)$:

$$\lim_{h \to a} \mathbf{F}(h) = (L_1, L_2, \ldots, L_n). \tag{2b}$$

Example. Let

$$\mathbf{F}(h) = \left(e^h, \frac{\sin h}{h}\right).$$

Then, as $h \to 0$,

$$\lim e^h = 1, \quad \lim \frac{\sin h}{h} = 1,$$

so that we find

$$\lim_{h \to 0} \mathbf{F}(h) = (1, 1).$$

Continuity

The vector function $\mathbf{F}$ is *continuous* at a if and only if each of its components is continuous there. This is equivalent to the ϵ, δ-requirement stated in this definition.

Definition. *The vector function* $\mathbf{F} = (f_1, f_2, \ldots, f_n)$ *is said to be continuous at* $h = a$ *if and only if, to each* $\epsilon > 0$, *there corresponds a* $\delta > 0$ *such that*

$$|\mathbf{F}(h) - \mathbf{F}(a)| < \epsilon \quad \text{when} \quad |h - a| < \delta. \tag{3}$$

The following inequalities enable us to show that this is equivalent to the continuity, at $h = a$, of every component of $\mathbf{F}$:

$$\begin{aligned} |f_i(h) - f_i(a)| &= \sqrt{[f_i(h) - f_i(a)]^2} \\ &\leq \sqrt{\sum_{j=1}^{n} [f_j(h) - f_j(a)]^2}, \\ |f_i(h) - f_i(a)| &\leq |\mathbf{F}(h) - \mathbf{F}(a)| \\ &\leq \sum_{j=1}^{n} |f_j(h) - f_j(a)|. \end{aligned} \tag{4}$$

The first of these inequalities holds because the summands under the radical are all nonnegative, so their sum is greater than or equal to any one of them. The third inequality is of the form

$$\sqrt{a^2 + b^2 + \cdots + k^2} \leq |a| + |b| + \cdots + |k|. \quad (5)$$

This can be proved directly [by squaring both sides of (5) then taking square roots] or the last inequality in (4) can be deduced from the triangle inequality

$$|\mathbf{V}_1 + \mathbf{V}_2 + \cdots + \mathbf{V}_n| \leq |\mathbf{V}_1| + |\mathbf{V}_2| + \cdots + |\mathbf{V}_n| \quad (6)$$

applied with vectors $\mathbf{V}_j$ whose components are all zero except for the jth component, which is

$$f_j(h) - f_j(a), \quad j = 1, 2, \ldots, n.$$

Now let us suppose that each component of $\mathbf{F}$ is a continuous function at $h = a$. Then, given $\epsilon > 0$ and j any integer from 1 through n, there corresponds δ_j such that

$$|f_j(h) - f_j(a)| < \epsilon/n \quad \text{when} \quad |h - a| < \delta_j, \quad j = 1, 2, \ldots, n.$$

If $\delta = \min\{\delta_1, \delta_2, \ldots, \delta_n\}$, then each term in the second summation in (4) is less than ϵ/n, so their sum is less than ϵ, and condition (3) is satisfied. Conversely, suppose that we had started with condition (3): Then the first inequality in (4) would give

$$|f_i(h) - f_i(a)| < \epsilon \quad \text{when} \quad |h - a| < \delta;$$

hence f_i is continuous at $h = a$ for any i from 1 through n.

Working rule. A vector function is continuous at a point of its domain at which *all* of its components are continuous.

Derivative of a vector function

We shall define the derivative of $\mathbf{F}$ by the same type of limit equation we use for scalar functions:

$$\mathbf{F}'(c) = \lim_{h \to 0} \frac{\mathbf{F}(c + h) - \mathbf{F}(c)}{h}, \quad (7)$$

provided that this limit exists. The following theorem provides a statement of the necessary and sufficient condition.

Theorem 1. *Let*

$$\mathbf{F} = (f_1, f_2, \ldots, f_n)$$

be a vector function whose component functions are defined in some neigbborhood of c. Then $\mathbf{F}$ *is differentiable at c if and only if each of its components is differentiable there. If this condition is met, then*

$$\mathbf{F}'(c) = \big(f_1'(c), f_2'(c), \ldots, f_n'(c)\big). \quad (8)$$

Proof. The difference quotient we need to consider is

$$\frac{\mathbf{F}(c + h) - \mathbf{F}(c)}{h} = \left(\frac{f_1(c + h) - f_1(c)}{h}, \ldots, \frac{f_n(c + h) - f_n(c)}{h}\right) \quad (9)$$

and the left-hand side of this equation has a limit if and only if each component on the right-hand side has a limit, as $h \to 0$. The ith component on the right-hand side has a limit if and only if f_i is differentiable at c. If each component is differentiable at c, then as $h \to 0$ and we pass to the limit in Eq. (9), we get Eq. (8). □

Problem 2. Where is the vector function

$$\mathbf{F} = [\sin t, \ln t, \tan^{-1}(3t)]$$

differentiable, and what is its derivative?

Solution. The first component, $\sin t$, is everywhere differentiable, and its derivative is $\cos t$. The second component, $\ln t$, is differentiable for $t > 0$, and its derivative is $1/t$. The third component, $\tan^{-1}(3t)$, is everywhere differentiable, and its derivative is

$$\frac{3}{(1 + 9t^2)}.$$

Hence the vector function is differentiable for $t > 0$, and its derivative is

$$\mathbf{F}'(t) = \left(\cos t, \frac{1}{t}, \frac{3}{1 + 9t^2}\right), \quad t > 0.$$

EXERCISES

In Exercises 1 through 5, find the derivative of the given vector function and state the domain of the derived function.

1. (e^{2t}, te^{-t})
2. $(\ln\sqrt{1+t}, \sqrt{1-t^2})$
3. $(\sin^{-1} 2t, \tan 3t, 1/t)$
4. $(\sec^{-1} 3x, \cosh 2x, \tanh 4x)$
5. $\left(\dfrac{2t-1}{2t+1}, \ln(1-4t^2)\right)$

14.2 VELOCITY AND ACCELERATION

We shall be interested in applications of vectors to problems in physics. For applications to *statics*, we need only a knowledge of the *algebra* of vectors. But for applications to *dynamics*, we also require a knowledge of the *calculus* of vectors. In this article we shall study motion of a particle in a plane, for simplicity, but most of the ideas can easily be extended to motion in space.

Position vector

Suppose the point P moves along a curve in the xy-plane, and suppose we know its position at any time t. This means that the motion of P is described by a pair of functions f and g:

$$x = f(t), \quad y = g(t). \tag{1}$$

The vector from the origin to P is customarily called the *position vector* of P, although it might be appropriate to call it the "radar" vector. This vector is a function of t given by

$$\mathbf{R} = \mathbf{i}x + \mathbf{j}y, \tag{2a}$$

or

$$\mathbf{R} = \mathbf{i}f(t) + \mathbf{j}g(t). \tag{2b}$$

Velocity vector

We now raise the question as to what physical meaning we might attach to the *derivative* of $\mathbf{R}$ with respect to t. Mathematically, we have already *defined* the derivative as

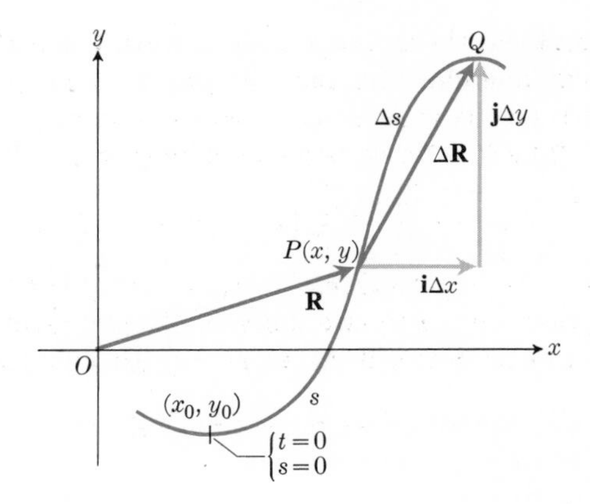

14.2

$$\frac{d\mathbf{R}}{dt} = \lim_{\Delta t\to 0} \frac{\Delta\mathbf{R}}{\Delta t}, \tag{3}$$

where $\mathbf{R}$ is given by (2a) and

$$\mathbf{R} + \Delta\mathbf{R} = \mathbf{i}(x + \Delta x) + \mathbf{j}(y + \Delta y). \tag{4}$$

Here $P(x, y)$ represents the position of the particle at time t, while $Q(x + \Delta x, y + \Delta y)$ gives its position at time $t + \Delta t$. By subtracting (2a) from (4), we obtain

$$\Delta\mathbf{R} = \mathbf{i}\,\Delta x + \mathbf{j}\,\Delta y, \tag{5}$$

which is the vector $\overrightarrow{PQ}$ in Fig. 14.2. The remaining calculations needed to give $d\mathbf{R}/dt$ proceed as follows:

$$\frac{\Delta\mathbf{R}}{\Delta t} = \mathbf{i}\frac{\Delta x}{\Delta t} + \mathbf{j}\frac{\Delta y}{\Delta t},$$

$$\lim_{\Delta t\to 0}\frac{\Delta\mathbf{R}}{\Delta t} = \lim_{\Delta t\to 0}\left(\mathbf{i}\frac{\Delta x}{\Delta t} + \mathbf{j}\frac{\Delta y}{\Delta t}\right)$$

$$= \mathbf{i}\lim_{\Delta t\to 0}\frac{\Delta x}{\Delta t} + \mathbf{j}\lim_{\Delta t\to 0}\frac{\Delta y}{\Delta t},$$

$$\frac{d\mathbf{R}}{dt} = \mathbf{i}\frac{dx}{dt} + \mathbf{j}\frac{dy}{dt}. \tag{6}$$

The result (6) is equivalent to what would be obtained by differentiating both sides of (2a) with respect to t, holding $\mathbf{i}$ and $\mathbf{j}$ constant. The geometric significance of (6) may be learned by calculating the direction and the magnitude of this vector:

$$\text{slope of } \frac{d\mathbf{R}}{dt} = \frac{\text{rise}}{\text{run}} = \frac{dy/dt}{dx/dt} = \frac{dy}{dx},$$

$$\text{magnitude of } \frac{d\mathbf{R}}{dt} = \left|\frac{d\mathbf{R}}{dt}\right| = \left|\mathbf{i}\frac{dx}{dt} + \mathbf{j}\frac{dy}{dt}\right| \tag{7}$$

$$= \sqrt{\left(\frac{dx}{dt}\right)^2 + \left(\frac{dy^2}{dt}\right)} = \left|\frac{ds}{dt}\right|.$$

Here s represents arc length along the curve measured from some starting point (x_0, y_0).

If we draw a vector equal to $d\mathbf{R}/dt$, placing its initial point at P, the resulting vector will

(a) be tangent to the curve at P, since its slope equals dy/dx, which is the same as the slope of the curve at P, and

(b) have magnitude $= |ds/dt|$, which gives the instantaneous speed of the particle at P.

Thus, physically, the vector $d\mathbf{R}/dt$, when drawn from P, is a suitable representation of the *velocity vector*, which has the same two properties (a) and (b).

We may now summarize by saying that if we differentiate the position vector

$$\mathbf{R} = \mathbf{i}x + \mathbf{j}y$$

with respect to time, the result gives the *velocity vector*

$$\mathbf{v} = \frac{d\mathbf{R}}{dt} = \mathbf{i}\frac{dx}{dt} + \mathbf{j}\frac{dy}{dt}.$$

It is customary to think of the velocity vector as being drawn at the point P.

Acceleration

The acceleration vector $\mathbf{a}$ is obtained from $\mathbf{v}$ by a further differentiation:

$$\mathbf{a} = \frac{d\mathbf{v}}{dt} = \mathbf{i}\frac{d^2x}{dt^2} + \mathbf{j}\frac{d^2y}{dt^2}. \tag{8}$$

For a particle of constant mass m moving under the action of an applied force $\mathbf{F}$, Newton's second law of motion states that

$$\mathbf{F} = m\mathbf{a}. \tag{9}$$

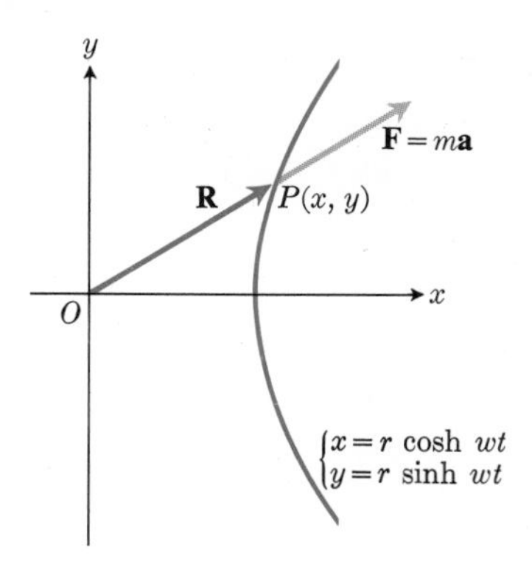

14.3

Since one ordinarily visualizes the force vector as being *applied at P*, it is customary to adopt the same viewpoint about the acceleration vector $\mathbf{a}$.

Example. A particle $P(x, y)$ moves on the hyperbola

$$x = r\cosh \omega t, \quad y = r \sinh \omega t, \tag{10}$$

where r and ω are positive constants. (See Fig. 14.3.) Then

$$\mathbf{R} = \mathbf{i}(r\cosh\omega t) + \mathbf{j}(r\sinh\omega t),$$

$$\mathbf{v} = \frac{d\mathbf{R}}{dt} = \mathbf{i}(\omega r\sinh\omega t) + \mathbf{j}(\omega r\cosh\omega t),$$

$$\mathbf{a} = \mathbf{i}(\omega^2 r\cosh\omega t) + \mathbf{j}(\omega^2 r\sinh\omega t)$$
$$= \omega^2\mathbf{R}.$$

This means that the force $\mathbf{F} = m\mathbf{a} = m\omega^2\mathbf{R}$ has a magnitude

$$m\omega^2|\mathbf{R}| = m\omega^2|\overrightarrow{OP}|,$$

which is directly proportional to the distance OP, and that its direction is the same as the direction of $\mathbf{R}$. Thus the force is directed away from O.

The next example illustrates how we obtain the path of motion by integrating Eq. (9) when the force $\mathbf{F}$ is a given function of time, and the initial position and initial velocity of the particle are given. In general the force $\mathbf{F}$ may depend upon the position of P as well as on the time, and the problem of integrating the differential equations so obtained is usually discussed in textbooks on that subject. (For example, see Martin and Reissner, *Elementary Differential Equations*, Addison-Wesley, 1961.)

Problem. (See Fig. 14.4.) The force acting on a particle P of mass m is given as a function of t by

$$\mathbf{F} = \mathbf{i}\cos t + \mathbf{j}\sin t.$$

If the particle starts at the point $(c, 0)$ with initial velocity $v_0\mathbf{j}$ perpendicular to the x-axis, find the curve it describes.

Solution. If we denote the position vector by

$$\mathbf{R} = \mathbf{i}x + \mathbf{j}y,$$

we may restate the problem as follows. Find $\mathbf{R}$ if

$$\mathbf{F} = m\frac{d^2\mathbf{R}}{dt^2} = \mathbf{i}\cos t + \mathbf{j}\sin t \tag{11}$$

and if, at $t = 0$,

$$\mathbf{R} = \mathbf{i}c, \quad \frac{d\mathbf{R}}{dt} = \mathbf{j}v_0. \tag{12}$$

In (11) we let $\mathbf{v} = d\mathbf{R}/dt$, and separate the variables to obtain

$$m\,d\mathbf{v} = (\mathbf{i}\cos t + \mathbf{j}\sin t)\,dt.$$

Integrating this, we have

$$m\mathbf{v} = m\frac{d\mathbf{R}}{dt} = \mathbf{i}\sin t - \mathbf{j}\cos t + \mathbf{C}_1, \tag{13}$$

where the constant of integration is a *vector* denoted by $\mathbf{C}_1$. The value of $\mathbf{C}_1$ may be found by using the initial velocity, the right-hand equation in (12), in Eq. (13), with $t = 0$:

$$m\mathbf{j}v_0 = -\mathbf{j} + \mathbf{C}_1,$$

$$\mathbf{C}_1 = (mv_0 + 1)\mathbf{j}.$$

Substituting this into (13), we have

$$m\frac{d\mathbf{R}}{dt} = \mathbf{i}\sin t + (mv_0 + 1 - \cos t)\mathbf{j}.$$

Another integration gives

$$m\mathbf{R} = -\mathbf{i}\cos t + \mathbf{j}(mv_0t + t - \sin t) + \mathbf{C}_2.$$

The initial condition $\mathbf{R} = \mathbf{i}c$ (Eq. 12) enables us to evaluate $\mathbf{C}_2$:

$$mc\mathbf{i} = -\mathbf{i} + \mathbf{C}_2, \quad \mathbf{C}_2 = \mathbf{i}(mc + 1),$$

so that the position vector $\mathbf{R}$ is given by

$$\mathbf{R} = \frac{1}{m}[\mathbf{i}(mc + 1 - \cos t) + \mathbf{j}(mv_0t + t - \sin t)].$$

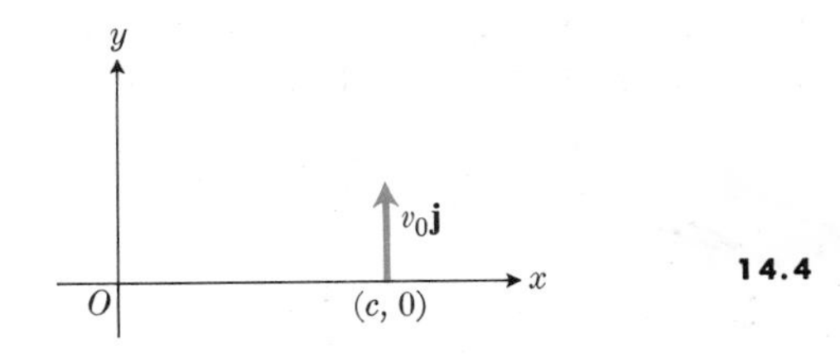

14.4

The parametric equations of the curve are found by equating components of this expression for $\mathbf{R}$ with

$$\mathbf{R} = \mathbf{i}x + \mathbf{j}y,$$

which gives

$$x = c + \frac{1 - \cos t}{m}, \quad y = v_0t + \frac{t - \sin t}{m}.$$

The foregoing equations for velocity and acceleration in two dimensions would be appropriate for describing the motion of a particle moving on a flat surface, such as a water bug skimming over the surface of a pond, or a hockey puck sliding on ice. But to describe the flight of a bumblebee or a rocket, we need three coordinates. Thus, if

$$\mathbf{R}(t) = \mathbf{i}x + \mathbf{j}y + \mathbf{k}z, \tag{14a}$$

where x, y, z are functions of t that are twice-differentiable, then the velocity of $P(x, y, z)$ is

$$\mathbf{v} = \frac{d\mathbf{R}}{dt} = \mathbf{i}\frac{dx}{dt} + \mathbf{j}\frac{dy}{dt} + \mathbf{k}\frac{dz}{dt}, \tag{14b}$$

and the acceleration is

$$\mathbf{a} = \mathbf{i}\frac{d^2x}{dt^2} + \mathbf{j}\frac{d^2y}{dt^2} + \mathbf{k}\frac{d^2z}{dt^2}. \tag{14c}$$

EXERCISES

In Exercises 1 through 8, $\mathbf{R} = \mathbf{i}x + \mathbf{j}y$ is the vector from the origin to the moving point $P(x, y)$ at time t. Find the velocity and acceleration vectors for any t. Also find these vectors and the speed at the particular instant given.

1. $\mathbf{R} = (a\cos\omega t)\mathbf{i} + (a\sin\omega t)\mathbf{j}$, a and ω being positive constants, $t = \pi/(3\omega)$
2. $\mathbf{R} = (2\cos t)\mathbf{i} + (3\sin t)\mathbf{j}$, $t = \pi/4$

3. $\mathbf{R} = (t+1)\mathbf{i} + (t^2 - 1)\mathbf{j}, \quad t = 2$
4. $\mathbf{R} = (\cos 2t)\mathbf{i} + (2 \sin t)\mathbf{j}, \quad t = 0$
5. $\mathbf{R} = e^t\mathbf{i} + e^{-2t}\mathbf{j}, \quad t = \ln 3$
6. $\mathbf{R} = (\sec t)\mathbf{i} + (\tan t)\mathbf{j}, \quad t = \pi/6$
7. $\mathbf{R} = (\cosh 3t)\mathbf{i} + (2 \sinh t)\mathbf{j}, \quad t = 0$
8. $\mathbf{R} = [\ln (t+1)]\mathbf{i} + t^2\mathbf{j}, \quad t = 1$

9. If the force that acts on a particle P of mass m is $F = mg\mathbf{j}$, where m and g are constants, and the particle starts from the origin with velocity

$$\mathbf{v}_0 = (v_0 \cos \alpha)\mathbf{i} + (v_0 \sin \alpha)\mathbf{j}$$

at time $t = 0$, find the vector $\mathbf{R} = \mathbf{i}x + \mathbf{j}y$ from the origin to P at time t.

10. Exercise 9 describes the motion of a projectile *in vacuo*. If the projectile encounters a resistance proportional to the velocity, the force is

$$\mathbf{F} = -mg\mathbf{j} - k\frac{d\mathbf{R}}{dt}.$$

Show that one integration of $\mathbf{F} = md^2\mathbf{R}/dt^2$ leads to the differential equation

$$\frac{d\mathbf{R}}{dt} + \frac{k}{m}\mathbf{R} = \mathbf{v}_0 - gt\mathbf{j}.$$

(This differential equation can in turn be integrated by means of the following device: Multiply both sides of the equation by $e^{(k/m)t}$. Then the left-hand side is the derivative of the product $\mathbf{R}e^{(k/m)t}$ and both sides can be integrated.)

Find the velocity $\mathbf{v}$ and acceleration $\mathbf{a}$ for the motion in Exercises 11 through 13. Also, find the angle θ between $\mathbf{v}$ and $\mathbf{a}$ at time $t = 0$.

11. $x = e^t \quad y = e^t \sin t, \quad z = e^t \cos t$
12. $x = \tan t, \quad y = \sinh 2t, \quad z = \operatorname{sech} 3t$
13. $x = \ln (t^2 + 1), \quad y = \tan^{-1} t, \quad z = \sqrt{t^2 + 1}$

14. The plane $z = 2x + 3y$ intersects the cylinder $x^2 + y^2 = 9$ in an ellipse.

(a) Express the position of a point $P(x, y, z)$ on this ellipse as a vector function $\mathbf{R} = \overrightarrow{OP} = \mathbf{R}(\theta)$, where θ is a measure of the dihedral angle between the xz-plane and the plane containing the z-axis and P.

(b) Using the equations of (a), find the velocity and acceleration of P, assuming that $d\theta/dt = \omega$ is constant.

14.3 TANGENTIAL VECTORS

As the point P moves along a given curve in the xy-plane, we may imagine its position as being specified by the length of arc s from some arbitrarily chosen reference point P_0 on the curve. The vector

$$\mathbf{R} = \mathbf{i}x + \mathbf{j}y$$

from O to $P(x, y)$ is therefore a function of s, and we shall now investigate the properties of $d\mathbf{R}/ds$. To this end, let P have coordinates (x, y) corresponding to the value s, while $Q(x + \Delta x, y + \Delta y)$ corresponds to $s + \Delta s$. Then

$$\frac{\Delta \mathbf{R}}{\Delta s} = \mathbf{i}\frac{\Delta x}{\Delta s} + \mathbf{j}\frac{\Delta y}{\Delta s} = \frac{\overrightarrow{PQ}}{\Delta s} \tag{1}$$

is a *vector* whose magnitude is chord PQ divided by arc PQ, and this approaches unity as $\Delta s \to 0$. Hence

$$\frac{d\mathbf{R}}{ds} = \lim_{\Delta s \to 0} \frac{\Delta \mathbf{R}}{\Delta s} \tag{2}$$

is a *unit* vector. The *direction* of this unit vector is the limiting direction approached by the direction of $\Delta\mathbf{R}/\Delta s$ as $\Delta s \to 0$. Now

$$\frac{\Delta \mathbf{R}}{\Delta s} = \frac{\overrightarrow{PQ}}{\Delta s}$$

has the same direction as $\overrightarrow{PQ}$ when (a) Δs is positive, or else it has the same direction as $\overrightarrow{QP}$ when (b) Δs is negative. Figure 14.5(a, b) illustrates these two cases and shows that in *either* case, $\Delta\mathbf{R}/\Delta s$ is directed along the chord through P and Q and that it points in the direction of increasing s [that is, upward to the right, although ΔR points downward and to the left in Fig. 14.5(b).] As $\Delta s \to 0$ and $Q \to P$, the direction of the chord through P and Q approaches the direction of the *tangent* to the curve at P. Thus the limiting direction of $\Delta\mathbf{R}/\Delta s$, in other words the *direction of* $d\mathbf{R}/ds$, is along the *tangent to the curve at* P, and its sense is that which points in the direction of increasing arc length s (that is, away from P_0 when s is positive, or toward P_0 when s is negative). Hence

$$\frac{d\mathbf{R}}{ds} = \mathbf{T} \tag{3}$$

is a *unit vector* which is *tangent to the curve at* P. (See Fig. 14.6.)

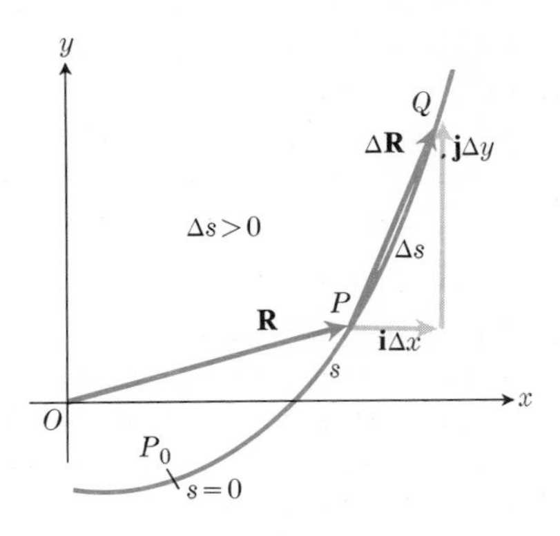

a|b

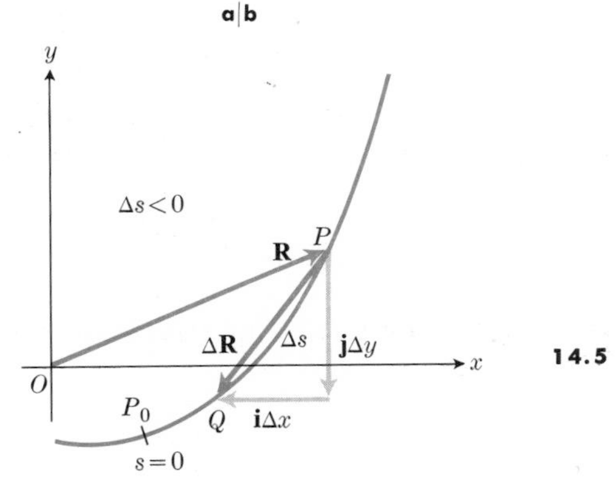

14.5

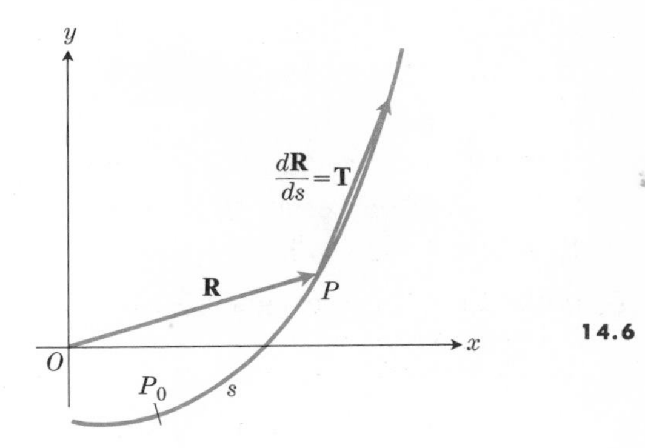

14.6

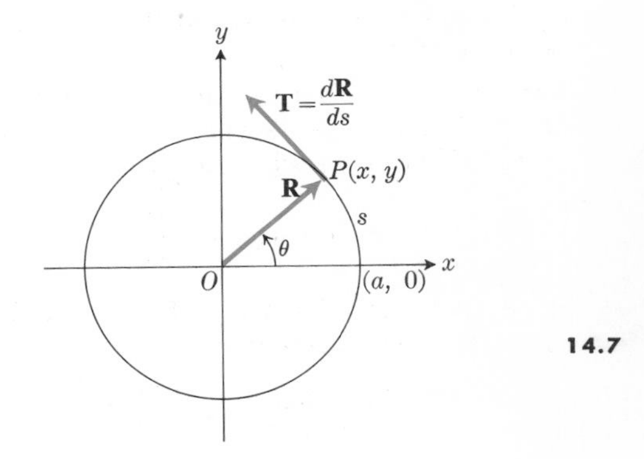

14.7

If we let $\Delta s \to 0$ in Eq. (1), we find that

$$\frac{d\mathbf{R}}{ds} = \mathbf{i}\,\frac{dx}{ds} + \mathbf{j}\,\frac{dy}{ds}, \tag{4}$$

and this may be used to find $\mathbf{T}$ at any point of a curve whose equation is given.

Problem 1. (See Fig. 14.7.) Find the unit vector $\mathbf{T}$ tangent to the circle

$$x = a\cos\theta, \quad y = a\sin\theta,$$

at any point $P(x, y)$.

Solution. From the equations of the curve, we have

$$dx = -a\sin\theta\,d\theta, \quad dy = a\cos\theta\,d\theta,$$

and

$$\begin{aligned} ds^2 = dx^2 + dy^2 &= a^2(\sin^2\theta + \cos^2\theta)\,d\theta^2 \\ &= a^2\,d\theta^2, \end{aligned}$$

so that

$$ds = \pm a\,d\theta.$$

If we measure arc length in the counterclockwise direction, with $s = 0$ at $(a, 0)$, s will be an increasing function of θ, so the $+$-sign should be taken: $ds = a\,d\theta$. Then

$$\begin{aligned} \mathbf{T} = \frac{d\mathbf{R}}{ds} &= \mathbf{i}\,\frac{dx}{ds} + \mathbf{j}\,\frac{dy}{ds} \\ &= \mathbf{i}\left(\frac{-a\sin\theta\,d\theta}{a\,d\theta}\right) + \mathbf{j}\left(\frac{a\cos\theta\,d\theta}{a\,d\theta}\right) \\ &= -\mathbf{i}\sin\theta + \mathbf{j}\cos\theta. \end{aligned}$$

Space curves and arc length

All that has been done above for two-dimensional motion in a plane can also be extended to three-dimensional motion in space. To this end, let $P(x, y, z)$ be a point whose position in space is given by the equations

$$x = f(t), \quad y = g(t), \quad z = h(t), \tag{5}$$

where f, g, and h are differentiable functions of t. As t varies continuously, the locus of P is a curve in space.

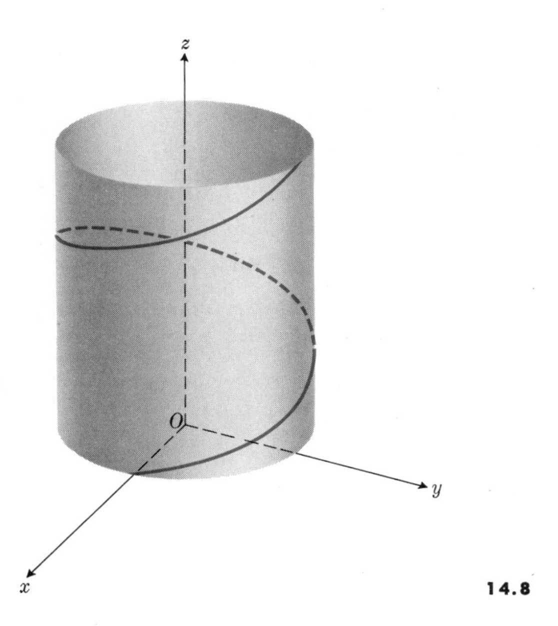

14.8

Example 1. The equations

$$x = a \cos \omega t, \quad y = a \sin \omega t, \quad z = bt, \tag{6}$$

where a, b, and ω are positive constants, represent a circular helix. (See Fig. 14.8.)

We now let P_0 be any fixed point on the space curve, and adopt a positive direction for measuring distance along the curve from P_0; for example, we may let P_0 be the position of P when $t = 0$ and let arc length be measured in the direction in which P first moves away from P_0 as t takes on positive values. Then the position of P on the curve becomes a function of the arc length s from P_0 to P. We can readily see that the vector

$$\mathbf{R} = \mathbf{i}x + \mathbf{j}y + \mathbf{k}z \tag{7}$$

from the origin to P is also a function of s, and we propose to discuss the geometrical significance of the derivative

$$\frac{d\mathbf{R}}{ds} = \mathbf{i}\frac{dx}{ds} + \mathbf{j}\frac{dy}{ds} + \mathbf{k}\frac{dz}{ds}. \tag{8}$$

If we calculate the derivative from the definition

$$\frac{d\mathbf{R}}{ds} = \lim_{\Delta s \to 0} \frac{\Delta \mathbf{R}}{\Delta s},$$

then we have (see Fig. 14.9)

$$\frac{\Delta \mathbf{R}}{\Delta s} = \begin{cases} \text{a vector of magnitude } \dfrac{\text{chord } PQ}{\text{arc } PQ} \text{ that} \\ \text{is directed along the secant line } PQ. \end{cases}$$

As $Q \to P$ and $\Delta s \to 0$, the direction of the secant line approaches the direction of the tangent to the curve at P, while the ratio of chord to arc approaches unity (for a "smooth" curve). Therefore the limit of $\Delta\mathbf{R}/\Delta s$ is a *unit* vector that is *tangent* to the curve at P and that points in the direction in which arc length increases along the curve. In other words, the vector $\mathbf{T}$, which is defined by the equation

$$\frac{d\mathbf{R}}{ds} = \mathbf{T}, \tag{9}$$

is a *unit* vector *tangent* to the space curve described by the endpoint P of the vector $\mathbf{R} = \overrightarrow{OP}$.

Example 2. From Eqs. (6) and (8), we find

$$\begin{aligned} \mathbf{T} &= \frac{d\mathbf{R}}{ds} = \mathbf{i}\frac{dx}{ds} + \mathbf{j}\frac{dy}{ds} + \mathbf{k}\frac{dz}{ds} \\ &= \mathbf{i}\left(-a\omega \sin \omega t \frac{dt}{ds}\right) + \mathbf{j}\left(a\omega \cos \omega t \frac{dt}{ds}\right) + \mathbf{k}\left(b\frac{dt}{ds}\right) \end{aligned}$$

as the unit vector tangent to the helix at any point P. To determine the magnitude of the scalar factor dt/ds, we use the fact that

$$|\mathbf{T}| = 1, \text{ and hence } \mathbf{T} \cdot \mathbf{T} = 1.$$

This leads to

$$\left(\frac{dt}{ds}\right)^2 (a^2\omega^2 + b^2) = 1,$$

so that

$$\frac{dt}{ds} = \pm \frac{1}{\sqrt{a^2\omega^2 + b^2}}.$$

Since dt/ds is a constant, we may as well agree to take it to be positive so that s is an increasing function of t. Then we have

$$\mathbf{T} = \frac{a\omega(-\mathbf{i} \sin \omega t + \mathbf{j} \cos \omega t) + b\mathbf{k}}{\sqrt{a^2\omega^2 + b^2}}.$$

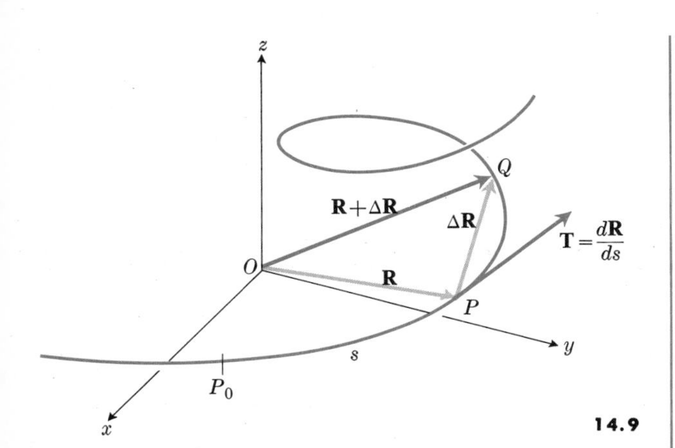

14.9

If we combine the results of Eqs. (8) and (9), we have

$$\mathbf{T} = \mathbf{i}\frac{dx}{ds} + \mathbf{j}\frac{dy}{ds} + \mathbf{k}\frac{dz}{ds}, \tag{10}$$

and since

$$\mathbf{T}\cdot\mathbf{T} = 1,$$

this means that

$$ds = \pm\sqrt{dx^2 + dy^2 + dz^2}. \tag{11}$$

The length of an arc of a curve may be calculated by computing ds from (11) and integrating between appropriate limits.

Example 3. For the helix of Examples 1 and 2, we have

$$ds = \sqrt{a^2\omega^2 + b^2}\,dt, \quad s = \sqrt{a^2\omega^2 + b^2}\int dt,$$

where appropriate limits of integration are to be supplied.

EXERCISES

In Exercises 1 through 5, $\mathbf{R} = \mathbf{i}x + \mathbf{j}y$ is the vector from the origin O to $P(x, y)$. For each of these motions, find the unit tangent vector $\mathbf{T} = d\mathbf{R}/ds$.

1. $\mathbf{R} = 2\mathbf{i}\cos t + 2\mathbf{j}\sin t$
2. $\mathbf{R} = e^t\mathbf{i} + t^2\mathbf{j}$
3. $\mathbf{R} = (\cos^3 t)\mathbf{i} + (\sin^3 t)\mathbf{j}$
4. $\mathbf{R} = \mathbf{i}x + \mathbf{j}x^2$
5. $\mathbf{R} = (\cos 2t)\mathbf{i} + (2\cos t)\mathbf{j}$

In Exercises 6 through 9, $\mathbf{R} = \mathbf{i}x + \mathbf{j}y + \mathbf{k}z$. Find the unit vector $\mathbf{T} = d\mathbf{R}/ds$ that is tangent to the given space curve. Take $ds/dt \geq 0$.

6. $x = 6\sin 2t, \quad y = 6\cos 2t, \quad z = 5t$
7. $x = e^t\cos t, \quad y = e^t\sin t, \quad z = e^t$
8. $x = 3\cosh 2t, \quad y = 3\sinh 2t, \quad z = 6t$
9. $x = 3t\cos t, \quad y = 3t\sin t, \quad z = 4t$

For the curves in Exercises 10 through 13, find the length of the curve between $t = 0$ and $t = \pi$.

10. The curve of Exercise 6
11. The curve of Exercise 7
12. The curve of Exercise 8
13. The curve of Exercise 9

14.4 CURVATURE AND NORMAL VECTORS

Our next step is to consider the rate of change of the unit tangent vector $\mathbf{T}$ as P moves along the curve. Of course the length of $\mathbf{T}$ is constant, always being equal to unity. But the direction of $\mathbf{T}$ changes, since it is tangent to the curve and this tangent changes direction from point to point unless the curve is a straight line.

Motion in a plane

We may specify the direction of $\mathbf{T}$ by means of the angle ϕ between the x-axis and $\mathbf{T}$ if the motion is in the xy-plane (see Fig. 14.10). The rate of change of this slope angle ϕ with respect to arc length s (measured in radians per unit of length) is taken as the

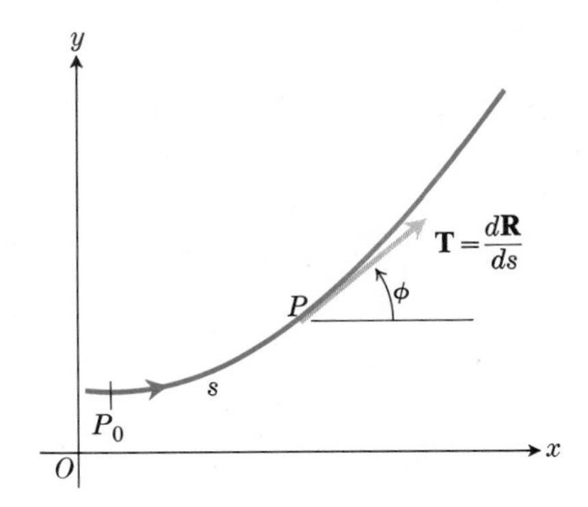

14.10

mathematical *definition of the curvature* of the curve at P. The Greek letter κ (kappa) is used to denote curvature, and its definition is given by the equation

$$\kappa = \left|\frac{d\phi}{ds}\right|, \tag{1}$$

where

$$\tan\phi = \frac{dy}{dx}, \quad ds = \pm\sqrt{dx^2 + dy^2}.$$

One may derive a formula for κ from the equations above in a straightforward manner:

$$\phi = \tan^{-1}\frac{dy}{dx}, \quad \frac{d\phi}{dx} = \frac{\dfrac{d^2y}{dx^2}}{1 + \left(\dfrac{dy}{dx}\right)^2},$$

and

$$\frac{ds}{dx} = \pm\sqrt{1 + \left(\frac{dy}{dx}\right)^2},$$

so that

$$\kappa = \left|\frac{d\phi}{ds}\right| = \left|\frac{d\phi/dx}{ds/dx}\right| = \frac{\left|\dfrac{d^2y}{dx^2}\right|}{\left[1 + \left(\dfrac{dy}{dx}\right)^2\right]^{3/2}}. \tag{2a}$$

We may arrive at a formula for κ in terms of dx/dy and d^2x/dy^2 if we use

$$\phi = \cot^{-1}\frac{dx}{dy} \quad \text{and} \quad \kappa = \left|\frac{d\phi}{ds}\right| = \left|\frac{d\phi/dy}{ds/dy}\right|.$$

The result, which corresponds to (2a), is

$$\kappa = \frac{\left|\dfrac{d^2x}{dy^2}\right|}{\left[1 + \left(\dfrac{dx}{dy}\right)^2\right]^{3/2}}. \tag{2b}$$

If the equation of the curve is given in parametric form:

$$x = f(t), \quad y = g(t),$$

then

$$\phi = \tan^{-1}\left(\frac{dy/dt}{dx/dt}\right),$$

and if we use

$$\kappa = \left|\frac{d\phi/dt}{ds/dt}\right|,$$

the calculations are as follows:

$$\frac{d\phi}{dt} = \frac{1}{1 + \left(\dfrac{dy/dt}{dx/dt}\right)^2} \frac{\dfrac{dx}{dt}\dfrac{d^2y}{dt^2} - \dfrac{dy}{dt}\dfrac{d^2x}{dt^2}}{\left(\dfrac{dx}{dt}\right)^2}$$

$$= \frac{\dot{x}\ddot{y} - \dot{y}\ddot{x}}{\dot{x}^2 + \dot{y}^2} \qquad \left[\dot{x} = \frac{dx}{dt},\ \ddot{x} = \frac{d^2x}{dt^2}\right]$$

and

$$\frac{ds}{dt} = \pm\sqrt{\dot{x}^2 + \dot{y}^2},$$

so that

$$\kappa = \frac{|\dot{x}\ddot{y} - \dot{y}\ddot{x}|}{[\dot{x}^2 + \dot{y}^2]^{3/2}}. \tag{2c}$$

On a straight line, ϕ is constant, so $d\phi/ds$ is zero and Eq. (2a) says that *the curvature of a straight line is zero.*

If we apply (2), or (1), to a circle of radius a, we find that the curvature of a circle is constant in magnitude. In fact, we see from Fig. 14.11 that both s and ϕ may be expressed in terms of the angle θ:

$$s = a\theta, \quad \phi = \theta + \frac{\pi}{2}, \tag{3}$$

so that

$$\kappa = \left|\frac{d\phi}{ds}\right| = \left|\frac{d\theta}{a\,d\theta}\right| = \frac{1}{a}.$$

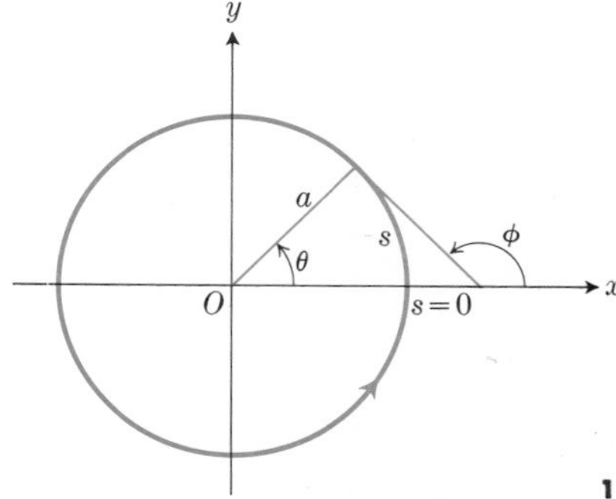

14.11

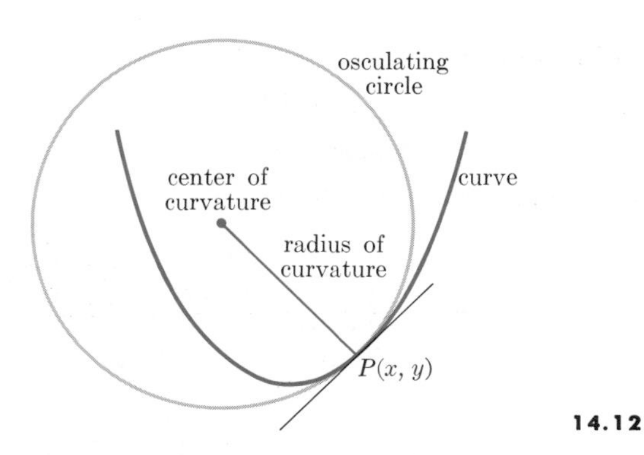

14.12

That is, the curvature of a circle is equal to the *reciprocal of its radius*. The smaller the circle, the greater its curvature. "Turning around on a dime" indicates a more rapid change of direction per unit of arc length than does turning around on a silver dollar!

Circle and radius of curvature

The circle that is tangent to a given curve at P, whose center lies on the concave side of the curve and which has the same curvature as the curve has at P, is called the *circle of curvature*. Its radius is $1/\kappa$. We define the *radius of curvature* at P to be ρ, where

$$\rho = \frac{1}{\kappa} = \frac{\left[1 + \left(\frac{dy}{dx}\right)^2\right]^{3/2}}{\left|\frac{d^2y}{dx^2}\right|}. \tag{4}$$

The center of the circle of curvature is called the center of curvature. The circle of curvature has its first and second derivatives equal respectively to the first and second derivatives of the curve itself at this point. For this reason it has a higher degree of contact with the curve at P than has any other circle, and so it is also called the *osculating* circle! (See Fig. 14.12.) Since velocity and acceleration involve only the first and second time-derivatives of the coordinates of P, it is natural to expect that the instantaneous velocity and acceleration of a particle moving on any curve may be expressed in terms of instantaneous velocity and acceleration of an associated particle moving on the osculating circle. This will be investigated in the next article.

Unit normal vector

We return once more to the question of the rate of change of the unit vector $\mathbf{T}$ as P moves along the curve. In terms of the slope angle ϕ, Fig. 14.13, we may write

$$\mathbf{T} = \mathbf{i} \cos \phi + \mathbf{j} \sin \phi, \tag{5}$$

and then the derivative

$$\frac{d\mathbf{T}}{d\phi} = -\mathbf{i} \sin \phi + \mathbf{j} \cos \phi \tag{6}$$

has magnitude

$$\left|\frac{d\mathbf{T}}{d\phi}\right| = \sqrt{\sin^2 \phi + \cos^2 \phi} = 1.$$

From Eqs. (5) and (6), we see that the inner product of $\mathbf{T}$ and $d\mathbf{T}/d\phi$ is zero:

$$\mathbf{T} \cdot \frac{d\mathbf{T}}{d\phi} = 0.$$

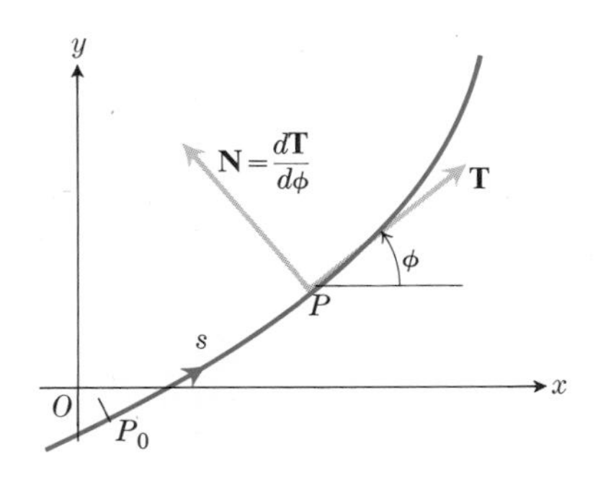

a | b

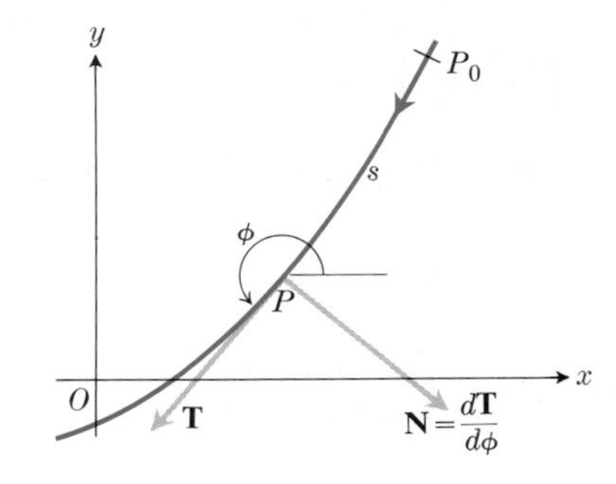

14.13

Therefore $d\mathbf{T}/d\phi$ is perpendicular to $\mathbf{T}$. In fact, we see from Eq. (6) that

$$\frac{d\mathbf{T}}{d\phi} = \mathbf{N}, \tag{7}$$

where $\mathbf{N}$ is the *unit normal* vector

$$\mathbf{N} = \mathbf{i} \cos (\phi + 90°) + \mathbf{j} \sin (\phi + 90°)$$
$$= -\mathbf{i} \sin \phi + \mathbf{j} \cos \phi$$

obtained by rotating the unit tangent vector $\mathbf{T}$ through 90° in the counterclockwise direction.

For a curve in 3-space, the direction of the unit tangent vector $\mathbf{T}$ is not determined by a single angle such as ϕ. Instead, we shall use arc length, s, as the parameter for the theoretical study of $\mathbf{T}$. In Article 14.5, we shall see that $\mathbf{T}$ and $d\mathbf{T}/ds$ are orthogonal vectors. If $d\mathbf{T}/ds$ is not the zero vector, we then use its direction to specify the *principal normal* to the curve. In the two-dimensional case, we have (from the chain rule)

$$\frac{d\mathbf{T}}{ds} = \frac{d\mathbf{T}}{d\phi}\frac{d\phi}{ds} = \mathbf{N}(\pm\kappa).$$

If we direct the curve so that ϕ is an *increasing function* of s, then

$$\frac{d\phi}{ds} = \kappa,$$

and

$$\frac{d\mathbf{T}}{ds} = \frac{d\mathbf{T}}{d\phi}\frac{d\phi}{ds} = \mathbf{N}\kappa. \tag{8}$$

To combine both the two-dimensional and three-dimensional curves in one equation, we drop the middle part of Eq. (8) and obtain the simple equation

$$\frac{d\mathbf{T}}{ds} = \mathbf{N}\kappa. \tag{9}$$

In Eq. (9), κ is the magnitude of $d\mathbf{T}/ds$:

$$\kappa = \left|\frac{d\mathbf{T}}{ds}\right|. \tag{10}$$

This number is called the *curvature*. Such a definition is consistent with Eq. (1) for a plane curve, and extends the concept of curvature to curves in space.

Equations (9) and (10) together define the unit principal normal vector $\mathbf{N}$, whenever $d\mathbf{T}/ds \neq \mathbf{0}$:

$$\mathbf{N} = \frac{d\mathbf{T}/ds}{|d\mathbf{T}/ds|}. \tag{11}$$

Example 1. For the helix in Examples 1 and 2 of Article 14.3, we have found

$$\mathbf{T} = \frac{a\omega(-\mathbf{i} \sin \omega t + \mathbf{j} \cos \omega t) + b\mathbf{k}}{\sqrt{a^2\omega^2 + b^2}} \tag{12}$$

and

$$\frac{ds}{dt} = \sqrt{a^2\omega^2 + b^2}.$$

Hence

$$\frac{d\mathbf{T}}{ds} = \frac{d\mathbf{T}/dt}{ds/dt} = \frac{-a\omega^2}{a^2\omega^2 + b^2}(\mathbf{i} \cos \omega t + \mathbf{j} \sin \omega t).$$

Comparing this with Eq. (10), we have

$$\kappa = \left|\frac{d\mathbf{T}}{ds}\right| = \frac{a\omega^2}{a^2\omega^2 + b^2}. \tag{13}$$

Two limiting cases of Eq. (13) are readily checked. First, if $b = 0$, then $z = 0$, and the curve reduces to a circle of radius a in the xy-plane, while Eq. (13) reduces to $\kappa = 1/a$. We recognize this as the correct relationship between the curvature and radius of a circle. Second, in the other limiting case, $a = 0$ gives $x = y = 0$ and $z = bt$ in Eq. (6) of Article 14.3. This tells us that the point moves along the z-axis and, in this case, Eq. (13) again gives us a correct answer, namely that $\kappa = 0$. Finally, in the general case, we note that the curvature of a circular helix is constant and less than the curvature of the circle which is the cross section of the cylinder (see Fig. 14.8). It is also worth noting that the *principal normal* in this case is the vector

$$\mathbf{N} = -(\mathbf{i} \cos \omega t + \mathbf{j} \sin \omega t)$$
$$= -\frac{\mathbf{i}x + \mathbf{j}y}{a}, \tag{14}$$

which is parallel to the vector from the point $(0, 0, z)$ on the z-axis to the point $P(x, y, z)$ on the helix.

Once the unit vectors $\mathbf{T}$ and $\mathbf{N}$ have been determined, it is a simple matter to define a third unit vector, perpendicular to both $\mathbf{T}$ and $\mathbf{N}$, by the equation

$$\mathbf{B} = \mathbf{T} \times \mathbf{N}. \tag{15}$$

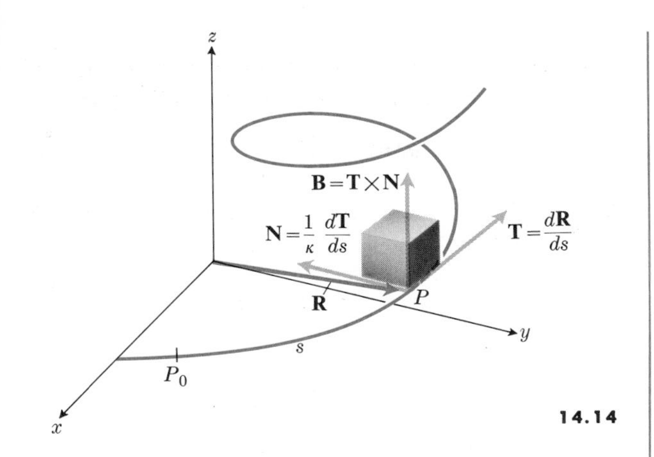

14.14

The vector **B** so defined may be thought of as lying in the plane normal to **T** at P and is called the *binormal* at P. These three unit vectors, **T**, **N**, and **B**, form a right-handed system of mutually orthogonal unit vectors which are useful in more thorough investigations of space curves (see Fig. 14.14). (Struik, *Differential Geometry*, Chapter 1, Addison-Wesley, 1961.)

EXERCISES

Find the curvature of each of the curves in Exercises 1 through 9.

1. $y = a \cosh (x/a)$
2. $y = \ln (\cos x)$
3. $y = e^{2x}$
4. $x = a \cos^3 t$, $y = a \sin^3 t$
5. $x = a (\cos \theta + \theta \sin \theta)$, $y = a (\sin \theta - \theta \cos \theta)$
6. $x = a(\theta - \sin \theta)$, $y = a(1 - \cos \theta)$
7. $x = \ln \sec y$
8. $x = \frac{1}{3}(y^2 + 2)^{3/2}$
9. $x = \frac{y^4}{4} + \frac{1}{8y^2}$

10. Find the equation of the osculating circle associated with the curve $y = e^x$ at the point (0, 1). By calculating dy/dx and d^2y/dx^2 at the point (0, 1) from the equation of this circle, verify that these derivatives have the same values there as do the corresponding derivatives for the curve $y = e^x$. Sketch the curve and the osculating circle.

11. Show that when x and y are considered as functions of arc length s, the unit vectors **T** and **N** may be expressed as follows:

$$\mathbf{T} = \mathbf{i}\frac{dx}{ds} + \mathbf{j}\frac{dy}{ds},$$

$$\mathbf{N} = -\mathbf{i}\frac{dy}{ds} + \mathbf{j}\frac{dx}{ds},$$

where $dx/ds = \cos \phi$, $dy/ds = \sin \phi$, and ϕ is the angle from the positive x-axis to the tangent line.

In Exercises 12 through 14, find the principal normal vector **N**, the curvature κ, and the unit binormal vector **B**.

12. The curve of Exercise 6, Article 14.3
13. The curve of Exercise 7, Article 14.3
14. The curve of Exercise 8, Article 14.3

15. Let $\mathbf{R} = \overrightarrow{OP}$ be the vector from the origin to a moving point P. Let **T** and **N** be the unit tangent and principal normal vectors, respectively, for the curve described by P. Express the velocity and acceleration vectors $d\mathbf{R}/dt$ and $d^2\mathbf{R}/dt^2$ in terms of their **T**- and **N**-components.

14.5 DIFFERENTIATION OF PRODUCTS OF VECTORS

If the components of a vector are differentiable functions of a scalar variable t, then we know that the vector is a differentiable function of t, and its derivative is obtained by differentiating the components (Article 14.1, Eq. 8).

It is also convenient to develop formulas for the derivative of the dot and the cross products of two vectors which are differentiable functions of t. Suppose, for example that

$$\begin{aligned} \mathbf{U} &= \mathbf{i}f_1(t) + \mathbf{j}g_1(t) + \mathbf{k}h_1(t), \\ \mathbf{V} &= \mathbf{i}f_2(t) + \mathbf{j}g_2(t) + \mathbf{k}h_2(t), \end{aligned} \tag{1}$$

where f, g, and h are differentiable functions of t. Then by the ordinary formulas for differentiating products of scalar functions, it is easy to verify that

$$\frac{d}{dt}(\mathbf{U}\cdot\mathbf{V}) = \frac{d\mathbf{U}}{dt}\cdot\mathbf{V} + \mathbf{U}\cdot\frac{d\mathbf{V}}{dt} \tag{2}$$

and

$$\frac{d}{dt}(\mathbf{U}\times\mathbf{V}) = \frac{d\mathbf{U}}{dt}\times\mathbf{V} + \mathbf{U}\times\frac{d\mathbf{V}}{dt}. \tag{3}$$

However, instead of verifying the identities in Eqs. (2) and (3) by verification of their components, it is instructive to think of how these equations might be established by direct appeal to the Δ-process for calculating derivatives. For example, let

$$\mathbf{W} = \mathbf{U}\times\mathbf{V},$$

where t has some specific value. Then give t an increment Δt and denote the new values of the vectors by $\mathbf{U} + \Delta\mathbf{U}$, and so on. This gives

$$\begin{aligned}\mathbf{W} + \Delta\mathbf{W} &= (\mathbf{U} + \Delta\mathbf{U})\times(\mathbf{V} + \Delta\mathbf{V}) \\ &= \mathbf{U}\times\mathbf{V} + \mathbf{U}\times\Delta\mathbf{V} \\ &\quad + \Delta\mathbf{U}\times\mathbf{V} + \Delta\mathbf{U}\times\Delta\mathbf{V},\end{aligned}$$

so that

$$\frac{\Delta\mathbf{W}}{\Delta t} = \mathbf{U}\times\frac{\Delta\mathbf{V}}{\Delta t} + \frac{\Delta\mathbf{U}}{\Delta t}\times\mathbf{V} + \frac{\Delta\mathbf{U}}{\Delta t}\times\Delta\mathbf{V}.$$

Now take limits as $\Delta t \to 0$, noting that

$$\lim \frac{\Delta\mathbf{W}}{\Delta t} = \frac{d\mathbf{W}}{dt}, \qquad \lim\frac{\Delta\mathbf{U}}{\Delta t} = \frac{d\mathbf{U}}{dt},$$

$$\lim \Delta\mathbf{V} = \lim\frac{\Delta\mathbf{V}}{\Delta t}\cdot\lim\Delta t = \mathbf{0},$$

so that

$$\frac{d\mathbf{W}}{dt} = \mathbf{U}\times\frac{d\mathbf{V}}{dt} + \frac{d\mathbf{U}}{dt}\times\mathbf{V},$$

which is equivalent to Eq. (3).

Equations (2) and (3) are just like the equations we have for the derivatives of products of scalar functions u and v, and indeed the proofs by the Δ-process look the same for vectors and for scalars. The only place we need to be careful is in a derivative involving a cross product; here it is essential that the relative order of the factors be preserved, since a reversal of order entails a change of sign.

Note. The formula for the derivative of the triple scalar product leads to an interesting identity regarding the derivative of a determinant of order three. Let

$$\begin{aligned}\mathbf{U} &= u_1\mathbf{i} + u_2\mathbf{j} + u_3\mathbf{k}, \\ \mathbf{V} &= v_1\mathbf{i} + v_2\mathbf{j} + v_3\mathbf{k}, \\ \mathbf{W} &= w_1\mathbf{i} + w_2\mathbf{j} + w_3\mathbf{k},\end{aligned} \tag{4}$$

where the components are differentiable functions of a scalar t. Then the left-hand side of the identity

$$\begin{aligned}&\frac{d}{dt}(\mathbf{U}\cdot\mathbf{V}\times\mathbf{W}) \\ &\quad = \frac{d\mathbf{U}}{dt}\cdot\mathbf{V}\times\mathbf{W} + \mathbf{U}\cdot\frac{d\mathbf{V}}{dt}\times\mathbf{W} + \mathbf{U}\cdot\mathbf{V}\times\frac{d\mathbf{W}}{dt}\end{aligned} \tag{5}$$

is equivalent to

$$\frac{d}{dt}\begin{vmatrix} u_1 & u_2 & u_3 \\ v_1 & v_2 & v_3 \\ w_1 & w_2 & w_3 \end{vmatrix}, \tag{6a}$$

and the right-hand side is equivalent to

$$\begin{vmatrix} \frac{du_1}{dt} & \frac{du_2}{dt} & \frac{du_3}{dt} \\ v_1 & v_2 & v_3 \\ w_1 & w_2 & w_3 \end{vmatrix} + \begin{vmatrix} u_1 & u_2 & u_3 \\ \frac{dv_1}{dt} & \frac{dv_2}{dt} & \frac{dv_3}{dt} \\ w_1 & w_2 & w_3 \end{vmatrix} + \begin{vmatrix} u_1 & u_2 & u_3 \\ v_1 & v_2 & v_3 \\ \frac{dw_1}{dt} & \frac{dw_2}{dt} & \frac{dw_3}{dt} \end{vmatrix}. \tag{6b}$$

The identity between (6a) and (6b) tells us that the derivative of a determinant of order three is the sum of three determinants obtained from the original determinant by differentiating one row of it at a time. This result may also be extended to determinants of any order n.

An interesting geometrical result is obtained by differentiating the identity

$$\mathbf{V}\cdot\mathbf{V} = |\mathbf{V}|^2 \tag{7a}$$

in the case where $\mathbf{V}$ is a vector of constant magnitude. Then $|\mathbf{V}|^2$ is a constant, so its derivative is zero, and we have

$$\mathbf{V}\cdot\frac{d\mathbf{V}}{dt} + \frac{d\mathbf{V}}{dt}\cdot\mathbf{V} = 0$$

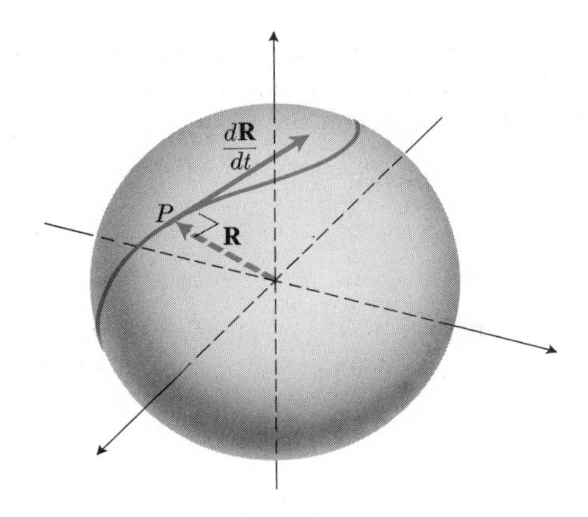

14.15

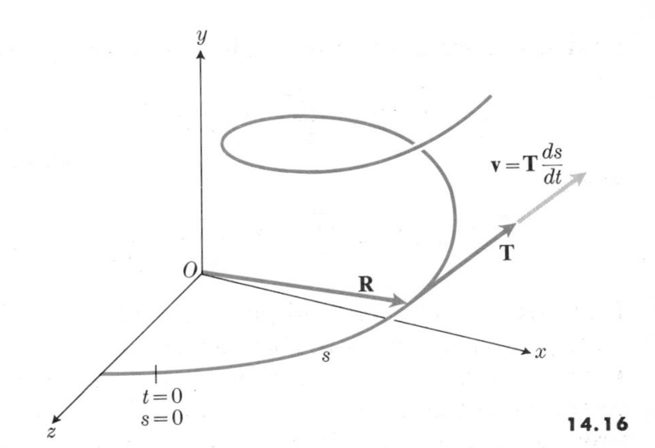

14.16

or, since the scalar product is commutative,

$$2\mathbf{V}\cdot\frac{d\mathbf{V}}{dt} = 0. \tag{7b}$$

Geometrically, this means that either $\mathbf{V}$ is zero, or that $d\mathbf{V}/dt$ is zero and hence $\mathbf{V}$ is constant in direction as well as in magnitude, or that $d\mathbf{V}/dt$ is perpendicular to $\mathbf{V}$. For example, if a point P moves about on the surface of a sphere, then the vector $\mathbf{R}$ from the center to P has constant magnitude, and $d\mathbf{R}/dt$, which in this case is the velocity vector of P, is perpendicular to $\mathbf{R}$ (see Fig. 14.15).

We can also use the foregoing results to show that the derivative of the unit tangent vector $\mathbf{T}$ is orthogonal to $\mathbf{T}$. Because $|\mathbf{T}| = 1$, we have

$$\mathbf{T}\cdot\mathbf{T} = 1,$$

so that, by the same kind of reasoning we used in going from Eq. (7a) to (7b), we can deduce that

$$\mathbf{T}\cdot\frac{d\mathbf{T}}{ds} = 0.$$

This validates our earlier statement that $d\mathbf{T}/ds$ is perpendicular to $\mathbf{T}$, so that the definition

$$\frac{d\mathbf{T}}{ds} = \mathbf{N}\kappa, \tag{8}$$

as given in Eq. (9), Article 14.4, with $\kappa = |d\mathbf{T}/ds|$, produces a vector $\mathbf{N}$ orthogonal to $\mathbf{T}$.

Tangential and normal components of the velocity and acceleration vectors

In mechanics, it is useful to be able to discuss the motion of a particle P in terms of its instantaneous speed ds/dt, its acceleration along its path d^2s/dt^2, and the curvature of its path. This is easy if we refer the velocity and acceleration vectors to the unit vectors $\mathbf{T}$ and $\mathbf{N}$.

In Article 14.2, we found the velocity vector is given by

$$\mathbf{v} = \frac{d\mathbf{R}}{dt}, \tag{9}$$

where $\mathbf{R} = \mathbf{i}x + \mathbf{j}y + \mathbf{k}z$ is the position vector $\overrightarrow{OP}$. We may also write this in the form

$$\mathbf{v} = \frac{d\mathbf{R}}{dt} = \frac{d\mathbf{R}}{ds}\,\frac{ds}{dt},$$

or

$$\mathbf{v} = \mathbf{T}\frac{ds}{dt}, \tag{10}$$

if we use the result of Eq. (9), Article 14.3. This is in keeping with our earlier remark that the velocity vector is tangent to the curve and has magnitude $|ds/dt|$. (See Fig. 14.16).

To obtain the acceleration vector, we differentiate Eq. (10) with respect to t:

$$\mathbf{a} = \frac{d\mathbf{v}}{dt} = \mathbf{T}\frac{d^2s}{dt^2} + \frac{ds}{dt}\,\frac{d\mathbf{T}}{dt}.$$

By Eq. (8), we have

$$\frac{d\mathbf{T}}{dt} = \frac{d\mathbf{T}}{ds}\frac{ds}{dt} = \mathbf{N}\kappa\frac{ds}{dt},$$

so that

$$\mathbf{a} = \mathbf{T}\frac{d^2s}{dt^2} + \mathbf{N}\kappa\left(\frac{ds}{dt}\right)^2. \tag{11}$$

Equation (11) expresses the acceleration vector in terms of its tangential and normal components. The *tangential component* d^2s/dt^2 is simply the derivative of the speed ds/dt of the particle in its path. The *normal component* is directed toward the concave side of the curve and has magnitude

$$\kappa\left(\frac{ds}{dt}\right)^2 = \frac{(ds/dt)^2}{\rho} = \frac{v^2}{\rho},$$

where v is the instantaneous speed of the particle and ρ is the radius of curvature of the path at the point in question. This explains why a large normal force, which must be supplied by friction between the tires and the roadway, is required to hold an automobile on a level road if it makes a sharp turn (small ρ) at a moderate speed or a moderate turn at high speed (large v^2). (See Sears, *Mechanics, Heat, and Sound*, p. 187, for a discussion of the banking of curves.)

If the particle is moving in a circle with *constant* speed $v = ds/dt$, then d^2s/dt^2 is zero, and the only acceleration is the normal acceleration v^2/p toward the center of the circle. If the speed is not constant, the acceleration vector $\mathbf{a}$ is the resultant of the tangential and normal components, as in Fig. 14.17.

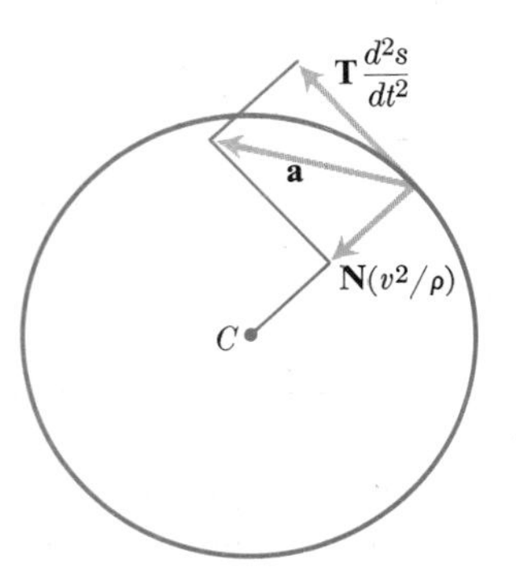

14.17

The following example illustrates how the tangential and normal components of velocity and acceleration may be computed when the equations of motion are known. In particular, it should be noted that the equation

$$|\mathbf{a}|^2 = a_x^2 + a_y^2 = a_T^2 + a_N^2 \tag{12}$$

is used to determine the normal component of acceleration

$$a_N = \sqrt{|\mathbf{a}|^2 - a_T^2}. \tag{13}$$

Example. The coordinates of a moving particle at time t are given by

$$x = \cos t + t\sin t, \quad y = \sin t - t\cos t.$$

Thus

$$\begin{aligned}\mathbf{v} &= \mathbf{i}\frac{dx}{dt} + \mathbf{j}\frac{dy}{dt}\\ &= \mathbf{i}[-\sin t + t\cos t + \sin t]\\ &\qquad + \mathbf{j}[\cos t + t\sin t - \cos t]\\ &= \mathbf{i}t\cos t + \mathbf{j}t\sin t\end{aligned}$$

and

$$\mathbf{a} = \frac{d\mathbf{v}}{dt} = \mathbf{i}[-t\sin t + \cos t] + \mathbf{j}[t\cos t + \sin t].$$

Now the tangential component of velocity is

$$\frac{ds}{dt} = |\mathbf{v}| = \sqrt{(t\cos t)^2 + (t\sin t)^2} = t,$$

and the tangential component of acceleration is

$$a_T = \frac{d^2s}{dt^2} = \frac{d}{dt}\left(\frac{ds}{dt}\right) = \frac{d}{dt}(t) = 1.$$

We use Eq. (13) to determine the normal component of acceleration:

$$\begin{aligned}a_N &= \sqrt{|\mathbf{a}|^2 - a_T^2}\\ &= \sqrt{(-t\sin t + \cos t)^2 + (t\cos t + \sin t)^2 - 1}\\ &= t.\end{aligned}$$

Here the tangential acceleration has constant magnitude and the normal acceleration starts with zero magnitude at $t = 0$ and increases with time. The equations of motion are the same as the parametric equations for the involute of a circle of unit radius with $\theta = t$ (see Fig. 14.18). We could also readily find the radius of

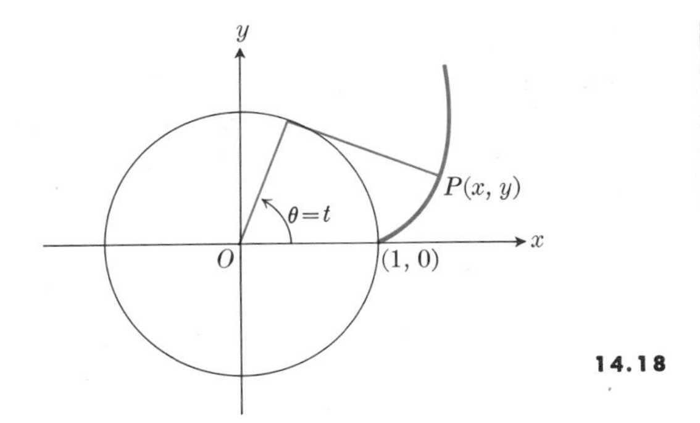

14.18

curvature without any complicated calculations from the equation $a_N = v^2/\rho$, since $v^2 = |\mathbf{v}|^2 = t^2$ and $a_N = t$ and hence

$$\rho = \frac{v^2}{a_N} = \frac{t^2}{t} = t.$$

Equations (10) and (11) can be used to find a formula for the curvature κ in terms of the velocity and acceleration. First, we compute the cross product of the velocity and acceleration vectors. We obtain

$$\mathbf{v} \times \mathbf{a} = \mathbf{T}\frac{ds}{dt} \times \left[\mathbf{T}\frac{d^2s}{dt^2} + \mathbf{N}\kappa\left(\frac{ds}{dt}\right)^2\right]$$

$$= \mathbf{T} \times \mathbf{N}\kappa\left(\frac{ds}{dt}\right)^3 \tag{14}$$

because we can apply the distributive law for the cross product, and $\mathbf{T} \times \mathbf{T} = \mathbf{0}$. Moreover, $\mathbf{T} \times \mathbf{N}$ is the unit binormal vector $\mathbf{B}$, as given by Eq. (15), Article 14.4. Therefore

$$\mathbf{v} \times \mathbf{a} = \mathbf{B}\kappa\left(\frac{ds}{dt}\right)^3. \tag{15}$$

Since $\mathbf{B}$ is a *unit* vector, the magnitude of $\mathbf{v} \times \mathbf{a}$ is

$$|\mathbf{v} \times \mathbf{a}| = \kappa\left|\frac{ds}{dt}\right|^3 = \kappa|\mathbf{v}|^3.$$

Finally, if $|\mathbf{v}| \neq 0$, we get (by division)

$$\kappa = \frac{|\mathbf{v} \times \mathbf{a}|}{|\mathbf{v}|^3}. \tag{16}$$

How do we use this equation? Given the motion in the form $\mathbf{R} = \mathbf{i}x + \mathbf{j}y + \mathbf{k}z$, we differentiate with respect to time to get $\mathbf{v}$, then differentiate again to get $\mathbf{a}$, and compute $\mathbf{v} \times \mathbf{a}$ using a determinant of order 3 in the usual way for cross products. We then divide the length of this vector by the cube of the length of $\mathbf{v}$.

Problem. Use Eq. (16) to find the curvature of the curve of the preceding example.

Solution. In that example, we found the velocity

$$\mathbf{v} = \mathbf{i}t\cos t + \mathbf{j}t\sin t,$$

and acceleration

$$\mathbf{a} = \mathbf{i}(-t\sin t + \cos t) + \mathbf{j}(t\cos t + \sin t).$$

Therefore

$$\mathbf{v} \times \mathbf{a} = \begin{vmatrix} \mathbf{i} & \mathbf{j} & \mathbf{k} \\ t\cos t & t\sin t & 0 \\ (-t\sin t + \cos t) & (t\cos t + \sin t) & 0 \end{vmatrix};$$

this leads to

$$\mathbf{v} \times \mathbf{a} = \mathbf{k}(t^2\cos^2 t + t\cos t\sin t + t^2\sin^2 t - t\sin t\cos t)$$
$$= \mathbf{k}t^2,$$

and

$$\kappa = \frac{|\mathbf{v} \times \mathbf{a}|}{|\mathbf{v}|^3} = \frac{t^2}{t^3} = \frac{1}{t}.$$

This result is valid for $t > 0$. If the curve and the motion also exist for $t < 0$, we should replace t by $|t|$ for $t < 0$.

EXERCISES

1. Derive Eq. (2) by the Δ-process.
2. Apply Eqs. (2) and (3) to $\mathbf{U} \cdot \mathbf{V}_1$ with $\mathbf{V}_1 = \mathbf{V} \times \mathbf{W}$ and thereby derive Eq. (5) for

$$\frac{d}{dt}[\mathbf{U} \cdot (\mathbf{V} \times \mathbf{W})].$$

3. If $\mathbf{F}(t) = \mathbf{i}f(t) + \mathbf{j}g(t) + \mathbf{k}h(t)$, where f, g, and h are functions of t which have derivatives of orders one, two, and three, show that

$$\frac{d}{dt}\left[\mathbf{F} \cdot \left(\frac{d\mathbf{F}}{dt} \times \frac{d^2\mathbf{F}}{dt^2}\right)\right] = \mathbf{F} \cdot \left(\frac{d\mathbf{F}}{dt} \times \frac{d^3\mathbf{F}}{dt^3}\right).$$

Explain why the answer contains just this one term rather than the three terms that one might expect.

4. With the book closed, derive vector expressions for the velocity and acceleration in terms of tangential and normal components. Check your derivations with those given in the text.

In Exercises 5 through 9, find the velocity and acceleration vectors, and then find the speed ds/dt and the tangential and normal components of acceleration.

5. $\mathbf{R} = \mathbf{i} \cosh 2t + \mathbf{j} \sinh 2t$
6. $\mathbf{R} = (2t + 3)\mathbf{i} + (t^2 - 1)\mathbf{j}$
7. $\mathbf{R} = a \cos \omega t\mathbf{i} + a \sin \omega t\mathbf{j}$,
   a and ω positive constants
8. $\mathbf{R} = \mathbf{i} \ln (t^2 + 1) + \mathbf{j}(t - 2 \tan^{-1} t)$
9. $\mathbf{R} = \mathbf{i}e^t \cos t + \mathbf{j}e^t \sin t$
10. Deduce from Eq. (11) that a particle will move in a straight line if the normal component of acceleration is identically zero.
11. Show that the radius of curvature of a plane curve is given by

$$\rho = \frac{(\dot{x}^2 + \dot{y}^2)}{\sqrt{\ddot{x}^2 + \ddot{y}^2 - \ddot{s}^2}},$$

where

$$\dot{x} = \frac{dx}{dt}, \quad \ddot{x} = \frac{d^2x}{dt^2}, \quad \ldots$$

and

$$\ddot{s} = \frac{d}{dt}(\sqrt{\dot{x}^2 + \dot{y}^2}).$$

12. If a particle moves in a curve with constant speed, show that the force is always directed along the normal.
13. If the force acting on a particle is at all times perpendicular to the direction of motion, show that the speed remains constant.

14.6 POLAR AND CYLINDRICAL COORDINATES

If the particle P moves on a plane curve whose equation is given in polar coordinates, it is convenient to express the velocity and acceleration vectors in terms of still a third set of unit vectors. We introduce **unit vectors**

$$\mathbf{u}_r = \mathbf{i} \cos \theta + \mathbf{j} \sin \theta, \qquad \mathbf{u}_\theta = -\mathbf{i} \sin \theta + \mathbf{j} \cos \theta, \tag{1}$$

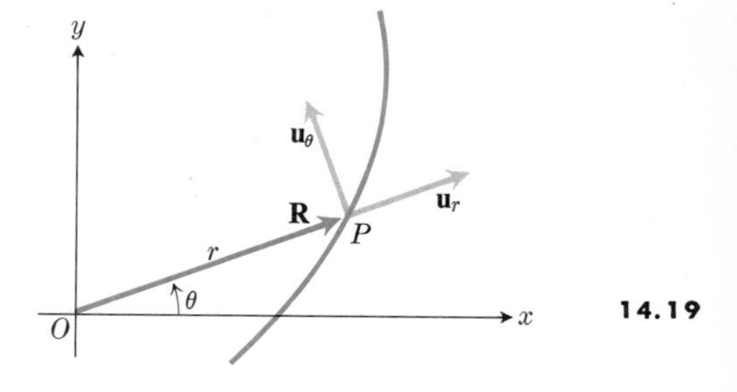

14.19

which point respectively along the radius vector $\overrightarrow{OP}$, and at right angles to $\overrightarrow{OP}$ and in the direction of increasing θ, as shown in Fig. 14.19. Then we find from (1) that

$$\frac{d\mathbf{u}_r}{d\theta} = -\mathbf{i} \sin \theta + \mathbf{j} \cos \theta = \mathbf{u}_\theta, \qquad \frac{d\mathbf{u}_\theta}{d\theta} = -\mathbf{i} \cos \theta - \mathbf{j} \sin \theta = -\mathbf{u}_r. \tag{2}$$

This says that the result of differentiating either one of the unit vectors $\mathbf{u}_r$ and $\mathbf{u}_\theta$ with respect to θ is equivalent to rotating that vector through 90° in the positive (counterclockwise) direction.

Since the vectors $\mathbf{R} = \overrightarrow{OP}$ and $r\mathbf{u}_r$ have the same direction, and the length of $\mathbf{R}$ is the absolute value of the polar coordinate r of $P(r, \theta)$, we have

$$\mathbf{R} = r\mathbf{u}_r. \tag{3}$$

To obtain the velocity, we must differentiate this with respect to t, remembering that both r and $\mathbf{u}_r$ may be variables. From (2) it is clear that

$$\frac{d\mathbf{u}_r}{dt} = \frac{d\mathbf{u}_r}{d\theta}\frac{d\theta}{dt} = \mathbf{u}_\theta \frac{d\theta}{dt}, \qquad \frac{d\mathbf{u}_\theta}{dt} = \frac{d\mathbf{u}_\theta}{d\theta}\frac{d\theta}{dt} = -\mathbf{u}_r \frac{d\theta}{dt}. \tag{4}$$

Hence

$$\mathbf{v} = \frac{d\mathbf{R}}{dt} = \mathbf{u}_r \frac{dr}{dt} + r \frac{d\mathbf{u}_r}{dt}$$

becomes

$$\mathbf{v} = \mathbf{u}_r \frac{dr}{dt} + \mathbf{u}_\theta r \frac{d\theta}{dt}. \tag{5}$$

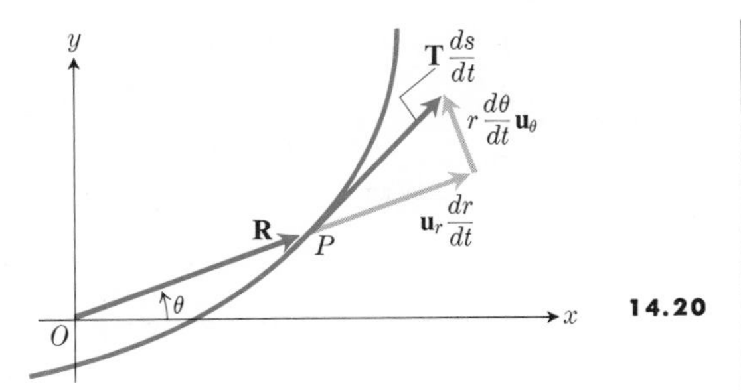

14.20

Of course this velocity vector is tangent to the curve at P and has magnitude

$$|\mathbf{v}| = \sqrt{(dr/dt)^2 + r^2(d\theta/dt)^2} = |ds/dt|.$$

In fact, if the three sides of the "differential triangle" of sides dr, r, $d\theta$, and ds are all divided by dt, the result will be a similar triangle having sides dr/dt, $r\,d\theta/dt$, and ds/dt, which illustrates the vector equation

$$\begin{aligned}\mathbf{v} &= \mathbf{T}\frac{ds}{dt}\\ &= \mathbf{u}_r\frac{dr}{dt} + \mathbf{u}_\theta r\frac{d\theta}{dt}.\end{aligned}$$

(See Fig. 14.20.)

The acceleration vector is found by differentiating the velocity vector in (5) as follows:

$$\begin{aligned}\mathbf{a} = \frac{d\mathbf{v}}{dt} &= \left(\mathbf{u}_r\frac{d^2r}{dt^2} + \frac{dr}{dt}\frac{d\mathbf{u}_r}{dt}\right)\\ &+ \left(\mathbf{u}_\theta r\frac{d^2\theta}{dt^2} + \mathbf{u}_\theta\frac{dr}{dt}\frac{d\theta}{dt} + \frac{d\mathbf{u}_\theta}{dt}r\frac{d\theta}{dt}\right).\end{aligned}$$

When Eqs. (4) are used to evaluate the derivatives of $\mathbf{u}_r$ and $\mathbf{u}_\theta$ and the components are separated, the result becomes

$$\begin{aligned}\mathbf{a} = \mathbf{u}_r&\left[\frac{d^2r}{dt^2} - r\left(\frac{d\theta}{dt}\right)^2\right]\\ &+ \mathbf{u}_\theta\left[r\frac{d^2\theta}{dt^2} + 2\frac{dr}{dt}\frac{d\theta}{dt}\right].\end{aligned} \tag{6}$$

Equations (5) and (6) apply to motion in the xy-plane. It is easy to modify them to apply to motion in 3-space. First, we need to add a term

$\mathbf{k}z$ to the right-hand side of Eq. (3),

$\mathbf{k}\dfrac{dz}{dt}$ to the velocity vector,

$\mathbf{k}\dfrac{d^2z}{dt^2}$ to the acceleration vector.

Thus we get

$$\mathbf{R} = r\mathbf{u}_r + \mathbf{k}z, \tag{7a}$$

$$\mathbf{v} = \mathbf{u}_r\frac{dr}{dt} + \mathbf{u}_\theta r\frac{d\theta}{dt} + \mathbf{k}\frac{dz}{dt}, \tag{7b}$$

$$\begin{aligned}\mathbf{a} = \mathbf{u}_r&\left[\frac{d^2r}{dt^2} - r\left(\frac{d\theta}{dt}\right)^2\right]\\ &+ \mathbf{u}_\theta\left[r\frac{d^2\theta}{dt^2} + 2\frac{dr}{dt}\frac{d\theta}{dt}\right] + \mathbf{k}\frac{d^2z}{dt^2}.\end{aligned} \tag{7c}$$

Equations (7) are particularly useful in connection with cylindrical coordinates. The three vectors $\mathbf{u}_r$, $\mathbf{u}_\theta$, and $\mathbf{k}$ are mutually orthogonal unit vectors that form a right-handed system:

$$\begin{aligned}\mathbf{u}_r \times \mathbf{u}_\theta &= \mathbf{k},\\ \mathbf{k} \times \mathbf{u}_r &= \mathbf{u}_\theta,\\ \mathbf{u}_\theta \times \mathbf{k} &= \mathbf{u}_r.\end{aligned} \tag{8}$$

Kepler's laws

If we assume that the motion is in accord with Newton's second law, $\mathbf{F} = m\mathbf{a}$, and that the only force acting on the particle is a gravitational attraction directed toward a fixed point O, of magnitude inversely proportional to the square of the distance of the particle from O, we can deduce Kepler's laws by integration of differential equations of motion. We get those equations from the physical assumptions

$$\mathbf{F} = m\mathbf{a} = m\frac{d^2\mathbf{R}}{dt^2} = -\frac{GmM}{|\mathbf{R}|^2}\frac{\mathbf{R}}{|\mathbf{R}|}, \tag{9}$$

where G is the gravitational constant, m the mass of the moving particle (or planet), and M the mass of the attracting object (or sun) at O. Equation (9) tells us that the acceleration vector $d^2\mathbf{R}/dt^2$ is proportional to the position vector $\mathbf{R}$. This is typical

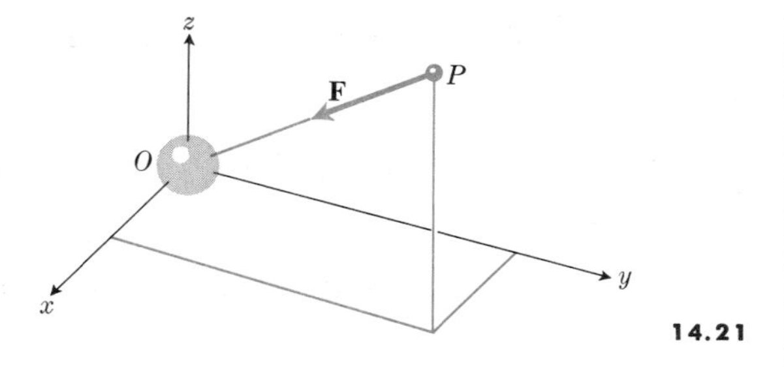

14.21

of a *central force field* (see Fig. 14.21), and it leads to the result

$$\mathbf{R} \times \frac{d^2\mathbf{R}}{dt^2} = \mathbf{0}, \tag{10}$$

because $\mathbf{A} \times \mathbf{B}$ is always equal to the zero vector if $\mathbf{B}$ is a scalar times $\mathbf{A}$. Now we can deduce that the left-hand side of Eq. (10) is the derivative of the cross product of $\mathbf{R}$ and $d\mathbf{R}/dt$. First we have

$$\frac{d}{dt}\left(\mathbf{R} \times \frac{d\mathbf{R}}{dt}\right) = \frac{d\mathbf{R}}{dt} \times \frac{d\mathbf{R}}{dt} + \mathbf{R} \times \frac{d^2\mathbf{R}}{dt^2}.$$

However, the right-hand side of this equation is equal to

$$\mathbf{0} + \mathbf{R} \times \frac{d^2\mathbf{R}}{dt^2},$$

because the cross product of any vector with itself is the zero vector. When the derivative of a vector is zero, each of its components (relative to a fixed coordinate system) is a constant, so the vector is a constant vector. Thus, Eq. (10) implies that

$$\mathbf{R} \times \frac{d\mathbf{R}}{dt} = \mathbf{C}. \tag{11}$$

From the geometric interpretation of the cross product, we conclude from Eq. (11) that both the position vector $\mathbf{R}$ and velocity vector $\mathbf{v} = d\mathbf{R}/dt$ lie in a plane perpendicular to the fixed vector $\mathbf{C}$. We choose a reference frame so that this plane, which includes the path of the particle, is the same as the xy-plane, and so that the unit vector in the direction of $\mathbf{C}$ is $\mathbf{k}$. We introduce polar coordinates in this plane, choosing as initial line $\theta = 0$, the direction of $\mathbf{R}$ when $|\mathbf{R}|$ is a *minimum*. In planetary motion, this corresponds to perihelion position of the planet (the position at which it is nearest to the sun). If we also measure the time t from the instant of passage through perihelion, we get the following initial values: when $t = 0$,

$$\begin{aligned} r &= r_0 \\ &= \text{minimum value of } r \Rightarrow \left(\frac{dr}{dt}\right)_0 = 0, \\ \theta &= 0, \\ |\mathbf{v}| &= v_0 = \left(r\frac{d\theta}{dt}\right)_0. \end{aligned} \tag{12}$$

We can now use Eqs. (3), (5), and (6) for $\mathbf{R}$, $\mathbf{v}$, and $\mathbf{a}$. Because $\mathbf{R} \times \mathbf{v}$ is constant (Eq. 11), we can evaluate that constant by forming the cross product at time $t = 0$:

$$\mathbf{C} = (\mathbf{R}_0) \times (\mathbf{v}_0) = \mathbf{k}\left(r^2\frac{d\theta}{dt}\right)_0 = \mathbf{k}(r_0v_0). \tag{13}$$

To get Eq. (13), cross the right-hand side of Eq. (3) and the right-hand side of Eq. (5), and recall that

$$\mathbf{u}_r \times \mathbf{u}_r = \mathbf{0} \quad \text{and} \quad \mathbf{u}_r \times \mathbf{u}_\theta = \mathbf{k}.$$

If we substitute this result for $\mathbf{C}$ in Eq. (11), we get

$$\mathbf{R} \times \frac{d\mathbf{R}}{dt} = \mathbf{k}(r_0v_0),$$

or

$$\mathbf{k}\left(r^2\frac{d\theta}{dt}\right) = \mathbf{k}(r_0v_0). \tag{14}$$

From Eq. (14) we get

$$r^2\frac{d\theta}{dt} = r_0v_0. \tag{15}$$

Kepler's second law of planetary motion follows from this, because the element of area in polar coordinates is

$$dA = \tfrac{1}{2}r^2\,d\theta$$

and Eq. (15) implies that

$$\frac{dA}{dt} = \tfrac{1}{2}r_0v_0 = \text{constant.}$$

This is the result: *The radius vector sweeps over area at a constant rate in a central force field.*

Our next goal is to deduce *Kepler's first law*, that the orbit is a conic section. To do this, we shall have to use the "inverse square" law, which we have not used so far. Also, to do this, we shall need to express r in terms of θ. This requires a rather lengthy sequence of calculations, and some substitutions that people have found to be very useful, if not altogether obvious. By equating the $\mathbf{u}_r$-components of acceleration from Eqs. (6) and (9), we get

$$\frac{d^2r}{dt^2} - r\left(\frac{d\theta}{dt}\right)^2 = -\frac{GM}{r^2}. \qquad (16)$$

We should like to integrate this equation with respect to t, but we are in no position to do so. We eliminate $d\theta/dt$ by substituting for it r_0v_0/r^2, from Eq. (15):

$$\frac{d^2r}{dt^2} - \frac{r_0^2v_0^2}{r^3} = -\frac{GM}{r^2}. \qquad (17)$$

It is possible to transform Eq. (17) into an integrable form by making the substitution of a new variable, say p, for dr/dt, and then using the chain rule to express dp/dt in terms of p and r:

$$\frac{dr}{dt} = p, \quad \frac{d^2r}{dt^2} = \frac{dp}{dt} = \frac{dp}{dr}\frac{dr}{dt} = p\frac{dp}{dr}.$$

Making this substitution for d^2r/dt^2 in Eq. (17) and separating the variables p and r, we get

$$p\,dp = \left(\frac{r_0^2v_0^2}{r^3} - \frac{GM}{r^2}\right)dr. \qquad (18)$$

Integrating both sides of Eq. (18) and multiplying by two, we get

$$p^2 = \left(\frac{dr}{dt}\right)^2 = -\frac{r_0^2v_0^2}{r^2} + \frac{2GM}{r} + c_1, \qquad (19)$$

where c_1 is a constant. We use the initial conditions

$$r = r_0 \quad \text{and} \quad \frac{dr}{dt} = 0 \quad \text{when } t = 0$$

to evaluate c_1:

$$c_1 = v_0^2 - \frac{2GM}{r_0}.$$

If this value of c_1 is substituted in Eq. (19), the result is

$$\left(\frac{dr}{dt}\right)^2 = v_0^2\left(1 - \frac{r_0^2}{r^2}\right) + 2GM\left(\frac{1}{r} - \frac{1}{r_0}\right). \qquad (20)$$

If we square both sides of Eq. (15) and divide the results into the corresponding sides of Eq. (20), we get

$$\frac{1}{r^4}\left(\frac{dr}{d\theta}\right)^2 = \frac{1}{r_0^2} - \frac{1}{r^2} + \frac{2GM}{r_0^2v_0^2}\left(\frac{1}{r} - \frac{1}{r_0}\right). \qquad (21)$$

For simplicity, let us take

$$h = \frac{GM}{r_0^2v_0^2}, \qquad (22a)$$

and rewrite Eq. (21) as

$$\frac{1}{r^4}\left(\frac{dr}{d\theta}\right)^2 = \frac{1}{r_0^2} - \frac{1}{r^2} + 2h\left(\frac{1}{r} - \frac{1}{r_0}\right).$$

This equation becomes more tractable if we substitute the reciprocal of r as a new variable, say u:

$$u = \frac{1}{r},$$

$$\frac{du}{d\theta} = -\frac{1}{r^2}\frac{dr}{d\theta}, \qquad \left(\frac{du}{d\theta}\right)^2 = \frac{1}{r^4}\left(\frac{dr}{d\theta}\right)^2.$$

Since a value $u = u_0$ would correspond to the initial value r_0 for r, we also substitute u_0 for $1/r_0$. These substitutions in Eq. (21) give

$$\begin{aligned}\left(\frac{du}{d\theta}\right)^2 &= u_0^2 - u^2 + 2hu - 2hu_0\\ &= (u_0 - h)^2 - (u - h)^2. \qquad (22b)\end{aligned}$$

When we take square roots in Eq. (22b) to solve for $du/d\theta$, we get

$$\frac{du}{d\theta} = \pm\sqrt{(u_0 - h)^2 - (u - h)^2}. \qquad (23)$$

The geometry is such that θ increases with time because

$$\frac{d\theta}{dt} = \frac{r_0v_0}{r^2} \quad \text{is positive.}$$

Also, because r starts from a *minimum* value at $t = 0$, it should not decrease, and so we find $dr/dt \geq 0$, at least for early positive values of t. Therefore

$$\frac{dr}{d\theta} \geq 0 \quad \text{and} \quad \frac{du}{d\theta} \leq 0,$$

and so we select the negative sign in Eq. (23) and get, after separation of variables,

$$\frac{-du}{\sqrt{(u_0 - h)^2 - (u - h)^2}} = d\theta. \tag{24}$$

The integral of Eq. (24) is

$$\cos^{-1}\left(\frac{u - h}{u_0 - h}\right) = \theta + c_2,$$

and the constant of integration c_2 is zero because

$$u = u_0 \quad \text{when} \quad \theta = 0,$$

and

$$\cos^{-1}(1) = 0 + c_2 = 0.$$

Therefore

$$\frac{u - h}{u_0 - h} = \cos\theta,$$

or

$$u = 1/r = h + (u_0 - h)\cos\theta.$$

A few more algebraic maneuvers produce the result

$$r = \frac{(1/h)}{1 + e\cos\theta}, \tag{25a}$$

with

$$\frac{1}{h} = \frac{r_0^2 v_0^2}{GM} \quad \text{and} \quad e = \frac{1}{r_0 h} - 1 = \frac{r_0 v_0^2}{GM} - 1. \tag{25b}$$

Equations (25) represent a conic section of eccentricity e with one focus at the origin. As a check on the algebra, we substitute $\theta = 0$ and $r = r_0$ and see if the results agree:

$$r_0 \stackrel{?}{=} \frac{(1/h)}{1 + e} = \frac{(r_0^2 v_0^2)/(GM)}{(r_0 v_0^2)/(GM)} = r_0 \qquad \text{(Yes)}.$$

Remark 1. For given values of r_0, M, and G, Eq. (25b) shows that the eccentricity depends on the velocity at perihelion. For a circular orbit, $e = 0$ and $v_0 = \sqrt{GM/r_0}$. For $0 < e < 1$, the orbit is an ellipse with one focus at O; for $e = 1$, the orbit is a parabola; and for $e > 1$, the orbit is a hyperbola.

Remark 2. Our integration of Eqs. (23) and (24) was based on the assumption that $dr/d\theta$ is nonnegative, which only holds for $0 \leq \theta \leq \pi$ in Eq. (25a). When $\theta = \pi$, Eq. (25a) gives the maximum value of r:

$$r_{\max} = \frac{(1/h)}{1 - e} = \frac{r_0}{2r_0 h - 1}, \tag{26a}$$

for an elliptical orbit with $e < 1$. The length of the major axis of the ellipse is

$$2a = r_0 + r_{\max} = \frac{2r_0^2 h}{2r_0 h - 1}, \tag{26b}$$

and the length of the minor axis is

$$2b = 2a\sqrt{1 - e^2}. \tag{26c}$$

The *period* [assuming that Eq. (25a) represents the motion for $\theta > \pi$ as well as for $0 < \theta < \pi$] can be found by using the rate at which the radius vector sweeps over the area:

$$\frac{dA}{dt} = \tfrac{1}{2} r_0 v_0. \tag{27}$$

If T is the period for one orbit, then the radius vector sweeps out the area of the ellipse, which is πab, in T units of time. Therefore

$$T = \frac{2\pi ab}{r_0 v_0} = \frac{2\pi a^2\sqrt{1 - e^2}}{r_0 v_0}, \tag{28}$$

and it is possible to deduce *Kepler's third law:*

$$\frac{T^2}{a^3} = \text{constant}\left(= \frac{4\pi^2}{GM}\right), \tag{29}$$

from Eq. (28) and the identities

$$r_0 h = \frac{a}{r_0}(2r_0 h - 1) \tag{30}$$

and

$$r_0^2 v_0^2 h = GM. \tag{31}$$

The reader is asked to verify some of these results in Exercise 13.

Table 1. Data on orbits of major planets

Planet	Mean distance a from sun†	Eccen-tricity e	Period T
Mercury	36.0	0.2056234	87.967 days
Venus	67.3	0.0067992	224.701 days
Earth	93.0	0.0167322	365.256 days
Mars	141.7	0.0933543	1.881 years
Jupiter	483.9	0.0484108	11.862 years
Saturn	887.1	0.0557337	29.458 years
Uranus	1785.0	0.0471703	84.015 years
Neptune	2797.0	0.0085646	164.788 years
Pluto	3670.0	0.2485200	247.697 years

† Millions of miles

Table 2. Data on some earth satellites

Name	Max. ht. (miles)	Min. ht. (miles)	Weight (lbs)	Orbit time (min.)
Sputnik I	560	145	184.00	96.2
Sputnik II	1056	150	1120.00	103.7
Explorer I	1587	219	30.80	114.5
Vanguard	2466	405	3.25	134.0
Explorer III	1741	117	31.00	115.7
Sputnik III	1168	150	2920.00	106.0
Explorer IV	1386	178	38.43	110.0

Table 3. Additional data

Gravitational constant

$$G = (6.673 \pm 0.003) \times 10^{-8} \frac{\text{cm}^3}{\text{gm} \cdot \text{sec}^2}$$

Sun's mass

$$2 \times 10^{33} \text{ gm}$$

Earth's mass

$$(5.976 \pm 0.005) \times 10^{27} \text{ gm}$$

Equatorial radius of earth

$$6378.388 \text{ km} = 3963.34 \text{ mi}$$

Polar radius of earth

$$6356.912 \text{ km} = 3949.99 \text{ mi}$$

$4\pi^2 = 39.4784\,176$ $\qquad$ $\log 4\pi^2 = 1.5963\,597$

1 year = 365.256 days $\qquad$ 1 mile = 1.609×10^5 cm

Table 1 contains some data about the orbits of the major planets in the solar system, Table 2 gives some data on some of the early artificial earth satellites,* and Table 3 contains some additional data related to the equations of motion of planets about the sun, and to artificial Earth satellites.

Here we might mention some statistics on the Syncom 3 satellite. The perigee of its initial transfer orbit was slightly less than 700 miles, and the apogee was about 23,675 miles. One day later its rocket engine was fired by remote control to boost the satellite into a nearly circular orbit over the equator. roughly 22,300 miles high. The goal had been a perfectly circular, *synchronous* orbit in which the satellite would appear stationary. As it was, the period of the orbit was 1436.158 min/day, which compares closely with the earth's period of 1436.068 min/day.

Circular orbits

If the eccentricity of an orbit is zero, the theory simplifies considerably, for then the orbit is a circle $r = (1/h) =$ constant. For earth satellites, even with elliptical orbits, the major axis of the ellipse is

$$2a = (\text{diameter of earth}) + (\text{perigee ht}) + (\text{apogee ht})$$

and the minor axis, $2b$, is $\geq$ (diameter of earth). (See Fig. 14.22.) So, for orbits near the earth's surface, one can use the approximation of a circular orbit of radius a,

$$a = (\text{earth's radius}) + \tfrac{1}{2}(\text{perigee ht} + \text{apogee ht}), \tag{32a}$$

or

$$a \approx 4000 \text{ mi} = 6440 \times 10^5 \text{ cm}. \tag{32b}$$

If we assume that the orbit is circular, and also that the gravitational force is given by the inverse square law, we can quickly rederive Kepler's second (area) and third (period) laws. Thus, we let

$$\mathbf{F} = m\mathbf{a} = \frac{-GMm}{r^2}\mathbf{u}_r, \tag{33}$$

* For more data, see articles on spacecraft in the Britannica's *Yearbooks*, or "Space Age Chronology" by Willy Ley in the 1968 *Information Please Almanac.*

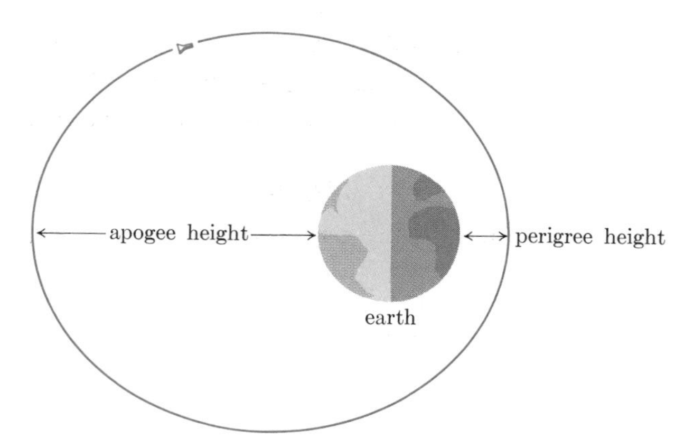

14.22 The orbit of an earth satellite: $2a$ = (diameter of earth) + (perigee height) + (apogee height).

with the following symbolism:

m = mass of orbiting particle,

M = mass of attracting object, considered as a point mass at the origin, and

r = distance from origin to orbiting particle.

We are now assuming that we know that the orbit is a circle, r = constant, so that

$$\frac{dr}{dt} = 0 \quad \text{and} \quad \frac{d^2r}{dt^2} = 0. \tag{34}$$

Because the force field satisfies Eqs. (33), (34), and (6), we have $\mathbf{F} \cdot \mathbf{u}_\theta = 0$, and hence

$$r\frac{d^2\theta}{dt^2} = 0,$$

which, together with r = constant $\neq 0$, implies

$$\frac{d\theta}{dt} = \omega = \text{constant}. \tag{35}$$

This means that the angular velocity is constant, which in turn implies that

$$\frac{dA}{dt} = \frac{r^2}{2}\frac{d\theta}{dt}$$

is also constant. This is Kepler's second law.

To deduce Kepler's third law for circular orbits, we next equate the radial components in $\mathbf{F} = m\mathbf{a}$, and have

$$\frac{-GMm}{r^2} = -mr\left(\frac{d\theta}{dt}\right)^2,$$

so that

$$\omega^2 r^3 = GM. \tag{36}$$

Because the velocity in orbit of the particle is

$$v = \omega r, \tag{37}$$

we can calculate the period T from the equation

$$T = \frac{2\pi r}{\omega r} = \frac{2\pi}{\omega}. \tag{38}$$

Therefore, if we square and use Eq. (36), we get

$$T^2 = \frac{4\pi^2}{\omega^2} = \frac{4\pi^2 r^3}{GM},$$

or

$$\frac{T^2}{r^3} = \frac{4\pi^2}{GM}, \tag{39}$$

which is Kepler's third law.

We also know that near the earth's surface,

$$\mathbf{a} = -g\mathbf{u}_r, \tag{40}$$

where g is the acceleration due to gravity (approximately 32.2 ft/sec^2, or 981 cm/sec^2). From the inverse square law, Eq. (33), we thus get

$$g = GM/r^2, \tag{41}$$

where M is the mass of the earth, r is the radius of the earth (treated as a sphere), and G is the gravitational constant.

Problem 1. Use Eq. (41) to find the mass of the earth, assuming that g, G, and r are known:

$$g = 981 \text{ cm/sec}^2,$$
$$r = 6440 \times 10^5 \text{ cm},$$
$$G = 6.670 \times 10^{-8} \text{ cm}^3/(\text{gm} \cdot \text{sec}^2).$$

Solution. From Eq. (41),

$$M = \frac{r^2 g}{G} = \frac{6.440^2 \times 9.81 \times 10^{18}}{6.670 \times 10^{-8}}$$
$$= 61.0 \times 10^{26} \text{ gm}.$$

This compares with the value 59.8×10^{26} gm given in Table 3. (One hardly knows whether this is good or bad agreement unless one has further experience in working with data of this nature. An obvious source of discrepancy is treating the earth as a sphere. The reader will undoubtedly think of others.) If we keep only one significant figure, we have agreement that $M = 6 \times 10^{27}$ gm, and this is approximately 6.6×10^{21} tons. (Approximately, 1000 gm = 1 kg = 2.205 lbs.)

Problem 2. Find the period for an earth satellite whose orbit is a circle of radius 4000 miles.

Solution. From Eqs. (39) and (41), we deduce that

$$T^2 = \frac{4\pi^2 r^3}{GM} = \frac{4\pi^2 r}{g}, \tag{42}$$

or

$$T = 2\pi\sqrt{\frac{r}{g}} = 2\pi\sqrt{\frac{4000 \times 5280}{32}} = 200\pi\sqrt{66} \text{ sec},$$
$$= 200\pi(8.124)/60 \text{ min} = 85.1 \text{ min}.$$

Remark 3. The 85-minute figure is less than any of the orbit times given in Table 2. One reason may be that data in Table 2 refer to the length of time it takes for a satellite to return to the same location over the earth. Meanwhile, of course, the earth has rotated approximately (85/1440) 360°. If this is interpreted to mean that the satellite travels this additional fraction of an orbit, then we should add about five minutes to the foregoing 85-minute figure to compensate for this effect. Also, observe that of the satellites listed in Table 2, Vanguard had the greatest orbit time and also the greatest maximum and minimum heights. Also, if one replaces the figure 4000 by 4410, for example, the period is multiplied by $\sqrt{441/400} = 21/20$, which adds five percent. Of course, for orbits that are not "near" the surface of the earth, one should use Eq. (39) with appropriate values of r and M [or Eq. (29), which is like Eq. (39) with a in place of r].

Problem 3. The purpose of the Syncom satellite program was to establish a circular orbit with period equal to the length of time for one rotation of the earth on its axis. This figure is given as 1436.068 min ≈ 86,164 sec. Find the radius of the appropriate orbit.

Solution. We use Eq. (39) with

$$G = 6.673 \times 10^{-8} \quad \text{and} \quad M = 5.976 \times 10^{27},$$

as given in Table 3, in cm-gm-sec units. Solving Eq. (39) for r, we get

$$r = \sqrt[3]{GMT^2/(4\pi^2)}.$$

We use a table of common logarithms for the calculations:

$$\begin{aligned}
\log G &= 2.82432 - 10\\
\log M &= 27.77641\\
2\log T &= 9.87066\\
\text{Total} &\quad 30.47139;\\
\log GMT^2 &= 30.47139\\
\log 4\pi^2 &= 1.59636\\
\log r^3 &= 28.87503;\\
\log r &= 9.62501,\\
r &= 4.2171 \times 10^9 \text{ cm} = 4.2171 \times 10^4 \text{ km},\\
&= 2.6505 \times 10^4 \text{ mi}.
\end{aligned}$$

If the distance from the center of the earth is 26,500 miles and the radius of the earth is 4000 miles, the height of the orbit above the earth's surface is 22,500 miles. (Whether the one percent discrepancy with the published figure of 22,300 miles is due to a computational error, or to other factors not taken into account, the author does not know. Perhaps one of the readers can supply the needed information. In any event, the whole mass of data that has been accumulating as a result of the satellite program is considered to further demonstrate the validity of Newton's theory of gravitational attraction, including the inverse square law.)

EXERCISES

1. With the book closed, derive vector expressions for velocity and acceleration in terms of components along and at right angles to the radius vector. Check your derivations with those given in the text.

In Exercises 2 through 6, find the velocity and acceleration vectors in terms of the unit vectors $\mathbf{u}_r$ and $\mathbf{u}_\theta$.

2. $r = a(1 - \cos\theta), \quad \dfrac{d\theta}{dt} = 3$

3. $r = a\sin 2\theta, \quad \dfrac{d\theta}{dt} = 2t$

4. $r = e^{a\theta}, \quad \dfrac{d\theta}{dt} = 2$

5. $r = a(1 + \sin t), \quad \theta = 1 - e^{-t}$
6. $r = 2 \cos 4t, \quad \theta = 2t$
7. If a particle moves in an ellipse whose polar equation is $r = c/(1 - e \cos \theta)$ and the force is directed toward the origin, show that the magnitude of the force is proportional to $1/r^2$.
8. Since the orbit of the Vanguard satellite (see Table 2) had major axis $2a = 2466 + 405 + 7913$ mi (approximately), Eq. (29), with M equal to earth's mass, should give the period. Compute, and compare with the figure given in Table 2.
9. Read the article *Orbit* in the Encyclopaedia Britannica. What dates does it give for Kepler's announcements of his first two laws? Of his third law?
10. In May 1965, the U.S.S.R. launched Proton I, weighing 26,900 lbs (at launch), with a perigee of 118 miles, an apogee of 390 miles, and a period of 92 minutes. Using the period $T = 5520$ sec and the relevant data for the mass of the earth and the gravitational constant G, compute the semimajor axis a of the orbit from Eq. (29), and compare $2a$ with the diameter of the earth plus the perigee and apogee heights.
11. Read the article *Spacecraft* in the most recent Yearbook of the Encyclopaedia Britannica (or other source). Report the perigee, apogee, and orbital period for at least one earth satellite as given in the article you read. (There are several in the 1966 Yearbook.)
12. In the Encyclopaedia Britannica (or elsewhere), read an article on the scientific work of Kepler. In what way was Kepler's work dependent on earlier work of Tycho Brahe? On contemporary work of Galileo?
13. From Eq. (25b), deduce that

$$1 - e^2 = (1 - e)(1 + e) = (2r_0h - 1)/(r_0h)^2,$$

and use the result to show that Eq. (29) follows from Eqs. (25b), (28), (30), and (31).
14. Without introducing coordinates, give a geometric argument for the validity of the equation

$$\frac{dA}{dt} = \frac{1}{2}\left|\mathbf{R} \times \frac{d\mathbf{R}}{dt}\right|,$$

where $\mathbf{R}$ is the position vector of a particle moving in a plane curve, and dA/dt is the rate at which that vector sweeps out area.
15. For what values of v_0 is the orbit of Eq. (25a) a parabola? A circle? An ellipse? A hyperbola?
16. Assuming that the earth's distance from the sun at perihelion is approximately 93,000,000 miles, and that the eccentricity of the earth's orbit about the sun is 0.0167, compute the velocity of the earth in its orbit at perihelion. [Use Eq. (25b).]

REVIEW QUESTIONS AND EXERCISES

1. Define the derivative of a vector function.
2. Develop formulas for the derivatives, with respect to θ, of the unit vectors $\mathbf{u}_r$ and $\mathbf{u}_\theta$.
3. Develop vector formulas for velocity and acceleration of a particle moving in a plane curve:
   (a) in terms of Cartesian coordinates,
   (b) in terms of polar coordinates,
   (c) in terms of distance traveled along the curve and unit vectors tangent and normal to the curve.
4. (a) Define curvature of a plane curve.
   (b) Define radius of curvature.
   (c) Define center of curvature.
   (d) Define osculating circle.
5. Develop a formula for the curvature of a curve whose parametric equations are $x = f(t)$, $y = g(t)$.
6. In what way does the curvature of a curve affect the acceleration of a particle moving along the curve? In particular, discuss the case of constant-speed motion along a curve.
7. State and derive Kepler's second law concerning motion in a central force field.
8. If a vector $\mathbf{V}$ is a differentiable function of t and $|\mathbf{V}| =$ constant, what do you know about $d\mathbf{V}/dt$?
9. Define arc length and curvature of a space curve.
10. For a space curve, explain how to find the unit tangent vector, unit principal normal, and unit binormal.
11. Express the vector $\mathbf{R} = \overrightarrow{OP}$ in terms of cylindrical coordinates and the unit vectors $\mathbf{u}_r$, $\mathbf{u}_\theta$, and $\mathbf{k}$.
12. Derive formulas for the velocity $\mathbf{v} = d\mathbf{R}/dt$ and acceleration $\mathbf{a} = d\mathbf{v}/dt$ in terms of cylindrical coordinates and the unit vectors $\mathbf{u}_r$, $\mathbf{u}_\theta$, and $\mathbf{k}$.

13. The volume of an ellipsoid of semiaxes a, b, c is $\frac{4}{3}\pi abc$. Apply this formula with

$$a = b = \text{equatorial radius of earth}$$

and

$$c = \text{polar radius of earth}$$

to find (approximately) the volume of the earth. Use this volume and the figure 5.522 gm/cm^3 for mean density to find the mass of the earth. Compare your answer with data given in the text.

MISCELLANEOUS EXERCISES

1. A particle moves in the xy-plane according to the time law

$$x = 1/\sqrt{1+t^2}, \quad y = t/\sqrt{1+t^2}.$$

(a) Compute the velocity vector and acceleration vector when $t = 1$.

(b) At what time is the speed of the particle a maximum?

2. A circular wheel with unit radius rolls along the x-axis uniformly, rotating one half-turn per second. The position of a point P on the circumference is given by the formula

$$\overrightarrow{OP} = \mathbf{R} = \mathbf{i}(\pi t - \sin \pi t) + \mathbf{j}(1 - \cos \pi t).$$

(a) Determine the velocity (*vector*) $\mathbf{v}$ and the acceleration (*vector*) $\mathbf{a}$ at time t.

(b) Determine the slopes (as functions of t) of the two straight lines PC and PQ joining P to the center C of the wheel and to the point Q that is topmost at the instant.

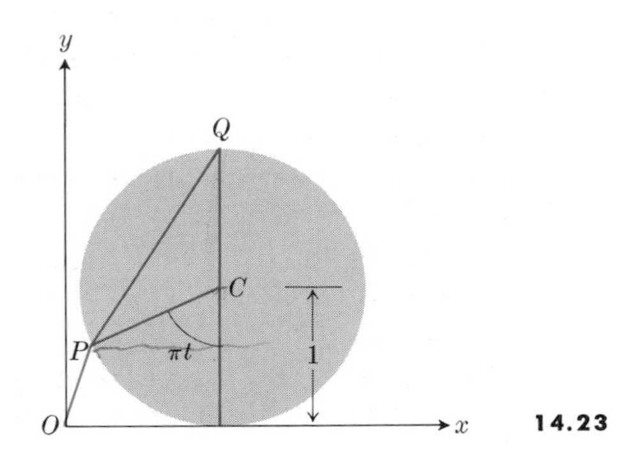

14.23

(c) Show that the directions of the vectors $\mathbf{v}$ and $\mathbf{a}$ can be expressed in terms of the straight lines described in (b).

3. The motion of a particle in the xy-plane is given by

$$\mathbf{R} = \mathbf{i}at \cos t + \mathbf{j}at \sin t.$$

Find the speed, and the tangential and normal components of the acceleration.

4. A particle moves in such a manner that the derivative of the position vector is always perpendicular to the position vector. Show that the particle moves on a circle with center at the origin.

5. The position of a point at time t is given by the formulas $x = e^t \cos t$, $y = e^t \sin t$.

(a) Show that $\mathbf{a} = 2\mathbf{v} - 2\mathbf{r}$.

(b) Show that the angle between the radius vector $\mathbf{r}$ and the acceleration vector $\mathbf{a}$ is constant, and find this angle.

6. Given the instantaneous velocity $\mathbf{v} = a\mathbf{i} + b\mathbf{j}$ and acceleration $\mathbf{a} = c\mathbf{i} + d\mathbf{j}$ of a particle at a point P on its path of motion, determine the curvature of the path at P.

7. Find the parametric equations, in terms of the parameter θ, of the locus of the center of curvature of the cycloid

$$x = a(\theta - \sin \theta), \quad y = a(1 - \cos \theta).$$

8. Find the point on the curve $y = e^x$ for which the radius of curvature is a minimum.

9. (a) Given a closed curve having the property that every line parallel to the x-axis or the y-axis has at most two points in common with the curve. Let

$$x = x(t), \quad y = y(t), \quad \alpha \le t \le \beta,$$

be equations of the curve. Prove that if dx/dt and dy/dt are continuous, then the area bounded by the curve is

$$\frac{1}{2}\left|\int_\alpha^\beta \left[x(t)\frac{dy}{dt} - y(t)\frac{dx}{dt}\right] dt\right|.$$

(b) Use the result of (a) to find the area inside the ellipse

$$x = a \cos \phi, \quad y = b \sin \phi, \quad 0 \le \phi \le 2\pi.$$

What does the answer become when $a = b$?

10. For the curve defined by the equations

$$x = \int_0^\theta \cos(\tfrac{1}{2}\pi t^2)\,dt, \quad y = \int_0^\theta \sin(\tfrac{1}{2}\pi t^2)\,dt,$$

calculate the curvature κ as a function of the length of arc s, where s is measured from $(0, 0)$.

11. The curve for which the length of the tangent intercepted between the point of contact and the y-axis is always equal to 1 is called the *tractrix*. Find its equation. Show that the radius of curvature at each point of the curve is inversely proportional to the length of the normal intercepted between the point on the curve and the y-axis. Calculate the length of arc of the tractrix, and find the parametric equations in terms of the length of arc.

12. Let $x = x(t)$, $y = y(t)$ be a closed curve. A constant length p is measured off along the normal to the curve. The extremity of this segment describes a curve which is called a *parallel curve* to the original curve. Find the length of arc, the radius of curvature, and the area enclosed by the parallel curve.

13. Given the curve represented by the parametric equations

$$x = 32t, \quad y = 16t^2 - 4.$$

(a) Calculate the radius of curvature of the curve at the point where $t = 3$.

(b) Find the length of the curve between the points where $t = 0$ and $t = 1$.

14. Find the velocity, acceleration, and speed of a particle whose position at time t is

$$x = 3\sin t, \quad y = 2\cos t.$$

Also find the tangential and normal components of the acceleration.

15. The position of a particle at time t is given by the equations

$$x = 1 + \cos 2t, \quad y = \sin 2t.$$

Find

(a) the normal and tangential components of acceleration at time t;

(b) the radius of curvature of the path;

(c) the equation of the path in polar coordinates, using the x-axis as the line $\theta = 0$ and the y-axis as the line $\theta = \pi/2$.

16. A particle moves so that its position at time t has the polar coordinates $r = t$, $\theta = t$. Find the velocity $\mathbf{v}$, the acceleration $\mathbf{a}$, and the curvature κ at any time t.

17. Find an expression for the curvature of the curve whose equation in polar coordinates is $r = f(\theta)$.

18. (a) Find the equation in *polar coordinates* of the curve

$$x = e^{2t}\cos t,$$
$$y = e^{2t}\sin t.$$

(b) Find the length of this curve from $t = 0$ to $t = 2\pi$.

19. Express the velocity vector in terms of $\mathbf{u}_r$ and $\mathbf{u}_\theta$ for a point moving in the xy-plane according to the law

$$\mathbf{r} = (t + 1)\mathbf{i} + (t - 1)\mathbf{j}.$$

20. The polar coordinates of a particle at time t are

$$r = e^{\omega t} + e^{-\omega t}, \quad \theta = t,$$

where ω is a constant. Find the acceleration vector when $t = 0$.

21. A slender rod, passing through the fixed point O, is rotating about O in a plane at the constant rate of 3 rad/min. An insect is crawling along the rod toward O at the constant rate of 1 in/min. Use polar coordinates in the plane, with point O as the origin, and assume that the insect starts at the point $r = 2$, $\theta = 0$.

(a) Find, in polar form, the vector velocity and vector acceleration of the insect when it is halfway to the origin.

(b) What will be the length of the path, in the plane, that the insect has traveled when it reaches the origin?

22. A smooth ball rolls inside a long hollow tube while the tube rotates with constant angular velocity ω about an axis perpendicular to the axis of the tube. Assuming no friction between the ball and the sides of the tube, show that the distance r from the axis of rotation to the ball satisfies the differential equation

$$d^2r/dt^2 - \omega^2 r = 0.$$

If at time $t = 0$ the ball is at rest (relative to the tube) at $r = a > 0$, find r as a function of t.

23. A particle P slides without friction along a coil spring having the form of a right circular helix. If the positive z-axis is taken downward, the cylindrical coordinates of P at time t are $r = a$, $z = b\theta$, where a and b are positive constants. If the particle starts at $r = a$, $\theta = 0$ with zero velocity and falls under gravity, the law of conservation of energy then tells us that its speed after it has fallen a vertical distance z is $\sqrt{2gz}$.

(a) Find the angular velocity $d\theta/dt$ when $\theta = 2\pi$.

(b) Express θ and z as functions of the time t.

(c) Determine the tangential and normal components of the velocity $d\mathbf{R}/dt$ and acceleration $d^2\mathbf{R}/dt^2$ as functions of t. Is there any component of acceleration in the direction of the binormal $\mathbf{B}$?

24. Suppose the curve in Exercise 23 is replaced by the conical helix

$$r = a\theta, \quad z = b\theta.$$

(a) Express the angular velocity $d\theta/dt$ as a function of θ.

(b) Express the distance that the particle travels along this helix as a function of θ.

25. Hold two of the three spherical coordinates ρ, ϕ, θ of point P in Fig. 12.20 constant while letting the other coordinate increase. Let $\mathbf{u}$, with subscript corresponding to the coordinate which is permitted to vary, denote the unit vector that points in the direction in which P starts to move under these conditions.

(a) Express the three unit vectors $\mathbf{u}_\rho$, $\mathbf{u}_\phi$, $\mathbf{u}_\theta$ which are obtained in this manner, in terms of ρ, ϕ, θ, and the unit vectors $\mathbf{i}$, $\mathbf{j}$, $\mathbf{k}$.

(b) Show that $\mathbf{u}_\rho \cdot \mathbf{u}_\phi = 0$.

(c) Show that $\mathbf{u}_\theta = \mathbf{u}_\rho \times \mathbf{u}_\phi$.

(d) Do the vectors $\mathbf{u}_\rho$, $\mathbf{u}_\phi$, $\mathbf{u}_\theta$ form a system of mutually orthogonal vectors? Is the system, in the order given, a right-handed or a left-handed system?

26. If the spherical coordinates ρ, ϕ, θ of a moving point P are differentiable functions of the time t, and $\mathbf{R} = \overrightarrow{OP}$ is the vector from the origin to P, express $\mathbf{R}$ and $d\mathbf{R}/dt$ in terms of ρ, ϕ, θ and their derivatives and the unit vectors $\mathbf{u}_\rho$, $\mathbf{u}_\phi$, $\mathbf{u}_\theta$ of Exercise 25.

27. Express $ds^2 = dx^2 + dy^2 + dz^2$ in terms of

(a) cylindrical coordinates r, θ, z,

(b) spherical coordinates ρ, ϕ, θ (see Exercise 26).

Interpret your results geometrically in terms of the sides and a diagonal of a rectangular box. Sketch.

28. Using the results of Exercise 27, find the lengths of the following curves between $\theta = 0$ and $\theta = \ln 8$.

(a) $z = r = ae^\theta$

(b) $\phi = \pi/6, \quad \rho = 2e^\theta$

29. Determine parametric equations giving x, y, z in terms of the parameter θ for the curve of intersection of the sphere $\rho = a$ and the plane $y + z = 0$, and find its length.

30. In Article 14.6, we found the velocity vector of a particle moving in a plane to be

$$\mathbf{v} = \frac{d\mathbf{R}}{dt} = \mathbf{i}\frac{dx}{dt} + \mathbf{j}\frac{dy}{dt} = \mathbf{u}_r\frac{dr}{dt} + \mathbf{u}_\theta\frac{r\,d\theta}{dt}.$$

(a) Express dx/dt and dy/dt in terms of dr/dt and $r\,d\theta/dt$ by computing $\mathbf{v} \cdot \mathbf{i}$ and $\mathbf{v} \cdot \mathbf{j}$.

(b) Express dr/dt and $r\,d\theta/dt$ in terms of dx/dt and dy/dt by computing $\mathbf{v} \cdot \mathbf{u}_r$ and $\mathbf{v} \cdot \mathbf{u}_\theta$.

31. The line through OA, A being the point $(1, 1, 1)$, is the axis of rotation of a rigid body that is rotating with a constant angular speed of 6 rad/sec. The rotation appears clockwise when we look towards the origin from A. Find the velocity vector of the point of the body that is at the position $(1, 3, 2)$ (see Problem 4, Article 5.5).

32. If a particle moves in the xy-plane with velocity $\mathbf{v}$ and acceleration $\mathbf{a}$, show that we have

$$|\mathbf{v} \times \mathbf{a}| = |\mathbf{v}|^3\,|\kappa|,$$

where κ is the curvature of the path traced out.

33. A curve is given by the parametric equations

$$x = e^t \sin 2t, \quad y = e^t \cos 2t, \quad z = 2e^t.$$

Let P_0 be the point where $t = 0$. Determine

(a) the direction cosines of the tangent, principal normal, and binormal at P_0;

(b) the curvature at P_0.

34. The *normal plane* to a space curve at any point P of the curve is defined as the plane through P that is perpendicular to the tangent vector. The *osculating plane* at P is the plane containing the tangent and the principal normal. Given the space curve whose vector equation is

$$\mathbf{r}(t) = t\mathbf{i} + t^2\mathbf{j} + t^3\mathbf{k},$$

find

(a) the equation of the normal plane at (1, 1, 1);

(b) the equation of the osculating plane at (1, 1, 1).

35. Given the curve whose vector is

$$\mathbf{r}(t) = (3t - t^3)\mathbf{i} + 3t^2\mathbf{j} + (3t + t^3)\mathbf{k},$$

compute the curvature.

36. Show that the length of the arc described by the endpoint of

$$\mathbf{R} = 3\cos t\mathbf{i} + 3\sin t\mathbf{j} + t^2\mathbf{k},$$

as t varies from 0 to 2, is $5 + \frac{9}{4}\ln 3$.

37. The curve whose vector equation is

$$\mathbf{r}(t) = 2\sqrt{t}\cos t\mathbf{i} + 3\sqrt{t}\sin t\mathbf{j} + \sqrt{1 - t}\,\mathbf{k}, \qquad 0 \leq t \leq 1,$$

lies on a quadric surface. Find the equation of this surface and describe it.

PARTIAL DIFFERENTIATION

CHAPTER 15

15.1 FUNCTIONS OF TWO OR MORE VARIABLES

There are many instances in science, engineering, and everyday life where a quantity is determined by a number of other quantities. For example, the volume of a right circular cone is

$$V = \tfrac{1}{3}\pi r^2 h, \tag{1}$$

where the volume is determined by numerical (constant) quantities, $\frac{1}{3}$ and π, and by variable quantities, radius r and altitude h. In the context of volume of a cone, the *domain* of the function described by Eq. (1) would be the region in E^2 for which both r and h are positive: $r > 0$, $h > 0$. But we notice that the right-hand side of the equation is meaningful for all real values of r and h, and we could use the equation to associate a real number V with every vector, or point, in E^2. In this sense, Eq. (1) produces a mapping from all of E^2 to E^1, and the range is the set of all real numbers, $-\infty < V < \infty$. This mapping is not linear, as are those that we have studied in Chapter 13. The purpose of this chapter is to study functions from E^n to E^1, usually for $n = 2$ or $n = 3$, for convenience.

For $n = 2$, we have the following definition of a *function*. The modification to $n > 2$ is easy.

Definition. *Let D be a set of points (x, y) in E^2. A mapping that assigns a unique number w to each point in D is called a function from D to E^1. The domain of the function is D, and the range is the set of numbers w that are images, under the mapping, of points in D. We indicate that w is the image, under the mapping f, of (x, y), by writing*

$$w = f(x, y). \tag{2}$$

We also say that w is the value of the function f at (x, y).

The notation

$$w = f(x, y, z) \tag{3}$$

similarly means that when values are assigned to the three variables x, y, and z, the value of w is uniquely determined. For example, if

$$\rho^2 = x^2 + y^2 + z^2, \tag{4}$$

as in the relationship between $\rho = |\overrightarrow{OP}|$ and the Cartesian coordinates of $P(x, y, z)$, the further condition

$$\rho \geq 0 \tag{5}$$

singles out a unique value of ρ when values are assigned to x, y, and z. This unique value is

$$\rho = +\sqrt{x^2 + y^2 + z^2}.$$

Continuity

A function $w = f(x, y)$ is said to be *continuous* at (x_0, y_0) if and only if

$$w \to w_0 = f(x_0, y_0) \quad \text{as} \quad (x, y) \to (x_0, y_0),$$

or, in other words, if and only if $|w - w_0|$ can be made arbitrarily small by making both $|x - x_0|$ and $|y - y_0|$ small. For instance, $w = xy$ is continuous at any (x_0, y_0), since

$$\begin{aligned} |w - w_0| &= |xy - x_0y_0| \\ &= |xy - xy_0 + xy_0 - x_0y_0| \\ &= |x(y - y_0) + y_0(x - x_0)| \end{aligned}$$

can be made arbitrarily small by making both $|x - x_0|$ and $|y - y_0|$ small.

On the other hand, the function defined by

$$w = \begin{cases} \dfrac{xy}{x^2 + y^2} & \text{when } (x, y) \neq (0, 0), \\ 0 & \text{when } (x, y) = (0, 0) \end{cases}$$

is not continuous at $(0, 0)$. We easily see this if we take

$$x = r\cos\theta, \quad y = r\sin\theta, \quad r \neq 0;$$

then we have

$$w = \sin\theta\cos\theta = \tfrac{1}{2}\sin 2\theta,$$

so that w takes all values between $-\frac{1}{2}$ and $+\frac{1}{2}$ as the point (x, y) or (r, θ), moves around the origin, no matter how small r may be. That is, we cannot make the w values stay close to zero by simply keeping the (x, y) values close to $(0, 0)$.

In most cases, the so-called elementary functions, which include algebraic expressions as well as trigonometric functions, logarithms, and exponentials, are continuous, except possibly where a denominator vanishes or the logarithm of zero is indicated.

EXERCISES

1. How close to the point $(0, 0)$ should one take the point (x, y) to make $|f(x, y) - f(0, 0)| < \epsilon$ if
   (a) $f(x, y) = x^2 + y^2$ and $\epsilon = 0.01$,
   (b) $f(x, y) = \dfrac{y}{x^2 + 1}$ and $\epsilon = 0.001$?

2. (a) How close to the point $(0, 0, 0)$ should one take the point (x, y, z) to make $|f(x, y, z) - f(0, 0, 0)| < \epsilon$
   if $f(x, y, z) = x^2 + y^2 + z^2$ and $\epsilon = 0.01$?
   If $f(x, y, z) = xyz$ and $\epsilon = 0.008$?
   (b) Is the function $f(x, y, z) = x^2 + y^2 + z^2$ continuous at $(0, 0, 0)$? Give reasons for your answer.

3. Let $f(x, y) = (x + y)/(x^2 + y)$ when $x^2 + y \neq 0$.
   (a) Is it possible to define $f(1, -1)$ in such a way that $f(x, y) \to f(1, -1)$ as $(x, y) \to (1, -1)$ along the line $x = 1$? Along the line $y = -1$?
   (b) Is it possible to define $f(1, -1)$ in such a way that f is continuous at $(1, -1)$? Give a reason for your answer.

15.2 THE DIRECTIONAL DERIVATIVE: SPECIAL CASES

For simplicity, consider a function of the two independent variables x and y, and denote the dependent variable by w. The equation

$$w = f(x, y) \tag{1}$$

may be interpreted as representing a surface in xyw-space. In particular, we might imagine that it represents the elevation of points on a hill above the plane $w = 0$ (see Fig. 15.1a).

As a second interpretation of the equation $w = f(x, y)$, we imagine a base region D in the xy-plane, where D is the *domain* of the function f. To each point in D, we may imagine that a marker is attached which bears the w-value associated with that point: $w = f(x, y)$. We can then bring some order into the system by constructing a *contour map*, consisting of a set of *contour curves* in D. Each contour curve consists of points (x, y) in D with equal w-values. We get the equations of contour

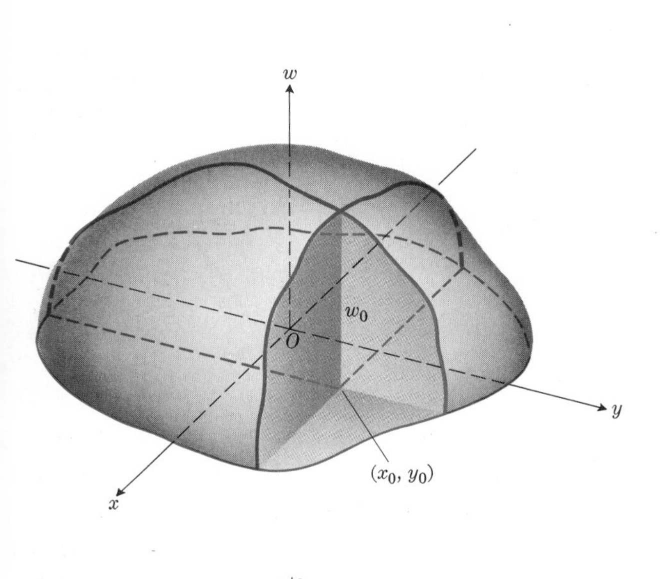

a | b

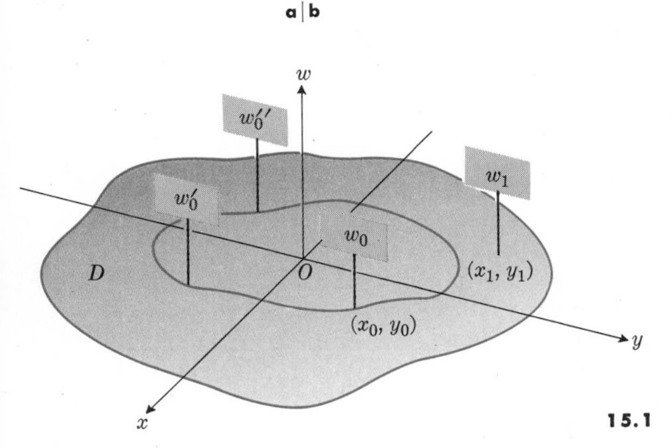

15.1

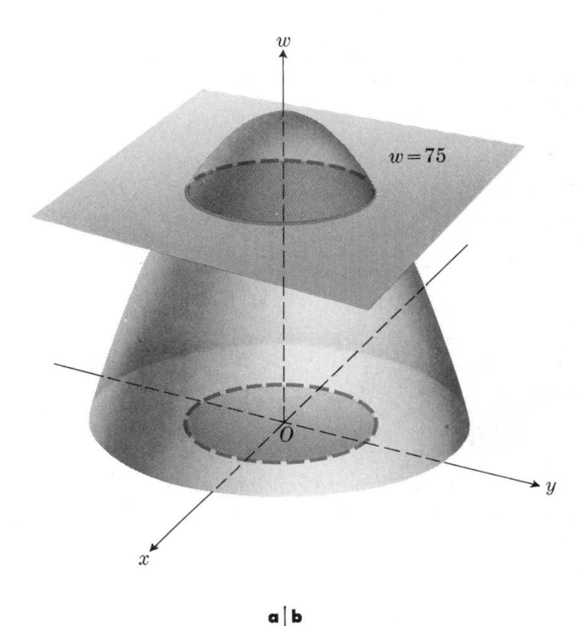

a | b

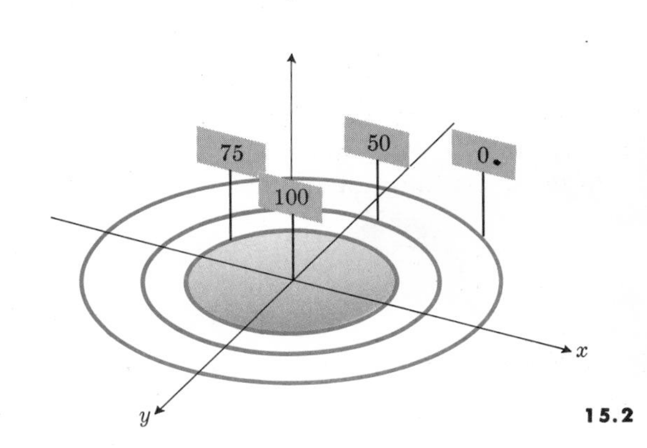

15.2

curves by setting $f(x, y) = w_0$, where w_0 is any number in the range of f. Figure 15.1(b) illustrates a contour curve related to the surface shown in Fig. 15.1(a), with $w_0 = w_0' = w_0''$.

Example. Figure 15.2 illustrates these two interpretations for

$$\begin{aligned} w &= 100 - x^2 - y^2, \\ D &= \{(x, y)\colon x^2 + y^2 \leq 100\}. \end{aligned} \tag{2}$$

Figure 15.2(a) shows the surface, which is a paraboloid of revolution, and indicates a cutting plane $w = 75$. The corresponding level curve in the xy-plane is the circle $x^2 + y^2 = 25$. This is the circle which, in Fig. 15.2(b), is marked $w = 75$. The second interpretation is particularly useful in certain engineering applications. For instance, Eq. (2) might represent the temperature w, in degrees centigrade, at each point (x, y) of a flat circular plate at some fixed instant of time. And if we were dealing with temperatures w inside a sphere, we might easily extend the second point of view and imagine a marker with the appropriate w-value attached to each point inside the sphere. The points having the same w-value would constitute an isothermal surface in space.

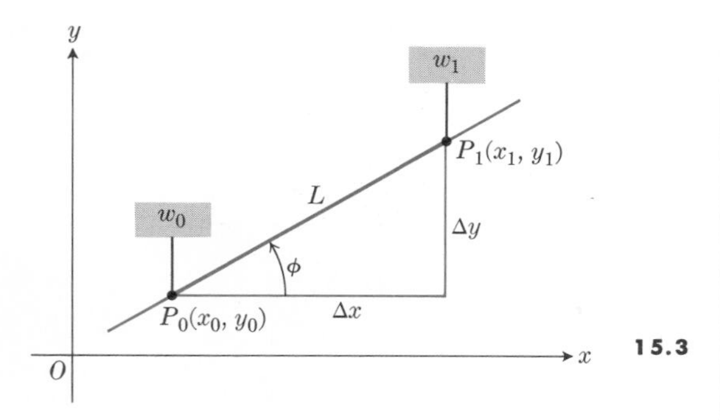

15.3

Suppose, now, that w is a function of x and y, defined for values of (x, y) in some region D of the xy-plane. Let $P_0(x_0, y_0)$ be any point of D and $P_1(x_1, y_1)$ a second point of D. Then the increment in w in going from w_0 at P_0 to w_1 at P_1 is

$$\Delta w = w_1 - w_0 = f(x_1, y_1) - f(x_0, y_0), \tag{3}$$

corresponding to

$$\Delta x = x_1 - x_0, \qquad \Delta y = y_1 - y_0.$$

Keeping P_0 fixed, suppose we require P_1 to approach it along some specific smooth curve in the xy-plane. In general, this curve would not be one of the level curves referred to above, but would instead cut across these level curves. To be definite, suppose P_1 approaches P_0 along a straight line L, making an angle ϕ with the x-axis (see Fig. 15.3). Then, if the limit

$$\frac{dw}{ds} = \lim_{\Delta s \to 0} \frac{\Delta w}{\Delta s} = \lim_{P_1 \to P_0} \frac{f(x_1, y_1) - f(x_0, y_0)}{\sqrt{\Delta x^2 + \Delta y^2}} \tag{4}$$

exists, its value is called the *directional derivative of* $w = f(x, y)$ at (x_0, y_0) in the direction of L. The adjective "directional" is used because the answer, Eq. (4), may depend not only on the function and the point P_0 but also on the *direction* from which P_1 approaches P_0. We shall investigate the general case of the directional derivative in more detail in Article 15.5. Now we shall just investigate two special cases: In the first case, P_1 approaches P_0 along the line $y = y_0$ parallel to the x-axis; in the second, P_1 approaches P_0 along the line $x = x_0$ parallel to the y-axis. These two cases are of interest in themselves, but they are also of interest because when we know the values of these directional derivatives, we may calculate the derivative for *any* other direction, provided the function f is *differentiable* at P_0. (The technical definition of differentiability is given later; it is roughly equivalent to the idea of the surface $w = f(x, y)$ being smooth enough to have a tangent plane at P_0.)

If P_1 approaches P_0 along the line $y = y_0$, we write

$$f_x(x_0, y_0) = \lim_{\Delta x \to 0} \frac{f(x_0 + \Delta x, y_0) - f(x_0, y_0)}{\Delta x}, \tag{5}$$

and call the resulting limit the *partial derivative of* $w = f(x, y)$ *with respect to* x at $P_0(x_0, y_0)$. From the definition (5), this is just the usual derivative with respect to x of the function $F(x) = f(x, y_0)$, obtained from $f(x, y)$ by holding the value of y constant. It measures the instantaneous rate of change, at P_0, of the function $w = f(x, y)$ per unit change in x. The notation $\partial w/\partial x$ is also used to denote the partial derivative of w with respect to x. If we delete the subscript 0 everywhere in (5), the result is the partial derivative at (x, y):

$$\frac{\partial w}{\partial x} = f_x(x, y) = \lim_{\Delta x \to 0} \frac{f(x + \Delta x, y) - f(x, y)}{\Delta x}. \tag{6}$$

To calculate such a partial derivative from the equation for w, we simply apply the rules for ordinary differentiation, treating y as a constant.

Note. In passing to the limit in Eq. (5) or (6), it is understood that Δx may be either positive or negative. If, on the other hand, we calculate the directional derivative in the direction of the *positive* x-axis, then Δx is restricted to positive values only and Eq. (5) is identical with Eq. (4), provided that we take

$$y_1 = y_0, \quad x_1 = x_0 + \Delta x, \quad \Delta x > 0, \quad \Delta y = 0.$$

The directional derivative and the partial derivative f_x differ in that in the directional derivative the point P_1 approaches P_0 always from the same side, while in f_x,

P_1 may approach P_0 either from the left or from the right. In certain "pathological" cases, a function may have a directional derivative from the right but not from the left or may have both directional derivatives but the two may fail to have the same magnitude. In either of these cases, the partial derivative f_x would not exist. If, however, f_x does exist, then it gives the directional derivative to the right, while $-f_x$ gives the directional derivative to the left (the change in sign is due to the fact that $\sqrt{\Delta x^2} = -\Delta x$ if Δx is negative). That is, if f_x exists at a point, then both the right- and left-directional derivatives exist at that point and have the same magnitude but opposite signs.

Example. If we interpret Eq. (2),

$$w = 100 - x^2 - y^2,$$

as temperature in degrees centigrade at (x, y), where x and y are in centimeters, then we have

$$\frac{\partial w}{\partial x} = -2x,$$

and at (3, 4), for example, the temperature is 75°C and

$$\frac{\partial w}{\partial x} = -6 \text{ (°C/cm)}.$$

That is, a small positive change in x would decrease the temperature at the rate of 6°C per cm change in x, while a negative change in x would increase w at the same rate.

The geometric interpretation of Eq. (5) (see Fig. 15.4) is that

$$f_x(x_0, y_0) = \left(\frac{\partial w}{\partial x}\right)_{(x_0, y_0)}$$

gives the slope at $P_0(x_0, y_0, w_0)$ of the curve

$$w = f(x, y_0)$$

in which the plane $y = y_0$ cuts the surface

$$w = f(x, y).$$

Thus, in Fig. 15.4, if x, y, and w are measured in the same units, then

$$\tan \alpha = \left(\frac{\partial w}{\partial x}\right)_{(x_0, y_0)} = f_x(x_0, y_0),$$

while similarly

$$\tan \beta = \left(\frac{\partial w}{\partial y}\right)_{(x_0, y_0)} = f_y(x_0, y_0).$$

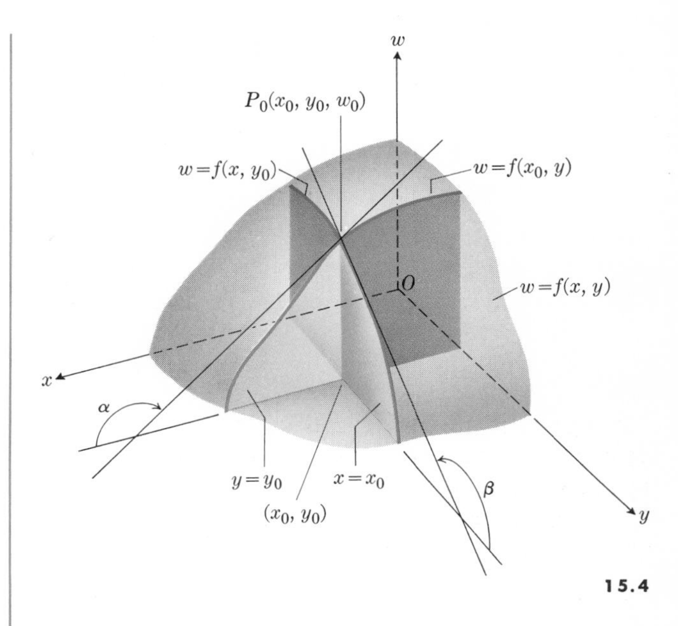

15.4

Here the partial derivative of $w = f(x, y)$ with respect to y is denoted either by $\partial w/\partial y$ or by $f_y(x, y)$, and we have the definitions (7):

$$\begin{aligned} f_y(x_0, y_0) &= \lim_{\Delta y \to 0} \frac{f(x_0, y_0 + \Delta y) - f(x_0, y_0)}{\Delta y}, \\ \frac{\partial w}{\partial y} &= f_y(x, y) \qquad (7) \\ &= \lim_{\Delta y \to 0} \frac{f(x, y + \Delta y) - f(x, y)}{\Delta y}. \end{aligned}$$

There are similar definitions for $\partial w/\partial x$, $\partial w/\partial y$, $\partial w/\partial z$, $\partial w/\partial u$, and $\partial w/\partial v$ when $w = f(x, y, z, u, v)$; in each case, we hold all but one of the variables constant while differentiating with respect to that one. The alternative subscript notation has the advantage of permitting us to exhibit the values of the variables where the derivative is to be evaluated; for example, $f_u(x_0, y_0, z_0, u_0, v_0)$ is the partial derivative of $w = f(x, y, z, u, v)$ with respect to u at $(x_0, y_0, z_0, u_0, v_0)$. This might also be denoted by

$$(\partial w/\partial u)_{(x_0, y_0, z_0, u_0, v_0)}$$

or, more simply, by $(\partial w/\partial u)_0$.

Problem 1. Three resistors of resistances R_1, R_2, and R_3 connected in parallel produce a resistance R given by

$$\frac{1}{R} = \frac{1}{R_1} + \frac{1}{R_2} + \frac{1}{R_3}.$$

Find $\partial R/\partial R_2$.

Solution. Treat R_1 and R_3 as constants and differentiate both sides of the equation implicitly with respect to R_2. Then

$$-\frac{1}{R^2}\frac{\partial R}{\partial R_2} = -\frac{1}{R_2^2}, \quad \text{or} \quad \frac{\partial R}{\partial R_2} = \left(\frac{R}{R_2}\right)^2.$$

Problem 2. If $w = (xy)^z$, find $\partial w/\partial z$.

Solution. Here, we treat x and y, and hence xy, as constant and apply the law

$$\frac{d(a^u)}{dz} = a^u \ln a \frac{du}{dz}.$$

Hence

$$\partial w/\partial z = (xy)^z \ln (xy).$$

EXERCISES

In Exercises 1 through 5, show two ways to represent the function $w = f(x, y)$,

(a) by sketching a surface in xyw-space, and

(b) by drawing a family of level curves,

$$f(x, y) = \text{constant}.$$

1. $f(x, y) = x$
2. $f(x, y) = y$
3. $f(x, y) = x^2 + y^2$
4. $f(x, y) = x^2 - y^2$
5. $f(x, y) = ye^x$

In Exercises 6 through 10, find $\partial w/\partial x$ and $\partial w/\partial y$.

6. $w = e^x \cos y$
7. $w = e^x \sin y$
8. $w = \tan^{-1} (y/x)$
9. $w = \ln \sqrt{x^2 + y^2}$
10. $w = \cosh (y/x)$

In Exercises 11 through 16, find the partial derivatives of the given function with respect to each variable.

11. $f(x, y, z, w) = x^2 e^{2y+3z} \cos (4w)$
12. $f(x, y, z) = z \sin^{-1} (y/x)$
13. $f(u, v, w) = \dfrac{u^2 - v^2}{v^2 + w^2}$
14. $f(r, \theta, z) = \dfrac{r(2 - \cos 2\theta)}{r^2 + z^2}$
15. $f(x, y, u, v) = \dfrac{x^2 + y^2}{u^2 + v^2}$
16. $f(x, y, r, s) = \sin 2x \cosh 3r + \sinh 3y \cos 4s$

In Exercises 17 and 18, A, B, C are the angles of a triangle and a, b, c are the respective opposite sides.

17. Express A (explicitly or implicitly) as a function of a, b, c and calculate $\partial A/\partial a$ and $\partial A/\partial b$.
18. Express a (explicitly or implicitly) as a function of A, b, B and calculate $\partial a/\partial A$ and $\partial a/\partial B$.

In Exercises 19 through 24, express the spherical coordinates ρ, ϕ, θ as functions of the Cartesian coordinates x, y, z and calculate the partial derivative.

19. $\partial\rho/\partial x$
20. $\partial\phi/\partial z$
21. $\partial\theta/\partial y$
22. $\partial\theta/\partial z$
23. $\partial\phi/\partial x$
24. $\partial\theta/\partial x$

In Exercises 25 through 27, let $\mathbf{R} = \mathbf{i}x + \mathbf{j}y + \mathbf{k}z$ be the vector from the origin to (x, y, z). Express x, y, z as functions of the spherical coordinates ρ, ϕ, θ, and calculate the partial derivative.

25. $\partial\mathbf{R}/\partial\rho$
26. $\partial\mathbf{R}/\partial\phi$
27. $\partial\mathbf{R}/\partial\theta$
28. Express the answers to Exercises 25 through 27 in terms of the unit vectors $\mathbf{u}_\rho$, $\mathbf{u}_\phi$, $\mathbf{u}_\theta$ discussed in Miscellaneous Exercise 25, Chapter 14.
29. In Fig. 15.4, let

$$\mathbf{R} = \mathbf{i}x + \mathbf{j}y + \mathbf{k}f(x, y)$$

be the vector from the origin to (x, y, w). What can you say about the direction of the vectors

(a) $\partial\mathbf{R}/\partial x$, (b) $\partial\mathbf{R}/\partial y$?

(c) Calculate the vector product

$$\mathbf{v} = \left(\frac{\partial\mathbf{R}}{\partial x}\right) \times \left(\frac{\partial\mathbf{R}}{\partial y}\right).$$

What can you say about the direction of this vector $\mathbf{v}$, with respect to the surface $w = f(x, y)$, at $P_0(x_0, y_0, w_0)$?

15.3 TANGENT PLANE AND NORMAL LINE

In Article 15.2, we saw that the partial derivatives

$$f_y(x_0, y_0) = \left(\frac{\partial w}{\partial y}\right)_{(x_0, y_0)}, \tag{1a}$$

$$f_x(x_0, y_0) = \left(\frac{\partial w}{\partial x}\right)_{(x_0, y_0)} \tag{1b}$$

give the slopes of the lines L_1 and L_2 which are tangent to the curves C_1 and C_2 in which the planes $x = x_0$ and $y = y_0$, respectively, intersect the surface $w = f(x, y)$. The lines L_1 and L_2 determine a plane. If the surface is sufficiently smooth near $P_0(x_0, y_0, w_0)$, this plane will be tangent to the surface at P_0.

Definition. Tangent plane. *Let $w = f(x, y)$ be the equation of a surface S. Let $P_0(x_0, y_0, w_0)$ be a point on the surface. Let T be a plane through P_0. Let $P(x, y, w)$ be any other point on S. If the measure of the angle between T and the line P_0P approaches zero as P approaches P_0, we say that T is tangent to S at P_0.*

The line through P_0 which is normal to this plane is called the *normal line* to the surface at P_0.

Remark 1. If a surface has a tangent plane at $P_0(x_0, y_0, w_0)$, the lines L_1 and L_2 tangent to the curves

$$C_1\colon w = f(x_0, y), \quad x = x_0,$$

and

$$C_2\colon w = f(x, y_0), \quad y = y_0,$$

must lie in this tangent plane. Since two intersecting lines determine a plane, the plane determined by L_1 and L_2 is the tangent plane, if there is one.

Remark 2. The curves C_1 and C_2, where the surface is cut by the planes $x = x_0$ and $y = y_0$, may be smooth enough to possess tangent lines L_1 and L_2 and the surface still not have a tangent plane at P_0. In other words, the plane determined by L_1 and L_2 may not, in fact, be tangent to the surface. This would happen, for example, if the curves cut from the surface by other planes, such as the planes $y - y_0 = \pm(x - x_0)$, either fail to have tangent lines, or have tangent lines L', L'' which don't lie

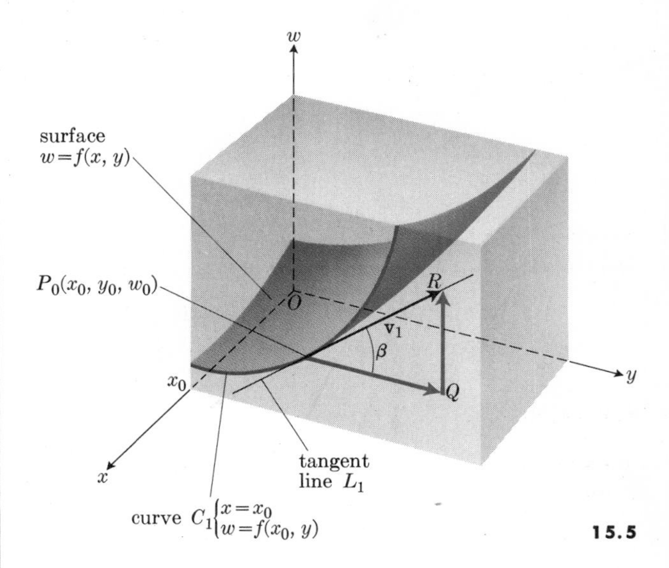

15.5

in the plane determined by L_1 and L_2. In the next article, we shall show that this does not happen if the partial derivatives $f_x(x, y)$, $f_y(x, y)$ exist in some rectangle centered at (x_0, y_0) and are continuous at (x_0, y_0).

We shall now assume that the surface does have a tangent plane and a normal line and see how we may find their equations. They may easily be written down once we have found a vector **N** perpendicular to the plane of L_1 and L_2. For such a vector **N**, we may use the cross product of vectors $\mathbf{v}_1$ and $\mathbf{v}_2$ along the lines L_1 and L_2. So we now consider how to find $\mathbf{v}_1$ and $\mathbf{v}_2$.

Figure 15.5 shows a portion of the curve C_1 cut from the surface by the plane $x = x_0$. Triangle P_0QR is a right triangle with hypotenuse P_0R lying along the tangent line L_1. The slope of L_1 is

$$\tan\beta = \frac{QR}{P_0Q} = f_y(x_0, y_0).$$

Therefore, if we take $P_0Q = 1$ y-unit, QR will be equal to $f_y(x_0, y_0)$ w-units. In terms of vectors, we may take

$$\overrightarrow{P_0Q} = 1\mathbf{j}, \qquad \overrightarrow{QR} = f_y(x_0, y_0)\mathbf{k},$$

and

$$\mathbf{v}_1 = \overrightarrow{P_0Q} + \overrightarrow{QR} = \mathbf{j} + f_y(x_0, y_0)\mathbf{k}, \tag{2a}$$

where **i**, **j**, and **k** are used to denote unit vectors along the x-, y-, and w-axes respectively.

Similarly, by considering the curve C_2 out from the surface by the plane $y = y_0$, we see that the vector

$$\mathbf{v}_2 = \mathbf{i} + f_x(x_0, y_0)\mathbf{k} \tag{2b}$$

is parallel to the line L_2.

For the normal vector **N**, we may therefore take

$$\mathbf{N} = \mathbf{v}_1 \times \mathbf{v}_2 = \begin{vmatrix} \mathbf{i} & \mathbf{j} & \mathbf{k} \\ 0 & 1 & f_y(x_0, y_0) \\ 1 & 0 & f_x(x_0, y_0) \end{vmatrix}$$

$$= \mathbf{i}f_x(x_0, y_0) + \mathbf{j}f_y(x_0, y_0) - \mathbf{k}. \tag{3a}$$

The equations of the tangent plane and normal line at $P_0(x_0, y_0, w_0)$ may now be written down at once.

Tangent plane:

$$A(x - x_0) + B(y - y_0) + C(w - w_0) = 0. \tag{3b}$$

Normal line:

$$(x, y, w) = (x_0, y_0, w_0) + t(A, B, C), \quad -\infty < t < \infty. \tag{3c}$$

The coefficients A, B, and C, determined from the *normal vector* (3a) are given by

$$A = f_x(x_0, y_0), \quad B = f_y(x_0, y_0), \quad C = -1. \tag{4}$$

If we denote the function by $z = f(x, y)$, with z in place of w, then we can rewrite Eqs. (3b, c) in vector form, as follows: We let $\mathbf{R} = \mathbf{i}x + \mathbf{j}y + \mathbf{k}z$ be the position vector of the point $P(x, y, z)$, and let $\mathbf{R}_0$ be the position vector of P_0. Then Eqs. (3a, b, c) take the following forms:

Normal vector:

$$\mathbf{N} = \mathbf{i}f_x(x_0, y_0) + \mathbf{j}f_y(x_0, y_0) - \mathbf{k}. \tag{5a}$$

Tangent plane:

$$\mathbf{N} \cdot (\mathbf{R} - \mathbf{R}_0) = 0. \tag{5b}$$

Normal line:

$$\mathbf{R} = \mathbf{R}_0 + t\mathbf{N}. \tag{5c}$$

EXERCISES

In Exercises 1 through 5, find the plane that is tangent to the given surface $z = f(x, y)$ at the given point P_0. Also find the line normal to the surface at P_0.

1. $z = x^2 + y^2, \quad (3, 4, 25)$
2. $z = \sqrt{9 - x^2 - y^2}, \quad (1, -2, 2)$
3. $z = x^2 - xy - y^2, \quad (1, 1, -1)$
4. $z = \tan^{-1}\dfrac{y}{x}, \quad \left(1, 1, \dfrac{\pi}{4}\right)$
5. $z = \dfrac{x}{\sqrt{x^2 + y^2}} \quad \left(3, -4, \dfrac{3}{5}\right)$
6. (a) If the equation of a surface is given in the form $x = f(y, z)$, what takes the place of Eq. (5a) for a vector **N** normal to the surface at a point $P_0(x_0, y_0, z_0)$? (b) Find the tangent plane and normal line to the surface
$$x = e^{2y-z}$$
at the point (1, 1, 2).
7. Show that there is a line on the cone
$$z^2 = 2x^2 + 4y^2$$
where the tangent plane is parallel to the plane
$$12x + 14y + 11z = 25.$$
Find the line and the tangent plane.
8. At each point of the curve of intersection of the paraboloid $z = x^2 + y^2$ and the plane $z = z_0$ (>0), a line is drawn normal to the paraboloid. Show that these lines generate a cone and find its vertex. Sketch the paraboloid and the associated cone.
9. The intersection of the surface $z = f(x, y)$ and the surface $z = g(x, y)$ is a curve C. Find a vector tangent to C at a point $P_0(x_0, y_0, z_0)$ on it. Express the result in terms of partial derivatives of f and g at P_0.
10. Apply the result of Exercise 9 to find a vector of length $\sqrt{3}$ tangent to the curve of intersection of the cone
$$z^2 = 4x^2 + 9y^2$$
and the plane
$$6x + 3y + 2z = 5$$
at the point $P_0(2, 1, -5)$.

15.4 APPROXIMATE VALUE OF Δw

We saw in the previous article, Eqs. (3b, 4), that if the surface $w = f(x, y)$ has a tangent plane at $P_0(x_0, y_0, w_0)$, then the equation of the tangent plane is

$$w_{\tan} - w_0 = f_x(x_0, y_0)(x - x_0) + f_y(x_0, y_0)(y - y_0), \tag{1a}$$

or

$$w_{\tan} = f(x_0, y_0) + A(x - x_0) + B(y - y_0), \tag{1b}$$

where

$$A = f_x(x_0, y_0), \quad B = f_y(x_0, y_0). \tag{1c}$$

Note that the right-hand side of Eq. (1b) is *linear* in both x and y. In Theorem 1 below, we shall show that under suitable hypotheses this formulation provides a good *linear approximation* to the function f near P_0. The right-hand side of Eq. (1b) is also called the *linearization* of f at P_0.

In these equations (x_0, y_0, w_0) are coordinates of a point on the surface, while $(x, y, w_{\tan})$ are coordinates of a point on the *tangent plane*. If we take

$$x = x_0 + \Delta x, \qquad y = y_0 + \Delta y$$

in Eq. (1a), and denote the change, $w_{\tan} - w_0$, by $\Delta w_{\tan}$, we have

$$\Delta w_{\tan} = f_x(x_0, y_0)\,\Delta x + f_y(x_0, y_0)\,\Delta y. \tag{2}$$

Note the similarity of Eq. (2) and the equation of the tangent-line approximation to the increment of a function of a *single* independent variable. With the notation f_x in place of f' for the derivative, we could write $\Delta w_{\tan} = f_x(x_0)\,\Delta x$ for the analog of Eq. (2) when $w = f(x)$. The right-hand side of Eq. (2) reduces to this expression when y is absent; or it reduces to $f_y(y_0)\,\Delta y$ if $w = f(y)$ and x is absent. The important thing to see is that Eq. (2) simply calls for the *addition* of the tangential increments caused by varying x alone or y alone; no interaction term is needed. We shall see that we can apply this way of thinking to functions of more than two independent variables.

Equation (2) tells us how much change is produced in w, corresponding to the changes Δx and Δy, when we move along the *tangent plane*. In this article we shall see that under suitable restrictions on the function f,

(i) the surface does have a tangent plane, and

(ii) the change in w *on the surface* $w = f(x, y)$ differs from $\Delta w_{\tan}$ by an amount $\epsilon_1\,\Delta x + \epsilon_2\,\Delta y$, where both ϵ_1 and ϵ_2 are small when Δx and Δy are small. We shall discuss the suitable restrictions on f and then prove the following theorem.

Theorem 1. *Let the function $w = f(x, y)$ be continuous and possess partial derivatives f_x, f_y throughout a region*

$$R: |x - x_0| < h, \quad |y - y_0| < k$$

of the xy-plane. Let f_x and f_y be continuous at (x_0, y_0). Let

$$\Delta w = f(x_0 + \Delta x, y_0 + \Delta y) - f(x_0, y_0). \tag{3}$$

Then

$$\Delta w = f_x(x_0, y_0)\,\Delta x + f_y(x_0, y_0)\,\Delta y + \epsilon_1\,\Delta x + \epsilon_2\,\Delta y, \tag{4}$$

where

$$\epsilon_1 \text{ and } \epsilon_2 \to 0 \quad \text{when} \quad \Delta x \text{ and } \Delta y \to 0. \tag{5}$$

Discussion. The region R is a rectangle, center at $A(x_0, y_0)$ and sides $2h$ by $2k$ (see Fig. 15.6). We shall restrict Δx and Δy to be so small that the points

$$(x_0, y_0), \quad (x_0 + \Delta x, y_0 + \Delta y),$$
$$(x_0 + \Delta x, y_0), \quad (x_0, y_0 + \Delta y)$$

all lie inside this rectangle R. The function f is assumed to be continuous and to have partial derivatives f_x and f_y throughout the rectangle R. In particular, these functions are well behaved at each

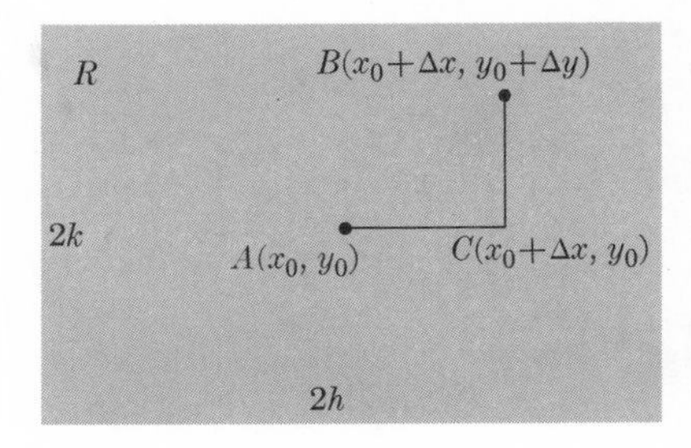

15.6

of the points listed and along the segment joining any pair of them. This is sufficient to show that the applications of the Mean Value Theorem in the following proof are valid.

Proof. The key to the proof is a double application of the Mean Value Theorem. The increment Δw is the change in f from $A(x_0, y_0)$ to

$$B(x_0 + \Delta x, y_0 + \Delta y)$$

in R. We resolve this into two parts, Δw_1 and Δw_2:

$$\Delta w_1 = f(x_0 + \Delta x, y_0) - f(x_0, y_0), \tag{6}$$

$$\Delta w_2 = f(x_0 + \Delta x, y_0 + \Delta y) - f(x_0 + \Delta x, y_0). \tag{7}$$

The first is the change in w from A to C, and the second is the change from C to B (see Fig. 15.6). Algebraically, we have

$$\begin{aligned}\Delta w_2 + \Delta w_1 &= f(x_0 + \Delta x, y_0 + \Delta y) - f(x_0 + \Delta x, y_0) \\ &\quad + f(x_0 + \Delta x, y_0) - f(x_0, y_0) \\ &= f(x_0 + \Delta x, y_0 + \Delta y) - f(x_0, y_0) \\ &= \Delta w.\end{aligned} \tag{8}$$

In Δw_1, we hold $y = y_0$ fixed and have an increment of a function of x that is continuous and differentiable. The Mean Value Theorem is therefore applicable and yields

$$\begin{aligned}\Delta w_1 &= f(x_0 + \Delta x, y_0) - f(x_0, y_0) \\ &= f_x(x_1, y_0)\,\Delta x\end{aligned} \tag{9}$$

for some x_1 between x_0 and $x_0 + \Delta x$.

Similarly, in Δw_2, we hold $x = x_0 + \Delta x$ and have an increment of a function of y that is continuous and differentiable. By the Mean Value Theorem,

$$\begin{aligned}\Delta w_2 &= f(x_0 + \Delta x, y_0 + \Delta y) - f(x_0 + \Delta x, y_0) \\ &= f_y(x_0 + \Delta x, y_1)\,\Delta y\end{aligned} \tag{10}$$

for some y_1 between y_0 and $y_0 + \Delta y$. Hence

$$\Delta w = f_x(x_1, y_0)\,\Delta x + f_y(x_0 + \Delta x, y_1)\,\Delta y \tag{11}$$

for some x_1 between x_0 and $x_0 + \Delta x$, and y_1 between y_0 and $y_0 + \Delta y$.

We now use the hypothesis that f_x and f_y are continuous at $P_0(x_0, y_0)$. This means that

$$f_x(x_1, y_0) \to f_x(x_0, y_0) \tag{12a}$$

and

$$f_y(x_0 + \Delta x, y_1) \to f_y(x_0, y_0) \tag{12b}$$

as Δx and Δy approach zero. Therefore we may write

$$f_x(x_1, y_0) = f_x(x_0, y_0) + \epsilon_1, \tag{13a}$$

$$f_y(x_0 + \Delta x, y_1) = f_y(x_0, y_0) + \epsilon_2, \tag{13b}$$

and ϵ_1, ϵ_2 both approach zero as Δx and Δy approach zero. Substituting (13a, b) into (11) gives the desired result:

$$\Delta w = [f_x(x_0, y_0) + \epsilon_1]\,\Delta x + [f_y(x_0, y_0) + \epsilon_2]\,\Delta y,$$

while (12a, b) guarantee that

$$\epsilon_1 \text{ and } \epsilon_2 \to 0 \quad \text{when} \quad \Delta x \text{ and } \Delta y \to 0. \ \square$$

Corollary. *Let $w = f(x, y)$ be continuous in a region R: $|x - x_0| < h$, $|y - y_0| < k$. Let f_x and f_y exist in R and be continuous at (x_0, y_0). Then the surface $w = f(x, y)$ has a tangent plane at $P_0(x_0, y_0, w_0)$, where $w_0 = f(x_0, y_0)$.*

Proof. Let $P(x, y, w)$ be any point, different from P_0, on the surface $w = f(x, y)$. Let $\mathbf{N}$ be the vector

$$\mathbf{N} = \mathbf{i}f_x(x_0, y_0) + \mathbf{j}f_y(x_0, y_0) - \mathbf{k}. \tag{14}$$

The cosine of the angle θ between $\overrightarrow{P_0P}$ and $\mathbf{N}$ is

$$\cos\theta = \frac{\overrightarrow{P_0P} \cdot \mathbf{N}}{|\overrightarrow{P_0P}|\,|\mathbf{N}|}. \tag{15}$$

Now

$$\begin{aligned}\overrightarrow{P_0P} &= \mathbf{i}(x - x_0) + \mathbf{j}(y - y_0) + \mathbf{k}(w - w_0) \\ &= \mathbf{i}\,\Delta x + \mathbf{j}\,\Delta y + \mathbf{k}\,\Delta w.\end{aligned} \tag{16}$$

Therefore

$$\begin{aligned}\cos\theta &= \frac{f_x(x_0, y_0)\,\Delta x + f_y(x_0, y_0)\,\Delta y - \Delta w}{|\overrightarrow{P_0P}|\,|\mathbf{N}|} \\ &= \frac{-\epsilon_1\,\Delta x - \epsilon_2\,\Delta y}{|\overrightarrow{P_0P}|\,|\mathbf{N}|}.\end{aligned} \tag{17}$$

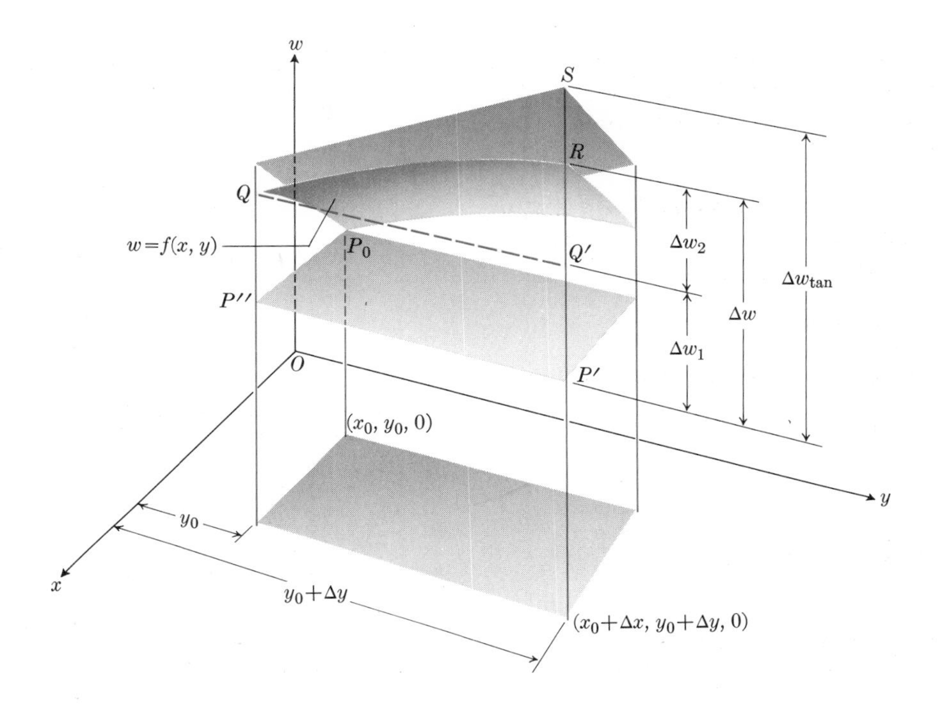

15.7

From (14) we see that $|\mathbf{N}| \geq 1$, and from (16) we see that

$$\frac{|\Delta x|}{|\overrightarrow{P_0P}|} \leq 1, \qquad \frac{|\Delta y|}{|\overrightarrow{P_0P}|} \leq 1.$$

Hence, from (17) and the previous theorem, it follows that

$$|\cos\theta| \leq |\epsilon_1| + |\epsilon_2| \to 0$$

as $P \to P_0$. Therefore, for any point $P \neq P_0$ on the surface, the angle between the vector $\overrightarrow{P_0P}$ and the vector $\mathbf{N}$ (Eq. 14) approaches 90°. This means that the plane through P_0, perpendicular to $\mathbf{N}$, is tangent to the surface at P_0. □

Remark 1. The significance of the theorem lies in Eqs. (4) and (5). It says that

$$\Delta w = \Delta w_{\tan} + \epsilon_1\,\Delta x + \epsilon_2\,\Delta y$$

consists of the part $\Delta w_{\tan}$, which is *linear* in Δx and Δy, plus the error terms $\epsilon_1\,\Delta x$ and $\epsilon_2\,\Delta y$ which are *products* of small terms when Δx and Δy are small.

Example. If

$$w = x^2 + y^2 = f(x, y),$$

then

$$\begin{aligned}\Delta w &= (x+\Delta x)^2 + (y+\Delta y)^2 - (x^2+y^2)\\ &= \underbrace{2x\,\Delta x + 2y\,\Delta y}_{} + \underbrace{(\Delta x)^2 + (\Delta y)^2}_{}\\ &= \qquad \Delta w_{\tan} \qquad + \epsilon_1\,\Delta x + \epsilon_2\,\Delta y.\end{aligned}$$

Since $f_x = 2x$ and $f_y = 2y$, the part $2x\,\Delta x + 2y\,\Delta y$ is the same as $\Delta w_{\tan}$. Then $\Delta w - \Delta w_{\tan}$, in this particular example, is simply $(\Delta x)^2 + (\Delta y)^2$. This agrees with Eq. (4) with $\epsilon_1 = \Delta x$ and $\epsilon_2 = \Delta y$, and both of these approach zero when Δx and Δy do.

Remark 2. Figure 15.7 represents a portion of the surface $w = f(x, y)$ near $P_0(x_0, y_0, w_0)$, together with a portion of the plane that is tangent to the surface at P_0. Points P_0, P', and P'' have the same elevation $w_0 = f(x_0, y_0)$ above the xy-plane. The change in w *on the surface* corresponds to $\Delta w = P'R$ and the change in w *on the tangent plane* corresponds to $\Delta w_{\tan} = P'S$. The change

$$\Delta w_1 = f(x_0 + \Delta x, y_0) - f(x_0, y_0),$$

which corresponds to $P''Q = P'Q'$ in the figure, is caused by changing x from x_0 to $x_0 + \Delta x$ while holding y constant, $y = y_0$. Then we hold x constant, $x = x_0 + \Delta x$, and we see that

$$\Delta w_2 = f(x_0 + \Delta x, y_0 + \Delta y) - f(x_0 + \Delta x, y_0)$$

is the change in w caused by changing y from y_0 to $y_0 + \Delta y$. This is represented by $Q'R$ in the figure. The total change in w is just the sum of these two.

Functions of more variables

An analogous result holds for a function of any finite number of independent variables. For a function of three variables,

$$w = f(x, y, z),$$

that is continuous and has partial derivatives f_x, f_y, f_z at and in some neighborhood of the point (x_0, y_0, z_0), and whose derivatives are continuous at the point, we have

$$\begin{aligned} \Delta w &= f(x_0 + \Delta x, y_0 + \Delta y, z_0 + \Delta z) \\ &\quad - f(x_0, y_0, z_0) \\ &= f_x \Delta x + f_y \Delta y + f_z \Delta z \\ &\quad + \epsilon_1 \Delta x + \epsilon_2 \Delta y + \epsilon_3 \Delta z, \end{aligned} \tag{18}$$

where

$$\epsilon_1, \epsilon_2, \epsilon_3 \to 0 \quad \text{when} \quad \Delta x, \Delta y, \Delta z \to 0.$$

In this formula, the partial derivatives f_x, f_y, f_z are to be evaluated at the point (x_0, y_0, z_0). This formula will be used in certain theoretical discussions below. It is also useful in making *approximations*, for we may approximate Δw by using only the part of Eq. (18) that is *linear* in Δx, Δy, and Δz. We shall denote this linear approximation to Δw by Δw_{lin}:

$$\begin{aligned} \Delta w_{\text{lin}} &= f_x(x_0, y_0, z_0)\, \Delta x + f_y(x_0, y_0, z_0)\, \Delta y \\ &\quad + f_z(x_0, y_0, z_0)\, \Delta z \\ &= \left(\frac{\partial w}{\partial x}\right)_0 \Delta x + \left(\frac{\partial w}{\partial y}\right)_0 \Delta y + \left(\frac{\partial w}{dz}\right)_0 \Delta z. \end{aligned} \tag{19a}$$

Equation (19a) can also be written in the alternative form

$$w_{\text{lin}} = w_0 + A(x - x_0) + B(y - y_0) + C(z - z_0), \tag{19b}$$

using the notations

$$A = \left(\frac{\partial w}{dx}\right)_0, \quad B = \left(\frac{\partial w}{\partial y}\right)_0, \quad C = \left(\frac{\partial w}{\partial z}\right)_0. \tag{19c}$$

The right-hand side of Eq. (19b) represents a *linear* function $L(x, y, z)$ that is called the *linearization* of f at P_0. It is relatively easy to work with, and has the property that

$$|f(x, y, z) - L(x, y, z)| \leq |\epsilon_1 \Delta x| + |\epsilon_2 \Delta y| + |\epsilon_3 \Delta z|,$$

in the notation of Eq. (18). If we let Δs be the distance $|PP_0|$, so that

$$\Delta s = \sqrt{(\Delta x)^2 + (\Delta y)^2 + (\Delta z)^2},$$

then $|\Delta x|$, $|\Delta y|$, and $|\Delta z|$ are all less than or equal to Δs. Therefore the difference between the *exact* value $f(x, y, z)$ and the *linear approximation* $L(x, y, z)$ is less than or equal to $\epsilon \cdot \Delta s$, where

$$\epsilon = |\epsilon_1| + |\epsilon_2| + |\epsilon_3| \text{ approaches zero as } \Delta s \to 0.$$

Equation (18) thus implies that under the hypotheses stated,

$$\lim_{\Delta s \to 0} \left| \frac{f(x, y, z) - L(x, y, z)}{\Delta s} \right| = 0. \tag{19d}$$

Note. The reader may prove the result (18) by treating Δw as the sum of the three increments

$$\Delta w_1 = f(x_0 + \Delta x, y_0, z_0) - f(x_0, y_0, z_0), \tag{20a}$$

$$\begin{aligned} \Delta w_2 &= f(x_0 + \Delta x, y_0 + \Delta y, z_0) \\ &\quad - f(x_0 + \Delta x, y_0, z_0), \end{aligned} \tag{20b}$$

$$\begin{aligned} \Delta w_3 &= f(x_0 + \Delta x, y_0 + \Delta y, z_0 + \Delta z) \\ &\quad - f(x_0 + \Delta x, y_0 + \Delta y, z_0), \end{aligned} \tag{20c}$$

and applying the Mean Value Theorem to each of these separately. Note that two coordinates remain constant and only one varies in each of these partial increments $\Delta w_1, \Delta w_2, \Delta w_3$. For example, in (20b), only y varies, since x is held equal to $x_0 + \Delta x$ and z is held equal to z_0. The function $f(x_0 + \Delta x, y, z_0)$, being a continuous function of y with a derivative f_y, is subject to the Mean Value Theorem, and we have

$$\Delta w_2 = f_y(x_0 + \Delta x, y_1, z_0)\, \Delta y$$

for some y_1 between y_0 and $y_0 + \Delta y$.

The fact that Δw_{lin} is a good approximation to Δw when the increments in the independent variables are small may be used to advantage in calculating, approximately, small changes in a function of two or more variables in a way that is precisely analogous to the way in which the tangent line is used to approximate the increment for a function $f(x)$ of a single variable.

Problem 1. Calculate Δw and $\Delta w_{\tan}$ for the function

$$w = x^2 + xy.$$

Solution. Here

$$f(x, y) = x^2 + xy, \quad f_x = 2x + y, \quad f_y = x,$$

so that

$$\Delta w_{\tan} = f_x \Delta x + f_y \Delta y = (2x + y)\Delta x + x \Delta y.$$

To find Δw, we calculate

$$\begin{aligned} w + \Delta w &= f(x + \Delta x, y + \Delta y) \\ &= (x + \Delta x)^2 + (x + \Delta x)(y + \Delta y) \\ &= x^2 + 2x\Delta x + (\Delta x)^2 + xy \\ &\quad + x\Delta y + y \Delta x + (\Delta x)(\Delta y) \end{aligned}$$

and then subtract

$$w = x^2 + xy$$

to obtain

$$\Delta w = (2x + y)\Delta x + x\Delta y + (\Delta x)^2 + (\Delta x)(\Delta y).$$

The difference,

$$\Delta w - \Delta w_{\tan} = (\Delta x)^2 + (\Delta x)(\Delta y),$$

is thus seen to consist of nonlinear terms in Δx and Δy. These approach zero faster than the linear terms.

Problem 2. Use the linear approximation

$$w \approx w_0 + \left(\frac{\partial w}{\partial x}\right)_0 \Delta x + \left(\frac{\partial w}{\partial y}\right)_0 \Delta y + \left(\frac{\partial w}{\partial z}\right)_0 \Delta z$$

to calculate

$$\sqrt{(0.98)^2 + (2.01)^2 + (1.94)^2}.$$

Solution. Let

$$w = \sqrt{x^2 + y^2 + z^2} = f(x, y, z),$$

$$x_0 = 1.00, \qquad y_0 = 2.00, \qquad z_0 = 2.00,$$

$$\Delta x = -0.02, \quad \Delta y = +0.01, \quad \Delta z = -0.06.$$

Then the square root we wish to calculate is

$$w_0 + \Delta w = f(x_0 + \Delta x, y_0 + \Delta y, z_0 + \Delta z),$$

where

$$w_0 = \sqrt{1 + 4 + 4} = 3$$

and

$$\begin{aligned} \Delta w &\approx \frac{\partial w}{\partial x}\Delta x + \frac{\partial w}{\partial y}\Delta y + \frac{\partial w}{\partial z}\Delta z \\ &= \frac{x\Delta x + y \Delta y + z\Delta z}{\sqrt{x^2 + y^2 + z^2}} \\ &= \frac{-0.02 + 0.02 - 0.12}{3} = -0.04. \end{aligned}$$

Therefore

$$w_0 + \Delta w \approx 3 - 0.04 = 2.96.$$

EXERCISES

1. Find $\Delta w_{\tan}$ and Δw if
$$w = x^2 - xy + y^2,$$
$$(x_0, y_0) = (1, -2), \quad \Delta x = 0.01, \quad \Delta y = -0.02.$$
2. Use Δw_{lin} to calculate by how much an error of 2% in each of the factors a, b, c may affect the product abc.
3. Find a reasonable approximation (two decimal places) to the value of
$$\sqrt{(3.01)^2 + (3.97)^2}.$$
4. Carry through the details of deriving Eq. (11)
5. The dimensions of a rectangular box are measured as 3, 4, and 12 in. If the measurements may be in error by ± 0.01, ± 0.01, and ± 0.03 in., respectively, calculate the length of the diagonal and estimate the possible error in this length.
6. A function $w = f(x, y)$ is said to be *differentiable* at $P(a, b)$ if there are constants M and N (possibly depending on f and P) such that
$$\Delta w = M\Delta x + N\Delta y + \alpha[|\Delta x| + |\Delta y|]$$
and
$$\alpha \to 0 \quad \text{as} \quad |\Delta x| + |\Delta y| \to 0.$$
Here
$$\Delta w = f(a + \Delta x, b + \Delta y) - f(a, b).$$
Prove that if f is differentiable at (a, b), then
$$M = f_x(a, b) \quad \text{and} \quad N = f_y(a, b).$$

7. If the function $w = f(x, y)$ is differentiable at $P(a, b)$ (see Exercise 6), prove it is continuous there.

In Exercises 8 through 10 find the *linearization* of the given function f at the indicated point P_0.

8. $f(x, y) = (2 + xy) \cos y, \quad P_0\left(2, \frac{\pi}{2}\right)$
9. $f(x, y, z) = x^2 y \exp z, \quad P_0(1, -2, \ln 2)$
10. $f(x, y, u, v) = (x^2 + y^2)/(u^2 - v^2)$, $P_0(1, -1, 0, 2)$
11. (a) Restate the limit results indicated in Eq. (19d) as a formal theorem,
    (b) Prove this theorem.
12. (a) State a limit result like Eq. (19d) for a function of two independent variables, $w = f(x, y)$.
    (b) Prove it.
13. Generalize the definition of the linearization of a function to apply to functions of n independent variables. State a limit theorem like Eq. (19d) that you think is probably true for suitable hypotheses on f. Include those hypotheses in your statement of the theorem.

15.5 THE DIRECTIONAL DERIVATIVE: GENERAL CASE

Let us return once more to the problem of determining the instantaneous rate of change at $P_0(x_0, y_0)$ of a function $w = f(x, y)$ measured in units of change of w per unit of distance along a ray L with vertex at P_0 and making an angle ϕ with the positive x-axis. Hold $P_0(x_0, y_0)$ fixed, and let $P_1(x_1, y_1)$ be a point on L and lying near P_0. (See Fig. 15.8.) Let

$$\Delta x = x_1 - x_0, \quad \Delta y = y_1 - y_0$$

both tend to zero in such a way that P_1 approaches P_0 along L. Then, for each position of P_1 on L, calculate the ratio

$$\frac{\Delta w}{\Delta s} = \frac{w_1 - w_0}{\sqrt{\Delta x^2 + \Delta y^2}} = \frac{f(x_1, y_1) - f(x_0, y_0)}{\sqrt{(x_1 - x_0)^2 + (y_1 - y_0)^2}},$$

which measures the average rate of change of w along L from P_0 to P_1. If this average rate of change has a limit as $P_1 \to P_0$, we denote the limit by dw/ds and call it the *directional derivative* of w at P_0 in the ϕ-direction. We shall now prove the following theorem.

Theorem 2. *Let $w = f(x, y)$ be continuous and possess partial derivatives f_x, f_y throughout some neighborhood of the point $P_0(x_0, y_0)$. Let f_x and f_y be continuous at P_0. Then the directional derivative at P_0 exists for any direction angle ϕ and is given by*

$$\frac{dw}{ds} = f_x(x_0, y_0) \cos \phi + f_y(x_0, y_0) \sin \phi. \tag{1}$$

Remark. Before we prove this result, it is worth noting that the special cases $\phi = 0, \pi/2, \pi, 3\pi/2$ respectively lead to $dw/ds = \partial f/\partial x, \partial f/\partial y, -\partial f/\partial x, -\partial f/\partial y$, in harmony with our earlier discussion.

Proof. To establish Eq. (1), we recall that Eq. (4) of the previous article is valid under the hypotheses we have made for the continuity of $f(x, y)$, $\partial f/\partial x$, and $\partial f/\partial y$. Hence

$$\frac{\Delta w}{\Delta s} = f_x(x_0, y_0) \frac{\Delta x}{\Delta s} + f_y(x_0, y_0) \frac{\Delta y}{\Delta s} + \epsilon_1 \frac{\Delta x}{\Delta s} + \epsilon_2 \frac{\Delta y}{\Delta s},$$

where

$$\epsilon_1 \text{ and } \epsilon_2 \to 0 \quad \text{as} \quad \Delta x \text{ and } \Delta y \to 0.$$

Now, if $P_1 \to P_0$ along L, or even along a smooth curve which is tangent to L at P_0, we have

$$\lim \frac{\Delta x}{\Delta s} = \frac{dx}{ds} = \cos \phi, \qquad \lim \frac{\Delta y}{\Delta s} = \frac{dy}{ds} = \sin \phi,$$

and hence

$$\frac{dw}{ds} = \lim \frac{\Delta w}{\Delta s} = f_x(x_0, y_0) \cos \phi + f_y(x_0, y_0) \sin \phi. \quad \square$$

Problem. Let $w = 100 - x^2 - y^2$. If one starts from the point $P_0(3, 4)$, in which direction should he go to make w increase most rapidly?

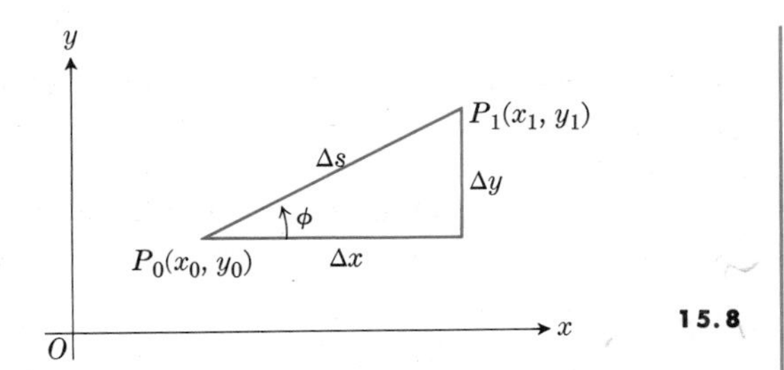

15.8

Solution. We have

$$f(x, y) = 100 - x^2 - y^2,$$
$$f_x(3, 4) = -6, \quad f_y(3, 4) = -8,$$
$$(dw/ds)_{(3,4)} = -6 \cos \phi - 8 \sin \phi.$$

To make w increase most rapidly, we seek the angle ϕ for which the function

$$F(\phi) = -6 \cos \phi - 8 \sin \phi$$

has a maximum. Since

$$F'(\phi) = 6 \sin \phi - 8 \cos \phi$$

is zero when

$$\tan \phi = \tfrac{4}{3}, \quad \sin \phi = \pm\tfrac{4}{5}, \quad \cos \phi = \pm\tfrac{3}{5},$$

while

$$F''(\phi) = 6 \cos \phi + 8 \sin \phi$$

is negative in case both $\sin \phi$ and $\cos \phi$ are negative, we observe that the maximum value of $F(\phi)$ is attained when

$$\cos \phi = -\tfrac{3}{5}, \qquad \sin \phi = -\tfrac{4}{5}.$$

It is readily seen that the geometrical meaning of this result is that w increases most rapidly if the ray L points from $P_0(3, 4)$ toward the origin. The derivative of w with respect to distance in this direction is found to be $+10$; that is, w increases at the instantaneous rate of 10 w-units per unit of length along this particular ray.

The notion of the directional derivative can easily be extended from the case of functions of two independent variables to the case of functions of three independent variables. To do this we consider the values of a function

$$w = f(x, y, z)$$

at $P_0(x_0, y_0, z_0)$ and at a nearby point $P_1(x_1, y_1, z_1)$ lying on a directed ray L through P_0. It will be convenient to specify the direction of L by means of a unit vector

$$\mathbf{u} = \mathbf{i} \cos \alpha + \mathbf{j} \cos \beta + \mathbf{k} \cos \gamma \tag{2a}$$

pointing in the same direction. If we let

$$x_1 - x_0 = \Delta x, \quad y_1 - y_0 = \Delta y, \quad z_1 - z_0 = \Delta z,$$

then

$$\overrightarrow{P_0P_1} = \mathbf{i}\,\Delta x + \mathbf{j}\,\Delta y + \mathbf{k}\,\Delta z.$$

The distance from P_0 to P_1 is

$$\Delta s = \sqrt{(\Delta x)^2 + (\Delta y)^2 + (\Delta z)^2}.$$

Since $\overrightarrow{P_0P_1}$ has the same direction as $\mathbf{u}$ and has length equal to Δs, it is clear that

$$\frac{\overrightarrow{P_0P_1}}{\Delta s} = \mathbf{u}.$$

That is,

$$\mathbf{i}\frac{\Delta x}{\Delta s} + \mathbf{j}\frac{\Delta y}{\Delta s} + \mathbf{k}\frac{\Delta z}{\Delta s} = \mathbf{i} \cos \alpha + \mathbf{j} \cos \beta + \mathbf{k} \cos \gamma,$$

or

$$\frac{\Delta x}{\Delta s} = \cos \alpha, \qquad \frac{\Delta y}{\Delta s} = \cos \beta, \qquad \frac{\Delta z}{\Delta s} = \cos \gamma.$$

These direction cosines of $\overrightarrow{P_0P_1}$ remain constant as P_1 approaches P_0 along L. Hence, in the limit as $\Delta s \to 0$, we also have

$$\frac{dx}{ds} = \cos \alpha, \qquad \frac{dy}{ds} = \cos \beta, \qquad \frac{dz}{ds} = \cos \gamma.$$

With these geometrical considerations out of the way, we now define the directional derivative of w at P_0 in the direction of $\mathbf{u}$ to be the limit, as P_1 approaches P_0 along L, of the average rate of change of w with respect to distance:

$$\frac{dw}{ds} = \lim_{\Delta s \to 0} \frac{\Delta w}{\Delta s}$$
$$= \lim_{P_1 \to P_0} \frac{f(x_1, y_1, z_1) - f(x_0, y_0, z_0)}{\sqrt{(x_1 - x_0)^2 + (y_1 - y_0)^2 + (z_1 - z_0)^2}}.$$

If f, f_x, f_y, and f_z are all continuous functions of x, y, z in some neighborhood of the point $P_0(x_0, y_0, z_0)$, then Eq. (18) of Article 14.4 applies, and we find that the directional derivative of $w = f(x, y, z)$ at P_0 in the direction $\mathbf{i} \cos \alpha + \mathbf{j} \cos \beta + \mathbf{k} \cos \gamma$ is

$$\begin{aligned} \frac{dw}{ds} &= f_x(x_0, y_0, z_0) \cos \alpha \\ &+ f_y(x_0, y_0, z_0) \cos \beta \\ &+ f_z(x_0, y_0, z_0) \cos \gamma. \end{aligned} \tag{2b}$$

This can be expressed as the dot product of the vector $\mathbf{u}$ of Eq. (2a) and the vector

$$\begin{aligned} \mathbf{v} = &\ \mathbf{i} f_x(x_0, y_0, z_0) \\ &+ \mathbf{j} f_y(x_0, y_0, z_0) \\ &+ \mathbf{k} f_z(x_0, y_0, z_0); \end{aligned} \tag{3}$$

that is,

$$\frac{dw}{ds} = \mathbf{u} \cdot \mathbf{v}. \tag{4}$$

This factorization separates the directional derivative into a part $\mathbf{u}$, which depends only on the *direction*, and a part $\mathbf{v}$, which depends only on the *function and the point P*. The vector $\mathbf{v}$ is called the *gradient* of f at P_0. It will be considered in detail in the next article. Equation (4) applies also in two dimensions as well as in three and includes Eq. (1) as a special case, with

$$\gamma = 90^\circ, \quad \cos \gamma = 0.$$

EXERCISES

In Exercises 1 through 4, find the directional derivative of the given function $f = f(x, y, z)$ at the given point, and in the direction of the given vector $\mathbf{A}$.

1. $f = e^x \cos (yz)$, $P_0(0, 0, 0)$, $\mathbf{A} = 2\mathbf{i} + \mathbf{j} - 2\mathbf{k}$.
2. $f = \ln \sqrt{x^2 + y^2 + z^2}$, $P_0(3, 4, 12)$, $\mathbf{A} = 3\mathbf{i} + 6\mathbf{j} - 2\mathbf{k}$.
3. $f = x^2 + 2y^2 + 3z^2$, $P_0(1, 1, 1)$, $\mathbf{A} = \mathbf{i} + \mathbf{j} + \mathbf{k}$.
4. $f = xy + yz + zx$, $P_0(1, -1, 2)$, $\mathbf{A} = 10\mathbf{i} + 11\mathbf{j} - 2\mathbf{k}$.
5. In which direction should one travel, starting from $P_0(1, 1, 0)$, to obtain the most rapid rate of decrease of the function
$$f = (x + y - 2)^2 + (3x - y - 6)^2?$$
6. The directional derivative of a given function $w = f(x, y)$ at $P_0(1, 2)$ in the direction toward $P_1(2, 3)$ is $+2\sqrt{2}$, and in the direction toward $P_2(1, 0)$ it is -3. What is the value of dw/ds at P_0 in the direction toward the origin?
7. Investigate the following graphical method of representing the directional derivative. Let $w = f(x, y)$ be a given function and let $P_0(x_0, y_0)$ be a given point. Through P_0, draw any ray making an angle θ with the positive x-direction, $0 \leq \theta \leq 2\pi$. On this directed line (or on its backward extension through P_0, if r is negative), mark the point Q such that the polar coordinates of Q relative to P_0 are (r, θ) with $r = (dw/ds)_0$. Show that the locus of Q is a circle of diameter
$$\sqrt{\left(\frac{\partial w}{\partial x}\right)_0^2 + \left(\frac{\partial w}{\partial y}\right)_0^2}.$$
Show that P_0 and
$$P_1(x_0 + (\partial w/\partial x)_0, \quad y_0 + (\partial w/\partial y)_0)$$
are opposite ends of one diameter of the circle. (This gives an easy method for constructing the circle. It is only necessary to locate P_0 and P_1 and then draw the circle. It is analogous to the Mohr circle in mechanics.)
8. Find the directional derivative of
$$f(x, y) = x \tan^{-1} y/x$$
at $(1, 1)$ in the direction of $\mathbf{A} = 2\mathbf{i} - \mathbf{j}$.
9. In which direction is the directional derivative of $f(x, y) = (x^2 - y^2)/(x^2 + y^2)$ at $(1, 1)$ equal to zero?

15.6 THE GRADIENT

In the previous article, we found that the directional derivative of a function $w = f(x, y, z)$ could be expressed as the dot product of a unit vector $\mathbf{u}$ specifying the direction and the vector $\mathbf{v}$ of Eq. (3). This latter vector depends only on the values of the partial derivatives of w at P_0, and is called the *gradient of w*. Two symbols are commonly used to denote the gradient, namely, grad w and ∇w, where ∇ is an inverted capital delta and is generally called *del*. The gradient is defined by the equation

$$\text{grad } w = \nabla w = \mathbf{i}\frac{\partial w}{\partial x} + \mathbf{j}\frac{\partial w}{\partial y} + \mathbf{k}\frac{\partial w}{\partial z}. \tag{1}$$

The del operator

$$\nabla = \mathbf{i}\frac{\partial}{\partial x} + \mathbf{j}\frac{\partial}{\partial y} + \mathbf{k}\frac{\partial}{\partial z} \tag{2}$$

is akin to, but somewhat more complex than, the familiar differentiation operator d/dx. When del operates on a differentiable function $w = f(x, y, z)$, it produces a vector, namely, the vector grad w (or grad f) given by Eq. (1).

In courses in advanced calculus and vector analysis, a detailed study is made of the operator ∇, including not only the operation of forming the gradient of a scalar function w but also the additional operations of forming the dot and cross products of the vector operator del with other vectors. (See Chapter 17.)*

It is our purpose in this article to develop some of the geometric properties of the gradient. The first property is the connection between the gradient and the directional derivative as developed in the previous article. Using ∇w to represent the gradient, we may write Eq. (2b), Article 15.5, in the form

$$\left(\frac{dw}{ds}\right)_0 = (\nabla w)_0 \cdot \mathbf{u}, \tag{3}$$

where the subscript 0 is used to indicate the fact that both ∇w and dw/ds are to be evaluated at the point $P_0(x_0, y_0, z_0)$.

From Eq. (3) and the geometric significance of the dot product of two vectors, we can say that

$$\begin{aligned}\left(\frac{dw}{ds}\right)_0 &= |(\nabla w)_0|\,|\mathbf{u}| \cos\theta \\ &= |(\nabla w)_0| \cos\theta, \end{aligned} \tag{4}$$

where θ is the angle between the vector $(\nabla w)_0$ and the unit vector $\mathbf{u}$. That is, $(dw/ds)_0$ is just the scalar

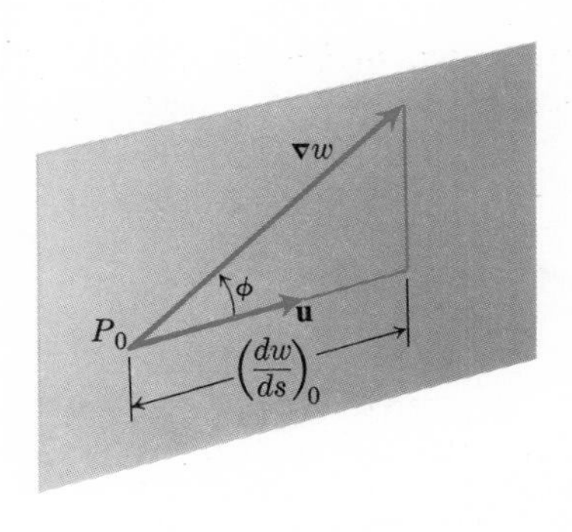

15.9

projection of grad w at P_0, onto the direction $\mathbf{u}$ (Fig. 15.9). Since this projection attains its maximum value when $\cos\theta = 1$ in Eq. (4), that is, when $\mathbf{u}$ and ∇w have the same direction, we can say that *the function $w = f(x, y, z)$ changes most rapidly in the direction given by the vector grad w itself. Moreover, the directional derivative in this direction is equal to the magnitude of the gradient.*

We may therefore characterize the gradient of the function $w = f(x, y, z)$ at the point $P_0(x_0, y_0, z_0)$ as a vector

(a) whose *direction* is that in which $(dw/ds)_0$ has its maximum value, and

(b) whose *magnitude* is equal to that maximum value of $(dw/ds)_0$.

A second interpretation of the gradient vector may be gained from the following considerations. The points at which the function $w = f(x, y, z)$ has the same value w_0 that it has at $P_0(x_0, y_0, z_0)$ will, in general, constitute a surface in space. The equation of this surface is

$$f(x, y, z) = w_0, \tag{5a}$$

$$f(x, y, z) - w_0 = 0, \tag{5b}$$

where w_0 is a constant. If w represents temperature, the surface given by Eq. (5) is an isothermal surface. If w represents electrical potential, then the surface is an equipotential surface. The fact that we now wish to establish is that *the gradient vector is normal to this isothermal or equipotential surface.* (See Fig. 15.10). To see that this is so, suppose we consider any curve C on the surface S of Eq. (5) and

---

* Also see Kaplan, *Advanced Calculus*, Chapters 3 and 5, Addison-Wesley, 1952.

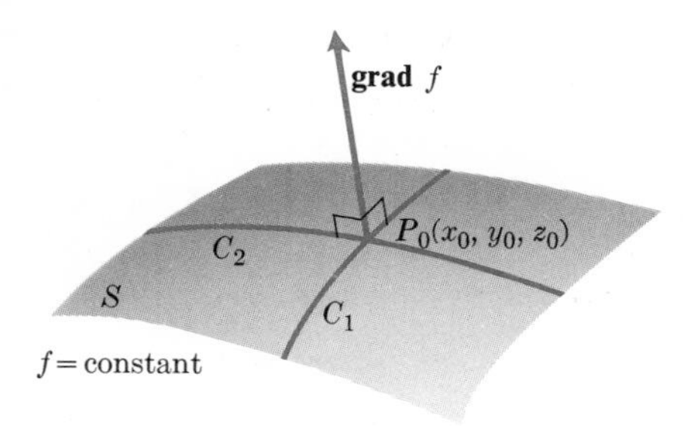

15.10

passing through P_0. Let us now calculate the directional derivative $(dw/ds)_0$ in the direction of the tangent to the curve C at P_0. This derivative is zero because w remains constant on C and hence $(\Delta w)_C = 0$, so that

$$\left(\frac{dw}{ds}\right)_0 = \lim \frac{\Delta w}{\Delta s} = 0.$$

If we compare this result with Eq. (4), we see that at any point P_0 where $(\nabla w)_0$ is not zero, $\cos\theta$ vanishes; that is,

$$(\nabla w)_0 \text{ is perpendicular to } \mathbf{u},$$

where $\mathbf{u}$ is a unit vector tangent to C at P_0. Now since C could be any curve on S through P_0, we have the result that $(\nabla w)_0$ *is normal to the surface* S *at* P_0. In the exceptional case where $(\nabla w)_0$ is the zero vector, it has no definite direction, but if we adopt the convention that the zero vector is orthogonal to every direction, then we may say without exception that the gradient of a function $f(x, y, z)$ at a point $P_0(x_0, y_0, z_0)$ is orthogonal to the surface

$$f(x, y, z) = \text{constant}$$

passing through that point.

Problem. Find the plane that is tangent to the surface

$$z = x^2 + y^2$$

at the point $P_0(1, -2, 5)$.

Solution. Let $w = f(x, y, z) = x^2 + y^2 - z$, so that the equation of the surface has the form

$$f(x, y, z) = \text{constant},$$

where the constant in this case is zero. Then the vector

$$(\text{grad}\, f)_0 = \left(\mathbf{i}\frac{\partial f}{\partial x} + \mathbf{j}\frac{\partial f}{\partial y} + \mathbf{k}\frac{\partial f}{\partial z}\right)_0$$
$$= (\mathbf{i}2x + \mathbf{j}2y - \mathbf{k})_{(1,-2,5)} = 2\mathbf{i} - 4\mathbf{j} - \mathbf{k}$$

is *normal* to the surface at P_0. But we recall that the equation of the plane through $P_0(x_0, y_0, z_0)$ normal to the vector

$$\mathbf{N} = A\mathbf{i} + B\mathbf{j} + C\mathbf{k}$$

is

$$A(x - x_0) + B(y - y_0) + C(z - z_0) = 0.$$

For the particular case at hand, we therefore have

$$2(x - 1) - 4(y + 2) - (z - 5) = 0,$$

or

$$2x - 4y - z = 5$$

as the equation of the tangent plane.

Note that the equation of the surface should be put into the form

$$f(x, y, z) = \text{constant}$$

to find the normal vector, grad f.

EXERCISES

In Exercises 1 through 7, find the electric intensity vector $\mathbf{E} = -\text{grad}\, V$ from the given potential function V, at the given point.

1. $V = x^2 + y^2 - 2z^2$, $(1, 1, 1)$
2. $V = 2z^3 - 3(x^2 + y^2)z$, $(1, 1, 1)$
3. $V = e^{-2y} \cos 2x$, $(\pi/4, 0, 0)$
4. $V = \ln \sqrt{x^2 + y^2}$, $(3, 4, 0)$
5. $V = (x^2 + y^2 + z^2)^{-(1/2)}$, $(1, 2, -2)$
6. $V = e^{3x+4y} \cos 5z$, $(0, 0, \pi/6)$
7. $V = \cos 3x \cos 4y \sinh 5z$, $(0, \pi/4, 0)$
8. Find equations of the line normal to the surface $z^2 = x^2 + y^2$ at $(3, 4, -5)$.
9. Lines are drawn through the origin and normal to the surface $xy + z = 2$.
   (a) Find equations of all such lines.
   (b) Find all points of intersection of these lines with the surface.

10. Find the locus of points on the surface

$$(y+z)^2+(z-x)^2 = 16,$$

where the normal is parallel to the yz-plane.

11. Find the tangent plane and normal to the hyperboloid

$$x^2+y^2-z^2 = 18 \quad \text{at} \quad (3, 5, -4).$$

12. In which direction should one travel, starting from the point $P_0(2, -1, 2)$, to obtain the most rapid rate of increase of the function

$$f = (x+y)^2+(y+z)^2+(z+x)^2?$$

What is the instantaneous rate of change of f per unit of distance in this direction?

13. Suppose cylindrical coordinates r, θ, z are introduced into a function $w = f(x, y, z)$ to yield $w = F(r, \theta, z)$. Show that the gradient may be expressed in terms of cylindrical coordinates and the unit vectors $\mathbf{u}_r$, $\mathbf{u}_\theta$, $\mathbf{k}$ as follows:

$$\nabla w = \mathbf{u}_r \frac{\partial w}{\partial r}+\frac{1}{r}\mathbf{u}_\theta \frac{\partial w}{\partial \theta}+\mathbf{k}\frac{\partial w}{\partial z}.$$

Hint. The component of ∇w in the direction of $\mathbf{u}_r$ is equal to the directional derivative dw/ds in that direction. But this is precisely $\partial w/\partial r$. Reason similarly for the components of ∇w in the directions of $\mathbf{u}_\theta$ and $\mathbf{k}$.

14. Express the gradient in terms of spherical coordinates and the appropriate unit vectors $\mathbf{u}_\rho$, $\mathbf{u}_\phi$, $\mathbf{u}_\theta$. Use a geometrical argument to determine the component of ∇w in each of these directions. (See hint for Exercise 13.)

15. (a) In the case of a function $w = f(x, y)$ of two independent variables, what is the expression for grad f?

(b) Find the direction in which the function

$$w = x^2+xy+y^2$$

increases most rapidly at the point $(-1, 1)$. What is the magnitude of dw/ds in this direction?

In Exercises 16 through 23, verify that the function V given in the exercise named satisfies Laplace's equation

$$\frac{\partial^2 V}{\partial x^2}+\frac{\partial^2 V}{\partial y^2}+\frac{\partial^2 V}{\partial z^2} = 0.$$

16. Exercise 1 17. Exercise 2 18. Exercise 3

19. Exercise 4 20. Exercise 5 21. Exercise 6

22. Exercise 7 23. Exercise 8

24. *Method of steepest descent.* Suppose it is desired to find a solution of the equation $f(x, y, z) = 0$. Let $P_0(x_0, y_0, z_0)$ be a first guess, and suppose

$$f(x_0, y_0, z_0) = f_0$$

is not zero. Let $(\nabla f)_0$ be the gradient vector normal to the surface $f(x, y, z) = f_0$ at P_0. If f_0 is positive, we want to decrease the value of f. The gradient points in the direction of most rapid increase, its negative in the direction of "steepest descent." We therefore take as next approximation

$$x_1 = x_0 - hf_x(x_0, y_0, z_0),$$
$$y_1 = y_0 - hf_y(x_0, y_0, z_0),$$
$$z_1 = z_0 - hf_z(x_0, y_0, z_0).$$

What value of h corresponds to making $\Delta f_{\text{lin}} = -f_0$? What change is suggested if f_0 is negative?

Note. The method could be applied to the problem of solving the simultaneous equations

$$2x+3y+4z = 5, \quad x^2+y^2+z^2 = 7, \quad xyz = 4$$

by writing

$$f(x, y, z) = (2x+3y+4z-5)^2 + (x^2+y^2+z^2-7)^2+(xyz-4)^2.$$

15.7 THE CHAIN RULE FOR PARTIAL DERIVATIVES

We recall that the formula

$$\frac{dy}{dt} = \frac{dy}{dx}\frac{dx}{dt}, \tag{1}$$

developed in an early chapter, is useful in connection with problems in related rates and with curves whose equations are given in parametric form. Equation (1) expresses the so-called "chain rule" for differentiating a function

$$y = f(x) \tag{2}$$

with respect to t, when x is a function of t,

$$x = g(t). \tag{3}$$

The result of substituting from Eq. (3) into Eq. (2) gives y as a function $F(t)$:

$$y = f[g(t)] = F(t). \tag{4}$$

Equation (1) tells us that the derivative of this function, namely $F'(t)$, can be found by calculating the terms on the right-hand side of the equation

$$F'(t) = f'(x)g'(t), \tag{5a}$$

or

$$F_t = f_x g_t, \tag{5b}$$

$$y_t = y_x x_t, \tag{5c}$$

where in (5a) we use primes to denote derivatives and in (5b, c) we use subscripts for the same purpose.

In this article, we shall extend this "chain rule" to cases in which functions of several variables and partial derivatives are involved. Suppose then that we have a function, say

$$w = f(x, y, z), \tag{6}$$

which has continuous partial derivatives

$$\frac{\partial w}{\partial x} = f_x, \quad \frac{\partial w}{\partial y} = f_y, \quad \frac{\partial w}{\partial z} = f_z \tag{7}$$

throughout some region R of xyz-space. Suppose we want to study the behavior of the function f along some curve C lying in R. Let the equations of C in terms of the parameter t be

$$x = x(t), \quad y = y(t), \quad z = z(t). \tag{8}$$

Such a situation arises, for example, in studying the pressure or density in a moving fluid. Then the equation which takes the place of Eq. (1) or (5) is

$$\frac{dw}{dt} = \frac{\partial w}{\partial x}\frac{dx}{dt} + \frac{\partial w}{\partial y}\frac{dy}{dt} + \frac{\partial w}{\partial z}\frac{dz}{dt}, \tag{9}$$

as we shall now show.

To prove Eq. (9), let t_0 be a value of t that corresponds to a point P_0 in R, and let Δt be an increment in t such that the point P that corresponds to $t_0 + \Delta t$ also lies in R. Let Δx, Δy, Δz, Δw denote the increments in x, y, z, w. Then by Eq. (18), Article 15.4, we may write

$$\frac{\Delta w}{\Delta t} = \left(\frac{\partial w}{\partial x}\right)_0 \frac{\Delta x}{\Delta t} + \left(\frac{\partial w}{\partial y}\right)_0 \frac{\Delta y}{\Delta t} + \left(\frac{\partial w}{\partial z}\right)_0 \frac{\Delta z}{\Delta t} + \epsilon_1 \frac{\Delta x}{\Delta t} + \epsilon_2 \frac{\Delta y}{\Delta t} + \epsilon_3 \frac{\Delta z}{\Delta t}, \tag{10}$$

where the subscript zero on the partial derivatives indicates that they are to be evaluated at P_0, and where

$$\epsilon_1, \epsilon_2, \epsilon_3 \to 0 \quad \text{as} \quad \Delta x, \Delta y, \Delta z \to 0.$$

Suppose we now let $\Delta t \to 0$ in Eq. (10) and assume that the curve C given by Eqs. (8) is such that the derivatives dx/dt, dy/dt, dz/dt all exist at t_0. Then

$$\Delta x, \Delta y, \Delta z \to 0 \quad \text{as} \quad \Delta t \to 0,$$

and the three terms in Eq. (10) that involve the epsilons go to zero, while the other terms give

$$\left(\frac{dw}{dt}\right)_0 = \left(\frac{\partial w}{\partial x}\right)_0 \left(\frac{dx}{dt}\right)_0 + \left(\frac{\partial w}{\partial y}\right)_0 \left(\frac{dy}{dt}\right)_0 + \left(\frac{\partial w}{\partial z}\right)_0 \left(\frac{dz}{dt}\right)_0,$$

which is Eq. (9) with all derivatives evaluated for $t = t_0$, $x = x_0$, $y = y_0$, and $z = z_0$.

Remark 1. Equation (9) can be written in the following vector form:

$$\frac{dw}{dt} = \nabla \mathbf{w} \cdot \mathbf{v}, \tag{11a}$$

where, if t is time, $\mathbf{v}$ is the velocity

$$\mathbf{v} = \mathbf{i}\frac{dx}{dt} + \mathbf{j}\frac{dy}{dt} + \mathbf{k}\frac{dz}{dt}. \tag{11b}$$

We can also use a matrix product to write Eq. (9) as

$$\frac{dw}{dt} = \begin{bmatrix} \dfrac{\partial w}{\partial x} & \dfrac{\partial w}{\partial y} & \dfrac{\partial w}{\partial x} \end{bmatrix} \begin{bmatrix} \dfrac{dx}{dt} \\ \dfrac{dy}{dt} \\ \dfrac{dz}{dt} \end{bmatrix}. \tag{11c}$$

There is no essential complication introduced by considering the behavior of the function w in Eq. (6) on a *surface* S lying in R. It takes, in general, *two* parameters to give the equations of a surface (for example, *latitude* and *longitude* on the surface of a sphere). Hence, consider the case where x, y, and z are functions of two parameters, say r and s.

$$x = x(r, s), \quad y = y(r, s), \quad z = z(r, s), \tag{12a}$$

and calculate

$$\frac{\partial w}{\partial r} = \lim_{\Delta r \to 0} \frac{\Delta w}{\Delta r} \tag{12b}$$

with s held constant. In this case, Eq. (10) is to be replaced by a similar equation with Δr in place of Δt throughout. When $\Delta r \to 0$ (s held constant), we have

$$\lim_{\Delta r \to 0} \frac{\Delta x}{\Delta r} = \frac{\partial x}{\partial r}$$

and two similar expressions with y and z in place of x. Thus, when $\Delta r \to 0$, the result is

$$\frac{\partial w}{\partial r} = \frac{\partial w}{\partial x}\frac{\partial x}{\partial r} + \frac{\partial w}{\partial y}\frac{\partial y}{\partial r} + \frac{\partial w}{\partial z}\frac{\partial z}{\partial r}. \tag{13}$$

A similar expression with s in place of r could also be derived for $\partial w/\partial s$.

More generally, we may consider a function

$$w = f(x, y, z, u, \ldots, v)$$

of any number of variables $x, y, z, u, \ldots, v$ and study the behavior of this function when these variables are related to any number of other variables $p, q, r, s, \ldots, t$ by equations

$$\begin{aligned} x &= x(p, q, r, s, \ldots, t), \\ y &= y(p, q, r, s, \ldots, t), \\ &\vdots \\ v &= v(p, q, r, s, \ldots, t). \end{aligned} \tag{14a}$$

Then suppose it is required to find

$$\partial w/\partial p, \quad \partial w/\partial q, \quad \partial w/\partial r, \quad \ldots, \quad \partial w/\partial t.$$

By the methods used above, we find

$$\frac{\partial w}{\partial p} = \frac{\partial w}{\partial x}\frac{\partial x}{\partial p} + \frac{\partial w}{\partial y}\frac{\partial y}{\partial p} + \frac{\partial w}{\partial z}\frac{\partial z}{\partial p} + \cdots + \frac{\partial w}{\partial v}\frac{\partial v}{\partial p}. \tag{14b}$$

In terms of the subscript notation for partial derivatives,

$$w_p = w_x x_p + w_y y_p + w_z z_p + \cdots + w_v v_p.$$

In matrix form, Eq. (14b) can be written

$$\frac{\partial w}{\partial p} = \begin{bmatrix} \dfrac{\partial w}{\partial x} & \dfrac{\partial w}{\partial y} & \dfrac{\partial w}{\partial z} & \cdots & \dfrac{\partial w}{\partial v} \end{bmatrix} \begin{bmatrix} \dfrac{\partial x}{\partial p} \\ \dfrac{\partial y}{\partial p} \\ \dfrac{\partial z}{\partial p} \\ \vdots \\ \dfrac{\partial v}{\partial p} \end{bmatrix}. \tag{14c}$$

If we replace p by q, or r, or s, . . . , or t on both sides of this equation, then we get $\partial w/\partial q$, or $\partial w/\partial r$, or $\partial w/\partial s$, . . . , or $\partial w/\partial t$. All these equations are indicated by the following matrix equation:

$$\begin{bmatrix} \dfrac{\partial w}{\partial p} & \dfrac{\partial w}{\partial q} & \cdots & \dfrac{\partial w}{\partial t} \end{bmatrix} = \begin{bmatrix} \dfrac{\partial w}{\partial x} & \dfrac{\partial w}{\partial y} & \cdots & \dfrac{\partial w}{\partial v} \end{bmatrix} \begin{bmatrix} \dfrac{\partial x}{\partial p} & \dfrac{\partial x}{\partial q} & \cdots & \dfrac{\partial x}{\partial t} \\ \dfrac{\partial y}{\partial p} & \dfrac{\partial y}{\partial q} & \cdots & \dfrac{\partial y}{\partial t} \\ \vdots & \vdots & & \vdots \\ \dfrac{\partial v}{\partial p} & \dfrac{\partial v}{\partial q} & \cdots & \dfrac{\partial v}{\partial t} \end{bmatrix}. \tag{14d}$$

To get the partial derivative in any column of the row vector on the left, take the inner product of the row vector

$$\text{grad } w = \begin{bmatrix} \dfrac{\partial w}{\partial x} & \dfrac{\partial w}{\partial y} & \cdots & \dfrac{\partial w}{\partial v} \end{bmatrix}$$

with the appropriate column in the matrix on the right.

Problem 1. Suppose that $w = r^2 \cos 2\theta$, where

$$x = r\cos\theta, \qquad y = r\sin\theta,$$
$$r = \sqrt{x^2+y^2}, \quad \theta = \tan^{-1}(y/x).$$

Find $\partial w/\partial x$ and $\partial w/\partial y$.

Solution. (The reader may wish to verify some of the following differentiation results with paper and pencil.)

$$\begin{bmatrix} \frac{\partial w}{\partial x} & \frac{\partial w}{\partial y} \end{bmatrix} = \begin{bmatrix} \frac{\partial w}{\partial r} & \frac{\partial w}{\partial \theta} \end{bmatrix} \begin{bmatrix} \frac{\partial r}{\partial x} & \frac{\partial r}{\partial y} \\ \frac{\partial \theta}{\partial x} & \frac{\partial \theta}{\partial y} \end{bmatrix}$$

$$= [2r\cos 2\theta \quad -2r^2 \sin 2\theta] \begin{bmatrix} \frac{x}{r} & \frac{y}{r} \\ -\frac{y}{r^2} & \frac{x}{r^2} \end{bmatrix}$$

$$= [2x\cos 2\theta + 2y\sin 2\theta \quad 2y\cos 2\theta - 2x\sin 2\theta].$$

Therefore

$$\frac{\partial w}{\partial x} = 2(x\cos 2\theta + y\sin 2\theta)$$
$$= 2r(\cos\theta\cos 2\theta + \sin\theta\sin 2\theta)$$
$$= 2r\cos(2\theta - \theta) = 2x,$$

and

$$\frac{\partial w}{\partial y} = 2(y\cos 2\theta - x\sin 2\theta)$$
$$= 2r(\sin\theta\cos 2\theta - \cos\theta\sin 2\theta)$$
$$= 2r\sin(\theta - 2\theta) = -2y.$$

Check. $w = r^2(\cos^2\theta - \sin^2\theta) = x^2 - y^2,$

so

$$\frac{\partial w}{\partial x} = 2x, \qquad \frac{\partial w}{\partial y} = -2y.$$

The chain rule for a change of variables is often used to transform a differential equation into a new form. Naturally we hope that the new form is easier to deal with than the original form. The following example illustrates such a situation.

Problem 2. Show that the change of variables from x and y to

$$r = y - ax, \qquad s = y + ax \tag{15a}$$

transforms the differential equation

$$\frac{\partial w}{\partial x} - a\frac{\partial w}{\partial y} = 0 \tag{15b}$$

into a form that is more easily solved, and solve it. (Here a is a constant.)

Solution. We imagine that $w = f(x, y)$ is transformed into $w = F(r, s)$ under the transformation represented by Eqs. (15a). If $a \neq 0$, those equations can be solved for x and y in terms of r and s:

$$x = \frac{1}{2a}(s - r), \qquad y = \frac{1}{2}(r + s).$$

In the exceptional case that $a = 0$, we would not make the suggested change of variables for reasons that will be discussed below. We proceed on the assumption that $a \neq 0$, and apply the chain rule:

$$\begin{bmatrix} \frac{\partial w}{\partial x} & \frac{\partial w}{\partial y} \end{bmatrix} = \begin{bmatrix} \frac{\partial w}{\partial r} & \frac{\partial w}{\partial s} \end{bmatrix} \begin{bmatrix} \frac{\partial r}{\partial x} & \frac{\partial r}{\partial y} \\ \frac{\partial s}{\partial x} & \frac{\partial s}{\partial y} \end{bmatrix}$$

$$= \begin{bmatrix} \frac{\partial w}{\partial r} & \frac{\partial w}{\partial s} \end{bmatrix} \begin{bmatrix} -a & 1 \\ a & 1 \end{bmatrix}$$

$$= \begin{bmatrix} -a\frac{\partial w}{\partial r} + a\frac{\partial w}{\partial s} & \frac{\partial w}{\partial r} + \frac{\partial w}{\partial s} \end{bmatrix}.$$

Therefore, from the definition of matrix equality,

$$\frac{\partial w}{\partial x} = -a\frac{\partial w}{\partial r} + a\frac{\partial w}{\partial s}, \qquad \frac{\partial w}{\partial y} = \frac{\partial w}{\partial r} + \frac{\partial w}{\partial s}. \tag{16a}$$

The result of substituting from Eqs. (16a) into the left-hand side of Eq. (15b) is $-2a(\partial w/\partial r)$, so the new equation is

$$-2a\frac{\partial w}{\partial r} = 0, \quad \text{or} \quad \frac{\partial w}{\partial r} = 0. \tag{16b}$$

But this equation is *easy* to solve! It simply requires that $w = F(r, s)$ be a constant when s is constant and r is allowed to vary. That is, w must be a function of s alone:

$$w = \phi(s) = \phi(y + ax).$$

Here $\phi(s)$ is *any* differentiable function of s whatever; for example,

$$\phi(s) = e^{2s} + \tan^{-1}(s^2) + \sqrt{s^2+4}$$

would be a suitable function. For this special case, we have

$$\begin{aligned} w &= \phi(y+ax) \\ &= e^{2y+2ax} + \tan^{-1}(y+ax)^2 + \sqrt{(y+ax)^2+4} \end{aligned}$$

as a function that satisfies the original partial differential equation.

Remark 2. If $a = 0$ in Problem 2, the original equation is $\partial w/\partial x = 0$. This is like Eq. (16b) with $r = y$, and the solution is $w = \phi(y)$.

Remark 3. In solving an *ordinary* differential equation (one that does not involve partial derivatives), we usually get a general solution that has one or more arbitrary constants of integration in it. Those can sometimes be evaluated from given initial conditions. Similarly, in solving a partial differential equation, we may (as above) get a general solution which involves one or more arbitrary functions corresponding to "constants of integration." These also can sometimes be determined from given conditions, called *initial* conditions or *boundary* conditions. Instead of specifying the value of w at a single point, these conditions might specify the values along some line or curve, as in the next problem.

Problem 3. In Problem 2 above, determine w if its values along the x-axis are given by $w = \sin x$, assuming that $a \neq 0$.

Solution. The general solution is

$$w = f(x, y) = \phi(y + ax),$$

as we determined in Problem 2. On the x-axis, $y = 0$, and we want

$$f(x, 0) = \phi(ax) = \sin x. \tag{17}$$

Temporarily, let $u = ax$ and $x = u/a$, and then Eq. (17) becomes

$$\phi(u) = \sin(u/a).$$

Now, just put $y + ax$ in place of u:

$$w = \phi(y+ax) = \sin\frac{y+ax}{a}. \tag{18}$$

Check. If w is given by Eq. (18), then

$$\frac{\partial w}{\partial x} = \cos\frac{y+ax}{a} \quad \text{and} \quad \frac{\partial w}{\partial y} = \frac{1}{a}\cos\frac{y+ax}{a}.$$

so that

$$\frac{\partial w}{\partial x} - a\frac{\partial w}{\partial y} = 0.$$

Also, when $y = 0$,

$$w = \sin\frac{0+ax}{a} = \sin x.$$

Therefore Eq. (18) satisfies the partial differential equation (15b) and the given initial conditions (17).

EXERCISES

In Exercises 1 through 3, find dw/dt

(a) by expressing w explicitly as a function of t and then differentiating, and

(b) by using the chain rule.

1. $w = x^2 + y^2 + z^2,$

   $x = e^t \cos t, \quad y = e^t \sin t, \quad z = e^t$

2. $w = \dfrac{xy}{x^2+y^2}, \quad x = \cosh t, \quad y = \sinh t$

3. $w = e^{2x+3y}\cos 4z,$

   $x = \ln t, \quad y = \ln(t^2+1), \quad z = t$

4. If $$w = \sqrt{x^2+y^2+z^2},$$

   $$x = e^r \cos s, \quad y = e^r \sin s, \quad z = e^s,$$

   find $\partial w/\partial r$ and $\partial w/\partial s$ by the chain rule and check your answer by using a different method.

5. If $$w = \ln(x^2 + y^2 + 2z),$$

   $$x = r + s, \quad y = r - s, \quad z = 2rs,$$

   find $\partial w/\partial r$ and $\partial w/\partial s$ by the chain rule and check your answer by a different method.

6. If a and b are constants and

   $$w = (ax+by)^3 + \tanh(ax+by) + \cos(ax+by),$$

   show that

   $$a\frac{\partial w}{\partial y} = b\frac{\partial w}{\partial x}.$$

7. If a and b are constants and $w = f(ax + by)$ is a differentiable function of $u = ax + by$, show that

$$a\frac{\partial w}{\partial y} = b\frac{\partial w}{\partial x}.$$

Hint. Apply the chain rule with u as the only independent variable in the first set of variables.

In Exercises 8 through 10, use the result of Exercise 7 to find a solution of the partial differential equation

$$a\frac{\partial w}{\partial y} = b\frac{\partial w}{\partial x}, \quad a \neq 0, b \neq 0,$$

that satisfies the initial conditions given.

8. $w = 5$ along the y-axis
9. $w = e^{-x}$ along the x-axis
10. $w = e^{-x}$ along the line $y = x$. (Here you may need to assume something more about a and b.)
11. Explain why each line of the form $ax + by = c$ is a contour line (or level line) for $w = f(ax + by)$. Would it be possible to find such a function f that satisfies the condition $w = \sin x$ along a line

$$ax + by = 5?$$

Why?

12. If $w = f[xy/(x^2 + y^2)]$ is a differentiable function of $u = xy/(x^2 + y^2)$, show that

$$x(\partial w/\partial x) + y(\partial w/\partial y) = 0.$$

(See the hint for Exercise 7.)

13. If $w = f(x + y, x - y)$ has continuous partial derivatives with respect to $u = x + y$, $v = x - y$, show that

$$\frac{\partial w}{\partial x}\frac{\partial w}{\partial y} = \left(\frac{\partial f}{\partial u}\right)^2 - \left(\frac{\partial f}{\partial v}\right)^2.$$

14. Verify the result given in Exercise 13, Article 15.6 by transforming the given expression on the right-hand side of the equation into **i**, **j**, **k**-components and replacing the cylindrical coordinates r, θ by Cartesian coordinates x, y, and making use of the chain rule for partial derivatives.
15. Verify the answer obtained in Exercise 14, Article 15.6 by transforming the expression you obtained in spherical coordinates back into Cartesian coordinates. Make use of the chain rule for partial derivatives.
16. If we substitute polar coordinates $x = r\cos\theta$ and $y = r\sin\theta$ in a function $w = f(x, y)$, show that

$$\frac{\partial w}{\partial r} = f_x\cos\theta + f_y\sin\theta,$$

$$\frac{1}{r}\frac{\partial w}{\partial \theta} = -f_x\sin\theta + f_y\cos\theta.$$

17. Using determinants, solve the equations given in Exercise 16 for f_x and f_y in terms of $(\partial w/\partial r)$ and $(\partial w/\partial \theta)$.
18. In connection with Exercise 16, show that

$$\left(\frac{\partial w}{\partial r}\right)^2 + \frac{1}{r^2}\left(\frac{\partial w}{\partial \theta}\right)^2 = f_x^2 + f_y^2.$$

15.8 THE TOTAL DIFFERENTIAL

The differential of a function

$$w = f(x, y, z) \tag{1}$$

is defined to be

$$dw = \frac{\partial w}{\partial x}dx + \frac{\partial w}{\partial y}dy + \frac{\partial w}{\partial z}dz. \tag{2}$$

The chain rule (Eq. 9, Article 15.7) tells us that we may formally divide both sides of Eq. (2) by dt to calculate dw/dt if x, y, z are differentiable functions of t. Or, if x, y, z are functions of the independent variables r, s and we want to calculate $\partial w/\partial r$, we hold s constant in calculating dx, dy, dz and divide both sides of Eq. (2) by dr, but write $\partial w/\partial r$, and so on, in place of dw/dr, and so on, to show that s has been held constant.

The separate terms

$$\frac{\partial w}{\partial x}dx, \qquad \frac{\partial w}{\partial y}dy, \qquad \frac{\partial w}{\partial z}dz$$

are sometimes called "partial differentials" of w with respect to x, y, z respectively. Then the sum of these partial differentials, Eq. (2), is called the *total differential* dw.

In general, the total differential of a function

$$w = F(x, y, z, u, \ldots, v)$$

is defined to be the sum of all its partial differentials:

$$dw = F_x\,dx + F_y\,dy + F_z\,dz + F_u\,du + \cdots + F_v\,dv.$$

If x, y, and z are independent variables in Eq. (1), then dx, dy, and dz are three *new independent* variables in Eq. (2). But in any problem involving increments we shall agree to take

$$dx = \Delta x, \quad dy = \Delta y, \quad dz = \Delta z, \tag{3}$$

so that we shall be able to use the differential dw as a good approximation to Δw (see Eq. 19a, Article 15.4). When x, y, z are *not* the independent variables, but are themselves given by equations such as

$$\begin{aligned} x &= x(t), & & & x &= x(r, s), \\ y &= y(t), & &\text{or} & y &= y(r, s), \\ z &= z(t), & & & z &= z(r, s), \end{aligned}$$

then in the first case we have

$$dx = x'(t)\,dt, \quad dy = y'(t)\,dt, \quad dz = z'(t)\,dt,$$

and in the second case,

$$\begin{aligned} dx &= \frac{\partial x}{\partial r}\,dr + \frac{\partial x}{\partial s}\,ds, \\ dy &= \frac{\partial y}{\partial r}\,dr + \frac{\partial y}{\partial s}\,ds, \\ dz &= \frac{\partial z}{\partial r}\,dr + \frac{\partial z}{\partial s}\,ds, \end{aligned} \tag{4}$$

if we are to be consistent.

Suppose we consider the second case in more detail. If we consider

$$w = f[x(r, s), y(r, s), z(r, s)] = F(r, s)$$

as a function of r and s, then instead of Eq. (2) we should have

$$dw = \frac{\partial w}{\partial r}\,dr + \frac{\partial w}{\partial s}\,ds, \tag{5}$$

where

$$\frac{\partial w}{\partial r} = F_r(r, s), \qquad \frac{\partial w}{\partial s} = F_s(r, s).$$

The question now arises as to whether or not the dw given by Eq. (2) is the same as the dw given by Eq. (5). The answer, which is "yes, they are the same," is a consequence of the chain rule for derivatives. For if we start with the dw given by Eq. (2) and into it substitute dx, dy, dz given by Eq. (4), then we obtain

$$\begin{aligned} dw &= \frac{\partial w}{\partial x}\left(\frac{\partial x}{\partial r}\,dr + \frac{\partial x}{\partial s}\,ds\right) \\ &\quad + \frac{\partial w}{\partial y}\left(\frac{\partial y}{\partial r}\,dr + \frac{\partial y}{\partial s}\,ds\right) + \frac{\partial w}{\partial z}\left(\frac{\partial z}{\partial r}\,dr + \frac{\partial z}{\partial s}\,ds\right) \\ &= \left(\frac{\partial w}{\partial x}\frac{\partial x}{\partial r} + \frac{\partial w}{\partial y}\frac{\partial y}{\partial r} + \frac{\partial w}{\partial z}\frac{\partial z}{\partial r}\right) dr \\ &\quad + \left(\frac{\partial w}{\partial x}\frac{\partial x}{\partial s} + \frac{\partial w}{\partial y}\frac{\partial y}{\partial s} + \frac{\partial w}{\partial z}\frac{\partial z}{\partial s}\right) ds, \end{aligned}$$

and the expressions in parentheses which here multiply dr and ds are the same as $\partial w/\partial r$ and $\partial w/\partial s$, respectively, by virtue of the chain rule for derivatives. Thus, starting with the expression for dw given by Eq. (2), we have transformed it into the expression for dw given by Eq. (5), thereby establishing the equivalence of the two.

It should be pointed out that in the case just discussed, where r and s are the *independent* variables, we are to treat dr and ds also as independent variables, but *not* dx, dy, and dz, which indeed are given by Eqs. (4). Thus, in a problem involving increments, we could for convenience take

$$dr = \Delta r \quad\text{and}\quad ds = \Delta s,$$

but we should *not* take

$$dx = \Delta x, \quad dy = \Delta y, \quad dz = \Delta z,$$

since we are bound by Eqs. (4). The differentials dx, dy, and dz will, however, usually be reasonably good *approximations* to the increments Δx, Δy, and Δz when Δr and Δs are small (Article 15.4).

Example 1. Consider the function

$$w = x^2 + y^2 + z^2,$$

with

$$x = r\cos s, \quad y = r\sin s, \quad z = r.$$

If we use Eq. (2), we have

$$dw = 2(x\,dx + y\,dy + z\,dz)$$

with

$$dx = \cos s\,dr - r\sin s\,ds,$$

$$dy = \sin s\,dr + r\cos s\,ds,$$

$$dz = dr,$$

and hence

$$\begin{aligned} dw &= 2(x\cos s + y\sin s + z)\,dr \\ &\quad + 2(-xr\sin s + yr\cos s)\,ds \\ &= 2(r\cos^2 s + r\sin^2 s + r)\,dr \\ &\quad + 2(-r^2\cos s\sin s + r^2\sin s\cos s)\,ds \\ &= 4r\,dr. \end{aligned}$$

On the other hand, if we first express w directly in terms of r and s, we obtain

$$w = r^2\cos^2 s + r^2\sin^2 s + r^2 = 2r^2,$$

from which we also obtain

$$dw = 4r\,dr.$$

Problem 1. Show that the slope at the point (x, y) of the plane curve whose equation is given implicitly by

$$F(x, y) = 0$$

is

$$\frac{dy}{dx} = \frac{-F_x(x, y)}{F_y(x, y)} \quad \text{if} \quad F_y(x, y) \neq 0.$$

Solution. Let $w = F(x, y)$ and consider the directional derivative of w at the point (x, y) in the direction of the tangent to the curve. If s denotes arc length along the curve, the directional derivative is

$$\frac{dF}{ds} = F_x\frac{dx}{ds} + F_y\frac{dy}{ds}.$$

But F is constant along the given curve, so that

$$\frac{dF}{ds} = \lim_{\Delta s \to 0}\frac{\Delta F}{\Delta s} = 0.$$

Therefore

$$F_x\frac{dx}{ds} + F_y\frac{dy}{ds} = 0,$$

and hence, if $F_y \neq 0$,

$$\frac{dy}{dx} = \frac{dy/ds}{dx/ds} = \frac{-F_x}{F_y}. \tag{6}$$

Remark 1. Equation (6) for dy/dx produces the same result we would get if we were to differentiate the equation $F(x, y) = 0$ *implicitly* with respect to x, *assuming* that the equation determines y as a differentiable function of x. The theorem that supports this assumption is called the *implicit function theorem*, and one part of the hypothesis of that theorem is that $F_y(x_0, y_0) \neq 0$. (For a complete statement of the theorem and its proof, see, for example, J. M. H. Olmsted, *Advanced Calculus*, Appleton-Century-Crofts, Inc., 1961, p. 326.)

Problem 2. Let $w = F(x, y, z)$ be *constant* along a curve C passing through $P_0(x_0, y_0, z_0)$. Let dx, dy, dz be such that the vector

$$d\mathbf{R} = \mathbf{i}\,dx + \mathbf{j}\,dy + \mathbf{k}\,dz$$

is tangent to C at P_0. Show that

$$dw = \text{grad } F \cdot d\mathbf{R} = 0.$$

Solution. By definition,

$$\begin{aligned} dw &= F_x\,dx + F_y\,dy + F_z\,dz \\ &= (\mathbf{i}F_x + \mathbf{j}F_y + \mathbf{k}F_z)\cdot(\mathbf{i}\,dx + \mathbf{j}\,dy + \mathbf{k}\,dz) \\ &= \text{grad } F \cdot d\mathbf{R}. \end{aligned}$$

Also, the directional derivative of w, at P_0, in the direction of $d\mathbf{R}$ is

$$\frac{dw}{ds} = \text{grad } F \cdot \mathbf{u},$$

where

$$\mathbf{u} = \frac{d\mathbf{R}}{|d\mathbf{R}|}$$

is a unit vector in the direction of $d\mathbf{R}$. Along C, w remains constant and $dw/ds = 0$. Therefore

$$\begin{aligned} dw &= \text{grad } F \cdot d\mathbf{R} \\ &= \text{grad } F \cdot \mathbf{u}\,|d\mathbf{R}| = \frac{dw}{ds}|d\mathbf{R}| = 0. \end{aligned}$$

Remark 2. Let $x = f(t)$, $y = g(t)$, $z = h(t)$ be parametric equations of the curve C of Problem 2. Then $d\mathbf{R}/dt$ is tangent to C, and $dw/dt = \nabla w \cdot (d\mathbf{R}/dt)$ is zero because w is constant on C. Hence ∇w is normal to C.

EXERCISES

1. Show that the formulas

(a) $d(u + v) = du + dv$,
(b) $d(uv) = v\,du + u\,dv$,
(c) $d\left(\frac{u}{v}\right) = \frac{v\,du - u\,dv}{v^2}$

are valid for total differentials, if u and v are independent variables or if they are functions of any number of independent variables, such as

$$u = u(x, y, \ldots, p), \quad v = v(x, y, \ldots, p).$$

2. Using differentials to approximate to increments, find the amount of material in a hollow rectangular box whose inside measurements are 5 ft long, 3 ft wide, and 2 ft deep, if the box is made of lumber which is $\frac{1}{2}$ in. thick and the box has no top.

3. The area of a triangle is $A = \frac{1}{2}ab \sin C$, where a and b are two sides of the triangle and C is the included angle. In surveying a particular triangular plot of land, a and b are measured to be 150 ft and 200 ft respectively, and C is read to be 60°. By how much (approximately) is the computed area in error if a and b are in error by $\frac{1}{2}$ ft each and C is in error by 2°?

4. (a) Given $x = r\cos\theta$, $y = r\sin\theta$, express dx and dy in terms of dr and $d\theta$.

(b) Solve the equations of (a) for dr and $d\theta$ in terms of dx and dy.

(c) In the answer to (b), suppose A and B are the coefficients of dx and dy in the expression for dr, that is,

$$dr = A\,dx + B\,dy.$$

Verify by direct computation that

$$A = \partial r/\partial x \quad \text{and} \quad B = \partial r/\partial y,$$

where $r^2 = x^2 + y^2$.

5. Given $x = f(u, v)$, $y = g(u, v)$. If these equations are considered as implicitly defining u and v as functions of x and y,

(a) express dx and dy in terms of du and dv,

(b) use determinants to solve the equations of (a) for du and dv in terms of dx and dy, and

(c) show that

$$\frac{\partial u}{\partial x} = \frac{g_v}{f_u g_v - f_v g_u}$$

wherever

$$f_u g_v - f_v g_u \neq 0.$$

6. (a) Given

$$x = \rho \sin\phi \cos\theta, \quad y = \rho \sin\phi \sin\theta, \quad z = \rho\cos\phi,$$

express dx, dy, dz in terms of $d\rho$, $d\phi$, $d\theta$.

(b) Solve the equations of (a) for $d\rho$ in terms of dx, dy, dz by the use of determinants.

(c) From your answers to (b), read off $\partial\rho/\partial x$, considering ρ, ϕ, θ as functions of x, y, z that are given implicitly by the equations of (a).

7. *Newton's method.* It is desired to find values of x and y that satisfy the pair of equations $f(x, y) = 0$ and $g(x, y) = 0$ simultaneously. Suppose it is found by trial and error or otherwise, that

$$u_0 = f(x_0, y_0) \quad \text{and} \quad v_0 = g(x_0, y_0)$$

are both small in absolute value. It is now desired to find "corrections" dx and dy such that

$$f(x_0 + dx, y_0 + dy) = g(x_0 + dx, y_0 + dy) = 0.$$

Using $u_0 + df$ and $v_0 + dg$ to approximate

$$f(x_0 + dx, y_0 + dy) \quad \text{and} \quad g(x_0 + dx, y_0 + dy),$$

respectively, determine approximate values of dx and dy. The procedure may be repeated with (x_0, y_0) replaced by $(x_1, y_1) = (x_0 + dx, y_0 + dy)$.

8. Generalize the method of Exercise 7 to the case of three equations in three unknowns:

$$f(x, y, z) = 0, \quad g(x, y, z) = 0, \quad h(x, y, z) = 0,$$

assuming that

$$u_0 = f(x_0, y_0, z_0),$$
$$v_0 = g(x_0, y_0, z_0),$$
$$w_0 = h(x_0, y_0, z_0)$$

are small in absolute value.

9. Suppose that $P_0(x_0, y_0, z_0)$ is a point on the surface S: $F(x, y, z) = 0$. Let dx and dy be arbitrary, except that at least one of them should be different from zero. Show that a dz can be found, provided

$F_z(x_0, y_0, z_0) \neq 0$, such that the vector

$$d\mathbf{R} = \mathbf{i}\,dx + \mathbf{j}\,dy + \mathbf{k}\,dz$$

is tangent to the surface S at P_0. Find an expression for such a dz. Also show that for such a vector $d\mathbf{R}$, dF is zero at P_0.

10. In Exercise 9, consider the equation $F(x, y, z) = 0$ as determining z implicitly as a function of x and y, say $z = \phi(x, y)$. Show that

$$\phi_x = -F_x/F_z, \qquad \phi_y = -F_y/F_z$$

at any point where $F_z \neq 0$.

11. *Ruled surfaces.* The surface $z = f(x, y)$ is said to have *rulings* if through the point $P_0(x_0, y_0, z_0)$ there is a straight line segment all of whose points are on the surface. This happens if through the point $(x_0, y_0, 0)$ in the xy-plane there is a line of points $(x_0 + h, y_0 + k, 0)$ such that along this line dz and Δz are equal. This happens if and only if

$$f(x_0 + h, y_0 + k) - f(x_0, y_0) = \left(\frac{\partial f}{\partial x}\right)_0 h + \left(\frac{\partial f}{\partial y}\right)_0 k,$$

when $h = \Delta x = dx$ and $k = \Delta y = dy$. Show that

(a) the surface $z = \sqrt{1 + xy}$ has rulings through the point $P_0(2, 4, 3)$, given by the conditions $h = k$ or $4h = k$,

(b) the surface $z = x^2 - y^2$ has rulings through any point $P_0(x_0, y_0, z_0)$, given by $h = \pm k$.

(See the article "Rulings" by C. S. Ogilvy in *The American Mathematical Monthly*, **59**, 1952, pp. 547–549.)

15.9 MAXIMA AND MINIMA OF FUNCTIONS OF TWO INDEPENDENT VARIABLES

In an early chapter, we learned how to use the differential calculus to solve max-min problems for functions $y = f(x)$ of a single independent variable. In this article, we shall extend the method to handle problems involving more than one independent variable.

As an illustration of the technique, we shall first discuss a geometrical application. Suppose it is required to find the high and low points on a smooth surface represented by the equation

$$z = f(x, y), \tag{1}$$

where the function f is defined, continuous, and has continuous partial derivatives with respect to x and y in some region R in the xy-plane. If there is a point (a, b) in R such that

$$f(x, y) \geq f(a, b) \tag{2}$$

for all points (x, y) sufficiently near to the point (a, b), then the function f is said to have a *local*, or *relative, minimum* at (a, b). If the inequality (2) holds for all points (x, y) in R, then f has an absolute minimum over R at (a, b). If the inequality in (2) is reversed, f then has a maximum (relative or absolute) at (a, b).

Suppose that the maximum (or minimum) value of f, over the region R, occurs at a point (a, b) that is not on the boundary of R, and suppose that both $\partial f/\partial x$ and $\partial f/\partial y$ exist at (a, b). Then the first *necessary condition* that must be satisfied is that

$$\frac{\partial f}{\partial x} = 0 \quad \text{and} \quad \frac{\partial f}{\partial y} = 0 \quad \text{at } (a, b),$$

as we shall now show. To begin with, the section of the surface (1) lying in the plane $y = b$ is simply the curve whose equation is

$$z = f(x, b), \qquad y = b,$$

and this curve has a high or low turning point at $x = a$ (see Fig. 15.11). Hence

$$\left(\frac{\partial z}{\partial x}\right)_{x=a,y=b} = 0.$$

Similarly, the curve

$$z = f(a, y), \qquad x = a,$$

in which the plane $x = a$ intersects the surface, also has a high or low turning point when $y = b$, so that

$$\left(\frac{\partial z}{\partial y}\right)_{x=a,y=b} = 0.$$

We shall not at this time enter into a detailed discussion of second-derivative tests for distinguish-

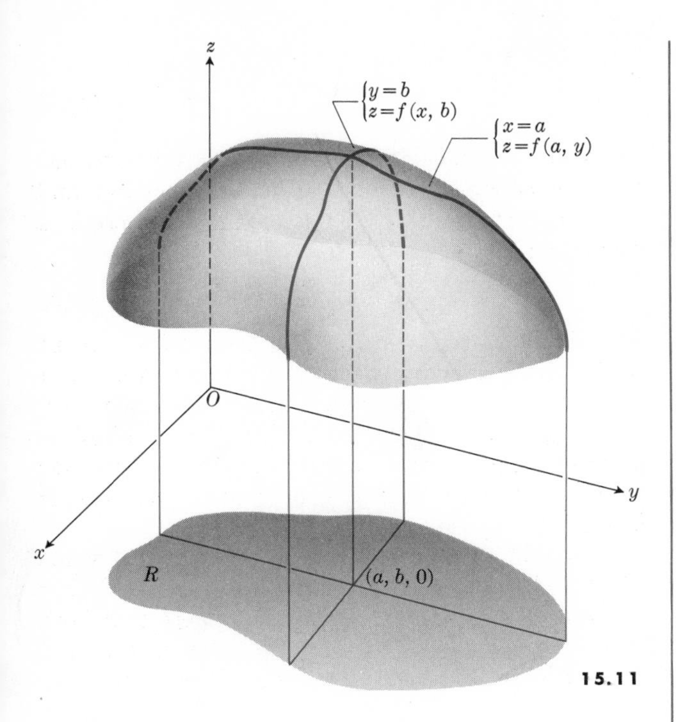

15.11

ing between maxima and minima. The fundamental principle used in deriving such a test is that the difference

$$D = f(x, y) - f(a, b)$$

should be nonnegative (that is, positive or zero) for all points (x, y) close to the point (a, b) in case of a minimum at (a, b), or nonpositive in the case of a maximum. One way to test for a maximum or a minimum is to take

$$x = a + h, \qquad y = b + k$$

and to examine the difference D for small values of h and k as in the problem that follows.

Problem. Find the high and low points on the surface

$$z = x^2 - xy + y^2 + 2x + 2y - 4 = f(x, y).$$

Solution. We apply the first necessary condition for a maximum or minimum of z, namely,

$$\frac{\partial z}{\partial x} = 0 \quad \text{and} \quad \frac{\partial z}{\partial y} = 0.$$

This leads to the simultaneous equations

$$2x - y = -2, \qquad -x + 2y = -2,$$

with solution

$$x = y = -2.$$

Thus, the point which we have been calling (a, b) is here the point $(-2, -2)$. The corresponding value of z is

$$f(-2, -2) = -8.$$

To examine the behavior of the difference

$$D = f(x, y) - f(-2, -2),$$

we let

$$x = -2 + h, \qquad y = -2 + k$$

and obtain

$$\begin{aligned} D &= f(-2 + h, -2 + k) - f(-2, -2) \\ &= h^2 - hk + k^2 \\ &= (h - k/2)^2 + 3k^2/4. \end{aligned}$$

This is readily seen to be positive for all values of h, k except $h = k = 0$. That is,

$$f(x, y) \geq f(-2, -2)$$

for all (x, y) different from $(-2, -2)$. Thus the surface has a *low* point at $(-2, -2, -8)$. The given function has an absolute minimum -8.

Remark. As with functions of a single independent variable, it is often possible to see that the function $z = f(x, y)$ has exactly one maximum (or minimum), that it occurs at an interior point of the domain of f, and that f everywhere possesses partial derivatives which must be zero at the critical point. No further test is then required. This is true, for instance, in the problem above.

EXERCISES

Examine the following surfaces for high and low points.

1. $z = x^2 + xy + y^2 + 3x - 3y + 4$
2. $z = x^2 + 3xy + 3y^2 - 6x + 3y - 6$
3. $z = 5xy - 7x^2 - y^2 + 3x - 6y + 2$
4. $z = 2xy - 5x^2 - 2y^2 + 4x + 4y - 4$

5. $z = x^2 + xy + 3x + 2y + 5$
6. $z = y^2 + xy - 2x - 2y + 2$
7. Sketch the surface $z = \sqrt{x^2 + y^2}$ over the region

$$\mathbf{R}: |x| \leq 1, \quad |y| \leq 1.$$

Find the high and low points of the surface over **R**. Discuss the existence, and the values of $\partial z/\partial x$ and $\partial z/\partial y$ at these points.

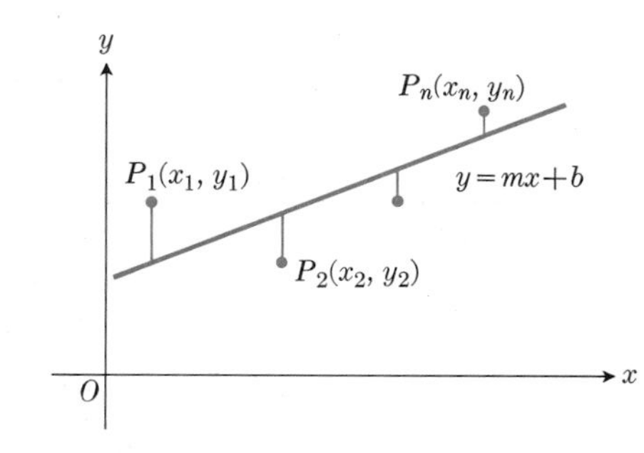

15.12

15.10 THE METHOD OF LEAST SQUARES

We next take up an important application of minimizing a function of two variables. It is the so-called "method of least squares," as applied to the problem of fitting a straight line

$$y = mx + b \tag{1}$$

to a set of experimentally observed points

$$(x_1, y_1), \quad (x_2, y_2), \quad \ldots, \quad (x_n, y_n).$$

(See Fig. 15.12.) Corresponding to each of the observed values of x, there are two values of y: the observed value y_{obs}, and the value predicted by the straight line $mx_{\text{obs}} + b$. We shall call the difference,

$$y_{\text{obs}} - (mx_{\text{obs}} + b), \tag{2}$$

a *deviation:*

$$\text{dev} = y_{\text{obs}} - mx_{\text{obs}} - b.$$

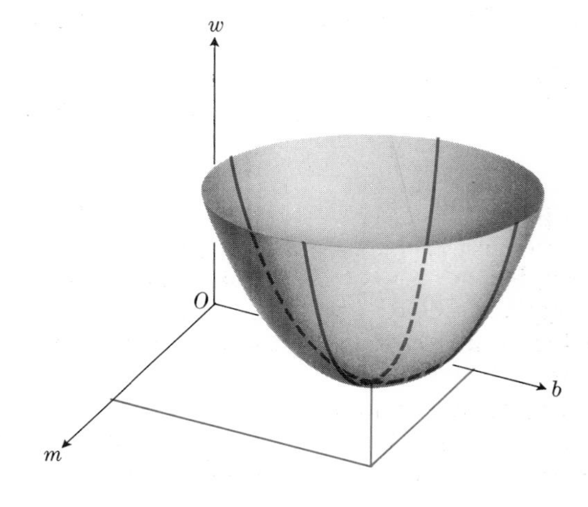

15.13

Each deviation measures the amount by which the predicted value of y falls short of the observed value. The set of all deviations

$$d_1 = y_1 - (mx_1 + b), \quad \ldots, \quad d_n = y_n - (mx_n + b) \tag{3}$$

gives a picture of the closeness of fit of the line, Eq. (1), to the observed data. The line is a perfect fit if and only if all of these deviations are zero. But in general no straight line will give a perfect fit. Then we are confronted with the problem of finding a line which fits *best* in some sense or other. Here is where the method of least squares comes in.

The method may be explained as follows. For a straight line which comes *close* to fitting all of the

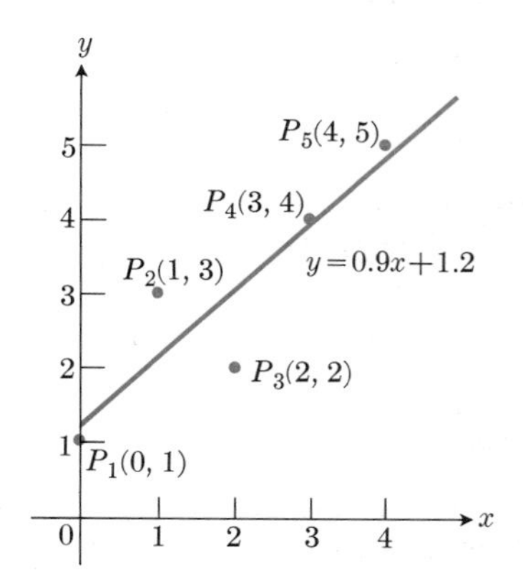

15.14

Table. Values for Problem 1.

x_{obs}	y_{obs}	dev	$(\text{dev})^2$
0	1	$1 - b$	$1 - 2b + b^2$
1	3	$3 - m - b$	$9 - 6b + b^2 - 6m + 2mb + m^2$
2	2	$2 - 2m - b$	$4 - 4b + b^2 - 8m + 4mb + 4m^2$
3	4	$4 - 3m - b$	$16 - 8b + b^2 - 24m + 6mb + 9m^2$
4	5	$5 - 4m - b$	$25 - 10b + b^2 - 40m + 8mb + 16m^2$

55 30

observed points, some of the deviations will probably be positive and some will probably be negative. But their squares will all be positive, and the expression

$$f(m, b) = (y_1 - mx_1 - b)^2 + (y_2 - mx_2 - b)^2 + \cdots + (y_n - mx_n - b)^2$$

counts a positive deviation $+d$ and a negative deviation $-d$ equally. This sum of squares of the deviations depends on the choice of m and b. It is never negative, and it can be zero only if m and b have values that produce a straight line that is a perfect fit.

Whether such a perfectly fitting line can be found or not, the method of least squares says: *Take as the line $y = mx + b$ of best fit that one for which the sum of squares of the deviations*

$$f(m, b) = d_1^2 + d_2^2 + \cdots + d_n^2$$

is a minimum. Thus we try to find the values of m and b where the surface

$$w = f(m, b)$$

in *mbw*-space has a low point (Fig. 15.13). To do this, we solve the equations

$$\frac{\partial f}{\partial m} = 0 \quad \text{and} \quad \frac{\partial f}{\partial b} = 0$$

simultaneously.

Problem 1. Find the straight line that best fits the points (0, 1), (1, 3), (2, 2), (3, 4), (4, 5), according to the method of least squares (see Fig. 15.14).

Solution. The sum of squares of the deviations is

$$f(m, b) = \sum (y_{obs} - mx_{obs} - b)^2,$$

where y_{obs} and x_{obs} are the observed (or given) coordinates of the points to be fitted by the line $y = mx + b$.

We list these, together with the deviations and their squares, in the table above. Then, as sum of squares, we have

$$f(m, b) = \sum (\text{dev})^2 = 55 - 30b + 5b^2 - 78m + 20mb + 30m^2;$$

$$\frac{\partial f}{\partial m} = -78 + 20b + 60m,$$

$$\frac{\partial f}{\partial b} = -30 + 10b + 20m.$$

The values of m and b for which f has a minimum must satisfy the simultaneous equations

$$\frac{\partial f}{\partial m} = 0, \quad 20b + 60m = 78,$$

$$\frac{\partial f}{\partial b} = 0, \quad 10b + 20m = 30.$$

The only solution is $m = 0.9$, $b = 1.2$. The "best-fitting" line (in the sense of least sum of squares of deviations) is therefore

$$y = 0.9x + 1.2.$$

To verify that these values of m and b do in fact correspond to a minimum, we let

$$m = 0.9 + h, \qquad b = 1.2 + k$$

and calculate

$$\Delta = f(0.9 + h, 1.2 + k) - f(0.9, 1.2).$$

Doing this algebraically, we find

$$\begin{aligned}\Delta f &= f(m+h, b+k) - f(m, b)\\ &= (-30 + 10b + 20m)k\\ &\quad + (-78 + 20b + 60m)h + 5k^2 + 20kh + 30h^2.\end{aligned}$$

The expressions in parentheses are $\partial f/\partial b$ and $\partial f/\partial m$, respectively, and these are zero if $m = 0.9$ and $b = 1.2$. Hence

$$\begin{aligned}f(0.9 + h, 1.2 + k) &- f(0.9, 1.2)\\ &= 5k^2 + 20kh + 30h^2\\ &= 5(k + 2h)^2 + 10h^2.\end{aligned}$$

This is greater than zero for all values of h and k other than $h = k = 0$. That is,

$$f(0.9 + h, 1.2 + k) \geq f(0.9, 1.2),$$

and we have found the values of m and b for which the function $f(m, b)$ is an absolute minimum.

It is customary to omit the details of testing the answer obtained in solving a problem in least squares. Indeed, it can be shown that for the case of fitting a straight line, the answer *always* corresponds to a minimum.

The method of least squares may also be applied to more complicated equations than that of a straight line, and the method has been widely extended.

EXERCISES

1. The observed points (x_i, y_i), $i = 1, 2, \ldots, n$, are to be fitted by a straight line $y = mx + b$ by the method of least squares. The sum of squares of the deviations is

$$f(m, b) = \sum_{i=1}^{n} (mx_i + b - y_i)^2.$$

(a) Show that the equations

$$\partial f/\partial b = 0 \quad \text{and} \quad \partial f/\partial m = 0$$

are equivalent to

$$\begin{aligned}m\left(\sum x_i\right) + nb &= \sum y_i,\\ m\left(\sum x_i^2\right) + b\left(\sum x_i\right) &= \sum x_i y_i,\end{aligned}$$

where all sums run from $i = 1$ to $i = n$.

(b) Express the solutions b, m of the equations of part (a) in terms of determinants.

In Exercises 2 through 4, apply the method of least squares to obtain the line $y = mx + b$ which best fits the three given points. [The computations can be systematized by making use of the results of Exercise 1(b).]

2. $(-1, 2)$, $(0, 1)$, $(3, -1)$
3. $(-2, 0)$, $(0, 2)$, $(2, 3)$
4. $(0, 0)$, $(1, 2)$, $(2, 3)$
5. If $y = mx + b$ is the best-fitting straight line, in the sense of least squares, show that the sum of deviations

$$\sum_{i=1}^{n} (y_i - mx_i - b)$$

is zero. (This means that positive and negative deviations cancel.)

6. Show that the point

$$(\bar{x}, \bar{y}) = \left[\frac{1}{n}\left(\sum_{i=1}^{n} x_i\right), \frac{1}{n}\left(\sum_{i=1}^{n} y_i\right)\right]$$

lies on the straight line $y = mx + b$ that is determined by the method of least squares. (This means that the "best-fitting" line passes through the center of gravity of the n points.)

15.11 MAXIMA AND MINIMA OF FUNCTIONS OF SEVERAL INDEPENDENT VARIABLES

In certain applications, particularly in statistics, it becomes necessary to find maximum or minimum values of a function

$$w = f(x, y, z, u, \ldots, v)$$

of several independent variables. If the given function has an extreme value at an interior point of the domain, say at

$$x = a, \quad y = b, \quad z = c, \quad \ldots, \quad v = e,$$

then by setting $y = b$, $z = c$, $\ldots$, $v = e$ we obtain a function of x alone,

$$F(x) = f(x, b, c, d, \ldots, e),$$

which has an extreme value at $x = a$. Hence, if f has a partial derivative with respect to x at $x = a$, $y = b, \ldots, v = e$, that partial derivative must be zero by virtue of the theory for max-min for functions $F(x)$ of a single independent variable. That is,

$$\frac{\partial f}{\partial x} = 0 \quad \text{at } (a, b, c, \ldots, e).$$

By similar reasoning, we arrive at the first necessary condition for extreme values of a function of several independent variables, namely,

$$\frac{\partial f}{\partial x} = 0, \quad \frac{\partial f}{\partial y} = 0, \quad \ldots, \quad \frac{\partial f}{\partial v} = 0 \quad \text{at } (a, b, \ldots, e).$$

The number of simultaneous equations $\partial f/\partial x = 0$, etc., which are thus obtained is precisely equal to the number of *independent variables* $x, y, \ldots, v$. Of course the solutions of this system of equations may correspond to maximum values of f, or to minimum values, or neither, in much the same way as occurs for solutions of the equation $dy/dx = 0$.

Occasionally, the problem arises of finding the extreme value of one function, say

$$w = f(x, y, z, u, \ldots, v),$$

subject to certain auxiliary restrictions that may be represented by such equations as

$$g(x, y, z, u, \ldots, v) = 0,$$
$$h(x, y, z, u, \ldots, v) = 0, \ldots$$

The theory behind such problems is discussed in most books on advanced calculus. (See Kaplan, *Advanced Calculus*, p. 128). Here we shall do no more than point out that the equations representing the side restrictions may be used to express some of the variables $x, y, z, \ldots, v$ in terms of the remaining ones before we take partial derivatives. This is done so that the variables that remain may be *independent*. After the next two problems, which show how this method works, we shall present a second method, that of Lagrange multipliers.

Problem 1. Find the minimum distance from the origin to the plane

$$2x + y - z = 5.$$

Solution. If $P(x, y, z)$ is any point on the plane, then the distance from the origin to P is

$$|\overrightarrow{OP}| = \sqrt{x^2 + y^2 + z^2},$$

and clearly this has a minimum wherever the function

$$f(x, y, z) = |\overrightarrow{OP}|^2 = x^2 + y^2 + z^2$$

does. (The latter is simpler to work with since it does not involve radicals.) But the three variables x, y, z are not all independent, since P is to lie on the plane

$$2x + y - z = 5.$$

If we solve this equation for z, we find

$$z = 2x + y - 5,$$

and we may treat x and y as independent variables and minimize the function

$$g(x, y) = x^2 + y^2 + (2x + y - 5)^2.$$

The necessary conditions

$$\frac{\partial g}{\partial x} = 0 \quad \text{and} \quad \frac{\partial g}{\partial y} = 0$$

lead to the equations

$$10x + 4y - 20 = 0, \qquad 4x + 4y - 10 = 0,$$

with solution $x = \frac{5}{3}$, $y = \frac{5}{6}$. The z-coordinate of the corresponding point P is $z = -\frac{5}{6}$, and thus we have found the point $(\frac{5}{3}, \frac{5}{6}, -\frac{5}{6})$ as the *only* point on the plane that satisfies the *necessary* conditions. That is, if the given problem has an answer, this is it. From our knowledge of solid geometry we know that the problem does possess an answer, hence we have found it. Of course, we may solve this same problem by strictly geometrical methods, but our purpose is not so much to solve this specific problem as it is to illustrate the *method* of solving such problems by use of partial differentiation. We may check our answer by noting that the vector from the origin to $P(\frac{5}{3}, \frac{5}{6}, -\frac{5}{6})$ is

$$\overrightarrow{OP} = \tfrac{5}{6}(2\mathbf{i} + \mathbf{j} - \mathbf{k}),$$

which is normal to the plane, as it should be.

Remark 1. The mechanics of setting $f_x = 0$, $f_y = 0$, and so forth may lead to a point not in the region where the function f is defined. Sometimes it is desirable to choose a different set of independent variables when this happens.

Problem 2. Find the minimum distance from the origin to the surface $x^2 - z^2 = 1$.

Solution. We seek to minimize

$$w = x^2 + y^2 + z^2,$$

where $x^2 = 1 + z^2$, or $z^2 = x^2 - 1$. If we eliminate z^2, we have

$$w = 2x^2 + y^2 - 1,$$

whose partial derivatives,

$$\frac{\partial w}{\partial x} = 4x, \qquad \frac{\partial w}{\partial y} = 2y,$$

are zero only at $x = 0$, $y = 0$. But then we run into trouble, for $z^2 = x^2 - 1 = -1$ means that z is imaginary when $x = 0$. In fact, the point $P(x, y, z)$ is on the surface $z^2 = x^2 - 1$ only when $|x| \geq 1$.

If we eliminate x^2, however, and express w as a function of y and z, then the function

$$w = 1 + y^2 + 2z^2$$

has partial derivatives

$$\frac{\partial w}{\partial y} = 2y, \qquad \frac{\partial w}{\partial z} = 4z,$$

which are both zero when $y = z = 0$. This leads to

$$x^2 = 1 + z^2 = 1, \quad x = \pm 1.$$

It is obvious from the expression $w = 1 + y^2 + 2z^2$ that $w \geq 1$ for all real values of y and z, since $y^2 + 2z^2 \geq 0$. Therefore the two points $(\pm 1, 0, 0)$ are nearer the origin than are any other points on the surface. By expressing w in terms of y and z as independent variables, we obtained variables which can take all real values

$$-\infty < y < \infty, \quad -\infty < z < \infty.$$

The surface is a two-sheeted hyperbolic cylinder with elements parallel to the y-axis (Fig. 15.15).

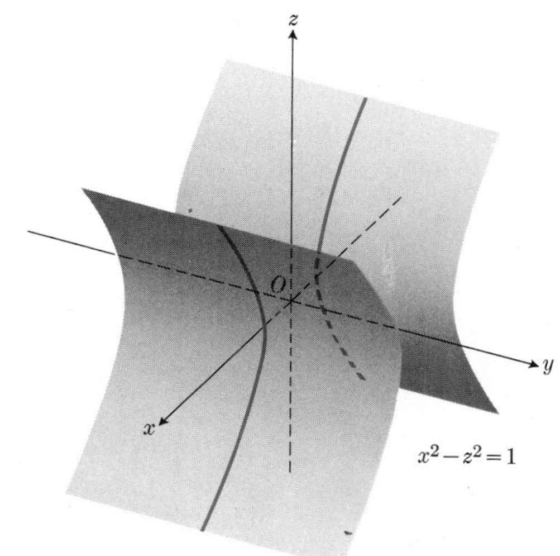

15.15

Method of Lagrange multipliers

In Problem 2, we wanted to minimize the function

$$f(x, y, z) = x^2 + y^2 + z^2,$$

subject to the side condition (or constraint) that

$$g(x, y, z) = x^2 - z^2 - 1 = 0.$$

The method of Lagrange multipliers applies to problems of this type. This is how it goes:

To minimize (or maximize) a function $f(x, y, z)$, subject to the constraint $g(x, y, z) = 0$, construct the auxiliary function

$$H(x, y, z, \lambda) = f(x, y, z) - \lambda g(x, y, z) \tag{1a}$$

and find values of x, y, z, λ for which the partial derivatives of H are all zero:

$$H_x = 0, \quad H_y = 0, \quad H_z = 0, \quad H_\lambda = 0. \tag{1b}$$

We shall discuss the theory behind the method after the next problem, which demonstrates how to use it.

Problem 3. Find the point on the plane

$$2x - 3y + 5z = 19$$

that is nearest the origin, using the method of Lagrange multipliers.

Solution. As before, the function to be minimized can be taken to be the square of the distance from the origin to $P(x, y, z)$:

$$f(x, y, z) = x^2 + y^2 + z^2. \tag{2a}$$

The constraint is

$$g(x, y, z) = 2x - 3y + 5z - 19 = 0. \tag{2b}$$

We let

$$H(x, y, z, \lambda) = x^2 + y^2 + z^2 - \lambda(2x - 3y + 5z - 19). \tag{2c}$$

Then

$$H_x = 2x - 2\lambda = 0, \quad H_y = 2y + 3\lambda = 0,$$
$$H_z = 2z - 5\lambda = 0, \tag{3a}$$

and

$$H_\lambda = -g(x, y, z) = -(2x - 3y + 5z - 19) = 0. \tag{3b}$$

From Eqs. (3a), we get

$$x = \lambda, \quad y = -\tfrac{3}{2}\lambda, \quad z = \tfrac{5}{2}\lambda, \tag{3c}$$

and when these are substituted in Eq. (3b), or (2b), we get

$$2\lambda + \tfrac{9}{2}\lambda + \tfrac{25}{2}\lambda = 19,$$

so

$$\lambda = 1. \tag{3d}$$

Substituting from Eq. (3d) in Eqs. (3c), we get the point $P_0 = (1, -\frac{3}{2}, \frac{5}{2})$. In this simple problem we can readily note that the vector $\overrightarrow{OP_0}$ is normal to the given plane, so $|\overrightarrow{OP_0}| = \frac{1}{2}\sqrt{38}$ is the minimum distance from the origin to the plane. (We could have found this answer using the normal vector

$$\mathbf{N} = 2\mathbf{i} - 3\mathbf{j} + 5\mathbf{k}$$

and no calculus.)

Discussion of the method and why it works

We pass now to the general situation. Assume that the equation representing the constraint

$$g(x, y, z) = 0 \tag{4a}$$

can be solved for z as a function of (x, y):

$$z = \phi(x, y), \tag{4b}$$

throughout some neighborhood of the point $P_0(x_0, y_0)$. Assume, further, that this point minimizes the function

$$w = f[x, y, \phi(x, y)] \tag{5}$$

that we get by substituting for z (Eq. 4b) into the function $f(x, y, z)$. For convenience, we shall use subscripts 1, 2, and 3 to denote partial derivatives of f and g with respect to the first, second, and third variables, respectively:

$$f_1(x, y, z) = f_x(x, y, z), \quad g_1(x, y, z) = g_x(x, y, z),$$
$$f_2(x, y, z) = f_y(x, y, z), \quad g_2(x, y, z) = g_y(x, y, z),$$
$$f_3(x, y, z) = f_z(x, y, z), \quad g_3(x, y, z) = g_z(x, y, z).$$

In particular, in the calculations that follow, we assume that

$$g_3[x, y, \phi(x, y)] \neq 0 \quad \text{near } P_0 \text{ and at } P_0.$$

We next differentiate both sides of Eq. (4a) implicitly with respect to x, holding y constant, and treating z as a differentiable function, as in Eq. (4b). This gives

$$g_1 + g_3 \frac{\partial z}{\partial x} = 0, \quad \text{or} \quad \frac{\partial z}{\partial x} = -\frac{g_1}{g_3}. \tag{6a}$$

Likewise, if we differentiate with respect to y, holding x constant, we get

$$g_2 + g_3 \frac{\partial z}{\partial y} = 0, \quad \text{or} \quad \frac{\partial z}{\partial y} = -\frac{g_2}{g_3}. \tag{6b}$$

In Eqs. (6a, b), the partial derivatives g_1, g_2, and g_3 are to be evaluated at $(x, y, \phi(x, y))$ for (x, y) in some suitably restricted neighborhood of $P_0(x_0, y_0)$.

Now we turn our attention to the function in Eq. (5), whose extreme value is assumed to occur at P_0. The necessary condition [assuming that the right-hand side of Eq. (5) is differentiable at P_0] for such a minimum or maximum is

$$f_1 + f_3 \frac{\partial \phi}{\partial x} = 0, \quad f_2 + f_3 \frac{\partial \phi}{\partial y} = 0, \quad \text{at } P_0. \tag{7}$$

Substituting from Eqs. (6a, b) for

$$\frac{\partial z}{\partial x} = \frac{\partial \phi}{\partial x} = \frac{-g_1}{g_3}, \qquad \frac{\partial z}{\partial y} = \frac{\partial \phi}{\partial y} = \frac{-g_2}{g_3},$$

we get

$$f_1 - f_3\left(\frac{g_1}{g_3}\right) = 0, \quad f_2 - f_3\left(\frac{g_2}{g_3}\right) = 0, \quad \text{at } P_0.$$

Thus at

$$(x_0, y_0, \phi(x_0, y_0)) = (x_0, y_0, z_0),$$

the following conditions hold:

$$f_1 = g_1\left(\frac{f_3}{g_3}\right), \quad f_2 = g_2\left(\frac{f_3}{g_3}\right), \tag{8a}$$

and, of course,

$$f_3 = g_3\left(\frac{f_3}{g_3}\right). \tag{8b}$$

Suppose, therefore, that we denote the ratio f_3/g_3 by λ. Then Eqs. (8a, b) can be combined into one vector equation:

$$\mathbf{i}f_1 + \mathbf{j}f_2 + \mathbf{k}f_3 = \lambda(\mathbf{i}g_1 + \mathbf{j}g_2 + \mathbf{k}g_3),$$

or

$$\nabla f = \lambda \nabla g \quad \text{at } (x_0, y_0, z_0), \tag{9}$$

and (x_0, y_0, z_0) is a point whose coordinates satisfy the constraint

$$g(x_0, y_0, z_0) = 0. \tag{10}$$

Equations (9, 10) are just the same as Eqs. (1a, b) in different form, because

$$\left.\begin{array}{l} H_x = f_x - \lambda g_x = 0 \\ H_y = f_y - \lambda g_y = 0 \\ H_z = f_z - \lambda g_z = 0 \end{array}\right\} \Leftrightarrow \nabla f = \lambda \nabla g,$$

and

$$H_\lambda = -g = 0 \Leftrightarrow g = 0.$$

Remark 2. We could make Eq. (9) seem plausible by a geometric argument. Let us imagine that we are traveling around on the surface S, whose equation is $g = 0$, noting the values of f as we go. In particular, we note that the *level surfaces* of f (which we shall call *iso-f surfaces*) intersect the surface S in curves along which f remains constant. To find either a maximum or a minimum of f on S, we should take a route that crosses these iso-f *curves* in a direction in which f-values increase or decrease; specifically, if we are searching for a maximum, we should look for a direction in which f-values increase, and if we are searching for a minimum, we should look for a direction in which f-values decrease. When we have once arrived at a maximum of f on S, there will be no direction in which we can travel to get to an iso-f curve with a larger f-value. Equation (9) says that at this point ∇f and ∇g must have the same direction (or *opposite* directions, if λ is *negative*). This means that S and the surface $f =$ constant are *tangent* at that point: For a slightly smaller f-value, there would be a curve of intersection of the iso-f surface and S, and for a larger f-value, there would be no intersection. For the maximum, there is just one point at which the surface S and the iso-f surface touch, and at that point it does indeed seem plausible that the two surfaces are tangent.

Remark 3. If there are two constraints, say

$$g(x, y, z) = 0 \quad \text{and} \quad h(x, y, z) = 0,$$

we introduce two Lagrange multipliers λ and μ, and work with the auxiliary function

$$H(x, y, z, \lambda, \mu) = f(x, y, z) - \lambda g(x, y, z) - \mu h(x, y, z).$$

We then treat x, y, z, λ, μ as five independent variables for H, and set the five first-order partial derivatives of H equal to zero:

$$H_x = 0, \quad H_y = 0, \quad H_z = 0, \quad H_\lambda = 0, \quad H_\mu = 0.$$

These results are equivalent to

$$\nabla f = \lambda \nabla g + \mu \nabla h \quad \text{at } (x_0, y_0, z_0), \tag{11a}$$

$$g(x_0, y_0, z_0) = 0 \quad \text{and} \quad h(x_0, y_0, z_0) = 0. \tag{11b}$$

The vector equation (11a) says that the gradient of f lies in the plane of the gradients of g and h at $Q_0 = (x_0, y_0, z_0)$, and this has a fairly simple geometrical interpretation. First, we know that ∇g is normal to the surface $g = 0$, and that ∇h is normal to the surface $h = 0$. The intersection of these two surfaces is usually a curve. On this curve, say C, we can think of x, y, and z as functions of one variable, for example, time (t) or arc length. Then $w = f(x, y, z)$ is a function of a single variable, say t, on C, and we want to find points where

$$\frac{dw}{dt} = 0.$$

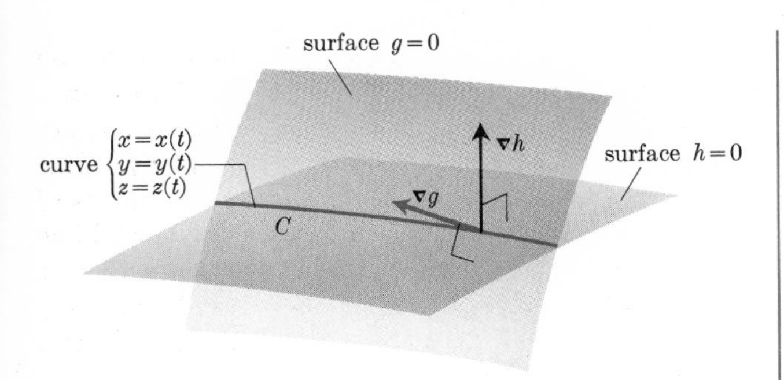

15.16 Vectors ∇g and ∇h are in a plane perpendicular to curve C because ∇g is normal to the surface $g = 0$, and ∇h is normal to the surface $h = 0$.

We also know, from the chain rule, that

$$\frac{dw}{dt} = \nabla f \cdot \mathbf{v},$$

where

$$\mathbf{v} = \mathbf{i}\frac{dx}{dt} + \mathbf{j}\frac{dy}{dt} + \mathbf{k}\frac{dz}{dt}$$

is a vector *tangent* to C. To make $dw/dt = 0$, we therefore want to find a point Q_0 on C where ∇f is orthogonal to the tangent vector. This means, however, that ∇f should lie in a plane that is perpendicular to C at Q_0. This is just the plane that contains the two vectors $(\nabla g)_0$ and $(\nabla h)_0$, so $(\nabla f)_0$ should be a linear combination of them. Equation (11a) expresses this relationship. Equations (11b) represent the constraints that put Q_0 on C.

Problem 4. The cone $z^2 = x^2 + y^2$ is cut by the plane $z = 1 + x + y$ in a curve C. Find the points on C that are nearest to, and farthest from, the origin.

Solution. The function $f(x, y, z) = x^2 + y^2 + z^2$ is to be a minimum (or maximum) subject to the constraints

$$g(x, y, z) = x^2 + y^2 - z^2 = 0, \tag{12a}$$

$$h(x, y, z) = 1 + x + y - z = 0. \tag{12b}$$

We use Eq. (11a). For the critical points, we have

$$2x\mathbf{i} + 2y\mathbf{j} + 2z\mathbf{k} = \lambda(2x\mathbf{i} + 2y\mathbf{j} - 2z\mathbf{k}) + \mu(\mathbf{i} + \mathbf{j} - \mathbf{k}),$$

or, because components must agree,

$$\left.\begin{aligned} 2x &= 2x\lambda + \mu \\ 2y &= 2y\lambda + \mu \end{aligned}\right\} \Rightarrow x - y = (x - y)\lambda$$

$$\left.\begin{aligned} 2y &= 2y\lambda + \mu \\ 2z &= -2z\lambda - \mu \end{aligned}\right\} \Rightarrow y + z = (y - z)\lambda$$

The equation $x - y = (x - y)\lambda$ is satisfied if $x = y$ or if $x \neq y$ and $\lambda = 1$. This latter case cannot apply: The case $\lambda = 1$ implies $y + z = y - z$, from which it follows that $z = 0$. From Eq. (12a), however, $z = 0$ gives $x^2 + y^2 = 0$, or $x = y = 0$. But the point $(0, 0, 0)$ is not on the plane $z = 1 + x + y$. Therefore $\lambda \neq 1$, and we must have $x = y$. The intersection of this plane with the plane $z = 1 + x + y$ is a line which cuts the cone in just two points. We find these points by substituting $y = x$ and $z = 1 + 2x$:

$$x^2 + x^2 = (1 + 2x)^2, \quad 2x^2 + 4x + 1 = 0,$$
$$x = -1 \pm \tfrac{1}{2}\sqrt{2}.$$

The points are

$$A = (-1 - \sqrt{1/2},\ -1 - \sqrt{1/2},\ -1 - \sqrt{2}), \tag{13a}$$

$$B = (-1 + \sqrt{1/2},\ -1 + \sqrt{1/2},\ -1 + \sqrt{2}). \tag{13b}$$

In this problem, we know that C is either an ellipse or a hyperbola. If it is an ellipse, we would conclude that B is the point on it nearest the origin, and A the point farthest from the origin. But if it is a hyperbola, then there is no point on it that is farthest away from the origin, and the points A and B are the points on the two branches that are nearest the origin. Exercise 10 asks you to think about these possibilities and decide between them. It seems obvious that the critical points should satisfy the condition $x = y$ because all three of the functions f, g, and h treat x and y alike.

EXERCISES

1. Find the point on the surface $z = xy + 1$ that is nearest the origin.
2. A rectangular box, open at the top, is to hold 256 in^3. Find the dimensions of the box for which the surface area is a minimum.

3. The base of a rectangular box costs three times as much per square foot as do the sides and top. Find the relative dimensions for the most economical box of given volume.

4. Find the equation of the plane through the point (2, 1, 1) that cuts off the least volume from the first octant. (Consider only those planes whose intercepts with the coordinate axes are positive.)

5. A pentagon is composed of a rectangle surmounted by an isosceles triangle. If the area is fixed, what are the dimensions for which the perimeter is a minimum?

6. A plane of the form

$$z = Ax + By + C$$

is to be "fitted" to the following points (x_i, y_i, z_i):

$$(0, 0, 0), \quad (0, 1, 1), \quad (1, 1, 1), \quad (1, 0, -1).$$

Find the plane that minimizes the sum of squares of the deviations

$$\sum_{i=1}^{4} (Ax_i + By_i + C - z_i)^2.$$

7. Consider the geometric argument in Remark 2 for the minimization problem of Problem 3.

(a) Describe the family of surfaces f = constant.

(b) Do all of the surfaces of (a) intersect the plane $2x - 3y + 5z = 19$?

(c) If a particular surface $f = k$ intersects the plane, what is the locus of intersection, geometrically? If $k' > k$, does the surface $f = k'$ also intersect the plane? In what kind of curve?

(d) Does the geometric plausibility argument apply to this example?

8. Use the method of Lagrange multipliers to find the points nearest to, and farthest from, the origin and lying on the curve $x^2 + 2xy + 3y^2 = 9$ in the xy-plane.

9. Use the method of Lagrange multipliers to find the points on the surface $x^2 + y^2 + z^2 = 25$, where the function $f(x, y, z) = x + 2y + 3z$ is:

(a) a minimum, (b) a maximum.

Comment on the geometric interpretation of $\nabla f = \lambda \nabla g$ at these points.

10. In the solution of Problem 4, two points A and B were located as candidates for maximum or minimum distances from the origin. Use cylindrical coordinates r, θ, z to express the equations of the cone, the plane $z = 1 + x + y$, and the cylinder that contains their curve of intersection and has elements parallel to the z-axis. Is this cylinder circular, elliptical, parabolic, or hyperbolic? (Consider its intersection with the xy-plane.) Express the distance from the origin to a point on the curve of intersection of the cone and the plane as a function of θ. Does it have a minimum? A maximum? What can you now say about the points A and B of Eqs. (13a, b) as solutions in Problem 4?

11. In Problem 4, the extrema for $f(x, y, z)$ on the cone $z^2 = x^2 + y^2$ and the plane $z = 1 + x + y$ were found to satisfy $y = x$ as well. Thus $z = 1 + 2x$, and the function to be made a maximum or minimum is $x^2 + y^2 + z^2 = x^2 + x^2 + (1 + 2x)^2$, which can also be written as $6(x + \frac{1}{3})^2 + \frac{1}{3}$. This is obviously a minimum when $x = -\frac{1}{3}$. But the point we get with

$$y = x, \quad z = 1 + 2x, \quad x = -1/3$$

is $(-\frac{1}{3}, -\frac{1}{3}, \frac{1}{3})$, which is not on the cone. What's wrong? (A sketch of the situation in the plane $y = x$ may throw some light on the question.)

12. In Problem 4, we can determine that for all points on the cone $z^2 = x^2 + y^2$, the square of the distance from the origin to $P(x, y, z)$ is $w = 2(x^2 + y^2)$, which is a function of two independent variables, x and y. But if P is also to be on the plane

$$z = 1 + x + y$$

as well as the cone, then

$$(1 + x + y)^2 = x^2 + y^2.$$

Show that these points have coordinates that satisfy the equation

$$2xy + 2x + 2y + 1 = 0.$$

Interpret this equation in two ways:

(a) as a curve in the xy-plane, and

(b) as a locus of points on a cylinder in 3-space.

Sketch the curve of (a) and find the point or points on it for which w is a minimum. Are there points on this curve for which w is a maximum? Use the information you now have to complete the discussion of Problem 4.

15.12 HIGHER-ORDER DERIVATIVES

Partial derivatives of the second order are denoted by such symbols as

$$\frac{\partial^2 f}{\partial x^2}, \quad \frac{\partial^2 f}{\partial y^2}, \quad \frac{\partial^2 f}{\partial x\,\partial y}, \quad \frac{\partial^2 f}{\partial y\,\partial x}$$

or by

$$f_{xx}, \quad f_{yy}, \quad f_{yx}, \quad f_{xy},$$

where these are defined by the equations

$$\frac{\partial^2 f}{\partial x^2} = \frac{\partial}{\partial x}\left(\frac{\partial f}{\partial x}\right), \qquad \frac{\partial^2 f}{\partial x\,\partial y} = \frac{\partial}{\partial x}\left(\frac{\partial f}{\partial y}\right),$$

and so forth.

Example. If $f(x, y) = x \cos y + ye^x$,

then
$$\frac{\partial f}{\partial x} = \cos y + ye^x,$$

$$\frac{\partial}{\partial y}\left(\frac{\partial f}{\partial x}\right) = -\sin y + e^x = \frac{\partial^2 f}{\partial y\,\partial x},$$

$$\frac{\partial}{\partial x}\left(\frac{\partial f}{\partial x}\right) = ye^x = \frac{\partial^2 f}{\partial x^2},$$

$$\frac{\partial}{\partial x}\left(\frac{\partial^2 f}{\partial x^2}\right) = ye^x = \frac{\partial^3 f}{\partial x^3},$$

$$\frac{\partial}{\partial y}\left(\frac{\partial^2 f}{\partial x^2}\right) = e^x = \frac{\partial^3 f}{\partial y\,\partial x^2},$$

and so on; while

$$\frac{\partial f}{\partial y} = -x \sin y + e^x,$$

$$\frac{\partial}{\partial x}\left(\frac{\partial f}{\partial y}\right) = -\sin y + e^x = \frac{\partial^2 f}{\partial x\,\partial y},$$

$$\frac{\partial}{\partial y}\left(\frac{\partial f}{\partial y}\right) = -x \cos y = \frac{\partial^2 f}{\partial y^2},$$

$$\frac{\partial}{\partial x}\left(\frac{\partial^2 f}{\partial x\,\partial y}\right) = e^x = \frac{\partial^3 f}{\partial x^2\,\partial y},$$

and so on.

The example shows how the order of differentiation is indicated by the notation. Thus, in calculating $\partial^2 f/\partial y\,\partial x$, we differentiate first with respect to x and then with respect to y. This might also be indicated by $(f_x)_y$, or f_{xy}, or by f_{12}. Now it is a remarkable fact that the so-called "mixed" second-order partial derivatives

$$\frac{\partial^2 f}{\partial y\,\partial x} \quad \text{and} \quad \frac{\partial^2 f}{\partial x\,\partial y}$$

are generally equal, just as they are seen to be equal in the Example. That is, we arrive at the same result whether we differentiate first with respect to x and then with respect to y or do the differentiation in the reverse order. The following theorem supports this assertion, under suitable hypotheses.

Theorem. *Let the function $w = f(x, y)$, together with the partial derivatives f_x, f_y, f_{xy}, and f_{yx}, be continuous in some neighborhood of a point $P(a, b)$. Then, at that point,*

$$\frac{\partial}{\partial x}\left(\frac{\partial f}{\partial y}\right) = \frac{\partial}{\partial y}\left(\frac{\partial f}{\partial x}\right).$$

Proof. Let (a, b) be a point in the interior of a rectangle R in the xy-plane such that f, f_x, f_y, f_{xy}, and f_{yx} are continuous functions throughout R. Then the fact that

$$f_{xy}(a, b) = f_{yx}(a, b) \tag{1}$$

may be proved by repeated application of the Mean Value Theorem. We let h and k be numbers such that the point $(a + h, b + k)$ also lies in the rectangle R, and consider the difference

$$\Delta = F(a + h) - F(a), \tag{2}$$

where we define $F(x)$ in terms of $f(x, y)$ by the equation

$$F(x) = f(x, b + k) - f(x, b). \tag{3}$$

We apply the Mean Value Thoerem to the function $F(x)$, and Eq. (2) becomes

$$\Delta = hF'(c_1), \tag{4}$$

where c_1 lies between a and $a + h$. From Eq. (3),

$$F'(x) = f_x(x, b + k) - f_x(x, b),$$

so Eq. (4) becomes

$$\Delta = h[f_x(c_1, b + k) - f_x(c_1, b)]. \tag{5}$$

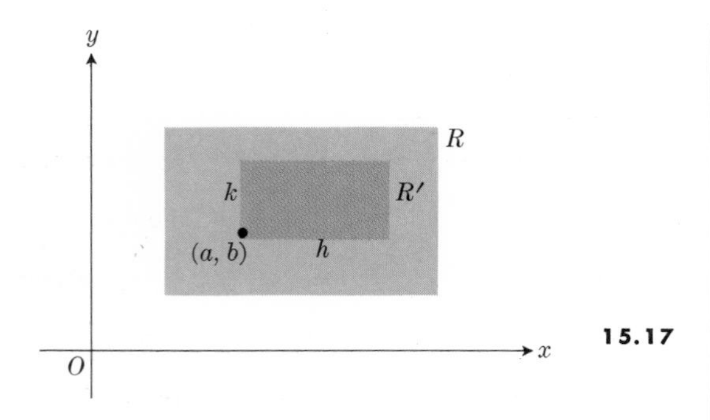

15.17

Now we apply the Mean Value Theorem to the function $g(y) = f_x(c_1, y)$ and have

$$g(b + k) - g(b) = kg'(d_1), \tag{6a}$$

or

$$f_x(c_1, b + k) - f_x(c_1, b) = kf_{xy}(c_1, d_1), \tag{6b}$$

for some d_1 between b and $b + k$. By substituting this into Eq. (5), we get

$$\Delta = hkf_{xy}(c_1, d_1), \tag{7}$$

for some point (c_1, d_1) in the rectangle R' whose vertices are the four points

$$(a, b), \quad (a + h, b), \quad (a + h, b + k), \quad (a, b + k).$$

(See Fig. 15.17.)

On the other hand, by substituting from Eq. (3) into Eq. (2), we may also write

$$\begin{aligned} \Delta &= f(a + h, b + k) - f(a + h, b) \\ &\quad - f(a, b + k) + f(a, b) \\ &= [f(a + h, b + k) - f(a, b + k)] \\ &\quad - [f(a + h, b) - f(a, b)] \\ &= \phi(b + k) - \phi(b), \end{aligned} \tag{8}$$

where

$$\phi(y) = f(a + h, y) - f(a, y). \tag{9}$$

The Mean Value Theorem applied to Eq. (8) now gives

$$\Delta = k\phi'(d_2), \tag{10}$$

for some d_2 between b and $b + k$. By Eq. (9),

$$\phi'(y) = f_y(a + h, y) - f_y(a, y). \tag{11}$$

Substituting from Eq. (11) into Eq. (10), we have

$$\Delta = k[f_y(a + h, d_2) - f_y(a, d_2)]. \tag{12}$$

Finally, we apply the Mean Value Theorem to the expression in brackets and get

$$\Delta = khf_{yx}(c_2, d_2), \tag{13}$$

for some c_2 between a and $a + h$.

A comparison of Eqs. (7) and (13) shows that

$$f_{xy}(c_1, d_1) = f_{yx}(c_2, d_2), \tag{14}$$

where (c_1, d_1) and (c_2, d_2) both lie in the rectangle R' (Fig. 15.17). Equation (14) is not quite the result we want, since it says only that the mixed derivative f_{xy} has the same value at (c_1, d_1) that the derivative f_{yx} has at (c_2, d_2). But the numbers h and k in our discussion may be made as small as we wish. The hypothesis that f_{xy} and f_{yx} are both continuous throughout R then means that

$$f_{xy}(c_1, d_1) = f_{xy}(a, b) + \epsilon_1$$

and

$$f_{yx}(c_2, d_2) = f_{yx}(a, b) + \epsilon_2,$$

where

$$\epsilon_1, \epsilon_2 \to 0 \quad \text{as} \quad h, k \to 0.$$

Hence, if we let h and $k \to 0$, we have

$$f_{xy}(a, b) = f_{yx}(a, b). \; \square$$

The proof just completed hinges on the consideration of the so-called "second difference" Δ given by Eq. (8). The reason for calling this a *second* difference is to be found from a closer examination of Eqs. (2) and (3). Note that both

$$F(a) = f(a, b + k) - f(a, b)$$

and

$$F(a + h) = f(a + h, b + k) - f(a + h, b)$$

are themselves differences ("first" differences), while

$$\Delta = F(a + h) - F(a)$$

is the difference between these first differences.

The introduction of such a *second* difference might occur to a mathematician who wishes to prove a theorem involving *second* derivatives, but the student should not feel he is expected to be able to reconstruct the proof from scratch.

An interesting by-product of the proof of the theorem is the fact that this second difference may, under the hypotheses of the theorem, be approximated by

$$\Delta \approx hkf_{xy}(a, b),$$

where the accuracy of the approximation depends on the size of h and k.

If we refer once more to the example at the beginning of this article, we note not only that

$$\frac{\partial^2 f}{\partial x\, \partial y} = \frac{\partial^2 f}{\partial y\, \partial x},$$

but also that

$$\frac{\partial^3 f}{\partial x^2\, \partial y} = \frac{\partial^3 f}{\partial y\, \partial x^2},$$

This equality may be derived from the theorem above as follows:

$$\begin{aligned}\frac{\partial^3 f}{\partial x^2\, \partial y} &= \frac{\partial}{\partial x}\left(\frac{\partial^2 f}{\partial x\, \partial y}\right) = \frac{\partial}{\partial x}\left(\frac{\partial^2 f}{\partial y\, \partial x}\right)\\ &= \frac{\partial}{\partial x}\left(\frac{\partial}{\partial y} f_x\right) = \frac{\partial}{\partial y}\left(\frac{\partial}{\partial x} f_x\right)\\ &= \frac{\partial}{\partial y}\left(\frac{\partial^2 f}{\partial x^2}\right) = \frac{\partial^3 f}{\partial y\, \partial x^2}.\end{aligned}$$

In fact, if all the partial derivatives that appear are continuous, the notation

$$\frac{\partial^{m+n} f}{\partial x^m\, \partial y^n}$$

may be used to denote the result of differentiating the function $f(x, y)$ m times with respect to x and n times with respect to y, the order in which these differentiations are performed being entirely arbitrary. For example, $\partial^5 f/(\partial x^2\, \partial y^3)$ is the result of five successive differentiations, two with respect to x and three with respect to y, such as f_{xyyxy}, where the latter notation means differentiation first with respect to x, then twice with respect to y, then again with respect to x, and finally with respect to y. (The subscripts are read from left to right.)

EXERCISES

1. If $w = \cos(x + y) + \sin(x - y)$, show that
$$(\partial^2 w/\partial x^2) = (\partial^2 w/\partial y^2).$$

2. If $w = \ln(2x + 2y) + \tan(2x - 2y)$, show that
$$(\partial^2 w/dx^2) = (\partial^2 w/\partial y^2).$$

3. If $w = f(x + y) + g(x - y)$, where $f(u)$ and $g(v)$ are twice-differentiable functions of $u = x + y$ and $v = x - y$, respectively, show that
$$\frac{\partial^2 w}{\partial x^2} = \frac{\partial^2 w}{\partial y^2} = f''(u) + g''(v).$$
Note. Exercises 1 and 2 are special examples of this result.

4. If c is a constant and
$$w = \sin(x + ct) + \cos(2x + 2ct),$$
show that
$$(\partial^2 w/\partial t^2) = c^2(\partial^2 w/dx^2).$$

5. If c is a constant and
$$w = 5\cos(3x + 3ct) - 7\sinh(4x - 4ct),$$
show that
$$(\partial^2 w/\partial t^2) = c^2(\partial^2 w/\partial x^2).$$

6. If c is a constant and
$$w = f(x + ct) + g(x - ct),$$
where $f(u)$ and $g(v)$ are twice-differentiable functions of $u = x + ct$ and $v = x - ct$, respectively, show that
$$\frac{\partial^2 w}{\partial t^2} = c^2\frac{\partial^2 w}{\partial x^2} = c^2[f''(u) + g''(v)].$$
Note. Exercises 4 and 5 are special cases. The equation
$$\frac{\partial^2 w}{\partial t^2} = c^2\frac{\partial^2 w}{\partial x^2}$$
describes the motion of a wave that travels with velocity c. See Sears, *Mechanics, Heat, and Sound*, Chapter 26.

In Exercises 7 through 13, verify that V satisfies Laplace's equation,

$$\frac{\partial^2 V}{\partial x^2}+\frac{\partial^2 V}{\partial y^2}+\frac{\partial^2 V}{\partial z^2}=0.$$

7. $V = x^2 + y^2 - 2z^2$
8. $V = 2z^3 - 3(x^2+y^2)z$
9. $V = e^{-2y}\cos 2x$
10. $V = \ln\sqrt{x^2+y^2}$
11. $V = (x^2+y^2+z^2)^{-1/2}$
12. $V = e^{3x+4y}\cos 5z$
13. $V = \cos 3x \cos 4y \sinh 5z$

In Exercises 14 through 17, verify that $w_{xy} = w_{yx}$.

14. $w = e^x \sinh y + \cos(2x - 3y)$
15. $w = \ln(2x+3y)$
16. $w = \tan^{-1} y/x$
17. $w = xy^2 + x^2y^3 + x^3y^4$
18. Let $f(x, y) = x^3y^2$. Following the notation in Eqs. (2) through (7),
    (a) find $F(x)$, c_1, $g(y)$ and thus verify Eq. (7) for this particular case, and
    (b) show that
$$f_{xy}(c_1, d_1) \to f_{xy}(a, b) \quad \text{as} \quad (h, k) \to (0, 0).$$
19. Use the function $f = x^3y^2$ and carry out the steps of Eqs. (8) to (13). In particular, find (c_2, d_2), and show that
$$f_{yx}(c_2, d_2) \to f_{xy}(a, b) \quad \text{as} \quad (h, k) \to (0, 0).$$

15.13 EXACT DIFFERENTIALS

We have seen that on numerous occasions the result of translating a physical problem into mathematical terms is a differential equation to be integrated. In the simplest cases, we may be able to separate the variables and write the differential equation in a form such as

$$f(x)\,dx = g(y)\,dy,$$

which we can solve if we are able to evaluate $\int f(x)\,dx$ and $\int g(y)\,dy$. Sometimes, however, we are not able to separate the variables, as in the equations

$$(x^2+y^2)\,dx + 2xy\,dy = 0 \tag{1a}$$

and

$$(x^2+y^2)\,dx - 2xy\,dy = 0. \tag{1b}$$

More generally, we may be led to a differential equation of the form

$$M(x, y)\,dx + N(x, y)\,dy = 0, \tag{2}$$

where $M(x, y)$ and $N(x, y)$ are functions of x and y. (In most cases, these functions will be continuous and have continuous partial derivatives with respect to x and y.) The differential equation (2) can be solved quite easily, provided that it is possible to find a function

$$w = f(x, y) \tag{3}$$

such that the left side of Eq. (2) is the total differential of w. For if such is the case, then Eq. (2) becomes

$$dw = 0,$$

and the solution of this equation is simply

$$w = C,$$

and the solution of Eq. (2) is

$$f(x, y) = C,$$

where C is an arbitrary constant.

Now expressions of the form given in Eq. (2) aren't always differentials of functions. In fact, of the two expressions in Eq. (1a, b), we shall see that the first is such a differential while the second is not.

The following terminology is commonly used: An expression

$$M(x, y)\,dx + N(x, y)\,dy \tag{4}$$

is called an *exact differential* if and only if a function $w = f(x, y)$ exists such that

$$df(x, y) = M(x, y)\,dx + N(x, y)\,dy. \tag{5}$$

If no such function exists, the expression is not an exact differential.

Three questions naturally present themselves:

(a) How can we tell whether a given expression is or is not an exact differential?

(b) If the expression is exact, how do we find the function $f(x, y)$ of which it is the differential?

(c) If the expression is not exact, how can we solve the differential equation (2)?

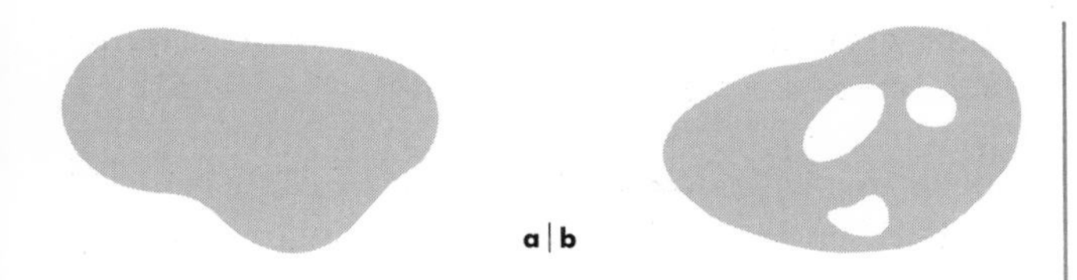

a| A simply connected region.

b| A connected region that is not simply connected.

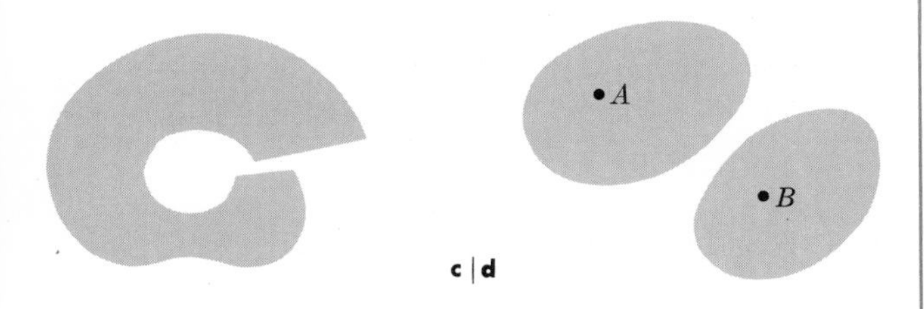

c| A simply connected region.

d| A region that is not connected. Note that there is no path from A to B that lies entirely inside the set.

15.18

In the proof of the theorem below, we shall give answers to the first two of these questions. Differential equations are discussed in more detail in Chapter 20.

The theorem introduces a hypothesis about the behavior of some functions in a *simply connected region* of the plane. This term refers to a set of points with the following three properties:

1. Each point of the set is an *interior* point of the set. (It can be the center of a small circle whose entire interior is in the set.) Technically, this means that the region is an *open* set in the plane.

2. Any two points of the set can be joined by a polygonal path, all of whose points are in the set. (This property makes the set a *connected* set.)

3. If C is any simple closed curve, all of whose points are in the set, then all points in the interior of C are also in the set.

Figure 15.18(a, c) show two simply connected regions; Fig. 15.18(b) shows a region that is connected but not simply connected; and Fig. 15.18(d) shows a region consisting of two disconnected components.

Theorem. *Let the functions $M(x, y)$ and $N(x, y)$ be continuous, and let them possess continuous partial derivatives M_x, M_y, N_x, N_y for all real values of x and y in some simply connected region G. Then a necessary and sufficient condition for*

$$M(x, y)\,dx + N(x, y)\,dy$$

to be an exact differential in G is that

$$\frac{\partial M}{\partial y} = \frac{\partial N}{\partial x}.$$

Proof. We shall first prove that the condition is necessary. Suppose the expression is an exact differential; that is, suppose a function $f(x, y)$ exists such that Eq. (5) is satisfied at all points in G. We also know that

$$df = \frac{\partial f}{\partial x}\,dx + \frac{\partial f}{\partial y}\,dy. \tag{6}$$

In Eqs. (5) and (6), dx and dy are independent variables and we may set either dx or dy equal to zero and keep the other one different from zero. Then the only way that both equations can hold is to have

$$\frac{\partial f}{\partial x} = M(x, y), \qquad \frac{\partial f}{\partial y} = N(x, y). \tag{7}$$

Now, by the theorem of Article 15.12, we know that if M and N are continuous and have continuous partial derivatives in G, then

$$\frac{\partial}{\partial y}\left(\frac{\partial f}{\partial x}\right) = \frac{\partial}{\partial x}\left(\frac{\partial f}{\partial y}\right),$$

so that the condition

$$\frac{\partial M}{\partial y} = \frac{\partial N}{\partial x} \tag{8}$$

is a necessary condition if Eq. (7) is to be satisfied.

Example 1. The criterion given by Eq. (8) is satisfied by the expression in Eq. (1a), where we have

$$M = x^2 + y^2, \quad N = 2xy.$$

Here

$$\frac{\partial M}{\partial y} = 2y = \frac{\partial N}{\partial x}.$$

We have not yet proved that this condition is sufficient, however, so we cannot at this point say that the expression is exact.

For Eq. (1b), we have

$$M = x^2 + y^2, \quad N = -2xy$$

and

$$\frac{\partial M}{\partial y} = 2y, \qquad \frac{\partial N}{\partial x} = -2y,$$

so that

$$\frac{\partial M}{\partial y} \neq \frac{\partial N}{\partial x}.$$

Since Eq. (8) is a necessary condition, we can say that the expression in (1b) is not an exact differential.

It is very easy to apply the criterion of Eq. (8), but we have not yet shown that it is a sufficient condition as well as a necessary one. But we shall now show that if

$$\frac{\partial M}{\partial y} = \frac{\partial N}{\partial x} \quad \text{in } G,$$

then there is a function $w = f(x, y)$ whose domain includes G and which is such that

$$df = M\,dx + N\,dy.$$

We shall establish this result by showing how to find the function $f(x, y)$ and this will also answer the second question previously raised.

From Eq. (6), we see that our sought-for function must have the property expressed by Eqs. (7):

$$\frac{\partial f}{\partial x} = M(x, y), \qquad \frac{\partial f}{\partial y} = N(x, y).$$

Integrating the first of these with respect to x, we find

$$f(x, y) = \int_x M(x, y)\,dx + g(y), \tag{9}$$

where $g(y)$ represents an unknown function of y that plays the role of an arbitrary constant of integration. The integral $\int_x M(x, y)\,dx$ is an ordinary indefinite integral with respect to x with y held constant during the integration. The function $f(x, y)$ produced by Eq. (9) satisfies the condition

$$\partial f/\partial x = M(x, y)$$

for any choice of the function $g(y)$, because $g(y)$ acts as a constant under partial differentiation with respect to x. To see how we may satisfy the second condition, $\partial f/\partial y = N(x, y)$, we differentiate both sides of Eq. (9) with respect to y, holding x fixed, and obtain

$$\frac{\partial f}{\partial y} = \frac{\partial}{\partial y} \int_x M(x, y)\,dx + \frac{\partial g(y)}{\partial y}. \tag{10}$$

Actually, since $g(y)$ is a function of y alone, we may write dg/gy instead of $\partial g/\partial y$. We then set $\partial f/\partial y$ from Eq. (10) equal to $N(x, y)$ and have

$$N(x, y) = \frac{\partial}{\partial y} \int_x M(x, y)\,dx + \frac{dg(y)}{dy},$$

or

$$\frac{dg(y)}{dy} = N(x, y) - \frac{\partial}{\partial y} \int_x M(x, y)\,dx. \tag{11}$$

We use this differential equation to determine $g(y)$ by simply integrating its right-hand member with respect to y and then substituting the result back into Eq. (9) to obtain our final answer, namely $f(x, y)$. Success of the method hinges on the fact that the expression on the right-hand side of Eq. (11) *is a function of y alone, provided that the condition*

$$\frac{\partial M(x, y)}{\partial y} = \frac{\partial N(x, y)}{\partial x}$$

is satisfied. [If the right-hand side of Eq. (11) depended on x as well as on y, it could not be equal to dg/dy, which involves only y.]

But how, in general, can we prove that the expression in question is independent of x? We can prove this by showing that its partial derivative with respect to x is identically zero. That is, we must calculate

$$\begin{aligned}\frac{\partial}{\partial x}\left(N(x, y) - \frac{\partial}{\partial y}\int_x M(x, y)\,dx\right) &= \frac{\partial N}{\partial x} - \frac{\partial^2}{\partial x\,\partial y}\int_x M(x, y)\,dx \\ &= \frac{\partial N}{\partial x} - \frac{\partial}{\partial y}\left(\frac{\partial}{\partial x}\int_x M(x, y)\,dx\right) \\ &= \frac{\partial N}{\partial x} - \frac{\partial}{\partial y}(M),\end{aligned}$$

which vanishes if it is true that

$$\partial N/\partial x = \partial M/\partial y,$$

as we have assumed. □

Example 2. We shall illustrate the method of finding the function $f(x, y)$ such that

$$df = (x^2 + y^2)\,dx + 2xy\,dy.$$

We set

$$M = x^2 + y^2, \qquad N = 2xy,$$

and note that the condition

$$\partial M/\partial y = \partial N/\partial x$$

is satisfied. We seek $f(x, y)$ such that

$$\frac{\partial f}{\partial x} = x^2 + y^2, \qquad \frac{\partial f}{\partial y} = 2xy.$$

Integrating the first of these with respect to x while holding y constant and adding $g(y)$ as our "constant of integration," we have

$$f(x, y) = \tfrac{1}{3}x^3 + y^2x + g(y).$$

Differentiating this with respect to y with x held constant and setting the result equal to $2xy$, we have

$$2xy = 2yx + \frac{dg}{dy}, \quad \text{or} \quad \frac{dg}{dy} = 0,$$

so that

$$g(y) = C$$

must be a pure constant. Hence

$$f(x, y) = \tfrac{1}{3}x^3 + y^2x + C.$$

Note that the method is very easy to apply!

EXERCISES

In Exercises 1 through 7, determine whether the given expression is or is not an exact differential. If the expression is the differential of a function $f(x, y)$, find f.

1. $2x(x^3 + y^3)\,dx + 3y^2(x^2 + y^2)\,dy$
2. $e^y\,dx + x(e^y + 1)\,dy$
3. $(2x + y)\,dx + (x + 2y)\,dy$
4. $(\cosh y + y \cosh x)\,dx + (\sinh x + x \sinh y)\,dy$
5. $(\sin y + y \sin x)\,dx + (\cos x + x \cos y)\,dy$
6. $(1 + e^x)\,dy + e^x(y - x)\,dx$
7. $(e^{x+y} + e^{x-y})(dx + dy)$

15.14 DERIVATIVES OF INTEGRALS

From the Fundamental Theorem of Integral Calculus, we know that if f is a continuous function on $a \le t \le b$, then

$$\frac{d}{dx}\int_a^x f(t)\,dt = f(x). \tag{1}$$

For example,

$$\frac{d}{dx}\int_0^x e^{-t^2}\,dt = e^{-x^2}$$

and

$$\frac{d}{dx}\int_1^x \frac{1}{t}\,dt = \frac{1}{x}, \quad x > 0.$$

Similarly, since

$$\int_x^b f(t)\,dt = -\int_b^x f(t)\,dt,$$

we have

$$\frac{d}{dx}\int_x^b f(t)\,dt = -f(x). \tag{2}$$

Equations (1) and (2) may be combined to give the following result.

Theorem. *Let f be continuous on $a \le t \le b$. Let u and v be differentiable functions of x such that $u(x)$ and $v(x)$ lie between a and b. Then*

$$\frac{d}{dx}\int_{u(x)}^{v(x)} f(t)\,dt = f[v(x)]\frac{dv}{dx} - f[u(x)]\frac{du}{dx}. \tag{3}$$

Proof. Let $F(u, v) = \int_u^v f(t)\,dt$. Then, by Eq. (1),

$$\partial F/\partial v = f(v), \tag{4a}$$

and by Eq. (2),

$$\partial F/\partial u = -f(u), \tag{4b}$$

provided that u and v lie between a and b. If u and v are differentiable functions of x, and $u(x)$, $v(x)$ are between a and b, then we may apply the chain rule:

$$\frac{dF}{dx} = \frac{\partial F}{\partial u}\frac{du}{dx} + \frac{\partial F}{\partial v}\frac{dv}{dx}. \tag{5}$$

The result of substituting from Eqs. (4a, b) into Eq. (5) is

$$\frac{dF}{dx} = f(v)\frac{dv}{dx} - f(u)\frac{du}{dx}.$$

This establishes Eq. (3). □

Problem. Verify Eq. (3) for

$$\frac{d}{dx}\int_x^{2x} \frac{1}{t}\,dt, \qquad x > 0.$$

Solution.

Let $F(u, v) = \int_u^v 1/t\,dt = \ln v - \ln u.$

Then $\dfrac{\partial F}{\partial u} = -\dfrac{1}{u}$ and $\dfrac{\partial F}{\partial v} = \dfrac{1}{v}.$

If $u = x$ and $v = 2x$,

then $du/dx = 1$, $dv/dx = 2$,

and

$$\frac{dF}{dx} = \frac{\partial F}{\partial u}\cdot\frac{du}{dx} + \frac{\partial F}{\partial v}\cdot\frac{dv}{dx}$$

$$= -\frac{1}{u} + \frac{2}{v}$$

$$= -\frac{1}{x} + \frac{2}{2x} = 0.$$

Alternatively,

$$F(x, 2x) = \int_x^{2x} \frac{1}{t}\,dt = \ln t\Big]_x^{2x} = \ln 2x - \ln x$$

$$= \ln\frac{2x}{x} = \ln 2, \qquad 0 < x$$

and

$$\frac{d}{dx}F(x, 2x) = \frac{d}{dx}(\ln 2) = 0.$$

EXERCISES

Find the derivatives with respect to x of the following integrals. Assume that $x > 0$.

1. $\displaystyle\int_x^{x^2} \frac{1}{t}\,dt$
2. $\displaystyle\int_x^{2x} \frac{1}{t^2}\,dt$
3. $\displaystyle\int_{2x}^{x^2} \frac{1}{t^2}\,dt$
4. $\displaystyle\int_0^{\sin^{-1}x} \frac{\sin t}{t}\,dt$
5. $\displaystyle\int_x^{x^2} \ln t\,dt$
6. If u and v are differentiable functions of x, and if $u(x) > 0$, show (by two different methods) that

$$\frac{d}{dx}\{[u(x)]^{v(x)}\} = u^v\left(\frac{v}{u}\cdot\frac{du}{dx} + \frac{dv}{dx}\cdot\ln u\right).$$

Find dy/dx in Exercises 7 through 10.

7. $y = (e^x)^x$
8. $y = (\cosh x)^{x^2}$
9. $y = (x^2 + 1)^{1/x}$
10. $y = (4x^2 + 4x + 3)^{\int_x^{x^2} \ln t\,dt}$

REVIEW QUESTIONS AND EXERCISES

1. Let $w = f(x, y)$ define a function of two independent variables for values of (x, y) in some region G of the xy-plane (the domain of f). Describe two geometrical ways of representing the function.
2. When is a function of two variables continuous at a point of its domain? Give an example, different from those in the text, of a function that is discontinuous at some point(s) of its domain.
3. Let $w = f(x, y)$. Define

$$\partial w/\partial x \quad\text{and}\quad \partial w/\partial y$$

at a point (x_0, y_0) in the domain of f.
4. When is a function of three variables continuous at a point of its domain? Give an example of a function of three independent variables that is continuous at some points of its domain and discontinuous at least one place in its domain. Give an example

of a function that is discontinuous at all points of a surface $F(x, y, z) = 0$; at all points of a line.

5. Define the directional derivative, at a point in its domain, of a function of three independent variables. Write a formula for the directional derivative in vector form. What is the analogous formula for a function of two variables?
6. Define *tangent plane* and *normal line to a surface S* at a point P_0 on S. Derive equations of the tangent plane and normal line in terms of the equations of the surface.
7. Write an expression for the tangent-plane approximation to the increment of a function of two independent variables. What is the corresponding expression approximating the increment of a function of three variables?
8. Define *gradient* of a scalar function. Give two properties of the gradient that you consider to be important.
9. State the chain rule for partial derivatives of functions of several variables.
10. Define the *total differential* of a function of several variables.
11. Outline a method for finding local maxima or minima of a function of two or three independent variables.
12. Outline the "method of least squares" as applied to fitting a straight line to a set of observations.
13. Describe the method of Lagrange multipliers as it applies to the problem of maximizing or minimizing a function $f(x, y, z)$

    (a) subject to a constraint

    $$g(x, y, z) = 0,$$

    (b) subject to two constraints,

    $$g(x, y, z) = 0 \quad \text{and} \quad h(x, y, z) = 0.$$

14. Sketch some of the contour curves for the function $f(x, y) = 2x + 3y$. Find points on the curve $C\colon x^2 + xy + y^2 = 3$ that

    (a) maximize f, (b) minimize f.

    Does it seem reasonable on geometric grounds that the contour curve of f at each one of these points is tangent to C at that point? Is it true?

MISCELLANEOUS EXERCISES

1. Let

    $$f(x, y) = (x^2 - y^2)/(x^2 + y^2)$$

    for $x^2 + y^2 \neq 0$. Is it possible to define the value of f at $x = 0$, $y = 0$ in such a way that the function would be continuous at $x = 0$, $y = 0$? Why?

2. Let the function $f(x, y)$ be defined by the relations

    $$f(x, y) = \frac{\sin^2(x - y)}{|x| + |y|} \quad \text{for } |x| + |y| \neq 0$$

    and $$f(0, 0) = 0.$$

    Is f continuous at $x = 0$, $y = 0$?

3. Prove that if $f(x, y)$ is defined for all x, y by

    $$\begin{aligned} f(x, y) &= 2xy/(x^2 + y^2) \quad \text{if} \quad (x, y) \neq (0, 0), \\ &= 0 \quad \text{if} \quad x = y = 0, \end{aligned}$$

    then

    (a) for any fixed x, $f(x, y)$ is a continuous function of y;

    (b) for any fixed y, $f(x, y)$ is a continuous function of x;

    (c) $f(x, y)$ is not continuous at $(0, 0)$;

    (d) $\partial f/\partial x$ and $\partial f/\partial y$ exist at $(0, 0)$ but are not continuous there.

    Note. This example shows that a function may possess partial derivatives at all points of a region, yet not be continuous in the region.

    (e) Contrast the case of a function of one variable, where the existence of a derivative implies continuity.

4. Let $f(x, y)$ be defined and continuous for all x, y (differentiability not assumed). Show that it is always possible to find arbitrarily many points

    $$(x_1, y_1), \quad (x_2, y_2), \quad \ldots, \quad (x_n, y_n)$$

    such that the function has the same value at each of them.

5. Find the first partial derivatives of the functions

    $$(\sin xy)^2 \quad \text{and} \quad \sin[(xy)^2].$$

6. Let α, β, γ be the direction angles of a line, and consider γ as a function of α and β. Find the value of $\partial\gamma/\partial\alpha$ when $\alpha = \pi/4$, $\beta = \pi/3$, $\gamma = \pi/3$.

7. Let (r, θ) and (x, y) be polar coordinates and Cartesian coordinates in the plane. Show geometrically why $\partial r/\partial x$ is not equal to $(\partial x/\partial r)^{-1}$ by appealing to the definitions of these derivatives.

8. Consider the surface whose equation is
$$x^3z + y^2x^2 + \sin(yz) + 54 = 0.$$
Give an equation of the tangent plane to the surface at the point $P(3, 0, -2)$, and give equations of the straight line through P normal to the surface. Determine direction cosines of the line.

9. (a) Sketch and name the surface
$$x^2 - y^2 + z^2 = 4.$$
(b) Find a vector that is normal to this surface at $(2, -3, 3)$.
(c) Find equations of the surface's tangent plane and normal line at $(2, -3, 3)$.

10. (a) Find an equation of the plane tangent to the surface
$$x^3 + xy^2 + y^3 + z^3 + 1 = 0$$
at the point $(-2, 1, 2)$.
(b) Find equations of the straight line perpendicular to the above plane at the point $(-2, 1, 2)$.

11. The directional derivative of a given function $w = f(x, y)$ at the point $P_0(1, 2)$ in the direction toward $P_1(2, 3)$ is $2\sqrt{2}$, and in the direction toward $P_2(1, 0)$ it is -3. Compute $\partial f/\partial x$ and $\partial f/\partial y$ at $P_0(1, 2)$, and compute the directional derivative dw/ds at $P_0(1, 2)$ in the direction toward $P_3(4, 6)$.

12. Let $z = f(x, y)$ have continuous first partial derivatives. Let C be any curve lying on the surface and passing through (x_0, y_0, z_0). Prove that the tangent line to C at (x_0, y_0, z_0) must lie wholly in the plane determined by the tangent lines to the curves C_x and C_y, where C_x is the curve of intersection of $y = y_0$ and $z = f(x, y)$, and C_y is the curve of intersection of $x = x_0$ and the surface.

13. Let $u = xyz$. Show that if x and y are the independent variables (so that u and z are functions of x and y), then $\partial u/\partial x = xy\,(\partial z/\partial x) + yz$; but that if x, y, and z are the independent variables, then $\partial u/\partial x = yz$.

14. Let $\mathbf{u} = u_1\mathbf{i} + u_2\mathbf{j} + u_3\mathbf{k}$ and $\mathbf{v} = v_1\mathbf{i} + v_2\mathbf{j} + v_3\mathbf{k}$ be given as constant unit vectors and let $f(x, y, z)$ be a given scalar function. Compute
(a) the directional derivative $D_{\mathbf{u}}f$, and
(b) the directional derivative $D_{\mathbf{v}}(D_{\mathbf{u}}f)$,
in terms of derivatives of f and the components of $\mathbf{u}$ and $\mathbf{v}$. Here $D_{\mathbf{u}}f$ denotes df/ds in the direction of $\mathbf{u}$.

15. Consider the function $w = xyz$.
(a) Compute the directional derivative of w at the point $(1, 1, 1)$ in the direction of the vector $\mathbf{i} + \mathbf{j} + \mathbf{k}$.
(b) Compute the largest value of the directional derivative of w at the point $(1, 1, 1)$.

16. The function $w = f(x, y)$ has, at the point $(1, 2)$, directional derivatives that are equal to $+2$ in the direction toward $(2, 2)$, and equal to -2 in the direction toward $(1, 1)$. What is its directional derivative at $(1, 2)$ in the direction toward $(4, 6)$?

17. Given the function $f(x, y, z) = x^2 + y^2 - 3z$, what is the maximum value of the directional derivative df/ds at the point $(1, 3, 5)$?

18. Given the function
$$(x - 1)^2 + 2(y + 1)^2 + 3(z - 2)^2 - 6.$$
Find the directional derivative of the function at the point $(2, 0, 1)$ in the direction of the vector $\mathbf{i} - \mathbf{j} + 2\mathbf{k}$.

19. Find the derivative of the function
$$f(x, y, z) = x^2 - 2y^2 + z^2$$
at the point $(3, 3, 1)$ in the direction of the vector $2\mathbf{i} + \mathbf{j} - \mathbf{k}$.

20. The two equations
$$e^u \cos v - x = 0, \quad e^u \sin v - y = 0$$
define u and v as functions of x and y, say
$$u = u(x, y) \quad \text{and} \quad v = v(x, y).$$
Show that the angle between the two vectors
$$\frac{\partial u}{\partial x}\mathbf{i} + \frac{\partial u}{\partial y}\mathbf{j}, \qquad \frac{\partial v}{\partial x}\mathbf{i} + \frac{\partial v}{\partial y}\mathbf{j}$$
is constant.

21. (a) Find a vector $\mathbf{N}(x, y, z)$ normal to the surface
$$z = \sqrt{x^2 + y^2} + (x^2 + y^2)^{3/2}$$
at the point (x, y, z) of the surface.
(b) Find the cosine of the angle γ between $\mathbf{N}(x, y, z)$ and the z-axis. Find the limit of $\cos\gamma$ as
$$(x, y, z) \to (0, 0, 0).$$

22. Find the locus of all points (a, b, c) in space for which the spheres

$$(x - a)^2 + (y - b)^2 + (z - c)^2 = 1$$

and

$$x^2 + y^2 + z^2 = 1$$

intersect orthogonally. (Their tangents are to be perpendicular at each point of intersection.)

23. (a) Find the gradient, at $P_0(1, -1, 3)$, of the function $x^2 + 2xy - y^2 + z^2$.

(b) Find the plane that is tangent to the surface

$$x^2 + 2xy - y^2 + z^2 = 7$$

at $P_0(1, -1, 3)$.

24. Find a unit vector normal to the surface $x^2 + y^2 = 3z$ at the point $(1, 3, \frac{10}{3})$.

25. In a flowing fluid, the density $\rho(x, y, z, t)$ depends on position and time. If $\mathbf{V} = \mathbf{V}(x, y, z, t)$ is the velocity of the fluid particle at the point (x, y, z) at time t, then

$$\frac{d\rho}{dt} = \mathbf{V} \cdot \nabla\rho + \frac{\partial\rho}{\partial t}$$

$$= V_1 \frac{\partial\rho}{\partial x} + V_2 \frac{\partial\rho}{\partial y} + V_3 \frac{\partial\rho}{\partial z} + \frac{\partial\rho}{\partial t},$$

where $\mathbf{V} = V_1\mathbf{i} + V_2\mathbf{j} + V_3\mathbf{k}$. Explain the physical and geometrical meaning of this relation.

26. Find a constant a such that at any point of intersection of the two spheres

$$(x - a)^2 + y^2 + z^2 = 3, \quad x^2 + (y - 1)^2 + z^2 = 1,$$

their tangent planes will be perpendicular to each other.

27. If the gradient of a function $f(x, y, z)$ is always parallel to the vector $x\mathbf{i} + y\mathbf{j} + z\mathbf{k}$, show that the function must assume the same value at the points $(0, 0, a)$, $(0, 0, -a)$.

28. Let $f(P)$ denote a function defined for points P in the plane; that is, to each point P there is attached a real number $f(P)$. Explain how one could introduce the notions of continuity and differentiability of the function and define the vector ∇f *without* introducing a coordinate system. If one introduces a polar coordinate system r, θ, $\mathbf{U}_r$, $\mathbf{U}_\theta$, what form does the vector $\nabla f(t, \theta)$ take?

29. Show that the directional derivative of

$$r = \sqrt{x^2 + y^2 + z^2}$$

equals unity in any direction at the origin, but that r does not have a gradient vector at the origin.

30. Let $\mathbf{R} = x\mathbf{i} + y\mathbf{j} + z\mathbf{k}$ and $r = |\mathbf{R}|$.

(a) From its geometrical interpretation, show that

$$\nabla r = \mathbf{R}/r.$$

(b) Show that $\nabla(r^n) = nr^{n-2}\mathbf{R}$.

(c) Find a function with gradient equal to $\mathbf{R}$.

(d) Show $\mathbf{R} \cdot d\mathbf{R} = r\,dr$.

(e) If $\mathbf{A}$ is a constant vector, show that $\nabla(\mathbf{A} \cdot \mathbf{R}) = \mathbf{A}$.

31. If θ is the polar coordinate in the xy-plane, find the direction and magnitude of $\nabla\theta$.

32. If r_1, r_2 are the distances from the point $P(x, y)$ on an ellipse to its foci, show that the equation

$$r_1 + r_2 = \text{constant},$$

satisfied by these distances, requires

$$\mathbf{U} \cdot \nabla(r_1 + r_2) = 0,$$

where $\mathbf{U}$ is a unit tangent to the curve. By geometrical interpretation, show that the tangent makes equal angles with the lines to the foci.

33. If A, B are fixed points and θ is the angle at $P(x, y, z)$ subtended by the line segment AB, show that $\nabla\theta$ is normal to the circle through A, B, P.

34. Find the general solution of the partial differential equations

(a) $af_x + bf_y = 0$, a, b, constants,

(b) $yf_x - xf_y = 0$.

Hint. Consider the geometrical meaning of the equations.

35. When y is eliminated from the two equations

$$z = f(x, y), \quad g(x, y) = 0,$$

the result is expressible in the form $z = h(x)$. Express the derivative $h'(x)$ in terms of $\partial f/\partial x$, $\partial f/\partial y$, $\partial g/\partial x$, $\partial g/dy$. Check your formula by computing $h(x)$ and $h'(x)$ explicitly in the example where

$$f(x, y) = x^2 + y^2, \quad g(x, y) = x^3 + y^2 - x.$$

36. Suppose the equation $F(x, y, z) = 0$ defines z as a function of x and y, say $z = f(x, y)$, with derivatives $\partial f/\partial x$ and $\partial f/\partial y$. Suppose also that the same equation $F(x, y, z) = 0$ defines x as a function of y and z, say $x = g(y, z)$, with derivatives $\partial g/\partial y$ and $\partial g/\partial z$. Prove that

$$\frac{\partial g}{\partial y} = -\frac{\partial f/\partial y}{\partial f/\partial x},$$

and also express $\partial g/\partial z$ in terms of $\partial f/\partial x$ and $\partial f/\partial y$.

37. Given $z = x \sin x - y^2$, $\cos y = y \sin z$; find dx/dz.

38. If

$$z = f\left(\frac{x-y}{y}\right),$$

show that $x(\partial z/\partial x) + y(\partial z/\partial y) = 0$.

39. If the substitution

$$u = \frac{x-y}{2}, \quad v = \frac{x+y}{2}$$

changes $f(u, v)$ into $F(x, y)$, express $\partial F/\partial x$ and $\partial F/\partial y$ in terms of the derivatives of $f(u, v)$ with respect to u and v.

40. Given $w = f(x, y)$, with $x = u + v$, $y = u - v$, show that

$$\frac{\partial^2 w}{\partial u\, \partial v} = \frac{\partial^2 w}{\partial x^2} - \frac{\partial^2 w}{\partial y^2}.$$

41. Suppose $f(x, y, z)$ is a function with continuous partial derivatives and suppose it satisfies

$$f(tx, ty, tz) = t^n f(x, y, z)$$

for every quadruple of numbers x, y, z, t (where n is a fixed integer). Show the identity

$$\frac{\partial f}{\partial x} x + \frac{\partial f}{\partial y} y + \frac{\partial f}{\partial z} z = nf.$$

Hint. Differentiate with respect to t and then set $t = 1$.

42. The substitution $u = x + y$, $v = xy^2$ changes the function $f(u, v)$ into $F(x, y)$. Express the partial derivative $\partial^2 F/\partial x \partial y$ in terms of x, y, and the partial derivatives of $f(u, v)$ with respect to u, v.

43. Given $z = u(x, y) \cdot e^{ax+by}$, where $u(x, y)$ is a function of x and y such that $\partial^2 u/\partial x \partial y = 0$, ($a$, b constants). Find values of a and b that make the expression $\partial^2 z/\partial x \partial y - \partial z/\partial x - \partial z/\partial y + z$ identically zero.

44. Introducing polar coordinates $x = r \cos \theta$, $y = r \sin \theta$ changes $f(x, y)$ into $g(r, \theta)$. Compute the value of the second derivative $\partial^2 g/\partial \theta^2$ at the point where $r = 2$ and $\theta = \pi/2$, given that

$$\partial f/\partial x = \partial f/\partial y = \partial^2 f/\partial x^2 = \partial^2 f/\partial y^2 = 1$$

at that point.

45. Let $w = f(u, v)$ be a function of u, v with continuous partial derivatives, where u, v in turn are functions of independent variables x, y, z with continuous partial derivatives. Show that if w is regarded as a function of x, y, z its gradient at any point (x_0, y_0, z_0) lies in a common plane with the gradients of

$$u = u(x, y, z) \quad \text{and} \quad v = v(x, y, z).$$

46. Show that if a function u has first derivatives that satisfy a relation of the form $F(u_x, y_y) = 0$, and if $\partial F/\partial u_x$ and $\partial F/\partial u_y$ are not both zero, then u also satisfies $u_{xx}u_{yy} - u_{xy}^2 = 0$.
Hint. Differentiate $F = 0$ with respect to x and y.

47. If $f(x, y) = 0$, find d^2y/dx^2.

48. If $F(x, y, z) = 0$, show that

$$\left(\frac{\partial x}{\partial y}\right)_z \left(\frac{\partial y}{\partial z}\right)_x \left(\frac{\partial z}{\partial x}\right)_y = -1.$$

Note. Here $(\partial x/\partial y)_z$ denotes the fact that z is held constant while we compute the partial derivative of x with respect to y and so forth.

49. If $f(x, y, z) = 0$ and $z = x + y$, find dz/dx.

50. The function $v(x, t)$ is defined for

$$0 \le x \le 1, \quad 0 \le t$$

and satisfies the partial differential equation

$$v_t = v_x(v - x) + av_{xx}$$

where a = constant > 0. It also satisfies the boundary conditions $v(0, t) = 0$, $v(1, t) = 1$. Suppose that for each fixed t, $v(x, t)$ is a strictly increasing function of x; that is, $v_x(x, t) > 0$. Show that v and t may be introduced as independent variables and x as dependent variable and find the partial differential equation satisfied by the function $x(v, t)$. Find also the region of definition of $x(v, t)$ and boundary values that it satisfies. By considering level curves, show geometrically why the assumption $v_x(x, t) > 0$ is necessary for the success of this transformation.

51. Let $f(x, y, z)$ be a function that is dependent only on $r = \sqrt{x^2 + y^2 + z^2}$; that is, $f(x, y, z) = g(r)$. Prove that if
$$f_{xx} + f_{yy} + f_{zz} = 0,$$
it follows that $f = (a/r) + b$, where a and b are constants.

52. A function $f(x, y)$, defined and differentiable for all x, y, is said to be homogeneous of degree n (a nonnegative integer) if $f(tx, ty) = t^n f(x, y)$ for all t, x, and y. For such a function prove

(a) $$x\frac{\partial f}{\partial x} + y\frac{\partial f}{\partial y} = nf(x, y),$$
and express this in vector form;

(b) $$x^2\frac{\partial^2 f}{\partial x^2} + 2xy\frac{\partial^2 f}{\partial x\,\partial y} + y^2\frac{\partial^2 f}{\partial y^2} = n(n-1)f,$$
if f has continuous second partial derivatives;

(c) a homogeneous function of degree zero is a constant.

53. Prove the Mean Value Theorem for functions of two variables:
$$f(x + h, y + k) - f(x, y)$$
$$= f_x(x + \theta h, y + \theta k)h + f_y(x + \theta h, y + \theta k)k$$
for some θ, $0 < \theta < 1$, with suitable assumptions about f. What assumptions?
Hint. Apply the Mean Value Theorem for functions of one variable to $F(t) = f(x + ht, y + kt)$.

54. Prove the theorem: If $f(x, y)$ is defined in a region R, and f_x, f_y exist and are bounded in R, then $f(x, y)$ is continuous in R. (The assumption of boundedness is essential.)

55. Using differentials, find a reasonable approximation to the value of
$$w = xy\sqrt{x^2 + y^2}$$
at $x = 2.98$, $y = 4.04$.

56. A flat circular plate has the shape of the region
$$x^2 + y^2 \leq 1.$$
The plate (including the boundary, where $x^2 + y^2 = 1$) is heated so that the temperature T at any point (x, y) is
$$T = x^2 + 2y^2 - x.$$
Locate the hottest and coldest points of the plate and find the temperature at each of these points.

57. The temperature T at any point (x, y, z) in space is $T = 400xyz^2$. Find the highest temperature on the surface of the unit sphere $x^2 + y^2 + z^2 = 1$.

58. For each of the following three surfaces, find all the values of x and y for which z is a maximum or minimum (if there are any). Give complete reasonings.
(a) $x^2 + y^2 + z^2 = 3$ (b) $x^2 + y^2 = 2z$
(c) $x^2 - y^2 = 2z$

59. Find the point(s) on the surface $xyz = 1$ whose distance from the origin is a minimum.

60. A closed rectangular box is to be made to hold a given volume V in^3. The cost of the material used in the box is a cents/in^2 for top and bottom, b cents/in^2 for front and back, c cents/in^2 for the remaining two sides. What dimensions make the total cost of materials a minimum?

61. Find the maximum value of the function $xye^{-(2x+3y)}$ in the first quadrant.

62. A surface is defined by $z = x^3 + y^3 - 9xy + 27$. Prove that the only possible maxima and minima of z occur at $(0, 0)$ or $(3, 3)$. Prove that $(0, 0)$ is neither a maximum nor a minimum. Determine whether $(3, 3)$ is a maximum or a minimum.

63. Given n positive numbers $a_1, a_2, \ldots, a_n$. Find the maximum value of the expression
$$a_1x_1 + a_2x_2 + \cdots + a_nx_n$$
if the variables $x_1, x_2, \ldots, x_n$ are restricted so that the sum of their squares is 1.

64. Find the minimum volume bounded by the planes $x = 0$, $y = 0$, $z = 0$ and a plane that is tangent to the ellipsoid $x^2/a^2 + y^2/b^2 + z^2/c^2 = 1$ at a point in the octant $x > 0$, $y > 0$, $z > 0$.

65. Among the points $P(x, y)$ on the contour curve $\phi(x, y) = 0$, it is desired to find one at which the function $f(x, y)$ has a (relative) maximum. Assuming that such a point p_0 exists, show that at p_0 the vectors ∇f and $\nabla\phi$ are parallel so that there is a number λ_0 such that $(\nabla f)_0 = \lambda_0(\nabla\phi)_0$. Explain geometrically by considering the contour curves of f and ϕ.

66. Let z be defined implicitly as a function of x and y by the equation $\sin(x + y) + \sin(y + z) = 1$. Compute $\partial^2 z/\partial x\partial y$ in terms of x, y, and z.

67. Given $z = xy^2 - y \sin x$, calculate the value of
$$y(\partial^2 z/\partial y\, \partial x) - \partial z/\partial x.$$

68. Let $w = z \tan^{-1}(x/y)$. Compute
$$\partial^2 w/\partial x^2 + \partial^2 w/\partial y^2 + \partial^2 w/\partial z^2.$$

69. Show that each given function satisfies the subsequent equation.

(a) $\log \sqrt{x^2 + y^2}$, $\quad f_{xx} + f_{yy} = 0$

(b) $\sqrt{(x^2 + y^2 + z^2)^{-1}}$, $\quad f_{xx} + f_{yy} + f_{zz} = 0$

(c) $\int_0^{x/2\sqrt{kt}} e^{-\sigma^2}\, d\sigma$, $\quad kf_{xx} - f_t = 0$ (k const.)

(d) $\phi(x + at) + \psi(x - at)$, $\quad f_{tt} = a^2 f_{xx}$

70. Consider the function defined by
$$f(x, y) = xy\frac{x^2 - y^2}{x^2 + y^2}, \quad (x, y) \neq (0, 0),$$
$$= \quad 0, \qquad (x, y) = (0, 0).$$
Find $f_{yx}(0, 0)$ and $f_{xy}(0, 0)$.

71. Is $2x(x^3 + y^3)\, dx + 3y^2(x^2 + y^2)\, dy$ the total differential df of a function $f(x, y)$? If so, find the function.

72. Find a function $f(x, y)$ whose differential is
$$df = (y/x + e^y)\, dx + (\ln x + 2y + xe^y)\, dy,$$
or else show that no such function exists.

73. Find a function $w = f(x, y)$ such that
$$\partial w/\partial x = 1 + e^x \cos y, \quad \partial w/\partial y = 2y - e^x \sin y,$$
or else explain why no such function exists.

74. In thermodynamics, the five quantities S, T, u, p, v, are such that any two of them may be considered independent variables, the others then being determined. They are connected by the differential relation $T\, dS = du + p\, dv$. Show that
$$\left(\frac{\partial S}{\partial v}\right)_T = \left(\frac{\partial p}{\partial T}\right)_v \quad \text{and} \quad \left(\frac{\partial v}{\partial S}\right)_p = \left(\frac{\partial T}{\partial p}\right)_S.$$

Note. Here the letter outside the parentheses indicates a quantity held constant during the partial differentiation.

Suggestion. For (a), show that $S\, dT + p\, dv$ is an exact differential because it equals $d(ST - u)$, and then apply the condition for exactness.

MULTIPLE INTEGRALS

CHAPTER 16

16.1 DOUBLE INTEGRALS

We shall show how to use the method of double integration to calculate the area or center of gravity of the region A (Fig. 16.1), which is bounded above by the curve $y = f_2(x)$, below by $y = f_1(x)$, on the left by the line $x = a$, and on the right by $x = b$. Before taking up this specific application, we shall first define what we mean by the double integral of a function $F(x, y)$ of two variables x and y. Then specific applications will follow at once by specializing the function $F(x, y)$ to be

$$\text{(a)} \quad F(x, y) = 1, \quad \text{or} \quad \text{(b)} \quad F(x, y) = y$$

for the calculation of

(a) the area, or

(b) the moment of the area about the x-axis.

The notation

$$\int_A \int F(x, y)\, dA \tag{1}$$

is used to denote the double integral, over the region A, of the function $F(x, y)$. We imagine that the region A is covered by a grid of lines parallel to the x- and y-axes. These lines divide the plane into small pieces of area

$$\Delta A = \Delta x\, \Delta y = \Delta y\, \Delta x, \tag{2}$$

some of which lie entirely within the given region, some entirely outside the region, and some of which are intersected by the boundary of the region. We shall disregard all those that lie outside the

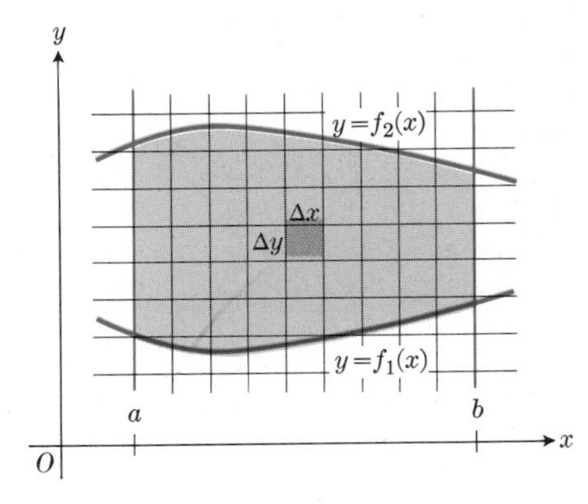

16.1

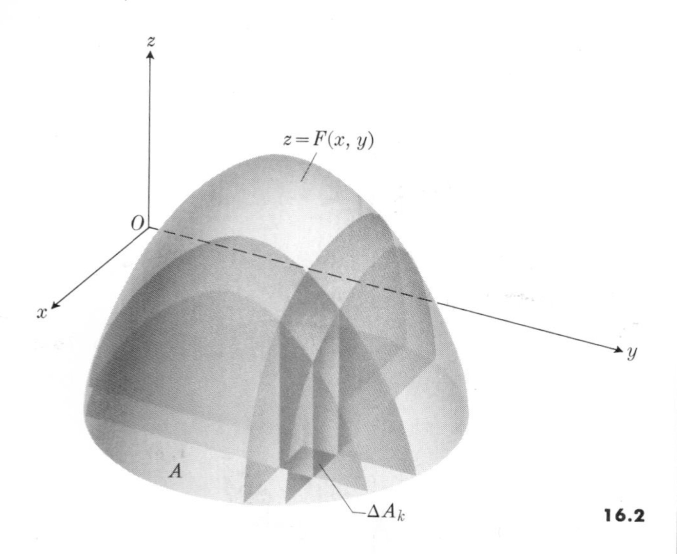

16.2

region and we may or may not take into account those which lie only partly inside, but in particular we shall consider all the pieces ΔA which lie completely inside. Suppose we number the elements ΔA lying inside the region

$$\Delta A_1, \Delta A_2, \ldots, \Delta A_n \tag{3}$$

in some order, and let (x_k, y_k) be any point lying in ΔA_k. We may then form the sum

$$S_n = \sum_{k=1}^{n} F(x_k, y_k)\, \Delta A_k. \tag{4}$$

If the function $F(x, y)$ is continuous throughout A, and if the curves which form the boundary of A are continuous and have a finite total length, then, as we refine the mesh width in such a way that Δx and Δy tend to zero (we may, for example, take $\Delta y = 2\,\Delta x$ and then make $\Delta x \to 0$), the limit

$$I = \lim_{\Delta A \to 0} \sum_{k=1}^{n} F(x_k, y_k)\, \Delta A_k \tag{5a}$$

exists, and it is this limit that is indicated by the notation in Eq. (1):

$$\int_A \int F(x, y)\, dA = \lim_{\Delta A \to 0} \sum_{k=1}^{n} F(x_k, y_k)\, \Delta A_k. \tag{5b}$$

The double integral (1) can also be interpreted as a volume, at least in the case where $F(x, y)$ is positive. Suppose, for example, that the region A is the base of a solid (Fig. 16.2) whose altitude above the point (x, y) is given by

$$z = F(x, y).$$

We can then see that the term

$$F(x_k, y_k)\, \Delta A_k$$

represents a reasonable approximation to the volume of that portion of the solid that rests on the base ΔA_k. The sum S_n in Eq. (4) then gives an approximation to the total volume of the solid, and the limit in Eq. (5) gives the exact volume.

The usefulness of the concept of the double integral should be apparent. Its usefulness as a working tool would be limited, however, if it were necessary to resort to the limit of sums, as in Eq. (5), to find numerical answers to specific problems. Fortunately, there is an alternative method for evaluating double integrals, namely, evaluating successive single integrals. In practice, the double integral in (1) is evaluated by calculating one or another of the *iterated* integrals

$$\begin{aligned} &\int_A \int F(x, y)\, dx\, dy, \\ &\int_A \int F(x, y)\, dy\, dx, \end{aligned} \tag{6}$$

which we shall explain below. Before doing so, however, we remark that a theorem of analysis (which we shall not prove here) asserts that the iterated integrals (6) are equal to each other and to the double integral (1), provided that the function $F(x, y)$ is continuous over A and the boundary of A is not too complicated. The necessary conditions are fulfilled in the examples and problems in this book. For more details, see Franklin, *A Treatise on Advanced Calculus*, Chapter XI.

Now we shall explain what we mean by the iterated integral

$$\int_A \int F(x, y)\, dy\, dx.$$

It is the result of

(a) integrating $\int F(x, y)\,dy$ with respect to y (with x held fixed) and evaluating the resulting integral between the limits $y = f_1(x)$ and $y = f_2(x)$, and then

(b) integrating the result of (a) with respect to x between the limits $x = a$ and $x = b$.

That is, we start with the innermost integral and perform successive integrations as follows:

$$\int_A \int F(x, y)\,dy\,dx = \int_a^b \left(\int_{f_1(x)}^{f_2(x)} F(x, y)\,dy\right) dx, \tag{7}$$

treating x as a constant while we perform the y-integration.

We can gain some insight into the geometrical significance of Eq. (7). We may again think of a solid with base covering the region A of the xy-plane and having altitude $z = F(x, y)$ at the point (x, y) of A. (Assume, for sake of simplicity, that F is positive.) Then imagine a slice cut from the solid by planes perpendicular to the x-axis at x and at $x + dx$. We may think of this as approximated by the differential of volume given by

$$dV = A(x)\,dx,$$

where $A(x)$ is the cross-sectional area cut from the solid by the plane at x (see Fig. 16.3). Now this cross-sectional area is given by the integral

$$A(x) = \int_{f_1(x)}^{f_2(x)} z\,dy = \int_{f_1(x)}^{f_2(x)} F(x, y)\,dy, \tag{8}$$

where x is held fixed. Note that the limits of integration depend on where the cutting plane is taken; that is, the y-limits are functions of x, specifically, the functions that represent the boundary curves. Finally, we see that the iterated integral in Eq. (7) is the same as

$$V = \int_a^b A(x)\,dx = \int_a^b \left(\int_{f_1(x)}^{f_2(x)} F(x, y)\,dy\right) dx.$$

Problem. Find the volume of the solid whose base is a triangle lying in the xy-plane, bounded by the x-axis, the line $y = x$, and the line $x = 1$, and whose top lies in the plane

$$z = x + y + 1.$$

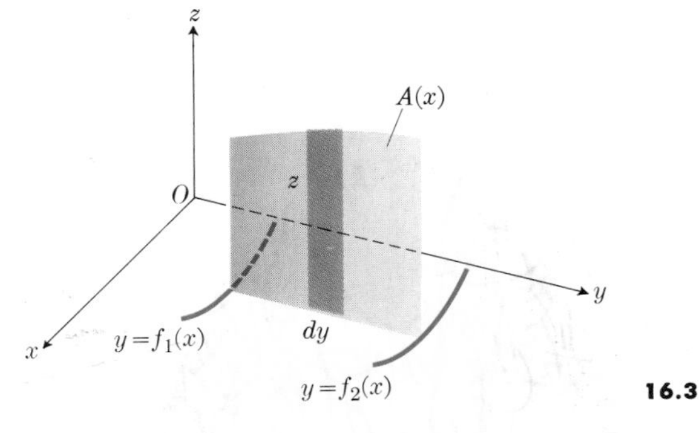

16.3

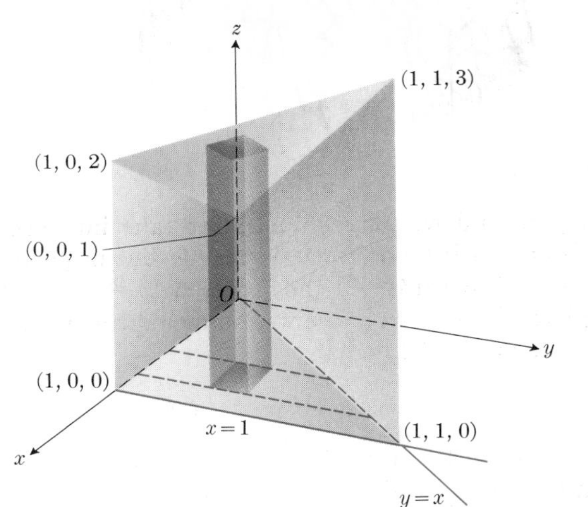

16.4 A prism with triangular base in the xy-plane.

Solution. (See Fig. 16.4.) The volume dV of a representative prism of altitude z and base $dy\,dx$ is

$$dV = (x + y + 1)\,dy\,dx.$$

For any x between 0 and 1, y may vary from $y = 0$ to $y = x$; hence

$$V = \int_0^1 \int_0^x (x + y + 1)\,dy\,dx$$

$$= \int_0^1 \left[xy + \frac{y^2}{2} + y\right]_{y=0}^{x} dx$$

$$= \int_0^1 \left(\frac{3x^2}{2} + x\right) dx = 1.$$

EXERCISES

Evaluate the double integrals in Exercises 1 through 4. Also sketch the region A over which the integration extends.

1. $\int_0^{\pi}\int_0^{x} x\sin y\,dy\,dx$
2. $\int_1^{\ln 8}\int_0^{\ln y} e^{x+y}\,dx\,dy$
3. $\int_0^{\pi}\int_0^{\sin x} y\,dy\,dx$
4. $\int_1^{2}\int_y^{y^2} dx\,dy$

Write an equivalent double integral with the order of integration reversed for each of Exercises 5 through 8. Check your answer by evaluating *both* double integrals.

5. $\int_0^{2}\int_1^{e^x} dy\,dx$
6. $\int_0^{1}\int_{\sqrt{y}}^{1} dx\,dy$
7. $\int_0^{\sqrt{2}}\int_{-\sqrt{4-2y^2}}^{\sqrt{4-2y^2}} y\,dx\,dy$
8. $\int_{-2}^{1}\int_{x^2+4x}^{3x+2} dy\,dx$
9. Find the volume of the solid whose base is the region in the xy-plane bounded by the parabola $y = 4 - x^2$ and the line $y = 3x$, and whose top is bounded by the plane $z = x + 4$.
10. The base of a solid is the region in the xy-plane bounded by the circle $x^2 + y^2 = a^2$, while the top of the solid is bounded by the paraboloid $az = x^2 + y^2$. Find the volume.
11. Evaluate
$$\int_0^2\int_y^2 \exp(x^2)\,dx\,dy$$
To do so, first sketch the region of integration and interchange the order of integration.
12. Evaluate
$$\int_0^{\pi}\int_x^{\pi}\frac{\sin y}{y}\,dy\,dx.$$
13. Evaluate
$$\int_{y_0}^{y_0+k}\int_{x_0}^{x_0+h} f_{yx}(x, y)\,dx\,dy.$$
14. Suppose that G is the interior of a circle with center at $(3, -4)$ and radius ϵ, where ϵ is a small positive number. Estimate the value of
$$\epsilon^{-2}\iint_G (x^2 + y^2 + 2)^{-2/3}\,dx\,dy$$
by a direct interpretation of the integral as defined in Eq. (5). What is the limit of the foregoing expression as $\epsilon \to 0$?

16.2 AREA BY DOUBLE INTEGRATION

The simplest application of double integration is that of finding the area of a region of the xy-plane. The area is given by either of the integrals

$$A = \iint dx\,dy = \iint dy\,dx, \tag{1}$$

with proper limits of integration to be supplied. We have already illustrated how this is done for such an area as the one shown in Fig. 16 1 when the integrations are carried out in the order of first y and then x; that is, where we take

$$A = \int_a^b \int_{f_1(x)}^{f_2(x)} dy\,dx. \tag{2}$$

If, however, the area is bounded on the left by the curve $x = g_1(y)$, on the right by $x = g_2(y)$, below by the line $y = c$, and above by the line $y = d$ (Fig. 16.5), then it is better to integrate first with respect to x [which may vary from $g_1(y)$ to $g_2(y)$] and then with respect to y. That is, we decide to take

$$A = \int_c^d \int_{g_1(y)}^{g_2(y)} dx\,dy. \tag{3}$$

The first integration, with respect to x, may be visualized as an adding together of all the representative elements $dA = dx\,dy$ that lie in a horizontal strip that extends from the curve $x = g_1(y)$ on the left to the curve $x = g_2(y)$ on the right. The evaluation of the integral in (3) gives

$$A = \int_c^d \int_{g_1(y)}^{g_2(y)} dx\,dy = \int_c^d \Big[x\Big]_{g_1(y)}^{g_2(y)} dy$$
$$= \int_c^d [g_2(y) - g_1(y)]\,dy.$$

This latter integral could have been written down at once, since it merely expresses the area as the limit of the sum of horizontal strips of area.

Problem 1. The integral

$$\int_0^1\int_{x^2}^{x} dy\,dx$$

represents the area of a region of the xy-plane. Sketch the region and express the same area as a double integral with the order of integration reversed.

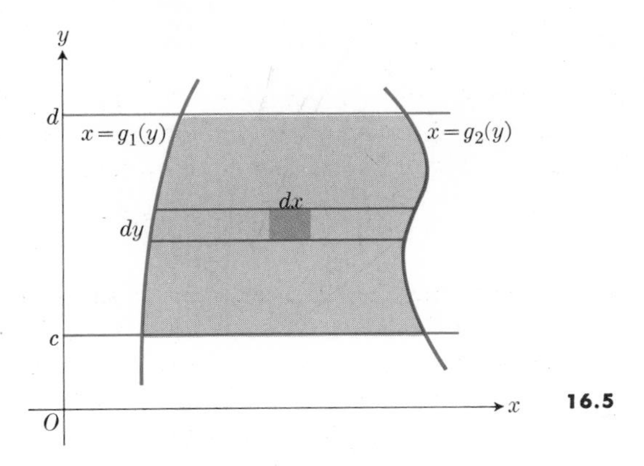

16.5

Solution. In the inner integral, y varies from the curve $y = x^2$ to the line $y = x$. This gives the area of a vertical strip between x and $x + dx$, for values of x from $x = 0$ to $x = 1$. The region of integration is shown in Fig. 16.6. If we integrate in the other order, taking the x integration first, then x varies from the line $x = y$ to the parabola $x = \sqrt{y}$ to fill out a horizontal strip between y and $y + dy$. These strips must then be added together for values of y from 0 to 1. Hence

$$A = \int_0^1 \int_y^{\sqrt{y}} dx\, dy.$$

As a check, we evaluate the area by both integrals and find

$$A = \int_0^1 \int_{x^2}^{x} dy\, dx = \int_0^1 (x - x^2)\, dx = \tfrac{1}{6}$$

and

$$A = \int_0^1 \int_y^{\sqrt{y}} dx\, dy = \int_0^1 (\sqrt{y} - y)\, dy = \tfrac{1}{6}.$$

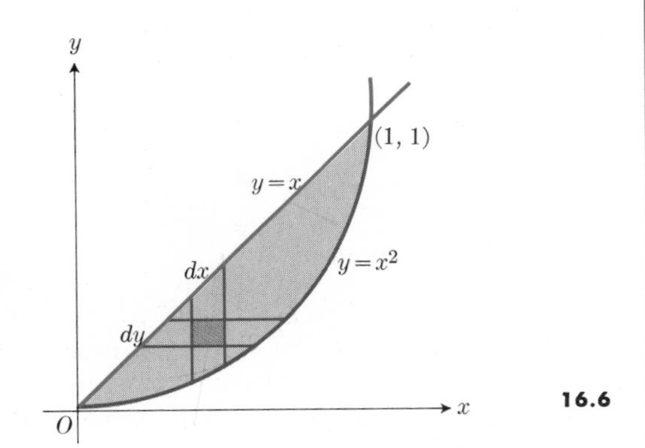

16.6

Problem 2. Find the area bounded by the parabola $y = x^2$ and the line $y = x + 2$.

Solution. The area to be found is shown in Fig. 16.7. We imagine a representative element of area

$$dA = dx\, dy = dy\, dx$$

lying in the region and ask ourselves what order of integration we should choose. We see that *horizontal* strips sometimes go from the line to the right branch of the parabola (if $1 \le y \le 4$) but sometimes go from the left branch of the parabola to its right side (if $0 \le y \le 1$). Thus integration in the order of first x and then y requires that the area be taken in two separate pieces, with the result to be given by

$$A = \int_0^1 \int_{-\sqrt{y}}^{\sqrt{y}} dx\, dy + \int_1^4 \int_{y-2}^{\sqrt{y}} dx\, dy.$$

On the other hand, *vertical* strips always go from the parabola as lower boundary up to the line, and the area is given by

$$A = \int_{-1}^{2} \int_{x^2}^{x+2} dy\, dx.$$

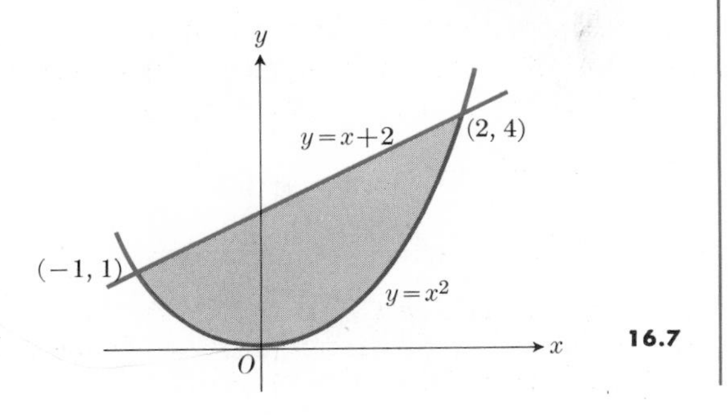

16.7

Clearly, this result is simpler and is the only one we would bother to write down in practice. Evaluation of this integral leads to the result

$$A = \int_{-2}^{2} y\Big]_{x^2}^{x+2} dx = \int_{-1}^{2} (x + 2 - x^2)\, dx = \tfrac{9}{2}.$$

EXERCISES

In these exercises, find the area of the region bounded by the given curves and lines, by double integration.

1. The coordinate axes and the line $x + y = a$
2. The x-axis, the curve $y = e^x$, and the lines $x = 0$, $x = 1$
3. The y-axis, the line $y = 2x$, and the line $y = 4$
4. The curve $y^2 + x = 0$, and the line $y = x + 2$
5. The curves $x = y^2$, $x = 2y - y^2$
6. The semicircle $y = \sqrt{a^2 - x^2}$, the lines $x = \pm a$, and the line $y = -a$
7. The parabola $x = y - y^2$ and the line $x + y = 0$

16.3 PHYSICAL APPLICATIONS

Let the representative element of mass dm in a mass that is continuously distributed over some region A of the xy-plane be

$$dm = \delta(x, y)\, dy\, dx = \delta(x, y)\, dA, \tag{1}$$

where $\delta = \delta(x, y)$ is the density at the point (x, y) of A (see Fig. 16.8). Then double integration may be used to calculate:

(a) the mass, $$M = \iint \delta(x, y)\, dA; \tag{2}$$

(b) the first moment of the mass with respect to the x-axis,

$$M_x = \iint y\, \delta(x, y)\, dA; \tag{3a}$$

(c) its first moment with respect to the y-axis,

$$M_y = \iint x\, \delta(x, y)\, dA. \tag{3b}$$

From (2) and (3) we get the coordinates of the center of mass:

$$\bar{x} = \frac{M_y}{M}, \quad \bar{y} = \frac{M_x}{M}.$$

Other moments of importance in physical application are the *moments of inertia* of the mass. These

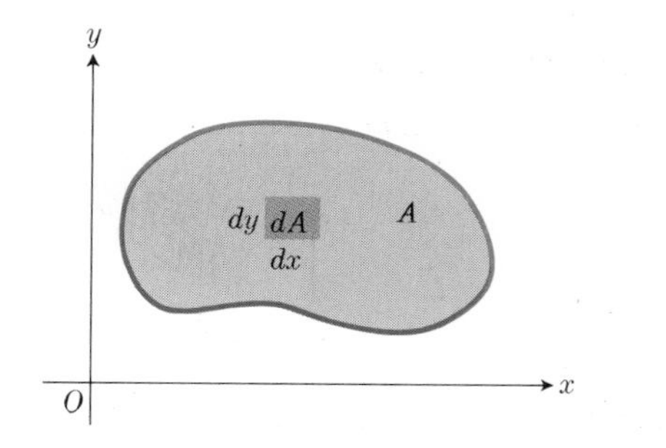

16.8

are the *second* moments that we get by using the squares instead of the first powers of the "lever arm" distances x and y. Thus the moment of inertia about the x-axis, denoted by I_x, is defined by

$$I_x = \iint y^2\, \delta(x, y)\, dA. \tag{4}$$

The moment of inertia about the y-axis is

$$I_y = \iint x^2\, \delta(x, y)\, dA. \tag{5}$$

Also of interest is the *polar moment of inertia* about the origin, I_0, given by

$$I_0 = \iint r^2\, \delta(x, y)\, dA. \tag{6}$$

Here $r^2 = x^2 + y^2$ is the square of the distance from the origin to the representative point (x, y) in the element of mass dm.

In all these integrals, the same limits of integration are to be supplied as would be called for if one were calculating only the area of A.

Remark 1. When a particle of mass m is rotating about an axis in a circle of radius r with angular velocity ω and linear velocity $v = \omega r$, its kinetic energy is

$$\tfrac{1}{2}mv^2 = \tfrac{1}{2}mr^2\omega^2.$$

If a system of particles of masses $m_1, m_2, \ldots, m_n$, at distances $r_1, r_2, \ldots, r_n$, respectively, from the axis of rotation, all rotate about the same axis with the same angular velocity ω, then the kinetic energy of the system of particles is

$$\text{K.E.} = \tfrac{1}{2}(m_1 v_1^2 + \cdots + m_n v_n^2)$$
$$= \tfrac{1}{2}\omega^2 \sum_{k=1}^{n} m_k r_k^2 = \tfrac{1}{2}\omega^2 I, \tag{7}$$

where

$$I = \sum_{k=1}^{n} m_k r_k^2 \tag{8}$$

is the *moment of inertia* of the system about the axis in question. This moment depends only on the magnitude m_k of the masses and their distances r_k from the axis. When a mass m is moving in a straight line with velocity v, its kinetic energy is $\frac{1}{2}mv^2$, and an amount of work equal to this must be expended to stop the object and bring it to rest. Similarly, when a system of mass is moving in a *rotational* motion (like a turning shaft), the kinetic energy it possesses is

$$\text{K.E.} = \tfrac{1}{2}I\omega^2, \tag{9}$$

and this amount of work is required to stop the rotating system. It is seen that I here plays the role that m plays in the case of motion in a straight line. In a sense, the moment of inertia of a large shaft is what makes it hard to start or to stop the rotation of the shaft, in the same way that the *mass* of an automobile is what makes it hard to start or to stop its motion.

If, instead of a system of discrete mass particles as in (7) and (8), we have a *continuous distribution* of mass in a fine wire, or in a thin film or plate over an area, or throughout a solid, then we may divide the total mass into small elements of mass Δm, such that if r represents the distance of some *one* point of the element Δm from an axis, then *all* points of that element are within a distance $r \pm \epsilon$ of the axis, where $\epsilon \to 0$ as the largest dimension of the elements $\Delta m \to 0$. Then we define the moment of inertia of the total mass about the axis in question to be

$$I = \lim_{\Delta m \to 0} \sum r^2 \, \Delta m = \int r^2 \, dm. \tag{10}$$

Thus, for example, the polar moment of inertia given by Eq. (6) is the moment of inertia with respect to a z-axis through O perpendicular to the xy-plane.

347

In addition to its importance in connection with the kinetic energy of rotating bodies, the concept of the moment of inertia plays an important part in the theory of the deflection of beams under transverse loading. Here the "stiffness factor" is given by EI, where E is Young's modulus, and I is the moment of inertia of a cross section of the beam with respect to a horizontal axis through its center of gravity. The greater the value of I, the stiffer the beam and the less it will deflect. This fact is exploited in so-called I-beams, where the flanges at the top and bottom of the beam are at relatively large distances from the center and hence correspond to large values of r^2 in Eq. (10), thereby contributing a larger amount to the moment of inertia than would be the case if the same mass were all distributed uniformly, say in a beam with a square cross section.

Remark 2. Moments are also of importance in statistics. The *first moment* is used in computing the mean (that is, average) value of a given set of data. The *second moment* (which corresponds to the moment of inertia) is used in computing the variance (σ^2) or standard deviation (σ). Third and fourth moments are also used for computing statistical quantities known as skewness and kurtosis. The tth moment is defined as

$$M_t = \sum_{k=1}^{n} m_k r_k^t.$$

Here r_k ranges over the values of the statistic under consideration (for example, r_k might represent height in quarter inches, or weight in ounces, or quiz grades in percentage points, and so on), while m_k is the number of individuals in the entire group whose "measurements" equal r_k. (For example, if 5 students get a grade of 75 on a quiz, then corresponding to $r_k = 75$ we would have $m_k = 5$). A table of values of m_k versus r_k is called a "frequency distribution," and one refers to M_t as the tth moment of this frequency distribution. The mean value $\bar{r}$ is defined by

$$\bar{r} = \frac{\sum m_k r_k}{\sum m_k} = \frac{M_1}{m}, \tag{11}$$

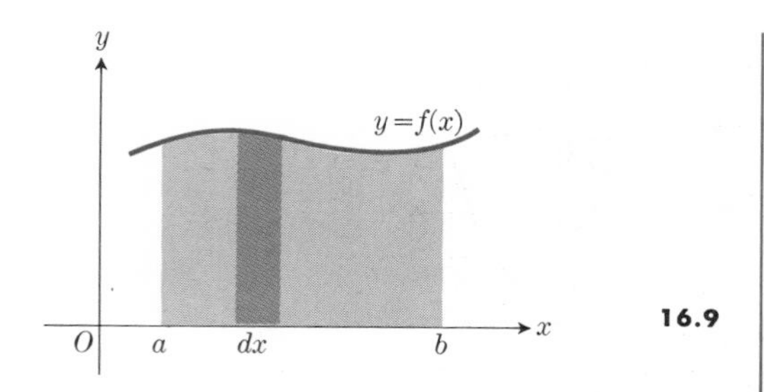

16.9

where M_1 is the first moment and $m = \sum m_k$ is the total number of individuals in the "population" under consideration. The *variance* σ^2 involves the second moment about the mean. It is defined by

$$\sigma^2 = \frac{\sum (r_k - \bar{r})^2 m_k}{\sum m_k}, \tag{12a}$$

where σ is the so-called *standard deviation*. Both the variance and standard deviation are measures of the way in which the r-values tend to bunch up close to $\bar{r}$ (small values of σ) or to be spread out (large values of σ). Algebraic manipulations with (12a) permit one also to write the variance in the alternative form

$$\sigma^2 = \frac{M_2}{m} - \bar{r}^2. \tag{12b}$$

Remark 3. There is a significant difference between the meaning attached to y in the formula

$$A = \int_a^b y\,dx, \tag{13}$$

meaning the area (Fig. 16.9) under a curve $y = f(x)$ from $x = a$ to $x = b$, and the meaning attached to y in the double integrals, Eqs. (2) through (6). In Eq. (13), one must replace y by $f(x)$ from the equation of the curve *before* integrating, because y means the ordinate of the point (x, y) *on* the curve $y = f(x)$. But in the case of the double integrals (2) through (6), one must *not* replace y by a function of x before integrating because the point (x, y) is in general a point of the element $dA = dy\,dx$, and both x and y are *independent* variables. The equations of the boundary curves of the region A enter only as the *limits of integration*. Thus:

1. In the case of single integrals, such as

$$A = \int_a^b y\,dx, \tag{14}$$

we do not integrate with respect to y and hence we must substitute for it.

2. In the case of *double* integrals, such as

$$I_x = \iint y^2\,\delta\,dy\,dx, \tag{15}$$

we *do* integrate with respect to y and therefore we do not substitute for it before performing the y-integration. The equations $y = f_1(x)$, $y = f_2(x)$ of the boundary curves are used as limits of integration and are only to be substituted after the y integration is performed.

Problem. A thin plate of uniform thickness and density covers the region of the xy-plane shown in Fig. 16.7. Find its moment of inertia I_y about the y-axis.

Solution. Integrating in the order of first y and then x, we have

$$\begin{aligned} I_y &= \int_{-1}^{2} \int_{x^2}^{x+2} x^2\,\delta\,dy\,dx \\ &= \delta \int_{-1}^{2} x^2 y\Big]_{y=x^2}^{y=x+2} dx \\ &= \delta \int_{-1}^{2} (x^3 + 2x^2 - x^4)\,dx = \tfrac{63}{20}\delta. \end{aligned}$$

The equation $\quad I_y = MR_y^2$

defines a number $\quad R_y = \sqrt{I_y/M}$,

called the *radius of gyration* with respect to the y-axis. It tells how far from the y-axis the entire mass M might be concentrated and still give the same I_y. In this example, the mass is

$$M = \int_{-1}^{2} \int_{x^2}^{x+2} \delta\,dy\,dx = \tfrac{9}{2}\delta.$$

Hence

$$R_y = \sqrt{I_y/M} = \sqrt{\tfrac{7}{10}}.$$

Note that the density δ in this problem is a constant and hence we were able to move it outside the integral

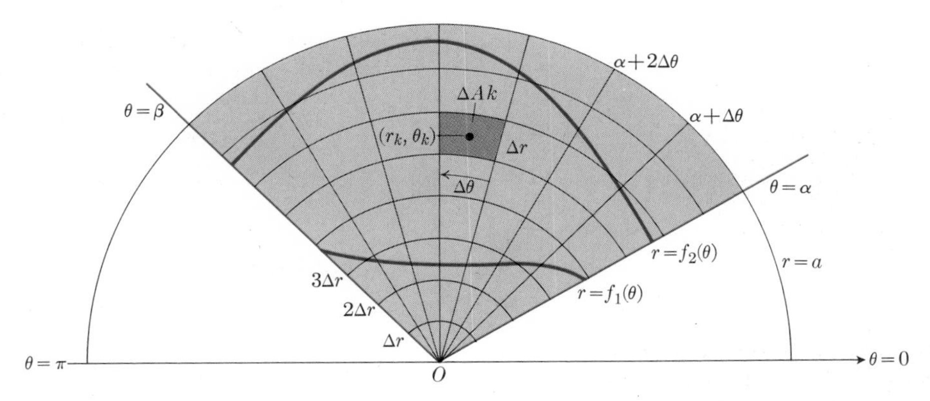

16.10. Subdivision of the region R: $0 \leq r \leq a$, $\alpha \leq \theta \leq \beta$, in polar coordinates: $\Delta A_k = r_k \,\Delta\theta\, \Delta r$.

signs. If the density had been given instead as some variable function of x and y, then we would have taken this into account, in both I_y and M, by simply substituting this function for δ before integrating.

EXERCISES

1. Find the center of gravity of the area of Exercise 1, Article 16.2.
2. Find the moment of inertia, about the x-axis, of the area of Exercise 2, Article 16.2. (For an area, take $\delta = 1$.)
3. Find the polar moment of inertia, about an axis through O perpendicular to the xy-plane, for the area of Exercise 3, Article 16.2.
4. Find the center of gravity of the area of Exercise 4, Article 16.2.
5. Find the moment of inertia about the x-axis of the area in Exercise 5, Article 16.2, if the density at (x, y) is $\delta = y + 1$.
6. Find the center of gravity of the area of Exercise 6, Article 16.2, if the density at (x, y) is $\delta = y + a$.
7. Find the moment of inertia, about the x-axis, of the area of Exercise 7, Article 16.2, if the density at (x, y) is $\delta = x + y$.
8. For any area in the xy-plane, show that its polar moment of inertia I_0 about an axis through O perpendicular to the xy-plane is equal to $I_x + I_y$.

16.4 POLAR COORDINATES

Let A be a region of the plane bounded by rays $\theta = \alpha$, $\theta = \beta$, and curves $r = f_1(\theta)$, $r = f_2(\theta)$, as in Fig. 16.10. Suppose that A is completely contained in the wedge

$$R\colon 0 \leq r \leq a, \quad \alpha \leq \theta \leq \beta.$$

Let m and n be positive integers and take

$$\Delta r = \frac{a}{m}, \qquad \Delta\theta = \frac{\beta - \alpha}{n}.$$

Now cover R by a grid of circular arcs with centers at O and radii Δr, $2\,\Delta r, \ldots, m\,\Delta r$, and rays through O along $\theta = \alpha, \alpha + \Delta\theta, \alpha + 2\,\Delta\theta, \ldots, \alpha + n\,\Delta\theta = \beta$. This grid partitions R into subregions of three types: (a) those exterior to A, (b) those interior to A, and (c) those that intersect the boundary of A. We henceforth ignore those of the first type, but we want to include all those of the second kind, and may include some, none, or all of those of the third kind. Those that are to be included may now be numbered in some order $1, 2, 3, \ldots, N$. In the kth subregion so included, let (r_k, θ_k) be the coordinates of its center.* We multiply the value of F at each of these centers by the area of the corresponding subregion and add the resulting products. That is,

---

* We mean the point halfway between the circular arcs, and on the ray that bisects them.

we consider the sum

$$S = \sum_{k=1}^{N} F(r_k, \theta_k) \cdot \Delta A_k$$
$$= \sum_{k=1}^{N} F(r_k, \theta_k) \cdot r_k \, \Delta\theta \, \Delta r, \tag{1}$$

since

$$\Delta A_k = r_k \, \Delta\theta \, \Delta r, \tag{2}$$

as we shall now see. The radius of the inner arc bounding ΔA_k is $r_k - \frac{1}{2}\Delta r$; of the outer arc, $r_k + \frac{1}{2}\Delta r$. Hence

$$\Delta A_k = \tfrac{1}{2}(r_k + \tfrac{1}{2}\Delta r)^2 \, \Delta\theta - \tfrac{1}{2}(r_k - \tfrac{1}{2}\Delta r)^2 \, \Delta\theta,$$

and simple algebra reduces this to Eq. (2).

We now imagine this process repeated over and over again with finer and finer grids, and consider the limit of the sums (1) as the diagonals of all subregions approach zero. If the function F is continuous, and the region A is bounded by continuous, rectifiable, curves, the sums approach as limit the double integral of F over A:

$$\lim_{N\to\infty} \sum_{k=1}^{N} F(r_k, \theta_k) r_k \, \Delta\theta \, \Delta r = \int_A \int F(r, \theta) \, dA. \tag{3}$$

This limit may also be computed from the iterated integral on the right below:

$$\int_A \int F(r, \theta) \, dA = \int_{\theta=\alpha}^{\beta} \int_{r=f_1(\theta)}^{f_2(\theta)} F(r, \theta) \, r \, dr \, d\theta. \tag{4}$$

The question naturally arises whether one might first set up the double integral in Cartesian coordinates and then change to polar coordinates. The answer is that in general we can.

Note. A rigorous treatment of the problem of changing the variables in a double integral may be found in Franklin, *A Treatise on Advanced Calculus*, p. 368. We shall here be content with citing the result and showing how it leads to Eq. (4) in the case of polar coordinates. In general, equations of the form

$$x = f(u, v), \quad y = g(u, v) \tag{5}$$

may be interpreted as mapping a region A of the xy-plane into a region G of the uv-plane. Then, under suitable restrictions on the functions f and g, the following equation gives the formula for changing from xy-coordinates to uv-coordinates in a double integral:

$$\int_A \int \phi(x, y) \, dx \, dy = \int_G \int \phi[f(u, v), g(u, v)] \frac{\partial(x, y)}{\partial(u, v)} \, du \, dv, \tag{6}$$

where the symbol $\partial(x, y)/\partial(u, v)$ denotes the so-called "Jacobian" of the transformation (5), and is defined by the determinant

$$\frac{\partial(x, y)}{\partial(u, v)} = \begin{vmatrix} \dfrac{\partial x}{\partial u} & \dfrac{\partial x}{\partial v} \\ \dfrac{\partial y}{\partial u} & \dfrac{\partial y}{\partial v} \end{vmatrix}. \tag{7}$$

(See also Exercise 10, Article 17.5.) In the case of polar coordinates, we have r and θ in place of u and v:

$$x = r\cos\theta, \quad y = r\sin\theta,$$

and

$$\frac{\partial(x, y)}{\partial(r, \theta)} = \begin{vmatrix} \cos\theta & -r\sin\theta \\ \sin\theta & r\cos\theta \end{vmatrix}$$
$$= r(\cos^2\theta + \sin^2\theta) = r.$$

Hence Eq. (6) becomes

$$\iint \phi(x, y) \, dx \, dy = \iint \phi(r\cos\theta, r\sin\theta) r \, dr \, d\theta, \tag{8}$$

which corresponds to Eq. (4).

In this connection, however, we must point out an important fact. The total area of a region is given by either one of the double integrals

$$A = \iint dx \, dy = \iint r \, dr \, d\theta, \tag{9}$$

with appropriate limits. This means, essentially, that the given region can be divided into pieces of area

$$dA_{xy} = dx \, dy$$

by lines parallel to the x- and y-axes, or that it can be divided into pieces of area

$$dA_{r\theta} = r \, dr \, d\theta$$

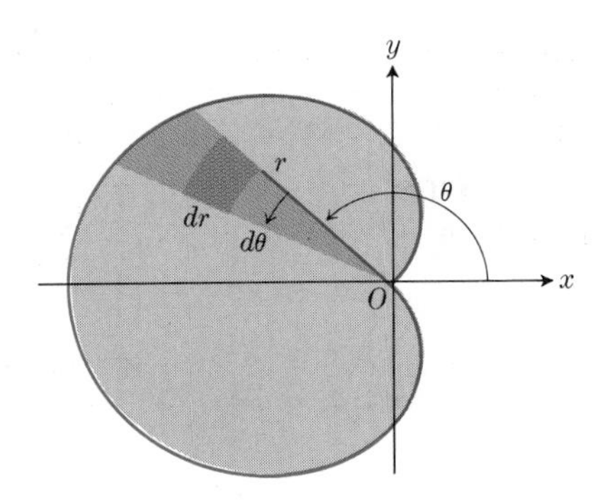

16.11

by radial lines and circular arcs, and that the total area can be found by adding together all of the elements of area of either type. But it is not to be expected that the *individual pieces* dA_{xy} and $dA_{r\theta}$ will be equal. In fact, an elementary calculation shows that

$$\begin{aligned} dA_{xy} &= dx\,dy \\ &= d(r\cos\theta)\,d(r\sin\theta) \\ &\neq r\,dr\,d\theta = dA_{r\theta}. \end{aligned}$$

When we say that integration can proceed with either Cartesian or polar coordinates, it is very much like saying that the two *sums* $1 + 4 = 5$ and $2 + 3 = 5$ are the same even though the individual terms of the two sums are not equal.

Problem. Find the moment of inertia, about the y-axis, of the area enclosed by the cardioid (Fig. 16.11)

$$r = a(1 - \cos\theta).$$

Solution. It is customary to take the density as unity whenever one is working with a geometrical area. Thus we have the integral

$$I_y = \int_A \int x^2\,dA,$$

with

$$x = r\cos\theta, \quad dA = r\,dr\,d\theta.$$

If we integrate first with respect to r, then r may vary from 0 to $a(1 - \cos\theta)$ for any θ between 0 and 2π. This permits the r-integration to extend over those elements of area lying in the wedge between the radius lines θ and $\theta + d\theta$. Next we integrate with respect to θ from 0 to 2π to allow these wedges to cover the entire area. Hence

$$\begin{aligned} I_y &= \int_0^{2\pi} \int_0^{a(1-\cos\theta)} r^3 \cos^2\theta\,dr\,d\theta \\ &= \int_0^{2\pi} \frac{a^4}{4} \cos^2\theta(1 - \cos\theta)^4\,d\theta. \end{aligned}$$

The evaluation of the integrals

$$\int_0^{2\pi} \cos^n\theta\,d\theta, \quad n = 2, 3, 4, 5, 6,$$

is made easier by use of the reduction formula

$$\int_0^{2\pi} \cos^n\theta\,d\theta = \frac{\cos^{n-1}\theta\sin\theta}{n}\Big]_0^{2\pi} + \frac{n-1}{n}\int_0^{2\pi} \cos^{n-2}\theta\,d\theta,$$

or, since $\sin\theta$ vanishes at both limits,

$$\int_0^{2\pi} \cos^n\theta\,d\theta = \frac{n-1}{n}\int_0^{2\pi} \cos^{n-2}\theta\,d\theta.$$

Thus

$$\int_0^{2\pi} \cos^2\theta\,d\theta = \frac{1}{2}\int_0^{2\pi} d\theta = \pi,$$

$$\int_0^{2\pi} \cos^3\theta\,d\theta = \frac{2}{3}\int_0^{2\pi} \cos\theta\,d\theta = \frac{2}{3}\sin\theta\Big]_0^{2\pi} = 0,$$

$$\int_0^{2\pi} \cos^4\theta\,d\theta = \frac{3}{4}\int_0^{2\pi} \cos^2\theta\,d\theta = \frac{3\pi}{4},$$

$$\int_0^{2\pi} \cos^5\theta\,d\theta = \frac{4}{5}\int_0^{2\pi} \cos^3\theta\,d\theta = 0,$$

$$\int_0^{2\pi} \cos^6\theta\,d\theta = \frac{5}{6}\int_0^{2\pi} \cos^4\theta\,d\theta = \frac{5\pi}{8}.$$

Therefore

$$\begin{aligned} I_y &= \frac{a^4}{4}\int_0^{2\pi} (\cos^2\theta - 4\cos^3\theta + 6\cos^4\theta \\ &\qquad - 4\cos^5\theta + \cos^6\theta)\,d\theta \\ &= \frac{a^4}{4}\left[1 + \frac{18}{4} + \frac{5}{8}\right]\pi = \frac{49\pi a^4}{32}. \end{aligned}$$

EXERCISES

Change each double integral in Exercises 1 through 6 to an equivalent double integral in terms of polar coordinates and then evaluate.

1. $\int_{-a}^{a} \int_{-\sqrt{a^2-x^2}}^{\sqrt{a^2-x^2}} dy\, dx$

2. $\int_0^a \int_0^{\sqrt{a^2-y^2}} (x^2 + y^2)\, dx\, dy$

3. $\int_0^{a/\sqrt{2}} \int_y^{\sqrt{a^2-y^2}} x\, dx\, dy$

4. $\int_0^{\infty} \int_0^{\infty} e^{-(x^2+y^2)}\, dx\, dy$

5. $\int_0^2 \int_0^x y\, dy\, dx$

6. $\int_0^{2a} \int_0^{\sqrt{2ax-x^2}} x^2\, dy\, dx$

7. By double integration, find the area that lies inside the cardioid $r = a(1 + \cos\theta)$ and outside the circle $r = a$.

8. Find the center of gravity of the area of Exercise 7.

9. Find the polar moment of inertia I_0 with respect to an axis through O perpendicular to the xy-plane, for the area of Exercise 7.

10. The base of a solid is the area of Exercise 7, and the solid is bounded at the top by the plane $z = x$. Find its volume.

11. Using double integration, find the total area enclosed by the lemniscate

$$r^2 = 2a^2 \cos 2\theta.$$

12. The base of a solid is the area of Exercise 11, and the solid is bounded at the top by the sphere $z = \sqrt{2a^2 - r^2}$. Find the volume.

16.5 TRIPLE INTEGRALS: VOLUME

Consider a region V, in xyz-space, completely contained within the box B bounded by the planes $x = a$, $x = b$, $y = c$, $y = d$, $z = e$, and $z = f$, with $a < b$, $c < d$, and $e < f$. Let $F(x, y, z)$ be a function whose domain includes V. Let m, n, p be positive integers, and let

$$\Delta x = \frac{b-a}{m}, \qquad \Delta y = \frac{d-c}{n}, \qquad \Delta z = \frac{f-e}{p}.$$

Divide B into $m \times n \times p$ subregions each with dimensions $\Delta x \times \Delta y \times \Delta z$, by planes

$$\begin{array}{llll} x = a, & a + \Delta x, & a + 2\,\Delta x, & \ldots, \quad a + m\,\Delta x; \\ y = c, & c + \Delta y, & c + 2\,\Delta y, & \ldots, \quad c + n\,\Delta y; \\ z = e, & e + \Delta z, & e + 2\,\Delta z, & \ldots, \quad e + p\,\Delta z. \end{array}$$

These subregions are of three kinds: (a) those interior to V, (b) those exterior to V, and (c) those intersecting the boundary of V. We include all those of type (a), exclude all those of type (b), and include some, none, or all those of type (c). Then we number the included subregions $1, 2, 3, \ldots, N$. Let (x_k, y_k, z_k) be a point in the kth subregion, multiply the value of F at that point by the volume ΔV_k of the subregion, and form the sum

$$S = \sum_{k=1}^{N} F(x_k, y_k, z_k)\, \Delta V_k$$

$$= \sum_{k=1}^{N} F(x_k, y_k, z_k)\, \Delta x\, \Delta y\, \Delta z. \tag{1}$$

Finally, suppose that the function F is continuous throughout V and on its boundary. Then, if the boundary of V is sufficiently "tame," the sums (1) have a limit as $\sqrt{(\Delta x)^2 + (\Delta y)^2 + (\Delta z)^2}$ approaches zero, and this limit is called the (Riemann) triple integral of F over V:

$$\iiint_V F\, dV = \lim \sum_{k=1}^{N} F(x_k, y_k, z_k)\, \Delta x\, \Delta y\, \Delta z. \tag{2}$$

Remark 1. There are many possible interpretations of (2). If $F(x, y, z) = 1$ for all points in V, the integral is just the volume of V. If $F(x, y, z) = x$, the integral is the first moment of the volume V with respect to the yz-plane. If $F(x, y, z)$ is the density at (x, y, z), then the integral is the mass in V. If $F(x, y, z)$ is the product of the density at (x, y, z) and the square of the distance from (x, y, z) to an axis L, then the integral is the moment of inertia of the mass with respect to L.

Remark 2. The triple integral is seldom evaluated directly from its definition as a limit. Instead, it is usually evaluated as an iterated integral. For

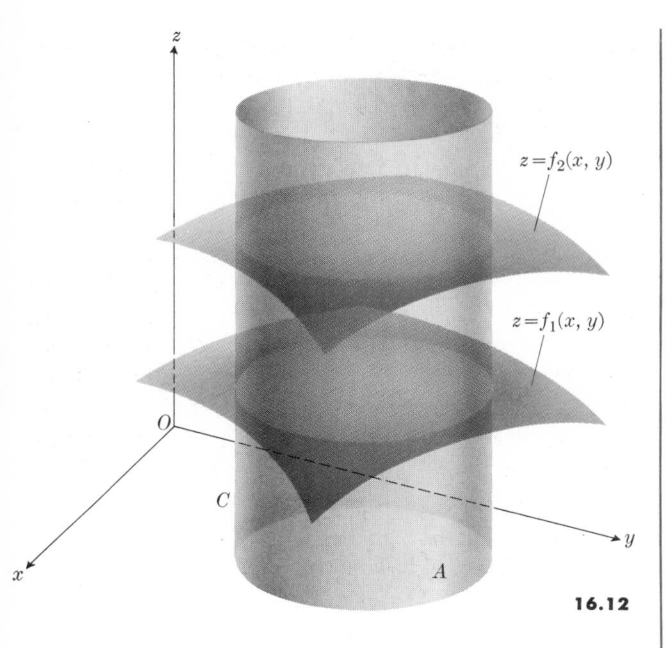

16.12

example, suppose V is bounded below by a surface $z = f_1(x, y)$, above by the surface $z = f_2(x, y)$, and laterally by a cylinder C with elements parallel to the z-axis (Fig. 16.12). Let A denote the region of the xy-plane enclosed by the cylinder C. (That is, A is the region covered by the orthogonal projection of the solid into the xy-plane.) Then the *volume* of the region V (which we shall also denote by V) can be found by evaluating the triply iterated integral

$$V = \int_A \int \int_{f_1(x,y)}^{f_2(x,y)} dz\, dy\, dx. \tag{3}$$

The z-limits of integration indicate that for every (x, y) in the region A, z may extend from the lower surface $z = f_1(x, y)$ to the upper surface $z = f_2(x, y)$. The y and x limits of integration have not been given explicitly in Eq. (3), but are indicated as extending over the region A. The problem of supplying these limits is precisely the problem we have previously considered in connection with *double* integrals. It is usually desirable to draw the xy-projection of the solid in order to see more easily what these limits are.

Sometimes it is not immediately clear that there is a cylinder such as the one shown in Fig. 16.12. In the following problem, two surfaces intersect along a curve in E^3, and it is the projection of this curve into the xy-plane that is the boundary of the region A. We get the equation of that boundary by eliminating z between the two equations $z = f_1(x, y)$ and $z = f_2(x, y)$, thereby obtaining an equation

$$f_1(x, y) = f_2(x, y) \tag{4}$$

that contains no z. Such an equation, interpreted as the equation of a surface in xyz-space, represents a cylinder with elements parallel to the z-axis. The curve of intersection of the two surfaces, and its projection into the xy-plane, both lie on the cylinder represented by Eq. (4). (See Fig. 16.13.) If we interpret Eq. (4) as an equation in the xy-plane, then it just represents the boundary of the region A.

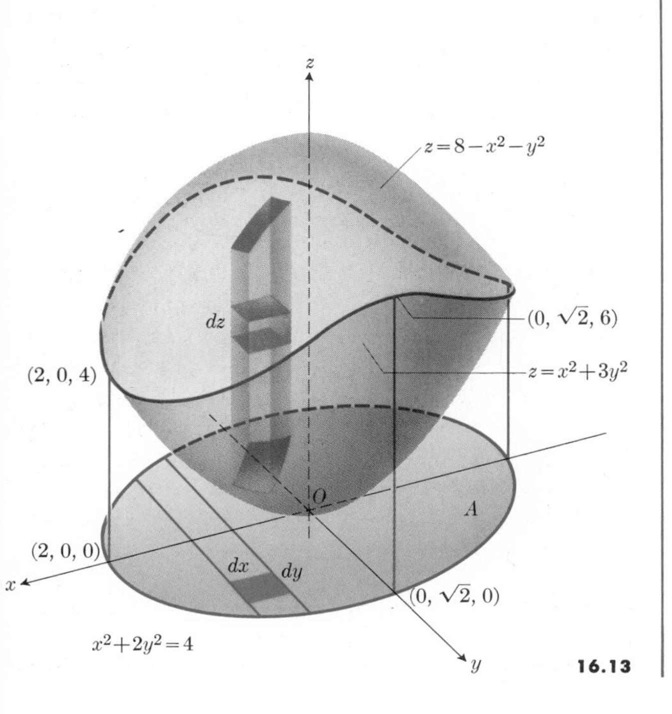

16.13

Problem. Find the volume enclosed between the two surfaces

$$z = 8 - x^2 - y^2$$

and

$$z = x^2 + 3y^2.$$

Solution. (See Fig. 16.13.) The two surfaces intersect on the elliptic cylinder

$$x^2 + 2y^2 = 4.$$

The volume projects into the region A in the xy-plane enclosed by the ellipse (Fig. 16.13) having this same equation. In the double integral with respect to y and x over this region A, if we integrate first with respect to y, holding x and dx fixed, we see that y varies from $-\sqrt{(4-x^2)/2}$ to $+\sqrt{(4-x^2)/2}$. Then x varies from -2 to $+2$. Thus we have

$$\begin{aligned} V &= \int_{-2}^{2} \int_{-\sqrt{(4-x^2)/2}}^{\sqrt{(4-x^2)/2}} \int_{x^2+3y^2}^{8-x^2-y^2} dz\, dy\, dx \\ &= \int_{-2}^{2} \int_{-\sqrt{(4-x^2)/2}}^{\sqrt{(4-x^2)/2}} (8 - 2x^2 - 4y^2)\, dy\, dx \\ &= \int_{-2}^{2} \left[2(8-2x^2)\sqrt{\frac{4-x^2}{2}} - \frac{8}{3}\left(\frac{4-x^2}{2}\right)^{3/2} \right] dx \\ &= \frac{4\sqrt{2}}{3} \int_{-2}^{2} (4-x^2)^{3/2}\, dx = 8\pi\sqrt{2}. \end{aligned}$$

EXERCISES

In these exercises, find the volume by triple integration.

1. The volume of the tetrahedron bounded by the plane

$$\frac{x}{a} + \frac{y}{b} + \frac{z}{c} = 1$$

and the coordinate planes (a, b, c positive)

2. The volume in the first octant bounded by the cylinder $x = 4 - y^2$ and the planes $z = y$, $x = 0$, $z = 0$
3. The volume bounded by the elliptic paraboloids $z = x^2 + 9y^2$, $z = 18 - x^2 - 9y^2$
4. The volume common to the two cylinders

$$x^2 + y^2 = a^2, \quad x^2 + z^2 = a^2$$

5. The volume of an ellipsoid of semiaxes a, b, c
6. The volume bounded below by the plane $z = 0$, laterally by the elliptic cylinder $x^2 + 4y^2 = 4$, and above the plane $z = x + 2$

16.6 CYLINDRICAL COORDINATES

Instead of using an element of volume

$$dV_{xyz} = dz\, dy\, dx \tag{1}$$

as we have done, we may use an element

$$dV_{r\theta z} = dz\, r\, dr\, d\theta. \tag{2}$$

Equation (2) may be conceived as giving the volume of an element having cross-sectional area $r\, dr\, d\theta$, such as that used with polar coordinates in Article 16.4, and altitude dz. Cylindrical coordinates, r, θ, z, are particularly useful in problems where there is an axis of symmetry of the solid. By proper choice of axes, this axis of symmetry may be taken to be the z-axis.

Problem. Find the center of gravity of a homogeneous solid hemisphere of radius a.

Solution. We may choose the origin at the center of the sphere and consider the hemisphere that lies above the xy-plane (see Fig. 16.14). The equation of the hemispherical surface is $z = \sqrt{a^2 - x^2 - y^2}$, or, in terms of cylindrical coordinates, $z = \sqrt{a^2 - r^2}$. By symmetry, we have $x = y = 0$. We then calculate $\bar{z}$:

$$\bar{z} = \frac{\iiint z\, dV}{\iiint dV} = \frac{\int_0^{2\pi} \int_0^a \int_0^{\sqrt{a^2-r^2}} z\, dz\, r\, dr\, d\theta}{\frac{2}{3}\pi a^3} = \frac{3a}{8}.$$

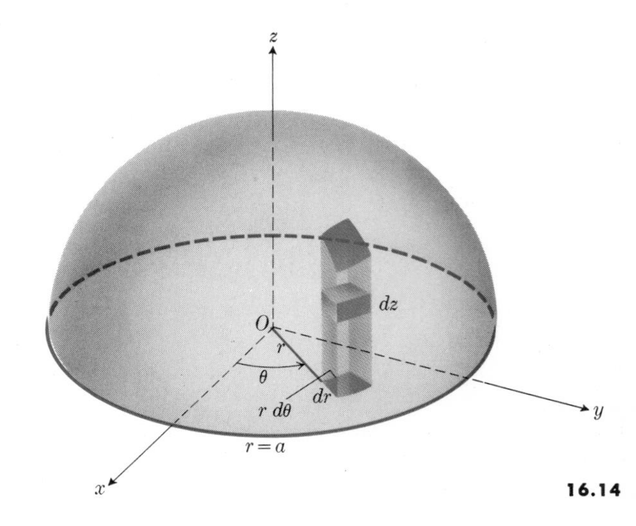

16.14

EXERCISES

20 all

Find the volumes in these exercises by triple integration.

1. The volume cut from the sphere $x^2 + y^2 + z^2 = 4a^2$ by the cylinder $x^2 + y^2 = a^2$
2. The volume bounded below by the paraboloid

$$z = x^2 + y^2$$

and above by the plane $z = 2y$
3. The volume bounded above by the sphere

$$x^2 + y^2 + z^2 = 2a^2$$

and below by the paraboloid $az = x^2 + y^2$
4. The volume in the first octant bounded by the cylinder $x^2 + y^2 = a^2$ and the planes $x = a$, $y = a$, $z = 0$, $z = x + y$

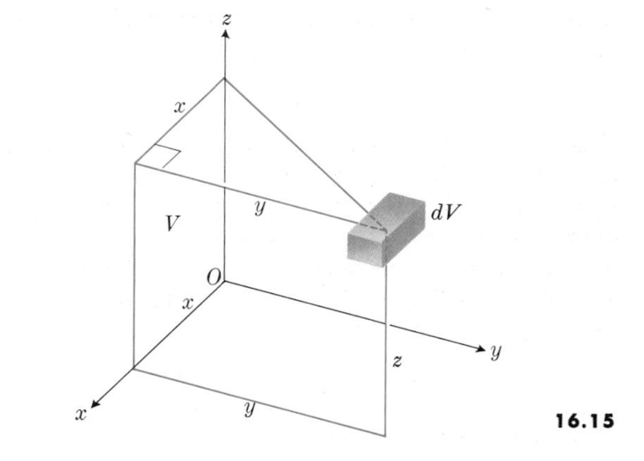

16.15

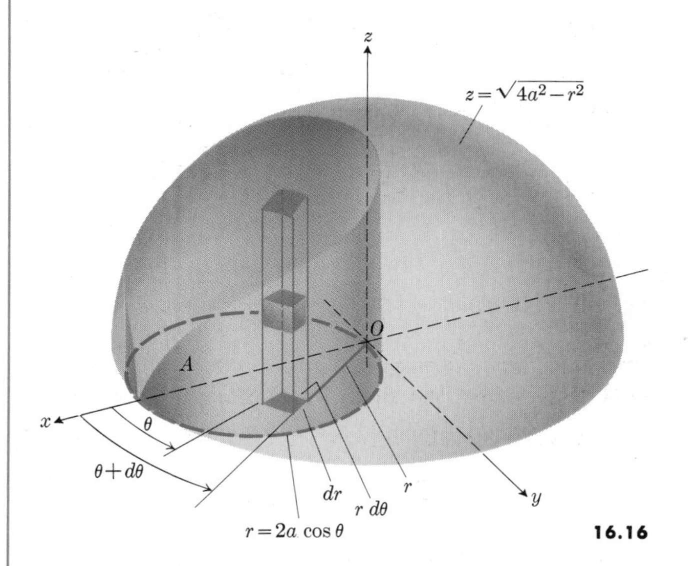

16.16

16.7 PHYSICAL APPLICATIONS OF TRIPLE INTEGRATION

The mass, center of gravity, and moments of inertia of a mass M distributed over a region V of xyz-space and having density $\delta = \delta(x, y, z)$ at the point (x, y, z) of V (see Fig. 16.15) are given by integrals of the type

$$M = \iiint \delta \, dV, \tag{1}$$

$$\bar{x} = \frac{\iiint x \, \delta \, dV}{\iiint \delta \, dV}, \tag{2}$$

$$I_z = \iiint (x^2 + y^2) \, \delta \, dV, \tag{3}$$

with similar integrals for $\bar{y}$, $\bar{z}$, I_x, and I_y. The integrals in Eqs. (1) through (3) may be evaluated as triple integrals with

$$dV = dz \, dy \, dx,$$

or, if it is more convenient to use cylindrical coordinates, with

$$dV = dz \, r \, dr \, d\theta.$$

Limits of integration are to be supplied so that the element of volume ranges over the volume V as discussed above.

Problem. A solid is bounded below by the xy-plane, above by the sphere $x^2 + y^2 + z^2 = 4a^2$, and laterally by the cylinder $r = 2a \cos\theta$. Find its moment of inertia I_z.

Solution. The solid lies in front of the yz-plane, as shown in Fig. 16.16. It projects orthogonally into the interior of the circle $r = 2a\cos\theta$ in the xy-plane. Taking an element of volume $dV = dz \, r \, dr \, d\theta$ enclosing a point r, θ, z, we have

$$dI_z = r^2 \, dV = dz \, r^3 \, dr \, d\theta.$$

When we integrate with respect to z, the lower limit is $z = 0$ (the plane) and the upper limit is

$$z = \sqrt{4a^2 - r^2}$$

(the sphere). This integral gives the moment of inertia of the elements dV in a prism extending from the xy-plane to the sphere and having its base inside the circle $r = 2a \cos \theta$. To add the moments of all the prisms, we examine this region by itself to see what the appropriate r- and θ-limits are. If we were finding only the *area* of this region, we would have

$$A = \int_{-\pi/2}^{\pi/2} \int_0^{2a \cos \theta} r\, dr\, d\theta.$$

In the present problem, we are finding the moment of inertia I_z of a solid standing on this area as a base, but *the same limits of integration* apply to the r- and θ-integrations. Therefore

$$\begin{aligned} I_z &= \int_{-\pi/2}^{\pi/2} \int_0^{2a \cos \theta} \int_0^{\sqrt{4a^2 - r^2}} dz\, r^3\, dr\, d\theta \\ &= \tfrac{64}{15} a^5 (\pi - \tfrac{26}{15}). \end{aligned}$$

EXERCISES

1. Find the moment of inertia about the x-axis for the volume of Exercise 1, Article 16.6.
2. Find the moment of inertia about the z-axis for the volume of Exercise 2, Article 16.6.
3. Find the x-coordinate of the center of gravity of the volume of Exercise 6, Article 16.6.
4. Find the center of gravity of the volume of Exercise 3, Article 16.6.
5. Use cylindrical coordinates to find the moment of inertia of a sphere of radius a and mass M about a diameter.
6. Find the volume generated by rotating the cardioid

$$r = a(1 - \cos \theta)$$

about the x-axis.
Hint. Use *double* integration. Rotate an area element dA around the x-axis to generate a volume element dV.

7. Find the moment of inertia about the x-axis for the volume of Exercise 6.
8. Find the moment of inertia of a right circular cone of base radius a, altitude h, and mass M about an axis through the vertex and parallel to the base.
9. Find the moment of inertia of a sphere of radius a and mass M with respect to a tangent line.
10. Find the center of gravity of that portion of the volume of the sphere $r^2 + z^2 = a^2$ that lies between the planes

$$\theta = -\frac{\pi}{4}, \qquad \theta = \frac{\pi}{4}.$$

11. A torus of mass M is generated by rotating a circle of radius a about an axis in its plane at distance b from the center (b greater than a). Find its moment of inertia about the axis of revolution.
12. Let V be a solid of mass M, and let R_0 be its radius of gyration with respect to a line L_0 through its centroid. Show that the moment of inertia with respect to a parallel line L_1 located R_0 units from L_0 is twice the moment of inertia with respect to L_0.

16.8 SPHERICAL COORDINATES

In a problem where there is symmetry with respect to a point, it may be convenient to choose that point as origin and to use spherical coordinates (Fig. 16.17). These are related to the Cartesian system by the equations

$$\begin{aligned} x &= \rho \sin \phi \cos \theta, \\ y &= \rho \sin \phi \sin \theta, \\ z &= \rho \cos \phi. \end{aligned} \tag{1}$$

If we give ρ, ϕ, and θ increments $d\rho$, $d\phi$, and $d\theta$, we are led to consider the volume element (Fig. 16.18)

$$\begin{aligned} dV_{\rho\phi\theta} &= d\rho \cdot \rho\, d\phi \cdot \rho \sin \phi\, d\theta \\ &= \rho^2 \sin \phi\, d\rho\, d\phi\, d\theta, \end{aligned} \tag{2}$$

and triple integrals of the form

$$\iiint F(\rho, \phi, \theta) \rho^2 \sin \phi\, d\rho\, d\phi\, d\theta. \tag{3}$$

Problem 1. Find the volume cut from the sphere $\rho = a$ by the cone $\phi = \alpha$.

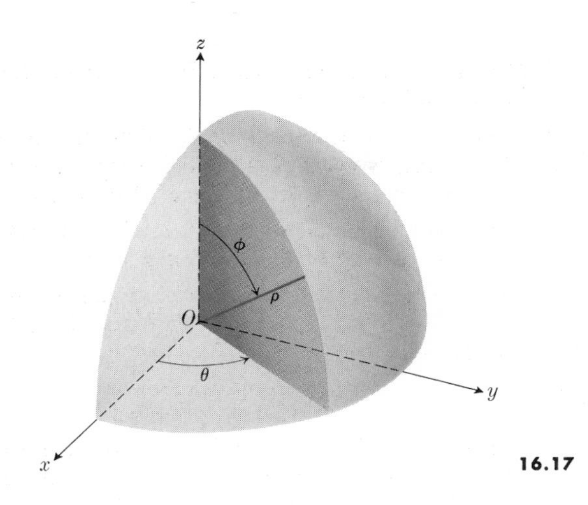

16.17

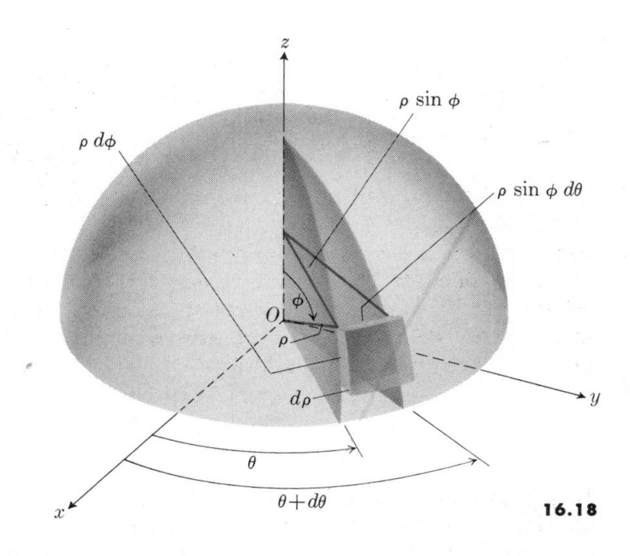

16.18

Solution. (See Fig. 16.19.) The volume is given by

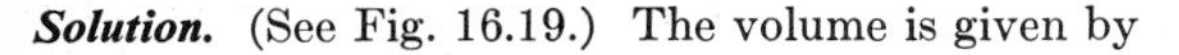

$$V = \int_0^{2\pi} \int_0^{\alpha} \int_0^{a} \rho^2 \sin \phi \, d\rho \, d\phi \, d\theta = \frac{2\pi a^3}{3} (1 - \cos \alpha).$$

As a check, we note that the special cases $\alpha = \pi/2$ and $\alpha = \pi$ correspond to the cases of a hemisphere and a sphere of volumes $2\pi a^3/3$ and $4\pi a^3/3$, respectively.

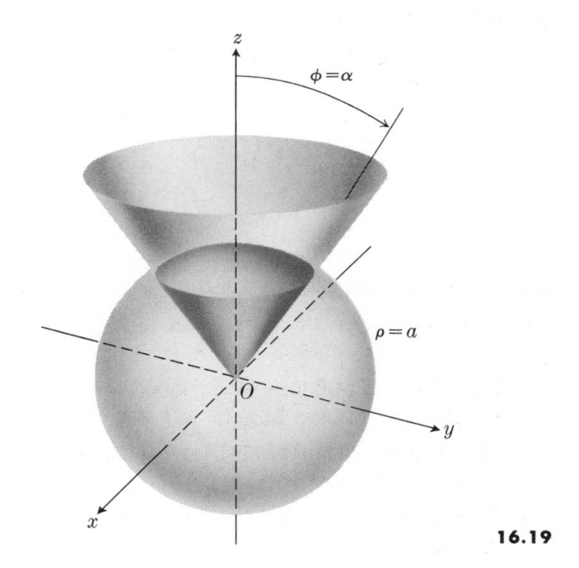

16.19

Gravitational attraction

Newton was interested in verifying (or disproving) a notion related to the fact that a solid spherical mass m, such as the sun, exerts a gravitational attraction on a particle of mass m' given by

$$F = G \frac{mm'}{r^2},$$

if the particle is at a distance r from its center (here, G is a constant). He had assumed in his earlier calculations that the mass of the sun could be considered as a point mass, but he wanted to know whether that assumption could in some way be justified mathematically. He now supposed instead that the sun is composed of a very large number of tiny masses, each to be treated as a point mass. If each of these tiny masses exerted a gravitational attraction according to the inverse-square law, he asked himself, would the particle be affected in the same way as it would by the attractive force of a body of the same total mass concentrated at its center? He found (as the next example will show) that the answer is "yes," if the attracting spherical body has uniform density, or, more realistically, if its density varies only with distance from the center, so that it remains constant in thin concentric spherical shells.

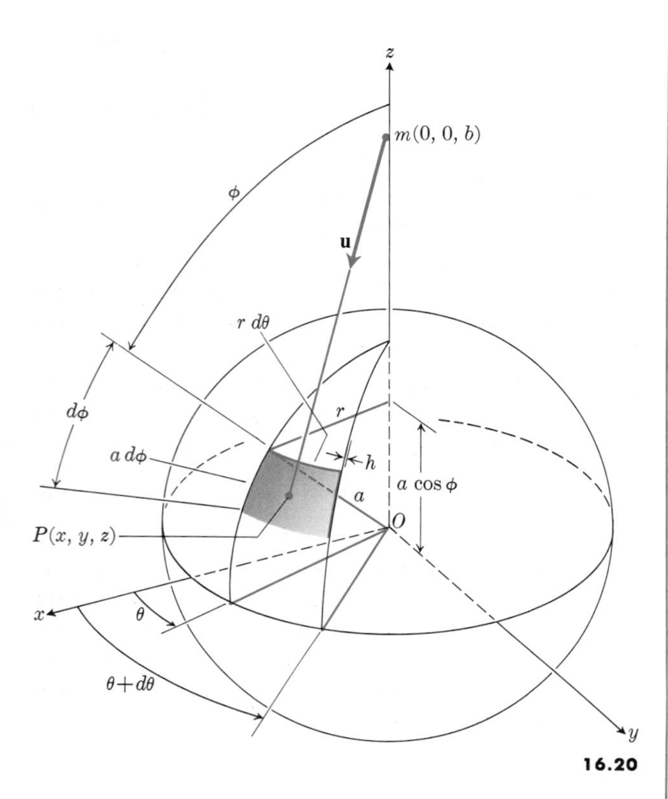

16.20

Problem 2. Find the gravitational attraction of a thin spherical shell, of inner radius a, thickness h, uniform density δ, on a particle of mass m at distance b from its center, $b > a$.

Solution. (See Fig. 16.20.) We choose the center of the spherical shell as origin, and the positive z-axis through the mass particle as indicated in the figure. Let D be the tiny volume element cut from the shell by cones of generating angles ϕ and $\phi + d\phi$, and by half-planes making dihedral angles θ and $\theta + d\theta$ with the xz-plane. The volume element is

$$dV = r\,d\theta \cdot a\,d\phi \cdot h, \tag{4a}$$

where

$$r = a \sin \phi. \tag{4b}$$

Because the dimensions of D are assumed to be very small, we consider that the element of mass in D is $dm = \delta\,dV$, located at the point $P(x, y, z)$ with

$$\begin{aligned} x &= r \cos \theta = a \sin \phi \cos \theta, \\ y &= r \sin \theta = a \sin \phi \sin \theta, \\ z &= a \cos \phi. \end{aligned} \tag{5}$$

The gravitational force that the element of mass dm at $P(x, y, z)$ exerts on the mass m at $(0, 0, b)$ is

$$d\mathbf{F} = \frac{Gm\delta\,dV}{x^2 + y^2 + (z - b)^2}\,\mathbf{u}, \tag{6a}$$

where $\mathbf{u}$ is a unit vector directed *from* $(0, 0, b)$ *toward* $P(x, y, z)$. Thus

$$\mathbf{u} = \frac{\mathbf{i}x + \mathbf{j}y + \mathbf{k}(z - b)}{\sqrt{x^2 + y^2 + (z - b)^2}}. \tag{6b}$$

To get the total gravitational attraction, we substitute from Eqs. (5) and (6b) into Eq. (6a), and integrate first with respect to θ from 0 to 2π, and then integrate that result with respect to ϕ from 0 to π. We can do this one component at a time. The **i**- and **j**-components are zero after integration because

$$\int_0^{2\pi} \cos \theta\,d\theta = 0, \qquad \int_0^{2\pi} \sin \theta\,d\theta = 0.$$

The resultant force is therefore

$$\mathbf{F} = Gm\mathbf{k} \int_{\phi=0}^{\pi} \int_{\theta=0}^{2\pi} \frac{\delta a^2 h(a \cos \phi - b) \sin \phi}{(a^2 + b^2 - 2ab \cos \phi)^{3/2}}\,d\theta\,d\phi. \tag{7a}$$

During the first integration, ϕ and $d\phi$ are held constant, and the density is constant, so the result is

$$\mathbf{F} = Gm\delta(2\pi a^2 h)\mathbf{k} \int_0^{\pi} \frac{(a \cos \phi - b) \sin \phi}{(a^2 + b^2 - 2ab \cos \phi)^{3/2}}\,d\phi. \tag{7b}$$

In Exercise 4, below, the reader is asked to complete the integration by making the substitutions

$$\begin{aligned} u &= a^2 + b^2 - 2ab \cos \phi, \qquad du = 2ab \sin \phi\,d\phi, \\ a \cos \phi &= \frac{a^2 + b^2 - u}{2b}, \qquad a \cos \phi - b = \frac{a^2 - b^2 - u}{2b}. \end{aligned} \tag{8}$$

The final result for the integral is

$$\frac{1}{4ab^2} \int_{(a-b)^2}^{(a+b)^2} [(a^2 - b^2)u^{-3/2} - u^{-1/2}]\,du = \frac{-2}{b^2}. \tag{9}$$

Substituting this for the integral in Eq. (7b), we get

$$\mathbf{F} = -\mathbf{k}\,\frac{(Gm)(4\pi a^2 h\delta)}{b^2}, \tag{10}$$

which is the same as the gravitational attraction that a point mass $4\pi a^2 h\delta$, the mass of the *thin* spherical shell, would exert if located at the origin (the center of the sphere).

EXERCISES

1. Find the center of gravity of the volume (which resembles a filled ice cream cone) that is bounded above by the sphere $\rho = a$ and below by the cone $\phi = \pi/6$.
2. Find the volume enclosed by the surface
$$\rho = a(1 - \cos\phi).$$
Compare with Exercise 6, Article 16.7.
3. Find the radius of gyration, with respect to a diameter, of a spherical shell of mass M bounded by the spheres $\rho = a$ and $\rho = 2a$ if the density is
$$\delta = \rho^2.$$
4. Show that the substitution of Eq. (8), or the equivalent substitution
$$\phi = \cos^{-1}\left(\frac{a^2 + b^2 - u}{2ab}\right),$$
leads to the result stated in Eq. (9).

Note. $\sqrt{(a - b)^2} = |a - b| = b - a,$

because $b > a.$

5. In Problem 2, replace the shell thickness h by $d\rho$, the radius a by ρ, and let the density $\delta = g(\rho)$ be a function of the spherical coordinate ρ. Suppose that these substitutions are made everywhere on the right-hand side of Eq. (7a), and that the result is then integrated with respect to ρ from 0 to a. (In other words, $\mathbf{F}$ is now the attraction of a solid spherical ball of radius a and density $\delta = g(\rho)$, depending only on ρ.) Show that these substitutions and those of Eqs. (8) result in
$$\mathbf{F} = -\mathbf{k}\frac{GmM}{b^2},$$
where M is the total mass of the solid spherical ball.

16.9 SURFACE AREA

Let G be a region of the xy-plane and let the function
$$z = f(x, y), \tag{1}$$
together with its first partial derivatives, be continuous in G. For simplicity, suppose that the surface represented by Eq. (1) has a normal $\mathbf{N}$ which is nowhere parallel to the xy-plane. The area of the surface (1) may then be computed in the following way.

Divide the region G into small rectangles, of dimensions $\Delta x \times \Delta y$, by a grid of lines parallel to the x- and y-axes. Project a typical rectangle of area
$$\Delta A = \Delta y\, \Delta x \tag{2}$$
vertically upward onto the surface and call the corresponding area on the surface ΔS. Along with ΔS, we also consider the area ΔP of a section of a plane tangent to the surface at some point of ΔS (see Fig. 16.21). More precisely, choose any point Q lying on ΔS and consider the plane tangent to the surface at Q. Then project the area ΔA, Eq. (2), vertically

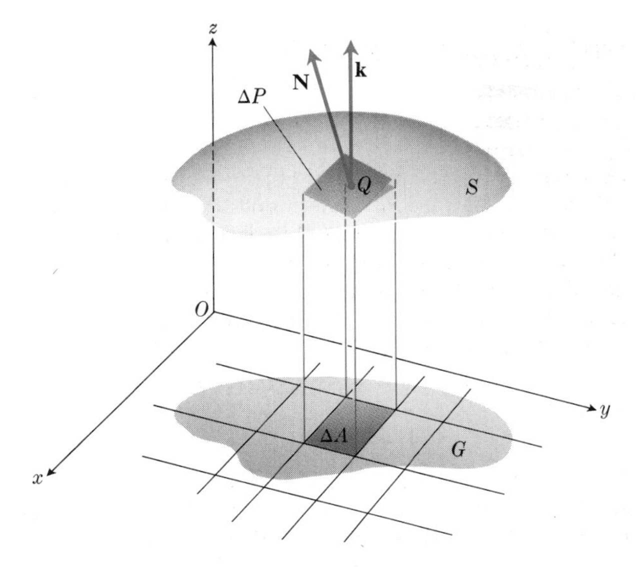

16.21 The area ΔP is the projection of ΔA in G onto the plane tangent to the surface S at the arbitrarily chosen point Q.

upward onto this tangent plane and denote the corresponding area in the tangent plane by ΔP. We shall use ΔP as an approximation to ΔS. If this procedure is carried out for all of the pieces of area ΔA lying in G and we add all the pieces of area ΔP obtained in this way, the result will approximate the total surface area S. That is,

$$S \approx \sum \Delta P. \tag{3}$$

The approximation becomes better as Δx and Δy approach zero. We take the limit of the sum in (3) as the *definition of the surface area S*, that is,

$$S = \lim_{\Delta x, \Delta y \to 0} \sum_G \Delta P. \tag{4}$$

Our problem now reduces to that of finding an analytic expression for ΔP as a function of x, y, Δx, and Δy. If we can do this, then we may expect that the limit in Eq. (4) can be expressed as a double integral over G.

Let $(x, y, 0)$, $(x + \Delta x, y, 0)$, $(x, y + \Delta y, 0)$, and $(x + \Delta x, y + \Delta y, 0)$ be the corners of the rectangle in G whose area ΔA projects onto the area ΔP in the tangent plane (see Fig. 16.22). For simplicity, let Q be the point $(x, y, f(x, y))$ on the surface (1). Let $\mathbf{u}$, $\mathbf{v}$ be vectors from Q forming two adjacent sides of the parallelogram whose area is ΔP. Then

$$\mathbf{u} = \mathbf{i}\,\Delta x + \mathbf{k} f_x(x, y)\,\Delta x,$$
$$\mathbf{v} = \mathbf{j}\,\Delta y + \mathbf{k} f_y(x, y)\,\Delta y.$$

A vector normal to the surface at Q, with magnitude equal to the area ΔP, is

$$\mathbf{N} = \mathbf{u} \times \mathbf{v} = \begin{vmatrix} \mathbf{i} & \mathbf{j} & \mathbf{k} \\ \Delta x & 0 & f_x(x, y)\,\Delta x \\ 0 & \Delta y & f_y(x, y)\,\Delta y \end{vmatrix}$$
$$= \Delta x\,\Delta y(-\mathbf{i} f_x(x, y) - \mathbf{j} f_y(x, y) + \mathbf{k}). \tag{5}$$

Therefore,

$$\Delta P = |\mathbf{u} \times \mathbf{v}|$$
$$= \Delta x\,\Delta y \sqrt{f_x^2(x, y) + f_y^2(x, y) + 1}. \tag{6}$$

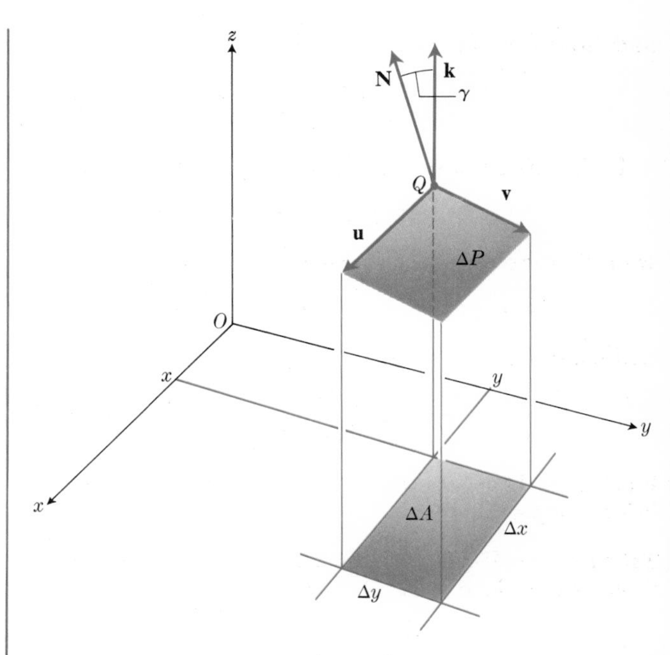

16.22 The area $\Delta P = |\mathbf{u} \times \mathbf{v}|$ is the projection of ΔA onto the tangent plane at Q: $\Delta P = \Delta A / \cos \gamma$.

Equations (4) and (6) together yield, for the area S,

$$S = \int_G \int \sqrt{\left(\frac{\partial f}{\partial x}\right)^2 + \left(\frac{\partial f}{\partial y}\right)^2 + 1}\; dx\, dy. \tag{7}$$

Other expressions for the area are easily found. If we apply Eq. (12a), Article 12.9, we have

$$(\mathbf{u} \times \mathbf{v}) \cdot \mathbf{k} = \text{area of projection of } \Delta P \text{ in } xy\text{-plane.}$$

This also agrees with Eq. (5) above, which gives

$$(\mathbf{u} \times \mathbf{v}) \cdot \mathbf{k} = \Delta x\,\Delta y = \Delta A. \tag{8a}$$

From the definition of the dot product, we also know that

$$(\mathbf{u} \times \mathbf{v}) \cdot \mathbf{k} = |\mathbf{u} \times \mathbf{v}|\,|\mathbf{k}| \cos \gamma = \Delta P \cos \gamma, \tag{8b}$$

where γ is the angle between $\mathbf{N} = \mathbf{u} \times \mathbf{v}$ and $\mathbf{k}$. Equations (8a) and (8b) give the result

$$\Delta P = \frac{\Delta A}{\cos \gamma}, \tag{9}$$

and from Eqs. (4) and (9) we have

$$S = \int_G\int \frac{dA}{\cos\gamma}. \tag{10}$$

If the equation of the surface is given in the form

$$F(x, y, z) = 0,$$

we may take as normal vector

$$\mathbf{N} = \text{grad } F = \mathbf{i}F_x + \mathbf{j}F_y + \mathbf{k}F_z, \tag{11a}$$

and compute $\cos\gamma$ in Eq. (10) from

$$\cos\gamma = \frac{\mathbf{N}\cdot\mathbf{k}}{|\mathbf{N}|\cdot|\mathbf{k}|}. \tag{11b}$$

Problem 1. Find the area of the paraboloid $z = x^2 + y^2$ below the plane $z = 1$.

Solution. (See Fig. 16.23.) The surface area S in question projects into the interior of the circle

$$x^2 + y^2 = 1,$$

in the xy-plane. This is the region denoted above by G. Here

$$z = f(x, y) = x^2 + y^2,$$

so that

$$\frac{\partial f}{\partial x} = 2x, \qquad \frac{\partial f}{\partial y} = 2y$$

and

$$S = \iint\limits_{x^2+y^2\le 1} \sqrt{4x^2 + 4y^2 + 1}\, dy\, dx. \tag{12}$$

Now the double integral (12) is easier to evaluate in polar coordinates, since the combination $x^2 + y^2$ may be replaced by r^2. Taking the element of area to be

$$dA = r\, dr\, d\theta$$

in place of $dy\, dx$, we thus have

$$S = \iint\limits_{r\le 1} \sqrt{4r^2 + 1}\, r\, dr\, d\theta$$

$$= \int_0^{2\pi}\int_0^1 \sqrt{4r^2 + 1}\, r\, dr\, d\theta = \frac{\pi}{6}(5\sqrt{5} - 1).$$

Remark 1. Equation (7) can easily be remembered as a modification or extension of the formula for

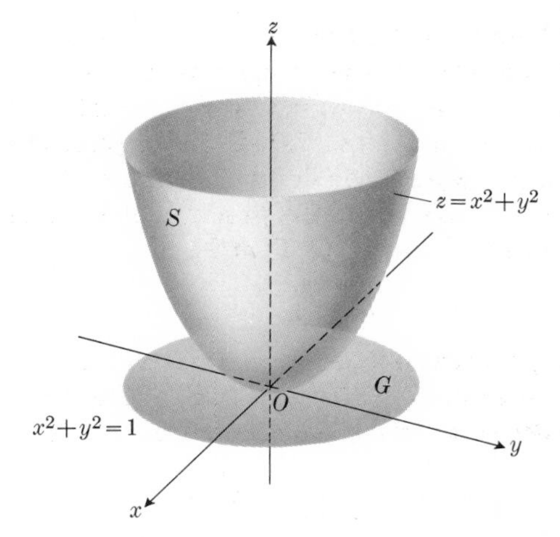

16.23

arc length in two dimensions:

$$\text{arc length} = \int_a^b \sqrt{1 + f_x^2}\, dx, \tag{13a}$$

$$\text{surface area} = \int_G\int \sqrt{1 + f_x^2 + f_y^2}\, dx\, dy. \tag{13b}$$

Remark 2. If the integrand in Eq. (13b) is hard to deal with, we can approximate the answer numerically. The idea here is to subdivide G into small pieces whose areas can be calculated, to multiply each of these by the value of the integrand

$$\sqrt{1 + f_x^2 + f_y^2}$$

at some point in the small piece, and then to add the results. The following problem illustrates how this can be done.

Problem 2. Estimate the area of that portion of the surface $z = \sin x + \sin y$ that lies over the square

$$G = \{(x, y): 0 \le x \le \pi/2,\ 0 \le y \le \pi/2\}.$$

Solution. Because

$$\frac{\partial z}{\partial x} = \cos x, \qquad \frac{\partial z}{\partial y} = \cos y,$$

we have the equation

$$S = \iint\limits_G \sqrt{1 + \cos^2 x + \cos^2 y}\, dx\, dy. \tag{14}$$

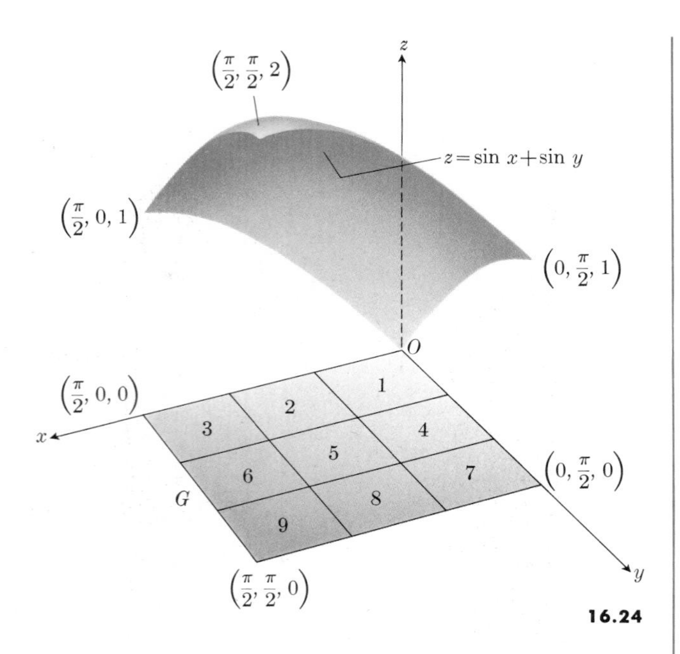

16.24

The integrand function

$$g(x, y) = (1 + \cos^2 x + \cos^2 y)^{1/2}$$

is a maximum at (0, 0) and a minimum at $(\pi/2, \pi/2)$:

$$g_{\max} = g(0, 0) = \sqrt{3} \approx 1.732$$
$$g_{\min} = g(\pi/2, \pi/2) = \sqrt{1} = 1.000.$$

We therefore know that if $A = (\pi/2)^2$ is the area of G, then

$$1.000A \le S \le 1.732A,$$

and one reasonable estimate might be the average:

$$S \approx \frac{1 + \sqrt{3}}{2} A \approx 1.366A \approx 1.366\left(\frac{9.86960}{4}\right)$$
$$\approx 3.370.$$

Another way of estimating S is to subdivide G into smaller squares and estimate the part of S above each of these, and then to add the results. Figure 16.24 shows G divided into nine squares, numbered 1 through 9. We estimate the integral by a sum

$$\sum_{k=1}^{9} g(x_k, y_k)\,\Delta A_k, \tag{15}$$

where the area of the kth small square is

$$\Delta A_k = \frac{\pi^2}{36} \approx \frac{9.86960}{36} \approx 0.27416,$$

and (x_k, y_k) is some point in or on the boundary of the kth subregion. The following table gives the results for one possible selection of these points.

k	(x_k, y_k)	$g(x_k, y_k)$
1	$(\frac{1}{6}\pi, 0)$	$\sqrt{1 + \frac{3}{4} + 1} = \sqrt{2.75} \approx 1.658$
2	$(\frac{1}{3}\pi, 0)$	$\sqrt{1 + \frac{1}{4} + 1} = \sqrt{2.25} = 1.500$
3	$(\frac{1}{2}\pi, 0)$	$\sqrt{1 + 0 + 1} = \sqrt{2.00} \approx 1.414$
4	$(\frac{1}{6}\pi, \frac{1}{6}\pi)$	$\sqrt{1 + \frac{3}{4} + \frac{3}{4}} = \sqrt{2.50} \approx 1.581$
5	$(\frac{1}{3}\pi, \frac{1}{6}\pi)$	$\sqrt{1 + \frac{1}{4} + \frac{3}{4}} = \sqrt{2.00} \approx 1.414$
6	$(\frac{1}{2}\pi, \frac{1}{6}\pi)$	$\sqrt{1 + 0 + \frac{3}{4}} = \sqrt{1.75} \approx 1.323$
7	$(\frac{1}{6}\pi, \frac{1}{3}\pi)$	$\sqrt{1 + \frac{3}{4} + \frac{1}{4}} = \sqrt{2.00} \approx 1.414$
8	$(\frac{1}{3}\pi, \frac{1}{3}\pi)$	$\sqrt{1 + \frac{1}{4} + \frac{1}{4}} = \sqrt{1.50} \approx 1.225$
9	$(\frac{1}{2}\pi, \frac{1}{3}\pi)$	$\sqrt{1 + 0 + \frac{1}{4}} = \sqrt{1.25} \approx 1.118$
		Sum $\approx$ 12.647

This gives as estimate for the surface area:

$$S \approx 12.647(9.86960/36) \approx 3.467.$$

Remark 3. In the foregoing problem, we could have chosen the points (x_k, y_k) in a variety of ways. In Exercise 10 below, you are asked which ways would guarantee an underestimate, and which ways an overestimate. We are not so concerned with pinning down the accuracy of the approximation as in illustrating how a double integral can be approximated.

Remark 4. In approximating the area of a surface, not all methods give reliable answers. We think that the following example is of interest.* It has to do with approximating the lateral surface area of a right circular cylinder of base radius a and altitude h. Imagine that a sequence of circles is drawn on the cylinder parallel to the base. These

---

* This example, which is due to H. A. Schwarz (1890), is discussed in Olmsted, *Advanced Calculus*, pp. 616–617.

divide the surface into bands, and we assume that there are m such bands, each of altitude h/m. We now pick out two adjacent bands, as in Fig. 16.25, and inscribe plane triangles in the manner suggested by the figure. More precisely, each of the three circles is divided into n congruent arcs and the endpoints of these arcs are the vertices of the triangles to be inscribed. The arcs on the top and bottom circles are vertically aligned, but on the circle in the middle, the arcs are all displaced through one-half an arc length. In the figure, O is the center of the circle containing the arc BC, so that OBC is a triangle lying in a plane parallel to the base of the cylinder. We choose 2θ as the measure of the vertex angle at O, and easily find that the length of the base BC is

$$|BC| = 2a \sin \theta = 2a \sin (\pi/n).$$

We focus attention next on the triangle ABC, whose area is to be used as an approximation to the corresponding part $ABEC$ of the cylinder. We can get the altitude AD of the triangle ABC by applying the theorem of Pythagoras to the triangle ADE, because $|AE| = h/m$ is just the distance between the circles drawn on the cylinder parallel to its base, and

$$|DE| = a - a \cos \theta = a(1 - \cos \theta).$$

Therefore

$$\begin{aligned} |AD|^2 &= |AE|^2 + |ED|^2 \\ &= \left(\frac{h}{m}\right)^2 + a^2 \left(1 - \cos \frac{\pi}{n}\right)^2. \end{aligned}$$

Thus the area of triangle ABC is

$$\begin{aligned} \text{area } (\triangle ABC) &= \tfrac{1}{2}|BC| \times |AD| \\ &= a \sin \frac{\pi}{n} \sqrt{\left(\frac{h}{m}\right)^2 + a^2 \left(1 - \cos \frac{\pi}{n}\right)^2}. \end{aligned}$$

In the band of the cylinder between two consecutive parallel circles, there are n triangles congruent to ABC with bases on the upper circle, and another n triangles with their bases on the lower circle, or $2n$ triangles altogether between the two

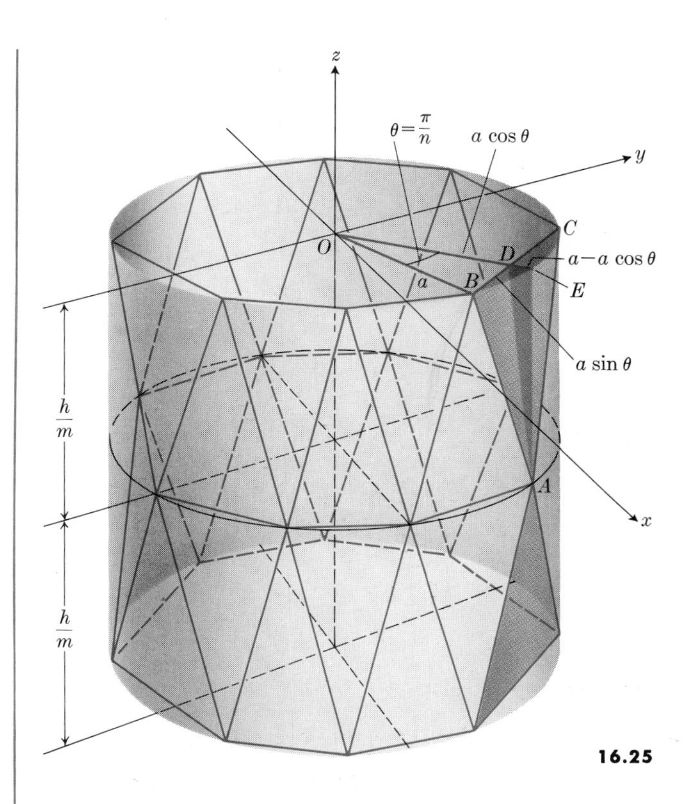

16.25

circles. In the m bands, there are $2nm$ such triangles. The sum of their areas is therefore

$$f(m, n) = 2nma \sin \frac{\pi}{n} \sqrt{\left(\frac{h}{m}\right)^2 + a^2 \left(1 - \cos \frac{\pi}{n}\right)^2}. \tag{16}$$

This function of the two variables m and n is an approximation to the lateral area of the surface of the cylinder, and we might naturally hope that its limit exists, as m and n tend to infinity, and that the limit is $2\pi ah$. But now comes a shock: The answer depends upon how m and n go to infinity.

Case 1. First, let $m \to \infty$, with n held fixed, and then let $n \to \infty$. If we do this, we get ∞ as the answer:

$$\lim_{n\to\infty} [\lim_{m\to\infty} f(m, n)] = \lim_{n\to\infty} (\infty) = \infty.$$

Case 2. Interchange the order, and let n go to ∞ first while holding m fixed, and then let m become infinite. Using the results

$$\lim_{n\to\infty} \frac{\sin(\pi/n)}{(\pi/n)} = 1 = \frac{1}{\pi} \lim_{n\to\infty} n \sin(\pi/n)$$

and

$$\lim_{n\to\infty} [1 - \cos(\pi/n)] = 0,$$

we get

$$\begin{aligned} \lim_{m\to\infty} [\lim_{n\to\infty} f(m, n)] &= \lim_{m\to\infty} 2m\pi a\sqrt{(h/m)^2 + 0} \\ &= 2\pi ah. \end{aligned}$$

Case 3. Suppose that we let m and n tend to infinity simultaneously, in some fixed ratio $m/n = c$, where c is a positive constant. Then, from Eq. (16),

$$\begin{aligned} f(m, n) &= f(cn, n) \\ &= 2cn^2 a \sin\frac{\pi}{n} \sqrt{\left(\frac{h}{cn}\right)^2 + a^2\left(1 - \cos\frac{\pi}{n}\right)^2} \\ &= 2ca\left(n \sin\frac{\pi}{n}\right) \\ &\quad \times \sqrt{\left(\frac{h}{c}\right)^2 + a^2\left[n\left(1 - \cos\frac{\pi}{n}\right)\right]^2} \end{aligned}$$

and

$$\lim_{n\to\infty} n \sin\frac{\pi}{n} = \pi, \quad \lim_{n\to\infty} \left[n\left(1 - \cos\frac{\pi}{n}\right)\right] = 0,$$

so that

$$\lim_{n\to\infty} f(cn, n) = 2ca\pi\sqrt{(h/c)^2 + 0} = 2\pi ah.$$

We therefore conclude that we could get the accepted answer for the lateral area of the cylinder by the second and third methods of taking limits, but not by the first. The trouble with the first method is that if $m \to \infty$ while n stays fixed, then each triangle like ABC approaches a triangle like EBC with nonzero area, and at the same time we get infinitely many circles each of which contains n triangles like EBC: Thus we get ∞ for the result. Our conclusion is that although we can approximate the length of a curve by inscribing polygons and taking a limit, we cannot be sure that we will get the correct area of a surface by inscribing polyhedra and taking a limit. For other, more striking paradoxes, see the article "Paradox Lost and Paradox Regained," by Edward Kasner and James R. Newman, in *The World of Mathematics,* Vol. 3 (especially pp. 1942–1945).

EXERCISES

1. Derive a formula to replace Eq. (7) for the area of a surface S whose equation is $F(x, y, z) = 0$ and where
   (a) the surface S is projected onto a region A in the xy-plane,
   (b) the surface S is projected onto a region B in the yz-plane,
   (c) the surface S is projected onto a region C in the zx-plane.
2. By integration, find the area of the triangle cut from the plane $x/a + y/b + z/c = 1$ by the coordinate planes. Check your answer by vector methods.
3. By integration, find the area of that portion of the surface of the sphere $x^2 + y^2 + z^2 = a^2$ that lies in the first octant.
4. Find the area of the surface of that portion of the sphere $x^2 + y^2 + z^2 = a^2$ that lies inside the cylinder $x^2 + y^2 = ax$.
5. In the preceding problem, find the area of that portion of the cylinder that lies inside the sphere. *Hint.* Project the area into the xz-plane. Or use single integration, $\int z\,ds$, where z is the altitude on the cylinder and ds is the element of arc length in the xy-plane.
6. Find the area of that portion of the sphere
$$x^2 + y^2 + z^2 = 2a^2$$
that is cut out by the upper nappe of the cone $x^2 + y^2 = z^2$.
7. Find the area cut from the plane $z = cx$ by the cylinder $x^2 + y^2 = a^2$.
8. Find the area of that portion of the cylinder
$$x^2 + z^2 = a^2$$
that lies between the planes $y = \pm a/2$, $x = \pm a/2$.

9. Find the area cut from the surface

$$az = y^2 - x^2$$

by the cylinder $x^2 + y^2 = a^2$.

10. In Problem 2 (see Fig. 16.24), suppose the 9 points (x_k, y_k) are all chosen at the lower left corners of the small squares. Would the corresponding sum in (15) be an overestimate or an underestimate for S? Answer the same question if the points are chosen at the upper right corners of the respective subregions. Without calculating either sum explicitly (unless you have ready access to a computer), show that their difference (overestimate minus underestimate) is

$$\frac{\pi^2}{36} \times \left[g(0, 0) + g\left(\frac{\pi}{6}, 0\right) + g\left(\frac{\pi}{3}, 0\right) + g\left(0, \frac{\pi}{6}\right) + g\left(0, \frac{\pi}{3}\right) - g\left(\frac{\pi}{2}, \frac{\pi}{6}\right) - g\left(\frac{\pi}{2}, \frac{\pi}{3}\right) - g\left(\frac{\pi}{2}, \frac{\pi}{2}\right) - g\left(\frac{\pi}{6}, \frac{\pi}{2}\right) - g\left(\frac{\pi}{3}, \frac{\pi}{2}\right)\right],$$

or approximately

$$(\pi^2/36)(8.048 - 5.882) = 0.594.$$

11. In the cylinder example discussed in Remark 4:

(a) Suppose that $m = 1$. What is $\lim f(1, n)$ as $n \to \infty$? Discuss the geometric significance.

(b) If $L_m = \lim f(m, n)$ as $n \to \infty$, does L_m increase, decrease, or remain constant as m increases? What is the geometric significance of this?

12. Suppose that the cylinder discussed in Remark 4 is replaced by a cone with base radius a and altitude h. Find a formula for $g(m, n)$, the sum of the areas of the inscribed triangles, following the same method. Discuss various limit possibilities as $m \to \infty$ and $n \to \infty$.

REVIEW QUESTIONS AND EXERCISES

1. Define the double integral of a function of two variables. What geometric interpretation may be given to the integral?
2. List four applications of multiple integration.
3. Define *moment of inertia* and *radius of gyration.*
4. How does a double integral in polar coordinates differ from a double integral in Cartesian coordinates? In what way are they alike?
5. What are the fundamental volume elements for triple integrals
   (a) in Cartesian coordinates,
   (b) in cylindrical coordinates,
   (c) in spherical coordinates?
6. How is surface area defined? Which formula of Article 16.9 is the most general one for computing surface area, in the sense that it includes many others as special cases?
7. How would you define $\iint_S F\, dS$, when S is a surface in space and F is a function defined at points on the surface? Illustrate when S is the hemisphere

$$z = \sqrt{1 - x^2 - y^2}, \quad x^2 + y^2 \leq 1,$$

and $F(x, y, z) = z$. What is the geometric interpretation of $\iint z\, dS$?

8. Suppose that D is a region in space bounded by a sphere of radius ϵ, with center at the point $(1, -2, 2)$. Evaluate

$$\lim_{\epsilon \to 0} \epsilon^{-3} \iiint_D (x^4 + y^2 - z^3)\, dx\, dy\, dz,$$

or explain why no such limit exists. (Very little calculation is required. Think.)

MISCELLANEOUS EXERCISES

1. Reverse the order of integration and evaluate

$$\int_0^4 \int_{-\sqrt{4-y}}^{(y-4)/2} dx\, dy.$$

2. Sketch the region over which the integral

$$\int_0^1 \int_{\sqrt{y}}^{2-\sqrt{y}} xy\, dx\, dy$$

is to be evaluated and find its value.

3. The integral

$$\int_{-1}^1 \int_{x^2}^1 dy\, dx$$

represents the area of a region of the xy-plane. Sketch the region and express the same area as a double integral with the order of integration reversed.

4. The base of a pile of sand covers the region in the xy-plane that is bounded by the parabola $x^2 + y = 6$ and the line $y = x$. The depth of the sand above the point (x, y) is x^2. Sketch the base of the sandpile and a representative element of volume dV, and find the volume of sand in the pile by double integration.

5. In setting up a double integral for the volume V under the paraboloid $z = x^2 + y^2$ and above a certain region R of the xy-plane, the following sum of iterated integrals is obtained:

$$V = \int_0^1 \left(\int_0^y (x^2 + y^2)\, dx\right) dy + \int_1^2 \left(\int_0^{2-y} (x^2 + y^2)\, dx\right) dy.$$

Sketch the region R in the xy-plane and express V as an iterated integral in which the order of integration is reversed.

6. By change of order of integration, show that this double integral can be reduced to a single integral:

$$\int_0^x du \int_0^u e^{m(x-t)} f(t)\, dt = \int_0^x (x - t)e^{m(x-t)} f(t)\, dt.$$

Similarly, it can be shown that

$$\int_0^x dv \int_0^v du \int_0^u e^{m(x-t)} f(t)\, dt = \int_0^x \frac{(x-t)^2}{2!} e^{m(x-t)} f(t)\, dt.$$

Evaluate integrals for the case $f(t) = \cos at$. (This example illustrates that such reductions usually make calculation easier.)

7. By changing the order of integration, show that

$$\int_0^1 f(x)\, dx \int_0^x \log (x - y) f(y)\, dy = \int_0^1 f(y)\, dy \int_y^1 \log (x - y) f(x)\, dx = \frac{1}{2} \int_0^1 \int_0^1 \log |x - y| f(x) f(y)\, dx\, dy.$$

(This example illustrates the fact that sometimes a multiple integral with variable limits may be changed into one with constant limits.)

8. Evaluate the integral

$$\int_0^\infty \frac{e^{-ax} - e^{-bx}}{x}\, dx.$$

Hint. Use the relation

$$\frac{e^{-ax} - e^{-bx}}{x} = \int_a^b e^{-xy}\, dy$$

to form a double integral, and evaluate it by change of the order of integration.

9. By double integration, find the center of gravity of that part of the area of the circle $x^2 + y^2 = a^2$ contained in the first quadrant.

10. Determine the centroid of the plane area that is given in polar coordinates by

$$0 \le r \le a, \quad -\alpha \le \theta \le \alpha.$$

11. Find the centroid of the area bounded by the lines $\theta = 0°$ and $\theta = 45°$, and by the circles $r = 1$ and $r = 2$.

12. By double integration, find the center of gravity of the area between the parabola $x + y^2 - 2y = 0$ and the line $x + 2y = 0$.

13. For a solid body of constant density, having its center of gravity at the origin, show that the moment of inertia about an axis parallel to Oz through (x_0, y_0) is equal to the moment of inertia about Oz plus $M(x_0^2 + z_0^2)$, where M is the mass of the body.

14. Find the moment of inertia of the angle section shown in the accompanying figure,

(a) with respect to the horizontal base,

(b) with respect to a horizontal line through its centroid.

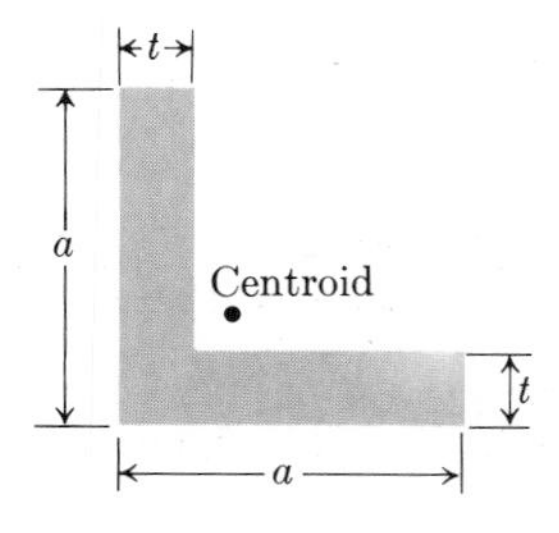

15. Consider a uniform elliptic lamina of semiaxes a, b, and suppose that its mass is M. Show that its moment of inertia about an axis in its plane through the center of the ellipse is

$$\tfrac{1}{4}M(a^2 \sin^2 \alpha + b^2 \cos^2 \alpha),$$

if the axis makes an angle α with the axis of length $2a$.

16. A counterweight of a flywheel has the form of the smaller segment cut from a circle of radius a by a chord at a distance b from the center ($b < a$). Find the area of this counterweight and its polar moment of inertia about the center of the circle.

17. The radius of gyration of a body with volume V is defined by $K = \sqrt{I/V}$, where K and its moment of inertia I are referred to the same axis. Consider an ellipse $(x^2/a^2) + (y^2/b^2) = 1$ revolving about the x-axis to generate an ellipsoid. Find the radius of gyration of the ellipsoid with respect to the x-axis.

18. Find the radii of gyration about $\theta = 0$ and $\theta = \pi/2$ for the area of a loop of the curve

$$r^2 = a^2 \cos 2\theta, \quad a > 0.$$

19. The hydrostatic pressure at a depth y in a fluid is wy. Taking the x-axis in the surface of the fluid and the y-axis vertically downward, consider a semicircular lamina, radius a, completely immersed with its bounding diameter horizontal, uppermost, and at a depth c. Show that the depth of the center of pressure is

$$\frac{3\pi a^2 + 32ac + 12\pi c^2}{4(4a + 3\pi c)}.$$

Note. The center of pressure is defined as the point where the entire hydrostatic force could be concentrated so as to produce the same first moment of force.

20. Show that

$$\iint \frac{\partial^2 F(x, y)}{\partial x\, \partial y}\, dx\, dy$$

over the rectangle $x_0 \le x \le x_1$, $y_0 \le y \le y_1$ is

$$F(x_1, y_1) - F(x_0, y_1) - F(x_1, y_0) + F(x_0, y_0).$$

Explain why

$$\lim \frac{F(x_1, y_1) - F(x_0, y_1) - F(x_1, y_0) + F(x_0, y_0)}{(x_1 - x_0)(y_1 - y_0)},$$

as $(x_1, y_1) \to (x_0, y_0)$, is equal to $F_{xy}(x_0, y_0)$, if F_{xy} is continuous in a neighborhood of (x_0, y_0).

21. Change the following double integral to an equivalent double integral in polar coordinates, and sketch the region of integration.

$$\int_{-a}^{a} \int_{0}^{\sqrt{a^2 - y^2}} x\, dx\, dy.$$

22. A customary method of evaluating the improper integral $I = \int_0^\infty e^{-x^2}\, dx$ is by considering its squared value

$$I^2 = \left(\int_0^\infty e^{-x^2}\, dx\right)\left(\int_0^\infty e^{-y^2}\, dy\right)$$
$$= \int_0^\infty \int_0^\infty e^{-(x^2+y^2)}\, dx\, dy.$$

Introduce polar coordinates in the last expression and show

$$I = \int_0^\infty e^{-x^2}\, dx = \frac{\sqrt{\pi}}{2}.$$

23. By transformation of variables $u = x - y$, $v = y$, show that

$$\int_0^\infty e^{-sx}\, dx \int_0^x f(x - y, y)\, dy$$
$$= \int_0^\infty \int_0^\infty e^{-s(u+v)} f(u, v)\, du\, dv.$$

24. How must a, b, c be chosen so that

$$\int_{-\infty}^{\infty} \int_{-\infty}^{\infty} e^{-(ax^2 + 2bxy + cy^2)}\, dx\, dy = 1?$$

Hint. Introduce the transformation

$$\xi = \alpha x + \beta y, \quad \eta = \gamma x + \delta y,$$

where $(\alpha\delta - \beta\gamma)^2 = ac - b^2;$

then $ax^2 + 2bxy + cy^2 = \xi^2 + \eta^2.$

25. Find the area enclosed by the lemniscate

$$r^2 = 2a^2 \cos 2\theta.$$

Also find the moment of inertia of this area about the y-axis.

26. Evaluate

$$\iint \frac{dx\, dy}{(1 + x^2 + y^2)^2},$$

(a) taken over one loop of the lemniscate

$$(x^2 + y^2) - (x^2 - y^2) = 0,$$

(b) taken over the triangle with vertices $(0, 0)$, $(2, 0)$, $(1, \sqrt{3})$.

Hint. Transform to polar coordinates.

27. Show, by transforming to polar coordinates, that

$$K(a) = \int_0^{a\sin\beta} dy \int_{y\cot\beta}^{\sqrt{a^2-y^2}} \ln(x^2+y^2)\,dx = a^2\beta\,(\ln a - \tfrac{1}{2}),$$

where $0 < \beta < \pi/2$. Changing the order of integration, what expression do you obtain?

28. Find the volume bounded by the cylinder $y = \cos x$ and the planes

$$z = y, \quad x = 0, \quad x = \pi/2, \quad z = 0.$$

29. Find the center of mass of the homogeneous pyramid whose base is the square enclosed by the lines $x = 1$, $x = -1$, $y = 1$, $y = -1$ in the plane $z = 0$, and whose vertex is at the point $(0, 0, 1)$.

30. Find the volume bounded above by the sphere

$$x^2 + y^2 + z^2 = 2a^2$$

and below by the paraboloid

$$az = x^2 + y^2.$$

31. Find the volume bounded by the surfaces

$$z = x^2 + y^2, \qquad z = \tfrac{1}{2}(x^2 + y^2 + 1).$$

32. By triple integration, determine the volume enclosed by the two surfaces

$$x = y^2 + z^2, \qquad x = 1 - y^2.$$

33. Find the moment of inertia, with respect to the z-axis, of a solid that is bounded below by the paraboloid $3az = x^2 + y^2$ and above by the sphere $x^2 + y^2 + z^2 = 4a^2$, if its density is constant.

34. By integration, find the volume in the ellipsoid

$$\frac{x^2}{a^2} + \frac{y^2}{b^2} + \frac{z^2}{c^2} = 1.$$

35. Evaluate the integral

$$\iiint |xyz|\,dx\,dy\,dz$$

taken throughout the ellipsoid

$$\frac{x^2}{a^2} + \frac{y^2}{b^2} + \frac{z^2}{c^2} \le 1.$$

Hint. Introduce new coordinates

$$x = a\xi, \quad y = b\eta, \quad z = c\zeta.$$

36. Two cylinders of radius a have their axes along the x- and y-axes respectively. Find the volume that they have in common.

37. The volume of a certain solid is given by the triple integral

$$\int_0^2 \left[\int_0^{\sqrt{2x-x^2}} \left(\int_{-\sqrt{4-x^2-y^2}}^{\sqrt{4-x^2-y^2}} dz\right) dy\right] dx.$$

(a) Describe the solid by giving the equations of all the surfaces that form its boundary.

(b) Express the volume as a triple integral in cylindrical coordinates. Give the limits of integration explicitly, but do not evaluate the integral.

38. A square hole of side $2b$ is cut symmetrically through a sphere of radius a $(a > b\sqrt{2})$. Find the volume removed.

39. A hole is bored through a sphere, the axis of the hole being a diameter of the sphere. The volume of the solid remaining is given by the integral

$$V = 2\int_0^{2\pi} \int_0^{\sqrt{3}} \int_0^{\sqrt{4-z^2}} r\,dr\,dz\,d\theta.$$

(a) By inspecting the given integral, determine the radius of the hole and the radius of the sphere.

(b) Calculate the numerical value of the integral.

40. Set up an equivalent triple integral in rectangular coordinates:

$$\int_0^{\pi/2} \int_1^{\sqrt{3}} \int_1^{\sqrt{4-r^2}} r^3 \sin\theta\cos\theta\, z^2\,dz\,dr\,d\theta.$$

Arrange the order so the first integration is with respect to z, the second with respect to y, and the last with respect to x.

41. Find the volume bounded by the plane $z = 0$, the cylinder $x^2 + y^2 = a^2$, and the cylinder $az = a^2 - x^2$.

42. Find the volume of that portion of the sphere $r^2 + z^2 = a^2$ that is inside the cylinder $r = a\sin\theta$. (Here r, θ, z are cylindrical coordinates.)

43. Find the moment of inertia about the z-axis of the volume that is bounded above by the sphere $\rho = a$ and below by the cone $\phi = \pi/3$. (Here ρ, ϕ, θ are spherical coordinates.)

44. Find the volume enclosed by the surface $\rho = a\sin\phi$, in spherical coordinates.

45. Find the moment of inertia of the solid of constant density δ bounded by two concentric spheres of radii a and b $(a < b)$, about a diameter.

46. Let S be a solid right circular cone with vertex at the origin, base in the plane $z = h$, and radius of base equal to a. Suppose that P is a particle of mass m at the origin. Let δ be the density, constant throughout S. Find the gravitational attraction of S for P. The force of gravitational attraction that the mass in the cone exerts on P is given by

$$\mathbf{F} = Gm \iiint \frac{\mathbf{u}\delta \, dV}{s^2},$$

where G is the gravitational constant, $\mathbf{u}$ is a unit vector in the direction from P toward the volume element dV, s^2 is the square of the distance from P to the volume element dV and the integration is extended throughout S.

47. The density at P, a point of a solid sphere of radius a and center O, is given to be

$$\rho_0[1 + \epsilon \cos \theta + \tfrac{1}{2}\epsilon^2(3 \cos \theta - 1)],$$

where θ is the angle OP makes with a fixed radius OQ, and ρ_0 and ϵ are constants. Find the average density of the sphere.

48. Find the area of the surface $y^2 + z^2 = 2x$ cut off by the plane $x = 1$.

49. Find the area cut from the plane $x + y + z = 1$ by the cylinder $x^2 + y^2 = 1$.

50. Find the area above the xy-plane cut from the cone

$$x^2 + y^2 = z^2$$

by the cylinder

$$x^2 + y^2 = 2ax.$$

51. Find the surface area of that portion of the sphere $r^2 + z^2 = a^2$ that is inside the cylinder $r = a \sin \theta$. (Here r, θ, z are cylindrical coordinates.)

52. The cylinder $x^2 + y^2 = 2x$ cuts out a portion of a surface S from the upper nappe of the cone

$$x^2 + y^2 = z^2.$$

Compute the value of the surface integral

$$\iint_S (x^4 - y^4 + y^2z^2 - z^2x^2 + 1)\, dS.$$

53. The sphere $x^2 + y^2 + z^2 = 25$ is cut by the plane $z = 3$, the smaller portion cut off forming a solid V which is bounded by a closed surface S_0 made up of two parts, the spherical part S_1 and the planar part S_2. If $\cos \alpha \mathbf{i} + \cos \beta \mathbf{j} + \cos \gamma \mathbf{k}$ is the unit outer normal of S_0, find the value of the surface integral

$$\iint_S (xz \cos \alpha + yz \cos \beta + \cos \gamma)\, dS$$

(a) If S is the spherical cap S_1,
(b) if S is the planar base S_2,
(c) if S is the complete boundary S_0.

54. Obtain the double integral expressing the surface area cut from the cylinder $z = a^2 - y^2$ by the cylinder $x^2 + y^2 = a^2$, and reduce this double integral to a definite single integral with respect to the variable y.

55. A square hole of side $2\sqrt{2}$ is cut symmetrically through a sphere of radius 2. Show that the area of the surface removed is

$$16\pi(\sqrt{2} - 1).$$

56. A torus-surface is generated by moving a sphere of unit radius whose center travels on a closed plane circle of radius 2. Calculate the area of this surface.

57. Calculate the area of the surface

$$(x^2 + y^2 + z^2)^2 = x^2 - y^2.$$

Hint. Use polar coordinates.

58. Calculate the area of the spherical part of the boundary of the region

$$x^2 + y^2 + z^2 = r^2,$$

where

$$x^2 + y^2 - rx \geq 0,$$
$$x^2 + y^2 + rx \geq 0.$$

Hint. Integrate first with respect to x and y.

59. Prove that the potential of a circular disk of mass m per unit area, and of radius a, at a point distant h from the center and on the normal to the disk through the center, is

$$2\pi m(\sqrt{(a^2 + h^2)} - h).$$

Note. The potential at a point P due to a mass Δm at Q is $\Delta m/r$, where r is the distance from P to Q.

60. The solid angle sustained by a surface Σ bounded by a closed curve is defined with respect to the origin as

$$\Omega = \left| \iint_{\Sigma} \frac{\cos \theta}{r^2}\, dS \right| ,$$

where the area element dS is located at the end of the position vector $\mathbf{R}$, θ is the angle between $\mathbf{R}$ and the normal to dS, and $r = |\mathbf{R}|$. Show that in Cartesian coordinates,

$$\Omega = \left| \iint_{\Sigma} \frac{x\, dy\, dz + y\, dz\, dx + z\, dx\, dy}{(x^2 + y^2 + z^2)^{3/2}} \right| .$$

61. Prove by direct integration that

$$\int_{-\infty}^{\infty} \int_{-\infty}^{\infty} \frac{dx\, dy}{(x^2 + y^2 + 1)^{3/2}} = 2\pi.$$

Interpret this integral as a solid angle sustained by a surface. What surface is this?

62. Show that the average distance of the points of the surface of a sphere of radius a from a point on the surface is $4a/3$.

VECTOR ANALYSIS

CHAPTER 17

17.1 INTRODUCTION: VECTOR FIELDS

Suppose that a certain region G of 3-space is occupied by a moving fluid: air, for example, or water. We may imagine that the fluid is made up of a large number of particles (probably an infinite number), and that at time t, the particle which is in position P at that instant has a velocity $\mathbf{v}$. If we stay at P and observe new particles that pass through it, we shall probably see that they have different velocities. This would surely be true, for example, in turbulent motions caused by high winds or stormy seas. Again, if we could take a picture of the velocities of particles at different places at the same instant, we would expect to find that these velocities vary from place to place. Thus the velocity at position P at time t is, in general, a function of both position and time:

$$\mathbf{v} = \mathbf{F}(x, y, z, t). \tag{1}$$

Equation (1) indicates that the velocity $\mathbf{v}$ is a vector function $\mathbf{F}$ of the four independent variables x, y, z, and t. Such functions have many applications, particularly in treatments of flows of material. In hydrodynamics, for example, if $\delta = \delta(x, y, z, t)$ is the *density* of the fluid at (x, y, z) at time t, and we take $\mathbf{F} = \mathbf{i}u + \mathbf{j}v + \mathbf{k}w$ to be the velocity expressed in terms of components, then we are able to derive the Euler partial differential equation of continuity of motion:

$$\frac{\partial \delta}{\partial t} + \frac{\partial(\delta u)}{\partial x} + \frac{\partial(\delta v)}{\partial y} + \frac{\partial(\delta w)}{\partial z} = 0.$$

(We are not prepared to deal with this subject* in any detail; we mention it as one important field in which the ideas presented in this chapter are applied.) Such functions are also applied in physics and electrical engineering; for example, in the study of propagation of electromagnetic waves. Also, much current research activity in applied mathematics has to do with such functions.

---

* Some of the laws of hydrodynamics can be found in the article "Hydromechanics" in the *Encyclopaedia Britannica*.

Steady-state flows

In this chapter, we shall deal only with those flows for which the velocity function, Eq. (1), does not depend on the time t. Such flows are called steady-state flows. They exemplify *vector fields.*

Definition. *If, to each point P in some region G, a vector $\mathbf{F}(P)$ is assigned, the collection of all such vectors is called a vector field.*

In addition to the vector fields that are associated with fluid flows, there are vector *force* fields that are associated with gravitational attraction, magnetic force fields, electric fields, and purely mathematical fields.

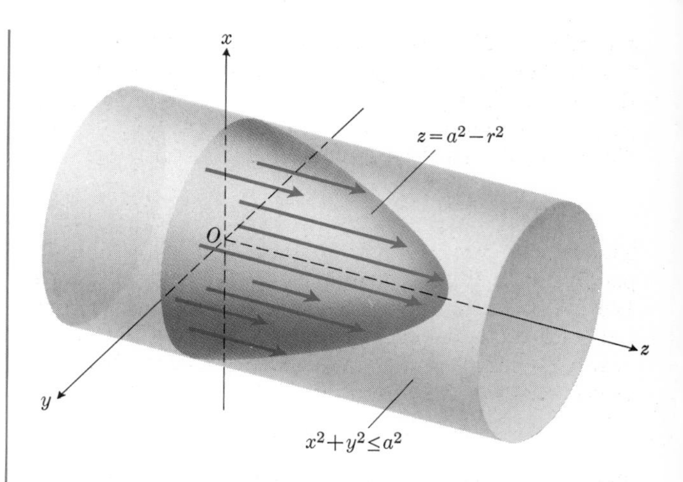

17.1 The flow of fluid in a long cylindrical pipe. The vectors $\mathbf{v} = (a^2 - r^2)\mathbf{k}$ inside the cylinder have their bases in the xy-plane and their tips on the paraboloid $z = a^2 - r^2$.

Problem 1. Imagine an idealized fluid flowing with steady-state flow in a long cylindrical pipe of radius a, so that particles at distance r from the long axis are moving parallel to the axis with speed $|\mathbf{v}| = a^2 - r^2$. Describe this field by a formula for $\mathbf{v}$.

Solution. Let the z-axis lie along the axis of the pipe, with positive direction in the direction of the flow. Then in the usual way, introduce a right-handed Cartesian coordinate system with unit vectors along the axes. By hypothesis, the flow has only a $\mathbf{k}$-component different from zero, so

$$\mathbf{v} = (a^2 - r^2)\mathbf{k} = (a^2 - x^2 - y^2)\mathbf{k}$$

for points inside the pipe. This vector field is not defined outside the cylinder $x^2 + y^2 = a^2$. If we were to draw the velocity vectors at all points in the disc

$$x^2 + y^2 \leq a^2, \qquad z = 0,$$

their tips would describe the surface

$$z = a^2 - r^2$$

(cylindrical coordinates) for $z \geq 0$. This velocity field is illustrated in Figure 17.1. Since this field does not depend on z, a similar figure would illustrate the flow field across any cross section of the pipe made by a plane perpendicular to its axis.

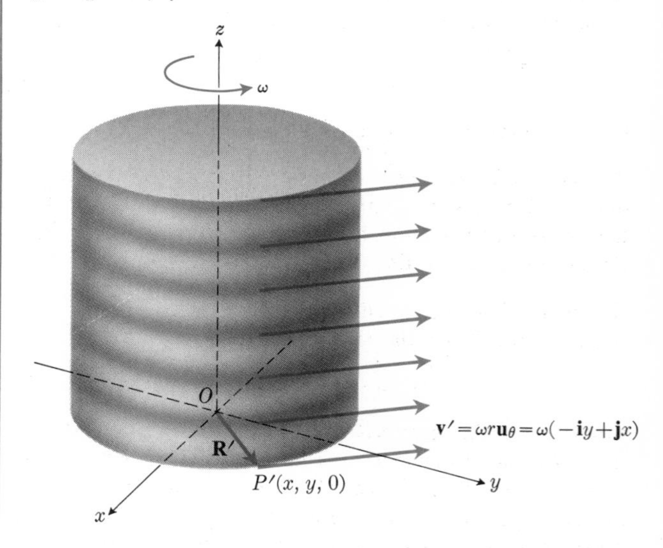

17.2 A steady flow parallel to the xy-plane, with constant angular velocity ω in the positive direction.

Problem 2. In another flow, a fluid is rotating about the z-axis with constant angular velocity ω. Hence every particle at a distance r from the z-axis and in a plane perpendicular to the z-axis traces a circle of radius r, and each such particle has constant speed $|\mathbf{v}| = \omega r$. Describe this field by writing an equation for the velocity at $P(x, y, z)$.

Solution. (See Fig. 17.2.) Each particle travels in a circle parallel to the xy-plane. Therefore it is convenient to begin by looking at the projection of such a circle onto this plane. The point $P(x, y, z)$ in space projects onto the image point $P'(x, y, 0)$, and the velocity vector $\mathbf{v}$ of a particle at P projects onto the velocity vector $\mathbf{v}'$ of a particle at P'. We assume that the motion is in the positive, or counterclockwise, direction, as indicated in the figure. The position vector of P' is $\mathbf{R}' = \mathbf{i}x + \mathbf{j}y$, and the vectors $-\mathbf{i}y + \mathbf{j}x$ and $\mathbf{i}y - \mathbf{j}x$ are both perpendicular to $\mathbf{R}'$. All three of these vectors have magnitude

$$\sqrt{x^2 + y^2} = r.$$

The velocity vector we want has magnitude ωr, is perpendicular to $\mathbf{R}'$, and points in the direction of motion. When x and y are both positive (that is, when the particle is in the first quadrant), the velocity should have a negative **i**-component and a positive **j**-component. The vector that has these properties is

$$\mathbf{v}' = \omega r \mathbf{u}_\theta = \omega(-\mathbf{i}y + \mathbf{j}x). \tag{2a}$$

This formula can be verified for P' in the other three quadrants as well; for example, in the third quadrant both x and y are negative, so Eq. (2a) gives a vector with a positive **i**-component and a negative **j**-component, which is correct. Also, because the motion of P is in a circle parallel to that described by P', and has the same velocity, we have

$$\mathbf{v} = \omega(-\mathbf{i}y + \mathbf{j}x) \tag{2b}$$

for any point in the fluid.

Problem 3. Imagine a fluid that has a velocity vector, at every point in space, that is the sum of a constant velocity vector parallel to the z-axis and a rotational velocity vector given by Eq. (2b). Describe the field.

Solution. Let the constant component parallel to the z-axis be $c\mathbf{k}$. Then the resultant field is

$$\mathbf{v} = \omega(-\mathbf{i}y + \mathbf{j}x) + c\mathbf{k}. \tag{3}$$

Problem 4. The gravitational force field induced at the point $P(x, y, z)$ in space by a mass M that is taken to lie at an origin is defined to be the force with which M would attract a particle of *unit* mass at P. Describe this field mathematically, assuming the inverse-square law.

Solution. Because we are now assuming that both M and the unit mass at P are *point* masses, we don't have to integrate anything; we just write down the force:

$$\mathbf{F} = \frac{GM(1)}{|\overrightarrow{OP}|^2}\mathbf{u}, \tag{4a}$$

where G is the gravitational constant, and

$$\mathbf{u} = -\overrightarrow{OP}/|\overrightarrow{OP}|$$

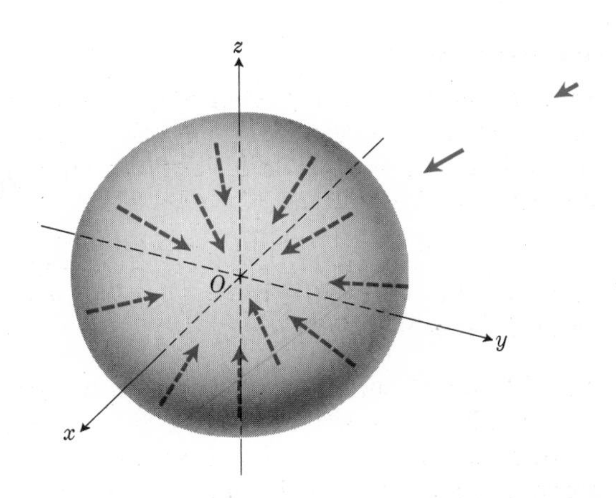

17.3 Some of the vectors of the gravitational field, Problem 4.

is a unit vector directed *from P toward O*. The position vector of P is $\overrightarrow{OP} = \mathbf{i}x + \mathbf{j}y + \mathbf{k}z$, so we find that

$$\mathbf{F} = \frac{-GM(\mathbf{i}x + \mathbf{j}y + \mathbf{k}z)}{(x^2 + y^2 + z^2)^{3/2}} \tag{4b}$$

gives the gravitational force field in question. Its graph would consist of infinitely many vectors, one starting from each point P (except the origin), and pointing straight toward the origin. If P is near the origin, the associated vector is longer than for points farther away from O. For points P on a ray through the origin, the **F**-vectors would lie along that same ray, and decrease in length in proportion to the square of the distance from O. Figure 17.3 is a partial representation of this field. As you look at the figure, however, you must also imagine that an **F**-vector is attached to *every* point $P \neq O$, and not just to those shown. At points on the surface of the sphere $|OP| = a$, the vectors all have the same length, and all point toward the center of the sphere.

Mathematically, a vector field $\mathbf{F}(x, y, z)$ need not be a velocity field or a force field. One easy way to construct another kind of vector field is to apply the gradient operator to a scalar function.

Problem 5. Suppose that the temperature T at each point $P(x, y, z)$ in some region of space is

$$T = 100 - x^2 - y^2 - z^2, \tag{5a}$$

and that $\mathbf{F}(x, y, z)$ is defined to be the gradient of T:

$$\mathbf{F} = \nabla T. \tag{5b}$$

Find this vector field and discuss some of its properties.

Solution. From Eq. (1) of Article 15.6, we have

$$\mathbf{F} = \nabla T = \text{grad } T = \mathbf{i}\frac{\partial T}{\partial x} + \mathbf{j}\frac{\partial T}{\partial y} + \mathbf{k}\frac{\partial T}{\partial z}$$

$$= -2x\mathbf{i} - 2y\mathbf{j} - 2z\mathbf{k}$$

$$= -2\mathbf{R}, \tag{5c}$$

where

$$\mathbf{R} = \overrightarrow{OP} = \mathbf{i}x + \mathbf{j}y + \mathbf{k}z$$

is the position vector of $P(x, y, z)$. This field is like a central force field, all vectors $\mathbf{F}$ being directed toward the origin. At points on a sphere with $|\overrightarrow{OP}|$ equal to a constant, the magnitude of the field vectors is a constant equal to twice the radius of the sphere. So, to represent the field, we could construct any sphere with center at O and draw a vector from any point P on the surface straight through O to the other side of the sphere. The collection of all such vectors, for points in the domain of the function T of Eq. (5a), constitutes the *gradient field* of this particular scalar function.

Remark. An *isothermal* surface for Eq. (5a) is any surface on which T is constant. For Problem 5, such a surface would be any sphere with center at the origin and radius $\sqrt{100 - T}$. Our calculation of $\mathbf{F} = \nabla T = -2\mathbf{R}$ has verified that the gradient of T at P is *normal* to the isothermal surface through P, because the diameter of such a spherical surface is always normal to the surface. [See Fig. 15.10(b) for the general picture of grad f normal to the surface $f =$ constant.]

EXERCISES

1. In Problem 1, where is the velocity (a) the greatest? (b) the least?

2. Suppose the density of the fluid is $\delta =$ constant at $P(x, y, z)$ in Problem 1. Explain why the double integral

$$\int_0^a \int_0^{2\pi} \delta(a^2 - r^2) r \, d\theta \, dr$$

represents the *mass transport* (amount of mass per unit of time) flowing across the surface

$$x^2 + y^2 \leq a^2, \qquad z = 0.$$

Evaluate the integral.

3. In Problem 2, the position vector of P is

$$\mathbf{R} = \mathbf{R}' + z\mathbf{k}.$$

Show that for the motion described, it is correct to say that

$$\frac{d\mathbf{R}}{dt} = \mathbf{v} = \frac{d\mathbf{R}'}{dt} = \mathbf{v}'.$$

4. Describe, in words, the motion of the fluid discussed in Problem 3. What is the path in space described by a particle of the fluid that goes through the point $A(a, 0, 0)$ at time $t = 0$? Prove your result by integrating the vector equation

$$\frac{d\mathbf{R}}{dt} = \omega(-\mathbf{i}y + \mathbf{j}x) + c\mathbf{k}.$$

You may find cylindrical coordinates helpful.

5. In Problem 4, suppose that the mass M is taken to be at the point (x_0, y_0, z_0) rather than at the origin. How should Eq. (4b) be modified to describe this new gravitational force field?

In Exercises 6 through 10, find the gradient fields $\mathbf{F}(x, y, z) = \nabla f$ for the given functions f.

6. $f(x, y, z) = x^2 \exp(2y + 3z)$
7. $f(x, y, z) = \ln(x^2 + y^2 + z^2)$
8. $f(x, y, z) = \tan^{-1}(xy/z)$
9. $f(x, y, z) = 2x - 3y + 5z$
10. $f(x, y, z) = (x^2 + y^2 + z^2)^{n/2}$

17.2 SURFACE INTEGRALS

In Article 16.9, we found how to calculate the area of a surface in space by projecting it onto one of the coordinate planes and integrating a suitable function over this shadow region. For a surface described by

$$z = f(x, y) \tag{1}$$

that projects into a region R in the xy-plane, we found the surface area to be

$$S = \iint_R \sqrt{1 + f_x^2 + f_y^2} \, dx \, dy. \tag{2}$$

The development there also showed that we could interpret the integrand

$$g(x, y) = \sqrt{1 + f_x^2 + f_y^2} \tag{3a}$$

as the amount by which we need to multiply the area of a small portion of R to obtain the area of the corresponding small portion of a tangent plane approximating the surface at $P(x, y, f(x, y))$. In such cases, it is customary to use differential notation and simply write

$$d\sigma = g(x, y)\, dA, \tag{3b}$$

where we consider $d\sigma$ to be an element of surface area in the tangent plane that approximates the corresponding portion $\Delta\sigma$ of the surface itself. This is precisely analogous to taking

$$ds = \sqrt{1 + f_x^2}\, dx$$

as an approximation of a portion Δs of arc length of a curve. Furthermore, just as $\int ds$ gives the length of a curve, so $\iint d\sigma$ gives the area of a surface. The dA in Eq. (3b) represents an element of area $dx\, dy$, or $r\, d\theta\, dr$, in the xy-plane, and Eq. (2) could be written as

$$\iint_{\Sigma} d\sigma = \iint_{R} g(x, y)\, dA. \tag{4}$$

The symbol $\sum$ under the integrals on the left-hand side in Eq. (4) symbolizes the surface in space over which we integrate, just as R on the right-hand side symbolizes that region of the xy-plane over which the actual calculations are made according to Eq. (2).

Exercise 7 of the Review Questions and Exercises, Chapter 16, asked you to formulate and use the concept of a surface integral of a slightly more general form than Eq. (4). This exercise has particular value because we often want to evaluate such an integral as

$$\iint_{\Sigma} h(x, y, z)\, d\sigma, \tag{5}$$

say, and understand its meaning and applications. For example, if h gives the charge density per unit area in some electrostatic field, then this integral could be interpreted as total charge on the surface $\sum$; or, if h represents the amount of fluid flowing in the direction of the normal to $\sum$ at $P(x, y, z)$ per unit area, per unit time, then the integral could be interpreted as the total fluid flow across $\sum$ per unit time. Many other interpretations are possible, some of which will be brought out in problems and exercises. You may encounter others in your studies in physics or engineering, or in other courses in mathematics.

Definition of the surface integral as a limit of sums

Although you should have no difficulty interpreting and using the surface integral indicated by (5), we ought first to give a mathematical definition of it. As in a discussion of surface area, therefore, we consider a surface consisting of those points $P(x, y, z)$ whose coordinates satisfy

$$z = f(x, y) \quad \text{for} \quad (x, y) \in R, \tag{6}$$

where R is a closed, bounded region of the xy-plane. We assume that f and its first partial derivatives f_x and f_y are continuous functions throughout R and on its boundary. We subdivide the region R into a finite number N of nonoverlapping subregions. We can do this, for example, by lines parallel to the y-axis spaced Δx apart, and lines parallel to the x-axis spaced Δy apart. When these subregions of R are projected vertically upward on the surface $\sum$ (see Fig. 17.4), they induce a subdivision of $\sum$ into N subregions. Let the areas of these subregions be numbered

$$\Delta\sigma_1, \Delta\sigma_2, \ldots, \Delta\sigma_N. \tag{7}$$

Next, we suppose that a point $P_i(x_i, y_i, z_i)$ is chosen on the surface in the ith subregion, and the product

$$h(x_i, y_i, z_i)\, \Delta\sigma_i$$

is formed for each i from 1 through N. Now consider the sum

$$\sum_{i=1}^{N} h(x_i, y_i, z_i)\, \Delta\sigma_i. \tag{8}$$

If such sums as (8) have a common limit L as the number N tends to infinity and the largest dimension of the subregions of R tends to zero, independently of the way in which the points P_i are selected within

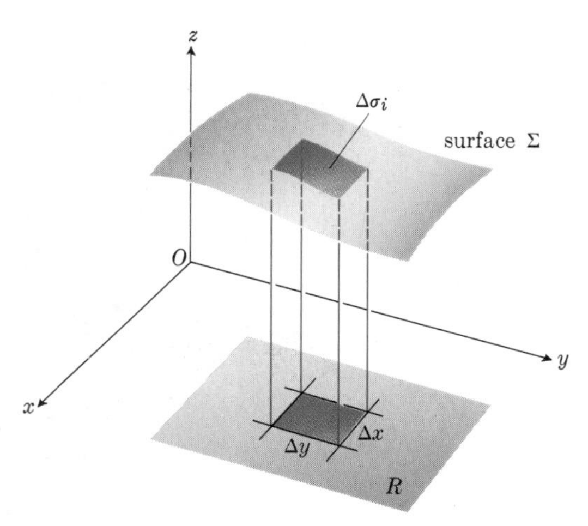

17.4 The surface $\sum$ in space projects onto the region R in the xy-plane. A subregion of $\sum$ with area $\Delta\sigma_i$ projects onto a subregion of R with area $\Delta x\,\Delta y$:

$$\iint_\Sigma h(x, y, z)\,d\sigma = \lim \sum h(x_i, y_i, z_i)\,\Delta\sigma_i.$$

the subregions $\Delta\sigma_i$ on the surface, then that limit is called the *surface integral* of h over $\sum$, and it is represented by the notation (5). Although we shall not prove it here, it is true that for surfaces $\sum$ of the type described, and for continuous functions h, the limit does exist as specified. We now turn to the practical matter of evaluating such surface integrals.

Evaluation of a surface integral

Continuing with the assumptions about $\sum$ and h just mentioned, we see that we can replace the area $\Delta\sigma_i$ in the sum (8), approximately, by

$$g(x_i, y_i)\,\Delta A_i,$$

where g is the function defined by Eq. (3a) and ΔA_i is the area of the ith subregion of R. For those subregions that are completely inside R, away from the boundary, we could take $\Delta A_i = \Delta x\,\Delta y$; but for subregions that are along the boundary, we might (at least in theory) compute ΔA_i itself by a separate double integration in the xy-plane. However, as $\Delta x \to 0$ and $\Delta y \to 0$, if the boundary of R is a curve of finite arc length, these broken subregions can in fact simply be ignored. We do this, and let N denote the number of those complete little rectangles that lie inside the boundary of R. The sum (8) is thus to be replaced by a sum of the form

$$\sum_{i=1}^{N} h(x_i, y_i, z_i)g(x_i, y_i)\,\Delta x\,\Delta y,$$

with $z_i = f(x_i, y_i)$, and g as given by Eq. (3a). Let Δx and Δy approach zero in this last sum: Then the approximations we have used for $\Delta\sigma_i$ become more and more accurate, and the limit is the same as

$$\iint_R h[x, y, f(x, y)]g(x, y)\,dx\,dy, \tag{9}$$

$$g(x, y) = \sqrt{1 + f_x^2 + f_y^2}.$$

As a practical procedure for evaluating the surface integral (5), replace z by its value $f(x, y)$ on the surface, and replace $d\sigma$ by

$$d\sigma = \sqrt{1 + f_x^2 + f_y^2}\,dA,$$

where dA is $dx\,dy$, $dy\,dx$, or $r\,d\theta\,dr$, and evaluate the resulting double integral over the region R in the xy-plane into which $\sum$ projects.

Problem 1. Evaluate $\iint z\,d\sigma$ over the hemisphere

$$z = \sqrt{a^2 - x^2 - y^2}, \qquad x^2 + y^2 \le a^2.$$

Solution

$$\frac{\partial z}{\partial x} = \frac{-x}{z}, \qquad \frac{\partial z}{\partial y} = \frac{-y}{z}.$$

Hence

$$\begin{aligned} d\sigma &= \sqrt{1 + (x^2/z^2) + (y^2/z^2)}\,dA \\ &= \frac{a}{z}\,dA, \end{aligned}$$

because $x^2 + y^2 + z^2 = a^2$ on $\sum$. Since the integrand is $z\,d\sigma = a\,dA$, we can omit substituting the radical expression for z, and simply get

$$\begin{aligned} \iint_\Sigma z\,d\sigma &= \iint_R a\,dA \\ &= a(\pi a^2) = \pi a^3. \end{aligned}$$

We skip the detailed evaluation of the double integral of dA over the interior of the circle $r = a$ in the plane, because we know it is just the area πa^2.

Problem 2. Evaluate $\iint (x^2 + y^2)\,d\sigma$ over the hemisphere $\sum$ of Problem 1.

Solution. From Problem 1, we have

$$d\sigma = \frac{a}{z}\,dA,$$

and so our integral is

$$\iint_{\Sigma} (x^2 + y^2)\,d\sigma = \iint_{R} (x^2 + y^2)(a/z)\,dA,$$

with

$$z = \sqrt{a^2 - x^2 - y^2}.$$

This integral is obviously a candidate for polar coordinates, with

$$x^2 + y^2 = r^2, \quad dA = r\,dr\,d\theta, \quad z = \sqrt{a^2 - r^2}.$$

Thus we obtain

$$\iint_{\Sigma} (x^2 + y^2)\,d\sigma = \int_0^{2\pi} \int_0^a \frac{ar^3\,dr\,d\theta}{(a^2 - r^2)^{1/2}}.$$

The r-integration is done using the substitutions

$$u = (a^2 - r^2)^{1/2}, \qquad u^2 = a^2 - r^2,$$
$$r^2 = a^2 - u^2, \qquad r\,dr = -u\,du,$$

so that

$$\int r^3(a^2 - r^2)^{-1/2}\,dr = -\int (a^2 - u^2)u^{-1}u\,du$$
$$= -a^2u + \tfrac{1}{3}u^3.$$

If $r = 0$, then $u = a$, and if $r = a$, then $u = 0$; and with another two or three simple steps, we get

$$2\pi a(a^3 - \tfrac{1}{3}a^3) = \tfrac{4}{3}\pi a^4$$

as the final answer.

Remark 1. There is a very easy way to check this result: If we were to integrate $x^2 + y^2$ over the entire sphere, we should get twice what we get for the integral over the top half. Also, because the sphere has so much symmetry, we see that

$$\iint x^2\,d\sigma = \iint y^2\,d\sigma = \iint z^2\,d\sigma$$
$$= \tfrac{1}{3}\iint (x^2 + y^2 + z^2)\,d\sigma,$$

if all integrals are extended over the entire surface of the sphere. But we can easily evaluate the last of these four integrals without any actual integra-

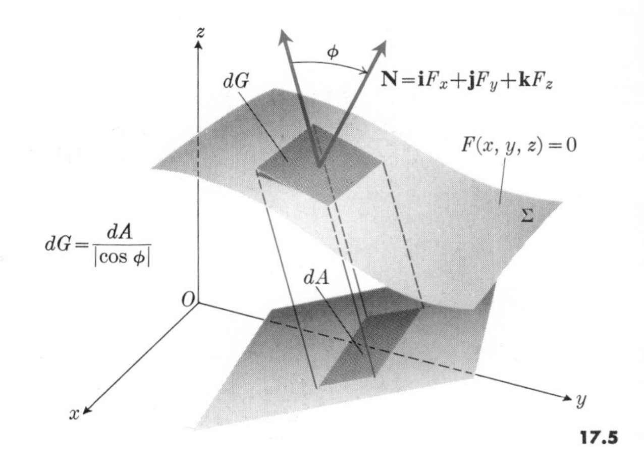

17.5

tion, because $x^2 + y^2 + z^2 = a^2$ is constant on the sphere: We have

$$\iint a^2\,d\sigma = a^2(4\pi a^2) = 4\pi a^4.$$

For Problem 2, we have to multiply this by $\frac{1}{3}$, then by 2 (for $x^2 + y^2$), and finally by $\frac{1}{2}$, because we want the value of the integral over a hemisphere. The result of this multiplication is $\frac{4}{3}\pi a^4$, which agrees with our earlier calculation.

Remark 2. If we refer back to Eqs. (10) and (11b), Article 16.9, we see that when the equation of the surface is $F(x, y, z) = 0$, we can let

$$\mathbf{N} = \mathbf{i}F_x + \mathbf{j}F_y + \mathbf{k}F_z$$

be a normal to $\sum$, and then take $d\sigma = dA/|\cos\phi|$, where ϕ is an angle between $\mathbf{N}$ and a vector normal to the plane onto which $\sum$ is projected. If we project $\sum$ into the xy-plane, we then take dA to be equal to $dx\,dy$, and ϕ to be the angle between $\mathbf{N}$ and $\mathbf{k}$, so that

$$d\sigma = (\sqrt{F_x^2 + F_y^2 + F_z^2}/|F_z|)\,dx\,dy. \tag{10}$$

(See Fig. 17.5.) To evaluate the surface integral

$$\iint_{\Sigma} h(x, y, z)\,d\sigma,$$

we use the equation of the surface to eliminate z in the integrand and in Eq. (10), thereby obtaining a function of x and y to be integrated, as usual, over the projection in the xy-plane. Two other equations can be derived from Eq. (10) by simply permuting the letters x, y, and z. These can be used whenever it is easier to work with the projection of $\sum$ into one of the other coordinate planes.

EXERCISES

In Exercises 1 through 3, explain briefly, with reference to the definition of the surface integral as a limit, why the stated result should be true if c is a constant and F and G are continuous functions.

1. $\iint\limits_{\Sigma} c\, d\sigma = c \times (\text{area of } \sum)$

2. $\iint\limits_{\Sigma} [F(x, y, z) + G(x, y, z)]\, d\sigma$
$$= \iint\limits_{\Sigma} F(x,y, z)\, a\sigma + \iint\limits_{\Sigma} G(x, y, z)\, d\sigma$$

3. $\iint\limits_{\Sigma} cF(x, y, z)\, d\sigma = c\iint\limits_{\Sigma} F(x, y, z)\, d\sigma$

4. Suppose that the surface of the hemisphere in Problem 1 is subdivided by arcs of great circles on which the spherical coordinate θ remains constant (meridians of longitude), and by circles parallel to the xy-plane on which ϕ remains constant (parallels of latitude). Let the angular spacings be $\Delta\theta$ and $\Delta\phi$, respectively. Express the integral of Problem 1 in the form
$$\lim_{\substack{\Delta\theta\to 0\\ \Delta\phi\to 0}} \sum F(\theta, \phi)\, \Delta\theta\, \Delta\phi = \iint F(\theta, \phi)\, d\theta\, d\phi,$$
with appropriate limits of integration, and evaluate.

In Exercises 5 and 6, let $h(x, y, z) = x + y + z$ and let $\sum$ be the portion of the plane $z = 2x + 3y$ for which $x \geq 0$, $y \geq 0$, $x + y \leq 2$.

5. Evaluate $\iint_{\Sigma} h\, d\sigma$ by projecting $\sum$ into the xy-plane. Sketch the projection.

6. Evaluate $\iint_{\Sigma} h\, d\sigma$ by projecting $\sum$ into the yz-plane. Sketch the projection and its boundaries.

In Exercises 7 through 10, you are asked to evaluate integrals of the form
$$\iint\limits_{\Sigma} \mathbf{F}\cdot\mathbf{n}\, d\sigma$$
for specific vector fields $\mathbf{F}$, given that $\sum$, lying in the first octant, is one-eighth of the sphere $x^2 + y^2 + z^2 = a^2$, and $\mathbf{n}$ is a unit vector normal to $\sum$ and pointing away from the origin. (Thus both $\mathbf{F}$ and $\mathbf{n}$ are vector functions of position on the sphere.)

7. $\mathbf{F} = \mathbf{n}$
8. $\mathbf{F} = -\mathbf{i}y + \mathbf{j}x$
9. $\mathbf{F} = z\mathbf{k}$
10. $\mathbf{F} = \mathbf{i}x + \mathbf{j}y$

17.3 LINE INTEGRALS

Suppose that C is a directed curve in space from A to B, and that $w = w(x, y, z)$ is a scalar function of position that is continuous in a region D that contains C. Figure 17.6 illustrates such a directed curve. It is the locus of points (x, y, z) such that
$$x = f(t), \quad y = g(t), \quad z = h(t), \quad t_A \leq t \leq t_B. \quad (1)$$

We assume that the functions f, g, h are continuous and have bounded and piecewise-continuous first derivatives on $[t_A, t_B]$. It is a theorem of higher mathematics that the object to be defined below, $\int_C w\, ds$, does not actually depend on the particular parameterization of C: All parameterizations satisfying the stated hypotheses on f, g, h give the same answer. Indeed, it is possible to define the integral

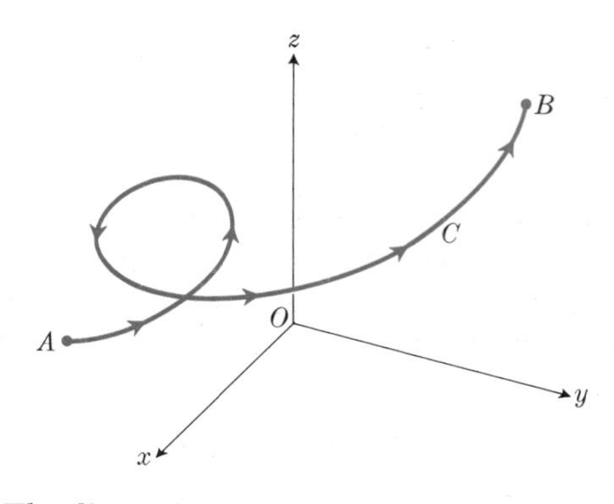

17.6 The directed curve C from A to B.

without reference to parametric equations, but evaluations of line integrals are almost always carried out in terms of some parameterization or other. So we proceed as follows:

We divide the interval $[t_A, t_B]$ into N subintervals of lengths

$$\Delta t_1, \Delta t_2, \ldots, \Delta t_N.$$

The points of subdivision also correspond to points on C which divide it into subarcs of lengths

$$\Delta s_1, \Delta s_2, \ldots, \Delta s_N.$$

Let $P_i(x_i, y_i, z_i)$ be an arbitrary point on the ith subarc, and form the sum

$$\sum_{i=1}^{N} w(x_i, y_i, z_i)\,\Delta s_i. \tag{2}$$

If these sums have a limit L, as $N \to \infty$ and the largest $\Delta t_i \to 0$, and if this limit is the same for *all* ways of subdividing $[t_A, t_B]$ and all choices of the points P_i, then we call this limit the *line integral* of w along C from A to B, and express it by the notation

$$\int_C w\,ds = \lim \sum_{i=1}^{N} w(x_i, y_i, z_i)\,\Delta s_i. \tag{3a}$$

It is also a theorem that under the stated hypotheses on w and on C, the line integral exists, and is the same as

$$\int_{t_A}^{t_B} w[f(t), g(t), h(t)]\sqrt{[f'(t)]^2 + [g'(t)]^2 + [h'(t)]^2}\,dt. \tag{3b}$$

Although we shall not prove this theorem, we shall show how line integrals are evaluated in practice, and illustrate some of their physical applications. Although Eq. (3b) looks quite complicated, it is just what one would naturally write if he simply substituted the parametric equations for x, y, z in w and used the standard formula for ds. In our first problem, we use a curve in the xy-plane for which the formula simplifies a bit.

Problem 1. Let C be the line segment from $A(0, 0)$ to $B(1, 1)$, and let $w = x + y^2$. Evaluate $\int_C w\,ds$ for two different parameterizations of C.

Solution 1. If we let

$$x = t \quad\text{and}\quad y = t, \quad 0 \le t \le 1,$$

then we get

$$\int_C w\,ds = \int_0^1 (t + t^2)\sqrt{1+1}\,dt$$

$$= \sqrt{2}\left[\frac{t^2}{2} + \frac{t^3}{3}\right]_0^1 = \frac{5\sqrt{2}}{6}.$$

Solution 2. As a second parameterization of the given segment of the line $y = x$, we let

$$x = \sin t \quad\text{and}\quad y = \sin t, \quad 0 \le t \le \pi/2,$$

and get

$$\int_C w\,ds = \int_0^{\pi/2} (\sin t + \sin^2 t)\sqrt{2\cos^2 t}\,dt$$

$$= \sqrt{2}\left[\frac{\sin^2 t}{2} + \frac{\sin^3 t}{3}\right]_0^{\pi/2} = \frac{5\sqrt{2}}{6}.$$

Remark 1. The line integral of Problem 1 can be interpreted as the area of the region R that lies in the plane $y = x$, above the plane $w = 0$, and under the surface $w = x + y^2$. We can, in fact, introduce an s-axis along the intersection of the planes $w = 0$ and $y = x$, as shown in Fig. 17.7. The s-coordinate of the point $P(x, x, 0)$ between $A(0, 0, 0)$ and $B(1, 1, 0)$ on C is

$$s = \sqrt{x^2 + x^2} = x\sqrt{2}.$$

Thus we can use the arc length s itself as parameter for C:

$$x = s/\sqrt{2}, \quad y = s/\sqrt{2}, \quad 0 \le s \le \sqrt{2}.$$

The upper boundary of R is the intersection of the surface $w = x + y^2$ and the plane $y = x$. In the sw-plane, this curve has equation

$$w = \frac{s}{\sqrt{2}} + \frac{s^2}{2},$$

and the area of R is given by

$$\int_C w\,ds = \int_0^{\sqrt{2}} \left(\frac{s}{\sqrt{2}} + \frac{s^2}{2}\right) ds$$

$$= \frac{s^2}{2\sqrt{2}} + \frac{s^3}{6}\Bigg]_0^{\sqrt{2}} = \frac{1}{\sqrt{2}} + \frac{2\sqrt{2}}{6} = \frac{5\sqrt{2}}{6}.$$

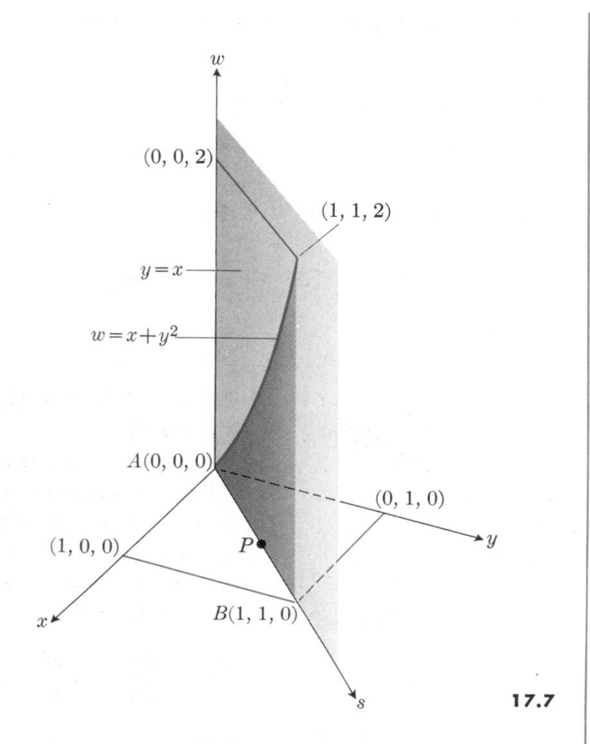

17.7

Whenever C is a *plane* curve, we can interpret the line integral $\int_C w\,ds$ as the area of a portion of a cylinder (or as the difference between areas above and below C, if w has both positive and negative values along C). Even if C is curved, one can at least think of measuring the distance s along C, starting with $s = 0$ at A and increasing to $s = l$ at B, where l is the length of arc of C from A to B. Because s increases along C, there is one and only one point on C for any value of $s \in [0, l]$, and we can imagine the parametric equations of C being given in terms of s; say,

$$x = x(s), \quad y = y(s), \quad z = z(s), \quad 0 \leq s \leq l.$$

Then, along C, $w = w[x(s), y(s), z(s)]$ is a function of s: $w = \phi(s)$. The way in which the line integral is defined as a limit of sums ensures that $\int_C w\,ds$ is just the integral of this function ϕ with respect to s:

$$\int_C w\,ds = \int_0^l \phi(s)\,ds. \tag{4}$$

Equation (4) also suggests that we can think of the cylinder as having the plane curve C as base, and the curve $w = \phi(s)$ as upper-boundary. We can then flatten the cylinder so that C becomes the segment $[0, l]$ of the s-axis in an sw-plane. Figure 17.8 illustrates this idea.

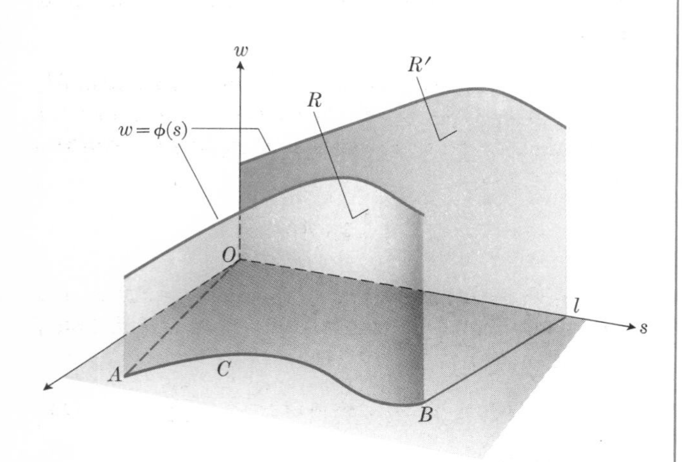

17.8 The region R above the curve C on the cylindrical surface maps onto the region R' in the sw-plane.

Line integrals and work

If the point of application of a force

$$\mathbf{F} = \mathbf{i}M(x, y, z) + \mathbf{j}N(x, y, z) + \mathbf{k}P(x, y, z) \tag{5}$$

moves along a curve C from a point $A(a_1, a_2, a_3)$ to a point $B(b_1, b_2, b_3)$, then the work done by the force is

$$W = \int_C \mathbf{F} \cdot d\mathbf{R}, \tag{6}$$

where

$$\mathbf{R} = \mathbf{i}x + \mathbf{j}y + \mathbf{k}z \tag{7}$$

is the vector from the origin to the point (x, y, z), and

$$d\mathbf{R} = \frac{d\mathbf{R}}{ds}\,ds = \mathbf{i}\,dx + \mathbf{j}\,dy + \mathbf{k}\,dz. \tag{8}$$

If we calculate the dot product of the vectors $\mathbf{F}$ and $d\mathbf{R}$ from Eqs. (5) and (8), then we may write Eq. (6)

in the alternative form

$$W = \int_C M\,dx + N\,dy + P\,dz, \tag{9}$$

where M, N, and P are functions of x, y, and z, and the subscript C on the integral refers to the curve C along which the integral is taken. Such an integral as Eq. (9) is also called a line integral (*curve integral* would perhaps be a more descriptive name). Because

$$\frac{d\mathbf{R}}{ds} = \mathbf{T} = \text{unit tangent vector},$$

Eq. (6) is just another way of saying that the work is the value of the line integral along C of the *tangential component* of the force field $\mathbf{F}$. This tangential component is a scalar function, say w, of position along C:

$$w = \mathbf{F} \cdot \mathbf{T} = w(x, y, z).$$

Therefore

$$W = \int_C \mathbf{F} \cdot d\mathbf{R} = \int_C \mathbf{F} \cdot \mathbf{T}\,ds = \int_C w(x, y, z)\,ds.$$

This is just the kind of line integral defined by Eq. (3a), but the form of the integral in Eq. (9) suggests another way of evaluating it.

To evaluate the integral in (9) we might express the equations of C in terms of a parameter t:

$$x = x(t), \quad y = y(t), \quad z = z(t), \tag{10}$$

such that the curve is described from A to B as t varies from a value t_1 to a value t_2. Then *all* quantities in the integral may be expressed in terms of one variable t and the result evaluated in the usual manner as a definite integral with respect to t from t_1 to t_2. In general, the value of the integral depends on the path C as well as on its endpoints.

Problem 2. Suppose the force is given by

$$\mathbf{F} = \mathbf{i}(x^2 - y) + \mathbf{j}(y^2 - z) + \mathbf{k}(z^2 - x),$$

and that its point of application moves from the origin O to the point A (1, 1, 1)

(a) along the straight line OA, and

(b) along the curve

$$x = t, \quad y = t^2, \quad z = t^3, \quad 0 \le t \le 1.$$

Find the work done in the two cases.

Solution. (a) Equations for the line OA are

$$x = y = z.$$

The integral to be evaluated is

$$W = \int_C (x^2 - y)\,dx + (y^2 - z)\,dy + (z^2 - x)\,dz,$$

which, for the path OA, becomes

$$W = \int_0^1 3(x^2 - x)\,dx = -\tfrac{1}{2}.$$

(b) Along the curve, we get

$$W = \int_0^1 2(t^4 - t^3)t\,dt + 3(t^6 - t)t^2\,dt = -\tfrac{29}{60}.$$

Now under certain conditions, the line integral between two points A and B is independent of the path C joining them. That is, the integral in Eq. (6) has the same value for any two paths C_1 and C_2 joining A and B. This happens when the force field $\mathbf{F}$ is a *gradient field*, that is, if and only if

$$\mathbf{F}(x, y, z) = \nabla f = \mathbf{i}\frac{\partial f}{\partial x} + \mathbf{j}\frac{\partial f}{\partial y} + \mathbf{k}\frac{\partial f}{\partial z},$$

for some differentiable function f. We state this as a formal theorem and prove the sufficiency and necessity of the conditions, with some interpolated remarks.

Theorem 1. *Let* $\mathbf{F}$ *be a vector field with components* M, N, P *that are continuous throughout some connected region* D. *Then a necessary and sufficient condition for the integral*

$$\int_A^B \mathbf{F} \cdot d\mathbf{R}$$

to be independent of the path joining the points A *and* B *in* D *is that there exist a differentiable function* f *such that*

$$\mathbf{F} = \nabla f = \mathbf{i}\frac{\partial f}{\partial x} + \mathbf{j}\frac{\partial f}{\partial y} + \mathbf{k}\frac{\partial f}{\partial z} \tag{11}$$

throughout D.

Proof. *Sufficiency.* First, we suppose that Eq. (11) is satisfied, and then consider A and B to be two points in D (see Fig. 17.9). Suppose that C is any piecewise

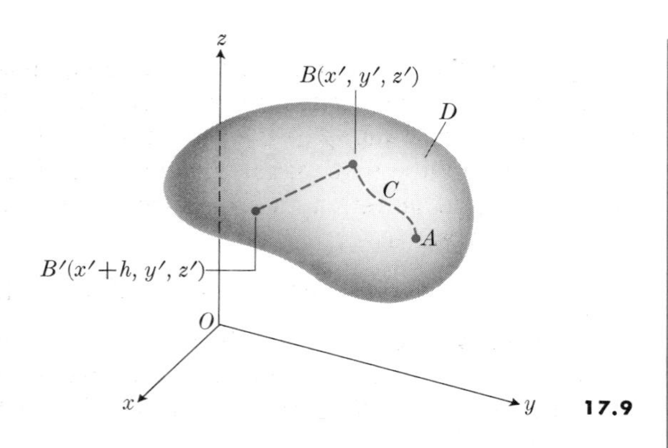

17.9

smooth curve joining A and B:

$$x = x(t), \quad y = y(t), \quad z = z(t), \quad t_1 \leq t \leq t_2.$$

Along C, $f = f[x(t), y(t), z(t)]$ is a function of t to which we may apply the chain rule to differentiate with respect to t:

$$\begin{aligned}\frac{df}{dt} &= \frac{\partial f}{\partial x}\frac{dx}{dt} + \frac{\partial f}{\partial y}\frac{dy}{dt} + \frac{\partial f}{\partial z}\frac{dz}{dt} \\ &= \nabla f \cdot \left(\mathbf{i}\frac{dx}{dt} + \mathbf{j}\frac{dy}{dt} + \mathbf{k}\frac{dz}{dt}\right) \\ &= \nabla f \cdot \frac{d\mathbf{R}}{dt}. \end{aligned} \tag{12a}$$

Because Eq. (11) holds, we also have

$$\mathbf{F}\cdot d\mathbf{R} = \nabla f \cdot d\mathbf{R} = \nabla f \cdot \frac{d\mathbf{R}}{dt}\,dt = \frac{df}{dt}\,dt. \tag{12b}$$

We now use this result to integrate $\mathbf{F}\cdot d\mathbf{R}$ along C from A to B:

$$\begin{aligned}\int_C \mathbf{F}\cdot d\mathbf{R} &= \int_{t_1}^{t_2} \frac{df}{dt}\,dt \\ &= \int_{t_1}^{t_2} \frac{d}{dt} f[x(t), y(t), z(t)]\,dt \\ &= f[x(t), y(t), z(t)]\Big]_{t_1}^{t_2} \\ &= f[x(t_2), y(t_2), z(t_2)] \\ &\quad - f[x(t_1), y(t_1), z(t_1)] \\ &= f(B) - f(A).\end{aligned}$$

Therefore, if $\mathbf{F} = \nabla f$, we have the result

$$\int_A^B \mathbf{F}\cdot d\mathbf{R} = \int_A^B \nabla f \cdot d\mathbf{R} = f(B) - f(A). \tag{13}$$

The value of the integral, $f(B) - f(A)$, does not depend on the path C at all. Equation (13) is the space analog of the First Fundamental Theorem of Integral Calculus (see Article 5.9):

$$\int_a^b f'(x)\,dx = f(b) - f(a).$$

The only difference is that we have $\nabla f \cdot d\mathbf{R}$ in place of $f'(x)\,dx$. This analogy also suggests that perhaps there is also a space analog of the Second Fundamental Theorem of Integral Calculus (again, see Article 5.9). In other words, if we define a function f by the rule

$$f(x', y', z') = \int_A^{(x',y',z')} \mathbf{F}\cdot d\mathbf{R}, \tag{14a}$$

perhaps it will be true that

$$\nabla f = \mathbf{F}. \tag{14b}$$

Equation (14b) is indeed true when the right-hand side of Eq. (14a) is path-independent, and the proof of this fact will complete our theorem.

Proof. *Necessity.* We now assume that the line integral in (14a) is path-independent, and prove that $\mathbf{F} = \nabla f$ for the function f defined by Eq. (14a). We first write $\mathbf{F}$ in terms of its $\mathbf{i}$-, $\mathbf{j}$-, and $\mathbf{k}$-components:

$$\mathbf{F}(x, y, z) = \mathbf{i}M(x, y, z) + \mathbf{j}N(x, y, z) + \mathbf{k}P(x, y, z), \tag{15}$$

and fix the points A and $B(x', y', z')$ in D. To establish Eq. (14b), we need to show that *the equalities*

$$\frac{\partial f}{\partial x} = M, \qquad \frac{\partial f}{\partial y} = N, \qquad \frac{\partial f}{\partial z} = P \tag{16}$$

hold at each point of D. In what follows, we either assume that D is an open set, so that all of its points are interior points, or we restrict our attention to interior points.

The point $B(x', y', z')$ is the center of some small sphere whose interoir lies entirely inside D. We let $h \neq 0$ be small enough so that all points on the ray from B to $B'(x'+h, y', z')$ lie in D (see Fig. 17.9), and consider the difference quotient

$$\frac{f(x'+h, y', z') - f(x', y', z')}{h} = \frac{1}{h}\int_B^{B'} \mathbf{F}\cdot d\mathbf{R}. \quad (17)$$

Since the integral does not depend on a particular path, we choose one convenient for our purpose:

$$x = x' + th, \quad y = y', \quad z = z', \quad 0 \le t \le 1,$$

along which neither y nor z varies, and along which $d\mathbf{R} = \mathbf{i}\,dx = \mathbf{i}h\,dt$. When this is substituted into Eq. (17), along with $\mathbf{F}$ from Eq. (15), we get

$$\begin{aligned}\frac{f(x'+h, y', z') - f(x', y', z')}{h} &= \frac{1}{h}\int_0^1 M(x'+ht, y', z')h\,dt \\ &= \int_0^1 M(x'+ht, y', z')\,dt. \quad (18)\end{aligned}$$

By hypothesis, $\mathbf{F}$ is continuous, so each component is a continuous function. Thus, given any $\epsilon > 0$, there is a $\delta > 0$ such that

$$|M(x'+ht, y', z') - M(x', y', z')| < \epsilon$$

when $\qquad |ht| < \delta.$

This implies that when $|h| < \delta$, the integral in Eq. (18) also differs from

$$\int_0^1 M(x', y', z')\,dt = M(x', y', z') \quad (19)$$

by less than ϵ. The equality in (19) follows from the fact that the integrand is a constant, and

$$\int_0^1 dt = 1.$$

Therefore, as $h \to 0$ in Eq. (18), the right-hand side has as limit $M(x', y', z')$, so the left-hand side must have the same limit. That, however, is just the partial derivative of f with respect to x at $B(x', y', z')$. Therefore

$$\partial f/\partial x = M \quad (20)$$

holds at each interior point of D.

Equation (20) is the first of the three equalities, Eqs. (16), that are needed to establish Eq. (14b). Proofs of the remaining two equalities in (16) are very similar, and you are asked to prove one of them in Exercise 7. One sets up a difference quotient like Eq. (17), but takes

$$B' = (x', y'+h, z') \quad \text{or} \quad B' = (x', y', z'+h).$$

In other words, from B, one integrates along a path parallel to the y-axis or parallel to the z-axis. On the former path, $d\mathbf{R} = \mathbf{j}h\,dt$, and on the latter, $d\mathbf{R} = \mathbf{k}h\,dt$. □

Problem 3. Find a function f such that

if $$\mathbf{F} = 2x\mathbf{i} + 2y\mathbf{j} + 2z\mathbf{k}, \quad (21a)$$

then $$\mathbf{F} = \nabla f. \quad (21b)$$

Solution. We might be lucky and guess

$$f(x, y, z) = x^2 + y^2 + z^2, \quad (22)$$

because $2x$, $2y$, $2z$ are its partial derivatives with respect to x, y, and z. But if we weren't so inspired, then we might try something like Eq. (14a). In the first place, the functions $2x$, $2y$, $2z$ are everywhere continuous, so the region D can be all of 3-space. Of course we don't know, until *after* we find that $\mathbf{F}$ is a gradient, that the integral in (14a) is path-independent, but we proceed on faith (or at least with hope). The choice of A is up to us, so we make life easy for ourselves by taking $A = (0, 0, 0)$. For the path of integration from A to $B(x', y', z')$, we take the line segment

$$x = x't, \quad y = y't, \quad z = z't, \quad 0 \le t \le 1,$$

along which

$$d\mathbf{R} = (x'\mathbf{i} + y'\mathbf{j} + z'\mathbf{k})\,dt$$

and

$$\begin{aligned}\mathbf{F}\cdot d\mathbf{R} &= (2xx' + 2yy' + 2zz')\,dt \\ &= (2x'^2 + 2y'^2 + 2z'^2)t\,dt.\end{aligned}$$

Therefore, when we substitute into Eq. (14a), we get

$$\begin{aligned}f(x', y', z') &= \int_{(0,0,0)}^{(x',y',z')} \mathbf{F}\cdot d\mathbf{R} \\ &= [x'^2 + y'^2 + z'^2]\int_0^1 2t\,dt \\ &= x'^2 + y'^2 + z'^2.\end{aligned}$$

If we delete the primes, this equation is identical with Eq. (22).

Remark 2. The upper limit of integration in Eq. (14a) is an arbitrary point in the domain of **F**, but we use (x', y', z') to designate it, rather than (x, y, z), because the latter is used for the running point that covers the arc C from A to B during the integration. After we have completed the computation of

$$f(x', y', z'),$$

we then delete the primes to express the result as $f(x, y, z)$. The analog in one dimension would be

$$\ln x' = \int_1^{x'} \frac{1}{x}\,dx.$$

We must be careful not to confuse the variable of integration x with the limit of integration x'. We distinguished between these two things in a slightly different manner in Eq. (1), Article 7.5, where we wrote

$$\ln x = \int_1^x \frac{1}{t}\,dt.$$

Our purpose there was the same as it is here, however: to maintain a notational difference between the variable *upper limit* and the *variable of integration*.

Remark 3. So far we have only the criterion

$$\mathbf{F} = \nabla f$$

for deciding whether

$$\int_A^B \mathbf{F} \cdot d\mathbf{R}$$

is path-independent. We shall discover another criterion in Eqs. (26) below. If we follow the method indicated by Eq. (14a) and illustrated in Problem 3 for a field **F** that is *not* path-independent, then we should discover, on trying to verify that $\mathbf{F} = \nabla f$, that it isn't so. The next problem illustrates exactly this situation.

Problem 4. Show that there is no function f such that

$$\mathbf{F} = \nabla f \quad \text{if} \quad \mathbf{F} = y\mathbf{i} - x\mathbf{j}.$$

Solution. Here's one way to show it: If there were such a function f, then

$$\frac{\partial f}{\partial x} = y, \quad \text{and} \quad \frac{\partial f}{\partial y} = -x,$$

from which we would get

$$\frac{\partial^2 f}{\partial y\,\partial x} = \frac{\partial(y)}{\partial y} = 1 \neq \frac{\partial^2 f}{\partial x\,\partial y} = \frac{\partial(-x)}{\partial x} = -1.$$

But we should have $f_{xy} = f_{yx}$, because

$$f_x = y \quad \text{and} \quad f_y = -x$$

are everywhere continuously differentiable. This contradiction shows that no such f exists.

Another method would be to compute $\int \mathbf{F} \cdot d\mathbf{R}$ between two points, say $A(0, 0, 0)$ and $B(1, 1, 0)$, along two different paths. If the answers turn out to be the same, we haven't proved a thing. But if they turn out to be different, then we know that **F** is not a gradient field. Exercise 5 asks you to do this for two specific paths.

A third method is to proceed blithely with Eq. (14a), and get a function f that satisfies Eq. (14a) for a particular path, but that fails to satisfy $\mathbf{F} = \nabla f$. Once again we would choose an origin $A = (0, 0, 0)$ and let $B = (x', y', z')$, and then integrate along the segment

$$x = x't, \quad y = y't, \quad z = z't, \quad 0 \le t \le 1.$$

We would get

$$\begin{aligned} \mathbf{F} \cdot d\mathbf{R} &= (y\mathbf{i} - x\mathbf{j}) \cdot (\mathbf{i}\,dx + \mathbf{j}\,dy + \mathbf{k}\,dz) \\ &= y\,dx - x\,dy \\ &= (y't)(x'\,dt) - (x't)(y'\,dt) \\ &= t(x'y' - x'y')\,dt = 0\,dt. \end{aligned}$$

Therefore Eq. (14a) produces

$$f(x', y', z') = 0 \quad \text{for all } (x', y', z').$$

This constant function obviously won't have a gradient equal to $y\mathbf{i} - x\mathbf{j}$. In Exercise 6 you are asked to explain why this also means that no other function exists whose gradient is the given **F**.

Exact differentials

When **F** is a force field such that the work integral from A to B is the same for all paths joining them, the field is said to be *conservative*. Theorem 1 therefore shows that a force field is *conservative* if and only if it is a *gradient* field:

$$\mathbf{F} \text{ is conservative} \Leftrightarrow \mathbf{F} = \nabla f. \tag{23}$$

This criterion for a conservative field can also be stated in terms of differentials: **F** is conservative if and only if the integrand in the work integral,

$$\mathbf{F} \cdot d\mathbf{R} = M\,dx + N\,dy + P\,dz, \tag{24a}$$

is an *exact differential*. By this we mean that there is a function f whose total differential is equal to the given integrand:

$$df = M\,dx + N\,dy + P\,dz. \tag{24b}$$

This is not a new result, but only a slightly different way of expressing Eq. (23). We understand this because

$$df = \nabla f \cdot d\mathbf{R} = \frac{\partial f}{\partial x}\,dx + \frac{\partial f}{\partial y}\,dy + \frac{\partial f}{\partial z}\,dz,$$

by definition of the differential df. To say that

$$M\,dx + N\,dy + P\,dz = \frac{\partial f}{\partial x}\,dx + \frac{\partial f}{\partial y}\,dy + \frac{\partial f}{\partial z}\,dz \tag{25a}$$

is the same as saying that

$$M\mathbf{i} + N\mathbf{j} + P\mathbf{k} = \frac{\partial f}{\partial x}\mathbf{i} + \frac{\partial f}{\partial y}\mathbf{j} + \frac{\partial f}{\partial z}\mathbf{k}, \tag{25b}$$

because either equation holds if and only if

$$M = \frac{\partial f}{\partial x}, \qquad N = \frac{\partial f}{\partial y}, \qquad P = \frac{\partial f}{\partial z}. \tag{25c}$$

In Article 15.13, we discussed exact differentials of functions $f(x, y)$ of two variables. The criterion for an exact differential stated in Theorem 1 of that article is easily extended to functions of three or more variables. For functions of three variables, it goes as follows.

Theorem 2. *Let $M(x, y, z)$, $N(x, y, z)$, and $P(x, y, z)$ be continuous, together with their first-order partial derivatives, for all real values of x, y, and z. Then a necessary and sufficient condition for the expression*

$$M\,dx + N\,dy + P\,dz$$

to be an exact differential is that the following equations all be satisfied:

$$\frac{\partial M}{\partial y} = \frac{\partial N}{\partial x}, \qquad \frac{\partial M}{\partial z} = \frac{\partial P}{\partial x}, \qquad \frac{\partial N}{\partial z} = \frac{\partial P}{\partial y}. \tag{26}$$

This theorem is a straightforward extention of Theorem 1, Article 15.13 from the two-dimensional to the three-dimensional case. We shall omit the proof, since it is similar to the proof of the earlier theorem.

Problem 5. Suppose

$$\mathbf{F} = \mathbf{i}(e^x \cos y + yz) + \mathbf{j}(xz - e^x \sin y) + \mathbf{k}(xy + z).$$

Is $\mathbf{F}$ conservative? If so, find f such that $\mathbf{F} = \nabla f$.

Solution. We apply the test of Eqs. (26) to the expression

$$\mathbf{F} \cdot d\mathbf{R} = (e^x \cos y + yz)\,dx + (xz - e^x \sin y)\,dy + (xy + z)\,dz.$$

We let

$$M = e^x \cos y + yz, \qquad N = xz - e^x \sin y, \qquad P = xy + z,$$

and calculate

$$\frac{\partial M}{\partial z} = y = \frac{\partial P}{\partial x}, \qquad \frac{\partial N}{\partial z} = x = \frac{\partial P}{\partial y},$$

$$\frac{\partial M}{\partial y} = -e^x \sin y + z = \frac{\partial N}{\partial x}.$$

The theorem tells us that there is a function $f(x, y, z)$ such that

$$\mathbf{F} \cdot d\mathbf{R} = df.$$

We find f by integrating the system of equations

$$\begin{aligned} \frac{\partial f}{\partial x} &= e^x \cos y + yz, \\ \frac{\partial f}{\partial y} &= xz - e^x \sin y, \\ \frac{\partial f}{\partial z} &= xy + z. \end{aligned} \tag{27}$$

We integrate the first of these with respect to x, holding y and z constant, and add an arbitrary function $g(y, z)$ as the "constant of integration"; we thus obtain

$$f(x, y, z) = e^x \cos y + xyz + g(y, z). \tag{28}$$

Next we differentiate this with respect to y and set it equal to $\partial f/\partial y$ as given by the second of Eqs. (27):

$$xz - e^x \sin y = -e^x \sin y + xz + \frac{\partial g}{\partial y},$$

or

$$\frac{\partial g(y, z)}{\partial y} = 0. \tag{29}$$

Integrating Eq. (29) with respect to y, holding z constant, and adding an arbitrary function $h(z)$ as constant of integration, we obtain

$$g(y, z) = h(z). \tag{30}$$

We substitute this into Eq. (28) and then calculate $\partial f/\partial z$, which we compare with the third of Eqs. (27). We find that

$$xy + z = xy + \frac{dh(z)}{dz},$$

or

$$\frac{dh(z)}{dz} = z,$$

so that

$$h(z) = \tfrac{1}{2}z^2 + C.$$

Hence we may write Eq. (28) as

$$f(x, y, z) = e^x \cos y + xyz + (z^2/2) + C.$$

Then

$$\mathbf{F} = \nabla f.$$

A function $f(x, y, z)$ which has the property that its gradient gives the force vector $\mathbf{F}$ is called a "potential" function. (Sometimes a minus sign is introduced. For example, the electric intensity of a field is the negative of the potential gradient in the field. See Sears, *Electricity and Magnetism*, p. 63.)

EXERCISES

1. In Problem 1, let C be given by
$$x = t^2, \quad y = t^2, \quad 0 \le t \le 1,$$
and evaluate $\int_C w\, ds$ for $w = x + y^2$.
2. In Problem 1, let C be given by $x = f(t) = y$, where $f(0) = 0$ and $f(1) = 1$. Show that if $f'(t)$ is continuous on $[0, 1]$, then $\int_C w\, ds = 5\sqrt{2}/6$, no matter what the particular function f may be.
3. Evaluate $\int \mathbf{F} \cdot d\mathbf{R}$ around the circle
$$x = \cos t, \quad y = \sin t, \quad z = 0, \quad 0 \le t \le 2\pi$$
for the force given in Problem 2.
4. In Problem 3, evaluate $\int \mathbf{F} \cdot d\mathbf{R}$ along a curve C lying on the sphere $x^2 + y^2 + z^2 = a^2$. Do you need to know anything more about C? Why?
5. Assume $\mathbf{F} = y\mathbf{i} - x\mathbf{j}$, as in Problem 4, and take $A = (0, 0, 0)$, $B = (1, 1, 0)$. Evaluate $\int \mathbf{F} \cdot d\mathbf{R}$ for
(a) $x = y = t, \quad 0 \le t \le 1,$
(b) $x = t, \quad y = t^2, \quad 0 \le t \le 1.$
Comment on the meaning of your answers.
6. In Problem 4, when we considered $\mathbf{F} = y\mathbf{i} - x\mathbf{j}$, we found a function $f(x', y', z') = 0$ which expresses the value of the integral
$$\int_{(0,0,0)}^{(x',y',z')} \mathbf{F} \cdot d\mathbf{R}$$
along the line segment from $(0, 0, 0)$ to an arbitrary point (x', y', z'). Using this result, and the first half of the proof of Theorem 1, prove that if $\mathbf{F}$ were a gradient field, say $\mathbf{F} = \nabla g$, then $g - f = $ constant. From this, show that no such g exists for the given $\mathbf{F}$.
7. Using the notations of Eqs. (14a) and (15), show that $\partial f/\partial y = N$ holds at each point of D if $\mathbf{F}$ is continuous and if the integral in (14a) is path-independent in D.
8. Let $\rho = (x^2 + y^2 + z^2)^{1/2}$. Show that
$$\nabla(\rho^n) = n\rho^{n-2}\mathbf{R},$$
where $\mathbf{R} = \mathbf{i}x + \mathbf{j}y + \mathbf{k}z$. Is there a value of n for which $\mathbf{F} = \nabla(\rho^n)$ represents the "inverse-square law" field? If so, what is this value of n?

In Exercises 9 through 13, find the work done by the given force $\mathbf{F}$ as the point of application moves from $(0, 0, 0)$ to $(1, 1, 1)$
(a) along the straight line $x = y = z$,
(b) along the curve $x = t$, $y = t^2$, $z = t^4$, and
(c) along the x-axis to $(1, 0, 0)$, then in a straight line to $(1, 1, 0)$, and from there in a straight line to $(1, 1, 1)$.

9. $\mathbf{F} = 2x\mathbf{i} + 3y\mathbf{j} + 4z\mathbf{k}$
10. $\mathbf{F} = \mathbf{i}x \sin y + \mathbf{j} \cos y + \mathbf{k}(x + y)$
11. $\mathbf{F} = \mathbf{i}(y + z) + \mathbf{j}(z + x) + \mathbf{k}(x + y)$
12. $\mathbf{F} = e^{y+2z}(\mathbf{i} + \mathbf{j}x + 2\mathbf{k}x)$
13. $\mathbf{F} = \mathbf{i}y \sin z + \mathbf{j}x \sin z + \mathbf{k}xy \cos z$

In Exercises 14 through 17, find a function $f(x, y, z)$ such that $\mathbf{F} = \operatorname{grad} f$ for the $\mathbf{F}$ given in the exercise named.

14. Exercise 9
15. Exercise 11
16. Exercise 12
17. Exercise 13

18. Prove that the line integral

$$\int (z^2 dx + 2y\, dy + 2xz\, dz)$$

is independent of the path of integration.

19. If $\mathbf{F} = y\mathbf{i} + x\mathbf{j}$, evaluate the line integral $\int_A^B \mathbf{F} \cdot d\mathbf{R}$ along the straight line from $A(1, 1, 1)$ to $B(3, 3, 3)$.

20. If $\mathbf{F} = \mathbf{i}x^2 + \mathbf{j}yz + \mathbf{k}y^2$, compute $\int_A^B \mathbf{F} \cdot d\mathbf{R}$, where $A = (0, 0, 0)$, $B = (0, 3, 4)$, along the straight line connecting these points.

21. Let C denote the plane curve whose vector equation is

$$\mathbf{r}(t) = e^t \cos t\mathbf{i} + e^t \sin t\mathbf{j}.$$

Evaluate the line integral

$$\int \frac{x\, dx + y\, dy}{(x^2 + y^2)^{3/2}}$$

along that arc of C from the point $(1, 0)$ to the point $(e^{2\pi}, 0)$.

22. If the density $\rho(x, y, z)$ of a fluid is a function of the pressure $p(x, y, z)$, and

$$\phi(x, y, z) = \int_{p_0}^{p} (dp/\rho),$$

where p_0 is constant, show that $\nabla\phi = \nabla p/\rho$.

23. If $\mathbf{F} = y\mathbf{i}$, show that the line integral $\int_A^B \mathbf{F} \cdot d\mathbf{R}$ along an arc AB in the xy-plane is equal to an area bounded by the x-axis, the arc, and the ordinates at A and B.

Remark. Despite similarity of appearance and identity of value, the integral of this problem and the integral of earlier calculus are conceptually distinct. The latter is a line integral for which the path lies along the x-axis.

24. The "curl" of a vector field

$$\mathbf{F} = \mathbf{i}f(x, y, z) + \mathbf{j}g(x, y, z) + \mathbf{k}h(x, y, z)$$

is defined to be del cross $\mathbf{F}$; that is,

$$\operatorname{curl} \mathbf{F} \equiv \nabla \times \mathbf{F} \equiv \begin{vmatrix} \mathbf{i} & \mathbf{j} & \mathbf{k} \\ \dfrac{\partial}{\partial x} & \dfrac{\partial}{\partial y} & \dfrac{\partial}{\partial z} \\ f & g & h \end{vmatrix},$$

or

$$\operatorname{curl} \mathbf{F} \equiv \mathbf{i}\left(\frac{\partial h}{\partial y} - \frac{\partial g}{\partial z}\right) + \mathbf{j}\left(\frac{\partial f}{\partial z} - \frac{\partial h}{\partial x}\right) + \mathbf{k}\left(\frac{\partial g}{\partial x} - \frac{\partial f}{\partial y}\right),$$

and the "divergence" of a vector field

$$\mathbf{V} = \mathbf{i}u(x, y, z) + \mathbf{j}v(x, y, z) + \mathbf{k}w(x, y, z)$$

is defined to be del dot $\mathbf{V}$; that is,

$$\operatorname{div} \mathbf{V} \equiv \nabla \cdot \mathbf{V} \equiv \frac{\partial u}{\partial x} + \frac{\partial v}{\partial y} + \frac{\partial w}{\partial z}.$$

If the components f, g, h of $\mathbf{F}$ are functions which possess continuous mixed partial derivatives

$$\frac{\partial^2 h}{\partial x\, \partial y}, \ldots,$$

show that

$$\operatorname{div} (\operatorname{curl} \mathbf{F}) = 0.$$

25. Assume the notation of Exercise 24.

(a) Prove that if ϕ is a scalar function of x, y, z, then

$$\operatorname{curl} (\operatorname{grad} \phi) = \nabla \times (\nabla\phi) = \mathbf{0}.$$

(b) State how you would express the condition that $\mathbf{F} \cdot d\mathbf{r}$ be an exact differential in terms of the vector field $\nabla \times \mathbf{F}$.

Prove the following results: If

$$\mathbf{r} = x\mathbf{i} + y\mathbf{j} + z\mathbf{k},$$

(c) $\operatorname{div} (\phi\mathbf{F}) \equiv \nabla \cdot (\phi\mathbf{F}) = \phi\nabla \cdot \mathbf{F} + \mathbf{F} \cdot \nabla\phi$,

(d) $\nabla \times (\phi\mathbf{F}) = \phi\nabla \times \mathbf{F} + (\nabla\phi) \times \mathbf{F}$,

(e) $\nabla \cdot (\mathbf{F}_1 \times \mathbf{F}_2) = \mathbf{F}_2 \cdot \nabla \times \mathbf{F}_1 - \mathbf{F}_1 \cdot \nabla \times \mathbf{F}_2$,

(f) $\nabla \cdot \mathbf{r} = 3$ and $\nabla \times \mathbf{r} = 0$.

17.4 TWO-DIMENSIONAL FIELDS: LINE INTEGRALS IN THE PLANE AND THEIR RELATION TO SURFACE INTEGRALS ON CYLINDERS

In this article, we turn our attention to two-dimensional vector fields of the form

$$\mathbf{F} = \mathbf{i}M(x, y) + \mathbf{j}N(x, y). \tag{1}$$

Figure 17.10 shows how such a two-dimensional vector field might look in space. In the figure, for example, $\mathbf{F}$ might represent a fluid flow in which each particle travels in a circle parallel to the xy-plane in such a way that all particles on a given line perpendicular to the xy-plane travel with the same velocity. Problem 1 provides another example, of an

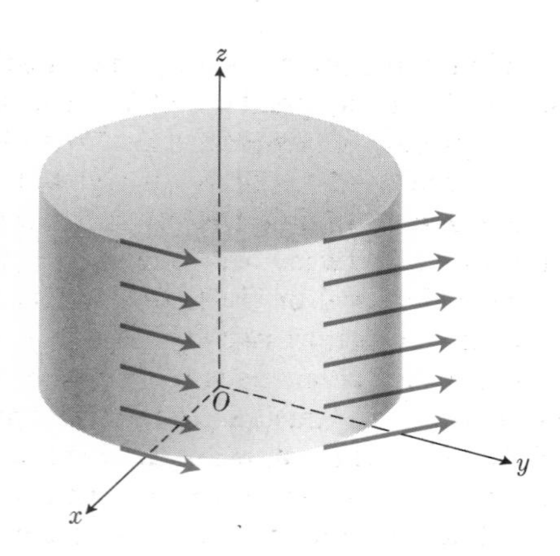

17.10 A two-dimensional vector field in space.

electric field with field strength

$$\mathbf{E} = \frac{\mathbf{i}x + \mathbf{j}y}{x^2 + y^2}. \tag{2}$$

Note that this formulation is like the right-hand side of Eq. (1) with

$$M(x, y) = \frac{x}{x^2 + y^2}, \qquad N(x, y) = \frac{y}{x^2 + y^2}.$$

The essential features of a two-dimensional field are (1) the vectors in $\mathbf{F}$ are all parallel to one plane, which we have taken to be the xy-plane, and (2) in every plane parallel to the xy-plane, the field is the same as it is in that plane. In Eq. (1), the field has everywhere a zero $\mathbf{k}$-component, which makes the vectors parallel to the xy-plane, and the $\mathbf{i}$- and $\mathbf{j}$-components do not depend on z, which makes them the same in all planes parallel to the xy-plane.

Problem 1. An infinitely long, thin, straight wire has a uniform electric charge density δ_0. Using Coulomb's law, find the electric field intensity around the wire due to this charge.

Solution. From a physics textbook,* we find that *Coulomb's law* is an inverse-square law. It says that the force acting on a positive test charge q_0 placed at a distance r from a positive *point charge* q is directed away from q and has magnitude $(4\pi\epsilon_0)^{-1}(qq_0)/r^2$, where ϵ_0 is a certain constant called the *permittivity*. For the charged wire, we have a distributed charge instead of a point charge, but we handle it in the familiar way: We replace this distributed charge by a large number of tiny elements, add, and take limits. More specifically, suppose that the wire runs along the z-axis from $-\infty$ to $+\infty$. Take a long but finite piece of the wire and divide it into a lot of small segments. One of these is indicated in Figure 17.11, with its center at $P(0, 0, z)$, and its length equal to Δz. We assume that Δz is so small that we can treat the charge $\delta_0\,\Delta z$ on this segment of the wire as a point charge at P. Now let $Q(a, b, c)$ be any point not on the z-axis, and let $\Delta\mathbf{F}$ denote the force at Q due to the point charge at P:

$$\Delta\mathbf{F} = \frac{\delta_0 q_0}{4\pi\epsilon_0}\,\frac{\Delta z\,\mathbf{u}}{a^2 + b^2 + (c - z)^2}, \tag{3a}$$

where $\mathbf{u}$ is a unit vector having the direction of $\overrightarrow{PQ}$:

$$\mathbf{u} = \frac{\mathbf{i}a + \mathbf{j}b + \mathbf{k}(c - z)}{[a^2 + b^2 + (c - z)^2]^{1/2}}. \tag{3b}$$

When we add the vector forces $\Delta\mathbf{F}$ for pieces of wire between $z = A$ and $z = B$ and take the limit as $\Delta z \to 0$, we get an integral of the form

$$\frac{\delta_0 q_0}{4\pi\epsilon_0}\int_A^B \frac{\mathbf{i}a + \mathbf{j}b + \mathbf{k}(c - z)}{[a^2 + b^2 + (c - z)^2]^{3/2}}\,dz. \tag{3c}$$

The denominator of the integrand behaves about like $|z|^3$ as $|z| \to \infty$, so the three component integrals in (3c) converge as $A \to -\infty$ and $B \to \infty$. As our final integral representation of the field, therefore, we get

$$\mathbf{F} = \frac{\delta_0 q_0}{4\pi\epsilon_0}\int_{-\infty}^{+\infty} \frac{\mathbf{i}a + \mathbf{j}b + \mathbf{k}(c - z)}{[a^2 + b^2 + (c - z)^2]^{3/2}}\,dz. \tag{4}$$

There are three separate integrals, but two are essentially the same except for the constant coefficients a and b, and the third is zero. It is easy to do the integration by making the substitutions

$$\sqrt{a^2 + b^2} = m,$$

$$z - c = m\tan\theta, \qquad \theta = tan^{-1}\left(\frac{z - c}{m}\right),$$

$$dz = m\sec^2\theta\,d\theta,$$

and observing that the limits for θ are from $-\pi/2$ to $\pi/2$.

---

* For example, Halliday and Resnick, *Physics for Students of Science and Engineering, Combined Edition*, 1962, John Wiley and Sons, Inc., p. 578.

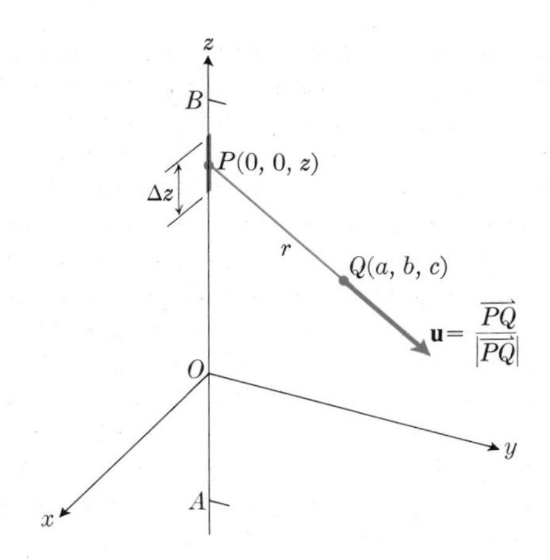

17.11 According to Coulombs' law, a charge $\delta_0\,\Delta z$ at $P(0, 0, z)$ produces a field $\Delta\mathbf{F}$ on a test charge q at $Q(a, b, c)$.

In Exercise 1 you are asked to finish these calculations and to show that the following result is correct:

$$\mathbf{F} = \frac{\delta_0 q_0}{2\pi\epsilon_0}\left(\frac{\mathbf{i}a + \mathbf{j}b}{a^2 + b^2}\right). \tag{5}$$

We observe that the result is a two-dimensional field that does not depend on $\mathbf{k}$ or on c. If we write the resultant field strength $\mathbf{E} = \mathbf{F}/q_0$ as a function of position (x, y, z) instead of (a, b, c), we have

$$\mathbf{E} = k_0\left(\frac{\mathbf{i}x + \mathbf{j}y}{x^2 + y^2}\right), \quad \text{where } k_0 = \frac{\delta_0}{2\pi\epsilon_0}. \tag{6}$$

Remark 1. There is also a fluid-flow interpretation for the field in Eq. (6) (or Eq. 2). To arrive at that interpretation, imagine a long, thin pipe running along the z-axis, and perforated with a very large number of little holes through which it supplies fluid at a constant rate. (It helps to be a bit vague about the actual physics of this: Don't try to be too literal-minded.) In other words, the z-axis is to be thought of as a *source* from which water flows radially outward and so produces a velocity field that is characterized by

$$\mathbf{v} = f(r)\mathbf{u}_r, \tag{7a}$$

where $\mathbf{u}_r$ is the usual unit vector associated with the cylindrical coordinates r, θ, z, and $f(r)$ is a function of r alone. Thus the velocity is all perpendicular to the z-axis, and it is independent of both θ and z. (These are our present interpretations of the phrases "at a constant rate" and "radially outward.") Consider, now, the amount of fluid that flows out through the cylinder $r = a$ between the planes $z = 0$ and $z = 1$ in a short interval of time, from t to $t + \Delta t$. According to the law expressed by Eq. (7a), every particle of water that is on the surface $r = a$ moves radially outward a distance Δr, which is approximately $f(a)\,\Delta t$. Thus the volume of fluid that crosses the boundary $r = a$ between $z = 0$ and $z = 1$ in this time interval is approximately the volume between the cylinders $r = a$ and $r = a + f(a)\,\Delta t$, or $2\pi a f(a)\,\Delta t$. If we multiply this by the density δ_1, we get the *mass* transported through a unit length of the cylinder $r = a$ in the interval from t to $t + \Delta t$:

$$\Delta m \approx \delta_1 2\pi a f(a)\,\Delta t. \tag{7b}$$

If we divide both sides of Eq. (7b) by Δt and take the limit as $\Delta t \to 0$, we get the *rate* at which fluid is flowing across the unit length of the cylinder $r = a$:

$$\frac{dm}{dt} = \delta_1 2\pi a f(a). \tag{7c}$$

For an incompressible fluid such as water, all fluid that flows across the cylinder $r = a$ flows across the cylinder $r = b$ as well (unless, of course, there are other sources or sinks between the two cylinders). Therefore, for the model under discussion, the rate of mass transport given by Eq. (7c) is independent of a, and its value for any radius $r = a$ is the same as for $r = 1$:

$$\delta_1 2\pi f(1) = \delta_1 2\pi a f(a). \tag{7d}$$

From Eq. (7d), we get

$$f(a) = f(1)/a \quad \text{for any } a > 0.$$

Writing r in place of a and substituting C for $f(1)$, we can rewrite the velocity field (7a) in the form

$$\mathbf{v} = (C/r)\mathbf{u}_r. \tag{8}$$

If we recall that the position vector in cylindrical coordinates is

$$\mathbf{R} = \overrightarrow{OP} = r\mathbf{u}_r + \mathbf{k}z$$

and that this must also be equal to $\mathbf{R} = \mathbf{i}x + \mathbf{j}y + \mathbf{k}z$, then we conclude that

$$r\mathbf{u}_r = \mathbf{i}x + \mathbf{j}y, \quad \text{or} \quad \mathbf{u}_r = \frac{\mathbf{i}x + \mathbf{j}y}{r},$$

where

$$r = \sqrt{x^2 + y^2}. \tag{9}$$

Therefore, Eqs. (6) and (8) describe the same field if $C = k_0$.

Remark 2. Instead of interpreting the two-dimensional vector fields as we have done in 3-space, we can interpret them simply as fields in the xy-plane itself. Then r in Eqs. (8) and (9) is just the distance from the origin to the point $P(x, y)$ in the plane, and the unit vector is

$$\mathbf{u}_r = (\mathbf{i}x + \mathbf{j}y)/r = \mathbf{i}\cos\theta + \mathbf{j}\sin\theta,$$

where $r > 0$ and θ measures the angle from the positive x-axis to the position vector $\overrightarrow{OP}$. Equation (8) then describes a vector field in the plane that is directed radially outward and whose strength decreases like $1/r$ as r increases. We still use the language of flow across boundaries, but in this interpretation the boundary would be a *curve* in the plane, rather than a unit length of a cylinder. Equation (7d) would be interpreted as saying that the amount of fluid flowing across the unit circle $r = 1$ per unit time is equal to the amount of fluid flowing across the circle $r = a$ in the same time. (This describes conditions after the flow has reached steady state, not during the transient phase.)

We can easily go back and forth between the two interpretations of two-dimensional fields, but henceforth we shall usually treat them as existing just in the xy-plane and ignore the fact that we can project the field onto any plane parallel to the xy-plane and thereby go to the 3-space view.

Problem 2. Given the velocity field $\mathbf{v} = (\mathbf{i}x + \mathbf{j}y)/r^2$, calculate the mass transport rate across the line segment AB joining the points $A(1, 0)$ and $B(0, 1)$.

Solution. Let δ denote the density factor by which we multiply the area of a region to get the mass of fluid in that region. (We assume the density to be constant.) Consider a segment of the line having length Δs, with its center at $P(x, y)$ on AB. In Remark 3 below, we see that the amount of fluid Δm that flows across the segment in time Δt is given, approximately, by

$$\Delta m \approx (\mathbf{v}\cdot\mathbf{n})(\Delta t\,\Delta s)\,\delta, \tag{10}$$

where $\mathbf{n}$ is a unit vector normal to the line AB at P, and pointing away from the origin:

$$\mathbf{n} = (\mathbf{i} + \mathbf{j})/\sqrt{2}.$$

If we divide both sides of Eq. (10) by Δt, then sum for all the pieces Δs of the segment AB, and take the limit as $\Delta s \to 0$, we get the *average rate* of mass transport across AB:

$$\frac{\Delta M}{\Delta t} \approx \int_{AB} \delta(\mathbf{v}\cdot\mathbf{n})\,ds.$$

Finally, letting $\Delta t \to 0$ and substituting for $\mathbf{v}$, $\mathbf{n}$, and ds, with

$$x = t, \quad y = 1 - t, \quad 0 \le t \le 1$$

as parameterization of the segment AB, we get as the *instantaneous* mass transport rate

$$\frac{dM}{dt} = \delta \int_0^1 \frac{x + y}{r^2\sqrt{2}} (\sqrt{2}\,dt)$$

$$= \delta \int_0^1 \frac{dt}{t^2 + (1 - t)^2} \tag{11a}$$

$$= \delta(\pi/2). \tag{11b}$$

[In Exercise 2 you are asked to verify that the integral in (11a) yields the result (11b).]

Remark 3. The terms on the right-hand side of Eq. (10) are explained this way (see Fig. 17.12): The quantity $\mathbf{v}\,\Delta t$ is, very nearly, the vector displacement of all particles of fluid that were on the segment Δs at time t; hence those particles have swept over a parallelogram with dimensions Δs and $|\mathbf{v}\,\Delta t|$. The $\mathbf{n}$-component of $\mathbf{v}\,\Delta t$ is the altitude h of this parallelogram. Its area is therefore approximately

$$(\mathbf{v}\,\Delta t)\cdot\mathbf{n} \text{ times } \Delta s,$$

and the mass of fluid that fills this parallelogram is what flows across the tiny segment Δs between t and $t + \Delta t$.

This same line of reasoning would apply to flows in the xy-plane in general. It leads to the result

$$\frac{dM}{dt} = \int_C \delta(\mathbf{v}\cdot\mathbf{n})\,ds, \tag{12}$$

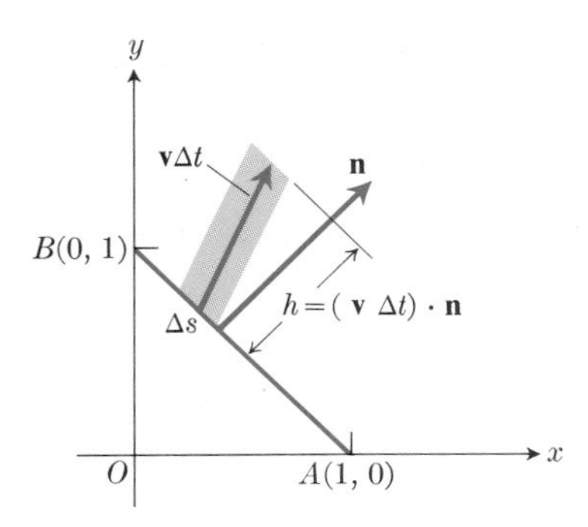

17.12 Fluid that flows across Δs in time Δt fills the parallelogram whose altitude is $h = (\mathbf{v}\,\Delta t) \cdot \mathbf{n}$.

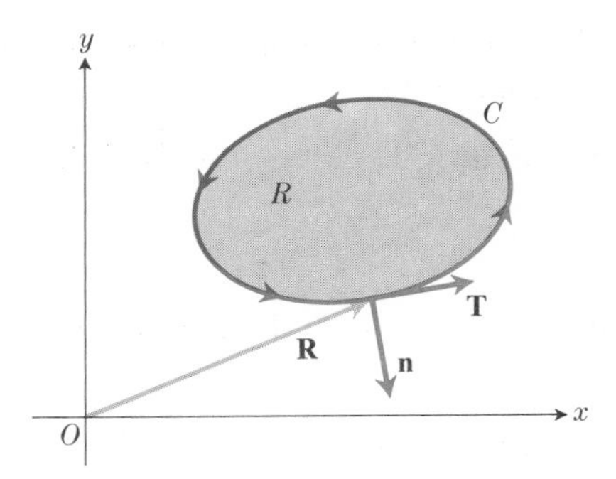

17.13 The position vector $\mathbf{R} = \mathbf{i}x + \mathbf{j}y$, the unit tangent vector $\mathbf{T} = d\mathbf{R}/ds$, and the unit normal vector $\mathbf{n} = \mathbf{T} \times \mathbf{k}$.

where dM/dt is the *rate* at which mass is being transported across the curve C, in the direction of the unit normal vector $\mathbf{n}$. One can interpret M as the amount of mass that has crossed C up to time t.

If the oppositely directed normal $\mathbf{n}' = -\mathbf{n}$ is substituted in place of $\mathbf{n}$ in the integral in Eq. (12), the sign of the answer changes. This just means that if flow in one direction across C is considered to be in the positive sense, then flow in the opposite direction is then considered to be negative. If C is a simple closed curve, we usually choose $\mathbf{n}$ to point outward. In Eqs. (11a, b), we chose the normal to point away from the origin; we got a positive answer because the flow in the first quadrant is generally upward and to the right for the given $\mathbf{v}$.

To simplify further discussion, we take

$$\delta \mathbf{v} = \mathbf{F}(x, y)$$

in Eq. (12), and call the resulting integral the *flux* of $\mathbf{F}$ across C:

$$\text{flux of } \mathbf{F} \text{ across } C = \int_C \mathbf{F} \cdot \mathbf{n}\, ds. \tag{13}$$

We shall use this terminology even when the field $\mathbf{F}$ has nothing to do with a fluid flow, but you may wish to keep the fluid-flow interpretation in mind too.

The curve C in Eq. (13) is to have a direction along it that is called the *positive* direction. For theoretical purposes, we find it convenient to use the arc length s, measured from some point on C arbitrarily chosen as origin, or from the starting point $s = 0$. Then s should *increase* as we proceed in the positive direction along C. (We assume that C is piecewise smooth enough to have a tangent.) Then the position vector $\mathbf{R} = \mathbf{i}x + \mathbf{j}y$ has derivative $d\mathbf{R}/ds = \mathbf{T}$, where $\mathbf{T}$ is a *unit tangent vector* pointing in the positive direction along C. (At a cusp or corner on C, the tangent does not exist, but this won't matter for most of our applications that involve integrals.) Because we often want the flux integral to represent flow outward from a region R bounded by a simple closed curve C, we choose the counterclockwise direction on C as positive, and choose the *outward*-pointing unit normal vector as $\mathbf{n}$. From Figure 17.13, we can see that, because

$$\mathbf{T} = \frac{dx}{ds}\mathbf{i} + \frac{dy}{ds}\mathbf{j}, \tag{14a}$$

for the indicated choice of $\mathbf{n}$ we should choose $\mathbf{n} = \mathbf{T} \times \mathbf{k}$, so that

$$\mathbf{n} = \frac{dy}{ds}\mathbf{i} - \frac{dx}{ds}\mathbf{j}. \tag{14b}$$

As a check, it is easy to see that the dot product of the vectors in Eqs. (14a) and (14b) is zero, that both have unit length, and that when $\mathbf{T}$ points upward and to the right, $\mathbf{n}$ has a positive $\mathbf{i}$-component and a negative $\mathbf{j}$-component. If we proceed around the curve C in the direction in which $\mathbf{T}$ points, with the interior toward our left, then $\mathbf{n}$, as given by Eq. (14b), points to our right, as it should.

We now make use of Eq. (14b) to write the flux integral. We assume that

$$\mathbf{F}(x, y) = \mathbf{i}M(x, y) + \mathbf{j}N(x, y).$$

Then it follows that

$$\begin{aligned}\text{flux across } C &= \int_C \mathbf{F} \cdot \mathbf{n}\, ds \\ &= \int_C \left(M\frac{dy}{ds} - N\frac{dx}{ds}\right) ds \\ &= \int_C (M\, dy - N\, dx). \qquad (15)\end{aligned}$$

The virtue of the final integral in Eq. (15) is this: It can be evaluated using *any* reasonable parameterization of C; we aren't restricted to using the arc length s, provided we integrate in the positive direction along C.

Problem 3. Find the flux of the field

$$\mathbf{F} = 2x\mathbf{i} - 3y\mathbf{j}$$

outward across the ellipse

$$x = \cos t, \quad y = 4 \sin t, \quad 0 \leq t \leq 2\pi.$$

Solution. By Eq. (15), for the flux we have

$$\begin{aligned}\int_C (M\, dy - N\, dx) &= \int_C (2x\, dy + 3y\, dx) \\ &= \int_0^{2\pi} (8 \cos^2 t - 12 \sin^2 t)\, dt \\ &= \int_0^{2\pi} [4(1 + \cos 2t) - 6(1 - \cos 2t)]\, dt \\ &= -4\pi.\end{aligned}$$

The negative answer just means that the net flux is inward.

Remark 4. The integral that we have just evaluated was set up as a flux integral, but it can also be interpreted as a work integral of the form

$$\int_C \mathbf{G} \cdot d\mathbf{R} = \int_C (3y\mathbf{i} + 2x\mathbf{j}) \cdot (\mathbf{i}\, dx + \mathbf{j}\, dy).$$

Conversely, any work integral in the plane can be reinterpreted as a flux integral of a related field. Exercise 7 asks you to supply the details.

EXERCISES

1. In Problem 1, complete the calculations that lead from Eq. (4) to Eq. (5).
2. Evaluate
$$\int_0^1 \frac{dt}{2t^2 - 2t + 1}$$
by changing it to
$$\int_0^1 \frac{2\, dt}{(2t - 1)^2 + 1}$$
and making a change of variables. You will thereby verify the answer given in Eq. (11b).
3. Find the rate of mass transport outward across the circle $r = a$ for the (velocity) flow field of Problem 2. [The calculations are trivial if you interpret Eq. (12) correctly.]
4. In a three-dimensional velocity field, let $\sum$ be a closed surface bounding a region D in its interior. Explain why
$$\iint_\Sigma \delta(\mathbf{v} \cdot \mathbf{n})\, d\sigma$$
can be interpreted as the rate of mass transport outward through $\sum$ if $\mathbf{n}$ is the outward-pointing unit vector normal to $\sum$.

In Exercises 5 and 6, use the result of Exercise 4 to find the rate of mass transport outward through the surface given, if the flow vector $\mathbf{F} = \delta\mathbf{v}$ is

(a) $\mathbf{F} = -\mathbf{i}y + \mathbf{j}x$,

(b) $\mathbf{F} = (x^2 + y^2 + z^2)^{-3/2}(\mathbf{i}x + \mathbf{j}y + \mathbf{k}z)$.

5. $\sum$ is the sphere $x^2 + y^2 + z^2 = a^2$.
6. $\sum$ is the closed cylinder $x^2 + y^2 = a^2$, $-h \leq z \leq h$, plus the end discs $z = \pm h$, $x^2 + y^2 \leq a^2$.
7. Suppose that C is a directed curve with unit tangent and normal vectors $\mathbf{T}$ and $\mathbf{n}$ related as in the text, and suppose that $\mathbf{F}$ and $\mathbf{G}$ are two-dimensional fields:
$$\begin{aligned}\mathbf{F}(x, y) &= \mathbf{i}M(x, y) + \mathbf{j}N(x, y), \\ \mathbf{G}(x, y) &= -\mathbf{i}N(x, y) + \mathbf{j}M(x, y).\end{aligned}$$
Show that
$$\int_C \mathbf{F} \cdot \mathbf{n}\, ds = \int_C \mathbf{G} \cdot \mathbf{T}\, ds.$$
Which integral represents work? Which represents flux?

17.5 GREEN'S THEOREM

This theorem asserts that under suitable conditions the line integral

$$\oint (M\,dx + N\,dy) \tag{1}$$

around a simple closed curve C in the xy-plane is equal to the double integral

$$\iint_R \left(\frac{\partial N}{\partial x} - \frac{\partial M}{\partial y}\right) dx\,dy \tag{2}$$

over the region R that lies inside C.

Notation. The symbol $\oint$ is only used when the curve C is *closed*.

Example 1. Let C be the circle $x = a\cos\theta$, $y = a\sin\theta$, $0 \le \theta \le 2\pi$, and let $M = -y$, $N = x$. Then (1) is

$$\oint_C (-y\,dx + x\,dy) = \int_0^{2\pi} a^2(\sin^2\theta + \cos^2\theta)\,d\theta = 2\pi a^2,$$

and the double integral (2) is

$$\iint_{x^2+y^2\le a^2} 2\,dx\,dy = 2\int_{\theta=0}^{2\pi}\int_0^a r\,dr\,d\theta = 2\pi a^2.$$

Both integrals equal twice the area inside the circle.

Green's theorem. *Let C be a simple closed curve in the xy-plane such that a line parallel to either axis cuts C in at most two points. Let M, N, $\partial N/\partial x$, and $\partial M/\partial y$ be continuous functions of (x, y) inside and on C. Let R be the region inside C. Then*

$$\oint_C M\,dx + N\,dy = \iint_R \left[\frac{\partial N}{\partial x} - \frac{\partial M}{\partial y}\right] dx\,dy. \tag{3}$$

Discussion. We indicate that the line integral on the left side of Eq. (3) is to be taken counterclockwise, the usual positive direction in the plane. We did this automatically in Example 1 by letting θ vary from 0 to 2π in the parametric representation of the circle. Figure 17.14 shows a curve C made up of two parts,

$$C_1\colon a \le x \le b, \quad y_1 = f_1(x),$$
$$C_2\colon b \ge x \ge a, \quad y_2 = f_2(x).$$

We use this notation in the proof.

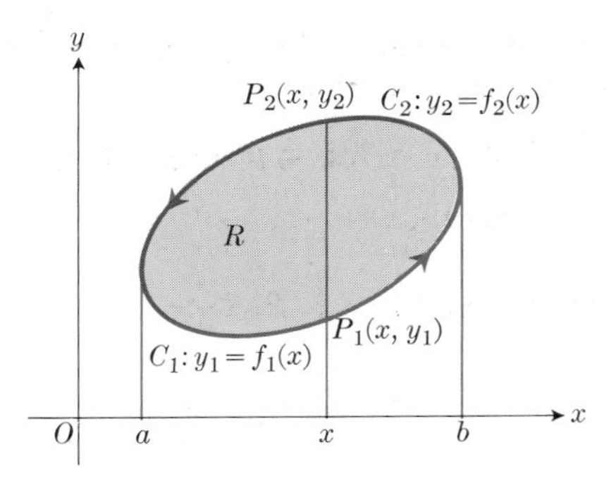

17.14 The boundary curve C, made up of $C_1\colon y = f_1(x)$ and $C_2\colon y = f_2(x)$.

Proof. Consider the double integral of $\partial M/\partial y$ over R in Fig. 17.14. For any x between a and b, we first integrate with respect to y from $y_1 = f_1(x)$ to $y_2 = f_2(x)$ and obtain

$$\int_{y_1}^{y_2} \frac{\partial M}{\partial y}\,dy = M(x, y)\Big]_{y=f_1(x)}^{y=f_2(x)}$$
$$= M(x, f_2(x)) - M(x, f_1(x)). \tag{4}$$

Next we integrate this with respect to x from a to b:

$$\int_a^b \int_{f_1(x)}^{f_2(x)} \frac{\partial M}{\partial y}\,dy\,dx$$
$$= \int_a^b \{M(x, f_1(x)) - M(x, f_1(x))\}\,dx$$
$$= -\int_b^a M(x, f_2(x))\,dx - \int_a^b M(x, f_1(x))\,dx$$
$$= -\int_{C_2} M\,dx - \int_{C_1} M\,dx$$
$$= -\oint_C M\,dx.$$

Therefore

$$\oint_C M\,dx = \iint_R \left(-\frac{\partial M}{\partial y}\right) dx\,dy. \tag{5}$$

Equation (5) is half the result we need for Eq. (3). In Exercise 1, the reader is asked to derive the other half, by integrating $\partial N/\partial x$ first with respect to x and

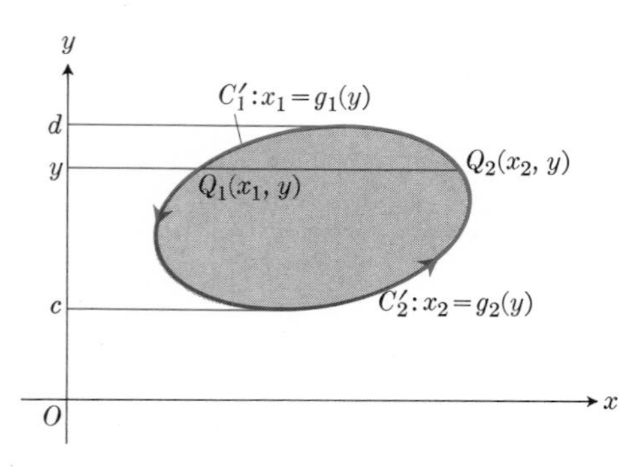

17.15 The boundary curve C', made up of C_1': $x_1 = g_1(y)$ and C_2': $x_2 = g_2(y)$.

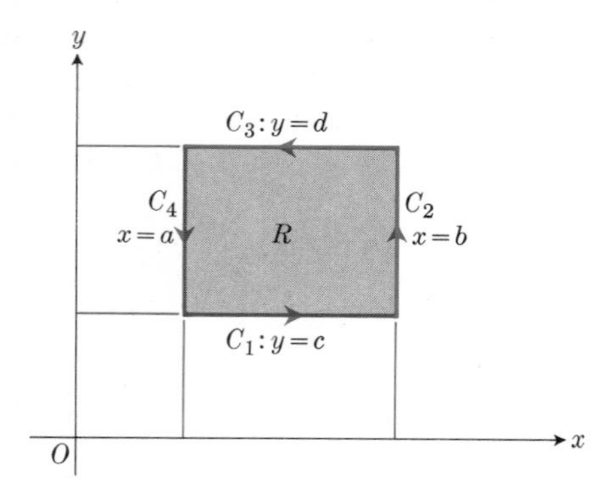

17.16 The rectangle made up of the four segments C_1, C_2, C_3, C_4.

then with respect to y, as suggested by Fig. 17.15. This shows the curve C of Fig. 17.14 reconsidered as now being composed of two directed parts,

$$C_1': c \leq y \leq d, \quad x = g_1(y),$$
$$C_2': d \geq y \geq c, \quad x = g_2(y).$$

The result of this double integration is expressed by

$$\oint_C N\,dy = \iint_R \frac{\partial N}{\partial x}\,dx\,dy. \tag{6}$$

Combining Eqs. (5) and (6), we get Eq. (3). □

Problem 1. Use Green's theorem to find the area enclosed by the ellipse

$$x = a\cos\theta, \quad y = b\sin\theta, \quad 0 \leq \theta \leq 2\pi.$$

Solution. If we take $M = -y$, $N = x$, as in Example 1, and apply Green's theorem, we obtain

$$\begin{aligned}\oint M\,dx + N\,dy &= \int_{\theta=0}^{2\pi} -y\,dx + x\,dy \\ &= \int_0^{2\pi} ab(\sin^2\theta + \cos^2\theta)\,d\theta \\ &= \int_0^{2\pi} ab\,d\theta = 2\pi ab,\end{aligned}$$

and

$$\begin{aligned}\iint_R \left(\frac{\partial N}{\partial x} - \frac{\partial M}{\partial y}\right) dx\,dy &= \iint_R 2\,dx\,dy \\ &= 2 \times (\text{area inside ellipse}).\end{aligned}$$

Therefore

$$\begin{aligned}\text{area inside ellipse} &= \tfrac{1}{2}\oint (-y\,dx + x\,dy) \\ &= \pi ab.\end{aligned}$$

Corollary to Green's theorem. *If C is a simple closed curve such that a line parallel to either axis cuts it in at most two points, then the area enclosed by C is equal to*

$$\tfrac{1}{2}\oint_C (x\,dy - y\,dx).$$

Proof. If we take $M = -y/2$, $N = x/2$, we obtain

$$\begin{aligned}&\oint_C (\tfrac{1}{2}x\,dy - \tfrac{1}{2}y\,dx) \\ &\quad = \iint_R (\tfrac{1}{2} + \tfrac{1}{2})\,dx\,dy \\ &\quad = \text{area of } R. \ \square\end{aligned} \tag{7}$$

Remark 1. Green's theorem may apply to curves and regions that don't meet all of the requirements stated in it. For example, C could be a rectangle, as shown in Fig. 17.16. Here C is considered as composed of four directed parts:

$$\begin{aligned}C_1&: y = c, \quad a \leq x \leq b, \\ C_2&: x = b, \quad c \leq y \leq d, \\ C_3&: y = d, \quad b \geq x \geq a, \\ C_4&: x = a, \quad d \geq y \geq c.\end{aligned}$$

The lines $x = a$ and $x = b$ intersect C in more than two points, and so do the boundaries $y = c$ and $y = d$.

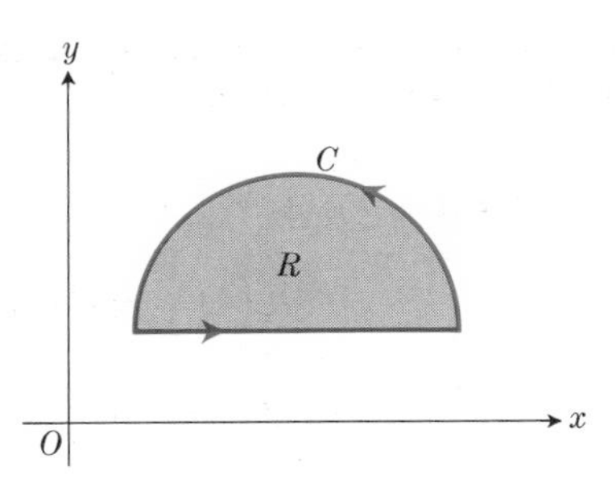

a|b

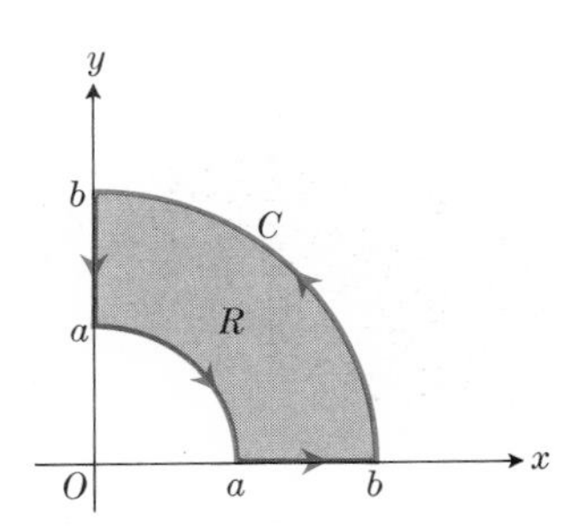

17.17

Proceeding as in the proof of Eq. (6), we have

$$\int_{y=c}^{d} \int_{x=a}^{b} \frac{\partial N}{\partial x}\, dx\, dy$$

$$= \int_{c}^{d} [N(b, y) - N(a, y)]\, dy$$

$$= \int_{c}^{d} N(b, y)\, dy + \int_{d}^{c} N(a, y)\, dy$$

$$= \int_{C_2} N\, dy + \int_{C_4} N\, dy. \tag{8}$$

Because y is constant along C_1 and C_3,

$$\int_{C_1} N\, dy = \int_{C_3} N\, dy = 0,$$

so we can add

$$\int_{C_1} N\, dy + \int_{C_3} N\, dy$$

to the right-hand side of Eq. (8) without changing the equality. Doing so, we have

$$\int_{c}^{d} \int_{a}^{b} \frac{\partial N}{\partial x}\, dx\, dy = \oint_C N\, dy. \tag{9}$$

Similarly we could show that

$$\int_{a}^{b} \int_{c}^{d} \frac{\partial M}{\partial y}\, dy\, dx = -\oint_C M\, dx. \tag{10}$$

Subtracting (10) from (9), we again arrive at

$$\oint_C M\, dx + N\, dy = \iint_R \left(\frac{\partial N}{\partial x} - \frac{\partial M}{\partial y}\right) dx\, dy.$$

Regions such as those shown in Fig. 17.17 can be handled with no greater difficulty. Equation (3) still applies. It also applies to the horseshoe-shaped region R shown in Fig. 17.18, as we see by putting together the regions R_1 and R_2 and their boundaries. Green's theorem applies to C_1, R_1 and to C_2, R_2 yielding

$$\int_{C_1} M\, dx + N\, dy = \iint_{R_1} \left(\frac{\partial N}{\partial x} - \frac{\partial M}{\partial y}\right) dx\, dy,$$

$$\int_{C_2} M\, dx + N\, dy = \iint_{R_2} \left(\frac{\partial N}{\partial x} - \frac{\partial M}{\partial y}\right) dx\, dy.$$

When we add, the line integral along the y-axis from b to a for C_1 cancels the integral over the same segment but in the opposite direction for C_2. Hence

$$\oint_C (M\, dx + N\, dy) = \iint_R \left(\frac{\partial N}{\partial x} - \frac{\partial M}{\partial y}\right) dx\, dy,$$

where C consists of the two segments of the x-axis

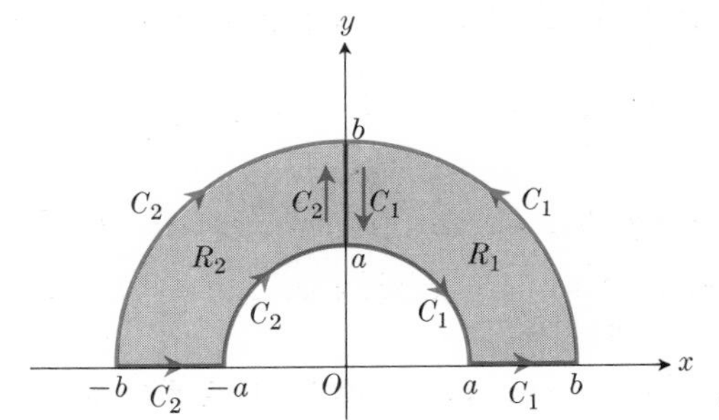

17.18

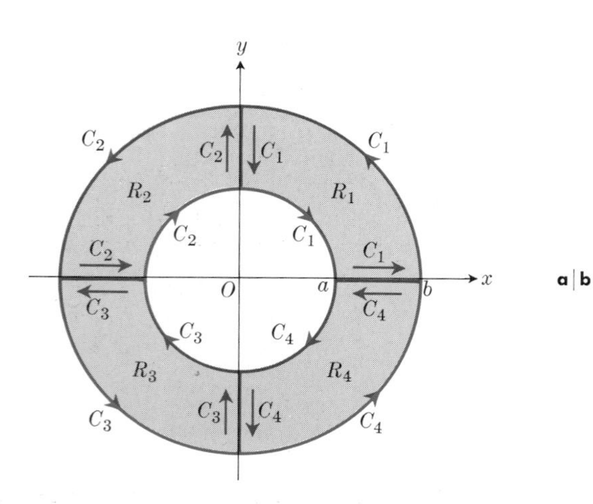

a | b

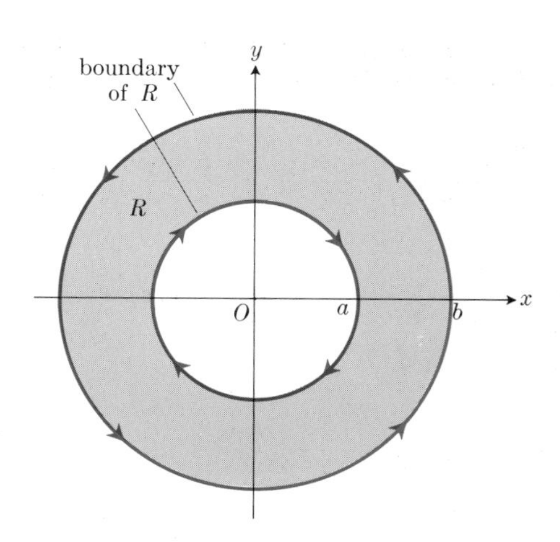

17.19

from $-b$ to $-a$ and from a to b, and of the two semicircles, and where R is the region inside C.

The device of adding line integrals over separate boundaries to build up an integral over a single boundary can be extended to any finite number of subregions. In Fig. 17.19(a), let C_1 be the boundary of the region R_1 in the first quadrant. Similarly for the other three quadrants: C_i is the boundary of the region R_i, $i = 1, 2, 3, 4$. By Green's theorem,

$$\oint_{C_i} M\,dx + N\,dy = \iint_{R_i} \left(\frac{\partial N}{\partial x} - \frac{\partial M}{\partial y}\right) dx\,dy. \tag{11}$$

We add Eqs. (11) for $i = 1, 2, 3, 4$, and get

$$\oint_{r=b} (M\,dx + N\,dy) + \oint_{r=a} (M\,dx + N\,dy) = \iint_{a \le r \le b} \left(\frac{\partial N}{\partial x} - \frac{\partial M}{\partial y}\right) dx\,dy. \tag{12}$$

Eq. (12) says that the double integral of

$$(\partial N/\partial x - \partial M/\partial y)\,dx\,dy$$

over the annular ring R is equal to the line integral of $M\,dx + N\,dy$ over the *entire* boundary of R, in that *direction* along the boundary that keeps the region R on one's left as one progresses. Figure 17.19(b) shows R and its boundary (two concentric circles) and the positive direction on the boundary.

Problem 2. Verify that Eq. (3) holds if

$$M = \frac{-y}{x^2 + y^2}, \quad N = \frac{x}{x^2 + y^2},$$

$$R = \{(x, y): h^2 \le x^2 + y^2 \le 1\},$$

where $0 < h < 1$.

Solution. (See Fig. 17.20.) The boundary of R consists of the circle C_1:

$$x = \cos\theta, \quad y = \sin\theta, \quad 0 \le \theta \le 2\pi,$$

around which we shall integrate counterclockwise, and the circle C_h:

$$x = h\cos\phi, \quad y = h\sin\phi, \quad 2\pi \ge \phi \ge 0,$$

around which we shall integrate in the clockwise direction. Note that the origin is not included in R, because

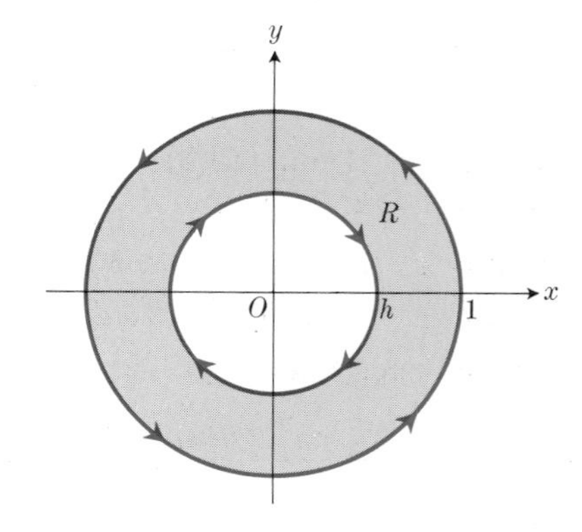

17.20

h is positive. For all $(x, y) \neq (0, 0)$, the functions M and N and their partial derivatives are continuous. Moreover

$$\frac{\partial M}{\partial y} = \frac{(x^2 + y^2)(-1) + y(2y)}{(x^2 + y^2)^2} = \frac{y^2 - x^2}{(x^2 + y^2)^2} = \frac{\partial N}{\partial x},$$

so

$$\iint_R \left(\frac{\partial N}{\partial x} - \frac{\partial M}{\partial y}\right) dx\,dy = \iint_R 0\,dx\,dy = 0.$$

The line integral is

$$\begin{aligned}\int_C M\,dx + N\,dy &= \oint_{C_1} \frac{x\,dy - y\,dx}{x^2 + y^2} + \oint_{C_2} \frac{x\,dy - y\,dx}{x^2 + y^2} \\ &= \int_0^{2\pi} (\cos^2\theta + \sin^2\theta)\,d\theta \\ &\quad + \int_{2\pi}^{0} \frac{h^2(\cos^2\phi + \sin^2\phi)\,d\phi}{h^2} \\ &= 2\pi - 2\pi = 0.\end{aligned}$$

Remark 2. In Problem 2, the functions M and N are discontinuous at $(0, 0)$, so we cannot immediately apply Green's theorem to C_1: $x^2 + y^2 = 1$, and all of the region inside it. We must delete the origin, which we did by excluding points inside C_h.

Remark 3. In Problem 2, we could replace the outer circle C_1 by an ellipse or any other simple closed curve Γ that lies outside C_h (for some positive h). The result would be

$$\oint_\Gamma (M\,dx + N\,dy) + \oint_{C_h} (M\,dx + N\,dy) = 0,$$

which leads to the conclusion

$$\oint_\Gamma \frac{x\,dy - y\,dx}{x^2 + y^2} = 2\pi.$$

This result is easily accounted for if we change to polar coordinates for Γ:

$$\begin{aligned} x &= r\cos\theta, \quad y = r\sin\theta, \\ dx &= -r\sin\theta\,d\theta + \cos\theta\,dr, \\ dy &= r\cos\theta\,d\theta + \sin\theta\,dr, \end{aligned}$$

$$\frac{x\,dy - y\,dx}{x^2 + y^2} = \frac{r^2(\cos^2\theta + \sin^2\theta)\,d\theta}{r^2} = d\theta;$$

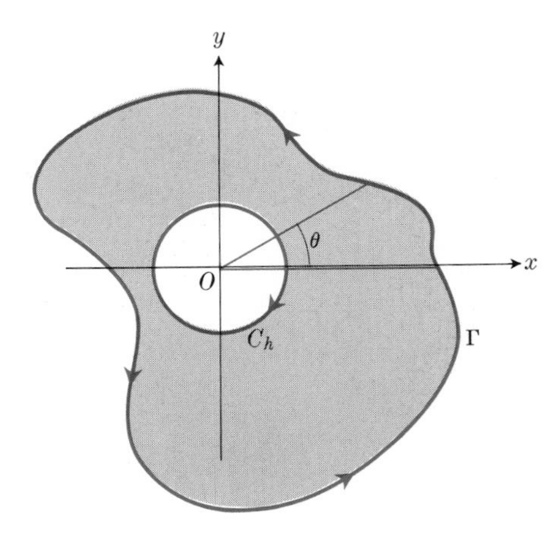

17.21 The region bounded by the circle C_h and the curve Γ.

and θ increases by 2π as we progress once around Γ counterclockwise (see Fig. 17.21).

Green's theorem in vector form

Let

$$\mathbf{F} = M\mathbf{i} + N\mathbf{j} + P\mathbf{k},$$

and

$$\mathbf{R} = x\mathbf{i} + y\mathbf{j}.$$

Then the left-hand side of Eq. (3) is given by

$$\oint_C \mathbf{F} \cdot d\mathbf{R} = \oint_C (M\,dx + N\,dy).$$

To express the right-hand side of Eq. (3) in vector form, we use the symbolic vector operator

$$\nabla = \mathbf{i}\frac{\partial}{\partial x} + \mathbf{j}\frac{\partial}{\partial y} + \mathbf{k}\frac{\partial}{\partial z}.$$

We met the del operator in Article 15.6, where we saw that if

$$w = f(x, y, z)$$

is a differentiable scalar function, then ∇w is the gradient of w:

$$\text{grad } w = \nabla w = \mathbf{i}\frac{\partial w}{\partial x} + \mathbf{j}\frac{\partial w}{\partial y} + \mathbf{k}\frac{\partial w}{\partial z}.$$

Other uses of the del operator are defined in Exercise 24, Article 17.3. In that exercise, the curl of a vector $\mathbf{F}$ is defined to be del cross $\mathbf{F}$. Hence, if $\mathbf{F} = M\mathbf{i} + N\mathbf{j} + P\mathbf{k}$, then we have

$$\text{curl } \mathbf{F} = \nabla \times \mathbf{F} = \begin{vmatrix} \mathbf{i} & \mathbf{j} & \mathbf{k} \\ \frac{\partial}{\partial x} & \frac{\partial}{\partial y} & \frac{\partial}{\partial z} \\ M & N & P \end{vmatrix}$$

$$= \mathbf{i}\left(\frac{\partial P}{\partial y} - \frac{\partial N}{\partial z}\right) + \mathbf{j}\left(\frac{\partial M}{\partial z} - \frac{\partial P}{\partial x}\right) + \mathbf{k}\left(\frac{\partial N}{\partial x} - \frac{\partial M}{\partial y}\right).$$

The component of curl $\mathbf{F}$ that is normal to the region R in the xy-plane is

$$(\nabla \times \mathbf{F}) \cdot \mathbf{k} = \frac{\partial N}{\partial x} - \frac{\partial M}{\partial y}.$$

Hence Green's theorem can be written in vector form as

$$\int_C \mathbf{F} \cdot d\mathbf{R} = \iint_R (\nabla \times \mathbf{F}) \cdot d\mathbf{A}, \tag{13}$$

where $d\mathbf{A} = \mathbf{k}\, dx\, dy$ is a vector normal to the region R and of magnitude $|d\mathbf{A}| = dx\, dy$. In words, Green's theorem states that the integral around C of the tangential component of $\mathbf{F}$ is equal to the integral, over the region R bounded by C, of the component of curl $\mathbf{F}$ that is normal to R; this integral, specifically, is the flux through R of curl $\mathbf{F}$. We shall later extend this result to more general curves and surfaces in a formulation that is known as Stokes's theorem.

There is a second, *normal*, vector form for Green's theorem. It involves the gradient operator ∇ in another form, one that produces the *divergence*. Now the integrand of the line integral of Eq. (13) is the *tangential* component of the field $\mathbf{F}$ because

$$\mathbf{F} \cdot d\mathbf{R} = \left(\mathbf{F} \cdot \frac{d\mathbf{R}}{ds}\right) ds = (\mathbf{F} \cdot \mathbf{T})\, ds.$$

As in Article 17.4, if we let

$$\mathbf{F} = \mathbf{i}M(x, y) + \mathbf{j}N(x, y),$$

and let $\mathbf{G}$ be the orthogonal field given by

$$\mathbf{G} = \mathbf{i}N(x, y) - \mathbf{j}M(x, y),$$

then it follows that

$$\mathbf{F} \cdot \mathbf{T} = \mathbf{G} \cdot \mathbf{n} = M\frac{dx}{ds} + N\frac{dy}{ds}$$

because

$$\mathbf{T} = \mathbf{i}\frac{dx}{ds} + \mathbf{j}\frac{dy}{ds}, \qquad \mathbf{n} = \mathbf{i}\frac{dy}{ds} - \mathbf{j}\frac{dx}{ds}.$$

Therefore Green's theorem, which says that

$$\int_C (M\, dx + N\, dy) = \iint_R \left[\frac{\partial N}{\partial x} - \frac{\partial M}{\partial y}\right] dx\, dy,$$

also says that

$$\int_C \mathbf{G} \cdot \mathbf{n}\, ds = \iint_R \nabla \cdot \mathbf{G}\, dx\, dy, \tag{14}$$

where

$$\nabla \cdot \mathbf{G} = \text{div } \mathbf{G} = \frac{\partial N}{\partial x} + \frac{\partial(-M)}{\partial y}.$$

In words, Eq. (14) says that the line integral of the *normal* component of any vector field $\mathbf{G}$ around the boundary of a region R in which $\mathbf{G}$ is continuous and has continuous partial derivatives is equal to the double integral of the *divergence* of $\mathbf{G}$ over R. In the next article, we shall extend this result to three-dimensional vector fields and discuss the physical interpretation of the divergence. For such vector fields

$$\mathbf{F}(x, y, z) = \mathbf{i}M(x, y, z) + \mathbf{j}N(x, y, z) + \mathbf{k}P(x, y, z),$$

the *divergence* is defined to be

$$\text{div } \mathbf{F} = \nabla \cdot \mathbf{F} = \frac{\partial M}{\partial x} + \frac{\partial N}{\partial y} + \frac{\partial P}{\partial z}. \tag{15}$$

EXERCISES

1. Supply the details necessary to establish Eq. (6).
2. Supply the steps necessary to establish Eq. (10).

In Exercises 3 through 7, C is the circle $x^2 + y^2 = a^2$ and R is C plus its interior. Verify that

$$\oint_C M\,dx + N\,dy = \iint_R \left(\frac{\partial N}{\partial x} - \frac{\partial M}{\partial y}\right) dx\,dy,$$

given the condition stated in each exercise.

3. $M = x,\ N = y$
4. $M = N = xy$
5. $M = -x^2y,\ N = xy^2$
6. $M = N = e^{x+y}$
7. $M = y,\ N = 0$

8. Suppose that

$$R = \{(x, y): 0 \le y \le \sqrt{a^2 - x^2},\ -a \le x \le a\},$$

and that C is the boundary of R.

(a) Sketch R and C.

(b) Write out the proof of Green's formula for this region.

Definition. *A region R is said to be simply connected if every simple closed curve lying in R can be continuously contracted to a point without its touching any part of the boundary of R. Examples are the interiors of circles, ellipses, cardioids, and rectangles; and, in three dimensions, the region between two concentric spheres. (The annular ring in Fig. 17.20 is not simply connected.)*

9. Show, by a geometric argument, that Green's formula, Eq. (3), holds for any simply connected region R whose boundary is a simple closed curve C, provided R can be decomposed into a finite number of nonoverlapping regions $R_1, R_2, \ldots, R_n$ with boundaries $C_1, C_2, \ldots, C_n$ of a type for which the formula (3) is true for each R_i and C_i, $i = 1, \ldots, n$.

10. Suppose R is a region in the xy-plane, C is its boundary, and the area of R is given by

$$A(R) = \oint_C \tfrac{1}{2}(x\,dy - y\,dx).$$

Suppose the equations $x = f(u, v)$, $y = g(u, v)$ map R and C in a continuous and one-to-one manner onto a region R' and curve C', respectively, in the uv-plane. Use Green's formula to show that

$$\iint_R dx\,dy = \iint_{R'} \begin{vmatrix} f_u & f_v \\ g_u & g_v \end{vmatrix} du\,dv = \iint_{R} \left(\frac{\partial f}{\partial u}\frac{\partial g}{\partial v} - \frac{\partial f}{\partial v}\frac{\partial g}{\partial u}\right) du\,dv.$$

Hint. Note that

$$\iint_R dx\,dy = \frac{1}{2}\int_C (x\,dy - y\,dx) = \frac{1}{2}\int_{C'} \left[f(u, v)\left(\frac{\partial g}{\partial u}du + \frac{\partial g}{\partial v}dv\right) - g(u, v)\left(\frac{\partial f}{\partial u}du + \frac{\partial f}{\partial v}dv\right)\right].$$

and apply Green's formula to C' and R'.

In Exercises 11 and 12, with C as given, evaluate the line integrals by applying Green's formula.

11. C is the triangle bounded by $x = 0$, $x + y = 1$, $y = 0$:

$$\int_C (y^2\,dx + x^2\,dy).$$

12. C is the boundary of $0 \le x \le \pi$, $0 \le y \le \sin x$:

$$\int_C (3y\,dx + 2x\,dy).$$

13. Rewrite Eq. (14) in nonvector notation for a vector field $\mathbf{F} = \mathbf{i}M(x, y) + \mathbf{j}N(x, y)$ in place of $\mathbf{G}$. (In other words, first write it in vector form with $\mathbf{F}$ in place of $\mathbf{G}$, and then translate the result into nonvector notation.)

17.6 DIVERGENCE THEOREM

This theorem states that under appropriate conditions, the triple integral

$$\iiint_D \operatorname{div} \mathbf{F}\,dV \tag{1}$$

is equal to the double integral

$$\iint_\Sigma \mathbf{F} \cdot d\boldsymbol{\sigma}. \tag{2}$$

Here $\mathbf{F} = \mathbf{i}M + \mathbf{j}N + \mathbf{k}P$, with M, N, and P continuous functions of (x, y, z) that have continuous first-order partial derivatives:

$$\operatorname{div} \mathbf{F} = \frac{\partial M}{\partial x} + \frac{\partial N}{\partial y} + \frac{\partial P}{\partial z};$$

$d\boldsymbol{\sigma} = \mathbf{n}\, d\sigma$ is a vector element of surface area directed along the unit outer normal vector $\mathbf{n}$; and $\sum$ is the surface enclosing the region D. We shall first show that (1) and (2) are equal if D is some convex region with no holes, such as the interior of a sphere, or a cube, or an ellipsoid, and if $\sum$ is a piecewise smooth surface. In addition, we assume that the projection of D into the xy-plane is a simply connected region R_{xy} and that any line perpendicular to the xy-plane at an interior point of R_{xy} intersects the surface S in at most two points, producing surfaces $\sum_1$ and $\sum_2$:

$$\textstyle\sum_1:\quad z_1 = f_1(x, y), \quad (x, y) \text{ in } R_{xy},$$
$$\textstyle\sum_2:\quad z_2 = f_2(x, y), \quad (x, y) \text{ in } R_{xy},$$

with $z_1 \leq z_2$. Similarly for the projection of D onto the other coordinate planes.

If we write the unit normal vector $\mathbf{n}$ in terms of its direction cosines, as

$$\mathbf{n} = \mathbf{i} \cos \alpha + \mathbf{j} \cos \beta + \mathbf{k} \cos \gamma,$$

then

$$\begin{aligned} \mathbf{F} \cdot d\boldsymbol{\sigma} &= \mathbf{F} \cdot (\mathbf{n}\, d\sigma) = (\mathbf{F} \cdot \mathbf{n})\, d\sigma \\ &= (M \cos \alpha + N \cos \beta + P \cos \gamma)\, d\sigma; \end{aligned} \tag{3}$$

and the divergence theorem states that

$$\begin{aligned} &\iiint_D \left(\frac{\partial M}{\partial x} + \frac{\partial N}{\partial y} + \frac{\partial P}{\partial z}\right) dx\, dy\, dz \\ &\quad = \iint_{\Sigma} (M \cos \alpha + N \cos \beta + P \cos \gamma)\, d\sigma. \end{aligned} \tag{4}$$

We see that both sides of Eq. (4) are additive with respect to M, N, and P, and that our task is to prove

$$\iiint \frac{\partial M}{\partial x}\, dx\, dy\, dz = \iint M \cos \alpha\, d\sigma, \tag{5a}$$

$$\iiint \frac{\partial N}{\partial y}\, dx\, dy\, dz = \iint N \cos \beta\, d\sigma, \tag{5b}$$

$$\iiint \frac{\partial P}{\partial z}\, dx\, dy\, dz = \iint P \cos \gamma\, d\sigma. \tag{5c}$$

We shall establish (5c) in detail.

Figure 17.22 illustrates the projection of D into the xy-plane. The surface $\sum$ consists of the *upper part*

$$\textstyle\sum_2:\quad z = f_2(x, y), \quad (x, y) \text{ in } R_{xy},$$

and the *lower part:*

$$\textstyle\sum_1:\quad z = f_1(x, y), \quad (x, y) \text{ in } R_{xy}.$$

On the surface $\sum_2$, the outer normal has a positive $\mathbf{k}$-component, and

$$\cos \gamma_2\, d\sigma_2 = dx\, dy \tag{6a}$$

is the projection of $d\sigma$ into R_{xy}. On the surface $\sum_1$, the outer normal has a negative $\mathbf{k}$-component, and

$$\cos \gamma_1\, d\sigma_1 = -dx\, dy. \tag{6b}$$

Therefore we can evaluate the surface integral on the right-hand side of Eq. (5c):

$$\begin{aligned} \iint P \cos \gamma\, d\sigma &= \iint_{\Sigma_2} P_2 \cos \gamma_2\, d\sigma_2 \\ &\qquad + \iint_{\Sigma_1} P_1 \cos \gamma_1\, d\sigma_1 \\ &= \iint_{R_{xy}} P(x, y, z_2)\, dx\, dy \\ &\qquad - \iint_{R_{xy}} P(x, y, z_1)\, dx\, dy \\ &= \iint_{R_{xy}} [P(x, y, z_2) - P(x, y, z_1)]\, dx\, dy \\ &= \iint_{R_{xy}} \left[\int_{z_1}^{z_2} \frac{\partial P}{\partial z}\, dz\right] dx\, dy \\ &= \iiint_D \frac{\partial P}{\partial z}\, dz\, dx\, dy. \end{aligned} \tag{7}$$

Thus we have established Eq. (5c). Proofs for (5a) and (5b) follow the same pattern; or, just permute x, y, z; M, N, P; α, β, γ, in order, and get those results from (5c) by renaming the axes. Finally, by

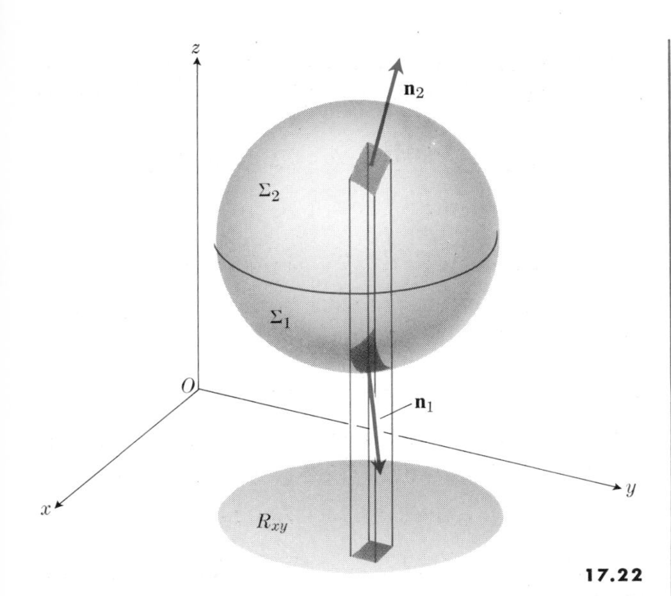

17.22

addition of (5a, b, c), we get Eq. (4):

$$\iiint_D \operatorname{div} \mathbf{F}\, dV = \iint_\Sigma \mathbf{F} \cdot d\boldsymbol{\sigma}. \tag{8}$$

Problem 1. Verify Eq. (8) for the sphere

$$x^2 + y^2 + z^2 = a^2$$

if

$$\mathbf{F} = \mathbf{i}x + \mathbf{j}y + \mathbf{k}z.$$

Solution

$$\operatorname{div} \mathbf{F} = \frac{\partial x}{\partial x} + \frac{\partial y}{\partial y} + \frac{\partial z}{\partial z} = 3,$$

so

$$\iiint_D \operatorname{div} \mathbf{F}\, dV = \iiint_D 3\, dV = 3(\tfrac{4}{3}\pi a^3) = 4\pi a^3.$$

To find a unit vector **n** normal to the surface $\sum$, we first write the equation of $\sum$ in the form $f(x, y, z) = 0$, and then (see Article 15.6) use

$$\mathbf{n} = \pm \frac{\operatorname{grad} f}{|\operatorname{grad} f|}.$$

Here

$$f(x, y, z) = x^2 + y^2 + z^2 - a^2;$$

the outer unit normal is

$$\mathbf{n} = \frac{2(x\mathbf{i} + y\mathbf{j} + z\mathbf{k})}{\sqrt{4(x^2 + y^2 + z^2)}} = \frac{x\mathbf{i} + y\mathbf{j} + z\mathbf{k}}{a},$$

and

$$\mathbf{F} \cdot \mathbf{n}\, d\sigma = \frac{x^2 + y^2 + z^2}{a}\, d\sigma = \frac{a^2}{a}\, d\sigma = a\, d\sigma,$$

because $x^2 + y^2 + z^2 = a^2$ on the surface. Therefore

$$\iint_\Sigma \mathbf{F} \cdot d\boldsymbol{\sigma} = \iint_\Sigma a\, d\sigma = a(4\pi a^2) = 4\pi a^3.$$

Problem 2. Show that Eq. (8) holds for the cube with faces in the planes

$$x = x_0, \quad x = x_0 + h,$$
$$y = y_0, \quad y = y_0 + h,$$
$$z = z_0, \quad z = z_0 + h,$$

where h is a positive constant.

Solution. We compute $\iint \mathbf{F} \cdot d\boldsymbol{\sigma}$ as the sum of the integrals over the six faces separately. We begin with the two faces perpendicular to the x-axis. For the face $x = x_0$ and the face $x = x_0 + h$, respectively, we have the first and second lines of the following table.

Range of integration	Outward unit normal	$\iint (\mathbf{F} \cdot \mathbf{n})\, d\sigma$
$y_0 \le y \le y_0 + h$, $z_0 \le z \le z_0 + h$	$-\mathbf{i}$	$-\iint M(x_0, y, z)\, dy\, dz$
$y_0 \le y \le y_0 + h$, $z_0 \le z \le z_0 + h$	$\mathbf{i}$	$+\iint M(x_0 + h, y, z)\, dy\, dz$

The sum of surface integrals over these two faces is

$$\begin{aligned}\iint (\mathbf{F} \cdot \mathbf{n})\, d\sigma &= \iint [M(x_0 + h, y, z) - M(x_0, y, z)]\, dy\, dz \\ &= \iint \left(\int_{x_0}^{x_0+h} \frac{\partial M}{\partial x}\, dx \right) dy\, dz \\ &= \iiint_D \frac{\partial M}{\partial x}\, dV.\end{aligned}$$

Similarly the sum of the surface integrals over the two faces perpendicular to the y-axis is equal to

$$\iiint (\partial N/\partial y)\, dV;$$

and the sum of the surface integrals over the other two faces is equal to $\iiint (\partial P/\partial z)\, dV$. Hence the surface integral over the six faces is equal to the sum of the three volume integrals, and Eq. (8) holds for the cube:

$$\iint_{\Sigma} \mathbf{F} \cdot \mathbf{n}\, d\sigma = \iiint_{D} \left(\frac{\partial M}{\partial x} + \frac{\partial N}{\partial y} + \frac{\partial P}{\partial z}\right) dV$$
$$= \iiint_{D} \operatorname{div} \mathbf{F}\, dV.$$

Remark 1. The divergence theorem can be extended to more complex regions that can be split up into a finite number of simple regions of the type discussed, and to regions that can be defined by certain limiting processes. For example, suppose D is the region between two concentric spheres, and $\mathbf{F}$ has continuously differentiable components throughout D and on the bounding surfaces. Split D by an equatorial plane and apply the divergence theorem to each half separately. The top half, D_1, is shown in Fig. 17.23. The surface that bounds D_1 consists of an outer hemisphere, a plane washer-shaped base, and an inner hemisphere. The divergence theorem says that

$$\iiint_{D_1} \operatorname{div} \mathbf{F}\, dV_1 = \iint_{\Sigma_1} \mathbf{F} \cdot \mathbf{n}_1\, d\sigma_1. \tag{9a}$$

The unit normal $\mathbf{n}_1$ that points outward from D_1 points away from the origin along the outer surface, points down along the flat base, and points toward the origin along the inner surface. Next apply the divergence theorem to D_2, as shown in Fig. 17.24:

$$\iiint_{D_2} \operatorname{div} \mathbf{F}\, dV_2 = \iint_{\Sigma_2} \mathbf{F} \cdot \mathbf{n}_2\, d\sigma_2. \tag{9b}$$

As we follow $\mathbf{n}_2$ over $\sum_2$, pointing outward from D_2, we see that $\mathbf{n}_2$ points upward along the flat surface in the xy-plane, points away from the origin on the outer sphere, and points toward the origin on the inner sphere. When we add (9a) and (9b), the surface integrals over the flat base cancel because of the opposite signs of $\mathbf{n}_1$ and $\mathbf{n}_2$. We thus arrive at the result

$$\iiint_{D} \operatorname{div} \mathbf{F}\, dV = \iint_{\Sigma} \mathbf{F} \cdot d\boldsymbol{\sigma}, \tag{10}$$

with D the region between the spheres, $\sum$ the boundary of D consisting of two spheres, and $d\boldsymbol{\sigma} = \mathbf{n}\, d\sigma$, where $\mathbf{n}$ is the unit normal to $\sum$ directed outward from D.

Problem 3. Verify Eq. (10) for the region

$$1 \le x^2 + y^2 + z^2 \le 4$$

if

$$\mathbf{F} = -\frac{\mathbf{i}x + \mathbf{j}y + \mathbf{k}z}{\rho^3}, \quad \rho = \sqrt{x^2 + y^2 + z^2}.$$

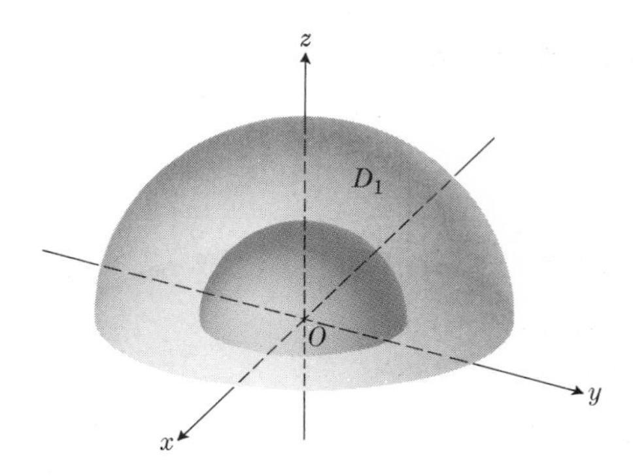

17.23 Upper half of the region between two spheres.

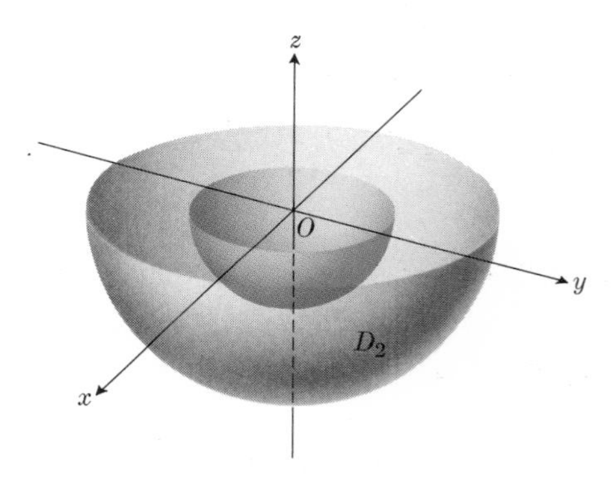

17.24 Lower half of the region between two spheres.

Solution. Observe that

$$\frac{\partial \rho}{\partial x} = \frac{x}{\rho},$$

and

$$\frac{\partial}{\partial x}(x\rho^{-3}) = \rho^{-3} - 3x\rho^{-4}\frac{\partial \rho}{\partial x} = \frac{1}{\rho^3} - \frac{3x^2}{\rho^5}.$$

Thus, throughout the region $1 \leq \rho \leq 2$, all functions considered are continuous, and

$$\operatorname{div} \mathbf{F} = \frac{-3}{\rho^3} + \frac{3}{\rho^5}(x^2 + y^2 + z^2) = -\frac{3}{\rho^3} + \frac{3\rho^2}{\rho^5} = 0.$$

Therefore

$$\iiint_D \operatorname{div} \mathbf{F}\, dV = 0. \tag{11}$$

On the outer sphere ($\rho = 2$), the positive unit normal is

$$\mathbf{n} = \frac{\mathbf{i}x + \mathbf{j}y + \mathbf{k}z}{\rho},$$

and

$$\mathbf{F} \cdot d\boldsymbol{\sigma} = \mathbf{F} \cdot \mathbf{n}\, d\sigma = -\frac{x^2 + y^2 + z^2}{\rho^4}\, d\sigma = -\frac{1}{\rho^2}\, d\sigma.$$

Hence

$$\iint_{\rho=2} \mathbf{F} \cdot d\sigma = -\tfrac{1}{4}\iint_{\rho=2} d\sigma$$

$$= -\tfrac{1}{4} \cdot 4\pi\rho^2 = -\pi\rho^2 = -4\pi. \tag{12a}$$

On the inner sphere ($\rho = 1$), the positive unit normal points toward the origin; its equation is

$$\mathbf{n} = \frac{-(\mathbf{i}x + \mathbf{j}y + \mathbf{k}z)}{\rho}.$$

Hence

$$\mathbf{F} \cdot \mathbf{n}\, d\sigma = +\frac{x^2 + y^2 + z^2}{\rho^4} = \frac{1}{\rho^2}\, d\sigma.$$

Thus

$$\iint_{\rho=1} \mathbf{F} \cdot d\sigma = \iint_{\rho=1} \frac{1}{\rho^2}\, d\sigma$$

$$= \frac{1}{\rho^2} \cdot 4\pi\rho^2 = 4\pi. \tag{12b}$$

The sum of (12a) and (12b) is the surface integral over the complete boundary of D:

$$-4\pi + 4\pi = 0,$$

which agrees with (11) as it should.

Remark 2. We can also conclude that if div $\mathbf{F}$ is continuous at a point Q, then

$$(\operatorname{div} \mathbf{F})_Q = \lim_{\rho \to 0} \left(\frac{3}{4\pi\rho^3} \iint_\Sigma \mathbf{F} \cdot \mathbf{n}\, d\sigma \right). \tag{13}$$

Here Σ is a sphere of radius ρ centered at Q. We take D to be the interior of such a sphere and apply the divergence theorem. The argument is this: If div $\mathbf{F}$ is continuous at Q, then its average value throughout V approaches the value at Q as $\rho \to 0$:

$$\lim_{\rho \to 0} \left(\frac{1}{\frac{4}{3}\pi\rho^3} \iiint_D \operatorname{div} \mathbf{F}\, dV \right) = (\operatorname{div} \mathbf{F})_Q.$$

We substitute

$$\iint_\Sigma \mathbf{F} \cdot \mathbf{n}\, d\sigma \quad \text{for} \quad \iiint_D \operatorname{div} \mathbf{F}\, dV$$

to get Eq. (13). Eq. (13) can be taken as the definition of div $\mathbf{F}$, and often is. Such a definition has the advantage of being coordinate-free. If one starts with Eq. (13) as a definition, he can then prove the divergence theorem and the coordinate representation of div $\mathbf{F}$ as

$$\frac{\partial F_x}{\partial x} + \frac{\partial F_y}{\partial y} + \frac{\partial F_z}{\partial z}.$$

However, we take Eq. (13) as a derived result, and see that it leads to the following interpretation of div $\mathbf{F}$. Suppose $\mathbf{v}$ is the velocity field of a moving fluid, and δ is the density. If $\mathbf{F} = \delta\mathbf{v}$, then $\mathbf{F}$ specifies the mass flow per unit of time, and $\mathbf{F} \cdot \mathbf{n}$ is the component of flow normal to Σ; hence

$$\iint_\Sigma \mathbf{F} \cdot \mathbf{n}\, d\sigma$$

gives the net rate of flow per unit time from the region bounded by Σ. If we divide this by the volume of that region (that is, by $\frac{4}{3}\pi\rho^3$, assuming that Σ is a sphere of radius ρ), then the result is the net outflow per unit of time, per unit volume. Thus $(\operatorname{div} \mathbf{F})_Q$ is the strength of the *source* at Q, if there is one. (A *sink* corresponds to a negative value of the divergence.) If we multiply $(\operatorname{div} \mathbf{F})_Q$ by ΔV, the result is approximately the rate of flow out from ΔV per unit time (say, in pounds per second).

For an incompressible fluid (δ = constant), if there are no sources or sinks in a region D, then the flow into D over its entire boundary just balances the flow out, so that the net flow is zero:

$$\iint \mathbf{F} \cdot d\boldsymbol{\sigma} = 0.$$

Equation (13) then leads to the conclusion that

$$\operatorname{div} \mathbf{F} = \operatorname{div} (\delta \mathbf{v}) = \delta \operatorname{div} \mathbf{v} = 0$$

at each interior point of D. In words, the divergence of the velocity of an incompressible fluid is zero in a region where there are no sources or sinks.

EXERCISES

In Exercises 1 through 5, verify the divergence theorem for the cube with center at the origin and faces in the planes $x = \pm 1$, $y = \pm 1$, $z = \pm 1$, and $\mathbf{F}$ as given.

1. $\mathbf{F} = 2\mathbf{i} + 3\mathbf{j} + 4\mathbf{k}$
2. $\mathbf{F} = \mathbf{i}x + \mathbf{j}y + \mathbf{k}z$
3. $\mathbf{F} = \mathbf{i}yz + \mathbf{j}xz + \mathbf{k}xy$
4. $\mathbf{F} = \mathbf{i}(x - y) + \mathbf{j}(y - z) + \mathbf{k}(x - y)$
5. $\mathbf{F} = \mathbf{i}x^2 + \mathbf{j}y^2 + \mathbf{k}z^2$

In Exercises 6 through 10, compute both

$$\iiint_D \operatorname{div} \mathbf{F}\, dV \quad \text{and} \quad \iint_\Sigma \mathbf{F} \cdot d\boldsymbol{\sigma}$$

directly. Compare the results with the divergence theorem expressed by Eq. (8), given that

$$\mathbf{F} = \mathbf{i}(x + y) + \mathbf{j}(y + z) + \mathbf{k}(z + x),$$

and given that Σ bounds the region D given in the exercise.

6. $0 \le z \le 4 - x^2 - y^2, \quad 0 \le x^2 + y^2 \le 4$
7. $-4 + x^2 + y^2 \le z \le 4 - x^2 - y^2,$
   $0 \le x^2 + y^2 \le 4$
8. $0 \le x^2 + y^2 \le 9, \quad 0 \le z \le 5$
9. $0 \le x^2 + y^2 + z^2 \le a^2$
10. $|x| \le 1, \quad |y| \le 1, \quad |z| \le 1$
11. A function f is said to be *harmonic* in a region D if, throughout D,
$$\frac{\partial^2 f}{\partial x^2} + \frac{\partial^2 f}{\partial y^2} + \frac{d^2 f}{\partial z^2} = 0.$$
Suppose f is harmonic throughout D, Σ is the boundary of D, $\mathbf{n}$ is the positive unit normal on Σ, and $\partial f/\partial n$ is the directional derivative of f in the direction of $\mathbf{n}$. Prove that
$$\iint_\Sigma \frac{\partial f}{\partial n}\, d\sigma = 0.$$
Hint. Let $\mathbf{F} = \operatorname{grad} f$.
12. Prove that if f is harmonic in D (see Exercise 11), then
$$\iint_\Sigma f \frac{\partial f}{\partial n}\, d\sigma = \iiint_D |\operatorname{grad} f|^2\, dV.$$
Hint. Let $\mathbf{F} = f \operatorname{grad} f$.
13. A function of two variables is harmonic in a region R of the xy-plane if
$$\frac{\partial^2 f}{\partial x^2} + \frac{\partial^2 f}{\partial y^2} = 0$$
throughout R. Let f be
(a) the real part, or (b) the imaginary part
of $(x + iy)^3$, where $i = \sqrt{-1}$ and x and y are real. Prove that f is harmonic in the entire xy-plane in both (a) and (b).

17.7 STOKES'S THEOREM

Stokes's theorem is an extension of Green's theorem in vector form to surfaces and curves in three dimensions. It says that the line integral

$$\oint \mathbf{F} \cdot d\mathbf{R} \tag{1}$$

is equal to the surface integral

$$\iint_\Sigma (\operatorname{curl} \mathbf{F}) \cdot d\boldsymbol{\sigma}, \tag{2}$$

under suitable restrictions (i) on the vector

$$\mathbf{F} = \mathbf{i}M + \mathbf{j}N + \mathbf{k}P, \tag{3}$$

(ii) on the simple closed curve C:

$$x = f(t), \quad y = g(t), \quad z = h(t), \quad 0 \leq t \leq 1, \tag{4}$$

(iii) on the surface

$$\sum: \phi(x, y, z) = 0 \tag{5}$$

bounded by C.

Example. Let $\sum$ be the hemisphere

$$z = \sqrt{4 - x^2 - y^2}, \quad 0 \leq x^2 + y^2 \leq 4, \tag{6}$$

lying above the xy-plane, with center at the origin. The boundary of this hemisphere is the circle C:

$$z = 0, \quad x^2 + y^2 = 4. \tag{7}$$

Let

$$\mathbf{F} = \mathbf{i}y - \mathbf{j}x. \tag{8}$$

The integrand in the line integral (1) is

$$\begin{aligned} \mathbf{F} \cdot d\mathbf{R} &= \mathbf{F} \cdot (\mathbf{i}\, dx + \mathbf{j}\, dy + \mathbf{k}\, dz) \\ &= y\, dx - x\, dy. \end{aligned} \tag{9}$$

By Green's theorem for *plane* curves and surfaces, we have

$$\begin{aligned} \oint_C \mathbf{F} \cdot dr &= \oint_C (y\, dx - x\, dy) \\ &= \iint_{x^2+y^2 \leq 4} -2\, dx\, dy \\ &= -8\pi. \end{aligned} \tag{10}$$

To evaluate the surface integral (2), we need to compute

$$\operatorname{curl} \mathbf{F} = \mathbf{i}\left(\frac{\partial P}{\partial y} - \frac{\partial N}{\partial z}\right) + \mathbf{j}\left(\frac{\partial M}{\partial z} - \frac{\partial P}{\partial x}\right) + \mathbf{k}\left(\frac{\partial N}{\partial x} - \frac{\partial M}{\partial y}\right), \tag{11}$$

where

$$M = y, \quad N = -x, \quad P = 0. \tag{12}$$

Substituting from (12) into (11), we get

$$\operatorname{curl} \mathbf{F} = -2\mathbf{k}. \tag{13}$$

The unit outer normal to the hemisphere of Eq. (6) is

$$\mathbf{n} = \frac{\mathbf{i}x + \mathbf{j}y + \mathbf{k}z}{\sqrt{x^2 + y^2 + z^2}} = \frac{\mathbf{i}x + \mathbf{j}y + \mathbf{k}z}{2}. \tag{14}$$

Therefore

$$\begin{aligned} \operatorname{curl} \mathbf{F} \cdot d\boldsymbol{\sigma} &= (\operatorname{curl} \mathbf{F} \cdot \mathbf{n})\, d\sigma \\ &= -z\, d\sigma. \end{aligned} \tag{15}$$

For element of surface area $d\sigma$ (Article 16.9), we use

$$d\sigma = \sqrt{1 + \left(\frac{\partial z}{\partial x}\right)^2 + \left(\frac{\partial z}{\partial y}\right)^2}\, dx\, dy, \tag{16}$$

with

$$\frac{\partial z}{\partial x} = \frac{-x}{\sqrt{4 - x^2 - y^2}} \quad \text{and} \quad \frac{\partial z}{\partial y} = \frac{-y}{\sqrt{4 - x^2 - y^2}},$$

or

$$\frac{\partial z}{\partial x} = \frac{-x}{z} \quad \text{and} \quad \frac{\partial z}{\partial y} = \frac{-y}{z}. \tag{17}$$

From (15), (16), and (17), we get

$$\operatorname{curl} \mathbf{F} \cdot d\boldsymbol{\sigma} = -z\, d\sigma = -z\sqrt{1 + \frac{x^2}{z^2} + \frac{y^2}{z^2}}\, dx\, dy. \tag{18}$$

Therefore

$$\iint_{\Sigma} \operatorname{curl} \mathbf{F} \cdot d\boldsymbol{\sigma} = \iint_{x^2+y^2 \leq 4} -2\, dx\, dy = -8\pi, \tag{19}$$

which agrees with the result found for the line integral in Eq. (10).

Remark 1. In this example, the surface integral (19) taken over the *hemisphere* turns out to have the same value as a surface integral taken over the *plane base* of that hemisphere. The underlying reason for this equality is that both surface integrals are equal to the line integral around the circle that is their common boundary.

Remark 2. In Stokes's theorem, we require that the surface be *orientable* and *simply connected*. By "simply connected," once again, we mean that any simple closed curve lying on the surface can be continuously contracted to a point (while staying on the surface) without its touching any part of the boundary of $\sum$. By "orientable," we mean that it is possible to consistently assign a unique direction, called *positive*, at each point of $\sum$, and that there exists a unit normal $\mathbf{n}$ pointing in this direction. As we move about over the surface $\sum$ without touching its boundary, the direction cosines of the unit vector $\mathbf{n}$ should vary continuously. Also, when we return to the starting position, $\mathbf{n}$ should return to its initial direction. This rules out such a surface as a Möbius

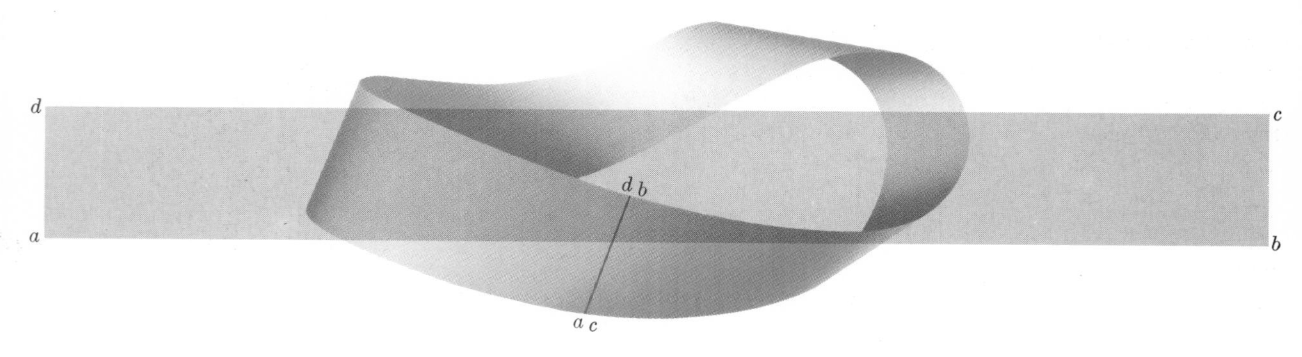

17.25 The construction of a Möbius strip.

strip, which can be constructed by taking a rectangular strip of paper $abcd$, giving the end bc a single twist to interchange the positions of the vertices b and c, and then pasting the ends of the strip together so as to bring vertices a and c together, and also b and d (see Fig. 17.25). The resulting surface is nonorientable because a unit normal vector (think of the shaft of a thumbtack) can be continuously moved around the surface without its touching the boundary of the surface, and in such a way that when it is returned to its initial position it will point in a direction exactly opposite to its initial direction.

Remark 3. We also want C to have a positive direction that is related to the positive direction on $\sum$. We imagine a simple closed curve Γ on $\sum$, near the boundary C, (see Fig. 17.26) and let $\mathbf{n}$ be normal to $\sum$ at some point inside Γ. We then assign to Γ a positive direction, the counterclockwise direction as viewed by an observer who is at the end of $\mathbf{n}$ and looking down. (Note that such a direction keeps the interior of Γ on the observer's left as he progresses around Γ. We could equally well have specified $\mathbf{n}$'s direction by this condition.) Now we move Γ about on $\sum$ until it touches and is tangent to C. The direction of the positive tangent to Γ at this point of common tangency we shall take to be the positive direction along C. It is a consequence of the orientability of $\sum$ that a consistent assignment of positive direction along C is induced by this process. The same positive direction is assigned all the way around C, no matter where on $\sum$ the process is begun. This would not be true of the (nonorientable) Möbius strip.

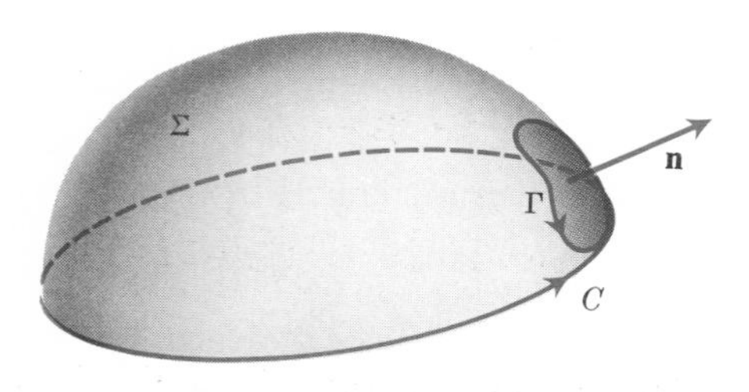

17.26 Orientation of the boundary of an oriented surface.

Stokes's theorem. *Let $\sum$ be a smooth, simply connected, orientable surface bounded by a simple closed curve C. Let*

$$\mathbf{F} = \mathbf{i}M + \mathbf{j}N + \mathbf{k}P, \tag{20}$$

where M, N, and P are continuous functions of (x, y, z), together with their first-order partial derivatives, throughout a region D containing S and C in its interior. Let $\mathbf{n}$ be a positive unit vector normal to S, and let the positive direction around C be the one induced by the positive orientation of S. Then

$$\oint_C \mathbf{F} \cdot d\mathbf{R} = \iint_S \operatorname{curl} \mathbf{F} \cdot d\boldsymbol{\sigma}, \tag{21a}$$

where

$$d\mathbf{R} = \mathbf{i}\,dx + \mathbf{j}\,dy + \mathbf{k}\,dz = \mathbf{T}\,ds \tag{21b}$$

and

$$d\boldsymbol{\sigma} = \mathbf{n}\,d\sigma = (\mathbf{i}\cos\alpha + \mathbf{j}\cos\beta + \mathbf{k}\cos\gamma)\,d\sigma. \tag{21c}$$

Proof for a polyhedral surface $\sum$. Let the surface $\sum$ be a polyhedral surface consisting of a finite number of plane regions. (This polyhedral surface $\sum$ might resemble a Buckminster Fuller geodesic dome.) We apply Green's theorem to each separate panel of $\sum$. There are two types of panels:

(1) those that are surrounded on all sides by other panels, and

(2) those that have one or more edges that are not adjacent to other panels.

Let Δ be part of the boundary of $\sum$ that consists of those edges of the type 2 panels that are not adjacent to other panels. In Fig. 17.27, the triangles ABE, BCE, and CDE represent a part of $\sum$, with $ABCD$ part of the boundary Δ. Applying Green's theorem to the three triangles in turn and adding the results, we get

$$\left(\oint_{ABE} + \oint_{BCE} + \oint_{CDE}\right) \mathbf{F} \cdot d\mathbf{R} = \left(\iint_{ABE} + \iint_{BCE} + \iint_{CDE}\right) \operatorname{curl} \mathbf{F} \cdot d\boldsymbol{\sigma}. \quad (22)$$

The three line integrals on the left-hand side of Eq. (22) combine into a single line integral taken around the periphery $ABCDE$, because the integrals along interior segments cancel in pairs. For example, the integral along the segment BE in triangle ABE is opposite in sign to the integral along the same segment in triangle EBC. Similarly for the segment CE. Hence (22) reduces to

$$\oint_{ABCDE} \mathbf{F} \cdot d\mathbf{R} = \iint_{ABCDE} \operatorname{curl} \mathbf{F} \cdot d\boldsymbol{\sigma}.$$

In general, when we apply Green's theorem to all the panels and add the results, we get

$$\oint_{\Delta} \mathbf{F} \cdot d\mathbf{R} = \iint_{\Sigma} \operatorname{curl} \mathbf{F} \cdot d\boldsymbol{\sigma}. \ \square \quad (23)$$

This is Stokes's theorem for a polyhedral surface $\sum$.

Remark 4. A rigorous proof of Stokes's theorem for more general surfaces is beyond the level of a beginning calculus course. (See, for example, Buck,

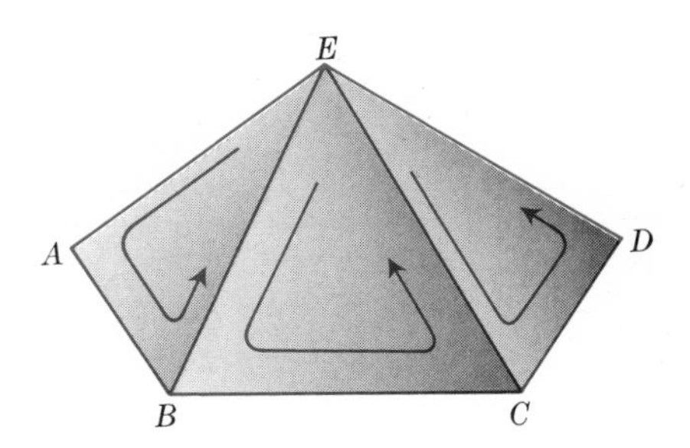

17.27

Advanced Calculus, p. 346, or Apostol, *Mathematical Analysis*, p. 335.) However, the following intuitive argument shows why one would expect Eq. (21a) to be true. Imagine a sequence of polyhedral surfaces

$$\textstyle\sum_1, \quad \sum_2, \quad \ldots,$$

and their corresponding boundaries $\Delta_1, \Delta_2, \ldots$ The surface $\sum_n$ should be constructed in such a way that its boundary Δ_n is inscribed in or tangent to C, the boundary of S, and so that the length of Δ_n approaches the length of C as $n \to \infty$. C needs to be rectifiable if this is to hold. The faces of $\sum_n$ might be polygonal regions, approximating pieces of S, and such that the area of $\sum_n$ approaches the area of S as $n \to \infty$. S also need to have finite area. Assuming that M, N, P, and their partial derivatives are continuous in a region D containing S and C, it is plausible to expect that

$$\oint_{\Delta_n} \mathbf{F} \cdot d\mathbf{R} \quad \text{approaches} \quad \oint_C \mathbf{F} \cdot d\mathbf{R}$$

and that

$$\iint_{\Sigma_n} \operatorname{curl} \mathbf{F} \cdot d\boldsymbol{\sigma}_n \quad \text{approaches} \quad \iint_S \operatorname{curl} \mathbf{F} \cdot d\boldsymbol{\sigma}$$

as $n \to \infty$. But if

$$\oint_{\Delta_n} \mathbf{F} \cdot d\mathbf{R} \to \oint_C \mathbf{F} \cdot d\mathbf{R} \quad (24a)$$

and

$$\iint_{\Sigma_n} \operatorname{curl} \mathbf{F} \cdot d\boldsymbol{\sigma} \to \iint_S \operatorname{curl} \mathbf{F} \cdot d\boldsymbol{\sigma}, \quad (24b)$$

and if the left-hand sides of (24a) and (24b) are equal by Stokes's theorem for polyhedra, we then have equality of their limits.

Problem 1. Let S be the portion of the paraboloid $z = 4 - x^2 - y^2$ that lies above the plane $z = 0$. Let C be their curve of intersection, and let

$$\mathbf{F} = \mathbf{i}(z - y) + \mathbf{j}(z + x) - \mathbf{k}(x + y).$$

Compute

$$\oint_C \mathbf{F} \cdot d\mathbf{R} \quad \text{and} \quad \iint_S \operatorname{curl} \mathbf{F} \cdot d\boldsymbol{\sigma}$$

and compare Eq. (21a).

Solution. The curve C lies in the xy-plane; it is the circle $x^2 + y^2 = 4$. We introduce a parameter θ such that

$$x = 2 \cos \theta, \quad y = 2 \sin \theta, \quad 0 \leq \theta \leq 2\pi.$$

Along C, we have

$$\mathbf{F} \cdot d\mathbf{R} = (z - y)\, dx + (z + x)\, dy - (x + y)\, dz,$$

or, since $z = 0$, $x = 2 \cos \theta$, $y = 2 \sin \theta$,

$$\begin{aligned} \mathbf{F} \cdot d\mathbf{R} &= -y\, dx + x\, dy \\ &\quad - 2 \sin \theta \cdot (-2 \sin \theta\, d\theta) \\ &\quad + 2 \cos \theta \cdot (2 \cos \theta\, d\theta) \\ &= 4\, d\theta \end{aligned}$$

and

$$\oint_C \mathbf{F} \cdot d\mathbf{R} = \int_0^{2\pi} 4\, d\theta = 8\pi.$$

For the surface integral, we compute

$$\operatorname{curl} \mathbf{F} = \begin{vmatrix} \mathbf{i} & \mathbf{j} & \mathbf{k} \\ \dfrac{\partial}{\partial x} & \dfrac{\partial}{\partial y} & \dfrac{\partial}{\partial z} \\ z - y & z + x & -x - y \end{vmatrix}$$

$$= -2\mathbf{i} + 2\mathbf{j} + 2\mathbf{k}.$$

For a positive unit normal on the surface

$$S\colon f(x, y, z) = z - 4 + x^2 + y^2 = 0,$$

we take

$$\mathbf{n} = \frac{\operatorname{grad} f}{|\operatorname{grad} f|} = \frac{2x\mathbf{i} + 2y\mathbf{j} + \mathbf{k}}{\sqrt{4x^2 + 4y^2 + 1}}.$$

The projection of S onto the xy-plane is the region

$$x^2 + y^2 \leq 4,$$

and for element of surface area $d\sigma$, we take

$$\begin{aligned} d\sigma &= \sqrt{\left(\frac{\partial z}{\partial x}\right)^2 + \left(\frac{\partial z}{\partial y}\right)^2 + 1}\; dx\, dy \\ &= \sqrt{4x^2 + 4y^2 + 1}\; dx\, dy. \end{aligned}$$

Thus

$$\begin{aligned} \iint_S \operatorname{curl} \mathbf{F} \cdot d\boldsymbol{\sigma} &= \iint_S \operatorname{curl} \mathbf{F} \cdot \mathbf{n}\, d\sigma \\ &= \iint_{x^2 + y^2 \leq 4} (-4x + 4y + 2)\, dx\, dy \qquad (\alpha) \\ &= \iint_{x^2 + y^2 \leq 4} 2\, dx\, dy = 8\pi, \qquad (\beta) \end{aligned}$$

where (β) follows from (α) because odd powers of x or y integrate to zero over the interior of the circle.

Remark 5. Stokes's theorem can also be extended to a surface S that has one or more holes in it (like a curved piece of Swiss cheese), in a way exactly analogous to Green's theorem: The surface integral over S of the *normal component* of curl $\mathbf{F}$ is equal to the line integral around all the boundaries of S (including boundaries of the holes) of the *tangential component* of $\mathbf{F}$, where the boundary curves are to be traced in the positive direction induced by the positive orientation of S.

Remark 6. Stokes's theorem provides the following vector interpretation for curl $\mathbf{F}$. As in the discussion of divergence, let $\mathbf{v}$ be the velocity field of a moving fluid, δ the density, and $\mathbf{F} = \delta\mathbf{v}$. Then

$$\oint_C \mathbf{F} \cdot \mathbf{T}\, ds$$

is a measure of the *circulation* of fluid around the closed curve C. By Stokes's theorem, this circulation is also equal to the flux of curl $\mathbf{F}$ through a surface S spanning C:

$$\oint_C \mathbf{F} \cdot d\mathbf{R} = \iint_S \operatorname{curl} \mathbf{F} \cdot d\boldsymbol{\sigma}.$$

Suppose we fix a point Q and a direction $\mathbf{u}$ at Q. Let C be a circle of radius ρ, with center at Q, whose plane is normal to $\mathbf{u}$. If curl $\mathbf{F}$ is continuous at Q, then the average value of the $\mathbf{u}$-component of curl $\mathbf{F}$

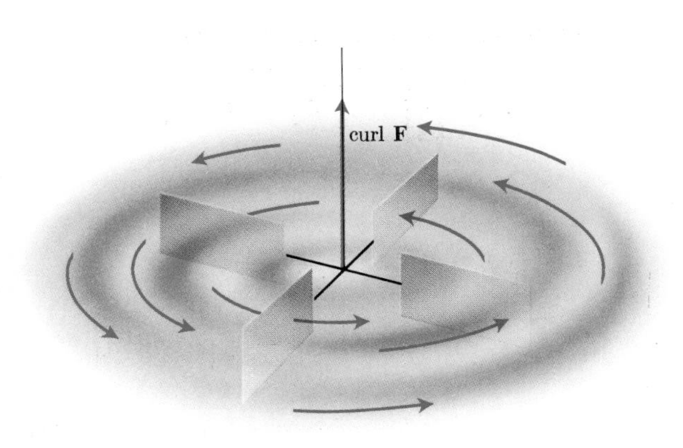

17.28 The paddle-wheel interpretation of curl **F**.

over the circular disk bounded by C approaches the **u**-component of curl **F** at Q as $\rho \to 0$:

$$(\text{curl } \mathbf{F} \cdot \mathbf{u})_Q = \lim_{\rho \to 0} \frac{1}{\pi \rho^2} \iint_S \text{curl } \mathbf{F} \cdot \mathbf{u} \, d\sigma. \qquad (25)$$

If we replace the double integral on the right-hand side of Eq. (25) by the circulation, we get

$$(\text{curl } \mathbf{F} \cdot \mathbf{u})_Q = \lim_{\rho \to 0} \frac{1}{\pi \rho^2} \oint_C \mathbf{F} \cdot d\mathbf{R}. \qquad (26)$$

The left-hand side of Eq. (26) is a maximum, at Q, when **u** has the same direction as curl **F**. When ρ is small, the right-hand side of Eq. (26) is approximately equal to

$$\frac{1}{\pi \rho^2} \oint_C \mathbf{F} \cdot d\mathbf{R},$$

which is the circulation around C divided by the area of the disk. Suppose that a small paddle wheel, of radius ρ, is introduced into the fluid at Q, with its axle directed along **u**. The circulation of the fluid around C will affect the rate of spin of the paddle wheel. The wheel will spin fastest when the circulation integral is maximized; therefore it will spin fastest when the axle of the paddle wheel points in the direction of curl **F**. (See Fig. 17.28.)

Problem 2. A fluid of constant density δ rotates around the z-axis with velocity $\mathbf{v} = \omega(\mathbf{j}x - \mathbf{i}y)$, where ω is a positive constant. If $\mathbf{F} = \delta\mathbf{v}$, find curl **F**, and comment.

Solution

$$\mathbf{F} = \delta\omega(\mathbf{j}x - \mathbf{i}y),$$

and

$$\text{curl } \mathbf{F} = \begin{vmatrix} \mathbf{i} & \mathbf{j} & \mathbf{k} \\ \dfrac{\partial}{\partial x} & \dfrac{\partial}{\partial y} & \dfrac{\partial}{\partial z} \\ -\delta\omega y & \delta\omega x & 0 \end{vmatrix} = 2\,\delta\omega\mathbf{k}.$$

The work done by a force equal to **F**, as the point of application moves around a circle C of radius ρ, is

$$\oint_C \mathbf{F} \cdot d\mathbf{R}.$$

If C is in a plane parallel to the xy-plane, Stokes's theorem gives the result

$$\oint_C \mathbf{F} \cdot d\mathbf{R} = \iint_S \text{curl } \mathbf{F} \cdot d\boldsymbol{\sigma}$$

$$= \iint 2\,\delta\omega\mathbf{k} \cdot \mathbf{k} \, dx \, dy$$

$$= (2\,\delta\omega)(\pi\rho^2).$$

We note that

$$(\text{curl } \mathbf{F}) \cdot \mathbf{k} = 2\,\delta\omega = \frac{1}{\pi\rho^2} \oint_C \mathbf{F} \cdot d\mathbf{r},$$

in agreement with Eq. (26) with $\mathbf{u} = \mathbf{k}$.

EXERCISES

In Exercises 1 through 4, verify the result of Stokes's theorem for the vector

$$\mathbf{F} = \mathbf{i}(y^2 + z^2) + \mathbf{j}(x^2 + z^2) + \mathbf{k}(x^2 + y^2)$$

for the given surface S and boundary C.

1. S: $z = \sqrt{1 - x^2}$, $-1 \le x \le 1$, $-2 \le y \le 2$,
   $y = 2$, $0 \le z \le \sqrt{1 - x^2}$, $-1 \le x \le 1$,
   $y = -2$, $0 \le z \le \sqrt{1 - x^2}$, $-1 \le x \le 1$;
   C: $z = 0$,
   $x = \pm 1$, $-2 \le y \le 2$,
   $y = \pm 2$, $-1 \le x \le 1$

2. The surface S is the surface of the upper half of the cube with one vertex at $(1, 1, 1)$, center at the origin, and edges parallel to the axes; the curve C is the intersection of S with the xy-plane.
3. The surface S is as in Exercise 2, with a hole cut out of the top face by the circular disk whose cylindrical coordinates satisfy

$$z = 1, \quad 0 \le r \le \cos\theta, \quad -\tfrac{1}{2}\pi \le \theta \le \tfrac{1}{2}\pi.$$

(The circle $z = 1$, $r = \cos\theta$ becomes part of the boundary of S.)
4. The surface S is the surface (excluding the face in the yz-plane) of a pyramid with vertices at the origin O and at $A(1, 0, 0)$, $B(0, 1, 0)$, and $D(0, 0, 1)$; the boundary curve C is the triangle OBD in the yz-plane.
5. Suppose $\mathbf{F} = \operatorname{grad}\phi$ is the gradient of a scalar function ϕ having continuous second-order partial derivatives

$$\frac{\partial^2\phi}{\partial x^2}, \frac{\partial^2\phi}{\partial x\,\partial y}, \ldots$$

throughout a simply connected region D that contains the surface S and its boundary C in the interior of D. What constant value does

$$\oint_C \mathbf{F}\cdot d\mathbf{R}$$

have in such circumstances? Explain.
6. Let $\phi = (x^2 + y^2 + z^2)^{-1/2}$. Let V be the spherical shell $1 \le x^2 + y^2 + z^2 \le 4$. Let $\mathbf{F} = \operatorname{grad}\phi$. If $1 < a < 2$ and C is the circle $z = 0$, $x^2 + y^2 = a^2$, show that

$$\oint_C \mathbf{F}\cdot d\mathbf{R} = 0$$

(a) by direct evaluation of the integral, and

(b) by applying Stokes's theorem with S the hemisphere $z = \sqrt{a^2 - x^2 - y^2}$, $x^2 + y^2 \le a^2$.
7. If the components of $\mathbf{F}$ have continuous second-order partial derivatives of all types, prove that

$$\operatorname{div}(\operatorname{curl}\mathbf{F}) = 0.$$

8. Use the result of Exercise 7 and the divergence theorem to show that

$$\iint_S \operatorname{curl}\mathbf{F}\cdot d\boldsymbol{\sigma} = 0$$

if the components of $\mathbf{F}$ have continuous second-order derivatives and S is a closed surface like a sphere, an ellipsoid, or a cube.
9. By Stokes's theorem, if S_1 and S_2 are two oriented surfaces having the same positively oriented curve C as boundary, then

$$\iint_{S_1} \operatorname{curl}\mathbf{F}\cdot d\boldsymbol{\sigma}_1 = \oint_C \mathbf{F}\cdot d\mathbf{r} = \iint_{S_2} \operatorname{curl}\mathbf{F}\cdot d\boldsymbol{\sigma}_2.$$

Deduce that

$$\iint_S \operatorname{curl}\mathbf{F}\cdot d\boldsymbol{\sigma}$$

has the same value for all oriented surfaces S that span C and that induce the same positive direction on C.
10. Use Stokes's theorem to deduce that if $\operatorname{curl}\mathbf{F} = \mathbf{0}$ throughout a simply connected region D, then

$$\int_{P_1}^{P_2} \mathbf{F}\cdot d\mathbf{R}$$

has the same value for all simple paths lying in D and joining P_1 and P_2.

REVIEW QUESTIONS AND EXERCISES

1. What is a vector field? Give an example of a two-dimensional vector field; of a three-dimensional field.
2. What is the velocity vector field for a fluid rotating about the x-axis if the angular velocity is a constant ω and the flow is counterclockwise as viewed by an observer at $(1, 0, 0)$ looking toward $(0, 0, 0)$?
3. Give examples of gradient fields

(a) in E^2, (b) in E^3.

State a property that gradient fields have and other fields do not.
4. If $\mathbf{F} = \nabla f$ is a gradient field, S is a level surface for f, and C is a curve on S, why is it true (or not true) that $\int_C \mathbf{F}\cdot d\mathbf{R} = 0$?
5. Suppose that Σ is a portion of a level surface S of a function $f(x, y, z)$. How could you select an orientation (if S is orientable) on S such that

$$\iint_\Sigma \nabla f\cdot\mathbf{n}\,d\sigma = \iint_\Sigma |\nabla f|\,d\sigma?$$

Why is the hypothesis that S is a level surface of f important?

6. If $f(x, y, z) = 2x - 3y + e^z$, and C is any smooth curve from $A(0, 0, 0)$ to $(1, 2, \ln 3)$, what is the value of $\int_C \nabla f \cdot d\mathbf{R}$? Why doesn't the answer depend on C?
7. Write a formula for a vector field $\mathbf{F}(x, y, z)$ such that $\mathbf{F} = f(\rho)\mathbf{R}$, where $\rho = |\mathbf{R}|$ and $\mathbf{R} = \mathbf{i}x + \mathbf{j}y + \mathbf{k}z$, if it is true that

   (a) $\mathbf{F}$ is directed radially outward from the origin and $|\mathbf{F}| = \rho^{-n}$,

   (b) $\mathbf{F}$ is directed toward the origin and $|\mathbf{F}| = 2$,

   (c) $\mathbf{F}$ is a gravitational attraction field for a mass M at the origin in which an inverse-*cube* law applies. (Don't worry about the possible nonexistence of such a field.)
8. Give one physical and one geometrical interpretation for a surface integral
$$\iint_{\Sigma} h\, d\sigma.$$
9. State both the normal form and the tangential form of Green's theorem in the plane.
10. State the divergence theorem and show that it applies to the region D described by $1 \leq |\mathbf{R}| \leq 2$, assuming that $\sum$ is the total boundary of this region, $\mathbf{n}$ is directed away from D at each point, and $\mathbf{F} = \nabla(1/|\mathbf{R}|)$, $\mathbf{R} = \mathbf{i}x + \mathbf{j}y + \mathbf{k}z$.

MISCELLANEOUS EXERCISES

In Exercises 1 through 4, describe the vector fields in words and with graphs.

1. $\mathbf{F} = x\mathbf{i} + y\mathbf{j} + z\mathbf{k}$
2. $\mathbf{F} = -x\mathbf{i} - y\mathbf{j} - z\mathbf{k}$
3. $\mathbf{F} = (x - y)\mathbf{i} + (x + y)\mathbf{j}$
4. $\mathbf{F} = (x\mathbf{i} + y\mathbf{j})/(x^2 + y^2)$

In Exercises 5 through 8, evaluate the surface integrals
$$\iint_{\Sigma} h\, d\sigma$$
for the given functions and surfaces.

5. The surface $\sum$ is the hemisphere $x^2 + y^2 + z^2 = a^2$, $z \geq 0$, and $h(x, y, z) = x + y$.
6. The surface $\sum$ is the portion of the plane $z = x + y$ for which $x \geq 0$, $y \geq 0$, $z \leq \pi$, and
$$h(x, y, z) = \sin z.$$
7. The surface $\sum$ is $z = 4 - x^2 - y^2$, $z \geq 0$; $h = z$.
8. The surface $\sum$ is the sphere $\rho = a$, and $h = z^2$. (You might do the upper and lower hemispheres separately and add; or use spherical coordinates and not project the surface.)

For the functions and surfaces given in Exercises 9 through 12, evaluate
$$\iint_{\Sigma} \frac{\partial f}{\partial n}\, d\sigma,$$
where $\partial f/\partial n$ is the directional derivative of f in the direction of the normal $\mathbf{n}$ in the sense specified.

9. The surface $\sum$ is the sphere $x^2 + y^2 + z^2 = a^2$, $\mathbf{n}$ is directed outward on $\sum$, and $f = x^2 + y^2 + z^2$.
10. The surface and normal are as in Exercise 9, and $f = (x^2 + y^2 + z^2)^{-1}$.
11. The surface $\sum$ is the portion of the plane
$$z = 2x + 3y$$
for which $x \geq 0$, $y \geq 0$, $z \leq 5$, $\mathbf{n}$ has a positive $\mathbf{k}$-component, and $f = x + y + z$.
12. The surface $\sum$ is the one-eighth of the sphere $x^2 + y^2 + z^2 = a^2$ that lies in the first octant, $\mathbf{n}$ is directed inward with respect to the sphere, and $f = \ln (x^2 + y^2 + z^2)^{1/2}$.

In Exercises 13 through 20, evaluate the line integrals
$$\int_C \mathbf{F} \cdot d\mathbf{R}$$
for the given fields $\mathbf{F}$ and paths C.

13. The field $\mathbf{F} = x\mathbf{i} + y\mathbf{j}$ and the circle C:
$$x = \cos t, \quad y = \sin t, \quad 0 \leq t \leq 2\pi$$
14. The field $\mathbf{F} = -y\mathbf{i} + x\mathbf{j}$ and C as in Exercise 13
15. The field $\mathbf{F} = (x - y)\mathbf{i} + (x + y)\mathbf{j}$ and C as in Exercise 13
16. The field $\mathbf{F} = x\mathbf{i} + y\mathbf{j} + z\mathbf{k}$ and the ellipse C, in which the plane $z = 2x + 3y$ cuts the cylinder $x^2 + y^2 = 12$, counterclockwise as viewed from the positive end of the z-axis looking toward the origin

17. The field $\mathbf{F} = \nabla(xy^2z^3)$ and C as in Exercise 16
18. The field $\mathbf{F} = \nabla\mathbf{x}(x\mathbf{i} + y\mathbf{j} + z\mathbf{k})$ and C as in Exercise 13
19. The field $\mathbf{F}$ as in Exercise 18 and C the line segment from the origin to the point (1, 2, 3)
20. The field $\mathbf{F}$ as in Exercise 17 and C the line segment from (1, 1, 1) to (2, 1, −1)
21. The First Law of Thermodynamics says that heat flows from a hotter body to a cooler body. In three-dimensional heat flow, the fundamental equation for the rate at which heat flows out of D is

$$\iint_\Sigma K\frac{\partial u}{\partial n}\,d\sigma = \iiint_D c\delta\frac{\partial u}{\partial t}\,dV. \tag{1}$$

The symbolism in this equation is as follows:

$u = u(x, y, z, t)$ the temperature at the point (x, y, z) at time t,

K the thermal conductivity coefficient,

δ the mass density,

c the specific heat coefficient. This is the amount of heat required to raise one unit of mass of the material of the body one degree,

Σ the boundary surface of the region D,

$\dfrac{\partial u}{\partial n}$ the directional derivative in the direction of the outward normal to Σ.

How is $\partial u/\partial n$ related to the *gradient* of the temperature? In which direction (described in words) does ∇u point? Why does the left-hand side of Eq. (1) appear to make sense as a measure of the rate of flow? Now look at the right-hand side of Eq. (1): If ΔV is a small volume element in D, what does $\delta\,\Delta V$ represent? If the temperature of this element changes by an amount Δu in time Δt, what is

(a) the amount, (b) the average rate

of change of heat in the element? In words, what does the right side of Eq. (1) represent physically? Is it reasonable to interpret Eq. (1) as saying that the rate at which heat flows out through the boundary of D is equal to the rate at which heat is being supplied from D?

22. Assuming Eq. (1), Exercise 21, and assuming that there is no heat source or sink in D, derive the equation

$$\nabla\cdot(K\nabla u) = c\delta\frac{\partial u}{\partial t} \tag{2}$$

as the equation that must be satisfied at each point in D.

Suggestion. Apply the divergence theorem to the left-hand side of Eq. (1), and make D be a sphere of radius ϵ; then let $\epsilon \to 0$.

23. Assuming the result of Exercise 22, and assuming that K, c, and δ are constants, deduce that the condition for steady-state temperature in D is Laplace's equation

$$\nabla^2 u = 0, \quad \text{or} \quad \text{div (grad } u) = 0.$$

In higher mathematics, the symbol Δ is used for the *Laplace operator:*

$$\Delta u = \frac{\partial^2 u}{\partial x^2} + \frac{\partial^2 u}{\partial y^2} + \frac{\partial^2 u}{\partial z^2}.$$

Thus, in this notation,

$$\Delta u = \nabla^2 u = \nabla\cdot\nabla u = \text{div (grad } u).$$

Using the divergence theorem, and assuming that the functions u and v and their first- and second-order partial derivatives are continuous in the regions considered, verify the formulas in Exercises 24 through 27. Assume that Σ is the boundary surface of the simply connected region D.

24. $$\iint_\Sigma u\nabla v\cdot d\boldsymbol{\sigma} = \iiint_D [u\,\Delta v + (\nabla u)\cdot(\nabla v)]\,dV$$

25. $$\iint_\Sigma \left(u\frac{\partial v}{\partial n} - v\frac{\partial u}{\partial n}\right)d\sigma = \iiint_D (u\,\Delta v - v\,\Delta u)\,dV$$

Suggestion. Use the result of Exercise 24 as given and in the form you get by interchanging u and v.

26. $$\iint_\Sigma u\frac{\partial u}{\partial n}\,d\sigma = \iiint_D [u\,\Delta u + |\nabla u|^2]\,dV$$

Suggestion. Use the result of Exercise 24 with $v = u$.

27. $$\iint_\Sigma \frac{\partial u}{\partial n}\,d\sigma = \iiint_D \Delta u\,dV$$

Suggestion. Use the result of Exercise 25 with $v = -1$.

28. A function u is *harmonic* in a region D if and only if it satisfies *Laplace's equation* $\Delta u = 0$ throughout D. Use the identity in Exercise 26 to deduce that if u is harmonic in D and either $u = 0$ or $\partial u/\partial n = 0$ at all points on the surface $\sum$ that is the boundary of D, then $\nabla u = \mathbf{0}$ throughout D, and therefore u is constant throughout D.

29. The result of Exercise 28 can be used to establish the uniqueness of solutions of Laplace's equation in D, provided that either (i) the value of u is prescribed at each point on $\sum$, or (ii) the value of $\partial u/\partial n$ is prescribed at each point of $\sum$. This is done by supposing that u_1 and u_2 are harmonic in D and that both satisfy the same boundary conditions, and then letting $u = u_1 - u_2$. Complete this uniqueness proof.

30. Exercises 21 through 23 deal with heat flow. Assume that K, c, and δ are constant and that the temperature $u = u(x, y, z)$ does not vary with time. Use the results of Exercises 23 and 27 to conclude that the net rate of outflow of heat through the surface $\sum$ is zero.

    Note. This result might apply, for example, to the region D between two concentric spheres if the inner one were maintained at 100° and the outer one at 0°, so that heat would flow into D through the inner surface and out through the outer surface at the same rate.

31. Let $\rho = (x^2 + y^2 + z^2)^{1/2}$. Show that

$$u = C_1 + C_2/\rho$$

is harmonic where $\rho > 0$, if C_1 and C_2 are constants. Find values of C_1 and C_2 so that the following boundary conditions are satisfied:

(a) $u = 100$ when $\rho = 1$, and $u = 0$ when $\rho = 2$,
(b) $u = 100$ when $\rho = 1$, and $\partial u/\partial n = 0$ when $\rho = 2$.

Note. Part (a) refers to a steady-state heat flow problem that is like the one discussed at the end of Exercise 30; part (b) refers to an insulated boundary on the sphere $\rho = 2$.

INFINITE SERIES

CHAPTER 18

18.1 INTRODUCTION AND DEFINITIONS

In Article 10.3, we became acquainted with Newton's method for finding a root of an equation $f(x) = 0$. The procedure was

(a) to guess a first approximation x_1, and

(b) to get a new approximation x_{n+1} from any current approximation x_n, by means of the formula

$$x_{n+1} = x_n - \frac{f(x_n)}{f'(x_n)}. \tag{1}$$

In theory, we might continue this procedure indefinitely, producing an infinite succession of approximations $x_1, x_2, x_3, \ldots$ Indeed, to each positive integer n there corresponds an nth approximation x_n. The set of all ordered pairs of the form (n, x_n), $n = 1, 2, \ldots$, is an example of a *sequence.*

There are many other ways of producing sequences. Consider, for example, a geometric progression like $1, \frac{1}{2}, \frac{1}{4}, \frac{1}{8}, \ldots$, where each term is one-half its predecessor. Suppose we number the terms $1, 2, 3, \ldots$ in order, as in the following table.

Number of term (index)	1	2	3	4	...	n	...
Value of term	1	$\frac{1}{2}$	$\frac{1}{4}$	$\frac{1}{8}$	...	$(\frac{1}{2})^{n-1}$	...

We now consider the set of all pairs of the form $(n, (\frac{1}{2})^{n-1})$, where the first element in each pair is the index that shows the ordinal number of that pair, and the second element is the value of the corresponding term of the geometric progression. This set is another example of a sequence, in agreement with the following general definition.

Definition. *A sequence is a function whose domain is the set of positive integers.*

We may denote the function by f. Then its value at n is $f(n)$. The sequence f is the set

$$\{(n, f(\text{n})) : n = 1, 2, 3, \ldots\};$$

that is, the set of all pairs $(n, f(n))$, with n a positive integer.

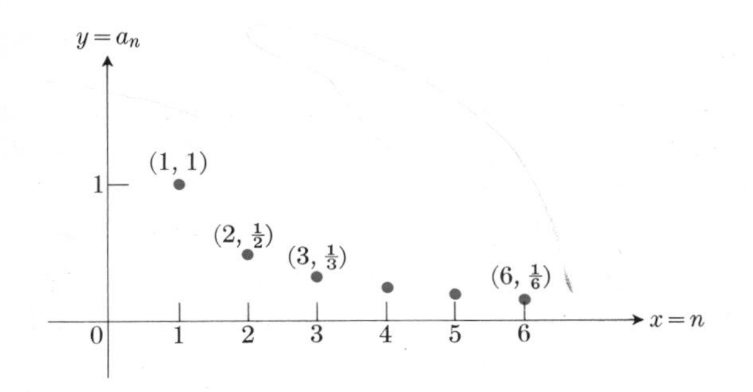

18.1 The graph of the sequence $\{(n, 1/n : n = 1, 2, \ldots\}$.

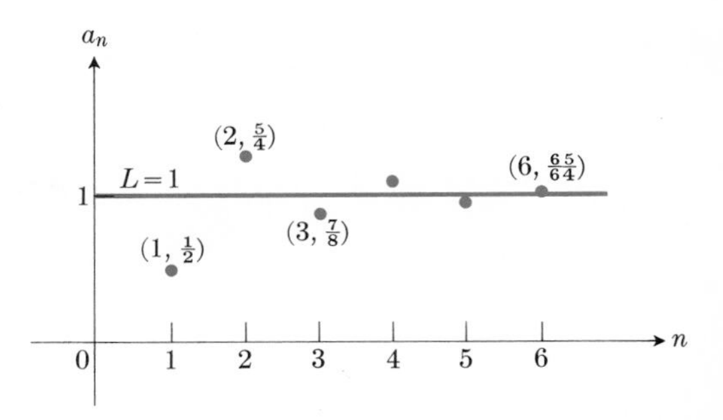

18.2 The graph of the sequence $\{1 + (-1)^n/2^n\}$, and the limit $L = 1$.

Example 1. The set $\{(n, 1/n) : n = 1, 2, 3, \ldots\}$ is a sequence whose value at n is $1/n$. (See Fig. 18.1.)

Notation. Since the domain of a sequence is always the same (the set of positive integers), it is customary to shorten the notation and just write $\{f(n)\}$ instead of $\{(n, f(n))\}$. Thus the sequence

$$\{(n, 1/n) : n = 1, 2, 3, \ldots\}$$

would be abbreviated simply $\{1/n\}$. Similarly $\{1/2^{n-1}\}$ would denote the sequence

$$\{(n, 1/2^{n-1}) : n = 1, 2, 3, \ldots\}.$$

We also write $\{a_n\}$ to denote the sequence whose ordinate is $y = a_n$ at the abscissa $x = n$.

Limit of a sequence

A sequence $\{(n, a_n)\}$ may have a different value a_n for each different value of n. But it may also happen that as n increases, the different a_n's tend to cluster around some fixed number L. If there is such a number L, with the property that $|L - a_n|$ is arbitrarily small for all sufficiently large values of the index n, we say that a_n *converges to L as limit*, and write

$$\lim_{n \to \infty} a_n = L.$$

By the phrase "$|L - a_n|$ is arbitrarily small for all sufficiently large values of n," we mean that to any positive number ϵ there corresponds an index N such that

$$|L - a_n| < \epsilon \quad \text{for all } n > N. \tag{2}$$

That is, all terms after the Nth lie within the distance ϵ from L. If no such limit exists, then we say that the given sequence *diverges*.

Remark 1. The sequence $\{(n, a_n)\}$ has the limit L if and only if its graph has the line $y = L$ as horizontal asymptote. For example, the x-axis ($y = 0$) is an asymptote of the graph of $\{(n, 1/n)\}$, and also of the graph of $\{(n, (\frac{1}{2})^{n-1})\}$. Both sequences converge to zero as $n \to \infty$. Again, the sequence

$$\left\{(n, a_n) : a_n = 1 + \frac{(-1)^n}{2^n}\right\}$$

converges to 1 as $n \to \infty$. (See Fig. 18.2.)

Remark 2. As we recalled above, Newton's method for finding a solution of an equation $f(x) = 0$ is to guess a first approximation x_1 and then to use the formula

$$x_{n+1} = x_n - \frac{f(x_n)}{f'(x_n)}$$

to generate, by iteration, a sequence $\{x_n\}$. In favorable circumstances, the sequence converges to a number L that satisfies the equation $f(L) = 0$.

Remark 3. Iteration is used as a method for solving many different types of problems, particularly with high-speed computing machines. The technique is to start with a first approximation, use that to compute a second approximation, the second to get a third, then a fourth, and so on. The process generates a sequence which we hope will converge to a solution of the problem.*

Infinite series

One very important way of creating a sequence $\{s_n\}$ is by addition. Suppose the numbers to be added are $u_1, u_2, u_3, \ldots, u_n, \ldots$ We let

$$\begin{aligned} s_1 &= u_1, \\ s_2 &= u_1 + u_2, \\ s_3 &= u_1 + u_2 + u_3, \\ &\vdots \\ s_n &= u_1 + u_2 + \cdots + u_n = \sum_{k=1}^{n} u_k. \end{aligned} \tag{3}$$

As n increases without bound, we are led to consider the symbol, or "set of marks,"

$$u_1 + u_2 + u_n + \cdots + u_n + \cdots, \tag{4a}$$

which we shall also denote by

$$\sum_{k=1}^{\infty} u_k. \tag{4b}$$

Such an expression is called an *infinite series.* One object in the present chapter is to see what meaning, if any, can be attached to an expression such as (4).

The sequence $\{s_n\}$, Eq. (3), is called the *sequence of partial sums* of the series (4a, b). The nth partial sum of the series is s_n. If s_n converges to a limit S as $n \to \infty$:

$$\lim_{n\to\infty} s_n = S, \tag{5}$$

then we say that the series *converges* and that its *sum* is S. We indicate this by writing

$$\sum_{k=1}^{\infty} u_k = \lim_{n\to\infty} \left(\sum_{k=1}^{n} u_k \right) = S. \tag{6}$$

On the other hand, if the sequence of partial sums diverges, we then say that the series *diverges.* In other words, the behavior of the series (convergence or divergence) is the behavior of its sequence of partial sums. Later in this chapter we shall study several methods for testing a given series in order to determine whether it converges or diverges. For the present, we shall want to use the following theorem.

Theorem 1. *A necessary condition for the convergence of an infinite series*

$$u_1 + u_2 + \cdots + u_n + \cdots$$

is that

$$\lim_{n\to\infty} u_n = 0. \tag{7}$$

Remark 4. The statement that the condition (7) is a "necessary condition for convergence" means that the condition must be satisfied if the series converges. Therefore it follows that when (7) is *not* satisfied, then the series *diverges.* This is frequently a useful test for divergence; namely, if the nth term of the series does not have a limit, or if it has a limit different from zero, as $n \to \infty$, then the series diverges. For example, the series

$$\frac{1}{2} + \frac{2}{3} + \frac{3}{4} + \frac{4}{5} + \cdots + \frac{n}{n+1} + \cdots$$

diverges, since

$$u_n = \frac{n}{n+1}$$

approaches 1 instead of 0 as $n \to \infty$. *Caution.* The theorem does *not* give a *sufficient* condition for convergence. That is, it does not follow that a series converges just because its nth term approaches zero as $n \to \infty$. The truth of this statement is illustrated by the series

$$\begin{aligned} &\tfrac{1}{2} + \underbrace{\tfrac{1}{4} + \tfrac{1}{4}}_{\text{2 terms}} + \underbrace{\tfrac{1}{8} + \tfrac{1}{8} + \tfrac{1}{8} + \tfrac{1}{8}}_{\text{4 terms}} \\ &+ \underbrace{\tfrac{1}{16} + \tfrac{1}{16} + \cdots + \tfrac{1}{16}}_{\text{8 terms}} + \cdots \\ &+ \underbrace{\frac{1}{2^k} + \frac{1}{2^k} + \cdots + \frac{1}{2^k}}_{2^{k-1}\text{ terms}} + \cdots \end{aligned} \tag{8}$$

---

* For a discussion of some specific examples, see the article "Feed it back," by Francis Scheid, in the *Mathematics Teacher,* Vol. LII, No. 4, April 1959, pp. 226–229.

Proof. Let

$$s_n = u_1 + u_2 + \cdots + u_n$$

denote the nth partial sum of the series. Suppose the series converges to S; that is, suppose

$$\lim_{n\to\infty} s_n = S.$$

Then, corresponding to any preassigned positive number ϵ, there is an index N such that all terms of the sequence $\{s_n\}$ after the Nth one lie between $S - \epsilon/2$ and $S + \epsilon/2$. Hence no two of them may differ by as much as ϵ. That is, if m and n are both greater than N, then

$$|s_n - s_m| < \epsilon.$$

In particular, this inequality holds if

$$m = n - 1 \quad \text{and} \quad n > N + 1.$$

But

$$\begin{aligned} s_n - s_{n-1} &= (u_1 + u_2 + \cdots + u_n) \\ &\quad - (u_1 + u_2 + \cdots + u_{n-1}) \\ &= u_n. \end{aligned}$$

Therefore

$$|u_n| < \epsilon \quad \text{when } n > N + 1.$$

Since ϵ was any positive number whatever, this means that

$$\lim_{n\to\infty} u_n = 0. \ \square$$

Remark 5. It is worth noting that the sum

$$u_1 + u_2 + \cdots + u_n$$

is always defined (if n is a positive integer and $u_1, u_2, \ldots, u_n$ are finite numbers). When the series

$$u_1 + u_2 + \cdots + u_n + \cdots$$

has a "sum," in the sense that its sequence of partial sums has a *limit,* the symbol "+" has acquired new meaning, as has the word "sum." We can now "add" infinitely many numbers, in certain cases, not by actual addition, but rather by the process of finding a *limit.* This is a typical illustration of the way in which mathematics generalizes familiar concepts to increase their usefulness.

Example 2. We shall illustrate the method of finding the "sum" of an infinite series for the case of the repeating decimal

$$0.3333\cdots = \frac{3}{10} + \frac{3}{100} + \frac{3}{1000} + \frac{3}{10{,}000} + \cdots.$$

Here

$$s_1 = \frac{3}{10},$$

$$s_2 = \frac{3}{10} + \frac{3}{10^2},$$

$$\vdots$$

$$s_n = \frac{3}{10} + \frac{3}{10^2} + \cdots + \frac{3}{10^n}.$$

We can obtain an explicit algebraic expression for s_n as follows. We multiply both sides of the equation for s_n by $\frac{1}{10}$ and obtain

$$\frac{1}{10} s_n = \frac{3}{10^2} + \frac{3}{10^3} + \cdots + \frac{3}{10^n} + \frac{3}{10^{n+1}}.$$

When we subtract this from s_n, we have

$$s_n - \frac{1}{10} s_n = \frac{3}{10} - \frac{3}{10^{n+1}} = \frac{3}{10}\left(1 - \frac{1}{10^n}\right).$$

Therefore

$$\frac{9}{10} s_n = \frac{3}{10}\left(1 - \frac{1}{10^n}\right),$$

or

$$s_n = \frac{3}{9}\left(1 - \frac{1}{10^n}\right).$$

Clearly, as n approaches ∞, $(\frac{1}{10})^n$ approaches 0, and

$$\lim_{n\to\infty} s_n = \tfrac{3}{9} = \tfrac{1}{3}.$$

We can therefore write the sum of the infinite series:

$$\frac{3}{10} + \frac{3}{10^2} + \frac{3}{10^3} + \cdots + \frac{3}{10^n} + \cdots = \frac{1}{3}.$$

The repeating-decimal illustration is a special case of a *geometric series.*

Definition. *A series of the form*

$$a + ar + ar^2 + ar^3 + \cdots + ar^{n-1} + \cdots \tag{9}$$

is called a geometric series. The ratio of any term to the one before it is r, if $a \neq 0$.

The sum of the first n terms of (9) is

$$s_n = a + ar + ar^2 + \cdots + ar^{n-1}. \tag{10}$$

Multiplying both sides of (10) by r gives

$$rs_n = ar + ar^2 + \cdots + ar^{n-1} + ar^n. \tag{11}$$

When we subtract (11) from (10), many terms cancel on the right-hand side, leaving

$$(1 - r)s_n = a(1 - r^n). \tag{12}$$

If $r \neq 1$, we may divide (12) by $(1 - r)$:

$$s_n = \frac{a(1 - r^n)}{1 - r}, \quad r \neq 1. \tag{13a}$$

On the other hand, if $r = 1$ in (10), we get

$$s_n = na, \quad r = 1. \tag{13b}$$

We are interested in the limit as $n \to \infty$ in Eqs. (13a) and (13b).

Clearly (13b), with $r = 1$, has no finite limit if $a \neq 0$. If $a = 0$, the series (9) is just

$$0 + 0 + 0 + \cdots,$$

and it converges to the sum zero.

If $r \neq 1$, we use (13a). In the right-hand side of (13a), n appears only in the expression r^n. This approaches zero as $n \to \infty$, if $|r| < 1$. Therefore

$$\lim_{n\to\infty} s_n = \lim_{n\to\infty} \frac{a(1 - r^n)}{1 - r}$$
$$= \frac{a}{1 - r}, \quad \text{if} \quad |r| < 1. \tag{14}$$

If $|r| > 1$, then $|r^n| \to \infty$, and (9) diverges if $a \neq 0$.

The last case is the one in which $r = -1$. Then

$$s_1 = a, \quad s_2 = a - a = 0, \quad s_3 = a, \quad s_4 = 0, \quad \ldots$$

If $a \neq 0$, then this sequence of partial sums has no limit as $n \to \infty$, and the series (9) diverges.

We have thus proved the following theorem.

Theorem 2. *If $|r| < 1$, then the geometric series*

$$a + ar + ar^2 + \cdots + ar^{n-1} + \cdots$$

converges to the sum $a/(1 - r)$. If $|r| \geq 1$, the series diverges unless $a = 0$. If $a = 0$, the series converges to the sum 0.

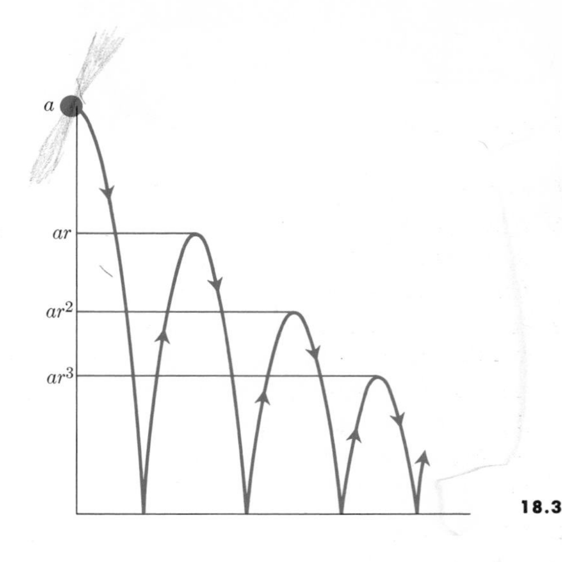

18.3

Problem. A ball is dropped from a point a feet above a flat surface. Each time the ball hits the ground after falling a distance h it rebounds a distance rh, where r is a positive fraction less than one. Find the total distance the ball travels.

Solution. (See Fig. 18.3.) The distance is given by the series

$$s = a + 2ar + 2ar^2 + 2ar^3 + \cdots.$$

The terms following the first one form a geometric series of sum $2ar/(1 - r)$. Hence the distance is

$$s = a + \frac{2ar}{1 - r} = a\frac{1 + r}{1 - r}.$$

For instance, if $a = 6$ ft and $r = \frac{2}{3}$, the distance is

$$s = 6\frac{1 + \frac{2}{3}}{1 - \frac{2}{3}} = 30 \text{ ft}.$$

EXERCISES

Determine which of the following sequences $\{a_n\}$ converge and which of them diverge. Find the limit of each sequence that converges.

1. $a_n = \dfrac{2n + 1}{1 - 3n}$
2. $a_n = \dfrac{n^2 - n}{2n^2 + n}$
3. $a_n = 1 + \dfrac{(-1)^n}{n}$
4. $a_n = \dfrac{1 + (-1)^n}{n}$
5. $a_n = 1 + (-1)^n$
6. $a_n = \sin n$

7. $a_n = \tanh n$

8. $a_n = \dfrac{\ln n}{n}$

9. $a_n = e^n$

10. $a_n = e^{-n}$

11. Prove that a given sequence $\{a_n\}$ cannot converge to two different limits L and L'.

Hint. Take $\epsilon = \frac{1}{2}|L - L'|$ in (2).

12. Suppose the sequence $\{a_n\}$ converges to the limit L and the sequence $\{b_n\}$ converges to L'. Show that

(a) $\{a_n + b_n\}$ converges to $L + L'$,

(b) $\{ca_n\}$ converges to cL,

(c) $\{a_n \cdot b_n\}$ converges to $L \cdot L'$,

(d) $\{a_n/b_n\}$ converges to L/L', if $L' \neq 0$.

13. (a) If each term of a given sequence $\{a_n\}$ is less than or equal to a constant M, that is,

$$\text{if} \quad a_n \leq M \quad \text{for all } n = 1, 2, 3, \ldots,$$

and if $\{a_n\}$ converges to L, show that L is also less than or equal to M.

(b) Give an example of a sequence $\{a_n\}$ such that $a_n \leq 1$ for all $n = 1, 2, 3, \ldots,$ but such that the sequence has no limit.

Find an explicit algebraic expression for the sum of the first n terms of each of the following series. Then compute the sum of the series, if the series converges.

14. $$\left(1 - \frac{1}{2}\right) + \left(\frac{1}{2} - \frac{1}{3}\right) + \left(\frac{1}{3} - \frac{1}{4}\right) + \cdots + \left(\frac{1}{n} - \frac{1}{n+1}\right) + \cdots$$

15. $$\frac{1}{1 \cdot 2} + \frac{2}{2 \cdot 3} + \frac{1}{3 \cdot 4} + \cdots + \frac{1}{n(n+1)} + \cdots$$

Hint. Compare with Exercise 14.

16. $$\ln \frac{1}{2} + \ln \frac{2}{3} + \ln \frac{3}{4} + \cdots + \ln \frac{n}{n+1} + \cdots$$

17. $$1 + \frac{1}{2} + \frac{1}{4} + \cdots + \frac{1}{2^{n-1}} + \cdots$$

Hint. Multiply s_n by $\frac{1}{2}$ and subtract the result from s_n.

18. $$2 + \frac{2}{3} + \frac{2}{9} + \frac{2}{27} + \cdots + \frac{2}{3^{n-1}} + \cdots$$

Hint. Compute $s_n - \frac{1}{3}s_n$.

19. $$5 - \frac{5}{2} + \frac{5}{4} - \frac{5}{8} + \cdots + \frac{(-1)^{n-1}5}{2^{n-1}} + \cdots$$

Hint. Compute $s_n + \frac{1}{2}s_n$.

20. $$\frac{9}{100} + \frac{9}{(100)^2} + \frac{9}{(100)^3} + \cdots + \frac{9}{(100)^n} + \cdots$$

21. A ball is dropped from a height of 4 ft. Each time it strikes the ground after falling from a height of h ft, it rebounds a distance $\frac{3}{4}h$ ft. Find the total distance traveled by the ball.

22. (a) Express the repeating decimal

$$0.234\ 234\ 234\ \ldots$$

as a ratio p/q of integers p and q.

(b) Is it true that *every* repeating decimal is a rational number p/q? Give a reason for your answer.

23. Express the decimal number

$$1.24\ 123\ 123\ 123\ \ldots$$

(which is a repeating decimal after the first three figures) as a rational number p/q.

24. (a) Show by long division that

$$\frac{1}{1+t} = 1 - t + t^2 - t^3 + \cdots + (-1)^n t^n + \frac{(-1)^{n+1} t^{n+1}}{1+t}.$$

(b) By integrating the equation of (a) with respect to t from 0 to x, show that

$$\ln(1+x) = x - \frac{x^2}{2} + \frac{x^3}{3} - \frac{x^4}{4} + \cdots + (-1)^n \frac{x^{n+1}}{n+1} + R,$$

where

$$R = (-1)^{n+1} \int_0^x \frac{t^{n+1}}{1+t}\, dt.$$

(c) If $x > 0$, show that

$$|R| \leq \int_0^x t^{n+1}\, dt = \frac{x^{n+2}}{n+2}.$$

Hint. As t varies from 0 to x, $1 + t \geq 1$.

(d) If $x = \frac{1}{2}$, how large should n be, in (c) above, to guarantee that $|R| < 0.001$?

(e) If $x = 1$, how large should n be, in (c) above, to guarantee that $|R| < 0.001$?

18.2 TESTS FOR CONVERGENCE OF A SERIES OF CONSTANTS

It is sometimes hard to apply the definition of convergence given in Article 18.1. This is true when we can't find any simple expression for the sum of the first n terms of a series as a function of n. In the present article, we shall learn several tests for determining whether or not a given series converges, tests that depend on the individual terms of the series rather than on their sums.

Consider a series of numbers

$$\sum_{k=1}^{\infty} u_k = u_1 + u_2 + u_3 + \cdots + u_n + \cdots. \qquad (1)$$

As in Article 18.1, we say that the series (1) converges if the sequence s_n of partial sums

$$\begin{aligned} s_1 &= u_1, \\ s_2 &= u_1 + u_2, \\ &\vdots \\ s_n &= u_1 + u_2 + \cdots + u_n \end{aligned} \qquad (2)$$

tends to a definite finite limit as n increases without bound. If the partial sums s_n do not have such a limit, then the series is said to diverge. Most of the series considered in this book belong to one or the other of the following two types:

1. *Positive series:* Those in which all terms of the series are positive numbers.

2. *Alternating series:* Those in which the terms are alternately positive and negative. For example,

$$a_1 - a_2 + a_3 - a_4 + \cdots, \qquad (3)$$

when each of the a's is positive.

Positive series

Suppose all the numbers u_k in (1) are positive. Then when we calculate the partial sums s_1, s_2, s_3, and so on, we see that each one is greater than its predecessor, since $s_{n+1} = s_n + u_{n+1}$. That is,

$$s_1 \leq s_2 \leq s_3 \leq \cdots \leq s_n \leq s_{n+1}. \qquad (4)$$

A sequence $\{s_n\}$ that has the property (4) is called an *increasing* sequence. Theorem 3 gives the cardinal principle governing increasing sequences.

Theorem 3. *Let s_1, s_2, s_3, ... be an increasing sequence of real numbers. Then one or the other of the following alternatives must hold.*

A. *There is a finite constant M such that all terms of the sequence are less than or equal to M. In this case the sequence possesses a definite finite limit L which is also less than or equal to M.*

B. *The sequence diverges to plus infinity; that is, the numbers in the sequence $\{s_n\}$ ultimately exceed any preassigned number, no matter how large.*

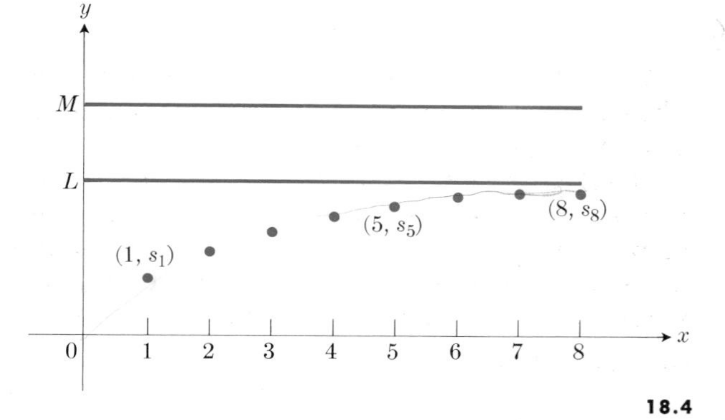

18.4

We shall not attempt to give a rigorous proof of this theorem, but we may gain an intuitive appreciation of the result by adopting the following graphical approach. Suppose we plot the points $(1, s_1)$, $(2, s_2)$, ..., (n, s_n) in the xy-plane (Fig. 18.4). Then if there is a line $y = M$ such that *none* of the points (n, s_n) lies above this line, it is intuitively clear that there is a *lowest* such line. That is, there is a line

$$y = L$$

such that none of the points lies above it but such that there are points (n, s_n) that lie above any *lower* line

$$y = L - \epsilon,$$

where ϵ is any positive number. Analytically, this means that the number L has the properties that

(a) $s_n \leq L$ for *all* values of n, and

(b) given any $\epsilon > 0$, there exists at least one integer N such that

$$s_N > L - \epsilon.$$

Then the fact that $\{s_n\}$ is an increasing sequence tells us further that

$$s_n \geq s_N > L - \epsilon \quad \text{for all } n \geq N.$$

This means that *all* the numbers s_n beyond the Nth one in the sequence lie within ϵ distance of L. This is precisely the condition for L to be the limit of the sequence s_n:

$$L = \lim_{n \to \infty} s_n.$$

Alternative B of the theorem comes into play when there are points (n, s_n) above any given line $y = M$, no matter how large M may be.

The theorem does not tell us how to find the limit L when it exists; it simply tells us whether or not there is such a limit. Briefly, it says that a *positive series either converges or else it becomes infinite.* Divergence by oscillation is ruled out.

Example 1. The series

$$\sum_{k=0}^{\infty} \frac{1}{k!} = 1 + \frac{1}{1!} + \frac{1}{2!} + \frac{1}{3!} + \cdots$$

converges because its terms are all less than or equal to the corresponding terms in the series

$$1 + \sum_{k=0}^{\infty} \frac{1}{2_k} = 1 + 1 + \frac{1}{2} + \frac{1}{2^2} + \cdots.$$

After its first term, the latter series is a simple geometric series of ratio $\frac{1}{2}$. The sum of terms after the first is $1/(1 - \frac{1}{2}) = 2$, so the sum of the series is 3. The partial sums of the first series are all less than 3. Hence the first series converges by virtue of our theorem about positive series, and its sum is less than or equal to 3. In fact, it is the series for e.

Example 2. The terms of the "harmonic" series

$$\sum_{k=1}^{\infty} \frac{1}{k} = 1 + \frac{1}{2} + \frac{1}{3} + \cdots$$

may be interpreted as representing the areas of rectangles, each of base unity and having altitudes equal

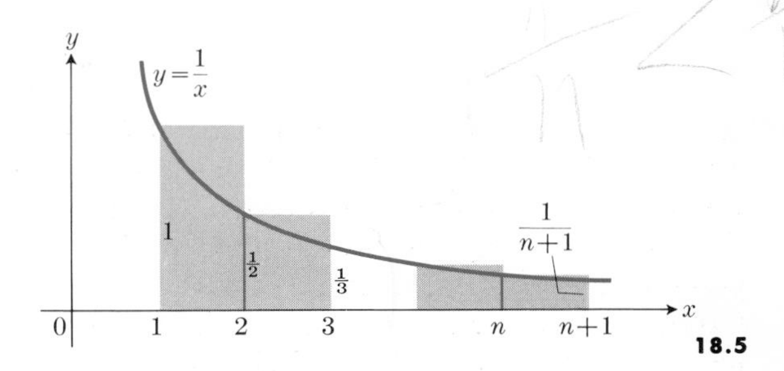

18.5

to $1, \frac{1}{2}, \frac{1}{3}, \ldots$, respectively. If we consider the sum of the first n terms of the series,

$$s_n = 1 + \frac{1}{2} + \frac{1}{3} + \cdots + \frac{1}{n},$$

we see (Fig. 18.5) that this represents the sum of the areas of n rectangles each of which is somewhat greater than the area under the corresponding portion of the curve $y = 1/x$. Therefore s_n is greater than the area under this curve between $x = 1$ and $x = n + 1$:

$$s_n > \int_1^{n+1} \frac{dx}{x} = \ln (n + 1).$$

By taking n sufficiently large, we can make

$$\ln (n + 1)$$

as large as we please:

$$\lim_{n \to \infty} \ln (n + 1) = +\infty.$$

Since $s_n > \ln (n + 1)$, this also means

$$\lim_{n \to \infty} s_n = +\infty.$$

Therefore the series

$$1 + \frac{1}{2} + \frac{1}{3} + \cdots + \frac{1}{n} + \cdots$$

diverges to plus infinity.

In the two examples just discussed we made two *comparisons*: first, between the given series and another series (Example 1), and second, between the given series and an integral (Example 2). We now state a theorem which embodies the so-called *comparison test* for positive series.

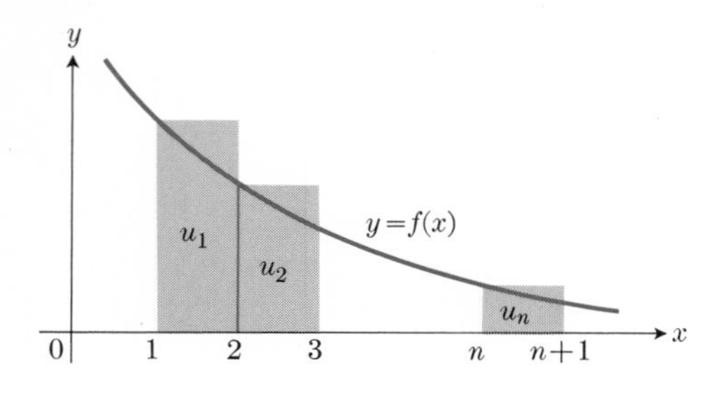

a | b

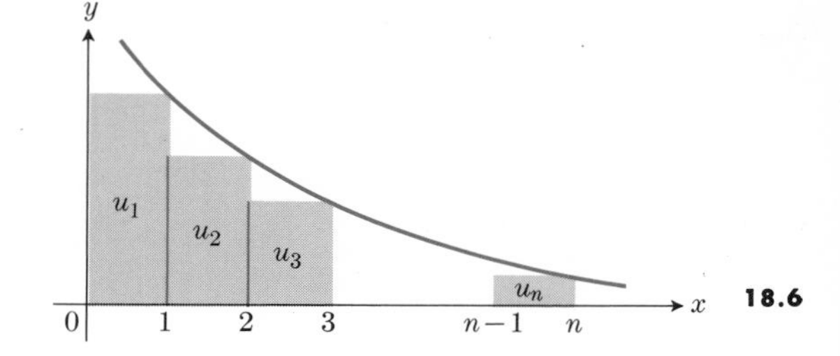

18.6

Theorem 4. *If each term of the positive series*

$$\sum_{k=1}^{\infty} u_k = u_1 + u_2 + u_3 + \cdots$$

is less *than the corresponding term of a known* convergent *series*

$$\sum_{k=1}^{\infty} c_k = c_1 + c_2 + c_3 + \cdots,$$

then $\sum u_k$ converges. But if each term of the series $\sum u_k$ is greater *than the corresponding term of a known* divergent *positive series*

$$\sum_{k=1}^{\infty} d_k = d_1 + d_2 + d_3 + \cdots,$$

then $\sum u_k$ diverges.

This theorem is an immediate consequence of Theorem 1.

When we compare a positive series with an integral, as we did above in Example 2, we are led to the *integral test,* which we now state as a theorem.

Theorem 5. *Suppose there is a decreasing continuous function $f(x)$ such that $f(n) = u_n$ is the nth term of the positive series*

$$u_1 + u_2 + \cdots + u_n + \cdots.$$

Then

$$\text{the series } \sum_{n=1}^{\infty} u_n \quad \text{and} \quad \text{the integral } \int_1^{\infty} f(x)\,dx$$

both converge or both diverge.

Proof. See Fig. 18.6(a). The rectangles of areas $u_1, u_2, \ldots, u_n$ enclose more area than that under the curve from $x = 1$ to $x = n + 1$; that is,

$$u_1 + u_2 + \cdots + u_n > \int_1^{n+1} f(x)\,dx.$$

In Fig. 18.6(b), the rectangles have been faced to the left instead of to the right. If we momentarily disregard the first rectangle, of area u_1, we see that

$$u_2 + u_3 + \cdots + u_n < \int_1^n f(x)\,dx.$$

If we include u_1, we have

$$u_1 + u_2 + \cdots + u_n < u_1 + \int_1^n f(x)\,dx.$$

Combining these results, we have

$$\int_1^{n+1} f(x)\,dx < u_1 + u_2 + \cdots + u_n < u_1 + \int_1^n f(x)\,dx.$$

If the integral $\int_1^{\infty} f(x)\,dx$ is finite, the right-hand inequality shows that the infinite series

$$\sum_{n=1}^{\infty} u_n$$

is also finite. But if $\int_1^{\infty} f(x)\,dx$ is infinite, then the left-hand inequality shows that the series is also infinite. Hence the series and the integral are both finite or both infinite. □

Example 3. The p-series

$$\frac{1}{1^p} + \frac{1}{2^p} + \frac{1}{3^p} + \cdots + \frac{1}{n^p} + \cdots$$

converges if $p > 1$ and diverges if $p \leq 1$. To prove this, let

$$f(x) = 1/x^p.$$

Then, if $p > 1$, we have

$$\int_1^\infty x^{-p}\,dx = \lim_{b\to\infty} \frac{x^{-p+1}}{-p+1}\Big]_1^b = \frac{1}{p-1},$$

which is finite. Hence the p-series converges if p is greater than one.

If $p = 1$, we have the *harmonic series*

$$1 + \frac{1}{2} + \frac{1}{3} + \cdots + \frac{1}{n} + \cdots,$$

which we already know diverges. Alternatively, by the integral test,

$$\int_1^\infty x^{-1}\,dx = \lim_{b\to\infty} \ln x\Big]_1^b = +\infty,$$

and since the integral diverges, the series does likewise.

Finally, if $p < 1$, then the terms of the p-series are greater than the corresponding terms of the divergent harmonic series. Hence the p-series diverges when $p < 1$, by the comparison test.

Remark 1. An integral may also be used to advantage to estimate the remainder R_n after n terms of a convergent series.

Example 4. Suppose we are interested in learning the numerical value of the series

$$\sum_{k=1}^{\infty} \frac{1}{k^2} = \frac{1}{1^2} + \frac{1}{2^2} + \frac{1}{3^2} + \cdots.$$

This is a p-series with $p = 2$, and hence we know that it converges. This means that the sequence of partial sums s_n,

$$s_n = \frac{1}{1^2} + \frac{1}{2^2} + \cdots + \frac{1}{n^2},$$

possesses a limit L. When n is "large enough," s_n is "close to" L. If we want to know L to a couple of decimal places, we might try to find an integer n such that the corresponding *finite* sum s_n differs from L by less, say, than 0.005. Then we would use this s_n in place of L, to two decimals. If we write

$$L = \sum_{k=1}^{\infty} \frac{1}{k^2} = \frac{1}{1^2} + \frac{1}{2^2} + \cdots + \frac{1}{n^2} + \frac{1}{(n+1)^2} + \cdots$$

and

$$s_n = \sum_{k=1}^{n} \frac{1}{k^2} = \frac{1}{1^2} + \frac{1}{2^2} + \cdots + \frac{1}{n^2},$$

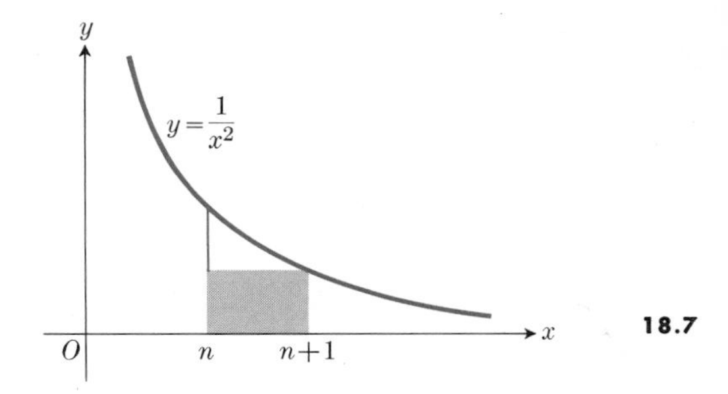

18.7

we see that the error R_n is

$$R_n = L - s_n = \frac{1}{(n+1)^2} + \cdots.$$

We may estimate this error by thinking of the area under the curve

$$y = 1/x^2$$

from $x = n$ to ∞. From Fig. 18.7, we see that

$$R_n < \int_n^\infty \frac{1}{x^2}\,dx = \frac{1}{n},$$

which tells us that by taking 200 terms of the series, we can be sure that the difference between the sum L of the entire series and the sum s_{200} of these 200 terms will be less than 0.005.

A somewhat closer estimate of R_n results from using the trapezoidal rule to approximate the area under the curve in Fig. 18.7. Let us write u_k for $1/k^2$ and consider the trapezoidal approximation

$$\begin{aligned} T_n &= \sum_{k=n}^{\infty} \tfrac{1}{2}(u_k + u_{k+1}) \\ &= \tfrac{1}{2}(u_n + u_{n+1}) + \tfrac{1}{2}(u_{n+1} + u_{n+2}) + \cdots \\ &= \tfrac{1}{2}u_n + u_{n+1} + u_{n+2} + \cdots = \tfrac{1}{2}u_n + R_n. \end{aligned}$$

Now since the curve $y = 1/x^2$ is concave upward, it is clear that

$$T_n > \int_n^\infty \frac{1}{x^2}\,dx = \frac{1}{n},$$

and we have

$$R_n = T_n - \tfrac{1}{2}u_n > \frac{1}{n} - \frac{1}{2n^2}.$$

We now know that

$$\frac{1}{n} > R_n > \frac{1}{n} - \frac{1}{2n^2},$$

and $L = s_n + R_n$ may be estimated as follows:

$$s_n + \frac{1}{n} > L > s_n + \frac{1}{n} - \frac{1}{2n^2}.$$

Thus, by using $s_n + 1/n$ in place of s_n to estimate L, we shall be making an error which is numerically less than $1/(2n^2)$. By taking $n \geq 10$, this error is then made less than 0.005. The difference in time required to compute the sum of 10 terms versus 200 terms is sufficiently great to make this sharper analysis of practical importance.

What we have done in this specific example may be done in any case where the graph of the function $y = f(x)$ is concave upward, as in Fig. 18.7. We find that

$$u_1 + u_2 + \cdots + u_n + \int_n^\infty f(x)\,dx \tag{5}$$

tends to overestimate the value of the series, but by an amount that is less than $u_n/2$.

Ratio test

The ratios

$$(u_2/u_1),\ (u_3/u_2),\ (u_4/u_3),\ \ldots,\ (u_{n+1}/u_n),\ \ldots$$

measure the rate of growth of the terms of a series. In a geometric series

$$1 + r + r^2 + r^3 + \cdots,$$

this rate of growth is a constant, r. We shall show that if the rate of growth in a positive series tends to a definite limit ρ as n becomes infinite:

$$\rho = \lim_{n\to\infty} \frac{u_{n+1}}{u_n},$$

then

(a) the series *converges* if $\rho < 1$,

(b) the series *diverges* if $\rho > 1$,

(c) the series may converge or it may diverge if $\rho = 1$.

(Remember that we are still speaking about *positive* series.)

To establish the validity of the ratio test, suppose we assume first that ρ is less than one. Let r be a number between ρ and one, for example,

$$r = (1 + \rho)/2.$$

Then the number ϵ defined by the equation

$$\epsilon = r - \rho$$

is positive. Since

$$\rho = \lim_{n\to\infty} \frac{u_{n+1}}{u_n},$$

we know that u_{n+1}/u_n must lie within a distance ϵ of ρ when n is large enough, say for all $n \geq N$. Then, in particular,

$$\frac{u_{n+1}}{u_n} < r = \rho + \epsilon \quad \text{when } n \geq N.$$

That is,

$$\begin{aligned} u_{N+1} &< ru_N, \\ u_{N+2} &< ru_{N+1} < r^2 u_N, \\ u_{N+3} &< ru_{N+2} < r^3 u_N, \\ &\vdots \\ u_{N+m} &< ru_{N+m-1} < r^m u_N, \end{aligned}$$

and

$$\begin{aligned} u_1 + u_2 + \cdots + u_N + u_{N+1} + u_{N+2} + \cdots \\ < u_1 + u_2 + \cdots + u_{N-1} \\ + u_N(1 + r + r^2 + \cdots). \end{aligned} \tag{6}$$

Since $|r|$ is less than one, the geometric series $1 + r + r^2 + \cdots$ converges, and the right-hand side of the inequality (6) is finite. Therefore the series on the left converges, by the comparison test.

Next, suppose ρ is greater than one. Then, from some index m on, we have

$$\frac{u_{n+1}}{u_n} \geq 1 \qquad (n \geq m),$$

or

$$u_{m+1} \geq u_m,\ u_{m+2} \geq u_{m+1} \geq u_m,\ \ldots$$

Since u_m is positive, the sum of the terms

$$\begin{aligned} u_m + u_{m+1} + u_{m+2} + \cdots + u_{m+q} \\ \geq u_m(1 + 1 + \cdots + 1) = (q+1)u_m \end{aligned}$$

tends to infinity as q does. Hence in this case the series diverges.

Finally, the two series

$$\sum_{n=1}^{\infty} \frac{1}{n} \quad \text{and} \quad \sum_{n=1}^{\infty} \frac{1}{n^2}$$

both have rate-of-growth ratios that approach one as limit. Since one of these series converges while the other diverges, we have specific examples which show that the ratio test cannot be used to distinguish between convergence and divergence in case $\rho = 1$.

Remark 2. The ratio test is also useful in estimating the truncation error which results from using

$$s_N = u_1 + u_2 + \cdots + u_N$$

as an approximation to the sum of a convergent series

$$S = u_1 + u_2 + \cdots + u_N + (u_{N+1} + \cdots).$$

Suppose that $n \leq N$, and we know that the rate-of-growth ratio lies between two constants r_1 and r_2, both of which are positive and less than one:

$$r_1 \leq \frac{u_{n+1}}{u_n} \leq r_2 \quad \text{for} \quad n \geq N;$$

then the inequalities

$$r_1 u_n \leq u_{n+1} \leq r_2 u_n, \quad n = N, N+1, N+2, \ldots,$$

enable us to deduce that

$$\begin{aligned} u_N(r_1 + r_1^2 + r_1^3 + \cdots) \\ &\leq u_{N+1} + u_{N+2} + u_{N+3} + \cdots \\ &\leq u_N(r_2 + r_2^2 + r_2^3 + \cdots). \end{aligned}$$

The two geometric series have sums

$$r_1 + r_1^2 + r_1^3 + \cdots = \frac{r_1}{1 - r_1},$$

$$r_2 + r_2^2 + r_2^3 + \cdots = \frac{r_2}{1 - r_2}.$$

Hence the error

$$R_N = u_{N+1} + u_{N+2} + u_{N+3} + \cdots$$

lies between

$$\frac{r_1 u_N}{1 - r_1} \quad \text{and} \quad \frac{r_2 u_N}{1 - r_2}.$$

That is,

$$\frac{r_1 u_N}{1 - r_1} \leq \sum_{k=1}^{\infty} u_k - \sum_{k=1}^{N} u_k \leq \frac{r_2 u_N}{1 - r_2} \tag{7a}$$

if

$$0 \leq r_1 \leq \frac{u_{n+1}}{u_n} \leq r_2 < 1, \quad \text{for } n \geq N. \tag{7b}$$

Problem 1. For what values of x does the series

$$x + \frac{x^3}{3} + \frac{x^5}{5} + \frac{x^7}{7} + \cdots \tag{8}$$

converge?

Solution. The nth term of the series (assuming the law of formation as indicated) is

$$u_n = \frac{x^{2n-1}}{2n - 1}.$$

We consider first the case where x is positive. Then the series is a positive series and the rate-of-growth ratio is

$$\frac{u_{n+1}}{u_n} = \frac{(2n - 1)}{(2n + 1)} x^2,$$

with limit

$$\rho = \lim_{n \to \infty} \frac{2n - 1}{2n + 1} x^2 = x^2.$$

The ratio test therefore tells us that the series converges if x is positive and less than one, and diverges if x is greater than one.

Since only odd powers of x occur in the series, we see that the series simply changes sign when x is replaced by $-x$. Therefore the series also converges for

$$-1 < x \leq 0,$$

and diverges for $x < -1$. The series converges to zero when $x = 0$.

We know, thus far, that the series

$$\text{converges for } |x| < 1,$$
$$\text{diverges for } |x| > 1,$$

but we don't know what happens when $|x| = 1$. To test at $x = 1$, we apply the integral test to the series

$$1 + \frac{1}{3} + \frac{1}{5} + \frac{1}{7} + \cdots + \frac{1}{2n - 1} + \cdots,$$

which we get by taking $x = 1$ in the series (8). The companion integral is

$$\int_1^\infty \frac{dx}{2x-1} = \tfrac{1}{2}\ln(2x-1)\Big]_1^\infty = \infty.$$

Hence the series diverges to $+\infty$ when $x = 1$, and it diverges to $-\infty$ when $x = -1$. We may therefore say that the only values of x for which the given series converges are $-1 < x < 1$.

EXERCISES

1. Prove that if $\sum u_k$ is a positive series and c is a positive constant, then $\sum u_k$ and $\sum (cu_k)$ both converge or both diverge.
2. What are the two alternatives, analogous to A and B of Theorem 1, for a sequence $\{s_n\}$ that is monotonically decreasing as n increases; that is, for which

$$s_1 \geq s_2 \geq s_3 \geq \cdots \geq s_n \geq s_{n+1} \geq \cdots ?$$

Draw a graph to illustrate what happens when all terms of the sequence are greater than or equal to some constant m.

3. (a) Show that the sequence of numbers

$$s_n = \left(1 + \frac{1}{n}\right)^n$$

is an increasing sequence and that each s_n is less than 3.

Hint. Expand $(1 + 1/n)^n$ by the binomial theorem and show that

$$\left(1+\frac{1}{n}\right)^n = 1 + 1 + \frac{\left(1-\frac{1}{n}\right)}{2!} + \frac{\left(1-\frac{1}{n}\right)\left(1-\frac{2}{n}\right)}{3!} + \cdots + \frac{\left(1-\frac{1}{n}\right)\left(1-\frac{2}{n}\right)\cdots\left(1-\frac{n-1}{n}\right)}{n!}.$$

(b) From these two statements, what conclusion can we draw about the existence of the limit of s_n as n becomes infinite?

4. Consider the sequence of numbers

$$1, 0, 1, 0, 1, 0, \ldots$$

formed according to the rule

$$s_{2m-1} = 1, \quad s_{2m} = 0.$$

Does this sequence have a limit as n becomes infinite? Give a reason for your answer.

In Exercises 5 through 19, determine whether the given series converge or diverge. In each case, give a reason for your answer.

5. $\displaystyle\sum_{n=1}^{\infty} \frac{1}{10n}$
6. $\displaystyle\sum_{n=1}^{\infty} \frac{n}{n+2}$
7. $\displaystyle\sum_{n=1}^{\infty} \frac{\sin^2 n}{2^n}$
8. $\displaystyle\sum_{n=1}^{\infty} \frac{n}{n^2+1}$
9. $\displaystyle\sum_{n=1}^{\infty} \frac{1}{1+\ln n}$
10. $\displaystyle\sum_{n=3}^{\infty} \frac{1}{n \ln n}$
11. $\displaystyle\sum_{n=0}^{\infty} \frac{1}{(2n+1)!}$
12. $\displaystyle\sum_{n=1}^{\infty} \frac{1}{n^2+1}$
13. $\displaystyle\sum_{n=1}^{\infty} \frac{\sqrt{n}}{n^2+1}$
14. $\displaystyle\sum_{n=1}^{\infty} \frac{n!}{n^n}$
15. $\displaystyle\sum_{n=1}^{\infty} \frac{(n+1)(n+2)}{n!}$
16. $\displaystyle\sum_{n=1}^{\infty} \frac{n^2}{2^n}$
17. $\displaystyle\sum_{n=0}^{\infty} \frac{(n+3)!}{3!n!3^n}$
18. $\displaystyle\sum_{n=1}^{\infty} \frac{2^n}{n^3+1}$
19. $\displaystyle\sum_{n=1}^{\infty} \frac{10^n}{n!}$

18.3 POWER SERIES EXPANSIONS OF FUNCTIONS

The rational operations of arithmetic are addition, subtraction, multiplication, and division. Using only these simple operations, we can evaluate any rational function of x. But other functions, such as $\ln x$, e^x, $\sin x$, and so on, cannot be evaluated so simply. Of course, these functions are so important that their values have been computed and the results printed in mathematical tables. But we may wonder how the tables were computed. The answer is: from power series.

Definition. *A power series is an expression of the form*

$$\sum_{k=0}^{\infty} a_k x^k = a_0 + a_1 x + a_2 x^2 + \cdots.$$

In this article we shall show how to use power series to represent a wide variety of functions.

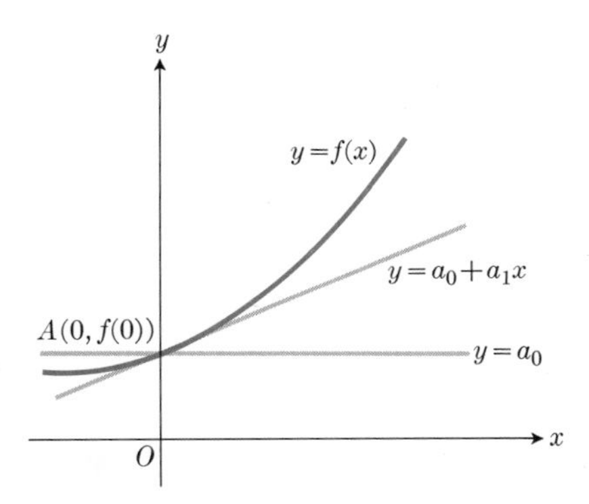

18.8 The tangent line $y = a_0 + a_1 x$ serves to approximate the function $y = f(x)$ for small values of x.

We shall concern ourselves with the problem of approximating a given function

$$y = f(x) \tag{1}$$

by a sequence of polynomials $f_n(x)$ of the form

$$f_n(x) = a_0 + a_1 x + a_2 x^2 + \cdots + a_n x^n. \tag{2}$$

In particular, since we expect to take larger and larger values of n, we shall be interested in making the approximation good for small values of x, since we want the terms that come far out in the series to be small. Hence we first focus our attention on a portion of the curve $y = f(x)$ in the neighborhood of the point $A(0, f(0))$ in Fig. 18.8. We then ask the following questions.

Question. What polynomial

$$y = f_0(x) = a_0,$$

of degree zero, gives the closest approximation to the given curve in the immediate neighborhood of the point A?

Answer. We want the horizontal line $y = a_0$ to pass through A. Hence we take $a_0 = f(0)$, which is the value of the given function at $x = 0$.

Question. What polynomial

$$y = a_0 + a_1 x,$$

of degree one, comes closest to fitting the given curve in the neighborhood of the point A?

Answer. The tangent line. That is, we want this approximating line to pass through A and to have the same slope there that the given curve has. This means that we should take

$$a_0 = f(0) \quad \text{and} \quad a_1 = f'(0).$$

Question. More generally, what polynomial

$$y = a_0 + a_1 x + a_2 x^2 + \cdots + a_n x^n,$$

of degree n, comes closest to fitting the given curve near A?

Answer. We take the approximating curve that passes through A and has the highest possible degree of contact with the given curve at A, in the sense that its derivatives of orders one, two, three, . . . , n match the corresponding derivatives of the given function there. Thus, if we write

$$f_n(x) = a_0 + a_1 x + a_2 x^2 + a_3 x^3 + \cdots + a_n x^n$$

for the approximating polynomial, then its derivatives are

$$\begin{aligned}
f_n'(x) &= a_1 + 2a_2 x + 3a_3 x^2 + \cdots + n a_n x^{n-1},\\
f_n''(x) &= 2a_2 + 3 \cdot 2a_3 x + \cdots + n(n-1) a_n x^{n-2},\\
&\ \vdots\\
f_n^{(n)}(x) &= n! a_n.
\end{aligned}$$

We substitute $x = 0$ on the right-hand side of each of the above equations and set the results, respectively, equal to the values

$$f(0), f'(0), f''(0), \ldots, f^{(n)}(0)$$

of the given function and its first n derivatives at $x = 0$. We then determine the coefficients

$$a_0, a_1, a_2, \ldots, a_n$$

of the approximating polynomial $y = f_n(x)$ that has the highest degree of contact with the given curve at $A(0, f(0))$:

$$\begin{aligned} a_0 &= f(0), \\ a_1 &= f'(0), \\ a_2 &= \frac{f''(0)}{2!}, \\ &\vdots \\ a_n &= \frac{f^{(n)}(0)}{n!}. \end{aligned}$$

That is, for our approximating polynomial of degree n, we take

$$f_n(x) = f(0) + f'(0)x + \frac{f''(0)}{2!}x^2 + \frac{f'''(0)}{3!}x^3 + \cdots + \frac{f^{(n)}(0)}{n!}x^n. \quad (3)$$

Problem 1. Compute the polynomial $f_n(x)$ given by Eq. 3 that would be used to approximate the exponential function $f(x) = e^x$ near $x = 0$.

Solution. We require $f(0), f'(0), \ldots, f^{(n)}(0)$. For the given function,

$$f(x) = e^x; \qquad f(0) = e^{\circ} = 1,$$

and

$$f'(x) = e^x; \qquad f'(0) = 1,$$
$$\vdots \qquad\qquad \vdots$$
$$f^{(n)}(x) = e^x; \qquad f^{(n)}(0) = 1.$$

Therefore we have

$$f_n(x) = 1 + x + \frac{x^2}{2!} + \frac{x^3}{3!} + \cdots + \frac{x^n}{n!}. \quad (4)$$

The question now naturally arises as to whether, for a fixed value of x, our approximating polynomials $f_n(x)$ converge to $f(x)$ as $n \to \infty$. Since $f_n(x)$ of Eq. (3) is the nth partial sum of the infinite series

$$f(0) + f'(0) \cdot x + \frac{f''(0)}{2!}x^2 + \cdots + \frac{f^{(n)}(0)}{n!}x^n + \cdots, \quad (5)$$

the question just posed is equivalent to inquiring whether the series (5) converges to $f(x)$ as sum. Of course, *every* one of our approximating polynomials has the correct value $f(0)$ at $x = 0$, but we are interested now in knowing also how far we may go away from $x = 0$ and still have convergence.

The series (5) is known as *Maclaurin's series* for the given function $f(x)$. If, instead of approximating to the function $f(x)$ for the values of x near zero, we are concerned with values of x near a, we write our approximating polynomials in powers of $x - a$:

$$f_n(x) = a_0 + a_1(x - a) + a_2(x - a)^2 + \cdots + a_n(x - a)^n.$$

When we now determine the coefficients

$$a_0, a_1, \ldots, a_n$$

so that the polynomial and its first n derivatives agree with the given function and its first n derivatives, at $x = a$, we are led to the series

$$f(a) + f'(a)(x - a) + \frac{f''(a)}{2!}(x - a)^2 + \frac{f'''(a)}{3!}(x - a)^3 + \cdots + \frac{f^{(n)}(a)}{n!}(x - a)^n + \cdots, \quad (6)$$

where $f^{(n)}(a)$ denotes the value of the nth derivative of the function $f(x)$ at $x = a$. This is known as the *Taylor series* expansion of $f(x)$ about $x = a$. Since this includes the Maclaurin series, (5), as a special case, [namely, Eq. (6) with $a = 0$], we shall henceforth concern ourselves with the Taylor series. Since Eq. (6) calls for derivatives of *all* orders at $x = a$, it is clear that a function cannot possess a Taylor series expansion about $x = a$ unless the function possesses finite derivatives of all orders at $x = a$. For instance, $f(x) = \ln x$ does not possess a Maclaurin series expansion $(a = 0)$, since the function itself (to say nothing of its derivatives) does not have a finite value at $x = 0$. On the other hand, it does possess a Taylor series expansion in powers of $(x - 1)$, since $\ln x$ and all its derivatives are finite at $x = 1$.

Problem 2. Find the Maclaurin series expansion for the function $f(x) = (1 + x)^m$.

Solution. We need the derivatives

$$\begin{aligned} f'(x) &= m(1+x)^{m-1}, \\ f''(x) &= m(m-1)(1+x)^{m-2}, \\ f'''(x) &= m(m-1)(m-2)(1+x)^{m-3}, \\ &\vdots \\ f^{(k)}(x) &= m(m-1)(m-2)\cdots(m-k+1)(1+x)^{m-k}. \end{aligned}$$

If we substitute the values of these derivatives at $x = 0$ into the basic Maclaurin series, (5):

$$f(0) + f'(0)x + \frac{f''(0)}{2!}x^2 + \cdots + \frac{f^{(k)}(0)}{k!}x^k + \cdots,$$

we then obtain

$$1 + mx + \frac{m(m-1)}{2!}x^2 + \cdots + \frac{m(m-1)(m-2)\cdots(m-k+1)}{k!}x^k + \cdots.$$

For a proof that this series converges to the given function $f(x) = (1 + x)^m$ when $|x| < 1$, see Olmsted, *Advanced Calculus*, pp. 423–425.

EXERCISES

In Exercises 1 through 6, use Eq. (3) to calculate the polynomial $f_n(x)$ that is associated with the given function $f(x)$.

1. e^{-x}
2. $\sin x$
3. $\cos x$
4. $\sin (x + \pi/4)$
5. $\sinh x$
6. $\cosh x$

In Exercises 7 through 13, use Eq. (6) to find the Taylor series expansion of the given function about the given value of a.

7. $f(x) = \ln x \quad a = 1$
8. $f(x) = \sqrt{x}, \quad a = 4$
9. $f(x) = 1/x, \quad a = -1$
10. $f(x) = \tan x, \quad a = \pi/4$ (Find three terms of the series.)
11. $f(x) = \ln \cos x, \quad a = \pi/3$ (Find three terms of the series.)
12. $f(x) = \sin x, \quad a = \pi/6$
13. $f(x) = \cos x, \quad a = -\pi/4$

18.4 TAYLOR'S THEOREM WITH REMAINDER

In the previous article, we raised some questions about convergence of the Taylor series. In this section, we give a theorem which answers some of these questions. Unfortunately, although several proofs of the theorem are known, none of them is easily motivated. The path we shall follow in arriving at the theorem starts with the simple formula

$$\int_a^b f'(t)\,dt = f(t)\Big]_a^b = f(b) - f(a). \tag{1}$$

We integrate the left side of (1) by parts, with

$$u = f'(t), \qquad dv = dt,$$
$$du = f''(t)\,dt, \qquad v = t - b,$$

where we introduce the constant of integration $-b$ for future convenience. Thus we have

$$\begin{aligned} \int_a^b f'(t)\,dt &= (t-b)f'(t)\Big]_a^b - \int_a^b (t-b)f''(t)\,dt \\ &= (b-a)f'(a) + \int_a^b (b-t)f''(t)\,dt. \end{aligned}$$

We perform a further integration by parts, this time with

$$u = f''(t), \qquad dv = (b-t)\,dt,$$
$$du = f'''(t)\,dt, \qquad v = -\frac{(b-t)^2}{2}.$$

Using this set of substitutions, we then arrive at the expression

$$\int_a^b (b-t)f''(t)\,dt = -\frac{(b-t)^2}{2}f''(t)\Big]_a^b + \int_a^b \frac{(b-t)^2}{2}f'''(t)\,dt = \frac{(b-a)^2}{2!}f''(a) + \int_a^b \frac{(b-t)^2}{2!}f'''(t)\,dt.$$

By continuing in this fashion, we find

$$\int_a^b f'(t)\,dt = (b-a)f'(a) + \frac{(b-a)^2}{2!}f''(a) + \frac{(b-a)^3}{3!}f'''(a) + \cdots + \frac{(b-a)^n}{n!}f^{(n)}(a) + \int_a^b \frac{(b-t)^n}{n!}f^{(n+1)}(t)\,dt.$$

Comparing this with Eq. (1), we have

$$f(b) = f(a) + (b-a)f'(a) + \frac{(b-a)^2}{2!}f''(a) + \cdots + \frac{(b-a)^n}{n!}f^{(n)}(a) + \int_a^b \frac{(b-t)^n}{n!}f^{(n+1)}(t)\,dt.$$

We may now replace b by x, and write

$$f(x) = f(a) + (x-a)f'(a) + \frac{(x-a)^2}{2!}f''(a) + \cdots + \frac{(x-a)^n}{n!}f^{(n)}(a) + R_n(x, a), \tag{2}$$

where

$$R_n(x, a) = \int_a^x \frac{(x-t)^n}{n!}f^{(n+1)}(t)\,dt. \tag{3}$$

The term $R_n(x, a)$ is called the *remainder* in the Taylor series expansion for $f(x)$. If the remainder is omitted in Eq. (2), the polynomial on the right-hand side of the equation gives only an approximation to the function on the left. The error in this approximation is what is measured by $R_n(x, a)$. Hence, to investigate convergence of the series, we need to investigate this remainder. First of all, the series converges to $f(x)$ if and only if

$$\lim_{n\to\infty} R_n(x, a) = 0. \tag{4}$$

When (4) holds, however, we may also write

$$f(x) = \sum_{k=0}^{\infty} \frac{f^{(k)}(a)}{k!}(x-a)^k,$$

since the series on the right converges, at x, to the value $f(x)$. It is often possible to show that the remainder does approach zero without evaluating the integral in (3) explicitly. (See Exercises 1 and 3 and Example 1 below.)

The application of integration by parts n times in succession requires, of course, that the integral in Eq. (3) exist. This will indeed be the case if the derivative of order $(n+1)$ of the function exists and is continuous in some closed interval that includes the range of integration $a \ldots x$ in (3).

Taylor's theorem. *Let f be a function that is continuous together with its first $n+1$ derivatives on an interval containing a and x. Then the value of the function at x is given by*

$$\begin{aligned} f(x) = f(a) &+ f'(a)(x-a) \\ &+ \frac{f''(a)}{2!}(x-a)^2 \\ &+ \frac{f'''(a)}{3!}(x-a)^3 + \cdots \\ &+ \frac{f^{(n)}(a)}{n!}(x-a)^n + R_n(x, a), \end{aligned}$$

where

$$R_n(x, a) = \int_a^x \frac{(x-t)^n}{n!}f^{(n+1)}(t)\,dt.$$

Example 1. Let $f(x) = e^x$. This function and all its derivatives are everywhere continuous, so Taylor's theorem may be applied with any convenient value of a. We select $a = 0$, since the values of f and its derivatives are easy to compute there. Taylor's theorem leads to

$$e^x = 1 + x + \frac{x^2}{2!} + \frac{x^3}{3!} + \cdots + \frac{x^n}{n!} + R_n(x, 0), \tag{5}$$

which contains the remainder

$$R_n(x, 0) = \int_0^x \frac{(x-t)^n}{n!} e^t\, dt. \tag{6}$$

To avoid unnecessary complications, we temporarily assume that x is positive. Then we may *estimate* the remainder (6) by observing that the integrand is positive for $0 < t < x$, and that $e^t < e^x < 3^x$. Therefore

$$|R_n(x, 0)| \le \int_0^x \frac{(x-t)^n}{n!} 3^x\, dt = 3^x \frac{x^{n+1}}{(n+1)!}$$

for $x > 0$. When x is negative we must replace x by $|x|$ on the right-hand side of this inequality. But for all real values of x, $-\infty < x < +\infty$, we have

$$|R_n(x, 0)| \le 3^{|x|} \frac{|x|^{n+1}}{(n+1)!}. \tag{7}$$

Now that we have this estimate of the remainder term, we are ready to discuss the convergence of the Maclaurin series (Taylor series with $a = 0$) for e^x. There are two separate cases to be considered, according to whether or not $|x| \le 1$.

Case 1. If $|x| \le 1$, then $3^{|x|} \le 3$, and (7) tells us that

$$|R_n(x, 0)| \le \frac{3}{(n+1)!} \quad \text{if} \quad |x| \le 1. \tag{8}$$

Hence $R_n(x, 0)$ certainly approaches zero as n tends to infinity, and the following series converges to e^x for any number x whose absolute value is less than or equal to 1:

$$e^x = 1 + x + \frac{x^2}{2!} + \frac{x^3}{3!} + \cdots + \frac{x^n}{n!} + \cdots, \quad |x| \le 1. \tag{9}$$

Using the estimate of the remainder given by (8), we may determine n (and thus the number of terms of the series to use) so as to guarantee a specified accuracy in computing the value of e^x. For example, we may calculate e by taking $x = 1$ in Eq. (9). If we do this, with $n = 3$, say, the error we make will be less than

$$3/4! = \tfrac{1}{8} = 0.125.$$

If we take $n = 7$, then the error is less than

$$3/8! < 0.75 \times 10^{-4},$$

and the approximation

$$e \approx 1 + 1 + \frac{1}{2!} + \frac{1}{3!} + \cdots + \frac{1}{7!}$$

is good to within one unit in the fourth decimal place. Since the remainder goes to zero as n tends to infinity, we also have the infinite series expansion for e:

$$e = 1 + 1 + \frac{1}{2!} + \frac{1}{3!} + \cdots. \tag{10}$$

The first two terms in the series (9) can be written as $1/0!$ and $x/1!$, provided that we adopt the definition

$$0! = 1.$$

This definition of $0!$ is commonly accepted, and with it the identity $n! = n(n-1)!$ is satisfied for all positive integers n, including $n = 1$. The series for e^x, Eq. (9), now takes the compact form

$$e^x = \sum_{k=0}^{\infty} \frac{x^k}{k!}. \tag{11}$$

So far, we have only established Eq. (11) for $|x| \le 1$, but we shall see that it is in fact valid for all real x.

Graphs of some of the approximating polynomials

$$f_n(x) = 1 + x + \frac{x^2}{2!} + \cdots + \frac{x^n}{n!} = \sum_{k=0}^{n} \frac{x^k}{k!},$$

for $-1 \le x \le 1$, are shown in Fig. 18.9 for $n = 1, 2$, and 3. All three approximations are tangent to the curve $y = e^x$ at $P_0(0, 1)$, and therefore lie close to that curve near P_0. But as $|x|$ increases, the parabola B is a better approximation than the straight line A; and the cubic curve C is better than either A or B. This is to be expected since the error (remainder term in the series) decreases as n increases.

Case 2. We now consider what happens to the remainder term, Eq. (6), as n increases, if $|x| > 1$. Then both $|x|^{n+1}$ and $(n+1)!$ increase indefinitely as n does, but the factorial increases so much faster (for large n) that the ratio approaches zero:

$$\lim_{n\to\infty} \frac{|x|^{n+1}}{(n+1)!} = 0. \tag{12}$$

(See Exercise 3 below.) For example, if $x = 5$, the ratio $5^{n+1}/(n+1)!$ increases with n until we get to $n = 4$.

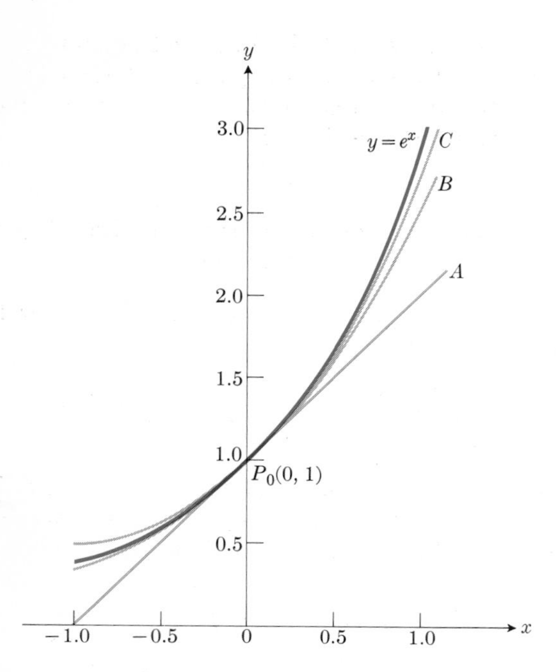

18.9 The graph of the function $y = e^x$, and graphs of three approximating polynomials, (A) a straight line, (B) a parabola, and (C) a cubic curve.

x	e^x	$1+x$	$1+x+\frac{x^2}{2!}$	$1+x+\frac{x^2}{2!}+\frac{x^3}{3!}$
0	1	1	1	1
0.1	1.1052	1.1	1.105	1.10517−
0.2	1.2214	1.2	1.220	1.22133+
0.3	1.3499	1.3	1.345	1.34950
0.4	1.4918	1.4	1.480	1.49067−
0.5	1.6487	1.5	1.625	1.64583+
0.6	1.8221	1.6	1.780	1.81600
0.7	2.0138	1.7	1.945	2.00217−
0.8	2.2255	1.8	2.120	2.20533+
0.9	2.4596	1.9	2.305	2.42650
1.0	2.7183	2.0	2.500	2.66667−
−0.1	0.90484	0.9	0.905	0.904833+
−0.2	0.81873	0.8	0.820	0.818667−
−0.3	0.74082	0.7	0.745	0.740500
−0.4	0.67032	0.6	0.680	0.669333+
−0.5	0.60653	0.5	0.625	0.604167−
−0.6	0.54881	0.4	0.580	0.544000
−0.7	0.49659	0.3	0.545	0.487833+
−0.8	0.44933	0.2	0.520	0.434667−
−0.9	0.40657	0.1	0.505	0.383500
−1.0	0.36788	0.0	0.500	0.333333+

But as n continues to increase, one unit at a time, the numerator is multiplied each time by 5, but the denominator is multiplied by 6, or 7, or 8, and so on, and the ratio ultimately approaches zero. The additional factor $3^{|x|}$ doesn't change as $n \to \infty$, so we have

$$\lim_{n\to\infty} 3^{|x|} \frac{|x|^{n+1}}{(n+1)!} = 0$$

and, from (7),

$$\lim_{n\to\infty} R_n(x, 0) = 0$$

for any real x, $-\infty < x < +\infty$. Therefore

$$e^x = \sum_{k=0}^{\infty} \frac{x^k}{k!}, \qquad -\infty < x < +\infty. \tag{13}$$

The table of values accompanying Fig. 18.9 shows e^x, $1 + x$, $1 + x + x^2/2!$, and $1 + x + x^2/2! + x^3/3!$ for values of x from -1 to $+1$ at intervals of 0.1.

The remainder term (other forms)

In the formula

$$f(x) = f(a) + f'(a)(x - a) + \frac{f''(a)}{2!}(x - a)^2 + \cdots + \frac{f^{(n)}(a)}{n!}(x - a)^n + R_n(x, a), \tag{14}$$

we have expressed the remainder as an integral:

$$R_n(x, a) = \int_a^x \frac{(x - t)^n}{n!} f^{(n+1)}(t)\, dt. \tag{15}$$

Other expressions for the remainder may be obtained by applying the following theorem to this integral.

Theorem 6. *Let $g(t)$ and $h(t)$ be continuous for*

$$a \le t \le b,$$

and suppose that $h(t)$ does not change sign in this interval. Then there exists a number c between a and b such that

$$\int_a^b g(t)h(t)\,dt = g(c)\int_a^b h(t)\,dt. \tag{16}$$

Proof. Let m and M be the least and greatest values of $g(t)$, respectively, for $t \in [a, b]$. Then

$$m \leq g(t_i) \leq M, \tag{17}$$

for any t_i between a and b inclusive. We form a subdivision

$$a = t_0 < t_1 < t_2 < \cdots < t_n = b$$

of the interval $[a, b]$, with

$$t_i - t_{i-1} = \Delta t = \frac{b-a}{n}, \qquad i = 1, \ldots, n,$$

and multiply each term in the inequalities (17) by $h(t_i)\,\Delta t$. Since the sign of $h(t)$ does not change over the interval, all the terms $h(t_1)\,\Delta t$ are of the same sign, say positive. [If they are all negative, replace $h(t)$ by $-h(t)$ in the argument that follows.] Then

$$mh(t_i)\,\Delta t \leq g(t_i)\,h(t_i)\,\Delta t \leq Mh(t_i)\,\Delta t,$$

and the order of inequality is preserved when we now sum on i from 1 to n; that is,

$$\sum_{i=1}^{n} mh(t_i)\,\Delta t \leq \sum_{i=1}^{n} g(t_i)h(t_i)\,\Delta t \leq \sum_{i=1}^{n} Mh(t_i)\,\Delta t. \tag{18}$$

Finally, we let n increase indefinitely and recall that the sums then become definite integrals. Hence (18) leads to

$$m\int_a^b h(t)\,dt \leq \int_a^b g(t)h(t)\,dt \leq M\int_a^b h(t)\,dt,$$

or

$$m \leq \frac{\int_a^b g(t)h(t)\,dt}{\int_a^b h(t)\,dt} \leq M. \tag{19}$$

Since

$$Q = \frac{\int_a^b g(t)h(t)\,dt}{\int_a^b h(t)\,dt} \tag{20}$$

is a number between the least and greatest values taken on by $g(t)$ for $t \in [a, b]$, and since $g(t)$ is assumed to be continuous, there is a number c between a and b such that

$$Q = g(c).$$

This leads at once to the result in Eq. (16) and completes the proof of the theorem. □

We now apply this theorem to the integral in Eq. (15), with

$$h(t) = \frac{(x-t)^n}{n!}, \qquad g(t) = f^{(n+1)}(t).$$

As t varies from a to x, $h(t)$ is continuous and does not change sign. Hence, if $f^{(n+1)}(t)$ is continuous, then Eq. (16) applies, and we have

$$\begin{aligned} R_n(x, a) &= f^{(n+1)}(c)\int_a^x \frac{(x-t)^n}{n!}\,dt \\ &= f^{(n+1)}(c)\,\frac{(x-a)^{n+1}}{(n+1)!}. \end{aligned}$$

This is known as *Lagrange's* form of the remainder:

$$R_n(x, a) = f^{(n+1)}(c)\,\frac{(x-a)^{n+1}}{(n+1)!}, \tag{21}$$

for c between a and x.*

In applying Eq. (21), we can ordinarily only estimate $f^{(n+1)}(c)$, since we do not know c itself exactly. But if we know that

$$m \leq f^{(n+1)}(t) \leq M$$

for

$$t \in [a, x] \quad \text{or} \quad t \in [x, a],$$

then we know from Eq. (21) that $R_n(x, a)$ *lies between*

$$m\,\frac{(x-a)^{n+1}}{(n+1)!} \quad \text{and} \quad M\,\frac{(x-a)^{n+1}}{(n+1)!}.$$

---

* Another proof of Taylor's theorem, leading directly to the Lagrange form of the remainder, is outlined in Exercise 74 (Extended Mean Value Theorem), in the Miscellaneous Exercises at the end of Chapter 4.

Note. There is another common form of remainder, known as *Cauchy's* form:

$$R_n(x, a) = \frac{(x - c')^n (x - a)}{n!} f^{(n+1)}(c'), \qquad (22)$$

for c' between a and x. This form may be obtained by applying Eq. (16) to Eq. (15) with

$$h(t) = 1, \qquad g(t) = \frac{(x - t)^n}{n!} f^{(n+1)}(t).$$

The Lagrange form, Eq. (21), is usually easier to apply, and in general gives satisfactory results. However, the Cauchy form is useful in proving convergence of certain series, such as the Maclaurin series for $\ln(1 + x)$ for $x \in [-1, 0]$.

Problem. Find the Maclaurin series expansion for $\sin x$, and show that the series converges to $\sin x$ for all finite x, $-\infty < x < +\infty$.

Solution. In Eq. (14), we take $f(x) = \sin x$, $a = 0$. Then

$$\begin{aligned}
f(x) &= \sin x, & f(0) &= 0 \\
f'(x) &= \cos x = \sin(x + \pi/2), & f'(0) &= 1 \\
f''(x) &= -\sin x = \sin(x + \pi), & f''(0) &= 0 \\
f'''(x) &= -\cos x = \sin(x + 3\pi/2), & f'''(0) &= -1 \\
&\vdots & &\vdots \\
f^{(k)}(x) &= \sin(x + k\pi/2), & f^{(k)}(0) &= \sin(k\pi/2).
\end{aligned}$$

The values of the function and its derivatives at $x = 0$ are given by the formula

$$f^{(k)}(0) = \sin(k\pi/2), \quad k = 0, 1, 2, \ldots, \qquad (23)$$

where the notation $f^{(k)}(0)$ means the value of the kth derivative of $f(x)$ at $x = 0$ and the zeroth derivative of the function means the function itself. If k is an even integer 0, 2, 4, . . . , then $\sin(k\pi/2)$ is zero. If k is one of the integers 1, 5, 9, 13, . . . , of the form $4m + 1$, then $\sin(k\pi/2)$ is plus one, while if k is one of the integers 3, 7, . . . , of the form $4m + 3$, then $\sin(k\pi/2)$ is minus one. Thus when we substitute these values (23) into the Taylor series formula, Eq. (14), with $a = 0$, we obtain

$$\sin x = x - \frac{x^3}{3!} + \frac{x^5}{5!} - \frac{x^7}{7!} + \cdots + \frac{(-1)^{n-1} x^{2n-1}}{(2n-1)!} + 0 \cdot x^{2n} + R_{2n}(x, 0). \qquad (24)$$

By Eq. (21), the remainder is

$$R_{2n}(x, 0) = \frac{x^{2n+1}}{(2n+1)!} \sin\left(c + \frac{(2n+1)\pi}{2}\right).$$

Even though the only thing we know about c is that it lies between 0 and x, we know that the sine never exceeds 1 in absolute value. Hence

$$|R_{2n}(x, 0)| \le \frac{|x|^{2n+1}}{(2n+1)!}. \qquad (25)$$

Now for any finite x whatever, no matter how large, the $(2n + 1)!$ in the denominator of (25) goes to infinity more rapidly than the $|x|^{2n+1}$ in the numerator, and

$$\lim_{n \to \infty} \frac{|x|^{2n-1}}{(2n+1)!} = 0. \qquad (26)$$

[See Exercise 3 for a discussion that establishes Eq. (26).] Therefore the sequence of partial sums $f_n(x)$ of the Maclaurin series

$$x - \frac{x^3}{3!} + \frac{x^5}{5!} - \frac{x^7}{7!} + \cdots$$

converges to $\sin x$ for any x, $-\infty < x < +\infty$, and we may write

$$\sin x = \sum_{k=1}^{\infty} \frac{(-1)^{k-1} x^{2k-1}}{(2k-1)!}, \quad -\infty < x < +\infty. \qquad (27)$$

Figure 18.10 shows graphs of $y = \sin x$ and the first two approximating polynomials

$$A\colon y = x, \quad B\colon y = x - \frac{x^3}{3!}.$$

The cubic approximation is in error by less than $x^5/120$, so it is almost indistinguishable from the sine curve from $x = -1$ to $x = +1$. However, it crosses the x-axis at $\pm\sqrt{6} \approx \pm 2.45$, whereas the sine curve crosses at $\pm\pi \approx \pm 3.14$. The table of data for Fig. 18.10 also shows that the cubic approximation is in error by less than 0.01 from $x = 0$ to $x = 1$. The values of $\sin x$ were taken from a four-place table, x in radians.

Analogous methods may be used to show that

$$\cos x = \sum_{k=0}^{\infty} \frac{(-1)^k x^{2k}}{(2k)!}, \quad -\infty < x < \infty. \qquad (28)$$

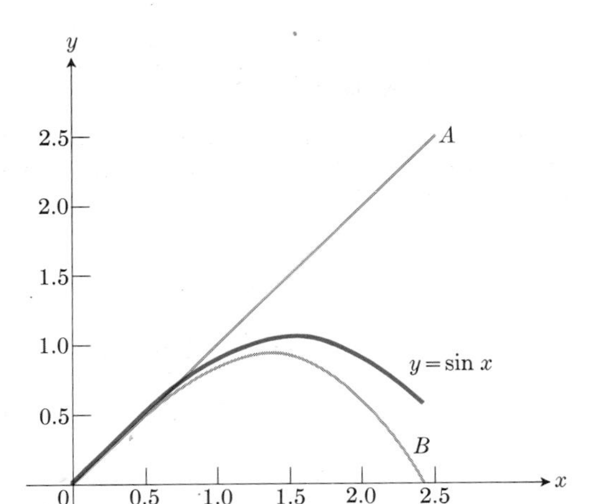

18.10 The graphs of $y = \sin x$ and of two approximating polynomials, $(A)\ y = x$ and $(B)\ y = x - x^3/6$. The three curves are symmetric with respect to the origin. (Note that the negative portions are not shown.)

x	$\sin x$	$x - \dfrac{x^3}{3!}$
0	0	0
0.1	0.0998	0.09983+
0.2	0.1987	0.19867−
0.3	0.2955	0.29550
0.4	0.3894	0.38933+
0.5	0.4794	0.47917−
0.6	0.5646	0.56400
0.7	0.6442	0.64283+
0.8	0.7174	0.71467−
0.9	0.7833	0.77850
1.0	0.8415	0.83333+
1.41	0.9871	0.94280−
1.5	0.9975	0.93750
1.57	1.0000−	0.92502
1.58	1.0000−	0.92261
1.6	0.9996	0.91733
2.45	0.63776	−0.00102

EXERCISES

1. In connection with Eq. (15), show that the remainder term $R_n(x, a)$ in Taylor's theorem satisfies the inequality
$$|R_n(x, a)| \leq M \frac{|x - a|^{n+1}}{(n+1)!}$$
if there is a positive constant M such that
$$|f^{(n+1)}(t)| \leq M$$
for all t between a and x inclusive.

2. Using series, calculate e to five decimal places.

3. We encounter series where the nth term involves the fraction $x^n/n!$, and it is important to know how this function of the two variables x and n behaves for fixed values of x as n becomes larger and larger. The following analysis can be used to show that when $n \geq 2$, $n!$ is greater than $e(n/e)^n$, where $e = 2.718...$ is the base of natural logarithms.

(a) $$\ln n! = \ln 2 + \ln 3 + \cdots + \ln n$$
can be represented as the sum of the areas of $n - 1$ rectangles, as indicated in Fig. 18.11. Note that the first rectangle has its base on the x-axis from 1 to 2 and has altitude equal to $\ln 2$, etc. Since each base is of unit length, the area of the rectangle has the same magnitude as its altitude. Show that
$$\ln n! > \int_1^n \ln x\, dx = n \ln n - n + 1, \quad n \geq 2.$$

(b) From the result of (a), show that
$$n! > e(n/e)^n, \quad n \geq 2.$$

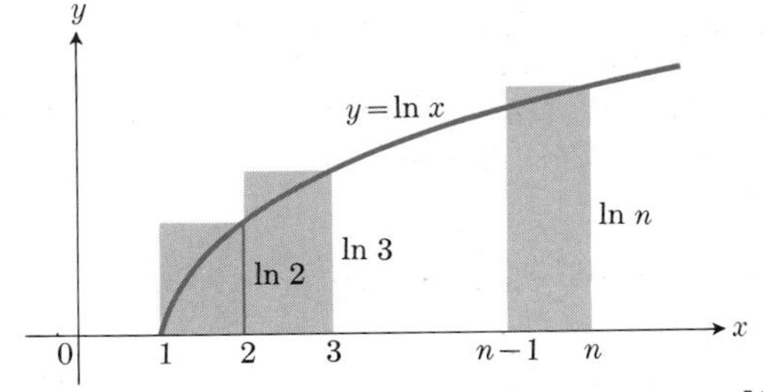

18.11

(c) Assume that x is any finite real number. Use the result of (b) to show that

$$\lim_{n\to\infty} (x^n/n!) = 0.$$

In particular, show that $|x^n/n!| < 2^{-n}$ when n is greater than 2 and also greater than $2|ex|$.

4. For what range of values of x can one replace $\sin x$ by $x - x^3/6$ with an error not greater than 5×10^{-4}?

5. If $\cos x$ is replaced by $1 - x^2/2$ and $|x| < 0.1$, what estimate can you give of the error?

6. For what range of values of x can one replace $\ln(1 + x)$ by x with an error not greater than one percent of the absolute value of x?

7. The approximation

$$\sqrt{1+x} = 1 + x/2$$

is used when $|x|$ is small. Give an estimate of the error if $|x| < 0.01$.

8. A log-log slide rule gives readings of e^h for

$$0.01 \le h \le 10.$$

When $0 \le h \le 0.01$, show that e^h may be replaced simply by $1 + h$ with an error not greater than six-tenths of one percent of h.

9. The quantity $\sqrt{e} = e^{0.5}$ is to be computed from the series

$$e^x = 1 + x + \frac{x^2}{2!} + \cdots + \frac{x^n}{n!} + R_n(x, 0).$$

How large should one choose n to guarantee that

$$|R_n(x, 0)| < 0.0005?$$

18.5 APPLICATION TO MAX-MIN THEORY FOR FUNCTIONS OF TWO INDEPENDENT VARIABLES

In this article, we propose to apply Taylor's theorem to study the behavior of a function

$$w = f(x, y),$$

when the point (x, y) is close to a point (a, b) at which the first-order partial derivatives vanish:

$$f_x(a, b) = f_y(a, b) = 0. \tag{1}$$

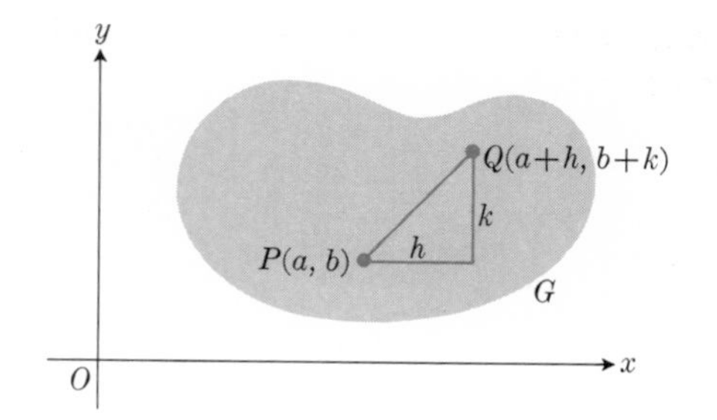

18.12

We shall assume that w is continuous, together with its first- and second-order partial derivatives, throughout some neighborhood G of the point $P(a, b)$ (see Fig. 18.12). In the discussion that follows, a, b, h, and k are held fixed. In addition, h and k are chosen as small, so the point

$$Q(a + h, b + k),$$

together with all points

$$\left.\begin{aligned} x &= a + ht, \\ y &= b + kt, \end{aligned}\right\} \quad 0 \le t \le 1, \tag{2}$$

on the line PQ, also lie in G. We consider the values of $w = f(x, y)$ along the line PQ of Eq. (2) as t varies from 0 to 1. Thus, if we let

$$F(t) = f(a + ht, b + kt), \tag{3}$$

we have

$$F(0) = f(a, b) \quad \text{when } t = 0,$$
$$F(1) = f(a + h, b + k) \quad \text{when } t = 1.$$

Now by Taylor's theorem, we have

$$F(t) = F(0) + tF'(0) + \frac{t^2}{2!} F''(t_1),$$

where t_1 is between 0 and t. In particular, taking $t = 1$,

$$F(1) = F(0) + F'(0) + \tfrac{1}{2}F''(t_1), \tag{4}$$

with $0 < t_1 < 1$.

The derivatives in Eq. (4) may be calculated from (3) by using the chain rule for partial derivatives. We have two variables, x and y, in the first set, and only one variable, t, in the second set, and they are related as in (2). Hence

$$F'(t) = \frac{\partial f}{\partial x}\frac{dx}{dt} + \frac{\partial f}{\partial y}\frac{dy}{dt},$$

and, since

$$\frac{dx}{dt} = h, \qquad \frac{dy}{dt} = k,$$

this means that

$$F'(t) = h\frac{\partial f}{\partial x} + k\frac{\partial f}{\partial y}. \tag{5}$$

The same chain rule may be applied to calculate

$$F''(t) = h\frac{\partial[\ ]}{\partial x} + k\frac{\partial[\ ]}{\partial y}, \tag{6}$$

where we have used the brackets to indicate the expression on the righthand side of Eq. (5); that is,

$$\begin{aligned} [\ ] &= h\frac{\partial f}{\partial x} + k\frac{\partial f}{\partial y}, \\ \frac{\partial[\ ]}{\partial x} &= h\frac{\partial^2 f}{\partial x^2} + k\frac{\partial^2 f}{\partial x\,\partial y}, \\ \frac{\partial[\ ]}{\partial y} &= h\frac{\partial^2 f}{\partial y\,\partial x} + k\frac{\partial^2 f}{\partial y^2}. \end{aligned} \tag{7}$$

When we substitute from (7) into (6), we have

$$F''(t) = h^2\frac{\partial^2 f}{\partial x^2} + 2hk\frac{\partial^2 f}{\partial x\,\partial y} + k^2\frac{\partial^2 f}{\partial y^2}. \tag{8}$$

Note 1. Equations (5) and (8) may be given the following interpretation in terms of operators. The operator d/dt operating on $F(t)$ is the same as the operator

$$h(\partial/\partial x) + k(\partial/\partial y)$$

operating on $f(x, y)$; and the operator d^2/dt^2 operating on $F(t)$ is the same as the operator

$$\left(h\frac{\partial}{\partial x} + k\frac{\partial}{\partial y}\right)^2 = h^2\frac{\partial^2}{\partial x^2} + 2hk\frac{\partial^2}{\partial x\,\partial y} + k^2\frac{\partial^2}{\partial y^2}$$

operating on $f(x, y)$. This may be extended, more generally, to say that

$$\frac{d^n}{dt^n}F(t) = \left(h\frac{\partial}{\partial x} + k\frac{\partial}{\partial y}\right)^n f(x, y),$$

where the term in parentheses on the right should be expanded by the binomial theorem and then made to operate on $f(x, y)$ with x and y as in Eq. 2.

If we now take $t = 0$ in Eq. (5) and $t = t_1$ in Eq. (8) and substitute the results into Eq. (4), we obtain

$$\begin{aligned} f(a + h,\, b + k) &= f(a, b) + (hf_x + kf_y)_{(a,b)} \\ &\quad + \tfrac{1}{2}(h^2 f_{xx} + 2hkf_{xy} + k^2 f_{yy})_{(a+t_1h,\, b+t_1k)}. \end{aligned} \tag{9}$$

Note 2. If we extend the Maclaurin series for $F(t)$ to more terms:

$$F(t) = F(0) + F'(0)\cdot t + \frac{F''(0)}{2!}t^2 + \cdots + \frac{F^{(n)}(0)}{n!}t^n + \cdots,$$

and then take $t = 1$:

$$F(1) = F(0) + F'(0) + \frac{F''(0)}{2!} + \cdots + \frac{F^{(n)}(0)}{n!} + \cdots,$$

we obtain the expansion

$$\begin{aligned} f(a + h,\, b + k) &= f(a, b) + (hf_x + kf_y)_{(a,b)} \\ &\quad + \frac{1}{2!}(h^2 f_{xx} + 2hkf_{xy} + k^2 f_{yy})_{(a,b)} \\ &\quad + \frac{1}{3!}(h^3 f_{xxx} + 3h^2kf_{xxy} + 3hk^2 f_{xyy} + k^3 f_{yyy})_{(a,b)} \\ &\quad + \cdots \\ &\quad + \frac{1}{n!}\left[\left(h\frac{\partial}{\partial x} + k\frac{\partial}{\partial y}\right)^n f\right]_{(a,b)} + \cdots. \end{aligned} \tag{10}$$

This expresses the value of the function $f(x, y)$ at

$$x = a + h, \qquad y = b + k$$

in terms of the values of the function and its partial derivatives at (a, b), and powers of

$$h = x - a \quad \text{and} \quad k = y - b.$$

This is the Taylor series expansion, about the point (a, b), of the function $f(x, y)$. Analogous formulas hold for functions of more independent variables.

Suppose, now, that we have found a point (a, b) where the first-order derivatives f_x and f_y are both zero, and we wish to determine whether the function $w = f(x, y)$ has or has not a maximum or a minimum at (a, b). We may then rewrite Eq. (9) in the form

$$\begin{aligned} f(a + h, b + k) &- f(a, b) \\ &= \tfrac{1}{2}(h^2 f_{xx} + 2hk f_{xy} + k^2 f_{yy})_{(a+t_1h,\, b+t_1k)} \\ &= \phi(t_1). \end{aligned}$$

Since a maximum or minimum value of w at (a, b) is reflected in the sign of $\phi(t_1)$, we are led to consider the sign of $\phi(t)$. The second-order derivatives which enter this expression are to be evaluated at a point on the line segment PQ (Fig. 18.12). This is not convenient for our purposes, since we do not know precisely where to take this point; that is, we do not know anything more about t_1 than that

$$0 \leq t_1 \leq 1.$$

However, since we are assuming that f_{xx}, f_{xy}, and f_{yy} are *continuous* throughout the region G, and since both h and k are assumed to be *small*, the values of these derivatives at $(a + t_1h, b + t_1k)$ are nearly the same as their values at (a, b). In particular, the *sign* of $\phi(t_1)$ is the same, for sufficiently small values of h and k, as the sign of

$$\phi(0) = h^2 f_{xx}(a, b) + 2hk f_{xy}(a, b) + k^2 f_{yy}(a, b), \tag{11}$$

if this quantity is not zero. We therefore have the following criteria.

1. The function $f(x, y)$ has a relative *minimum* at (a, b), provided that

$$f_x(a, b) = f_y(a, b) = 0,$$

and that $\phi(0)$, Eq. (11), is *positive* for all sufficiently small values of h and k. (We exclude, of course, the case where $h = k = 0$).

2. The function $f(x, y)$ has a relative *maximum* at (a, b), provided that

$$f_x(a, b) = f_y(a, b) = 0,$$

and that $\phi(0)$, Eq. (11), is *negative* for all sufficiently small values of h and k. (We again exclude $h = k = 0$.)

3. The function $f(x, y)$ has a *saddle point* at (a, b), provided that

$$f_x(a, b) = f_y(a, b) = 0,$$

and that $\phi(0)$, Eq. (11), is positive for some and negative for other small values of h and k. (For further discussion, see Exercises 1 and 3 below.)

Example. The function

$$f(x, y) = x^2 + xy + y^2 + x - 4y + 5$$

has partial derivatives

$$f_x = 2x + y + 1, \qquad f_y = x + 2y - 4,$$

which vanish at $(-2, 3)$. The second partial derivatives are all constant:

$$f_{xx} = 2, \qquad f_{xy} = 1, \qquad f_{yy} = 2,$$

and the expression whose sign determines whether f has a maximum, a minimum, or a saddle point at $(-2, 3)$ is

$$2h^2 + 2hk + 2k^2.$$

If we multiply this by 2, we have

$$4h^2 + 4hk + 4k^2 = (2h + k)^2 + 3k^2,$$

which is the sum of two nonnegative terms and is zero only when $h = k = 0$. Hence, the function has a relative *minimum* at $(-2, 3)$. In fact, this is its absolute minimum, since

$$f(-2 + h, 3 + k) \geq f(-2, 3)$$

for *all* h and k.

EXERCISES

1. (a) Let $A = f_{xx}(a, b)$, $B = f_{xy}(a, b)$, $C = f_{yy}(a, b)$, and show that the expression for $\phi(0)$, Eq. (11), becomes the same as

$$A\phi(0) = (Ah + Bk)^2 + (AC - B^2)k^2$$

when multiplied by A.

(b) Suppose $f_x(a, b) = f_y(a, b) = 0$ and $A \neq 0$. Use the result of (a) to show that $f(x, y)$ has

(1) a relative minimum at (a, b) if

$$AC - B^2 > 0 \quad \text{and} \quad A > 0,$$

(2) a relative maximum at (a, b) if

$$AC - B^2 > 0 \quad \text{and} \quad A < 0,$$

(3) a saddle point at (a, b) if $AC - B^2 < 0$.

2. Test the following surfaces for maxima, minima, and saddle points.
   (a) $z = x^2 + y^2 - 2x + 4y + 6$
   (b) $z = x^2 - y^2 - 2x + 4y + 6$
   (c) $z = x^2 - 2xy + 2y^2 - 2x + 2y + 1$
   (d) $z = x^2 + 2xy$
   (e) $z = 3 + 2x + 2y - 2x^2 - 2xy - y^2$
   (f) $z = x^3 - y^3 - 2xy + 6$
   (g) $z = x^3 + y^3 + 3x^2 - 3y^2 - 8$
3. In Eq. (11), let

$$h = c\cos\alpha, \quad k = c\sin\alpha, \quad c > 0,$$

and show that

$$\phi(0) = c^2\frac{d^2f}{ds^2},$$

where d^2f/ds^2 is the second-order directional derivative of f at (a, b) in the direction of the unit vector $\mathbf{u} = \mathbf{i}\cos\alpha + \mathbf{j}\sin\alpha$. Express the three criteria for maxima, minima, and saddle points at the end of Article 18.5 in terms of d^2f/ds^2. (Similar criteria also apply in higher-dimensional problems.)
4. (a) Taking $a + h = x$, $b + k = y$, and $a = b = 0$ in Eq. (10), obtain the series through the terms of second degree in x and y, for the function

$$f(x, y) = e^x \cos y.$$

(b) Obtain the series for (a) more simply by multiplying the series for e^x by the series for $\cos y$.
5. Write out explicitly, through the terms of second degree, the Taylor series for a function $f(x, y, z)$, in powers of $(x - a)$, $(y - b)$, $(z - c)$, and the values of f and its partial derivatives at (a, b, c).

18.6 COMPUTATIONS

The Taylor series expansion

$$f(x) = f(a) + f'(a)(x - a) + \frac{f''(a)}{2!}(x - a)^2 + \cdots + \frac{f^{(n)}(a)}{n!}(x - a)^n + R_n(x, a) \tag{1}$$

expresses the value of the function at x in terms of its value and the values of its derivatives at a, plus a remainder term which we hope is so small that it may safely be omitted. In applying series to numerical computations, it is therefore *necessary* that a be chosen so that $f(a), f'(a), f''(a), \ldots$ are known. In dealing with the trigonometric functions, for example, one might take

$$a = 0, \ \pm\pi/6, \ \pm\pi/4, \ \pm\pi/3, \ \pm\pi/2,$$

and so on. It is also clear that it is *desirable* to choose the value of a near the value of x for which the function is to be computed, to make $x - a$ small, and thereby to ensure that the terms of the series decrease rapidly as n increases.

Problem 1. What value of a might one choose in the Taylor series (1) to compute

(a) $\cos 5°$, (b) $\sin 35°$?

Solution. (a) In the first case, one would probably choose $a = 0$. If we substitute

$$f(x) = \cos x \quad \text{and} \quad a = 0$$

in Eq. (1), we find that the series for $\cos x$ in powers of $x - 0$ is

$$\cos x = 1 - \frac{x^2}{2!} + \frac{x^4}{4!} - \cdots + (-1)^n\frac{x^{2n}}{(2n)!} + 0 \cdot x^{2n+1} + R_{2n+1}(x, 0). \tag{2}$$

Since the derivatives of $\cos x$ are $\pm\sin x$ or $\pm\cos x$, which never exceed one in absolute value, the remainder with $|f^{(2n+1)}(c)| \le 1$, leads to the estimate

$$|R_{2n+1}(x, 0)| \le \frac{|x|^{2n+2}}{(2n+2)!};$$

this tends to zero as $n \to \infty$, for all $|x| < \infty$. In particular, to calculate $\cos 5°$, we take

$$x = \frac{5\pi}{180} = 0.08726\ 65 \quad \text{(radian measure!)}$$

and see that

$$\left|R_5\left(\frac{5\pi}{180}, 0\right)\right| \le \frac{(0.088)^6}{6!} < \frac{(0.1)^6}{6!} < 10^{-8}.$$

Hence we may calculate cos 5° accurately to seven decimal places by using the approximation

$$\cos x \approx 1 - \frac{x^2}{2} + \frac{x^4}{24} \qquad \left(x = \frac{5\pi}{180}\right).$$

(b) In calculating sin 35°, we could again choose $a = 0$ and use the series

$$\sin x = x - \frac{x^3}{3!} + \frac{x^5}{5!} - \cdots + (-1)^n \frac{x^{2n+1}}{(2n+1)!} + 0 \cdot x^{2n+2} + R_{2n+2}(x, 0), \tag{3}$$

or we could choose $a = \pi/6$ (which corresponds to 30°) and use the series

$$\begin{aligned}\sin x = \sin\frac{\pi}{6} + \cos\frac{\pi}{6}\left(x - \frac{\pi}{6}\right) &- \sin\frac{\pi}{6}\frac{(x-\pi/6)^2}{2!} - \cos\frac{\pi}{6}\frac{(x-\pi/6)^3}{3!} + \cdots \\ &+ \sin\left(\frac{\pi}{6} + n\frac{\pi}{2}\right)\frac{(x-\pi/6)^n}{n!} + R_n\left(x, \frac{\pi}{6}\right).\end{aligned}$$

The remainder in the series (3) satisfies the inequality

$$|R_{2n+2}(x, 0)| \le \frac{|x|^{2n+3}}{(2n+3)!}, \tag{4}$$

which tends to zero as n becomes infinite, no matter how large $|x|$ may be. We could therefore calculate sin 35° by placing

$$x = \frac{35\pi}{180} = 0.61086\ 53$$

in the approximation

$$\sin x \approx x - \frac{x^3}{6} + \frac{x^5}{120} - \frac{x^7}{5040},$$

with an error no greater than $(0.5)10^{-7}$, since

$$\left|R_8\left(\frac{35\pi}{180}, 0\right)\right| \le \frac{(0.62)^9}{9!} < \frac{0.016}{360{,}000} < (0.5)10^{-7}.$$

By using the series with $a = \pi/6$, we could obtain equal accuracy with a smaller exponent n, but at the expense of introducing $\cos(\pi/6) = \sqrt{3}/2$ as one of the coefficients. In this series, with $a = \pi/6$, we would take

$$x = 35\pi/180,$$

but the quantity that appears raised to the various powers is

$$x - \frac{\pi}{6} = \frac{5\pi}{180} = 0.08726\ 65,$$

which decreases rapidly when raised to high powers.

As a matter of fact, various trigonometric identities may be used, such as

$$\sin\left(\frac{\pi}{2} - x\right) = \cos x,$$

to enable one to calculate quite easily the sine or cosine of any number by means of the series in Eq. (2) or (3). This method of finding the sine or cosine of a number is used in modern high-speed computers. It is more efficient for these large machines to calculate these trigonometric functions from series than it is for them to "read" tables.

Computation of logarithms

Tables of natural logarithms are computed from series. The starting point is the series for $\ln(1 + x)$ in powers of x:

$$\ln(1+x) = x - \frac{x^2}{2} + \frac{x^3}{3} - \cdots + (-1)^{n-1}\frac{x^n}{n} + \cdots.$$

This series may be found directly from the Taylor series expansion, Eq. (1), with $a = 0$. It may also be obtained by integrating the geometric series for $1/(1 + t)$ from $t = 0$ to $t = x$:

$$\int_0^x \frac{dt}{1+t} = \int_0^x (1 - t + t^2 - t^3 + \cdots)\, dt,$$

$$\ln(1+t)\Big]_0^x = t - \frac{t^2}{2} + \frac{t^3}{3} - \frac{t^4}{4} + \cdots\Big]_0^x,$$

$$\ln(1+x) = x - \frac{x^2}{2} + \frac{x^3}{3} - \frac{x^4}{4} + \cdots. \tag{5}$$

The expansion (5) is valid for $|x| < 1$, since then the remainder $R_n(x, 0)$ approaches zero as $n \to \infty$, as we shall now see.

The remainder is given by the integral of the remainder in the geometric series; that is,

$$R_n(x, 0) = \int_0^x \frac{(-1)^n t^n}{1+t}\,dt. \tag{6}$$

We now suppose that $|x| < 1$. For every t between 0 and x inclusive, we have

$$|1 + t| \geq 1 - |x|$$

and

$$|(-1)^n t^n| = |t|^n,$$

so that

$$\left|\frac{(-1)^n t^n}{1+t}\right| \leq \frac{|t|^n}{1 - |x|}.$$

Therefore

$$\begin{aligned} |R_n(x, 0)| &\leq \int_0^{|x|} \frac{t^n}{1 - |x|}\,dt \\ &= \frac{1}{n+1}\,\frac{|x|^{n+1}}{1 - |x|}. \end{aligned} \tag{7}$$

When $n \to \infty$, the right-hand side of the inequality in (7) approaches zero, and so must the left-hand side. Thus (5) holds for $|x| < 1$.

If we replace x by $-x$, we obtain

$$\ln(1 - x) = -x - \frac{x^2}{2} - \frac{x^3}{3} - \cdots - \frac{x^n}{n} - \cdots, \tag{8}$$

which is also valid for $|x| < 1$. When we subtract (8) from (5), we get

$$\ln\frac{1+x}{1-x} = 2\left(x + \frac{x^3}{3} + \frac{x^5}{5} + \cdots + \frac{x^{2k-1}}{2k-1} + \cdots\right) \tag{9}$$

for $|x| < 1$. Equation (9) may be used to compute the natural logarithm of any positive number y (see Fig. 18.13) by taking

$$y = \frac{1+x}{1-x} \quad \text{or} \quad x = \frac{y-1}{y+1}.$$

But the series converges most rapidly for values of x near zero, or when $(1 + x)/(1 - x)$ is near one.

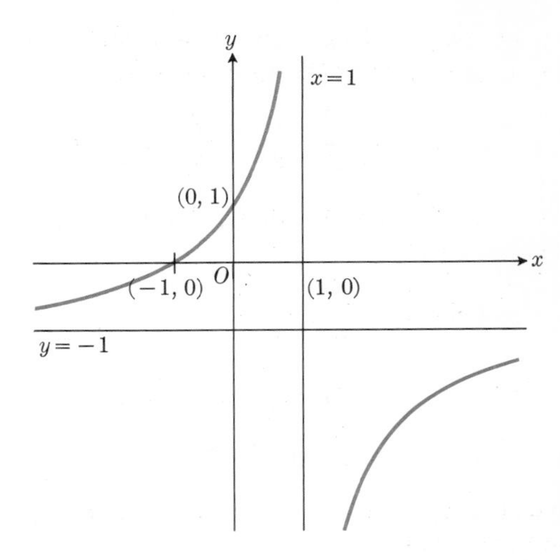

18.13

For this reason, the logarithms of the numbers

$$\frac{2}{1}, \frac{3}{2}, \frac{4}{3}, \frac{5}{4}, \ldots, \frac{N+1}{N}$$

are ordinarily computed first in forming a table of natural logarithms of the integers. Then it is a matter of simple arithmetic to compute

$$\begin{aligned} \ln 3 &= \ln \tfrac{2}{1} + \ln \tfrac{3}{2}, \\ \ln 4 &= \ln 3 + \ln \tfrac{4}{3}, \\ \ln 5 &= \ln 4 + \ln \tfrac{5}{4}, \\ &\vdots \\ \ln(N+1) &= \ln N + \ln \frac{N+1}{N}. \end{aligned}$$

For this purpose, one may solve the equation

$$\frac{1+x}{1-x} = \frac{N+1}{N}$$

for

$$x = \frac{1}{2N+1}.$$

This may be substituted into Eq. (9), which becomes

$$\begin{aligned} &\ln\frac{N+1}{N} \\ &= 2\left(\frac{1}{2N+1} + \frac{1}{3(2N+1)^3} + \frac{1}{5(2N+1)^5} + \cdots\right). \end{aligned} \tag{10}$$

For example, to calculate $\ln 2$, we take $N = 1$, $x = \frac{1}{3}$, and calculate

$$\begin{aligned} x &= 0.33333 \quad 33, \\ \tfrac{1}{3}x^3 &= 0.01234 \quad 57, \\ \tfrac{1}{5}x^5 &= 0.00082 \quad 30, \\ \tfrac{1}{7}x^7 &= 0.00006 \quad 53, \\ \tfrac{1}{9}x^9 &= 0.00000 \quad 56, \\ \tfrac{1}{11}x^{11} &= 0.00000 \quad 05, \\ \hline \text{Sum} &\quad 0.34657 \quad 34; \\ \ln 2 &\approx 0.69314 \quad 68. \end{aligned}$$

Computation of π

Archimedes (287–212 B.C.) gave the approximation

$$3\tfrac{1}{7} > \pi > 3\tfrac{10}{71}$$

in the third century B.C. A French mathematician, Vieta (1540–1603), gave the formula for $2/\pi$ as

$$\sqrt{\tfrac{1}{2} \times \sqrt{(\tfrac{1}{2} + \tfrac{1}{2}\sqrt{\tfrac{1}{2}}) \times \sqrt{(\tfrac{1}{2} + \tfrac{1}{2}\sqrt{\tfrac{1}{2} + \tfrac{1}{2}\sqrt{\tfrac{1}{2}}}) \times \cdots}}},$$

which Turnbull* calls "the first actual formula for the time-honoured number π." Other interesting formulas for π include the following†:

$$\frac{\pi}{4} = \cfrac{1}{1 + \cfrac{1^2}{2 + \cfrac{3^2}{2 + \cfrac{5^2}{2 + \cdots}}}},$$

credited to Lord Brouncker, an Irish peer;

$$\frac{\pi}{4} = \frac{2 \times 4 \times 4 \times 6 \times 6 \times 8 \times \cdots}{3 \times 3 \times 5 \times 5 \times 7 \times 7 \times \cdots},$$

discovered by the English mathematician Wallis; and

$$\frac{\pi}{4} = 1 - \frac{1}{3} + \frac{1}{5} - \frac{1}{7} + \cdots,$$

known as Leibniz's formula. All these formulas involve limits. For example, Vieta's formula may be expressed in terms of the sequences $\{x_n\}$ and $\{s_n\}$ defined as follows:

$$\begin{aligned} x_1 &= \sqrt{\tfrac{1}{2}}, \\ x_2 &= \sqrt{\tfrac{1}{2} + \tfrac{1}{2}x_1}, \\ x_3 &= \sqrt{\tfrac{1}{2} + \tfrac{1}{2}x_2}, \\ &\vdots \\ x_{n+1} &= \sqrt{\tfrac{1}{2} + \tfrac{1}{2}x_n} \end{aligned}$$

and

$$\begin{aligned} s_1 &= x_1, \\ s_2 &= x_1\sqrt{x_2} = x_1 x_2^{1/2}, \\ s_3 &= x_1\sqrt{x_2\sqrt{x_3}} = x_1 x_2^{1/2} x_3^{1/4}, \\ s_4 &= x_1\sqrt{x_2\sqrt{x_3\sqrt{x_4}}} = x_1 x_2^{1/2} x_3^{1/4} x_4^{1/8}, \\ &\vdots \\ s_{n+1} &= x_1 x_2^{1/2} x_3^{1/4} \ldots x_n^{1/2^n}. \end{aligned}$$

Vieta's formula says that

$$\lim_{n\to\infty} s_n = \frac{2}{\pi}.$$

Similarly, we could interpret the other formulas as expressing certain limits in terms of π. However, we now turn our attention to the series for $\tan^{-1} x$, since it leads to the Leibniz formula and others from which π has been computed to a great many decimal places.

Since

$$\tan^{-1} x = \int_0^x \frac{dt}{1 + t^2},$$

we integrate the geometric series (with remainder)

$$\frac{1}{1+t^2} = 1 - t^2 + t^4 - t^6 + \cdots + (-1)^n t^{2n} + \frac{(-1)^{n+1} t^{2n+2}}{1+t^2}. \tag{11}$$

Thus

$$\tan^{-1} x = x - \frac{x^3}{3} + \frac{x^5}{5} - \frac{x^7}{7} + \cdots + (-1)^n \frac{x^{2n+1}}{2n+1} + R,$$

* *World of Mathematics*, Vol. 1, p. 121.
† *World of Mathematics*, Vol. 1, p. 138.

where $$R = \int_0^x \frac{(-1)^{n+1}t^{2n+2}}{1+t^2}\,dt.$$

The denominator of the integrand is greater than or equal to 1; hence

$$|R| \leq \int_0^{|x|} t^{2n+2}\,dt = \frac{|x|^{2n+3}}{2n+3}.$$

If $|x| \leq 1$, then the right-hand side of this inequality approaches zero as $n \to \infty$. Therefore R also approaches zero, and we have

$$\tan^{-1} x = \sum_{n=0}^{\infty} \frac{(-1)^n x^{2n+1}}{2n+1},$$

or

$$\tan^{-1} x = x - \frac{x^3}{3} + \frac{x^5}{5} - \frac{x^7}{7} + \cdots, \quad |x| \leq 1. \tag{12}$$

In 1961 this equation was used to compute π to more than 100,000 decimal places on an IBM 7090 computer.

Various trigonometric identities are useful if one wishes to use Eq. (12) to calculate π. For example, if

$$\alpha = \tan^{-1} \tfrac{1}{2} \quad \text{and} \quad \beta = \tan^{-1} \tfrac{1}{3},$$

then

$$\tan(\alpha + \beta) = \frac{\tan\alpha + \tan\beta}{1 - \tan\alpha\tan\beta} = \frac{\frac{1}{2}+\frac{1}{3}}{1-\frac{1}{6}} = 1 = \tan\frac{\pi}{4}$$

and

$$\frac{\pi}{4} = \alpha + \beta = \tan^{-1}\tfrac{1}{2} + \tan^{-1}\tfrac{1}{3}. \tag{13}$$

Now Eq. (12) may be used with $x = \frac{1}{2}$ to evaluate $\tan^{-1}\frac{1}{2}$, and with $x = \frac{1}{3}$ to evaluate $\tan^{-1}\frac{1}{3}$. The sum of these results, multiplied by 4, gives π. One identity that has been used in place of Eq. (13) for the calculation* of π is

$$\frac{\pi}{4} = 4\tan^{-1}\frac{1}{5} - \tan^{-1}\frac{1}{239}. \tag{14}$$

* For details, see Wrench and Shanks, "Calculation of π to 100,000 Decimals," *Mathematics of Computation*, **16,** No. 77, Jan. 1962, pp. 76–99.

Remark. We should mention the two types of numerical error that occur in computing with series. On the one hand, there is the so-called *truncation error,* which is the remainder $R_n(x, a)$ and consists of the sum of the infinite number of terms in the series that follow the term $(x - a)^n f^{(n)}(a)/n!$. This is the only error we have discussed so far. On the other hand, there is the so-called *round-off error* that enters in calculating the sum of the finite number of terms

$$f(a) + f'(a)(x-a) + \cdots + \frac{f^{(n)}(a)(x-a)^n}{n!}$$

when we approximate each of these terms by a decimal number with only a finite number of decimal places. For example, taking 0.3333 in place of $\frac{1}{3}$ introduces a round-off error equal to $10^{-4}/3$. There is likely to be a round-off error associated with each term in our computations, some of these being positive and some negative. In highly accurate computations, such as the fifteen-place tables of functions published by the Works Progress Administration, it is important to control both the truncation error and the round-off errors. *Truncation* errors can be reduced by taking more terms of the series; *round-off* errors can be reduced by taking more decimal places.

EXERCISES

In Exercises 1 through 8, use a suitable series to calculate the indicated quantity to three decimal places. In each case, show that the remainder term does not exceed 5×10^{-4}.

1. $\cos 31°$
2. $\tan 46°$
3. $\sqrt[3]{9}$
4. $(36)^{-1/5}$
5. $\cosh 0.5$
6. $\sinh 0.5$
7. $\ln 1.25$
8. $\tan^{-1} 1.02$
9. Show that the ordinate of the catenary

$$y = a\cosh(x/a)$$

deviates from the ordinate of the parabola

$$x^2 = 2a(y - a)$$

by less than $0.003\,|a|$ over the range $|x/a| \leq \frac{1}{3}$.

In Exercises 10 and 11, use series to evaluate the integrals to three decimal places.

10. $\int_0^{0.1} \frac{\sin x}{x}\,dx$ 11. $\int_0^{0.1} e^{-x^2}\,dx$

12. Construct a table of natural logarithms $\ln N$ for $N = 1, 2, 3, \ldots, 10$ by the method discussed in connection with Eq. (10), but taking advantage of the relationships

$$\ln 4 = 2\ln 2, \quad \ln 6 = \ln 2 + \ln 3,$$
$$\ln 8 = 3\ln 2, \quad \ln 9 = 2\ln 3,$$
$$\ln 10 = \ln 2 + \ln 5$$

to reduce the job to the calculation of relatively few logarithms by series. In fact, you may use $\ln 2$ as given in the text and calculate $\ln \frac{3}{2}$, $\ln \frac{5}{4}$, and $\ln \frac{7}{6}$ by series, and then combine these in suitable ways to get the logarithms of numbers N from 1 to 10.

13. Use Eqs. (12, 14) to calculate π to three decimals.

14. (a) Show that the integral

$$\int_0^x \frac{dt}{1-t^2}$$

is equal to the integral

$$\int_0^x \left(1 + t^2 + t^4 + \cdots + t^{2n} + \frac{t^{2n+2}}{1-t^2}\right) dt;$$

or, in other words, that

$$\tanh^{-1} x = x + \frac{x^3}{3} + \frac{x^5}{5} + \cdots + \frac{x^{2n+1}}{2n+1} + R,$$

where

$$R = \int_0^x \frac{t^{2n+2}}{1-t^2}\,dt.$$

(b) Show that R in (a) is not greater than

$$\frac{1}{1-x^2}\cdot\frac{|x|^{2n+3}}{2n+3}, \quad \text{if } x^2 < 1.$$

15. (a) Differentiate the identity

$$\frac{1}{1-x} = 1 + x + x^2 + \cdots + x^n + \frac{x^{n+1}}{1-x}$$

to obtain the expansion

$$\frac{1}{(1-x)^2} = 1 + 2x + 3x^2 + \cdots + nx^{n-1} + R.$$

(b) Prove that if $|x| < 1$, then $R \to 0$ as $n \to \infty$.

(c) In one throw of two dice, the probability of getting a score of 7 is $p = \frac{1}{6}$. If the dice are thrown repeatedly, the probability that a 7 will appear for the first time at the nth throw is $q^{n-1}p$, where $q = 1 - p = \frac{5}{6}$. The expected number of throws until a 7 first appears is $\sum_{n=1}^{\infty} nq^{n-1}p$. Evaluate this series numerically.

(d) In applying statistical quality control to an industrial operation, an engineer inspects items taken at random from the assembly line. He classifies each item sampled as "good" or "bad." If the probability of a good item is p and of a bad item is $q = 1 - p$, the probability that the first bad item he finds is the nth inspected is $p^{n-1}q$. The average number inspected up to and including the first bad item found is $\sum_{k=1}^{\infty} np^{n-1}q$. Evaluate this series, assuming $0 < p < 1$.

16. In probability theory, a random variable X may assume the values $1, 2, 3, \ldots,$ with probabilities $p_1, p_2, p_3, \ldots,$ where p_k is the probability that X is equal to k, $(k = 1, 2, \ldots)$. It is customary to assume $p_k \geq 0$ and $\sum_{k=1}^{\infty} p_k = 1$. The *expected value* of X denoted by $E(X)$ is defined as $\sum_{k=1}^{\infty} kp_k$, provided that this series converges. In each of the following cases, show that $\sum p_k = 1$ and find $E(X)$ if it exists. [*Hint.* See Exercise 15.]

(a) $p_k = 2^{-k}$ (b) $p_k = \frac{5^{k-1}}{6^k}$

(c) $p_k = \frac{1}{k(k+1)} = \frac{1}{k} - \frac{1}{k+1}$

18.7 INDETERMINATE FORMS

The indeterminate form 0/0

In considering the ratio of two functions $f(x)$ and $g(x)$, we sometimes wish to know the value of

$$\lim_{x\to a} \frac{f(x)}{g(x)} \tag{1}$$

at a point a where both $f(x)$ and $g(x)$ vanish. In such a case we are led to the meaningless expression $0/0$ if we substitute $x = a$ in the numerator and denominator of the fraction.

Example 1. Both x and $\sin x$ are zero at $x = 0$. Therefore we cannot just put $x = 0$ in the numerator and denominator if we want to find the limit

$$\lim_{x\to 0} \frac{\sin x}{x},$$

which we know (from earlier investigations) is 1. This is also the derivative of $\sin x$ at $x = 0$. Indeed, the derivative

$$f'(a) = \lim_{x\to a} \frac{f(x) - f(a)}{x - a}$$

was our first example of the indeterminate form $0/0$, since both the numerator and the denominator of the fraction $[f(x) - f(a)]/(x - a)$ approach zero when x approaches a.

Suppose the functions f and g both possess Taylor series expansions in powers of $x - a$:

$$f(x) = f(a) + f'(a)\cdot(x - a) + \frac{f''(a)}{2!}(x - a)^2 + \cdots, \tag{2a}$$

$$g(x) = g(a) + g'(a)\cdot(x - a) + \frac{g''(a)}{2!}(x - a)^2 + \cdots, \tag{2b}$$

that converge to $f(x)$ and $g(x)$, respectively, in some interval $|x - a| < \delta$, where δ is a positive number. Then the series (2a, b) may be used to calculate the limit (1), provided that the limit exists.

Problem 1

$$\lim_{x\to 1} \frac{\ln x}{x - 1}.$$

Solution. Let $f(x) = \ln x$, $g(x) = x - 1$. The Taylor series for $f(x)$, with $a = 1$, is found as follows:

$$\begin{aligned} f(x) &= \ln x, & f(1) &= \ln 1 = 0,\\ f'(x) &= 1/x, & f'(1) &= 1,\\ f''(x) &= -1/x^2, & f''(1) &= -1, \end{aligned}$$

so that

$$\ln x = 0 + (x - 1) - \tfrac{1}{2}(x - 1)^2 + \cdots.$$

Hence

$$\frac{\ln x}{x - 1} = 1 - \tfrac{1}{2}(x - 1) + \cdots$$

and

$$\lim_{x\to 1} \frac{\ln x}{x - 1} = \lim_{x\to 1} [1 - \tfrac{1}{2}(x - 1) + \cdots] = 1.$$

Problem 2

$$\lim_{x\to 0} \frac{\sin x - \tan x}{x^3}.$$

Solution. The Maclaurin series for $\sin x$ and $\tan x$, to terms in x^5, are

$$\sin x = x - \frac{x^3}{3!} + \frac{x^5}{5!} - \cdots,$$

$$\tan x = x + \frac{x^3}{3} + \frac{2x^5}{15} + \cdots.$$

Hence

$$\sin x - \tan x = -\frac{x^3}{2} - \frac{x^5}{8} - \cdots,$$

and

$$\lim_{x\to 0} \frac{\sin x - \tan x}{x^3} = \lim_{x\to 0}\left(-\frac{1}{2} - \frac{x^2}{8} - \cdots\right) = -\frac{1}{2}.$$

It is sometimes more convenient to apply the following theorem instead of resorting to the use of series to evaluate the limit (1).

Theorem 7. **l'Hôpital's rule.** *If $f(a) = g(a) = 0$, and if the limit of the ratio $f'(t)/g'(t)$ as t approaches a exists, then*

$$\lim_{t\to a} \frac{f(t)}{g(t)} = \lim_{t\to a} \frac{f'(t)}{g'(t)}. \tag{3}$$

Remark 1. The primes denote derivatives with respect to t and it is tacitly assumed when we write $f'(t)/g'(t)$ that both $f'(t)$ and $g'(t)$ exist and that $g'(t)$ is not zero when t is different from, but sufficiently near to, $t = a$. Note particularly that $f'(t)/g'(t)$ is the derivative of the numerator divided by the derivative of the denominator, which is not the same as the derivative of the fraction $f(t)/g(t)$.

Proof. We shall establish Eq. (3) for the case where $t \to a+$. The method needs only minor modifications, such as reversing certain inequalities, to apply

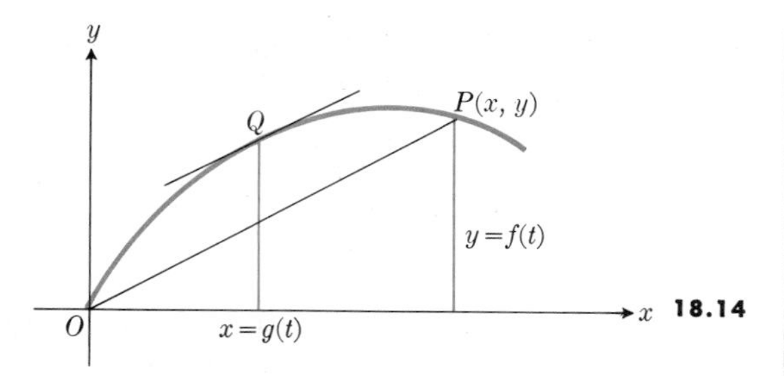

18.14

to $t \to a-$. The combination of these two cases then establishes the result.

Suppose, then, that $f(a) = g(a) = 0$ and that

(a) $f(t)$ and $g(t)$ are continuous functions of t for all t in the closed interval $[a, b]$,

(b) $f'(t)$ and $g'(t)$ exist for each t in the open interval (a, b), and

(c) $g'(t) \neq 0$ for all t in the open interval (a, b).

Now consider any t in the open interval (a, b). (See Fig. 18.14.) The fraction $F(t) = f(t)/g(t)$ can be interpreted as the slope of the chord OP joining the point $O(g(a), f(a))$ and the point $P(g(t), f(t))$ on the curve C represented by the parametric equations

$$x = g(t), \quad y = f(t), \quad a \leq t \leq b.$$

Our hypotheses are such that we may conclude that there is a point Q on the arc OP where the tangent to the curve is parallel to the chord. But the slope of the tangent at Q is

$$\left(\frac{dy}{dx}\right)_Q = \left(\frac{dy/dt}{dx/dt}\right)_Q = \frac{f'(t_1)}{g'(t_1)},$$

where t_1 is the value of t that corresponds to the point Q on the curve. We therefore have the result that if t is any number in the open interval (a, b), then there is at least one number t_1 between a and t such that

$$\frac{f(t)}{g(t)} = \frac{f'(t_1)}{g'(t_1)}. \tag{4}$$

[Note that the hypothesis $g'(t) \neq 0$ for $a < t < b$ assures us that the denominator on the right does not vanish.]

Now as $t \to a+$, t_1 does likewise, and

$$\lim_{t \to a+} \frac{f(t)}{g(t)} = \lim_{t_1 \to a+} \frac{f'(t_1)}{g'(t_1)} = \lim_{t \to a+} \frac{f'(t)}{g'(t)},$$

provided that the last limit above exists. This establishes the theorem for the case in which t approaches a from above. The case in which t approaches a from below is proved in an analogous manner, and the two cases together give us the result stated as l'Hôpital's rule. □

From the geometric interpretation given in Fig. 18.14, it is clear that when the limit

$$\lim_{t \to a} \frac{f(t)}{g(t)} = \lim_{P \to O} \text{ (slope of } OP)$$

exists, it is the same as the slope of the tangent to the curve at O. If $dx/dt = g'(t)$ did not vanish at $t = a$, we could simply write this slope as $f'(a)/g'(a)$ and have

$$\lim_{t \to a} \frac{f(t)}{g(t)} = \frac{f'(a)}{g'(a)}.$$

Equation (3) does better than this, however. Since it was established on the assumption that $g'(t)$ is different from zero in an *open* interval (a, b), it is valid whether or not $g'(a)$ is zero. Thus, if

$$f'(a) = g'(a) = 0,$$

then we are led to another indeterminate form $0/0$ and may again apply l'Hôpital's rule:

$$\lim_{t \to a} \frac{f'(t)}{g'(t)} = \lim_{t \to a} \frac{f''(t)}{g''(t)},$$

provided that the limit on the right exists.

In practice, the functions we deal with in this book are such as to satisfy the hypotheses laid down in deriving l'Hôpital's rule. We apply the method by proceeding to differentiate the numerator and denominator separately so long as we still get the form $0/0$ at $t = a$. As soon as one or the other of these derivatives is *different* from zero at $t = a$, we stop differentiating! It may happen that *one* of the derivatives is zero at $t = a$ and the other one not. Then the limit of the fraction is either zero or infinity, accordingly as the zero derivative occurs in the numerator or in the denominator. If we arrive at a limit that can be evaluated by some special method that we recognize, then we may use that method instead of continuing with l'Hôpital's.

Example 2

$$\lim_{t\to 0} \frac{\sqrt{1+t}-(1+t/2)}{t^2} \qquad [=\tfrac{0}{0}]$$

$$= \lim_{t\to 0} \frac{\frac{1}{2}(1+t)^{-1/2}-\frac{1}{2}}{2t} \qquad [\text{still } = \tfrac{0}{0}]$$

$$= \lim_{t\to 0} \frac{-\frac{1}{4}(1+t)^{-3/2}}{2} = -\frac{1}{8}.$$

Note that one obtains the same result by making use of the binomial series

$$(1+t)^{1/2} = 1+\tfrac{1}{2}t-\tfrac{1}{8}t^2+\cdots.$$

The indeterminate forms ∞/∞ and $\infty\cdot 0$

In more advanced textbooks it is proved that if $f(t)\to\infty$ and $g(t)\to\infty$ as $t\to a$, then

$$\lim_{t\to a}\frac{f(t)}{g(t)} = \lim_{t\to a}\frac{f'(t)}{g'(t)},$$

provided that the limit on the right exists.* This simply says that l'Hôpital's rule applies to the indeterminate form ∞/∞ as well as to the form $0/0$. The form $\infty\cdot 0$ must be reduced to one or the other of these two. In the notation $t\to a$, a may either be finite or infinite.

Problem 3

$$\lim_{t\to\infty}\frac{t^2+t}{2t^2+1}.$$

Solution. This assumes the form ∞/∞ as $t\to\infty$.

(a) We apply l'Hôpital's rule and have

$$\lim_{t\to\infty}\frac{t^2+t}{2t^2+1} = \lim_{t\to\infty}\frac{2t+1}{4t} \qquad \left[\text{still } = \frac{\infty}{\infty}\right]$$

$$= \lim_{t\to\infty}\frac{2}{4} = \frac{1}{2}.$$

* For more details, see "L'Hôpital's Rule," by A. E. Taylor, *American Mathematical Monthly*, **59**, 20–24, 1952.

(b) Note that this could also have been evaluated by writing $t = 1/h$ and letting $h\to 0$:

$$\lim_{t\to\infty}\frac{t^2+t}{2t^2+1} = \lim_{h\to 0}\frac{\dfrac{1}{h^2}+\dfrac{1}{h}}{\dfrac{2}{h^2}+1} = \lim_{h\to 0}\frac{1+h}{2+h^2} = \frac{1}{2}.$$

(c) Another way is to divide both the numerator and denominator by t^2 (the highest power of t that occurs in either numerator or denominator) and observe that $1/t$ and $1/t^2$ approach zero as t increases without bound:

$$\lim_{t\to\infty}\frac{t^2+t}{2t^2+1} = \lim_{t\to\infty}\frac{1+\dfrac{1}{t}}{2+\dfrac{1}{t^2}} = \frac{1+0}{2+0} = \frac{1}{2}.$$

This is actually the same as the method given in (b).

The indeterminate form $\infty-\infty$

The differences

$$n-n=0, \qquad n-n^2=n(1-n),$$
$$n^2-n=n(n-1)$$

behave entirely differently as $n\to\infty$: But all become formally $\infty-\infty$, and illustrate why such an expression is called an indeterminate form.

It may be possible, by preliminary algebraic manipulations, to reduce an expression $F(t)-G(t)$ that becomes $\infty-\infty$ as t approaches a to a form $f(t)/g(t)$ that becomes $0/0$ or ∞/∞. Although l'Hôpital's rule is not to be applied to the original form, it may be applied to either of the forms ∞/∞ or $0/0$.

Problem 4

$$\lim_{x\to 0}\left(\frac{1}{\sin x}-\frac{1}{x}\right).$$

Solution. If $\qquad x\to 0+,$

then $\qquad \sin x\to 0+$ and $1/\sin x\to+\infty,$

while $\qquad 1/x\to+\infty.$

Thus the expression $(1/\sin x)-(1/x)$ formally becomes $+\infty-(+\infty)$, which is indeterminate. On the other hand, if $x\to 0-$, then

$$1/\sin x\to-\infty, \quad\text{and}\quad 1/x\to-\infty,$$

so that $(1/\sin x) - (1/x)$ becomes $-\infty + \infty$, which is also indeterminate. But we may also write

$$\frac{1}{\sin x} - \frac{1}{x} = \frac{x - \sin x}{x \sin x} = \frac{x - \left[x - \frac{x^3}{3!} + \frac{x^5}{5!} - \cdots\right]}{x \cdot \left[x - \frac{x^3}{3!} + \frac{x^5}{5!} - \cdots\right]}$$

$$= \frac{x^3\left[\frac{1}{3!} - \frac{x^2}{5!} + \cdots\right]}{x^2\left[1 - \frac{x^2}{3!} + \cdots\right]} = x\,\frac{\frac{1}{3!} - \frac{x^2}{5!} + \cdots}{1 - \frac{x^2}{3!} + \cdots}.$$

Therefore

$$\lim_{x\to 0}\left(\frac{1}{\sin x} - \frac{1}{x}\right) = \lim_{x\to 0}\left[x\,\frac{\frac{1}{3!} - \frac{x^2}{5!} + \cdots}{1 - \frac{x^2}{3!} + \cdots}\right] = 0.$$

In fact, from the series expressions above we can see that if $|x|$ is small, then

$$\frac{1}{\sin x} - \frac{1}{x} \approx x \cdot \frac{1}{3!} = \frac{x}{6}, \quad \text{or} \quad \csc x \approx \frac{1}{x} + \frac{x}{6}.$$

The indeterminate forms 0^0, 1^∞, and ∞^0

The expressions 0^0, 1^∞, and ∞^0 symbolize the situations that arise when we consider the limit, as t approaches a, of a function of the form

$$y = f(t)^{g(t)},$$

where
$$f(a) = g(a) = 0; \tag{a}$$
or
$$f(a) = 1 \quad \text{and} \quad \lim_{t\to a} g(t) = \infty; \tag{b}$$
or
$$g(a) = 0 \quad \text{and} \quad \lim_{t\to a} f(t) = \infty. \tag{c}$$

In any such case, the natural logarithm of y becomes an indeterminate form of the type $0 \cdot (-\infty)$ or $\infty \cdot 0$. The method of dealing with such cases is illustrated by the following problem.

Problem 5. $\lim_{h\to 0} (1 + h)^{1/h}$.

Solution. Let $y = (1 + h)^{1/h}$,

so that

$$\ln y = \ln (1 + h)^{1/h} = \frac{1}{h} \ln (1 + h) = \frac{\ln (1 + h)}{h}.$$

This would assume the form

$$\frac{\ln (1 + 0)}{0} = \frac{0}{0},$$

if we were to substitute $h = 0$. By l'Hôpital's rule,

$$\lim_{h\to 0} \frac{\ln (1 + h)}{h} = \lim_{h\to 0} \frac{1/(1 + h)}{1} = 1;$$

that is,

$$\ln y \to 1 \quad \text{as} \quad h \to 0,$$

and, since $\ln y$ is a continuous function of y, which means that

$$\lim_{h\to 0} \ln y = \ln (\lim_{h\to 0} y) = 1,$$

we therefore have $\lim_{h\to 0} y = e^1 = e$.

Remark 2. Many people feel that the expression 1^∞ cannot represent an indeterminate form. They say that one to any finite power is one, hence one to the power infinity is also one. But note, in the example above, that the base $1 + h$ is a variable which is not exactly equal to 1 except in the *limit* as $h \to 0$. There is a difference, in other words, between

$$1^{1/h} \quad \text{and} \quad (1 + h)^{1/h},$$

even though both expressions formally become 1^∞ as $h \to 0$. If the reader is still skeptical, let him write out a few terms of the binomial expansion of $(1 + 1/n)^n$ for some large value of n, say $n = 10{,}000$. The result will be a fairly good approximation to $\lim_{h\to 0} (1 + h)^{1/h}$ (since $h = 1/n = 0.0001$ when $n = 10{,}000$).

EXERCISES

Evaluate the following limits by using series, or l'Hôpital's rule, or otherwise.

1. $\lim_{t\to 0} \dfrac{1 - \cos t - \frac{1}{2}t^2}{t^4}$
2. $\lim_{h\to 0} \dfrac{\sqrt{4 + h} - 2}{h}$
3. $\lim_{x\to 0} \dfrac{\sqrt[3]{1+x} - (1+x/3)}{x^2}$
4. $\lim_{x\to 0} \dfrac{\sinh x - \sin x}{x^3}$
5. $\lim_{x\to 0} \dfrac{e^x - (1 + x)}{x^2}$
6. $\lim_{h\to 0} \dfrac{(\sin h)/h - \cos h}{h^2}$

7. $\lim_{z\to 0} \dfrac{\sin(z^2) - \sin^2 z}{z^4}$

8. $\lim_{n\to\infty} \dfrac{n^3 + 3n^2}{2n^3 - n}$

9. $\lim_{x\to\pi/2} \left(x - \dfrac{\pi}{2}\right) \tan 3x$

10. $\lim_{x\to 0+} x \ln x$

11. $\lim_{x\to+\infty} \dfrac{\ln x}{x}$

12. $\lim_{x\to+\infty} \dfrac{x^{100}}{e^x}$

13. $\lim_{x\to+\infty} x \sin \dfrac{1}{x}$

14. $\lim_{x\to+\infty} (x - \sqrt{x^2 + x})$

15. $\lim_{x\to 1+} \left(\dfrac{1}{x-1} - \dfrac{1}{\sqrt{x-1}}\right)$

16. $\lim_{x\to 0} (1 - 2x)^{3/x}$

17. $\lim_{x\to+\infty} \left(1 + \dfrac{2}{x}\right)^x$

18. $\lim_{x\to+\infty} (x + e^x)^{2/x}$

19. $\lim_{x\to 0+} (x + \sin x)^{\tan x}$

20. $\lim_{x\to 0+} x^x$

21. $\lim_{x\to 1-} x^{1/(1-x^2)}$

22. $\lim_{x\to 0+} (\cos\sqrt{x})^{1/x}$

23. $\lim_{x\to+\infty} x^{1/x}$

24. $\lim_{x\to+\infty} (\cosh x)^{1/x}$

25. (a) Prove that $\left(\int_0^x e^{t^2}\, dt\right) \to +\infty$ as $x \to +\infty$.

(b) Find $\lim_{x\to\infty} x \int_0^x e^{t^2 - x^2}\, dt$.

26. Find values of r and s such that

$$\lim_{x\to 0} (x^{-3} \sin 3x + rx^{-2} + s) = 0.$$

18.8 FOURIER SERIES

We have seen how the Taylor series is a natural outgrowth of the problem of approximating to a given function $f(x)$ by means of polynomials. This approximation is based on getting a close fit to $f(x)$ near some particular point $x = a$. The Taylor series usually converges in an interval about a (in fact, in an interval having a as center), but any one of its partial sums is a polynomial which can only be expected to come close to fitting $f(x)$ in a rather restricted neighborhood of a. We might say that the Taylor series does a good job *locally*, or *in the small*. In many important applications, however, it is desirable to approximate to a function $f(x)$ over a fairly wide interval, or *in the large*. For such purposes, the *Fourier series* is often used. Whereas the power series uses powers of x as its fundamental elements, the Fourier series uses sines and cosines as basic components. Suppose, for instance, that a function $f(x)$ is to be approximated on the interval $0 \le x \le 2\pi$ by a trigonometric polynomial

$$\begin{aligned}\phi_n(x) = a_0 &+ (a_1 \cos x + b_1 \sin x) \\ &+ (a_2 \cos 2x + b_2 \sin 2x) + \cdots \\ &+ (a_n \cos nx + b_n \sin nx),\end{aligned}$$

$$\phi_n(x) = \sum_{k=0}^{n} (a_k \cos kx + b_k \sin kx). \tag{1}$$

What choice of the coefficients

$$a_0, a_1, \ldots, a_n, b_1, \ldots, b_n$$

will make $\phi_n(x)$ the "best possible" approximation to $f(x)$? Of course, the words "best possible" need some further interpretation. Suppose we require that $f(x)$ and $\phi_n(x)$ shall give the same answer when integrated with respect to x from 0 to 2π. We also require that when $f(x)$ and $\phi_n(x)$ are both multiplied by $\cos kx$ or $\sin kx$, $(k = 1, \ldots, n)$, and the results integrated from 0 to 2π, the integrals shall be the same. That is, we shall impose upon $\phi_n(x)$ the $2n + 1$ conditions

$$\int_0^{2\pi} \phi_n(x)\, dx = \int_0^{2\pi} f(x)\, dx,$$

$$\int_0^{2\pi} \phi_n(x) \cos kx\, dx = \int_0^{2\pi} f(x) \cos kx\, dx, \quad k = 1, 2, \ldots, n, \tag{2}$$

$$\int_0^{2\pi} \phi_n(x) \sin kx\, dx = \int_0^{2\pi} f(x) \sin kx\, dx, \quad k = 1, 2, \ldots, n.$$

We may ask whether or not these conditions can be met. The right-hand sides of Eqs. (2) are integrals that depend on the particular function f; but the lefthand sides of the equations involve only the constants $a_0, a_1, \ldots, a_n, b_1, \ldots, b_n$, as we shall now see.

From Eq. (1), we have

$$\int_0^{2\pi} \phi_n(x)\, dx = 2\pi a_0,$$

since all the cosine and sine terms give zero when integrated from 0 to 2π. Next, suppose we multiply both sides of Eq. (1) by $\cos x$ and integrate from 0 to 2π. The only nonzero term we get from the right-hand side of the equation is

$$\int_0^{2\pi} a_1 \cos^2 x \, dx = \pi a_1.$$

This results from the fact that

$$\int_0^{2\pi} \cos px \cos qx \, dx = \int_0^{2\pi} \cos px \sin mx \, dx = \int_0^{2\pi} \sin px \sin qx \, dx = 0,$$

provided that p, q, and m are integers and p is not equal to q. Similarly, if we multiply Eq. (1) by $\sin x$ and integrate from 0 to 2π, the only nonzero term remaining on the right is

$$\int_0^{2\pi} b_1 \sin^2 x \, dx = \pi b_1.$$

Proceeding in similar fashion with

$$\cos 2x, \sin 2x, \ldots, \cos nx, \sin nx,$$

we obtain each time only one nonzero term, and that is the one which has a sine-squared or a cosine-squared term. These results are all summarized by saying that

$$\int_0^{2\pi} \phi_n(x) \cos kx \, dx = \pi a_k, \quad k = 1, 2, \ldots, n,$$

$$\int_0^{2\pi} \phi_n(x) \sin kx \, dx = \pi b_k, \quad k = 1, 2, \ldots, n;$$

$$\int_0^{2\pi} \phi_n(x) \, dx = 2\pi a_0.$$

If we now require that Eqs. (2) be satisfied, we see that the coefficients must be

$$a_0 = \frac{1}{2\pi} \int_0^{2\pi} f(x) \, dx,$$

$$a_k = \frac{1}{\pi} \int_0^{2\pi} f(x) \cos kx \, dx, \quad k = 1, 2, \ldots, n, \tag{3}$$

$$b_k = \frac{1}{\pi} \int_0^{2\pi} f(x) \sin kx \, dx, \quad k = 1, 2, \ldots, n.$$

Clearly these conditions can be met if the required integrals exist. If we let n tend to infinity, we obtain an infinite series in Eq. (1), and if the coefficients are determined as in Eqs. (3), then this series is the *Fourier series* for $f(x)$.

Remark 1. The Fourier series also has the property that the coefficients as given by Eqs. (3) are precisely those that one should choose to minimize the integral of the square of the error in approximating $f(x)$ by $\phi_n(x)$. That is,

$$\int_0^{2\pi} [f(x) - \phi_n(x)]^2 \, dx$$

is minimized by choosing $a_0, a_1, \ldots, a_n, b_1, \ldots, b_n$ as in Eqs. (3).

Remark 2. If we approximate the function $f(x)$ by one trigonometric polynomial $\phi_n(x)$, using Eqs. (3), and then decide to take instead a second approximation $\phi_N(x)$ using more terms $(N > n)$, we may simply add more terms to $\phi_n(x)$ without changing any of the coefficients $a_0, \ldots, b_n$ used in the first approximation.

Remark 3. Fourier series can be used to represent some functions which cannot be expanded in power series. For example, a step function such as

$$f(x) = \begin{cases} 1, & \text{if } 0 < x < \pi, \\ 2, & \text{if } \pi < x < 2\pi \end{cases} \tag{4}$$

can be represented by a Fourier series. At $x = \pi$, the Fourier series compromises between the left-hand limit of $f(x)$ and its right-hand limit and gives a value $\frac{3}{2}$, which is simply the average of the two values on either side of the discontinuity. The reason that a Fourier series may be used to represent some functions which cannot be represented by power series is that the Fourier series depends on the existence of certain *integrals,* whereas the power series depends on the existence of *derivatives.* A function must be very "smooth" to possess derivatives, but it may be fairly "rough" and still be integrable. (For more details, see Franklin, *A Treatise on Advanced Calculus,* Chapter XIV.)

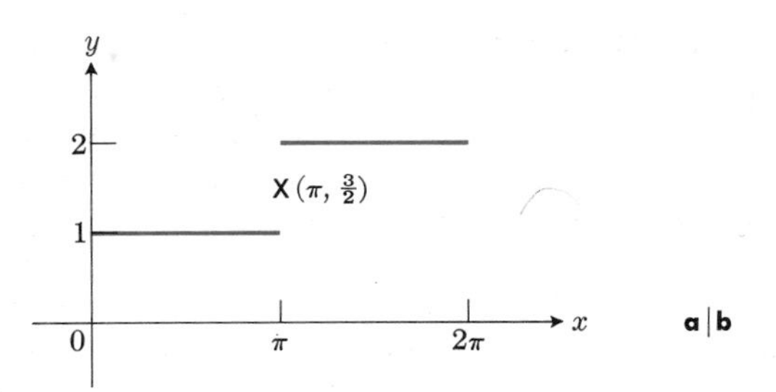

a|b

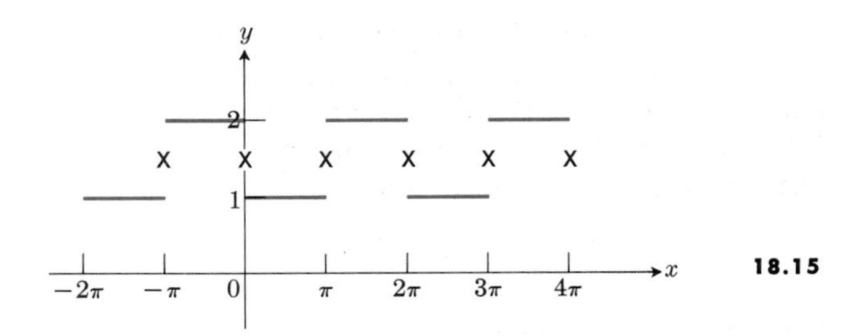

18.15

Remark 4. If a function is to be represented by a Fourier series for $0 \leq x \leq L$ instead of for

$$0 \leq x \leq 2\pi,$$

one may substitute $t = (2\pi x)/L$ and work with the Fourier series in terms of t. If the interval is $a \leq x \leq b$, one may first take $x' = x - a$ and then $t = (2\pi x')/(b - a)$.

Example. As an illustration of the method to be followed in finding the Fourier series for a given function $f(x)$, we shall analyze the step function in Eq. (4). We find

$$a_0 = \frac{1}{2\pi}\int_0^{2\pi} f(x)\,dx$$

$$= \frac{1}{2\pi}\left(\int_0^{\pi} 1\,dx + \int_{\pi}^{2\pi} 2\,dx\right) = \frac{3}{2},$$

$$a_k = \frac{1}{\pi}\int_0^{2\pi} f(x)\cos kx\,dx$$

$$= \frac{1}{\pi}\left(\int_0^{\pi} \cos kx\,dx + \int_{\pi}^{2\pi} 2\cos kx\,dx\right)$$

$$= \frac{1}{\pi}\left(\frac{\sin kx}{k}\Big]_0^{\pi} + \frac{2\sin kx}{k}\Big]_{\pi}^{2\pi}\right) = 0, \qquad k \geq 1,$$

$$b_k = \frac{1}{\pi}\int_0^{2\pi} f(x)\sin kx\,dx$$

$$= \frac{1}{\pi}\left(\int_0^{\pi} \sin kx\,dx + \int_{\pi}^{2\pi} 2\sin kx\,dx\right)$$

$$= \frac{1}{\pi}\left(-\frac{\cos kx}{k}\Big]_0^{\pi} - \frac{2\cos kx}{k}\Big]_{\pi}^{2\pi}\right)$$

$$= \frac{\cos k\pi - 1}{k\pi} = \frac{(-1)^k - 1}{k\pi};$$

or,

$$a_0 = 2/3, \qquad a_1 = a_2 = \cdots = 0,$$
$$b_1 = -2/\pi, \qquad b_2 = 0, \qquad b_3 = -2/3\pi,$$
$$b_4 = 0, \qquad b_5 = -2/5\pi, \qquad b_6 = 0, \quad \ldots,$$

and the Fourier series is

$$\frac{3}{2} - \frac{2}{\pi}\left(\sin x + \frac{\sin 3x}{3} + \frac{\sin 5x}{5} + \cdots\right).$$

We see that all the sine terms vanish when $x = \pi$, leaving $\frac{3}{2}$ as the value of the series at this point. We also note that the series gives the value $\frac{3}{2}$ at $x = 0$ and $x = 2\pi$ as well as at $x = \pi$. In fact, all terms in the Fourier series are periodic, of period 2π. Such a series thus has the same value at $x + 2\pi$ or $x - 2\pi$ as it has at x. The present series represents the function shown in Fig. 18.15(a), between 2π and 0, but the series repeats this same pattern between -2π and 0, between 2π and 4π, and so on to the left and to the right. This extended function will then be seen to have jumps at $x = n\pi$, $n = 0, \pm 1, \pm 2, \ldots$ and the value $\frac{3}{2}$ is the average of the values of the function immediately to either side of one of these points of discontinuity. The Fourier series converges to the periodic function shown in Fig. 18.15(b).

EXERCISES

Find the Fourier series associated with the following functions. Sketch each function.

1. $f(x) = 1, \quad 0 < x < \pi,$
   $\quad\;\; = 0, \quad \pi < x < 2\pi$

2. $f(x) = -1, \quad 0 < x < \pi,$
   $\quad\;\; = 1, \quad \pi < x < 2\pi$

3. $f(x) = x, \quad 0 < x < 2\pi$
4. $f(x) = x, \quad 0 < x < \pi,$
   $= 2\pi - x, \quad \pi < x < 2\pi$
5. $f(x) = x, \quad 0 < x < \pi,$
   $= x - 2\pi, \quad \pi < x < 2\pi$

18.9 CONVERGENCE OF POWER SERIES: ABSOLUTE CONVERGENCE

Problem 1, Article 18.2 showed how the ratio test may be applied to find the region of convergence of a power series. We must remember that the ratio test applies only to *positive* series. For a negative series, we simply factor out a minus sign from every term. This doesn't work, however, for a series such as

$$1 + 2x + 3x^2 + 4x^3 + \cdots + nx^{n-1} + \cdots, \tag{1}$$

which becomes an *alternating* series when x is negative. To handle such series, we introduce the concept of *absolute* convergence.

Definition. *A series*

$$\sum_{k=1}^{\infty} u_k = u_1 + u_2 + \cdots$$

is said to be absolutely convergent if the corresponding series of absolute values

$$\sum_{k=1}^{\infty} |u_k| = |u_1| + |u_2| + \cdots$$

converges.

One reason that absolute convergence is important appears in the following theorem.

Theorem 8. *If a series converges absolutely, then it also converges when the absolute value signs are removed.*

Proof. Let the sum of the first n terms of $\sum u_k$ be denoted as usual by s_n, and let the sum of the first n terms of $\sum |u_k|$ be denoted by p_n. Suppose that

$$\lim_{n\to\infty} p_n = p$$

exists. We want to prove that $\lim s_n$ also exists. The following trick is effective: Consider the sum

$$q_n = (u_1 + |u_1|) + (u_2 + |u_2|) + \cdots + (u_n + |u_n|) = s_n + p_n. \tag{2}$$

Each term of the series $\sum (u_k + |u_k|)$ is greater than or equal to zero, so the sequence of numbers $\{q_n\}$ is a monotone increasing sequence. On the other hand, q_n is less than or equal to

$$(|u_1| + |u_1|) + (|u_2| + |u_2|) + \cdots + (|u_n| + |u_n|);$$

in other words,

$$q_n \leq 2p_n. \tag{3}$$

Further, since $\{p_n\}$ is a monotone increasing sequence whose limit is p, we have

$$p_n \leq p \quad \text{for all } n.$$

The inequality (3) therefore tells us that $q_n \leq 2p$, and alternative (A) of Theorem 3, Article 18.2 holds with $M = 2p$. Therefore

$$\lim_{n\to\infty} q_n = q$$

exists. Hence, from Eq. (2), $s_n = q_n - p_n$ also has a limit as n becomes infinite:

$$\lim_{n\to\infty} s_n = \lim_{n\to\infty} q_n - \lim_{n\to\infty} p_n = q - p. \; \square$$

We also repeat the following theorem from Article 18.1.

Theorem 9. *If a series $\sum u_k$ converges, then*

$$\lim_{n\to\infty} u_n = 0. \tag{4}$$

Remark. In determining the interval of convergence of a power series, we shall apply the ratio test to the series of *absolute values*. Thus, if

$$\rho = \lim_{n\to\infty} \left| \frac{u_{n+1}}{u_n} \right|,$$

the series

(a) converges absolutely if $\rho < 1$,
(b) diverges if $\rho > 1$, and
(c) may either converge or diverge if $\rho = 1$.

The only case for which the introduction of absolute value signs requires further discussion is case (b), $\rho > 1$. Here the terms do not get smaller in magnitude as n increases, since there is then an index N such that

$$|u_{n+1}/u_n| \geq 1 \quad \text{for all } n \geq N.$$

This, in turn, implies

$$|u_N| \leq |u_{N+1}| \leq |u_{N+2}| \leq |u_{N+3}| \leq \cdots,$$

so that the nth term does not approach zero as n becomes infinite. Hence the series diverges (with or without absolute value signs) whenever it happens that $\rho > 1$.

Theorem 10. *If a power series*

$$\sum_{k=0}^{\infty} a_k x^k = a_0 + a_1 x + a_2 x^2 + \cdots \tag{5}$$

converges for $x = c$ ($c \neq 0$), then it converges absolutely for all $|x| < |c|$. If the series diverges for $x = d$, then it diverges for all $|x| > |d|$.

Proof. Suppose the series

$$\sum_{k=0}^{\infty} a_k c^k \tag{6}$$

converges. Then, by Theorem 9,

$$\lim_{n\to\infty} a_n c^n = 0.$$

Hence there is an index N such that

$$|a_n c^n| < 1 \quad \text{for all } n \geq N.$$

That is,

$$|a_n| < \frac{1}{|c|^n} \quad \text{for } n \geq N. \tag{7}$$

Now take any x such that $|x| < |c|$, and consider

$$|a_0| + |a_1 x| + \cdots + |a_{N-1}x^{N-1}| + |a_N x^N| + |a_{N+1}x^{N+1}| + \cdots.$$

There is only a finite number of terms prior to $|a_N x^N|$ and their sum is finite. Starting with $|a_N x^N|$ and beyond, the terms are less than

$$\left|\frac{x}{c}\right|^N + \left|\frac{x}{c}\right|^{N+1} + \left|\frac{x}{c}\right|^{N+2} + \cdots, \tag{8}$$

by virtue of the inequality (7). But the series in (8) is a geometric series with ratio $r = |x/c|$, which is less than 1 since $|x| < |c|$. Hence the series (8) converges, and thus the original series (5) converges absolutely. This argument establishes the first half of the theorem.

The second half of the theorem involves nothing new. For if the series diverges at $x = d$ and converges at a value x_0 with $|x_0| > |d|$, we may take $c = x_0$ in the first half of the theorem and conclude that the series converges absolutely at d. But the series cannot both converge absolutely and diverge at one and the same time. Hence if it diverges at d, it diverges for all $|x| > |d|$. □

It is a consequence of Theorem 10 that a power series behaves in one or another of the following ways.

1. It converges only at $x = 0$.
2. It converges absolutely for all finite x,

$$-\infty < x < \infty.$$

3. There is a positive number c such that the series converges absolutely for $|x| < c$ and diverges for $|x| > c$. In this case, the series may converge or diverge at either or both of the points $x = c$, $x = -c$, the endpoints of the interval of convergence.

Problem. Find the interval of convergence for the series

$$\sum_{n=1}^{\infty} nx^{n-1}.$$

Solution. First we test for absolute convergence by calculating

$$\rho = \lim_{n\to\infty} \left|\frac{(n+1)x^n}{nx^{n-1}}\right| = |x|.$$

By the ratio test, the series converges absolutely if $|x| < 1$ and diverges if $|x| > 1$. When x is either $+1$ or -1, the nth term does not approach zero as n becomes infinite. Hence the series diverges at $x = \pm 1$. The interval of convergence is $-1 < x < 1$.

EXERCISES

In these exercises, find the interval in which the given power series converges absolutely.

1. $\sum_{n=0}^{\infty} x^n$
2. $\sum_{n=0}^{\infty} \frac{x^n}{n+1}$
3. $\sum_{n=1}^{\infty} n^2 x^n$
4. $\sum_{n=1}^{\infty} \frac{nx^n}{2^n}$
5. $\sum_{n=0}^{\infty} \frac{x^n}{n!}$
6. $\sum_{n=1}^{\infty} \frac{(-1)^{n-1} x^{2n-1}}{(2n-1)!}$
7. $\sum_{n=1}^{\infty} (-1)^{n-1} \frac{(x-1)^n}{n}$
8. $\sum_{n=1}^{\infty} \frac{n^2}{2^n} (x+2)^n$
9. $\sum_{n=1}^{\infty} \frac{2^n}{n(n+1)} (x-3)^{2n-1}$
10. $\sum_{n=0}^{\infty} \frac{(-1)^n x^{2n}}{2^{2n}(n!)^2}$

18.10 ALTERNATING SERIES: CONDITIONAL CONVERGENCE

When some of the terms of a series $\sum u_k$ are positive and some are negative, *the series converges if* $\sum |u_k|$ *converges*. Thus we may apply any of our tests for convergence of positive series, provided that we apply them to the series of absolute values. But we do not know, when the series of absolute values *diverges*, whether the *original* series diverges or not. If it converges, but not absolutely, we say that it *converges conditionally*.

We discuss one simple case of series with mixed signs, that of an alternating series of the form

$$a_1 - a_2 + a_3 - a_4 + \cdots + (-1)^{n-1} a_n + \cdots, \quad (1)$$

where the a's are all positive.

Theorem 11. *If the following three conditions are all satisfied, then the series does converge.*

(a) *The series* $\sum u_k$ *is strictly alternating.*

(b) *The nth term tends to zero as n becomes infinite.*

(c) *Each term is numerically less than, or at most equal to, its predecessor.*

Proof. Suppose these three conditions are satisfied. Then we may write

$$u_k = (-1)^{k-1} a_k$$

and put the series in the form (1) with $a_{n+1} \le a_n$. If n is an even integer, say $n = 2m$, then the sum of the first n terms is

$$\begin{aligned} s_{2m} &= (a_1 - a_2) + (a_3 - a_4) + \cdots \\ &\quad + (a_{2m-1} - a_{2m}) \\ &= a_1 - (a_2 - a_3) - (a_4 - a_5) - \cdots \\ &\quad - (a_{2m-2} - a_{2m-1}) - a_{2m}. \end{aligned}$$

The first way of writing s_{2m} exhibits it as the sum of m nonnegative terms, since each expression in parentheses is positive or zero. Hence $s_{2m+2} \ge s_{2m}$ and the sequence $\{s_{2m}\}$ is monotonically increasing. The second way of expressing s_{2m} shows, on the other hand, that

$$s_{2m} \le a_1.$$

Hence we have a sequence $\{s_{2m}\}$ that is monotone increasing and is bounded above. Such a sequence has a limit, say

$$\lim_{m \to \infty} s_{2m} = L. \quad (2)$$

If we now form the sum of an odd number of terms, say

$$\begin{aligned} s_{2m+1} &= a_1 - a_2 + a_3 - \cdots \\ &\quad + a_{2m-1} - a_{2m} + a_{2m+1} \\ &= s_{2m} + a_{2m+1}, \end{aligned}$$

the fact that the nth term approaches zero as n becomes infinite means that

$$\lim_{m \to \infty} a_{2m+1} = 0.$$

Hence, as $m \to \infty$,

$$\lim s_{m+1} = \lim s_{2m} + \lim a_{2m+1} = L. \quad (3)$$

Finally, we may combine Eqs. (2) and (3) and say simply

$$\lim_{n \to \infty} s_n = L,$$

where n may be either even or odd. □

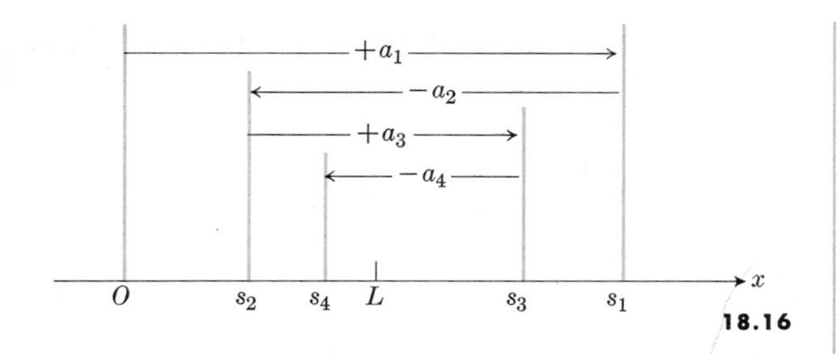

18.16

The following graphical interpretation of the partial sums gives us added insight into the way in which an alternating series converges to its limit L when the three conditions of the theorem are satisfied. (See Fig. 18.16.) Starting from the origin O on a scale of real numbers, we lay off the positive distance

$$s_1 = a_1.$$

To find the point corresponding to

$$s_2 = a_1 - a_2$$

we must back up a distance equal to a_2. Since $a_2 \leq a_1$, we do not back up any farther than O at most. Next we go forward a distance a_3 and mark the point corresponding to

$$s_3 = a_1 - a_2 + a_3.$$

Since $a_3 \leq a_2$, we go forward by an amount which is no greater than the previous backward step; that is, s_3 is less than or equal to s_1. We continue in this seesaw fashion, backing up or going forward as the signs in the series demand. But each forward or backward step is shorter than (or at most the same size as) the preceding step, because $a_{n+1} \leq a_n$. And since the nth term approaches zero as n increases, the size of step we take forward or backward gets smaller and smaller. We thus oscillate across the limit L, but the amplitude of oscillation continually decreases and approaches zero as its limit. The even-numbered partial sums $s_2, s_4, s_6, \ldots, s_{2m}$ continually increase toward L, while the odd-numbered sums $s_1, s_3, s_5, \ldots, s_{2m+1}$ continually decrease toward L. The limit L is between any two successive sums s_n and s_{n+1} and hence differs from either of them by an amount not greater than a_{n+1}.

Example 1. The series

$$1 - \frac{1}{2} + \frac{1}{3} - \frac{1}{4} + \cdots + \frac{(-1)^{n-1}}{n} + \cdots$$

satisfies the three conditions of the theorem.

(a) It is alternating.

(b) The nth term is $(-1)^{n-1}/n$, which tends to zero as n increases.

(c) The terms decrease in absolute value, since

$$\frac{1}{n+1} < \frac{1}{n}.$$

Hence this series converges. But it converges conditionally and not absolutely, since the series of absolute values is

$$1 + \frac{1}{2} + \frac{1}{3} + \cdots + \frac{1}{n} + \cdots,$$

which we know diverges.

Remark. Suppose that when we compute the numerical value of an alternating series that satisfies the three conditions of the theorem and converges to a limit L, we stop after computing the sum s_n of the first n terms. Then the error, which is represented by the remainder

$$\begin{aligned} R_n &= L - s_n \\ &= (-1)^n(a_{n+1} - a_{n+2} + a_{n+3} - \cdots), \end{aligned}$$

has the same sign, $(-1)^n$, as the next term, and its absolute value is no greater than the absolute value of that term. That is, $L = s_n + R_n$, R_n has the same sign as $\pm a_{n+1}$, and $|R_n| \leq a_{n+1}$. This follows at once from the fact that

$$\begin{aligned} R_n = (-1)^n[a_{n+1} &- (a_{n+2} - a_{n+3}) \\ &- (a_{n+4} - a_{n+5}) - \cdots] \\ = (-1)^n[(a_{n+1} &- a_{n+2}) + (a_{n+3} - a_{n+4}) + \cdots]; \end{aligned}$$

note that the expression in brackets is positive (second form) but not greater than a_{n+1} (first form). Thus, in Example 1, if we stop with

$$s_{10} = 1 - \tfrac{1}{2} + \tfrac{1}{3} - \cdots - \tfrac{1}{10},$$

then we know that the exact value of the series is $L = s_{10} + R_{10}$, where R_{10} is positive (because the next term is $+\frac{1}{11}$) and where $|R_{10}| < \frac{1}{11}$.

Example 2. We can now show that the series

$$\sum_{k=1}^{\infty} \frac{x^k}{k} = x + \frac{x^2}{2} + \frac{x^3}{3} + \frac{x^4}{4} + \cdots \tag{4}$$

converges for $-1 \leq x < 1$. To find this interval of convergence, we let $u_n = x_n/n$ and use the ratio test:

$$\rho = \lim_{n\to\infty} \left|\frac{u_{n+1}}{u_n}\right| = \lim_{n\to\infty} \left|\frac{x^{n+1}}{n+1}\frac{n}{x^n}\right| = |x|.$$

Hence the series converges absolutely if $|x| < 1$, and it diverges if $|x| > 1$. When $x = +1$, the series becomes

$$1 + \tfrac{1}{2} + \tfrac{1}{3} + \tfrac{1}{4} + \cdots,$$

which diverges to $+\infty$. When $x = -1$, the series becomes

$$-(1 - \tfrac{1}{2} + \tfrac{1}{3} - \tfrac{1}{4} + \cdots),$$

which is just the negative of the convergent alternating series considered in Example 1. Hence the series (4) converges for $-1 \leq x < 1$, and diverges for all other x.

EXERCISES

Find the interval of absolute convergence for each of the following series. If the interval is finite, determine whether the series converges or diverges at the endpoints.

1. $\sum_{n=0}^{\infty} \frac{x^{2n+1}}{2n+1}$
2. $\sum_{n=0}^{\infty} (-1)^n \frac{x^{2n+1}}{2n+1}$
3. $\sum_{n=1}^{\infty} \frac{(x-2)^n}{n^2}$
4. $\sum_{n=1}^{\infty} (-1)^{n-1} \frac{(x+1)^n}{n}$
5. $\sum_{n=1}^{\infty} \frac{(-1)^{n-1}(x-2)^n}{n \cdot 2^n}$
6. $\sum_{n=0}^{\infty} (-2)^n (n+1)(x-1)^n$
7. $\sum_{k=0}^{\infty} \frac{(3x+6)^k}{k!}$
8. $\sum_{n=0}^{\infty} n!x^n$
9. $\sum_{n=1}^{\infty} (-1)^n \frac{(2n-2)!}{2^{2n-2}(n-1)!(n-1)!} \frac{x^{2n-1}}{2n-1}$
10. $\sum_{n=1}^{\infty} \frac{(x+3)^{n-1}}{n}$

REVIEW QUESTIONS AND EXERCISES

1. Define "sequence," "series," "sequence of partial sums of a series."
2. Define convergence
   (a) of a sequence, (b) of an infinite series.
3. Which of the following statements are true, and which are false?
   (a) If a sequence does not converge, then it diverges.
   (b) If a sequence $\{n, f(n)\}$ does not converge, then $f(n)$ tends to infinity as n does.
   (c) If a series does not converge, then its nth term does not approach zero as n tends to infinity.
   (d) If the nth term of a series does not approach zero as n tends to infinity, then the series diverges.
   (e) If a sequence $\{n, f(n)\}$ converges, then there is a number L such that $f(n)$ lies within one unit of L
      (i) for all values of n,
      (ii) for all but a finite number of values of n.
   (f) If all partial sums of a series are less than some constant L, then the series converges.
   (g) If a series converges, then its partial sums s_n are bounded (that is, $m \leq s_n \leq M$ for some constants m and M).
4. List three tests for convergence (or divergence) of an infinite series.
5. Under what circumstances do you know that a bounded sequence converges?
6. Define absolute convergence and conditional convergence. Give examples of series that are
   (a) absolutely convergent,
   (b) conditionally convergent.
7. State Taylor's theorem with remainder, giving two different expressions for the remainder.
8. It can be shown (though not very simply) that the function f defined by

$$f(x) = \begin{cases} 0 & \text{when } x = 0, \\ e^{-1/x^2} & \text{when } x \neq 0 \end{cases}$$

is everywhere continuous, together with its derivatives of all orders. At zero, the derivatives are all equal to zero.

(a) Write the Taylor series expansion of f in powers of x.

(b) What is the remainder $R_n(x, 0)$ for this function? Does the Taylor series for f converge to $f(x)$ at some value of x different from zero? Give a reason for your answer.

9. If a Taylor series in powers of $x - a$ is to be used for the numerical evaluation of a function, what is necessary or desirable in the choice of a?

10. Write the Taylor series in powers of $x - a$ and $y - b$ for a function f of two variables x and y, about the point (a, b).

11. List two methods that may be useful in finding

$$\lim_{x \to a} \frac{f(x)}{g(x)} \quad \text{if} \quad f(a) = g(a) = 0.$$

Illustrate both methods with a single example.

12. What test is usually used to find the interval of convergence of a power series? Does this test also work at the end points of the interval? Illustrate with examples.

13. What test is usually used to decide whether or not a given alternating series converges? Give examples of convergent and divergent alternating series.

MISCELLANEOUS EXERCISES

1. Find explicitly the nth partial sum of the series

$$\sum_{n=2}^{\infty} \ln\left(1 - \frac{1}{n^2}\right),$$

and thereby determine whether or not the series converges.

2. Evaluate

$$\sum_{k=2}^{\infty} \frac{1}{k^2 - 1}$$

by finding the nth partial sum and taking the limit as n becomes infinite.

3. Prove that the sequence $\{x_n\}$ and the series

$$\sum_{k=1}^{\infty} (x_{k+1} - x_k)$$

both converge or both diverge.

4. In an attempt to find a root of the equation $x = f(x)$, a first approximation x_1 is estimated from the graphs of $y = x$ and $y = f(x)$. Then $x_2, x_3, \ldots, x_n, \ldots$ are computed successively from the formula

$$x_n = f(x_{n-1}).$$

If the points $x_1, x_2, \ldots, x_n, \ldots$ all lie on an interval $[a, b]$ on which $f(x)$ has a derivative such that $|f'(x)| < M < 1$, show that the sequence $\{x_n\}$ converges to a root of the given equation.

5. Assuming $|x| > 1$, show that

$$\frac{1}{1 - x} = -\frac{1}{x} - \frac{1}{x^2} - \frac{1}{x^3} - \cdots.$$

6. (a) Find the expansion in powers of x of

$$\frac{x^2}{1 + x}.$$

(b) Does the series expansion of $x^2/(1 + x)$ in powers of x converge when $x = 2$? Give a brief reason.

7. Obtain the Maclaurin series expansion for $\sin^{-1} x$ by integrating the series for $(1 - t^2)^{-1/2}$ from 0 to x. Find the intervals of convergence of these series.

8. Obtain the Maclaurin series for

$$\ln\left(x + \sqrt{x^2 + 1}\right) = \sinh^{-1} x$$

by integrating the series for $(1 + t^2)^{-1/2}$ from 0 to x. Find the intervals of convergence of these two series.

9. Obtain the first four terms in the Maclaurin series for $e^{\sin x}$ by substituting the series for $y = \sin x$ in the series for e^y.

10. Assuming $|x| > 1$, obtain the expansions

$$\tan^{-1} x = \frac{\pi}{2} - \frac{1}{x} + \frac{1}{3x^3} - \frac{1}{5x^5} + \cdots, \quad x > 1,$$

$$\tan^{-1} x = -\frac{\pi}{2} - \frac{1}{x} + \frac{1}{3x^3} - \frac{1}{5x^5} + \cdots, \quad x < -1$$

by integrating the series

$$\frac{1}{1 + t^2} = \frac{1}{t^2} \cdot \frac{1}{1 + (1/t^2)}$$
$$= \frac{1}{t^2} - \frac{1}{t^4} + \frac{1}{t^6} - \frac{1}{t^8} + \cdots$$

from $x(> 1)$ to $+\infty$ or from $-\infty$ to $x(< -1)$.

11. (a) Obtain the Maclaurin series, through the term in x^6, for $\ln(\cos x)$ by substituting the series for $y = 1 - \cos x$ in the series for $\ln(1 - y)$.
(b) Use the result of (a) to estimate

$$\int_0^{0.1} \ln(\cos x)\, dx$$

to five decimal places.

12. Compute

$$\int_0^1 \frac{\sin x}{x}\, dx$$

to three decimal places.

13. Compute $\int_0^1 e^{-x^2}\, dx$ to three decimal places.

14. Expand the function $f(x) = \sqrt{1 + x^2}$ in powers of $(x - 1)$, obtaining three nonvanishing terms.

15. Expand the function $f(x) = 1/(1 - x)$ in powers of $(x - 2)$, and find the interval of convergence.

16. If $\tan x$ has the Maclaurin expansion

$$a_0 + a_1x + a_2x^2 + a_3x^3 + \cdots,$$

find the first three nonvanishing terms.

17. Determine the Taylor series expansion of the function $f(x) = 1/(x + 1)$ in powers of $(x - 3)$.

18. Expand $\cos x$ in powers of $(x - \pi/3)$.

19. Find the first three terms of the Taylor series expansion of the function $1/x$ about the point π.

20. Let f and g be functions satisfying the following conditions:

(a) $f(0) = 1$, (b) $g(0) = 0$,
(c) $f'(x) = g(x)$, $g'(x) = f(x)$.

Find $f(1)$, accurate to three decimal places.

21. Suppose

$$f(x) = \sum_{n=0}^{\infty} a_nx^n.$$

Prove that

(a) if $f(x)$ is an even function, then

$$a_1 = a_3 = a_5 = \cdots = 0;$$

(b) if $f(x)$ is an odd function, then

$$a_0 = a_2 = a_4 = \cdots = 0.$$

22. Show that the function $f(x) = e^{(e^x)}$ can be expanded into a power series in x, and find the first four terms (up to x^3).

23. Using a suitable series, complete the following four-place table of the function $f(x) = \arctan(x)$:

x	0.01	0.02	0.03	0.04	0.05	0.06	0.10
$f(x)$	0.0100	0.0200	0.0300	0.0400			

24. Give a quantitative estimate of the error involved in using $x - x^2/2$ as an approximation to $\ln(1 + x)$ for values of x between 0 and 0.2 inclusive.

25. If $(1 + x)^{1/3}$ is replaced by $1 + x/3$, and if $0 \le x \le \frac{1}{10}$, what estimate can be given for the error?

26. By considering the quotient of the power series for numerator and denominator, find

$$\lim_{x \to 0} \frac{\ln(1 - x) - \sin x}{1 - \cos^2 x}.$$

27. Find the limit of $[(\sin x)]^{1/x^2}$ as x approaches zero.

28. (a) Find the Fourier series of period 2π generated by the function

$$f(x) = \begin{cases} -\pi, & \text{if } -\pi < x < 0, \\ x, & \text{if } 0 < x < \pi. \end{cases}$$

(b) Take $x = 0$ in your result in (a) to find the sum of the series

$$\frac{1}{1^2} + \frac{1}{3^2} + \frac{1}{5^2} + \frac{1}{7^2} + \cdots + \frac{1}{(2n - 1)^2} + \cdots.$$

(The Fourier series converges to the "average value" $-\pi/2$ at $x = 0$.)

29. Does the series $\sum_{n=1}^{\infty} \operatorname{sech} n$ converge or diverge? Why?

30. Does the series $\sum_{n=1}^{\infty} (-1)^n \tanh n$ converge or diverge? Why?

31. If $a_n > 0$ and $\sum_{n=1}^{\infty} a_n$ converges, prove that $\sum_{n=1}^{\infty} (1/a_n)$ diverges.

32. Establish the convergence or divergence of the series whose nth terms are as given.

(a) $\dfrac{1}{\ln(n + 1)}$ (b) $\dfrac{n}{2(n + 1)(n + 2)}$

(c) $\dfrac{\sqrt{n + 1} - \sqrt{n}}{\sqrt{n}}$ (d) $\dfrac{1}{n(\ln n)^2}$, $n \ge 2$

(e) $\dfrac{1+(-2)^{n-1}}{2^n}$ (f) $\dfrac{n}{1000n^2+1}$

(g) $\dfrac{e^n}{n!}$ (h) $\dfrac{1}{n\sqrt{n^2+1}}$

(i) $\dfrac{1}{n^{1+1/n}}$ (j) $\dfrac{1\cdot 3\cdot 5\cdots(2n-1)}{2\cdot 4\cdot 6\cdots(2n)}$

(k) $\dfrac{n^2}{n^3+1}$ (l) $\dfrac{n+1}{n!}$

33. Find the sum of the convergent series

$$\sum_{n=1}^{\infty}\frac{1}{(n+1)(n+2)}.$$

34. (a) Suppose $a_1, a_2, a_3, \ldots, a_n$ are positive numbers satisfying the following conditions:

(1) $a_1 \geq a_2 \geq a_3 \cdots$,

(2) the series $a_2 + a_4 + a_8 + a_{16} + \cdots$ diverges.

Show that the series

$$\frac{a_1}{1}+\frac{a_2}{2}+\frac{a_3}{3}+\cdots$$

is divergent.

(b) Use the result above to show that the series

$$\sum_{n=2}^{\infty}\frac{1}{n\ln n}$$

is divergent.

35. Let $a_n \neq 1$ and $a_n > 0$, and suppose that $\sum a_n$ converges.

(a) Show that $\sum a_n^2$ converges.

(b) Does $\sum a_n/(1-a_n)$ converge?

(c) Does $\sum \ln(1+a_n)$ converge?

Justify your conclusions.

36. Show that the series

$$\sum_{n=2}^{\infty}\frac{1}{n(\ln n)^k}$$

is convergent for $k > 1$.

37. Find the interval of convergence for the following series and test for convergence or divergence at the end points if the interval is finite.

(a) $1+\dfrac{x+2}{3\cdot 1}+\dfrac{(x+2)^2}{3^2\cdot 2}+\cdots+\dfrac{(x+2)^n}{3^n\cdot n}+\cdots$

(b) $1+\dfrac{(x-1)^2}{2!}+\dfrac{(x-1)^4}{4!}+\cdots+\dfrac{(x-1)^{2n-2}}{(2n-2)!}+\cdots$

(c) $\displaystyle\sum_{n=1}^{\infty}\frac{x^n}{n^n}$ (d) $\displaystyle\sum_{n=1}^{\infty}\frac{n!x^n}{n^n}$

(e) $\displaystyle\sum_{n=0}^{\infty}\frac{n+1}{2n+1}\,\frac{(x+3)^n}{2^n}$

(f) $\displaystyle\sum_{n=0}^{\infty}\frac{n+1}{2n+1}\,\frac{(x-2)^n}{3^n}$

(g) $\displaystyle\sum_{n=1}^{\infty}\frac{(-1)^{n-1}(x-1)^n}{n^2}$

(h) $\displaystyle\sum_{n=1}^{\infty}\frac{x^n}{n}$

38. Determine *all* the values of x for which the following series converge.

(a) $\displaystyle\sum_{n=1}^{\infty}\frac{(x-2)^{3n}}{n!}$ (b) $\displaystyle\sum_{n=1}^{\infty}\frac{2^n(\sin x)^n}{n^2}$

(c) $\displaystyle\sum_{n=1}^{\infty}\frac{1}{n}\left(\frac{x-1}{x}\right)^n$

39. A function is defined by the power series

$$y = 1+\frac{1}{6}x^3+\frac{1}{180}x^6+\cdots + \frac{1\cdot 4\cdot 7\cdots(3n-2)}{(3n)!}x^{3n}+\cdots.$$

(a) Find the interval of convergence of the series.

(b) Show that there exist two constants a and b such that the function so defined satisfies a differential equation of the form $y'' = x^a y + b$.

40. (a) Show that the series

$$1-\int_1^2\frac{dx}{x}+\frac{1}{2}-\int_2^3\frac{dx}{x}+\frac{1}{3} -\int_3^4\frac{dx}{x}+\frac{1}{4}-\int_4^5\frac{dx}{x}+\frac{1}{5}-\cdots$$

is convergent.

(b) If S is the sum of the series in (a), deduce that

$$\lim_{n\to\infty}\left(1+\frac{1}{2}+\frac{1}{3}+\cdots+\frac{1}{n}-\ln n\right) = S.$$

41. (a) Does the series

$$\sum_{n=1}^{\infty} \frac{1}{[1 + (1/n)]^n}$$

converge or diverge? Why?

(b) Does the series

$$\sum_{n=1}^{\infty} \frac{1}{[1 + (1/n)]^{n^2}}$$

converge or diverge? Why?

42. If $a_n > 0$ and the series $\sum_{n=1}^{8} a_n$ converges, prove that

$$\sum_{n=1}^{\infty} \frac{a_n}{1 + a_n}$$

converges.

43. If $1 > a_n > 0$ and $\sum_{n=1}^{\infty} a_n$ converges, prove that

$$\sum_{n=1}^{\infty} \ln (1 - a_n)$$

converges.

Hint. First show that $|\ln (1 - a_n)| \leq a_n/(1 - a_n)$, and then apply the answer to Exercise 35(b).

44. An infinite product, indicated by

$$\prod_{n=1}^{\infty} (1 + a_n),$$

is said to converge if the series $\sum_{n=1}^{\infty} \ln (1 + a_n)$ converges. (The series is the natural logarithm of the product.) Prove that the product converges if every $a_n > -1$ and if $\sum_{n=1}^{\infty} |a_n|$ converges.

Hint. Show that

$$|\ln (1 + a_n)| \leq |a_n|/(1 - |a_n|) < 2|a_n|$$

when $|a_n| < \frac{1}{2}$.

COMPLEX NUMBERS AND FUNCTIONS

CHAPTER 19

19.1 INVENTED NUMBER SYSTEMS

In this chapter we shall discuss complex numbers. These are expressions of the form $a + ib$, where a and b are "real" numbers and i is a symbol for $\sqrt{-1}$. Unfortunately, the words "real" and "imaginary" have connotations which somehow place $\sqrt{-1}$ in a less favorable position than $\sqrt{2}$ in our minds. As a matter of fact, a good deal of imagination, in the sense of *inventiveness*, has been required to construct the *real* number system which forms the basis of the calculus we have studied thus far. In this article, we shall review the various stages of this process of invention. The further invention of a complex number system will not then seem so strange. It is fitting for us to study this system, since modern engineering has found in it a language convenient for expressing vibratory motion, harmonic oscillation, damped vibrations, alternating currents, and other wave phenomena.

The earliest stage of development of man's number consciousness was his recognition of the *counting numbers* 1, 2, 3, . . . , to which we now refer as the *positive integers.** Certain simple arithmetical operations can be performed with these numbers without getting outside the system. That is, the system of positive integers is *closed* with respect to the operations of *addition* and *multiplication*. By this we mean that if m and n are any positive integers, then

$$m + n = p \quad \text{and} \quad mn = q \tag{1}$$

are also positive integers. Given the two positive integers on the *left-hand side* of either equation in (1) we can find the positive integer on the right. More than this, we may sometimes specify the positive integers m and p and find a positive integer n such that $m + n = p$. For instance, $3 + n = 7$ can be *solved* when the only numbers we know are the positive integers. But the equation $7 + n = 3$ cannot be solved unless we enlarge our number system. Man therefore used his *imagination* and invented the number concepts which we denote by zero and

---

* Also called *natural numbers*.

the *negative* integers. In a civilization which recognizes all the integers

$$\ldots, -3, -2, -1, 0, 1, 2, 3, \ldots, \tag{2}$$

an educated man may always find the missing integer which solves the equation $m + n = p$ when he is given two of the integers in the equation.

Suppose our educated man also knows how to multiply any two integers of the set in (2). If, in Eq. (1), he is given m and q, he discovers that sometimes he can find n and sometimes he can't. If his *imagination* is still in good working order he may be inspired to invent still more numbers and introduce fractions, which are just ordered pairs m/n of integers m and n. The number zero has special properties that may bother him for awhile, but he ultimately discovers that it is handy to have all ratios of integers m/n, excluding only those having zero in the denominator. This system, called the set of *rational numbers*, is now rich enough for him to perform the so-called *rational operations* of

addition; subtraction,
multiplication; division,

on any two numbers in the system, *except that he cannot divide by zero.*

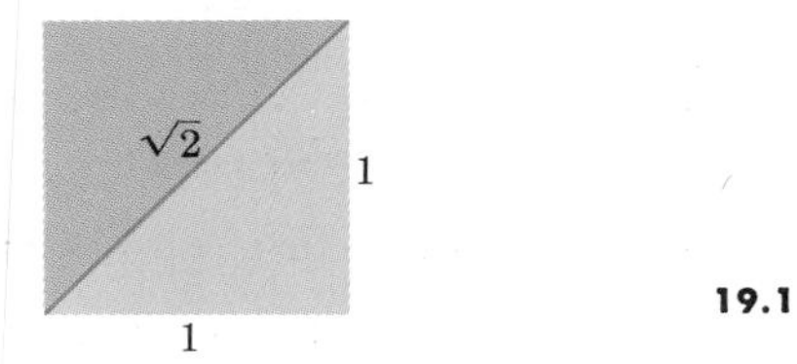

19.1

The geometry of the unit square (Fig. 19.1) and the Pythagorean Theorem showed that man could construct a geometrical line segment which, in terms of some basic unit of length, has length equal to $\sqrt{2}$. Thus man could solve the equation

$$x^2 = 2$$

by means of a geometrical construction. But then he discovered that the line segment representing $\sqrt{2}$ and the line segment representing the unit of length 1 were incommensurable quantities. This means that their ratio $\sqrt{2}/1$ cannot be expressed as the ratio of two *integral* multiples of some other, presumably more fundamental, unit of length. That is, man could not solve the equation $x^2 = 2$ in terms of his rational numbers.

Note. If there is a rational number whose square is equal to 2, we may find integers p and q such that they have no common factor other than unity, and such that

$$p^2 = 2q^2. \tag{3}$$

Since p and q are integers, p must then be even, say $p = 2p_1$, where p_1 is an integer. This in turn leads to $2p_1^2 = q^2$, which then says that q must also be even, say $q = 2q_1$, where q_1 is also an integer. But this is contrary to our choice of p and q as integers having no common factor other than unity. Hence there is no rational number whose square is 2.

Our educated man *could,* however, get a *sequence* of rational numbers

$$\frac{1}{1}, \frac{7}{5}, \frac{41}{29}, \frac{239}{169}, \ldots \tag{4}$$

whose squares form a sequence

$$\frac{1}{1}, \frac{49}{25}, \frac{1681}{841}, \frac{57{,}121}{28{,}561}, \ldots \tag{5}$$

which converges to 2 as its *limit.* This time his *imagination* suggested to him that he needed the concept of a *limit of a sequence* of rational numbers. If we accept the fact that a monotone increasing sequence which is bounded above possesses a limit, and observe that the sequence in (4) has these properties, then we want it to have a limit L. This would also mean, from (5), that $L^2 = 2$, and hence L is *not* one of our rational numbers. If to the *rational* numbers we further add the *limits* of all bounded monotone increasing sequences, we arrive at the system of all "real" numbers. The word *real* is placed in quotes because there is nothing which is either "more real" or "less real" about this system than there is about any other well-defined mathematical system.

Man's *imagination* was called upon at many stages in the development of the real number system from the system of positive integers. In fact, the art of invention was needed at least three times in constructing the three systems we have discussed so far:

1. The *first invented* system; the set of *all integers,* as constructed from the counting numbers.

2. The *second invented* system; the set of *all rational numbers m/n,* as constructed from the integers.

3. The *third invented* system; the set of *all real numbers x,* as constructed from the rational numbers.

These invented systems form a hierarchy in which each system contains the previous system. Each system is also richer than its predecessor in that it permits more operations to be performed without going outside the system. We may express this in algebraic terms.

1. In the system of all integers, we can solve all equations of the form

$$x + a = 0, \tag{6}$$

where a may be an integer.

2. In the system of all rational numbers, we can solve all equations of the form

$$ax + b = 0 \tag{7}$$

provided a and b are rational numbers and $a \neq 0$.

3. In the system of all real numbers, we can solve all the Eqs. (6) and (7) and, in addition, all quadratic equations

$$ax^2 + bx + c = 0 \tag{8a}$$

having

$$a \neq 0 \quad \text{and} \quad b^2 - 4ac \geq 0. \tag{8b}$$

Every student of algebra is familiar with the formula that gives the solutions of (8), namely,

$$x = \frac{-b \pm \sqrt{b^2 - 4ac}}{2a}, \tag{9}$$

and with the further fact that when the discriminant, $d = b^2 - 4ac$, is *negative,* the solutions in (9) do *not* belong to any of the systems discussed so far. In fact, the very simple quadratic equation

$$x^2 + 1 = 0 \tag{10}$$

is impossible to solve if the only number systems that can be used are the three invented systems so far described.

Thus we come to the *fourth invented* system, the set of all complex numbers $a \times ib$. We could, in fact, dispense entirely with the symbol i and use a notation such as (a, b). We would then speak simply of a pair of real numbers a and b. Since, under algebraic operations, the numbers a and b are treated somewhat differently, it is essential to keep the *order* straight. We therefore might say that *the complex number system consists of all ordered pairs of real numbers (a, b) subject to certain laws,* laws which we shall set forth below. We shall use both the (a, b)-notation and the notation $a + ib$. We call a the "real part" and b the "imaginary part" of (a, b).

Definitions

Equality. *Two complex numbers (a, b) and (c, d) are equal if and only if $a = c$ and $b = d$:*

$$a + ib = c + id$$

if and only if

$$a = c \quad \textit{and} \quad b = d.$$

Addition. *The sum of the two complex numbers (a, b) and (c, d) is the complex number $(a + c, b + d)$:*

$$(a + ib) + (c + id) = (a + c) + i(b + d).$$

Multiplication. *The product of two complex numbers (a, b) and (c, d) is the complex number $(ac - bd, ad + bc)$:*

$$(a + ib)(c + id) = (ac - bd) + i(ad + bc),$$

and the product of a real number c and the complex number (a, b) is the complex number (ac, bc):

$$c(a + ib) = ac + i(bc).$$

The set of all complex numbers (a, b) in which the second number is zero has all the properties of the set of ordinary real numbers a. For example, addition and multiplication of $(a, 0)$ and $(c, 0)$ give

$$(a, 0) + (c, 0) = (a + c, 0),$$
$$(a, 0) \cdot (c, 0) = (ac, 0),$$

which are numbers of the same type with imaginary part equal to zero. Also, if we multiply a real number $(a, 0)$ with a complex number (c, d), we get

$$(a, 0) \cdot (c, d) = (ac, ad) = a(c, d).$$

In particular, the complex number $(0, 0)$ plays the role of zero in the complex number system, and the complex number $(1, 0)$ plays the role of unity.

The number pair $(0, 1)$, which has real part equal to zero and imaginary part equal to one, has the property that its square,

$$(0, 1)(0, 1) = (-1, 0),$$

has real part equal to minus one and imaginary part equal to zero. Therefore, in the system of complex numbers (a, b), there is a number $x = (0, 1)$ whose square can be added to unity $= (1, 0)$ to produce zero $= (0, 0)$; that is,

$$(0, 1)^2 + (1, 0) = (0, 0).$$

The equation

$$x^2 + 1 = 0$$

therefore has a solution $x = (0, 1)$ in this new number system.

The student is probably more familiar with the $a + ib$ notation than he is with the notation (a, b). And since our laws of algebra for the ordered pairs enable us to write

$$(a, b) = (a, 0) + (0, b) = a(1, 0) + b(0, 1)$$

and since $(1, 0)$ behaves like unity and $(0, 1)$ behaves like a square root of minus one, we need not hesitate to write $a + ib$ in place of (a, b). The i associated with b is like a tracer element which tags the imaginary part of $a + ib$; we can pass at will from the realm of ordered pairs (a, b) to the realm of expressions $a + ib$, and conversely. But there is nothing less "real" about the symbol $(0, 1)(= i)$ than there is about the symbol $(1, 0)(= 1)$, once we have learned the laws of algebra in the system of complex numbers (a, b).

EXERCISES

1. When complex numbers are denoted by ordered pairs of real numbers, the product $(a, b) \cdot (c, d)$ can be found as follows. First, write a and b on one line, and write c and d beneath them. Then:

   For the *real* part of the product, multiply the numbers a and c in the first column and from their product subtract the product of the numbers b and d in the second column.

   For the *imaginary* part of the product, cross-multiply and add the products ad and bc.

   Apply this method to find the following products.

   (a) $(2, 3) \cdot (4, -2)$ (b) $(2, -1) \cdot (-2, 3)$
   (c) $(-1, -2) \cdot (2, 1)$

   Note. This is the way in which complex numbers are multiplied on modern computing machinery.

2. Solve the following equations for the real numbers x and y.

   (a) $(3 + 4i)^2 - 2(x - iy) = x + iy$

   (b) $\left(\dfrac{1 + i}{1 - i}\right)^2 + \dfrac{1}{z + iy} = 1 + i$

   (c) $(3 - 2i)(x + iy) = 2(x - 2iy) + 2i - 1$

19.2 THE ARGAND DIAGRAM

To reduce any rational combination of complex numbers to a single complex number, we need only apply the laws of elementary algebra and replace i^2 by -1 whenever it appears. Of course, we cannot divide by the complex number $(0, 0) = 0 + i0$. But if $a + ib \neq 0$, then we proceed as follows when we have a division:

$$\frac{c + id}{a + ib} = \frac{(c + id)(a - ib)}{(a + ib)(a - ib)} = \frac{(ac + bd) + i(ad - bc)}{a^2 + b^2}.$$

The result is a complex number $x + iy$, with

$$x = \frac{ac + bd}{a^2 + b^2}, \qquad y = \frac{ad - bc}{a^2 + b^2},$$
$$a^2 + b^2 \neq 0$$

[since $a + ib = (a, b) \neq (0, 0)$]. The number $a - ib$ that we have used as a multiplier to clear the i out of the denominator is called the *complex conjugate of* $a + ib$. It is customary to use $\bar{z}$ (read "z bar") to denote the complex conjugate of z:

$$z = a + ib, \qquad \bar{z} = a - ib.$$

Thus we multiplied the numerator and denominator of the complex fraction $(c + id)/(a + ib)$ by the complex conjugate of the denominator. This procedure will always replace the denominator by a real number.

There are two geometric representations of the complex number $z = x + iy$:

(a) as the point $P(x, y)$ in the xy-plane, or

(b) as the vector $\overrightarrow{OP}$ from the origin to P.

In either representation, the x-axis is called the *axis of reals* and the y-axis, the *imaginary axis*. Either representation is called an *Argand diagram*. (See Fig. 19.2.)

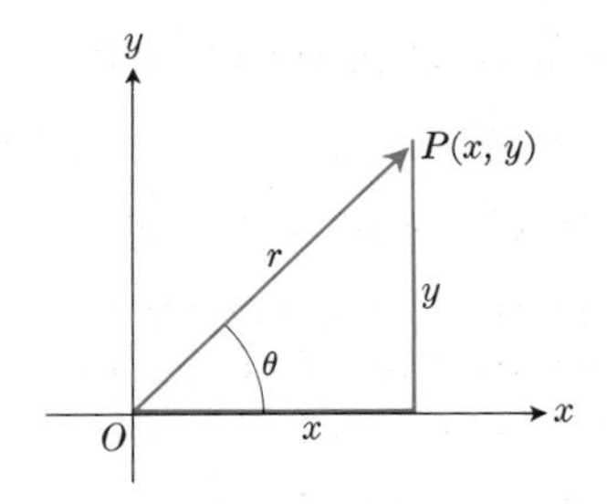

19.2 An Argand diagram representing $z = x + iy$ as (a) the point $P(x, y)$, and (b) the vector $\overrightarrow{OP}$.

In terms of its polar coordinates (r, θ), we have

$$x = r\cos\theta, \qquad y = r\sin\theta,$$

and

$$z = x + iy = r(\cos\theta + i\sin\theta). \tag{1}$$

We define the *absolute value* of a complex number $x + iy$ to be the length r of a vector $\overrightarrow{OP}$ from the origin to $P(x, y)$. We denote the absolute value by vertical bars, thus:

$$|x + iy| = \sqrt{x^2 + y^2}. \tag{2a}$$

Since we can always choose the polar coordinates r and θ so that $r \geq 0$, we have

$$r = |x + iy|. \tag{2b}$$

The polar angle θ is called the *argument* of z and written $\theta = \arg z$. Of course, any integral multiple of 2π may be added to θ to produce another appropriate angle. The *principal value* of the argument will, in this book, be taken to be that value of θ for which $-\pi < \theta \leq +\pi$.

The following equation gives a useful formula connecting a complex number z, its conjugate $\bar{z}$, and its absolute value $|z|$:

$$z \cdot \bar{z} = |z|^2. \tag{2c}$$

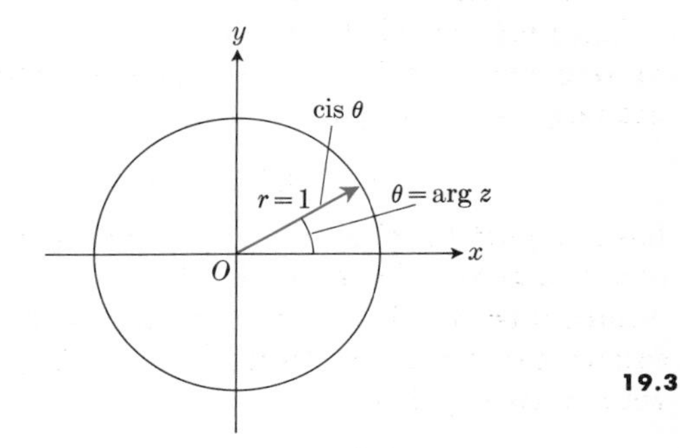

19.3

We shall show in a later section how $\cos\theta + i\sin\theta$ can be expressed very conveniently as $e^{i\theta}$. But for the present, let us just introduce the abbreviation

$$\operatorname{cis}\theta = \cos\theta + i\sin\theta. \tag{3}$$

Since $\operatorname{cis}\theta$ is what we get from Eq. (1) by taking $r = 1$, we can say that $\operatorname{cis}\theta$ is represented by a *unit* vector that makes an angle θ with the positive x-axis (see Fig. 19.3).

Example 1

$$\operatorname{cis} 0 = \cos 0 + i\sin 0 = 1,$$

$$\operatorname{cis}\frac{\pi}{4} = \cos\frac{\pi}{4} + i\sin\frac{\pi}{4} = \frac{1+i}{\sqrt{2}},$$

$$\operatorname{cis}\frac{3\pi}{2} = \cos\frac{3\pi}{2} + i\sin\frac{3\pi}{2} = -i.$$

The complex-valued function cis θ has some interesting properties. For example, we shall show that

$$\text{cis}\,\theta_1 \cdot \text{cis}\,\theta_2 = \text{cis}\,(\theta_1 + \theta_2), \tag{4a}$$

$$(\text{cis}\,\theta)^{-1} = \text{cis}\,(-\theta), \tag{4b}$$

$$\frac{\text{cis}\,\theta_1}{\text{cis}\,\theta_2} = \text{cis}\,(\theta_1 - \theta_2). \tag{4c}$$

To prove the first of these, we simply multiply

$$\begin{aligned}\text{cis}\,\theta_1 \cdot \text{cis}\,\theta_2 &= (\cos\theta_1 + i\sin\theta_1)(\cos\theta_2 + i\sin\theta_2)\\ &= (\cos\theta_1\cos\theta_2 - \sin\theta_1\sin\theta_2)\\ &\quad + i(\sin\theta_1\cos\theta_2 + \cos\theta_1\sin\theta_2).\end{aligned}$$

From trigonometry, we recognize the expressions in parentheses to be

$$\cos(\theta_1 + \theta_2) = \cos\theta_1\cos\theta_2 - \sin\theta_1\sin\theta_2,$$

$$\sin(\theta_1 + \theta_2) = \sin\theta_1\cos\theta_2 + \cos\theta_1\sin\theta_2,$$

which give us

$$\begin{aligned}\text{cis}\,\theta_1 \cdot \text{cis}\,\theta_2 &= \cos(\theta_1 + \theta_2) + i\sin(\theta_1 + \theta_2)\\ &= \text{cis}\,(\theta_1 + \theta_2)\end{aligned}$$

and establishes (4a). In particular,

$$\text{cis}\,\theta \cdot \text{cis}\,(-\theta) = \text{cis}\,(\theta - \theta) = \text{cis}\,0 = 1,$$

whence

$$\text{cis}\,(-\theta) = \frac{1}{\text{cis}\,\theta},$$

which establishes (4b). Finally, we may combine Eqs. (4a, b) and write

$$\begin{aligned}\frac{\text{cis}\,\theta_1}{\text{cis}\,\theta_2} &= (\text{cis}\,\theta_1)(\text{cis}\,\theta_2)^{-1}\\ &= \text{cis}\,\theta_1 \cdot \text{cis}\,(-\theta_2) = \text{cis}\,(\theta_1 - \theta_2).\end{aligned}$$

These properties of cis θ lead to interesting geometrical interpretations of the product and quotient of two complex numbers in terms of the vectors that represent them.

Product. Let

$$z_1 = r_1\,\text{cis}\,\theta_1, \qquad z_2 = r_2\,\text{cis}\,\theta_2, \tag{5}$$

so that

$$|z_1| = r_1, \quad \arg z_1 = \theta_1,$$

$$|z_2| = r_2, \quad \arg z_2 = \theta_2. \tag{6}$$

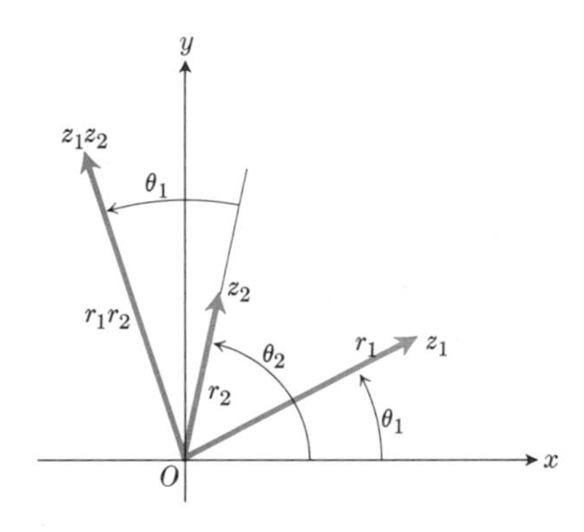

19.4

Then

$$z_1z_2 = r_1\,\text{cis}\,\theta_1 \cdot r_2\,\text{cis}\,\theta_2 = r_1r_2\,\text{cis}\,(\theta_1 + \theta_2),$$

and hence

$$\begin{aligned}|z_1z_2| &= r_1r_2 = |z_1|\cdot|z_2|,\\ \arg(z_1z_2) &= \theta_1 + \theta_2 = \arg z_1 + \arg z_2.\end{aligned} \tag{7}$$

Thus the product of two complex numbers is represented by a vector whose length is the product of the lengths of the two factors r and whose argument is the sum of their arguments θ (see Fig. 19.4). In particular, a vector may be rotated in the counterclockwise direction through an angle θ by simply multiplying it by cis θ. Multiplication by i rotates 90°, by -1 rotates 180°, by $-i$ rotates 270°, and so forth.

Example 2. Let

$$z_1 = 1 + i, \qquad z_2 = \sqrt{3} - i.$$

We plot these complex numbers in an Argand diagram, Fig. 19.5, from which we read off the polar representations

$$z_1 = \sqrt{2}\,\text{cis}\,\frac{\pi}{4}, \qquad z_2 = 2\,\text{cis}\left(-\frac{\pi}{6}\right).$$

Then

$$\begin{aligned}z_1z_2 &= 2\sqrt{2}\,\text{cis}\left(\frac{\pi}{4} - \frac{\pi}{6}\right)\\ &= 2\sqrt{2}\,\text{cis}\,\frac{\pi}{12}\\ &= 2\sqrt{2}\,(\cos 15° + i\sin 15°)\\ &\approx 2.73 + 0.73\,i.\end{aligned}$$

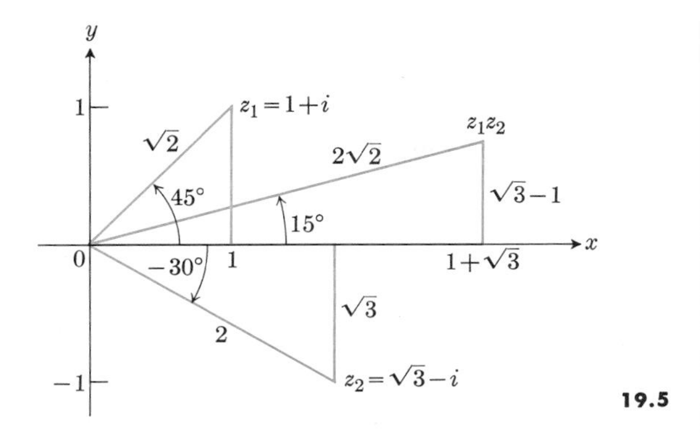

19.5

Quotient. Suppose $r_2 \neq 0$ in Eqs. (5). Then

$$\frac{z_1}{z_2} = \frac{r_1 \operatorname{cis} \theta_1}{r_2 \operatorname{cis} \theta_2} = \frac{r_1}{r_2} \operatorname{cis} (\theta_1 - \theta_2).$$

Hence

$$\left|\frac{z_1}{z_2}\right| = \frac{r_1}{r_2} = \frac{|z_1|}{|z_2|},$$

$$\arg (z_1/z_2) = \theta_1 - \theta_2 = \arg z_1 - \arg z_2.$$

That is, when taking a quotient we divide lengths and subtract angles. In Example 2, if we divide z_1 by z_2, we would thus have

$$\begin{aligned}\frac{1+i}{\sqrt{3}-i} &= \frac{\sqrt{2} \operatorname{cis} (\pi/4)}{2 \operatorname{cis} (-\pi/6)} = \frac{\sqrt{2}}{2} \operatorname{cis} \frac{5\pi}{12} \\ &\approx 0.707 (\cos 75° + i \sin 75°) \\ &\approx 0.183 + 0.683\, i.\end{aligned}$$

Powers. If n is a positive integer, we may apply the product formulas, Eqs. (7), to find

$$z^n = z \cdot z \cdot z \cdot \cdots \cdots z \quad (n \text{ factors})$$

when

$$z = r \operatorname{cis} \theta.$$

Doing so, we obtain

$$\begin{aligned}(r \operatorname{cis} \theta)^n &= r^n \operatorname{cis} (\theta + \theta + \cdots + \theta) \quad (n \text{ summands}) \\ &= r^n \operatorname{cis} n\theta. \qquad (8)\end{aligned}$$

In terms of vectors, we see that the length $r = |z|$ is raised to the nth power and the angle $\theta = \arg z$ is multiplied by n.

In particular, if we take $r = 1$ in Eq. (8), we obtain *De Moivre's theorem:*

$$(\cos \theta + i \sin \theta)^n = \cos n\theta + i \sin n\theta. \qquad (9)$$

If we expand the left-hand side of this equation by the binomial theorem and reduce it to the standard form $a + ib$, we then obtain formulas for $\cos n\theta$ and for $\sin n\theta$ as polynomials of degree n in $\cos \theta$ and $\sin \theta$.

Example 3. If we take $n = 3$ in Eq. (9), we have

$$(\cos \theta + i \sin \theta)^3 = \cos 3\theta + i \sin 3\theta.$$

The left-hand side of this equation is

$$\cos^3 \theta + 3\, i \cos^2 \theta \sin \theta - 3 \cos \theta \sin^2 \theta - i \sin^3 \theta.$$

The real part of this must equal $\cos 3\theta$ and the imaginary part must equal $\sin 3\theta$; hence

$$\begin{aligned}\cos 3\theta &= \cos^3 \theta - 3 \cos \theta \sin^2 \theta, \\ \sin 3\theta &= 3 \cos^2 \theta \sin \theta - \sin^3 \theta.\end{aligned}$$

Roots. If $z = r \operatorname{cis} \theta$ is a complex number different from zero and n is a positive integer, then there are precisely n different complex numbers $w_0, w_1, \ldots, w_{n-1}$, each of which is an nth root of z. To see this, we let

$$w = \rho \operatorname{cis} \alpha$$

be an nth root of $z = r \operatorname{cis} \theta$, so that

$$w^n = z,$$

or

$$\rho^n \operatorname{cis} n\alpha = r \operatorname{cis} \theta. \qquad (10)$$

Then

$$\rho = \sqrt[n]{r} \qquad (11)$$

is the real, positive nth root of r. As regards the angle: Although we cannot say that $n\alpha$ and θ must be equal, we can say that they may differ only by an integral multiple of 2π. That is,

$$n\alpha = \theta + 2k\pi, \qquad k = 0, \pm 1, \pm 2, \ldots \qquad (12)$$

Therefore

$$\alpha = \frac{\theta}{n} + k \frac{2\pi}{n}.$$

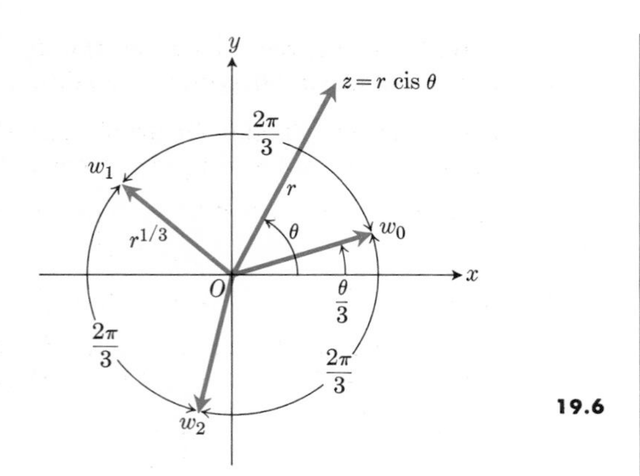

19.6

Hence all nth roots of $z = r \text{ cis } \theta$ are given by

$$\sqrt[n]{r \text{ cis } \theta} = \sqrt[n]{r} \text{ cis } \left(\frac{\theta}{n} + k\frac{2\pi}{n}\right), \qquad (13)$$
$$k = 0, \pm 1, \pm 2, \ldots$$

Remark 1. It might appear that there are infinitely many different answers corresponding to the infinitely many possible values of k. But one readily sees that $k = n + m$ gives the same answer as $k = m$ in Eq. (13). Thus we need only take n consecutive values for k to obtain all the different nth roots of z. For convenience, we may take

$$k = 0, 1, 2, \ldots, n - 1.$$

It is worth noting that all the nth roots of $r \text{ cis } \theta$ lie on a circle centered at the origin O and having radius equal to the real, positive nth root of r. One of them has argument $\alpha = \theta/n$. The others are uniformly spaced around the circumference of the circle, each being separated from its neighbors by an angle equal to $2\pi/n$. Figure 19.6 illustrates the placement of the three cube roots w_0, w_1, w_2 of the complex number $z = r \text{ cis } \theta$.

Problem. Find the four fourth roots of -16.

Solution. As our first step, we plot the given number in an Argand diagram, Fig. 19.7, and determine its polar representation $r \cos \theta$. Here

$$z = -16, \qquad r = +16, \qquad \theta = \pi.$$

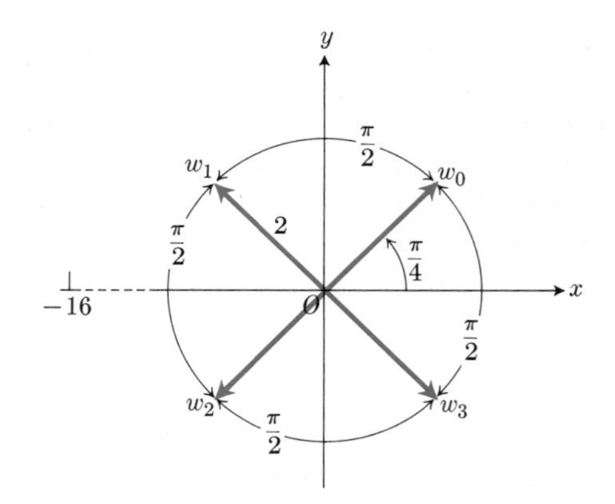

19.7

One of the fourth roots of $16 \text{ cis } \pi$ is $2 \text{ cis } (\pi/4)$. We obtain others by successive additions of $2\pi/4 = \pi/2$ to the argument of this first one. Hence

$$\sqrt[4]{16 \text{ cis } \pi} = 2 \text{ cis } \left(\frac{\pi}{4}, \frac{3\pi}{4}, \frac{5\pi}{4}, \frac{7\pi}{4}\right),$$

and the four roots are

$$w_0 = 2\left(\cos\frac{\pi}{4} + i \sin\frac{\pi}{4}\right) = \sqrt{2}(1 + i),$$

$$w_1 = 2\left(\cos\frac{3\pi}{4} + i \sin\frac{3\pi}{4}\right) = \sqrt{2}(-1 + i),$$

$$w_2 = 2\left(\cos\frac{5\pi}{4} + i \sin\frac{5\pi}{5}\right) = \sqrt{2}(-1 - i),$$

$$w_3 = 2\left(\cos\frac{7\pi}{4} + i \sin\frac{7\pi}{4}\right) = \sqrt{2}(1 - i).$$

Remark 2. The reader may well say that the invention of $\sqrt{-1}$ is all well and good and leads to a number system which is richer than the real number system alone. But where will this process end? Are we also going to invent still more systems so as to obtain $\sqrt[4]{-1}$, $\sqrt[6]{-1}$, and so on? By now it should be clear that this is not necessary. These numbers are already expressible in terms of the complex number system $a + ib$. In fact, the *Fundamental Theorem of Algebra* (which is rather difficult to prove; we only state it here) says that every polynomial equation of the form

$$a_0 z^n + a_1 z^{n-1} + a_2 z^{n-2} + \cdots + a_{n-1} z + a_n = 0$$

in which the coefficients $a_0, a_1, \ldots, a_n$ are any complex numbers, whose degree n is greater than or equal to one, and whose leading coefficient a_0 is not zero, possesses precisely n roots in the complex number system, provided that each multiple root of multiplicity m is counted as m roots.

EXERCISES

1. Show by an Argand diagram that the law for addition of complex numbers is the same as the parallelogram law for addition of vectors.
2. How may the following complex numbers be obtained from $z = x + iy$ geometrically? Sketch.

   (a) $\bar{z}$ (b) $(\overline{-z})$ (c) $-z$ (d) $1/z$

3. Show that the conjugate of the sum (product, or quotient) of two complex numbers z_1 and z_2 is the same as the sum (product, or quotient) of their conjugates.
4. (a) Extend the results of Exercise 3 to show that
$$f(\bar{z}) = \overline{f(z)}$$
if
$$f(z) = a_0 z^n + a_1 z^{n-1} + \cdots + a_{n-1} z + a_n$$
is a polynomial with real coefficients $a_0, a_1, \ldots, a_n$.
(b) If z is a root of the equation $f(z) = 0$, where $f(z)$ is a polynomial with real coefficients as in (a), show that the conjugate $\bar{z}$ is also a root of the equation. *Hint.* Let $f(z) = u + iv = 0$; then both u and v are zero. Now make use of the fact that
$$f(\bar{z}) = \overline{f(z)} = u - iv.$$
5. Show that $|\bar{z}| = |z|$.
6. If z and $\bar{z}$ are equal, what can you say about the location of the point z in the complex plane?
7. Let $R(z)$, $I(z)$ denote respectively the real and imaginary parts of z. Show that the following equations hold.

   (a) $z + \bar{z} = 2R(z)$ (b) $z - \bar{z} = 2iI(z)$
   (c) $|R(z)| \leq |z|$
   (d) $|z_1 + z_2|^2 = |z_1|^2 + |z_2|^2 + 2R(z_1\bar{z}_2)$
   (e) $|z_1 + z_2| \leq |z_1| + |z_2|$

8. Show that the distance between the two points z_1 and z_2 in an Argand diagram is equal to $|z_1 - z_2|$.

In Exercises 9 through 13, indicate graphically the locus of points $z = x + iy$ that satisfy the given conditions.

9. (a) $|z| = 2$ (b) $|z| < 2$ (c) $|z| > 2$
10. $|z - 1| = 2$
11. $|z + 1| = 1$
12. $|z + 1| = |z - 1|$
13. $|z + i| = |z - 1|$

Express the answers to Exercises 14 through 17 in the form r cis θ, with $r \geq 0$ and $-\pi < \theta \leq \pi$. Sketch.

14. $(1 + \sqrt{-3})^2$
15. $\dfrac{1+i}{1-i}$
16. $\dfrac{1 + i\sqrt{3}}{1 - i\sqrt{3}}$
17. $(2 + 3i)(1 - 2i)$
18. Use De Moivre's theorem to express $\cos 4\theta$ and $\sin 4\theta$ as polynomials in $\cos\theta$ and $\sin\theta$.
19. Find the three cube roots of unity.
20. Find the two square roots of i.
21. Find the three cube roots of $-8i$.
22. Find the six sixth roots of 64.
23. Find the four roots of the equation
$$z^4 - 2z^2 + 4 = 0.$$
24. Find the six roots of the equation
$$z^6 + 2z^3 + 2 = 0.$$
25. Find all roots of the equation
$$x^4 + 4x^2 + 16 = 0.$$
26. Solve
$$x^4 + 1 = 0.$$

19.3 THE COMPLEX VARIABLE

A set S of complex numbers $z = x + iy$ may be represented by points in an Argand diagram. For instance, S might be all complex z's for which $|z| \leq 1$. The corresponding points in the Argand diagram, or in the z-plane as it is called, then would be all points inside or on the circumference of a circle of radius 1 centered at O. (Fig. 19.8 shows the region $|z| < 1$ and circle $|z| = 1$.) During a discussion, we might wish to consider the symbol z as representing

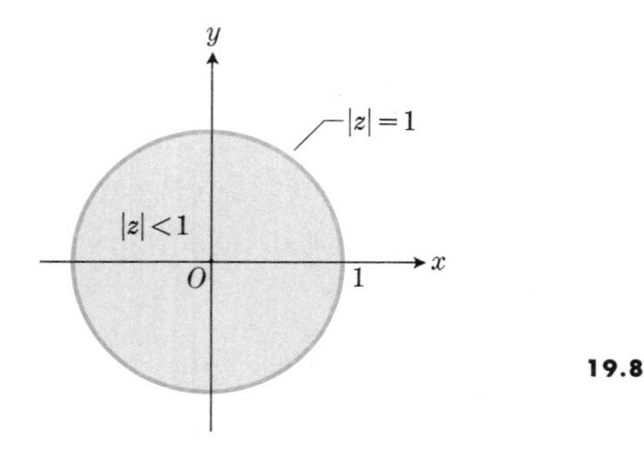

19.8

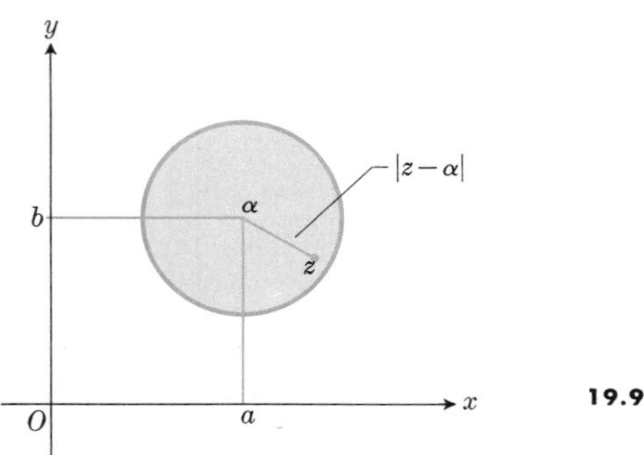

19.9

any one of the complex numbers in the set S. We would then think of z as a variable whose domain is S. Or, we might wish to consider a moving point which starts at time $t = 0$ from the point z_0 and moves continuously along some path in the z-plane as t increases to a second value, say $t = 1$. We would again consider the complex number z associated with this moving point (x, y) by the equation $z = x + iy$ to be a variable. This time it is a dependent variable, since its value depends on the value of t.

We say that the complex variable $z = x + iy$ approaches the *limit* $\alpha = a + ib$ if the *distance* between z and α approaches zero. (See Fig. 19.9.) That is,

$$z \longrightarrow \alpha \quad \text{if and only if} \quad |z - \alpha| \longrightarrow 0. \tag{1}$$

If we imagine z to be a function of time, we would say, that

$$\lim_{t \to 1} z = \alpha, \tag{2}$$

for example, provided it is true that

$$\lim_{t \to 1} |z - \alpha| = 0. \tag{3}$$

One way of interpreting Eq. (3) is to say that if it holds, then one may prescribe as small a circle about α as center as one pleases, and the point $z = x + iy$ will be inside that circle for all values of t sufficiently close to 1. That is, $|z - \alpha|$ is small when $|t - 1|$ is small. Since

$$\begin{aligned} |z - \alpha| &= |(x + iy) - (a + ib)| \\ &= |(x - a) + i(y - b)| \\ &= \sqrt{(x - a)^2 + (y - b)^2} \\ &\leq |x - a| + |y - b|, \end{aligned}$$

while both

$$|x - a| \leq |z - \alpha| \quad \text{and} \quad |y - b| \leq |z - \alpha|,$$

we see that

$$z \longrightarrow \alpha \quad \text{if and only if} \quad x \longrightarrow a \quad \text{and} \quad y \longrightarrow b. \tag{4}$$

This follows because both $|x - a|$ and $|y - b|$ are small when $|z - \alpha|$ is small; and, conversely, if both $|x - a|$ and $|y - b|$ are small, then $|z - \alpha|$ is also small.

Function

We say that w is a single-valued function of z on a domain S and write

$$w = f(z) \quad \text{for all } z \text{ in } S \tag{5}$$

if to each z in the set S there corresponds a complex number $w = u + iv$. For instance, S may be the set of all complex numbers and

$$w = z^2. \tag{6}$$

Then Eq. (6) produces a complex number $w = u + iv$ for each point $z = x + iy$ in the z-plane:

$$\begin{aligned} u + iv = (x + iy)^2 &= x^2 + 2ixy + i^2y^2 \\ &= (x^2 - y^2) + i(2xy). \end{aligned} \tag{7}$$

One way of representing such a function graphically is by a technique called *mapping*. For example, we may use Eq. (7) to map a vertical line $x = a$ into the w-plane. (See Fig. 19.10.) We would then ask this question: What does the image point w do as

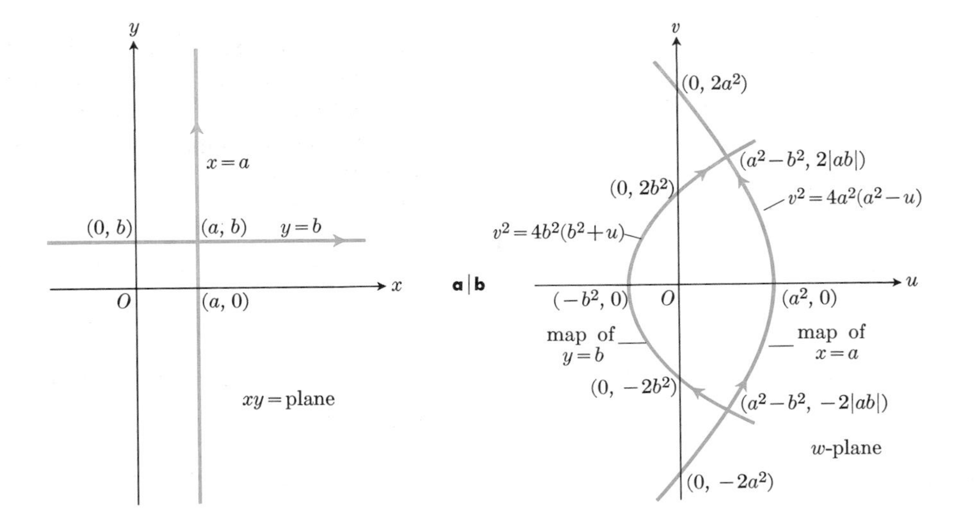

19.10

the z-point traverses the line $x = a$ from $y = -\infty$ to $y = +\infty$? To find out, we separate the real and imaginary parts of Eq. (7) and obtain

$$\begin{aligned} u &= x^2 - y^2 = a^2 - y^2, \\ v &= 2xy = 2ay. \end{aligned} \tag{8a}$$

These equations are parametric equations (in terms of the parameter y) of the parabola

$$v^2 = 4a^2(a^2 - u). \tag{8b}$$

For the case illustrated in Fig. 19.10, a is positive, and hence $v = 2ay$ has the same sign as y. Therefore, as the z-point moves up the line $x = a$ from $y = -\infty$ to $y = +\infty$, the w-point moves upward along the parabola (8b) in the direction indicated by the arrows in Fig. 19.10(b).

Similarly, the map in the w-plane of the line $y = b$ has parametric equations

$$\left.\begin{aligned} u &= x^2 - b^2, \\ v &= 2bx, \end{aligned}\right\} \quad -\infty < x < +\infty, \tag{9a}$$

which represent the parabola

$$v^2 = 4b^2(b^2 + u). \tag{9b}$$

Once again, see Fig. 19.10(b).

It is easily seen from (8a, b) that the line $x = -a$ maps into the same parabola (8b) as does the line $x = a$; but this time the parabola is described in the opposite sense as y varies from $-\infty$ to $+\infty$. Similarly, the line $y = -b$ maps into the same parabola as does the line $y = b$. These phenomena are to be expected, since the point z and the point $-z$ both map into the same point $w = (-z)^2$ in the w-plane.

Continuity

A function $w = f(z)$ that is defined throughout some neighborhood of the point $z = \alpha$ is said to be *continuous* at α if

$$|f(z) - f(\alpha)| \to 0 \quad \text{as} \quad |z - \alpha| \to 0. \tag{10}$$

Expressed in the language of mapping, the conditions in (10) say simply that the point w, which is the image of z, is near the point β, which is the image of α, when z is near α. That is, the image points are close to each other when the original points are. Another way of saying it is that when a small circle C is prescribed in the w-plane with β as its center, then it is possible to describe a circle C' in the z-plane with center at α and radius sufficiently small so that whenever z is inside C' the image point w is guaranteed to be inside C. (See Fig. 19.11.)

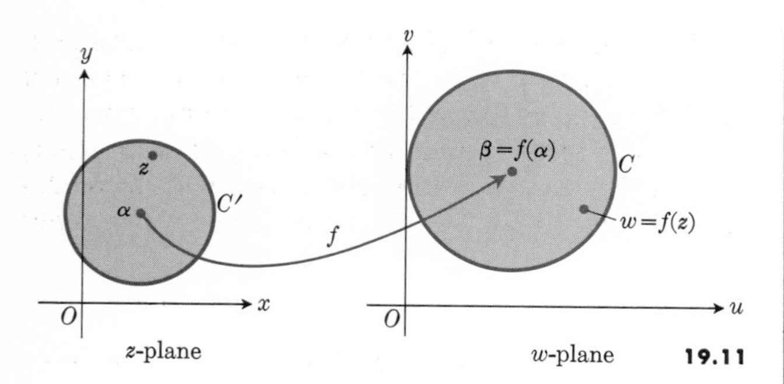

19.11

Problem. Show that $f(z) = z^2$ is continuous at any point $z = \alpha$.

Solution. We observe that

$$\begin{aligned}|f(z) - f(\alpha)| &= |z^2 - \alpha^2| \\ &= |(z-\alpha)(z+\alpha)| = |z-\alpha| \cdot |z+\alpha|.\end{aligned}$$

Now as $z \to \alpha$, we have

$$\lim_{z \to \alpha} (z + \alpha) = 2\alpha,$$

so that

$$\begin{aligned}\lim_{z\to\alpha} |f(z) - f(\alpha)| &= \lim_{z\to\alpha} |(z-\alpha)| \cdot |(z+\alpha)| \\ &= |2\alpha| \lim_{z\to\alpha} |z-\alpha| = 0,\end{aligned}$$

and hence the condition (10) is satisfied.

EXERCISES

1. For the function $w = z^2$ discussed in the text, sketch the maps in the w-plane of the following figures in the z-plane.

   Hint. Use polar coordinates.

   (a) $|z| = 1, \quad 0 \leq \arg z \leq \pi$
   (b) $|z| = 2, \quad \pi/2 \leq \arg z \leq \pi$
   (c) $\arg z = \pi/4$
   (d) $|z| < 1, \quad -\pi < \arg z \leq 0$
   (e) the x-axis
   (f) the y-axis

2. Show that the two parabolas in the w-plane in Fig. 19.10 intersect orthogonally if neither a nor b is zero.

3. Show that the function $w = z^3$ maps the wedge $0 \leq \arg z \leq \pi/3$ in the z-plane onto the upper half of the w-plane. Use polar coordinates and sketch.

4. Show that $f(z) = z^3$ is continuous at $z = \alpha$ for any α.

5. Show that $f(z) = 1/z$ is continuous at $z = \alpha$ if $\alpha \neq 0$.

19.4 DERIVATIVES

The derivative of a function $w = f(z)$ is defined in the same way as the derivative of a real-valued function of the real variable x. Namely, the derivative at $z = \alpha$ is

$$f'(\alpha) = \lim_{z\to\alpha} \frac{f(z) - f(\alpha)}{z - \alpha}, \tag{1}$$

provided that the limit exists. By saying that the limit in (1) exists, we mean, of course, that there is some complex number, which we have called $f'(\alpha)$, such that

$$\left| f'(\alpha) - \frac{f(z) - f(\alpha)}{z - \alpha} \right| \to 0 \quad \text{as} \quad |z - \alpha| \to 0. \tag{2}$$

Since z may approach α from *any direction* (how does this differ from the real variable case?), the existence of such a limit imposes a rather strong restriction on the function $w = f(z)$.

Example. The rather simple function

$$w = \bar{z} = f(z), \tag{3}$$

where

$$z = x + iy, \qquad \bar{z} = x - iy,$$

can be shown to have no derivative at any point. For if we take $\alpha = a + ib$, then

$$\begin{aligned}\frac{f(z) - f(\alpha)}{z - \alpha} &= \frac{\bar{z} - \bar{\alpha}}{z - \alpha} \\ &= \frac{(x-a) - i(y-b)}{(x-a) + i(y-b)}.\end{aligned} \tag{4}$$

Now from among the many different ways in which z might approach α, we shall single out for special attention the following two:

(a) z approaches α along the line $y = b$;
(b) z approaches α along the line $x = a$.

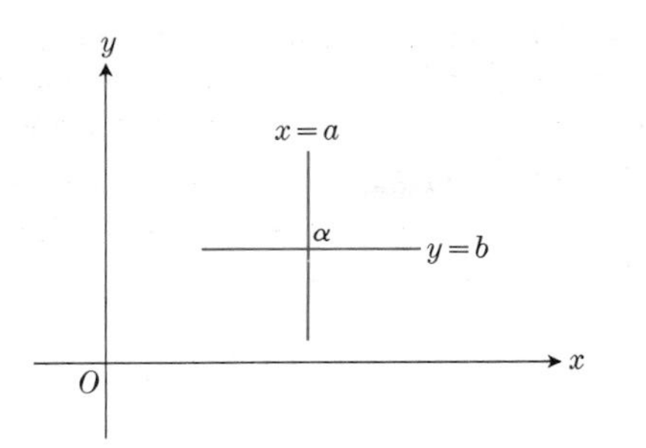

19.12

(See Fig. 19.12.) In the first case, consequently, we take $y = b$ and let $x \to a$. Then Eq. (4) becomes

$$y = b:\quad \frac{f(z) - f(\alpha)}{z - \alpha} = \frac{x - a}{x - a} = +1,$$

and hence

$$\lim_{\substack{x \to a \\ y = b}} \frac{f(z) - f(\alpha)}{z - \alpha} = +1. \tag{5a}$$

In the second case, we take $x = a$ and let $y \to b$. This time Eq. (4) becomes

$$x = a:\quad \frac{f(z) - f(\alpha)}{z - \alpha} = \frac{-i(y - b)}{i(y - b)} = -1,$$

and hence

$$\lim_{\substack{x = a \\ y \to b}} \frac{f(z) - f(\alpha)}{z - \alpha} = -1. \tag{5b}$$

Since these two different paths along which z may approach α lead to two different answers, there is no single complex number which we can call $f'(\alpha)$ in Eq. (1) or (2). That is, the function $w = \bar{z}$ does not possess a derivative. *Question.* Is the function $w = \bar{z}$ *continuous* for some, none, or all values of z? Can you justify your answer?

In terms of the Δ-notation, we may write

$$z - \alpha = \Delta z, \quad f(z) - f(\alpha) = \Delta w,$$

and say that $w = f(z)$ has a derivative

$$\frac{dw}{dz} = \lim_{\Delta z \to 0} \frac{\Delta w}{\Delta z}, \tag{6}$$

provided that the limit exists and is independent of the manner in which Δz approaches 0.

The formulas for differentiating sums, products, quotients, and powers are the same for the complex variable z as for the real variable x. In other words, if c is any complex constant, n is a positive integer, and if $f(z)$ and $g(z)$ are functions which have derivatives at $z = \alpha$, then the following formulas apply at $z = \alpha$.

Derivatives

i. $\dfrac{d}{dz} c = 0$ ii. $\dfrac{d}{dz} cf(z) = c \dfrac{d}{dz} f(z)$

iii. $\dfrac{d}{dz} [f(z) + g(z)] = \dfrac{df(z)}{dz} + \dfrac{dg(z)}{dz}$

iv. $\dfrac{d}{dz} [f(z)g(z)] = f(z) \dfrac{dg(z)}{dz} + g(z) \dfrac{df(z)}{dz}$

v. $\dfrac{d}{dz} \dfrac{f(z)}{g(z)} = \dfrac{g(z) \dfrac{df(z)}{dz} - f(z) \dfrac{dg(z)}{dz}}{[g(z)]^2}, \quad g(z) \neq 0$

vi. (a) $\dfrac{d}{dz} [f(z)]^n = n[f(z)]^{n-1} \dfrac{df(z)}{dz}$

(b) $\dfrac{d}{dz} z^n = nz^{n-1}$

These formulas may all be derived rather easily if the Δ-notation if one first proves that the limit of a sum, product, or quotient of two complex functions is the sum, product, or quotient of their limits (always, of course, excluding division by zero), provided that the individual limits exist. We shall omit these proofs and content ourselves here with a simple specific illustration.

Problem. Show that $d(z^3)/dz = 3z^2$.

Solution. Let $w = z^3$. Then

$$w + \Delta w = z^3 + 3z^2 \Delta z + 3z (\Delta z)^2 + (\Delta z)^3,$$

$$\frac{\Delta w}{\Delta z} = 3z^2 + 3z (\Delta z) + (\Delta z)^2,$$

and

$$\left| \frac{\Delta w}{\Delta z} - 3z^2 \right| = |3z + \Delta z| \cdot |\Delta z| \to 0 \tag{7}$$

as $\Delta z \to 0$. Therefore

$$\lim_{\Delta z \to 0} \frac{\Delta w}{\Delta z} = 3z^2;$$

that is,

$$\frac{d(z^3)}{dz} = 3z^2.$$

Note that it makes no difference *how* Δz approaches zero in Eq. (7). When $|\Delta z|$ is small, the whole right-hand side of the equation is small.

EXERCISES

In Exercises 1 through 3, find the derivative with respect to z of the given function at the given point z_0.

1. $\dfrac{z+1}{z-1}$, $z_0 = 1 + i$
2. $z^3 + 3z^2 + 3z + 2$, $z_0 = -1 + 2i$
3. $\sqrt{z^2 + 1}$, $z_0 = (1 + i)/\sqrt{2}$

Note. Here we get *two* answers, depending on our choice of the square root.

4. Use the definition in Eq. (1) to find $f'(\alpha)$ if

$$f(z) = 1/z \quad \text{and} \quad \alpha \neq 0.$$

19.5 CAUCHY-RIEMANN DIFFERENTIAL EQUATIONS

Suppose the complex function

$$w = u + iv = f(z)$$

is differentiable at the point $\alpha = a + ib$. Then, by making z approach α, once along the line $y = b$ (that is, taking $\Delta y = 0$ and making Δx approach 0) and then along the line $x = a$ (that is, taking $\Delta x = 0$ and making Δy approach 0), we quickly learn that the equations

$$\frac{\partial u}{\partial x} = \frac{\partial v}{\partial y} \quad \text{and} \quad \frac{\partial v}{\partial x} = -\frac{\partial u}{\partial y} \tag{1}$$

must be satisfied at the point (a, b).

Remark. To show this, let us assume that dw/dz does exist at $z = \alpha$. We then have

$$f'(\alpha) = \lim_{\substack{\Delta y = 0 \\ \Delta x \to 0}} \frac{\Delta u + i\,\Delta v}{\Delta x + i\,\Delta y} = \lim_{\Delta x \to 0} \left(\frac{\Delta u}{\Delta x} + i\frac{\Delta v}{\Delta x}\right)$$
$$= \left(\frac{\partial u}{\partial x} + i\frac{\partial v}{\partial x}\right)_{z=\alpha}, \tag{2a}$$

and also

$$f'(\alpha) = \lim_{\substack{\Delta x = 0 \\ \Delta y \to 0}} \frac{\Delta u + i\,\Delta v}{\Delta x + i\,\Delta y} = \lim_{\Delta y \to 0} \left(\frac{\Delta u}{i\,\Delta y} + \frac{\Delta v}{\Delta y}\right)$$
$$= \left(\frac{1}{i}\frac{\partial u}{\partial y} + \frac{\partial v}{\partial y}\right)_{z=\alpha}$$
$$= \left(-i\frac{\partial u}{\partial y} + \frac{\partial v}{\partial y}\right)_{z=\alpha}. \tag{2b}$$

We need now only equate the real and imaginary parts of these two expressions for $f'(\alpha)$, Eqs. (2a) and (2b), to obtain the results in Eq. (1).

The relationships in Eqs. 1 connect the four partial derivatives of u and v with respect to x and y. They are known as the *Cauchy-Riemann* differential equations. We have just shown that they must be satisfied at any point where $w = f(z)$ has a derivative. Thus we cannot, in general, specify the functions $u = u(x, y)$ and $v = v(x, y)$ independently and then hope that the resulting function $w = u + iv$ is differentiable with respect to $z = x + iy$. However, if we take functions which do satisfy the Cauchy-Riemann equations and which, in addition, have *continuous* partial derivatives u_x, u_y, v_x, v_y, then it is true (but we shall not prove it here) that the resulting function $w = u + iv$ is differentiable with respect to z. In a sense, this says that if the derivatives as calculated along the *two* directions $x = a$ and $y = b$ are equal, and if the partial derivatives $u_x, \ldots$ are *continuous*, then one will also get the same answer for $f'(\alpha)$ along *all* directions.

If a function $w = f(z)$ has a derivative at every point of some region G in the z-plane, then the function is said to be *analytic* in G. If a function fails to have a derivative at one point α but does have a derivative everywhere else in a region G, we still say that it is analytic in G *except* at α and say that α is a *singular* point of the function. Thus, for example, a rational function $f(z)/g(z)$, where $f(z)$ and $g(z)$ are polynomials, is analytic everywhere except at those points where the denominator is zero. For all points where $g(z) \neq 0$, the function has a derivative

$$\frac{g(z)f'(z) - f(z)g'(z)}{[g(z)]^2}.$$

Problem. Show that the real and imaginary parts of the function $w = 1/z$ satisfy the Cauchy-Riemann equations at all points where $z \neq 0$.

Solution. Let

$$w = u + iv = \frac{1}{x + iy} = \frac{x - iy}{x^2 + y^2},$$

so that

$$u = \frac{x}{x^2 + y^2}, \quad v = \frac{-y}{x^2 + y^2}; \quad x^2 + y^2 \neq 0.$$

Then we find, by calculating the partial derivatives, that

$$\frac{\partial u}{\partial x} = \frac{y^2 - x^2}{(x^2 + y^2)^2} = \frac{\partial v}{\partial y},$$

$$\frac{\partial v}{\partial x} = \frac{2xy}{(x^2 + y^2)^2} = -\frac{\partial u}{\partial y},$$

so that the Cauchy-Riemann equations are satisfied at all points where $x^2 + y^2 \neq 0$, that is, where $z \neq 0$.

EXERCISES

Find the real and imaginary parts of the functions

$$w = f(z), \quad w = u + iv, \quad z = x + iy,$$

and show that they satisfy the Cauchy-Riemann equations.

1. z^2 2. z^3 3. z^4 4. $1/z^2$, $z \neq 0$

5. If the partial derivatives of first and second order of the real and imaginary parts $u = u(x, y)$, $v = v(x, y)$ of an analytic function $w = f(z)$ are continuous, show that

$$\frac{\partial^2 u}{\partial x^2} + \frac{\partial^2 u}{\partial y^2} = 0, \quad \frac{\partial^2 v}{\partial x^2} + \frac{\partial^2 v}{\partial y^2} = 0.$$

6. Verify that the equations in Exercise 5 are satisfied by the real and imaginary parts of these functions.

(a) z (b) z^2 (c) z^3

19.6 COMPLEX SERIES

The simplest functions of the complex variable $z = x + iy$ are polynomials and rational functions (ratios of polynomials) in z. These have been briefly discussed above. We might now ask whether or not it is possible to make useful definitions of some other elementary functions, such as $\sin z$, $\cos z$, e^z, $\cosh z$, and so on. It is hard for us to imagine $\sin (2 + 3i)$, for example, if we try to think of $2 + 3i$ as meaning an angle. In fact, we may have had trouble thinking of the meaning of the simpler expression $\sin 2$. Of course, in the latter case, we ask ourselves "2 what?" and then answer "2 radians," and then we can translate this into degrees and look it up in a table. But how was the table itself constructed? Why, by means of the series

$$\sin x = x - \frac{x^3}{3!} + \frac{x^5}{5!} - \frac{x^7}{7!} + \cdots. \tag{1}$$

To use the series to calculate $\sin 2$, for example, we don't need to think of radians at all, but just take $x = 2$ as a pure number. To be sure, certain basic trigonometric identities for $\sin (x + y)$, $\cos (x + y)$, and so on are also used in constructing tables, but the series in Eq. (1) is the basic thing.

Now, we may ask ourselves, why can't we go ahead and define $\sin z = \sin (x + iy)$ by a series like (1) but having z in place of x? The answer is, of course, we can do this if we want to and if the series converges. We are therefore led to investigate power series in z, such as the power series

$$\sin z = z - \frac{z^3}{3!} + \frac{z^5}{5!} - \frac{z^7}{7!} + \cdots. \tag{2}$$

When y is zero, then $z = x + iy = x$, and the series in (2) is the same as the series in (1). Thus it would not be inconsistent for us to extend the domain of definition of $\sin z$ from the real axis into the complex plane by means of the series.

Convergence

We say that a power series

$$\sum_{n=0}^{\infty} a_n z^n = a_0 + a_1 z + a_2 z^2 + \cdots \tag{3}$$

converges at a point z if the sequence of partial sums

$$\begin{aligned} s_0 &= a_0, \\ s_1 &= a_0 + a_1 z, \\ &\vdots \\ s_n &= a_0 + a_1 z + \cdots + a_n z^n \end{aligned} \tag{4}$$

tends to a limit as n becomes infinite. If we separate s_n into its real and imaginary parts,

$$s_n = u_n(x, y) + i v_n(x, y), \tag{5}$$

then

$$s_n \to u + iv$$

if and only if

$$u_n \to u \quad \text{and} \quad v_n \to v. \tag{6}$$

This follows from the fact that

$$\begin{aligned} |s_n - (u + iv)| &= |(u_n - u) + i(v_n - v)| \\ &= \sqrt{(u_n - u)^2 + (v_n - v)^2} \end{aligned}$$

approaches zero if and only if

$$u_n - u \to 0 \quad \text{and} \quad v_n - v \to 0.$$

We say that the series (3) *converges absolutely* if and only if the corresponding series of absolute values

$$\sum_{n=0}^{\infty} |a_n z^n| = |a_0| + |a_1 z| + |a_2 z^2| + \cdots \tag{7}$$

converges. Since this is a series of nonnegative real numbers, we are already familiar with certain tests (the comparison test, integral test, and ratio test) for determining whether or not it converges. Then, if (7) does converge, the following theorem tells us that (3) also converges.

Theorem 1. *If a series $\sum_{n=0}^{\infty} a_n z^n$ converges absolutely, then it converges.*

Proof. Separate each term of the series into its real and imaginary parts, say

$$a_k z^k = c_k + i d_k, \quad k = 0, 1, 2, \ldots, \tag{8}$$

where c_k and d_k are real. Then if the series (7) converges, we use the facts that

$$\begin{aligned} |c_k| &\le \sqrt{c_k^2 + d_k^2} = |a_k z^k|, \\ |d_k| &\le \sqrt{c_k^2 + d_k^2} = |a_k z^k|, \end{aligned}$$

and the comparison test for nonnegative real series, to tell us that

$$\sum |c_k| \quad \text{and} \quad \sum |d_k|$$

both converge. But from this we may also conclude that the two series without absolute value signs,

$$\sum c_k \quad \text{and} \quad \sum d_k,$$

both converge. Hence

$$\sum (c_k + i d_k) = \sum c_k + i \sum d_k$$

also converges. □

Problem. Show that the series for $\sin z$ in Eq. (2) converges for all complex numbers z, $|z| < \infty$.

Solution. We test (2) for *absolute* convergence and examine

$$\sum_{k=1}^{\infty} \left| \frac{(-1)^{k-1} z^{2k-1}}{(2k-1)!} \right|. \tag{9}$$

As is usual when we have a power series, we apply the ratio test. We take

$$U_n = \left| \frac{(-1)^{n-1} z^{2n-1}}{(2n-1)!} \right|$$

and calculate

$$\frac{U_{n+1}}{U_n} = \frac{|z^2|}{2n(2n+1)}.$$

For *any* fixed z such that $|z| < \infty$, we then find

$$\lim_{n \to \infty} \frac{U_{n+1}}{U_n} = \lim_{n \to \infty} \frac{|z^2|}{2n(2n+1)} = 0.$$

Since this limit is less than unity, series (9) converges. That is, series (2) converges *absolutely*. Hence, by our theorem, it also converges without the absolute value signs.

EXERCISES

Find the regions in which the series in Exercises 1 through 5 converge absolutely. Sketch the regions of absolute convergence.

1. $1 + z + z^2 + z^3 + \cdots$
2. $1 - 2z + 3z^2 - 4z^3 + \cdots$
3. $(z - 1) - \dfrac{(z-1)^2}{2} + \dfrac{(z-1)^3}{3} - \dfrac{(z-1)^4}{4} + \cdots$
4. $(z + i) + \dfrac{(z+i)^2}{2} + \dfrac{(z+i)^3}{3} + \dfrac{(z+i)^4}{4} + \cdots$
5. $1 - \dfrac{(z+1)}{2} + \dfrac{(z+1)^2}{4} - \dfrac{(z+1)^3}{8} + \dfrac{(z+1)^4}{16} - \cdots$
6. Show that the series $\sum z^k/k!$ converges absolutely for all $|z| < \infty$.
7. Show that the series $\sum(-1)^k z^{2k}/(2k)!$ converges absolutely for all $|z| < \infty$.
8. Show that the series $\sum (-1)^{k-1} z^k/k$ converges absolutely for $|z| < 1$.
9. Use the series (2) to compute sin 2 to three decimals.

19.7 CERTAIN ELEMENTARY FUNCTIONS

In addition to defining sin z, we may also define other functions by means of series. Thus, extending formulas which we developed in the case of real-valued functions of x, we define the following functions by the series we get when we substitute z for x:

$$e^z = 1 + z + \frac{z^2}{2!} + \frac{z^3}{3!} + \cdots = \sum_{k=0}^{\infty} \frac{z}{k!}, \quad (1)$$

$$\sin z = z - \frac{z^3}{3!} + \frac{z^5}{5!} - \frac{z^7}{7!} + \cdots = \sum_{k=1}^{\infty} \frac{(-1)^{k-1} z^{2k-1}}{(2k-1)!}, \quad (2)$$

$$\cos z = 1 - \frac{z^2}{2!} + \frac{z^4}{4!} - \frac{z^6}{6!} + \cdots = \sum_{k=0}^{\infty} \frac{(-1)^k z^{2k}}{(2k)!}, \quad (3)$$

$$\tan^{-1} z = z - \frac{z^3}{3} + \frac{z^5}{5} - \frac{z^7}{7} + \cdots = \sum_{k=1}^{\infty} \frac{(-1)^{k-1} z^{2k-1}}{2k-1}, \quad (4)$$

$$\ln(1 + z) = z - \frac{z^2}{2} + \frac{z^3}{3} - \frac{z^4}{4} + \cdots = \sum_{k=1}^{\infty} \frac{(-1)^{k-1} z^k}{k}. \quad (5)$$

It is easy to show by the ratio test that the first three of these series converge absolutely for $|z| < \infty$. The last two converge absolutely if z is inside the unit circle $|z| < 1$, and diverge if $|z| > 1$. We realize that the inverse tangent function is multiple-valued, and the series in (4) gives the so-called *principal value* of $\tan^{-1} z$. We shall show that the logarithm of a complex number is also multiple-valued, and the series in (5) gives the principal value of $\ln(1 + z)$ when $|z| < 1$.

It is a basic theorem in the theory of functions of a complex variable that a power series $\sum a_k z^k$ either

(a) converges only at $z = 0$, or

(b) converges inside a circle $|z| < R$, or

(c) converges for all z, $|z| < \infty$.

The second case occurs when the function represented by the power series is analytic everywhere inside the circle $|z| < R$, but has a singularity on the circle $|z| = R$. In this case, the theorem tells us that the *largest* circle throughout which the series will converge has radius R equal to the distance from $z = 0$ to the nearest singular point of the function. Thus, for example,

$$\frac{1}{1 + z^2} = 1 - z^2 + z^4 - z^6 + \cdots \quad (6)$$

converges in a circle of radius $R = 1$, since the singularities of the function occur at $z = \pm i$

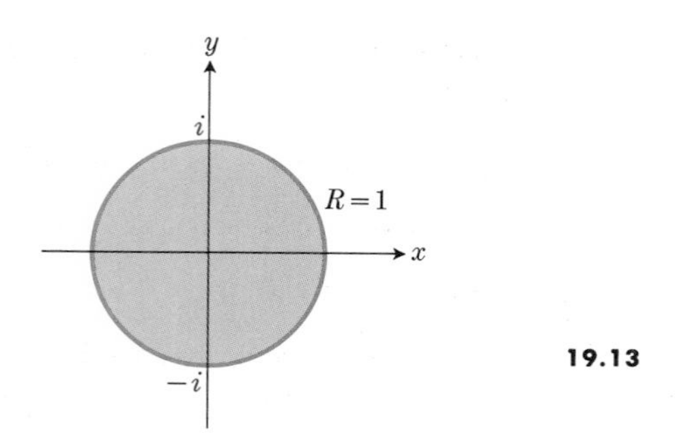

19.13

which are at distance $R = 1$ from the origin. (See Fig. 19.13.) We call the circle $|z| = 1$ the "circle of convergence," by which we mean that the series converges for all z *inside* this circle and diverges for any z *outside* it. Behavior *on* the circle of convergence constitutes a difficult problem which we are not prepared to discuss.

Certain basic properties of the elementary functions can be shown to extend from the domain of real variables x to complex variables z by algebraic manipulation with series. For example, we can prove the following theorem by substituting the series definitions of the various terms.

Theorem 2. $$e^{z_1} \cdot e^{z_2} = e^{z_1+z_2}. \tag{7}$$

Proof. We multiply the series for e^{z_1} by the series for e^{z_2} and collect terms of like degree. By the *degree* of a term

$$z_1^p z_2^q,$$

we mean simply the sum of the exponents $p + q$. We shall focus our attention on all those terms in the product that are of a certain degree, say n. Then we may safely ignore all powers

$$z_1^{n+1},\ z_1^{n+2},\ \ldots \qquad \text{and} \qquad z_2^{n+1},\ z_2^{n+2},\ \ldots,$$

since these would only enter in terms of degree greater than n. Now we want all terms of the form $z_1^k z_2^{n-k}$, with $k = 0, 1, 2, \ldots, n$. When we multiply

$$\left[e^{z_1} = 1 + z_1 + \frac{1}{2!}z_1^2 + \frac{1}{3!}z_1^3 + \cdots + \frac{1}{n!}z_1^n + \cdots\right] \times \left[e^{z_2} = 1 + z_2 + \frac{1}{2!}z_2^2 + \frac{1}{3!}z_2^3 + \cdots + \frac{1}{n!}z_2^n + \cdots\right],$$

the following terms are of degree n:

$$\frac{1}{n!}z_1^n \cdot 1 + \frac{1}{(n-1)!}z_1^{n-1} \cdot z_2 + \frac{1}{(n-2)!}z_1^{n-2} \cdot \frac{z_2^2}{2!} + \cdots + \frac{1}{(n-k)!}z_1^{n-k} \cdot \frac{1}{k!}z_2^k + \cdots + 1 \cdot \frac{1}{n!}z_2^n$$
$$= \sum_{k=0}^{n} \frac{1}{(n-k)!\,k!} z_1^{n-k} z_2^k. \tag{8}$$

On the other hand, the terms of degree n in the series for

$$e^{z_1+z_2} = 1 + (z_1 + z_2) + \frac{1}{2!}(z_1 + z_2)^2 + \cdots + \frac{1}{n!}(z_1 + z_2)^n + \cdots$$

are in the one binomial expression

$$\frac{1}{n!}(z_1+z_2)^n = \frac{1}{n!}\left[z_1^n + nz_1^{n-1}z_2 + \frac{n(n-1)}{2!}z_1^{n-2}z_2^2 + \cdots + \frac{n(n-1)\cdots(n-k+1)}{k!}z_1^{n-k}z_2^k + \cdots + z_2^n\right].$$

By noting that

$$\frac{n(n-1)\cdots(n-k+1)}{k!} \cdot \frac{(n-k)!}{(n-k)!} = \frac{n!}{k!(n-k)!},$$

we may also write this in summation form as

$$\frac{1}{n!}(z_1 + z_2)^n = \frac{1}{n!}\sum_{k=0}^{n} \frac{n!}{k!(n-k)!} z_1^{n-k} z_2^k = \sum_{k=0}^{n} \frac{1}{(n-k)!\,k!} z_1^{n-k} z_2^k. \tag{9}$$

A comparison of Eqs. (8) and (9) shows us that we have precisely the same terms of degree n in the product of the series for e^{z_1} and e^{z_2} that we have in the series for $e^{z_1+z_2}$. The above argument is valid for any positive integer n. For $n = 0$, the terms of degree zero are

$$z_1^0 \cdot z_2^0 = 1 \qquad \text{and} \qquad (z_1 + z_2)^0 = 1,$$

respectively, and these again are equal. Thus the terms of degree $n (n = 0, 1, 2, 3, \ldots)$ are the same in the product of the series for e^{z_1} and e^{z_2} as they are in the series for $e^{z_1+z_2}$; that is,

$$(\text{series for } e^{z_1}) \cdot (\text{series for } e^{z_2}) = (\text{series for } e^{z_1+z_2}). \; \square$$

Note. It is not always true that one may multiply two series together and then arrange the terms in groups which are then to be added in the most convenient way. For instance, in multiplying the series for e^{z_1} and e^{z_2}, we might have wanted to write the answer as

$$\begin{aligned} 1 \cdot &\left(1 + z_2 + \frac{z_2^2}{2!} + \cdots\right) \\ &+ z_1\left(1 + z_2 + \frac{z_2^2}{2!} + \cdots\right) + \cdots \\ &+ \frac{z_1^n}{n!}\left(1 + z_2 + \frac{z_2^2}{2!} + \cdots\right) + \cdots, \end{aligned}$$

where we agree to add all terms not containing z_1, then all terms containing only the first power of z_1, then all terms containing only the second power of z_1, and so on. This certainly isn't convenient to do if we are trying to show that the result is the same as

$$1 + (z_1 + z_2) + \frac{(z_1 + z_2)^2}{2!} + \frac{(z_1 + z_2)^3}{3!} + \cdots,$$

and conceivably it might lead to a different answer. However, it is shown in more advanced courses in analysis that when both the series in a product are *absolutely convergent*, then it is permissible to arrange the terms in any way one wishes, provided one ultimately takes all terms into account. The series for e^z does converge absolutely for all values of z. Hence the series for e^{z_1} and e^{z_2} satisfy the requirements of this theorem, and it is permissible to arrange the terms according to ascending degree, as we have done above.

By similar operations with power series, one can show that

$$\begin{aligned} \sin(z_1 + z_2) &= \sin z_1 \cos z_2 + \cos z_1 \sin z_2, \\ \cos(z_1 + z_2) &= \cos z_1 \cos z_2 - \sin z_1 \sin z_2, \\ \sin^2 z + \cos^2 z &= 1. \end{aligned}$$

One of the most famous results involving the elementary complex functions is the formula

$$e^{iz} = \cos z + i \sin z, \tag{10}$$

which is known as *Euler's formula.* To establish Eq. (10), we simply substitute iz in place of z in the series (1). The various powers of i that enter can all be reduced to one of the four numbers

$$i, \quad -1, \quad -i, \quad +1$$

by observing that

$$\begin{aligned} i^2 &= -1, \\ i^3 &= i^2 \cdot i = -i, \\ i^4 &= (i^2)^2 = (-1)^2 = +1, \\ i^5 &= i^4 \cdot i = i, \end{aligned}$$

and so on. In fact, if n is any integer, then

$$i^{4n} = +1,\ i^{4n+1} = i,\ i^{4n+2} = -1,\ i^{4n+3} = -i.$$

Thus we have

$$\begin{aligned} e^{iz} &= 1 + iz + \frac{(iz)^2}{2!} + \frac{(iz)^3}{3!} + \frac{(iz)^4}{4!} + \frac{(iz)^5}{5!} + \cdots \\ &= \left(1 - \frac{z^2}{2!} + \frac{z^4}{4!} - \frac{z^6}{6!} + \cdots\right) \\ &\quad + i\left(z - \frac{z^3}{3!} + \frac{z^5}{5!} - \frac{z^7}{7!} + \cdots\right) \\ &= \cos z + i \sin z, \end{aligned}$$

where we have recognized the series for $\sin z$ and $\cos z$, Eqs. (2) and (3).

If we make use of Euler's formula (10) and the companion equation

$$e^{-iz} = \cos z - i \sin z \tag{11}$$

which results from replacing z by $-z$ in (10), we may express the trigonometric functions of z in terms of exponentials. For example, if we add Eqs. (10) and (11) and then divide by two, we obtain

$$\cos z = \frac{1}{2}(e^{iz} + e^{-iz}). \tag{12a}$$

On the other hand, if we subtract (11) from (10), we obtain

$$\sin z = \frac{1}{2i}(e^{iz} - e^{-iz}). \tag{12b}$$

The other trigonometric functions of z are defined in the usual way in terms of $\sin z$ and $\cos z$: For example,

$$\tan z = \frac{\sin z}{\cos z} = \frac{1}{i}\frac{e^{iz} - e^{-iz}}{e^{iz} + e^{-iz}}.$$

The usual trigonometric identities can be established by expressing the trigonometric functions as exponentials and then making use of the basic identity (7).

Example. Suppose we wish to show that

$$\cos^2 z + \sin^2 z = 1.$$

We simply square both sides of Eqs. (12a, b) and add:

$$\begin{aligned}\cos^2 z + \sin^2 z &= \tfrac{1}{4}(e^{2iz} + 2e^0 + e^{-2iz}) \\ &\quad - \tfrac{1}{4}(e^{2iz} - 2e^0 + e^{-2iz}) \\ &= \tfrac{1}{4}(4e^0) = 1.\end{aligned}$$

Equations (12a, b) show the very intimate relationship between the circular functions and the hyperbolic functions: If, as in the real case, we define

$$\cosh z = \tfrac{1}{2}(e^z + e^{-z}) = 1 + \frac{z^2}{2!} + \frac{z^4}{4!} + \frac{z^6}{6!} + \cdots,$$

$$\sinh z = \tfrac{1}{2}(e^z - e^{-z}) = z + \frac{z^3}{3!} + \frac{z^5}{5!} + \frac{z^7}{7!} + \cdots,$$

then Eqs. (12a, b) say that

$$\cos z = \cosh iz, \tag{13a}$$

$$i \sin z = \sinh iz. \tag{13b}$$

These relationships explain the similarity in form between the identities of circular trigonometry, such as

$$\cos^2 z + \sin^2 z = 1, \tag{14a}$$

and the corresponding identities of hyperbolic trigonometry, such as

$$\cosh^2 u - \sinh^2 u = 1. \tag{14b}$$

In fact, any identity in circular functions produces a corresponding identity in hyperbolic functions, provided $\sin^2$ is replaced by $-\sinh^2$. The presence of the minus sign is a consequence of the i in Eq. (13b).

EXERCISES

1. Take $z = (1 + i)/\sqrt{2}$ in Eq. (1), and calculate an approximation to $e^{(1+i)\sqrt{2}}$ by using the first four terms of the series.
2. Express the sines and cosines in the identity
$$\sin(A + B) = \sin A \cos B + \cos A \sin B$$
in terms of exponentials and thereby show that this identity is a consequence of the exponential law $e^{z_1} \cdot e^{z_2} = e^{z_1 + z_2}$.
3. By differentiating the appropriate series, Eqs. (1) through (5), term by term, demonstrate these results.
(a) $\dfrac{de^z}{dz} = e^z$ (b) $\dfrac{d \sin z}{dz} = \cos z$
(c) $\dfrac{d \cos z}{dz} = -\sin z$ (d) $\dfrac{d \tan^{-1} z}{dz} = \dfrac{1}{1 + z^2}$
(e) $\dfrac{d \ln(1 + z)}{dz} = \dfrac{1}{1 + z}$
4. (a) Show that $y = e^{i\omega x}$ and $y = e^{-i\omega x}$ are solutions of the differential equation
$$\frac{d^2y}{dx^2} + \omega^2 y = 0.$$
(b) Show that $y = e^{(a+ib)x}$ and $y = e^{(a-ib)x}$ are solutions of the differential equation
$$\frac{d^2y}{dx^2} - 2a\frac{dy}{dx} + (a^2 + b^2)y = 0.$$

(c) Assuming that a and b in (b) are real, show that both

$$e^{ax}\cos bx = R(e^{(a+ib)x})$$

and

$$e^{ax}\sin bx = I(e^{(a+ib)x})$$

are solutions of the given differential equation.

5. Find a value of m such that $y = e^{mx}$ is a solution of each differential equation.

(a) $\dfrac{d^2y}{dx^2} + 2\dfrac{dy}{dx} + 5y = 0$

(b) $\dfrac{d^4y}{dx^4} + 4y = 0$ (c) $\dfrac{d^3y}{dx^3} - 8y = 0$

6. Show that the point (x, y) describes a unit circle with angular velocity ω if $z = x + iy = e^{i\omega t}$ and ω is a real constant.

7. If x and y are real, show that $|e^{x+iy}| = e^x$.

8. Show that the real and imaginary parts of $w = e^z$ satisfy the Cauchy-Riemann equations.

9. Show by reference to the appropriate series that

(a) $\sin(iz) = i\sinh z$, (b) $\cos(iz) = \cosh z$.

10. Show, by reference to the results of Exercises 2 and 9, that

$$\sin(x+iy) = \sin x\cosh y + i\cos x\sinh y,$$

and find a value of z such that $\sin z = 2$.

11. Show that the real and imaginary parts of $w = \sin z$, Exercise 10, satisfy the Cauchy-Riemann equations.

12. Show that

$$|\sin(x+iy)| = \sqrt{\sin^2 x + \sinh^2 y}$$

if x and y are real.

13. Let $z = x + iy$, $w = u + iv$, $w = \sin z$. Show that the line segment $y =$ constant, $-\pi < x \leq +\pi$, in the z-plane maps into the ellipse

$$\frac{u^2}{\cosh^2 y} + \frac{v^2}{\sinh^2 y} = 1$$

in the w-plane.

14. Show that the only complex roots $z = x + iy$ (x and y real) of the equation $\sin z = 0$ are at points on the real axis ($y = 0$) at which $\sin x = 0$.

15. Show that

$$\cosh(x+iy) = \cosh x\cos y + i\sinh x\sin y.$$

16. Show that

$$|\cosh(x+iy)| = \sqrt{\cos^2 y + \sinh^2 x}$$

if x and y are real.

17. If x and y are real, and $\cosh(x+iy) = 0$, show that $x = 0$ and $\cos y = 0$.

18. What is the image in the w-plane of a line $x =$ constant in the z-plane if

(a) $w = e^z$, (b) $w = \sin z$?

Sketch.

19. Integrate $$\int e^{(a+ib)x}\,dx,$$

and, by equating real and imaginary parts, obtain formulas for

$$\int e^{ax}\cos bx\,dx \quad\text{and}\quad \int e^{ax}\sin bx\,dx.$$

19.8 LOGARITHMS

In the previous article, we mentioned that the logarithm (as an inverse of the exponential) is a multiple-valued function of z. The multiple-valuedness of this function is introduced by the polar angle θ. We see this when we write

$$z = r\operatorname{cis}\theta,$$

or, as we may do in view of Euler's formula,

$$z = re^{i\theta}, \tag{1}$$

and ask that

$$\begin{aligned}\log_e z &= \log_e r + \log_e e^{i\theta}\\ &= \ln r + i\theta.\end{aligned} \tag{2}$$

Then the angle θ may be given its principal value θ_0, $-\pi < \theta_0 \leq \pi$, which leads to the *principal value* of $\log z$:

$$\ln z = \ln r + i\theta_0, \quad -\pi < \theta_0 \leq \pi. \tag{3}$$

But many other values of θ will still give the same z in Eq. (1), namely,

$$\theta = \theta_0 + 2n\pi, \quad n = 0, \pm 1, \pm 2, \ldots, \tag{4}$$

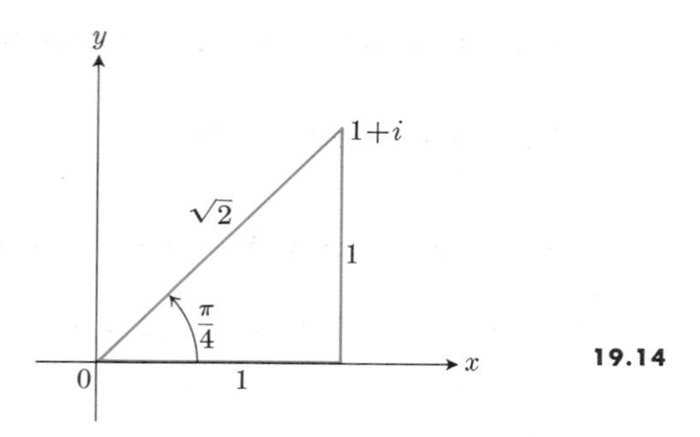

19.14

and each of these values of θ gives rise, in turn, to a value of $\log z$ (we shall omit writing the base e henceforth):

$$\log z = \ln r + i(\theta_0 + 2n\pi). \tag{5a}$$

In terms of the principal value, we have

$$\log z = \ln z + 2n\pi i, \tag{5b}$$

so that all the infinitely many different values of $\log z$ differ from the principal value by an integral multiple of $2\pi i$.

Problem. Find all values of $\log (1 + i)$.

Solution. From Fig. 19.14, we see that the complex number $1 + i$ has polar coordinates $r = \sqrt{2}$, $\theta_0 = \pi/4$. Hence

$$1 + i = \sqrt{2}\, e^{i(\pi/4+2n\pi)},$$

and

$$\log (1 + i) = \ln \sqrt{2} + i(\pi/4 + 2n\pi),$$
$$n = 0, \pm 1, \pm 2, \ldots$$

EXERCISES

1. Find the principal value of $\log z$ for the following complex numbers z.

   (a) $2 - 2i$ (b) $\sqrt{3} + i$ (c) -4

   (d) $+4$ (e) $2i$ (f) $\dfrac{1+i}{1-i}$

2. Find *all* values of $\log z$ for the following complex numbers z.

   (a) 2 (b) -2 (c) $2i$

   (d) $-2i$ (e) $i - \sqrt{3}$

3. Express $w = \tan z$ in terms of exponentials; then solve for z in terms of w and thereby show that

$$\tan^{-1} w = \frac{1}{2i} \log \frac{1 + iw}{1 - iw}.$$

4. (a) Show that $\sin^{-1} z = -i \log (iz + \sqrt{1 - z^2})$.
   (b) Find $\sin^{-1} 3$.

5. Sketch the maps in the w-plane of the following loci in the z-plane, under the mapping function $w = \ln z$.

   (a) $|z| = \text{constant}, \quad -\pi < \arg z \leq +\pi$
   (b) $\arg z = \text{constant}, \quad 0 < |z| < +\infty$

REVIEW QUESTIONS AND EXERCISES

1. Define the system of complex numbers.
2. Define the concepts of equality, addition, multiplication, and division of complex numbers.
3. Is the system of complex numbers closed under the operations of addition, subtraction, multiplication, division (by numbers $\neq 0$), and raising to powers [including complex exponents; see Article 6.7, Eqs. (3 a, b)]? (A system is said to be *closed* under an operation $\otimes$ if $a \otimes b$ is in the system whenever a and b are.)
4. How may the complex number $a + ib$ be represented graphically in an Argand diagram?
5. On an Argand diagram, illustrate how the absolute values and arguments of the product and quotient of two complex numbers z_1 and z_2 are related to the absolute values and the arguments of z_1 and z_2.
6. State De Moivre's theorem, and explain how it may be used to find expressions for $\cos n\theta$ and $\sin n\theta$ as polynomials in $\cos \theta$ and $\sin \theta$.
7. Prove that if n is an even positive integer, then $\cos n\theta$ may be expressed as a polynomial, with integral coefficients, in $\cos^2 \theta$.

   Example. $\cos 2\theta = 2 \cos^2 \theta - 1.$

8. Using an Argand diagram, explain how to find the n complex nth roots of any complex number $a + ib$.
9. On an Argand diagram, illustrate how the conjugate and the reciprocal of a complex number $a + ib$ are related to that number.
10. If the conjugate of a complex number is equal to the number, what else can you conclude about the number?
11. If z is a complex number such that $z = -\bar{z}$, what more can you conclude about z?
12. What is the locus of the complex variable z if

    (a) $|z - \alpha| = k$, (b) $|z - \alpha| < k$,
    (c) $|z - \alpha| > k$,

    when $\alpha = a + ib$ is a given complex number and k is a positive real constant?
13. Define the concept of continuity of a function of a complex variable $w = f(z)$. Define the derivative of f at $\alpha = a + ib$.
14. What are the Cauchy-Riemann equations, and when are they known to be satisfied?
15. Define convergence of a series of complex numbers.
16. How do we define

$$e^z, \quad \sin z, \quad \log z, \quad \tan z$$

    for complex $z = x + iy$?
17. How would you define z^α for complex z and α? Illustrate for

$$z = 1 + i, \quad \alpha = 2i.$$

MISCELLANEOUS EXERCISES

1. Let $z = 2 - 2i$, $i = \sqrt{-1}$.

    (a) Plot the points z, $\bar{z}$, and z^2.

    (b) Plot the three complex cube roots of z^2 (that is, $z^{2/3}$).
2. Express the following complex numbers in the form $re^{i\theta}$ with $r \geq 0$ and $-\pi < \theta \leq \pi$. Sketch.

    (a) $(1 + i)(1 - i\sqrt{3})^2$

    (b) $\sqrt[3]{2 - 2i}$ (three answers)
3. Express the following complex numbers in the form $a + bi$.

    (a) the four 4th roots of $-16i$

    (b) $\sin^{-1}(5)$
4. (a) Solve the equation $z^4 + 16 = 0$, obtaining four distinct roots.

    (b) Express the five roots of the equation

$$z^5 + 32 = 0$$

    in polar form.
5. (a) Find all complex numbers z such that

$$z^4 + 1 + i\sqrt{3} = 0.$$

    (b) Express the number $e^{2+\pi i/4}$ in the form $a + bi$.
6. Plot the complex number $2 - 2\sqrt{3}i$ in an Argand diagram and find

    (a) its two square roots,

    (b) the principal value of its logarithm.
7. Find values of r and θ such that $3 + 4i = re^{i\theta}$.
8. If $z = 3 - 3i$, find all values of $\log z$.
9. Find a complex number $a + ib$ that will satisfy the equation

$$e^{a+ib} = 1 - i\sqrt{3}.$$

10. Express the following in the form $a + bi$.

    (a) $(-1 - i)^{1/3}$ (Write down all the roots.)

    (b) $\ln(3 + i\sqrt{3})$ (c) $e^{2+\pi i}$
11. Let $f(z) = \bar{z} =$ the conjugate of z.

    (a) Study the behavior of the quotient

$$\frac{f(z) - f(z_1)}{z - z_1}$$

    when $z \to z_1$ along straight lines of slope m.

    (b) From the results of (a), what can you conclude about the derivative of $f(z)$?
12. If

$$f(z) = \sum_{n=0}^{\infty} a_n z^n$$

    and

$$f(\bar{z}) = \overline{f(z)} \quad \text{for } |z| < R,$$

    show that the a_n are real.

13. If
$$f(z) = \sum_{n=0}^{\infty} a_n z^n$$
and
$$f(\bar{z}) = f(z) \quad \text{for } |z| < R,$$
show that $f(z)$ is a constant.

14. Show that
$$\frac{d^n}{d\theta^n}(\cos\theta + i\sin\theta) = i^n(\cos\theta + i\sin\theta).$$

15. Indicate graphically the locus of points $z = x + iy$ that satisfy each given condition.

(a) $R(z) > 0$
(b) $I(z - i) \le 0$
(c) $\left|\dfrac{z - i}{z + i}\right| < 1$
(d) $|e^z| \ge 1$
(e) $|\sin z| \le 1$

16. If $u(x, y)$, $v(x, y)$ are the real and imaginary parts of an analytic function of $z = x + iy$, show that the family of curves $u =$ *constant* is orthogonal to the family $v =$ *constant* at every point of intersection where $f'(z) \neq 0$.

17. Let
$$u(x, y) + iv(x, y) = (x + iy)^2.$$
Sketch the curves $u(x, y) = a$, $v(x, y) = b$, for the cases $a = 1, 0, -1$, and $b = 1, 0, -1$. Show that the locus $u = 0$ is not orthogonal to the locus $v = 0$.

18. Verify that the real and imaginary parts of the following functions satisfy Laplace's equation $\phi_{xx} + \phi_{yy} = 0$.

(a) $\sin z$
(b) $\ln z$ $(z \neq 0)$
(c) e^z
(d) $\cosh z$

19. Find all solutions of the following equations.

(a) $z^5 = 32$
(b) $e^z = -2$
(c) $\cos z = 10$
(d) $\tanh z = 2$
(e) $z = i^i$
(f) $z^3 + 3z^2 + 3z + 9 = 0$

DIFFERENTIAL EQUATIONS

CHAPTER 20

20.1 INTRODUCTION

A differential equation is an equation that involves one or more derivatives, or differentials. Differential equations are classified by

(a) *type:* ordinary or partial,

(b) *order:* the order of the highest-order derivative that occurs in the equation,

(c) *degree:* the exponent of the highest power of the highest-order derivative, after the equation has been cleared of fractions and radicals in the dependent variable and its derivatives.

For example,

$$\left(\frac{d^3y}{dx^3}\right)^2 + \left(\frac{d^2y}{dx^2}\right)^5 + \frac{y}{x^2+1} = e^x \tag{1}$$

is an ordinary differential equation, of order three and degree two.

Only "ordinary" derivatives occur when the dependent variable y is a function of a single independent variable x. On the other hand, if the dependent variable y is a function of two or more independent variables, say if

$$y = f(x, t)$$

where x and t are independent variables, then partial derivatives of y may occur. For example,

$$\frac{\partial^2 y}{\partial t^2} = a^2 \frac{\partial^2 y}{\partial x^2} \tag{2}$$

is a partial differential equation, of order two and degree one. (This is the one-dimensional "wave-equation." A systematic treatment of partial differential equations lies beyond the scope of this book. For a discussion of partial differential equations, including the wave equation and solutions of associated physical problems, see Kaplan, *Advanced Calculus*, Chapter 10.)

Many physical problems lead to differential equations when formulated in mathematical terms. In Article 7.10, for example, we discussed two differential equations from the field of electrical engineering:

$$\frac{dQ}{dt} = -kQ \tag{3}$$

and, for an RL-series circuit,

$$L\frac{di}{dt} + Ri = E. \tag{4}$$

Again, in Article 12.1, we solved the system of differential equations

$$m\frac{d^2x}{dt^2} = 0, \qquad m\frac{d^2y}{dt^2} = -mg \tag{5}$$

that arose from the formulation of the laws of motion of a projectile (neglecting air resistance). Indeed, one of the chief sources of differential equations in mechanics is Newton's second law:

$$\mathbf{F} = \frac{d}{dt}(m\mathbf{v}), \tag{6}$$

where $\mathbf{F}$ is the resultant of the forces acting on a body of mass m, and $\mathbf{v}$ is its velocity.

The following situation is a typical example of an application in the field of radiochemistry. Suppose that at time t, there are x, y, and z grams, respectively, of three radioactive substances A, B, and C which have the following properties. A, through radioactive decomposition, transforms into B at a rate proportional to the amount of A present. B, in turn, transforms into C at a rate which is proportional to the amount of B present. Finally, C transforms back into A at a rate which is proportional to the amount of C present. (See Fig. 20.1.) If we call the proportionality factors k_1, k_2, and k_3, respectively, we have the following system of differential equations:

$$\begin{aligned}\frac{dx}{dt} &= -k_1x + k_3z,\\ \frac{dy}{dt} &= k_1x - k_2y, \\ \frac{dz}{dt} &= k_2y - k_3z.\end{aligned} \tag{7}$$

The first of these simply says that the amount of A present at time t, namely x, is decreasing at a rate k_1x through transformation into substance B, but is gaining at a rate k_3z from substance C. The other two equations have similar meanings. It is an immediate consequence of Eqs. (7) that

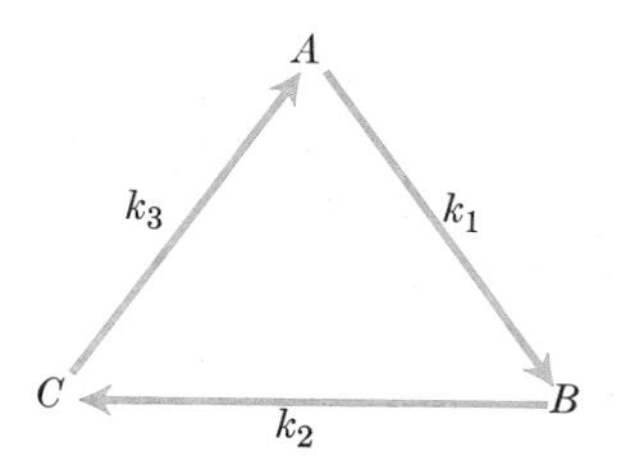

20.1

$$\frac{dx}{dt} + \frac{dy}{dt} + \frac{dz}{dt} = 0. \tag{8}$$

From Eq. (8), in turn, we readily deduce that

$$x + y + z = \text{constant}.$$

In other words, in the hypothetical case under consideration, the total amount of substances A, B, and C remains constant.

20.2 SOLUTIONS

A function $y = f(x)$ is said to be a *solution* of a differential equation if the latter is identically satisfied when y and its derivatives are replaced throughout by $f(x)$ and its corresponding derivatives. For example, if c_1 and c_2 are any constants, then

$$y = c_1 \cos x + c_2 \sin x \tag{1a}$$

is a solution of the differential equation

$$\frac{d^2y}{dx^2} + y = 0. \tag{1b}$$

A physical problem that translates into a differential equation usually involves additional conditions not expressed by the differential equation itself. In mechanics, for example, the initial position and velocity of the moving body are usually prescribed, as well as the forces. The differential equation, or equations, of motion will usually have solutions in which certain arbitrary constants occur. [Note the constants in Eq. (1a), for example.] However, it will be necessary to assign specific values to these arbitrary constants in order to meet the prescribed initial conditions. (See the problems and examples worked out in Articles 7.10 and 12.1.)

A differential equation of order n will in general possess a solution involving n arbitrary constants. (There is a more precise mathematical theorem, which we shall neither state nor prove.) This solution is called the *general* solution. Once this general solution is known, it is only a matter of algebra to determine specific values of the constants if initial conditions are also prescribed. Hence we shall devote our attention to the problem of finding the general solutions of certain types of differential equations.

The subject of differential equations is a complex one. The interested reader may consult textbooks on differential equations* or on advanced calculus† for more exhaustive treatments than we shall present here. The field is also the subject of current research by several eminent mathematicians, and nobody "knows all the answers."

Throughout this chapter, only *ordinary* differential equations will be considered. In outline, the next nine articles cover the following topics.

1. First-order equations
   (a) variables separable
   (b) homogeneous
   (c) linear
   (d) exact differentials
2. Special types of second-order equations
3. Linear equations with constant coefficients
   (a) homogeneous
   (b) nonhomogeneous

The reader will note that several new terms occur in this list; these will be defined in the appropriate places.

* See, for example, Martin and Reissner, *Elementary Differential Equations*, Addison-Wesley, 1961.

† See, for example, Kaplan, *Advanced Calculus*, Addison-Wesley, 1952.

20.3 FIRST-ORDER EQUATIONS WITH VARIABLES SEPARABLE

A first-order differential equation can be solved by integration if it is possible to collect all y terms with dy and all x terms with dx. That is, if it is possible to write the equation in the form

$$f(y)\,dy + g(x)\,dx = 0,$$

then the general solution is

$$\int f(y)\,dy + \int g(x)\,dx = C,$$

where C is an arbitrary constant.

Problem. Solve the equation

$$(x+1)\frac{dy}{dx} = x(y^2+1).$$

Solution. We change to differential form, separate the variables, and integrate:

$$(x+1)\,dy = x(y^2+1)\,dx,$$

$$\frac{dy}{y^2+1} = \frac{x\,dx}{x+1},$$

$$\tan^{-1} y = x - \ln|x+1| + C.$$

EXERCISES

Separate the variables and solve the following differential equations.

1. $x(2y-3)\,dx + (x^2+1)\,dy = 0$
2. $x^2(y^2+1)\,dx + y\sqrt{x^3+1}\,dy = 0$
3. $\dfrac{dy}{dx} = e^{x-y}$
4. $\sqrt{2xy}\,\dfrac{dy}{dx} = 1$
5. $\sin x\,\dfrac{dx}{dy} + \cosh 2y = 0$
6. $\ln x\,\dfrac{dx}{dy} = \dfrac{x}{y}$
7. $xe^y\,dy + \dfrac{x^2+1}{y}\,dx = 0$
8. $y\sqrt{2x^2+3}\,dy + x\sqrt{4-y^2}\,dx = 0$
9. $\sqrt{1+x^2}\,dy + \sqrt{y^2-1}\,dx = 0$
10. $x^2y\,\dfrac{dy}{dx} = (1+x)\csc y.$

20.4 FIRST-ORDER HOMOGENEOUS EQUATIONS

A differential equation that can be put into the form

$$\frac{dy}{dx} = F\left(\frac{y}{x}\right) \tag{1}$$

is said to be *homogeneous*. Such an equation can be solved by introducing a new dependent variable

$$v = \frac{y}{x}. \tag{2}$$

Then

$$y = vx, \qquad \frac{dy}{dx} = v + x\frac{dv}{dx},$$

and (1) becomes

$$v + x\frac{dv}{dx} = F(v). \tag{3}$$

Equation (3) can be solved by separation of variables:

$$\frac{dx}{x} + \frac{dv}{v - F(v)} = 0. \tag{4}$$

After (4) is solved, the solution of the original equation is obtained when we replace v by y/x.

Problem. Show that the equation

$$(x^2 + y^2)\,dx + 2xy\,dy = 0$$

is homogeneous, and solve it.

Solution. From the given equation, we have

$$\frac{dy}{dx} = -\frac{x^2 + y^2}{2xy} = -\frac{1 + (y/x)^2}{2(y/x)}.$$

This has the form of Eq. (1), with

$$F(v) = -\frac{1 + v^2}{2v}, \qquad \text{where } v = \frac{y}{x}.$$

Then Eq. (4) becomes

$$\frac{dx}{x} + \frac{dv}{v + \dfrac{1 + v^2}{2v}} = 0, \quad \text{or} \quad \frac{dx}{x} + \frac{2v\,dv}{1 + 3v^2} = 0.$$

The solution of this is

$$\ln|x| + \tfrac{1}{3}\ln(1 + 3v^2) = \tfrac{1}{3}\ln C,$$

or

$$x^3(1 + 3v^2) = \pm C.$$

In terms of y and x, the solution is

$$x(x^2 + 3y^2) = C.$$

EXERCISES

Show that the following equations are homogeneous, and solve.

1. $(x + y)\,dy + (x - y)\,dx = 0$
2. $x^2\,dy + (y^2 - xy)\,dx = 0$
3. $(xe^{y/x} + y)\,dx - x\,dy = 0$
4. $(x^2 + y^2)\,dy - y^2\,dx = 0$
5. $\dfrac{dy}{dx} = \dfrac{-x + \sqrt{x^2 + y^2}}{y}$
6. If every member of a family of curves is a solution of the differential equation
$$M(x, y)\,dx + N(x, y)\,dy = 0,$$
while every member of a second family of curves is a solution of the related equation
$$N(x, y)\,dx - M(x, y)\,dy = 0,$$
then each curve of the one family is orthogonal to every curve of the other family. Each family is said to be a family of *orthogonal trajectories* of the other. Find the family of solutions of the differential equation
$$\frac{dy}{dx} = \frac{y^2 - x^2}{2xy}$$
and their orthogonal trajectories. Sketch both families of curves.

20.5 FIRST-ORDER LINEAR EQUATIONS

The complexity of a differential equation depends primarily on the way in which the *dependent* variable and its derivatives occur. Of particular importance are those equations that are linear. To find whether or not a differential equation is linear, we compute the degree of each of its terms, adding the exponents of the dependent variable, and of any of its derivatives that occur in the term. Thus, for example, the term (d^2y/dx^2) is of the first degree, while the term $y(dy/dx)$ is of the second degree because we must add 1 for the exponent of y, and 1 for the exponent of dy/dx. If every term of a differential equation is of degree zero or degree one, then the equation is linear.

A linear differential equation of first order can always be put into the standard form

$$\frac{dy}{dx} + Py = Q, \tag{1}$$

where P and Q are functions of x. One method for solving Eq. (1) is to find a function $\rho = \rho(x)$ such that if the equation is multiplied by ρ, the left-hand side becomes the derivative of the product ρy. That is, we multiply (1) by ρ:

$$\rho \frac{dy}{dx} + \rho P y = \rho Q, \tag{1'}$$

and then try to impose on ρ the condition that

$$\rho \frac{dy}{dx} + \rho P y = \frac{d}{dx}(\rho y). \tag{2}$$

When we expand the right-hand side of (2) and cancel terms, we obtain

$$\frac{d\rho}{dx} = \rho P \tag{3}$$

as the condition to be satisfied by ρ. In Eq. (3), $P = P(x)$ is a known function, so we can separate the variables and solve for ρ:

$$\frac{d\rho}{\rho} = P\,dx, \qquad \ln|\rho| = \int P\,dx + \ln C,$$

$$\rho = \pm C e^{\int P\,dx}. \tag{4}$$

Since we do not require the most general function ρ, we may take $\pm C = 1$ in (4) and use

$$\rho = e^{\int P\,dx}. \tag{5}$$

This function is called an *integrating factor* for Eq. (1). With its help, (1′) becomes

$$\frac{d}{dx}(\rho y) = \rho Q,$$

the solution of which is

$$\rho y = \int \rho Q\,dx + C. \tag{6}$$

Since ρ is given by (5), and since P and Q are known from the given differential equation (1), we have in Eqs. (5) and (6) a summary of all that is required to solve (1).

Problem 1. $$\frac{dy}{dx} + y = e^x.$$

Solution.

$$P = 1, \qquad Q = e^x,$$

$$\rho = e^{\int dx} = e^x,$$

$$e^x y = \int e^{2x}\,dx + C = \tfrac{1}{2}e^{2x} + C,$$

$$y = \tfrac{1}{2}e^x + Ce^{-x}.$$

Problem 2. $$x\frac{dy}{dx} - 3y = x^2.$$

Solution. We put the equation in standard form:

$$\frac{dy}{dx} - \frac{3}{x}y = x,$$

and then read off

$$P = -\frac{3}{x}, \qquad Q = x.$$

Hence

$$\rho = e^{\int -(3/x)\,dx} = e^{-3\ln x} = \frac{1}{e^{3\ln x}} = \frac{1}{x^3},$$

and

$$\frac{1}{x^3}y = \int \frac{x}{x^3}\,dx + C = -\frac{1}{x} + C,$$

$$y = -x^2 + Cx^3.$$

Remark 1. Note that whenever $\int P\,dx$ involves logarithms, as in the last example, it is profitable to simplify the expression for $e^{\int P\,dx}$ before substituting into Eq. (6). We make use of the properties of the logarithmic and exponential functions:

$$e^{\ln A} = A,$$

$$e^{m\ln A} = A^m,$$

$$e^{n+m\ln A} = A^m e^n.$$

Remark 2. A differential equation which is linear in y and dy/dx may also be separable, or homogeneous. In such cases, we have a choice of methods of solution. Observe also that an equation which is linear in x and dx/dy can be solved by the technique of this article; one need only interchange the roles of x and y in Eqs. (1), (5), and (6).

EXERCISES

1. $\frac{dy}{dx} + 2y = e^{-x}$
2. $2\frac{dy}{dx} - y = e^{x/2}$
3. $x\frac{dy}{dx} + 3y = \frac{\sin x}{x^2}$
4. $x\,dy + y\,dx = \sin x\,dx$
5. $x\,dy + y\,dx = y\,dy$
6. $(x-1)^3\frac{dy}{dx} + 4(x-1)^2 y = x + 1$
7. $\cosh x\,dy + (y\sinh x + e^x)\,dx = 0$
8. $e^{2y}\,dx + 2(xe^{2y} - y)\,dy = 0$
9. $(x - 2y)\,dy + y\,dx = 0$
10. $(y^2 + 1)\,dx + (2xy + 1)\,dy = 0$

20.6 FIRST-ORDER EQUATIONS WITH EXACT DIFFERENTIALS

An equation that can be written in the form

$$M(x, y)\,dx + N(x, y)\,dy = 0, \tag{1}$$

and having the property that

$$\frac{\partial M}{\partial y} = \frac{\partial N}{\partial x}, \tag{2}$$

is said to be *exact*, because its left-hand side is an exact differential. The technique of solving an exact equation consists in finding a function $f(x, y)$ such that

$$df = M\,dx + N\,dy. \tag{3}$$

Then (1) becomes

$$df = 0$$

and the solution is

$$f(x, y) = C,$$

where C is an arbitrary constant. The method of finding $f(x, y)$ to satisfy (3) has been discussed and illustrated in Article 15.13.

We remark that mathematicians have proved that every first-order differential equation

$$P(x, y)\,dx + Q(x, y)\,dy = 0$$

can be made exact by multiplication by a suitable *integrating factor* $\rho(x, y)$. Such an integrating factor has the property that

$$\frac{\partial}{\partial y}[\rho(x, y)P(x, y)] = \frac{\partial}{\partial x}[\rho(x, y)Q(x, y)].$$

It is not easy to determine ρ from this equation. However, one can often recognize certain combinations of differentials that can be made exact by "ingenious devices."

Problem. $x\,dy - y\,dx = xy^2\,dx.$

Solution. The combination $x\,dy - y\,dx$ may "ring a bell" in our memories and cause us to recall the formula

$$d\left(\frac{u}{v}\right) = \frac{v\,du - u\,dv}{v^2} = -\frac{u\,dv - v\,du}{v^2}.$$

Therefore we might divide the given equation by x^2, or change signs and divide by y^2. Clearly, the latter approach will be more profitable, so we proceed as follows:

$$x\,dy - y\,dx = xy^2\,dx,$$

$$\frac{y\,dx - x\,dy}{y^2} = -x\,dx,$$

$$d\left(\frac{x}{y}\right) + x\,dx = 0, \qquad \frac{x}{y} + \frac{x^2}{2} = C.$$

The same result would be obtained if we wrote our equation in the form

$$(xy^2 + y)\,dx - x\,dy = 0$$

and multiplied by the integrating factor $1/y^2$. This would give

$$\left(x + \frac{1}{y}\right)dx - \frac{x}{y^2}\,dy = 0,$$

which is exact since

$$\frac{\partial}{\partial y}\left(x + \frac{1}{y}\right) = \frac{\partial}{\partial x}\left(-\frac{x}{y^2}\right).$$

EXERCISES

1. $(x + y)\,dx + (x + y^2)\,dy = 0$
2. $(2xe^y + e^x)\,dx + (x^2 + 1)e^y\,dy = 0$
3. $(2xy + y^2)\,dx + (x^2 + 2xy - y)\,dy = 0$

4. $(x+\sqrt{y^2+1})\,dx-\left(y-\dfrac{xy}{\sqrt{y^2+1}}\right)dy=0$

5. $x\,dy-y\,dx+x^3\,dx=0$

6. $x\,dy-y\,dx=(x^2+y^2)\,dx$

7. $(x^2+x-y)\,dx+x\,dy=0$

8. $\left(e^x+\ln y+\dfrac{y}{x}\right)dx+\left(\dfrac{x}{y}+\ln x+\sin y\right)dy=0$

9. $\left(\dfrac{y^2}{1+x^2}-2y\right)dx+(2y\tan^{-1}x-2x+\sinh y)\,dy=0$

10. $dy+\dfrac{y-\sin x}{x}\,dx=0.$

20.7 SPECIAL TYPES OF SECOND-ORDER EQUATIONS

Certain types of second-order differential equations, of which the general form is

$$F\left(x,\,y,\,\frac{dy}{dx},\,\frac{d^2y}{dx^2}\right)=0, \tag{1}$$

can be reduced to first-order equations by a suitable change of variables.

Type 1. Equations with dependent variable missing. If Eq. (1) has the special form

$$F\left(x,\,\frac{dy}{dx},\,\frac{d^2y}{dx^2}\right)=0, \tag{2}$$

then we can reduce it to a first-order equation by substituting

$$p=\frac{dy}{dx},\qquad \frac{d^2y}{dx^2}=\frac{dp}{dx}.$$

Then Eq. (2) takes the form

$$F\left(x,\,p,\,\frac{dp}{dx}\right)=0,$$

which is of the first order in p. If this can be solved for p as a function of x, say

$$p=\phi(x,C_1),$$

then we can find y by an additional integration:

$$y=\int(dy/dx)\,dx=\int p\,dx=\int\phi(x,C_1)\,dx+C_2.$$

Note. For an example, see the solution of the differential equation

$$\frac{d^2y}{dx^2}=\frac{w}{H}\sqrt{1+\left(\frac{dy}{dx}\right)^2}$$

in Article 8.6.

Type 2. Equations with independent variable missing. If Eq. (1) does not contain x explicitly but has the form

$$F\left(y,\,\frac{dy}{dx},\,\frac{d^2y}{dx^2}\right)=0, \tag{3}$$

then the substitutions to use are

$$p=\frac{dy}{dx},\qquad \frac{d^2y}{dx^2}=p\,\frac{dp}{dy}.$$

Then Eq. (3) takes the form

$$F\left(y,\,p,\,p\,\frac{dp}{dy}\right)=0,$$

which is of order 1 in p. Its solution gives p in terms of y; a further integration gives the solution of Eq. (3).

Problem. $\dfrac{d^2y}{dx^2}+y=0.$

Solution. Let

$$\frac{dy}{dx}=p,\qquad \frac{d^2y}{dx^2}=\frac{dp}{dx}=\frac{dp}{dy}\,\frac{dy}{dx}=\frac{dp}{dy}\,p,$$

and proceed as follows:

$$p\,\frac{dp}{dy}+y=0,\quad\text{or}\quad p\,dp+y\,dy=0;$$

hence

$$\frac{p^2}{2}+\frac{y^2}{2}=\frac{C_1^2}{2},\quad\text{or}\quad p=\pm\sqrt{C_1^2-y^2}.$$

But $p=dy/dx$, and therefore

$$\frac{dy}{dx}=\pm\sqrt{C_1^2-y^2},\quad\text{or}\quad \frac{dy}{\sqrt{C_1^2-y^2}}=\pm dx.$$

Noting the trigonometric differential, we have

$$\sin^{-1}(y/C_1)=\pm(x+C_2),$$

or

$$y=C_1\sin[\pm(x+C_2)]=\pm C_1\sin(x+C_2).$$

Since C_1 is arbitrary, there is no need for the $\pm$-sign, and we have $y=C_1\sin(x+C_2)$ as the general solution.

EXERCISES

1. $\frac{d^2y}{dx^2} + \frac{dy}{dx} = 0$
2. $\frac{d^2y}{dx^2} + y\frac{dy}{dx} = 0$
3. $\frac{d^2y}{dx^2} + x\frac{dy}{dx} = 0$
4. $x\frac{d^2y}{dx^2} + \frac{dy}{dx} = 0$
5. $\frac{d^2y}{dx^2} - y = 0$
6. $\frac{d^2y}{dx^2} + \omega^2 y = 0, \quad \omega$ a constant $\neq 0$
7. A body of mass m is suspended from one end of a spring whose other end is attached to a rigid support. The body is allowed to come to rest, and is then pulled down an additional slight amount A and released. Find its motion.

   Hint. Assume Newton's second law of motion and Hooke's law, which says that the tension in the spring is proportional to the amount it is stretched. Let x denote the displacement of the body at time t, measured from the equilibrium position. Then

   $$m\frac{d^2x}{dt^2} = -kx,$$

   where k, the "spring constant," is the proportionality factor in Hooke's law.
8. A man suspended from a parachute falls through space under the pull of gravity. If air resistance produces a retarding force proportional to the man's velocity and he starts from rest at time $t = 0$, find the distance he falls in time t.

20.8 LINEAR EQUATIONS WITH CONSTANT COEFFICIENTS

An equation of the form

$$\frac{d^n}{dx^n}y + a_1\frac{d^{n-1}}{dx^{n-1}}y + a_2\frac{d^{n-2}}{dx^{n-2}}y + \cdots + a_{n-1}\frac{d}{dx}y + a_ny = F(x), \tag{1}$$

which is linear in y and its derivatives, is called a *linear* equation of order n. If $F(x)$ is identically zero, the equation is said to be *homogeneous*, otherwise it is called nonhomogeneous. The equation is linear even though the coefficients $a_1, a_2, \ldots, a_n$ are functions of x. However, we shall consider only the case where these coefficients are *constants*.

It is convenient to introduce the symbol D to represent the operation of differentiation with respect to x. That is, we write $Df(x)$ to mean the same thing as $(d/dx)f(x)$. Furthermore, we define powers of D to mean taking successive derivatives:

$$D^2f(x) = D\{Df(x)\} = \frac{d^2f(x)}{dx^2},$$

$$D^3f(x) = D\{D^2f(x)\} = \frac{d^3f(x)}{dx^3},$$

and so on. A polynomial in D is to be interpreted as an operator which, when applied to $f(x)$, produces a linear combination of f and its successive derivatives. For example,

$$\begin{aligned}(D^2 + D - 2)f(x) &= D^2f(x) + Df(x) - 2f(x)\\ &= \frac{d^2f(x)}{dx^2} + \frac{df(x)}{dx} - 2f(x).\end{aligned}$$

Such a polynomial in D is called a "linear differential operator," and may be denoted by the single letter L. If L_1 and L_2 are two such linear operators, their sum and product are defined by the equations

$$\begin{aligned}(L_1 + L_2)f(x) &= L_1f(x) + L_2f(x),\\ L_1L_2f(x) &= L_1\{L_2f(x)\}.\end{aligned}$$

Linear differential operators that are polynomials in D with constant coefficients satisfy basic algebraic laws. This makes it possible to treat them like ordinary polynomials so far as addition, multiplication, and factorization are concerned. Thus

$$\begin{aligned}(D^2 + D - 2)f(x) &= (D + 2)(D - 1)f(x)\\ &= (D - 1)(D + 2)f(x).\end{aligned}$$

Since this is true for any twice-differentiable function f, we also write the equality between operators:

$$\begin{aligned}D^2 + D - 2 &= (D + 2)(D - 1)\\ &= (D - 1)(D + 2).\end{aligned}$$

20.9 HOMOGENEOUS LINEAR SECOND-ORDER DIFFERENTIAL EQUATIONS WITH CONSTANT COEFFICIENTS

Now suppose we wish to solve a differential equation of order two, say

$$\frac{d^2y}{dx^2} + 2a\frac{dy}{dx} + by = 0, \tag{1}$$

where a and b are constants. In operator notation, this becomes

$$(D^2 + 2aD + b)y = 0. \tag{1$'$}$$

In association with this differential equation, we consider the algebraic equation

$$r^2 + 2ar + b = 0, \tag{1$''$}$$

which we get by replacing D by r and suppressing y. This is called the *characteristic equation* of the differential equation. Suppose the roots of (1″) are r_1 and r_2. Then

$$r^2 + 2ar + b = (r - r_1)(r - r_2)$$

and

$$D^2 + 2aD + b = (D - r_1)(D - r_2).$$

Hence Eq. (1′) is equivalent to

$$(D - r_1)(D - r_2)y = 0. \tag{2}$$

If we now let

$$(D - r_2)y = u, \tag{3a}$$

$$(D - r_1)u = 0, \tag{3b}$$

we can solve Eq. (1′) in two steps. From Eq. (3b), which is separable, we find

$$u = C_1e^{r_1x}.$$

We substitute this into (3a), which becomes

$$(D - r_2)y = C_1e^{r_1x},$$

or

$$\frac{dy}{dx} - r_2y = C_1e^{r_1x}.$$

This equation is linear; its integrating factor is

$$\rho = e^{-r_2x};$$

its solution (see Article 20.5) is

$$e^{-r_2x}y = C_1\int e^{(r_1-r_2)x}\,dx + C_2. \tag{4}$$

At this point we must consider two cases, $r_1 \neq r_2$ and $r_1 = r_2$.

Case 1. If $r_1 \neq r_2$, the evaluation of the integral in Eq. (4) leads to

$$e^{-r_2x}y = \frac{C_1}{r_1 - r_2}e^{(r_1-r_2)x} + C_2,$$

or

$$y = \frac{C_1}{r_1 - r_2}e^{r_1x} + C_2e^{r_2x}.$$

Since C_1 is an arbitrary constant, so is $C_1/(r_1 - r_2)$, and the solution of Eq. (2) can be written simply as

$$y = C_1e^{r_1x} + C_2e^{r_2x}, \quad \text{if} \quad r_1 \neq r_2. \tag{5}$$

Case 2. If $\quad r_1 = r_2,$

then $\quad e^{(r_1-r_2)x} = e^0 = 1,$

and Eq. (4) reduces to

$$e^{-r_2x}y = C_1x + C_2,$$

or

$$y = (C_1x + C_2)e^{r_2x}, \quad \text{if} \quad r_1 = r_2. \tag{6}$$

Problem 1. $\quad \dfrac{d^2y}{dx^2} + \dfrac{dy}{dx} - 2y = 0.$

Solution. The equation $r^2 + r - 2 = 0$ has roots

$$r_1 = 1, \qquad r_2 = -2.$$

Hence, by Eq. (5), the solution of the differential equation is

$$y = C_1e^x + C_2e^{-2x}.$$

Problem 2. $\quad \dfrac{d^2y}{dx^2} + 4\dfrac{dy}{dx} + 4y = 0.$

Solution.

$$r^2 + 4r + 4 = (r + 2)^2,$$

$$r_1 = r_2 = -2,$$

$$y = (C_1x + C_2)e^{-2x}.$$

Imaginary roots

If the coefficients a and b in Eq. (1) are real, the roots of the characteristic Eq. (1″) will either be real, or will be a pair of complex conjugate numbers

$$r_1 = \alpha + i\beta, \qquad r_2 = \alpha - i\beta. \tag{7}$$

If $\beta \neq 0$, then Eq. (5) applies, with the result

$$\begin{aligned} y &= c_1 e^{(\alpha + i\beta)x} + c_2 e^{(\alpha - i\beta)x} \\ &= e^{\alpha x}[c_1 e^{i\beta x} + c_2 e^{-i\beta x}]. \end{aligned} \tag{8}$$

By Euler's formula (Eq. 10, Article 19.7),

$$e^{i\beta x} = \cos \beta x + i \sin \beta x,$$

$$e^{-i\beta x} = \cos \beta x - i \sin \beta x.$$

Hence Eq. (8) may be replaced by

$$y = e^{\alpha x}[(c_1 + c_2) \cos \beta x + i(c_1 - c_2) \sin \beta x]. \tag{9}$$

Finally, if we introduce new arbitrary constants

$$C_1 = c_1 + c_2 \quad \text{and} \quad C_2 = i(c_1 - c_2),$$

Eq. (9) takes the form

$$y = e^{\alpha x}[C_1 \cos \beta x + C_2 \sin \beta x]. \tag{9'}$$

The arbitrary constants C_1 and C_2 in (9′) will be real, provided that the constants c_1 and c_2 in (9) are complex conjugates:

$$c_1 = \tfrac{1}{2}(C_1 - iC_2), \qquad c_2 = \tfrac{1}{2}(C_1 + iC_2).$$

To solve a problem in which the roots of the characteristic equation are complex conjugates, we simply write down the answer by Eq. (9′).

Problem 3. $\dfrac{d^2y}{dx^2} + 2\dfrac{dy}{dx} + 2y = 0.$

Solution. The equation $r^2 + 2r + 2 = 0$ has roots

$$r_1 = -1 + i, \qquad r_2 = -1 - i.$$

Hence in Eq. (9′), we take

$$\alpha = -1, \qquad \beta = 1$$

and obtain

$$y = e^{-x}[C_1 \cos x + C_2 \sin x].$$

Problem 4. $\dfrac{d^2y}{dx^2} + \omega^2 y = 0, \quad \omega \neq 0.$

Solution. The equation $r^2 + \omega^2 = 0$ has roots

$$r_1 = i\omega, \qquad r_2 = -i\omega.$$

Hence we take $\alpha = 0$, $\beta = \omega$ in Eq. (9′), and obtain

$$y = C_1 \cos \omega x + C_2 \sin \omega x.$$

EXERCISES

1. $\dfrac{d^2y}{dx^2} + 2\dfrac{dy}{dx} = 0$
2. $\dfrac{d^2y}{dx^2} + 5\dfrac{dy}{dx} + 6y = 0$
3. $\dfrac{d^2y}{dx^2} + 6\dfrac{dy}{dx} + 5y = 0$
4. $\dfrac{d^2y}{dx^2} - 2\dfrac{dy}{dx} - 3y = 0$
5. $\dfrac{d^2y}{dx^2} + \dfrac{dy}{dx} + y = 0$
6. $\dfrac{d^2y}{dx^2} - 4\dfrac{dy}{dx} + 4y = 0$
7. $\dfrac{d^2y}{dx^2} + 6\dfrac{dy}{dx} + 9y = 0$
8. $\dfrac{d^2y}{dx^2} - 6\dfrac{dy}{dx} + 10y = 0$
9. $\dfrac{d^2y}{dx^2} - 2\dfrac{dy}{dx} + 4y = 0$
10. $\dfrac{d^2y}{dx^2} - 10\dfrac{dy}{dx} + 16y = 0$

20.10 NONHOMOGENEOUS LINEAR SECOND-ORDER DIFFERENTIAL EQUATIONS WITH CONSTANT COEFFICIENTS

In Article 20.9, we learned how to solve the homogeneous equation

$$\frac{d^2y}{dx^2} + 2a\frac{dy}{dx} + by = 0. \tag{1}$$

We are now in a position to describe a method for solving the nonhomogeneous equation

$$\frac{d^2y}{dx^2} + 2a\frac{dy}{dx} + by = F(x). \tag{2}$$

To solve Eq. (2), we first obtain the general solution of the related homogeneous equation (1) that is obtained by replacing $F(x)$ by zero. Denote this solution by

$$y_h = C_1 u_1(x) + C_2 u_2(x), \tag{3}$$

where C_1 and C_2 are arbitrary constants and $u_1(x)$, $u_2(x)$ are functions of one or more of the following forms: e^{rx}, xe^{rx}, $e^{\alpha x} \cos \beta x$, $e^{\alpha x} \sin \beta x$.

Now we might, by inspection or by inspired guesswork, be able to discover *one* particular function $y = y_p(x)$ which satisfies Eq. (2). In this case, we would be able to solve Eq. (2) completely, as

$$y = y_h(x) + y_p(x).$$

Problem 1. $$\frac{d^2y}{dx^2} + 2\frac{dy}{dx} - 3y = 6.$$

Solution. The quantity y_h satisfies

$$\frac{d^2y_h}{dx^2} + 2\frac{dy_h}{dx} - 3y_h = 0.$$

The characteristic equation is

$$r^2 + 2r - 3 = 0,$$

and its roots are

$$r_1 = -3, \qquad r_2 = 1.$$

Hence

$$y_h = C_1e^{-3x} + C_2e^x.$$

Now to find a particular integral of the original equation, observe that y equal to a constant would do, provided that $-3y = 6$. Hence $y_p = -2$ is one particular solution. The complete solution is

$$y = y_p + y_h = -2 + C_1e^{-3x} + C_2e^x.$$

Variation of parameters

Fortunately, there is a general method for finding the solution of the nonhomogeneous Eq. (2) once the general solution of the corresponding homogeneous equation is known. The method is known as the method of *variation of parameters*. It consists in replacing the constants C_1 and C_2 in Eq. (3) by functions $v_1 = v_1(x)$ and $v_2 = v_2(x)$, and requiring (in a way to be explained) that the resulting expression satisfy Eq. (2). There are two functions to be determined, and requiring that Eq. (2) be satisfied is only one of the necessary conditions. As a second condition, we also require that

$$v_1'u_1 + v_2'u_2 = 0. \tag{4}$$

Then we have

$$y = v_1u_1 + v_2u_2, \qquad \frac{dy}{dx} = v_1u_1' + v_2u_2',$$

$$\frac{d^2y}{dx^2} = v_1u_1'' + v_2u_2'' + v_1'u_1' + v_2'u_2'.$$

If we substitute these expressions into the left-hand side of Eq. (2), we obtain

$$v_1\left(\frac{d^2u_1}{dx^2} + 2a\frac{du_1}{dx} + bu_1\right)$$
$$+ v_2\left(\frac{d^2u_2}{dx^2} + 2a\frac{du_2}{dx} + bu_2\right) + v_1'u_1' + v_2'u_2' = F(x).$$

The two terms in parentheses are zero, since u_1 and u_2 are solutions of the homogeneous equation (1). Hence Eq. (2) is satisfied if, in addition to Eq. (4), we require that

$$v_1'u_1' + v_2'u_2' = F(x). \tag{5}$$

Equations (4) and (5) together may be solved for the unknown functions v_1 and v_2. Then v_1 and v_2 can be found by integration. In applying the method, we can work directly from Eqs. (4) and (5); it is not necessary to rederive them.

Problem 1′. $$\frac{d^2y}{dx^2} + 2\frac{dy}{dx} - 3y = 6.$$

Solution. $u_1(x) = e^{-3x}, \qquad u_2(x) = e^x.$

$$v_1'e^{-3x} + v_2'e^x = 0, \qquad v_1'(-3e^{-3x}) + v_2'e^x = 6.$$

By Cramer's rule (Appendix I),

$$v_1' = \frac{\begin{vmatrix} 0 & e^x \\ 6 & e^x \end{vmatrix}}{\begin{vmatrix} e^{-3x} & e^x \\ -3e^{-3x} & e^x \end{vmatrix}} = -\tfrac{3}{2}e^{3x},$$

$$v_2' = \frac{\begin{vmatrix} e^{-3x} & 0 \\ -3e^{-3x} & 6 \end{vmatrix}}{\begin{vmatrix} e^{-3x} & e^x \\ -3e^{-3x} & e^x \end{vmatrix}} = \tfrac{3}{2}e^{-x}.$$

Hence

$$v_1 = \int -\tfrac{3}{2}e^{3x}\,dx = -\tfrac{1}{2}e^{3x} + c_1,$$

$$v_2 = \int \tfrac{3}{2}e^{-x}\,dx = -\tfrac{3}{2}e^{-x} + c_2,$$

and

$$\begin{aligned} y &= v_1u_1 + v_2u_2 \\ &= (-\tfrac{1}{2}e^{3x} + c_1)e^{-3x} + (-\tfrac{3}{2}e^{-x} + c_2)e^x \\ &= -2 + c_1e^{-3x} + c_2e^x. \end{aligned}$$

EXERCISES

1. $\dfrac{d^2y}{dx^2} + \dfrac{dy}{dx} = x$
2. $\dfrac{d^2y}{dx^2} + y = \tan x$
3. $\dfrac{d^2y}{dx^2} + y = \sin x$
4. $\dfrac{d^2y}{dx^2} + 2\dfrac{dy}{dx} + y = e^x$
5. $\dfrac{d^2y}{dx^2} + 2\dfrac{dy}{dx} + y = e^{-x}$
6. $\dfrac{d^2y}{dx^2} - y = x$
7. $\dfrac{d^2y}{dx^2} - y = e^x$
8. $\dfrac{d^2y}{dx^2} - y = \sin x$
9. $\dfrac{d^2y}{dx^2} + 4\dfrac{dy}{dx} + 5y = 10$
10. $\dfrac{d^2y}{dx^2} + 4\dfrac{dy}{dx} + 5y = x + 2$

20.11 HIGHER-ORDER LINEAR DIFFERENTIAL EQUATIONS WITH CONSTANT COEFFICIENTS

The methods of Articles 20.9 and 20.10 can be extended to equations of higher order. The characteristic algebraic equation associated with the differential equation

$$(D^n + a_1 D^{n-1} + \cdots + a_{n-1}D + a_n)y = F(x) \tag{1}$$

is

$$r^n + a_1 r^{n-1} + \cdots + a_{n-1}r + a_n = 0. \tag{2}$$

If its roots $r_1, r_2, \ldots, r_n$ are all distinct, the solution of the homogeneous equation obtained by replacing $F(x)$ by zero in Eq. (1) is

$$y_h = c_1 e^{r_1 x} + c_2 e^{r_2 x} + \cdots + c_n e^{r_n x}.$$

Pairs of complex conjugate roots $\alpha \pm i\beta$ can be grouped together, and the corresponding part of y_h can be written in terms of the functions

$$e^{\alpha x} \cos \beta x \quad \text{and} \quad e^{\alpha x} \sin \beta x.$$

If the roots of Eq. (2) are not all distinct, then the portion of y_h that corresponds to a root r of multiplicity m is to be replaced by

$$(C_1 x^{m-1} + C_2 x^{m-2} + \cdots + C_m)e^{rx}.$$

Note that the polynomial in parentheses contains m arbitrary constants.

Problem. $$\frac{d^4y}{dx^4} - 3\frac{d^3y}{dx^3} + 3\frac{d^2y}{dx^2} - \frac{dy}{dx} = 0.$$

Solution

$$r^4 - 3r^3 + 3r^2 - r = r(r-1)^3.$$

The roots of the characteristic equation are

$$r_1 = 0, \qquad r_2 = r_3 = r_4 = 1.$$

The solution is $y = C_1 + (C_2 x^2 + C_3 x + C_4)e^x$.

Variation of parameters

If the general solution of the homogeneous equation is

$$y_h = C_1 u_1 + C_2 u_2 + \cdots + C_n u_n,$$

then

$$y = v_1 u_1 + v_2 u_2 + \cdots + v_n u_n$$

will be a solution of the nonhomogeneous Eq. (1) if and only if

$$\begin{aligned}
v_1' u_1 &+ v_2' u_2 &+ \cdots &+ v_n' u_n &= 0,\\
v_1' u_1' &+ v_2' u_2' &+ \cdots &+ v_n' u_n' &= 0,\\
&\vdots\\
v_1' u_1^{(n-2)} &+ v_2' u_2^{(n-2)} &+ \cdots &+ v_n' u_n^{(n-2)} &= 0,\\
v_1' u_1^{(n-1)} &+ v_2' u_2^{(n-1)} &+ \cdots &+ v_n' u_n^{(n-1)} &= F(x).
\end{aligned}$$

These equations may be solved for $v_1', v_2', \ldots, v_n'$ by Cramer's rule, and the results integrated to give $v_1, v_2, \ldots, v_n$.

EXERCISES

1. $\dfrac{d^3y}{dx^3} - 3\dfrac{d^2y}{dx^2} + 2\dfrac{dy}{dx} = 0$
2. $\dfrac{d^3y}{dx^3} - y = 0$
3. $\dfrac{d^4y}{dx^4} - 4\dfrac{d^2y}{dx^2} + 4y = 0$
4. $\dfrac{d^4y}{dx^4} - 16y = 0$
5. $\dfrac{d^4y}{dx^4} + 16y = 0$
6. $\dfrac{d^3y}{dx^3} - 3\dfrac{dy}{dx} + 2y = e^x$
7. $\dfrac{d^4y}{dx^4} - 4\dfrac{d^3y}{dx^3} + 6\dfrac{d^2y}{dx^2} - 4\dfrac{dy}{dx} + y = 7$
8. $\dfrac{d^4y}{dx^4} + y = x + 1$

20.12 VIBRATIONS

Suppose that a spring of natural length L has its upper end fastened to a rigid support at A (Fig. 20.2). A weight W of mass m is suspended from the spring. The weight stretches the spring to a length $L + s$ when the system is allowed to come to rest in a new equilibrium position. By Hooke's law, the tension in the spring is ks, where k is the so-called spring constant. The force of gravity pulling down on the weight is $W = mg$. Equilibrium requires that

$$ks = mg. \tag{1}$$

Suppose now that the weight is pulled down an additional amount a beyond the equilibrium position, and released. We shall discuss its motion.

Let x, with positive direction downward, denote the displacement of the weight away from equilibrium at any time t after the motion has started. Then the forces acting on the weight are

$$+mg \quad \text{(due to gravity)},$$

$$-k(s + x) \quad \text{(due to the spring tension)}.$$

By Newton's second law, the resultant of these forces is also equal to

$$m\,\frac{d^2x}{dt^2}.$$

Therefore

$$m\,\frac{d^2x}{dt^2} = mg - ks - kx. \tag{2}$$

By Eq. (1), $mg - ks = 0$, so that (2) becomes

$$m\,\frac{d^2x}{dt^2} + kx = 0. \tag{3}$$

In addition to the differential equation (3), the motion satisfies the initial conditions

$$\text{at} \quad t = 0, \quad x = a \quad \text{and} \quad \frac{dx}{dt} = 0. \tag{4}$$

Let $\omega = \sqrt{k/m}$. Then Eq. (3) becomes

$$\frac{d^2x}{dt^2} + \omega^2 x = 0, \quad \text{or} \quad (D^2 + \omega^2)x = 0,$$

where

$$D = d/dt.$$

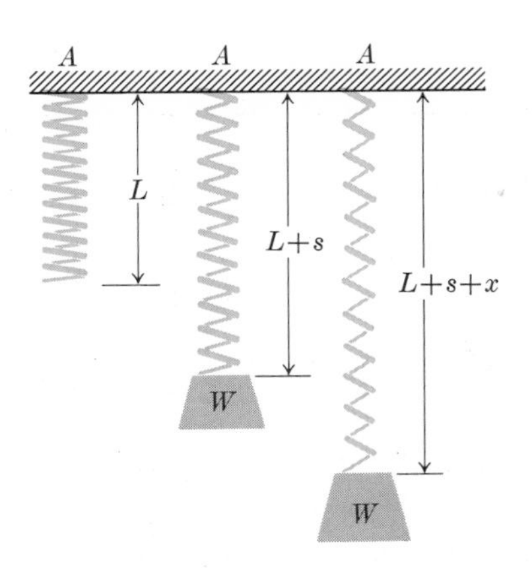

20.2

The roots of the characteristic equation

$$r^2 + \omega^2 = 0$$

are complex conjugates $r = \pm\omega i$. Hence

$$x = c_1 \cos \omega t + c_2 \sin \omega t \tag{5}$$

is the general solution of the differential equation. To accommodate the initial conditions, we also compute

$$\frac{dx}{dt} = -c_1\omega \sin \omega t + c_2\omega \cos \omega t$$

and then substitute from (4). This yields

$$a = c_1, \qquad 0 = c_2\omega.$$

Therefore

$$c_1 = a, \qquad c_2 = 0,$$

and

$$x = a \cos \omega t \tag{6}$$

describes the motion of the weight. Equation (6) represents simple harmonic motion of amplitude a and period $T = 2\pi/\omega$.

The two terms on the right-hand side of Eq. (5) can be combined into a single term by making use of the trigonometric identity

$$\sin(\omega t + \phi) = \cos \omega t \sin \phi + \sin \omega t \cos \phi.$$

This identity can be applied to Eq. (5) if and only if we take

$$c_1 = C \sin \phi, \qquad c_2 = C \cos \phi, \tag{7a}$$

where

$$C = \sqrt{c_1^2 + c_2^2} \quad \text{and} \quad \phi = \tan^{-1} \frac{c_1}{c_2}. \tag{7b}$$

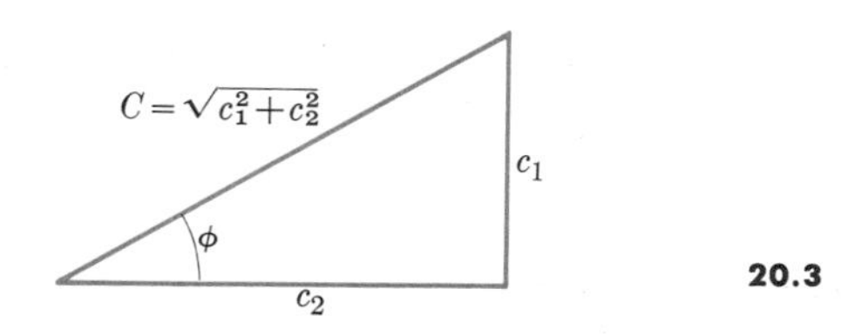

20.3

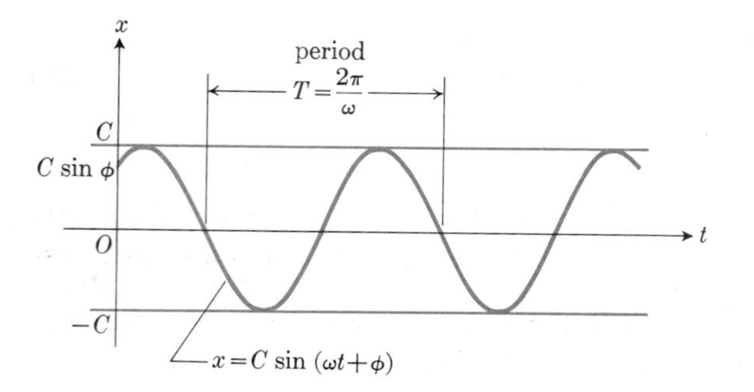

20.4 Representation of a vibration without damping.

(See Fig. 20.3.) Then Eq. (5) can be written in the alternative form

$$x = C \sin(\omega t + \phi). \tag{8}$$

Here C and ϕ may be taken as two new arbitrary constants, replacing the two constants c_1 and c_2 of Eq. (5). Equation (8) represents simple harmonic motion of amplitude C and period $T = 2\pi/\omega$. The angle $\omega t + \phi$ is called the *phase angle*, and ϕ may be interpreted as the initial value of the phase angle. A graph of Eq. (8) is given in Fig. 20.4.

Equation (3) assumes that there is no friction in the system. Let's now consider the case where the motion of the weight is retarded by a friction force $c(dx/dt)$ proportional to the velocity, where c is a positive constant. Then the differential equation is

$$m\frac{d^2x}{dt^2} = -kx - c\frac{dx}{dt},$$

or

$$\frac{d^2x}{dt^2} + 2b\frac{dx}{dt} + \omega^2 x = 0, \tag{9}$$

where

$$2b = \frac{c}{m} \quad \text{and} \quad \omega = \sqrt{\frac{k}{m}}.$$

If we introduce the operator $D = d/dt$, Eq. (9) then becomes

$$(D^2 + 2bD + \omega^2)x = 0.$$

The characteristic equation is

$$r^2 + 2br + \omega^2 = 0$$

with roots

$$r = -b \pm \sqrt{b^2 - \omega^2}. \tag{10}$$

Three cases now present themselves, depending upon the relative sizes of b and ω.

Case 1. If $b = \omega$, then the two roots (10) are equal, and the solution of (9) is

$$x = (c_1 + c_2 t)e^{-\omega t}. \tag{11}$$

As time goes on, x approaches zero. The motion is not oscillatory.

Case 2. If $b > \omega$, then the roots (10) are both real but unequal, and

$$x = c_1 e^{r_1 t} + c_2 e^{r_2 t}, \tag{12}$$

where

$$r_1 = -b + \sqrt{b^2 - \omega^2},$$
$$r_2 = -b - \sqrt{b^2 - \omega^2}.$$

Here again the motion is not oscillatory. Both r_1 and r_2 are negative, and x approaches zero as time goes on.

Case 3. If $b < \omega$, let

$$\omega^2 - b^2 = \alpha^2.$$

Then

$$r_1 = -b + \alpha i, \qquad r_2 = -b - \alpha i$$

and

$$x = e^{-bt}(c_1 \cos \alpha t + c_2 \sin \alpha t). \tag{13a}$$

If we introduce the substitutions (7), we may also write Eq. (13a) in the equivalent form

$$x = Ce^{-bt} \sin(\alpha t + \phi). \tag{13b}$$

This equation represents damped vibratory motion. It is analogous to simple harmonic motion, of period $T = 2\pi/\alpha$, except that the amplitude is not

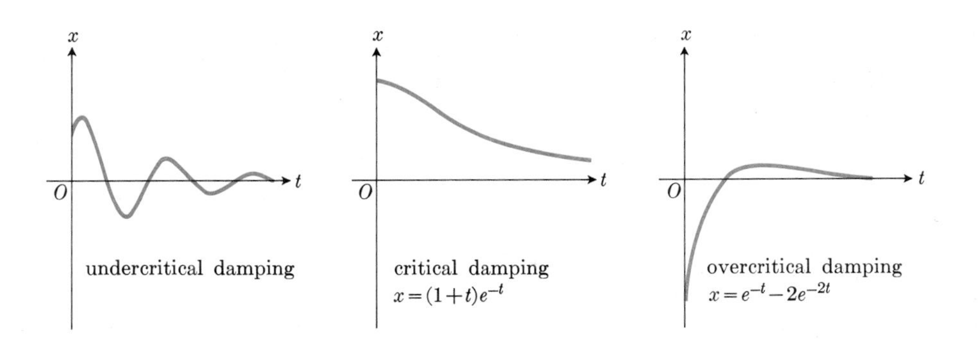

20.5

constant but is given by Ce^{-bt}. Since this tends to zero as t increases, the vibrations tend to die out as time goes on. Observe, however, that Eq. (13b) reduces to Eq. (8) in the absence of friction. The effect of friction is twofold:

1. The quantity $b = c/(2m)$ appears as a coefficient in the exponential *damping factor* e^{-bt}. The larger b is, the more quickly do the vibrations tend to become unnoticeable.

2. The period $T = 2\pi/\alpha$ is greater than the period $T_0 = 2\pi/\omega$ in the friction-free system.

Curves representing solutions of Eq. (9) in typical cases are depicted in Figs. 20.4 and 20.5. The size of b, relative to ω, determines the kind of solution, and b also determines the rate of damping. It is therefore customary to say that there is

(a) critical damping if $b = \omega$,
(b) overcritical damping if $b > \omega$,
(c) undercritical damping if $0 < b < \omega$,
(d) no damping if $b = 0$.

EXERCISES

1. Suppose the motion of the weight in Fig. 20.2 is governed by the differential equation (3). Find the motion if $x = x_0$ and $dx/dt = v_0$ at $t = 0$. Express the answer in two equivalent forms (Eqs. 5 and 8).

2. A 5-lb weight is suspended from the lower end of a spring whose upper end is attached to a rigid support. The weight extends the spring by 2 in. If, after the weight has come to rest in its new equilibrium position, it is struck a sharp blow which starts it downward with a velocity of 4 ft/sec, find its subsequent motion, assuming there is no friction.

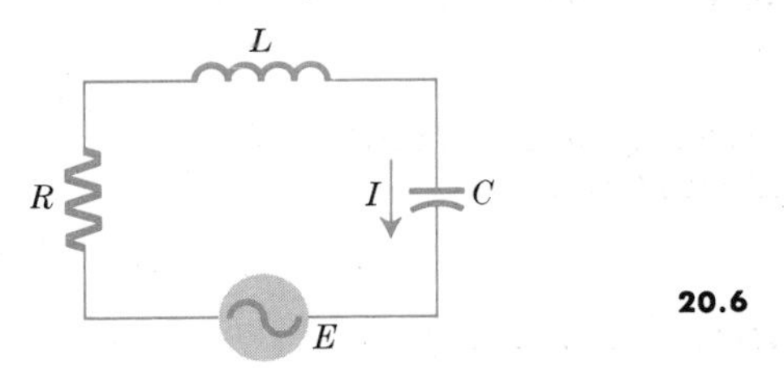

20.6

3. A simple electrical circuit contains a capacitor of capacity C farads, a coil of inductance L henrys, a resistance of R ohms, and a generator which produces an electromotive force of E volts, in series. If the current intensity at time t at some point of the circuit is I amperes, the differential equation governing the current I is

$$L\frac{d^2I}{dt^2} + R\frac{dI}{dt} + \frac{1}{C}I = \frac{dE}{dt}.$$

Find I as a function of t if

(a) $R = 0,\quad 1/(LC) = \omega^2,\quad E = \text{constant},$
(b) $R = 0,\quad 1/(LC) = \omega^2,\quad E = A \sin \alpha t,$
$\alpha = \text{constant} \neq \omega,$
(c) $R = 0,\quad 1/(LC) = \omega^2,\quad E = A \sin \omega t,$
(d) $R = 50,\quad L = 5,\quad C = 9 \times 10^{-6},$
$E = \text{constant}.$

4. A simple pendulum of length l makes an angle θ with the vertical. As it swings back and forth, its motion (neglecting friction) is governed by the differential equation

$$\frac{d^2\theta}{dt^2} = -\frac{g}{l}\sin\theta,$$

where g (the acceleration due to gravity, $g \approx 32$ ft/sec^2) is a constant. Solve the differential equation of motion, under the assumption that θ is sufficiently small that $\sin\theta$ may be replaced by θ without appreciable error. Assume that $\theta = \theta_0$ and $d\theta/dt = 0$ when $t = 0$.

5. A circular disc of mass m and radius r is suspended by a thin wire attached to the center of one of its flat faces. If the disc is twisted through an angle θ, torsion in the wire tends to turn the disc back in the opposite direction. The differential equation that governs the motion is

$$\frac{1}{2}mr^2\frac{d^2\theta}{dt^2} = -k\theta,$$

where k is the coefficient of torsion of the wire. Find the motion if $\theta = \theta_0$ and $d\theta/dt = v_0$ at $t = 0$.

6. A cylindrical spar buoy, diameter 1 foot, weight 100 lb, floats partially submerged in an upright position. When it is depressed slightly from its equilibrium position and released, it bobs up and down, its motion being governed by the differential equation

$$\frac{100}{g}\frac{d^2x}{dt^2} = -16\pi x - c\frac{dx}{dt}.$$

Here $c\,(dx/dt)$ is the frictional resistance of the water. Find c if the period of oscillation is observed to be 1.6 sec.

7. Suppose the upper end of the spring in Fig. 20.2 is attached, not to a rigid support at A, but to a member which itself undergoes up and down motion given by a function of the time t, say $y = f(t)$. If the positive direction of y is downward, then the differential equation of motion is

$$m\frac{d^2x}{dt^2} + kx = kf(t).$$

Let $x = x_0$ and $dx/dt = 0$ when $t = 0$, and solve for x

(a) if $f(t) = A\sin\alpha t$ and $\alpha \neq \sqrt{k/m}$,

(b) if $f(t) = A\sin\alpha t$ and $\alpha = \sqrt{k/m}$.

20.13 POISSON PROBABILITY DISTRIBUTION

In the study of emission of particles from a radioactive substance, it is found that the number X of particles emitted during a fixed time interval is not precisely predictable. In fact, X is a variable that can assume only the values 0, 1, 2, . . . Some of these values are more likely to be observed than others. It is useful to introduce the symbol $P_n(t)$ to denote the *probability* that n particles will be emitted in a t-second interval. It is possible to arrive at a formula for $P_n(t)$ if we assume the following postulates.

Postulate 1. *The number of particles emitted in any time interval is independent of the number emitted in any other nonoverlapping time interval.*

Postulate 2. *The probability that a single particle is emitted in an interval $(t, t + h)$ is*

$$\lambda \cdot h + o(h),$$

where λ is a constant (for a particular substance) and $o(h)$, read "small oh of h," denotes any function having the property

$$\lim_{h \to 0} \frac{o(h)}{h} = 0. \tag{1}$$

Postulate 3. *The probability that more than one particle will be emitted in $(t, t + h)$ is $o(h)$.*

A variable that satisfies these three postulates will be called a *Poisson* random variable. The significances of the postulates are as follows.

First. Independence of numbers of particles emitted in nonoverlapping time intervals means that the probability of r particles in one interval and s in another is the product of the respective probabilities:

$$P \text{ (r particles emitted in 1st interval)} \\ \text{times} \\ P \text{ (s particles emitted in 2nd interval).*}$$

Second. The statement

$$P \text{ (single particle emitted in interval } (t, t+h)) \\ = \lambda \cdot h + o(h)$$

means that the probability is approximately proportional to the length h of the time interval. The approximation λh is better the smaller h is, the error $o(h)$ being composed of terms (like h^2 or $h \sin h$, for instance) which are themselves small compared to h, since

$$\frac{o(h)}{h} \to 0 \quad \text{as} \quad h \to 0.$$

Third. The probability that more than one particle is emitted in a very short time interval $(t, t+h)$ is $o(h)$, and is small compared to h.

From Postulates 2 and 3, it follows that the probability of one *or more* particles in the interval $(t, t+h)$ is also $\lambda h + o(h)$, because

$$P(\text{1 or more particles}) = P(\text{1 particle}) \\ + P(\text{more than 1 particle})$$

and

$$P(\text{1 particle}) = \lambda h + o_1(h),$$

$$P(\text{more than 1 particle}) = o_2(h).$$

Hence

$$P(\text{1 or more particles}) = \lambda h + o_1(h) + o_2(h) \\ = \lambda h + o(h), \tag{2}$$

since the sum of two functions $o_1(h)$ and $o_2(h)$ having the property that

$$\lim_{h\to 0} \frac{o_1(h)}{h} = 0, \qquad \lim_{h\to 0} \frac{o_2(h)}{h} = 0$$

is a function having the same property:

$$\lim_{h\to 0} \frac{o_1(h) + o_2(h)}{h} = \lim_{h\to 0} \frac{o_1(h)}{h} + \lim_{h\to 0} \frac{o_2(h)}{h} = 0.$$

From the postulates for a Poisson random variable we may deduce the following system of differential equations:

$$\frac{dP_0(t)}{dt} = -\lambda P_0(t), \tag{3a}$$

$$\frac{dP_n(t)}{dt} = -\lambda P_n(t) + \lambda P_{n-1}(t), \\ n = 1, 2, 3, \ldots \tag{3b}$$

Demonstration. We first establish (3a). Consider

$$\begin{aligned} P_0(t+h) &= P(\text{0 emissions in time } t+h) \\ &= P(\text{0 in time } t) \times P(\text{0 in time } (t, t+h)) \\ &= P_0(t) \cdot \{1 - P(\text{1 or more in } (t, t+h))\} \\ &= P_0(t) \cdot \{1 - \lambda h - o(h)\}. \end{aligned}$$

Therefore

$$P_0(t+h) - P_0(t) = -P_0(t) \cdot [\lambda h + o(h)],$$

$$\frac{P_0(t+h) - P_0(t)}{h} = -P_0(t) \cdot \left(\lambda + \frac{o(h)}{h}\right),$$

$$\begin{aligned} \frac{dP_0(t)}{dt} &= \lim_{h\to 0} \frac{P_0(t+h) - P_0(t)}{h} \\ &= -P_0(t) \cdot (\lambda + 0) = -\lambda P_0(t). \end{aligned}$$

Thus (3a) holds.

To establish (3b), consider the probability of n emissions in time $(0, t+h)$, where n is a positive integer. Consider the nonoverlapping intervals of time $(0, t)$ and $(t, t+h)$. The only possibilities for n particles in $(0, t+h)$ are:

(a) n particles in $(0, t)$ and 0 in $(t, t+h)$,

(b) $(n-1)$ particles in $(0, t)$ and 1 in $(t, t+h)$,

(c) $(n-2)$ particles in $(0, t)$ and 2 in $(t, t+h)$,

(d) $(n-3)$ particles in $(0, t)$ and 3 in $(t, t+h)$, . . .

* The notation P(1 or more particles) is an abbreviation for "probability that 1 or more particles are emitted."

Since these are mutually exclusive events, their probabilities add, and

$$P_n(t+h) = P_n(t)\cdot P_0(h) + P_{n-1}(t)\cdot P_1(h) + P_{n-2}(t)\cdot P_2(h) + \cdots + P_0(t)\cdot P_n(h). \quad (4)$$

Now for any $t \geq 0$,

$$P_0(t) + P_1(t) + P_2(t) + \cdots = 1, \quad (5)$$

because it is certain (that is, the probability is 1) that the number of particles emitted in time t is one of the integers 0, or 1, or 2, or 3, . . . , and since these possibilities are mutually exclusive, their probabilities therefore add. Hence in particular, if $t = h$ is small, we may replace t by h in (5) and use the fact (Eq. 2) that

$$P(\text{1 or more particles}) = P_1(h) + P_2(h) + \cdots = \lambda h + o(h)$$

to obtain

$$P_0(h) = 1 - \lambda h - o(h). \quad (6)$$

Hence Eq. (4) becomes

$$P_n(t+h) = P_n(t)\cdot[1 - \lambda h - o(h)] + P_{n-1}(t)\cdot[\lambda h + o_1(h)] + o_2(h), \quad (7)$$

since

$$P_{n-2}(t)\cdot P_2(h) + P_{n-3}(t)\cdot P_3(h) + \cdots + P_0(t)\cdot P_n(h)$$
$$\leq 1\cdot P_2(h) + 1\cdot P_3(h) + \cdots + 1\cdot P_n(h) = o(h).$$

Therefore, from Eq. (7),

$$P_n(t+h) - P_n(t) = -P_n(t)\cdot[\lambda h + o(h)] + P_{n-1}(t)\cdot[\lambda h + o_1(h)] + o_2(h)$$

and we want to find the limit

$$\frac{dP_n}{dt} = \lim_{h\to 0} \frac{P_n(t+h) - P_n(t)}{h}.$$

But this limit is the same as the limit of the expression

$$-P_n(t)\cdot\left(\lambda + \frac{o(h)}{h}\right) + P_{n-1}(t)\cdot\left(\lambda + \frac{o_1(h)}{h}\right) + \frac{o_2(h)}{h}$$

as $h \to 0$. Therefore

$$\frac{dP_n}{dt} = -\lambda P_n(t) + \lambda P_{n-1}(t). \ \square$$

The probabilities $P_0(t), P_1(t), \ldots$ satisfy the system of differential equations (3a, b) and the initial conditions

$$P_0(0) = 1,$$
$$P_1(0) = P_2(0) = \cdots = P_n(0) = 0, \quad n > 1. \quad (8)$$

That is, the number of particles emitted in 0 time is zero, with probability one. The solution of Eq. (3a), with $P_0(0) = 1$, is

$$P_0(t) = e^{-\lambda t}.$$

If we substitute this into Eq. (3b), with $n = 1$, we have

$$\frac{dP_1(t)}{dt} = -\lambda P_1(t) + e^{-\lambda t},$$

or

$$\frac{dP_1(t)}{dt} + \lambda P_1(t) = e^{-\lambda t}.$$

This equation is linear in $P_1(t)$, with integrating factor $e^{\lambda t}$. It is equivalent to

$$\frac{d}{dt}[e^{\lambda t}P_1(t)] = \lambda.$$

Hence

$$e^{\lambda t}P_1(t) = \lambda t + C_1.$$

But when $t = 0$, $P_1(0) = 0$, and thus $C_1 = 0$. Therefore

$$P_1(t) = \lambda t e^{-\lambda t}.$$

We can now substitute this into (3b) with $n = 2$ and obtain

$$\frac{dP_2(t)}{dt} + \lambda P_2(t) = \lambda(\lambda t)e^{-\lambda t}.$$

Again the equation is linear, with integrating factor $e^{\lambda t}$, and

$$\frac{d}{dt}[e^{\lambda t}P_2(t)] = e^{\lambda t}\cdot\lambda(\lambda t)e^{-\lambda t} = \lambda^2 t.$$

Hence

$$e^{\lambda t}P_2(t) = \frac{\lambda^2 t^2}{2} + C_2.$$

But $P_2(0) = 0$, and thus $C_2 = 0$; therefore

$$P_2(t) = \frac{(\lambda t)^2}{2!}\, e^{-\lambda t}.$$

Proceeding in this fashion, one discovers that

$$P_3(t) = \frac{(\lambda t)^3}{3!}\, e^{-\lambda t}, \quad P_4(t) = \frac{(\lambda t)^4}{4!}\, e^{-\lambda t}, \quad \ldots$$

The general term is

$$P_n(t) = e^{-\lambda t}\,\frac{(\lambda t)^n}{n!}, \qquad n = 0, 1, 2, \ldots \tag{9}$$

The set of numbers 0, 1, 2, 3, . . . with corresponding probabilities given by Eq. (9) is called a *Poisson probability distribution*. Figure 20.7 shows the graph of a Poisson distribution with $\lambda t = 2$. The viewpoint to adopt when considering a Poisson distribution is, in most instances, to imagine λt as fixed, and then Eq. (9) gives the probability of observing n occurrences of the event in question, for all possible different values of n.

The probability of zero emissions is $P_0(t) = e^{-\lambda t}$. Observe that this probability tends to zero as t increases: it becomes very improbable that *no* emissions occur over any very long interval of time.

Do the probabilities found in Eq. (9) add up to 1, as Eq. (4) requires? Let's try it and see:

$$\begin{aligned} P_0(t) + P_1(t) + \cdots &= e^{-\lambda t}\left(1 + \frac{\lambda t}{1!} + \frac{(\lambda t)^2}{2!} + \cdots\right) \\ &= e^{-\lambda t}\sum_{n=0}^{\infty}\frac{(\lambda t)^n}{n!} \\ &= e^{-\lambda t}\cdot e^{\lambda t} \\ &= e^0 = 1. \qquad \text{(Yes)} \end{aligned}$$

Interpretation of λ

The *mean* value of a random variable X is obtained by taking the weighted average of the possible values X may assume, each weighted according to the

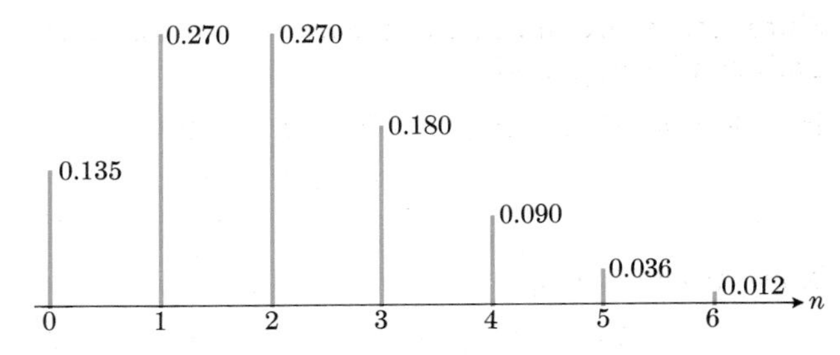

20.7 Poisson distribution with $\lambda t = 2$.

probability of its occurring. Equations (9) give these probabilities for a Poisson random variable.

Possible values of X	Probabilities of these possibilities	(Possible values) × (probability)
0	$e^{-\lambda t}$	$0 \cdot e^{-\lambda t}$
1	$e^{-\lambda t}\cdot(\lambda t)$	$1 \cdot e^{-\lambda t}\cdot(\lambda t)$
2	$e^{-\lambda t}\cdot(\lambda t)^2/2!$	$2 \cdot e^{-\lambda t}\cdot(\lambda t)^2/2!$
3	$e^{-\lambda t}\cdot(\lambda t)^3/3!$	$3 \cdot e^{-\lambda t}\cdot(\lambda t)^3/3!$
⋮	⋮	⋮
n	$e^{-\lambda t}(\lambda t)^n/n!$	$n \cdot e^{-\lambda t}(\lambda t)^n/n!$
⋮	⋮	⋮

Thus, the mean value of the Poisson random variable is

$$\begin{aligned} &e^{-\lambda t}\left[0 + 1(\lambda t) + 2\,\frac{(\lambda t)^2}{2!} + 3\,\frac{(\lambda t)^3}{3!} + \cdots + n\,\frac{(\lambda t)^n}{n!} + \cdots\right] \\ &= e^{-\lambda t}(\lambda t)\left[1 + (\lambda t) + \frac{(\lambda t)^2}{2!} + \cdots + \frac{(\lambda t)^{n-1}}{(n-1)!} + \cdots\right] \\ &= e^{-\lambda t}(\lambda t)e^{\lambda t} = \lambda t. \end{aligned}$$

Therefore λt is the *mean* number of particles emitted in time $(0, t)$. In particular, the mean number emitted in unit time, $t = 1$, is λ. In two units of time, it is 2λ, and so on. The mean number in t units of time is proportional to t, and λ is the proportionality factor.

Remark. Many phenomena have been found that satisfy the postulates for a Poisson random variable, or nearly do.* Among these are: the number of telephone calls coming into a central office during a particular hour of the day, the number of splices in a certain manufactured tape (per linear foot), the number of surface flaws (per square foot) in plating sheets of metal, the number of failures of electron tubes (per hour) in a given airborne instrumentation device, the number of flying bomb hits in the south of London during World War II (on regions of area $= \frac{1}{4}$ square kilometer). In some of these cases, the "t" in Eqs. (8) should be interpreted as so many units of length, or of area, rather than of time; but the formulas are correct with such interpretations and with λ equal to the mean number of occurrences of the event in question per unit of t.

Problem. Suppose there are 500 misprints in a book of 750 pages. Assuming that the number of misprints per page is (approximately) a Poisson random variable, find

(a) the probability that a page selected at random has no misprints,

(b) the probability of more than one misprint on a given page.

Solution. Let one page be the unit, $t = 1$. The mean number of misprints per page is $500/750 = \frac{2}{3}$. So we apply (9) with $\lambda = \frac{2}{3}$.

$$\begin{aligned} P_n(t) &= \text{probability of } n \text{ misprints on } t \text{ pages} \\ &= e^{-(2/3)t}\frac{(\frac{2}{3}t)^n}{n!}, \qquad n = 0, 1, 2, \ldots \end{aligned}$$

(a) The probability of no misprints on a single page is $P_n(t)$, with $n = 0$, $t = 1$; or

$$P_0(1) = e^{-2/3} \approx 0.514.$$

(b) The probability of more than one misprint on a single page is $\sum_{n=2}^{\infty} P_n(t)$, with $t = 1$. This is also equal to $1 - P_0(t) - P_1(t)$, or

$$1 - e^{-2/3}[1 + \tfrac{2}{3}] \approx 0.143.$$

---

* *References.* W. Feller, *An Introduction to Probability Theory and Its Applications*, 3rd edition, Wiley, 1968, p. 159. B. W. Lindgren and G. W. McElrath, *Introduction to Probability and Statistics*, Macmillan, 1959, pp. 54–59.

EXERCISES

---

1. Substitute $n = 3$ and $P_2(t) = e^{-\lambda t}(\lambda t)^2/2!$ into Eq. (3b) and solve for $P_3(t)$, subject to the appropriate initial condition.

2. Assuming that Eq. (9) is correct for some integer $n - 1$, so that

$$P_{n-1}(t) = \frac{e^{-\lambda t}(\lambda t)^{n-1}}{(n-1)!}$$

solve Eq. (3b) for $P_n(t)$, subject to the appropriate initial condition. [In other words, show that Eq. (9) is valid for all integers $n \geq 0$ by the method of mathematical induction.]

3. Sketch the graphs of $y = P_n(t)$ as functions of t, $t \geq 0$, for $n = 2$ and $\lambda = 2$, 1, and $\frac{1}{2}$. Find the maximum, minimum, and inflection points.

4. Prove directly that $e^{-\lambda t}(\lambda t)^n/n!$ is never greater than 1 if $\lambda > 0$, $t \geq 0$, and n is an integer ≥ 0. This result must be true, of course, since $P_n(t)$ is a probability and hence is never greater than 1.

5. In the problem of the book of 750 pages and 500 misprints, what is the probability

   (a) that a chapter of 15 pages has no misprints?

   (b) that a section of 6 pages has one or more misprints?

6. Assume that a certain manufacturer of automobile radiators has found that the number of minor defects in his radiators is (approximately) a Poisson random variable with mean equal to 0.02. If 20 radiators are selected at random from his production line, what is the probability that there will be

   (a) no minor defects in the entire group of 20 radiators,

   (b) one or more minor defects in the group of 20 radiators,

   (c) no minor defects in the first radiator of the 20?

7. A baker finds that the number of raisins per loaf in his raisin bread is (approximately) a Poisson random variable with mean equal to 200 raisins per loaf. Suppose that each loaf is sliced into 20 slices of uniform size.

   (a) What is the mean number of raisins per slice?

   (b) What is the probability of getting five or fewer raisins in a slice?

8. The first term of a certain Poisson probability distribution is $P_0(t) = 0.135$. Find

(a) $P_1(t)$, (b) $P_3(t)$, (c) $P_0(t/2)$, (d) $P_1(2t)$.

9. Assume that the number of α-particles registered by a Geiger counter in a certain experiment is a Poisson random variable. Suppose that the mean number registered per 10-second interval of time is eight. What is the probability that there will be exactly four registered in a given interval of five seconds?

10. Let $Q_n(t, t+h)$ be the probability that the nth particle is emitted between times t and $t+h$. Then explain why it is true that if $t+h > t > 0$, it is true that

$$Q_n(t, t+h) = P_{n-1}(t) \cdot P_1(h) + P_{n-2}(t) \cdot P_2(h) + \cdots + P_0(t) \cdot P_n(h);$$

and from this show that

$$Q_n(t, t+h) = \frac{he^{-\lambda t}\lambda^n t^{n-1}}{(n-1)!} + o(h).$$

11. Using the result in Exercise 10, show that

$$Q_n(a, b) = \int_a^b \frac{e^{-\lambda t}\lambda^n t^{n-1}}{(n-1)!}\, dt, \qquad b > a > 0.$$

Hint. Let p be a positive integer, $\Delta t = h = (b-a)/p$, and let

$$\begin{aligned} Q_n(a, b) &= Q_n(a, a+h) + Q_n(a+h, a+2h) + \cdots + Q_n(b-h, b) \\ &= \sum_{i=1}^{p} f(t_i)\,\Delta t + \frac{b-a}{h}\, o(h), \end{aligned}$$

with

$$f(t) = \frac{e^{-\lambda t}\lambda^n t^{n-1}}{(n-1)!} \quad \text{and} \quad t_i = a + (i-1)h.$$

Then let $p \to \infty$.

12. Let

$$F(n) = \int_0^\infty e^{-\lambda t}\lambda^n t^{n-1}\, dt = \int_0^\infty e^{-u} u^{n-1}\, du.$$

(a) Integrate by parts and show that

$$F(n) = (n-1)F(n-1).$$

(b) Show directly that $F(1) = 1$.

(c) From (a) and (b), show that

$$F(2) = 1, \qquad F(3) = 2F(2) = 2!,$$

and in general that $F(n) = (n-1)!$.

(d) Using the result of (c) and of Exercise 11, show that $Q_n(0, \infty) = 1$ for any integer $n \geq 1$. What does this mean in terms of probability?

REVIEW QUESTIONS AND EXERCISES

1. List some differential equations (having physical interpretations) that you have come across in your courses in chemistry, physics, or engineering; or look at some in the articles on differential equations, dynamics, electromagnetic waves, hydromechanics, quantum mechanics, or thermodynamics in the *Encyclopaedia Britannica.*
2. How are differential equations classified?
3. What is meant by a "solution" of a differential equation?
4. Review methods for solving ordinary first-order and first-degree differential equations

   (a) when the variables are separable,
   (b) when the equation is homogeneous,
   (c) when the equation is linear in one variable,
   (d) when the equation is exact.

   Illustrate each with an example.
5. Review methods of solving second-order equations

   (a) with dependent variable missing,
   (b) with independent variable missing.

   Illustrate each with an example.
6. Review methods for solving linear differential equations with constant coefficients

   (a) in the homogeneous case,
   (b) in the nonhomogeneous case.

   Illustrate each with an example.
7. If an external force F acts on a system whose mass varies with time, Newton's law of motion is

$$\frac{d(mv)}{dt} = F + (v+u)\frac{dm}{dt}.$$

In this equation, m is the mass of the system at time t, v is its velocity, and $v + u$ is the velocity of the mass that is entering (or leaving) the system at the rate dm/dt. Suppose that a rocket of initial mass m_0 starts from rest, but is driven upward by firing some of its mass directly backward at the constant rate of $dm/dt = -b$ units per second and at constant speed relative to the rocket $u = -c$. The only external force acting on the rocket is $F = -mg$, due to gravity. Under these assumptions, show that the height of the rocket above the ground at the end of t seconds (t small compared to m_0/b) is

$$y = c\left(t + \frac{m_0 - bt}{b} \ln \frac{m_0 - bt}{m_0}\right) - \frac{1}{2} gt^2.$$

(See Martin and Reissner, *Elementary Differential Equations*, 2nd edition, Addison-Wesley, 1961, pp. 26 ff.)

MISCELLANEOUS EXERCISES

Solve the following differential equations.

1. $y \ln y \, dx + (1 + x^2) \, dy = 0$
2. $\dfrac{dy}{dx} = \dfrac{y^2 - y - 2}{x^2 + x}$
3. $e^{x+2y} \, dy - e^{y-2x} \, dx = 0$
4. $\sqrt{1 + (dy/dx)^2} = ky$
5. $y \, dy = \sqrt{1 + y^4} \, dx$
6. $(2x + y) \, dx + (x - 2y) \, dy = 0$
7. $\dfrac{dy}{dx} = \dfrac{x^2 + y^2}{2xy}$
8. $x \dfrac{dy}{dx} = y + \sqrt{x^2 + y^2}$
9. $x \, dy = \left(y + x \cos^2 \dfrac{y}{x}\right) dx$
10. $x(\ln y - \ln x) \, dy = y(1 + \ln y - \ln x) \, dx$
11. $x \, dy + (2y - x^2 - 1) \, dx = 0$
12. $\cos y \, dx + (x \sin y - \cos^2 y) \, dy = 0$
13. $\cosh x \, dy - (y + \cosh x) \sinh x \, dx = 0$
14. $(x + 1) \, dy + (2y - x) \, dx = 0$
15. $(1 + y^2) \, dx + (2xy + y^2 + 1) \, dy = 0$
16. $(x^2 + y) \, dx + (e^y + x) \, dy = 0$
17. $(x^2 + y^2) \, dx + (2xy + \cosh y) \, dy = 0$
18. $(e^x + \ln y) \, dx + \dfrac{x + y}{y} \, dy = 0$
19. $x(1 + e^y) \, dx + \frac{1}{2}(x^2 + y^2)e^y \, dy = 0$
20. $\left(\sin x + \tan^{-1} \dfrac{y}{x}\right) dx - (y - \ln \sqrt{x^2 + y^2}) \, dy = 0$
21. $\dfrac{d^2y}{dx^2} - 2y \dfrac{dy}{dx} = 0$
22. $\dfrac{d^2x}{dy^2} + 4x = 0$
23. $\dfrac{d^2y}{dx^2} = 1 + \left(\dfrac{dy}{dx}\right)^2$
24. $\dfrac{d^2x}{dy^2} = 1 - \left(\dfrac{dx}{dy}\right)^2$
25. $x^2 \dfrac{d^2y}{dx^2} + x \dfrac{dy}{dx} = 1$
26. $\dfrac{d^2y}{dx^2} - 4 \dfrac{dy}{dx} + 3y = 0$
27. $\dfrac{d^3y}{dx^3} - 2 \dfrac{d^2y}{dx^2} + \dfrac{dy}{dx} = 0$
28. $\dfrac{d^2y}{dx^2} + 4y = \sec 2x$
29. $\dfrac{d^2y}{dx^2} - \dfrac{dy}{dx} - 2y = e^{2x}$
30. $\dfrac{d^2y}{dx^2} - 2 \dfrac{dy}{dx} + 5y = e^{-x}$
31. Find the *general solution* of the differential equation $4x^2y'' + 4xy' - y = 0$, given that there is a particular solution of the form $y = x^c$ for some constant c.
32. Show that the only curves having constant curvature are circles and straight lines.
33. Find the orthogonal trajectories of the family of curves $x^2 = Cy^3$.

    Caution. The differential equation should not contain the arbitrary constant C.
34. Find the orthogonal trajectories of the family of circles $(x - C)^2 + y^2 = C^2$.
35. Find the orthogonal trajectories of the family of parabolas $y^2 = 4C(C - x)$.
36. The equation $d^2y/dt^2 + 100y = 0$ represents a simple harmonic motion. Find the general solution of the equation and determine the constants of integration if $y = 10$, $dy/dt = 50$, when $t = 0$. Find the period and the amplitude of the motion.

DETERMINANTS AND LINEAR EQUATIONS

APPENDIX I

A.1 INTRODUCTION

At various places in this book some knowledge of determinants has been assumed. This topic is part of algebra, and might have been included in the chapter on linear algebra. But since some readers will already have enough familiarity with the subject to meet their needs, and since others may only wish to study it briefly and learn the operational rules, we include it here as an appendix. We shall develop nearly all the theory of determinants that the student needs for this book and for other elementary courses in mathematics (for example, in differential equations), in science, and in engineering. This appendix is almost entirely independent of the rest of the book, so that the student can read it whenever he finds he needs to know something about the subject.

The square array of numbers displayed in Eq. (1) is called *an $n \times n$ matrix*, or a *square matrix of order n*. Because it takes a lot of space to write it out *in extenso*, we also abbreviate it as $A = [a_{ij}]$. By convention, the *first* subscript, i, tells the number of the *row* in the square array in which the element a_{ij} is located, and the second subscript, j, tells the number of the *column* in which it is located. Because A has n rows and n columns, the indices i and j range from 1 through n inclusive; we indicate this by writing

$$A = [a_{ij}], \quad 1 \leq i \leq n, 1 \leq j \leq n,$$

as the complete abbreviation for

$$A = \begin{bmatrix} a_{11} & a_{12} & \dots & a_{1n} \\ a_{21} & a_{22} & \dots & a_{2n} \\ \dots & \dots & \dots & \dots \\ a_{n1} & a_{n2} & \dots & a_{nn} \end{bmatrix}. \tag{1}$$

The square brackets that enclose the array in (1) are sometimes replaced by large parentheses, or by a pair of vertical lines before and after the array; e.g.,

$$\begin{pmatrix} a_{11} & a_{12} \\ a_{21} & a_{22} \end{pmatrix} \quad \text{and} \quad \begin{Vmatrix} a_{11} & a_{12} \\ a_{21} & a_{22} \end{Vmatrix}$$

represent the same 2×2 matrix. However, if only one vertical bar appears before and after the array, the object so denoted is the *determinant* of the matrix. The *determinant* of any square matrix of numbers is a *single number*, called det A or det$[a_{ij}]$, and denoted by the expressions

$$\det A = \det [a_{ij}] = \begin{vmatrix} a_{11} & a_{12} & \cdots & a_{1n} \\ a_{21} & a_{22} & \cdots & a_{2n} \\ \cdots & \cdots & \cdots & \cdots \\ a_{n1} & a_{n2} & \cdots & a_{nn} \end{vmatrix}. \tag{2}$$

There is little occasion to talk about the case $n = 1$, but in that special case the vertical bars that signify the determinant look like absolute value signs; however, context makes the meaning clear. Specifically, for $n = 1$, the determinant of a 1×1 matrix is just the element a_{11} and not its absolute value.

In the next article, we first indicate briefly how determinants of order two arise naturally in the solution of pairs of simultaneous linear equations in two unknowns. Then we go on to discuss third-order determinants, and introduce some technical vocabulary which we then use in developing the theory for nth-order determinants. It is now old-fashioned (although still standard usage) to employ such terminology as "a determinant of order n" in speaking of an array such as (2); it is more appropriate to adhere strictly to the notion that a determinant is NOT the array, but the single number that we assign to that array as its value, and to govern the terminology accordingly. However, since we shall restrict our theoretical discussion to just the theory of determinants, we shall use the standard, older terminology. Thus, for example, when we speak of interchanging two rows of a determinant, we are abusing the language; we should speak of interchanging two rows of the corresponding *matrix*. But the important point is not this fact about the language, but rather the theorem that tells what effect that operation has on the determinant of the matrix: It changes the sign. (See Theorem 4, Article A.5.)

A.2 DETERMINANTS AND LINEAR EQUATIONS

Determinants originate in connection with the problem of solving simultaneous linear equations. For example, if we wish to find the point of intersection of two straight lines

$$a_1x + b_1y = c_1, \tag{1a}$$

$$a_2x + b_2y = c_2, \tag{1b}$$

we try to find a pair of numbers x and y that satisfies both equations simultaneously. There are several methods of doing this, but one fairly common method is to find equations equivalent to the given equations but having the property that the unknowns (x and y) are separated. This may be achieved, for example, by multiplying the first equation by b_2, the second equation by $-b_1$, and adding, to obtain the equation

$$(a_1b_2 - a_2b_1)x = c_1b_2 - c_2b_1; \tag{2a}$$

and by multiplying the first equation by $-a_2$, the second by a_1, and adding, to obtain the equation

$$(a_1b_2 - a_2b_1)y = a_1c_2 - a_2c_1. \tag{2b}$$

Equations (2a) and (2b) must be satisfied by any pair of numbers x and y that satisfies the original equations (1). They have the added merit of being easy to solve, since each equation contains only one of the unknowns. Thus, we find that the only pair of numbers which can satisfy Eqs. (1) is

$$x = \frac{c_1b_2 - c_2b_1}{a_1b_2 - a_2b_1}, \qquad y = \frac{a_1c_2 - a_2c_1}{a_1b_2 - a_2b_1}, \tag{3}$$

provided that

$$a_1b_2 - a_2b_1 \neq 0. \tag{4}$$

By direct substitution from (3) back into (1) we find, conversely, that these values of x and y do indeed satisfy both of the original equations. We shall verify that this is so for the first equation and leave to the reader the computation for the second. Substituting for x and y from (3) into (1a), we find

$$a_1x + b_1y = a_1\left(\frac{c_1b_2 - c_2b_1}{a_1b_2 - a_2b_1}\right) + b_1\left(\frac{a_1c_2 - a_2c_1}{a_1b_2 - a_2b_1}\right)$$

$$= \frac{a_1c_1b_2 - a_1c_2b_1 + b_1a_1c_2 - b_1a_2c_1}{a_1b_2 - a_2b_1}$$

$$= \frac{c_1(a_1b_2 - a_2b_1)}{a_1b_2 - a_2b_1} = c_1.$$

In the exceptional case where $a_1b_2 - a_2b_1$ vanishes, the left-hand sides of Eqs. (2a) and (2b) are both zero and the Eqs. (1) are inconsistent unless

$$c_1b_2 - c_2b_1 = a_1c_2 - a_2c_1 = 0$$

as well. In this case we have a pair of straight lines having equal slopes, and one of two things happens:

(a) The straight lines coincide and there are infinitely many number-pairs (x, y) that satisfy the equations, or else

(b) the lines are parallel and there are no number-pairs (x, y) that satisfy the equations simultaneously.

Example 1. If the equations are

$$2x + 3y = 5, \quad 4x + 6y = 10,$$

then x may have any value; and provided $y = (5 - 2x)/3$, the point (x, y) will lie on both lines, since they are really not distinct. But if the equations are

$$2x + 3y = 5, \quad 4x + 6y = 11,$$

then the lines are parallel, and we say that the equations are inconsistent because no pair of numbers x and y can satisfy both equations simultaneously.

Let us, then, rule out the case of parallel or coincident lines and discuss the results obtained in (3) when condition (4) is satisfied. The expression $a_1b_2 - a_2b_1$ is defined to be the expanded value of the *determinant*

$$D_2 = \begin{vmatrix} a_1 & b_1 \\ a_2 & b_2 \end{vmatrix} \tag{5a}$$

which we get by writing the coefficients of x and y from Eqs. (1) in the order indicated in (5a). Such an expression is called a *determinant of order two,* since it contains two rows and two columns. To evaluate such a determinant, we multiply the diagonal elements in the upper-left and lower-right corners to get a_1b_2, then multiply the elements on the other diagonal to get a_2b_1, and finally subtract the latter product from the former:

$$D_2 = a_1b_2 - a_2b_1. \tag{5b}$$

Example 2

$$\begin{vmatrix} 3 & 2 \\ 7 & 5 \end{vmatrix} = 3 \cdot 5 - 7 \cdot 2 = 15 - 14 = 1.$$

Example 3

$$\begin{vmatrix} 4 & -3 \\ 2 & -6 \end{vmatrix} = 4 \cdot (-6) - 2 \cdot (-3) = -24 + 6 = -18.$$

Now the equations

$$a_1x + b_1y = c_1, \quad a_2x + b_2y = c_2$$

have the solutions (3), which may be expressed by determinants as

$$x = \frac{\begin{vmatrix} c_1 & b_1 \\ c_2 & b_2 \end{vmatrix}}{\begin{vmatrix} a_1 & b_1 \\ a_2 & b_2 \end{vmatrix}}, \qquad y = \frac{\begin{vmatrix} a_1 & c_1 \\ a_2 & c_2 \end{vmatrix}}{\begin{vmatrix} a_1 & b_1 \\ a_2 & b_2 \end{vmatrix}}. \tag{6}$$

In the equations for both x and y, the denominator is the determinant of the coefficients, arranged in order, from the given equations. The numerator in the equation for x is a determinant which may be obtained from the denominator determinant by suppressing from the latter the coefficients of x and writing the constants c_1 and c_2 in their places. The numerator in the equation for y may similarly be obtained from the denominator determinant by suppressing from the latter the coefficients of y and writing the constants c_1 and c_2 in their places. In applying this rule, it is assumed that the denominator determinant is not zero. If it is zero, one has the case where there is no solution if the equations are inconsistent, or there are infinitely many solutions if the straight lines coincide. If the denominator determinant is zero, Eqs. (6) are not to be used.

Problem. Solve the simultaneous equations

$$2x - 3y = 8,$$

$$3x + y = 1.$$

Solution

$$x = \frac{\begin{vmatrix} 8 & -3 \\ 1 & 1 \end{vmatrix}}{\begin{vmatrix} 2 & -3 \\ 3 & 1 \end{vmatrix}} = \frac{8 \cdot 1 - 1 \cdot (-3)}{2 \cdot 1 - 3 \cdot (-3)} = \frac{8 + 3}{2 + 9} = 1,$$

$$y = \frac{\begin{vmatrix} 2 & 8 \\ 3 & 1 \end{vmatrix}}{\begin{vmatrix} 2 & -3 \\ 3 & 1 \end{vmatrix}} = \frac{2 \cdot 1 - 3 \cdot 8}{2 \cdot 1 - 3 \cdot (-3)} = \frac{-22}{11} = -2.$$

EXERCISES

Expand the following second-order determinants.

1. $\begin{vmatrix} 3 & 4 \\ 2 & 5 \end{vmatrix}$
2. $\begin{vmatrix} 0 & 3 \\ -1 & 7 \end{vmatrix}$
3. $\begin{vmatrix} 5 & 7 \\ 6 & 2 \end{vmatrix}$
4. $\begin{vmatrix} a & b \\ 2a & 2b \end{vmatrix}$

Solve the sets of simultaneous equations in Exercises 5 and 6 by means of determinants. In both cases, check your answers by substituting them back into the original equations.

5. $2x + 3y = 5, \quad 3x - y = 2$
6. $4x - 3y = 6, \quad 3x - 2y = 5$
7. Show that the two lines
$$a_1x + b_1y = c_1, \quad a_2x + b_2y = c_2$$
are parallel or coincide if and only if the determinant D_2, Eq. (5 a, b), is zero.
8. Show that all the determinants appearing in Eq. (6) are zero if the equation $a_2x + b_2y = c_2$ is a multiple of the equation $a_1x + b_1y = c_1$; that is, if there is a constant k such that
$$a_2 = ka_1, \quad b_2 = kb_1, \quad c_2 = kc_1.$$
9. Suppose the elements of the determinant
$$F = \begin{vmatrix} u_1 & v_1 \\ u_2 & v_2 \end{vmatrix}$$
are differentiable functions of x. Show that
$$\frac{dF}{dx} = \begin{vmatrix} \dfrac{du_1}{dx} & \dfrac{dv_1}{dx} \\ u_2 & v_2 \end{vmatrix} + \begin{vmatrix} u_1 & v_1 \\ \dfrac{du_2}{dx} & \dfrac{dv_2}{dx} \end{vmatrix}.$$
In particular, verify the result for the case
$$F = \begin{vmatrix} \cos x & -\sin x \\ \sin x & \cos x \end{vmatrix}.$$
10. Solve the equations
$$x' \cos \alpha - y' \sin \alpha = x, \quad x' \sin \alpha + y' \cos \alpha = y$$
for x' and y' in terms of x and y by means of determinants.

A.3 DETERMINANTS OF ORDER THREE

A third-order determinant

$$D_3 = \begin{vmatrix} a_1 & b_1 & c_1 \\ a_2 & b_2 & c_2 \\ a_3 & b_3 & c_3 \end{vmatrix} \tag{1}$$

contains nine elements arranged in three horizontal rows and three vertical columns. Each element has an "address" in the determinant given by its row number r and its column number s. Thus, in (1), the element b_3 is in the third row and second column; that is, $r = 3$, $s = 2$ for b_3. Similarly, for the element c_1, $r = 1$ and $s = 3$. We have used different letters a, b, and c to denote the different columns in (1) and subscripts 1, 2, and 3 on these letters to denote the rows.

To obtain the expanded value of the determinant of the third order as in (1), one forms *all possible products* $a_ib_jc_k$, where the subscripts i, j, and k form some permutation of the numbers 1, 2, and 3. There are precisely six possible products of this kind: $a_1b_2c_3$, $a_1b_3c_2$, $a_2b_1c_3$, $a_2b_3c_1$, $a_3b_1c_2$, and $a_3b_2c_1$. One must next attach an algebraic sign to each of these six products according to this rule: The sign is to be plus if the subscripts i, j, and k in $a_ib_jc_k$ form an *even permutation* of the integers 1, 2, 3; and the sign is to be minus if i, j, k is an *odd permutation* of 1, 2, 3. This terminology will now be explained.

When the numbers 1, 2, 3 are arranged in their normal order, we say there are no inversions among them. If, on the other hand, they are arranged in some other order, such as 3, 1, 2, we say that we now have a *permutation* of them, and the permutation is "even" or "odd" according as the number of *inversions* (defined below) is an even integer (0, 2, 4, . . .) or an odd integer (1, 3, 5, . . .), respectively.

The number of inversions in any permutation is determined as follows. If, in a given permutation, i is any integer that is followed by a smaller integer, we say that there is an *inversion* relative to i and the smaller integer. The number of integers following i and smaller than i gives the total number of inversions relative to i. The total number of inversions relative to all the integers of a permutation may be called the *index* of that permutation. In the permutation

$$5 \quad 2 \quad 1 \quad 4 \quad 3,$$

for example, we may count the following:

relative to	*the number of inversions is*
5	4
2	1
1	0
4	1
3	0.

Thus the index of the given permutation is

$$4 + 1 + 1 = 6,$$

and the permutation is said to be *even* because 6 is an even integer.

Let us now examine in turn each of the six products in the expansion of D_3, count the inversions in its subscripts, and then affix a plus sign if this number is even and a minus sign if the number is odd.

Product	Subscripts	Index	Signed product
$a_1b_2c_3$	1 2 3	0	$+a_1b_2c_3$
$a_1b_3c_2$	1 3 2	1	$-a_1b_3c_2$
$a_2b_1c_3$	2 1 3	1	$-a_2b_1c_3$
$a_2b_3c_1$	2 3 1	2	$+a_2b_3c_1$
$a_3b_1c_2$	3 1 2	2	$+a_3b_1c_2$
$a_3b_2c_1$	3 2 1	3	$-a_3b_2c_1$

We are now in a position to complete the definition of a third-order determinant. We form all possible signed products $(\pm)\ a_ib_jc_k$, where i, j, k is a permutation of 1, 2, 3 in some order and the sign is plus or minus according as the number of inversions in i, j, k is even or odd. The expanded value of the determinant is the algebraic sum of the signed products:

$$D_3 = \begin{vmatrix} a_1 & b_1 & c_1 \\ a_2 & b_2 & c_2 \\ a_3 & b_3 & c_3 \end{vmatrix}$$

$$= a_1b_2c_3 + a_2b_3c_1 + a_3b_1c_2 - a_3b_2c_1 - a_2b_1c_3 - a_1b_3c_2. \tag{2}$$

The expansion (2) can also be expressed in several ways as the sum of three second-order *minor* determinants, each multiplied by an element of a row or column of D_3. For example, it is a simple matter to verify by direct computation that the expanded value in (2) is precisely the same as

$$D_3 = \begin{vmatrix} a_1 & b_1 & c_1 \\ a_2 & b_2 & c_2 \\ a_3 & b_3 & c_3 \end{vmatrix}$$

$$= a_1 \begin{vmatrix} b_2 & c_2 \\ b_3 & c_3 \end{vmatrix} - b_1 \begin{vmatrix} a_2 & c_2 \\ a_3 & c_3 \end{vmatrix} + c_1 \begin{vmatrix} a_2 & b_2 \\ a_3 & b_3 \end{vmatrix}. \tag{3}$$

Here, we say that we have expanded D_3 by minors according to elements of its first row. The *minor* of the element in the rth row and the sth column of a determinant of any order n is the determinant of order $n - 1$ obtained by suppressing all the elements in the rth row and sth column in the original array. The minor of a_1 is found by covering up the first row and first column in D_3, and the result, which we shall call A_1, is

$$A_1 = \begin{vmatrix} b_2 & c_2 \\ b_3 & c_3 \end{vmatrix} = b_2c_3 - b_3c_2.$$

Similarly, the minor of b_1 is found by covering up the first row and second column of D_3, and the result, denoted by $-B_1$, is

$$-B_1 = \begin{vmatrix} a_2 & c_2 \\ a_3 & c_3 \end{vmatrix} = a_2c_3 - a_3c_2.$$

The *negative* of this minor is called the *cofactor* of b_1 in D_3, and we arrange to call the minor $-B_1$ so that the cofactor is $+B_1$. In general, the cofactor of the element in the rth row and sth column of a determinant is related to its minor according to the law

$$\text{cofactor} = (-1)^{r+s}\ \text{minor}. \tag{4}$$

Using cofactors instead of minors in the expansion (3), with A_1, B_1, and C_1 denoting the cofactors of a_1, b_1, and c_1, respectively, we have

$$D_3 = a_1A_1 + b_1B_1 + c_1C_1. \tag{5}$$

The determinant is the sum of the products of the elements of the first row by their corresponding cofactors.

It may also be verified by comparison with the expansion in (2) that the determinant is the sum of the products of the elements of *any* row (or column) by their corresponding cofactors. The cofactor of an element is simply its *signed minor*, the sign being + or − according to the law in Eq. (4). The sign may also be obtained from a checkerboard arrangement of + and − signs. One starts with + in the upper left corner and changes sign in going from any square to an adjacent square, as follows:

$$\begin{vmatrix} + & - & + \\ - & + & - \\ + & - & + \end{vmatrix}.$$

Problem 1. Evaluate

$$D = \begin{vmatrix} 2 & 2 & 1 \\ -1 & 0 & -1 \\ 3 & -1 & 3 \end{vmatrix}.$$

Solution. We exploit the presence of small numbers and the zero in the second row and expand the determinant by cofactors of the elements of this row.

$$D = -(-1)\begin{vmatrix} 2 & 1 \\ -1 & 3 \end{vmatrix} + 0\begin{vmatrix} 2 & 1 \\ 3 & 3 \end{vmatrix} - (-1)\begin{vmatrix} 2 & 2 \\ 3 & -1 \end{vmatrix}$$

$$= +(6+1) + 0 + (-2-6) = -1.$$

Actually, we need not have written the cofactor of zero, since it contributes nothing to the answer.

Problem 2. Show that the equation

$$\begin{vmatrix} x & y & 1 \\ -1 & 2 & 1 \\ 1 & 0 & 1 \end{vmatrix} = 0$$

represents a straight line which passes through the points $(-1, 2)$ and $(1, 0)$.

Solution. Expanding the determinant by cofactors of the elements of its first row and equating the result with zero, we get

$$x\begin{vmatrix} 2 & 1 \\ 0 & 1 \end{vmatrix} - y\begin{vmatrix} -1 & 1 \\ 1 & 1 \end{vmatrix} + 1\begin{vmatrix} -1 & 2 \\ 1 & 0 \end{vmatrix} = 0,$$

or

$$2x + 2y - 2 = 0,$$

or

$$x + y = 1.$$

This is the equation of a straight line. Clearly, by substitution, the given points satisfy the equation.

Another fairly convenient device for evaluating a third-order determinant consists in repeating the first two columns of the determinant to the right of it, and then taking the sum of the products along diagonals parallel to the main diagonal minus the products along the diagonals that run up from left to right. This is illustrated schematically in the accompanying diagram. A word of caution: This method does not work for determinants of order 4 or higher!

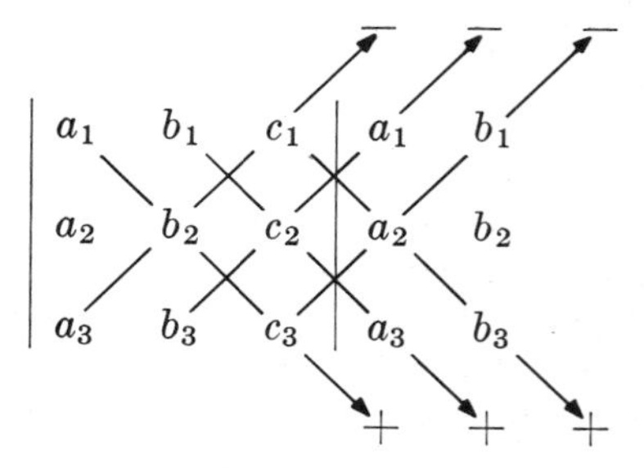

Example

$$\begin{vmatrix} 3 & 2 & 1 \\ 2 & 3 & 1 \\ 1 & 2 & 3 \end{vmatrix}\begin{matrix} 3 & 2 \\ 2 & 3 \\ 1 & 2 \end{matrix} = (27+2+4) - (3+6+12) = 12.$$

EXERCISES

1. Find the index of the permutation 4 1 3 2.
2. Verify Eq. (3) by expanding the second-order determinants on the right and comparing the result with Eq. (2).
3. Find the *cofactor* of the element in the second row and first column of the determinant

$$\begin{vmatrix} 2 & 3 & 4 \\ 3 & 2 & -1 \\ 4 & 3 & 7 \end{vmatrix}.$$

4. Show that the third-order determinant, Eq. (2), is equal to the sum of the products of the elements of the second column by their corresponding cofactors.

5. Evaluate the determinant

$$\begin{vmatrix} 2 & 2 & 1 \\ 1 & 3 & 2 \\ 3 & 1 & -1 \end{vmatrix}$$

by two different methods.

Evaluate these third-order determinants.

6. $\begin{vmatrix} 1 & 3 & -2 \\ 2 & -1 & 1 \\ -2 & 2 & 3 \end{vmatrix}$ 7. $\begin{vmatrix} 2 & 0 & 2 \\ 0 & 3 & -3 \\ -3 & -2 & 0 \end{vmatrix}$

A.4 DETERMINANTS OF ORDER n

To denote a determinant of order n, it is advantageous to use a slightly different notation from that introduced above. Namely, we shall not use different letters of the alphabet to represent the elements in the different columns, but shall instead use double subscripts and write a_{ij} to denote the element in the ith row, jth column of the determinant:

$$D_n = \begin{vmatrix} a_{11} & a_{12} & a_{13} & \cdots & a_{1n} \\ a_{21} & a_{22} & a_{23} & \cdots & a_{2n} \\ \cdots & \cdots & \cdots & \cdots & \cdots \\ a_{n1} & a_{n2} & a_{n3} & \cdots & a_{nn} \end{vmatrix}. \tag{1}$$

The expanded value of D_n is the sum of all signed products

$$(\pm) a_{1i_1} a_{2i_2} a_{3i_3} \cdots a_{ni_n}, \tag{2}$$

where $i_1, i_2, \ldots, i_n$ is a permutation of the integers $1, 2, 3, \ldots, n$ in some order and the sign is plus or minus, according as the permutation is even or odd.

For a determinant of order four, for example, we shall have

$$D_4 = \sum (\pm) a_{1i_1} a_{2i_2} a_{3i_3} a_{4i_4},$$

where i_1, i_2, i_3, i_4 is to range over all permutations of 1, 2, 3, 4. There are $4! = 24$ such permutations, and accordingly the determinant D_4 will be the sum of 24 signed products.

In general, a determinant of order n is the sum of $n!$ signed products. Since this number increases rapidly with n, it is desirable to develop equivalent ways of expanding determinants, but ways that will be less laborious than the direct application of the definition. This is the purpose behind the discussion which follows.

EXERCISE

Write out the 24 permutations of the numbers 1, 2, 3, 4. What is the index of the permutation 4 3 2 1? In the expansion of the determinant D_4, what sign is to be attached to the product $a_{14}a_{23}a_{32}a_{41}$?

A.5 PROPERTIES OF DETERMINANTS OF ORDER n

Theorem 1. *If all elements of any row of D_n are zero, then the expanded value of the determinant is zero.*

Proof. Each of the signed products contains one factor from each row (because the first subscripts include all the integers $1, 2, \ldots, n$), so that each product is zero under the given hypothesis. Hence their sum is also zero; that is, $D_n = 0$. □

Theorem 2. *If D_n and D_n' are two determinants of the nth order which differ only in that the elements in some one row of D_n' are k times the corresponding elements in D_n, then the expanded value of D_n' is k times the value of D_n.*

(The proof is similar to the proof of Theorem 1, and we omit it.) This result permits us to factor out any common factor from a row of a determinant and to write the factor in front of the new determinant. For example:

$$\begin{vmatrix} a_{11} & a_{12} & a_{13} \\ ka_{21} & ka_{22} & ka_{23} \\ a_{31} & a_{32} & a_{33} \end{vmatrix} = k \begin{vmatrix} a_{11} & a_{12} & a_{13} \\ a_{21} & a_{22} & a_{23} \\ a_{31} & a_{32} & a_{33} \end{vmatrix}.$$

Theorem 3. *If D_n' is obtained from D_n by interchanging two adjacent rows of D_n, then $D_n' = -D_n$.*

Proof. Suppose D_n' is obtained from D_n by interchanging its kth and $(k+1)$th rows. Then, denoting the elements by a_{ij}' and a_{ij}, respectively, we have

$$a_{ij}' = a_{ij} \quad \text{if} \quad i \neq k,\ i \neq k+1,$$

while

$$a_{kj}' = a_{k+1,j} \quad a_{k+1,j}' = a_{kj}, \quad j = 1, 2, \ldots, n.$$

By definition, D_n is the sum of signed products

$$\pm a_{1i_1}a_{2i_2}a_{3i_2} \cdots a_{ni_n}, \tag{1}$$

while D_n' is the sum of similar products with primes throughout. Let P be a typical product in the expansion of D_n:

$$P = \pm a_{1i_1}a_{2i_2} \cdots a_{ki_k}a_{k+1,i_{k+1}} \cdots a_{ni_n}, \tag{2}$$

where the sign is determined by the number of inversions in the permutation $i_1, i_2, \ldots, i_n$. If we interchange a_{ki_k} and $a_{k+1,i_{k+1}}$ in (2), we obtain a typical product in the expansion of D_n', except that the sign is now to be determined by the number of inversions in the permutation

$$\underbrace{i_1, i_2, \ldots, i_{k-1},}_{Q'} \quad \underbrace{i_{k+1}, i_k,}_{R'} \quad \underbrace{i_{k+2}, \ldots, i_n.}_{S'} \tag{3}$$

Next, we shall show that the number of inversions in the permutation (3) differs by unity from the number of inversions in

$$\underbrace{i_1, i_2, \ldots, i_{k-1},}_{Q} \quad \underbrace{i_k, i_{k+1},}_{R} \quad \underbrace{i_{k+2}, \ldots, i_n,}_{S} \tag{4}$$

which determines the sign in (2).

We have indicated the permutations in (3) and (4) as composed of three segments. Suppose we examine the number of inversions in (3) term by term, and record the total number q' relative to elements in Q' that are followed by smaller elements in Q', R', and S'. Similarly, we record the numbers of inversions r' and s' from the segments R' and S'. We can then formulate the total number of inversions in (3) as

$$I' = q' + r' + s'.$$

We repeat the process for the permutation (4) and obtain the total number of inversions

$$I = q + r + s,$$

where the letters have meanings which are clear by analogy. However, $q = q'$ and $s = s'$, and we need only look more closely at the relation between r and r'. The number r' consists of two parts:

$$r' = r_1' + r_2',$$

where r_1' is the number of inversions caused by the occurrence in R' of numbers that are greater than numbers in S', and where r_2 is 0 or 1, depending on whether $i_{k+1} < i_k$ or $i_{k+1} > i_k$. Defining r_1 analogously, we write

$$r = r_1 + r_2$$

and observe that $r_1 = r_1'$, since the segments R and S contain the same integers as do the segments R' and S'. However,

$$r_2 \neq r_2',$$

but rather

$$r_2 = 1 \quad \text{if} \quad i_k > i_{k+1}; \quad r_2 = 0 \quad \text{if} \quad i_k < i_{k+1}.$$

That is,

$$r_2 = r_2' \pm 1.$$

This leads to the end result that

$$I = I' \pm 1,$$

so that one of the permutations (3) or (4) is even and the other is odd. Therefore every product P in D_n undergoes a change in sign in becoming a product P' in D_n'. The sum of all the products P is therefore the negative of the sum of the products P'; that is,

$$D_n = -D_n', \quad \text{or} \quad D_n' = -D_n. \ \square$$

To illustrate the relationship $I' = I \pm 1$, let us consider an example. For instance, suppose we interchange the 3 and the 7 in the permutation

$$\underbrace{8, 1, 4,}_{Q} \quad \underbrace{3, 7,}_{R} \quad \underbrace{6, 2, 5}_{S}$$

and thus obtain

$$\underbrace{8, 1, 4,}_{Q'} \quad \underbrace{7, 3,}_{R'} \quad \underbrace{6, 2, 5.}_{S'}$$

We may record the number of inversions relative to a given number directly beneath it as follows:

$$I = \overbrace{\underset{\displaystyle q}{\begin{matrix}8, & 1, & 4,\\ 7+0+2\end{matrix}}}^{Q} + \overbrace{\underset{\displaystyle r}{\begin{matrix}3, & 7,\\ 1+3\end{matrix}}}^{R} + \overbrace{\underset{\displaystyle s}{\begin{matrix}6, & 2, & 5\\ 2+0+0\end{matrix}}}^{S},$$

$$I' = \overbrace{\underset{\displaystyle q'}{\begin{matrix}8, & 1, & 4,\\ 7+0+2\end{matrix}}}^{Q'} + \overbrace{\underset{\displaystyle r'}{\begin{matrix}7, & 3,\\ 4+1\end{matrix}}}^{R'} + \overbrace{\underset{\displaystyle s'}{\begin{matrix}6, & 2, & 5\\ 2+0+0\end{matrix}}}^{S'}.$$

Thus

$$q = q' = 7 + 0 + 2 = 9,$$
$$s = s' = 2 + 0 + 0 = 2,$$

but

$$r = 1 + 3 = 4 \quad \text{while} \quad r' = 4 + 1 = 5;$$

hence

$$I' = I + 1.$$

When we interchanged the 3 and the 7 we added one inversion to the total, namely, the inversion caused by having 7 ahead of 3. All other inversions were unaltered; hence we increased the index of the permutation by one, from $I = 15$ to $I' = 16$.

Theorem 4. *If D_n' is obtained from D_n by interchanging any pair of rows, then*

$$D_n' = -D_n.$$

Proof. By Theorem 3, the theorem is true if the two rows in question are adjacent. Suppose, then, in the general case, that the two rows are not adjacent but that there are m intervening rows. Call the two rows to be interchanged α and β. By successively interchanging α and an *adjacent* row, m times, we may bring α adjacent to β. This changes the sign of the determinant m times. Next, interchange α and β and then continue to move β one row at a time, interchanging it with the m rows mentioned above, until it comes into the position originally occupied by α. This changes the sign $(m + 1)$ times and results in the determinant D_n' in which the rows α and β of D_n have been interchanged. Altogether, the sign has been changed $m + (m + 1) = 2m + 1$ times; that is,

$$D_n' = (-1)^{2m+1} D_n.$$

But $2m + 1$ is odd for any integer m; hence

$$D_n' = -D_n. \; \square$$

Theorem 5. *If two rows of D_n are identical, then*

$$D_n = 0.$$

Proof. Interchange the two identical rows and call the result D_n'. Then $D_n' = -D_n$, by Theorem 4. But the interchange of two *identical* rows leaves D_n unaltered; that is, $D_n' = D_n$. Therefore

$$D_n = -D_n, \qquad 2D_n = 0;$$

hence $D_n = 0$. $\square$

Corollary. *If two rows of D_n are proportional, then*

$$D_n = 0.$$

Proof. Suppose the elements of row α are all k times the corresponding elements of row β. Then by Theorem 2, the determinant is k times a determinant whose α and β rows are identical; and this is zero, by Theorem 5. $\square$

Example 1

$$\begin{vmatrix} 1 & 3 & -2 \\ 2 & 4 & 5 \\ -4 & -12 & 8 \end{vmatrix} = -4 \begin{vmatrix} 1 & 3 & -2 \\ 2 & 4 & 5 \\ 1 & 3 & -2 \end{vmatrix} = 0.$$

The -4 may be factored out of the third row. Then the first and third rows are identical, and the result is zero by Theorem 5.

Theorem 6. *Suppose each element a_{kj} in the kth row of D_n is a sum*

$$a_{kj} = b_{kj} + c_{kj}, \quad j = 1, 2, \ldots, n.$$

Let D_n' be obtained from D_n by replacing the elements a_{kj} of the kth row by b_{kj}, and let D_n'' be obtained from D_n by replacing the elements of the kth row by c_{kj}. Then

$$D_n = D_n' + D_n''.$$

Proof. Each product in the expansion of D_n contains a term

$$a_{ki_k} = b_{ki_k} + c_{ki_k}$$

as a factor, and each of these products leads to a product in the expansion of D'_n plus a product in the expansion of D''_n. Adding these (signed) products leads to the result stated. □

Example 2

$$\begin{vmatrix} a_{11} & a_{12} & a_{13} \\ b_{21}+c_{21} & b_{22}+c_{22} & b_{23}+c_{23} \\ a_{31} & a_{32} & a_{33} \end{vmatrix} = \begin{vmatrix} a_{11} & a_{12} & a_{13} \\ b_{21} & b_{22} & b_{23} \\ a_{31} & a_{32} & a_{33} \end{vmatrix} + \begin{vmatrix} a_{11} & a_{12} & a_{13} \\ c_{21} & c_{22} & c_{23} \\ a_{31} & a_{32} & a_{33} \end{vmatrix}.$$

Theorem 7. *The value of a determinant of order n is unaltered if to each element of any row is added k times the corresponding element of some other row, where k is any constant.*

Proof. Let D_n be the given determinant and let D'_n be the result of adding k times the rth row of D_n to its tth row ($r \neq t$). Then, by Theorem 6, D'_n is the sum of two determinants, one of which is D_n and the other of which is zero, by the corollary to Theorem 5. □

Example 3

$$\begin{vmatrix} a_1 & b_1 & c_1 \\ a_2 & b_2 & c_2 \\ a_3+ka_1 & b_3+kb_1 & c_3+kc_1 \end{vmatrix} = \begin{vmatrix} a_1 & b_1 & c_1 \\ a_2 & b_2 & c_2 \\ a_3 & b_3 & c_3 \end{vmatrix} + \begin{vmatrix} a_1 & b_1 & c_1 \\ a_2 & b_2 & c_2 \\ ka_1 & kb_1 & kc_1 \end{vmatrix}.$$

The second determinant on the right is zero, since its third row is proportional to its first row.

Theorem 8. *Let D_n be a determinant of order n and let D'_n be obtained by taking the 1st, 2nd, . . . , nth rows, respectively, of D_n as the 1st, 2nd, . . . , nth columns, respectively, of D'_n. Then*

$$D'_n = D_n.$$

Proof. If we denote the elements in the ith row and jth column of D_n and D'_n, respectively, by a_{ij} and a'_{ij}, the relation between D'_n and D_n is expressed by

$$a'_{ij} = a_{ji}, \quad i, j = 1, 2, \ldots, n.$$

By definition,

$$D'_n = \sum \pm a'_{1j_1} a'_{2j_2} \ldots a'_{nj_n},$$

where $j_1, j_2, \ldots, j_n$ is a permutation of $1, 2, \ldots, n$. In terms of the elements a_{ij} of D_n, this is the same as

$$D'_n = \sum \pm a_{j_1 1} a_{j_2 2} \ldots a_{j_n n},$$

and each signed product

$$\pm a_{j_1 1} a_{j_2 2} \ldots a_{j_n n} \tag{5}$$

corresponds to one and only one signed product

$$\pm a_{1i_1} a_{2i_2} \ldots a_{ni_n} \tag{6}$$

in the expansion of D_n, as we show in the following manner. Since $j_1, j_2, \ldots, j_n$ is a permutation of $1, 2, \ldots, n$, we may rearrange the factors in the product (5) so as to have the first subscripts appearing in the natural order $1, 2, 3, \ldots, n$. This induces a permutation $i_1, i_2, \ldots, i_n$ in the second subscripts and gives the product (6), except that there is at least a possibility that the signs in (5) and in (6) may not agree. The fact is, however, that the signs do agree, since the permutation $j_1, j_2, \ldots, j_n$ in (5) and the related permutation $i_1, i_2, \ldots, i_n$ in (6) are both even or both odd, as we shall prove below. Let us accept this as true for the present; then D'_n is the sum of all signed products (5) and D_n is the sum of the equal signed products (6); that is,

$$D'_n = D_n. \ \square$$

Let us now prove the statement that the permutations $j_1, \ldots, j_n$ and $i_1, \ldots, i_n$ above are both even or both odd. To this end, we consider the two sequences of double subscripts

$$P': \quad (j_1 1), (j_2 2), \ldots, (j_n n) \tag{7}$$

and

$$P: \quad (1i_1), (2i_2), \ldots, (ni_n). \tag{8}$$

In P', the second subscripts are in their natural order 1, 2, ..., n, so that there are no inversions in their order. Hence the total number T' of inversions in P', counting inversions in the first subscripts j_1, $j_2, \ldots, j_n$ plus inversions in the second subscripts, is the same as the index I' of the permutation j_1, $j_2, \ldots, j_n$. Similarly, the total number T of inversions in P, counting inversions in the first subscripts 1, 2, ..., n plus inversions in the second subscripts i_1, $i_2, \ldots, i_n$, is equal to the index I of the permutation $i_1, i_2, \ldots, i_n$. Now (8) may be obtained from (7) by a sequence of transpositions of adjacent terms. Each such transposition changes the number of inversions in the first subscripts by ± 1 and also changes the number of inversions in the second subscripts by ± 1, and hence changes the sum total of inversions by 2, 0, or -2, but always by an *even* number. Thus T and T' differ by an even integer, so that I and I' also differ by an even integer. Thus the permutations are both even or both odd as was asserted.

Example 4. The term $a_{21}a_{32}a_{13}$ has its second subscripts arranged in natural order. The index of the permutation 2, 3, 1 of its first subscripts is 2. If we rearrange the factors so as to put the first subscripts in natural order, we have $a_{13}a_{21}a_{32}$. Then the second subscripts form the permutation 3, 1, 2, which also has index 2.

Remark. The import of Theorem 8 is that each theorem about determinants which contains the word "row" in its statement is true when the word "column" is substituted everywhere for the word row. For example, Theorem 5 is true if stated for columns instead of rows: "If two columns of a determinant are identical, the determinant is zero," for we may, by Theorem 8, interchange rows and columns in the original determinant. (In some theorems a second interchange of rows and columns may be required to establish the end result.)

Theorem 9. *If D has the special form*

$$D = \begin{vmatrix} a_{11} & 0 & 0 & \ldots & 0 \\ a_{21} & a_{22} & a_{23} & \ldots & a_{2n} \\ \ldots & \ldots & \ldots & \ldots & \ldots \\ a_{n1} & a_{n2} & a_{n3} & \ldots & a_{nn} \end{vmatrix}, \tag{9}$$

in which all elements after the first in its first row are zero, then D is a_{11} times its minor; that is,

$$D = a_{11} \begin{vmatrix} a_{22} & a_{23} & \ldots & a_{2n} \\ \ldots & \ldots & \ldots & \ldots \\ a_{n2} & a_{n3} & \ldots & a_{nn} \end{vmatrix}. \tag{10}$$

Proof. In the expansion for D,

$$D = \sum \pm a_{1i_1}a_{2i_2} \ldots a_{ni_n},$$

each product is zero except possibly those in which $i_1 = 1$. Hence

$$D = a_{11}\left(\sum \pm a_{2i_2} \ldots a_{ni_n}\right),$$

and the signed products in the sum are precisely the same as the signed products in the expansion of the minor of a_{11}. □

Corollary. *The value of a triangular determinant*

$$D = \begin{vmatrix} a_{11} & 0 & 0 & \ldots & 0 \\ a_{21} & a_{22} & 0 & \ldots & 0 \\ \ldots & \ldots & \ldots & \ldots & \ldots \\ a_{n1} & a_{n2} & a_{n3} & \ldots & a_{nn} \end{vmatrix} \tag{11}$$

in which all elements above the main diagonal are zero, is simply the product of the elements on this diagonal; that is,

$$D = a_{11}a_{22} \ldots a_{nn}.$$

It should of course be noted that Theorem 9 has a companion theorem for the case where the elements below a_{11} in the first column are zero. Also, the value of a triangular determinant which is zero everywhere *below* the main diagonal is just the product of the diagonal elements.

Theorem 10. *If D has the special form*

$$D = \begin{vmatrix} a_{11} & a_{12} & \cdots & a_{1n} \\ \cdots & \cdots & \cdots & \cdots \\ 0 & 0 & a_{ij} & 0 \\ \cdots & \cdots & \cdots & \cdots \\ a_{n1} & a_{n2} & \cdots & a_{nn} \end{vmatrix}, \tag{12}$$

in which all elements of the ith row are zero except for the element a_{ij} in the jth column, then

$$D = (-1)^{i+j} a_{ij} D_{ij}, \tag{13}$$

where D_{ij} is the minor of a_{ij} obtained by covering up the ith row and jth column in D.

Proof. By $i - 1$ successive row transpositions followed by $j - 1$ column transpositions, the element a_{ij} of D may be brought into the upper left corner of a determinant D' having the special form described in Theorem 9. The minor of this element in D' is the same as the minor D_{ij} of the element a_{ij} in D. Hence

$$D' = a_{ij} D_{ij}.$$

But, by Theorem 4 and its analog for columns,

$$D' = (-1)^{(i-1)+(j-1)} D, \quad \text{or} \quad D = (-1)^{i+j} D',$$

so that

$$D = (-1)^{i+j} a_{ij} D_{ij}. \ \square$$

Of course, a companion theorem for columns is also true. The results may be summarized by saying that whenever all but one of the elements of a row or column of a determinant are zero, the order of the determinant may be reduced by unity by simply multiplying the element times its signed minor, the sign being given by $(-1)^{i+j}$, where i and j are the row and column numbers of the element in question.

Example 5. Exploiting the first row of the fourth-order determinant below, we have

$$\begin{vmatrix} 0 & 2 & 0 & 0 \\ 3 & 4 & 0 & 0 \\ -2 & -7 & 4 & 0 \\ 6 & 8 & -1 & 5 \end{vmatrix} = -2 \begin{vmatrix} 3 & 0 & 0 \\ -2 & 4 & 0 \\ 6 & -1 & 5 \end{vmatrix} = -120.$$

The 2 in the first row and second column must be multiplied by $(-1)^{1+2}$ times its minor. The minor is a triangular determinant whose value is simply

$$(3)(4)(5) = 60.$$

In the next example, we illustrate how Theorem 7 and the corresponding theorem for columns are used to transform a given determinant into the special form in which all but one of the elements of a row or column are zero. The order of the determinant is then reduced by one and the process repeated if desired. Note that several applications of Theorem 7 may be combined in one step, provided the same base row (or column) is used to modify several other rows (or columns). In every case, the base row itself is *not* changed. That is, if we multiply the first row by 3 and add it to the second row, we thereby get a new second row, but we do not change the first row when we do this. Of course we may now, if we wish, use a new base row to alter the first row.

Example 6. In the fourth-order determinant, Eq. (14), we use the first row to obtain zeros in all but one place in the fourth column. We denote the second row by R_2, and so forth, and replace

$$\begin{aligned} R_2 &\quad \text{by} \quad R_2 + R_1, \\ R_3 &\quad \text{by} \quad R_3 + R_1, \\ R_4 &\quad \text{by} \quad R_4 + 3R_1, \end{aligned}$$

and do not change R_1.

$$D = \begin{vmatrix} 1 & 3 & 2 & -1 \\ 2 & -1 & 3 & 1 \\ -1 & 2 & 1 & 1 \\ -2 & -5 & 2 & 3 \end{vmatrix} = \begin{vmatrix} 1 & 3 & 2 & -1 \\ 3 & 2 & 5 & 0 \\ 0 & 5 & 3 & 0 \\ 1 & 4 & 8 & 0 \end{vmatrix}. \tag{14}$$

This determinant is equal to (the element -1 in its first row and fourth column) $\times$ $(-1)^{4+1}$ $\times$ (the minor obtained by deleting the first row and fourth column); that is,

$$D = (-1)(-1)^{4+1} \begin{vmatrix} 3 & 2 & 5 \\ 0 & 5 & 3 \\ 1 & 4 & 8 \end{vmatrix}.$$

This third-order determinant can be evaluated at once, or we may further transform it to get another zero in its first column by adding -3 times its third row to its first row. Thus

$$D = \begin{vmatrix} 0 & -10 & -19 \\ 0 & 5 & 3 \\ 1 & 4 & 8 \end{vmatrix} = (1)(-1)^{3+1} \begin{vmatrix} -10 & -19 \\ 5 & 3 \end{vmatrix}$$
$$= +(-30 + 95) = 65.$$

EXERCISES

1. Following the notation used in the proof of Theorem 3, calculate the numbers q, r, s and q', r', s' relative to the permutations

$$\underbrace{5,\ 1,}_{Q}\ \underbrace{3,\ 6,}_{R}\ \underbrace{2,\ 4}_{S} \quad \text{and} \quad \underbrace{5,\ 1,}_{Q'}\ \underbrace{6,\ 3,}_{R'}\ \underbrace{2,\ 4.}_{S'}$$

Apply Theorem 5 to Exercises 2 and 3.

2. Show that the determinantal equation

$$\begin{vmatrix} x^2 & x & 1 \\ 4 & 2 & 1 \\ 9 & -3 & 1 \end{vmatrix} = 0$$

has roots $x = 2$, $x = -3$.

3. Let $P_1(x_1, y_1)$, $P_2(x_2, y_2)$, and $P_3(x_3, y_3)$ be the vertices of a triangle. Show that the area of the triangle is the absolute value of

$$\frac{1}{2}\begin{vmatrix} x_1 & y_1 & 1 \\ x_2 & y_2 & 1 \\ x_3 & y_3 & 1 \end{vmatrix}.$$

Hint. Write the coordinates of P_2 and P_3 as

$$x_2 = x_1 + a\cos\alpha, \qquad y_2 = y_1 + a\sin\alpha,$$
$$x_3 = x_1 + b\cos\beta, \qquad y_3 = y_1 + b\sin\beta,$$

where $a = P_1P_2$, $b = P_1P_3$, while α and β are angles from the horizontal to P_1P_2 and P_1P_3, respectively. Sketch.

Evaluate the determinants, Exercises 4 and 5.

4. $\begin{vmatrix} 1 & 3 & -1 & 2 \\ 2 & 1 & 3 & 1 \\ -1 & 2 & -1 & 3 \\ -2 & 1 & 2 & -3 \end{vmatrix}$. 5. $\begin{vmatrix} 4 & 3 & -2 & 7 \\ 8 & 1 & -4 & 6 \\ 6 & 2 & -3 & 11 \\ 10 & 4 & -5 & -8 \end{vmatrix}$.

6. Show that the determinant

$$\begin{vmatrix} 1 & x_1 & x_1^2 \\ 1 & x_2 & x_2^2 \\ 1 & x_3 & x_3^2 \end{vmatrix}$$

is equal to the product

$$(x_3 - x_1)(x_3 - x_2)(x_2 - x_1).$$

Hint. Use Theorem 5.

7. Generalize the result of Exercise 6 and show that

$$\begin{vmatrix} 1 & x_1 & x_1^2 & \cdots & x_1^{n-1} \\ 1 & x_2 & x_2^2 & \cdots & x_2^{n-1} \\ 1 & x_3 & x_3^2 & \cdots & x_3^{n-1} \\ \cdots & & \cdots\ \cdots & & \cdots \\ 1 & x_n & x_n^2 & \cdots & x_n^{n-1} \end{vmatrix} = \prod_{i>j} (x_i - x_j),$$

where the $\prod$ symbol denotes the product of all factors of the form $(x_i - x_j)$ that can be formed with $i > j$, i and j ranging between 1 and n inclusive; that is, $n \geq i > j \geq 1$. There are $[n(n-1)]/2$ such factors.

A.6 EXPANSION BY COFACTORS

In the expansion of a determinant

$$D = \begin{vmatrix} a_{11} & a_{12} & \cdots & a_{1n} \\ a_{21} & a_{22} & \cdots & a_{2n} \\ \cdots & \cdots & \cdots & \cdots \\ a_{n1} & a_{n2} & \cdots & a_{nn} \end{vmatrix} \tag{1}$$

as the sum of all signed products

$$D = \sum \pm\, a_{1i_1}a_{2i_2}\cdots a_{ni_n}, \tag{2}$$

we may arrange the products in blocks from which common factors may be removed. Thus, for example, we may collect all those products for which $i_1 = 1$ in (2), then all terms in which $i_1 = 2$, and so forth. The result may be written in the form

$$D = a_{11}A_{11} + a_{12}A_{12} + \cdots + a_{1n}A_{1n}, \quad (3)$$

where A_{1j} is called the cofactor of $a_{1j}; j = 1, \ldots, n$. More generally, we could single out the kth row of D instead of its first row, and write the expression in the form

$$D = a_{k1}A_{k1} + a_{k2}A_{k2} + \cdots + a_{kn}A_{kn}, \quad (4)$$

which is called the expansion of D by cofactors of the elements of the kth row.

We may interpret Eq. (4) as follows. If we were to take all elements $a_{k2}, a_{k3}, \ldots, a_{kn}$ after the first, in the kth row of D, to be zero, then by (4), the resulting determinant D' would have the value

$$D' = a_{k1}A_{k1};$$

that is,

$$D' = \begin{vmatrix} a_{11} & a_{12} & \cdots & a_{1n} \\ \cdots & \cdots & \cdots & \cdots \\ a_{k1} & 0 & \cdots & 0 \\ \cdots & \cdots & \cdots & \cdots \\ a_{n1} & a_{n2} & \cdots & a_{nn} \end{vmatrix} = a_{k1}A_{k1}. \quad (5)$$

In general, the right-hand side of (4) may be interpreted as the sum of n determinants of the type exhibited in (5), in which all elements of the kth row are replaced by zeros except for one element a_{kj}; that is,

$$D = \begin{vmatrix} a_{11} & a_{12} & a_{13} & \cdots & a_{1n} \\ \cdots & \cdots & \cdots & \cdots & \cdots \\ a_{k1} & a_{k2} & a_{k3} & \cdots & a_{kn} \\ \cdots & \cdots & \cdots & \cdots & \cdots \\ a_{n1} & a_{n2} & a_{n3} & \cdots & a_{nn} \end{vmatrix} = \begin{vmatrix} a_{11} & a_{12} & \cdots & a_{1n} \\ \cdots & \cdots & \cdots & \cdots \\ a_{k1} & 0 & \cdots & 0 \\ \cdots & \cdots & \cdots & \cdots \\ a_{n1} & a_{n2} & \cdots & a_{nn} \end{vmatrix} + \begin{vmatrix} a_{11} & a_{12} & \cdots & a_{1n} \\ \cdots & \cdots & \cdots & \cdots \\ 0 & a_{k2} & \cdots & 0 \\ \cdots & \cdots & \cdots & \cdots \\ a_{n1} & a_{n2} & \cdots & a_{nn} \end{vmatrix} + \cdots + \begin{vmatrix} a_{11} & a_{12} & \cdots & a_{1n} \\ \cdots & \cdots & \cdots & \cdots \\ 0 & 0 & \cdots & a_{kn} \\ \cdots & \cdots & \cdots & \cdots \\ a_{n1} & a_{n2} & \cdots & a_{nn} \end{vmatrix}. \quad (6)$$

This may also be interpreted as a generalization of Theorem 6 of the preceding article, and may in fact be obtained by repeated application of that theorem.

Now each of the determinants in (6) which has all but one element zero in its kth row may be evaluated by means of Theorem 10 of the preceding article. These results are now summarized in the following theorem.

Theorem 11. *The sum of the products of the elements of any row (or column) by their respective cofactors is equal to the determinant. The cofactor of the element in the kth row and jth column is* $(-1)^{k+j}$ *times its minor.*

In expanding a determinant by cofactors, it may be convenient first to transform the determinant into a form in which all elements, except one, of some row (or column) are zero. This method was illustrated in the preceding article in connection with Theorem 10.

On the other hand, if we wish to avoid these preliminary transformations, we may do so at the expense of using all the elements of a row times their respective cofactors. Thus a fourth-order determinant reduces to four third-order determinants, in general.

Problem. Show that

$$\begin{vmatrix} x & y & 1 \\ x_1 & y_1 & 1 \\ x_2 & y_2 & 1 \end{vmatrix} = 0$$

is the equation of the straight line through the points (x_1, y_1) and (x_2, y_2), provided that the two points are distinct.

Solution. If the determinant is expanded by elements of its first row, the result is a linear equation of the form $Ax + By + C = 0$, which represents a straight line since one, at least, of A and B is not zero. Furthermore, if one substitutes (x_1, y_1) for (x, y) in the determinant, the first two rows are identical and the equation is satisfied. Hence (x_1, y_1) does lie on the line. The same is true of (x_2, y_2).

Theorem 12. *If the elements of any row (or column) are multiplied by the cofactors of the corresponding elements of some other row (or column) and the products added, the result is zero.*

Proof. The result may be interpreted as the expansion of a determinant having two identical rows (or columns); this expansion is therefore equal to zero, by a previous theorem. □

We shall illustrate this interpretation for a third-order determinant. Let us start with any determinant

$$D = \begin{vmatrix} a_{11} & a_{12} & a_{13} \\ a_{21} & a_{22} & a_{23} \\ a_{31} & a_{32} & a_{33} \end{vmatrix}$$

and expand it by cofactors of any row, say the second, to be specific. Then

$$D = a_{21}A_{21} + a_{22}A_{22} + a_{23}A_{23}. \tag{7}$$

Suppose we now replace a_{21}, a_{22}, and a_{23} by other numbers, say

$$c_1,\ c_2, \text{ and } c_3,$$

respectively, in (7), with the result

$$D' = c_1A_{21} + c_2A_{22} + c_3A_{23}. \tag{8}$$

We may interpret this as the expansion by cofactors relative to the second row in

$$D' = \begin{vmatrix} a_{11} & a_{12} & a_{13} \\ c_1 & c_2 & c_3 \\ a_{31} & a_{32} & a_{33} \end{vmatrix}, \tag{9}$$

since the cofactors of c_1, c_2, and c_3 in D' are precisely the original cofactors A_{21}, A_{22}, A_{23}, respectively. In particular, if c_1, c_2, and c_3 are taken to be the elements of either the first or the third row of D, then (8) is the sum of products of elements of that row by the cofactors of corresponding elements of a different row, while simultaneously the determinant D' in (9) then has two identical rows. Hence $D' = 0$, as stated in the theorem.

Theorems 11 and 12 form the basis for the solution by determinants, of n simultaneous linear equations in n unknowns, which we shall discuss in the next article.

EXERCISES

1. Suppose the elements of the determinant

$$F = \begin{vmatrix} u_1 & v_1 & w_1 \\ u_2 & v_2 & w_2 \\ u_3 & v_3 & w_3 \end{vmatrix}$$

are differentiable functions of x. Prove that dF/dx is equal to

$$\begin{vmatrix} \frac{du_1}{dx} & \frac{dv_1}{dx} & \frac{dw_1}{dx} \\ u_2 & v_2 & w_2 \\ u_3 & v_3 & w_3 \end{vmatrix} + \begin{vmatrix} u_1 & v_1 & w_1 \\ \frac{du_2}{dx} & \frac{dv_2}{dx} & \frac{dw_2}{dx} \\ u_3 & v_3 & w_3 \end{vmatrix} + \begin{vmatrix} u_1 & v_1 & w_1 \\ u_2 & v_2 & w_2 \\ \frac{du_3}{dx} & \frac{dv_3}{dx} & \frac{dw_3}{dx} \end{vmatrix}.$$

State the result in words as a rule for differentiating a determinant of order three. Can you generalize the result to a determinant of any order n?

2. Write out the sum of the products of the elements of the second row times the cofactors of the corresponding elements of the third row in the determinant

$$\begin{vmatrix} 2 & 3 & -1 \\ -2 & -3 & 2 \\ 3 & 2 & 1 \end{vmatrix}.$$

A.7 SOLUTION OF SIMULTANEOUS LINEAR EQUATIONS

In this article we shall work only with three equations in three unknowns to simplify the writing, but the argument used may be applied to n equations in n unknowns. We shall accordingly denote the unknowns by x_1, x_2, and x_3, since such a notation lends itself to generalization. The equations may then be represented by

$$\begin{aligned} a_{11}x_1 + a_{12}x_2 + a_{13}x_3 &= c_1, \\ a_{21}x_1 + a_{22}x_2 + a_{23}x_3 &= c_2, \\ a_{31}x_1 + a_{32}x_2 + a_{33}x_3 &= c_3. \end{aligned} \tag{1}$$

The coefficients of the unknowns may be used to form the determinant

$$D = \begin{vmatrix} a_{11} & a_{12} & a_{13} \\ a_{21} & a_{22} & a_{23} \\ a_{31} & a_{32} & a_{33} \end{vmatrix}, \tag{2}$$

which is called *the determinant of the system.* In this determinant, the first row contains the coefficients, in order, of the unknowns x_1, x_2, x_3 in the first equation. Similarly, each row in D corresponds to one equation of the system, and each column of D corresponds to one of the unknowns.

The solution of the equations (1) by determinants is given by the following theorem, which is known as "Cramer's rule."

Theorem 13. *If the determinant D of the coefficients in a system of n linear equations in n unknowns is not zero, then the equations have a unique solution. In the solution, each unknown may be expressed as a fraction of two determinants with denominator D and with numerator obtained from D by replacing the column of coefficients of the unknown in question by the constants $c_1, c_2, \ldots, c_n$.*

Proof. Suppose x_1, x_2, x_3 are three numbers that satisfy the system (1). If we multiply these equations by constants and add to get new equations, then the same three numbers must satisfy these equations also. Now we choose multipliers in the following way.

Suppose, first, that we wish to find x_1. We calculate the cofactors A_{11}, A_{21}, and A_{31} of the elements in the first column of D. Then we multiply the first equation by A_{11}, the second by A_{21}, and the third by A_{31}, and add. The resulting new equation is

$$\begin{aligned} &(a_{11}A_{11} + a_{21}A_{21} + a_{31}A_{31})x_1 \\ &\qquad + (a_{12}A_{11} + a_{22}A_{21} + a_{23}A_{31})x_2 \\ &\qquad + (a_{13}A_{11} + a_{23}A_{21} + a_{33}A_{31})x_3 \\ &\quad = c_1A_{11} + c_2A_{21} + c_3A_{31}. \end{aligned} \tag{3}$$

In this equation, the coefficient that multiplies x_1 is the sum of products of the elements of the first column of D by their cofactors, which by Theorem 11 of the previous article is just D. But the coefficient that multiplies x_2 is the sum of products of the elements of the second column of D by the cofactors of the corresponding elements of the first column of D, and this is zero by Theorem 12. Similarly, the coefficient of x_3 is zero. Hence (3) contains only x_1, and has the form

$$x_1 \begin{vmatrix} a_{11} & a_{12} & a_{13} \\ a_{21} & a_{22} & a_{23} \\ a_{31} & a_{32} & a_{33} \end{vmatrix} = \begin{vmatrix} c_1 & a_{12} & a_{13} \\ c_2 & a_{22} & a_{23} \\ c_3 & a_{32} & a_{33} \end{vmatrix}, \tag{4}$$

where we have simply written out the determinant D on the left, while the right-hand sides of (4) and (3) are equal, as can be seen at once if the determinant in (4) is expanded by cofactors according to the elements of its first column. Hence, if the determinant D is not zero, then we may divide to find the value that x_1 must have if x_1, x_2, x_3 are to satisfy the system (1):

$$x_1 = \frac{\begin{vmatrix} c_1 & a_{12} & a_{13} \\ c_2 & a_{22} & a_{23} \\ c_3 & a_{32} & a_{33} \end{vmatrix}}{\begin{vmatrix} a_{11} & a_{12} & a_{13} \\ a_{21} & a_{22} & a_{23} \\ a_{31} & a_{32} & a_{33} \end{vmatrix}}.$$

The same argument may be repeated for each of the other unknowns: We simply multiply the equations of the system (1) by the cofactors in D of the elements in the column of D that corresponds to the unknown in question, and then add the results to obtain a new equation. In the equation thus obtained, all terms drop out except for the one unknown and the value of the unknown is then found to be precisely the kind of fraction described in the theorem. □

Problem 1. Solve the system

$$\begin{aligned} x + y + z &= 1, \\ 2x - y + z &= 0, \\ x + 2y - z &= 4. \end{aligned}$$

Solution. The determinant of the system is

$$D = \begin{vmatrix} 1 & 1 & 1 \\ 2 & -1 & 1 \\ 1 & 2 & -1 \end{vmatrix}.$$

Suppose we are interested in finding z. We then calculate the cofactors of the elements of the third column of D:

$$\begin{vmatrix} 2 & -1 \\ 1 & 2 \end{vmatrix} = 5, \quad -\begin{vmatrix} 1 & 1 \\ 1 & 2 \end{vmatrix} = -1, \quad \begin{vmatrix} 1 & 1 \\ 2 & -1 \end{vmatrix} = -3.$$

Multiplying the first equation by 5, the second by -1, and the third by -3, and adding, we obtain

$$(5 - 2 - 3)x + (5 + 1 - 6)y + (5 - 1 + 3)z = 5 - 12,$$

or

$$7z = -7, \qquad z = -1.$$

The method described shows that if the original equations have a solution, then it must be given by Cramer's rule. It does not, however, show that these values of the unknowns actually do satisfy the original equations. This may be done by substituting the answers back into the original equations.

Note. Consider the first equation in (1). If we write it in the form

$$a_{11}x_1 + a_{12}x_2 + a_{13}x_3 - c_1 = 0,$$

then substitute the expressions for x_1, x_2, and x_3 in terms of determinants, and finally clear of fractions, the left-hand side becomes

$$a_{11}\begin{vmatrix} c_1 & a_{12} & a_{13} \\ c_2 & a_{22} & a_{23} \\ c_3 & a_{32} & a_{33} \end{vmatrix} + a_{12}\begin{vmatrix} a_{11} & c_1 & a_{13} \\ a_{21} & c_2 & a_{23} \\ a_{31} & c_3 & a_{33} \end{vmatrix}$$
$$+ a_{13}\begin{vmatrix} a_{11} & a_{12} & c_1 \\ a_{21} & a_{22} & c_2 \\ a_{31} & a_{32} & c_3 \end{vmatrix} - c_1\begin{vmatrix} a_{11} & a_{12} & a_{13} \\ a_{21} & a_{22} & a_{23} \\ a_{31} & a_{32} & a_{33} \end{vmatrix}. \tag{5}$$

In each of the first two determinants, we move the column of constants c_1, c_2, c_3 to the right, changing signs twice in the first determinant and once in the second. The result may then be identified as the expansion, by cofactors relative to the first row, of the fourth-order determinant

$$\begin{vmatrix} a_{11} & a_{12} & a_{13} & c_1 \\ a_{11} & a_{12} & a_{13} & c_1 \\ a_{21} & a_{22} & a_{23} & c_2 \\ a_{31} & a_{32} & a_{33} & c_3 \end{vmatrix}. \tag{6}$$

This is zero, since its first two rows are identical. A determinant similar to the one in (6) also results when the expressions for the unknowns are substituted into the remaining equations in (1), except that the first row becomes the same as the third or fourth row respectively. Thus the expressions given by Cramer's rule give the unique solution of the system of equations when $D \neq 0$.

Example. To illustrate Cramer's rule, we solve this system by determinants:

$$\begin{aligned} 2x + y - z &= 0, \\ x - y + z &= 6, \\ x + 2y + z &= 3. \end{aligned}$$

The determinant of the coefficients is

$$D = \begin{vmatrix} 2 & 1 & -1 \\ 1 & -1 & 1 \\ 1 & 2 & 1 \end{vmatrix} = \begin{vmatrix} 2 & 1 & -1 \\ 3 & 0 & 0 \\ 3 & 3 & 0 \end{vmatrix} = -9.$$

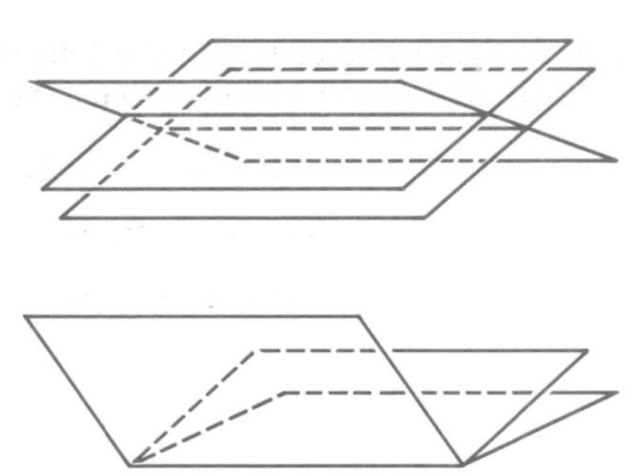

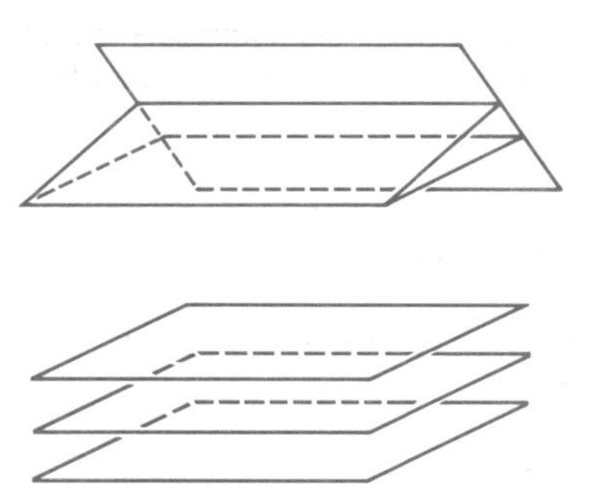

I.1

Since $D \neq 0$, we may solve by determinants. Thus

$$x = \frac{\begin{vmatrix} 0 & 1 & -1 \\ 6 & -1 & 1 \\ 3 & 2 & 1 \end{vmatrix}}{D} = \frac{-18}{-9} = 2,$$

$$y = \frac{\begin{vmatrix} 2 & 0 & -1 \\ 1 & 6 & 1 \\ 1 & 3 & 1 \end{vmatrix}}{D} = \frac{9}{-9} = -1,$$

$$z = \frac{\begin{vmatrix} 2 & 1 & 0 \\ 1 & -1 & 6 \\ 1 & 2 & 3 \end{vmatrix}}{D} = \frac{-27}{-9} = 3.$$

As a check, we quickly verify that $x = 2$, $y = -1$, $z = 3$ actually do satisfy the original three equations.

One naturally wonders if the original equations are solvable when the determinant of coefficients is zero. A complete answer will not be given here, but is given in more advanced algebra courses in which the subject of matrices is studied. We can gain some idea of the complications that may arise by considering the case of three equations in three unknowns. Each such equation represents a plane in space, and the determinant of coefficients vanishes if

(a) any two of the planes are parallel or coincident, or

(b) the line of intersection of two of the planes is parallel to, or lies in, the third plane.

A solution of the system of simultaneous equations corresponds to a point on all three planes. If the cases (a) or (b) occur, there may be no point at all that lies on all three planes, or there may [for example, if the two planes in (a) are identical] be infinitely many points on all three planes. In particular, there are infinitely many solutions if the three planes all intersect along one straight line. (See Fig. I.1.)

In any case where $D = 0$, one should attempt to solve the system of equations by the method of "successive elimination" illustrated below. In fact, this method is usually more economical than the solution by determinants, even in those cases where D does not vanish.

Problem 2. Solve

$$\begin{aligned} 2x - y + z &= 4, \\ x + 3y + 2z &= 12, \\ 3x + 2y + 3z &= 16. \end{aligned}$$

Solution. The determinant of coefficients is

$$D = \begin{vmatrix} 2 & -1 & 1 \\ 1 & 3 & 2 \\ 3 & 2 & 3 \end{vmatrix} = \begin{vmatrix} 2 & -1 & 1 \\ 3 & 2 & 3 \\ 3 & 2 & 3 \end{vmatrix} = 0.$$

Multiply the second equation by 2 and by 3, respectively, and subtract these two results from the first and third equations to eliminate x from them. The three new equations are

$$\begin{aligned} x + 3y + 2z &= 12, \\ -7y - 3z &= -20, \\ -7y - 3z &= -20. \end{aligned}$$

Here the third equation is superfluous, since it contains no information not already given by the second equation. Thus we need only solve

$$x + 3y = 12 - 2z,$$
$$7y = 20 - 3z.$$

In these, let z have any value whatever. Then we have a solution, provided we take

$$y = \frac{20 - 3z}{7}$$

and

$$x = 12 - 2z - 3y, \quad \text{or} \quad x = \frac{24 - 5z}{7}.$$

By substituting into the original equations, we easily verify that

$$x = \frac{24 - 5z}{7}, \qquad y = \frac{20 - 3z}{7}, \qquad z \text{ arbitrary},$$

do satisfy the equations. For example,

$$z = -5, \quad x = 7, \quad y = 5;$$
$$z = 2, \quad x = 2, \quad y = 2$$

are particular solutions.

EXERCISES

Solve these two systems of simultaneous equations by Cramer's rule.

1. $x + y + z = 2,$
   $2x - y + z = 0,$
   $x + 2y - z = 4$

2. $2x + y - z = 2,$
   $x - y + z = 7,$
   $x + 2y + z = 4$

3. By multiplying the following equations by appropriate factors and adding, eliminate x, y, and t.

$$x + y + z + t = 6,$$
$$x - 2y + 3z + t = 2,$$
$$3x + y - z + 2t = 6,$$
$$2x + 3y + 2z - t = 2$$

4. For what value of k may the following set of equations fail to have a unique solution?

$$x + y - z = 3,$$
$$kx - y + 2z = 5,$$
$$x + 2y - z = 4$$

A.8 HOMOGENEOUS LINEAR EQUATIONS

There is one case in particular where one rather hopes that the determinant of the coefficients will be zero. That is in the case of homogeneous equations:

$$\begin{aligned} a_{11}x_1 + a_{12}x_2 + a_{13}x_3 &= 0, \\ a_{21}x_1 + a_{22}x_2 + a_{23}x_3 &= 0, \\ a_{31}x_1 + a_{32}x_2 + a_{33}x_3 &= 0, \end{aligned} \tag{1}$$

where $x_1 = x_2 = x_3 = 0$ always gives a solution regardless of the coefficients a_{ij}. Since this solution contains no distinguishing information that is characteristic of the particular system of equations, it is referred to as the *trivial solution*. The question of interest in such a case is whether or not a *nontrivial* solution can be found. By Cramer's rule, it follows that the answer is in the negative if $D \neq 0$. On the other hand, if $D = 0$, it is true that nontrivial solutions do exist, as we shall prove in the next theorem.

Theorem 14. *A necessary and sufficient condition for existence of nontrivial solutions of a system of n homogeneous linear equations in n unknowns is that the determinant of coefficients be zero.*

Proof *(by induction on n).* If $n = 1$, the system of equations is just $a_{11}x_1 = 0$, and this implies that $a_{11} = 0$ or $x_1 = 0$. Therefore, if the "determinant" a_{11} is not zero, then only the trivial solution $x_1 = 0$ exists. But if the determinant a_{11} is zero, then x_1 may have any value; in particular, $x_1 = 1$ gives a nontrivial solution.

Suppose now that n is a positive integer and the theorem is known to be true for the smaller integer $n - 1$. We then proceed as follows to reduce the problem to the case of $n - 1$ equations in $n - 1$

unknowns. Let the given system of n equations in n unknowns be

$$\begin{aligned} a_{11}x_1 + a_{12}x_2 + \cdots + a_{1n}x_n &= 0, \\ a_{21}x_1 + a_{22}x_2 + \cdots + a_{2n}x_n &= 0, \\ &\vdots \\ a_{n1}x_1 + a_{n2}x_2 + \cdots + a_{nn}x_n &= 0. \end{aligned} \tag{E}$$

Case 1. If it happens that

$$a_{11} = a_{21} = \cdots = a_{n1} = 0,$$

so that x_1 doesn't actually appear in any of equations (E), we may take

$$x_1 = 1, \quad x_2 = x_3 = \cdots = x_n = 0$$

and obtain a nontrivial solution of (E).

Case 2. Suppose that x_1 does appear in (E) with a nonzero coefficient. If necessary, we could rearrange the equations to have a nonzero coefficient of x_1 in the first equation. So, with no loss in generality, suppose $a_{11} \neq 0$. Then we replace (E) by the equivalent set of equations (E′):

$$\begin{aligned} x_1 + b_{12}x_2 + b_{13}x_3 + \cdots + b_{1n}x_n &= 0, \\ b_{22}x_2 + b_{23}x_3 + \cdots + b_{2n}x_n &= 0, \\ &\vdots \\ b_{n2}x_2 + b_{n3}x_3 + \cdots + b_{nn}x_n &= 0. \end{aligned} \tag{E′}$$

We divide the first equation of (E) by a_{11}, thus getting the first equation of (E′) with leading coefficient 1. We multiply this equation by a_{k1} and subtract the result from the kth equation of (E), thus obtaining the kth equation of (E′); $k = 2, 3, \ldots, n$. These operations are reversible, and hence any solution of the system (E) is a solution of (E′), and conversely. Thus the determinant D' of the coefficients of the system (E′) may be obtained from the determinant D of the system (E) by dividing the first row by a_{11}, then subtracting a_{k1} times the new first row from the kth row; $k = 2, 3, \ldots, n$. Therefore

$$D' = \frac{D}{a_{11}}.$$

Hence $D' = 0$ if and only if $D = 0$.

But if we expand

$$D' = \begin{vmatrix} 1 & b_{12} & b_{13} & \cdots & b_{1n} \\ 0 & b_{22} & b_{23} & \cdots & b_{2n} \\ 0 & b_{32} & b_{33} & \cdots & b_{3n} \\ \vdots & & & & \\ 0 & b_{n2} & b_{n3} & \cdots & b_{nn} \end{vmatrix}$$

by cofactors of its first column, we see that it is equal to B_{11}, the cofactor of the element in its upper left corner:

$$B_{11} = \begin{vmatrix} b_{22} & b_{23} & \cdots & b_{2n} \\ b_{32} & b_{33} & \cdots & b_{3n} \\ \cdots & \cdots & \cdots & \cdots \\ b_{n2} & b_{n3} & \cdots & b_{nn} \end{vmatrix}.$$

This is also the determinant of the $n - 1$ equations in $n - 1$ unknowns that we get by temporarily ignoring the first equation of the system (E′).

We now apply our inductive hypothesis to this system of $n - 1$ equations. By our assumption, the theorem is true for $n - 1$ equations in $n - 1$ unknowns. Therefore, if $D = 0$, then $B_{11} = 0$, and the last $n - 1$ equations have a solution in which $x_2, x_3, \ldots, x_n$ are not all zero. From this we get a nontrivial solution of the full system (E′), and hence of (E), by solving the first equation for x_1 in terms of the numbers $x_2, x_3, \ldots, x_n$:

$$x_1 = -b_{12}x_2 - b_{13}x_3 - \cdots - b_{1n}x_n.$$

Therefore, if $D = 0$, the system has a nontrivial solution.

Conversely, if $D \neq 0$, then $B_{11} \neq 0$, and the only solution of the last $n - 1$ equations of (E′) is

$$x_2 = x_3 = \cdots = x_n = 0.$$

To satisfy the first equation of (E′), we must then also take $x_1 = 0$. That is, if $D \neq 0$, then the system has only the trivial solution. □

Corollary. *A system of m homogeneous linear equations in n unknowns always has a nontrivial solution if $m < n$.*

Proof. We augment the given set of equations by adjoining $n - m$ equations of the form

$$0 \cdot x_1 + 0 \cdot x_2 + \cdots + 0 \cdot x_n = 0.$$

The determinant of coefficients of the new set of n equations in n unknowns has one or more rows of zeros; hence its value is zero. Therefore it has nontrivial solutions. □

Problem 1. Solve the system

$$x + 3y + 2z = 0,$$
$$2x - y + z = 0,$$
$$3x + 2y + 3z = 0.$$

Solution. The coefficient of x in the first equation is already 1. We subtract twice the first equation from the second, and three times the first from the third. The new system is

$$x + 3y + 2z = 0,$$
$$-7y - 3z = 0,$$
$$-7y - 3z = 0.$$

The last two equations are identical: $7y + 3z = 0$. Infinitely many numbers satisfy this equation. We may, for example, let $z = a$, where a is chosen arbitrarily, so long as we then take

$$y = -3a/7.$$

The original system requires that

$$x = -3y - 2z = \frac{9a}{7} - 2a = \frac{-5a}{7}.$$

In particular, if we take $a = 1$, we have a solution

$$x = -\tfrac{5}{7}, \quad y = -\tfrac{3}{7}, \quad z = 1.$$

Another solution, with $a = 7$, is

$$x = -5, \quad y = -3, \quad z = 7.$$

Every solution in this case is proportional to this solution; that is,

$$x = -5k, \quad y = -3k, \quad z = 7k$$

represents the general solution of the system. Assigning different values to k, $-\infty < k < +\infty$, corresponds to letting the point (x, y, z) trace out the line of intersection of the three given planes.

Problem 2. Solve the system

$$x + 2y - 3z + 4t = 0,$$
$$2x - y + z - 2t = 0.$$

Solution. We subtract twice the first equation from the second to obtain

$$x + 2y - 3z + 4t = 0,$$
$$-5y + 7z - 10t = 0.$$

In this system, or an augmented system with two more equations of the form

$$0 \cdot z + 0 \cdot t = 0, \quad 0 \cdot t = 0,$$

we may take

$$t = a \quad (a \text{ arbitrary}),$$
$$z = b \quad (b \text{ arbitrary}),$$
$$y = \frac{7z - 10t}{5} = -2a + \frac{7}{5}b,$$
$$x = -2y + 3z - 4t = \frac{b}{5}.$$

Here we have four equations and only two unknowns, so it is not surprising that there are two "degrees of freedom" evidenced by the two arbitrary numbers a and b. If we take $a = 0$, $b = 5$, we get a solution

$$x_1 = 1, \quad y_1 = 7, \quad z_1 = 5, \quad t_1 = 0.$$

If we take $a = 1$, $b = 0$, we get another solution,

$$x_2 = 0, \quad y_2 = -2, \quad z_2 = 0, \quad t_2 = 1.$$

The solution corresponding to $a = 1$, $b = 5$ is

$$x_3 = x_2 + x_1, \quad y_3 = y_2 + y_1,$$

and so on.

For the following important problem, and for Exercises 5 through 13 at the end of this appendix, you need to know how to differentiate the exponential function e^u. (See formula XX, Article 7.8.)

Problem 3. For what values of r is

$$x = Ae^{rt}, \qquad y = Be^{rt}$$

a nontrivial solution of the pair of differential equations

$$\frac{dx}{dt} = x - 2y, \qquad \frac{dy}{dt} = y - 2x?$$

Solution. Substituting the expressions given for x and y into the differential equations produces

$$Are^{rt} = Ae^{rt} - 2Be^{rt}, \qquad Bre^{rt} = Be^{rt} - 2Ae^{rt}.$$

Dividing by e^{rt} and collecting terms, we get

$$(r-1)A + 2B = 0, \qquad 2A + (r-1)B = 0.$$

These are two homogeneous linear equations for A and B. The determinant is

$$D = \begin{vmatrix} r-1 & 2 \\ 2 & r-1 \end{vmatrix} = (r-1)^2 - 4.$$

If $D \neq 0$, then the equations have only the trivial solution $A = B = 0$. To get a nontrivial solution, it is necessary that $D = 0$, or $r = 1 \pm 2 = 3$ or -1.

Case 1. Consider $r = 3$. Then the two equations for A and B are identical:

$$2A + 2B = 0, \quad \text{or} \quad B = -A.$$

Therefore one solution is $x = Ae^{3t}$, $y = -Ae^{3t}$.

Case 2. Consider $r = -1$. Then the equations for A and B become

$$-2A + 2B = 0, \qquad 2A - 2B = 0,$$

or $B = A$. Therefore another solution is

$$x = Ae^{-t}, \qquad y = Ae^{-t}.$$

EXERCISES

1. For what value of k do the following equations possess a nontrivial solution?

$$2x + 3y - 4z = 0,$$
$$x + ky + 3z = 0,$$
$$3x + ky - 2z = 0$$

2. Find at least one nontrivial solution of the equations

$$x + y - z = 0,$$
$$3x - y - z = 0,$$
$$x - 3y + z = 0,$$

or else show that there is no nontrivial solution.

3. Suppose the determinant of the coefficients of the system of equations (1) is zero. Show that

$$x_1 = kA_{11}, \quad x_2 = kA_{12}, \quad x_3 = kA_{13}$$

is a solution of the system for any k, where A_{11} is the cofactor of a_{11}, and so on. Could cofactors of elements of other rows than the first be used? Give a reason for your answer.

4. Verify that the system of equations

$$2x - y + z = 0,$$
$$x + 2y - z = 0,$$
$$4x + 3y - z = 0$$

has nontrivial solutions. Find some of them by using the results of Exercise 3.

The sets of simultaneous differential equations in Exercises 5 through 12 possess nonzero solutions of the form

$$x = Ae^{rt}, \qquad y = Be^{rt}$$

for certain values of the constants A, B, and r. Find them.

5. $\dfrac{dx}{dt} = 2x + 2y, \quad \dfrac{dy}{dt} = x + y$

6. $\dfrac{dx}{dt} = 3x + y, \quad \dfrac{dy}{dt} = -x + y$

7. $\dfrac{dx}{dt} = x + y, \quad \dfrac{dy}{dt} = x + y$

8. $\dfrac{dx}{dt} = 2x + 3y, \quad \dfrac{dy}{dt} = -x - 2y$

9. $\dfrac{dx}{dt} + x - 2y = 0, \quad \dfrac{dy}{dt} - 2x + y = 0$

10. $\dfrac{dx}{dt} + 2x - y = 0, \quad \dfrac{dy}{dt} - x + 2y = 0$

11. $\dfrac{dx}{dt} + 2x = 0, \quad \dfrac{dy}{dt} - 4x - 2y = 0$

12. $\dfrac{dx}{dt} = 3x + 4y, \quad \dfrac{dy}{dt} + x + 2y = 0$

13. If $x = A_1e^{r_1t}$, $y = B_1e^{r_1t}$ is one solution of any of the sets of differential equations of Exercises 5 through 12, and $x = A_2e^{r_2t}$, $y = B_2e^{r_2t}$ is another solution, prove that

$$x = A_1e^{r_1t} + A_2e^{r_2t}, \qquad y = B_1e^{r_1t} + B_2e^{r_2t}$$

is also a solution.

REVIEW QUESTIONS AND EXERCISES

1. What methods are available for evaluating a determinant?

2. What is the effect on a determinant if

 (a) two rows are interchanged?

 (b) three rows are permuted cyclically (the first one to be moved goes to the position of the second, the second to the position of the third, the third back to the position of the first)?

 (c) each element of a row is divided by 3?

 (d) each element of a column is subtracted from the corresponding element of a different column?

3. Under what conditions do n linear equations in n unknowns have a unique solution? When they fail to have a unique solution, they may have no solution or more than one solution. Discuss the possibilities for $n = 3$.

4. Define "homogeneous linear equations." When does a system of n homogeneous linear equations in n unknowns have nontrivial solutions? When do they have a unique solution, and what is that solution?

5. State Cramer's rule. Illustrate for three equations in three unknowns.

MISCELLANEOUS EXERCISES

In Exercises 1 through 10, solve the system of equations by Cramer's rule, or by the method of elimination, or show that the system has no solution.

1. $2x + 3y = 5,\quad 4x - y = 3$

2. $3x + y - z = 2,\quad x - 2y + z = 0,\quad 4x - y + z = 3$

3. $x - y + 2z = 2,\quad x + 3y + z = 1,\quad -7x + y + 4z = 0$

4. $x + y + z = 1,\quad x - y = 4,\quad x + y - 2z = 4$

5. $x + 2y + 3z = 0,\quad x - 5y + 3z = 0,\quad 3x - z = 0$

6. $2x + y = 0,\quad -x + 2y + z = 2,\quad x - 2y + 5z = 5$

7. $x - y + 2z = 2,\quad 3x - 2y + 4z = 5,\quad 2y - 3z + 2 = 0$

8. $2x - y + 3z = 6,\quad 4x + y = 9,\quad 7x + 2y - z = 15$

9. $x + y + 3z = 6,\quad x - 3y - 3z = -4,\quad 5x - 3y + 3z = 8$

10. $x - y - 3z - t = 1,\quad 2x + 4y - 2t = 2,\quad 3x + 4y - 2z = 0,\quad x + 2z - 3t = 3$

11. Show that the three points $P_1(x_1, y_1)$, $P_2(x_2, y_2)$, $P_3(x_3, y_3)$ are collinear if and only if

$$\begin{vmatrix} x_1 & y_1 & 1 \\ x_2 & y_2 & 1 \\ x_3 & y_3 & 1 \end{vmatrix} = 0.$$

12. Let $P_1(x_1, y_1)$, $P_2(x_2, y_2)$, $P_3(x_3, y_3)$ be vertices of a triangle. Let

$$\begin{vmatrix} x & y & 1 \\ x_2 & y_2 & 1 \\ x_3 & y_3 & 1 \end{vmatrix} \equiv Ax + By + C.$$

Let $b = P_2P_3$ and let h be the length of the altitude from P_1 to the base P_2P_3.

(a) Show that $b = \sqrt{A^2 + B^2}$.

(b) Show that

$$h = \left|\frac{1}{b} D\right|, \quad \text{where} \quad D = \begin{vmatrix} x_1 & y_1 & 1 \\ x_2 & y_2 & 1 \\ x_3 & y_3 & 1 \end{vmatrix}.$$

(c) Deduce that the area of the triangle is therefore $\frac{1}{2}|D|$.

Note. Compare G. M. Petersen, "Area of a Triangle and the Determinant," *Amer. Math. Monthly*, **62**, 1955, p. 249.

(d) Find the area of the triangle with vertices $P_1(1, -1)$, $P_2(3, 1)$, $P_3(2, 4)$.

13. Show that

$$\begin{vmatrix} 1 & 1 & 1 \\ a & b & c \\ a^3 & b^3 & c^3 \end{vmatrix} = (b - a)(c - a)(c - b)(a + b + c).$$

14. Let u, v, and w be functions of x with

$$u' = \frac{du}{dx}, \qquad u'' = \frac{d^2u}{dx^2}, \qquad v' = \frac{dv}{dx}, \qquad \ldots.$$

If

$$A = \begin{vmatrix} u & v & w \\ u' & v' & w' \\ u'' & v'' & w'' \end{vmatrix},$$

show that

$$\frac{dA}{dx} = \begin{vmatrix} u & v & w \\ u' & v' & w' \\ u''' & v''' & w''' \end{vmatrix}.$$

15. Without expanding, show that

$$\begin{vmatrix} b+c & a & 1 \\ c+a & b & 1 \\ b+a & c & 1 \end{vmatrix} = 0.$$

16. Reduce the following to an array with only four nonzero entries and thus evaluate the determinant

$$\begin{vmatrix} -1 & 1 & 1 & 1 \\ 1 & -1 & 1 & 1 \\ 1 & 1 & -1 & 1 \\ 1 & 1 & 1 & -1 \end{vmatrix}.$$

17. Solve the following system of equations.

$$x + y + z = a + b + c,$$
$$bx + cy + az = ab + bc + ca,$$
$$b^2x + c^2y + a^2z = ca^2 + ab^2 + bc^2,$$

where a, b, c are three given numbers, different from each other.

18. Show that

$$F(x) = \begin{vmatrix} A+x & a+x & a+x \\ b+x & B+x & a+x \\ b+x & b+x & C+x \end{vmatrix}$$

defines a linear function of x, and that

$$F(0) = \frac{af(b) - bf(a)}{a - b},$$

where

$$f(x) = (A - x)(B - x)(C - x).$$

19. If

$$\Delta_1 = \begin{vmatrix} a_{11} & a_{12} \\ a_{21} & a_{22} \end{vmatrix} \quad \text{and} \quad \Delta_2 = \begin{vmatrix} b_{11} & b_{12} \\ b_{21} & b_{22} \end{vmatrix},$$

show that

$$\begin{vmatrix} c_{11} & c_{12} \\ c_{21} & c_{22} \end{vmatrix} = \Delta_1\Delta_2,$$

where $c_{ij} = a_{i1}b_{1j} + a_{i2}b_{2j}$

20. (a) If three lines

$$a_1x + b_1y + c_1 = 0,$$
$$a_2x + b_2y + c_2 = 0,$$
$$a_3x + b_3y + c_3 = 0$$

pass through a point, show that $\Delta = 0$, where

$$\Delta = \begin{vmatrix} a_1 & b_1 & c_1 \\ a_2 & b_2 & c_2 \\ a_3 & b_3 & c_3 \end{vmatrix}.$$

(b) If $\Delta = 0$, do the lines necessarily pass through a point?

Hint. Consider the case of three parallel lines.

21. If

$$a_1x + b_1y + c_1 = 0,$$
$$a_2x + b_2y + c_2 = 0,$$
$$a_3x + b_3y + c_3 = 0$$

have a simultaneous solution, show that constants K_1, K_2, and K_3 not all zero exist such that

$$K_1(a_1x + b_1y + c_1) + K_2(a_2x + b_2y + c_2) + K_3(a_3x + b_3y + c_3) \equiv 0 \quad \text{for all } x \text{ and } y.$$

Is the converse true?

Hint. For the converse, consider the equations

$$x + y - 1 = 0,$$
$$x + y - 2 = 0,$$
$$x + y - 3 = 0.$$

APPENDICES II & III

APPENDIX II: FORMULAS FROM ELEMENTARY MATHEMATICS

A. Algebra

1. *Laws of exponents*

$$a^m a^n = a^{m+n},$$
$$(ab)^m = a^m b^m,$$
$$(a^m)^n = a^{mn},$$
$$a^{m/n} = \sqrt[n]{a^m}.$$

If $a \neq 0$,

$$\frac{a^m}{a^n} = a^{m-n},$$
$$a^0 = 1,$$
$$a^{-m} = \frac{1}{a^m}.$$

2. *Zero*

$a \cdot 0 = 0 \cdot a = 0$ for any finite number a.

If $a \neq 0$,

$$\frac{0}{a} = 0, \quad 0^a = 0, \quad a^0 = 1.$$

Division by zero is not defined.

3. *Fractions*

$$\frac{a}{b} + \frac{c}{d} = \frac{ad + bc}{bd}, \qquad \frac{a}{b} \cdot \frac{c}{d} = \frac{ac}{bd},$$

$$\frac{a/b}{c/d} = \frac{a}{b} \cdot \frac{d}{c}, \qquad \frac{-a}{b} = -\frac{a}{b} = \frac{a}{-b}.$$

$$\frac{(a/b) + (c/d)}{(e/f) + (g/h)} = \frac{(a/b) + (c/d)}{(e/f) + (g/h)} \cdot \frac{bdfh}{bdfh} = \frac{(ad + bc)fh}{(eh + fg)bd}.$$

4. *Binomial Theorem, for n any positive integer*

$$(a + b)^n = a^n + na^{n-1}b + \frac{n(n-1)}{1 \cdot 2} a^{n-2}b^2 + \frac{n(n-1)(n-2)}{1 \cdot 2 \cdot 3} a^{n-3}b^3 + \cdots + nab^{n-1} + b^n.$$

5. *Proportionality factor.* If y is proportional to x, or y varies directly as x, then $y = kx$ for some constant k, called the *proportionality factor.*

6. *Remainder Theorem and Factor Theorem.* If the polynomial $f(x)$ is divided by $x - r$ until a remainder is obtained that is independent of x, then this remainder R is equal to $f(r)$:

$$R = f(r).$$

In particular, $x - r$ is a *factor* of $f(x)$ if and only if r is a *root* of the equation $f(x) = 0$.

7. *Quadratic formula.* If $a \neq 0$, the roots of the equation $ax^2 + bx + c = 0$ are given by the formula

$$x = \frac{-b \pm \sqrt{b^2 - 4ac}}{2a}.$$

B. Geometry

Notation

A = area, B = area of base,

C = circumference,

S = lateral or surface area,

V = volume.

1. *Triangle* $A = \frac{1}{2}bh.$ (Fig. 1)

2. *Similar triangles*

$$\frac{a'}{a} = \frac{b'}{b} = \frac{c'}{c}. \quad \text{(Fig. 2)}$$

3. *Theorem of Pythagoras*

$$c^2 = a^2 + b^2. \quad \text{(Fig. 3)}$$

4. *Parallelogram*

$$A = bh. \quad \text{(Fig. 4)}$$

5. *Trapezoid*

$$A = \tfrac{1}{2}(a + b)h. \quad \text{(Fig. 5)}$$

6. *Circle*

$$A = \pi r^2, \qquad C = 2\pi r. \quad \text{(Fig. 6)}$$

7. *Any cylinder or prism with parallel bases*

$$V = Bh. \quad \text{(Fig. 7)}$$

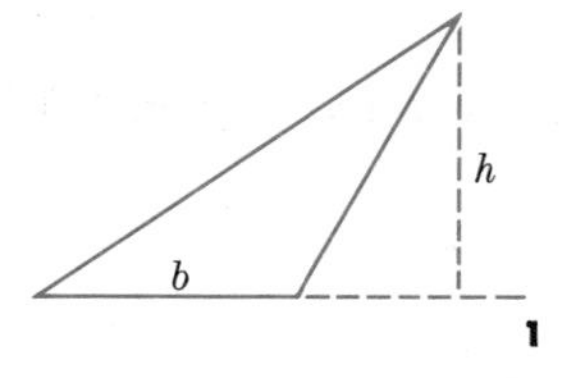

1

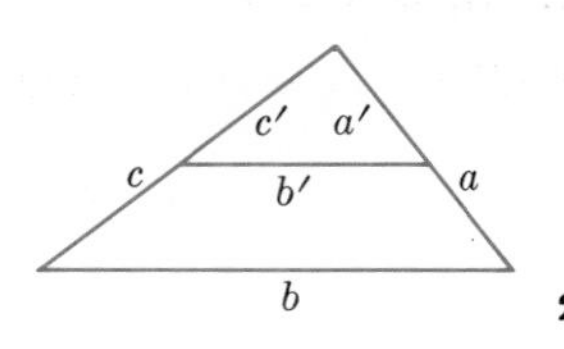

2

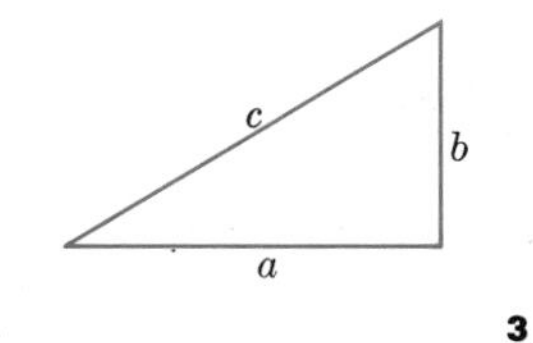

3

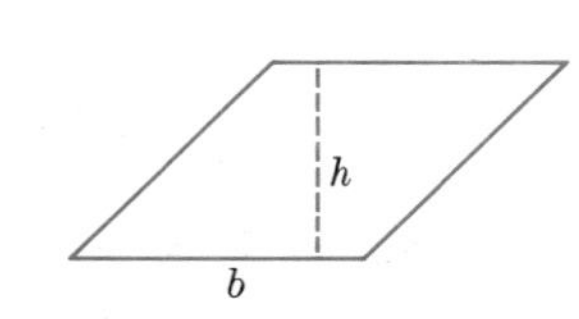

4

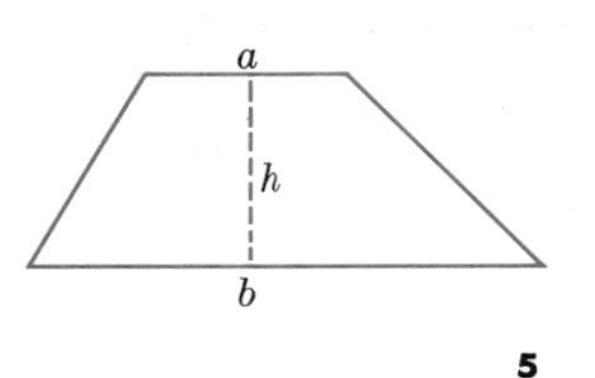

5

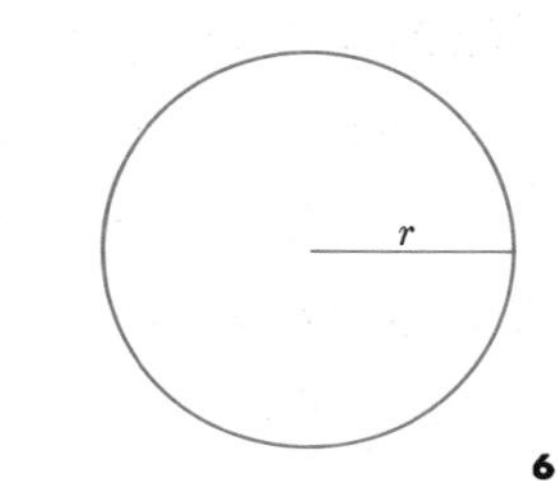

6

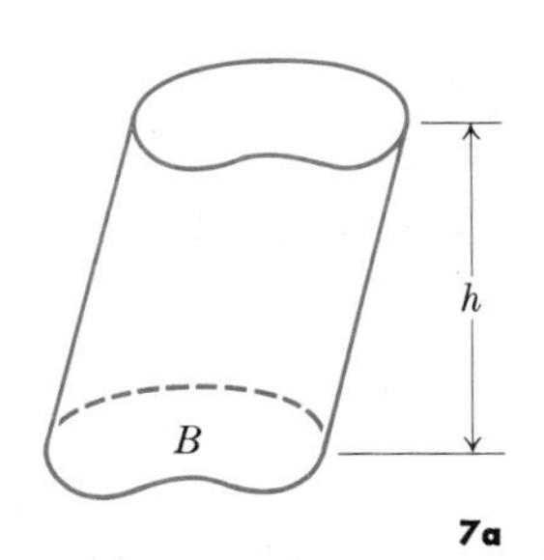

7a

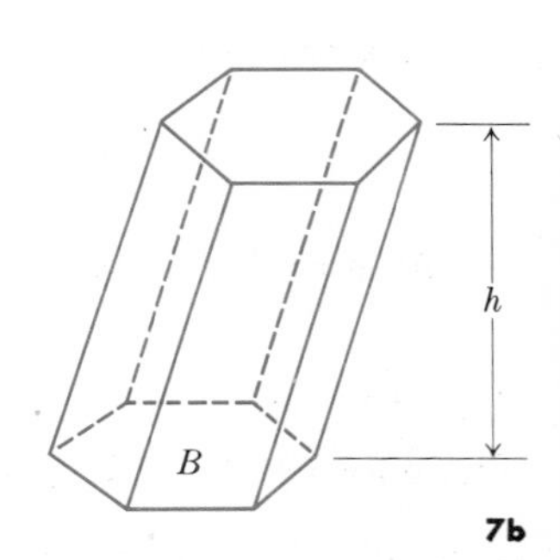

7b

8. *Right circular cylinder*

$$V = \pi r^2 h, \qquad S = 2\pi rh. \quad \text{(Fig. 8)}$$

9. *Any cone or pyramid*

$$V = \tfrac{1}{3}Bh. \quad \text{(Fig. 9)}$$

10. *Right circular cone*

$$V = \tfrac{1}{3}\pi r^2 h, \qquad S = \pi rs. \quad \text{(Fig.10)}$$

11. *Sphere*

$$V = \tfrac{4}{3}\pi r^3, \qquad A = 4\pi r^2. \quad \text{(Fig. 11)}$$

C. Trigonometry

1. *Definitions and fundamental identities* (Fig. 12)

$$\sin\theta = \frac{y}{r} = \frac{1}{\csc\theta},$$

$$\cos\theta = \frac{x}{r} = \frac{1}{\sec\theta},$$

$$\tan\theta = \frac{y}{x} = \frac{1}{\cot\theta},$$

$$\sin(-\theta) = -\sin\theta, \quad \cos(-\theta) = \cos\theta,$$

$$\sin^2\theta + \cos^2\theta = 1,$$

$$\sin 2\theta = 2\sin\theta\cos\theta, \quad \cos 2\theta = \cos^2\theta - \sin^2\theta,$$

$$\sin(\alpha+\beta) = \sin\alpha\cos\beta + \cos\alpha\sin\beta,$$

$$\cos(\alpha+\beta) = \cos\alpha\cos\beta - \sin\alpha\sin\beta.$$

2. *Angles and sides of triangle*

Law of cosines:

$$a^2 = b^2 + c^2 - 2bc\cos A.$$

Law of sines:

$$\frac{\sin A}{a} = \frac{\sin B}{b} = \frac{\sin C}{c}.$$

$$\text{Area} = \tfrac{1}{2}bc\sin A = \tfrac{1}{2}ac\sin B = \tfrac{1}{2}ab\sin C.$$

D. Sets

The notation $A = \{x: \ldots\}$

is equivalent to the statement, "A is the set of all elements x such that . . ."

The *union* of A and B contains every element that is in A, or in B, or in both A and B. It is denoted by

$$A \cup B.$$

The *intersection* of A and B contains every element that is in both A and B. It is denoted by $A \cap B$.

The *empty* set is denoted by $\emptyset$. If A and B have no elements in common, then

$$A \cap B = \emptyset.$$

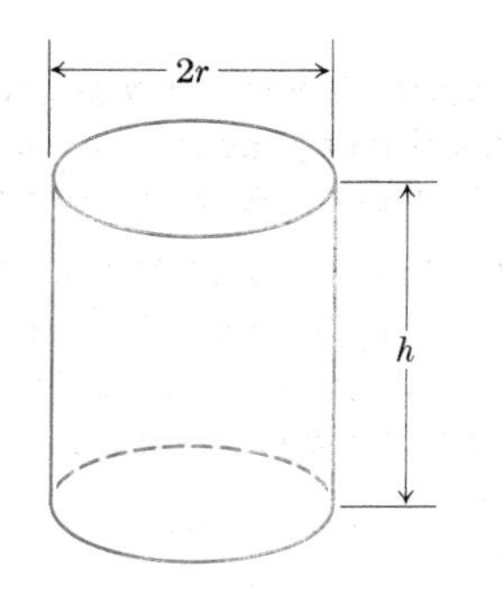

8

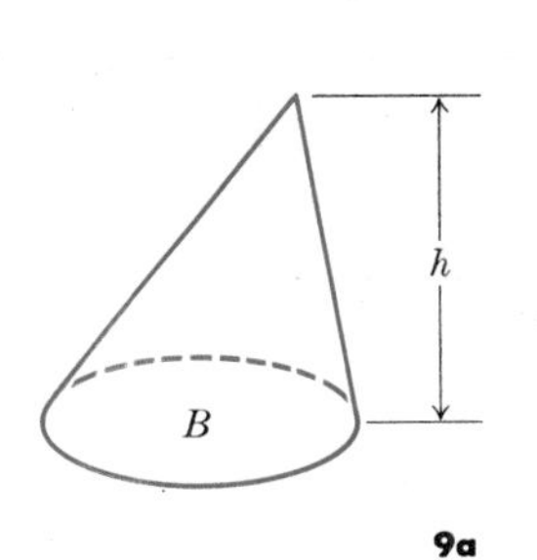

9a

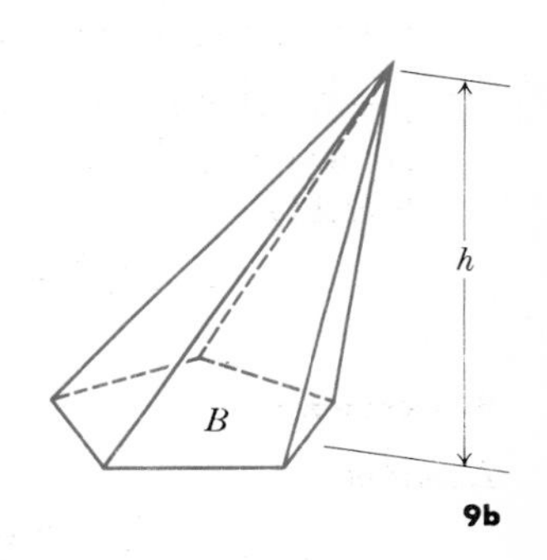

9b

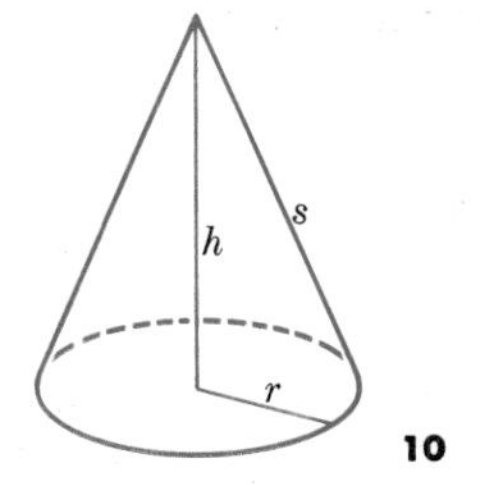

10

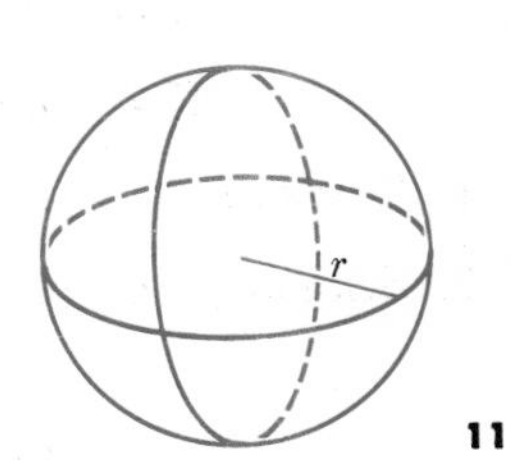

11

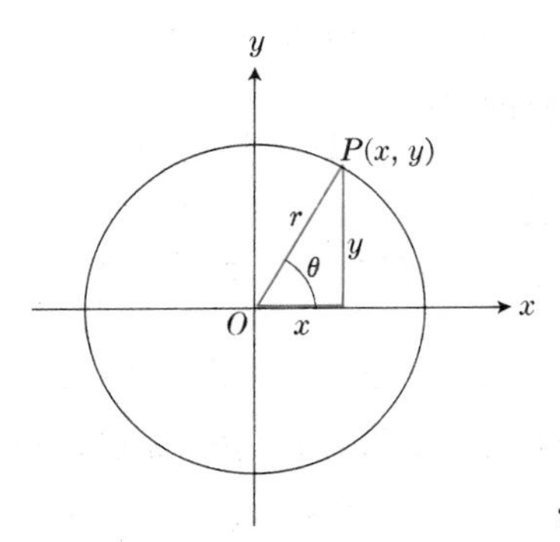

12

APPENDIX III: TABLES OF FUNCTIONS

Table 1. Natural trigonometric functions

Angle					Angle				
Degree	Radian	Sine	Cosine	Tangent	Degree	Radian	Sine	Cosine	Tangent
0°	0.000	0.000	1.000	0.000					
1°	0.017	0.017	1.000	0.017	46°	0.803	0.719	0.695	1.036
2°	0.035	0.035	0.999	0.035	47°	0.820	0.731	0.682	1.072
3°	0.052	0.052	0.999	0.052	48°	0.838	0.743	0.669	1.111
4°	0.070	0.070	0.998	0.070	49°	0.855	0.755	0.656	1.150
5°	0.087	0.087	0.996	0.087	50°	0.873	0.766	0.643	1.192
6°	0.105	0.105	0.995	0.105	51°	0.890	0.777	0.629	1.235
7°	0.122	0.122	0.993	0.123	52°	0.908	0.788	0.616	1.280
8°	0.140	0.139	0.990	0.141	53°	0.925	0.799	0.602	1.327
9°	0.157	0.156	0.988	0.158	54°	0.942	0.809	0.588	1.376
10°	0.175	0.174	0.985	0.176	55°	0.960	0.819	0.574	1.428
11°	0.192	0.191	0.982	0.194	56°	0.977	0.829	0.559	1.483
12°	0.209	0.208	0.978	0.213	57°	0.995	0.839	0.545	1.540
13°	0.227	0.225	0.974	0.231	58°	1.012	0.848	0.530	1.600
14°	0.244	0.242	0.970	0.249	59°	1.030	0.857	0.515	1.664
15°	0.262	0.259	0.966	0.268	60°	1.047	0.866	0.500	1.732
16°	0.279	0.276	0.961	0.287	61°	1.065	0.875	0.485	1.804
17°	0.297	0.292	0.956	0.306	62°	1.082	0.883	0.469	1.881
18°	0.314	0.309	0.951	0.325	63°	1.100	0.891	0.454	1.963
19°	0.332	0.326	0.946	0.344	64°	1.117	0.899	0.438	2.050
20°	0.349	0.342	0.940	0.364	65°	1.134	0.906	0.423	2.145
21°	0.367	0.358	0.934	0.384	66°	1.152	0.914	0.407	2.246
22°	0.384	0.375	0.927	0.404	67°	1.169	0.921	0.391	2.356
23°	0.401	0.391	0.921	0.424	68°	1.187	0.927	0.375	2.475
24°	0.419	0.407	0.914	0.445	69°	1.204	0.934	0.358	2.605
25°	0.436	0.423	0.906	0.466	70°	1.222	0.940	0.342	2.748
26°	0.454	0.438	0.899	0.488	71°	1.239	0.946	0.326	2.904
27°	0.471	0.454	0.891	0.510	72°	1.257	0.951	0.309	3.078
28°	0.489	0.469	0.883	0.532	73°	1.274	0.956	0.292	3.271
29°	0.506	0.485	0.875	0.554	74°	1.292	0.961	0.276	3.487
30°	0.524	0.500	0.866	0.577	75°	1.309	0.966	0.259	3.732
31°	0.541	0.515	0.857	0.601	76°	1.326	0.970	0.242	4.011
32°	0.559	0.530	0.848	0.625	77°	1.344	0.974	0.225	4.332
33°	0.576	0.545	0.839	0.649	78°	1.361	0.978	0.208	4.705
34°	0.593	0.559	0.829	0.675	79°	1.379	0.982	0.191	5.145
35°	0.611	0.574	0.819	0.700	80°	1.396	0.985	0.174	5.671
36°	0.628	0.588	0.809	0.727	81°	1.414	0.988	0.156	6.314
37°	0.646	0.602	0.799	0.754	82°	1.431	0.990	0.139	7.115
38°	0.663	0.616	0.788	0.781	83°	1.449	0.993	0.122	8.144
39°	0.681	0.629	0.777	0.810	84°	1.466	0.995	0.105	9.514
40°	0.698	0.643	0.766	0.839	85°	1.484	0.996	0.087	11.43
41°	0.716	0.656	0.755	0.869	86°	1.501	0.998	0.070	14.30
42°	0.733	0.669	0.743	0.900	87°	1.518	0.999	0.052	19.08
43°	0.750	0.682	0.731	0.933	88°	1.536	0.999	0.035	28.64
44°	0.768	0.695	0.719	0.966	89°	1.553	1.000	0.017	57.29
45°	0.785	0.707	0.707	1.000	90°	1.571	1.000	0.000	

Table 2. Exponential functions

x	e^x	e^{-x}
0.00	1.0000	1.0000
0.05	1.0513	0.9512
0.10	1.1052	0.9048
0.15	1.1618	0.8607
0.20	1.2214	0.8187
0.25	1.2840	0.7788
0.30	1.3499	0.7408
0.35	1.4191	0.7047
0.40	1.4918	0.6703
0.45	1.5683	0.6376
0.50	1.6487	0.6065
0.55	1.7333	0.5769
0.60	1.8221	0.5488
0.65	1.9155	0.5220
0.70	2.0138	0.4966
0.75	2.1170	0.4724
0.80	2.2255	0.4493
0.85	2.3396	0.4274
0.90	2.4596	0.4066
0.95	2.5857	0.3867
1.0	2.7183	0.3679
1.1	3.0042	0.3329
1.2	3.3201	0.3012
1.3	3.6693	0.2725
1.4	4.0552	0.2466
1.5	4.4817	0.2231
1.6	4.9530	0.2019
1.7	5.4739	0.1827
1.8	6.0496	0.1653
1.9	6.6859	0.1496
2.0	7.3891	0.1353
2.1	8.1662	0.1225
2.2	9.0250	0.1108
2.3	9.9742	0.1003
2.4	11.023	0.0907

x	e^x	e^{-x}
2.5	12.182	0.0821
2.6	13.464	0.0743
2.7	14.880	0.0672
2.8	16.445	0.0608
2.9	18.174	0.0550
3.0	20.086	0.0498
3.1	22.198	0.0450
3.2	24.533	0.0408
3.3	27.113	0.0369
3.4	29.964	0.0334
3.5	33.115	0.0302
3.6	36.598	0.0273
3.7	40.447	0.0247
3.8	44.701	0.0224
3.9	49.402	0.0202
4.0	54.598	0.0183
4.1	60.340	0.0166
4.2	66.686	0.0150
4.3	73.700	0.0136
4.4	81.451	0.0123
4.5	90.017	0.0111
4.6	99.484	0.0101
4.7	109.95	0.0091
4.8	121.51	0.0082
4.9	134.29	0.0074
5	148.41	0.0067
6	403.43	0.0025
7	1096.6	0.0009
8	2981.0	0.0003
9	8103.1	0.0001
10	22026	0.00005

Table 3. Natural logarithms of numbers

n	$\log_e n$	n	$\log_e n$	n	$\log_e n$
0.0	*	4.5	1.5041	9.0	2.1972
0.1	7.6974	4.6	1.5261	9.1	2.2083
0.2	8.3906	4.7	1.5476	9.2	2.2192
0.3	8.7960	4.8	1.5686	9.3	2.2300
0.4	9.0837	4.9	1.5892	9.4	2.2407
0.5	9.3069	5.0	1.6094	9.5	2.2513
0.6	9.4892	5.1	1.6292	9.6	2.2618
0.7	9.6433	5.2	1.6487	9.7	2.2721
0.8	9.7769	5.3	1.6677	9.8	2.2824
0.9	9.8946	5.4	1.6864	9.9	2.2925
1.0	0.0000	5.5	1.7047	10	2.3026
1.1	0.0953	5.6	1.7228	11	2.3979
1.2	0.1823	5.7	1.7405	12	2.4849
1.3	0.2624	5.8	1.7579	13	2.5649
1.4	0.3365	5.9	1.7750	14	2.6391
1.5	0.4055	6.0	1.7918	15	2.7081
1.6	0.4700	6.1	1.8083	16	2.7726
1.7	0.5306	6.2	1.8245	17	2.8332
1.8	0.5878	6.3	1.8405	18	2.8904
1.9	0.6419	6.4	1.8563	19	2.9444
2.0	0.6931	6.5	1.8718	20	2.9957
2.1	0.7419	6.6	1.8871	25	3.2189
2.2	0.7885	6.7	1.9021	30	3.4012
2.3	0.8329	6.8	1.9169	35	3.5553
2.4	0.8755	6.9	1.9315	40	3.6889
2.5	0.9163	7.0	1.9459	45	3.8067
2.6	0.9555	7.1	1.9601	50	3.9120
2.7	0.9933	7.2	1.9741	55	4.0073
2.8	1.0296	7.3	1.9879	60	4.0943
2.9	1.0647	7.4	2.0015	65	4.1744
3.0	1.0986	7.5	2.0149	70	4.2485
3.1	1.1314	7.6	2.0281	75	4.3175
3.2	1.1632	7.7	2.0142	80	4.3820
3.3	1.1939	7.8	2.0541	85	4.4427
3.4	1.2238	7.9	2.0669	90	4.4998
3.5	1.2528	8.0	2.0794	95	4.5539
3.6	1.2809	8.1	2.0919	100	4.6052
3.7	1.3083	8.2	2.1041		
3.8	1.3350	8.3	2.1163		
3.9	1.3610	8.4	2.1282		
4.0	1.3863	8.5	2.1401		
4.1	1.4110	8.6	2.1518		
4.2	1.4351	8.7	2.1633		
4.3	1.4586	8.8	2.1748		
4.4	1.4816	8.9	2.1861		

* Subtract 10 from $\log_e n$ entries for $n < 1.0$.

ANSWERS TO EXERCISES

Chapter 1

Article 1.2

1 through 15. If P is the point (x, y), the other points are $Q(x, -y)$, $R(-x, y)$, $S(-x, -y)$, $T(y, x)$.

16. 45°
17. 2
19. −1
20. No, because the line joining the given points crosses the y-axis at $(0, \frac{1}{2})$

Article 1.3

1. (3, 1), 2, 3, $\sqrt{13}$
2. (1, 4), −2, −3, $\sqrt{13}$
3. (2, 2), 3, −3, $\sqrt{18}$
4. (−1, −1), −3, 3, $\sqrt{18}$
5. (1, −1), 2, 3, $\sqrt{13}$
6. (−1, 2), −2, −3, $\sqrt{13}$
7. (−1, 2), 2, −4, $\sqrt{20}$
8. (−3, −2), −2, 4, $\sqrt{20}$
9. (−1, −2), 2, 1, $\sqrt{5}$
10. (−3, −1), −2, −1, $\sqrt{5}$
11. (a) $x^2 + y^2 = 25$
    (b) $x^2 + y^2 - 10x = 0$
    (c) $x^2 + y^2 - 6x - 8y = 0$
    (d) $(x - h)^2 + (y - k)^2 = 25$
12. $B(3, -3)$
13. $A(u + h, v + k)$
14. $B(0, 11)$

Article 1.4

1. $-3, \frac{1}{3}$
2. $3, -\frac{1}{3}$
3. $-\frac{1}{3}, 3$
4. $-\frac{1}{2}, 2$
5. $-1, 1$
6. $0, \infty$
7. $0, \infty$
8. $-3, \frac{1}{3}$
9. $-1, 1$
10. $\infty, 0$
11. $\frac{2}{3}, -\frac{3}{2}$
12. $y/x, -x/y$
13. $0, \infty$
14. $\infty, 0$
15. $-b/a, a/b$
16. Rectangle
17. Parallelogram
18. Rectangle
19. Parallelogram
20. No
21. (1, 2)
23. Yes, they lie on a line because the slope of BA is equal to the slope of AC.
24. No, because the slope of AC is 1 but the slope of CB is 3.
25. Yes, because $m_{AB} = m_{BC} = m_{CD} = 2$
26. No
27. Yes
28. $\left(\frac{x_1 + x_2}{2}, \frac{y_1 + y_2}{2}\right)$
29. Center $(\frac{3}{2}, \frac{3}{2})$, radius $\sqrt{8.5}$

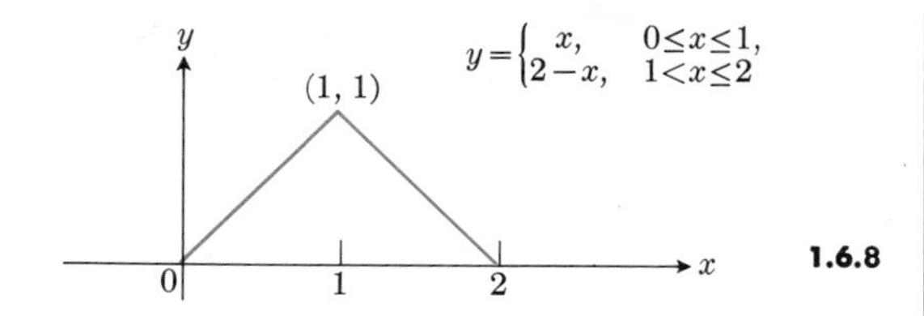

1.6.8

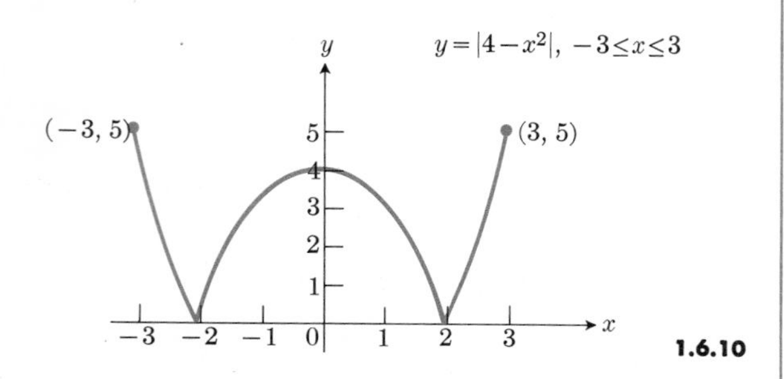

1.6.10

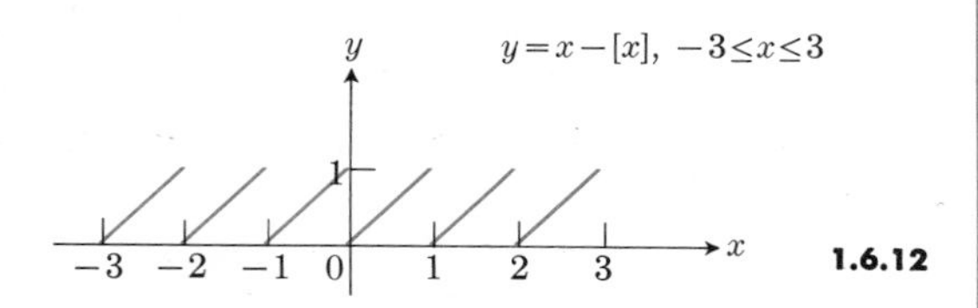

1.6.12

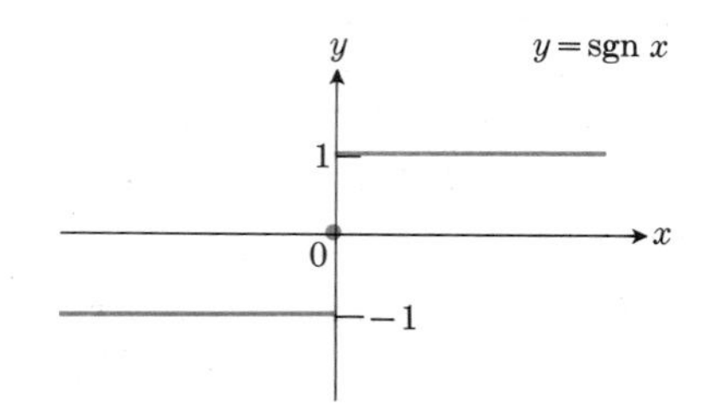

a|b

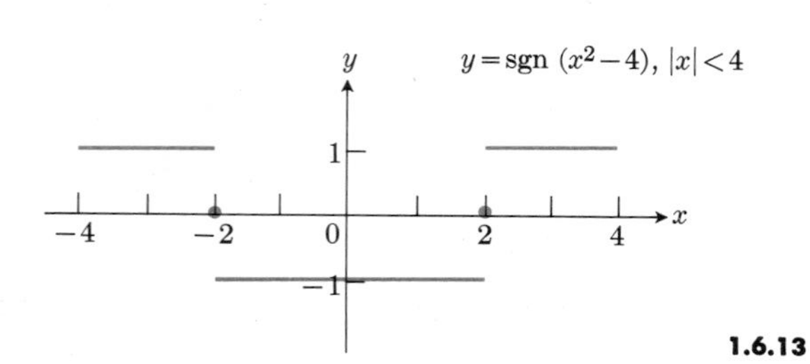

1.6.13

Article 1.5

1. $2y = 3x$
2. $y = 1$
3. $x = 1$
4. $3x + 4y + 2 = 0$
5. $x = -2$
6. $x + y = 4$
7. $x/a + y/b = 1$
8. $y = 0$
9. $x = 0$
10. $x + y = 1$
11. 3
12. $\frac{3}{2}$
13. -1
14. 2
15. $\frac{1}{2}$
16. $-\frac{3}{4}$
17. $\frac{4}{3}$
18. $\frac{1}{2}$
19. $-b/a$
20. $-x_1/y_1$
21. $x + 2y = 5$
22. (a) $x - 2y + 6 = 0$ (b) $(\frac{2}{5}, \frac{16}{5})$ (c) $6/\sqrt{5}$
23. $y - 4 = \sqrt{3}(x - 1)$
24. $180° - \tan^{-1} 2$
26. (a) $F = 1.8C + 32$ (b) $F = C$ at $-40°$
28. (a) Let $D = Ax_1 + By_1 + C$. If D and B agree in sign, then $BD > 0$, so that

$$\frac{DB}{B^2} = y_1 + \frac{Ax_1 + C}{B} > 0, \qquad y_1 > \frac{-(Ax_1 + B)}{C}.$$

Hence P_1 is above L.

(b) The inequalities are reversed, and P_1 is below L, if D and B have opposite signs.

(c) If $B = 0$, then we may assume $A \neq 0$. It then follows that P_1 is to the right of L when D and A agree in sign, but P_1 is to the left of L when D and A have opposite signs, because

$$AD/A^2 = x_1 + (C/A)$$

has the same sign as AD.

29. (a) 1 (b) $\frac{7}{5}$ (c) $\frac{9}{5}$ (d) 0

Article 1.6

1. $x = \pm\sqrt{y}, \quad -\infty < x < \infty, \quad y \geq 0$
2. $x = \dfrac{y+1}{y-1}, \quad x \neq 1, \quad y \neq 1$
3. $x = \pm\sqrt{\dfrac{y}{1-y}}, \quad -\infty < x < \infty, \quad 0 \leq y < 1$
4. $x = \dfrac{y^2}{1-y^2}, \quad x < -1$ or $x \geq 0,\ y \geq 0,\ y \neq 1$
5. $x = \dfrac{y \pm \sqrt{y^2+4}}{2}, \quad x \neq 0, \ -\infty < y < \infty$

6. $y = \dfrac{-x \pm \sqrt{12 - 3x^2}}{2}$, $\begin{cases} -2 \le x \le 2, \\ -2 \le y \le 2 \end{cases}$

7. $\frac{9}{2}$, 0, $\dfrac{1}{x^2} + x$, $x^2 + 2x\,\Delta x + (\Delta x)^2 + \dfrac{1}{x + \Delta x}$

8. See Fig. 1.6.8

9. (a) $-2 < x < 2$ (b) $x \le -2$ or $x \ge 2$
   (c) $-2 \le x \le 4$ (d) $0 < x \le 2$

10. See Fig. 1.6.10

11. $|1 - x| = 1 - x$ if $x \le 1$,
    $|1 - x| = x - 1$ if $x \ge 1$

12. See Fig. 1.6.12.

13. See Fig. 1.6.13.

14. Only (b) and (d)

15. Yes: $D_f = D_g = D_h = (-\infty, \infty)$, and, for all x, $f(x) = g(x) = h(x)$.

17. The functions r, s, and v are equal because their domains are the same and, for all x,

$$r(x) = s(x) = v(x).$$

18. All except g, whose domain does not include $x = 1$.

19. See Fig. 1.6.19.

20. (a) The independent variables are l, w, h.
    (b) V
    (c) The domain of each is $\{x: x > 0\}$.
    (d) $\{V: V > 0\}$

21. (a) $h = 0$
    (b) $\Delta f(0, 0) = 0$, $\Delta f(-2, 4) = 0$, $\Delta f(0.0001, -0.0001) = -10^{-8}$
    (c) $-5 < \Delta f < 7$

22. $A \cup B = (a, b)$

23. Yes, this is a deleted neighborhood of c. $A^- \cup B^- = (a, c) \cup (c, b)$.

24. One point. The intersection is not an open interval and hence it is not a neighborhood of zero. It is the single number zero.

Article 1.7

1. $D_f = \{x: -\infty < x < \infty\}$, $D_g = \{x: x \ge 1\}$, $D_{f+g} = D_{f \cdot g} = D_{g/f} = D_g$, $D_{f/g} = \{x: x > 1\}$

2. $D_f = \{x: x \ne 2\}$, $D_g = \{x: x > 1\}$,
   $D_{f+g} = D_{f \cdot g} = D_{f/g} = D_{g/f} = D_f \cap D_g = \{x: x > 1, x \ne 2\}$

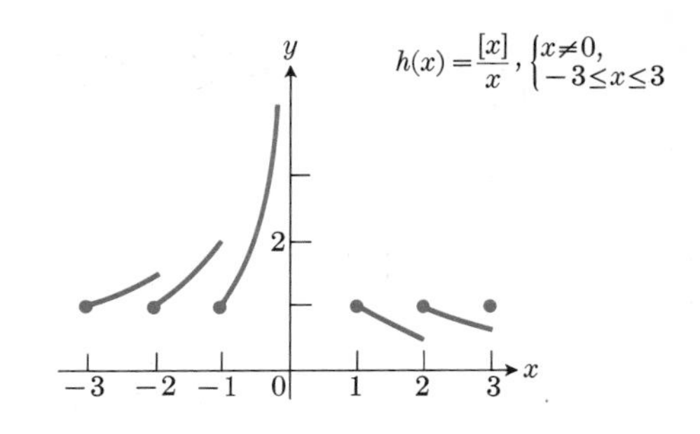

a|b

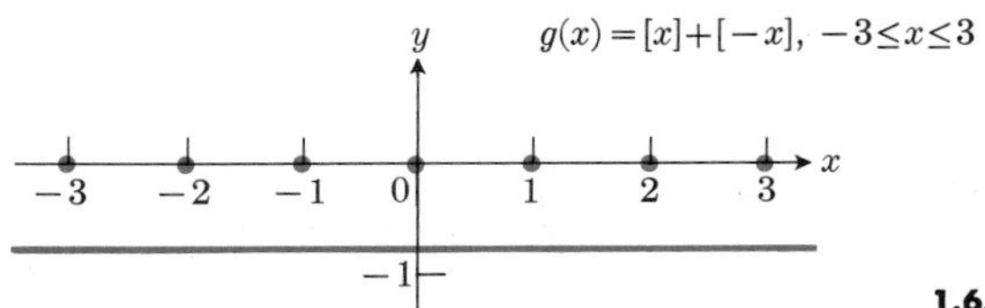

1.6.19

3. $D_f = \{x: x \ge 0\}$, $D_g = \{x: x \ge -1\}$, $D_{f+g} = D_{f \cdot g} = D_{f/g} = D_f$, $D_{g/f} = \{x: x > 0\}$

4. The graph of g is the graph of f moved three units to the right.

5. $D_g = [a - c, b - c]$. Move the graph of f to the left by c units to get the graph of g. When the sine curve over $[-\pi, \pi]$ is moved $\pi/2$ units to the left, the result is the cosine curve over $[-3\pi/2, \pi/2]$. This agrees with the identity $\sin(x + \pi/2) = \cos x$.

6. (a) $D_p = [-2\pi, 2\pi]$, $R_p = [-1, 1]$
   (b) $D_r = [-\pi/3, \pi/3]$, $R_r = [-1, 1]$
   (c) $D_q = [-\pi, \pi]$, $R_q = [-1, 1]$

7. See Fig. 1.7.7, and parts (a), (b), (c) on next page.

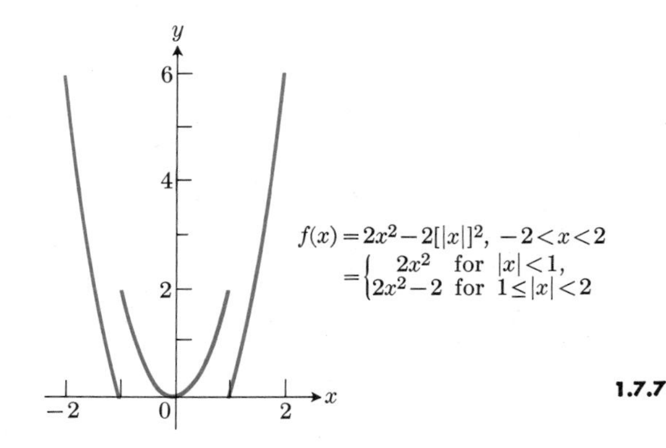

1.7.7

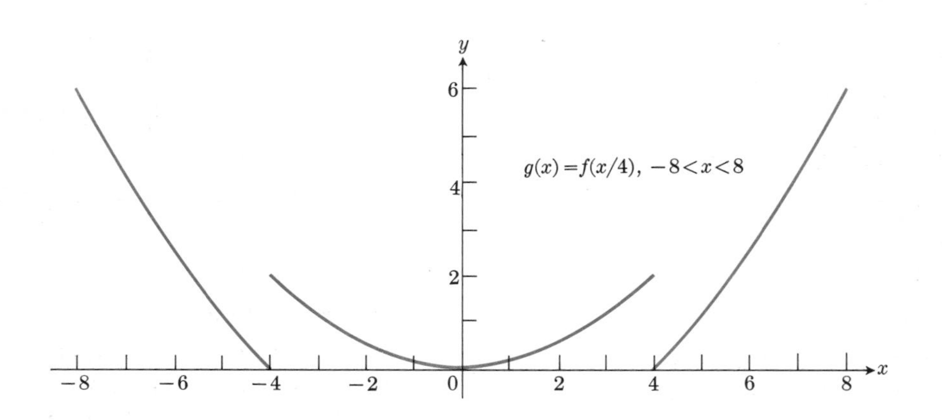

1.7.7a

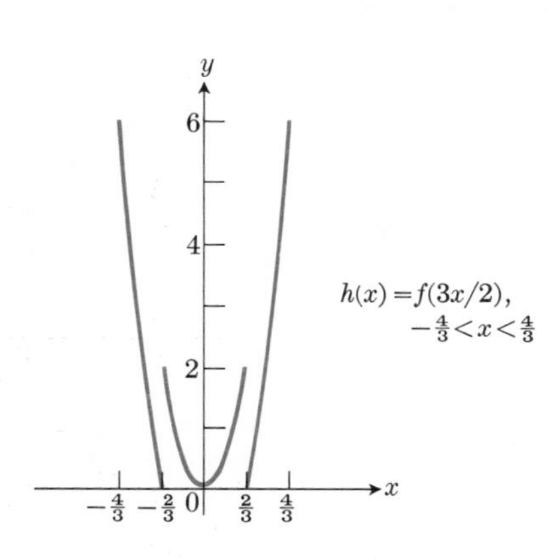

1.7.7b

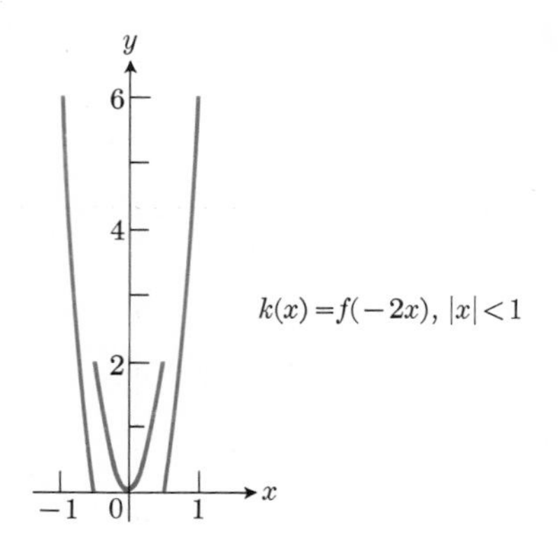

1.7.7c

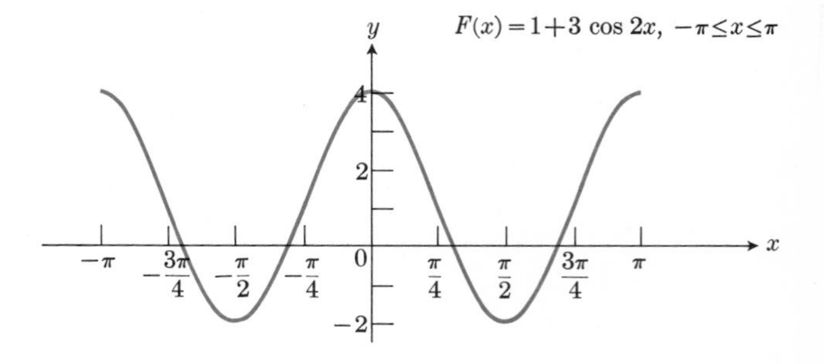

a|b

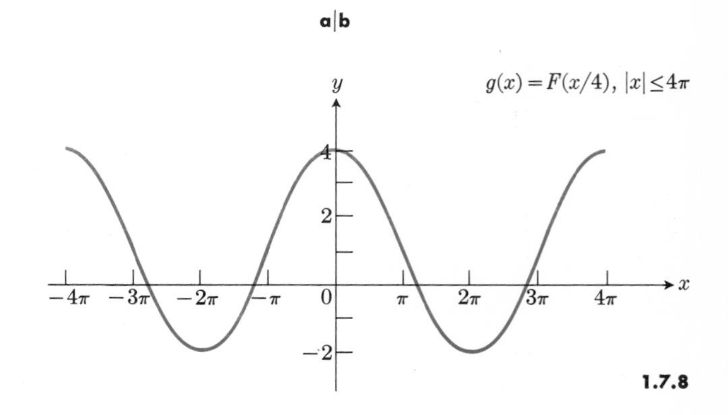

1.7.8

7. (Continued.)
   (a) $D_g = (-8, 8), \quad R_g = [0, 6)$
   See Fig. 1.7.7(a).
   (b) $D_h = (-\frac{4}{3}, \frac{4}{3}), \quad R_h = [0, 6)$
   See Fig. 1.7.7(b).
   (c) $D_k = (-1, 1), \quad R_k = [0, 6)$
   See Fig. 1.7.7(c).

8. (a) $D_f = [-\pi, \pi], \quad R_f = [-1, 2]$
   (b) $D_g = [-4\pi, 4\pi], \quad R_g = [-2, 4]$
   (c) $D_h = [-3\pi/2, \pi/2], \quad R_h = [-2, 4]$
   (d) $D_H = [-\pi, \pi], \quad R_H = [-4, 2]$

   See Fig. 1.7.8 for graph of $F(x)$ and the functions of parts (a) and (b).

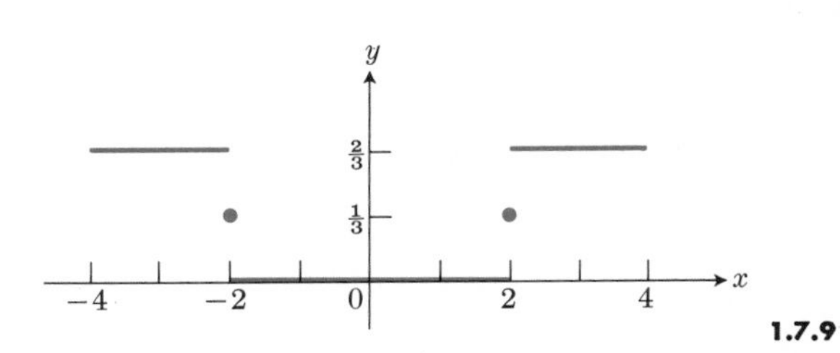

1.7.9

9. $R_f = \{0, \frac{1}{3}, \frac{2}{3}\}$
   See Fig. 1.7.9 above.

10. (a) $D_g = [-1, 1], \quad R_g = [0, \frac{3}{2}]$
    (b) $D_h = [-\frac{3}{2}, \frac{3}{2}], \quad R_h = [0, 1]$
    (c) $D_F = [-\frac{1}{2}, \frac{1}{2}], \quad R_F = [0, 3]$
    (d) $D_H = [-1, 1], \quad R_H = [-1, 0]$

11. See Fig. 1.7.11 below.

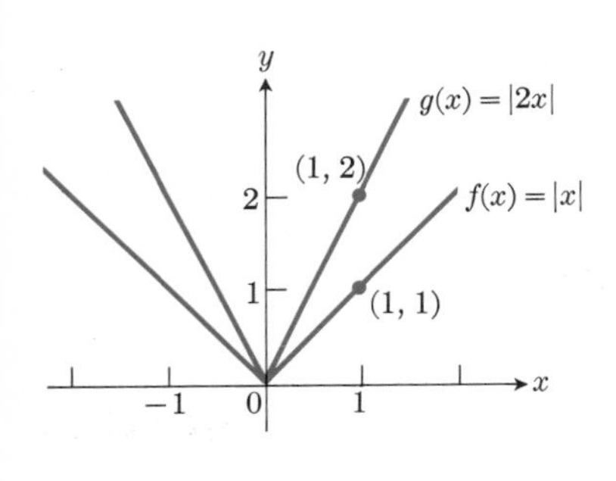

|a|b|

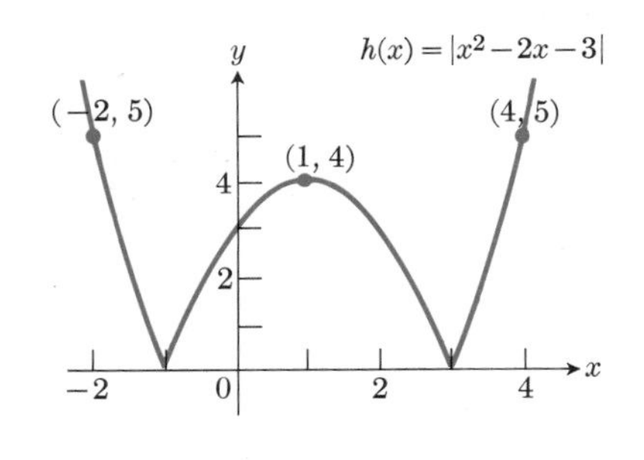

|c|

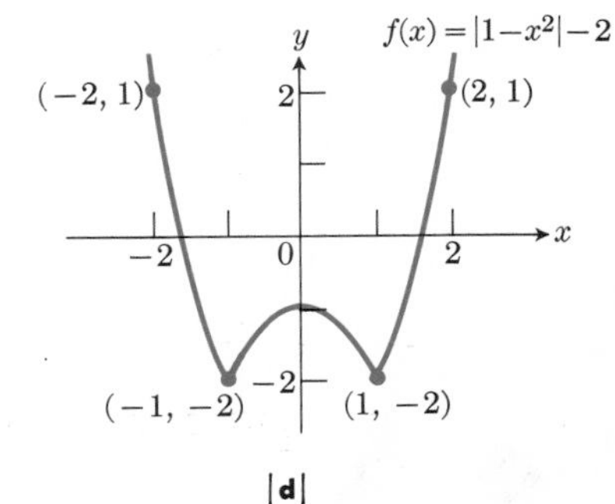

|d|

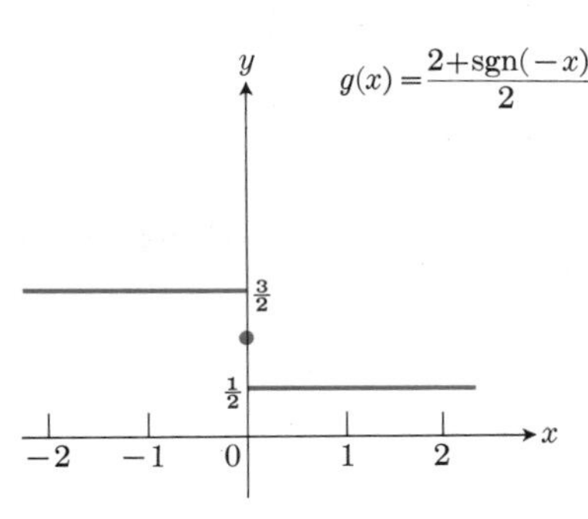
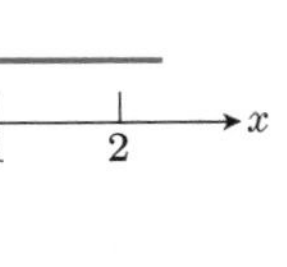

|e|

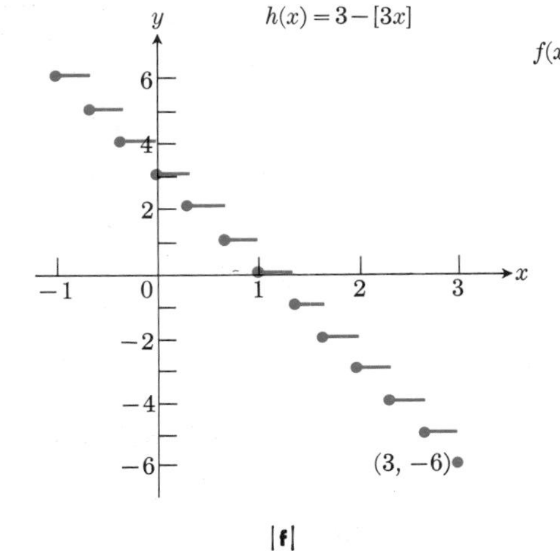

|f|

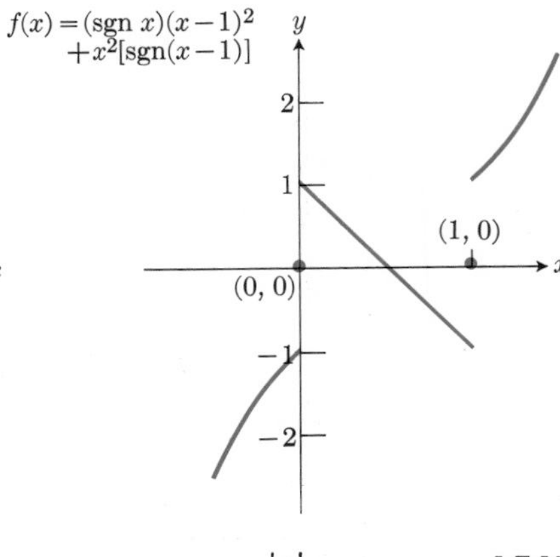

|g|

1.7.11

Article 1.8

1. $m(h) = 2, \quad m(4) = 2$

3. $m(h) = \dfrac{-1}{x_1(x_1 + h)}, \quad m(\frac{1}{4}) = -\frac{4}{5}$

5. $m(h) = \dfrac{|x_1 + h| - |x_1|}{h}, \quad m(\frac{1}{2}) = -1$

7. True 9. False 11. True

13. $L = 4, \quad \epsilon = 0.6$ 15. $L = 6, \quad \epsilon \geq \frac{3}{4}$

Article 1.9

15.

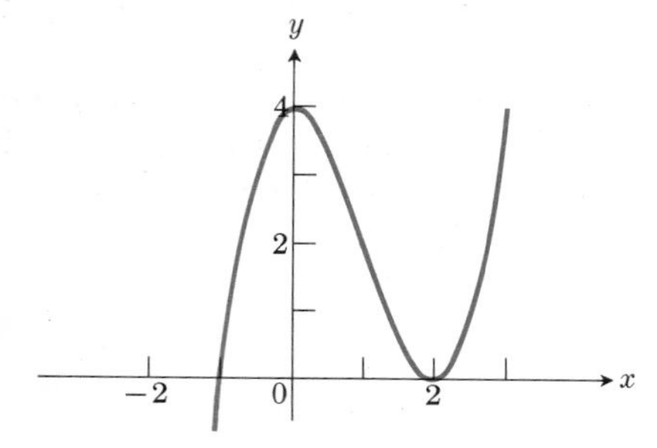

Article 1.10

1. $2x$
2. $3x^2$
3. 2
4. $2x - 1$
5. $-1/x^2$
6. $-2/x^3$
7. $-2/(2x+1)^2$
8. $1/(x+1)^2$
9. $4x - 1$
10. $3x^2 - 12$
11. $4x^3$
12. $2ax + b$
13. $1 + 1/x^2$
14. $a - b/x^2$
15. $1/\sqrt{2x}$
16. $1/(2\sqrt{x} + 1)$
17. $1/\sqrt{2x + 3}$
18. $-1/(2x\sqrt{x})$
19. $-1/(2x + 3)^{3/2}$
20. $x/\sqrt{x^2 + 1}$
21. $R_{f'} = \{-1, 1\}$, same graphs

Article 1.11

2. $4t + 5$ 3. $gt + v_0$ 4. 4
5. $2t - 3$ 6. $-2 - 2t$ 7. $8t + 12$
8. $2t - 4$ 9. $-4t$ 10. $64 - 32t$
12. 8,000 gal/min, 10,000 gal/min
13. (a) $2\pi r + \pi\,\Delta r$ (b) $2\pi r$
14. $4\pi r^2$ 15. πh^2

Miscellaneous exercises

1. (b) $-\frac{3}{2}, \ \frac{2}{3}, \ -\frac{3}{2}, \ \frac{2}{3}, \ 0, \ \infty$
   (c) Yes, $ABCD$ (d) Yes, AEB (e) Yes, CD
   (f) $2y + 3x = 26, \quad 2y + 3x = 0,$
   $3y - 2x + 13 = 0, \quad y = 6, \quad x = 2$
   (g)

L	AB	CD	AD	CE	BD
y-intercept	13	0	$-\frac{13}{3}$	6	none
x-intercept	$\frac{26}{3}$	0	$\frac{13}{2}$	none	2

2. (a) $3y + 2x = -7$ (b) $\sqrt{13}$
3. (a) No (b) No
   (c) Outside (d) $(-2, -\frac{9}{7})$
4. $4y + 3x = 0, \quad x = 0$
5. $((x_1 + x_2)/2, \ (y_1 + y_2)/2)$
7. (a) $-a/b$ (b) $(0, -c/b)$
   (c) $(-c/a, 0)$ (d) $ay - bx = 0$
8. Min $= 6$, max $= 23$. S is the triangle with vertices at $(0, 2)$, $(2, 0)$, and $(1, 5)$, plus its interior. The lines $3x + 4y = k$, $k \geq 0$, first meet S at the vertex $(2, 0)$, if we start with $k = 0$ and let k increase, and they leave at the vertex $(1, 5)$ when k increases through 23.
9. Four. Centers: $(0, \sqrt{5} - 2)$, $(0, -\sqrt{5} - 2)$, $(-\sqrt{5} - 1, 1)$, $(\sqrt{5} - 1, 1)$
   Radii (in same order): $(3\sqrt{2} - \sqrt{10})/2$, $(3\sqrt{2} + \sqrt{10})/2$, $(\sqrt{10} + \sqrt{2})/2$, $(\sqrt{10} - \sqrt{2})/2$
10. $|b' - b|/\sqrt{1 + m^2}$
11. Straight line through point of intersection of the two given lines
12. $(\frac{17}{18}, \frac{23}{6})$ 13. $x + 5y = \pm 5\sqrt{2}$
14. $x = \sqrt{(y + 2)/(y - 1)}, \quad y \leq -2, \ y > 1$
15. $A = \pi r^2, \quad C = 2\pi r, \quad A = C^2/4\pi$
16. (a) $x < -1, x > -1, \quad y < 0, y > 0$
    (b) All x, $\quad 0 < y \leq 1$
    (c) $x \geq 0, \quad 0 < y \leq 1$
17. $-5 < x < 3$
18. For $y \leq 4$, $x = y - 2$. For $y \geq 4$, $x = (y + 2)/3$.
19. $y > 2, \quad x = (y + 2)/2$

20. $x = \dfrac{y \pm \sqrt{y^2 - 4}}{2}$, $|y| \geq 2$

21. (a) -3, -4, $x_1^2 + 2x_1 - 3$,
$x_1^2 + 2x_1\,\Delta x + (\Delta x)^2 + 2x_1 + 2\,\Delta x - 3$

22. (c) $(\frac{1}{2}, -\frac{3}{4})$

24. $m(a, b) = (a + b - |a - b|)/2$

26. Take $\phi_k(x) = g_k(x)/g_k(x_k)$ where $g_k(x)$ is the product

$$(x - x_1)(x - x_2)\cdots(x - x_n)$$

with the factor $(x - x_k)$ deleted.

27. $f(d) = g(b)$

29. (a) $1/(1 - x)$ (b) $x/(x + 1)$
(c) x (d) $1 - x$

30. (a) $2/(x + 1)^2$ (b) $3\sqrt{x}/2$
(c) $\frac{1}{3}x^{-2/3}$

31. (a) $2x - 3$ (b) $-1/(3x^2) + 2$
(c) $1/(2\sqrt{t - 4})$

32. (a) 6 (b) (0, 2)

33. (a) $f(0) = 0$, $f(-1) = 1$, $f(1/x) = 2/(1 - x)$

(b) $\dfrac{-2}{(x - 1)(x + \Delta x - 1)}$ (c) $\dfrac{-2}{(x - 1)^2}$

34. $P(\frac{45}{8}, \frac{2025}{4})$

35. $v = 180 - 32t$, $t = \frac{45}{8}$

36. $t = 1$ sec, $s = 16$ ft

37. (a) $-\dfrac{1}{V(V + \Delta V)}$ (b) $-\frac{1}{4}$

38. 28 in^3/sec

41. (a) $\sqrt{x}$ (b) $1/\sqrt{x}$ (c) $\sqrt{(2 - x)/(1 + x)}$
(d) $(x^2 - 16)^{1/2}$ (e) $(x^2 - 16)^{-1/4}$ (f) x

Chapter 2

Article 2.1

1. $L = -1$, $0 < |x - 2| < \epsilon/3$
2. $L = 7$, N any deleted neighborhood of -1
3. $L = 2$, $0 < |x - 2| < 2\epsilon/(2 + \epsilon)$
4. $L = 4$, $0 < |x - 2| < \epsilon$
5. $L = 6$, $0 < |x - 5| < \epsilon$
6. $L = -5$, $0 < |x + 3| < 2\epsilon/3$
7. $L = -1/9$, $0 < |x - 3| < 27\epsilon/(1 + 9\epsilon)$
8. $h_3 = \max\{h_1, h_2\}$, $k_3 = \min\{k_1, k_2\}$
9. $$|t^2 + t - 12| = |(t - 3)(t + 4)| < \epsilon$$
if $|t + 4| < 8$ and $|t - 3| < \epsilon/8$,

thus if $0 < |t - 3| < \delta$ and $\delta \leq \min\{1, \epsilon/8\}$. Take $\epsilon = 0.1$ or 0.01 for parts (a) and (b).

10. Yes, because $|\sin(1/x)| \leq 1$, so $|f(x)| \leq |x| < \epsilon$ if $0 < |x| < \delta$ and $\delta \leq \epsilon$
11. No, because $\operatorname{sgn} x \to -1$, or $+1$, if $x \to 0$ through negative, or through positive, values. There is no number L that $\operatorname{sgn} x$ stays close to when x is near zero.

Article 2.2

1. 4	2. 0	3. 4	4. 4	5. 2
6. 0	7. 5	8. −1	9. 4	

10. Apply Theorem 3 twice, first with
$$f(x) = p(x) + r(x) \quad \text{and} \quad g(x) = s(x),$$
second with
$$f(x) = p(x) \quad \text{and} \quad g(x) = r(x).$$

11. 1

12. (a) $$\left|\frac{1}{x} - \frac{1}{3}\right| = \frac{|3 - x|}{3|x|} < \epsilon$$
if $|3x| \geq 3$ and $|x - 3| < 3\epsilon$.
Both are satisfied when x is in
$$N = \{x : 0 < |x - 3| < 3\epsilon \quad \text{and} \quad x \geq 1\}.$$
(b) When x is near c, $|cx|$ is near c^2 and is $>c^2/2$. Then
$$\left|\frac{1}{x} - \frac{1}{c}\right| = \frac{|c - x|}{|cx|} < \frac{2}{c^2}|c - x| < \epsilon$$
if both
$$|cx| > \frac{c^2}{2} \quad \text{and} \quad |x - c| < c^2\frac{\epsilon}{2},$$
or if x is in
$$N = \left\{x : 0 < |x - c| < c^2\frac{\epsilon}{2} \quad \text{and} \quad cx > \frac{c^2}{2}\right\}.$$

13. 25 14. 0 15. 1 16. -8 17. 1

18. 2, because

$$|\sqrt{x}-2| = \frac{|x-4|}{\sqrt{x}+2} \leq \frac{1}{2}|x-4| < \epsilon$$

if $|x-4| < 2\epsilon$

19. Let $f(x) = -g(x) = \sin(1/x)$.

20. Let $f(x) = g(x) = \operatorname{sgn} x$.

Article 2.3

1. (a) -10 (b) -20

2. 1 3. 18 4. 27 5. 36

6. 81 7. 2 8. 4 9. 4

10. 1 11. -1 12. -15

13. (a) $\frac{2}{5}$ (b) $\frac{2}{15}$ 14. $\frac{5}{4}$ 15. 10

16. 0 17. $\frac{1}{10}$ 18. No limit

19. No limit 20. $-\frac{1}{2}$

21. Let $f(x) = g(x) = 2 + \operatorname{sgn} x$.

22. $1/x^2$ most rapidly, $1/\sqrt{x}$ least rapidly

24. $(1-\cos\theta) < \theta^2/2$ for all $\theta \neq 0$. Not true for $\theta = 0$.

25. $\frac{1}{2}$ 26. $\frac{3}{5}$ 27. $\frac{2}{7}$ 28. 1

29. 2 30. $\cos x$

Article 2.4

1. $\frac{2}{5}$ 2. $\frac{3}{10}$ 3. ∞ 4. 0 5. 0

6. 5 7. ∞ 8. 5 9. $\frac{1}{2}$ 10. 0

11. $\pm\infty$, the sign being the same as that of the quotient of the coefficients of the highest powers in f and g

17. (a, b) Yes, because

$$g(x) \geq \tfrac{7}{4} \quad \text{for all } x$$

(c) No

19. 2

20. Because the degree of $g(x)$ is greater than the degree of $f(x)$, $g(x)$ is not the zero polynomial and $g(x) \neq 0$ for sufficiently large values of x. Hence

$$f(x) + g(x) = g(x)[1 + f(x)/g(x)].$$

By Exercise 10, $f(x)/g(x) \to 0$ as $x \to \infty$. For some real numbers b and M_1, $g(x) > b$ for $x > M_1$; and for some real number M_2, $-\frac{1}{2} < f(x)/g(x) < \frac{1}{2}$ when $x > M_2$. Let $M = \max\{M_1, M_2\}$ and get $f(x)+g(x)$ between $\frac{1}{2}g(x)$ and $\frac{3}{2}g(x)$ for $x > M$. Therefore $f(x) + g(x) > -\frac{3}{2}|b|$ for $x > M$.

Article 2.5

5. $$s_n = \frac{(n-1)n(2n-1)b^3}{6n^3} \to \frac{b^3}{3} \quad \text{as} \quad n \to \infty$$

6. $N = 1 + (b^3/\epsilon)$. Yes, because the area lies in $[s_n, S_n]$.

8. $$s_n = (b^3/3)[1-(1/n)][1-(1/2n)].$$

When n is replaced by $n+1$, both factors

$$a_n = 1 - (1/n) \quad \text{and} \quad b_n = 1-(1/2n)$$

increase because

$$a_{n+1} - a_n = \frac{1}{n(n+1)} > 0$$

and

$$b_{n+1} - b_n = \frac{1}{2n(n+1)} > 0.$$

Therefore $s_{n+1} > s_n$. Geometrically this means that the inscribed rectangles fit more closely as n increases. The argument is similar for $S_n = (b^3/3)A_nB_n$ with

$$A_n = 1 + (1/n) > A_{n+1}$$

and

$$B_n = 1 + (1/2n) > B_{n+1},$$

so that $S_n > S_{n+1}$.

9. $c = \sqrt{(a^2+ab+b^2)/3} \in (a, b)$

10. If $F(x) = x^3/3$, then $F'(x) = x^2$.

11. $$s_n = mh^2(n-1)n/2 = m(b^2/2)[1-(1/n)],$$
$$\lim s_n = b^2/2 \quad \text{as} \quad n \to \infty$$

12. $$S_n = mh^2n(n+1)/2 = m(b^2/2)[1+(1/n)],$$
$$\lim S_n = b^2/2 \quad \text{as} \quad n \to \infty$$

13. $$T_n = (s_n + S_n)/2 = mb^2/2 = \lim T_n$$

14. $$S_n - T_n = \tfrac{1}{2}(S_n - s_n) = mb^2/2n < \epsilon$$
if $$n > mb^2/2\epsilon$$

15. (a) $G'(x) = mx$
(b) Area $= (mb^2/2) - (ma^2/2) = G(b) - G(a)$

Miscellaneous exercises

1. $\frac{1}{2}$ 2. 0 3. $\frac{3}{7}$

4. $\frac{1}{3}$ 5. No limit 6. $2a$

7. 0 8. $2x$ 9. $1/(2\sqrt{x})$, $x > 0$

10. $-1/x^2$ 11. ∞ 12. $\frac{1}{2}$

13. $-\frac{1}{10}$ 14. No limit 20. $\frac{3}{4}$

21. 0 22. $\frac{1}{2}$ 23. (a) 0 (b) 1

24. 1 25. -1 26. -1 27. 0

28. 0 29. ∞ 30. 4 31. 4

32. (a) 0 (b) $-\frac{1}{3}$

(c) $f(-1/x) = -(x^2+x)/(5x^2+7x+2)$, $x \neq 0, -1, -\frac{2}{5}$,

$f(0) = -\frac{1}{5}$,

$1/f(x) = (2x^2-7x+5)/(x-1)$, $x \neq 1, \frac{5}{2}$.

33. Only one line if $c = 1$; two parallel lines if $c = -\frac{2}{3}$. If $c \neq 1$ or $-\frac{2}{3}$, the intersection point is

$$\left(\frac{c+1}{3c+2}, \frac{-1}{15c+10}\right),$$

and the limit is $(\frac{2}{5}, -\frac{1}{25})$.

34. (a) 0 (b) $\frac{1}{2}$

35. δ less than or equal to the smaller of 1 and $\epsilon^2/3$

36. $M = 1 + 1/\epsilon$

Chapter 3

Article 3.1

1. $v = 2t - 4$, $a = 2$
2. $v = 6t^2 - 10t + 4$, $a = 12t - 10$
3. $v = gt + v_0$, $a = g$
4. $v = 4 - 2t$, $a = -2$
5. $v = 8t + 12$, $a = 8$
6. $y' = 4x^3 - 21x^2 + 4x$, $y'' = 12x^2 - 42x + 4$
7. $y' = 15(x^2 - x^4)$, $y'' = 30(x - 2x^3)$
8. $y' = 8(x-1)$, $y'' = 8$
9. $y' = x^3 - x^2 + x - 1$, $y'' = 3x^2 - 2x + 1$
10. $y' = 8(x^3 - x)$, $y'' = 8(3x^2 - 1)$
11. $y' = 2x^3 - 3x - 1$, $y'' = 6x^2 - 3$
12. $y' = 21(x^6 - x^2 + 2x)$, $y'' = 42(3x^5 - x + 1)$
13. $y' = 5x^4 - 2x$, $y'' = 20x^3 - 2$
14. $y' = 2x + 1$, $y'' = 2$
15. $y' = 12x + 13$, $y'' = 12$
16. (a) 400 ft (b) 96 ft/sec
17. $y = 8x - 5$
18. $y = 4x \pm 2$. Least slope is $+1$, at $(0, 0)$.
19. $(-1, 27)$, $(2, 0)$
20. $a = b = 1$, $c = 0$
21. $a = -3$, $b = 2$, $c = 1$
22. $c = \frac{1}{4}$

Article 3.2

1. $x^2 - x + 1$
2. $(x-1)^2(x+2)^3(7x+2)$
3. $10x(x^2+1)^4$
4. $12(x^2-1)(x^3-3x)^3$
5. $-2(2x^2+3x-1)(x+1)(x^2+1)^{-4}$
6. $-2(x^2+x+1)/(x^2-1)^2$
7. $-19/(3x-2)^2$
8. $-4(x+1)(x-1)^{-3}$
9. $(1-t^2)/(t^2+1)^2$
10. $6(2t+3)^2$
11. $-2(2t-1)(t^2-t)^{-3}$
12. $(t^2+2t)(t+1)^{-2}$
13. $(2-6t^2)/(3t^2+1)^2$
14. $2(t-t^{-3})$
15. $3(2t+3)(t^2+3t)^2$

Article 3.3

1. $-x/y$
2. $\dfrac{1}{y(x+1)^2}$
3. $-(2x+y)/x$
4. $-(2xy+y^2)/(x^2+2xy)$
5. $3x^2/(2y)$
6. $-(y/x)^{1/3}$
7. $-(y/x)^{1/2}$
8. $\dfrac{y-3x^2}{3y^2-x}$
9. $\dfrac{y}{x} - (x+y)^2$
10. $(x^2+1)^{-3/2}$
11. $(2x^2+1)(x^2+1)^{-1/2}$
12. $\dfrac{x^4-1}{x^3y}$
13. $\dfrac{1-2y}{2x+2y-1}$
14. $\frac{1}{2}x^{-1/2} + \frac{1}{3}x^{-2/3} + \frac{1}{4}x^{-3/4}$
15. $\dfrac{2x}{y(x^2+1)^2}$
16. $\dfrac{2x^3-3x^2-3y^2}{6xy-2y^3}$
17. $5(3x+7)^4/(2y^2)$
18. $(10x^2+30x-8)(x+5)^3(x^2-2)^2$
19. $-y^2/x^2$
20. $(6x+15)(x^2+5x)^2$
21. $(2x-1)/(2y)$
22. $\dfrac{x(y^2-1)}{y(1-x^2)}$
23. $-\dfrac{x^2+9}{3x^2(x^2+3)^{2/3}}$
25. (a) $-x/y$ (b) $-1/y^3$
26. (a) $-x^2/y^2$ (b) $-2x/y^5$
27. (a) $-x^{-1/3}y^{1/3}$ (b) $\frac{1}{3}x^{-4/3}y^{-1/3}$
28. (a) $-y(x+2y)^{-1}$ (b) $2(x+2y)^{-3}$
30. (a) $7x - 4y = 2$ (b) $4x + 7y = 29$
31. (a) $3x - 4y = 25$ (b) $4x + 3y = 0$
32. (a) $y = 3x + 6$ (b) $x + 3y = 8$
33. (a) $x = 3y$ (b) $3x + y = 10$
34. (a) $3x - 4y = 10$ (b) $4x + 3y = 30$

35. (b) The function $y = x^{1/3}$ has no derivative at $x = 0$, but the y-axis is tangent to its graph there. The slope of $y = x^3$ is zero, and the x-axis is tangent to its graph at (0, 0).

37. For $0 \leq x < 1$, $x^n \to 0$ as $n \to \infty$. The functions $f_n(1) = 1$ for all n, so the limit is 1. Yes, there is a limit for each $x \in [0, 1]$, with $g(x) = 0$ for $x \in [0, 1)$ and $g(1) = 1$. The graph of g is the semiclosed interval [0, 1), and the single point (1, 1).

38. All five are differentiable on (0, 1). The defining equation makes sense for $(-\infty, \infty)$, $n = 1, 3, 10, \frac{1}{3}$, and for $x \in [0, \infty)$, $n = \frac{1}{10}$. Those with fractional exponents fail to have derivatives at $x = 0$.

39. For $n = 1$, f_1 is its own inverse. For $n = 3$ and $n = \frac{1}{3}$, f_3 and $f_{1/3}$ are inverses. The inverse of $f_{1/10}$ is a function with domain $[0, \infty)$, but it is not f_{10}. The inverse of f_{10} on $(-\infty, \infty)$ is not a function because $(-x)^{10} = x^{10}$.

Article 3.4

1. (a) $(2x + 2)\,\Delta x + (\Delta x)^2$
   (b) $(2x + 2)\,\Delta x$
   (c) $(\Delta x)^2$

2. (a) $(4x + 4)\,\Delta x + 2\,(\Delta x)^2$
   (b) $(4x + 4)\,\Delta x$
   (c) $2\,(\Delta x)^2$

3. (a) $(3x^2 - 1)\,\Delta x + 3x\,(\Delta x)^2 + (\Delta x)^3$
   (b) $(3x^2 - 1)\,\Delta x$
   (c) $3x\,(\Delta x)^2 + (\Delta x)^3$

4. (a) $4x^3\,\Delta x + 6x^2\,(\Delta x)^2 + 4x\,(\Delta x)^3 + (\Delta x)^4$
   (b) $4x^3\,\Delta x$
   (c) $6x^2\,(\Delta x)^2 + 4x\,(\Delta x)^3 + (\Delta x)^4$

5. (a) $-\dfrac{\Delta x}{x(x + \Delta x)}$ (b) $-\dfrac{\Delta x}{x^2}$ (c) $\dfrac{(\Delta x)^2}{x^2(x + \Delta x)}$

6. 2.13 7. 0.51 8. 1.96

9. 0.494 10. 0.58 12. Figure

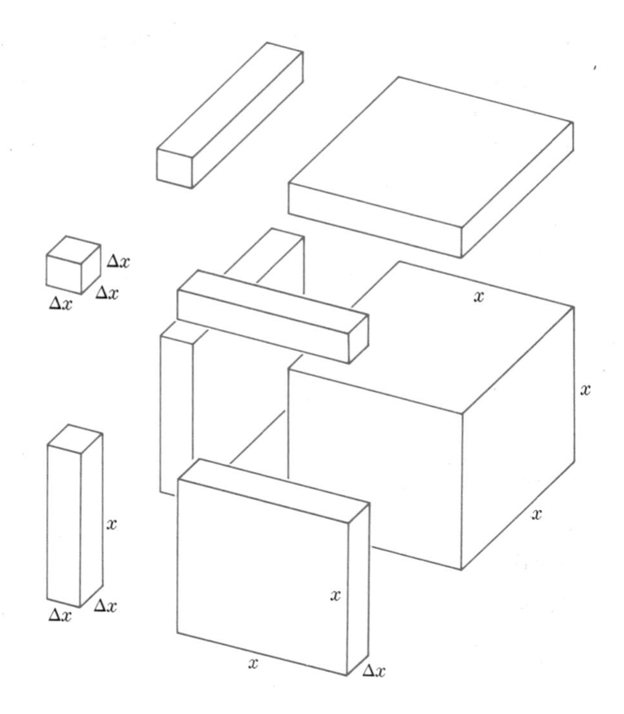

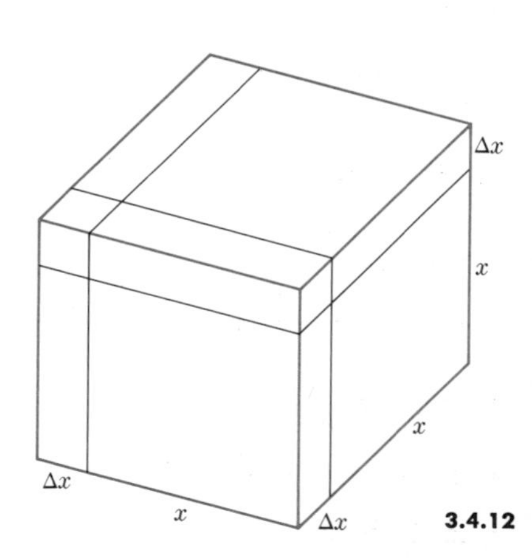

3.4.12

Article 3.5

1. $f_1(x) = 2\sqrt{x} + 1$, $x \geq 0$,
   $f_2(x) = \sqrt{2x + 1}$, $x \geq -\frac{1}{2}$

2. $f_1(x) = 1/(1 + x)$, $x \neq 0, x \neq -1$,
   $f_2(x) = 1 + (1/x)$, $x \neq 0, x \neq -1$

3. $f_1(x) = \sqrt{(1/x) - 1}$, $0 < x \leq 1$,
   $f_2(x) = 1/\sqrt{x - 1}$, $x > 1$

4. $f_1(x) = \sin(1/x)$, $x \neq 0$,
   $f_2(x) = \csc x$, $x \neq n\pi$, $n = 0, \pm 1, \pm 2, \ldots$

5. $f_1(x) = \cos(1/\sqrt{x})$, $x \geq 1/(4\pi^2)$,
   $f_2(x) = 1/\sqrt{\cos x}$, $|x| < \pi/2$ or $3\pi/2 < |x| \leq 2\pi$

6. $f_1(x) = |\sin x|, \quad f_2(x) = \sin |x|, \quad x \in (-\infty, \infty)$
7. $f_1(x) = 2|x| + 4, \quad f_2(x) = |2x + 4|, \quad x \in (-\infty, \infty)$
8. $f_1(x) = \frac{1}{2}(\tan x + |\tan x|), \quad x \in (-\pi/2, \pi/2),$
   $f_2(x) = \tan[(x + |x|)/2], \quad x \in (-\infty, \pi/2)$
9. $f_1(x) = f_2(x) = x, \quad x \in (-\infty, \infty)$
10. $f_1(x) = x, \quad x \in [0, \infty),$
    $f_2(x) = |x|, \quad x \in (-\infty, \infty)$
12.

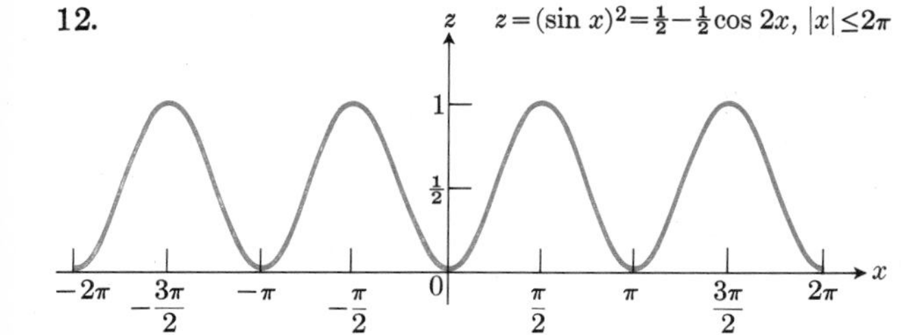

Article 3.6

1. $9y = (x - 1)^2$
2. $y^2 = x^3$
3. $y = \left(\dfrac{x}{1 + x}\right)^2$
4. $y = \dfrac{x^2}{1 - x}$
5. -3
6. $8x + 10$
7. $\dfrac{1}{2(x + 1)^2}$
8. $\dfrac{4x}{3(x^2 + 1)^{1/3}}$
9. $18x + \dfrac{1}{3x^2}$
10. $12\,\dfrac{(3x + 2)^4 - 1}{(3x + 2)^3}$

Article 3.8

1. $(3x^2 - 6x + 5)\,dx$
2. $\dfrac{9x}{2y}(3x^2 + 1)^{1/2}\,dx$
3. $-\dfrac{2xy + y^2}{2xy + x^2}\,dx$
4. $2(1 - x^2)(1 + x^2)^{-2}\,dx$
5. $\dfrac{1 - 2x^2}{\sqrt{1 - x^2}}\,dx$
6. $(6 - 2x - x^2)(x^2 - 2x + 4)^{-2}\,dx$
7. $3(1 - x)^2(2x - 1)(2 - 3x)^{-2}\,dx$
8. $(2 - 2x + x^2)(1 - x)^{-2}\,dx$
9. 12.04
10. 9.2
11. 2.03
12. 0.5013
13. 79.69
14. (c) $5x - 3y = 8$
15. (c) $15x - y = 20$
16. (c) $11y = x + 16$
17. (c) $9x + y = 6$
18. (c) $4x + 2y = 9$
19. $-4t^3(t^2 - 1)^{-3}$
20. $2(t + 1)^3(t - 1)^{-3}$

Article 3.9

1. (a) $x = -1$ (b) $x = 1$ or 3
2. Continuous
3. Take $\delta \le \epsilon$
4. Max = 1, min = 0
5. Neither a maximum nor a minimum since

$$f(x_2) > f(x_1) \quad \text{if} \quad 0 < x_1 < x_2 < 1.$$

Hence we can always find x_2 such that

$$f(x_2) > f(x_1),$$

so $f(x_1)$ is *not* the maximum. This is true for any x_1 in the given domain. The same argument shows that no x_2 in the domain can give a minimum.

6. Take $a = 0$, $b = 1$, and $N = 0$ in Theorem 15 (ii).
7. Continuous but not differentiable
9. 7
10. $$|\sin x - \sin a| \le 2\left|\sin\frac{x - a}{2}\right| \le |x - a|$$

Miscellaneous exercises

1. $-4(x^2 - 4)^{-3/2}$
2. $(5 - 2x - y)/(x + 2y)$
3. $-y/(x + 2y)$
4. $(3x^2 + 4y)/(9y^2 - 4x)$
5. $-\dfrac{2xy}{x^2 + 2y}$
6. $\dfrac{-2(x + 1)(x^2 + 2x + 2)}{x^3(x + 2)^3}$
7. $(1 - x^2)^{-3/2}$
8. $\dfrac{-(2x + 3y^2)}{3y(2x + y)}$
9. $(x + 1)^{-2}$
10. $(2x + 1)^{-1/2}$
11. $(3x^3 - 2a^2x)/\sqrt{x^2 - a^2}$
12. $-4(2x - 1)^{-2}$
13. $2x(1 - x^2)^{-2}$
14. $3(x^2 + x + 1)^2(2x + 1)$
15. $-\dfrac{2x + 1}{2}(x^2 + x - 1)^{-3/2}$
16. $2x + 1$
17. $\dfrac{4x + 5}{2}\sqrt{2x^2 + 5x}$
18. $-(9/2)(4x + 5)(2x^2 + 5x)^{-5/2}$
19. $-\dfrac{y}{x}\left(\dfrac{2y\sqrt{xy} + 1}{4y\sqrt{xy} + 1}\right)$
20. $\dfrac{y - 2x}{2y - x}$

21. $-(y/x)^{1/3}$

22. $-\sqrt{y/x}$

23. $-y/x$

24. $-y/x$

25. $-\frac{1}{2}\left(\frac{x+2y+y^2}{x+2y+xy}\right)$

26. $\frac{x^2-2x-1}{2(1-x)^{1/2}(1+x^2)^{3/2}}$

27. $\frac{1}{2y(x+1)^2}$

28. $\frac{12x-2xy-y^2}{x^2+2xy-12y}$

29. $-\frac{y+2}{x+3}$

30. 1

31. $\frac{4x+8}{\sqrt{4x^2+16x+15}}$

32. $\frac{2t(1+t^2)^2}{(1-t^2)}$

33. $2x\left[1+\frac{(1-x^2)}{(1+x^2)^3}\right]$

34. $m = 1, y = x$

35. $y + 2x = 6$

36. $c = 4$

37. $m = 2, y - 2x = -5$

38. $(2, 0)$ and $(-1, 27)$

39. (a) $10(x^2+2x)^4(x+1)$

(b) $\frac{3t-1}{\sqrt{3t^2-2t}}$

(c) $r[(r^2+5)^{-1/2}+(r^2-5)^{-1/2}]$

(d) $\frac{4x}{(x^2+1)^2}$

40. $27y + x = 28$

41. $y + x = 3$

42. $(2x+3)^{-1/2}$

43. $-6(2-3x)$, $f(x) = (2-3x)^2$

44. $-\frac{8}{5}$

45. $6\pi\,(\Delta r)^2$. It is the volume of a shell around the can with thickness Δr.

46. $\pi x(20-x)$

47. $x_1 = 40$, $p = 4(=20¢)$

49. $(x-1)(2x-1)$

50. $(2x+1)/9$

51. 56 ft/sec

52. $-\frac{12}{5}$

53. (a) $h + 2k = 5$

(b) $h = -4$, $k = \frac{9}{2}$, $r = 5\sqrt{5}/2$

54. $2\sqrt{x^2+1}$

55. $\frac{2}{3(2y+1)(2x+1)\sqrt{x^2+x}}$

56. $2x\sqrt{3x^4-1}$

57. $\frac{3}{(x+1)^2}\sin\left[\left(\frac{2x-1}{x+1}\right)^2\right]$

58. $y' = 2x - 3$, $y'' = 2$

59. $a = -32$

60. $\frac{3(x-2)}{(2x-3)^{3/2}}$

61. $-\frac{3}{32}$

62. $y' = 2$, $y'' = -2$

64. (a) $3(2x-1)^{-5/2}$ (b) $-162(3x+2)^{-4}$
(c) $6a$

66. $\Delta y = 0.92$, principal part $= 0.9$

67. $\sqrt{26} \approx 5.1$,
$\sqrt[3]{26} \approx 2.963$, $y = x^{1/3}$, $x = 27$, $\Delta x = -1$

68. 0.9999

69. 14 ± 0.044 ft

70. (a) $\frac{x(x+2)}{(x+1)^2}\,dx$ (b) $\frac{x}{y}\,dx$ (c) $\frac{-y}{x+2y}\,dx$

72. $\delta = \epsilon/4$. The function is uniformly continuous for $-2 \le x \le 2$.

75. $m = \frac{1}{2}$, $C = \sqrt{2}$

Chapter 4

Article 4.1

1. Falling, $x < \frac{1}{2}$; rising, $x > \frac{1}{2}$. Low point, $(\frac{1}{2}, \frac{3}{4})$.
2. Rising, $x < -1$; falling, $-1 < x < 2$; rising, $x > 2$. High point at $(-1, \frac{3}{2})$, low at $(2, -3)$.
3. Rising, $x < 0$; falling, $0 < x < 1$; rising $x > 1$. High point at $(0, 3)$, low at $(1, 2)$.
4. Rising, $x < -3$; falling, $-3 < x < 3$; rising, $x > 3$. High point at $(-3, 90)$, low at $(3, -18)$.
5. Falling, $x < -2$; rising, $-2 < x < 0$; falling, $0 < x < 2$; rising, $x > 2$. Low points at $(\pm 2, 0)$, high at $(0, 16)$.
6. Rising, $x < 0$; falling, $0 < x < 2$; rising, $x > 2$. High point at $(0, 4)$, low at $(2, 0)$.
7. Rising, $x < -4$; falling, $-4 < x < -1$, $-1 < x < 2$; rising, $x > 2$. High point at $(-4, -12)$, low at $(2, 0)$.
8. Falling, $x < -4$; rising, $-4 < x < 2$; falling, $x > 2$. Low point at $(-4, -\frac{1}{12})$. As $x \to 2$, $y \to \infty$.
9. Rising, $-2 < x < 0$; falling, $0 < x < 2$. High point at $(0, 2)$, low points at $(\pm 2, 0)$.
10. Falling, $x < -2$; rising, $x > 2$. Low points at $(\pm 2, 0)$. No curve for $|x| < 2$.
11. Rising for all x. Zero slope at $(0, 0)$.

Article 4.2

1. $\dfrac{dA}{dt} = 2\pi r \dfrac{dr}{dt}$
2. $\dfrac{dV}{dt} = 4\pi r^2 \dfrac{dr}{dt}$
3. $\dfrac{8}{5\pi}$ ft/min
5. $\dfrac{ax + by}{\sqrt{x^2 + y^2}}$ ft/sec
6. $\dfrac{25}{9\pi}$ ft/min
7. $\dfrac{10}{\sqrt{21}} = 2.2$ ft/sec
8. Increasing 33.7 ft/sec
9. $\dfrac{125}{144\pi}$ ft/min
10. $\dfrac{dy}{dt} = -x$. Clockwise
11. 8 ft/sec toward the lamppost. Decreasing 3 ft/sec
12. Increasing 8.75 psi/sec
13. 1500 ft/sec
14. 20 ft/sec
15. Thickness decreasing at the rate of $5/(72\pi)$ in/min
Area decreasing at the rate of $\frac{10}{3}$ in^2/min
16. $260/\sqrt{37}$ mi/hr

Article 4.4

M = high turning point
m = low turning point
I = inflection point

1. (a) $x > 2$ (b) $x < 2$ (c) Always
(d) Never $m(2, -1)$
2. (a) Always (b) Never (c) $x < -1$
(d) $x > -1$. Asymptote $x = -1$. (There is also a horizontal asymptote $y = 1$.)
3. (a) $-1 < x < 1$ (b) $x < 1$, $x > 1$
(c) $x < 0$
(d) $x > 0$, $M(1, 6)$, $m(-1, 2)$, $I(0, 4)$
4. (a) $x < -2$, $x > 3$ (b) $-2 < x < 3$
(c) $x > \frac{1}{2}$
(d) $x < \frac{1}{2}$, $M(-2, \frac{22}{3})$, $m(3, -\frac{27}{2})$, $I(\frac{1}{2}, -\frac{37}{12})$
5. (a) $x < -2$, $x > 2$ (b) $-2 < x < 2$, $x \neq 0$
(c) $x > 0$ (d) $x < 0$, $M(-2, -4)$, $m(2, 4)$
Asymptote $x = 0$ (also $y = x$).
6. (a) $x < -1$, $-1 < x < 0$
(b) $0 < x < 1$, $x > 1$ (c) $|x| > 1$
(d) $|x| < 1$, $M(0, -1)$. Asymptotes $x = \pm 1$. (There is also a horizontal asymptote $y = 1$.)
9. $M(-1, 7)$
10. $M(-2, 28)$, $m(2, -4)$, $I(0, 12)$
11. $m(2, 0)$
12. $M(0, 2)$, $m(2, -2)$, $I(1, 0)$
13. $(1, -1)$
14. $(16, -3)$, $(-16, 1)$
15. $(2, 0)$, $(1, \pm 1)$
16. $(3, 1)$
19. Concave downward, because $f''(x) = -1/x^2$ is negative
21. (a) Three times: near $x = -1.2, -0.4, +0.8$
(b) Once: near $x = -1.8$
(c) Once: near $x = 1.2$

Article 4.5

M = local maximum, m = local minimum

1. $M(\frac{1}{2}, \frac{1}{4})$, $m_1(0, 0)$, $m_2(1, 0)$
2. $M(1/\sqrt{3}, 2/3\sqrt{3})$, $m_1(0, 0)$, $m_2(1, 0)$
3. $M_1(-2, 6)$, $M_2(\frac{1}{2}, \frac{1}{4})$, $M_3(2, 2)$, $m_1(0, 0)$, $m_2(1, 0)$
4. $m(-1, -\frac{1}{2})$, $M(1, \frac{1}{2})$
5. $f'(x) = (1 + |x|)^{-2}$. There are no critical points.
6. $m(0, 0)$
7. No maxima; minima at $(n, 0)$ for $n = 0, \pm 1, \pm 2, \ldots$
8. Each point in $[-2\pi, 2\pi]$ is a critical point. Local maxima at $(x, 1)$ for $x \in \{(-2\pi, -\pi) \cup (0, \pi)\}$, and local minima at $(x, -1)$ for $x \in \{(-\pi, 0) \cup (\pi, 2\pi)\}$
9. $m(\frac{1}{2}, 4)$
10. On $[-2, 3]$, $M_1(-2, 8)$, $M_2(3, 27)$ (the absolute maximum), and $m(0, 0)$. There is no absolute maximum on $[-2, 3)$.

Article 4.6

2. $(r\sqrt{2}) \times (r/\sqrt{2})$
3. 32
4. $\frac{5}{3} \times \frac{14}{3} \times \frac{35}{3}$ (in.)
5. Use one-half of the fence parallel to the river.
6. $\sqrt{A/3} \times \sqrt{A/3} \times \frac{1}{2}\sqrt{A/3}$, where A is the given amount of material
7. $4 \times 4 \times 2$ (in.)
8. 4
9. 9 in. wide and 18 in. high
10. $V_{\max} = \dfrac{2\pi h^3}{9\sqrt{3}}$, h the given hypotenuse

11. $(100 + c)/2$

12. $0.58L$

13. (a) 16 (b) -54 (c) -1
The derivative $f'(x)$ is equal to zero when $2x^3 = a$, and then $f''(x) = +6$ if $a = 0$, or $f''(x) = +2$ if $a \neq 0$.

14. (a) $a = -3,\ b = -9$ (b) $a = -3,\ b = -24$

15. (a) Square, $\dfrac{4L}{4 + 3\sqrt{3}}$; triangle, $\dfrac{3\sqrt{3}L}{4 + 3\sqrt{3}}$
(b) Use it all for the square.

16. $(\frac{1}{2}, -\frac{1}{2})$ and $(-\frac{1}{2}, \frac{1}{2})$

17. (a) $(c - \frac{1}{2}, \sqrt{c - \frac{1}{2}})$ (b) $(0, 0)$

18. $\frac{32}{81}\pi r^3$

19. $\frac{4}{9}\pi r^3\sqrt{3}$

21. Width, $\dfrac{2r}{\sqrt{3}}$; depth, $2r\sqrt{\frac{2}{3}}$

22. Width of beam = radius of log.

23. $\dfrac{x}{c - x} = \sqrt[3]{\dfrac{a}{b}}$, where x is the distance from the source of strength a.

24. Height of rectangle is $(4 + \pi)/8$ times diameter of semicircle.

25. $4/\pi$

26. Diameter = alt. of cylinder = $\sqrt[3]{\dfrac{3V}{\pi}}$,
V = total volume

27. (a) Unity (b) Zero

31. Before tax: Production = 7,500; price = \$1.25, profit = \$5,425. After tax: Production = 7,000; price = \$1.30, profit = \$4,700. To maximize his profit he should absorb 5¢ of the tax and add 5¢ to the price.

Article 4.7

4. If three of the roots of $f(x) = 0$, $x \in (a, b)$, are $x_1 < x_2 < x_3$, then, by Rolle's Theorem,
$$f'(c_1) = 0 \quad \text{and} \quad f'(c_2) = 0$$
for some
$$c_1 \in (x_1, x_2) \quad \text{and} \quad c_2 \in (x_2, x_3).$$
Now apply Rolle's Theorem to f' on $[c_1, c_2]$. *Generalization.* If $f(x) = 0$ has n roots in $[a, b]$, then $f'(x) = 0$ has $n - 1$ roots, $f''(x) = 0$ has $n - 2$ roots, ..., and $f^{(n-1)}(x) = 0$ has at least one root in (a, b).

5. The maximum of $\sin x$ is at $x = \pi/2$, so $c = \pi/2$.

Article 4.8

1. $\frac{1}{2}$ 2. $\sqrt{3}$ 3. $\frac{8}{27}$ 4. 1 5. $\frac{3}{2}$

10. (b) For $1 < c < 2$, we have $2 < c^2 + 1 < 5$, so $\frac{1}{2} > f'(c) > \frac{1}{5}$. Therefore $(\pi/4) + 0.2 < f(2) < (\pi/4) + 0.5$.

Article 4.9

1. $4.9 - \epsilon$, with $1/1000 < \epsilon < \sqrt{6}/2304$

3. (a) 2.1, 2.0975 (b) 1.5, 1.4375
(c) 1.9, 1.8975 (d) 2.325, 2.2986

5. (a) 0 (b) $f'(a)$ (c) 0 (d) $\frac{1}{2}f''(a)$

Miscellaneous exercises

1. (a) $x < \frac{9}{2}$
(b) $x > \frac{9}{2}$
(c) No values
(d) All values

2. (a) $x > 3,\ x < \frac{1}{3}$
(b) $\frac{1}{3} < x < 3$
(c) $x > \frac{5}{3}$
(d) $x < \frac{5}{3}$

3. (a) $x < 3$
(b) $x > 3$
(c) $0 < x < 2$
(d) $x < 0,\ x > 2$

4. (a) $|x| > \frac{1}{2}$
(b) $|x| < \frac{1}{2}$
(c) $x > 0$
(d) $x < 0$

5. (a) $x > \sqrt[3]{2}$
(b) $x < \sqrt[3]{2}$
(c) $x > 0,\ x < -\sqrt[3]{4}$
(d) $-\sqrt[3]{4} < x < 0$

6. (a) $x < 0,\ x > 2$
(b) $0 < x < 2$
(c) $x \neq 0$
(d) Never

7. (a) $x < 0$
(b) $x > 0$
(c) $x \neq 0$
(d) Never

8. (a) $x \neq -1$
(b) Never
(c) $x < -1$
(d) $x > -1$

9. (a) $x \neq 0$
(b) Never
(c) $x < 0$
(d) $x > 0$

10. (a) $-1 < x < 0,\ x > 1$
(b) $x < -1,\ 0 < x < 1$
(c) $x < -1/\sqrt{3},\ x > 1/\sqrt{3}$
(d) $|x| < 1/\sqrt{3}$

11. (a) $x < -2b/a,\ x > 0$
(b) $-2b/a < x < 0$
(c) $x > -b/a$
(d) $x < -b/a$

12. (a) $x < 1,\ x > 2$
(b) $1 < x < 2$
(c) $x > \frac{3}{2}$
(d) $x < \frac{3}{2}$

13. (a) $x < -1, \quad x > \frac{1}{3}$
(b) $-1 < x < \frac{1}{3}$
(c) $x > -\frac{1}{3}$
(d) $x < -\frac{1}{3}$

14. (a) $0 < x < 4$
(b) $x < 0, \quad x > 4$
(c) $x < 2$
(d) $x > 2$

15. (a) At $x = 1$; because y' goes from + to −
(b) At $x = 3$; because y' goes from − to +

17. If $v = k/\sqrt{s}$, then $dv/dt = -k/2s^2$.

18. $k^2/2$ 19. $\frac{1}{4}$ in/min 20. $(3/400\pi)$ ft/min

21. $dr/dt = -(3/400)$ ft/min
$dA/dt = -\frac{6}{5}\pi$ ft²/min
$\Delta r \approx -(3/4000)$ ft; $\Delta A \approx -(3\pi/25)$ ft²

22. (a) Approx. 606 mi/hr (b) Approx. 0.83 mi

23. $\frac{1}{2}\sqrt{3}$ 24. $-480\sqrt{2}/7$

25. Yes, it will fill, because $dy/dt > 0.007$ for $y \leq 10$.

26. $a = -gR^2/s^2$ 28. $(\sqrt{3}/30)\, r$, increasing

29. ≈ 18 mi/hr 30. $(8/9\pi)$ ft/min

31. $\frac{20}{3}$ and $\frac{40}{3}$ 32. 10

33. 18 and 18, no

34. $a = 1, \quad b = -3, \quad c = -9, \quad d = 5$

35. $\frac{1}{2}$ 36. $r = 25$ ft

37. $t = 1, \quad s = 16$ 38. $r = 4, \quad h = 4$

39. Diameter = height = 4 in.

40. $4\sqrt{3}$ 41. 276 (approx.)

42. (a) $(2, \pm\sqrt{3})$, (b, c) $(1, 0)$

43. (a) $2\frac{1}{12}$ mi from A (b) $2\frac{1}{12}$ mi from A (c) at B

44. (a) True (b) False (c) True (d) False

45. Each $= (P - b)/2$

46. Each $= \tan^{-1}(4k/b^2)$ 48. $m = \frac{1}{4}$

50. Sides are 110 yd, radii of semicircular ends are $110/\pi$ yd.

57. 6×18 ft 58. Approx. 16.4 in.

59. $(h^{2/3} + w^{2/3})^{3/2}$ ft

60. $v = \sqrt[n]{\dfrac{a}{(n-1)b}}$ 61. $r = \sqrt{A}, \quad \theta = 2$ rad

62. Approx. 1.94 gal

63. (a) A decreasing 0.04π in²/sec
(b) $t = (5b - 3a)/(a^2 - b^2)$

64. $t = (R - cr)/(ca - b)$, where $c = \sqrt{a/b}$

66. (a) True
(b) Not necessarily true:

$$f(x) = g(x) = 5 + x + x^3 \quad \text{for} \quad |x| \leq 1, a = 0.$$

67. $x = \dfrac{1}{n}\displaystyle\sum_{i=1}^{n} c_i$ 68. $m = \frac{5}{7}$

70. Doesn't apply; derivative doesn't exist at $x = 0$.

72. $c = \pm\left(\dfrac{b^{2/3} + a^{1/3}b^{1/3} + a^{2/3}}{3}\right)^{3/2}$

73. Root -0.67

75. (a) $x = (c - b)/(2e)$ (b) $P = \frac{1}{2}(b + c)$
(c) $-a + (c - b)^2/(4e)$
(d) $P = \frac{1}{2}(b + t + c)$; that is, he adds $\frac{1}{2}t$ to his previous price.

Chapter 5

Article 5.2

1. $y = \dfrac{x^3}{3} + x + C$ 2. $y = -\dfrac{1}{x} + \dfrac{x^2}{2} + C$

3. $y^2 = x^2 + C$ 4. $3y^{1/2} = x^{3/2} + C$

5. $y^{2/3} = x^{2/3} + C$ 6. $-1/y = x^2 + C$

7. $y = x^3 - x^2 + 5x + C$ 8. $s = t^3 + 2t^2 - 6t + C$

9. $r = (2z + 1)^4/8 + C$

10. $-u^{-1} = 2v^4 - 4v^{-2} + C$

11. $x^{1/2} = 4t + C$

12. $y = 4t^3/3 + 4t - t^{-1} + C$

13. $y = z^3/3 - z^{-1} + C$ 14. $x^2 + 3x + C$

15. $\dfrac{x^3}{3} - \dfrac{2x^{3/2}}{3} + C$ 16. $\dfrac{(3x - 1)^{235}}{705} + C$

17. $\dfrac{-3(2 - 7t)^{5/3}}{35} + C$ 18. $\dfrac{2(2 + 5y)^{3/2}}{15} + C$

19. $\dfrac{-1}{9x + 6} + C$ 20. $-3\sqrt{1 - r^2} + C$

21. $\dfrac{(2x^2 + 1)^{3/2}}{6} + C$ 22. $\frac{1}{2}(1 + 2t^3)^{1/3} + C$

23. $\frac{1}{2}\sqrt{2y^2 + 1} + C$ 24. $\frac{2}{3}x^{3/2} + 2x^{1/2} + C$

25. $\frac{3}{4}(z^2 + 2z + 2)^{2/3} + C$

Article 5.3

1. $s = t^3 + s_0$
2. $s = t^2 + t + s_0$
3. $s = \frac{1}{3}(t+1)^3 - \frac{1}{3} + s_0$
4. $s = \frac{1}{5}t^5 + \frac{2}{3}t^3 + t + s_0$
5. $s = -(t+1)^{-1} + 1 + s_0$
6. $s = s_0 + t\sqrt{2gs_0} + \frac{1}{2}gt^2$
7. $v = gt + v_0,$ $s = \frac{1}{2}gt^2 + v_0t + s_0$
8. $v = \frac{1}{2}t^2 + v_0,$ $s = \frac{1}{6}t^3 + v_0t + s_0$
9. $v = \frac{3}{8}(2t+1)^{4/3} + v_0 - \frac{3}{8},$ $s = \frac{9}{112}(2t+1)^{7/3} + (v_0 - \frac{3}{8})t + s_0 - \frac{9}{112}$
10. $v = -\frac{1}{4}(2t+1)^{-2} + v_0 + \frac{1}{4},$ $s = \frac{1}{8}(2t+1)^{-1} + (v_0 + \frac{1}{4})t + s_0 - \frac{1}{8}$
11. $v = \frac{1}{5}t^5 + \frac{2}{3}t^3 + t + v_0,$ $s = \frac{1}{30}t^6 + \frac{1}{6}t^4 + \frac{1}{2}t^2 + v_0t + s_0$
13. $4\sqrt{y} = x^2 + 4$
14. $-y^{-1} = x^2 - 2$
15. $(3y+10)^2 = (1+x^2)^3$
16. $(1+y^2)^{-1/2} = \dfrac{1}{\sqrt{2}} - 4x$

Article 5.4

9. $\cos^2 A + \sin^2 A = \cos 0 = 1$
10. $\sin 0 = 0$
11. $\tan(A - B) = \dfrac{\tan A - \tan B}{1 + \tan A \tan B}$
12. $\tan(A + B) = \dfrac{\tan A + \tan B}{1 - \tan A \tan B}$
14. Cosine and secant are even. Sine, tangent, cotangent, cosecant are odd.
15. $\cos 2A = \cos^2 A - \sin^2 A,$ $\sin 2A = 2 \sin A \cos A$

Article 5.5

1. 1 2. 1 3. 2 4. $\frac{1}{3}$ 5. $\frac{5}{3}$
6. $\frac{1}{2}$ 7. 0 8. 0 9. $\frac{1}{2}$ 10. $\frac{2}{3}$
11. $\frac{3}{5}$ 12. 1 13. 2π 14. -2 15. $\frac{1}{2}$
16. 1 17. 2 18. $\cos a$ 19. $-\sin a$
20. $\pi/4$
21. $3\cos(3x+4)$
22. $x\cos x + \sin x$
23. $(x\cos x - \sin x)/x^2$
24. $-5\sin 5x$
25. $3x^2\cos 3x + 2x\sin 3x$
26. $-\sin 2x/(2+\cos 2x)^{1/2}$
27. 0
28. $6\sin 3x/\cos^2 3x$
29. $6\cos 2x + 8\sin 2x$
30. $-24\sin 2x\cos 2x$ or $-12\sin 4x$
31. $2(\cos^2 x - \sin^2 x)$ or $2\cos 2x$
32. $-\cos x/\sin^2 x$
33. $-3\sin 6x$
34. $\dfrac{-1}{\sin^2 x}$
35. $\dfrac{\sin 2y + 2y\sin 2x}{\cos 2x - 2x\cos 2y}$
36. $(-\sin 8x)/y$
40. $-\frac{1}{3}\cos 3x + C$
41. $\frac{1}{2}\sin(2x+4) + C$
42. $-\frac{1}{4}\cos(2x^2) + C$
43. $2\sin\sqrt{x} + C$
44. $-\frac{1}{2}\cos 2t + C$
45. $\frac{1}{3}\sin(3\theta - 1) + C$
46. $\frac{4}{3}\sin 3y + C$
47. $\sin^2 z + C$
48. $\frac{1}{3}\sin^3 x + C$
49. $-\frac{1}{6}\cos^3 2y + C$
50. $\frac{1}{3}\sin 3t - \frac{1}{9}\sin^3 3t + C$
51. $1/\cos x + C$
52. $-1/\sin x + C$
53. $\frac{2}{9}(2+\sin 3t)^{3/2} + C$
54. $\sqrt{2 - \cos 2t} + C$
55. $\dfrac{1}{2}\sin^4\dfrac{y}{2} + C$
56. $3\sec\dfrac{z-1}{3} + C$
57. $-\frac{1}{2}\cos^3(2x/3) + C$
58. $\frac{1}{5}(1+\sin 2t)^{5/2} + C$
59. $-\frac{3}{2}\cos 2x + \frac{4}{3}\sin 3x + C$
60. $\frac{1}{3}(\sin^3 t - \cos^3 t) + C$

Article 5.6

1. 1.75, 2.25
2. 0.25, 1.25
3. $\pi/(2\sqrt{2}) \approx 1.12,$ $(\pi/2)(1 + 1/\sqrt{2}) \approx 2.68$
4. 0.6345, 0.7595
5. 4.146, 6.146

Article 5.7

6. 1

Article 5.8

1. 12 2. 4 3. $8\frac{2}{3}$
4. 2 5. $\frac{1}{15}$ 6. $57\frac{1}{3}$
7. 10 8. $4\frac{1}{3}$ 9. 1
10. $\frac{2}{9}$ 11. $\frac{1}{2}$ 12. $10\frac{2}{3}$
13. 2 14. $\frac{2}{3}$ 15. $\frac{4}{3}$
16. $\frac{2}{3}$ 17. $\pi/6$ 18. (b) $\pi a^2/4$

Article 5.9

1. (a) $c_k = \frac{1}{2}(x_k + x_{k-1})$

(b) $c_k = \pm\sqrt[3]{\dfrac{x_{k-1}^2 + x_{k-1}x_k + x_k^2}{3}},$

the sign being determined by the interval x_{k-1}, x_k.

(c) $c_k = \sqrt[3]{\dfrac{x_{k-1}^3 + x_{k-1}^2 x_k + x_{k-1}x_k^2 + x_k^3}{4}}$

(d) $c_k = \left(\dfrac{\sqrt{x_{k-1}} + \sqrt{x_k}}{2}\right)^2$

4. $S_n = b^4/4$ 5. $S_n = 2(\sqrt{b} - 1)$

6. 8 7. $\frac{7}{3}$ 8. $\frac{8}{3}$ 9. $\frac{13}{3}$

10. 2 11. 0 12. $\sqrt{2} - 1$ 13. $\frac{1}{2}$

14. $\pi/2$ 15. π/ω 16. $\frac{2}{9}$

17. (a) $A = \lim_{n\to\infty} \sum_{k=1}^{n} (18 - 2x_k^2)\,\Delta x_k$

$(x_1 = -3,\ x_{n+1} = 3)$

(b) $A = \int_{-3}^{3} (18 - 2x^2)\,dx = 72$

18. 0.692 19. $\sqrt{1 + x^2}$

20. $1/x$ 21. $-\sqrt{1 - x^2}$

22. $1/(1 + x^2)$ 23. $2\cos(4x^2)$

24. $2x/(1 + \sqrt{1 - x^2})$ 25. $-\cos x/(2 + \sin x)$

Article 5.10 (I = integral, A = approximation)

2. I = 2, A = 2 3. I = 4, A = 4.25

4. I = $4\sqrt{2}/3$, A $\approx$ 1.82 5. I = $\frac{1}{2}$, A $\approx$ 0.53

Article 5.12

1. $\int_a^b f(t)\,dt = -\int_b^a f(t)\,dt,$

$\int_a^c f(t)\,dt = \int_a^b f(t)\,dt + \int_b^c f(t)\,dt,$

$\frac{d}{dx}\int_a^x f(t)\,dt = f(x),$

$\int_a^b f(t)\,dt = f(c)\cdot(b - a),$

$\int_a^b f(t)\,dt = F(b) - F(a)$

2. (a) $\int_0^1 x\,dx + \int_1^2 \sin(\pi x)\,dx = \frac{1}{2} - \frac{2}{\pi}$

(b) $\int_0^1 (1 - x)^{1/2}\,dx + \int_1^2 (7x - 6)^{-1/3}\,dx = \frac{55}{42}$

3. (a) $F'(\frac{1}{2}) = \frac{1}{2}$, $F'(\frac{3}{2}) = -1$
(b) $F'(\frac{1}{2}) = \sqrt{\frac{1}{2}}$, $F'(\frac{3}{2}) = (\frac{9}{2})^{-1/3}$
(c) $F'(\frac{1}{2}) = \frac{2}{3}$, $F'(\frac{3}{2}) = -\frac{2}{5}$

Miscellaneous exercises

1. $x^2y + 2 = cy$
2. $3\sqrt{y + 1} = (1 + x)^{3/2} + c$
3. $y^3 + 3y = x^3 - 3x + c$
4. $3(x^2 - y^2) + 4(x^{3/2} + y^{3/2}) = c$
5. $(2 + x)^{-1} = -(3 - y)^{-1} + c$
6. (a) $y = \frac{1}{3}(x^2 - 4)^{3/2} + 3$ (b) $y^2(1 - x^2) = 1$
7. Yes, $y = x$ is such a curve.
8. $y = x^3 + 2x - 4$ 9. $(2\sqrt{2}/3)b^{3/4}$
10. (a) $v = \frac{2}{3}t^{3/2} - 2\sqrt{t} + 2$
(b) $s = \frac{4}{15}t^{5/2} - \frac{4}{3}t^{3/2} + 2t + 5$
11. $v = t^2 + 3t + 4$, $\Delta s = 61\frac{1}{3}$
12. -8 ft/sec
[*Note.* $d^2x/dt^2 = v\,(dv/dx)$, where $v = dx/dt$.]
13. $h(10) = h(0) = 5$
14. $(1 + y')(1 + y'^2) + y''(y - x) = 0$
15. (a) $k = 38.72$ ft/sec² (b) 25 ft
16. $3y = (1 + x^2)^{3/2} - 7$ 17. $h = 96t - 16t^2$, 144 ft
19. (a) ≈ 1.76 (b) ≈ 1.73
20. (a) 4 ft/sec (b) $\frac{64}{3}$ ft
21. $(x^2/2) - x^{-1} + c$ 22. $\frac{1}{3}(y^2 + 1)^{3/2} + c$
23. $-\frac{1}{8}(1 + t^{4/3})^{-6} + c$ 24. $\frac{4}{3}(1 + \sqrt{u})^{3/2} + c$
25. $-\frac{3}{5}(7 - 5r)^{1/3} + c$ 26. $\frac{1}{4}\sin 4x + c$
27. $\frac{1}{9}\sin^3 3x + c$ 28. $2\sqrt{\sin x} + c$
29. $\frac{1}{2}\sin(2x - 1) + c$ 30. $-\frac{1}{4}\sqrt{25 - 4y^2} + c$
31. $-\sqrt{2/t} + c$ 32. $\frac{1}{3}(x^3 - 2x^{3/2}) + c$
33. $\frac{1}{3}(2 - 3x)^{-1} + c$ 34. $2\sin(1 - 2x)$
35. $-\csc^2 x$ 36. $10\sec^2 5x \tan 5x$
37. $20\sin^3 5x\cos 5x$
38. $\dfrac{\sin x\cos x\,(\sin x - \cos x)}{(\sin^3 x + \cos^3 x)^{2/3}}$
39. 0 42. $r\sqrt{2}/2$ ft 44. $|a|$
45. Yes; no; applies to identities, but not to equations
46. 480π mi/hr 47. $|3a\omega\sin\theta\cos\theta|$
48. $f(x) = x/\sqrt{x^2 + 1}$

49. (a) $\frac{10}{3}$ (b) $\frac{5}{3}$ (c) $\frac{1}{2}$
50. $dy/dx = -2x, \quad d^2y/dx^2 = -2$
51. $\int_0^1 f(x)\,dx$
52. (a) $\frac{1}{16}$ (b) $\frac{2}{3}$ (c) $2/\pi$
53. (a) $\sqrt{1+x^2} - x$ (b) $x_1 f(x_1)$
54. $d^2y/dx^2 = 4y$

Chapter 6

Article 6.2

1. (a) $\lim_{\Delta y \to 0} \sum_a^b (f(y) - g(y))\,\Delta y$

   (b) $\int_a^b (f(y) - g(y))\,dy$

2. (a) $\frac{4}{3}$ (b) $\frac{13}{12}$ (c) $\frac{32}{3}$ (d) $\frac{32}{3}$ (e) $\frac{1}{6}$ (f) $\frac{4}{3}$ (g) $\frac{128}{15}$
3. $\sqrt{2} - 1$
4. $\sqrt[3]{16} \approx 2.52$
5. $\frac{1}{6}$

Article 6.3

1. 6 ft
2. $5\frac{1}{6}$ ft
3. $3\frac{2}{3}$ ft
4. 1 ft
5. 6 ft
6. 8 ft
7. $2\sqrt{2}$ ft
8. $2\frac{2}{3}$ ft
9. $6 - \sin 2 \approx 5.09$ (ft)
10. $1 + \cos 2 \approx 0.58$ (ft)
11. $2g$
12. $4\frac{29}{30}$ ft
13. $3\frac{1}{6}$ ft
15. 6.45 ft, 3.22 ft/sec

Article 6.4

1. $\dfrac{8\pi}{3}$
2. $\dfrac{\pi^2}{2}$
3. $\dfrac{\pi}{30}$
4. $\dfrac{56\pi}{15}$
5. $\dfrac{8\pi}{3}$
6. $\dfrac{2\pi}{3}$
7. $\dfrac{256\pi}{5}$
8. $\dfrac{48\pi}{5}$
9. (a) $\frac{1}{3}\pi r^2 h$ (b) $\frac{2}{3}\pi h^2 r$
10. (a) $\pi/6$ (b) $2\pi/15$
11. (a) 8π (b) $\frac{512}{15}\pi$ (c) $\frac{256}{5}\pi$ (d) $\frac{1088}{15}\pi$ (e) $\frac{128}{3}\pi$
12. (a) $\pi h^2(a - \frac{1}{3}h)$

    (b) $0.1/12\pi$ (ft/sec) $= 6/\pi$ (in/min)
13. $\frac{4}{3}\pi ab^2$
14. $\frac{512}{15}\pi$
15. $\frac{16}{3}a^3$
16. $\frac{8}{3}a^3$
17. $\frac{8}{3}a^3$
18. $\frac{1}{16}\pi\sqrt{3}$
19. 15,990 ft^3
20. $2\pi^2 a^2 b$

Article 6.6

1. 12
2. $\frac{8}{27}(10\sqrt{10} - 1)$
3. $\frac{14}{3}$
4. $\frac{53}{6}$
5. $\frac{123}{32}$
6. $\frac{4}{27}(10\sqrt{10} - 1)$
7. $\frac{1}{8}a\pi^2$
8. 12
9. $\frac{21}{2}$

Article 6.7

1. $\frac{12}{5}\pi a^2$
2. $\frac{99}{2}\pi$
3. $\frac{1}{27}\pi(10\sqrt{10} - 1)$
4. $\frac{1}{6}\pi(17\sqrt{17} - 1)$
5. $\frac{1823}{18}\pi$
6. $\frac{253}{20}\pi$
7. $\frac{2}{3}\pi(26\sqrt{26} - 2\sqrt{2})$
8. $56\pi\sqrt{3}/5$

Article 6.8

1. (a) $2/\pi$ (b) 0
2. (a) $\frac{1}{2}$ (b) $\frac{1}{2}$
3. $\frac{49}{12}$
4. $\frac{1}{2}$
5. $\alpha \dfrac{a+b}{2} + \beta$
6. $\pi a/2$
7. $4a/\pi$
8. $8a^2/3$
9. $2a^2$

Article 6.9

1. $(4a/3\pi,\ 4a/3\pi)$
2. $(0,\ \frac{2}{5}h^2)$
3. $\left(\dfrac{2a}{3(4-\pi)},\ \dfrac{2a}{3(4-\pi)}\right)$
4. $\left(\dfrac{\pi}{2},\ \dfrac{\pi}{8}\right)$
5. $(\frac{2}{5},\ 1)$
6. $\frac{3}{7}h$
7. $\frac{3}{5}h$
8. On the axis, $\frac{3}{4}h$ from vertex
9. On the axis, $\frac{3}{5}h$ from vertex
10. On the axis, $\frac{1}{2}h$ from vertex
11. $(0,\ \frac{1}{4}\pi r)$

Article 6.10

1. $(0,\ \frac{2}{5}c^2)$
2. $(\frac{16}{105},\ \frac{8}{15})$
3. $(0,\ \frac{12}{5})$
4. $(1,\ -\frac{3}{5})$

5. $(\frac{3}{5}, 1)$
6. On the axis, $\frac{3}{4}h$ from the vertex
7. $(0, \frac{8}{3})$
8. $(\frac{4}{5}, 0)$
9. On the axis, $\frac{2}{3}h$ from the vertex
10. $\left(-r, \dfrac{3}{\pi + 2} r\right)$
11. $\dfrac{17\sqrt{17} - 1}{12}$
12. $\left(\dfrac{2r}{\pi}, \dfrac{2r}{\pi}\right)$

Article 6.11

2. $(0, 2r/\pi)$
3. $2\pi(\pi - 2)r^2$
4. $\dfrac{3\pi + 4}{3} \pi r^3$
5. $\dfrac{3\pi + 4}{3} \dfrac{\pi r^3}{\sqrt{2}}$
6. $\pi\sqrt{2}(\pi + 2)r^2$
7. $2r^3/3$
8. $(4 + 3\pi)r^3/6$
9. $(4 + 3\pi)r^3/(6\sqrt{2})$

Article 6.12

1. 375 lb
2. $111\frac{1}{9}$ lb
3. $1666\frac{2}{3}$ lb
4. $116\frac{2}{3}$ lb
5. $41\frac{2}{3}$ lb
6. 975 tons
7. 8450 ton-ft

Article 6.13

1. $c = 5$ lb/in, 80 inch-lb
2. $c = \dfrac{10}{\sin 10°}$ lb, 78.5 inch-lb
3. $\frac{1}{2}k$, where k is the proportionality factor
4. $\frac{1}{3}k$, k as in Problem 3
5. $\frac{1}{2}mg\ R$
6. 1944 ft-lb
8. (Work done *on* the gas) = —(Work done *by* the gas) = $3164\frac{1}{16}$ ft-lb
9. $200\pi/3$ ft-tons
10. 27.2 ft-tons

Miscellaneous exercises

1. $\frac{9}{2}$
2. $6\frac{3}{4}$
3. 1
4. $(7 - 4\sqrt{2})/2$
5. $\frac{9}{2}$
6. $\frac{32}{3}$
7. 18
8. $3^5/2^3$
9. $a^2/6$
10. 14.95 correct value
11. 8
12. $\pi/4$
13. $\frac{13}{3}$
14. $\frac{9}{8}$
15. $\frac{64}{3}$
16. (a) $0 \leq x < 2$ (b) $2 < x < 3$ (c) $\frac{29}{2}$ ft
17. (a) $\frac{3}{7}$ (b) $\frac{3}{8}$ (c) $y = 3/(2\sqrt{1 + 3x})$ (d) $\frac{3}{7}$
18. $f(x) = \sqrt{(2x - a)/\pi}$
19. (a) $3^5\pi/5$ (b) $189\pi/2$ (c) $81\pi/2$ (d) $2^9\pi/15$ (e) $2^7\pi/3$
20. $f(x) = \pm\sqrt{(2x + 1)/\pi}$
21. $32\pi/3$
22. (a) $2\pi a^3$ (b) $(\frac{16}{15})\pi a^3$ (c) $(\frac{8}{5})\pi a^3$
23. $V = 2\pi \int_0^2 x^2/\sqrt{x^3 + 8}\, dx = (8\pi/3)(2 - \sqrt{2})$
24. $(\pi/15)(88\sqrt{2} + 107)$
25. $112\pi a^3/15$
26. $V = hs^2$
27. $(\frac{8}{3})r^3$
28. $\pi^2/4$
29. $28\pi/3$
30. $72\pi/35$
31. $2\sqrt{3}a^3$
32. $\frac{19}{3}$
33. $424\pi/15$
34. $\frac{27}{20}$
35. $153\pi/40$
36. $(a/3)(3\sqrt{3} - 1)$
37. (a) $(\frac{2}{3})b^2$ (b) $(\frac{2}{3})b$
38. (a) $f(x) = a \pm x\sqrt{A^2 - 1}$, $|A| \geq 1$ (b) No
39. (a) $\bar{v} = 72$ (b) $82\frac{2}{3}$
40. Vol. 401π, average area 40.1π
41. $(\frac{9}{10}, \frac{9}{5})$
42. $(\frac{3}{2}, \frac{12}{5})$
43. $(1, \frac{12}{5})$
44. $(\frac{3}{4}, \frac{3}{10})$
46. (a) $(5a/7, 0)$ (b) $(2a/3, 0)$
47. (a) $\bar{x} = \bar{y} = \dfrac{4(a^2 + ab + b^2)}{3\pi(a + b)}$ (b) $(2a/\pi, 2a/\pi)$
48. (a) $(2a/5, 2a/5)$ (b) $(0, 15\pi a/256)$
49. 5 in., $7\frac{1}{9}$ in., $4\frac{8}{9}$ in.
50. $7000/3$ lb
51. $5 \times 10^6/3$ lb
52. (a) $2h/3$ below surface (b) $a + h(4a + 3h)/(6a + 4h)$ below surface
53. $504d_1 + 72d_2$
54. $Mah + (h^2/a)$
55. $k(b - a)/ab$
56. $wr[1 - r/2R]$ ft-lb, R = radius of earth in feet
57. $(320w/3)(15\pi + 8)$ ft-lb

Chapter 7

Article 7.1

13. $6x \sec^2 (3x^2)$
14. $-2 \sin x \tan (\cos x) \sec^2 (\cos x)$

15. $-3 \csc^2 (3x + 5)$
16. $-\frac{1}{2} (\cot x)^{-1/2} \csc^2 x$
17. 0
18. $\sec^3 x \tan x$
19. $3(1 + \sec^2 3x)$
20. $6 \sec^2 2x$
21. $-\tan^2 x$
22. $4 \sec^2 x \tan x$
23. $-2 \csc x (\csc x + \cot x)^2$
24. $12 \sin 3x \cos 3x$, or $6 \sin 6x$
25. $\frac{1}{2}(1 - \cos 2x)$, or $\sin^2 x$
26. $\frac{1}{2}(1 + \cos 2x)$, or $\cos^2 x$
27. $-\cos^3 (x/3)$
28. $-3 \cos \frac{x}{2} \sin \frac{x}{2}$, or $-\frac{3}{2} \sin x$
29. $\frac{1}{2}x \sin \frac{x}{2}$
30. $\dfrac{2x}{\cos y + 2 \cos 2y}$
31. $3(\sin^2 x \cos x - \cos 3x)$
32. $-5 \cos^3 5x$
33. $-\dfrac{\cos^2 (xy) + y}{x}$
34. 10 when $x = -\sin^{-1} \frac{4}{5}$
35. max $y = 3/2$ when $x = \pi/6$, min $y = 1$ when $x = 0$ or $\pi/2$
38. 1.6π mi/sec
39. 120°
40. $-\frac{1}{3} \cos 3t + C$
41. $\frac{1}{2} \tan 2\theta + C$
42. $\frac{1}{4} \tan^4 x + C$
43. $\frac{1}{3} \sec^3 x + C$
44. $2 \sec (x/2) + C$
45. $\frac{1}{2}y + \frac{1}{4} \sin 2y + C$
46. $\tan \theta + C$
47. $-3 \cot (\theta/3) + C$
48. $\frac{3}{2}$
49. $(\pi/4)(4 - \pi)$

Article 7.2

1. $\frac{\sqrt{3}}{2}, \frac{\sqrt{3}}{3}, \frac{2\sqrt{3}}{3}, 2$
2. $\frac{\sqrt{3}}{2}, -\sqrt{3}, -2, \frac{2\sqrt{3}}{3}$
3. π
4. $\pi/2$
5. $-\pi/3$
8. (a) 0.735 (b) 0.6 (c) 0.96 (d) $\pi/3$ (e) $2\pi/3$ (f) $\pi/6$

Article 7.3

5. $\dfrac{1}{\sqrt{4 - x^2}}$
6. $\dfrac{1}{9 + x^2}$
7. $\dfrac{1}{|x|\sqrt{25x^2 - 1}}$
8. $\dfrac{-2}{\sqrt{1 - 4x^2}}$
9. $\dfrac{4}{4 + x^2}$
10. $\dfrac{1}{(x + 1)\sqrt{x}}$
11. $\dfrac{1}{x^2 + 1}$
12. $\sin^{-1} x$
13. $(\sin^{-1} x)^2$
14. $\cos^{-1} 2x$
15. $\sqrt{b(a + b)}$ ft
16. $\pi/6$
17. $\pi/2$
18. $-\pi/12$
19. $-\pi/6$
20. $\frac{1}{2} \sin^{-1} 2x + C$
21. $\pi/6$

Article 7.4

1. 0.1826
2. 0.3369
3. 0.6938
4. -0.6956
5. (a) Circumscribed (b) Inscribed
6. (a) 0.0392 (b) -0.0408
7. $\ln 2 \approx \frac{2}{3}$
8. $0.18226 < \ln 1.2 < 0.18267$

Article 7.5

1. $\dfrac{2x + 2}{x^2 + 2x}$
2. $\dfrac{3}{x} (\ln x)^2$
3. $-\tan x$
4. $\sec x$
5. $\dfrac{1}{x} + \dfrac{x}{x^2 + 1}$
6. $\dfrac{1}{x} + \dfrac{1}{2(x + 2)}$
7. $\ln x$
8. $3x^2 \ln (2x) + x^2$
9. $\dfrac{1}{1 - x^2}$
10. $\dfrac{1}{x(1 + x^3)}$
11. $\dfrac{2}{x(2 + 3x)}$
12. $-\tan^{-1} \dfrac{x}{2}$
13. $\dfrac{\tan^{-1} x}{x^2}$
14. $3 (\ln x)^2 + (\ln x)^3$
15. $2 \cos (\ln x)$
16. $\sec^{-1} x$
17. $\ln (a^2 + x^2)$
18. $\dfrac{1}{x \ln x}$
19. $\frac{1}{2} \ln |2x + 3| + C$
20. $-\frac{1}{3} \ln |2 - 3x| + C$
21. $\frac{1}{8} \ln (4x^2 + 1) + C$
22. $\ln (2 - \cos x) + C$
23. $\ln |\sin x| + C$
24. $2x - 5 \ln |x| + C$
25. $x - \ln |x + 1| + C$
26. $-\frac{1}{3} \ln |4 - x^3| + C$
27. $-\frac{1}{2} \ln |1 - x^2| + C$
28. $2 \ln (1 + \sqrt{x}) + C$
29. $\frac{1}{3} (\ln x)^3 + C$
30. $\dfrac{-1}{4x + 6} + C$

Article 7.6

1. $4a$
2. $2b/3$
3. $3a/2$
4. $-2a$
5. $2a - 2b$
6. $2a + b$
7. $2b - 3a$
8. $2a + 2b$
9. $2b - a$
10. $(3b - a)/2$

Article 7.7

2. No inflection point because $y'' = 1/x$ is positive when x is positive.
3. Slope at $(x, y) = y$. Concave up because $y'' = y' = y$ is positive.
7. $\log_{10} N = \dfrac{\ln N}{\ln 10}$

Article 7.8

1. (a) x (b) x (c) x^{-2}
(d) $-x^2$ (e) $1/x$ (f) $-x$
(g) $1/x$ (h) x (i) $2x$
(j) x^2 (k) $x - x^2$ (l) $-2x + 2 \ln x$
(m) xe^x (n) x/y^2

2. $(x^2 + 2x)e^x$
3. $13e^{2x} \cos 3x$
4. $1/(1 + e^x)$
5. $\frac{1}{2}(e^x + e^{-x})$
6. $\frac{1}{2}(e^x - e^{-x})$
7. $4/(e^x + e^{-x})^2$
8. $e^{\sin^{-1} x}/\sqrt{1 - x^2}$
9. $-4xe^{-2x}$
10. $27x^2e^{3x}$
11. xe^{ax}
12. $-2xe^{-x^2}$
13. $2e^{-x^2}(x - x^3)$
14. $e^x\left(\ln x + \dfrac{1}{x}\right)$
15. $\dfrac{e^x}{1 + e^{2x}}$
16. $2/\sqrt{e^{4x} - 1}$
17. $\frac{2}{3}e^{2x} \sec (x + 3y) - \frac{1}{3}$
18. $-e^{1/x}/x^2$
19. $(\cos^2 y)(e^x + 1/x)$
20. Start adding the terms: $(h) + (h/2) + (h/4) + \cdots$. The sum is always in $[h, 2h)$, and we can make it as close as we wish to either h or $2h$ by taking only one term, or by taking a very large number of terms. (The geometric series has the property that the sum of all the terms following any term is equal to that single term.) Because $h < 1 < 2h$, we can get sums that are less than 1 or greater than 1. Our aim is to come very close to 1, by judicious choice of summands. We let $f(n)$ be the cumulated sum after n steps, with $f(1) = h$. If, at any step,
$$1 - \epsilon < f(n) < 1 + \epsilon,$$
then we stop. Also keep $f(n) < 1 + \epsilon$ at all times. If $f(n) \le 1 - \epsilon$, and the last summand used in $f(n)$ is $h/2^m$, then the next candidate is $h/2^{m+1}$, and we test $f(n) + h/2^{m+1}$. If this number is less than $1 + \epsilon$, then we call it $f(n + 1)$. If it is too large, we discard $h/2^{m+1}$ and try $f(n) + h/2^{m+2}$, and so on until we reach the first new summand $h/2^{m+k}$ for which $f(n + 1) = f(n) + h/2^{m+k} < 1 + \epsilon$. We continue, with $n + 1$ in place of n. These operations can be programmed.

22. (b) $y = 3e^{-2t}$
24. $\frac{1}{2}e^{2x} + C$
25. $\frac{1}{2}e^{x^2} + C$
26. $e^{\sin x} + C$
27. $3e^{x/3} + C$
28. $-\frac{4}{3}e^{-3x} + C$
29. $\frac{1}{2} \ln |1 + 2e^x| + C$
30. $\ln 2$
31. $\ln |e^x + e^{-x}| + C$
33. $p \ge 90$
34. $b \ge 40$

Article 7.9

1. $\dfrac{y(2x + 1)}{2x(x + 1)}$
2. $\dfrac{-2y}{3(x^2 - 1)}$
3. $\dfrac{y(4x^3 + 6x^2 + x + 3)}{3x(x + 1)(x^2 + 1)}$
4. $\dfrac{y}{3}\left[\dfrac{1}{x} + \dfrac{1}{x + 1} + \dfrac{1}{x - 2} - \dfrac{2x}{x^2 + 1} - \dfrac{2}{2x + 3}\right]$
5. $y\left[\dfrac{\sin x}{x} + \cos x \cdot \ln x\right]$
6. $y[1 + \sec^2 x \cdot \ln (\sin x)]$
7. $y(\sec x \tan x) \ln 2$
8. $(2y \ln x)/x$
11. $\frac{3}{8}$
12. $(3^{1.2} - 1)/\ln 3 \approx 2.49$
13. $(4 \ln 4)^{-1} \approx 0.180$
14. $24/(50 \ln 5) \approx 0.298$
15. $(2 \ln 4)^{-1} \approx 0.360$
18. (a) 2 (b) $\frac{5}{3}$ (c) -2 (d) -2
19. (a) 2 (b) $\frac{1}{2}$ (c) $\frac{4}{3}$ (d) $\frac{2}{5}$
20. $\log_{1.5} 2 = \dfrac{\ln 2}{\ln 1.5}$
21. 12
24. (a) 1.30103, 2.30103, 9.30103 − 10, 8.30103 − 10
(b) 2.99574, 5.29833, 8.39056 − 10, 6.08797 − 10

25. (a) 1.46 (b) 0.68 (c) 2.47
26. (a) 2.02 (b) −1.36
27. (a) $3^{\tan x}\sec^2 x \ln 3$
(b) $y\left[\frac{1}{x}\ln(x^2+1)+\frac{2x}{x^2+1}\ln x\right]$
(c) $-2t(\ln 2)2^{-t^2}$ (d) $e^{-2\theta}(1-2\theta)$

Article 7.10

1. $$\frac{mv_0}{k}[1-e^{-kt/m}],$$
where k is a constant of proportionality.
2. Q_0e^{-kt}
3. N_0e^{kt}
4. 1, 3
5. −1, 2
6. −5, −1
7. ±3, ±2

Miscellaneous exercises

1. $2\cos 2x$
2. $-8\sin(2x+\pi/4)$
3. $2\cos x + 2\cos 2x$
4. $(1+\cos x)^{-1}$
5. $-\csc y$
6. $\cos^2 y$
7. $-2\cos(x/2)$
8. $-\cos x$
9. $1-\cos x$
10. $\frac{x\cos x-\sin x}{x^2}$, $\begin{cases} y\to 1 \text{ as } x\to 0,\\ y\to 0 \text{ as } x\to\infty\end{cases}$
11. $\frac{x\cos x-2\sin x}{x^3}$, $\begin{cases} y\to \infty \text{ as } x\to 0,\\ y\to 0 \text{ as } x\to\infty\end{cases}$
12. $\sin\frac{1}{x}-\frac{1}{x}\cos\frac{1}{x}$, $\begin{cases} y\to 0 \text{ as } x\to 0,\\ y\to 1 \text{ as } x\to\infty\end{cases}$
13. $2x\sin\frac{1}{x}-\cos\frac{1}{x}$, $\begin{cases} y\to 0 \text{ as } x\to 0,\\ y\to \infty \text{ as } x\to\infty\end{cases}$
14. $(2/\pi)\cos^2(\pi y/2)$
15. (a) $\frac{1}{2}(e^x-e^{-x})$ (b) $\frac{1}{2}(e^x+e^{-x})$
(c) $4(e^x+e^{-x})^{-2}$
16. $e^{-x}(1-x)$
17. (a) $1+\ln x$ (b) $\frac{1}{2}x^{-1/2}(2+\ln x)$
18. (a) $x^{-2}(1-\ln x)$ (b) $\frac{1}{2}x^{-3/2}(2-\ln x)$
(c) $x^{-3}(1-2\ln x)$
19. $e^{-x}(2\cos 2x-\sin 2x)$
20. (a) $2-e^{2x}$ (b) $(2-e^{2x})e^{(2x-\frac{1}{2}e^{2x})}$
21. (a) $1-e^x$ (b) $(1-e^x)e^{(x-e^x)}$
22. $(1+x^2)^{-1/2}$
23. $(1-x^2)^{-1}$
24. $2x/(x^2+4)$
25. $\tan^{-1}(x/2)+2x/(x^2+4)$
26. $8(e^{2x}+e^{-2x})^{-2}$
27. $2\sec x$
28. $xe^{2x}(3x\cos 3x+2x\sin 3x+2\sin 3x)$
29. $2x(1-x^4)^{-1/2}-(2x^2+1)e^{x^2}$
30. $\frac{4+x^3}{x(1+x^3)}+\frac{2}{3}x^{-1/3}7^{x^{2/3}}\ln 7$
31. $2x(2-x)(x^2+2)^{1-x}-(x^2+2)^{2-x}\ln(x^2+2)$
32. $x^{-1}e^{-x}(1-x\ln x)$
33. $1/x(x^2+1)$
34. $3/(3x-4)$
35. $(6x+4)/(3x^2+4x)$
36. $(\ln x)^2(\ln x+3)$
37. $3(1+\ln x)$
38. $x^2(1+3\ln x)$
39. 1
40. $(x^2+2x)e^x$
41. e^x
42. $1/x\ln x$
43. (a) $(1-x^2)/(1+x^2)^2$
(b) $\frac{2}{3}(x^2+x-1)x^{-2/3}(x-2)^{-2/3}(x^2+1)^{-4/3}$
44. $f(0)=1$,
$f'(0)=5$,
$f(0.01)\approx 1.05$
45. (a) $a^{x^2-x}(2x-1)\ln a$
(b) $(1+e^x)^{-1}$
(c) $x^x(1+\ln x)$
(d) $x^{(1-2x)/x}(1-\ln x)$
48. $x=1+\sqrt{2}$
49. $\frac{d^2y}{dx^2}=4(e^x+e^{-x})^{-2}$
50. $y=\ln|2x-9|$
52. 0.88
53. $E=E_0e^{-t/40}$, $t=92.1$ sec
54. (b) $L=(e^{x_1}-e^{-x_1})/2$
55. $\pi^2/2$
56. $\frac{y}{y_1}=\left(\frac{x}{x_1}\right)^2$
57. $y=1-\frac{1}{2}(e^x+e^{-x})$
58. $t(1/4)=1/\pi$; it will not reach $x=1/2$, because $t=(1/\pi)\tan(\pi x)$.
59. $1+4\ln\frac{3}{2}\approx 2.62$
60. $2\sqrt{2}$
61. $y=2e^x/(2-e^x)$
62. $N=N_0e^{0.02t}$, $t=34.7$ yr
64. $(\pm a/\sqrt{2}, e^{-1/2})$
65. $3\pi/8$
66. $\pi\sqrt{3}$

67. (a) $\sec x$ (b) $\ln(1+\sqrt{2})$

68. $(64a^2 + b^2 \ln 3)/8a$

69. $\pi(e^4 - 1)/2 \approx 84.2$

70. $(e - e^{-1})a^2 \approx 2.35a^2$

71. $y^2 = 4x$

72. (a) $\ln 0.5 = -0.693$
(b) $\ln 3 = 1.099$ (from tables)

73. (a) $y' = \ln x$ (b) 1.3

74. (a) $-\frac{1}{3}\ln|4-3x| + c$ (b) $5\ln|x-3| + c$

75. (a) $\frac{1}{2}\ln 3 = 0.550$ (b) $\frac{1}{6}$

76. (a) $x + \ln x + C$ (b) $-\frac{1}{4}\ln|2x+1| + x/2 + C$

77. $\ln 0.1 = -2.3026$, $\ln 0.25 = -1.3864$,
$\ln 10 = 2.3026$, $\ln 20 = 2.9958$

78. $\pi \ln 4 \approx 4.35$

79. $28\pi/3 \approx 29.4$

80. (a) $\frac{1}{3}\ln 2 \approx 0.23$ (b) $\frac{1}{2}(e^8 - e^{-1}) \approx 1490$
(c) $\ln 3 \approx 1.10$ (d) $\frac{1}{2}\ln 26 \approx 1.63$ (e) e

81. 585 lb

82. $2000\pi(\frac{9}{10})^t$ ft^3

84. (a) No, x could equal 4.
(b) Yes, consider the graph of $y = (\ln x)/x$ for $x > 0$.

90. $e - 1$

Chapter 8

Article 8.2

6–11. Values of sinh u, cosh u, tanh u, coth u, sech u, csch u are listed in the order given.

6. $-\frac{3}{4}, \frac{5}{4}, -\frac{3}{5}, -\frac{5}{3}, \frac{4}{5}, -\frac{4}{3}$
7. $\pm\frac{8}{15}, \frac{17}{15}, \pm\frac{8}{17}, \pm\frac{17}{8}, \frac{15}{17}, \pm\frac{15}{8}$
8. $-\frac{7}{24}, \frac{25}{24}, -\frac{7}{25}, -\frac{25}{7}, \frac{24}{25}, -\frac{24}{7}$
9. $\frac{12}{5}, \frac{13}{5}, \frac{12}{13}, \frac{13}{12}, \frac{5}{13}, \frac{5}{12}$
10. $\pm\frac{4}{3}, \frac{5}{3}, \pm\frac{4}{5}, \pm\frac{5}{4}, \frac{3}{5}, \pm\frac{3}{4}$
11. Same as 9.

Article 8.3

2. $3\cosh 3x$
3. $10\cosh 5x \sinh 5x = 5\sinh 10x$
4. 0
5. $2\,\text{sech}^2\, 2x$
6. $-\sec^2 x\, \text{csch}^2 (\tan x)$
7. $-3\,\text{sech}^3 x \tanh x$
8. $-\text{csch}(x/4)\coth(x/4)$
9. $\sec^2 x\, \text{sech}\, y$
10. $\frac{1}{2}\sinh(2x+1) + C$
11. $\ln\cosh x + C$
12. $-\frac{1}{3}\,\text{sech}^3 x + C$
13. $\tanh x + C$
14. $\ln\cosh x + C$
15. $x - \tanh x + C$
16. $2\cosh\sqrt{x} + C$
17. $\frac{1}{2}x + \frac{1}{12}\sinh 6x + C$
18. $2\sqrt{2}\cosh(x/2) + C$
19. $u/2$
21. (a) 115 ft (b) 16 lb

Article 8.5

2. $\cosh^{-1} x = \ln(x + \sqrt{x^2 - 1})$
6. $\dfrac{2}{\sqrt{1+4x^2}}$
7. $-\text{csc}\, x$
8. $\sec x$ if $\tan x > 0$, $-\sec x$ if $\tan x < 0$
9. $-\text{csc}\, x$
10. $-2\csc 2x$ if $\cos 2x > 0$, $2\csc 2x$ if $\cos 2x < 0$
11. $\frac{1}{2}\sinh^{-1} 2x + C$
12. $\sinh^{-1}\dfrac{x}{2} + C$
13. 0.5493
14. -0.5493
15. $-\frac{1}{2}\,\text{csch}^{-1}\dfrac{|x|}{2} + C = -\frac{1}{2}\sinh^{-1}\dfrac{2}{|x|} + C$

Article 8.6

1. $a\sinh\dfrac{x_1}{a}$
3. $\dfrac{\pi a}{2}\left(2x_1 + a\sinh\dfrac{2x_1}{a}\right)$
4. $\bar{x} = 0$, $\bar{y} = \frac{1}{2}y_1 + \dfrac{x_1}{2}\,\text{csch}\,\dfrac{x_1}{a}$
5. $\dfrac{\pi a^2}{4}\left(2x_1 + a\sinh\dfrac{2x_1}{a}\right)$
6. (b) $\dfrac{dx}{ds} = \dfrac{a}{\sqrt{s^2+a^2}}$, $\dfrac{dy}{ds} = \dfrac{s}{\sqrt{a^2+s^2}}$.
7. (c) $a = 25$ ft, dip ≈ 4.6 ft (d) $H = 50$ lb

Miscellaneous exercises

3. $\lim PQ = 0$
4. $\sinh x = \pm\frac{3}{4}$, $\tanh x = \pm\frac{3}{5}$
5. $\cosh x = \frac{41}{9}$, $\tanh x = -\frac{40}{41}$
9. $y = 1$
12. $3\sinh 6x$
13. $\sec^2 x/(2\tanh y\, \text{sech}^2 y)$
14. $-(\cosh y \coth y)/\sqrt{1-x^2}$
15. $\tanh y \tan x$
16. $(1+y^2)/(1-x^2)$
17. $x^{-1}\,\text{sech}^2(\ln x) = 4x(x^2+1)^{-2}$

18. $y \operatorname{csch}(\ln y)$
19. $3e^{3x}\cosh(\tan^{-1} e^{3x})/(1+e^{6x})$
20. $|\sec x|$
21. $-(\cosh y + \sinh 2x)/(2y + x\sinh y)$
22. $-e^{-\theta} + C$
23. $(2\theta - e^{-2\theta})/4$
24. $(\cosh^3 x - 3\cosh x)/3 + C$
25. $(e^{3x} + 3e^{-x})/6 + C$
25. $\ln(\cosh x) + C$
27. $\frac{1}{2}\tanh^{-1}(\frac{1}{2})$
28. $\frac{1}{2}(\coth^{-1}\frac{5}{2} - \coth^{-1}\frac{3}{2})$
29. $\sinh^{-1}(e^t) + C$
30. $-\tanh^{-1}(\cos x) + C$
31. $\cosh^{-1}(\tan\theta) + C$
36. (a) $\lim v = \sqrt{mg/k}$, where k = resistance proportionality factor; m = mass of body; g = acceleration due to gravity
    (b) $s = (m/k)\ln\cosh(t\sqrt{gk/m})$
37. $\ln 2$
38. $\ln[2/(1+\sqrt{2})]$

Chapter 9

Article 9.1

1. $\frac{1}{3}(2x+3)^{3/2} + C$
2. $\frac{1}{3}\ln|3x+5| + C$
3. $-1/(4x+14) + C$
4. $\frac{1}{2}\ln|x^2+2x+3| + C$
5. $-\ln|2+\cos x| + C$
6. $\frac{1}{8}\tan^4 2x + C$
7. $-\frac{1}{4}\sqrt{1-4x^2} + C$
8. $\frac{1}{2}(x^{4/3} - 1)^{3/2} + C$
9. $\dfrac{-1}{12(3x^2+4)^2} + C$
10. $\frac{2}{9}(x^3+5)^{3/2} + C$
11. $\frac{2}{3}(x^3+5)^{1/2} + C$
12. $\frac{1}{8}\ln(4x^2+1) + C$
13. $\frac{1}{2}e^{2x} + C$
14. $-e^{\cos x} + C$
15. $-\frac{1}{3}e^{-3x} + C$
16. $2e^{\sqrt{x+1}} + C$
17. $-\frac{1}{3}\cos^3 x + C$
18. $-\frac{1}{2}\csc^2 x + C$
19. $-\frac{1}{4}\cot^4 x + C$
20. $\frac{1}{6}\tan^2 3x + C$
21. $\frac{1}{2}\ln|e^{2x} - e^{-2x}| + C$
22. $-\frac{1}{6}\cos^3 2x + C$
23. $-\frac{1}{4}(1+\cos\theta)^4 + C$
24. $-\frac{1}{2}e^{-t^2} + C$
25. $\ln|\sin x| + C$
26. $\ln|1+\sin x| + C$
27. $\frac{1}{3}\sec^3 x + C$
28. $-2\sqrt{1+\cos\theta} + C$
29. $\frac{1}{3}e^{\tan 3x} + C$
30. $-\frac{1}{3}(4-\sin 2t)^{3/2} + C$
31. $-\frac{1}{2}(\csc 2x + \cot 2x) + C$
32. $x - \frac{1}{2}\sin 2x + C$
33. $-\sqrt{1+\cot 2t} + C$
34. $\frac{1}{3}e^{3x} + C$
35. $\frac{1}{2}e^{\tan^{-1}2t} + C$
36. $-\frac{1}{2}e^{-x^2} + C$
37. $3^x/(\ln 3) + C$
38. $10^{2x}/(2\ln 10) + C$
39. (a) -1, $\frac{1}{2}(\ln x)^2 + C$ (b) 2, $\frac{1}{3}e^{x^3} + C$
    (c) $-\frac{1}{2}$, $-2\cos\sqrt{x} + C$

Article 9.2

1. $-\frac{2}{3}(1+\cos t)^{3/2} + C$
2. $\ln|2-\cos\theta| + C$
3. $\frac{1}{2}\ln|1+\tan 2x| + C$
4. $-\frac{1}{3}\ln|\cos 3x| + C$
5. $\sin x - \frac{1}{3}\sin^3 x + C$
6. $\frac{1}{4}\tan 4\theta - \theta + C$
7. (a) $\frac{1}{5}\cos^5 x - \frac{1}{3}\cos^3 x + C$
    (b) $\cos x + \sec x + C$
8. $\dfrac{\sec^n x}{n} + C$ if $n \neq 0$,
    $\ln|\sec x| + C$ if $n = 0$
9. $\dfrac{\tan^{n+1} x}{n+1} + C$ if $n \neq -1$,
    $\ln|\tan x| + C$ if $n = -1$
10. $\dfrac{\sin^{n+1} x}{n+1} + C$ if $n \neq -1$,
    $\ln|\sin x| + C$ if $n = -1$
11. $\dfrac{-\cos^{n+1} x}{n+1} + C$ if $n \neq -1$,
    $-\ln|\cos x| + C$ if $n = -1$
12. $\frac{1}{9}\sin^3 3x + C$
13. $-\frac{1}{8}\cos^4 2x + C$
14. $\frac{1}{12}\sec^4 3x + C$
15. $\frac{1}{9}\tan^3 3x + \frac{1}{3}\tan 3x + C$
16. $\frac{1}{2}\sin 2x - \frac{1}{6}\sin^3 2x + C$
17. $\frac{1}{4}\tan^2 2x + \frac{1}{2}\ln|\cos 2x| + C$
18. $\frac{1}{3}\sec^3 x - \sec x + C$
19. $\frac{1}{3}\cos^3 x - \cos x + C$
20. $-1/(1+\sin x) + C$
21. $\ln|2+\tan x| + C$
22. $-\sin t - \csc t + C$
23. $\ln|1+e^x| + C$
24. $\dfrac{(\ln ax)^{n+1}}{n+1} + C$ if $n \neq -1$, $\ln|\ln ax| + C$ if $n = -1$
25. $\ln|\ln 3x| + C$
26. $-\ln|\sin x| - \frac{1}{2}\cot^2 x + C$
27. $-\frac{1}{6}\csc^3 2t + C$
28. $-\cot x - \frac{1}{3}\cot^3 x + C$

29. $-\dfrac{\cot^{n-1} ax}{(n-1)a} - \int \cot^{n-2} ax\, dx,$

$-\dfrac{\cot^3 3x}{9} + \dfrac{\cot 3x}{3} + x + C$

30. $\ln \cosh u + C$

31. $x - \frac{1}{3} \tanh 3x + C$

32. $\frac{1}{5} \tan^{-1} (\sinh 5x) + C$

33. $(1/2a)[\frac{1}{2} \sinh (2at) - at] + C$

34. $\frac{1}{2} \sinh^2 u + C$, or $\frac{1}{2} \cosh^2 u + C$

Article 9.3

1. $\frac{1}{3} \sin^3 x - \frac{1}{5} \sin^5 x + C$
2. $\cos x + \sec x + C$
3. $\frac{1}{2}t - \frac{1}{8} \sin 4t + C$
4. $\frac{1}{2}\theta + \frac{1}{12} \sin 6\theta + C$
5. $\dfrac{3}{8}x - \dfrac{1}{4a} \sin 2ax + \dfrac{1}{32a} \sin 4ax + C$
6. $\frac{1}{8}y - \frac{1}{32} \sin 4y + C$
7. $\tan x + C$
8. $-\frac{1}{3} \cot^3 x - \cot x + C$
9. $-\frac{1}{6} \csc^3 2t + C$
10. $\frac{5}{16}x - \frac{1}{4} \sin 2x + \frac{3}{64} \sin 4x + \frac{1}{48} \sin^3 2x + C$
11. $\frac{1}{6} \sinh^3 2x + \frac{1}{10} \sinh^5 2x + C$
12. $\frac{1}{3} (\cosh 3t + \operatorname{sech} 3t) + C$
13. $(1/a)(\tanh ax - \frac{1}{3} \tanh^3 ax) + C$
14. $\frac{3}{5} \tanh^{5/3} u + C$
15. $\frac{1}{5} \ln (\cosh 5z) - \frac{1}{10} \tanh^2 5z + C$

Article 9.4

1. $0.2a$
2. $\frac{1}{2} \sin^{-1} 2x + C$
3. $\dfrac{\pi a^2}{4}$
4. $\sin^{-1} \dfrac{x-1}{2} + C$
5. $\frac{1}{2} \ln |\sec 2t + \tan 2t| + C$
6. $\ln (1 + \sqrt{2}) \approx 0.88$
7. $\pi/6$
8. $\sqrt{4 + x^2} + C$
9. $\frac{1}{2} \ln 2$
10. $\pi/8$
11. $\frac{1}{4} \ln 3 = 0.2747$
12. $-\ln |\csc u + \cot u| + C$
13. $\dfrac{1}{a} \ln \left| \dfrac{x}{a + \sqrt{a^2 + x^2}} \right| + C$
14. $-\sqrt{4 - x^2} + \sin^{-1} \dfrac{x}{2} + C$
15. $\dfrac{1}{\sqrt{5}} \sin^{-1} (x\sqrt{5/2}) + C$
16. $\cos^{-1} \left(\dfrac{\cos \theta}{\sqrt{2}} \right) + C$
17. $\dfrac{1}{|a|} \sec^{-1} \left| \dfrac{x}{a} \right| + C$
18. $\dfrac{1}{a} \ln \left| \dfrac{x}{a + \sqrt{a^2 - x^2}} \right| + C$
19. $\dfrac{1}{a^2} \left(\dfrac{x}{\sqrt{a^2 - x^2}} \right) + C$
20. $\dfrac{1}{2a^3} \left(\tan^{-1} \dfrac{x}{a} + \dfrac{ax}{a^2 + x^2} \right) + C$

Article 9.5

1. $\pi/8$
2. $\sqrt{x^2 - 2x + 5} + \ln |x - 1 + \sqrt{x^2 - 2x + 5}| + C$
3. $2 \sin^{-1} (x - 1) - \sqrt{2x - x^2} + C$
4. $\sqrt{x^2 - 4x + 3} + \ln |x - 2 + \sqrt{x^2 - 4x + 3}| + C$
5. $2 \sin^{-1} \dfrac{x-2}{3} - \sqrt{5 + 4x - x^2} + C$
6. $\ln |x - 1 + \sqrt{x^2 - 2x - 8}| + C$
7. $\sqrt{8 + 2x - x^2} + C$
8. $\sqrt{x^2 + 4x + 5} - 2 \ln |x + 2 + \sqrt{x^2 + 4x + 5}| + C$
9. $\frac{1}{2} \ln (x^2 + 4x + 5) - 2 \tan^{-1} (x + 2) + C$
10. $\frac{1}{4} \ln (4x^2 + 4x + 5) + \frac{1}{2} \tan^{-1} (x + \frac{1}{2}) + C$

Article 9.6

1. $\frac{1}{6} \ln |(x + 5)^5(x - 1)| + C$
2. $\frac{3}{4} \ln |x - 3| + \frac{1}{4} \ln |x + 1| + C$
3. $\frac{2}{3} \ln |x + 5| + \frac{1}{3} \ln |x - 1| + C$
4. $x - 2 \ln |x + 1| - \dfrac{1}{x + 1} + C$
5. $\ln \left| \dfrac{x}{x + 1} \right| + \dfrac{1}{x + 1} + C$
6. $\dfrac{1}{4} \ln \dfrac{(x + 1)^2}{x^2 + 1} + \dfrac{1}{2} \tan^{-1} x + C$

7. $\frac{1}{2}\ln\frac{x^2}{x^2+x+1}-\frac{1}{\sqrt{3}}\tan^{-1}\frac{2x+1}{\sqrt{3}}+C$

8. $\frac{1}{3}\ln\frac{2+\cos\theta}{1-\cos\theta}+C$

9. $\ln\frac{1+e^t}{2+e^t}+C$

10. $\frac{1}{2}\left(\tan^{-1}x+\frac{x}{1+x^2}\right)+C$

11. $x-\frac{3}{2}\tan^{-1}x+\frac{1}{2}\frac{x}{1+x^2}+C$

Article 9.7

1. $\frac{x^2}{2}\ln x-\frac{x^2}{4}+C$

2. $\frac{x^{n+1}}{n+1}\ln ax-\frac{x^{n+1}}{(n+1)^2}+C$

3. $\frac{x^2+1}{2}\tan^{-1}x-\frac{1}{2}x+C$

4. $x\sin^{-1}ax+\frac{1}{a}\sqrt{1-a^2x^2}+C$

5. $\frac{1}{a^2}\sin ax-\frac{1}{a}x\cos ax+C$

6. $\frac{2x}{a^2}\cos ax-\frac{2}{a^3}\sin ax+\frac{x^2}{a}\sin ax+C$

7. $\frac{x}{a}\tan ax-\frac{1}{a^2}\ln|\sec ax|+C$

8. $\frac{e^{ax}}{a^2+b^2}(a\sin bx-b\cos bx)+C$

9. $(x/2)[\sin(\ln x)-\cos(\ln x)]+C$

10. $(x/2)[\sin(\ln x)+\cos(\ln x)]+C$

11. $x\ln(a^2+x^2)-2x+2a\tan^{-1}(x/a)+C$

12. $(x/2)\sin(2x+1)+\frac{1}{4}\cos(2x+1)+C$

13. $\left(\frac{x^2}{2}-\frac{1}{4}\right)\sin^{-1}x+\frac{x}{4}\sqrt{1-x^2}+C$

14. $\frac{4\pi-3\sqrt{3}}{6}$

15. $\frac{4\pi-3\sqrt{3}}{3}$

16. $\frac{x^3}{3}\tan^{-1}x-\frac{x^2}{6}+\frac{1}{6}\ln(x^2+1)+C$

17. (b) $\frac{1}{32}x^4[8(\ln x)^2-4\ln x+1]+C$

18. (b) $\pi/16$

19. (b) $(x^3-3x^2+6x-6)e^x+C$

20. (b) $\frac{1}{2}(\sec x\tan x+\ln|\sec x+\tan x|)+C$

(c) $\frac{1}{2}(x\sqrt{x^2+a^2}+a^2\ln|x+\sqrt{x^2+a^2}|)+C$

21. $P(x)=1-3x^2$

Article 9.8

1. 2

2. 1

3. $\tan x+\sec x+C$

4. $\pi/3\sqrt{3}$

5. $\frac{4}{\sqrt{3}}\tan^{-1}\left(\sqrt{3}\tan\frac{x}{2}\right)-x+C$

6. $-\frac{2}{5}$

7. 0

8. π

Article 9.9

1. $-x+4\sqrt{x}-4\ln|1+\sqrt{x}|+C$

2. $\frac{2\sqrt{x}}{b}-\frac{2a}{b^2}\ln|a+b\sqrt{x}|+C$

3. $\frac{2(x+a)^{3/2}}{105}(15x^2-12ax+8a^2)+C$

4. $2\sqrt{x+a}-2\sqrt{b-a}\tan^{-1}\sqrt{\frac{x+a}{b-a}}+C$

5. $\frac{b-a}{2}\left(\ln\left|\frac{u-1}{u+1}\right|-\frac{2u}{u^2-1}\right)+C,$

where $u=\sqrt{\frac{a+x}{b+x}}$

6. $\frac{1}{2}\ln|x+\sqrt{x^2+a^2}|+\frac{x\sqrt{a^2+x^2}-x^2}{2a^2}+C$

7. $(3\sqrt{2}-4)/2$

8. $\frac{1}{nc}\ln\left|\frac{x^n}{ax^n+c}\right|+C$

9. $2+2\ln(\frac{3}{2})$

10. $3\ln|\sqrt[3]{x}-1|+C$

11. $-1+4\ln(\frac{3}{2})$

12. $4\ln\left|\frac{\sqrt[4]{x}}{1-\sqrt[4]{x}}\right|+C$

13. $\frac{3}{2}y^{2/3}+\ln|1+y^{1/3}|-\frac{1}{2}\ln|y^{2/3}-y^{1/3}+1|$
$-\sqrt{3}\tan^{-1}\frac{2y^{1/3}-1}{\sqrt{3}}+C$

14. $\frac{2}{9}(z^3-2)\sqrt{z^3+1}+C$

15. $\frac{2}{3}(\sqrt{t^3-1}-\tan^{-1}\sqrt{t^3-1})+C$

16. $\ln(\frac{4}{3})$

17. $2+4\ln(\frac{2}{3})$

18. $-\frac{1}{2}\ln|\sin x+\cos^2 x|$
$$-\frac{1}{2\sqrt{5}}\ln\left|\frac{2\sin x-1-\sqrt{5}}{2\sin x-1+\sqrt{5}}\right|+C$$

19. $\frac{1}{3}\ln|x+1|-\frac{1}{6}\ln|x^2-x+1|$
$$+\frac{1}{\sqrt{3}}\tan^{-1}\frac{2x-1}{\sqrt{3}}+C$$

20. $-2y^{1/2}-3y^{1/3}-6y^{1/6}-6\ln|y^{1/6}-1|+C$

Article 9.10

1. $\pi/2$
2. 2
3. 6
4. 1000
5. 4
6. $\pi/2$
7. $\frac{1}{2}$
8. 1000
9. Diverges
10. Converges
11. Converges
12. Diverges
13. Converges
14. Converges
15. Diverges
16. Diverges
17. Converges
18. Diverges
19. Converges

Article 9.11

1. $\frac{8}{3}$, $\frac{8}{3}$, $\frac{8}{3}$
2. $2\pi/3$, $(1+2\sqrt{2})\pi/6$, 2
3. $2\pi/3$, 2.28
4. $5\pi/9$, 1.8, $\pi/\sqrt{3}$
5. 1.86, 1.85
6. 0.78, 0.785

Miscellaneous exercises

1. $2\sqrt{1+\sin x}+C$
2. $(\sin^{-1}x)^2/2+C$
3. $(\tan^2 x)/2+C$
4. $\tan x+\sec x+C$
5. $(2x^{3/2})/3+C$
6. $2\sin\sqrt{x}+C$
7. $\ln|x+1+\sqrt{x^2+2x+2}|+C$
8. $\ln|(x-2)(x-3)/(x-1)^2|+C$
9. $(x^2-2x+2)e^x+C$
10. $\frac{1}{2}(\ln|x+\sqrt{x^2+1}|+x\sqrt{1+x^2})+C$
11. $\tan^{-1}(e^t)+C$
12. $\tan^{-1}(e^x)+C$
13. $2\sqrt{x}-2\ln(1+\sqrt{x})+C$
14. $\frac{4}{3}(\sqrt{x}-2)\sqrt{1+\sqrt{x}}+C$
15. $\frac{9}{25}(t^{5/3}+1)^{5/3}+C$
16. $\ln|\ln\sin x|+C$
17. $\ln\dfrac{\sqrt{1+e^t}-1}{\sqrt{1+e^t}+1}+C$
18. $t+2\ln(1+\sqrt{1-e^{-t}})+C$
19. $e^{\sec x}+C$
20. $\tan^{-1}(\sin x)+C$
21. $\sin^{-1}(x-1)+C$
22. $\cot^{-1}(\cos x)+C$
23. $\frac{1}{2}\ln|1+\sin 2t|+C$
24. $\ln|\tan x|+C$
25. $(-2\cos x)/\sqrt{1+\sin x}+C$
26. $(2\cos x)/\sqrt{1-\sin x}+C$
27. $x/(a^2\sqrt{a^2-x^2})+C$
28. $x/(a^2\sqrt{a^2+x^2})+C$
29. $\dfrac{1}{3}\ln\dfrac{1-\cos x}{4-\cos x}+C$
30. $\frac{3}{10}(2e^x-3)(1+e^x)^{2/3}+C$
31. $\displaystyle\sum_{k=0}^{m}\frac{(-1)^k}{k!(m-k)!}\ln|x+k|+C$
32. $\dfrac{1}{12}\ln\dfrac{(x-1)^2(x^2-x+1)}{(x+1)^2(x^2+x+1)}$
$$-\frac{\sqrt{3}}{6}\tan^{-1}\left(\frac{x\sqrt{3}}{1-x^2}\right)+C$$
33. $\frac{1}{3}[(2y^3+1)^{-1}+\ln|2y^3/(2y^3+1)|]+C$
34. $\frac{2}{3}x^{3/2}-x+2x^{1/2}-2\ln(1+x^{1/2})+C$
35. $\frac{1}{2}[\ln x^2-\ln(1+x^2)-x^2(1+x^2)^{-1}]+C$
36. $(x-1)\ln\sqrt{x-1}-(x/2)+C$
37. $\ln|e^x-1|-x+C$
38. $\dfrac{1}{4}\ln\left|\dfrac{1+\tan\theta}{1-\tan\theta}\right|+\dfrac{1}{2}\theta+C$
39. $x^{-1}+2\ln|1-x^{-1}|+C$
40. $\frac{1}{2}\ln|(x+3)^3(x+1)^{-1}|+C$
41. $-\dfrac{1}{2}\dfrac{1}{e^{2u}-1}+C$
42. $\ln|x/\sqrt{x^2+4}|+C$
43. $\frac{1}{3}\tan^{-1}[(5x+4)/3]+C$
44. $\sqrt{x^2-a^2}-a\tan^{-1}(\sqrt{x^2-a^2}/a)+C$
45. $e^x(\cos 2x+2\sin 2x)/5+C$
46. $\ln x-2\ln(1+3\sqrt{x})+C$
47. $\ln|x|-3\ln|1+\sqrt[3]{x}|+C$
48. $\ln|\sin\theta|-\frac{1}{2}\ln(1+\sin^2\theta)+C$
49. $\frac{1}{15}(8-4z^2+3z^4)(1+z^2)^{1/2}+C$
50. $\frac{3}{8}(e^{2t}-3)(1+e^{2t})^{1/3}+C$
51. $\frac{5}{2}(1+x^{4/5})^{1/2}+C$
52. $x\tan x+\ln|\cos x|+C$
53. $\frac{1}{4}[(2x^2-1)\sin^{-1}x+x\sqrt{1-x^2}]+C$

54. $\frac{x^2}{2}+\frac{4}{3}\ln|x+2|+\frac{2}{3}\ln|x-1|+C$

55. $x+\ln|(x-1)/x|+C$

56. $\ln|x-1|-(x-1)^{-1}+C$

57. $\frac{2}{3}(3e^{2x}-6e^x-1)^{1/2}$
$+\frac{1}{\sqrt{3}}\ln|e^x-1+\sqrt{e^{2x}-2e^x-\frac{1}{3}}|+C$

58. $\frac{3}{2}(x^2+2x-3)^{1/3}+C$ 59. $\tan^{-1}(2\sqrt{y^2+y})+C$

60. $-\sqrt{a^2-x^2}/a^2x+C$

61. $\frac{3}{8}\sin^{-1}x+\frac{1}{8}(5x-2x^3)\sqrt{1-x^2}+C$

62. $x\ln(x+\sqrt{1+x^2})-\sqrt{1+x^2}+C$

63. $x\tan x-(x^2/2)+\ln|\cos x|+C$

64. $-(\tan^{-1}x)/x+\ln|x|-\ln\sqrt{1+x^2}+C$

65. $\frac{x^2}{4}+\frac{x\sin 2x}{4}+\frac{\cos 2x}{8}+C$

66. $-x^2\cos x+2x\sin x+2\cos x+C$

67. $\frac{x^2}{4}-\frac{x\sin 2x}{4}-\frac{\cos 2x}{8}+C$

68. $\frac{1}{2}\tan^{-1}t-\frac{1}{2\sqrt{3}}\tan^{-1}\left(\frac{t}{\sqrt{3}}\right)+C$

69. $\frac{1}{12}[4u-3\ln(1+e^{2u})+\ln(3+e^{2u})]+C$

70. $\frac{1}{4}(x^2-4)\ln(x+2)-x^2/8+x/2+C$

71. $(x^2+1)e^x+C$

72. $x\sec^{-1}x-\ln(|x|+\sqrt{x^2-1})+C$

73. $-2x^{-2}+2x^{-1}+\ln|x/(x+2)|+C$

74. $\frac{1}{16}\ln|(x^2-4)/(x^2+4)|+C$

75. $2-\sqrt{2}\ln(1+\sqrt{2})$ 76. $\ln|\cot x|+C$

77. $-\frac{1}{2}\frac{1}{e^{2u}+1}+C$

78. $x-2\sqrt{x}+\frac{4}{\sqrt{3}}\tan^{-1}\left(\frac{1+2\sqrt{x}}{\sqrt{3}}\right)+C$

79. $\ln|\tan t-2|-\ln|\tan t-1|+C$

80. $-t+\sqrt{2}\tan^{-1}(\sqrt{2}\tan t)+C$

81. $\frac{1}{\sqrt{2}}\tan^{-1}\left(\frac{\tan x}{\sqrt{2}}\right)+C$

82. $e^t\sin(e^t)+\cos(e^t)+C$

83. $x\ln\sqrt{x^2+1}-x+\tan^{-1}x+C$

84. $\frac{1}{2}x^2\ln(x^3+x)-\frac{3}{4}x^2+\frac{1}{2}\ln(x^2+1)+C$

85. $\frac{1}{2}(x^2-1)e^{x^2}+C$

86. $\ln|\sin x+\sqrt{3+\sin^2 x}|+C$

87. $\sin^{-1}(3^{-1/2}\tan x)+C$

88. $(x^2-2)\cos(1-x)+2x\sin(1-x)+C$

89. $\tan x-\sec x+C$

90. $3^{-1/2}[\ln|1+2\sin x|$
$-\ln|2+\sin x+\sqrt{3}\cos x|]+C$

91. $-\frac{1}{2}\cot x\csc x+\frac{1}{4}\ln\frac{1-\cos x}{1+\cos x}+C$

92. $\frac{1}{2}\tan^2 x+\ln|\cos x|+C$

93. $x(\sin^{-1}x)^2-2x+2\sqrt{1-x^2}\sin^{-1}x+C$

94. $\frac{1}{54}(9x^2-1)\ln|3x+1|-x^2/12+x/18+C$

95. $\frac{1}{2}\ln(x^2+1)+\frac{1}{2}(x^2+1)^{-1}+C$

96. $-\frac{2}{3}(x+2)\sqrt{1-x}+C$

97. $\frac{1}{15}(3x-1)(2x+1)^{3/2}+C$

98. $x\ln(x+\sqrt{x^2-1})-\sqrt{x^2-1}+C$

99. $x\ln(x-\sqrt{x^2-1})+\sqrt{x^2-1}+C$

100. $\frac{1}{2}\ln|t-\sqrt{1-t^2}|-\frac{1}{2}\sin^{-1}t+C$

101. $x-e^{-x}\tan^{-1}(e^x)-\frac{1}{2}\ln(1+e^{2x})+C$

102. $\frac{1}{2}(x-x^2)^{1/2}-\frac{1}{2}(1-2x)\sin^{-1}\sqrt{x}+C$

103. $x\ln(x+\sqrt{x})-x+\sqrt{x}-\ln(1+\sqrt{x})+C$

104. $(x+1)\tan^{-1}\sqrt{x}-\sqrt{x}+C$

105. $x\ln(x^2+x)+\ln(x+1)-2x+C$

106. $(x+\frac{1}{2})\ln(\sqrt{x}+\sqrt{1+x})-\frac{1}{2}\sqrt{x^2+x}+C$

107. $2\sqrt{x}\sin\sqrt{x}+2\cos\sqrt{x}+C$

108. $-2\sqrt{x}\cos\sqrt{x}+2\sin\sqrt{x}+C$

109. $(x+2)\tan^{-1}\sqrt{x+1}-\sqrt{x+1}+C$

110. $\frac{1}{4}(\sin^{-1}x)^2+\frac{1}{2}x\sqrt{1-x^2}\sin^{-1}x-x^2/4+C$

111. $\frac{1}{4}x^2-\frac{1}{8}x\sin 4x-\frac{1}{32}\cos 4x+C$

112. $\sec x-\tan x+x+C$

113. $\ln(\sqrt{e^{2t}+1}-1)-t+C$

114. $\ln[(1-\sin x)^{1/2}(1+\sin x)^{-1/18}(2-\sin x)^{-4/9}]$
$+1/(6-3\sin x)+C$

115. $\dfrac{1}{ac} \ln \left| \dfrac{e^{ct}}{a + be^{ct}} \right| + C$

116. $-2 \sin^{-1} \left[\dfrac{\cos (x/2)}{\cos (\alpha/2)} \right] + C$

117. $\dfrac{1}{\sqrt{2}} \ln \left| \csc \left(x - \dfrac{\pi}{4} \right) - \cot \left(x - \dfrac{\pi}{4} \right) \right| + C$

118. $x \ln \sqrt{1 + x^2} - x + \tan^{-1} x + C$

119. $x \ln (2x^2 + 4) + 2\sqrt{2} \tan^{-1} (x/\sqrt{2}) - 2x + C$

120. $-\frac{1}{3}(1 - x^2)^{1/2}(2 + x^2) + C$

121. $\ln |2 + \ln x| + C$ 122. $x - \tan x + C$

123. $\frac{1}{3} \ln |x + 1| - \frac{1}{6} \ln (x^2 - x + 1) + \dfrac{1}{\sqrt{3}} \tan^{-1} \dfrac{2x - 1}{\sqrt{3}} + C$

124. $\frac{4}{21}(e^x + 1)^{3/4}(3e^x - 4) + C$

125. $\tan^{-1} (e^x) + C$

126. $2e^{\sqrt{t}}(\sqrt{t} - 1) + C$

127. $-2\sqrt{x + 1} \cos \sqrt{x + 1} + 2 \sin \sqrt{x + 1} + C$

128. $-2\sqrt{1 - x} \sin \sqrt{1 - x} - 2 \cos \sqrt{1 - x} + C$

129. $\ln 2$ 130. $2\sqrt{2} - 2$ 131. $2/\pi$

132. $e - 1$ 133. $\pi/4$ 134. $2 \ln 2 - 1$

135. $\pi/2$ 136. $1/2$ 137. -1

139. e^{-4} 140. $4p^2(2\sqrt{2} - 1)/3$

141. Area $(\pi - 2)r^2/4$, centroid $2r\sqrt{2}/(3\pi - 6)$ from center.

142. $[1/(2e - 4), (e^2 - 3)/(4e - 8)] \approx (0.7, 1.5)$

143. $2\pi^2 kr^2$

144. $w(3e^8 - 20e^4 + 9)/4 \approx 123{,}000$ lb

145. Only V is finite; integrals for A and for S diverge.

146. $A = e^{-1}$, $V = \pi/(2e^2)$,
$S = \pi[e^{-1}\sqrt{1 + e^{-2}} + \ln (e^{-1} + \sqrt{e^{-2} + 1})]$

147. (0.58, 0.93) 148. (1.53, 0.696)

149. About 2 150. About 23

151. About 6.8 152. About 175

153. $2\pi[\sqrt{2} + \ln (1 + \sqrt{2})] \approx 14.4$

154. About 3.82

155. (0, 0.6) for arch $|x| \leq \pi/2$

156. 2π 157. $6a$

158. $\ln (2 + \sqrt{3}) \approx 1.32$ 159. $12\pi a^2/5$

160. $(2a/5, 2a/5)$ for first-quadrant arch

161. Converges, compare with $\int_1^\infty x^{-3/2}\, dx$.

162. (c) $1 - e^{-e^b}$
(d) 1

164. Converges

165. $x = a \cos kt$ 166. $\frac{1}{2}$

Chapter 10

Article 10.1

Note. The answer for each exercise in this article is accompanied by an appropriately numbered figure.

1. (a) About the x-axis
   (b) No curve for $0 < x < 2$
   (c) (0, 0), (2, 0) (e) ∞
2. (a) About the x-axis
   (b) No curve for $0 < x \leq 2$
   (c) (0, 0) (d) $y = \pm 1$, $x = 2$ (e) ∞

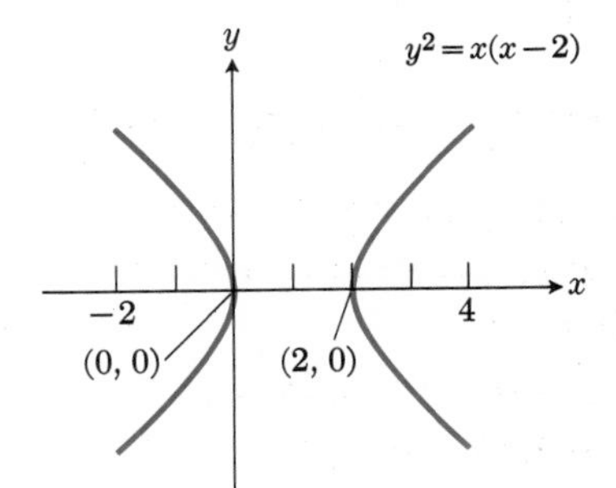

10.1.1

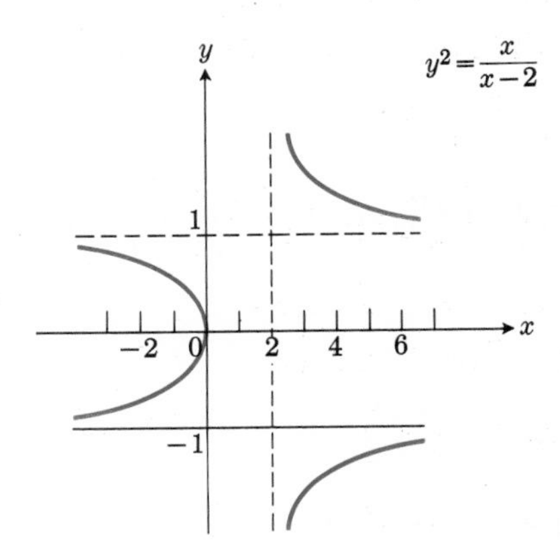

10.1.2

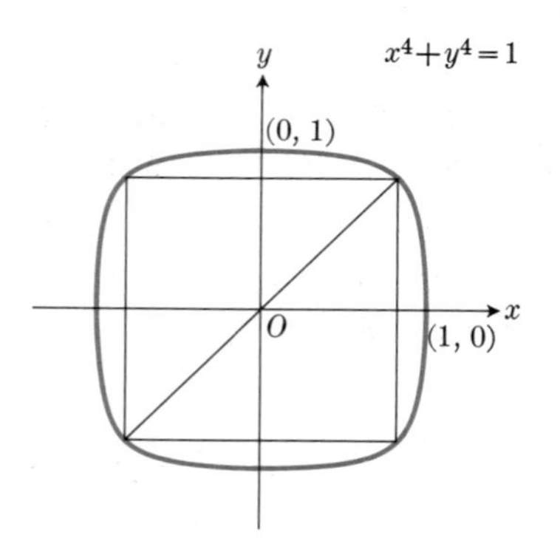

10.1.3

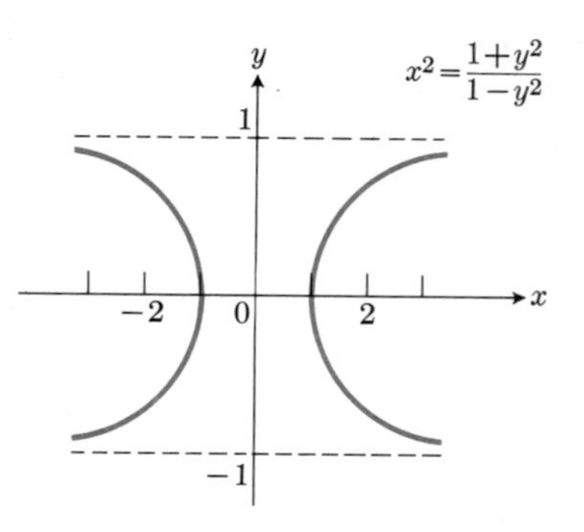

10.1.4

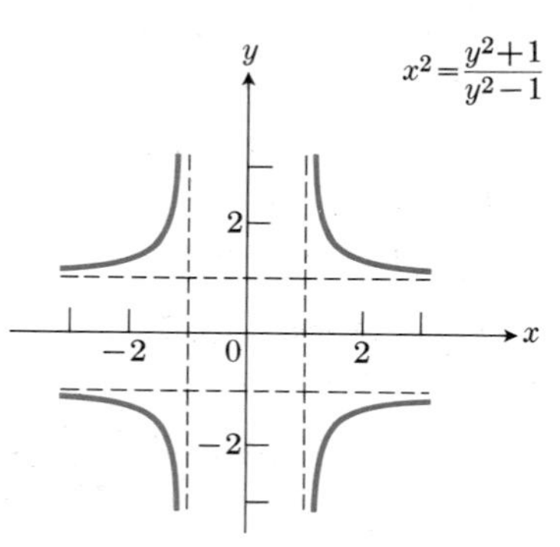

10.1.5

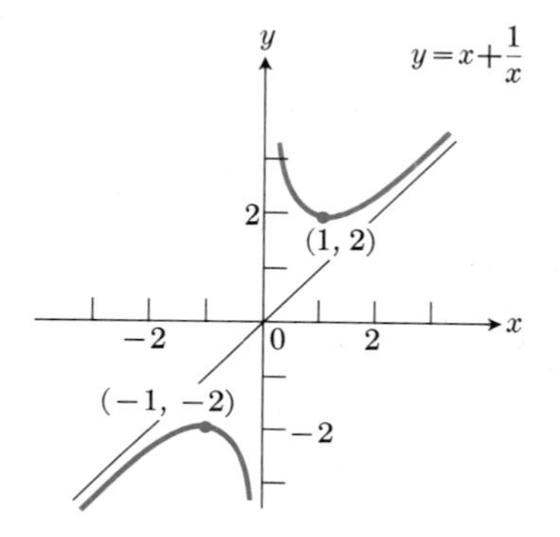

10.1.6

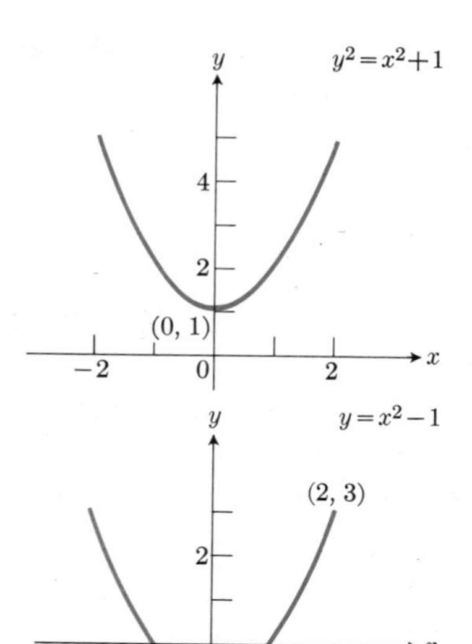

10.1.7

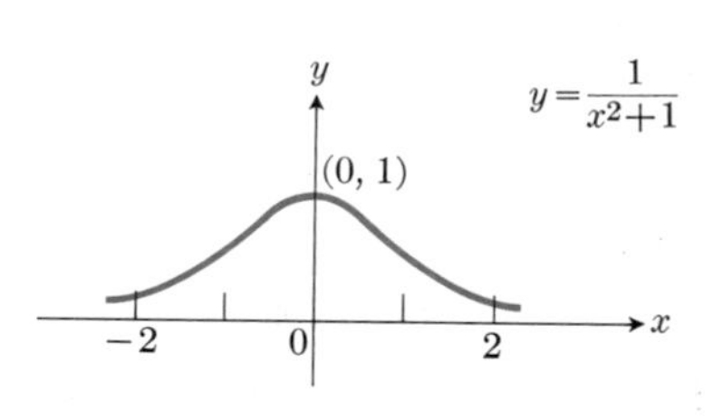

10.1.8

10.1.9

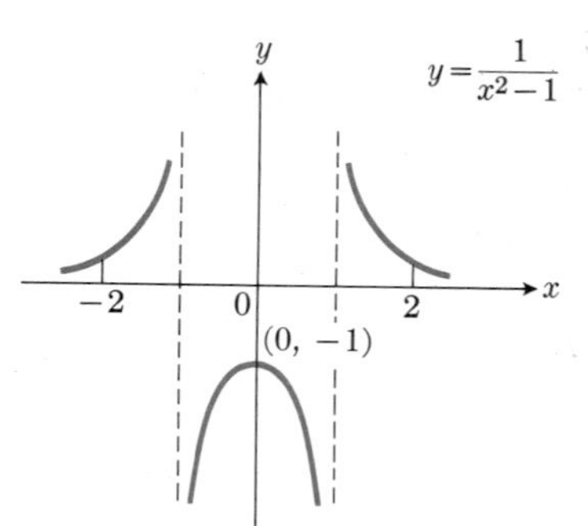

10.1.10

3. (a) About both axes, the origin, and the line $y = x$
   (b) Curve is inside the square $|x| \leq 1, \quad |y| \leq 1$
   (c) $(0, \pm 1), \quad (\pm 1, 0)$
   (e) Slope is 0 at $(0, \pm 1)$, ∞ at $(\pm 1, 0)$
4. (a) About both axes, and the origin
   (b) No curve for $|y| \geq 1$. No curve for $|x| < 1$
   (c) $(\pm 1, 0)$ (d) $y = \pm 1$ (e) ∞
5. (a) About both axes, the origin, and $y = x$
   (b) No curve for $|y| \leq 1$. No curve for $|x| \leq 1$
   (c) No intercepts (d) $x = \pm 1, \quad y = \pm 1$
6. (a) About the origin (b) No curve for $|y| < 2$
   (c) No intercepts
   (d) $x = 0$; $y = x$ is also an asymptote.
   (e) Low point at (1, 2), high point at (—1, —2)
7. (a) About the y-axis
   (b) Curve is confined to $y \geq 1$
   (c) (0, 1) (e) 0
8. (a) About the y-axis (b) $0 < y \leq 1$
   (c) (0, 1) (d) $y = 0$ (e) 0
9. (a) About the y-axis (b) $y \geq -1$
   (c) $(0, -1), \quad (\pm 1, 0)$ (e) $0, \quad \pm 2$
10. (a) About the y-axis
    (b) No curve for $-1 < y \leq 0$
    (c) $(0, -1)$ (d) $x = \pm 1, \quad y = 0$ (e) 0

Article 10.2

1. $x + y + 1 = 0$
2. $x + 2y = 7$
3. $6x - y = 14, \quad y - 6x = 18$
4. $2x + y = \pm 3$
7. Yes; $(-2, -8)$
8. $(-1, -1)$. No, because

$$2x_1^3 - 3x_1^2 + 5 = 0,$$

and this equation has only one real root.

9. $2x + 3y = \pm 12$
10. $x + y = 1, \quad 4x + y = -2$
11. $(-\frac{13}{4}, \frac{17}{16})$
15. ± 16
16. $\pm 1/\sqrt{3}$
17. $A = \tan^{-1} \frac{5}{3}, \quad B = \tan^{-1} 5, \quad C = \tan^{-1} \frac{10}{11}$
21. (a) 45° (b) $\tan^{-1} 3$
    (c) $\tan^{-1} (5/\sqrt{3})$ (d) $\tan^{-1} (\frac{14}{3})$

Article 10.3

1. 0.618
2. 0.682
3. 1.165
4. 1.189
5. —1.189
6. 3
7. $x_1 = x_2 = x_3$, etc.
9. $m_1 = (q - 1)/q, \quad m_2 = 1/q$
12. Yes, they may get worse.

Article 10.4

1. $x - y = 1$
2. $x^2 + y^2 - 6x + 5 = 0$
3. $(x^2 + y^2)^2 + 8(y^2 - x^2) = 0$
4. $3x^2 + 3y^2 + 2xy - 4x - 4y = 0$
5. $3x^2 + 4y^2 - 20x + 12 = 0$
6. $5x^2 - 4y^2 = 20$ (right half only)
7. $x^2 = 10 - 2y + 6|y - 1|$
8. $x^2 + y^2 - 4x - 6y + 4 = 0,$

$$\text{or } \begin{cases} x^2 = 4y + 4 & \text{when } y \geq 1, \\ x^2 = 16 - 8y & \text{when } y < 1 \end{cases}$$

9. $(2, 2), \quad \sqrt{5}$
10. $\dfrac{|Ax_1 + By_1 + C|}{\sqrt{A^2 + B^2}}$

Article 10.5

1. $x^2 + y^2 - 4y = 0$
2. $x^2 + y^2 + 4x = 5$
3. $x^2 + y^2 - 6x + 8y = 0$
4. $x^2 + y^2 - 2x - 2y = 0$
5. $x^2 + y^2 + 4x + 2y = 1$
6. $x^2 + y^2 + 8x - 4y + 4 = 0$
7. (0, 1), 2
8. (—1, 0), 3
9. $(-1, 0), \quad 2/\sqrt{3}$
10. $(-\frac{1}{4}, -\frac{1}{4}), \quad 1/(2\sqrt{2})$
11. (—1, 2), 0
12. No real locus, radius becomes imaginary.
13. $x^2 + y^2 - 4x - 4y = 5$
14. $5x^2 + 5y^2 + 10x - 10y + 1 = 0$
15. $(x - 0.7)^2 + (y - 1.3)^2 = 12.58$
17. $x^2 + y^2 + 2x + 2y - 23 = 0$
18. Circle, center (—2, 3), radius 4
19. Center (1, 2), radius $\sqrt{2}$
    Distance of (0.1, 3.1) from center is $\sqrt{2.02}$.
    Point is outside the circle.
20. Center (—2, 4), radius $2\sqrt{5}$
21. Center (—1, 1), radius 5

24. (a) $C_1 = -2h$, $C_2 = -2k$, $C_3 = h^2 + k^2 - r^2$

(b) $h = -\dfrac{C_1}{2}$, $k = -\dfrac{C_2}{2}$, $r = \sqrt{\dfrac{C_1^2 + C_2^2 - 4C_3}{4}}$

Article 10.6

1. $x^2 = 8y$, $y = -2$
2. $y^2 = -8x$, $x = 2$
3. $(x + 2)^2 = 4(y - 3)$, $y = 2$
4. $(y - 3)^2 = -4x$, $x = 1$
5. $(y - 1)^2 = 12(x + 3)$, $x = -6$
6. $(x - 1)^2 = 12(y + 3)$, $y = -6$
7. $y^2 = 8(x - 2)$, $F(4, 0)$
8. $(x - 1)^2 = -8(y + 2)$, $F(1, -4)$
9. $(y - 1)^2 = -16(x + 3)$, $F(-7, 1)$
10. $(x + 2)^2 = 4(y + 2)$, $F(-2, -1)$
11. $(y - 1)^2 = 4x$, $F(1, 1)$
12. $x^2 = -4(y - 1)$, $F(0, 0)$
13. $V(1, 1)$, $x = 1$, $F(1, -1)$, $y = 3$
14. $V(-4, -3)$, $x = -4$, $F(-4, -\frac{5}{2})$, $y = -\frac{7}{2}$
15. $V(2, 0)$, $y = 0$, $F(1, 0)$, $x = 3$
16. $V(0, -\frac{1}{2})$, $x = 0$, $F(0, \frac{3}{2})$, $y = -\frac{5}{2}$
17. $V(-1, -1)$, $x = -1$, $F(-1, 0)$, $y = -2$
18. $V(\frac{1}{4}, -\frac{1}{2})$, $y = -\frac{1}{2}$, $F(0, -\frac{1}{2})$, $x = \frac{1}{2}$
19. $V(2, 1)$, $y = 1$, $F(\frac{29}{16}, 1)$, $x = \frac{35}{16}$
20. $V(2, -3)$, $y = -3$, $F(\frac{3}{2}, -3)$, $x = \frac{5}{2}$
21. $V(2, -2)$, $x = 2$, $F(2, -\frac{4}{3})$, $y = -\frac{8}{3}$
22. $V(-3, -1)$, $y = -1$, $F(-\frac{21}{8}, -1)$, $x = -\frac{27}{8}$
23. Points above parabola $x^2 = 8y$
26. $(x - y)^2 - 8(x + y) + 16 = 0$
27. $b^2 y = 4hx(b - x)$
28. (a) $7y^2 - 3y + 6x - 16 = 0$
(b) $7x^2 - 9x - 6y - 4 = 0$
30. The product is zero if and only if at least one factor is zero. Locus is the line $2x + y - 3 = 0$, the circle $x^2 + y^2 - 4 = 0$, and the parabola $x^2 - 8y = 0$.

Article 10.7

1. $x^2/12 + y^2/16 = 1$, $e = \frac{1}{2}$
2. $x^2/25 + y^2/16 = 1$, $e = \frac{3}{5}$
3. $9x^2 + 5(y - 2)^2 = 45$, $e = \frac{2}{3}$
4. $(x + 3)^2/12 + y^2/16 = 1$, $e = \frac{1}{2}$
5. $(x - 2)^2 + 10(y - 2)^2 = 10$, $e = 3/\sqrt{10}$
6. $9(x - 1)^2 + 4(y - 4)^2 = 36$, $F(1, 4 \pm \sqrt{5})$
7. $(2, -3)$, $(2, 2)$ and $(2, -8)$, $(2, 1)$ and $(2, -7)$
9. $x^2 + 2y^2 - 4y = 0$
10. $e = \frac{3}{4}$, $y = \pm\frac{16}{3}$
11. $\frac{4}{3}\pi ab^2$
12. (a) $\int_0^a \sqrt{a^2 - x^2}\, dx$ ($= \frac{1}{4}\pi a^2$ for $\frac{1}{4}$ of a circle)
(b) $\dfrac{b}{a}\int_0^a \sqrt{a^2 - x^2}\, dx$. Area of ellipse $= \pi ab$.
14. $2b^2/a$
15. $5x^2 + 9y^2 = 180$
16. $A = C = 0$, $B = -4$
18. Interior of ellipse $9x^2 + 16y^2 = 144$
19. Graph consists of parabola $x^2 + 4y = 0$, straight line $2x - y - 3 = 0$, circle $x^2 + y^2 - 25 = 0$, ellipse
$$x^2 + 4y^2 - 4 = 0.$$
20. One factor must be positive, the other negative. Graph consists of region between circle $x^2 + y^2 = 1$ and ellipse
$$9x^2 + 4y^2 = 36.$$

Article 10.8

2. $C(2, -3)$, $V(2 \pm 2, -3)$, $F(2 \pm \sqrt{13}, -3)$
$A: 3(x - 2) = \pm 2(y + 3)$
3. $C(2, -3)$, $V(2 \pm 3, -3)$, $F(2 \pm \sqrt{13}, -3)$
$A: 2(x - 2) = \pm 3(y + 3)$
4. $C(2, -3)$, $V(2, -3 \pm \frac{1}{2})$, $F(2, -3 \pm \sqrt{13}/6)$
$A: 2(y + 3) = \pm 3(x - 2)$
5. $C(-2, 1)$, $V(-2 \pm 2, 1)$, $F(-2 \pm 3, 1)$
$A: \sqrt{5}(x + 2) = \pm 2(y - 1)$
6. $C(0, 2)$, $V(\pm 1, 2)$, $F(\pm\sqrt{5}, 2)$
$A: 2x = \pm(y - 2)$
7. $C(2, 0)$, $V(2 \pm 2, 0)$, $F(2 \pm \sqrt{5}, 0)$
$A: x - 2 = \pm 2y$
8. $C(2, 1)$, $V(2, 1 \pm 2)$, $F(2, 1 \pm 3)$
$A: 2(x - 2) = \pm\sqrt{5}(y - 1)$
10. $\frac{4}{3}\pi(b^4/a)$
12. $x^2 - 3y^2 + 12y = 9$
13. $4x^2 - 5y^2 - 8x + 60y + 4 = 0$

Article 10.9

2. $15x^2 + 13y^2 - 2\sqrt{3}xy - 14\sqrt{3}y + 18x = 33,\ 60°$

7. $2y'^2 = 1$

8. $4x'^2 + 2y'^2 = 19$

9. $5y'^2 - x'^2 = 10$

10. $5x'^2 - 3y'^2 = 7$

Article 10.10

1. Ellipse
2. Hyperbola
3. Parabola
4. Circle
5. Hyperbola
6. Parabola
7. Hyperbola
8. Ellipse
9. Ellipse
10. After rotation of axes to a form with $B' = 0$, we have $A' = 1/a^2$, $C' = 1/b^2$. Then $B'^2 - 4A'C' = -4/(a^2b^2)$; hence

$$\pi ab = 2\pi/\sqrt{4A'C' - B'^2} = 2\pi/\sqrt{4AC - B^2}.$$

12. If $C = -A$, then $A + C = A' + C' = 0$ and $C' = -A'$. Hence $A' = C' = 0$ if $\tan 2\alpha = -2A/B$, by Eqs. (6), Article 10.9.

Article 10.11

3. Line L and the point D do not exist.
4. $(1 - e^2)x^2 + y^2 - 2pe^2x - p^2e^2 = 0$

Miscellaneous exercises

1. (a) A. Symmetric about the x-axis
   B. $0 \le x \le 4,\quad -2 \le y \le 2$
   C. $(0, 0),\quad (4, 0)$
   D. None
   E. ∞ at both intercepts

   (b) A. Symmetric about the x-axis
   B. $x \le 0,\quad x \ge 4,\quad$ all y
   C. $(0, 0),\quad (4, 0)$
   D. Asymptotic to $y = \pm(x - 2)$
   E. ∞ at both intercepts

   (c) A. Symmetric about the x-axis
   B. $0 \le x < 4,\quad$ all y
   C. $(0, 0)$
   D. Asymptotic to $x = 4$
   E. ∞ at $(0, 0)$

2. (a) A. None
   B. All $y,\quad |x| > 0$
   C. $(-1, 0)$
   D. $y = x,\quad x = 0$
   E. 3

   (b) A. Symmetric about the x-axis
   B. All $y,\quad -1 \le x < 0,\quad x > 0$
   C. $(-1, 0)$
   D. $x = 0$
   E. ∞

   (c) A. None
   B. All $y,\quad |x| > 0$
   C. $(-1, 0)$
   D. $x = 0$
   E. -3

3. (a) A. None
   B. All $y,\quad$ all x
   C. $(0, 0),\quad (-1, 0),\quad (2, 0)$
   D. None
   E. -2 at $(0, 0)$, $\quad$ 3 at $(-1, 0)$, $\quad$ 6 at $(2, 0)$

   (b) A. Symmetric about the x-axis
   B. All $y,\quad -1 \le x \le 0,\quad x \ge 2$
   C. $(0, 0),\quad (-1, 0),\quad (2, 0)$
   D. None
   E. ∞ at all intercepts

4. (a) A. Symmetric about the y-axis
   B. All $x,\quad 0 < y \le 2$
   C. $(0, 2)$
   D. $y = 0$
   E. 0

   (b) A. Symmetric about the y-axis
   B. $y < 0,\quad y \ge 2,\quad x^2 \ne 4$
   C. $(0, 2)$
   D. $y = 0,\quad x = \pm 2$
   E. 0

   (c) A. Symmetric about the origin
   B. $|y| \le 2,\quad$ all x
   C. $(0, 0)$
   D. $y = 0$
   E. 2

5. (a) A. Symmetric about the origin
   B. $|y| \ge 2,\quad |x| > 0$
   C. None
   D. $y = x,\quad x = 0$
   E. ———

   (b) A. About $(1, 2)$
   B. $y \le 0,\quad y \ge 4,\quad x \ne 1$
   C. $(0, 0)$
   D. $x = 1,\quad y = x + 1$
   E. 0

   Remark: Translate axes to new origin at $(1, 2)$; becomes same as 5(a).

6. (a) A. Symmetric about both axes and the origin
   B. All y, $|x| \geq 1$ and $x = 0$
   C. $(0, 0)$, $(-1, 0)$, $(1, 0)$
   D. None
   E. ∞ at $(\pm 1, 0)$. No derivative exists at the isolated point $(0, 0)$.
   (b) A. Symmetric about the x-axis
   B. $x \leq 1$, $x > 2$, $y^2 \neq 1$
   C. $(0, \pm\sqrt{2}/2)$, $(1, 0)$
   D. $y = \pm 1$, $x = 2$
   E. $\mp\sqrt{2}/8$ at $(0, \pm\sqrt{2}/2)$, ∞ at $(1, 0)$
7. A. None
   B. $y \leq 4 - 2\sqrt{3}$, $y \geq 4 + 2\sqrt{3}$, $x^2 \neq 1$
   C. $(0, 8)$, $(2, 0)$
   D. $x = \pm 1$, $y = 0$
   E. -4 at $(0, 8)$, $\frac{4}{3}$ at $(2, 0)$
8. A. Symmetric about the y-axis
   B. $y \leq -1$, $y > 1$, $x^2 \neq 1$
   C. $(0, -1)$
   D. $x = \pm 1$, $y = 1$
   E. 0
9. A. Symmetric about the origin
   B. $|x| \leq 2$, $|y| \leq 2$
   C. $(0, \pm\sqrt{3})$, $(\pm\sqrt{3}, 0)$
   D. None
   E. $-\frac{1}{2}$ at $(0, \pm\sqrt{3})$, -2 at $(\pm\sqrt{3}, 0)$
10. $x^2 + xy + y^2 = 3k^2$
12. $a = c = d = e = 1$, $b = f = g = 0$
14. $y + 2x = 6$
16. $y = 6x - 14$, $y = 6x + 18$
18. (*b*) 1:1
21. $5x^2 + 5y^2 = 4x + 8y$
23. $x = 2pm$
24. $Q(4p^3/y_1^2, -4p^2/y_1)$
25. $(\pm\frac{8}{9}\sqrt{3}, \frac{4}{3})$
27. $(-2, 1)$, $(2, -1)$
28. $27x - 3y + 56 = 0$, $27x - 3y - 104 = 0$
30. $r = 2.56$, center $(4, 1.61)$. Results are approximate and are based on 1.55 as positive root of
$$x^4 + 4x - 12 = 0.$$
31. $(-\frac{16}{5}, \frac{53}{10})$
32. (a) A conic passing through the points of intersection of the three lines, and tangent to the lines $L_1 + hL_3 = 0$, $L_2 + kL_3 = 0$, $hL_2 + kL_1 = 0$
   (b) $3x^2 + 3y^2 - 8x - 16y + 20 = 0$
   (c) $2y = 3x^2 - 4x + 4$
33. 10^7 mi
34. $\pi/2$
35. $y^2 = 2x - 7$
36. $(5, 10\sqrt{10}/3)$
38. $3x^2 + 3y^2 - 16y + 16 = 0$, circle
40. Locus consists of circle $x^2 + y^2 = 9$ and hyperbola $y^2 - x^2 = 9$.
41. No. For points inside the circle, $x^2 + (y^2 - 9)$ is negative while $x^2 - (y^2 - 9)$ is positive, so their product cannot equal 1.
43. $5x^2 + 9y^2 - 30x = 0$
44. (a) $F_1P = 5$, $F_2P = \sqrt{5}$
   (b) Outside, since
$$OF_1 + OF_2 = 8 > F_1P + F_2P = 5 + \sqrt{5}$$
45. $2ab$
47. $x^2 + 4xy + 5y^2 = 1$
48. $(x + 3)^2/36 + (y - 1)^2/20 = 1$
50. $x^2 - y^2 = 4$
51. Center at $(\frac{1}{2}\sqrt{2}, 0)$, $e = \sqrt{2}$
52. *Hint.* Eliminate λ^2 and obtain the differential equation
$$xy(y')^2 + (x^2 - y^2 - c^2)y' - xy = 0$$
for both families. The product of the roots $y' = m_1$ and $y' = m_2$ is $m_1m_2 = -1$.
53. Square, area $= 2$
55. (a) Center $(2, -1)$, foci $((4 \pm 3\sqrt{3})/2, -1)$, major axis 6, minor axis 3, $e = \sqrt{3}/2$
   (b) Center $(-2, -\frac{3}{2})$, foci $(-2 \pm \sqrt{13}, -\frac{3}{2})$, major axis $= \sqrt{13}$, conjugate axis $= \sqrt{39}$, $e = 2$, asymptotes: $2y \pm 2\sqrt{3}x + (3 \pm 4\sqrt{3}) = 0$
57. (b) Bounded. Both $9x^2 + 4y^2 - 36$ and $4x^2 + 9y^2 - 36$ exceed one when $4(x^2 + y^2) > 37$
58. Foci $(\pm\sqrt{p - q}, 0)$
59. $\sqrt{2}$
60. On a branch of a hyperbola with the rifle and target at its foci
62. Center $(1, -2)$, foci $(1, -2 \pm \sqrt{13})$, asymptotes $2(y + 2) = \pm 3(x - 1)$
63. $9x^2 - y^2 = 9$, or $9y^2 - x^2 = 9$
65. (a) Center $(3, 2)$, $e = \sqrt{\frac{11}{12}}$ (b) $(3, 0)$

66. $x^2 + y^2 - 6x - 15y + 9 = 0$

67. With $A(0, 0)$ and $B(c, 0)$, locus is left branch of hyperbola
$$3x^2 - y^2 - 4cx + c^2 = 0.$$

68. $2y'^2 - 2a\sqrt{2}x' + a^2 = 0$

Chapter 11

Article 11.1

1. (a) $(3, \frac{1}{4}\pi + 2n\pi)$, $(-3, -\frac{3}{4}\pi + 2n\pi)$,
(b) $(-3, \frac{1}{4}\pi + 2n\pi)$, $(3, -\frac{3}{4}\pi + 2n\pi)$,
(c) $(3, -\frac{1}{4}\pi + 2n\pi)$, $(-3, \frac{3}{4}\pi + 2n\pi)$,
(d) $(-3, -\frac{1}{4}\pi + 2n\pi)$, $(3, \frac{3}{4}\pi + 2n\pi)$,
$n = 0, \pm 1, \pm 2, \ldots$

3. (a) $(3/\sqrt{2}, 3/\sqrt{2})$ (b) $(-3/\sqrt{2}, -3/\sqrt{2})$
(c) $(3/\sqrt{2}, -3/\sqrt{2})$ (d) $(-3/\sqrt{2}, 3/\sqrt{2})$

5. $(2, \frac{3}{4}\pi)$ is the same as $(-2, -\frac{1}{4}\pi)$, and $2 \sin(-\frac{1}{2}\pi) = -2$.

6. $(\frac{1}{2}, \frac{3}{2}\pi) = (-\frac{1}{2}, \frac{1}{2}\pi)$ and $-\sin \frac{1}{6}\pi = -\frac{1}{2}$

7. If (r_1, θ_1) satisfies $r_1 = \cos\theta_1 + 1$, then the same point has coordinates $(-r_1, \theta_1 + \pi)$ which satisfy the second equation. Also, if (r_2, θ_2) satisfies the second equation, then the coordinates $(-r_2, \theta_2 + \pi)$ of the same point satisfy the first equation.

8. $(\pm a, 15°)$, $(\pm a, 75°)$

9. The origin and $(a/\sqrt{2}, 45°)$

10. The origin and $(a + a/\sqrt{2}, -\frac{1}{4}\pi)$, $(a - a/\sqrt{2}, \frac{3}{4}\pi)$

11. The origin and $(\frac{8}{5}a, \sin^{-1}\frac{3}{5})$

12. The origin, $(a, \pm\frac{1}{2}\pi)$, $(\frac{1}{2}a, \pm\frac{2}{3}\pi)$, $(0.22a, \pm 141°.3)$

Article 11.2

1. $r(3\cos\theta + 4\sin\theta) = 5$ 2. $x^2 + y^2 = 2ax$

11. (a) $(\frac{3}{2}a, \pm 60°)$, $(0, 180°)$
(b) $(2a, 0°)$, $(\frac{1}{2}a, \pm 120°)$

12. $30°$

Article 11.3

1. $r = a \sin 2\theta$; a rose of 4 leaves

2. $r = 2a \sin\theta \tan\theta$;
$x = r\cos\theta = 2a\sin^2\theta \to 2a$ as $\theta \to \pi/2$

3. $r = 2a\sin\theta$

4. $r = a(1 - \cos\theta)$, $r = -a(1 + \cos\theta)$. (Both equations represent the same curve.)

5. $r\cos(\theta - \alpha) = p$

6. $r(1 - \cos\theta) = 2a$, $r(1 + \cos\theta) = -2a$. (Both equations represent the same curve.)

7. $r^2 = \cos 2\theta$ 8. $x^2 + y^2 = 4x$

9. $x^2 + y^2 = 6y$ 10. $(x^2 + y^2)^3 = 4x^2y^2$

11. $(x^2 + y^2)^2 = 2a^2(x^2 - y^2)$

12. $3x^2 - y^2 + 32x + 64 = 0$

13. $x^2 + y^2 - ay = a\sqrt{x^2 + y^2}$

15. (a) $p = |c|/\sqrt{a^2 + b^2}$, $\cos\beta = a/\sqrt{a^2 + b^2}$, $\sin\beta = b/\sqrt{a^2 + b^2}$
(b) $p = 3$, $\beta = 30°$

16. $r = 4/[1 - \sin(\theta - \frac{1}{6}\pi)]$

17. $(50, 0°)$, $r = 45/(4 + 5\cos\theta)$

18. $r = a(1 + \cos\theta)$

19. (a) $\max = \sqrt{a^2 + b^2}$, $\min = \sqrt{|a^2 - b^2|}$
(b) $\max = a\sqrt{5}$, $\min = a\sqrt{3}$
$\max = a\sqrt{2}$, $\min = 0$
$\max = a\sqrt{5}/2$, $\min = a\sqrt{3}/2$

20. ± 1 21. $(\frac{2}{3}, 0°)$ 22. $\pm\sqrt{e^2 - 1}$

Article 11.4

2. When $\tan\psi_2 = -1/\tan\psi_1$

3. $(2, \pm 60°)$, $90°$

4. $(\frac{3}{2}a, \pm 60°)$, $(0, 180°)$ 5. $8a$

6. $(\frac{4}{5}a, 0)$ 7. $4\pi a^2(2 - \sqrt{2})$

8. $2a$. The curve is the cardioid $r = \frac{1}{2}a(1 - \cos\theta)$.

9. $\frac{1}{3}a[(4 + \pi^2)^{3/2} - 8]$ 10. $\frac{1}{8}a(4\pi - 3\sqrt{3})$

11. $4\pi^2a^2$

Article 11.5

1. $\frac{3}{2}\pi a^2$ 2. πa^2 3. $2a^2$

4. $\frac{1}{3}a^2(3\sqrt{3} - \pi)$ 5. $\frac{9}{2}\pi a^2$

6. $\frac{1}{2}a^2(\pi - 2)$ 7. πa^2 9. $(\frac{5}{6}a, 0)$

10. $4a/3\pi$ from center

Review questions and exercises

6. (a) $e = \frac{7}{93} \approx 0.0754$
(b) $r = ke/(1 - e\cos\theta) \approx 4623/(1 - 0.0754\cos\theta)$

Miscellaneous exercises

15. $\pi/3$ at $(a, \pi/6)$ and $(a, 5\pi/6)$
16. $\pi/4$ at $(a, 0)$ and (a, π)
17. 0 at $(a\sqrt{2}, \pi/4)$
18. $\pi/2$ at $r = 0$
19. $\pi/4$ at $r = 0$, and $2\tan^{-1}(\sqrt{3}/6)$ at $(a/2, \theta_i)$, $\theta_i = \pm\pi/3, \pm 2\pi/3$
20. $\pi/3$ at $(\sqrt{2}, \pm\pi/6)$ and $(\sqrt{2}, \pm 5\pi/6)$
21. $r = 2/(1 + \cos\theta)$
22. $r(b\cos\theta + a\sin\theta) = ab$
23. $r = -2a\cos\theta$
24. $r[1 + \cos(\theta - \frac{1}{4}\pi)] = 2a$
25. $r = 8/(3 - \cos\theta)$
26. $r = 3/(1 + 2\sin\theta)$
27. $(0.75a, -3.44°)$, approximately
29. $\pi/4$
30. $\pi/2$
31. -1
32. $\pi/2$
33. (b) $xy = 1$
(c) $\pi/2$
34. (a) $(1/4)\tan\alpha$
(b) $\sec\alpha$
35. (a) $32\pi a^2/5$
(b) $\bar{y} = 0, \quad \bar{x} = 5a\int_0^\pi \sin^6\frac{\theta}{2}\cos\frac{\theta}{2}\cos\theta\,d\theta$
(c) $\bar{x} = -50a/63$
38. $P = 3\pi a/2$
39. $2a^2$
40. $(3\pi - 8)a^2/2$
41. a^2
42. $9a^2\pi/2$
43. $3\pi a^2/2$
44. πa^2
45. πa^2
46. $4a^2$
47. $a^2(\pi + 2)$
48. $5\pi a^2/4$

Chapter 12

Article 12.1

1. $\alpha_1 = \frac{1}{2}\sin^{-1}0.8 = 26°34'$
$\alpha_2 = 90° - \alpha_1 = 63°26'$
5. $x = e^t, \quad y = -(7 + e^{2t})/2$
6. $y = (1 - t)^{-1}, \quad x = -\ln|1 - t|$
7. $x = \sin t, \quad y = 1 + (t - \sin t\cos t)/2, \quad t \in [0, \pi/2]$
$x = 1, \quad y = t + 1 - \pi/4, \quad t > \pi/2$

Article 12.2

1. $x^2 + y^2 = 1$
2. $x = 1 - 2y^2, \quad |y| \le 1$
3. $x^2 - y^2 = 1, \quad x > 0$
4. $\dfrac{(x-2)^2}{16} + \dfrac{(y-3)^2}{4} = 1$
5. $y = x^2 - 6x$
6. $4x^2 - y^2 = 4, \quad x \ge 1, y \ge 0$
7. $(x - 2)(y - 2) + 1 = 0, \quad x > 2, \quad y < 2$
8. $y = x^2 - 2x + 5, \quad x \ge 1$
9. $2(x + y) = (x - y)^2$
10. $\dfrac{(x-3)^2}{4} + \dfrac{(y-4)^2}{9} = 1, \quad x > 3$
11. $x = \dfrac{-at}{\sqrt{t^2+1}}, \quad y = \dfrac{a}{\sqrt{t^2+1}}$
12. $x = a\tanh\theta, \quad y = a\,\text{sech}\,\theta$
13. $x = a\cos\dfrac{s}{a}, \quad y = a\sin\dfrac{s}{a}$
14. $x = a\sinh^{-1}\dfrac{s}{a}, \quad y = \sqrt{a^2 + s^2}$
15. $x = a\cos\phi + a\phi\sin\phi, \quad y = a\sin\phi - a\phi\cos\phi$
16. $x = (a + b)\cos\theta - b\cos\left(\dfrac{a+b}{b}\theta\right),$
$y = (a + b)\sin\theta - b\sin\left(\dfrac{a+b}{b}\theta\right)$
17. $x = (a - b)\cos\theta + b\cos\left(\dfrac{a-b}{b}\theta\right),$
$y = (a - b)\sin\theta - b\sin\left(\dfrac{a-b}{b}\theta\right)$
18. $8a$
21. (a) $x = 2a\cot\theta, \quad y = 2a\sin^2\theta$
(b) $y = \dfrac{8a^3}{x^2 + 4a^2}$

Article 12.3

1. $\mathbf{i} - 4\mathbf{j}$
2. $-\mathbf{i} + \mathbf{j}$
3. $-2\mathbf{i} - 3\mathbf{j}$
4. $\mathbf{0}$
5. $\frac{1}{2}\sqrt{3}\mathbf{i} + \frac{1}{2}\mathbf{j}$
6. $\frac{1}{2}\sqrt{3}\mathbf{i} - \frac{1}{2}\mathbf{j}$

7. $\frac{3}{5}\mathbf{i} - \frac{4}{5}\mathbf{j}$

8. $\dfrac{\mathbf{i} + 4\mathbf{j}}{\sqrt{17}}$

9. $(-4\mathbf{i} + \mathbf{j})/\sqrt{17}$

10. $\mathbf{i} \cos\theta + \mathbf{j} \sin\theta$

11. $\sqrt{2}$, 45°

12. $\sqrt{13}$, $\tan^{-1}(-1.5) = -56°.3$

13. 2, 30°

14. $\sqrt{13}$, $180° - \tan^{-1}(1.5) = 123°.7$

15. 13, $\tan^{-1}\frac{12}{5} = 67°.4$

16. 13, $180° + \tan^{-1}\frac{12}{5} = 247°.4$

Article 12.4

1. Straight line parallel to z-axis
2. Straight line 5 units above and parallel to the line $y = x$ in the xy-plane
3. Circle in the plane $z = -2$, center $(0, 0, -2)$, radius 2
4. Ellipse in the yz-plane, center $(0, 0, 0)$, semi-axes a and b
5. Circle in the plane $z = 3$, radius 2, center $(0, 0, 3)$.
6. Straight line in the plane $\theta = \pi/6$, making an angle of 45° with the plane $z = 0$.
7. A right circular helix (Fig. 14.8) wound on a cylinder of radius 3
8. A right conical helix wound on a right circular cone
9. A semicircle of radius 5, center at the origin, lying in the plane $\theta = \pi/4$
10. The intersection of the sphere $\rho = 5$ and the cone $\phi = \pi/4$ is a circle of radius $\rho \sin\phi = 5/\sqrt{2}$.
11. The plane $\theta = \pi/4$ and the cone $\phi = \pi/4$ intersect in a straight line through the origin.
12. A semicircle in the yz-plane, center $(0, 0, 2)$ on the z-axis, radius 2.
13. $\rho = 2$, $r^2 + z^2 = 4$
14. $\rho = 4\cos\phi$, $r^2 + z^2 = 4z$
15. $z^2 = x^2 + y^2$, $\phi = \pi/4$, $\phi = 3\pi/4$
16. $x^2 + y^2 + z^2 = 6z$, $r^2 + z^2 = 6z$
17. Plane $x = 0$ and half-space where $x > 0$
18. Spheres $\rho = 3$, $\rho = 5$, and shell between them
19. On or inside sphere $\rho = 5$ but outside or on cylinder $r = 2$
20. Wedge with two intersecting plane boundaries $\theta = 0$, $\theta = \pi/4$, and curved boundary part of cone $\phi = \pi/4$
21. Interior and boundary of elliptic cylinder

$$4x^2 + 9y^2 = 36$$

Article 12.5

1. $(-2, 0, 2)$, $\sqrt{8}$

2. $(-\frac{1}{4}, -\frac{1}{4}, -\frac{1}{4})$, $\sqrt{75}/4$

3. $(0, 0, a)$, $|a|$

4. $(0, -\frac{1}{3}, \frac{1}{3})$, $\sqrt{29}/3$

5. (a) $\sqrt{y^2 + z^2}$ (b) $\sqrt{z^2 + x^2}$
   (c) $\sqrt{x^2 + y^2}$ (d) $|z|$

6. (a) $x^2 + y^2 + (z - 4)^2 = 4$
   (b) $4x^2 + 4y^2 + 3(z - \frac{3}{2})^2 = 27$
   (c) $5(z - \frac{3}{2})^2 - 4x^2 - 4y^2 = 5$

7. 3 8. 7 9. 9 10. 11

11. $(4\mathbf{i} + 3\mathbf{j} + 12\mathbf{k})/13$

12. $(2\mathbf{i} + 2\mathbf{j} + 4\mathbf{k})/3$

Article 12.6

1. No, we cannot conclude that $\mathbf{B}_1 = \mathbf{B}_2$. All we can say is that $\mathbf{B}_1$ and $\mathbf{B}_2$ have the same projection on $\mathbf{A}$ when all three vectors start from the same initial point.

2. (a) $\dfrac{\mathbf{A} \cdot \mathbf{B}}{\mathbf{A} \cdot \mathbf{A}}\mathbf{A}$ (b) $\frac{7}{45}\mathbf{A}$

3. $\angle A = \angle C = 71°.1$, $\angle B = 37°.8$

4. $(3, 3, 0)$

5. $\frac{13}{15}$

6. $\cos^{-1}(1/\sqrt{3}) = 54°.7$

7. $\cos^{-1}(\sqrt{6}/3) = 35°.3$

8. $\cos^{-1}(\frac{13}{45}) = 73°.2$

9. Two, 45°, 135°

13. $w(z_1 - z_2)$

16. $c_1 = -\frac{14}{25}$, $c_2\mathbf{B}' = \frac{23}{25}(4\mathbf{i} + 3\mathbf{j})$

Article 12.7

1. $-\mathbf{i} - 3\mathbf{j} + 4\mathbf{k}$

2. $c(2\mathbf{i} + \mathbf{j} + \mathbf{k})$, c = scalar

3. $2\sqrt{6}$

4. $\sqrt{6}/2$

5. $c(\mathbf{i} - \mathbf{j})$, c = scalar

6. $\pm[4/(3\sqrt{2}), 1/(3\sqrt{2}), 1/(3\sqrt{2})]$

7. $11/\sqrt{107}$

8. (a) $60°.6$

(b) Yes, because the angle between the two planes is equal to the angle between their normals, and this angle is neither $0°$ nor $180°$.

(c) $\mathbf{j} + \mathbf{k}$

Article 12.8

1. $(9, -5, 12)$
2. $x = 1 + t, \quad y = 2 + t, \quad z = -1 - t,$
$x - 1 = y - 2 = -(z + 1)$
4. (a) The complement of the angle between the line and a normal to the plane
(b) $22°.4$
5. $3x + y + z = 5$
6. $|D_2 - D_1|/7$
8. $7x - 5y - 4z = 6$
9. Plane through the three points
10. $x - 2y + z = 6$
11. $x - y + z = 0$
12. $x + 6y + z = 16$
13. 3
14. (b) Yes
16. $17x - 26y + 11z = -40$
17. $2x - y + z = 5$
18. $1/\sqrt{75}, 5/\sqrt{75}, 7/\sqrt{75}$
19. Locus is the set of points in the half-space that lies on the side of the plane toward which **N** points.
20. $\cos\alpha(x - x_0) + \cos\beta(y - y_0) + \cos\gamma(z - z_0) = 0$

Article 12.9

1. $-2\mathbf{C} = -6\mathbf{i} + 8\mathbf{j} - 24\mathbf{k},$
$-22\mathbf{A} = -88\mathbf{i} + 176\mathbf{j} - 22\mathbf{k}$
2. 245
3. (a) $\mathbf{A} \times \mathbf{B} = 15\mathbf{i} + 10\mathbf{j} + 20\mathbf{k},$
$(\mathbf{A} \times \mathbf{B}) \times \mathbf{C} = 200\mathbf{i} - 120\mathbf{j} - 90\mathbf{k}$
(b) $(\mathbf{A} \cdot \mathbf{C})\mathbf{B} - (\mathbf{B} \cdot \mathbf{C})\mathbf{A} = 56\mathbf{B} + 22\mathbf{A}$
$= 200\mathbf{i} - 120\mathbf{j} - 90\mathbf{k}$
5. $a = (\mathbf{A} \times \mathbf{B}) \cdot \mathbf{D}, \quad b = -(\mathbf{A} \times \mathbf{B}) \cdot \mathbf{C}$
6. $\frac{2}{3}$
10. $7 - x = \dfrac{3y - 19}{12} = \dfrac{4 - 3z}{30}$
11. (a) $S(0, 9, -3)$
(b) Area $PQRS = |-29\mathbf{i} - \mathbf{j} + 11\mathbf{k}|$
$= \sqrt{963} \approx 31.03$
(c) 11, 29, 1

Article 12.10

1. Circular cylinder of radius a, axis along Oz
2. Hyperbolic cylinder, axis along Oy
3. Circular cylinder of radius a, one element being the z-axis
4. Cylinder with elements parallel to Oz. Cross sections are cardioids.
5. Circular cylinder of radius 2, one element being the x-axis
6. Elliptic cylinder, one element being the y-axis

Article 12.11

1. Paraboloid of revolution, vertex $(-2, 3, -13)$, and opening upward
2. Sphere of radius 4, center $(-2, 3, 0)$
3. Ellipsoid, center $(0, 1, 0)$, $a = 2, b = 1, c = 2$
4. Ellipsoid, center $(0, 1, 0)$, $a = 2, b = c = 1$
5. Sphere, center $(0, 1, 0)$, radius 1
6. One-sheeted hyperboloid of revolution, axis of symmetry parallel to Oy, center $(-2, -3, 0)$
7. Two-sheeted hyperboloid of revolution, axis parallel to Ox, center $(-2, -3, 0)$
8. Parabolic cylinder, one element being Oy
9. Rotate the xy-axes through $45°$ and have
$$z^2 + 2y'^2 = 2x'^2.$$
Elliptic cone with vertex O, axis along Ox'
10. Rotate xy-axes $45°$; $z = 2(x'^2 - y'^2)$, hyperbolic paraboloid ("saddle")
11. Paraboloid of revolution obtained by rotating the parabola $z = x^2$ about the z-axis
12. Rotate the line $z = x$ about the z-axis, right circular cone
13. Rotate $z^2 = x$ about the z-axis.
14. Elliptic cone, vertex O, axis Oz
15. Elliptic cone, vertex O, axis Ox
16. Right circular cone, vertex O, axis Oy
17. $(x - 1)^2 + 4(y + 1)^2 - (z - 2)^2 = 1,$
one-sheeted hyperboloid, center $(1, -1, 2)$,
axis $x - 1 = y + 1 = 0$
18. $(z - 2)^2 = (x - 1)^2 + 4(y + 1)^2$. Elliptic cone, vertex $(1, -1, 2)$, axis $x - 1 = y + 1 = 0$

19. Elliptic cylinder, axis along Oy
20. Elliptic paraboloid, vertex $(1, 0, 0)$, axis along Ox
21. Plane, $z = x$ 22. Plane, $z = y$
23. Ruled surface generated by rotating about Oz a line parallel to the xy-plane and passing through Oz, line's distance above the xy-plane is $z = \sin\theta$.
24. Ruled surface generated by rotating about Oz a line parallel to the xy-plane and passing through Oz, line's distance above the xy-plane is $z = \cosh\theta$.
25. (a) $A(z_1) = \pi ab(1 - z_1^2/c^2)$
(b) $V = \frac{4}{3}\pi abc$
27. (a) $\pi abh(1 + h^2/3c^2)$
(b) $A_0 = \pi ab$, $A_h = \pi ab(1 + h^2/c^2)$, $V = \frac{1}{3}h(2A_0 + A_h)$
28. Vertex $(0, y_1, cy_1^2/b^2)$, focus $\left(0, y_1, \frac{cy_1^2}{b^2} - \frac{a^2}{4c}\right)$
29. In any plane θ = constant, the equation $\rho = F(\phi)$ may be considered as the polar equation of a plane curve C having polar coordinates ρ and ϕ. Since the space equation is independent of θ, the surface is generated by rotating the curve C around the z-axis.
30. A sphere, center $\rho = a/2$, $\phi = 0$, $\theta = 0$ and radius $a/2$
31. A cardioid of revolution

Miscellaneous exercises

1. $x = (1 - t)^{-1}$, $y = 1 - \ln(1 - t)$, $t < 1$
2. $x = \tan^{-1}(t + 1)$,
$y = (t + 1)\tan^{-1}(t + 1) - \frac{1}{4}\pi - \frac{1}{2}\ln(1 + t + \frac{1}{2}t^2)$
3. $x = e^t$, $y = e^{e^t} - e$
4. $x = 3 - 3\cos 2t$, $y = 4 + 2\sin 2t$
5. $x = t - \sin t$, $y = 1 - \cos t$
6. $y = e^t$, $x = 2\sqrt{1 + e^t} - 2\sqrt{2} - 2\coth^{-1}\sqrt{1 + e^t} + 2\coth^{-1}\sqrt{2}$
7. $x = \sinh^{-1} t$, $y = t\sinh^{-1} t - \sqrt{1 + t^2} + 1$
8. $x = 2 + 2\sinh(t/2)$, $y = 2t - 4 + 4\cosh(t/2)$
9. $x = 4\sin t$, $y = 4\cos t$
10. $x = 2 - e^{-t}$, $y = 2t + e^{-t}$
12. $x = 8t - 2\sin 2t$, $y = 4 - 2\cos 2t$
13. $x = 2a\sin^2\theta\tan\theta$, $y = 2a\sin^2\theta$
14. $x = \cot\theta \pm a\cos\theta$, $y = \pm a\sin\theta$
15. (a) $3\pi a^2$ (b) $8a$
(c) $64\pi a^2/3$ (d) πa, $5a/6$
17. 0
24. $\frac{3}{11}(3\mathbf{i} - \mathbf{j} + \mathbf{k})$ 25. $-\frac{1}{10}$
26. $\mathbf{C} = \dfrac{\mathbf{A}\cdot\mathbf{B}}{\mathbf{B}\cdot\mathbf{B}}\mathbf{B}$, $\mathbf{D} = \dfrac{(\mathbf{B}\cdot\mathbf{B})\mathbf{A} - (\mathbf{A}\cdot\mathbf{B})\mathbf{B}}{\mathbf{B}\cdot\mathbf{B}}$
27. $7\sqrt{2}/10$ 29. $\mathbf{j} - \mathbf{k}$
30. $5\mathbf{i} + 7\mathbf{j} + \mathbf{k}$ 31. $(\mathbf{j} + \mathbf{k})/\sqrt{2}$
33. $(10\mathbf{i} - 2\mathbf{j} - 6\mathbf{k})/\sqrt{35}$ 35. $\cos^{-1}(3/\sqrt{35})$
36. $\mathbf{B} = \dfrac{1}{|\mathbf{A}|^2}(d\mathbf{A} + \mathbf{C}\times\mathbf{A})$
38. $z = 3$, $x - \sqrt{3}\,y + 2\sqrt{3} - 1 = 0$
39. $(1, -2, -1)$, $\dfrac{x - 1}{-5} = \dfrac{y + 2}{3} = \dfrac{z + 1}{4}$
40. $25/\sqrt{38}$ 42. $(\frac{11}{9}, \frac{26}{9}, -\frac{7}{9})$
43. $3x + 3y + 3z - 8 = 0$
44. $2x - y + 2z - 8 = 0$
45. (a) $2x + 7y + 2z + 10 = 0$ (b) $9/5\sqrt{57}$
46. $2x + 2y + z = 5$
49. (a) 1 (b) $-10\mathbf{i} - 2\mathbf{j} - 12\mathbf{k}$
50. $\frac{1}{3}$, $\cos\theta = -\sqrt{2}/3$
56. Hyperboloid of two sheets, center $(2, -1, -1)$
57. $(x - 1)^2 + (y - 2)^2 + (z - 3)^2 = 3$
58. $z^6 - y = 0$, $y = (x^2 + z^2)^3$

Chapter 13

Article 13.1

5. $|\mathbf{A}| = \sqrt{6}$, $|\mathbf{B}| = \sqrt{3}$, $\mathbf{A}\cdot\mathbf{B} = 0$, $\cos\theta = 0$, $\text{proj}_{\mathbf{B}}(\mathbf{A}) = \mathbf{0}$.
The component of $\mathbf{A} \perp \mathbf{B} = \mathbf{A}$ agrees with geometric intuition.
6. $\text{Proj}_{\mathbf{C}}(\mathbf{B}) = (0, 0, \frac{1}{2}, \frac{1}{2})$,
component of $\mathbf{B} \perp \mathbf{C} = (1, -1, -\frac{1}{2}, \frac{1}{2})$
7. $\text{Proj}_{\mathbf{C}}(\mathbf{A}) = (0, 0, 1, 1)$, $\text{proj}_{\mathbf{C}}(\mathbf{B}) = (0, 0, \frac{1}{2}, \frac{1}{2})$, $\text{proj}_{\mathbf{C}}(\mathbf{A} + \mathbf{B}) = (0, 0, \frac{3}{2}, \frac{3}{2}) = \text{proj}_{\mathbf{C}}(\mathbf{A}) + \text{proj}_{\mathbf{C}}(\mathbf{B})$.
Yes.
9. $f_1(x) = \frac{1}{2}$, $f_2(x) = x - \frac{1}{2}$
10. $f_1(x) = \frac{3}{4}x$, $f_2(x) = x^2 - \frac{3}{4}x$
11. $f_1(x) = \frac{5}{6}x^2$, $f_2(x) = x^3 - \frac{5}{6}x^2$

12. $(a, b, -c)$. Reflection in the xy-plane changes the sign of z-component.
13. $(a, -b, c)$. Reflection in the xz-plane changes the sign of the y-component.
14. $(-a, b, c)$. Reflection in the yz-plane changes the sign of the x-component.
15. (b, a, c). Reflection in the plane $y = x$ interchanges the x- and y-components.

Article 13.2

1. $\begin{bmatrix} 1 & -1 \\ -1 & 2 \end{bmatrix}$ 2. $\begin{bmatrix} -1 & 2 \\ 1 & -1 \end{bmatrix}$ 3. $\begin{bmatrix} -1 & 2 \\ 2 & -3 \end{bmatrix}$

4. Same as number 3. 10. $\det(c\mathbf{A}) = c^n \det(\mathbf{A})$.

Article 13.3

1. $\mathbf{E} = \begin{bmatrix} 1 & 0 & \frac{3}{5} \\ 0 & 1 & \frac{6}{5} \end{bmatrix}$, $\mathbf{P} = \begin{bmatrix} \frac{1}{5} & \frac{2}{5} \\ \frac{2}{5} & -\frac{1}{5} \end{bmatrix}$

2. $\mathbf{E} = \begin{bmatrix} 1 & 0 \\ 0 & 1 \end{bmatrix}$, $\mathbf{P} = \begin{bmatrix} \frac{2}{3} & -\frac{1}{3} \\ -\frac{1}{3} & \frac{2}{3} \end{bmatrix}$

3. $\mathbf{E} = \begin{bmatrix} 1 & 0 \\ 0 & 1 \end{bmatrix}$, $\mathbf{P} = \begin{bmatrix} -1 & 1 \\ 1 & 0 \end{bmatrix}$

4. $\mathbf{E} = \begin{bmatrix} 1 & 0 & 1 \\ 0 & 1 & 0 \\ 0 & 0 & 0 \end{bmatrix}$, $\mathbf{P} = \begin{bmatrix} 1 & 0 & 0 \\ 0 & 1 & 0 \\ -1 & 0 & 1 \end{bmatrix}$

5. $\mathbf{E} = \begin{bmatrix} 1 & 0 & 0 \\ 0 & 1 & 0 \\ 0 & 0 & 1 \end{bmatrix}$, $\mathbf{P} = \begin{bmatrix} 1 & -a & a^2 - b \\ 0 & 1 & -a \\ 0 & 0 & 1 \end{bmatrix}$

6. $\mathbf{E} = \begin{bmatrix} 1 & 0 & 0 \\ 0 & 1 & 0 \\ 0 & 0 & 1 \end{bmatrix}$, $\mathbf{P} = \begin{bmatrix} 1 & 0 & 0 \\ \frac{1}{2} & \frac{1}{2} & 0 \\ -\frac{5}{2} & -\frac{1}{2} & 0 \end{bmatrix}$

7. The element in the ith row and jth column of $\mathbf{IC}$ is
$$\sum_{k=1}^{m} \delta_{ik}c_{kj} = \delta_{ii}c_{ij} = c_{ij} \quad \text{for } 1 \leq i \leq m, 1 \leq j \leq p.$$

8. $\mathbf{A}(\mathbf{BC}) = (\mathbf{AB})\mathbf{C} = \begin{bmatrix} 6 & 0 & -6 & -12 \\ 1 & -1 & 2 & -2 \end{bmatrix}$

9. $\mathbf{BA} = \begin{bmatrix} -2 & -1 & 0 \\ 5 & 0 & -1 \\ 0 & 5 & 2 \end{bmatrix}$. There are three rows and three columns. No, $(\mathbf{BA})\mathbf{C} \neq \mathbf{B}(\mathbf{AC})$, because $\mathbf{C}$ is a 2×4 matrix and neither $\mathbf{AC}$ nor $(\mathbf{BA})\mathbf{C}$ exists.

10. Augmented matrix: $\left[\begin{array}{ccc|c} 1 & -2 & 3 & 4 \\ 2 & 1 & -4 & -1 \\ 0 & 1 & 2 & 3 \end{array}\right]$

Echelon form: $\left[\begin{array}{ccc|c} 1 & 0 & 0 & \frac{8}{5} \\ 0 & 1 & 0 & \frac{3}{5} \\ 0 & 0 & 1 & \frac{6}{5} \end{array}\right]$, $x = \frac{8}{5}$, $y = \frac{3}{5}$, $z = \frac{6}{5}$

12. $x = \frac{4}{5}$, $y = \frac{3}{5}$, $z = -\frac{13}{5}$

13. $\mathbf{A}^{-1} = \begin{bmatrix} 0 & \frac{1}{5} & \frac{1}{5} \\ -1 & -1 & 1 \\ -1 & -\frac{2}{5} & \frac{3}{5} \end{bmatrix}$, $\mathbf{X} = \mathbf{A}^{-1}\mathbf{B} = \begin{bmatrix} \frac{8}{5} \\ 5 \\ \frac{9}{5} \end{bmatrix}$

15. $\operatorname{adj} \mathbf{A} = \begin{bmatrix} 4 & 3 & 5 \\ -2 & 1 & 5 \\ 2 & -1 & 5 \end{bmatrix}$, $\mathbf{A}^{-1} = \frac{1}{10} \operatorname{adj} \mathbf{A}$.

Article 13.4

1. $x\mathbf{V}_1 + y\mathbf{V}_2 + z\mathbf{V}_3$, with
$(x, y, z) = \frac{1}{5}(-3 - 2t, 1 - t, 5t)$, $t \in (-\infty, \infty)$
2. $\mathbf{V}_1 = \frac{1}{2}\mathbf{W}_1 - \frac{1}{2}\mathbf{W}_2$, $\mathbf{V}_2 = \frac{1}{2}\mathbf{W}_1 + \frac{1}{2}\mathbf{W}_2$
3. $\mathbf{V}_1 = \frac{1}{2}\mathbf{W}_1 = (\frac{1}{2}, \frac{1}{2}, \frac{1}{2}, \frac{1}{2})$, $\mathbf{V}_2 = \frac{1}{6}\sqrt{3}(1, 1, 1, -3)$
4. $\mathbf{X} = \frac{1}{5}(s - 2t, 2s + t, 5s, 5t)$, s, t real
5. In the answer to Exercise 4, take $s = 0$, $t = 1$ for one new vector and take $s = 1$, $t = 0$ for another. Basis $= \{\mathbf{A}, \mathbf{B}, \mathbf{V}_1, \mathbf{V}_2\}$, with
$$\mathbf{V}_1 = (-\tfrac{2}{5}, \tfrac{1}{5}, 0, 1), \quad \mathbf{V}_2 = (\tfrac{1}{5}, \tfrac{2}{5}, 1, 0)$$
orthogonal to each other and to $\mathbf{A}$ and $\mathbf{B}$.
7. Linearly independent 8. Linearly dependent
9. Linearly dependent 10. Linearly independent
13. $\mathbf{V}_1 = \mathbf{W}_1$, $\mathbf{V}_2 = \frac{1}{3}\mathbf{W}_1 + \frac{1}{3}\mathbf{W}_2$,
$\mathbf{V}_3 = \mathbf{W}_1 + \frac{2}{11}\mathbf{W}_2 + \frac{1}{11}\mathbf{W}_3$

15. (b) $\mathbf{Y} = s(-\frac{1}{2}, \frac{3}{2}, 1, 0) + t(0, -2, 0, 1)$, $s \in (-\infty, \infty)$, $t \in (-\infty, \infty)$.

16. (1, 1), (0, 1), or (1, 0), (0, 1), or any two linearly independent vectors in E^2

17. (1, −1, 2), (0, 1, −2)

18. (−1, 1, 0, 2), (1, 0, 1, −2)

19. (0, −1, 0, 1), (1, 0, 2, 0)

20. (0, 1), (1, 0)

21. $\begin{bmatrix}1\\0\end{bmatrix}, \begin{bmatrix}1\\1\end{bmatrix}$ 22. $\begin{bmatrix}1\\0\end{bmatrix}, \begin{bmatrix}-1\\1\end{bmatrix}$ 23. $\begin{bmatrix}1\\0\end{bmatrix}, \begin{bmatrix}0\\1\end{bmatrix}$

24. $\begin{bmatrix}0\\1\\1\end{bmatrix}, \begin{bmatrix}1\\0\\1\end{bmatrix}$ 25. $\begin{bmatrix}1\\2\\0\\1\end{bmatrix}, \begin{bmatrix}-2\\1\\1\\0\end{bmatrix}$

26. (0, 2, 1)

27. (−1, 0, −2, 1), (−2, 1, −5, 0)

28. (a) **0** (b) Space spanned by (−1, −1, 1)

29. (a) They are linearly independent, by Theorem 5, and there are n of them. (See Remark 6.)
(b) $x_j = (\mathbf{B} \cdot \mathbf{V}_j)/(\mathbf{V}_j \cdot \mathbf{V}_j)$, $j = 1, 2, \ldots, n$

30. $$\mathbf{B} = \frac{a+2c}{5}\mathbf{V}_1 + \frac{-2a+c}{5}\mathbf{V}_2 + \frac{b+4d}{17}\mathbf{V}_3 + \frac{-4b+d}{17}\mathbf{V}_4$$

31. (a) $\mathbf{V}_1 = (1, 0, -1, 1)$, $\mathbf{V}_2'' = (\frac{1}{3}, 1, \frac{2}{3}, \frac{1}{3})$,
$\mathbf{W}_1 = (1, 0, 0, -1)$, $\mathbf{W}_2'' = (-\frac{1}{2}, 1, -1, -\frac{1}{2})$
(b) $x = a + \frac{1}{3}b$, $y = b$, $z = c - \frac{1}{2}d$, $u = d$
(c) $a = \frac{1}{3}$, $b = \frac{1}{5}$, $c = 2$, $d = \frac{4}{5}$,
$x = \frac{2}{5}$, $y = \frac{1}{5}$, $z = \frac{8}{5}$, $u = \frac{4}{5}$

Article 13.5

1. (a) $\begin{bmatrix}0\\3\end{bmatrix}$ (b) $\begin{bmatrix}3\\-5\end{bmatrix}$

3. (b) $\mathbf{A} = \begin{bmatrix}-1 & 1\\1 & 0\\1 & 0\end{bmatrix}$ 4. $\mathbf{T} = \begin{bmatrix}1 & -3\\-1 & 3\\1 & -1\end{bmatrix}$

Miscellaneous exercises

1. (a) $\begin{bmatrix}2 & 1\\1 & 0\\0 & -2\end{bmatrix}$ (b) $\begin{bmatrix}0\\1\end{bmatrix}$

(c) $|\mathbf{B}|^2 = 1 = \mathbf{B}^T\mathbf{B}$,

$$\mathbf{A}^T\mathbf{A} = \begin{bmatrix}5 & 2 & -2\\2 & 1 & 0\\-2 & 0 & 4\end{bmatrix}, \quad (\mathbf{A}^T\mathbf{A})\mathbf{X} = \begin{bmatrix}-1\\0\\2\end{bmatrix}$$

2. $$\mathbf{B}^T\mathbf{A}^T = \left[\sum_{k=1}^{n} b_{ki}a_{jk}\right] = \left[\sum_{k=1}^{n} a_{jk}b_{ki}\right] = (\mathbf{AB})^T.$$

The sums inside the brackets are the elements in the ith row and jth column.

3. $$\mathbf{A}^T\mathbf{A} = \mathbf{I} \Leftrightarrow \sum_{k=1}^{n} a_{ki}a_{kj} = \delta_{ij} = \begin{cases}0 & \text{if } i \neq j,\\ 1 & \text{if } i = j,\end{cases}$$
$$\mathbf{A}\mathbf{A}^T = \mathbf{I} \Leftrightarrow \sum_{k=1}^{n} a_{ik}a_{jk} = \delta_{ij}.$$

The first sum is the inner product of the ith and jth columns of **A**, and the second sum is the inner product of the ith and jth rows of **A**.

5. Let $\mathbf{B} = \mathbf{A}^T$. Then $\mathbf{B}^T = \mathbf{A}$, and
$$\mathbf{B}\mathbf{B}^T = \mathbf{A}^T\mathbf{A} = \mathbf{I}$$
and
$$\mathbf{B}^T\mathbf{B} = \mathbf{A}\mathbf{A}^T = \mathbf{I}$$
if **A** is orthogonal.

6. $$|T(\mathbf{X})|^2 = [T(\mathbf{X})]^T[T(\mathbf{X})] = \mathbf{X}^T(\mathbf{A}^T\mathbf{A})\,\mathbf{X} = \mathbf{X}^T\mathbf{I}\mathbf{X} = \mathbf{X}^T\mathbf{X} = |\mathbf{X}|^2$$

7. Let **A** and **B** be orthogonal matrices, and let $\mathbf{C} = \mathbf{AB}$. Then
$$\mathbf{C}^T\mathbf{C} = \mathbf{B}^T\mathbf{A}^T\mathbf{A}\mathbf{B} = \mathbf{B}^T\mathbf{I}\mathbf{B} = \mathbf{B}^T\mathbf{B} = \mathbf{I},$$
$$\mathbf{C}\mathbf{C}^T = \mathbf{A}\mathbf{B}\mathbf{B}^T\mathbf{A}^T = \mathbf{A}\mathbf{I}\mathbf{A}^T = \mathbf{A}\mathbf{A}^T = \mathbf{I},$$
so **C** is also orthogonal.

9. $\begin{bmatrix}2 & -2\\-2 & 0\end{bmatrix}$ 10. $\begin{bmatrix}1 & 1\\1 & -1\end{bmatrix}$ 11. $\begin{bmatrix}1 & 0\\0 & 1\end{bmatrix}$

12. $\begin{bmatrix}1 & 0\\0 & -4\end{bmatrix}$ 13. $\begin{bmatrix}0 & 1\\1 & 0\end{bmatrix}$ 14. $\begin{bmatrix}2 & -\frac{3}{2}\\-\frac{3}{2} & 5\end{bmatrix}$

15. $$\mathbf{Q} = \mathbf{X}^T\mathbf{A}\mathbf{X} = (\mathbf{P}\mathbf{X}')^T\mathbf{A}(\mathbf{P}\mathbf{X}') = (\mathbf{X}')^T\mathbf{P}^T\mathbf{A}\mathbf{P}(\mathbf{X}') = (\mathbf{X}')^T\mathbf{B}(\mathbf{X}'),$$
with $\mathbf{B} = \mathbf{P}^T\mathbf{A}\mathbf{P}$

16. Yes. $\mathbf{P}$ is orthogonal for any choice of θ, and $\mathbf{B} = \mathbf{P}^T\mathbf{AP}$ is diagonal if and only if

$$b \cos 2\theta = [(c - a)/2] \sin 2\theta.$$

If $b = 0$, let $\theta = 0$. If $b \neq 0$, let

$$\theta = \cot^{-1} [(c - a)/2b].$$

The matrix $\mathbf{A}$ is as in Miscellaneous Exercise 8.

17.

$$\mathbf{P}^T\mathbf{AP} = \begin{bmatrix} \mathbf{P}_1^T \\ \mathbf{P}_2^T \\ \mathbf{P}_3^T \end{bmatrix} \times \mathbf{A}[\mathbf{P}_1 \quad \mathbf{P}_2 \quad \mathbf{P}_3]$$

$$= \begin{bmatrix} \mathbf{P}_1^T \\ \mathbf{P}_2^T \\ \mathbf{P}_3^T \end{bmatrix} \times [\mathbf{AP}_1 \quad \mathbf{AP}_2 \quad \mathbf{AP}_3]$$

$$= \begin{bmatrix} \mathbf{P}_1^T \\ \mathbf{P}_2^T \\ \mathbf{P}_3^T \end{bmatrix} [d_1\mathbf{P}_1 \quad d_2\mathbf{P}_2 \quad d_3\mathbf{P}_3]$$

$$= \begin{bmatrix} d_1\mathbf{P}_1^T\mathbf{P}_1 & d_2\mathbf{P}_1^T\mathbf{P}_2 & d_3\,\mathbf{P}_1^T\mathbf{P}_3 \\ d_1\mathbf{P}_2^T\mathbf{P}_1 & d_2\mathbf{P}_2^T\mathbf{P}_2 & d_3\mathbf{P}_2^T\mathbf{P}_3 \\ d_1\mathbf{P}_3^T\mathbf{P}_1 & d_2\mathbf{P}_3^T\mathbf{P}_2 & d_3\mathbf{P}_3^T\mathbf{P}_3 \end{bmatrix}$$

$$= \begin{bmatrix} d_1 & 0 & 0 \\ 0 & d_2 & 0 \\ 0 & 0 & d_3 \end{bmatrix}$$

18. $[\mathbf{X} \neq \mathbf{0}, \mathbf{AX} = d\mathbf{X}] \Rightarrow [(\mathbf{A} - d\mathbf{I})\mathbf{X} = \mathbf{0}, \mathbf{X} \neq 0]$
$\Rightarrow \det(\mathbf{A} - d\mathbf{I}) = 0$

19. $d_1 = 6$; $\mathbf{X}_1 = s(3, 1)$, $s \neq 0$, $s \in (-\infty, \infty)$,
$d_2 = -4$, $\mathbf{X}_2 = t(1, -3)$, $t \in (-\infty, \infty)$, $t \neq 0$

20. (a) $\mathbf{P}^T = \mathbf{P}$ and $\mathbf{P}^2 = \mathbf{I}$. $\mathbf{P}^T\mathbf{AP} = \begin{bmatrix} 6 & 0 \\ 0 & -4 \end{bmatrix}$

(b) $\mathbf{X} = \begin{bmatrix} x \\ y \end{bmatrix} = \mathbf{PX}' = \dfrac{1}{\sqrt{10}} \begin{bmatrix} 3 & 1 \\ 1 & -3 \end{bmatrix} \begin{bmatrix} x' \\ y' \end{bmatrix}$,

or

$$x = \frac{3x' + y'}{\sqrt{10}}, \quad y = \frac{x' - 3y'}{\sqrt{10}}, \quad d_1 = 6, \quad d_2 = -4$$

(c) Hyperbolas for $k \neq 0$, two straight lines if $k = 0$

21.

$$D^m = \begin{bmatrix} k_1^m & 0 & \cdots & 0 \\ 0 & k_2^m & \cdots & 0 \\ \vdots & & & \\ 0 & 0 & & k_n^m \end{bmatrix}$$

for each positive integer m.

22. Proof by induction on r: true for $r = 1$. Assume true for $r - 1$, then

$$\begin{aligned} \mathbf{A}^r = \mathbf{A}^{r-1}\mathbf{A} &= [\mathbf{P}(\mathbf{D}^{r-1})\mathbf{P}^{-1}]\mathbf{PDP}^{-1} \\ &= \mathbf{PD}^{r-1}(\mathbf{P}^{-1}\mathbf{P})\mathbf{DP}^{-1} \\ &= \mathbf{PD}^{r-1}\mathbf{IDP}^{-1} = \mathbf{PD}^r\mathbf{P}^{-1}. \end{aligned}$$

Same method for the other half.

23. Combine the results of Exercises 21 and 22:

$$\mathbf{A}^r = \mathbf{PD}^r\mathbf{P}^{-1}.$$

For the matrices $\mathbf{A}$ and $\mathbf{P}$ given,

$$\mathbf{D}^3 = \begin{bmatrix} 6^3 & 0 \\ 0 & (-4)^3 \end{bmatrix} = \begin{bmatrix} 216 & 0 \\ 0 & -64 \end{bmatrix},$$

and

$$\mathbf{P}^{-1} = \mathbf{P} = \mathbf{P}^T;$$

$$\mathbf{A}^3 = \begin{bmatrix} 188 & 84 \\ 84 & -36 \end{bmatrix}.$$

24. Rotation through 2θ, $-\theta$

25. $$\begin{aligned} \det \mathbf{AB} &= (ae + bg)(cf + dh) - (ce + dg)(af + bh) \\ &= adeh - bceh - adfg + bcfg \\ &= (ad - bc)(eh - fg) \\ &= (\det \mathbf{A})(\det \mathbf{B}). \end{aligned}$$

26. Let $G = \{\mathbf{A}, \mathbf{B}\}$. G is closed under multiplication because $\mathbf{AA} = \mathbf{A}$, $\mathbf{AB} = \mathbf{B}$, $\mathbf{BA} = \mathbf{B}$, $\mathbf{BB} = \mathbf{A}$. The associative law holds for all multiplication of square $n \times n$ matrices (here $n = 2$). $\mathbf{A}$ is the identity matrix. Inverses are $\mathbf{A}^{-1} = \mathbf{A}$ and $\mathbf{B}^{-1} = \mathbf{B}$, in G. So G is a group under multiplication.

27. (Key steps only.) Closure: $\det \mathbf{A} = \pm 1$ and $\det \mathbf{B} = \pm 1$ imply

$$\det(\mathbf{AB}) = (\det \mathbf{A})(\det \mathbf{B}) = \pm 1,$$

so $\mathbf{AB} \in G$. Associative law true for any $n \times n$ matrices, hence true in G. The identity matrix $\mathbf{I}$ has $\det \mathbf{I} = 1$, and hence $\mathbf{I} \in G$. The inverse of $\mathbf{A}$ is [adj **A**]/det **A**, and det $(\mathbf{A}\mathbf{A}^{-1}) = \det \mathbf{I} = 1$ implies

$$\det \mathbf{A}^{-1} = 1/(\det \mathbf{A}) = 1/(\pm 1) = \pm 1,$$

and $\mathbf{A}^{-1} \in G$ when $\mathbf{A} \in G$.

28. H is the set of all matrices like **P** in Exercise 15, where **P** is an orthogonal matrix and

$$\det \mathbf{P} = \cos^2\theta + \sin^2\theta = 1.$$

If **A** and **B** belong to H, then so does **AB**, because **AB** is orthogonal and det (**AB**) = 1. The associative law holds in H, and the identity matrix belongs to H. Finally, if $\mathbf{A} \in H$, then $\mathbf{A}^T = \mathbf{A}^{-1} \in H$, because $\det(\mathbf{A}^T) = \det(\mathbf{A}) = 1$ and $\mathbf{A}^T$ is orthogonal. Thus H is a group. But H does not contain matrices of determinant -1 (like **B** of Exercise 26) which are in G. So H is a proper subset of G.

29. For an orthogonal matrix **A**,

$$\det(\mathbf{A}\mathbf{A}^T) = \det(\mathbf{I}) = 1.$$

Also

$$\det(\mathbf{A}^T) = \det(\mathbf{A}),$$
$$\det(\mathbf{A}\mathbf{A}^T) = (\det \mathbf{A})(\det \mathbf{A}^T) = (\det \mathbf{A})^2 = 1,$$

so $\det \mathbf{A} = \pm 1$. The proof that the 2×2 orthogonal matrices of determinant 1 form a group under multiplication is like that given in the answer to Exercise 28. Finally, if

$$\mathbf{A} = \begin{bmatrix} a & b \\ c & d \end{bmatrix}$$

is an orthogonal matrix with $\det \mathbf{A} = 1$, then $ad - bc = 1$, $a^2 + b^2 = 1$, $c^2 + d^2 = 1$, and $ac + bd = ab + cd = 0$. These equations imply $d = a$, $c = -b$, and $a = \cos\theta$, $b = \sin\theta$, for some θ [locate (a, b) on unit circle]. Thus **A** represents a rotation.

30. **B** is orthogonal but not a rotation, because

$$\det \mathbf{B} = -1.$$

The change of coordinates $\mathbf{X} = \mathbf{B}\mathbf{X}'$ represents a reflection across the line $y = x$. If **A** is orthogonal and $\det \mathbf{A} = 1$, then **A** represents a rotation, by Exercise 29. If $\det \mathbf{A} = -1$, let $\mathbf{C} = \mathbf{B}\mathbf{A}$. Then **C** is orthogonal and $\det \mathbf{C} = (\det \mathbf{A})(\det \mathbf{B}) = 1$, so **C** is a rotation and $\mathbf{B}\mathbf{C} = \mathbf{B}^2\mathbf{A} = \mathbf{I}\mathbf{A} = \mathbf{A}$.

31. (a) Project parallel to $\mathbf{V}_1$, $\mathbf{V}_2$.

(c) $\det([\mathbf{W}_1\ \mathbf{W}_2]) = \det(\mathbf{A}[\mathbf{V}_1\ \mathbf{V}_2])$
$= (\det \mathbf{A})(\det[\mathbf{V}_1\ \mathbf{V}_2])$

Chapter 14

Article 14.1

1. $(2e^{2t}, e^{-t} - te^{-t}),\quad t \in (-\infty, \infty)$

2. $\left(\dfrac{1}{2 + 2t}, \dfrac{-t}{(1 - t^2)^{1/2}}\right),\quad |t| < 1$

3. $(2/\sqrt{1 - 4t^2}, 3\sec^2 3t, -t^{-2}),\quad 0 < |t| < \frac{1}{2}$

4. $\left(\dfrac{3}{|x|(9x^2 - 1)^{1/2}}, 2\sinh 2x, 4\,\mathrm{sech}^2\, 4x\right),\quad |x| > \frac{1}{3}$

5. $(4/(2t + 1)^2, -8t/(1 - 4t^2)),\quad |t| < \frac{1}{2}$

Article 14.2

1. $\mathbf{v} = -a\omega \sin \omega t\mathbf{i} + a\omega \cos wt\mathbf{j}$,
$\mathbf{a} = -a\omega^2 \cos \omega t\mathbf{i} - a\omega^2 \sin \omega t\mathbf{j} = -\omega^2\mathbf{R}$;
when $\omega t = \pi/3$,

$$\mathbf{v} = a\omega(-\tfrac{1}{2}\sqrt{3}\mathbf{i} + \tfrac{1}{2}\mathbf{j}),\ \text{speed} = a\omega,$$
$$\mathbf{a} = -a\omega^2(\tfrac{1}{2}\mathbf{i} + \tfrac{1}{2}\sqrt{3}\mathbf{j})$$

2. $\mathbf{v} = -2\sin t\mathbf{i} + 3\cos t\mathbf{j}$, $\quad \mathbf{a} = -2\cos t\mathbf{i} - 3\sin t\mathbf{j}$;
at $t = \pi/4$,

$$\mathbf{v} = -\sqrt{2}\mathbf{i} + \tfrac{3}{2}\sqrt{2}\mathbf{j},\quad \text{speed} = \sqrt{6.5},$$
$$\mathbf{a} = -\sqrt{2}\mathbf{i} - \tfrac{3}{2}\sqrt{2}\mathbf{j}$$

3. $\mathbf{v} = \mathbf{i} + 2t\mathbf{j}$, $\quad \mathbf{a} = 2\mathbf{j}$; $\quad$ at $t = 2$,

$$\mathbf{v} = \mathbf{i} + 4\mathbf{j},\quad \mathbf{a} = 2\mathbf{j},\quad \text{speed} = \sqrt{17}$$

4. $\mathbf{v} = -2\sin 2t\mathbf{i} + 2\cos t\mathbf{j}$,
$\mathbf{a} = -4\cos 2t\mathbf{i} - 2\sin t\mathbf{j}$; $\quad$ at $t = 0$,

$$\mathbf{v} = 2\mathbf{j},\quad \mathbf{a} = -4\mathbf{i},\quad \text{speed} = 2$$

5. $\mathbf{v} = e^t\mathbf{i} - 2e^{-2t}\mathbf{j}$, $\quad \mathbf{a} = e^t\mathbf{i} + 4e^{-2t}\mathbf{j}$; $\quad$ at $t = \ln 3$,

$$\mathbf{v} = 3\mathbf{i} - \tfrac{2}{9}\mathbf{j},\quad \mathbf{a} = 3\mathbf{i} + \tfrac{4}{9}\mathbf{j},\quad \text{speed} = \tfrac{1}{9}\sqrt{733}$$

6. $\mathbf{v} = \sec t \tan t\mathbf{i} + \sec^2 t\mathbf{j}$,
$\mathbf{a} = (\sec^3 t + \sec t \tan^2 t)\mathbf{i} + 2\sec^2 t \tan t\mathbf{j}$;
at $t = \pi/6$,

$$\mathbf{v} = \tfrac{2}{3}\mathbf{i} + \tfrac{4}{3}\mathbf{j},\quad \mathbf{a} = (10\mathbf{i} + 8\mathbf{j})/3\sqrt{3},\quad \text{speed} = \tfrac{1}{3}\sqrt{20}$$

7. $\mathbf{v} = 3\sinh 3t\mathbf{i} + 2\cosh t\mathbf{j}$,
$\mathbf{a} = 9\cosh 3t\mathbf{i} + 2\sinh t\mathbf{j}$; $\quad$ at $t = 0$,

$$\mathbf{v} = 2\mathbf{j},\quad \mathbf{a} = 9\mathbf{i},\quad \text{speed} = 2$$

8. $\mathbf{v} = \dfrac{\mathbf{i}}{t + 1} + 2t\mathbf{j}$, $\quad \mathbf{a} = \dfrac{-\mathbf{i}}{(t + 1)^2} + 2\mathbf{j}$;
at $t = 1$,
$\mathbf{v} = \frac{1}{2}\mathbf{i} + 2\mathbf{j}$, $\quad \mathbf{a} = -\frac{1}{4}\mathbf{i} + 2\mathbf{j}$, $\quad$ speed $= \frac{1}{2}\sqrt{17}$

9. $\mathbf{R} = -\frac{1}{2}gt^2\mathbf{j} + \mathbf{v}_0 t = tv_0 \cos\alpha\,\mathbf{i} + (tv_0 \sin\alpha - \frac{1}{2}gt^2)\mathbf{j}$

11. $\mathbf{v} = e^t[\mathbf{i} + \mathbf{j}(\cos t + \sin t) + \mathbf{k}(\cos t - \sin t)]$,
$\mathbf{a} = e^t[\mathbf{i} + 2\mathbf{j}\cos t - 2\mathbf{k}\sin t]$,
$\theta = \cos^{-1}(\sqrt{15}/5) = 39°.2$

12. $\mathbf{v} = \mathbf{i}\sec^2 t + 2\mathbf{j}\cosh 2t - 3\mathbf{k}\,\mathrm{sech}\, 3t \tanh 3t$,
$\mathbf{a} = 2\mathbf{i}\sec^2 t \tan t + 4\mathbf{j}\sinh 2t + 9\mathbf{k}\,\mathrm{sech}\, 3t(\tanh^2 3t - \mathrm{sech}^2 3t)$,
$\theta = 90°$

13. $\mathbf{v} = \dfrac{2t}{t^2+1}\mathbf{i} + \dfrac{1}{t^2+1}\mathbf{j} + \dfrac{t}{\sqrt{t^2+1}}\mathbf{k}$,
$\mathbf{a} = \dfrac{2(1-t^2)}{(1+t^2)^2}\mathbf{i} - \dfrac{2t}{(1+t^2)^2}\mathbf{j} + \dfrac{\mathbf{k}}{(t^2+1)^{3/2}}$,
$\theta = 90°$

14. $\mathbf{R} = (3\cos\theta)\mathbf{i} + (3\sin\theta)\mathbf{j} + (6\cos\theta + 9\sin\theta)\mathbf{k}$,
$\mathbf{v} = -3\omega(\sin\theta)\mathbf{i} + 3\omega(\cos\theta)\mathbf{j} + 3\omega(3\cos\theta - 2\sin\theta)\mathbf{k}$,
$\mathbf{a} = -\omega^2\mathbf{R}$

Article 14.3

1. $-\mathbf{i}\sin t + \mathbf{j}\cos t$
2. $\dfrac{e^t\mathbf{i} + 2t\mathbf{j}}{\sqrt{e^{2t} + 4t^2}}$
3. $-\cos t\mathbf{i} + \sin t\mathbf{j}$
4. $\dfrac{\mathbf{i} + 2x\mathbf{j}}{\sqrt{1+4x^2}}$
5. $-\dfrac{2\cos t\mathbf{i} + \mathbf{j}}{\sqrt{1 + 4\cos^2 t}}$
6. $\mathbf{T} = \frac{1}{13}(12\cos 2t\mathbf{i} - 12\sin 2t\mathbf{j} + 5\mathbf{k})$
7. $\mathbf{T} = \sqrt{\frac{1}{3}}[(\cos t - \sin t)\mathbf{i} + (\cos t + \sin t)\mathbf{j} + \mathbf{k}]$
8. $\mathbf{T} = (\mathbf{i}\tanh 2t + \mathbf{j} + \mathbf{k}\,\mathrm{sech}\, 2t)/\sqrt{2}$
9. $\mathbf{T} = \dfrac{3(\cos t - t\sin t)\mathbf{i} + 3(\sin t + t\cos t)\mathbf{j} + 4\mathbf{k}}{\sqrt{9t^2 + 25}}$
10. 13π
11. $\sqrt{3}(e^\pi - 1)$
12. $3\sqrt{2}\sinh 2\pi$
13. $\dfrac{\pi\sqrt{9\pi^2+25}}{2} + \dfrac{25}{6}\sinh^{-1}\dfrac{3\pi}{5}$

Article 14.4

1. $\dfrac{1}{a}\mathrm{sech}^2\dfrac{x}{a}$
2. $|\cos x|$
3. $\dfrac{4e^{2x}}{(1+4e^{4x})^{3/2}}$
4. $\dfrac{1}{3a|\sin t\cos t|}$
5. $\dfrac{1}{|a\theta|}$
6. $\dfrac{1}{4a}\left|\csc\dfrac{\theta}{2}\right|$
7. $|\cos y|$
8. $\dfrac{2}{\sqrt{y^2 + 2(y^2+1)^2}}$
9. $\dfrac{|48y^5|}{(4y^6+1)^2}$
10. $(x+2)^2 + (y-3)^2 = 8$.
For the circle
$$y' = -\frac{x+2}{y-3}, \qquad y'' = \frac{-8}{(y-3)^3}.$$
Both of these are equal to $+1$ at $(0, 1)$.
12. $\mathbf{N} = -\mathbf{i}\sin 2t - \mathbf{j}\cos 2t$, $\kappa = \frac{24}{169}$,
$\mathbf{B} = \frac{1}{13}(5\cos 2t\mathbf{i} - 5\sin 2t\mathbf{j} - 12\mathbf{k})$
13. $\mathbf{N} = \dfrac{-\mathbf{i}}{\sqrt{2}}(\sin t + \cos t) + \dfrac{\mathbf{j}}{\sqrt{2}}(\cos t - \sin t)$,
$\kappa = \sqrt{2}\,e^{-t}/3$,
$\mathbf{B} = [\mathbf{i}(\sin t - \cos t) - \mathbf{j}(\sin t + \cos t) + 2\mathbf{k}]/\sqrt{6}$
14. $\mathbf{N} = \mathbf{i}\,\mathrm{sech}\, 2t - \mathbf{k}\tanh 2t$, $\kappa = \frac{1}{6}\mathrm{sech}^2 2t$,
$\mathbf{B} = (-\mathbf{i}\tanh 2t + \mathbf{j} - \mathbf{k}\,\mathrm{sech}\, 2t)/\sqrt{2}$
15. $d\mathbf{R}/dt = \mathbf{T}(ds/dt)$,
$d^2\mathbf{R}/dt^2 = \mathbf{T}(d^2s/dt^2) + \mathbf{N}\kappa(ds/dt)^2$

Article 14.5

5. $\mathbf{v} = 2\mathbf{i}\sinh 2t + 2\mathbf{j}\cosh 2t$,
$\mathbf{a} = 4\mathbf{i}\cosh 2t + 4\mathbf{j}\sinh 2t$,
$$\frac{ds}{dt} = |\mathbf{v}| = 2\sqrt{\sinh^2 2t + \cosh^2 2t} = 2\sqrt{\cosh 4t}.$$
$$a_T = \frac{d^2s}{dt^2} = \frac{4\sinh 4t}{\sqrt{\cosh 4t}},$$
$$a_N = \sqrt{|\mathbf{a}|^2 - a_T^2} = 4\sqrt{\mathrm{sech}\, 4t}$$
6. $a_T = \dfrac{2t}{\sqrt{1+t^2}}$, $a_N = \dfrac{2}{\sqrt{1+t^2}}$
7. $a_T = 0$, $a_N = \omega^2 a$
8. $a_T = 0$, $a_N = \dfrac{2}{t^2+1}$
9. $a_T = \sqrt{2}e^t$, $a_N = \sqrt{2}e^t$

Article 14.6

2. $\mathbf{v} = 3a \sin \theta \mathbf{u}_r + 3a(1 - \cos \theta)\mathbf{u}_\theta$,
$\mathbf{a} = 9a(2 \cos \theta - 1)\mathbf{u}_r + 18a \sin \theta \mathbf{u}_\theta$

3. $\mathbf{v} = 4at \cos 2\theta \mathbf{u}_r + 2at \sin 2\theta \mathbf{u}_\theta$,
$\mathbf{a} = (4a \cos 2\theta - 20at^2 \sin 2\theta)\mathbf{u}_r + (2a \sin 2\theta + 16at^2 \cos 2\theta)\mathbf{u}_\theta$

4. $\mathbf{v} = 2ae^{a\theta}\mathbf{u}_r + 2e^{a\theta}\mathbf{u}_\theta$,
$\mathbf{a} = 4e^{a\theta}(a^2 - 1)\mathbf{u}_r + 8ae^{a\theta}\mathbf{u}_\theta$

5. $\mathbf{v} = a \cos t\mathbf{u}_r + ae^{-t}(1 + \sin t)\mathbf{u}_\theta$,
$\mathbf{a} = -a[\sin t + e^{-2t}(1 + \sin t)]\mathbf{u}_r + ae^{-t}[2 \cos t - 1 - \sin t]\mathbf{u}_\theta$

6. $\mathbf{v} = -8 \sin 4t\mathbf{u}_r + 4 \cos 4t\mathbf{u}_\theta$,
$\mathbf{a} = -40 \cos 4t\mathbf{u}_r - 32 \sin 4t\mathbf{u}_\theta$

8. $T = 8042$ sec (approx) $= 134$ min 2 sec

10. $2a = 13.5046 \times 10^8$ cm $= 8391.3$ mi (compared with 8421 mi)

14. In an increment of time Δt, the tip of the position vector moves about $(d\mathbf{R}/dt)\,\Delta t$, and $\mathbf{R}$ sweeps over the interior of a triangle with one side $\mathbf{R}$ and the other side $\mathbf{v}\,\Delta t$. Call the area of this triangle ΔA, and then divide by Δt and take the limit as $\Delta t \to 0$.

15. Let $a = \sqrt{GM/r_0}$, $b = \sqrt{2GM/r_0}$. Parabola, $v_0 = b$; circle, $v_0 = a$; ellipse, $a < v_0 < b$; hyperbola, $v_0 > b$

16. $v_0 = 3.01 \times 10^6$ cm/sec $= 18.7$ mi/sec

Miscellaneous exercises

1. (a) $\mathbf{v} = 2^{-3/2}(-\mathbf{i} + \mathbf{j})$, $\mathbf{a} = 2^{-5/2}(\mathbf{i} - 3\mathbf{j})$
(b) $t = 0$

2. (a) $\mathbf{v} = \pi(\mathbf{i}(1 - \cos \pi t) + \mathbf{j} \sin \pi t)$,
$\mathbf{a} = \pi^2(\mathbf{i} \sin \pi t + \mathbf{j} \cos \pi t)$
(b) Slope of $PC = \cot \pi t$,
slope of $PQ = \csc \pi t + \cot \pi t$
(c) Slope of $\mathbf{v}$ = slope of PQ,
slope of $\mathbf{a}$ = slope of PC

3. Speed $= a\sqrt{1 + t^2}$,
$a_t = at/\sqrt{1 + t^2}$,
$a_n = a(t^2 + 2)/\sqrt{1 + t^2}$

5. (b) $\pi/2$

6. $|ad - bc|\,(a^2 + b^2)^{-3/2}$, provided $a^2 + b^2 \neq 0$

7. $x = a\theta + a \sin \theta$, $y = -a(1 - \cos \theta)$

8. $(-\frac{1}{2} \ln 2, 1/\sqrt{2})$

9. (b) πab, πa^2

10. $\kappa = \pi s$

11. $y = \pm\sqrt{1 - x^2} \pm \ln \dfrac{1 - \sqrt{1 - x^2}}{x} + C.$
Set $C = 0$. Then
$x = e^{-s}$ [$s = -\ln x$, measured from (1, 0)],
$y = \pm\sqrt{1 - e^{-2s}} \pm \ln (e^s - \sqrt{e^{2s} - 1})$.

12. $$s = s_0 + 2\pi p,$$
$$A = A_0 + s_0 p + \pi p^2, \quad R = R_0 + p,$$
where s_0, A_0, R_0 are the length of arc, area, and radius of curvature of the original curve.

13. (a) $320\sqrt{10}$ (b) $16[\sqrt{2} + \ln (\sqrt{2} + 1)]$

14. $\mathbf{v} = 3 \cos t\mathbf{i} - 2 \sin t\mathbf{j}$,
$\mathbf{a} = -3 \sin t\mathbf{i} - 3 \cos t\mathbf{j}$,
speed $= (4 + 5 \cos^2 t)^{1/2}$,
$a_t = -5 \sin t \cos t\,(4 + 5 \cos^2 t)^{-1/2}$,
$a_n = 6(4 + 5 \cos^2 t)^{-1/2}$

15. (a) $a_t = 0$, $a_n = 4$ (b) 1 (c) $r = 2 \cos \theta$

16. $\mathbf{v} = (\cos t - t \sin t)\mathbf{i} + (\sin t + t \cos t)\mathbf{j}$,
$\mathbf{a} = -(t \cos t + 2 \sin t)\mathbf{i} - (t \sin t - 2 \cos t)\mathbf{j}$,
$\kappa = (t^2 + 2)/(t^2 + 1)^{3/2}$

17. $\kappa = |f^2 + 2f'^2 - ff''|/(f^2 + f'^2)^{3/2}$

18. (a) $r = e^{2\theta}$ (b) $\frac{1}{2}\sqrt{5}(e^{4\pi} - 1)$

19. $\sqrt{2}\mathbf{u}_r$

20. $2(\omega^2 - 1)\mathbf{i}$

21. (a) $\mathbf{v} = -\mathbf{u}_r + 3\mathbf{u}_\theta$, $\mathbf{a} = -9\mathbf{u}_r - 6\mathbf{u}_\theta$
(b) $\sqrt{37} + \frac{1}{6} \ln (6 + \sqrt{37})$

22. $r = a \cosh \omega t$

23. (a) $2\sqrt{\dfrac{\pi bg}{a^2 + b^2}}$

(b) $\theta = \dfrac{b}{a^2 + b^2}\,(\frac{1}{2}gt^2)$, $z = \dfrac{b^2}{a^2 + b^2}\,(\frac{1}{2}gt^2)$

(c) $$\frac{d\mathbf{R}}{dt} = gt\,\frac{b}{\sqrt{a^2 + b^2}}\,\mathbf{T},$$
$$\frac{d^2\mathbf{R}}{dt^2} = g\,\frac{b}{\sqrt{a^2 + b^2}}\,\mathbf{T} + (gt)^2\,\frac{ab^2}{(a^2 + b^2)^2}\,\mathbf{N}$$

There is never any component of acceleration in the direction of the binormal.

24. (a) $\dfrac{d\theta}{dt} = \sqrt{\dfrac{2gb\theta}{a^2 + b^2 + a^2\theta^2}}$

(b) $\frac{1}{2}\{\theta\sqrt{a^2 + b^2 + a^2\theta^2} + [(a^2 + b^2)/a] \sinh^{-1}\,(a\theta)/\sqrt{a^2 + b^2}\}$

25. (a) $\mathbf{u}_\rho = \mathbf{i}\sin\phi\cos\theta + \mathbf{j}\sin\phi\sin\theta + \mathbf{k}\cos\phi$,
$\mathbf{u}_\phi = \mathbf{i}\cos\phi\cos\theta + \mathbf{j}\cos\phi\sin\theta - \mathbf{k}\sin\phi$,
$\mathbf{u}_\theta = -\mathbf{i}\sin\theta + \mathbf{j}\cos\theta$
(d) Yes, they form a right-handed system of mutually orthogonal vectors because of (b) and (c).

26. $\mathbf{R} = \rho\mathbf{u}_\rho$, $\dfrac{d\mathbf{R}}{dt} = \mathbf{u}_\rho\dfrac{d\rho}{dt} + \mathbf{u}_\phi\rho\dfrac{d\phi}{dt} + \mathbf{u}_\theta\rho\sin\phi\dfrac{d\theta}{dt}$

27. (a) $ds^2 = dr^2 + r^2\,d\theta^2 + dz^2$
(b) $ds^2 = d\rho^2 + \rho^2\,d\phi^2 + \rho^2\sin^2\phi\,d\theta^2$

28. (a) $7\sqrt{3}\,a$ (b) $7\sqrt{5}$

29. $x = a\cos\theta/\sqrt{1+\sin^2\theta}$,
$y = a\sin\theta/\sqrt{1+\sin^2\theta}$,
$z = -a\sin\theta/\sqrt{1+\sin^2\theta}$

30. (a) $\dfrac{dx}{dt} = \cos\theta\dfrac{dr}{dt} - \sin\theta\dfrac{r\,d\theta}{dt}$,
$\dfrac{dy}{dt} = \sin\theta\dfrac{dr}{dt} + \cos\theta\dfrac{r\,d\theta}{dt}$
(b) $\dfrac{dr}{dt} = \cos\theta\dfrac{dx}{dt} + \sin\theta\dfrac{dy}{dt}$,
$\dfrac{r\,d\theta}{dt} = -\sin\theta\dfrac{dx}{dt} + \cos\theta\dfrac{dy}{dt}$

31. $2\sqrt{3}(\mathbf{i} + \mathbf{j} - 2\mathbf{k})$

33. (a) $\mathbf{t}(\frac{2}{3}, \frac{1}{3}, \frac{2}{3})$, $\mathbf{n}(1/\sqrt{5}, -2/\sqrt{5}, 0)$,
$\mathbf{b}(4/3\sqrt{5}, 2/3\sqrt{5}, -5/3\sqrt{5})$
(b) $2\sqrt{5}/9$

34. (a) $x + 2y + 3z = 6$ (b) $3x - 3y + z = 1$

35. $\frac{1}{3}(t^2+1)^{-2}$

37. $\dfrac{x^2}{4} + \dfrac{y^2}{9} + \dfrac{z^2}{1} = 1$, an ellipsoid

Chapter 15

Article 15.1

1. Let $r = \sqrt{x^2+y^2}$. (a) $r < 0.1$ (b) $r < 0.001$

2. (a) Let $\rho = \sqrt{x^2+y^2+z^2}$;
$\rho < 0.1$, $\rho < 0.2\sqrt{3}$
(b) Yes, because
$|f(x, y, z) - f(0, 0, 0)| = x^2 + y^2 + z^2 = \rho^2$
is less than any positive ϵ if $\rho < \sqrt{\epsilon}$.

3. (a) Along line $x = 1$, $f(x, y) = f(1, y) \to 1$
Along line $y = -1$, $f(x, y) = f(x, -1) \to \frac{1}{2}$
(b) No, because different limits are approached as $(x, y) \to (1, -1)$ in different ways. [See answer to (a).]

Article 15.2

6. $\dfrac{\partial w}{\partial x} = e^x\cos y$, $\dfrac{\partial w}{\partial y} = -e^x\sin y$

7. $\dfrac{\partial w}{\partial x} = e^x\sin y$, $\dfrac{\partial w}{\partial y} = e^x\cos y$

8. $\dfrac{\partial w}{\partial x} = \dfrac{-y}{x^2+y^2}$, $\dfrac{\partial w}{\partial y} = \dfrac{x}{x^2+y^2}$

9. $\dfrac{\partial w}{\partial x} = \dfrac{x}{x^2+y^2}$ $\dfrac{\partial w}{\partial y} = \dfrac{y}{x^2+y^2}$

10. $\dfrac{\partial w}{\partial x} = -\dfrac{y}{x^2}\sinh\dfrac{y}{x}$, $\dfrac{\partial w}{\partial y} = \dfrac{1}{x}\sinh\dfrac{y}{x}$

11. $f_x = 2xe^{2y+3z}\cos(4w)$,
$f_y = 2f$,
$f_z = 3f$,
$f_w = -4x^2e^{2y+3z}\sin(4w)$

12. $f_x = \dfrac{-yz}{\sqrt{x^4 - x^2y^2}}$,
$f_y = \dfrac{|x|z}{x\sqrt{x^2-y^2}}$,
$f_z = \sin^{-1}(y/x)$

13. $f_u = \dfrac{2u}{v^2+w^2}$,
$f_v = \dfrac{-2v(u^2+w^2)}{(v^2+w^2)^2}$,
$f_w = \dfrac{-2w(u^2-v^2)}{(v^2+w^2)^2}$

14. $f_r = \dfrac{(z^2-r^2)(2-\cos 2\theta)}{(r^2+z^2)^2}$,
$f_\theta = \dfrac{2r\sin 2\theta}{r^2+z^2}$,
$f_z = \dfrac{-2rz(2-\cos 2\theta)}{(r^2+z^2)^2}$

15. $f_x = \dfrac{2x}{u^2 + v^2}$, $f_y = \dfrac{2y}{u^2 + v^2}$, $f_u = \dfrac{-2u(x^2 + y^2)}{(u^2 + v^2)^2}$, $f_v = \dfrac{-2v(x^2 + y^2)}{(u^2 + v^2)^2}$

16. $f_x = 2 \cos 2x \cosh 3r$, $f_y = 3 \cosh 3y \cos 4s$, $f_r = 3 \sin 2x \sinh 3r$, $f_s = -4 \sinh 3y \sin 4s$

17. By the law of cosines, $\cos A = (b^2 + c^2 - a^2)/(2bc)$.

$$\frac{\partial A}{\partial a} = \frac{a}{bc \sin A}, \quad \frac{\partial A}{\partial b} = \frac{c^2 - a^2 - b^2}{2b^2c \sin A}.$$

18. By the law of sines, $a = b \sin A \csc B$.

$$\frac{\partial a}{\partial A} = b \cos A \csc B, \quad \frac{\partial a}{\partial B} = -b \sin A \csc B \cot B.$$

19. $\dfrac{x}{\rho} = \dfrac{x}{\sqrt{x^2 + y^2 + z^2}}$

20. $-\dfrac{x^2 + y^2}{\rho^3 \sin \phi} = -\dfrac{\sqrt{x^2 + y^2}}{x^2 + y^2 + z^2}$

21. $\dfrac{x}{x^2 + y^2}$

22. 0

23. $\dfrac{xz}{\rho^3 \sin \phi} = \dfrac{xz}{(x^2 + y^2 + z^2)\sqrt{x^2 + y^2}}$

24. $\dfrac{-y_2}{x^2 + y^2}$

25. $\mathbf{i} \sin \phi \cos \theta + \mathbf{j} \sin \phi \sin \theta + \mathbf{k} \cos \phi$

26. $\rho \cos \phi\, (\mathbf{i} \cos \theta + \mathbf{j} \sin \theta) - \mathbf{k}\rho \sin \phi$

27. $\rho \sin \phi(-\mathbf{i} \sin \theta + \mathbf{j} \cos \theta)$

28. $\dfrac{\partial \mathbf{R}}{\partial \rho} = \mathbf{u}_\rho$, $\dfrac{\partial \mathbf{R}}{\partial \phi} = \rho \mathbf{u}_\phi$, $\dfrac{\partial \mathbf{R}}{\partial \theta} = \rho \sin \phi \mathbf{u}_\theta$

29. (a) $\partial \mathbf{R}/\partial x$ is tangent to the curve in which the surface $w = f(x, y)$ and the plane $y = y_0$ intersect.
(b) $\partial \mathbf{R}/\partial y$ is tangent to the curve in which the surface $w = f(x, y)$ and the plane $x = x_0$ intersect.
(c) $\mathbf{v} = -\mathbf{i}f_x(x_0, y_0) - \mathbf{j}f_y(x_0, y_0) + \mathbf{k}$ is normal to the surface at (x_0, y_0, w_0).

Article 15.3

1. $6x + 8y = z + 25$, $\dfrac{x - 3}{6} = \dfrac{y - 4}{8} = \dfrac{z - 25}{-1}$
2. $x - 2y + 2z = 9$, $2x = -y = z$
3. $x - 3y = z - 1$, $x - 1 = \dfrac{1 - y}{3} = -z - 1$
4. $x - y + 2z = \dfrac{\pi}{2}$, $x - 1 = 1 - y = \dfrac{z - \pi/4}{2}$
5. $16x + 12y = 125z - 75$,
$\dfrac{x - 3}{16} = \dfrac{y + 4}{12} = \dfrac{3 - 5z}{625}$
6. (a) $\mathbf{N} = c(-\mathbf{i} + \mathbf{j}f_y + \mathbf{k}f_z)$, c scalar
(b) $x - 2y + z = 1$, $2x = 3 - y = 2(z - 1)$
7. Line: $z = \dfrac{-11x}{6} = \dfrac{-22y}{7}$,
plane: $12x + 14y + 11z = 0$
8. Vertex: $(0, 0, z_0 + \frac{1}{2})$
9. $\begin{vmatrix} \mathbf{i} & \mathbf{j} & \mathbf{k} \\ f_x & f_y & -1 \\ g_x & g_y & -1 \end{vmatrix}$
10. $\pm\sqrt{3/1105}\,[3\mathbf{i} + 14\mathbf{j} - 30\mathbf{k}]$

Article 15.4

1. $\Delta w_{\tan} = 0.14$, $\Delta w = 0.1407$
2. 6%
3. 4.98
5. (13 ± 0.03) in.
8. $L(x, y) = -(2 + \pi)(y - \pi/2)$
9. $L(x, y, z) = -4 - 8(x - 1) + 2(y + 2) - 4(z - \ln 2)$
10. $L(x, y, u, v) = \frac{1}{2}(-1 - x + y + v)$
11. The answer is given in the following theorem.

Theorem. *Let the function $w = f(x, y, z)$ be continuous and possess partial derivatives f_x, f_y, f_z in some region*

$$G = \{(x, y, z) : |x - x_0| < h, |y - y_0| < h, |z - z_0| < h\}.$$

Let

$$L(x, y, z) = f(x_0, y_0, z_0) + f_x(x_0, y_0, z_0)(x - x_0) + f_y(x_0, y_0, z_0)(y - y_0) + f_z(x_0, y_0, z_0)(z - z_0)$$

be the linearization of f at $P_0(x_0, y_0, z_0)$. If the partial derivatives f_x, f_y, f_z are continuous at P_0, then

$$\lim_{\Delta s \to 0} \left| \frac{f(x, y, z) - L(x, y, z)}{\Delta s} \right| = 0,$$

where

$$\Delta s = \sqrt{(x - x_0)^2 + (y - y_0)^2 + (z - z_0)^2}.$$

Proof *(incomplete outline).* Apply the Mean Value Theorem for functions of one variable to the right-hand sides of Eqs. (20a, b, c) and get

$$\begin{aligned}\Delta w = \Delta w_1 + \Delta w_2 + \Delta w_3 &= f_x(x_1, y_0, z_0)\,\Delta x \\ &\quad + f_y(x_0 + \Delta x, y_1, z_0)\,\Delta y \\ &\quad + f_z(x_0 + \Delta x, \\ &\qquad y_0 + \Delta y, z_1)\,\Delta z, \\ &= [f_x(x_0, y_0, z_0) + \epsilon_1]\,\Delta x \\ &\quad + [f_y(x_0, y_0, z_0) + \epsilon_2]\,\Delta y \\ &\quad + [f_z(x_0, y_0, z_0) + \epsilon_3]\,\Delta z,\end{aligned}$$

with $\epsilon_1, \epsilon_2, \epsilon_3 \to 0$ as $\Delta s \to 0$.

12. As given in the theorem above, with all z-terms omitted.
13. As given in the theorem above, with (x, y, z) replaced by $(x_1, x_2, \ldots, x_n)$ and (x_0, y_0, z_0) replaced by $(a_1, a_2, \ldots, a_n)$, and with partial derivatives indicated by $f_1, f_2, \ldots, f_n$.

Article 15.5

1. $\frac{2}{3}$ 2. $\frac{9}{1183}$ 3. $4\sqrt{3}$ 4. $\frac{43}{15}$
5. In the direction of $3\mathbf{i} - \mathbf{j}$
6. $-7/\sqrt{5}$ 8. $\dfrac{\pi - 3}{2\sqrt{5}} \approx 0.032$ 9. $\pm(\mathbf{i} + \mathbf{j})$

Article 15.6

1. $-2\mathbf{i} - 2\mathbf{j} + 4\mathbf{k}$ 2. $6\mathbf{i} + 6\mathbf{j}$ 3. $2\mathbf{i}$
4. $(-3\mathbf{i} - 4\mathbf{j})/25$ 5. $(\mathbf{i} + 2\mathbf{j} - 2\mathbf{k})/27$
6. $(3\sqrt{3}\mathbf{i} + 4\sqrt{3}\mathbf{j} + 5\mathbf{k})/2$ 7. $5\mathbf{k}$
8. $\dfrac{x-3}{3} = \dfrac{y-4}{4} = \dfrac{z+5}{5}$
9. (a) $x = y = z, \quad -x = -y = z, \quad x = y = 0$
   (b) $(1, 1, 1)$ and $(-2, -2, -2)$, $(2, 2, -2)$ and $(-1, -1, 1)$, $(0, 0, 2)$
10. Two lines: $x = z = -y \pm 4$
11. Tangent plane: $3x + 5y + 4z = 18$,
    normal line $\dfrac{x-3}{3} = \dfrac{y-5}{5} = \dfrac{z+4}{4}$
12. In the direction of $\operatorname{grad} f = 10\mathbf{i} + 4\mathbf{j} + 10\mathbf{k}$. Max $(df/ds) = |\operatorname{grad} f| = \sqrt{216}$
14. $\operatorname{grad} w = \mathbf{u}_\rho \dfrac{\partial w}{\partial \rho} + \mathbf{u}_\phi \dfrac{1}{\rho}\dfrac{\partial w}{\partial \phi} + \mathbf{u}_\theta \dfrac{1}{\rho \sin\phi}\dfrac{\partial w}{\partial \theta}$
15. (a) $\mathbf{i}\dfrac{\partial f}{\partial x} + \mathbf{j}\dfrac{\partial f}{\partial y}$
    (b) In the direction given by the vector $\operatorname{grad} f = -\mathbf{i} + \mathbf{j}$; $\left|\dfrac{dw}{ds}\right| = \sqrt{2}$ in this direction.
24. $h = f_0/(f_x^2 + f_y^2 + f_z^2)_0$

Article 15.7

1. $2e^t[x(\cos t - \sin t) + y(\sin t + \cos t) + z] = 4e^{2t}$
2. $\dfrac{y(y^2 - x^2)\sinh t + x(x^2 - y^2)\cosh t}{(x^2 + y^2)^2} = \operatorname{sech}^2 2t$
3. $e^{2x+3y}\left(\dfrac{(8t^2 + 2)\cos 4z}{t(t^2+1)} - 4\sin 4z\right) = 2t(t^2+1)^2[(4t^2+1)\cos 4t - 2t(t^2+1)\sin 4t]$
4. $\dfrac{\partial w}{\partial r} = e^{2r}/\sqrt{e^{2r} + e^{2s}}, \quad \dfrac{\partial w}{\partial s} = e^{2s}/\sqrt{e^{2r} + e^{2s}}$
5. $\dfrac{\partial w}{\partial r} = \dfrac{\partial w}{\partial s} = \dfrac{2}{r+s}$
8. $w = 5$
9. $w = \exp[-(ax + by)/a]$
10. $w = \exp[(-ax + by)/(a + b)]$
11. If $ax + by = c$, then $f(ax + by) = f(c)$ is the same number for all points on the line. Thus w is constant on the line. It would not be possible to satisfy $w = \sin x$ along the line $ax + by = 5$ unless x were constant on the line; that is, unless b were equal to zero. However, if $b = 0$, then $w = \sin x$ is a solution.
17. $f_x = \dfrac{\partial w}{\partial r}\cos\theta - \dfrac{1}{r}\dfrac{\partial w}{\partial \theta}\sin\theta$,
    $f_y = \dfrac{\partial w}{\partial r}\sin\theta + \dfrac{1}{r}\dfrac{\partial w}{\partial \theta}\cos\theta$

Article 15.8

2. $\frac{47}{24}$ ft^3 3. Approx. 340 ft^2
4. (a) $dx = \cos\theta\,dr - r\sin\theta\,d\theta$,
   $dy = \sin\theta\,dr + r\cos\theta\,d\theta$
   (b) $dr = \cos\theta\,dx + \sin\theta\,dy$,
   $d\theta = -\dfrac{\sin\theta}{r}\,dx + \dfrac{\cos\theta}{r}\,dy$

(c) $\dfrac{\partial r}{\partial x} = \cos\theta = \dfrac{x}{r}, \quad \dfrac{\partial r}{\partial y} = \sin\theta = \dfrac{y}{r}$

5. (a) $dx = f_u\,du + f_v\,dv, \quad dy = g_u\,du + g_v\,dv$

(b) $du = \dfrac{\begin{vmatrix} dx & f_v \\ dy & g_v \end{vmatrix}}{\begin{vmatrix} f_u & f_v \\ g_u & g_v \end{vmatrix}}, \quad dv = \dfrac{\begin{vmatrix} f_u & dx \\ g_u & dy \end{vmatrix}}{\begin{vmatrix} f_u & f_v \\ g_u & g_v \end{vmatrix}}$

6. (a) $dx = \sin\phi\cos\theta\,d\rho + \rho\cos\phi\cos\theta\,d\phi - \rho\sin\phi\sin\theta\,d\theta,$

$dy = \sin\phi\sin\theta\,d\rho + \rho\cos\phi\sin\theta\,d\phi + \rho\sin\phi\cos\theta\,d\theta,$

$dz = \cos\phi\,d\rho - \rho\sin\phi\,d\phi$

(b)

$$d\rho = \frac{\begin{vmatrix} dx & \rho\cos\phi\cos\theta & -\rho\sin\phi\sin\theta \\ dy & \rho\cos\phi\sin\theta & \rho\sin\phi\cos\theta \\ dx & -\rho\sin\phi & 0 \end{vmatrix}}{\begin{vmatrix} \sin\phi\cos\theta & \rho\cos\phi\cos\theta & -\rho\sin\phi\sin\theta \\ \sin\phi\sin\theta & \rho\cos\phi\sin\theta & \rho\sin\phi\cos\theta \\ \cos\phi & -\rho\sin\phi & 0 \end{vmatrix}}$$

(c) $\dfrac{\partial\rho}{\partial x} = \sin\phi\cos\theta = \dfrac{x}{\rho}$

7. $df = -u_0$ and $dg = -v_0$ provided

$$dx = \left(\frac{vf_y - ug_y}{f_xg_y - g_xf_y}\right)_0, \quad dy = \left(\frac{ug_x - vf_x}{f_xg_y - g_xf_y}\right)_0.$$

8.

$$dx = -\frac{\begin{vmatrix} u_0 & f_y & f_z \\ v_0 & g_y & g_z \\ w_0 & h_y & h_z \end{vmatrix}}{\begin{vmatrix} f_x & f_y & f_z \\ g_x & g_y & g_z \\ h_x & h_y & h_z \end{vmatrix}}$$

, with similar expressions for dy and dz.

Article 15.9

1. Low, $(-3, 3, -5)$
2. Low, $(15, -8, -63)$
3. High, $(-8, -23, 59)$
4. High, $(\frac{2}{3}, \frac{4}{3}, 0)$
5. $(-2, 1, 3)$ is a saddle point.
6. $(-2, 2, 2)$ is a saddle point.
7. Low, $(0, 0, 0)$; high, $(\pm 1, \pm 1, \sqrt{2})$. The partial derivatives do not exist at $(0, 0)$. They exist but are not zero at $(\pm 1, \pm 1)$.

Article 15.10

1. (b)

$$m = \frac{\begin{vmatrix} \sum y_i & n \\ \sum x_iy_i & \sum x_i \end{vmatrix}}{D}, \quad b = \frac{\begin{vmatrix} \sum x_i & \sum y_i \\ \sum x_i^2 & \sum x_iy_i \end{vmatrix}}{D},$$

where

$$D = \begin{vmatrix} \sum x_i & n \\ \sum x_i^2 & \sum x_i \end{vmatrix}.$$

2. $26y + 19x = 30$
3. $12y = 9x + 20$
4. $6y = 9x + 1$

Article 15.11

1. $(0, 0, 1)$
2. Square bottom, 8 in. by 8 in., 4 in. deep
3. Square bottom, $\sqrt[3]{V/2}$ by $\sqrt[3]{V/2}$, with depth $= \sqrt[3]{4V}$, where V is the given volume
4. $x + 2y + 2z = 6$
5. h = altitude of triangle $= \sqrt{\dfrac{A}{6 + 3\sqrt{3}}}$

   $2x$ = width of rectangle $= 2\sqrt{3}\,h$

   y = altitude of rectangle $= (1 + \sqrt{3})h$
6. $z = -\frac{1}{2}x + \frac{3}{2}y - \frac{1}{4}$
7. (a) Spheres, center at origin
   (b) No
   (c) Circle or point of tangency. Yes. In a circle.
   (d) Yes
8. $(-3.621, 1.5)$ and $(3.621, -1.5)$ for maximum, $(0.621, 1.5)$ and $(-0.621, -1.5)$ for minimum
9. (a) $(-5/\sqrt{14}, -10/\sqrt{14}, -15/\sqrt{14}),\ f = -5\sqrt{14}$
   (b) $(5/\sqrt{14}, 10/\sqrt{14}, 15/\sqrt{14}),\ f = 5\sqrt{14}$
   Geometric interpretation: Planes f = constant are tangent to the sphere for $f = \pm 5\sqrt{14}$.

10. Cone $|z| = r$, plane

$$z = 1 + r(\cos\theta + \sin\theta) = 1 + r\sqrt{2}\sin(\theta + \pi/4),$$

curve of intersection lies on the hyperbolic cylinder

$$r = g(\theta) = 1/[\pm 1 - \sqrt{2}\sin(\theta + \pi/4)],$$

$|OP| = \sqrt{2}\,r$ with $r = g(\theta)$. The minimum value of $|g(\theta)|$ is $1/(1+\sqrt{2}) = \sqrt{2} - 1$, and

$$|OB| = \sqrt{2}(\sqrt{2} - 1) = 2 - \sqrt{2} = \sqrt{(6 - 4\sqrt{2})}.$$

B is the point on the cone nearest the origin; A is the point nearest the origin on the other branch of the hyperbola; there is no point farthest from the origin.

11. If $y = x$, $z = 1 + 2x$, and $z^2 = x^2 + y^2$, then

$$(1 + 2x)^2 = 2x^2 \quad \text{and} \quad x = -1 \pm \sqrt{\tfrac{1}{2}},$$

so x is not an independent *variable.* (We have no right to differentiate with respect to x if x is restricted to the set $\{-1 + \sqrt{\frac{1}{2}}, -1 - \sqrt{\frac{1}{2}}\}$.)

12. (a) $2xy + 2x + 2y + 1 = 0$ is the hypberbola

$$2(x + 1)(y + 1) = 1$$

in the xy-plane.

(b) In 3-space, $\{(x, y, z): 2xy + 2x + 2y + 1 = 0\}$ is a hyperbolic cylinder. Minimum for

$$y = x = -1 + \sqrt{\tfrac{1}{2}};$$

no maximum

Article 15.12

14. $e^x \cosh y + 6\cos(2x - 3y)$
15. $-6/(2x + 3y)^2$ 16. $(y^2 - x^2)(x^2 + y^2)^{-2}$
17. $2y + 6xy^2 + 12x^2y^3$
18. (a) $F(x) = (2bk + k^2)x^3$

(b) $c_1 = \pm\sqrt{a^2 + ah + \frac{1}{3}h^2}$ ($\pm$-sign depending on the signs of a and $a + h$)

(c) $g(y) = 3c_1^2y^2 = (3a^2 + 3ah + h^2)y^2$

(d) $d_1 = b + k/2$

19. $d_2 = b + k/2$, $c_2 = \pm\sqrt{a^2 + ah + \frac{1}{3}h^2}$ ($\pm$-sign depending on the signs of a and $a + h$)

Article 15.13

1. Exact, $f = (2x^5 + 5x^2y^3 + 3y^5)/5 + C$
2. Not exact
3. Exact, $f = x^2 + xy + y^2 + C$
4. Exact, $f = x\cosh y + y\sinh x + C$
5. Not exact
6. Exact, $f = e^x(y - x + 1) + y + C$
7. Not exact

Article 15.14

1. $1/x$ 2. $-1/(2x^2)$
3. $(4 - x)/(2x^3)$ 4. $x/(\sqrt{1 - x^2}\sin^{-1} x)$
5. $(4x - 1)\ln x$ 7. $2xe^{x^2}$
8. $y(x^2\tanh x + 2x\ln\cosh x)$
9. $y\left(\dfrac{2}{x^2 + 1} - \dfrac{\ln(x^2 + 1)}{x^2}\right)$
10. $y\left\{\dfrac{8x + 4}{4x^2 + 4x + 3}[(2x^2 - x)\ln x - x^2 + x] + (4x - 1)\ln x \ln(4x^2 + 4x + 3)\right\}$

Miscellaneous exercises

1. No, $\lim_{x\to 0} f(x, 0) = 1$, $\lim_{x\to 0} f(x, x) = 0$.
2. Yes
4. If f is constant, then f has the same value for all points and there is nothing more to prove. Next, suppose f is not constant. Then there are points Q_1 and Q_2 such that $f(Q_1) \neq f(Q_2)$. Let m be any number between $f(Q_1)$ and $f(Q_2)$. Let C be a circular arc joining Q_1 and Q_2. Since f is continuous along this arc it takes all values between $f(Q_1)$ and $f(Q_2)$. Hence there is a point P_1 on C such that $f(P_1) = m$. Repeat the argument with a different circular arc; there is a point P_2 on it such that $f(P_2) = f(P_1) = m$. There are infinitely many circular arcs joining Q_1 and Q_2, and on each of them is a point P such that and $f(P) = m$.
5. $\dfrac{\partial}{\partial x}(\sin xy)^2 = y\sin 2xy$, $\dfrac{\partial}{\partial y}(\sin xy)^2 = x\sin 2xy$;

$$\frac{\partial}{\partial x}\sin(xy)^2 = 2xy^2\cos(xy)^2,$$

$$\frac{\partial}{\partial y}\sin(xy)^2 = 2x^2y\cos(xy)^2$$

6. $-2/\sqrt{3}$

8. $54(x-3)+2y-27(z+2)=0$,
$(x-3)/54 = y/2 = -(z+2)/27$,
$(54/\sqrt{3649}, 2/\sqrt{3649}, -27/\sqrt{3649})$

9. (a) Hyperboloid of one sheet
(b) $2\mathbf{i}+3\mathbf{j}+3\mathbf{k}$
(c) $2(x-2)+3(y+3)+3(z-3)=0$,
$(x-2)/2 = (y+3)/3 = (z-3)/3$

10. (a) $13(x+2)-(y-1)+12(z-2)=0$
(b) $(x+2)/13 = 1-y = (z-2)/12$

11. $\dfrac{\partial f}{\partial x}=1,\quad \dfrac{\partial f}{\partial y}=3,\quad \dfrac{dw}{ds}=3$

14. (a) $D_u f = f_x u_1 + f_y u_2 + f_z u_3$
(b) $D_v(D_u f) = f_{xx}u_1v_1 + f_{yy}u_2v_2 + f_{zz}u_3v_3 + f_{xy}(u_1v_2+u_2v_1) + f_{xz}(u_1v_3+u_3v_1) + f_{yz}(u_2v_3+u_3v_2)$

15. $\dfrac{dw}{ds}=\sqrt{3}$, max $=\sqrt{3}$ 16. $(f_x = f_y = 2)\ \dfrac{df}{ds} = \frac{14}{5}$

17. 7 18. $-14/\sqrt{6}$ 19. $-\sqrt{\frac{2}{3}}$

21. (a) $\mathbf{N}(x, y, z) = [x(1+3x^2+3y^2)\mathbf{i} + y(1+3x^2+3y^2)\mathbf{j} - \sqrt{x^2+y^2}\,\mathbf{k}]/\sqrt{x^2+y^2}$
(b) $[(1+3x^2+3y^2)^2+1]^{-1/2},\quad 1/\sqrt{2}$

22. $a^2+b^2+c^2=2$

23. (a) $4\mathbf{j}+6\mathbf{k}$ (b) $4(y+1)+6(z-3)=0$

24. $(2\mathbf{i}+6\mathbf{j}-3\mathbf{k})/7$ 26. $\pm\sqrt{3}$

28. $\nabla f = \mathbf{u}_r(\partial f/\partial r) + \mathbf{u}_\theta(1/r)(\partial f/\partial\theta)$

30. (c) $(x^2+y^2+z^2)/2$ 31. $\theta+(\pi/2),\quad 1/r$

34. (a) $f=\phi(bx-ay)$
(b) $f=\phi(x^2+y^2)$, ϕ arbitrary in each case

35. $h'(x) = f_x(x, y) + f_y(x, y)[-g_x(x, y)/g_y(x, y)]$

36. $\partial g/\partial z = 1/(\partial f/\partial x)$

37. $\dfrac{dx}{dz} = \dfrac{\sin y + \sin z - 2y^2\cos z}{(\sin y+\sin z)(\sin x + x\cos x)}$

39. $\dfrac{\partial F}{\partial x} = \dfrac{1}{2}\left(\dfrac{\partial f}{\partial u}+\dfrac{\partial f}{\partial v}\right),\quad \dfrac{\partial F}{\partial y} = \dfrac{1}{2}\left(\dfrac{\partial f}{\partial v}-\dfrac{\partial f}{\partial u}\right)$

42. $\dfrac{\partial^2 f}{\partial u^2} + (y^2+2xy)\dfrac{\partial^2 f}{\partial u\partial v} + 2xy^3\dfrac{\partial^2 f}{\partial v^2} + 2y\dfrac{\partial f}{\partial v}$

43. $a=b=1$ 44. 2

47. $\dfrac{d^2y}{dx^2} = -\dfrac{f_y^3}{f_{xx}f_y^2 - 2f_{xy}f_xf_y + f_{yy}f_x^2}$

49. $dz/dx = (f_y - f_x)/(f_y+f_z)$

50. $x_t = x - v + ax_{vv}/x_v^2$,
$0\le v\le 1,\quad t\ge 0,\quad x(0, t)=0,\quad x(1, t)=1$

55. 60.44

56. Min $(\frac{1}{2}, 0)$, $T=-\frac{1}{4}$,
max $(-\frac{1}{2}, \pm\frac{1}{2}\sqrt{3})$, $T=2\frac{1}{4}$

57. 50

58. (a) Max $z=\sqrt{3}$ at $(0, 0)$, min $z=-\sqrt{3}$ at $(0, 0)$
(b) Min $z=0$ at $(0, 0)$
(c) None

59. $(1, 1, 1)$, $(1, -1, -1)$,
$(-1, 1, -1)$, $(-1, -1, 1)$

60. Length, $\left(\dfrac{c^2V}{ab}\right)^{1/3}$; depth, $\left(\dfrac{b^2V}{ca}\right)^{1/3}$;
height, $\left(\dfrac{a^2V}{bc}\right)^{1/3}$

61. $e^{-2}/6$ 62. Minimum

63. $\left(\sum_{i=1}^{n} a_i^2\right)^{1/2}$ 64. $\left(\dfrac{\sqrt{3}}{2}\right)abc$

66. $\partial^2 z/\partial x\partial y = \cos^{-3}(y+z)[\cos^2(y+z)\sin(x+y) + \sin(y+z)\cos^2(x+y)]$

67. y^2 68. $2xyz(x^2+y^2)^{-2}$

70. $f_{yx}(0, 0) = -1,\quad f_{xy}(0, 0) = 1$

71. $f(x, y) = \frac{1}{5}(2x^5+5x^2y^3+3y^5)+C$

72. $f(x, y) = y\ln x + xe^y + y^2 + C$

73. $w = f(x, y) = x + e^x\cos y + y^2 + C$

Chapter 16

Article 16.1

1.

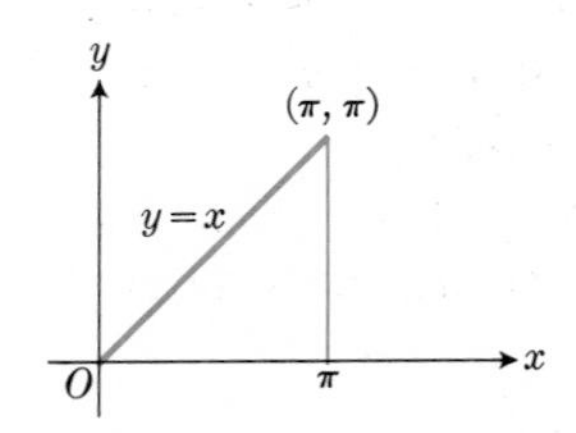

Answer. $(4+\pi^2)/2$

2.

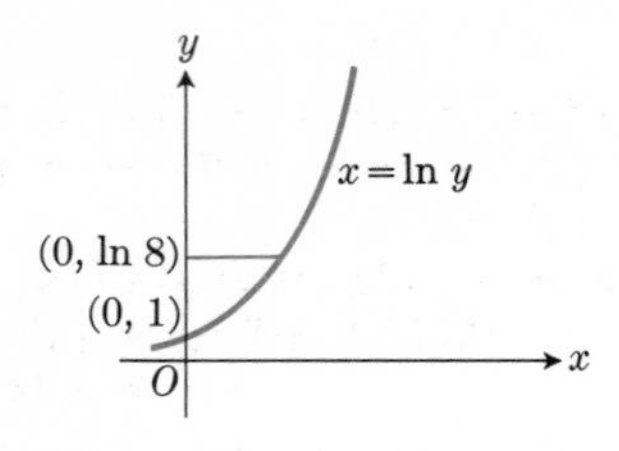

Answer. $8 \ln 8 - 16 + e$

3.

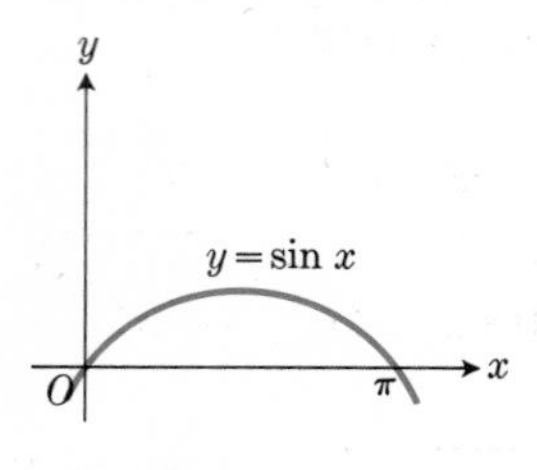

Answer. $\pi/4$

4.

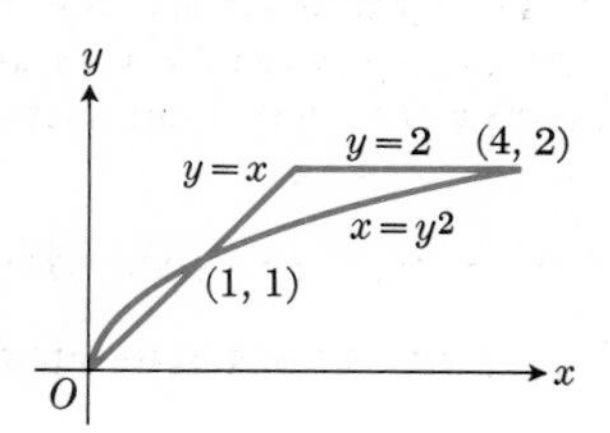

Answer. $\frac{5}{6}$

5. $\int_1^{e^2} \int_{\ln y}^{2} dx\,dy = e^2 - 3$ 6. $\int_0^1 \int_0^{x^2} dy\,dx = \frac{1}{3}$

7. $\int_{-2}^{2} \int_0^{\sqrt{(4-x^2)/2}} y\,dy\,dx = \frac{8}{3}$

8. $\int_{-4}^{5} \int_{(y-2)/3}^{-2+\sqrt{y+4}} dx\,dy = \frac{9}{2}$

9. $V = \int_{-4}^{1} \int_{3x}^{4-x^2} (x+4)\,dy\,dx = \frac{625}{12}$

10. $V = 4\int_0^a \int_0^{\sqrt{a^2-x^2}} \frac{x^2+y^2}{a}\,dy\,dx = \frac{\pi a^3}{2}$

11. $\frac{1}{2}[e^4 - 1]$ 12. 2

13. $f(x_0 + h, y_0 + k) - f(x_0 + h, y_0) - f(x_0, y_0 + k) + f(x_0, y_0)$

14. Approximately $\epsilon^{-2}\pi\epsilon^2(9 + 16 + 2)^{-2/3}$. Limit, $\pi/9$

Article 16.2

1. $a^2/2$ 2. $e - 1$ 3. 4 4. $\frac{9}{2}$
5. $\frac{1}{3}$ 6. $a^2(\pi + 4)/2$ 7. $\frac{4}{3}$

Article 16.3

1. $(a/3, a/3)$ 2. $I_x = (e^3 - 1)/9$
3. $\frac{104}{3}$ 4. $(-\frac{8}{5}, -\frac{1}{2})$
5. $\frac{1}{6}$ 6. $\left(0, a\frac{8 + 3\pi}{40 + 12\pi}\right)$
7. $\frac{64}{105}$

Article 16.4

1. $\int_0^{2\pi} \int_0^a r\,dr\,d\theta = \pi a^2$ 2. $\int_0^{\pi/2} \int_0^a r^3\,dr\,d\theta = \frac{\pi a^4}{8}$

3. $\int_0^{\pi/4} \int_0^a r^2 \cos\theta\,dr\,d\theta = \frac{a^3\sqrt{2}}{6}$

4. $\int_0^{\pi/2} \int_0^{\infty} e^{-r^2} r\,dr\,d\theta = \frac{\pi}{4}$

5. $\int_0^{\pi/4} \int_0^{2\sec\theta} r^2 \sin\theta\,dr\,d\theta = \frac{4}{3}$

6. $\int_0^{\pi/2} \int_0^{2a\cos\theta} r^3 \cos^2\theta\,dr\,d\theta = \frac{5\pi a^4}{8}$

9. $I_0 = a^4(320 + 81\pi)/96$

10. $a^3(15\pi + 32)/24$ 11. $2a^2$

12. $(3\pi + 20 - 16\sqrt{2})2\sqrt{2}\,a^3/9$

Article 16.5

1. $\frac{1}{6}|abc|$ 2. 4 3. 27π
4. $16a^3/3$ 5. $\frac{4}{3}\pi abc$ 6. 4π

Article 16.6

1. $4\pi a^3(8 - 3\sqrt{3})/3$ 2. $\pi/2$
3. $(8\sqrt{2} - 7)\pi a^3/6$ 4. $a^3/3$

Article 16.7

1. $64\pi a^5\left(\frac{4}{15} - \frac{17\sqrt{3}}{160}\right) = 16.7a^5$

2. $2\pi/3$ 3. $\bar{x} = \frac{1}{2}$

4. $\bar{x} = \bar{y} = 0,\quad \bar{z} = 7a/2(8\sqrt{2} - 7)$

5. $\frac{2}{5}Ma^2$

6. $8\pi a^2/3$

7. $64\pi a^5/35$

8. $\frac{3}{20}(a^2 + 4h^2)M$

9. $\frac{7}{5}Ma^2$

10. $\bar{y} = \bar{z} = 0,\quad \bar{x} = \frac{3}{8}\sqrt{2}\,a$

11. $M(3a^2 + 4b^2)/4$

Article 16.8

1. $\bar{x} = \bar{y} = 0,\quad \bar{z} = \dfrac{3a}{16(2 - \sqrt{3})} = 0.7a$

2. $8\pi a^3/3$

3. $a\sqrt{\frac{1270}{651}} \approx 1.4a$

Article 16.9

1. (a) $\displaystyle\int_A\int \frac{\sqrt{F_x^2 + F_y^2 + F_z^2}}{|F_z|}\,dx\,dy$

(b) $\displaystyle\int_B\int \frac{\sqrt{F_x^2 + F_y^2 + F_z^2}}{|F_x|}\,dy\,dz$

(c) $\displaystyle\int_C\int \frac{\sqrt{F_x^2 + F_y^2 + F_z^2}}{|F_y|}\,dx\,dz$

2. $\frac{1}{2}\sqrt{b^2c^2 + a^2c^2 + a^2b^2}$

3. $\pi a^2/2$

4. $2a^2(\pi - 2)$

5. $4a^2$

6. $2\pi a^2(2 - \sqrt{2})$

7. $\pi a^2\sqrt{1 + c^2}$

8. $2\pi a^2/3$

9. $(5\sqrt{5} - 1)\pi a^2/6$

10. The lower left corners give an overestimate, the upper right corners an underestimate.

11. (a) $2\pi ah$
(b) $L_m = 2\pi ah$ remains constant as m increases. The inscribed polygon, in each cross section, approaches its boundary circle as $n \to \infty$, and the triangles in each band approach the corresponding lateral area on the cylinder. The sum is just the total lateral area, independent of m.

Miscellaneous exercises

1. $\displaystyle\int_{-2}^{0}\int_{2x+4}^{4-x^2} dy\,dx = \frac{4}{3}$

2. $\frac{1}{5}$

3. $\displaystyle\int_0^1\int_{-\sqrt{y}}^{\sqrt{y}} dx\,dy$

4. 31.25

5. $\displaystyle\int_0^1\int_x^{2-x} (x^2 + y^2)\,dy\,dx = \frac{4}{3}$

6. (a) $(a^2 + m^2)^{-2}$
$\times \{(m^2 - a^2)\cos ax - 2am\sin ax$
$+ [(a^2 - m^2) + mx(a^2 + m^2)]e^{mx}\}$
(b) $(a^2 + m^2)^{-3}$
$\times \{m(3a^2 - m^2)\cos ax + a(3m^2 - a^2)\sin ax$
$+ [(m/2)(a^2 + m^2)^2x^2 + (a^4 - m^4)x$
$+ m(m^2 - 3a^2)]e^{mx}\}$

8. $\ln(b/a)$

9. $(4a/3\pi,\ 4a/3\pi)$

10. $(2a\sin\alpha/3\alpha,\ 0)$

11. $(28\sqrt{2}/9\pi,\ 28\sqrt{2}(\sqrt{2} - 1)/9\pi)$
(Cartesian coordinates)

12. $(-\frac{12}{5},\ 2)$

14. (a) $\frac{1}{3}(a^3t + at^3 - t^4)$ (b) 0

16. $A = a^2\cos^{-1}(b/a) - b\sqrt{a^2 - b^2}$,
$I_0 = (a^4/2)\cos^{-1}(b/a) - b\sqrt{a^2 - b^2}\,(a^2 + 2b^2)/6$

17. $b\sqrt{2/5}$

18. (a) $(a/2)\sqrt{\pi/4 - \frac{2}{3}}$ (b) $(a/2)\sqrt{\pi/4 + \frac{2}{3}}$

20. When the rectangle is small, the integrand stays near its value at (x_0, y_0), and the value of the integral is approximately the area of the rectangle times that number; in other words,

$$(x_1 - x_0)\,(y_1 - y_0)F_{xy}(x_0, y_0).$$

Now divide by the area of the rectangle and take the limit.

21. $\displaystyle\int_{-\pi/2}^{\pi/2}\int_0^a r^2\cos\theta\,dr\,d\theta$

24. $a > 0,\quad c > 0,\quad ac - b^2 = \pi^2$

25. $A = 2a^2,\quad I_y = (3\pi + 8)a^4/12$

26. (a) $(\pi - 2)/4$ (b) $(\sqrt{3}/4)\tan^{-1}\frac{4}{3}$

27. $\displaystyle K(a) = \int_0^{a\cos\beta} dx\int_0^{x\tan\beta}\ln(x^2 + y^2)\,dy + \int_{a\cos\beta}^{a} dx\int_0^{\sqrt{a^2-x^2}}\ln(x^2 + y^2)\,dy$

28. $\pi/8$

29. $(0, 0, \frac{1}{4})$

30. $(4\pi a^3/3)(\sqrt{2} - \frac{7}{8})$

31. $\pi/4$

32. $\pi/\sqrt{2}$

33. $49\pi a^5/15$

34. $\frac{4}{3}\pi abc$

35. $a^2b^2c^2/6$

36. $16a^3/3$

37. $x^2 + y^2 + z^2 = 4,\quad x^2 + y^2 = 2x,\quad y = 0,$

$$\int_0^{\pi/2}\int_0^{2\cos\theta}\int_{-\sqrt{4-r^2}}^{\sqrt{4-r^2}} dz\,r\,dr\,d\theta$$

38. $\frac{4\pi a^2}{3} + \frac{16}{3}\left[\frac{b}{2}(3a^2 - b^2)\sin^{-1}\frac{b}{\sqrt{a^2-b^2}} + \frac{b^2}{2}\sqrt{a^2-2b^2} - a^3\tan^{-1}\frac{a}{\sqrt{a^2-2b^2}}\right]$

39. (a) Radius of sphere, 2; of hole, 1 (b) $4\pi\sqrt{3}$

40. $\int_0^1 \int_{\sqrt{1-x^2}}^{\sqrt{3-x^2}} \int_1^{\sqrt{4-x^2-y^2}} xyz^2\,dz\,dy\,dx + \int_1^{\sqrt{3}} \int_0^{\sqrt{3-x^2}} \int_1^{\sqrt{4-x^2-y^2}} xyz^2\,dz\,dy\,dx$

41. $3\pi a^3/4$ 42. $2a^3(3\pi - 4)/9$ 43. $\pi a^5/12$

44. $\pi^2 a^3/4$ 45. $8\pi\,\delta(b^5 - a^5)/15$

46. $6GmMa^{-2}[1 - h(a^2 + h^2)^{-1/2}]$, where $M = \frac{1}{3}\pi a^2 h\delta$ = mass of cone

47. $\rho_0(1 - \frac{1}{2}\epsilon^2)$ 48. $2\pi(\sqrt{3} - 1)$

49. $\pi\sqrt{3}$ 50. $\pi\sqrt{2}\,a^2$

51. $2a^2(\pi - 2)$ 52. $\pi\sqrt{2}$

53. (a) 144π (b) -16π (c) 128π

54. $4\int_0^a \sqrt{a^2 - y^2}\,\sqrt{4y^2 + 1}\,dy$

55. $16\pi(\sqrt{2} - 1)$

56. $8\pi^2$ 57. $\pi^2/2$ 58. $8r^2$ 61. $z = 1$

Chapter 17

Article 17.1

1. (a) $x = 0,\ y = 0$ (b) $x^2 + y^2 = a^2$
2. $\pi a^4/2$ 4. Helix
5. Replace x, y, z by $x - x_0, y - y_0, z - z_0$.
6. $\mathbf{F}(x, y, z) = \exp(2y + 3z)[2x\mathbf{i} + 2x^2\mathbf{j} + 3x^2\mathbf{k}]$
7. $\mathbf{F}(x, y, z) = 2(x^2 + y^2 + z^2)^{-1}[x\mathbf{i} + y\mathbf{j} + z\mathbf{k}]$
8. $\mathbf{F}(x, y, z) = (x^2 + y^2 + z^2)^{-1}[yz\mathbf{i} + xz\mathbf{j} - xy\mathbf{k}]$
9. $\mathbf{F}(x, y, z) = 2\mathbf{i} - 3\mathbf{j} + 5\mathbf{k}$
10. $\mathbf{F}(x, y, z) = n(x^2 + y^2 + z^2)^{(n-2)/2}[x\mathbf{i} + y\mathbf{j} + z\mathbf{k}]$

Article 17.2

4. πa^3 5. $\frac{28}{3}\sqrt{14}$
6. $\frac{28}{3}\sqrt{14}$ 7. $\pi a^2/2$
8. 0 9. $\pi a^4/6$
10. $\pi a^4/3$

Article 17.3

1. $\frac{5}{6}\sqrt{2}$ 3. π 4. 0
5. (a) 0 (b) $-\frac{1}{3}$ 8. $n = -1$
9. (a) $\frac{9}{2}$ (b) $\frac{9}{2}$ (c) $\frac{9}{2}$
10. (a) 2.143 (b) 2.538 (c) $2 + \sin 1$
11. (a) 3 (b) 3 (c) 3
12. (a) e^3 (b) e^3 (c) e^3
13. (a) $\sin 1 = 0.8415$ (b) 0.8415 (c) 0.8415
14. $f(x, y, z) = x^2 + \frac{3}{2}y^2 + 2z^2 + C$
15. $f(x, y, z) = xy + yz + zx + C$
16. $f(x, y, z) = xe^{y+2z} + C$
17. $f(x, y, z) = xy\sin z + C$
19. 8 20. 24 21. $-e^{-2\pi} + 1$ 25. (b) $x\mathbf{F} = \mathbf{0}$

Article 17.4

2. $\pi/2$ 3. $2\pi\delta/a$
5. (a) 0 (b) 4π 6. (a) 0 (b) 4π

Article 17.5

3. 0 4. 0 5. $\pi a^4/2$ 6. 0
7. $-\pi a^2$ 11. 0 12. -2
13. $\int_C M(x, y)\,dy - N(x, y)\,dx = \iint_R (\partial M/\partial x + \partial N/\partial y)\,dx\,dy$

Article 17.6

1. 0 2. 24 3. 0 4. 16 5. 0
6. 24π 7. 48π 8. 135π 9. $4\pi a^3$ 10. 24

Article 17.7

1. 0 2. 0 3. $-\pi/4$ 4. 0
5. Zero, because curl grad $\phi = \mathbf{0}$.

Miscellaneous exercises

5. 0 6. π 7. $\frac{\pi}{60}(17^{5/2} - 41)$
8. $4\pi a^4/3$ 9. $8\pi a^3$ 10. $-8\pi a$
11. $-\frac{25}{3}$ 12. $\pi a/2$ 13. 0
14. 2π 15. 2π 16. $36\pi\sqrt{14}$
17. 0 18. 0 19. 0 20. 0
31. (a) $C_1 = -100,\ C_2 = 200$ (b) $C_1 = 100,\ C_2 = 0$

Chapter 18

Article 18.1

1. Converges to $-\frac{2}{3}$
2. Converges to $\frac{1}{2}$
3. Converges to 1
4. Converges to 0
5. Diverges
6. Diverges
7. Converges to 1
8. Converges to 0
9. Diverges
10. Converges to 0

13. (b) $a_n = \sin n$ is an example, $a_n = -n$ is another example.

14. $s_n = 1 - \dfrac{1}{n+1}$, sum $= 1$

15. Same as Exercise 14. Note that

$$\frac{1}{n\cdot(n+1)} = \frac{1}{n} - \frac{1}{n+1}.$$

16. $s_n = \ln \dfrac{1}{n+1}$, diverges

17. $s_n = 2 - \dfrac{1}{2^{n-1}}$, sum $= 2$

18. $s_n = 3 - \dfrac{1}{3^{n-1}}$, sum $= 3$

19. $s_n = \dfrac{10}{3}\left[1 + \dfrac{(-1)^{n+1}}{2^n}\right]$, sum $= \dfrac{10}{3}$

20. $s_n = \dfrac{1}{11}\left(1 - \dfrac{1}{100^n}\right)$, sum $= \dfrac{1}{11}$

21. 28 ft

22. (a) $\frac{234}{999}$
(b) Yes, because it is of the form

$$p\cdot 10^{-k} + p\cdot 10^{-2k} + \cdots = p/(10^k - 1),$$

where p and $10^k - 1 = q$ are integers.

23. $\dfrac{124}{100} + \dfrac{123}{99{,}900} = \dfrac{123{,}999}{99{,}900}$

24. (d) $n = 6$ (e) $n = 998$

Article 18.2

2. If the sequence is bounded below, then it has a definite finite limit. Otherwise, it diverges to minus infinity.
3. The limit exists and is less than or equal to three.
4. No, diverges, by oscillation
5. Diverges, by the integral test
6. The nth term approaches 1. Series diverges to plus infinity, by comparison with $\sum 1/n$.
7. Converges, by comparison with $\sum 1/2^n$
8. Diverges, by integral test
9. Diverges, by comparison with $\sum 1/n$
10. Diverges, by integral test
11. Converges, by ratio test
12. Converges, by comparison with $\sum 1/n^2$
13. Converges, by comparison with $\sum 1/n^{3/2}$

14–17. Converge, by ratio test

18. Diverges, by ratio test
19. Converges, by ratio test

Article 18.3

1. $1 - x + \dfrac{x^2}{2!} - \dfrac{x^3}{3!} + \cdots + \dfrac{(-1)^n x^n}{n!}$

2. $x - \dfrac{x^3}{3!} + \dfrac{x^5}{5!} - \dfrac{x^7}{7!} + \cdots + \dfrac{(-1)^m x^{2m+1}}{(2m+1)!}$

3. $1 - \dfrac{x^2}{2!} + \dfrac{x^4}{4!} - \dfrac{x^6}{6!} + \cdots + \dfrac{(-1)^m x^{2m}}{(2m)!}$

4. $\dfrac{\sqrt{2}}{2}\left(1 + x - \dfrac{x^2}{2!} - \dfrac{x^3}{3!} + \dfrac{x^4}{4!} + \dfrac{x^5}{5!} - \dfrac{x^6}{6!} - \cdots + \dfrac{x^{4m}}{(4m)!} + \dfrac{x^{4m+1}}{(4m+1)!} - \dfrac{x^{4m+2}}{(4m+2)!} - \dfrac{x^{4m+3}}{(4m+3)!}\right)$

5. $x + \dfrac{x^3}{3!} + \dfrac{x^5}{5!} + \dfrac{x^7}{7!} + \cdots + \dfrac{x^{2m+1}}{(2m+1)!}$

6. $1 + \dfrac{x^2}{2!} + \dfrac{x^4}{4!} + \cdots + \dfrac{x^{2m}}{(2m)!}$

7. $(x-1) - \dfrac{(x-1)^2}{2} + \dfrac{(x-1)^3}{3} - \dfrac{(x-1)^4}{4} + \cdots$

8. $2 + \dfrac{1}{2}\dfrac{(x-4)}{2} - \dfrac{1}{2\cdot 4}\dfrac{(x-4)^2}{2^3} + \dfrac{1\cdot 3}{2\cdot 4\cdot 6}\dfrac{(x-4)^3}{2^5} - \dfrac{1\cdot 3\cdot 5}{2\cdot 4\cdot 6\cdot 8}\dfrac{(x-4)^4}{2^7} + \cdots$

9. $-1 - (x+1) - (x+1)^2 - (x+1)^3 - \cdots$

10. $1 + 2(x - \pi/4) + 2(x - \pi/4)^2 + \cdots$

11. $-\ln 2 - \sqrt{3}(x - \pi/3) - 2(x - \pi/3)^2 - \cdots$

12. $\dfrac{1}{2} + \dfrac{\sqrt{3}}{2}\left(x - \dfrac{\pi}{6}\right) - \dfrac{1}{2}\dfrac{(x - \pi/6)^2}{2!} - \dfrac{\sqrt{3}}{2}\dfrac{(x-\pi/6)^3}{3!} + \cdots$

13. $\frac{\sqrt{2}}{2}\left[1+\left(x+\frac{\pi}{4}\right)-\frac{(x+\pi/4)^2}{2!}-\frac{(x+\pi/4)^3}{3!}+\cdots\right]$

Article 18.4

4. $|x| < \sqrt[5]{0.06} \approx 0.57$
5. $|R_3(x, 0)| < 10^{-4}/24$
6. $-26+\sqrt{675} \leq x \leq 0.02$
7. $|R_1(x, 0)| < \frac{x^2}{8(0.99)^{3/2}} < 1.3 \times 10^{-5}$
9. $n \geq 4$

Article 18.5

2. (a) Minimum at $x = 1, \quad y = -2$
   (b) Saddle point at $x = 1, \quad y = 2$
   (c) Minimum at $x = 1, \quad y = 0$
   (d) Saddle point at $x = 0, \quad y = 0$
   (e) Maximum at $x = 0, \quad y = 1$
   (f) Saddle point at $(0, 0, 6)$, relative maximum at $(-\frac{2}{3}, \frac{2}{3}, 6+\frac{8}{27})$
   (g) Saddle points at $(0, 0, -8)$, and $(-2, 2, -8)$, relative maximum at $(-2, 0, -4)$, and relative minimum at $(0, 2, -12)$
3. If $f_x(a, b) = f_y(a, b) = 0$, then $f(x, y)$ has at the point $x = a, y = b$:
   (a) a relative minimum, if d^2f/ds^2 is positive in all directions emanating from (a, b),
   (b) a relative maximum, if d^2f/ds^2 is negative in all directions emanating from (a, b),
   (c) a saddle point, if d^2f/ds^2 is positive in some directions and negative in other directions emanating from (a, b).
4. $1+x+\frac{x^2-y^2}{2!}+\cdots$
5. $$\begin{aligned} f(x, y, z) = {} & f(a, b, c)+f_x(a, b, c)\cdot(x-a) \\ & +f_y(a, b, c)\cdot(y-b) \\ & +f_z(a, b, c)\cdot(z-c) \\ & +\frac{1}{2!}[f_{xx}(a, b, c)\cdot(x-a)^2 \\ & +f_{yy}(a, b, c)\cdot(y-b)^2 \\ & +f_{zz}(a, b, c)\cdot(z-c)^2 \\ & +2f_{xy}(a, b, c)\cdot(x-a)(y-b) \\ & +2f_{xz}(a, b, c)\cdot(x-a)(z-c) \\ & +2f_{yz}(a, b, c)\cdot(y-b)(z-c)]+\cdots \end{aligned}$$

Article 18.6

1. 0.85717
2. 1.0355
3. 2.08008
4. 0.48836
5. 1.1276
6. 0.5211
7. 0.22314
8. $\pi/4+0.01 = 0.795$
10. 0.09994
11. 0.09967
15. (c) 6 (d) $1/q$
16. (a) $E(X) = 2$ (b) $E(X) = 6$
    (c) The series for $E(X)$ diverges.

Article 18.7

1. $-\frac{1}{24}$
2. $\frac{1}{4}$
3. $-\frac{1}{9}$
4. $\frac{1}{3}$
5. $\frac{1}{2}$
6. $\frac{1}{3}$
7. $\frac{1}{3}$
8. $\frac{1}{2}$
9. $-\frac{1}{3}$
10. 0
11. 0
12. 0
13. 1
14. $-\frac{1}{2}$
15. $+\infty$
16. e^{-6}
17. e^2
18. e^2
19. 1
20. 1
21. $e^{-1/2}$
22. $e^{-1/2}$
23. 1
24. e
25. (b) $\frac{1}{2}$
26. $r = -3, \quad s = \frac{9}{2}$

Article 18.8

1. $$\frac{1}{2}+\sum_{n=1}^{\infty}\frac{1-\cos n\pi}{n\pi}\sin nx = \frac{1}{2}+\frac{2}{\pi}\left(\sin x+\frac{\sin 3x}{3}+\frac{\sin 5x}{5}+\cdots\right)$$
2. $-\frac{4}{\pi}\left(\sin x+\frac{\sin 3x}{3}+\frac{\sin 5x}{5}+\cdots\right)$
3. $\pi-2\left(\sin x+\frac{\sin 2x}{2}+\frac{\sin 3x}{3}+\cdots\right)$
4. $\frac{\pi}{2}-\frac{4}{\pi}\left(\cos x+\frac{\cos 3x}{3^2}+\frac{\cos 5x}{5^2}+\cdots\right)$
5. $2\left(\sin x-\frac{\sin 2x}{2}+\frac{\sin 3x}{3}-\frac{\sin 4x}{4}+\cdots\right)$

Article 18.9

1. $-1 < x < 1$
2. $-1 < x < 1$
3. $-1 < x < 1$
4. $-2 < x < 2$
5. $-\infty < x < \infty$
6. $-\infty < x < \infty$
7. $0 < x < 2$
8. $-4 < x < 0$
9. $|x-3| < 1/\sqrt{2}$
10. $-\infty < x < \infty$

Article 18.10

1. $|x| < 1$, diverges at both $x = \pm 1$
2. $|x| < 1$, converges at both $x = \pm 1$
3. $|x - 2| \leq 1$
4. $|x + 1| < 1$, converges at $x = 0$, diverges at -2
5. $|x - 2| < 2$, diverges at $x = 0$, converges at $x = 4$
6. $|x - 1| < \frac{1}{2}$, diverges at both $x = \frac{1}{2}$, $x = \frac{3}{2}$
7. $-\infty < x < \infty$
8. Converges only at $x = 0$
9. $|x| < 1$, converges at both $x = \pm 1$
10. $|x + 3| < 1$, converges at $x = -4$, diverges at -2

Miscellaneous exercises

1. $s_n = \ln[(n + 1)/(2n)]$, series converges to $-\ln 2$.
2. $s_n = \frac{1}{2}[\frac{3}{2} - 1/n - 1/(n + 1)]$, limit $= \frac{3}{4}$
6. (a) $\sum_{n=2}^{\infty} (-1)^n x^n$
   (b) No, because it will converge in an interval symmetric about $x = 0$, and it cannot converge when $x = -1$.
7. $(1 - t^2)^{-1/2} = \sum_{k=0}^{\infty} 2^{-2k}(k!)^{-2}(2k)!\, t^{2k}$, $\quad -1 < t < 1$;
   $\sin^{-1} x = \sum_{k=0}^{\infty} 2^{-2k}(k!)^{-2}(2k)!\, x^{2k+1}/(2k + 1)$, $\quad -1 \leq x \leq 1$
8. $(1 + t^2)^{-1/2} = \sum_{k=0}^{\infty} (-1)^k 2^{-2k}(k!)^{-2}(2k)!\, t^{2k}$, $\quad -1 \leq t \leq 1$;
   $\sinh^{-1} x = \sum_{k=0}^{\infty} \dfrac{(-1)^k 2^{-2k}(k!)^{-2}(2k)!\, x^{2k+1}}{2k + 1}$, $\quad -1 \leq x \leq 1$
9. $e^{\sin x} = 1 + x + \dfrac{x^2}{2!} - \dfrac{3x^4}{4!} - \cdots$
11. (a) $\ln \cos x = -\dfrac{x^2}{2} - \dfrac{x^4}{12} - \dfrac{x^6}{45} - \cdots$
    (b) -0.00017
12. 0.946
13. 0.747
14. $\sqrt{2}[1 + (x - 1)/2 + (x - 1)^2/8 + \cdots]$
15. $\sum_{n=0}^{\infty} (-1)^{n+1}(x - 2)^n$, $\quad 1 < x < 3$
16. $\tan x = x + \dfrac{x^3}{3} + \dfrac{2x^5}{15} + \cdots$
17. $\sum_{n=0}^{\infty} \dfrac{(-1)^n (x - 3)^n}{4^{n+1}}$
18. $\cos x = \dfrac{1}{2} \sum_{n=0}^{\infty} (-1)^n (x - \pi/3)^{2n}/(2n)!$
    $+ \dfrac{\sqrt{3}}{2} \sum_{n=0}^{\infty} (-1)^n (x - \pi/3)^{2n+1}/(2n + 1)!$
19. $\sum_{n=0}^{\infty} \dfrac{(-1)^n (x - \pi)^n}{\pi^{n+1}}$
20. 1.543
22. $e^{e^x} = e\left(1 + x + \dfrac{2x^2}{2!} + \dfrac{5x^3}{3!} + \dfrac{15x^4}{4!} + \dfrac{52x^5}{5!} + \cdots\right)$
23. $\tan^{-1} 0.05 = 0.0500$, $\tan^{-1} 0.06 = 0.0599$, $\tan^{-1} 0.1 = 0.0997$
24. 0.0027
25. -0.0011
26. $-\infty$
27. $e^{-1/6}$
28. (a) $-\dfrac{\pi}{4} - \dfrac{2}{\pi} \sum_{n=0}^{\infty} \dfrac{\cos(2n + 1)x}{(2n + 1)^2}$
    $+ \sum_{n=1}^{\infty} \left[\dfrac{3 \sin(2n - 1)x}{2n - 1} - \dfrac{\sin 2nx}{2n}\right]$
    (b) $\pi^2/8$
29. Converges, by the integral test
30. Diverges, since nth term doesn't approach zero
32. (a) Diverges (b) Diverges
    (c) Diverges (d) Converges
    (e) Diverges (f) Diverges
    (g) Converges (h) Converges
    (i) Diverges (j) Diverges
    (k) Diverges (l) Converges
33. $\frac{1}{2}$
35. (b) Converges (c) Converges, by comparison test
37. (a) $-5 \leq x < 1$ (b) $-\infty < x < \infty$
    (c) $-\infty < x < \infty$ (d) $-e < x < e$
    (e) $1 < x < 5$ (f) $-1 < x < 5$
    (g) $0 \leq x \leq 2$ (h) $-1 \leq x < 1$
38. (a) $-\infty < x < \infty$
    (b) $(6n - 1)\pi/6 \leq x \leq (6n + 1)\pi/6$, where $n = 0, \pm 1, \pm 2, \ldots$
    (c) $x \geq \frac{1}{2}$

39. (a) $-\infty < x < \infty$
(b) $y'' = xy \quad (a = 1, b = 0)$

41. (a) Diverges, because the nth term approaches e^{-1}, and not zero, as $n \to \infty$
(b) Converges, by comparison with $\sum e^{-n}$

Chapter 19

Article 19.1

1. (a) $(14, 8)$ (b) $(-1, 8)$ (c) $(0, -5)$
2. (a) $(-\frac{7}{3}, -24)$ (b) $(\frac{2}{5}, -\frac{1}{5})$ (c) $(-1, 0)$

Article 19.2

2. (a) By reflecting z across the real axis
(b) By reflecting z across the imaginary axis
(c) By reflecting z in the origin
(d) By reflecting z in the real axis and then multiplying the length of the vector by $1/|z|^2$

6. On the real axis

9. (a) Points on the circle $x^2 + y^2 = 4$
(b) Points inside the circle $x^2 + y^2 = 4$
(c) Points outside the circle $x^2 + y^2 = 4$

10. Points on a circle of radius 2, center $(1, 0)$

11. Points on a circle of radius 1, center $(-1, 0)$

12. Points on the y-axis

13. Points on the line $y = -x$

14. $4 \text{ cis } (2\pi/3)$ 15. $1 \text{ cis } (\pi/2)$ 16. $1 \text{ cis } (2\pi/3)$

17. $\sqrt{65} \text{ cis } (-\tan^{-1} 0.125) = \sqrt{65} \text{ cis } (-7°7'.5)$

18. $\cos 4\theta = \cos^4 \theta - 6 \cos^2 \theta \sin^2 \theta + \sin^4 \theta$,
$\sin 4\theta = 4 \sin \theta \cos \theta (\cos^2 \theta - \sin^2 \theta)$

19. $1, \quad \text{cis } (2\pi/3) = (-1 + i\sqrt{3})/2$,
$\text{cis } (-2\pi/3) = (-1 - i\sqrt{3})/2$

20. $\text{cis } (\pi/4, 5\pi/4) = \pm\text{cis } (\pi/4) = \pm(1 + i)/\sqrt{2}$

21. $2i, \quad 2 \text{ cis } (-5\pi/6) = -\sqrt{3} - i$,
$2 \text{ cis } (-\pi/6) = \sqrt{3} - i$

22. $2 \text{ cis } (0°, 60°, 120°, 180°, 240°, 300°)$; that is,

$$2, \quad 1 + i\sqrt{3}, \quad -1 + i\sqrt{3}, \quad -2, \quad -1 - i\sqrt{3}, \quad 1 - i\sqrt{3}.$$

23. $\pm\sqrt{2} \text{ cis } (\pm 30°)$; that is,

$$\frac{\sqrt{3} + i}{\sqrt{2}}, \quad -\frac{\sqrt{3} + i}{\sqrt{2}}, \quad \frac{\sqrt{3} - i}{\sqrt{2}}, \quad \frac{-\sqrt{3} + i}{\sqrt{2}}$$

24. $\sqrt[6]{2} \text{ cis } (45°, 165°, 285°, 75°, 195°, 315°)$

25. $2 \text{ cis } (60°, 120°, 240°, 300°)$, or

$$1 \pm i\sqrt{3}, \quad -1 \pm i\sqrt{3}$$

26. $\text{cis } (45°, 135°, 225°, 315°)$, or

$$(1 \pm i)/\sqrt{2}, \quad (-1 \pm i)/\sqrt{2}$$

Article 19.3

1. (a) $|w| = 1, \quad 0 \leq \arg w \leq 2\pi$
(b) $|w| = 4, \quad \pi \leq \arg w \leq 2\pi$
(c) $\arg w = \pi/2$
(d) $|w| < 1, \quad -2\pi < \arg w \leq 0$
(e) The u-axis, $u \geq 0$
(f) The u-axis, $u \leq 0$

Article 19.4

1. $-2/(z_0 - 1)^2 = +2$ 2. $3(z_0 + 1)^2 = -12$
3. $z_0/\sqrt{z_0^2 + 1} = 2^{-1/4} \text{ cis } (\pi/8, 9\pi/8)$

Article 19.5

1. $u = x^2 - y^2, \quad v = 2xy$
2. $u = x^3 - 3xy^2, \quad v = 3x^2y - y^3$
3. $u = x^4 - 6x^2y^2 + y^4, \quad v = 4x^3y - 4xy^3$
4. $u = (x^2 - y^2)/(x^2 + y^2)^2, \quad v = -2xy/(x^2 + y^2)^2$

Article 19.6

1. $|z| < 1$ 2. $|z| < 1$
3. $|z - 1| < 1$ 4. $|z + i| < 1$
5. $|z + 1| < 2$ 9. 0.909

Article 19.7

1. $1.59 + 1.32i$
5. (a) $-1 \pm 2i$ (b) $1 \pm i, \quad -1 \pm i$
(c) $2, \quad -1 \pm i\sqrt{3}$

18. (a) The circle $u^2 + v^2 = e^{2x}$, described once in the counterclockwise direction each time y increases through a 2π-interval of values; for instance, $-\pi < y < \pi$.
(b) The branch of the hyperbola

$$\frac{u^2}{\sin^2 x} - \frac{v^2}{\cos^2 x} = 1$$

(x not a multiple of $\pi/2$) on which u and $\sin x$ have the same sign. The lines

$$x = n\pi, \quad n = 0, \pm 1, \ldots,$$

map into the v-axis. The lines

$$x = \pi/2 + 2n\pi, \quad n = 0, \pm 1, \ldots,$$

map into the u-axis, $u \geq 1$. The lines

$$x = -\pi/2 + 2n\pi, \quad n = 0, \pm 1, \ldots,$$

map into the u-axis, $u \leq -1$.

19. $$\int e^{ax} \cos bx\, dx = \frac{e^{ax}}{a^2 + b^2}(a \cos bx + b \sin bx) + C,$$

$$\int e^{ax} \sin bx\, dx = \frac{e^{ax}}{a^2 + b^2}(a \sin bx - b \cos bx) + C$$

Article 19.8

1. (a) $\ln 2^{3/2} - i(\pi/4)$ (b) $\ln 2 + i(\pi/6)$
   (c) $\ln 4 + i\pi$ (d) $\ln 4$
   (e) $\ln 2 + i(\pi/2)$ (f) $i(\pi/2)$
2. (a) $\ln 2 + 2n\pi i$ (b) $\ln 2 + (2n + 1)\pi i$
   (c) $\ln 2 + (2n + \frac{1}{2})\pi i$ (d) $\ln 2 + (2n - \frac{1}{2})\pi i$
   (e) $\ln 2 + (2n + \frac{5}{6})\pi i$, $n = 0, \pm 1, \pm 2, \ldots$
4. (b) $\left(\frac{\pi}{2} + 2n\pi\right) - i \ln (3 + 2\sqrt{2})$
5. (a) Locus is a straight line segment joining the points $(\ln |z|, -\pi)$ and $(\ln |z|, +\pi)$.
   (b) Locus is the line $v = \arg z$, parallel to the real axis.

Miscellaneous exercises

2. (a) $r = 4\sqrt{2}$, $\theta = -\tan^{-1}(2 + \sqrt{3})$
   (b) $r = \sqrt{2}$, $\theta = -\pi/12, 7\pi/12, 15\pi/12$
3. (a) $\pm(\sqrt{2 + \sqrt{2}} - i\sqrt{2 - \sqrt{2}})$,
   $\pm(\sqrt{2 - \sqrt{2}} + i\sqrt{2 + \sqrt{2}})$
   (b) $\sin^{-1} 5 = 2n\pi + \pi/2 - i \ln (5 + \sqrt{24})$, $n = 0, \pm 1, \pm 2, \ldots$
4. (a) $\pm\sqrt{2}(1 + i)$, $\pm\sqrt{2}(1 - i)$
   (b) $2e^{i(2n\pi + \pi)/5}$, $n = 0, 1, 2, 3, 4$
5. (a) $2^{1/4} e^{i\pi(2 + 3n)/6}$, $n = 0, 1, 2, 3$
   (b) $e^2/\sqrt{2} + i(e^2/\sqrt{2})$
6. (a) $\pm(\sqrt{3} - i)$ (b) $\ln 4 - i\pi/3$
7. $r = 5$, $\theta = \tan^{-1} \frac{4}{3}$
8. $\ln (3\sqrt{2}) + i(2n\pi - \pi/4)$, n any integer
9. $\ln 2 - i\pi/3$
10. (a) $2^{1/6} \operatorname{cis}\left(\frac{5\pi}{12} + \frac{2n\pi}{3}\right)$, n any integer
    (b) $\ln (3\sqrt{2}) + i(2n\pi + \pi/4)$, n any integer
    (c) $-e^2$
11. (b) Derivative doesn't exist.
19. (a) $2e^{i2n\pi/5}$, $n = 0, 1, 2, 3, 4$
    (b) $\ln 2 + (2n + 1)\pi i$, n any integer
    (c) $2n\pi \pm i \cosh^{-1}(10)$, or $2n\pi + i \ln (10 \pm 3\sqrt{11})$, n any integer
    (d) $\ln \sqrt{3} + (2n + 1)\pi i/2$, n any integer
    (e) $e^{-(4n+1)\pi/2}$, n any integer
    (f) -3, $\pm i\sqrt{3}$

Chapter 20

Article 20.3

1. $(x^2 + 1)(2y - 3) = C$
2. $\frac{2}{3}\sqrt{x^3 + 1} + \frac{1}{2} \ln (y^2 + 1) = C$
3. $e^y = e^x + C$
4. $y^{3/2} = 3\sqrt{x/2} + C$
5. $\sinh 2y - 2 \cos x = C$
6. $\frac{1}{2}(\ln |x|)^2 = \ln |y| + C$
7. $(y - 1)e^y + \frac{x^2}{2} + \ln |x| = C$
8. $\sqrt{2x^2 + 3} - 2\sqrt{4 - y^2} = C$
9. $\cosh^{-1} y + \sinh^{-1} x = C$
10. $-y \cos y + \sin y = -x^{-1} + \ln |x| + C$

Article 20.4

1. $2 \tan^{-1} \frac{y}{x} + \ln (x^2 + y^2) = C$
2. $\frac{x}{y} = \ln |x| + C$; or $y = 0$
3. $\ln |x| + e^{-y/x} = C$
4. $\ln |y| + \frac{2}{\sqrt{3}} \tan^{-1} \frac{2y - x}{x\sqrt{3}} = C$
5. $y^2 = 2cx + c^2$
6. (a) $x^2 + y^2 = Cx$ (b) $x^2 + y^2 = Cy$

Article 20.5

1. $y = e^{-x} + Ce^{-2x}$
2. $y = \frac{1}{2}(x + C)e^{x/2}$
3. $x^3y = C - \cos x$
4. $xy = C - \cos x$
5. $x = \dfrac{y}{2} + \dfrac{C}{y}$
6. $(x-1)^4 y = \dfrac{x^3}{3} - x + C$
7. $y \cosh x = C - e^x$
8. $xe^{2y} = y^2 + C$
9. $xy = y^2 + C$
10. $x = (C - y)/(1 + y^2)$

Article 20.6

1. $\dfrac{x^2}{2} + xy + \dfrac{y^3}{3} = C$
2. $e^x + e^y(x^2 + 1) = C$
3. $x^2y + xy^2 - \dfrac{y^2}{2} = C$
4. $x^2 + 2x\sqrt{y^2+1} - y^2 = C$
5. $2y + x^3 = Cx$
6. $y = x \tan(x + C)$
7. $y = x(C - x - \ln|x|)$
8. $e^x + x \ln y + y \ln x - \cos y = C$
9. $y^2 \tan^{-1} x - 2xy + \cosh y = C$
10. $xy + \cos x = C$

Article 20.7

1. $y = C_1e^{-x} + C_2$
2. $y = C$, or $y = -2a \tan(ax + C)$,
   or $y = 2a \tanh(ax + C)$,
   or $y = 2/(x + C)$
3. $y = C_1 \int e^{-x^2/2}\,dx + C_2$. Note that the integral appearing here could be evaluated as an infinite series, giving
$$y = C_1\left[x - \frac{x^3}{2\cdot 3} + \frac{x^5}{2\cdot 4\cdot 5} - \frac{x^7}{2\cdot 4\cdot 6\cdot 7} + \cdots\right] + C_2.$$
4. $y = C_1 \ln|x| + C_2$
5. $y = C_1 \sinh(x + C_2)$
6. $y = C_1 \sin(\omega x + C_2)$
7. $x = A \sin(\sqrt{k/m}\,t + \pi/2) = A \cos(\sqrt{k/m}\,t)$
8. $s = \dfrac{m^2g}{k^2}\left[\dfrac{kt}{m} + e^{-kt/m} - 1\right]$,
   where m is man's mass, g is the acceleration due to gravity, and k is the factor of proportionality in the air resistance $-kv$ (v being velocity)

Article 20.9

1. $y = C_1 + C_2e^{-2x}$
2. $y = C_1e^{-2x} + C_2e^{-3x}$
3. $y = C_1e^{-x} + C_2e^{-5x}$
4. $y = C_1e^{3x} + C_2e^{-x}$
5. $y = e^{-x/2}[C_1 \cos(x\sqrt{3}/2) + C_2 \sin(x\sqrt{3}/2)]$
6. $y = (C_1 + C_2x)e^{2x}$
7. $y = (C_1 + C_2x)e^{-3x}$
8. $y = e^{3x}(C_1 \cos x + C_2 \sin x)$
9. $y = e^x(C_1 \cos\sqrt{3}x + C_2 \sin\sqrt{3}x)$
10. $y = C_1e^{2x} + C_2e^{8x}$

Article 20.10

1. $y = \dfrac{x^2}{2} - x + C_1 + C_2e^{-x}$
2. $y = \cos x \sinh^{-1}(\tan x) + C_1 \cos x + C_2 \sin x$
3. $y = -\frac{1}{2}x \cos x + C_1 \cos x + C_2 \sin x$
4. $y = \frac{1}{4}e^x + e^{-x}(C_1 + C_2x)$
5. $y = e^{-x}(C_1 + C_2x + \frac{1}{2}x^2)$
6. $y = C_1e^x + C_2e^{-x} - x$
7. $y = C_1e^x + C_2e^{-x} + \frac{1}{2}xe^x$
8. $y = C_1e^x + C_2e^{-x} - \frac{1}{2} \sin x$
9. $y = e^{-2x}(C_1 \cos x + C_2 \sin x) + 2$
10. $y = e^{-2x}(C_1 \cos x + C_2 \sin x) + \frac{1}{5}x + \frac{6}{25}$

Article 20.11

1. $y = C_1 + C_2e^x + C_3e^{2x}$
2. $y = C_1e^x + e^{-x/2}[C_2 \cos(x\sqrt{3}/2) + C_3 \sin(x\sqrt{3}/2)]$
3. $y = (C_1 + C_2x)e^{\sqrt{2}x} + (C_3 + C_4x)e^{-\sqrt{2}x}$
4. $y = C_1e^{2x} + C_2e^{-2x} + C_3 \cos 2x + C_4 \sin 2x$
5. $y = e^{\sqrt{2}x}(C_1 \cos\sqrt{2}x + C_2 \sin\sqrt{2}x) + e^{-\sqrt{2}x}(C_3 \cos\sqrt{2}x + C_4 \sin\sqrt{2}x)$
6. $y = e^x\left(\dfrac{x^2}{6} + C_1x + C_2\right) + C_3e^{-2x}$
7. $y = (C_1 + C_2x + C_3x^2 + C_4x^3)e^x + 7$
8. $y = e^{x/\sqrt{2}}\left(C_1 \cos\dfrac{x}{\sqrt{2}} + C_2 \sin\dfrac{x}{\sqrt{2}}\right) + e^{-x/\sqrt{2}}\left(C_3 \cos\dfrac{x}{\sqrt{2}} + C_4 \sin\dfrac{x}{\sqrt{2}}\right) + x + 1$

Article 20.12

1. $x = x_0 \cos \omega t + \dfrac{v_0}{\omega} \sin \omega t = \sqrt{x_0^2 + \dfrac{v_0^2}{\omega^2}} \sin(\omega t + \phi)$,

   where $\omega = \sqrt{k/m}$, $\phi = \tan^{-1}(\omega x_0/v_0)$
2. $x = 0.288 \sin(13.9t)$; x in ft, t in sec.
3. (a) $I = C_1 \cos \omega t + C_2 \sin \omega t$

   (b) $I = C_1 \cos \omega t + C_2 \sin \omega t + \dfrac{A\alpha}{L(\omega^2 - \alpha^2)} \cos \alpha t$

   (c) $I = C_1 \cos \omega t + C_2 \sin \omega t + \dfrac{A}{2L} t \sin \omega t$

   (d) $I = e^{-5t}(C_1 \cos 149t + C_2 \sin 149t)$
4. $\theta = \theta_0 \cos(\sqrt{g/l}\, t)$
5. $\theta = \theta_0 \cos \omega t + (v_0/\omega) \sin \omega t$, where $\omega = \sqrt{2k/mr^2}$
6. $c = 5.1$ (lb sec/ft)
7. (a) $x = C_1 \cos \omega t + C_2 \sin \omega t + \dfrac{\omega^2 A}{\omega^2 - \alpha^2} \sin \alpha t$,

   where $\omega = \sqrt{k/m}$

   (b) $x = C_1 \cos \omega t + C_2 \sin \omega t - \dfrac{\omega A}{2} t \cos \omega t$,

   where $\omega = \sqrt{k/m}$

Article 20.13

5. (a) $e^{-10} = 0.000045$ (b) $1 - e^{-4} = 0.9817$
6. (a) 0.67 (b) 0.33 (c) 0.98
7. (a) 10 (b) $e^{-10} \sum_{n=0}^{5} \dfrac{(10)^n}{n!} \approx 0.067$
8. (a) 0.270 (b) 0.180 (c) 0.368 (d) 0.073
9. 0.195

Miscellaneous exercises

1. $\ln(c \ln y) = -\tan^{-1} x$
2. $y = [cx^3 + 2(x+1)^3][(x+1)^3 - cx^3]^{-1}$
3. $y = \ln(c - \frac{1}{3}e^{-3x})$
4. $ky + \sqrt{k^2y^2 - 1} = ce^{kx}$
5. $y^2 + \sqrt{y^4 + 1} = ce^{2x}$ 6. $x^2 + xy - y^2 = c$
7. $y^2 = x^2 + Cx$ 8. $y + \sqrt{x^2 + y^2} = cx^2$
9. $y = x \tan^{-1}(\ln cx)$ 10. $y = xe^{\sqrt{2 \ln cx}}$
11. $y = (x^2 + 2)/4 + cx^{-2}$
12. $y - x \sec y = c$
13. $y = \cosh x \ln(c \cosh x)$
14. $y = \frac{1}{6}(2x^3 + 3x^2 + c)(x+1)^{-2}$
15. $y^3 + 3xy^2 + 3(x + y) = c$
16. $x^3 + 3xy + 3e^y = c$
17. $x^3 + 3xy^2 + 3 \sinh y = c$
18. $e^x + x \ln y + y = c$
19. $x^2 + e^y(x^2 + y^2 - 2y + 2) = c$
20. $2x \tan^{-1}(y/x) + y \ln(x^2 + y^2) - 2 \cos x - (y+1)^2 = c$
21. $y = C$, or $y = a \tan(ax + C)$,

    or $y = -a \tanh(ax + C)$,

    or $y = -1/(x + C)$
22. $x = C_1 \cos 2y + C_2 \sin 2y$
23. $y = -\ln|\cos(x + c)| + d$
24. $x = \pm(y + C)$, or $x + C_2 = \ln \cosh(y + C_1)$
25. $y = \frac{1}{2}(\ln cx)^2 + d$ 26. $y = c_1e^x + c_2e^{3x}$
27. $y = c_1 + (c_2x + c_3)e^x$
28. $y = c_1 \sin 2x + c_2 \cos 2x + (x/2) \sin 2x + \frac{1}{4} \cos 2x \ln \cos 2x$
29. $y = c_1e^{-x} + (c_2 + x/3)e^{2x}$
30. $y = e^x(c_1 \cos 2x + c_2 \sin 2x) + \frac{1}{8}e^{-x}$
31. $y = c_1\sqrt{x} + c_2/\sqrt{x}$ 33. $\frac{3}{2}x^2 + y^2 = d$
34. $x^2 + (y - d)^2 = d^2$ 35. $y^2 = d^2 + 2\,dx$
36. $y = 10 \cos 10t + 5 \sin 10t$

Appendix 1

Article A.1

1. 7 2. 3 3. −32 4. 0
5. $x = 1$, $y = 1$ 6. $x = 3$, $y = 2$
10. $x' = x \cos \alpha + y \sin \alpha$,

    $y' = -x \sin \alpha + y \cos \alpha$

Article A.3

1. 4 3. −9 5. −4 6. −33 7. 6

Article A.4

1. Index of 4 3 2 1 is $3 + 2 + 1 = 6$.

   Sign of $a_{14}\, a_{23}\, a_{32}\, a_{41}$ is +.

Article A.5

1. $q = 4, \quad r = 3, \quad s = 0$
   $q' = 4, \quad r' = 4, \quad s' = 0$

4. 118

5. 0

Article A.6

1. Differentiate the elements of any one row of the determinant and leave the other rows unaltered. The sum of the n determinants which can be obtained in this manner is the derivative of the original nth-order determinant.

2. $-2(3) - 3(-2) + 2(0) = 0$

Article A.7

1. $x = \frac{6}{7}, \quad y = \frac{10}{7}, \quad z = -\frac{2}{7}$

2. $x = 3, \quad y = -1, \quad z = 3$

3. Multiply the equations respectively by -14, -7, $+9$, and -3, and add to obtain $-50z = -50$.

4. -2

Article A.8

1. 16.5

2. $x = y = a, \quad z = 2a, \quad a$ arbitrary

4. $x = k, \quad y = -3k, \quad z = -5k, \quad k$ arbitrary

5. $r = 0, \quad A = -B; \quad r = 3, \quad A = 2B$

6. $r = 2, \quad A = -B$

7. $r = 0, \quad A = -B; \quad r = 2, \quad A = B$

8. $r = 1, \quad A = -3B; \quad r = -1, \quad A = -B$

9. $r = 1, \quad A = B; \quad r = -3, \quad A = -B$

10. $r = -1, \quad A = B; \quad r = -3, \quad A = -B$

11. $r = 2, \quad A = 0, \quad B$ arbitrary;
    $r = -2, \quad A = -B$

12. $r = 2, \quad A = -4B; \quad r = -1, \quad A = -B$

Miscellaneous exercises

1. $x = y = 1$

2. $x = \frac{5}{7}, \quad y = \frac{6}{7}, \quad z = 1$

3. $\frac{14}{33}, -\frac{2}{33}, \frac{25}{33}$

4. $3, -1, -1$

5. $x = y = z = 0$

6. $-\frac{1}{6}, \frac{2}{6}, \frac{7}{6}$

7. $1, -1, 0$

8. $2, 1, 1$

9. No solution

10. $0, 0, 0, -1$

12. (d) 4

16. -16

17. $x = a, \quad y = b, \quad z = c$

20. (b) No

21. The converse is false.

INDEX

Note. Numbers in parentheses refer to exercises on the pages indicated.

A

B

C

D

E

F

J

K

L

M

N

Q

R

S

T

ABCDE698

70. $$\int \frac{dx}{b+c\sin ax} = \frac{-2}{a\sqrt{b^2-c^2}}\tan^{-1}\left[\sqrt{\frac{b-c}{b+c}}\tan\left(\frac{\pi}{4}-\frac{ax}{2}\right)\right]+C, \quad b^2>c^2$$

71. $$\int \frac{dx}{b+c\sin ax} = \frac{-1}{a\sqrt{c^2-b^2}}\ln\left|\frac{c+b\sin ax+\sqrt{c^2-b^2}\cos ax}{b+c\sin ax}\right|+C, \quad b^2<c^2$$

72. $$\int \frac{dx}{1+\sin ax} = -\frac{1}{a}\tan\left(\frac{\pi}{4}-\frac{ax}{2}\right)+C$$

73. $$\int \frac{dx}{1-\sin ax} = \frac{1}{a}\tan\left(\frac{\pi}{4}+\frac{ax}{2}\right)+C$$

74. $$\int \frac{dx}{b+c\cos ax} = \frac{2}{a\sqrt{b^2-c^2}}\tan^{-1}\left[\sqrt{\frac{b-c}{b+c}}\tan\frac{ax}{2}\right]+C, \quad b^2>c^2$$

75. $$\int \frac{dx}{b+c\cos ax} = \frac{1}{a\sqrt{c^2-b^2}}\ln\left|\frac{c+b\cos ax+\sqrt{c^2-b^2}\sin ax}{b+c\cos ax}\right|+C, \quad b^2<c^2$$

76. $$\int \frac{dx}{1+\cos ax} = \frac{1}{a}\tan\frac{ax}{2}+C$$

77. $$\int \frac{dx}{1-\cos ax} = -\frac{1}{a}\cot\frac{ax}{2}+C$$

78. $$\int x\sin ax\,dx = \frac{1}{a^2}\sin ax-\frac{x}{a}\cos ax+C$$

79. $$\int x\cos ax\,dx = \frac{1}{a^2}\cos ax+\frac{x}{a}\sin ax+C$$

80. $$\int x^n\sin ax\,dx = -\frac{x^n}{a}\cos ax+\frac{n}{a}\int x^{n-1}\cos ax\,dx$$

81. $$\int x^n\cos ax\,dx = \frac{x^n}{a}\sin ax-\frac{n}{a}\int x^{n-1}\sin ax\,dx$$

82. $$\int \tan ax\,dx = -\frac{1}{a}\ln|\cos ax|+C$$

83. $$\int \cot ax\,dx = \frac{1}{a}\ln|\sin ax|+C$$

84. $$\int \tan^2 ax\,dx = \frac{1}{a}\tan ax-x+C$$

85. $$\int \cot^2 ax\,dx = -\frac{1}{a}\cot ax-x+C$$

86. $$\int \tan^n ax\,dx = \frac{\tan^{n-1}ax}{a(n-1)}-\int\tan^{n-2}ax\,dx, \quad n\neq 1$$

87. $$\int \cot^n ax\,dx = -\frac{\cot^{n-1}ax}{a(n-1)}-\int\cot^{n-2}ax\,dx, \quad n\neq 1$$

88. $$\int \sec ax\,dx = \frac{1}{a}\ln|\sec ax+\tan ax|+C$$

89. $$\int \csc ax\,dx = -\frac{1}{a}\ln|\csc ax+\cot ax|+C$$

90. $$\int \sec^2 ax\,dx = \frac{1}{a}\tan ax+C$$

91. $$\int \csc^2 ax\,dx = -\frac{1}{a}\cot ax+C$$

92. $$\int \sec^n ax\,dx = \frac{\sec^{n-2}ax\tan ax}{a(n-1)}+\frac{n-2}{n-1}\int\sec^{n-2}ax\,dx, \quad n\neq 1$$

93. $$\int \csc^n ax\,dx = -\frac{\csc^{n-2}ax\cot ax}{a(n-1)}+\frac{n-2}{n-1}\int\csc^{n-2}ax\,dx, \quad n\neq 1$$

94. $$\int \sec^n ax\tan ax\,dx = \frac{\sec^n ax}{na}+C, \quad n\neq 0$$

95. $$\int \csc^n ax\cot ax\,dx = -\frac{\csc^n ax}{na}+C, \quad n\neq 0$$

96. $$\int \sin^{-1}ax\,dx = x\sin^{-1}ax+\frac{1}{a}\sqrt{1-a^2x^2}+C$$

97. $$\int \cos^{-1}ax\,dx = x\cos^{-1}ax-\frac{1}{a}\sqrt{1-a^2x^2}+C$$

98. $$\int \tan^{-1}ax\,dx = x\tan^{-1}ax-\frac{1}{2a}\ln(1+a^2x^2)+C$$

99. $$\int x^n\sin^{-1}ax\,dx = \frac{x^{n+1}}{n+1}\sin^{-1}ax-\frac{a}{n+1}\int\frac{x^{n+1}\,dx}{\sqrt{1-a^2x^2}}, \quad n\neq -1$$

100. $$\int x^n\cos^{-1}ax\,dx = \frac{x^{n+1}}{n+1}\cos^{-1}ax+\frac{a}{n+1}\int\frac{x^{n+1}\,dx}{\sqrt{1-a^2x^2}}, \quad n\neq -1$$

101. $$\int x^n\tan^{-1}ax\,dx = \frac{x^{n+1}}{n+1}\tan^{-1}ax-\frac{a}{n+1}\int\frac{x^{n+1}\,dx}{1+a^2x^2}, \quad n\neq -1$$

102. $$\int e^{ax}\,dx = \frac{1}{a}e^{ax}+C$$

103. $$\int b^{ax}\,dx = \frac{1}{a}\frac{b^{ax}}{\ln b}+C, \quad b>0, \quad b\neq 1$$

104. $$\int xe^{ax}\,dx = \frac{e^{ax}}{a^2}(ax-1)+C$$

105. $$\int x^ne^{ax}\,dx = \frac{1}{a}x^ne^{ax}-\frac{n}{a}\int x^{n-1}e^{ax}\,dx$$

106. $$\int x^nb^{ax}\,dx = \frac{x^nb^{ax}}{a\ln b}-\frac{n}{a\ln b}\int x^{n-1}b^{ax}\,dx, \quad b>0, \quad b\neq 1$$